POWER ELECTRONICS HANDBOOK

Blackburn College

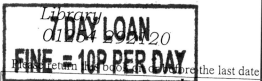

Library
01254 292120

Please return this book on or before the last date

0 8 OCT 2007

2 0 OCT 2008

- 3 NOV 2008

SHORT LOAN

7 DAY LOAN ONLY

NO RENEWAL

POWER ELECTRONICS HANDBOOK

DEVICES, CIRCUITS, AND APPLICATIONS

Edited by

Muhammad H. Rashid
Ph.D., Fellow IEE, Fellow IEEE
Professor and Director
Department of Electrical and Computer Engineering
University of West Florida
Pensacola, Florida

AMSTERDAM • BOSTON • HEIDELBERG • LONDON • NEW YORK • OXFORD
PARIS • SAN DIEGO • SAN FRANCISCO • SINGAPORE • SYDNEY • TOKYO
Academic Press is an imprint of Elsevier

ELSEVIER

Academic Press is an imprint of Elsevier
30 Corporate Drive, Suite 400, Burlington, MA 01803, USA
525 B Street, Suite 1900, San Diego, California 92101-4495, USA
84 Theobald's Road, London WC1X 8RR, UK

This book is printed on acid-free paper. ∞

Copyright © 2007, Elsevier Inc. All rights reserved.

No part of this publication may be reproduced or transmitted in any form or by any means, electronic or mechanical, including photocopy, recording, or any information storage and retrieval system, without permission in writing from the publisher.

Permissions may be sought directly from Elsevier's Science & Technology Rights Department in Oxford, UK: phone: (+44) 1865 843830, fax: (+44) 1865 853333, E-mail: permissions@elsevier.com. You may also complete your request on-line via the Elsevier homepage (http://elsevier.com), by selecting "Support & Contact" then "Copyright and Permission" and then "Obtaining Permissions."

Library of Congress Cataloging-in-Publication Data

Application submitted

British Library Cataloguing-in-Publication Data

A catalogue record for this book is available from the British Library.

ISBN 13: 978-0-12-088479-7
ISBN 10: 0-12-088479-8

For information on all Academic Press publications
visit our Web site at www.books.elsevier.com

Printed in the United States of America
07 08 09 10   9 8 7 6 5 4 3 2 1

Working together to grow
libraries in developing countries

www.elsevier.com | www.bookaid.org | www.sabre.org

ELSEVIER   BOOK AID International   Sabre Foundation

Dedication

To those who promote power electronics and inspire students for finding applications for the benefits of the people and the environment in the global community

Table of Contents

Chapter 1	Introduction	1
	Philip T. Krein *Department of Electrical and Computer Engineering* *University of Illinois* *Urbana, Illinois, USA*	
Chapter 2	The Power Diode	15
	Ali I. Maswood *School of EEE* *Nanyang Technological University* *Nanyang Avenue, Singapore*	
Chapter 3	Power Bipolar Transistors	27
	Marcelo Godoy Simoes *Engineering Division* *Colorado School of Mines* *Golden, Colorado, USA*	
Chapter 4	The Power MOSFET	41
	Issa Batarseh *School of Electrical Engineering and Computer Science* *University of Central Florida* *4000 Central Florida Blvd.* *Orlando, Florida, USA*	
Chapter 5	Insulated Gate Bipolar Transistor	71
	S. Abedinpour and K. Shenai *Department of Electrical Engineering and Computer Science* *University of Illinois at Chicago* *851, South Morgan Street (M/C 154)* *Chicago, Illinois, USA*	

Chapter 6	Thyristors	89

Angus Bryant
Department of Engineering
University of Warwick
Coventry CV4 7AL, UK

Enrico Santi
Department of Electrical Engineering
University of South Carolina
Columbia, South Carolina, USA

Jerry Hudgins
Department of Electrical Engineering
University of Nebraska
Lincoln, Nebraska, USA

Patrick Palmer
Department of Engineering
University of Cambridge
Trumpington Street
Cambridge CB2 1PZ, UK

Chapter 7	Gate Turn-off Thyristors	115

Muhammad H. Rashid
Electrical and Computer Engineering
University of West Florida
11000 University Parkway
Pensacola, Florida, USA

Chapter 8	MOS Controlled Thyristors (MCTs)	123

S. Yuvarajan
Department of Electrical Engineering
North Dakota State University
P.O. Box 5285
Fargo, North Dakota, USA

Chapter 9	Static Induction Devices	133

Bogdan M. Wilamowski
Alabama Microelectronics Science and Technology Center
Auburn University
Alabama, USA

Chapter 10	Diode Rectifiers	145

Yim-Shu Lee and Martin H. L. Chow
Department of Electronic and Information Engineering
The Hong Kong Polytechnic
University Hung Hom
Hong Kong

Table of Contents

Chapter 11	Single-phase Controlled Rectifiers	179

José Rodríguez, Pablo Lezana,
Samir Kouro, and
Alejandro Weinstein
Department of Electronics
Universidad Técnica Federico
Santa María, Valparaíso, Chile

Chapter 12	Three-phase Controlled Rectifiers	201

Juan W. Dixon
Department of Electrical Engineering
Pontificia Universidad Católica de Chile
Vicuña Mackenna 4860
Santiago, Chile

Chapter 13	DC–DC Converters	245

Dariusz Czarkowski
Department of Electrical and Computer Engineering
Polytechnic University
Brooklyn, New York, USA

Chapter 14	DC/DC Conversion Technique and Twelve Series Luo-converters	261

Fang Lin Luo
School of EEE, Block S1
Nanyang Technological University
Nanyang Avenue, Singapore

Hong Ye
School of Biological Sciences, Block SBS
Nanyang Technological University
Nanyang Avenue, Singapore

Chapter 15	Inverters	353

José R. Espinoza
Departamento de Ingeniería Eléctrica, of. 220
Universidad de Concepción
Casilla 160-C, Correo 3
Concepción, Chile

Chapter 16	Resonant and Soft-switching Converters	405

S. Y. (Ron) Hui and Henry S. H. Chung
Department of Electronic Engineering
City University of Hong Kong
Tat Chee Avenue, Kowloon
Hong Kong

Chapter 17	Multilevel Power Converters	451

Surin Khomfoi and Leon M. Tolbert
The University of Tennessee
Department of Electrical and Computer Engineering
Knoxville, Tennessee, USA

Chapter 18	AC–AC Converters	483

A. K. Chattopadhyay
Electrical Engg. Department
Bengal Engineering & Science University
Shibpur, Howrah, India

Chapter 19	Power Factor Correction Circuits	517

Issa Batarseh and Huai Wei
School of Electrical Engineering and Computer Science
University of Central Florida
4000 Central Florida Blvd.
Orlando, Florida, USA

Chapter 20	Gate Drive Circuitry for Power Converters	543

Irshad Khan
University of Cape Town
Department of Electrical Engineering
Cape Town, South Africa

Chapter 21	Power Electronics in Capacitor Charging Applications	559

William C. Dillard
Archangel Systems, Incorporated
1635 Pumphrey Avenue Auburn
Alabama, USA

Chapter 22	Electronic Ballasts	565

J. Marcos Alonso
University of Oviedo
DIEECS - Tecnologia Electronica
Campus de Viesques s/n
Edificio de Electronica
33204 Gijon, Asturias, Spain

Chapter 23	Power Supplies	593

Y. M. Lai
Department of Electronic and Information Engineering
The Hong Kong Polytechnic University
Hong Kong

Table of Contents

Chapter 24	Uninterruptible Power Supplies	619

Adel Nasiri
Power Electronics and Motor Drives Laboratory
University of Wisconsin-Milwaukee
3200 North Cramer Street
Milwaukee, Wisconsin, USA

Chapter 25	Automotive Applications of Power Electronics	635

David J. Perreault
Massachusetts Institute of Technology
Laboratory for Electromagnetic and Electronic Systems
77 Massachusetts Avenue, 10-039
Cambridge, Massachusetts, USA

Khurram Afridi
Techlogix, 800 West Cummings Park
1925, Woburn, Massachusetts, USA

Iftikhar A. Khan
Delphi Automotive Systems
2705 South Goyer Road
MS D35 Kokomo
Indiana, USA

Chapter 26	Solar Power Conversion	661

Lana Chaar
Electrical Engineering Department
American University in Dubai
P. O. Box 28282
Dubai, UAE

Chapter 27	Power Electronics for Renewable Energy Sources	673

C. V. Nayar, S. M. Islam,
H. Dehbonei, and K. Tan
Department of Electrical & Computer Engineering
Curtin University of Technology
GPO Box U1987, Perth
Western Australia, Australia

H. Sharma
Research Institute for Sustainable Energy
Murdoch University
Perth, Western Australia, Australia

Chapter 28	Fuel-cell Power Electronics for Distributed Generation	717

S. K. Mazumder
Department of Electrical and Computer Engineering
Director Laboratory for Energy and
Switching-Electronics Systems (LESES)
University of Illinois
Chicago, Illinois, USA

| Chapter 29 | Wind Turbine Applications | 737 |

Juan M. Carrasco,
Eduardo Galván, and
Ramón Portillo
Department of Electronic Engineering
Engineering School, Seville University
Spain

| Chapter 30 | HVDC Transmission | 769 |

Vijay K. Sood
Hydro-Quebec (IREQ)
1800 Lionel Boulet
Varennes, Quebec, Canada

| Chapter 31 | Flexible AC Transmission Systems | 797 |

E. H. Watanabe, M. Aredes, G. Santos Jr.,
F. K. de Araújo Lima, and R. F. da Silva Dias
Electrical Engineering Department
COPPE/Federal University of Rio de Janeiro
Brazil, South America

P. G. Barbosa
Electrical Engineering Department
Federal University of Juiz de Fora
Brazil, South America

| Chapter 32 | Drives Types and Specifications | 823 |

Yahya Shakweh
Technical Director
FKI Industrial Drives & Controls, England, UK

| Chapter 33 | Motor Drives | 857 |

M. F. Rahman
School of Electrical Engineering and Telecommunications
The University of New South Wales
Sydney, New South Wales, Australia

D. Patterson
Northern Territory Centre for Energy Research
Faculty of Technology
Northern Territory University
Darwin, Northern Territory, Australia

A. Cheok
Department of Electrical and Computer Engineering
National University of Singapore
10 Kent Ridge Crescent
Singapore

R. Betz
Department of Electrical and Computer Engineering
University of Newcastle
Callaghan, New South Wales, Australia

Table of Contents xiii

| Chapter 34 | Control Methods for Switching Power Converters | 935 |

J. Fernando Silva and
Sónia Ferreira Pinto
Instituto Superior Técnico, DEEC,
A.C. Energia, Laboratório de Máquinas Eléctricas e Electrónica de Potência
Centro de Automática da Universidade Técnica de Lisboa
AV. Rorisco Pais 1
Lisboa, Portugal

| Chapter 35 | Fuzzy Logic in Electric Drives | 999 |

Ahmed Rubaai
Department of Electrical Engineering
Howard University
Washington, D.C., USA

| Chapter 36 | Artificial Neural Network Applications in Power Electronics and Electrical Drives | 1015 |

B. Karanayil and M. F. Rahman
School of Electrical Engineering and Telecommunications
The University of New South Wales
Sydney, New South Wales, Australia

| Chapter 37 | DSP-based Control of Variable Speed Drives | 1031 |

Hamid A. Toliyat
Electrical and Computer Engineering Department
Texas A&M University
3128 Tamus
216g Zachry Engineering Center
College Station, Texas, USA

Mehdi Abolhassani
Black & Decker (US) Inc.
701 E Joppa Rd., TW100
Towson, Maryland, USA

Peyman Niazi
Maxtor Co.
333 South St., Shrewsbury
Massachusetts, USA

Lei Hao
Wavecrest Laboratories
1613 Star Batt Drive
Rochester Hills, Michigan, USA

| Chapter 38 | Power Quality | 1053 |

S. Mark Halpin and Angela Card
Department of Electrical and Computer Engineering
Auburn University
Alabama, USA

Chapter 39	Active Filters	1067

Luis Morán
Electrical Engineering Dept.
Universidad de Concepción
Concepción, Chile

Juan Dixon
Electrical Engineering Dept.
Universidad Católica de Chile
Santiago, Chile

Chapter 40	EMI Effects of Power Converters	1103

Andrzej M. Trzynadlowski
University of Nevada
Electrical Engineering Dept.
260 Reno, Nevada, USA

Chapter 41	Computer Simulation of Power Electronics and Motor Drives	1121

Michael Giesselmann, P. E.
Center for Pulsed Power and Power Electronics
Department of Electrical and Computer Engineering
Texas Tech University, Lubbock
Texas, USA

Chapter 42	Packaging and Smart Power Systems	1147

Douglas C. Hopkins
Dir.—Electronic Power and Energy Research Laboratory
University at Buffalo
332 Bonner Hall
Buffalo, New York, USA

Index	1159

Preface

Introduction

The purpose of *Power Electronics Handbook* second edition is to provide an up-to-date reference that is both concise and useful for engineering students and practicing professionals. It is designed to cover a wide range of topics that make up the field of power electronics in a well-organized and highly informative manner. The Handbook is a careful blend of both traditional topics and new advancements. Special emphasis is placed on practical applications, thus, this Handbook is not a theoretical one, but an enlightening presentation of the usefulness of the rapidly growing field of power electronics. The presentation is tutorial in nature in order to enhance the value of the book to the reader and foster a clear understanding of the material.

The contributors to this Handbook span the globe, with fifty-four authors from twelve different countries, some of whom are the leading authorities in their areas of expertise. All were chosen because of their intimate knowledge of their subjects, and their contributions make this a comprehensive state-of-the-art guide to the expanding field of power electronics and its applications covering:

- the characteristics of modern power semiconductor devices, which are used as switches to perform the power conversions from ac–dc, dc–dc, dc–ac, and ac–ac;
- both the fundamental principles and in-depth study of the operation, analysis, and design of various power converters; and
- examples of recent applications of power electronics.

Power Electronics Backgrounds

The first electronics revolution began in 1948 with the invention of the silicon transistor at Bell Telephone Laboratories by Bardeen, Bratain, and Shockley. Most of today's advanced electronic technologies are traceable to that invention, and modern microelectronics has evolved over the years from these silicon semiconductors. The second electronics revolution began with the development of a commercial thyristor by the General Electric Company in 1958. That was the beginning of a new era of *power electronics*. Since then, many different types of power semiconductor devices and conversion techniques have been introduced.

The demand for energy, particularly in electrical forms, is ever-increasing in order to improve the standard of living. *Power electronics* helps with the efficient use of electricity, thereby reducing power consumption. Semiconductor devices are used as switches for power conversion or processing, as are solid state electronics for efficient control of the amount of power and energy flow. Higher efficiency and lower losses are sought for devices for a range of applications, from microwave ovens to high-voltage dc transmission. New devices and power electronic systems are now evolving for even more efficient control of power and energy.

Power electronics has already found an important place in modern technology and has revolutionized control of power and energy. As the voltage and current ratings and switching characteristics of power semiconductor devices keep improving, the range of applications continues to expand in areas such as lamp controls, power supplies to motion control, factory automation, transportation, energy storage, multi-megawatt industrial drives, and electric power transmission and distribution. The greater efficiency and tighter control features of power electronics are becoming attractive for applications in motion control by replacing the earlier electro-mechanical and electronic systems. Applications in power transmission include high-voltage dc (VHDC) converter stations, flexible ac transmission system (FACTS), and static-var compensators. In power distribution these include dc-to-ac conversion, dynamic filters, frequency conversion, and Custom Power System.

Almost all new electrical or electromechanical equipment, from household air conditioners and computer power supplies to industrial motor controls, contain power electronic circuits and/or systems. In order to keep up, working engineers involved in control and conversion of power and energy into applications ranging from several hundred voltages at a fraction of an ampere for display devices to about 10,000 V at high-voltage dc transmission, should have a working knowledge of power electronics.

Organization

The Handbook starts with an introductory chapter and moves on to cover topics on power semiconductor devices, power converters, applications, and peripheral issues. The book is organized into six areas, the first of which includes Chapters 2 to 9 on operation and characterizations of power semiconductor devices: Power Diode, Thyristor, Gate Turn-off Thyristor (GTO), Power Bipolar Transistor (BJT), Power MOSFET, Insulated Gate Bipolar Transistor, MOS Controlled Thyristor (MCT), and Static Induction Devices.

The next topic area includes Chapters 10 to 20 covering various types of power converters, the principles of operation, and the methods for the analysis and design of power converters. This also includes gate drive circuits and control methods for power converters. The next 13 chapters 21 to 33 cover applications in power supplies, electronics ballasts, renewable energy soruces, HVDC transmission, VAR compensation, and capacitor charging. Power Electronics in Capacitor Charging Applications, Electronic Ballasts, Power Supplies, Uninterruptible Power Supplies, Automotive Applications of Power Electronics, Solar Power Conversion, Power Electronics for Renewable Energy Sources, Fuel-cell Power Electronics for Distributed Generation, Wind Turbine Applications, HVDC Transmission, Flexible AC Transmission Systems, Drives Types and Specifications, Motor Drives.

The following four chapters 34 to 37 focus on the Operation, Theory, and Control Methods of Motor Drives, and Automotive Systems. We then move on to three chapters 38 to 40 on Power Quality Issues, Active Filters, and EMI Effects of Power Converters and two chapters 41 to 42 on Computer Simulation, Packaging and Smart Power Systems.

Locating Your Topic

A table of contents is presented at the front of the book, and each chapter begins with its own table of contents. The reader should look over these tables of contents to become familiar with the structure, organization, and content of the book.

Audience

The Handbook is designed to provide both students and practicing engineers with answers to questions involving the wide spectrum of power electronics. The book can be used as a textbook for graduate students in electrical or systems engineering, or as a reference book for senior undergraduate students and for engineers who are interested and involved in operation, project management, design, and analysis of power electronics equipment and motor drives.

Acknowledgments

This Handbook was made possible through the expertise and dedication of outstanding authors from throughout the world. I gratefully acknowledge the personnel at Academic Press who produced the book, including Jane Phelan. In addition, special thanks are due to Joel D. Claypool, the executive editor for this book.

Finally, I express my deep appreciation to my wife, Fatema Rashid, who graciously puts up with my publication activities.

Muhammad H. Rashid, Editor-in-Chief

1
Introduction

Philip T. Krein, Ph.D.
Department of Electrical and Computer Engineering, University of Illinois, Urbana, Illinois, USA

1.1 Power Electronics Defined 1
1.2 Key Characteristics 2
   1.2.1 The Efficiency Objective – The Switch • 1.2.2 The Reliability Objective – Simplicity and Integration
1.3 Trends in Power Supplies 4
1.4 Conversion Examples 4
   1.4.1 Single-Switch Circuits • 1.4.2 The Method of Energy Balance
1.5 Tools for Analysis and Design 7
   1.5.1 The Switch Matrix • 1.5.2 Implications of Kirchhoff's Voltage and Current Laws • 1.5.3 Resolving the Hardware Problem – Semiconductor Devices • 1.5.4 Resolving the Software Problem – Switching Functions • 1.5.5 Resolving the Interface Problem – Lossless Filter Design
1.6 Summary 13
   References 13

1.1 Power Electronics Defined[1]

It has been said that people do not use electricity, but rather they use communication, light, mechanical work, entertainment, and all the tangible benefits of both energy and electronics. In this sense, electrical engineering as a discipline is much involved in energy conversion and information. In the general world of electronics engineering, the circuits engineers design and use are intended to convert information. This is true of both analog and digital circuit design. In radio frequency applications, energy and information are sometimes on more equal footing, but the main function of any circuit is information transfer.

What about the conversion and control of electrical energy itself? Energy is a critical need in every human endeavor. The capabilities and flexibility of modern electronics must be brought to bear to meet the challenges of reliable, efficient energy. It is essential to consider how electronic circuits and systems can be applied to the challenges of energy conversion and management. This is the framework of *power electronics*, a discipline defined in terms of *electrical energy conversion, applications,* and *electronic devices*. More specifically,

DEFINITION *Power electronics* involves the study of electronic circuits intended to control the flow of electrical energy. These circuits handle power flow at levels much higher than the individual device ratings.

Rectifiers are probably the most familiar examples of circuits that meet this definition. Inverters (a general term for dc–ac converters) and dc–dc converters for power supplies are also common applications. As shown in Fig. 1.1, power electronics represents a median point at which the topics of energy systems, electronics, and control converge and combine [1]. Any useful circuit design for an energy application must address issues of both devices and control, as well as of the energy itself. Among the unique aspects of power electronics are its emphasis on large semiconductor devices, the application of magnetic devices for energy storage, special control methods that must be applied to nonlinear systems, and its fundamental place as a vital component of today's energy systems. In any study of electrical engineering, power electronics must be placed on a level with digital, analog, and radio-frequency electronics to reflect the distinctive design methods and unique challenges.

Applications of power electronics are expanding exponentially. It is not possible to build practical computers, cell phones, cars, airplanes, industrial processes, and a host of

---

[1] Portions of this chapter are from P. T. Krein, *Elements of Power Electronics*. New York: Oxford University Press, 1998. Copyright © 1998, Oxford University Press. Used by permission.

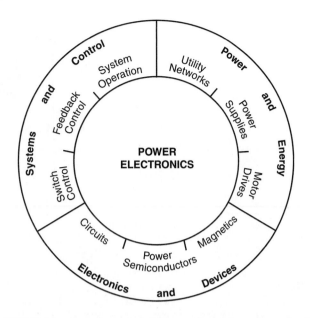

FIGURE 1.1 Control, energy, and power electronics are interrelated.

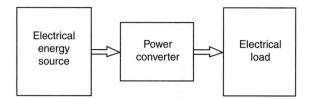

FIGURE 1.2 General system for electric power conversion. (From Reference [2], copyright © 1998, Oxford University Press, Inc.; used by permission.)

other everyday products without power electronics. Alternative energy systems such as wind generators, solar power, fuel cells, and others require power electronics to function. Technology advances such as hybrid vehicles, laptop computers, microwave ovens, plasma displays, and hundreds of other innovations were not possible until advances in power electronics enabled their implementation. While no one can predict the future, it is certain that power electronics will be at the heart of fundamental energy innovations.

The history of power electronics [2–5] has been closely allied with advances in electronic devices that provide the capability to handle high power levels. Since about 1990, devices have become so capable that a transition is being made from a "device-driven" field to an "applications-driven" field. This transition has been based on two factors: advanced semiconductors with suitable power ratings exist for almost every application of wide interest; and the general push toward miniaturization is bringing advanced power electronics into a growing variety of products. While the devices continue to improve, their development now tends to follow innovative applications.

1.2 Key Characteristics

All power electronic circuits manage the flow of electrical energy between an electrical source and a load. The parts in a circuit must direct electrical flows, not impede them. A general power conversion system is shown in Fig. 1.2. The function of the power converter in the middle is to control the energy flow between a source and a load. For our purposes, the power converter will be implemented with a power electronic circuit. Since a power converter appears between a source and a load, any energy used within the converter is lost to the overall system. A crucial point emerges: to build a power converter, we should consider only lossless components. A realistic converter design must approach 100% efficiency.

A power converter connected between a source and a load also affects system reliability. If the energy source is perfectly reliable (it is on all the time), then a failure in the converter affects the user (the load) just as if the energy source had failed. An unreliable power converter creates an unreliable system. To put this in perspective, consider that a typical American household loses electric power only a few minutes a year. Energy is available 99.999% of the time. A converter must be better than this to prevent system degradation. An ideal converter implementation will not suffer any failures over its application lifetime. Extreme high reliability can be a more difficult objective than high efficiency.

1.2.1 The Efficiency Objective – The Switch

A circuit element as simple as a light switch reminds us that the extreme requirements in power electronics are not especially novel. Ideally, when a switch is on, it has zero voltage drop and will carry any current imposed on it. When a switch is off, it blocks the flow of current regardless of the voltage across it. The *device power*, the product of the switch voltage and current, is identically zero at all times. A switch therefore controls energy flow with no loss. In addition, reliability is also high. Household light switches perform over decades of use and perhaps 100,000 operations. Unfortunately, a mechanical light switch does not meet all practical needs. A switch in a power supply must often function 100,000 times each second. Even the best mechanical switch will not last beyond a few million cycles. Semiconductor switches (without this limitation) are the devices of choice in power converters.

A circuit built from ideal switches will be lossless. As a result, switches are the main components of power converters, and many people equate power electronics with the study of switching power converters. Magnetic transformers and lossless storage elements such as capacitors and inductors are also valid components for use in power converters. The complete concept, shown in Fig. 1.3, illustrates a *power electronic system*. Such a system consists of an electrical energy source, an

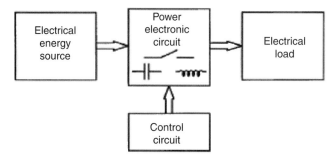

FIGURE 1.3 A basic power electronic system. (From Reference [2], copyright © 1998, Oxford University Press, Inc.; used by permission.)

electrical load, a *power electronic circuit*, and a control function. The power electronic circuit contains switches, lossless energy storage elements, and magnetic transformers. The controls take information from the source, the load, and the designer, and then determine how the switches operate to achieve the desired conversion. The controls are built up with conventional low-power analog and digital electronics.

Switching devices are selected based on their *power handling rating* – the product of their voltage and current ratings – rather than on power dissipation ratings. This is in contrast to other applications of electronics, in which power dissipation ratings dominate. For instance, a typical stereo receiver performs a conversion from ac line input to audio output. Most audio amplifiers do not use the techniques of power electronics, and the semiconductor devices do not act as switches. A commercial 100 W amplifier usually is designed with transistors big enough to dissipate the full 100 W. The semiconductor devices are used primarily to reconstruct the audio information rather than to manipulate the energy flows. The sacrifice in energy is large – a home theater amplifier often functions at less than 10% energy efficiency. In contrast, emerging *switching amplifiers* do use the techniques of power electronics. They provide dramatic efficiency improvements. A home theater system implemented with switching amplifiers can exceed 90% energy efficiency in a smaller, cooler package. The amplifiers can even be packed inside the loudspeaker.

Switches can reach extreme power levels, far beyond what might be expected for a given size. Consider the following examples.

EXAMPLE 1.1 The NTP30N20 is a metal oxide semiconductor field effect transistor (MOSFET) with a drain current rating of 30 A, a maximum drain source breakdown voltage of 200 V, and rated power dissipation of up to 200 W under ideal conditions. Without a heat sink, however, the device can handle less than 2.5 W of dissipation. For power electronics purposes, the power handling rating is 30 A × 200 V = 6 kW. Several manufacturers have developed controllers for domestic refrigerators, air conditioners, and high-end machine tools based on this device and its relatives. The second part of the definition of power electronics in Section 1.1 points out that the circuits handle power at levels much higher than that of the ratings of individual devices. Here a device is used to handle 6000 W – as compared with its individual rating of no more than 200 W. The ratio 30:1 is high, but not unusual in power electronics contexts. In contrast, the same ratio in a conventional audio amplifier is close to unity.

EXAMPLE 1.2 The IRGPS60B120KD is an insulated gate bipolar transistor (IGBT) – a relative of the bipolar transistor that has been developed specifically for power electronics – rated for 1200 V and 120 A. Its power handling rating is 144 kW. This is sufficient to control an electric or hybrid car.

1.2.2 The Reliability Objective – Simplicity and Integration

High-power applications lead to interesting issues. In an inverter, the semiconductors often manipulate 30 times their power dissipation capability or more. This implies that only about 3% of the power being controlled is lost. A small design error, unexpected thermal problem, or minor change in layout could alter this somewhat. For instance, if the loss turns out to be 4% rather than 3%, the device stresses are 33% higher, and quick failure is likely to occur. The first issue for reliability in power electronic circuits is that of managing device voltage, current, and power dissipation levels to keep them well within rating limits. This can be challenging when power handling levels are high.

The second issue for reliability is simplicity. It is well established in electronics design that the more parts there are in a system, the more likely it is to fail. Power electronic circuits tend to have few parts, especially in the main energy flow paths. Necessary operations must be carried out through shrewd use of these parts. Often, this means that sophisticated control strategies are applied to seemingly simple conversion circuits.

The third issue for reliability is integration. One way to avoid the reliability-complexity tradeoff is to integrate multiple components and functions on a single substrate. A microprocessor, for example, might contain more than a million gates. All interconnections and signals flow within a single chip, and the reliability is nearly to that of a single part. An important parallel trend in power electronic devices involves the integrated module [6]. Manufacturers seek ways to package several switching devices, with their interconnections and protection components, together as a unit. Control circuits for converters are also integrated as much as possible to keep the reliability high. The package itself becomes a fourth issue for reliability, and one that is a subject of active research. Many semiconductor packages include small bonding wires that can be susceptible to thermal or vibration damage.

The small geometries tend to enhance electromagnetic interference among the internal circuit components.

1.3 Trends in Power Supplies

Two distinct trends drive electronic power supplies, one of the major classes of power electronic circuits. At the high end, microprocessors, memory chips, and other advanced digital circuits require increasing power levels and increasing performance at very low voltage. It is a challenge to deliver 100 A or more efficiently at voltages that can be less than 1 V. These types of power supplies are asked to deliver precise voltages even though the load can change by an order of magnitude in a few nanoseconds.

At the other end is the explosive growth of portable devices with rechargeable batteries. The power supplies for these devices, for televisions, and for many other consumer products must be cheap and efficient. Losses in low-cost power supplies are a problem today; often low-end power supplies and battery chargers draw energy even when their load is off. It is increasingly important to use the best possible power electronics design techniques for these supplies to save energy while minimizing the costs. Efficiency standards such as the EnergyStar® program place increasingly stringent requirements on a wide range of low-end power supplies.

In the past, bulky "linear" power supplies were designed with transformers and rectifiers from the ac line frequency to provide low level dc voltages for electronic circuits. Late in the 1960s, use of dc sources in aerospace applications led to the development of power electronic dc–dc conversion circuits for power supplies. In a well-designed power electronics arrangement today, called a *switch-mode power supply*, an ac source from a wall outlet is rectified without direct transformation. The resulting high dc voltage is converted through a dc–dc converter to the 3, 5, and 12 V, or other level required. Switch-mode power supplies to continue to supplant linear supplies across the full spectrum of circuit applications. A personal computer commonly requires three different 5 V supplies, a 3.3 V supply, two 12 V supplies, a −12 V supply, a 24 V supply, and a separate converter for 1 V delivery to the microprocessor. This does not include supplies for the video display or peripheral devices. Only a switch-mode supply can support such complex requirements with acceptable costs.

Switch-mode supplies often take advantage of MOSFET semiconductor technology. Trends toward high reliability, low cost, and miniaturization have reached the point at which a 5 V power supply sold today might last 1,000,000 h (more than a century), provide 100 W of output in a package with volume less than 15 cm³, and sell for a price approaching US$ 0.10 per watt. This type of supply brings an interesting dilemma: the ac line cord to plug it in takes up more space than the power supply itself. Innovative concepts such as integrating a power supply within a connection cable will be used in the future.

Device technology for power supplies is also being driven by expanding needs in the automotive and telecommunications industries as well as in markets for portable equipment. The automotive industry is making a transition to higher voltages to handle increasing electric power needs. Power conversion for this industry must be cost effective, yet rugged enough to survive the high vibration and wide temperature range to which a passenger car is exposed. Global communication is possible only when sophisticated equipment can be used almost anywhere. This brings a special challenge, because electrical supplies are neither reliable nor consistent throughout much of the world. While in North America voltage swings in the domestic ac supply are often ±5% around a nominal value, in many developing nations the swing can be ±25% – when power is available. Power converters for communications equipment must tolerate these swings, and must also be able to make use of a wide range of possible backup sources. Given the enormous size of worldwide markets for telephones and consumer electronics, there is a clear need for flexible-source equipment. Designers are challenged to obtain maximum performance from small batteries, and to create equipment with minimal energy requirements.

1.4 Conversion Examples

1.4.1 Single-Switch Circuits

Electrical energy sources take the form of dc voltage sources at various values, sinusoidal ac sources, polyphase sources, and many others. A power electronic circuit might be asked to transfer energy between two different dc voltage levels, between an ac source and a dc load, or between sources at different frequencies. It might be used to adjust an output voltage or power level, drive a nonlinear load, or control a load current. In this section, a few basic converter arrangements are introduced and energy conservation provides a tool for analysis.

EXAMPLE 1.3 Consider the circuit shown in Fig. 1.4. It contains an ac source, a switch, and a resistive load. It is a simple but complete power electronic system.

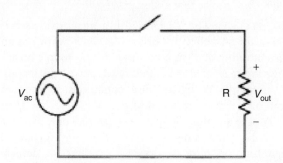

FIGURE 1.4 A simple power electronic system. (From Reference [2], copyright © 1998, Oxford University Press, Inc.; used by permission.)

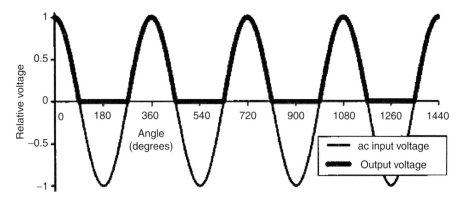

FIGURE 1.5 Input and output waveforms for Example 1.4.

Let us assign a (somewhat arbitrary) control scheme to the switch. What if the switch is turned on whenever $V_{ac} > 0$, and turned off otherwise? The input and output voltage waveforms are shown in Fig. 1.5. The input has a time average of 0, and root-mean-square (RMS) value equal to $V_{peak}/\sqrt{2}$, where V_{peak} is the maximum value of V_{ac}. The output has a nonzero average value given by

$$\langle v_{out}(t) \rangle = \frac{1}{2\pi} \left(\int_{-\pi/2}^{\pi/2} V_{peak} \cos\theta \, d\theta + \int_{\pi/2}^{3\pi/2} 0 \, d\theta \right) \quad (1.1)$$

$$= \frac{V_{peak}}{\pi} = 0.3183 V_{peak}$$

and an RMS value equal to $V_{peak}/2$. Since the output has nonzero dc voltage content, the circuit can be used as an ac–dc converter. To make it more useful, a low-pass filter would be added between the output and the load to smooth out the ac portion. This filter needs to be lossless, and will be constructed from only inductors and capacitors.

The circuit in Example 1.3 acts as a half-wave rectifier with a resistive load. With the hypothesized switch action, a diode can substitute for the ideal switch. The example confirms that a simple switching circuit can perform power conversion functions. But, notice that a diode is not, in general, the same as an ideal switch. A diode places restrictions on the current direction, while a true switch would not. An ideal switch allows control over whether it is on or off, while a diode's operation is constrained by circuit variables.

Consider a second half-wave circuit, now with a series L–R load, shown in Fig. 1.6.

EXAMPLE 1.4 A series diode L–R circuit has ac voltage source input. This circuit operates much differently than the half-wave rectifier with resistive load. A diode will be on if forward biased, and off if reverse biased. In this circuit, an off diode will give current of zero. Whenever

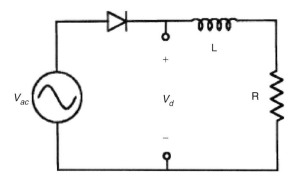

FIGURE 1.6 Half-wave rectifier with L–R load for Example 1.5.

the diode is on, the circuit is the ac source with L–R load. Let the ac voltage be $V_0 \cos(\omega t)$. From Kirchhoff's Voltage Law (KVL),

$$V_0 \cos(\omega t) = L \frac{di}{dt} + Ri$$

Let us assume that the diode is initially off (this assumption is arbitrary, and we will check it as the example is solved). If the diode is off, the diode current $i = 0$, and the voltage across the diode will be v_{ac}. The diode will become forward-biased when v_{ac} becomes positive. The diode will turn on when the input voltage makes a zero-crossing in the positive direction. This allows us to establish initial conditions for the circuit: $i(t_0) = 0$, $t_0 = -\pi/(2\omega)$. The differential equation can be solved in a conventional way to give

$$i(t) = V_0 \left[\frac{\omega L}{R^2 + \omega^2 L^2} \exp\left(\frac{-t}{\tau} - \frac{\pi}{2\omega\tau} \right) \right.$$
$$+ \frac{R}{R^2 + \omega^2 L^2} \cos(\omega t)$$
$$\left. + \frac{\omega L}{R^2 + \omega^2 L^2} \sin(\omega t) \right] \quad (1.2)$$

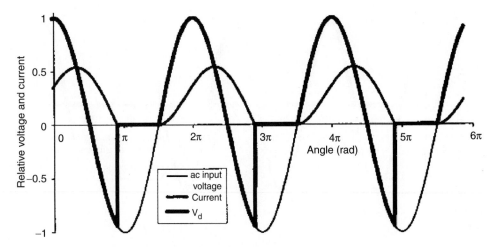

FIGURE 1.7 Input and output waveforms for Example 1.5.

where τ is the time constant L/R. What about diode turn off? One first guess might be that the diode turns off when the voltage becomes negative, but this is not correct. From the solution, the current is not zero when the voltage first becomes negative. If the switch attempts to turn off, it must drop the inductor current to zero instantly. The derivative of current in the inductor, di/dt, would become negative infinite. The inductor voltage $L(di/dt)$ similarly becomes negative infinite – and the devices are destroyed. What really happens is that the falling current allows the inductor to maintain forward bias on the diode. The diode will turn off only when the *current* reaches zero. A diode has definite properties that determine the circuit action, and both the voltage and current are relevant. Figure 1.7 shows the input and output waveforms for a time constant τ equal to about one-third of the ac waveform period.

1.4.2 The Method of Energy Balance

Any circuit must satisfy conservation of energy. In a lossless power electronic circuit, energy is delivered from source to load, possibly through an intermediate storage step. The energy flow must balance over time such that the energy drawn from the source matches that delivered to the load. The converter in Fig. 1.8 serves as an example of how the method of energy balance can be used to analyze circuit operation.

EXAMPLE 1.5 The switches in the circuit of Fig. 1.8 are controlled cyclically to operate in alternation: when the left switch is on, the right one is off, and so on. What does the circuit do if each switch operates half the time? The inductor and capacitor have large values.

When the left switch is on, the source voltage V_{in} appears across the inductor. When the right switch is on,

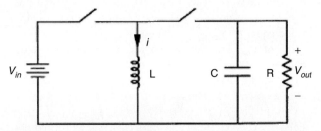

FIGURE 1.8 Energy transfer switching circuit for Example 1.5. (From Reference [2], copyright © 1998, Oxford University Press, Inc.; used by permission.)

the output voltage V_{out} appears across the inductor. If this circuit is to be a useful converter, we want the inductor to receive energy from the source, then deliver it to the load without loss. Over time, this means that energy does not build up in the inductor (instead it flows through on average). The power into the inductor therefore must equal the power out, at least over a cycle. Therefore, the *average* power in should equal the *average* power out of the inductor. Let us denote the inductor current as i. The input is a constant voltage source. Since L is large, this constant voltage source will not be able to change the inductor current quickly, and we can assume that the inductor current is also constant. The average power into L over the cycle period T is

$$P_{in} = \frac{1}{T} \int_0^{T/2} V_{in}\, i\, dt = \frac{V_{in}\, i}{2} \qquad (1.3)$$

For the average power out of L, we must be careful about current directions. The current *out* of the inductor will

1 Introduction

have a value $-i$. The average output power is

$$P_{out} = \frac{1}{T}\int_{T/2}^{T} -iV_{out}\,dt = -\frac{V_{out}\,i}{2} \qquad (1.4)$$

For this circuit to be useful as a converter, there is net energy flow from the source to the load over time. The power conservation relationship $P_{in} = P_{out}$ requires that $V_{out} = -V_{in}$.

The method of energy balance shows that when operated as described in the example, the circuit of Fig. 1.8 serves as a polarity reverser. The output voltage magnitude is the same as that of the input, but the output polarity is negative with respect to the reference node. The circuit is often used to generate a negative supply for analog circuits from a single positive input level. Other output voltage magnitudes can be achieved at the output if the switches alternate at unequal times.

If the inductor in the polarity reversal circuit is moved instead to the input, a step-up function is obtained. Consider the circuit of Fig. 1.9 in the following example.

EXAMPLE 1.6 The switches of Fig. 1.9 are controlled cyclically in alternation. The left switch is on for two-third of each cycle, and the right switch for the remaining one-third of each cycle. Determine the relationship between V_{in} and V_{out}. The inductor's energy should not build up when the circuit is operating normally as a converter. A power balance calculation can be used to relate the input and output voltages. Again, let i be the inductor current. When the left switch is on, power is injected into the inductor. Its average value is

$$P_{in} = \frac{1}{T}\int_{0}^{2T/3} V_{in}\,i\,dt = \frac{2V_{in}\,i}{3} \qquad (1.5)$$

Power leaves the inductor when the right switch is on. Care must be taken with respect to polarities, and the current should be set negative to represent output power.

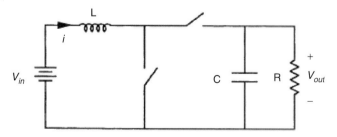

FIGURE 1.9 Switching converter Example 1.6. (From Reference [2], copyright © 1998, Oxford University Press, Inc.; used by permission.)

The result is

$$\begin{aligned}P_{out} &= \frac{1}{T}\int_{2T/3}^{T} -(V_{in}-V_{out})i\,dt \\ &= -\frac{V_{in}\,i}{3} + \frac{V_{out}\,i}{3}\end{aligned} \qquad (1.6)$$

When the input and output power are equated,

$$\frac{2V_{in}\,i}{3} = -\frac{V_{out}\,i}{3} + \frac{V_{out}\,i}{3}, \quad \text{and} \quad 3V_{in} = V_{out} \qquad (1.7)$$

and the output voltage is found to be triple the input. Many seasoned engineers find the dc–dc step-up function of Fig. 1.9 to be surprising. Yet Fig. 1.9 is just one example of such action. Others (including flyback circuits related to Fig. 1.8) are used in systems ranging from CRT electron guns to spark ignitions for automobiles.

The circuits in the preceding examples have few components, provide useful conversion functions, and are efficient. If the switching devices are ideal, each circuit is lossless. Over the history of power electronics, development has tended to flow around the discovery of such circuits: a circuit with a particular conversion function is discovered, analyzed, and applied. As the circuit moves from laboratory testing to a complete commercial product, control, and protection functions are added. The power portion of the circuit remains close to the original idea. The natural question arises as to whether a systematic approach to conversion is possible. Can we start with a desired function and design an appropriate converter, rather than starting from the converter and working backwards toward the application? What underlying principles can be applied to design and analysis? In this introductory chapter, a few of the key concepts are introduced. Keep in mind that while many of the circuits look deceptively simple, all are nonlinear systems with unusual behavior.

1.5 Tools for Analysis and Design

1.5.1 The Switch Matrix

The most readily apparent difference between a power electronic circuit and other types of electronic circuits is the switch action. In contrast to a digital circuit, the switches do not indicate a logic level. Control is effected by determining the times at which switches should operate. Whether there is just one switch or a large group, there is a complexity limit: if a converter has m inputs and n outputs, even the densest possible collection of switches would have a single switch between each input line and each output line. The $m \times n$ switches in the circuit can be arranged according to their connections. The pattern suggests a matrix, as shown in Fig. 1.10.

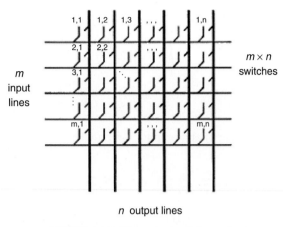

FIGURE 1.10 The general switch matrix.

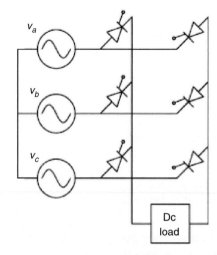

FIGURE 1.12 Three-phase bridge rectifier circuit, a 3 × 2 switch matrix.

Power electronic circuits fall into two broad classes:

1. **Direct switch matrix circuits.** In these circuits, energy storage elements are connected to the matrix only at the input and output terminals. The storage elements effectively become part of the source or the load. A rectifier with an external low-pass filter is an example of a direct switch matrix circuit. In the literature, these circuits are sometimes called *matrix converters*.
2. **Indirect switch matrix circuits**, also termed **embedded converters**. These circuits, like the polarity-reverser example, have energy storage elements connected *within* the matrix structure. There are usually very few storage elements. Indirect switch matrix circuits are most commonly analyzed as a cascade connection of direct switch matrix circuits with the storage in between.

The switch matrices in realistic applications are small. A 2 × 2 switch matrix, for example, covers all possible cases with a single-port input source and a two-terminal load. The matrix is commonly drawn as the *H-bridge* shown in Fig. 1.11. A more complicated example is the three-phase bridge rectifier shown in Fig. 1.12. There are three possible inputs, and the two terminals of the dc circuit provide outputs, which gives

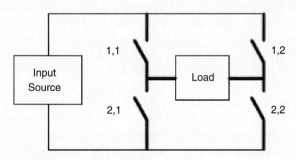

FIGURE 1.11 H-bridge configuration of a 2 × 2 switch matrix.

a 3 × 2 switch matrix. In a personal computer power supply, there are commonly five separate dc loads, and the switch matrix is 2 × 10. Very few practical converters have more than 24 switches, and most designs use fewer than 12.

A switch matrix provides a way to organize devices for a given application. It also helps to focus the effort into three major task areas. Each of these areas must be addressed effectively in order to produce a useful power electronic system.

- The "**Hardware**" Task – Build a switch matrix. This involves the selection of appropriate semiconductor switches and the auxiliary elements that drive and protect them.
- The "**Software**" Task – Operate the matrix to achieve the desired conversion. All operational decisions are implemented by adjusting switch timing.
- The "**Interface**" Task – Add energy storage elements to provide the filters or intermediate storage necessary to meet the application requirements. Unlike most filter applications, lossless filters with simple structures are required.

In a rectifier or other converter, we must choose the electronic parts, how to operate them, and how best to filter the output to satisfy the needs of the load.

1.5.2 Implications of Kirchhoff's Voltage and Current Laws

A major challenge of switch circuits is their capacity to "violate" circuit laws. Consider first the simple circuits of Fig. 1.13. The circuit of Fig. 1.13a is something we might try for ac–dc conversion. This circuit has problems. Kirchhoff's Voltage Law (KVL) tells us that the "sum of voltage drops around a closed loop is zero." However, with the switch closed,

1 Introduction

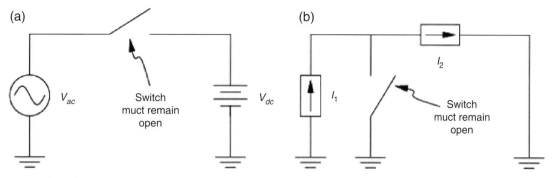

FIGURE 1.13 Hypothetical power converters: (a) possible ac–dc converter and (b) possible dc–dc converter. (From [2], copyright © 1998, Oxford University Press Inc.; used by permission.)

the sum of voltages around the loop is not zero. In reality, this is not a valid result. Instead, a very large current will flow and cause a large $I \cdot R$ drop in the wires. KVL *will* be satisfied by the wire voltage drop, but a fire or, better yet, fuse action, might result. There is, however, nothing that would prevent an operator from trying to close the switch. KVL, then, implies a crucial restriction: a switch matrix must not attempt to interconnect unequal voltage sources directly. Notice that a wire, or dead short, can be thought of as a voltage source with $V = 0$, so KVL is a generalization for avoiding shorts across an individual voltage source.

A similar constraint holds for Kirchhoff's Current Law (KCL). The law states that "currents into a node must sum to zero." When current sources are present in a converter, we must avoid any attempts to violate KCL. In Fig. 1.13b, if the current sources are different and if the switch is opened, the sum of the currents into the node will not be zero. In a real circuit, high voltages will build up and cause an arc to create another current path. This situation has real potential for damage, and a fuse will not help. As a result, KCL implies the restriction that a switch matrix must not attempt to interconnect unequal current sources directly. An open circuit can be thought of as a current source with $I = 0$, so KCL applies to the problem of opening an individual current source.

In contrast to conventional circuits, in which KVL and KCL are automatically satisfied, switches do not "know" KVL or KCL. If a designer forgets to check, and accidentally shorts two voltages or breaks a current source connection, some problem or damage will result. On the other hand, KVL and KCL place necessary constraints on the operating strategy of a switch matrix. In the case of voltage sources, switches must not act to create short-circuit paths among unlike sources. In the case of KCL, switches must act to provide a path for currents. These constraints drastically reduce the number of valid switch operating conditions in a switch matrix, and lead to manageable operating design problems.

When energy storage is included, there are interesting implications of the current law restrictions. Figure 1.14 shows two "circuit law problems." In Fig. 1.14a, the voltage source will cause the inductor current to ramp up indefinitely, since

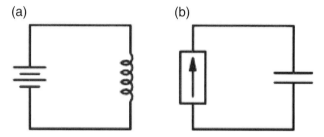

FIGURE 1.14 Short-term KVL and KCL problems in energy storage circuits: (a) an inductor cannot sustain dc voltage indefinitely and (b) a capacitor cannot sustain dc current indefinitely.

$V = L\, di/dt$. We might consider this to be a "KVL problem," since the long-term effect is similar to shorting the source. In Fig. 1.14b, the current source will cause the capacitor voltage to ramp towards infinity. This causes a "KCL problem;" eventually, an arc will be formed to create an additional current path, just as if the current source had been opened. Of course, these connections are not problematic if they are only temporary. However, it should be evident that an inductor will not support dc voltage, and a capacitor will not support dc current. On average over an extended time interval, the voltage across an inductor must be zero, and the current into a capacitor must be zero.

1.5.3 Resolving the Hardware Problem – Semiconductor Devices

A switch is either on or off. An ideal switch, when on, will carry any current in any direction. When off, it will never carry current, no matter what voltage is applied. It is entirely lossless, and changes from its on-state to its off-state instantaneously. A *real switch* can only approximate an ideal switch. Those aspects of real switches that differ from the ideal include the following:

- limits on the amount and direction of on-state current;
- a nonzero on-state voltage drop (such as a diode forward voltage);

- some level of leakage current when the device is supposed to be off;
- limitations on the voltage that can be applied when off; and
- operating speed. The duration of transition between the on- and off-states can be important.

The degree to which the properties of an ideal switch must be met by a real switch depends on the application. For example, a diode can easily be used to conduct dc current; the fact that it conducts only in one direction is often an advantage, not a weakness.

Many different types of semiconductors have been applied in power electronics. In general, these fall into three groups:

- Diodes, which are used in rectifiers, dc–dc converters, and in supporting roles.
- Transistors, which in general are suitable for control of single-polarity circuits. Several types of transistors are applied to power converters. The most recent type, the IGBT, is unique to power electronics and has good characteristics for applications such as inverters.
- Thyristors, which are multi-junction semiconductor devices with latching behavior. Thyristors in general can be switched with short pulses, and then maintain their state until current is removed. They act only as switches. The characteristics are especially well suited to controllable rectifiers, although thyristors have been applied to all power conversion applications.

Some of the features of the most common power semiconductors are listed in Table 1.1. The table shows a wide variety of speeds and rating levels. As a rule, faster speeds apply to lower ratings. For each device type, cost tends to increase both for faster devices and for devices with higher power-handling capacity.

Conducting direction and blocking behavior are fundamentally tied to the device type, and these basic characteristics constrain the choice of device for a given conversion function. Consider again a diode. It carries current in only one direction and always blocks current in the other. Ideally, the diode exhibits no forward voltage drop or off-state leakage current. Although it lacks many features of an ideal switch, the ideal diode is an important switching device. Other real devices operate with polarity limits on current and voltage and have corresponding ideal counterparts. It is convenient to define a special type of switch to represent this behavior: the *restricted switch*.

DEFINITION A *restricted switch* is an ideal switch with the addition of restrictions on the direction of current

TABLE 1.1 Semiconductor devices used in power electronics

Device type	Characteristics of power devices
Diode	Current ratings from under 1 A to more than 5000 A. Voltage ratings from 10 V to 10 kV or more. The fastest power devices switch in less than 20 ns, while the slowest require 100 µs or more. The function of a diode applies in rectifiers and dc–dc circuits.
BJT	(Bipolar junction transistor) Conducts collector current (in one direction) when sufficient base current is applied. Power device current ratings from 0.5 to 500 A or more; voltages from 30 to 1200 V. Switching times from 0.5 to 100 µs. The function applies to dc–dc circuits; combinations with diodes are used in inverters. Power BJTs are being supplanted by FETs and IGBTs.
FET	(Field effect transistor) Conducts drain current when sufficient gate voltage is applied. Power FETs (nearly always enhancement-mode MOSFETs) have a parallel connected reverse diode by virtue of their construction. Ratings from about 0.5 A to about 150 A and 20 V up to 1000 V. Switching times are fast, from 50 ns or less up to 200 ns. The function applies to dc–dc conversion, where the FET is in wide use, and to inverters.
IGBT	(Insulated gate bipolar transistor) A special type of power FET that has the function of a BJT with its base driven by an FET. Faster than a BJT of similar ratings, and easy to use. Ratings from 10 A to more than 600 A, with voltages of 600 to 2500 V. The IGBT is popular in inverters from about 1 to 200 kW or more. It is found almost exclusively in power electronics applications.
SCR	(Silicon controlled rectifier) A thyristor that conducts like a diode after a gate pulse is applied. Turns off only when current becomes zero. Prevents current flow until a pulse appears. Ratings from 10 A to more than 5000 A, and from 200 V up to 6 kV. Switching requires 1 to 200 µs. Widely used for controlled rectifiers. The SCR is found almost exclusively in power electronics applications, and is the most common member of the thyristor family.
GTO	(Gate turn-off thyristor) An SCR that can be turned off by sending a negative pulse to its gate terminal. Can substitute for BJTs in applications where power ratings must be very high. The ratings approach those of SCRs, and the speeds are similar as well. Used in inverters rated above about 100 kW.
TRIAC	A semiconductor constructed to resemble two SCRs connected in reverse parallel. Ratings from 2 to 50 A and 200 to 800 V. Used in lamp dimmers, home appliances, and hand tools. Not as rugged as many other device types, but very convenient for many ac applications.
MCT	(MOSFET controlled thyristor) A special type of SCR that has the function of a GTO with its gate driven from an FET. Much faster than conventional GTOs, and easier to use. These devices and relatives such as the IGCT (integrated gate controlled thyristor) are supplanting GTOs in some application areas.

1 Introduction

TABLE 1.2 The types of restricted switches

Action	Device	Quadrants	Restricted switch symbol	Device symbol
Carries current in one direction, blocks in the other (forward-conducting reverse-blocking)	Diode	(upper-left quadrant)	─▷╂─	─▷╂─
Carries or blocks current in one direction (forward-conducting forward-blocking)	BJT	(upper-right quadrant)	─▷╳─	(BJT symbol)
Carries in one direction or blocks in both directions (forward-conducting bidirectional-blocking)	GTO	(upper half)	─▷╳─	─▷╳─
Carries in both directions, but blocks only in one direction (bidirectional-carrying forward-blocking)	FET	(right half)	─▷╳─	(FET symbol with body diode)
Fully bidirectional	Ideal switch	(all four quadrants)	─▷╳─	─╱─

flow and voltage polarity. The *ideal diode* is one example of a restricted switch.

The diode always permits current flow in one direction, while blocking flow in the other. It therefore represents a *forward-conducting reverse-blocking* (FCRB) restricted switch, and operates in one quadrant on a graph of device current vs. voltage. This FCRB function is automatic – the two diode terminals provide all the necessary information for switch action. Other restricted switches require a third *gate* terminal to determine their state. Consider the polarity possibilities given in Table 1.2. Additional functions such as bidirectional-conducting reverse-blocking can be obtained by reverse connection of one of the five types in the table.

The quadrant operation shown in the table indicates polarities. For example, the current in a diode will be positive when on and the voltage will be negative when off. This means diode operation is restricted to the single quadrant comprising the upper vertical (current) axis and the left horizontal (voltage) axis. The other combinations appear in the table. Symbols for restricted switches can be built up by interpreting the diode's triangle as the current-carrying direction and the bar as the blocking direction. The five types can be drawn as in Table 1.2. These symbols are used infrequently, but are valuable for showing the polarity behavior of switching devices. A circuit drawn with restricted switches represents an idealized power converter.

Restricted switch concepts guide the selection of devices. For example, consider an inverter intended to deliver ac load current from a dc voltage source. A switch matrix built to perform this function must be able to manipulate ac current and dc voltage. Regardless of the physical arrangement of the matrix, we would expect bidirectional-conducting forward-blocking switches to be useful for this conversion. This is a correct result: modern inverters operating from dc voltage sources are built with FETs, or with IGBTs arranged with reverse-parallel diodes. As new power devices are introduced to the market, it is straightforward to determine what types of converters will use them.

1.5.4 Resolving the Software Problem – Switching Functions

The physical $m \times n$ switch matrix can be associated with a mathematical $m \times n$ *switch state matrix*. Each element of this matrix, called a *switching function*, shows whether the corresponding physical device is on or off.

DEFINITION A *switching function*, $q(t)$, has a value of 1 when the corresponding physical switch is on and 0 when it is off. Switching functions are discrete-valued functions of time, and control of switching devices can be represented with them.

Figure 1.15 shows a typical switching function. It is periodic, with period T, representing the most likely repetitive switch action in a power converter. For convenience, it is drawn on a relative time scale that begins at 0 and draws out the square

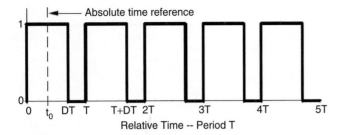

FIGURE 1.15 A generic switching function with period T, duty ratio D, and time reference t_0.

wave period by period. The actual timing is arbitrary, so the center of the first pulse is defined as a specified time t_0 in the figure. In many converters, the switching function is generated as an actual control voltage signal that might drive the gate of either a MOSFET or some other semiconductor switching device.

The timing of switch action is the only alternative for control of a power converter. Since switch action can be represented with a discrete-valued switching function, timing can be represented within the switching function framework. Based on Fig. 1.15, a generic switching function can be characterized completely with three parameters:

1. The *duty ratio*, D, is the fraction of time during which the switch is on. For control purposes, the pulse width can be adjusted to achieve a desired result. We can term this adjustment process *pulse-width modulation* (PWM), perhaps the most important process for implementing control in power converters.
2. The frequency $f_{switch} = 1/T$ (with radian frequency $\omega = 2\pi f_{switch}$) is most often constant, although not in all applications. For control purposes, frequency can be adjusted. This strategy is sometimes used in low-power dc–dc converters to manage wide load ranges. In other converters, frequency control is unusual because the operating frequency is often dictated by the application.
3. The time delay t_0 or phase $\varphi_0 = \omega t_0$. Rectifiers often make use of *phase control* to provide a range of adjustment. A few specialized ac–ac converter applications use phase modulation.

With just three parameters to vary, there are relatively few possible ways to control any power electronic circuit. Dc–dc converters usually rely on duty ratio adjustment (PWM) to alter their behavior. Phase control is common in controlled rectifier applications. Many types of inverters use PWM.

Switching functions are powerful tools for the general representation of converter action [7]. The most widely used control approaches derive from averages of switching functions [2, 8]. Their utility comes from their application in writing circuit equations. For example, in the boost converter of Fig. 1.9, the loop and node equations change depending on which switch is acting at a given moment. The two possible circuit configurations each have distinct equations. Switching functions allow them to be combined. By assigning switching functions $q_1(t)$ and $q_2(t)$ to the left and right switching devices, respectively, we obtain

$$q_1\left(V_{in} - L\frac{di_L}{dt} = 0\right),$$

$$q_1\left(C\frac{dv_C}{dt} + \frac{v_C}{R} = 0\right), \quad \text{left switch on} \quad (1.8)$$

$$q_2\left(V_{in} - L\frac{di_L}{dt} = v_C\right),$$

$$q_2\left(C\frac{dv_C}{dt} + \frac{v_C}{R} = i_L\right), \quad \text{right switch on} \quad (1.9)$$

Because the switches alternate, and the switching functions must be 0 or 1, these sets of equations can be combined to give

$$V_{in} - L\frac{di_L}{dt} = q_2 v_C, \quad C\frac{dv_C}{dt} + \frac{v_C}{R} = q_2 i_L \quad (1.10)$$

The combined expressions are simpler and easier to analyze than the original equations.

For control purposes, the average of equations such as (1.10) often proceeds with the replacement of switching functions q with duty ratios d. The discrete time action of a switching function thus will be represented by an average duty cycle parameter. Switching functions, the advantages gained by averaging, and control approaches such as PWM are discussed at length in several chapters in this handbook.

1.5.5 Resolving the Interface Problem – Lossless Filter Design

Lossless filters for power electronic applications are sometimes called smoothing filters [9]. In applications in which dc outputs are of interest, such filters are commonly implemented as simple low-pass LC structures. The analysis is facilitated because in most cases the residual output waveform, termed ripple, has a known shape. Filter design for rectifiers or dc–dc converters is a question of choosing storage elements large enough to keep ripple low, but not so large that the whole circuit becomes unwieldy or expensive.

Filter design is more challenging when ac outputs are desired. In some cases, this is again an issue of low-pass filter design. In many applications, low-pass filters are not adequate to meet low noise requirements. In these situations, active filters can be used. In power electronics, the term *active filter* refers to lossless switching converters that actively inject or remove energy moment-by-moment to compensate for distortion. The circuits (discussed elsewhere in this handbook) are not related to the linear active filter op-amp circuits used in analog signal processing. In ac cases, there is a continuing opportunity for innovation in filter design.

1.6 Summary

Power electronics is the study of electronic circuits for the control and conversion of electrical energy. The technology is a critical part of our energy infrastructure, and is a key driver for a wide range of uses of electricity. It is becoming increasingly important as an essential tool for efficient, convenient energy conversion, and management. For power electronics design, we consider only those circuits and devices that, in principle, introduce no loss and achieve near-perfect reliability. The two key characteristics of high efficiency and high reliability are implemented with switching circuits, supplemented with energy storage. Switching circuits can be organized as switch matrices. This facilitates their analysis and design.

In a power electronic system, the three primary challenges are the hardware problem of implementing a switch matrix, the software problem of deciding how to operate that matrix, and the interface problem of removing unwanted distortion and providing the user with the desired clean power source. The hardware is implemented with a few special types of power semiconductors. These include several types of transistors, especially MOSFETs and IGBTs, and several types of thyristors, especially SCRs and GTOs. The software problem can be represented in terms of switching functions. The frequency, duty ratio, and phase of switching functions are available for operational purposes. The interface problem is addressed by means of lossless filter circuits. Most often, these are lossless LC passive filters to smooth out ripple or reduce harmonics. Active filter circuits also have been applied to make dynamic corrections in power conversion waveforms.

Improvements in devices and advances in control concepts have led to steady improvements in power electronic circuits and systems. This is driving tremendous expansion of their application. Personal computers, for example, would be unwieldy and inefficient without power electronic dc supplies. Portable communication devices and laptop computers would be impractical. High-performance lighting systems, motor controls, and a wide range of industrial controls depend on power electronics. Strong growth is occurring in automotive applications, in dc power supplies for communication systems, in portable devices, and in high-end converters for advanced microprocessors. In the near future, power electronics will be the enabler for alternative and renewable energy resources. During the next generation, we will reach a time when almost all electrical energy is processed through power electronics somewhere in the path from generation to end use.

References

1. J. Motto, ed., *Introduction to Solid State Power Electronics*. Youngwood, PA: Westinghouse, 1977.
2. P. T. Krein, *Elements of Power Electronics*. New York: Oxford University Press, 1998.
3. T. M. Jahns and E. L. Owen, "Ac adjustable-speed drives at the millenium: how did we get here?" in *Proc. IEEE Applied Power Electronics Conf.*, 2000, pp. 18–26.
4. C. C. Herskind and W. McMurray, "History of the static power converter committee," *IEEE Trans. Industry Applications*, vol. IA-20, no. 4, pp. 1069–1072, July 1984.
5. E. L. Owen, "Origins of the inverter," *IEEE Industry Applications Mag.*, vol. 2, p. 64, January 1996.
6. J. D. Van Wyk and F. C. Lee, "Power electronics technology at the dawn of the new millennium – status and future," in *Rec., IEEE Power Electronics Specialists Conf.*, 1999, pp. 3–12.
7. P. Wood, *Switching Power Converters*. New York: Van Nostrand Reinhold, 1981.
8. R. Erickson, *Fundamentals of Power Electronics*. New York: Chapman and Hall, 1997.
9. P. T. Krein and D. C. Hamill, "Smoothing circuits," in J. Webster (ed.), *Wiley Encyclopedia of Electrical and Electronics Engineering*. New York: John Wiley, 1999.

2
The Power Diode

Ali I. Maswood, Ph.D.
School of EEE
Nanyang Technological University, Nanyang Avenue, Singapore

2.1 Diode as a Switch 15
2.2 Properties of PN Junction 15
2.3 Common Diode Types 17
2.4 Typical Diode Ratings 17
   2.4.1 Voltage Ratings • 2.4.2 Current Ratings
2.5 Snubber Circuits for Diode 19
2.6 Series and Parallel Connection of Power Diodes 19
2.7 Typical Applications of Diodes 22
2.8 Standard Datasheet for Diode Selection 24
    References 25

2.1 Diode as a Switch

Among all the static switching devices used in power electronics (PE), the power diode is perhaps the simplest. Its circuit symbol is shown in Fig. 2.1. It is a two terminal device, and terminal A is known as the anode whereas terminal K is known as the cathode. If terminal A experiences a higher potential compared to terminal K, the device is said to be forward biased and a current called forward current (I_F) will flow through the device in the direction as shown. This causes a small voltage drop across the device (<1 V), which in ideal condition is usually ignored. On the contrary, when a diode is reverse biased, it does not conduct and a practical diode do experience a small current flowing in the reverse direction called the leakage current. Both the forward voltage drop and the leakage current are ignored in an ideal diode. Usually in PE applications a diode is considered to be an ideal static switch.

The characteristics of a practical diode show a departure from the ideals of zero forward and infinite reverse impedance, as shown in Fig. 2.2a. In the forward direction, a potential barrier associated with the distribution of charges in the vicinity of the junction, together with other effects, leads to a voltage drop. This, in the case of silicon, is in the range of 1 V for currents in the normal range. In reverse, within the normal operating range of voltage, a very small current flows which is largely independent of the voltage. For practical purposes, the static characteristics is often represented by Fig. 2.2b.

In the figure, the forward characteristic is expressed as a threshold voltage V_o and a linear incremental or slope resistance, r. The reverse characteristic remains the same over the range of possible leakage currents irrespective of voltage within the normal working range.

2.2 Properties of PN Junction

From the forward and reverse biased condition characteristics, one can notice that when the diode is forward biased, current rises rapidly as the voltage is increased. Current in the reverse biased region is significantly small until the breakdown voltage of the diode is reached. Once the applied voltage is over this limit, the current will increase rapidly to a very high value limited only by an external resistance.

DC diode parameters. The most important parameters are the followings:

- *Forward voltage,* V_F is the voltage drop of a diode across A and K at a defined current level when it is forward biased.
- *Breakdown voltage,* V_B is the voltage drop across the diode at a defined current level when it is beyond reverse biased level. This is popularly known as avalanche.
- *Reverse current* I_R is the current at a particular voltage, which is below the breakdown voltage.

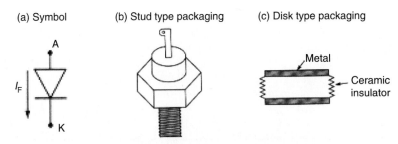

FIGURE 2.1 Power diode: (a) symbol; (b) and (c) types of packaging.

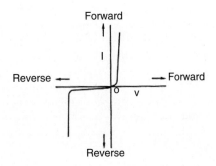

FIGURE 2.2a Typical static characteristic of a power diode (forward and reverse have different scale).

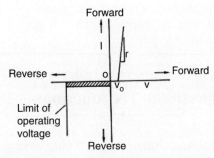

FIGURE 2.2b Practical representation of the static characteristic of a power diode.

AC diode parameters. The commonly used parameters are the followings:

- *Forward recovery time, t_{FR}* is the time required for the diode voltage to drop to a particular value after the forward current starts to flow.
- *Reverse recovery time t_{rr}* is the time interval between the application of reverse voltage and the reverse current dropped to a particular value as shown in Fig. 2.3. Parameter t_a is the interval between the zero crossing of the diode current to when it becomes I_{RR}. On the other hand, t_b is the time interval from the maximum reverse recovery current to approximately 0.25 of I_{RR}. The ratio of the two parameters t_a and t_b is known as the softness factor (SF). Diodes with abrupt recovery characteristics are used for high frequency switching.

In practice, a design engineer frequently needs to calculate the reverse recovery time. This is in order to evaluate the possibility of high frequency switching. As a thumb rule, the lower t_{RR} the faster the diode can be switched.

$$t_{rr} = t_a + t_b \quad (2.1)$$

If t_b is negligible compared to t_a which is a very common case, then the following expression is valid:

$$t_{RR} = \sqrt{\frac{2Q_{RR}}{(di/dt)}}$$

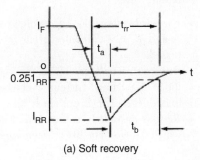
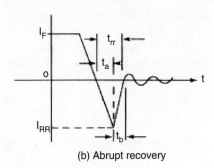

FIGURE 2.3 Diode reverse recovery with various softness factors.

from which the reverse recovery current

$$I_{RR} = \sqrt{\frac{di}{dt} 2 Q_{RR}}$$

where Q_{RR} is the storage charge and can be calculated from the area enclosed by the path of the recovery current.

EXAMPLE 2.1 The manufacturer of a selected diode gives the rate of fall of the diode current $di/dt = 20$ A/μs, and its reverse recovery time $t_{rr} = 5$ μs. What value of peak reverse current do you expect?

SOLUTION. The peak reverse current is given as:

$$I_{RR} = \sqrt{\frac{di}{dt} 2 Q_{RR}}$$

The storage charge Q_{RR} is calculated as:

$$Q_{RR} = \frac{1}{2} \frac{di}{dt} t_{rr}^2 = 1/2 \times 20 \,\text{A/μs} \times (5 \times 10^{-6})^2 = 50 \,\text{μC}$$

Hence,

$$I_{RR} = \sqrt{20 \frac{\text{A}}{\text{μs}} \times 2 \times 50 \,\text{μC}} = 44.72 \,\text{A}$$

- **Diode capacitance**, C_D is the net diode capacitance including the junction (C_J) plus package capacitance (C_P).

In high-frequency pulse switching, a parameter known as transient thermal resistance is of vital importance since it indicates the instantaneous junction temperature as a function of time under constant input power.

2.3 Common Diode Types

Depending on their applications, diodes can be segregated into the following major divisions:

Small signal diode: They are perhaps the most widely used semiconductor devices used in wide variety of applications. In general purpose applications, they are used as a switch in rectifiers, limiters, capacitors, and in wave-shaping. Some common diode parameters a designer needs to know are the forward voltage, reverse breakdown voltage, reverse leakage current, and the recovery time.

Silicon rectifier diode: These are the diodes, which have high forward current carrying capability, typically up to several hundred amperes. They usually have a forward resistance of only a fraction of an ohm while their reverse resistance is in the mega-ohm range. Their primary application is in power conversion, like in power supplies, UPS, rectifiers/inverters, etc.

In case of current exceeding the rated value, their case temperature will rise. For stud-mounted diodes, their thermal resistance is between 0.1 and 1°C/W.

Zener diode: Its primary applications are in the voltage reference or regulation. However, its ability to maintain a certain voltage depends on its temperature coefficient and the impedance. The voltage reference or regulation applications of zener diodes are based on their avalanche properties. In the reverse biased mode, at a certain voltage the resistance of these devices may suddenly drop. This occurs at the zener voltage V_X, a parameter the designer knows beforehand.

Figure 2.4 shows a circuit using a zener diode to control a reference voltage of a linear power supply. Under normal operating condition, the transistor will transmit power to the load (output) circuit. The output power level will depend on the transistor base current. A very high base current will impose a large voltage across the zener and it may attain zener voltage V_X, when it will crush and limit the power supply to the load.

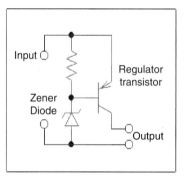

FIGURE 2.4 Voltage regulator with a zener diode for reference.

Photo diode: When a semiconductor junction is exposed to light, photons generate hole–electron pairs. When these charges diffuse across the junction, they produce photocurrent. Hence this device acts as a source of current, which increases with the intensity of light.

Light emitting diode (**LED**): Power diodes used in PE circuits are high power versions of the commonly used devices employed in analog and digital circuits. They are manufactured in wide varieties and ranges. The current rating can be from a few amperes to several hundreds while the voltage rating varies from tens of volts to several thousand volts.

2.4 Typical Diode Ratings

2.4.1 Voltage Ratings

For power diodes, a given datasheet has two voltage ratings. One is the repetitive peak inverse voltage (V_{RRM}), the other is the non-repetitive peak inverse voltage. The non-repetitive voltage (V_{RM}) is the diodes capability to block a reverse voltage that may occur occasionally due to a overvoltage surge.

Repetitive voltage on the other hand is applied on the diode in a sustained manner. To understand this, let us look at the circuit in Fig. 2.5.

EXAMPLE 2.2 Two equal source voltages of 220 V peak and phase shifted from each other by 180° are supplying a common load as shown. (a) Show the load voltage; (b) describe when diode D1 will experience V_{RRM}; and (c) determine the V_{RRM} magnitude considering a safety factor of 1.5.

SOLUTION. (a) The input voltage, load voltage, and the voltage across D1 when it is not conducting (V_{RRM}) are shown in Fig. 2.5b.

(b) Diode D1 will experience V_{RRM} when it is not conducting. This happens when the applied voltage V1 across it is in the negative region (from 70 to 80 ms as shown in the figure) and consequently the diode is reverse biased. The actual ideal voltage across it is the peak value of the two input voltages i.e. $220 \times 2 = 440$ V. This is because when D1 is not conducting, D2 conducts. Hence in addition V_{an}, V_{bn} is also applied across it since D2 is practically shorted.

(c) The $V_{RRM} = 440$ V is the value in ideal situation. In practice, higher voltages may occur due to stray circuit inductances and/or transients due to the reverse recovery of the diode. They are hard to estimate. Hence, a design engineer would always use a safety factor to cater to these overvoltages. Hence, one should use a diode with a $220 \times 2 \times 1.5 = 660$ V rating.

2.4.2 Current Ratings

Power diodes are usually mounted on a heat sink. This effectively dissipates the heat arising due to continuous conduction. Hence, current ratings are estimated based on temperature rise considerations. The datasheet of a diode normally specifies three different current ratings. They are (1) the average current, (2) the rms current, and (3) the peak current. A design engineer must ensure that each of these values is not exceeded. To do that, the actual current (average, rms, and peak) in the circuit must be evaluated either by calculation, simulation, or measurement. These values must be checked against the ones given in the datasheet for that selected diode. The calculated values must be less than or equal to the datasheet values. The following example shows this technique.

EXAMPLE 2.3 The current waveform passing through a diode switch in a switch mode power supply application is shown in Fig. 2.6. Find the average, rms, and the peak current.

SOLUTION. The current pulse duration is shown to be 0.2 ms within a period of 1 ms and with a peak amplitude of 50 A. Hence the required currents are:

$$I_{average} = 50 \times \frac{0.2}{1} = 10 \text{ A}$$

$$I_{rms} = \sqrt{50^2 \times \frac{0.2}{1}} = 22.36 \text{ A}$$

$$I_{peak} = 50 \text{ A}$$

Sometimes, a surge current rating and its permissible duration is also given in a datasheet. For protection of diodes and other semiconductor devices, a fast acting fuse is required. These fuses are selected based on their I^2t rating which is normally specified in a datasheet for a selected diode.

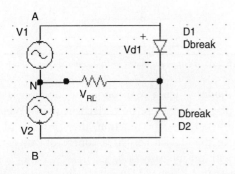

FIGURE 2.5a The circuit.

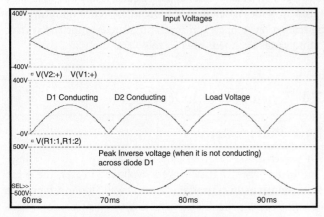

FIGURE 2.5b The waveforms.

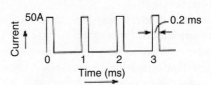

FIGURE 2.6 The current waveform.

2.5 Snubber Circuits for Diode

Snubber circuits are essential for diodes used in switching circuits. It can save a diode from overvoltage spikes, which may arise during the reverse recovery process. A very common snubber circuit for a power diode consists of a capacitor and a resistor connected in parallel with the diode as shown in Fig. 2.7.

When the reverse recovery current decreases, the capacitor by virtue of its property will try to hold the voltage across it, which, approximately, is the voltage across the diode. The resistor on the other hand will help to dissipate some of the energy stored in the inductor, which forms the I_{RR} loop. The dv/dt across a diode can be calculated as:

$$\frac{dv}{dt} = \frac{0.632 \times V_S}{\tau} = \frac{0.632 \times V_S}{R_S \times C_S} \quad (2.2)$$

where V_S is the voltage applied across the diode.

Usually the dv/dt rating of a diode is given in the manufacturers datasheet. Knowing dv/dt and the R_S, one can choose the value of the snubber capacitor C_S. The R_S can be calculated from the diode reverse recovery current:

$$R_S = \frac{V_S}{I_{RR}} \quad (2.3)$$

The designed dv/dt value must always be equal or lower than the dv/dt value found from the datasheet.

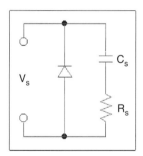

FIGURE 2.7 A typical snubber circuit.

2.6 Series and Parallel Connection of Power Diodes

For specific applications, when the voltage or current rating of a chosen diode is not enough to meet the designed rating, diodes can be connected in series or parallel. Connecting them in series will give the structure a high voltage rating that may be necessary for high-voltage applications. However, one must ensure that the diodes are properly matched especially in terms of their reverse recovery properties. Otherwise, during reverse recovery there may be a large voltage imbalances between the

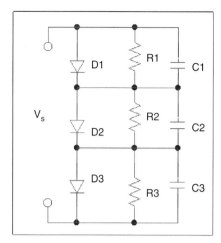

FIGURE 2.8 Series connected diodes with necessary protection.

series connected diodes. Additionally, due to the differences in the reverse recovery times, some diodes may recover from the phenomenon earlier than the other causing them to bear the full reverse voltage. All these problems can effectively be overcome by connecting a bank of a capacitor and a resistor in parallel with each diode as shown in Fig. 2.8.

If a selected diode cannot match the required current rating, one may connect several diodes in parallel. In order to ensure equal current sharing, the designer must choose diodes with the same forward voltage drop properties. It is also important to ensure that the diodes are mounted on similar heat sinks and are cooled (if necessary) equally. This will affect the temperatures of the individual diodes, which in turn may change the forward characteristics of diode.

Tutorial 2.1 Reverse Recovery and Overvoltages

Figure 2.9 shows a simple switch mode power supply. The switch (1-2) is closed at $t = 0$ s. When the switch is open, a freewheeling current $I_F = 20$ A flows through the load (RL), freewheeling diode (DF), and the large load circuit inductance (LL). The diode reverse recovery current is 20 A and it then decays to zero at the rate of 10 A/µs. The load is rated at 10 Ω and the forward on-state voltage drop is neglected.

(a) Draw the current waveform during the reverse recovery (I_{RR}) and find its time (t_{rr}).
(b) Calculate the maximum voltage across the diode during this process (I_{RR}).

Solution. (a) A typical current waveform during reverse recovery process is shown in Fig. 2.10 for an ideal diode.

When the switch is closed, the steady-state current is, $I_{SS} = 200 \text{ V}/10 \text{ Ω} = 20$ A, since under steady-state condition, the inductor is shorted. When the switch is open, the reverse recovery current flows in the right-hand side

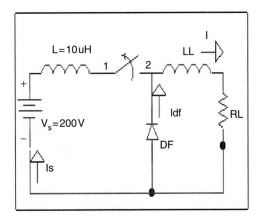

FIGURE 2.9 A simple switch mode power supply with freewheeling diode.

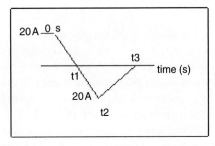

FIGURE 2.10 Current through the freewheeling diode during reverse recovery.

loop consisting of the LL, RL, and DF. The load inductance, LL is assumed to be shorted. Hence, when the switch is closed, the loop equation is:

$$V = L\frac{di_S}{dt}$$

from which

$$\frac{di_S}{dt} = \frac{V}{L} = \frac{200}{10} = 20 \text{ A}/\mu\text{s}$$

At the moment the switch is open, the same current keeps flowing in the right-hand side loop. Hence,

$$\frac{di_d}{dt} = -\frac{di_S}{dt} = -20 \text{ A}/\mu\text{s}$$

from time zero to time t_1 the current will decay at a rate of 20 A/s and will be zero at $t_1 = 20/20 = 1\,\mu\text{s}$. The reverse recovery current starts at this point and, according to the given condition, becomes 20 A at t_2. From this point on, the rate of change remains unchanged at 20 A/μs. Period $t_2 - t_1$ is found as:

$$t_2 - t_1 = \frac{20 \text{ A}}{20 \text{ A}/\mu\text{s}} = 1\,\mu\text{s}$$

From t_2 to t_3, the current decays to zero at the rate of 20 A/μs. The required time:

$$t_3 - t_2 = \frac{20 \text{ A}}{10 \text{ A}/\mu\text{s}} = 2\,\mu\text{s}$$

Hence the actual *reverse recovery time*: $t_{rr} = t_3 - t_1 = (1 + 1 + 2) - 1 = 3\,\mu\text{s}$.

(b) The diode experiences the maximum voltage just when the switch is open. This is because both the source voltage 200 V and the newly formed voltage due to the change in current through the inductor L. The voltage across the diode:

$$V_D = -V + L\frac{di_S}{dt} = -200 + (10 \times 10^{-6})(-20 \times 10^6) = -400\,\text{V}$$

Tutorial 2.2 Ideal Diode Operation, Mathematical Analysis, and PSPICE Simulation

This tutorial illustrates the operation of a diode circuit. Most of the PE applications operate at a relative high voltage, and in such cases, the voltage drop across the power diode usually is small. It is quite often justifiable to use the ideal diode model. An ideal diode has a zero conduction drop when it is forward biased and has zero current when it is reverse biased. The explanation and the analysis presented below is based on the ideal diode model.

Circuit Operation A circuit with a single diode and an RL load is shown in Fig. 2.11. The source V_S is an alternating sinusoidal source. If $V_S = E\sin(\omega t)$, then V_S is positive when $0 < \omega t < \pi$, and V_S is negative when $\pi < \omega t < 2\pi$. When V_S starts becoming positive, the diode starts conducting and the positive source keeps the diode in conduction till ωt reaches π radians. At that instant, defined by $\omega t = \pi$ radians, the current through the circuit is not zero and there is some energy stored in the inductor. The voltage across an inductor is positive when the current through it is increasing and becomes negative when the current through it tends to fall. When the

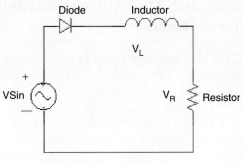

FIGURE 2.11 Circuit diagram.

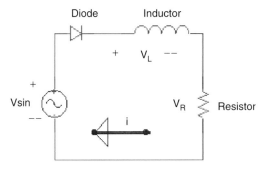

FIGURE 2.12 Current increasing, $0 < \omega t < \pi/2$.

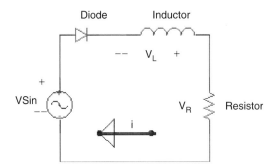

FIGURE 2.13 Current decreasing, $\pi/2 < \omega t < \pi$.

voltage across the inductor is negative, it is in such a direction as to forward bias the diode. The polarity of voltage across the inductor is as shown in Fig. 2.12 or 2.13.

When V_S changes from a positive to a negative value, there is current through the load at the instant $\omega t = \pi$ radians and the diode continues to conduct till the energy stored in the inductor becomes zero. After that the current tends to flow in the reverse direction and the diode blocks conduction. The entire applied voltage now appears across the diode.

Mathematical Analysis An expression for the current through the diode can be obtained as shown in the equations. It is assumed that the current flows for $0 < \omega t < \beta$, where $\beta > \pi$, when the diode conducts, the driving function for the differential equation is the sinusoidal function defining the source voltage. During the period defined by $\beta < \omega t < 2\pi$, the diode blocks current and acts as an open switch. For this period, there is no equation defining the behavior of the circuit. For $0 < \omega t < \beta$, Eq. (2.4) applies.

$$L\frac{di}{dt} + R \times i = E \times \sin(\theta), \text{ where } -0 \leq \theta \leq \beta \quad (2.4)$$

$$L\frac{di}{dt} + R \times i = 0 \quad (2.5)$$

$$\omega L \frac{di}{d\theta} + R \times i = 0 \quad (2.6)$$

$$i(\theta) = A \times e^{-R\theta/\omega L} \quad (2.7)$$

Given a linear differential equation, the solution is found out in two parts. The homogeneous equation is defined by Eq. (2.5). It is preferable to express the equation in terms of the angle θ instead of "t." Since $\theta = \omega t$, we get that $d\theta = \omega \cdot dt$. Then Eq. (2.5) gets converted to Eq. (2.6). Equation (2.7) is the solution to this homogeneous equation and is called the complementary integral.

The value of constant A in the complimentary solution is to be evaluated later. The particular solution is the steady-state response and Eq. (2.8) expresses the particular solution. The steady-state response is the current that would flow in steady state in a circuit that contains only the source, resistor, and inductor shown in the circuit, the only element missing being the diode. This response can be obtained using the differential equation or the Laplace transform or the ac sinusoidal circuit analysis. The total solution is the sum of both the complimentary and the particular solution and it is shown in Eq. (2.9). The value of A is obtained using the initial condition. Since the diode starts conducting at $\omega t = 0$ and the current starts building up from zero, $i(0) = 0$. The value of A is expressed by Eq. (2.10).

Once the value of A is known, the expression for current is known. After evaluating A, current can be evaluated at different values of ωt, starting from $\omega t = \pi$. As ωt increases, the current would keep decreasing. For some values of ωt, say β, the current would be zero. If $\omega t > \beta$, the current would evaluate to a negative value. Since the diode blocks current in the reverse direction, the diode stops conducting when ωt reaches. Then an expression for the average output voltage can be obtained. Since the average voltage across the inductor has to be zero, the average voltage across the resistor and average voltage at the cathode of the diode are the same. This average value can be obtained as shown in Eq. (2.11).

$$i(\theta) = \left(\frac{E}{Z}\right) \sin(\omega t - \alpha) \quad (2.8)$$

where

$$\alpha = a\tan\left(\frac{\omega l}{R}\right) \text{ and } Z^2 = R^2 + \omega l^2$$

$$i(\theta) = A \times e^{(-R\theta/\omega L)} + \frac{E}{Z}\sin(\theta - \alpha) \quad (2.9)$$

$$A = \left(\frac{E}{Z}\right) \sin(\alpha) \quad (2.10)$$

Hence, the average output voltage:

$$V_{OAVG} = \frac{E}{2\pi} \int_0^\beta \sin\theta \cdot d\theta = \frac{E}{2\pi} \times [1 - \cos(\beta)] \quad (2.11)$$

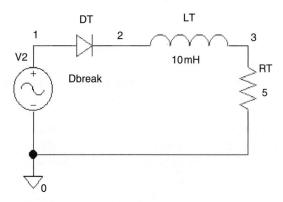

FIGURE 2.14 PSPICE model to study an R–L diode circuit.

```
.MODEL Dbreak D(IS=10N N=1 BV=1200
   IBV=10E-3 VJ=0.6)
.TRAN 10 uS 100 mS 60 mS 100 uS
.PROBE
.OPTIONS (ABSTOL=1N RELTOL=.01 VNTOL=1MV)
.END
```

The diode is described using the MODEL statement. The TRAN statement simulates the transient operation for a period of 100 ms at an interval of 10 ms. The OPTIONS statement sets limits for tolerances. The output can be viewed on the screen because of the PROBE statement. A snapshot of various voltages/currents is shown in Fig. 2.15.

From Fig. 2.15, it is evident that the current lags the source voltage. This is a typical phenomenon in any inductive circuit and is associated with the energy storage property of the inductor. This property of the inductor causes the current to change slowly, governed by the time constant $\tau = \tan^{-1}(\omega l/R)$. Analytically, this is calculated by the expression in Eq. (2.8).

2.7 Typical Applications of Diodes

A. In rectification

Four diodes can be used to fully rectify an ac signal as shown in Fig. 2.16. Apart from other rectifier circuits, this topology does not require an input transformer. However, they are used for isolation and protection. The direction of the current is decided by two diodes conducting at any given time. The direction of the current through the load is always the same. This rectifier topology is known as the full bridge rectifier.

PSPICE Simulation For simulation using PSPICE, the circuit used is shown in Fig. 2.14. Here the nodes are numbered. The ac source is connected between the nodes 1 and 0. The diode is connected between the nodes 1 and 2 and the inductor links the nodes 2 and 3. The resistor is connected from the node 3 to the reference node, that is, node 0. The circuit diagram is shown in Fig. 2.14.

The PSPICE program in textform is presented below.

```
*Half-wave Rectifier with RL Load
*An exercise to find the diode current
VIN 1 0 SIN(0 100 V 50 Hz)
D1 1 2 Dbreak
L1 2 3 10 mH
R1 3 0 5 Ohms
```

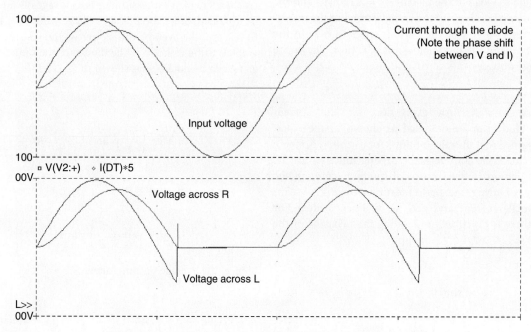

FIGURE 2.15 Voltage/current waveforms at various points in the circuit.

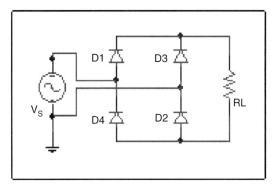
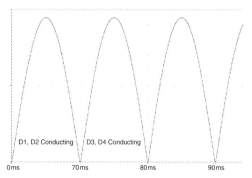

FIGURE 2.16 Full bridge rectifier and its output dc voltage.

The average rectifier output voltage:

$$V_{dc} = \frac{2V_m}{\pi},$$ where V_m is the peak input voltage

The rms rectifier output voltage:

$$V_{rms} = \frac{V_m}{\sqrt{2}}$$

This rectifier is twice as efficient as compared to a single phase one.

B. For voltage clamping

Figure 2.17 shows a voltage clamper. The negative pulse of the sinusoidal input voltage charges the capacitor to its maximum value in the direction shown. After charging, the capacitor cannot discharge, since it is open circuited by the diode. Hence the output voltage:

$$V_o = V_c + V_i = V_m(1 + \sin(\omega t))$$

The output voltage is clamped between zero and $2V_m$.

C. As voltage multiplier

Connecting diode in a predetermined manner, an ac signal can be doubled, tripled, and even quadrupled. This is shown in Fig. 2.18. As evident, the circuit will yield a dc voltage equal to $2V_m$. The capacitors are alternately charged to the maximum value of the input voltage.

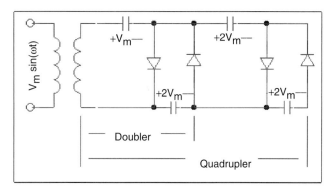

FIGURE 2.18 Voltage doubler and quadrupler circuit.

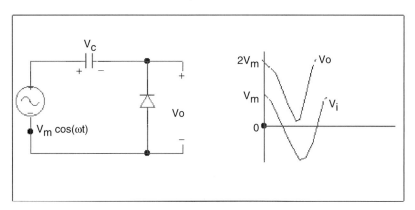

FIGURE 2.17 Voltage clamping with diode.

2.8 Standard Datasheet for Diode Selection

In order for a designer to select a diode switch for specific applications, the following tables and standard test results can be used. A power diode is primarily chosen based on forward current (I_F) and the peak inverse (V_{RRM}) voltage. For example, the designer chooses the diode type V30 from the table in Fig. 2.19 because it closely matches their calculated values of I_F and V_{RRM} without going over. However, if for some reason only the V_{RRM} matches but the calculated value of I_F comes higher, one should go for diode H14, and so on. Similar concept is used for V_{RRM}.

General-Use Rectifier Diodes
Glass Molded Diodes

$I_{F(AV)}$ (A)	V_{RRM}(V) / Type	50	100	200	300	400	500	600	800	1000	1300	1500
0.4	V30	-	-	-	-	-	-	-	yes	yes	yes	yes
1.0	H14	-	yes	yes	yes	yes	yes	yes	yes	yes	-	-
1.1	V06	-	-	yes	-	yes	-	yes	yes	-	-	-
1.3		-	-	yes	-	yes	-	yes	yes	-	-	-
2.5	U05	-	yes	yes	-	yes	-	yes	yes	-	-	-
3.0	U15	-	yes	yes	-	yes	-	yes	yes	-	-	-

FIGURE 2.19 Table of diode selection based on average forward current, $I_{F(AV)}$ and peak inverse voltage, V_{RRM} (courtesy of Hitachi semiconductors).

ABSOLUTE MAXIMUM RATINGS

Item	Type		V30J	V30L	V30M	V30N
Repetitive Peak Reverse Voltage	V_{RRM}	V	800	1000	1300	1500
Non-Repetitive Peak Reverse Voltage	V_{RSM}	V	1000	1300	1600	1800
Average Forward Current	$I_{F(AV)}$	A	0.4 (Single-phase half sine wave 180° conduction, TL = 100°C, Lead length = 10 mm)			
Surge(Non-Repetitive) Forward Current	I_{FSM}	A	30 (Without PIV, 10 ms conduction, Tj = 150°C start)			
I^2t Limit Value	I^2t	A^2s	3.6 (Time = 2 ~ 10 ms, I = RMS value)			
Operating Junction Temperature	T_j	°C	−50 ~ +150			
Storage Temperature	T_{stg}	°C	−50 ~ +150			

Notes (1) Lead Mounting: Lead temperature 300°C max. to 3.2 mm from body for 5 sec. max.
(2) Mechanical strength: Bending 90° × 2 cycles or 180° × 1 cycle, Tensile 2 kg, Twist 90° × l cycle.

CHARACTERISTICS (T_L=25°C)

Item	Symbols	Units	Min.	Typ.	Max.	Test Conditions
Peak Reverse Current	I_{RRM}	μA	–	0.6	10	All class Rated V_{RRM}
Peak Forward Voltage	V_{FM}	V	–	–	1.3	I_{FM}=0.4 Ap, Single-phase half sine wave 1 cycle
Reverse Recovery Time	t_{rr}	μs	–	3.0	–	I_F=2 mA, V_R=−15 V
Steady State Thermal Impedance	$R_{th(j-a)}$	°C/W	–	–	80	Lead length = 10 mm
	$R_{th(j-1)}$		–	–	50	

FIGURE 2.20 Details of diode characteristics for diode *V30* selected from Fig. 2.19.

In addition to the above mentioned diode parameters, one should also calculate parameters like the peak forward voltage, reverse recovery time, case and junction temperatures, etc. and check them against the datasheet values. Some of these datasheet values are provided in Fig. 2.20 for the selected diode V30. Figures 2.21–2.23 give the standard experimental relationships between voltages, currents, power, and case temperatures for our selected V30 diode. These characteristics help a designer to understand the safe operating area for the diode, and to make a decision whether or not to use a snubber or a heat sink. If one is particularly interested in the actual reverse recovery time measurement, the circuit given in Fig. 2.24 can be constructed and experimented upon.

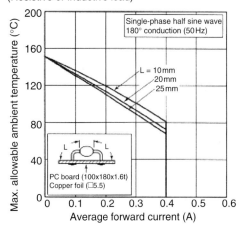

FIGURE 2.23 Maximum allowable case temperature with variation of average forward current.

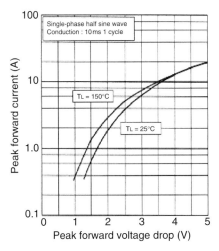

FIGURE 2.21 Variation of peak forward voltage drop with peak forward current.

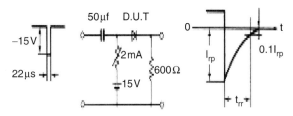

FIGURE 2.24 Reverse recovery time (t_{rr}) measurement.

References

1. N. Lurch, *Fundamentals of Electronics*, 3rd ed., John Wiley & Sons Ltd., New York, 1981.
2. R. Tartar, *Solid-State Power Conversion Handbook*, John Wiley & Sons Ltd., New York, 1993.
3. R.M. Marston, *Power Control Circuits Manual*, Newnes circuits manual series. Butterworth Heinemann Ltd., New York, 1995.
4. Internet information on "Hitachi Semiconductor Devices," http://semiconductor.hitachi.com.
5. International rectifier, *Power Semiconductors Product Digest*, 1992/93.
6. Internet information on, "Electronic Devices & SMPS Books," http://www.smpstech.com/books/booklist.htm.

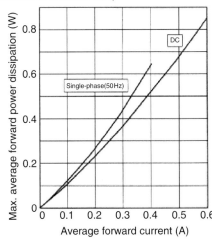

FIGURE 2.22 Variation of maximum forward power dissipation with average forward current.

3
Power Bipolar Transistors

Marcelo Godoy Simoes, Ph.D.
Engineering Division, Colorado School of Mines, Golden, Colorado, USA

3.1 Introduction 27
3.2 Basic Structure and Operation.. 28
3.3 Static Characteristics 29
3.4 Dynamic Switching Characteristics... 32
3.5 Transistor Base Drive Applications... 33
3.6 SPICE Simulation of Bipolar Junction Transistors 36
3.7 BJT Applications.. 37
    Further Reading 39

3.1 Introduction

The first transistor was discovered in 1948 by a team of physicists at the Bell Telephone Laboratories and soon became a semiconductor device of major importance. Before the transistor, amplification was achieved only with vacuum tubes. Even though there are now integrated circuits with millions of transistors, the flow and control of all the electrical energy still require single transistors. Therefore, power semiconductors switches constitute the heart of modern power electronics. Such devices should have larger voltage and current ratings, instant turn-on and turn-off characteristics, very low voltage drop when fully on, zero leakage current in blocking condition, ruggedness to switch highly inductive loads which are measured in terms of safe operating area (SOA) and reverse-biased second breakdown (ES/b), high temperature and radiation withstand capabilities, and high reliability. The right combination of such features restrict the devices suitability to certain applications. Figure 3.1 depicts voltage and current ranges, in terms of frequency, where the most common power semiconductors devices can operate.

The plot gives actually an overall picture where power semiconductors are typically applied in industries: high voltage and current ratings permit applications in large motor drives, induction heating, renewable energy inverters, high voltage DC (HVDC) converters, static VAR compensators, and active filters, while low voltage and high-frequency applications concern switching mode power supplies, resonant converters, and motion control systems, low frequency with high current and voltage devices are restricted to cycloconverter-fed and multimegawatt drives.

Power-npn or -pnp bipolar transistors are used to be the traditional component for driving several of those industrial applications. However, insulated gate bipolar transistor (IGBT) and metal oxide field effect transistor (MOSFET) technology have progressed so that they are now viable replacements for the bipolar types. Bipolar-npn or -pnp transistors still have performance areas in which they may be still used, for example they have lower saturation voltages over the operating temperature range, but they are considerably slower, exhibiting long turn-on and turn-off times. When a bipolar transistor is used in a totem-pole circuit the most difficult design aspects to overcome are the based drive circuitry. Although bipolar transistors have lower input capacitance than that of MOSFETs and IGBTs, they are current driven. Thus, the drive circuitry must generate high and prolonged input currents.

The high input impedance of the IGBT is an advantage over the bipolar counterpart. However, the input capacitance is also high. As a result, the drive circuitry must rapidly charge and discharge the input capacitor of the IGBT during the transition time. The IGBTs low saturation voltage performance is analogous to bipolar power-transistor performance, even over the operating-temperature range. The IGBT requires a −5 to 10 V gate–emitter voltage transition to ensure reliable output switching.

The MOSFET gate and IGBT are similar in many areas of operation. For instance, both devices have high input impedance, are voltage-driven, and use less silicon than the bipolar power transistor to achieve the same drive performance. Additionally, the MOSFET gate has high input capacitance, which places the same requirements on the gate-drive circuitry as the IGBT employed at that stage. The IGBTs

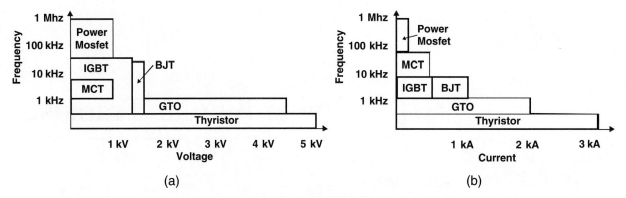

FIGURE 3.1 Power semiconductor operating regions; (a) voltage vs frequency and (b) current vs frequency.

outperform MOSFETs when it comes to conduction loss vs supply-voltage rating. The saturation voltage of MOSFETs is considerably higher and less stable over temperature than that of IGBTs. For such reasons, during the 1980s, the insulated gate bipolar transistor took the place of bipolar junction transistors (BJTs) in several applications. Although the IGBT is a cross between the bipolar and MOSFET transistor, with the output switching and conduction characteristics of a bipolar transistor, but voltage-controlled like a MOSFET, early IGBT versions were prone to latch up, which was largely eliminated. Another characteristic with some IGBT types is the negative temperature coefficient, which can lead to thermal runaway and making the paralleling of devices hard to effectively achieve. Currently, this problem is being addressed in the latest generations of IGBTs.

It is very clear that a categorization based on voltage and switching frequency are two key parameters for determining whether a MOSFET or IGBT is the better device in an application. However, there are still difficulties in selecting a component for use in the crossover region, which includes voltages of 250–1000 V and frequencies of 20–100 kHz. At voltages below 500 V, the BJT has been entirely replaced by MOSFET in power applications and has been also displaced in higher voltages, where new designs use IGBTs. Most of regular industrial needs are in the range of 1–2 kV blocking voltages, 200–500 A conduction currents, and switching speed of 10–100 ns. Although on the last few years, new high voltage projects displaced BJTs towards IGBT, and it is expected to see a decline in the number of new power system designs that incorporate BJTs, there are still some applications for BJTs; in addition the huge built-up history of equipments installed in industries make the BJT yet a lively device.

3.2 Basic Structure and Operation

The bipolar junction transistor (BJT) consists of a three-region structure of n-type and p-type semiconductor materials, it

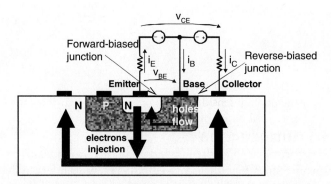

FIGURE 3.2 Structure of a planar bipolar junction transistor.

can be constructed as npn as well as pnp. Figure 3.2 shows the physical structure of a planar npn BJT. The operation is closely related to that of a junction diode where in normal conditions the pn junction between the base and collector is forward-biased ($V_{BE} > 0$) causing electrons to be injected from the emitter into the base. Since the base region is thin, the electrons travel across arriving at the reverse-biased base–collector junction ($V_{BC} < 0$) where there is an electric field (depletion region). Upon arrival at this junction the electrons are pulled across the depletion region and draw into the collector. These electrons flow through the collector region and out the collector contact. Because electrons are negative carriers, their motion constitutes positive current flowing into the external collector terminal. Even though the forward-biased base–emitter junction injects holes from base to emitter they do not contribute to the collector current but result in a net current flow component into the base from the external base terminal. Therefore, the emitter current is composed of those two components: electrons destined to be injected across the base–emitter junction, and holes injected from the base into the emitter. The emitter current is exponentially related to the base–emitter voltage by the equation:

$$i_E = i_{E0}(e^{V_{BE}/\eta V_T} - 1) \tag{3.1}$$

where i_E is the saturation current of the base–emitter junction which is a function of the doping levels, temperature, and the area of the base–emitter junction, V_T is the thermal voltage Kt/q, and η is the emission coefficient. The electron current arriving at the collector junction can be expressed as a fraction α of the total current crossing the base–emitter junction

$$i_C = \alpha i_E \quad (3.2)$$

Since the transistor is a three terminals device, i_E is equal to $i_C + i_B$, hence the base current can be expressed as the remaining fraction

$$i_B = (1 - \alpha) i_E \quad (3.3)$$

The collector and base currents are thus related by the ratio

$$\frac{i_C}{i_B} = \frac{\alpha}{1 - \alpha} = \beta \quad (3.4)$$

The values of α and β for a given transistor depend primarily on the doping densities in the base, collector, and emitter regions, as well as on the device geometry. Recombination and temperature also affect the values for both parameters. A power transistor requires a large blocking voltage in the off state and a high current capability in the on state, and a vertically oriented four layers structures as shown in Fig. 3.3 is preferable because it maximizes the cross-sectional area through which the current flows, enhancing the on-state resistance and power dissipation in the device. There is an intermediate collector region with moderate doping, the emitter region is controlled so as to have an homogenous electrical field.

Optimization of doping and base thickness are required to achieve high breakdown voltage and amplification capabilities.

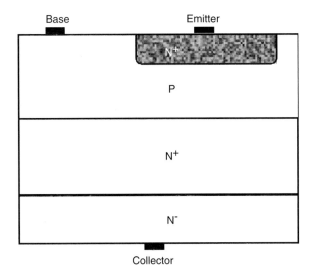

FIGURE 3.3 Power transistor vertical structure.

Power transistors have their emitters and bases interleaved to reduce parasitic ohmic resistance in the base current path and also improving the device for second breakdown failure. The transistor is usually designed to maximize the emitter periphery per unit area of silicon, in order to achieve the highest current gain at a specific current level. In order to ensure those transistors have the greatest possible safety margin, they are designed to be able to dissipate substantial power and, thus, have low thermal resistance. It is for this reason, among others, that the chip area must be large and that the emitter periphery per unit area is sometimes not optimized. Most transistor manufacturers use aluminum metalization, since it has many attractive advantages, among these are ease of application by vapor deposition and ease of definition by photolithography. A major problem with aluminum is that only a thin layer can be applied by normal vapor deposition techniques. Thus, when high currents are applied along the emitter fingers, a voltage drop occurs along them, and the injection efficiency on the portions of the periphery that are furthest from the emitter contact is reduced. This limits the amount of current each finger can conduct. If copper metalization is substituted for aluminum, then it is possible to lower the resistance from the emitter contact to the operating regions of the transistors (the emitter periphery).

From a circuit point of view, the Eqs. (3.1)–(3.4) are used to relate the variables of the BJT input port (formed by base (B) and emitter (E)) to the output port (collector (C) and emitter (E)). The circuit symbols are shown in Fig. 3.4. Most of the power electronics applications use npn transistor because electrons move faster than holes, and therefore, npn transistors have considerable faster commutation times.

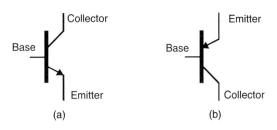

FIGURE 3.4 Circuit symbols: (a) npn transistor and (b) pnp transistor.

3.3 Static Characteristics

Device static ratings determine the maximum allowable limits of current, voltage, and power dissipation. The absolute voltage limit mechanism is concerned to the avalanche such that thermal runaway does not occur. Forward current ratings are specified at which the junction temperature does not exceed a rated value, so leads and contacts are not evaporated. Power dissipated in a semiconductor device produces

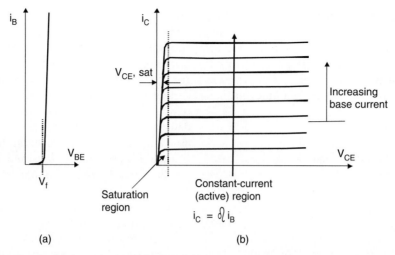

FIGURE 3.5 Family of current–voltage characteristic curves: (a) base–emitter input port and (b) collector–emitter output port.

a temperature rise and are related to the thermal resistance. A family of voltage–current characteristic curves is shown in Fig. 3.5. Figure 3.5a shows the base current i_B plotted as a function of the base–emitter voltage V_{BE} and Fig. 3.5b depicts the collector current i_C as a function of the collector–emitter voltage V_{CE} with i_B as the controlling variable.

Figure 3.5 shows several curves distinguished each other by the value of the base current. The active region is defined where flat, horizontal portions of voltage–current curves show "constant" i_C current, because the collector current does not change significantly with V_{CE} for a given i_B. Those portions are used only for small signal transistor operating as linear amplifiers. Switching power electronics systems on the other hand require transistors to operate in either the saturation region where V_{CE} is small or in the cut off region where the current is zero and the voltage is uphold by the device. A small base current drives the flow of a much larger current between collector and emitter, such gain called beta (Eq. (3.4)) depends upon temperature, V_{CE} and i_C. Figure 3.6 shows current gain increase with increased collector voltage; gain falls off at both high and low current levels.

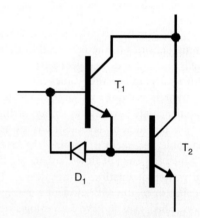

FIGURE 3.7 Darlington connected BJTs.

High voltage BJTs typically have low current gain, and hence Darlington connected devices, as indicated in Fig. 3.7 are commonly used. Considering gains β_1 and β_2 for each one of those transistors, the Darlington connection will have an increased gain of $\beta_1 + \beta_2 + \beta_1\beta_2$, diode D_1 speedsup the turn-off process, by allowing the base driver to remove the stored charge on the transistor bases.

Vertical structure power transistors have an additional region of operation called quasi-saturation, indicated in the characteristics curve of Fig. 3.8. Such feature is a consequence of the lightly doped collector drift region where the collector–base junction supports a low reverse bias. If the transistor enters in the hard-saturation region the on-state power dissipation is minimized, but has to be traded off with the fact that in quasi-saturation the stored charges are smaller. At high collector currents beta gain decreases with increased temperature and with quasi-saturation operation such negative feedback allows careful device paralleling. Two mechanisms on microelectronic level determine the fall off in beta, namely

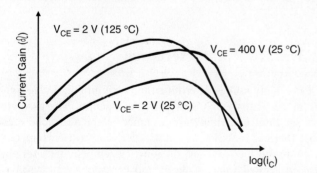

FIGURE 3.6 Current gain depends on temperature, V_{CE} and i_C.

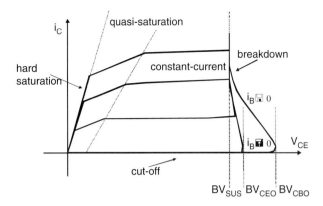

FIGURE 3.8 Voltage–current characteristics for a vertical power transistor.

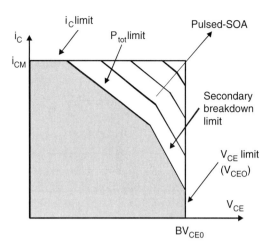

FIGURE 3.9 Forward-bias safe operating area (FBSOA).

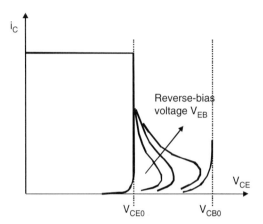

FIGURE 3.10 Reverse-bias safe operating area (RBSOA).

conductivity modulation and emitter crowding. One can note that there is a region called primary breakdown due to conventional avalanche of the C–B junction and the attendant large flow of current. The BV_{SUS} is the limit for primary breakdown, it is the maximum collector–emitter voltage that can be sustained across the transistor when it is carrying high collector current. The BV_{SUS} is lower than BV_{CEO} or BV_{CBO} which measure the transistor's voltage standoff capability when the base current is zero or negative. The bipolar transistor have another potential failure mode called second breakdown, which shows as a precipitous drop in the collector–emitter voltage at large currents. Because the power dissipation is not uniformly spread over the device but it is rather concentrated on regions make the local gradient of temperature can rise very quickly. Such thermal runaway brings hot spots which can eventually melt and recrystallize the silicon resulting in the device destruction. The key to avoid second breakdown is to (1) keep power dissipation under control, (2) use a controlled rate of change of base current during turn-off, (3) use of protective snubbers circuitry, and (4) positioning the switching trajectory within the safe operating area (SOA) boundaries.

In order to describe the maximum values of current and voltage, to which the BJT should be subjected two diagrams, are used: the forward-bias safe operating area (FBSOA) given in Fig. 3.9 and the reverse-bias safe operating area (RBSOA) shown in Fig. 3.10. In the FBSOA the current I_{CM} is the maximum current of the device, there is a boundary defining the maximum thermal dissipation and a margin defining the second breakdown limitation. Those regions are expanded for switching mode operation. Inductive load generates a higher peak energy at turn-off than its resistive counterpart. It is then possible to have a secondary breakdown failure if RBSOA is exceeded. A reverse base current helps the cut off characteristics expanding RBSOA. The RBSOA curve shows that for voltages below V_{CEO} the safe area is independent of reverse bias voltage V_{EB} and is only limited by the device collector current, whereas above V_{CEO} the collector current must be under control depending upon the applied reverse-bias voltage, in addition temperature effects derates the SOA. Ability for the transistor to switch high currents reliably is thus determined by its peak power handling capabilities. This ability is dependent upon the transistor's current and thermal density throughout the active region. In order to optimize the SOA capability, the current density and thermal density must be low. In general, it is the hot spots occurring at the weakest area of the transistor that will cause a device to fail due to second breakdown phenomena. Although a wide base width will limit the current density across the base region, good heat sinking directly under the collector will enable the transistor to withstand high peak power. When the power and heat are spread over a large silicon area, all of these destructive tendencies are held to a minimum, and the transistor will have the highest SOA capability.

When the transistor is on, one can ignore the base current losses and calculate the power dissipation on the on state (conduction losses) by Eq. (3.5). Hard saturation minimizes collector–emitter voltage, decreasing on-state losses.

$$P_{ON} = I_C V_{CE(sat)} \quad (3.5)$$

3.4 Dynamic Switching Characteristics

Switching characteristics are important to define the device velocity in changing from conduction (on) to blocking (off) states. Such transition velocity is of paramount importance also because most of the losses are due to high-frequency switching. Figure 3.11 shows typical waveforms for a resistive load. Index "r" refers to the rising time (from 10 to 90% of maximum value), for example t_{ri} is the current rise time which depends upon the base current. The falling time is indexed by "f"; the parameter t_{fi} is the current falling time, i.e. when the transistor is blocking such time corresponds to crossing from the saturation to the cut off state. In order to improve t_{fi} the base current for blocking must be negative and the device must be kept in quasi-saturation to minimize the stored charges. The delay time is denoted by t_d, corresponding to the time to discharge the capacitance of base–emitter junction, which can be reduced with a larger current base with high slope. Storage time (t_s) is a very important parameter for BJT transistor, it is the required time to neutralize the carriers stored in the collector and base. Storage time and switching losses are key points to deal extensively with bipolar power transistors. Switching losses occur at both turn-on and turn-off and for high frequency operation the rising and falling times for voltage and current transitions play important role as indicated by Fig. 3.12.

A typical inductive load transition is indicated in Fig. 3.13. The figure indicates a turn-off transition. Current and voltage are interchanged at turn-on and an approximation based upon straight line switching intervals (resistive load) gives the

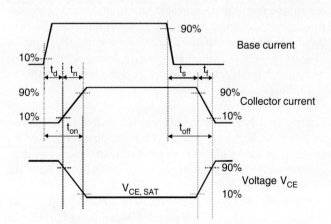

FIGURE 3.11 Resistive load dynamic response.

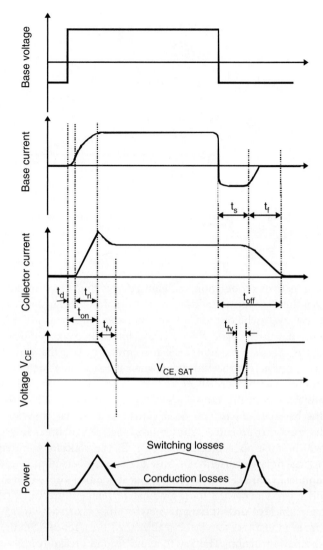

FIGURE 3.12 Inductive load switching characteristics.

switching losses by Eq. (3.6).

$$P_s = \frac{V_S I_M}{2} \tau f_s \quad (3.6)$$

where τ is the period of the switching interval, and V_S and I_M are the maximum voltage and current levels as shown in Fig. 3.10.

Most advantageous operation is achieved when fast transitions are optimized. Such requirement minimizes switching losses. Therefore, a good bipolar drive circuit highly influences the transistor performance. A base drive circuit should provide a high forward base drive current (I_{B1}) as indicated in Fig. 3.14 to ensure the power semiconductor turn-on quickly. Base drive current should keep the BJT fully saturated to minimize forward conduction losses, but a level I_{B2} would maintain the transistor in quasi-saturation avoiding excess of charges

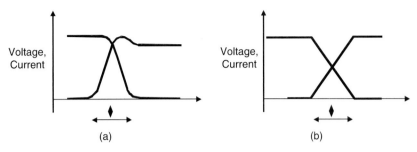

FIGURE 3.13 Turn-off voltage and current switching transition: (a) inductive load and (b) resistive load.

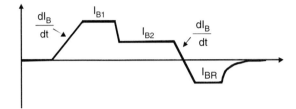

FIGURE 3.14 Recommended base current for BJT driving.

in base. Controllable slope and reverse current I_{BR} sweeps out stored charges in the transistor base, speeding up the device turn-off.

3.5 Transistor Base Drive Applications

A plethora of circuits have been suggested to successfully command transistors for operating in power electronics switching systems. Such base drive circuits try to satisfy the following requirements: supply the right collector current, adapt the base current to the collector current, and extract a reverse current from base to speed up the device blocking. A good base driver reduces the commutation times and total losses, increasing efficiency and operating frequency. Depending upon the grounding requirements between the control and the power circuits, the base drive might be isolated or non-isolated types. Fig. 3.15 shows a non-isolated circuit. When T_1 is switched on T_2 is driven and diode D_1 is forward-biased providing a reverse-bias keeping T_3 off. The base current I_B is positive and saturates the power transistor T_P. When T_1 is switched off, T_3 switches on due to the negative path provided by R_3, and $-V_{CC}$, providing a negative current for switching off the power transistor T_P.

When a negative power supply is not provided for the base drive, a simple circuit like Fig. 3.16 can be used in low power applications (step per motors, small dc–dc converters, relays, pulsed circuits). When the input signal is high, T_1 switches on and a positive current goes to T_P keeping the capacitor charged with the zener voltage, when the input signal goes low

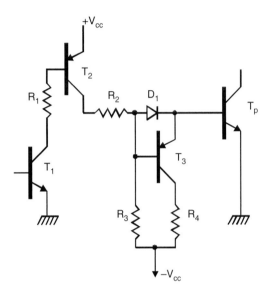

FIGURE 3.15 Non-isolated base driver.

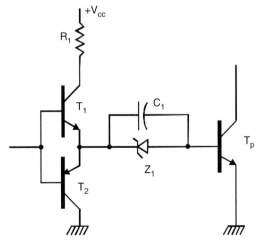

FIGURE 3.16 Base command without negative power supply.

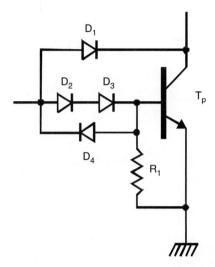

FIGURE 3.17 Antisaturation diodes (Baker's clamp) improve power transistor storage time.

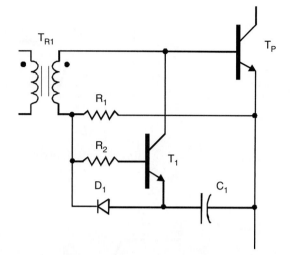

FIGURE 3.18 Isolated base drive circuit.

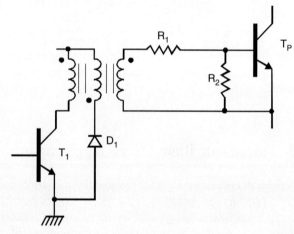

FIGURE 3.19 Transformer coupled base drive with tertiary winding transformer.

T_2 provides a path for the discharge of the capacitor, imposing a pulsed negative current from the base–emitter junction of T_P.

A combination of large reverse base drive and antisaturation techniques may be used to reduce storage time to almost zero. A circuit called Baker's clamp may be employed as illustrated in Fig. 3.17. When the transistor is on, its base is two diode drops below the input. Assuming that diodes D_2 and D_3 have a forward-bias voltage of about 0.7 V, then the base will be 1.4 V below the input terminal. Due to diode D_1 the collector is one diode drop, or 0.7 V below the input. Therefore, the collector will always be more positive than the base by 0.7 V, staying out of saturation, and because collector voltage increases, the gain β also increases a little bit. Diode D_4 provides a negative path for the reverse base current. The input base current can be supplied by a driver circuit similar to the one discussed in Fig. 3.15.

Several situations require ground isolation, off-line operation, floating transistor topology, in addition safety needs may call for an isolated base drive circuit. Numerous circuits have been demonstrated in switching power supplies isolated topologies, usually integrating base drive requirements with their power transformers. Isolated base drive circuits may provide either constant current or proportional current excitation. A very popular base drive circuit for floating switching transistor is shown in Fig. 3.18. When a positive voltage is impressed on the secondary winding (V_S) of T_{R1} a positive current flows into the base of the power transistor T_P which switches on (resistor R_1 limits the base current). The capacitor C_1 is charged by ($V_S-V_{D1}-V_{BE}$) and T_1 is kept blocked because the diode D_1 reverse biases T_1 base–emitter. When V_S is zipped off, the capacitor voltage V_C brings the emitter of T_1 to a negative potential with respect to its base. Therefore, T_1 is excited so as to switch on and start pulling a reverse current from T_P base. Another very effective circuit is shown in Fig. 3.19 with a minimum number of components. The base transformer has a tertiary winding which uses the energy stored in the transformer to generate the reverse base current during the turn-off command. Other configurations are also possible by adding to the isolated circuits the Baker clamp diodes, or zener diodes with paralleled capacitors.

Sophisticated isolated base drive circuits can be used to provide proportional base drive currents where it is possible to control the value of β, keeping it constant for all collector currents leading to shorter storage time. Figure 3.20 shows one of the possible ways to realize a proportional base drive circuit. When transistor T_1 turns on, the transformer T_{R1} is in negative saturation and the power transistor T_P is off. During the time that T_1 is on, a current flows through winding N_1, limited by resistor R_1, storing energy in the transformer, holding

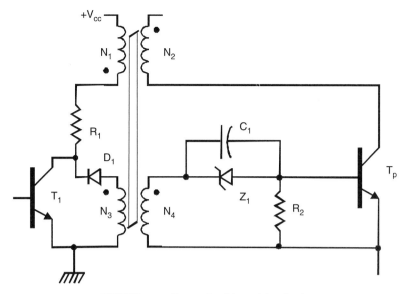

FIGURE 3.20 Proportional base drive circuit.

it into saturation. When the transistor T_1 turns off, the energy stored in N_1 is transferred to winding N_4, pulling the core from negative to positive saturation. The windings N_2 and N_4 will withstand as a current source, the transistor T_P will stay on and the gain β will be imposed by the turns ratio given by Eq. (3.7).

$$\beta = \frac{N_4}{N_2} \qquad (3.7)$$

In order to use the proportional drive given in Fig. 3.20 careful design of the transformer must be done, so as to have flux balanced which will keep core under saturation. The transistor gain must be somewhat higher than the value imposed by the transformer turns ratio, which requires cautious device matching.

The most critical portion of the switching cycle occurs during transistor turn-off, since normally reverse base current is made very large in order to minimize storage time, such conditions may avalanche the base–emitter junction leading to destruction. There are two options to prevent this from happening: turning off the transistor at low values of collector–emitter voltage (which is not practical in most of the applications) or reducing collector current with rising collector voltage, implemented by RC protective networks called snubbers. Therefore, an RC snubber network can be used to divert the collector current during the turn-off improving the RBSOA in addition the snubber circuit dissipates a fair amount of switching power relieving the transistor. Figure 3.21 shows a turn-off snubber network; when the power transistor is off, the capacitor C is charged through diode D_1. Such collector current flows temporarily into the capacitor as the collector-voltage rises; as the power transistor turns on,

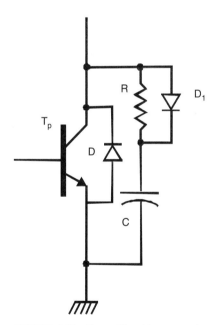

FIGURE 3.21 Turn-off snubber network.

the capacitor discharges through the resistor R back into the transistor.

It is not possible to fully develop all the aspects regarding simulation of BJT circuits. Before giving an example some comments are necessary regarding modeling and simulation of BJT circuits. There are a variety of commercial circuit simulation programs available on the market, extending from a set of functional elements (passive components, voltage controlled current sources, semiconductors) which can be used to model devices, to other programs with the possibility of implementation of algorithm relationships. Those streams are called subcircuit (building auxiliary circuits around a SPICE primitive) and mathematical (deriving models from internal device physics) methods. Simulators can solve circuit equations exactly, given models for the non-linear transistors, and predict the analog behavior of the node voltages and currents in continuous time. They are costly in computer time and such programs have not been written to usually serve the needs of designing power electronic circuits, rather for designing low-power and low-voltage electronic circuits. Therefore, one has to decide which approach should be taken for incorporating BJT power transistor modeling, and a trade-off between accuracy and simplicity must be considered. If precise transistor modeling are required subcircuit oriented programs should be used. On the other hand, when simulation of complex power electronic system structures, or novel power electronic topologies are devised, the switch modeling should be rather simple, by taking in consideration fundamental switching operations, and a mathematical oriented simulation program should be used.

3.6 SPICE Simulation of Bipolar Junction Transistors

SPICE is a general-purpose circuit program that can be applied to simulate electronic and electrical circuits and predict the circuit behavior. SPICE was originally developed at the Electronics Research Laboratory of the University of California, Berkeley (1975), the name stands for: Simulation Program for Integrated Circuits Emphasis. A circuit must be specified in terms of element names, element values, nodes, variable parameters, and sources. SPICE can do several types of circuit analyses:

- Non-linear dc analysis, calculating the dc transference.
- Non-linear transient analysis: calculates signals as a function of time.
- Linear ac analysis: computes a bode plot of output as a function of frequency.
- Noise analysis.
- Sensitivity analysis.
- Distortion analysis.
- Fourier analysis.
- Monte-Carlo analysis.

In addition, PSpice has analog and digital libraries of standard components such as operational amplifiers, digital gates, flip-flops. This makes it a useful tool for a wide range of analog and digital applications. An input file, called source file, consists of three parts: (1) data statements, with description of the components and the interconnections, (2) control statements, which tells SPICE what type of analysis to perform on the circuit, and (3) output statements, with specifications of what outputs are to be printed or plotted. Two other statements are required: the title statement and the end statement. The title statement is the first line and can contain any information, while the end statement is always .END. This statement must be a line be itself, followed by a carriage return. In addition, there are also comment statements, which must begin with an asterisk (*) and are ignored by SPICE. There are several model equations for BJTs.

SPICE has built-in models for the semiconductor devices, and the user need to specify only the pertinent model parameter values. The model for the BJT is based on the integral-charge model of Gummel and Poon. However, if the Gummel–Poon parameters are not specified, the model reduces to piecewise-linear Ebers-Moll model as depicted in Fig. 3.22. In either case, charge-storage effects, ohmic resistances, and a current-dependent output conductance may be included. The forward gain characteristics is defined by the parameters I_S and B_F, the reverse characteristics by I_S and B_R. Three ohmic resistances R_B, R_C, and R_E are also included. The two diodes are modeled by voltage sources, exponential equations of Shockley can be transformed into logarithmic ones. A set of device model parameters is defined on a separate .MODEL card and assigned a unique model name. The device element cards in SPICE then reference the model name. This scheme lessens the need to specify all of the model parameters on each device element card. Parameter values are defined by appending the parameter name, as given below for each model type, followed by an equal sign and the parameter value. Model parameters that are not given a value are assigned the default values given below for each model type. As an example, the

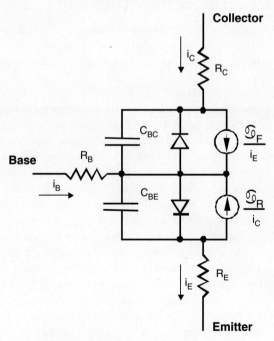

FIGURE 3.22 Ebers–Moll transistor model.

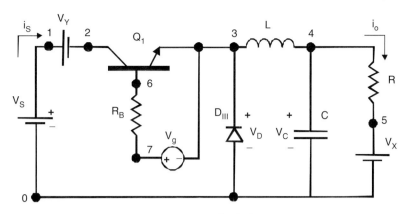

FIGURE 3.23 BJT buck chopper.

model parameters for the 2N2222A NPN transistor is given below:

```
.MODEL Q2N2222A NPN (IS=14.34F XTI=3 EG=1.11
VAF= 74.03 BF=255.9 NE=1.307 ISE=14.34F IKF=.2847
XTB=1.5 BR=6.092 NC=2 ISC=0 IKR=0 RC=1 CJC=7.306P
MJC=.3416 VJC=.75 FC=.5 CJE=22.01P MJE=.377 VJE=.75
TR=46.91N TF=411.1P ITF=.6 VTF=1.7 XTF=3 RB=10)
```

Figure 3.23 shows a BJT buck chopper. The dc input voltage is 12 V, the load resistance R is 5 Ω, the filter inductance L is 145.84 μH, and the filter capacitance C is 200 μF. The chopping frequency is 25 kHz and the duty cycle of the chopper is 42% as indicated by the control voltage statement (V_G). The listing below plots the instantaneous load current (I_O), the input current (I_S), the diode voltage (V_D), the output voltage (V_C), and calculate the Fourier coefficients of the input current (I_S). It is suggested for the careful reader to have more details and enhancements on using SPICE for simulations on specialized literature and references.

```
*SOURCE
VS  1  0    DC   12V
VY  1  2    DC   0V        ;Voltage source to measure
                            input current
VG  7  3    PULSE 0V 30V 0.1NS 0/1Ns 16.7US 40US)
*CIRCUIT
RB  7  6    250            ;Transistor base
                            resistance
R   5  0    5
L   3  4    145.8UH
C   5  0    200UF IC=3V    ;Initial voltage
VX  4  5    DC   0V        ;Source to inductor
                            current
DM  0  3    DMOD           ;Freewheeling diode
.MODEL DMOD D(IS=2.22E-15 BV=1200V CJO=0 TT=0)
Q1  2  6  3  3  2N6546 ;BJT switch
.MODEL 2N6546 NPN (IS=6.83E-14 BF=13 CJE=1PF
                    CJC=607.3PF TF=26.5NS)
*ANALYSIS
.TRAN   2US 2.1MS   2MS UIC ;Transient analysis
.PROBE                      ;Graphics post-processor
.OPTIONS    ABSTOL=1.00N RELTOL=0.01 VNTOL=0.1
                            ITL5=40000
.FOUR   25KHZ I (VY)        ;Fourier analysis
.END
```

3.7 BJT Applications

Bipolar junction power transistors are applied to a variety of power electronic functions, switching mode power supplies, dc motor inverters, PWM inverters just to name a few. To conclude the present chapter, three applications are next illustrated.

A flyback converter is exemplified in Fig. 3.24. The switching transistor is required to withstand the peak collector voltage at turn-off and peak collector currents at turn-on. In order to limit the collector voltage to a safe value, the duty cycle must be kept relatively low, normally below 50%, i.e. 6. < 0.5. In practice, the duty cycle is taken around 0.4, which limits the peak collector. A second design factor which the transistor must meet is the working collector current at turn-on, dependent on the primary transformer-choke peak current, the primary-to-secondary turns ratio, and the output load current. When the transistor turns on the primary current builds up in the primary winding, storing energy, as the transistor turns off, the diode at the secondary winding is forward biasing, releasing such stored energy into the output capacitor and load. Such transformer operating as a coupled inductor is actually defined as a transformer-choke. The design of the transformer-choke of the flyback converter must be done carefully to avoid saturation because the operation is unidirectional on the B–H characteristic curve.

Therefore, a core with a relatively large volume and air gap must be used. An advantage of the flyback circuit is the simplicity by which a multiple output switching power supply may

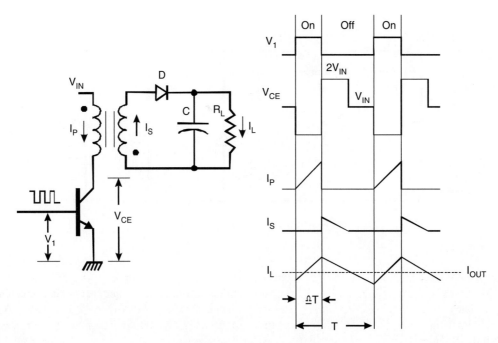

FIGURE 3.24 Flyback converter.

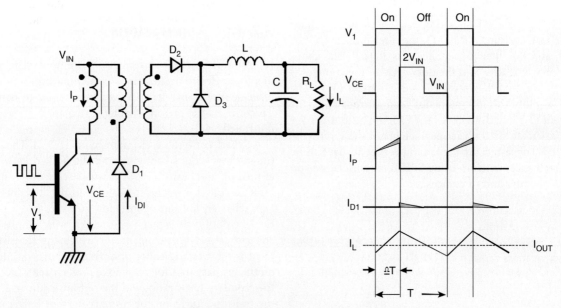

FIGURE 3.25 Isolated forward converter.

be realized. This is because the isolation element acts as a common choke to all outputs, thus only a diode and a capacitor are needed for an extra output voltage.

Figure 3.25 shows the basic forward converter and its associated waveforms. The isolation element in the forward converter is a pure transformer which should not store energy, and therefore, a second inductive element L is required at the output for proper and efficient energy transfer. Notice that the primary and secondary windings of the transformer have the same polarity, i.e. the dots are at the same winding ends. When the transistor turns on, current builds up in the primary winding. Because of the same polarity of the transformer secondary winding, such energy is forward transferred to the output and also stored in inductor L through diode D_2 which is forward-biased. When the transistor turns off, the transformer winding voltage reverses, back-biasing diode D_2

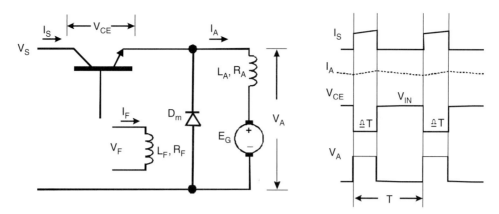

FIGURE 3.26 Chopper-fed dc drive.

and the flywheel diode D_3 is forward-biased, conducting current in the output loop and delivering energy to the load through inductor L. The tertiary winding and diode D, provide transformer demagnetization by returning the transformer magnetic energy into the output dc bus. It should be noted that the duty cycle of the switch must be kept below 50%, so that when the transformer voltage is clamped through the tertiary winding, the integral of the volt-seconds between the input voltage and the clamping level balances to zero. Duty cycles above 50%, i.e. $6 > 0.5$, will upset the volt-seconds balance, driving the transformer into saturation, which in turn produces high collector current spikes that may destroy the switching transistor. Although the clamping action of the tertiary winding and the diode limit the transistor peak collector voltage to twice the dc input, care must be taken during construction to couple the tertiary winding tightly to the primary (bifilar wound) to eliminate voltage spikes caused by leakage inductance.

Chopper drives are connected between a fixed-voltage dc source and a dc motor to vary the armature voltage. In addition to armature voltage control, a dc chopper can provide regenerative braking of the motors and can return energy back to the supply. This energy-saving feature is attractive to transportation systems as mass rapid transit (MRT), chopper drives are also used in battery electric vehicles. A dc motor can be operated in one of the four quadrants by controlling the armature or field voltages (or currents). It is often required to reverse the armature or field terminals in order to operate the motor in the desired quadrant. Figure 3.26 shows a circuit arrangement of a chopper-fed dc separately excited motor. This is a one-quadrant drive, the waveforms for the armature voltage, load current, and input current are also shown. By varying the duty cycle, the power flow to the motor (and speed) can be controlled.

Further Reading

1. B. K. Bose, *Power Electronics and Ac Drives*, Prentice-Hall, Englewood Cliffs, NJ; 1986.
2. G. C. Chryssis, *High Frequency Switching Power Supplies: Theory and Design*, McGraw-Hill, NY; 1984
3. N. Mohan, T. M. Undeland, and W. P. Robbins, *Power Electronics: Converters, Applications, and Design*, John Wiley & Sons, NY; 1995.
4. M. H. Rashid, *Power Electronics: Circuits, Devices, and Applications*, Prentice-Hall, Englewood Cliffs, NJ; 1993.
5. B.W. Williams, *Power Electronics: Devices, Drivers and Applications*, John Wiley & Sons, NY; 1987.

4
The Power MOSFET

Issa Batarseh, Ph.D.
School of Electrical Engineering and Computer Science, University of Central Florida, 4000 Central Florida Blvd., Orlando, Florida, USA

4.1 Introduction 41
4.2 Switching in Power Electronic Circuits 42
4.3 General Switching Characteristics 44
    4.3.1 The Ideal Switch • 4.3.2 The Practical Switch
4.4 The Power MOSFET 48
    4.4.1 MOSFET Structure • 4.4.2 MOSFET Regions of Operation • 4.4.3 MOSFET Switching Characteristics • 4.4.4 MOSFET PSPICE Model • 4.4.5 MOSFET Large-signal Model • 4.4.6 Current MOSFET Performance
4.5 Future Trends in Power Devices 68
    References 69

4.1 Introduction

In this chapter, an overview of power MOSFET (metal oxide semiconductor field effect transistor) semiconductor switching devices will be given. The detailed discussion of the physical structure, fabrication, and physical behavior of the device and packaging is beyond the scope of this chapter. The emphasis here will be given on the terminal i–v switching characteristics of the available device, turn-on, and turn-off switching characteristics, PSPICE modeling and its current, voltage, and switching limits. Even though, most of today's available semiconductor power devices are made of silicon or germanium materials, other materials such as gallium arsenide, diamond, and silicon carbide are currently being tested.

One of the main contributions that led to the growth of the power electronics field has been the unprecedented advancement in the semiconductor technology, especially with respect to switching speed and power handling capabilities. The area of power electronics started by the introduction of the silicon controlled rectifier (SCR) in 1958. Since then, the field has grown in parallel with the growth of the power semiconductor device technology. In fact, the history of power electronics is very much connected to the development of switching devices and it emerged as a separate discipline when high power bipolar junction transistors (BJTs) and MOSFETs devices where introduced in the 1960s and 1970s. Since then, the introduction of new devices has been accompanied with dramatic improvement in power rating and switching performance. Because of their functional importance, drive complexity, fragility, cost, power electronic design engineer must be equipped with the thorough understanding of the device operation, limitation, drawbacks, and related reliability and efficiency issues.

In the 1980s, the development of power semiconductor devices took an important turn when new process technology was developed that allowed the integration of MOS and BJT technologies on the same chip. Thus far, two devices using this new technology have been introduced: integrated gate bipolar transistor (IGBT) and MOS controlled thyristor (MCT). Many of the IC processing methods and equipment have been adopted for the development of power devices. However, unlike microelectronic IC's which process information, power devices IC's process power, hence, their packaging and processing techniques are quite different. Power semiconductor devices represent the "heart" of modern power electronics, with two major desirable characteristics of power semiconductor devices that guided their development are: the *switching speed* and *power handling capabilities.*

The improvement of semiconductor processing technology along with manufacturing and packaging techniques has allowed power semiconductor development for high voltage and high current ratings and fast turn-*on* and turn-*off* characteristics. Today, switching devices are manufactured with amazing power handling capabilities and switching speeds as will be shown later. The availability of different devices with different switching speed, power handling capabilities, size, cost, ... etc. make it possible to cover many power

Copyright © 2007, 2001, Elsevier Inc.
All rights reserved.

electronics applications. As a result, trade-offs are made when it comes to selecting power devices.

4.2 Switching in Power Electronic Circuits

As stated earlier, the heart of any power electronic circuit is the semiconductor-switching network. The question arises here is do we have to use switches to perform electrical power conversion from the source to the load? The answer of course is no, there are many circuits which can perform energy conversion without switches such as linear regulators and power amplifiers. However, the need for using semiconductor devices to perform conversion functions is very much related to the converter efficiency. In power electronic circuits, the semiconductor devices are generally operated as switches, i.e. either in the *on*-state or the *off*-state. This is unlike the case in power amplifiers and linear regulators where semiconductor devices operate in the linear mode. As a result, very large amount of energy is lost within the power circuit before the processed energy reaches the output. The need to use semiconductor switching devices in power electronic circuits is their ability to control and manipulate very large amounts of power from the input to the output with a relatively very low power dissipation in the switching device. Hence, resulting in a very high efficient power electronic system.

Efficiency is considered as an important figure of merit and has significant implications on the overall performance of the system. Low efficient power systems means large amounts of power being dissipated in a form of heat, resulting in one or more of the following implications:

1. Cost of energy increases due to increased consumption.
2. Additional design complications might be imposed, especially regarding the design of device heat sinks.
3. Additional components such as heat sinks increase cost, size, and weight of the system, resulting in low power density.
4. High power dissipation forces the switch to operate at low switching frequency, resulting in limited bandwidth, slow response, and most importantly, the size and weight of magnetic components (inductors and transformers), and capacitors remain large. Therefore, it is always desired to operate switches at very high frequencies. But, we will show later that as the switching frequency increases, the average switching power dissipation increases. Hence, a trade-off must be made between reduced size, weight, and cost of components vs reduced switching power dissipation, which means inexpensive low switching frequency devices.
5. Reduced component and device reliability.

For more than forty years, it has been shown that in order to achieve high efficiency, switching (mechanical or electrical) is the best possible way to accomplish this. However, unlike mechanical switches, electronic switches are far more superior because of their speed and power handling capabilities as well as reliability.

We should note that the advantages of using switches don't come at no cost. Because of the nature of switch currents and voltages (square waveforms), normally high order harmonics are generated in the system. To reduce these harmonics, additional input and output filters are normally added to the system. Moreover, depending on the device type and power electronic circuit topology used, driver circuit control and circuit protection can significantly increase the complexity of the system and its cost.

EXAMPLE 4.1 The purpose of this example is to investigate the efficiency of four different power circuits whose functions are to take in power from 24 volts dc source and deliver a 12 volts dc output to a 6 Ω resistive load. In other words, the tasks of these circuits are to serve as *dc transformer* with a ratio of 2:1. The four circuits are shown in Fig. 4.1a–d representing voltage divider circuit, zener-regulator, transistor linear regulator, and switching circuit, respectively. The objective is to calculate the efficiency of those four power electronic circuits.

SOLUTION.

Voltage Divider DC Regulator The first circuit is the simplest forming a voltage divider with $R = R_L = 6\,\Omega$ and $V_o = 12$ volts. The efficiency defined as the ratio of the average load power, P_L, to the average input power, P_{in}

$$\eta = \frac{P_L}{P_{in}}\%$$
$$= \frac{R_L}{R_L + R}\% = 50\%$$

In fact efficiency is simply $V_o/V_{in}\%$. As the output voltage becomes smaller, the efficiency decreases proportionally.

Zener DC Regulator Since the desired output is 12 V, we select a zener diode with zener breakdown $V_Z = 12$ V. Assume the zener diode has the *i–v* characteristic shown in Fig. 4.1e since $R_L = 6\,\Omega$, the load current, I_L, is 2 A. Then we calculate R for $I_Z = 0.2$ A (10% of the load current). This results in $R = 5.27\,\Omega$. Since the input power is $P_{in} = 2.2\,\text{A} \times 24\,\text{V} = 52.8\,\text{W}$ and the output power is $P_{out} = 24$ W. The efficiency of

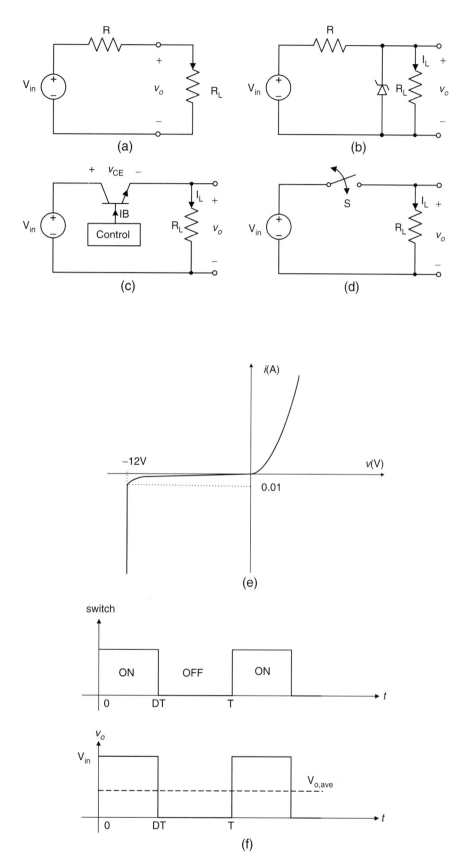

FIGURE 4.1 (a) Voltage divider; (b) zener regulator; (c) transistor regulator; (d) switching circuit; (e) *i–v* zener diode characteristics; and (f) switching waveform for S and corresponding output waveform.

the circuit is given by,

$$\eta = \frac{24\,\text{W}}{52.8\,\text{W}}\%$$
$$= 45.5\%$$

Transistor DC Regulator It is clear from Fig. 4.1c that for $V_o = 12\,\text{V}$, the collector emitter voltage must be around 12 V. Hence, the control circuit must provide base current, I_B to put the transistor in the active mode with $V_{CE} \approx 12\,\text{V}$. Since the load current is 2 A, then collector current is approximately 2 A (assume small I_B). The total power dissipated in the transistor can be approximated by the following equation:

$$P_{diss} = V_{CE}I_C + V_{BE}I_B$$
$$\approx V_{CE}I_C \approx 12 \times 2 = 24\,\text{watts}$$

Therefore, the efficiency of the circuit is 50%.

Switching DC Regulator Let us consider the switching circuit of Fig. 4.1d by assuming the switch is ideal and periodically turned *on* and *off* is shown in Fig. 4.1f. The output voltage waveform is shown in Fig. 4.1f. Even though the output voltage is not constant or pure dc, its average value is given by,

$$V_{o,ave} = \frac{1}{T}\int_0^{T_0} V_{in}\,dt = V_{in}D$$

where D is the duty ratio equals the ratio of the on-time to the switching period. For $V_{o,ave} = 12\,\text{V}$, we set $D = 0.5$, i.e. the switch has a duty cycle of 0.5 or 50%. In case, the average output power is 48 W and the average input power is also 48 W, resulting in 100% efficiency! This is of course because we assumed the switch is ideal. However, let us assume a BJT switch is used in the above circuit with $V_{CE,sat} = 1\,\text{V}$ and I_B is small, then the average power loss across the switch is approximately 2 W, resulting in overall efficiency of 96%. Of course the switching circuit given in this example is over simplified, since the switch requires additional driving circuitry that was not shown, which also dissipates some power. But still, the example illustrates that high efficiency can be acquired by switching power electronic circuits when compared to the efficiency of linear power electronic circuits. Also, the difference between the linear circuit in Figs. 4.1b and c and the switched circuit of Fig. 4.1d is that the power delivered to the load in the later case in pulsating between zero and 96 watts. If the application calls for constant power delivery with little output voltage ripple, then an LC filter must be added to smooth out the output voltage.

The final observation is regarding what is known as load regulation and line regulation. The line regulation is defined as the ratio between the change in output voltage, ΔV_o, with respect to the change in the input voltage ΔV_{in}. This is a very important parameter in power electronics since the dc input voltage is obtained from a rectified line voltage that normally changes by $\pm 20\%$. Therefore, any off-line power electronics circuit must have a limited or specified range of line regulation. If we assume the input voltage in Figs. 4.1a,b is changed by 2 V, i.e. $\Delta V_{in} = 2\,\text{V}$, with R_L unchanged, the corresponding change in the output voltage ΔV_o is 1 V and 0.55 V, respectively. This is considered very poor line regulation. Figures 4.1c,d have much better line and load regulations since the closed-loop control compensate for the line and load variations.

4.3 General Switching Characteristics

4.3.1 The Ideal Switch

It is always desired to have the power switches perform as close as possible to the ideal case. Device characteristically speaking, for a semiconductor device to operate as an ideal switch, it must possess the following features:

1. No limit on the amount of current (known as forward or reverse current) the device can carry when in the conduction state (*on*-state);
2. No limit on the amount of the device-voltage ((known as forward or reverse blocking voltage) when the device is in the non-conduction state (*off*-state);
3. Zero *on*-state voltage drop when in the conduction state;
4. Infinite *off*-state resistance, i.e. zero leakage current when in the non-conduction state; and
5. No limit on the operating speed of the device when changes states, i.e. zero rise and fall times.

The switching waveforms for an ideal switch is shown in Fig. 4.2, where i_{sw} and v_{sw} are the current through and the voltage across the switch, respectively.

Both during the switching and conduction periods, the power loss is zero, resulting in a 100% efficiency, and with no switching delays, an infinite operating frequency can be achieved. In short, an ideal switch has infinite speed, unlimited power handling capabilities, and 100% efficiency. It must be noted that it is not surprising to find semiconductor-switching devices that can almost, for all practical purposes, perform as ideal switches for number of applications.

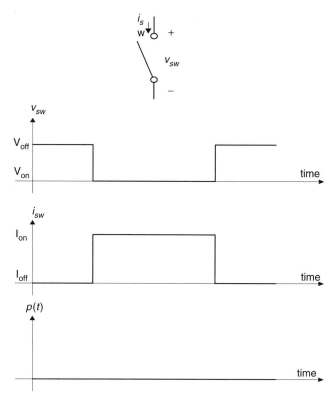

FIGURE 4.2 Ideal switching current, voltage, and power waveforms.

4.3.2 The Practical Switch

The practical switch has the following switching and conduction characteristics:

1. Limited power handling capabilities, i.e. limited conduction current when the switch is in the *on*-state, and limited blocking voltage when the switch is in the *off*-state.
2. Limited switching speed that is caused by the finite turn-*on* and turn-*off* times. This limits the maximum operating frequency of the device.
3. Finite *on*-state and *off*-state resistance's i.e. there exists forward voltage drop when in the *on*-state, and reverse current flow (leakage) when in the off-state.
4. Because of characteristics 2 and 3 above, the practical switch experiences power losses in the on and the off states (known as conduction loss), and during switching transitions (known as switching loss). Typical switching waveforms of a practical switch are shown in Fig. 4.3a.

The average switching power and conduction power losses can be evaluated from these waveforms. We should point out the exact practical switching waveforms vary from one device to another device, but Fig. 4.3a is a reasonably good representation. Moreover, other issues such as temperature dependence, power gain, surge capacity, and over voltage capacity must be considered when addressing specific devices for specific applications. A useful plot that illustrates how switching takes place from *on* to *off* and vice versa is what is called *switching trajectory*, which is simply a plot of i_{sw} vs v_{sw}. Figure 4.3b shows several switching trajectories for the ideal and practical cases under resistive loads.

EXAMPLE 4.2 Consider a linear approximation of Fig. 4.3a as shown in Fig. 4.4a: (a) Give a possible circuit implementation using a power switch whose switching waveforms are shown in Fig. 4.4a, (b) Derive the expressions for the instantaneous switching and conduction power losses and sketch them, (c) Determine the total average power dissipated in the circuit during one switching frequency, and (d) The maximum power.

SOLUTION. (a) First let us assume that the turn-on time, t_{on}, and turn-off time, t_{off}, the conduction voltage V_{ON}, and the leakage current, I_{OFF}, are part of the switching characteristics of the switching device and have nothing to do with circuit topology.

When the switch is *off*, the blocking voltage across the switch is V_{OFF} that can be represented as a dc voltage source of value V_{OFF} reflected somehow across the switch during the *off*-state. When the switch is *on*, the current through the switch equals I_{ON}, hence, a dc current is needed in series with the switch when it is in the *on*-state. This suggests that when the switch turns *off* again, the current in series with the switch must be diverted somewhere else (this process is known as *commutation*). As a result, a second switch is needed to carry the main current from the switch being investigated when it's switched *off*. However, since i_{sw} and v_{sw} are linearly related as shown in Fig. 4.4b, a resistor will do the trick and a second switch is not needed. Figure 4.4 shows a one-switch implementation with S, the switch and R represents the switched-load.

(b) The instantaneous current and voltage waveforms during the transition and conduction times are given as follows

$$i_{sw}(t) = \begin{cases} \dfrac{t}{t_{ON}}(I_{ON} - I_{OFF}) + I_{OFF} & 0 \leq t \leq t_{ON} \\ I_{ON} & t_{ON} \leq t \leq T_s - t_{OFF} \\ -\dfrac{t - T_s}{t_{OFF}}(I_{ON} - I_{OFF}) + I_{OFF} & T_s - t_{OFF} \leq t \leq T_s \end{cases}$$

$$v_{sw}(t) = \begin{cases} -\dfrac{V_{OFF} - V_{ON}}{t_{ON}} \times (t - t_{ON}) + V_{ON} & 0 \leq t \leq t_{ON} \\ V_{ON} & t_{ON} \leq t \leq T_s - t_{OFF} \\ \dfrac{V_{OFF} - V_{ON}}{t_{OFF}} \times (t - (T_s - t_{OFF})) + V_{ON} & T_s - t_{OFF} \leq t \leq T_s \end{cases}$$

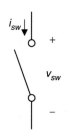
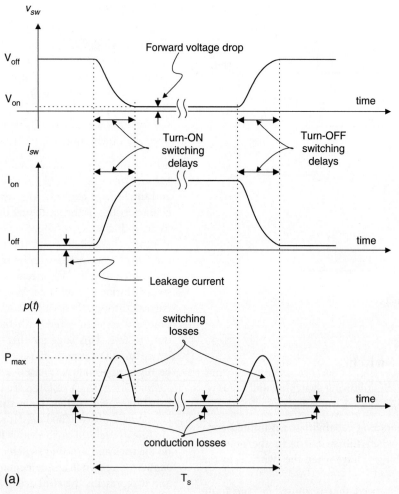
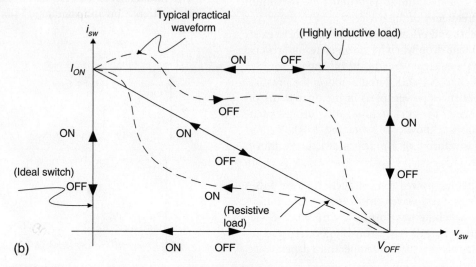

FIGURE 4.3 (a) Practical switching current, voltage, and power waveforms and (b) switching trajectory.

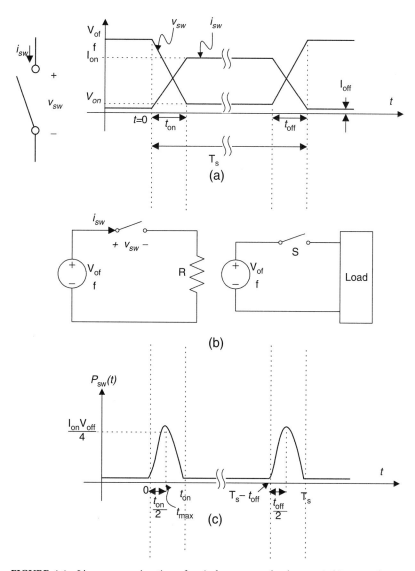

FIGURE 4.4 Linear approximation of typical current and voltage switching waveforms.

It can be shown that if we assume $I_{ON} \gg I_{OFF}$ and $V_{OFF} \gg V_{ON}$, then the instantaneous power, $p(t) = i_{sw} v_{sw}$ can be given as follows,

$$p(t) = \begin{cases} -\dfrac{V_{OFF} I_{ON}}{t_{ON}^2}(t - t_{ON})\, t & 0 \leq t \leq t_{ON} \\ V_{ON} I_{ON} & t_{ON} \leq t \leq T_s - t_{OFF} \\ -\dfrac{V_{OFF} I_{ON}}{t_{OFF}^2}(t - (T_s - t_{OFF}))(t - T_s) & T_s - t_{OFF} \leq t \leq T_s \end{cases}$$

Figure 4.4c shows a plot of the instantaneous power where the maximum power during turn-on and off is $V_{OFF} I_{ON}/4$.

(c) The total average dissipated power is given by

$$P_{ave} = \dfrac{1}{T_s}\int_0^{T_s} p(t)\,dt = \dfrac{1}{T_s}\left[\int_0^{t_{ON}} -\dfrac{V_{OFF} I_{ON}}{t_{ON}^2}(t - t_{ON})\, t \, dt \right.$$

$$+ \int_{t_{ON}}^{T_s - t_{OFF}} V_{ON} I_{ON}\, dt$$

$$\left. + \int_{T_s - t_{OFF}}^{T_s} -\dfrac{V_{OFF} I_{ON}}{t_{OFF}^2}(t - (T_s - t_{OFF}))(t - T_s)\, dt \right]$$

The evaluation of the above integral gives

$$P_{ave} = \frac{V_{OFF} I_{ON}}{T_s}\left(\frac{t_{ON}+t_{OFF}}{6}\right)$$
$$+ \frac{V_{ON} I_{ON}}{T_s}(T_s - t_{OFF} - t_{ON})$$

The first expression represents the total switching loss, whereas the second expression represents the total conduction loss over one switching cycle. We notice that as the frequency increases, the average power increases linearly. Also the power dissipation increases with the increase in the forward conduction current and the reverse blocking voltage.

(d) The maximum power occurs at the time when the first derivative of $p(t)$ during switching is set to zero, i.e.

$$\left.\frac{dp(t)}{dt}\right|_{t=t_{max}} = 0$$

solve the above equation for t_{max}, we obtain values at turn on and off, respectively,

$$t_{max} = \frac{t_{rise}}{2}$$

$$t_{max} = T - \frac{t_{fall}}{2}$$

Solving for the maximum power, we obtain

$$P_{max} = \frac{V_{off} I_{on}}{4}$$

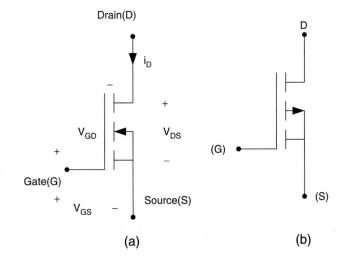

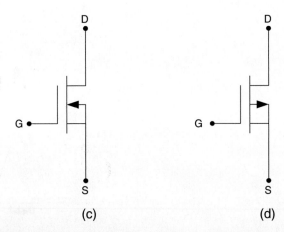

FIGURE 4.5 Device symbols: (a) n-channel enhancement-mode; (b) p-channel enhancement-mode; (c) n-channel depletion-mode; and (d) p-channel depletion-mode.

4.4 The Power MOSFET

Unlike the bipolar junction transistor (BJT), the MOSFET device belongs to the *Unipolar Device family, since it uses only the majority carriers in conduction*. The development of the metal oxide semiconductor technology for microelectronic circuits opened the way for developing the power metal oxide semiconductor field effect transistor (MOSFET) device in 1975. Selecting the most appropriate device for a given application is not an easy task, requiring knowledge about the device characteristics, its unique features, innovation, and engineering design experience. Unlike low power (signal devices), power devices are more complicated in structure, driver design, and understanding of their operational i–v characteristics. This knowledge is very important for power electronics engineer to design circuits that will make these devices close to ideal. The device symbol for a p- and n-channel enhancement and depletion types are shown in Fig. 4.5. Figure 4.6 shows the i–v characteristics for the n-channel enhancement-type MOSFET. It is the fastest power switching device with switching frequency more than 1 MHz, with voltage power ratings up to 1000 V and current rating as high as 300 A.

MOSFET regions of operations will be studied shortly.

4.4.1 MOSFET Structure

Unlike the lateral channel MOSFET devices used in many IC technology in which the gate, source, and drain terminals are located in the same surface of the silicon wafer, power MOSFET use vertical channel structure in order to increase the device power rating [1]. In the vertical channel structure, the source and drain are in opposite side of the silicon wafer. Figure 4.7a shows vertical cross-sectional view for a power MOSFET. Figure 4.7b shows a more simplified representation. There are several discrete types of the vertical structure power MOSFET available commercially today such as V-MOSFET,

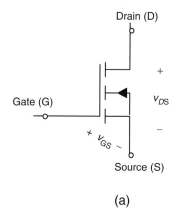

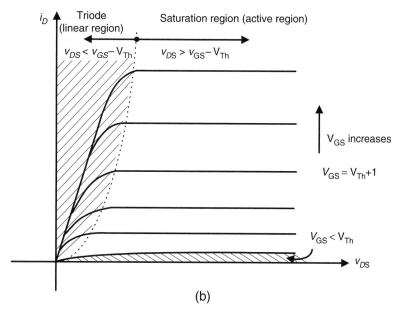

FIGURE 4.6 (a) n-Channel enhancement-mode MOSFET and (b) its i_D vs v_{DS} characteristics.

U-MOSFET, D-MOSFET, and S-MOSFET [1, 2]. The P–N junction between p-base (also referred to as body or bulk region) and the n-drift region provide the forward voltage blocking capabilities. The source metal contact is connected directly to the p-base region through a break in the n$^+$ source region in order to allow for a fixed potential to p-base region during the normal device operation. When the gate and source terminal are set the same potential ($V_{GS} = 0$), no channel is established in the p-base region, i.e. the channel region remain unmodulated. The lower doping in the n-drift region is needed in order to achieve higher drain voltage blocking capabilities. For the drain–source current, I_D, to flow, a conductive path must be established between the n$^+$ and n$^-$ regions through the p-base diffusion region.

A. On-state Resistance When the MOSFET is in the on-state (triode region), the channel of the device behaves like a constant resistance, $R_{DS(on)}$, that is linearly proportional to the change between v_{DS} and i_D as given by the following relation:

$$R_{DS(ON)} = \frac{\partial v_{DS}}{\partial i_D}\bigg|_{V_{GS}=Constant} \quad (4.1)$$

The total conduction (on-state) power loss for a given MOSFET with forward current I_D and on-resistance $R_{DS(on)}$ is given by,

$$P_{on,diss} = I_D^2 R_{DS(on)} \quad (4.2)$$

The value of $R_{DS(on)}$ can be significant and it varies between tens of milliohms and a few ohms for low-voltage and high-voltage MOSFETS, respectively. The *on*-state resistance is an important data sheet parameter, since it determines the forward voltage drop across the device and its total power losses.

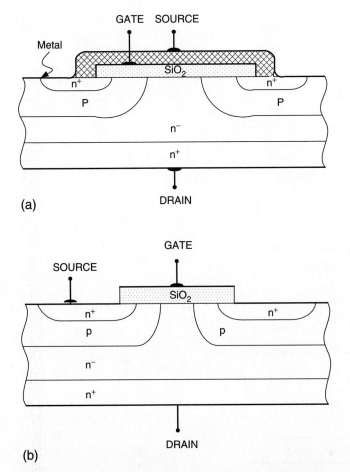

FIGURE 4.7 (a) Vertical cross-sectional view for a power MOSFET and (b) simplified representation.

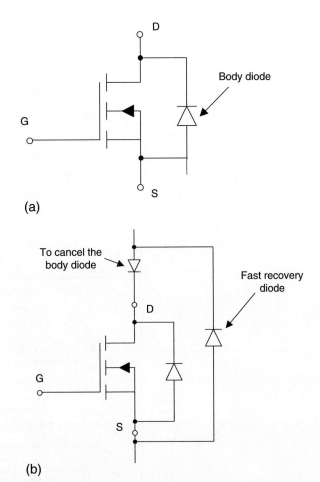

FIGURE 4.8 (a) MOSFET Internal body diode and (b) implementation of a fast body diode.

Unlike the current-controlled bipolar device, which requires base current to allow the current to flow in the collector, the power MOSFET device is a voltage-controlled unipolar device and requires only a small amount of input (gate) current. As a result, it requires less drive power than the BJT. However, it is a non-latching current like the BJT i.e. a gate source voltage must be maintained. Moreover, since only majority carriers contribute to the current flow, MOSFETs surpass all other devices in switching speed with switching speeds exceeding a few megahertz. Comparing the BJT and the MOSFET, the BJT has higher power handling capabilities and smaller switching speed, while the MOSFET device has less power handling capabilities and relatively fast switching speed. The MOSFET device has higher *on*-state resistor than the bipolar transistor. Another difference is that the BJT parameters are more sensitive to junction temperature when compared to the MOSFET, and unlike the BJT, MOSFET devices do not suffer from second breakdown voltages, and sharing current in parallel devices is possible.

B. Internal Body Diode The modern power MOSFET has an internal diode called a body diode connected between the source and the drain as shown in Fig. 4.8a. This diode provides a reverse direction for the drain current, allowing a bi-directional switch implementation. Even though the MOSFET body diode has adequate current and switching speed ratings, in some power electronic applications that require the use of ultra-fast diodes, an external fast recovery diode is added in anti-parallel fashion after blocking the body diode by a slow recovery diode as shown in Fig. 4.8b.

C. Internal Capacitors Another important parameter that effect the MOSFET's switching behavior is the parasitic capacitance between the device's three terminals, namely, gate-to-source, C_{gs}, gate-to-drain, C_{gd}, and drain-to-source (C_{ds}) capacitance as shown in Fig. 4.9a. Figure 4.9b shows the physical representation of these capacitors. The values of these capacitances are non-linear and a function of device structure, geometry, and bias voltages. During turn on, capacitor C_{gd} and C_{gs} must be charged through the gate, hence, the design

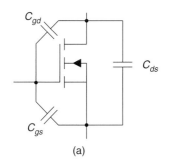

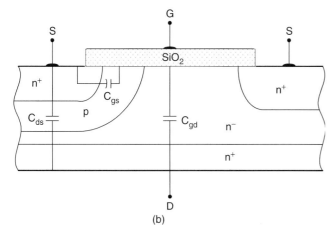

FIGURE 4.9 (a) Equivalent MOSFET representation including junction capacitances and (b) representation of this physical location.

of the gate control circuit must take into consideration the variation in this capacitance (Fig. 4.9b). The largest variation occurs in the gate-to-drain capacitance as the drain-to-gate voltage varies. The MOSFET parasitic capacitance are given in terms of the device data sheet parameters C_{iss}, C_{oss}, and C_{rss} as follows,

$$C_{gd} = C_{rss}$$

$$C_{gs} = C_{iss} - C_{rss}$$

$$C_{ds} = C_{oss} - C_{rss}$$

where C_{rss} = small-signal reverse transfer capacitance.
 C_{iss} = small-signal input capacitance with the drain and source terminals are shorted.
 C_{oss} = small-signal output capacitance with the gate and source terminals are shorted.

The MOSFET capacitances C_{gs}, C_{gd}, and C_{ds} are non-linear and function of the dc bias voltage. The variations in C_{oss} and C_{iss} are significant as the drain-to-source and gate-to-source voltages cross zero, respectively. The objective of the drive circuit is to charge and discharge the gate-to-source and gate-to-drain parasitic capacitance to turn *on* and *off* the device, respectively.

In power electronics, the aim is to use power-switching devices to operate at higher and higher frequencies. Hence, size and weight associated with the output transformer, inductors, and filter capacitors will decrease. As a result, MOSFETs are used extensively in power supply design that requires high switching frequencies including switching and resonant mode power supplies and brushless dc motor drives. Because of its large conduction losses, its power rating is limited to a few kilowatts. Because of its many advantages over the BJT devices, modern MOSFET devices have received high market acceptance.

4.4.2 MOSFET Regions of Operation

Most of the MOSFET devices used in power electronics applications are of the n-channel, enhancement-type like that which is shown in Fig. 4.6a. For the MOSFET to carry drain current, a channel between the drain and the source must be created. This occurs when the gate-to-source voltage exceeds the device threshold voltage, V_{Th}. For $v_{GS} > V_{Th}$, the device can be either in the triode region, which is also called "constant resistance" region, or in the saturation region, depending on the value of v_{DS}. For given v_{GS}, with small $v_{DS}(v_{DS} < v_{GS} - V_{Th})$, the device operates in the triode region(saturation region in the BJT), and for larger $v_{DS}(v_{DS} > v_{GS} - V_{Th})$, the device enters the saturation region (active region in the BJT). For $v_{GS} < V_{Th}$, the device turns off, with drain current almost equals zero. Under both regions of operation, the gate current is almost zero. This is why the MOSFET is known as a voltage-driven device, and therefore, requires simple gate control circuit.

The characteristic curves in Fig. 4.6b show that there are three distinct regions of operation labeled as triode region, saturation region, and cut-off-region. When used as a switching device, only triode and cut-off regions are used, whereas, when it is used as an amplifier, the MOSFET must operate in the saturation region, which corresponds to the active region in the BJT.

The device operates in the cut-off region (off-state) when $v_{GS} < V_{Th}$, resulting in no induced channel. In order to operate the MOSFET in either the triode or saturation region, a channel must first be induced. This can be accomplished by applying gate-to-source voltage that exceeds V_{Th}, i.e.

$$v_{GS} > V_{Th}$$

Once the channel is induced, the MOSFET can either operate in the triode region (when the channel is continuous with no pinch-off, resulting in the drain current proportioned to the channel resistance) or in the saturation region (the channel pinches off, resulting in constant I_D). The gate-to-drain bias voltage (v_{GD}) determines whether the induced channel enter pinch-off or not. This is subject to the following restriction.

For triode mode of operation, we have

$$v_{GD} > V_{Th}$$

$$v_{GD} < V_{Th}$$

And for the saturation region of operation,

Pinch-off occurs when $v_{GD} = V_{Th}$.

In terms of v_{DS}, the above inequalities may be expressed as follows:

1. For triode region of operation

$$v_{DS} < v_{GS} - V_{Th} \quad \text{and} \quad v_{GS} > V_{Th} \quad (4.3)$$

2. For saturation region of operation

$$v_{DS} > v_{GS} - V_{Th} \quad \text{and} \quad v_{GS} > V_{Th} \quad (4.4)$$

3. For cut-off region of operation

$$v_{GS} < V_{Th} \quad (4.5)$$

It can be shown that drain current, i_D, can be mathematically approximated as follows:

$$i_D = K[2(v_{GS} - V_{Th})v_{DS} - v_{DS}^2] \quad \text{Triode Region} \quad (4.6)$$

$$i_D = K(v_{GS} - V_{Th})^2 \quad \text{Saturation Region} \quad (4.7)$$

where, $K = \dfrac{1}{2}\mu_n C_{OX}\left(\dfrac{W}{L}\right)$

μ_n = electron mobility
C_{OX} = oxide capacitance per unit area
L = length of the channel
W = width of the channel.

Typical values for the above parameters are given in the PSPICE model discussed later. At the boundary between the saturation (active) and triode regions, we have,

$$v_{DS} = v_{GS} - V_{Th} \quad (4.8)$$

Resulting in the following equation for i_D,

$$i_D = kv_{DS}^2 \quad (4.9)$$

The input transfer characteristics curve for i_D and v_S. v_{GS} is when the device is operating in the saturation region shown in Fig. 4.10.

The large signal equivalent circuit model for a n-channel enhancement-type MOSFET operating in the saturation mode

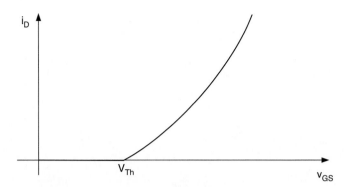

FIGURE 4.10 Input transfer characteristics for a MOSFET device when operating in the saturation region.

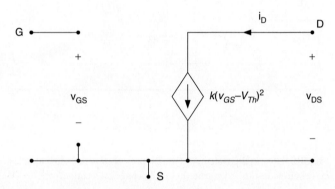

FIGURE 4.11 Large signal equivalent circuit model.

is shown in Fig. 4.11. The drain current is represented by a current source as the function of V_{Th} and v_{GS}.

If we assume that once the channel is pinched-off, the drain–source current will no longer be constant but rather depends on the value of v_{DS} as shown in Fig. 4.12. The increased value of v_{DS} results in reduced channel length, resulting in a phenomenon known as channel-length modulation [3, 4]. If the v_{DS}–i_D lines are extended as shown in Fig. 4.12, they all intercept the v_{DS}-axis at a single point labeled $-1/\lambda$, where λ is a positive constant MOSFET parameter. The term $(1 + \lambda v_{DS})$ is added to the i_D equation in order to account for the increase in i_D due to the channel-length modulation, i.e. i_D is given by,

$$i_D = k(v_{GS} - V_{Th})^2(1 + \lambda v_{DS}) \quad \text{Saturation Region} \quad (4.10)$$

From the definition of the r_o given in Eq. (4.1), it is easy to show the MOSFET output resistance which can be expressed as follows:

$$r_o = \dfrac{1}{\lambda k(v_{GS} - V_{Th})} \quad (4.11)$$

If we assume the MOSFET is operating under small signal condition, i.e. the variation in v_{GS} on i_D vs v_{GS} is in the

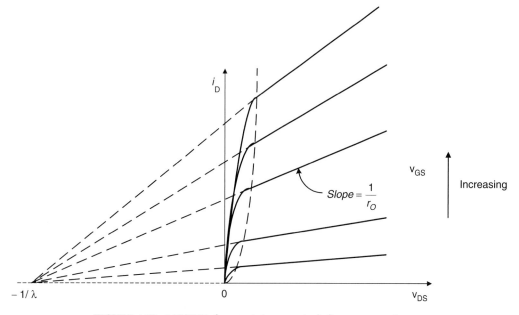

FIGURE 4.12 MOSFET characteristics curve including output resistance.

neighborhood of the dc operating point Q at i_D and v_{GS} as shown in Fig. 4.13. As a result, the i_D current source can be represented by the product of the slope g_m and v_{GS} as shown in Fig. 4.14.

4.4.3 MOSFET Switching Characteristics

Since the MOSFET is a majority carrier transport device, it is inherently capable of a high frequency operation [5–8]. But still the MOSFET has two limitations:

1. High input gate capacitances.
2. Transient/delay due to carrier transport through the drift region.

As stated earlier the input capacitance consists of two components: the gate-to-source and gate-to-drain capacitances. The input capacitances can be expressed in terms of the device

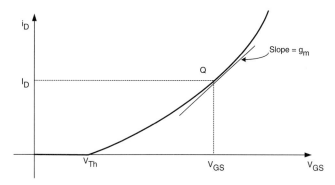

FIGURE 4.13 Linearized i_D vs v_{GS} curve with operating dc point (Q).

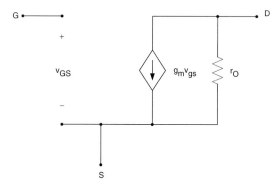

FIGURE 4.14 Small-signal equivalent circuit including MOSFET output resistance.

junction capacitances by applying Miller theorem to Fig. 4.15a. Using Miller theorem, the total input capacitance, C_{in}, seen between the gate-to-source is given by,

$$C_{in} = C_{gs} + (1 + g_m R_L) C_{gd} \qquad (4.12)$$

The frequency response of the MOSFET circuit is limited by the charging and discharging times of C_{in}. Miller effect is inherent in any feedback transistor circuit with resistive load that exhibits a feedback capacitance from the input and output. The objective is to reduce the feedback gate-to-drain resistance. The output capacitance between the drain-to-source, C_{ds}, does not affect the turn-on and turn-off MOSFET switching characteristics. Figure 4.16 shows how C_{gd} and C_{gs} vary under increased drain-source, v_{Ds}, voltage.

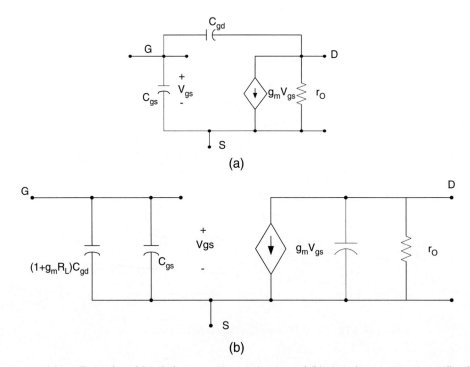

FIGURE 4.15 (a) Small-signal model including parasitic capacitances and (b) equivalent circuit using Miller theorem.

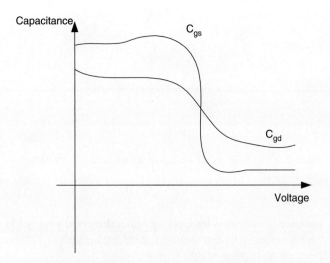

FIGURE 4.16 Variation of C_{gd} and C_{gs} as a function of v_{DS}.

In power electronics applications, the power MOSFETs are operated at high frequencies in order to reduce the size of the magnetic components. In order to reduce the switching losses, the power MOSFETs are maintained in either the on-state (conduction state) or the off-state (forward blocking) state.

It is important we understand the internal device behavior; therefore, the parameters that govern the device transition from the on-state and off-states. To investigate the on and off switching characteristics, we consider the simple power electronic circuit shown in Fig. 4.17a under inductive load.

The fly back diode D is used to pick up the load current when the switch is off. To simplify the analysis we will assume the load inductance is L_0 large enough so that the current through it is constant as shown in Fig. 4.17b.

A. Turn-on Analysis Let us assume initially the device is off, the load current, I_0, flows through D as shown in the Fig. 4.18a $v_{GG} = 0$. The voltage $v_{DS} = V_{DD}$ and $i_G = i_D$. At $t = t_0$, the voltage v_{GG} is applied as shown in Fig. 4.19a. The voltage across C_{GS} starts charging through R_G. The gate–source voltage, v_{GS}, controls the flow of the drain-to-source current i_D. Let us assume that for $t_0 \leq t < t_1$, $v_{GS} < V_{Th}$, i.e. the MOSFET remains in the cut-off region with $i_D = 0$, regardless of v_{DS}. The time interval (t_1, t_0) represents the delay turn-on time needed to change C_{GS} from zero to V_{Th}. The expression for the time interval $\Delta t_{10} = t_1 - t_0$ can be obtained as follows:

The gate current is given by,

$$i_G = \frac{v_{GG} - v_{GS}}{R_G}$$

$$= i_{C_{GS}} + i_{C_{GD}}$$

$$= C_{GS}\frac{dv_{GS}}{dt} - C_{GD}\frac{d(v_G - v_D)}{dt} \quad (4.13)$$

where v_G and v_D are gate-to-ground and drain-to-ground voltages, respectively.

From Eqs. (4.13) and (4.14), we obtain,

$$\frac{V_{GG} - v_{GS}}{R_G} = (C_{GS} + C_{GD})\frac{dv_{GS}}{dt} \quad (4.15)$$

Solving Eq. (4.15) for $v_{GS}(t)$ for $t > t_0$ with $v_{GS}(t_0) = 0$, we obtain,

$$v_{GS}(t) = V_{GG}(1 - e^{(t-t_0)/\tau}) \quad (4.16)$$

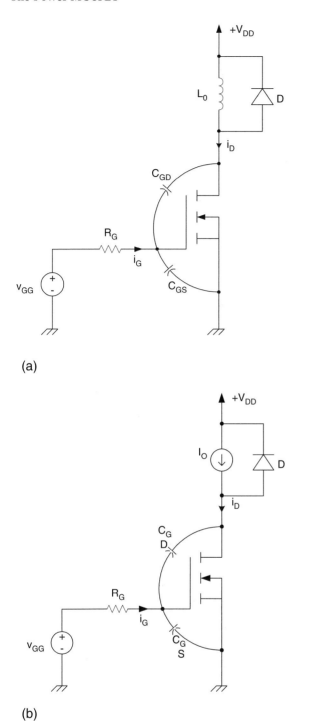

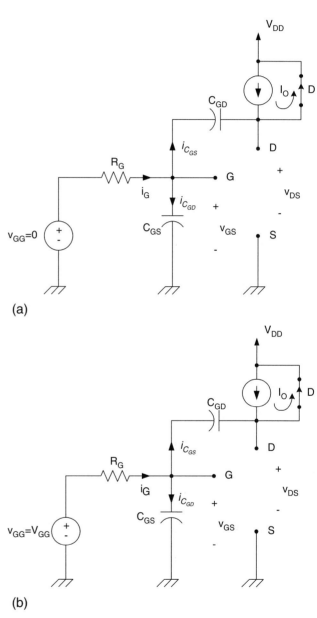

FIGURE 4.17 (a) Simplified equivalent circuit used to study turn-on and turn-off characteristics of the MOSFET and (b) simplified equivalent circuit.

Since we have $v_G = v_{GS}$, $v_D = +V_{DD}$, then i_G is given by

$$i_G = C_{GS}\frac{dv_{GS}}{dt} + C_{GD}\frac{dv_{GS}}{dt} = (C_{GS} + C_{GD})\frac{dv_{GS}}{dt} \quad (4.14)$$

FIGURE 4.18 Equivalent modes: (a) MOSFET is in the off-state for $t < t_0$, $v_{GG} = 0$, $v_{DS} = V_{DD}$, $i_G = 0$, $i_D = 0$; (b) MOSFET in the off-state with $v_{GS} < V_{Th}$ for $t_1 > t > t_0$; (c) $v_{GS} > V_{Th}$, $i_D < I_0$ for $t_1 < t < t_2$; (d) $v_{GS} > V_{Th}$, $i_D = I_0$ for $t_2 \leq t < t_3$; and (e) $V_{GS} > V_{Th}$, $i_D = I_o$ for $t_3 \leq t < t_4$.

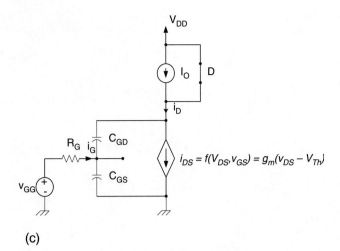

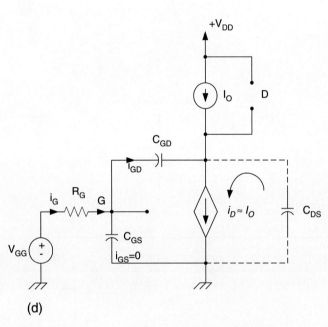

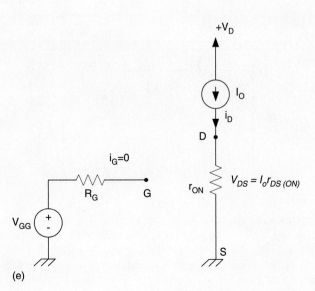

FIGURE 4.18 *continued*

where,

$$\tau = R_G(C_{GS} + C_{GD})$$

The gate current, i_G, is given by,

$$i_G = \frac{v_{GG} - v_{GS}}{R_G}$$

$$i_G = \frac{V_{GG}}{R_G}e^{-(t-t_0)/\tau} \quad (4.17)$$

As long as $v_{GS} < V_{Th}$, i_D remains zero. At $t = t_1$, v_{GS} reaches V_{Th} causing the MOSFET to start conducting. Waveforms for i_G and v_{GS} are shown in Fig. 4.19. The time interval $(t_1 - t_0)$ is given by,

$$\Delta t_{10} = t_1 - t_0 = -\tau \ln\left(1 - \frac{V_{Th}}{V_{GG}}\right)$$

Δt_{10} represents the first delay interval in the turn-on process.

For $t > t_1$ with $v_{GS} > V_{Th}$, the device starts conducting and its drain current is given as a function of v_{GS} and V_{Th}. In fact i_D starts flowing exponentially from zero as shown in Fig. 4.19d. Assume the input transfer characteristics for the MOSFET is limited as shown in Fig. 4.20 with slope of g_m that is given by

$$g_m = \frac{(\partial i_D/\partial v_{GS})}{I_D} = \frac{2\sqrt{I_{DSS}I_D}}{V_{Th}} \quad (4.18)$$

The drain current can be approximately given as follows:

$$i_D(t) = g_m(v_{GS} - V_{Th}) \quad (4.19)$$

As long as $i_D(t) < I_0$, D remains on and $v_{DS} = V_{DD}$ as shown in Fig. 4.18c.

The equation for $v_{GS}(t)$ remains the same as in Eq. (4.16), hence, Eq. (4.19) results in $i_D(t)$ given by,

$$i_D(t) = g_m(V_{GG} - V_{Th}) - g_m V_{GG} e^{-(t-t_1)/\tau} \quad (4.20)$$

The gate current continues to decrease exponentially as shown in Fig. 4.19c. At $t = t_2$, i_D reaches its maximum value of I_0, turning D off. The time interval $\Delta t_{21} = (t_2 - t_1)$ is obtained from Eq. (4.20) by setting $i_D(t_2) = I_0$.

$$\Delta t_{21} = \tau \ln\frac{g_m V_{GG}}{g_m(V_{GG} - V_{Th}) - I_0} \quad (4.21)$$

For $t > t_2$, the diode turns off and $i_D \approx I_0$ as shown in Fig. 4.18d. Since the drain current is nearly a constant, then

4 The Power MOSFET

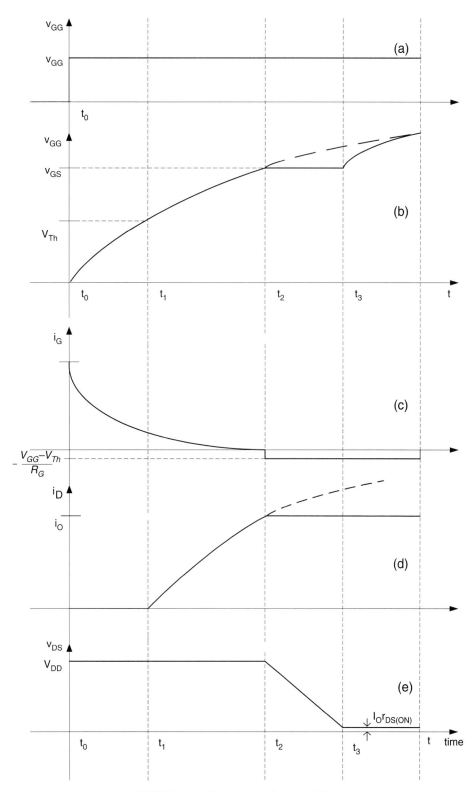

FIGURE 4.19 Turn-on waveform switiching.

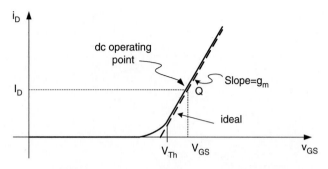

FIGURE 4.20 Input transfer characteristics.

the gate–source voltage is also constant according to the input transfer characteristic of the MOSFET, i.e.

$$i_D = g_m(v_{GS} - V_{Th}) \approx I_0 \quad (4.22)$$

Hence,

$$v_{GS}(t) = \frac{I_0}{g_m} + V_{Th} \quad (4.23)$$

At $t = t_2$, $i_G(t)$ is given by,

$$i_G(t_2) = \frac{V_{GG} - v_{GS}(t_2)}{V_{Th}} = \frac{V_{GG} - (I_0/g_m) - V_{Th}}{V_{Th}} \quad (4.24)$$

Since the time constant τ is very small, it is safe to assume $v_{GS}(t_2)$ reaches its maximum, i.e.

$$v_{GS}(t_2) \approx V_{GG}$$

and

$$i_G(t_2) \approx 0$$

For $t_2 \leq t < t_3$, the diode turns off the load current I_0 (drain current i_D), which starts discharging the drain-to-source capacitance.

Since v_{GS} is constant, the entire gate current flows through C_{GD}, resulting in the following relation,

$$i_G(t) = i_{C_{GD}}$$
$$= C_{GD}\frac{d(v_G - v_D)}{dt}$$

Since v_G is constant and $v_s = 0$, we have

$$i_G(t) = -C_{GD}\frac{dv_{DS}}{dt}$$
$$= -\frac{V_{GG} - V_{Th}}{R_G}$$

Solving for $v_{DS}(t)$ for $t > t_2$, with $v_{DS}(t_2) = V_{DD}$, we obtain

$$v_{DS}(t) = -\frac{V_{GG} - V_{Th}}{R_G C_{GD}}(t - t_2) + V_{DD} \quad \text{For } t > t_2 \quad (4.25)$$

This is a linear discharge of C_{GD} as shown in Fig. 4.19e

The time interval $\Delta t_{32} = (t_3 - t_2)$ is determined by assuming that at $t = t_3$, the drain-to-source voltage reaches its minimum value determined by its on resistance, $v_{DS(ON)}$ i.e. $v_{DS(ON)}$ is given by,

$$v_{DS(ON)} \approx I_0 r_{DS(ON)}$$
$$= \text{constant}$$

For $t > t_3$, the gate current continues to charge C_{GD} and since v_{DS} is constant, v_{GS} starts charging at the same rate as in interval $t_0 \leq t < t_1$, i.e.

$$v_{GS}(t) = V_{GG}(1 - e^{-(t-t_3)/\tau})$$

The gate voltage keeps increasing exponentially until $t = t_3$ when it reaches V_{GG}, at which $i_G = 0$ and the device fully turns on as shown in Fig. 4.18e.

The equivalent circuit model when the MOSFET is completely turned on is for $t > t_1$. At this time, the capacitors C_{GS} and C_{GD} are charged with V_{GG} and $(I_0 r_{ds(ON)} - V_{GG})$, respectively.

The time interval $\Delta t_{32} = (t_3 - t_2)$ is obtained by evaluating v_{DS} at $t = t_3$ as follows

$$v_{DS}(t_3) = -\frac{V_{GG} - V_{Th}}{R_G C_{GD}}(t_3 - t_2) + V_{DD} \quad (4.26)$$
$$= I_0 r_{DS(ON)}$$

Hence, $\Delta t_{32} = (t_3 - t_2)$ is given by,

$$\Delta t_{32} = t_3 - t_2 = R_G C_{GD}\frac{(V_{DD} - I_D r_{DS(ON)})}{V_{GG} - V_{Th}} \quad (4.27)$$

The total delay in turning on the MOSFET is given by

$$t_{ON} = \Delta t_{10} + \Delta t_{21} + \Delta t_{32} \quad (4.28)$$

Notice the MOSFET sustains high voltage and current simultaneously during intervals Δt_{21} and Δt_{32}. This results in large power dissipation during turn on, that contributes to the overall switching losses. The smaller the R_G, the smaller Δt_{21} and Δt_{32} become.

4 The Power MOSFET

B. Turn-off Characteristics To study the turn-off characteristic of the MOSFET, we will consider Fig. 4.17b again by assuming the MOSFET is ON and in steady state at $t > t_0$ with the equivalent circuit of Fig. 4.18e. Therefore, at $t = t_0$ we have the following initial conditions.

$$v_{DS}(t_0) = I_D r_{DS(ON)}$$
$$v_{GS}(t_0) = V_{GG}$$
$$i_{DS}(t_0) = I_0 \quad (4.29)$$
$$i_G(t_0) = 0$$
$$v_{C_{GS}}(t_0) = V_{GG}$$
$$v_{C_{GD}}(t_0) = V_{GG} - I_0 r_{DS(ON)}$$

At $t = t_0$, the gate voltage, $v_{GG}(t)$ is reduced to zero as shown in Fig. 4.21a. The equivalent circuit at $t > t_0$ is shown in Fig. 4.22a.

If we assume the drain-to-source voltage remains constant, C_{GS} and C_{GD} are discharging through R_G as governed by the following relations

$$i_G = \frac{-v_G}{R_G} = i_{C_{GS}} + i_{C_{GD}}$$
$$= C_{GS}\frac{dv_{GS}}{dt} + C_{GD}\frac{dv_{GD}}{dt}$$

Since v_{DS} is assumed constant, then i_G becomes,

$$i_G = \frac{-v_{GS}}{R_G}$$
$$= (C_{GS} + C_{GD})\frac{dv_{GS}}{dt} \quad (4.30)$$

Hence, evaluating for v_{GS} for $t \geq t_0$, we obtain

$$v_{GS}(t) = v_{GS}(t_0)e^{-(t-t_0)/\tau} \quad (4.31)$$

where,

$$v_{GS}(t_0) = v_{GG}$$
$$\tau = (C_{GS} + C_{GD})R_G$$

As v_{GS} continues to decrease exponentially, drawing current from C_{GD} will reach a constant value at which drain current is fixed, i.e. $I_D = I_0$. From the input transfer characteristics, the value of v_{GS} when $I_D = I_0$ is given by,

$$v_{GS} = \frac{I_0}{g_m} + V_{Th} \quad (4.32)$$

The time interval $\Delta t_{10} = t_1 - t_0$ can be obtained easily by setting Eq. (4.31) to (4.32) at $t = t_1$.

The gate current during the $t_2 \leq t < t_1$ is given by

$$i_G = -\frac{V_{GG}}{R_G} - e^{-(t-t_0)/\tau} \quad (4.33)$$

Since, for $t_2 - t_1$, the gate-to-source voltage is constant and equals $v_{GS}(t_1) = (I_0/g_m) + V_{Th}$ as shown in Fig. 4.21b, then the entire gate current is being drawn from C_{GD}, hence,

$$i_G = C_{GD}\frac{dv_{GD}}{dt} = C_{GD}\frac{d(v_{GS} - v_{DS})}{dt} = -C_{GD}\frac{dv_{DS}}{dt}$$
$$= \frac{v_{GS}(t_1)}{R_G} = \frac{1}{R_G}\left(\frac{I_0}{g_m} + V_{Th}\right)$$

Assuming i_G constant at its initial value at $t = t_1$, i.e.

$$i_G = \frac{v_{GS}(t_1)}{R_G} = \frac{1}{R_G}\left(\frac{I_0}{g_m} + V_{Th}\right)$$

Integrating both sides of the above equation from t_1 to t with $v_{DS}(t_1) = -v_{DS(ON)}$, we obtain,

$$v_{DS}(t) = v_{DS(ON)} + \frac{1}{R_G C_{GD}}\left(\frac{I_0}{g_m} + V_{Th}\right)(t - t_1) \quad (4.34)$$

hence, v_{DS} charges linearly until it reaches V_{DD}.

At $t = t_2$, the drain-to-source voltage becomes equal to V_{DD}, forcing D to turn on as shown in Fig. 4.22c.

The drain-to-source current is obtained from the transfer characteristics and given by

$$i_{DS}(t) = g_m(v_{GS} - V_{Th})$$

where $v_{GS}(t)$ is obtained from the following equation

$$i_G = -\frac{v_{GS}}{R_G} = (C_{GS} + C_{GD})\frac{dv_{GS}}{dt} \quad (4.35)$$

Integrate both sides from t_2 to t with $v_{GS}(t_2) = (I_0/g_m) + V_{Th}$, we obtain the following expression for $v_{GS}(t)$,

$$v_{GS}(t) = \left(\frac{I_0}{g_m} + V_{Th}\right)e^{-(t-t_2)/\tau} \quad (4.36)$$

Hence the gate current and drain-to-source current are given by,

$$i_G(t) = \frac{-1}{R_G}\left(\frac{I_0}{g_m} + V_{Th}\right)e^{-(t-t_2)/\tau} \quad (4.37)$$

$$i_{DS}(t) = g_m V_{Th}(e^{-(t-t_2)/\tau} - 1) + I_0 e^{-(t-t_2)/\tau} \quad (4.38)$$

The time interval between $t_2 \leq t < t_3$ is obtained by evaluating $v_{GS}(t_3) = V_{Th}$, at which the drain current becomes

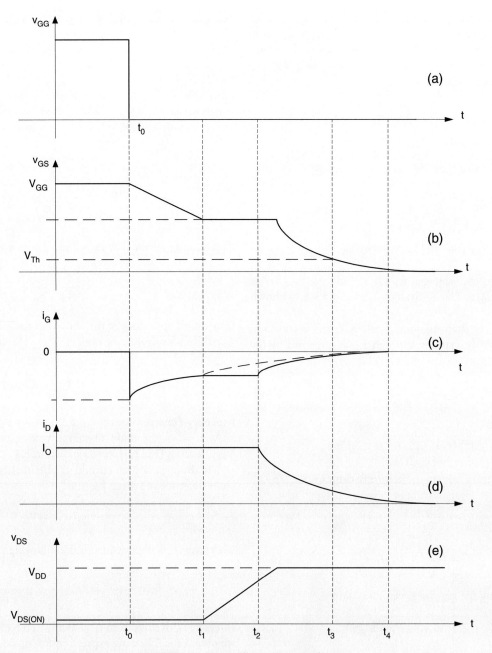

FIGURE 4.21 Turn-off switching waveforms.

approximately zero and the MOSFET turn off hence,

$$v_{GS}(t_3) = V_{Th}$$
$$= \left(\frac{I_0}{g_m} + V_{Th}\right) e^{-(t_3-t_2)/\tau}$$

Solving for $\Delta t_{32} = t_3 - t_2$ we obtain,

$$\Delta t_{32} = t_3 - t_2 = \tau \ln\left(1 + \frac{I_0}{V_{Th}g_m}\right) \quad (4.39)$$

For $t > t_3$, the gate voltage continues to decrease exponentially to zero, at which the gate current becomes zero and C_{GD} charges to $-V_{DD}$. Between t_3 and t_4, I_D discharges to zero as shown in the equivalent circuit Fig. 4.22d.

The total turn-off time for the MOSFET is given by,

$$t_{off} = \Delta t_{10} + \Delta t_{21} + \Delta t_{32} + \Delta t_{43}$$
$$\approx \Delta t_{21} + \Delta t_{32} \quad (4.40)$$

The time intervals that most effect the power dissipation are Δt_{21} and Δt_{32}. It is clear that in order to reduce

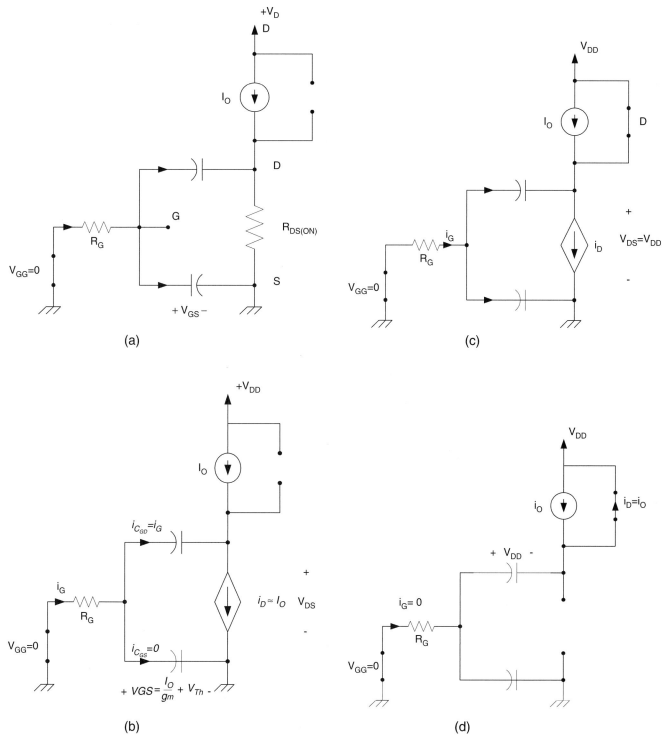

FIGURE 4.22 Equivalent circuits: (a) $t_0 \leq t < t_1$; (b) $t_1 \leq t < t_2$; (c) $t_2 \leq t < t_3$; and (d) $t_3 \leq t < t_4$.

the MOSFET t_{on} and t_{off} times, the gate–drain capacitance must be reduced. Readers are encouraged to see the reference by Baliga for detailed discussion on the turn-on and turn-off characteristics of the MOSFET and to explore various fabrication methods.

C. Safe Operation Area The safe operation area (SOA) of a device provides the current and voltage limits. The device must handle to avoid destructive failure. Typical SOA for a MOSFET device is shown in Fig. 4.23. The maximum current limit while the device is on is determined by the maximum

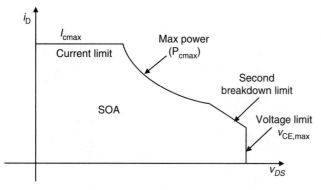

FIGURE 4.23 Safe operation area for MOSFET.

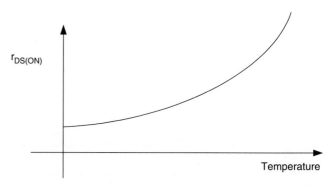

FIGURE 4.25 The on-state resistance as a fraction of temperature.

power dissipation

$$P_{diss,ON} = I_{DS(ON)} R_{DS(ON)}$$

As the drain–source voltage starts increasing, the device starts leaving the on-state and enters the saturation (linear) region. During the transition time, the device exhibits large voltage and current simultaneously. At higher drain–source voltage values that approach the avalanche breakdown it is observed that power MOSFET suffers from second breakdown phenomenon. The second breakdown occurs when the MOSFET is in the blocking state (off) and a further increase in v_{DS} will cause a sudden drop in the blocking voltage. The source of this phenomenon in MOSFET is caused by the presence of a parasitic n-type bipolar transistor as shown in Fig. 4.24.

The inherent presence of the body diode in the MOSFET structure makes the device attractive to application in which bi-directional current flow is needed in the power switches.

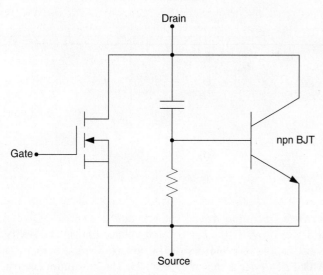

FIGURE 4.24 MOSFET equivalent circuit including the parasitic BJT.

Today's commercial MOSFET devices have excellent high operating temperatures. The effect of temperature is more prominent on the on-state resistance as shown in Fig. 4.25.

As the on-state resistance increases, the conduction losses also increase. This large $v_{DS(ON)}$ limits the use of the MOSFET in high voltage applications. The use of silicon carbide instead of silicon has reduced $v_{DS(ON)}$ by many folds.

As the device technology keeps improving in terms of improving switch speeds, increased power handling capabilities, it is expected that the MOSFET will continue to replace BJTs in all types of power electronics systems.

4.4.4 MOSFET PSPICE Model

The PSPICE simulation package has been used widely by electrical engineers as an essential software tool for circuit design. With the increasing number of devices available in the market place, PSPICE allows for the accurate extraction and understanding of various device parameters and their variation effect on the overall design prior to their fabrication. Today's PSPICE library is rich with numerous commercial MOSFET models. This section will give a brief overview of how the MOSFET model is implemented in PSPICE. A brief overview of the PSPICE modeling of the MOSFET device will be given here.

A. PSPICE Static Model There are four different types of MOSFET models that are also known as *levels*. The simplest MOSFET model is called LEVEL1 model and is shown in Fig. 4.26 [9, 10].

LEVEL2 model uses the same parameters as LEVEL1, but it provides a better model for Ids by computing the model coefficients KP, VTO, LAMBDA, PHI, and GAMMA directly from the geometrical, physical, and technological parameters [10]. LEVEL3 is used to model the short-channel devices and LEVEL4 represents the Berkeley Short-channel IGFET model (BSIM-model).

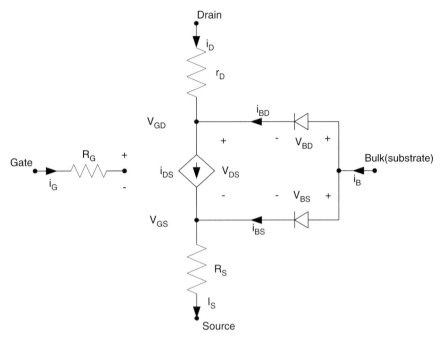

FIGURE 4.26 PSPICE LEVEL1 MOSFET static model.

The triode region, $v_{GS} > V_{Th}$ and $v_{DS} < v_{GS}$ and $v_{DS} < v_{GS} - V_{Th}$ the drain current is given by,

$$i_D = \frac{K_P}{2} \frac{W}{L - 2X_{jl}} \left(v_{GS} - V_{Th} - \frac{v_{DS}}{2} \right) v_{DS}(1 + \lambda v_{DS}) \tag{4.41}$$

In the saturation (linear) region, where $v_{GS} > V_{Th}$ and $v_{DS} > v_{GS} - V_{Th}$, the drain current is given by

$$I_D = \frac{K_P}{2} \frac{W}{L - 2X_{jl}} (V_{GS} - V_{Th})^2 (1 + \lambda V_{DS}) \tag{4.42}$$

where K_P is the transconductance and X_{jl} is the lateral diffusion.

The threshold voltage, V_{Th}, is given by,

$$V_{Th} = V_{T0} + \partial \left(\sqrt{2\phi_p - V_{BS}} - \sqrt{2\phi_p} \right) \tag{4.43}$$

where,

V_{T0} = Zero-bias threshold voltage.
δ = Body-effect parameter.
ϕ_p = Surface inversion potential.

Typically, $X_{ij} \ll L$ and $\lambda \approx 0$.

The term $(1 + \lambda V_{DS})$ is included in the model as empirical connection to model the effect of the output conductance when the MOSFET is operating in triode region. λ is known as the channel-length modulation parameter.

When the bulk and source terminals are connected together, i.e. $V_{BS} = 0$, the device threshold voltage equals the zero-bias threshold voltage, i.e.

$$V_{Th} = V_{T0}$$

V_{T0} is positive for the n-channel enhancement-mode devices and negative for the n-channel depletion-mode devices.

The parameters K_P, V_{T0}, δ, ϕ are electrical parameters that can be either specified directly in the MODEL statement under the Pspice keywords KP, VTO, GAMMA, and PHI, respectively, as shown in Table 4.1. They also can be calculated when the geometrical and physical parameters are known. The two-substrate currents that flow from the bulk to the source, I_{BS} and from the bulk to the drain, I_{BD} are simply diode currents, which are given by,

$$I_{BS} = I_{SS} \left(e^{-(V_{BS}/V_T)} - 1 \right) \tag{4.44}$$

$$I_{BD} = I_{DS} \left(e^{-(V_{BD}/V_T)} - 1 \right) \tag{4.45}$$

where I_{SS} and I_{DS} are the substrate source and substrate drain saturation currents. These currents are considered equal and given as I_S in the MODEL statement with a default value of 10^{-14} A. Where the equation symbols and their corresponding PSPICE parameter names are shown in Table 4.1.

In PSPICE, a MOSFET device is described by two statements: the first statement start with the letter M and the

TABLE 4.1 PSPICE MOSFET parameters

Symbol	Name	Description	Default	Units
(a) Device dc and parasitic parameters				
Level	LEVEL	Model type (1, 2, 3, or 4)	1	–
V_{TO}	VTO	Zero-bias threshold voltage	0	V
λ	LAMDA	Channel-length modulation [1,2]*	0	v^{-1}
γ	GAMMA	Body-effect (bulk) threshold parameter	0	$v^{-1/2}$
Φ_ρ	PHI	Surface inversion potential	0.6	V
η	ETA	Static feedback[3]	0	–
κ	KAPPA	Saturation field factor[3]	0.2	–
μ_0	UO	Surface mobility	600	$cm^2/V \cdot s$
I_s	IS	Bulk saturation current	10^{-14}	A
J_s	JS	Bulk saturation current/area	0	A/m^2
J_{SSW}	JSSW	Bulk saturation current/length	0	A/m
N	N	Bulk emission coefficient n	1	–
P_B	PB	Bulk junction voltage	0.8	V
P_{BSW}	PBSW	Bulk sidewall diffusion voltage	PB	V
R_D	RD	Drain resistance	0	Ω
R_S	RS	Source resistance	0	Ω
R_G	RG	Gate resistance	0	Ω
R_B	RB	Bulk resistance	0	Ω
R_{ds}	RDS	Drain–source shunt resistance	α	Ω
R_{sh}	RSH	Drain and source diffusion sheet resistance	0	Ω/m^2
(b) Device process and dimensional parameters				
N_{sub}	NSUB	Substrate doping density	None	cm^{-3}
W	W	Channel width	DEFW	m
L	L	Channel length	DEFL	m
W_D	WD	Lateral Diffusion width	0	m
X_{jl}	LD	Lateral Diffusion length	0	m
K_p	KP	Transconductance coefficient	20μ	A/v^2
t_{OX}	TOX	Oxide thickness	10^{-7}	m
N_{SS}	NSS	Surface-state density	None	cm^{-2}
N_{FS}	NFS	Fast surface-state density	0	cm^{-2}
N_A	NSUB	Substrate doping	0	cm^{-3}
T_{PG}	TPG	Gate material	1	–
		+1 Opposite of substrate	–	–
		−1 Same as substrate	–	–
		0 Aluminum	–	–
X_j	XJ	Metallurgical junction depth[2,3]	0	m
μ_0	UO	Surface mobility	600	$cm^2/V \cdot s$
U_c	UCRIT	Mobility degradation critical field[2]	10^4	V/cm
U_e	UEXP	Mobility degradation exponent[2]	0	–
U_t	VMAX	Maximum drift velocity of carriers[2]	0	m/s
N_{eff}	NEFF	Channel charge coefficient[2]	1	–
δ	DELTA	Width effect on threshold[2,3]	0	–
θ	THETA	Mobility modulation[3]	0	–
(c) Device capacitance parameters				
C_{BD}	CBD	Bulk-drain zero-bias capacitance	0	F
C_{BS}	CBS	Bulk-source zero-bias capacitance	0	F
C_j	CJ	Bulk zero-bias bottom capacitance	0	F/m^2
C_{jsw}	CJSW	Bulk zero-bias perimeter capacitance/length	0	F/m
M_j	MJ	Bulk bottom grading coefficient	0.5	–
M_{jsw}	MJSW	Bulk sidewall grading coefficient	0.33	–
F_C	FC	Bulk forward-bias capacitance coefficient	0.5	–
C_{GSO}	CGSO	Gate–source overlap capacitance/channel width	0	F/m
C_{GDO}	CGDO	Gate–drain overlap capacitance/channel width	0	F/m
C_{GBO}	CGBO	Gate-bulk overlap capacitance/channel length	0	F/m
X_{QC}	XQC	Fraction of channel charge that associates with drain[1,2]	0	–
K_F	KF	Flicker noise coefficient	0	–
α_F	AF	Flicker noise exponent	0	–

*These numbers indicate that this parameter is available in this level number, otherwise it is available in all levels.

second statement starts with .Model that defines the model used in the first statement. The following syntax is used:

```
M<device_name><Drain_node_number>
<Gate_node_number>
<Source_node_number><Substrate_node number>
<Model_name>
* [<param_1>=<value_1><param_2>=<value_2>....]
.MODEL <Model_name><type_name>
[(<param_1>=<value_1>
<param_2>=<value_2>.....]
```

where the starting letter "M" in M<device_name> statement indicates that the device is a MOSFET and <device_name> is a user specified label for the given device, the <Model_name> is one of the hundreds of device models specified in the PSPICE library, <Model_name> the same name specified in the device name statement, <type_name> is either NMOS of PMOS, depending on whether the device is n-channel or p-channel MOS, respectively, that follows by optional list of parameter types and their values. The length L and the width W and other parameters can be specified in the M<device_name>, in the .MODEL or .OPTION statements. User may select not to include any value, and PSPICE will use the specified default values in the model. For normal operation (physical construction of the MOS devices), the source and bulk substrate nodes must be connected together. In all the PSPICE library files, a default parameter values for L, W, AS, AD, PS, PD, NRD, and NDS are included, hence, user should not specify such values in the device "M" statement or in the OPTION statement.

The power MOSFET device PSPICE models include relatively complete static and dynamic device characteristics given in the manufacturing data sheet. In general, the following effects are specified in a given PSPICE model: dc transfer curves, on-resistance, switching delays, and gate drive characteristics and reverse-mode "body-diode" operation. The device characteristics that are not included in the model are noise, latch-ups, maximum voltage, and power ratings. Please see OrCAD Library Files.

EXAMPLE 4.3 Let us consider an example of using IRF MOSFET that was connected as shown in Fig. 4.27.

It was decided that the device should have a blocking voltage (V_{DSS}) of 600 V and drain current, i_d, of 3.6 A. The device selected is IRF CC30 with case TO220. This device is listed in PSPICE library under model number IRFBC30 as follows:

```
*Library of Power MOSFET Models
*Copyright OrCAD, Inc. 1998 All Rights Reserved.
*
*$Revision: 1.24 $
*$Author: Rperez $
*$Date: 19 October 1998 10:22:26 $
*
. Model IRFBC30 NMOS NMOS
```

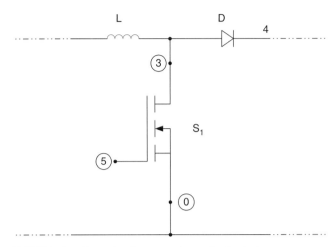

FIGURE 4.27 Example of a power electronic circuit that uses a power MOSFET.

The PSPICE code for the MOS device labeled S1 used in Fig. 4.27 is given by,

```
MS1 3 5 0 0 IRFBC30
.MODEL IRFBC30
.Model IRFBC30 NMOS(Level=3 Gamma=0 Delta=0
 Eta=0 Theta=0 Kappa=0.2 Vmax=0 Xj=0
+ Tox=100n Uo=600 Phi=.6 Rs=5.002m Kp=20.43u
  W=.35 L=2u Vto=3.625
+ Rd=1.851 Rds=2.667MEG Cbd=790.1p Pb=.8 Mj=.5
  Fc=.5 Cgso=1.64n
+ Cgdo=123.9p Rg=1.052 Is=720.2p N=1 Tt=685)
* Int'l Rectifier pid=IRFCC30 case=TO220
```

4.4.5 MOSFET Large-signal Model

The equivalent circuit of Fig. 4.28 includes five device parasitic capacitances. The capacitors C_{GB}, C_{GS}, C_{GD}, represent the charge-storage effect between the gate terminal and the bulk, source, and drain terminals, respectively. These are non-linear two-terminal capacitors expressed as function of W, L, C_{ox}, V_{GS}, V_{T0}, V_{DS}, and C_{GBO}, C_{GSO}, C_{GDO}, where the capacitors C_{GBO}, C_{GSO}, C_{GDO} are outside the channel region, known as overlap capacitances, that exist between the gate electrode and the other three terminals, respectively. Table 4.1 shows the list of PSPICE MOSFET capacitance parameters and their default values. Notice that the PSPICE overlap capacitors keywords (C_{GBO}, C_{GSO}, C_{GDO}) are proportional either to the MOSFET width or length of the channel as follows:

$$C_{GBO} = \frac{C_{GBO}}{L}$$

$$C_{GSO} = \frac{C_{GSO}}{W} \quad (4.46)$$

$$C_{GDO} = \frac{C_{GDO}}{W}$$

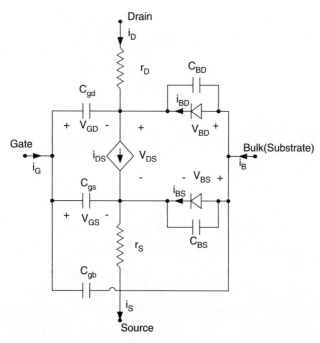

FIGURE 4.28 Large-signal model for the n-channel MOSFET.

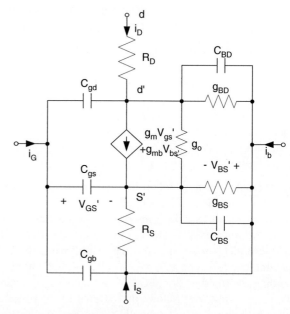

FIGURE 4.29 Small-signal equivalent circuit model for MOSFET.

In the triode region, $v_{GS} > v_{DS} - V_{Th}$, the terminal capacitors are given by,

$$C_{GS} = L_W C_{OX}\left[1 - \left(\frac{v_{GS} - v_{DS} - V_{Th}}{2(v_{GS} - V_{Th}) - V_{DS}}\right)^2\right] + C_{GSO}$$

$$C_{GD} = L_W C_{OX}\left[1 - \left(\frac{v_{GS} - V_{Th}}{2(v_{GS} - V_{Th}) - v_{DS}}\right)^2\right] + C_{GDO}$$

$$C_{GB} = C_{GB0}L \tag{4.47}$$

In the saturation (linear) region, we have

$$C_{GS} = \frac{2}{3}L_W C_{OX} + C_{GSO}$$

$$C_{GB} = C_{GB0}L \tag{4.48}$$

$$C_{GD} = C_{GD0}$$

where C_{OX} is the per-unit-area oxide capacitance given by,

$$C_{OX} = \frac{K_{OX}E_0}{T_{OX}}$$

K_{OX} = Oxide's relative dielectric constant.
E_0 = Free space dielectric constant equals 8.854×10^{-12} F/m.
T_{OX} = Oxide's thickness layer given as T_{OX} in Table 4.1.

Finally, the diffusion and junction region capacitances between the bulk-to-channel (drain and source) are modeled by C_{BD} and C_{BS} across the two diodes. Because for almost all power MOSFETs, the bulk and source terminals are connected together and at zero potential, diodes D_{BD} and D_{BS} don't have forward bias, resulting in very small conductance values, i.e. small diffusion capacitances. The small-signal model for MOSFET devices is given in Fig. 4.29.

EXAMPLE 4.4 Figure 4.30a shows an example of a soft-switching power factor connection circuit that has two MOSFETs. Its PSPICE simulation waveforms are shown in Fig. 4.30b.

Table 4.2 shows the PSPICE code for Fig. 4.30a.

4.4.6 Current MOSFET Performance

The current focus of MOSFET technology development is much more broad than power handling capacity and switching speed; the size, packaging, and cooling of modern MOSFET technology is a major focus. Of course, the development of higher power and efficiency is still paramount, but as modern electronics have become increasingly smaller, the packaging and cooling of power circuits has become more important. It has been indicated by manufacturers that many of their modern MOSFETs are not limited by their semiconductor, but by the packaging. If the MOSFET cannot properly disperse heat, the device will become overheated, which will lead to failure.

An example of modern MOSFET technology is the DirectFET surface mounted MOSFET manufactured by International Rectifier. Part number IRF6662, for example, can handle 47 A at 100 V, while consuming a board space of

4 *The Power MOSFET*

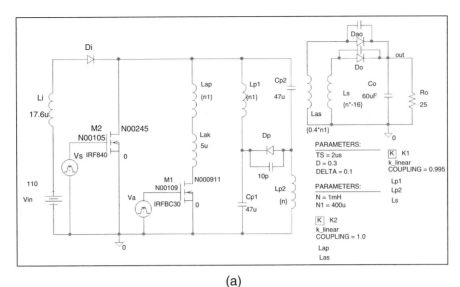

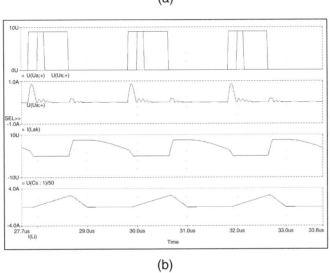

FIGURE 4.30 (a) Example of power electronic circuit and (b) PSPICE simulation waveforms.

5 × 6 mm, and being only 0.6 mm thick. This switch is efficient at frequencies greater than 1 MHz, and the packaging can dissipate over 50% more heat than traditional surface mounted MOSFETs of similar power ratings. The power density of this switch is many times the power density of similarly rated devices made by International Rectifier in the past. One major factor in the performance gain of this product line is dual-sided cooling. By designing the package to mount to the board through a large contact patch, and by using materials with high heat conductivity, the switch has a very high surface area vs volume ratio, which allows for the heat to be dissipated through the top heat sink as well as through the circuit board.

Another example of manufacturers that are focusing on packaging and cooling to increase the performance of their products is Vishay's PolarPAK and PowerPAK. These devices have a 65% smaller board surface area than traditional SO-8 packages. Also, the thermal conductivity of the package is 88%

TABLE 4.2 PSPICE MOSFET capacitance parameters and their default values for Fig. 4.30a

```
* source ZVT-ZCS
D_Do                         N00111  OUT      Dbreak
V_Vs                         N00105  0    DC  0  AC  0  PULSE  0  9  0  0  0  {D*Ts}  {Ts}
L_Ls                         0   N00111   {n*.16}
Kn_K1                        L_Lp1  L_Lp2  L_Ls    0.995
C_Co                         OUT   0    60uF   IC=50
V_Vin                        N00103  0    110
L_Li                         N00103  N00099    17.6u   IC=0
V_Va                         N00109  0   DC  0  AC  0  PULSE  0  9  {-Delta*Ts/1.1}  0  0  {2.0*Delta*Ts}
{Ts}
D_Dp                         N00121  N00169   Dbreak
C_C7                         N00111  OUT      30p
R_Ro                         OUT   0    25
C_C8                         N00143  OUT      10p
D_Dao                        N00143  OUT      Dbreak
D_Di                         N00099  N00245   Dbreak
L_Lp2                        N00121  0     {n}    IC=0
C_C9                         N00169  N00121   10p
L_Las                        N00143  0     {0.4*n1}
Kn_K2                        L_Lap  L_Las    1.0
L_Lp1                        N00245  N00169    {n}    IC=0
L_Lap                        N00245  N000791   {n1}
C_Cp2                        N00245  N00121    47u   IC=170
C_Cp1                        N00169  0     47u   IC=170
L_Lak                        N000791  N000911    5u   IC=0
M_M1                         N000911  N00109   0   0   IRFBC30
M_M2                         N00245  N00105   0   0   IRF840
.PARAM   D=0.3   DELTA=0.1   N1=400u   N=1mH   TS=2us
```

```
                **** MOSFET MODEL PARAMETERS
*****************************************************
                IRFBC30          IRF840
                NMOS             NMOS
LEVEL           3                3
L               2.000000E-06     2.000000E-06
W               .35              .68
VTO             3.625            3.879
KP              20.430000E-06    20.850000E-06
GAMMA           0                0
PHI             .6               .6
LAMBDA          0                0
RD              1.851            .6703
RS              5.002000E-03     6.382000E-03
RG              1.052            .6038
RDS             2.667000E+06     2.222000E+06
IS              720.200000E-12   56.030000E-12
JS              0                0
PB              .8               .8
PBSW            .8               .8
CBD             790.100000E-12   1.415000E-09
CJ              0                0
CJSW            0                0
TT              685.000000E-09   710.000000E-09
CGSO            1.640000E-09     1.625000E-09
CGDO            123.900000E-12   133.400000E-12
CGBO            0                0
TOX             100.000000E-09   100.000000E-09
XJ              0                0
UCRIT           10.000000E+03    10.000000E+03
DELTA           0                0
ETA             0                0
DIOMOD          1                1
VFB             0                0
LETA            0                0
WETA            0                0
U0              0                0
TEMP            0                0
VDD             0                0
XPART           0                0
```

greater than traditional devices. The PolarPAK device increases the performance by cooling the part from the top and the bottom of the package. These advances in packaging and cooling have allowed the devices to have power densities greater than 250 W/mm^3 as well, while maintaining high efficiencies into the megahertz.

Another important characteristic of any solid-state device is the expected service life. For MOSFETs, manufacturers have indicated that the mean time before failure (MTBF) approximately decreases by 50% for every 10°C that the operational temperature increases. For this reason, the current

Examples of modern MOSFETs

Device type	Rated voltage	Rated current	Frequency limit	Rated power	Footprint mm^2
High voltage	1000 V	6.1 A	1 MHz	6 kW	310
High voltage	600 V	40 A	1 MHz	24 kW	320
High power	100 V	180 A	500 kHz	18 kW	310
High current	40 V	280 A	1 MHz	11 kW	310
High efficiency	30 V	40 A	2 MHz	1.2 kW	31.5
High efficiency	30 V	60 A	2 MHz	1.8 kW	36
High efficiency	100 V	47 A	2 MHz	4.7 kW	30.9
High freq. – low power	10 V	0.7 A	200 MHz	7 W	21

advancement in cooling and packaging has a direct effect on the longevity of the components. While there are definite increases in device longevity every year, the easiest way to have a large impact on the life of the device is to keep the temperature down.

As development continues, MOSFETs will become smaller, more efficient, higher power density, and higher frequency of operation. As such, MOSFETs will continue to expand into applications that typically use other forms of power switches.

4.5 Future Trends in Power Devices

As stated earlier, depending on the applications, the power range processed in power electronic range is very wide, from hundreds of milliwatts to hundreds of megawatts, therefore, it is very difficult to find a single switching device type to cover all power electronic applications. Today's available power devices have tremendous power and frequency rating range as well as diversity. Their forward current ratings range from a few amperes to a few kiloamperes, blocking voltage rating ranges from a few volts to a few of kilovolts, and switching frequency ranges from a few hundred of hertz to a few megahertz as illustrated in Table 4.3. This table illustrates the relative comparison between available power semiconductor devices. We only give relative comparison because there is no straightforward technique that gives ranking of these devices. As we accumulate this table, devices are still being developed very rapidly with higher current, voltage ratings, and switching frequency.

TABLE 4.3 Comparison of power semiconductor devices

Device type	Year made available	Rated voltage	Rated current	Rated frequency	Rated power	Forward voltage
Thyristor (SCR)	1957	6 kV	3.5 kA	500 Hz	100's MW	1.5–2.5 V
Triac	1958	1 kV	100 A	500 Hz	100's kW	1.5–2 V
GTO	1962	4.5 kV	3 kA	2 kHz	10's MW	3–4 V
BJT (Darlington)	1960s	1.2 kV	800 A	10 kHz	1 MW	1.5–3 V
MOSFET	1976	500 V	50 A	1 MHz	100 kW	3–4 V
IGBT	1983	1.2 kV	400 A	20 kHz	100's kW	3–4 V
SIT		1.2 kV	300 A	100 kHz	10's kW	10–20 V
SITH		1.5 kV	300 A	10 kHz	10's kW	2–4 V
MCT	1988	3 kV	2 kV	20–100 kHz	10's MW	1–2 V

It is expected that improvement in power handling capabilities and increasing frequency of operation of power devices will continue to drive the research and development in semiconductor technology. From power MOSFET to power MOS-IGBT and to power MOS-controlled thyristors, power rating has consistently increased by a factor of 5 from one type to another. Major research activities will focus on obtaining new device structure based on MOS-BJT technology integration to rapidly increase power ratings. It is expected that the power MOS-BJT technology will capture more than 90% of the total power transistor market.

The continuing development of power semiconductor technology has resulted in power systems with driver circuit, logic and control, device protection, and switching devices being designed and fabricated on a single-chip. Such power IC modules are called "smart power" devices. For example, some of today's power supplies are available as IC's for use in low-power applications. No doubt the development of smart power devices will continue in the near future, addressing more power electronic applications.

References

1. B. Jayant Baliga, *Power Semiconductor Devices*, 1996.
2. L. Lorenz, M. Marz, and H. Amann, "Rugged Power MOSFET- A milestone on the road to a simplified circuit engineering," SIEMENS application notes on S-FET application, 1998.
3. M. Rashid, *Microelectronics*, Thomson-Engineering, 1998.
4. Sedra and Smith, *Microelectronic Circuits*, 4th Edition, Oxford Series, 1996.
5. Ned Mohan, Underland, and Robbins, *Power Electronics: Converters, Applications and Design*, 2nd Edition. John Wiley. 1995.
6. R. Cobbold, *Theory and Applications of Field Effect Transistor*, John Wiley, 1970.
7. R.M. Warner and B.L. Grung, *MOSFET: Theory and Design*, Oxford, 1999.
8. *Power FET's and Their Application*, Prentice-Hall, 1982.
9. J. G. Gottling, *Hands on pspice*, Houghton Mifflin Company, 1995.
10. G. Massobrio and P. Antognetti, *Semiconductor Device Modeling with PSPICE*, McGraw-Hill, 1993.

5
Insulated Gate Bipolar Transistor

S. Abedinpour, Ph.D. and
K. Shenai, Ph.D.
Department of Electrical Engineering and Computer Science, University of Illinois at Chicago, 851, South Morgan Street (M/C 154), Chicago, Illinois, USA

5.1 Introduction 71
5.2 Basic Structure and Operation... 72
5.3 Static Characteristics 74
5.4 Dynamic Switching Characteristics... 76
    5.4.1 Turn-on Characteristics • 5.4.2 Turn-off Characteristics • 5.4.3 Latch-up of Parasitic Thyristor
5.5 IGBT Performance Parameters.. 78
5.6 Gate Drive Requirements 80
    5.6.1 Conventional Gate Drives • 5.6.2 New Gate Drive Circuits • 5.6.3 Protection
5.7 Circuit Models 82
    5.7.1 Input and Output Characteristics • 5.7.2 Implementing the IGBT Model into a Circuit Simulator
5.8 Applications... 85
    Further Reading 87

5.1 Introduction

The insulated gate bipolar transistor (IGBT), which was introduced in early 1980s, is becoming a successful device because of its superior characteristics. IGBT is a three-terminal power semiconductor switch used to control the electrical energy. Many new applications would not be economically feasible without IGBTs. Prior to the advent of IGBT, power bipolar junction transistors (BJT) and power metal oxide field effect transistors (MOSFET) were widely used in low to medium power and high-frequency applications, where the speed of gate turn-off thyristors was not adequate. Power BJTs have good on-state characteristics but have long switching times especially at turn-off. They are current-controlled devices with small current gain because of high-level injection effects and wide base width required to prevent reach-through breakdown for high blocking voltage capability. Therefore, they require complex base drive circuits to provide the base current during on-state, which increases the power loss in the control electrode.

On the other hand power MOSFETs are voltage-controlled devices, which require very small current during switching period and hence have simple gate drive requirements. Power MOSFETs are majority carrier devices, which exhibit very high switching speeds. But the unipolar nature of the power MOSFETs causes inferior conduction characteristics as the voltage rating is increased above 200 V. Therefore their on-state resistance increases with increasing breakdown voltage. Furthermore, as the voltage rating increases the inherent body diode shows inferior reverse recovery characteristics, which leads to higher switching losses.

In order to improve the power device performance, it is advantageous to have the low on-state resistance of power BJTs with an insulated gate input like that of a power MOSFET. The Darlington configuration of the two devices shown in Fig. 5.1 has superior characteristics as compared to the two discrete devices. This hybrid device could be gated like a power MOSFET with low on-state resistance because the majority of the output current is handled by the BJT. Because of the low current gain of BJT, a MOSFET of equal size is required as a driver. A more powerful approach to obtain the maximum benefits of the MOS gate control and bipolar current conduction is to integrate the physics of MOSFET and BJT within the same semiconductor region. This concept gave rise to the commercially available IGBTs with superior on-state characteristics, good switching speed and excellent safe operating area. Compared to power MOSFETs the absence of the integral body diode can be considered as an advantage or disadvantage depending on the switching speed and current requirements. An external fast recovery diode or a diode in the same package

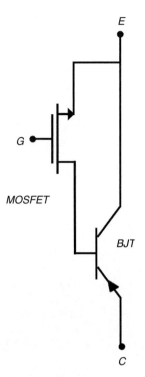

FIGURE 5.1 Hybrid Darlington configuration of MOSFET and BJT.

can be used for specific applications. The IGBTs are replacing MOSFETs in high-voltage applications with lower conduction losses. They have on-state voltage and current density comparable to a power BJT with higher switching frequency. Although they exhibit fast turn-on, their turn-off is slower than a MOSFET because of current fall time. The IGBTs have considerably less silicon area than similar rated power MOSFETs. Therefore by replacing power MOSFETs with IGBTs, the efficiency is improved and cost is reduced. IGBT is also known as conductivity modulated FET (COMFET), insulated gate transistor (IGT), and bipolar-mode MOSFET.

As soft switching topologies offer numerous advantages over the hard switching topologies, their use is increasing in the industry. By the use of soft-switching techniques, IGBTs can operate at frequencies up to hundreds of kilohertz. The IGBTs behave differently under soft switching condition as opposed to hard switching conditions. Therefore, the device tradeoffs involved in soft switching circuits are different than those in hard switching case. Application of IGBTs in high power converters subjects them to high-transient electrical stress such as short circuit and turn-off under clamped inductive load and therefore robustness of IGBTs under stress conditions is an important requirement. Traditionally, there has been limited interaction between device manufacturers and power electronic circuit designers. Therefore, shortcomings of device reliability are observed only after the devices are used in actual circuits. This significantly slows down the process of power electronic system optimization. The development

time can be significantly reduced if all issues of device performance and reliability are taken into consideration at the design stage. As high stress conditions are quite frequent in circuit applications, it is extremely cost efficient and pertinent to model the IGBT performance under these conditions. However, development of the model can follow only after the physics of device operation under stress conditions imposed by the circuit is properly understood. Physically based process and device simulations are a quick and cheap way of optimizing the IGBT. The emergence of mixed mode circuit simulators in which semiconductor carrier dynamics is optimized within the constraints of circuit level switching is a key design tool for this task.

5.2 Basic Structure and Operation

The vertical cross section of a half cell of one of the parallel cells of an n-channel IGBT shown in Fig. 5.2 is similar to that of a double diffused power MOSFET (DMOS) except for a p^+ layer at the bottom. This layer forms the IGBT collector and a pn junction with n^- drift region, where conductivity modulation occurs by injecting minority carriers into the drain drift region of the vertical MOSFET. Therefore, the current density is much greater than a power MOSFET and the forward voltage drop is reduced. The p^+ substrate, n^- drift layer, and p^+ emitter constitute a BJT with a wide base region and hence small current gain. The device operation can be explained by a BJT with its base current controlled by the voltage applied to the MOS gate. For simplicity, it is assumed that the emitter terminal is connected to the ground potential. By applying a negative voltage to the collector, the pn junction between the p^+ substrate

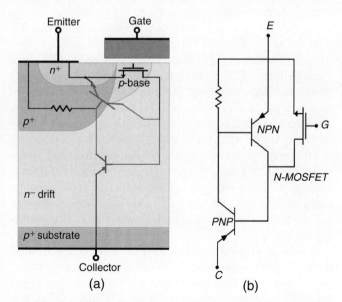

FIGURE 5.2 IGBT: (a) half-cell vertical cross section and (b) equivalent circuit model.

and the n⁻ drift region is reverse biased which prevents any current flow and the device is in its reverse blocking state. If the gate terminal is kept at ground potential but a positive potential is applied to the collector, the pn junction between the p-base and n⁻ drift region is reverse biased. This prevents any current flow and the device is in its forward blocking state until the open base breakdown of the pnp transistor is reached.

When a positive potential is applied to the gate and exceeds the threshold voltage required to invert the MOS region under the gate an n channel is formed, which provides a path for electrons to flow into the n⁻ drift region. The pn junction between the p$^+$ substrate and n⁻ drift region is forward biased and holes are injected into the drift region. The electrons in the drift region recombine with these holes to maintain space charge neutrality and the remaining holes are collected at the emitter, causing a vertical current flow between the emitter and collector. For small values of collector potential and a gate voltage larger than the threshold voltage the on-state characteristics can be defined by a wide base power BJT. As the current density increases, the injected carrier density exceeds the low doping of the base region and becomes much larger than the background doping. This conductivity modulation decreases the resistance of the drift region, and therefore IGBT has a much greater current density than a power MOSFET with reduced forward voltage drop. The base–collector junction of the pnp BJT cannot be forward biased, and therefore this transistor will not operate in saturation. But when the potential drop across the inversion layer becomes comparable to the difference between the gate voltage and threshold voltage, channel pinch-off occurs. The pinch-off limits the electron current and as a result the holes injected from the p$^+$ layer. Therefore, base current saturation causes the collector current to saturate.

Typical forward characteristics of an IGBT as a function of gate potential and IGBT transfer characteristics are shown in Fig. 5.3. The transfer characteristics of IGBT and MOSFET are similar. The IGBT is in the off-state if the gate–emitter potential is below the threshold voltage. For gate voltages greater than the threshold voltage, the transfer curve is linear over most of the drain current range. Gate-oxide breakdown and the maximum IGBT drain current limit the maximum gate–emitter voltage.

To turn-off the IGBT, gate is shorted to the emitter to remove the MOS channel and the base current of the pnp transistor. The collector current is suddenly reduced because the electron current from channel is removed. Then the excess carriers in the n⁻ drift region decay by electron–hole recombination, which causes a gradual collector current decay. In order to keep the on-state voltage drop low, the excess carrier lifetime must be kept large. Therefore, similar to the other minority carrier devices there is a tradeoff between on-state losses and faster turn-off switching times. In the punch-through (PT) IGBT structure of Fig. 5.4 the switching time is reduced by use of a heavily doped n buffer layer in the drift region near the collector. Because of much higher doping density in the buffer layer, the injection efficiency of the collector junction and the minority carrier lifetime in the base region is reduced. The smaller excess carrier lifetime in the buffer layer sinks the excess holes. This speeds up the removal of holes from the drift region and therefore decreases the turn-off time. Non-punch-through (NPT) IGBTs have higher carrier lifetimes and low doped shallow collector region, which affect their electrical characteristics. In order to prevent punch through, NPT IGBTs have a thicker drift region, which results in a higher base transit time. Therefore in NPT structure carrier lifetime is kept more than that of a PT structure, which causes conductivity modulation of the drift region and reduces the on-state voltage drop.

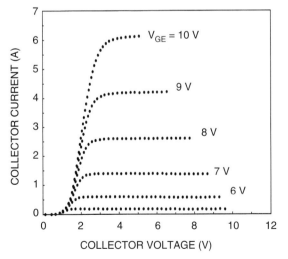 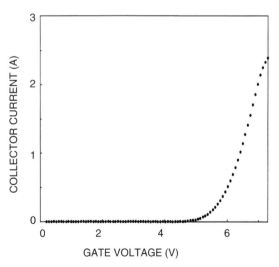

FIGURE 5.3 IGBT: (a) forward characteristics and (b) transfer characteristics.

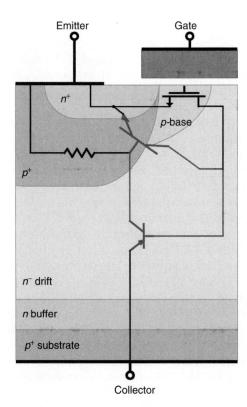

FIGURE 5.4 Punch-through (PT) IGBT structure.

5.3 Static Characteristics

In the IGBT structure of Fig. 5.2, if a negative voltage is applied to the collector, the junction between the p^+ substrate and n^- drift region becomes reverse biased. The drift region is lightly doped and the depletion layer extends principally into the drift region. An open base transistor exists between the p^+ substrate, n^- drift region, and the p-base region. The doping concentration (N_D) and thickness of the n^- drift region (W_D) are designed to avoid the breakdown of this structure. The width of the drift region affects the forward voltage drop and therefore, should be optimized for a desired breakdown voltage. The thickness of the drift region (W_D) is chosen equal to the sum of one diffusion length (L_p) and the width of the depletion layer at maximum applied voltage (V_{max}).

$$W_D = \sqrt{\frac{2\varepsilon_s V_{max}}{qN_D}} + L_P \quad (5.1)$$

When the gate is shorted to the emitter, no channel exists under the gate. Therefore, if a positive voltage is applied to the collector the junction between the p-base and n^- drift region is reverse biased and only a small leakage current flows through IGBT. Similar to a MOSFET the depletion layer extends into the p-base and n^- drift region. The p-base doping concentration, which also controls the threshold voltage is chosen to avoid punch through of the p-base to n^+ emitter. In ac circuit applications, which require identical forward and reverse blocking capability the drift region thickness of the symmetrical IGBT shown in Fig. 5.2 is designed by use of Eq. (5.1) to avoid reach through of the depletion layer to the junction between the p^+ collector and the n^- drift region. When IGBT is used in dc circuits, which do not require reverse blocking capability a highly doped n buffer layer is added to the drift region near the collector junction to form a PT IGBT. In this structure, the depletion layer occupies the entire drift region and the n buffer layer prevents reach through of the depletion layer to the p^+ collector layer. Therefore the required thickness of the drift region is reduced, which reduces the on-state losses. But the highly doped n buffer layer and p^+ collector layer degrade the reverse blocking capability to a very low value. Therefore on-state characteristics of a PT IGBT can be optimized for a required forward blocking capability while the reverse blocking capability is neglected.

When a positive voltage is applied to the gate of an IGBT, an MOS channel is formed between the n^+ emitter and the n^- drift region. Therefore a base current is provided for the parasitic pnp BJT. By applying a positive voltage between the collector and emitter electrodes of an n type IGBT, minority carriers (holes) are injected into the drift region. The injected minority carriers reduce the resistivity of the drift region and reduce the on-state voltage drop resulting in a much higher current density compared to a power MOSFET.

If the shorting resistance between the base and emitter of the npn transistor is small, the n^+ emitter p-base junction does not become forward biased and therefore the parasitic npn transistor is not active and can be deleted from the equivalent IGBT circuit. The analysis of the forward conduction characteristics of an IGBT is possible by the use of two equivalent circuit approaches. The model based on a PiN rectifier in series with a MOSFET, shown in Fig. 5.5b is easy to analyze and gives a reasonable understanding of the IGBT operation. But this model does not account for the hole current component flowing into the p-base region. The junction between the p-base and the n^- drift region is reverse biased. This requires that the free carrier density be zero at this junction, and therefore results in a different boundary condition for IGBT compared to those for PiN rectifier. The IGBT conductivity modulation in the drift region is identical to the PiN rectifier near the collector junction, but it is less than a PiN rectifier near the p-base junction. Therefore, the model based on a bipolar pnp transistor driven by a MOSFET in Fig. 5.5a gives a more complete description of the conduction characteristics.

Analyzing the IGBT operation by the use of these models shows that IGBT has one diode drop due to the parasitic diode. Below the diode knee voltage, there is negligible current flow due to the lack of minority carrier injection from the collector. Also by increasing the applied voltage between the gate and emitter, the base of the internal bipolar transistor is supplied by more base current, which results in an increase in the collector

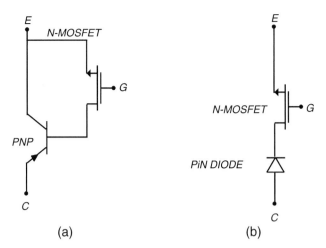

FIGURE 5.5 IGBT equivalent circuits: (a) BJT/MOSFET and (b) PiN/MOSFET.

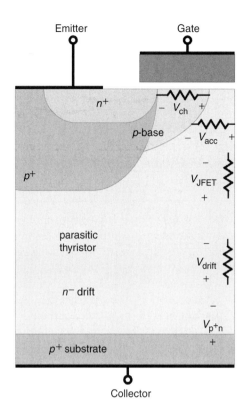

FIGURE 5.6 Components of on-state voltage drop within the IGBT structure.

current. The IGBT current shows saturation due to the pinch-off of the MOS channel. This limits the input base current of the bipolar transistor. The MOS channel of the IGBT reverse biases the collector–base junction and forces the bipolar pnp transistor to operate in its active region. The drift region is in high-level injection at the required current densities and wider n^- drift region results in higher breakdown voltage.

Because of the very low gain of the pnp BJT, the driver MOSFET in the equivalent circuit of the IGBT carries a major portion of the total collector current. Therefore, the IGBT on-state voltage drop as is shown in Fig. 5.6 consists of voltage drop across the collector junction, drift region, and MOSFET portion. The low value of the drift region conductivity modulation near the p-base junction causes a substantial drop across the junction field effect transistor (JFET) resistance of the MOSFET (V_{JFET}) in addition to the voltage drop across the channel resistance (V_{ch}) and the accumulation layer resistance (V_{acc}).

$$V_{CE(on)} = V_{p^+n} + V_{drift} + V_{MOSFET} \quad (5.2)$$

$$V_{MOSFET} = V_{ch} + V_{JFET} + V_{acc} \quad (5.3)$$

When the lifetime in the n^- drift region is large, the gain of the pnp bipolar transistor is high and its collector current is much larger than the MOSFET current and therefore, the voltage drop across the MOSFET component of IGBT is a small fraction of the total voltage drop. When lifetime control techniques are used to increase the switching speed, the current gain of the bipolar transistor is reduced and a greater portion of the current flows through the MOSFET channel and therefore the voltage drop across the MOSFET increases. In order to decrease the resistance of the MOSFET current path, trench IGBTs can be used as shown in Fig. 5.7. Extending the trench gate below the p-base and n^- drift region junction forms a channel between the n^+ emitter and the n^- drift region. This eliminates the JFET and accumulation layer resistance

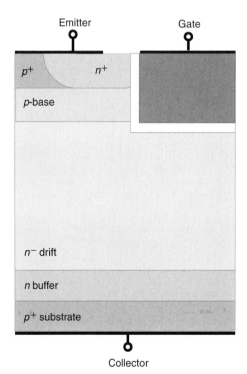

FIGURE 5.7 Trench IGBT structure.

and therefore reduces the voltage drop across the MOSFET component of IGBT, which results in a superior conduction characteristics. By the use of trench structure, the IGBT cell density and latching current density are also improved.

5.4 Dynamic Switching Characteristics

5.4.1 Turn-on Characteristics

The switching waveforms of an IGBT in a clamped inductive circuit are shown in Fig. 5.8. The L/R time constant of the inductive load is assumed to be large compared to the switching frequency and therefore, can be considered as a constant current source I_{on}. The IGBT turn-on switching performance is dominated by its MOS structure. During $t_{d(on)}$, the gate current charges the constant input capacitance with a constant slope until the gate–emitter voltage reaches the threshold voltage $V_{GE(th)}$ of the device. During t_{ri}, load current is transferred from the diode into the device and increases to its steady-state value.

The gate voltage rise time and IGBT transconductance determine the current slope and results as t_{ri}. When the gate–emitter voltage reaches $V_{GE(Ion)}$, which will support the steady-state collector current, collector–emitter voltage starts to decrease. After this there are two distinct intervals, during IGBT turn-on. In the first interval, the collector to emitter voltage drops rapidly as the gate–drain capacitance C_{gd} of the MOSFET portion of IGBT discharges. At low collector–emitter voltage C_{gd} increases. A finite time is required for high-level injection conditions to set in the drift region. The pnp transistor portion of IGBT has a slower transition to its on-state than the MOSFET. The gate voltage starts rising again only after the transistor comes out of its saturation region into the linear region, when complete conductivity modulation occurs and the collector–emitter voltage reaches its final on-state value.

5.4.2 Turn-off Characteristics

Turn-off begins by removing the gate–emitter voltage. Voltage and current remain constant until the gate voltage reaches $V_{GE(Ion)}$, required to maintain the collector steady-state current as shown in Fig. 5.9. After this delay time ($t_{d(off)}$) the collector voltage rises, while the current is held constant. The gate resistance determines the rate of collector voltage rise. As the MOS channel turns off, collector current decreases sharply during t_{fi1}. The MOSFET portion of IGBT determines the turn-off delay time $t_{d(off)}$ and the voltage rise time t_{rv}. When the collector voltage reaches the bus voltage, the freewheeling diode starts to conduct.

However the excess stored charge in the n$^-$ drift region during on-state conduction, must be removed for the device to turn-off. The high minority carrier concentration stored in the n$^-$ drift region supports the collector current after the MOS channel is turned off. Recombination of the minority carriers in the wide base region gradually decreases the collector current and results in a current tail. Since there is no access to the base of the pnp transistor, the excess minority carriers cannot be removed by reverse biasing the gate. The t_{fi2} interval is long because the excess carrier lifetime in this region is normally kept high to reduce the on-state voltage drop. Since the collector–emitter voltage has reached the bus voltage in this interval, a significant power loss occurs which increases with frequency. Therefore, the current tail limits the IGBT operating frequency and there is a tradeoff between the on-state losses and faster switching times. For an on-state current of I_{on}, the magnitude of the current tail, and the time required for the collector current to decrease to 10% of its on-state value, turn-off (t_{off}) time, are approximated as:

$$I_c(t) = \alpha_{pnp} I_{on} e^{-t/\tau_{HL}} \tag{5.4}$$

$$t_{off} = \tau_{HL} \ln(10 \alpha_{pnp}) \tag{5.5}$$

where

$$\alpha_{pnp} = \operatorname{sec} h\left(\frac{l}{L_a}\right) \tag{5.6}$$

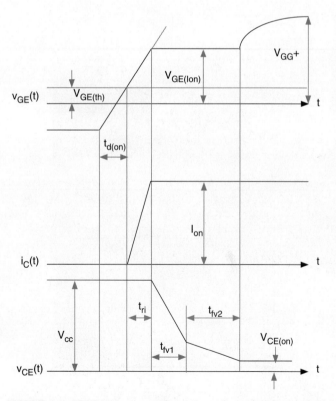

FIGURE 5.8 IGBT turn-on waveforms in a clamped inductive load circuit.

is the gain of the bipolar pnp transistor, l is the undepleted base width, and L_a is the ambipolar diffusion length and it

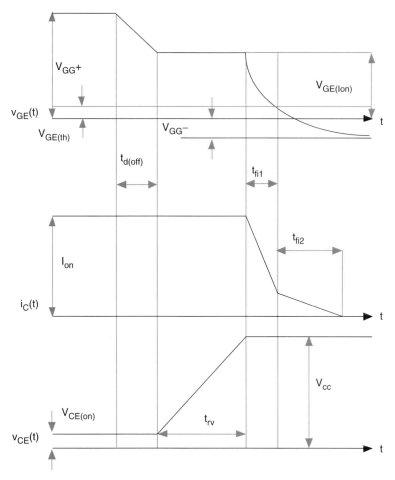

FIGURE 5.9 Switching waveforms during IGBT clamped inductive load turn-off.

is assumed that the high level lifetime (τ_{HL}) is independent of the minority carrier injection during the collector current decay.

Lifetime-control techniques are used to reduce the lifetime (τ_{HL}) and the gain of the bipolar transistor (α_{pnp}). As a result, the magnitude of the current tail and t_{off} decrease. But the conductivity modulation decreases, which increases the on-state voltage drop in the drift region. Therefore, higher speed IGBTs have a lower current rating. Thermal diffusion of impurities such as gold and platinum introduces recombination centers, which reduce the lifetime. The device can also be irradiated with high-energy electrons to generate recombination centers. Electron irradiation introduces a uniform distribution of defects, which results in reduction of lifetime in the entire wafer and affects the conduction properties of the device. Another method of lifetime control is proton implantation, which can place defects at a specific depth. Therefore, it is possible to have a localized control of lifetime to improve the tradeoff between the on-state voltage and switching speed of the device. The turn-off loss can be minimized by curtailing the current tail as a result of speeding up the recombination process in the portion of the drift region, which is not swept by the reverse bias.

5.4.3 Latch-up of Parasitic Thyristor

A portion of minority carriers injected into the drift region from the collector of an IGBT flows directly to the emitter terminal. The negative charge of electrons in the inversion layer attracts the majority of holes and generates the lateral component of hole current through the p-type body layer as shown in Fig. 5.10. This lateral current flow develops a voltage drop across the spreading resistance of the p-base region, which forward biases the base–emitter junction of the npn parasitic BJT. By designing a small spreading resistance, the voltage drop is lower than the built-in potential and therefore the parasitic thyristor between the p^+ collector region, n^- drift region, p-base region, and n^+ emitter does not latch-up. Larger values of on-state current density produce a larger voltage drop, which causes injection of electrons from the emitter region into the p-base region and hence turns on the npn transistor. When this occurs the pnp transistor will turn-on,

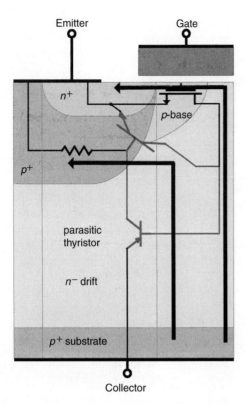

FIGURE 5.10 On-state current flow paths in an IGBT structure.

therefore the parasitic thyristor will latch-up and the gate loses control over the collector current.

Under dynamic turn-off conditions the magnitude of the lateral hole current flow increases and latch-up can occur at lower on-state currents compared to the static condition. The parasitic thyristor latches up when the sum of the current gains of the npn and pnp transistors exceeds one. When the gate voltage is removed from IGBT with a clamped inductive load, its MOSFET component turns off and reduces the MOSFET current to zero very rapidly. As a result the drain–source voltage rises rapidly and is supported by the junction between the n^- drift region and the p-base region. The drift region has a lower doping and therefore the depletion layer extends more in the drift region. As a result the current gain of the pnp transistor portion, α_{pnp} increases and a greater portion of the injected holes into the drift region will be collected at the junction of p-base and n^- drift regions. Therefore, the magnitude of the lateral hole current increases, which increases the lateral voltage drop. As a result the parasitic thyristor will latch-up even if the on-state current is less than the static latch-up value.

Reducing the gain of the npn or pnp transistors can prevent the parasitic thyristor latch-up. A reduction in the gain of the pnp transistor increases the IGBT on-state voltage drop. Therefore in order to prevent the parasitic thyristor latch-up, it is better to reduce the gain of the npn transistor component of IGBT. Reduction of carrier lifetime, use of buffer layer, and use of deep p^+ diffusion improve the latch-up immunity of IGBT. But inadequate extent of the p^+ region may fail to prevent the device from latch-up. Also care should be taken that the p^+ diffusion does not extend into the MOS channel because this causes an increase in the MOS threshold voltage.

5.5 IGBT Performance Parameters

The IGBTs are characterized by certain performance parameters. The manufacturers specify these parameters, which are described below, in the IGBT data sheet. The important ratings of IGBTs are values, which establish either a minimum or maximum limiting capability or limiting condition. The IGBTs cannot be operated beyond the maximum or minimum rating's value, which are determined for a specified operating point and environment condition.

Collector–Emitter blocking voltage (BV_{CES}): This parameter specifies the maximum off-state collector–emitter voltage when the gate and emitter are shorted. Breakdown is specified at a specific leakage current and varies with temperature by a positive temperature coefficient.

Emitter–Collector blocking voltage (BV_{ECS}): This parameter specifies the reverse breakdown of the collector–base junction of the pnp transistor component of IGBT.

Gate–Emitter voltage (V_{GES}): This parameter determines the maximum allowable gate–emitter voltage, when collector is shorted to emitter. The thickness and characteristics of the gate-oxide layer determine this voltage. The gate voltage should be limited to a much lower value to limit the collector current under fault conditions.

Continuous collector current (I_C): This parameter represents the value of the dc current required to raise the junction to its maximum temperature, from a specified case temperature. This rating is specified at a case temperature of 25°C and maximum junction temperature of 150°C. Since normal operating condition cause higher case temperatures, a plot is given to show the variation of this rating with case temperature.

Peak collector repetitive current (I_{CM}): Under transient conditions, the IGBT can withstand higher peak currents compared to its maximum continuous current, which is described by this parameter.

Maximum power dissipation (P_D): This parameter represents the power dissipation required to raise the junction temperature to its maximum value of 150°C, at a case temperature of 25°C. Normally a plot is provided to show the variation of this rating with temperature.

Junction temperature (T_j): Specifies the allowable range of the IGBT junction temperature during its operation.

Clamped inductive load current (I_{LM}): This parameter specifies the maximum repetitive current that IGBT can turn-off under a clamped inductive load. During IGBT turn-on, the reverse recovery current of the freewheeling diode in parallel with the inductive load increases the IGBT turn-on switching loss.

Collector–Emitter leakage current (I_{CES}): This parameter determines the leakage current at the rated voltage and specific temperature when the gate is shorted to emitter.

Gate–Emitter threshold voltage ($V_{GE(th)}$): This parameter specifies the gate–emitter voltage range, where the IGBT is turned on to conduct the collector current. The threshold voltage has a negative temperature coefficient. Threshold voltage increases linearly with gate-oxide thickness and as the square root of the p-base doping concentration. Fixed surface charge at the oxide–silicon interface and mobile ions in the oxide shift the threshold voltage.

Collector–Emitter saturation voltage ($V_{CE(SAT)}$): This parameter specifies the collector–emitter forward voltage drop and is a function of collector current, gate voltage, and temperature. Reducing the resistance of the MOSFET channel and JFET region, and increasing the gain of the pnp bipolar transistor can minimize the on-state voltage drop. The voltage drop across the MOSFET component of IGBT, which provides the base current of the pnp transistor is reduced by a larger channel width, shorter channel length, lower threshold voltage, and wider gate length. Higher minority carrier lifetime and a thin n-epi region cause high carrier injection and reduce the voltage drop in the drift region.

Forward transconductance (g_{FE}): Forward transconductance is measured with a small variation on the gate voltage, which linearly increases the IGBT collector current to its rated current at 100°C. The transconductance of an IGBT is reduced at currents much higher than its thermal handling capability. Therefore, unlike the bipolar transistors, the current handling capability of IGBTs is limited by thermal consideration and not by its gain. At higher temperatures, the transconductance starts to decrease at lower collector currents. Therefore, these features of transconductance protects the IGBT under short circuit operation.

Total gate charge (Q_G): This parameter helps to design a suitable size gate drive circuit and approximately calculate its losses. Because of the minority carrier behavior of device, the switching times cannot be approximately calculated by the use of gate charge value. This parameter varies as a function of the gate–emitter voltage.

Turn-on delay time (t_d): It is defined as the time between 10% of gate voltage and 10% of the final collector current.

Rise time (t_r): It is the time required for the collector current to increase to 90% of its final value from 10% of its final value.

Turn-off delay time ($t_{d(off)}$): It is the time between 90% of gate voltage and 10% of final collector voltage.

Fall time (t_f): It is the time required for the collector current to drop from 90% of its initial value to 10% of its initial value.

Input capacitance (C_{ies}): It is the measured gate–emitter capacitance when collector is shorted to emitter. The input capacitance is the sum of the gate–emitter and the miller capacitance. The gate–emitter capacitance is much larger than the miller capacitance.

Output capacitance (C_{oes}): It is the capacitance between collector and emitter when gate is shorted to the emitter, which has the typical pn junction voltage dependency.

Reverse transfer capacitance (C_{res}): It is the miller capacitance between gate and collector, which has a complex voltage dependency.

Safe operating area (SOA): The safe operating area determines the current and voltage boundary within which the IGBT can be operated without destructive failure. At low currents the maximum IGBT voltage is limited by the open base transistor breakdown. The parasitic thyristor latch-up limits the maximum collector current at low voltages. The IGBTs immune to static latch-up may be vulnerable to dynamic latch-up. Operation in short circuit and inductive load switching are conditions that would subject an IGBT to a combined voltage and current stress. Forward biased safe operating area (FBSOA) is defined during the turn-on transient of the inductive load switching when both electron and hole current flow in the IGBT in the presence of high voltage across the device. The reverse biased safe operating area (RBSOA) is defined during the turn-off transient, where only hole current flows in the IGBT with high voltage across it.

If the time duration of simultaneous high voltage and high current is long enough, the IGBT failure will occur because of thermal breakdown. But if this time duration is short, the temperature rise due to power dissipation will not be enough to cause thermal breakdown. Under this condition the avalanche breakdown occurs at voltage levels lower than the breakdown voltage of the device. Compared to the steady-state forward blocking condition the much larger charge in the drift region causes a higher electric field and narrower depletion region at the p-base and n^- drift junction. Under RBSOA conditions there is no electron in the space charge region, and therefore there is a larger increase in electric field than the FBSOA condition.

The IGBT SOA is indicated in Fig. 5.11. Under short-switching times the rectangular SOA shrinks by increase in

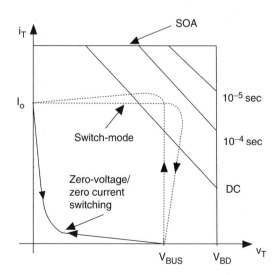

FIGURE 5.11 IGBT safe operating area (SOA).

the duration of on-time. Thermal limitation is the reason for smaller SOA and the lower limit is set by dc operating conditions. The device switching loci under hard switching (dashed lines) and zero voltage or zero current switching (solid lines) is also indicated in Fig. 5.11. The excursion is much wider for switch-mode hard-switching applications than for the soft-switching case, and therefore a much wider SOA is required for hard-switching applications. Presently IGBTs are optimized for hard-switching applications. In soft-switching applications the conduction losses of IGBT can be optimized at the cost of smaller SOA. In this case the p-base doping can be adjusted to result in a much lower threshold voltage and hence forward voltage drop. But in hard-switching applications, the SOA requirements dominate over forward voltage drop and switching time. Therefore, the p-base resistance should be reduced, which causes a higher threshold voltage. As a result, the channel resistance and forward voltage drop will increase.

5.6 Gate Drive Requirements

The gate drive circuit acts as an interface between the logic signals of the controller and the gate signals of the IGBT, which reproduces the commanded switching function at a higher power level. Non-idealities of the IGBT such as finite voltage and current rise and fall times, turn-on delay, voltage and current overshoots, and parasitic components of the circuit cause differences between the commanded and real waveforms. Gate drive characteristics affect the IGBT non-idealities. The MOSFET portion of the IGBT drives the base of the pnp transistor and therefore the turn-on transient and losses is greatly affected by the gate drive.

Due to lower switching losses, soft-switched power converters require gate drives with higher power ratings. The IGBT gate drive must have sufficient peak current capability to provide the required gate charge for zero current switching and zero voltage switching. The delay of the input signal to the gate drive should be small compared to the IGBT switching period and therefore, the gate drive speed should be designed properly to be able to use the advantages of faster switching speeds of the new generation IGBTs.

5.6.1 Conventional Gate Drives

The first IGBT gate drives used fixed passive components and were similar to MOSFET gate drives. Conventional gate drive circuits use a fixed gate resistance for turn-on and turn-off as shown in Fig. 5.12. The turn-on gate resistor R_{gon} limits the maximum collector current during turn-on, and the turn-off gate resistor R_{goff} limits the maximum collector–emitter voltage. In order to decouple the dv_{ce}/dt and di_c/dt control, an external capacitance C_g can be used at the gate, which increases the time constant of the gate circuit and reduces the di_c/dt as shown in Fig. 5.13. But C_g does not affect the dv_{ce}/dt

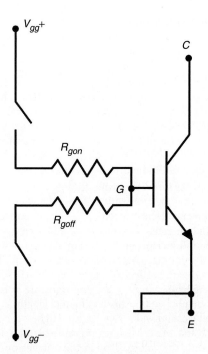

FIGURE 5.12 Gate drive circuit with independent turn-on and turn-off resistors.

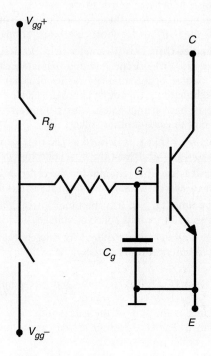

FIGURE 5.13 External gate capacitor for decoupling dv_{ce}/dt and di_c/dt during switching transient.

transient, which occurs during the miller plateau region of the gate voltage.

5.6.2 New Gate Drive Circuits

In order to reduce the delay time required for the gate voltage to increase from V_{gg-} to $V_{ge}(th)$, the external gate capacitor can be introduced in the circuit only after V_{ge} reaches $V_{ge}(th)$ as is shown in Fig. 5.14, where the collector current rise occurs. The voltage tail during turn-on transient is not affected by this method. In order to prevent shoot through caused by accidental turn-on of IGBT due to noise, a negative gate voltage is required during off-state. Low gate impedance reduces the effect of noise on gate.

During the first slope of the gate voltage turn-on transient, the rate of charge supply to the gate determines the collector current slope. During the miller effect zone of the turn-on transient the rate of charge supply to the gate determines the collector voltage slope. Therefore, the slope of the collector current, which is controlled by the gate resistance, strongly affects the turn-on power loss. Reduction in switching power loss requires low gate resistance. But the collector current slope also determines the amplitude of the conducted electromagnetic interference (EMI) during turn-on switching transient. Lower EMI generation requires higher values of gate resistance. Therefore, in conventional gate drive circuits by selecting an optimum value for R_g, there is a tradeoff between lower switching losses and lower EMI generation.

But the turn-off switching of IGBT depends on the bipolar characteristics. Carrier lifetime determines the rate at which the minority carriers stored in the drift region recombine. The charge removed from the gate during turn-off has small influence on minority carrier recombination. The tail current and di/dt during turn-off, which determine the turn-off losses, depend mostly on the amount of stored charge and the minority carriers lifetime. Therefore, the gate drive circuit has a minor influence on turn-off losses of the IGBT, while it affects the turn-on switching losses.

The turn-on transient is improved by use of the circuit shown in Fig. 5.15. The additional current source increases the gate current during the tail voltage time, and therefore reduces the turn-on loss. The initial gate current is determined by $V_{gg}+$ and R_{gon}, which are chosen to satisfy device electrical specifications and EMI requirements. After the collector current reaches its maximum value, the miller effect occurs and the controlled current source is enabled to increase the gate current to increase the rate of collector voltage fall. This reduces the turn-on switching loss. Turn-off losses can only be reduced during the miller effect and MOS turn-off portion of the turn-off transient, by reducing the gate resistance. But this increases the rate of change of collector voltage, which strongly affects the IGBT latching current and RBSOA. During the turn-off

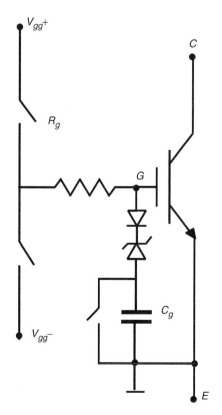

FIGURE 5.14 A circuit for reducing the turn-on delay.

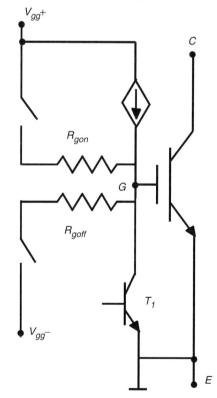

FIGURE 5.15 Schematic circuit of an IGBT gate drive circuit.

period, the turn-off gate resistor R_{goff} determines the maximum rate of collector voltage change. After the device turns off, turning on transistor T_1 prevents the spurious turn-on of IGBT by preventing the gate voltage to reach the threshold voltage.

5.6.3 Protection

Gate drive circuits can also provide fault protection of IGBT in the circuit. The fault protection methods used in IGBT converters are different from their gate turn-off thyristor (GTO) counterparts. In a GTO converter, a crowbar is used for protection and as a result there is no current limiting. When the short circuit is detected the control circuit turns on all the GTO switches in the converter, which results in the opening of a fuse or circuit breaker on the dc input. Therefore, series di/dt snubbers are required to prevent rapid increase of the fault current and the snubber inductor has to be rated for large currents in the fault condition. But IGBT has an important ability to intrinsically limit the current under over-current and short circuit fault conditions. However, the value of the fault current can be much larger than the nominal IGBT current. Therefore, IGBT has to be turned off rapidly after the fault occurs. The magnitude of the fault current depends on the positive gate bias voltage V_{gg+}. A higher V_{gg+} is required to reduce conduction loss in the device, but this leads to larger fault currents. In order to decouple the tradeoff limitation between conduction loss and fault current level, a protection circuit can reduce the gate voltage when a fault occurs. But this does not limit the peak value of the fault current, and therefore, a fast fault detection circuit is required to limit the peak value of the fault current. Fast integrated sensors in the gate drive circuit are essential for proper IGBT protection.

Various methods have been studied to protect IGBTs under fault conditions. One of the techniques uses a capacitor to reduce the gate voltage when the fault occurs. But depending on the initial condition of the capacitor and its value the IGBT current may reduce to zero and then turned on again. Another method is to softly turn-off the IGBT after the fault and to reduce the over-voltage due to di_c/dt. Therefore the over-voltage on IGBT caused by the parasitic inductance is limited while turning off large currents. The most common method of IGBT protection is the collector voltage monitoring or desat detection. The monitored parameter is the collector–emitter voltage, which makes fault detection easier compared to measuring the device current. But voltage detection can be activated only after the complete turn-on of IGBT. If the fault current increases slowly due to large fault inductance, the fault detection is difficult because the collector–emitter voltage will not change significantly. In order to determine whether the current that is being turned off is over-current or nominal current, the miller voltage plateau level can be used. This method can be used to initiate soft turn-off and reduce the over-voltage during over-currents.

Special sense IGBTs have been introduced at low power levels with a sense terminal to provide a current signal proportional to the IGBT collector current. A few active device cells are used to mirror the current carried by the other cells. But unfortunately, sense IGBTs are not available at high power levels and there are problems related to the higher conduction losses in the sense device. The most reliable method to detect an over-current fault condition is to introduce a current sensor in series with the IGBT. The additional current sensor makes the power circuit more complex and may lead to parasitic bus inductance, which results in higher over-voltages during turn-off.

After the fault occurs, the IGBT has to be safely turned off. Due to large di_c/dt during turn-off, the over-voltage can be very large. Therefore, many techniques have been investigated to obtain soft turn-off. The most common method is to use large turn-off gate resistor when the fault occurs. Another method to reduce the turn-off over-voltage is to lower the fault current level by reducing the gate voltage before initiating the turn-off. A resistive voltage divider can be used to reduce the gate voltage during fault turn-off. For example, the gate voltage reduction can be obtained by turning on simultaneously R_{goff} and R_{gon} in the circuit of Fig. 5.12. Another method is to switch a capacitor into the gate and rapidly discharge the gate during the occurrence of a fault. To prevent the capacitor from charging back up to the nominal on-state gate voltage, a large capacitor should be used, which may cause a rapid gate discharge. Also a zener can be used in the gate to reduce the gate voltage after a fault occurs. But the slow transient behavior of the zener leads to large initial peak fault current. The power dissipation during a fault determines the time duration that the fault current can flow in the IGBT without damaging it. Therefore, the IGBT fault endurance capability is improved by the use of fault current limiting circuits to reduce the power dissipation in the IGBT under fault conditions.

5.7 Circuit Models

High-quality IGBT model for circuit simulation is essential for improving the efficiency and reliability in the design of power electronic circuits. Conventional models for power semiconductor devices simply described an abrupt or linear switching behavior and a fixed resistance during the conduction state. Low switching frequencies of power circuits made it possible to use these approximate models. But moving to higher switching frequencies to reduce the size of a power electronic system requires high-quality power semiconductor device models for circuit simulation.

The n-channel IGBT consists of a pnp bipolar transistor whose base current is provided by an n-channel MOSFET, as is shown in Fig. 5.1. Therefore, the IGBT behavior is determined by the physics of the bipolar and MOSFET devices.

Several effects dominate the static and dynamic device characteristics. The influence of these effects on low-power semiconductor device is negligible and therefore they cannot be described by standard device models. The conventional circuit models were developed to describe the behavior of low power devices, and therefore were not adequate to be modified for IGBT. The reason is that the bipolar transistor and MOSFET in the IGBT have a different behavior compared to their low-power counterparts and have different structures.

The present available models have different levels of accuracy at the expense of speed. Circuit issues such as switching losses and reliability are strongly dependent on the device and require accurate device models. But simpler models are only adequate for system oriented issues such as the behavior of an electric motor driven by a pulse width modulation (PWM) converter. Finite element models have high accuracy, but are slow and require internal device structure details. Macro models are fast but have low accuracy, which depends on the operating point. Recently commercial circuit simulators have introduced one-dimensional physics-based models, which offer a compromise between the finite element models and macro models. The Hefner model and the Kraus model are such examples that have been implemented in Saber and there has been some effort to implement them in PSPICE. The Hefner model depends on the redistribution of charge in the drift region during transients. The Kraus model depends on the extraction of charge from the drift region by the electric field and emitter back injection.

The internal BJT of the IGBT has a wide base, which is lightly doped to support the depletion region to have high blocking voltages. The excess carrier lifetime in the base region is low to have fast turn-off. But low power bipolar transistors have high excess carrier lifetime in the base, narrow base, and high current gain. A finite base transit time is required for a change in the injected base charge to change the collector current. Therefore, quasi-static approximation cannot be used at high speeds and the transport of carriers in the base should be described by ambipolar transport theory.

5.7.1 Input and Output Characteristics

The bipolar and MOSFET components of a symmetric IGBT are shown in Fig. 5.16. The components between the emitter (e), base (b), and collector (c) terminals correspond to the bipolar transistor and those between gate (g), source (s), and drain (d) are associated with MOSFET. The combination of the drain–source and gate–drain depletion capacitances is identical to the base–collector depletion capacitance, and therefore they are shown for the MOSFET components. The gate-oxide capacitance of the source overlap (C_{oxs}) and source metallization capacitance (C_m) form the gate–source capacitance (C_{gs}). When the MOSFET is in its linear region the gate-oxide capacitance of the drain overlap (C_{oxd}) forms the gate–drain capacitance (C_{gd}). In the saturation region of

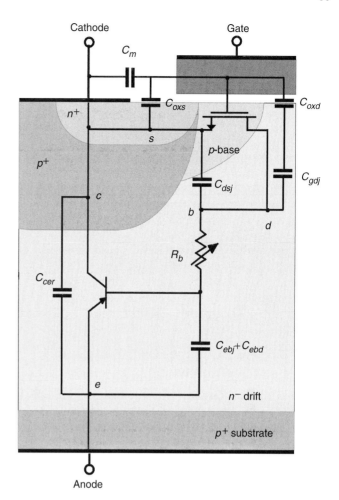

FIGURE 5.16 Symmetric IGBT half cell.

MOSFET the equivalent series connection of gate–drain overlap oxide capacitance and the depletion capacitance of the gate–drain overlap (C_{gdj}) forms the gate–drain miller capacitance. The gate–drain depletion width and the drain–source depletion width are voltage dependent, which has the same effect on the corresponding capacitances.

The most important capacitance in IGBT is the capacitance between the input terminal (g) and output terminal (a), because the switching characteristics is affected by this feedback.

$$C_{ga}\frac{dQ_g}{dv_{ga}} = C_{ox}\frac{dv_{ox}}{dv_{ga}} \quad (5.7)$$

C_{ox} is determined by the oxide thickness and device area. The accumulation, depletion, and inversion states below the gate cause different states of charge and therefore different capacitance values.

The stored charge in the lightly doped wide base of the bipolar component of IGBT causes switching delays and switching losses. The standard quasi-static charge description is not adequate for IGBT because it assumes that the charge distribution is a function of the IGBT terminal voltage. But the stored charge density ($P(x,t)$) changes with time and position and therefore the ambipolar diffusion equation must be used to describe the charge variation.

$$\frac{dP(x,t)}{dt} = -\frac{P(x,t)}{\tau_a} + D_a \frac{d^2 P(x,t)}{dx^2} \qquad (5.8)$$

The slope of the charge carrier distribution determines the sum of electron and hole currents. The non-quasi-static behavior of the stored charge in the base of the bipolar component of IGBT results in the collector–emitter redistribution capacitance (C_{cer}). This capacitance dominates the output capacitance of IGBT during turn-off and describes the rate of change of base–collector depletion layer with the rate of change of base–collector voltage. But the base–collector displacement current is determined by the gate–drain (C_{gdj}) and drain–source (C_{dsj}) capacitance of the MOSFET component.

5.7.2 Implementing the IGBT Model into a Circuit Simulator

Usually a netlist is used in a circuit simulator such as Saber to describe an electrical circuit. Each component of the circuit is defined by a model template with the component terminal connection and the model parameters values. While Saber libraries provide some standard component models, the models can be generated by implementing the model equations in a defined saber template. Electrical component models of IGBT are defined by the current through each component element as a function of component variables, such as terminal and internal node voltages and explicitly defined variables. The circuit simulator uses the Kirchhoff's current law to solve for electrical component variables such that the total current into each node is equal to zero, while satisfying the explicitly defined component variables needed to describe the state of the device.

The IGBT circuit model is generated by defining the currents between terminal nodes as a non-linear function of component variables and their rate of change. An IGBT circuit model is shown in Fig. 5.17. Compared to Fig. 5.16, the bipolar transistor is replaced by the two base and collector current sources. There is a distributed voltage drop due to diffusion and drift in the base regions. The drift terms in the ambipolar diffusion equation depends on base and collector currents. Therefore, both of these currents generate the resistive voltage drop V_{ae} and R_b is placed at the emitter-terminal in the IGBT circuit model. The capacitance of the emitter–base junction (C_{eb}) is implicitly defined by the emitter–base voltage as a function of base charge. I_{ceb} is the emitter–base capacitor current which defines the rate of change of the base charge. The current through the collector–emitter redistribution capacitance (I_{ccer}) is part of the collector current, which in contrast

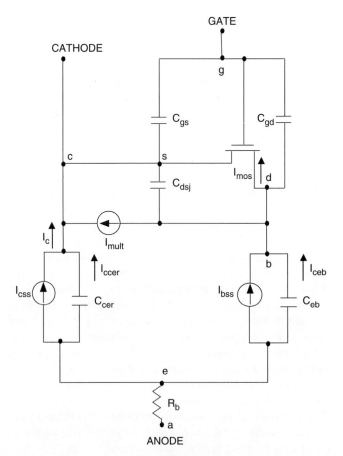

FIGURE 5.17 IGBT circuit model.

to I_{css} depends on the rate of change of the base–emitter voltage. I_{bss} is part of the base current that does not flow through C_{eb} and does not depend on rate of change of base–collector voltage.

Impact ionization causes carrier multiplication in the high electric field of the base–collector depletion region. This carrier multiplication generates an additional base–collector current component (I_{mult}), which is proportional to I_c, I_{mos}, and the multiplication factor. The resulting Saber IGBT model should be able to describe accurately the experimental results for the range of static and dynamic conditions where IGBT operates. Therefore, the model can be used to describe the steady-state and dynamic characteristics under various circuit conditions.

The present available models have different levels of accuracy at the expense of speed. Circuit issues such as switching losses and reliability are strongly dependent on the device and require accurate device models. But simpler models are adequate for system oriented issues such as the behavior of an electric motor driven by a PWM converter. Finite element models have high accuracy, but are slow and require internal device structure details. Macro models are fast but have low accuracy, which depends on the operating point. Recently commercial circuit simulators have introduced one-dimensional physics-based models, which offer a compromise between the finite element models and macro models.

5.8 Applications

Power electronics evolution is a result of the evolution of power semiconductor devices. Applications of power electronics are still expanding in industrial and utility systems. A major challenge in designing power electronic systems is a simultaneous operation at high power and high-switching frequency. The advent of IGBTs has revolutionized power electronics by extending the power and frequency boundary. During the last decade, the conduction and switching losses of IGBTs has been reduced in the process of transition from the first to the third generation IGBTs. The improved charcteristics of the IGBTs have resulted in higher switching speed and lower energy losses. High voltage IGBTs are expected to take the place of high voltage GTO thyristor converters in the near future. To advance the performance beyond the third generation IGBTs, the fourth generation devices will require exploiting fine-line lithographic technology and employing the trench technology used to produce power MOSFETs with very low on-state resistance. Intelligent IGBT or intelligent power module (IPM) is an attractive power device integrated with circuits to protect against over-current, over-voltage, and over-heat. The main application of IGBT is for use as a switching component in inverter circuits, which are used in both power supply and motor-drive applications. The advantages of using IGBT in these converters are simplicity and modularity of the converter, simple gate drive, elimination of snubber circuits due to the square SOA, lower switching loss, improved protection characteristics in case of over-current and short circuit fault, galvanic isolation of the modules, and simpler mechanical construction of the power converter. These advantages have made the IGBT the preferred switching device in the power range below 1 MW.

Power supply applications of IGBTs include uninterruptible power supplies (UPS) as is shown in Fig. 5.18, constant voltage, constant frequency power supplies, induction heating systems, switch mode power supplies, welders (Fig. 5.19), cutters, traction power supplies, and medical equipment (CT, X-ray). Low noise operation, small size, low cost, and high accuracy are chracteristics of the IGBT converters in these applications. Examples of motor-drive applications include variable voltage, variable frequency inverter as is shown in Fig. 5.20. The IGBTS have been recently introduced at high voltage and current levels, which has enabled their use in high power converters utilized for medium voltage motor drives. The improved characteristics of the IGBTs have introduced power converters in megawatt power applications such as traction drives. One of the critical issues in realizing high power

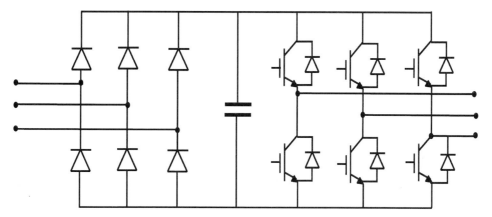

FIGURE 5.18 Constant voltage, constant frequency inverter (UPS).

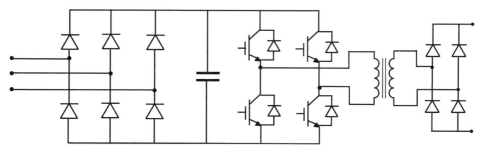

FIGURE 5.19 IGBT welder.

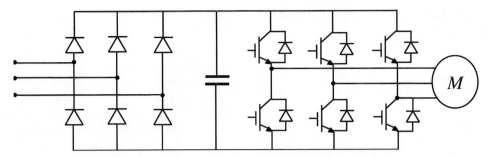

FIGURE 5.20 Variable voltage, variable frequency inverter (PWM).

converters is the reliability of the power switches. The devices used in these applications must be robust and capable of withstanding faults long enough for a protection scheme to be activated. The hard switching voltage source power converter is the most commonly used topology. In this switch-mode operation, the switches are subjected to high switching stresses and high switching power loss that increases linearly with the switching frequency of the PWM. The resulting switching loci in the v_t–i_t plane is shown by the dotted lines in Fig. 5.11. Because of simultaneous large switch voltage and large switch current, the switch must be capable of withstanding high switching stresses with a large SOA. The requirement of being able to withstand large stresses results in design compromises in other characteristics of the power semiconductor device. Often forward voltage drop and switching speed are sacrificed for enhanced short circuit capability. Process parameters of the IGBT such as threshold voltage, carrier lifetime, and the device thickness can be varied to obtain various combinations of SOA, on-state voltage, and switching time. However, there is very little overlap in the optimum combination for more than one performance parameter. Therefore, improved performance in one parameter is achieved at the cost of other parameters.

In order to reduce the size, the weight, and the cost of circuit components used in a power electronics converter very high-switching frequencies of the order of few megahertz are being contemplated. In order to be able to increase the switching frequency, the problems of switch stresses, switching losses, and the EMI associated with switch-mode applications need to be overcome. Use of soft-switching converters reduces the problems of high dv/dt and high di/dt by the use of external inductive and capacitive components to shape the switching trajectory of device. The device switching loci resulting from soft switching is shown in Fig. 5.11, where significant reduction in switching stress can be noticed. The traditional snubber circuits achieves this goal without the added control complexity, but the power dissipation in these snubber circuits can be large and limit the switching frequency of the converter. Also passive components significantly add to the size, weight, and cost of the converter at high power levels. Soft switching uses lossless resonant circuits, which overcomes the problem of power loss in the snubber circuit, but increases the conduction loss. Resonant transition circuits eliminate the problem of high peak device stress in the soft-switched converters. The main drawback of these circuits is the increased control complexity required to obtain the resonant switching transition. The large number of circuit variables that have to be sensed in such power converters can affect their reliability. Short circuit capability no longer being the primary concern, designers can push the performance envelope for their circuits until the device becomes the limiting factor once again.

The transient response of the conventional volts/hertz induction motor drive is sluggish, because both torque and flux are functions of stator voltage and frequency. Use of vector or field oriented control methods makes the performance of the induction motor drive almost identical to that of a separately excited dc motor. Therefore, the transient response is like a dc machine, where torque and flux can be controlled in a decoupled manner. Vector controlled induction motors with shaft encoders or speed sensors have been widely applied in combination with voltage source PWM inverters using IGBT modules. According to the specification of the new products, vector controlled induction motor drive systems ranging from kilowatts to megawatts provide a broad range of speed control, constant torque operation, and high starting torque.

Because of their simple gate drives and modular packaging, IGBTs lead to simpler construction of power electronic circuits. This feature has lead to a trend to standardize and modularize power electronic circuits. Simplification of the overall system design and construction and significant cost reduction are the main implications of this approach. With these goals the power electronics building block (PEBB) program has been introduced, where the entire power electronic converter system is reduced to a single block. Similar modular power electronic blocks are commercially available at low power levels in the form of power integrated circuits. At higher power levels, these blocks have been realized in the form of intelligent power modules and power blocks. But these high power modules do not encompass the entire power electronic systems like motor drives and UPS. The aim of the PEBB program is to realize the whole power handling system within standardized blocks. A PEBB is a universal power processor that changes any electrical power input to any desired form of voltage, current, and frequency output. A PEBB is a single package with a multi-function controller that replaces the complex power electronic circuits with a single device and therefore reduces the development and design costs of the complex power circuits and simplifies the development and design of large electric power systems.

The applications of power electronics are varied and various applications have their own specific design requirement. There is a wide choice of available power devices. Because of physical, material, and design limitations, none of the presently available devices behave as an ideal switch, which should block arbitrarily large forward and reverse voltages with zero current in the off-state, conduct arbitrarily large currents with zero voltage drop in the on-state, and have negligible switching time and power loss. Therefore, power electronic circuits should be designed by considering the capabilities and limitations of available devices. Traditionally there has been limited interaction between device manufacturers and circuit designers. Therefore, manufacturers have been fabricating generic power semiconductor devices with inadequate consideration of the specific applications where the devices are used. The diverse nature of power electronics does not allow the use of generic power semiconductor devices in all applications as it leads to non-optimal systems. Therefore, the devices and circuits need to be optimized at the application level. Soft-switching topologies offer numerous advantages over conventional hard-switching applications such as reduced switching stress and EMI, and higher switching speed at reduced power loss. The IGBTs behave dissimilarly in the two circuit conditions. As a result, devices optimized for hard switching conditions do not necessarily give the best possible performance when used in soft switching circuits. In order to extract maximum system performance, it is necessary to develop IGBTs suited for specific applications. These optimized devices need to be manufacturable and cost effective in order to be commercially viable.

Further Reading

1. Adler, M. S., Owyang, K. W., Baliga, B. J., and Kokosa, R. A., "The evolution of power device technology," *IEEE Trans. Electron. Devices* **ED-31**: 1570–1591 (1984).
2. Akagi, H., "The state-of-the-art of power electronics in Japan," *IEEE Trans. Power Electron.* **13**: 345–356 (1998).
3. Baliga, B. J., Adler, M. S., Love, R. P., Gray, P. V., and Zommer, N., "The insulated gate transistor: a new three-terminal MOS controlled bipolar power device," *IEEE Trans. Electron. Devices* **ED-31**: 821–828 (1984).
4. Baliga B. J., *Power Semiconductor Devices*, PWS Publishing, Boston, MA, 1996.
5. Blaabjerg, F. and Pedersen, J. K., "An optimum drive and clamp circuit design with controlled switching for a snubberless PWM-VSI-IGBT inverterleg," in *IEEE Power Electronics Specialists Conference Records*, pp. 289–297, 1992.
6. Chokhawala, R. and Castino, G., "IGBT fault current limiting circuits," in *IEEE Industry Applications Society Annual Meeting Records*, pp. 1339–1345, 1993.
7. Clemente, S. *et al.*, *IGBT Characteristics*, IR Applications note AN-983A.
8. Divan, D. M. and Skibinski, G., "Zero-switching-loss inverters for high power applications," *IEEE Trans. Industry Applications* **25**: 634–643 (1989).
9. Elasser, A., Parthasarathy, V., and Torrey, D., "A study of the internal device dynamics of punch-through and non punch-through IGBTs under zero-current switching," *IEEE Trans. Power Electron.* **12**: 21–35 (1997).
10. Ghandi, S. K., *Semiconductor Power Devices*, John Wiley & Sons, NY, 1977.
11. Hefner, A. R., "An improved understanding for the transient operation of the insulated gate bipolar transistor (IGBT)," *IEEE Trans. Power Electron.* **5**: 459–468 (1990).
12. Hefner, A. R. and Blackburn, D. L., "An analytical model for the steady-state and transient characteristics of the power insulated gate bipolar transistor," *Solid-State Electron.* **31**: 1513–1532 (1988).
13. Hefner, A. R., "An investigation of the drive circuit requirements for the power insulated gate bipolar transistor (IGBT)," *IEEE Trans. Power Electron.* **6**: 208–219 (1991).
14. Jahns, T.M. "Designing intelligent muscle into industrial motion control," in *Industrial Electronics Conference Records*, pp. 1–14, 1989.
15. John, V., Suh, B. S., and Lipo, T. A., "Fast clamped short circuit protection of IGBTs," in *IEEE Applied Power Electronics Conference Records*, pp. 724–730, 1998.
16. Kassakian, J. G., Schlecht, M. F., and Verghese, G. C., *Principles of Power Electronics*, Addison Wesley, Reading, MA, 1991.
17. Kraus, R. and Hoffman, K., "An analytical model of IGBTs with low emitter efficiency," in *ISPSD'93*, pp. 30–34.
18. Lee, H. G., Lee, Y. H., Suh, B. S., and Lee, J. W., "A new intelligent gate control scheme to drive and protect high power IGBTs," in *European Power Electronics Conference Records*, pp. 1.400–1.405, 1997.
19. Licitra, C., Musumeci, S., Raciti, A., Galluzzo, A. U., and Letor, R., "A new driving circuit for IGBT devices," *IEEE Trans. Power Electron.* **10**: 373–378 (1995).
20. McMurray, W., "Resonant snubbers with auxiliary switches," *IEEE Trans. Industry Applications* **29**: 355–362 (1993).
21. Mohan, N., Undeland, T., and Robbins, W., *Power Electronics – Design, Converters and Applications*, John Wiley & Sons, NY, 1996.
22. Penharkar, S. and Shenai, K., "Zero voltage switching behavior of punchthrough and nonpunchthrough insulated gate bipolar transistors (IGBTs)," *IEEE Trans. Electron. Devices* **45**: 1826–1835 (1998).
23. Powerex IGBTMOD and intellimod – *Intelligent Power Modules Applications and Technical Data Book*, 1994.
24. Sze, S. M., *Physics of Semiconductor Devices*, John Wiley & Sons, NY, 1981.
25. Sze, S. M., *Modern Semiconductor Device Physics*, John Wiley & Sons, NY, 1998.
26. Trivedi, M., Pendharkar, S., and Shenai, K., "Switching charcteristics of IGBTs and MCTs in power converters," *IEEE Trans. Electron. Devices* **43**: 1994–2003 (1996).
27. Trivedi, M. and Shenai, K., "Modeling the turn-off of IGBTs in hard- and soft-switching applications," *IEEE Trans. Electron. Devices* **44**: 887–893 (1997).

28. Trivedi, M. and Shenai, K., "Internal dynamics of IGBT under zero-voltage and zero-current switching conditions," *IEEE Trans. Electron. Devices* **46**: 1274–1282 (1999).
29. Trivedi, M. and Shenai, K., "Failure mechanisms of IGBTs under short-circuit and clamped inductive switching stress," *IEEE Trans. Power Electron.* **14**: 108–116 (1999).
30. Undeland, T., Jenset, F., Steinbakk, A., Ronge, T., and Hernes, M., "A snubber configuration for both power transistor and GTO PWM inverters," in *IEEE Power Electronics Specialists Conference Records*, pp. 42–53, 1984.
31. Venkatesan, V., Eshaghi, M., Borras, R., and Deuty, S., "IGBT turn-off characteristics explained through measurements and device simulation," in *IEEE Applied Power Electronics Conference Records*, pp. 175–178, 1997.
32. Widjaja, I., Kurnia, A., Shenai, K., and Divan, D., "Switching dynamics of IGBTs in soft-switching converters," *IEEE Trans. Electron. Devices* **42**: 445–454 (1995).

6
Thyristors

Angus Bryant, Ph.D.
Department of Engineering, University of Warwick, Coventry CV4 7AL, UK

Enrico Santi, Ph.D.
Department of Electrical Engineering, University of South Carolina, Columbia, South Carolina, USA

Jerry Hudgins, Ph.D.
Department of Electrical Engineering, University of Nebraska, Lincoln, Nebraska, USA

Patrick Palmer, Ph.D.
Department of Engineering, University of Cambridge, Trumpington Street, Cambridge CB2 1PZ, UK

6.1	Introduction	89
6.2	Basic Structure and Operation	90
6.3	Static Characteristics	92
	6.3.1 Current–Voltage Curves for Thyristors • 6.3.2 Edge and Surface Terminations • 6.3.3 Packaging	
6.4	Dynamic Switching Characteristics	95
	6.4.1 Cathode Shorts • 6.4.2 Anode Shorts • 6.4.3 Amplifying Gate • 6.4.4 Temperature Dependencies	
6.5	Thyristor Parameters	99
6.6	Types of Thyristors	101
	6.6.1 SCRs and GTOs • 6.6.2 MOS-controlled Thyristors • 6.6.3 Static Induction Thyristors • 6.6.4 Optically Triggered Thyristors • 6.6.5 Bi-directional Thyristors	
6.7	Gate Drive Requirements	106
	6.7.1 Snubber Circuits • 6.7.2 Gate Circuits	
6.8	PSpice Model	109
6.9	Applications	110
	6.9.1 DC–AC Utility Inverters • 6.9.2 Motor Control • 6.9.3 VAR Compensators and Static Switching Systems • 6.9.4 Lighting Control Circuits	
	Further Reading	114

6.1 Introduction

Thyristors are usually three-terminal devices that have four layers of alternating *p*-type and *n*-type material (i.e. three *p–n* junctions) comprising its main power handling section. In contrast to the linear relation which exists between load and control currents in a transistor, the thyristor is bistable. The control terminal of the thyristor, called the gate (*G*) electrode, may be connected to an integrated and complex structure as a part of the device. The other two terminals, called the anode (*A*) and cathode (*K*), handle the large applied potentials (often of both polarities) and conduct the major current through the thyristor. The anode and cathode terminals are connected in series with the load to which power is to be controlled.

Thyristors are used to approximate ideal closed (no voltage drop between anode and cathode) or open (no anode current flow) switches for control of power flow in a circuit. This differs from low-level digital switching circuits that are designed to deliver two distinct small voltage levels while conducting small currents (ideally zero). Thyristor circuits must have the capability of delivering large currents and be able to withstand large externally applied voltages. All thyristor types are controllable in switching from a forward-blocking state (positive potential applied to the anode with respect to the cathode, with correspondingly little anode current flow) into a forward-conduction state (large forward anode current flowing, with a small anode–cathode potential drop). Most thyristors have the characteristic that after switching from a forward-blocking state into the forward-conduction state, the gate signal can be removed and the thyristor will remain in its forward-conduction mode. This property is termed "latching" and is an important distinction between thyristors and other types of power electronic devices. Some thyristors are also controllable in switching from forward-conduction back to a forward-blocking state. The particular design of a thyristor will determine its controllability and often its application.

Thyristors are typically used at the highest energy levels in power conditioning circuits because they are designed to handle the largest currents and voltages of any device technology (systems approximately with voltages above 1 kV or currents above 100 A). Many medium-power circuits (systems operating at less than 1 kV or 100 A) and particularly low-power circuits (systems operating below 100 V or several amperes) generally make use of power bipolar transistors, power metal oxide semiconductor field effect transistors (MOSFETs) or insulated gate bipolar transistors (IGBTs) as the main switching elements because of the relative ease in controlling them. IGBT technology, however, continues to improve and multiple silicon die are commonly packaged together in a module. These modules are replacing thyristors in applications operating up to 3 kV that require controllable turn-off because of easier gate-drive requirements. Power diodes are used throughout all levels of power conditioning circuits and systems for component protection and wave shaping.

A thyristor used in some ac power circuits (50 or 60 Hz in commercial utilities or 400 Hz in aircraft) to control ac power flow can be made to optimize internal power loss at the expense of switching speed. These thyristors are called phase-control devices, because they are generally turned from a forward-blocking into a forward-conducting state at some specified phase angle of the applied sinusoidal anode–cathode voltage waveform. A second class of thyristors is used in association with dc sources or in converting ac power at one amplitude and frequency into ac power at another amplitude and frequency, and must generally switch on and off relatively quickly. A typical application for the second class of thyristors is in converting a dc voltage or current into an ac voltage or current. A circuit that performs this operation is often called an inverter, and the associated thyristors used are referred to as inverter thyristors.

There are four major types of thyristors: (i) the silicon-controlled rectifier (SCR); (ii) the gate turn-off thyristor (GTO) and its close relative the integrated gate commutated thyristor (IGCT); (iii) the MOS-controlled thyristor (MCT) and its various forms; and (iv) the static induction thyristor (SITh). MCTs are so-named because many parallel enhancement mode, MOSFET structures of one charge type are integrated into the thyristor for turn-on and many more MOSFETs of the other charge type are integrated into the thyristor for turn-off. A SITh or field-controlled thyristor (FCTh), has essentially the same construction as a power diode with a gate structure that can pinch-off anode current flow. Although MCTs, derivative forms of the MCT and SITHs have the advantage of being essentially voltage-controlled devices (i.e. little control current is required for turn-on or turn-off, and therefore require simplified control circuits attached to the gate electrode), they are currently only found in niche applications such as pulse power. Detailed discussion of variations of MCTs and SITHs, as well as additional references on these devices are discussed by Hudgins in [1]. Other types of thyristors include the Triac (a pair of anti-parallel SCRs integrated together to form a bi-directional current switch) and the programmable unijunction transistor (PUT).

The SCRs and GTOs are designed to operate at all power levels. These devices are primarily controlled using electrical signals (current), though some types are made to be controlled using optical energy (photons) for turn-on. Subclasses of SCRs and GTOs are reverse conducting types and symmetric structures that block applied potentials in the reverse and forward polarities. Other variations of GTOs are the gate-commutated turn-off thyristor (GCT), commonly available as the IGCT, and the bi-directional controlled thyristor (BCT). Most power converter circuits incorporating thyristors make use of SCRs, GTOs, or IGCTs, and hence the chapter will focus on these devices, though the basics of operation are applicable to all thyristor types.

All power electronic devices must be derated (e.g. power dissipation levels, current conduction, voltage blocking, and switching frequency must be reduced), when operating above room temperature (defined as approximately 25°C). Bipolar-type devices have thermal runaway problems, in that if allowed to conduct unlimited current, these devices will heat up internally causing more current to flow, thus generating more heat, and so forth until destruction. Devices that exhibit this behavior are *pin* diodes, bipolar transistors, and thyristors.

Almost all power semiconductor devices are made from silicon (Si). Research and development continues in developing other types of devices in silicon carbide (SiC), gallium nitride (GaN), and related material systems. However, the physical description and general behavior of thyristors is unimportant to the semiconductor material system used, though the discussion and any numbers cited in the chapter will be associated with Si devices.

6.2 Basic Structure and Operation

Figure 6.1 shows a conceptual view of a typical thyristor with the three p–n junctions and the external electrodes labeled. Also shown in the figure is the thyristor circuit symbol used in electrical schematics.

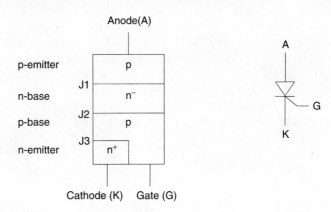

FIGURE 6.1 Simple cross section of a typical thyristor and the associated electrical schematic symbols.

A high-resistivity region, n-base, is present in all thyristors. It is this region, the n-base and associated junction, J_2 of Fig. 6.1, which must support the large applied forward voltages that occur when the switch is in its off- or forward-blocking state (non-conducting). The n-base is typically doped with impurity phosphorous atoms at a concentration of 10^{13} to 10^{14} cm^{-3}. The n-base can be tens to hundreds of micrometer thick to support large voltages. High-voltage thyristors are generally made by diffusing aluminum or gallium into both surfaces to create p-doped regions forming deep junctions with the n-base. The doping profile of the p-regions ranges from about 10^{15} to 10^{17} cm^{-3}. These p-regions can be up to tens of micrometer thick. The cathode region (typically only a few micrometer thick) is formed by using phosphorous atoms at a doping density of 10^{17} to 10^{18} cm^{-3}.

The higher the forward-blocking voltage rating of the thyristor, the thicker the n-base region must be. However, increasing the thickness of this high-resistivity region results in slower turn-on and turn-off (i.e. longer switching times and/or lower frequency of switching cycles because of more stored charge during conduction). For example, a device rated for a forward-blocking voltage of 1 kV will, by its physical construction, switch much more slowly than one rated for 100 V. In addition, the thicker high-resistivity region of the 1 kV device will cause a larger forward voltage drop during conduction than the 100 V device carrying the same current. Impurity atoms, such as platinum or gold, or electron irradiation are used to create charge-carrier recombination sites in the thyristor. The large number of recombination sites reduces the mean carrier lifetime (average time that an electron or hole moves through the Si before recombining with its opposite charge-carrier type). A reduced carrier lifetime shortens the switching times (in particular the turn-off or recovery time) at the expense of increasing the forward-conduction drop. There are other effects associated with the relative thickness and layout of the various regions that make up modern thyristors, but the major tradeoff between forward-blocking voltage rating and switching times, and between forward-blocking voltage rating and forward-voltage drop during conduction should be kept in mind. (In signal-level electronics an analogous trade-off appears as a lowering of amplification (gain) to achieve higher operating frequencies, and is often referred to as the gain-bandwidth product.)

Operation of thyristors is as follows. When a positive voltage is applied to the anode (with respect to cathode), the thyristor is in its forward-blocking state. The center junction, J_2 (see Fig. 6.1) is reverse biased. In this operating mode the gate current is held to zero (open circuit). In practice, the gate electrode is biased to a small negative voltage (with respect to the cathode) to reverse bias the GK-junction J_3 and prevent charge-carriers from being injected into the p-base. In this condition only thermally generated leakage current flows through the device and can often be approximated as zero in value (the actual value of the leakage current is typically many orders of magnitude lower than the conducted current in the on-state). As long as the forward applied voltage does not exceed the value necessary to cause excessive carrier multiplication in the depletion region around J_2 (avalanche breakdown), the thyristor remains in an off-state (forward-blocking). If the applied voltage exceeds the maximum forward-blocking voltage of the thyristor, it will switch to its on-state. However, this mode of turn-on causes non-uniformity in the current flow, is generally destructive, and should be avoided.

When a positive gate current is injected into the device, J_3 becomes forward biased and electrons are injected from the n-emitter into the p-base. Some of these electrons diffuse across the p-base and get collected in the n-base. This collected charge causes a change in the bias condition of J_1. The change in bias of J_1 causes holes to be injected from the p-emitter into the n-base. These holes diffuse across the n-base and are collected in the p-base. The addition of these collected holes in the p-base acts the same as gate current. The entire process is regenerative and will cause the increase in charge carriers until J_2 also becomes forward biased and the thyristor is latched in its on-state (forward-conduction). The regenerative action will take place as long as the gate current is applied in sufficient amount and for a sufficient length of time. This mode of turn-on is considered to be the desired one as it is controlled by the gate signal.

This switching behavior can also be explained in terms of the two-transistor analog shown in Fig. 6.2. The two transistors are regeneratively coupled so that if the sum of their forward current gains (α's) exceeds unity, each drives the other into saturation. Equation 6.1 describes the condition necessary for the thyristor to move from a forward-blocking state into the forward-conduction state. The forward current gain (expressed as the ratio of collector current to emitter current) of the pnp transistor is denoted by α_p, and that of the npn as α_n. The α's are current dependent and increase slightly as the current increases. The center junction J_2 is reverse biased under forward applied voltage (positive, v_{AK}). The associated electric field in the depletion region around the junction can result

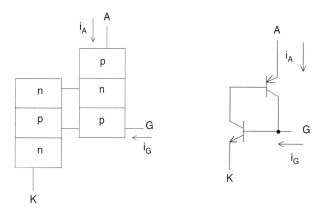

FIGURE 6.2 Two-transistor behavioral model of a thyristor.

in significant carrier multiplication, denoted as a multiplying factor M on the current components, I_{co} and i_G.

$$i_A = \frac{MI_{co} + M\alpha_n i_G}{1 - M(\alpha_n + \alpha_p)} \quad (6.1)$$

In the forward-blocking state, the leakage current I_{co} is small, both α's are small, and their sum is less than unity. Gate current increases the current in both transistors, increasing their α's. Collector current in the *npn* transistor acts as base current for the *pnp*, and analogously, the collector current of the *pnp* acts as base current driving the *npn* transistor. When the sum of the two α's equals unity, the thyristor switches to its on-state (latches). This condition can also be reached, without any gate current, by increasing the forward applied voltage so that carrier multiplication ($M \gg 1$) at J_2 increases the internal leakage current, thus increasing the two α's. A third way to increase the α's exists by increasing the device (junction) temperature. Increasing the temperature causes a corresponding increase in the leakage current I_{co} to the point where latching can occur. The typical manifestation of this temperature dependence is to cause an effective lowering of the maximum blocking voltage that can be sustained by the thyristor.

Another way to cause a thyristor to switch from forward-blocking to forward-conduction exists. Under a forward applied voltage, J_2 is reverse biased while the other two junctions are forward-biased in the blocking mode. The reverse-biased junction of J_2 is the dominant capacitance of the three and determines the displacement current that flows. If the rate of increase in the applied v_{AK} (dv_{AK}/dt) is sufficient, it will cause a significant displacement current through the J_2 capacitance. This displacement current can initiate switching similar to an externally applied gate current. This dynamic phenomenon is inherent in all thyristors and causes there to be a limit (dv/dt) to the time rate of applied v_{AK} that can be placed on the device to avoid uncontrolled switching. Alterations to the basic thyristor structure can be produced that increase the dv/dt limit and will be discussed in Section 6.4.

Once the thyristor has moved into forward-conduction, any applied gate current is superfluous. The thyristor is latched and, for SCRs, cannot be returned to a blocking mode by using the gate terminal. Anode current must be commutated away from the SCR for a sufficient time to allow stored charge in the device to recombine. Only after this recovery time has occurred, can a forward voltage be reapplied (below the dv/dt limit of course) and the SCR again be operated in a forward-blocking mode. If the forward voltage is reapplied before sufficient recovery time has elapsed, the SCR will move back into forward-conduction. For GTOs and IGCTs, a large applied reverse gate current (typically in the range of 10–50% of the anode current for GTOs, and 100% of the anode current for IGCTs) applied for a sufficient time can remove enough charge near the GK junction to cause it to turn-off. This interrupts the base current to the *pnp* transistor, leaving the *pnp* open-base, causing thyristor turn-off. This is similar in principle to use negative base current to quickly turn-off a traditional bipolar transistor.

6.3 Static Characteristics

6.3.1 Current–Voltage Curves for Thyristors

A plot of the anode current (i_A) as a function of anode–cathode voltage (v_{AK}) is shown in Fig. 6.3. The forward-blocking mode is shown as the low-current portion of the graph (solid curve around operating point "1"). With zero gate current and positive v_{AK}, the forward characteristic in the off- or blocking-state is determined by the center junction J_2, which is reverse biased. At operating point "1" very little current flows (I_{co} only) through the device. However, if the applied voltage exceeds the forward-blocking voltage, the thyristor switches to its on- or conducting-state (shown as operating point "2") because of carrier multiplication (M in Eq. (6.1)). The effect of gate current is to lower the blocking voltage at which switching takes place. The thyristor moves rapidly along the negatively-sloped portion of the curve until it reaches a stable operating point determined by the external circuit (point "2"). The portion of the graph indicating forward-conduction shows the large values of i_A that may be conducted at relatively low values of v_{AK}, similar to a power diode.

As the thyristor moves from forward-blocking to forward-conduction, the external circuit must allow sufficient anode current to flow to keep the device latched. The minimum anode current that will cause the device to remain in forward-conduction as it switches from forward-blocking is called the

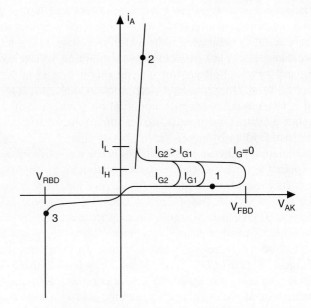

FIGURE 6.3 Static characteristic i–v curve typical of thyristors.

latching current I_L. If the thyristor is already in forward-conduction and the anode current is reduced, the device can move its operating mode from forward-conduction back to forward-blocking. The minimum value of anode current necessary to keep the device in forward-conduction after it has been operating at a high anode current value is called the holding current I_H. The holding current value is lower than the latching current value as indicated in Fig. 6.3.

The reverse thyristor characteristic, quadrant III of Fig. 6.3, is determined by the outer two junctions (J_1 and J_3), which are reverse biased in this operating mode (applied v_{AK} is negative). Symmetric thyristors are designed so that J_1 will reach reverse breakdown due to carrier multiplication at an applied reverse potential near the forward breakdown value (operating point "3" in Fig. 6.3). The forward- and reverse-blocking junctions are usually fabricated at the same time with a very long diffusion process (10–50 h) at high temperatures (>1200°C). This process produces symmetric blocking properties. Wafer edge termination processing causes the forward-blocking capability to be reduced to about 90% of the reverse-blocking capability. Edge termination is discussed below. Asymmetric devices are made to optimize forward-conduction and turn-off properties, and as such reach reverse breakdown at a lower voltage than that applied in the forward direction. This is accomplished by designing the asymmetric thyristor with a much thinner n-base than is used in symmetric structures. The thin n-base leads to improved properties such as lower forward drop and shorter switching times. Asymmetric devices are generally used in applications when only forward voltages (positive, v_{AK}) are to be applied (including many inverter designs).

The form of the gate-to-cathode i–v characteristic of SCRs, GTOs and IGCTs is similar to that of a diode. With positive gate bias, the gate–cathode junction is forward biased and permits the flow of a large current in the presence of a low voltage drop. When negative gate voltage is applied to an SCR, the gate–cathode junction is reverse biased and prevents the flow of current until the avalanche breakdown voltage is reached. In a GTO or IGCT, a negative gate voltage is applied to provide a low impedance path for anode current to flow out of the device instead of out the cathode. In this way the cathode region (base–emitter junction of the equivalent npn transistor) turns off, thus pulling the equivalent npn transistor out of conduction. This causes the entire thyristor to return to its blocking state. The problem with the GTO and IGCT is that the gate-drive circuitry is typically required to sink 10–50% (for the GTO) or 100% (for the IGCT) of the anode current to achieve turn-off.

6.3.2 Edge and Surface Terminations

Thyristors are often made with planar diffusion technology to create the cathode region. Formation of these regions creates cylindrical curvature of the metallurgical gate–cathode junction. Under reverse bias, the curvature of the associated depletion region results in electric field crowding along the curved section of the p^+ diffused region. The field crowding seriously reduces the breakdown potential below that expected for the bulk semiconductor. A floating field ring, an extra p diffused region with no electrical connection at the surface, is often added to modify the electric field profile and thus reduce it to a value below or at the field strength in the bulk. An illustration of a single floating field ring is shown in Fig. 6.4. The spacing, W, between the main anode region and the field ring is critical. Multiple rings can also be employed to further modify the electric field in high-voltage rated thyristors.

Another common method for altering the electric field at the surface is by using a field plate as shown in cross section in Fig. 6.5. By forcing the potential over the oxide to be the same as at the surface of the p^+ region, the depletion region can be extended so that the electric field intensity is reduced near the curved portion of the diffused p^+ region. A common practice is to use field plates with floating field rings to obtain optimum breakdown performance.

High-voltage thyristors are made from single wafers of Si and must have edge terminations other than floating field rings or field plates to promote bulk breakdown and limit leakage

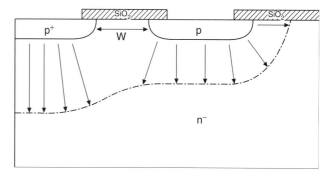

FIGURE 6.4 Cross section showing a floating field ring to decrease the electric field intensity near the curved portion of the main anode region (left-most p^+ region).

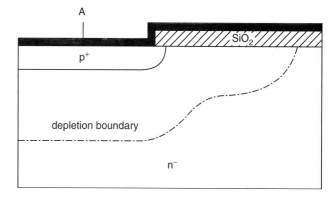

FIGURE 6.5 Cross section showing a field plate used to reduce the electric field intensity near the curved portion of the p^+-region (anode).

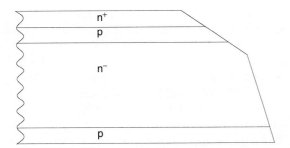

FIGURE 6.6 Cross section of a thyristor showing the negative bevel (upper p–n^- and p–n^+ junctions) and positive bevel (lower p–n^- junction) used for edge termination of large-area devices.

current at the surface. Controlled bevel angles can be created using lapping and polishing techniques during production of large-area thyristors. Two types of bevel junctions can be created: (i) a positive bevel defined as one in which the junction area decreases when moving from the highly-doped to the lightly-doped side of the depletion region and (ii) a negative bevel defined as one in which the junction area increases when moving from the highly-doped to the lightly-doped side of the depletion region. In practice, the negative bevel must be lapped at an extremely shallow angle to reduce the surface field below the field intensity in the bulk. All positive bevel angles between 0 and 90° result in a lower surface field than in the bulk. Figure 6.6 shows the use of a positive bevel for the J_1 junction and a shallow negative bevel for the J_2 and J_3 junctions on a thyristor cross section to make maximum use of the Si area for conduction and still reduce the surface electric field. Further details of the use of beveling, field plates, and field rings can be found in Ghandi [2] and Baliga [3].

6.3.3 Packaging

Thyristors are available in a wide variety of packages, from small plastic ones for low-power (i.e. TO-247), to stud-mount packages for medium-power, to press-pack (also called flat-pack) for the highest power devices. The press-packs must be mounted under pressure to obtain proper electrical and thermal contact between the device and the external metal electrodes. Special force-calibrated clamps are made for this purpose. Large-area thyristors cannot be directly attached to the large copper pole piece of the press-pack because of the difference in the coefficient of thermal expansion (CTE), hence the use of a pressure contact for both anode and cathode. Figure 6.7 shows typical thyristor stud-mount and press-pack packages.

Many medium power thyristors are appearing in modules where a half- or full-bridge (and associated anti-parallel diodes) is put together in one package. A power module package should have five characteristics:

i) electrical isolation of the baseplate from the semiconductor;
ii) good thermal performance;
iii) good electrical performance;
iv) long life/high reliability; and
v) low cost.

Electrical isolation of the baseplate from the semiconductor is necessary in order to contain both halves of a phase leg in one package as well as for convenience (modules switching different phases can be mounted on one heatsink) and safety (heatsinks can be held at ground potential).

Thermal performance is measured by the maximum temperature rise in the Si die at a given power dissipation level with a fixed heat sink temperature. The lower the die temperature, the better the package. A package with a low thermal resistance from junction-to-sink can operate at higher power densities for the same temperature rise or lower temperatures for the same power dissipation than a more thermally resistive package. While maintaining low device temperature is generally preferable, temperature variation affects majority carrier and bipolar devices differently. Roughly speaking, in a bipolar device such as a thyristor, switching losses increase and

FIGURE 6.7 Examples of thyristor packaging: stud-mount (left) and press-pack/capsule (right).

6 Thyristors

TABLE 6.1 Thermal conductivity of thyristor package materials

Material	Thermal conductivity (W/m·K) at 300 K
Silicon	150
Copper (baseplate and pole pieces)	390–400
AlN substrate	170
Al_2O_3 (Alumina)	28
Aluminum (Al)	220
Tungsten (W)	167
Molybdenum (Mo)	138
Metal matrix composites (MMC)	170
Thermal grease (heatsink compound)	0.75
60/40 solder (Pb/Sn eutectic)	50
95/5 solder (Pb/Sn high temperature)	35

TABLE 6.2 CTE for thyristor package materials

Material	CTE (μm/m·K) at 300 K
Silicon	4.1
Copper (baseplate and pole pieces)	17
AlN substrate	4.5
Al_2O_3 (Alumina)	6.5
Tungsten (W)	4.6
Molybdenum (Mo)	4.9
Aluminum (Al)	23
Metal matrix composites (MMC)	5–20
60/40 solder (Pb/Sn eutectic)	25

conduction losses decrease with increasing temperature. In a majority carrier device, such as a MOSFET, conduction losses increase with increasing temperature. The thermal conductivity of typical materials used in thyristor packages is shown in Table 6.1.

Electrical performance refers primarily to the stray inductance in series with the die, as well as the capability of mounting a low-inductance bus to the terminals. Another problem is the minimization of capacitive cross-talk from one switch to another, which can cause an abnormal on-state condition by charging the gate of an off-state switch, or from a switch to any circuitry in the package (as would be found in a hybrid power module). Capacitive coupling is a major cause of electromagnetic interference (EMI). As the stray inductance of the module and the bus sets a minimum switching loss for the device – because the switch must absorb the stored inductive energy – it is very important to minimize inductance within the module. Reducing the parasitic inductance reduces the high-frequency ringing during transients that is another cause of radiated electromagnetic interference. Since stray inductance can cause large peak voltages during switching transients, minimizing it helps to maintain the device within its safe area of operation.

Long life and high reliability are primarily attained through minimization of thermal cycling, minimization of ambient temperature, and proper design of the transistor stack. Thermal cycling fatigues material interfaces because of coefficient of thermal expansion (CTE) mismatch between dissimilar materials. As the materials undergo temperature variation, they expand and contract at different rates which stresses the interface between the layers and can cause interface deterioration (e.g. cracking of solder layers or wire debonding). Chemical degradation processes such as dendrite growth and impurity migration are accelerated with increasing temperature, so keeping the absolute temperature of the device low, as well as minimizing the temperature changes to which it is subjected is important. Typical CTE values for common package materials are given in Table 6.2.

Low cost is achieved in a variety of ways. Both manufacturing and material costs must be taken into account when designing a power module. Materials that are difficult to machine or process, even if they are relatively cheap in raw form (molybdenum, for example), should be avoided. Manufacturing processes that lower yield also drive up costs. In addition, a part that is very reliable can reduce future costs by reducing the need for repair and replacement.

The basic half-bridge module has three power terminals: plus, minus and phase. Advanced modules differ from traditional high power commercial modules in several ways. The baseplate is metallized aluminum nitride (AlN) ceramic rather than the typical 0.125" thick nickel-plated copper baseplate with a soldered metallized ceramic substrate for electrical isolation. This AlN baseplate stack provides a low thermal resistance from die to heatsink. The copper terminal power busses are attached by solder to the devices in a wirebond-free, low-inductance, low-resistance, device interconnect configuration. The balance of the assembly is typical for module manufacturing with attachment of shells, use of dielectric gels, and hard epoxies and adhesive to seal the finished module. Details of the thermal performance of modules and advanced modules can be found in Beker et al. [4] and Godbold et al. [5].

6.4 Dynamic Switching Characteristics

The time rate of rise of anode current (di/dt) during turn-on and the time rate of rise of anode–cathode voltage (dv/dt) during turn-off are important parameters to control for ensuring proper and reliable operation. All thyristors have maximum limits for di/dt and dv/dt that must not be exceeded. Devices capable of conducting large currents in the on-state, are necessarily made with large surface areas through which the current flows. During turn-on, localized areas of a device (near the gate region) begin to conduct current. The initial turn-on of an SCR is shown in Fig. 6.8. The cross section illustrates how injected gate current flows to the nearest cathode region, causing this portion of the npn transistor to begin conducting. The pnp transistor then follows the npn into conduction such that anode current begins flowing only in a small portion of the

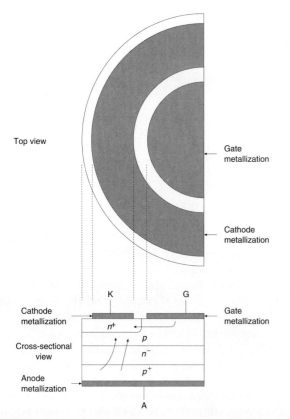

FIGURE 6.8 Top view and associated cross section of gate–cathode periphery showing initial turn-on region in a center-fired thyristor.

applications, the entire cathode region is never fully in conduction. Prevention of di/dt failure can be accomplished if the rate of increase of conduction area exceeds the di/dt rate such that the internal junction temperature does not exceed a specified critical temperature (typically approximately 350°C). This critical temperature decreases as the blocking voltage increases. Adding series inductance to the thyristor to limit di/dt below its maximum usually causes circuit design problems.

Another way to increase the di/dt rating of a device is to increase the amount of gate–cathode periphery. Inverter SCRs (so-named because of their use in high-frequency power converter circuits that convert dc to ac, i.e. invert) are designed so that there is a large amount of gate edge adjacent to a significant amount of cathode edge. A top surface view of two typical gate–cathode patterns, found in large thyristors, is shown in Fig. 6.9. An inverter SCR often has a stated maximum di/dt limit of approximately 2000 A/μs. This value has been shown to be conservative [6], and by using excessive gate current under certain operating conditions, an inverter SCR can be operated reliably at 10,000 A/μs–20,000 A/μs.

A GTO takes the interdigitation of the gate and cathode to the extreme (Fig. 6.9, left). In Fig. 6.10 a cross section of a GTO shows the amount of interdigitation. A GTO often has cathode islands that are formed by etching the Si. A metal plate can be placed on the top to connect the individual cathodes into a large arrangement of electrically parallel cathodes. The gate metallization is placed so that the gate surrounding each cathode is electrically in parallel as well. This construction not only allows high di/dt values to be reached, as in an inverter SCR, but also provides the capability to turn-off the anode current by shunting it away from the individual cathodes and out of the gate electrode upon reverse biasing of the gate. During turn-off, current is decreasing while voltage across the device is increasing. If the forward voltage becomes too high while sufficient current is still flowing, then the device will drop back into its conduction mode instead of completing its turn-off cycle. Also, during turn-off, the power dissipation can

cathode region. If the local current density becomes too large (in excess of several thousand amperes per square centimeter), then self-heating will damage the device. Sufficient time (referred to as plasma spreading time) must be allowed for the entire cathode area to begin conducting before the localized currents become too high. This phenomenon results in a maximum allowable rate of rise of anode current in a thyristor and is referred to as a di/dt limit. In many high-frequency

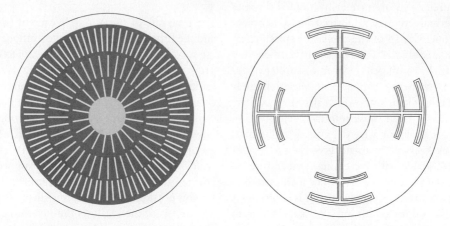

FIGURE 6.9 Top view of typical interdigitated gate–cathode patterns used for thyristors.

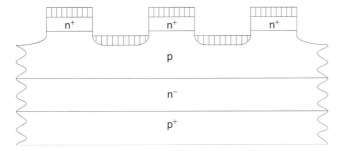

FIGURE 6.10 Cross section of a GTO showing the cathode islands and interdigitation with the gate (p-base).

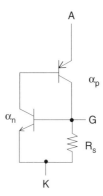

FIGURE 6.11 Two-transistor equivalent circuit showing the addition of a resistive shunt path for anode current.

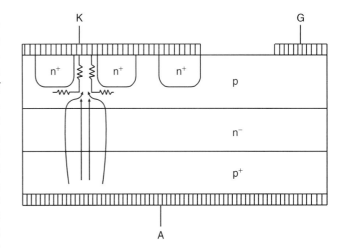

FIGURE 6.12 Cross section showing cathode shorts and the resulting resistive shunt path for anode current.

become excessive if the current and voltage are simultaneously too large. Both of these turn-off problems can damage the device as well as other portions of the circuit.

Another switching problem that occurs is associated primarily with thyristors, though other power electronic devices suffer some degradation of performance from the same problem. This problem is that thyristors can self-trigger into a forward-conduction mode from a forward-blocking mode if the rate of rise of forward anode–cathode voltage is too large. This triggering method is due to displacement current through the associated junction capacitances (the capacitance at J_2 dominates because it is reverse biased under forward applied voltage). The displacement current contributes to the leakage current I_{co}, shown in Eq. (6.1). The SCRs, GTOs and IGCTs, therefore, have a maximum dv/dt rating that should not be exceeded (typical values are 100–1000 V/μs). Switching into a reverse-conducting from a reverse-blocking state, due to an applied reverse dv/dt, is not possible because the values of the reverse α's of the equivalent transistors can never be made large enough to cause the necessary feedback (latching) effect. An external capacitor is often placed between the anode and cathode of the thyristor to help control the dv/dt experienced. Capacitors and other components that are used to form such protection circuits, known as snubbers, may be found in all power semiconductor devices.

6.4.1 Cathode Shorts

As the temperature in the thyristor increases above 25°C, the minority carrier lifetime and the corresponding diffusion lengths in the n- and p-bases increase. This leads to an increase in the α's of the equivalent transistors. Discussion of the details of the minority carrier diffusion length and its role in determining the current gain factor α can be found in Sze [7]. Referring to Eq. (6.1), it is seen that a lower applied bias will give a carrier multiplication factor M sufficient to switch the device from forward-blocking into conduction, because of this increase of the α's with increasing temperature. Placing a shunt resistor in parallel with the base–emitter junction of the equivalent npn transistor (shown in Fig. 6.11) will result in an effective current gain, α_{neff}, that is lower than α_n, as given by Eq. (6.2), where v_{GK} is the applied gate–cathode voltage, R_s is the equivalent lumped value for the distributed current shunting structure, and the remaining factors form the appropriate current factor based on the applied bias and characteristics of the gate–cathode junction. The shunt current path is implemented by providing intermittent shorts, called cathode shorts, between the p-base (gate) region and the n^+-emitter (cathode) region in the thyristor as illustrated in Fig. 6.12. The lumped shunt resistance value is in the range of 1–15 Ω as measured from gate to cathode.

$$\alpha_{neff} = \alpha_n \left(\frac{1}{1 + (v_{GK}\alpha_n)/(R_s i_0\ exp(qv_{GK}/kT))} \right) \quad (6.2)$$

Low values of anode current (e.g. those associated with an increase in temperature under forward-blocking conditions) will flow through the shunt path to the cathode contact, bypassing the n^+-emitter and keeping the device out of its forward-conduction mode. As the anode current becomes large, the potential drop across the shunt resistance will be sufficient to forward bias the gate–cathode junction, J3, and bring the thyristor into forward-conduction. The cathode shorts also provide a path for displacement current to flow without forward biasing J3. The *dv/dt* rating of the thyristor is thus improved as well as the forward-blocking characteristics by using cathode shorts. However, the shorts do cause a lowering of cathode current handling capability because of the loss of some of the cathode area (n^+-region) to the shorting pattern, an increase in the necessary gate current to obtain switching from forward-blocking to forward-conduction, and an increase in complexity of manufacturing of the thyristor. The loss of cathode area due to the shorting-structure is from 5 to 20%, depending on the type of thyristor. By careful design of the cathode short windows to the *p*-base, the holding current can be made lower than the latching current. This is important so that the thyristor will remain in forward-conduction when used with varying load impedances.

6.4.2 Anode Shorts

A further increase in forward-blocking capability can be obtained by introducing anode shorts in addition to the cathode shorts. This reduces α_p in a similar manner that cathode shorts reduce α_n. An illustration of this is provided in Fig. 6.13. In this structure both J1 and J3 are shorted (anode and cathode shorts), so that the forward-blocking capability of the thyristor is completely determined by the avalanche breakdown characteristics of J2. Anode shorts will result in the complete loss of reverse-blocking capability and is only suitable for thyristors used in asymmetric circuit applications.

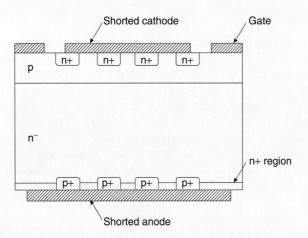

FIGURE 6.13 Cross section showing integrated cathode and anode shorts.

6.4.3 Amplifying Gate

The cathode-shorting structure will reduce the gate sensitivity dramatically. To increase this sensitivity and yet retain the

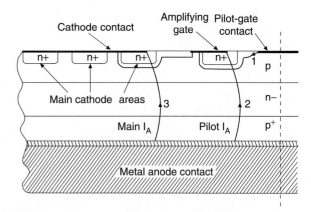

FIGURE 6.14 Cross section showing the amplifying gate structure in a thyristor.

benefits of the cathode-shorts, a structure called an amplifying gate (or regenerative gate) is used, as shown in Fig. 6.14 (and Fig. 6.9, right). When the gate current (1) is injected into the *p*-base through the pilot-gate contact, electrons are injected into the *p*-base by the n^+-emitter with a given emitter injection efficiency. These electrons traverse through the *p*-base (the time taken for this process is called the transit time) and accumulate near the depletion region. This negative charge accumulation leads to injection of holes from the anode. The device then turns on after a certain delay, dictated by the *p*-base transit time, and the pilot anode current (2) begins to flow through a small region near the pilot-gate contact as shown in Fig. 6.14.

This flow of pilot anode current corresponds to the initial sharp rise in the anode current waveform (phase I), as shown in Fig. 6.15. The device switching then goes into phase II, during which the anode current remains fairly constant, suggesting that the resistance of the region has reached its lower limit. This is due to the fact that the pilot anode current (2) takes a finite time to traverse through the *p*-base laterally and become the gate current for the main cathode area. The n^+-emitters start to inject electrons which traverse the *p*-base vertically and after a certain finite time (transit time of the *p*-base) reach the depletion region. The total time taken by the lateral traversal of pilot anode current and the electron transit time across the *p*-base is the reason for observing this characteristic phase II interval. The width of the phase II interval is comparable to the switching delay, suggesting that the *p*-base transit time is of primary importance. Once the main cathode region turns on, the resistance of the device decreases and the anode current begins to rise again (transition from phase II to III). From this time onward in the switching cycle, the plasma spreading velocity will dictate the rate at which the conduction area will increase. The current density during phase I and II can be quite large, leading to a considerable increase in the local temperature and device failure. The detailed effect of the amplifying gate on the anode current rise will only be noticed

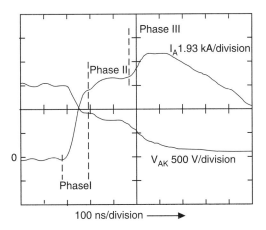

FIGURE 6.15 Turn-on waveforms showing the effect of the amplifying gate in the anode current rise.

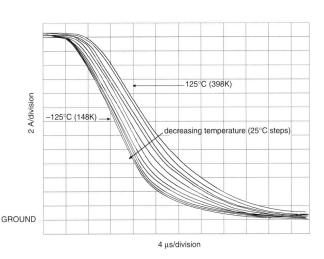

FIGURE 6.16 Temperature effect on the anode current tail during turn-off.

at high levels of *di/dt* (in the range of 1000 A/μs), shown in Fig. 6.15. It can be concluded that the amplifying gate will increase gate sensitivity at the expense of some *di/dt* capability, as demonstrated by Sankaran [8]. This lowering of *di/dt* capability can be somewhat off-set by an increase in gate–cathode interdigitation as previously discussed.

6.4.4 Temperature Dependencies

The forward-blocking voltage of an SCR has been shown to be reduced from 1350 V at 25°C to 950 V at −175°C in a near linear fashion [8]. Above 25°C, the forward-blocking capability is again reduced, due to changes in the minority carrier lifetime which cause the leakage current to increase and the associated breakover voltage to decrease. Several dominant physical parameters associated with semiconductor devices are sensitive to temperature variations, causing their dependent device characteristics to change dramatically. The most important of these parameters are: (i) the minority carrier lifetimes (which control the high-level injection lifetimes); (ii) the hole and electron mobilities; (iii) the impact ionization collision cross sections; and (iv) the free-carrier concentrations (primarily the ionized impurity-atom concentration). Almost all of the impurity atoms are ionized at temperatures above 0°C, and so further discussion of the temperature effects on ionization is not relevant for normal operation. As the temperature increases above 25°C, the following trends are observed: the carrier lifetimes increase, giving longer recovery times and greater switching losses; the carrier mobilities are reduced, increasing the on-state voltage drop; and at very high temperatures, the intrinsic carrier concentration becomes sufficiently high that the depletion layer will not form and the device cannot switch off. A more detailed discussion of these physical parameters is beyond the scope of this article, but references are listed for those persons interested in pursuing relevant information about temperature effects.

It is well known that charge carrier recombination events are more efficient at lower temperatures. This shows up as a larger potential drop during forward-conduction and a shorter recovery time during turn-off. A plot of the anode current during turn-off, at various temperatures, for a typical GTO is shown in Fig. 6.16.

An approximate relation between the temperature and the forward drop across the *n*-base of a thyristor is discussed in detail by Herlet [10] and Hudgins *et al.* [11]. Temperature dependent equations relating the anode current density, J_A and the applied anode–cathode voltage V_{AK} are also given in Reference [11]; these include the junction potential drops in the device, the temperature dependence of the bandgap energy, and the *n*-base potential drop. Data from measurements at forward current densities of approximately 100 A/cm^2 on a GTO rated for 1 kV symmetric blocking gives forward voltage drops of 1.7 V at −50°C and 1.8 V at 150°C.

6.5 Thyristor Parameters

Understanding of a thyristor's maximum ratings and electrical characteristics is required for proper application. Use of a manufacturer's data sheet is essential for good design practice. *Ratings* are maximum or minimum values that set limits on device capability. A measure of device performance under specified operating conditions is a *characteristic* of the device. A summary of some of the maximum ratings which must be considered when choosing a thyristor for a given application is provided in Table 6.3. Thyristor types shown in parentheses indicate a maximum rating unique to that device. Both forward and reverse repetitive and non-repetitive voltage ratings must be considered, and a properly rated device must be chosen so that the maximum voltage ratings are never exceeded. In most cases, either forward or reverse voltage

TABLE 6.3 Thyristor maximum ratings specified by manufacturers

Symbol	Description
V_{RRM}	Peak repetitive reverse voltage
V_{RSM}	Peak non-repetitive reverse voltage (transient)
$V_{R(DC)}$	DC reverse blocking voltage
V_{DRM}	Peak repetitive forward off-state voltage
V_{DSM}	Peak non-repetitive forward off-state voltage (transient)
$V_{D(DC)}$	DC forward-blocking voltage
$I_{T(RMS)}, I_{F(RMS)}$	RMS forward on-state current
$I_{T(AV)}, I_{F(AV)}$	Average forward on-state current at specified case or junction temperature
$I_{TSM}, I_{F(TSM)}$	Peak one-cycle surge on-state current (values specified at 60 and 50 Hz)
I_{TGQ} (GTO)	Peak controllable current
$I^2 t$	Non-repetitive pulse overcurrent capability ($t = 8.3$ ms for a 60 Hz half cycle)
P_T	Maximum power dissipation
di/dt	Critical rate of rise of on-state current at specified junction temperature, gate current and forward-blocking voltage
P_{GM} (P_{FGM} for GTO)	Peak gate power dissipation (forward)
P_{RGM} (GTO)	Peak gate power dissipation (reverse)
$P_{G(AV)}$	Average gate power dissipation
V_{FGM}	Peak forward gate voltage
V_{RGM}	Peak reverse gate voltage
I_{FGM}	Peak forward gate current
I_{RGM} (GTO)	Peak reverse gate current
T_{STG}	Storage temperature
T_j	Junction operating temperature
V_{RMS}	Voltage isolation (modules)

transients in excess of the non-repetitive maximum ratings result in destruction of the device. The maximum root mean square (RMS) or average current ratings given are usually those which cause the junction to reach its maximum rated temperature. Because the maximum current will depend upon the current waveform and upon thermal conditions external to the device, the rating is usually shown as a function of case temperature and conduction angle. The peak single half-cycle surge-current rating must be considered, and in applications where the thyristor must be protected from damage by overloads, a fuse with an $I^2 t$ rating smaller than the maximum rated value for the device must be used. Maximum ratings for both forward and reverse gate voltage, current and power also must not be exceeded.

The maximum rated operating junction temperature T_J must not be exceeded, since device performance, in particular voltage-blocking capability, will be degraded. Junction temperature cannot be measured directly but must be calculated from a knowledge of steady-state thermal resistance $R_{\Theta(J-C)}$, and the average power dissipation. For transients or surges, the transient thermal impedance ($Z_{\Theta(J-C)}$) curve must be used (provided in manufacturer's data sheets). The maximum average power dissipation P_T is related to the maximum rated operating junction temperature and the case temperature by the steady-state thermal resistance. In general, both the maximum dissipation and its derating with increasing case temperature are provided.

The number and type of thyristor characteristics specified varies widely from one manufacturer to another. Some characteristics are given only as typical values of minima or maxima, while many characteristics are displayed graphically. Table 6.4 summarizes some of the typical characteristics provided as maximum values. The maximum value means that the manufacturer guarantees that the device will not exceed the value given under the specified operating or switching conditions. A minimum value means that the manufacturer guarantees that the device will perform at least, as well as the characteristic given under the specified operating or switching conditions. Thyristor types shown in parenthesis indicate a characteristic unique to that device. Gate conditions of both voltage and current to ensure either non-triggered or triggered device operation are included. The turn-on and turn-off transients of the thyristor are characterized by switching times like the

TABLE 6.4 Typical thyristor characteristic maximum and minimum values specified by manufacturers

Symbol	Description
V_{TM}, V_{FM}	Maximum on-state voltage drop (at specified junction temperature and forward current)
I_{DRM}	Maximum forward off-state current (at specified junction temperature and forward voltage)
I_{RRM}	Maximum reverse off-state current (at specified junction temperature and reverse voltage)
dv/dt	Minimum critical rate of rise of off-state voltage at specified junction temperature and forward-blocking voltage level
V_{GT}	Maximum gate trigger voltage (at specified temperature and forward applied voltage)
V_{GD}, V_{GDM}	Maximum gate non-trigger voltage (at specified temperature and forward applied voltage)
I_{GT}	Maximum gate trigger current (at specified temperature and forward applied voltage)
T_{gt} (GTO)	Maximum turn-on time (under specified switching conditions)
T_q	Maximum turn-off time (under specified switching conditions)
t_D	Maximum turn-on delay time (for specified test)
$R_{\Theta(J-C)}$	Maximum junction-to-case thermal resistance
$R_{\Theta(C-S)}$	Maximum case-to-sink thermal resistance (interface lubricated)

turn-off time listed in Table 6.4. The turn-on transient can be divided into three intervals: (i) gate-delay interval; (ii) turn-on of initial area; and (iii) spreading interval. The gate-delay interval is simply the time between application of a turn-on pulse at the gate and the time the initial cathode area turns on. This delay decreases with increasing gate drive current and is of the order of a few microseconds. The second interval, the time required for turn-on of the initial area, is quite short, typically less than 1 μs. In general, the initial area turned on is a small percentage of the total useful device area. After the initial area turns on, conduction spreads (spreading interval or plasma spreading time) throughout the device in tens of microseconds for high-speed or thyristors. The plasma spreading time may take up to hundreds of microseconds in large-area phase-control devices.

Table 6.5 lists many of the thyristor parameters that appear as listed values or as information on graphs. The definition of each parameter and the test conditions under which they are measured are given in the table as well.

6.6 Types of Thyristors

In recent years, most development effort has gone into continued integration of the gating and control electronics into thyristor modules, and the use of MOS-technology to create gate structures integrated into the thyristor itself. Many variations of this theme are being developed and some technologies should rise above the others in the years to come. Further details concerning most of the following discussion of thyristor types can be found in [1].

6.6.1 SCRs and GTOs

The highest power handling devices continue to be bipolar thyristors. High powered thyristors are large diameter devices, some well in excess of 100 mm, and as such have a limitation on the rate of rise of anode current, a di/dt rating. The depletion capacitances around the p–n junctions, in particular the center junction J_2, limit the rate of rise in forward voltage that can be applied even after all the stored charge, introduced during conduction, is removed. The associated displacement current under application of forward voltage during the thyristor blocking state sets a dv/dt limit. Some effort in improving the voltage hold-off capability and over-voltage protection of conventional SCRs is underway by incorporating a lateral high resistivity region to help dissipate the energy during breakover. Most effort, though, is being placed in the further development of high performance GTOs and IGCTs because of their controllability and to a lesser extent in optically triggered structures that feature gate circuit isolation.

High voltage GTOs with symmetric blocking capability require thick n-base regions to support the high electric field. The addition of an n+ buffer layer next to the p+-anode allows high voltage forward-blocking and a low forward voltage drop during conduction because of the thinner n-base required. Cylindrical anode shorts have been incorporated to facilitate excess carrier removal from the n-base during turn-off and still retain the high blocking capability. This device structure can control 200 A, operating at 900 Hz, with a 6 kV hold-off. Some of the design tradeoff between the n-base width and turn-off energy losses in these structures have been determined. A similar GTO incorporating an n^+-buffer layer and a *pin* structure has been fabricated that can control up to 1 kA (at a forward drop of 4 V) with a forward blocking capability of 8 kV. A reverse conducting GTO has been fabricated that can block 6 kV in the forward direction, interrupt a peak current of 3 kA and has a turn-off gain of about 5.

The IGCT is a modified GTO structure. It is designed and manufactured so that it commutates all of the cathode current away from the cathode region and diverts it out of the gate contact. The IGCT is similar to a GTO in structure except that it always has a low-loss n-buffer region between the n-base and p-emitter. The IGCT device package is designed to result in a very low parasitic inductance and is integrated with a specially designed gate-drive circuit. The gate drive contains all the necessary di/dt and dv/dt protection; the only connections required are a low-voltage power supply for the gate drive and an optical signal for controlling the gate. The specially designed gate drive and ring-gate package circuit allows the IGCT to be operated without a snubber circuit, and to switch with a higher anode di/dt than a similar GTO. At blocking voltages of 4.5 kV and higher the IGCT provides better performance than a conventional GTO. The speed at which the cathode current is diverted to the gate (di_{GQ}/dt) is directly related to the peak snubberless turn-off capability of the IGCT. The gate drive circuit can sink current for turn-off at di_{GQ}/dt values in excess of 7000 A/μs. This hard gate drive results in a low charge storage time of about 1 μs. The low storage time and the fail-short mode makes the IGCT attractive for high-power, high-voltage series applications; examples include high-power converters in excess of 100 MVA, static vol-ampere reactive (VAR) compensators and converters for distributed generation such as wind power.

6.6.2 MOS-controlled Thyristors

The cross section of the *p*-type MCT unit cell is given in Fig. 6.17. When the MCT is in its forward-blocking state and a negative gate–anode voltage is applied, an inversion layer is formed in the n-doped material that allows holes to flow laterally from the p-emitter (p-channel FET source) through the channel to the p-base (p-channel FET drain). This hole flow is the base current for the *npn* transistor. The n-emitter then injects electrons which are collected in the n-base, causing the p-emitter to inject holes into the n-base so that the *pnp* transistor is turned on and latches the MCT. The MCT is brought

TABLE 6.5 Symbols and definitions of major thyristor parameters

Symbol	Name	Definition
R_θ	Thermal resistance	Specifies the degree of temperature rise per unit of power, measuring junction temperature from a specified external point. Defined when junction power dissipation results in steady-state thermal flow.
$R_{\theta(J-A)}$	Junction-to-ambient thermal resistance	The steady-state thermal resistance between the junction and ambient.
$R_{\theta(J-C)}$	Junction-to-case thermal resistance	The steady-state thermal resistance between the junction and case surface.
$R_{\theta(J-S)}$	Junction-to-sink thermal resistance	The steady-state thermal resistance between the junction and the heatsink mounting surface.
$R_{\theta(C-S)}$	Contact thermal resistance	The steady-state thermal resistance between the surface of the case and the heatsink mounting surface.
Z_θ	Transient thermal impedance	The change of temperature difference between two specified points or regions at the end of a time interval divided by the step function change in power dissipation at the beginning of the same interval causing the change of temperature difference.
$Z_{\theta(J-A)}$	Junction-to-ambient transient thermal impedance	The transient thermal impedance between the junction and ambient.
$Z_{\theta(J-C)}$	Junction-to-case transient thermal impedance	The transient thermal impedance between the junction and the case surface.
$Z_{\theta(J-S)}$	Junction-to-sink transient thermal impedance	The transient thermal impedance between the junction and the heatsink mounting surface.
T_A	Ambient temperature	It is the temperature of the surrounding atmosphere of a device when natural or forced-air cooling is used, and is not influenced by heat dissipation of the device.
T_S	Sink temperature	The temperature at a specified point on the device heatsink.
T_C	Case temperature	The temperature at a specified point on the device case.
T_J	Junction temperature	The device junction temperature rating. Specifies the maximum and minimum allowable operation temperatures.
T_{STG}	Storage temperature	Specifies the maximum and minimum allowable storage temperatures (with no electrical connections).
V_{RRM}	Peak reverse blocking voltage	Within the rated junction temperature range, and with the gate terminal open circuited, specifies the repetitive peak reverse anode to cathode voltage applicable on each cycle.
V_{RSM}	Transient peak reverse-blocking voltage	Within the rated junction temperature range, and with the gate terminal open circuited, specifies the non-repetitive peak reverse anode to cathode voltage applicable for a time width equivalent to less than 5 ms.
$V_{R(DC)}$ SCR only	dc reverse-blocking voltage	Within the rated junction temperature range, and with the gate terminal open-circuited, specifies the maximum value for dc anode to cathode voltage applicable in the reverse direction.
V_{DRM}	Peak forward-blocking voltage	Within the rated junction temperature range, and with the gate terminal open circuited (SCR), or with a specified reverse voltage between the gate and cathode (GTO), specifies the repetitive peak off-state anode to cathode voltage applicable on each cycle. This does not apply for transient off-state voltage application.
V_{DSM}	Transient peak forward-blocking voltage	Within the rated junction temperature range, and with the gate terminal open circuited (SCR), or with a specified reverse voltage between the gate and cathode (GTO), specifies the non-repetitive off-state anode to cathode voltage applicable for a time width equivalent to less than 5 ms. This gives the maximum instantaneous value for non-repetitive transient off-state voltage.
$V_{D(DC)}$	dc forward-blocking voltage	Within the rated junction temperature range, and with the gate terminal open circuited (SCR), or with a specified reverse voltage between the gate and cathode (GTO), specifies the maximum value for dc anode to cathode voltage applicable in the forward direction.
dv/dt	Critical rate of rise of off-state voltage $dv/dt = (0.632 V_D)/\tau V_D$ is specified off-state voltage τ is time constant for exponential	At the maximum rated junction temperature range, and with the gate terminal open circuited (SCR), or with a specified reverse voltage between the gate and cathode (GTO), this specifies the maximum rate of rise of off-state voltage that will not drive the device from an off-state to an on-state when an exponential off-state voltage of specified amplitude is applied to the device.
V_{TM}	Peak on-state voltage	At specified junction temperature, and when on-state current (50 or 60 Hz, half sine wave of specified peak amplitude) is applied to the device, indicates peak value for the resulting voltage drop.

TABLE 6.5—Contd.

Symbol	Term	Description
$I_{T(RMS)}$	RMS on-state current	At specified case temperature, indicates the RMS value for on-state current that can be continuously applied to the device.
$I_{T(AV)}$	Average on-state current	At specified case temperature, and with the device connected to a resistive or inductive load, indicates the average value for forward-current (sine half wave, commercial frequency) that can be continuously applied to the device.
I_{TSM}	Peak on-state current	Within the rated junction temperature range, indicates the peak-value for non-repetitive on-state current (sine half wave, 50 or 60 Hz). This value indicated for one cycle, or as a function of a number of cycles.
$I^2 t$	Current-squared time	The maximum, on-state, non-repetitive short-time thermal capacity of the device and is helpful in selecting a fuse or providing a coordinated protection scheme of the device in the equipment. This rating is intended specifically for operation less than one half cycle of a 180° (degree) conduction angle sinusoidal wave form. The off-state blocking capability cannot be guaranteed at values near the maximum $I^2 t$.
di/dt	Critical rate of Rise of on-state current	At specified case temperature, specified off-state voltage, specified gate conditions, and at a frequency of less than 60 Hz, indicates the maximum rate of rise of on-state current which the thyristor will withstand when switching from an off-state to an on-state, when using recommended gate drive.
I_{RRM}	Peak reverse leakage current	At maximum rated junction temperature, indicates the peak value for reverse current flow when a voltage (sine half wave, 50 or 60 Hz, and having a peak value as specified for repetitive peak reverse-voltage rating) is applied in a reverse direction to the device.
I_{DRM}	Peak forward-leakage current	At maximum rated junction temperature, indicates the peak value for off-state current flow when a voltage (sine half wave, 50 or 60 Hz, and having a peak value for repetitive off-state voltage rating) is applied in a forward direction to the device. For a GTO, a reverse voltage between the gate and cathode is specified.
P_{GM} (SCR) P_{GFM} (GTO)	Peak gate power dissipation peak gate forward power dissipation	Within the rated junction temperature range, indicates the peak value for maximum allowable power dissipation over a specified time period, when the device is in forward-conduction between the gate and cathode.
$P_{G(AV)}$	Average gate power dissipation	Within the rated junction temperature range, indicates the average value for maximum allowable power dissipation when the device is forward-conducting between the gate and cathode.
P_{GRM} GTO only	Peak gate reverse power dissipation	Within the rated junction temperature range, indicates the peak value for maximum allowable power dissipation in the reverse direction between the gate and cathode, over a specified time period.
$P_{GR(AV)}$ GTO only	Average gate reverse power dissipation	Within the rated junction temperature range, indicates the average value for maximum allowable power dissipation in the reverse direction between the gate and cathode.
I_{GFM}	Peak forward gate current	Within the rated junction temperature range, indicates the peak value for forward current flow between the gate and cathode.
I_{GRM} GTO only	Peak reverse gate current	Within the rated junction temperature range, indicates peak value for reverse current that can be conducted between the gate and cathode.
V_{GRM}	Peak reverse gate voltage	Within the rated junction temperature range, indicates the peak value for reverse voltage applied between the gate and cathode.
V_{GFM}	Peak forward gate voltage	Within the rated junction temperature range, indicates the peak value for forward voltage applied between the gate and cathode.
I_{GT}	Gate current to trigger	At a junction temperature of 25°C, and with a specified off-voltage, and a specified load resistance, indicates the minimum gate dc current required to switch the thyristor from an off-state to an on-state.
V_{GT}	Gate voltage to trigger	At a junction temperature of 25°C, and with a specified off-state voltage, and a specified load resistance, indicates the minimum dc gate voltage required to switch the thyristor from an off-state to an on-state.
V_{GDM} SCR Only	Non-triggering gate voltage	At maximum rated junction temperature, and with a specified off-state voltage applied to the device, indicates the maximum dc gate voltage which will not switch the device from an off-state to an on-state.
I_{TGQ} GTO only	Gate controlled turn-off current	Under specified conditions, indicates the instantaneous value for on-current usable in gate control, specified immediately prior to device turn-off.

continued

R_θ	Thermal resistance	Specifies the degree of temperature rise per unit of power, measuring junction temperature from a specified external point. Defined when junction power dissipation results in steady-state thermal flow.
t_{on} SCR only	Turn-on time	At specified junction temperature, and with a peak repetitive off-state voltage of half rated value, followed by device turn-on using specified gate current, and when specified on-state current of specified di/dt flows, indicated as the time required for the applied off-state voltage to drop to 10% of its initial value after gate current application. Delay time is the term used to define the time required for applied voltage to drop to 90% of its initial value following gate-current application. The time required for the voltage level to drop from 90 to 10% of its initial value is referred to as rise time. The sum of both these defines turn-on time.
T_q SCR Only	Turn-off time	Specified at maximum rated junction temperature. Device set up to conduct on-state current, followed by application of specified reverse anode-cathode voltage to quench on-state current, and then increasing the anode-cathode voltage at a specified rate of rise as determined by circuit conditions controlling the point where the specified off-state voltage is reached. Turn-off time defines the minimum time which the device will hold its off-state, starting from the time on-state current reached zero until the time forward voltage is again applied (i.e. applied anode–cathode voltage becomes positive again).
t_{gt} GTO only	Turn-on time	When applying forward current to the gate, indicates the time required to switch the device from an off-state to an on-state.
t_{qt} GTO only	Turn-off time	When applying reverse current to the gate, indicates the time required to switch the device from an on-state to an off-state.

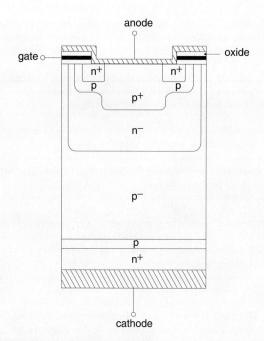

FIGURE 6.17 Cross section of unit cell of a *p*-type MCT.

out of conduction by applying a positive gate–anode voltage. This signal creates an inversion layer that diverts electrons in the *n*-base away from the *p*-emitter and into the heavily doped *n*-region at the anode. This *n*-channel FET current amounts to a diversion of the *pnp* transistor base current so that its base–emitter junction turns off. Holes are then no longer available for collection by the *p*-base. The elimination of this hole current (*npn* transistor base current) causes the *npn* transistor to turn-off. The remaining stored charge recombines and returns the MCT to its blocking state.

The seeming variability in fabrication of the turn-off FET structure continues to limit the performance of MCTs, particularly current interruption capability, though these devices can handle two to five times the conduction current density of IGBTs. Numerical modeling and its experimental verification show that ensembles of cells are sensitive to current filamentation during turn-off. All MCT device designs suffer from the problem of current interruption capability. Turn-on is relatively simple, by comparison; both the turn-on and conduction properties of the MCT approach the one-dimensional thyristor limit.

Other variations on the MCT structure have been demonstrated, namely the emitter switched thyristor (EST) and the dual-gate emitter switched thyristor (DG-EST) [12]. These comprise integrated lateral MOSFET structures which connect a floating thyristor *n*-emitter region to an *n*+ thyristor cathode region. The MOS channels are in series with the floating *n*-emitter region, allowing triggering of the thyristor with electrons from the *n*-base and interruption of the current to initiate turn-off. The DG-EST behaves as a dual-mode device, with the two gates allowing an IGBT mode to operate during switching and a thyristor mode to operate in the on-state.

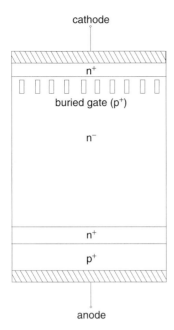

FIGURE 6.18 Cross section of a SITh or FCT.

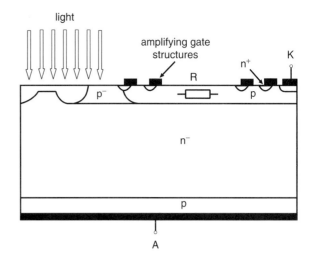

FIGURE 6.19 Cross section of a light-triggered thyristors (LTT).

6.6.3 Static Induction Thyristors

A SITh or FCTh has a cross section similar to that shown in Fig. 6.18. Other SITh configurations have surface gate structures. The device is essentially a *pin* diode with a gate structure that can pinch-off anode current flow. High power SIThs have a sub-surface gate (buried-gate) structure to allow larger cathode areas to be utilized, and hence larger current densities are possible.

Planar gate devices have been fabricated with blocking capabilities of up to 1.2 kV and conduction currents of 200 A, while step-gate (trench-gate) structures have been produced that are able to block up to 4 kV and conduct 400 A. Similar devices with a "Verigrid" structure have been demonstrated that can block 2 kV and conduct 200 A, with claims of up to 3.5 kV blocking and 200 A conduction. Buried-gate devices that block 2.5 kV and conduct 300 A have also been fabricated. Recently there has been a resurgence of interest in these devices for fabrication in SiC.

6.6.4 Optically Triggered Thyristors

Optically gated thyristors have traditionally been used in power utility applications where series stacks of devices are necessary to achieve the high voltages required. Isolation between gate drive circuits for circuits such as static VAR compensators and high voltage dc to ac inverters (for use in high voltage dc (HVDC) transmission) have driven the development of this class of devices, which are typically available in ratings from 5 to 8 kV. The cross section is similar to that shown in Fig. 6.19, showing the photosensitive region and the amplifying gate structures. Light-triggered thyristors (LTTs) may also integrate over-voltage protection.

One of the most recent devices can block 6 kV forward and reverse, conduct 2.5 kA average current, maintain a *di/dt* capability of 300 A/µs and a *dv/dt* capability of 3000 V/µs, with a required trigger power of 10 mW. An integrated light triggered and light quenched SITh has been produced that can block 1.2 kV and conduct up to 20 A (at a forward drop of 2.5 V). This device is an integration of a normally off buried-gate static induction photo-thyristor and a normally off *p*-channel darlington surface-gate static induction phototransistor. The optical trigger and quenching power required is less than 5 and 0.2 mW, respectively.

6.6.5 Bi-directional Thyristors

The BCT is an integrated assembly of two anti-parallel thyristors on one Si wafer. The intended applications for this switch are VAR compensators, static switches, soft starters and motor drives. These devices are rated up to 6.5 kV blocking. Crosstalk between the two halves has been minimized. A cross section of the BCT is shown in Fig. 6.20. Note that each surface has a cathode and an anode (opposite devices). The small gate–cathode periphery necessarily restricts the BCT to low-frequency applications because of its *di/dt* limit.

Low-power devices similar to the BCT, but in existence for many years, are the diac and triac. A simplified cross section of a diac is shown in Fig. 6.21. A positive voltage applied to the anode with respect to the cathode forward biases J_1, while reverse biasing J_2. J_4 and J_3 are shorted by the metal contacts. When J_2 is biased to breakdown, a lateral current flows in the p_2 region. This lateral flow forward biases the edge of J_3, causing carrier injection. The result is that the device switches into its thyristor mode and latches. Applying a reverse voltage causes the opposite behavior at each junction, but

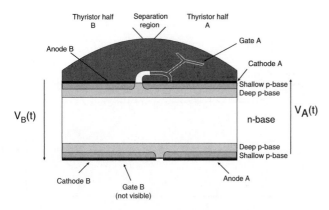

FIGURE 6.20 Cross section of a bi-directional control thyristor (BCT).

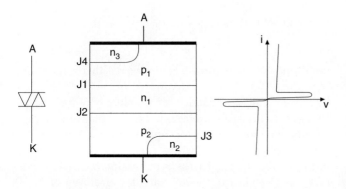

FIGURE 6.21 Cross section and i–v plot of a diac.

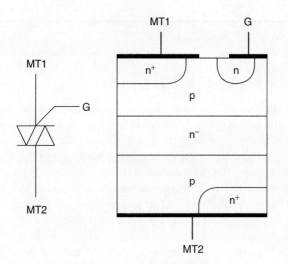

FIGURE 6.22 Cross section of a triac.

with the same result. Figure 6.21 also shows the i–v plot for a diac.

The addition of a gate connection, to form a triac, allows the breakover to be controlled at a lower forward voltage. Figure 6.22 shows the structure for the triac. Unlike the diac, this is not symmetrical, resulting in differing forward and reverse breakover voltages for a given gate voltage. The device is fired by applying a gate pulse of the same polarity relative to MT1 as that of MT2.

6.7 Gate Drive Requirements

6.7.1 Snubber Circuits

To protect a thyristor, from a large di/dt during turn-on and a large dv/dt during turn-off, a snubber circuit is needed. A general snubber topology is shown in Fig. 6.23. The turn-on snubber is made by inductance L_1 (often $L1$ is stray inductance only). This protects the thyristor from a large di/dt during the turn-on process. The auxiliary circuit made by R_1 and D_1 allows the discharging of L_1 when the thyristor is turned off. The turn-off snubber is made by resistor R_2 and capacitance C_2. This circuit protects a GTO from large dv/dt during the turn-off process. The auxiliary circuit made by D_2 and R_2 allows the discharging of C_2 when the thyristor is turned on. The circuit of capacitance C_2 and inductance L_1 also limits the value of dv/dt across the thyristor during forward-blocking. In addition, L_1 protects the thyristor from reverse over-currents. R_1 and diodes D_1, D_2 are usually omitted in ac circuits with converter-grade thyristors. A similar second set of L, C and R may be used around this circuit in HVDC applications.

6.7.2 Gate Circuits

It is possible to turn on a thyristor by injecting a current pulse into its gate. This process is known as gating, triggering or firing the thyristor. The most important restrictions are on the maximum peak and duration of the gate pulse current.

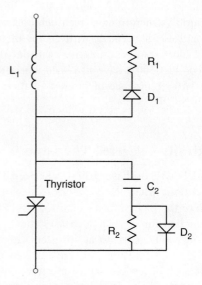

FIGURE 6.23 Turn-on (top elements) and turn-off (bottom elements) snubber circuits for thyristors.

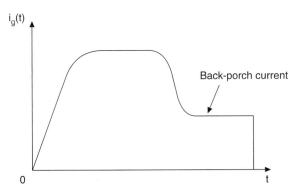

FIGURE 6.24 Gate current waveform showing large initial current followed by a suitable back-porch value.

In order to allow a fast turn-on, and a correspondingly large anode di/dt during the turn-on process, a large gate current pulse is supplied during the initial turn-on phase with a large di_G/dt. The gate current is kept on, at lower value, for some times after the thyristor turned on in order to avoid unwanted turn-off of the device; this is known as the "back-porch" current. A shaped gate current waveform of this type is shown in Fig. 6.24.

Figure 6.25 shows typical gate i–v characteristics for the maximum and minimum operating temperatures. The dashed line represents the minimum gate current and corresponding gate voltage needed to ensure that the thyristor will be triggered at various operating temperatures. It is also known as the locus of minimum firing points. On the data sheet it is possible find a line representing the maximum operating power of the thyristor gating internal circuit. The straight line, between V_{GG} and V_{GG}/R_G, represents the current voltage characteristic of the equivalent trigger circuit. If the equivalent trigger circuit line intercepts the two gate i–v characteristics for the maximum and minimum operating temperatures between where they intercept the dashed lines (minimum gate current to trigger and maximum gate power dissipation), then the trigger circuit is able to turn-on the thyristor at any operating temperature without destroying or damaging the device.

In order to keep the power circuit and the control circuit electrically unconnected, the gate signal generator and the gate of the thyristor are often connected through a transformer. There is a transformer winding for each thyristor, and in this way unwanted short circuits between devices are avoided. A general block diagram of a thyristor gate-trigger circuit is shown in Fig. 6.26. This application is for a standard bridge configuration often used in power converters.

Another problem can arise if the load impedance is high, particularly if the load is inductive and the supply voltage is low. In this situation, the latching current may not be reached during the trigger pulse. A possible solution to this problem could be the use of a longer current pulse. However, such a solution is not attractive because of the presence of the isolation transformer. An alternative solution is the generation of a series of short pulses that last for the same duration as a single long pulse. A single short pulse, a single long pulse and a series of short pulses are shown in Fig. 6.27. Reliable gating of the thyristor is essential in many applications.

There are many gate trigger circuits that use optical isolation between the logic-level electronics and a drive stage (typically MOSFETs) configured in a push–pull output. The dc power supply voltage for the drive stage is provided

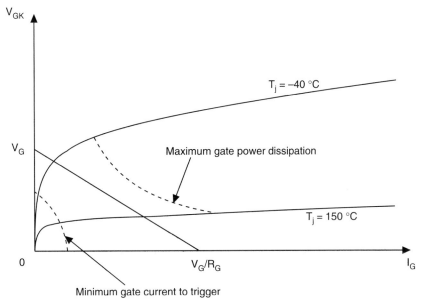

FIGURE 6.25 Gate i–v curve for a typical thyristor.

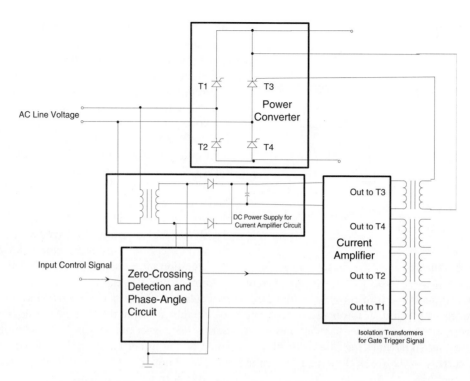

FIGURE 6.26 Block diagram of a transformer-isolated gate drive circuit.

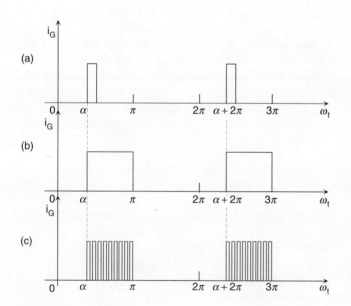

FIGURE 6.27 Multiple gate pulses used as an alternative to one long current pulse.

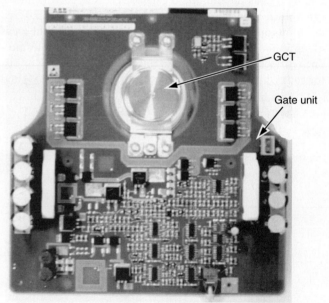

FIGURE 6.28 Typical layout of an IGCT gate drive.

through transformer isolation. Many device manufacturers supply drive circuits available on printed circuit (PC) boards or diagrams of suggested circuits.

IGCT gate drives consist of an integrated module to which the thyristor is connected via a low-inductance mounting; an example is given in Fig. 6.28. Multiple MOSFETs and capacitors connected in parallel may be used to source or sink the necessary currents to turn the device on or off. Logic in the module controls the gate drive from a fiber-optic trigger input, and provides diagnostic feedback from a fiber-optic output. A simple power supply connection is also required.

6.8 PSpice Model

Circuit simulators such as Spice and PSpice are widely used as tools in the design of power systems. For this purpose equivalent circuit models of thyristors have been developed. A variety of models have been proposed with varying degrees of complexity and accuracy. Frequently the simple two-transistor model described in Section 6.2 is used in PSpice. This simple structure, however, cannot model the appropriate negative-differential-resistance (NDR) behavior as the thyristor moves from forward-blocking to forward-conduction. Few other models for conventional thyristors have been reported. A PSpice model for a GTO has been developed by Tsay et al. [13], which captures much of thyristor behavior, such as the static i–v curve shown in Fig. 6.3, dynamic characteristics (turn-on and turn-off times), device failure modes (e.g. current crowding due to excessive di/dt at turn-on and spurious turn-on due to excessive dv/dt at turn-off), and thermal effects. Specifically, three resistors are added to the two-transistor model to create the appropriate behavior.

The proposed two-transistor, three-resistor model (2T-3R) is shown in Fig. 6.29. This circuit exhibits the desired NDR behavior. Given the static i–v characteristics for an SCR or GTO, it is possible to obtain similar curves from the model by choosing appropriate values for the three resistors and for the forward current gains α_p and α_n of the two transistors. The process of curve fitting can be simplified by keeping in mind that resistor R_1 tends to affect the negative slope of the i–v characteristic, resistor R_2 tends to affect the value of the holding current I_H and resistor R_3 tends to affect the value of the forward breakdown voltage V_{FBD}. When modeling thyristors with cathode or anode shorts, as described in Section 6.4, the presence of these shorts determines the values of R_1 and R_2, respectively. In the case of a GTO or IGCT, an important device characteristic is the so-called turn-off gain $K_{off}=I_A/|I_G|$, i.e. the ratio of the anode current to the negative gate current required to turn-off the device. An approximate formula relating the turn-off gain to the α's of the two transistors is given by,

$$K_{off} = \frac{\alpha_n}{\alpha_n + \alpha_p - 1} \quad (6.3)$$

The ability of this model to predict dynamic effects depends on the dynamics included in the transistor models. If transistor junction capacitances are included, it is possible to model the dv/dt limit of the thyristor. Too high a value of dv_{AK}/dt will cause significant current to flow through the J_2 junction capacitance. This current acts like gate current and can turn on the device.

This model does not accurately represent spatial effects such as current crowding at turn-on (the di/dt limit), when only part of the device is conducting, and, in the case of a GTO, current crowding at turn-off, when current is extracted from the gate to turn-off the device. Current crowding is caused by the location of the gate connection with respect to the conducting area of the thyristor and by the magnetic field generated by the changing conduction current. To model these effects, Tsay et al. [13] propose a multi-cell circuit model, in which the device is discretized in a number of conducting cells, each having the structure of Fig. 6.29. This model, shown in Fig. 6.30, takes into account the mutual inductive coupling, the delay in the gate turn-off signal due to positions of the cells relative to the gate connection, and non-uniform gate- and cathode-contact resistance. In particular, the RC delay circuits (series R with a shunt C tied to the cathode node) model the time delays between the gate triggering signals due to the position of the cell with respect to the gate connection; coupled inductors, M, model magnetic coupling between cells; resistors, R_{KC}, model non-uniform contact resistance; and resistors, R_{GC}, model gate contact resistances. The various circuit elements in the model can be estimated from device geometry and measured electrical characteristics. The choice of the number of cells is a tradeoff between accuracy and complexity. Example values of the RC delay network, R_{GC}, R_{KC}, and M are given in Table 6.6.

Other GTO thyristor models have been developed which offer improved accuracy at the expense of increased complexity. The model by Tseng et al. [14] includes charge storage

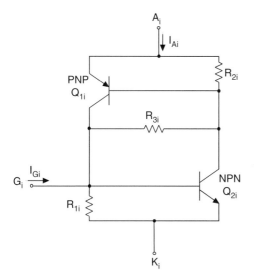

FIGURE 6.29 A two-transistor, three-resistor model for SCRs and GTOs.

TABLE 6.6 Element values for each cell of a multi-cell GTO model

Model component	Symbol	Value
Delay resistor	R	1 μΩ
Delay capacitor	C	1 nF
Mutual coupling inductance	M	10 nH
Gate contact resistance	R_{GC}	1 mΩ
Cathode contact resistance	R_{KC}	1 mΩ

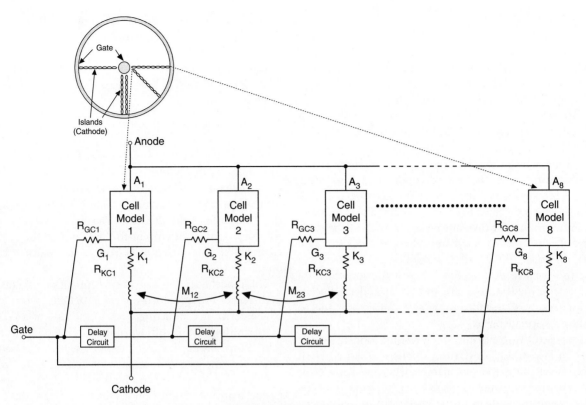

FIGURE 6.30 Thyristor multi-cell circuit model containing eight cells.

effects in the *n*-base, and its application to a multi-cell model, as in Fig. 6.30, has been demonstrated successfully. Models for the IGCT, based on the lumped-charge approach [15] and the Fourier-based solution of the ambipolar diffusion equation (ADE) [16], have also been developed.

6.9 Applications

The most important application of thyristors is for line-frequency phase-controlled rectifiers. This family includes several topologies, of which one of the most important is used to construct HVDC transmission systems. A single-phase controlled rectifier is shown in Fig. 6.31.

The use of thyristors instead of diodes allows the average output voltage to be controlled by appropriate gating of the thyristors. If the gate signals to the thyristors were continuously applied, the thyristors in Fig. 6.31 would behave as diodes. If no gate currents are supplied they behave as open circuits. Gate current can be applied any time (phase delay) after the forward voltage becomes positive. Using this phase-control feature, it is possible to produce an average dc output voltage less than the average output voltage obtained from an uncontrolled diode rectifier.

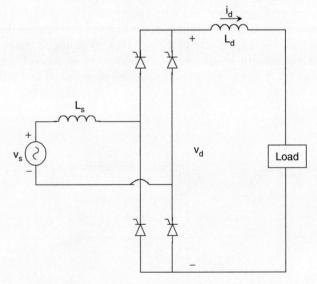

FIGURE 6.31 Single-phase controlled rectifier circuit.

6.9.1 DC–AC Utility Inverters

Three-phase converters can be made in different ways, according to the system in which they are employed. The basic circuit used to construct these topologies – the three-phase controlled rectifier – is shown in Fig. 6.32.

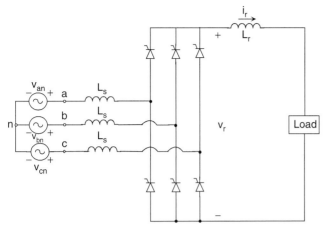

FIGURE 6.32 A three-phase controlled bridge circuit used as a basic topology for many converter systems.

Starting from this basic configuration, it is possible to construct more complex circuits in order to obtain high-voltage or high-current outputs, or just to reduce the output ripple by constructing a multi-phase converter. One of the most important systems using the topology shown in Fig. 6.32 as a basic circuit is the HVDC system represented in Fig. 6.33. This system is made by two converters, a transmission line, and two ac systems. Each converter terminal is made of two poles. Each pole is made by two six-pulse line-frequency converters connected through Δ-Y and Y-Y transformers in order to obtain a twelve-pulse converter and a reduced output ripple. The filters are required to reduce the current harmonics generated by the converter.

When a large amount of current and relatively low voltage is required, it is possible to connect in parallel, using a specially designed inductor, two six-pulse line-frequency converters connected through Δ-Y and Y-Y transformers. The special inductor is designed to absorb the voltage between the two converters, and to provide a pole to the load. This topology is shown in Fig. 6.34. This configuration is often known as a twelve-pulse converter. Higher pulse numbers may also be found.

6.9.2 Motor Control

Another important application of thyristors is in motor control circuits. Historically thyristors have been used heavily in traction, although most new designs are now based on IGBTs. Such motor control circuits broadly fall into four types: i) chopper control of a dc motor from a dc supply; ii) single- or three-phase converter control of a dc motor from an ac supply; iii) inverter control of an ac synchronous or induction machine from a dc supply and iv) cycloconverter control of an ac machine from an ac supply. An example of a GTO chopper is given in Fig. 6.35. L_1, R_1, D_1, and C_1 are the turn-on snubber; R_2, D_2, and C_2 are the turn-off snubber; finally R_3 and D_3 form the snubber for the freewheel diode D_3. A thyristor cycloconverter is shown in Fig. 6.36; the waveforms show the fundamental component of the output voltage for one phase. Three double converters are used to produce a three-phase

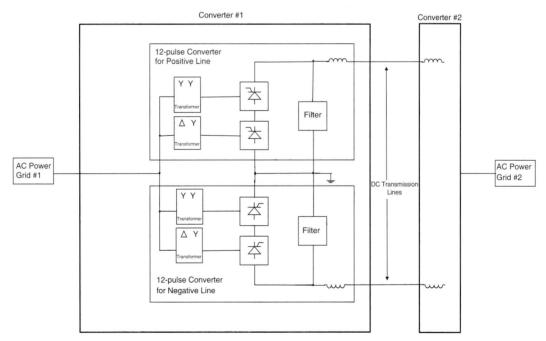

FIGURE 6.33 A HVDC transmission system.

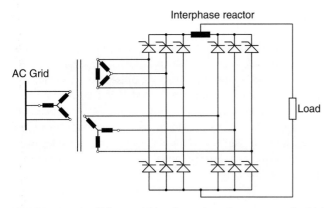

FIGURE 6.34 Parallel connection of two six-pulse converters for high current applications.

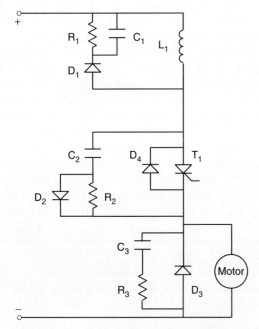

FIGURE 6.35 GTO chopper for dc motor control.

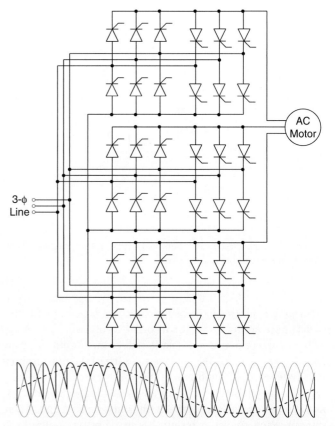

FIGURE 6.36 Cycloconverter for control of large ac machines.

variable-frequency, variable voltage sinusoidal output for driving ac motors. However, the limited frequency range (less than a third of the line frequency) restricts the application to large, low-speed machines at high powers.

A single- or three-phase thyristor converter (controlled rectifier) may be used to provide a variable dc supply for controlling a dc motor. Such a converter may also be used as the front end of a three-phase induction motor drive. The variable voltage, variable frequency motor drive requires a dc supply, which is supplied by the thyristor converter. The drive may use a square-wave or PWM voltage source inverter (VSI), or a current source inverter (CSI). Figure 6.37 shows a square-wave or PWM VSI with a controlled rectifier on the input side. The switch block inverter may be made of thyristors (usually GTOs or IGCTs) for high power, although most new designs now use IGBTs. Low-power motor controllers generally use IGBT inverters.

In motor control, thyristors are also used in CSI topologies. When the motor is controlled by a CSI, a controlled rectifier is also needed on the input side. Figure 6.38 shows a typical CSI inverter. The capacitors are needed to force the current in the thyristors to zero at each switching event. This is not needed when using GTOs. This inverter topology does not need any additional circuitry to provide the regenerative braking (energy recovery when slowing the motor). Historically, two back-to-back connected line-frequency thyristor converters have been employed to allow bi-directional power flow, and thus regenerative braking. Use of anti-parallel GTOs with symmetric blocking capability, or the use of diodes in series with each asymmetric GTO, reduces the number of power devices needed, but greatly increases the control complexity.

6.9.3 VAR Compensators and Static Switching Systems

Thyristors are also used to switch capacitors or inductors in order to control the reactive power in the system. Such arrangements may also be used in phase-balancing circuits for

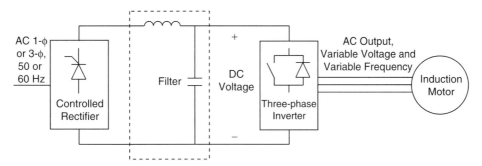

FIGURE 6.37 PWM or square-wave inverter with a controlled rectifier input.

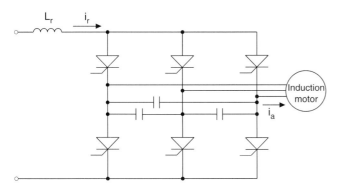

FIGURE 6.38 CSI on the output section of a motor drive system using capacitors for power factor correction.

balancing the load fed from a three-phase supply. Examples of these circuits are shown in Fig. 6.39. These circuits act as static VAR controllers. The topology represented on the left of Fig. 6.39 is called a thyristor-controlled inductor (TCI) and it acts as a variable inductor where the inductive VAR supplied can be varied quickly. Because the system may require either inductive or capacitive VAR compensation, it is possible to connect a bank of capacitors in parallel with a TCI. The topology shown on the right of Fig. 6.39 is called a thyristor-switched capacitor (TSC). Capacitors can be switched out by blocking the gate pulse of all thyristors in the circuit. The problem of this topology is the voltage across the capacitors at the thyristor turn-off. At turn-on the thyristor must be gated at the instant of the maximum ac voltage to avoid large over-currents. Many recent SVCs have used GTOs.

A similar application of thyristors is in solid-state fault current limiters and circuit breakers. In normal operation, the thyristors are continuously gated. However, under fault conditions they are switched rapidly to increase the series impedance in the load and to limit the fault current. Key advantages are the flexibility of the current limiting, which is independent of the location of the fault and the change in load impedance, the reduction in fault level of the supply, and a smaller voltage sag during a short-circuit fault.

A less important application of thyristors is as a static transfer switch, used to improve the reliability of uninterruptible power supplies (UPS) as shown in Fig. 6.40. There are two modes of using the thyristors. The first leaves the load permanently connected to the UPS system and in case of emergency disconnects the load from the UPS and connects it directly to the power line. The second mode is opposite to the first one. Under normal conditions the load is permanently connected to the power line, and in event of a line outage, the load is disconnected from the power line and connected to the UPS system.

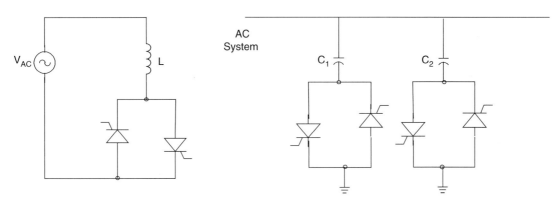

FIGURE 6.39 Per phase TCI and TSC system.

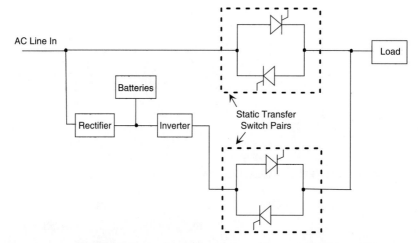

FIGURE 6.40 Static transfer switch used in an UPS system.

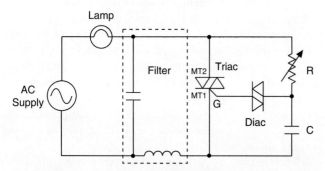

FIGURE 6.41 Basic dimmer circuit used in lighting control.

6.9.4 Lighting Control Circuits

An important circuit used in lighting control is the dimmer, based on a triac and shown in Fig. 6.41. The R–C network applies a phase shift to the gate voltage, delaying the triggering of the triac. Varying the resistance, controls the firing angle of the triac and therefore the voltage across the load. The diac is used to provide symmetrical triggering for the positive and negative half-cycles, due to the non-symmetrical nature of the triac. This ensures symmetrical waveforms and elimination of even harmonics. An L–C filter is often used to reduce any remaining harmonics.

Further Reading

1. J.L. Hudgins, "A review of modern power semiconductor electronic devices," *Microelectronics Journal*, vol. 24, pp. 41–54, Jan. 1993.
2. S.K. Ghandi, *Semiconductor Power Devices – Physics of Operation and Fabrication Technology*, New York, John Wiley and Sons, 1977, pp. 63–84.
3. B.J. Baliga, *Power Semiconductor Devices*, Boston, PWS Publishing, 1996, pp. 91–110.
4. B. Beker, J.L. Hudgins, J. Coronati, B. Gillett, and S. Shekhawat, "Parasitic parameter extraction of PEBB module using VTB technology," *IEEE IAS Ann. Mtg. Rec.*, pp. 467–471, Oct. 1997.
5. C.V. Godbold, V.A. Sankaran, and J.L. Hudgins, "Thermal analysis of high power modules," *IEEE Trans. PEL*, vol. 12, no. 1, pp. 3–11, Jan. 1997.
6. J.L. Hudgins and W.M. Portnoy, "High di/dt pulse switching of thyristors," *IEEE Tran. PEL*, vol. 2, pp. 143–148, April 1987.
7. S.M. Sze, *Physics of Semiconductor Devices*, 2nd ed., New York, John Wiley and Sons, 1984, pp. 140–147.
8. V.A. Sankaran, J.L. Hudgins, and W.M. Portnoy, "Role of the amplifying gate in the turn-on process of involute structure thyristors," *IEEE Tran. PEL*, vol. 5, no. 2, pp. 125–132, April 1990.
9. S. Menhart, J.L. Hudgins, and W.M. Portnoy, "The low temperature behavior of thyristors," *IEEE Tran. ED*, vol. 39, pp. 1011–1013, April 1992.
10. A. Herlet, "The forward characteristic of silicon power rectifiers at high current densities," *Solid-State Electron.*, vol. 11, no. 8, pp. 717–742, 1968.
11. J.L. Hudgins, C.V. Godbold, W.M. Portnoy, and O.M. Mueller, "Temperature effects on GTO characteristics," *IEEE IAS Annual Mtg. Rec.*, pp. 1182–1186, Oct. 1994.
12. P.R. Palmer and B.H. Stark, "A PSPICE model of the DG-EST based on the ambipolar diffusion equation," *IEEE PESC Rec.*, pp. 358–363, June 1999.
13. C.L. Tsay, R. Fischl, J. Schwartzenberg, H. Kan, and J. Barrow, "A high power circuit model for the gate turn off thyristor," *IEEE IAS Annual Mtg. Rec.*, pp. 390–397, Oct. 1990.
14. K.J. Tseng and P.R. Palmer, "Mathematical model of gate-turn-off thyristor for use in circuit simulations," *IEE Proc.-Electr. Power Appl.*, vol. 141, no. 6, pp. 284–292, Nov. 1994.
15. X. Wang, A. Caiafa, J. Hudgins, and E. Santi, "Temperature effects on IGCT performance," *IEEE IAS Annual Mtg. Rec.*, Oct. 2003.
16. X. Wang, A. Caiafa, J.L. Hudgins, E. Santi, and P.R. Palmer, "Implementation and validation of a physics-based circuit model for IGCT with full temperature dependencies," *IEEE PESC Rec.*, pp. 597–603, June 2004.

7
Gate Turn-off Thyristors

Muhammad H. Rashid, Ph.D.
Electrical and Computer Engineering, University of West Florida, 11000 University Parkway, Pensacola, Florida 32514-5754, USA

7.1 Introduction 115
7.2 Basic Structure and Operation.. 115
7.3 GTO Thyristor Models 116
7.4 Static Characteristics 117
   7.4.1 On-state Characteristics • 7.4.2 Off-state Characteristics • 7.4.3 Rate of Rise of Off-state Voltage (dv_T/dt) • 7.4.4 Gate Triggering Characteristics
7.5 Switching Phases.. 118
7.6 SPICE GTO Model.. 120
7.7 Applications.. 121
   References 121

7.1 Introduction

A gate turn-off thyristor (known as a GTO) is a three terminal power semiconductor device. GTOs belong to a thyristor family having a four-layer structure. GTOs also belong to a group of power semiconductor devices that have the ability for full control of on- and off-states via the control terminal (gate). To fully understand the design, development and operation of the GTO, it is easier to compare with the conventional thyristor. Like a conventional thyristor, applying a positive gate signal to its gate terminal can turn-on to a GTO. Unlike a standard thyristor, a GTO is designed to turn-off by applying a negative gate signal.

GTOs are of two types: asymmetrical and symmetrical. The asymmetrical GTOs are the most common type on the market. This type of GTOs is normally used with a anti-parallel diode and hence high reverse blocking capability is not available. The reverse conducting is accomplished with an anti-parallel diode integrated onto the same silicon wafer. The symmetrical type of GTOs has an equal forward and reverse blocking capability.

7.2 Basic Structure and Operation

The symbol of a GTO is shown in Fig. 7.1a. A high degree of interdigitation is required in GTOs in order to achieve efficient turn-off. The most common design employs the cathode area separated into multiple segments (cathode fingers) arranged in concentric rings around the device center. The internal structure is shown in Fig. 7.1b. A common contact disc pressed against the cathode fingers connects the fingers together. It is important that all the fingers turns off simultaneously, otherwise the current may be concentrated into a fewer fingers which are likely to be damaged due to over heating.

The high level of gate interdigitation also results in a fast turn-on speed and a high di/dt performance of the GTOs. The most remote part of a cathode region is not more than 0.16 mm from a gate edge and hence the whole GTO can conduct within about 5 µs with sufficient gate drive and the turn-on losses can be reduced. However, the interdigitation reduces the available emitter area so the low frequency average current rating is less than for a standard thyristor with an equivalent diameter.

The basic structure of a GTO consists of a four-layer-PNPN semiconductor device, which is very similar in construction to a thyristor. It has several design features which allow it to be turned on and off by reversing the polarity of the gate signal. The most important differences are that the GTO has long narrow emitter fingers surrounded by gate electrodes and no cathode shorts.

The turn-on mode is similar to a standard thyristor. The injection of the hole current from the gate forward biases the cathode p-base junction causing electron emission from the cathode. These electrons flow to the anode and induce hole injection by the anode emitter. The injection of holes and electrons into the base regions continues until charge

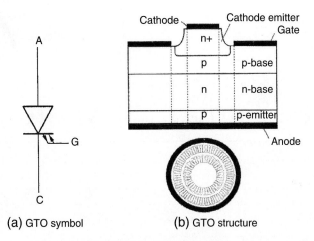

FIGURE 7.1 GTO structure.

multiplication effects bring the GTO into conduction. This is shown in Fig. 7.2a. As with a conventional thyristor only the area of cathode adjacent to the gate electrode is turned on initially, and the remaining area is brought into conduction by plasma spreading. However, unlike the thyristor, the GTO consists of many narrow cathode elements, heavily interdigitated with the gate electrode, and therefore the initial turned-on area is very large and the time required for plasma spreading is small. The GTO, therefore, is brought into conduction very rapidly and can withstand a high turn-on di/dt.

In order to turn-off a GTO, the gate is reversed biased with respect to the cathode and holes from the anode are extracted from the p-base. This is shown in Fig. 7.2b. As a result a voltage drop is developed in the p-base region, which eventually reverse biases the gate cathode junction cutting off the injection of electrons. As the hole extraction continues, the p-base is further depleted, thereby squeezing the remaining conduction area. The anode current then flows through the most remote areas from the gate contacts, forming high current density filaments. This is the most crucial phase of the turn-off process in GTOs, because high density filaments leads to localized heating which can cause device failure unless these filaments are extinguished quickly. An application of higher negative gate voltage may aid in extinguishing the filaments rapidly. However, the breakdown voltage of the gate-cathode junction limits this method.

When the excess carrier concentration is low enough for carrier multiplication to cease, the device reverts to the forward blocking condition. At this point although the cathode current has stopped flowing, anode-to-gate current continues to flow supplied by the carriers from n-base region stored charge. This is observed as a tail current that decays exponentially as the remaining charge concentration is reduced by recombination process. The presence of this tail current with the combination of high GTO off-state voltage produces substantial power losses. During this transition period, the electric field in the n-base region is grossly distorted due to the presence of the charge carriers and may result in premature avalanche breakdown. The resulting impact ionization can cause device failure. This phenomenon is known as "dynamic avalanche." The device regains its steady-state blocking characteristics when the tail current diminishes to leakage current level.

7.3 GTO Thyristor Models

One-dimensional two-transistor model of GTOs is shown in Fig. 7.3. The device is expected to yield the turn-off gain g given by

$$A_g = \frac{I_A}{I_G} = \frac{\alpha_{npn}}{\alpha_{pnp} + \alpha_{npn} - 1} \qquad (7.1)$$

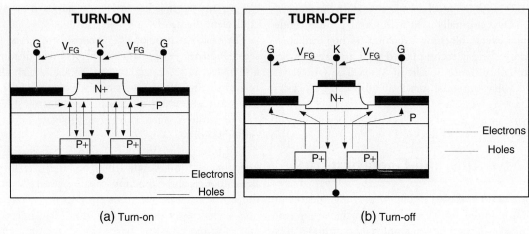

FIGURE 7.2 Turn-on and turn-off of GTOs.

7 Gate Turn-off Thyristors

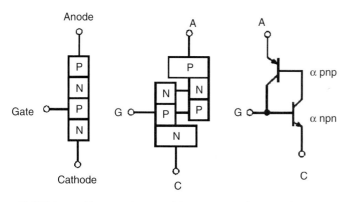

FIGURE 7.3 Two-transistor model representing the GTO thyristor.

where I_A is the anode current and I_G the gate current at turn-off, and α_{npn} and α_{pnp} are the common-base current gains in the NPN and PNP transistors sections of the device. For a non-shorted device, the charge is drawn from the anode and regenerative action commences, but the device does not latch on (remain on when the gate current is removed) until

$$\alpha_{npn} + \alpha_{pnp} \geq 1 \qquad (7.2)$$

This process takes a short period while the current and the current gains increase until they satisfy Eq. (7.2). For anode-shorted devices, the mechanism is similar but the anode short impairs the turn-on process by providing a base–emitter short thus reducing the PNP transistor gain, which is shown in Fig. 7.4. The composite PNP gain of the emitter-shorted structure is given as follows

$$\alpha_{pnp}(\text{composite}) = \alpha_{pnp}\left(\frac{1 - V_{be}}{R_{Sanode}}\right) \qquad (7.3)$$

where V_{be} = emitter base voltage (generally 0.6 V for injection of carriers), and R_S is the anode short resistance. The anode emitter injects when the voltage around it exceed 0.06 V, and therefore the collector current of the NPN transistor flowing through the anode shorts influences turn-on.

The GTO remains in a transistor state if the load circuit limits the current through the shorts.

7.4 Static Characteristics

7.4.1 On-state Characteristics

In the on-state the GTO operates in a similar manner to the thyristor. If the anode current remains above the holding current level then positive gate drive may be reduced to zero and the GTO will remain in conduction. However, as a result of the turn-off ability of the GTO, it does posses a higher holding current level than the standard thyristor, and in addition, the cathode of the GTO thyristor is sub-divided into small finger elements to assist turn-off. Thus, if the GTO thyristor anode current transiently dips below the holding current level, localized regions of the device may turn-off, thus forcing a high anode current back into the GTO at a high rate of rise of anode current after this partial turn-off. This situation could be potentially destructive. It is recommended, therefore, that the positive gate drive is not removed during conduction but is held at a value $I_{G(ON)}$, where $I_{G(ON)}$ is greater than the maximum critical trigger current (I_{GT}) over the expected operating temperature range of the GTO thyristor.

Figure 7.5 shows the typical on-state V–I characteristics for a 4000 A, 4500 V GTO from Dynex range of GTOs [1] at junction temperatures of 25°C and 125°C. The curves can be approximated to a straight line of the form

$$V_{TM} = V_0 + IR_0 \qquad (7.4)$$

where V_0 = voltage intercept, models the voltage across the cathode and anode forward biased junctions and R_0 = on state resistance. When average and RMS values of on-state current (I_{TAV}, I_{TRMS}) are known, then the on-state power dissipation P_{ON} can be determined using V_0 and R_0. That is,

$$P_{ON} = V_0 I_{TAV} + R_0 I_{TRMS}^2 \qquad (7.5)$$

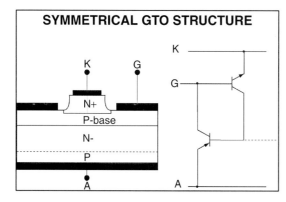
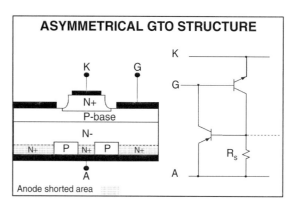

FIGURE 7.4 Two-transistor models of GTO structures.

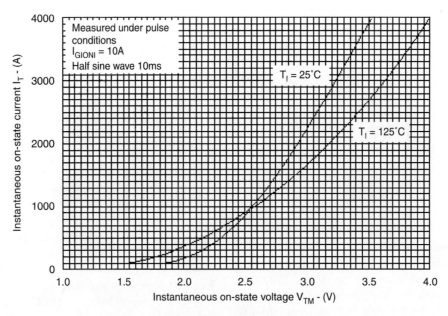

FIGURE 7.5 V–I characteristics of GTO [see data sheet in Ref. 1].

7.4.2 Off-state Characteristics

Unlike the standard thyristor, the GTO does not include cathode emitter shorts to prevent non-gated turn-on effects due to *dv/dt* induced forward biased leakage current. In the off-state of the GTO, steps should, therefore, be taken to prevent such potentially dangerous triggering. This can be accomplished by either connecting the recommended value of resistance between the gate and the cathode (R_{GK}) or by maintaining a small reverse bias on the gate contact ($V_{RG} = -2$ V). This will prevent the cathode emitter becoming forward biased and therefore sustain the GTO thyristor in the off state.

The peak off-state voltage is a function of resistance R_{GK}. This is shown in Fig. 7.6. Under ordinary operating conditions, GTOs are biased with a negative gate voltage of around −15 V supplied from the gate drive unit during the off-state interval. Nevertheless, provision of R_{GK} may be desirable design practice in the event of the gate-drive failure for any reason ($R_{GK} < 1.5 \Omega$ is recommended for a large GTO). R_{GK} dissipates energy and hence adds to the system losses.

7.4.3 Rate of Rise of Off-state Voltage (dv_T/dt)

The rate of rise of off-state voltage (dv_T/dt) depends on the resistance R_{GK} connected between the gate and the cathode and the reverse bias applied between the gate and the cathode. This relationship is shown in Fig. 7.7.

7.4.4 Gate Triggering Characteristics

The gate trigger current (I_{GT}) and the gate trigger voltage (V_{GT}) are both dependent on junction temperature T_j as shown in Fig. 7.8. During the conduction state of the GTO a certain value of gate current must be supplied and this value should be larger than the I_{GT} at the lowest junction temperature at which the GTO operates. In dynamic conditions the specified I_{GT} is not sufficient to trigger the GTO switching from higher voltage and high *di/dt*. In practice a much high peak gate current I_{GM} (in order of ten times I_{GT}) at T_j min is recommended to obtain good turn-on performance.

7.5 Switching Phases

The switching process of a GTO thyristor goes through four operating phases (a) turn-on, (b) on-state, (c) turn-off, and (d) off-state.

Turn-on: A GTO has a highly interdigited gate structure with no regenerative gate. Thus it requires a large initial gate trigger pulse. A typical turn-on gate pulse and its important parameters are shown in Fig. 7.9. A minimum and maximum values of I_{GM} can be derived from the device data sheet. A value of di_g/dt is given in device characteristics of the data sheet, against turn-on time. The rate of rise of gate current, di_g/dt will affect the device turn-on losses. The duration of

7 Gate Turn-off Thyristors

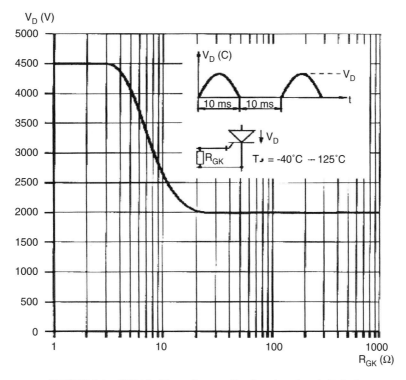

FIGURE 7.6 GTO blocking voltage vs. R_{GK} [see data sheet in Ref. 1].

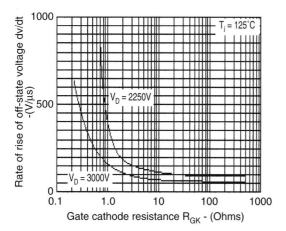

FIGURE 7.7 dV_D/dt vs. R_{GK} [see data sheet in Ref. 1].

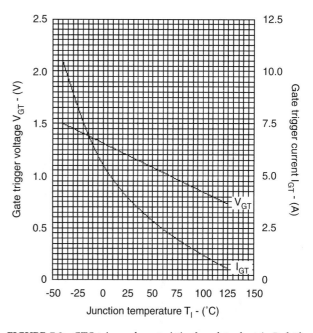

FIGURE 7.8 GTO trigger characteristics [see data sheet in Ref. 1].

the I_{GM} pulse should not be less than half the minimum on time given in data sheet ratings. A longer period will be required if the anode current di/dt is low such that I_{GM} is maintained until a sufficient level of anode current is established.

On-state: Once the GTO is turned on, forward gate current must be continued for the whole of the conduction period. Otherwise, the device will not remain in conduction during the on-state period. If large negative di/dt or anode current reversal occurs in the circuit during the on-state, then higher values of I_G may be required. Much lower values of I_G are, however, required when the device has heated up.

Turn-off: The turn-off performance of a GTO is greatly influenced by the characteristics of the gate turn-off circuit. Thus the characteristics of the turn-off circuit must match

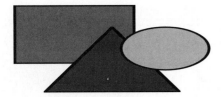

FIGURE 7.9 A typical turn-on gate pulse [see data sheet in Ref. 2].

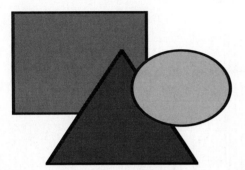

FIGURE 7.10 Anode and gate currents during turn-off [see data sheet in Ref. 2].

with the device requirements. Figure 7.10 shows the typical anode and gate currents during the turn-off. The gate turn-off process involves the extraction of the gate charge, the gate avalanche period and the anode current decay. The amount of the charge extraction is a device parameter and its value is not significantly affected by the external circuit conditions. The initial peak turn-off current and turn-off time, which are important parameters of the turning-off process, depend on the external circuit components. The device data sheet gives typical values for I_{GQ}.

The turn-off circuit arrangement of a GTO is shown in Fig. 7.11. The turn-off current gain of a GTO is low, typically 6–15. Thus, for a GTO with a turn-off gain of 10, it will require a turn-off gate current of 10 A to turn-off an on-state of 100 A. A charged capacitor C is normally used to provide the required turn-off gate current. Inductor L limits the turn-off di/dt of the gate current through the circuit formed by R_1, R_2, SW_1, and L. The gate circuit supply voltage V_{GS} should be selected to give the required value of V_{GQ}. The values of R_1 and R_2 should also be minimized.

Off-state period: During the off-state period, which begins after the fall of the tail current to zero, the gate should

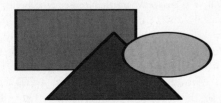

FIGURE 7.11 Turn-off circuit [see data sheet in Ref. 2].

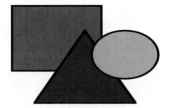

FIGURE 7.12 Gate-cathode resistance, R_{GK} [see data sheet in Ref. 2].

ideally remain reverse biased. This reverse bias ensures maximum blocking capability and dv/dt rejection. The reverse bias can be obtained either by keeping SW_1 closed during the whole off-state period or via a higher impedance circuit SW_2 and R_3 provided a minimum negative voltage exits. This higher impedance circuit SW_2 and R_3 must sink the gate leakage current.

In case of a failure of the auxiliary supplies for the gate turn-off circuit, the gate may be in reverses biased condition and the GTO may not be able to block the voltage. To ensure blocking voltage of the device is maintained, then a minimum gate-cathode resistance (R_{GK}) should be applied as shown in Fig. 7.12. The value of R_{GK} for a given line voltage can be derived from the data sheet.

7.6 SPICE GTO Model

A GTO may be modelled with two transistors shown in Fig. 7.3. However, a GTO model [3] consisting of two thyristors, which are connected in parallel, yield improved on-state, turn-on, and turn-off characteristics. This is shown in Fig. 7.13 with four transistors.

When the anode to cathode voltage, V_{AK} is positive and there is no gate voltage, the GTO model will be in the off-state like a standard thyristor. If a positive voltage (V_{AK}) is applied to the anode with respect to the cathode and no gate

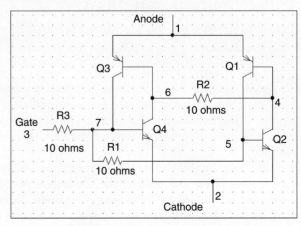

FIGURE 7.13 Four-transistor GTO model.

pulse applied, $I_{B1} = I_{B2} = 0$ and therefore $I_{C1} = I_{C2} = 0$. Thus no anode current will flow, $I_A = I_K = 0$.

When a small voltage is applied to the gate, then I_{B2} is non-zero and therefore both $I_{C1} = I_{C2} = 0$ are non-zero. Thus the internal circuit will conduct and there will a current flow from the anode to the cathode.

When a negative gate pulse is applied to the GTO model, the PNP junction near to the cathode will behave as a diode. The diode will be reverse biased since the gate voltage is negative with to the cathode. Therefore the GTO will stop conduction.

When the anode-to-cathode voltage is negative, that is, the anode voltage is negative with respect to the cathode, the GTO model will act like a reverse biased diode. This is because the PNP transistor will see a negative voltage at the emitter and the NPN transistor will see a positive voltage at the emitter. Therefore both transistors will be in the off-state and hence the GTO will not conduct. The SPICE sub-circuit description of the GTO model will be as follows

.SUBCIRCUIT	1		2	3	; GTO Sub-circuit definition
*Terminal			anode	cathode gate	
Q1	5	4	1	DMOD1	PNP ; PNP transistor with model DMOD1
Q3	7	6	1	DMOD1	PNP
Q2	4	5	2	DMOD2	NPN ; PNP transistor with model DMOD2
Q4	6	7	2	DMOD2	NPN
R1	7	5	10 ohms		
R2	6	4	10 ohms		
R3	3	7	10 ohms		
.MODEL	DMOD1			PNP	; Model statement for a PNP transistor
.MODEL	DMOD2			NPN	; Model statement for an NPN transistor
.ENDS					; End of sub-circuit definition

7.7 Applications

GTO thyristors find many applications such as in motor drives, induction heating [4], distribution lines [5], pulsed power [6], and Flexible AC transmission systems [7, 8].

References

1. Dynex semiconductor: Data GTO data-sheets web-site: http://www.dynexsemi.com/products/power_search.cgi
2. Westcode semiconductor: Data GTO data-sheets web-site: http://www.westcode.com/ws-gto.html
3. El-Amin, I.M.A. "GTO PSPICE Model and its applications," *The Fourth Saudi Engineering Conference*, November 1995, vol. III, pp. 271–7.
4. Busatto, G., Iannuzzo, F., and Fratelli, L., "PSPICE model for GTOs," *Proceedings of Symposium on Power Electronics Electrical Drives. Advanced Machine Power Quality. SPEEDAM Conference.* Sorrento, Italy, 3–5 June 1998, vol. 1, pp. 2/5–10.
5. Malesani, L. and Tenti, P. "Medium-frequency GTO inverter for induction heating applications," *Second European Conference on Power Electronics and Applications. EPE. Proceedings*, Grenoble, France, 22–24, September 1987, vol. 1, pp. 271–6.
6. Souza, L.F.W., Watanabe, E.H., and Aredes, M.A. "GTO controlled series capacitor for distribution lines," *International Conference on Large High Voltage Electric Systems.* CIGRE'98, 1998.
7. Chamund, D.J. "Characterisation of 3.3 kV asymmetrical thyristor for pulsed power application," *IEE Symposium Pulsed Power 2000* (Digest No.00/053) pp. 35/1–4, London, UK, 3–4 May 2000.
8. Moore, P. and Ashmole, P. "Flexible AC transmission systems: 4. Advanced FACTS controllers," *Power Engineering Journal*, vol. 12, no. 2, pp. 95–100, April 1998.

8
MOS Controlled Thyristors (MCTs)

S. Yuvarajan, Ph.D.
Department of Electrical Engineering, North Dakota State University, P.O. Box 5285, Fargo, North Dakota, USA

8.1 Introduction 123
8.2 Equivalent Circuit and Switching Characteristics 124
    8.2.1 Turn-on and Turn-off
8.3 Comparison of MCT and other Power Devices 125
8.4 Gate Drive for MCTs 126
8.5 Protection of MCTs 126
    8.5.1 Paralleling of MCTs • 8.5.2 Overcurrent Protection
8.6 Simulation Model of an MCT 127
8.7 Generation-1 and Generation-2 MCTs 127
8.8 N-channel MCT 127
8.9 Base Resistance-controlled Thyristor 127
8.10 MOS Turn-off Thyristor 128
8.11 Applications of PMCT 128
    8.11.1 Soft-switching • 8.11.2 Resonant Converters
8.12 Conclusions 130
8.13 Appendix 130
    References 130

8.1 Introduction

The efficiency, capacity, and ease of control of power converters depend mainly on the power devices employed. Power devices, in general, belong to either bipolar-junction type or field-effect type and each one has its advantages and disadvantages. The silicon controlled rectifier (SCR), also known as a thyristor, is a popular power device that has been used over the past several years. It has a high current density and a low forward voltage drop, both of which make it suitable for use in large power applications. The inability to turn-off through the gate and the low switching speed are the main limitations of an SCR. The gate turn-off (GTO) thyristor was proposed as an alternative to SCR. However, the need for a higher gate turn-off current limited its application.

The power MOSFET has several advantages such as high input impedance, ease of control, and higher switching speeds. Lower current density and higher forward drop limited the device to low-voltage and low-power applications. An effort to combine the advantages of bipolar junction and field-effect structures has resulted in hybrid devices such as the insulated gate bipolar transistor (IGBT) and the MOS controlled thyristor (MCT). While an IGBT is an improvement over a bipolar junction transistor (BJT) using a MOSFET to turn-on and turn-off current, an MCT is an improvement over a thyristor with a pair of MOSFETs to turn-on and turn-off current. The MCT overcomes several of the limitations of the existing power devices and promises to be a better switch for the future. While there are several devices in the MCT family with distinct combinations of channel and gate structures [1], one type, called the P-channel MCT, has been widely reported and is discussed here. Because the gate of the device is referred to with respect to the anode rather than the cathode, it is sometimes referred to as a complementary MCT (C-MCT) [2]. Harris Semiconductors (Intersil) originally made the MCTs, but the MCT division was sold to Silicon Power Corporation (SPCO), which has continued the development of MCTs.

8.2 Equivalent Circuit and Switching Characteristics

The SCR is a 4-layer *pnpn* device with a control gate, and applying a positive gate pulse turns it on when it is forward-biased. The regenerative action in the device helps to speed up the turn-on process and to keep it in the "ON" state even after the gate pulse is removed. The MCT uses an auxiliary MOS device (PMOSFET) to turn-on and this simplifies the gate control. The turn-on has all the characteristics of a power MOSFET. The turn-off is accomplished using

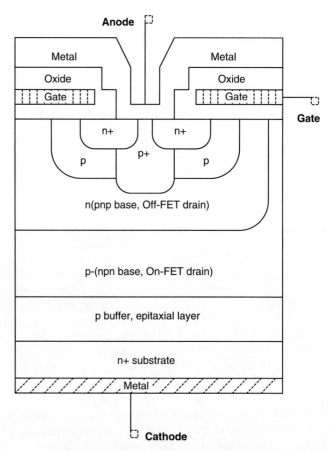

FIGURE 8.2 Cross section of an MCT unit cell.

another MOSFET (NMOSFET), which essentially diverts the base current of one of the BJTs and breaks the regeneration.

The transistor-level equivalent circuit of a P-channel MCT and the circuit symbol are shown in Fig. 8.1. The cross section of a unit cell is shown in Fig. 8.2. The MCT is modeled as an SCR merged with a pair of MOSFETs. The SCR consists of the bipolar junction transistors (BJTs) Q_1 and Q_2, which are interconnected to provide regenerative feedback such that the transistors drive each other into saturation. Of the two MOSFETs, the PMOS located between the collector and emitter of Q_2 helps to turn the SCR on, and the NMOS located across the base-emitter junction of Q_2 turns it off. In the actual fabrication, each MCT is made up of a large number (~100,000) cells, each of which contains a wide-base *npn* transistor and a narrow-base *pnp* transistor. While each *pnp* transistor in a cell is provided with an N-channel MOSFET across its emitter and base, only a small percentage (~4%) of *pnp* transistors are provided with P-channel MOSFETs across their emitters and collectors. The small percentage of PMOS cells in an MCT provides just enough current for turn-on and the large number of NMOS cells provide plenty of current for turn-off.

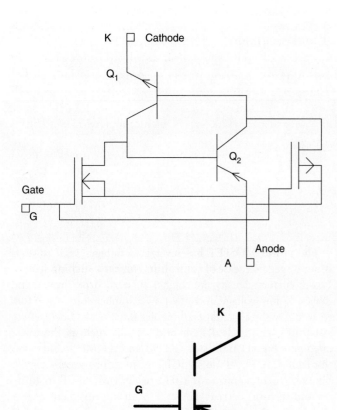

FIGURE 8.1 Equivalent circuit and symbol of an MCT.

8.2.1 Turn-on and Turn-off

When the MCT is in the forward blocking state, it can be turned on by applying a negative pulse to its gate with respect to the anode. The negative pulse turns on the PMOSFET (On-FET) whose drain current flows through the base-emitter junction of Q_1 *(npn)* thereby turning it on. The regenerative action within $Q_1 - Q_2$ turns the MCT on into full conduction within a very short time and maintains it even after the gate pulse is removed. The MCT turns on without a plasma-spreading phase giving a high dI/dt capability and ease of overcurrent protection. The on-state resistance of an MCT is slightly higher than that of an equivalent thyristor because of the degradation of the injection efficiency of the N^+ emitter/p-base junction. Also, the peak current rating of an MCT is much higher than its average or rms current rating.

An MCT will remain in the "ON" state until the device current is reversed or a turn-off pulse is applied to its gate. Applying a positive pulse to its gate turns off a conducting MCT. The positive pulse turns on the NMOSFET (Off-FET), thereby diverting the base current of Q_2 *(pnp)* away to the anode of the MCT and breaking the latching action of the SCR. This stops the regenerative feedback within the SCR and turns the MCT off. All the cells within the device are to be turned off at the same time to avoid a sudden increase in current density. When the Off-FETs are turned on, the SCR section is heavily shorted and this results in a high dV/dt rating for the MCT. The highest current that can be turned off with the application of a gate bias is called the "maximum controllable current." The MCT can be gate controlled if the device current is less than the maximum controllable current. For smaller device currents, the width of the turn-off pulse is not critical. However, for larger currents, the gate pulse has to be wider and more often has to occupy the entire off-period of the switch.

8.3 Comparison of MCT and Other Power Devices

An MCT can be compared to a power MOSFET, a power BJT, and an IGBT of similar voltage and current ratings. The operation of the devices is compared under on-state, off-state, and transient conditions. The comparison is simple and very comprehensive.

The current density of an MCT is ≈70% higher than that of an IGBT having the same total current [2]. During its on-state, an MCT has a lower conduction drop compared to other devices. This is attributed to the reduced cell size and the absence of emitter shorts present in the SCR within the MCT. The MCT also has a modest negative temperature coefficient at

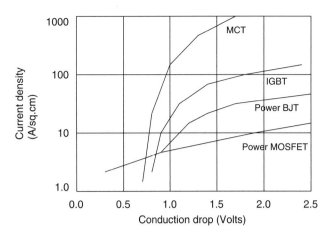

FIGURE 8.3 Comparison of forward drop for different devices.

lower currents with the temperature coefficient turning positive at larger current [2]. Figure 8.3 shows the conduction drop as a function of current density. The forward drop of a 50-A MCT at 25 °C is around 1.1 V, while that for a comparable IGBT is over 2.5 V. The equivalent voltage drop calculated from the value of $r_{DS}(ON)$ for a power MOSFET will be much higher. However, the power MOSFET has a much lower delay time (30 ns) compared to that of an MCT (300 ns). The turn-on of a power MOSFET can be so much faster than an MCT or an IGBT therefore, the switching losses would be negligible compared to the conduction losses. The turn-on of an IGBT is intentionally slowed down to control the reverse recovery of the freewheeling diode used in inductive switching circuits [3].

The MCT can be manufactured for a wide range of blocking voltages. Turn-off speeds of MCTs are supposed to be higher as initially predicted. The turn-on performance of Generation-2 MCTs are reported to be better compared to Generation-1 devices. Even though the Generation-1 MCTs have higher turn-off times compared to IGBTs, the newer ones with higher radiation (hardening) dosage have comparable turn-off times. At present, extensive development activity in IGBTs has resulted in high-speed switched mode power supply (SMPS) IGBTs that can operate at switching speeds ≈150 kHz [4]. The turn-off delay time and the fall time for an MCT are much higher compared to a power MOSFET, and they are found to increase with temperature [2]. Power MOSFETs becomes attractive at switching frequencies above 200 kHz, and they have the lowest turn-off losses among the three devices.

The turn-off safe operating area (SOA) is better in the case of an IGBT than an MCT. For an MCT, the full switching current is sustainable at ≈50 to 60% of the breakdown voltage rating, while for an IGBT it is about 80%. The use of capacitive snubbers becomes necessary to shape the turn-off locus of an MCT. The addition of even a small capacitor improves the SOA considerably.

8.4 Gate Drive for MCTs

The MCT has a MOS gate similar to a power MOSFET or an IGBT and hence it is easy to control. In a PMCT, the gate voltage must be applied with respect to its anode. A negative voltage below the threshold of the On-FET must be applied to turn on the MCT. The gate voltage should fall within the specified steady-state limits in order to give a reasonably low delay time and to avoid any gate damage due to overvoltage [3]. Similar to a GTO, the gate voltage rise-time has to be limited to avoid hot spots (current crowding) in the MCT cells. A gate voltage less than −5 V for turn-off and greater than 10 V for turn-on ensures proper operation of the MCT. The latching of the MCT requires that the gate voltage be held at a positive level in order to keep the MCT turned off.

Because the peak-to-peak voltage levels required for driving the MCT exceeds those of other gate-controlled devices, the use of commercial drivers is limited. The MCT can be turned on and off using a push–pull pair with discrete NMOS–PMOS devices, which, in turn, are driven by commercial integrated circuits (ICs). However, some drivers developed by MCT manufacturers are not commercially available [3].

A Baker's clamp push–pull can also be used to generate gate pulses of negative and positive polarity of adjustable width for driving the MCT [5–7]. The Baker's clamp ensures that the push–pull transistors will be in the quasi-saturated state prior to turn-off and this results in a fast switching action. Also, the negative feedback built into the circuit ensures satisfactory operation against variations in load and temperature. A similar circuit with a push–pull transistor pair in parallel with a pair of power BJTs is available [8]. An intermediate section, with a BJT that is either cut off or saturated, provides −10 and +15 V through potential division.

8.5 Protection of MCTs

8.5.1 Paralleling of MCTs

Similar to power MOSFETs, MCTs can be operated in parallel. Several MCTs can be paralleled to form larger modules with only slight derating of the individual devices provided the devices are matched for proper current sharing. In particular, the forward voltage drops of individual devices have to be matched closely.

8.5.2 Overcurrent Protection

The anode-to-cathode voltage in an MCT increases with its anode current and this property can be used to develop a protection scheme against overcurrent [5, 6]. The gate pulses to the MCT are blocked when the anode current and hence the anode-to-cathode voltage exceeds a preset value. A Schmitt trigger comparator is used to allow gate pulses to the MCT when it is in the process of turning on, during which time the anode voltage is relatively large and decreasing.

8.5.2.1 Snubbers

As with any other power device, the MCT is to be protected against switching-induced transient voltage and current spikes by using suitable snubbers. The snubbers modify the voltage and current transients during switching such that the switching trajectory is confined within the safe operating area (SOA). When the MCT is operated at high frequencies, the snubber increases the switching loss due to the delayed voltage and current responses. The power circuit of an MCT chopper including an improved snubber circuit is shown in Fig. 8.4 [5, 7]. The turn-on snubber consists of L_s and D_{LS} and the turn-off snubber consists of R_s, C_s, and D_{CS}. The series-connected turn-on snubber reduces the rate of change of the anode current dI_A/dt. The MCT does not support V_s until the current through the freewheeling diode reaches zero at turn-on. The turn-off snubber helps to reduce the peak power and the total power dissipated by the MCT by reducing the voltage across the MCT when the anode current decays to zero. The analysis and design of the snubber and the effect of the snubber on switching loss and electromagnetic interference are given in References [5] and [7]. An alternative snubber configuration for the two MCTs in an ac–ac converter has also been reported [8]. This snubber uses only one capacitor and one inductor for both the MCT switches (PMCT and NMCT) in a power-converter leg.

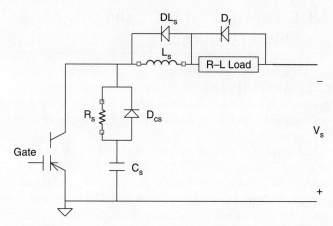

FIGURE 8.4 An MCT chopper with turn-on and turn-off snubbers.

8.6 Simulation Model of an MCT

The operation of power converters can be analyzed using PSPICE and other simulation software. As it is a new device, models of MCTs are not provided as part of the simulation libraries. However, an appropriate model for the MCT would be helpful in predicting the performance of novel converter topologies and in designing the control and protection circuits. Such a model must be simple enough to keep the simulation time and effort at a minimum, and must represent most of the device properties that affect the circuit operation. The PSPICE models for Harris PMCTs are provided by the manufacturer and can be downloaded from the internet. However, a simple model presenting most of the characteristics of an MCT is available [9, 10]. It is derived from the transistor-level equivalent circuit of the MCT by expanding the SCR model already reported the literature. The improved model [10] is capable of simulating the breakover and breakdown characteristics of an MCT and can be used for the simulation of high-frequency converters.

8.7 Generation-1 and Generation-2 MCTs

The Generation-1 MCTs were commercially introduced by Harris Semiconductors in 1992. However, the development of Generation-2 MCTs is continuing. In Gen-2 MCTs, each cell has its own turn-on field-effect transistor (FET). Preliminary test results on Generation-2 devices and a comparison of their performance with those of Generation-1 devices and highspeed IGBTs are available [11, 12]. The Generation-2 MCTs have a lower forward drop compared to the Generation-1 MCTs. They also have a higher dI/dt rating for a given value of capacitor used for discharge. During hard switching, the fall time and the switching losses are lower for the Gen-2 MCTs. The Gen-2 MCTs have the same conduction loss characteristics as Gen-1 with drastic reductions in turn-off switching times and losses [13].

Under zero-current switching conditions, Gen-2 MCTs have negligible switching losses [13]. Under zero-voltage switching, the turn-off losses in a Gen-2 device are one-half to one-fourth (depending on temperature and current level) the turn-off losses in Gen-1 devices. In all soft-switching applications, the predominant loss, namely, the conduction loss, reduces drastically allowing the use of fewer switches in a module.

8.8 N-channel MCT

The PMCT discussed in this chapter uses an NMOSFET for turn-off and this results in a higher turn-off current capability.

The PMCT can only replace a P-channel IGBT and inherits all the limitations of a P-channel IGBT. The results of a 2D simulation show that the NMCT can have a higher controllable current [13]. It is reported that NMCT versions of almost all Harris PMCTs have been fabricated for analyzing the potential for a commercial product [3]. The NMCTs are also being evaluated for use in zero-current soft-switching applications. However, the initial results are not quite encouraging in that the peak turn-off current of an NMCT is one-half to one-third of the value achievable in a PMCT. It is hoped that the NMCTs will eventually have a lower switching loss and a larger SOA as compared to PMCTs and IGBTs.

8.9 Base Resistance-controlled Thyristor [14]

The base resistance-controlled thyristor (BRT) is another gate-controlled device that is similar to the MCT but with a different structure. The Off-FET is not integrated within the p-base region but is formed within the n-base region. The diverter region is a shallow p-type junction formed adjacent to the p-base region of the thyristor. The fabrication process is simpler for this type of structure. The transistor level equivalent circuit of a BRT is shown in Fig. 8.5.

The BRT will be in the forward blocking state with a positive voltage applied to the anode and with a zero gate bias. The forward blocking voltage will be equal to the breakdown voltage of the open-base pnp transistor. A positive gate bias turns on the BRT. At low current levels, the device behaves similarly to an IGBT. When the anode current increases, the operation changes to thyristor mode resulting

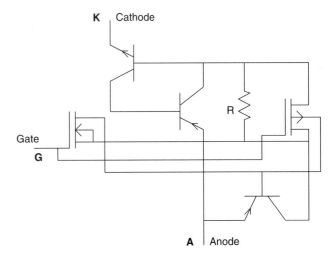

FIGURE 8.5 Equivalent circuit of base resistance-controlled thyristor (BRT).

in a low forward drop. Applying a negative voltage to its gate turns off the BRT. During the turn-off process, the anode current is diverted from the N^+ emitter to the diverter. The BRT has a current tail during turn-off that is similar to an MCT or an IGBT.

8.10 MOS Turn-off Thyristor [15]

The MOS turn-off (MTO) thyristor or the MTOT is a replacement for the GTO and it requires a much smaller gate drive. It is more efficient than a GTO, it can have a maximum blocking voltage of about 9 kV, and it will be used to build power converters in the 1–20 MVA range. Silicon Power Corporation (SPCO) manufactures the device.

The transistor-level equivalent circuit of the MTOT (hybrid design) and the circuit symbol are shown in Fig. 8.6. Applying a current pulse at the turn-on gate (G1), as with a conventional GTO, turns on the MTOT. The turn-on action, including regeneration, is similar to a conventional SCR. Applying a positive voltage pulse to the turn-off gate (G2), as with an MCT, turns off the MTOT. The voltage pulse turns on the FET, thereby shorting the emitter and base of the *npn* transistor and breaking the regenerative action. The MTOT is a faster switch than a GTO in that it is turned off with a reduced storage time compared to a GTO. The disk-type construction allows double-side cooling.

8.11 Applications of PMCT

The MCTs have been used in various applications, some of which are in the area of ac-dc and ac-ac conversion where the input is 60 Hz ac. Variable power factor operation was achieved using the MCTs as a force-commutated power switch [5]. The power circuit of an ac voltage controller capable of operating at a leading, lagging, and unity power factor is shown in Fig. 8.7. Because the switching frequency is low, the switching losses are negligible. Because the forward drop is low, the conduction losses are also small. The MCTs are also used in circuit breakers.

8.11.1 Soft-switching

The MCT is intended for high-frequency switching applications where it is supposed to replace a MOSFET or an IGBT. Similar to a Power MOSFET or an IGBT, the switching losses will be high at high switching frequencies. The typical characteristics of an MCT during turn-on and turn-off under hard switching (without snubber) are shown in Fig. 8.8. During turn-on and turn off, the device current and voltage take a finite time to reach their steady-state values. Each time the device changes state, there is a short period during which the voltage and current variations overlap. This results in a transient power loss that contributes to the average power loss.

Soft-switching converters are being designed primarily to enable operation at higher switching frequencies. In these

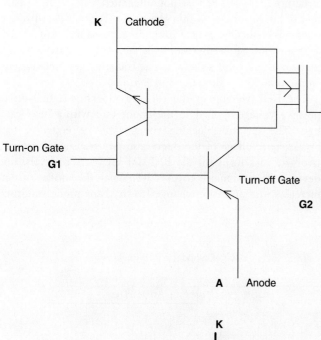

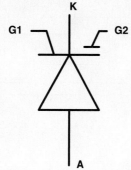

FIGURE 8.6 Equivalent circuit and symbol of a MOS turn-off (MTO) thyristor.

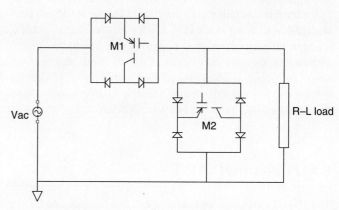

FIGURE 8.7 Power circuit of MCT ac voltage controller.

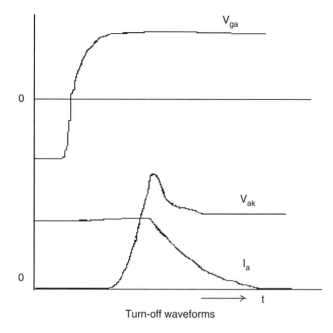

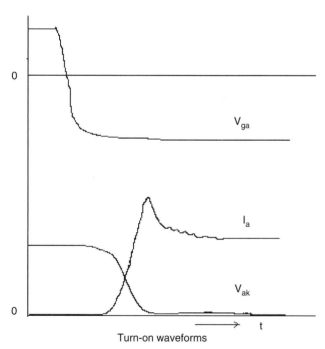

FIGURE 8.8 The MCT turn-off and turn-on waveforms under hard switching.

converters, the power devices switch at zero voltage or zero current, thereby eliminating the need for a large safe operating area (SOA) and at the same time eliminating the switching losses entirely. The MCT converters will outperform IGBT and power MOSFET converters in such applications by giving the highest possible efficiency. In soft-switching applications, the MCT will have only conduction loss, which is low and is close to that in a power diode with similar power ratings [12]. The Generation-1 MCTs did not turn on rapidly in the vicinity of zero anode-cathode voltage and this posed a problem in softswitching applications of an MCT. However, Generation-2 MCTs have enhanced dynamic characteristics under zero voltage soft switching [16]. In an MCT, the PMOS On-FET together with the *pnp* transistor constitute a *p*-IGBT. An increase in the number of turn-on cells (decrease in the on-resistance of the *p*-IGBT) and an enhancement of their distribution across the MCT active area enable the MCT to turn on at a very low transient voltage allowing zero voltage switching (ZVS). During zero voltage turn-on, a bipolar device such as the MCT takes more time to establish conductivity modulation. Before the device begins to conduct fully, a voltage spike appears, thus causing a modest switching loss [12]. Reducing the tail-current amplitude and duration by proper circuit design can minimize the turn-off losses in softswitching cases.

8.11.2 Resonant Converters

Resonant and quasi-resonant converters are known for their reduced switching loss [17]. Resonant converters with zero current switching are built using MCTs and the circuit of one such, a buck-converter, is shown in Fig. 8.9. The resonant commutating network consisting of L_r, C_r, auxiliary switch T_r, and diode D_r enables the MCT to turn off under zero current. The MCT must be turned off during the conduction period of D_Z. Commutating switch T_r must be turned off when the resonant current reaches zero.

A resonant dc link circuit with twelve parallel MCTs has been reported [18]. In this circuit, the MCTs switch at zero-voltage instants. The elimination of the switching loss allows operation at higher switching frequencies, which in turn increases the power density and offers better control of the spectral content. The use of MCTs with the same forward drop provides good current sharing.

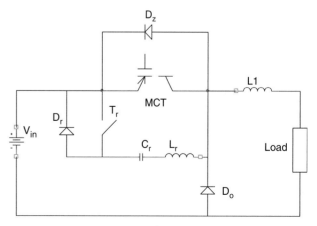

FIGURE 8.9 Power circuit of MCT resonant buck converter.

The MCTs are also used in ac-resonant-link converters with pulse density modulation (PDM) [19]. The advantages of the PDM converter, such as zero-voltage switching, combined with those of the MCT make the PDM converter a suitable candidate for many ac–ac converter applications. In an ac–ac PDM converter, a low-frequency ac voltage is obtained by switching the high-frequency ac link at zero-crossing voltages. Two MCTs with reverse-connected diodes form a bidirectional switch that is used in the circuit. A single capacitor was used as a simple snubber for both MCTs in the bidirectional switch.

8.12 Conclusions

The MCT is a power switch with a MOS gate for turn-on and turn-off. It is derived from a thyristor by adding the features of a MOSFET. It has several advantages compared to modern devices like the power MOSFET and the IGBT. In particular, the MCT has a low forward drop and a higher current density which are required for high-power applications. The characteristics of Generation-2 MCTs are better than those of Generation-1 MCTs. The switching performance of Generation-2 MCTs is comparable to the IGBTs. At one time, SPCO was developing both PMCTs and NMCTs. The only product that is currently under the product list of SPCO is the voltage/current controlled Solidtron, which is a discharge switch utlizing an n-type MCT. The device features a high current and high dI/dt capability and is used in capacitor discharge applications. The data on Solidtron can be obtained at: http://www.siliconpower.com/Solidtron/Solid_home.htm.

Acknowledgment

The author is grateful to Ms. Jing He and Mr. Rahul Patil for their assistance in collecting the reference material for this chapter.

8.13 Appendix

The following is a summary of the specifications on a 600 V/150 A PMCT made by SPC:

Peak Off-state Voltage, V_{DRM}	−600 V
Peak Reverse Voltage, V_{RRM}	+40V
Continuous Cathode Current, ($T = +90°C$), I_{K90}	150 A
Non-repetitive Peak Cathode Current, I_{KSM}	5000 A
Peak Controllable Current, I_{KC}	300 A
Gate to Anode Voltage (Continuous), V_{GA}	±15V
Gate to Anode Voltage (Peak), V_{GAM}	±20 V
Rate of Change of Voltage ($V_{GA} = 15$ V), dV/dt	10 kV/μs
Rate of Change of Current, di/dt	80 kA/μ
Peak Off-state Blocking Current (I_{DRM}) ($V_{KA} = -600$ V $V_{GA} = +15$ V, $Tc = +25°C$)v	200 μA
On-state Voltage (V_{TM}) ($I_K = 100$ A, $V_{GA} = -10V Tc = +25°C$)	1.3 V

References

1. V. A. K. Temple, "MOS-Controlled Thyristors — A new class of power devices," *IEEE Trans. on Electron Devices* **33**: 1609–1618 (1986).
2. T. M. Jahns, R. W. A. A. De Doncker, J. W. A. Wilson, V. A. K. Temple, and D. L. Waltrus, "Circuit utilization characteristics of MOS-Controlled Thyristors," *IEEE Trans. on Industry Applications* **27**:3, 589–597 (May/June 1991).
3. Harris Semiconductor, *MCT/IGBTs/Diodes Databook*, 1995.
4. P. Holdman and F. Lotuka, "SMPS IGBTs — High switching frequencies allow efficient switchers," *PCIM Power Electronics Systems* **25**:2, 38–2 (February 1999).
5. D. Quek, Design of Protection and Control Strategies for Low-loss MCT Power Converters, Ph.D. Thesis, North Dakota State University, July 1994.
6. D. Quek and S. Yuvarajan, "A novel gate drive for the MCT incorporating overcurrent protection," *Proc. of IEEE IAS Annual Meeting* 1994, pp. 1297–1302.
7. S. Yuvarajan, R. Nelson, and D. Quek, "A study of the effects of snubber on switching loss and EMI in an MCT converter," *Proc. of IEEE IAS Annual Meeting* 1994, pp. 1344–1349.
8. T. C. Lee, M. E. Elbuluk, and D. S. Zinger, "Characterization and snubbing of a bidirectional MCT switch in a resonant ac link converter," *IEEE Trans. Industry Applications* **31**:5, 978–985 (Sept./Oct. 1995).
9. S. Yuvarajan and D. Quek, "A PSPICE model for the MOS Controlled Thyristor," *IEEE Trans, on Industrial Electronics* **42**:5, 554–558 (Oct. 1995).
10. G. L. Arsov and L. P. Panovski, "An improved PSPICE model for the MOS-Controlled Thyristor," *IEEE Trans. Industrial Electronics* **46**:2, 473–477 (April 1999).
11. P. D. Kendle, V. A. K. Temple, and S. D. Arthur, "Switching comparison of Generation-1 and Generation-2 P-MCTs and ultrafast N-IGBTs," *Proc. of IEEE IAS Annual Meeting* 1993, pp. 1286–1292.
12. E. Yang, V. Temple, and S. Arthur, "Switching loss of Gen-1 and Gen-2 P-MCTs in soft-switching circuits," *Proc. of IEEE APEC* 1995, pp. 746–754.
13. Q. Huang, G. A. J. Amaratunga, E. M. Sankara Narayanan, and W. I. Milne, "Analysis of n-channel MOS-Controlled Thyristors," *IEEE Trans. Electron Devices* **38**:7, 1612–1618 (1991).
14. B. Jayant Baliga, *Power Semiconductor Devices,* PWS Publishing Co., Boston, 1996.
15. R. Rodrigues, A. Huang, and R. De Doncker, "MTO Thyristor Power Switches," *Proc. of PCIM'97 Power Electronics Conference* 1997, pp. 4-1–4-12.

16. R. W. A. A. De Doncker, T. M. Jahns, A. V. Radun, D. L. Waltrus, and V. A. K. Temple, "Characteristics of MOS-Controlled Thyristors under zero voltage soft-switching conditions," *IEEE Trans. Industry Applications* **28**:2, 387–394 (March/April 1992).
17. A. Dmowski, R. Bugyi, and P. Szewczyk, "Design of a buck converter with zero-current turn-off MCT," *Proc. IEEE IAS Annual Meeting* 1994, pp. 1025–1030.
18. H.-R. Chang and A. V. Radun, "Performance of 500 V, 450 A Parallel MOS Controlled Thyristors (MCTs) in a resonant dc-link circuit," *Proc. IEEE IAS Annual Meeting* 1990, pp. 1613–1617.
19. M. E. Elbuluk, D. S. Zinger, and T. Lee, "Performance of MCT's in a current-regulated ac/ac PDM converter," *IEEE Trans. Power Electronics* **11**:1, 49–56 (January 1996).

Static Induction Devices

Bogdan M. Wilamowski, Ph.D.
Alabama Microelectronics Science and Technology Center, Auburn University, Alabama, USA

	Summary	133
9.1	Introduction	133
9.2	Theory of Static Induction Devices	133
9.3	Characteristics of Static Induction Transistor	134
9.4	Bipolar Mode Operation of SI devices (BSIT)	136
9.5	Emitters for Static Induction Devices	136
9.6	Static Induction Diode	137
9.7	Lateral Punch-through Transistor	137
9.8	Static Induction Transistor Logic	137
9.9	BJT Saturation Protected by SIT	138
9.10	Static Induction MOS Transistor	138
9.11	Space Charge Limiting Load (SCLL)	140
9.12	Power MOS Transistors	140
9.13	Static Induction Thyristor	142
9.14	Gate Turn Off Thyristor	142
	References	142

Summary

Several devices from the static induction family such as: static induction transistor (SIT), static induction diode (SID), static induction thyristor, lateral punch-through transistor (LPTT), static induction transistor logic (SITL), static induction MOS transistor (SIMOS), and space charge limiting load (SCLL) are described. The theory of operation of static induction devices is given for both a current controlled by a potential barrier and a current controlled by space charge. The new concept of a punch-through emitter (PTE), which operates with majority carrier transport, is presented.

9.1 Introduction

Static induction devices were invented in 1975 by J. Nishizawa [1] and for many years Japan was the only country where static induction family devices were successfully fabricated. Static induction transistor can be considered as a short channel junction field effect transistor (JFET) device operating in pre-punch-through region. The number of devices in this family is growing with time. The SIT can operate with the power over 100 kW at 100 kHz; above 150 W at 3 GHz. [2]. These devices may operate upto THz frequencies [3, 4]. Static induction transistor logic had 100 times smaller switching energy than its I^2L competitor [5, 6]. Static induction thyristor has many advantages over the traditional silicon controlled rectifier (SCR), and SID exhibits high switching speed, large reverse voltage, and low forward voltage drops [7].

9.2 Theory of Static Induction Devices

The cross section of the SIT is shown in Fig. 9.1, while its characteristics are shown in Fig. 9.2. An induced electrostatically potential barrier controls the current in static induction devices. The derivations of formulas will be done for an n-channel device, but the obtained results, with a little modification can also be applied to p-channel devices. For a small electrical field existing in the vicinity of the potential barrier, the drift and diffusion current can be approximated by

$$J_n = -qn(x)\mu_n \frac{d\varphi(x)}{dx} + qD_n \frac{dn(x)}{dx} \qquad (9.1)$$

Copyright © 2007, 2001, Elsevier Inc.
All rights reserved.

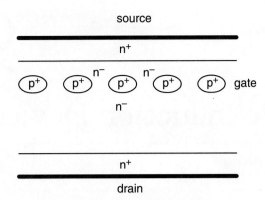

FIGURE 9.1 Cross section of the static induction transistor.

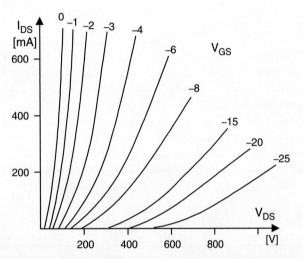

FIGURE 9.2 Characteristics of the early SIT design [1].

where $D_n = \mu_n V_T$ and $V_T = kT/q$. By multiplying both sides of the equation by $\exp(-\varphi(x)/V_T)$ and rearranging

$$J_n \exp\left(-\frac{\varphi(x)}{V_T}\right) = qD_n \frac{d}{dx}\left[n(x)\exp\left(-\frac{\varphi(x)}{V_T}\right)\right] \quad (9.2)$$

Integrating from x_1 to x_2 one can obtain

$$J_n = qD_n \frac{n(x_2)\exp(-\varphi(x_2)/V_T) - n(x_1)\exp(-\varphi(x_1)/V_T)}{\int_{x_1}^{x_2} \exp(-\varphi(x)/V_T)\, dx} \quad (9.3)$$

With the following boundary conditions

$$\varphi(x_1) = 0; \quad n(x_1) = N_S$$
$$\varphi(x_2) = V_D; \quad n(x_2) = N_D \quad (9.4)$$

Eq. (9.3) reduces to

$$J_n = \frac{qD_n N_S}{\int_{x_1}^{x_2} \exp(-\varphi(x)/V_T)\, dx} \quad (9.5)$$

Note that the above equations derived for SIT can also be used to find current in any devices controlled by a potential barrier, such as a bipolar transistor, MOS transistor operation in subthreshold mode, or in a Schottky diode.

9.3 Characteristics of Static Induction Transistor

Samples of the potential distribution in the SI devices are shown in Fig. 9.3 [7]. The vicinity of the potential barrier can be approximated using parabolic formulas (Fig. 9.4) along and across the channel [8, 9].

$$\varphi(x) = \Phi\left[1 - \left(2\frac{x}{L} - 1\right)^2\right] \quad (9.6)$$

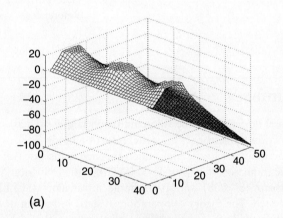

(a)

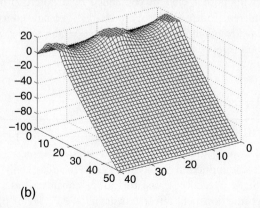

(b)

FIGURE 9.3 Potential distribution in SIT: (a) view from the source side and (b) view from the drain side.

9 Static Induction Devices

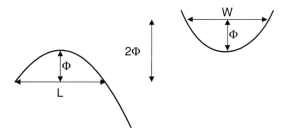

FIGURE 9.4 Potential distribution in the vicinity of the barrier approximated by parabolic shapes.

$$\varphi(y) = \Phi\left[1 - \left(2\frac{y}{W} - 1\right)^2\right] \quad (9.7)$$

Integrating Eq. (9.5) first along the channel and then across the channel, yields a very simple formula for drain currents in n-channel SITs

$$I_D = qD_pN_SZ\frac{W}{L}\exp\left(\frac{\Phi}{V_T}\right) \quad (9.8)$$

where Φ is the potential barrier height in reference to the source potential, N_S is the electron concentration at the source, W/L ratio describes the shape of the potential saddle in vicinity of the barrier, and Z is the length of the source strip.

Since barrier height Φ can be a linear function of gate and drain voltages; therefore,

$$I_D = qD_pN_SZ\frac{W}{L}\exp\bigl(a(V_{GS} + bV_{DS} + \Phi_0)/V_T\bigr) \quad (9.9)$$

The above equation describes characteristics of SIT for small current range. For large current levels, the device current is controlled by the space charge of moving carriers. In the one-dimensional case, the potential distribution is described by the Poisson equation:

$$\frac{d^2\varphi}{dx^2} = -\frac{\rho(x)}{\varepsilon_{Si}\varepsilon_0} = \frac{I_{DS}}{A v(x)} \quad (9.10)$$

where A is the effective device cross section and $v(x)$ is carrier velocity. For a small electrical field $v(x) = \mu E(x)$ and the solution of Eq. (9.10) is

$$I_{DS} = \frac{9}{8}V_{DS}^2 \mu \varepsilon_{Si}\varepsilon_0 \frac{A}{L^3} \quad (9.11)$$

and for a large electrical field $v(x) = \text{const}$ and Eq. (9.10) results in:

$$I_{DS} = 2V_{DS}v_{sat}\varepsilon_{Si}\varepsilon_0 \frac{A}{L^2} \quad (9.12)$$

where L is the channel length and $v_{sat} \approx 10^{11}\,\mu\text{m/s}$ is the carrier saturation velocity. In practical devices, the current–voltage relationship is described by an exponential relationship, Eq. (9.9), for small currents, a quadratic relationship, Eq. (9.11), and finally for large voltages by an almost linear relationship, Eq. (9.12). Static induction transistor characteristics drawn in linear and logarithmic scales are shown in Figs. 9.5 and 9.6, respectively.

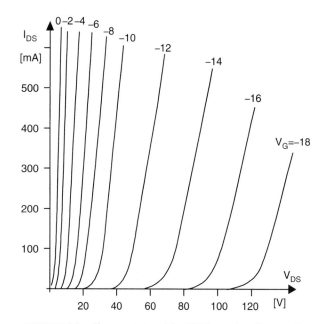

FIGURE 9.5 Characteristics of the SIT drawn in a linear scale.

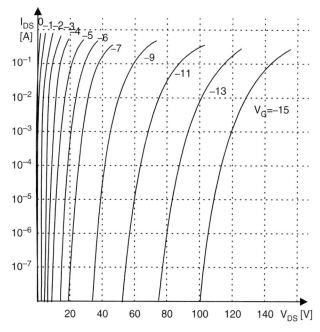

FIGURE 9.6 Characteristics of the SIT drawn in a logarithmic scale.

9.4 Bipolar Mode Operation of SI devices (BSIT)

The bipolar mode of operation of SIT was first reported in 1976 by Nishizawa and Wilamowski [5, 6]. Several complex theories for the bipolar mode of operation were developed [10–14], but actually the simple formula given by Eq. (9.5) works well not only for the typical mode of the SIT operation, but also for the bipolar mode of the SIT operation. Furthermore, the same formula works very well for the classical bipolar transistors. Typical characteristics of the SI transistor operating in normal and in bipolar modes are shown in Figs. 9.7 and 9.8.

A potential barrier controls the current in the SIT and it is given by

$$J_n = \frac{qD_n N_S}{\int_{x_1}^{x_2} \exp\left(-\frac{\varphi(x)}{V_T}\right) dx} \quad (9.13)$$

where $\varphi(x)$ is the profile of the potential barrier along the channel.

For example, in the case of npn bipolar transistors, the potential distribution across the base in reference to emitter potential at the reference impurity level $N_E = N_S$ is described by:

$$\varphi(x) = V_T \ln\left(\frac{N_B(x) N_S}{n_i^2}\right) \exp\left(-\frac{V_{BE}}{V_T}\right) \quad (9.14)$$

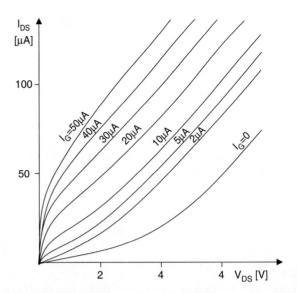

FIGURE 9.8 Small size SIT transistor characteristics, operating in both the normal and bipolar modes, $I_D = f(V_{DS})$ with I_G as a parameter.

After inserting Eq. (9.14) into Eq. (9.13), one can obtain the well-known equation for electron current injected into the base

$$J_n = \frac{qD_n n_i^2}{\int_{x_1}^{x_2} N_B(x) dx} \exp\left(\frac{V_{BE}}{V_T}\right) \quad (9.15)$$

If Eq. (9.13) is valid for SIT and BJT, then one may assume that it is also valid for the bipolar mode of operation of the SIT transistor. This is a well-known equation for the collector current in the bipolar transistor, but this time it was derived using the concept of the current flow through the potential barrier.

9.5 Emitters for Static Induction Devices

One of the disadvantages of the SIT is the relatively flat shape of the potential barrier (Fig. 9.9a). This leads to slow, diffusion-based transport of carriers in the vicinity of the potential barrier. The carrier transit time can be estimated using the formula:

$$t_{transit} = \frac{l_{eff}^2}{D} \quad (9.16)$$

where l_{eff} is the effective length of the channel and $D = \mu V_T$ is the diffusion constant. In the case of a traditional SIT transistor, this channel length is about 2 μm while in the case of SIT transistors with sharper barriers (Fig. 9.9b) the channel length is reduced to about 0.2 μm. The corresponding transient times are 2 ns and 20 ps, respectively.

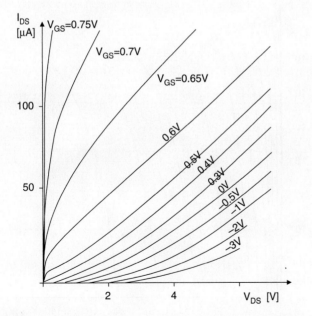

FIGURE 9.7 Small size SIT transistor characteristics, operating in both the normal and bipolar modes, $I_D = f(V_{DS})$ with V_{GS} as a parameter.

9 Static Induction Devices

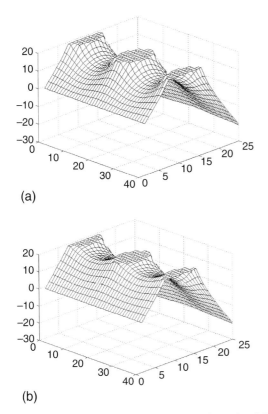

FIGURE 9.9 Potential distributions in SIT: (a) traditional and (b) with sharp potential barrier.

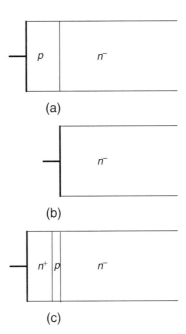

FIGURE 9.10 Various structures of emitters: (a) *p–n* junction including heterostructure with SiGe materials; (b) Schottky junction; and (c) punch-through emitter (in normal operation condition the *p* region is depleted from carriers).

Potential distributions shown in Fig. 9.3 are valid for SIT with an emitter made of a traditional *p–n* junction. A much narrower potential barrier can be obtained when other type of emitter is being used. There are two well-known emitters: (1) *p–n* junction (Fig. 9.10a) and (2) Schottky junction (Fig. 9.10b). For silicon devices, *p–n* junctions have a forward voltage drop of 0.7–0.8 V, while Schottky emitters have 0.2–0.3 V only. Since the Schottky diode is a majority carrier device, the carrier storage effect is negligible.

Another interesting emitter structure is shown in Fig. 9.10c. This emitter has all the advantages of the Schottky diode, with majority carrier injection, even though it is fabricated out of *p–n* junctions.

The concept of static induction devices can be used independently of the type of emitter shown in Fig. 9.10. With Schottky type and punch-through type emitters, the potential barrier is much narrower and this results in faster response time and larger current gain in the bipolar mode of operation.

9.6 Static Induction Diode

The bipolar mode of operation of SIT can also be used to obtain diodes with low forward voltage drop and negligible carrier storage effect [10, 11, 13, 15]. A static induction diode (SID) can be obtained by shorting a gate to the emitter of the SIT [16, 17]. Such diode has all the advantages of the SIT such as thermal stability and short switching time. The cross section of such diode is shown in Fig. 9.11.

The quality of the SID can be further improved with more sophisticated emitters (Figs. 9.10b and c). The SI diode with Schottky emitter was described by Wilamowski in 1983 [18] (Fig. 9.12). A similar structure was later published by Baliga [19].

9.7 Lateral Punch-through Transistor

Fabrications of SI transistors usually require very sophisticated technology. It is much simpler to fabricate a lateral punch-through transistor, which operates on the same principle and has similar characteristics [20] (Fig. 9.14). The cross section of the LPTT is shown in Fig. 9.13.

9.8 Static Induction Transistor Logic

The static induction transistor logic (SITL) was proposed by Nishizawa and Wilamowski [5, 6]. This logic circuit has almost 100 times better power-delay product than its I^2L competitor. Such drastic improvement of the power-delay product is possible because the SITL structure has a significantly smaller junction parasitic capacitance and also the voltage swing

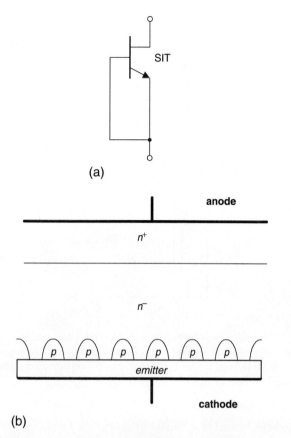

FIGURE 9.11 Static induction diode: (a) circuit diagram and (b) cross section.

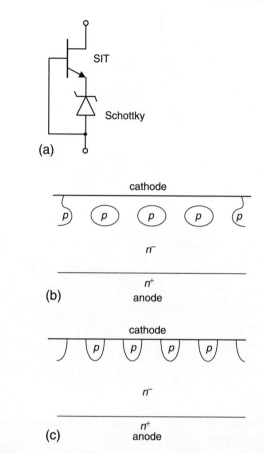

FIGURE 9.12 Schottky diode with enlarged breakdown voltages: (a) circuit diagram; (b) and (c) two cross section of possible implementation.

is reduced. Figs. 9.15 and 9.16 illustrate the concept of SITL. Measured characteristics of *n*-channel transistor of the static induction logic are shown in Fig. 9.17.

9.9 BJT Saturation Protected by SIT

The SI transistor can also be used instead of a Schottky diode to protect a bipolar junction transistor against saturation [21]. This leads to faster switching time. The concept is shown in Figs. 9.18 and 9.19. Note that this approach is advantageous to the solution with Schottky diode, since it does not require additional area on a chip and it does not introduce additional capacitance between the base and the collector. The base collector capacitance is always enlarged by the Miller effect and this leads to slower switching in the case of the solution with the Schottky diode.

9.10 Static Induction MOS Transistor

The punch-through transistor with MOS-controlled gate was described in 1983 [22, 23]. In the structure in Fig. 9.20a current

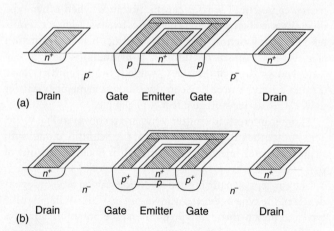

FIGURE 9.13 Structures of the lateral punch-through transistors: (a) simple and (b) with sharper potential barrier.

can flow in a similar fashion as in the lateral punch-through transistor [20]. In this mode of operation, carriers are moving far from the surface with a velocity close to the saturation velocity. The real advantage of such structure is the very low gate capacitance.

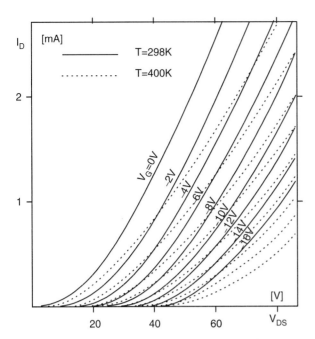

FIGURE 9.14 Characteristics of lateral punch-through transistor.

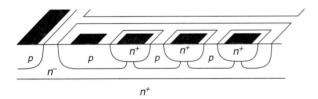

FIGURE 9.15 Cross section of SIT logic.

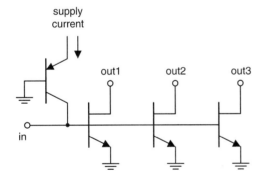

FIGURE 9.16 Diagrams of SIT logic.

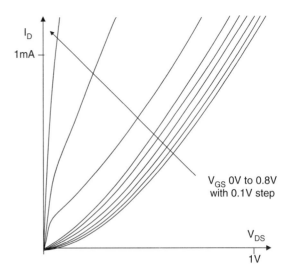

FIGURE 9.17 Measured characteristics of n-channel transistor of the logic circuit of Fig. 9.16.

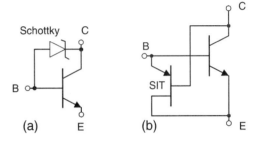

FIGURE 9.18 Protection of bipolar transistor against deep saturation: (a) using Schottky diode and (b) using SIT.

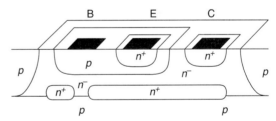

FIGURE 9.19 Cross sections of bipolar transistors protected against deep saturation using SIT.

Another implementation of static induction MOS transistor (SIMOS) is shown in Fig. 9.21. The buried p^+ layer is connected to the substrate, which has a large negative potential. As a result, the potential barrier is high and the emitter–drain current cannot flow. The punch-through current may start to flow when the positive voltage is applied to the gate and in this way the potential barrier is lowered. The p-implant layer is depleted and due to the high horizontal electrical field under the gate there is no charge accumulation under this gate. Such a transistor has several advantages over the traditional MOS transistor.

1. The gate capacitance is very small, since there is no accumulation layer under the gate.
2. Carriers are moving with a velocity close to saturation velocity.
3. Much lower substrate doping and the existing depletion layer lead to much smaller drain capacitance.

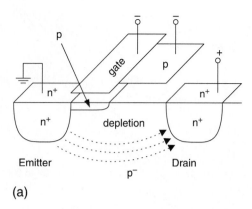

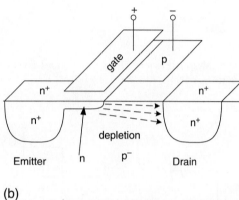

FIGURE 9.20 MOS controlled punch-through transistor: (a) transistor in the punch-through mode for the negative gate potential and (b) transistor in the on-state for the positive gate potential.

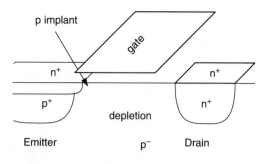

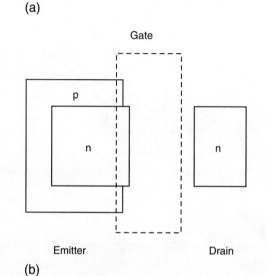

FIGURE 9.21 Static induction MOS structure: (a) cross section and (b) top view.

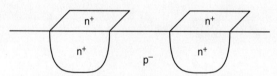

FIGURE 9.22 Space charge limiting load (SCLL).

The device operates in a similar fashion as MOS transistor in subthreshold conditions, but this process occurs at much higher current levels. Such "bipolar mode" of operation may have many advantages in VLSI applications.

9.11 Space Charge Limiting Load (SCLL)

Using the concept of the space charge limited current flow it is possible to fabricate very large resistors on a very small area. Moreover these resistors have a very small parasitic capacitance. For example, a 50 kΩ resistor requires only several square μm when 2 μm feature size technology is used [7].

Depending on the value of the electrical field, the device current is described by the following two equations. For a small electrical field $v(x) = \mu E(x)$

$$I_{DS} = \frac{9}{8} V_{DS}^2 \mu \varepsilon_{Si} \varepsilon_0 \frac{A}{L^3} \qquad (9.17)$$

For a large electrical field $v(x) = \text{const}$

$$I_{DS} = 2 V_{DS} v_{sat} \varepsilon_{Si} \varepsilon_0 \frac{A}{L^2} \qquad (9.18)$$

Moreover these resistors, which are based on the space charge limit flow, have a very small parasitic capacitance.

9.12 Power MOS Transistors

Power MOS transistors are being used for fast switching power supplies and for switching power converters. They can be driven with relatively small power and switching frequencies could be very high. High switching frequencies lead to compact circuit implementations with small inductors and small capacitances. Basically only two technologies, DMOS and VMOS, are used for power MOS devices as shown in Figs. 9.23 and 9.24.

9 *Static Induction Devices*

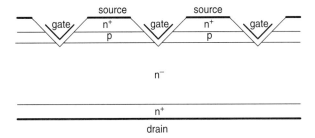

FIGURE 9.23 Cross section of the VMOS transistor.

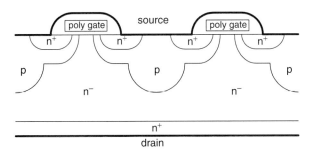

FIGURE 9.24 Cross section of the DMOS transistor.

A more popular structure is the DMOS shown in Fig. 9.24. This structure also uses the SIT concept. Note that for large drain voltages the *n*-region is depleted from carriers and statically induced electrical field in the vicinity of the virtual drain is significantly reduced. As a result this transistor may withstand much larger drain voltages and also the effect of channel length modulation is significantly reduced. The later effect leads to larger output resistances of the transistor. Therefore, the drain current is less sensitive to drain voltage variations. The structure in Fig. 9.24 can be considered as a composition of the MOS transistor and the SIT transistor as is shown in Fig. 9.25.

The major disadvantage of power MOS transistors is relatively large drain series resistance and much smaller transconductance in comparison to bipolar transistors. Both of these parameters can be improved dramatically by a simple change of the type of drain. In the case of *n*-channel device from *n*-type to *p*-type drain. This way the integrated structure is being built where its equivalent diagram consists MOS transistor integrated with bipolar transistor. Such structure has β times larger transconductance (β is the current gain of bipolar transistor) and much smaller series resistance due to the conductivity modulation effect caused by holes injected into lightly doped drain region. Such device is known as insulated gate bipolar transistors (IGBT) shown in Fig. 9.26. Their main disadvantage is large switching time limited primarily by poor switching performance of bipolar transistor. Another difficulty is related to a possible latch-up action of four layer $n^+pn^-p^+$ structure. This undesired effect could be suppressed by using heavily doped p^+ region in the base of NPN structure, which leads to significant reduction of the current gain of this parasitic transistor. The gain of other PNP transistor must be kept large so the transconductance of the entire device is large too. The IGBT transistor has breakdown voltages up to 1500 V, turn-off times are in range 0.1–0.5 μs. They may operate with currents above 100 A with a forward voltage drop about 3 V.

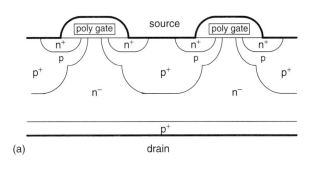

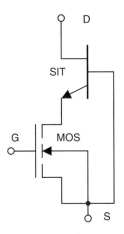

FIGURE 9.25 Equivalent diagram with MOS and SIT of the structure of Fig. 9.24.

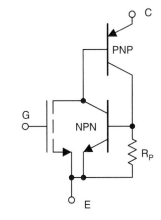

FIGURE 9.26 Insulated gate bipolar transistor (IGBT): (a) cross section and (b) equivalent diagram.

9.13 Static Induction Thyristor

There are several special semiconductor devices dedicated to high power applications. The most popular is thyristor known also as silicon control rectifier (SCR). This device has a four layer structure Fig. 9.27a and it can be considered as two transistors *npn* and *pnp* connected as shown in Fig. 9.27b.

In normal mode of operation (anode has positive potential) only one junction is reverse-biased and it can be represented by capacitance C. A spike of anode voltage can therefore get through capacitor C and it can trigger SCR. This behavior is not acceptable in practical application and therefore a different device structure is being used as is shown in Fig. 9.28. Note that by shorting gate to cathode by resistor R it is much more difficult to trigger the *npn* transistor by spike of anode voltage. This way rapid change of anode voltages is not able to trigger thyristor. Therefore this structure has very large dv/dt parameter.

When NPN transistor is replaced with SI transistor parameters of a thyristor can be significantly improved. For example, with breaking voltage in the range of 5 kV and current of 600 A the switching on time can be as short as 100 ns and dv/dt parameter can be as large as 50 kV/s [15, 24].

Most of the SCRs sold in the market comprise an integrated structure composed of two or more thyristors. This structure

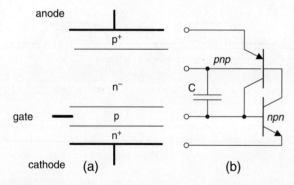

FIGURE 9.27 Silicon control rectifier: (a) cross section and (b) equivalent diagram.

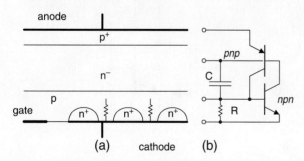

FIGURE 9.28 Silicon control rectifier with larger dv/dt parameter: (a) cross section and (b) equivalent diagram.

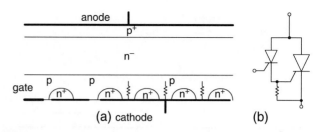

FIGURE 9.29 Integrated structure of silicon control rectifier: (a) cross section and (b) equivalent diagram.

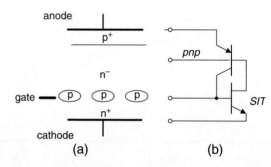

FIGURE 9.30 GTO–SIT: (a) cross section and (b) equivalent diagram.

has both large dv/dt and di/dt parameters. This structure consists of internal thyristor which significantly amplifies the gate signal.

One can notice that the classical thyristor as shown in Fig. 9.27 can be turned off by the gate voltage while integrated SCR shown in Fig. 9.29 can be only turned off by reducing anode current to zero. Most of the SCRs sold in the market have an integrated structure composed of two or more thyristors. This structure has both large dv/dt and di/dt parameters.

9.14 Gate Turn Off Thyristor

For the dc operation it is important to have a thyristor which can be turned off by the gate voltage. Such thyristor has a structure similar to the one shown in Fig. 9.27. It is important, however, to have significantly different current gains β for *pnp* and *npn* transistors. The current gain of *npn* transistor should be as large as possible and the current gain of *pnp* transistor should be small. The product of β_{npn} and β_{pnp} should be larger than one. This can be easily implemented using SI structure as shown in Fig. 9.30.

References

1. Nishizawa J., Terasaki T., and Shibata J., "Field-Effect Transistor versus Analog Transistor (Static Induction Transistor)," *IEEE Trans. on Electron Devices*, vol. **22**, No. 4, pp. 185–197, April 1975.

2. Tatsude M., Yamanaka E., and Nishizawa J., "High-Frequency High-Power Static Induction Transistor," *IEEE Industry Application Magazine*, pp. 40–45, March/April 1995.
3. Nishizawa J., Plotka P., and Kurabayashi T., "Ballistic and Tunneling GaAs Static Induction Transistors: Nano-Devices for THz Electronics," *IEEE Trans. on Electron Devices*, vol. **49**, No. 7, pp. 1102–1111, 2002.
4. Nishizawa J., Suto K., and Kurabayashi T., "Recent Advance in Tetrahertz Wave and Material Basis," *Russian Physics Journal*, vol. **46**, No. 6, pp. 615–622, 2003.
5. Nishizawa J. and Wilamowski B. M., "Integrated Logic – State Induction Transistor Logic," *International Solid State Circuit Conference*, Philadelphia USA, pp. 222–223, 1977.
6. Nishizawa J. and Wilamowski B. M., "Static Induction Logic – A Simple Structure with Very Low Switching Energy and Very High Packing Density," *International Conference on Solid State Devices*, Tokyo, Japan, pp. 53–54, 1976 and *Journal of Japanese Soc. Appl. Physics*, vol. **16**, No. 1, pp. 158–162, 1977.
7. Wilamowski B. M., "High Speed, High Voltage, and Energy Efficient Static Induction Devices," *12 Symposium of Static Induction Devices – SSID'99*, (invited speech) Tokyo, Japan, pp. 23–28, April 23, 1999.
8. Plotka P. and Wilamowski B. M., "Interpretation of Exponential Type Drain Characteristics of the SIT," *Solid-State Electronics*, vol. **23**, pp. 693–694, 1980.
9. Plotka P. and Wilamowski B. M., "Temperature Properties of the Static Induction Transistor," *Solid-State Electronics*, vol. **24**, pp. 105–107, 1981.
10. Kim C. W., Kimura M., Yano K., Tanaka, A., and Sukegawa, T., "Bipolar-Mode Static Induction Transistor: Experiment and Two-Dimensional Analysis," *IEEE Trans. on Electron Devices*, vol. **37**, No. 9, pp. 2070–2075, September 1990.
11. Nakamura Y., Tadano H., Takigawa M., Igarashi I., and Nishizawa J., "Experimental Study on Current Gain of BSIT," *IEEE Trans. on Electron Devices*, vol. **33**, No. 6, pp. 810–815, June 1986.
12. Nishizawa J., Ohmi T., and Chen H. L., "Analysis of Static Characteristics of a Bipolar-Mode SIT (BSIT)," *IEEE Trans. on Electron Devices*, vol. **29**, No. 8, pp. 1233–1244, August 1982.
13. Yano K., Henmi I., Kasuga M., and Shimizu A., "High-Power Rectifier Using the BSIT Operation," *IEEE Trans. on Electron Devices*, vol. **45**, No. 2, pp. 563–565, February 1998.
14. Yano K., Masahito M., Moroshima H., Morita J., Kasuga M., and Shimizu A., "Rectifier Characteristics Based on Bipolar-Mode SIT Operation," *IEEE Electron Device Letters*, vol. **15**, No. 9, pp. 321–323, September 1994.
15. Hironaka R., Watanabe M., Hotta E., and Okino A. "Performance of Pulsed Power Generator using High Voltage Static Induction Thyristor," *IEEE Trans. on Plasma Science*, vol. **28**, No. 5, pp. 1524–1527, 2000.
16. Yano K., Honarkhah S., and Salama A., "Lateral SOI Static Induction Rectifiers", *Proc. of 2001 Int. Symp. on Power Semiconductor Devices*, Osaka, pp. 247–250, 2001.
17. Yano K., Hattori N., Yamamoto Y., and Kasuga M., "Impacts of Channel Implantation on Performance of Static Shielding Diodes and Static Induction Rectifiers," *Proc. of 2001 Int. Symp. on Power Semiconductor Devices*, Osaka, pp. 219–222, 2001.
18. Wilamowski B. M., "Schottky Diodes with High Breakdown voltage," *Solid-State Electronics*, vol. **26**, No. 5, pp. 491–493, 1983.
19. Baliga B. J., "The Pinch Rectifier: A Low Forward-Drop, High-Speed Power Diode," *IEEE Electron Device Letters*, vol. **5**, pp. 194–196, 1984.
20. Wilamowski B. M. and Jaeger R. C., "The Lateral Punch-Through Transistor," *IEEE Electron Device Letters*, vol. **3**, No. 10, pp. 277–280, 1982.
21. Wilamowski B. M., Mattson R. H., and Staszak Z. J., "The SIT saturation protected bipolar transistor," *IEEE Electron Device Letters*, vol. **5**, pp. 263–265, 1984.
22. Wilamowski B. M., "The Punch-Through Transistor with MOS Controlled Gate," *Phys. Status Solidi (a)*, vol. **79**, pp. 631–637, 1983.
23. Wilamowski B. M., Jaeger R. C., and Fordemwalt J. N., "Buried MOS Transistor with Punch-Through," *Solid State Electronics*, vol. **27**, No. 8/9, pp. 811–815, 1984.
24. Shimizu N., Sekiya T., Iida K., Imanishi Y., Kimura M., and Nishizawa J., "Over 55kV/us, dv/dt turnoff characteristics of 4kV-Static Induction Thyristor for Pulsed Power Applications," *Proc. of 2004 Int. Symp. on Power Semiconductor Devices*, Kitakyushu, Japan, pp. 281–284, 2004.

10
Diode Rectifiers

Yim-Shu Lee and Martin H. L. Chow
Department of Electronic and Information Engineering, The Hong Kong Polytechnic, University Hung Hom, Hong Kong

10.1 Introduction 145
10.2 Single-phase Diode Rectifiers 145
    10.2.1 Single-phase Half-wave Rectifiers • 10.2.2 Single-phase Full-wave Rectifiers • 10.2.3 Performance Parameters • 10.2.4 Design Considerations
10.3 Three-phase Diode Rectifiers 150
    10.3.1 Three-phase Star Rectifiers • 10.3.2 Three-phase Bridge Rectifiers • 10.3.3 Operation of Rectifiers with Finite Source Inductance
10.4 Poly-phase Diode Rectifiers 155
    10.4.1 Six-phase Star Rectifier • 10.4.2 Six-phase Series Bridge Rectifier • 10.4.3 Six-phase Parallel Bridge Rectifier
10.5 Filtering Systems in Rectifier Circuits 158
    10.5.1 Inductive-input DC Filters • 10.5.2 Capacitive-input DC Filters
10.6 High-frequency Diode Rectifier Circuits 162
    10.6.1 Forward Rectifier Diode, Flywheel Diode, and Magnetic-reset Clamping Diode in a Forward Converter • 10.6.2 Flyback Rectifier Diode and Clamping Diode in a Flyback Converter • 10.6.3 Design Considerations • 10.6.4 Precautions in Interpreting Simulation Results
    Further Reading 177

10.1 Introduction

This chapter describes the application and design of diode rectifier circuits. It covers single-phase rectifier circuits, three-phase rectifier circuits, poly-phase rectifier circuits, and high-frequency rectifier circuits. The objectives of this chapter are:

- To enable the readers to understand the operation of typical rectifier circuits.
- To enable the readers to appreciate the different qualities of rectifiers required for different applications.
- To enable the reader to design practical rectifier circuits.

The high-frequency rectifier waveforms given are obtained from PSPICE simulations, which take into account the secondary effects of stray and parasitic components. In this way, the waveforms can closely resemble the real ones. These waveforms are particularly useful to help designers determine the practical voltage, current, and other ratings of high-frequency rectifiers.

10.2 Single-phase Diode Rectifiers

There are two types of single-phase diode rectifier that convert a single-phase ac supply into a dc voltage, namely, single-phase half-wave rectifiers and single-phase full-wave rectifiers. In the following subsections, the operations of these rectifier circuits are examined and their performances are analyzed and compared in a tabulated form. For the sake of simplicity, the diodes are considered to be ideal, i.e. they have zero forward voltage drop and reverse recovery time. This assumption is generally valid for the case of diode rectifiers which use the mains, a low-frequency source, as the input, and when the forward voltage drop is small compared with the peak voltage of the mains. Furthermore, it is assumed that the load is purely resistive such that the load voltage and the load current have similar waveforms. In Section 10.5, Filtering Systems in Rectifiers, the effects of inductive load and capacitive load on a diode rectifier are considered in detail.

10.2.1 Single-phase Half-wave Rectifiers

The simplest single-phase diode rectifier is the single-phase half-wave rectifier. A single-phase half-wave rectifier with resistive load is shown in Fig. 10.1. The circuit consists of only one diode that is usually fed with a secondary transformer as shown. During the positive half-cycle of the transformer secondary voltage, diode D conducts. During the negative half-cycle, diode D stops conducting. Assuming that the transformer has zero internal impedance and provides perfect sinusoidal voltage on its secondary winding, the voltage and current waveforms of resistive load R and the voltage waveform of diode D are shown in Fig. 10.2.

By observing the voltage waveform of diode D in Fig. 10.2, it is clear that the peak inverse voltage (PIV) of diode D is equal to V_m during the negative half-cycle of the transformer secondary voltage. Hence the peak repetitive reverse voltage (V_{RRM}) rating of diode D must be chosen to be higher than V_m to avoid reverse breakdown. In the positive half-cycle of the transformer secondary voltage, diode D has a forward current which is equal to the load current, therefore the peak repetitive forward current (I_{FRM}) rating of diode D must be chosen to be higher than the peak load current, $V_m = R$, in practice. In addition, the transformer has to carry a dc current that may result in a dc saturation problem of the transformer core.

10.2.2 Single-phase Full-wave Rectifiers

There are two types of single-phase full-wave rectifier, namely, full-wave rectifiers with center-tapped transformer and bridge rectifiers. A full-wave rectifier with a center-tapped transformer is shown in Fig. 10.3. It is clear that each diode, together with the associated half of the transformer, acts as a half-wave rectifier. The outputs of the two half-wave rectifiers are combined to produce full-wave rectification in the load. As far as the transformer is concerned, the dc currents of the two half-wave rectifiers are equal and opposite, such that there is no dc current for creating a transformer core saturation problem. The voltage and current waveforms of the full-wave rectifier are shown in Fig. 10.4. By observing diode voltage waveforms v_{D1} and v_{D2} in Fig. 10.4, it is clear that the PIV of the diodes is equal to $2V_m$ during their blocking state. Hence the V_{RRM} rating of the diodes must be chosen to be higher than $2V_m$ to avoid reverse breakdown. (Note that, compared with the half-wave rectifier shown in Fig. 10.1, the full-wave rectifier has twice the dc output voltage, as shown in Section 10.2.4.) During its conducting state, each diode has a forward current which is equal to the load current, therefore the I_{FRM} rating of these diodes must be chosen to be higher than the peak load current, $V_m = R$, in practice.

Employing four diodes instead of two, a bridge rectifier as shown in Fig. 10.5 can provide full-wave rectification without using a center-tapped transformer. During the positive half-cycle of the transformer secondary voltage, the current flows to the load through diodes D_1 and D_2. During the negative half-cycle, D_3 and D_4 conduct. The voltage and current waveforms of the bridge rectifier are shown in Fig. 10.6 As with the full-wave rectifier with center-tapped transformer, the I_{FRM} rating of the employed diodes must be chosen to be higher than the peak load current, $V_m = R$. However, the PIV of the diodes is reduced from $2V_m$ to V_m during their blocking state.

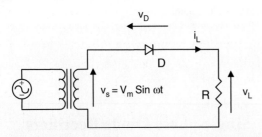

FIGURE 10.1 A single-phase half-wave rectifier with resistive load.

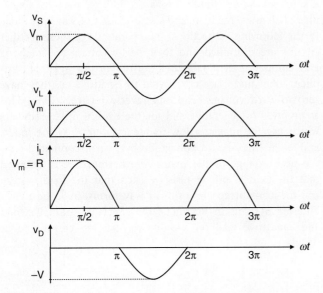

FIGURE 10.2 Voltage and current waveforms of the half-wave rectifier with resistive load.

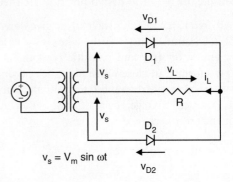

FIGURE 10.3 Full-wave rectifier with center-tapped transformer.

10 Diode Rectifiers

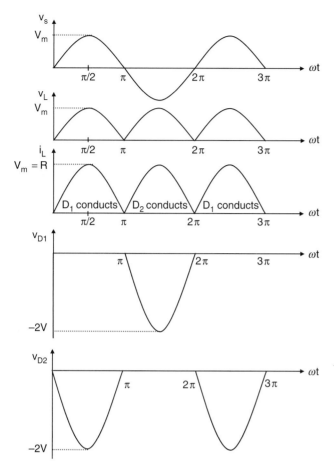

FIGURE 10.4 Voltage and current waveforms of the full-wave rectifier with center-tapped transformer.

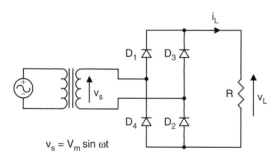

$v_s = V_m \sin \omega t$

FIGURE 10.5 Bridge rectifier.

10.2.3 Performance Parameters

In this subsection, the performance of the rectifiers mentioned above will be evaluated in terms of the following parameters.

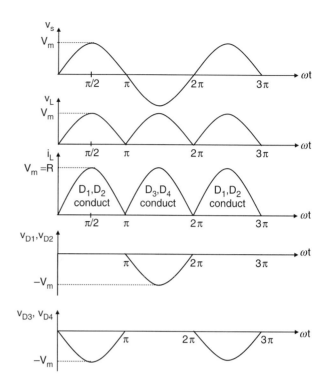

FIGURE 10.6 Voltage and current waveforms of the bridge rectifier.

10.2.3.1 Voltage Relationships

The average value of the load voltage v_L is V_{dc} and it is defined as

$$V_{dc} = \frac{1}{T} \int_0^T v_L(t)\, dt \qquad (10.1)$$

In the case of a half-wave rectifier, Fig. 10.2 indicates that load voltage $v_L(t) = 0$ for the negative half-cycle. Note that the angular frequency of the source $\omega = 2\pi = T$, and Eq. (10.1) can be re-written as

$$V_{dc} = \frac{1}{2\pi} \int_0^\pi V_m \sin \omega t\, d(\omega t) \qquad (10.2)$$

Therefore,

$$\text{Half-wave} \quad V_{dc} = \frac{V_m}{\pi} = 0.318 V_m \qquad (10.3)$$

In the case of a full-wave rectifier, Figs. 10.4 and 10.6 indicate that $v_L(t) = V_m |\sin \omega t|$ for both the positive and negative half-cycles. Hence Eq. (10.1) can be re-written as

$$V_{dc} = \frac{1}{\pi} \int_0^\pi V_m \sin \omega t\, d(\omega t) \qquad (10.4)$$

Therefore,

$$\text{Full-wave} \quad V_{dc} = \frac{2V_m}{\pi} = 0.636 V_m \qquad (10.5)$$

The root-mean-square (rms) value of load voltage v_L is V_L, which is defined as

$$V_L = \left[\frac{1}{T}\int_0^T v_L^2(t)dt\right]^{1/2} \tag{10.6}$$

In the case of a half-wave rectifier, $v_L(t) = 0$ for the negative half-cycle, therefore Eq. (10.6) can be re-written as

$$V_L = \sqrt{\frac{1}{2\pi}\int_0^\pi (V_m \sin \omega t)^2 d(\omega t)} \tag{10.7}$$

or

$$\text{Half-wave} \quad V_L = \frac{V_m}{2} = 0.5V_m \tag{10.8}$$

In the case of a full-wave rectifier, $v_L(t) = V_m|\sin \omega t|$ for both the positive and negative half-cycles. Hence Eq. (10.6) can be re-written as

$$V_L = \sqrt{\frac{1}{\pi}\int_0^\pi (V_m \sin \omega t)^2 d(\omega t)} \tag{10.9}$$

or

$$\text{Full-wave} \quad V_L = \frac{V_m}{\sqrt{2}} = 0.707V_m \tag{10.10}$$

The result of Eq. (10.10) is as expected because the rms value of a full-wave rectified voltage should be equal to that of the original ac voltage.

10.2.3.2 Current Relationships

The average value of load current i_L is I_{dc} and because load R is purely resistive it can be found as

$$I_{dc} = \frac{V_{dc}}{R} \tag{10.11}$$

The rms value of load current i_L is I_L and it can be found as

$$I_L = \frac{V_L}{R} \tag{10.12}$$

In the case of a half-wave rectifier, from Eq. (10.3)

$$\text{Half-wave} \quad I_{dc} = \frac{0.318V_m}{R} \tag{10.13}$$

and from Eq. (10.8)

$$\text{Half-wave} \quad I_L = \frac{0.5V_m}{R} \tag{10.14}$$

In the case of a full-wave rectifier, from Eq. (10.5)

$$\text{Full-wave} \quad I_{dc} = \frac{0.636V_m}{R} \tag{10.15}$$

and from Eq. (10.10)

$$\text{Full-wave} \quad I_L = \frac{0.707V_m}{R} \tag{10.16}$$

10.2.3.3 Rectification Ratio

The rectification ratio, which is a figure of merit for comparing the effectiveness of rectification, is defined as

$$\sigma = \frac{P_{dc}}{P_L} = \frac{V_{dc}I_{dc}}{V_L I_L} \tag{10.17}$$

In the case of a half-wave diode rectifier, the rectification ratio can be determined by substituting Eqs. (10.3), (10.13), (10.8), and (10.14) into Eq. (10.17).

$$\text{Half-wave} \quad \sigma = \frac{(0.318V_m)^2}{(0.5V_m)^2} = 40.5\% \tag{10.18}$$

In the case of a full-wave rectifier, the rectification ratio is obtained by substituting Eqs. (10.5), (10.15), (10.10), and (10.16) into Eq. (10.17).

$$\text{Full-wave} \quad \sigma = \frac{(0.636V_m)^2}{(0.707V_m)^2} = 81\% \tag{10.19}$$

10.2.3.4 Form Factor

The form factor (FF) is defined as the ratio of the root-mean-square value (heating component) of a voltage or current to its average value,

$$\text{FF} = \frac{V_L}{V_{dc}} \quad \text{or} \quad \frac{I_L}{I_{dc}} \tag{10.20}$$

In the case of a half-wave rectifier, the FF can be found by substituting Eqs. (10.8) and (10.3) into Eq. (10.20).

$$\text{Half-wave} \quad \text{FF} = \frac{0.5V_m}{0.318V_m} = 1.57 \tag{10.21}$$

In the case of a full-wave rectifier, the FF can be found by substituting Eqs. (10.16) and (10.15) into Eq. (10.20).

$$\text{Full-wave} \quad \text{FF} = \frac{0.707V_m}{0.636V_m} = 1.11 \tag{10.22}$$

10.2.3.5 Ripple Factor

The ripple factor (RF), which is a measure of the ripple content, is defined as

$$\text{RF} = \frac{V_{ac}}{V_{dc}} \quad (10.23)$$

where V_{ac} is the effective (rms) value of the ac component of load voltage v_L.

$$V_{ac} = \sqrt{V_L^2 - V_{dc}^2} \quad (10.24)$$

Substituting Eq. (10.24) into Eq. (10.23), the RF can be expressed as

$$\text{RF} = \sqrt{\left(\frac{V_L}{V_{dc}}\right)^2 - 1} = \sqrt{\text{FF}^2 - 1} \quad (10.25)$$

In the case of a half-wave rectifier,

$$\text{Half-wave} \quad \text{RF} = \sqrt{1.57^2 - 1} = 1.21 \quad (10.26)$$

In the case of a full-wave rectifier,

$$\text{Full-wave} \quad \text{RF} = \sqrt{1.11^2 - 1} = 0.482 \quad (10.27)$$

10.2.3.6 Transformer Utilization Factor

The transformer utilization factor (TUF), which is a measure of the merit of a rectifier circuit, is defined as the ratio of the dc output power to the transformer volt–ampere (VA) rating required by the secondary winding,

$$\text{TUF} = \frac{P_{dc}}{V_s I_s} = \frac{V_{dc} I_{dc}}{V_s I_s} \quad (10.28)$$

where V_s and I_s are the rms voltage and rms current ratings of the secondary transformer.

$$V_s = \frac{V_m}{\sqrt{2}} = 0.707 V_m \quad (10.29)$$

The rms value of the transformer secondary current I_s is the same as that of the load current I_L. For a half-wave rectifier, I_s can be found from Eq. (10.14).

$$\text{Half-wave} \quad I_s = \frac{0.5 V_m}{R} \quad (10.30)$$

For a full-wave rectifier, I_s is found from Eq. (10.16).

$$\text{Full-wave} \quad I_s = \frac{0.707 V_m}{R} \quad (10.31)$$

Therefore, the TUF of a half-wave rectifier can be obtained by substituting Eqs. (10.3), (10.13), (10.29), and (10.30) into Eq. (10.28).

$$\text{Half-wave} \quad \text{TUF} = \frac{0.318^2}{0.707 \times 0.5} = 0.286 \quad (10.32)$$

The poor TUF of a half-wave rectifier signifies that the transformer employed must have a 3.496 (1/0.286) VA rating in order to deliver 1 W dc output power to the load. In addition, the transformer secondary winding has to carry a dc current that may cause magnetic core saturation. As a result, half-wave rectifiers are used only when the current requirement is small.

In the case of a full-wave rectifier with center-tapped transformer, the circuit can be treated as two half-wave rectifiers operating together. Therefore, the transformer secondary VA rating, $V_s I_s$, is double that of a half-wave rectifier, but the output dc power is increased by a factor of four due to higher the rectification ratio as indicated by Eqs. (10.5) and (10.15). Therefore, the TUF of a full-wave rectifier with center-tapped transformer can be found from Eq. (10.32)

$$\text{Full-wave} \quad \text{TUF} = \frac{4 \times 0.318^2}{2 \times 0.707 \times 0.5} = 0.572 \quad (10.33)$$

In the case of a bridge rectifier, it has the highest TUF in single-phase rectifier circuits because the currents flowing in both the primary and secondary windings are continuous sinewaves. By substituting Eqs. (10.5), (10.15), (10.29), and (10.31) into Eq. (10.28), the TUF of a bridge rectifier can be found.

$$\text{Bridge} \quad \text{TUF} = \frac{0.636^2}{(0.707)^2} = 0.81 \quad (10.34)$$

The transformer primary VA rating of a full-wave rectifier is equal to that of a bridge rectifier since the current flowing in the primary winding is also a continuous sinewave.

10.2.3.7 Harmonics

Full-wave rectifier circuits with resistive load do not produce harmonic currents in their transformers. In half-wave rectifiers, harmonic currents are generated. The amplitudes of the harmonic currents of a half-wave rectifier with resistive load, relative to the fundamental, are given in Table 10.1. The extra loss caused by the harmonics in the resistive loaded rectifier circuits is often neglected because it is not high compared with other losses. However, with non-linear loads, harmonics can cause appreciable loss and other problems such as poor power factor and interference.

TABLE 10.1 Harmonic percentages of a half-wave rectifier with resistive load

Harmonic	2nd	3rd	4th	5th	6th	7th	8th
%	21.2	0	4.2	0	1.8	0	1.01

10.2.4 Design Considerations

In a practical design, the goal is to achieve a given dc output voltage. Therefore, it is more convenient to put all the design parameters in terms of V_{dc}. For example, the rating and turns ratio of the transformer in a rectifier circuit can be easily determined if the rms input voltage to the rectifier is in terms of the required output voltage V_{dc}.

Denote the rms value of the input voltage to the rectifier as V_s, which is equal to $0.707 V_m$. Based on this relation and Eq. (10.3), the rms input voltage to a half-wave rectifier is found as

$$\text{Half-wave} \quad V_s = 2.22 V_{dc} \quad (10.35)$$

Similarly, from Eqs. (10.5) and (10.29), the rms input voltage per secondary winding of a full-wave rectifier is found as

$$\text{Full-wave} \quad V_s = 1.11 V_{dc} \quad (10.36)$$

Another important design parameter is the V_{RRM} rating of the diodes employed.

In the case of a half-wave rectifier, from Eq. (10.3),

$$\text{Half-wave} \quad V_{RRM} = V_m = \frac{V_{dc}}{0.318} = 3.14 V_{dc} \quad (10.37)$$

In the case of a full-wave rectifier with center-tapped transformer, from Eq. (10.5),

$$\text{Full-wave} \quad V_{RRM} = 2V_m = \frac{2V_{dc}}{0.636} = 3.14 V_{dc} \quad (10.38)$$

In the case of a bridge rectifier, also from Eq. (10.5),

$$\text{Bridge} \quad V_{RRM} = V_m = \frac{V_{dc}}{0.636} = 1.57 V_{dc} \quad (10.39)$$

It is important to evaluate the I_{FRM} rating of the employed diodes in rectifier circuits.

In the case of a half-wave rectifier, from Eq. (10.13),

$$\text{Half-wave} \quad I_{FRM} = \frac{V_m}{R} = \frac{I_{dc}}{0.318} = 3.41 I_{dc} \quad (10.40)$$

In the case of full-wave rectifiers, from Eq. (10.15),

$$\text{Full-wave} \quad I_{FRM} = \frac{V_m}{R} = \frac{I_{dc}}{0.636} = 1.57 I_{dc} \quad (10.41)$$

The important design parameters of basic single-phase rectifier circuits with resistive loads are summarized in Table 10.2.

10.3 Three-phase Diode Rectifiers

It has been shown in Section 10.2 that single-phase diode rectifiers require a rather high transformer VA rating for a given dc output power. Therefore, these rectifiers are suitable only for low to medium power applications. For power output higher than 15 kW, three-phase or poly-phase diode rectifiers should be employed. There are two types of three-phase diode rectifier that convert a three-phase ac supply into a dc voltage, namely, star rectifiers and bridge rectifiers. In the following subsections,

TABLE 10.2 Important design parameters of basic single-phase rectifier circuits with resistive load

	Half-wave rectifier	Full-wave rectifier with center-tapped transformer	Full-wave bridge rectifier
Peak repetitive reverse voltage V_{RRM}	$3.14 V_{dc}$	$3.14 V_{dc}$	$1.57 V_{dc}$
RMS input voltage per transformer leg V_s	$2.22 V_{dc}$	$1.11 V_{dc}$	$1.11 V_{dc}$
Diode average current $I_{F(AV)}$	$1.00 I_{dc}$	$0.50 I_{dc}$	$0.50 I_{dc}$
Peak repetitive forward current I_{FRM}	$3.14 I_{F(AV)}$	$1.57 I_{F(AV)}$	$1.57 I_{F(AV)}$
Diode rms current $I_{F(RMS)}$	$1.57 I_{dc}$	$0.785 I_{dc}$	$0.785 I_{dc}$
Form factor of diode current $I_{F(RMS)}/I_{F(AV)}$	1.57	1.57	1.57
Rectification ratio	0.405	0.81	0.81
Form factor	1.57	1.11	1.11
Ripple factor	1.21	0.482	0.482
Transformer rating primary VA	$2.69 P_{dc}$	$1.23 P_{dc}$	$1.23 P_{dc}$
Transformer rating secondary VA	$3.49 P_{dc}$	$1.75 P_{dc}$	$1.23 P_{dc}$
Output ripple frequency f_r	$1 f_i$	$2 f_i$	$2 f_i$

10 Diode Rectifiers

the operations of these rectifiers are examined and their performances are analyzed and compared in tabulated form. For the sake of simplicity, the diodes and the transformers are considered to be ideal, i.e. the diodes have zero forward voltage drop and reverse current, and the transformers possess no resistance and no leakage inductance. Furthermore, it is assumed that the load is purely resistive, such that the load voltage and the load current have similar waveforms. In Section 10.5 Filtering Systems in Rectifier Circuits, the effects of inductive load and capacitive load on a diode rectifier are considered in detail.

10.3.1 Three-phase Star Rectifiers

10.3.1.1 Basic Three-phase Star Rectifier Circuit

A basic three-phase star rectifier circuit is shown in Fig. 10.7. This circuit can be considered as three single-phase half-wave rectifiers combined together. Therefore it is sometimes referred to as a three-phase half-wave rectifier. The diode in a particular phase conducts during the period when the voltage on that phase is higher than that on the other two phases. The voltage waveforms of each phase and the load are shown in Fig. 10.8. It is clear that, unlike the single-phase rectifier circuit, the conduction angle of each diode is $2\pi/3$, instead of π. This circuit finds uses where the required dc output voltage is relatively low and the required output current is too large for a practical single-phase system.

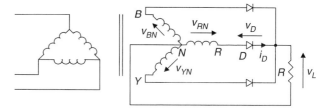

FIGURE 10.7 Three-phase star rectifier.

Taking phase R as an example, diode D conducts from $\pi/6$ to $5\pi/6$. Therefore, using Eq. (10.1) the average value of the output can be found as

$$V_{dc} = \frac{3}{2\pi} \int_{\pi/6}^{5\pi/6} V_m \sin\theta \, d\theta \tag{10.42}$$

or

$$V_{dc} = V_m \frac{3}{\pi} \frac{\sqrt{3}}{2} = 0.827 V_m \tag{10.43}$$

Similarly, using Eq. (10.6), the rms value of the output voltage can be found as

$$V_L = \sqrt{\frac{3}{2\pi} \int_{\pi/6}^{5\pi/6} (V_m \sin\theta)^2 \, d\theta} \tag{10.44}$$

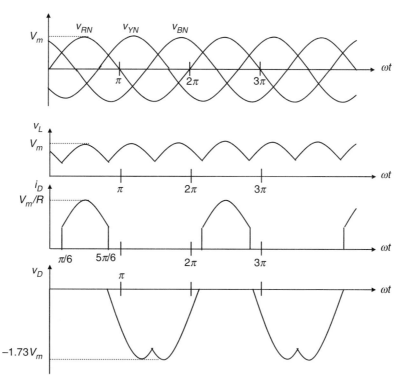

FIGURE 10.8 Waveforms of voltage and current of the three-phase star rectifier shown in Fig. 10.7.

TABLE 10.3 Important design parameters of the three-phase rectifier circuits with the resistive load

	Three-phase star rectifier	Three-phase double-star rectifier with inter-phase transformer	Three-phase bridge rectifier
Peak repetitive reverse voltage V_{RRM}	$2.092V_{dc}$	$1.06V_{dc}$	$1.05V_{dc}$
RMS input voltage per transformer leg V_s	$0.855V_{dc}$	$0.855V_{dc}$	$0.428V_{dc}$
Diode average current $I_{F(AV)}$	$0.333I_{dc}$	$0.167I_{dc}$	$0.333I_{dc}$
Peak repetitive forward current I_{FRM}	$3.63I_{F(AV)}$	$3.15I_{F(AV)}$	$3.14I_{F(AV)}$
Diode rms current $I_{F(RMS)}$	$0.587I_{dc}$	$0.293I_{dc}$	$0.579I_{dc}$
Form factor of diode current $I_{F(RMS)}/I_{F(AV)}$	1.76	1.76	1.74
Rectification ratio	0.968	0.998	0.998
Form factor	1.0165	1.0009	1.0009
Ripple factor	0.182	0.042	0.042
Transformer rating primary VA	$1.23P_{dc}$	$1.06P_{dc}$	$1.05P_{dc}$
Transformer rating secondary VA	$1.51P_{dc}$	$1.49P_{dc}$	$1.05P_{dc}$
Output ripple frequency f_r	$3f_i$	$6f_i$	$6f_i$

or

$$V_L = V_m \sqrt{\frac{3}{2\pi}\left(\frac{\pi}{3}+\frac{\sqrt{3}}{4}\right)} = 0.84V_m \quad (10.45)$$

In addition, the rms current in each transformer secondary winding can also be found as

$$I_s = I_m \sqrt{\frac{1}{2\pi}\left(\frac{\pi}{3}+\frac{\sqrt{3}}{4}\right)} = 0.485I_m \quad (10.46)$$

where $I_m = V_m/R$.

Based on the relationships stated in Eqs. (10.43), (10.45), and (10.46), all the important design parameters of the three-phase star rectifier can be evaluated, as listed in Table 10.3, which is given at the end of Subsection 10.3.2. Note that, as with a single-phase half-wave rectifier, the three-phase star rectifier shown in Fig. 10.7 has direct currents in the secondary windings that can cause a transformer core saturation problem. In addition, the currents in the primary do not sum to zero. Therefore it is preferable not to have star-connected primary windings.

10.3.1.2 Three-phase Inter-star Rectifier Circuit

The transformer core saturation problem in the three-phase star rectifier can be avoided by a special arrangement in its secondary windings, known as zig-zag connection. The modified circuit is called the three-phase inter-star or zig-zag rectifier circuit, as shown in Fig. 10.9. Each secondary phase voltage is obtained from two equal-voltage secondary windings (with a phase displacement of $\pi/3$) connected in series so that the dc magnetizing forces due to the two secondary windings on any limb are equal and opposite. At the expense of extra secondary windings (increasing the transformer secondary rating factor from 1.51 to 1.74 VA/W), this circuit connection eliminates the effects of core saturation and reduces the transformer primary rating factor to the minimum of 1.05 VA/W. Apart from transformer ratings, all the design parameters of this circuit are the same as those of a three-phase star rectifier (therefore not separately listed in Table 10.3). Furthermore, a star-connected primary winding with no neutral connection

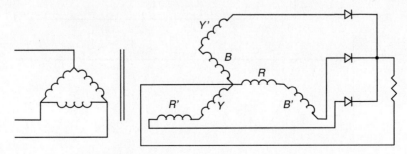

FIGURE 10.9 Three-phase inter-star rectifier.

10.3.1.3 Three-phase Double-star Rectifier with Inter-phase Transformer

This circuit consists essentially of two three-phase star rectifiers with their neutral points interconnected through an inter-phase transformer or reactor (Fig. 10.10). The polarities of the corresponding secondary windings in the two interconnected systems are reversed with respect to each other, so that the rectifier output voltage of one three-phase unit is at a minimum when the rectifier output voltage of the other unit is at a maximum as shown in Fig. 10.11. The function of the inter-phase transformer is to cause the output voltage v_L to be the average of the rectified voltages $v1$ and $v2$ as shown in Fig. 10.11. In addition, the ripple frequency of the output voltage is now six times that of the mains and therefore the component size of the filter (if there is any) becomes smaller. In a balanced circuit, the output currents of two three-phase units flowing in opposite directions in the inter-phase transformer winding will produce no dc magnetization current. Similarly, the dc magnetization currents in the secondary windings of two three-phase units cancel each other out.

By virtue of the symmetry of the secondary circuits, the three primary currents add up to zero at all times. Therefore, a star primary winding with no neutral connection would be equally permissible.

10.3.2 Three-phase Bridge Rectifiers

Three-phase bridge rectifiers are commonly used for high power applications because they have the highest possible transformer utilization factor for a three-phase system. The circuit of a three-phase bridge rectifier is shown in Fig. 10.12.

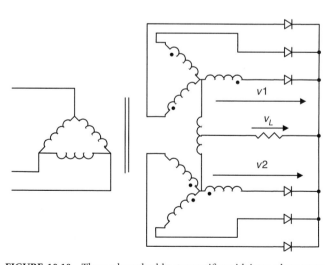

FIGURE 10.10 Three-phase double-star rectifier with inter-phase transformer.

The diodes are numbered in the order of conduction sequences and the conduction angle of each diode is $2\pi/3$.

The conduction sequence for diodes is 12, 23, 34, 45, 56, and 61. The voltage and the current waveforms of the three-phase bridge rectifier are shown in Fig. 10.13. The line voltage is 1.73 times the phase voltage of a three-phase star-connected source. It is permissible to use any combination of star- or delta-connected primary and secondary windings because the currents associated with the secondary windings are symmetrical.

Using Eq. (10.1) the average value of the output can be found as

$$V_{dc} = \frac{6}{2\pi} \int_{\pi/3}^{2\pi/3} \sqrt{3} V_m \sin\theta \, d\theta \qquad (10.47)$$

or

$$V_{dc} = V_m \frac{3\sqrt{3}}{\pi} = 1.654 V_m \qquad (10.48)$$

Similarly, using Eq. (10.6), the rms value of the output voltage can be found as

$$V_L = \sqrt{\frac{9}{\pi} \int_{\pi/3}^{2\pi/3} (V_m \sin\theta)^2 \, d\theta} \qquad (10.49)$$

or

$$V_L = V_m \sqrt{\frac{3}{2} + \frac{9\sqrt{3}}{4\pi}} = 1.655 V_m \qquad (10.50)$$

In addition, the rms current in each transformer secondary winding can also be found as

$$I_s = I_m \sqrt{\frac{2}{\pi}\left(\frac{\pi}{6} + \frac{\sqrt{3}}{4}\right)} = 0.78 I_m \qquad (10.51)$$

and the rms current through a diode is

$$I_D = I_m \sqrt{\frac{1}{\pi}\left(\frac{\pi}{6} + \frac{\sqrt{3}}{4}\right)} = 0.552 I_m \qquad (10.52)$$

where $I_m = 1.73 V_m/R$.

Based on Eqs. (10.48), (10.50), (10.51), and (10.52), all the important design parameters of the three-phase star rectifier can be evaluated, as listed in Table 10.3. The dc output voltage is slightly lower than the peak line voltage or 2.34 times the rms phase voltage. The V_{RRM} rating of the employed diodes is 1.05 times the dc output voltage, and the I_{FRM} rating of the employed diodes is 0.579 times the dc output current. Therefore, this three-phase bridge rectifier is very efficient and

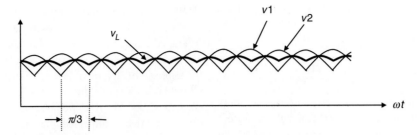

FIGURE 10.11 Voltage waveforms of the three-phase double-star rectifier.

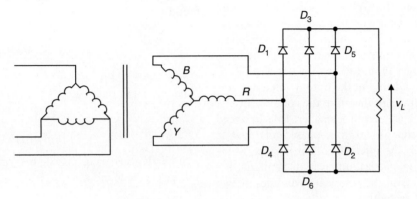

FIGURE 10.12 Three-phase bridge rectifier.

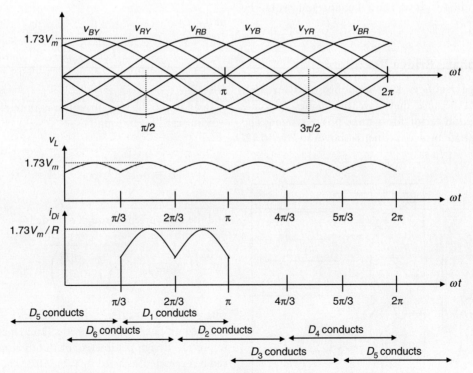

FIGURE 10.13 Voltage and current waveforms of the three-phase bridge rectifier.

popular wherever both dc voltage and current requirements are high. In many applications, no additional filter is required because the output ripple voltage is only 4.2%. Even if a filter is required, the size of the filter is relatively small because the ripple frequency is increased to six times the input frequency.

10.3.3 Operation of Rectifiers with Finite Source Inductance

It has been assumed in the preceding sections that the commutation of current from one diode to the next takes place instantaneously when the inter-phase voltage assumes the necessary polarity. In practice this is hardly possible, because there are finite inductances associated with the source. For the purpose of discussing the effects of the finite source inductance, a three-phase star rectifier with transformer leakage inductances is shown in Fig. 10.14, where l_1, l_2, l_3 denote the leakage inductances associated with the transformer secondary windings.

Refer to Fig. 10.15. At the time when v_{YN} is about to become larger than v_{RN}, due to leakage inductance l_1, the current in D_1 cannot fall to zero immediately. Similarly, due to the leakage inductance l_2, the current in D_2 cannot increase immediately

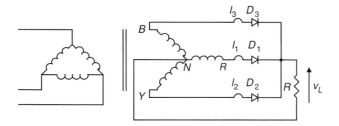

FIGURE 10.14 Three-phase star rectifier with the transformer leakage inductances.

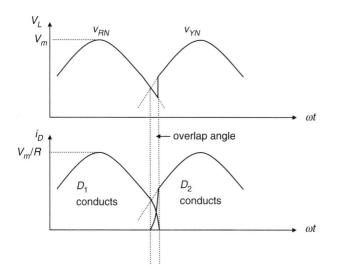

FIGURE 10.15 Waveforms during commutation in Fig. 10.14.

to the full value. The result is that both the diodes conduct for a certain period, which is called the overlap (or commutation) angle. The overlap reduces the rectified voltage v_L as shown in the upper voltage waveform of Fig. 10.15. If all the leakage inductances are equal, i.e. $l_1 = l_2 = l_3 = l_c$, then the amount of reduction of dc output voltage can be estimated as $mf_i l_c I_{dc}$, where m is the ratio of the lowest-ripple frequency to the input frequency.

For example, for a three-phase star rectifier operating from a 60-Hz supply with an average load current of 50 A, the amount of reduction of the dc output voltage is 2.7 V if the leakage inductance in each secondary winding is 300 µH.

10.4 Poly-phase Diode Rectifiers

10.4.1 Six-phase Star Rectifier

A basic six-phase star rectifier circuit is shown in Fig. 10.16. The six-phase voltages on the secondary are obtained by means of a center-tapped arrangement on a star-connected three-phase winding. Therefore, it is sometimes referred to as a three-phase full-wave rectifier. The diode in a particular phase conducts during the period when the voltage on that phase is higher than that on the other phases. The voltage waveforms of each phase and the load are shown in Fig. 10.17. It is clear that, unlike the three-phase star rectifier circuit, the conduction angle of each diode is $\pi/3$, instead of $2\pi/3$. Currents flow in only one rectifying element at a time, resulting in a low average current, but a high peak to an average current ratio in the diodes and poor transformer secondary utilization. Nevertheless, the dc currents in the secondary of the six-phase star rectifier cancel in the secondary windings like a full-wave rectifier and, therefore, core saturation is not encountered. This six-phase star circuit is attractive in applications which require a low ripple factor and a common cathode or anode for the rectifiers.

By considering the output voltage provided by v_{RN} between $\pi/3$ and $2\pi/3$, the average value of the output voltage can be

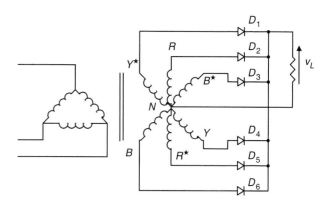

FIGURE 10.16 Six-phase star rectifier.

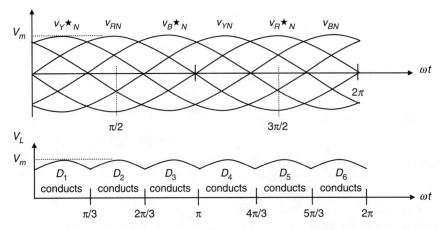

FIGURE 10.17 Voltage waveforms of the six-phase star rectifier.

found as

$$V_{dc} = \frac{6}{2\pi} \int_{\pi/3}^{2\pi/3} V_m \sin\theta \, d\theta \quad (10.53)$$

or

$$V_{dc} = V_m \frac{6}{\pi} \frac{1}{2} = 0.955 V_m \quad (10.54)$$

Similarly, the rms value of the output voltage can be found as

$$V_L = \sqrt{\frac{6}{2\pi} \int_{\pi/3}^{2\pi/3} (V_m \sin\theta)^2 \, d\theta} \quad (10.55)$$

or

$$V_L = V_m \sqrt{\frac{6}{2\pi}\left(\frac{\pi}{6} + \frac{\sqrt{3}}{4}\right)} = 0.956 V_m \quad (10.56)$$

In addition, the rms current in each transformer secondary winding can also be found as

$$I_s = I_m \sqrt{\frac{1}{2\pi}\left(\frac{\pi}{6} + \frac{\sqrt{3}}{4}\right)} = 0.39 I_m \quad (10.57)$$

where $I_m = V_m/R$.

Based on the relationships stated in Eqs. (10.55), (10.56), and (10.57), all the important design parameters of the six-phase star rectifier can be evaluated, as listed in Table 10.4 (given at the end of Subsection 10.4.3).

10.4.2 Six-phase Series Bridge Rectifier

The star- and delta-connected secondaries have an inherent $\pi/6$-phase displacement between their output voltages. When a star- and a delta-connected bridge rectifier are connected

TABLE 10.4 Important design parameters of the six-phase rectifier circuits with resistive load

	Six-phase star rectifier	Six-phase series bridge rectifier	Six-phase parallel bridge rectifier (with inter-phase transformer)
Peak repetitive reverse voltage V_{RRM}	$2.09 V_{dc}$	$0.524 V_{dc}$	$1.05 V_{dc}$
RMS input voltage per transformer leg V_s	$0.74 V_{dc}$	$0.37 V_{dc}$	$0.715 V_{dc}$
Diode average current $I_{F(AV)}$	$0.167 I_{dc}$	$0.333 I_{dc}$	$0.167 I_{dc}$
Peak repetitive forward current I_{FRM}	$6.28 I_{F(AV)}$	$3.033 I_{F(AV)}$	$3.14 I_{F(AV)}$
Diode rms current $I_{F(RMS)}$	$0.409 I_{dc}$	$0.576 I_{dc}$	$0.409 I_{dc}$
Form factor of diode current $I_{F(RMS)}/I_{F(AV)}$	2.45	1.73	2.45
Rectification ratio	0.998	1.00	1.00
Form factor	1.0009	1.00005	1.00005
Ripple factor	0.042	0.01	0.01
Transformer rating primary VA	$1.28 P_{dc}$	$1.01 P_{dc}$	$1.01 P_{dc}$
Transformer rating secondary VA	$1.81 P_{dc}$	$1.05 P_{dc}$	$1.05 P_{dc}$
Output ripple frequency f_r	$6 f_i$	$12 f_i$	$12 f_i$

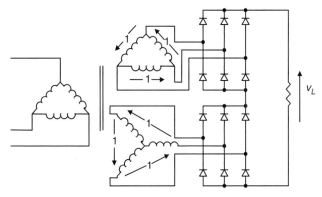

FIGURE 10.18 Six-phase series bridge rectifier.

in series as shown in Fig. 10.18, the combined output voltage will have a doubled ripple frequency (12 times that of the mains). The ripple of the combined output voltage will also be reduced from 4.2% (for each individual bridge rectifier) to 1%. The combined bridge rectifier is referred to as a six-phase series bridge rectifier.

In the six-phase series bridge rectifier shown in Fig. 10.18, let V_m^* be the peak voltage of the delta-connected secondary. The peak voltage between the lines of the star-connected secondary is also V_m^*. The peak voltage across the load, denoted as V_m, is equal to $2V_m^* \times \cos(\pi/12)$ or $1.932 V_m^*$ because there is $\pi/6$-phase displacement between the secondaries. The ripple frequency is twelve times the mains frequency. The average value of the output voltage can be found as

$$V_{dc} = \frac{12}{2\pi} \int_{5\pi/12}^{7\pi/12} V_m \sin\theta \, d\theta \qquad (10.58)$$

or

$$V_{dc} = V_m \frac{12}{\pi} \frac{\sqrt{3}-1}{2\sqrt{2}} = 0.98862 V_m \qquad (10.59)$$

The rms value of the output voltage can be found as

$$V_L = \sqrt{\frac{12}{2\pi} \int_{5\pi/12}^{7\pi/12} (V_m \sin\theta)^2 \, d\theta} \qquad (10.60)$$

or

$$V_L = V_m \sqrt{\frac{12}{2\pi} \left(\frac{\pi}{12} + \frac{1}{4}\right)} = 0.98867 V_m \qquad (10.61)$$

The rms current in each transformer secondary winding is

$$I_s = I_m \sqrt{\frac{4}{\pi} \left(\frac{\pi}{12} + \frac{1}{4}\right)} = 0.807 I_m \qquad (10.62)$$

The rms current through a diode is

$$I_s = I_m \sqrt{\frac{2}{\pi} \left(\frac{\pi}{12} + \frac{1}{4}\right)} = 0.57 I_m \qquad (10.63)$$

where $I_m = V_m/R$.

Based on Eqs. (10.59), (10.61), (10.62), and (10.63), all the important design parameters of the six-phase series bridge rectifier can be evaluated, as listed in Table 10.4 (given at the end of Subsection 10.4.3).

10.4.3 Six-phase Parallel Bridge Rectifier

The six-phase series bridge rectifier described above is useful for high output voltage applications. However, for high output current applications, the six-phase parallel bridge rectifier (with an inter-phase transformer) shown in Fig. 10.19 should be used.

The function of the inter-phase transformer is to cause the output voltage v_L to be the average of the rectified voltages $v1$ and $v2$ as shown in Fig. 10.20. As with the six-phase series

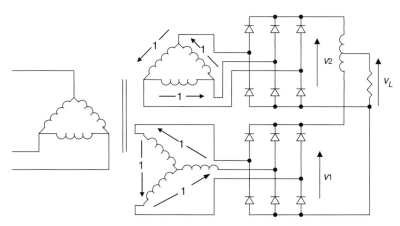

FIGURE 10.19 Six-phase parallel bridge rectifier.

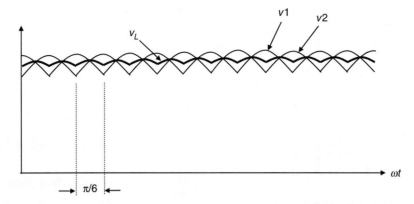

FIGURE 10.20 Voltage waveforms of the six-phase bridge rectifier with inter-phase transformer.

bridge rectifier, the output ripple frequency of the six-phase parallel bridge rectifier is also 12 times that of the mains. Further filtering on the output voltage is usually not required. Assuming a balanced circuit, the output currents of two three-phase units (flowing in opposite directions in the inter-phase transformer winding) produce no dc magnetization current.

All the important design parameters of the six-phase parallel rectifiers with inter-phase transformer are also listed in Table 10.4.

10.5 Filtering Systems in Rectifier Circuits

Filters are commonly employed in rectifier circuits for smoothing out the dc output voltage of the load. They are classified as inductor-input dc filters and capacitor-input dc filters. Inductor-input dc filters are preferred in high-power applications because more efficient transformer operation is obtained due to the reduction in the form factor of the rectifier current. Capacitor-input dc filters can provide volumetrically efficient operation, but they demand excessive turn-on and repetitive surge currents. Therefore, capacitor-input dc filters are suitable only for lower-power systems where close regulation is usually achieved by an electronic regulator cascaded with the rectifier.

10.5.1 Inductive-input DC Filters

The simplest inductive-input dc filter is shown in Fig. 10.21a. The output current of the rectifier can be maintained at a steady value if the inductance of L_f is sufficiently large ($\omega L_f \gg R$). The filtering action is more effective in heavy load conditions than in light load conditions. If the ripple attenuation is not sufficient even with large values of inductance, an L-section filter as shown in Fig. 10.21b can be used for further filtering. In practice, multiple L-section filters can also be employed if the requirement on the output ripple is very stringent.

For a simple inductive-input dc filter shown in Fig. 10.21a, the ripple is reduced by the factor

$$\frac{v_o}{v_L} = \frac{R}{\sqrt{R^2 + (2\pi f_r L_f)^2}} \quad (10.64)$$

where v_L is the ripple voltage before filtering, v_o is the ripple voltage after filtering, and f_r is the ripple frequency.

For the inductive-input dc filter shown in Fig. 10.21b, the amount of reduction in the ripple voltage can be estimated as

$$\frac{v_o}{v_L} = \left| \frac{1}{1 - (2\pi f_r)^2 L_f C_f} \right| \quad (10.65)$$

where f_r is the ripple frequency, if $R \gg 1/2\pi f_r C_f$.

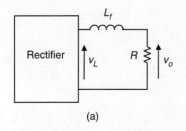
(a)

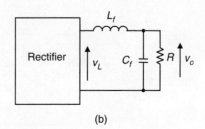
(b)

FIGURE 10.21 Inductive-input dc filters.

10.5.1.1 Voltage and Current Waveforms of Full-wave Rectifier with Inductor-input DC Filter

Figure 10.22 shows a single-phase full-wave rectifier with an inductor-input dc filter. The voltage and current waveforms are illustrated in Fig. 10.23.

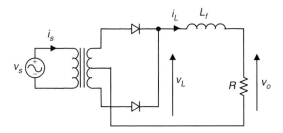

FIGURE 10.22 A full-wave rectifier with inductor-input dc filter.

When the inductance of L_f is infinite, the current through the inductor and the output voltage are constant. When inductor L_f is finite, the current through the inductor has a ripple component, as shown by the dotted lines in Fig. 10.23. If the input inductance is too small, the current decreases to zero (becoming discontinuous) during a portion of the time between the peaks of the rectifier output voltage. The minimum value of inductance required to maintain a continuous current is known as the critical inductance L_C.

10.5.1.2 Critical inductance L_C

In the case of single-phase full-wave rectifiers, the critical inductance can be found as

$$\text{Full-wave} \quad L_C = \frac{R}{6\pi f_i} \quad (10.66)$$

where f_i is the input mains frequency.

In the case of poly-phase rectifiers, the critical inductance can be found as

$$\text{Poly-phase} \quad L_C = \frac{R}{3\pi m \left(m^2 - 1\right) f_i} \quad (10.67)$$

where m is ratio of the lowest ripple frequency to the input frequency, e.g. $m = 6$ for a three-phase bridge rectifier.

10.5.1.3 Determining the Input Inductance for a Given Ripple Factor

In practice, the choice of the input inductance depends on the required ripple factor of the output voltage. The ripple voltage of a rectifier without filtering can be found by means of Fourier Analysis. For example, the coefficient of the nth harmonic component of the rectified voltage v_L shown in Fig. 10.22 can be expressed as:

$$v_{L_n} = \frac{-4V_m}{\pi \left(n^2 - 1\right)} \quad (10.68)$$

where $n = 2, 4, 8, \ldots$ etc.

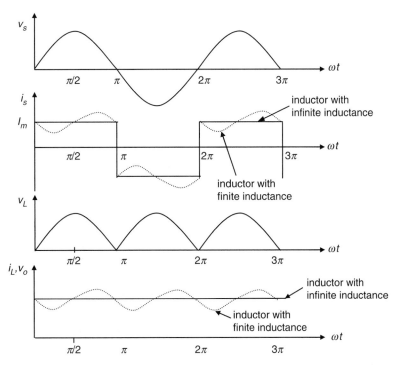

FIGURE 10.23 Voltage and current waveforms of full-wave rectifier with inductor-input dc filter.

The dc component of the rectifier voltage is given by Eq. (10.5). Therefore, in addition to Eq. (10.27), the ripple factor can also be expressed as

$$\text{RF} = \sqrt{2 \sum_{n=2,4,8,} \left(\frac{1}{n^2 - 1}\right)^2} \qquad (10.69)$$

Considering only the lowest-order harmonic ($n = 2$), the output ripple factor of a simple inductor-input dc filter (without C_f) can be found, from Eqs. (10.64) and (10.69), as

$$\text{Filtered RF} = \frac{0.4714}{\sqrt{1 + (4\pi f_i L_f / R)^2}} \qquad (10.70)$$

10.5.1.4 Harmonics of the Input Current

In general, the total harmonic distortion (THD) of an input current is defined as

$$\text{THD} = \sqrt{\left(\frac{I_s}{I_{s1}}\right)^2 - 1} \qquad (10.71)$$

where I_s is the rms value of the input current and I_{s1} and the rms value of the fundamental component of the input current. The THD can also be expressed as

$$\text{THD} = \sqrt{\sum_{n=2,3,4,} \left(\frac{I_{sn}}{I_{s1}}\right)^2} \qquad (10.72)$$

where I_{sn} is the rms value of the nth harmonic component of the input current.

Moreover, the input power factor is defined as

$$\text{PF} = \frac{I_{s1}}{I_s} \cos\phi \qquad (10.73)$$

where ϕ is the displacement angle between the fundamental components of the input current and voltage.

Assume that inductor L_f of the circuit shown in Fig. 10.22 has an infinitely large inductance. The input current is then a square wave. This input current contains undesirable higher harmonics that reduce the input power factor of the system. The input current can be easily expressed as

$$i_s = \frac{4I_m}{\pi} \sum_{n=1,3,5,} \frac{1}{n} \sin 2n\pi f_i t \qquad (10.74)$$

The rms values of the input current and its fundamental component are I_m and $4I_m/(\pi\sqrt{2})$ respectively. Therefore, the THD of the input current of this circuit is 0.484. Since the displacement angle $\phi = 0$, the power factor is $4/(\pi\sqrt{2}) = 0.9$.

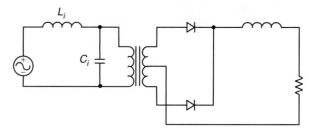

FIGURE 10.24 Rectifier with input ac filter.

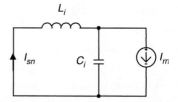

FIGURE 10.25 Equivalent circuit for input ac filter.

The power factor of the circuit shown in Fig. 10.22 can be improved by installing an ac filter between the source and the rectifier, as shown in Fig. 10.24.

Considering only the harmonic components, the equivalent circuit of the rectifier given in Fig. 10.24 can be found as shown in Fig. 10.25. The rms value of the nth harmonic current appearing in the supply can then be obtained using the current-divider rule,

$$I_{sn} = \left|\frac{1}{1 - (2n\pi f_i)^2 L_i C_i}\right| I_{rn} \qquad (10.75)$$

where I_{rn} is the rms value of the nth harmonic current of the rectifier.

Applying Eq. (10.73) and knowing $I_{rn}/I_{r1} = 1/n$ from Eq. (10.74), the THD of the rectifier with input filter shown in Fig. 10.24 can be found as

$$\text{Filtered THD} = \sqrt{\sum_{n=3,5} \frac{1}{n^2} \left|\frac{1}{1 - (2n\pi f_i)^2 L_i C_i}\right|^2} \qquad (10.76)$$

The important design parameters of typical single-phase and three-phase rectifiers with inductor-input dc filter are listed in Table 10.5. Note that, in a single-phase half-wave rectifier, a freewheeling diode is required to be connected across the input of the dc filters such that the flow of load current can be maintained during the negative half-cycle of the supply voltage.

10.5.2 Capacitive-input DC Filters

Figure 10.26 shows a full-wave rectifier with capacitor-input dc filter. The voltage and current waveforms of this rectifier

10 Diode Rectifiers

TABLE 10.5 Important design parameters of typical rectifier circuits with inductor-input dc filter

	Full-wave rectifier with center-tapped transformer	Full-wave bridge rectifier	Three-phase star rectifier	Three-phase bridge rectifier	Three-phase double-star rectifier with inter-phase transformer
Peak repetitive reverse voltage V_{RRM}	$3.14 V_{dc}$	$1.57 V_{dc}$	$2.09 V_{dc}$	$1.05 V_{dc}$	$2.42 V_{dc}$
RMS input voltage per transformer leg V_s	$1.11 V_{dc}$	$1.11 V_{dc}$	$0.885 V_{dc}$	$0.428 V_{dc}$	$0.885 V_{dc}$
Diode average current $I_{F(AV)}$	$0.5 I_{dc}$	$0.5 I_{dc}$	$0.333 I_{dc}$	$0.333 I_{dc}$	$0.167 I_{dc}$
Peak repetitive forward current I_{FRM}	$2.00 I_{F(AV)}$	$2.00 I_{F(AV)}$	$3.00 I_{F(AV)}$	$3.00 I_{F(AV)}$	$3.00 I_{F(AV)}$
Diode rms current $I_{F(RMS)}$	$0.707 I_{dc}$	$0.707 I_{dc}$	$0.577 I_{dc}$	$0.577 I_{dc}$	$0.289 I_{dc}$
Form factor of diode current $I_{F(RMS)}/I_{F(AV)}$	1.414	1.414	1.73	1.73	1.73
Transformer rating primary VA	$1.11 P_{dc}$	$1.11 P_{dc}$	$1.21 P_{dc}$	$1.05 P_{dc}$	$1.05 P_{dc}$
Transformer rating secondary VA	$1.57 P_{dc}$	$1.11 P_{dc}$	$1.48 P_{dc}$	$1.05 P_{dc}$	$1.48 P_{dc}$
Output ripple frequency f_r	$2 f_i$	$2 f_i$	$3 f_i$	$6 f_i$	$6 f_i$
Ripple component V_r at					
(a) fundamental,	$0.667 V_{dc}$	$0.667 V_{dc}$	$0.250 V_{dc}$	$0.057 V_{dc}$	$0.057 V_{dc}$
(b) second harmonic,	$0.133 V_{dc}$	$0.133 V_{dc}$	$0.057 V_{dc}$	$0.014 V_{dc}$	$0.014 V_{dc}$
(c) third harmonic of the ripple frequency	$0.057 V_{dc}$	$0.057 V_{dc}$	$0.025 V_{dc}$	$0.006 V_{dc}$	$0.006 V_{dc}$

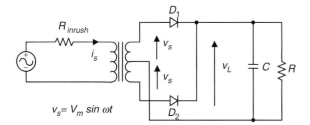

FIGURE 10.26 Full-wave rectifier with capacitor-input dc filter.

are shown in Fig. 10.27. When the instantaneous voltage of the secondary winding v_s is higher than the instantaneous value of capacitor voltage v_L, either D_1 or D_2 conducts, and the capacitor C is charged up from the transformer. When the instantaneous voltage of the secondary winding v_s falls below the instantaneous value of capacitor voltage v_L, both the diodes are reverse biased and the capacitor C is discharged through load resistance R. The resulting capacitor voltage v_L varies between a maximum value of V_m and a minimum value of $V_m - V_{r(pp)}$ as shown in Fig. 10.27. ($V_{r(pp)}$ is the peak-to-peak ripple voltage.) As shown in Fig. 10.27, the conduction angle θ_c of the diodes becomes smaller when the output-ripple voltage decreases. Consequently, the power supply and the diodes suffer from high repetitive surge currents. An LC ac filter, as shown in Fig. 10.24, may be required to improve the input power factor of the rectifier.

In practice, if the peak-to-peak ripple voltage is small, it can be approximated as

$$V_{r(pp)} = \frac{V_m}{f_r RC} \quad (10.77)$$

where f_r is the output ripple frequency of the rectifier.

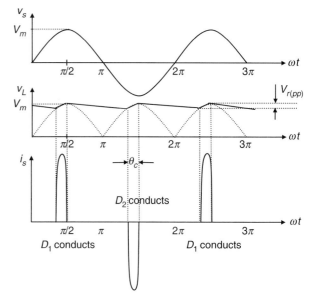

FIGURE 10.27 Voltage and current waveforms of the full-wave rectifier with capacitor-input dc filter.

Therefore, the average output voltage V_{dc} is given by

$$V_{dc} = V_m \left(1 - \frac{1}{2 f_r RC} \right) \quad (10.78)$$

The rms output ripple voltage V_{ac} is approximately given by

$$V_{ac} = \frac{V_m}{2\sqrt{2} f_r RC} \quad (10.79)$$

The ripple factor RF can be found from

$$\text{RF} = \frac{1}{\sqrt{2}\left(2f_r RC - 1\right)} \quad (10.80)$$

10.5.2.1 Inrush Current

The resistor R_{inrush} in Fig. 10.26 is used to limit the inrush current imposed on the diodes during the instant when the rectifier is being connected to the supply. The inrush current can be very large because capacitor C has zero charge initially. The worst case occurs when the rectifier is connected to the supply at its maximum voltage. The worst-case inrush current can be estimated from

$$I_{inrush} = \frac{V_m}{R_{sec} + R_{ESR}} \quad (10.81)$$

where R_{sec} is the equivalent resistance looking from the secondary transformer and R_{ESR} is the equivalent series resistance (ESR) of the filtering capacitor. Hence the employed diode should be able to withstand the inrush current for a half cycle of the input voltage. In other words, the *Maximum Allowable Surge Current* (I_{FSM}) rating of the employed diodes must be higher than the inrush current. The equivalent resistance associated with the transformer windings and the filtering capacitor is usually sufficient to limit the inrush current to an acceptable level. However, in cases where the transformer is omitted, e.g. the rectifier of an off-line switch-mode supply, resistor R_{inrush} must be added for controlling the inrush current.

Consider as an example, a single-phase bridge rectifier, which is to be connected to a 120-V–60-Hz source (without transformer). Assume that the I_{FSM} rating of the diodes is 150 A for an interval of 8.3 ms. If the ESR of the filtering capacitor is zero, the value of the resistor for limiting inrush current resistance can be estimated to be 1.13 Ω using Eq. (10.81).

10.6 High-frequency Diode Rectifier Circuits

In high-frequency converters, diodes perform various functions, such as rectifying, flywheeling, and clamping. One special quality a high-frequency diode must possess is a fast switching speed. In technical terms, it must have a short reverse recovery time and a short forward recovery time.

The reverse recovery time of a diode may be understood as the time a forwardly conducting diode takes to recover to a blocking state when the voltage across it is suddenly reversed (which is known as forced turn-off). The temporary short circuit during the reverse recovery period may result in large reverse current, excessive ringing, and large power dissipation, all of which are highly undesirable.

The forward recovery time of a diode may be understood as the time a non-conducting diode takes to change to the fully-on state when a forward current is suddenly forced into it (which is known as forced turn-on). Before the diode reaches the fully-on state, the forward voltage drop during the forward recovery time can be significantly higher than the normal on-state voltage drop. This may cause voltage spikes in the circuit.

It should be interesting to note that, as far as circuit operation is concerned, a diode with a long reverse recovery time is similar to a diode with a large parasitic capacitance. A diode with a long forward recovery time is similar to a diode with a large parasitic inductance. (Spikes caused by the slow forward recovery of diodes are often wrongly thought to be caused by leakage inductance.) Comparatively, the adverse effect of a long reverse recovery time is much worse than that of a long forward recovery time.

Among commonly used diodes, the Schottky diode has the shortest forward and reverse recovery times. Schottky diodes are therefore most suitable for high-frequency applications. However, Schottky diodes have relatively low reverse breakdown voltage (normally lower than 200 V) and large leakage current. If, due to these limitations, Schottky diodes cannot be used, ultra-fast diodes should be used in high-frequency converter circuits.

Using the example of a forward converter, the operations of a forward rectifier diode, a flywheel diode, and a clamping diode will be studied in Subsection 10.6.1. Because of the difficulties encountered in the full analyses taking into account parasitic/stray/leakage components, PSpice simulations are extensively used here to study the following:

- The idealized operation of the converter.
- The adverse effects of relatively slow rectifiers (e.g. the so-called ultra-fast diodes, which are actually much slower than Schottky diodes).
- The improvement achievable by using high-speed rectifiers (Schottky diodes).
- The effects of leakage inductance of the transformer.
- The use of snubber circuits to reduce ringing.
- The operation of a practical converter with snubber circuits.

Using the example of a flyback converter, the operations of a flyback rectifier diode and a clamping diode will also be studied in Subsection 10.6.2.

The design considerations for high-frequency diode rectifier circuits will be discussed in Subsection 10.6.3. Some precautions which must be taken in the interpretation of computer simulation results are briefed in Subsection 10.6.4.

10.6.1 Forward Rectifier Diode, Flywheel Diode, and Magnetic-reset Clamping Diode in a Forward Converter

10.6.1.1 Ideal Circuit

Figure 10.28 shows the basic circuit of a forward converter. Figure 10.29 shows the idealized steady-state waveforms for continuous-mode operation (the current in L_1 being continuous). These waveforms are obtained from PSpice simulations, based on the following assumptions:

- Rectifier diode D_R, flywheel diode D_F, and magnetic-reset clamping diode D_M are ideal diodes with infinitely fast switching speed.
- Electronic switch $M1$ is an idealized MOS switch with infinitely fast switching speed and

    On-state resistance = 0.067 Ω

    Off-state resistance = 1 MΩ

It should be noted that PSpice does not allow a switch to have zero on-state resistance and infinite off-state resistance.
- Transformer T_1 has a coupling coefficient of 0.99999999. PSpice does not accept a coupling coefficient of 1.
- The switching operation of the converter has reached a steady state.

Referring to the circuit shown in Fig. 10.28 and the waveforms shown in Fig. 10.29, the operation of the converter can be explained as follows:

1. For $0 < t < DT$ (D is the duty cycle of the MOS switch M_1 and T is the switching period of the converter. M_1 is turned on when V1(VPULSE) is 15 V, and turned off when V1(VPULSE) is 0 V).

The switch M_1 is turned on at $t = 0$.
The voltage at node 3, denoted as $V(3)$, is

$$V(3) = 0 \quad \text{for} \quad 0 < t < DT \quad (10.82)$$

The voltage induced at node 6 of the secondary winding L_S is

$$V(6) = V_{IN} (N_s/N_P) \quad (10.83)$$

This voltage drives a current $I(DR)$ (current through rectifier diode D_R) into the output circuit to produce the output voltage V_o. The rate of increase of $I(DR)$ is given by

$$\frac{d I(DR)}{dt} = \left[V_{IN} \frac{N_S}{N_P} - V_o \right] \frac{1}{L_1} \quad (10.84)$$

where V_o is the dc output voltage of the converter.
The flywheel diode D_F is reversely biased by $V(9)$, the voltage at node 9.

$$V(9) = V_{IN} (N_S/N_P) \quad \text{for} \quad 0 < t < DT \quad (10.85)$$

The magnetic-rest clamping diode D_M is reversely biased by the negative voltage at node 100. Assuming that L_M and L_P have the same number of turns, we have

$$V(100) = -V_{IN} \quad \text{for } 0 < t < DT \quad (10.86)$$

A magnetizing current builds up linearly in L_P. This magnetizing current reaches the maximum value of $(V_{IN}DT)/L_P$ at $t = DT$.

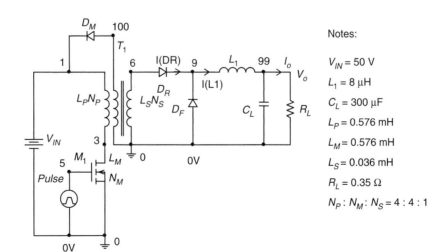

FIGURE 10.28 Basic circuit of forward converter.

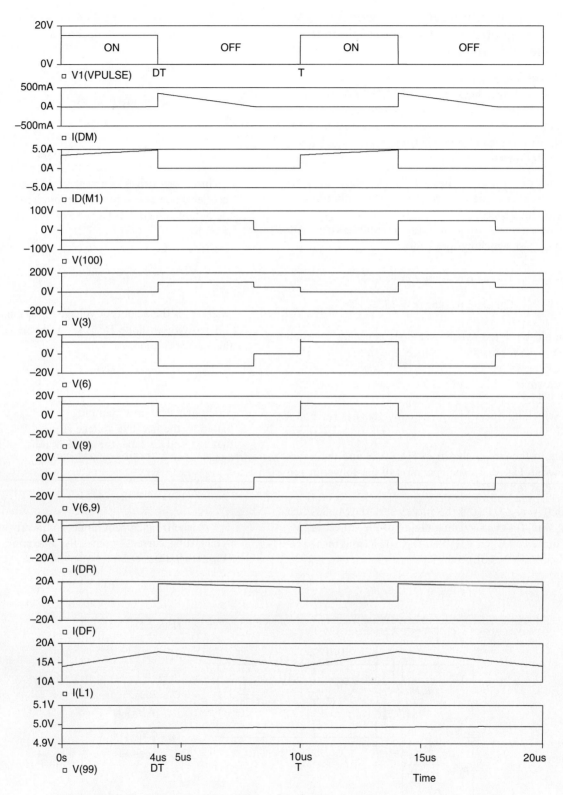

FIGURE 10.29 Idealized steady-state waveforms of forward converter for continuous-mode operation.

2. For $DT < t < 2DT$

   The switch M_1 is turned off at $t = DT$.

   The collapse of magnetic flux induces a back emf in L_M, which is equal to L_P, to turn-on the clamping diode D_M. The magnetizing current in L_M drops (from the maximum value of $(V_{IN}DT)/L_P$, as mentioned above) at the rate of V_{IN}/L_P. It reaches zero at $t = 2DT$.

   The back emf induced across L_P is equal to V_{IN}. The voltage at node 3 is

   $$V(3) = 2V_{IN} \quad \text{for } DT < t < 2DT \quad (10.87)$$

   The back emf across L_S forces D_R to stop conducting.

   The inductive current in L_1 forces the flywheel diode D_F to conduct. $I(L1)$ (current through L_1) falls at the rate of

   $$\frac{dI(L1)}{dt} = \frac{-V_o}{L_1} \quad (10.88)$$

   The voltage across D_R, denoted as $V(6,9)$ (the voltage at node 6 with respect to node 9), is

   $$V(DR) = V(6,9)$$
   $$= -V_{IN}(N_S/N_P) \quad \text{for } DT < t < 2DT \quad (10.89)$$

3. For $2DT < t < T$

   D_M stops conducting at $t = 2DT$. The voltage across L_M then falls to zero.

   The voltage across L_P is zero.

   $$V(3) = V_{IN} \quad (10.90)$$

   The voltage across L_S is also zero.

   $$V(6) = 0 \quad (10.91)$$

   Inductive current $I(L1)$ continues to fall at the rate of

   $$\frac{dI(L1)}{dt} = \frac{-V_o}{L_1} \quad (10.92)$$

   The switching cycle restarts when the switch M_1 is turned on again at $t = T$.

From the waveforms shown in Fig. 10.29, the following useful information (for continuous-mode operation) can be found:

- The output voltage V_o is equal to the average value of $V(9)$.

$$V_o = D\frac{N_S}{N_P}V_{IN} \quad (10.93)$$

- The maximum current in the forward rectifying diode D_R and flywheel diode D_F is

$$I(DR)_{max} = I(DF)_{max}$$
$$= I_o + \frac{1}{2}\frac{V_o}{L_1}(1-D)T \quad (10.94)$$

where $V_o = DV_{IN}(N_S/N_P)$ and I_o is the output loading current.

- The maximum reverse voltage of D_R and D_F is

$$V(DR)_{max} = V(DF)_{max}$$
$$= V(6,9)_{max} = V_{IN}\frac{N_S}{N_P} \quad (10.95)$$

- The maximum reverse voltage of D_M is

$$V(DM)_{max} = V_{IN} \quad (10.96)$$

- The maximum current in D_M is

$$I(DM)_{max} = DT\frac{V_{IN}}{L_P} \quad (10.97)$$

- The maximum current in the switch $M1$, denoted as $ID(M1)$, is

$$ID(M1)_{max} = \frac{N_S}{N_P}I(DR)_{max} + I(DM)_{max}$$
$$= \frac{N_S}{N_P}\left[I_o + \frac{1}{2}\frac{V_o}{L_1}(1-D)T\right] + DT\frac{V_{IN}}{L_P} \quad (10.98)$$

It should, however, be understood that, due to the non-ideal characteristics of practical components, the idealized waveforms shown in Fig. 10.29 cannot actually be achieved in the real world. In the following, the effects of non-ideal diodes and transformers will be examined.

10.6.1.2 Circuit Using Ultra-fast Diodes

Figure 10.30 shows the waveforms of the forward converter (circuit given in Fig. 10.28) when ultra-fast diodes are used as D_M, D_R, and D_F. (Note that ultra-fast diodes are actually much slower than Schottky diodes.) The waveforms are obtained by PSpice simulations, based on the following assumptions:

- D_M is an MUR460 ultra-fast diode. D_R and D_F are MUR1560 ultra-fast diodes.
- M_1 is an IRF640 MOS transistor.
- Transformer T_1 has a coupling coefficient of 0.99999999 (which may be assumed to be 1).
- The switching operation of the converter has reached a steady state.

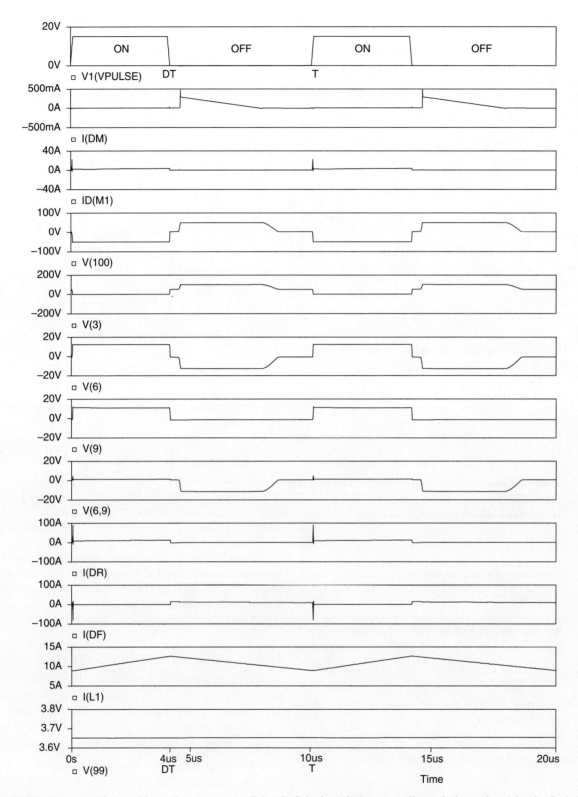

FIGURE 10.30 Waveforms of forward converter using "ultra-fast" diodes (which are actually much slower than Schottky diodes).

It is observed that a large spike appears in the current waveforms of diodes D_R and D_F (denoted as $I(DR)$ and $I(DF)$ in Fig. 10.30) whenever the MOS transistor M_1 is turned on. This is due to the relatively slow reverse recovery of the flywheel diode D_F. During the reverse recovery time, the positive voltage suddenly appearing across L_S (which is equal to $V_{IN}(N_S/N_P)$) drives a large transient current through D_R and D_F. This current spike results in large current stress and power dissipation in D_R, D_F, and M_1.

A method of reducing the current spikes is to use Schottky diodes as D_R and D_F, as described below.

10.6.1.3 Circuit Using Schottky Diodes

In order to reduce the current spikes caused by the slow reverse recovery of rectifiers, Schottky diodes are now used as D_R and D_F. The assumptions made here are (referring to the circuit shown in Fig. 10.28):

- D_R and D_F are MBR2540 Schottky diodes.
- D_M is an MUR460 ultra-fast diode.
- M_1 is an IRF640 MOS transistor.
- Transformer T_1 has a coupling coefficient of 0.99999999.
- The switching operation of the converter has reached a steady state.

The new simulated waveforms are given in Fig. 10.31. It is found that, by employing Schottky diodes as D_R and D_F, the amplitudes of the current spikes in $ID(M1)$, $I(DR)$, and $I(DF)$ can be reduced to practically zero. This solves the slow-speed problem of ultra-fast diodes.

10.6.1.4 Circuit with Practical Transformer

The simulation results given above in Figs. 10.29–10.31 (for the forward converter circuit shown in Fig. 10.28) are based on the assumption that transformer $T1$ has effectively no leakage inductance (with coupling coefficient $K = 0.99999999$). It is, however, found that when a practical transformer (having a slightly lower K) is used, severe ringings occur. Figure 10.32 shows some simulation results to demonstrate this phenomenon, where the following assumptions are made:

- D_R and D_F are MBR2540 Schottky diodes. D_M is an MUR460 ultra-fast diode.
- M_1 is an IRF640 MOS transistor.
- Transformer T_1 has a practical coupling coefficient of 0.996.
- The effective winding resistance of L_P is 0.1 Ω. The effective winding resistance of L_M is 0.4 Ω. The effective winding resistance of L_S is 0.01 Ω.
- The effective series resistance of the output filtering capacitor is 0.05 Ω.
- The switching operation of the converter has reached a steady state.

The resultant waveforms shown in Fig. 10.32 indicate that there are large voltage and current ringings in the circuit. These ringings are caused by the resonant circuits formed by the leakage inductance of the transformer and the parasitic capacitances of diodes and transistor.

A practical converter may therefore need snubber circuits to damp these ringings, as described below.

10.6.1.5 Circuit with Snubber Across the Transformer

In order to suppress the ringing voltage caused by the resonant circuit formed by transformer leakage inductance and the parasitic capacitance of the MOS switch, a snubber circuit, shown as R_1 and C_1 in Fig. 10.33, is now connected across the primary winding of transformer T_1. The new waveforms are shown in Fig. 10.34. Here the drain-to-source voltage waveform of the MOS transistor, $V(3)$, is found to be acceptable. However, there are still large ringing voltages across the output rectifiers ($V(6,9)$ and $V(9)$).

In order to damp the ringing voltages across the output rectifiers, additional snubber circuits across the rectifiers may therefore also be required in a practical circuit, as described below.

10.6.1.6 Practical Circuit

Figure 10.35 shows a practical forward converter with snubber circuits added also to rectifiers (R_2C_2 for D_R and R_3C_3 for D_F) to reduce the voltage ringing. Figures 10.36 and 10.37 show the resultant voltage and current waveforms. Figure 10.36 is for continuous-mode operation ($R_L = 0.35$ Ω), where $I(L1)$ (current in L_1) is continuous. Figure 10.37 is for discontinuous-mode operation ($R_L = 10$ Ω), where $I(L1)$ becomes discontinuous due to an increased value of R_L. These waveforms are considered to be acceptable.

The design considerations of diode rectifier circuits in high-frequency converters will be discussed later in Subsection 10.6.3.

10.6.2 Flyback Rectifier Diode and Clamping Diode in a Flyback Converter

10.6.2.1 Ideal Circuit

Figure 10.38 shows the basic circuit of a flyback converter. Due to its simple circuit, this type of converter is widely used in low-cost low-power applications. Discontinuous-mode operation (meaning that the magnetizing current in the transformer falls to zero before the end of each switching cycle) is often used because it offers the advantages of easy control and low diode reverse-recovery loss. Figure 10.39 shows the idealized steady-state waveforms for discontinuous-mode operation. These waveforms are obtained from PSpice simulations, based on

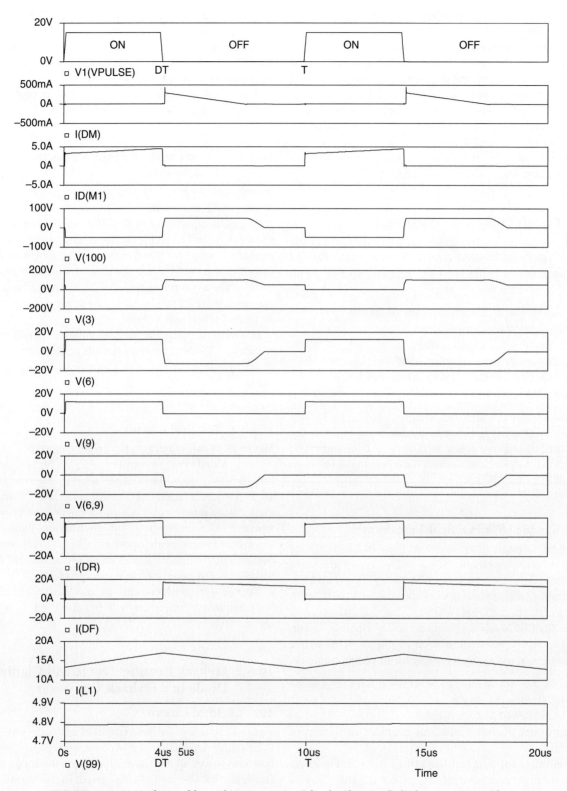

FIGURE 10.31 Waveforms of forward converter using Schottky (fast-speed) diodes as output rectifiers.

10 Diode Rectifiers 169

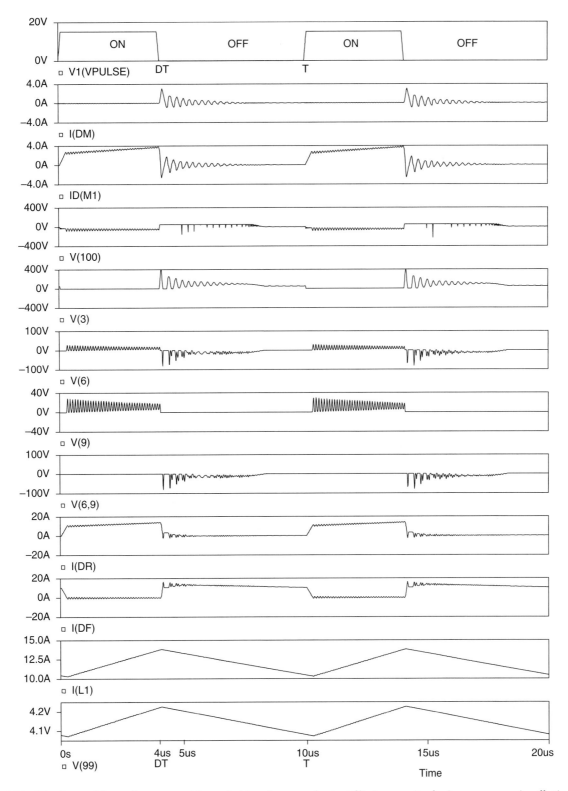

FIGURE 10.32 Waveforms of forward converter with practical transformer and output filtering capacitor having non-zero series effective resistance.

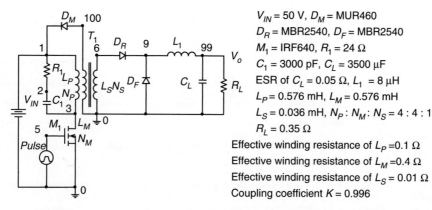

FIGURE 10.33 Forward converter with snubber circuit ($R_1 C_1$) across transformer.

the following assumptions:

- D_R is an idealized rectifier diode with infinitely fast switching speed.
- M_1 is an idealized MOS switch with infinitely fast switching speed and

$$\text{On-state resistance} = 0.067\ \Omega$$

$$\text{Off-state resistance} = 1\ \text{M}\Omega$$

- Transformer T_1 has a coupling coefficient of 0.99999999.
- The switching operation of the converter has reached a steady state.

Referring to the circuit shown in Fig. 10.38 and the waveforms shown in Fig. 10.39, the operation of the converter can be explained as follows:

1. For $0 < t < DT$
   The switch M_1 is turned on at $t = 0$.

$$V(3) = 0 \quad \text{for } 0 < t < DT$$

The current in M_1, denoted as $ID(M1)$, increases at the rate of

$$\frac{d\,ID(M1)}{dt} = \frac{V_{IN}}{L_P} \tag{10.99}$$

The output rectifier D_R is reversely biased.

2. For $DT < t < (D + D_2)T$
   The switching M_1 is turned off at $t = DT$.
   The collapse of magnetic flux induces a back emf in L_S to turn-on the output rectifier D_R. The initial amplitude of the rectifier current $I(DR)$, which is also denoted as $I(LS)$, can be found by equating the energy stored in the primary-winding current $I(LP)$ just before $t = DT$ to the energy stored in the secondary-winding current $I(LS)$ just after $t = DT$:

$$\frac{1}{2} L_P [I(LP)]^2 = \frac{1}{2} L_S [I(LS)]^2 \tag{10.100}$$

$$\frac{1}{2} L_P \left[\frac{V_{IN}}{L_P} DT \right]^2 = \frac{1}{2} L_S [I(LS)]^2 \tag{10.101}$$

$$I(LS) = \sqrt{\frac{L_P}{L_S}} \frac{V_{IN}}{L_P} DT \tag{10.102}$$

$$I(LS) = \frac{N_P}{N_S} \frac{V_{IN}}{L_P} DT \tag{10.103}$$

The amplitude of $I(LS)$ falls at the rate of

$$\frac{dI(LS)}{dt} = \frac{-V_o}{L_S} \tag{10.104}$$

and $I(LS)$ falls to zero at $t = (D + D_2)T$. Since $D_2 V_o = V_{IN}(N_S/N_P)D$

$$D_2 = \frac{V_{IN}}{V_o} \frac{N_S}{N_P} D \tag{10.105}$$

D_2 is effectively the duty cycle of the output rectifier D_R.

3. For $(D + D_2)T < t < T$
   The output rectifier D_R is off.
   The output capacitor C_L provides the output current to the load R_L.
   The switching cycle restarts when the switch M_1 is turned on again at $t = T$.

From the waveforms shown in Fig. 10.39, the following information (for discontinuous-mode operation)

10 Diode Rectifiers 171

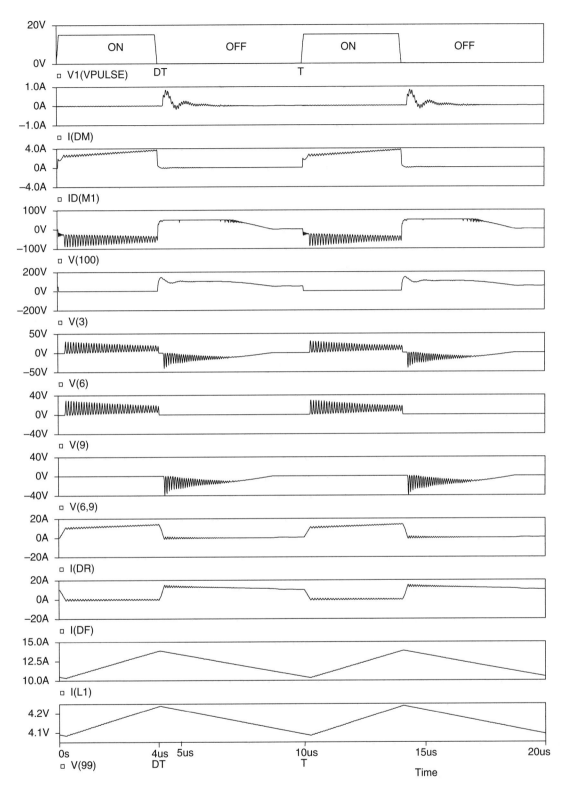

FIGURE 10.34 Waveforms of forward converter with snubber circuit across the transformer.

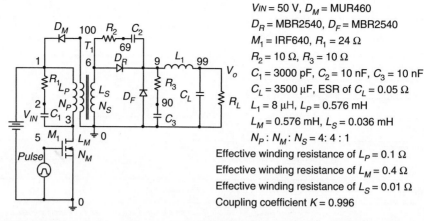

FIGURE 10.35 Practical forward converter with snubber circuits across the transformer and rectifiers.

can be obtained:

- The maximum value of the current in the switch M_1 is

$$ID(M1)_{max} = \frac{V_{IN}}{L_P} DT \qquad (10.106)$$

- The maximum value of the current in the output rectifier D_R is

$$I(DR)_{max} = \frac{N_P}{N_S} \frac{V_{IN}}{L_P} DT \qquad (10.107)$$

- The output voltage V_o can be found by equating the input energy to the output energy within a switching cycle.

$V_{IN} \times$ [Charge taken from V_{IN} in a switching cycle]
$= \frac{V_o^2}{R_L} T$

$$V_{IN} \left[\frac{1}{2} DT \frac{DT}{L_P} V_{IN} \right] = \frac{V_o^2}{R_L} T \qquad (10.108)$$

$$V_o = \sqrt{\frac{R_L T}{2 L_P}} D V_{IN} \qquad (10.109)$$

- The maximum reverse voltage of D_R, $V(6,9)$ (which is the voltage at node 6 with respect to node 9), is

$$V(DR)_{max} = V(6,9)_{max} = V_{IN} \frac{N_S}{N_P} + V_o \qquad (10.110)$$

10.6.2.2 Practical Circuit

When a practical transformer (with leakage inductance) is used in the flyback converter circuit shown in Fig. 10.38, there will be large ringings. In order to reduce these ringings to practically acceptable levels, snubber and clamping circuits have to be added. Figure 10.40 shows a practical flyback converter circuit where a resistor–capacitor snubber ($R_2 C_2$) is used to damp the ringing voltage across the output rectifier D_R, and a resistor–capacitor-diode clamping ($R_1 C_1 D_S$) is used to clamp the ringing voltage across the switch M_1. What the diode D_S does here is to allow the energy stored by the current in the leakage inductance to be converted to the form of a dc voltage across the clamping capacitor C_1. The energy transferred to C_1 is then dissipated slowly in the parallel resistor R_1, without ringing problems.

The simulated waveforms of the flyback converter (circuit given in Fig. 10.40) for discontinuous-mode operation are shown in Fig. 10.41, where the following assumptions are made:

- D_R and D_S are MUR460 ultra-fast diodes.
- M_1 is an IRF640 MOS transistor.
- Transformer T_1 has a practical coupling coefficient of 0.992.
- The effective winding resistance of L_P is 0.025 Ω. The effective winding resistance of L_S is 0.1 Ω.
- The effective series resistance of the output filtering capacitor C_L is 0.05 Ω.
- The switching operation of the converter has reached a steady state.

The waveforms shown in Fig. 10.40 are considered to be acceptable.

10.6.3 Design Considerations

In the design of rectifier circuits, it is necessary for the designer to determine the voltage and current ratings of the diodes. The idealized waveforms and expressions for the maximum diode voltages and currents given under the heading of "Ideal circuit" above (for both forward and flyback converters) are a good starting point. However, when parasitic/stray components are also considered, the simulation results given under

10 Diode Rectifiers

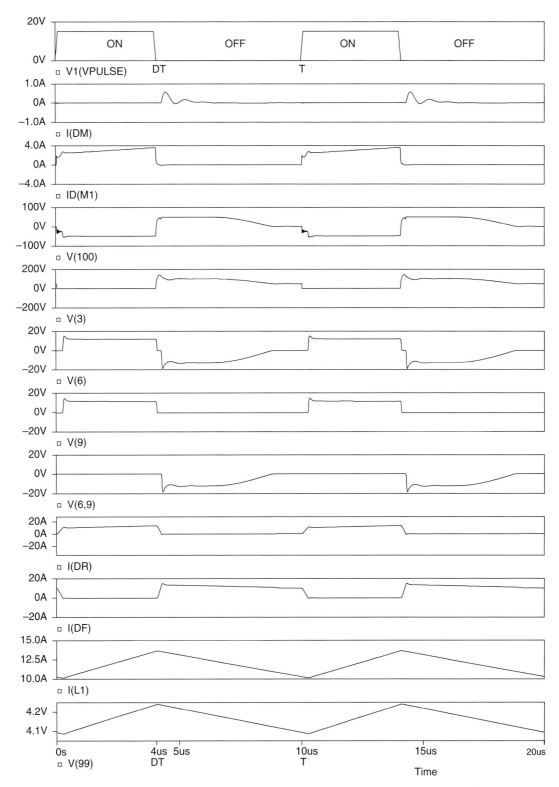

FIGURE 10.36 Waveforms of practical forward converter for continuous-mode operation.

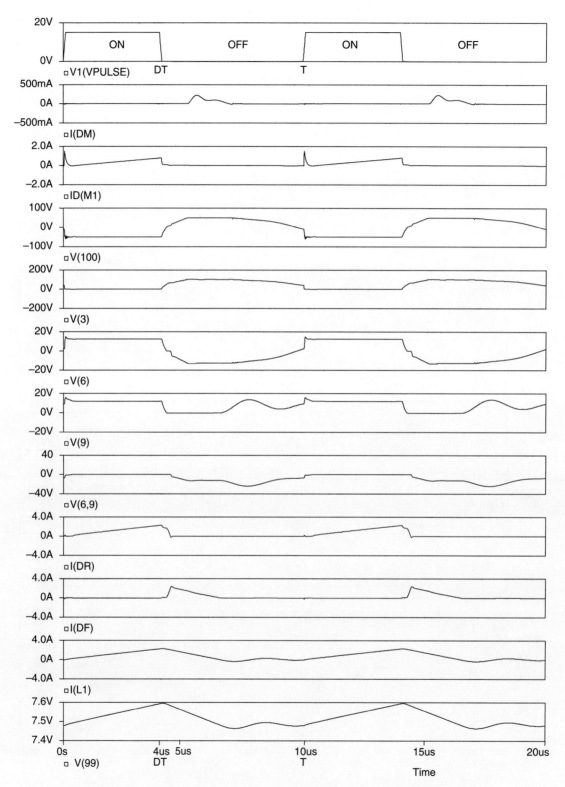

FIGURE 10.37 Waveforms of practical forward converter for discontinuous-mode operation.

10 Diode Rectifiers

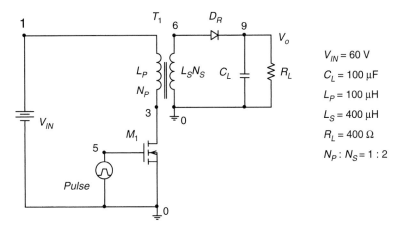

FIGURE 10.38 Basic circuit of flyback converter.

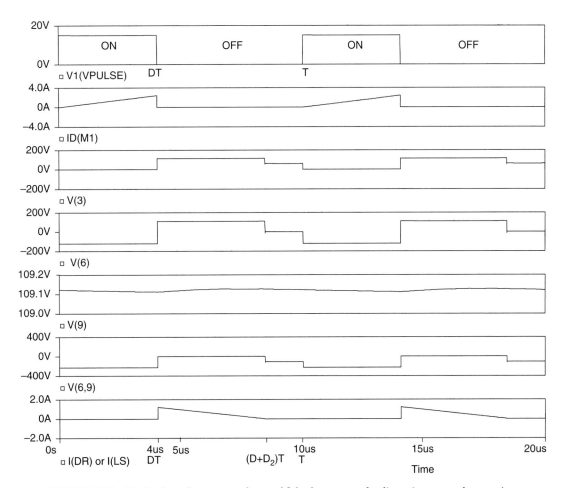

FIGURE 10.39 Idealized steady-state waveforms of flyback converter for discontinuous-mode operation.

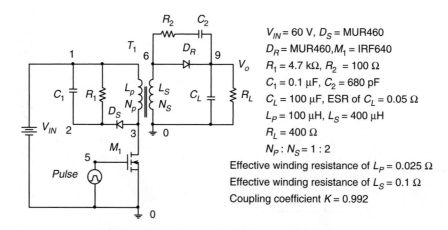

FIGURE 10.40 Practical flyback converter circuit.

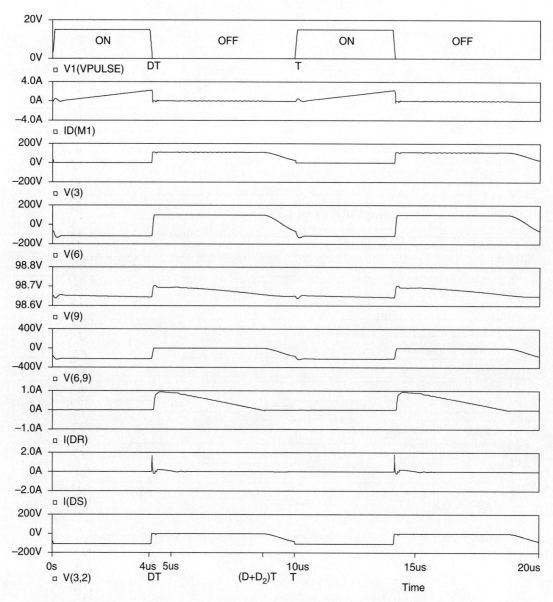

FIGURE 10.41 Waveforms of practical flyback converter for discontinuous-mode operation.

"Practical circuit" are much more useful for the determination of the voltage and current ratings of the high-frequency rectifier diodes.

Assuming that the voltage and current ratings have been determined, proper diodes can be selected to meet the requirements. The following are some general guidelines on the selection of diodes:

- For low-voltage applications, Schottky diodes should be used because they have very fast switching speed and low forward voltage drop. If Schottky diodes cannot be used, either because of their low reverse breakdown voltage or because of their large leakage current (when reversely biased), ultra-fast diodes should be used.
- The reverse breakdown-voltage rating of the diode should be reasonably higher (e.g. 10 or 20% higher) than the maximum reverse voltage, the diode is expected to encounter under the worst-case condition. However, an overly-conservative design (using a diode with much higher breakdown voltage than necessary) would result in a lower rectifier efficiency, because a diode having a higher reverse-voltage rating would normally have a larger voltage drop when it is conducting.
- The current rating of the diode should be substantially higher than the maximum current the diode is expected to carry during normal operation. Using a diode with a relatively large current rating has the following advantages:
  - It reduces the possibility of damage due to transients caused by start-up, accidental short circuit, or random turning on and off of the converter.
  - It reduces the forward voltage drop because the diode is operated in the lower current region of the V–I characteristic.

In some of the "high-efficiency" converter circuits, the current rating of the output rectifier can be many times larger than the actual current expected in the rectifier. In this way, a higher efficiency is achieved at the expense of a larger silicon area.

In the design of R–C snubber circuits for rectifiers, it should be understood that a larger C (and a smaller R) will give better damping. However, a large C (and a small R) will result in a large switching loss (which is equal to $0.5CV^2f$). As a guideline, a capacitor with five to ten times the junction capacitance of the rectifier may be used as a starting point for iterations. The value of the resistor should be chosen to provide a slightly underdamped operating condition.

10.6.4 Precautions in Interpreting Simulation Results

In using the simulated waveforms as references for design purposes, attention should be paid to the following:

- The voltage/current spikes that appear in the practically measured waveforms may not appear in the simulated waveforms. This is due to the lack of a model in the computer simulation to simulate unwanted coupling among the practical components.
- Most of the computer models of diodes, including those used in the simulations given above, do not take into account the effects of the forward recovery time. (The forward recovery time is not even mentioned in most manufacturers' data sheets.) However, it is also interesting to note that in most cases the effect of the forward recovery time of a diode is masked by that of the effective inductance in series with the diode (e.g. the leakage inductance of a transformer).

Further Reading

1. *Rectifier Applications Handbook*, 3rd ed., Phoenix, Ariz.: Motorola, Inc., 1993.
2. M. H. Rashid, *Power Electronics: Circuits, Devices, and Applications*, 2nd ed., Englewood Cliffs, NJ: Prentice Hall, Inc., 1993.
3. Y.-S. Lee, *Computer-Aided Analysis and Design of Switch-Mode Power Supplies*, New York: Marcel Dekker, Inc., 1993.
4. J. W. Nilsson, *Introduction to PSpice Manual, Electric Circuits Using OrCAD Release 9.1*, 4th ed., Upper Saddle River, NJ: Prentice Hall, Inc., 2000.
5. J. Keown, *OrCAD PSpice and Circuit Analysis*, 4th ed., Upper Saddle River, NJ: Prentice Hall, Inc., 2001.

11
Single-phase Controlled Rectifiers

José Rodríguez, Ph.D.,
Pablo Lezana, Samir
Kouro, and Alejandro
Weinstein
*Department of Electronics,
Universidad Técnica Federico
Santa María, Valparaíso, Chile*

11.1 Introduction 179
11.2 Line-commutated Single-phase Controlled Rectifiers 179
   11.2.1 Single-phase Half-wave Rectifier • 11.2.2 Bi-phase Half-wave Rectifier
   • 11.2.3 Single-phase Bridge Rectifier • 11.2.4 Analysis of the Input Current • 11.2.5 Power
   Factor of the Rectifier • 11.2.6 The Commutation of the Thyristors • 11.2.7 Operation in the
   Inverting Mode • 11.2.8 Applications
11.3 Unity Power Factor Single-phase Rectifiers 188
   11.3.1 The Problem of Power Factor in Single-phase Line-commutated Rectifiers
   • 11.3.2 Standards for Harmonics in Single-phase Rectifiers • 11.3.3 The Single-phase Boost
   Rectifier • 11.3.4 Voltage Doubler PWM Rectifier • 11.3.5 The PWM Rectifier in Bridge
   Connection • 11.3.6 Applications of Unity Power Factor Rectifiers
   References 199

11.1 Introduction

This chapter is dedicated to single-phase controlled rectifiers, which are used in a wide range of applications. As shown in Fig. 11.1, single-phase rectifiers can be classified into two big categories:

(i) Topologies working with low switching frequency, also known as line commutated or phase controlled rectifiers.
(ii) Circuits working with high switching frequency, also known as power factor correctors (PFCs).

Line-commutated rectifiers with diodes, covered in a previous chapter of this handbook, do not allow the control of power being converted from ac to dc. This control can be achieved with the use of thyristors. These controlled rectifiers are addressed in the first part of this chapter.

In the last years, increasing attention has been paid to the control of current harmonics present at the input side of the rectifiers, originating from a very important development in the so-called PFC. These circuits use power transistors working with high switching frequency to improve the waveform quality of the input current, increasing the power factor. High power factor rectifiers can be classified in regenerative and non-regenerative topologies and they are covered in the second part of this chapter.

11.2 Line-commutated Single-phase Controlled Rectifiers

11.2.1 Single-phase Half-wave Rectifier

The single-phase half-wave rectifier uses a single thyristor to control the load voltage as shown in Fig. 11.2. The thyristor will conduct, on-state, when the voltage v_T is positive and a firing current pulse i_G is applied to the gate terminal. The control of the load voltage is performed by delaying the firing pulse by an angle α. The firing angle α is measured from the position where a diode would naturally conduct. In case of Fig. 11.2 the angle α is measured from the zero-crossing point of the supply voltage v_s. The load in Fig. 11.2 is resistive and therefore the current i_d has the same waveform of the load voltage. The thyristor goes to the non-conducting condition, off-state, when the load voltage, and consequently the current, reaches a negative value.

The load average voltage is given by

$$V_{d\alpha} = \frac{1}{2\pi} \int_\alpha^\pi V_{max} \sin(\omega t) d(\omega t) = \frac{V_{max}}{2\pi}(1 + \cos\alpha) \quad (11.1)$$

where V_{max} is the supply peak voltage. Hence, it can be seen from Eq. (11.1) that changing the firing angle α controls

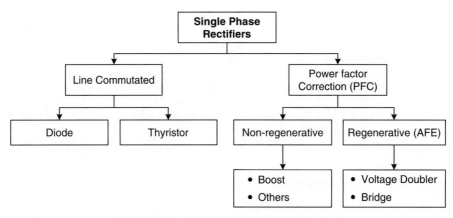

FIGURE 11.1 Single-phase rectifier classification.

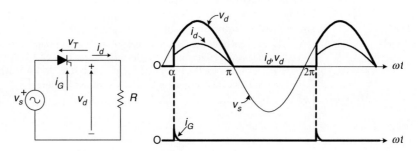

FIGURE 11.2 Single-thyristor rectifier with resistive load.

both the load average voltage and the amount of transferred power.

Figure 11.3a shows the rectifier waveforms for an R–L load. When the thyristor is turned on, the voltage across the inductance is

$$v_L = v_s - v_R = L \frac{di_d}{dt} \quad (11.2)$$

where v_R is the voltage in the resistance R, given by $v_R = R \cdot i_d$. If $v_s - v_R > 0$, from Eq. (11.2) holds that the load current increases its value. On the other hand, i_d decreases its value when $v_s - v_R < 0$. The load current is given by

$$i_d(\omega t) = \frac{1}{\omega L} \int_\alpha^{\omega t} v_L d\theta \quad (11.3)$$

Graphically, Eq. (11.3) means that the load current i_d is equal to zero when $A_1 = A_2$, maintaining the thyristor in conduction state even when $v_s < 0$.

When an inductive–active load is connected to the rectifier, as illustrated in Fig. 11.3b, the thyristor will be turned on if the firing pulse is applied to the gate when $v_s > E_d$. Again, the thyristor will remain in the on-state until $A_1 = A_2$. When the thyristor is turned off, the load voltage will be $v_d = E_d$.

11.2.2 Bi-phase Half-wave Rectifier

The bi-phase half-wave rectifier, shown in Fig. 11.4, uses a center-tapped transformer to provide two voltages v_1 and v_2. These two voltages are 180° out of phase with respect to the mid-point neutral N. In this scheme, the load is fed via thyristors T_1 and T_2 during each positive cycle of voltages v_1 and v_2, respectively, while the load current returns via the neutral N.

As illustrated in Fig. 11.4, thyristor T_1 can be fired into the on-state at any time while voltage $v_{T1} > 0$. The firing pulses are delayed by an angle α with respect to the instant where diodes would conduct. Also the current paths for each conduction state are presented in Fig. 11.4. Thyristor T_1 remains in the on-state until the load current tends to a negative value. Thyristor T_2 is fired into the on-state when $v_{T2} > 0$, which corresponds in Fig. 11.4 to the condition when $v_2 > 0$.

The mean value of the load voltage with resistive load is determined by

$$V_{di\alpha} = \frac{1}{\pi} \int_\alpha^\pi V_{max} \sin(\omega t) d(\omega t) = \frac{V_{max}}{\pi}(1 + \cos \alpha) \quad (11.4)$$

The ac supply current is equal to $i_{T1}(N_2/N_1)$ when T_1 is in the on-state and $-i_{T2}(N_2/N_1)$ when T_2 is in the on-state, where N_2/N_1 is the transformer turns ratio.

11 Single-phase Controlled Rectifiers

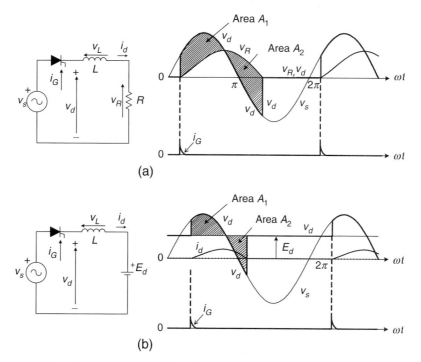

FIGURE 11.3 Single-thyristor rectifier with: (a) resistive-inductive load and (b) active load.

The effect of the load time constant $T_L = L/R$, on the normalized load current $i_d(t)/i_R(t)$ for a firing angle $\alpha = 0°$ is shown in Fig. 11.5. The ripple in the load current reduces as the load inductance increases. If the load inductance $L \to \infty$, then the current is perfectly filtered.

11.2.3 Single-phase Bridge Rectifier

Figure 11.6a shows a fully controlled bridge rectifier, which uses four thyristors to control the average load voltage. In addition, Fig. 11.6b shows the half-controlled bridge rectifier which uses two thyristors and two diodes.

The voltage and current waveforms of the fully controlled bridge rectifier for a resistive load are illustrated in Fig 11.7. Thyristors T_1 and T_2 must be fired on simultaneously during the positive half-wave of the source voltage v_s, to allow the conduction of current. Alternatively, thyristors T_3 and T_4 must be fired simultaneously during the negative half-wave of the source voltage. To ensure simultaneous firing, thyristors T_1 and T_2 use the same firing signal. The load voltage is similar to the voltage obtained with the bi-phase half-wave rectifier. The input current is given by

$$i_S = i_{T1} - i_{T4} \quad (11.5)$$

and its waveform is shown in Fig. 11.7.

Figure 11.8 presents the behavior of the fully controlled rectifier with resistive–inductive load (with $L \to \infty$). The high load inductance generates a perfectly filtered current and the rectifier behaves like a current source. With continuous load current, thyristors T_1 and T_2 remain in the on-state beyond the positive half-wave of the source voltage v_s. For this reason, the load voltage v_d can have a negative instantaneous value. The firing of thyristors T_3 and T_4 has two effects:

(i) they turn-off thyristors T_1 and T_2 and
(ii) after the commutation, they conduct the load current.

This is the main reason why this type of converters are called "naturally commutated" or "line commutated" rectifiers. The supply current i_S has the square waveform, as shown in Fig. 11.9, for continuous conduction. In this case, the average load voltage is given by

$$V_{di\alpha} = \frac{1}{\pi} \int_{\alpha}^{\pi+\alpha} V_{max} \sin(\omega t) d(\omega t) = \frac{2V_{max}}{\pi} \cos \alpha \quad (11.6)$$

11.2.4 Analysis of the Input Current

Considering a very high inductive load, the input current in a bridge-controlled rectifier is filtered and presents a square waveform. In addition, the input current i_s is shifted by the firing angle α with respect to the input voltage v_s, as shown in Fig. 11.9a. The input current can be expressed as a Fourier series, where the amplitude of the different harmonics are

$$I_{smax,n} = \frac{4}{\pi} \frac{I_d}{n} \quad (n = 1, 3, 5, \ldots) \quad (11.7)$$

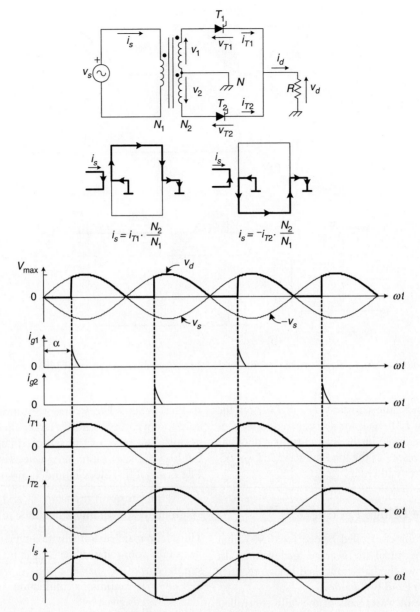

FIGURE 11.4 Bi-phase half-wave rectifier.

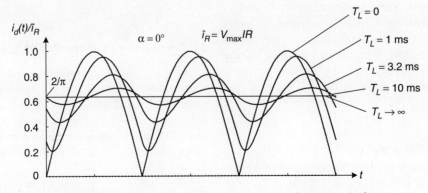

FIGURE 11.5 Effect of the load time constant over the current ripple.

11 Single-phase Controlled Rectifiers

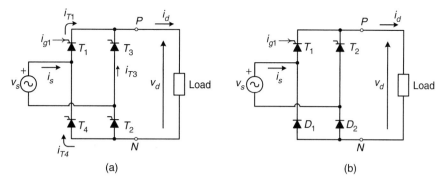

FIGURE 11.6 Single-phase bridge rectifier: (a) fully controlled and (b) half-controlled.

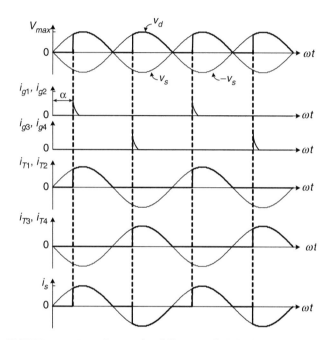

FIGURE 11.7 Waveforms of a fully controlled bridge rectifier with resistive load.

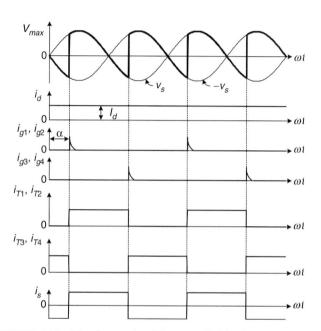

FIGURE 11.8 Waveforms of a fully controlled bridge rectifier with resistive–inductive load ($L \to \infty$).

where n is the harmonic order. The root mean square (rms) value of each harmonic can be expressed as

$$I_{sn} = \frac{I_{smax,n}}{\sqrt{2}} = \frac{2\sqrt{2}}{\pi}\frac{I_d}{n} \qquad (11.8)$$

Thus, the rms value of the fundamental current i_{s1} is

$$I_{s1} = \frac{2\sqrt{2}}{\pi} I_d = 0.9 I_d \qquad (11.9)$$

It can be observed from Fig. 11.9a that the displacement angle ϕ_1 of the fundamental current i_{s1} corresponds to the firing angle α. Figure 11.9b shows that in the harmonic spectrum of the input current, only odd harmonics are present with

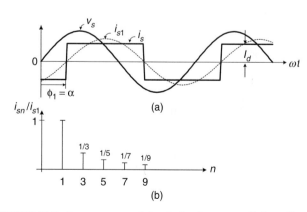

FIGURE 11.9 Input current of the single-phase controlled rectifier in bridge connection: (a) waveforms and (b) harmonics spectrum.

decreasing amplitude while the frequency increases. Finally the rms value of the input current i_s is

$$I_s = I_d \qquad (11.10)$$

The total harmonic distortion (THD) of the input current can be determined by

$$\text{THD} = \frac{\sqrt{I_s^2 - I_{s1}^2}}{I_{s1}} 100 = 48.4\% \qquad (11.11)$$

11.2.5 Power Factor of the Rectifier

The displacement factor of the fundamental current, obtained from Fig. 11.9a is

$$\cos \phi_1 = \cos \alpha \qquad (11.12)$$

In the case of non-sinusoidal currents, the active power delivered by the sinusoidal single-phase supply is

$$P = \frac{1}{T} \int_0^T v_s(t) i_s(t) dt = V_s I_{s1} \cos \phi_1 \qquad (11.13)$$

where V_s is the rms value of the single-phase voltage v_s.

The apparent power is given by

$$S = V_s I_s \qquad (11.14)$$

The power factor (PF) is defined by

$$\text{PF} = \frac{P}{S} \qquad (11.15)$$

Substitution from Eqs. (11.12), (11.13), and (11.14) in Eq. (11.15) yields

$$\text{PF} = \frac{I_{s1}}{I_s} \cos \alpha \qquad (11.16)$$

This equation shows clearly that due to the non-sinusoidal waveform of the input current, the power factor of the rectifier is negatively affected both by the firing angle α and by the distortion of the input current. In effect, an increase in the distortion of the current produces an increase in the value of I_s in Eq. (11.16), which deteriorates the power factor.

11.2.6 The Commutation of the Thyristors

Until now, the current commutation between thyristors has been considered to be instantaneous. This condition is not valid in real cases due to the presence of the line inductance L, as shown in Fig. 11.10a. During the commutation, the current through the thyristors cannot change instantaneously, and for this reason, during the commutation angle μ, all four thyristors are conducting simultaneously. Therefore, during the commutation, the following relationship for the load voltage holds

$$v_d = 0 \quad \alpha \leq \omega t \leq \alpha + \mu \qquad (11.17)$$

The effect of the commutation on the supply current, voltage waveforms, and the thyristor current waveforms can be observed in Fig. 11.10b.

During the commutation, the following expression holds

$$L \frac{di_s}{dt} = v_s = V_{max} \sin(\omega t) \quad \alpha \leq \omega t \leq \alpha + \mu \qquad (11.18)$$

Integrating Eq. (11.18) over the commutation interval yields

$$\int_{-I_d}^{I_d} di_s = \frac{V_{max}}{L} \int_{\alpha/\omega}^{\alpha + \mu/\omega} \sin(\omega t) dt \qquad (11.19)$$

From Eq. (11.19), the following relationship for the commutation angle μ is obtained

$$\cos(\alpha + \mu) = \cos \alpha - \frac{2\omega L}{V_{max}} I_d \qquad (11.20)$$

Equation (11.20) shows that an increase of the line inductance L or an increase of the load current I_d increases the commutation angle μ. In addition, the commutation angle is affected by the firing angle α. In effect, Eq. (11.18) shows that with different values of α, the supply voltage v_s has a different instantaneous value, which produces different di_s/dt, thereby affecting the duration of the commutation.

Equation (11.17) and the waveform of Fig. 11.10b show that the commutation process reduces the average load voltage $V_{d\alpha}$. When the commutation is considered, the expression for the average load voltage is given by

$$V_{d\alpha} = \frac{1}{\pi} \int_{\alpha + \mu}^{\pi + \alpha} \sin(\omega t) d(\omega t) = \frac{V_{max}}{\pi} [\cos(\alpha + \mu) + \cos \alpha] \qquad (11.21)$$

Substituting Eq. (11.20) into Eq. (11.21) yields

$$V_{d\alpha} = \frac{2}{\pi} V_{max} \cos \alpha - \frac{2\omega L}{\pi} I_d \qquad (11.22)$$

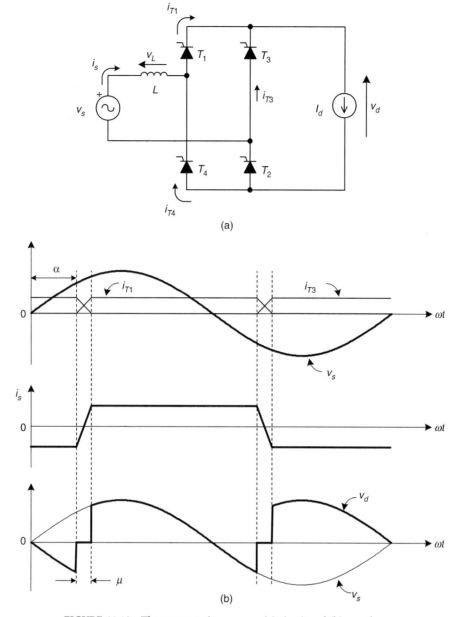

FIGURE 11.10 The commutation process: (a) circuit and (b) waveforms.

11.2.7 Operation in the Inverting Mode

When the angle $\alpha > 90°$, it is possible to obtain a negative average load voltage. In this condition, the power is fed back to the single-phase supply from the load. This operating mode is called inverter or inverting mode, because the energy is transferred from the dc to the ac side. In practical cases, this operating mode is obtained when the load configuration is as shown in Fig. 11.11a. It must be noticed that this rectifier allows unidirectional load current flow.

Figure 11.11b shows the waveform of the load voltage with the rectifier in the inverting mode, neglecting the source inductance L.

Section 11.2.6 described how supply inductance increases the conduction interval of the thyristors by the angle μ. As shown in Fig. 11.11c, the thyristor voltage v_{T1} has a negative value during the extinction angle γ, defined by

$$\gamma = 180 - (\alpha + \mu) \tag{11.23}$$

To ensure that the outgoing thyristor will recover its blocking capability after the commutation, the extinction angle should satisfy the following restriction

$$\gamma > \omega t_q \tag{11.24}$$

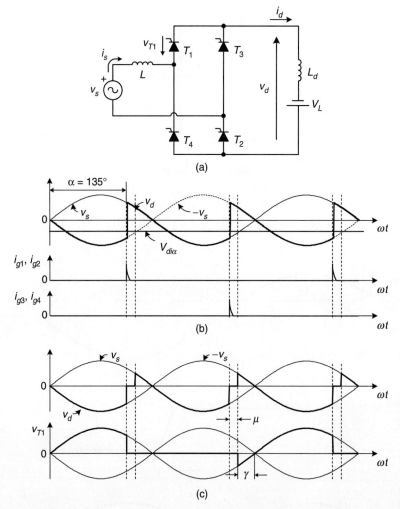

FIGURE 11.11 Rectifier in the inverting mode: (a) circuit; (b) waveforms neglecting source inductance L; and (c) waveforms considering L.

where ω is the supply frequency and t_q is the thyristor turn-off time. Considering Eqs. (11.23) and (11.24) the maximum firing angle is, in practice,

$$\alpha_{max} = 180 - \mu - \gamma \qquad (11.25)$$

If the condition of Eq. (11.25) is not satisfied, the commutation process will fail, originating destructive currents.

11.2.8 Applications

Important application areas of controlled rectifiers include uninterruptible power supplies (UPS), for feeding critical loads. Figure 11.12 shows a simplified diagram of a single-phase UPS configuration, typically rated for <10 kVA. A fully controlled or half-controlled rectifier is used to generate the dc voltage for the inverter. In addition, the input rectifier acts as a battery charger. The output of the inverter is filtered before it is fed to the load. The most important operation modes of the UPS are:

(i) Normal mode. In this case the line voltage is present. The critical load is fed through the rectifier-inverter scheme. The rectifier keeps the battery charged.
(ii) Outage mode. During a loss of the ac main supply, the battery provides the energy for the inverter.
(iii) Bypass operation. When the load demands an overcurrent to the inverter, the static bypass switch is turned on and the critical load is fed directly from the mains.

The control of low power dc motors is another interesting application of controlled single-phase rectifiers. In the circuit of Fig. 11.13, the controlled rectifier regulates the armature voltage and consequently controls the motor current i_d in order to produce a required torque.

This configuration allows only positive current flow in the load. However the load voltage can be both positive

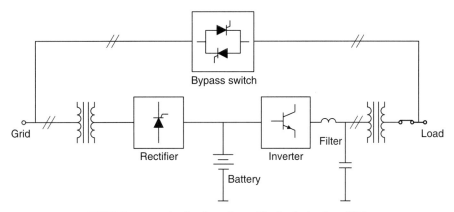

FIGURE 11.12 Application of a rectifier in single-phase UPS.

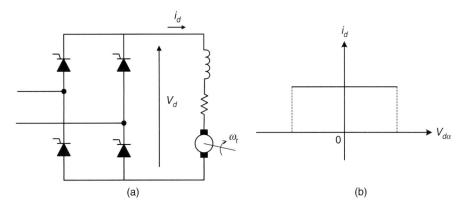

FIGURE 11.13 Two-quadrant dc drive: (a) circuit and (b) quadrants of operation.

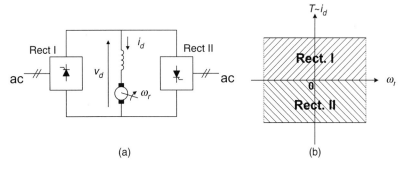

FIGURE 11.14 Single-phase dual-converter drive: (a) connection and (b) four-quadrant operation.

and negative. For this reason, this converter works in the two-quadrant mode of operation in the plane i_d vs $V_{d\alpha}$.

A better performance can be obtained with two rectifiers in back-to-back connection at the dc terminals, as shown in Fig. 11.14a. This arrangement, known as a dual converter configuration, allows four-quadrant operation of the drive. Rectifier I provides positive load current i_d, while rectifier II provides negative load current. The motor can work in forward powering, forward braking (regenerating), reverse powering, and reverse braking (regenerating). These operating modes are shown in Fig. 11.14b, where the torque T vs the rotor speed ω_r is illustrated.

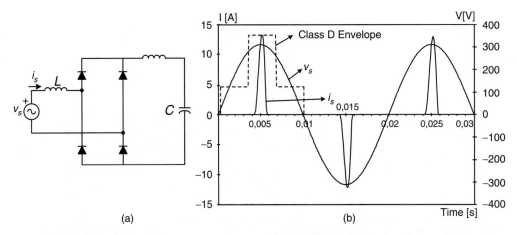

FIGURE 11.15 Single-phase rectifier: (a) circuit and (b) waveforms of the input voltage and current.

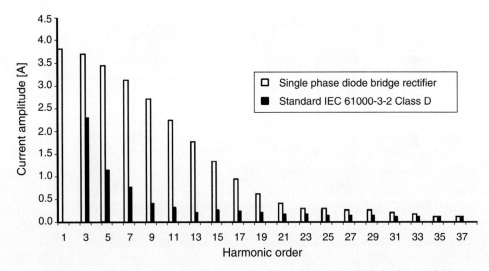

FIGURE 11.16 Input current harmonics produced by a single-phase diode bridge rectifier.

11.3 Unity Power Factor Single-phase Rectifiers

11.3.1 The Problem of Power Factor in Single-phase Line-commutated Rectifiers

The main disadvantages of the classical line-commutated rectifiers are that (i) they produce a lagging displacement factor with respect to the voltage of the utility; (ii) they generate an important amount of input current harmonics.

These aspects have a negative influence on both power factor and power quality. In the last several years, the massive use of single-phase power converters has increased the problems of power quality in electrical systems. In effect, modern commercial buildings have 50% and even up to 90% of the demand, originated by non-linear loads, which are composed mainly by rectifiers [1]. Today it is not unusual to find rectifiers with total harmonic distortion of the current $THD_i > 40\%$, originating severe overloads in conductors and transformers.

Figure 11.15a shows a single-phase rectifier with a capacitive filter, used in much of today's low-power equipment. The input current produced by this rectifier is illustrated in Fig. 11.15b, it appears highly distorted due to the presence of the filter capacitor. This current has a harmonic content shown in Fig. 11.16 and Table 11.1, with a $THD_i = 197\%$.

The rectifier in Fig. 11.15 has a very low power factor of $PF = 0.45$, due mainly to its large harmonic content.

TABLE 11.1 Harmonic content of the current of Fig. 11.15

n	3	5	7	9	11	13	15	17	19	21
I_n/I_1 [%]	96.8	90.5	81.7	71.0	59.3	47.3	35.7	25.4	16.8	10.6

11.3.2 Standards for Harmonics in Single-phase Rectifiers

The relevance of the problems originated by harmonics in single-phase line-commutated rectifiers has motivated some agencies to introduce restrictions to these converters. The IEC 61000-3-2 Class D International Standard establishes limits to all low-power single-phase equipment having an input current with a "special wave shape" and an active input power $P \leq 600$ W. Class D equipment has an input current with a special wave shape contained within the envelope given in Fig. 11.15b. This class of equipment must satisfy certain harmonic limits, shown in Fig. 11.16. It is clear that a single-phase line-commutated rectifier with the parameters shown in Fig. 11.15a is not able to comply with the standard IEC 61000-3-2 Class D. The standard can be satisfied only by adding huge passive filters, which increases the size, weight, and cost of the rectifier. This standard has been the motivation for the development of active methods to improve the quality of the input current and, consequently, the power factor.

11.3.3 The Single-phase Boost Rectifier

One of the most important high power factor rectifiers, from a theoretical and conceptual point of view, is the so-called single-phase boost rectifier, shown in Fig. 11.17a, which is obtained from a classical non-controlled bridge rectifier, with the addition of transistor T, diode D, and inductor L.

11.3.3.1 Working Principle, Basic Concepts

In boost rectifiers, the input current $i_s(t)$ is controlled by changing the conduction state of transistor T. When transistor T is in the on-state, the single-phase power supply is short-circuited through the inductance L, as shown in Fig. 11.17b; the diode D avoids the discharge of the filter capacitor C through the transistor. The current of the inductance i_L is given by the following equation

$$\frac{di_L}{dt} = \frac{v_L}{L} = \frac{|v_s|}{L} \quad (11.26)$$

Due to the fact that $|v_s| > 0$, the on-state of transistor T always produces an increase in the inductance current i_L and consequently an increase in the absolute value of the source current i_s.

When transistor T is turned off, the inductor current i_L cannot be interrupted abruptly and flows through diode D, charging capacitor C. This is observed in the equivalent circuit of Fig. 11.17c. In this condition, the behavior of the inductor current is described by

$$\frac{di_L}{dt} = \frac{v_L}{L} = \frac{|v_s| - v_o}{L} \quad (11.27)$$

If $v_o > |v_s|$, which is an important condition for the correct behavior of the rectifier, then $|v_s| - v_o < 0$, and this means that in the off-state the inductor current decreases its instantaneous value.

11.3.3.2 Continuous Conduction Mode (CCM)

With an appropriate firing pulse sequence is applied to transistor T, the waveform of the input current i_s can be controlled to follow a sinusoidal reference, as can be observed in the positive half-wave of i_s in Fig. 11.18. This figure shows the reference

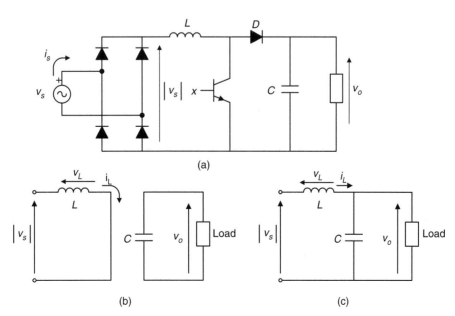

FIGURE 11.17 Single-phase boost rectifier: (a) power circuit and equivalent circuit for transistor T in; (b) on-state; and (c) off-state.

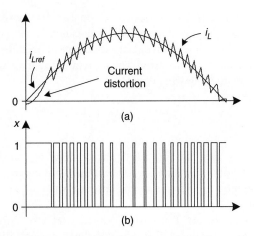

FIGURE 11.18 Behavior of the inductor current i_L: (a) waveforms and (b) transistor T gate drive signal x.

inductor current i_{Lref}, the inductor current i_L and the gate drive signal x for transistor T. Transistor T is on when $x = 1$ and it is off when $x = 0$.

Figure 11.18 clearly shows that the on- (off-) state of transistor T produces an increase (decrease) in the inductor current i_L.

Note that for low values of v_s the inductor does not have enough energy to increase the current value, for this reason it presents a distortion in their current waveform as shown in Fig. 11.18a.

Figure 11.19 presents a block diagram of the control system for the boost rectifier, which includes a proportional-integral (PI) controller, to regulate the output voltage v_o. The reference value i_{Lref} for the inner current control loop is obtained from the multiplication between the output of the voltage controller and the absolute value $|v_s(t)|$. A hysteresis controller provides a fast control for the inductor current i_L, resulting in a practically sinusoidal input current i_s.

Typically, the output voltage v_o should be at least 10% higher than the peak value of the source voltage $v_s(t)$, in order to assure good dynamic control of the current. The control works with the following strategy: a step increase in the reference voltage v_{oref} will produce an increase in the voltage error $v_{oref} - v_o$ and an increase of the output of the PI controller, which originates an increase in the amplitude of the reference

current i_{Lref}. The current controller will follow this new reference and will increase the amplitude of the sinusoidal input current i_s, which will increase the active power delivered by the single-phase power supply, producing finally an increase in the output voltage v_o.

Figure 11.20a shows the waveform of the input current i_s and the source voltage v_s. The ripple of the input current can be reduced by shortening the hysteresis width δ. The tradeoff for this improvement is an increase in the switching frequency, which is proportional to the commutation losses of the transistors. For a given hysteresis width δ, a reduction of inductance L also produces an increase in the switching frequency. As can be seen, the input current presents a third-harmonic component. This harmonic is generated by the second-harmonic component present in v_o, which is fed back through the voltage (PI) controller and multiplied by the sinusoidal waveform, generating a third-harmonic component on i_{Lref}. This harmonic contamination can be avoided by filtering the v_o measurement with a lowpass filter or a bandstop filter around $2\omega_s$. The input current obtained using the measurement filter is shown in Fig. 11.20b. Figure 11.20d confirms the reduction of the third-harmonic component.

However, in both cases, a drastic reduction in the harmonic content of the input current i_s can be observed in the frequency spectrum of Figs. 11.20c and 11.20d. This current fulfills the restrictions established by standard IEC 61000-3-2. The total harmonic distortion of the current in Fig. 11.20a is THD = 7.46%, while the THD of the current of Fig. 11.20b is 4.83%, in both cases a very high power factor, over 0.99, is reached.

Figure 11.21 shows the dc voltage control loop dynamic behavior for step changes in the load. An increase in the load, at $t = 0.3$ [s], produces an initial reduction of the output voltage v_o, which is compensated by an increase in the input current i_s. At $t = 0.6$ [s] a step decrease in the load is applied. The dc voltage controller again adjusts the supply current in order to balance the active power.

11.3.3.3 Discontinuous Conduction Mode (DCM)

This PFC method is based on an active current waveform-shaping principle. There are two different approaches considering fixed and variable switching frequency, both operating principles are illustrated in Fig. 11.22.

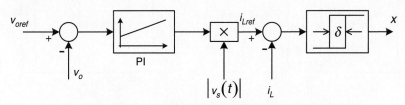

FIGURE 11.19 Control system of the boost rectifier.

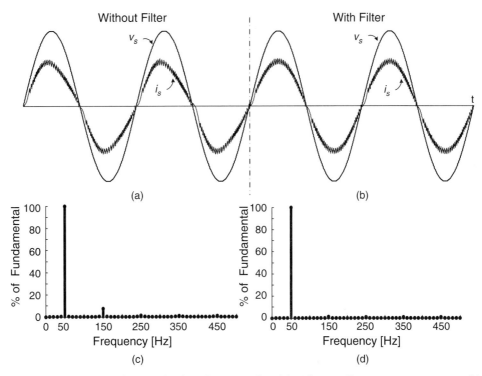

FIGURE 11.20 Input current and voltage of the single-phase boost rectifier: (a) without a filter on v_o measurement; (b) with a filter on v_o measurement; frequency spectrum; (c) without filter; and (d) with filter.

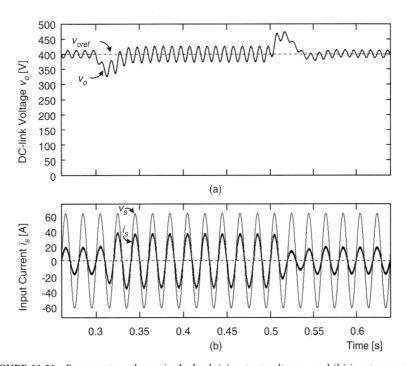

FIGURE 11.21 Response to a change in the load: (a) output-voltage v_o and (b) input current i_s.

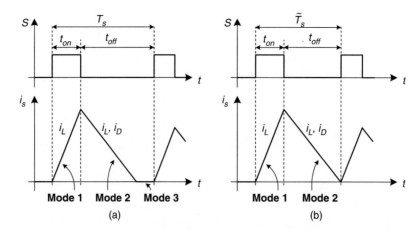

FIGURE 11.22 Boost DCM operating principle: (a) with fixed switching frequency and (b) with variable switching frequency.

a) DCM with Fixed Switching Frequency

The current shaping strategy is achieved by combining three different conduction modes, performed over a fixed switching period T_s. At the beginning of each period the power semiconductor is turned on. During the on-state, shown in Fig. 11.23a, the power supply is short-circuited through the rectifier diodes, the inductor L, and the boost switch T. Hence, the inductor current i_L increases at a rate proportional to the instantaneous value of the supply voltage. As a result, during the on-state, the average supply current i_s is proportional to the supply voltage v_s which yields to power factor correction.

When the switch is turned off, the current flows to the load trough diode D, as shown in Fig. 11.23b. The instantaneous current value decreases (since the load voltage v_o is higher than the supply peak voltage) at a rate proportional to the difference between the supply and load voltage. Finally, the last mode, illustrated in Fig. 11.23c, corresponds to the time in which the current reaches zero value, completing the switching period T_s. Therefore, the supply current is not proportional to the voltage source during the whole control period, introducing distortion and undesirable EMI in comparison to CCM.

The duty cycle $D = t_{on}/T_s$ is determined by the control loop, in order to obtain the desired output power and to ensure operation in DCM, i.e. to reach zero current before the new switching cycle starts. The control strategy can be implemented with analog circuitry as shown in Fig. 11.24, or digitally with modern computing devices. Generally, the duty cycle is controlled with a slow control loop, maintaining the output voltage and duty cycle constant over a half-source cycle.

A qualitative example of the supply voltage and current obtained using DCM is illustrated in Fig. 11.25.

b) DCM with Variable Switching Frequency

The operating principle is similar to the one used in the previous case, the main difference is that mode 3 is avoided by switching the transistor again to the on-state, immediately after the inductor current reaches zero value. This reduces

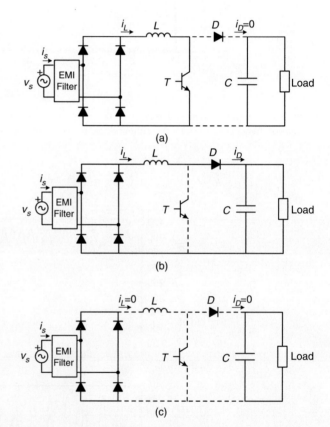

FIGURE 11.23 Boost DCM equivalent circuits: (a) mode 1: transistor on, inductor current increasing; (b) mode 2: transistor off, inductor current decreasing; and (c) mode 3: transistor off, inductor current reaches zero.

the current distortion, with the tradeoff of introducing variable switching frequency (T_s is variable) and consequently lower-order harmonic content.

Both CCM and DCM achieve an improvement in the power factor. The DCM is more efficient since reverse-recovery losses

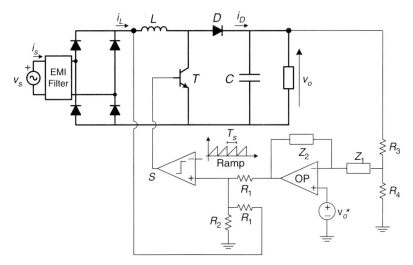

FIGURE 11.24 Boost DCM control circuit with fixed switching frequency.

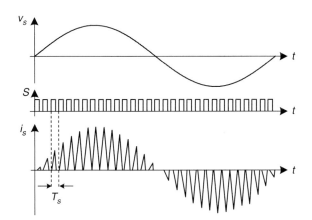

FIGURE 11.25 Boost DCM waveforms: supply voltage, transistor control signal, and supply current.

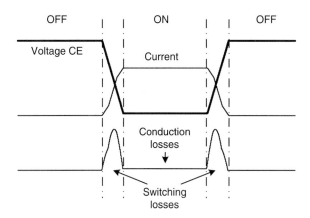

FIGURE 11.26 Conduction and switching losses on a power switch.

of the boost diode are eliminated, however this mode introduces high-current ripple and considerable distortion and usually an important fifth-order harmonic is obtained. Therefore boost-DCM applications are limited to 300 W power levels, to meet standards and regulations. The DCM with variable switching frequency reduces this harmonic content, at expends of a wide distributed current spectrum and all related design problems.

11.3.3.4 Resonant Structures for the Boost Rectifiers

An important issue in power electronics is the power losses in power semiconductors. These losses can be classified in two groups: conduction losses and switching losses, as shown in Fig. 11.26.

The conduction losses are produced by the current through the semiconductor juncture, so these losses are unavoidable. However, the switching losses, which are produced while the power semiconductors work in linear state during the transition from on- to off- state or from off- to on-state, can be reduced or even eliminated, if the switch (transition) occurs when: (a) the current across the power semiconductor is zero; (b) the voltage between the power terminals of the power semiconductor is zero.

This operation mode is used in the so-called *resonant* or *soft-switched* converters, which are discussed in detail in a different chapter of this handbook.

Resonant operation can also be used with the boost converter topology. In order to produce this condition, topology of Fig. 11.17 needs to be modified, by including reactive components and additional semiconductors.

In Fig. 11.27 a resonant structure for zero current switching (ZCS) [2] is shown. As can be seen, additional resonant inductors (L_{r1}, L_{r2}), capacitors (C_r), diodes (D_{r1}, D_{r2}), and power switch (S_r) have been included.

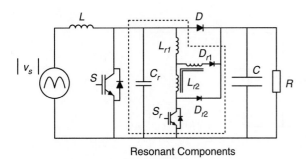

FIGURE 11.27 Boost rectifier with ZCS.

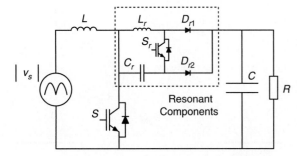

FIGURE 11.28 Boost rectifier with ZVS.

In a similar way, in Fig. 11.28 a resonant structure for zero voltage switching (ZVS) [3] is shown. Once again, additional inductance (L_r), capacitor (C_r), and power switch (S_r) are added, note however, that diode D has been replaced by two "resonant diodes," D_{r1} and D_{r2}.

In both cases, the ZVS or ZCS condition is reached through a proper control of S_r. Other resonant topologies are described in the literature [4–6] with similar behavior.

11.3.3.5 Bridgeless Boost Rectifier

The bridgeless boost rectifier [7] is shown in Fig. 11.29a. This rectifier replaces the input diode rectifier by a combination of two boost rectifiers which work alternately: (a) when v_s is positive, T_1 and D_1 operate as boost rectifier 1 (Fig. 11.29b); (b) when v_s is negative, T_2 and D_2 operate as boost rectifier 2 (Fig. 11.29c);

This topology reduces the conduction losses of the rectifier [8, 9], but requires a slightly more complex control scheme, also EMI and EMC aspects must be considered.

11.3.4 Voltage Doubler PWM Rectifier

Figure 11.30a shows the power circuit of the voltage doubler pulse width modulated (PWM) rectifier, which uses two transistors T_1 and T_2 and two filter capacitors C_1 and C_2. The transistors are switched complementary to control the waveform of the input current i_s and the output dc voltage v_o. Capacitor voltages V_{C1} and V_{C2} must be higher than the peak value of the input voltage v_s to ensure the control of the input current.

The equivalent circuit of this rectifier with transistor T_1 in the on-state is shown in Fig. 11.30b. For this case, the inductor voltage dynamic equation is

$$v_L = L\frac{di_s}{dt} = v_s(t) - V_{C1} < 0 \qquad (11.28)$$

Equation (11.28) means that under this conduction state, current $i_s(t)$ decreases its value.

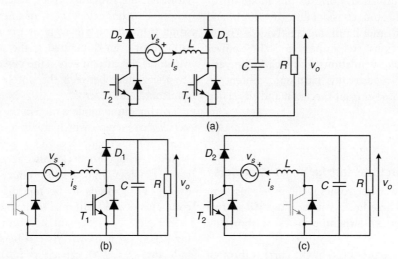

FIGURE 11.29 (a) Power circuit of bridgeless boost rectifier; equivalent circuit when; (b) $v_s > 0$; and (c) $v_s < 0$.

11 Single-phase Controlled Rectifiers

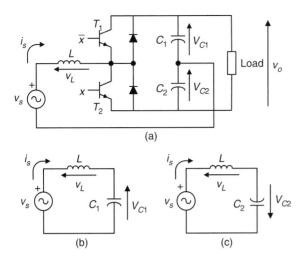

FIGURE 11.30 Voltage doubler rectifier: (a) power circuit; (b) equivalent circuit with T_1 on; and (c) equivalent circuit with T_2 on.

On the other hand, the equivalent circuit of Fig. 11.30c is valid when transistor T_2 is in the conduction state, resulting in the following expression for the inductor voltage

$$v_L = L\frac{di_s}{dt} = v_s(t) + V_{C2} > 0 \quad (11.29)$$

hence, for this condition, the input current $i_s(t)$ increases.

Therefore the waveform of the input current can be controlled by switching appropriately transistors T_1 and T_2 in a similar way as shown in Fig. 11.18a for the single-phase boost converter. Figure 11.31 shows a block diagram of the control system for the voltage doubler rectifier, which is very similar to the control scheme of the boost rectifier. This topology can present an unbalance in the capacitor voltages V_{C1} and V_{C2}, which will affect the quality of the control. This problem is solved by adding to the actual current value i_s an offset signal proportional to the capacitor's voltage difference.

Figure 11.32 shows the waveform of the input current. The ripple amplitude of this current can be reduced by decreasing the hysteresis width of the controller.

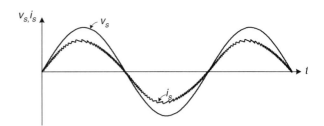

FIGURE 11.32 Waveform of the input current in the voltage doubler rectifier.

11.3.5 The PWM Rectifier in Bridge Connection

Figure 11.33a shows the power circuit of the fully controlled single-phase PWM rectifier in bridge connection, which uses four transistors with antiparallel diodes to produce a controlled dc voltage v_o. Using a bipolar PWM switching strategy, this converter may have two conduction states: (i) Transistors T_1 and T_4 in the on-state and T_2 and T_3 in the off-state; (ii) Transistors T_2 and T_3 in the on-state and T_1 and T_4 in the off-state.

In this topology, the output voltage v_o must be higher than the peak value of the ac source voltage v_s, to ensure a proper control of the input current.

Figure 11.33b shows the equivalent circuit with transistors T_1 and T_4 on. In this case, the inductor voltage is given by

$$v_L = L\frac{di_s}{dt} = v_s(t) - V_0 < 0 \quad (11.30)$$

Therefore, in this condition a reduction of the inductor current i_s is produced.

Figure 11.33c shows the equivalent circuit with transistors T_2 and T_3 on. Here, the inductor voltage has the following expression

$$v_L = L\frac{di_s}{dt} = v_s(t) + V_0 > 0 \quad (11.31)$$

which means an increase in the instantaneous value of the input current i_s.

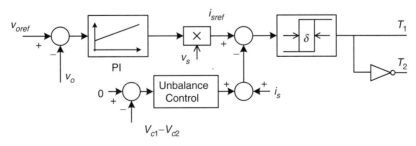

FIGURE 11.31 Control system of the voltage doubler rectifier.

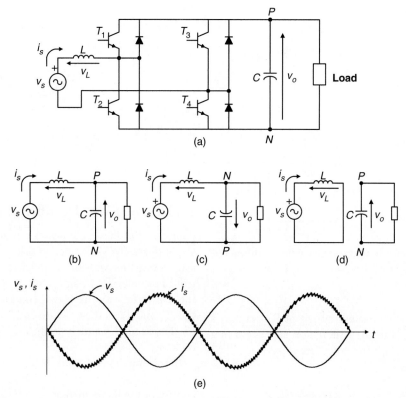

FIGURE 11.33 Single-phase PWM rectifier in bridge connection: (a) power circuit; (b) equivalent circuit with T_1 and T_4 on; (c) equivalent circuit with T_2 and T_3 on; (d) equivalent circuit with T_1 and T_3 or T_2 and T_4 on; and (e) waveform of the input current during regeneration.

Finally, Fig. 11.33d shows the equivalent circuit with transistors T_1 and T_3 or T_2 and T_4 are in the on-state. In this case, the input voltage source is short-circuited through inductor L, which yields

$$v_L = L\frac{di_s}{dt} = v_s(t) + V_0 > 0 \quad (11.32)$$

Equation (11.32) implies that the current value will depend on the sign of v_s.

The waveform of the input current i_s can be controlled by appropriately switching transistors T_1–T_4 or T_2–T_3, originating a similar shape to the one shown in Fig. 11.18a for the single-phase boost rectifier.

The control strategy for the rectifier is similar to the one illustrated in Fig. 11.31, for the voltage doubler topology. The quality of the input current obtained with this rectifier is the same as presented in Fig. 11.32 for the voltage doubler configuration.

The input current waveform can be slightly improved if the state of Fig. 11.33d is used. This can be done by replacing the hysteresis current control with a more complex linear control plus a three-level PWM modulator. This method reduces the semiconductor switching frequency and provides a more defined current spectrum.

Finally, it must be said that one of the most attractive characteristics of the fully controlled PWM converter in bridge connection and the voltage doubler is their regeneration capability. In effect, these rectifiers can deliver power from the load to the single-phase supply, operating with sinusoidal current and a high power factor of PF > 0.99. Figure 11.33e shows that during regeneration, the input current i_s is 180° out of phase with respect to the supply voltage v_s, which means operation with power factor PF ≈ −1 (PF is approximately 1 because of the small harmonic content in the input current).

11.3.6 Applications of Unity Power Factor Rectifiers

11.3.6.1 Boost Rectifier Applications

The single-phase boost rectifier has become the most popular topology for power factor correction (PFC) in general purpose power supplies. To reduce the costs, the complete control system shown in Fig. 11.19 and the gate drive circuit of the power transistor have been included in a single integrated circuit (IC), like the UC3854 [10] or MC33262, shown in Fig. 11.34.

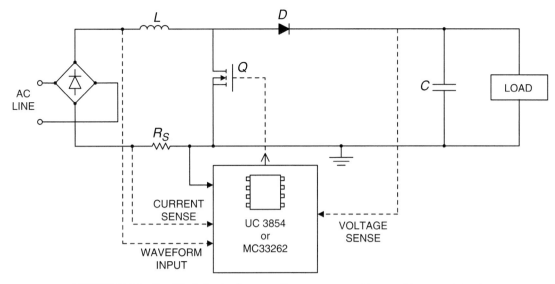

FIGURE 11.34 Simplified circuit of a power factor corrector with control integrated circuit.

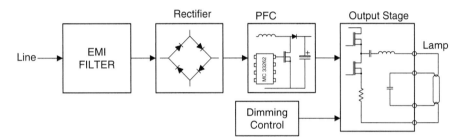

FIGURE 11.35 Functional block diagram of electronic ballast with power factor correction.

Today there is increased interest in developing high-frequency electronic ballasts to replace the classical electromagnetic ballast present in fluorescent lamps. These electronic ballasts require an ac–dc converter. To satisfy the harmonic current injection from electronic equipment and to maintain a high power quality, a high power factor rectifier can be used, as shown in Fig. 11.35 [11].

11.3.6.2 Voltage Doubler PWM Rectifier

The development of low-cost compact motor drive systems is a very relevant topic, particularly in the low-power range. Figure 11.36 shows a low-cost converter for low-power induction motor drives. In this configuration, a three-phase induction motor is fed through the converter from a single-phase power supply. Transistors T_1, T_2 and capacitors C_1, C_2

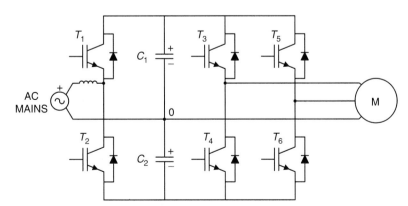

FIGURE 11.36 Low-cost induction motor drive.

constitute the voltage doubler single-phase rectifier, which controls the dc link voltage and generates sinusoidal input current, working with close-to-unity power factor [12]. On the other hand, transistors T_3, T_4, T_5, and T_6 and capacitors C_1 and C_2 constitute the power circuit of an asymmetric inverter that supplies the motor. An important characteristic of the power circuit shown in Fig. 11.36 is the capability of regenerating power to the single-phase mains.

11.3.6.3 PWM Rectifier in Bridge Connection

Distortion of the input current in the line-commutated rectifiers with capacitive filtering is particularly critical in the UPS fed from motor-generator sets. In effect, due to the higher value of the generator impedance, the current distortion can originate an unacceptable distortion on the ac voltage, which affects the behavior of the whole system. For this reason, it is very attractive to use rectifiers with low distortion in the input current.

Figure 11.37 shows the power circuit of a single-phase UPS, which has a PWM rectifier in bridge connection at the input side. This rectifier generates a sinusoidal input current and controls the charge of the battery [13].

Perhaps the most typical and widely accepted area of application of high power factor single-phase rectifiers is in locomotive drives [14]. An essential prerequisite for proper operation of voltage source three-phase inverter drives in modern locomotives is the use of four-quadrant line-side converters, which ensure motoring and braking of the drive, with reduced harmonics in the input current. Figure 11.38 shows a simplified power circuit of a typical drive for a locomotive connected to a single-phase power supply [14], which includes a high power factor rectifier at the input.

Finally, Fig. 11.39 shows the main circuit diagram of the 300 series Shinkansen train [15]. In this application, ac power from the overhead catenary is transmitted through a transformer to single-phase PWM rectifiers, which provide the dc voltage for the inverters. The rectifiers are capable of controlling the input ac current in an approximate sine waveform and in phase with the voltage, achieving power factor close to unity on powering and on regenerative braking. Regenerative braking produces energy savings and an important operational flexibility.

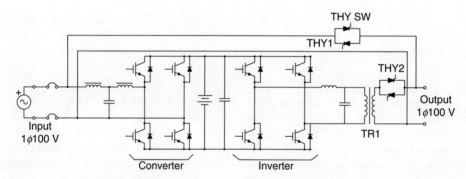

FIGURE 11.37 Single-phase UPS with PWM rectifier.

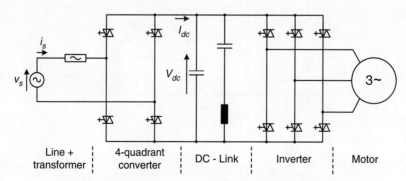

FIGURE 11.38 Typical power circuit of an ac drive for locomotive.

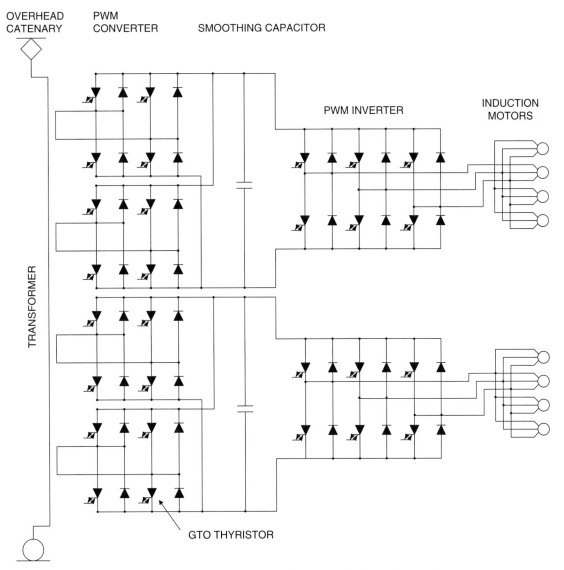

FIGURE 11.39 Main circuit diagram of 300 series Shinkansen locomotives.

Acknowledgment

The authors gratefully acknowledge the valuable contribution of Dr. Rubén Peña, and support provided by the Millennium Science Initiative (ICM) from Mideplan, Chile.

References

1. R. Dwyer and D. Mueller, "Selection of transformers for comercial buildings," in Proc. of IEEE/IAS 1992 Annual Meeting, U.S.A., Oct 1992, pp. 1335–1342.
2. D. C. Martins, F. J. M. de Seixas, J. A. Brilhante, and I. Barbi, "A family of dc-to-dc PWM converters using a new ZVS commutation cell," in Proc. IEEE PESC'93, 1993, pp. 524–530.
3. J. Bassett, "New, zero voltage switching, high frequency boost converter topology for power factor correction," in Proc. INTELEC'95, 1995, pp. 813–820.
4. R. Streit and D. Tollik, "High efficiency telecom rectifier using a novel soft-switched boost-based input current shaper," in Proc. INTELEC'91, 1991, pp. 720–726.
5. Y. Jang and M. M. Jovanovic´, "A new, soft-switched, high-power-factor boost converter with IGBTs," presented at the INTELEC'99, 1999, Paper 8-3.
6. M. M. Jovanovic´, "A technique for reducing rectifier reverse-recovery-related losses in high-voltage, high-power boost converters," in Proc. IEEE APEC'97, 1997, pp. 1000–1007.
7. D. M. Mitchell, "AC-DC converter having an improved power factor," U.S. Patent 4 412 277, Oct 25, 1983.

8. A. F. de Souza and I. Barbi, "A new ZVS-PWM unity power factor rectifier with reduced conduction losses," IEEE Trans. Power Electron, Vol. 10, No. 6, Nov 1995, pp. 746–752.
9. A. F. de Souza and I. Barbi, "A new ZVS semiresonant power factor rectifier with reduced conduction losses," IEEE Trans. Ind. Electron, Vol. 46, No. 1, Feb 1999, pp. 82–90.
10. P. Todd, "UC3854 controlled power factor correction circuit design," Application Note U-134, Unitrode Corp.
11. J. Adams, T. Ribarich, and J. Ribarich, "A new control IC for dimmable high-frequency electronic ballast," IEEE Applied Power Electronics Conference APEC'99, USA,1999, pp. 713–719.
12. C. Jacobina, M. Beltrao, E. Cabral, and A. Nogueira, "Induction motor drive system for low-power applications," IEEE Transactions on Industry Applications, Vol. 35, No. 1. Jan/Feb 1999, pp. 52–60.
13. K. Hirachi, H. Yamamoto, T. Matsui, S. Watanabe, and M. Nakaoka, "Cost-effective practical developments of high-performance 1kVA UPS with new system configurations and their specific control implementations," European Conference on Power Electronics EPE 95, Spain 1995, pp. 2035–2040.
14. K. Hückelheim and Ch. Mangold, "Novel 4-quadrant converter control method," European Conference on Power Electronics EPE 89, Germany 1989, pp. 573–576.
15. T. Ohmae and K. Nakamura, "Hitachi's role in the area of power electronics for transportation," Proc. of the IECON'93. Hawai, Nov 1993, pp. 714–718.

12
Three-phase Controlled Rectifiers

Juan W. Dixon, Ph.D.
Department of Electrical Engineering, Pontificia Universidad Católica de Chile Vicuña Mackenna 4860, Santiago, Chile

12.1 Introduction 201
12.2 Line-commutated Controlled Rectifiers 201
    12.2.1 Three-phase Half-wave Rectifier • 12.2.2 Six-pulse or Double Star Rectifier • 12.2.3 Double Star Rectifier with Interphase Connection • 12.2.4 Three-phase Full-wave Rectifier or Graetz Bridge • 12.2.5 Half Controlled Bridge Converter • 12.2.6 Commutation • 12.2.7 Power Factor • 12.2.8 Harmonic Distortion • 12.2.9 Special Configurations for Harmonic Reduction • 12.2.10 Applications of Line-commutated Rectifiers in Machine Drives • 12.2.11 Applications in HVDC Power Transmission • 12.2.12 Dual Converters • 12.2.13 Cycloconverters • 12.2.14 Harmonic Standards and Recommended Practices
12.3 Force-commutated Three-phase Controlled Rectifiers 221
    12.3.1 Basic Topologies and Characteristics • 12.3.2 Operation of the Voltage Source Rectifier • 12.3.3 PWM Phase-to-phase and Phase-to-neutral Voltages • 12.3.4 Control of the *DC* Link Voltage • 12.3.5 New Technologies and Applications of Force-commutated Rectifiers
    Further Reading 242

12.1 Introduction

Three-phase controlled rectifiers have a wide range of applications, from small rectifiers to large high voltage direct current (HVDC) transmission systems. They are used for electrochemical processes, many kinds of motor drives, traction equipment, controlled power supplies and many other applications. From the point of view of the commutation process, they can be classified into two important categories: *line-commutated controlled rectifiers (thyristor rectifiers)* and *force-commutated pulse width modulated (PWM) rectifiers*.

12.2 Line-commutated Controlled Rectifiers

12.2.1 Three-phase Half-wave Rectifier

Figure 12.1 shows the three-phase half-wave rectifier topology. To control the load voltage, the half-wave rectifier uses three common-cathode thyristor arrangement. In this figure, the power supply and the transformer are assumed ideal. The thyristor will conduct (*ON* state), when the anode-to-cathode voltage v_{AK} is positive and a firing current pulse i_G is applied to the gate terminal. Delaying the firing pulse by an angle α controls the load voltage. As shown in Fig. 12.2, the firing angle α is measured from the crossing point between the phase supply voltages. At that point, the anode-to-cathode thyristor voltage v_{AK} begins to be positive. Figure 12.3 shows that the possible range for gating delay is between $\alpha = 0°$ and $\alpha = 180°$, but because of commutation problems in actual situations, the maximum firing angle is limited to around 160°. As shown in Fig. 12.4, when the load is resistive, current i_d has the same waveform of the load voltage. As the load becomes more and more inductive, the current flattens and finally becomes constant. The thyristor goes to the non-conducting condition (*OFF* state) when the following thyristor is switched *ON*, or the current, tries to reach a negative value.

With the help of Fig. 12.2, the load average voltage can be evaluated, and is given by:

$$V_D = \frac{V_{MAX}}{\frac{2}{3}\pi} \int_{-\pi/3+\alpha}^{\pi/3+\alpha} \cos \omega t \cdot d(\omega t)$$

$$= V_{MAX} \frac{\sin \frac{\pi}{3}}{\frac{\pi}{3}} \cdot \cos \alpha \approx 1.17 \cdot V_{f-N}^{rms} \cdot \cos \alpha \quad (12.1)$$

where V_{MAX} is the secondary phase-to-neutral peak voltage, V_{f-N}^{rms} its root mean square (*rms*) value and ω is the angular frequency of the main power supply. It can be seen from Eq. (12.1) that the load average voltage V_D is modified by

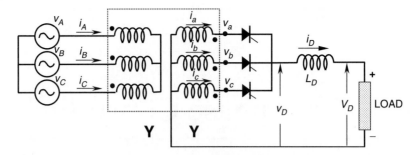

FIGURE 12.1 Three-phase half-wave rectifier.

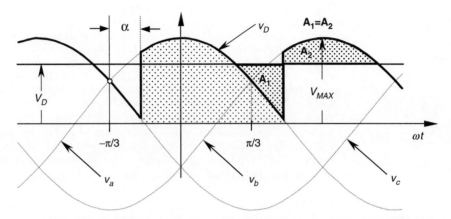

FIGURE 12.2 Instantaneous dc voltage v_D, average dc voltage V_D, and firing angle α.

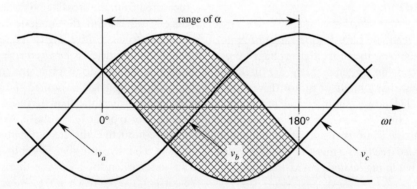

FIGURE 12.3 Possible range for gating delay in angle α.

changing firing angle α. When $\alpha < 90°$, V_D is positive and when $\alpha > 90°$, the average dc voltage becomes negative. In such a case, the rectifier begins to work as an inverter and the load needs to be able to generate power reversal by reversing its dc voltage.

The ac currents of the half-wave rectifier are shown in Fig. 12.5. This drawing assumes that the dc current is constant (very large L_D). Disregarding commutation overlap, each valve conducts during 120° per period. The secondary currents (and thyristor currents) present a dc component that is undesirable, and makes this rectifier not useful for high power applications.

The primary currents show the same waveform, but with the dc component removed. This very distorted waveform requires an input filter to reduce harmonics contamination.

The current waveforms shown in Fig. 12.5 are useful for designing the power transformer. Starting from:

$$VA_{prim} = 3 \cdot V^{rms}_{(prim)f-N} \cdot I^{rms}_{prim}$$
$$VA_{sec} = 3 \cdot V^{rms}_{(sec)f-N} \cdot I^{rms}_{sec} \quad (12.2)$$
$$P_D = V_D \cdot I_D$$

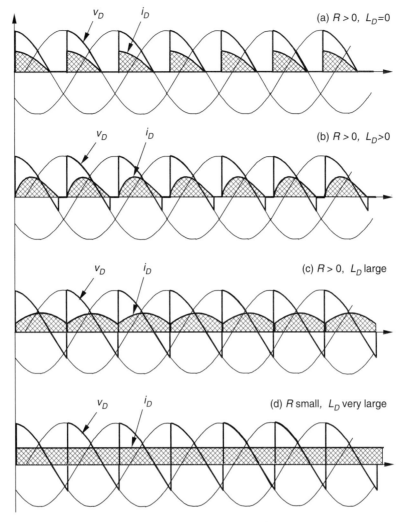

FIGURE 12.4 DC current waveforms.

where VA_{prim} and VA_{sec} are the ratings of the transformer for the primary and secondary side respectively. Here P_D is the power transferred to the dc side. The maximum power transfer is with $\alpha = 0°$ (or $\alpha = 180°$). Then, to establish a relation between ac and dc voltages, Eq. (12.1) for $\alpha = 0°$ is required:

$$V_D = 1.17 \cdot V^{rms}_{(sec)f-N} \tag{12.3}$$

and:

$$V_D = 1.17 \cdot a \cdot V^{rms}_{(prim)f-N} \tag{12.4}$$

where a is the secondary to primary turn relation of the transformer. On the other hand, a relation between the currents is also possible to obtain. With the help of Fig. 12.5:

$$I^{rms}_{sec} = \frac{I_D}{\sqrt{3}} \tag{12.5}$$

$$I^{rms}_{prim} = a \cdot \frac{I_D \sqrt{2}}{3} \tag{12.6}$$

Combining Eqs. (12.2) to (12.6), it yields:

$$\begin{aligned} VA_{prim} &= 1.21 \cdot P_D \\ VA_{sec} &= 1.48 \cdot P_D \end{aligned} \tag{12.7}$$

Equation (12.7) shows that the power transformer has to be oversized 21% at the primary side, and 48% at the secondary side. Then, a special transformer has to be built for this rectifier. In terms of average VA, the transformer needs to be 35% larger that the rating of the dc load. The larger rating of the secondary with respect to primary is because the secondary carries a dc component inside the windings. Furthermore, the transformer is oversized because the circulation of current harmonics does not generate active power. Core saturation, due to

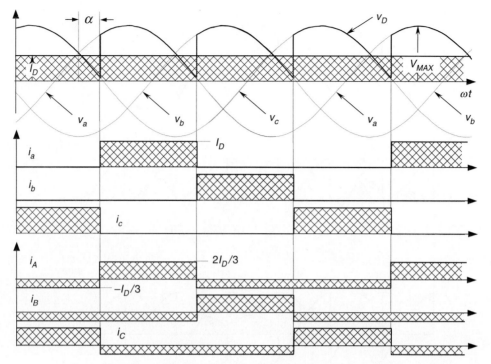

FIGURE 12.5 *AC* current waveforms for the half-wave rectifier.

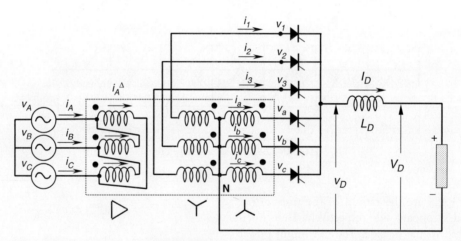

FIGURE 12.6 Six-pulse rectifier.

the *dc* components inside the secondary windings, also needs to be taken in account for iron oversizing.

12.2.2 Six-pulse or Double Star Rectifier

The thyristor side windings of the transformer shown in Fig. 12.6 form a six-phase system, resulting in a six-pulse starpoint (midpoint connection). Disregarding commutation overlap, each valve conducts only during 60° per period. The direct voltage is higher than that from the half-wave rectifier

and its average value is given by:

$$V_D = \frac{V_{MAX}}{\frac{\pi}{3}} \int_{-\pi/6+\alpha}^{\pi/6+\alpha} \cos \omega t \cdot d(\omega t)$$

$$= V_{MAX} \frac{\sin \frac{\pi}{6}}{\frac{\pi}{6}} \cdot \cos \alpha \approx 1.35 \cdot V_{f-N}^{rms} \cdot \cos \alpha \quad (12.8)$$

The *dc* voltage ripple is also smaller than the one generated by the half-wave rectifier, due to the absence of the third harmonic with its inherently high amplitude. The smoothing

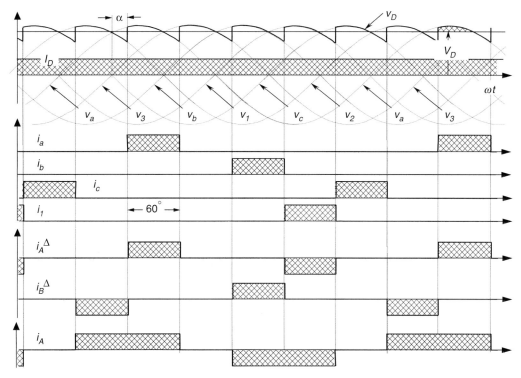

FIGURE 12.7 AC current waveforms for the six-pulse rectifier.

reactor L_D is also considerably smaller than the one needed for a three-pulse (half-wave) rectifier.

The ac currents of the six-pulse rectifier are shown in Fig. 12.7. The currents in the secondary windings present a dc component, but the magnetic flux is compensated by the double star. As can be observed, only one valve is fired at a time and then this connection in no way corresponds to a parallel connection. The currents inside the delta show a symmetrical waveform with 60° conduction. Finally, due to the particular transformer connection shown in Fig. 12.6, the source currents also show a symmetrical waveform, but with 120° conduction.

Evaluation of the the rating of the transformer is done in similar fashion to the way the half-wave rectifier is evaluated:

$$VA_{prim} = 1.28 \cdot P_D$$
$$VA_{sec} = 1.81 \cdot P_D \qquad (12.9)$$

Thus the transformer must be oversized 28% at the primary side and 81% at the secondary side. In terms of size it has an average apparent power of 1.55 times the power P_D (55% oversized). Because of the short conducting period of the valves, the transformer is not particularly well utilized.

12.2.3 Double Star Rectifier with Interphase Connection

This topology works as two half-wave rectifiers in parallel, and is very useful when high dc current is required. An optimal way to reach both good balance and elimination of harmonics is through the connection shown in Fig. 12.8. The two rectifiers are shifted by 180°, and their secondary neutrals are connected through a middle-point autotransformer called "interphase transformer." The interphase transformer is connected between the two secondary neutrals and the middle point at the load return. In this way, both groups operate in parallel. Half the direct current flows in each half of the interphase transformer, and then its iron core does not become saturated. The potential of each neutral can oscillate independently, generating an almost triangular voltage waveform (v_T) in the interphase transformer, as shown in Fig. 12.9. As this converter work like two half-wave rectifiers connected in parallel, the load average voltage is the same as in Eq. (12.1):

$$V_D \approx 1.17 \cdot V_{f-N}^{rms} \cdot \cos \alpha \qquad (12.10)$$

where V_{f-N}^{rms} is the phase-to-neutral rms voltage at the valve side of the transformer (secondary).

The Fig. 12.9 also shows the two half-wave rectifier voltages, related to their respective neutrals. Voltage v_{D1} represents the potential between the common cathode connection and the neutral N1. The voltage v_{D2} is between the common cathode connection and N2. It can be seen that the two instantaneous voltages are shifted, which gives as a result, a voltage v_D that is smoother than v_{D1} and v_{D2}.

Figure 12.10 shows how v_D, v_{D1}, v_{D2}, and v_T change when the firing angle changes from $\alpha = 0°$ to 180°.

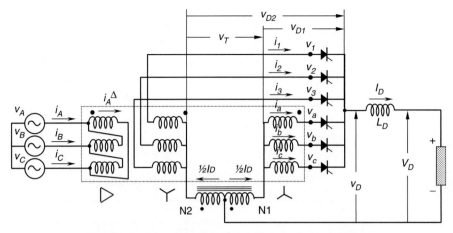

FIGURE 12.8 Double star rectifier with interphase transformer.

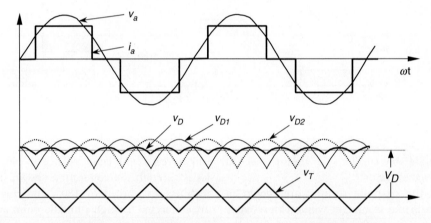

FIGURE 12.9 Operation of the interphase connection for $\alpha = 0°$.

The transformer rating in this case is:

$$VA_{prim} = 1.05 \cdot P_D$$
$$VA_{sec} = 1.48 \cdot P_D \qquad (12.11)$$

And the average rating power will be $1.26\,P_D$, which is better than the previous rectifiers (1.35 for the half-wave rectifier and 1.55 for the six-pulse rectifier). Thus the transformer is well utilized. Figure 12.11 shows ac current waveforms for a rectifier with interphase transformer.

12.2.4 Three-phase Full-wave Rectifier or Graetz Bridge

Parallel connection via interphase transformers permits the implementation of rectifiers for high current applications. Series connection for high voltage is also possible, as shown in the full-wave rectifier of Fig. 12.12. With this arrangement, it can be seen that the three common cathode valves generate a positive voltage with respect to the neutral, and the three common anode valves produce a negative voltage. The result is a dc voltage, twice the value of the half-wave rectifier. Each half of the bridge is a three-pulse converter group. This bridge connection is a two-way connection and alternating currents flow in the valve-side transformer windings during both half periods, avoiding dc components into the windings, and saturation in the transformer magnetic core. These characteristics make the so-called Graetz bridge the most widely used line-commutated thyristor rectifier. The configuration does not need any special transformer and works as a six-pulse rectifier. The series characteristic of this rectifier produces a dc voltage twice the value of the half-wave rectifier. The load average voltage is given by:

$$V_D = \frac{2 \cdot V_{MAX}}{\frac{2}{3}\pi} \int_{-\pi/3+\alpha}^{\pi/3+\alpha} \cos \omega t \cdot d(\omega t)$$

$$= 2 \cdot V_{MAX} \frac{\sin \frac{\pi}{3}}{\frac{\pi}{3}} \cdot \cos \alpha \approx 2.34 \cdot V_{f-N}^{rms} \cdot \cos \alpha \quad (12.12)$$

12 Three-phase Controlled Rectifiers

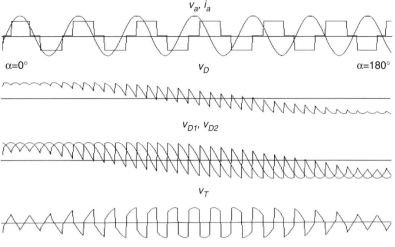

FIGURE 12.10 Firing angle variation from $\alpha = 0°$ to $180°$.

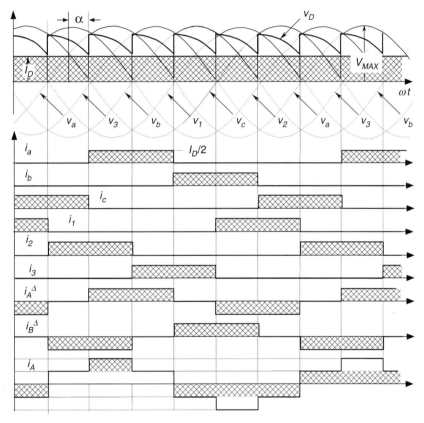

FIGURE 12.11 AC current waveforms for the rectifier with interphase transformer.

or

$$V_D = \frac{3 \cdot \sqrt{2} \cdot V_{f-f}^{sec}}{\pi} \cos\alpha \approx 1.35 \cdot V_{f-f}^{sec} \cdot \cos\alpha \quad (12.13)$$

where V_{MAX} is the peak phase-to-neutral voltage at the secondary transformer terminals, V_{f-N}^{rms} its *rms* value, and V_{f-f}^{sec} the *rms* phase-to-phase secondary voltage, at the valve terminals of the rectifier.

Figure 12.13 shows the voltages of each half-wave bridge of this topology, v_D^{pos} and v_D^{neg}, the total instantaneous *dc* voltage v_D, and the anode-to-cathode voltage v_{AK} in one of the bridge thyristors. The maximum value of v_{AK} is $\sqrt{3} \cdot V_{MAX}$,

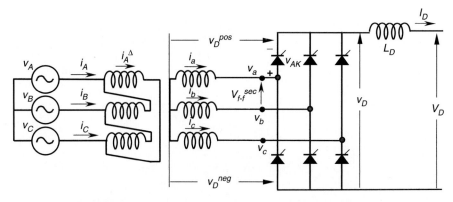

FIGURE 12.12 Three-phase full-wave rectifier or Graetz bridge.

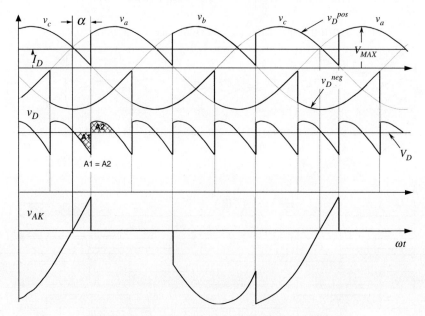

FIGURE 12.13 Voltage waveforms for the Graetz bridge.

which is the same as that of the half-wave converter and the interphase transformer rectifier. The double star rectifier presents a maximum anode-to-cathode voltage of two times V_{MAX}. Figure 12.14 shows the currents of the rectifier, which assumes that L_D is large enough to keep the *dc* current smooth. The example is for the same ΔY transformer connection shown in the topology of Fig. 12.12. It can be noted that the secondary currents do not carry any *dc* component, thereby avoiding overdesign of the windings and transformer saturation. These two figures have been drawn for a firing angle, α of approximately 30°. The perfect symmetry of the currents in all windings and lines is one of the reasons why this rectifier is the most popular of its type. The transformer rating in this case is

$$VA_{prim} = 1.05 \cdot P_D$$
$$VA_{sec} = 1.05 \cdot P_D \quad (12.14)$$

As can be noted, the transformer needs to be oversized only 5%, and both primary and secondary windings have the same rating. Again, this value can be compared with the previous rectifier transformers: $1.35 P_D$ for the half-wave rectifier, $1.55 P_D$ for the six-pulse rectifier, and $1.26 P_D$ for the interphase transformer rectifier. The Graetz bridge makes excellent use of the power transformer.

12.2.5 Half Controlled Bridge Converter

The fully controlled three-phase bridge converter shown in Fig. 12.12 has six thyristors. As already explained here, this circuit operates as a rectifier when each thyristor has a firing angle $\alpha < 90°$ and functions as an inverter for $\alpha > 90°$. If inverter operation is not required, the circuit may be simplified by replacing three controlled rectifiers with power diodes, as in Fig. 12.15a. This simplification is economically attractive

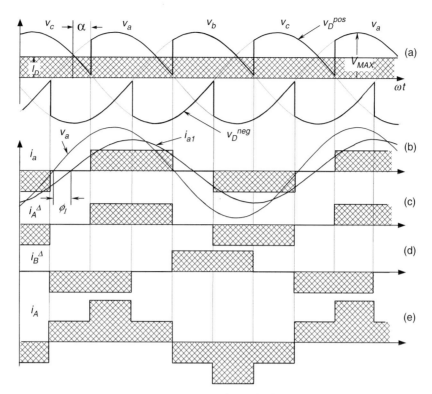

FIGURE 12.14 Current waveforms for the Graetz bridge.

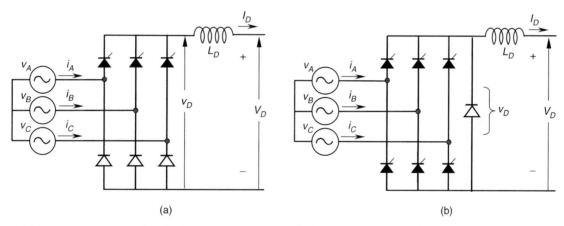

FIGURE 12.15 One-quadrant bridge converter circuits: (a) half-controlled bridge and (b) free-wheeling diode bridge.

because diodes are considerably less expensive than thyristors and they do not require firing angle control electronics.

The half controlled bridge, or "semiconverter," is analyzed by considering it as a phase-controlled half-wave circuit in series with an uncontrolled half-wave rectifier. The average dc voltage is given by the following equation

$$V_D = \frac{3 \cdot \sqrt{2} \cdot V_{f-f}^{sec}}{2\pi}(1 + \cos\alpha) \quad (12.15)$$

Then, the average voltage V_D never reaches negative values. The output voltage waveforms of half-controlled bridge are similar to those of a fully controlled bridge with a free-wheeling diode. The advantage of the free-wheeling diode connection, shown in Fig. 12.15b is that there is always a path for the dc current, independent of the status of the ac line and of the converter. This can be important if the load is inductive–resistive with a large time constant, and there is an interruption in one or more of the line phases. In such a case, the load current could commutate to the free-wheeling diode.

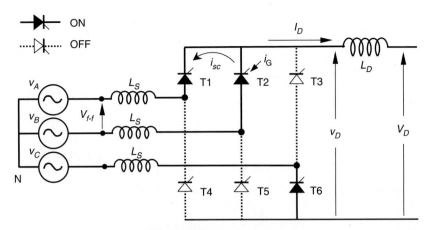

FIGURE 12.16 Commutation process.

12.2.6 Commutation

The description of the converters in the previous sections was based upon assumption that the commutation was instantaneous. In practice this is not possible, because the transfer of current between two consecutive valves in a commutation group takes a finite time. This time, called *overlap time*, depends on the phase-to-phase voltage between the valves participating in the commutation process, and the line inductance L_S between the converter and power supply. During the overlap time, two valves conduct, and the phase-to-phase voltage drops entirely on the inductances L_S. Assuming the *dc* current I_D to be smooth and with the help of Fig. 12.16, the following relation is deduced

$$2L_S \cdot \frac{di_{sc}}{dt} = \sqrt{2} \cdot V_{f-f} \sin \omega t = v_A - v_B \quad (12.16)$$

where i_{sc} is the current in the valve being fired during the commutation process (thyristor T2 in Fig. 12.16). This current can be evaluated, and it yields

$$i_{sc} = -\frac{\sqrt{2}}{2L_S} \cdot V_{f-f} \frac{\cos \omega t}{\omega} + C \quad (12.17)$$

The constant "C" is evaluated through initial conditions at the instant when T2 is ignited. In terms of angle, when $\omega t = \alpha$

$$\text{when } \omega t = \alpha, \quad i_{sc} = 0$$

$$\therefore C = \frac{V_{f-f}^{sec}}{\sqrt{2} \cdot \omega L_S} \cos \alpha \quad (12.18)$$

Replacing Eq. (12.18) in (12.17):

$$i_{sc} = \frac{V_{f-f}}{\sqrt{2} \cdot \omega L_S} \cdot (\cos \alpha - \cos \omega t) \quad (12.19)$$

Before commutation, the current I_D was carried by thyristor T1 (see Fig. 12.16). During the commutation time, the load current I_D remains constant, i_{sc} returns through T1, and T1 is automatically switched off when the current i_{sc} reaches the value of I_D. This happens because thyristors cannot conduct in reverse direction. At this moment, the overlap time lasts and the current I_D is then conducted by T2. In terms of angle, when $\omega t = \alpha + \mu$, $i_{sc} = I_D$, where μ is defined as the "*overlap angle*." Replacing this final condition in Eq. (12.19) yields

$$I_D = \frac{V_{f-f}^{sec}}{\sqrt{2} \cdot \omega L_S} \cdot [\cos \alpha - \cos (\alpha + \mu)] \quad (12.20)$$

To avoid confusion in a real analysis, it has to be remembered that V_{f-f} corresponds to the secondary voltage in case of transformer utilization. For this reason, the abbreviation "sec" has been added to the phase-to-phase voltage in Eq. (12.20).

During commutation, two valves conduct at a time, which means that there is an instantaneous short circuit between the two voltages participating in the process. As the inductances of each phase are the same, the current i_{sc} produces the same voltage drop in each L_S, but with opposite sign because this current flows in reverse direction and with opposite slope in each inductance. The phase with the higher instantaneous voltage suffers a voltage drop $-\Delta v$ and the phase with the lower voltage suffers a voltage increase $+\Delta v$. This situation affects the *dc* voltage V_C, reducing its value an amount ΔV_{med}. Figure 12.17 shows the meanings of Δv, ΔV_{med}, μ, and i_{sc}.

The area ΔV_{med} showed in Fig. 12.17, represents the loss of voltage that affects the average voltage V_C, and can be evaluated through the integration of Δv during the overlap angle μ. The voltage drop Δv can be expressed as

$$\Delta v = \left(\frac{v_A - v_B}{2}\right) = \frac{\sqrt{2} \cdot V_{f-f}^{sec} \sin \omega t}{2} \quad (12.21)$$

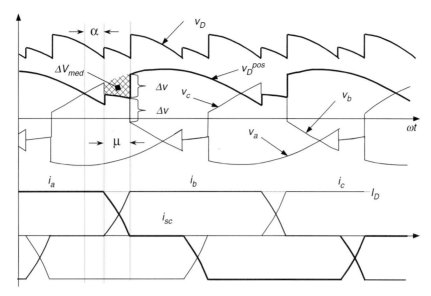

FIGURE 12.17 Effect of the overlap angle on the voltages and currents.

Integrating Eq. (12.21) into the corresponding period (60°) and interval (μ) and starting at the instant when the commutation begins (α)

$$\Delta V_{med} = \frac{3}{\pi} \cdot \frac{1}{2} \int_{\alpha}^{\alpha+\mu} \sqrt{2} \cdot V_{f-f}^{sec} \sin \omega t \cdot d\omega t \quad (12.22)$$

$$\Delta V_{med} = \frac{3 \cdot V_{f-f}^{sec}}{\pi \cdot \sqrt{2}} [\cos \alpha - \cos(\alpha + \mu)] \quad (12.23)$$

Subtracting ΔV_{med} in Eq. (12.13)

$$V_D = \frac{3 \cdot \sqrt{2} \cdot V_{f-f}^{sec}}{\pi} \cos \alpha - \Delta V_{med} \quad (12.24)$$

$$V_D = \frac{3 \cdot \sqrt{2} \cdot V_{f-f}^{sec}}{2\pi} [\cos \alpha + \cos(\alpha + \mu)] \quad (12.25)$$

or

$$V_D = \frac{3 \cdot \sqrt{2} \cdot V_{f-f}^{sec}}{\pi} \left[\cos\left(\alpha + \frac{\mu}{2}\right) \cos \frac{\mu}{2} \right] \quad (12.26)$$

Equations (12.20) and (12.25) can be written as a function of the primary winding of the transformer, if there is any transformer.

$$I_D = \frac{a \cdot V_{f-f}^{prim}}{\sqrt{2} \cdot \omega L_S} \cdot [\cos \alpha - \cos(\alpha + \mu)] \quad (12.27)$$

$$V_D = \frac{3 \cdot \sqrt{2} \cdot a \cdot V_{f-f}^{prim}}{2\pi} [\cos \alpha + \cos(\alpha + \mu)] \quad (12.28)$$

where $a = V_{f-f}^{sec}/V_{f-f}^{prim}$. With Eqs. (12.27) and (12.28) one gets

$$V_D = \frac{3 \cdot \sqrt{2}}{\pi} \cdot a \cdot V_{f-f}^{prim} \cos \alpha - \frac{3 I_D \omega L_S}{\pi} \quad (12.29)$$

Equation (12.29) allows a very simple equivalent circuit of the converter to be made, as shown in Fig. 12.18. It is important to note that the equivalent resistance of this circuit is not real, because it does not dissipate power.

From the equivalent circuit, regulation curves for the rectifier under different firing angles are shown in Fig. 12.19. It should be noted that these curves correspond only to an ideal situation, but helps in understanding the effect of voltage drop Δv on dc voltage. The commutation process and the overlap angle also affects the voltage v_a and anode-to-cathode thyristor voltage, as shown in Fig. 12.20.

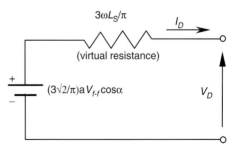

FIGURE 12.18 Equivalent circuit for the converter.

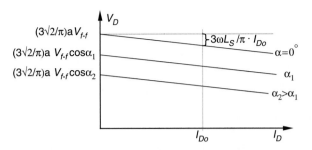

FIGURE 12.19 DC voltage regulation curves for rectifier operation.

12.2.7 Power Factor

The displacement factor of the fundamental current, obtained from Fig. 12.14 is

$$\cos\phi_1 = \cos\alpha \tag{12.30}$$

In the case of non-sinusoidal current, the active power delivered per phase by the sinusoidal supply is

$$P = \frac{1}{T}\int_0^T v_a(t)i_a(t)dt = V_a^{rms} I_{a1}^{rms} \cos\phi_1 \tag{12.31}$$

where V_a^{rms} is the *rms* value of the voltage v_a and I_{a1}^{rms} the *rms* value of i_{a1} (fundamental component of i_a). Analog relations can be obtained for v_b and v_c.

The apparent power per phase is given by

$$S = V_a^{rms} I_a^{rms} \tag{12.32}$$

The power factor is defined by

$$PF = \frac{P}{S} \tag{12.33}$$

By substituting Eqs. (12.30), (12.31), and (12.32) into Eq. (12.33), the power factor can be expressed as follows

$$PF = \frac{I_{a1}^{rms}}{I_a^{rms}} \cos\alpha \tag{12.34}$$

This equation shows clearly that due to the non-sinusoidal waveform of the currents, the power factor of the rectifier is negatively affected by both the firing angle α and the distortion of the input current. In effect, an increase in the distortion of the current produces an increase in the value of I_a^{rms} in Eq. (12.34), which deteriorates the power factor.

12.2.8 Harmonic Distortion

The currents of the line-commutated rectifiers are far from being sinusoidal. For example, the currents generated from the Graetz rectifier (see Fig. 12.14b) have the following harmonic content

$$i_A = \frac{2\sqrt{3}}{\pi} I_D (\cos\omega t - \frac{1}{5}\cos 5\omega t + \frac{1}{7}\cos 7\omega t$$
$$- \frac{1}{11}\cos 11\omega t + \ldots) \tag{12.35}$$

Some of the characteristics of the currents, obtained from Eq. (12.35) include: (i) the absence of triple harmonics; (ii) the presence of harmonics of order $6k \pm 1$ for integer values of k; (iii) those harmonics of orders $6k+1$ are of positive sequence; (iv) those of orders $6k-1$ are of negative sequence; (v) the *rms* magnitude of the fundamental frequency is

$$I_1 = \frac{\sqrt{6}}{\pi} I_D \tag{12.36}$$

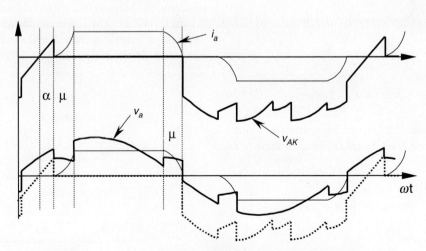

FIGURE 12.20 Effect of the overlap angle on v_a and on thyristor voltage v_{AK}.

and (vi) the *rms* magnitude of the nth harmonic is

$$I_n = \frac{I_1}{n} \quad (12.37)$$

If either, the primary or the secondary three-phase windings of the rectifier transformer are connected in delta, the *ac* side current waveforms consist of the instantaneous differences between two rectangular secondary currents 120° apart as shown in Fig. 12.14e. The resulting Fourier series for the current in phase "a" on the primary side is

$$i_A = \frac{2\sqrt{3}}{\pi} I_D (\cos \omega t + \frac{1}{5} \cos 5\omega t - \frac{1}{7} \cos 7\omega t$$
$$- \frac{1}{11} \cos 11\omega t + \ldots) \quad (12.38)$$

This series differs from that of a star connected transformer only by the sequence of rotation of harmonic orders $6k \pm 1$ for odd values of k, i.e. the 5th, 7th, 17th, 19th, etc.

12.2.9 Special Configurations for Harmonic Reduction

A common solution for harmonic reduction is through the connection of passive filters, which are tuned to trap a particular harmonic frequency. A typical configuration is shown in Fig. 12.21.

However, harmonics also can be eliminated using special configurations of converters. For example, 12-pulse configuration consists of two sets of converters connected as shown

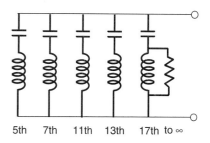

FIGURE 12.21 Typical passive filter for one phase.

in Fig. 12.22. The resultant *ac* current is given by the sum of the two Fourier series of the star connection (Eq. (12.35)) and delta connection transformers (Eq. (12.38))

$$i_A = 2\left(\frac{2\sqrt{3}}{\pi}\right) I_D (\cos \omega t - \frac{1}{11} \cos 11\omega t + \frac{1}{13} \cos 13\omega t$$
$$- \frac{1}{23} \cos 23\omega t + \ldots) \quad (12.39)$$

The series only contains harmonics of order $12k \pm 1$. The harmonic currents of orders $6k \pm 1$ (with k odd), i.e. 5th, 7th, 17th, 19th, etc. circulate between the two converter transformers but do not penetrate the *ac* network.

The resulting line current for the 12-pulse rectifier, shown in Fig. 12.23, is closer to a sinusoidal waveform than previous line currents. The instantaneous *dc* voltage is also smoother with this connection.

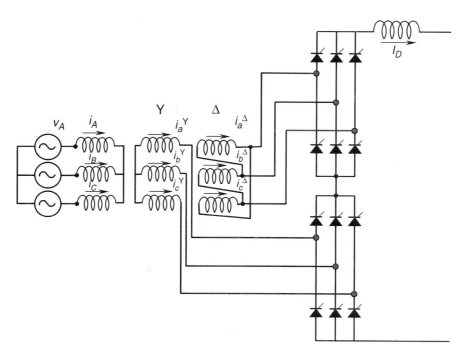

FIGURE 12.22 12-pulse rectifier configuration.

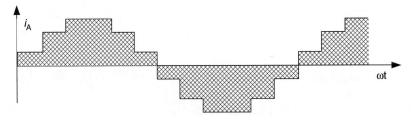

FIGURE 12.23 Line current for the 12-pulse rectifier.

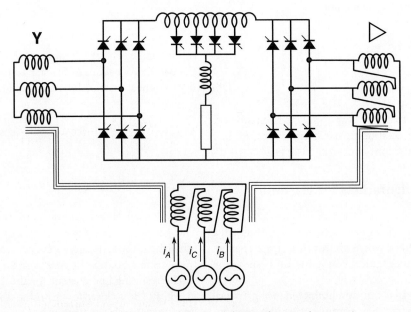

FIGURE 12.24 *DC* ripple reinjection technique for 48-pulse operation.

Higher pulse configuration using the same principle is also possible. The 12-pulse rectifier was obtained with a 30° phase shift between the two secondary transformers. The addition of further, appropriately shifted, transformers in parallel provides the basis for increasing pulse configurations. For instance, 24-pulse operation is achieved by means of four transformers with 15° phase shift, and 48-pulse operation requires eight transformers with 7.5° phase shift (transformer connections in zig-zag configuration).

Although theoretically possible, pulse numbers above 48 are rarely justified due to the practical levels of distortion found in the supply voltage waveforms. Further, the converter topology becomes more and more complicated.

An ingenious and very simple way to reach high pulse operation is shown in Fig. 12.24. This configuration is called *dc ripple reinjection*. It consists of two parallel converters connected to the load through a multistep reactor. The reactor uses a chain of thyristor-controlled taps, which are connected to symmetrical points of the reactor. By firing the thyristors located at the reactor at the right time, high-pulse operation is reached. The level of pulse operation depends on the number of thyristors connected to the reactor. They multiply the basic level of operation of the two converters. The example, is Fig. 12.24, shows a 48-pulse configuration, obtained by the multiplication of basic 12-pulse operation by four reactor thyristors. This technique also can be applied to series connected bridges.

Another solution for harmonic reduction is the utilization of active power filters. Active power filters are special pulse width modulated (PWM) converters, able to generate the harmonics the converter requires. Figure 12.25 shows a current controlled shunt active power filter.

12.2.10 Applications of Line-commutated Rectifiers in Machine Drives

Important applications for line-commutated three-phase controlled rectifiers, are found in machine drives. Figure 12.26 shows a *dc* machine control implemented with a six-pulse rectifier. Torque and speed are controlled through the armature current I_D, and excitation current I_{exc}. Current I_D is adjusted with V_D, which is controlled by the firing angle α through Eq. (12.12). This *dc* drive can operate in two quadrants: positive and negative *dc* voltage. This two-quadrant

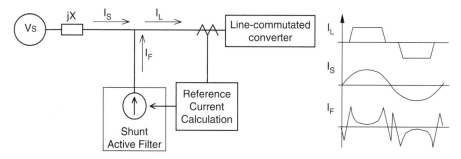

FIGURE 12.25 Current controlled shunt active power filter.

operation allows regenerative braking when $\alpha > 90°$ and $I_{exc} < 0$.

The converter of Fig. 12.26 can also be used to control synchronous machines, as shown in Fig. 12.27. In this case, a second converter working in the inverting mode operates the machine as self-controlled synchronous motor. With this second converter, the synchronous motor behaves like a *dc* motor but has none of the disadvantages of mechanical commutation. This converter is not line commutated, but machine commutated.

The nominal synchronous speed of the motor on a 50 or 60 Hz *ac* supply is now meaningless and the upper speed limit is determined by the mechanical limitations of the rotor construction. There is disadvantage that the rotational *emfs* required for load commutation of the machine side converter are not available at standstill and low speeds. In such a case, auxiliary force commutated circuits must be used.

The line-commutated rectifier controls the torque of the machine through firing angle α. This approach gives direct torque control of the commutatorless motor and is analogous to the use of armature current control shown in Fig. 12.26 for the converter-fed *dc* motor drive.

Line-commutated rectifiers are also used for speed control of wound rotor induction motors. Subsynchronous and supersynchronous static converter cascades using a naturally commutated *dc* link converter, can be implemented. Figure 12.28 shows a supersynchronous cascade for a wound rotor induction motor, using a naturally commutated *dc* link converter.

In the supersynchronous cascade shown in Fig. 12.28, the right hand bridge operates at slip frequency as a rectifier or inverter, while the other operates at network frequency as an inverter or rectifier. Control is difficult near synchronism when slip frequency *emfs* are insufficient for natural commutation and special circuit configuration employing forced commutation or devices with a self-turn-off capability are necessary for the passage through synchronism. This kind of supersynchronous cascade works better with cycloconverters.

12.2.11 Applications in HVDC Power Transmission

High voltage direct current (HVDC) power transmission is the most powerful application for line-commutated converters that exist today. There are power converters with ratings in excess of 1000 MW. Series operation of hundreds of valves can be found in some HVDC systems. In high power and long distance applications, these systems become more economical than conventional *ac* systems. They also have some other advantages compared with *ac* systems:

1. they can link two *ac* systems operating unsynchronized or with different nominal frequencies, that is 50 ↔ 60 Hz;

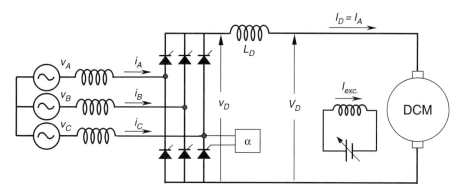

FIGURE 12.26 *DC* machine drive with a six-pulse rectifier.

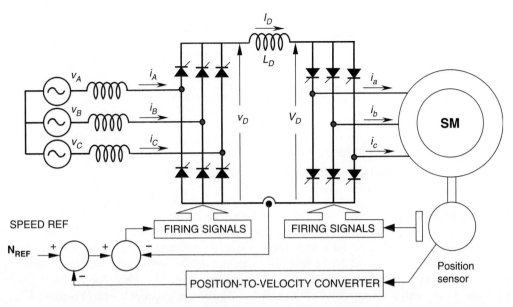

FIGURE 12.27 Self-controlled synchronous motor drive.

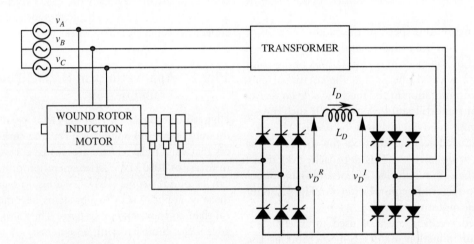

FIGURE 12.28 Supersynchronous cascade for a wound rotor induction motor.

2. they can help in stability problems related with subsynchronous resonance in long *ac* lines;
3. they have very good dynamic behavior and can interrupt short-circuit problems very quickly;
4. if transmission is by submarine or underground cable, it is not practical to consider *ac* cable systems exceeding 50 km, but *dc* cable transmission systems are in service whose length is in hundreds of kilometers and even distances of 600 km or greater have been considered feasible;
5. reversal of power can be controlled electronically by means of the delay firing angles α; and
6. some existing overhead *ac* transmission lines cannot be increased. If overbuilt with or upgraded to *dc* transmission can substantially increase the power transfer capability on the existing right-of-way.

The use of HVDC systems for interconnections of asynchronous systems is an interesting application. Some continental electric power systems consist of asynchronous networks such as those for the East, West, Texas, and Quebec networks in North America, and islands loads such as that for the Island of Gotland in the Baltic Sea make good use of the HVDC interconnections.

Nearly all HVDC power converters with thyristor valves are assembled in a converter bridge of 12-pulse configuration, as shown in Fig. 12.29. Consequently, the *ac* voltages applied to each six-pulse valve group which makes up the 12-pulse

12 Three-phase Controlled Rectifiers

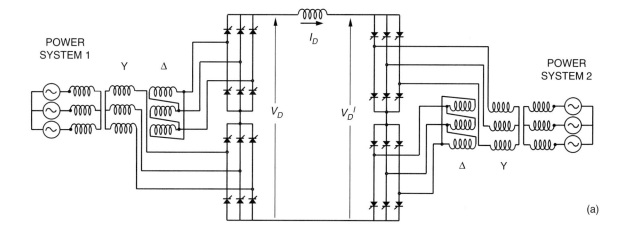

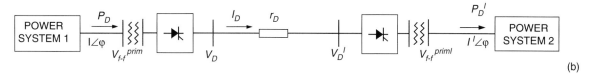

FIGURE 12.29 Typical HVDC power system: (a) detailed circuit and (b) unilinear diagram.

valve group have a phase difference of 30° which is utilized to cancel the ac side, 5th and 7th harmonic currents and dc side, 6th harmonic voltage, thus resulting in a significant saving in harmonic filters.

Some useful relations for HVDC systems include:

a) Rectifier Side:

$$P_D = V_D \cdot I_D = \sqrt{3} \cdot V_{f-f}^{prim} \cdot I_{line}^{rms} \cos\varphi \quad (12.40)$$

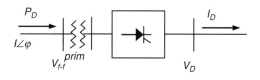

$$I_P = I \cos\varphi$$

$$I_Q = I \sin\varphi$$

$$\therefore P_D = V_D \cdot I_D = \sqrt{3} \cdot V_{f-f}^{prim} \cdot I_P \quad (12.41)$$

$$I_P = \frac{V_D \cdot I_D}{\sqrt{3} \cdot V_{f-f}^{prim}} \quad (12.42)$$

$$I_P = \frac{a^2\sqrt{3} \cdot V_{f-f}^{prim}}{4\pi \cdot \omega L_S} [\cos 2\alpha - \cos 2(\alpha + \mu)] \quad (12.43)$$

$$I_Q = \frac{a^2\sqrt{3} \cdot V_{f-f}^{prim}}{4\pi \cdot \omega L_S} [\sin 2(\alpha + \mu) - \sin 2\alpha - 2\mu] \quad (12.44)$$

$$I_P = I_D \frac{a\sqrt{6}}{\pi} \left[\frac{\cos\alpha + \cos(\alpha + \mu)}{2} \right] \quad (12.45)$$

Fundamental secondary component of I

$$I = \frac{a\sqrt{6}}{\pi} I_D \quad (12.46)$$

Replacing Eq. (12.46) in (12.45)

$$I_P = I \cdot \left[\frac{\cos\alpha + \cos(\alpha + \mu)}{2} \right] \quad (12.47)$$

as $I_P = I \cos\varphi$, it yields Fig. 12.30a

$$\cos\varphi = \left[\frac{\cos\alpha + \cos(\alpha + \mu)}{2} \right] \quad (12.48)$$

b) Inverter Side:

The same equations are applied for the inverter side, but the firing angle α is replaced by γ, where γ is

$$\gamma = 180° - (\alpha + \mu) \quad (12.49)$$

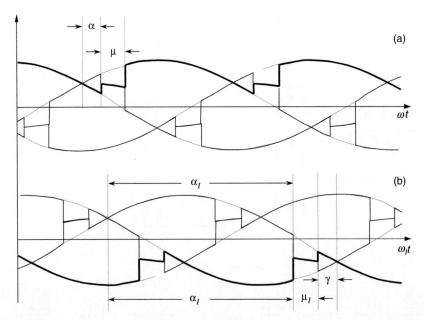

FIGURE 12.30 Definition of angle γ for inverter side: (a) rectifier side and (b) inverter side.

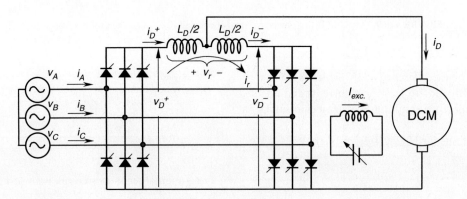

FIGURE 12.31 Dual converter in a four-quadrant dc drive.

As reactive power always goes in the converter direction, Eq. (12.44) for inverter side becomes (Fig. 12.30b)

$$I_{QI} = -\frac{a_I^2\sqrt{3} \cdot V_{f-f_I}^{prim}}{4\pi \cdot \omega_I L_I}[\sin 2(\gamma + \mu_I) - \sin 2\gamma - 2\mu_I] \quad (12.50)$$

12.2.12 Dual Converters

In many variable-speed drives, four-quadrant operation is required, and three-phase dual converters are extensively used in applications up to 2 MW level. Figure 12.31 shows a three-phase dual converter, where two converters are connected back-to-back.

In the dual converter, one rectifier provides the positive current to the load and the other the negative current. Due to the instantaneous voltage differences between the output voltages of the converters, a circulating current flows through the bridges. The circulating current is normally limited by circulating reactor, L_D, as shown in Fig. 12.31. The two converters are controlled in such a way that if α^+ is the delay angle of the positive current converter, the delay angle of the negative current converter is $\alpha^- = 180° - \alpha^+$.

Figure 12.32 shows the instantaneous dc voltages of each converter, v_D^+ and v_D^-. Despite the average voltage V_D is the same in both the converters, their instantaneous voltage differences shown as voltage v_r, are producing the circulating current i_r, which is superimposed with the load currents i_D^+ and i_D^-.

To avoid the circulating current i_r, it is possible to implement a "circulating current free" converter if a dead time of a few milliseconds is acceptable. The converter section, not required to supply current, remains fully blocked. When a current reversal is required, a logic switch-over system determines

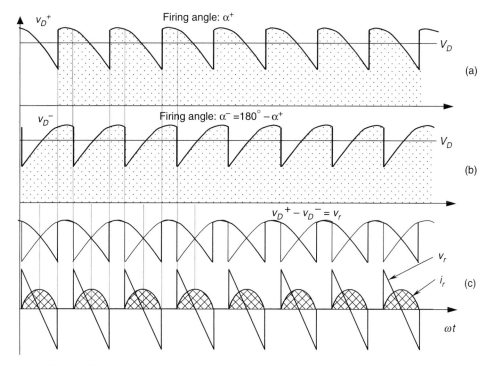

FIGURE 12.32 Waveform of circulating current: (a) instantaneous *dc* voltage from positive converter: (b) instantaneous *dc* voltage from negative converter; and (c) voltage difference between v_D^+ and v_D^-, v_r, and circulating current i_r.

at first the instant at which the conducting converter's current becomes zero. This converter section is then blocked and the further supply of gating pulses to it is prevented. After a short safety interval (dead time), the gating pulses for the other converter section are released.

12.2.13 Cycloconverters

A different principle of frequency conversion is derived from the fact that a dual converter is able to supply an *ac* load with a lower frequency than the system frequency. If the control signal of the dual converter is a function of time, the output voltage will follow this signal. If this control signal value alters sinusoidally with the desired frequency, then the waveform depicted in Fig. 12.33a consists of a single-phase voltage with a large harmonic current. As shown in Fig. 12.33b, if the load is inductive, the current will present less distortion than voltage.

The cycloconverter operates in all four quadrants during a period. A pause (dead time) at least as small as the time required by the switch-over logic occurs after the current reaches zero, that is, between the transfer to operation in the quadrant corresponding to the other direction of current flow.

Three single-phase cycloconverters may be combined to build a three-phase cycloconverter. The three-phase cycloconverters find an application in low-frequency, high-power requirements. Control speed of large synchronous motors in the low-speed range is one of the most common applications of three-phase cycloconverters. Figure 12.34 is a diagram for this application. They are also used to control slip frequency in wound rotor induction machines, for supersynchronous cascade (Scherbius system).

12.2.14 Harmonic Standards and Recommended Practices

In view of the proliferation of the power converter equipment connected to the utility system, various national and international agencies have been considering limits on harmonic current injection to maintain good power quality. As a consequence, various standards and guidelines have been established that specify limits on the magnitudes of harmonic currents and harmonic voltages.

The Comité Européen de Normalisation Electrotechnique (CENELEC), International Electrical Commission (IEC), and West German Standards (VDE) specify the limits on the voltages (as a percentage of the nominal voltage) at various harmonics frequencies of the utility frequency, when the equipment-generated harmonic currents are injected into a network whose impedances are specified.

According with Institute of Electrical and Electronic Engineers-519 standards (IEEE), Table 12.1 lists the limits on the harmonic currents that a user of power electronics equipment and other non-linear loads is allowed to inject into the

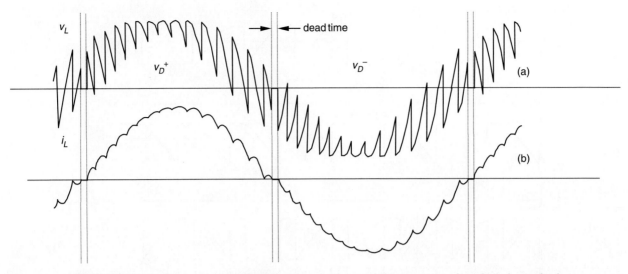

FIGURE 12.33 Cycloconverter operation: (a) voltage waveform and (b) current waveform for inductive load.

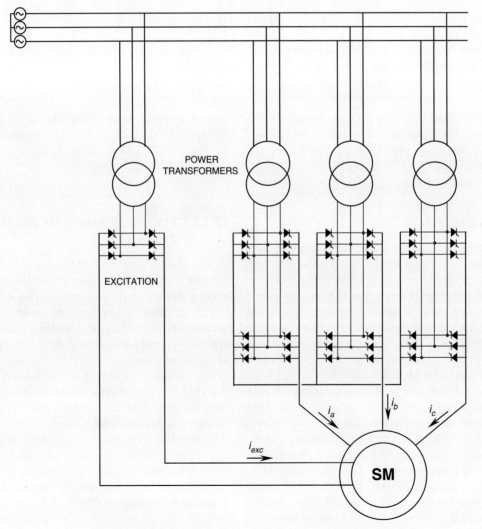

FIGURE 12.34 Synchronous machine drive with a cycloconverter.

# 12 Three-phase Controlled Rectifiers

TABLE 12.1 Harmonic current limits in percent of fundamental

Short circuit current (pu)	$h < 11$	$11 < h < 17$	$17 < h < 23$	$23 < h < 35$	$35 < h$	THD
<20	4.0	2.0	1.5	0.6	0.3	5.0
20–50	7.0	3.5	2.5	1.0	0.5	8.0
50–100	10.0	4.5	4.0	1.5	0.7	12.0
100–1000	12.0	5.5	5.0	2.0	1.0	15.0
>1000	15.0	7.0	6.0	2.5	1.4	20.0

TABLE 12.2 Harmonic voltage limits in percent of fundamental

Voltage level	2.3–69 kV	69–138 kV	>138 kV
Maximum for individual harmonic	3.0	1.5	1.0
Total harmonic distortion (THD)	5.0	2.5	1.5

utility system. Table 12.2 lists the quality of voltage that the utility can furnish the user.

In Table 12.1, the values are given at the point of connection of non-linear loads. The THD is the total harmonic distortion given by Eq. (12.51) and h is the number of the harmonic.

$$THD = \frac{\sqrt{\sum_{h=2}^{\infty} I_h^2}}{I_1} \qquad (12.51)$$

The total current harmonic distortion allowed in Table 12.1 increases with the value of short circuit current.

The total harmonic distortion in the voltage can be calculated in a manner similar to that given by Eq. (12.51). Table 12.2 specifies the individual harmonics and the THD limits on the voltage that the utility supplies to the user at the connection point.

12.3 Force-commutated Three-phase Controlled Rectifiers

12.3.1 Basic Topologies and Characteristics

Force-commutated rectifiers are built with semiconductors with gate-turn-off capability. The gate-turn-off capability allows full control of the converter, because valves can be switched ON and OFF whenever is required. This allows the commutation of the valves, hundreds of times in one period which is not possible with line-commutated rectifiers, where thyristors are switched ON and OFF only once a cycle. This feature has the following advantages: (a) the current or voltage can be modulated (PWM), generating less harmonic contamination; (b) power factor can be controlled and even it can be made leading; and (c) they can be built as voltage source or current source rectifiers; (d) the reversal of power in thyristor rectifiers is by reversal of voltage at the dc link. Instead, force-commutated rectifiers can be implemented for both, reversal of voltage or reversal of current.

There are two ways to implement force-commutated three-phase rectifiers: (a) as a current source rectifier, where power reversal is by dc voltage reversal; and (b) as a voltage source rectifier, where power reversal is by current reversal at the dc link. Figure 12.35 shows the basic circuits for these two topologies.

12.3.2 Operation of the Voltage Source Rectifier

The voltage source rectifier is by far the most widely used, and because of the duality of the two topologies showed in Fig. 12.35, only this type of force-commutated rectifier will be explained in detail.

The voltage source rectifier operates by keeping the dc link voltage at a desired reference value, using a feedback control loop as shown in Fig. 12.36. To accomplish this task, the dc link voltage is measured and compared with a reference V_{REF}. The error signal generated from this comparison is used to switch the six valves of the rectifier ON and OFF. In this way, power can come or return to the ac source according with the dc link voltage requirements. The voltage V_D is measured at the capacitor C_D.

When the current I_D is positive (rectifier operation), the capacitor C_D is discharged, and the error signal ask the control block for more power from the ac supply. The control block takes the power from the supply by generating the appropriate PWM signals for the six valves. In this way, more current flows from the ac to the dc side and the capacitor voltage is recovered. Inversely, when I_D becomes negative (inverter operation), the capacitor C_D is overcharged and the error signal ask the control to discharge the capacitor and return power to the ac mains.

The PWM control can manage not only the active power, but also the reactive power, allowing this type of rectifier to correct power factor. In addition, the ac current waveforms can be maintained as almost sinusoidal, which reduces harmonic contamination to the mains supply.

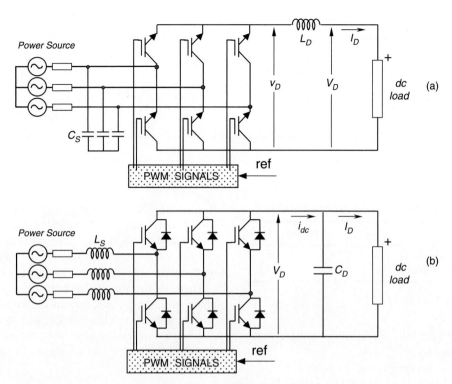

FIGURE 12.35 Basic topologies for force-commutated PWM rectifiers: (a) current source rectifier and (b) voltage source rectifier.

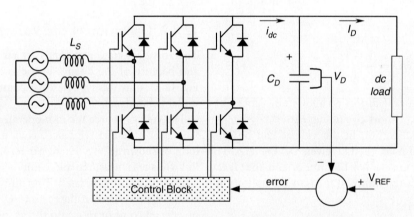

FIGURE 12.36 Operation principle of the voltage source rectifier.

The PWM consists of switching the valves ON and OFF, following a pre-established template. This template could be a sinusoidal waveform of voltage or current. For example, the modulation of one phase could be as the one shown in Fig. 12.37. This PWM pattern is a periodical waveform whose fundamental is a voltage with the same frequency of the template. The amplitude of this fundamental, called V_{MOD} in Fig. 12.37, is also proportional to the amplitude of the template.

To make the rectifier work properly, the PWM pattern must generate a fundamental V_{MOD} with the same frequency as the power source. Changing the amplitude of this fundamental and its phase shift with respect to the mains, the rectifier can be controlled to operate in the four quadrants: leading power factor rectifier, lagging power factor rectifier, leading power factor inverter, and lagging power factor inverter. Changing the pattern of modulation, as shown in Fig. 12.38, modifies the magnitude of V_{MOD}. Displacing the PWM pattern changes the phase shift.

The interaction between V_{MOD} and V (source voltage) can be seen through a phasor diagram. This interaction permits understanding of the four-quadrant capability of this rectifier. In Fig. 12.39, the following operations are displayed: (a) rectifier at unity power factor; (b) inverter at unity power factor; (c) capacitor (zero power factor); and (d) inductor (zero power factor).

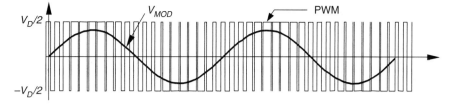

FIGURE 12.37 A PWM pattern and its fundamental V_{MOD}.

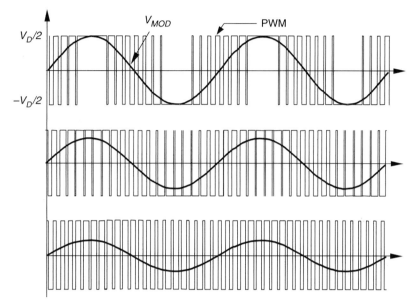

FIGURE 12.38 Changing V_{MOD} through the PWM pattern.

In Fig. 12.39, I_S is the *rms* value of the source current i_s. This current flows through the semiconductors in the way shown in Fig. 12.40. During the positive half cycle, the transistor T_N, connected at the negative side of the *dc* link is switched ON, and the current i_s begins to flow through T_N (i_{Tn}). The current returns to the mains and comes back to the valves, closing a loop with another phase, and passing through a diode connected at the same negative terminal of the *dc* link. The current can also go to the *dc* load (inversion) and return through another transistor located at the positive terminal of the *dc* link. When the transistor T_N is switched OFF, the current path is interrupted and the current begins to flow through the diode D_P, connected at the positive terminal of the *dc* link. This current, called i_{Dp} in Fig. 12.39, goes directly to the *dc* link, helping in the generation of the current i_{dc}. The current i_{dc} charges the capacitor C_D and permits the rectifier to produce *dc* power. The inductances L_S are very important in this process, because they generate an induced voltage which allows the conduction of the diode D_P. Similar operation occurs during the negative half cycle, but with T_P and D_N (see Fig. 12.40).

Under inverter operation, the current paths are different because the currents flowing through the transistors come mainly from the *dc* capacitor, C_D. Under rectifier operation, the circuit works like a boost converter and under inverter, it works as a buck converter.

To have full control of the operation of the rectifier, their six diodes must be polarized negatively at all values of instantaneous *ac* voltage supply. Otherwise diodes will conduct, and the PWM rectifier will behave like a common diode rectifier bridge. The way to keep the diodes blocked is to ensure a *dc* link voltage higher than the peak *dc* voltage generated by the diodes alone, as shown in Fig. 12.41. In this way, the diodes remain polarized negatively, and they will conduct only when at least one transistor is switched ON, and favorable instantaneous *ac* voltage conditions are given. In the Fig. 12.41, V_D represents the capacitor *dc* voltage, which is kept higher than the normal diode-bridge rectification value v_{BRIDGE}. To maintain this condition, the rectifier must have a control loop like the one displayed in Fig. 12.36.

12.3.3 PWM Phase-to-phase and Phase-to-neutral Voltages

The PWM waveforms shown in the preceding figures are voltages measured between the middle point of the *dc* voltage and the corresponding phase. The phase-to-phase PWM voltages

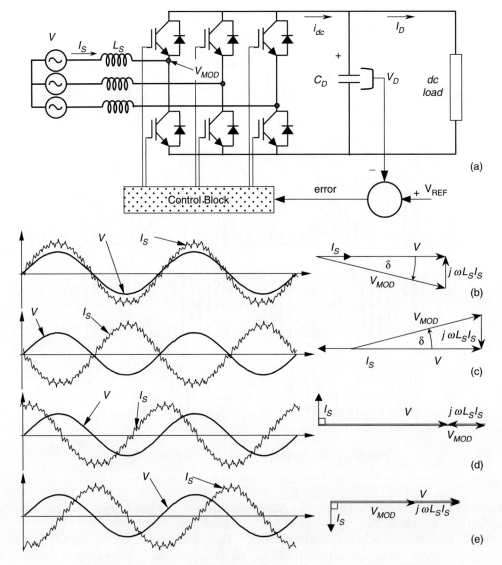

FIGURE 12.39 Four-quadrant operation of the force-commutated rectifier: (a) the PWM force-commutated rectifier; (b) rectifier operation at unity power factor; (c) inverter operation at unity power factor; (d) capacitor operation at zero power factor; and (e) inductor operation at zero power factor.

can be obtained with the help of Eq. (12.52), where the voltage V_{PWM}^{AB} is evaluated.

$$V_{PWM}^{AB} = V_{PWM}^{A} - V_{PWM}^{B} \qquad (12.52)$$

where V_{PWM}^{A} and V_{PWM}^{B} are the voltages measured between the middle point of the dc voltage, and the phases a and b respectively. In a less straightforward fashion, the phase-to-neutral voltage can be evaluated with the help of Eq. (12.53).

$$V_{PWM}^{AN} = \frac{1}{3}(V_{PWM}^{AB} - V_{PWM}^{CA}) \qquad (12.53)$$

where V_{PWM}^{AN} is the phase-to-neutral voltage for phase a, and V_{PWM}^{jk} is the phase-to-phase voltage between phase j and phase k. Figure 12.42 shows the PWM patterns for the phase-to-phase and phase-to-neutral voltages.

12.3.4 Control of the DC Link Voltage

Control of the dc link voltage requires a feedback control loop. As already explained in Section 12.3.2, the dc voltage V_D is compared with a reference V_{REF}, and the error signal "e" obtained from this comparison is used to generate a template waveform. The template should be a sinusoidal waveform with the same frequency of the mains supply. This template is used to produce the PWM pattern and allows controlling

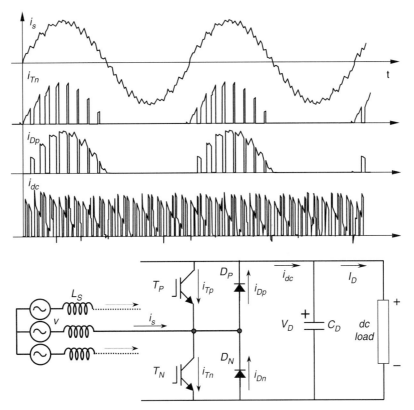

FIGURE 12.40 Current waveforms through the mains, the valves, and the *dc* link.

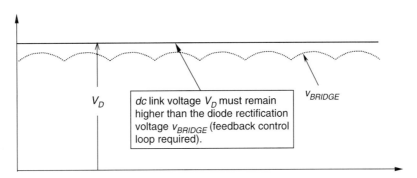

FIGURE 12.41 The *DC* link voltage condition for the operation of the PWM rectifier.

the rectifier in two different ways: (1) as a voltage-source current-controlled PWM rectifier or (2) as a voltage-source voltage-controlled PWM rectifier. The first method controls the input current, and the second controls the magnitude and phase of the voltage V_{MOD}. The current-controlled method is simpler and more stable than the voltage-controlled method, and for these reasons it will be explained first.

12.3.4.1 Voltage-source Current-controlled PWM Rectifier

This method of control is shown in the rectifier in Fig. 12.43. Control is achieved by measuring the instantaneous phase currents and forcing them to follow a sinusoidal current reference template, I_ref. The amplitude of the current reference template, I_{MAX} is evaluated using the following equation

$$I_{MAX} = G_C \cdot e = G_C \cdot (V_{REF} - v_D) \quad (12.54)$$

where G_C is shown in Fig. 12.43 and represents a controller such as PI, P, Fuzzy, or other. The sinusoidal waveform of the template is obtained by multiplying I_{MAX} with a sine function, with the same frequency of the mains, and with the desired phase-shift angle φ, as shown in Fig. 12.43. Further, the template must be synchronized with the power supply. After that,

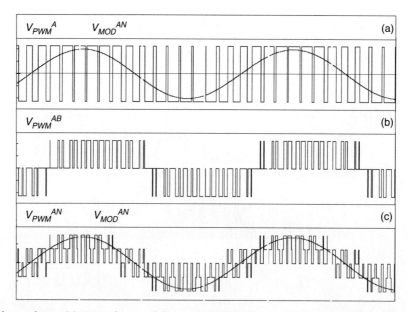

FIGURE 12.42 PWM phase voltages: (a) PWM phase modulation; (b) PWM phase-to-phase voltage; and (c) PWM phase-to-neutral voltage.

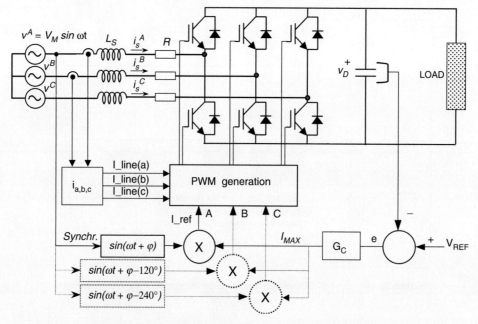

FIGURE 12.43 Voltage-source current-controlled PWM rectifier.

the template has been created and is ready to produce the PWM pattern.

However, one problem arises with the rectifier because the feedback control loop on the voltage V_C can produce instability. Then it becomes necessary to analyze this problem during rectifier design. Upon introducing the voltage feedback and the G_C controller, the control of the rectifier can be represented in a block diagram in Laplace dominion, as shown in Fig. 12.44. This block diagram represents a linearization of the system around an operating point, given by the *rms* value of the input current, I_S.

The blocks $G_1(s)$ and $G_2(s)$ in Fig. 12.44 represent the transfer function of the rectifier (around the operating point) and the transfer function of the *dc* link capacitor C_D respectively.

$$G_1(s) = \frac{\Delta P_1(s)}{\Delta I_S(s)} = 3 \cdot (V \cos \varphi - 2RI_S - L_S I_S s) \quad (12.55)$$

$$G_2(s) = \frac{\Delta V_D(s)}{\Delta P_1(s) - \Delta P_2(s)} = \frac{1}{V_D \cdot C_D \cdot s} \quad (12.56)$$

where $\Delta P_1(s)$ and $\Delta P_2(s)$ represent the input and output power of the rectifier in Laplace dominion, V the *rms* value of

12 Three-phase Controlled Rectifiers

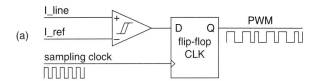

FIGURE 12.44 Closed-loop rectifier transfer function.

the mains voltage supply (phase-to-neutral), I_S the input current being controlled by the template, L_S the input inductance, and R the resistance between the converter and power supply. According to stability criteria, and assuming a *PI* controller, the following relations are obtained

$$I_S \leq \frac{C_D \cdot V_D}{3K_P \cdot L_S} \qquad (12.57)$$

$$I_S \leq \frac{K_P \cdot V \cdot \cos\varphi}{2R \cdot K_P + L_S \cdot K_I} \qquad (12.58)$$

These two relations are useful for the design of the current-controlled rectifier. They relate the values of *dc* link capacitor, *dc* link voltage, *rms* voltage supply, input resistance and inductance, and input power factor, with the *rms* value of the input current, I_S. With these relations the proportional and integral gains K_P and K_I can be calculated to ensure the stability of the rectifier. These relations only establish limitations for rectifier operation, because negative currents always satisfy the inequalities.

With these two stability limits satisfied, the rectifier will keep the *dc* capacitor voltage at the value of V_{REF} (*PI* controller), for all load conditions, by moving power from the *ac* to the *dc* side. Under inverter operation, the power will move in the opposite direction.

Once the stability problems have been solved and the sinusoidal current template has been generated, a modulation method will be required to produce the PWM pattern for the power valves. The PWM pattern will switch the power valves to force the input currents *I_line* to follow the desired current template *I_ref*. There are many modulation methods in the literature, but three methods for voltage-source current-controlled rectifiers are the most widely used ones: *periodical sampling* (PS), *hysteresis band* (HB), and *triangular carrier* (TC).

The PS method switches the power transistors of the rectifier during the transitions of a square wave clock of fixed frequency: the *periodical sampling* frequency. In each transition, a comparison between *I_ref* and *I_line* is made, and corrections take place. As shown in Fig. 12.45a, this type of control is very simple to implement: only a comparator and a D-type flip-flop are needed per phase. The main advantage of this method is that the minimum time between switching

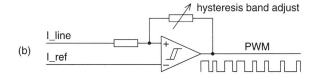

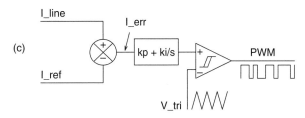

FIGURE 12.45 Modulation control methods: (a) periodical sampling; (b) hysteresis band; and (c) triangular carrier.

transitions is limited to the period of the sampling clock. This characteristic determines the maximum switching frequency of the converter. However, the average switching frequency is not clearly defined.

The HB method switches the transistors when the error between *I_ref* and *I_line* exceeds a fixed magnitude: the *hysteresis band*. As it can be seen in Fig. 12.45b, this type of control needs a single comparator with hysteresis per phase. In this case the switching frequency is not determined, but its maximum value can be evaluated through the following equation

$$f_S^{max} = \frac{V_D}{4h \cdot L_S} \qquad (12.59)$$

where h is the magnitude of the hysteresis band.

The TC method, shown in Fig. 12.45c, compares the error between *I_ref* and *I_line* with a triangular wave. This triangular wave has fixed amplitude and frequency and is called the *triangular carrier*. The error is processed through a proportional-integral (PI) gain stage before comparison with the TC takes place. As can be seen, this control scheme is more complex than PS and HB. The values for k_p and k_i determine the transient response and steady-state error of the TC method. It has been found empirically that the values for k_p and k_i shown in Eqs. (12.60) and (12.61) give a good dynamic performance under several operating conditions.

$$k_p = \frac{L_s \cdot \omega_c}{2 \cdot V_D} \quad (12.60)$$

$$k_i = \omega_c \cdot K_P \quad (12.61)$$

where L_S is the total series inductance seen by the rectifier, ω_c is the TC frequency, and V_D is the dc link voltage of the rectifier.

In order to measure the level of distortion (or undesired harmonic generation) introduced by these three control methods, Eq. (12.62) is defined

$$\%Distortion = \frac{100}{I_{rms}} \sqrt{\frac{1}{T} \int_T (I\_line - I\_ref)^2 \, dt} \quad (12.62)$$

In Eq. (12.62), the term I_{rms} is the effective value of the desired current. The term inside the square root gives the rms value of the error current, which is undesired. This formula measures the percentage of error (or distortion) of the generated waveform. This definition considers the ripple, amplitude, and phase errors of the measured waveform, as opposed to the THD, which does not take into account offsets, scalings, and phase shifts.

Figure 12.46 shows the current waveforms generated by the three aforementioned methods. The example uses an average switching frequency of 1.5 kHz. The PS is the worst, but its digital implementation is simpler. The HB method and TC

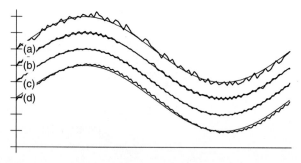

FIGURE 12.46 Waveforms obtained using 1.5 kHz switching frequency and $L_S = 13$ mH: (a) PS method; (b) HB method; (c) TC method ($K_P + K_I$); and (d) TC method (K_P only).

with PI control are quite similar, and the TC with only proportional control gives a current with a small phase shift. However, Fig. 12.47 shows that the higher the switching frequency, the closer the results obtained with the different modulation methods. Over 6 kHz of switching frequency, the distortion is very small for all methods.

12.3.4.2 Voltage-source Voltage-controlled PWM Rectifier

Figure 12.48 shows a one-phase diagram from which the control system for a voltage-source voltage-controlled rectifier is derived. This diagram represents an equivalent circuit of the fundamentals, that is, pure sinusoidal at the mains side and pure dc at the dc link side. The control is achieved by creating a sinusoidal voltage template V_{MOD}, which is modified in amplitude and angle to interact with the mains voltage V. In this way the input currents are controlled without measuring them. The template V_{MOD} is generated using the differential equations that govern the rectifier.

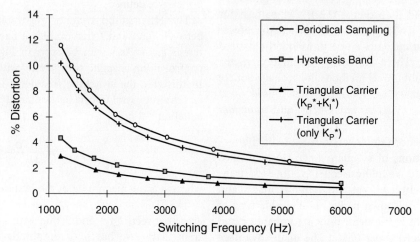

FIGURE 12.47 Distortion comparison for a sinusoidal current reference.

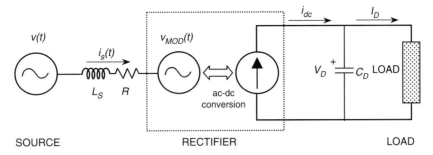

FIGURE 12.48 One-phase fundamental diagram of the voltage source rectifier.

The following differential equation can be derived from Fig. 12.48

$$v(t) = L_S \frac{di_s}{dt} + Ri_s + v_{MOD}(t) \quad (12.63)$$

Assuming that $v(t) = V\sqrt{2} \sin \omega t$, then the solution for $i_s(t)$, to acquire a template V_{MOD} able to make the rectifier work at constant power factor, should be of the form

$$i_s(t) = I_{max}(t) \sin(\omega t + \varphi) \quad (12.64)$$

Equations (12.63), (12.64), and $v(t)$ allows a function of time able to modify V_{MOD} in amplitude and phase that will make the rectifier work at a fixed power factor. Combining these equations with $v(t)$ yields

$$v_{MOD}(t) = \left[V\sqrt{2} + X_S I_{max} \sin \varphi - \left(R I_{max} + L_S \frac{dI_{max}}{dt} \right) \cos \varphi \right] \sin \omega t$$
$$- \left[X_S I_{max} \cos \varphi + \left(R I_{max} + L_S \frac{dI_{max}}{dt} \right) \sin \varphi \right] \cos \omega t$$
$$(12.65)$$

Equation (12.65) provides a template for V_{MOD}, which is controlled through variations of the input current amplitude I_{max}. Substituting the derivatives of I_{max} into Eq. (12.65) make sense, because I_{max} changes every time the *dc* load is modified. The term X_S in Eq. (12.65) is ωL_S. This equation can also be written for unity power factor operation. In such a case, $\cos \varphi = 1$ and $\sin \varphi = 0$.

$$v_{MOD}(t) = \left(V\sqrt{2} - RI_{max} - L_S \frac{dI_{max}}{dt} \right) \sin \omega t$$
$$- X_S I_{max} \cos \omega t \quad (12.66)$$

With this last equation, a unity power factor, voltage source, voltage-controlled PWM rectifier can be implemented as shown in Fig. 12.49. It can be observed that Eqs. (12.65) and (12.66) have an *in-phase* term with the mains supply ($\sin \omega t$) and an *in-quadrature* term ($\cos \omega t$). These two terms allow the template V_{MOD} to change in magnitude and phase so as to have full unity power factor control of the rectifier.

Compared with the control block of Fig. 12.43, in the voltage-source voltage-controlled rectifier of Fig. 12.49, there is no need to sense the input currents. However, to ensure stability limits as good as the limits of the current-controlled rectifier, the blocks "$-R-sL_S$" and "$-X_S$" in Fig. 12.49, have to emulate and reproduce exactly the real values of R, X_S, and L_S of the power circuit. However, these parameters do not remain constant, and this fact affects the stability of this system, making it less stable than the system showed in Fig. 12.43. In theory, if the impedance parameters are reproduced exactly, the stability limits of this rectifier are given by the same equations as used for the current-controlled rectifier seen in Fig. 12.43 (Eqs. (12.57) and (12.58)).

Under steady-state, I_{max} is constant, and Eq. (12.66) can be written in terms of phasor diagram, resulting in Eq. (12.67). As shown in Fig. 12.50, different operating conditions for the unity power factor rectifier can be displayed with this equation

$$\vec{V}_{MOD} = \vec{V} - R\vec{I}_S - jX_S\vec{I}_S \quad (12.67)$$

With the sinusoidal template V_{MOD} already created, a modulation method to commutate the transistors will be required. As in the case of current-controlled rectifier, there are many methods to modulate the template, with the most well known the so-called *sinusoidal pulse width modulation* (SPWM), which uses a TC to generate the PWM as shown in Fig. 12.51. Only this method will be described in this chapter.

In this method, there are two important parameters to define: the amplitude modulation ratio or modulation index m, and the frequency modulation ratio p. Definitions are given by

$$m = \frac{V_{MOD}^{max}}{V_{TRIANG}^{max}} \quad (12.68)$$

$$p = \frac{f_T}{f_S} \quad (12.69)$$

where V_{MOD}^{max} and V_{TRIANG}^{max} are the amplitudes of V_{MOD} and V_{TRIANG} respectively. On the other hand, f_S is the frequency of

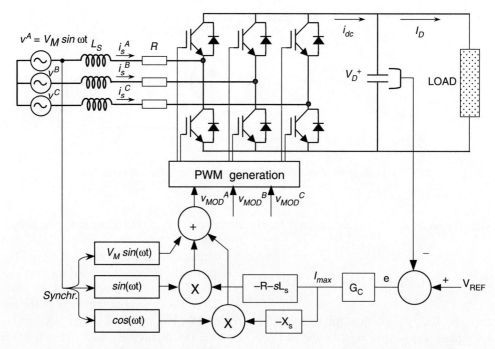

FIGURE 12.49 Implementation of the voltage-controlled rectifier for unity power factor operation.

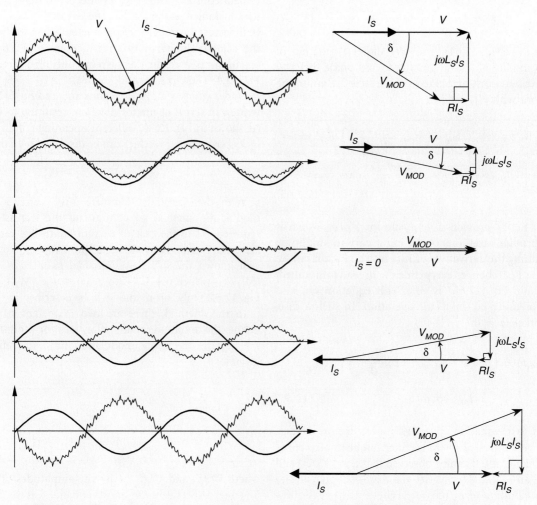

FIGURE 12.50 Steady-state operation of the unity power factor rectifier under different load conditions.

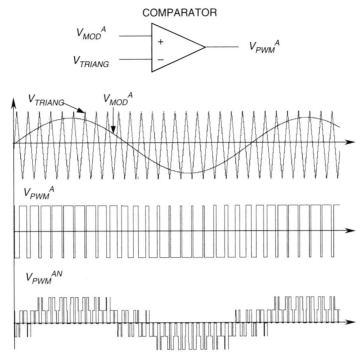

FIGURE 12.51 Sinusoidal modulation method based on TC.

the mains supply and f_T the frequency of the TC. In Fig. 12.51, $m = 0.8$ and $p = 21$. When $m > 1$ overmodulation is defined.

The modulation method described in Fig. 12.51 has a harmonic content that changes with p and m. When $p < 21$, it is recommended that synchronous PWM be used, which means that the TC and the template should be synchronized. Furthermore, to avoid subharmonics, it is also desired that p be an integer. If p is an odd number, even harmonics will be eliminated. If p is a multiple of 3, then the PWM modulation of the three phases will be identical. When m increases, the amplitude of the fundamental voltage increases proportionally, but some harmonics decrease. Under overmodulation ($m > 1$), the fundamental voltage does not increase linearly, and more harmonics appear. Figure 12.52 shows the harmonic spectrum

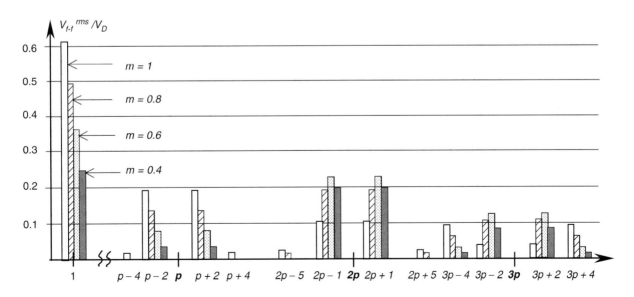

FIGURE 12.52 Harmonic spectrum for SPWM modulation.

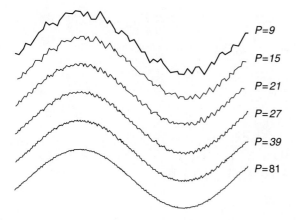

FIGURE 12.53 Current waveforms for different values of p.

of the three-phase PWM voltage waveforms for different values of m, and $p = 3k$ where k is an odd number.

Due to the presence of the input inductance L_S, the harmonic currents that result are proportionally attenuated with the harmonic number. This characteristic is shown in the current waveforms of Fig. 12.53, where larger p numbers generate cleaner currents. The rectifier that originated the currents of Fig. 12.53 has the following characteristics: $V_D = 450\,V_{dc}$, $V_{f-f}^{rms} = 220\,V_{ac}$, $L_S = 2\,\text{mH}$, and input current $I_S = 80\,A_{rms}$. It can be observed that with $p > 21$ the current distortion is quite small. The value of $p = 81$ in Fig. 12.53 produces an almost pure sinusoidal waveform, and it means 4860 Hz of switching frequency at 60 Hz or only 4.050 Hz in a rectifier operating in a 50 Hz supply. This switching frequency can be managed by MOSFETs, IGBTs, and even Power Darlingtons. Then a number $p = 81$, is feasible for today's low and medium power rectifiers.

12.3.4.3 Voltage-source Load-controlled PWM Rectifier

A simple method of control for small PWM rectifiers (up to 10–20 kW) is based on direct control of the *dc* current. Figure 12.54 shows the schematic of this control system. The fundamental voltage V_{MOD} modulated by the rectifier is produced by a fixed and unique PWM pattern, which can be carefully selected to eliminate most undesirable harmonics. As the PWM does not change, it can be stored in a permanent digital memory (ROM).

The control is based on changing the power angle δ between the mains voltage V and fundamental PWM voltage V_{MOD}. When δ changes, the amount of power flow transferred from the *ac* to the *dc* side also changes. When the power angle is negative (V_{MOD} lags V), the power flow goes from the *ac* to the *dc* side. When the power angle is positive, the power flows in the opposite direction. Then, the power angle can be controlled through the current I_D. The voltage V_D does not need to be sensed, because this control establishes a stable *dc* voltage operation for each *dc* current and power angle. With these characteristics, it is possible to find a relation between I_D and δ so as to obtain constant *dc* voltage for all load conditions.

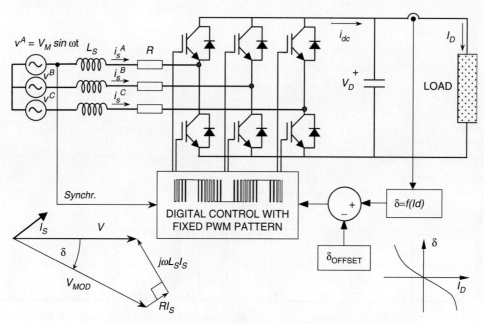

FIGURE 12.54 Voltage-source load-controlled PWM rectifier.

This relation is given by

$$I_D = f(\delta) = \frac{V\left(\cos\delta - \frac{\omega L_S}{R}\sin\delta - 1\right)}{R\left[1 + \left(\frac{\omega L_S}{R}\right)^2\right]} \qquad (12.70)$$

From Eq. (12.70) a plot and a reciprocal function $\delta = f(I_D)$ is obtained to control the rectifier. The relation between I_D and δ allows for leading power factor operation and null regulation. The leading power factor operation is shown in the phasor diagram of Fig. 12.54.

The control scheme of the voltage-source load-controlled rectifier is characterized by the following: (i) there are neither input current sensors nor dc voltage sensor; (ii) it works with a fixed and predefined PWM pattern; (iii) it presents very good stability; (iv) its stability does not depend on the size of the dc capacitor; (v) it can work at leading power factor for all load conditions; and (vi) it can be adjusted with Eq. (12.70) to work at zero regulation. The drawback appears when R in Eq. (12.70) becomes negligible, because in such a case the control system is unable to find an equilibrium point for the dc link voltage. That is why this control method is not applicable to large systems.

12.3.5 New Technologies and Applications of Force-commutated Rectifiers

The additional advantages of force-commutated rectifiers with respect to line-commutated rectifiers, make them better candidates for industrial requirements. They permit new applications such as rectifiers with harmonic elimination capability (active filters), power factor compensators, machine drives with four-quadrant operation, frequency links to connect 50 Hz with 60 Hz systems, and regenerative converters for traction power supplies. Modulation with very fast valves such as IGBTs permit almost sinusoidal currents to be obtained. The dynamics of these rectifiers is so fast that they can reverse power almost instantaneously. In machine drives, current source PWM rectifiers, like the one shown in Fig. 12.35a, can be used to drive dc machines from the three-phase supply. Four-quadrant applications using voltage-source PWM rectifiers, are extended for induction machines, synchronous machines with starting control, and special machines such as brushless-dc motors. Back-to-back systems are being used in Japan to link power systems with different frequencies.

12.3.5.1 Active Power Filter

Force-commutated PWM rectifiers can work as active power filters. The voltage-source current-controlled rectifier has the capability to eliminate harmonics produced by other polluting loads. It only needs to be connected as shown in Fig. 12.55.

The current sensors are located at the input terminals of the power source and these currents (instead of the rectifier currents) are forced to be sinusoidal. As there are polluting loads in the system, the rectifier is forced to deliver the harmonics that loads need, because the current sensors do not allow the harmonics going to the mains. As a result, the rectifier currents become distorted, but an adequate dc capacitor C_D can keep the dc link voltage in good shape. In this way the rectifier can do its duty, and also eliminate harmonics to the source. In addition, it also can compensate power factor and unbalanced load problems.

12.3.5.2 Frequency Link Systems

Frequency link systems permit power to be transferred form one frequency to another one. They are also useful for linking unsynchronized networks. Line-commutated converters are widely used for this application, but they have some drawbacks that force-commutated converters can eliminate. For example, the harmonic filters requirement, the poor power factor, and the necessity to count with a synchronous compensator when generating machines at the load side are absent. Figure 12.56 shows a typical line-commutated system in which a 60 Hz load is fed by a 50 Hz supply. As the 60 Hz side needs excitation to commutate the valves, a synchronous compensator has been required.

In contrast, an equivalent system with force-commutated converters is simpler, cleaner, and more reliable. It is implemented with a dc voltage-controlled rectifier, and another identical converter working in the inversion mode. The power factor can be adjusted independently at the two ac terminals, and filters or synchronous compensators are not required. Figure 12.57 shows a frequency link system with force-commutated converters.

12.3.5.3 Special Topologies for High Power Applications

High power applications require series- and/or parallel-connected rectifiers. Series and parallel operation with force-commutated rectifiers allow improving the power quality because harmonic cancellation can be applied to these topologies. Figure 12.58 shows a series connection of force-commutated rectifiers, where the modulating carriers of the valves in each bridge are shifted to cancel harmonics. The example uses sinusoidal PWM that are with TC shifted.

The waveforms of the input currents for the series connection system are shown in Fig. 12.59. The frequency modulation ratio shown in this figure is for $p = 9$. The carriers are shifted by 90° each, to obtain harmonics cancellation. Shifting of the carriers δ_T depends on the number of converters in series (or in parallel), and is given by

$$\delta_T = \frac{2\pi}{n} \qquad (12.71)$$

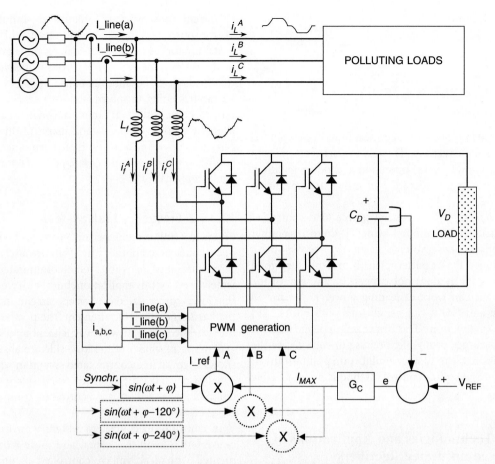

FIGURE 12.55 Voltage-source rectifier with harmonic elimination capability.

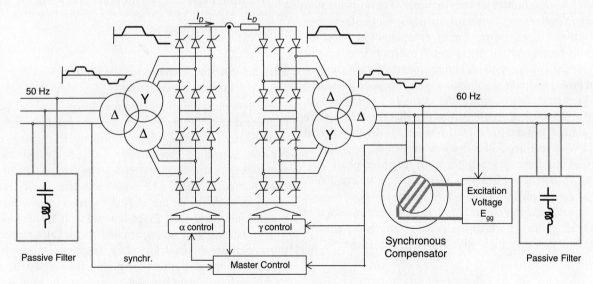

FIGURE 12.56 Frequency link systems with line-commutated converters.

12 Three-phase Controlled Rectifiers

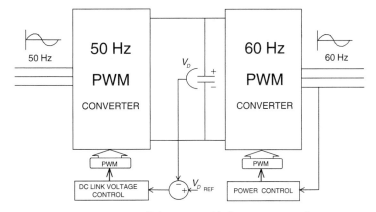

FIGURE 12.57 Frequency link systems with force-commutated converters.

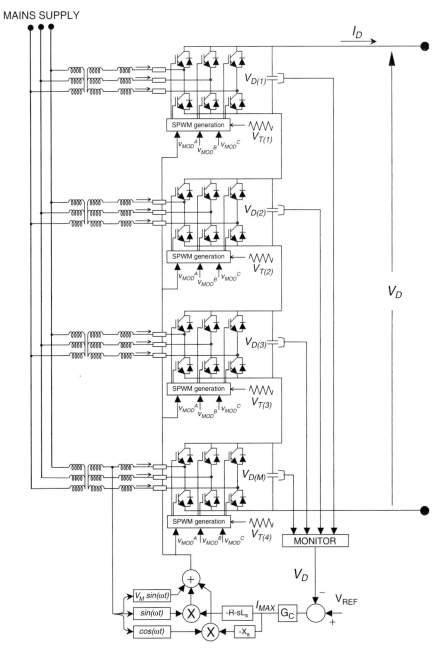

FIGURE 12.58 Series connection system with force-commutated rectifiers.

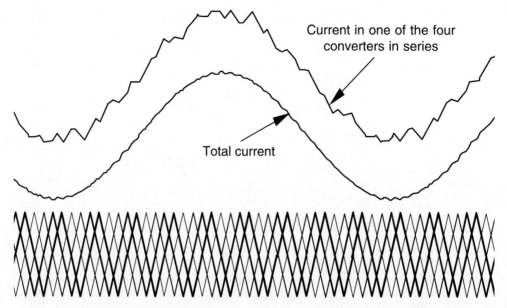

FIGURE 12.59 Input currents and carriers of the series connection system of Fig. 12.58.

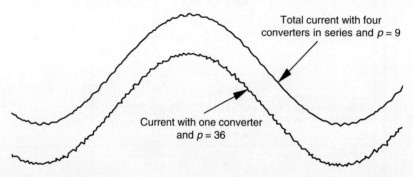

FIGURE 12.60 Four converters in series and $p = 9$ compared with one converter and $p = 36$.

where n is the number of converters in series or in parallel. It can be observed that despite the low value of p, the total current becomes quite clean and clear, better than the current of one of the converters in the chain.

The harmonic cancellation with series- or parallel-connected rectifiers, using the same modulation but the carriers shifted, is quite effective. The resultant current is better with n converters and frequency modulation $p = p_1$ than with one converter and $p = n \cdot p_1$. This attribute is verified in Fig. 12.60, where the total current of four converters in series with $p = 9$ and carriers shifted, is compared with the current of only one converter and $p = 36$. This technique also allows for the use of valves with slow commutation times, such as high power GTOs. Generally, high power valves have low commutation times and hence the parallel and/or series options remain very attractive.

Another special topology for high power was implemented for Asea Brown Boveri (ABB) in Bremen. A 100 MW power converter supplies energy to the railways at $16^{2/3}$ Hz. It uses basic "H" bridges like the one shown in Fig. 12.61, connected to the load through power transformers. These transformers are connected in parallel at the converter side, and in series at the load side.

The system uses SPWM with TCs shifted, and depending on the number of converters connected in the chain of bridges, the voltage waveform becomes more and more sinusoidal. Figure 12.62 shows a back-to-back system using a chain of 12 "H" converters connected as showed in Fig. 12.61b.

The ac voltage waveform obtained with the topology of Fig. 12.62 is displayed in Fig. 12.63. It can be observed that the voltage is formed by small steps that depend on the number of converters in the chain (12 in this case). The current is almost perfectly sinusoidal.

Figure 12.64 shows the voltage waveforms for different number of converters connected in the bridge. It is clear that the larger the number of converters, the better the voltage.

Another interesting result with this converter is that the ac voltages become modulated by both PWM and amplitude

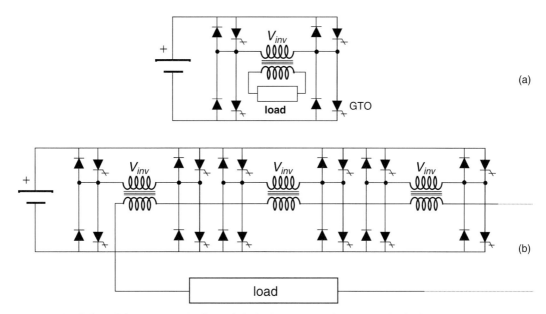

FIGURE 12.61 The "H" modulator: (a) one bridge and (b) bridges connected in series at load side through isolation transformers.

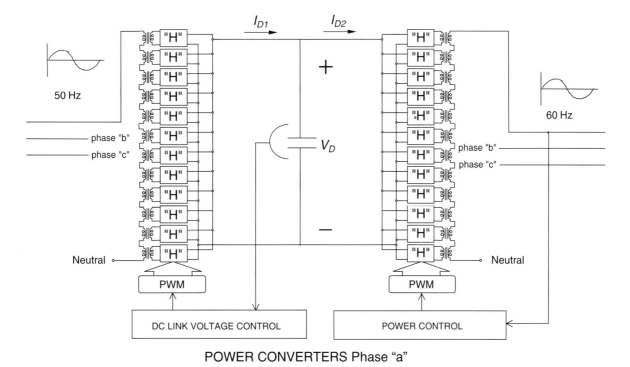

FIGURE 12.62 Frequency link with force-commutated converters and sinusoidal voltage modulation.

modulation (AM). This is because when the pulse modulation changes, the steps of the amplitude change. The maximum number of steps of the resultant voltage is equal to the number of converters. When the voltage decreases, some steps disappear, and then the AM becomes a discrete function. Figure 12.65 shows the AM of the voltage.

12.3.5.4 Machine Drives Applications

One of the most important applications of force-commutated rectifiers is in machine drives. Line-commutated thyristor converters have limited applications because they need excitation to extinguish the valves. This limitation do not allow the use of line-commutated converters in induction machine drives.

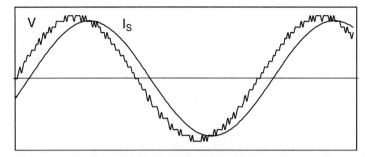

FIGURE 12.63 Voltage and current waveforms with 12 converters.

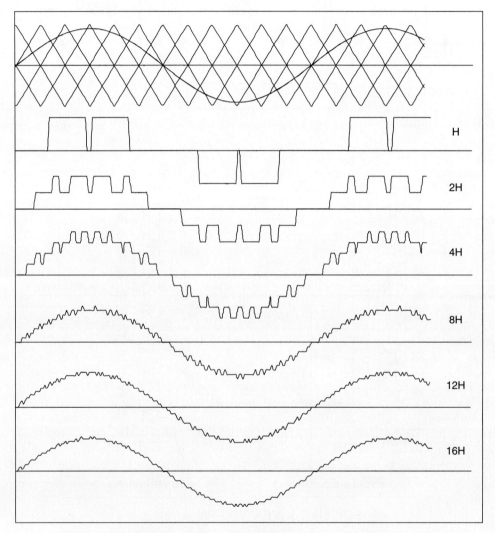

FIGURE 12.64 Voltage waveforms with different numbers of "H" bridges in series.

On the other hand, with force-commutated converters four-quadrant operation is achievable. Figure 12.66 shows a typical frequency converter with a force-commutated rectifier–inverter link. The rectifier side controls the *dc* link, and the inverter side controls the machine. The machine can be a synchronous, brushless *dc*, or induction machine. The reversal of both speed and power are possible with this topology. At the rectifier side, the power factor can be controlled, and even with an inductive load such as an induction machine, the source can "see" the load as capacitive or resistive. Changing the frequency of the inverter controls the machine speed, and the torque is controlled through the stator currents and

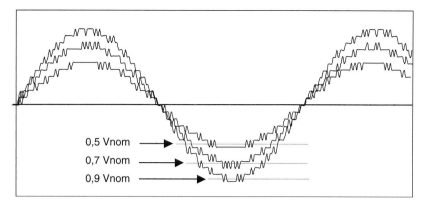

FIGURE 12.65 Amplitude modulation of the "H" bridges of Fig. 12.62.

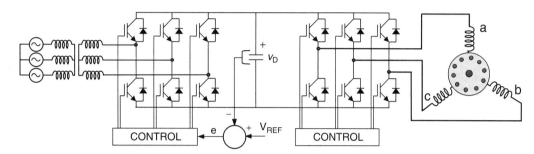

FIGURE 12.66 Frequency converter with force-commutated converters.

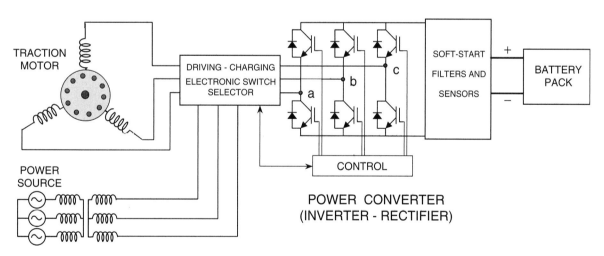

FIGURE 12.67 Electric bus system with regenerative braking and battery charger.

torque angle. The inverter will become a rectifier during regenerative braking, which is possible by making slip negative in an induction machine, or by making the torque angle negative in synchronous and brushless *dc* machines.

A variation of the drive of Fig. 12.66 is found in electric traction applications. Battery powered vehicles use the inverter as a rectifier during regenerative braking, and sometimes the inverter is also used as a battery charger. In this case, the rectifier can be fed by a single-phase or three-phase system.

Figure 12.67 shows a battery-powered electric bus system. This system uses the power inverter of the traction motor as rectifier for two purposes: regenerative braking and as a battery charger fed by a three-phase power source.

12.3.5.5 Variable Speed Power Generation

Power generation at 50 or 60 Hz requires constant speed machines. In addition, induction machines are not currently

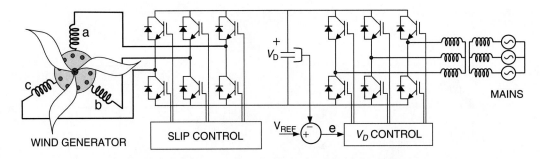

FIGURE 12.68 Variable-speed constant-frequency wind generator.

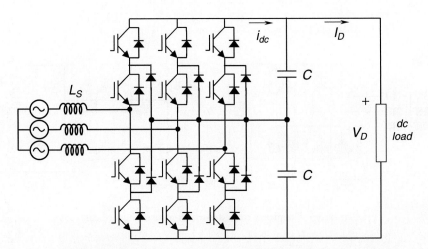

FIGURE 12.69 Voltage-source rectifier using three-level converter.

used in power plants because of magnetization problems. With the use of frequency-link force-commutated converters, variable-speed constant-frequency generation becomes possible even with induction generators. The power plant in Fig. 12.68 shows a wind generator implemented with an induction machine, and a rectifier–inverter frequency link connected to the utility. The *dc* link voltage is kept constant with the converter located at the mains side. The converter connected at the machine side controls the slip of the generator and adjusts it according to the speed of wind or power requirements. The utility is not affected by the power factor of the generator, because the two converters keep the $\cos\varphi$ of the machine independent of the mains supply. The converter at the mains side can even be adjusted to operate at leading power factor.

Variable-speed constant-frequency generation also can be used in either hydraulic or thermal plants. This allows for optimal adjustment of the efficiency-speed characteristics of the machines. In many places, wound rotor induction generators working as variable speed synchronous machines are being used as constant frequency generators. They operate in hydraulic plants that are able to store water during low demand periods. A power converter is connected at the slip rings of the generator. The rotor is then fed with variable frequency excitation. This allows the generator to generate at different speeds around the synchronous rotating flux.

12.3.5.6 Power Rectifiers Using Multilevel Topologies

Almost all voltage source rectifiers already described are two-level configurations. Today, multilevel topologies are becoming very popular, mainly three-level converters. The most popular three-level configuration is called diode clamped converter, which is shown in Fig. 12.69. This topology is today the standard solution for high power steel rolling mills, which uses back-to-back three-phase rectifier–inverter link configuration. In addition, this solution has been recently introduced in high power downhill conveyor belts which operate almost permanently in the regeneration mode or rectifier operation. The more important advantage of three-level rectifiers is that voltage and current harmonics are reduced due to the increased number of levels.

Higher number of levels can be obtained using the same diode clamped strategy, as shown in Fig. 12.70, where only one phase of a general approach is displayed. However, this topology becomes more and more complex with the increase of number of levels. For this reason, new topologies are being

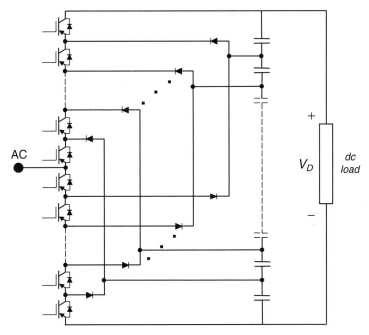

FIGURE 12.70 Multilevel rectifier using diode clamped topology.

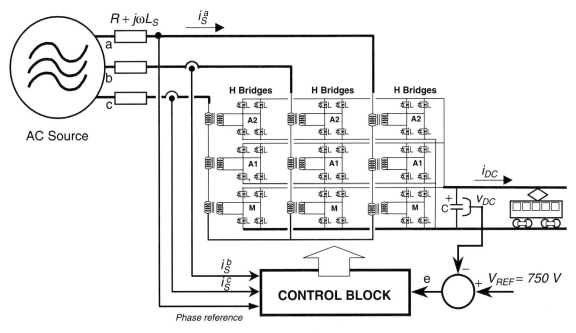

FIGURE 12.71 27-Level rectifier for railways, using H-bridges scaled in power of three.

studied to get a large number of levels with less power transistors. One example of such of these topologies is the multistage, 27-level converter shown in Fig. 12.71. This special 27-level, four-quadrant rectifier, uses only three H-bridges per phase with independent input transformers for each H-bridge. The transformers allow galvanic isolation and power escalation to get high quality voltage waveforms, with *THD* of less than 1%.

The power scalation consists on increasing the voltage rates of each transformer making use of the "three-level" characteristics of H-bridges. Then, the number of levels is optimized when transformers are scaled in power of three. Some advantages of this 27-level topology are: (a) only one of the three H-bridges, called "main converter," manages more than 80% of the total active power in each phase and (b) this main

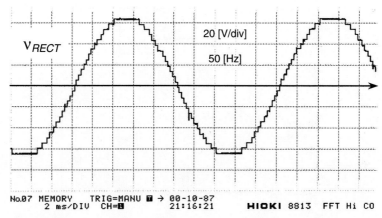

FIGURE 12.72 *AC* voltage waveform generated by the 27-level rectifier.

converter switches at fundamental frequency reducing the switching losses at a minimum value. The rectifier of Fig. 12.71 is a current-controlled voltage source type, with a conventional feedback control loop, which is being used as a rectifier in a subway substation. It includes fast reversal of power and the ability to produce clean *ac* and *dc* waveforms with negligible ripple. This rectifier can also compensate power factor and eliminate harmonics produced by other loads in the *ac* line. Figure 12.72 shows the *ac* voltage waveform obtained with this rectifier from an experimental prototype. If one more H-bridge is added, 81 levels are obtained, because the number of levels increases according with $N = 3^k$, where N is the number of levels or voltage steps and k the number of H-bridges used per phase.

Many other high-level topologies are under study but this matter is beyond the main topic of this chapter.

Further Reading

1. G. Möltgen, "Line Commutated Thyristor Converters," Siemens Aktiengesellschaft, Berlin-Munich, Pitman Publishing, London, 1972.
2. G. Möltgen, "Converter Engineering, and Introduction to Operation and Theory," John Wiley and Sons, New York, 1984.
3. K. Thorborg, "Power Electronics," Prentice-Hall International (UK) Ltd., London, 1988.
4. M. H. Rashid, "Power Electronics, Circuits Devices and Applications," Prentice-Hall International Editions, London, 1992.
5. N. Mohan, T. M. Undeland, and W. P. Robbins, "Power Electronics: Converters, Applications, and Design," John Wiley and Sons, New York 1989.
6. J. Arrillaga, D. A. Bradley, and P. S. Bodger, "Power System Harmonics," John Wiley and Sons, New York, 1989.
7. J. M. D. Murphy and F. G. Turnbull, "Power Electronic Control of AC Motors," Pergamon Press, 1988.
8. M. E. Villablanca and J. Arrillaga, "Pulse Multiplication in Parallel Convertors by Multitap Control of Interphase Reactor," IEE Proceedings-B, Vol. 139, No 1; January 1992, pp. 13–20.
9. D. A. Woodford, "HVDC Transmission," Professional Report from Manitoba HVDC Research Center, Winnipeg, Manitoba, March 1998.
10. D. R. Veas, J. W. Dixon, and B. T. Ooi, "A Novel Load Current Control Method for a Leading Power Factor Voltage Source PEM Rectifier," IEEE Transactions on Power Electronics, Vol. 9, No 2, March 1994, pp. 153–159.
11. L. Morán, E. Mora, R. Wallace, and J. Dixon, "Performance Analysis of a Power Factor Compensator which Simultaneously Eliminates Line Current Harmonics," IEEE Power Electronics Specialists Conference, PESC'92, Toledo, España, June 29 to July 3, 1992.
12. P. D. Ziogas, L. Morán, G. Joos, and D. Vincenti, "A Refined PWM Scheme for Voltage and Current Source Converters," IEEE-IAS Annual Meeting, 1990, pp. 977–983.
13. W. McMurray, "Modulation of the Chopping Frequency in DC Choppers and PWM Inverters Having Current Hysteresis Controllers," IEEE Transaction on Ind. Appl., Vol. IA-20, July/August 1984, pp. 763–768.
14. J. W. Dixon and B. T. Ooi, "Indirect Current Control of a Unity Power Factor Sinusoidal Current Boost Type Three-Phase Rectifier," IEEE Transactions on Industrial Electronics, Vol. 35, No 4, November 1988, pp. 508–515.
15. L. Morán, J. Dixon, and R. Wallace "A Three-Phase Active Power Filter Operating with Fixed Switching Frequency for Reactive Power and Current Harmonic Compensation," IEEE Transactions on Industrial Electronics, Vol. 42, No 4, August 1995, pp. 402–408.
16. M. A. Boost and P. Ziogas, "State-of-the-Art PWM Techniques, a Critical Evaluation," IEEE Transactions on Industry Applications, Vol. 24, No 2, March/April 1988, pp. 271–280.
17. J. W. Dixon and B. T. Ooi, "Series and Parallel Operation of Hysteresis Current-Controlled PWM Rectifiers," IEEE Transactions on Industry Applications, Vol. 25, No 4, July/August 1989, pp. 644–651.

18. B. T. Ooi, J. W. Dixon, A. B. Kulkarni, and M. Nishimoto, "An integrated AC Drive System Using a Controlled-Current PWM Rectifier/Inverter Link," IEEE Transactions on Power Electronics, Vol. 3, No 1, January 1988, pp. 64–71.
19. M. Koyama, Y. Shimomura, H. Yamaguchi, M. Mukunoki, H. Okayama, and S. Mizoguchi, "Large Capacity High Efficiency Three-Level GCT Inverter System for Steel Rolling Mill Drivers," Proceedings of the 9th European Conference on Power Electronics, EPE 2001, Austria, CDROM.
20. J. Rodríguez, J. Dixon, J. Espinoza, and P. Lezana, "PWM Regenerative Rectifiers: State of the Art," IEEE Transactions on Industrial Electronics, Vol. 52, No 4, January/February 2005, pp. 5–22.
21. J. Dixon and L. Morán, "A Clean Four-Quadrant Sinusoidal Power Rectifier, Using Multistage Converters for Subway Applications," IEEE Transactions on Industrial Electronics, Vol. 52, No 5, May–June 2005, pp. 653–661.

13
DC–DC Converters

Dariusz Czarkowski
Department of Electrical and Computer Engineering, Polytechnic University, Brooklyn, New York, USA

13.1 Introduction 245
13.2 DC Choppers 246
13.3 Step-down (Buck) Converter 247
    13.3.1 Basic Converter • 13.3.2 Transformer Versions of Buck Converter
13.4 Step-up (Boost) Converter 250
13.5 Buck–Boost Converter 251
    13.5.1 Basic Converter • 13.5.2 Flyback Converter
13.6 Ćuk Converter 252
13.7 Effects of Parasitics 253
13.8 Synchronous and Bidirectional Converters 254
13.9 Control Principles 255
13.10 Applications of DC–DC Converters 258
    Further Reading 259

13.1 Introduction

Modern electronic systems require high quality, small, lightweight, reliable, and efficient power supplies. Linear power regulators, whose principle of operation is based on a voltage or current divider, are inefficient. They are limited to output voltages smaller than the input voltage. Also, their power density is low because they require low-frequency (50 or 60 Hz) line transformers and filters. Linear regulators can, however, provide a very high quality output voltage. Their main area of application is at low power levels as low drop-out voltage (LDO) regulators. Electronic devices in linear regulators operate in their active (linear) modes. At higher power levels, switching regulators are used. Switching regulators use power electronic semiconductor switches in *on* and *off* states. Since there is a small power loss in those states (low voltage across a switch in the *on* state, zero current through a switch in the *off* state), switching regulators can achieve high energy conversion efficiencies. Modern power electronic switches can operate at high frequencies. The higher the operating frequency, the smaller and lighter the transformers, filter inductors, and capacitors. In addition, dynamic characteristics of converters improve with increasing operating frequencies. The bandwidth of a control loop is usually determined by the corner frequency of the output filter. Therefore, high operating frequencies allow for achieving a faster dynamic response to rapid changes in the load current and/or the input voltage.

High-frequency electronic power processors are used in dc–dc power conversion. The functions of dc–dc converters are:

- to convert a dc input voltage V_S into a dc output voltage V_O;
- to regulate the dc output voltage against load and line variations;
- to reduce the ac voltage ripple on the dc output voltage below the required level;
- to provide isolation between the input source and the load (isolation is not always required);
- to protect the supplied system and the input source from electromagnetic interference (EMI);
- to satisfy various international and national safety standards.

The dc–dc converters can be divided into two main types: hard-switching pulse width modulated (PWM) converters, and resonant and soft-switching converters. This chapter deals with the former type of dc–dc converters. The PWM converters have been very popular for the last three decades. They are widely used at all power levels. Topologies and properties of PWM converters are well understood and described in literature. Advantages of PWM converters include low component count, high efficiency, constant frequency operation, relatively simple control and commercial availability of integrated circuit controllers, and ability to achieve high conversion ratios for both step-down and step-up application. A disadvantage of PWM dc–dc converters is that PWM rectangular voltage and current waveforms cause turn-on and turn-off losses in semiconductor devices which limit practical operating frequencies to a megahertz range. Rectangular waveforms also inherently generate EMI.

This chapter starts from a section on dc choppers which are used primarily in dc drives. The output voltage of dc choppers is controlled by adjusting the *on* time of a switch which in turn adjusts the width of a voltage pulse at the output. This is so called pulse-width modulation (PWM) control. The dc choppers with additional filtering components form PWM dc–dc converters. Four basic dc–dc converter topologies are presented in Sections 13.3–13.6: buck, boost, buck–boost, and Ćuk converters. Popular isolated versions of these converters are also discussed. Operation of converters is explained under ideal component and semiconductor device assumptions. Section 13.7 discusses effects of non-idealities in PWM converters. Section 13.8 presents topologies for increased efficiency at low output voltage and for bidirectional power flow. Section 13.9 reviews control principles of PWM dc–dc converters. Two main control schemes, voltage-mode control and current-mode control, are described. Summary of application areas of PWM dc–dc converters is given in Section 13.10. Finally, a list of modern textbooks on power electronics is provided. These books are excellent resources for deeper exploration of the area of dc–dc power conversion.

13.2 DC Choppers

A step-down dc chopper with a resistive load is shown in Fig. 13.1a. It is a series connection of a dc input voltage source V_S, controllable switch S, and load resistance R. In most cases, switch S has a unidirectional voltage blocking capabilities and unidirectional current conduction capabilities. Power electronic switches are usually implemented with power MOSFETs, IGBTs, MCTs, power BJTs, or GTOs. If an antiparallel diode is used or embedded in a switch, a switch exhibits a bidirectional current conduction property. Figure 13.1b depicts waveforms in a step-down chopper. The switch is being operated with a duty ratio D defined as a ratio of the switch *on* time to the sum of the *on* the *off* times. For a constant frequency operation,

$$D \equiv \frac{t_{on}}{t_{on} + t_{off}} = \frac{t_{on}}{T} \quad (13.1)$$

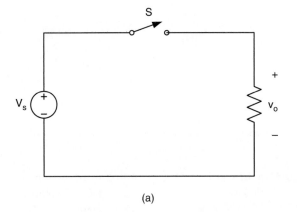

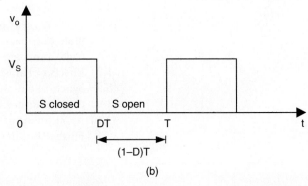

FIGURE 13.1 DC chopper with resistive load: (a) circuit diagram and (b) output voltage waveform.

where $T = 1/f$ is the period of the switching frequency f. The average value of the output voltage is

$$V_O = DV_S \quad (13.2)$$

and can be regulated by adjusting duty ratio D. The average output voltage is always smaller than the input voltage, hence, the name of the converter.

The dc step-down choppers are commonly used in dc drives. In such a case, the load is presented as a series combination of inductance L, resistance R, and back emf E as shown in Fig. 13.2a. To provide a path for a continuous inductor current flow when the switch is in the *off* state, an antiparallel diode D must be connected across the load. Since the chopper of Fig. 13.2a provides a positive voltage and a positive current to the load, it is called a first-quadrant chopper. The load voltage and current are graphed in Fig. 13.2b under assumptions that the load current never reaches zero and the load time constant $\tau = L/R$ is much greater than the period T. Average values of the output voltage and current can be adjusted by changing the duty ratio D.

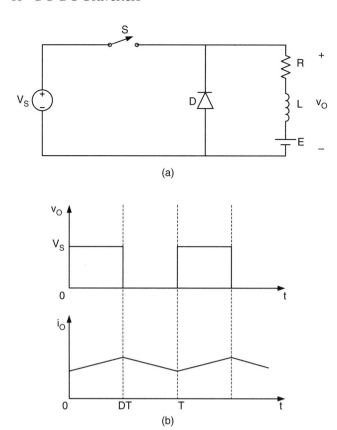

FIGURE 13.2 DC chopper with RLE load: (a) circuit diagram and (b) waveforms.

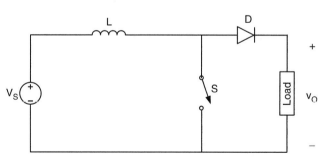

FIGURE 13.3 The dc step-up chopper.

The dc choppers can also provide peak output voltages higher than the input voltage. Such a step-up configuration is presented in Fig. 13.3. It consists of dc input source V_S, inductor L connected in series with the source, switch S connecting the inductor to ground, and a series combination of diode D and load. If the switch operates with a duty ratio D, the output voltage is a series of pulses of duration $(1-D)T$ and amplitude $V_S/(1-D)$. Neglecting losses, the average value of the output voltage is V_S. To obtain an average value of the output voltage greater than V_S, a capacitor must be connected in parallel with the load. This results in a topology of a boost dc–dc converter that is described in Section 13.4.

13.3 Step-down (Buck) Converter

13.3.1 Basic Converter

The step-down dc–dc converter, commonly known as a buck converter, is shown in Fig. 13.4a. It consists of dc input voltage source V_S, controlled switch S, diode D, filter inductor L, filter capacitor C, and load resistance R. Typical waveforms in the converter are shown in Fig. 13.4b under assumption that the inductor current is always positive. The state of the converter in which the inductor current is never zero for any period of time is called the continuous conduction mode (CCM). It can be seen from the circuit that when the switch S is commanded to the *on* state, the diode D is reverse biased. When the switch S is off, the diode conducts to support an uninterrupted current in the inductor.

The relationship among the input voltage, output voltage, and the switch duty ratio D can be derived, for instance, from the inductor voltage v_L waveform (see Fig. 13.4b). According to Faraday's law, the inductor volt–second product over a period of steady-state operation is zero. For the buck converter

$$(V_S - V_O)DT = -V_O(1-D)T \qquad (13.3)$$

Hence, the dc voltage transfer function, defined as the ratio of the output voltage to the input voltage, is

$$M_V = \frac{V_O}{V_S} = D \qquad (13.4)$$

It can be seen from Eq. (13.4) that the output voltage is always smaller than the input voltage.

The dc–dc converters can operate in two distinct modes with respect to the inductor current i_L. Figure 13.4b depicts the CCM in which the inductor current is always greater than zero. When the average value of the input current is low (high R) and/or the switching frequency f is low, the converter may enter the discontinuous conduction mode (DCM). In the DCM, the inductor current is zero during a portion of the switching period. The CCM is preferred for high efficiency and good utilization of semiconductor switches and passive components. The DCM may be used in applications with special control requirements, since the dynamic order of the converter is reduced (the energy stored in the inductor is zero at the beginning and at the end of each switching period). It is uncommon to mix these two operating modes because of different control algorithms. For the buck converter, the value

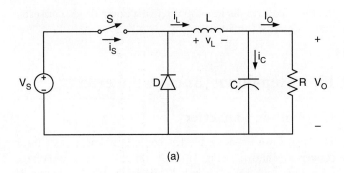

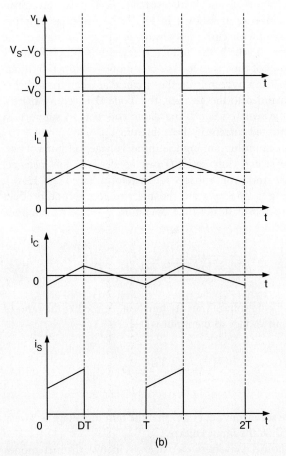

FIGURE 13.4 Buck converter: (a) circuit diagram and (b) waveforms.

of the filter inductance that determines the boundary between CCM and DCM is given by

$$L_b = \frac{(1-D)R}{2f} \quad (13.5)$$

For typical values of $D = 0.5$, $R = 10\,\Omega$, and $f = 100\,\text{kHz}$, the boundary is $L_b = 25\,\mu\text{H}$. For $L > L_b$, the converter operates in the CCM.

The filter inductor current i_L in the CCM consists of a dc component I_O with a superimposed triangular ac component.

Almost all of this ac component flows through the filter capacitor as a current i_c. Current i_c causes a small voltage ripple across the dc output voltage V_O. To limit the peak-to-peak value of the ripple voltage below certain value V_r, the filter capacitance C must be greater than

$$C_{min} = \frac{(1-D)V_O}{8V_r L f^2} \quad (13.6)$$

At $D = 0.5$, $V_r/V_O = 1\%$, $L = 25\,\mu\text{H}$, and $f = 100\,\text{kHz}$, the minimum capacitance is $C_{min} = 25\,\mu\text{F}$.

Equations (13.5) and (13.6) are the key design equations for the buck converter. The input and output dc voltages (hence, the duty ratio D), and the range of load resistance R are usually determined by preliminary specifications. The designer needs to determine values of passive components L and C, and of the switching frequency f. The value of the filter inductor L is calculated from the CCM/DCM condition using Eq. (13.5). The value of the filter capacitor C is obtained from the voltage ripple condition Eq. (13.6). For the compactness and low conduction losses of a converter, it is desirable to use small passive components. Equations (13.5) and (13.6) show that it can be accomplished by using a high switching frequency f. The switching frequency is limited, however, by the type of semiconductor switches used and by switching losses. It should be also noted that values of L and C may be altered by effects of parasitic components in the converter, especially by the equivalent series resistance of the capacitor. The issue of parasitic components in dc–dc converters is discussed in Section 13.7.

13.3.2 Transformer Versions of Buck Converter

In many dc power supplies, a galvanic isolation between the dc or ac input and the dc output is required for safety and reliability. An economical mean of achieving such an isolation is to employ a transformer version of a dc–dc converter. High-frequency transformers are of a small size and weight and provide high efficiency. Their turns ratio can be used to additionally adjust the output voltage level. Among buck-derived dc–dc converters, the most popular are: forward converter, push–pull converter, half-bridge converter, and full-bridge converter.

A. Forward Converter

The circuit diagram of a forward converter is depicted in Fig. 13.5. When the switch S is *on*, diode D_1 conducts and diode D_2 is *off*. The energy is transferred from the input, through the transformer, to the output filter. When the switch is *off*, the state of diodes D_1 and D_2 is reversed. The dc voltage transfer function of the forward converter is

$$M_V = \frac{D}{n} \quad (13.7)$$

where $n = N_1/N_2$.

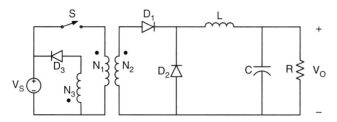

FIGURE 13.5 Forward converter.

In the forward converter, the energy-transfer current flows through the transformer in one direction. Hence, an additional winding with diode D_3 is needed to bring the magnetizing current of the transformer to zero. This prevents transformer saturation. The turns ratio N_1/N_3 should be selected in such a way that the magnetizing current decreases to zero during a fraction of the time interval when the switch is *off*.

Equations (13.5) and (13.6) can be used to design the filter components. The forward converter is very popular for low power applications. For medium power levels, converters with bidirectional transformer excitation (push–pull, half-bridge, and full-bridge) are preferred due to better utilization of magnetic components.

B. Push–Pull Converter

The PWM dc–dc push–pull converter is shown in Fig. 13.6. The switches S_1 and S_2 operate shifted in phase by $T/2$ with the same duty ratio D. The duty ratio must be smaller than 0.5. When switch S_1 is *on*, diode D_1 conducts and diode D_2 is *off*. Diode states are reversed when switch S_2 is *on*. When both controllable switches are *off*, the diodes are *on* and share equally the filter inductor current. The dc voltage transfer function of the push–pull converter is

$$M_V = \frac{2D}{n} \quad (13.8)$$

where $n = N_1/N_2$. The boundary value of the filter inductor is

$$L_b = \frac{(1-2D)R}{4f} \quad (13.9)$$

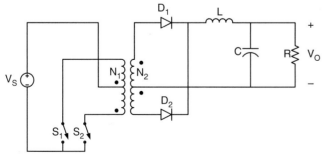

FIGURE 13.6 Push–pull converter.

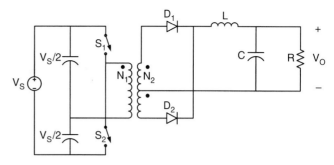

FIGURE 13.7 Half-bridge converter.

The filter capacitor can be obtained from

$$C_{min} = \frac{(1-2D)V_O}{32 V_r L f^2} \quad (13.10)$$

C. Half-bridge Converter

Figure 13.7 shows the dc–dc half-bridge converter. The operation of the PWM half-bridge converter is similar to that of the push–pull converter. In comparison to the push–pull converter, the primary of the transformer is simplified at the expense of two voltage-sharing input capacitors. The half-bridge converter dc voltage transfer function is

$$M_V \equiv \frac{V_D}{V_S} = \frac{D}{n} \quad (13.11)$$

where $D \leq 0.5$. Equations (13.9) and (13.10) apply to the filter components.

D. Full-bridge Converter

Comparing the PWM dc–dc full-bridge converter of Fig. 13.8 to the half-bridge converter, it can be seen that the input capacitors have been replaced by two controllable switches. The controllable switches are operated in pairs. When S_1 and S_4 are *on*, voltage V_S is applied to the primary of the transformer and diode D_1 conducts, With S_2 and S_3 *on*, there is voltage $-V_S$ across the primary transformer and diode D_2

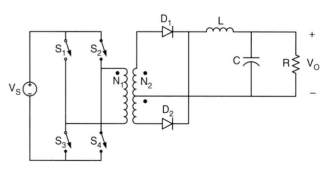

FIGURE 13.8 Full-bridge converter.

is *on*. With all controllable switches *off*, both diodes conduct, similarly as in the push–pull and half-bridge converters. The dc voltage transfer function of the full-bridge converter is

$$M_V \equiv \frac{V_O}{V_S} = \frac{2D}{n} \quad (13.12)$$

where $D \leq 0.5$. Values of filter components can be obtained from Eqs. (13.9) and (13.10).

It should be stressed that the full-bridge topology is a very versatile one. With different control algorithms, it is very popular in dc–ac conversion (square-wave and PWM single-phase inverters). It is also used in four-quadrant dc drives.

13.4 Step-up (Boost) Converter

Figure 13.9a depicts a step-up or a PWM boost converter. It is comprised of dc input voltage source V_S, boost inductor L, controlled switch S, diode D, filter capacitor C, and load resistance R. The converter waveforms in the CCM are presented in Fig. 13.9b. When the switch S is in the *on* state, the current in the boost inductor increases linearly. The diode D is *off* at the time. When the switch S is turned *off*, the energy stored in the inductor is released through the diode to the input RC circuit.

Using the Faraday's law for the boost inductor

$$V_S DT = (V_O - V_S)(1 - D)T \quad (13.13)$$

from which the dc voltage transfer function turns out to be

$$M_V \equiv \frac{V_O}{V_S} = \frac{1}{1 - D} \quad (13.14)$$

As the name of the converter suggests, the output voltage is always greater than the input voltage.

The boost converter operates in the CCM for $L > L_b$ where

$$L_b = \frac{(1 - D)^2 DR}{2f} \quad (13.15)$$

For $D = 0.5$, $R = 10\,\Omega$, and $f = 100\,\text{kHz}$, the boundary value of the inductance is $L_b = 6.25\,\mu\text{H}$.

As shown in Fig. 13.9b, the current supplied to the output RC circuit is discontinuous. Thus, a larger filter capacitor is required in comparison to that in the buck-derived converters to limit the output voltage ripple. The filter capacitor must provide the output dc current to the load when the diode D

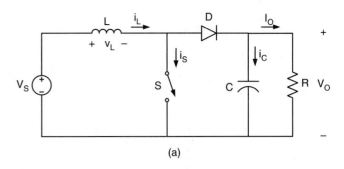

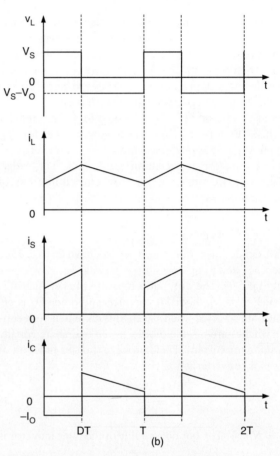

FIGURE 13.9 Boost converter: (a) circuit diagram and (b) waveforms.

is *off*. The minimum value of the filter capacitance that results in the voltage ripple V_r is given by

$$C_{min} = \frac{DV_O}{V_r R f} \quad (13.16)$$

At $D = 0.5$, $V_r/V_O = 1\%$, $R = 10\,\Omega$, and $f = 100\,\text{kHz}$, the minimum capacitance for the boost converter is $C_{min} = 50\,\mu\text{F}$.

The boost converter does not have a popular transformer (isolated) version.

13.5 Buck–Boost Converter

13.5.1 Basic Converter

A non-isolated (transformerless) topology of the buck–boost converter is shown in Fig. 13.10a. The converter consists of dc input voltage source V_S, controlled switch S, inductor L, diode D, filter capacitor C, and load resistance R. With the switch *on*, the inductor current increases while the diode is maintained *off*. When the switch is turned *off*, the diode provides a path for the inductor current. Note the polarity of the diode which results in its current being *drawn from* the output.

The buck–boost converter waveforms are depicted in Fig. 13.10b. The condition of a zero volt–second product for the inductor in steady state yields

$$V_S D T = -V_O (1 - D) T \quad (13.17)$$

Hence, the dc voltage transfer function of the buck–boost converter is

$$M_V \equiv \frac{V_O}{V_S} = -\frac{D}{1-D} \quad (13.18)$$

The output voltage V_O is negative with respect to the ground. Its magnitude can be either greater or smaller (equal at $D = 0.5$) than the input voltage as the converter's name implies.

The value of the inductor that determines the boundary between the CCM and DCM is

$$L_b = \frac{(1-D)^2 R}{2f} \quad (13.19)$$

The structure of the output part of the converter is similar to that of the boost converter (reversed polarities being the only difference). Thus, the value of the filter capacitor can be obtained from Eq. (13.16).

13.5.2 Flyback Converter

A PWM flyback converter is a very practical isolated version of the buck–boost converter. The circuit of the flyback converter is presented in Fig. 13.11a. The inductor of the buck–boost converter has been replaced by a flyback transformer. The input dc source V_S and switch S are connected in series with the primary transformer. The diode D and the RC output circuit are connected in series with the secondary of the flyback transformer. Figure 13.11b shows the converter with a simple flyback transformer model. The model includes a magnetizing inductance L_m and an ideal transformer with a turns ratio $n = N_1/N_2$. The flyback transformer leakage inductances and losses are neglected in the model. It should be noted that leakage inductances, although not important from the principle of operation point of view, affect adversely switch and diode transitions. Snubbers are usually required in flyback converters.

Refer to Fig. 13.11b for the converter operation. When the switch S is *on*, the current in the magnetizing inductance increases linearly. The diode D is *off* and there is no current in the ideal transformer windings. When the switch is turned *off*, the magnetizing inductance current is diverted into the ideal transformer, the diode turns *on*, and the transformed

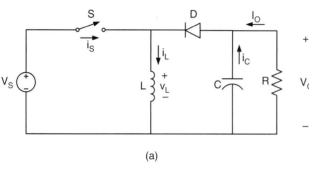

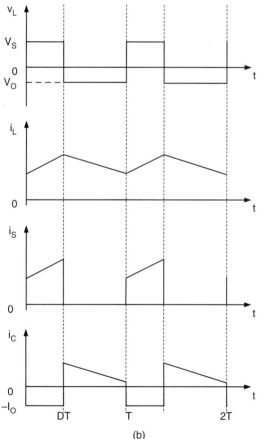

FIGURE 13.10 Buck–boost converter: (a) circuit diagram and (b) waveforms.

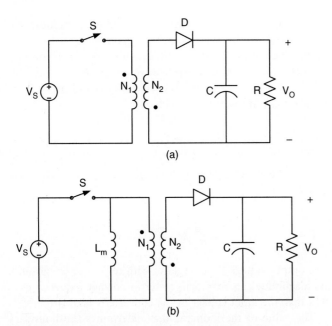

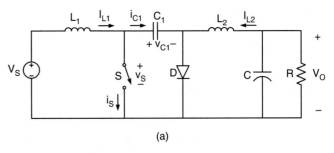

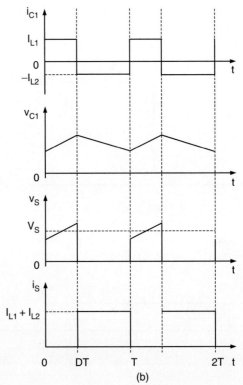

FIGURE 13.11 Flyback converter: (a) circuit diagram and (b) circuit with a transformer model showing the magnetizing inductance L_m.

magnetizing inductance current is supplied to the RC load. The dc voltage transfer function of the flyback converter is

$$M_V \equiv \frac{V_O}{V_S} = \frac{D}{n(1-D)} \qquad (13.20)$$

It differs from the buck–boost converter voltage transfer function by the turns ratio factor n. A positive sign has been obtained by an appropriate coupling of the transformer windings.

Unlike in transformer buck-derived converters, the magnetizing inductance L_m of the flyback transformer is an important design parameter. The value of the magnetizing inductance that determines the boundary between the CCM and DCM is given by

$$L_{mb} = \frac{n^2(1-D)^2 R}{2f} \qquad (13.21)$$

The value of the filter capacitance can be calculated using Eq. (13.16).

13.6 Ćuk Converter

The circuit of the Ćuk converter is shown in Fig. 13.12a. It consists of dc input voltage source V_S, input inductor L_1, controllable switch S, energy transfer capacitor C_1, diode D, filter inductor L_2, filter capacitor C, and load resistance R.

FIGURE 13.12 Ćuk converter: (a) circuit diagram and (b) waveforms.

An important advantage of this topology is a continuous current at both the input and the output of the converter. Disadvantages of the Ćuk converter include a high number of reactive components and high current stresses on the switch, the diode, and the capacitor C_1. Main waveforms in the converter are presented in Fig. 13.12b. When the switch is *on*, the diode is *off* and the capacitor C_1 is discharged by the inductor L_2 current. With the switch in the *off* state, the diode conducts currents of the inductors L_1 and L_2 whereas capacitor C_1 is charged by the inductor L_1 current.

To obtain the dc voltage transfer function of the converter, we shall use the principle that the average current through a capacitor is zero for steady-state operation. Let us assume that inductors L_1 and L_2 are large enough that their ripple current can be neglected. Capacitor C_1 is in steady state if

$$I_{L2}DT = I_{L1}(1-D)T \qquad (13.22)$$

For a lossless converter

$$P_S = V_S I_{L1} = -V_O I_{L2} = P_O \quad (13.23)$$

Combining these two equations, the dc voltage transfer function of the Ćuk converter is

$$M_V \equiv \frac{V_O}{V_S} = -\frac{D}{1-D} \quad (13.24)$$

This voltage transfer function is the same as that for the buck–boost converter.

The boundaries between the CCM and DCM are determined by

$$L_{b1} = \frac{(1-D)R}{2Df} \quad (13.25)$$

for L_1 and

$$L_{b2} = \frac{(1-D)R}{2f} \quad (13.26)$$

for L_2.

The output part of the Ćuk converter is similar to that of the buck converter. Hence, the expression for the filter capacitor C is

$$C_{min} = \frac{(1-D)V_O}{8V_r L_2 f^2} \quad (13.27)$$

The peak-to-peak ripple voltage in the capacitor C_1 can be estimated as

$$V_{r1} = \frac{DV_O}{C_1 R f} \quad (13.28)$$

A transformer (isolated) version of the Ćuk converter can be obtained by splitting capacitor C_1 and inserting a high-frequency transformer between the split capacitors.

13.7 Effects of Parasitics

The analysis of converters in Sections 13.2 through 13.6 has been performed under ideal switch, diode, and passive component assumptions. Non-idealities or parasitics of practical devices and components may, however, greatly affect some performance parameters of dc–dc converters. In this section, effects of parasitics on output voltage ripple, efficiency, and voltage transfer function of converters will be illustrated.

A more realistic model of a capacitor than just a capacitance C, consists of a series connection of capacitance C and resistance r_C. The resistance r_C is called an equivalent series resistance (ESR) of the capacitor and is due to losses in the dielectric and physical resistance of leads and connections. Recall Eq. (13.6) which provided a value of the filter capacitance in a buck converter that limits the peak-to-peak output voltage ripple to V_r. The equation was derived under an assumption that the entire triangular ac component of the inductor current flows through a capacitance C. It is, however, closer to reality to maintain that this triangular component flows through a series connection of capacitance C and resistance r_C.

The peak-to-peak ripple voltage is independent of the voltage across the filter capacitor and is determined only by the ripple voltage of the ESR if the following condition is satisfied,

$$C \geq C_{min} = \max\left\{\frac{1-D_{min}}{2r_C f}, \frac{D_{max}}{2r_C f}\right\} \quad (13.29)$$

If condition (13.29) is satisfied, the peak-to-peak ripple voltage of the buck and forward converters is

$$V_r = r_C \Delta i_{Lmax} = \frac{r_C V_O(1-D_{min})}{fL} \quad (13.30)$$

For push–pull, half-bridge, and full-bridge converters,

$$C \geq C_{min} = \max\left\{\frac{0.5-D_{min}}{2r_C f}, \frac{D_{max}}{2r_C f}\right\} \quad (13.31)$$

where $D_{max} \leq 0.5$. If condition (13.31) is met, the peak-to-peak ripple voltage V_r of these converters is given by

$$V_r = r_C \Delta i_{Lmax} = \frac{r_C V_O(0.5-D_{min})}{fL} \quad (13.32)$$

Waveforms of voltage across the ESR V_{rC}, voltage across the capacitance V_C, and total ripple voltage V_r are depicted in Fig. 13.13 for three values of the filter capacitances. For the case of the top graph in Fig. 13.13, the peak-to-peak value of V_r is higher than the peak-to-peak value of V_{rC} because $C < C_{min}$. Middle and bottom graphs in Fig. 13.13 show the waveforms for $C = C_{min}$ and $C > C_{min}$, respectively. For both these cases, the peak-to-peak voltages of V_r and V_{rC} equal to each other.

Note that when the resistance r_C sets the ripple voltage V_r, the minimum value of inductance L is determined either by the boundary between the CCM and DCM according to Eq. (13.5) (buck and forward converters) or Eq. (13.9) (push–pull, half-bridge, and full-bridge converters), or by the voltage ripple condition (13.30) or (13.32).

In buck–boost and boost converters, the peak-to-peak capacitor current I_{Cpp} is equal to the peak-to-peak diode current and is given by

$$I_{Cpp} = \frac{I_O}{1-D} \quad (13.33)$$

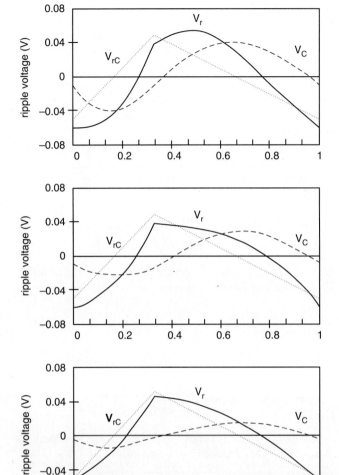

FIGURE 13.13 Voltage ripple waveforms V_{rC}, V_C, and V_r for a buck converter at $V_O = 12\,\text{V}$, $f = 100\,\text{kHz}$, $L = 40\,\mu\text{H}$, $r_C = 0.05\,\Omega$, and various values of C: $C = 33\,\mu\text{F}$ (top graph), $C = C_{min} = 65\,\mu\text{F}$ (middle graph), and $C = 100\,\mu\text{F}$ (bottom graph).

under condition that the inductor current ripple is much lower than the average value of the inductor current. The peak-to-peak voltage across the ESR is

$$V_{rC} = r_C I_{Cpp} = \frac{r_C I_O}{1-D} \quad (13.34)$$

Assuming that the total ripple voltage V_r is approximately equal to the sum of the ripple voltages across the ESR and the capacitance, the maximum value of the peak-to-peak ripple voltage across the capacitance is

$$V_{Cmax} \approx V_r - V_{rC} \quad (13.35)$$

Finally, by analogy to Eq. (13.16), when the ESR of the filter capacitor is taken into account in the boost-type output filter, the filter capacitance should be greater than

$$C_{min} = \frac{DV_O}{V_{Cmax} R f} \quad (13.36)$$

Parasitic resistances, capacitances, and voltage sources affect also an energy conversion efficiency of dc–dc converters. The efficiency η is defined as a ratio of output power to the input power

$$\eta \equiv \frac{P_O}{P_S} = \frac{V_O I_O}{V_S I_S} \quad (13.37)$$

Efficiencies are usually specified in percent. Let us consider the boot converter as an example. Under low ripple assumption, the boost converter efficiency can be estimated as

$$\eta = \frac{R(1-D)^2}{R(1-D)^2(1+(V_D/V_O)+fC_oR)+r_L+Dr_S+(1-D)r_D+D(1-D)r_C} \quad (13.38)$$

where V_D is the forward conduction voltage drop of the diode, C_o is the output capacitance of the switch, r_L is the ESR of the inductor, and r_D is the forward *on* resistance of the diode. The term fC_oR in Eq. (13.38) represents switching losses in the converter. Other terms account for conduction losses. Losses in a dc–dc converter also contribute to a decrease in the dc voltage transfer function. The non-ideal dc voltage transfer function M_{Vn} is a product of the ideal one and the efficiency

$$M_{Vn} = \eta M_V \quad (13.39)$$

Sample graphs for the boost converter that correspond to Eqs. (13.38) and (13.39) are presented in Fig. 13.14.

13.8 Synchronous and Bidirectional Converters

It can be observed in Eq. (13.38) that the forward voltage of a diode V_D contributes to a decrease in efficiency. This contribution is especially significant in low output voltage power supplies, e.g. 3.3 V power supplies for microprocessors or power supplies for portable telecommunication equipment. Even with a Schottky diode, which has V_D in the range of 0.4 V, the power loss in the diode can easily exceed 10% of the total power delivered to the load. To reduce conduction losses in the diode, a low on-resistance switch can be added in parallel as shown in Fig. 13.15 for a buck converter. The input switch and the switch parallel to the diode must be

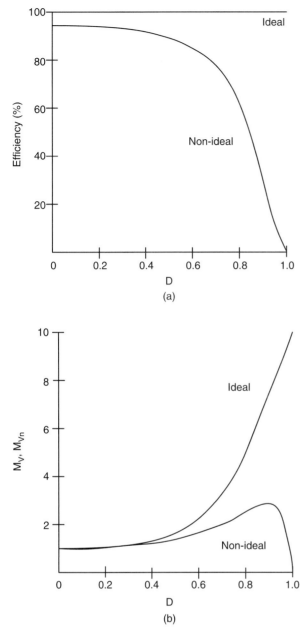

FIGURE 13.14 Effects of parasitics on characteristics of a boost converter: (a) efficiency and (b) dc voltage transfer function.

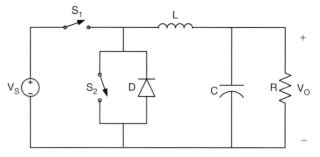

FIGURE 13.15 Synchronous buck converter.

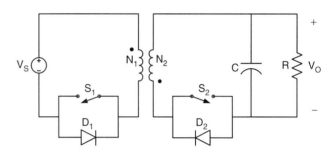

FIGURE 13.16 Bidirectional flyback converter.

turned on and off alternately. The arrangement of Fig. 13.15 is called a synchronous converter or a synchronous rectifier. Modern low-voltage MOSFETs have *on* resistances of only several milliohms. Hence, a synchronous converter may exhibit higher efficiency than a conventional one at output currents as large as tens of amperes. The efficiency is increased at an expense of more complicated driving circuitry for the switches. In particular, a special can must be exercised to avoid having both switches *on* at the same time as this would short the input voltage source. Since power semiconductor devices usually have longer turn-off times than turn-on times, a dead time (sometimes called a blanking time) must be introduced in PWM driving signals.

The parallel combination of a controllable switch and a diode is also used in converters which allow for a current flow in both directions: from the input source to the load and from the load back to the input source. Such converters are called bidirectional power flow or simply bidirectional converters. As an example, a flyback bidirectional converter is shown in Fig. 13.16. It contains unipolar voltage and bidirectional current switch–diode combinations at both primary and secondary of the flyback transformer. When the primary switch and secondary diode operate, the current flows from the input source to the load. The converter current can also flow from the output to the input through the secondary switch and primary diode. Bidirectional arrangements can be made for buck and boost converters. A bidirectional buck converter operates as a boost converter when the current flow is from the output to the input. A bidirectional boost converter operates as a buck converter with a reversed current flow. If for any reason (for instance to avoid the DCM) the controllable switches are driven at the same time, they must be driven alternately with a sufficient dead time to avoid a shot-through current.

13.9 Control Principles

A dc–dc converter must provide a regulated dc output voltage under varying load and input voltage conditions.

The converter component values are also changing with time, temperature, pressure, etc. Hence, the control of the output voltage should be performed in a closed-loop manner using principles of negative feedback. Two most common closed-loop control methods for PWM dc–dc converters, namely, the voltage-mode control and the current-mode control, are presented schematically in Fig. 13.17.

In the voltage-mode control scheme shown in Fig. 13.17a, the converter output voltage is sensed and subtracted from an external reference voltage in an error amplifier. The error amplifier produces a control voltage that is compared to a constant-amplitude sawtooth waveform. The comparator produces a PWM signal which is fed to drivers of controllable switches in the dc–dc converter. The duty ratio of the PWM signal depends on the value of the control voltage. The frequency of the PWM signal is the same as the frequency of the sawtooth waveform. An important advantage of the voltage-mode control is its simple hardware implementation and flexibility.

The error amplifier in Fig. 13.17a reacts fast to changes in the converter output voltage. Thus, the voltage-mode control provides good load regulation, that is, regulation against variations in the load. Line regulation (regulation against variations in the input voltage) is, however, delayed because changes in the input voltage must first manifest themselves in the converter output before they can be corrected. To alleviate this problem, the voltage-mode control scheme is sometimes augmented by so-called voltage feedforward path. The feedforward path affects directly the PWM duty ratio according to variations in the input voltage. As will be explained below, the input voltage feedforward is an inherent feature of current-mode control schemes.

The current-mode control scheme is presented in Fig. 13.17b. An additional inner control loop feeds back an inductor current signal. This current signal, converted into its voltage analog, is compared to the control voltage. This modification of replacing the sawtooth wavefrom of the voltage-mode control scheme by a converter current signal

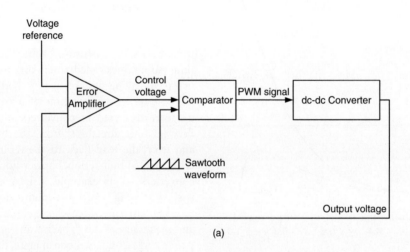

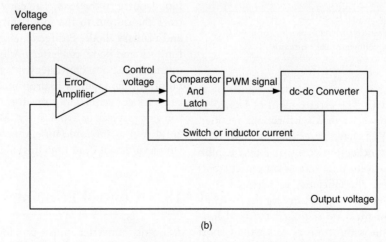

FIGURE 13.17 Main control schemes for dc–dc converters: (a) voltage-mode control and (b) current-mode control.

significantly alters the dynamic behavior of the converter. The converter takes on some characteristics of a current source. The output current in PWM dc–dc converters is either equal to the *average* value of the output inductor current (buck-derived and Ċuk converters) or is a product of an *average* inductor current and a function of the duty ratio. In practical implementations of the current-mode control, it is feasible to sense the peak inductor current instead of the average value. Since the peak inductor current is equal to the peak switch current, the latter can be used in the inner loop which often simplifies the current sensor. Note that the peak inductor (switch) current is proportional to the input voltage. Hence, the inner loop of the current-mode control naturally accomplishes the input voltage feedforward technique. Among several current-mode control versions, the most popular is the constant-frequency one which requires a clock signal. Advantages of the current-mode control include: input voltage feedforward, limit on the peak switch current, equal current sharing in modular converters, and reduction in the converter dynamic order. The main disadvantage of the current-mode control is its complicated hardware which includes a need to compensate the control voltage by ramp signals (to avoid converter instability).

Among other control methods of dc–dc converters, a hysteretic (or bang-bang) control is very simple for hardware implementation. The hysteretic control results, however, in variable frequency operation of semiconductor switches. Generally, a constant switching frequency is preferred in power electronic circuits for easier elimination of electromagnetic interference and better utilization of magnetic components.

Application specific integrated circuits (ASICs) are commercially available that contain main elements of voltage- or current-mode control schemes. On a single 14 or 16-pin chip, there is error amplifier, comparator, sawtooth generator or sensed current input, latch, and PWM drivers. The switching frequency is usually set by an external RC network and can be varied from tens of kilohertz to a few megahertz. The controller has an oscillator output for synchronization with other converters in modular power supply systems. A constant voltage reference is generated on the chip as well. Additionally, the ASIC controller may be equipped in various diagnostic and protection features: current limiting, overvoltage and undervoltage protection, soft start, dead time in case of multiple PWM outputs, and duty ratio limiting. In several dc–dc converter topologies, e.g. buck and buck–boost, neither control terminal of semiconductor switches is grounded (so-called high-side switches). The ASIC controllers are usually designed for a particular topology and their PWM drivers may be able to drive high-side switches in low voltage applications. In high voltage applications, external PWM drivers must be used. External PWM drivers are also used for switches with high input capacitances. To take a full advantage of the input–output isolation in transformer versions of dc–dc converters, such an isolation must be also provided in the control loop. Signal transformers or optocouplers are used for isolating feedback signals.

Dynamic characteristic of closed-loop dc–dc converters must fulfill certain requirements. To simply analysis, these requirements are usually translated into desired properties of the open loop. The open loop should provide a sufficient (typically, at least 45°) phase margin for stability, high bandwidth (about one-tenth of the switching frequency) for good transient response, and high gain (several tens of decibels) at low frequencies for small steady-state error.

The open loop dynamic characteristics are shaped by compensating networks of passive components around the error amplifier. Second or third order RC networks are commonly used. Since the converter itself is a part of the control loop, the design of compensating networks requires a knowledge of small-signal characteristics of the converter. There are several methods of small-signal characterization of PWM dc–dc converters. The most popular methods provide average models of converters under high switching frequency assumption. The averaged models are then linearized at an operating point to obtain small-signal transfer functions. Among analytical averaging methods, state-space averaging has been popular since late 1970s. Circuit-based averaging is usually performed using PWM switch or direct replacement of semiconductor switches by controlled current and voltage sources. All these methods can take into account converter parasitics.

The most important small-signal characteristic is the *control-to-output* transfer function T_p. Other converter characteristics that are investigated include the *input-to-output* (or *line-to-output*) voltage transfer function, also called the *open-loop dynamic line regulation* or the *audio susceptibility*, which describes the input–output disturbance transmission; the *open-loop input impedance*; and the *open-loop dynamic load regulation*. Buck-derived, boost, and buck–boost converters are second order dynamic systems; the Ċuk converter is a fourth-order system. Characteristics of buck and buck-derived converters are similar to each other. Another group of converters with similar small-signal characteristics is formed by boost, buck–boost, and flyback converters. Among parasitic components, the ESR of the filter capacitor r_C introduces additional dynamic terms into transfer functions. Other parasitic resistances usually modify slightly the effective value of the load resistance. Sample characteristics below are given for non-zero r_C, neglecting other parasitics.

The control-to-output transfer function of the forward converter is

$$T_p(s) \equiv \frac{v_o(s)}{d(s)}\bigg|_{v_s(s)=0} = \frac{V_I R r_C}{nL(R + r_C)}$$

$$\times \frac{s + (1/Cr_C)}{s^2 + s(CRr_C + L/LC(R + r_C)) + R/(LC(R + r_C))}$$
(13.40)

It can be seen that this transfer function has two poles and one zero. The zero is due to the filter capacitor ESR. Buck-derived converters can be easily compensated for stability with second-order controllers.

The control-to-output transfer function of the boost converter is given by

$$T_p(s) = -\frac{V_O r_C}{(1-D)(R+r_C)} \times \frac{[s+(1/Cr_C)][s-((1-D)^2 R)/L]}{s^2 + s[((1-D)^2 CRr_C + L)/(LC(R+r_C))] + [((1-D)^2 R)/(LC(R+r_C))]} \quad (13.41)$$

The zero $-(1-D)^2 R/L$ is located in the right half of the s-plane. Therefore, the boost converter (as well as buck–boost and flyback converters) is a non-minimum phase system. Non-minimum phase dc–dc converters are typically compensated with third-order controllers. Step-by-step procedures for a design of compensating networks are usually given by manufacturers of ASIC controllers in application notes.

The final word of this section is on the behavior of dc–dc converters in distributed power supply systems. An important feature of closed-loop regulated dc–dc converters is that they exhibit a negative input resistance. As the load voltage is kept constant by the controller, the output power changes with the load. With slow load changes, an increase (decrease) in the input voltage results in a decrease (increase) in the input power. This negative resistance property must be carefully examined during the system design to avoid resonances.

13.10 Applications of DC–DC Converters

Step-down choppers find most of their applications in high-performance dc drive systems, e.g. electric traction, electric vehicles, and machine tools. The dc motors with their winding inductances and mechanical inertia act as filters resulting in high-quality armature currents. The average output voltage of step-down choppers is a linear function of the switch duty ratio. Step-up choppers are used primarily in radar and ignition systems. The dc choppers can be modified for two-quadrant and four-quadrant operation. Two-quadrant choppers may be a part of autonomous power supply system that contain battery packs and such renewable dc sources as photovoltaic arrays, fuel cells, or wind turbines. Four-quadrant choppers are applied in drives in which regenerative breaking of dc motors is desired, e.g. transportation systems with frequent stops. The dc choppers with inductive outputs serve as inputs to current-driven inverters.

An addition of filtering reactive components to dc choppers results in PWM dc–dc converters. The dc–dc converters can be viewed as dc transformers that deliver to the load as dc voltage or current at a different level than the input source. This dc transformation is performed by electronic switching means, not by electromagnetic means like in conventional transformers. Output voltages of dc–dc converters range from a volt for special VLSI circuits to tens of kilovolts in X-ray lamps. The most common output voltages are: 3.3 V for modern microprocessors, 5 and 12 V for logic circuits, 48 V for telecommunication equipment, and 270 V for main dc bus on airplanes. Typical input voltages include 48 V, 170 V (the peak value of a 120 V rms line), and 270 V.

Selection of a topology of dc–dc converters is determined not only by input/output voltages, which can be additionally adjusted with the turns ratio in isolated converters, but also by power levels, voltage and current stresses of semiconductor switches, and utilization of magnetic components. The low part-count flyback converter is popular in low power applications (up to 200 W). Its main deficiencies are the large size of the flyback transformer core and high voltage stress on the semiconductor switch. The forward converter is also a single switch converter. Since its core size requirements are smaller, it is popular in low/medium (up to several hundreds of watts) power applications. Disadvantages of the forward converter are in a need for demagnetizing winding and in a high voltage stress on the semiconductor switch. The push–pull converter is also used at medium power levels. Due to bidirectional excitation, the transformer size is small. An advantage of the push–pull converter is also a possibility to refer driving terminals of both switches to the ground which greatly simplifies the control circuitry. A disadvantage of the push–pull converter is a potential core saturation in a case of asymmetry. The half-bridge converter has similar range of applications as the push–pull converter. There is no danger of transformer saturation in the half-bridge converter. It requires, however, two additional input capacitors to split in half the input dc source. The full-bridge converter is used at high (several kilowatts) power and voltage levels. The voltage stress on power switches is limited to the input voltage source value. A disadvantage of the full-bridge converter is a high number of semiconductor devices.

The dc–dc converters are building blocks of distributed power supply systems in which a common dc bus voltage is converted to various other voltage according to requirements of particular loads. Such distributed dc systems are common in space stations, ships and airplanes, as well as in computer and telecommunication equipment. It is expected that modern portable wireless communication and signal processing systems will use variable supply voltages to minimize power consumption and extend battery life. Low output voltage converters in these applications utilize the synchronous rectification arrangement.

Another big area of dc–dc converter applications is related to the utility ac grid. For critical loads, if the utility grid fails, there must be a backup source of energy, e.g. a battery pack. This need for continuous power delivery gave rise to various types of uninterruptible power supplies (UPSs). The dc–dc converters are used in UPSs to adjust the level of a rectified grid voltage to that of the backup source. Since during normal operation, the energy flows from the grid to the backup source and during emergency conditions the backup source must supply the load, bidirectional dc–dc converters are often used. The dc–dc converters are also used in dedicated battery chargers.

Power electronic loads, especially those with front-end rectifiers, pollute the ac grid with odd harmonics. The dc–dc converters are used as intermediate stages, just after a rectifier and before the load-supplying dc–dc converter, for shaping the input ac current to improve power factor and decrease the harmonic content. The boost converter is especially popular in such power factor correction (PFC) applications. Another utility grid related application of dc–dc converters is in interfaces between ac networks and dc renewable energy sources such as fuel cells and photovoltaic arrays.

In isolated dc–dc converters, multiple outputs are possible with additional secondary windings of transformers. Only one output is regulated with a feedback loop. Other outputs depend on the duty ratio of the regulated one and on their loads. A multiple-output dc–dc converter is a convenient solution in application where there is a need for one closely regulated output voltage and for one or more non-critical other output voltage levels.

Further Reading

1. R. P. Severns and G. Bloom, *Modern DC-to-DC Switchmode Power Converter Circuits*, New York: Van Nostrand Reinhold Company, 1985.
2. D. W. Hart, *Introduction to Power Electronics*, Englewood Cliffs, NJ: Prentice Hall, 1997.
3. P. T. Krein, *Elements of Power Electronics*, New York: Oxford University Press, 1998.
4. A. I. Pressman, *Switching Power Supply Design, 2nd Ed.*, New York: McGraw-Hill, 1998.
5. A. M. Trzynadlowski, *Introduction to Modern Power Electronics*, New York: Wiley Interscience, 1998.
6. R. Erickson and D. Maksimovic, *Fundamentals of Power Electronics, 2nd Ed.*, Norwell, MA: Kluwer Academic, 2001.
7. M. H. Rashid, *Power Electronics Circuits, Devices, and Applications 3rd Ed.*, Upper Saddle River, NJ: Pearson Prentice Hall, 2003.
8. N. Mohan, T. M. Undeland, and W. P. Robbins, *Power Electronics: Converters, Applications and Design, 3rd Ed.*, New York: John Wiley & Sons, 2003.

14

DC/DC Conversion Technique and Twelve Series Luo-converters

Fang Lin Luo, Ph.D.
School of EEE, Block S1, Nanyang Technological University, Nanyang Avenue, Singapore

Hong Ye, Ph.D.
School of Biological Sciences, Block SBS, Nanyang Technological University, Nanyang Avenue, Singapore

14.1 Introduction 262
14.2 Fundamental, Developed, Transformer-type, and Self-lift Converters 263
  14.2.1 Fundamental Topologies • 14.2.2 Developed Topologies • 14.2.3 Transformer-type Topologies • 14.2.4 Seven (7) Self-lift DC/DC Converters • 14.2.5 Tapped Inductor (Watkins–Johnson) Converters
14.3 Voltage-lift Luo-converters 271
  14.3.1 Positive Output Luo-converters • 14.3.2 Simplified Positive Output (S P/O) Luo-converters • 14.3.3 Negative Output Luo-converters
14.4 Double Output Luo-converters 284
14.5 Super-lift Luo-converters 288
  14.5.1 P/O Super-lift Luo-converters • 14.5.2 N/O Super-lift Luo-converters • 14.5.3 P/O Cascade Boost-converters • 14.5.4 N/O Cascade Boost-converters
14.6 Ultra-lift Luo-converters 299
  14.6.1 Continuous Conduction Mode • 14.6.2 Discontinuous Conduction Mode
14.7 Multiple-quadrant Operating Luo-converters 301
  14.7.1 Forward Two-quadrant DC/DC Luo-converter • 14.7.2 Two-quadrant DC/DC Luo-converter in Reverse Operation • 14.7.3 Four-quadrant DC/DC Luo-converter
14.8 Switched-capacitor Multi-quadrant Luo-converters 306
  14.8.1 Two-quadrant Switched-capacitor DC/DC Luo-converter • 14.8.2 Four-quadrant Switched-capacitor DC/DC Luo-converter
14.9 Multiple-lift Push–Pull Switched-capacitor Luo-converters. 315
  14.9.1 P/O Multiple-lift Push–Pull Switched-capacitor DC/DC Luo-converter • 14.9.2 N/O Multiple-lift Push–Pull Switched-capacitor DC/DC Luo-converter
14.10 Switched-inductor Multi-quadrant Operation Luo-converters 318
  14.10.1 Two-quadrant Switched-inductor DC/DC Luo-converter in Forward Operation • 14.10.2 Two-quadrant Switched-inductor DC/DC Luo-converter in Reverse Operation • 14.10.3 Four-quadrant Switched-inductor DC/DC Luo-converter
14.11 Multi-quadrant ZCS Quasi-resonant Luo-converters 323
  14.11.1 Two-quadrant ZCS Quasi-resonant Luo-converter in Forward Operation • 14.11.2 Two-quadrant ZCS Quasi-resonant Luo-Converter in Reverse Operation • 14.11.3 Four-quadrant ZCS Quasi-resonant Luo-converter
14.12 Multi-quadrant ZVS Quasi-resonant Luo-converters 327
  14.12.1 Two-quadrant ZVS Quasi-resonant Luo-converter in Forward Operation • 14.12.2 Two-quadrant ZVS Quasi-resonant Luo-converter in Reverse Operation • 14.12.3 Four-quadrant ZVS Quasi-resonant Luo-converter
14.13 Synchronous-rectifier DC/DC Luo-converters 331
  14.13.1 Flat Transformer Synchronous-rectifier DC/DC Luo-converter • 14.13.2 Double Current SR DC/DC Luo-converter with Active Clamp Circuit • 14.13.3 Zero-current-switching Synchronous-rectifier DC/DC Luo-converter • 14.13.4 Zero-voltage-switching Synchronous-rectifier DC/DC Luo-converter
14.14 Multiple-element Resonant Power Converters 335
  14.14.1 Two Energy-storage Elements Resonant Power Converters • 14.14.2 Three Energy-storage Elements Resonant Power Converters • 14.14.3 Four Energy-storage Elements Resonant Power Converters • 14.14.4 Bipolar Current and Voltage Sources
14.15 Gate Control Luo-resonator 342
14.16 Applications 343
  14.16.1 5000 V Insulation Test Bench • 14.16.2 MIT 42/14 V–3 KW DC/DC Converter • 14.16.3 IBM 1.8 V/200 A Power Supply
14.17 Energy Factor and Mathematical Modeling for Power DC/DC Converters 345
  14.17.1 Pumping Energy (*PE*) • 14.17.2 Stored Energy (*SE*) • 14.17.3 Energy Factor (*EF*) • 14.17.4 Time Constant τ and Damping Time Constant τ_d • 14.17.5 Mathematical Modeling for Power DC/DC Converters • 14.17.6 Buck Converter with Small Energy Losses ($r_L = 1.5\,\Omega$) • 14.17.7 A Super-lift Luo-converter in CCM
  Further Reading 350

14.1 Introduction

DC/DC converters are widely used in industrial applications and computer hardware circuits. DC/DC conversion technique has been developed very quickly. Since 1920s there have been more than 500 DC/DC converters' topologies developed. Professor Luo and Dr. Ye have systematically sorted them in six generations in 2001. They are the first-generation (classical) converters, second-generation (multi-quadrant) converters, third-generation (switched-component) converters, fourth-generation (soft-switching) converters, fifth-generation (synchronous-rectifier) converters and sixth-generation (multi-element resonant power) converters.

The first-generation converters perform in a single quadrant mode with low power range (up to around 100 W), such as buck converter, boost converter and buck–boost converter. Because of the effects of parasitic elements, the output voltage and power transfer efficiency of all these converters are restricted.

The voltage-lift (VL) technique is a popular method that is widely applied in electronic circuit design. Applying this technique effectively overcomes the effects of parasitic elements and greatly increases the output voltage. Therefore, these DC/DC converters can convert the source voltage into a higher output voltage with high power efficiency, high power density, and a simple structure.

The VL converters have high voltage transfer gains, which increase in arithmetical series stage-by-stage. Super-lift (SL) technique is more powerful to increase the converters voltage transfer gains in geometric series stage-by-stage. Even higher, ultra-lift (UL) technique is most powerful to increase the converters voltage transfer gain.

The second-generation converters perform in two- or four-quadrant operation with medium output power range (say hundreds watts or higher). Because of high power conversion, these converters are usually applied in industrial applications with high power transmission. For example, DC motor drives with multi-quadrant operation. Since most of second-generation converters are still made of capacitors and inductors, they are large.

The third-generation converters are called switched-component DC/DC converters, and made of either inductor or capacitors, which are so-called switched-inductor and switched-capacitors. They usually perform in two- or four-quadrant operation with high output power range (say thousands watts). Since they are made of only inductor or capacitors, they are small.

Switched-capacitor (SC) DC/DC converters are made of only switched-capacitors. Since switched-capacitors can be integrated into power semiconductor integrated circuits (IC) chips, they have limited size and work in high switching frequency. They have been successfully employed in the inductorless DC/DC converters and opened the way to build the converters with high power density. Therefore, they have drawn much attention from the research workers and manufacturers. However, most of these converters in the literature perform single-quadrant operation. Some of them work in the push–pull status. In addition, their control circuit and topologies are very complex, especially, for the large difference between input and output voltages.

Switched-inductor (SI) DC/DC converters are made of only inductor, and have been derived from four-quadrant choppers. They usually perform multi-quadrant operation with very simple structure. The significant advantage of these converters is its simplicity and high power density. No matter how large the difference between the input and output voltages, only one inductor is required for each SI DC/DC converter. Therefore, they are widely required for industrial applications.

The fourth-generation converters are called soft-switching converters. Soft-switching technique involves many methods implementing resonance characteristics. Popular method is resonant-switching. There are three main groups: zero-current-switching (ZCS), zero-voltage-switching (ZVS), and zero-transition (ZT) converters. They usually perform in single quadrant operation in the literature. We have developed this technique in two- and four-quadrant operation with high output power range (say thousands watts).

Multi-quadrant ZCS/ZVS/ZT converters implement ZCS/ZVS technique in four-quadrant operation. Since switches turn on and off at the moment that the current/voltage is equal to zero, the power losses during switching on and off become zero. Consequently, these converters have high power density and transfer efficiency. Usually, the repeating frequency is not very high and the converters work in a mono-resonance frequency, the components of higher order harmonics is very low. Using fast fourier transform (FFT) analysis, we obtain that the total harmonic distortion (THD) is very small. Therefore, the electromagnetic interference (EMI) is weaker, electromagnetic sensitivity (EMS) and electromagnetic compatibility (EMC) are reasonable.

The fifth-generation converters are called synchronous rectifier (SR) DC/DC Converters. Corresponding to the development of the microelectronics and computer science, the power supplies with low output voltage (5 V, 3.3 V, and $1.8 \sim 1.5$ V) and strong output current (30 A, 50 A, 100 A up to 200 A) are widely required in industrial applications and computer peripheral equipment. Traditional diode bridge rectifiers are not available for this requirement. Many prototypes of SR DC/DC converters with soft-switching technique have been developed. The SR DC/DC converters possess the technical feathers with very low voltage and strong current and high power transfer efficiency η (90%, 92% up to 95%) and high power density (22–25 W/in^3).

The sixth-generation converters are called multi-element resonant power converters (RPCs). There are eight topologies of 2-E RPC, 38 topologies of 3-E RPC, and 98 topologies of 4-E RPC. The RPCs have very high current transfer gain, purely harmonic waveform, low power losses and EMI since

14 DC/DC Conversion Technique and 12 Series Luo-converters

they are working in resonant operation. Usually, the sixth-generation RPCs used in large power industrial applications with high output power range (say thousands watts).

The DC/DC converter family tree is shown in Fig. 14.1.

Professor F. L. Luo and Dr. H. Ye have devoted in the subject area of DC/DC conversion technique for a long time and harvested outstanding achievements. They have created **twelve** (12) series converters namely **Luo-converters** and more knowledge which are listed below:

- Positive output Luo-converters;
- Negative output Luo-converters;
- Double output Luo-converters;
- Positive/Negative output super-lift Luo-converters;
- Ultra-lift Luo-converter;
- Multiple-quadrant Luo-converters;
- Switched capacitor multi-quadrant Luo-converters;
- Multiple-lift push-pull switched-capacitor Luo-converters;
- Switched-inductor multi-quadrant Luo-converters;
- Multi-quadrant ZCS quasi-resonant Luo-converters;
- Multi-quadrant ZVS quasi-resonant Luo-converters;
- Synchronous-rectifier DC/DC Luo-converters;
- Multi-element resonant power converters;
- Energy factor and mathematical modeling for power DC/DC converters.

All of their research achievements have been published in the international top-journals and conferences. Many experts, including Prof. Rashid of West Florida University, Prof. Kassakian of MIT, and Prof. Rahman of Memorial University of Newfoundland are very interested in their work, and acknowledged their outstanding achievements.

In this handbook, we only show the circuit diagram and list a few parameters of each converter for readers, such as the output voltage and current, voltage transfer gain and output voltage variation ratio, and the discontinuous condition and output voltage.

After a well discussion of steady-state operation, we prepare one section to investigate the dynamic transient process of DC/DC converters. Energy storage in DC/DC converters have been paid attention long time ago, but it was not well investigated and defined. Professor Fang Lin Luo and Dr. Hong Ye have theoretically defined it and introduced new parameters: energy factor (EF) and other variables. They have also fundamentally established the mathematical modeling and discussed the characteristics of all power DC/DC converters. They have successfully solved the traditional problems.

In this chapter, the input voltage is V_I or V_1 and load voltage is V_O or V_2. Pulse width modulated (PWM) pulse train has repeating frequency f, the repeating period is $T = 1/f$. Conduction duty is k, the switching-on period is kT, and switching-off period is $(1-k)T$. All average values are in capital letter, and instantaneous values in small letter, e.g. V_1 and $v_1(t)$ or v_1. The variation ratio of the free-wheeling diode's current is ζ. Voltage transfer gain is M and power transfer efficiency is η.

14.2 Fundamental, Developed, Transformer-type, and Self-lift Converters

The first-generation converters are called classical converters which perform in a single-quadrant mode and in low. Historically, the development of the first generation converters covers very long time. Many prototypes of these converters have been created. We can sort them in six categories:

- Fundamental topologies: buck converter, boost converter, and buck–boost converter.
- Developed topologies: positive output Luo-converter, negative output Luo-converter, double output Luo-converter, Cúk-converter, and single-ended primary inductance converter (SEPIC).
- Transformer-type topologies: forward converter, push–pull converter, fly-back converter, half-bridge converter, bridge converter, and ZETA.
- Voltage-lift topologies: self-lift converters, positive output Luo-converters, negative output Luo-converters, double output Luo-converters.
- Super-lift topologies: positive/negative output super-lift Luo-converters, positive/negative output cascade boost-converters.
- Ultra-lift topologies: ultra-lift Luo-converter.

14.2.1 Fundamental Topologies

Buck converter is a step-down converter, which is shown in Fig. 14.2a, the equivalent circuits during switch-on and -off periods are shown in Figs. 14.2b and c. Its output voltage and output current are

$$V_2 = kV_1 \tag{14.1}$$

and

$$I_2 = \frac{1}{k}I_1 \tag{14.2}$$

This converter may work in discontinuous mode if the frequency f is small, conduction duty k is small, inductance L is small, and load current is high.

Boost converter is a step-up converter, which is shown in Fig. 14.3a, the equivalent circuits during switch-on and -off periods are shown in Figs. 14.3b and c. Its output voltage and current are

$$V_2 = \frac{1}{1-k}V_1 \tag{14.3}$$

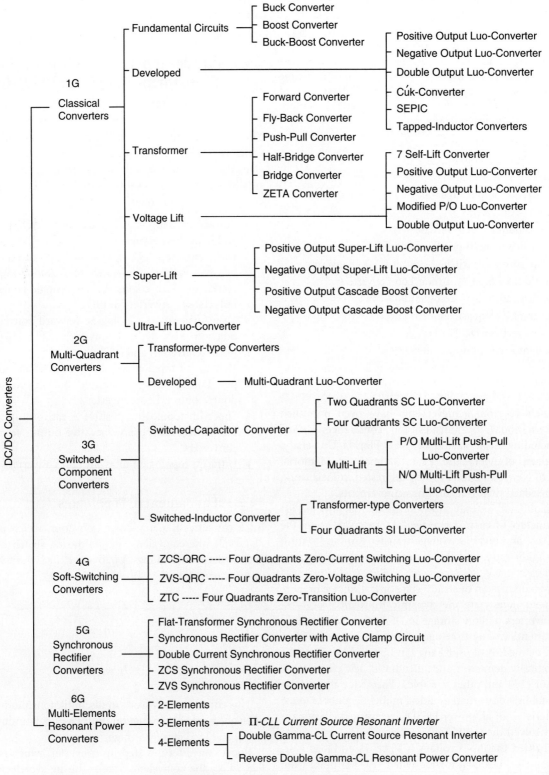

FIGURE 14.1 DC/DC converter family tree.

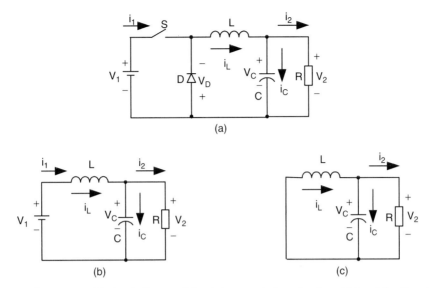

FIGURE 14.2 Buck converter: (a) circuit diagram; (b) switch-on equivalent circuit; and (c) switch-off equivalent circuit.

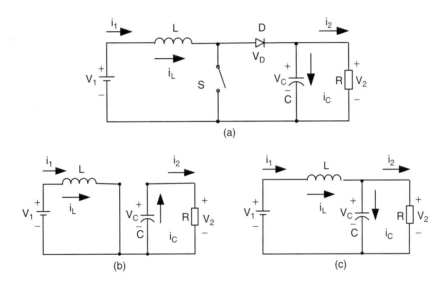

FIGURE 14.3 Boost converter: (a) circuit diagram; (b) switch-on equivalent circuit; and (c) switch-off equivalent circuit.

and

$$I_2 = (1-k)I_1 \quad (14.4)$$

The output voltage is higher than the input voltage. This converter may work in discontinuous mode if the frequency f is small, conduction duty k is small, inductance L is small, and load current is high.

Buck–boost converter is a step–down/up converter, which is shown in Fig. 14.4a, the equivalent circuits during switch-on and -off periods are shown in Figs. 14.4b and c. Its output voltage and current are

$$V_2 = \frac{k}{1-k}V_1 \quad (14.5)$$

and

$$I_2 = \frac{1-k}{k}I_1 \quad (14.6)$$

When k is greater than 0.5, the output voltage can be higher than the input voltage. This converter may work in

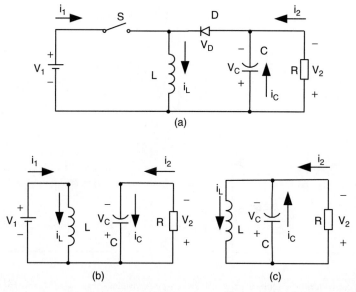

FIGURE 14.4 Buck-boost converter: (a) circuit diagram; (b) switch-on equivalent circuit; and (c) switch-off equivalent circuit.

discontinuous mode if the frequency f is small, conduction duty k is small, inductance L is small, and load current is high.

14.2.2 Developed Topologies

For convenient applications, all developed converters have output voltage and current as

$$V_2 = \frac{k}{1-k} V_1 \qquad (14.7)$$

and

$$I_2 = \frac{1-k}{k} I_1 \qquad (14.8)$$

Positive output (P/O) Luo-converter is a step-down/up converter, and is shown in Fig. 14.5. This converter may work in discontinuous mode if the frequency f is small, k is small, and inductance L is small.

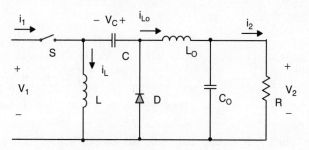

FIGURE 14.5 Positive output Luo-converter.

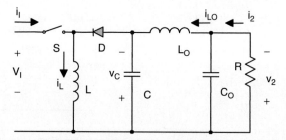

FIGURE 14.6 Negative output Luo-converter.

Negative output (N/O) Luo-converter is shown in Fig. 14.6. This converter may work in discontinuous mode if the frequency f is small, k is small, inductance L is small, and load current is high.

Double output Luo-converter is a double output step-down/up converter, which is derived from **P/O Luo-converter** and **N/O Luo-converter**. It has two conversion paths and two output voltages V_{O+} and V_{O-}. It is shown in Fig. 14.7. If the components are carefully selected the output voltages and currents (concentrate the absolute value) obtained are

$$V_2^+ = |V_2^-| = \frac{k}{1-k} V_1 \qquad (14.9)$$

and

$$I_2^+ = \frac{1-k}{k} I_1^+ \quad \text{and} \quad I_2^- = \frac{1-k}{k} I_1^- \qquad (14.10)$$

When k is greater than 0.5, the output voltage can be higher than the input voltage. This converter may work in discontinuous mode if the frequency f is small, k is small, inductance L is small, and load current is high.

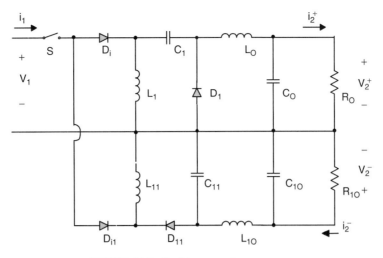

FIGURE 14.7 Double output Luo-converter.

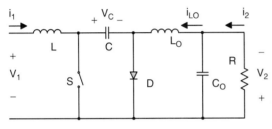

FIGURE 14.8 Cúk converter.

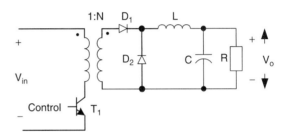

FIGURE 14.10 Forward converter.

Cúk-converter is a negative output step-down/up converter, which is derived from boost and buck converters. It is shown in Fig. 14.8.

Single-ended primary inductance converter is a positive output step-down/up converter, which is derived from boost converters. It is shown in Fig. 14.9.

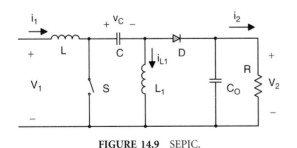

FIGURE 14.9 SEPIC.

14.2.3 Transformer-type Topologies

All transformer-type converters have transformer(s) to isolate the input and output voltages. Therefore, it is easy to obtain the high or low output voltage by changing the turns ratio N, the positive or negative polarity by changing the winding direction, and multiple output voltages by setting multiple secondary windings.

Forward converter is a step-up/down converter, which is shown in Fig. 14.10. The transformer turns ratio is N (usually $N > 1$). If the transformer has never been saturated during operation, it works as a buck converter. The output voltage and current are

$$V_O = kNV_I \qquad (14.11)$$

and

$$I_O = \frac{1}{kN} I_I \qquad (14.12)$$

This converter may work in discontinuous mode if the frequency f is small, conduction duty k is small, inductance L is small, and load current is high.

To avoid the saturation of transformer applied in forward converters, a tertiary winding is applied. The corresponding circuit diagram is shown in Fig. 14.11.

To obtain multiple output voltages we can set multiple secondary windings. The corresponding circuit diagram is shown in Fig. 14.12.

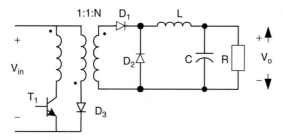

FIGURE 14.11 Forward converter with tertiary winding.

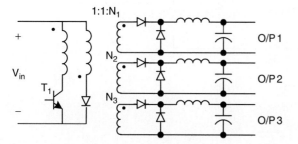

FIGURE 14.12 Forward converter with multiple secondary windings.

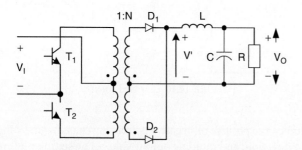

FIGURE 14.13 Push–pull converter.

Push–pull converter is a step-up/down converter, which is shown in Fig. 14.13. It is not necessary to set the tertiary winding. The transformer turns ratio is N (usually $N > 1$). If the transformer has never been saturated during operation, it works as a buck converter with the conduction duty cycle $k < 0.5$. The output voltage and current are

$$V_O = 2kNV_I \tag{14.13}$$

and

$$I_O = \frac{1}{2kN}I_I \tag{14.14}$$

This converter may work in discontinuous mode if the frequency f is small, conduction duty k is small, inductance L is small, and load current is high.

Fly-back converter is a high step-up converter, which is shown in Fig. 14.14. The transformer turns ratio is N (usually $N > 1$). It effectively uses the transformer leakage inductance

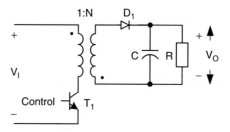

FIGURE 14.14 Fly-back converter.

in fly-back operation to obtain high surge voltage induced, then get high output voltage. It works likely in buck–boost operation as a buck–boost converter. Its output voltage and current are

$$V_O = \frac{kN}{1-k}V_I \tag{14.15}$$

and

$$I_O = \frac{1-k}{kN}I_I \tag{14.16}$$

Half-bridge converter is a step-up converter, which is shown in Fig. 14.15. There are two switches and one double secondary coils transformer required. The transformer turns ratio is N. It works as a half-bridge rectifier (half of V_I inputs to primary winding) plus a buck converter circuit in secondary side. The conduction duty cycle k is set in $0.1 < k < 0.5$. Its output voltage and current are

$$V_O = 2kN\frac{V_I}{2} = kNV_I \tag{14.17}$$

and

$$I_O = \frac{1}{kN}I_I \tag{14.18}$$

This converter may work in discontinuous mode if the frequency f is small, conduction duty k is small, inductance L is small, and load current is high.

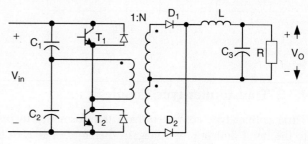

FIGURE 14.15 Half-bridge converter.

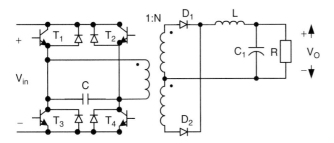

FIGURE 14.16 Bridge converter.

Bridge converter is a step-up converter, which is shown in Fig. 14.16. There are four switches and one double secondary coils transformer required. The transformer turns ratio is N. It works as a full-bridge rectifier (full V_1 inputs to primary winding) plus a buck-converter circuit in secondary side. The conduction duty cycle k is set in $0.1 < k < 0.5$. Its output voltage and current are

$$V_O = 2kNV_I \tag{14.19}$$

and

$$I_O = \frac{1}{2kN} I_I \tag{14.20}$$

ZETA (zeta) converter is a step-up converter, which is shown in Fig. 14.17. The transformer turns ratio is N. The transformer functions as a inductor (L_1) plus a buck–boost converter plus a low-pass filter (L_2–C_2). Its output voltage and current are

$$V_O = \frac{k}{1-k} NV_I \tag{14.21}$$

and

$$I_O = \frac{1-k}{kN} I_I \tag{14.22}$$

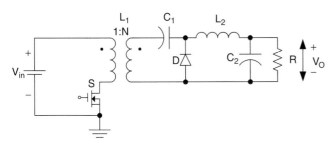

FIGURE 14.17 ZETA (zeta) converter.

14.2.4 Seven (7) Self-lift DC/DC Converters

Because of the effect of the parasitic elements, the voltage conversion gain is limited. Especially, when the conduction duty k is towards unity, the output voltage is sharply reduced.

Voltage-lift technique is a popular method used in electronic circuit design. Applying this technique can effectively overcome the effect of the parasitic elements, and largely increase the voltage transfer gain. In this section, we introduce seven self-lift converters which are working in continuous mode.

- Positive output (P/O) self-lift Luo-converter;
- Reverse P/O self-lift Luo-converter;
- Negative output (N/O) self-lift Luo-converter;
- Reverse N/O self-lift Luo-converter;
- Self-lift Cúk-converter;
- Self-lift SEPIC;
- Enhanced self-lift Luo-converter.

All self-lift converters (except enhanced self-lift circuit) have the output voltage and current to be

$$V_O = \frac{1}{1-k} V_I \tag{14.23}$$

and

$$I_O = (1-k) I_I \tag{14.24}$$

The voltage transfer gain in continuous mode is

$$M_S = \frac{V_O}{V_I} = \frac{I_I}{I_O} = \frac{1}{1-k} \tag{14.25}$$

P/O self-lift Luo-converter is shown in Fig. 14.18. The variation ratio of the output voltage v_O in continuous conduction mode (CCM) is

$$\varepsilon = \frac{\Delta v_O/2}{V_O} = \frac{k}{8M_S} \frac{1}{f^2 C_O L_2} \tag{14.26}$$

Reverse P/O self-lift Luo-converter is shown in Fig. 14.19. The variation ratio of the output voltage v_O in CCM is

$$\varepsilon = \frac{\Delta v_O/2}{V_O} = \frac{k}{16M_S} \frac{1}{f^2 C_O L_2} \tag{14.27}$$

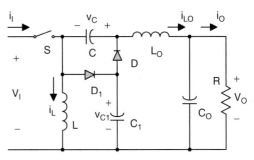

FIGURE 14.18 P/O self-lift Luo-converter.

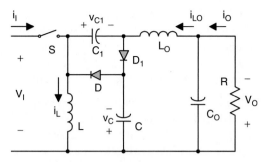

FIGURE 14.19 Reverse P/O self-lift Luo-converter.

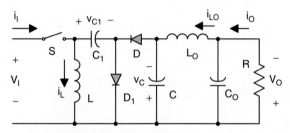

FIGURE 14.20 N/O self-lift Luo-converter.

N/O self-lift Luo-converter is shown in Fig. 14.20. The variation ratio of the output voltage v_O in CCM is

$$\varepsilon = \frac{\Delta v_O/2}{V_O} = \frac{k}{128} \frac{1}{f^3 L_O C_1 C_O R} \quad (14.28a)$$

Reverse N/O self-lift Luo-converter is shown in Fig. 14.21. The variation ratio of the output voltage v_O in CCM is

$$\varepsilon = \frac{\Delta v_O/2}{V_O} = \frac{k}{128} \frac{1}{f^3 L_O C_1 C_O R} \quad (14.28b)$$

Self-lift Cúk-converter is shown in Fig. 14.22. The variation ratio of the output voltage v_O in CCM is

$$\varepsilon = \frac{\Delta v_O/2}{V_O} = \frac{k}{128} \frac{1}{f^3 L_O C_1 C_O R} \quad (14.28c)$$

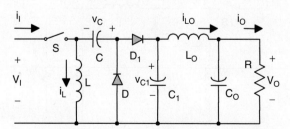

FIGURE 14.21 Reverse N/O self-lift Luo-converter.

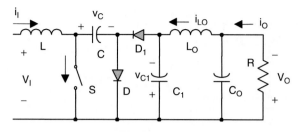

FIGURE 14.22 Self-lift Cúk-converter.

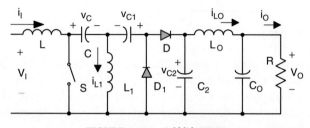

FIGURE 14.23 Self-lift SEPIC.

Self-lift SEPIC is shown in Fig. 14.23. The variation ratio of the output voltage v_O in CCM is

$$\varepsilon = \frac{\Delta v_O/2}{V_O} = \frac{k}{128} \frac{1}{f^3 L_O C_1 C_O R} \quad (14.28d)$$

Enhanced self-lift Luo-converter is shown in Fig. 14.24. Its output voltage and current are

$$V_O = \frac{2-k}{1-k} V_I \quad (14.29)$$

and

$$I_O = \frac{1-k}{2-k} I_I \quad (14.30)$$

The voltage transfer gain in continuous mode is

$$M_S = \frac{V_O}{V_I} = \frac{I_I}{I_O} = \frac{1}{1-k} + 1 = \frac{2-k}{1-k} \quad (14.31)$$

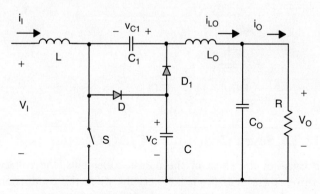

FIGURE 14.24 Enhanced self-lift Luo-converter.

TABLE 14.1 The circuit diagrams of the tapped inductor fundamental converters

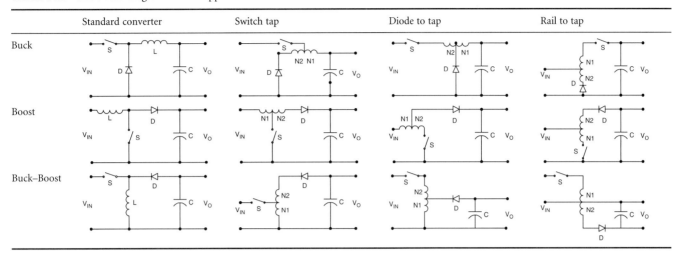

The variation ratio of the output voltage v_O in CCM is as in Eq. (14.26)

$$\varepsilon = \frac{\Delta v_O/2}{V_O} = \frac{k}{8M_S} \frac{1}{f^2 C_O L_2}$$

14.2.5 Tapped Inductor (Watkins–Johnson) Converters

Tapped inductor (Watkins–Johnson) converters have been derived from fundamental converters, which circuit diagrams are shown in Table 14.1. The voltage transfer gains are shown in Table 14.2. Here the tapped inductor ratio is $n = n1/(n1 + n2)$.

14.3 Voltage-lift Luo-converters

Voltage-lift (**VL**) technique is very popular for electronic circuit design. Professor Luo and Dr. Ye have successfully applied this technique in the design of DC/DC converters, and created

TABLE 14.2 The voltage transfer gains of the tapped inductor fundamental converters

Converter	No tap	Switched to tap	Diode to tap	Rail to tap
Buck	k	$\dfrac{k}{n+k(1-n)}$	$\dfrac{nk}{1+k(n-1)}$	$\dfrac{k-n}{k(1-n)}$
Boost	$\dfrac{1}{1-k}$	$\dfrac{n+k(1-n)}{n(1-k)}$	$\dfrac{1+k(n-1)}{1-k}$	$\dfrac{n-k}{n(1-k)}$
Buck–Boost	$\dfrac{k}{1-k}$	$\dfrac{k}{n(1-k)}$	$\dfrac{nk}{1-k}$	$\dfrac{k}{1-k}$

a number of up-to-date converters. There are three series of Luo-converters introduced in this section:

- Positive output Luo-converters;
- Simplified positive output Luo-converters;
- Negative output Luo-converters.

14.3.1 Positive Output Luo-converters

Positive output (P/O) Luo-converters perform the voltage conversion from positive to positive voltages using the voltage lift technique. They work in the first-quadrant with large voltage amplification. Their voltage transfer gains are high. Five circuits are introduced in the literature. They are:

- Elementary circuit;
- Self-lift circuit;
- Re-lift circuit;
- Triple-lift circuit;
- Quadruple-lift circuit.

Further lift circuits can be derived from the above circuits. In all P/O Luo-Converters, we define normalized inductance $L = L_1 L_2/(L_1 + L_2)$ and normalized impedance $z_N = R/fL$.

P/O Luo-converter elementary circuit is shown in Fig. 14.25a. The equivalent circuits during switch-on and -off periods are shown in Figs. 14.25b and c. Its output voltage and current are

$$V_O = \frac{k}{1-k} V_I$$

and

$$I_O = \frac{1-k}{k} I_I$$

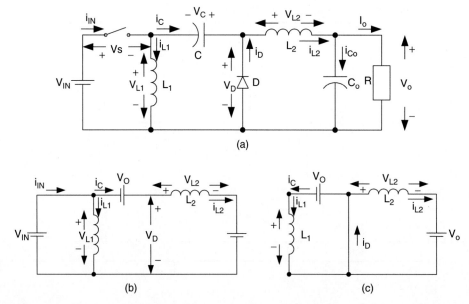

FIGURE 14.25 P/O Luo-converter elementary circuit; (a) circuit diagram; (b) switch on; and (c) switch off.

The voltage transfer gain in continuous mode is

$$M_E = \frac{V_O}{V_I} = \frac{I_I}{I_O} = \frac{k}{1-k} \quad (14.32)$$

The variation ratio of the output voltage v_O in CCM is

$$\varepsilon = \frac{\Delta v_O/2}{V_O} = \frac{k}{16 M_E} \frac{1}{f^2 C_O L_2} \quad (14.33)$$

This converter may work in discontinuous conduction mode if the frequency f is small, conduction duty k is small, inductance L is small, and load current is high. The condition for discontinuous conduction mode (DCM) is

$$M_E \leq k\sqrt{\frac{z_N}{2}} \quad (14.34)$$

The output voltage in DCM is

$$V_O = k(1-k)\frac{R}{2fL}V_I \quad \text{with} \quad \sqrt{\frac{R}{2fL}} \geq \frac{1}{1-k} \quad (14.35)$$

P/O Luo-converter self-lift circuit is shown in Fig. 14.26a. The equivalent circuits during switch-on and -off periods are shown in Figs. 14.26b and c. Its output voltage and current are

$$V_O = \frac{1}{1-k}V_I$$

and

$$I_O = (1-k)I_I$$

The voltage transfer gain in continuous mode is

$$M_S = \frac{V_O}{V_I} = \frac{I_I}{I_O} = \frac{1}{1-k} \quad (14.36)$$

The variation ratio of the output voltage v_O in CCM is

$$\varepsilon = \frac{\Delta v_O/2}{V_O} = \frac{k}{16 M_S} \frac{1}{f^2 C_O L_2} \quad (14.37)$$

This converter may work in discontinuous conduction mode if the frequency f is small, conduction duty k is small, inductance L is small, and load current is high. The condition for DCM is

$$M_S \leq \sqrt{k}\sqrt{\frac{z_N}{2}} \quad (14.38)$$

The output voltage in DCM is

$$V_O = \left[1 + k^2(1-k)\frac{R}{2fL}\right]V_I \quad \text{with} \quad \sqrt{k}\sqrt{\frac{R}{2fL}} \geq \frac{1}{1-k} \quad (14.39)$$

P/O Luo-converter re-lift circuit is shown in Fig. 14.27a. The equivalent circuits during switch-on and -off periods are shown in Figs. 14.27b and c. Its output voltage and current are

$$V_O = \frac{2}{1-k}V_I$$

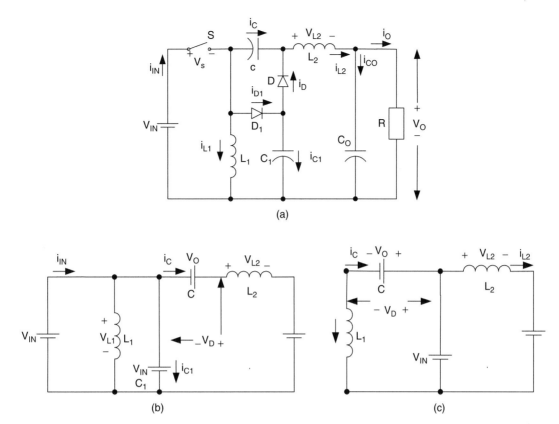

FIGURE 14.26 P/O Luo-converter self-lift circuit: (a) circuit diagram; (b) switch on; and (c) switch off.

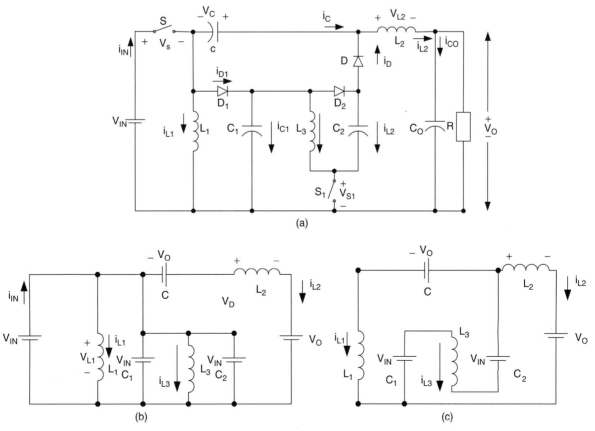

FIGURE 14.27 P/O Luo-converter re-lift circuit: (a) circuit diagram; (b) switch on; and (c) switch off.

and

$$I_O = \frac{1-k}{2} I_I$$

The voltage transfer gain in CCM is

$$M_R = \frac{V_O}{V_I} = \frac{I_I}{I_O} = \frac{2}{1-k} \quad (14.40)$$

The variation ratio of the output voltage v_O in CCM is

$$\varepsilon = \frac{\Delta v_O/2}{V_O} = \frac{k}{16 M_R} \frac{1}{f^2 C_O L_2} \quad (14.41)$$

This converter may work in discontinuous conduction mode if the frequency f is small, conduction duty k is small, inductance L is small, and load current is high. The condition for DCM is

$$M_R \leq \sqrt{k z_N} \quad (14.42)$$

The output voltage in DCM is

$$V_O = \left[2 + k^2(1-k)\frac{R}{2fL}\right] V_I \quad \text{with} \quad \sqrt{k}\sqrt{\frac{R}{fL}} \geq \frac{2}{1-k} \quad (14.43)$$

P/O Luo-converter triple-lift circuit is shown in Fig. 14.28a. The equivalent circuits during switch-on and -off periods are shown in Figs. 14.28b and c. Its output voltage and current are

$$V_O = \frac{3}{1-k} V_I$$

and

$$I_O = \frac{1-k}{3} I_I$$

The voltage transfer gain in CCM is

$$M_T = \frac{V_O}{V_I} = \frac{I_I}{I_O} = \frac{3}{1-k} \quad (14.44)$$

The variation ratio of the output voltage v_O in CCM is

$$\varepsilon = \frac{\Delta v_O/2}{V_O} = \frac{k}{16 M_T} \frac{1}{f^2 C_O L_2} \quad (14.45)$$

This converter may work in discontinuous conduction mode if the frequency f is small, conduction duty k is small, inductance L is small, and load current is high. The condition for DCM is

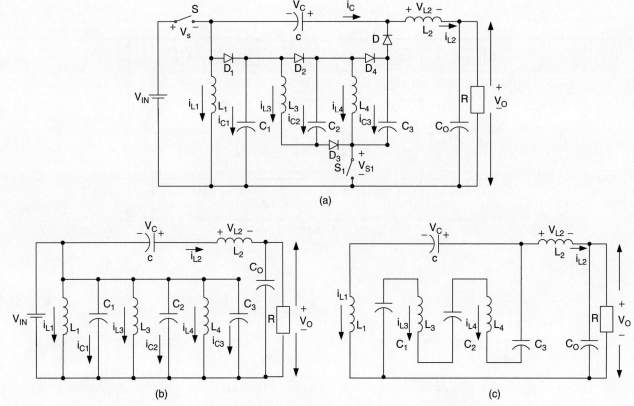

FIGURE 14.28 P/O Luo-converter triple-lift circuit: (a) circuit diagram; (b) switch on; and (c) switch off.

$$M_T \leq \sqrt{\frac{3kz_N}{2}} \quad (14.46)$$

The output voltage in DCM is

$$V_O = \left[3 + k^2(1-k)\frac{R}{2fL}\right]V_I \quad \text{with} \quad \sqrt{k}\sqrt{\frac{3R}{2fL}} \geq \frac{3}{1-k} \quad (14.47)$$

P/O Luo-converter quadruple-lift circuit is shown in Fig. 14.29a. The equivalent circuits during switch-on and -off periods are shown in Figs. 14.29b and c. Its output voltage and current are

$$V_O = \frac{4}{1-k}V_I$$

and

$$I_O = \frac{1-k}{4}I_I$$

The voltage transfer gain in CCM is

$$M_Q = \frac{V_O}{V_I} = \frac{I_I}{I_O} = \frac{4}{1-k} \quad (14.48)$$

The variation ratio of the output voltage v_O in CCM is

$$\varepsilon = \frac{\Delta v_O/2}{V_O} = \frac{k}{16M_Q}\frac{1}{f^2 C_O L_2} \quad (14.49)$$

This converter may work in discontinuous conduction mode if the frequency f is small, conduction duty k is small, inductance L is small, and load current is high. The condition for DCM is

$$M_Q \leq \sqrt{2kz_N} \quad (14.50)$$

The output voltage in DCM is

$$V_O = \left[4 + k^2(1-k)\frac{R}{2fL}\right]V_I \quad \text{with} \quad \sqrt{k}\sqrt{\frac{2R}{fL}} \geq \frac{4}{1-k} \quad (14.51)$$

Summary for all P/O Luo-converters:

$$M = \frac{V_O}{V_I} = \frac{I_I}{I_O}; \quad L = \frac{L_1 L_2}{L_1 + L_2}; \quad z_N = \frac{R}{fL}; \quad R = \frac{V_O}{I_O}$$

To write common formulas for all circuits parameters, we define that subscript $j = 0$ for the elementary circuit, $j = 1$

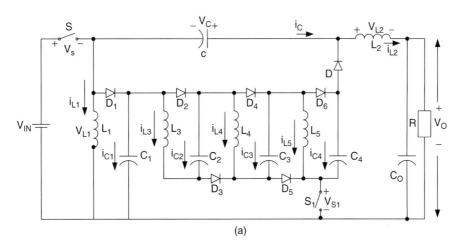

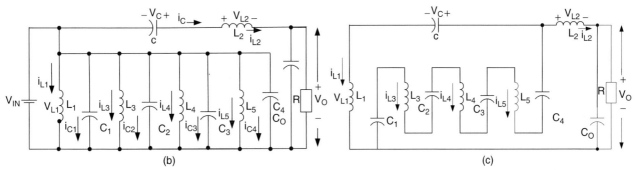

FIGURE 14.29 P/O Luo-converter quadruple-lift circuit: (a) circuit diagram; (b) switch on; and (c) switch off.

for the self-lift circuit, $j = 2$ for the re-lift circuit, $j = 3$ for the triple-lift circuit, $j = 4$ for the quadruple-lift circuit, and so on. The voltage transfer gain is

$$M_j = \frac{k^{h(j)}[j + h(j)]}{1 - k} \quad (14.52)$$

The variation ratio of the output voltage is

$$\varepsilon_j = \frac{\Delta v_O/2}{V_O} = \frac{k}{16 M_j} \frac{1}{f^2 C_O L_2} \quad (14.53)$$

The condition for discontinuous conduction mode is

$$\frac{k^{[1+h(j)]}}{M_j^2} \frac{j + h(j)}{2} z_N \geq 1 \quad (14.54)$$

The output voltage in discontinuous conduction mode is

$$V_{O-j} = \left\{ j + k^{[2-h(j)]} \frac{1 - k}{2} z_N \right\} V_I \quad (14.55)$$

where

$$h(j) = \begin{cases} 0 & \text{if } j \geq 1 \\ 1 & \text{if } j = 0 \end{cases} \quad (14.56)$$

is the **Hong** function.

14.3.2 Simplified Positive Output (S P/O) Luo-converters

Carefully check P/O Luo-converters, we can see that there are two switches required from re-lift circuit. In order to use only one switch in all P/O Luo-converters, we modify the circuits. In this section we introduce following four circuits:

- Simplified self-lift circuit;
- Simplified re-lift circuit;
- Simplified triple-lift circuit;
- Simplified quadruple-lift circuit.

Further lift circuits can be derived from the above circuits. In all S P/O Luo-converters, we define normalized impedance $z_N = R/fL$.

S P/O Luo-converter self-lift circuit is shown in Fig. 14.30a. The equivalent circuits during switch-on and -off periods are shown in Figs. 14.30b and c. Its output voltage and current are

$$V_O = \frac{1}{1 - k} V_I$$

and

$$I_O = (1 - k) I_I$$

The voltage transfer gain in CCM is

$$M_S = \frac{V_O}{V_I} = \frac{I_I}{I_O} = \frac{1}{1 - k} \quad (14.57)$$

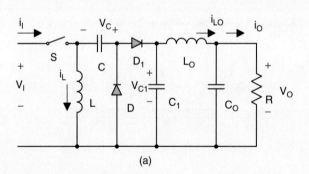

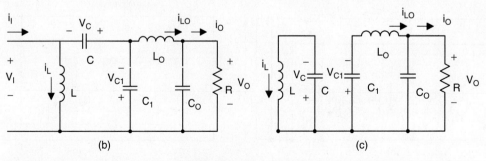

FIGURE 14.30 S P/O Luo-converter self-lift circuit: (a) circuit diagram; (b) switch on; and (c) switch off.

The variation ratio of the output voltage v_O in CCM is

$$\varepsilon = \frac{\Delta v_O/2}{V_O} = \frac{k}{128} \frac{1}{f^3 L_O C_1 C_O R} \quad (14.58)$$

This converter may work in discontinuous conduction mode if the frequency f is small, conduction duty k is small, inductance L is small, and load current is high. The condition for DCM is

$$M_S \leq \sqrt{k}\sqrt{\frac{z_N}{2}} \quad (14.59)$$

The output voltage in DCM is

$$V_O = \left[1 + k^2(1-k)\frac{R}{2fL}\right] V_I \quad \text{with} \quad \sqrt{k}\sqrt{\frac{R}{2fL}} \geq \frac{1}{1-k} \quad (14.60)$$

S P/O Luo-converter re-lift circuit is shown in Fig. 14.31a. The equivalent circuits during switch-on and -off periods are shown in Figs. 14.31b and c. Its output voltage and current are

$$V_O = \frac{2}{1-k} V_I$$

and

$$I_O = \frac{1-k}{2} I_I$$

The voltage transfer gain in CCM is

$$M_R = \frac{V_O}{V_I} = \frac{I_I}{I_O} = \frac{2}{1-k} \quad (14.61)$$

The variation ratio of the output voltage v_O in CCM is

$$\varepsilon = \frac{\Delta v_O/2}{V_O} = \frac{k}{128} \frac{1}{f^3 L_O C_1 C_O R} \quad (14.62)$$

This converter may work in discontinuous conduction mode if the frequency f is small, conduction duty k is small, inductance L is small, and load current is high. The condition for DCM is

$$M_R \leq \sqrt{k z_N} \quad (14.63)$$

The output voltage in DCM is

$$V_O = \left[2 + k^2(1-k)\frac{R}{2fL}\right] V_I \quad \text{with} \quad \sqrt{k}\sqrt{\frac{R}{fL}} \geq \frac{2}{1-k} \quad (14.64)$$

S P/O Luo triple-lift circuit is shown in Fig. 14.32a. The equivalent circuits during switch-on and -off periods are shown in Figs. 14.32b and c. Its output voltage and current are

$$V_O = \frac{3}{1-k} V_I$$

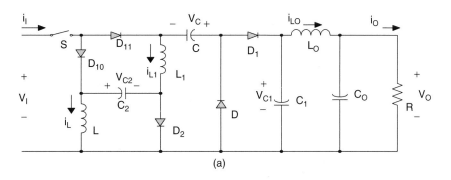

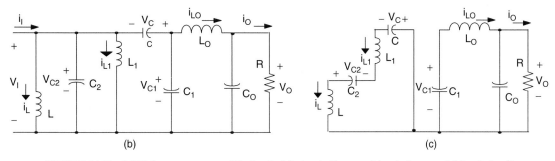

FIGURE 14.31 S P/O Luo-converter re-lift circuit: (a) circuit diagram; (b) switch on; and (c) switch off.

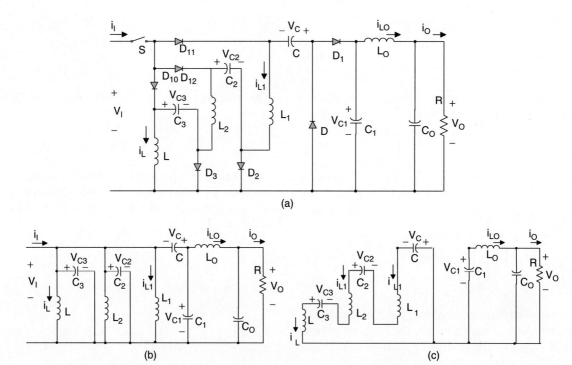

FIGURE 14.32 S P/O Luo-converter triple-lift circuit: (a) circuit diagram; (b) switch on; and (c) switch off.

and

$$I_O = \frac{1-k}{3} I_I$$

The voltage transfer gain in CCM is

$$M_T = \frac{V_O}{V_I} = \frac{I_I}{I_O} = \frac{3}{1-k} \quad (14.65)$$

The variation ratio of the output voltage v_O in CCM is

$$\varepsilon = \frac{\Delta v_O/2}{V_O} = \frac{k}{128} \frac{1}{f^3 L_O C_1 C_O R} \quad (14.66)$$

This converter may work in discontinuous conduction mode if the frequency f is small, conduction duty k is small, inductance L is small, and load current is high. The condition for DCM is

$$M_T \leq \sqrt{\frac{3kz_N}{2}} \quad (14.67)$$

The output voltage in DCM is

$$V_O = \left[3 + k^2(1-k)\frac{R}{2fL}\right] V_I \quad \text{with} \quad \sqrt{k}\sqrt{\frac{3R}{2fL}} \geq \frac{3}{1-k} \quad (14.68)$$

S P/O Luo quadruple-lift circuit is shown in Fig. 14.33a. The equivalent circuits during switch-on and -off periods are shown in Figs. 14.33b and c. Its output voltage and current are

$$V_O = \frac{4}{1-k} V_I$$

and

$$I_O = \frac{1-k}{4} I_I$$

The voltage transfer gain in CCM is

$$M_Q = \frac{V_O}{V_I} = \frac{I_I}{I_O} = \frac{4}{1-k} \quad (14.69)$$

The variation ratio of the output voltage v_O in CCM is

$$\varepsilon = \frac{\Delta v_O/2}{V_O} = \frac{k}{128} \frac{1}{f^3 L_O C_1 C_O R} \quad (14.70)$$

This converter may work in discontinuous conduction mode if the frequency f is small, conduction duty k is small, inductance L is small, and load current is high. The condition for DCM is

$$M_Q \leq \sqrt{2kz_N} \quad (14.71)$$

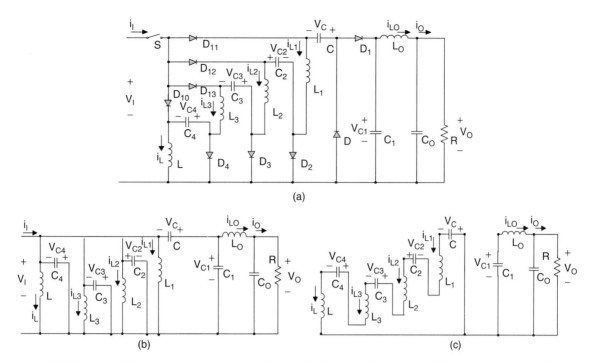

FIGURE 14.33 S P/O Luo-converter quadruple-lift circuit: (a) circuit diagram; (b) switch on; and (c) switch off.

The output voltage in DCM is

$$V_O = \left[4 + k^2(1-k)\frac{R}{2fL}\right]V_I \quad \text{with} \quad \sqrt{k}\sqrt{\frac{2R}{fL}} \geq \frac{4}{1-k}$$
(14.72)

Summary for all S P/O Luo-converters:

$$M = \frac{V_O}{V_I} = \frac{I_I}{I_O}; \quad z_N = \frac{R}{fL}; \quad R = \frac{V_O}{I_O}$$

To write common formulas for all circuits parameters, we define that subscript $j = 1$ for the self-lift circuit, $j = 2$ for the re-lift circuit, $j = 3$ for the triple-lift circuit, $j = 4$ for the quadruple-lift circuit, and so on. The voltage transfer gain is

$$M_j = \frac{j}{1-k}$$
(14.73)

The variation ratio of the output voltage is

$$\varepsilon_j = \frac{\Delta v_O/2}{V_O} = \frac{k}{128}\frac{1}{f^3 L_O C_1 C_O R}$$
(14.74)

The condition for discontinuous mode is

$$M_j \leq \sqrt{\frac{jkz_N}{2}}$$
(14.75)

The output voltage in discontinuous mode is

$$V_{O-j} = \left[j + k^2(1-k)\frac{z_N}{2}\right]V_I$$
(14.76)

14.3.3 Negative Output Luo-converters

Negative output (N/O) Luo-converters perform the voltage conversion from positive to negative voltages using the voltage-lift technique. They work in the third-quadrant with large voltage amplification. Their voltage transfer gains are high. Five circuits are introduced in the literature. They are:

- Elementary circuit;
- Self-lift circuit;
- Re-lift circuit;
- Triple-lift circuit;
- Quadruple-lift circuit.

Further lift circuits can be derived from above circuits. In all N/O Luo-converters, we define normalized impedance $z_N = R/fL$.

N/O Luo-converter elementary circuit is shown in Fig. 14.34a. The equivalent circuits during switch-on and -off periods are shown in Figs. 14.34b and c. Its output voltage and current (the absolute value) are

$$V_O = \frac{k}{1-k}V_I$$

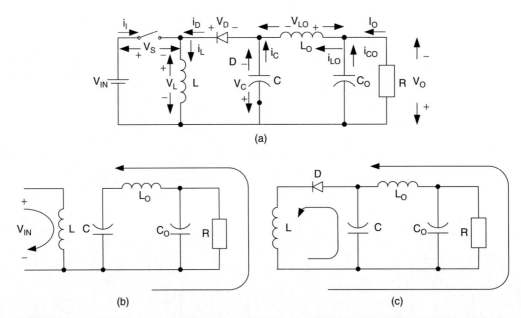

FIGURE 14.34 N/O Luo-converter elementary circuit: (a) circuit diagram; (b) switch on; and (c) switch off.

and

$$I_O = \frac{1-k}{k} I_I$$

When k is greater than 0.5, the output voltage can be higher than the input voltage.

The voltage transfer gain in CCM is

$$M_E = \frac{V_O}{V_I} = \frac{I_I}{I_O} = \frac{k}{1-k} \quad (14.77)$$

The variation ratio of the output voltage v_O in CCM is

$$\varepsilon = \frac{\Delta v_O/2}{V_O} = \frac{k}{128} \frac{1}{f^3 CC_O L_O R} \quad (14.78)$$

This converter may work in discontinuous conduction mode if the frequency f is small, conduction duty k is small, inductance L is small, and load current is high. The condition for DCM is

$$M_E \leq k\sqrt{\frac{z_N}{2}} \quad (14.79)$$

The output voltage in DCM is

$$V_O = k(1-k)\frac{R}{2fL} V_I \quad \text{with} \quad \sqrt{\frac{R}{2fL}} \geq \frac{1}{1-k} \quad (14.80)$$

N/O Luo-converter self-lift circuit is shown in Fig. 14.35a. The equivalent circuits during switch-on and -off periods are shown in Figs. 14.35b and c. Its output voltage and current (the absolute value) are

$$V_O = \frac{1}{1-k} V_I$$

and

$$I_O = (1-k) I_I$$

The voltage transfer gain in CCM is

$$M_S = \frac{V_O}{V_I} = \frac{I_I}{I_O} = \frac{1}{1-k} \quad (14.81)$$

The variation ratio of the output voltage v_O in CCM is

$$\varepsilon = \frac{\Delta v_O/2}{V_O} = \frac{k}{128} \frac{1}{f^3 CC_O L_O R} \quad (14.82)$$

This converter may work in discontinuous conduction mode if the frequency f is small, conduction duty k is small, inductance L is small, and load current is high. The condition for DCM is

$$M_S \leq \sqrt{k}\sqrt{\frac{z_N}{2}} \quad (14.83)$$

The output voltage in DCM is

$$V_O = \left[1 + k^2(1-k)\frac{R}{2fL}\right] V_I \quad \text{with} \quad \sqrt{k}\sqrt{\frac{R}{2fL}} \geq \frac{1}{1-k} \quad (14.84)$$

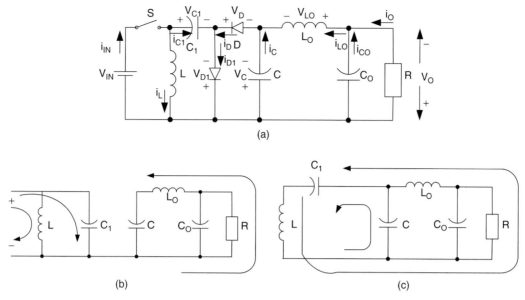

FIGURE 14.35 N/O Luo-converter self-lift circuit: (a) circuit diagram; (b) switch on; and (c) switch off.

N/O Luo-converter re-lift circuit is shown in Fig. 14.36a. The equivalent circuits during switch-on and -off periods are shown in Figs. 14.36b and c. Its output voltage and current (the absolute value) are

$$V_O = \frac{2}{1-k} V_I$$

and

$$I_O = \frac{1-k}{2} I_I$$

The voltage transfer gain in CCM is

$$M_R = \frac{V_O}{V_I} = \frac{I_I}{I_O} = \frac{2}{1-k} \quad (14.85)$$

The variation ratio of the output voltage v_O in CCM is

$$\varepsilon = \frac{\Delta v_O/2}{V_O} = \frac{k}{128} \frac{1}{f^3 C C_O L_O R} \quad (14.86)$$

This converter may work in discontinuous conduction mode if the frequency f is small, conduction duty k is small, inductance L is small, and load current is high. The condition for DCM is

$$M_R \leq \sqrt{k z_N} \quad (14.87)$$

The output voltage in DCM is

$$V_O = \left[2 + k^2(1-k)\frac{R}{2fL}\right] V_I \quad \text{with} \quad \sqrt{k}\sqrt{\frac{R}{fL}} \geq \frac{2}{1-k} \quad (14.88)$$

N/O Luo-converter triple-lift circuit is shown in Fig. 14.37a. The equivalent circuits during switch-on and -off periods are shown in Figs. 14.37b and c. Its output voltage and current (the absolute value) are

$$V_O = \frac{3}{1-k} V_I$$

and

$$I_O = \frac{1-k}{3} I_I$$

The voltage transfer gain in CCM is

$$M_T = \frac{V_O}{V_I} = \frac{I_I}{I_O} = \frac{3}{1-k} \quad (14.89)$$

The variation ratio of the output voltage v_O in CCM is

$$\varepsilon = \frac{\Delta v_O/2}{V_O} = \frac{k}{128} \frac{1}{f^3 C C_O L_O R} \quad (14.90)$$

This converter may work in discontinuous conduction mode if the frequency f is small, conduction duty k is small,

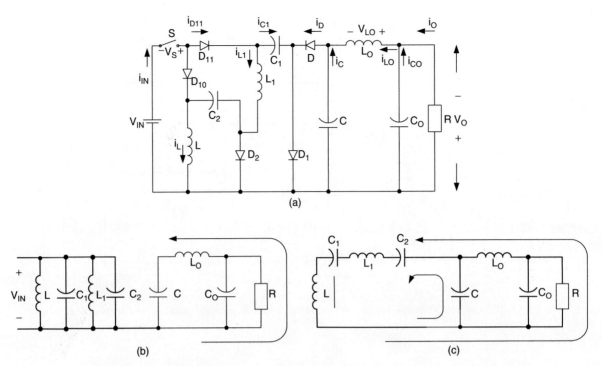

FIGURE 14.36 N/O Luo-converter re-lift circuit: (a) circuit diagram; (b) switch on; and (c) switch off.

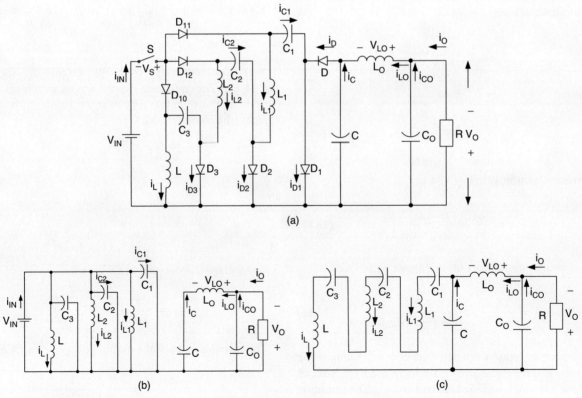

FIGURE 14.37 N/O Luo-converter triple-lift circuit: (a) circuit diagram; (b) switch on; and (c) switch off.

inductance L is small, and load current is high. The condition for DCM is

$$M_T \leq \sqrt{\frac{3kz_N}{2}} \quad (14.91)$$

The output voltage in DCM is

$$V_O = \left[3 + k^2(1-k)\frac{R}{2fL}\right]V_I \quad \text{with} \quad \sqrt{k}\sqrt{\frac{3R}{2fL}} \geq \frac{3}{1-k} \quad (14.92)$$

N/O Luo-converter quadruple-lift circuit is shown in Fig. 14.38a. The equivalent circuits during switch-on and -off periods are shown in Figs. 14.38b and c. Its output voltage and current (the absolute value) are

$$V_O = \frac{4}{1-k}V_I$$

and

$$I_O = \frac{1-k}{4}I_I$$

The voltage transfer gain in CCM is

$$M_Q = \frac{V_O}{V_I} = \frac{I_I}{I_O} = \frac{4}{1-k} \quad (14.93)$$

The variation ratio of the output voltage v_O in CCM is

$$\varepsilon = \frac{\Delta v_O/2}{V_O} = \frac{k}{128}\frac{1}{f^3 CC_O L_O R} \quad (14.94)$$

This converter may work in discontinuous conduction mode if the frequency f is small, conduction duty k is small, inductance L is small, and load current is high. The condition for DCM is

$$M_Q \leq \sqrt{2kz_N} \quad (14.95)$$

The output voltage in DCM is

$$V_O = \left[4 + k^2(1-k)\frac{R}{2fL}\right]V_I \quad \text{with} \quad \sqrt{k}\sqrt{\frac{2R}{fL}} \geq \frac{4}{1-k} \quad (14.96)$$

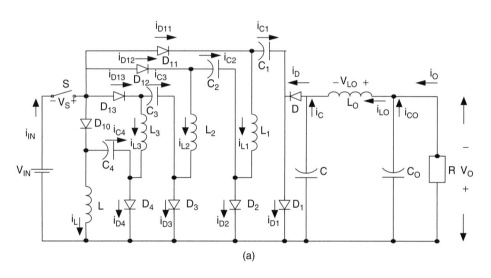

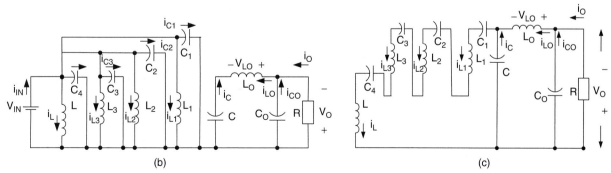

FIGURE 14.38 N/O Luo-converter quadruple-lift circuit: (a) circuit diagram; (b) switch on; and (c) switch off.

Summary for all N/O Luo-converters:

$$M = \frac{V_O}{V_I} = \frac{I_I}{I_O}; \quad z_N = \frac{R}{fL}; \quad R = \frac{V_O}{I_O}$$

To write common formulas for all circuits parameters, we define that subscript $j = 0$ for the elementary circuit, $j = 1$ for the self-lift circuit, $j = 2$ for the re-lift circuit, $j = 3$ for the triple-lift circuit, $j = 4$ for the quadruple-lift circuit, and so on. The voltage transfer gain is

$$M_j = \frac{k^{h(j)}[j + h(j)]}{1 - k} \quad (14.97)$$

The variation ratio of the output voltage is

$$\varepsilon = \frac{\Delta v_O/2}{V_O} = \frac{k}{128} \frac{1}{f^3 CC_O L_O R} \quad (14.98)$$

The condition for discontinuous conduction mode is

$$\frac{k^{[1+h(j)]}}{M_j^2} \frac{j + h(j)}{2} z_N \geq 1 \quad (14.99)$$

The output voltage in discontinuous conduction mode is

$$V_{O-j} = \left\{ j + k^{[2-h(j)]} \frac{1-k}{2} z_N \right\} V_I \quad (14.100)$$

where

$$h(j) = \begin{cases} 0 & \text{if } j \geq 1 \\ 1 & \text{if } j = 0 \end{cases}$$

is the **Hong** function.

14.4 Double Output Luo-converters

Double output (D/O) Luo-converters perform the voltage conversion from positive to positive and negative voltages simultaneously using the voltage-lift technique. They work in the first- and third-quadrants with high voltage transfer gain. There are five circuits introduced in this section:

- D/O Luo-converter elementary circuit;
- D/O Luo-converter self-lift circuit;
- D/O Luo-converter re-lift circuit;
- D/O Luo-converter triple-lift circuit;
- D/O Luo-converter quadruple-lift circuit.

Further lift circuits can be derived from above circuits. In all D/O Luo-converters, each circuit has two conversion paths – positive conversion path and negative conversion path. The positive path likes P/O Luo-converters, and the negative path likes N/O Luo-converters. We define normalized impedance $z_{N+} = R/fL$ for positive path, and normalized impedance $z_{N-} = R_1/fL_{11}$. We usually purposely select $R = R_1$ and $L = L_{11}$, so that we have $z_N = z_{N+} = z_{N-}$.

D/O Luo-converter elementary circuit is shown in Fig. 14.7. Its output voltages and currents (absolute values) are

$$V_{O+} = |V_{O-}| = \frac{k}{1-k} V_I$$

$$I_{O+} = \frac{1-k}{k} I_{I+}$$

and

$$I_{O-} = \frac{1-k}{k} I_{I-}$$

When k is greater than 0.5, the output voltage can be higher than the input voltage.

The voltage transfer gain in CCM is

$$M_E = \frac{V_{O+}}{V_I} = \frac{|V_{O-}|}{V_I} = \frac{k}{1-k} \quad (14.101)$$

The variation ratio of the output voltage v_{O+} in CCM is

$$\varepsilon_+ = \frac{\Delta v_{O+}/2}{V_{O+}} = \frac{k}{16 M_E} \frac{1}{f^2 C_O L_2} \quad (14.102)$$

The variation ratio of the output voltage v_{O-} in CCM is

$$\varepsilon_- = \frac{\Delta v_{O-}/2}{V_{O-}} = \frac{k}{128} \frac{1}{f^3 C_{11} C_{10} L_{12} R_1} \quad (14.103)$$

This converter may work in discontinuous conduction mode if the frequency f is small, conduction duty k is small, inductance L is small, and load current is high. The condition for DCM is

$$M_E \leq k \sqrt{\frac{z_N}{2}} \quad (14.104)$$

The output voltages in DCM are

$$V_O = V_{O+} = |V_{O-}| = k(1-k)\frac{z_N}{2} V_I \quad \text{with} \quad \sqrt{\frac{z_N}{2}} \geq \frac{1}{1-k} \quad (14.105)$$

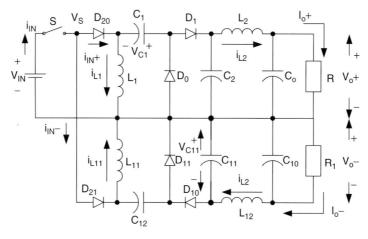

FIGURE 14.39 Double output Luo-converter self-lift circuit.

D/O Luo-converter self-lift circuit is shown in Fig. 14.39. Its output voltages and currents (absolute values) are

$$V_{O+} = |V_{O-}| = \frac{1}{1-k} V_I$$

$$I_{O+} = (1-k) I_{I+}$$

and

$$I_{O-} = (1-k) I_{I-}$$

The voltage transfer gain in CCM is

$$M_S = \frac{V_{O+}}{V_I} = \frac{|V_{O-}|}{V_I} = \frac{1}{1-k} \quad (14.106)$$

The variation ratio of the output voltage v_{O+} in CCM is

$$\varepsilon_+ = \frac{\Delta v_{O+}/2}{V_{O+}} = \frac{k}{128} \frac{1}{f^3 L_2 C_O C_2 R} \quad (14.107)$$

The variation ratio of the output voltage v_{O-} in CCM is

$$\varepsilon_- = \frac{\Delta v_{O-}/2}{V_{O-}} = \frac{k}{128} \frac{1}{f^3 C_{11} C_{10} L_{12} R_1} \quad (14.108)$$

This converter may work in discontinuous conduction mode if the frequency f is small, conduction duty k is small, inductance L is small, and load current is high. The condition for DCM is

$$M_S \leq \sqrt{k}\sqrt{\frac{z_N}{2}} \quad (14.109)$$

The output voltages in DCM are

$$V_O = V_{O+} = |V_{O-}| = \left[1 + k^2(1-k)\frac{z_N}{2}\right] V_I$$

with $\quad \sqrt{\dfrac{k z_N}{2}} \geq \dfrac{1}{1-k} \quad (14.110)$

D/O Luo-converter re-lift circuit is shown in Fig. 14.40. Its output voltages and currents (absolute values) are

$$V_{O+} = |V_{O-}| = \frac{2}{1-k} V_I$$

$$I_{O+} = \frac{1-k}{2} I_{I+}$$

and

$$I_{O-} = \frac{1-k}{2} I_{I-}$$

The voltage transfer gain in CCM is

$$M_R = \frac{V_{O+}}{V_I} = \frac{|V_{O-}|}{V_I} = \frac{2}{1-k} \quad (14.111)$$

The variation ratio of the output voltage v_{O+} in CCM is

$$\varepsilon_+ = \frac{\Delta v_{O+}/2}{V_{O+}} = \frac{k}{128} \frac{1}{f^3 L_2 C_O C_2 R} \quad (14.112)$$

The variation ratio of the output voltage v_{O-} in CCM is

$$\varepsilon_- = \frac{\Delta v_{O-}/2}{V_{O-}} = \frac{k}{128} \frac{1}{f^3 C_{11} C_{10} L_{12} R_1} \quad (14.113)$$

This converter may work in discontinuous conduction mode if the frequency f is small, conduction duty k is small,

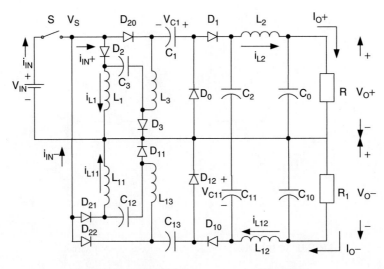

FIGURE 14.40 D/O Luo-converter re-lift circuit.

inductance L is small, and load current is high. The condition for DCM is

$$M_R \leq \sqrt{kz_N} \quad (14.114)$$

The output voltages in DCM are

$$V_O = V_{O+} = |V_{O-}| = \left[2 + k^2(1-k)\frac{z_N}{2}\right]V_I$$

with $\sqrt{kz_N} \geq \dfrac{2}{1-k} \quad (14.115)$

D/O Luo-converter triple-lift circuit is shown in Fig. 14.41. Its output voltages and currents (absolute values) are

$$V_{O+} = |V_{O-}| = \frac{3}{1-k}V_I$$

$$I_{O+} = \frac{1-k}{3}I_{I+}$$

and

$$I_{O-} = \frac{1-k}{3}I_{I-}$$

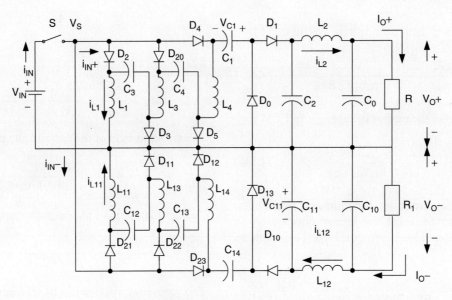

FIGURE 14.41 D/O Luo-converter triple-lift circuit.

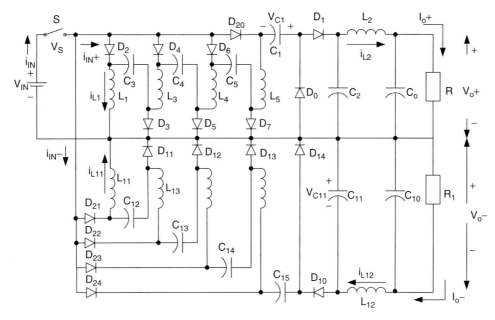

FIGURE 14.42 D/O Luo-converter quadruple-lift circuit.

The voltage transfer gain in CCM is

$$M_T = \frac{V_{O+}}{V_I} = \frac{|V_{O-}|}{V_I} = \frac{3}{1-k} \quad (14.116)$$

The variation ratio of the output voltage v_{O+} in CCM is

$$\varepsilon_+ = \frac{\Delta v_{O+}/2}{V_{O+}} = \frac{k}{128} \frac{1}{f^3 L_2 C_O C_2 R} \quad (14.117)$$

The variation ratio of the output voltage v_{O-} in CCM is

$$\varepsilon_- = \frac{\Delta v_{O-}/2}{V_{O-}} = \frac{k}{128} \frac{1}{f^3 C_{11} C_{10} L_{12} R_1} \quad (14.118)$$

This converter may work in discontinuous conduction mode if the frequency f is small, conduction duty k is small, inductance L is small, and load current is high. The condition for DCM is

$$M_T \leq \sqrt{\frac{3kz_N}{2}} \quad (14.119)$$

The output voltages in DCM are

$$V_O = V_{O+} = |V_{O-}| = \left[3 + k^2(1-k)\frac{z_N}{2}\right] V_I$$

with $\sqrt{\dfrac{3kz_N}{2}} \geq \dfrac{3}{1-k} \quad (14.120)$

D/O Luo-converter quadruple-lift circuit is shown in Fig. 14.42. Its output voltages (absolute values) are

$$V_{O+} = |V_{O-}| = \frac{4}{1-k} V_I$$

$$I_{O+} = \frac{1-k}{4} I_{I+}$$

and

$$I_{O-} = \frac{1-k}{4} I_{I-}$$

The voltage transfer gain in CCM is

$$M_Q = \frac{V_{O+}}{V_I} = \frac{|V_{O-}|}{V_I} = \frac{4}{1-k} \quad (14.121)$$

The variation ratio of the output voltage v_{O+} in CCM is

$$\varepsilon_+ = \frac{\Delta v_{O+}/2}{V_{O+}} = \frac{k}{128} \frac{1}{f^3 L_2 C_O C_2 R} \quad (14.122)$$

The variation ratio of the output voltage v_{O-} in CCM is

$$\varepsilon_- = \frac{\Delta v_{O-}/2}{V_{O-}} = \frac{k}{128} \frac{1}{f^3 C_{11} C_{10} L_{12} R_1} \quad (14.123)$$

This converter may work in discontinuous conduction mode if the frequency f is small, conduction duty k is small,

inductance L is small, and load current is high. The condition for DCM is

$$M_Q \leq \sqrt{2kz_N} \qquad (14.124)$$

The output voltages in DCM are

$$V_O = V_{O+} = |V_{O-}| = \left[4 + k^2(1-k)\frac{z_N}{2}\right]V_I$$

with $\sqrt{2kz_N} \geq \dfrac{4}{1-k}$ (14.125)

Summary for all D/O Luo-converters:

$$M = \frac{V_{O+}}{V_I} = \frac{|V_{O-}|}{V_I}; \quad L = \frac{L_1 L_2}{L_1 + L_2}; \quad L = L_{11}; \quad R = R_1;$$

$$z_{N+} = \frac{R}{fL}; \quad z_{N-} = \frac{R_1}{fL_{11}}$$

so that

$$z_N = z_{N+} = z_{N-}$$

To write common formulas for all circuits parameters, we define that subscript $j = 0$ for the elementary circuit, $j = 1$ for the self-lift circuit, $j = 2$ for the re-lift circuit, $j = 3$ for the triple-lift circuit, $j = 4$ for the quadruple-lift circuit, and so on. The voltage transfer gain is

$$M_j = \frac{k^{h(j)}[j + h(j)]}{1-k} \qquad (14.126)$$

The variation ratio of the output voltage v_{O+} in CCM is

$$\varepsilon_{+j} = \frac{\Delta v_{O+}/2}{V_{O+}} = \frac{k}{128}\frac{1}{f^3 L_2 C_O C_2 R} \qquad (14.127)$$

The variation ratio of the output voltage v_{O-} in CCM is

$$\varepsilon_{-j} = \frac{\Delta v_{O-}/2}{V_{O-}} = \frac{k}{128}\frac{1}{f^3 C_{11} C_{10} L_{12} R_1} \qquad (14.128)$$

The condition for DCM is

$$\frac{k^{[1+h(j)]}}{M_j^2}\frac{j + h(j)}{2}z_N \geq 1 \qquad (14.129)$$

The output voltage in DCM is

$$V_{O-j} = \left\{j + k^{[2-h(j)]}\frac{1-k}{2}z_N\right\}V_I \qquad (14.130)$$

where

$$h(j) = \begin{cases} 0 & \text{if } j \geq 1 \\ 1 & \text{if } j = 0 \end{cases}$$

is the **Hong** function.

14.5 Super-lift Luo-converters

Voltage-lift (VL) technique has been successfully applied in DC/DC converter's design. However, the output voltage of all VL converters increases in arithmetic progression stage-by-stage. Super-lift (SL) technique is more powerful than VL technique. The output voltage of all SL converters increases in geometric progression stage-by-stage. All super-lift converters are outstanding contributions in DC/DC conversion technology, and invented by Professor Luo and Dr. Ye in 2000–2003. There are four series SL Converters introduced in this section:

1. Positive output (P/O) super-lift Luo-converters;
2. Negative output (N/O) super-lift Luo-converters;
3. Positive output (P/O) cascade boost-converter;
4. Negative output (N/O) cascade boost-converter;

14.5.1 P/O Super-lift Luo-converters

There are several sub-series of P/O super-lift Luo-converters:

- Main series;
- Additional series;
- Enhanced series;
- Re-enhanced series;
- Multi-enhanced series.

We only introduce three circuits of main series and additional series.

P/O SL Luo-converter elementary circuit is shown in Fig. 14.43a. The equivalent circuits during switch on and switch off are shown in Figs. 14.43b and c. Its output voltage and current are

$$V_O = \frac{2-k}{1-k}V_I$$

and

$$I_O = \frac{1-k}{2-k}I_I$$

The voltage transfer gain is

$$M_E = \frac{V_O}{V_I} = \frac{2-k}{1-k} \qquad (14.131)$$

The variation ratio of the output voltage v_O is

$$\varepsilon = \frac{\Delta v_O/2}{V_O} = \frac{k}{2RfC_2} \qquad (14.132)$$

P/O SL Luo-converter re-lift circuit is shown in Fig. 14.44a. The equivalent circuits during switch on and switch off are

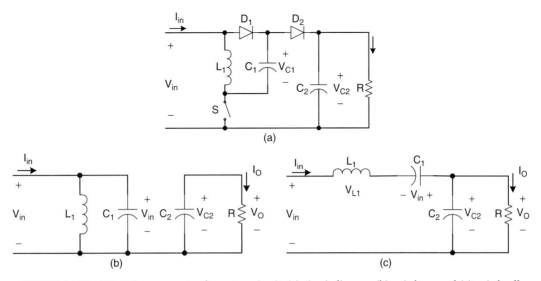

FIGURE 14.43 P/O SL Luo-converter elementary circuit: (a) circuit diagram; (b) switch on; and (c) switch off.

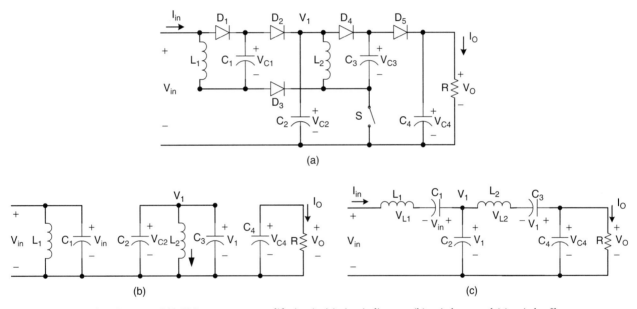

FIGURE 14.44 P/O SL Luo-converter re-lift circuit: (a) circuit diagram; (b) switch on; and (c) switch off.

shown in Figs. 14.44b and c. Its output voltage and current are

$$V_O = \left(\frac{2-k}{1-k}\right)^2 V_I$$

and

$$I_O = \left(\frac{1-k}{2-k}\right)^2 I_I$$

The voltage transfer gain is

$$M_R = \frac{V_O}{V_I} = \left(\frac{2-k}{1-k}\right)^2 \qquad (14.133)$$

The variation ratio of the output voltage v_O is

$$\varepsilon = \frac{\Delta v_O/2}{V_O} = \frac{k}{2RfC_4} \qquad (14.134)$$

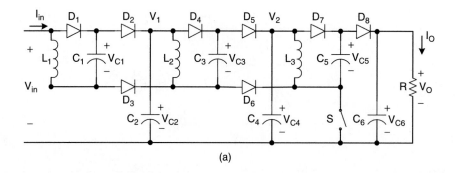

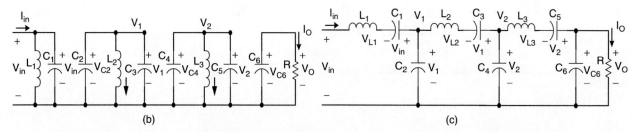

FIGURE 14.45 P/O SL Luo-converter triple-lift circuit: (a) circuit diagram; (b) switch on; and (c) switch off.

P/O SL Luo-converter triple-lift circuit is shown in Fig. 14.45a. The equivalent circuits during switch on and switch off are shown in Figs. 14.45b and c. Its output voltage and current are

$$V_O = \left(\frac{2-k}{1-k}\right)^3 V_I$$

and

$$I_O = \left(\frac{1-k}{2-k}\right)^3 I_I$$

The voltage transfer gain is

$$M_T = \frac{V_O}{V_I} = \left(\frac{2-k}{1-k}\right)^3 \quad (14.135)$$

The variation ratio of the output voltage v_O is

$$\varepsilon = \frac{\Delta v_O/2}{V_O} = \frac{k}{2RfC_6} \quad (14.136)$$

P/O SL Luo-converter additional circuit is shown in Fig. 14.46a. The equivalent circuits during switch on and switch off are shown in Figs. 14.46b and c. Its output voltage and current are

$$V_O = \frac{3-k}{1-k} V_I$$

and

$$I_O = \frac{1-k}{3-k} I_I$$

The voltage transfer gain is

$$M_A = \frac{V_O}{V_I} = \frac{3-k}{1-k} \quad (14.137)$$

The variation ratio of the output voltage v_O is

$$\varepsilon = \frac{\Delta v_O/2}{V_O} = \frac{k}{2RfC_{12}} \quad (14.138)$$

P/O SL Luo-converter additional re-lift circuit is shown in Fig. 14.47a. The equivalent circuits during switch on and switch off are shown in Figs. 14.47b and c. Its output voltage and current are

$$V_O = \frac{2-k}{1-k}\frac{3-k}{1-k} V_I$$

and

$$I_O = \frac{1-k}{2-k}\frac{1-k}{3-k} I_I$$

The voltage transfer gain is

$$M_{AR} = \frac{V_O}{V_I} = \frac{2-k}{1-k}\frac{3-k}{1-k} \quad (14.139)$$

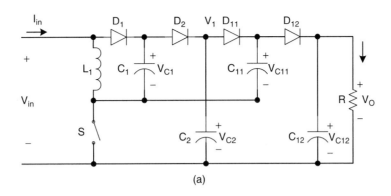

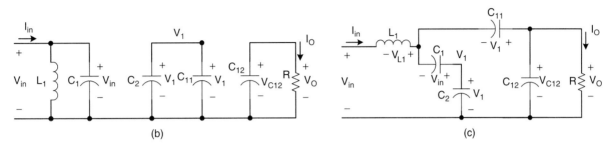

FIGURE 14.46 P/O SL Luo-converter additional circuit: (a) circuit diagram; (b) switch on; and (c) switch off.

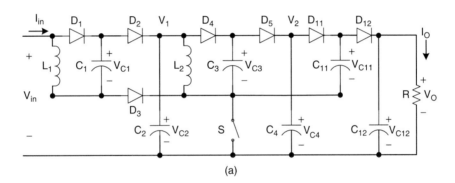

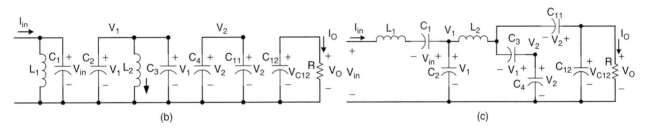

FIGURE 14.47 P/O SL Luo-converter additional re-lift circuit: (a) circuit diagram; (b) switch on; and (c) switch off.

The variation ratio of the output voltage v_O is

$$\varepsilon = \frac{\Delta v_O/2}{V_O} = \frac{k}{2RfC_{12}} \quad (14.140)$$

P/O SL Luo-converter additional triple-lift circuit is shown in Fig. 14.48a. The equivalent circuits during switch on and switch off are shown in Figs. 14.48b and c. Its output voltage and current are

$$V_O = \left(\frac{2-k}{1-k}\right)^2 \frac{3-k}{1-k} V_I$$

and

$$I_O = \left(\frac{1-k}{2-k}\right)^2 \frac{1-k}{3-k} I_I$$

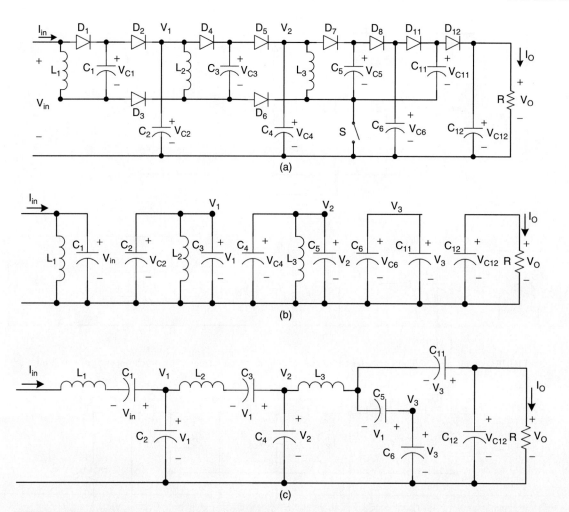

FIGURE 14.48 P/O SL Luo-converter additional triple-lift circuit: (a) circuit diagram; (b) switch on; and (c) switch off.

The voltage transfer gain is

$$M_{AT} = \frac{V_O}{V_I} = \left(\frac{2-k}{1-k}\right)^2 \frac{3-k}{1-k} \quad (14.141)$$

The variation ratio of the output voltage v_O is

$$\varepsilon = \frac{\Delta v_O/2}{V_O} = \frac{k}{2RfC_{12}} \quad (14.142)$$

14.5.2 N/O Super-lift Luo-converters

There are several subseries of N/O Super-lift Luo-converters:

- Main series;
- Additional series;
- Enhanced series;
- Re-enhanced series;
- Multi-enhanced series.

We only introduce three circuits of main series and additional series.

N/O SL Luo-converter elementary circuit is shown in Fig. 14.49. Its output voltage and current are

$$V_O = \left[\frac{2-k}{1-k} - 1\right] = \frac{1}{1-k} V_I$$

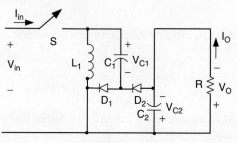

FIGURE 14.49 N/O SL Luo-converter elementary circuit.

and
$$I_O = (1-k)I_I$$

The voltage transfer gain is
$$M_E = \frac{V_O}{V_I} = \frac{1}{1-k} \quad (14.143)$$

The variation ratio of the output voltage v_O is
$$\varepsilon = \frac{\Delta v_O/2}{V_O} = \frac{k}{2RfC_2} \quad (14.144)$$

N/O SL Luo-converter re-lift circuit is shown in Fig. 14.50. Its output voltage and current are
$$V_O = \left[\left(\frac{2-k}{1-k}\right)^2 - 1\right]V_I$$

and
$$I_O = \frac{I_I}{\left((2-k)/(1-k)\right)^2 - 1}$$

The voltage transfer gain is
$$M_R = \frac{V_O}{V_I} = \left(\frac{2-k}{1-k}\right)^2 - 1 \quad (14.145)$$

The variation ratio of the output voltage v_O is
$$\varepsilon = \frac{\Delta v_O/2}{V_O} = \frac{k}{2RfC_4} \quad (14.146)$$

N/O SL Luo-converter triple-lift circuit is shown in Fig. 14.51. Its output voltage and current are
$$V_O = \left[\left(\frac{2-k}{1-k}\right)^3 - 1\right]V_I$$

and
$$I_O = \frac{I_I}{\left((2-k)/(1-k)\right)^3 - 1}$$

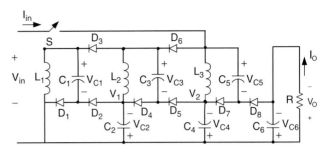

FIGURE 14.51 N/O SL Luo-converter triple-lift circuit.

The voltage transfer gain is
$$M_T = \frac{V_O}{V_I} = \left(\frac{2-k}{1-k}\right)^3 - 1 \quad (14.147)$$

The variation ratio of the output voltage v_O is
$$\varepsilon = \frac{\Delta v_O/2}{V_O} = \frac{k}{2RfC_6} \quad (14.148)$$

N/O SL Luo-converter additional circuit is shown in Fig. 14.52. Its output voltage and current are
$$V_O = \left[\frac{3-k}{1-k} - 1\right]V_I = \frac{2}{1-k}V_I$$

and
$$I_O = \frac{1-k}{2}I_I$$

The voltage transfer gain is
$$M_A = \frac{V_O}{V_I} = \frac{3-k}{1-k} - 1 = \frac{2}{1-k} \quad (14.149)$$

The variation ratio of the output voltage v_O is
$$\varepsilon = \frac{\Delta v_O/2}{V_O} = \frac{k}{2RfC_{12}} \quad (14.150)$$

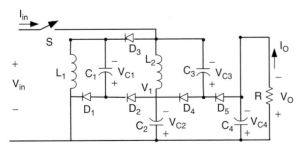

FIGURE 14.50 N/O SL Luo-converter re-lift circuit.

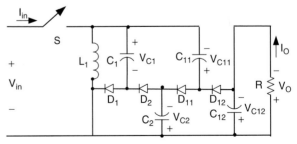

FIGURE 14.52 N/O SL Luo-converter additional circuit.

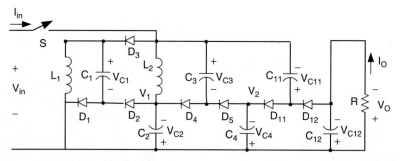

FIGURE 14.53 N/O SL Luo-converter additional re-lift circuit.

NO SL Luo-converter additional re-lift circuit is shown in Fig. 14.53. Its output voltage and current are

$$V_O = \left[\frac{2-k}{1-k}\frac{3-k}{1-k} - 1\right]V_I$$

and

$$I_O = \frac{I_I}{\left((2-k)/(1-k)\right)\left((3-k)/(1-k)\right) - 1}$$

The voltage transfer gain is

$$M_{AR} = \frac{V_O}{V_I} = \frac{2-k}{1-k}\frac{3-k}{1-k} - 1 \quad (14.151)$$

The variation ratio of the output voltage v_O is

$$\varepsilon = \frac{\Delta v_O/2}{V_O} = \frac{k}{2RfC_{12}} \quad (14.152)$$

N/O SL Luo-converter additional triple-lift circuit is shown in Fig. 14.54. Its output voltage and current are

$$V_O = \left[\left(\frac{2-k}{1-k}\right)^2 \frac{3-k}{1-k} - 1\right]V_I$$

and

$$I_O = \frac{I_I}{\left((2-k)/(1-k)\right)^2 \left((3-k)/(2-k)\right) - 1}$$

The voltage transfer gain is

$$M_{AT} = \frac{V_O}{V_I} = \left(\frac{2-k}{1-k}\right)^2 \frac{3-k}{1-k} - 1 \quad (14.153)$$

The variation ratio of the output voltage v_O is

$$\varepsilon = \frac{\Delta v_O/2}{V_O} = \frac{k}{2RfC_{12}} \quad (14.154)$$

14.5.3 P/O Cascade Boost-converters

There are several subseries of P/O cascade boost-converters (CBC):

- Main series;
- Additional series;
- Double series;
- Triple series;
- Multiple series.

We only introduce three circuits of main series and additional series.

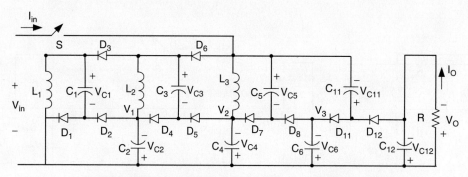

FIGURE 14.54 N/O SL Luo-converter additional triple-lift circuit.

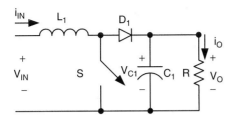

FIGURE 14.55 P/O CBC elementary circuit.

P/O CBC elementary circuit is shown in Fig. 14.55. Its output voltage and current are

$$V_O = \frac{1}{1-k} V_I$$

and

$$I_O = (1-k) I_I$$

The voltage transfer gain is

$$M_E = \frac{V_O}{V_I} = \frac{1}{1-k} \tag{14.155}$$

The variation ratio of the output voltage v_O is

$$\varepsilon = \frac{\Delta v_O/2}{V_O} = \frac{k}{2RfC_1} \tag{14.156}$$

P/O CBC two-stage circuit is shown in Fig. 14.56. Its output voltage and current are

$$V_O = \left(\frac{1}{1-k}\right)^2 V_I$$

and

$$I_O = (1-k)^2 I_I$$

The voltage transfer gain is

$$M_2 = \frac{V_O}{V_I} = \left(\frac{1}{1-k}\right)^2 \tag{14.157}$$

The variation ratio of the output voltage v_O is

$$\varepsilon = \frac{\Delta v_O/2}{V_O} = \frac{k}{2RfC_2} \tag{14.158}$$

P/O CBC three-stage circuit is shown in Fig. 14.57. Its output voltage and current are

$$V_O = \left(\frac{1}{1-k}\right)^3 V_I$$

and

$$I_O = (1-k)^3 I_I$$

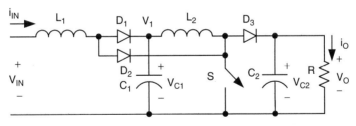

FIGURE 14.56 P/O CBC two-stage circuit.

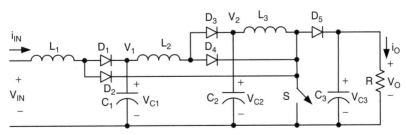

FIGURE 14.57 P/O CBC three-stage circuit.

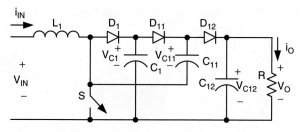

FIGURE 14.58 P/O CBC additional circuit.

The voltage transfer gain is

$$M_3 = \frac{V_O}{V_I} = \left(\frac{1}{1-k}\right)^3 \quad (14.159)$$

The variation ratio of the output voltage v_O is

$$\varepsilon = \frac{\Delta v_O/2}{V_O} = \frac{k}{2RfC_3} \quad (14.160)$$

P/O CBC additional circuit is shown in Fig. 14.58. Its output voltage and current are

$$V_O = \frac{2}{1-k}V_I$$

and

$$I_O = \frac{1-k}{2}I_I$$

The voltage transfer gain is

$$M_A = \frac{V_O}{V_I} = \frac{2}{1-k} \quad (14.161)$$

The variation ratio of the output voltage v_O is

$$\varepsilon = \frac{\Delta v_O/2}{V_O} = \frac{k}{2RfC_{12}} \quad (14.162)$$

P/O CBC additional two-stage circuit is shown in Fig. 14.59. Its output voltage and current are

$$V_O = 2\left(\frac{1}{1-k}\right)^2 V_I$$

and

$$I_O = \frac{(1-k)^2}{2}I_I$$

The voltage transfer gain is

$$M_{A2} = \frac{V_O}{V_I} = 2\left(\frac{1}{1-k}\right)^2 \quad (14.163)$$

The variation ratio of the output voltage v_O is

$$\varepsilon = \frac{\Delta v_O/2}{V_O} = \frac{k}{2RfC_{12}} \quad (14.164)$$

P/O CBC additional three-stage circuit is shown in Fig. 14.60. Its output voltage and current are

$$V_O = 2\left(\frac{1}{1-k}\right)^3 V_I$$

and

$$I_O = \frac{(1-k)^3}{2}I_I$$

The voltage transfer gain is

$$M_{A3} = \frac{V_O}{V_I} = 2\left(\frac{1}{1-k}\right)^3 \quad (14.165)$$

The variation ratio of the output voltage v_O is

$$\varepsilon = \frac{\Delta v_O/2}{V_O} = \frac{k}{2RfC_{12}} \quad (14.166)$$

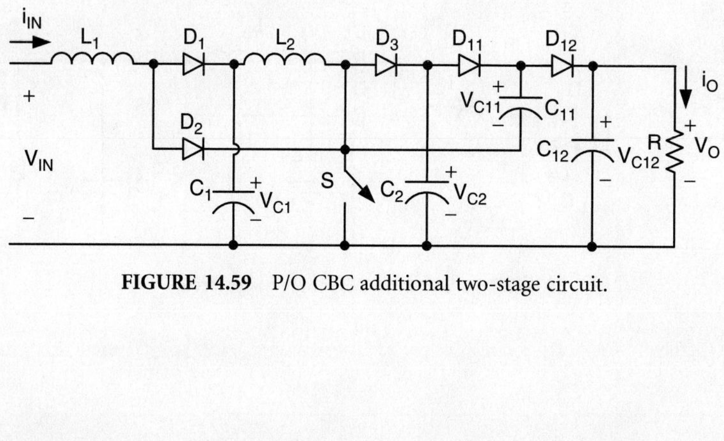

FIGURE 14.59 P/O CBC additional two-stage circuit.

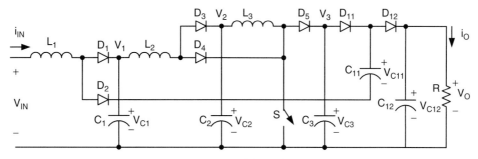

FIGURE 14.60 P/O CBC additional three-stage circuit.

14.5.4 N/O Cascade Boost-converters

There are several subseries of N/O CBC:

- Main series;
- Additional series;
- Double series;
- Triple series;
- Multiple series.

We only introduce three circuits of main series and additional series.

N/O CBC elementary circuit is shown in Fig. 14.61. Its output voltage and current are

$$V_O = \left[\frac{1}{1-k} - 1\right] V_I = \frac{k}{1-k} V_I$$

and

$$I_O = \frac{1-k}{k} I_I$$

The voltage transfer gain is

$$M_E = \frac{V_O}{V_I} = \frac{k}{1-k} \quad (14.167)$$

The variation ratio of the output voltage v_O is

$$\varepsilon = \frac{\Delta v_O/2}{V_O} = \frac{k}{2RfC_1} \quad (14.168)$$

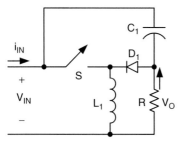

FIGURE 14.61 N/O CBC elementary circuit.

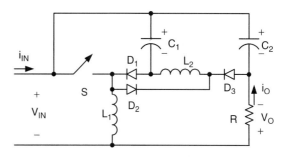

FIGURE 14.62 N/O CBC two-stage circuit.

N/O CBC two-stage circuit is shown in Fig. 14.62. Its output voltage and current are

$$V_O = \left[\left(\frac{1}{1-k}\right)^2 - 1\right] V_I$$

and

$$I_O = \frac{I_I}{\left(1/(1-k)\right)^2 - 1}$$

The voltage transfer gain is

$$M_2 = \frac{V_O}{V_I} = \left(\frac{1}{1-k}\right)^2 - 1 \quad (14.169)$$

The variation ratio of the output voltage v_O is

$$\varepsilon = \frac{\Delta v_O/2}{V_O} = \frac{k}{2RfC_2} \quad (14.170)$$

N/O CBC three-stage circuit is shown in Fig. 14.63. Its output voltage and current are

$$V_O = \left[\left(\frac{1}{1-k}\right)^3 - 1\right] V_I$$

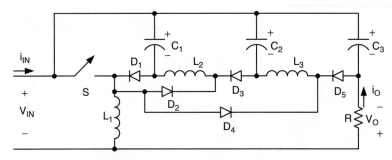

FIGURE 14.63 N/O CBC three-stage circuit.

and

$$I_O = \frac{I_I}{\left(1/(1-k)\right)^3 - 1}$$

The voltage transfer gain is

$$M_3 = \frac{V_O}{V_I} = \left(\frac{1}{1-k}\right)^3 - 1 \quad (14.171)$$

The variation ratio of the output voltage v_O is

$$\varepsilon = \frac{\Delta v_O/2}{V_O} = \frac{k}{2RfC_3} \quad (14.172)$$

N/O CBC additional circuit is shown in Fig. 14.64. Its output voltage and current are

$$V_O = \left[\frac{2}{1-k} - 1\right]V_I = \frac{1+k}{1-k}V_I$$

and

$$I_O = \frac{1-k}{1+k}I_I$$

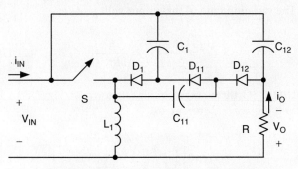

FIGURE 14.64 N/O CBC additional circuit.

The voltage transfer gain is

$$M_A = \frac{V_O}{V_I} = \frac{1+k}{1-k} \quad (14.173)$$

The variation ratio of the output voltage v_O is

$$\varepsilon = \frac{\Delta v_O/2}{V_O} = \frac{k}{2RfC_{12}} \quad (14.174)$$

N/O CBC additional two-stage circuit is shown in Fig. 14.65. Its output voltage and current are

$$V_O = \left[2\left(\frac{1}{1-k}\right)^2 - 1\right]V_I$$

and

$$I_O = \frac{I_I}{2\left(1/(1-k)\right)^2 - 1}$$

The voltage transfer gain is

$$M_{A2} = \frac{V_O}{V_I} = 2\left(\frac{1}{1-k}\right)^2 - 1 \quad (14.175)$$

The variation ratio of the output voltage v_O is

$$\varepsilon = \frac{\Delta v_O/2}{V_O} = \frac{k}{2RfC_{12}} \quad (14.176)$$

N/O CBC additional three-stage circuit is shown in Fig. 14.66. Its output voltage and current are

$$V_O = \left[2\left(\frac{1}{1-k}\right)^3 - 1\right]V_I$$

and

$$I_O = \frac{I_I}{2\left(1/(1-k)\right)^3 - 1}$$

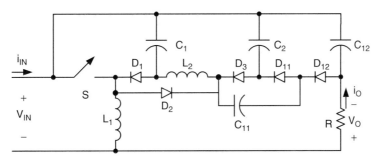

FIGURE 14.65 N/O CBC additional two-stage circuit.

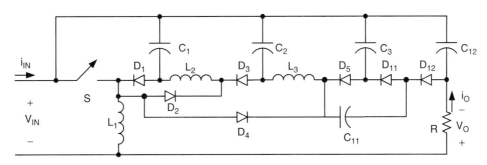

FIGURE 14.66 N/O CBC additional three-stage circuit.

The voltage transfer gain is

$$M_{A3} = \frac{V_O}{V_I} = 2\left(\frac{1}{1-k}\right)^3 - 1 \quad (14.177)$$

The variation ratio of the output voltage v_O is

$$\varepsilon = \frac{\Delta v_O/2}{V_O} = \frac{k}{2RfC_{12}} \quad (14.178)$$

14.6 Ultra-lift Luo-converters

Ultra-lift (UL) Luo-converter performs very high voltage transfer gain conversion. Its voltage transfer gain is the product of those of VL Luo-converter and SL Luo-converter.

We know that the gain of P/O VL Luo-converters (as in Eq. (14.52)) is

$$M = \frac{V_O}{V_I} = \frac{k^{h(n)}[n+h(n)]}{1-k}$$

where n is the stage number, $h(n)$ (as in Eq. (14.56)) is the Hong function.

$$h(n) = \begin{cases} 1 & n = 0 \\ 0 & n > 0 \end{cases}$$

(from Eq. (14.32)) $n = 0$ for the elementary circuit with the voltage transfer gain

$$M_E = \frac{V_O}{V_I} = \frac{k}{1-k}$$

The voltage transfer gain of P/O SL Luo-converters is

$$M = \frac{V_O}{V_I} = \left(\frac{j+2-k}{1-k}\right)^n \quad (14.179)$$

where n is the stage number, j is the multiple-enhanced number. $n = 1$ and $j = 0$ for the elementary circuit with gain (as in Eq. (14.131))

$$M_E = \frac{V_O}{V_I} = \frac{2-k}{1-k}$$

The circuit diagram of UL Luo-converter is shown in Fig. 14.67a, which consists of one switch S, two inductors L_1 and L_2, two capacitors C_1 and C_2, three diodes, and the load R. Its switch-on equivalent circuit is shown in Fig. 14.67b. Its switch-off equivalent circuit for the continuous conduction mode is shown in Fig. 14.67c and switch-off equivalent circuit for the discontinuous conduction mode is shown in Fig. 14.67d.

14.6.1 Continuous Conduction Mode

Referring to Figs. 14.67b and c, we have got the current i_{L1} increases with the slope $+V_I/L_1$ during switch on, and

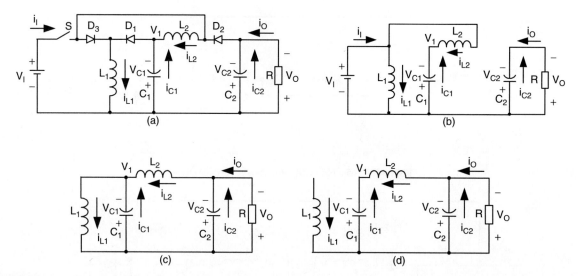

FIGURE 14.67 Ultra-lift (UL) Luo-converter: (a) circuit diagram; (b) switch on; (c) switch off in CCM; and (d) switch off in DCM.

decreases with the slope $-V_1/L_1$ during switch off. In the steady state, the current increment is equal to the decrement in a whole period T. The relation below is obtained

$$kT\frac{V_I}{L_1} = (1-k)T\frac{V_1}{L_1} \qquad (14.180)$$

Thus,

$$V_{C1} = V_1 = \frac{k}{1-k}V_I \qquad (14.181)$$

The current i_{L2} increases with the slope $+(V_I - V_1)/L_2$ during switch on, and decreases with the slope $-(V_1 - V_O)/L_2$ during switch off. In the steady state, the current increment is equal to the decrement in a whole period T. We obtain the relation below

$$kT\frac{V_I + V_1}{L_2} = (1-k)T\frac{V_O - V_1}{L_2} \qquad (14.182)$$

$$V_O = V_{C2} = \frac{2-k}{1-k}V_1 = \frac{k}{1-k}\frac{2-k}{1-k}V_I = \frac{k(2-k)}{(1-k)^2}V_I \qquad (14.183)$$

$$I_O = \frac{(1-k)^2}{k(2-k)}I_I \qquad (14.184)$$

The voltage transfer gain is

$$M = \frac{V_O}{V_I} = \frac{k(2-k)}{(1-k)^2} = \frac{k}{1-k}\frac{2-k}{1-k} = M_{E-VL} \times M_{E-SL} \qquad (14.185)$$

From Eq. (14.185) we can see that the voltage transfer gain of UL Luo-converter is very high which is the product of those of VL Luo-converter and SL Luo-converter. We list the transfer gains of various converters in Table 14.3 for reference.

TABLE 14.3 Comparison of various converters gains

k	0.2	0.33	0.5	0.67	0.8	0.9
Buck	0.2	0.33	0.5	0.67	0.8	0.9
Boost	1.25	1.5	2	3	5	10
Buck–Boost	0.25	0.5	1	2	4	9
VL Luo-converter	0.25	0.5	1	2	4	9
SL Luo-converter	2.25	2.5	3	4	6	11
UL Luo-converter	0.56	1.25	3	8	24	99

The variation of inductor current i_{L1} is

$$\Delta i_{L1} = kT\frac{V_I}{L_1} \qquad (14.186)$$

and its variation ratio is

$$\xi_1 = \frac{\Delta i_{L1}/2}{I_{L1}} = \frac{k(1-k)^2 TV_I}{2L_1 I_2} = \frac{k(1-k)^2 TR}{2L_1 M} = \frac{(1-k)^4 TR}{2(2-k)fL_1} \qquad (14.187)$$

The variation of inductor current i_{L2} is

$$\Delta i_{L2} = \frac{kTV_I}{(1-k)L_2} \qquad (14.188)$$

and its variation ratio is

$$\xi_2 = \frac{\Delta i_{L2}/2}{I_{L2}} = \frac{kTV_I}{2L_2 I_2} = \frac{kTR}{2L_2 M} = \frac{(1-k)^2 TR}{2(2-k)fL_2} \qquad (14.189)$$

The variation of capacitor voltage v_{C1} is

$$\Delta v_{C1} = \frac{\Delta Q_{C1}}{C_1} = \frac{kTI_{L2}}{C_1} = \frac{kTI_O}{(1-k)C_1} \quad (14.190)$$

and its variation ratio is

$$\sigma_1 = \frac{\Delta v_{C1}/2}{V_{C1}} = \frac{kTI_O}{2(1-k)V_1C_1} = \frac{k(2-k)}{2(1-k)^2 fC_1 R} \quad (14.191)$$

The variation of capacitor voltage v_{C2} is

$$\Delta v_{C2} = \frac{\Delta Q_{C2}}{C_2} = \frac{kTI_O}{C_2} \quad (14.192)$$

and its variation ratio is

$$\varepsilon = \sigma_2 = \frac{\Delta v_{C2}/2}{V_{C2}} = \frac{kTI_O}{2V_O C_2} = \frac{k}{2fC_2 R} \quad (14.193)$$

From the analysis and calculations, we can see that all variations are very small. A design example is that $V_I = 10$ V, $L_1 = L_2 = 1$ mH, $C_1 = C_2 = 1\,\mu$F, $R = 3000\,\Omega$, $f = 50$ kHz, and conduction duty cycle k varies from 0.1 to 0.9. We then obtain the output voltage variation ratio ε, which is less than 0.003. The output voltage is very smooth DC voltage nearly no ripple.

14.6.2 Discontinuous Conduction Mode

Referring to Fig. 14.67d, we have got the current i_{L1} decreases to zero before $t = T$, i.e. the current becomes zero before next time the switch turns on. The DCM operation condition is defined as

$$\xi \geq 1$$

or

$$\xi_1 = \frac{k(1-k)^2 TR}{2L_1 M} = \frac{(1-k)^4 TR}{2(2-k)fL_1} \geq 1 \quad (14.194)$$

The normalized impedance Z_N is,

$$Z_N = \frac{R}{fL_1} \quad (14.195)$$

We define the filling factor m to describe the current exists time. For DCM operation, $0 < m \leq 1$,

$$m = \frac{1}{\xi_1} = \frac{2L_1 G}{k(1-k)^2 TR} = \frac{2(2-k)}{(1-k)^4 Z_N} \quad (14.196)$$

$$kT\frac{V_I}{L_1} = (1-k)mT\frac{V_1}{L_1}$$

Thus,

$$V_{C1} = V_1 = \frac{k}{(1-k)m} V_I \quad (14.197)$$

We finally obtain the relation below

$$kT\frac{V_I + V_1}{L_2} = (1-k)T\frac{V_O - V_1}{L_2} \quad (14.198)$$

$$V_O = V_{C2} = \frac{2-k}{1-k} V_1 = \frac{k(2-k)}{m(1-k)^2} V_I \quad (14.199)$$

The voltage transfer gain in DCM is higher than that in CCM.

$$M_{DCM} = \frac{V_O}{V_I} = \frac{k(2-k)}{m(1-k)^2} = \frac{M_{CCM}}{m} \quad \text{with } m < 1 \quad (14.200)$$

14.7 Multiple-quadrant Operating Luo-converters

Multiple-quadrant operating converters are the second-generation converters. These converters usually perform between two voltage sources: V_1 and V_2. Voltage source V_1 is proposed positive voltage and voltage V_2 is the load voltage. In the investigation both voltages are proposed constant voltage. Since V_1 and V_2 are constant values, voltage transfer gain is constant. Our interesting research will concentrate the working current, minimum conduction duty k_{min}, and the power transfer efficiency η.

Multiple-quadrant operating Luo-converters are the second-generation converters and they have three modes:

- Two-quadrant DC/DC Luo-converter in forward operation;
- Two-quadrant DC/DC Luo-converter in reverse operation;
- Four-quadrant DC/DC Luo-converter.

The two-quadrant DC/DC Luo-converter in forward operation has been derived from the positive output Luo-converter. It performs in the first-quadrant Q_I and the second-quadrant Q_{II} corresponding to the DC motor forward operation in motoring and regenerative braking states.

The two-quadrant DC/DC Luo-converter in reverse operation has been derived from the N/O Luo-converter. It performs in the third-quadrant Q_{III} and the fourth-quadrant Q_{IV} corresponding to the DC motor reverse operation in motoring and regenerative braking states.

The four-quadrant DC/DC Luo-converter has been derived from the double output Luo-converter. It performs four-quadrant operation corresponding to the DC motor forward and reverse operation in motoring and regenerative braking states.

In the following analysis the input source and output load are usually constant voltages as shown, V_1 and V_2. Switches S_1 and S_2 in this diagram are power metal oxide semiconductor field effect transistor (MOSFET) devices, and they are driven by a PWM switching signal with repeating frequency f and conduction duty k. In this paper the switch repeating period is $T = 1/f$, so that the switch-on period is kT and switch-off period is $(1 − k)T$. The equivalent resistance is R for each inductor. During switch-on the voltage drop across the switches and diodes are V_S and V_D respectively.

14.7.1 Forward Two-quadrant DC/DC Luo-converter

Forward Two-quadrant (F 2Q) Luo-converter is shown in Fig. 14.68. The source voltage (V_1) and load voltage (V_2) are usually considered as constant voltages. The load can be a battery or motor back electromotive force (EMF). For example, the source voltage is 42 V and load voltage is +14 V. There are two modes of operation:

1. Mode A (Quadrant I): electrical energy is transferred from source side V_1 to load side V_2;
2. Mode B (Quadrant II): electrical energy is transferred from load side V_2 to source side V_1.

Mode A: The equivalent circuits during switch-on and -off periods are shown in Figs. 14.69a and b. The typical output voltage and current waveforms are shown in Fig. 14.69c. We have the output current I_2 as

$$I_2 = \frac{1-k}{k} I_1 \tag{14.201}$$

and

$$I_2 = \frac{V_1 - V_S - V_D - V_2((1-k)/k)}{R\left((k/(1-k)) + ((1-k)/k)\right)} \tag{14.202}$$

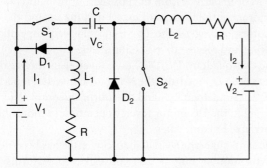

FIGURE 14.68 Forward two-quadrant operating Luo-converter.

The minimum conduction duty k corresponding to $I_2 = 0$ is

$$k_{min} = \frac{V_2}{V_1 + V_2 - V_S - V_D} \tag{14.203}$$

The power transfer efficiency is

$$\eta_A = \frac{P_O}{P_I} = \frac{V_2 I_2}{V_1 I_1}$$

$$= \frac{1}{1 + ((V_S + V_D)/V_2)(k/(1-k)) + (RI_2/V_2)\left[1 + ((1-k)/k)^2\right]} \tag{14.204}$$

The variation ratio of capacitor voltage v_C is

$$\rho = \frac{\Delta v_C/2}{V_C} = \frac{(1-k)I_2}{2fC(V_1 - RI_2(1/(1-k)))} \tag{14.205}$$

The variation ratio of inductor current i_{L1} is

$$\xi_1 = \frac{\Delta i_{L1}/2}{I_{L1}} = k\frac{V_1 - V_S - RI_1}{2fL_1 I_1} \tag{14.206}$$

The variation ratio of inductor current i_{L2} is

$$\xi_2 = \frac{\Delta i_{L2}/2}{I_{L2}} = k\frac{V_1 - V_S - RI_1}{2fL_2 I_2} \tag{14.207}$$

The variation ratio of diode current i_{D2} is

$$\zeta_{D2} = \frac{\Delta i_{D2}/2}{I_{L1} + I_{L2}} = k\frac{V_1 - V_S - RI_1}{2fL(I_1 + I_2)} = k^2\frac{V_1 - V_S - RI_1}{2fLI_1} \tag{14.208}$$

If the diode current becomes zero before S_1 switch on again, the converter works in discontinuous region. The condition is

$$\zeta_{D2} = 1, \quad \text{i.e.} \quad k^2 = \frac{2fLI_1}{V_1 - V_S - RI_1} \tag{14.209}$$

Mode B: The equivalent circuits during switch-on and -off periods are shown in Figs. 14.70a and b. The typical output voltage and current waveforms are shown in Fig. 14.70c. We have the output current I_1 as

$$I_1 = \frac{1-k}{k} I_2 \tag{14.210}$$

and

$$I_1 = \frac{V_2 - (V_1 + V_S + V_D)((1-k)/k)}{R\left((k/(1-k)) + ((1-k)/k)\right)} \tag{14.211}$$

The minimum conduction duty k corresponding to $I_1 = 0$ is

$$k_{min} = \frac{V_1 + V_S + V_D}{V_1 + V_2 + V_S + V_D} \tag{14.212}$$

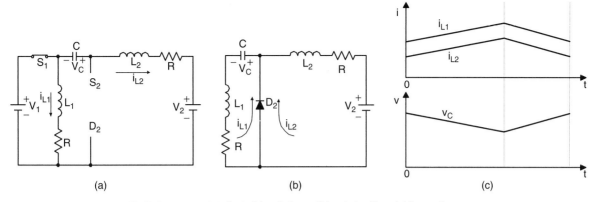

FIGURE 14.69 Mode A: (a) switch on; (b) switch off; and (c) waveforms.

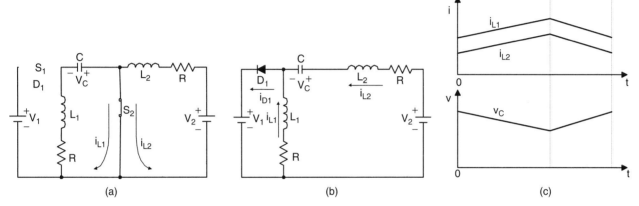

FIGURE 14.70 Mode B: (a) switch on; (b) switch off; and (c) waveforms.

The power transfer efficiency

$$\eta_B = \frac{P_O}{P_I} = \frac{V_1 I_1}{V_2 I_2}$$

$$= \frac{1}{1 + ((V_S + V_D)/V_1) + (RI_1/V_1)[1 + ((1-k)/k)^2]}$$

(14.213)

The variation ratio of capacitor voltage v_C is

$$\rho = \frac{\Delta v_C/2}{V_C} = \frac{kI_1}{2fC[(V_2/(1-k)) - V_1 - RI_1(k/(1-k)^2)]}$$

(14.214)

The variation ratio of inductor current i_{L1} is

$$\xi_1 = \frac{\Delta i_{L1}/2}{I_{L1}} = k\frac{V_2 - V_S - RI_2}{2fL_1 I_1}$$

(14.215)

The variation ratio of inductor current i_{L2} is

$$\xi_2 = \frac{\Delta i_{L2}/2}{I_{L2}} = k\frac{V_2 - V_S - RI_2}{2fL_2 I_2}$$

(14.216)

The variation ratio of diode current i_{D1} is

$$\zeta_{D1} = \frac{\Delta i_{D2}/2}{I_{L1} + I_{L2}} = k\frac{V_2 - V_S - RI_2}{2fL(I_1 + I_2)} = k^2 \frac{V_2 - V_S - RI_2}{2fLI_2}$$

(14.217)

If the diode current becomes zero before S_2 switch on again, the converter works in discontinuous region. The condition is

$$\zeta_{D1} = 1, \quad \text{i.e.} \quad k^2 = \frac{2fLI_2}{V_2 - V_S - RI_2}$$

(14.218)

14.7.2 Two-quadrant DC/DC Luo-converter in Reverse Operation

Reverse two-quadrant operating (R 2Q) Luo-converter is shown in Fig. 14.71, and it consists of two switches with two passive diodes, two inductors and one capacitor. The source voltage (V_1) and load voltage (V_2) are usually considered as constant voltages. The load can be a battery or motor back EMF. For example, the source voltage is 42 V and load voltage

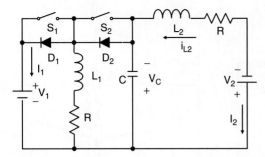

FIGURE 14.71 Reverse two-quadrant operating Luo-converter.

is -14 V. There are two modes of operation:

1. Mode C (Quadrant III): electrical energy is transferred from source side V_1 to load side $-V_2$;
2. Mode D (Quadrant IV): electrical energy is transferred from load side $-V_2$ to source side V_1.

Mode C: The equivalent circuits during switch-on and -off periods are shown in Figs. 14.72a and b. The typical output voltage and current waveforms are shown in Fig. 14.72c. We have the output current I_2 as

$$I_2 = \frac{1-k}{k} I_1 \quad (14.219)$$

and

$$I_2 = \frac{V_1 - V_S - V_D - V_2((1-k)/k)}{R\left[(1/(k(1-k))) + ((1-k)/k)\right]} \quad (14.220)$$

The minimum conduction duty k corresponding to $I_2 = 0$ is

$$k_{min} = \frac{V_2}{V_1 + V_2 - V_S - V_D} \quad (14.221)$$

The power transfer efficiency is

$$\eta_C = \frac{P_O}{P_I} = \frac{V_2 I_2}{V_1 I_1}$$

$$= \frac{1}{1 + ((V_S + V_D)/V_2)(k/(1-k)) + (RI_2/V_2)[1 + (1/(1-k))^2]} \quad (14.222)$$

The variation ratio of capacitor voltage v_C is

$$\rho = \frac{\Delta v_C/2}{V_C} = \frac{k I_2}{2 f C \left[(k/(1-k))V_1 - ((RI_2)/(1-k)^2)\right]} \quad (14.223)$$

The variation ratio of inductor current i_{L1} is

$$\xi_1 = \frac{\Delta i_{L1}/2}{I_{L1}} = k \frac{V_1 - V_S - R I_1}{2 f L_1 I_1} \quad (14.224)$$

The variation ratio of inductor current i_{D2} is

$$\zeta_{D2} = \xi_1 = \frac{\Delta i_{D2}/2}{I_{L1}} = k \frac{V_1 - V_S - R I_1}{2 f L_1 I_1} \quad (14.225)$$

The variation ratio of inductor current i_{L2} is

$$\xi_2 = \frac{\Delta i_{L2}/2}{I_2} = \frac{k}{16 f^2 C L_2} \quad (14.226)$$

If the diode current becomes zero before S_1 switch on again, the converter works in discontinuous region. The condition is

$$\zeta_{D2} = 1, \quad \text{i.e.} \quad k = \frac{2 f L_1 I_1}{V_1 - V_S - R I_1} \quad (14.227)$$

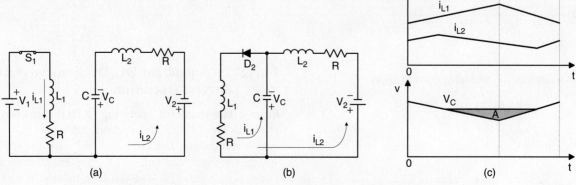

FIGURE 14.72 Mode C: (a) switch on; (b) switch off; and (c) waveforms.

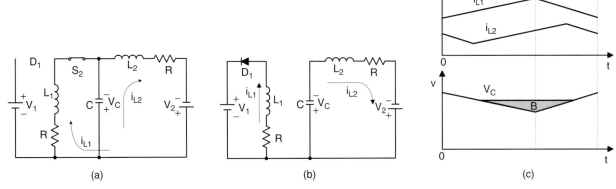

FIGURE 14.73 Mode D: (a) switch on; (b) switch off; and (c) waveforms.

Mode D: The equivalent circuits during switch-on and -off periods are shown in Figs. 14.73a and b. The typical output voltage and current waveforms are shown in Fig. 14.73c. We have the output current I_1 as

$$I_1 = \frac{1-k}{k} I_2 \qquad (14.228)$$

and

$$I_1 = \frac{V_2 - (V_1 + V_S + V_D)((1-k)/k)}{R[(1/(k(1-k))) + (k/(1-k))]} \qquad (14.229)$$

The minimum conduction duty k corresponding to $I_1 = 0$ is

$$k_{min} = \frac{V_1 + V_S + V_D}{V_1 + V_2 + V_S + V_D} \qquad (14.230)$$

The power transfer efficiency is

$$\eta_D = \frac{P_O}{P_I} = \frac{V_1 I_1}{V_2 I_2}$$
$$= \frac{1}{1 + ((V_S + V_D)/V_1) + (RI_1/V_1)[(1/(1-k)^2) + (k/(1-k))^2]} \qquad (14.231)$$

The variation ratio of capacitor voltage v_C is

$$\rho = \frac{\Delta v_C/2}{V_C} = \frac{kI_1}{2fC[((1-k)/k)V_1 + ((RI_1)/(k(1-k)))]} \qquad (14.232)$$

The variation ratio of inductor current i_{L1} is

$$\xi_1 = \frac{\Delta i_{L1}/2}{I_{L1}} = (1-k)\frac{V_2 - V_S - RI_2}{2fL_1 I_1} \qquad (14.233)$$

And the variation ratio of inductor current i_{D1} is

$$\zeta_{D1} = \xi_1 = \frac{\Delta i_{D1}/2}{I_{L1}} = k\frac{V_2 - V_S - RI_2}{2fL_1 I_2} \qquad (14.234)$$

The variation ratio of inductor current i_{L2} is

$$\xi_2 = \frac{\Delta i_{L2}/2}{I_2} = \frac{1-k}{16f^2 CL_2} \qquad (14.235)$$

If the diode current becomes zero before S_2 switch on again, the converter works in discontinuous region. The condition is

$$\zeta_{D1} = 1, \quad \text{i.e.} \quad k = \frac{2fL_1 I_2}{V_2 - V_S - RI_2} \qquad (14.236)$$

14.7.3 Four-quadrant DC/DC Luo-converter

Four-quadrant DC/DC Luo-converter is shown in Fig. 14.74, which consists of two switches with two passive diodes, two inductors, and one capacitor. The source voltage (V_1) and load voltage (V_2) are usually considered as constant voltages. The load can be a battery or motor back EMF. For example, the source voltage is 42 V and load voltage is ±14 V. There are four modes of operation:

1. Mode A (Quadrant I): electrical energy is transferred from source side V_1 to load side V_2;
2. Mode B (Quadrant II): electrical energy is transferred from load side V_2 to source side V_1;
3. Mode C (Quadrant III): electrical energy is transferred from source side V_1 to load side $-V_2$;
4. Mode D (Quadrant IV): electrical energy is transferred from load side $-V_2$ to source side V_1.

Each mode has two states: "on" and "off." Usually, each state is operating in different conduction duty k. The switches

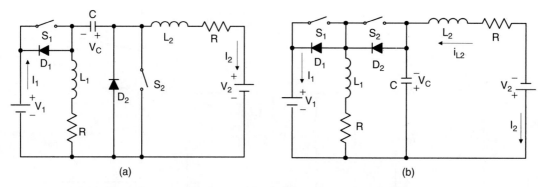

FIGURE 14.74 Four-quadrant operating Luo-converter: (a) circuit 1 and (b) circuit 2.

are the power MOSFET devices. The circuit 1 in Fig. 14.74 implements Modes A and B, and the circuit 2 in Fig. 14.74 implements Modes C and D. Circuits 1 and 2 can changeover by auxiliary switches (not in the figure).

Mode A: During state-**on** switch S_1 is closed, switch S_2 and diodes D_1 and D_2 are not conducted. In this case inductor currents i_{L1} and i_{L2} increase, and $i_1 = i_{L1} + i_{L2}$. During state-**off** switches S_1, S_2, and diode D_1 are off and diode D_2 is conducted. In this case current i_{L1} flows via diode D_2 to charge capacitor C, in the meantime current i_{L2} is kept to flow through load battery V_2. The free-wheeling diode current $i_{D2} = i_{L1} + i_{L2}$. Mode A implements the characteristics of the buck–boost conversion.

Mode B: During state-**on** switches S_2 is closed, switch S_1 and diodes D_1 and D_2 are not conducted. In this case inductor current i_{L2} increases by biased V_2, inductor current i_{L1} increases by biased V_C. Therefore capacitor voltage V_C reduces. During state-**off** switches S_1, S_2, and diode D_2 are not on, and only diode D_1 is on. In this case source current $i_1 = i_{L1} + i_{L2}$ which is a negative value to perform the regenerative operation. Inductor current i_{L2} flows through capacitor C, it is charged by current i_{L2}. After capacitor C, i_{L2} then flows through the source V_1. Inductor current i_{L1} flows through the source V_1 as well via diode D_1. Mode B implements the characteristics of the boost conversion.

Mode C: During state-**on** switch S_1 is closed, switch S_2 and diodes D_1 and D_2 are not conducted. In this case inductor currents i_{L1} and i_{L2} increase, and $i_1 = i_{L1}$. During state-**off** switches S_1, S_2, and diode D_1 are off and diode D_2 is conducted. In this case current i_{L1} flows via diode D_2 to charge capacitor C and the load battery V_2 via inductor L_2. The free-wheeling diode current $i_{D2} = i_{L1} = i_C + i_2$. Mode C implements the characteristics of the buck–boost conversion.

Mode D: During state-**on** switches S_2 is closed, switch S_1 and diodes D_1 and D_2 are not conducted. In this case inductor current i_{L1} increases by biased V_2, inductor current i_{L2} decreases by biased $(V_2 - V_C)$. Therefore capacitor voltage V_C reduces. Current $i_{L1} = i_{C-on} + i_2$. During state-**off** switches S_1, S_2, and diode D_2 are not on, and only diode D_1 is on. In this case source current $i_1 = i_{L1}$ which is a negative value to perform the regenerative operation. Inductor current i_2 flows through capacitor C that is charged by current i_2, i.e. $i_{C-off} = i_2$. Mode D implements the characteristics of the boost conversion.

Summary: The switch status is shown in Table 14.4. The operation of all modes A, B, C, and D is same to the description in Sections 14.7.1 and 14.7.2.

14.8 Switched-capacitor Multi-quadrant Luo-converters

Switched-component converters are the third-generation converters. These converters are made of only inductor

TABLE 14.4 Switch's status (the blank status means OFF)

Switch or diode	Mode A (QI)		Mode B (QII)		Mode C (QIII)		Mode D (QIV)	
	State-on	State-off	State-on	State-off	State-on	State-off	State-on	State-off
Circuit	Circuit 1				Circuit 2			
S_1	ON				ON			
D_1			ON					ON
S_2			ON				ON	
D_2		ON				ON		

14 DC/DC Conversion Technique and 12 Series Luo-converters

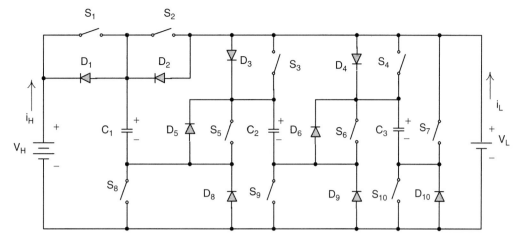

FIGURE 14.75 Two-quadrant switched-capacitor DC/DC Luo-converter.

or capacitors. They usually perform in the systems between two voltage sources: V_1 and V_2. Voltage source V_1 is proposed positive voltage and voltage V_2 is the load voltage that can be positive or negative. In the investigation both voltages are proposed constant voltage. Since V_1 and V_2 are constant values, so that voltage transfer gain is constant. Our interesting research will concentrate on the working current and the power transfer efficiency η. The resistance R of the capacitors and inductor has to be considered for the power transfer efficiency η calculation.

Reviewing the papers in the literature, we can find that almost of the papers investigating the switched-component converters are working in single-quadrant operation. Professor Luo and colleagues have developed this technique into multi-quadrant operation. We describe these in this and next sections.

Switched-capacitor multi-quadrant Luo-converters are the third-generation converters, and they are made of only capacitors. Because these converters implement voltage-lift and current-amplification techniques, they have the advantages of high power density, high power transfer efficiency, and low EMI. They have two modes:

- Two-quadrant switched-capacitor DC/DC Luo-converter;
- Four-quadrant switched-capacitor DC/DC Luo-converter.

The two-quadrant switched-capacitor DC/DC Luo-converter in forward operation has been derived for the energy transmission of a dual-voltage system in two-quadrant operation. The both, source and load voltages are positive polarity. It performs in the first-quadrant Q_I and the second-quadrant Q_{II} corresponding to the DC motor forward operation in motoring and regenerative braking states.

The four-quadrant switched-capacitor DC/DC Luo-converter has been derived for the energy transmission of a dual-voltage system in four-quadrant operation. The source voltage is positive and load voltage can be positive or negative polarity. It performs four-quadrant operation corresponding to the DC motor forward and reverse operation in motoring and regenerative braking states.

From the analysis and calculation, the conduction duty k does not affect the power transfer efficiency. It affects the input and output power in a small region. The maximum output power corresponds at $k = 0.5$.

14.8.1 Two-quadrant Switched-capacitor DC/DC Luo-converter

This converter is shown in Fig. 14.75. It consists of nine switches, seven diodes, and three capacitors. The high source voltage V_H and low load voltage V_L are usually considered as constant voltages, e.g. the source voltage is 48 V and load voltage is 14 V. There are two modes of operation:

- Mode A (Quadrant I): electrical energy is transferred from V_H side to V_L side;
- Mode B (Quadrant II): electrical energy is transferred from V_L side to V_H side.

Each mode has two states: "on" and "off." Usually, each state is operating in different conduction duty k. The switching period is T where $T = 1/f$, where f is the switching frequency. The switches are the power MOSFET devices. The parasitic resistance of all switches is r_S. The equivalent resistance of all capacitors is r_C and the equivalent voltage drop of all diodes is V_D. Usually we select the three capacitors having same capacitance $C = C_1 = C_2 = C_3$. Some reference data are useful: $r_S = 0.03\,\Omega$, $r_C = 0.02\,\Omega$, and $V_D = 0.5$ V, $f = 5$ kHz, and $C = 5000\,\mu$F. The switch's status is shown in Table 14.5.

For Mode A, state-**on** is shown in Fig. 14.76a: switches S_1 and S_{10} are closed and diodes D_5 and D_5 are conducted. Other switches and diodes are open. In this case capacitors C_1, C_2, and C_3 are charged via the circuit V_H–S_1–C_1–D_5–C_2–D_6–C_3–S_{10}, and the voltage across capacitors

TABLE 14.5 Switch's status (the blank status means OFF)

Switch or diode	Mode A		Mode B	
	State-on	State-off	State-on	State-off
S_1	ON			
D_1				ON
S_2, S_3, S_4		ON	ON	
D_2, D_3, D_4				ON
S_5, S_6, S_7				ON
D_5, D_6	ON			
S_8, S_9			ON	
S_{10}	ON		ON	
D_8, D_9, D_{10}		ON		

The variation of the voltage across capacitor C_1 is:

$$\Delta v_{C1} = \frac{k(V_H - 3V_{C1} - 2V_D)}{fCR_{AN}}$$

$$= \frac{2.4k(1-k)(V_H - 3V_L - 5V_D)}{(2.4 + 0.6k)fCR_{AN}} \quad (14.237)$$

After calculation,

$$V_{C1} = \frac{k(V_H - 2V_D) + 2.4(1-k)(V_L + V_D)}{2.4 + 0.6k} \quad (14.238)$$

The average output current is

$$I_L = \frac{3}{T}\int_{kT}^{T} i_{C1}(t)dt \approx 3(1-k)\frac{V_{C1} - V_L - V_D}{R_{AF}} \quad (14.239)$$

The average input current is

$$I_H = \frac{1}{T}\int_0^{kT} i_{C1}(t)dt \approx k\frac{V_H - 3V_{C1} - 2V_D}{R_{AN}} \quad (14.240)$$

Therefore, we have $3I_H = I_L$.
Output power is

$$P_O = V_L I_L = 3(1-k)V_L \frac{V_{C1} - V_L - V_D}{R_{AF}} \quad (14.241)$$

C_1, C_2, and C_3 is increasing. The equivalent circuit resistance is $R_{AN} = (2r_S + 3r_C) = 0.12\,\Omega$, and the voltage deduction is $2V_D = 1$ V. State-**off** is shown in Fig. 14.76b: switches S_2, S_3, and S_4 are closed and diodes D_8, D_9, and D_{10} are conducted. Other switches and diodes are open. In this case capacitor $C_1(C_2$ and $C_3)$ is discharged via the circuit $S_2(S_3$ and $S_4)$–V_L–$D_8(D_9$ and $D_{10})$–$C_1(C_2$ and $C_3)$, and the voltage across capacitor $C_1(C_2$ and $C_3)$ is decreasing. Mode A implements the **current-amplification technique**. The voltage and current waveforms are shown in Fig. 14.76c. All three capacitors are charged in series during state-on. The input current flows through three capacitors and the charges accumulated on the three capacitors should be the same. These three capacitors are discharged in parallel during state-off. Therefore, the output current is amplified by three times.

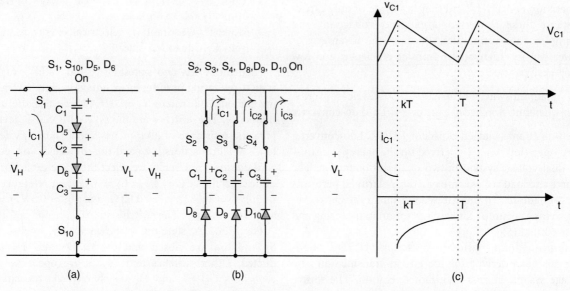

FIGURE 14.76 Mode A operation: (a) state-on; (b) state-off; and (c) voltage and current waveforms.

Input power

$$P_I = V_H I_H = kV_H \frac{V_H - 3V_{C1} - V_D}{R_{AN}} \quad (14.242)$$

The transfer efficiency is

$$\eta_A = \frac{P_O}{P_I} = \frac{1-k}{k} \frac{3V_L}{V_H} \frac{V_{C1} - V_L - V_D}{V_H - 3V_{C1} - V_D} \frac{R_{AN}}{R_{AF}} = \frac{3V_L}{V_H} \quad (14.243)$$

For Mode B, state-**on** is shown in Fig. 14.77a: switches S_8, S_9, and S_{10} are closed and diodes D_2, D_3, and D_4 are conducted. Other switches and diodes are off. In this case all three capacitors are charged via each circuit V_L–D_2(and D_3, D_4)–C_1(and C_2, C_3)–S_8(and S_9, S_{10}), and the voltage across three capacitors are increasing. The equivalent circuit resistance is $R_{BN} = r_S + r_C$ and the voltage deduction is V_D in each circuit. State-**off** is shown in Fig. 14.77b: switches S_5, S_6, and S_7 are closed and diode D_1 is on. Other switches and diodes are open. In this case all capacitors is discharged via the circuit V_L–S_7–C_3–S_6–C_2–S_5–C_1–D_1–V_H, and the voltage across all capacitors is decreasing. Mode B implements the **voltage-lift technique**. The voltage and current waveforms are shown in Fig. 14.77c. All three capacitors are charged in parallel during state-on. The input voltage is applied to the three capacitors symmetrically, so that the voltages across these three capacitors should be same. They are discharged in series during state-off. Therefore, the output voltage is lifted by three times.

The variation of the voltage across capacitor C is:

$$\Delta v_{C1} = \frac{k(1-k)[4(V_L - V_D) - V_H]}{fCR_{BN}} \quad (14.244)$$

After calculation

$$V_{C1} = k(V_L - V_D) + \frac{1-k}{3}(V_H - V_L + V_D) \quad (14.245)$$

The average input current is

$$I_L = \frac{1}{T}\left[3\int_0^{kT} i_{C1}(t)dt + \int_{kT}^{T} i_{C1}(t)dt\right]$$

$$\approx 3k\frac{V_L - V_{C1} - V_D}{R_{BN}} + (1-k)\frac{3V_{C1} + V_L - V_H - V_D}{R_{BF}} \quad (14.246)$$

The average output current is

$$I_H = \frac{1}{T}\int_{kT}^{T} i_{C1}(t)dt \approx (1-k)\frac{3V_{C1} + V_L - V_H - V_D}{R_{BF}} \quad (14.247)$$

From this formula, we have $4I_H = I_L$.

Input power is

$$P_I = V_L I_L$$
$$= V_L\left[3k\frac{V_L - V_C - V_D}{R_{BN}} + (1-k)\frac{3V_C + V_L - V_H - V_D}{R_{BF}}\right] \quad (14.248)$$

Output power is

$$P_O = V_H I_H = V_H(1-k)\frac{3V_C + V_L - V_H - V_D}{R_{BF}} \quad (14.249)$$

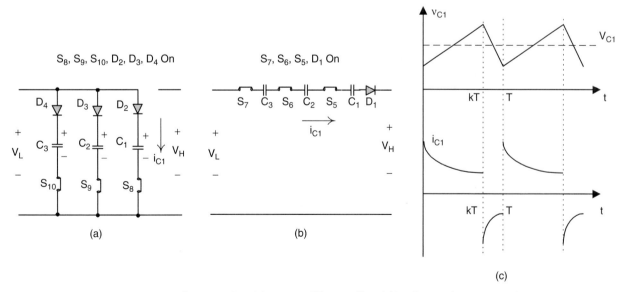

FIGURE 14.77 Mode B operation: (a) state-on; (b) state-off; and (c) voltage and current waveforms.

The efficiency is

$$\eta_B = \frac{P_O}{P_I} = \frac{V_H}{4V_L} \quad (14.250)$$

14.8.2 Four-quadrant Switched-capacitor DC/DC Luo-converter

Four-quadrant switched-capacitor DC/DC Luo-converter is shown in Fig. 14.78. Since it performs the *voltage-lift technique*, it has a simple structure with four-quadrant operation. This converter consists of eight switches and two capacitors. The source voltage V_1 and load voltage V_2 (e.g. a battery or DC motor back EMF) are usually constant voltages. In this paper they are supposed to be ± 21 V and ± 14 V. Capacitors C_1 and C_2 are same and $C_1 = C_2 = 2000\,\mu$F. The circuit equivalent resistance $R = 50$ mΩ. Therefore, there are four modes of operation for this converter:

1. Mode A: energy is converted from source to positive voltage load; the first-quadrant operation, Q_I;
2. Mode B: energy is converted from positive voltage load to source; the second-quadrant operation, Q_{II};
3. Mode C: energy is converted from source to negative voltage load; the third-quadrant operation, Q_{III};
4. Mode D: energy is converted from negative voltage load to source; the fourth-quadrant operation, Q_{IV}.

The first-quadrant (Mode A) is so called the forward motoring (Forw. Mot.) operation. V_1 and V_2 are positive, and I_1 and I_2 are positive as well. The second-quadrant (Mode B) is so called the forward regenerative (Forw. Reg.) braking operation. V_1 and V_2 are positive, and I_1 and I_2 are negative. The third-quadrant (Mode C) is so-called the reverse motoring (Rev. Mot.) operation. V_1 and I_1 are positive, and V_2 and I_2 are negative. The fourth-quadrant (Mode D) is so-called the reverse regenerative (Rev. Reg.) braking operation. V_1 and I_2 are positive, and I_1 and V_2 are negative.

Each mode has two conditions: $V_1 > V_2$ and $V_1 < V_2$ (or $|V_2|$ for Q_{III} and Q_{IV}). Each condition has two states: "on" and

TABLE 14.6 Switch's status (mentioned switches are not open)

Quadrant No. and mode	Condition	State ON	State OFF	Source side	Load side
QI, Mode A Forw. Mot.	$V_1 > V_2$	$S_{1,4,6,8}$	$S_{2,4,6,8}$	V_1+	V_2+
	$V_1 < V_2$	$S_{1,4,6,8}$	$S_{2,4,7}$	I_1+	I_2+
QII, Mode B Forw. Reg.	$V_1 > V_2$	$S_{2,4,6,8}$	$S_{1,4,7}$	V_1+	V_2+
	$V_1 < V_2$	$S_{2,4,6,8}$	$S_{1,4,6,8}$	I_1-	I_2-
QIII, Mode C Rev. Mot.	$V_1 > |V_2|$	$S_{1,4,6,8}$	$S_{3,5,6,8}$	V_1+	V_2-
	$V_1 < |V_2|$	$S_{1,4,6,8}$	$S_{3,5,7}$	I_1+	I_2-
QIV Mode D Rev. Reg.	$V_1 > |V_2|$	$S_{3,5,6,8}$	$S_{1,4,7}$	V_1+	V_2-
	$V_1 < |V_2|$	$S_{3,5,6,8}$	$S_{1,4,6,8}$	I_1-	I_2+

"off." Usually, each state is operating in various conduction duty k for different currents. As usual, the efficiency of all SC DC/DC converters is independent from the conduction duty cycle k. The switching period is T where $T = 1/f$. The switch status is shown in Table 14.6.

As usual, the transfer efficiency only relies on the ratio of the source and load voltages, and it is independent on R, C, f, and k. We select $k = 0.5$ for our description. Other values for the reference are $f = 5$ kHz, $V_1 = 21$ V, $V_2 = 14$ V, and total $C = 4000\,\mu$F, $R = 50$ mΩ.

For **Mode A1**, condition $V_1 > V_2$ is shown in Fig. 14.78a. Since $V_1 > V_2$, two capacitors C_1 and C_2 are connected in parallel. During switch-**on** state, switches S_1, S_4, S_6, and S_8 are closed and other switches are open. In this case, capacitors $C_1//C_2$ are charged via the circuit V_1–S_1–$C_1//C_2$–S_4, and the voltage across capacitors C_1 and C_2 is increasing. During switch-**off** state, switch S_2, S_4, S_6, and S_8 are closed and other switches are open. In this case capacitors $C_1//C_2$ are discharged via the circuit S_2–V_2–S_4–$C_1//C_2$, and the voltage across capacitors C_1 and C_2 is decreasing. Capacitors C_1 and C_2 transfer the energy from the source to the load.

The average capacitor voltage

$$V_C = kV_1 + (1-k)V_2 \quad (14.251)$$

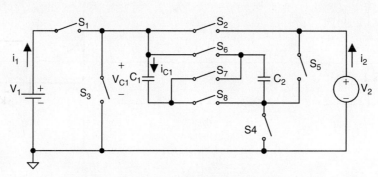

FIGURE 14.78 Four-quadrant sc DC/DC Luo-converter.

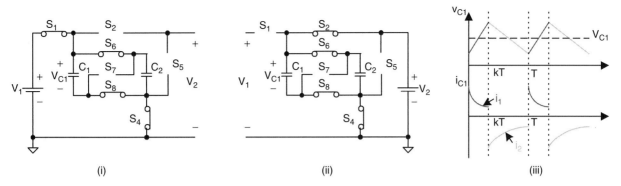

FIGURE 14.78a Mode A1 (QI): forward motoring with $V_1 > V_2$: (i) switch on: S_1, S_4, S_6, and S_8 on; (ii) switch off: S_2, S_4, S_6, and S_8, on; and (iii) waveforms.

The average current is

$$I_2 = \frac{1}{T}\int_{kT}^{T} i_C(t)dt \approx (1-k)\frac{V_C - V_2}{R} \qquad (14.252)$$

and

$$I_1 = \frac{1}{T}\int_{0}^{kT} i_C(t)dt \approx k\frac{V_1 - V_C}{R} \qquad (14.253)$$

The transfer efficiency is

$$\eta_{A1} = \frac{P_O}{P_I} = \frac{1-k}{k}\frac{V_2}{V_1}\frac{V_C - V_2}{V_1 - V_C} = \frac{V_2}{V_1} \qquad (14.254)$$

For **Mode A2**, condition $V_1 < V_2$ is shown in Fig. 14.78b. Since $V_1 < V_2$, two capacitors C_1 and C_2 are connected in parallel during switch on and in series during switch off. This is so-called the *voltage-lift technique*. During switch-**on** state, switches S_1, S_4, S_6, and S_8 are closed and other switches are open. In this case, capacitors $C_1 // C_2$ are charged via the circuit V_1–S_1–C_1//C_2–S_4, and the voltage across capacitors C_1 and C_2 is increasing. During switch-**off** state, switches S_2, S_4, and S_7 are closed and other switches are open. In this case, capacitors C_1 and C_2 are discharged via the circuit S_2–V_2–S_4–C_1–S_7–C_2, and the voltage across capacitor C_1 and C_2 is decreasing. Capacitors C_1 and C_2 transfer the energy from the source to the load.

The average capacitor voltage is

$$V_C = \frac{0.5V_1 + V_2}{2.5} = 11.2 \qquad (14.255)$$

The average current is

$$I_2 = \frac{1}{T}\int_{kT}^{T} i_C(t)dt \approx (1-k)\frac{2V_C - V_2}{R} \qquad (14.256)$$

and

$$I_1 = \frac{1}{T}\int_{0}^{kT} i_C(t)dt \approx k\frac{V_1 - V_C}{R} \qquad (14.257)$$

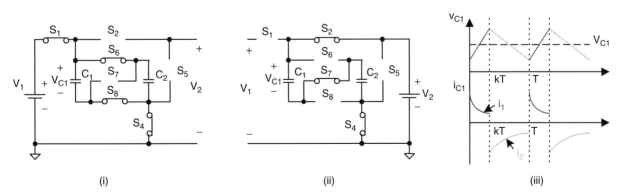

FIGURE 14.78b Mode A2 (QI): forward motoring with $V_1 < V_2$: (i) switch on: S_1, S_4, S_6, and S_8, on; (ii) switch off: S_2, S_4, and S_7, on; and (iii) waveforms.

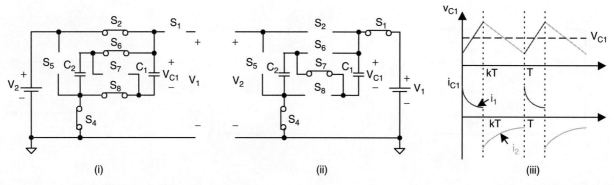

FIGURE 14.78c Mode B1 (QII): forward regenerative braking with $V_1 > V_2$: (i) switch on: S_2, S_4, S_6, and S_8, on; (ii) switch off; S_1, $S_4(S_5)$, and S_7 on; and (iii) waveforms.

The transfer efficiency is

$$\eta_{A2} = \frac{P_O}{P_I} = \frac{1-k}{k} \frac{V_2}{V_1} \frac{2V_C - V_2}{V_1 - V_C} = \frac{V_2}{2V_1} \quad (14.258)$$

For **Mode B1**, condition $V_1 > V_2$ is shown in Fig. 14.78c. Since $V_1 > V_2$, two capacitors C_1 and C_2 are connected in parallel during switch on and in series during switch off. The *voltage-lift technique* is applied. During switch-**on** state, switches S_2, S_4, S_6, and S_8 are closed. In this case, capacitors $C_1 /\!/ C_2$ are charged via the circuit V_2–S_2–$C_1 /\!/ C_2$–S_4, and the voltage across capacitors C_1 and C_2 is increasing. During switch-**off** state, switches S_1, S_4, and S_7 are closed. In this case, capacitors C_1 and C_2 are discharged via the circuit S_1–V_1–S_4–C_2–S_7–C_1, and the voltage across capacitor C_1 and C_2 is decreasing. Capacitors C_1 and C_2 transfer the energy from the load to the source. Therefore, we have $I_2 = 2I_1$.

The average capacitor voltage is

$$V_C = \frac{0.5V_2 + V_1}{2.5} = 11.2 \quad (14.259)$$

The average current is

$$I_1 = \frac{1}{T} \int_{kT}^{T} i_C(t)dt \approx (1-k)\frac{2V_C - V_1}{R} \quad (14.260)$$

and

$$I_2 = \frac{1}{T} \int_{0}^{kT} i_C(t)dt \approx k\frac{V_2 - V_C}{R} \quad (14.261)$$

The transfer efficiency is

$$\eta_{B1} = \frac{P_O}{P_I} = \frac{1-k}{k} \frac{V_1}{V_2} \frac{2V_C - V_1}{V_2 - V_C} = \frac{V_1}{2V_2} \quad (14.262)$$

For **Mode B2**, condition $V_1 < V_2$ is shown in Fig. 14.78d. Since $V_1 < V_2$, two capacitors C_1 and C_2 are connected in parallel. During switch-**on** state, switches S_2, S_4, S_6, and S_8 are closed. In this case, capacitors $C_1 /\!/ C_2$ are charged via the circuit V_2–S_2–$C_1 /\!/ C_2$–S_4, and the voltage across capacitors C_1 and C_2

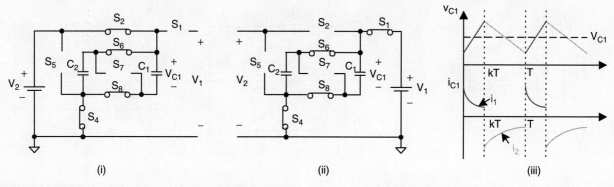

FIGURE 14.78d Mode B2 (QII): forward regenerative braking with $V_1 < V_2$: (i) switch on: S_2, S_4, S_6, and S_8, on; (ii) switch off: S_1, $S_4(S_5)$, S_6, and S_8 on; and (iii) waveforms.

is increasing. During switch-**off** state, switches S_1, S_4, S_6, and S_8 are closed. In this case capacitors $C_1 /\!/ C_2$ is discharged via the circuit S_1–V_1–S_4–$C_1 /\!/ C_2$, and the voltage across capacitors C_1 and C_2 is decreasing. Capacitors C_1 and C_2 transfer the energy from the load to the source. Therefore, we have $I_2 = I_1$.

The average capacitor voltage is

$$V_C = kV_2 + (1-k)V_1 \tag{14.263}$$

The average current is

$$I_1 = \frac{1}{T}\int_{kT}^{T} i_C(t)dt \approx (1-k)\frac{V_C - V_1}{R} \tag{14.264}$$

and

$$I_2 = \frac{1}{T}\int_{0}^{kT} i_C(t)dt \approx k\frac{V_2 - V_C}{R} \tag{14.265}$$

The transfer efficiency is

$$\eta_{B2} = \frac{P_O}{P_I} = \frac{1-k}{k}\frac{V_1}{V_2}\frac{V_C - V_1}{V_2 - V_C} = \frac{V_1}{V_2} \tag{14.266}$$

For **Mode C1**, condition $V_1 > |V_2|$ is shown in Fig. 14.78e. Since $V_1 > |V_2|$, two capacitors C_1 and C_2 are connected in parallel. During switch-**on** state, switches S_1, S_4, S_6, and S_8 are closed. In this case, capacitors $C_1 /\!/ C_2$ are charged via the circuit V_1–S_1–$C_1 /\!/ C_2$–S_4, and the voltage across capacitors C_1 and C_2 is increasing. During switch-**off** state, switches S_3, S_5, S_6, and S_8 are closed. Capacitors C_1 and C_2 are discharged via the circuit S_3–V_2–S_5–$C_1 /\!/ C_2$, and the voltage across capacitors C_1 and C_2 is decreasing. Capacitors C_1 and C_2 transfer the energy from the source to the load. We have $I_1 = I_2$.

The average capacitor voltage is

$$V_C = kV_1 + (1-k)|V_2| \tag{14.267}$$

The average current (absolute value) is

$$I_2 = \frac{1}{T}\int_{kT}^{T} i_C(t)dt \approx (1-k)\frac{V_C - |V_2|}{R} \tag{14.268}$$

and the average input current is

$$I_1 = \frac{1}{T}\int_{0}^{kT} i_C(t)dt \approx k\frac{V_1 - V_C}{R} \tag{14.269}$$

The transfer efficiency is

$$\eta_{C1} = \frac{P_O}{P_I} = \frac{1-k}{k}\frac{|V_2|}{V_1}\frac{V_C - |V_2|}{V_1 - V_C} = \frac{|V_2|}{V_1} \tag{14.270}$$

For **Mode C2**, condition $V_1 < |V_2|$ is shown in Fig. 14.78f. Since $V_1 < |V_2|$, two capacitors C_1 and C_2 are connected in parallel during switch on and in series during switch off, applying the *voltage-lift technique*. During switch-**on** state, switches S_1, S_4, S_6, and S_8, are closed. Capacitors C_1 and C_2 are charged via the circuit V_1–S_1–$C_1 /\!/ C_2$–S_4, and the voltage across capacitors C_1 and C_2 is increasing. During switch-**off** state, switches S_3, S_5, and S_7 are closed. Capacitors C_1 and C_2 is discharged via the circuit S_3–V_2–S_5–C_1–S_7–C_2, and the voltage across capacitor C_1 and C_2 is decreasing. Capacitors C_1 and C_2 transfer the energy from the source to the load. We have $I_1 = 2I_2$.

The average capacitor voltage is

$$V_C = \frac{0.5V_1 + |V_2|}{2.5} = 11.2 \tag{14.271}$$

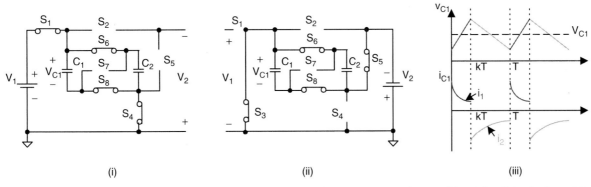

FIGURE 14.78e Mode C1 (QIII): reverse motoring with $V_1 > |V_2|$: (i) switch on: S_1, S_4, S_6, and S_8 on; (ii) switch off: S_3, S_5, S_6, and S_8 on; and (iii) waveforms.

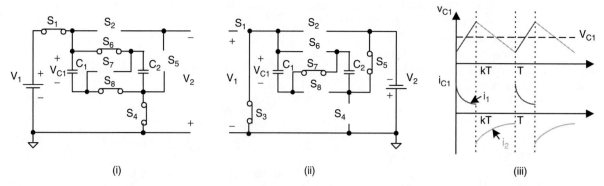

FIGURE 14.78f Mode C2 (QIII): reverse motoring with $V_1 < |V_2|$: (i) switch on: S_1, S_4, S_6, and S_8, on; (ii) switch off: S_3, S_5, and S_7, on; and (iii) waveforms.

The average currents are

$$I_2 = \frac{1}{T}\int_{kT}^{T} i_C(t)dt \approx (1-k)\frac{2V_C - |V_2|}{R} \quad (14.272)$$

and

$$I_1 = \frac{1}{T}\int_{0}^{kT} i_C(t)dt \approx k\frac{V_1 - V_C}{R} \quad (14.273)$$

The transfer efficiency is

$$\eta_{C2} = \frac{P_O}{P_I} = \frac{1-k}{k}\frac{|V_2|}{V_1}\frac{2V_C - |V_2|}{V_1 - V_C} = \frac{|V_2|}{2V_1} \quad (14.274)$$

For **Mode D1**, condition $V_1 > |V_2|$ is shown in Fig. 14.78g. Since $V_1 > |V_2|$, two capacitors C_1 and C_2 are connected in parallel during switch on and in series during switch off, applying the *voltage-lift technique*. During switch-**on** state, switches S_3, S_5, S_6, and S_8 are closed. In this case, capacitors $C_1 /\!/ C_2$ are charged via the circuit V_2–S_3–$C_1 /\!/ C_2$–S_5, and the voltage across capacitors C_1 and C_2 is increasing. During switch-**off** state, switches S_1, S_4, and S_7 are closed. Capacitors C_1 and C_2 are discharged via the circuit S_1–V_1–S_4–C_2–S_7–C_1, and the voltage across capacitor C_1 and C_2 is decreasing. Capacitors C_1 and C_2 transfer the energy from the load to the source. We have $I_2 = 2I_1$.

The average capacitor voltage is

$$V_C = \frac{0.5|V_2| + V_1}{2.5} = 11.2 \quad (14.275)$$

The average currents are

$$I_1 = \frac{1}{T}\int_{kT}^{T} i_C(t)dt \approx (1-k)\frac{2V_C - V_1}{R} \quad (14.276)$$

and

$$I_2 = \frac{1}{T}\int_{0}^{kT} i_C(t)dt \approx k\frac{|V_2| - V_C}{R} \quad (14.277)$$

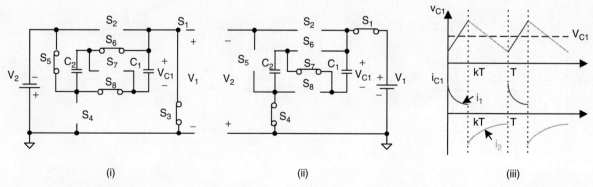

FIGURE 14.78g Mode D1 (QIV): reverse regenerative braking with $V_1 > |V_2|$: (i) switch on: S_3, S_4, S_6, and S_8, on; (ii) switch off: S_1, S_4, and S_7 on; and (iii) waveforms.

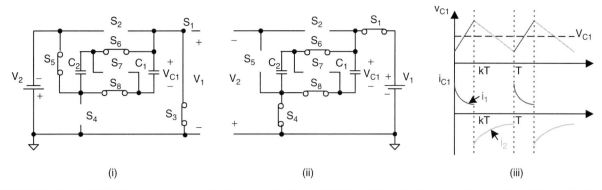

FIGURE 14.78h Mode D2 (QIV): reverse regenerative braking with $V_1 < |V_2|$: (i) switch on: S_3, S_5, S_6, and S_8, on; (ii) switch off: S_1, S_4, S_6, and S_8 on; and (iii) waveforms.

The transfer efficiency is

$$\eta_{D1} = \frac{P_O}{P_I} = \frac{1-k}{k} \frac{V_1}{|V_2|} \frac{2V_C - V_1}{|V_2| - V_C} = \frac{V_1}{2|V_2|} \quad (14.278)$$

For **Mode D2**, condition $V_1 < |V_2|$ is shown in Fig. 14.78h. Since $V_1 < |V_2|$, two capacitors C_1 and C_2 are connected in parallel. During switch-**on** state, switches S_3, S_5, S_6, and S_8 are closed. In this case, capacitors $C_1 /\!/ C_2$ are charged via the circuit V_2–S_3–$C_1 /\!/ C_2$–S_5, and the voltage across capacitors C_1 and C_2 is increasing. During switch-**off** state, switches S_1, S_4, S_6, and S_8 are closed. Capacitors C_1 and C_2 are discharged via the circuit S_1–V_1–S_4–$C_1 /\!/ C_2$, and the voltage across capacitors C_1 and C_2 is decreasing. Capacitors C_1 and C_2 transfer the energy from the load to the source. We have $I_2 = I_1$.

The average capacitor voltage is

$$V_C = k|V_2| + (1-k)V_1 \quad (14.279)$$

The average currents are

$$I_1 = \frac{1}{T} \int_{kT}^{T} i_C(t)dt \approx (1-k)\frac{V_C - V_1}{R} \quad (14.280)$$

and

$$I_2 = \frac{1}{T} \int_{0}^{kT} i_C(t)dt \approx k\frac{|V_2| - V_C}{R} \quad (14.281)$$

The transfer efficiency is

$$\eta_{D2} = \frac{P_O}{P_I} = \frac{1-k}{k} \frac{V_1}{|V_2|} \frac{V_C - V_1}{|V_2| - V_C} = \frac{V_1}{|V_2|} \quad (14.282)$$

14.9 Multiple-lift Push–Pull Switched-capacitor Luo-converters

Micro-power-consumption technique requires high power density DC/DC converters and power supply source. Voltage-lift (VL) technique is a popular method to apply in electronic circuit design. Since switched-capacitor can be integrated into power integrated circuit (IC) chip, its size is small. Combining switched-capacitor and VL techniques the DC/DC converters with small size, high power density, high voltage transfer gain, high power efficiency, and low EMI can be constructed. This section introduces a new series DC/DC converters – multiple-lift push–pull switched-capacitor DC/DC Luo-converters. There are two subseries:

- P/O multiple-lift (ML) push–pull (PP) switched-capacitor (SC) DC/DC Luo-converter;
- N/O multiple-lift push–pull switched-capacitor DC/DC Luo-converter.

14.9.1 P/O Multiple-lift Push–Pull Switched-capacitor DC/DC Luo-converter

P/O ML-PP SC DC/DC Luo-converters have several subseries:

- Main series;
- Additional series;
- Enhanced series;
- Re-enhanced series;
- Multiple-enhanced series.

We only introduce three circuits of main series and additional series in this section.

P/O ML-PP SC Luo-converter elementary circuit is shown in Fig. 14.79a. Its output voltage and current are

$$V_O = 2V_I$$

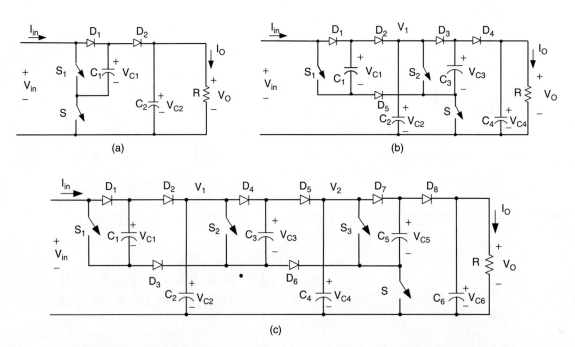

FIGURE 14.79 P/O ML-PP SC Luo-converter: (a) elemental; (b) re-lift; and (c) triple-lift circuits.

and

$$I_O = \frac{1}{2}I_I$$

The voltage transfer gain is

$$M_E = \frac{V_O}{V_I} = 2 \qquad (14.283)$$

P/O ML-PP SC Luo-converter re-lift circuit is shown in Fig. 14.79b. Its output voltage and current are

$$V_O = 4V_I$$

and

$$I_O = \frac{1}{4}I_I$$

The voltage transfer gain is

$$M_R = 4 \qquad (14.284)$$

P/O ML-PP SC Luo-converter triple-lift circuit is shown in Fig. 14.79c. Its output voltage and current are

$$V_O = 8V_I$$

and

$$I_O = \frac{1}{8}I_I$$

The voltage transfer gain is

$$M_T = 8 \qquad (14.285)$$

P/O ML-PP SC Luo-converter additional circuit is shown in Fig. 14.80a. Its output voltage and current are

$$V_O = 3V_I$$

and

$$I_O = \frac{1}{3}I_I$$

The voltage transfer gain is

$$M_A = \frac{V_O}{V_I} = 3 \qquad (14.286)$$

P/O ML-PP SC Luo-converter additional re-lift circuit is shown in Fig. 14.80b. Its output voltage and current are

$$V_O = 6V_I$$

and

$$I_O = \frac{1}{6}I_I$$

The voltage transfer gain is

$$M_{AR} = \frac{V_O}{V_I} = 6 \qquad (14.287)$$

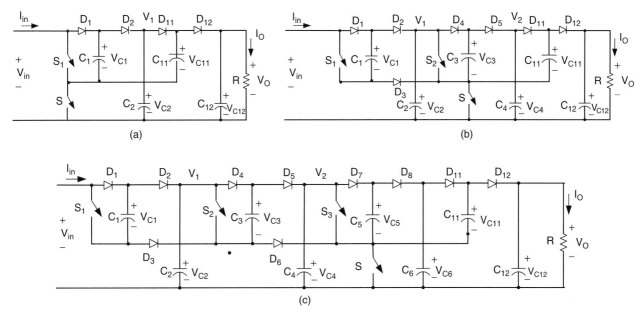

FIGURE 14.80 P/O ML-PP SC Luo-converter re-lift circuit: (a) additional; (b) re-lift; and (c) triple-lift circuits.

P/O ML-PP SC Luo-converter additional triple-lift circuit is shown in Fig. 14.80c. Its output voltage and current are

$$V_O = 12 V_I$$

and

$$I_O = \frac{1}{12} I_I$$

The voltage transfer gain is

$$M_{AT} = \frac{V_O}{V_I} = 12 \qquad (14.288)$$

14.9.2 N/O Multiple-lift Push–Pull Switched-capacitor DC/DC Luo-converter

N/O ML-PP SC DC/DC Luo-converters have several subseries:

- Main series;
- Additional series;
- Enhanced series;
- Re-enhanced series;
- Multiple-enhanced series.

We only introduce three circuits of main series and additional series in this section.

N/O ML-PP SC Luo-converter elementary circuit is shown in Fig. 14.81a. Its output voltage and current are

$$V_O = V_I$$

and

$$I_O = I_I$$

The voltage transfer gain is

$$M_E = \frac{V_O}{V_I} = 1 \qquad (14.289)$$

N/O ML-PP SC Luo-converter re-lift circuit is shown in Fig. 14.81b. Its output voltage and current are

$$V_O = 3 V_I$$

and

$$I_O = \frac{1}{3} I_I$$

The voltage transfer gain is

$$M_R = 3 \qquad (14.290)$$

N/O ML-PP SC Luo-converter triple-lift circuit is shown in Fig. 14.81c. Its output voltage and current are

$$V_O = 7 V_I$$

and

$$I_O = \frac{1}{7} I_I$$

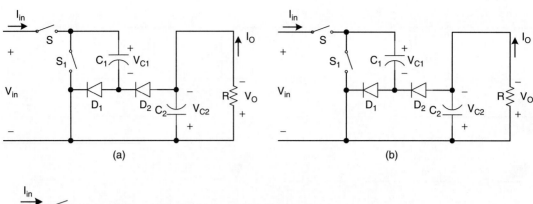

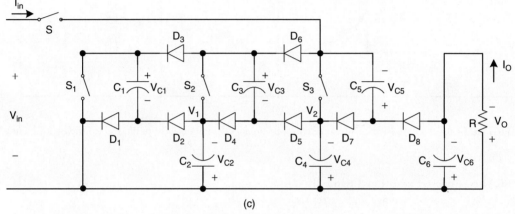

FIGURE 14.81 N/O ML-PP SC Luo-converter: (a) elemental; (b) re-lift; and (c) triple-lift circuits.

The voltage transfer gain is

$$M_T = 7 \quad (14.291)$$

N/O ML-PP SC Luo-converter additional circuit is shown in Fig. 14.82a. Its output voltage and current are

$$V_O = 2V_I$$

and

$$I_O = \frac{1}{2}I_I$$

The voltage transfer gain is

$$M_A = \frac{V_O}{V_I} = 2 \quad (14.292)$$

N/O ML-PP SC Luo-converter additional re-lift circuit is shown in Fig. 14.82b. Its output voltage and current are

$$V_O = 5V_I$$

and

$$I_O = \frac{1}{5}I_I$$

The voltage transfer gain is

$$M_{AR} = \frac{V_O}{V_I} = 5 \quad (14.293)$$

N/O ML-PP SC Luo-converter additional triple-lift circuit is shown in Fig. 14.82c. Its output voltage and current are

$$V_O = 11V_I$$

and

$$I_O = \frac{1}{11}I_I$$

The voltage transfer gain is

$$M_{AT} = \frac{V_O}{V_I} = 11 \quad (14.294)$$

14.10 Switched-inductor Multi-quadrant Operation Luo-converters

Switched-capacitor converters usually have many switches and capacitors, especially for the system with high ratio between

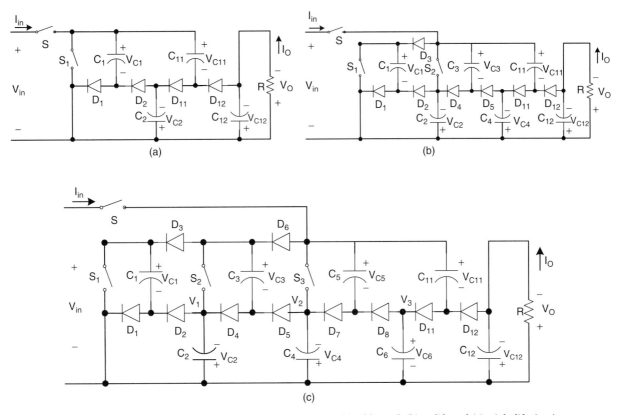

FIGURE 14.82 N/O ML-PP SC Luo-converter re-lift circuit: (a) additional; (b) re-lift; and (c) triple-lift circuits.

source and load voltages. Switched-inductor converter usually has only one inductor even if it works in single-, two-, and/or four-quadrant operation. Simplicity is the main advantage of all switched inductor converters.

Switched-inductor multi-quadrant Luo-converters are the third-generation converters, and they are made of only inductor. These converters have been derived from chopper circuits. They have three modes:

- Two-quadrant switched-inductor DC/DC Luo-converter in forward operation;
- Two-quadrant switched-inductor DC/DC Luo-converter in reverse operation;
- Four-quadrant switched-inductor DC/DC Luo-converter.

The two-quadrant switched-inductor DC/DC Luo-converter in forward operation has been derived for the energy transmission of a dual-voltage system. The both, source and load voltages are positive polarity. It performs in the first-quadrant Q_I and the second-quadrant Q_{II} corresponding to the DC motor forward operation in motoring and regenerative braking states.

The two-quadrant switched-inductor DC/DC Luo-converter in reverse operation has been derived for the energy transmission of a dual-voltage system. The source voltage is positive and load voltage is negative polarity. It performs in the third-quadrant Q_{III} and the fourth-quadrant Q_{IV} corresponding to the DC motor reverse operation in motoring and regenerative braking states.

The four-quadrant switched-inductor DC/DC Luo-converter has been derived for the energy transmission of a dual-voltage system. The source voltage is positive and load voltage can be positive or negative polarity. It performs four-quadrant operation corresponding to the DC motor forward and reverse operation in motoring and regenerative braking states.

14.10.1 Two-quadrant Switched-inductor DC/DC Luo-converter in Forward Operation

Forward operation (F) 2Q SI Luo-converter is shown in Fig. 14.83, and it consists of two switches with two passive diodes, two inductors, and one capacitor. The source voltage (V_1) and load voltage (V_2) are usually considered as constant voltages. The load can be a battery or motor back EMF. For example, the source voltage is 42 V and load voltage is +14 V. There are two modes of operation:

1. Mode A (QI): electrical energy is transferred from source side V_1 to load side V_2;

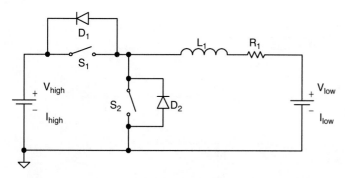

FIGURE 14.83 Switched-inductor QI and II DC/DC Luo-converter.

2. Mode B (QII): electrical energy is transferred from load side V_2 to source side V_1.

Mode A: The equivalent circuits during switch-on and -off periods are shown in Figs. 14.84a and b. The typical output voltage and current waveforms are shown in Fig. 14.84c.

We have the average inductor current I_L as

$$I_L = \frac{kV_1 - V_2}{R} \quad (14.295)$$

The variation ratio of the inductor current i_L is

$$\zeta = \frac{\Delta i_L/2}{I_L} = \frac{k(1-k)V_1}{kV_1 - V_2} \frac{R}{2fL} \quad (14.296)$$

The power transfer efficiency is

$$\eta_A = \frac{P_O}{P_I} = \frac{V_2}{kV_1} \quad (14.297)$$

The boundary between continuous and discontinuous regions is defined as

$$\zeta \geq 1 \quad \text{i.e.}$$

$$\frac{k(1-k)V_1}{kV_1 - V_2} \frac{R}{2fL} \geq 1 \quad \text{or} \quad k \leq \frac{V_2}{V_1} + k(1-k)\frac{R}{2fL} \quad (14.298)$$

Average inductor current I_L in discontinuous region is

$$I_L = \frac{V_1}{V_2 + RI_L} \frac{V_1 - V_2 - RI_L}{2fL} k^2 \quad (14.299)$$

The power transfer efficiency is

$$\eta_{A-dis} = \frac{P_O}{P_I} = \frac{V_2}{V_2 + RI_L} \quad \text{with} \quad k \leq \frac{V_2}{V_1} + k(1-k)\frac{R}{2fL} \quad (14.300)$$

Mode B: The equivalent circuits during switch-on and -off periods are shown in Figs. 14.85a and b. The typical output voltage and current waveforms are shown in Fig. 14.85c.

The average inductor current I_L is

$$I_L = \frac{V_2 - (1-k)V_1}{R} \quad (14.301)$$

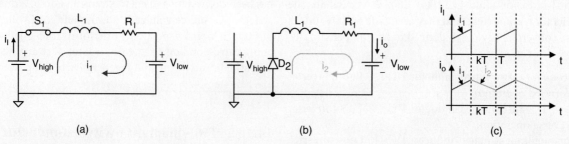

FIGURE 14.84 Mode A of F 2Q SI Luo-converter: (a) state-on: S_1 on; (b) state-off: D_2 on, S_1 off; and (c) input and output current waveforms.

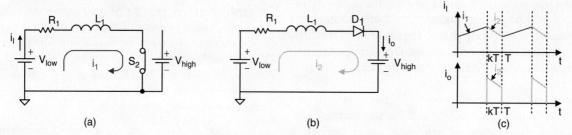

FIGURE 14.85 Mode B of F 2Q SI Luo-converter: (a) state-on: S_2 on; (b) state-off: D_1 on, S_2 off; and (c) input and output current waveforms.

The variation ratio of the inductor current i_L is

$$\zeta = \frac{\Delta i_L/2}{I_L} = \frac{k(1-k)V_1}{V_2 - (1-k)V_1} \frac{R}{2fL} \quad (14.302)$$

The power transfer efficiency

$$\eta_B = \frac{P_O}{P_I} = \frac{(1-k)V_1}{V_2} \quad (14.303)$$

The boundary between continuous and discontinuous regions is defined as

$$\zeta \geq 1, \quad \text{i.e.}$$

$$\frac{k(1-k)V_1}{V_2-(1-k)V_1} \frac{R}{2fL} \geq 1 \quad \text{or} \quad k \leq \left(1 - \frac{V_2}{V_1}\right) + k(1-k)\frac{R}{2fL} \quad (14.304)$$

Average inductor current I_L in discontinuous region is

$$I_L = \frac{V_1}{V_1 - V_2 + RI_L} \frac{V_2 - RI_L}{2fL} k^2 \quad (14.305)$$

The power transfer efficiency is

$$\eta_{B-dis} = \frac{P_O}{P_I} = \frac{V_2 - RI_L}{V_2}$$

with $\quad k \leq \left(1 - \frac{V_2}{V_1}\right) + k(1-k)\frac{R}{2fL} \quad (14.306)$

14.10.2 Two-quadrant Switched-inductor DC/DC Luo-converter in Reverse Operation

Reverse operation (R) 2Q SI Luo-converter is shown in Fig. 14.86, and it consists of two switches with two passive diodes, two inductors, and one capacitor. The source voltage (V_1) and load voltage (V_2) are usually considered as constant

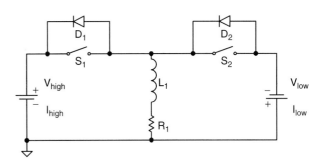

FIGURE 14.86 Switched-inductor QIII and IV DC/DC Luo-converter.

voltages. The load can be a battery or motor back EMF. For example, the source voltage is 42 V and load voltage is −14 V. There are two modes of operation:

1. Mode C (QIII): electrical energy is transferred from source side V_1 to load side $-V_2$;
2. Mode D (QIV): electrical energy is transferred from load side $-V_2$ to source side V_1.

Mode C: The equivalent circuits during switch-on and -off periods are shown in Figs. 14.87a and b. The typical output voltage and current waveforms are shown in Fig. 14.87c.

We have the average inductor current I_L as

$$I_L = \frac{kV_1 - (1-k)V_2}{R} \quad (14.307)$$

The variation ratio of the inductor current i_L is

$$\zeta = \frac{\Delta i_L/2}{I_L} = \frac{k(1-k)(V_1 + V_2)}{kV_1 - (1-k)V_2} \frac{R}{2fL} \quad (14.308)$$

The power transfer efficiency is

$$\eta_C = \frac{P_O}{P_I} = \frac{(1-k)V_2}{kV_1} \quad (14.309)$$

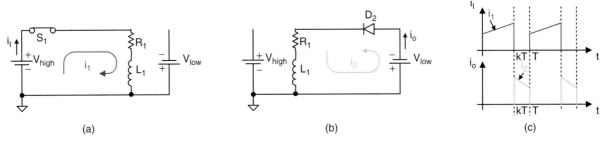

FIGURE 14.87 Mode C of F 2Q SI Luo-converter: (a) state-on; S_1 on; (b) state-off: D_2 on, S_1 off; and (c) input and output current waveforms.

The boundary between continuous and discontinuous regions is defined as

$$\zeta \geq 1, \quad \text{i.e.}$$

$$\frac{k(1-k)(V_1+V_2)}{kV_1-(1-k)V_2}\frac{R}{2fL} \geq 1 \quad \text{or} \quad k \leq \frac{V_2}{V_1+V_2} + k(1-k)\frac{R}{2fL} \tag{14.310}$$

Average inductor current I_L in discontinuous region is

$$I_L = \frac{V_1+V_2}{V_2+RI_L}\frac{V_1-RI_L}{2fL}k^2 \tag{14.311}$$

The power transfer efficiency is

$$\eta_{C-dis} = \frac{P_O}{P_I} = \frac{V_2}{V_1}\frac{V_1-RI_L}{V_2+RI_L}$$

with $\quad k \leq \dfrac{V_2}{V_1+V_2} + k(1-k)\dfrac{R}{2fL} \tag{14.312}$

Mode D: The equivalent circuits during switch-on and -off periods are shown in Figs. 14.88a and b. The typical output voltage and current waveforms are shown in Fig. 14.88c.

The average inductor current I_L is

$$I_L = \frac{kV_2-(1-k)V_1}{R} \tag{14.313}$$

The variation ratio of the inductor current i_L is

$$\zeta = \frac{\Delta i_L/2}{I_L} = \frac{k(1-k)(V_1+V_2)}{kV_2-(1-k)V_1}\frac{R}{2fL} \tag{14.314}$$

The power transfer efficiency is

$$\eta_D = \frac{P_O}{P_I} = \frac{(1-k)V_1}{kV_2} \tag{14.315}$$

The boundary between continuous and discontinuous regions is defined as

$$\zeta \geq 1, \quad \text{i.e.}$$

$$\frac{k(1-k)(V_1+V_2)}{kV_2-(1-k)V_1}\frac{R}{2fL} \geq 1 \quad \text{or} \quad k \leq \frac{V_1}{V_1+V_2} + k(1-k)\frac{R}{2fL} \tag{14.316}$$

Average inductor current I_L in discontinuous region is

$$I_L = \int_0^{t_4} = \frac{V_1+V_2}{V_1+RI_L}\frac{V_2-RI_L}{2fL}k^2 \tag{14.317}$$

The power transfer efficiency is

$$\eta_{D-dis} = \frac{P_O}{P_I} = \frac{V_1}{V_2}\frac{V_2-RI_L}{V_1+RI_L}$$

with $\quad k \leq \dfrac{V_1}{V_1+V_2} + k(1-k)\dfrac{R}{2fL} \tag{14.318}$

14.10.3 Four-quadrant Switched-inductor DC/DC Luo-converter

Switched-inductor DC/DC converters successfully overcome the disadvantage of switched-capacitor converters. Usually, only one inductor is required for each converter with one- or two- or four-quadrant operation, no matter how large the difference between the input and output voltage is. Therefore, switched-inductor converter has very simple topology and circuit. Consequently, it has high power density. This paper introduces a *switched-inductor four-quadrant DC/DC Luo-converter*.

This converter, shown in Fig. 14.89, consists of three switches, two diodes, and only one inductor L. The source voltage V_1 and load voltage V_2 (e.g. a battery or DC motor back EMF) are usually constant voltages. R is the equivalent resistance of the circuit, it is usually small. In this paper, $V_1 > |V_2|$,

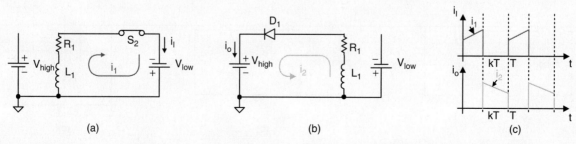

FIGURE 14.88 Mode D of F 2Q SI Luo-converter: (a) state-on: S_2 on; (b) state-off: D_1 on, S_2 off; and (c) input and output waveforms.

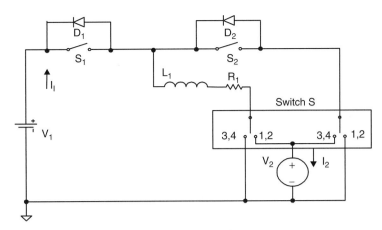

FIGURE 14.89 Four-quadrant switched-inductor DC/DC Luo-converter.

they are supposed as +42 V and ±14 V, respectively. Therefore, there are four-quadrants (modes) of operation:

1. Mode A: energy is transferred from source to positive voltage load; the first-quadrant operation, Q_I;
2. Mode B: energy is transferred from positive voltage load to source; the second-quadrant operation, Q_{II};
3. Mode C: energy is transferred from source to negative voltage load; the third-quadrant operation, Q_{III};
4. Mode C: energy is transferred from negative voltage load to source; the fourth-quadrant operation, Q_{IV}.

The first-quadrant is so-called the forward motoring (Forw. Mot.) operation. V_1 and V_2 are positive, and I_1 and I_2 are positive as well. The second-quadrant is so-called the forward regenerative (Forw. Reg.) braking operation. V_1 and V_2 are positive, and I_1 and I_2 are negative. The third-quadrant is so-called the reverse motoring (Rev. Mot.) operation. V_1 and I_1 are positive, and V_2 and I_2 are negative. The fourth-quadrant is so-called the reverse regenerative (Rev. Reg.) braking operation. V_1 and I_2 are positive, and I_1 and V_2 are negative. Each mode has two states: "on" and "off." Usually, each state is operating in different conduction duty k. The switching period is T, where $T = 1/f$. The switch status is shown in Table 14.7.

Mode A is shown in Fig. 14.84. During switch-**on** state, switch S_1 is closed. In this case the source voltage V_1 supplies the load V_2 and inductor L, inductor current i_L increases.

During switch-**off** state, diode D_2 is on. In this case current i_L flows through the load V_2 via the free-wheeling diode D_2, and it decreases.

Mode B is shown in Fig. 14.85. During switch-**on** state, switch S_2 is closed. In this case the load voltage V_2 supplies the inductor L, inductor current i_L increases. During switch-**off** state, diode D_1 is on, current i_L flows through the source V_1 and load V_2 via the diode D_1, and it decreases.

Mode C is shown in Fig. 14.87. During switch-**on** state, switch S_1 is closed. The source voltage V_1 supplies the inductor L, inductor current i_L increases. During switch-**off** state, diode D_2 is on. Current i_L flows through the load V_2 via the free-wheeling diode D_2, and it decreases.

Mode D is shown in Fig. 14.88. During switch-**on** state, switch S_2 is closed. The load voltage V_2 supplies the inductor L, inductor current i_L increases. During switch-**off** state, diode D_1 is on. Current i_L flows through the source V_1 via the diode D_1, and it decreases.

All description of the Modes A, B, C, and D is same as in Sections 14.10.1 and 14.10.2.

14.11 Multi-quadrant ZCS Quasi-resonant Luo-converters

Soft-switching converters are the fourth-generation converters. These converters are made of only inductor or capacitors. They usually perform in the systems between two voltage sources: V_1 and V_2. Voltage source V_1 is proposed positive voltage and voltage V_2 is the load voltage that can be positive or negative. In the investigation, both voltages are proposed constant voltage. Since V_1 and V_2 are constant value, the voltage transfer gain is constant. Our interesting research will concentrate on the working current and the power transfer efficiency η. The resistance R of the inductor has to be considered for the power transfer efficiency η calculation.

Reviewing the papers in the literature, we can find that most of the papers investigating the switched-component converters

TABLE 14.7 Switch's status (mentioned switches are not off)

Q no.	State	S_1	D_1	S_2	D_2	S_3	Source	Load
Q_I, Mode A	ON	ON				ON 1/2	V_1+	V_2+
Forw. Mot.	OFF				ON	ON 1/2	I_1+	I_2+
Q_{II}, Mode B	ON			ON		ON 1/2	V_1+	V_2+
Forw. Reg.	OFF		ON			ON 1/2	I_1-	I_2-
Q_{III}, Mode C	ON	ON				ON 3/4	V_1+	V_2-
Rev. Mot.	OFF				ON	ON 3/4	I_1+	I_2-
Q_{IV}, Mode D	ON			ON		ON 3/4	V_1+	V_2-
Rev. Reg.	OFF		ON			ON 3/4	I_1-	I_2+

are working in single-quadrant operation. Professor Luo and colleagues have developed this technique into multi-quadrant operation. We describe these in this section and the next sections.

Multi-quadrant ZCS quasi-resonant Luo-converters are the fourth-generation converters. Because these converters implement ZCS technique, they have the advantages of high power density, high power transfer efficiency, low EMI, and reasonable EMC. They have three modes:

- Two-quadrant ZCS quasi-resonant DC/DC Luo-converter in forward operation;
- Two-quadrant ZCS quasi-resonant DC/DC Luo-converter in reverse operation;
- Four-quadrant ZCS quasi-resonant DC/DC Luo-converter.

The two-quadrant ZCS quasi-resonant DC/DC Luo-converter in forward operation is derived for the energy transmission of a dual-voltage system. Both, the source and load voltages are positive polarity. It performs in the first-quadrant Q_I and the second-quadrant Q_{II} corresponding to the DC motor forward operation in motoring and regenerative braking states.

The two-quadrant ZCS quasi-resonant DC/DC Luo-converter in reverse operation is derived for the energy transmission of a dual-voltage system. The source voltage is positive and load voltage is negative polarity. It performs in the third-quadrant Q_{III} and the fourth-quadrant Q_{IV} corresponding to the DC motor reverse operation in motoring and regenerative braking states.

The four-quadrant ZCS quasi-resonant DC/DC Luo-converter is derived for the energy transmission of a dual-voltage system. The source voltage is positive, and load voltage can be positive or negative polarity. It performs four-quadrant operation corresponding to the DC motor forward and reverse operation in motoring and regenerative braking states.

14.11.1 Two-quadrant ZCS Quasi-resonant Luo-converter in Forward Operation

Since both voltages are low, this converter is designed as a ZCS quasi-resonant converter (ZCS-QRC). It is shown in Fig. 14.90. This converter consists of one main inductor L and

TABLE 14.8 Switch's status (the blank status means off)

Switch or diode	Mode A (QI)		Mode B (QII)	
	State-on	State-off	State-on	State-off
S_1	ON			
D_1				ON
S_2			ON	
D_2		ON		

two switches with their auxiliary components. A switch S_a is used for two-quadrant operation. Assuming the main inductance is sufficiently large, the current i_L is constant. The source voltage V_1 and load voltage V_2 are usually constant, $V_1 = 42$ V and $V_2 = 14$ V. There are two modes of operation:

1. Mode A (Quadrant I): electrical energy is transferred from V_1 side to V_2 side, switch S_a links to D_2;
2. Mode B (Quadrant II): electrical energy is transferred from V_2 side to V_1 side, switch S_a links to D_1.

Each mode has two states: "on" and "off." The switch status of each state is shown in Table 14.8.

Mode A is a ZCS buck converter. The equivalent circuit, current, and voltage waveforms are shown in Fig. 14.91. There are four time regions for the switching on and off period. The conduction duty cycle is $k = (t_1 + t_2)$ when the input current

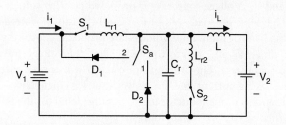

FIGURE 14.90 Two-quadrant (QI+QII) DC/DC ZCS quasi-resonant Luo-converter.

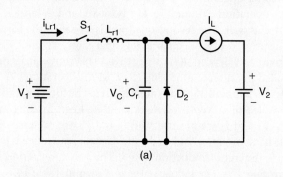

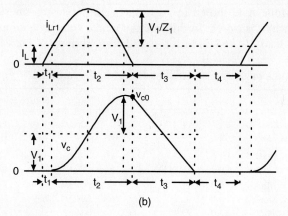

FIGURE 14.91 Mode A operation: (a) equivalent circuit and (b) waveforms.

flows through the switch S_1 and inductor L. The whole period is $T = (t_1 + t_2 + t_3 + t_4)$. Some formulas are listed below

$$\omega_1 = \frac{1}{\sqrt{L_{r1}C_r}}; \quad Z_1 = \sqrt{\frac{L_{r1}}{C_r}}; \quad i_{1-peak} = I_L + \frac{V_1}{Z_1} \quad (14.319)$$

$$t_1 = \frac{I_L L_{r1}}{V_1}; \quad \alpha_1 = \sin^{-1}\left(\frac{I_L Z_1}{V_1}\right) \quad (14.320)$$

$$t_2 = \frac{1}{\omega_1}(\pi + \alpha_1); \quad v_{CO} = V_1(1 + \cos\alpha_1) \quad (14.321)$$

$$t_3 = \frac{v_{CO}C_r}{I_L}; \quad \frac{I_L V_2}{V_1} = \frac{t_1 + t_2}{T}\left(I_L + \frac{V_1}{Z_1}\frac{\cos\alpha_1}{\pi/2 + \alpha_1}\right) \quad (14.322)$$

$$t_4 = \frac{V_1(t_1 + t_2)}{V_2 I_L}\left(I_L + \frac{V_1}{Z_1}\frac{\cos\alpha_1}{\pi/2 + \alpha_1}\right) - (t_1 + t_2 + t_3); \quad (14.323)$$

$$k = \frac{t_1 + t_2}{t_1 + t_2 + t_3 + t_4}; \quad T = t_1 + t_2 + t_3 + t_4; \quad f = 1/T \quad (14.324)$$

Mode B is a ZCS boost converter. The equivalent circuit, current, and voltage waveforms are shown in Fig. 14.92. There are four time regions for the switching on and off period. The conduction duty cycle is $k = (t_1 + t_2)$, but the output current only flows through the source V_1 in the period t_4. The whole period is $T = (t_1 + t_2 + t_3 + t_4)$. Some formulas are listed below

$$\omega_2 = \frac{1}{\sqrt{L_{r2}C_r}}; \quad Z_2 = \sqrt{\frac{L_{r2}}{C_r}}; \quad i_{2-peak} = I_L + \frac{V_1}{Z_2} \quad (14.325)$$

$$t_1 = \frac{I_L L_{r2}}{V_1}; \quad \alpha_2 = \sin^{-1}\left(\frac{I_L Z_2}{V_1}\right) \quad (14.326)$$

$$t_2 = \frac{1}{\omega_2}(\pi + \alpha_2); \quad v_{CO} = -V_1 \cos\alpha_2 \quad (14.327)$$

$$t_3 = \frac{(V_1 - v_{CO})C_r}{I_L}; \quad \frac{I_L V_2}{V_1} = \frac{t_4}{T}I_L \quad (14.328)$$

$$\frac{V_2}{V_1} = \frac{t_4}{T} = \frac{t_4}{t_1 + t_2 + t_3 + t_4}; \quad t_4 = \frac{t_1 + t_2 + t_3}{(V_1/V_2) - 1} \quad (14.329)$$

$$k = \frac{t_1 + t_2}{t_1 + t_2 + t_3 + t_4}; \quad T = t_1 + t_2 + t_3 + t_4; \quad f = 1/T \quad (14.330)$$

14.11.2 Two-quadrant ZCS Quasi-resonant Luo-converter in Reverse Operation

Two-quadrant ZCS quasi-resonant Luo-converter in reverse operation is shown in Fig. 14.93. It is a new soft-switching technique with two-quadrant operation, which effectively reduces the power losses and largely increases the power transfer efficiency. It consists of one main inductor L and two switches with their auxiliary components. A switch S_a is used for two-quadrant operation. Assuming the main inductance L is sufficiently large, the current i_L is constant. The source voltage V_1 and load voltage V_2 are usually constant, e.g. $V_1 = 42\,\text{V}$ and $V_2 = -28\,\text{V}$. There are two modes of operation:

1. Mode C (Quadrant III): electrical energy is transferred from V_1 side to $-V_2$ side, switch S_a links to D_2;
2. Mode D (Quadrant IV): electrical energy is transferred from $-V_2$ side to V_1 side, switch S_a links to D_1.

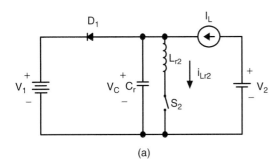

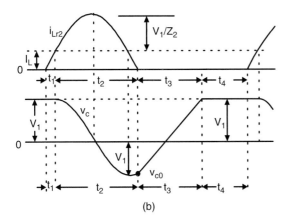

FIGURE 14.92 Mode B operation: (a) equivalent circuit and (b) waveforms.

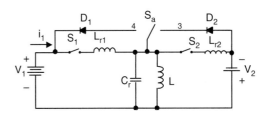

FIGURE 14.93 Two-quadrant (QIII+IV) DC/DC ZCS quasi-resonant Luo-converter.

Each mode has two states: "on" and "off." The switch status of each state is shown in Table 14.9.

Mode C is a ZCS buck–boost converter. The equivalent circuit, current, and voltage waveforms are shown in Fig. 14.94. There are four time regions for the switching on and off period. The conduction duty cycle is $kT = (t_1 + t_2)$ when the input current flows through the switch S_1 and the main inductor L. The whole period is $T = (t_1 + t_2 + t_3 + t_4)$. Some formulas are listed below

$$\omega_1 = \frac{1}{\sqrt{L_{r1} C_r}}; \quad Z_1 = \sqrt{\frac{L_{r1}}{C_r}}; \quad i_{1-peak} = I_L + \frac{V_1}{Z_1} \tag{14.331}$$

TABLE 14.9 Switch's status (the blank status means off)

Switch or diode	Mode C (QIII)		Mode D (QIV)	
	State-on	State-off	State-on	State-off
S_1	ON			
D_1				ON
S_2			ON	
D_2		ON		

$$t_1 = \frac{I_L L_{r1}}{V_1 + V_2}; \quad \alpha_1 = \sin^{-1}\left(\frac{I_L Z_1}{V_1 + V_2}\right) \tag{14.332}$$

$$t_2 = \frac{1}{\omega_1}(\pi + \alpha_1); \quad v_{CO} = (V_1 - V_2) + V_1 \sin(\pi/2 + \alpha_1)$$
$$= V_1(1 + \cos \alpha_1) - V_2 \tag{14.333}$$

$$t_3 = \frac{(v_{CO} + V_2)C_r}{I_L} = \frac{V_1(1 + \cos \alpha_1)C_r}{I_L};$$

$$I_1 = \frac{t_1 + t_2}{T}\left(I_L + \frac{V_1 + V_2}{Z_1}\frac{\cos \alpha_1}{\pi/2 + \alpha_1}\right); \quad I_2 = \frac{t_4}{T}I_L \tag{14.334}$$

$$t_4 = \frac{V_1(t_1 + t_2)}{V_2 I_L}\left(I_L + \frac{V_1 + V_2}{Z_1}\frac{\cos \alpha_1}{\pi/2 + \alpha_1}\right) \tag{14.335}$$

$$k = \frac{t_1 + t_2}{t_1 + t_2 + t_3 + t_4}; \quad T = t_1 + t_2 + t_3 + t_4; \quad f = 1/T \tag{14.336}$$

Mode D is a cross ZCS buck–boost converter. The equivalent circuit, current, and voltage waveforms are shown in Fig. 14.95. There are four time regions for the switching on and off period. The conduction duty cycle is $kT = (t_1 + t_2)$, but the output current only flows through the source V_1 in

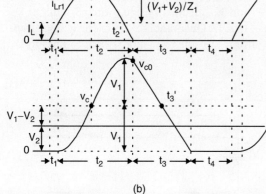

FIGURE 14.94 Mode C operation: (a) equivalent circuit and (b) waveforms.

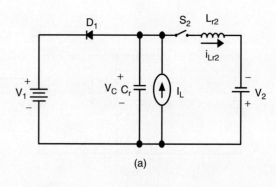

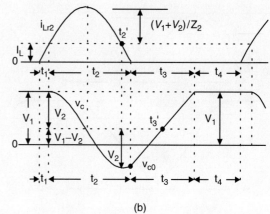

FIGURE 14.95 Mode D operation: (a) equivalent circuit and (b) waveforms.

the period t_4. The whole period is $T = (t_1 + t_2 + t_3 + t_4)$. Some formulas are listed below

$$\omega_2 = \frac{1}{\sqrt{L_{r2}C_r}}; \quad Z_2 = \sqrt{\frac{L_{r2}}{C_r}}; \quad i_{2-peak} = I_L + \frac{V_2}{Z_2} \tag{14.337}$$

$$t_1 = \frac{I_L L_{r2}}{V_1 + V_2}; \quad \alpha_2 = \sin^{-1}\left(\frac{I_L Z_2}{V_2 + V_2}\right) \tag{14.338}$$

$$t_2 = \frac{1}{\omega_2}(\pi + \alpha_2); \quad v_{CO} = (V_1 - V_2) - V_2 \sin(\pi/2 + \alpha_2)$$
$$= V_1 - V_2(1 + \cos\alpha_2) \tag{14.339}$$

$$t_3 = \frac{(V_1 - v_{CO})C_r}{I_L} = \frac{V_2(1 + \cos\alpha_2)C_r}{I_L};$$

$$I_2 = \frac{t_1 + t_2}{T}\left(I_L + \frac{V_1 + V_2}{Z_2}\frac{\cos\alpha_2}{\pi/2 + \alpha_2}\right); \quad I_1 = \frac{t_4}{T}I_L \tag{14.340}$$

$$t_4 = \frac{V_2(t_1 + t_2)}{V_1 I_L}\left(I_L + \frac{V_1 + V_2}{Z_2}\frac{\cos\alpha_2}{\pi/2 + \alpha_2}\right) \tag{14.341}$$

$$k = \frac{t_1 + t_2}{t_1 + t_2 + t_3 + t_4}; \quad T = t_1 + t_2 + t_3 + t_4; \quad f = 1/T \tag{14.342}$$

14.11.3 Four-quadrant ZCS Quasi-resonant Luo-converter

Four-quadrant ZCS quasi-resonant Luo-converter is shown in Fig. 14.96. Circuit 1 implements the operation in quadrants I and II, circuit 2 implements the operation in quadrants III and IV. Circuit 1 and 2 can be converted to each other by auxiliary switch. Each circuit consists of one main inductor L and two switches. A switch S_a is used for four-quadrant operation. Assuming that the main inductance L is sufficiently large,

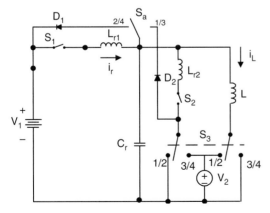

FIGURE 14.96 Four-quadrant DC/DC ZCS quasi-resonant Luo-converter.

the current i_L remains constant. The source and load voltages are usually constant, e.g. $V_1 = 42$ V and $V_2 = \pm 28$ V [7–9]. There are four modes of operation:

1. Mode A (Quadrant I): electrical energy is transferred from V_1 side to V_2 side, switch S_a links to D_2;
2. Mode B (Quadrant II): electrical energy is transferred from V_2 side to V_1 side, switch S_a links to D_1;
3. Mode C (Quadrant III): electrical energy is transferred from V_1 side to $-V_2$ side, switch S_a links to D_2;
4. Mode D (Quadrant IV): electrical energy is transferred from $-V_2$ side to V_1 side, switch S_a links to D_1.

Each mode has two states: "on" and "off." The switch status of each state is shown in Table 14.10.

The operation of Mode A, B, C, and D is same as in the previous Sections 14.11.1 and 14.11.2.

14.12 Multi-quadrant ZVS Quasi-resonant Luo-converters

Multi-quadrant ZVS quasi-resonant Luo-converters are the fourth-generation converters. Because these converters implement ZCS technique, they have the advantages of high power density, high power transfer efficiency, low EMI, and reasonable EMC. They have three modes:

- Two-quadrant ZVS quasi-resonant DC/DC Luo-converter in forward operation;
- Two-quadrant ZVS quasi-resonant DC/DC Luo-converter in reverse operation;
- Four-quadrant ZVS quasi-resonant DC/DC Luo-converter.

The two-quadrant ZVS quasi-resonant DC/DC Luo-converter in forward operation is derived for the energy transmission of a dual-voltage system. Both, the source and load voltages are positive polarity. It performs in the first-quadrant Q_I and the second-quadrant Q_{II} corresponding to the DC motor forward operation in motoring and regenerative braking states.

The two-quadrant ZVS quasi-resonant DC/DC Luo-converter in reverse operation is derived for the energy transmission of a dual-voltage system. The source voltage is positive and load voltage is negative polarity. It performs in the third-quadrant Q_{III} and the fourth-quadrant Q_{IV} corresponding to the DC motor reverse operation in motoring and regenerative braking states.

The four-quadrant ZVS quasi-resonant DC/DC Luo-converter is derived for the energy transmission of a dual-voltage system. The source voltage is positive, and load voltage can be positive or negative polarity. It performs four-quadrant operation corresponding to the DC motor forward and reverse operation in motoring and regenerative braking states.

TABLE 14.10 Switch's status (the blank status means off)

Circuit//switch or diode	Mode A (QI)		Mode B (QII)		Mode C (QIII)		Mode D (QIV)	
	State-on	State-off	State-on	State-off	State-on	State-off	State-on	State-off
Circuit	Circuit 1				Circuit 2			
S_1	ON				ON			
D_1				ON				ON
S_2			ON				ON	
D_2		ON				ON		

14.12.1 Two-quadrant ZVS Quasi-resonant DC/DC Luo-converter in Forward Operation

Two-quadrant ZVS quasi-resonant Luo-converter in forward operation is shown in Fig. 14.97. It consists of one main inductor L and two switches with their auxiliary components.

Assuming the main inductance L is sufficiently large, the current i_L is constant. The source voltage V_1 and load voltage V_2 are usually constant, e.g. $V_1 = 42$ V and $V_2 = 14$ V. There are two modes of operation:

1. Mode A (Quadrant I): electrical energy is transferred from V_1 side to V_2 side;
2. Mode B (Quadrant II): electrical energy is transferred from V_2 side to V_1 side.

Each mode has two states: "on" and "off." The switch status of each state is shown in Table 14.11.

Mode A is a ZVS buck converter shown in Fig. 14.98. There are four time regions for the switching on and off period.

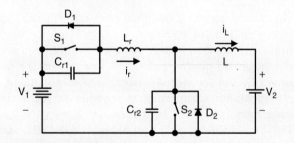

FIGURE 14.97 Two-quadrant (QI+QII) DC/DC ZVS quasi-resonant Luo-converter.

TABLE 14.11 Switch's status (the blank status means off)

Switch	Mode A (QI)		Mode B (QII)	
	State-on	State-off	State-on	State-off
S_1	ON			
D_1				ON
S_2			ON	
D_2		ON		

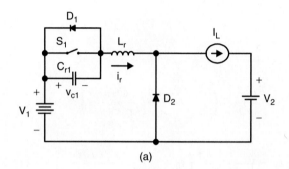

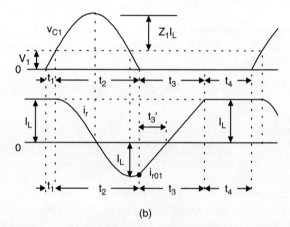

FIGURE 14.98 Mode A operation: (a) equivalent circuit and (b) waveforms.

The conduction duty cycle is $kT = (t_3 + t_4)$ when the input current flows through the switch S_1 and the main inductor L. The whole period is $T = (t_1 + t_2 + t_3 + t_4)$. Some relevant formulas are listed below

$$\omega_1 = \frac{1}{\sqrt{L_r C_{r1}}}; \quad Z_1 = \sqrt{\frac{L_r}{C_{r1}}}; \quad v_{c1-peak} = V_1 + Z_1 I_L \quad (14.343)$$

$$t_1 = \frac{V_1 C_{r1}}{I_L}; \quad \alpha_1 = \sin^{-1}\left(\frac{V_1}{Z_1 I_L}\right) \quad (14.344)$$

$$t_2 = \frac{1}{\omega_1}(\pi + \alpha_1); \quad i_{rO1} = -I_L \cos\alpha_1 \quad (14.345)$$

$$t_3 = \frac{(I_L - i_{rO1})L_r}{V_1}; \quad I_1 = \frac{I_L V_2}{V_1} = \frac{1}{T}\int_{t_3}^{t_4} i_r\, dt \approx \frac{1}{T}(I_L t_4) = \frac{t_4}{T} I_L \tag{14.346}$$

$$t_4 = \frac{t_1 + t_2 + t_3}{(V_1/V_2) - 1} \tag{14.347}$$

$$k = \frac{t_3 + t_4}{t_1 + t_2 + t_3 + t_4}; \quad T = t_1 + t_2 + t_3 + t_4; \quad f = 1/T \tag{14.348}$$

Mode B is a ZVS boost converter shown in Fig. 14.99. There are four time regions for the switching on and off period. The conduction duty cycle is $kT = (t_3 + t_4)$, but the output current only flows through the source V_1 in the period $(t_1 + t_2)$. The whole period is $T = (t_1 + t_2 + t_3 + t_4)$. Some relevant formulas are listed below

$$\omega_2 = \frac{1}{\sqrt{L_r C_{r2}}}; \quad Z_2 = \sqrt{\frac{L_r}{C_{r2}}}; \quad v_{C2-peak} = V_1 + Z_2 I_L \tag{14.349}$$

$$t_1 = \frac{V_1 C_{r2}}{I_L}; \quad \alpha_2 = \sin^{-1}\left(\frac{V_1}{Z_2 I_L}\right) \tag{14.350}$$

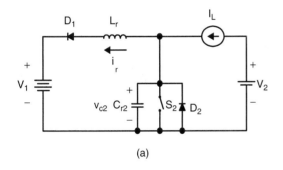

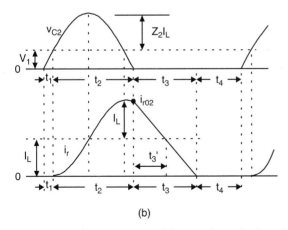

FIGURE 14.99 Mode B operation: (a) equivalent circuit and (b) waveforms.

$$t_2 = \frac{1}{\omega_2}(\pi + \alpha_2); \quad i_{rO2} = I_L(1 + \cos\alpha_2) \tag{14.351}$$

$$t_3 = \frac{i_{rO2} L_r}{V_1};$$

$$I_1 = \frac{I_L V_2}{V_1} = \frac{1}{T}\int_{t_1}^{t_3} i_r\, dt \approx \frac{1}{T}[I_L(t_2 + t_3)] = \frac{t_2 + t_3}{T} I_L;$$

$$\text{or} \quad \frac{V_2}{V_1} = \frac{1}{T}(t_2 + t_3) = \frac{t_2 + t_3}{t_1 + t_2 + t_3 + t_4} \tag{14.352}$$

$$t_4 = \left(\frac{V_1}{V_2} - 1\right)(t_2 + t_3) - t_1; \tag{14.353}$$

$$k = \frac{t_3 + t_4}{t_1 + t_2 + t_3 + t_4}; \quad T = t_1 + t_2 + t_3 + t_4; \quad f = 1/T \tag{14.354}$$

14.12.2 Two-quadrant ZVS Quasi-resonant DC/DC Luo-converter in Reverse Operation

Two-quadrant ZVS quasi-resonant Luo-converter in reverse operation is shown in Fig. 14.100. It consists of one main inductor L and two switches with their auxiliary components. Assuming the main inductance L is sufficiently large, the current i_L is constant. The source voltage V_1 and load voltage V_2 are usually constant, e.g. $V_1 = +42$ V and $V_2 = -28$ V. There are two modes of operation:

1. Mode C (Quadrant III): electrical energy is transferred from V_1 side to $-V_2$ side;
2. Mode D (Quadrant IV): electrical energy is transferred from $-V_2$ side to V_1 side.

Each mode has two states: "on" and "off." The switch status of each state is shown in Table 14.12.

Mode C is a ZVS buck–boost converter shown in Fig. 14.101. There are four time regions for the switching on and off period. The conduction duty cycle is $kT = (t_3 + t_4)$ when the input current flows through the switch S_1 and the main inductor L. The whole period is $T = (t_1 + t_2 + t_3 + t_4)$.

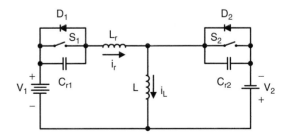

FIGURE 14.100 Two-quadrant (QIII+IV) DC/DC ZVS quasi-resonant Luo-converter.

TABLE 14.12 Switch's status (the blank status means off)

Switch	Mode C (QIII)		Mode D (QIV)	
	State-on	State-off	State-on	State-off
S_1	ON			
D_1				ON
S_2			ON	
D_2		ON		

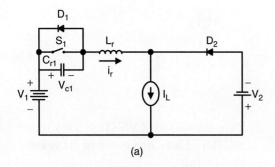

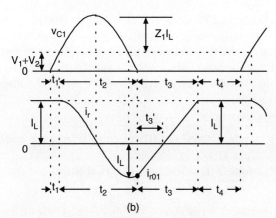

FIGURE 14.101 Mode C operation: (a) equivalent circuit and (b) waveforms.

Some formulas are listed below

$$\omega_1 = \frac{1}{\sqrt{L_r C_{r1}}}; \quad Z_1 = \sqrt{\frac{L_r}{C_{r1}}}; \quad v_{c1-peak} = V_1 + V_2 + Z_1 I_L \quad (14.355)$$

$$t_1 = \frac{(V_1+V_2)C_{r1}}{I_L}; \quad \alpha_1 = \sin^{-1}\left(\frac{V_1+V_2}{Z_1 I_L}\right) \quad (14.356)$$

$$t_2 = \frac{1}{\omega_1}(\pi + \alpha_1); \quad i_{rO1} = -I_L \sin(\pi/2 + \alpha_1) \quad (14.357)$$

$$t_3 = \frac{(I_L - i_{rO1})L_r}{V_1 + V_2} = \frac{I_L(1+\cos\alpha_1)L_r}{V_1+V_2};$$

$$I_1 = \frac{I_L V_2}{V_1} = \frac{1}{T}\int_{t_3}^{t_4} i_r\, dt \approx \frac{1}{T}(I_L t_4) = \frac{t_4}{T}I_L \quad (14.358)$$

$$t_4 = \frac{t_1+t_2+t_3}{(V_1/V_2)-1}; \quad I_2 = \frac{1}{T}\int_{t_1}^{t_3}(I_L - i_r)dt \approx \frac{t_1+t_2+t_3}{T}I_L \quad (14.359)$$

$$k = \frac{t_3+t_4}{t_1+t_2+t_3+t_4}; \quad T = t_1+t_2+t_3+t_4; \quad f = 1/T \quad (14.360)$$

Mode D is a cross ZVS buck–boost converter shown in Fig. 14.102. There are four time regions for the switching on and off period. The conduction duty cycle is $kT = (t_3 + t_4)$, but the output current only flows through the source V_1 in the period $(t_1 + t_2)$. The whole period is $T = (t_1 + t_2 + t_3 + t_4)$. Some formulae are listed below

$$\omega_2 = \frac{1}{\sqrt{L_r C_{r2}}}; \quad Z_2 = \sqrt{\frac{L_r}{C_{r2}}}; \quad v_{C2-peak} = V_1 + V_2 + Z_2 I_L \quad (14.361)$$

$$t_1 = \frac{(V_1+V_2)C_{r2}}{I_L}; \quad \alpha_2 = \sin^{-1}\left(\frac{V_1+V_2}{Z_2 I_L}\right) \quad (14.362)$$

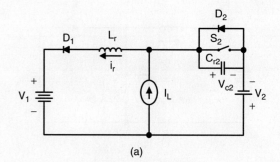

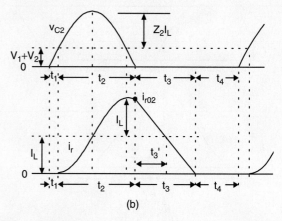

FIGURE 14.102 Mode D operation: (a) equivalent circuit and (b) waveforms.

$$t_2 = \frac{1}{\omega_2}(\pi + \alpha_2); \quad i_{rO2} = I_L[1 + \sin(\pi/2 + \alpha_2)] \quad (14.363)$$

$$t_3 = \frac{i_{rO2} L_r}{V_1 + V_2} = \frac{I_L(1 + \cos\alpha_2) L_r}{V_1 + V_2};$$

$$I_1 = \frac{1}{T}\int_{t_1}^{t_3} i_r dt \approx \frac{t_1 + t_2 + t_3}{T} I_L;$$

$$I_2 = \frac{1}{T}\int_{t_3}^{t_4} i_r dt \approx \frac{1}{T}(I_L t_4) = \frac{t_4}{T} I_L;$$

$$\frac{V_2}{V_1} = \frac{1}{T}(t_1 + t_2 + t_3) = \frac{t_1 + t_2 + t_3}{t_1 + t_2 + t_3 + t_4} \quad (14.364)$$

$$t_4 = \left(\frac{V_1}{V_2} - 1\right)(t_1 + t_2 + t_3) \quad (14.365)$$

$$k = \frac{t_3 + t_4}{t_1 + t_2 + t_3 + t_4}; \quad T = t_1 + t_2 + t_3 + t_4; \quad f = 1/T \quad (14.366)$$

14.12.3 Four-quadrant ZVS Quasi-resonant DC/DC Luo-converter

Four-quadrant ZVS quasi-resonant Luo-converter is shown in Fig. 14.103. Circuit 1 implements the operation in quadrants I and II, circuit 2 implements the operation in quadrants III and IV. Circuit 1 and 2 can be converted to each other by auxiliary switch. Each circuit consists of one main inductor L and two switches. Assuming that the main inductance L is sufficiently large, the current i_L is constant. The source and load voltages are usually constant, e.g. $V_1 = 42$ V and $V_2 = \pm 28$ V. There are four modes of operation:

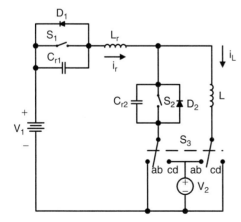

FIGURE 14.103 Four-quadrant DC/DC ZVS quasi-resonant Luo-converter.

- Mode A (Quadrant I): electrical energy is transferred from V_1 side to V_2 side;
- Mode B (Quadrant II): electrical energy is transferred from V_2 side to V_1 side;
- Mode C (Quadrant III): electrical energy is transferred from V_1 side to $-V_2$ side;
- Mode D (Quadrant IV): electrical energy is transferred from $-V_2$ side to V_1 side.

Each mode has two states: "on" and "off." The switch status of each state is shown in Table 14.13.

The description of Modes A, B, C, and D is same as in the previous Sections 14.12.1 and 14.12.2.

14.13 Synchronous-rectifier DC/DC Luo-converters

Synchronous-rectifier (SR) DC/DC converters are called the fifth-generation converters. The development of the microelectronics and computer science requires the power supplies with low output voltage and strong current. Traditional diode bridge rectifiers are not available for this requirement. Soft-switching technique can be applied in SR DC/DC converters. We have created few converters with very low voltage (5 V, 3.3 V, and $1.8 \sim 1.5$ V) and strong current (30 A, 60 A up to 200 A) and high power transfer efficiency (86%, 90% up to 93%). In this section, few new circuits, different from the ordinary SR DC/DC converters, are introduced:

- Flat transformer synchronous-rectifier DC/DC Luo-converter;
- Double current synchronous-rectifier DC/DC Luo-converter with active clamp circuit;
- Zero-current-switching synchronous-rectifier DC/DC Luo-converter;
- Zero-voltage-switching synchronous-rectifier DC/DC Luo-converter.

14.13.1 Flat Transformer Synchronous-rectifier DC/DC Luo-converter

Flat transformer **SR DC/DC Luo-converter** is shown in Fig. 14.104. The switches S_1, S_2, and S_3 are the low-resistance MOSFET devices with very low resistance R_S (7–8 mΩ). Since we use a flat transformer, the leakage inductance L_m and resistance R_L are small. Other parameters are $C = 1\,\mu$F, $L_m = 1$ nH, $R_L = 2$ mΩ, $L = 5\,\mu$H, $C_O = 10\,\mu$F. The input voltage is $V_1 = 30$ VDC and output voltage is V_2, the output current is I_O. The transformer term's ratio is $N = 12:1$. The repeating period is $T = 1/f$ and conduction duty is k. There are four working modes.

TABLE 14.13 Switch's status (the blank status means off)

Circuit//switch or diode	Mode A (QI)		Mode B (QII)		Mode C (QIII)		Mode D (QIV)	
	State-on	State-off	State-on	State-off	State-on	State-off	State-on	State-off
Circuit	Circuit 1				Circuit 2			
S_1	ON				ON			
D_1			ON					ON
S_2			ON				ON	
D_2		ON				ON		

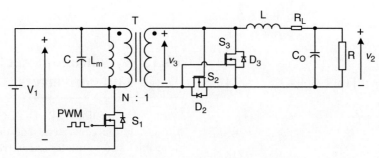

FIGURE 14.104 Flat transformer SR Luo-converter.

The natural resonant frequency is

$$\omega = \frac{1}{\sqrt{L_m C}} \quad (14.367)$$

The intervals are

$$t_1 = \frac{L_m}{V_1} \frac{I_O}{N}; \quad t_2 \approx kT; \quad (14.368)$$

$$t_3 = \sqrt{L_m C} \left[\frac{\pi}{2} + \frac{V_1}{\sqrt{V_1^2 + \frac{L_m}{C}\left(\frac{I_O}{N}\right)^2}} \right]; \quad t_4 \approx (1-k)T \quad (14.369)$$

Average output voltage V_2 and input current I_1 are

$$V_2 = \frac{kV_1}{N} - \left(R_L + R_S + \frac{L_m}{TN^2}\right)I_O; \quad I_1 = k\frac{I_O}{N} \quad (14.370)$$

The power transfer efficiency is

$$\eta = \frac{V_2 I_O}{V_1 I_1} = 1 - \frac{R_L + R_S + (L_m/TN^2)}{kV_1/N} I_O \quad (14.371)$$

When we set the frequency $f = 150$–200 kHz, we obtained the $V_2 = 1.8$ V, $N = 12$, $I_O = 0$–30 A, Volume $= 2.5$ in^3. The average power transfer efficiency is 92.3% and the maximum power density (PD) is 21.6 W/in^3.

14.13.2 Double Current SR DC/DC Luo-converter with Active Clamp Circuit

The converter in Fig. 14.104 resembles a half-wave rectifier. **Double current (DC) SR DC/DC Luo-converter** with active clamp circuit is shown in Fig. 14.105. The switches S_1–S_4 are the low-resistance MOSFET devices with very low resistance R_S (7–8 mΩ). Since S_3 and S_4 plus L_1 and L_2 form a double current circuit and S_2 plus C is the active clamp circuit, this converter resembles a full-wave rectifier and obtains strong output current. Other parameters are $C = 1$ μF, $L_m = 1$ nH, $R_L = 2$ mΩ, $L = 5$ μH, $C_O = 10$ μF. The input voltage is $V_1 = 30$ VDC and output voltage is V_2, the output current is I_O. The transformer term's ratio is $N = 12 : 1$. The repeating period is $T = 1/f$ and conduction duty is k. There are four working modes.

The natural resonant frequency is

$$\omega = \frac{1}{\sqrt{L_m C}}; \quad V_C = \frac{k}{1-k} V_1 \quad (14.372)$$

The interval of t_1 is

$$t_1 = \frac{L_m}{V_1} \frac{I_O}{N}; \quad t_2 \approx kT; \quad (14.373)$$

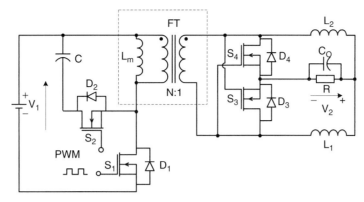

FIGURE 14.105 Double current SR Luo-converter.

$$t_3 = \sqrt{L_m C} \left[\frac{\pi}{2} + \frac{V_1}{\sqrt{V_1^2 + \frac{L_m}{C}\left(\frac{I_O}{N}\right)^2}} \right]; \quad t_4 \approx (1-k)T$$
(14.374)

Average output voltage V_2 and input current I_1 are

$$V_2 = \frac{kV_1}{N} - \left(R_L + R_S + \frac{L_m}{TN^2}\right)I_O; \quad I_1 = k\frac{I_O}{N} \quad (14.375)$$

The power transfer efficiency

$$\eta = \frac{V_2 I_O}{V_1 I_1} = 1 - \frac{R_L + R_S + (L_m/TN^2)}{kV_1/N} I_O \quad (14.376)$$

When we set the frequency $f = 200\text{--}250$ kHz, we obtained the $V_2 = 1.8$ V, $N = 12$, $I_O = 0\text{--}35$ A, Volume $= 2.5$ in^3. The average power transfer efficiency is 94% and the maximum power density (PD) is 25 W/in^3.

14.13.3 Zero-current-switching Synchronous-rectifier DC/DC Luo-converter

Since the power loss across the main switch S_1 is high in **DC SR DC/DC Luo-converter**, we designed **ZCS SR DC/DC Luo-converter** shown in Fig. 14.106. This converter is based on the **DC SR DC/DC Luo-converter** plus ZCS technique. It employs a double core flat transformer.

The ZCS resonant frequency is

$$\omega_r = \frac{1}{\sqrt{L_r C_r}} \quad (14.377)$$

The normalized impedance is

$$Z_r = \sqrt{\frac{L_r}{C_r}} \quad \text{and} \quad \alpha = \sin^{-1}\left(\frac{I_1 Z_r}{V_1}\right) \quad (14.378)$$

The intervals are

$$t_1 = \frac{I_1 L_r}{V_1}; \quad t_2 = \frac{1}{\omega_r}(\pi + \alpha); \quad (14.379)$$

$$t_3 = \frac{V_1(1 + \cos\alpha)C_r}{I_1};$$

$$t_4 = \frac{V_1(t_1 + t_2)}{V_2 I_1}\left(I_L + \frac{V_1}{Z_r}\frac{\cos\alpha}{\pi/2 + \alpha}\right) - (t_1 + t_2 + t_3)$$
(14.380)

Average output voltage V_2 and input current I_1 are

$$V_2 = \frac{kV_1}{N} - \left(R_L + R_S + \frac{L_m}{TN^2}\right)I_O; \quad I_1 = k\frac{I_O}{N} \quad (14.381)$$

The power transfer efficiency

$$\eta = \frac{V_2 I_O}{V_1 I_1} = 1 - \frac{R_L + R_S + (L_m/TN^2)}{kV_1/N} I_O \quad (14.382)$$

When we set the $V_1 = 60$ V and frequency $f = 200\text{--}250$ kHz, we obtained the $V_2 = 1.8$ V, $N = 12$, $I_O = 0\text{--}60$ A, Volume $= 4$ in^3. The average power transfer efficiency is 94.5% and the maximum power density (PD) is 27 W/in^3.

14.13.4 Zero-voltage-switching Synchronous-rectifier DC/DC Luo-converter

ZVS SR DC/DC Luo-converter is shown in Fig. 14.107. This converter is based on the **DC SR DC/DC Luo-converter** plus ZVS technique. It employs a double core flat transformer.

The ZVS resonant frequency is

$$\omega_r = \frac{1}{\sqrt{L_r C_r}} \quad (14.383)$$

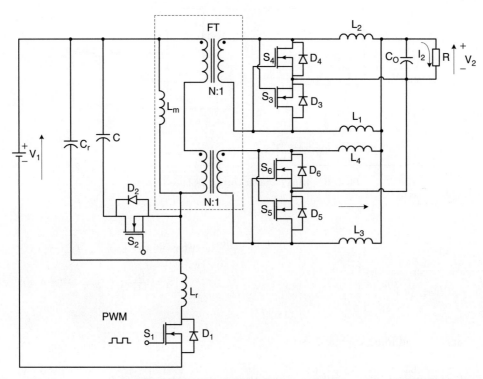

FIGURE 14.106 ZCS DC SR Luo-converter.

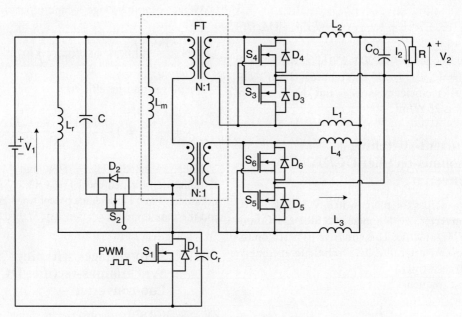

FIGURE 14.107 ZVS DC SR Luo-converter.

The normalized impedance is

$$Z_r = \sqrt{\frac{L_r}{C_r}}; \quad \alpha = \sin^{-1}\left(\frac{V_1}{Z_r I_1}\right) \qquad (14.384)$$

The intervals are

$$t_1 = \frac{V_1 C_r}{I_1}; \quad t_2 = \frac{1}{\omega_r}(\pi + \alpha); \qquad (14.385)$$

$$t_3 = \frac{I_1(1+\cos\alpha)L_r}{V_1}; \quad t_4 = \frac{t_1 + t_2 + t_3}{(V_1/V_2) - 1} \qquad (14.386)$$

Average output voltage V_2 and input current I_1 are

$$V_2 = \frac{kV_1}{N} - \left(R_L + R_S + \frac{L_m}{TN^2}\right)I_O; \quad I_1 = k\frac{I_O}{N} \qquad (14.387)$$

The power transfer efficiency

$$\eta = \frac{V_2 I_O}{V_1 I_1} = 1 - \frac{R_L + R_S + (L_m/TN^2)}{kV_1/N} I_O \qquad (14.388)$$

When we set the $V_1 = 60$ V and frequency $f = 200$–250 kHz, we obtained the $V_2 = 1.8$ V, $N = 12$, $I_O = 0$–60 A, Volume $= 4$ in^3. The average power transfer efficiency is 94.5% and the maximum power density (PD) is 27 W/in^3.

14.14 Multiple-element Resonant Power Converters

Multiple energy-storage elements resonant power converters (x-Element RPC) are the sixth-generation converters. According to the transferring, power becomes higher and higher, traditional methods are hardly satisfied to deliver large power from source to final actuators with high efficiency. In order to reduce the power losses during the conversion process the sixth-generation converters – multiple energy-storage elements resonant power converters (x-Element RPC) – are created. They can be sorted into two main groups:

- DC/DC resonant converters;
- DC/AC resonant inverters.

Both groups converters consist of multiple energy-storage elements: two elements, three elements, or four elements. These energy-storage elements are passive parts: inductors and capacitors. They can be connected in series or parallel in various methods. In full statistics, the circuits of the multiple energy-storage elements converters are:

- 8 topologies of 2-element RPC;
- 38 topologies of 3-element RPC;
- 98 topologies of 4-element (2L-2C) RPC.

How to investigate the large quantity converters is a vital task. This problem was addressed in the last decade of last century. Unfortunately, much attention was not paid to it. This generation converters were not well discussed, only limited number of papers was published in the literature.

14.14.1 Two Energy-storage Elements Resonant Power Converters

The 8 topologies of 2-element RPC are shown in Fig. 14.108. These topologies have simple circuit structure and least components. Consequently, they can transfer the power from source to end-users with higher power efficiency and lower power losses.

Usually, the 2-Element RPC has very narrow response frequency bands, which is defined as the frequency width between the two half-power points. The working point must be selected in the vicinity of the natural resonant frequency $\omega_0 = 1/\sqrt{LC}$. Another drawback is that the transferred waveform is usually not a perfect sinusoidal, i.e. the output waveform THD is not zero.

Since total power losses are mainly contributed by the power losses across the main switches. As resonant conversion technique, the 2-Element RPC has high power transferring efficiency.

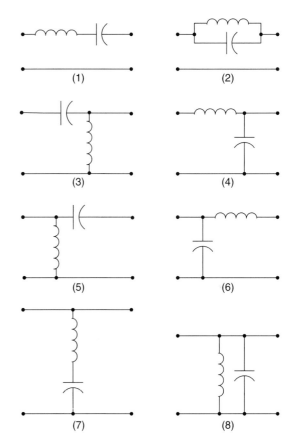

FIGURE 14.108 8 topologies of 2-element RPC.

14.14.2 Three Energy-storage Elements Resonant Power Converters

The 38 topologies of 3-element RPC are shown in Fig. 14.109. These topologies have one more component when compared to the 2-element RPC topologies. Consequently, they can transfer the power from source to end-users with higher lower power and lower power transfer efficiency.

Usually, the 3-element RPC has a much wider response frequency bands, which is defined as the frequency width between the two half-power points. If the circuit is a low-pass filter, the frequency bands can cover the frequency range from 0 to the natural resonant frequency $\omega_0 = 1/\sqrt{LC}$. The working point can be selected from a much wider frequency width which is lower than the natural resonant frequency $\omega_0 = 1/\sqrt{LC}$.

Another advantage, better than the 2-element RPC topologies, is that the transferred waveform can usually be a perfect sinusoidal, i.e. the output waveform THD is nearly zero. As well-known, mono-frequency waveform transferring operation has very low EMI.

14.14.3 Four Energy-storage Elements Resonant Power Converters

The 98 topologies of 4-element (2L-2C) RPC are shown in Fig. 14.110. If no restriction such as 2L-2C for 4-element RPC,

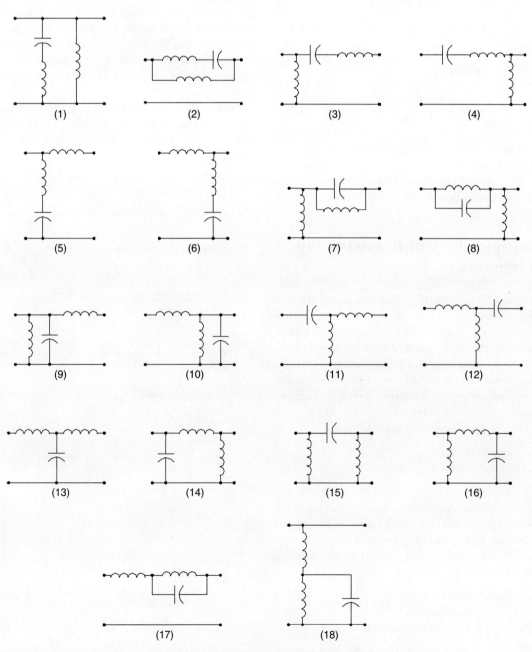

FIGURE 14.109 38 topologies of 3-element RPC.

14 DC/DC Conversion Technique and 12 Series Luo-converters

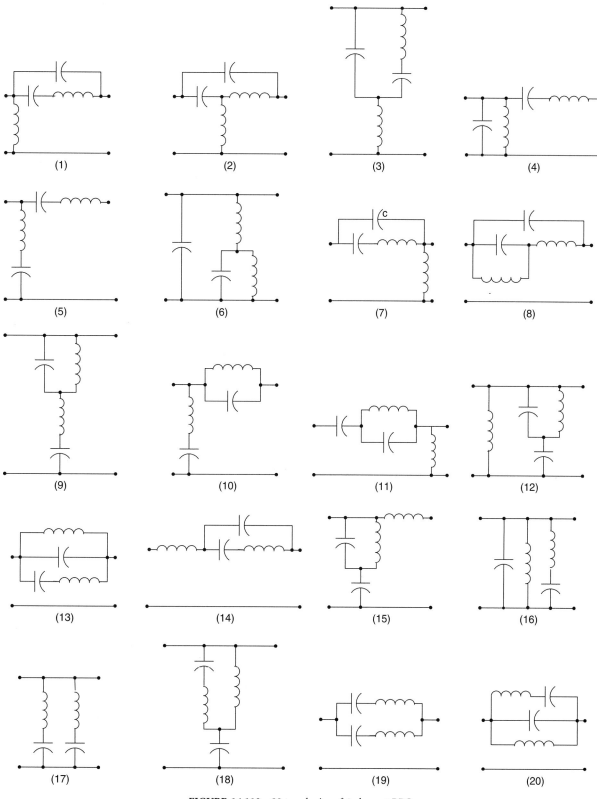

FIGURE 14.110 98 topologies of 4-element RPC.

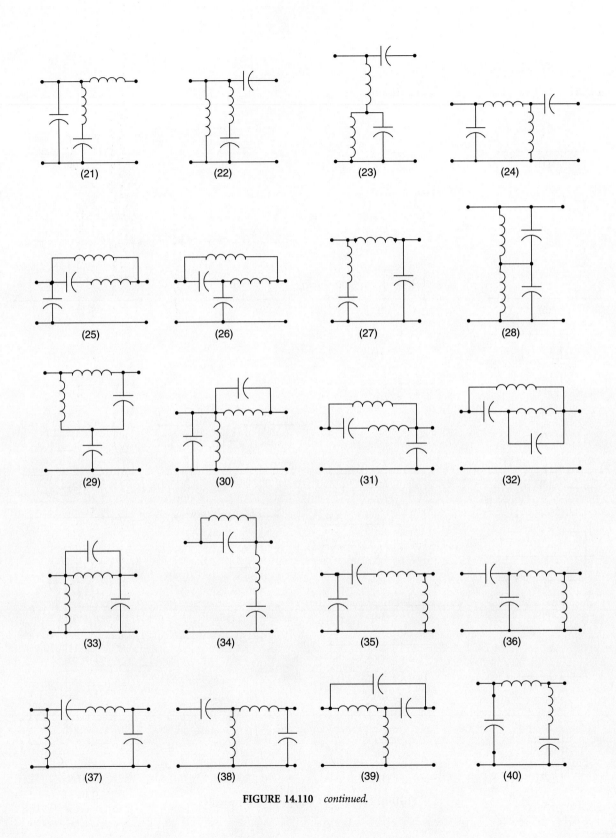

FIGURE 14.110 continued.

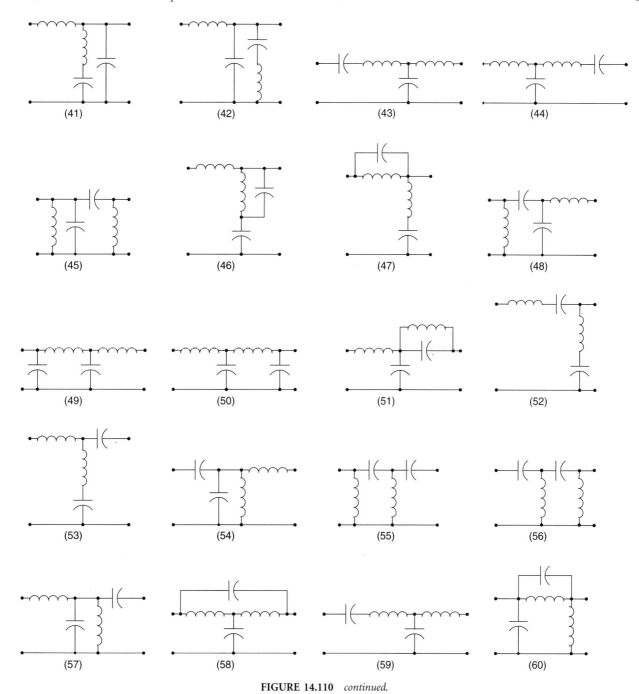

FIGURE 14.110 *continued.*

the number of the topologies of 4-element RPC can be much larger. Although these topologies have comparably complex circuit structure, they can still transfer the power from source to end-users with higher power efficiency and lower power losses.

Usually, the 4-element RPC has a wide response frequency bands, which is defined as the frequency width between the two half-power points. If the circuit is a low-pass filter, the frequency bands can cover the frequency range from 0 to the high half-power point which is definitely higher that the natural resonant frequency $\omega_0 = 1/\sqrt{LC}$. The working point can be selected from a wide area across (lower and higher than) the natural resonant frequency $\omega_0 = 1/\sqrt{LC}$. Another advantage is that the transferred waveform is usually a perfect sinusoidal, i.e. the output waveform THD is very close to zero. As well-known, mono-frequency-waveform

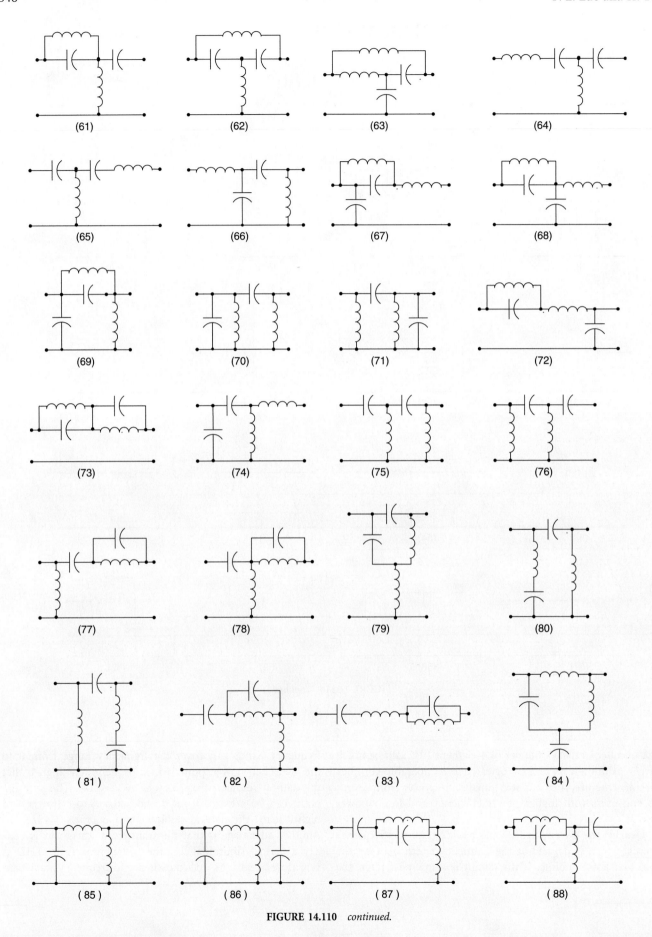

FIGURE 14.110 *continued.*

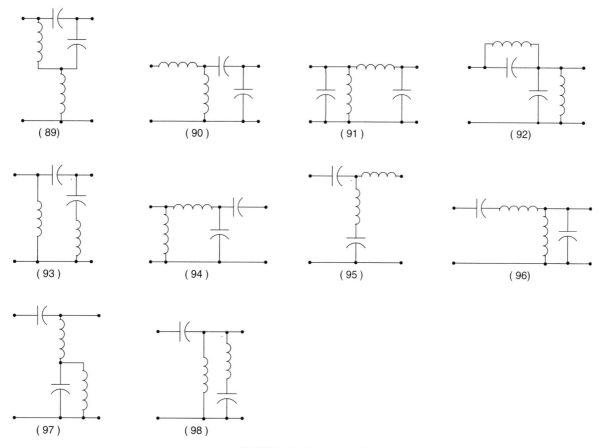

FIGURE 14.110 *continued.*

transferring operation has very low EMI and reasonable EMS and EMC.

14.14.4 Bipolar Current and Voltage Sources

Depending on different applications, resonant network can be low-pass filter, high-pass filter, or band-pass filter. For large power transferring, low-pass filter is usually employed. In this case, inductors are arranged in series arms and capacitors are arranged in shunt arms. If the first component is inductor, only voltage source can be applied since inductor current is continuous. Vice versa, if the first component is capacitor, only current source can be applied since capacitor voltage is continuous.

14.14.4.1 Bipolar Voltage Source

A bipolar voltage source using single voltage source is shown in Fig. 14.111. Since only voltage source is applied, there are four switches applied alternatively switching on or off to supply positive and negative voltage to the network. In the figure, the load is a resistance R.

The circuit of this voltage source is likely a four-quadrant operational chopper. The conduction duty cycle for each switch is 50%. For safety reason, the particular circuitry design has to consider some small gap between the turn-over (commutation) operation to avoid the short-circuit incidence.

The repeating frequency is theoretically not restricted. For industrial applications, the operating frequency is usually arranged in the range between 10 kHz and 5 MHz, depending on the application conditions.

14.14.4.2 Bipolar Current Source

A bipolar current voltage source using single voltage sources is shown in Fig. 14.112. To obtain stable current, the voltage source is connected in series by a large inductor. There are four switches applied alternatively switching on or off to supply positive and negative current to the network. In the figure, the load is a resistance R.

The circuit of this current source is likely a two-quadrant operational chopper. The conduction duty cycle for each

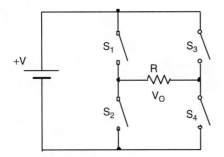

FIGURE 14.111 A bipolar voltage source using single voltage source.

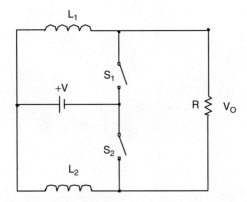

FIGURE 14.112 A bipolar current voltage source using single voltage source.

switch is 50%. For safety reason, the particular circuitry design has to consider some small gap between the turnover (commutation) operation to avoid the short-circuit incidence.

The repeating frequency is theoretically not restricted. For industrial applications, the operating frequency is usually arranged in the range between 10 kHz and 5 MHz, depending on the application conditions.

14.15 Gate Control Luo-resonator

Luo-resonator is shown in Fig. 14.113. It generates the PWM pulse train to drive the static switch S. Luo-resonator is a high efficiency and simple structure circuit with easily adjusting frequency f and conduction duty k. It consists of three operational amplifiers (**OA**) named OA1-3 and auxiliary. These three 741-type OA's are integrated in a chip TL074 (which contains four OA's). Two potentiometers are applied to adjust the frequency f and conduction duty k. The voltage waveforms are shown in Fig. 14.114.

Type-741 OA can work at the power supply $\pm 3 - \pm 18\,\text{V}$ that are marked $V+$, G, and $V-$ with $|V-| = V+$. OA2 in Fig. 14.113 acts as the integration operation, its output V_C is a triangle waveform with regulated frequency $f = 1/T$ controlled by potentiometer R_4. OA1 acts as a resonant operation, its output V_B is a square-waveform with the frequency f. OA3 acts as a comparator, its output V_D is a square-waveform pulse train with regulated conduction duty k controlled by R_7.

Firstly, assuming the voltage $V_B = V+$ at $t = 0$ and feeds positively back to OA1 via R_2. This causes the OA1's output voltage maintained at $V_B = V+$. In the meantime, V_B inputs to OA2 via R_4, the output voltage V_C of OA2, therefore, decreases towards $V-$ with the slope $1/R_4C$. Voltage V_C feeds negatively back to OA1 via R_3. Voltage V_A at point A changes from $(mV+)$ to 0 in the period of $2mR_4C$. Usually, R_3 is set slightly smaller than R_2, the ratio is defined as $m = R_3/R_2$. Thus, voltage V_A intends towards negative. It causes the OA1's output voltage $V_B = V-$ at $t = 2mR_4C$ and voltage V_A jumps to $mV-$. Vice versa, the voltage $V_B = V-$ at $t = 2mR_4C$ and feeds positively back to OA1 via R_2. This causes the OA1's output voltage maintained at $V_B = V-$. In the meantime, V_B inputs to OA2 via R_4, the output voltage V_C of OA2, therefore, increases towards $V+$ with the slope $1/R_4C$. Voltage V_C feeds negatively back to OA1 via R_3. Voltage V_A at point A changes from $(mV-)$ to 0 in the period of $2mR_4C$. Thus, voltage V_A intends towards positive. It causes the OA1's output voltage $V_B = V+$ at $t = 4mR_4C$ and voltage V_A jumps to $mV+$.

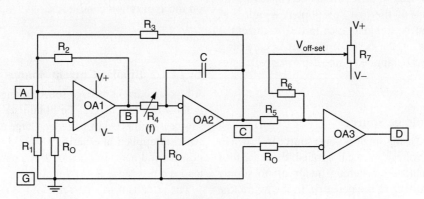

FIGURE 14.113 Luo-resonator.

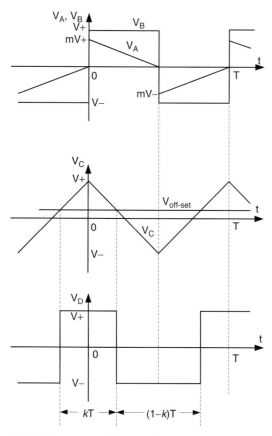

FIGURE 14.114 Voltage waveforms of Luo-resonator.

Then V_C inputs to OA3 and compares with shift signal $V_{off\text{-}set}$ regulated by the potentiometer R_7 via R_6. When $V_{off\text{-}set} = 0$, OA3 yields its output voltage V_D as a pulse train with conduction duty $k = 0.5$. Positive $V_{off\text{-}set}$ shifts the zero-cross point of voltage V_C downwards, hence, OA3 yields its output voltage V_D as a pulse train with conduction duty $k > 0.5$. Vice versa, negative $V_{off\text{-}set}$ shifts the zero-cross point of voltage V_C upwards, hence, OA3 yields its output voltage V_D as a pulse train with conduction duty $k < 0.5$ as shown in Fig. 14.114. Conduction duty k is controlled by $V_{off\text{-}set}$ via the potentiometer R_7.

The calculation formulas are

$$m = \frac{R_3}{R_2} \qquad (14.389)$$

$$f = \frac{1}{4mR_4C} \qquad (14.390)$$

$$k = 0.5 + \frac{R_5 V_{off\text{-}set}}{2R_6 V+} \qquad (14.391)$$

This PWM pulse train V_D is applied to the DC/DC converter switch such as a transistor, MOSFET, or IGBT via a coupling circuit.

A design example: A Luo-resonator was designed as shown in Fig. 14.113 with the component values of $R_0 = 10\,\text{k}\Omega$; $R_1 = R_2 = R_5 = 100\,\text{k}\Omega$, $R_3 = R_6 = 95\,\text{k}\Omega$; $R_4 = 510\,\Omega$–$5.1\,\text{k}\Omega$ $R_7 = 20\,\text{k}\Omega$; and $C = 5.1\,\text{nF}$. The results are $m = 0.95$, frequency $f = 10$–$100\,\text{kHz}$ and conduction duty $k = 0$–1.0.

14.16 Applications

The DC/DC conversion technique has been rapidly developed and has been widely applied in industrial applications and computer peripheral equipment. Three examples are listed below:

- 5000 V insulation test bench;
- MIT 42/14 V DC/DC converter;
- IBM 1.8 V/200 A power supply.

14.16.1 5000 V Insulation Test Bench

Insulation test bench is the necessary equipment for semiconductor manufacturing organizations. An adjustable DC voltage power supply is the heart of this equipment. Traditional method to obtain the adjustable high DC voltage is a diode rectifier via a setting up transformer. It is costly and larger in size with poor efficiency.

Using a positive output super-lift Luo-converter triple-lift circuit, which is shown in Fig. 14.115. This circuit is small, effective, and low cost. The output voltage can be determined by

$$V_O = \left(\frac{2-k}{1-k}\right)^3 V_{in} \qquad (14.392)$$

The conduction duty cycle k is only adjusted in the range 0–0.8 to carry out the output voltage in the range of 192–5184 V.

The experimental results are listed in Table 14.14. The measured data verified the advantages of this power supply.

14.16.2 MIT 42/14 V–3 KW DC/DC Converter

MIT 42/14 V–3 KW DC/DC converter was requested to transfer 3 kW energy between two battery sources with 42 and 14 V. The circuit diagram is shown in Fig. 14.116. This is a two-quadrant zero-voltage-switching (ZVS) quasi-resonant-converter (QRC). The current in low voltage side can be up to 250 A. This is a typical low voltage strong current converter. It is easier to carry out by ZVS-QRC.

This converter consists of two sources V_1 and V_2, one main inductor L, two main switches S_1 and S_2, two reverse-paralleled

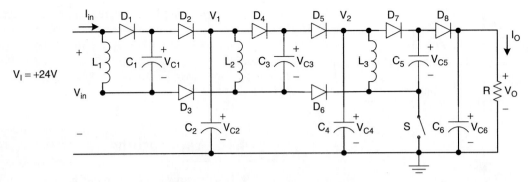

FIGURE 14.115 5000 V Insulation test bench.

TABLE 14.14 The experimental results of the 5000 V test bench

Conduction duty, k	0	0.1	0.2	0.3	0.4	0.5	0.6	0.7	0.8	0.82
Output voltage V_O (V)	192	226	273	244	455	648	1029	1953	5184	6760

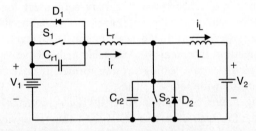

FIGURE 14.116 MIT 42/14 V-3 kW DC/DC converter.

diodes D_1 and D_2, one resonant inductor L_r and two resonant capacitors C_{r1} and C_{r2}. The working condition is selected

$$V_1 = 42\,\text{V}; \quad V_2 = 14\,\text{V}$$
$$L = 470\,\mu\text{H}; \quad C_{r1} = C_{r2} = C_r = 1\,\mu\text{F}$$
$$L_r = \begin{cases} 1\,\mu\text{H} & \text{normal operation} \\ 9\,\mu\text{H} & \text{low current operation} \end{cases}$$

Therefore,

$$\omega_O = \frac{1}{\sqrt{L_r C_r}} = 10^6\,\text{rad/s} \quad (14.393)$$

$$Z_O = \sqrt{\frac{L_r}{C_r}} = 1\,\Omega \text{ (normal operation)} \quad (14.394)$$

$$\alpha = \sin^{-1}\frac{V_1}{Z_O I_2} \quad (14.395)$$

It is easy to keep the quasi-resonance when the working current $I_2 > 50$ A. If the working current is too low, the resonant inductor will take large value to guarantee the quasi-resonance state. This converter performs two-quadrant operation:

- Mode A (Quadrant I): energy transferred from V_1 side to V_2 side;
- Mode B (Quadrant II): energy transferred from V_2 side to V_1 side.

Assuming the working current is $I_2 = 100$ A and the converter works in Mode A, following calculations are obtained

$$\omega_O = \frac{1}{\sqrt{L_r C_r}} = 10^6\,\text{rad/s}$$

$$Z_O = \sqrt{\frac{L_r}{C_r}} = 1\,\Omega$$

$$\alpha = \sin^{-1}\frac{V_1}{Z_O I_2} = 24.83° \quad (14.396)$$

$$t_1 = \frac{V_1 C_r}{I_2} = 0.42\,\mu\text{s} \quad (14.397)$$

$$t_2 = \frac{\pi + \alpha}{\omega_O} = 3.58\,\mu\text{s} \quad (14.398)$$

$$t_3 = \frac{1 + \cos\alpha}{V_1} I_2 L_r = \frac{1 + 0.908}{42} 100 \times 10^{-6} = 4.54\,\mu\text{s} \quad (14.399)$$

$$t_4 = \frac{t_1 + t_2 + t_3}{V_1/V_2 - 1} = \frac{0.42 + 3.58 + 4.54}{2} = 4.27\,\mu\text{s} \quad (14.400)$$

14 DC/DC Conversion Technique and 12 Series Luo-converters

TABLE 14.15 The experimental test results of MIT 42V/14 converter (with the condition: $L_r = 1\,\mu\text{H}$, $C_{r1} = C_{r2} = 1\,\mu\text{F}$)

Mode	f(kHz)	I_1 (A)	I_2 (A)	I_L (A)	P_1 (W)	P_2 (W)	η (%)	PD(W/in^3)
A	78	77.1	220	220	3239	3080	95.1	23.40
A	80	78.3	220	220	3287	3080	93.7	23.58
A	82	81	220	220	3403	3080	90.5	24.01
B	68	220	69.9	220	3080	2939	95.3	22.28
B	70	220	68.3	220	3080	2871	93.2	22.04
B	72	220	66.6	220	3080	2797	90.8	21.77

$$T = t_1 + t_2 + t_3 + t_4 = 0.42 + 3.58 + 4.54 + 4.27 = 12.81\,\mu\text{s} \quad (14.401)$$

$$f = \frac{1}{T} = \frac{1}{12.81} = 78.06\text{ kHz} \quad (14.402)$$

$$k = \frac{t_3 + t_4}{T} = \frac{4.54 + 4.27}{12.81} = 0.688 \quad (14.403)$$

The volume of this converter is 270 in^3. The experimental test results in full power 3 kW are listed in Table 14.15. From the tested data, a high power density 22.85 W/in^3 and a high efficiency 93% are obtained. Because of soft-switching operation, the EMI is low and EMS and EMC are reasonable.

14.16.3 IBM 1.8 V/200 A Power Supply

This equipment is suitable for IBM new generation computer with power supply 1.8 V/200 A. This is a ZCS SR DC/DC Luo-converter, and is shown in Fig. 14.117. This converter is based on the double-current synchronous-rectifier DC/DC converter plus ZCS technique. It employs a hixaploid-core flat-transformer with the turns ratio $N = 1/12$. It has six-unit ZCS synchronous-rectifier double-current DC/DC converter. The six primary coils are connected in series, and six secondary circuits are connected in parallel. Each unit has particular input voltage V_{in} to be about 33 V, and can offer 1.8 V/35 A individually. Total output current is 210 A. The equivalent primary full current is $I_1 = 14.5$ A and equivalent primary load voltage is $V_2 = 200$ V. The ZCS natural resonant frequency is

$$\omega_O = \frac{1}{\sqrt{L_r C_r}} \quad (14.404)$$

$$Z_O = \sqrt{\frac{L_r}{C_r}} \quad (14.405)$$

$$\alpha = \sin^{-1}\frac{Z_O I_1}{V_1} \quad (14.406)$$

The main power supply is from public utility board (PUB) via a diode rectifier. Therefore V_1 is nearly 200 V, and the each unit input voltage V_{in} is about 33 V. Other calculation formulas are

$$t_1 = \frac{I_1 L_r}{V_1} \quad (14.407)$$

$$t_2 = \frac{\pi + \alpha}{\omega_O} \quad (14.408)$$

$$t_3 = \frac{1 + \cos\alpha}{I_1} V_1 C_r \quad (14.409)$$

$$t_4 = \frac{V_1(t_1 + t_2)}{I_1 V_2}\left(I_1 + \frac{V}{Z_O}\frac{\cos\alpha}{\pi/2 + \alpha}\right) - (t_1 + t_2 + t_3) \quad (14.410)$$

$$T = t_1 + t_2 + t_3 + t_4 \quad (14.411)$$

$$f = \frac{1}{T} \quad (14.412)$$

$$k = \frac{t_1 + t_2}{T} \quad (14.413)$$

Real output voltage and input current are

$$V_O = kNV_1 - \left(R_L + R_S + \frac{L_m}{T}N^2\right)I_O \quad (14.414)$$

$$I_{in} = kNI_O \quad (14.415)$$

The efficiency is

$$\eta = \frac{V_O I_O}{V_{in} I_{in}} = 1 - \frac{R_L + R_S + (L_m/T)N^2}{kNV_{in}}I_O \quad (14.416)$$

The commercial unit of this power supply works in voltage closed loop control with inner current closed loop to keep the output voltage constant. Applying frequency is arranged in the band of 200–250 kHz. Whole volume of the power supply is 14 in^3. The transfer efficiency is about 88–92% and power density is about 25.7 W/in^3.

14.17 Energy Factor and Mathematical Modeling for Power DC/DC Converters

We have well discussed the various DC/DC converters operating in steady state in previous sections. We will investigate the transient process of DC/DC converters. Furthermore, we define a series of new parameters such as energy factor (*EF*) and so on to establish the mathematical modeling of all power DC/DC converters.

Energy storage in power DC/DC converters has been paid attention long time ago. Unfortunately, there is no clear

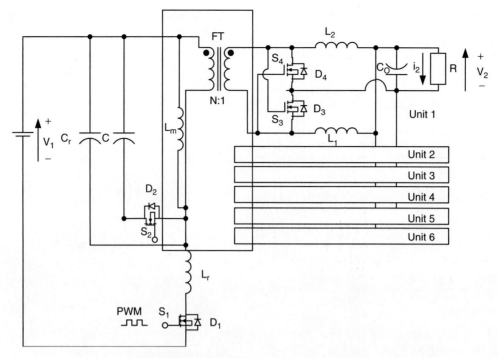

FIGURE 14.117 IBM 1.8 V/200 A power supply.

concept to describe the phenomena and reveal the relationship between the stored energy and the characteristics of power DC/DC converters. We have theoretically defined a new concept – energy factor (*EF*) and researched the relations between *EF* and the mathematical modeling of power DC/DC converters. Energy factor is a new concept in power electronics and conversion technology, which thoroughly differs from the traditional concepts such as power factor (*PF*), power transfer efficiency (η), total harmonic distortion (*THD*), and ripple factor (*RF*). Energy factor and the other sub-sequential parameters can illustrate the system stability, reference response, and interference recovery. This investigation is very helpful for system design and DC/DC converters characteristics foreseeing.

14.17.1 Pumping Energy (*PE*)

All power DC/DC converters have pumping circuit to transfer the energy from the source to some energy storage passive elements, e.g. inductors and capacitors. The *PE* is used to count the input energy in a switching period *T*. Its calculation formula is

$$PE = \int_0^T P_{in}(t)dt = \int_0^T V_1 i_1(t)dt = V_1 I_1 T \quad (14.417)$$

where

$$I_1 = \int_0^T i_1(t)dt$$

is the average value of the input current if the input voltage V_1 is constant. Usually the input average current I_1 depends on the conduction duty cycle.

14.17.2 Stored Energy (*SE*)

The stored energy in an inductor is

$$W_L = \frac{1}{2}LI_L^2 \quad (14.418)$$

The stored energy across a capacitor is

$$W_C = \frac{1}{2}CV_C^2 \quad (14.419)$$

Therefore, if there are n_L inductors and n_C capacitors the total stored energy in a DC/DC converter is

$$SE = \sum_{j=1}^{n_L} W_{Lj} + \sum_{j=1}^{n_C} W_{Cj} \quad (14.420)$$

Capacitor–inductor stored energy ratio (*CIR*) – Most power DC/DC converters consist of inductors and capacitors.

Therefore, we can define the capacitor–inductor stored energy ratio (**CIR**).

$$CIR = \frac{\sum_{j=1}^{n_C} W_{Cj}}{\sum_{j=1}^{n_L} W_{Lj}} \quad (14.421)$$

Energy losses (**EL**) – Usually, most analysis applied in DC/DC converters is assuming no power losses, i.e. the input power is equal to the output power, $P_{in} = P_O$ or $V_1 I_1 = V_2 I_2$, so that pumping energy is equal to output energy in a period, T.

Particularly, power losses always exist during the conversion process. They are caused by the resistance of the connection cables, resistance of the inductor and capacitor wire, and power losses across the semiconductor devices (diode, IGBT, MOSFET, and so on). We can sort them as the resistance power losses P_r, passive element power losses P_e, and device power losses P_d. The total power losses are P_{loss}.

$$P_{loss} = P_r + P_e + P_d$$

and

$$P_{in} = P_O + P_{loss} = P_O + P_e + P_e + P_d = V_2 I_2 + P_e + P_e + P_d$$

Therefore,

$$EL = P_{loss} \times T = (P_r + P_e + P_d) T$$

The energy losses (**EL**) is in a period T,

$$EL = \int_0^T P_{loss} dt = P_{loss} T \quad (14.422)$$

14.17.3 Energy Factor (**EF**)

As described in previous section the input energy in a period T is the pumping energy $PE = P_{in} \times T = V_{in} I_{in} \times T$. We now define the **EF**, that is the ratio of the SE over the pumping energy

$$EF = \frac{SE}{PE} = \frac{SE}{V_1 I_1 T} = \frac{\sum_{j=1}^m W_{Lj} + \sum_{j=1}^n W_{Cj}}{V_1 I_1 T} \quad (14.423)$$

Energy factor is a very important factor of a power DC/DC converter. It is usually independent from the conduction duty cycle k, and proportional to the switching frequency f (inversely proportional to the period T) since the pumping energy PE is proportional to the switching period T.

14.17.4 Time Constant τ and Damping Time Constant τ_d

The **time constant** τ of a power DC/DC converter is a new concept to describe the transient process of a DC/DC converter. If no power losses in the converter, it is defined

$$\tau = \frac{2T \times EF}{1 + CIR}\left(1 + CIR\frac{1-\eta}{\eta}\right) \quad (14.424)$$

The **damping time constant** τ_d of a power DC/DC converter is new concept to describe the transient process of a DC/DC converter. If no power losses, it is defined

$$\tau_d = \frac{2T \times EF}{1 + CIR}\frac{CIR}{\eta + CIR(1-\eta)} \quad (14.425)$$

The **time constants ratio** ξ of a power DC/DC converter is new concept to describe the transient process of a DC/DC converter. If no power losses, it is defined

$$\xi = \frac{\tau_d}{\tau} = \frac{CIR}{\eta\left(1 + CIR\frac{1-\eta}{\eta}\right)^2} \quad (14.426)$$

14.17.5 Mathematical Modeling for Power DC/DC Converters

The mathematical modeling for all power DC/DC converters is

$$G(s) = \frac{M}{1 + s\tau + s^2 \tau \tau_d} \quad (14.427)$$

where M is the voltage transfer gain: $M = V_O / V_{in}$, τ is the time constant in Eq. (14.424), τ_d the damping time constant in Eq. (14.425), $\tau_d = \xi\tau$. Using this mathematical model of power DC/DC converters, it is significantly easy to describe the characteristics of power DC/DC converters. In order to verify this theory, few converters are investigated to demonstrate the characteristics of power DC/DC converters and applications of the theory.

14.17.6 Buck Converter with Small Energy Losses ($r_L = 1.5\,\Omega$)

A buck converter shown in Fig. 14.118 has the components values: $V_1 = 40\,V$, $L = 250\,\mu H$ with resistance $r_L = 1.5\,\Omega$, $C = 60\,\mu F$, $R = 10\,\Omega$, the switching frequency $f = 20\,kHz$ ($T = 1/f = 50\,\mu s$) and conduction duty cycle $k = 0.4$. This converter is stable and works in CCM.

Therefore, we have got the voltage transfer gain $M = 0.35$, i.e. $V_2 = V_C = MV_1 = 0.35 \times 40 = 14\,V$. $I_L = I_2 = 1.4\,A$, $P_{loss} = I_L^2 \times r_L = 1.4^2 \times 1.5 = 2.94\,W$, and $I_1 = 0.564\,A$. The parameter EF and others are listed below

$$PE = V_1 I_1 T = 40 \times 0.564 \times 50\,\mu = 1.128\,mJ;$$

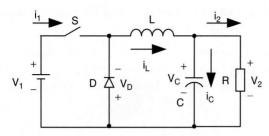

FIGURE 14.118 A buck converter.

$$W_L = \frac{1}{2}LI_L^2 = 0.5 \times 250\,\mu \times 1.4^2 = 0.245 \text{ mJ}$$

$$W_C = \frac{1}{2}CV_C^2 = 0.5 \times 60\,\mu \times 14^2 = 5.88 \text{ mJ};$$

$$SE = W_L + W_C = 0.245 + 5.88 = 6.125 \text{ mJ}$$

$$EF = \frac{SE}{PE} = \frac{6.125}{1.128} = 5.43; \quad CIR = \frac{W_C}{W_L} = \frac{5.88}{0.245} = 24$$

$$EL = P_{loss} \times T = 2.94 \times 50 = 0.147 \text{ mJ};$$

$$\eta = \frac{P_O}{P_O + P_{loss}} = 0.87$$

$$\tau = \frac{2T \times EF}{1 + CIR}\left(1 + CIR\frac{1-\eta}{\eta}\right) = 99.6\,\mu s;$$

$$\tau_d = \frac{2T \times EF}{1 + CIR}\frac{CIR}{\eta + CIR(1-\eta)} = 130.6\,\mu s;$$

$$\xi = \frac{\tau_d}{\tau} = \frac{CIR}{\eta\left(1 + CIR\frac{1-\eta}{\eta}\right)^2} = 1.31 \gg 0.25$$

By cybernetic theory, since the damping time constant τ_d is larger than the time constant τ, the corresponding ratio ξ is $1.31 \gg 0.25$. The output voltage has heavy oscillation with high overshot. The corresponding transfer function is

$$G(s) = \frac{M}{1 + s\tau + s^2\tau\tau_d} = \frac{M/\tau\tau_d}{(s+s_1)(s+s_2)} \quad (14.428)$$

where

$$s_1 = \sigma + j\omega \quad \text{and} \quad s_2 = \sigma - j\omega$$

with

$$\sigma = \frac{1}{2\tau_d} = \frac{1}{261.2\,\mu s} = 3833 \text{ Hz}$$

and

$$\omega = \frac{\sqrt{4\tau\tau_d - \tau^2}}{2\tau\tau_d} = \frac{\sqrt{52{,}031 - 9920}}{26{,}015.5} = \frac{205.2}{26{,}015.5\mu} = 7888 \text{ rad/s}$$

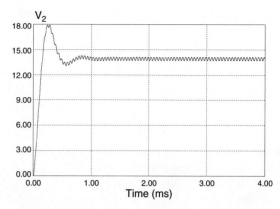

FIGURE 14.119 Buck converter unit-step response.

The unit-step function response is

$$v_2(t) = 14\left[1 - e^{-(t/0.000261)}(\cos 7888t - 0.486 \sin 7888t)\right] \text{ V} \quad (14.429)$$

The unit-step function response (transient process) has oscillation progress with damping factor σ and frequency ω. The simulation result is shown in Fig. 14.119.

The impulse interference response is

$$\Delta v_2(t) = 0.975 U e^{-(t/0.000261)} \sin 7888t$$

where U is the interference signal. The impulse response (interference recovery process) has oscillation progress with damping factor σ and frequency ω. The simulation result is shown in Fig. 14.120.

In order to verify the analysis, calculation and simulation results, we constructed a test rig with same conditions. The corresponding experimental results are shown in Figs. 14.121 and 14.122.

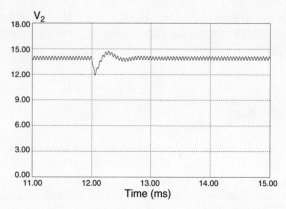

FIGURE 14.120 Buck converter impulse response.

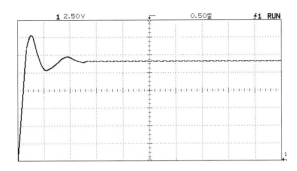

FIGURE 14.121 Unit-step response (test).

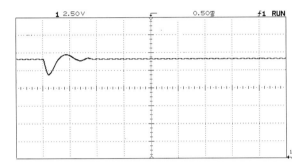

FIGURE 14.122 Impulse response (test).

14.17.7 A Super-lift Luo-converter in CCM

Figure 14.123 shows a super-lift Luo-converter with the conduction duty $k = 0.5$. The components values are $V_1 = 20$ V, $f = 50$ kHz ($T = 20\,\mu$s), $L = 100\,\mu$H with resistance $r_L = 0.12\,\Omega$, $C_1 = 2500\,\mu$F, $C_2 = 800\,\mu$F, and $R = 10\,\Omega$. This converter is stable and works in CCM.

Therefore, we have got the voltage transfer gain $M = 2.863$, i.e. the output voltage $V_2 = V_{C2} = 57.25$ V. $V_{C1} = V_1 = 20$ V, $I_1 = 14.145$ A, $I_2 = 5.725$ A, $I_L = 11.45$ A, and $P_{loss} = I_L^2 \times r_L = 11.45^2 \times 0.12 = 15.73$ W. The parameter EF and others are listed below

$$PE = V_1 I_1 T = 20 \times 17.175 \times 20\,\mu = 6.87 \text{ mJ};$$

$$W_L = \frac{1}{2}LI_L^2 = 0.5 \times 100\,\mu \times 11.45^2 = 6.555 \text{ mJ};$$

$$W_{C1} = \frac{1}{2}C_1 V_{C1}^2 = 0.5 \times 2500\,\mu \times 20^2 = 500 \text{ mJ};$$

$$W_{C2} = \frac{1}{2}C_2 V_{C2}^2 = 0.5 \times 800\,\mu \times 57.25^2 = 1311 \text{ mJ}$$

$$SE = W_L + W_{C1} + W_{C2} = 6.555 + 500 + 1311 = 1817.6 \text{ mJ};$$

$$EF = \frac{SE}{PE} = \frac{1817.6}{6.87} = 264.6;$$

$$CIR = \frac{W_{C1} + W_{C2}}{W_L} = \frac{1811}{6.555} = 276.3$$

$$EL = P_{loss} T = 15.73 \times 20 = 0.3146 \text{ mJ};$$

$$\eta = \frac{P_O}{P_O + P_{loss}} = \frac{327.76}{343.49} = 0.9542$$

$$\tau = \frac{2T \times EF}{1 + CIR}(1 + CIR\frac{1-\eta}{\eta})$$

$$= \frac{40\,\mu \times 264.6 \times 13.26}{277.3} = 506\,\mu\text{s}$$

$$\tau_d = \frac{2T \times EF}{1 + CIR} \frac{CIR}{\eta + CIR(1-\eta)}$$

$$= \frac{40 \times 264.6 \times 20.3}{277.3} = 775\,\mu\text{s}$$

By cybernetic theory, since the damping time constant τ_d is much larger than the time constant τ, the corresponding ratio $\xi = 775/506 = 1.53 \gg 0.25$. The output voltage has heavy oscillation with high overshot. The transfer function of this converter has two poles ($-s_1$ and $-s_2$) that are located in the left-hand half plane (LHHP).

$$G(s) = \frac{M}{1 + s\tau + s^2\tau\tau_d} = \frac{M/\tau\tau_d}{(s + s_1)(s + s_2)} \quad (14.430)$$

where

$$s_1 = \sigma + j\omega \quad \text{and} \quad s_2 = \sigma - j\omega$$

with

$$\sigma = \frac{1}{2\tau_d} = \frac{1}{1.55 \text{ ms}} = 645 \text{ Hz}$$

$$\omega = \frac{\sqrt{4\tau\tau_d - \tau^2}}{2\tau\tau_d} = \frac{\sqrt{16,86,400 - 2,95,936}}{8,43,200}$$

$$= \frac{1197.2}{8,43,200\,\mu} = 1398 \text{ rad/s}$$

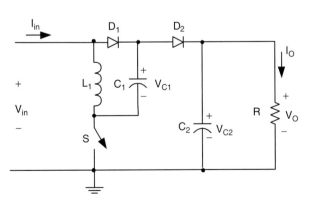

FIGURE 14.123 A super-lift Luo-converter.

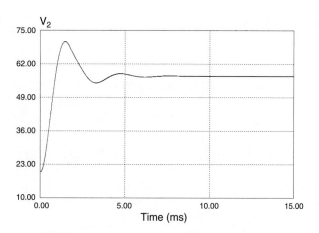

FIGURE 14.124 SL Luo-converter unit-step response.

The unit-step function response is

$$v_2(t) = 57.25\left[1 - e^{-(t/0.00155)}(\cos 1398t - 0.461\sin 1398t)\right]\text{V}$$

The unit-step function response (transient process) has oscillation progress with damping factor σ and frequency ω. The simulation is shown in Fig. 14.124.

The impulse interference response is

$$\Delta v_2(t) = 0.923 U e^{-(t/0.00155)} \sin 1398t$$

where U is the interference signal. The impulse response (interference recovery process) has oscillation progress with damping factor σ and frequency ω, and is shown in Fig. 14.125.

In order to verify the analysis, calculation and simulation results, we constructed a test rig with same conditions. The corresponding test results are shown in Figs. 14.126 and 14.127.

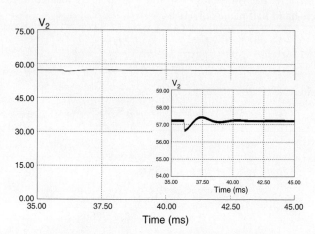

FIGURE 14.125 SL Luo-converter impulse response.

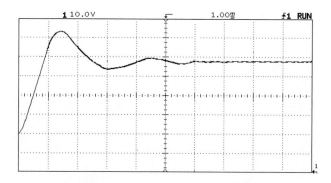

FIGURE 14.126 SL Luo-converter unit-step response (test).

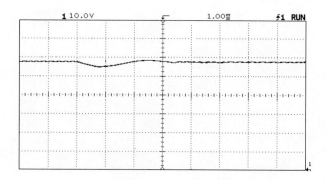

FIGURE 14.127 SL Luo-converter impulse response (test).

Further Reading

1. Luo F. L. and Ye H. "Advanced DC/DC Converters" **CRC Press LLC**, Boca Raton, Florida 07030, USA, 2004. **ISBN: 0-8493-1956-0**.
2. Luo F. L., Ye H., and Rashid M. H. "Digital Power Electronics and Applications" **Elsevier Academic Press**, Burlington, Massachusetts 01803, USA, June 2005. **ISBN: 0-1208-8757-6**.
3. Luo F. L. and Ye H. "Essential DC/DC Converters" **Taylor and Francis Group LLC**, Boca Raton, Florida 07030, USA, October 2005. **ISBN: 0-8493-7238-0**.
4. Luo F. L. "Positive Output Luo-Converters, Voltage Lift Technique" **IEE Proceedings on Electric Power Applications**, Vol. 146, No. 4, July 1999, pp. 415–432.
5. Luo F. L. "Negative Output Luo–Converters, Voltage Lift Technique" **IEE Proceedings on Electric Power Applications**, Vol. 146, No. 2, July 1999, pp. 208–224.
6. Luo F. L. "Double Output Luo–Converters, Advanced Voltage Lift Technique" **IEE Proceedings on Electric Power Applications**, Vol. 147, No. 6, November 2000, pp. 618–633.
7. Luo F. L. "Re–Lift Converter: Design, Test, Simulation and Stability Analysis" **IEE Proceedings on Electric Power Applications**, Vol. 145, No. 4, July 1998, pp. 315–325.
8. Luo F. L., Ye H., and Rashid M. H. "Chapter 14: DC/DC Conversion Techniques and Nine Series Luo–Converters" of "Power Electronics Handbook" (Edited by M. H. Rashid) **Academic Press**, San Diego, USA, August 2001, pp. 335–406. **ISBN: 0-12-581650-2**.
9. Rashid M. H. "Power Electronics: Circuits, Devices and Applications" Second edition, **Prentice-Hall**, USA, 1993.

10. Mohan N., Undeland T. M., and Robbins W. P. "Power Electronics: Converters, Applications and Design" **John Wiley & Sons**, New York, 1995.
11. Severns R. P. and Bloom G. "Modern DC-to-DC Switchmode Power Converter Circuits" **Van Nostrand Reinhold Company**, New York, 1985.
12. Kularatna N. "Power Electronics Design Handbook" **John Wiley & Sons**, New York, 1985.
13. Luo F. L. and Ye H. "Advanced Multi-Quadrant Operation DC/DC Converters" **Taylor and Francis Group LLC**, Boca Raton, Florida 07030, USA, June 2005. **ISBN: 0-8493-7239-9**.
14. Luo F. L. "Neural Network Control for Synchronous Rectifier DC/DC Converter" **Proceedings of the International Conference ICARCV'2000**, Singapore, 5–8 December 2000 (CD ROM).
15. Luo F. L., Ye H., and Rashid M. H. "Two-Quadrant DC/DC ZCS Quasi-Resonant Luo-Converter" **Proceedings of the IEEE International Conference IPEMC'2000**, Beijing, China, 15–18 August 2000, pp. 272–277.
16. Luo F. L., Ye H., and Rashid M. H. "Two-Quadrant DC/DC ZVS Quasi-Resonant Luo-Converter" **Proceedings of the IEEE International Conference IPEMC'2000**, Beijing, China, 15–18 August 2000, pp. 1132–1137.
17. Luo F. L. "Negative Output Luo–Converters – Voltage Lift Technique" **Proceedings of the Second World Energy System International Conference WES'98**, Toronto, Canada, 19–22 May 1998, pp. 253–260.
18. Luo F. L. and Ye H. "Ultra-Lift Luo-Converter" **IEE-EPA Proceedings**, Vol. 152, No. 1, January 2005, pp. 27–32.
19. Luo F. L. and Ye H. "Positive Output Cascade Boost Converters" **IEE-EPA Proceedings**, Vol. 151, No. 5, September 2004, pp. 590–606.
20. Luo F. L. and Ye H. "Positive Output Multiple-Lift Push-Pull Switched-Capacitor Luo-Converters" **IEEE-Transactions on Industrial Electronics**, Vol. 51, No. 3, June 2004, pp. 594–602.
21. Luo F. L. and Ye H. "Negative Output Super-Lift Converters" **IEEE-Transactions on Power Electronics**, Vol. 18, No. 5, September 2003, pp. 1113–1121.
22. Luo F. L. and Ye H. "Positive Output Super-Lift Converters" **IEEE-Transactions on Power Electronics**, Vol. 18, No. 1, January 2003, pp. 105–113.
23. Luo F. L. and Ye H. "Investigation and Verification of a Cascade Double Γ-CL Current Source Resonant Inverter" **IEE-EPA Proceedings**, Vol. 149, No. 5, September 2002, pp. 369–378.
24. Luo F., Ye H., and Rashid M. H. "Multiple Quadrant Operation Luo-Converters" **IEE-EPA Proceedings**, Vol. 149, No. 1, January 2002, pp. 9–18.
25. Luo F. L. "Six Self-Lift DC/DC Converters: Voltage Lift Technique" **IEEE-Transactions on Industrial Electronics**, Vol. 48, No. 6, December 2001, pp. 1268–1272.
26. Luo F. L. "Seven Self-Lift DC/DC Converters: Voltage Lift Technique" **IEE-EPA Proceedings**, Vol. 148, No. 4, July 2001, pp. 329–338.
27. Luo F. L., Ye H., and Rashid M. H. "Four-Quadrant Operating Luo-Converters" **Proceedings of the IEEE International Conference PESC'2000**, Galway, Ireland, 18–23 June 2000, pp. 1047–1052.
28. Luo F. L. "42/14 V Two-Quadrant DC/DC Soft-Switching Converter" **Proceedings of the IEEE International Conference PESC'2000**, Galway, Ireland, 18–23 June 2000, pp. 143–148.
29. Luo F. L. and Ye H. "Two-Quadrant Switched Capacitor Converter" **Proceedings of the 13th Chinese Power Supply Society IAS Annual Meeting**, Shenzhen, China, 15–18 November 1999, pp. 164–168.
30. Luo F. L. and Ye H. "Four-Quadrant Switched Capacitor Converter" **Proceedings of the 13th Chinese Power Supply Society IAS Annual Meeting**, Shenzhen, China, 15–18 November 1999, pp. 513–518.
31. Luo F. L., Ye H., and Rashid M. H. "Switched Capacitor Four-Quadrant Luo-Converter" **Proceedings of the IEEE-IAS Annual Meeting, IAS'99**, Phoenix, Arizona, USA, 3–7 October 1999, pp. 1653–1660.
32. Luo F. L., Ye H., and Rashid M. H. "Switched Inductor Four-Quadrant Luo-Converter" **Proceedings of the IEEE-IAS Annual Meeting, IAS'99**, Phoenix, Arizona, USA, 3–7 October 1999, pp. 1631–1638.
33. Luo F. L. "Four-Quadrant DC/DC ZCS Quasi-Resonant Luo-Converter" Accepted for publication by IEE International Conference IPEC'2001, Singapore, 14–19 May 2001.
34. Luo F. L. "Four-Quadrant DC/DC ZVS Quasi-Resonant Luo-Converter" Accepted for publication by IEE International Conference IPEC'2001, Singapore, 14–19 May 2001.
35. Gao Y. and Luo F. L. "Theoretical Analysis on performance of a 5V/12V Push-Pull Switched Capacitor DC/DC Converter" Accepted for publication by IEE International Conference IPEC'2001, Singapore, 14–19 May 2001.
36. Luo F. L. and Chua L. M. "Fuzzy Logic Control for Synchronous Rectifier DC/DC Converter" **Proceedings of the IASTED International Conference ASC'2000**, Banff, Alberta, Canada, 24–26 July 2000, pp. 24–28.
37. Luo F. L. "Luo-Converters – Voltage Lift Technique" **Proceedings of the IEEE Power Electronics Special Conference IEEE-PESC'98**, Fukuoka, Japan, 14–22 May 1998, pp. 1483–1489.
38. Luo F. L. "Luo-Converters, A Series of New DC-DC Step-Up (Boost) Conversion Circuits" **Proceedings of the IEEE International Conference PEDS'97**, 26–29 May 1997, Singapore, pp. 882–888.
39. Luo F. L. "Re-Lift Circuit, A New DC-DC Step-Up (Boost) Converter" **IEE – Electronics Letters**, Vol. 33, No. 1, 2 January 1997, pp. 5–7.
40. Luo F. L., Lee W. C., and Lee G. B., "Self-Lift Circuit, A New DC-DC Converter" **Proceedings of the 3rd National Undergraduate Research Programme (NURP), Congress 97**, Singapore, 13 September 1997, pp. 31–36.
41. Luo F. L. "DSP-Controlled PWM L-Converter Used for PM DC Motor Drives" **Proceedings of the IEEE International Conference SISCTA'97**, Singapore, 29–30 July 1997, pp. 98–102.
42. Luo F. L. "Luo-Converters, New DC-DC Step-Up Converters" **Proceedings of the IEE International Conference ISIC-97**, Singapore, 10–12 September 1997, pp. 227–230.
43. Luo F. L. and Ye H. "Synchronous and Resonant DC/DC Conversion Technology, Energy Factor and Mathematical Modeling" **Taylor and Francis Group LLC**, Boca Raton, Florida 07030, USA, October 2005. **ISBN: 0-8493-7237-2**.
44. Luo F. L. and Ye H. "Chapter 11 (32): D/A and A/D Converters" of Volume 2 of "Electrical Engineering Handbook" (Edited by R. C. Dorf) Third edition, **CRC Press LLC**, Boca Raton, Florida 07030, USA, September 2004. **ISBN: 0-8493-7339-5 (0-8493-2774-0)**.
45. Maksimovic D. and Cuk S. "A General Approach to Synthesis and Analysis of Quasi-Resonant Converters" **IEEE Transactions on PE**, Vol. 6, No. 1, January 1991, pp. 127–140.

46. Middlebrook R. D. and Cuk S. "Advances in Switched-Mode Power Conversion" Vols. I and II, **TESLAco**, Pasadena, CA, 1981.
47. Liu Y. and Sen P. C., "New Class-E DC-DC Converter Topologies with Constant Switching Frequency" **IEEE-IA Transactions**, Vol. 32, No. 4, July/August 1996, pp. 961–969.
48. Redl R., Molnar B., and Sokal N. O. "Class-E Resonant DC-DC Power Converters: Analysis of Operations, and Experimental Results at 1.5 MHz" **IEEE Transactions, Power Electronics**, Vol. 1, April 1986, pp. 111–119.
49. Kazimierczuk M. K. and Bui X. T. "Class-E DC-DC Converters with an Inductive Impedance Inverter" **IEEE Transactions, Power Electronics**, Vol. 4, July 1989, pp. 124–135.
50. Massey R. P. and Snyder E. C. "High Voltage Single-Ended DC-DC Converter" **IEEE PESC, 1977 Record**, pp. 156–159
51. Jozwik J. J. and Kazimerczuk M. K. "Dual Sepic PWM Switching-Mode DC/DC Power Converter" **IEEE Transactions on Industrial Electronics**, Vol. 36, No. 1, 1989, pp. 64–70.
52. Martins D. C. "Application of the Zeta Converter in Switch-Mode Power Supplies" **Proc. of IEEE APEC'93, USA**, pp. 214–220.
53. Kassakian J. G., Wolf H-C., Miller J. M. and Hurton C. J. "Automotive electrical systems, circa 2005" **IEEE Spectrum**, August 1996, pp. 22–27.
54. Wang J., Dunford W. G., and Mauch K. "Some Novel Four-Quadrant DC-DC Converters" **Proc. Of IEEE-PESC'98**, Fukuoka, Japan, pp. 1475–1482.
55. Luo F. L. "Double Output Luo-Converters" **Proceedings of the IEE International Conference IPEC'99**, Singapore, 24–26 May 1999, pp. 647–652.
56. Ye H. and Luo F. L. "Luo-Converters, A Series of New DC-DC Step-Up Conversion Circuits" **International Journal <Power Supply Technologies and Applications>**, Vol. 1, No. 1, April 1998, Xi'an, China, pp. 30–39.
57. Ye H. and Luo F. L. "Advanced Voltage Lift Technique – Negative Output Luo-Converters" **International Journal <Power Supply Technologies and Applications>**, Vol. 1, No. 3, August 1998, Xi'an, China, pp. 152–168.
58. Luo F. L. "Negative Output Luo-Converters – Voltage Lift Technique" **Proceedings of the Second world energy system International Conference WES'98**, Toronto, Canada, 19–22 May 1998, pp. 253–260.
59. Luo F. L. "Luo-Converters – Voltage Lift Technique" **Proceedings of the IEEE Power Electronics Special Conference IEEE-PESC'98**, Fukuoka, Japan, 14–22 May 1998, pp. 1483–1489.
60. Luo F. L. and Ye H. "Two-Quadrant DC/DC Converter with Switched Capacitors" **Proceedings of the International Conference IPEC'99**, Singapore, 24–26 May 1999, pp. 641–646.
61. Midgley D. and Sigger M. "Switched-capacitors in power control" **IEE Proc.**, Vol. 121, July 1974, pp. 703–704.
62. Cheong S. V., Chung H., and Ioinovici A. "Inductorless DC-DC Converter with high Power Density" **IEEE Trans. on Industrial Electronics**, Vol. 41, No. 2, April 1994, pp. 208–215.
63. Midgley D. and Sigger M. "Switched-capacitors in power control" **IEE Proc.**, Vol. 121, July 1974, pp. 703–704.
64. Chung H., Hui S. Y. R., and Tang S. C. "A low-profile Switched-capacitor-based DC/DC Converter" **Proc. of AUPEC'97**, October 1997, pp. 73–78.
65. Ngo K. D. T. and Webster R. "Steady-state Analysis and Design of a Switched-Capacitor DC-DC Converter" **IEEE Transactions on ANES**, Vol. 30, No. 1, January 1994, pp. 92–101.
66. Harris W. S. and Ngo K. D. T. "Power Switched-Capacitor DC-DC Converter: Analysis and Design" **IEEE Transactions on ANES**, Vol. 33, No. 2, April 1997, pp. 386–395.
67. Tse C. K., Wong S. C., and Chow M. H. L. "On Lossless Switched-Capacitor Power Converters" **IEEE Trans. on PE**, Vol. 10, No. 3, May 1995, pp. 286–291.
68. Luo F. L. and Ye H. "Energy Factor and Mathematical Modeling for Power DC/DC Converters" **IEE-EPA Proceedings**, Vol. 152, No. 2, March 2005, pp. 191–198.
69. Mak O. C., Wong Y. C., and Ioinovici A. "Step-up DC Power Supply Based on a Switched-capacitor Circuit" **IEEE Trans. on IE**, Vol. 42, No. 1, February 1995, pp. 90–97.
70. Mak O. C. and Ioinovici A. "Switched-capacitor Inverter with High Power Density and Enhanced Regulation Capability" **IEEE Trans. on CAS-I**, Vol. 45, No. 4, April 1998, pp. 336–347.
71. Luo F. L. and Ye H. "Switched Inductor Two-Quadrant DC/DC Converter with Fuzzy Logic Control" **Proceedings of IEEE International Conference PEDS'99**, Hong Kong, 26–29 July 1999, pp. 773–778.
72. Luo F. L. and Ye H. "Switched Inductor Two-Quadrant DC/DC Converter with Neural Network Control" **Proceedings of IEEE International Conference PEDS'99**, Hong Kong, 26–29 July 1999, pp. 1114–1119.
73. Liu K. H. and Lee F. C. "Resonant Switches – A Unified Approach to Improved Performances of Switching Converters," **International Telecommunications Energy Conference (INTELEC) Proc.**, New Orleans, LA, USA, 4–7 November 1984, pp. 344–351.
74. Liu K. H. and Lee F. C. "Zero-Voltage Switching Techniques in DC/DC Converter Circuits" **Power Electronics Specialist's Conf. (PESC) Record**, June 1986, Vancouver, Canada, pp. 58, 70.
75. Martinez Z. R. and Ray B. "Didirectional DC/DC Power Conversion Using Constant-Frequency Multi-Resonant Topology," **Applied Power Electronics Conf. (APEC) Proc.**, Orlando, FL, USA, 13–14 February 1994, pp. 991–997.
76. Pong M. H., Ho W. C., and Poon N. K. "Soft Switching Converter with Power Limiting Feature" **IEE-EPA Proceedings**, Vol. 146, No. 1, January 1999, pp. 95–102.
77. Gu W. J. and Harada K. "A Novel Self-Excited Forward DC-DC Converter with Zero-Voltage-Switched Resonant Transitions using a Saturable Core" **IEEE-PE Transactions**, Vol. 10, No. 2, March 1995, pp. 131–141.
78. Cho J. G., Sabate J. A., Hua G., and Lee F. C. "Zero-Voltage and Zero-Current-Switching Full Bridge PWM Converter for High Power Applications" **IEEE-PE Transactions**, Vol. 11, No. 4, July 1996, pp. 622–628.
79. Poon N. K. and Pong M. H. "Computer Aided Design of a Crossing Current Resonant Converter (XCRC)" **Proceedings of IECON'94**, Bologna, Italy, 5–9 September 1994, pp. 135–140.
80. Kassakian J. G., Schlecht M. F., and Verghese G. C. "Principles of Power Electronics" **Addison-Wesley**, New York, 1991, p. 214.

15
Inverters

José R. Espinoza, Ph.D.
Departamento de Ingeniería Eléctrica, of. 220, Universidad de Concepción, Casilla 160-C, Correo 3, Concepción, Chile

15.1 Introduction 353
15.2 Single-phase Voltage Source Inverters 355
    15.2.1 Half-bridge VSI • 15.2.2 Full-bridge VSI
15.3 Three-phase Voltage Source Inverters 363
    15.3.1 Sinusoidal PWM • 15.3.2 Square-wave Operation of Three-phase VSIs • 15.3.3 Sinusoidal PWM with Zero Sequence Signal Injection • 15.3.4 Selective Harmonic Elimination in Three-phase VSIs • 15.3.5 Space-vector (SV)-based Modulating Techniques • 15.3.6 DC Link Current in Three-phase VSIs • 15.3.7 Load-phase Voltages in Three-phase VSIs
15.4 Current Source Inverters 371
    15.4.1 Carrier-based PWM Techniques in CSIs • 15.4.2 Square-wave Operation of Three-phase CSIs • 15.4.3 Selective Harmonic Elimination in Three-phase CSIs • 15.4.4 Space-vector-based Modulating Techniques in CSIs • 15.4.5 DC Link Voltage in Three-phase CSIs
15.5 Closed-loop Operation of Inverters 379
    15.5.1 Feedforward Techniques in Voltage Source Inverters • 15.5.2 Feedforward Techniques in Current Source Inverters • 15.5.3 Feedback Techniques in Voltage Source Inverters • 15.5.4 Feedback Techniques in Current Source Inverters
15.6 Regeneration in Inverters 386
    15.6.1 Motoring Operating Mode in Three-phase VSIs • 15.6.2 Regenerative Operating Mode in Three-phase VSIs • 15.6.3 Regenerative Operating Mode in Three-phase CSIs
15.7 Multistage Inverters 390
    15.7.1 Multicell Topologies • 15.7.2 Voltage Source-based Multilevel Topologies • 15.7.3 Current Source-based Multilevel Topologies
    Further Reading 402

15.1 Introduction

The main objective of static power converters is to produce an ac output waveform from a dc power supply. These are the types of waveforms required in adjustable speed drives (ASDs), uninterruptible power supplies (UPSs), static var compensators, active filters, flexible ac transmission systems (FACTSs), and voltage compensators, which are only a few applications. For sinusoidal ac outputs, the magnitude, frequency, and phase should be controllable. According to the type of ac output waveform, these topologies can be considered as voltage-source inverters (VSIs), where the independently controlled ac output is a voltage waveform. These structures are the most widely used because they naturally behave as voltage sources as required by many industrial applications, such as ASDs, which are the most popular application of inverters (Fig. 15.1a). Similarly, these topologies can be found as current-source inverters (CSIs), where the independently controlled ac output is a current waveform. These structures are still widely used in medium-voltage industrial applications, where high-quality voltage waveforms are required.

Static power converters, specifically inverters, are constructed from power switches and the ac output waveforms are therefore made up of discrete values. This leads to the generation of waveforms that feature fast transitions rather than smooth ones. For instance, the ac output voltage produced by the VSI of a three-level ASD is a, Pulse Width Modulation (PWM) type of waveform (Fig. 15.1c). Although this waveform is not sinusoidal as expected (Fig. 15.1b), its fundamental component behaves as such. This behavior should be ensured by a modulating technique that controls

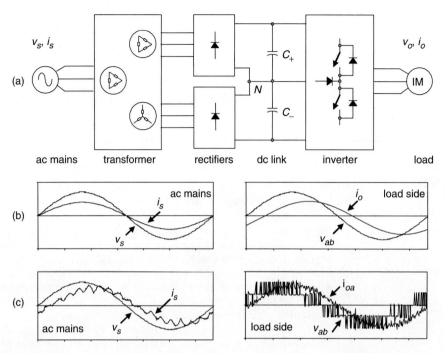

FIGURE 15.1 A three-level adjustable speed drive scheme and associated waveforms: (a) the electrical power conversion topology; (b) the ideal input (ac mains) and output (load) waveforms; and (c) the actual input (ac mains) and output (load) waveforms.

the amount of time and the sequence used to switch the power valves on and off. The modulating techniques most used are the carrier-based technique (e.g. sinusoidal pulsewidth modulation, SPWM), the space-vector (SV) technique, and the selective-harmonic-elimination (SHE) technique.

The discrete shape of the ac output waveforms generated by these topologies imposes basic restrictions on the applications of inverters. The VSI generates an ac output voltage waveform composed of discrete values (high dv/dt); therefore, the load should be inductive at the harmonic frequencies in order to produce a smooth current waveform. A capacitive load in the VSIs will generate large current spikes. If this is the case, an inductive filter between the VSI ac side and the load should be used. On the other hand, the CSI generates an ac output current waveform composed of discrete values (high di/dt); therefore, the load should be capacitive at the harmonic frequencies in order to produce a smooth voltage waveform. An inductive load in CSIs will generate large voltage spikes. If this is the case, a capacitive filter between the CSI ac side and the load should be used.

A three-level voltage waveform is not recommended for medium-voltage ASDs due to the high dv/dt that would apply to the motor terminals. Several negative side effects of this approach have been reported (bearing and isolation problems). As alternatives, to improve the ac output waveforms in VSIs are the multistage topologies (multilevel and multicell). The basic principle is to construct the required ac output waveform from various voltage levels, which achieves medium-voltage waveforms at reduced dv/dt. Although these topologies are well developed in ASDs, they are also suitable for static var compensators, active filters, and voltage compensators. Specialized modulating techniques have been developed to switch the higher number of power valves involved in these topologies. Among others, the carrier-based (SPWM) and SV-based techniques have been naturally extended to these applications.

In many applications, it is required to take energy from the ac side of the inverter and send it back into the dc side. For instance, whenever ASDs need to either brake or slow down the motor speed, the kinetic energy is sent into the voltage dc link (Fig. 15.1a). This is known as the regenerative operating mode and, in contrast to the motoring mode, the dc link current direction is reversed due to the fact that the dc link voltage is fixed. If a capacitor is used to maintain the dc link voltage (as in standard ASDs) the energy must either be dissipated or fed back into the distribution system, otherwise, the dc link voltage gradually increases. The first approach requires the dc link capacitor be connected in parallel with a resistor, which must be properly switched only when the energy flows from the motor into the dc link. A better alternative is to feed back such energy into the distribution system. However, this alternative requires a reversible-current topology connected between the distribution system and the dc link capacitor. A modern approach to such a requirement is to use the active front-end rectifier technologies, where the regeneration mode is a natural operating mode of the system.

15 Inverters

In this chapter, single- and three-phase inverters in their voltage and current source alternatives will be reviewed. The dc link will be assumed to be a perfect dc, either voltage or current source that could be fixed as the dc link voltage in standard ASDs, or variable as the dc link current in some medium-voltage current source drives. Specifically, the topologies, modulating techniques and control aspects oriented to standard applications, are analyzed. In order to simplify the analysis, the inverters are considered lossless topologies, which are composed of ideal power valves. Nevertheless, some practical non-ideal conditions are also considered.

15.2 Single-phase Voltage Source Inverters

Single-phase VSI can be found as half-bridge and full-bridge topologies. Although, the power range they cover is the low one, they are widely used in power supplies, single-phase UPSs, and currently to form high-power static power topologies, such as the multicell configurations that are reviewed in Section 15.7. The main features of both approaches are reviewed and presented in the following.

15.2.1 Half-bridge VSI

Figure 15.2 shows the power topology of a half-bridge VSI, where two large capacitors are required to provide a neutral point N, such that each capacitor maintains a constant voltage $v_i/2$. Because the current harmonics injected by the operation of the inverter are low-order harmonics, a set of large capacitors (C_+ and C_-) is required. It is clear that both switches S_+ and S_- cannot be on simultaneously because a short circuit across the dc link voltage source v_i would be produced. There are two defined (states 1 and 2) and one undefined (state 3) switch state as shown in Table 15.1. In order to avoid the short circuit across the dc bus and the undefined ac output-voltage condition, the modulating technique should always ensure that at any instant either the top or the bottom switch of the inverter leg is on.

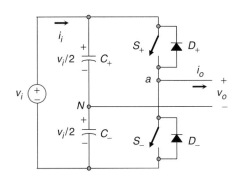

FIGURE 15.2 Single-phase half-bridge VSI.

TABLE 15.1 Switch states for a half-bridge single-phase VSI

State	State #	v_o	Components conducting	
S_+ is on and S_- is off	1	$v_i/2$	S_+	if $i_o > 0$
			D_+	if $i_o < 0$
S_- is on and S_+ is off	2	$-v_i/2$	D_-	if $i_o > 0$
			S_-	if $i_o < 0$
S_+ and S_- are all off	3	$-v_i/2$	D_-	if $i_o > 0$
			$v_i/2$	D_+	if $i_o < 0$

Figure 15.3 shows the ideal waveforms associated with the half-bridge inverter shown in Fig. 15.2. The states for the switches S_+ and S_- are defined by the modulating technique, which in this case is a carrier-based PWM.

A. The Carrier-based Pulse Width Modulation (PWM) Technique

As mentioned earlier, it is desired that the ac output voltage, $v_o = v_{aN}$, follow a given waveform (e.g. sinusoidal) on a continuous basis by properly switching the power valves. The carrier-based PWM technique fulfills such a requirement as it defines the on- and off-states of the switches of one leg of a VSI by comparing a modulating signal v_c (desired ac output voltage) and a triangular waveform v_Δ (carrier signal). In practice, when $v_c > v_\Delta$ the switch S_+ is on and the switch S_- is off; similarly, when $v_c < v_\Delta$ the switch S_+ is off and the switch S_- is on.

A special case is when the modulating signal v_c is a sinusoidal at frequency f_c and amplitude $\hat{v}_c$, and the triangular signal v_Δ is at frequency f_Δ and amplitude $\hat{v}_\Delta$. This is the sinusoidal PWM (SPWM) scheme. In this case, the modulation index m_a (also known as the amplitude-modulation ratio) is defined as

$$m_a = \frac{\hat{v}_c}{\hat{v}_\Delta} \quad (15.1)$$

and the normalized carrier frequency m_f (also known as the frequency-modulation ratio) is

$$m_f = \frac{f_\Delta}{f_c} \quad (15.2)$$

Figure 15.3e clearly shows that the ac output voltage $v_o = v_{aN}$ is basically a sinusoidal waveform plus harmonics, which features: (a) the amplitude of the fundamental component of the ac output voltage $\hat{v}_{o1}$ satisfying the following expression:

$$\hat{v}_{o1} = \hat{v}_{aN1} = \frac{v_i}{2} m_a \quad (15.3)$$

for $m_a \leq 1$, which is called the linear region of the modulating technique (higher values of m_a leads to overmodulation that

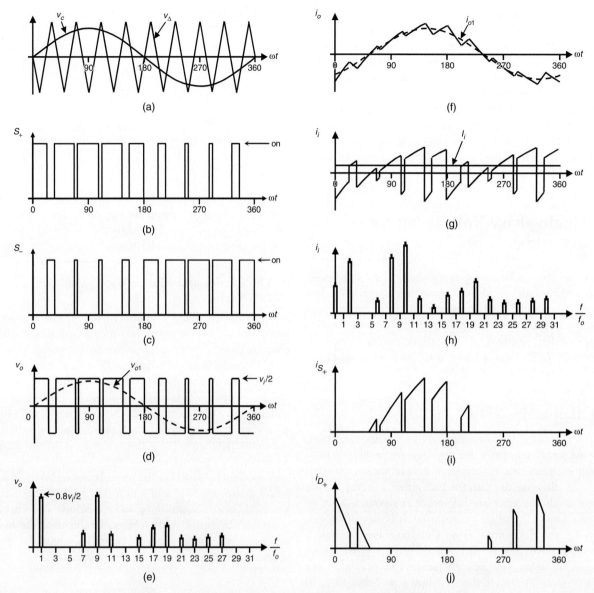

FIGURE 15.3 The half-bridge VSI. Ideal waveforms for the SPWM ($m_a = 0.8$, $m_f = 9$): (a) carrier and modulating signals; (b) switch S_+ state; (c) switch S_- state; (d) ac output voltage; (e) ac output voltage spectrum; (f) ac output current; (g) dc current; (h) dc current spectrum; (i) switch S_+ current; and (j) diode D_+ current.

will be discussed later); (b) for odd values of the normalized carrier frequency m_f the harmonics in the ac output voltage appear at normalized frequencies f_h centered around m_f and its multiples, specifically,

$$h = l\, m_f \pm k \quad l = 1, 2, 3, \ldots \quad (15.4)$$

where $k = 2, 4, 6, \ldots$ for $l = 1, 3, 5, \ldots$; and $k = 1, 3, 5, \ldots$ for $l = 2, 4, 6, \ldots$; (c) the amplitude of the ac output voltage harmonics is a function of the modulation index m_a and is independent of the normalized carrier frequency m_f for $m_f > 9$; (d) the harmonics in the dc link current (due to the modulation) appear at normalized frequencies f_p centered around the normalized carrier frequency m_f and its multiples, specifically,

$$p = l\, m_f \pm k \pm 1 \quad l = 1, 2, \ldots \quad (15.5)$$

where $k = 2, 4, 6, \ldots$ for $l = 1, 3, 5, \ldots$; and $k = 1, 3, 5, \ldots$ for $l = 2, 4, 6, \ldots$. Additional important issues are: (a) for small values of m_f ($m_f < 21$), the carrier signal v_Δ and the signal v_c should be synchronized to each other (m_f integer), which is required to hold the previous features; if this is not the case, subharmonics will be present in the ac output voltage;

(b) for large values of m_f ($m_f > 21$), the subharmonics are negligible if an asynchronous PWM technique is used, however, due to potential very low-order subharmonics, its use should be avoided; finally (c) in the overmodulation region ($m_a > 1$) some intersections between the carrier and the modulating signal are missed, which leads to the generation of low-order harmonics but a higher fundamental ac output voltage is obtained; unfortunately, the linearity between m_a and $\hat{v}_{o1}$ achieved in the linear region does not hold in the overmodulation region, moreover, a saturation effect can be observed (Fig. 15.4).

The PWM technique allows an ac output voltage to be generated that tracks a given modulating signal. A special case is the SPWM technique (the modulating signal is a sinusoidal) that provides, in the linear region, an ac output voltage that varies linearly as a function of the modulation index, and the harmonics are at well-defined frequencies and amplitudes. These features simplify the design of filtering components. Unfortunately, the maximum amplitude of the fundamental ac voltage is $v_i/2$ in this operating mode. Higher voltages are obtained by using the overmodulation region ($m_a > 1$); however, low-order harmonics appear in the ac output voltage. Very large values of the modulation index ($m_a > 3.24$) lead to a totally square ac output voltage that is considered as the square-wave modulating technique.

B. Square-wave Modulating Technique

Both switches S_+ and S_- are on for one half-cycle of the ac output period. This is equivalent to the SPWM technique with an infinite modulation index m_a. Figure 15.5 shows the following: (a) the normalized ac output voltage harmonics are

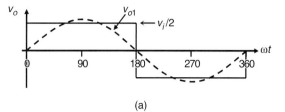

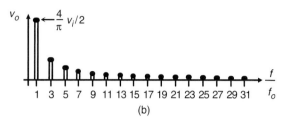

FIGURE 15.5 The half-bridge VSI. Ideal waveforms for the square-wave modulating technique: (a) ac output voltage and (b) ac output voltage spectrum.

at frequencies $h = 3, 5, 7, 9, \ldots$, and for a given dc link voltage; (b) the fundamental ac output voltage features an amplitude given by

$$\hat{v}_{o1} = \hat{v}_{aN1} = \frac{4}{\pi} \frac{v_i}{2} \tag{15.6}$$

and the harmonics feature an amplitude given by

$$\hat{v}_{oh} = \frac{\hat{v}_{o1}}{h} \tag{15.7}$$

It can be seen that the ac output voltage cannot be changed by the inverter. However, it could be changed by controlling the dc link voltage v_i. Other modulating techniques that are applicable to half-bridge configurations (e.g., selective harmonic elimination) are reviewed here as they can easily be extended to modulate other topologies.

C. Selective Harmonic Elimination

The main objective is to obtain a sinusoidal ac output voltage waveform where the fundamental component can be adjusted arbitrarily within a range and the intrinsic harmonics selectively eliminated. This is achieved by mathematically generating the exact instant of the turn-on and turn-off of the power valves. The ac output voltage features odd half- and quarter-wave symmetry; therefore, even harmonics are not present ($v_{oh} = 0$, $h = 2, 4, 6, \ldots$). Moreover, the phase voltage waveform ($v_o = v_{aN}$ in Fig. 15.2), should be chopped N times per half-cycle in order to adjust the fundamental and eliminate $N - 1$ harmonics in the ac output voltage waveform. For instance, to eliminate the third and fifth harmonics

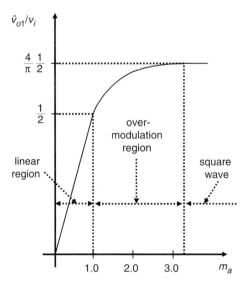

FIGURE 15.4 Normalized fundamental ac component of the output voltage in a half-bridge VSI SPWM modulated.

and to perform fundamental magnitude control ($N = 3$), the equations to be solved are the following:

$$\cos(1\alpha_1) - \cos(1\alpha_2) + \cos(1\alpha_3) = (2 + \pi\hat{v}_{o1}/v_i)/4$$

$$\cos(3\alpha_1) - \cos(3\alpha_2) + \cos(3\alpha_3) = 1/2 \quad (15.8)$$

$$\cos(5\alpha_1) - \cos(5\alpha_2) + \cos(5\alpha_3) = 1/2$$

where the angles α_1, α_2 and α_3 are defined as shown in Fig. 15.6a. The angles are found by means of iterative algorithms as no analytical solutions can be derived. The angles α_1, α_2, and α_3 are plotted for different values of $\hat{v}_{o1}/v_i$ in Fig. 15.7a. The general expressions to eliminate an even $N-1$ ($N - 1 = 2, 4, 6, \ldots$) number of harmonics are

$$-\sum_{k=1}^{N}(-1)^k \cos(\alpha_k) = \frac{(2 + \pi\hat{v}_{o1})/v_i}{4}$$

$$-\sum_{k=1}^{N}(-1)^k \cos(n\alpha_k) = \frac{1}{2} \quad \text{for } n = 3, 5, \ldots, 2N-1$$

(15.9)

where $\alpha_1, \alpha_2, \ldots, \alpha_N$ should satisfy $\alpha_1 < \alpha_2 < \cdots < \alpha_N < \pi/2$. Similarly, to eliminate an odd number of harmonics, for instance the third, fifth, and seventh, and to perform the

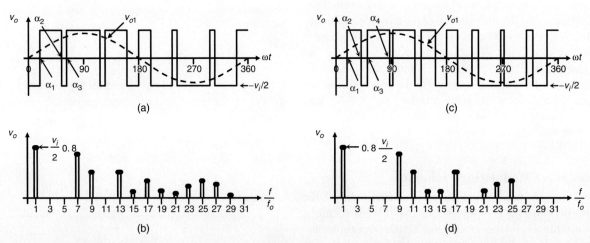

FIGURE 15.6 The half-bridge VSI. Ideal waveforms for the SHE technique: (a) ac output voltage for third and fifth harmonic elimination; (b) spectrum of (a); (c) ac output voltage for third, fifth, and seventh harmonic elimination; and (d) spectrum of (c).

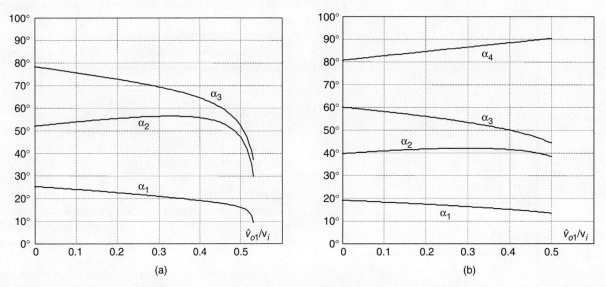

FIGURE 15.7 Chopping angles for SHE and fundamental voltage control in half-bridge VSIs: (a) third and fifth harmonic elimination and (b) third, fifth, and seventh harmonic elimination.

fundamental magnitude control ($N - 1 = 3$), the equations to be solved are:

$$\cos(1\alpha_1) - \cos(1\alpha_2) + \cos(1\alpha_3) - \cos(1\alpha_4) = (2 - \pi\hat{v}_{o1}/v_i)/4$$
$$\cos(3\alpha_1) - \cos(3\alpha_2) + \cos(3\alpha_3) - \cos(3\alpha_4) = 1/2$$
$$\cos(5\alpha_1) - \cos(5\alpha_2) + \cos(5\alpha_3) - \cos(5\alpha_4) = 1/2$$
$$\cos(7\alpha_1) - \cos(7\alpha_2) + \cos(7\alpha_3) - \cos(7\alpha_4) = 1/2$$

(15.10)

where the angles α_1, α_2, α_3, and α_4 are defined as shown in Fig. 15.6b. The angles α_1, α_2, and α_3 are plotted for different values of $\hat{v}_{o1}/v_i$ in Fig. 15.7b. The general expressions to eliminate an odd $N - 1$ ($N - 1 = 3, 5, 7, \ldots$) number of harmonics are given by

$$-\sum_{k=1}^{N}(-1)^k \cos(n\alpha_k) = \frac{(2 - \pi\hat{v}_{o1})/v_i}{4}$$

$$-\sum_{k=1}^{N}(-1)^k \cos(n\alpha_k) = \frac{1}{2} \quad \text{for } n = 3, 5, \ldots, 2N - 1$$

(15.11)

where $\alpha_1, \alpha_2, \ldots, \alpha_N$ should satisfy $\alpha_1 < \alpha_2 < \cdots < \alpha_N < \pi/2$.

To implement the SHE modulating technique, the modulator should generate the gating pattern according to the angles as shown in Fig. 15.7. This task is usually performed by digital systems that normally store the angles in look-up tables.

D. DC Link Current

The split capacitors are considered a part of the inverter and therefore an instantaneous power balance cannot be considered due to the storage energy components (C_+ and C_-). However, if a lossless inverter is assumed, the average power absorbed in one period by the load must be equal to the average power supplied by the dc source. Thus, we can write

$$\int_0^T v_i(t) \cdot i_i(t) \cdot dt = \int_0^T v_o(t) \cdot i_o(t) \cdot dt \qquad (15.12)$$

where T is the period of the ac output voltage. For an inductive load and a relatively high switching frequency, the load current i_o is nearly sinusoidal and therefore, only the fundamental component of the ac output voltage provides power to the load. On the other hand, if the dc link voltage remains constant $v_i(t) = V_i$, Eq. (15.12) can be simplified to

$$\int_0^T i_i(t) \cdot dt = \frac{1}{V_i} \int_0^T \sqrt{2}V_{o1}\sin(\omega t) \cdot \sqrt{2}I_o\sin(\omega t - \phi) \cdot dt = I_i$$

(15.13)

where V_{o1} is the fundamental rms ac output voltage, I_o is the rms load current, ϕ is an arbitrary inductive load power factor, and I_i is the dc link current that can be further simplified to

$$I_i = \frac{V_{o1}}{V_i} I_o \cos(\phi) \qquad (15.14)$$

15.2.2 Full-bridge VSI

Figure 15.8 shows the power topology of a full-bridge VSI. This inverter is similar to the half-bridge inverter; however, a second leg provides the neutral point to the load. As expected, both switches S_{1+} and S_{1-} (or S_{2+} and S_{2-}) cannot be on simultaneously because a short circuit across the dc link voltage source v_i would be produced. There are four defined (states 1, 2, 3, and 4) and one undefined (state 5) switch state as shown in Table 15.2.

The undefined condition should be avoided so as to be always capable of defining the ac output voltage always. In order to avoid the short circuit across the dc bus and the undefined ac output voltage condition, the modulating technique should ensure that either the top or the bottom switch of each leg is on at any instant. It can be observed that the ac output voltage can take values up to the dc link value v_i, which is twice that obtained with half-bridge VSI topologies.

Several modulating techniques have been developed that are applicable to full-bridge VSIs. Among them are the PWM (bipolar and unipolar) techniques.

A. Bipolar PWM Technique

States 1 and 2 (Table 15.2) are used to generate the ac output voltage in this approach. Thus, the ac output voltage waveform features only two values, which are v_i and $-v_i$. To generate the

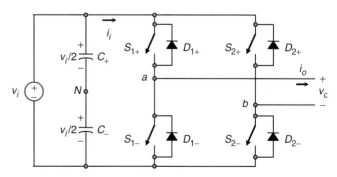

FIGURE 15.8 Single-phase full-bridge VSI.

TABLE 15.2 Switch states for a full-bridge single-phase VSI

State	State #	v_{aN}	v_{bN}	v_o	Components conducting	
S_{1+} and S_{2-} are on and S_{1-} and S_{2+} are off	1	$v_i/2$	$-v_i/2$	v_i	S_{1+} and S_{2-}	if $i_o > 0$
					D_{1+} and D_{2-}	if $i_o < 0$
S_{1-} and S_{2+} are on and S_{1+} and S_{2-} are off	2	$-v_i/2$	$v_i/2$	$-v_i$	D_{1-} and D_{2+}	if $i_o > 0$
					S_{1-} and S_{2+}	if $i_o < 0$
S_{1+} and S_{2+} are on and S_{1-} and S_{2-} are off	3	$v_i/2$	$v_i/2$	0	S_{1+} and D_{2+}	if $i_o > 0$
					D_{1+} and S_{2+}	if $i_o < 0$
S_{1-} and S_{2-} are on and S_{1+} and S_{2+} are off	4	$-v_i/2$	$-v_i/2$	0	D_{1-} and S_{2-}	if $i_o > 0$
					S_{1-} and D_{2-}	if $i_o < 0$
S_{1-}, S_{2-}, S_{1+}, and S_{2+} are all off	5	$-v_i/2$	$v_i/2$	v_i	D_{1-} and D_{2+}	if $i_o > 0$
		$v_i/2$	$-v_i/2$	$-v_i$	D_{1+} and D_{2-}	if $i_o < 0$

states, a carrier-based technique can be used as in half-bridge configurations (Fig. 15.3), where only one sinusoidal modulating signal has been used. It should be noted that the on-state in switch S_+ in the half-bridge corresponds to both switches S_{1+} and S_{2-} being in the on-state in the full-bridge configuration. Similarly, S_- in the on-state in the half-bridge corresponds to both switches S_{1-} and S_{2+} being in the on-state in the full-bridge configuration. This is called bipolar carrier-based SPWM. The ac output voltage waveform in a full-bridge VSI is basically a sinusoidal waveform that features a fundamental component of amplitude $\hat{v}_{o1}$ that satisfies the expression

$$\hat{v}_{o1} = \hat{v}_{ab1} = v_i m_a \quad (15.15)$$

in the linear region of the modulating technique ($m_a \leq 1$), which is twice that obtained in the half-bridge VSI. Identical conclusions can be drawn for the frequencies and the amplitudes of the harmonics in the ac output voltage and dc link current, and for operations at smaller and larger values of odd m_f (including the overmodulation region ($m_a > 1$)), than in half-bridge VSIs, but considering that the maximum ac output voltage is the dc link voltage v_i. Thus, in the overmodulation region the fundamental component of amplitude $\hat{v}_{o1}$ satisfies the expression

$$v_i < \hat{v}_{o1} = \hat{v}_{ab1} < \frac{4}{\pi} v_i \quad (15.16)$$

B. Unipolar PWM Technique

In contrast to the bipolar approach, the unipolar PWM technique uses the states 1, 2, 3, and 4 (Table 15.2) to generate the ac output voltage. Thus, the ac output voltage waveform can instantaneously take one of the three values, namely v_i, $-v_i$, and 0. To generate the states, a carrier-based technique can be used as shown in Fig. 15.9, where two sinusoidal modulating signals (v_c and $-v_c$) are used. The signal v_c is used to generate v_{aN}, and $-v_c$ is used to generate v_{bN}; thus $v_{bN1} = -v_{aN1}$. On the other hand, $v_{o1} = v_{aN1} - v_{bN1}, = 2 \cdot v_{aN1}$; thus $\hat{v}_{o1} = 2 \cdot \hat{v}_{aN1} = m_a \cdot v_i$ This is called unipolar carrier-based SPWM.

Identical conclusions can be drawn for the amplitude of the fundamental component and harmonics in the ac output voltage and dc link current, and for operations at smaller and larger values of m_f (including the overmodulation region ($m_a > 1$)) than in full-bridge VSIs modulated by the bipolar SPWM. However, because the phase voltages (v_{aN} and v_{bN}) are identical but 180° out of phase, the output voltage ($v_o = v_{ab} = v_{aN} - v_{bN}$) will not contain even harmonics. Thus, if m_f is taken even, the harmonics in the ac output voltage appear at normalized odd frequencies f_h centered around twice the normalized carrier frequency m_f and its multiples. Specifically,

$$h = l\,m_f \pm k \quad l = 2, 4, \ldots \quad (15.17)$$

where $k = 1, 3, 5, \ldots$ and the harmonics in the dc link current appear at normalized frequencies f_p centered around twice the normalized carrier frequency m_f and its multiples. Specifically,

$$p = l\,m_f \pm k \pm 1 \quad l = 2, 4, \ldots \quad (15.18)$$

where $k = 1, 3, 5, \ldots$ This feature is considered to be an advantage because it allows the use of smaller filtering components to obtain high-quality voltage and current waveforms while using the same switching frequency as in VSIs modulated by the bipolar approach.

C. Selective Harmonic Elimination

In contrast to half-bridge VSIs, this approach is applied in a per-line fashion for full-bridge VSIs. The ac output voltage features odd half- and quarter-wave symmetry; therefore, even harmonics are not present ($\hat{v}_{oh} = 0$, $h = 2, 4, 6, \ldots$). Moreover, the ac output voltage waveform ($v_o = v_{ab}$ in Fig. 15.8), should feature N pulses per half-cycle in order to adjust the fundamental component and eliminate $N - 1$ harmonics. For instance, to eliminate the third, fifth, and the seventh harmonics and to perform fundamental component magnitude

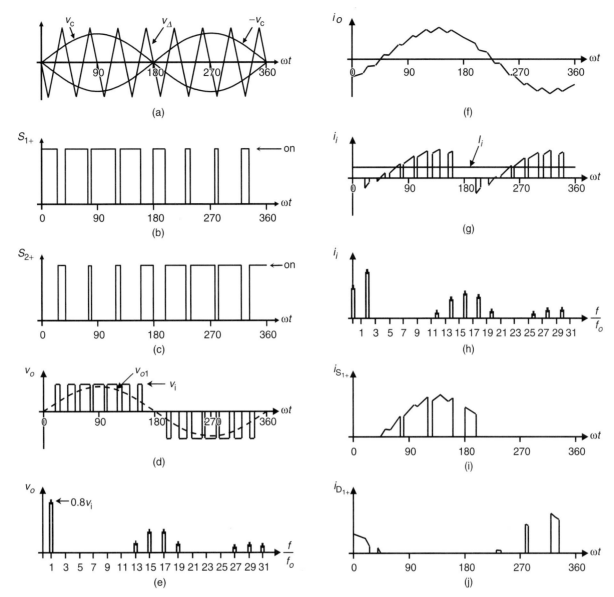

FIGURE 15.9 The full-bridge VSI. Ideal waveforms for the unipolar SPWM ($m_a = 0.8$, $m_f = 8$): (a) carrier and modulating signals; (b) switch S_{1+} state; (c) switch S_{2+} state; (d) ac output voltage; (e) ac output voltage spectrum; (f) ac output current; (g) dc current; (h) dc current spectrum; (i) switch S_{1+} current; and (j) diode D_{1+} current.

control ($N = 4$), the equations to be solved are:

$$\cos(1\alpha_1) - \cos(1\alpha_2) + \cos(1\alpha_3) - \cos(1\alpha_4) = \pi\hat{v}_{o1}/(v_i 4)$$
$$\cos(3\alpha_1) - \cos(3\alpha_2) + \cos(3\alpha_3) - \cos(3\alpha_4) = 0$$
$$\cos(5\alpha_1) - \cos(5\alpha_2) + \cos(5\alpha_3) - \cos(5\alpha_4) = 0$$
$$\cos(7\alpha_1) - \cos(7\alpha_2) + \cos(7\alpha_3) - \cos(7\alpha_4) = 0$$
(15.19)

where the angles α_1, α_2, α_3, and α_4 are defined as shown in Fig. 15.10a. The angles α_1, α_2, α_3, and α_4 are plotted for different values of $\hat{v}_{o1}/v_i$ in Fig. 15.11a. The general expressions to eliminate an arbitrary $N - 1$ ($N - 1 = 3, 5, 7, \ldots$) number of harmonics are given by

$$-\sum_{k=1}^{N}(-1)^k \cos(n\alpha_k) = \frac{\pi}{4}\left(\frac{\hat{v}_{o1}}{v_i}\right)$$
$$-\sum_{k=1}^{N}(-1)^k \cos(n\alpha_k) = 0 \quad \text{for } n = 3, 5, \ldots, 2N - 1$$
(15.20)

where $\alpha_1, \alpha_2, \ldots, \alpha_N$ should satisfy $\alpha_1 < \alpha_2 < \cdots < \alpha_N < \pi/2$.

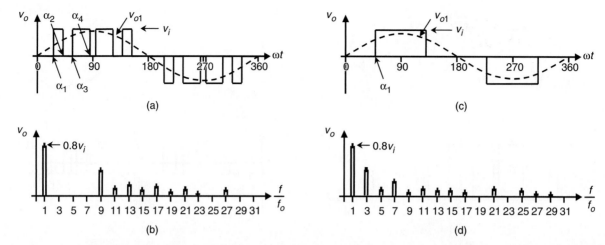

FIGURE 15.10 The half-bridge VSI. Ideal waveforms for the SHE technique: (a) ac output voltage for third, fifth, and seventh harmonic elimination; (b) spectrum of (a); (c) ac output voltage for fundamental control; and (d) spectrum of (c).

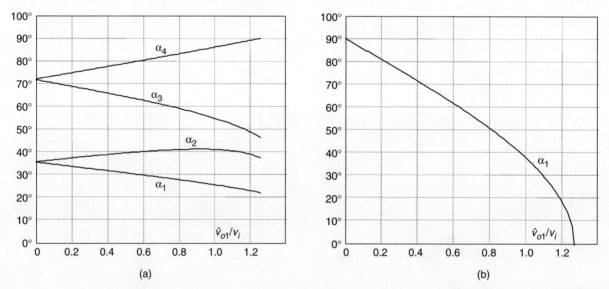

FIGURE 15.11 Chopping angles for SHE and fundamental voltage control in half-bridge VSIs: (a) fundamental control and third, fifth, and seventh harmonic elimination and (b) fundamental control.

Figure 15.10c shows a special case where only the fundamental ac output voltage is controlled. This is known as output control by voltage cancellation, which derives from the fact that its implementation is easily attainable by using two phase-shifted square-wave switching signals as shown in Fig. 15.12. The phase-shift angle becomes $2 \cdot \alpha_1$ (Fig. 15.11b). Thus, the amplitude of the fundamental component and harmonics in the ac output voltage are given by

$$\hat{v}_{oh} = \frac{4}{\pi} v_i \frac{(-1)^{(h-1)/2}}{h} \cos(h\alpha_1) \quad h = 1, 3, 5, \ldots \quad (15.21)$$

It can also be observed in Fig. 15.12c that for $\alpha_1 = 0$ square-wave operation is achieved. In this case, the fundamental ac output voltage is given by

$$\hat{v}_{o1} = \frac{4}{\pi} v_i \quad (15.22)$$

where the fundamental load voltage can be controlled by the manipulation of the dc link voltage.

D. DC Link Current

Due to the fact that the inverter is assumed lossless and constructed without storage energy components, the instantaneous power balance indicates that,

$$v_i(t) \cdot i_i(t) = v_o(t) \cdot i_o(t) \quad (15.23)$$

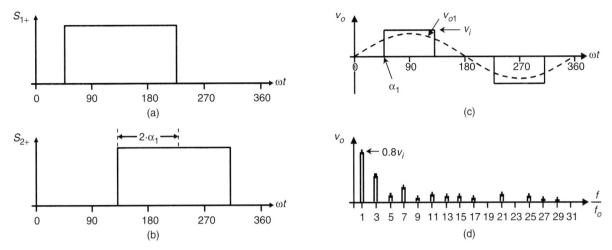

FIGURE 15.12 The full-bridge VSI. Ideal waveforms for the output control by voltage cancellation: (a) switch S_{1+} state; (b) switch S_{2+} state; (c) ac output voltage; and (d) ac output voltage spectrum.

For inductive load and relatively high switching frequencies, the load current i_o is nearly sinusoidal. As a first approximation, the ac output voltage can also be considered sinusoidal. On the other hand, if the dc link voltage remains constant $v_i(t) = V_i$, Eq. (15.23) can be simplified to

$$i_i(t) = \frac{1}{V_i}\sqrt{2}V_{o1}\sin(\omega t) \cdot \sqrt{2}I_o\sin(\omega t - \phi) \quad (15.24)$$

where V_{o1} is the fundamental rms ac output voltage, I_o is the rms load current, and ϕ is an arbitrary inductive load power factor. Thus, the dc link current can be further simplified to

$$i_i(t) = \frac{V_{o1}}{V_i}I_o\cos(\phi) - \frac{V_{o1}}{V_i}I_o\cos(2\omega t - \phi) \quad (15.25)$$

The preceding expression reveals an important issue, that is, the presence of a large second-order harmonic in the dc link current (its amplitude is similar to the dc link current). This second harmonic is injected back into the dc voltage source, thus its design should consider it in order to guarantee a nearly constant dc link voltage. In practical terms, the dc voltage source is required to feature large amounts of capacitance, which is costly and demands space, both undesired features, especially in medium- to high-power supplies.

15.3 Three-phase Voltage Source Inverters

Single-phase VSIs cover low-range power applications and three-phase VSIs cover medium- to high-power applications. The main purpose of these topologies is to provide a three-phase voltage source, where the amplitude, phase, and frequency of the voltages should always be controllable. Although most of the applications require sinusoidal voltage waveforms (e.g. ASDs, UPSs, FACTS, var compensators), arbitrary voltages are also required in some emerging applications (e.g. active filters, voltage compensators).

The standard three-phase VSI topology is shown in Fig. 15.13 and the eight valid switch states are given in Table 15.3. As in single-phase VSIs, the switches of any leg

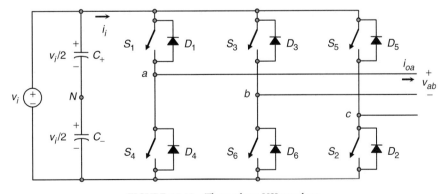

FIGURE 15.13 Three-phase VSI topology.

TABLE 15.3 Valid switch states for a three-phase VSI

State	State #	v_{ab}	v_{bc}	v_{ca}	Space vector
S_1, S_2, and S_6 are on and S_4, S_5, and S_3 are off	1	v_i	0	$-v_i$	$\vec{v}_1 = 1 + j0.577$
S_2, S_3, and S_1 are on and S_5, S_6, and S_4 are off	2	0	v_i	$-v_i$	$\vec{v}_2 = j1.155$
S_3, S_4, and S_2 are on and S_6, S_1, and S_5 are off	3	$-v_i$	v_i	0	$\vec{v}_3 = -1 + j0.577$
S_4, S_5, and S_3 are on and S_1, S_2, and S_6 are off	4	$-v_i$	0	v_i	$\vec{v}_4 = -1 - j0.577$
S_5, S_6, and S_4 are on and S_2, S_3, and S_1 are off	5	0	$-v_i$	v_i	$\vec{v}_5 = -j1.155$
S_6, S_1, and S_5 are on and S_3, S_4, and S_2 are off	6	v_i	$-v_i$	0	$\vec{v}_6 = 1 - j0.577$
S_1, S_3, and S_5 are on and S_4, S_6, and S_2 are off	7	0	0	0	$\vec{v}_7 = 0$
S_4, S_6, and S_2 are on and S_1, S_3, and S_5 are off	8	0	0	0	$\vec{v}_8 = 0$

of the inverter (S_1 and S_4, S_3 and S_6, or S_5 and S_2) cannot be switched on simultaneously because this would result in a short circuit across the dc link voltage supply. Similarly, in order to avoid undefined states in the VSI, and thus undefined ac output line voltages, the switches of any leg of the inverter cannot be switched off simultaneously as this will result in voltages that will depend upon the respective line current polarity.

Of the eight valid states, two of them (7 and 8 in Table 15.3) produce zero ac line voltages. In this case, the ac line currents freewheel through either the upper or lower components. The remaining states (1 to 6 in Table 15.3) produce non-zero ac output voltages. In order to generate a given voltage waveform, the inverter moves from one state to another. Thus the resulting ac output line voltages consist of discrete values of voltages that are v_i, 0, and $-v_i$ for the topology shown in Fig. 15.13. The selection of the states in order to generate the given waveform is done by the modulating technique that should ensure the use of only the valid states.

15.3.1 Sinusoidal PWM

This is an extension of the one introduced for single-phase VSIs. In this case and in order to produce 120° out-of-phase load voltages, three modulating signals that are 120° out-of-phase are used. Figure 15.14 shows the ideal waveforms of three-phase VSI SPWM. In order to use a single carrier signal and preserve the features of the PWM technique, the normalized carrier frequency m_f should be an odd multiple of 3. Thus, all phase voltages (v_{aN}, v_{bN}, and v_{cN}) are identical, but 120° out-of-phase without even harmonics; moreover, harmonics at frequencies, a multiple of 3, are identical in amplitude and phase in all phases. For instance, if the ninth harmonic in phase aN is

$$v_{aN9}(t) = \hat{v}_9 \sin(9\omega t) \quad (15.26)$$

the ninth harmonic in phase bN will be

$$v_{bN9}(t) = \hat{v}_9 \sin\left[9(\omega t - 120°)\right]$$
$$= \hat{v}_9 \sin(9\omega t - 1080°) = \hat{v}_9 \sin(9\omega t) \quad (15.27)$$

Thus, the ac output line voltage $v_{ab} = v_{aN} - v_{bN}$ will not contain the ninth harmonic. Therefore, for odd multiple of 3 values of the normalized carrier frequency m_f, the harmonics in the ac output voltage appear at normalized frequencies f_h centered around m_f and its multiples, specifically, at

$$h = l\, m_f \pm k \quad l = 1, 2, \ldots \quad (15.28)$$

where $l = 1, 3, 5, \ldots$ for $k = 2, 4, 6, \ldots$ and $l = 2, 4, \ldots$ for $k = 1, 5, 7, \ldots$ such that h is not a multiple of 3. Therefore, the harmonics will be at $m_f \pm 2$, $m_f \pm 4, \ldots$, $2m_f \pm 1$, $2m_f \pm 5, \ldots$, $3m_f \pm 2$, $3m_f \pm 4, \ldots$, $4m_f \pm 1$, $4m_f \pm 5, \ldots$. For nearly sinusoidal ac load current, the harmonics in the dc link current are at frequencies given by

$$h = l\, m_f \pm k \pm 1 \quad l = 1, 2, \ldots \quad (15.29)$$

where $l = 0, 2, 4, \ldots$ for $k = 1, 5, 7, \ldots$ and $l = 1, 3, 5, \ldots$ for $k = 2, 4, 6, \ldots$ such that $h = l \cdot m_f \pm k$ is positive and not a multiple of 3. For instance, Fig. 15.14h shows the sixth harmonic ($h = 6$), which is due to $h = 1 \cdot 9 - 2 - 1 = 6$.

The identical conclusions can be drawn for the operation at small and large values of m_f as for the single-phase configurations. However, because the maximum amplitude of the fundamental phase voltage in the linear region ($m_a \leq 1$) is $v_i/2$, the maximum amplitude of the fundamental ac output line voltage is $\sqrt{3}v_i/2$. Therefore, one can write

$$\hat{v}_{ab1} = m_a \sqrt{3}\frac{v_i}{2} \quad 0 < m_a \leq 1 \quad (15.30)$$

To further increase the amplitude of the load voltage, the amplitude of the modulating signal $\hat{v}_c$ can be made higher than the amplitude of the carrier signal $\hat{v}_\Delta$, which leads to overmodulation. The relationship between the amplitude of the fundamental ac output line voltage and the dc link voltage becomes non-linear as in single-phase VSIs. Thus, in the overmodulation region, the line voltages range is

$$\sqrt{3}\frac{v_i}{2} < \hat{v}_{ab1} = \hat{v}_{bc1} = \hat{v}_{ca1} < \frac{4}{\pi}\sqrt{3}\frac{v_i}{2} \quad (15.31)$$

15.3.2 Square-wave Operation of Three-phase VSIs

Large values of m_a in the SPWM technique lead to full overmodulation. This is known as square-wave operation as illustrated in Fig. 15.15, where the power valves are on for 180°.

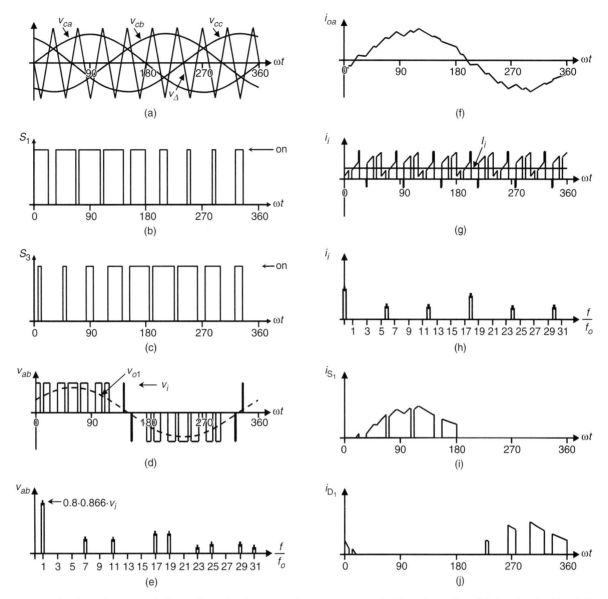

FIGURE 15.14 The three-phase VSI. Ideal waveforms for the SPWM ($m_a = 0.8$, $m_f = 9$): (a) carrier and modulating signals; (b) switch S_1 state; (c) switch S_3 state; (d) ac output voltage; (e) ac output voltage spectrum; (f) ac output current; (g) dc current; (h) dc current spectrum; (i) switch S_1 current; and (j) diode D_1 current.

In this operation mode, the VSI cannot control the load voltage except by means of the dc link voltage v_i. This is based on the fundamental ac line-voltage expression

$$\hat{v}_{ab\,1} = \frac{4}{\pi}\sqrt{3}\frac{v_i}{2} \qquad (15.32)$$

The ac line output voltage contains the harmonics f_h, where $h = 6 \cdot k \pm 1$ ($k = 1, 2, 3, \ldots$) and they feature amplitudes that are inversely proportional to their harmonic order (Fig. 15.15d). Their amplitudes are

$$\hat{v}_{ab\,h} = \frac{1}{h}\frac{4}{\pi}\sqrt{3}\frac{v_i}{2} \qquad (15.33)$$

15.3.3 Sinusoidal PWM with Zero Sequence Signal Injection

The restriction for m_a ($m_a \leq 1$) can be relaxed if a zero sequence signal is added to the modulating signals before they are compared to the carrier signal. Figure 15.16 shows the block diagram of the technique. Clearly, the addition of

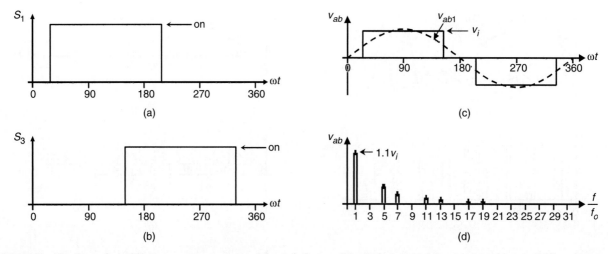

FIGURE 15.15 The three-phase VSI. Square-wave operation: (a) switch S_1 state; (b) switch S_3 state; (c) ac output voltage; and (d) ac output voltage spectrum.

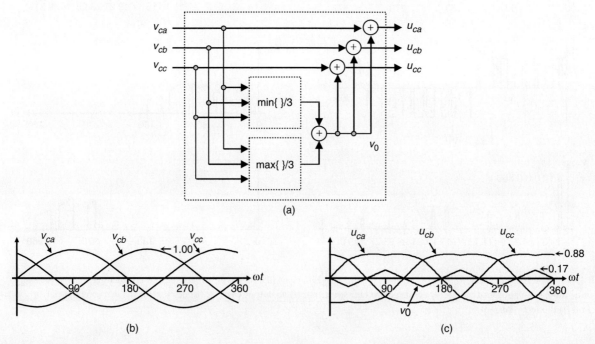

FIGURE 15.16 Zero sequence signal generator ($m_a = 1.0$, $m_f = 9$): (a) block diagram; (b) modulating signals; and (c) zero sequence and modulating signals with zero sequence injection.

the zero sequence reduces the peak amplitude of the resulting modulating signals (u_{ca}, u_{cb}, u_{cc}), while the fundamental components remain unchanged. This approach expands the range of the linear region as it allows the use of modulation indexes m_a up to $2/\sqrt{3}$ without getting into the overmodulating region.

The maximum amplitude of the fundamental phase voltage in the linear region ($m_a \leq 2/\sqrt{3}$) is $v_i/2$, thus, the maximum amplitude of the fundamental ac output line voltage is v_i. Therefore, one can write

$$\hat{v}_{ab1} = m_a \sqrt{3} \frac{v_i}{2} \quad \left(0 < m_a \leq 2/\sqrt{3}\right) \quad (15.34)$$

Figure 15.17 shows the ideal waveforms of a three-phase VSI SPWM with zero injection for $m_a = 0.8$.

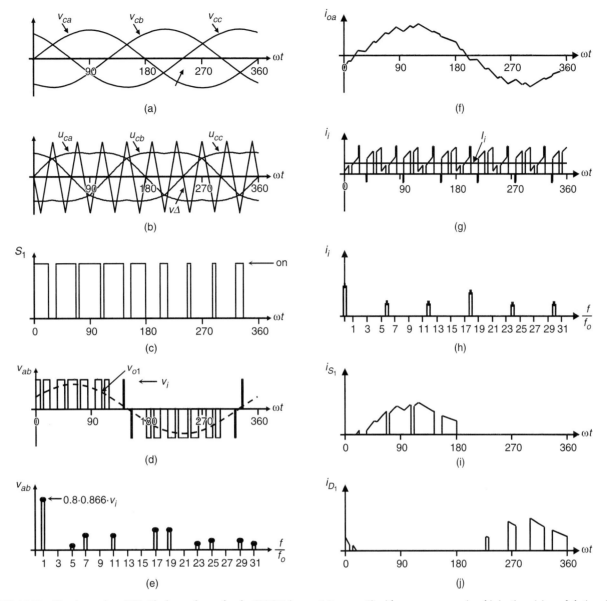

FIGURE 15.17 The three-phase VSI. Ideal waveforms for the SPWM ($m_a = 0.8$, $m_f = 9$) with zero sequence signal injection: (a) modulating signals; (b) carrier and modulating signals with zero sequence signal injection; (c) switch S_1 state; (d) ac output voltage; (e) ac output voltage spectrum; (f) ac output current; (g) dc current; (h) dc current spectrum; (i) switch S_1 current; and (j) diode D_1 current.

15.3.4 Selective Harmonic Elimination in Three-phase VSIs

As in single-phase VSIs, the SHE technique can be applied to three-phase VSIs. In this case, the power valves of each leg of the inverter are switched so as to eliminate a given number of harmonics and to control the fundamental phase-voltage amplitude. Considering that in many applications, the required line output voltages should be balanced and 120° out of phase, the harmonics multiples of 3 ($h = 3, 9, 15, \ldots$), which could be present in the phase voltages (v_{aN}, v_{bN}, and v_{cN}), will not be present in the load voltages (v_{ab}, v_{bc}, and v_{ca}). Therefore, these harmonics are not required to be eliminated, thus the chopping angles are used to eliminate only the harmonics at frequencies $h = 5, 7, 11, 13, \ldots$ as required.

The expressions to eliminate a given number of harmonics are the same as those used in single-phase inverters. For instance, to eliminate the fifth and seventh harmonics and perform fundamental magnitude control ($N = 3$), the equations

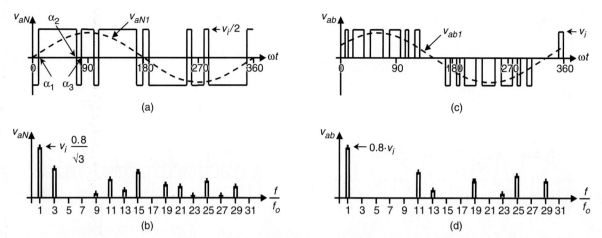

FIGURE 15.18 The three-phase VSI. Ideal waveforms for the SHE technique: (a) phase voltage v_{aN} for fifth and seventh harmonic elimination; (b) spectrum of (a); (c) line voltage v_{ab} for fifth and seventh harmonic elimination; and (d) spectrum of (c).

to be solved are:

$$\cos(1\alpha_1) - \cos(1\alpha_2) + \cos(1\alpha_3) = (2 + \pi \hat{v}_{aN1}/v_i)/4$$
$$\cos(5\alpha_1) - \cos(5\alpha_2) + \cos(5\alpha_3) = 1/2$$
$$\cos(7\alpha_1) - \cos(7\alpha_2) + \cos(7\alpha_3) = 1/2$$
(15.35)

where the angles α_1, α_2, and α_3 are defined as shown in Fig. 15.18a and plotted in Fig. 15.19. Figure 15.18b shows that the third, ninth, fifteenth, ... harmonics are all present in the phase voltages; however, they are not in the line voltages (Fig. 15.18d).

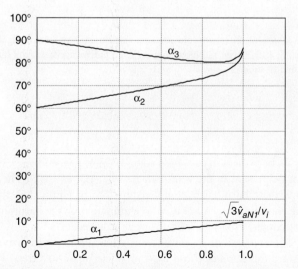

FIGURE 15.19 Chopping angles for SHE and fundamental voltage control in three-phase VSIs: fifth and seventh harmonic elimination.

15.3.5 Space-vector (SV)-based Modulating Techniques

At present, the control strategies are implemented in digital systems, and therefore digital modulating techniques are also available. The SV-based modulating technique is a digital technique in which the objective is to generate PWM load line voltages that are on average equal to given load line voltages. This is done in each sampling period by properly selecting the switch states from the valid ones of the VSI (Table 15.3) and by proper calculation of the period of times they are used. The selection and calculation times are based upon the SV transformation.

A. Space-vector Transformation

Any three-phase set of variables that add up to zero in the stationary abc frame can be represented in a complex plane by a complex vector that contains a real (α) and an imaginary (β) component. For instance, the vector of three-phase line-modulating signals $\mathbf{v_c^{abc}} = [v_{ca} v_{cb} v_{cc}]^T$ can be represented by the complex vector $\vec{\mathbf{v}}_\mathbf{c} = \mathbf{v_c^{\alpha\beta}} = [v_{c\alpha} v_{c\beta}]^T$ by means of the following transformation:

$$v_{c\alpha} = \frac{2}{3}[v_{ca} - 0.5(v_{cb} + v_{cc})] \quad (15.36)$$

$$v_{c\beta} = \frac{\sqrt{3}}{3}(v_{cb} - v_{cc}) \quad (15.37)$$

If the line-modulating signals $\mathbf{v_c^{abc}}$ are three balanced sinusoidal waveforms that feature an amplitude $\hat{v}_c$ and an angular frequency ω, the resulting modulating signals in the $\alpha\beta$ stationary frame become a vector $\vec{\mathbf{v}}_\mathbf{c} = \mathbf{v_c^{\alpha\beta}}$ of fixed module $\hat{v}_c$, which rotates at frequency ω (Fig. 15.20). Similarly, the SV transformation is applied to the line voltages of the eight states of the

15 Inverters

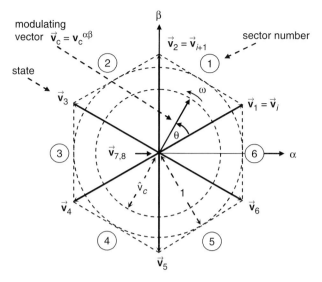

FIGURE 15.20 The space-vector representation.

VSI normalized with respect to v_i (Table 15.3), which generates the eight space vectors ($\vec{v}_i$, $i = 1, 2, \ldots, 8$) in Fig. 15.20. As expected, $\vec{v}_1$ to $\vec{v}_6$ are non-null line-voltage vectors and $\vec{v}_7$ and $\vec{v}_8$ are null line-voltage vectors.

The objective of the SV technique is to approximate the line-modulating signal space vector $\vec{v}_c$ with the eight space vectors ($\vec{v}_i$, $i = 1, 2, \ldots, 8$) available in VSIs. However, if the modulating signal $\vec{v}_c$ is laying between the arbitrary vectors $\vec{v}_i$ and $\vec{v}_{i+1}$, only the nearest two non-zero vectors ($\vec{v}_i$ and $\vec{v}_{i+1}$) and one zero SV ($\vec{v}_z = \vec{v}_7$ or $\vec{v}_8$) should be used. Thus, the maximum load line voltage is maximized and the switching frequency is minimized. To ensure that the generated voltage in one sampling period T_s (made up of the voltages provided by the vectors $\vec{v}_i$, $\vec{v}_{i+1}$, and $\vec{v}_z$ used during times T_i, T_{i+1}, and T_z) is on average equal to the vector $\vec{v}_c$ the following expression should hold:

$$\vec{v}_c \cdot T_s = \vec{v}_i \cdot T_s + \vec{v}_{i+1} \cdot T_{i+1} + \vec{v}_z \cdot T_z \quad (15.38)$$

The solution of the real and imaginary parts of Eq. (15.37) for a line-load voltage that features an amplitude restricted to $0 \le \hat{v}_c \le 1$ gives

$$T_i = T_s \cdot \hat{v}_c \cdot \sin(\pi/3 - \theta) \quad (15.39)$$

$$T_{i+1} = T_s \cdot \hat{v}_c \cdot \sin(\theta) \quad (15.40)$$

$$T_z = T_s - T_i - T_{i+1} \quad (15.41)$$

The preceding expressions indicate that the maximum fundamental line-voltage amplitude is unity as $0 \le \theta \le \pi/3$. This is an advantage over the SPWM technique which achieves a $\sqrt{3}/2$ maximum fundamental line-voltage amplitude in the linear operating region. Although, the space vector modulation (SVM) technique selects the vectors to be used and their respective on-times, the sequence in which they are used, the selection of the zero space vector, and the normalized sampled frequency remain undetermined.

For instance, if the modulating line-voltage vector is in sector 1 (Fig. 15.20), the vectors $\vec{v}_1$, $\vec{v}_2$, and $\vec{v}_z$ should be used within a sampling period by intervals given by T_1, T_2, and T_z, respectively. The question that remains is whether the sequence (i) $\vec{v}_1 - \vec{v}_2 - \vec{v}_z$, (ii) $\vec{v}_z - \vec{v}_1 - \vec{v}_2 - \vec{v}_z$, (iii) $\vec{v}_z - \vec{v}_1 - \vec{v}_2 - \vec{v}_1 - \vec{v}_z$, (iv) $\vec{v}_z - \vec{v}_1 - \vec{v}_2 - \vec{v}_z - \vec{v}_2 - \vec{v}_1 - \vec{v}_z$, or any other sequence should actually be used. Finally, the technique does not indicate whether $\vec{v}_z$ should be $\vec{v}_7$, $\vec{v}_8$, or a combination of both.

B. Space-vector Sequences and Zero Space-vector Selection

The sequence to be used should ensure load line-voltages that feature quarter-wave symmetry in order to reduce unwanted harmonics in their spectra (even harmonics). Additionally, the zero SV selection should be done in order to reduce the switching frequency. Although there is not a systematic approach to generate a SV sequence, a graphical representation shows that the sequence $\vec{v}_i$, $\vec{v}_{i+1}$, $\vec{v}_z$ (where $\vec{v}_z$ is alternately chosen among $\vec{v}_7$ and $\vec{v}_8$) provides high performance in terms of minimizing unwanted harmonics and reducing the switching frequency.

C. The Normalized Sampling Frequency

The normalized carrier frequency m_f in three-phase carrier-based PWM techniques is chosen to be an odd integer number multiple of 3 ($m_f = 3 \cdot n$, $n = 1, 3, 5, \ldots$). Thus, it is possible to minimize parasitic or non-intrinsic harmonics in the PWM waveforms. A similar approach can be used in the SVM technique to minimize uncharacteristic harmonics. Hence, it is found that the normalized sampling frequency f_{sn} should be an integer multiple of 6. This is due to the fact that in order to produce symmetrical line voltages, all the sectors (a total of 6) should be used equally in one period. As an example, Fig. 15.21 shows the relevant waveforms of a VSI SVM for $f_{sn} = 18$ and $\hat{v}_c = 0.8$. Figure 15.21 confirms that the first set of relevant harmonics in the load line voltage are at f_{sn} which is also the switching frequency.

15.3.6 DC Link Current in Three-phase VSIs

Due to the fact that the inverter is assumed to be lossless and constructed without storage energy components, the instantaneous power balance indicates that

$$v_i(t) \cdot i_i(t) = v_{ab}(t) \cdot i_a(t) + v_{bc}(t) \cdot i_b(t) + v_{ca}(t) \cdot i_c(t) \quad (15.42)$$

where $i_a(t)$, $i_b(t)$, and $i_c(t)$ are the phase-load currents as shown in Fig. 15.22. If the load is balanced and inductive, and a relatively high switching frequency is used, the load currents become nearly sinusoidal balanced waveforms. On the other

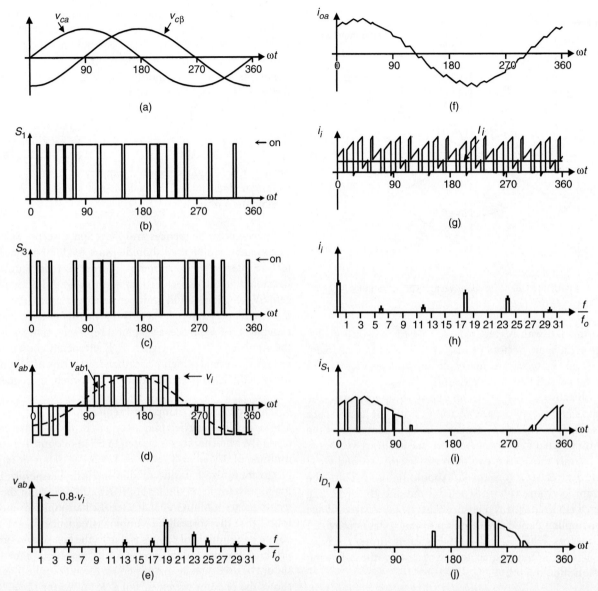

FIGURE 15.21 The three-phase VSI. Ideal waveforms for space-vector modulation ($\hat{v}_c = 0.8$, $f_{sn} = 18$): (a) modulating signals; (b) switch S_1 state; (c) switch S_3 state; (d) ac output voltage; (e) ac output voltage spectrum; (f) ac output current; (g) dc current; (h) dc current spectrum; (i) switch S_1 current; and (j) diode D_1 current.

hand, if the ac output voltages are considered sinusoidal and the dc link voltage is assumed constant $v_i(t) = V_i$, Eq. (15.42) can be simplified to

$$i_i(t) = \frac{1}{V_i} \left\{ \begin{array}{l} \sqrt{2}V_{o1}\sin(\omega t) \cdot \sqrt{2}I_o\sin(\omega t - \phi) \\ +\sqrt{2}V_{o1}\sin(\omega t - 120°) \cdot \sqrt{2}I_o\sin(\omega t - 120° - \phi) \\ +\sqrt{2}V_{o1}\sin(\omega t - 240°) \cdot \sqrt{2}I_o\sin(\omega t - 240° - \phi) \end{array} \right\}$$
(15.43)

where V_{o1} is the fundamental rms ac output line voltage, I_o is the rms load-phase current, and ϕ is an arbitrary inductive load power factor. Hence, the dc link current expression can be further simplified to

$$i_i(t) = 3\frac{V_{o1}}{V_i}I_o\cos(\phi) = \sqrt{3}\frac{V_{o1}}{V_i}I_l\cos(\phi) \quad (15.44)$$

where $I_l = \sqrt{3}I_o$ is the rms load line current. The resulting dc link current expression indicates that under harmonic-free load voltages, only a clean dc current should be expected in the dc bus and, compared to single-phase VSIs, there is no presence of second harmonic. However, as the ac load line voltages contain harmonics around the normalized sampling

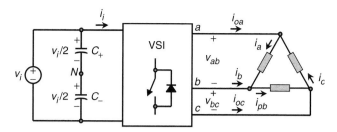

FIGURE 15.22 Phase-load currents definition in a delta-connected load.

frequency f_{sn}, the dc link current will contain harmonics but around f_{sn} as shown in Fig. 15.21h.

15.3.7 Load-phase Voltages in Three-phase VSIs

The load is sometimes wye-connected and the phase-load voltages v_{an}, v_{bn}, and v_{cn} may be required (Fig. 15.23). To obtain them, it should be considered that the line-voltage vector is

$$\begin{bmatrix} v_{ab} \\ v_{bc} \\ v_{ca} \end{bmatrix} = \begin{bmatrix} v_{an} - v_{bn} \\ v_{bn} - v_{cn} \\ v_{cn} - v_{an} \end{bmatrix} \quad (15.45)$$

which can be written as a function of the phase-voltage vector $[v_{an} v_{bn} v_{cn}]^T$ as

$$\begin{bmatrix} v_{ab} \\ v_{bc} \\ v_{ca} \end{bmatrix} = \begin{bmatrix} 1 & -1 & 0 \\ 0 & 1 & -1 \\ -1 & 0 & 1 \end{bmatrix} \begin{bmatrix} v_{an} \\ v_{bn} \\ v_{cn} \end{bmatrix} \quad (15.46)$$

Expression (15.46) represents a linear system where the unknown quantity is the vector $[v_{an} v_{bn} v_{cn}]^T$. Unfortunately,

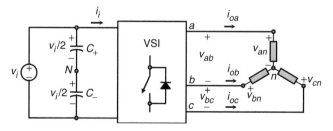

FIGURE 15.23 Phase-load voltages definition in a wye-connected load.

the system is singular as the rows add up to zero (line voltages add up to zero), therefore, the phase-load voltages cannot be obtained by matrix inversion. However, if the phase-load voltages add up to zero, Eq. (15.46) can be rewritten as

$$\begin{bmatrix} v_{ab} \\ v_{bc} \\ 0 \end{bmatrix} = \begin{bmatrix} 1 & -1 & 0 \\ 0 & 1 & -1 \\ 1 & 1 & 1 \end{bmatrix} \begin{bmatrix} v_{an} \\ v_{bn} \\ v_{cn} \end{bmatrix} \quad (15.47)$$

which is not singular and hence,

$$\begin{bmatrix} v_{an} \\ v_{bn} \\ v_{cn} \end{bmatrix} = \begin{bmatrix} 1 & -1 & 0 \\ 0 & 1 & -1 \\ 1 & 1 & 1 \end{bmatrix}^{-1} \begin{bmatrix} v_{ab} \\ v_{bc} \\ 0 \end{bmatrix} = \frac{1}{3} \begin{bmatrix} 2 & 1 & 1 \\ -1 & 1 & 1 \\ -1 & -2 & 1 \end{bmatrix} \begin{bmatrix} v_{ab} \\ v_{bc} \\ 0 \end{bmatrix} \quad (15.48)$$

that can be further simplified to

$$\begin{bmatrix} v_{an} \\ v_{bn} \\ v_{cn} \end{bmatrix} = \frac{1}{3} \begin{bmatrix} 2 & 1 \\ -1 & 1 \\ -1 & -2 \end{bmatrix} \begin{bmatrix} v_{ab} \\ v_{bc} \end{bmatrix} \quad (15.49)$$

The final expression for the phase-load voltages is only a function of v_{ab} and v_{bc}, which is due to fact that the last row in Eq. (15.46) is chosen to be only ones. Figure 15.24 shows the line- and phase-voltages obtained using Eq. (15.49).

15.4 Current Source Inverters

The main objective of these static power converters is to produce an ac output current waveforms from a dc current power supply. For sinusoidal ac outputs, its magnitude, frequency, and phase should be controllable. Due to the fact that the ac line currents i_{oa}, i_{ob}, and i_{oc} (Fig. 15.25) feature high di/dt, a capacitive filter should be connected at the ac terminals in inductive load applications (such as ASDs). Thus, nearly sinusoidal load voltages are generated that justifies the use of these topologies in medium-voltage industrial applications, where high-quality voltage waveforms are required. Although single-phase CSIs can in the same way as three-phase CSIs topologies, be developed under similar principles, only three-phase applications are of practical use and are analyzed below.

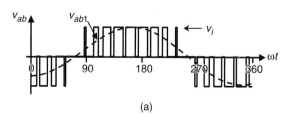
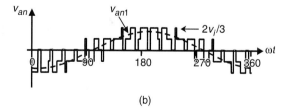

FIGURE 15.24 The three-phase VSI. Line- and phase-load voltages: (a) line-load voltage v_{ab}; and (b) phase-load voltage v_{an}.

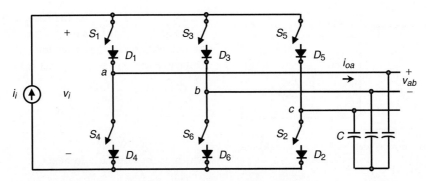

FIGURE 15.25 Three-phase CSI topology.

In order to properly gate the power switches of a three-phase CSI, two main constraints must always be met: (a) the ac side is mainly capacitive, thus, it must not be short-circuited; this implies that, at most one top switch (1, 3, or 5 (Fig. 15.25)) and one bottom switch (4, 6, or 2 (Fig. 15.25)) should be closed at any time; and (b) the dc bus is of the current-source type and thus it cannot be opened; therefore, there must be at least one top switch (1, 3, or 5) and one bottom switch (4, 6, or 2) closed at all times. Note that both constraints can be summarized by stating that at any time, only one top switch and one bottom switch must be closed.

There are nine valid states in three-phase CSIs. The states 7, 8, and 9 (Table 15.4) produce zero ac line currents. In this case, the dc link current freewheels through either the switches S_1 and S_4, switches S_3 and S_6, or switches S_5 and S_2. The remaining states (1 to 6 in Table 15.4) produce non-zero ac output line currents. In order to generate a given set of ac line current waveforms, the inverter must move from one state to another. Thus, the resulting line currents consist of discrete values of current, which are i_i, 0, and $-i_i$. The selection of the states in order to generate the given waveforms is done by the modulating technique that should ensure the use of only the valid states.

There are several modulating techniques that deal with the special requirements of CSIs and can be implemented online. These techniques are classified into three categories: (a) the carrier-based; (b) the SHE-based; and (c) the SV-based techniques. Although they are different, they generate gating signals that satisfy the special requirements of CSIs. To simplify the analysis, a constant dc link-current source is considered ($i_i = I_i$).

15.4.1 Carrier-based PWM Techniques in CSIs

It has been shown that the carrier-based PWM techniques that were initially developed for three-phase VSIs can be extended to three-phase CSIs. The circuit shown in Fig. 15.26 obtains the gating pattern for a CSI from the gating pattern developed for a VSI. As a result, the line current appears to be identical to the line voltage in a VSI for similar carrier and modulating signals.

It is composed of a *switching pulse generator*, a *shorting pulse generator*, a *shorting pulse distributor*, and a *switching and shorting pulse combinator*. The circuit basically produces the gating signals ($\mathbf{s} = [s_1 \ldots s_6]^T$) according to a carrier i_Δ and three modulating signals $\mathbf{i}_c^{abc} = [i_{ca}\ i_{cb}\ i_{ca}]^T$. Therefore, any set of modulating signals which when combined result in a sinusoidal line-to-line set of signals, will satisfy the requirement for a sinusoidal line current pattern. Examples of such a modulating signals are the standard sinusoidal, sinusoidal with third harmonic injection, trapezoidal, and deadband waveforms.

The first component of this stage (Fig. 15.26) is the *switching pulse generator*, where the signals s_a^{123} are generated according to:

$$s_a^{123} = \begin{cases} \text{HIGH} = 1 & \text{if } \mathbf{i}_c^{abc} > v_c \\ \text{LOW} = 0 & \text{otherwise} \end{cases} \tag{15.50}$$

TABLE 15.4 Valid switch states for a three-phase CSI

State	State #	i_{oa}	i_{ob}	i_{oc}	Space vector
S_1 and S_2 are on and S_3, S_4, S_5, and S_6 are off	1	i_i	0	$-i_i$	$\vec{i}_1 = 1 + j0.577$
S_2 and S_3 are on and S_4, S_5, S_6, and S_1 are off	2	0	i_i	$-i_i$	$\vec{i}_2 = j1.155$
S_3 and S_4 are on and S_5, S_6, S_1, and S_2 are off	3	$-i_i$	i_i	0	$\vec{i}_3 = -1 + j0.577$
S_4 and S_5 are on and S_6, S_1, S_2, and S_3 are off	4	$-i_i$	0	i_i	$\vec{i}_4 = -1 - j0.577$
S_5 and S_6 are on and S_1, S_2, S_3, and S_4 are off	5	0	$-i_i$	i_i	$\vec{i}_5 = -j1.155$
S_6 and S_1 are on and S_2, S_3, S_4, and S_5 are off	6	i_i	$-i_i$	0	$\vec{i}_6 = 1 - j0.577$
S_1 and S_4 are on and S_2, S_3, S_5, and S_6 are off	7	0	0	0	$\vec{i}_7 = 0$
S_3 and S_6 are on and S_1, S_2, S_4, and S_5 are off	8	0	0	0	$\vec{i}_8 = 0$
S_5 and S_2 are on and S_6, S_1, S_3, and S_4 are off	9	0	0	0	$\vec{i}_9 = 0$

15 Inverters

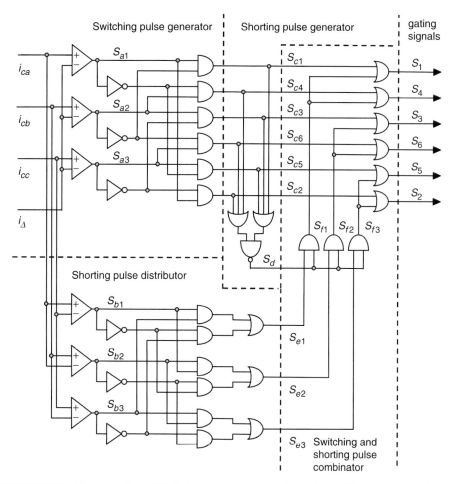

FIGURE 15.26 The three-phase CSI. Gating pattern generator for analog on-line carrier-based PWM.

The outputs of the *switching pulse generator* are the signals s_c, which are basically the gating signals of the CSI without the shorting pulses. These are necessary to freewheel the dc link current i_i when zero ac output currents are required. Table 15.5 shows the truth table of s_c for all combinations of their inputs s_a^{123}. It can be clearly seen that at most one top switch and one bottom switch is on, which satisfies the first constraint of the gating signals as stated before.

In order to satisfy the second constraint, the shorting pulse ($s_d = 1$) ($s_d = 1$) is generated (*shorting pulse generator* (Fig. 15.26)) the top switches ($s_{c1} = s_{c3} = s_{c5} = 0$) or none of the bottom switches ($s_{c4} = s_{c6} = s_{c2} = 0$) are gated. Then, this pulse is added (using OR gates) to only one leg of the CSI (either to the switches 1 and 4, 3 and 6, or 5 and 2) by means of the *switching and shorting pulse combinator* (Fig. 15.26). The signals generated by the *shorting pulse generator* s_e^{123} ensure that: (a) only one leg of the CSI is shorted, as only one of the signals is HIGH at any time; and (b) there is an even distribution of the shorting pulse, as s_e^{123} is high for 120° in each period. This ensures that the rms currents are equal in all legs.

TABLE 15.5 Truth table for the switching pulse generator stage (Fig. 15.26)

s_{a1}	s_{a2}	s_{a3}	Top switches			Bottom switches		
			s_{c1}	s_{c3}	s_{c5}	s_{c4}	s_{c6}	s_{c2}
0	0	0	0	0	0	0	0	0
0	0	1	0	0	1	0	1	0
0	1	0	0	1	0	1	0	0
0	1	1	0	0	1	1	0	0
1	0	0	1	0	0	0	0	1
1	0	1	1	0	0	0	1	0
1	1	0	0	1	0	0	0	1
1	1	1	0	0	0	0	0	0

Figure 15.27 shows the relevant waveforms if a triangular carrier i_Δ and sinusoidal modulating signals i_c^{abc} are used in combination with the gating pattern generator circuit (Fig. 15.26); this is SPWM in CSIs. It can be observed that some of the waveforms (Fig. 15.27) are identical to those

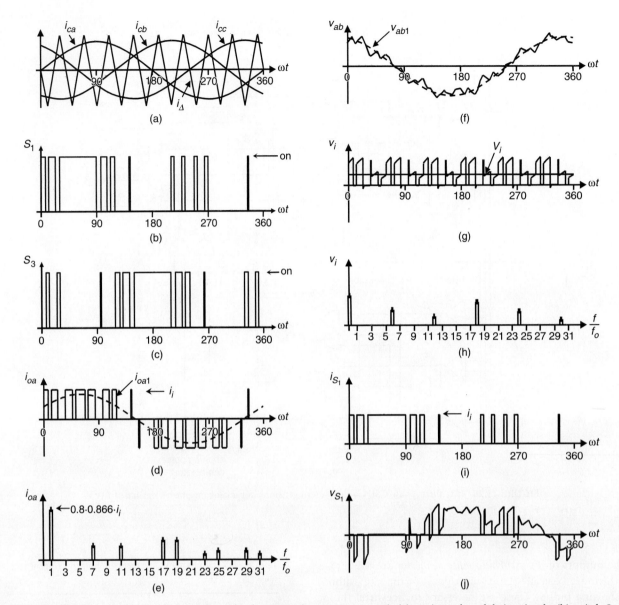

FIGURE 15.27 The three-phase CSI. Ideal waveforms for the SPWM ($m_a = 0.8$, $m_f = 9$): (a) carrier and modulating signals; (b) switch S_1 state; (c) switch S_3 state; (d) ac output current; (e) ac output current spectrum; (f) ac output voltage; (g) dc voltage; (h) dc voltage spectrum; (i) switch S_1 current; and (j) switch S_1 voltage.

obtained in three-phase VSIs, where a SPWM technique is used (Fig. 15.15). Specifically: (i) the load line voltage (Fig. 15.15d) in the VSI is identical to the load line current (Fig. 15.27d) in the CSI; and (ii) the dc link current (Fig. 15.15g) in the VSI is identical to the dc link voltage (Fig. 15.27g) in the CSI.

This brings up the duality issue between both the topologies when similar modulation approaches are used. Therefore, for odd multiples of 3 values of the normalized carrier frequency m_f, the harmonics in the ac output current appear at normalized frequencies f_h centered around m_f and its multiples, specifically, at

$$h = l\, m_f \pm k \quad l = 1, 2, \ldots \tag{15.51}$$

where $l = 1, 3, 5, \ldots$ for $k = 2, 4, 6, \ldots$ and $l = 2, 4, \ldots$ for $k = 1, 5, 7, \ldots$ such that h is not a multiple of 3. Therefore, the harmonics will be at $m_f \pm 2$, $m_f \pm 4, \ldots$, $2m_f \pm 1$, $2m_f \pm 5, \ldots$, $3m_f \pm 2$, $3m_f \pm 4, \ldots$, $4m_f \pm 1$, $4m_f \pm 5, \ldots$. For nearly sinusoidal ac load voltages, the harmonics in the dc link voltage are at frequencies given by

$$h = l\, m_f \pm k \pm 1 \quad l = 1, 2, \ldots \tag{15.52}$$

where $l = 0, 2, 4, \ldots$ for $k = 1, 5, 7, \ldots$ and $l = 1, 3, 5, \ldots$ for $k = 2, 4, 6, \ldots$ such that $h = l \cdot m_f \pm k$ is positive and not a multiple of 3. For instance, Fig. 15.27h shows the sixth harmonic ($h = 6$), which is due to $h = 1 \cdot 9 - 2 - 1 = 6$. Identical conclusions can be drawn for the small and large values of m_f in the same way as for three-phase VSI configurations. Thus, the maximum amplitude of the fundamental ac output line current is $\hat{i}_{oa1} = \sqrt{3} i_i/2$ and therefore one can write

$$\hat{i}_{oa1} = m_a \frac{\sqrt{3}}{2} i_i \quad 0 < m_a \leq 1 \quad (15.53)$$

To further increase the amplitude of the load current, the overmodulation approach can be used. In this region, the fundamental line currents range in

$$\frac{\sqrt{3}}{2} i_i < \hat{i}_{oa1} = \hat{i}_{ob1} = \hat{i}_{oc1} < \frac{4}{\pi} \frac{\sqrt{3}}{2} i_i \quad (15.54)$$

To further test the gating signal generator circuit (Fig. 15.26), a sinusoidal set with third and ninth harmonic injection modulating signals are used. Figure 15.28 shows the relevant waveforms.

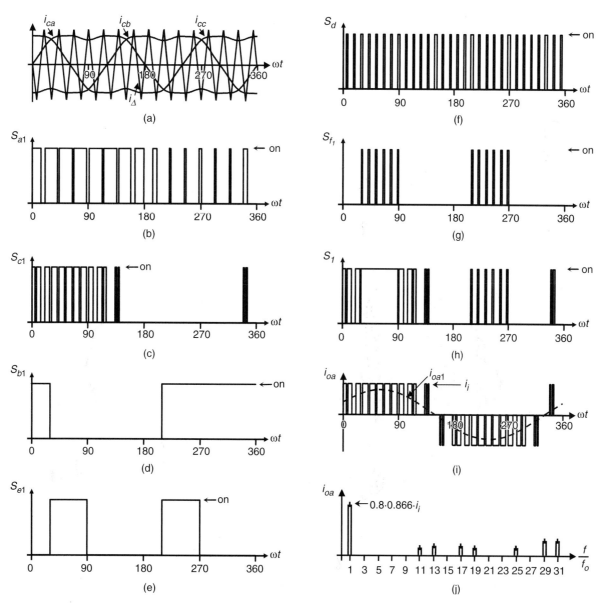

FIGURE 15.28 Gating pattern generator. Waveforms for third and ninth harmonic injection PWM ($m_a = 0.8$, $m_f = 15$): signals as described in Fig. 15.26.

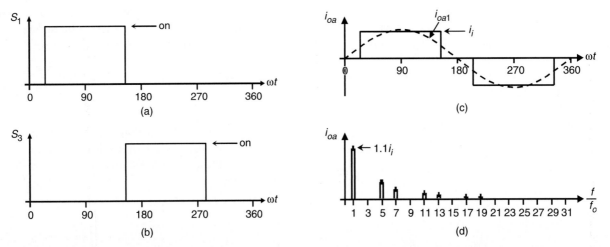

FIGURE 15.29 The three-phase CSI. Square-wave operation: (a) switch S_1 state; (b) switch S_3 state; (c) ac output current; and (d) ac output current spectrum.

15.4.2 Square-wave Operation of Three-phase CSIs

As in VSIs, large values of m_a in the SPWM technique lead to full overmodulation. This is known as square-wave operation. Figure 15.29 depicts this operating mode in a three-phase CSI, where the power valves are on for 120°. As presumed, the CSI cannot control the load current except by means of the dc link current i_i. This is due to the fact that the fundamental ac line current expression is

$$\hat{i}_{oa1} = \frac{4}{\pi} \frac{\sqrt{3}}{2} i_i \quad (15.55)$$

The ac line current contains the harmonics f_h, where $h = 6 \cdot k \pm 1$ ($k = 1, 2, 3, \ldots$), and they feature amplitudes that is inversely proportional to their harmonic order (Fig. 15.29d). Thus,

$$\hat{i}_{oah} = \frac{1}{h} \frac{4}{\pi} \frac{\sqrt{3}}{2} i_i \quad (15.56)$$

The duality issue among both the three-phase VSI and CSI should be noted especially in terms of the line-load waveforms. The line-load voltage produced by a VSI is identical to the load line current produced by the CSI when both are modulated using identical techniques. The next section will show that this also holds for SHE-based techniques.

15.4.3 Selective Harmonic Elimination in Three-phase CSIs

The SHE-based modulating techniques in VSIs define the gating signals such that a given number of harmonics are eliminated and the fundamental phase-voltage amplitude is controlled. If the required line output voltages are balanced and 120° out-of-phase, the chopping angles are used to eliminate only the harmonics at frequencies $h = 5, 7, 11, 13, \ldots$ as required.

The circuit shown in Fig. 15.30 uses the gating signals s_a^{123} developed for a VSI and a set of synchronizing signals $\mathbf{i}_c^{abc}$ to obtain the gating signals $\mathbf{s}$ for a CSI. The synchronizing signals $\mathbf{i}_c^{abc}$ are sinusoidal balanced waveforms that are synchronized with the signals s_a^{123} in order to symmetrically distribute the shorting pulse and thus generate symmetrical gating patterns. The circuit ensures line current waveforms as the line voltages in a VSI. Therefore, any arbitrary number of harmonics can be eliminated and the fundamental line current can be controlled in CSIs. Moreover, the same chopping angles obtained for VSIs can be used in CSIs.

For instance, to eliminate the fifth and seventh harmonics, the chopping angles are shown in Fig. 15.31, which are identical to that obtained for a VSI using Eq. (15.9). Figure 15.32 shows that the line current does not contain the fifth and the seventh harmonics as expected. Hence, any number of harmonics can be eliminated in three-phase CSIs by means of the circuit (Fig. 15.30) without the hassle of how to satisfy the gating signal constrains.

15.4.4 Space-vector-based Modulating Techniques in CSIs

The objective of the SV-based modulating technique is to generate PWM load line currents that are on average equal to given load line currents. This is done digitally in each sampling period by properly selecting the switch states from the valid ones of the CSI (Table 15.4) and the proper calculation of the period of times they are used. As in VSIs, the selection and time calculations are based upon the space-vector transformation.

15 Inverters

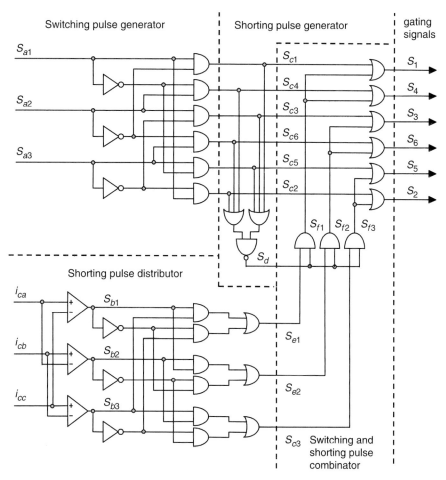

FIGURE 15.30 The three-phase CSI. Gating pattern generator for SHE PWM techniques.

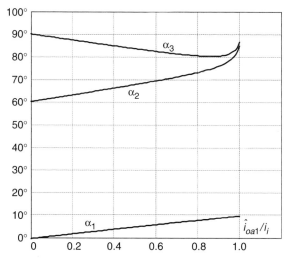

FIGURE 15.31 Chopping angles for SHE and fundamental current control in three-phase CSIs: fifth and seventh harmonic elimination.

A. Space-vector Transformation in CSIs

Similarly to VSIs, the vector of three-phase line-modulating signals $\vec{i}_c^{abc} = [i_{ca}\, i_{cb}\, i_{cc}]^T$ can be represented by the complex vector $\vec{i}_c = \vec{i}_c^{\alpha\beta} = [i_{c\alpha}\, i_{c\beta}]^T$ by means of Eqs. (15.36) and (15.37). For three-phase balanced sinusoidal modulating waveforms, which feature an amplitude $\hat{i}_c$ and an angular frequency ω, the resulting modulating signals complex vector $\vec{i}_c = \vec{i}_c^{\alpha\beta}$ becomes a vector of fixed module $\hat{i}_c$, which rotates at frequency ω (Fig. 15.33). Similarly, the SV transformation is applied to the line currents of the nine states of the CSI normalized with respect to i_i, which generates nine space vectors ($\vec{i}_i$, $i = 1, 2, \ldots, 9$ in Fig. 15.33). As expected, $\vec{i}_1$ to $\vec{i}_6$ are non-null line current vectors and $\vec{i}_7$, $\vec{i}_8$, and $\vec{i}_9$ are null line current vectors.

The SV technique approximates the line-modulating signal space vector $\vec{i}_c$ by using the nine space vectors ($\vec{i}_i$, $i = 1, 2, \ldots, 9$) available in CSIs. If the modulating signal vector $\vec{i}_c$ is between the arbitrary vectors $\vec{i}_i$ and $\vec{i}_{i+1}$, then $\vec{i}_i$ and $\vec{i}_{i+1}$ combined with one zero SV ($\vec{i}_z = \vec{i}_7$ or $\vec{i}_8$ or $\vec{i}_9$) should be used to generate $\vec{i}_c$. To ensure that the generated current in

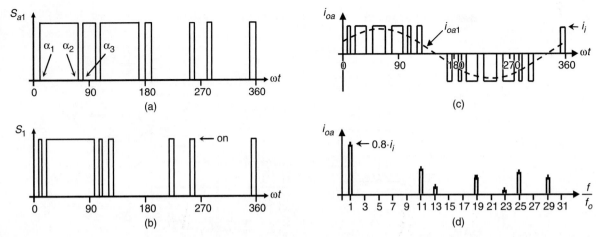

FIGURE 15.32 The three-phase CSI. Ideal waveforms for the SHE technique: (a) VSI gating pattern for fifth and seventh harmonic elimination; (b) CSI gating pattern for fifth and seventh harmonic elimination; (c) line current i_{oa} for fifth and seventh harmonic elimination; and (d) spectrum of (c).

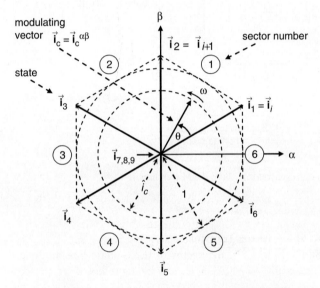

FIGURE 15.33 The space-vector representation in CSIs.

B. Space-vector Sequences and Zero Space-vector Selection

Although there is no systematic approach to generate a SV sequence, a graphical representation shows that the sequence $\vec{i}_i$, $\vec{i}_{i+1}$, $\vec{i}_z$ (where the chosen $\vec{i}_z$ depends upon the sector) provides high performance in terms of minimizing unwanted harmonics and reducing the switching frequency. To obtain the zero SV that minimizes the switching frequency, it is assumed that $\mathbf{I_c}$ is in Sector ②. Then Fig. 15.34 shows all

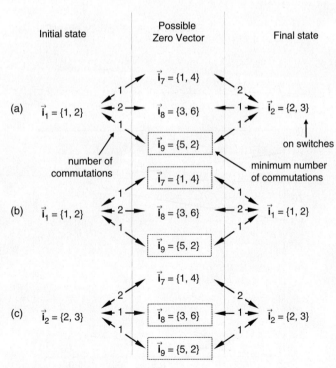

FIGURE 15.34 Possible state transitions in Sector ② involving a zero SV: (a) transition: $\vec{i}_1 \Leftrightarrow \vec{i}_z \Leftrightarrow \vec{i}_2$ or $\vec{i}_2 \Leftrightarrow \vec{i}_z \Leftrightarrow \vec{i}_1$; (b) transition: $\vec{i}_1 \Leftrightarrow \vec{i}_z \Leftrightarrow \vec{i}_1$; and (c) transition: $\vec{i}_2 \Leftrightarrow \vec{i}_z \Leftrightarrow \vec{i}_2$.

one sampling period T_s (made up of the currents provided by the vectors $\vec{i}_i$, $\vec{i}_{i+1}$, and $\vec{i}_z$ used during times T_i, T_{i+1}, and T_z) is on average equal to the vector $\vec{i}_c$, the following expressions should hold:

$$T_i = T_s \cdot \hat{i}_c \cdot \sin(\pi/3 - \theta) \quad (15.57)$$

$$T_{i+1} = T_s \cdot \hat{i}_c \cdot \sin(\theta) \quad (15.58)$$

$$T_z = T_s - T_i - T_{i+1} \quad (15.59)$$

where $0 \leq \hat{i}_c \leq 1$. Although, the SVM technique selects the vectors to be used and their respective on-times, the sequence in which they are used, the selection of the zero space vector, and the normalized sampled frequency remain undetermined.

TABLE 15.6 Zero SV for minimum switching frequency in CSI and sequence $\vec{i}_i, \vec{i}_{i+1}, \vec{i}_z$

Sector	$\vec{i}_i$	$\vec{i}_{i+1}$	$\vec{i}_z$
①	$\vec{i}_6$	$\vec{i}_1$	$\vec{i}_7$
②	$\vec{i}_1$	$\vec{i}_2$	$\vec{i}_9$
③	$\vec{i}_2$	$\vec{i}_3$	$\vec{i}_8$
④	$\vec{i}_3$	$\vec{i}_4$	$\vec{i}_7$
⑤	$\vec{i}_4$	$\vec{i}_5$	$\vec{i}_9$
⑥	$\vec{i}_5$	$\vec{i}_6$	$\vec{i}_8$

the possible transitions that could be found in Sector ②. It can be seen that the zero vector $\vec{i}_9$ should be chosen to minimize the switching frequency. Table 15.6 gives a summary of the zero space vector to be used in each sector in order to minimize the switching frequency. However, should be noted that Table 15.6 is valid only for the sequence $\vec{i}_i$, $\vec{i}_{i+1}$, $\vec{i}_z$. Another sequence will require reformulating the zero space-vector selection algorithm.

C. The Normalized Sampling Frequency

As in VSIs modulated by a SV approach, the normalized sampling frequency f_{sn} should be an integer multiple of 6 to minimize uncharacteristic harmonics. As an example, Fig. 15.35 shows the relevant waveforms of a CSI SVM for $f_{sn} = 18$ and $\hat{i}_c = 0.8$. Figure 15.35 also shows that the first set of relevant harmonics load line current are at f_{sn}.

15.4.5 DC Link Voltage in Three-phase CSIs

An instantaneous power balance indicates that

$$v_i(t) \cdot i_i(t) = v_{an}(t) \cdot i_{oa}(t) + v_{bn}(t) \cdot i_{ob}(t) + v_{cn}(t) \cdot i_{oc}(t) \quad (15.60)$$

where $v_{an}(t)$, $v_{bn}(t)$, and $v_{cn}(t)$ are the phase filter voltages as shown in Fig. 15.36. If the filter is large enough and a relatively high switching frequency is used, the phase voltages become nearly sinusoidal balanced waveforms. On the other hand, if the ac output currents are considered sinusoidal and the dc link current is assumed constant $i_i(t) = I_i$, Eq. (15.60) can be simplified to

$$v_i(t) = \frac{1}{I_i} \left\{ \begin{array}{l} \sqrt{2}V_{on}\sin(\omega t) \cdot \sqrt{2}I_{o1}\sin(\omega t - \phi) \\ +\sqrt{2}V_{on}\sin(\omega t - 120°) \cdot \sqrt{2}I_{o1}\sin(\omega t - 120° - \phi) \\ +\sqrt{2}V_{on}\sin(\omega t - 240°) \cdot \sqrt{2}I_{o1}\sin(\omega t - 240° - \phi) \end{array} \right\} \quad (15.61)$$

where V_{on} is the rms ac output phase voltage, I_{o1} is the rms fundamental line current, and ϕ is an arbitrary filter-load angle. Hence, the dc link voltage expression can be further simplified to the following:

$$v_i(t) = 3\frac{I_{o1}}{I_i}V_{on}\cos(\phi) = \sqrt{3}\frac{I_{o1}}{I_i}V_o\cos(\phi) \quad (15.62)$$

where $V_o = \sqrt{3}V_{on}$ is the rms load line voltage. The resulting dc link voltage expression indicates that the first line-current harmonic I_{o1} generates a clean dc current. However, as the load line currents contain harmonics around the normalized sampling frequency f_{sn}, the dc link current will contain harmonics but around f_{sn} as shown in Fig. 15.35h. Similarly, in carrier-based PWM techniques, the dc link current will contain harmonics around the carrier frequency m_f (Fig. 15.27).

In practical implementations, a CSI requires a dc current source that should behave as a constant (as required by PWM CSIs) or variable (as square-wave CSIs) current source. Such current sources should be implemented as separate units and they are described earlier in this book.

15.5 Closed-loop Operation of Inverters

Inverters generate variable ac waveforms from a dc power supply to feed, for instance, ASDs. As the load conditions usually change, the ac waveforms should be adjusted to these new conditions. Also, as the dc power supplies are not ideal and the dc quantities are not fixed, the inverter should compensate for such variations. Such adjustments can be done automatically by means of a closed-loop approach. Inverters also provide an alternative to changing the load operating conditions (i.e. speed in an ASD).

There are two alternatives for closed-loop operation the feedback and the feedforward approaches. It is known that the feedback approach can compensate for both the perturbations (dc power variations) and the load variations (load torque changes). However, the feedforward strategy is more effective in mitigating perturbations as it prevents its negative effects at the load side. These cause-effect issues are analyzed in three-phase inverters in the following, although similar results are obtained for single-phase VSIs.

15.5.1 Feedforward Techniques in Voltage Source Inverters

The dc link bus voltage in VSIs is usually considered a constant voltage source v_i. Unfortunately, and due to the fact that most practical applications generate the dc bus voltage by means of a diode rectifier (Fig. 15.37), the dc bus voltage contains low-order harmonics such as the sixth, twelfth, ... (due to six-pulse

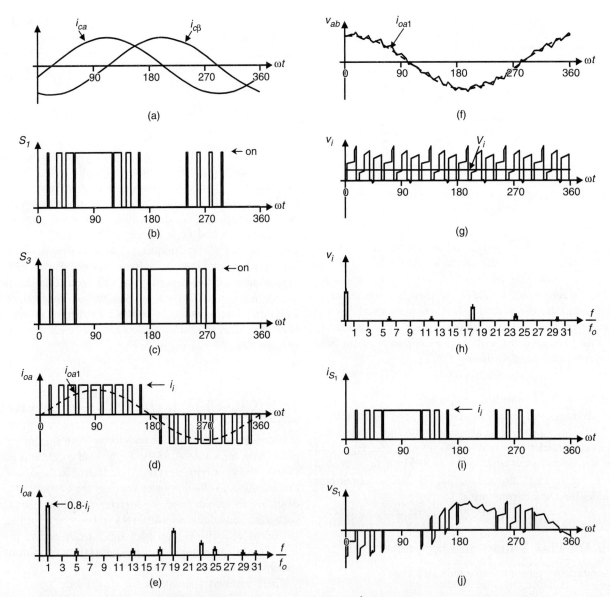

FIGURE 15.35 The three-phase CSI. Ideal waveforms for space-vector modulation ($\hat{i}_c = 0.8$, $f_{sn} = 18$): (a) modulating signals; (b) switch S_1 state; (c) switch S_3 state; (d) ac output current; (e) ac output current spectrum; (f) ac output voltage; (g) dc voltage; (h) dc voltage spectrum; (i) switch S_1 current; and (j) switch S_1 voltage.

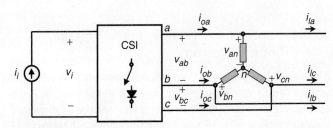

FIGURE 15.36 Phase-voltage definition in a wye-connected filter.

diode rectifiers), and the second if the ac voltage supply features an unbalance, which is usually the case. Additionally, if the three-phase load is unbalanced, as in UPS applications, the dc input current in the inverter i_i also contains the second harmonic, which in turn contributes to the generation of a second voltage harmonic in the dc bus.

The basic principle of feedforward approaches is to sense the perturbation and then modify the input in order to compensate for its effect. In this case, the dc link voltage should be sensed and the modulating technique should accordingly be modified. The fundamental *ab* line voltage in a VSI SPWM

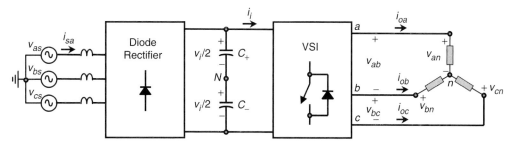

FIGURE 15.37 Three-phase VSI topology with a diode-based front-end rectifier.

can be written as

$$v_{ab1}(t) = \left\{ \frac{v_{ca1}(t)}{\hat{v}_\Delta} - \frac{v_{cb1}(t)}{\hat{v}_\Delta} \right\} \frac{\sqrt{3}}{2} v_i(t) \quad \hat{v}_\Delta > \hat{v}_{ca1}, \hat{v}_{cb1}$$
(15.63)

where $\hat{v}_\Delta$ is the carrier signal peak, $\hat{v}_{ca1}$ and $\hat{v}_{cb1}$ are the modulating signal peaks, and $v_{ca}(t)$ and $v_{ca}(t)$ are the modulating signals. If the dc bus voltage v_i varies around a nominal V_i value, then the fundamental line voltage varies proportionally; however, if the carrier signal peak $\hat{v}_\Delta$ is redefined as

$$\hat{v}_\Delta = \hat{v}_{\Delta m} \frac{v_i(t)}{V_i}$$
(15.64)

where $\hat{v}_{\Delta m}$ is the carrier signal peak (Fig. 15.38), then the resulting fundamental ab line voltage in a VSI SPWM is

$$v_{ab1}(t) = \left\{ \frac{v_{ca1}(t)}{\hat{v}_{\Delta m}} - \frac{v_{cb1}(t)}{\hat{v}_{\Delta m}} \right\} \frac{\sqrt{3}}{2} V_i$$
(15.65)

where, clearly, the result does not depend upon the variations of the dc bus voltage.

Figure 15.39 shows the waveforms generated by the SPWM under a severe dc bus voltage variation (a second harmonic has been added manually to a constant V_i). As a consequence, the ac line voltage generated by the VSI is distorted as it contains

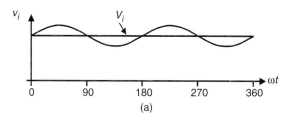
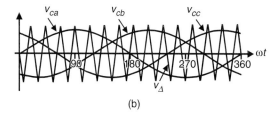
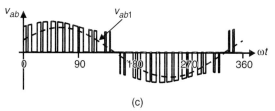
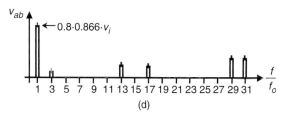

FIGURE 15.39 The three-phase VSI. Waveforms for regular SPWM ($m_a = 0.8$, $m_f = 9$): (a) dc bus voltage; (b) carrier and modulating signals; (c) ac output voltage; and (d) ac output voltage spectrum.

low-order harmonics (Fig. 15.39e). These operating conditions may not be acceptable in standard applications such as ASDs because the load will draw distorted three-phase currents as well. The feedforward loop performance is illustrated in Fig. 15.40. As expected, the carrier signal is modified so as to compensate for the dc bus voltage variation (Fig. 15.40b). This is probed by the spectrum of the ac line voltage that does not

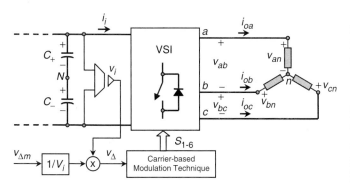

FIGURE 15.38 The three-phase VSI. Feedforward control technique to reject dc bus voltage variations.

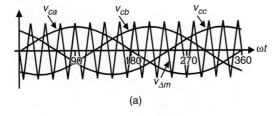

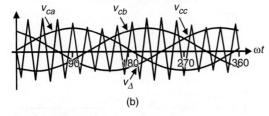

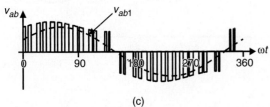

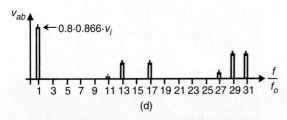

FIGURE 15.40 The three-phase VSI. Waveforms for SPWM including a feedforward loop ($m_a = 0.8$, $m_f = 9$): (a) carrier and modulating signals; (b) modified carrier and modulating signals; (c) ac output voltage; and (d) ac output voltage spectrum.

contain low-order harmonics (Fig. 15.40e). It should be noted that $\hat{v}_\Delta > \hat{v}_{ca1}, \hat{v}_{cb1}$; therefore, the compensation capabilities are limited by the required ac line voltage.

The performance of the feedforward approach depends upon the frequency of the harmonics present in the dc bus voltage and the carrier signal frequency. Fortunately, the relevant unwanted harmonics to be found in the dc bus voltage are the second, due to unbalanced supply voltages, and/or the sixth as the dc bus voltage is generated by means of a six-pulse diode rectifier. Therefore, a carrier signal featuring a 15-pu frequency is found to be sufficient to properly compensate for dc bus voltage variations.

Unbalanced loads generate a dc input current i_i that contains a second harmonic, which contributes to the dc bus voltage variation. The previous feedforward approach can compensate for such perturbation and maintain balanced ac load voltages.

Digital techniques can also be modified in order to compensate for dc bus voltage variations by means of a feedforward approach. For instance, the SVM techniques indicate that the on-times of the vectors $\vec{v}_i$, $\vec{v}_{i+1}$, and $\vec{v}_z$ are

$$T_i = T_s \cdot \hat{v}_c \cdot \sin(\pi/3 - \theta) \quad (15.66)$$

$$T_{i+1} = T_s \cdot \hat{v}_c \cdot \sin(\theta) \quad (15.67)$$

$$T_z = T_s - T_i - T_{i+1} \quad (15.68)$$

respectively, where $\hat{v}_c$ is the amplitude of the desired ac line voltage, as shown in Fig. 15.18. By redefining this quantity to

$$0 \leq \hat{v}_c = \hat{v}_{cm} \frac{V_i}{v_i(t)} \leq 1 \quad (15.69)$$

where V_i is the nominal dc bus voltage and $v_i(t)$ is the actual dc bus voltage. Thus, the on-times become

$$T_i = T_s \cdot \hat{v}_{cm} \frac{V_i}{v_i(t)} \cdot \sin(\pi/3 - \theta) \quad (15.70)$$

$$T_{i+1} = T_s \cdot \hat{v}_{cm} \frac{V_i}{v_i(t)} \cdot \sin(\theta) \quad (15.71)$$

$$T_z = T_s - T_i - T_{i+1} \quad (15.72)$$

where $\hat{v}_{cm}$ is the desired maximum ac line voltage. The previous expressions account for dc bus voltage variations and behave as a feedforward loop as it needs to sense the perturbation in order to be implemented. The previous expressions are valid for the linear region, thus $\hat{v}_c$ is restricted to $0 \leq \hat{v}_c \leq 1$, which indicates that the compensation is indeed limited.

15.5.2 Feedforward Techniques in Current Source Inverters

The duality principle between the voltage and the current source inverters indicates that, as described previously, the feedforward approach can be used for CSIs as well as for VSIs. Therefore, low-order harmonics present in the dc bus current can be compensated for before they appear at the load side. This can be done for both analog-based (e.g. carrier-based) and digital-based (e.g. space-vector) modulating techniques.

15.5.3 Feedback Techniques in Voltage Source Inverters

Unlike the feedforward approach, the feedback techniques correct the input to the system (gating signals) depending upon the deviation of the output to the system (e.g. ac load line currents in VSIs). Another important difference is that feedback techniques need to sense the controlled variables. In general, the controlled variables (output to the system) are chosen according to the control objectives. For instance, in ASDs, it is usually necessary to keep the motor line currents equal to a given set of sinusoidal references. Therefore, the controlled variables become the ac line currents. There are several alternatives to implement feedback techniques in VSIs, and three of them are discussed in the following.

A. Hysteresis Current Control

The main purpose here is to force the ac line current to follow a given reference. The status of the power valves S_1 and S_4 are changed whenever the actual i_{oa} current goes beyond a given reference $i_{oa,ref} \pm \Delta i/2$. Figure 15.41 shows the hysteresis current controller for phase a. Identical controllers are used in phase b and c. The implementation of this controller is simple as it requires an operational amplifier (op-amp) operating in the hysteresis mode, thus the controller and modulator are combined in one unit.

Unfortunately, there are several drawbacks associated with the technique itself. First, the switching frequency cannot be predicted as in carrier-based modulators and therefore the harmonic content of the ac line voltages and currents becomes random (Fig. 15.42d). This could be a disadvantage when designing the filtering components. Second, as three-phase loads do not have the neutral connected as in ASDs, the load currents add up to zero. This means that only two ac line currents can be controlled independently at any given instant. Therefore, one of the hysteresis controllers is redundant at a given time. This explains why the load current goes beyond the limits and introduces limit cycles (Fig. 15.42a). Finally, although the ac load currents add up to zero, the controllers cannot ensure that all load line currents feature a zero dc component in one load cycle.

B. Linear Control of VSIs

Proportional and proportional-integrative controllers can also be used in VSIs. The main purpose is to generate the modulating signals v_{ca}, v_{cb}, and v_{cc} in a closed-loop fashion as depicted in Fig. 15.43. The modulating signals can be used by a carrier-based technique such as the SPWM (as depicted in Fig. 15.43) or by space vector modulation. Because the load line currents add up to zero, the load line current references must add up to zero. Thus, the abc/αβγ transformation can be used to reduce to two controllers the overall implementation scheme as the γ component is always zero. This avoids limit cycles in the ac load currents.

The transformation of a set of variables in the stationary abc frame $\mathbf{x}^{abc}$ into a set of variables in the stationary αβ frame $\mathbf{x}^{\alpha\beta}$ is given by

$$\mathbf{x}^{\alpha\beta} = \frac{2}{3} \begin{bmatrix} 1 & -1/2 & -1/2 \\ 0 & \sqrt{3}/2 & -\sqrt{3}/2 \end{bmatrix} \mathbf{x}^{abc} \qquad (15.73)$$

The selection of the controller (P, PI,...) is done according to the control procedures such as steady-state error, settling time, overshoot, and so forth. Figure 15.44 shows the relevant waveforms of a VSI SPWM controlled by means of a PI controller as shown in Fig. 15.43.

Although it is difficult to prove that no limit cycles are generated, the ac line current appears very much sinusoidal. Moreover, the ac line voltage generated by the VSI preserves the characteristics of such waveforms generated by SPWM modulators. This is confirmed by the harmonic spectrum

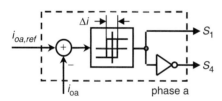

FIGURE 15.41 The three-phase VSI. Hysteresis current control (phase a).

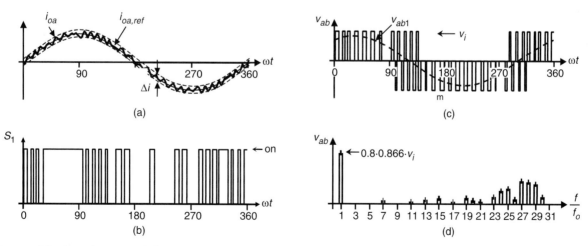

FIGURE 15.42 The three-phase VSI. Ideal waveforms for hysteresis current control: (a) actual ac load current and reference; (b) switch S_1 state; (c) ac output voltage; and (d) ac output voltage spectrum.

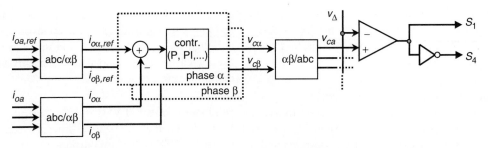

FIGURE 15.43 The three-phase VSI. Feedback control based on linear controllers.

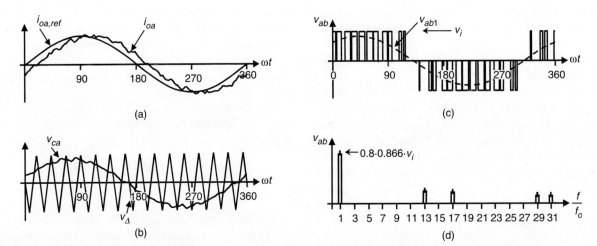

FIGURE 15.44 The three-phase VSI. Ideal waveforms for a PI controller in a feedback loop ($m_a = 0.8$, $m_f = 15$): (a) actual ac load current and reference; (b) carrier and modulating signals; (c) ac output voltage; and (d) ac output voltage spectrum.

shown in Fig. 15.44d, where the first set of characteristic harmonics are around the normalized carrier frequency $m_f = 15$.

However, an error between the actual i_{oa} and the ac line current reference $i_{oa,ref}$ can be observed (Fig. 15.44a). This error is inherent to linear controllers and cannot be totally eliminated, but it can be minimized by increasing the gain of the controller. However, the noise in the circuit is also increased, which could deteriorate the overall performance of the control scheme. The inherent presence of the error in this type of controllers is due to the fact that the controller needs a sinusoidal error to generate sinusoidal modulating signals v_{ca}, v_{cb}, and v_{cc}, as required by the modulator. Therefore, an error must exist between the actual and the ac line current references.

Nevertheless, as current-controlled VSIs are actually the inner loops in many control strategies, their inherent errors are compensated by the outer loop. This is the case of ASDs, where the outer speed loop compensates the inner current loops. In general, if the outer loop is implemented with dc quantities (such as speed), it can compensate the ac inner loops (such as ac line currents). If it is mandatory that a zero steady-state error be achieved with the ac quantities, then a stationary (abc frame) to rotating (dq frame) transformation is a valid alternative to use.

C. Linear Control of VSIs in a Rotating Frame

The rotating dq transformation allows ac three-phase circuits to be operated as if they were dc circuits. This is based upon a mathematical operation, that is the transformation of a set of variables in the stationary abc frame $\mathbf{x}^{abc}$ into a set of variables in the rotating dq0 frame $\mathbf{x}^{dq0}$. The transformation is given by

$$\mathbf{x}^{dq} = \frac{2}{3} \begin{bmatrix} \sin(\omega t) & \sin(\omega t - 2\pi/3) & \sin(\omega t - 4\pi/3) \\ \cos(\omega t) & \cos(\omega t - 2\pi/3) & \cos(\omega t - 4\pi/3) \\ 1/\sqrt{2} & 1/\sqrt{2} & 1/\sqrt{2} \end{bmatrix} \mathbf{x}^{abc} \quad (15.74)$$

where ω is the angular frequency of the ac quantities. For instance, the current vector given by

$$\mathbf{i}^{abc} = \begin{bmatrix} i_a \\ i_b \\ i_c \end{bmatrix} = \begin{bmatrix} I\sin(\omega t - \varphi) \\ I\sin(\omega t - 2\pi/3 - \varphi) \\ I\sin(\omega t - 4\pi/3 - \varphi) \end{bmatrix} \quad (15.75)$$

15 Inverters

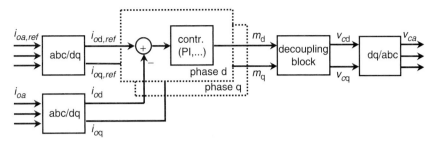

FIGURE 15.45 The three-phase VSI. Feedback control based on dq0 transformation.

becomes the vector

$$\mathbf{i}^{dq0} = \begin{bmatrix} i_d \\ i_q \\ i_0 \end{bmatrix} = \begin{bmatrix} I\cos(\varphi) \\ -I\sin(\varphi) \\ 0 \end{bmatrix} \quad (15.76)$$

where I and φ are the amplitude and phase of the line currents, respectively. It can be observed that: (a) the zero component i_0 is always zero as the three-phase quantities add up to zero; and (b) the d and q components i_d, i_q are dc quantities. Thus, linear controllers should help to achieve zero steady-state error. The control strategy shown in Fig. 15.45 is an alternative where the zero-component controller has been eliminated due to fact that the line currents at the load side add up to zero.

The controllers in Fig. 15.45 include an integrator that generates the appropriate dc outputs m_d and m_q even if the actual and the line current references are identical. This ensures that the zero steady-state error is achieved. The decoupling block in Fig. 15.45 is used to eliminate the cross-coupling effect generated by the dq0 transformation and to allow an easier design of the parameters of the controllers.

The dq0 transformation requires the intensive use of multiplications and trigonometric functions. These operations can readily be done by means of digital microprocessors. Also, analog implementations would indeed be involved.

15.5.4 Feedback Techniques in Current Source Inverters

Duality indicates that CSIs should be controlled as equally as VSIs except that the voltages become currents and the currents become voltages. Thus, hysteresis, linear and dq linear-based control strategies are also applicable to CSIs; however, the controlled variables are the load voltages instead of the load line currents.

For instance, the linear control of a CSI based on a dq transformation is depicted in Fig. 15.46. In this case, a passive balanced load is considered. In order to show that zero steady-state error is achieved, the per phase equations of the converter are written as

$$C\frac{d}{dt}\mathbf{v}_p^{abc} = \mathbf{i}_o^{abc} - \mathbf{i}_l^{abc} \quad (15.77)$$

$$L\frac{d}{dt}\mathbf{i}_l^{abc} = \mathbf{v}_p^{abc} - R\mathbf{i}_l^{abc} \quad (15.78)$$

the ac line currents are in fact imposed by the modulator and they satisfy

$$\mathbf{i}_o^{abc} = i_i \mathbf{i}_c^{abc} \quad (15.79)$$

Replacing Eq. (15.79) into the model of the converter Eqs. (15.77) and (15.78), using the dq0 transformation and assuming null zero component, the model of the converter becomes

$$\frac{d}{dt}\mathbf{v}_p^{dq} = -\mathbf{W}\mathbf{v}_p^{dq} + \frac{i_i}{C}\mathbf{i}_c^{dq} - \frac{1}{C}\mathbf{i}_l^{dq} \quad (15.80)$$

$$\frac{d}{dt}\mathbf{i}_l^{dq} = -\mathbf{W}\mathbf{i}_l^{dq} + \frac{1}{L}\mathbf{v}_p^{dq} - \frac{R}{L}\mathbf{i}_l^{dq} \quad (15.81)$$

where $\mathbf{W}$ is given by

$$\mathbf{W} = \begin{bmatrix} 0 & -\omega \\ \omega & 0 \end{bmatrix} \quad (15.82)$$

A first approximation is to assume that the decoupling block is not there; in other words, $\mathbf{i}_c^{dq} = \mathbf{m}^{dq}$. On the other hand, the model of the controllers can be written as

$$\mathbf{m}^{dq} = k\left\{\mathbf{v}_{p,ref}^{dq} - \mathbf{v}_p^{dq}\right\} + \frac{1}{T}\int_{-\infty}^{t}\left(\mathbf{v}_{p,ref}^{dq} - \mathbf{v}_p^{dq}\right)dt \quad (15.83)$$

where k and T are the proportional and integrative gains of the PI controller that are chosen to achieve a desired dynamic response. Combining the model of the controllers and the model of the converter in dq coordinates and using

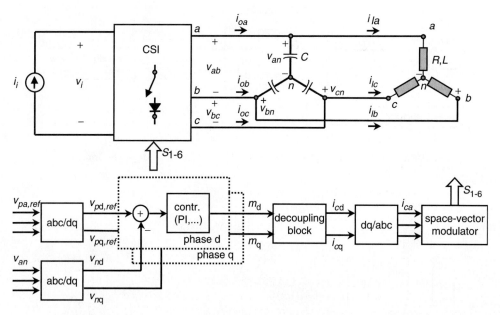

FIGURE 15.46 The three-phase CSI. Feedback control based on dq0 transformation.

the Laplace transform, the following relationship between the reference and actual load-phase voltages is found:

$$\mathbf{v}_\mathbf{p}^{dq} = \frac{i_i}{C} \left\{ sk + \frac{1}{T} \right\} \left\{ s\mathbf{I} + \mathbf{W} + \frac{R}{L}\mathbf{I} \right\} \times$$

$$\left[\left\{ s\mathbf{I} + \mathbf{W} + \frac{R}{L}\mathbf{I} \right\} \left\{ s^2\mathbf{I} + s\left(\mathbf{W} + \frac{i_i}{C}k\mathbf{I}\right) + \frac{i_i}{CT}\mathbf{I} \right\} \right.$$

$$\left. + \frac{s}{LC}\mathbf{I} \right]^{-1} \mathbf{v}_{\mathbf{p},\mathbf{ref}}^{dq} \qquad (15.84)$$

Finally, in order to prove that the zero steady-state error is achieved for step inputs in either the d or q component of the load-phase voltage reference, the previous expression is evaluated in $s = 0$. This results in the following:

$$\mathbf{v}_\mathbf{p}^{dq} = \frac{i_i}{C} \left\{ \frac{1}{T} \right\} \left\{ \mathbf{W} + \frac{R}{L}\mathbf{I} \right\} \left[\left\{ \mathbf{W} + \frac{R}{L}\mathbf{I} \right\} \left\{ \frac{i_i}{CT}\mathbf{I} \right\} \right]^{-1} \mathbf{v}_{\mathbf{p},\mathbf{ref}}^{dq}$$

$$= \mathbf{v}_{\mathbf{p},\mathbf{ref}}^{dq} \qquad (15.85)$$

As expected, the actual and reference values are identical. Finally, the relationship in Eq. (15.84) is a matrix that is not diagonal. This means that both the actual and the reference load-phase voltages are coupled. In order to obtain a decoupled control, the decoupling block in Fig. 15.46 should be properly chosen.

15.6 Regeneration in Inverters

Industrial applications are usually characterized by a power flow that goes from the ac distribution system to the load. This is, for example, the case of an ASD operating in the motoring mode. In this instance, the active power flows from the dc side to the ac side of the inverter. However, there are an important number of applications in which the load may supply power to the system. Moreover, this could be an occasional condition as well as a normal operating condition. This is known as the regenerative operating mode. For example, when an ASD reduces the speed of an electrical machine this can be considered a transient condition. Downhill belt conveyors in mining applications can be considered as a normal operating condition. In order to simplify the notation, it could be said that an inverter operates in the motoring mode when the power flows from the dc to the ac side, and in the regenerative mode when the power flows from the ac to the dc side.

15.6.1 Motoring Operating Mode in Three-phase VSIs

This is the case where the power flows from the dc side to the ac side of the inverter. Figure 15.47 shows a simplified scheme of an ASD where the motor has been modeled by three RLe branches, where the sources $\mathbf{e}^{abc}$ are the back-emf. Because the ac line voltages applied by the inverter are imposed by the pulsewidth modulation technique being used, they can be adjusted according to specific requirements. In particular, Fig. 15.48 shows the relevant waveforms in steady state for the

15 Inverters

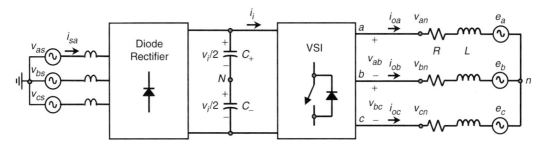

FIGURE 15.47 Three-phase VSI topology with a diode-based front-end rectifier.

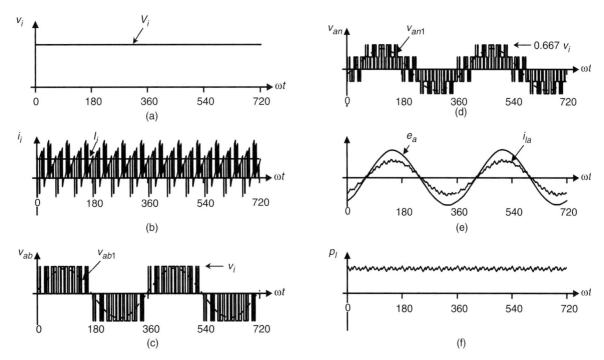

FIGURE 15.48 The ASD based on a VSI. Motoring mode: (a) dc bus voltage; (b) dc bus current; (c) ac line-load voltage; (d) ac phase-load voltage; (e) motor line current and back-emf; and (f) shaft power.

motoring operating mode of the ASD. To simplify the analysis, a constant dc bus voltage $v_i = V_i$ has been considered.

It can be observed that: (i) the dc bus current i_i features a dc value I_i that is positive; and (ii) the motor line current is in phase with the back-emf. Both features confirm that the active power flows from the dc source to the motor. This is also confirmed by the shaft power plot (Fig. 15.48f), which is obtained as:

$$p_l(t) = e_a(t)i_{la}(t) + e_b(t)i_{lb}(t) + e_c(t)i_{lc}(t) \quad (15.86)$$

15.6.2 Regenerative Operating Mode in Three-phase VSIs

The back-emf sources e^{abc} are functions of the machine speed and as such they ideally change just as the speed changes. The regeneration operating mode can be achieved by properly modifying the ac line voltages applied to the machine. This is done by the speed outer loop that could be based on a scalar (e.g. V/f) or vectorial (e.g. field-oriented) control strategy. As indicated earlier, there are two cases of regenerative operating modes.

A. Occasional Regenerative Operating Mode

This mode is required during transient conditions such as in occasional braking of electrical machines (ASDs). Specifically, the speed needs to be reduced and the kinetic energy is taken into the dc bus. Because the motor line voltage is imposed by the VSI, the speed reduction should be done in such a way that the motor line currents do not exceed the maximum values. This boundary condition will limit the ramp-down speed to a minimum, but shorter braking times will require a mechanical braking system.

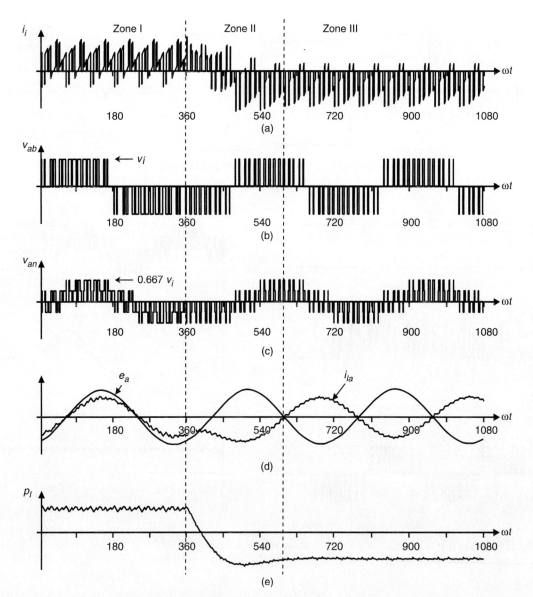

FIGURE 15.49 The ASD based on a VSI. Motoring to regenerative operating mode transition: (a) dc bus current; (b) ac line motor voltage; (c) ac phase motor voltage; (d) motor line current and back-emf; and (e) shaft power.

Figure 15.49 shows a transition from the motoring to regenerative operating mode for an ASD as shown in Fig. 15.47. Here, a stiff dc bus voltage has been used. Zone I in Fig. 15.49 is the motoring mode, Zone II is a transition condition, and Zone III is the regeneration mode. The line voltage is adjusted dynamically to obtain nominal motor line currents during regeneration (Fig. 15.49d). Zone III clearly shows that the shaft power gets reversed.

Occasional regeneration means that the drive rarely goes into this operating mode. Therefore, such energy can be: (a) left uncontrolled or (b) burned in resistors that are paralleled to the dc bus. The first option is used in low- to medium-power applications that use diode-based front-end rectifiers. Therefore, the dc bus current flows into the dc bus capacitor and the dc bus voltage rises accordingly to

$$\Delta v_i = \frac{1}{C} I_i \Delta t \qquad (15.87)$$

where Δv_i is the dc bus voltage variation, C is the dc bus voltage capacitor, I_i is the average dc bus current during regeneration, and Δt is the duration of the regeneration operating mode. Usually, the drives have the capacitor C designed to allow a 10% overvoltage in the dc bus.

The second option uses burning resistors R_R that are paralleled in the dc bus as shown in Fig. 15.50 by means of the

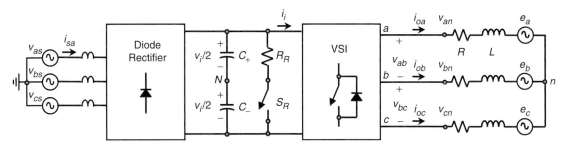

FIGURE 15.50 The ASD based on a VSI. Burning resistor strategy.

switch S_R. A closed-loop strategy based on the actual dc bus voltage modifies the duty cycle of the turn-on/turn-off of the switch S_R in order to keep such voltage under a given reference. This alternative is used when the energy recovered by the VSI would result in an acceptable dc bus voltage variation if an uncontrolled alternative is used.

There are some special cases where the regeneration operating mode is frequently used. For instance, electrical shovels in mining companies have repetitive working cycles and ≈15% of the energy is sent back into the dc bus. In this case, a valid alternative is to send back the energy into the ac distribution system.

The schematic shown in Fig. 15.51 is capable of taking the kinetic energy and sending it into the ac grid. As reviewed earlier, the regeneration operating mode reverses the polarity of the dc current i_i, and because the diode-based front-end converter cannot take negative currents, a thyristor-based front-end converter is added. Similarly to the burning-resistor approach, a closed-loop strategy based on the actual dc bus voltage v_i modifies the commutation angle α of the thyristor rectifier in order to keep such voltage under a given reference.

B. Regenerative Operating Mode as Normal operating Mode

Fewer industrial applications are capable of returning energy into the ac distribution system on a continuous basis. For instance, mining companies usually transport their product downhill for a few kilometers before processing it. In such cases, the drive maintains the transportation belt conveyor at constant speed and takes the kinetic energy. Due to the large amount of energy and the continuous operating mode, the drive should be capable of taking the kinetic energy, transforming it into electrical energy, and sending it into the ac distribution system. This would make the drive a generator that would compensate for the active power required by other loads connected to the electrical grid.

The schematic shown in Fig. 15.52 is a modern alternative for adding regeneration capabilities to the VSI-based drive on a continuous basis. In contrast to the previous alternatives, this scheme uses a VSI topology as an active front-end converter, which is generally called voltage-source rectifier (VSR). The VSR operates in two quadrants, that is, positive dc voltages and positive/negative dc currents as reviewed earlier. This feature makes it a perfect match for ASDs based on a VSI. Some of the advantages of using a VSR topology are: (i) the ac supply

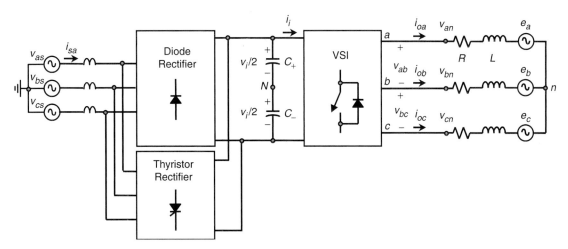

FIGURE 15.51 The ASD based on a VSI. Diode-thyristor-based front-end rectifier with regeneration capabilities.

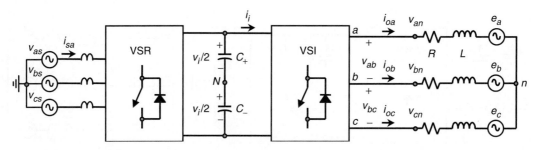

FIGURE 15.52 The ASD based on a VSI. Active front-end rectifier with regeneration capabilities.

current can be as sinusoidal as required (by increasing the switching frequency of the VSR or the ac line inductance); (ii) the operation can be done at a unity displacement power factor in both motoring and regenerative operating modes; and (iii) the control of the VSR is done in both motoring and regenerative operating modes by a single dc bus voltage loop.

15.6.3 Regenerative Operating Mode in Three-phase CSIs

There are drives where the motor side converter is a CSI. This is usually the case where near sinusoidal motor voltages are needed instead of the PWM type of waveform generated by VSIs. This is normally the case for medium-voltage applications. Such inverters require a dc current source that is constructed by means of a controlled rectifier.

Figure 15.53 shows a CSI-based ASD where the dc current source is generated by means of a thyristor-based rectifier in combination with a dc link inductor L_{dc}. In order to maintain a constant dc link current $i_i = I_i$, the thyristor-based rectifier adjusts the commutation angle α by means of a closed-loop control strategy. Assuming a constant dc link current, the regenerating operating mode is achieved when the dc link voltage v_i reverses its polarity. This can be done by modifying the PWM pattern applied to the CSI as in the VSI-based drive. To maintain the dc link current constant, the thyristor-based rectifier also reverses its dc link voltage v_r. Fortunately, the thyristor rectifier operates in two quadrant, that is, positive dc link currents and positive/ negative dc link voltages.

Thus, no additional equipment is required to include regeneration capabilities in CSI-based drives.

Similarly, an active front-end rectifier could be used to improve the overall performance of the thyristor-based rectifier. A PWM current-source rectifier (CSR) could replace the thyristor-based rectifier with the following added advantages: (i) the ac supply current can be as sinusoidal as required (e.g. by increasing the switching frequency of the CSR); (ii) the operation can be done at a unity displacement power factor in both motoring and regenerative operating modes; and (iii) the control of the CSR is done in both motoring and regenerative operating modes by a single dc bus current loop.

15.7 Multistage Inverters

The most popular three-phase voltage source inverter (VSI) consists of a six-switch topology (Fig. 15.54a). The topology can generate a three-phase set of ac line voltages such that each line voltage v_{ab} (Fig. 15.54b) features a fundamental ac line voltage v_{ab1} and unwanted harmonics Fig. 15.54c. The fundamental ac line voltage is usually required as a sinusoidal waveform at variable amplitude and frequency, and the unwanted harmonics are located at high frequencies. These requirements are met by means of a modulating technique as shown earlier. Among the applications in low-voltage ranges of six-switch VSIs are the adjustable speed drives (ASDs). The range is in low voltages due to: (a) the high dv/dt present in the PWM ac line voltages (Fig. 15.54b), which will be unacceptable

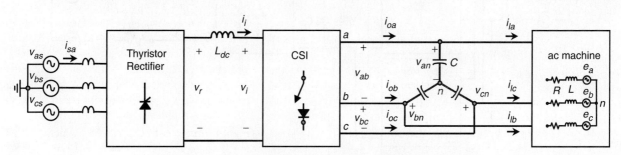

FIGURE 15.53 The ASD based on a CSI. Thyristor-based rectifier.

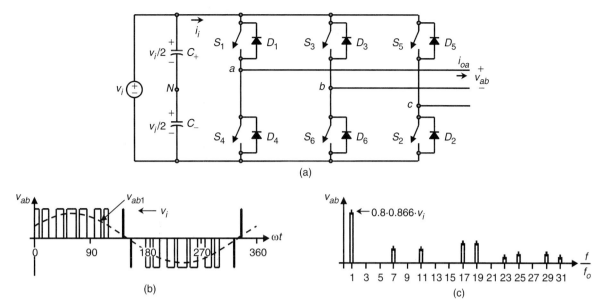

FIGURE 15.54 Six-switch voltage source inverter ($m_f = 9$, $m_a = 0.8$): (a) power topology; (b) ac output voltage; and (c) ac output voltage spectrum.

in the medium- to high-voltage ranges and (b) the load power would be shared only among six switches. This may require paralleling and series-connected power valves, an option usually avoided as symmetrical sharing of the power is not natural in these arrangements.

Two solutions are available to generate near-sinusoidal voltage waveforms while using six-switch topologies. The first is a topology based on a CSI in combination with a capacitive filter. The second solution is a topology based on a VSI including an inductive or inductive/capacitive filter at the load terminals. Although both alternatives generate near-sinusoidal voltage waveforms, both continue sharing the load power only among six power valves.

Solutions based on multistage voltage source topologies have been proposed. They provide medium voltages at the ac terminals while keeping low dv/dts and a large number of power valves that symmetrically share the total load power. The multistage VSIs can be classified in multicell and multilevel topologies.

15.7.1 Multicell Topologies

The goal is to develop a new structure with improved performance based on standard structures that are known as cells. For instance, Fig. 15.55a shows a cell featuring a three-phase input and a single-phase output. The front-end converter is a six-diode-based rectifier, and a single-phase VSI generates a single-phase ac voltage v_o. Figure 15.55b and c shows characteristic waveforms where a sinusoidal unipolar PWM ($m_f = 6$, $m_a = 0.8$) has been used to modulate the inverter.

Standard cells are meant to be used at low voltages, thus they can use standard components that are less expensive and widely available. The new structure should generate near-sinusoidal ac load voltages, draw near-sinusoidal ac line currents, and more importantly the load voltages should feature moderate dv/dts.

Figure 15.56 shows a multicell converter that generates a three-phase output voltage out of a three-phase ac distribution system. The structure uses three standard cells (as shown in Fig. 15.55) connected in series to form one phase; thus the phase-load voltages are the sum of the single-phase voltages generated by each cell. For instance, the phase voltage a is given by

$$v_{an} = v_{o11} + v_{o21} + v_{o31} \qquad (15.88)$$

In order to maximize the load-phase voltages, the ac voltages generated by the cells should feature identical fundamental components. On the other hand, each cell generates a PWM voltage waveform at the ac side, which contains unwanted voltage harmonics. If a carrier-based modulating technique is used, the harmonics generated by each cell are at well-defined frequencies (Fig. 15.55c). Some of these harmonics are not present in the phase-load voltage if the carrier signals of each cell are properly phase shifted.

In fact, Fig. 15.57 shows the voltages generated by cells c_{11}, c_{21}, and c_{31}, which are v_{o11}, v_{o21}, and v_{o31}, respectively, and form the load-phase voltage a. They are generated using the unipolar SPWM approach, that is, one modulating signal v_{ca} and three carrier signals $v_{\Delta 1}$, $v_{\Delta 2}$, and $v_{\Delta 3}$ that are used by cells c_{11}, c_{21}, and c_{31}, respectively (Fig. 15.57a). The carrier signals have a normalized frequency m_f, which ensures an m_f switching frequency in each power valve and the lowest unwanted set of harmonics $\approx 2 \cdot m_f$ (m_f even) in the ac cell

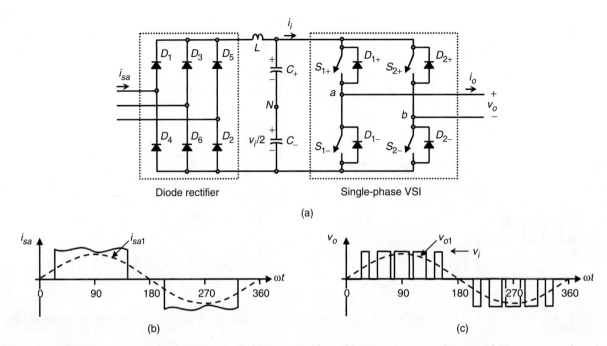

FIGURE 15.55 Three-phase-input single-phase output cell: (a) power topology; (b) ac input current, phase a; and (c) ac output voltage ($m_f = 6$, $m_a = 0.8$).

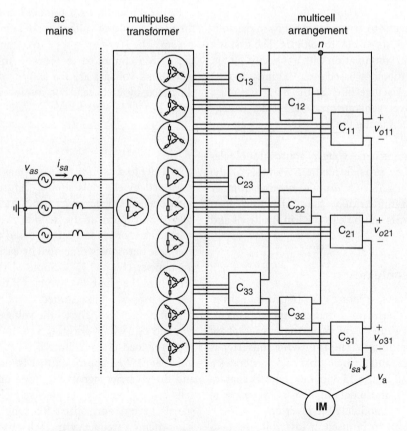

FIGURE 15.56 Multistage converter based on a multicell arrangement.

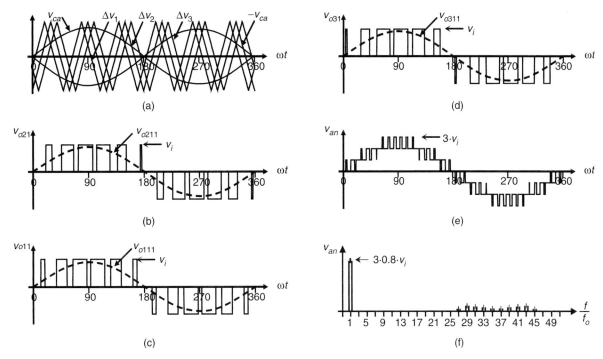

FIGURE 15.57 Multicell topology. Cell voltages in phase a using a unipolar SPWM ($m_f = 6$, $m_a = 0.8$): (a) modulating and carrier signals; (b) cell c_{11} ac output voltage; (c) cell c_{21} ac output voltage; (d) cell c_{31} ac output voltage; (e) phase a load voltage; and (f) phase a load-voltage spectrum.

voltages v_{o11}, v_{o21}, and v_{o31}. More importantly, the carrier signals are $\psi = 60°$ out-of-phase, which ensures the lowest unwanted set of voltage harmonics $\approx 6 \cdot m_f$ in the load-phase voltage v_{an}, that is, the lowest set of harmonics in Fig. 15.57f is $6 \cdot m_f = 6 \cdot 6 = 36$.

This can be explained as follows. The voltage harmonics present in the PWM voltage of each cell are at $l \cdot m_f \pm k$, $l = 2, 4, \ldots$ (where $k = 1, 3, 5, \ldots$); for instance, for $m_f = 6$, the first set of harmonics is at 12 ± 1, 12 ± 3, ... in all cells. Because the cells in one phase use carrier signals that are $60°$ out-of-phase, all the voltage harmonics $\approx l \cdot m_f$ in all cells are $l \cdot 60°$ out-of-phase. Therefore, for $l = 2$, the cell c_{11} generates the harmonics $l \cdot m_f \pm k = 2 \cdot m_f \pm k$ at a given phase φ, the cell c_{21} generates the harmonics $2 \cdot m_f \pm k$ at a phase $\varphi + l \cdot 60° = \varphi + 2 \cdot 60° = \varphi + 120° = \varphi - 240°$, and the cell c_{21} generates the harmonics $2 \cdot m_f \pm k$ at a phase $\varphi - l \cdot 60° = \varphi - 2 \cdot 60° = \varphi - 120° = \varphi + 240°$; thus, if the voltages have identical amplitudes, the harmonics $\approx 2 \cdot m_f$ add up to zero. Similarly, for $l = 4$, the cell c_{11} generates the harmonics $l \cdot m_f \pm k = 4 \cdot m_f \pm k$ at a given phase φ, the cell c_{21} generates the harmonics $4 \cdot m_f \pm k$ at a phase $\varphi + l \cdot 60° = \varphi + 4 \cdot 60° = \varphi + 240° = \varphi - 120°$, and the cell c_{21} generates the harmonics $4 \cdot m_f \pm k$ at a phase $\varphi - l \cdot 60° = \varphi - 4 \cdot 60° = \varphi - 240° = \varphi + 120°$; thus, if the voltages have identical amplitudes, the harmonics $\approx 4 \cdot m_f$ add up to zero. However, for $l = 6$, the cell c_{11} generates the harmonics $l \cdot m_f \pm k = 6 \cdot m_f \pm k$ at a given phase φ, the cell c_{21} generates the harmonics $6 \cdot m_f \pm k$ at a phase $\varphi + l \cdot 60° = \varphi + 6 \cdot 60° = \varphi + 360° = \varphi$, and the cell c_{21} generates the harmonics $6 \cdot m_f \pm k$ at a phase $\varphi - l \cdot 60° = \varphi - 6 \cdot 60° = \varphi - 360° = \varphi$; thus, if the voltages have identical amplitudes, the harmonics $\approx 6 \cdot m_f$ become triplicated rather than cancelled out.

In general, due to the fact that $n_c = 3$, cells are connected in series in each phase, n_c carriers are required, which should be $\psi = 180°/n_c$ out-of-phase. The number of cells per phase n_c depends on the required phase voltage. For instance, a 600 V dc cell generates an ac voltage of $\approx 600/\sqrt{2} = 424$ V. Then three cells connected in series generate a phase voltage of $3 \cdot 424 = 1.27$ kV, which in turn generates a $1.27 \cdot \sqrt{3} = 2.2$ kV line-to-line voltage.

Phases b and c are generated similarly to phase a. However, the modulating signals v_{cb} and v_{cc} should be $120°$ out-of-phase. In order to use identical carrier signals in phases b and c, the carrier-normalized frequency m_f should be a multiple of 3. Thus, three modulating signals and n_c carrier signals are required to generate three phase voltages by means of a multicell approach, where n_c depends upon the required load line voltage and the dc bus voltage of each cell.

The ac supply current of each cell is a six-pulse type of current as shown in Fig. 15.58, which feature harmonics at $6 \cdot k \pm 1$ ($k = 1, 2, \ldots$). Similarly to the load side, the ac supply currents of each cell are combined so as to achieve high-performance overall supply currents. Because the front-end converter of

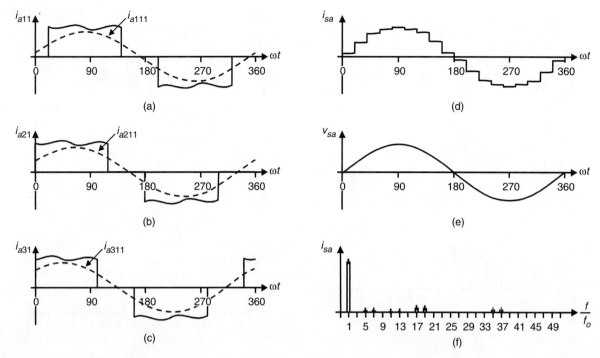

FIGURE 15.58 Multicell topology. Ac input current, phase a: (a) cell c_{11}; (b) cell c_{21}; (c) cell c_{31}; (d) overall supply current; (e) supply phase voltage; and (f) overall supply current spectrum.

each cell is a six-pulse diode rectifier, a multipulse approach is used. This is based on the natural harmonic cancellation when, for instance, a wye to delta/wye transformer is used to form an $N = 12$-pulse configuration from two six-pulse diode rectifiers. In this case, the fifth and seventh harmonics are cancelled out because the supply voltages applied to each six-pulse rectifier become 30° out-of-phase. In general, to form an $N = 6 \cdot n_s$ pulse configuration, n_s set of supply voltages that should be $60°/n_s$ out-of-phase is required. This would ensure the first set of unwanted current harmonics at $6 \cdot n_s \pm 1$.

The configuration depicted in Fig. 15.56 contains $n_c = 9$ cells, and a transformer capable of providing $n_s = 9$ sets of three-phase voltages that should be $60°/n_s = 60°/9$ out-of-phase to form an $N = 6 \cdot n_s = 6 \cdot 9 = 54$-pulse configuration is required. Although this alternative would provide a near-sinusoidal overall supply current, a fewer number of pulses are also acceptable that would reduce the transformer complexity. An $N = 18$-pulse configuration usually satisfies all the requirements. In the example, this configuration can be achieved by means of a transformer with $n_c = 9$ isolated secondaries; however, only $n_s = 3$ set of three-phase voltages that are $60°/n_s = 60°/3 = 20°$ out-of-phase are generated (Fig. 15.56). The configuration of the transformer restricts the connection of the cells in groups of three as shown in Fig. 15.56. In this case, the fifth, seventh, eleventh, and thirteenth harmonics are cancelled out and thus the first set of harmonics in the supply currents are the seventeenth and the nineteenth. Figure 15.58d shows the resulting supply current that is near-sinusoidal and Fig. 15.58f shows the corresponding spectrum. The fifth, seventh, eleventh, and thirteenth harmonics are still there, which is due to the fact that the ac input currents in each cell are not exactly the six-pulse type of waveforms as seen in Fig. 15.58a, b, and c. This is mainly because: (i) the dc link in the cells contains a small inductor L, which does not smooth out sufficiently the dc bus current (Fig. 15.55a) and (ii) the transformer leakage inductance (or added line inductance) smoothes out the edges of the current, which also contributes to the reactive power required by the cells. This last effect is not shown in Fig. 15.58a, b, and c.

15.7.2 Voltage Source-based Multilevel Topologies

The six-switch VSI is usually called a two-level VSI due to the fact that the inverter phase voltages v_{aN}, v_{bN}, and v_{cN} (Fig. 15.54a) are instantaneously either $v_i/2$ or $-v_i/2$. In other words, the phase voltages can take one of the two voltage levels. Multilevel topologies provide an alternative to these voltages to take one value out of N levels. For instance, Fig. 15.59 shows an $N = 3$-level topology, where the values of the inverter phase voltage are either $v_i/2$, 0, or $-v_i/2$ (Fig. 15.60d). An interesting problem is how to obtain the gating pattern for the 12 switches required in an $N = 3$-level topology. There are several modulating techniques to overcome this problem,

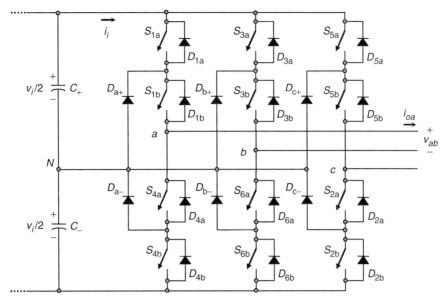

FIGURE 15.59 Three-phase three-level VSI topology.

which can be classified as analog (e.g. carrier-based) and digital (SV-based). Both approaches have to deal with the valid switch states of the inverter.

A. Valid Switch States in a Three-level VSI

The easiest way of obtaining the valid switch states is to analyze each phase separately. Phase a contains the switches S_{1a}, S_{1b}, S_{4a}, and S_{4b}, which cannot be on simultaneously because a short circuit across the dc bus would be produced, and cannot be off simultaneously because an undefined phase voltage v_{aN} would be produced. A summary of the valid switch combinations is given in Table 15.7. It is important to note that all valid switch combinations satisfy the condition that switch S_{1a} state is always the opposite to switch S_{4a} state, and that switch S_{1b} state is always the opposite to switch S_{4b} state. Any other switch-state combination would result in an undefined inverter phase a voltage because it will depend upon the load-phase current i_{oa} polarity. The switch states for phases b and c are identical to that of phase a; moreover, because they are paralleled, they can operate in an independent manner.

B. The SPWM Technique in Three-level VSIs

The main objective is to generate the appropriate 12 gating signals so as to obtain fundamental inverter phase voltages equal to a given set of modulating signals. Specifically, the SPWM in three-level inverters uses a sinusoidal set of modulating signals (v_{ca}, v_{cb}, and v_{cc} for phases a, b, and c, respectively) and $N - 1 = 2$ triangular type of carrier signals ($v_{\Delta 1}$ and $v_{\Delta 2}$) as illustrated in Fig. 15.60a. The best results are obtained if the carrier signals are in-phase and feature an odd normalized frequency (e.g. $m_f = 15$). According to Fig. 15.60a, switch S_{1a} is either turned on if $v_{ca} > v_{\Delta 1}$ or off if $v_{ca} < v_{\Delta 1}$, and switch S_{1b} is either turned on if $v_{ca} > v_{\Delta 2}$ or off if $v_{ca} < v_{\Delta 2}$. Additionally, the switch S_{4a} status is obtained as the opposite to switch S_{1a}, and the switch S_{4b} status is obtained as the opposite to switch S_{1b}. In order to use the same set of carrier signals to generate the gating signals for phases b and c, the normalized frequency of the carrier signal m_f should be a multiple of 3. Thus, the possible values are $m_f = 3, 9, 15, 21, \ldots$

Figure 15.60 shows the relevant waveforms for a three-level inverter modulated by means of a SPWM technique ($m_f = 15$, $m_a = 0.8$). Specifically, Fig. 15.60d shows the inverter phase voltage, which is clearly a three-level type of voltage, and Fig. 15.60f shows the load line voltage, which shows that the step voltages are at most $v_i/2$. More importantly, the inverter phase voltage (Fig. 15.60e) contains harmonics at $l \cdot m_f \pm k$ with $l = 1, 3, \ldots$ and $k = 0, 2, 4, \ldots$ and at $l \cdot m_f \pm k$ with $l = 2, 4, \ldots$ and $k = 1, 3, \ldots$ For instance, the first set of harmonics ($l = 1$, $m_f = 15$) are at 15, 15 ± 2, 15 ± 4, ...

TABLE 15.7 Valid switch states for a three-level VSI, phase a

s_{1a}	s_{1b}	s_{4a}	s_{4b}	v_o	Components conducting	
1	1	0	0	$v_i/2$	S_{1a}, S_{1b}	if $i_{oa} > 0$
					D_{1a}, D_{1b}	if $i_{oa} < 0$
0	1	1	0	0	S_{1b}, D_{a+}	if $i_{oa} > 0$
					S_{4a}, D_{a-}	if $i_{oa} < 0$
0	0	1	1	$-v_i/2$	D_{4a}, D_{4b}	if $i_{oa} > 0$
					S_{4a}, S_{4b}	if $i_{oa} < 0$

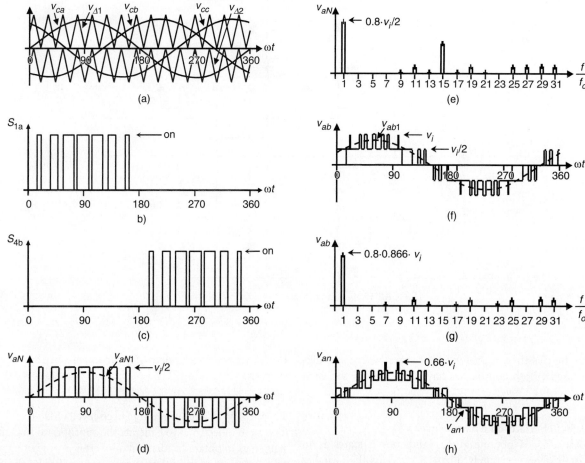

FIGURE 15.60 Three-level VSI topology. Relevant waveforms using a SPWM ($m_f = 15$, $m_a = 0.8$): (a) modulating and carrier signals; (b) switch S_{1a} status; (c) switch S_{4b} status; (d) inverter phase a voltage; (e) inverter phase a voltage spectrum; (f) load line voltage; (g) load line voltage spectrum; and (h) load phase a voltage.

The inverter line voltage (Fig. 15.60g) contains harmonics at $l \cdot m_f \pm k$ with $l = 1, 3, \ldots$ and $k = 2, 4, \ldots$ and at $l \cdot m_f \pm k$ with $l = 2, 4, \ldots$ and $k = 1, 3, \ldots$. For instance, the first set of harmonics in the line voltages ($l = 1$, $m_f = 15$) are at 15 ± 2, $15 \pm 4, \ldots$.

All the other features of carrier-based PWM techniques also apply in multilevel inverters. For instance, (I) the fundamental component of the inverter phase voltages satisfies

$$\hat{v}_{aN1} = \hat{v}_{bN1} = \hat{v}_{cN1} = m_a \frac{v_i}{2} \quad 0 < m_a \leq 1 \quad (15.89)$$

and thus the line voltages satisfy

$$\hat{v}_{ab1} = \hat{v}_{bc1} = \hat{v}_{ca1} = m_a \sqrt{3} \frac{v_i}{2} \quad 0 < m_a \leq 1 \quad (15.90)$$

where $0 < m_a \leq 1$ the linear operating region. To further increase the amplitude of the load voltages, the overmodulation operating region can be used by further increasing the modulating signal amplitudes ($m_a > 1$), where the line voltages range in

$$\sqrt{3} \frac{v_i}{2} < \hat{v}_{ab1} = \hat{v}_{bc1} = \hat{v}_{ca1} < \frac{4}{\pi} \sqrt{3} \frac{v_i}{2} \quad (15.91)$$

Also, (II) the modulating signals could be improved by adding a third harmonic (zero sequence), which will increase the linear region up to $m_a = 1.15$. This results in a maximum fundamental line-voltage component equal to v_i; (III) a non-sinusoidal set of modulating signals could also be used by the modulating technique. This is the case where nonsinusoidal line voltages are required as in active filter applications; and (IV) because of the two quadrants operation of VSIs, the multilevel inverter could equally be used in applications where the active power flow goes from the dc to the ac side or from the ac to the dc side.

15 Inverters

In general, for an N-level inverter modulated by means of a carrier-based technique, the following conclusions can be drawn:

(a) three modulating signals $120°$ out of phase and $N-1$ carrier signals are required;
(b) the phase voltages in the inverters have a peak value of $v_i/(N-1)$;
(c) the phase voltages in the inverters are discrete waveforms constructed from the values

$$\frac{v_i}{2}, \frac{v_i}{2} - \frac{v_i}{N-1}, \frac{v_i}{2} - \frac{2 \cdot v_i}{N-1}, \ldots, -\frac{v_i}{2} \quad (15.92)$$

(d) the maximum voltage step in the line voltages is

$$\frac{v_i}{N-1} \quad (15.93)$$

for instance, an $N = 5$-level inverter requires four carrier signals, the discrete values of the phase voltages are: $v_i/2$, $v_i/4$, 0, $-v_i/4$, and $-v_i/2$, and the maximum step voltage at the load side is $v_i/4$. Key waveforms are shown in Fig. 15.61.

One of the drawbacks of the multilevel inverter is that the dc link capacitors should be equal. Unfortunately, this is not a natural operating condition mainly due to the fact that the currents required by the inverter in the dc bus are not symmetrical and therefore the capacitors will not equally share the total dc supply voltage v_i. To overcome this problem, two alternatives are developed later on.

C. The Space-vector Modulation in Three-level VSIs

Digital techniques are naturally extended to multilevel inverters. In fact, the SV modulating technique can be applied using the same principles used in two-level inverters. However, the higher number of voltage levels increases the complexity of the practical implementation of the technique. For instance, in $N = 3$-level inverters, each leg allows $N = 3$ different switch combinations as indicated in Table 15.7. Therefore, there are $N^3 = 27$ total valid switch combinations, which generate $N^3 = 27$ load line voltages that are represented by $N^3 = 27$ space vectors ($\vec{v}_1, \vec{v}_2, \ldots, \vec{v}_{27}$) in Fig. 15.62. For instance, $\vec{v}_2 = 0.5 + j0.866$ is due to the line voltages $v_{ab} = 0.5$, $v_{bc} = 0.5$, $v_{ca} = -1.0$ in pu. Thus, although the principle of operation is the same, the SV digital algorithm will have to deal with a higher number of states N^3. Moreover, because some space vectors (e.g. $\vec{v}_{13}$ and $\vec{v}_{14}$ in Fig. 15.62) produce the same load-voltage terminals, the algorithm will have to decide between the two based on additional criteria and that of the basic SV-approach. Clearly, as the number of level increases, the algorithm becomes more and more elaborate. However, the benefits are not evident as the number of level increases. The maximum number of levels used in practical applications is five. This is based on a compromise between the complexity of the implementation and the benefits of the resulting waveforms.

D. DC Link Voltage Balancing Issues

Figure 15.59 shows a three-level inverter and the ideal waveforms are shown in Fig. 15.60, which assume an even distribution of the voltage across the dc link capacitors. This even distribution is not naturally achieved and could be overcome by supplying both capacitors from independent supplies or properly gating the power valves of the inverter in order to minimize the unbalance.

Figure 15.63 shows an ASD based on a three-level VSI, where the dc link capacitors are feed from two different sources. This approach is being commercially used as it ensures a robust balanced dc link voltage distribution and operates with a high-performance type of ac mains current.

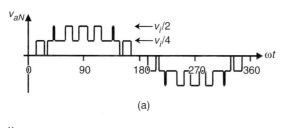

(a)

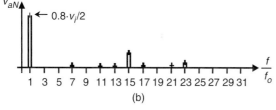

(b)

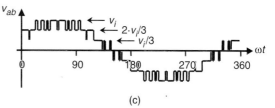

(c)

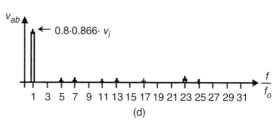

(d)

FIGURE 15.61 Five-level VSI topology. Relevant waveforms using a SPWM ($m_f = 15$, $m_a = 0.8$): (a) inverter phase a voltage; (b) inverter phase a voltage spectrum; (c) load line voltage; and (d) load line voltage spectrum.

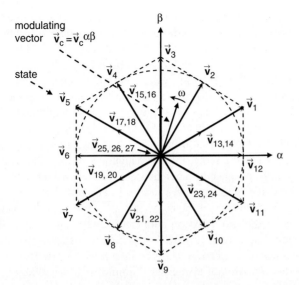

FIGURE 15.62 The space-vector representation in a three-level VSI.

Indeed, for a N level inverter, $N - 1$ independent dc voltage supplies are required that could be provided by $N - 1$ six-pulse rectifiers feed from an $N - 1$-pulse transformer. Therefore, the ac main currents is a $N - 1$ level type of waveform.

This approach cannot be used when the inverter does not feature dc link voltage supplies. This is the case of static power reactive power compensators and static power active filters. In this case, the proper gating of the power valves becomes the only choice to keep and balance the dc link voltages. Figure 15.64a shows this case where the current added by the inverter i_o^{abc} provides the reactive power and current harmonics such that the ac mains current i_s^{abc} features a given power factor.

The SPWM modulating technique could be used as in Fig. 15.60; however, the zero level of the carriers δ is left as a manipulable variable Fig. 15.64b. In fact, it is used to control the difference of the upper and lower capacitor voltages $\Delta v_i = v_{i1} - v_{i2}$. A closed loop alternative is depicted in Fig. 15.64c to manipulate δ. The modulating signals $\mathbf{v}_c^{abc}$ are left to control the reactive power and current harmonics injected into the ac mains by regulating the currents $\mathbf{i}_o^{abc}$ and keep the total dc link voltage $v_i = v_{i1} + v_{i2}$ equal to a reference. Both loops are not included in Fig. 15.64c.

15.7.3 Current Source-based Multilevel Topologies

Duality is found in many aspects related to voltage and current source inverters. Perhaps, the most evident is the duality in terms of modulating techniques. Thus, current source based multilevel topologies are available as well. As expected, all the benefits and all the drawbacks found in voltage source topologies should be found in current source topologies.

Figure 15.65 shows a three-level $N = 3$ current source topology, which is formed by paralleling two standard six-switches topologies. The main goal is to share evenly the ac current $\mathbf{i}_o^{abc}$ among the two topologies ($\mathbf{i}_o^{abc}/2 = \mathbf{i}_{o1}^{abc} = \mathbf{i}_{o2}^{abc}$). This should be ensured by having equal dc link currents ($i_{i1} = i_{i2}$). Similarly to voltages source based mutlilevel topologies, this could be achieved by using either two independent dc link currents or by properly gating the power valves. Both alternatives are reviewed later on.

A. The SPWM Technique in Three-level CSIs

As in three-level VSIs, the main objective is to generate the appropriate 12 gating signals so as to obtain fundamental inverter line currents equal to a given set of modulating signals. Specifically, the SPWM in three-level inverters uses a sinusoidal set of modulating signals (i_{ca}, i_{cb}, and i_{cc} for phases a, b, and c, respectively) and $N - 1 = 2$ triangular type of carrier signals ($i_{\Delta 1}$ and $i_{\Delta 2}$) as illustrated in Fig. 15.66a and 15.66e. The best results are obtained if the carrier signals are

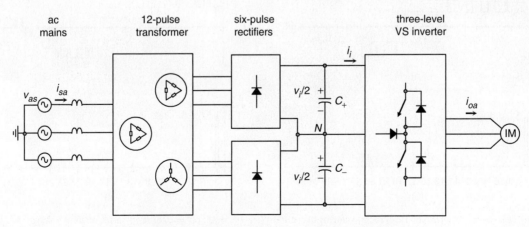

FIGURE 15.63 ASD based on a three-phase three-level VSI topology.

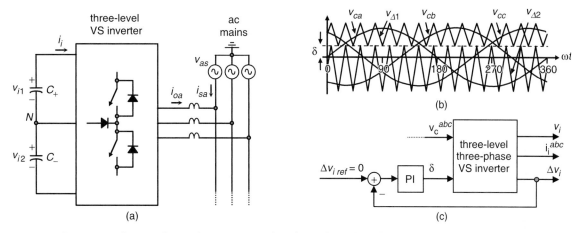

FIGURE 15.64 Reactive power and current harmonics compensator based on a three-phase three-level VSI topology: (a) power topology; (b) carrier and modulating signals; and (c) δ closed loop scheme.

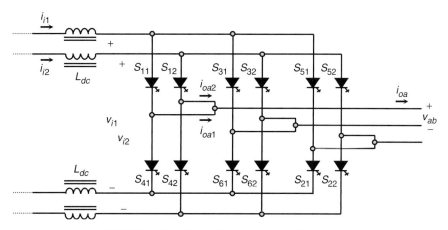

FIGURE 15.65 Three-phase three-level CSI topology.

180° out of phase and feature an odd normalized frequency (e.g. $m_f = 15$). In order to use the same set of carrier signals to generate the gating signals for phases b and c, the normalized frequency of the carrier signal m_f should be a multiple of 3. Thus, the possible values are $m_f = 3, 9, 15, 21, \ldots$.

Figure 15.66 shows the relevant waveforms for a three-level inverter modulated by means of a SPWM technique ($m_f = 15$, $m_a = 0.8$). Specifically, Fig. 15.66b and 15.66f show the gating signals obtained as described earlier in this chapter. The inverter line currents shown in 15.66c and 15.66g feature spectra shown in 15.66d and 15.66h, respectively. As expected, the inverter line currents contain harmonics at $l \cdot m_f \pm k$ with $l = 1, 3, \ldots$ and $k = 2, 4, \ldots$ and at $l \cdot m_f \pm k$ with $l = 2, 4, \ldots$ and $k = 1, 3, \ldots$. For instance, the first set of harmonics in the line currents ($l=1$, $m_f = 15$) are at 15 ± 2, $15 \pm 4, \ldots$.

The total inverter line current is shown in Fig. 15.67a, and features the first set of unwanted harmonics around $2m_f$ Fig. 15.67b. This becomes the first advantage of using a multilevel topology as the filtering component requirements become more relaxed. All the other features of carrier-based PWM techniques also apply in current source multilevel inverters. For instance: (I) the fundamental component of the line currents satisfy

$$\hat{i}_{oa1} = \hat{i}_{ob1} = \hat{i}_{oc1} = m_a \frac{\sqrt{3}}{2}(i_{i1} + i_{i2}) \quad 0 < m_a \leq 1 \quad (15.94)$$

where $0 < m_a \leq 1$ is the linear operating region. Also: (II) to further increase the amplitude of the load currents, a zero sequence signal could be injected to the modulating signals, in this case

$$\hat{i}_{oa1} = \hat{i}_{ob1} = \hat{i}_{oc1} = m_a \frac{\sqrt{3}}{2}(i_{i1} + i_{i2}) \quad 0 < m_a \leq 2/\sqrt{3} \quad (15.95)$$

the overmodulation operating region can be used by further increasing the modulating signal amplitudes ($m_a > 2/\sqrt{3}$),

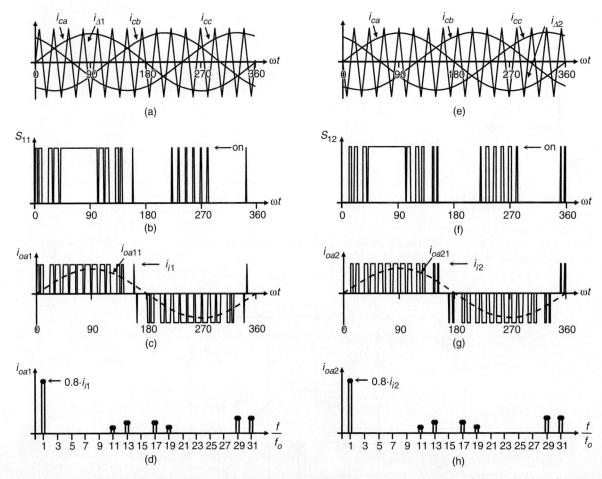

FIGURE 15.66 Three-level CSI topology. Relevant waveforms using a SPWM ($m_f = 15$, $m_a = 0.8$): (a) modulating signals and carrier signal 1; (b) switch S_{11} status; (c) inverter 1 linea current; (d) inverter 1 linea current spectrum; (e) modulating signals and carrier signal 2; (f) switch S_{12} status; (g) inverter 2 linea current; and (h) inverter 2 linea current spectrum.

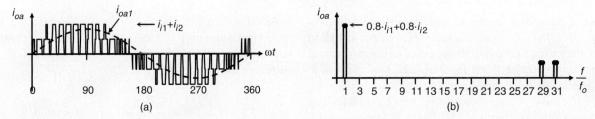

FIGURE 15.67 Three-level CSI topology. Relevant waveforms using a SPWM ($m_f = 15$, $m_a = 0.8$): (a) total inverter line current and (b) total inverter line current spectrum.

where the line currents range in

$$(i_{i1} + i_{i2}) < \hat{i}_{oa1} = \hat{i}_{ob1} = \hat{i}_{oc1} < \frac{4}{\pi}(i_{i1} + i_{i2}) \quad (15.96)$$

Also: (III) a nonsinusoidal set of modulating signals could also be used by the modulating technique. This is the case where nonsinusoidal line currents are required as in active filter applications; and (IV) because of the two quadrants operation of CSIs, the multilevel inverter could equally be used in applications where the active power flow goes from the dc to the ac side or from the ac to the dc side. In general, for an N-level inverter modulated by means of a carrier-based technique, three modulating signals 120° out-of-phase and $N-1$ carrier signals are required and the line currents in the inverters have a peak value of $i_i/(N-1)$.

15 Inverters

One of the drawbacks of the multilevel inverter is that the dc link capacitors cannot be supplied by a single dc voltage source. This is due to the fact that the currents required by the inverter in the dc bus are not symmetrical and therefore the capacitors will not equally share the dc supply voltage v_i. To overcome this problem, two alternatives are developed later on.

B. DC Link Voltage Balancing Issues

Figure 15.65 shows a three-level inverter and the ideal waveforms are shown in Fig. 15.66 and Fig. 15.67, which assume equal dc link currents, $i_{i1} = i_{i2}$. This even distribution is not naturally achieved and could be overcome by supplying the dc link inductors from independent supplies or properly gating the power valves of the inverter in order to minimize the unbalance.

Figure 15.68 shows an ASD based on a three-level current source inverter, where the dc link inductors are feed from two different sources. Unlike the VS topology, the scheme needs a closed loop control strategy to keep constant the dc link currents and equal to a given reference. This is achieved in commercial units by using either phase-controlled rectifiers or PWM rectifiers. Nevertheless, the multipulse transformer required to provide isolated dc link currents improves the ac mains current as in the VS multilevel topology.

This approach cannot be used when the inverter is not feed from an external power supply. This is the case of static series voltage compensators. In this case, the proper gating of the power valves becomes the only choice to keep and balance the dc link currents. Figure 15.64a shows this case where the voltage added by the inverter $n\mathbf{v}_o^{abc}$ compensates the sags and/or swells present in the ac mains in order to provide a constant voltage to the load.

The SPWM modulating technique could be used as in Fig. 15.66 and Fig. 15.67; however, the peak amplitude of one triangular is amplified in the factor $1 + \delta$ and the peak amplitude of other triangular is amplified in the factor $1 - \delta$, where δ is left as a manipulable variable Fig. 15.69b. In fact, δ is used to control the difference of the dc link currents $\Delta i_i = i_{i1} - i_{i2}$.

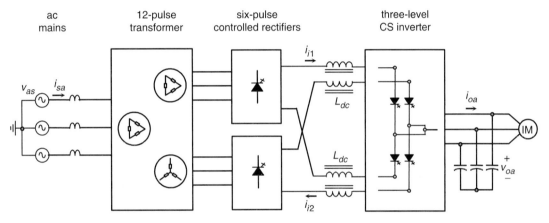

FIGURE 15.68 ASD based on a three-phase three-level CSI topology.

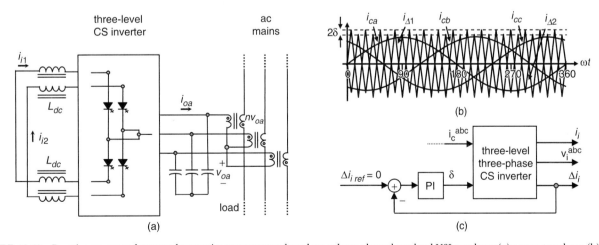

FIGURE 15.69 Reactive power and current harmonics compensator based on a three-phase three-level VSI topology: (a) power topology; (b) carrier and modulating signals; and (c) δ closed loop scheme.

A closed loop alternative is depicted in Fig. 15.69c to manipulate δ. The modulating signals $\mathbf{i}_c^{abc}$ are left to control the series injected voltage into the ac mains by regulating the voltages $\mathbf{v}_o^{abc}$ and keep the total dc link current $i_i = i_{i1} + i_{i2}$ equal to a reference. Both loops are not included in Fig. 15.69c.

Acknowledgment

The author is grateful for the financial support from the Chilean Fund for Scientific and Technological Development (FONDECYT) through project 105 0958.

Further Reading

Inverters Applications

1. Chih-Yi Huang, Chao-Peng Wei, Jung-Tai Yu, and Yeu-Jent Hu, "Torque and current control of induction motor drives for inverter switching frequency reduction," *IEEE Trans. Industrial Electronics*, **52**: (5), 1364–1371 (2005).
2. J. Rodriguez, L. Moran, J. Pontt, J. Espinoza, R. Diaz, and E. Silva, "Operating experience of shovel drives for mining applications," *IEEE Trans. Industry Applications*, **40**: (2), 664–671 (2004).
3. J. Espinoza, L. Morán, and J. Guzmán "Multi-level three-phase current source inverter based AC drive for high performance applications," *Conf. Rec. PESC'05*, Recife, Brazil, June 2005.
4. G. Joós and J. Espinoza, "Three phase series var compensation based on a voltage controlled current source inverter with supplemental modulation index control," *IEEE Trans. Power Electronics*, **15**: (3), 587–598 (1999).
5. P. Jain, J. Espinoza, and H. Jin, "Performance of a single-stage UPS system for single-phase trapezoidal-shaped ac voltage supplies," *IEEE Trans. Power Electronics*, **13**: (5), 912–923 (1998).
6. H. Akagi, "The state-of-the-art of power electronics in Japan," *IEEE Trans. Power Electronics*, **13**: (2), 345–356 (1998).
7. T. Wu and T. Yu, "Off-line applications with single-stage converters," *IEEE Trans. Industry Applications*, **44**: (5), 638–647 (1997).
8. H. Akagi, "New trends in active filters for power conditioning," *IEEE Trans. Industry Applications*, **32**: (6), 1312–1322 (1996).
9. J. Espinoza and G. Joós, "A current source inverter induction motor drive system with reduced losses," *IEEE Trans. Industry Applications*, **34**: (4), 796–805 (1998).
10. M. Ryan, W. Brumsickle, and R. Lorenz, "Control topology options for single-phase UPS inverters," *IEEE Trans. Industry Applications*, **33**: (2), 493–501 (1997).
11. A. Jungreis and A. Kelly, "Adjustable speed drive for residential applications," *IEEE Trans. Industry Applications*, **31**: (6), 1315–1322 (1995).
12. K. Rajashekara, "History of electrical vehicles in General Motors," *IEEE Trans. Industry Applications*, **30**: (4), 897–904 (1994).
13. B. Bose, "Power electronics and motion control – Technology status and recent trends," *IEEE Trans. Industry Applications*, **29**: (5), 902–909 (1993).
14. S. Bhowmik and R. Spée, "A guide to the application-oriented selection of ac/ac converter topologies," *IEEE Trans. Power Electronics*, **8**: (2), 156–163 (1993).

Current Source Inverters

15. J. Espinoza, L. Morán, and N. Zargari "Multi-level three-phase current source inverter based series voltage compensator," *Conf. Rec. PESC'05*, Recife, Brazil, June (2005).
16. M. Pande, H. Jin, and G. Joós, "Modulated integral control technique for compensating switch delays and nonideal dc buses in voltage-source inverters," *IEEE Trans. Industrial Electronics*, **44**: (2), 182–190 (1997).
17. J. Espinoza and G. Joós, "Current-source converter on-line pattern generator switching frequency minimization," *IEEE Trans. Industry Applications*, **44**: (2), 198–206 (1997).
18. G. Joós, G. Moschopoulos, and P. Ziogas, "A high performance current source inverter," *IEEE Trans. Power Electronics*, **8**: (4), 571–579 (1993).
19. Poh Chiang Loh and D.G. Holmes, "Analysis of multiloop control strategies for LC/CL/LCL-filtered voltage-source and current-source inverters," *IEEE Trans. Industry Applications*, **41**: (2), 644–654 (2005).
20. M. Salo and H. Tuusa, "Vector-controlled PWM current-source-inverter-fed induction motor drive with a new stator current control method," *IEEE Trans. Industrial Electronics*, **52**: (2), 523–531 (2005).
21. Dong Shen and P.W. Lehn, "Modeling, analysis, and control of a current source inverter-based STATCOM," *IEEE Trans. on Power Delivery*, **17**: (1), 248–253 (2002).
22. A. Bendre, I. Wallace, J. Nord, and G. Venkataramanan, "A current source PWM inverter with actively commutated SCRs," *IEEE Trans. Power Electronics*, **17**: (4), 461–468 (2002).
23. B.M. Han and S.I. Moon, "Static reactive-power compensator using soft-switching current-source inverter," *IEEE Trans. Industrial Electronics*, **48**: (6), 1158–1165 (2001).
24. D.N. Zmood and D.G. Holmes, "Improved voltage regulation for current-source inverters," *IEEE Trans. Industry Applications*, **37**: (4), 1028–1036 (2001).

Modulating Techniques and Control Strategies

25. J. Espinoza and G. Joós, "DSP implementation of output voltage reconstruction in CSI based converters," *IEEE Trans. Industrial Electronics*, **45**: (6), 895–904 (1998).
26. M. Kazmierkowski and L. Malesani, "Current control techniques for three-phase voltage-source PWM converters: A survey," *IEEE Trans. Industrial Electronics*, **45**: (5), 691–703 (1998).
27. A. Tilli and A. Tonielli, "Sequential design of hysteresis current controller for three-phase inverter," *IEEE Trans. Industrial Electronics*, **45**: (5), 771–781 (1998).
28. D. Chung, J. Kim, and S. Sul, "Unified voltage modulation technique for real-time three-phase power conversion," *IEEE Trans. Industry Applications*, **34**: (2), 374–380 (1998).
29. L. Malesani, P. Mattavelli, and P. Tomasin, "Improved constant-frequency hysteresis current control of VSI inverters with simple feed-forward bandwidth prediction," *IEEE Trans. Industry Applications*, **33**: (5), 1194–1202 (1997).
30. M. Rahman, T. Radwin, A. Osheiba, and A. Lashine, "Analysis of current controllers for voltage-source inverter," *IEEE Trans. Industrial Electronics*, **44**: (4), 477–485 (1997).
31. A. Trzynadlowski, R. Kirlin, and S. Legowski, "Space vector PWM technique with minimum switching losses and a variable pulse rate," *IEEE Trans. Industrial Electronics*, **44**: (2), 173–181 (1997).

32. S. Tadakuma, S. Tanaka, H. Naitoh, and K. Shimane, "Improvement of robustness of vector-controlled induction motors using feedforward and feedback control," *IEEE Trans. Power Electronics*, **12**: (2), 221–227 (1997).
33. J. Holtz and B. Beyer, "Fast current trajectory tracking control based on synchronous optimal pulse width modulation," *IEEE Trans. Industry Applications*, **31**: (5), 1110–1120 (1995).
34. J. Espinoza, G. Joós, and P. Ziogas, "Voltage controlled current source inverters," *Conf. Rec. IECON'92*, San Diego CA, USA, pp. 512–517, November (1992).
35. Fei Wang, "Sine-triangle versus space-vector modulation for three-level PWM voltage-source inverters," *IEEE Trans. Industry Applications*, **38**: (2), 500–506 (2002).
36. K.K. Tse, Henry Shu-Hung Chung; S.Y. Ron Hui, and H.C. So, "A comparative study of carrier-frequency modulation techniques for conducted EMI suppression in PWM converters," *IEEE Trans. Industrial Electronics*, **49**: (3), 618–627 (2002).
37. K.L. Shi and H. Li, "Optimized PWM strategy based on genetic algorithms," *IEEE Trans. Industrial Electronics*, **52**: (5), 1558–1561 (2005).

Overmodulation

38. A. Hava, S. Sul, R. Kerkman, and T. Lipo, "Dynamic overmodulation characteristics of triangle intersection PWM methods," *IEEE Trans. Industry Applications*, **35**: (4), 896–907 (1999).
39. A. Hava, R. Kerkman, and T. Lipo, "Carrier-based PWM-VSI overmodulation strategies: Analysis, comparison, and design," *IEEE Trans. Power Electronics*, **13**: (4), 674–689 (1998).
40. Bon-Ho Bae and Seung-Ki Sul, "A novel dynamic overmodulation strategy for fast torque control of high-saliency-ratio AC motor," *IEEE Trans. Industry Applications*, **41**: (4), 1013–1019 (2005).
41. Hee-Jhung Park and Myung-Joong Youn, "A new time-domain discontinuous space-vector PWM technique in overmodulation region," *IEEE Trans. Industrial Electronics*, **50**: (2), 349–355 (2003).
42. S.K. Mondal, B.K. Bose, V. Oleschuk, and J.O.P. Pinto, "Space vector pulse width modulation of three-level inverter extending operation into overmodulation region," *IEEE Trans. Power Electronics*, **18**: (2), 604–611 (2003).
43. A.M. Khambadkone and J. Holtz, "Compensated synchronous PI current controller in overmodulation range and six-step operation of space-vector-modulation-based vector-controlled drives," *IEEE Trans. Industrial Electronics*, **49**: (3), 574–580 (2002).
44. G. Narayanan and V.T. Ranganathan, "Extension of operation of space vector PWM strategies with low switching frequencies using different overmodulation algorithms," *IEEE Trans. Power Electronics*, **17**: (5), 788–798 (2002).
45. A.R. Bakhshai, G. Joos, P.K. Jain, and Hua Jin, "Incorporating the overmodulation range in space vector pattern generators using a classification algorithm," *IEEE Trans. Power Electronics*, **15**: (1), 83–91 (2000).

Selective Harmonic Elimination

46. S. Bowe and S. Grewal, "Novel space-vector-based harmonic elimination inverter control," *IEEE Trans. Industry Applications*, **36**: (2), 549–557 (2000).
47. L. Li, D. Czarkowski, Y. Liu, and P. Pillay, "Multilevel selective harmonic elimination PWM technique in series-connected voltage inverters," *IEEE Trans. Industry Applications*, **36**: (1), 160–170 (2000).
48. H. Karshenas, H. Kojori, and S. Dewan, "Generalized techniques of selective harmonic elimination and current control in current source inverters/converters," *IEEE Trans. Power Electronics*, **10**: (5), 566–573 (1995).
49. H. Patel and R. Hoft, "Generalized techniques of harmonic elimination and voltage control in thyristor inverters, Part I-Harmonic elimination," *IEEE Trans. Industry Applications*, **IA-9**: (3), 310–317 (1973).
50. J.R. Wells, B.M. Nee, P.L. Chapman, and P.T. Krein, "Selective harmonic control: a general problem formulation and selected solutions," *IEEE Trans. Power Electronics*, **20**: (6), 1337–1345 (2005).
51. M.J. Newman, D.G. Holmes, J.G. Nielsen, and F. Blaabjerg, "A dynamic voltage restorer (DVR) with selective harmonic compensation at medium voltage level," *IEEE Trans. Industry Applications*, **41**: (6), 1744–1753 (2005).
52. J.R. Espinoza, G. Joos, J.I. Guzman, L.A. Moran, and R.P. Burgos, "Selective harmonic elimination and current/voltage control in current/voltage-source topologies: a unified approach," *IEEE Trans. Industrial Electronics*, **48**: (1), 71–81 (2001).

Effects of PWM-type of Voltage Waveforms

53. N. Aoki, K. Satoh, and A. Nabae, "Damping circuit to suppress motor terminal overvoltage and ringing in PWM inverter-fed ac motor drive systems with long motor leads," *IEEE Trans. Industry Applications*, **35**: (5), 1015–1020 (1999).
54. D. Rendusara and P. Enjeti, "An improved inverter output filter configuration reduces common and differential modes dv/dt at the motor terminals in PWM drive systems," *IEEE Trans. Power Electronics*, **13**: (6), 1135–1153 (1998).
55. S. Chen and T. Lipo, "Bearing currents and shaft voltages of an induction motor under hard- and soft-switching inverter excitation," *IEEE Trans. Industry Applications*, **34**: (5), 1042–1048 (1998).
56. A. von Jouanne, H. Zhang, and A. Wallace, "An evaluation of mitigation techniques for bearing currents, EMI and overvoltages in ASD applications," *IEEE Trans. Industry Applications*, **34**: (5), 1113–1122 (1998).
57. H. Akagi, and T. Doumoto, "A passive EMI filter for preventing high-frequency leakage current from flowing through the grounded inverter heat sink of an adjustable-speed motor drive system," *IEEE Trans. Industry Applications*, **41**: (5), 1215–1223 (2005).

Multilevel Structures

58. L. Tolbert and T. Habetler, "Novel multilevel inverter carrier-based PWM method," *IEEE Trans. Industry Applications*, **35**: (5), 1098–1107 (1999).
59. G. Walker and G. Ledwich, "Bandwidth considerations for multilevel converters," *IEEE Trans. Power Electronics*, **15**: (1), 74–81 (1999).
60. Y. Liang and C. Nwankpa, "A new type of STATCOM based on cascading voltage-source inverters with phase-shifted unipolar SPWM," *IEEE Trans. Industry Applications*, **35**: (5), 1118–1123 (1999).

61. N. Schibli, T. Nguyen, and A. Rufer, "A three-phase multilevel converter for high-power induction motors," *IEEE Trans. Power Electronics*, **13**: (5), 978–986 (1998).
62. J. Lai, and F. Peng, "Multilevel converters – A new breed of power converters," *IEEE Trans. Industry Applications*, **32**: (3), 509–517 (1997).
63. R.M Tallam, R. Naik, and T.A. Nondahl, "A carrier-based PWM scheme for neutral-point voltage balancing in three-level inverters," *IEEE Trans. Industry Applications*, **41**: (6), 1734–1743 (2005).
64. M.A. Perez, J.R. Espinoza, J.R. Rodriguez, and P. Lezana, "Regenerative medium-voltage AC drive based on a multicell arrangement with reduced energy storage requirements," *IEEE Trans. Industrial Electronics*, **52**: (1), 171–180 (2005).

Regeneration

65. P. Verdelho and G. Marques, "DC voltage control and stability analysis of PWM-voltage-type reversible rectifiers," *IEEE Trans. Industrial Electronics*, **45**: (2), 263–273 (1998).
66. J. Espinoza, G. Joós, and A. Bakhshai, "Non-linear control and stabilization of PWM current source rectifiers in the regeneration mode," *Conf. Rec. APEC'97*, Atlanta GA, USA, pp. 902–908, February (1997).
67. M. Hinkkanen and J. Luomi, "Stabilization of regenerating-mode operation in sensorless induction motor drives by full-order flux observer design," *IEEE Transaction on Industrial Electronics*, **51**: (6), 1318–1328 (2004).
68. T. Tanaka, S. Fujikawa, and S. Funabiki, "A new method of damping harmonic resonance at the DC link in large-capacity rectifier-inverter systems using a novel regenerating scheme," *IEEE Trans. Industry Applications*, **38**: (4), 1131–1138 (2002).
69. J. Rodriguez, J. Pontt, E. Silva, J. Espinoza, and M. Perez, "Topologies for regenerative cascaded multilevel inverters," *Conf. Rec. PESC'03*, Acapulco, Mexico, pp. 519–524, June (2003).

16
Resonant and Soft-switching Converters

S. Y. (Ron) Hui and
Henry S. H. Chung
Department of Electronic Engineering, City University of Hong Kong, Tat Chee Avenue, Kowloon, Hong Kong

16.1 Introduction 405
16.2 Classification 407
16.3 Resonant Switch 407
    16.3.1 ZC Resonant Switch • 16.3.2 ZV Resonant Switch
16.4 Quasi-resonant Converters... 408
    16.4.1 ZCS-QRCs • 16.4.2 ZVS-QRC • 16.4.3 Comparisons between ZCS and ZVS
16.5 ZVS in High Frequency Applications 412
    16.5.1 ZVS with Clamped Voltage • 16.5.2 Phase-shifted Converter with Zero Voltage Transition
16.6 Multi-resonant Converters (MRC) 415
16.7 Zero-voltage-transition (ZVT) Converters 421
16.8 Non-dissipative Active Clamp Network... 421
16.9 Load Resonant Converters 421
    16.9.1 Series Resonant Converters • 16.9.2 Parallel Resonant Converters • 16.9.3 Series–Parallel Resonant Converter
16.10 Control Circuits for Resonant Converters 425
    16.10.1 QRCs and MRCs • 16.10.2 Phase-shifted, ZVT FB Circuit
16.11 Extended-period Quasi-resonant (EP-QR) Converters 427
    16.11.1 Soft-switched DC–DC Flyback Converter • 16.11.2 A ZCS Bidirectional Flyback DC–DC Converter
16.12 Soft-switching and EMI Suppression 434
16.13 Snubbers and Soft-switching for High Power Devices 435
16.14 Soft-switching DC–AC Power Inverters... 436
    16.14.1 Resonant (Pulsating) DC Link Inverter • 16.14.2 Active-clamped Resonant DC Link Inverter • 16.14.3 Resonant DC Link Inverter with Low Voltage Stress [49] • 16.14.4 Quasi-resonant Soft-switched Inverter [50] • 16.14.5 Resonant Pole Inverter (RPI) and Auxiliary Resonant Commutated Pole Inverter (ARCPI)
    References.. 448

16.1 Introduction

Advances in power electronics in the last few decades have led to not just improvements in power devices, but also new concepts in converter topologies and control. In the 1970s, conventional pulse width modulated (PWM) power converters were operated in a switched mode operation. Power switches have to cut off the load current within the turn-on and turn-off times under the hard switching conditions. Hard switching refers to the stressful switching behavior of the power electronic devices. The switching trajectory of a hard-switched power device is shown in Fig. 16.1. During the turn-on and turn-off processes, the power device has to withstand high voltage and current simultaneously, resulting in

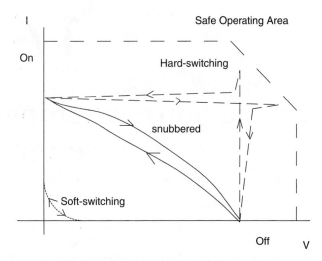

FIGURE 16.1 Typical switching trajectories of power switches.

high switching losses and stress. Dissipative passive snubbers are usually added to the power circuits so that the dv/dt and di/dt of the power devices could be reduced, and the switching loss and stress be diverted to the passive snubber circuits. However, the switching loss is proportional to the switching frequency, thus limiting the maximum switching frequency of the power converters. Typical converter switching frequency was limited to a few tens of kilo-Hertz (typically 20–50 kHz) in early 1980s. The stray inductive and capacitive components in the power circuits and power devices still cause considerable transient effects, which in turn give rise to electromagnetic interference (EMI) problems. Figure 16.2 shows ideal switching waveforms and typical practical waveforms of the switch voltage. The transient ringing effects are major causes of EMI.

In the 1980s, lots of research efforts were diverted towards the use of resonant converters. The concept was to incorporate resonant tanks in the converters to create oscillatory (usually sinusoidal) voltage and/or current waveforms so that zero-voltage switching (ZVS) or zero-current switching (ZCS) conditions can be created for the power switches. The reduction of switching loss and the continual improvement of power switches allow the switching frequency of the resonant converters to reach hundreds of kilo-Hertz (typically 100–500 kHz). Consequently, the size of magnetic components can be reduced and the power density of the converters increased. Various forms of resonant converters have been proposed and developed. However, most of the resonant converters suffer several problems. When compared with the conventional PWM converters, the resonant current and the voltage of resonant converters have high peak values, leading to higher conduction loss and higher V and I rating requirements for the power devices. Also, many resonant converters require frequency modulation (FM) for output regulation. Variable switching frequency operation makes the filter design and control more complicated.

In late 1980s and throughout 1990s, further improvements have been made in converter technology. New generations of soft-switched converters that combine the advantages of conventional PWM converters and resonant converters have been developed. These soft-switched converters have switching waveforms similar to those of conventional PWM converters except that the rising and falling edges of the waveforms are "smoothed" with no transient spikes. Unlike the resonant converters, new soft-switched converters usually utilize the resonance in a controlled manner. Resonance is allowed to occur just before and during the turn-on and turn-off processes so as to create ZVS and ZCS conditions. Other than that, they behave just like conventional PWM converters. With simple modifications, many customized control integrated circuits (ICs) designed for conventional converters can be employed for soft-switched converters. Because the switching loss and stress have been reduced, soft-switched converter can be operated at the very high frequency (typically 500 kHz to a few Mega-Hertz). Soft-switching converters also provide an effective solution to suppress EMI and have been applied to DC–DC, AC–DC, and DC–AC converters. This chapter covers the basic technology of resonant and soft-switching converters. Various forms of soft-switching techniques such as ZVS, ZCS, voltage clamping, zero-voltage transition methods, etc. are addressed. The emphasis is placed on the basic operating principle and practicality of the converters without using much mathematical analysis.

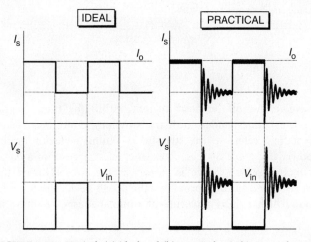

FIGURE 16.2 Typical: (a) ideal and (b) practical switching waveforms.

16.2 Classification

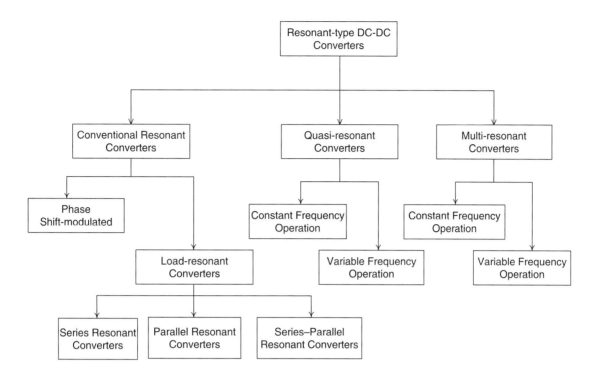

16.3 Resonant Switch

Prior to the availability of fully controllable power switches, thyristors were the major power devices used in power electronic circuits. Each thyristor requires a commutation circuit, which usually consists of a LC resonant circuit, for forcing the current to zero in the turn-off process [1]. This mechanism is in fact a type of zero-current turn-off process. With the recent advancement in semiconductor technology, the voltage and current handling capability, and the switching speed of fully controllable switches have significantly been improved. In many high power applications, controllable switches such as gate turn-offs (GTOs) and insulated gate bipolar transistors (IGBTs) have replaced thyristors [2, 3]. However, the use of resonant circuit for achieving ZCS and/or ZVS [4–8] has also emerged as a new technology for power converters. The concept of resonant switch that replaces conventional power switch is introduced in this section.

A resonant switch is a sub-circuit comprising a semiconductor switch S and resonant elements, L_r and C_r [9–11]. The switch S can be implemented by a unidirectional or bidirectional switch, which determines the operation mode of the resonant switch. Two types of resonant switches [12], including zero-current (ZC) resonant switch and zero-voltage (ZV) resonant switches, are shown in Figs. 16.3 and 16.4, respectively.

16.3.1 ZC Resonant Switch

In a ZC resonant switch, an inductor L_r is connected in series with a power switch S in order to achieve zero-current switching (ZCS). If the switch S is a unidirectional switch, the switch current is allowed to resonate in the positive half cycle only. The resonant switch is said to operate in *half-wave* mode. If a diode is connected in anti-parallel with the unidirectional switch, the switch current can flow in both directions. In this case, the resonant switch can operate in *full-wave* mode. At turn-on, the switch current will rise slowly from zero. It will then oscillate, because of the resonance between L_r and C_r. Finally, the switch can be commutated at the next zero current duration. The objective of this type of switch is to shape the switch current waveform during conduction time in order to create a zero-current condition for the switch to turn off [13].

16.3.2 ZV Resonant Switch

In a ZV resonant switch, a capacitor C_r is connected in parallel with the switch S for achieving zero-voltage switching (ZVS).

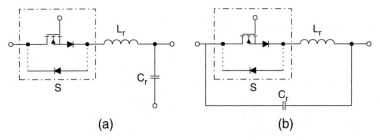

FIGURE 16.3 Zero-current (ZC) resonant switch.

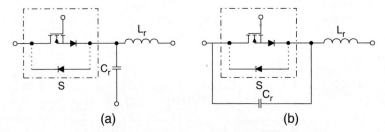

FIGURE 16.4 Zero-voltage (ZV) resonant switch.

If the switch S is a unidirectional switch, the voltage across the capacitor C_r can oscillate freely in both positive and negative half-cycle. Thus, the resonant switch can operate in *full-wave* mode. If a diode is connected in anti-parallel with the unidirectional switch, the resonant capacitor voltage is clamped by the diode to zero during the negative half-cycle. The resonant switch will then operate in *half-wave* mode. The objective of a ZV switch is to use the resonant circuit to shape the switch voltage waveform during the off time in order to create a zero-voltage condition for the switch to turn on [13].

16.4 Quasi-resonant Converters

Quasi-resonant converters (QRCs) can be considered as a hybrid of resonant and PWM converters. The underlying principle is to replace the power switch in PWM converters with the resonant switch. A large family of conventional converter circuits can be transformed into their resonant converter counterparts. The switch current and/or voltage waveforms are forced to oscillate in a quasi-sinusoidal manner, so that ZCS and/or ZVS can be achieved. Both ZCS-QRCs and ZVS-QRCs have *half-wave* and *full-wave* mode of operations [8–10, 12].

16.4.1 ZCS-QRCs

A ZCS-QRC designed for *half-wave* operation is illustrated with a buck type DC–DC converter. The schematic is shown in Fig. 16.5a. It is formed by replacing the power switch in conventional PWM buck converter with the ZC resonant switch in Fig. 16.3a. The circuit waveforms in steady state are shown in Fig. 16.5b. The output filter inductor L_f is sufficiently large so that its current is approximately constant. Prior to turning the switch on, the output current I_o freewheels through the output diode D_f. The resonant capacitor voltage V_{Cr} equals zero. At t_0, the switch is turned on with ZCS. A quasi-sinusoidal current I_S flows through L_r and C_r, the output filter, and the load. S is then softly commutated at t_1 with ZCS again. During and after the gate pulse, the resonant capacitor voltage V_{Cr} rises and then decays at a rate depending on the output current. Output voltage regulation is achieved by controlling the switching frequency. Operation and characteristics of the converter depend mainly on the design of the resonant circuit $L_r C_r$. The following parameters are defined: voltage conversion ratio M, characteristic impedance Z_r, resonant frequency f_r, normalized load resistance r, normalized switching frequency γ.

$$M = \frac{V_o}{V_i} \quad (16.1a)$$

$$Z_r = \sqrt{\frac{L_r}{C_r}} \quad (16.1b)$$

$$f_r = \frac{1}{2\pi\sqrt{L_r C_r}} \quad (16.1c)$$

$$r = \frac{R_L}{Z_r} \quad (16.1d)$$

$$\gamma = \frac{f_s}{f_r} \quad (16.1e)$$

It can be seen from the waveforms that if $I_o > V_i/Z_r$, I_S will not come back to zero naturally and the switch will have to be

16 Resonant and Soft-switching Converters

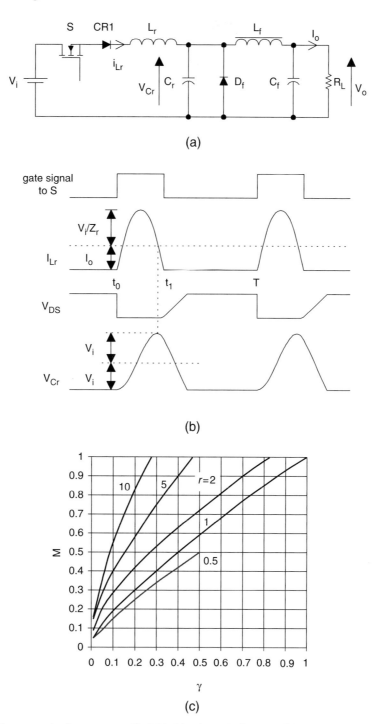

FIGURE 16.5 Half-wave, quasi-resonant buck converter with ZCS: (a) schematic diagram; (b) circuit waveforms; and (c) relationship between M and γ.

forced off, thus resulting in turn-off losses. The relationships between M and γ at different r are shown in Fig. 16.5c. It can be seen that M is sensitive to the load variation. At light load conditions, the unused energy is stored in C_r, leading to an increase in the output voltage. Thus, the switching frequency has to be controlled, in order to regulate the output voltage.

If an anti-parallel diode is connected across the switch, the converter will be operating in *full-wave* mode. The circuit schematic is shown in Fig. 16.6a. The circuit waveforms in steady state are shown in Fig. 16.6b. The operation is similar to the one in *half-wave* mode. However, the inductor current is allowed to reverse through the anti-parallel diode and the duration for the resonant stage is lengthened. This permits excess energy in the resonant circuit at light loads to be transferred back to the voltage source V_i. This significantly reduces the dependence of V_o on the output load. The relationships between M and γ at different r are shown in Fig. 16.6c. It can be seen that M is insensitive to load variation.

By replacing the switch in the conventional converters, a family of QRC [9] with ZCS is shown in Fig. 16.7.

16.4.2 ZVS-QRC

In these converters, the resonant capacitor provides a zero-voltage condition for the switch to turn on and off.

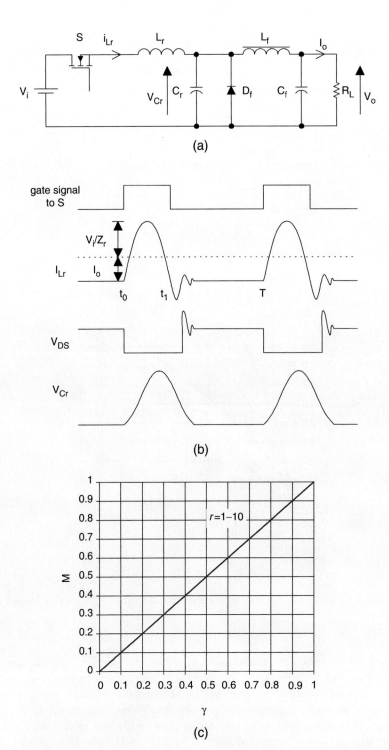

FIGURE 16.6 Full-wave, quasi-resonant buck converter with ZCS: (a) schematic diagram; (b) circuit waveforms; and (c) relationship between M and γ.

16 Resonant and Soft-switching Converters

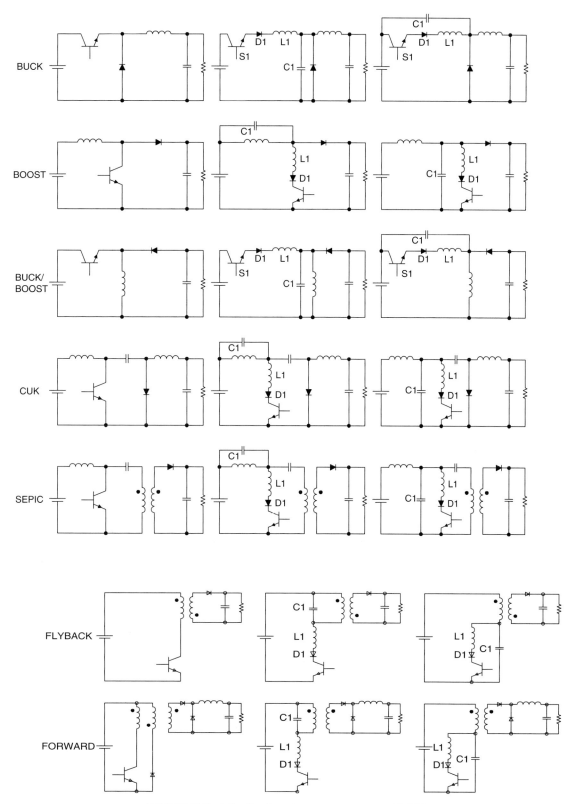

FIGURE 16.7 A family of quasi-resonant converter with ZCS.

A quasi-resonant buck converter designed for *half-wave* operation is shown in Fig. 16.8a – using a ZV resonant switch in Fig. 16.4b. The steady-state circuit waveforms are shown in Fig. 16.8b. Basic relations of ZVS-QRCs are given in Eqs. (16.1a–e). When the switch S is turned on, it carries the output current I_o. The supply voltage V_i reverse biases the diode D_f. When the switch is zero-voltage (ZV) turned off, the output current starts to flow through the resonant capacitor C_r. When the resonant capacitor voltage V_{Cr} is equal to V_i, D_f turns on. This starts the resonant stage. When V_{Cr} equals zero, the anti-parallel diode turns on. The resonant capacitor is shorted and the source voltage is applied to the resonant inductor L_r. The resonant inductor current I_{Lr} increases linearly until it reaches I_o. Then D_f turns off. In order to achieve ZVS, S should be triggered during the time when the anti-parallel diode conducts. It can be seen from the waveforms that the peak amplitude of the resonant capacitor voltage should be greater or equal to the input voltage (i.e. $I_o Z_r > V_{in}$). From Fig. 16.8c, it can be seen that the voltage conversion ratio is load-sensitive. In order to regulate the output voltage for different loads r, the switching frequency should also be changed accordingly.

ZVS converters can be operated in *full-wave* mode. The circuit schematic is shown in Fig. 16.9a. The circuit waveforms in steady state are shown in Fig. 16.9b. The operation is similar to *half-wave* mode of operation, except that V_{Cr} can swing between positive and negative voltages. The relationships between M and γ at different r are shown in Fig. 16.9c.

Comparing Fig. 16.8c with Fig. 16.9c, it can be seen that M is load-insensitive in *full-wave* mode. This is a desirable feature. However, as the series diode limits the direction of the switch current, energy will be stored in the output capacitance of the switch and will dissipate in the switch during turn on. Hence, the *full-wave* mode has the problem of capacitive turn-on loss, and is less practical in high frequency operation. In practice, ZVS-QRCs are usually operated in *half-wave* mode rather than *full-wave* mode.

By replacing the ZV resonant switch in the conventional converters, various ZVS-QRCs can be derived. They are shown in Fig. 16.10.

16.4.3 Comparisons between ZCS and ZVS

ZCS can eliminate the switching losses at turn off and reduce the switching losses at turn on. As a relatively large capacitor is connected across the output diode during resonance, the converter operation becomes insensitive to the diode's junction capacitance. When power MOSFETs are zero-current switched on, the energy stored in the device's capacitance will be dissipated. This capacitive turn-on loss is proportional to the switching frequency. During turn on, considerable rate of change of voltage can be coupled to the gate drive circuit through the Miller capacitor, thus increasing switching loss and noise. Another limitation is that the switches are under high current stress, resulting in higher conduction loss. However, it should be noted that ZCS is particularly effective in reducing switching loss for power devices (such as IGBT) with large tail current in the turn-off process.

ZVS eliminates the capacitive turn-on loss. It is suitable for high-frequency operation. For single-ended configuration, the switches could suffer from excessive voltage stress, which is proportional to the load. It will be shown in Section 16.5 that the maximum voltage across switches in half-bridge and full-bridge configurations is clamped to the input voltage.

For both ZCS and ZVS, output regulation of the resonant converters can be achieved by variable frequency control. ZCS operates with constant on-time control, while ZVS operates with constant off-time control. With a wide input and load range, both techniques have to operate with a wide switching frequency range, making it not easy to design resonant converters optimally.

16.5 ZVS in High Frequency Applications

By the nature of the resonant tank and ZCS, the peak switch current in resonant converters is much higher than that in the square-wave counterparts. In addition, a high voltage will be established across the switch in the off state after the resonant stage. When the switch is switched on again, the energy stored in the output capacitor will be discharged through the switch, causing a significant power loss at high frequencies and high voltages. This switching loss can be reduced by using ZVS.

ZVS can be viewed as square-wave power utilizing a constant off-time control. Output regulation is achieved by controlling the on time or switching frequency. During the off time, the resonant tank circuit traverses the voltage across the switch from zero to its peak value and then back to zero again. At that ZV instant, the switch can be reactivated. Apart from the conventional single-ended converters, some other examples of converters with ZVS are illustrated in the following section.

16.5.1 ZVS with Clamped Voltage

The high voltage stress problem in the single-switch configuration with ZVS can be avoided in half-bridge (HB) and full-bridge (FB) configurations [14–17]. The peak switch voltage can be clamped to the dc supply rail, and thus reducing the switch voltage stress. In addition, the series transformer leakage and circuit inductance can form parts of the resonant path. Therefore, these parasitic components, which are undesirable in hard-switched converter become useful components in ZVS ones. Figures 16.11 and 16.12 show the ZVS HB and FB circuits, respectively, together with the circuit waveforms.

16 Resonant and Soft-switching Converters

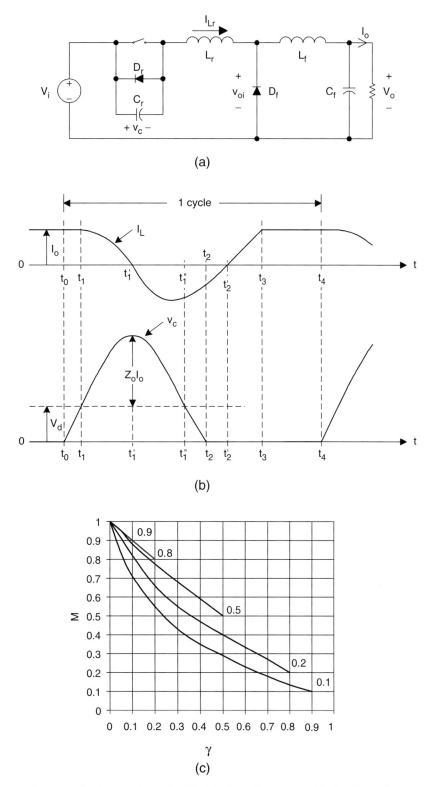

FIGURE 16.8 Half-wave, quasi-resonant buck converter with ZVS: (a) schematic diagram; (b) circuit waveforms; and (c) relationship between M and γ.

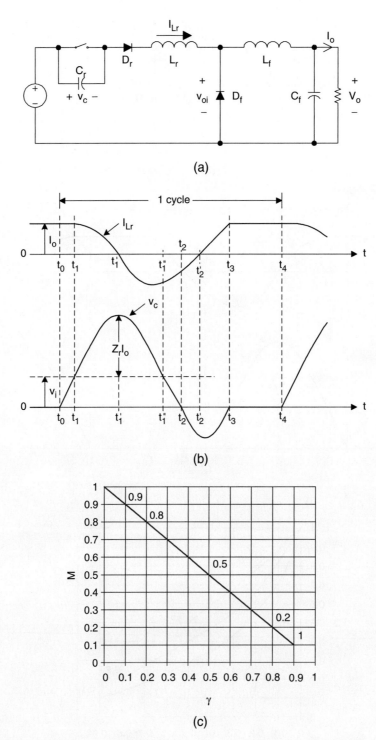

FIGURE 16.9 Full-wave, quasi-resonant buck converter with ZVS: (a) schematic diagram; (b) circuit waveforms; and (c) relationship between M and γ.

The resonant capacitor is equivalent to the parallel connection of the two capacitors ($C_r/2$) across the switches. The off-state voltage of the switches will not exceed the input voltage during resonance because they will be clamped to the supply rail by the anti-parallel diode of the switches.

16.5.2 Phase-shifted Converter with Zero Voltage Transition

In a conventional FB converter, the two diagonal switch pairs are driven alternatively. The output transformer is fed with an

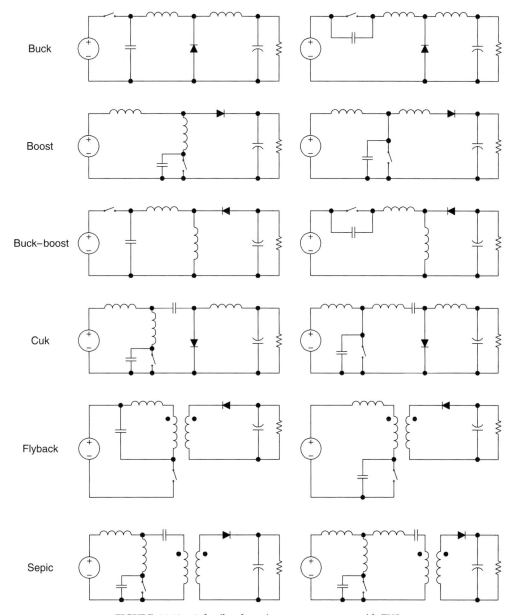

FIGURE 16.10 A family of quasi-resonant converter with ZVS.

ac rectangular voltage. By applying a phase-shifting approach, a deliberate delay can be introduced between the gate signals to the switches [18]. The circuit waveforms are shown in Fig. 16.13. Two upper or lower switches can be conducting (either through the switch or the anti-parallel diode), yet the applied voltage to the transformer is zero. This zero-voltage condition appears in the interval $[t_1, t_2]$ of V_{pri} in Fig. 16.13. This operating stage corresponds to the required off time for that particular switching cycle. When the desired switch is turned off, the primary transformer current flows into the switch output capacitance causing the switch voltage to resonate to the opposite input rail. Effects of the parasitic circuit components are used advantageously to facilitate the resonant transitions. This enables a ZVS condition for turning on the opposite switch. Thus, varying the phase shift controls the effective duty cycle and hence the output power. The resonant circuit is necessary to meet the requirement of providing sufficient inductive energy to drive the capacitors to the opposite bus rail. The resonant transition must be achieved within the designed transition time.

16.6 Multi-resonant Converters (MRC)

The ZCS- and ZVS-QRCs optimize the switching condition for either the active switch or the output diode only, but not

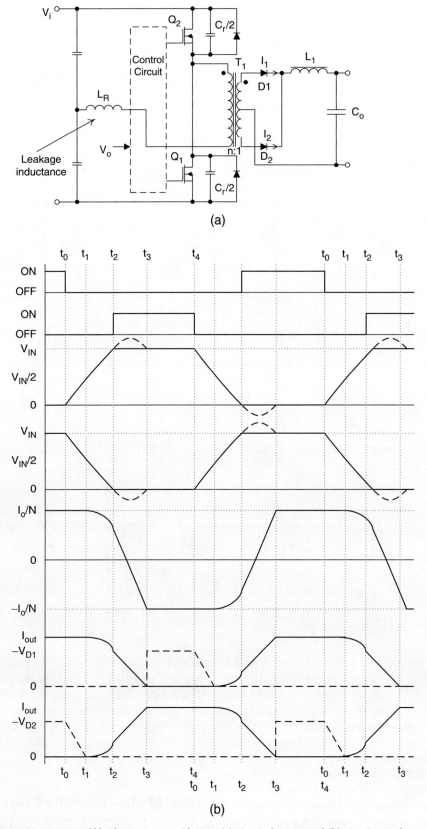

FIGURE 16.11 Half-bridge converter with ZVS: (a) circuit diagram and (b) circuit waveforms.

16 Resonant and Soft-switching Converters

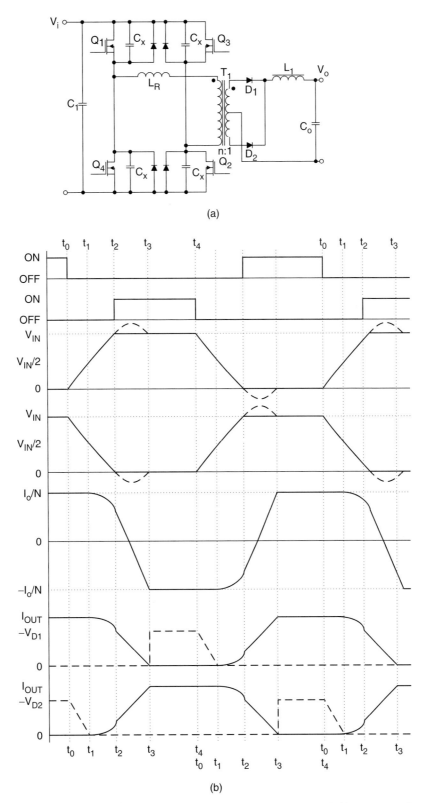

FIGURE 16.12 Full-bridge converter with ZVS: (a) circuit schematics and (b) circuit waveforms.

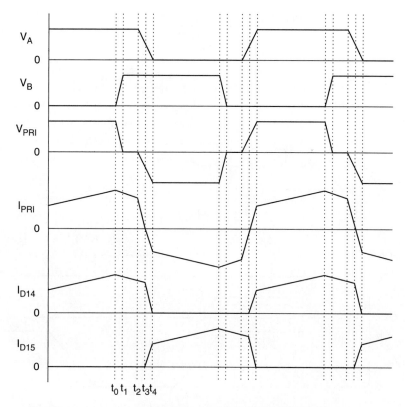

FIGURE 16.13 Circuit waveforms of the phase-shifted, ZVT FB converter.

for both of them simultaneously. Multi-resonant switch concept, which is an extension of the concept of the resonant switch, has been developed to overcome such limitation. The zero-current multi-resonant (ZC-MR) and zero-voltage multi-resonant (ZV-MR) switches [12, 17] are shown in Fig. 16.14.

The multi-resonant circuits incorporate all major parasitic components, including switch output capacitance, diode junction capacitance, and transformer leakage inductance into the resonant circuit. In general, ZVS (*half-wave* mode) is more favorable than ZCS in DC–DC converters for high-frequency operation because the parasitic capacitance of the active switch and the diode will form a part of the resonant circuit.

An example of a buck ZVS-MRC is shown in Fig. 16.15. Depending on the ratio of the resonant capacitance C_D/C_S, two possible topological modes, namely mode I and mode II, can be operated [19]. The ratio affects the time at which the voltages across the switch S and the output diode D_f become zero. Their waveforms are shown in Figs. 16.16a and b, respectively. If diode voltage V_D falls to zero earlier than the switch

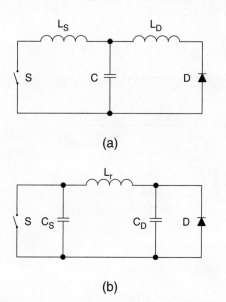

FIGURE 16.14 Multi-resonant switches: (a) ZC-MR switch and (b) ZV-MR switch.

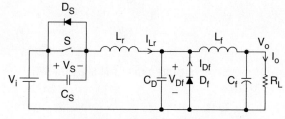

FIGURE 16.15 Buck ZVS-MRC.

16 Resonant and Soft-switching Converters

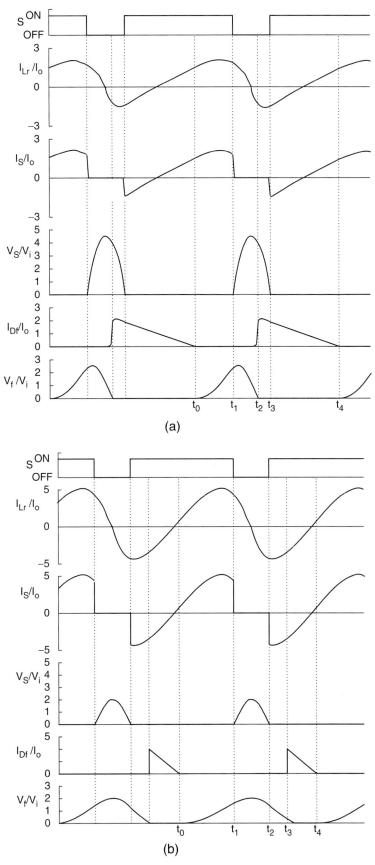

FIGURE 16.16 Possible modes of the buck ZVS-MRC: (a) mode I and (b) mode II.

voltage V_S, the converter will follow mode I. Otherwise, the converter will follow mode II.

Instead of having one resonant stage, there are three in this converter. The mode I operation in Fig. 16.16a is described first. Before the switch S is turned on, the output diode D_f is conducting and the resonant inductor current I_{Lr} is negative (flowing through the anti-parallel diode of S). S is then turned on with ZVS. The resonant inductor current I_{Lr} increases linearly and D_f is still conducting. When I_{Lr} reaches the output current I_o, the first resonant stage starts. The resonant circuit is formed by the resonant inductor L_r and the capacitor C_D across the output diode. This stage ends when S is turned off with ZVS. Then, a second resonant stage starts. The resonant circuit consists of L_r, C_D, and the capacitor across the switch C_s. This stage ends when the output diode becomes forward biased. A third resonant stage will then start. L_r and C_s form the resonant circuit. This stage ends and completes one operation cycle when the diode C_s becomes forward biased.

The only difference between mode I and mode II in Fig. 16.16b is in the third resonant stage, in which the resonant circuit is formed by L_r and C_D. This stage ends when D_f becomes forward biased. The concept of the multi-resonant switch can be applied to conventional converters [19–21]. A family of MRCs are shown in Fig. 16.17.

Although the variation of the switching frequency for regulation in MRCs is smaller than that of QRCs, a wide-band frequency modulation is still required. Hence, the optimal design of magnetic components and the EMI filters in MRCs is not easy. It would be desirable to have a constant switching frequency operation. In order to operate the MRCs with constant switching frequency, the diode in Fig. 16.14 can be replaced with an active switch S_2 [22]. A constant-frequency multi-resonant (CF-MR) switch is shown in Fig. 16.18. The output voltage is regulated by controlling the on-time of the two switches. This concept can be illustrated with the buck converter as shown in Fig. 16.19, together with the gate drive waveforms and operating stages. S_1 and S_2 are turned on during the time when currents flow through the anti-parallel diodes of S_1 and S_2. This stage ends when S_2 is turned off with ZVS. The first resonant stage is then started. L_r and C_{S2} form the resonant circuit. A second resonant stage begins. L_r resonates with C_{S1} and C_{S2}. The voltage across S_1 oscillates to zero. When I_{Lr} becomes negative, S_1 will be turned on with ZVS. Then, L_r resonates with C_{S2}. S_2 will be turned on when current flows through D_{S2}. As the output voltage is the average voltage across S_2, output voltage regulation is achieved by controlling the conduction time of S_2.

All switches in MRCs operate with ZVS, which reduces the switching losses and switching noise and eliminates the oscillation due to the parasitic effects of the components (such as the junction capacitance of the diodes). However, all switches are under high current and voltage stresses, resulting in an increase in the conduction loss.

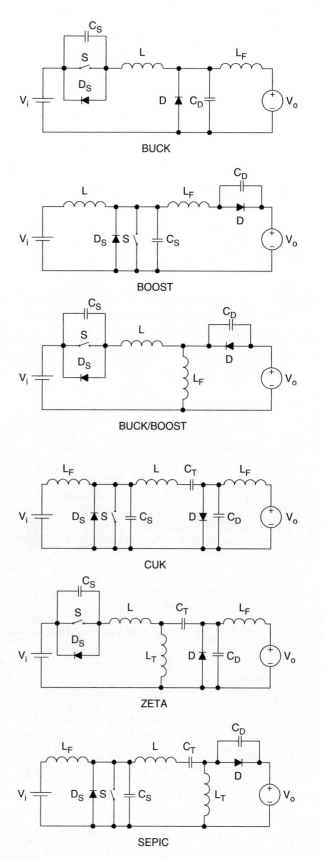

FIGURE 16.17 Use of the multi-resonant switch in conventional PWM converters.

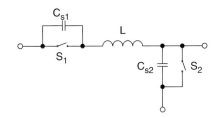

FIGURE 16.18 Constant frequency multi-resonant switch.

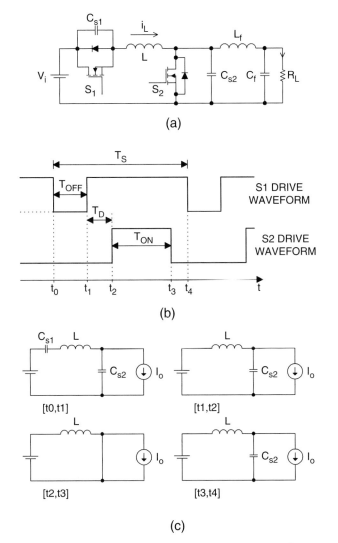

FIGURE 16.19 Constant frequency buck MRC: (a) circuit schematics; (b) gate drive waveforms; and (c) operating stages.

16.7 Zero-voltage-transition (ZVT) Converters

By introducing a resonant circuit in parallel with the switches, the converter can achieve ZVS for both power switch and diode without significantly increasing their voltage and current stresses [23]. Figure 16.20a shows a buck type ZVT-PWM converter and Fig. 16.20b shows the associated waveforms. The converter consists of a main switch S and an auxiliary switch S_1. It can be seen that the voltage and current waveforms of the switches are square-wave-like except during turn-on and turn-off switching intervals, where ZVT takes place. The main switch and the output diode are under ZVS and are subjected to low voltage and current stresses. The auxiliary switch is under ZCS, resulting in low switching loss.

The concept of ZVT can be extended to other PWM circuits by adding the resonant circuit. Some basic ZVT-PWM converters are shown in Fig. 16.21.

16.8 Non-dissipative Active Clamp Network

The active-clamp circuit can utilize the transformer leakage inductance energy and can minimize the the turn-off voltage stress in the isolated converters. The active clamp circuit provides a means of achieving ZVS for the power switch and reducing the rate of change of the diode's reverse recovery current. An example of a flyback converter with active clamp is shown in Fig. 16.22a and the circuit waveforms are shown in Fig. 16.22b. Clamping action is obtained by using a series combination of an active switch (i.e. S_2) and a large capacitor so that the voltage across the main switch (i.e. S_1) is clamped to a minimum value. S_2 is turned on with ZVS. However, S_2 is turned off with finite voltage and current, and has turn-off switching loss. The clamp-mode ZVS-MRCs is discussed in [24–26].

16.9 Load Resonant Converters

Load resonant converters (LRCs) have many distinct features over conventional power converters. Due to the soft commutation of the switches, no turn-off loss or stress is present. LRCs are specially suitable for high-power applications because they allow high-frequency operation for equipment size/weight reduction, without sacrificing the conversion efficiency and imposing extra stress on the switches. Basically, LRCs can be divided into three different configurations, namely series resonant converters, parallel resonant converters, and series–parallel resonant converters.

16.9.1 Series Resonant Converters

Series resonant converters (SRCs) have their load connected in series with the resonant tank circuit, which is formed by L_r and C_r [15, 27–29]. The half-bridge configuration is shown in Fig. 16.23. When the resonant inductor current i_{Lr} is positive, it flows through T_1 if T_1 is on; otherwise it flows through the diode D_2. When i_{Lr} is negative, it flows through T_2 if T_2 is on;

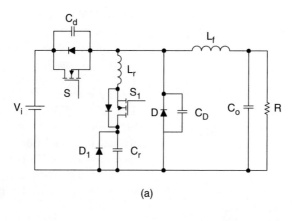

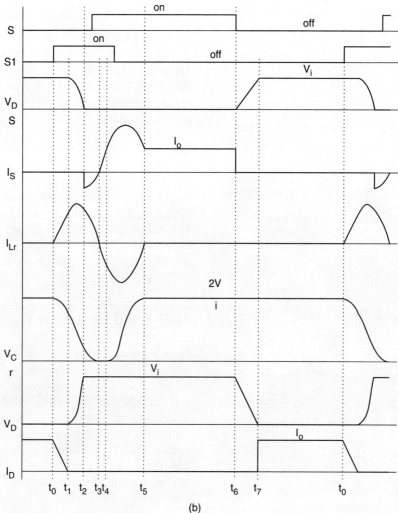

FIGURE 16.20 Buck ZVT-PWM converter: (a) circuit schematics and (b) waveforms.

otherwise it flows through the diode D_1. In the steady-state symmetrical operation, both the active switches are operated in a complementary manner. Depending on the ratio between the switching frequency ω_S and the converter resonant frequency ω_r, the converter has several possible operating modes.

A. Discontinuous Conduction Mode (DCM) with $\omega_S < 0.5\omega_r$

Figure 16.24a shows the waveforms of i_{Lr} and the resonant capacitor voltage v_{Cr} in this mode of operation. From 0 to t_1, T_1 conducts. From t_1 to t_2, the current in T_1 reverses its direction. The current flows through D_1 and back to the

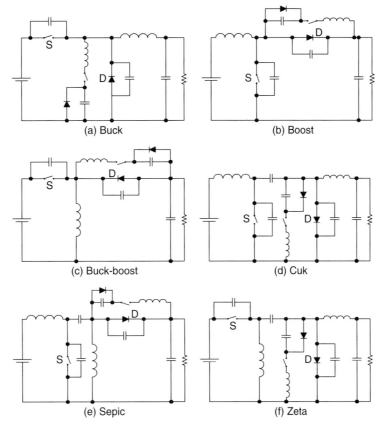

FIGURE 16.21 Conventional ZVT-PWM converters.

supply source. From t_2 to t_3, all switches are in the off state. From t_3 to t_4, T_2 conducts. From t_4 to t_5, the current in T_2 reverses its direction. The current flows through D_2 and back to the supply source. T_1 and T_2 are switched on under ZCS condition and they are switched off under zero-current and zero-voltage conditions. However, the switches are under high current stress in this mode of operation and thus have higher conduction loss.

B. Continuous Conduction Mode (CCM) with $0.5\omega_r < \omega_S < \omega_r$

Figure 16.24b shows the circuit waveforms. From 0 to t_1, i_{Lr} transfers from D_2 to T_1. T_1 is switched on with finite switch current and voltage, resulting in turn-on switching loss. Moreover, the diodes must have good reverse recovery characteristics in order to reduce the reverse recovery current. From t_1 to t_2, D_1 conducts and T_1 is turned off softly with zero voltage and zero current. From t_2 to t_3, T_2 is switched on with finite switch current and voltage. At t_3, T_2 is turned off softly and D_2 conducts until t_4.

C. Continuous Conduction Mode (CCM) with $\omega_r < \omega_S$

Figure 16.24c shows the circuit waveforms. From 0 to t_1, i_{Lr} transfers from D_1 to T_1. Thus, T_1 is switched on with zero current and zero voltage. At t_1, T_1 is switched off with finite voltage and current, resulting in turn-off switching loss. From t_1 to t_2, D_2 conducts. From t_2 to t_3, T_2 is switched on with zero current and zero voltage. At t_3, T_2 is switched off. i_{Lr} transfers from T_2 to D_1. As the switches are turned on with ZVS, lossless snubber capacitors can be added across the switches.

The following parameters are defined: voltage conversion ratio M, characteristic impedance Z_r, resonant frequency f_r, normalized load resistance r, normalized switching frequency γ.

$$M = nV_o/V_{in} \qquad (16.2a)$$

$$Z_r = \sqrt{L_r/C_r} \qquad (16.2b)$$

$$f_r = 1/\left(2\pi\sqrt{L_rC_r}\right) \qquad (16.2c)$$

$$r = n^2 R_L/Z_r \qquad (16.2d)$$

$$\gamma = f_s/f_r \qquad (16.2e)$$

$$M = 1\left/\sqrt{(\gamma - 1/\gamma)^2/(r^2 + 1)}\right. \qquad (16.2f)$$

The relationships between M and γ for different value of r are shown in Fig. 16.25. The boundary between CCM and DCM

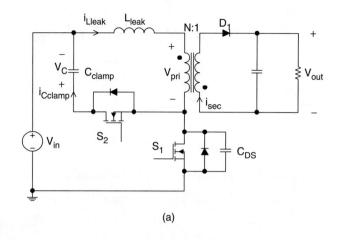

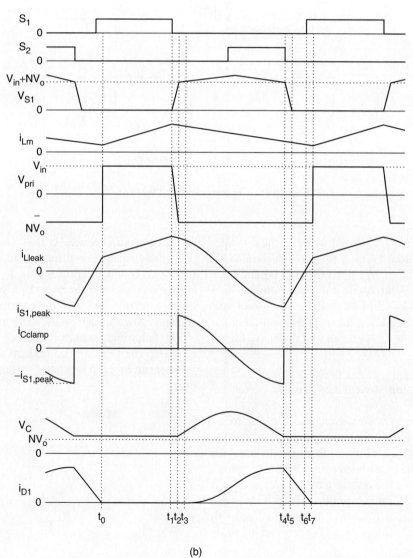

FIGURE 16.22 Active-clamp flyback converter: (a) circuit schematics and (b) circuit waveforms.

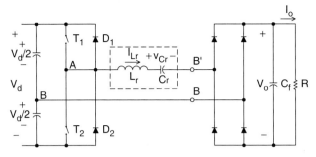

FIGURE 16.23 SRC half-bridge configuration.

is at $r = 1.27\gamma$. When the converter is operating in DCM and $0.2 < \gamma < 0.5$, $M = 1.27 \, r\gamma$.

The SRC has the following advantages. Transformer saturation can be avoided since the series capacitor can block the dc component. The light load efficiency is high because the device current and conduction loss are low. However, the major disadvantages are that there is difficulty in regulating the output voltage under light load and no load conditions. Moreover, the output dc filter capacitor has to carry high ripple current, which could be a major problem in low-output voltage and high-output current applications [29].

16.9.2 Parallel Resonant Converters

Parallel resonant converters (PRCs) have their load connected in parallel with the resonant tank capacitor C_r [27–30]. The half-bridge configuration is shown in Fig. 16.26. SRC behaves as a current source, whereas the PRC acts as a voltage source. For voltage regulation, PRC requires a smaller operating frequency range than the SRC to compensate for load variation.

A. Discontinuous Conduction Mode (DCM)

The steady-state waveforms of the resonant inductor current i_{Lr} and the resonant capacitor voltage v_{Cr} are shown in Fig. 16.27a. Initially both i_{Lr} and v_{Cr} are zero. From 0 to t_2, T_1 conducts and is turned on with zero current. When i_{Lr} is less than the output current I_o, i_{Lr} increases linearly from 0 to t_1 and the output current circulates through the diode bridge. From t_1 to t_3, L_r resonates with C_r. Starting from t_2, i_{Lr} reverses its direction and flows through D_1. T_1 is then turned off with zero current and zero voltage. From t_3 to t_4, v_{Cr} decreases linearly due to the relatively constant value of I_o. At t_4, when v_{Cr} equals zero, the output current circulates through the diode bridge again. Both i_{Lr} and v_{Cr} will stay at zero for an interval. From t_5 to t_9, the above operations will be repeated for T_2 and D_2. The output voltage is controlled by adjusting the time interval of $[t_4, t_5]$.

B. Continuous Conduction Mode $\omega_S < \omega_r$

This mode is similar to the operation in the DCM, but with a higher switching frequency. Both i_{Lr} and v_{Cr} become continuous. The waveforms are shown in Fig. 16.27b. The switches T_1 and T_2 are hard turned on with finite voltage and current and are soft turned off with ZVS.

C. Continuous Conduction Mode $\omega_S > \omega_r$

If the switching frequency is higher than ω_r (Fig. 16.27c), the anti-parallel diode of the switch will be turned on before the switch is triggered. Thus, the switches are turned on with ZVS. However, the switches are hard turned off with finite current and voltage.

The parameters defined in Eq. (16.2) are applicable. The relationships between M and γ for various values of r are shown in Fig. 16.28. During the DCM (i.e. $\gamma < 0.5$), M is in linear relationship with γ. Output voltage regulation can be achieved easily. The output voltage is independent on the output current. The converter shows a good voltage source characteristics. It is also possible to step up and step down the input voltage.

The PRC has the advantages that the load can be short-circuited and the circuit is suitable for low-output voltage, high-output current applications. However, the major disadvantage of the PRC is the high device current. Moreover, since the device current do not decrease with the load, the efficiency drops with a decrease in the load [29].

16.9.3 Series–Parallel Resonant Converter

Series–Parallel Resonant Converter (SPRC) combines the advantages of the SRC and PRC. The SPRC has an additional capacitor or inductor connected in the resonant tank circuit [29–31]. Figure 16.29a shows an LCC-type SPRC, in which an additional capacitor is placed in series with the resonant inductor. Figure 16.29b shows an LLC-type SPRC, in which an additional inductor is connected in parallel with the resonant capacitor in the SRC. However, there are many possible combinations of the resonant tank circuit. Detailed analysis can be found in [31].

16.10 Control Circuits for Resonant Converters

Since the 1985s, various control integrated circuits (ICs) for resonant converters have been developed. Some common ICs for different converters are described in this section.

16.10.1 QRCs and MRCs

Output regulations in many resonant-type converters, such as QRCs and MRCs, are achieved by controlling the

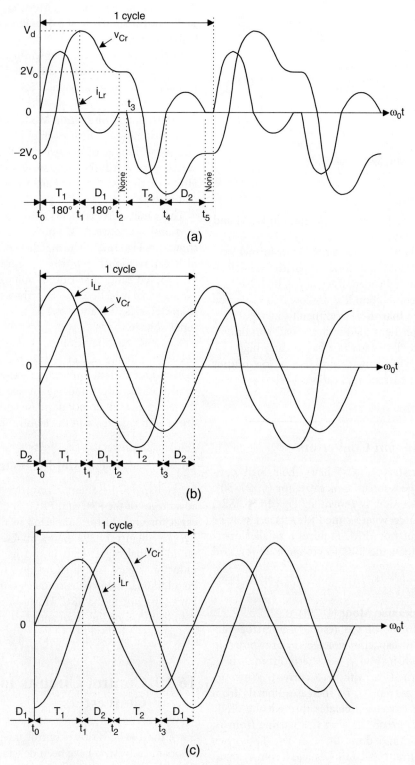

FIGURE 16.24 Circuit waveforms under different operating conditions: (a) $\omega_S < 0.5\,\omega_r$; (b) $0.5\,\omega_r < \omega_S < \omega$; and (c) $\omega_r < \omega_S$.

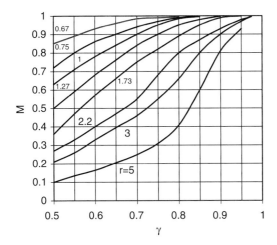

FIGURE 16.25 M vs γ in SRC.

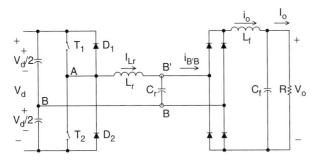

FIGURE 16.26 PRC half-bridge configuration.

switching frequency. ZCS applications require controlled switch-on times while ZVS applications require controlled switch-off times. The fundamental control blocks in the IC include an error amplifier, voltage controlled oscillator (VCO), one shot generator with a zero wave-crossing detection comparator, and an output stage to drive the active switch. Typical ICs include UC1861–UC1864 for ZVS applications and UC1865–UC1868 for ZCS applications [32]. Figure 16.30 shows the controller block diagram of UC1864.

The maximum and minimum switching frequencies (i.e. f_{max} and f_{min}) are controlled by the resistors R_{range} and R_{min} and the capacitor C_{vco}. f_{max} and f_{min} can be expressed as

$$f_{max} = \frac{3.6}{(R_{range}//R_{min})C_{VCO}} \quad \text{and} \quad f_{min} = \frac{3.6}{R_{min}C_{VCO}} \tag{16.3}$$

The frequency range Δf is then equal to

$$\Delta f = f_{max} - f_{min} = \frac{3.6}{R_{range}C_{VCO}} \tag{16.4}$$

The frequency range of the ICs is from 10 kHz to 1 MHz. The output frequency of the oscillator is controlled by the error amplifier (E/A) output. An example of a ZVS-MR forward converter is shown in Fig. 16.31.

16.10.2 Phase-shifted, ZVT FB Circuit

The UCC3895 is a phase shift PWM controller that can generate a phase shifting pattern of one half-bridge with respect to the other. The application diagram is shown in Fig. 16.32.

The four outputs "OUTA," "OUTB," "OUTC," and "OUTD" are used to drive the MOSFETs in the full-bridge. The dead time between "OUTA" and "OUTB" is controlled by "DELAB" and the dead time between "OUTC" and "OUTD" is controlled by "DELCD." Separate delays are provided for the two half-bridges to accommodate differences in resonant capacitor charging currents. The delay in each set is approximated by

$$t_{DELAY} = \frac{25 \times 10^{-12} R_{DEL}}{0.75(V_{CS} - V_{ADS}) + 0.5} + 25\,ns \tag{16.5}$$

where R_{DEL} is the resistor value connected between "DELAB" or "DELCD" to ground.

The oscillator period is determined by R_T and C_T. It is defined as

$$t_{OSC} = \frac{5R_T C_T}{48} + 120\,ns \tag{16.6}$$

The maximum operating frequency is 1 MHz. The phase shift between the two sets of signals is controlled by the ramp voltage and the error amplifier output having a 7 MHz bandwidth.

16.11 Extended-period Quasi-resonant (EP-QR) Converters

Generally, resonant and quasi-resonant converters operate with frequency control. The extended-period quasi-resonant converters proposed by Barbi [33] offer a simple solution to modify existing hard-switched converters into soft-switched ones with constant frequency operation. This makes both output filter design and control simple. Figure 16.33 shows a standard hard-switched boost type PFC converter. In this hard-switched circuit, the main switch SW1 could be subject to significant switching stress because the reverse recovery current of the diode D_F could be excessive when SW1 is turned on. In practice, a small saturable inductor may be added in series with the power diode D_F in order to reduce the di/dt of the reverse-recovery current. In addition, an optional R–C snubber may be added across SW1 to reduce the dv/dt of SW1. These extra reactance components can in fact be used in the EP-QR circuit to achieve soft switching, as shown in Fig. 16.34. The resonant components L_r and C_r are of small

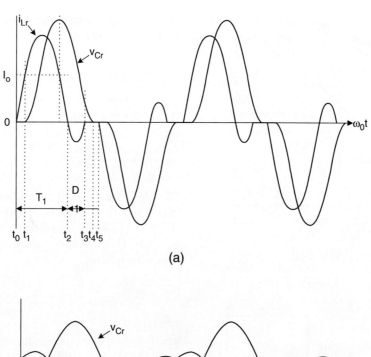

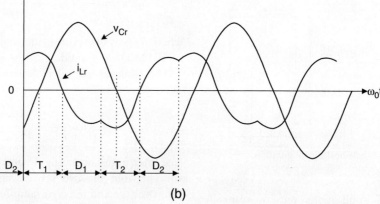

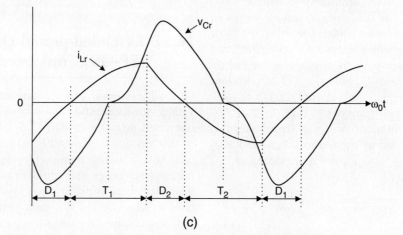

FIGURE 16.27 Circuit waveforms under different operating conditions: (a) discontinuous conduction mode; (b) continuous conduction mode $\omega_S < \omega_r$; and (c) continuous conduction mode $\omega_S > \omega_r$.

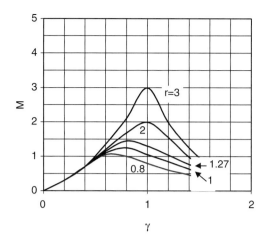

FIGURE 16.28 M vs γ in PRC.

values and can come from the snubber circuits of a standard hard-switched converter. Thus, the only additional component is the auxiliary switch Q2. The small resonant inductor is put in series with the main switch SW1 so that SW1 can be switched on under ZC condition and the di/dt problem of the reverse-recovery current be eliminated. The resonant capacitor C_r is used to store energy for creating condition for soft switching. Q2 is used to control the resonance during the main switch transition. It should be noted that all power devices including SW1, Q1 and main power diode D_F are turned on and off under ZV and/or ZC conditions. Therefore, the large di/dt problem due to the reverse recovery of the power diode can be eliminated. The soft-switching method is an effective technique for EMI suppression.

Together with power factor correction technique, soft-switching converters offer a complete solution to meet EMI regulations for both conducted and radiated EMI. The operation of the EP-QR boost PFC circuit [34, 35] can be described in six modes as shown in Fig. 16.35. The corresponding idealized waveforms are included in Fig. 16.36.

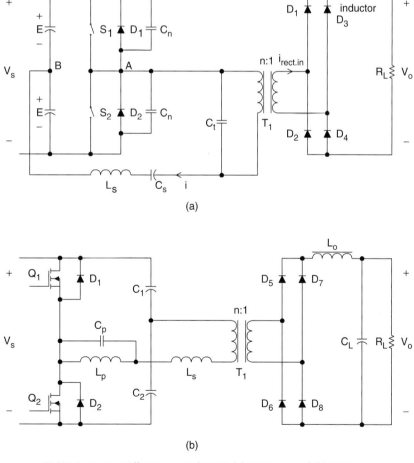

FIGURE 16.29 Different types of SPRC: (a) LCC-type and (b) LLC-type.

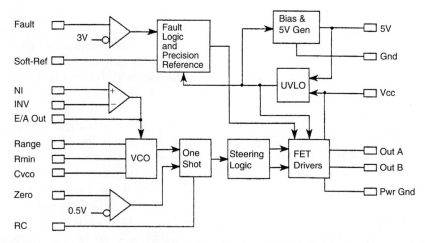

FIGURE 16.30 Controller block diagram of UC1864 (Courtesy of Unitrode Corp. and Texas Instruments).

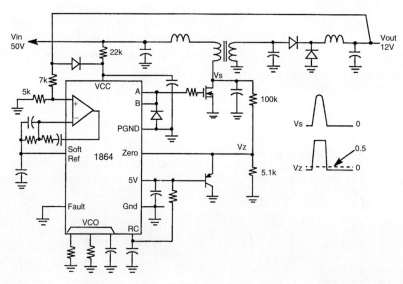

FIGURE 16.31 ZV-MR forward converter (Courtesy of Unitrode Corp. and Texas Instruments).

A. Circuit Operation

Interval I: (t_0–t_1) Due to the resonant inductor L_r which limits the di/dt of the switch current, switch SW1 is turned on at zero-current condition with a positive gating signal V_{GS1} to start a switching cycle at $t = t_0$. Current in D_F is diverted to inductor L_r. Because D_F is still conducting during this short period, D_{S2} is still reverse biased and is thus not conducting. The equivalent circuit topology for the conducting paths is shown in Fig. 16.35a. Resonant switch Q_2 remains off in this interval.

Interval II: (t_1–t_a) When D_F regains its blocking state, D_{S2} becomes forward biased. The first half of the resonance cycle occurs and resonant capacitor C_r starts to discharge and current flows in the loop C_r–Q_2–L_r–SW_1. The resonance half-cycle stops at time $t = t_a$ because D_{S2} prevents the loop current i_{Cr} from flowing in the opposition direction. The voltage across C_r is reversed at the end of this interval. The equivalent circuit is shown in Fig. 16.35b.

Interval III: (t_a–t_b) Between t_a and t_b, current in L_F and L_r continues to build up. This interval is the extended-period for the resonance during which energy is pumped into L_r. The corresponding equivalent circuit is showed in Fig. 16.35c.

Interval IV: (t_b–t_2) Figure 16.35d shows the equivalent circuit for this operating mode. Before SW_1 is turned off, the second half of the resonant cycle needs to take place in order that a zero-voltage condition can be created for the turn-off process of SW_1. The second half of the resonant cycle starts when auxiliary switch Q_2 is turned on at $t = t_b$. Resonant current then flows through the loop L_r–Q_2–C_r–anti-parallel diode

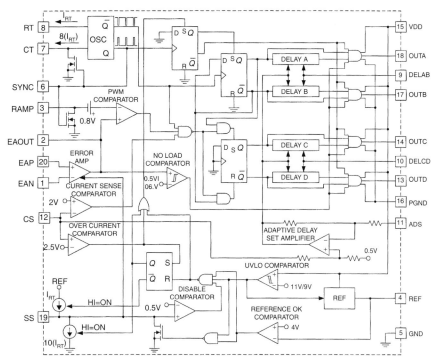

FIGURE 16.32 Application diagram of UCC3895 (Courtesy of Unitrode Corp. and Texas Instruments).

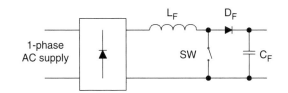

FIGURE 16.33 Boost-type AC–DC power factor correction circuit.

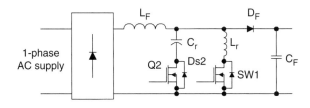

FIGURE 16.34 EP-QR boost-type AC–DC power factor correction circuit.

of SW_1. This current is limited by L_r and thus Q_2 is turned on under zero-current condition. Since the anti-parallel diode of SW_1 is conducting, the voltage across SW_1 is clamped to the on-state voltage of the anti-parallel diode. SW_1 can therefore be turned off at (near) zero-voltage condition before $t = t_2$ at which the second half of the resonant cycle ends.

Interval V: (t_2–t_3) During this interval, the voltage across C_r is less than the output voltage V_o. Therefore D_F is still reverse biased. Inductor current I_s flows into C_r until V_{Cr} reaches V_o at $t = t_3$. The equivalent circuit is represented in Fig. 16.35e.

Interval VI: (t_3–t_4) During this period, the resonant circuit is not in action and the inductor current I_s charges the output capacitor C_F via D_F, as in the case of a classical boost-type PFC circuit. C_r is charged to V_o, therefore Q_2 can be turned off at zero-voltage and zero-current conditions. Figure 16.35f shows the equivalent topology of this operating mode.

In summary, SW1, Q2, and D_F are fully soft-switched. Since the two resonance half-cycles take place within a closed loop outside the main inductor, the high resonant pulse will not occur in the inductor current, thus making a well-established averaged current mode control technique applicable for such QR circuit. For full soft-switching in the turn-off process, the resonant components need to be designed so that the peak resonant current exceeds the maximum value of the inductor current. Typical measured switching waveforms and trajectories of SW1, Q2 and D_F are shown in Figs. 16.37, 16.38, and 16.39, respectively.

B. Design Procedure

Given: Input AC voltage = V_s (V)
Peak AC voltage = $V_s(max)$ (V)
Nominal output DC voltage = V_o (V)
Switching frequency = f_{sw} (Hz)
Output power = P_o (W)

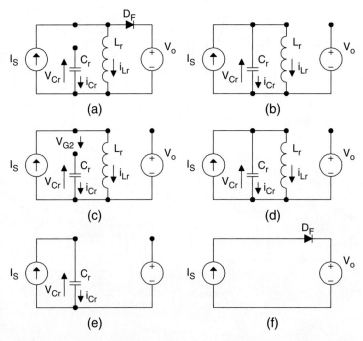

FIGURE 16.35 Operating modes of EP-QR boost-type AC–DC power factor correction circuit.

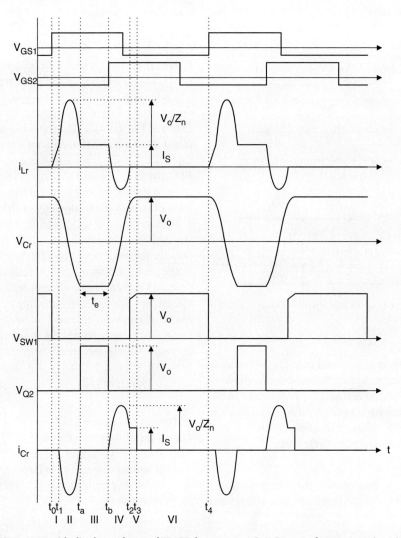

FIGURE 16.36 Idealized waveforms of EP-QR boost-type AC–DC power factor correction circuit.

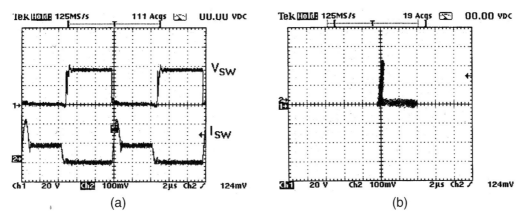

FIGURE 16.37 (a) Drain-source voltage and current of SW1 and (b) switching locus of SW1.

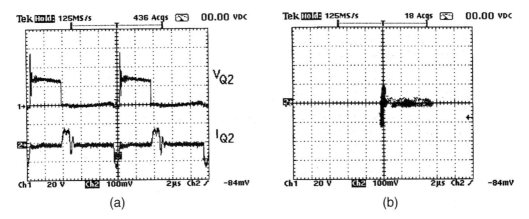

FIGURE 16.38 (a) Drain-source voltage and current of Q2 and (b) switching locus of Q2.

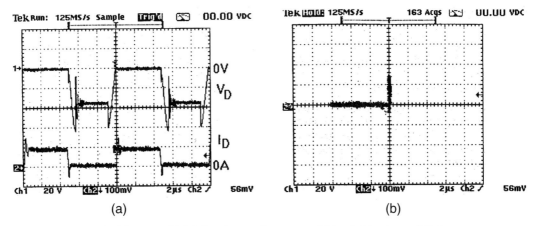

FIGURE 16.39 (a) Diode voltage and current and (b) switching locus of diode.

Input current ripple = ΔI (A)
Output voltage ripple = ΔV (V)

(I) Resonant tank design:
Step 1: Because the peak resonant current must be greater than the peak inductor current (same as peak input line current) in order to achieve soft-turn-off, it is necessary to determine the peak input current $I_s(max)$. Assuming lossless AC–DC power conversion, $I_s(max)$ can be estimated from the following equation

$$I_{s(max)} \approx \frac{2V_o I_o}{V_{s(max)}} \qquad (16.7)$$

where $I_o = P_o/V_o$ is the maximum output current.
Step 2: Soft-switching criterion is

$$Z_r \le \frac{V_o}{I_{s(max)}} \qquad (16.8)$$

where $Z_r = \sqrt{\frac{L_r}{C_r}}$ is the impedance of the resonant tank.

For a chosen resonant frequency f_r, L_r, and C_r can be obtained from:

$$2\pi f_r = \frac{1}{\sqrt{L_r C_r}} \qquad (16.9)$$

(II) Filter component design:
The minimum conversion ratio is

$$M_{(min)} = \frac{V_o}{V_{s(max)}}$$
$$= \frac{1}{1 - \left(\frac{f_{sw}}{f_r} + \frac{t_e}{T_{sw}}\right)} \qquad (16.10)$$

where $T_{sw} = 1/f_{sw}$ and t_e is the extended period. From Eq. (16.10), minimum t_e can be estimated.

The turn-on period of the SW1 is

$$T_{on}(sw1) = t_e + 1/f_r \qquad (16.11)$$

Inductor value L is obtained from:

$$L \ge \left(\frac{T_{on(sw1)}}{\Delta I}\right) V_{s(max)} \qquad (16.12)$$

The filter capacitor value C can be determined from:

$$C\left(\frac{\Delta V}{\frac{T_s}{\pi} \sin^{-1}\left(\frac{I_o}{I_{s(max)}}\right)}\right) = I_o \qquad (16.13)$$

where $T_s = 1/f_s$ is the period of the AC voltage supply frequency.

16.11.1 Soft-switched DC–DC Flyback Converter

A simple approach that can turn an existing hard-switched converter design into a soft-switched one is shown in Fig. 16.40. The key advantage of the proposal is that many well proven and reliable hard-switched converter designs can be kept. The modification required is the addition of a simple circuit (consisting an auxiliary winding, a switch, and a small capacitor) to an existing isolated converter [36]. This principle, which is the modified version of the EPQR technique for isolated converter, is demonstrated in an isolated soft-switched flyback converter with multiple outputs. Other advantageous features of the proposal are:

- All switches and diodes of the converter are 'fully' soft-switched, i.e. soft-switched at both turn-on and turn-off transitions under zero-voltage and/or zero-current conditions.
- The leakage inductance of each winding in the flyback transformer forms part of the resonant circuit for achieving ZVS and ZCS of all switches and diodes.
- The control technique is simply PWM-based as in standard hard-switched converters.
- The soft-switched technique is a proven method for EMI reduction [37].

16.11.2 A ZCS Bidirectional Flyback DC–DC Converter

A bi-directional flyback dc–dc converter that uses one auxiliary circuit for both power flow directions is proposed in Fig. 16.41 [38]. The methodology is based on extending the unidirectional soft-switched flyback converter in [36] and replacing the output diode with a controlled switch, which acts as either a rectifier [39] or a power control switch in the corresponding power flow direction. An auxiliary circuit that consists of a winding in the coupled inductor, a switch, and a capacitor converts the hard-switched design into a soft-switched one. The operation is the same as [36] in the forward mode. An extended-period resonant stage [34] is introduced when the power control switch is on. Conversely, in the reverse mode, a complete resonant stage is initiated before the main switch is off. In both the power flow operations, the leakage inductance of the coupled inductor is used to create zero-current switching conditions for all switches.

16.12 Soft-switching and EMI Suppression

A family of EP-QR converters are displayed in Fig. 16.42. Their radiated EMI emission have been compared with that from their hard-switched counterparts [37]. Figures 16.43a, b show the conducted EMI emission from a hard-switched flyback

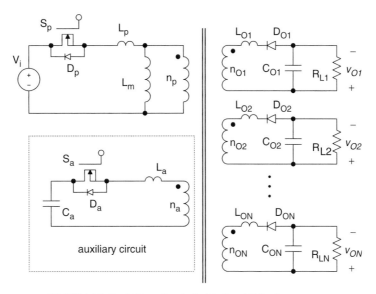

FIGURE 16.40 Fully soft-switched isolated flyback converter.

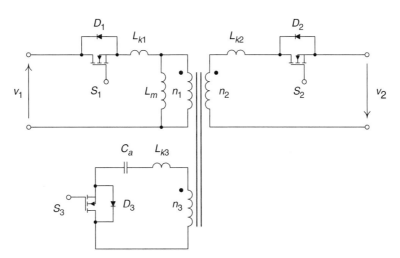

FIGURE 16.41 Bidirectional soft-switched isolated flyback converter.

converter and a soft-switched one, respectively. Their radiated EMI emissions are included in Fig. 16.44. Both converters are tested at an output power of 50 W. No special filtering or shielding measures have been taken during the measurement. It is clear from the measurements that soft-switching is an effective means to EMI suppression.

16.13 Snubbers and Soft-switching for High Power Devices

Today, most of the medium power (up to 200 kVA) and medium voltage (up to 800 V) inverter are hard-switched. Compared with low-power switched mode power supplies, the high voltage involved in the power inverters makes the dv/dt, di/dt, and the switching stress problems more serious. In addition, the reverse recovery of power diodes in the inverter leg may cause very sharp current spike, leading to severe EMI problem. It should be noted that some high power devices such as GTO thyristors do not have a square safe operating area (SOA). It is therefore essential that the switching stress they undergone must be within their limits. Commonly used protective measures are to use snubber circuits for protecting high power devices.

Among various snubbers, two snubber circuits are most well-known for applications in power inverters. They are the Undeland snubber [40] (Fig. 16.45) and McMurray snubber circuits [41] (Fig. 16.46). The Undeland snubber is an asymmetric snubber circuit with one turn-on inductor and one turn-off capacitor. The turn-off snubber capacitor C_s is clamped by another capacitor C_c. At the end of each switching

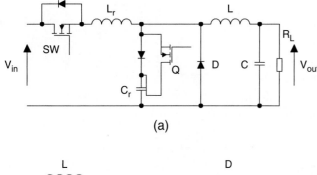

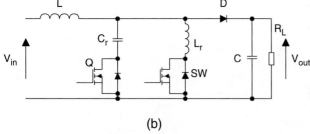

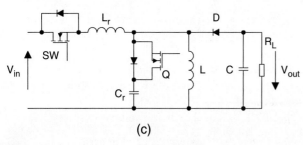

FIGURE 16.42 A family of EP-QR converters: (a) buck converter; (b) boost converter; and (c) flyback converter.

cycle, the snubber energy is dumped into C_c and then discharged into the dc bus via a discharge resistor. In order to reduce the snubber loss, the discharge resistor can be replaced by a switched mode circuit. In this way, the Undeland snubber can become a snubber with energy recovery. The McMurray snubber is symmetrical. Both the turn-off snubber capacitors share current in parallel during turn off. The voltage transient is limited by the capacitor closest to the turning-off device because the stray inductance to the other capacitor will prevent instantaneous current sharing. The turn-on inductors require mid-point connection. Snubber energy is dissipated into the snubber resistor. Like the Undeland snubber, the McMurray snubber can be modified into an energy recovery snubber. By using an energy recovery transformer as shown in Fig. 16.47, this snubber becomes a regenerative one. Although other regenerative circuits have been proposed, their complexity makes them unattractive in industrial applications. Also, they do not necessarily solve the power diode reverse-recovery problems.

Although the use of snubber circuits can reduce the switching stress in the power devices, the switching loss is actually damped into the snubber resistors unless regenerative snubbers are used. The switching loss is still a limiting factor to the high frequency operation of power inverters. However, the advent of soft-switching techniques opens a new way to high-frequency inverter operation. Because the switching trajectory of a soft-switched switch is close to the voltage and current axis, faster power electronic devices with smaller SOAs can in principle be used. In general, both ZVS and ZCS can reduce switching loss in high-power power switches. However, for power switches with tail currents, such as IGBT, ZCS is more effective than ZVS.

16.14 Soft-switching DC–AC Power Inverters

Soft-switching technique not only offers a reduction in switching loss and thermal requirement, but also allows the possibility of high frequency and snubberless operation. Improved circuit performance and efficiency, and reduction of EMI emission can be achieved. For zero-voltage switching (ZVS) inverter applications, two major approaches which enable

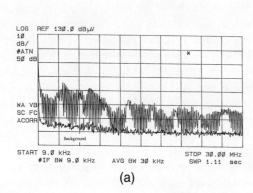

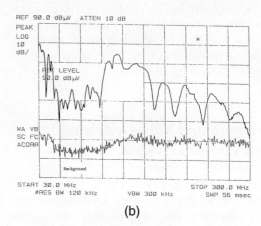

FIGURE 16.43 (a) Conducted EMI from hard-switched flyback converter and (b) radiated EMI from hard-switched flyback converter.

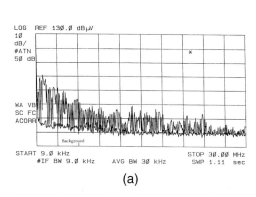
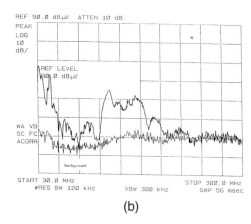

FIGURE 16.44 (a) Conducted EMI from soft-switched flyback converter and (b) radiated EMI from soft-switched flyback converter.

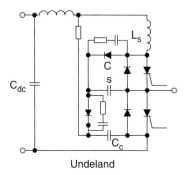

FIGURE 16.45 Undeland snubber.

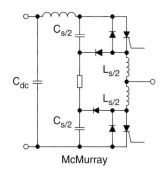

FIGURE 16.46 McMurray snubber.

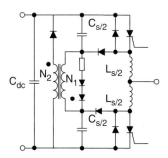

FIGURE 16.47 McMurray snubber with energy recovery.

inverters to be soft-switched have been proposed. The first approach pulls the dc link voltage to zero momentarily so that the inverter's switches can be turned on and off with ZVS. Resonant dc link and quasi-resonant inverters belong to this category. The second approach uses the resonant pole idea. By incorporating the filter components into the inverter operation, resonance condition and thus zero voltage/current conditions can be created for the inverter switches.

In this section, the following soft-switched inverters are described.

Approach 1: Resonant dc link inverters

1. Resonant (pulsating) dc link inverters
2. Actively-clamped resonant dc link inverters
3. Resonant inverters with minimum voltage stress
4. Quasi-resonant soft-switched inverter
5. Parallel resonant dc link inverter.

Approach 2: Resonant pole inverters

1. Resonant pole inverters
2. Auxiliary resonant pole inverters
3. Auxiliary resonant commutated pole inverters.

Type 1 is the resonant dc link inverter [42–44] which sets the dc link voltage into oscillation so that the zero-voltage instants are created periodically for ZVS. Despite the potential advantages that this soft-switching approach can offer, a recent review on existing resonant link topologies for inverters [45] concludes that the resonant dc link system results in an increase in circuit complexity and the frequency spectrum is restricted by the need of using integral pulse density modulation (IPDM) when compared with a standard hard-switched inverter. In addition, the peak pulsating link voltage of resonant link inverters is twice the dc link voltage in a standard hard-switched inverter. Although clamp circuits (Type 2) can be used to limit the peak voltage to 1.3–1.5 per unit [44], power devices with higher than normal voltage ratings have to be used.

Circuits of Type 3–5 employ a switched mode front stage circuit which pulls the dc link voltage to zero momentarily whenever inverter switching is required. This soft-switching approach does not cause extra voltage stress to the inverter and hence the voltage rating of the power devices is only 1 per unit. As ZVS conditions can be created at any time, there is virtually no restriction in the PWM strategies. Therefore, well established PWM schemes developed in the last two decades can be employed. In some ways, this approach is similar to some dc-side commutation techniques proposed in the past for thyristor inverters [46, 47], although these dc-side commutation techniques were used for turning off thyristors in the inverter bridge and were not primarily developed for soft-switching.

Circuits of Type 6–8 retain the use of a constant dc link voltage. They incorporate the use of the resonant components and/or filter components into the inverter circuit operation. This approach is particularly useful for inverter applications in which output filters are required. Examples include uninterruptible power supplies (UPS) and inverters with output filters for motor drives. The LC filter components can form the auxiliary resonant circuits that create the soft-switching conditions. However, these tend to have high power device count and require complex control strategy.

16.14.1 Resonant (Pulsating) DC Link Inverter

Resonant DC link converter for DC–AC power conversion was proposed in 1986 [42]. Instead of using a nominally constant DC link voltage, a resonant circuit is added to cause the DC link voltage to be pulsating at a high frequency. This resonant circuit theoretically creates periodic zero-voltage duration at which the inverter switches can be turned on or off. Figure 16.48 shows the schematics of the pulsating link inverter. Typical dc link voltage, inverter's phase voltage and the line voltages are shown in Fig. 16.49. Because the inverter switching can only occur at zero voltage duration, integral pulse density modulation (IPDM) has to be adopted in the switching strategy.

Analysis of the resonant dc link converter can be simplified by considering that the inverter system is highly inductive. The equivalent circuit is shown in Fig. 16.50.

The link current I_x may vary with the changing load condition, but can be considered constant during the short resonant cycle. If switch S is turned on when the inductor current is I_{Lo}, the resonant dc link voltage can be expressed as

$$V_c(t) = V_s + e^{-\alpha t}\left[-V_s \cos(\omega t) + \omega L I_M \sin(\omega t)\right] \quad (16.14)$$

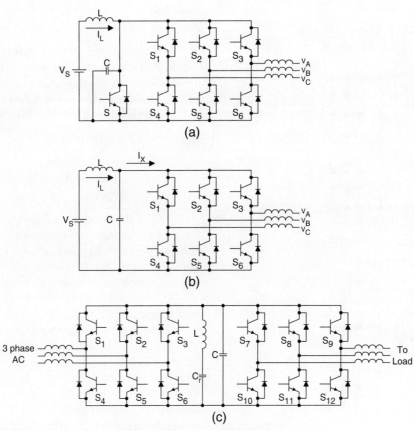

FIGURE 16.48 Resonant-link inverters.

16 Resonant and Soft-switching Converters

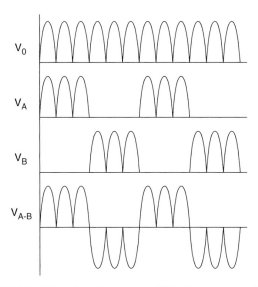

FIGURE 16.49 Typical dc link voltage (V_0), phase voltages (V_A, V_B), and line voltage (V_{AB}) of resonant link inverters.

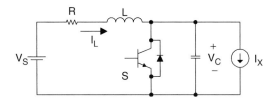

FIGURE 16.50 Equivalent circuit of resonant link inverter.

and inductor current i_L is

$$i_L(t) \approx I_x + e^{-\alpha t}\left[I_M \cos(\omega t) + \frac{V_s}{\omega L}\sin(\omega t)\right] \quad (16.15)$$

where $\quad \alpha = \dfrac{R}{2L} \quad (16.16)$

$$\omega_o = (LC)^{-0.5} \quad (16.17)$$

$$\omega = \left(\omega_o^2 - \alpha^2\right)^{0.5} \quad (16.18)$$

and $\quad I_M = I_{Lo} - I_x \quad (16.19)$

The resistance in the inductor could affect the resonant behavior because it dissipates some energy. In practice, ($i_L - I_x$) has to be monitored when S is conducting. S can be turned on when ($i_L - I_x$) equal to a desired value. The objective is to ensure that the dc link voltage can be resonated to zero voltage level in the next cycle.

The pulsating dc link inverter has the following advantages:

- Reduction of switching loss.
- Snubberless operation.

- High switching frequency (> 18 kHz) operation becomes possible, leading to the reduction of acoustic noise in inverter equipment.
- Reduction of heatsink requirements and thus improvement of power density.

This approach has the following limitations:

- The peak dc pulsating link voltage (2.0 per unit) is higher than the nominal dc voltage value of a conventional inverter. This implies that power devices and circuit components of higher voltage ratings must be used. This could be a serious drawback because power components of higher voltage ratings are not only more expensive, but usually have inferior switching performance than their low-voltage counterparts.
- Although voltage clamp can be used to reduce the peak dc link voltage, the peak voltage value is still higher than normal and the additional clamping circuit makes the control more complicated.
- Integral pulse-density modulation has to be used. Many well-established PWM techniques cannot be employed.

Despite these advantages, this resonant converter concept has paved the way for other soft-switched converters to develop.

16.14.2 Active-clamped Resonant DC Link Inverter

In order to solve the high voltage requirement in the basic pulsating dc link inverters, active clamping techniques (Fig. 16.51) have been proposed. The active clamp can reduce the per-unit peak voltage from 2.0 to about 1.3–1.5 [44]. It has been reported that operating frequency in the range of 60–100 kHz has been achieved [48] with an energy efficiency of 97% for a 50 kVA drive system.

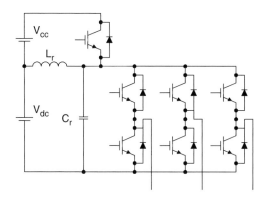

FIGURE 16.51 Active-clamp resonant link inverter.

The design equations for active-clamped resonant link inverter are

$$T_L = \frac{1}{f_L} = 2\sqrt{L_r C_r}\left(\cos^{-1}(1-k) + \frac{\sqrt{k(2-k)}}{k-1}\right) \quad (16.20)$$

where T_L is the minimum link period, f_L is the maximum link frequency, and k is the clamping ratio. For the active-clamped resonant inverter, k is typically 1.3–1.4 per unit.

The rate of rise of the current in the clamping device is

$$\frac{di}{dt} = \frac{(k-1)V_s}{L_r} \quad (16.21)$$

The peak clamping current required to ensure that the dc bus return to zero volt is

$$I_{sp} = V_s\sqrt{\frac{k(2-k)C_r}{L_r}} \quad (16.22)$$

In summary, resonant (pulsating) dc link inverters offer significant advantages such as:

- High switching frequency operation.
- Low dv/dt for power devices.
- ZVS with reduced switching loss.
- Suitable for 1–250 kW.
- Rugged operation with few failure mode.

16.14.3 Resonant DC Link Inverter with Low Voltage Stress [49]

A resonant dc link inverter with low voltage stress is shown in Fig. 16.52. It consists of a front-end resonant converter that can pull the dc link voltage down just before any inverter switching. This resonant dc circuit serves as an interface between the dc power supply and the inverter. It essentially retains all the advantages of the resonant (pulsating) dc link inverters. But it offers extra advantages such as:

- No increase in the dc link voltage when compared with conventional hard-switched inverter. That is, the dc link voltage is 1.0 per unit.
- The zero voltage condition can be created at any time. The ZVS is not restricted to the periodic zero-voltage instants as in resonant dc link inverter.
- Well-established PWM techniques can be employed.
- Power devices of standard voltage ratings can be used.

The timing program and the six operating modes of this resonant circuit are as shown in Figs. 16.53 and 16.54, respectively.

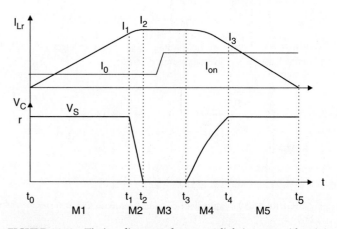

FIGURE 16.53 Timing diagram of resonant link inverter with minimum voltage stress.

(1) *Normal mode:*
This is the standard PWM inverter mode. The resonant inductor current $i_{Lr}(t)$ and the resonant voltage $V_{cr}(t)$ are given by

$$i_{Lr}(t) = 0$$

$$v_{Cr}(t) = V_s$$

where V_s is the nominal dc link voltage.

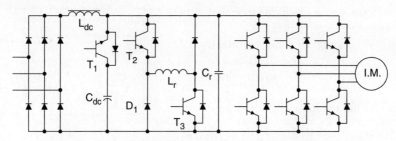

FIGURE 16.52 Resonant DC link inverter with low voltage stress.

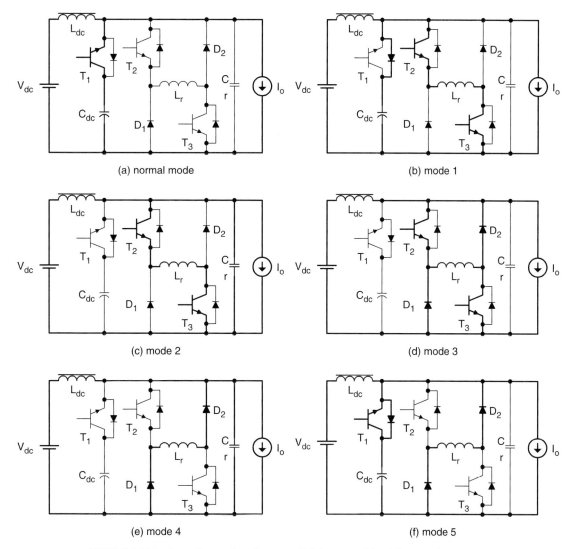

FIGURE 16.54 Operating modes of resonant link inverter with minimum voltage stress.

(2) *Mode 1 (initiating mode): (t_0-t_1)*
At t_0, mode 1 begins by switching on T2 and T3 on with zero current. $i_{Lr}(t)$ increases linearly with a di/dt of V_s/L_r. If $i_{Lr}(t)$ is equal to the initialized current I_i, T1 is zero-voltage turned off. If $(I_s-I_o) < I_i$, then the initialization is ended when $i_{Lr}(t)$ is equal to I_i, where I_s is the current flowing into the dc inductor L_{dc}. If $(I_s-I_o) > I_i$, then this mode continues until $i_{Lr}(t)$ is equal to (I_s-I_o). The equations in this interval are

$$i_{Lr}(t) = \frac{V_s}{L_r} t$$

$$v_{Cr}(t) = V_s$$

$$i_{Lr}(t_1) = \frac{V_s}{L_r} t_1 = I_i$$

(3) *Mode 2 (Resonant mode): (t_1-t_2)*
After T1 is turned off under ZVS condition, resonance between L_r and C_r occurs. $V_{cr}(t)$ decreases from V_s to 0. At t_2, $i_{Lr}(t)$ reaches the peak value in this interval. The equations are:

$$i_{Lr}(t) = \frac{V_s}{Z_r} \sin(\omega_r t) + [I_1 + (I_o - I_s)] \cos(\omega_r t) - (I_o - I_s)$$

$$V_{Cr}(t) = -V_s \cos(\omega_r t) - [I_1 + (I_o - I_s)] Z_r \sin(\omega_r t)$$

$$i_{Lr}(t_2) = I_2 = I_{Lr,peak}$$

$$V_{Cr}(t_2) = 0$$

where

$$Z_r = \sqrt{\frac{L_r}{C_r}} \quad \text{and} \quad \omega_r = \frac{1}{\sqrt{L_r C_r}}$$

(4) **Mode 3 (Freewheeling mode):** $(t_2 - t_3)$
The resonant inductor current flows through two freewheeling paths (T2-Lr-D2 and T3-D1-Lr). This duration is the zero voltage period created for ZVS of the inverter, and should be longer than the minimum on and off times of the inverter's power switches.

$$i_{Lr}(t) = I_2$$

$$v_{Cr}(t) = 0$$

(5) **Mode 4 (Resonant mode):** $(t_3 - t_4)$
This mode begins when T2 and T3 are switched off under ZVS. The second half of the resonance between L_r and C_r starts again. The capacitor voltage $V_{cr}(t)$ increases back from 0 to V_s and is clamped to V_s. The relevant equations in this mode are

$$i_{Lr}(t) = [I_2 - (I_{on} - I_s)]\cos(\omega_r t) - (I_{on} - I_s)$$

$$V_{Cr}(t) = [I_2 - (I_{on} - I_s)]Z_r \sin(\omega_r t)$$

$$i_{Lr}(t_4) = I_3$$

$$V_{Cr}(t_4) = V_s$$

where I_{on} is the load current after the switching state.

(6) **Mode 5 (Discharging mode):** $(t_4 - t_5)$
In this period, T1 is switched on under ZV condition because $V_{cr}(t) = V_s$. The inductor current decreases linearly. This mode finishes when $i_{Lr}(t)$ becomes zero.

$$i_{Lr}(t) = -\frac{V_s}{L_r}t + I_3$$

$$v_{Cr}(t) = V_s$$

$$i_{Lr}(t_5) = 0$$

16.14.4 Quasi-resonant Soft-switched Inverter [50]

(A) Circuit Operation
Consider an inverter fed by a dc voltage source vs a front-stage interface circuit shown in Fig. 16.55, can be added between the dc voltage source and the inverter. The front-stage circuit consists of a quasi-resonant circuit in which the first half of the resonance cycle is set to occur to create the zero-voltage condition whenever inverter switching is needed. After inverter switching has been completed, the second half of the resonance cycle takes place so that the dc link voltage is set back to its normal level. To avoid excessive losses in the resonant circuit, a small capacitor C_{r1} is normally used to provide the dc link voltage whilst the large smoothing dc link capacitor C_1 is isolated from the resonant circuit just before the zero-voltage

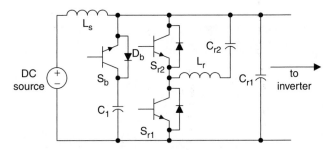

FIGURE 16.55 Quasi-resonant circuit for soft-switched inverter.

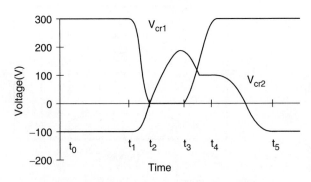

FIGURE 16.56 Typical waveforms for V_{cr1} and V_{cr2}.

duration. This method avoids the requirement for pulling the dc voltage of the bulk capacitor to zero.

The period for this mode is from t_0 to t_1 in Fig. 16.56. In this mode, switch S_b is turned on and switches S_{r1} and S_{r2} are turned off. The inverter in Fig. 16.55 works like a conventional dc link inverter. In this mode, $V_{cr1} = V_{c1}$. The voltage across switch S_b is zero. Before an inverter switching takes place, when switch S_{r1} is triggered at t_1 to discharge C_{r1}. This operating mode ends at t_2 when V_{cr1} approaches zero. The equivalent circuit in this mode is shown in Fig. 16.57a. The switch S_b must be turned off at zero voltage when switch S_{r1} is triggered. After S_{r1} is triggered, C_{r1} will be discharged via the loop C_{r1}, C_{r2}, L_r, and S_{r1}. Under conditions of $V_{cr2} \leq 0$ and $C_{r1} \leq C_{r2}$, the energy stored in C_{r1} will be transferred to C_{r2} and V_{cr1} falls to zero in the first half of the resonant cycle in the equivalent circuit of Fig. 16.57a. V_{cr1} will be clamped to zero by the freewheel diodes in the inverter bridge and will not become negative. Thus, V_{cr1} can be pulled down to zero for zero-voltage switching. When the current in inductor L_r becomes zero, switch S_{r1} can be turned off at zero current.

Inverter switching can take place in the period from t_2 to t_3 in which V_{cr1} remains zero. This period must be longer than the turn-on and turn-off times of the switches. When inverter switching has been completed, it is necessary to reset the voltage of capacitor C_{r1}. The equivalent circuit in this mode is shown in Fig. 16.57b. The current in inductor L_r reaches zero at t_3. Due to the voltage in V_{cr2} and the presence

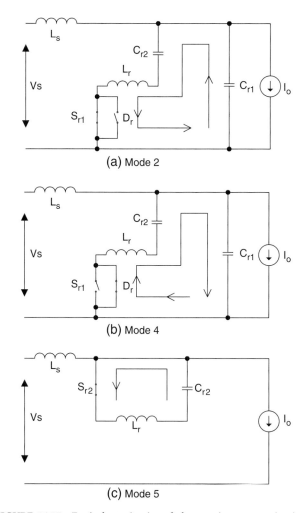

FIGURE 16.57 Equivalent circuits of the quasi-resonant circuit for different modes.

of diode D_r, this current then flows in the opposite direction. C_{r1} will be recharged via L_r, C_{r2}, C_{r1}, and D_r. The diode D_r turns off when the current in L_r becomes zero. V_{cr1} will not go beyond 1 per unit because C_{r1} is clamped to supply voltage by diode D_b. The switch S_b can be turned on again at zero-voltage condition when V_{cr1} returns to normal dc supply voltage. After D_r turns off, V_{cr2} may not be zero. Some positive residual capacitor voltage remains in C_2 at t_4, as shown in Fig. 16.56. In case V_{cr2} is positive, V_{cr1} cannot be pulled down to zero again in the next switching cycle. Therefore, S_{r2} should be triggered after t_4 to reverse the residual voltage in C_{r2}. At time t_5, S_{r2} turns off at zero-current condition and V_{cr2} is now reversed to negative. The equivalent circuit in this mode is shown in Fig. 16.57c. When $V_{cr2} \leq 0$ and $C_{r1} \leq C_{r2}$, V_{cr1} can be pulled down to zero again before the next inverter switching. The operation can then be repeated in next switching cycle.

(B) Design Considerations

(1) C_{r1} and C_{r2}

The criterion for getting zero capacitor voltages V_{cr1} is:

$$(C_{r1} - C_{r2})V_s + 2C_{r2}V_{o2} - \Delta I \pi \sqrt{L_r C_e} \leq 0 \quad (16.23)$$

where

- V_{o1} is the initial voltage of C_{r1};
- V_{o2} is the initial voltage of C_{r2};
- i_{L0} is the initial current of inductor L_r;
- $\Delta I = I_o - I_s$, which is the difference between load current and supply current. It is assumed to be a constant within a resonant cycle;
- R_r is the equivalent resistance in the resonant circuit.
- $C_e = \dfrac{C_{r1} C_{r2}}{C_{r1} + C_{r2}}$

When $\Delta I \geq 0$, the above criterion is always true under conditions of:

$$C_{r1} \leq C_{r2}, \quad V_{o2} \leq 0$$

The criterion for recharging voltage V_{cr1} to 1 per unit dc link voltage is:

$$\frac{2C_{r2}}{C_{r1} + C_{r2}} V_{o2} - \frac{\Delta I}{C_{r1} + C_{r2}} \pi \sqrt{L_r C_e} \geq V_s \quad (16.24)$$

(2) Inductor L_r

The inductor L_r should be small so that the dc link voltage can be decreased to zero quickly. However, a small L_r could result in large peak resonant current and therefore requirement of power devices with large current pulse ratings. An increase in the inductance of L_r can limit the peak current in the quasi-resonant circuit. Because the resonant frequency depends on both the inductor and the capacitor, therefore, the selection of L_r can be considered together with the capacitors C_{r1} and C_{r2} and with other factors such as the current ratings of power devices, the zero-voltage duration and the switching frequency required in the soft-switching circuit.

(3) Triggering instants of the switches

The correct triggering instants for the switches are essential for the successful operation of this soft-switched inverter. For the inverter switches, the triggering instants are determined from a PWM modulation. Let T_s be the time at which the inverter switches change states. To get the zero-voltage inverter switching, switch S_{r1} should be turned on half resonant cycle before the inverter switching instant. The turn-on instant of S_{r1}, which is t_1 in Fig. 16.56, can be written as:

$$t_1 = T_s - \frac{\pi}{\omega} \quad (16.25)$$

where $\omega = \sqrt{\omega_0^2 - \alpha^2}$, $\alpha = \frac{R_r}{2L_r}$, $\omega_0 = \sqrt{\frac{1}{L_r C_e}}$. The switch S_b is turned off at t_1.

S_{r1} may be turned off during its zero current period when diode D_r is conducting. For easy implementation, its turn-off time can be selected as $T_s + \pi/\omega$. Because the dc link voltage can be pulled down to zero in less than half resonant cycle, T_s should occur between t_2 and t_3.

At time t_3 (the exact instant depends on the ΔI), the diode D_r turns on in the second half of the resonant cycle to recharge C_{r1}. At t_4, V_{cr1} reaches 1 per unit and diode D_b clamps V_{cr1} to 1 per unit. The switch S_b can be turned on again at t_4, which is half resonant cycle after the start of t_3:

$$t_4 \approx t_3 + \frac{\pi}{\omega} \qquad (16.26)$$

As t_3 cannot be determined accurately, a voltage sensor in principle can be used to provide information for t_4 so that S_b can be turned on to reconnect C_1 to the inverter. In practice, however, S_b can be turned on a few microseconds (longer than $T_s + \pi/\omega$) after t_2 without using a voltage sensor (because it is not critical for S_b to be on exactly at the moment V_{cr} reaches the nominal voltage). As for switch S_{r2}, it can be turned on a few microseconds after t_4. It will be turned off half of resonant cycle ($\pi\sqrt{L_r C_{r2}}$) in the L_r–C_{r2} circuit later. In practice, the timing of S_{r1}, S_{r2}, and S_b can be adjusted in a simple tuning procedure for a given set of parameters. Figure 16.58 shows the measured gating signals of S_{r1}, S_{r2}, and S_b with the dc link voltage V_{cr1} in a 20 kHz switching inverter. Figures 16.59 and 16.60 show the measured waveforms of I_s, I_o, and V_{cr1} under no load condition and loaded condition, respectively.

(C) Control of Quasi-resonant Soft-switched Inverter Using Digital Time Control (DTC) [51]

Based on the zero-average-current error (ZACE) control concept, a digital time control method has been developed for a current-controlled quasi-resonant soft-switched inverter. The basic ZACE concept is shown in Fig. 16.61. The current error is obtained from the difference of a reference current and the sensed current. The idea is to make the areas of each transition (A1 and A2) equal. If the switching frequency is significantly greater than the fundamental frequency of the reference signal,

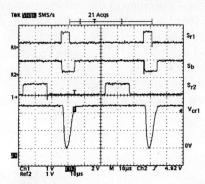

FIGURE 16.58 Gating signals for S_{r1}, S_{r2}, and S_b with V_{cr1}.

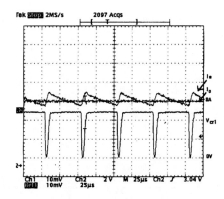

FIGURE 16.59 Typical I_s, I_o, and V_{cr1} under no-load condition.

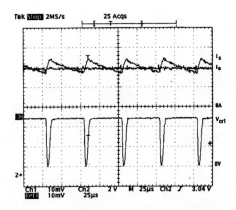

FIGURE 16.60 Typical I_s, I_o, and V_{cr1} under loaded condition.

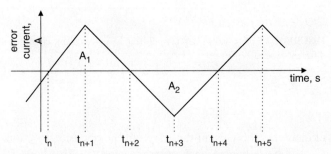

FIGURE 16.61 Zero-average-current error (ZACE) control concept.

the rising and falling current segments can be assumed to be linear. The following simplified equation can be established.

$$\Delta t_{n+1} = t_{n+1} - t_n \qquad (16.27)$$

The control algorithm for the inverter is

$$\Delta t_{n+3} = \Delta t_{n+2} + D\left[\frac{T_{sw}}{2} - (t_{n+2} - t_n)\right] \qquad (16.28)$$

where $D = \dfrac{\Delta t_{n+2}}{t_{n+2} - t_n}$ and $T_{sw} = t_{n+4} - t_n$.

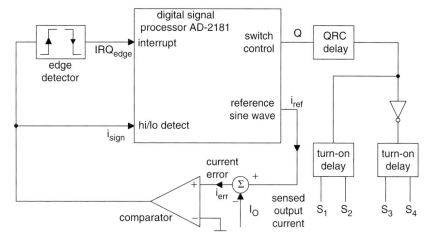

FIGURE 16.62 Implementation of DTC.

The schematic of a digital signal processor (DSP) based controller for the DTC method is shown in Fig. 16.62. The duty cycle can be approximated from the reference sine wave by level shifting and scaling it between 0 and 1. The time $t_{n+2} - t_n$ is the sum of Δt_{n+1} and Δt_{n+2}. These data provide information for the calculation of the next switching time Δt_{n+1}.

The switches are triggered by the changing edge of the switch control Q. Approximate delays are added to the individual switching signals for both the inverter switches and the quasi-resonant switches. Typical gating waveforms are shown in Fig. 16.63. The use of the quasi-resonant soft-switched inverter is a very effective way in suppressing switching transient and EMI emission. Figures 16.64a,b show the inverter switch voltage waveforms of a standard hard-switched inverter and a quasi-resonant soft-switched inverter, respectively. It is clear that the soft-switched waveform has much less transient than the hard-switched waveform.

16.14.5 Resonant Pole Inverter (RPI) and Auxiliary Resonant Commutated Pole Inverter (ARCPI)

The resonant pole inverter integrates the resonant components with the output filter components L_f and C_f. The load

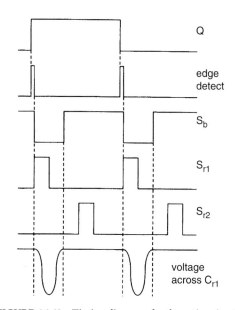

FIGURE 16.63 Timing diagrams for the gating signals.

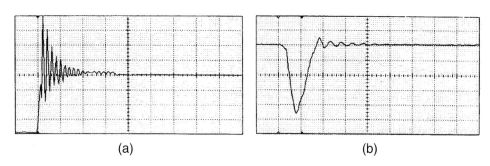

FIGURE 16.64 (a) Typical switch voltage under hard turn-off and (b) typical switch voltage under soft turn-off.

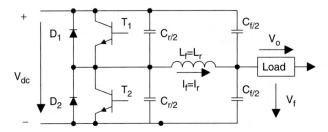

FIGURE 16.65 One leg of a resonant pole inverter.

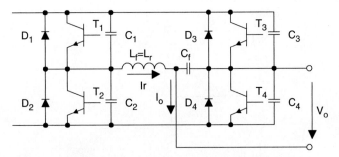

FIGURE 16.66 Single-phase resonant pole inverter.

is connected to the mid-point of the dc bus capacitors as shown in Fig. 16.65. It should however be noted that the RPI can be described as a resonant inverter. Figure 16.66 shows a single-phase RPI. Its operation can be described with the timing diagram in Fig. 16.67. The operating modes are included in Fig. 16.68. The RPI provides soft-switching for all power switches. But it has two disadvantages. First, the power devices have to be switched continuously at the resonant frequency determined by the resonant components. Second, the power devices in the RPI circuit require a 2.2–2.5 p.u. current turn-off capability.

An improved version of the RPI is the auxiliary resonant commutated pole inverter (ARCPI). The ARCPI for one inverter leg is shown in Fig. 16.69. Unlike the basic RPI, the ARCPI allows the switching frequency to be controlled. Each of the primary switches is closely paralleled with a snubber capacitor to ensure ZV turn off. Auxiliary switches are connected in series with an inductor, ensuring that they operate under ZC conditions. For each leg, an auxiliary circuit comprising two extra switches A1 and A2, two freewheeling diodes, and a resonant inductor L_r is required. This doubles the number of power switches when compared with hard-switched inverters. Figure 16.70 shows the three-phase ARCPI system. Depending on the load conditions, three commutation modes are generally needed. The commutation methods at low and high current are different. This makes the control of the ARCPI very complex. The increase in control and circuit complexity represents a considerable cost penalty [52, 53].

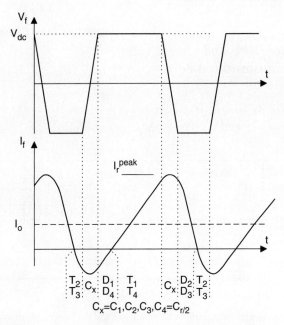

FIGURE 16.67 Timing diagram for a single-phase resonant pole inverter.

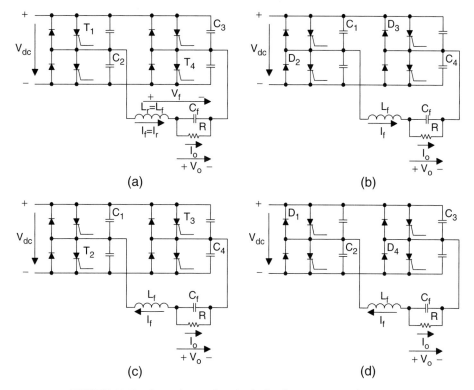

FIGURE 16.68 Operating modes of a single-phase resonant pole inverter.

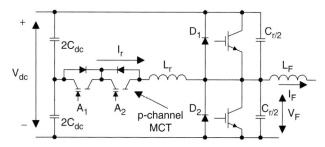

FIGURE 16.69 Improved resonant pole inverter leg.

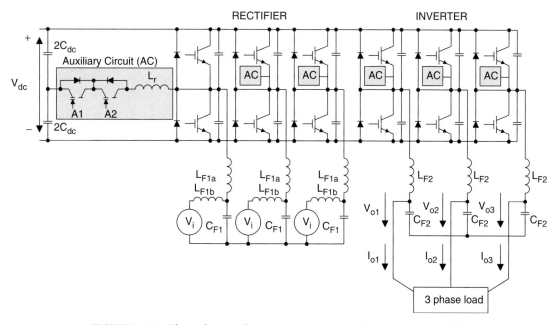

FIGURE 16.70 Three-phase auxiliary resonant commutated pole inverter (ARCPI).

References

1. R. M. Davis, *Power Diode and Thyristor Circuits*, IEE Monograph, Series 7, Herts: Peregrinus, 1971.
2. N. Mohan, T. Undeland, and W. Robbins, *Power Electronics, Converters, Applications, and Design*, Hoboken, NJ: John Wiley and Sons, 1995.
3. M. H. Rashid, *Power Electronics, Circuits, Devices, and Applications*, Upper Saddle River, NJ: Pearson/Prentice Hall, 2004.
4. P. Vinciarelli, "Forward Converter Switching at Zero Current," U.S. Patent 4,416,959, November 1983.
5. F. C. Lee, *High-Frequency Resonant, Quasi-Resonant and Multi-Resonant Converters*, Blacksburg, VA: Virginia Power Electronics Center, 1991.
6. F. C. Lee, *High-Frequency Resonant and Soft-Switching Converters*, Virginia Power Electronics Center, 1991.
7. K. Kit Sum, *Recent Developments in Resonant Converters*, Ventura, CA, Intertec Communication Press, 1988.
8. R. E. Tarter, *Solid-State Power Conversion Handbook*, New York Wiley-Interscience publication, 1993.
9. K. H. Liu and F. C. Lee, "Resonant Switches – A Unified Approach to Improve Performances of Switching Converters," in *Proc. Int. Telecomm. Energy Conf.*, 1984, pp. 344–351.
10. K. H. Liu, R. Oruganti, and F. C. Lee, "Resonant Switches – Topologies and Characteristics," in *Proc. IEEE Power Electron. Spec. Conf.*, 1985, pp. 62–67.
11. K. D. T. Ngo, "Generalized of Resonant Switches and Quasi-Resonant DC-DC Converters," in *Proc. IEEE Power Electron. Spec. Conf.*, 1986, pp. 58–70.
12. F. C. Lee, "High-frequency Quasi-Resonant and Multi-Resonant Converter Technologies," in *Proc. IEEE Int. Conf. Ind. Electron.*, 1988, pp. 509–521.
13. K. H. Liu and F. C. Lee, "Zero-Voltage Switching Techniques in DC/DC Converter Circuits," in *Proc. IEEE Power Electron. Spec. Conf.*, 1986, pp. 58–70.
14. M. Jovanovic, W. Tabisz, and F. C. Lee, "Zero-voltage Switching Technique in High Frequency Off-line Converters," in *Proc. Applied Power Electron. Conf. and Expo.*, 1988, pp. 23–32.
15. R. Steigerwald, "A Comparison of Half-Bridge Resonant Converter Topologies," *IEEE Trans. Power Electron.*, vol. 3, no. 2, April. 1988, pp. 174–182.
16. O. D. Patterson and D.M. Divan, "Pseudo-resonant Full-Bridge DC/DC Converter," in *Proc. IEEE Power Electron. Spec. Conf.*, 1987, pp. 424–430.
17. C. P. Henze, H. C. Martin, and D. W. Parsley, "Zero-Voltage Switching in High Frequency Power Converters Using Pulse Width Modulation," in *Proc. IEEE Applied Power Electron. Conf. and Expo.*, 1988, pp. 33–40.
18. General Electric Company, "Full-Bridge Lossless Switching Converters," U.S. Patent 4,864,479, 1989.
19. W. A. Tabisz and F. C. Lee, "DC Analysis and Design of Zero-Voltage Switched Multi-Resonant Converters," in *Proc. IEEE Power Electron. Spec. Conf.*, 1989, pp. 243–251.
20. W. A. Tabisz and F. C. Lee, "Zero-Voltage-Switching Multi-Resonant Technique – a Novel Approach to Improve Performance of High Frequency Quasi-Resonant Converters," *IEEE Trans. Power Electron.*, vol. 4, no. 4, October 1989, pp. 450–458.
21. M. Jovanovic and F. C. Lee, "DC Analysis of Half-Bridge Zero-Voltage-Switched Multi-Resonant Converter," *IEEE Trans. Power Electron.*, vol. 5, no. 2, April 1990, pp. 160–171.
22. R. Farrington, M. M. Jovanovic, and F. C. Lee, "Constant-Frequency Zero-Voltage-Switched Multi-resonant converters: Analysis, Design, and Experimental Results," in *Proc. IEEE Power Electron. Spec. Conf.*, 1990, pp. 197–205.
23. G. Hua, C. S. Leu, and F. C. Lee, "Novel zero-voltage-transition PWM converters," in *Proc. IEEE Power Electron. Spec. Conf.*, 1992, pp. 55–61.
24. R. Watson, G. Hua, and F. C. Lee, "Characterization of an Active Clamp Flyback Topology for Power Factor Correction Applications," *IEEE Trans. Power Electron.*, vol. 11, no. 1, January 1996, pp. 191–198.
25. R. Watson, F. C. Lee, and G. Hua, "Utilization of an Active-Clamp Circuit to Achieve Soft Switching in Flyback Converters," *IEEE Trans. Power Electron.*, vol. 11, no. 1, January 1996, pp. 162–169.
26. B. Carsten, "Design Techniques for Transformer Active Reset Circuits at High Frequency and Power Levels," in *Proc. High Freq. Power Conversion Conf.*, 1990, pp. 235–245.
27. S. D. Johnson, A. F. Witulski, and E. W. Erickson, "A Comparison of Resonant Technologies in High Voltage DC Applications," in *Proc. IEEE Appl. Power Electron. Conf.*, 1987, pp. 145–166.
28. A. K. S. Bhat and S. B. Dewan, "A Generalized Approach for the Steady State Analysis of Resonant Inverters," *IEEE Trans. Ind. Appl.*, vol. 25, no. 2, March 1989, pp. 326–338.
29. A. K. S. Bhat, "A Resonant Converter Suitable for 650V DC Bus Operation," *IEEE Trans. Power Eletcron.*, vol. 6, no. 4, October 1991, pp. 739–748.
30. A. K. S. Bhat and M. M. Swamy, "Loss Calculations in Transistorized Parallel Converters Operating Above Resonance," *IEEE Trans. Power Electron.*, vol. 4, no. 4, July 1989, pp. 449–458.
31. A. K. S. Bhat, "A Unified Approach for the Steady-State Analysis of Resonant Converter," *IEEE Trans. Ind. Electron.*, vol. 38, no. 4, August 1991, pp. 251–259.
32. *Applications Handbook*, Unitrode Corporation, 1999.
33. Barbi, J. C. Bolacell, D. C. Martins, and F.B. Libano, "Buck Quasi-resonant Converter Operating at Constant Frequency: Analysis, Design and Experimentation," PESC'89, pp. 873–880.
34. K. W. E. Cheng and P. D. Evans, "A Family of Extended-period Circuits for Power Supply Applications using High Conversion Frequencies," EPE'91, pp. 4.225–4.230.
35. S. Y. R. Hui, K. W. E. Cheng, and S. R. N. Prakash, "A Fully Soft-switched Extended-period Quasi-resonant Power Correction Circuit," *IEEE Transactions on Power Electronics*, vol. 12, no. 5, September 1997, pp. 922–930.
36. H. S. H. Chung, S. Y. R. Hui, and W. H. Wang, "A Zero-Current-Switching PWM Flyback Converter with a Simple Auxiliary Switch," *IEEE Transactions on Power Electronics*, vol. 14, no. 2, March 1999, pp. 329–342.
37. H. S. H. Chung and S. Y. R. Hui, "Reduction of EMI in Power Converters Using fully Soft-switching Technique," *IEEE Transactions on Electromagnetics*, vol. 40, no. 3, August 1998, pp. 282–287.
38. H. S. H. Chung, W. L. Cheung, and K. S. Tang, "A ZCS Bidirectional Flyback Converter," *IEEE Transactions on Power Electronics*, vol. 19, no. 6, November 2004, pp. 1426–1434.

39. M. T. Zhang, M. M. Jovanovic, and F. C. Lee, "Design Considerations and Performance Evaluation of Synchronous Rectification in Flyback Converter," *IEEE Transactions on Power Electronics*, vol. 13, no. 3, May 1998, pp. 538–546.
40. T. Undeland, "Snubbers for Pulse Width Modulated Bridge Converters with Power Transistors or GTOs," IPEC, Tokyo, Conference proceedings, vol. 1, 1983, pp. 313–323.
41. McMurray, "Efficient snubbers for voltage source GTO inverters," *IEEE Transactions on Power Electronics*, vol. 2, no. 3, July 1987, pp. 264–272.
42. D. M. Divan, "The resonant dc link converter-A new concept in static power conversion," *IEEE Trans. IA*, vol. 25, no. 2, 1989, pp. 317–325.
43. Jin-Sheng Lai and B. K. Bose, "An Induction Motor Drive Using an Improved High Frequency Resonant DC Link Inverter," *IEEE Trans. on Power Electronics*, vol. 6, no. 3, 1991, pp. 504–513.
44. D. M. Divan and G. Skibinski, "Zero-Switching-Loss Inverters for High-Power Applications," *IEEE Trans. IA*, vol. 25, no. 4, 1989, pp. 634–643.
45. S. J. Finney, T. C. Green, and B. W. Williams, "Review of Resonant Link Topologies for Inverters," *IEE Proc. B*, vol. 140, no. 2, 1993, pp. 103–114.
46. S. B. Dewan and D. L. Duff, "Optimum Design of an Input Commutated Inverter for AC Motor Control," *IEEE Trans. on Industry Gen. Applications*, vol. IGA-5, no. 6, November/December 1969, pp. 699–705.
47. V.R. Stefanov and P. Bhagwat, "A Novel DC Side Commutated Inverter," *IEEE PESC Record*, 1980.
48. A. Kurnia, H. Cherradi, and D. Divan, "Impact of IGBT behavior on design optimization of soft switching inverter topologies," *IEEE IAS Conf. Record*, 1993, pp. 140–146.
49. Y. C. Jung, J. G. Cho, and G. H. Cho, "A New Zero Voltage Switching Resonant DC-Link Inverter with Low Voltage Stress," *Proc. of IEEE Industrial Electronics Conference*, 1991, pp. 308–13.
50. S. Y. R. Hui, E. S. Gogani and J. Zhang, "Analysis of a Quasi-resonant Circuit for Soft-switched Inverters," *IEEE Transactions on Power Electronics*, vol. 11, no. 1, January, 1999, pp. 106–114.
51. D. M. Baker, V. G. Ageliidis, C. W. Meng, and C. V. Nayar, "Integrating the Digital Time Control Algorithm with DC-Bus 'Notching' Circuit Based Soft-Switching Inverter," *IEE Proceedings-Electric Power Applications*, vol. 146, no. 5, September 1999, pp. 524–529.
52. F. C. Lee and D. Borojevic, "Soft-switching PWM Converters and Inverters," Tutorial notes, PESC'94.
53. D. M. Divan and R. W. De Doncker, "Hard and Soft-switching Voltage Source Inverters", Tutorial notes, PESC'94.

17
Multilevel Power Converters

Surin Khomfoi and
Leon M. Tolbert, Ph.D.
The University of Tennessee,
Department of Electrical and
Computer Engineering,
Knoxville, Tennessee, USA

17.1 Introduction 451
17.2 Multilevel Power Converter Structures... 452
  17.2.1 Cascaded H-bridges • 17.2.2 Diode-clamped Multilevel Inverter •
  17.2.3 Flying-capacitor Multilevel Inverter • 17.2.4 Other Multilevel Inverter Structures
17.3 Multilevel Converter PWM Modulation Strategies 459
  17.3.1 Multilevel Carrier-based PWM • 17.3.2 Multilevel Space Vector PWM •
  17.3.3 Selective Harmonic Elimination
17.4 Multilevel Converter Design Example.. 470
  17.4.1 Interface with Electrical System • 17.4.2 Number of Levels and Voltage Rating of
  Active Devices • 17.4.3 Number and Voltage Rating of Clamping Diodes • 17.4.4 Current
  Rating of Active Devices • 17.4.5 Current Rating of Clamping Diodes • 17.4.6 DC Link
  Capacitor Specifications
17.5 Fault Diagnosis in Multilevel Converters 478
17.6 Renewable Energy Interface 478
17.7 Conclusion 480
     References 480

17.1 Introduction

Numerous industrial applications have begun to require higher power apparatus in recent years. Some medium voltage motor drives and utility applications require medium voltage and megawatt power level. For a medium voltage grid, it is troublesome to connect only one power semiconductor switch directly. As a result, a multilevel power converter structure has been introduced as an alternative in high power and medium voltage situations. A multilevel converter not only achieves high power ratings, but also enables the use of renewable energy sources. Renewable energy sources such as photovoltaic, wind, and fuel cells can be easily interfaced to a multilevel converter system for a high power application [1–3].

The concept of multilevel converters has been introduced since 1975 [4]. The term multilevel began with the three-level converter [5]. Subsequently, several multilevel converter topologies have been developed [6–13]. However, the elementary concept of a multilevel converter to achieve higher power is to use a series of power semiconductor switches with several lower voltage dc sources to perform the power conversion by synthesizing a staircase voltage waveform. Capacitors, batteries, and renewable energy voltage sources can be used as the multiple dc voltage sources. The commutation of the power switches aggregate these multiple dc sources in order to achieve high voltage at the output, however, the rated voltage of the power semiconductor switches depends only upon the rating of the dc voltage sources to which they are connected.

A multilevel converter has several advantages over a conventional two-level converter that uses high switching frequency pulse width modulation (PWM). The attractive features of a multilevel converter can be briefly summarized as follows.

- *Staircase waveform quality:* Multilevel converters not only can generate the output voltages with very low distortion, but also can reduce the dv/dt stresses, therefore electromagnetic compatibility (EMC) problems can be reduced.
- *Common-mode (CM) voltage:* Multilevel converters produce smaller CM voltage, therefore, the stress in the bearings of a motor connected to a multilevel motor drive can be reduced. Furthermore, CM voltage can be eliminated by using advanced modulation strategies such as that proposed in [14].
- *Input current:* Multilevel converters can draw input current with low distortion.
- *Switching frequency:* Multilevel converters can operate at both fundamental switching frequency and high switching frequency PWM. It should be noted that lower

Copyright © 2007, 2001, Elsevier Inc.
All rights reserved.

switching frequency usually means lower switching loss and higher efficiency.

Unfortunately, multilevel converters do have some disadvantages. One particular disadvantage is the greater number of power semiconductor switches needed. Although lower voltage rated switches can be utilized in a multilevel converter, each switch requires a related gate drive circuit. This may cause the overall system to be more expensive and complex.

Plentiful multilevel converter topologies have been proposed during the last two decades. Contemporary research has engaged novel converter topologies and unique modulation schemes. Moreover, three different major multilevel converter structures have been reported in the literature: cascaded H-bridges converter with separate dc sources, diode clamped (neutral clamped), and flying capacitors (capacitor clamped). Moreover, abundant modulation techniques and control paradigms have been developed for multilevel converters such as sinusoidal pulse width modulation (SPWM), selective harmonic elimination (SHE-PWM), space vector modulation (SVM), and others. In addition, many multilevel converter applications focus on industrial medium-voltage motor drives [11, 15], utility interface for renewable energy systems [16], flexible ac transmission system (FACTS) [17], and traction drive systems [18].

This chapter reviews state of the art of multilevel power converter technology. Fundamental multilevel converter structures and modulation paradigms are discussed including the pros and cons of each technique. Particular concentration is addressed in modern and more practical industrial applications of multilevel converters. A procedure for calculating the required ratings for the active switches, clamping diodes, and dc link capacitors including a design example are described. Finally, the possible future developments of multilevel converter technology are noted.

17.2 Multilevel Power Converter Structures

As previously mentioned, three different major multilevel converter structures have been applied in industrial applications: cascaded H-bridges converter with separate dc sources, diode clamped, and flying capacitors. Before continuing discussion in this topic, it should be noted that the term *multilevel converter* is utilized to refer to a power electronic circuit that could operate in an inverter or rectifier mode. The multilevel inverter structures are the focus of this chapter, however, the illustrated structures can be implemented for rectifying operation as well.

17.2.1 Cascaded H-bridges

A single-phase structure of an m-level cascaded inverter is illustrated in Fig. 17.1. Each separate dc source (SDCS) is connected to a single-phase full-bridge or H-bridge inverter.

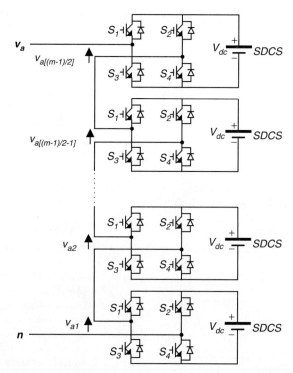

FIGURE 17.1 Single-phase structure of a multilevel cascaded H-bridges inverter.

Each inverter level can generate three different voltage outputs, $+V_{dc}$, 0, and $-V_{dc}$ by connecting the dc source to the ac output by different combinations of the four switches, S_1, S_2, S_3, and S_4. To obtain $+V_{dc}$, switches S_1 and S_4 are turned on, whereas $-V_{dc}$ can be obtained by turning on switches S_2 and S_3. By turning on S_1 and S_2 or S_3 and S_4, the output voltage is Zero. The ac outputs of each of the different full-bridge inverter levels are connected in series such that the synthesized voltage waveform is the sum of the inverter outputs. The number of output phase voltage levels m in a cascade inverter is defined by $m = 2s + 1$, where s is the number of separate dc sources. An example phase voltage waveform for an 11-level cascaded H-bridge inverter with five SDCSs and five full bridges is shown in Fig. 17.2. The phase voltage $v_{an} = v_{a1} + v_{a2} + v_{a3} + v_{a4} + v_{a5}$.

For a stepped waveform such as the one depicted in Fig. 17.2 with s steps, the Fourier Transform for this waveform follows [14, 18]

$$V(\omega t) = \frac{4V_{dc}}{\pi} \sum_n [\cos(n\theta_1) + \cos(n\theta_2) + \cdots + \cos(n\theta_s)]$$

$$\times \frac{\sin(n\omega t)}{n}, \quad \text{where } n = 1, 3, 5, 7, \ldots \quad (17.1)$$

From Eq. (17.1), the magnitudes of the Fourier coefficients when normalized with respect to V_{dc} are as follows

$$H(n) = \frac{4}{\pi n}[\cos(n\theta_1) + \cos(n\theta_2) + \cdots$$

$$+ \cos(n\theta_s)], \quad \text{where } n = 1, 3, 5, 7, \ldots \quad (17.2)$$

17 Multilevel Power Converters

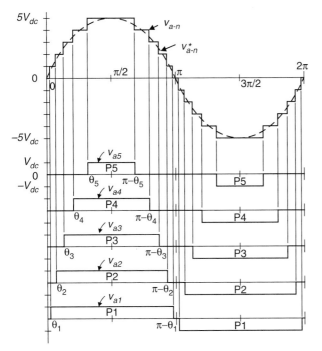

FIGURE 17.2 Output phase voltage waveform of an 11-level cascade inverter with five separate dc sources.

The conducting angles, $\theta_1, \theta_2, \ldots, \theta_s$, can be chosen such that the voltage total harmonic distortion is minimum. Generally, these angles are chosen so that predominant lower frequency harmonics, 5th, 7th, 11th, and 13th harmonics are eliminated [19]. More details on harmonic elimination techniques will be presented in the next section.

Multilevel cascaded inverters have been proposed for such applications as static var generation, an interface with renewable energy sources, and for battery-based applications.

Three-phase cascaded inverters can be connected in wye, as shown in Fig. 17.3, or in delta. Peng has demonstrated a prototype multilevel cascaded static var generator connected in parallel with the electrical system that could supply or draw reactive current from an electrical system [20–23]. The inverter could be controlled to either regulate the power factor of the current drawn from the source or the bus voltage of the electrical system where the inverter was connected. Peng [20] and Joos [24] have also shown that a cascade inverter can be directly connected in series with the electrical system for static var compensation. Cascaded inverters are ideal for connecting renewable energy sources with an ac grid, because of the need for separate dc sources, which is the case in applications such as photovoltaics or fuel cells.

Cascaded inverters have also been proposed for use as the main traction drive in electric vehicles, where several batteries or ultracapacitors are well suited to serve as SDCSs [18, 25]. The cascaded inverter could also serve as a rectifier/charger for the batteries of an electric vehicle while the vehicle was connected to an ac supply as shown in Fig. 17.3. Additionally, the cascade inverter can act as a rectifier in a vehicle that uses regenerative braking.

Manjrekar has proposed a cascade topology that uses multiple dc levels, which instead of being identical in value are multiples of each other [26, 27]. He also uses a combination of fundamental frequency switching for some of the levels and PWM switching for part of the levels to achieve the output voltage waveform. This approach enables a wider diversity of output voltage magnitudes, however, it also results in unequal voltage and current ratings for each of the levels and loses the advantage of being able to use identical modular units for each level.

The main advantages and disadvantages of multilevel cascaded H-bridge converters are as follows [28, 29].

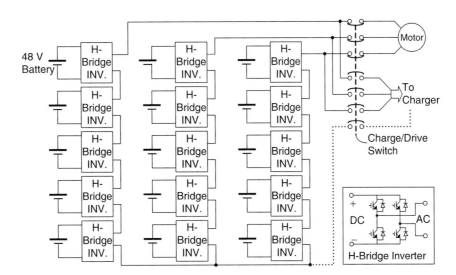

FIGURE 17.3 Three-phase wye-connection structure for electric vehicle motor drive and battery charging.

Advantages:

- The number of possible output voltage levels is more than twice the number of dc sources ($m = 2s + 1$).
- The series of H-bridges makes for modularized layout and packaging. This will enable the manufacturing process to be done more quickly and cheaply.

Disadvantages:

- Separate dc sources are required for each of the H-bridges. This will limit its application to products that already have multiple SDCSs readily available.

Another kind of cascaded multilevel converter with transformers using standard three-phase bi-level converters has been proposed [14]. The circuit is shown in Fig. 17.4. The converter uses output transformers to add different voltages. In order for the converter output voltages to be added up, the outputs of the three converters need to be synchronized with a separation of 120° between each phase. For example, obtaining a three-level voltage between outputs a and b, the output voltage can be synthesized by $V_{ab} = V_{a1-b1} + V_{b1-a2} + V_{a2-b2}$. An isolated transformer is used to provide voltage boost. With three converters synchronized, the voltages $V_{a1-b1}, V_{b1-a2}, V_{a2-b2}$, are all in phase, thus, the output level can be tripled [1].

The advantage of the cascaded multilevel converters with transformers using standard three-phase bi-level converters is that the three converters are identical and thus control is more simple. However, the three converters need separate DC sources, and a transformer is needed to add up the output voltages.

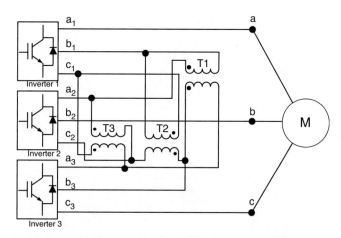

FIGURE 17.4 Cascaded multilevel converter with transformers using standard three-phase bi-level converters.

17.2.2 Diode-clamped Multilevel Inverter

The neutral point converter proposed by Nabae, Takahashi, and Akagi in 1981 was essentially a three-level diode-clamped inverter [5]. In the 1990s, several researchers published articles that have reported experimental results for four-, five-, and six-level diode-clamped converters for uses such as static var compensation, variable speed motor drives, and high-voltage system interconnections [17–30]. A three-phase six-level diode-clamped inverter is shown in Fig. 17.5. Each of the three phases of the inverter shares a common dc bus,

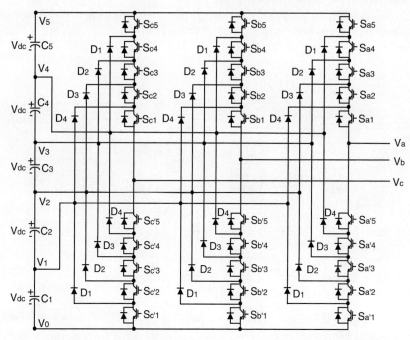

FIGURE 17.5 Three-phase six-level structure of a diode-clamped inverter.

TABLE 17.1 Diode-clamped six-level inverter voltage levels and corresponding switch states

Voltage V_{a0}	Switch state									
	S_{a5}	S_{a4}	S_{a3}	S_{a2}	S_{a1}	$S_{a'5}$	$S_{a'4}$	$S_{a'3}$	$S_{a'2}$	$S_{a'1}$
$V_5 = 5V_{dc}$	1	1	1	1	1	0	0	0	0	0
$V_4 = 4V_{dc}$	0	1	1	1	1	1	0	0	0	0
$V_3 = 3V_{dc}$	0	0	1	1	1	1	1	0	0	0
$V_2 = 2V_{dc}$	0	0	0	1	1	1	1	1	0	0
$V_1 = 1V_{dc}$	0	0	0	0	1	1	1	1	1	0
$V_0 = 0$	0	0	0	0	0	1	1	1	1	1

which has been subdivided by five capacitors into six levels. The voltage across each capacitor is V_{dc}, and the voltage stress across each switching device is limited to V_{dc} through the clamping diodes. Table 17.1 lists the output voltage levels possible for one phase of the inverter with the negative dc rail voltage V_0 as a reference. State condition 1 means the switch is on, and 0 means the switch is off. Each phase has five complementary switch pairs such that turning on one of the switches of the pair requires that the other complementary switch be turned off. The complementary switch pairs for phase leg a are $(S_{a1}, S_{a'1})$, $(S_{a2}, S_{a'2})$, $(S_{a3}, S_{a'3})$, $(S_{a4}, S_{a'4})$, and $(S_{a5}, S_{a'5})$. Table 17.1 also shows that in a diode-clamped inverter, the switches that are on for a particular phase leg are always adjacent and in series. For a six-level inverter, a set of five switches is on at any given time.

Figure 17.6 shows one of the three line–line voltage waveforms for a six-level inverter. The line voltage V_{ab} consists of a phase-leg a voltage and a phase-leg b voltage. The resulting line voltage is an 11-level staircase waveform. This means that an m-level diode-clamped inverter has an m-level output phase voltage and a $(2m - 1)$-level output line voltage.

Although each active switching device is required to block only a voltage level of V_{dc}, the clamping diodes require different ratings for reverse voltage blocking. Using phase a of Fig. 17.5 as an example, when all the lower switches $S_{a'1}$

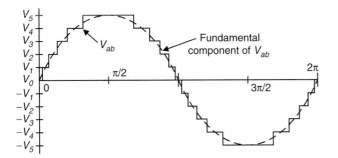

FIGURE 17.6 Line voltage waveform for a six-level diode-clamped inverter.

through $S_{a'5}$ are turned on, D_4 must block four voltage levels, or $4V_{dc}$. Similarly, D_3 must block $3V_{dc}$, D_2 must block $2V_{dc}$, and D_1 must block V_{dc}. If the inverter is designed such that each blocking diode has the same voltage rating as the active switches, D_n will require n diodes in series; consequently, the number of diodes required for each phase would be $(m-1) \times (m-2)$. Thus, the number of blocking diodes is quadratically related to the number of levels in a diode-clamped converter [28].

One application of the multilevel diode-clamped inverter is an interface between a high-voltage dc transmission line and an ac transmission line [29]. Another application would be as a variable speed drive for high-power medium-voltage (2.4–13.8 kV) motors as proposed in [3, 6, 19, 28–30]. Static var compensation is an additional function which several authors have proposed for the diode-clamped converter. The main advantages and disadvantages of multilevel diode-clamped converters are as follows [1–3].

Advantages:

- All of the phases share a common dc bus, which minimizes the capacitance requirements of the converter. For this reason, a back-to-back topology is not only possible but also practical for uses such as a high-voltage back-to-back inter-connection or an adjustable speed drive.
- The capacitors can be pre-charged as a group.
- Efficiency is high for fundamental frequency switching.

Disadvantages:

- Real power flow is difficult for a single inverter because the intermediate dc levels will tend to overcharge or discharge without precise monitoring and control.
- The number of clamping diodes required is quadratically related to the number of levels, which can be cumbersome for units with a high number of levels.

17.2.3 Flying-capacitor Multilevel Inverter

Meynard and Foch introduced a flying-capacitor-based inverter in 1992 [31]. The structure of this inverter is similar to that of the diode-clamped inverter except that instead of using clamping diodes, the inverter uses capacitors in their place. The circuit topology of the flying-capacitor multilevel inverter is shown in Fig. 17.7. This topology has a ladder structure of dc side capacitors, where the voltage on each capacitor differs from that of the next capacitor. The voltage increment between two adjacent capacitor legs gives the size of the voltage steps in the output waveform.

One advantage of the flying-capacitor-based inverter is that it has redundancies for inner voltage levels, in other words, two or more valid switch combinations can synthesize an output voltage. Table 17.2 shows a list of all the combinations of phase voltage levels that are possible for the six-level circuit shown

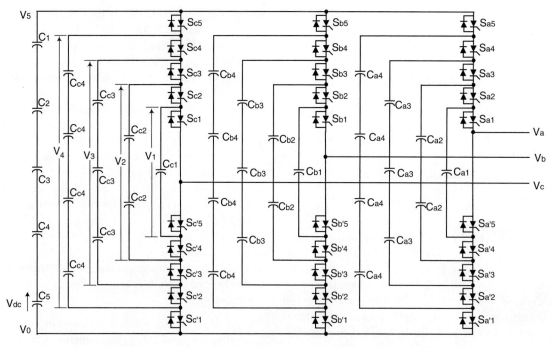

FIGURE 17.7 Three-phase six-level structure of a flying-capacitor inverter.

in Fig. 17.7. Unlike the diode-clamped inverter, the flying-capacitor inverter does not require all of the switches that are on (conducting) be in a consecutive series. Moreover, the flying-capacitor inverter has *phase* redundancies, whereas the diode-clamped inverter has only *line–line* redundancies [2, 3, 32]. These redundancies allow a choice of charging/discharging specific capacitors and can be incorporated in the control system for balancing the voltages across the various levels.

In addition to the $(m-1)$ dc link capacitors, the m-level flying-capacitor multilevel inverter will require $(m-1) \times (m-2)/2$ auxiliary capacitors per phase if the voltage rating of the capacitors is identical to that of the main switches. One application proposed in the literature for the multilevel flying capacitor is static var generation [2, 3]. The main advantages and disadvantages of multilevel flying-capacitor converters are as follows [2, 3].

Advantages:

- Phase redundancies are available for balancing the voltage levels of the capacitors.
- Real and reactive power flow can be controlled.
- The large number of capacitors enables the inverter to ride through short duration outages and deep voltage sags.

Disadvantages:

- Control is complicated to track the voltage levels for all of the capacitors. Also, precharging all of the capacitors to the same voltage level and startup are complex.
- Switching utilization and efficiency are poor for real power transmission.
- The large number of capacitors are both more expensive and bulky than clamping diodes in multilevel diode-clamped converters. Packaging is also more difficult in inverters with a high number of levels.

17.2.4 Other Multilevel Inverter Structures

Besides the three basic multilevel inverter topologies previously discussed, other multilevel converter topologies have been proposed, however, most of these are "hybrid" circuits that are combinations of two of the basic multilevel topologies or slight variations to them. Additionally, the combination of multilevel power converters can be designed to match with a specific application based on the basic topologies. In the interest of completeness, some of these will be identified and briefly described.

A. Generalized Multilevel Topology

Existing multilevel converters such as diode-clamped and capacitor-clamped multilevel converters can be derived from the generalized converter topology called P2 topology proposed by Peng [33] as illustrated in Fig. 17.8. The generalized multilevel converter topology can balance each voltage level by itself regardless of load characteristics, active or reactive power conversion, and without any assistance from other circuits at any number of levels automatically. Thus, the topology

TABLE 17.2 Flying-capacitor six-level inverter redundant voltage levels and corresponding switch states

Voltage V_{a0}	S_{a5}	S_{a4}	S_{a3}	S_{a2}	S_{a1}	$S_{a'5}$	$S_{a'4}$	$S_{a'3}$	$S_{a'2}$	$S_{a'1}$
$V_{a0} = 5V_{dc}$ (no redundancies)										
$5V_{dc}$	1	1	1	1	1	0	0	0	0	0
$V_{a0} = 4V_{dc}$ (4 redundancies)										
$5V_{dc} - V_{dc}$	1	1	1	1	0	0	0	0	0	1
$4V_{dc}$	0	1	1	1	1	1	0	0	0	0
$5V_{dc} - 4V_{dc} + 3V_{dc}$	1	0	1	1	1	0	1	0	0	0
$5V_{dc} - 3V_{dc} + 2V_{dc}$	1	1	0	1	1	0	0	1	0	0
$5V_{dc} - 2V_{dc} + V_{dc}$	1	1	1	0	1	0	0	0	1	0
$V_{a0} = 3V_{dc}$ (5 redundancies)										
$5V_{dc} - 2V_{dc}$	1	1	1	0	0	0	0	0	1	1
$4V_{dc} - V_{dc}$	0	1	1	1	0	1	0	0	0	1
$3V_{dc}$	0	0	1	1	1	1	1	0	0	0
$5V_{dc} - 4V_{dc} + 3V_{dc} - V_{dc}$	1	0	1	1	0	0	1	0	0	1
$5V_{dc} - 3V_{dc} + V_{dc}$	1	1	0	0	1	0	0	1	1	0
$4V_{dc} - 2V_{dc} + V_{dc}$	0	1	1	0	1	1	0	0	1	0
$V_{a0} = 2V_{dc}$ (6 redundancies)										
$5V_{dc} - 3V_{dc}$	1	1	0	0	0	0	0	1	1	1
$5V_{dc} - 4V_{dc} + V_{dc}$	1	0	0	0	1	0	1	1	1	0
$4V_{dc} - 2V_{dc}$	0	1	1	0	0	1	0	0	1	1
$4V_{dc} - 3V_{dc} + V_{dc}$	0	1	0	0	1	1	0	1	1	0
$3V_{dc} - V_{dc}$	0	0	1	1	0	1	1	0	0	1
$3V_{dc} - 2V_{dc} + V_{dc}$	0	0	1	0	1	1	1	0	1	0
$2V_{dc}$	0	0	0	1	1	1	1	1	0	0
$V_{a0} = V_{dc}$ (4 redundancies)										
$5V_{dc} - 4V_{dc}$	1	0	0	0	0	0	1	1	1	1
$4V_{dc} - 3V_{dc}$	0	1	0	0	0	1	0	1	1	1
$3V_{dc} - 2V_{dc}$	0	0	1	0	0	1	1	0	1	1
$2V_{dc} - V_{dc}$	0	0	0	1	0	1	1	1	0	1
V_{dc}	0	0	0	0	1	1	1	1	1	0
$V_{a0} = 0$ (no redundancies)										
0	0	0	0	0	0	1	1	1	1	1

provides a complete multilevel topology that embraces the existing multilevel converters in principle.

Figure 17.8 shows the P2 multilevel converter structure per phase leg. Each switching device, diode, or capacitor's voltage is $1V_{dc}$, for instance, $1/(m-1)$ of the dc-link voltage. Any converter with any number of levels, including the conventional bi-level converter can be obtained using this generalized topology [1, 33].

B. Mixed-level Hybrid Multilevel Converter

To reduce the number of separate dc sources for high-voltage, high-power applications with multilevel converters, diode-clamped or capacitor-clamped converters could be used to replace the full-bridge cell in a cascaded converter [34]. An example is shown in Fig. 17.9. The nine-level cascade converter incorporates a three-level diode-clamped converter as the cell. The original cascaded H-bridge multilevel converter requires four separate dc sources for one phase leg and twelve for a three-phase converter. If a five-level converter replaces the full-bridge cell, the voltage level is effectively doubled for each cell. Thus, to achieve the same nine voltage levels for each phase, only two separate dc sources are needed for one phase leg and six for a three-phase converter. The configuration has mixed-level hybrid multilevel units because it embeds multilevel cells as the building block of the cascade converter. The advantage of the topology is it needs less separate dc sources. The disadvantage for the topology is its control will be complicated due to its hybrid structure.

C. Soft-switched Multilevel Converter

Some soft-switching methods can be implemented for different multilevel converters to reduce the switching loss and

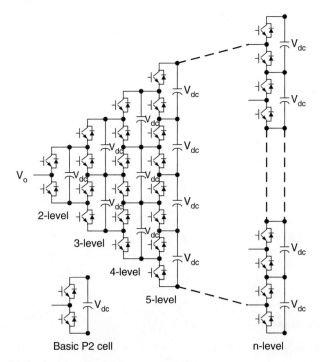

FIGURE 17.8 Generalized P2 multilevel converter topology for one phase leg.

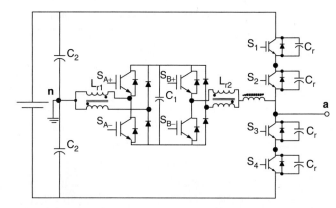

FIGURE 17.10 Zero-voltage-switching capacitor-clamped inverter circuit.

to increase efficiency. For the cascaded converter, because each converter cell is a bi-level circuit, the implementation of soft switching is not at all different from that of conventional bi-level converters. For capacitor-clamped or diode-clamped converters, soft-switching circuits have been proposed with different circuit combinations. One of the soft-switching circuits is a zero-voltage-switching type which includes auxiliary resonant commutated pole (ARCP), coupled inductor with zero-voltage transition (ZVT), and their combinations [1, 35] as shown in Fig. 17.10.

D. Back-to-back Diode-clamped Converter

Two multilevel converters can be connected in a back-to-back arrangement and then the combination can be connected to the electrical system in a series–parallel arrangement as shown in Fig. 17.11. Both the current demanded from the utility and the voltage delivered to the load can be controlled at the same time. This series–parallel active power filter has been referred to as a universal power conditioner [36–42] when used on electrical distribution systems and as a universal power flow controller [43–47] when applied at the transmission level. Previously, Lai and Peng [29] proposed the back-to-back diode-clamped topology shown in Fig. 17.12 for use as a high-voltage dc inter-connection between two asynchronous ac systems or as a rectifier/inverter for an adjustable speed drive for high-voltage motors. The diode-clamped inverter has been chosen over the other two basic multilevel circuit topologies for use in a universal power conditioner for the following reasons:

- All six phases (three on each inverter) can share a common dc link. Conversely, the cascade inverter requires that each dc level be separate, and this is not conducive to a back-to-back arrangement.
- The multilevel flying-capacitor converter also shares a common dc link however, each phase leg requires several additional auxiliary capacitors. These extra capacitors would add substantially to the cost and the size of the conditioner.

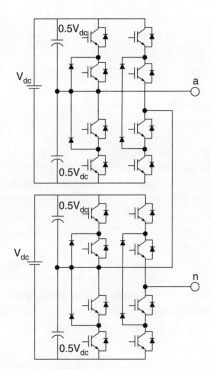

FIGURE 17.9 Mixed-level hybrid unit configuration using the three-level diode-clamped converter as the cascaded converter cell to increase the voltage levels.

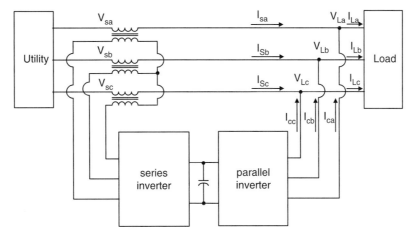

FIGURE 17.11 Series–parallel connection to electrical system of two back-to-back inverters.

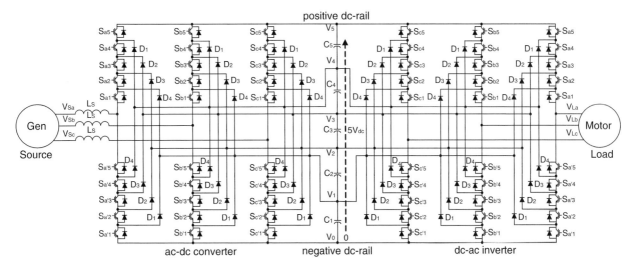

FIGURE 17.12 Six-level diode-clamped back-to-back converter structure.

Because a diode-clamped converter acting as a universal power conditioner will be expected to compensate for harmonics and/or operate in low amplitude modulation index regions, a more sophisticated higher-frequency switch control than the fundamental frequency switching method will be needed. For this reason, multilevel space vector and carrier-based PWM approaches are compared in the next section, as well as novel carrier-based PWM methodologies.

17.3 Multilevel Converter PWM Modulation Strategies

Pulse width modulation strategies used in a conventional inverter can be modified to use in multilevel converters. The advent of the multilevel converter PWM modulation methodologies can be classified according to switching frequency as illustrated in Fig. 17.13. The three multilevel PWM methods most discussed in the literature have been multilevel carrier-based PWM, selective harmonic elimination, and multilevel space vector PWM, all are extensions of traditional two-level PWM strategies to several levels. Other multilevel PWM methods have been used to a much lesser extent by researchers; therefore, only the three major techniques will be discussed in this chapter.

17.3.1 Multilevel Carrier-based PWM

Several different two-level, multilevel carrier-based PWM techniques have been extended by previous authors as a means for controlling the active devices in a multilevel converter. The most popular and easiest technique to implement uses several triangle carrier signals and one reference, or modulation, signal per phase. Figure 17.14 illustrates three major carrier-based techniques used in a conventional inverter that can be

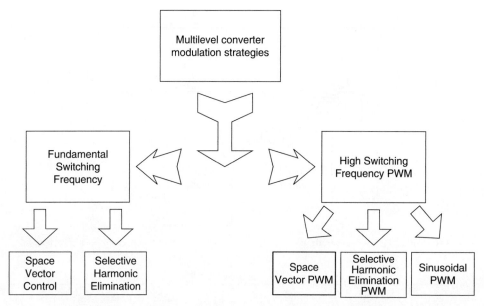

FIGURE 17.13 Classification of PWM multilevel converter modulation strategies.

applied in a multilevel inverter: sinusoidal PWM (SPWM), third harmonic injection PWM (THPWM), and space vector PWM (SVM). SPWM is a very popular method in industrial applications.

In order to achieve better dc link utilization at high modulation indices, the sinusoidal reference signal can be injected by a third harmonic with a magnitude equal to 25% of the fundamental, its line–line output voltage is shown in Fig. 17.14b. As can be seen in Fig. 17.14b and c, the reference signals have some margin at unity amplitude modulation index. Obviously, the dc utilization of THPWM and SVM are better than SPWM in the linear modulation region. The dc utilization means the ratio of the output fundamental voltage to the dc link voltage. Other interesting carrier-based multilevel PWM are subharmonic PWM (SH-PWM) and switching frequency optimal PWM (SFO-PWM). In addition, some particular aspects of these carrier-based methods are also discussed as follows.

A. Subharmonic PWM

Carrara [48] extended SH-PWM to multiple levels as follows: for an m-level inverter, $m-1$ carriers with the same frequency f_c and the same amplitude A_c are disposed such that the bands they occupy are contiguous. The reference waveform has peak-to-peak amplitude A_m, a frequency f_m, and its zero centered in the middle of the carrier set. The reference is continuously compared with each of the carrier signals. If the reference is greater than a carrier signal, then the active device corresponding to that carrier is switched on, and if the reference is less than a carrier signal, then the active device corresponding to that carrier is switched off.

In multilevel inverters, the amplitude modulation index, m_a, and the frequency ratio, m_f, are defined as

$$m_a = \frac{A_m}{(m-1) \cdot A_c} \quad (17.3)$$

$$m_f = \frac{f_c}{f_m} \quad (17.4)$$

Figure 17.15 shows a set of carriers ($m_f = 21$) for a six-level diode-clamped inverter and a sinusoidal reference, or modulation waveform with an amplitude modulation index of 0.8. The resulting output voltage of the inverter is also shown in Fig. 17.15.

B. Switching Frequency Optimal PWM

Another carrier-based method that was extended to multi-level applications by Menzies is termed SFO-PWM, and it is similar to SH-PWM except that a zero sequence (triplen harmonic) voltage is added to each of the carrier waveforms [49]. This method takes the instantaneous average of the maximum and minimum of the three reference voltages (V_a^*, V_b^*, V_c^*) and subtracts this value from each of the individual reference voltages, i.e.

$$V_{offset} = \frac{\max(V_a^*, V_b^*, V_c^*) + \min(V_a^*, V_b^*, V_c^*)}{2} \quad (17.5)$$

$$V_{aSFO}^* = V_a^* - V_{offset} \quad (17.6)$$

$$V_{bSFO}^* = V_b^* - V_{offset} \quad (17.7)$$

$$V_{cSFO}^* = V_c^* - V_{offset} \quad (17.8)$$

17 Multilevel Power Converters

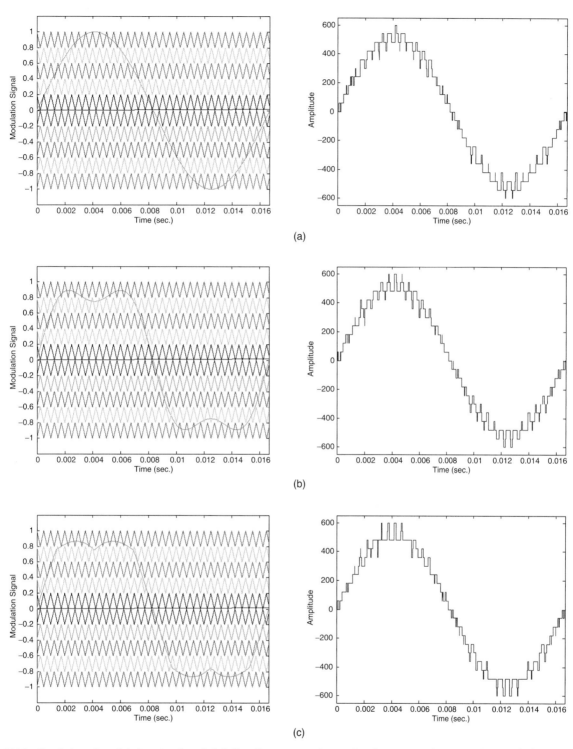

FIGURE 17.14 Simulation of modulation signals and their line–line output voltage using five separate dc sources (60 V each dc source) cascaded multilevel inverter with three major conventional carrier-based PWM techniques at unity modulation index and 2 kHz switching frequency: (a) SPWM; (b) THPWM; and (c) SVM.

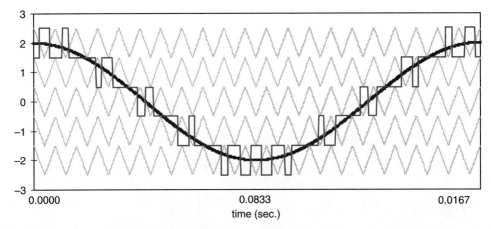

FIGURE 17.15 Multilevel carrier-based SH-PWM showing carrier bands, modulation waveform, and inverter output waveform ($m = 6$, $m_f = 21$, $m_a = 0.8$).

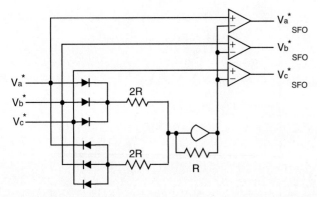

FIGURE 17.16 Analog circuit for zero sequence addition in SFO-PWM.

The addition of this triplen offset voltage centers all of the three reference waveforms in the carrier band, which is equivalent to using space vector PWM [50, 51]. The analog equivalent of Eqs. (17.5–17.8) is shown in Fig. 17.16 [52]. The SFO-PWM is illustrated in Fig. 17.17 for the same reference voltage waveform that was used in Fig. 17.15. The resulting output voltage of the inverter is also shown in Fig. 17.17. The SFO-PWM technique enables the modulation index to be increased by 15% before the overmodulation region is reached.

For the SH-PWM and SFO-PWM techniques shown in Figs. 17.15 and 17.17, the top and bottom switches are switched much more often than the intermediate devices. Methods to balance and/or reduce the device switchings without an adverse effect on a multilevel inverter's output voltage total harmonic distortion would be beneficial. Methods to do just that are developed in [53]. A novel method to balance device switchings for all of the levels in a diode clamped inverter has been demonstrated for SH-PWM and SFO-PWM by varying the frequency for the different triangle wave carrier bands as shown in Fig. 17.18 [53].

C. Modulation Index Effect on Level Utilization

For low amplitude modulation indices, a multilevel inverter will not make use of all of its levels and at very low modulation indices it operates as if it is a traditional two-level inverter. Figure 17.19 shows two simulation results of what the output voltage waveform looks like at amplitude modulation indices

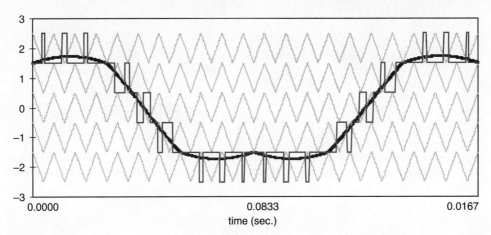

FIGURE 17.17 Multilevel carrier-based SFO-PWM showing carrier bands, modulation waveform, and inverter output waveform ($m = 6$, $m_f = 21$, $m_a = 0.8$).

17 Multilevel Power Converters   463

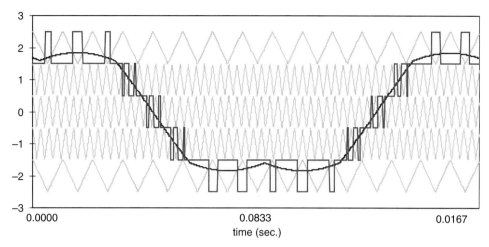

FIGURE 17.18 SFO-PWM where carriers have different frequencies ($m_a = 0.85$, $m_f = 15$ for Band$_2$, Band$_{-2}$; $m_f = 55$ for Band$_1$, Band$_{-1}$; Band$_0$, $\varphi = 0.10$ rad).

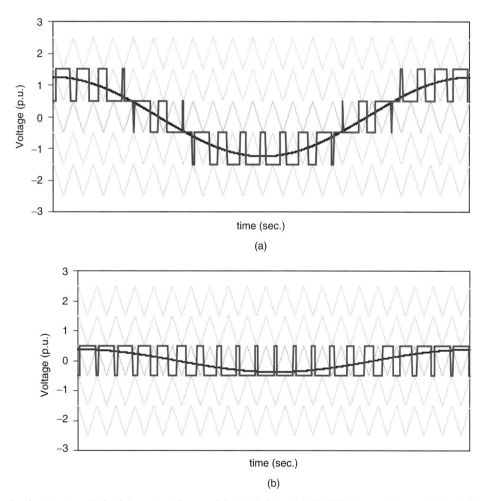

FIGURE 17.19 Level reduction in a six-level inverter at low modulation indices: (a) SH-PWM, $m = 6, m_a = 0.5$ and (b) SH-PWM, $m = 6$, $m_a = 0.15$.

of 0.5 and 0.15. Figure 17.19a shows how the bottom and top switches (S_{a1}–$S_{a'1}$, S_{a5}–$S_{a'5}$ in Fig. 17.5) go unused for amplitude modulation indices less than 0.6 in a six-level inverter. Figure 17.19b shows how only the middle switches (S_{a3}–$S_{a'3}$ in Fig. 17.5) change states when a six-level inverter is operated at an amplitude modulation index less than 0.2. The output waveform in Fig. 17.19b appears to be that of a traditional two-level inverter rather than a multilevel inverter.

The minimum modulation index m_{amin} for which a multilevel inverter controlled with SH-PWM makes use of all of its levels, m, is

$$m_{amin} = \frac{m-3}{m-1} \quad (17.9)$$

Table 17.3 lists the minimum modulation index where a multilevel inverter uses all its constituent levels for both SH-PWM and SFO-PWM techniques. Table 17.3 also shows that the maximum modulation index before pulse dropping (overmodulation) occurs is 1.000 for SH-PWM and 1.155 for SFO-PWM. As shown in Table 17.3, when a multilevel inverter operates at modulation indices much less than 1.000, not all of its levels are involved in the generation of the output voltage and simply remain in an unused state until the modulation index increases sufficiently. The table also shows that level usage is more likely to suffer to a greater extent as the number of levels in the inverter increases.

One way to make use of the multiple levels, even during low modulation periods, is to take advantage of the redundant output voltage states by rotating level usage in the inverter after each modulation cycle. This will reduce the switching stresses on some of the inner levels by making use of those outer voltage levels that otherwise would go unused.

As previously mentioned, diode-clamped inverters have redundant line–line voltage states for low modulation indices, but have no phase redundancies [54]. For an output voltage

TABLE 17.3 Modulation index ranges without level reduction (Min) or pulse dropping because of overmodulation (Max)

Levels	SH-PWM		SFO-PWM	
	Min	Max	Min	Max
3	0.000	1.000	0.000	1.155
4	0.333	1.000	0.385	1.155
5	0.500	1.000	0.578	1.155
6	0.600	1.000	0.693	1.155
7	0.667	1.000	0.770	1.155
8	0.714	1.000	0.825	1.155
9	0.750	1.000	0.866	1.155
10	0.778	1.000	0.898	1.155
11	0.800	1.000	0.924	1.155
12	0.818	1.000	0.945	1.155
13	0.833	1.000	0.962	1.155

TABLE 17.4 Six-level inverter line–line voltage redundancies

max(i,j,k)−min(i,j,k)	Number of distinct states	Number of redundancies per distinct state	Total number of states
0	1	5	6
1	6	4	30
2	12	3	48
3	18	2	54
4	24	1	48
5	30	0	30
Total	91	–	216

state (i, j, k) in an m-level diode-clamped inverter, the number of redundant states available is given by

$$N_{\substack{redundancies\\available}} = m - 1 - \left[\max(i,j,k) - \min(i,j,k)\right] \quad (17.10)$$

As the modulation index decreases, more redundant states are available. Table 17.4 shows the number of distinct and redundant line–line voltage states available in a six-level inverter for different output voltages.

In the next section, a carrier-based method is given that uses line–line redundancies in a diode-clamped inverter operating at a low modulation index so that active device usage is more balanced among the levels.

D. Increasing Switching Frequency at Low Modulation Indices

For amplitude modulation indices less than 0.5, the level usage in odd-level inverters can be sufficiently rotated so that the switching frequency can be doubled and still keep the thermal losses within the limits of the device. For inverters with an even number of levels, the modulation index at which frequency doubling can be accomplished varies with the levels as shown in Table 17.5. This increase in switching frequency enables the inverter to compensate for higher frequency harmonics and will yield a waveform that more closely tracks a reference.

As an example of how to accomplish this doubling of inverter frequency, an analysis of a seven-level diode-clamped inverter with an amplitude modulation index of 0.4 is conducted. During one cycle, the reference waveform is centered in the upper three carrier bands, and during the next cycle, the reference waveform is centered in the lower three carrier bands as shown in Fig. 17.20. This technique enables half of the switches to "rest" every other cycle and not incur any switching losses. With this method, the switching frequency (or carrier frequency f_c in the case of multilevel inverters) can effectively be doubled to $2f_c$, but the switches will have the same thermal losses as if they were switching at f_c but every cycle.

TABLE 17.5 Increased switching frequency possible at lower modulation indices

Inverter levels	Modulation index, m_a		Frequency multiplier
	Min	Max	
3	0.000	0.500	2X
4	0.000	0.333	3X
5	0.250	0.500	2X
	0.000	0.250	4X
6	0.200	0.400	2X
	0.000	0.200	5X
7	0.333	0.500	2X
	0.167	0.333	3X
	0.000	0.167	6X
8	0.285	0.428	2X
	0.142	0.285	3X
	0.000	0.142	7X
9	0.25	0.500	2X
	0.125	0.250	4X
	0.000	0.125	8X
10	0.333	0.444	2X
	0.222	0.333	3X
	0.111	0.222	4X
	0.000	0.111	9X
11	0.333	0.500	2X
	0.200	0.333	3X
	0.000	0.200	5X
12	0.272	0.454	2X
	0.181	0.272	3X
	0.090	0.181	5X
	0.000	0.090	11X
13	0.333	0.500	2X
	0.250	0.333	3X
	0.167	0.250	4X
	0.0833	0.167	6X
	0.000	0.0833	12X

three phases are moved from one carrier set to the next set at the same time. In the case of frequency doubling, all three phases add or subtract the following number of states (or levels) every other reference cycle

$$h_a(j+1) = h_a(j) + (-1)^j \cdot \frac{\lfloor m-1 \rfloor}{2} \quad (17.11)$$

At modulation indices closer to zero, the switching frequency can be increased even more. This is possible because the reference waveform can be rotated among the carrier bands for a few cycles before returning to a previous set of switches for use. The switches are allowed to "rest" for a few cycles and thus are able to absorb higher losses during the cycle that they are in use. Table 17.5 shows the possible increased switching frequencies available at lower amplitude modulation indices for several different inverter levels.

Some additional switching loss is associated with the redundant switchings of the three phases at the end of each modulation cycle when rotating among carrier bands. For instance, for Fig. 17.20 each of the three phases in the seven-level inverter will have three switch pairs change states at the end of every reference cycle. Compared to the switching loss associated with just the normal PWM switchings, however, this redundant switching loss is quite small, typically less than 5% of the total switching loss.

Figures 17.21 and 17.22 illustrate two different methods of rotating the reference waveform among three different regions (top, middle, and bottom) for modulation indices less than 0.333 in a seven-level inverter to enable the carrier frequency to be increased by a factor of three. The method shown in Fig. 17.21 is preferred over that shown in Fig. 17.22 because of less redundant state switching. The method in Fig. 17.21 requires only four redundant state switchings for every three reference cycles, whereas the method in Fig. 17.22 requires eight redundant switchings for every three reference cycles. In general for any multilevel inverter regardless of the number of levels or number of rotation regions, using the preferred reference rotation method will

This method is possible only for three-wire systems because the diode-clamped inverter has line–line redundancies and no phase redundancies. This means that at the discontinuity where the reference moves from one carrier band set to another, the transition has to be synchronized such that all

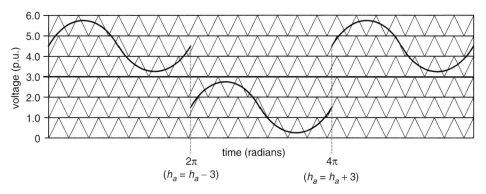

FIGURE 17.20 Reference rotation among carrier bands at low modulation indices ($m_a < 0.5$).

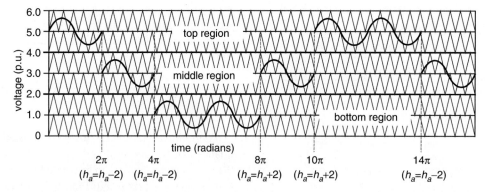

FIGURE 17.21 Preferred method of reference rotation among carrier bands with 3× carrier frequency at very low modulation indices.

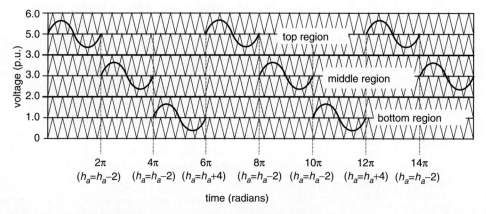

FIGURE 17.22 Alternate method of reference rotation among carrier bands with 3× carrier frequency at very low modulation indices.

have 1/2 of the redundant switching losses that the alternate method would have.

Unlike the diode-clamped inverter, the cascaded H-bridges inverter has phase redundancies in addition to the aforementioned line–line redundancies. Phase redundancies are much easier to exploit than line–line redundancies because the output voltage in each phase of a three-phase inverter can be generated independently of the other two phases when only phase redundancies are used. A method was given in [18] that makes use of these phase redundancies in a cascaded inverter so that each active device's duty cycle is balanced over $(m-1)/2$ modulation waveform cycles regardless of the modulation index. The same pulse rotation technique used for fundamental frequency switching of cascade inverters was used but with a PWM output voltage waveform [55], which is a much more effective means of controlling a driven motor at low speeds than continuing to do fundamental frequency switching. The effect of this control is that the output waveform can have a high switching frequency but the individual levels can still switch at a constant switching frequency of 60 Hz, if desired.

17.3.2 Multilevel Space Vector PWM

Choi [56] was the first author to extend the two-level space vector PWM technique to more than three levels for the diode-clamped inverter. Figure 17.23 shows what the space vector d–q plane looks like for a six-level inverter. Figure 17.24 represents the equivalent dc link of a six-level inverter as a multiplexer that connects each of the three output phase voltages to one of the dc link voltage tap points [57]. Each integral point on the space vector plane represents a particular three-phase output voltage state of the inverter. For instance, the point (3, 2, 0) on the space vector plane means, that with respect to ground, a phase is at $3V_{dc}$, b phase is at $2V_{dc}$, and c phase is at $0V_{dc}$. The corresponding connections between the dc link and the output lines for the six-level inverter are also shown in Fig. 17.24 for the point (3, 2, 0). An algebraic way to represent the output voltages in terms of the switching states and dc link capacitors is described in the following [58]. For $n = m-1$, where m is the number of levels in the inverter

$$V_{abc0} = H_{abc}V_c \qquad (17.12)$$

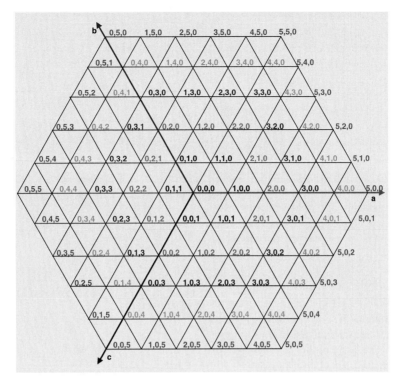

FIGURE 17.23 Voltage space vectors for a six-level inverter.

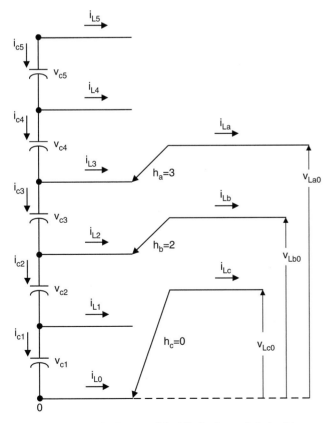

FIGURE 17.24 Multiplexer model of diode-clamped six-level inverter.

where,

$$V_c = \begin{bmatrix} V_{c1} & V_{c2} & V_{c3} & \ldots & V_{cn} \end{bmatrix}^T;$$

$$H_{abc} = \begin{bmatrix} h_{a1} & h_{a2} & h_{a3} & \ldots & h_{an} \\ h_{b1} & h_{b2} & h_{b3} & \ldots & h_{bn} \\ h_{c1} & h_{c2} & h_{c3} & \ldots & h_{cn} \end{bmatrix}; \quad V_{abc0} = \begin{bmatrix} V_{a0} \\ V_{b0} \\ V_{c0} \end{bmatrix}$$

and

$$h_{aj} = \sum_j^n \delta(h_a - j)$$

where h_a is the switch state and j is an integer from 0 to n, and where $\delta(x) = 1$ if $x \geq 0$, $\delta(x) = 0$ if $x < 0$.

Besides the output voltage state, the point (3, 2, 0) on the space vector plane can also represent the switching state of the converter. Each integer indicates how many upper switches in each phase leg are on for a diode-clamped converter. As an example, for $h_a = 3$, $h_b = 2$, $h_c = 0$, the H_{abc} matrix for this particular switching state of a six-level inverter would be

$$H_{abc} = \begin{bmatrix} 0 & 0 & 1 & 1 & 1 \\ 0 & 0 & 0 & 1 & 1 \\ 0 & 0 & 0 & 0 & 0 \end{bmatrix}$$

Redundant switching states are those states for which a particular output voltage can be generated by more than one switch combination. Redundant states are possible at lower modulation indices, or at any point other than those on the outermost hexagon shown in Fig. 17.23. Switch state (3, 2, 0) has redundant states (4, 3, 1) and (5, 4, 2). Redundant switching states differ from each other by an identical integral value, i.e. (3, 2, 0) differs from (4, 3, 1) by (1, 1, 1) and from (5, 4, 2) by (2, 2, 2).

For an output voltage state (x, y, z) in an m-level diode-clamped inverter, the number of redundant states available is given by $m - 1 - \max(x, y, z)$. As the modulation index decreases (or the voltage vector in the space vector plane gets closer to the origin), more redundant states are available. The number of possible zero states is equal to the number of levels, m. For a six-level diode-clamped inverter, the zero voltage states are (0, 0, 0), (1, 1, 1), (2, 2, 2), (3, 3, 3), (4, 4, 4), and (5, 5, 5).

The number of possible switch combinations is equal to the cube of the level (m^3). For this six-level inverter, there are 216 possible switching states. The number of distinct or unique states for an m-level inverter can be given by

$$m^3 - (m-1)^3 = \left[6 \sum_{n=1}^{m-1} n \right] + 1 \qquad (17.13)$$

Therefore, the number of redundant switching states for an m-level inverter is $(m-1)^3$. Table 17.6 summarizes the available redundancies and distinct states for a six-level diode-clamped inverter.

In two-level PWM, a reference voltage is tracked by selecting the two nearest voltage vectors and a zero vector and then by calculating the time required to be at each of these three vectors such that their sum equals the reference vector. In multilevel PWM, generally the nearest three triangle vertices, V_1, V_2, and V_3, to a reference point V^* are selected so as to minimize the harmonic components of the output line–line voltage [59]. The respective time duration, T_1, T_2, and T_3, required of these vectors is then solved from the following equations

$$\vec{V}_1 T_1 + \vec{V}_2 T_2 + \vec{V}_3 T_3 = V^* T_s \qquad (17.14)$$

$$T_1 + T_2 + T_3 = T_s \qquad (17.15)$$

where T_s is the switching period. Equation (17.14) actually represents two equations, one with the real part of the terms and one with the imaginary part of the terms

$$V_{1d} T_1 + V_{2d} T_2 + V_{3d} T_3 = V_d^* T_s \qquad (17.16)$$

$$V_{1q} T_1 + V_{2q} T_2 + V_{3q} T_3 = V_q^* T_s \qquad (17.17)$$

Equations (17.15) – (17.18) can then be solved for T_1, T_2, and T_3 as follows:

$$\begin{bmatrix} T_1 \\ T_2 \\ T_3 \end{bmatrix} = \begin{bmatrix} V_{1d} & V_{2d} & V_{3d} \\ V_{1q} & V_{2q} & V_{3q} \\ 1 & 1 & 1 \end{bmatrix}^{-1} \begin{bmatrix} V_d^* T_s \\ V_q^* T_s \\ T_s \end{bmatrix} \qquad (17.18)$$

Others have proposed space vector methods that did not use the nearest three vectors, but these methods generally add complexity to the control algorithm. Figure 17.25 shows what a sinusoidal reference voltage (circle of points) and the inverter output voltages look like in the d–q plane.

Redundant switch levels can be used to help manage the charge on the dc link capacitors [60]. Generalizing from

TABLE 17.6 Line–line redundancies of six-level three-phase diode-clamped inverter

Redundancies	Distinct states	Redundant states	Unique state coordinates: (a, b, c) where $0 \leq a, b, c \leq 5$
5	1	5	(0,0,0)
4	6	24	(0,0,1),(0,1,0),(1,0,0),(1,0,1),(1,1,0),(0,0,1)
3	12	36	(p,0,2),(p,2,0),(0,p,2),(2,p,0),(0,2,p,),(2,0,p) where $p \leq 2$
2	18	36	(0,3,p),(3,0,p),(p,3,0),(p,0,3),(3,p,0),(0,p,3) where $p \leq 3$
1	24	24	(0,4,p),(4,0,p),(p,4,0),(p,0,4),(4,p,0),(0,p,4) where $p \leq 4$
0	30	0	(0,5,p),(5,0,p),(p,5,0),(p,0,5),(5,p,0),(0,p,5) where $p \leq 5$
Total	91	125	216 total states

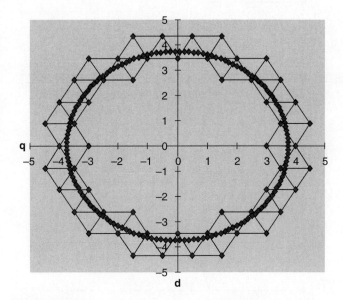

FIGURE 17.25 Sinusoidal reference and inverter output voltage states in d–q plane.

Fig. 17.24, the equations for the currents through the dc link capacitors can be given as

$$i_{cn} = -i_{Ln} \quad (17.19)$$

and

$$i_{c(n-j)} = -i_{L(n-j)} + i_{c(n-j+1)}, \quad \text{where } j = 1, 2, 3, \ldots, n-1 \quad (17.20)$$

The dc link currents for $h_a = 3$, $h_b = 2$, $h_c = 0$ would be $i_{c5} = i_{c4} = 0$, $i_{c3} = -i_a$, $i_{c2} = -i_a - i_b$, $i_{c1} = -i_a - i_b$. To see how redundant states affect the dc link currents, consider the two redundant states for (3, 2, 0). In state (4, 3, 1), the dc link currents would be $i_{c5} = 0$, $i_{c4} = -i_a$, $i_{c3} = -i_a - i_b$, $i_{c2} = -i_a - i_b$, $i_{c1} = -i_a - i_b - i_c = 0$; and for the state (5, 4, 2), the dc link currents would be $i_{c5} = -i_a$, $i_{c4} = -i_a - i_b$, $i_{c3} = -i_a - i_b$, $i_{c2} = i_{c1} = -i_a - i_b - i_c = 0$.

From this example, one can see that the choice of redundant switching states can be used to determine which capacitors will be charged/discharged or unaffected during the switching period. While this control is helpful in balancing the individual dc voltages across the capacitors that make up the dc link, this method is quite complicated in selecting which of the redundant states to use. Constant use of redundant switching states also results in a higher switching frequency and lower efficiency of the inverter because of the extra switchings. Recently, optimized space vector switching sequences for multilevel inverters have been proposed in [61].

17.3.3 Selective Harmonic Elimination

A. Fundamental Switching Frequency

The selective harmonic elimination method is also called fundamental switching frequency method based on the harmonic elimination theory proposed by Patel [62, 63]. A typical 11-level multilevel converter output with fundamental frequency switching scheme is shown in Fig. 17.2. The Fourier series expansion of the output voltage waveform as shown in Fig. 17.2 is expressed in Eqs. (17.1) and (17.2).

The conducting angles, $\theta_1, \theta_2, \ldots, \theta_s$, can be chosen such that the voltage total harmonic distortion is a minimum. Normally, these angles are chosen so as to cancel the predominant lower frequency harmonics [19].

For the 11-level case in Fig. 17.2, the 5th, 7th, 11th, and 13th harmonics can be eliminated with the appropriate choice of the conducting angles. One degree of freedom is used so that the magnitude of the fundamental waveform corresponds to the reference waveform's amplitude or modulation index, m_a, which is defined as V_L^*/V_{Lmax}. V_L^* is the amplitude command of the inverter for a sine wave output phase voltage, and V_{Lmax} is the maximum attainable amplitude of the converter,

i.e. $V_{Lmax} = s \cdot V_{dc}$. The equations from Eq. (17.2) will now be as follows

$$\cos(5\theta_1) + \cos(5\theta_2) + \cos(5\theta_3) + \cos(5\theta_4) + \cos(5\theta_5) = 0$$
$$\cos(7\theta_1) + \cos(7\theta_2) + \cos(7\theta_3) + \cos(7\theta_4) + \cos(7\theta_5) = 0$$
$$\cos(11\theta_1) + \cos(11\theta_2) + \cos(11\theta_3) + \cos(11\theta_4) + \cos(11\theta_5) = 0$$
$$\cos(13\theta_1) + \cos(13\theta_2) + \cos(13\theta_3) + \cos(13\theta_4) + \cos(13\theta_5) = 0$$
$$\cos(\theta_1) + \cos(\theta_2) + \cos(\theta_3) + \cos(\theta_4) + \cos(\theta_5) = 5m_a$$
$$(17.21)$$

The above equations are non-linear transcendental equations that can be solved by an iterative method such as the Newton–Raphson method. For example, using a modulation index of 0.8 obtains: $\theta_1 = 6.57°$, $\theta_2 = 18.94°$, $\theta_3 = 27.18°$, $\theta_4 = 45.14°$, $\theta_5 = 62.24°$. Thus, if the inverter output is symmetrically switched during the positive half cycle of the fundamental voltage to $+V_{dc}$ at 6.57°, $+2V_{dc}$ at 18.94°, $+3V_{dc}$ at 27.18°, $+4V_{dc}$ at 45.14°, and $+5V_{dc}$ at 62.24°, and similarly in the negative half cycle to $-V_{dc}$ at 186.57°, $-2V_{dc}$ at 198.94°, $-3V_{dc}$ at 207.18°, $-4V_{dc}$ at 225.14°, $-5V_{dc}$ at 242.24°, the output voltage of the 11-level inverter will not contain the 5th, 7th, 11th, and 13th harmonic components [18]. Other methods to solve these equations include using genetic algorithms [64] and resultant theory [65–67].

Practically, the precalculated switching angles are stored as the data in memory (look-up table). Therefore, a microcontroller could be used to generate the PWM gate drive signals.

B. Selective Harmonic Elimination PWM

In order to achieve a wide range of modulation indexes with minimized THD for the synthesized waveforms, a generalized selective harmonic modulation method [68, 69] was proposed, which is called virtual stage PWM [64]. An output waveform is shown in Fig. 17.26. The virtual stage PWM is a combination of unipolar programmed PWM and the fundamental frequency switching scheme. The output waveform of unipolar programmed PWM is shown in Fig. 17.27. When unipolar

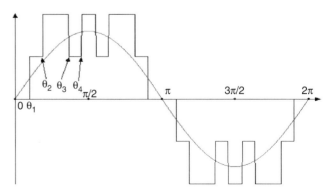

FIGURE 17.26 Output waveform of virtual stage PWM control.

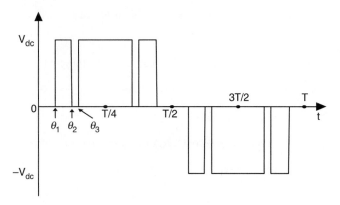

FIGURE 17.27 Unipolar switching output waveform.

programmed PWM is employed on a multilevel converter, typically one dc voltage is involved, where the switches connected to the dc voltage are switched "on" and "off" several times per fundamental cycle. The switching pattern decides what the output voltage waveform looks like.

For fundamental switching frequency method, the number of switching angles is equal to the number of dc sources. However, for the virtual stage PWM method, the number of switching angles is not equal to the number of dc voltages. For example, in Fig. 17.26, only two dc voltages are used, whereas there are four switching angles.

Bipolar programmed PWM and unipolar programmed PWM could be used for modulation indices, too low for the applicability of the multilevel fundamental frequency switching method. Virtual stage PWM can also be used for low modulation indices. Virtual stage PWM will produce output waveforms with a lower THD most of the time [64]. Therefore, virtual stage PWM provides another alternative to bipolar programmed PWM and unipolar programmed PWM for low modulation index control.

The major difficulty for selective harmonic elimination methods, including the fundamental switching frequency method and the virtual stage PWM method, is to solve the transcendental Eq. (17.21) for switching angles. Newton's method can be used to solve Eq. (17.21), but it needs good initial guesses, and solutions are not guaranteed. Therefore, Newton's method is not feasible to solving equations for large number of switching angles if good initial guesses are not available [70].

Recently, the resultant method has been proposed in [65–67] to solve the transcendental equations for switching angles. The transcendental equations characterizing the harmonic content can be converted into polynomial equations. Elimination resultant theory has been employed to determine the switching angles to eliminate specific harmonics, such as the 5th, 7th, 11th, and 13th. However, as the number of dc voltages or the number of switching angles increases, the degrees of the polynomials in these equations become bulky.

To conquer this problem, the fundamental frequency switching angle computation is solved by Newton's method. The initial guess can be provided by the results of lower order transcendental equations by the resultant method [70].

17.4 Multilevel Converter Design Example

The objective of this section is to give a general idea how to design a multilevel converter in a specific application. Different applications for multilevel converters might have different specification requirements. Therefore, the multilevel universal power conditioner (MUPC) is utilized to demonstrate as the design example in this section.

Multilevel diode clamped converters can be designed where different levels have unequal voltage and current ratings; however, this approach would lose the advantage of being able to use identical, modular units for each leg of the inverter. The method used in this chapter to specify a back-to-back diode clamped converter for use as a universal power conditioner is for all voltage levels and legs in each of the two inverters to be the same. (The current ratings in the series inverter may be different than those in the parallel inverter.) This approach also allows the control system to extend the frequency range of the inverter by exploiting the additional voltage redundancies available at lower modulation indices as discussed in [71].

17.4.1 Interface with Electrical System

Figure 17.28 illustrates the proposed electrical system connection topology for two diode-clamped inverters connected back-to-back and sharing a common dc bus. One inverter interfaces with the electrical system by means of a parallel connection through output inductors L_{PI}. The other inverter interfaces with the electrical system through a set of single-phase transformers in a series fashion. The primaries of the transformers are inserted in series with each of the three phase conductors supplied from a utility. The secondaries of the transformers are connected in an ungrounded wye and to the output of the series inverter. By having two inverters, this arrangement allows both the source voltage and the load current to be compensated independently of each other [71, 72]. With only a single inverter, regulating the load voltage and source current at the same time would not be possible.

The voltage injected into the electrical system by the series inverter compensates for deviations in the source voltage such that a regulated distortion-free waveform is supplied to the load. The parallel inverter injects current into the electrical system to compensate for current harmonics and/or reactive current demanded by the load such that the current drawn from the utility is in phase with the source voltage and contains no harmonic components.

17 Multilevel Power Converters

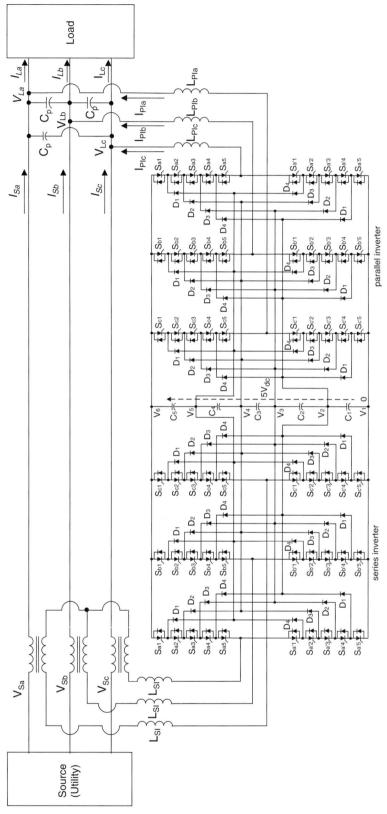

FIGURE 17.28 Electrical system connection of multilevel diode-clamped power conditioner.

17.4.2 Number of Levels and Voltage Rating of Active Devices

In a multilevel inverter, determining the number of levels will be one of the most important factors because this affects many of the other sizing factors and control techniques. Tradeoffs, in specifying the number of levels that the power conditioner will need and the advantages and complexity of having multiple voltage levels available, are the primary differences that set a multilevel filter apart from a single level filter.

As a starting point, known is the nominal RMS voltage rating, V_{nom}, of the electrical system to which the diode-clamped power conditioner will be connected. The dc link voltage must be at least as high as the amplitude of the nominal line–neutral voltage at the point of connection, or $\sqrt{2} \cdot V_{nom}$.

The parallel inverter must be able to inject currents by imposing a voltage across the parallel inductors, L_{PI}, that is the difference between the load voltage V_L and parallel inverter output voltage V_{PI}. The most difficult time to impose a voltage across the inductors is when the load voltage waveform is at its maximum or minimum. Simulation results have shown that the amplitude of the desired load voltage V_{nom} should not be more than 70% of the overall dc link voltage for the parallel inverter to have sufficient margin to inject appropriate compensation currents. Without this margin, complete compensation of reactive currents may not be possible. This margin can be incorporated into a design factor for the inverter. Because, the dc link voltage and the voltage at the connection point, both can vary, the design factor used in the rating selection process incorporates these elements as well as the small voltage drops that occur in the inverters during active device conduction.

The product of the number of the active devices in series $(m - 1)$ and the voltage rating of the devices V_{dev} must then be such that

$$V_{\substack{device \\ rating}} \cdot (m - 1) \geq \sqrt{2} \cdot V_{nom} \cdot D_{\substack{design \\ factor}} \qquad (17.22)$$

The minimum number of levels and the voltage rating of the active devices (IGBTs, GTOs, power MOSFETs, etc.) are inversely related to each other. More levels in the inverter will lower the required voltage device rating of individual devices, or looking at it another way, a higher voltage rating of the devices will enable a fewer minimum number of levels to be used.

Increasing the number of levels does not affect the total voltage blocking capability of the active devices in each phase leg because lower device ratings can be used. Some of the benefits of using more than the minimum required number of levels in a diode-clamped inverter are as follows:

1. Voltage stress across each device is lower. Both active devices and dc link capacitors could be used that have lower voltage ratings (which sometimes are much cheaper and have greater availability).
2. The inverter will have a lower EMI because the dv/dt during each switching will be lower.
3. The output of the waveform will have more steps, or degrees of freedom, which enables the output waveform to closely track a reference waveform.
4. Lower individual device switching frequency will achieve the same results as an inverter with a fewer number of levels and higher device switching frequency. Or the switching frequency can be kept the same as that in an inverter with a fewer number of levels to achieve a better waveform.

The drawbacks of using more than the required minimum number of levels are as follows:

1. Six active device control signals (one for each phase of the parallel inverter and the series inverter) are needed for each hardware level of the inverter – i.e. $6 \cdot (m - 1)$ control signals. Additional levels require more computational resources and add complexity to the control.
2. If the blocking diodes used in the inverter have the same rating as the active devices, their number increases dramatically because $6 \cdot (m-2) \cdot (m-1)$ diodes would be required for the back-to-back structure.

Considering the tradeoffs between the number of levels and the voltage rating of the devices will generally lead the designer to choose an appropriate value for each.

17.4.3 Number and Voltage Rating of Clamping Diodes

As shown in the previous section, $6 \cdot (m-1) \cdot (m-2)$ clamping diodes are required for an m-level back-to-back converter if the diodes have the same voltage rating as the active devices. As discussed in Section 17.2, the voltage rating of each series of clamping diodes is designated by the subscript of the diode shown in Fig. 17.28. For instance, D_4 must block $4V_{dc}$, D_3 must block $3V_{dc}$, and so on.

If diodes that have higher voltage ratings than the active devices are available, then the number of diodes required can be reduced accordingly. When considering diodes of different ratings, the minimum number of clamping diodes per phase leg of the inverter is $2 \cdot (m - 2)$ and for the complete back-to-back converter, $12 \cdot (m - 2)$. Unlike the active devices, additional levels do *not* enable a decrease in the voltage rating of the clamping diodes. In each phase leg, note that the voltage rating of each pair of diodes adds up to the overall dc link voltage $(m - 1) \cdot V_{dc}$. Considering the six-level converter in Fig. 17.28, connected to voltage level V_5 are the anode of D_1 and the cathode of D_4. D_1 must be able to block V_{dc}, and D_4

must block $4V_{dc}$, the sum of their voltage blocking capabilities is $5V_{dc}$. For voltage level V_4, the anode of D_2 and the cathode of D_3 are connected together to this point. Again, the sum of their voltage blocking capability is $2V_{dc} + 3V_{dc} = 5V_{dc}$. The same is true for the other intermediate voltage levels. Therefore, the total voltage blocking capability per phase of an m-level converter is $(m-2) \cdot (m-1) \cdot V_{dc}$ and for the back-to-back converter,

$$V_{\substack{clamp \\ total}} = 6 \cdot (m-2) \cdot (m-1) \cdot V_{dc} \quad (17.23)$$

Each additional level added to the converter will require an additional $6 \cdot (m-1) \cdot V_{dc}$ in voltage blocking capabilities. From this, one can see that unnecessarily adding more than the required number of voltage levels can quickly become cost prohibitive.

17.4.4 Current Rating of Active Devices

In order to determine the required current rating of the active switching devices for the parallel and series portions of the back-to-back converter shown in Fig. 17.28, the maximum apparent power that each inverter will either supply or draw from the electrical system must be known. These ratings will largely depend on the compensation objectives and to what limits they are specified to maintain. Of the three voltage compensation objectives (voltage sag, unbalanced voltages, voltage harmonics), the greatest power demands of the series inverter will almost always occur during voltage sag conditions. For the parallel inverter, generally the reactive power compensation demands will dominate the design of the converter, as opposed to harmonic current compensation.

For this analysis, balanced voltage sag conditions will be considered in the specification of the power ratings of the two inverters. A one-line diagram circuit is shown in Fig. 17.29 for the converter and electrical system represented in Fig. 17.28. Equations can be developed for the apparent power required of each of the inverters based on the three phase rated apparent load power S_{Lnom}, rated line–line load voltage V_{Lnom}, and line–line source voltage V_S [73].

A. Series Inverter Power Rating
First, the rating of the series inverter will be considered. The voltage V_{SI} across the series transformer shown in Fig. 17.29, is given by the vector equation

$$\vec{V}_{SI} = \vec{V}_L - \vec{V}_S \quad (17.24)$$

The apparent power delivered from the series converter can then be given as

$$\vec{S}_{SI} = \left(\vec{V}_L - \vec{V}_S\right) \cdot \vec{I}_S^* \quad (17.25)$$

where $\vec{I}_S^*$ is the conjugate of $\vec{I}_S$, the source current.

If the load voltage V_L is regulated such that it is in phase with the source voltage V_S, then Eq. (17.24) can be rewritten as an algebraic equation

$$V_{SI} = V_L - V_S \quad (17.26)$$

Assuming that the back-to-back converter is lossless, the entire real power P_L drawn by the load must be supplied by the utility source, $P_S = P_L$. If the source current is regulated such that it is in phase with the source voltage, then

$$P_S = V_S \cdot I_S = P_L \quad (17.27)$$

Combining Eqs. (17.25)–(17.27), the real power delivered from the series converter is

$$P_{SI} = (V_L - V_S) \cdot I_S \quad (17.28)$$

Multiplying and dividing the right side of Eq. (17.28) by V_s yields

$$P_{SI} = \left(\frac{V_L}{V_S} - \frac{V_S}{V_S}\right) \cdot I_S \cdot V_S \quad (17.29)$$

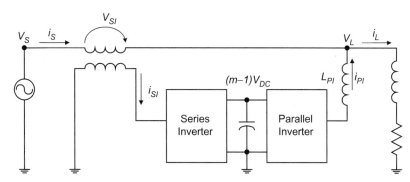

FIGURE 17.29 One-line diagram of a MUPC connected to the electrical system.

Substituting Eq. (17.27) into Eq. (17.29) produces the following equation for the rated apparent power of the series inverter

$$S_{SI} = P_{SI} = P_L \cdot \left(\frac{V_L}{V_S} - 1\right) \qquad (17.30)$$

Choosing the rated load power S_L and the rated load voltage V_L as bases, Fig. 17.30 shows the apparent power S_{SI} in per unit that the series inverter must provide as a function of the source voltage V_S. Each of the curves in Fig. 17.30 is for loads of different power factors. As shown in Fig. 17.30, the apparent power that the series inverter has to transfer is proportional to the power factor of the load [73].

Figure 17.30a shows that for voltage sags less than 50% of nominal, the series inverter would have to be rated to transfer more power than the rated load power, which in most applications would not be practical. Figure 17.30b shows that for sags that are small in magnitude, the series inverter would have a rating much less than that of the rated load power. For example, for a 20% voltage sag ($V_s = 0.8 \cdot V_{nom}$), the required power rating of the series inverter is only 25% of the rated load.

When considering selection of the active devices for the series inverter, as shown in Fig. 17.30, the magnitude of the voltage sag to be compensated will play a large role in determining the current rating required. The formula for determining the current rating of each of the devices as a function of the minimum source voltage to be compensated, $\min(V_s)$, is given in Eq. (17.31)

$$I_{SI \atop device \atop rating} = \frac{S_{Lnom} \cdot ((V_{nom})/(V_{nom} - \min(V_S)) - 1)}{\sqrt{3} \cdot V_{nom}} \cdot D_{safety \atop factor} \qquad (17.31)$$

The safety factor, or design factor, in Eq. (17.31) should be chosen to allow for future growth in the load supplied by the utility and compensated by the power conditioner.

B. Parallel Inverter Power Rating

The power rating of the parallel inverter will now be considered. From Fig. 17.29, the apparent power delivered to the electrical system by the parallel inverter can be expressed as

$$\vec{S}_{PI} = \vec{S}_L - \vec{V}_L \cdot \vec{I}_S^* = (P_L + jQ_L) - V_L I_S \qquad (17.32)$$

because the source current I_s and load voltage V_L are controlled such that they are in phase with the source voltage [73]. Multiplying and dividing the second term of Eq. (17.32) by V_S and substituting Eq. (17.27) yields the following

$$V_L \cdot I_S = \frac{V_L}{V_S} \cdot V_S I_S = \frac{V_L}{V_S} \cdot P_L \qquad (17.33)$$

Substituting Eq. (17.33) into Eq. (17.32) and combining like terms yields

$$\vec{S}_{PI} = P_L \cdot \left(1 - \frac{V_L}{V_S}\right) + jQ_L \qquad (17.34)$$

Figure 17.31 shows the apparent power S_{PI} in per unit that the parallel inverter must provide as a function of the source voltage V_S for loads of different power factors. Because the power transferred for voltage dips to less than 50% of nominal is predominantly real power, the parallel inverter would have to have an extraordinarily high rating if the conditioner were designed to compensate for such large voltage sags, just like the series inverter. From Fig. 17.31b, one can see that for a voltage sag to 50% of nominal, the parallel inverter has to draw a current I_{PI} equal to that drawn by the rated load, I_L. Unlike the series inverter, however, the dominant factor in determining the power rating of the parallel inverter is the load power factor if the conditioner is designed to compensate for only marginal voltage sags as shown in Fig. 17.31b.

If the design of the universal power conditioner is to compensate for voltage sags to less than 50% of nominal voltage, then Eq. (17.31) should be used to determine the current rating of the parallel inverter. If the design of the conditioner is for marginal voltage sags (to 70% of nominal voltage) and the MUPC will be applied to a customer load that has a power factor of less than 0.9, then the following equation is more suited for calculating the current rating of the parallel inverter's active devices

$$I_{PI \atop device \atop rating} = \frac{Q_{Lnom}}{\sqrt{3} \cdot V_{nom}} \cdot D_{design \atop factor} = \frac{S_{Lnom} \cdot \left(1 - (p.f.)^2\right)^{1/2}}{\sqrt{3} \cdot V_{nom}} \cdot D_{design \atop factor} \qquad (17.35)$$

One common design for the parallel inverter in a universal power conditioner is for the inverter to have a current rating equal to that of the rated load current I_L.

17.4.5 Current Rating of Clamping Diodes

When a multilevel inverter outputs an intermediate voltage level, not 0 or $(m-1) \cdot V_{dc}$, only one clamping diode in each phase leg conducts current at any instant in time whereas half of the active switches are conducting at all times. The diode which is conducting current is determined by the intermediate dc voltage level which is connected to the output phase conductor and by the direction of the current flowing, positive or negative. For instance, when a phase leg of the series inverter in Fig. 17.28 is connected to level V_4, then diode D_2 conducts for current flowing from the inverter to the electrical system, and diode D_3 conducts for current flowing into the inverter from the electrical system.

17 *Multilevel Power Converters* 475

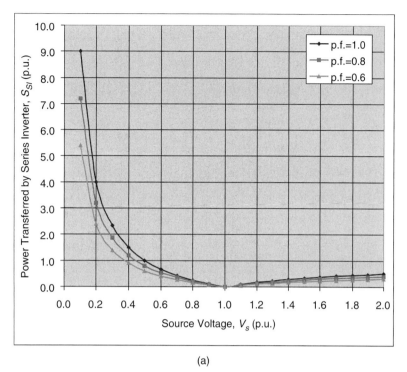

(a)

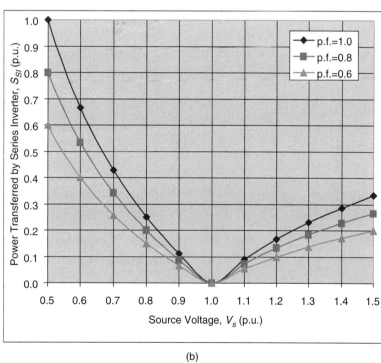

(b)

FIGURE 17.30 Apparent power requirements of series inverter during voltage sags.

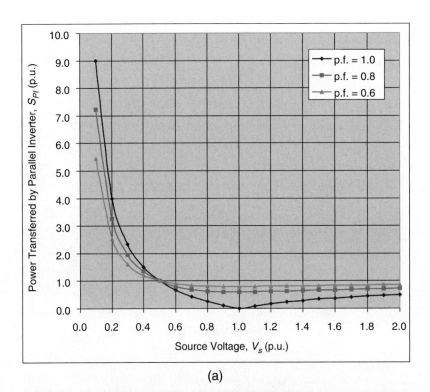

(a)

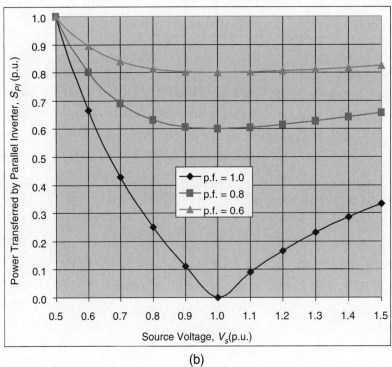

(b)

FIGURE 17.31 Apparent power requirements of parallel inverter during voltage sags.

This example illustrates that for current flowing out of an inverter, only the clamping diodes in the top half of a phase leg will conduct, and for current flowing into an inverter, only the clamping diodes in the bottom half of the phase leg will conduct. In all likelihood the current waveforms will be odd symmetric. These facts alone enable the *average* current rating for the clamping diodes to be at most one half that of the active devices. The clamping diodes should all have a *pulse or short time* current rating equal to the amplitude of the maximum compensation current that the inverter is expected to conduct. Generally, this is equal to $\sqrt{2}$ times the value calculated in Eq. (17.31) or (17.35) for the series and parallel inverters, respectively.

The average current that flows through each clamping diode is dependent on currents i_{SI} and i_{PI}, the modulation index, and the control of the voltage level outputs of the inverter. Because all of these are widely varying attributes in a power conditioner, an explicit formula for determining their ratings would be difficult at best. Nonetheless, for the assumption that each clamping diode conducts an equal amount of current and that each level of the inverter is "on" for an equal duration of time, their average current ratings for the series and parallel inverter could be found from the following equations

$$I_{SI \atop clamping \atop diode} = \frac{I_{SI \atop device \atop rating}}{2 \cdot (m-1)} \quad (17.36)$$

or

$$I_{PI \atop clamping \atop diode} = \frac{I_{PI \atop device \atop rating}}{2 \cdot (m-1)} \quad (17.37)$$

17.4.6 DC Link Capacitor Specifications

Unipolar capacitors can be used for the dc link capacitors. Just like the voltage rating of the active devices in Eq. (17.22), the sum of the voltage ratings of the dc link capacitors should be greater than or equal to the overall dc link voltage which is equal to the right side of Eq. (17.22). The design factor in this case would include the dc link voltage ripple plus any safety factor, the designer feels is necessary to maintain the capacitors within their safe operating range.

The capacitance of each capacitor in the dc link is determined by the equation

$$C_n = \frac{\Delta q_n}{\Delta V_n} \quad (17.38)$$

where $n = 1, 2, 3, \ldots, m-1$, Δq_n is the change in charge, and ΔV_n is the change in voltage over a specified period.

The required capacitance of the dc link and the voltage ripple are inversely related to each other. An increase in the capacitance will decrease the amount of ripple in the dc voltage. By assuming that each level has the same voltage V_{dc} across it,

$$\Delta V_n = \%V_{ripple} \cdot V_{dc} \quad (17.39)$$

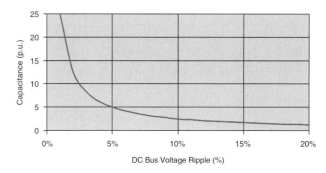

FIGURE 17.32 Capacitance required as a function of the maximum voltage ripple on dc bus.

Figure 17.32 shows a graph of the required capacitance as a function of the maximum permissible voltage ripple on the dc link. The graph indicates that an unnecessarily strict tolerance on the voltage ripple of the dc bus will result in extraordinarily large capacitor values. For this reason, the maximum voltage ripple is normally chosen to be in the 5–10% range.

The current that flows through the capacitor determines the change in charge Δq_n for a capacitor C_n. This current is a function of what input and output voltage states the inverter progresses through each cycle, and will largely be dependent on the control method implemented by the series and parallel inverter in maintaining the voltage on the dc link. In addition, the current waveforms i_{PI} and i_{SI} also will depend largely on the system conditions, in other words, the type of compensation that the converter is conducting. Although the current that flows through each capacitor C_n that makes up the dc link will be different, for the reasons mentioned previously in Section 17.4.2, normally each of the capacitors will be identically sized such that Eq. (17.38) can then be rewritten as

$$C_n = C_{dc} = \frac{\max(\Delta q_n)}{\%V_{ripple} \cdot V_{dc}} \quad (17.40)$$

Suppose a UPFC is connected to a distribution system with a voltage of 13.8 kV line–line (7970 V line–ground). From Eq. (17.22), $V_{device \atop rating} \cdot (m-1) \geq \sqrt{2} \cdot 7970 \cdot 1.5 = 16,900\,\text{V}$. If 3300 V IGBTs are chosen for the design, then the number of levels m would be 6. The next lower rating of available IGBTs is 2500 V, and use of these devices would require 7 levels. Because of the added complexity and computational burden of 7 levels, the design with 6 levels of 3300 V IGBTs is chosen.

A 13.8 kV line–line ac waveform from an inverter requires a minimum dc link voltage of approximately 11.3 kV. The nominal dc voltage for each level would be approximately 2000 V.

TABLE 17.7 Back-to-back MUPC clamping diode ratings

Per unit voltage rating	Blocking voltage required	Voltage rating of diode used	Number of diodes per leg	Total number per phase
$1V_{dc}$	3000 V	3000 V	1	2
$2V_{dc}$	6000 V	3000 V	2	4
$3V_{dc}$	9000 V	3000 V	3	6
$4V_{dc}$	12,000 V	3000 V	4	8

For a design factor of 1.5, the design voltage for each level of the inverter would be approximately 3000 V.

From Eq. (17.23), the minimum total voltage blocking capability for a back-to-back converter would be, $V_{clamp\ total} = 6 \cdot (6 - 2) \cdot (6 - 1) \cdot 3000\,V = 360\,kV$. Each phase of the converter will require the blocking voltages shown in Table 17.7.

The current rating of the active devices in the series inverter is found from Eq. (17.31)

$$I_{SI\ device\ rating} = \frac{20\,MVA \cdot ((1/1 - 0.3) - 1)}{\sqrt{3} \cdot 13{,}800\,V} \cdot 1.5 = 540\,A$$

The current rating of the active devices in the parallel inverter is found from Eq. (17.35):

$$I_{PI\ device\ rating} = \frac{20\,MVA \cdot (1 - 0.85^2)^{1/2}}{\sqrt{3} \cdot 13{,}800\,V} \cdot 1.5 = 661\,A$$

Use of 3300 V, 800 A IGBTs would be sufficient for both the series and parallel inverter.

The current rating of the clamping diodes in the series inverter is found from Eq. (17.36)

$$I_{SI\ clamping\ diode} = \frac{540\,A}{2 \cdot (6 - 1)} = 54\,A$$

Likewise, the current rating of the clamping diodes in the parallel inverter is found from Eq. (17.37)

$$I_{PI\ clamping\ diode} = \frac{660\,A}{2 \cdot (6 - 1)} = 66\,A$$

Use of 3000 V, 75 A diodes would be sufficient for both the series and parallel inverter.

17.5 Fault Diagnosis in Multilevel Converters

Since a multilevel converter is normally used in medium to high power applications, the reliability of the multilevel converter system is very important. For instance industrial drive applications in manufacturing plants are dependent upon induction motors and their inverter systems for process control. Generally, the conventional protection systems are passive devices such as fuses, overload relays, and circuit breakers to protect the inverter systems and the induction motors. The protection devices will disconnect the power sources from the multilevel inverter system whenever a fault occurs, stopping the operated process. Downtime of manufacturing equipment can add up to be thousands or hundreds of thousands of dollars per hour, therefore fault detection and diagnosis is vital to a company's bottom line. In order to maintain continuous operation for a multilevel inverter system, knowledge of fault behaviors, fault prediction, and fault diagnosis are necessary. Faults should be detected as soon as possible after they occur, because if a motor drive runs continuously under abnormal conditions, the drive or motor may quickly fail.

The possible structure for a fault diagnosis system is illustrated in Fig. 17.33. The system is composed of four major states: feature extraction, neural network classification, fault diagnosis, and switching pattern calculation with gate signal output. The feature extraction performs the voltage input signal transformation, with rated signal values as important features, and the output of the transformed signal is transferred to the neural network classification. The networks are trained with both normal and abnormal data for the multilevel inverter drive (MLID) thus, the output of this network is nearly 0 and 1 as binary code. The binary code is sent to the fault diagnosis to decode the fault type and its location. Then, the switching pattern is calculated to reconfigure the multilevel inverter.

Switching patterns and the modulation index of other active switches can be adjusted to maintain voltage and current in a balanced condition after reconfiguration recovers from a fault. The MLID can continuously operate in a balanced condition, of course, the MLID will not be able to operate at its rated power. Therefore, the MLID can operate in balanced condition at reduced power after the fault occurs until the operator locates and replaces the damaged switch [74].

17.6 Renewable Energy Interface

Multilevel converters can be used to interface with renewable energy and/or distributed energy resources because several batteries, fuel cells, solar cells, wind turbines, and micro turbines can be connected through a multilevel converter to

17 Multilevel Power Converters

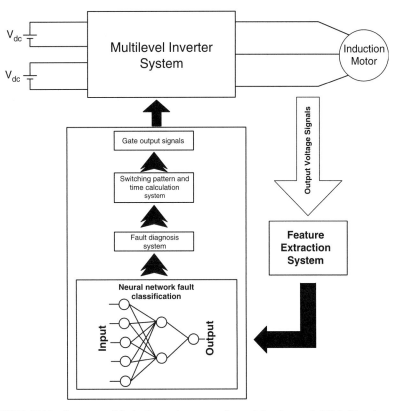

FIGURE 17.33 Structure of fault diagnosis system of a multilevel cascaded H-bridges inverter.

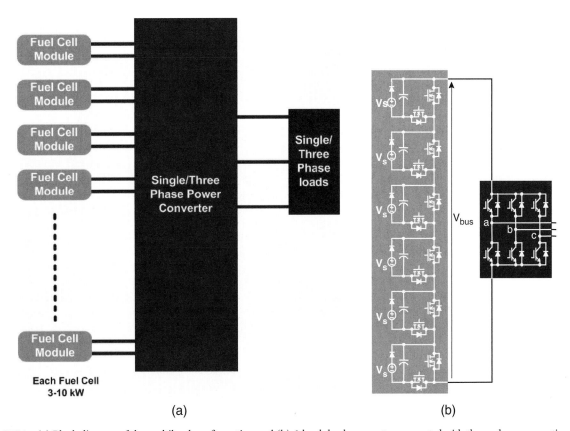

FIGURE 17.34 (a) Block diagram of the multilevel configuration and (b) 6-level dc–dc converter connected with three-phase conventional inverter.

supply a load or the ac grid without voltage balancing problems. Nevertheless, the intrinsic characteristics of renewable energy sources might have some trouble with their energy source utilization, for instance, fuel cell energy sources have some problems associated with their V–I characteristics. The static V–I characteristic of fuel cells illustrates more than 30% difference in the output voltage between no load and full load condition. This unavoidable decrease, caused by internal losses reduces fuel cell utilization factor. Therefore, a multilevel dc–dc converter might be used to overcome the problem as shown in Fig. 17.34. To overcome the fuel cell voltage drop, either voltage regulators have to be connected at the fuel cell outputs or fuel cell voltages have to be monitored and the control signals have to be modified accordingly.

Five different approaches for integrating numerous fuel cell modules have been evaluated and compared with respect to cost, control complexity, ease of modularity, and fault tolerance in [75]. In addition, the optimum fuel cell utilization technique with a multilevel dc–dc converter has been proposed in [76, 77].

17.7 Conclusion

This chapter has demonstrated the state of the art of multilevel power converter technology. Fundamental multilevel converter structures and modulation paradigms including the pros and cons of each technique have been discussed. Most of the chapter focus has addressed modern and more practical industrial applications of multilevel converters. A procedure for calculating the required ratings for the active switches, clamping diodes, and dc link capacitors with a design example has been described. The possible future enlargements of multilevel converter technology such as fault diagnosis system and renewable energy sources have been noted. It should be noted that this chapter could not cover all multilevel power converter related applications, however the basic principles of different multilevel converters have been discussed methodically. The main objective of this chapter is to provide a general notion to readers who are interested in multilevel power converters and their applications.

References

1. J. Rodriguez, J. S. Lai, and F. Z. Peng, "Multilevel Inverters: Survey of Topologies, Controls, and Applications," *IEEE Transactions on Industry Applications*, vol. 49, no. 4, Aug. 2002, pp. 724–738.
2. J. S. Lai and F. Z. Peng, "Multilevel Converters – A New Breed of Power Converters," *IEEE Transactions on Industry Applications*, vol. 32, May/June 1996, pp. 509–517.
3. L. M. Tolbert, F. Z. Peng, and T. Habetler, "Multilevel Converters for Large Electric drives," *IEEE Transactions on Industry Applications*, vol. 35, Jan./Feb. 1999, pp. 36–44.
4. R. H. Baker and L. H. Bannister, "Electric Power Converter," U.S. Patent 3 867 643, Feb. 1975.
5. A. Nabae, I. Takahashi, and H. Akagi, "A New Neutral-point Clamped PWM Inverter," *IEEE Transactions on Industry Applications*, vol. IA-17, Sept./Oct. 1981, pp. 518–523.
6. R. H. Baker, "Bridge Converter Circuit," U.S. Patent 4 270 163, May 1981.
7. P. W. Hammond, "Medium Voltage PWM Drive and Method," U.S. Patent 5 625 545, Apr. 1977.
8. F. Z. Peng and J. S. Lai, "Multilevel Cascade Voltage-source Inverter with Separate DC Source," U.S. Patent 5 642 275, June 24, 1997.
9. P. W. Hammond, "Four-quadrant AC-AC Drive and Method," U.S. Patent 6 166 513, Dec. 2000.
10. M. F. Aiello, P. W. Hammond, and M. Rastogi, "Modular Multi-level Adjustable Supply with Series Connected Active Inputs," U.S. Patent 6 236 580, May 2001.
11. M. F. Aiello, P. W. Hammond, and M. Rastogi, "Modular Multi-Level Adjustable Supply with Parallel Connected Active Inputs," U.S. Patent 6 301 130, Oct. 2001.
12. J. P. Lavieville, P. Carrere, and T. Meynard, "Electronic Circuit for Converting Electrical Energy and a Power Supply Installation Making Use Thereof," U.S. Patent 5 668 711, Sept. 1997.
13. T. Meynard, J.-P. Lavieville, P. Carrere, J. Gonzalez, and O. Bethoux, "Electronic Circuit for Converting Electrical Energy," U.S. Patent 5 706 188, Jan. 1998.
14. E. Cengelci, S. U. Sulistijo, B. O. Woom, P. Enjeti, R. Teodorescu, and F. Blaabjerg, "A New Medium Voltage PWM Inverter Topology for Adjustable Speed Drives," in *Conf. Rec. IEEE-IAS Annu. Meeting*, St. Louis, MO, Oct. 1998, pp. 1416–1423.
15. M. F. Escalante, J. C. Vannier, and A. Arzande, "Flying Capacitor Multilevel Inverters and DTC Motor Drive Applications," *IEEE Transactions on Industry Electronics*, vol. 49, no. 4, Aug. 2002, pp. 809–815.
16. L. M. Tolbert and F. Z. Peng, "Multilevel Converters as a Utility Interface for Renewable Energy Systems," in *Proceedings of 2000 IEEE Power Engineering Society Summer Meeting*, pp. 1271–1274.
17. L. M. Tolbert, F. Z. Peng, and T. G. Habetler, "A Multilevel Converter-based Universal Power Conditioner," *IEEE Transactions on Industry Applications*, vol. 36, no. 2, Mar./Apr. 2000, pp. 596–603.
18. L. M. Tolbert, F. Z. Peng, and T. G. Habetler, "Multilevel Inverters for Electric Vehicle Applications," *IEEE Workshop on Power Electronics in Transportation*, Oct. 22–23 1998, Dearborn, Michigan, pp. 1424–1431.
19. R. W. Menzies and Y. Zhuang, "Advanced Static Compensation Using a Multilevel GTO Thyristor Inverter," *IEEE Transactions on Power Delivery*, vol. 10, no. 2, Apr. 1995, pp. 732–738.
20. F. Z. Peng, J. S. Lai, J. W. McKeever, and J. VanCoevering, "A Multilevel Voltage-Source Inverter with Separate DC Sources for Static Var Generation," *IEEE Transactions on Industry Applications*, vol. 32, no. 5, Sept. 1996, pp. 1130–1138.
21. F. Z. Peng and J. S. Lai, "Dynamic Performance and Control of a Static Var Generator Using Cascade Multilevel Inverters," *IEEE Transactions on Industry Applications*, vol. 33, no. 3, May 1997, pp. 748–755.
22. F. Z. Peng, J. W. McKeever, and D. J. Adams, "A Power Line Conditioner Using Cascade Multilevel Inverters for Distribution Systems,"

Conference Record – IEEE Industry Applications Society 32nd Annual Meeting, 1997, pp. 1316–1321.
23. F. Z. Peng, J. W. McKeever, and D. J. Adams, "Cascade Multilevel Inverters for Utility Applications," *Proceedings of 23rd International Conference on Industrial Electronics, Control, and Instrumentation*, 1997, pp. 437–442.
24. G. Joos, X. Huang, and B. T. Ooi, "Direct-Coupled Multi-level Cascaded Series VAR Compensators," *Conference Record – IEEE Industry Applications Society 32nd Annual Meeting*, 1997, pp. 1608–1615.
25. L. M. Tolbert, F. Z. Peng, T. Cunnyngham, and J. N. Chiasson, "Charge Balance Control Schemes for Multilevel Converter in Hybrid Electric Vehicles," *IEEE Transactions on Industrial Electronics*, vol. 49, no. 5, Oct. 2002, pp. 1058–1065.
26. M. D. Manjrekar and T. A. Lipo, "A Hybrid Multilevel Inverter Topology for Drive Applications," *IEEE Applied Power Electronics Conference*, 1998, pp. 523–529.
27. M. D. Manjrekar and T. A. Lipo, "A Generalized Structure of Multi-level Power Converter," *IEEE Conference on Power Electronics, Drives, and Energy Systems*, 1998, Australia, pp. 62–67.
28. C. Hochgraf, R. Lasseter, D. Divan, and T. A. Lipo, "Comparison of Multilevel Inverters for Static Var Compensation," *Conference Record – IEEE Industry Applications Society 29th Annual Meeting*, 1994, pp. 921–928.
29. J. S. Lai and F. Z. Peng, "Multilevel Converters – A New Breed of Power Converters," *IEEE Transactions on Industry Applications*, vol. 32, no. 3, May 1996, pp. 509–517.
30. K. Corzine and Y. Familiant, "A New Cascaded Multilevel H-Bridge Drive," *IEEE Transactions on Power Electronics*, vol. 17, no. 1, Jan. 2002, pp. 125–131.
31. T. A. Meynard and H. Foch, "Multi-Level Conversion: High Voltage Choppers and Voltage-Source Inverters," *IEEE Power Electronics Specialists Conference*, 1992, pp. 397–403.
32. G. Sinha and T. A. Lipo, "A New Modulation Strategy for Improved DC Bus Utilization in Hard and Soft Switched Multilevel Inverters," *IECON*, 1997, pp. 670–675.
33. F. Z. Peng, "A Generalized Multilevel Converter Topology with Self Voltage Balancing," *IEEE Transactions on Industry Applications*, vol. 37, Mar./Apr. 2001, pp. 611–618.
34. W. A. Hill and C. D. Harbourt, "Performance of Medium Voltage Multilevel Converters," in *Conf. Rec. IEEE-IAS Annu. Meeting*, Oct. 1999, Phoenix, AZ, pp. 1186–1192.
35. B. M. Song and J. S. Lai, "A Multilevel Soft-switching Inverter with Inductor Coupling," *IEEE Transactions on Industry. Applications*, vol. 37, Mar./Apr. 2001, pp. 628–636.
36. H. Fujita and H. Akagi, "The Unified Power Quality Conditioner: The Integration of Series- and Shunt-Active Filters," *IEEE Transactions on Power Electronics*, vol. 13, no. 2, March 1998, pp. 315–322.
37. S.-J. Jeon and G.-H. Cho, "A Series-Parallel Compensated Uninterruptible Power Supply with Sinusoidal Input Current and Sinusoidal Output Voltage," *IEEE Power Electronics Specialists Conference*, 1997, pp. 297–303.
38. F. Kamran and T. G. Habetler, "A Novel On-Line UPS with Universal Filtering Capabilities," *IEEE Power Electronics Specialists Conference*, 1995, pp. 500–506.
39. F. Kamran and T. G. Habetler, "Combined Deadbeat Control of a Series-Parallel Converter Combination Used as a Universal Power Filter," *IEEE Transactions on Power Electronics*, vol. 13, no. 1, Jan. 1998, pp. 160–168.
40. L. Moran and G. Joos, "Principles of Active Power Filters," *IEEE Industry Applications Society Annual Meeting*, Oct. 1998, Tutorial Course Notes.
41. S. Muthu and J. M. S. Kim, "Steady-State Operating Characteristics of Unified Active Power Filters," *IEEE Applied Power Electronics Conference*, 1997, pp. 199–205.
42. A. van Zyl, J. H. R. Enslin, and R. Spee, "A New Unified Approach to Power Quality Management," *IEEE Transactions on Power Electronics*, vol. 11, no. 5, Sept. 1996, pp. 691–697.
43. Y. Chen, B. Mwinyiwiwa, Z. Wolanski, and B.-T. Ooi, "Unified Power Flow Controller (UPFC) Based on Chopper Stabilized Multi-level Converter," *IEEE Power Electronics Specialists Conference*, 1997, pp. 331–337.
44. J. H. R. Enslin, J. Zhao, and R. Spee, "Operation of the Unified Power Flow Controller as Harmonic Isolator," *IEEE Transactions on Power Electronics*, vol. 11, no. 6, Sept. 1996, pp. 776–784.
45. H. Fujita, Y. Watanabe, and H. Akagi, "Control and Analysis of a Unified Power Flow Controller," *IEEE Power Electronics Specialists Conference*, 1998, pp. 805–811.
46. L. Gyugyi, "Dynamic Compensation of AC Transmission Lines by Solid-State Synchronous Voltage Sources," *IEEE Transactions on Power Delivery*, vol. 9, no. 2, Apr. 1994, pp. 904–911.
47. L. Gyugi, C. D. Schauder, S. L. Williams, T. R. Reitman, D. R. Torgerson, and A. Edris, "The Unified Power Flow Controller: A New Approach to Power Transmission Control," IEEE *Transactions on Power Delivery*, vol. 10, no. 2, April 1995, pp. 1085–1093.
48. G. Carrara, S. Gardella, M. Marchesoni, R. Salutari, and G. Sciutto, "A New Multilevel PWM Method: A Theoretical Analysis," *IEEE Transactions on Power Electronics*, vol. 7, no. 3, July 1992, pp. 497–505.
49. R. W. Menzies, P. Steimer, and J. K. Steinke, "Five-Level GTO Inverters for Large Induction Motor Drives," *IEEE Transactions on Industry Applications*, vol. 30, no. 4, July 1994, pp. 938–944.
50. S. Halasz, G. Csonka, and A. A. M. Hassan, "Sinusoidal PWM Techniques with Additional Zero-Sequence Harmonics," *Proceedings of 20th International Conference on Industrial Electronics, Control, and Instrumentation*, 1994, pp. 85–90.
51. D. G. Holmes, "The Significance of Zero Space Vector Placement for Carrier Based PWM Schemes," *Conference Record – IEEE Industry Applications Society 30th Annual Meeting*, 1995, pp. 2451–2458.
52. S. Bhattacharya, D. G. Holmes, and D. M. Divan, "Optimizing Three Phase Current Regulators for Low Inductance Loads," *Conference Record – IEEE Industry Applications Society 30th Annual Meeting*, 1995, pp. 2357–2364.
53. L. M. Tolbert and T. G. Habetler, "Novel Multilevel Inverter Carrier-based PWM Method," *IEEE Transactions on Industry Applications*, vol. 25, no. 5, Sep./Oct., 1999, pp. 1098–1107.
54. F. Z. Peng and J. S. Lai, "A Static Var Generator Using a Staircase Waveform Multilevel Voltage-Source Converter," *PCIM/Power Quality Conference*, Sept. 1994, Dallas, Texas, pp. 58–66.
55. L. M. Tolbert, F. Z. Peng, and T. G. Habetler, "Multilevel PWM Methods at Low Modulation Indices,"*IEEE Transactions on Power Electronics*, vol. 15, no. 4, July 2000, pp. 719–725.
56. N. S. Choi, J. G. Cho, and G. H. Cho, "A General Circuit Topology of Multilevel Inverter," *IEEE Power Electronics Specialists Conference*, 1991, pp. 96–103.
57. G. Sinha and T. A. Lipo, "A Four-Level Inverter Drive with Passive Front End," *IEEE Power Electronics Specialists Conference*, 1997, pp. 590–596.

58. G. Sinha and T. A. Lipo, "A Four-Level Rectifier-Inverter System for Drive Applications," *Conference Record – IEEE Industry Applications Society 31st Annual Meeting*, 1996, pp. 980–987.
59. H. L. Liu, N. S. Choi, and G. H. Cho, "DSP Based Space Vector PWM for Three-Level Inverter with DC-Link Voltage Balancing," *Proceedings of the IECON '91 17th International Conference on Industrial Electronics, Control, and Instrumentation*, 1991, pp. 197–203.
60. G. Sinha and T. A. Lipo, "A New Modulation Strategy for Improved DC Bus Utilization in Hard and Soft Switched Multilevel Inverters," *IECON*, 1997, pp. 670–675.
61. B. P. McGrath, D. G. Holmes, and T. Lipo, "Optimized Space Vector Switching Sequences for Multilevel Inverters," *IEEE Transactions on Power Electronics*, vol. 18, no. 6, Nov. 2003, pp. 1293–1301.
62. H. S. Patel and R. G. Hoft, "Generalized Harmonic Elimination and Voltage Control in Thyristor Converters: Part I – Harmonic Elimination," *IEEE Transactions on Industry Applications*, vol. 9, May/June 1973, pp. 310–317.
63. H. S. Patel and R. G. Hoft, "Generalized Harmonic Elimination and Voltage Control in Thyristor Converters: Part II –Voltage Control Technique," *IEEE Transactions on Industry Applications*, vol. 10, Sept./Oct. 1974, pp. 666–673.
64. K. J. McKenzie, "Eliminating Harmonics in a Cascaded H-bridges Multilevel Converter using Resultant Theory, Symmetric Polynomials, and Power Sums," Master Thesis, The University of Tennessee, 2004.
65. J. N. Chiasson, L. M. Tolbert, K. J. McKenzie, and Z. Du, "A Complete Solution to the Harmonic Elimination Problem," *IEEE Transactions on Power Electronics*, vol. 19, no. 2, Mar. 2004, pp. 491–499.
66. J. N. Chiasson, L. M. Tolbert, K. J. McKenzie, and Z. Du, "Control of a Multilevel Converter Using Resultant Theory,"*IEEE Transactions on Control System Theory*, vol. 11, no. 3, May 2003, pp. 345–354.
67. J. N. Chiasson, L. M. Tolbert, K. J. McKenzie, and Z. Du, "A New Approach to Solving the Harmonic Elimination Equations for a Multilevel Converter," *IEEE Industry Applications Society Annual Meeting*, Oct. 12–16, 2003, Salt Lake City, Utah, pp. 640–645.
68. L. Li, D. Czarkowski, Y. Liu, and P. Pillay, "Multilevel Selective Harmonic Elimination PWM Technique in Series-Connected Voltage Converters," *IEEE Transactions on Industry Applications*, vol. 36, no. 1, Jan.–Feb. 2000, pp. 160–170.
69. S. Sirisukprasert, J. S. Lai, and T. H. Liu, "Optimum Harmonic Reduction with a Wide Range of Modulation Indexes for Multilevel Converters," *IEEE Transactions on Industrial Electronics*, vol. 49, no. 4, Aug. 2002, pp. 875–881.
70. Z. Du, "Active Harmonic Elimination in Multilevel Converters," Ph.D. Dissertation, The University of Tennessee, 2005, pp. 33–36.
71. L. M. Tolbert, "New Multilevel Carrier-Based Pulse Width Modulation Techniques Applied to a Diode-Clamped Converter for Use as a Universal Power Conditioner," Ph.D. Dissertation, Georgia Institute of Technology, 1999, pp. 76–103.
72. F. Kamran and T. G. Habetler, "Combined Deadbeat Control of a Series-Parallel Converter Combination Used as a Universal Power Filter," *IEEE Transactions on Power Electronics*, vol. 13, no. 1, Jan. 1998, pp. 160–168
73. S. J. Jeon and G.-H. Cho, "A Series-Parallel Compensated Uninterruptible Power Supply with Sinusoidal Input Current and Sinusoidal Output Voltage," *IEEE Power Electronics Specialists Conference*, 1997, pp. 297–303.
74. S. Khomfoi and L. M. Tolbert, "Fault Diagnosis System for a Multilevel Inverters Using a Neural Network," *IEEE Industrial Electronics Conference*, Nov. 6–10, 2005, Raleigh, North Carolina, pp. 1455–1460.
75. B. Ozpineci, Z. Du, L. M. Tolbert, D. J. Adams, and D. Collins, "Integrating Multiple Solid Oxide Fuel Cell Modules," *IEEE Industrial Electronics Conference*, Nov. 2–6, 2003, Roanoke Virginia, pp. 1568–1573.
76. B. Ozpineci, L. M. Tolbert, and Z. Du, "Optimum Fuel Cell Utilization with Multilevel Inverters," *IEEE Power Electronics Specialists Conference*, June 20–25, 2004, Aachen, Germany, pp. 4798–4802.
77. B. Ozpineci, Z. Du, L. M. Tolbert, and G. J. Su, "Optimum Fuel Cell Utilization with Multilevel DC-DC Converters," *IEEE Applied Power Electronics Conference*, Feb. 22–26, 2004, Anaheim, California, pp. 1572–1576.

18
AC–AC Converters

A. K. Chattopadhyay, Ph.D.
Electrical Engg. Department, Bengal Engineering & Science University, Shibpur, Howrah, India

18.1 Introduction 483
18.2 Single-phase AC–AC Voltage Controller 484
   18.2.1 Phase-controlled Single-phase AC Voltage Controller • 18.2.2 Single-phase AC–AC Voltage Controller with On/Off Control
18.3 Three-phase AC–AC Voltage Controllers 488
   18.3.1 Phase-controlled Three-phase AC Voltage Controllers • 18.3.2 Fully Controlled Three-phase Three-wire AC Voltage Controller
18.4 Cycloconverters 493
   18.4.1 Single-phase to Single-phase Cycloconverter • 18.4.2 Three-phase Cycloconverters • 18.4.3 Cycloconverter Control Scheme • 18.4.4 Cycloconverter Harmonics and Input Current Waveform • 18.4.5 Cycloconverter Input Displacement/Power Factor • 18.4.6 Effect of Source Impedance • 18.4.7 Simulation Analysis of Cycloconverter Performance • 18.4.8 Power Quality Issues • 18.4.9 Forced Commutated Cycloconverter
18.5 Matrix Converter 503
   18.5.1 Operation and Control of the Matrix Converter • 18.5.2 Commutation and Protection Issues in a Matrix Converter
18.6 High Frequency Linked Single-phase to Three-phase Matrix Converters 509
   18.6.1 High Frequency Integral-pulse Cycloconverter [48] • 18.6.2 High Frequency Phase-controlled Cycloconverter [49]
18.7 Applications of AC–AC Converters 510
   18.7.1 Applications of AC Voltage Controllers • 18.7.2 Applications of Cycloconverters • 18.7.3 Applications of Matrix Converters
   References 513

18.1 Introduction

A power electronic ac–ac converter, in generic form, accepts electric power from one system and converts it for delivery to another ac system with waveforms of different amplitude, frequency, and phase. They may be single- or three-phase types depending on their power ratings. The ac–ac converters employed to vary the rms voltage across the load at constant frequency are known as *ac voltage controllers or ac regulators.* The voltage control is accomplished either by (i) *phase control* under natural commutation using pairs of silicon controlled rectifiers (SCRs) or triacs or (ii) by *on/off control* under forced commutation/ self-commutation using fully controlled self-commutated switches like gate turn-off thyristors (GTOs), power transistors, integrated gate bipolar transistor (IGBTs), MOS controlled thyristors (MCTs), integrated gate commutated thyristor (IGCTs), etc. The ac–ac power converters in which ac power at one frequency is directly converted to ac power at another frequency without any intermediate dc conversion link (as in the case of inverters) are known as *cycloconverters*, the majority of which use naturally commutated SCRs for their operation when the maximum output frequency is limited to a fraction of the input frequency. With rapid advancements of fast-acting fully controlled switches, forced commutated cycloconverters, or recently developed *matrix converters* with bi-directional on/off control switches provide independent control of the magnitude and the frequency of the generated output voltage as well as sinusoidal modulation of output voltage and current.

While typical applications of ac voltage controllers include lighting and heating control, online transformer tap changing, soft-starting and speed control of pump and fan drives, the cycloconverters are mainly used for high power low speed large ac motor drives for application in cement kilns, rolling mills, and ship propellers. The power circuits, control methods and the operation of the ac voltage controllers, cycloconverters, and matrix converters are introduced in this chapter. A brief review is also made regarding their applications.

18.2 Single-phase AC–AC Voltage Controller

The basic power circuit of a single-phase ac–ac voltage controller, as shown in Fig. 18.1a, comprises a pair of SCRs connected back-to-back (also known as inverse-parallel or anti-parallel) between the ac supply and the load. This connection provides a *bi-directional full-wave symmetrical* control and the SCR pair can be replaced by a triac (Fig. 18.1b) for low-power applications. Alternate arrangements are as shown in Fig. 18.1c with two diodes and two SCRs to provide a common cathode connection for simplifying the gating circuit without needing isolation, and in Fig. 18.1d with one SCR and four diodes to reduce the device cost but with increased device conduction loss. An SCR and diode combination, known as *thyrode controller*, as shown in Fig. 18.1e provides a *uni-directional half-wave asymmetrical voltage* control with device economy, but introduces dc component and more harmonics and thus is not so practical to use except for very low power heating load.

With *phase control*, the switches conduct the load current for a chosen period of each input cycle of voltage and with *on/off* control, the switches connect the load either for a few cycles of input voltage and disconnect it for the next few cycles (*integral cycle control*) or the switches are turned on and off several times within alternate half cycles of input voltage (*ac chopper or pulse width modulated (PWM) ac voltage controller*).

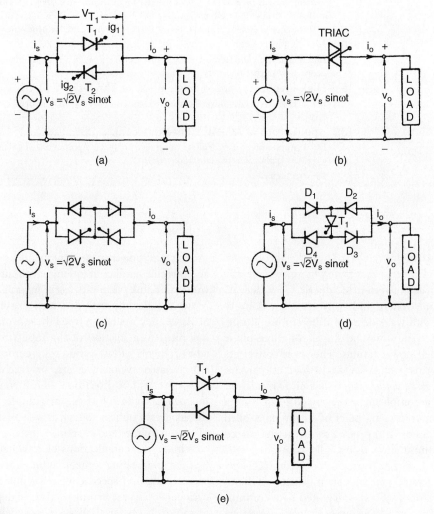

FIGURE 18.1 Single-phase ac voltage controllers: (a) full wave, two SCRs in inverse-parallel; (b) full wave with triac; (c) full wave with two SCRs and two diodes; (d) full wave with four diodes and one SCR; and (e) half wave with one SCR and one diode in anti-parallel.

18.2.1 Phase-controlled Single-phase AC Voltage Controller

For a full wave, symmetrical phase control, the SCRs T_1 and T_2 in Fig. 18.1a are gated at α and $\pi + \alpha$, respectively from the zero crossing of the input voltage and by varying α, the power flow to the load is controlled through voltage control in alternate half cycles. As long as one SCR is carrying current, the other SCR remains reverse biased by the voltage drop across the conducting SCR. The principle of operation in each half cycle is similar to that of the controlled half-wave rectifier, and one can use the same approach for analysis of the circuit.

Operation with R-load: Figure 18.2 shows the typical voltage and current waveforms for the single-phase bi-directional phase-controlled ac voltage controller of Fig. 18.1a with a resistive load. The output voltage and current waveforms have half-wave symmetry and so no dc component.

If $v_s = \sqrt{2} V_s \sin \omega t$ is the source voltage, the rms output voltage with T_1 triggered at α can be found from the half-wave symmetry as

$$V_o = \left[\frac{1}{\pi} \int_\alpha^\pi 2 V_s^2 \sin^2 \omega t \, d(\omega t) \right]^{\frac{1}{2}} = V_s \left[1 - \frac{\alpha}{\pi} + \frac{\sin 2\alpha}{2\pi} \right]^{\frac{1}{2}} \tag{18.1}$$

Note that V_o can be varied from V_s to 0 by varying α from 0 to π.

$$\text{The rms value of load current, } I_o = \frac{V_o}{R} \tag{18.2}$$

$$\text{The input power factor} = \frac{P_o}{VA} = \frac{V_o}{V_s} = \left[1 - \frac{\alpha}{\pi} + \frac{\sin 2\alpha}{2\pi} \right]^{\frac{1}{2}} \tag{18.3}$$

$$\text{The average SCR current, } I_{A,\text{SCR}} = \frac{1}{2\pi R} \int_\alpha^\pi \sqrt{2} V_s \sin \omega t \, d(\omega t) \tag{18.4}$$

Since each SCR carries half the line current, the rms current in each SCR is

$$I_{o,\text{SCR}} = \frac{I_o}{\sqrt{2}} \tag{18.5}$$

Operation with RL Load: Figure 18.3 shows the voltage and current waveforms for the controller in Fig. 18.1a with RL load. Due to the inductance, the current carried by the SCR T_1 may not fall to zero at $\omega t = \pi$ when the input voltage goes negative and may continue till $\omega t = \beta$, the extinction angle, as shown. The conduction angle,

$$\theta = \beta - \alpha \tag{18.6}$$

of the SCR depends on the firing delay angle α and the load impedance angle ϕ.

The expression for the load current $I_o(\omega t)$ when conducting from α to β can be derived in the same way as that used for a phase-controlled rectifier in a discontinuous mode by solving the relevant Kirchoff's voltage equation:

$$i_o(\omega t) = \frac{\sqrt{2} V}{Z} [\sin(\omega t - \phi) - \sin(\alpha - \phi) e^{(\alpha - \omega t)/\tan \phi}], \quad \alpha < \omega t < \beta \tag{18.7}$$

where $Z = (R^2 + \omega^2 L^2)^{\frac{1}{2}}$ = load impedance and ϕ = load impedance angle = $\tan^{-1}(\omega L/R)$

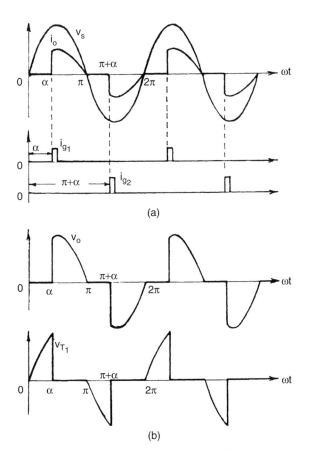

FIGURE 18.2 Waveforms for single-phase ac full-wave voltage controller with R-load.

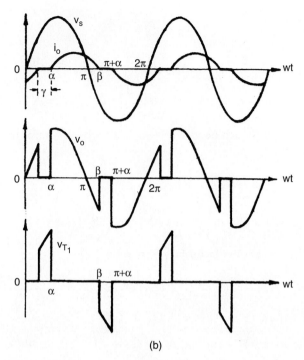

(b)

FIGURE 18.3 Typical waveforms of single-phase ac voltage controller with an RL load.

The angle β, when the current i_o falls to zero, can be determined from the following transcendental equation resulted by putting $i_o(\omega t = \beta) = 0$ in Eq. (18.7)

$$\sin(\beta - \phi) = \sin(\alpha - \phi) - \sin(\alpha - \phi) e^{(\alpha - \beta)/\tan\phi} \quad (18.8)$$

From Eqs. (18.6) and (18.8) one can obtain a relationship between θ and α for a given value of ϕ as shown in Fig. 18.4 which shows that as α is increased, the conduction angle θ decreases and the rms value of the current decreases.

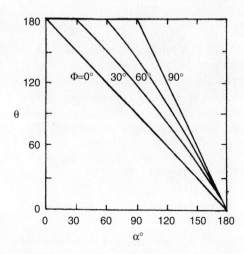

FIGURE 18.4 θ vs α curves for single-phase ac voltage controller with RL load.

The rms output voltage

$$V_o = \left[\frac{1}{\pi}\int_\alpha^\beta 2 V_s^2 \sin^2 \omega t \, d(\omega t)\right]^{\frac{1}{2}}$$

$$= \frac{V_s}{\pi}\left[\beta - \alpha + \frac{\sin 2\alpha}{2} - \frac{\sin 2\beta}{2}\right]^{\frac{1}{2}} \quad (18.9)$$

V_o can be evaluated for two possible extreme values of $\phi = 0$ when $\beta = \pi$, and $\phi = \pi/2$ when $\beta = 2\pi - \alpha$ and the envelope of the voltage control characteristics for this controller is shown in Fig. 18.5.

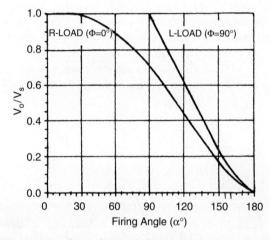

FIGURE 18.5 Envelope of control characteristics of a single-phase ac voltage controller with RL load.

The rms SCR current can be obtained from Eq. (18.7) as:

$$I_{o,SCR} = \left[\frac{1}{2\pi}\int_\alpha^\beta i_o^2 d(\omega t)\right]^{\frac{1}{2}} \quad (18.10)$$

The rms load current, $I_o = \sqrt{2}\, I_{o,SCR} \quad (18.11)$

The average value of SCR current, $I_{A,SCR} = \frac{1}{2\pi}\int_\alpha^\beta i_o d(\omega t)$

$$(18.12)$$

Gating Signal Requirements: For the inverse-parallel SCRs as shown in Fig. 18.1a, the gating signals of SCRs must be isolated from one another since there is no common cathode. For R-load, each SCR stops conducting at the end of each half cycle and under this condition, single short pulses may be used for gating as shown in Fig. 18.2. With RL load, however, this single short pulse gating is not suitable as shown in Fig. 18.6. When SCR T_2 is triggered at $\omega t = \pi + \alpha$, SCR T_1 is still conducting due to the load inductance. By the time the SCR T_1 stops conducting at β, the gate pulse for SCR T_2 has already

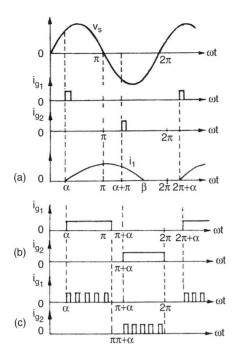

FIGURE 18.6 Single-phase full-wave controller with RL load: gate pulse requirements.

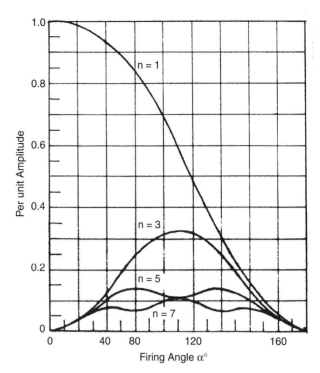

FIGURE 18.7 Harmonic content as a function of the firing angle for a single-phase voltage controller with RL load.

ceased and T_2 will fail to turn on resulting the converter to operate as a single-phase rectifier with conduction of T_1 only. This necessitates the application of a sustained gate pulse either in the form of a continuous signal for the half cycle period which increases the dissipation in SCR gate circuit and a large isolating pulse transformer or better a *train of pulses* (*carrier frequency gating*) to overcome these difficulties.

Operation with $\alpha < \phi$: If $\alpha = \phi$, then from Eq.(18.8),

$$\sin(\beta - \phi) = \sin(\beta - \alpha) = 0 \quad (18.13)$$

and $\quad \beta - \alpha = \theta = \pi \quad (18.14)$

As the conduction angle θ cannot exceed π and the load current must pass through zero, the control range of the firing angle is $\phi \leq \alpha \leq \pi$. With narrow gating pulses and $\alpha < \phi$, only one SCR will conduct resulting in a rectifier action as shown. Even with a train of pulses, if $\alpha < \phi$, the changes in the firing angle will not change the output voltage and current but both the SCRs will conduct for the period π with T_1 becoming on at $\omega t = \pi$ and T_2 at $\omega t + \pi$.

This *dead zone* ($\alpha = 0$ to ϕ) whose duration varies with the load impedance angle ϕ is not a desirable feature in closed-loop control schemes. An alternative approach to the phase control with respect to the input voltage zero crossing has been visualized in which the firing angle is defined with respect to the instant when it is the load current, not the input voltage, that reaches zero, this angle being called *the hold-off angle (γ) or the control angle* (as marked in Fig. 18.3). This method needs sensing the load current – which may otherwise be required anyway in a closed-loop controller for monitoring or control purposes.

Power Factor and Harmonics: As in the case of phase-controlled rectifiers, the important limitations of the phase-controlled ac voltage controllers are the poor power factor and the introduction of harmonics in the source currents. As seen from Eq.(18.3), the input power factor depends on α and as α increases, the power factor decreases.

The harmonic distortion increases and the quality of the input current decreases with increase of firing angle. The variations of low-order harmonics with the firing angle as computed by Fourier analysis of the voltage waveform of Fig. 18.2 (with R-load) are shown in Fig. 18.7. Only odd harmonics exist in the input current because of half-wave symmetry.

18.2.2 Single-phase AC–AC Voltage Controller with On/Off Control

Integral Cycle Control: As an alternative to the phase control, the method of integral cycle control or burst-firing is used for heating loads. Here, the switch is turned on for a time t_n with n integral cycles and turned off for a time t_m with m integral cycles (Fig. 18.8). As the SCRs or triacs used here are turned on at the zero crossing of the input voltage and turn off occurs at zero current, supply harmonics and radio

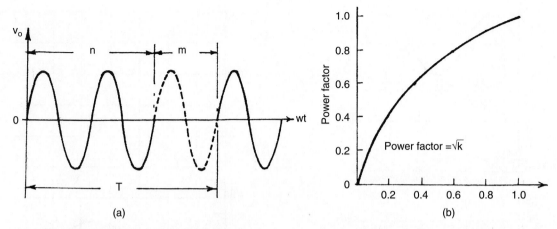

FIGURE 18.8 Integral cycle control: (a) typical load voltage waveforms and (b) power factor with the duty cycle k.

frequency interference are very low. However, sub-harmonic frequency components may be generated which are undesirable as they may set up sub-harmonic resonance in the power supply system, cause lamp flicker and may interfere with the natural frequencies of motor loads causing shaft oscillations.

For sinusoidal input voltage, $v = \sqrt{2}V_s \sin \omega t$, the rms output voltage,

$$V_o = V_s\sqrt{k} \quad \text{where } k = n/(n+m) = \text{duty cycle} \quad (18.15)$$

and V_s = rms phase voltage

$$\text{The power factor} = \sqrt{k} \quad (18.16)$$

which is poorer for lower values of the duty cycle k.

PWM AC Chopper: As in the case of controlled rectifier, the performance of ac voltage controllers can be improved in terms of harmonics, quality of output current, and input power factor by PWM control in PWM ac choppers, the circuit configuration of one such single phase unit being shown in Fig. 18.9. Here, fully controlled switches S_1 and S_2 connected in anti-parallel are turned on and off many times during the

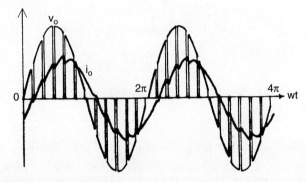

FIGURE 18.10 Typical output voltage and current waveforms of a single-phase PWM ac chopper.

positive and negative half cycles of the input voltage, respectively. S_1' and S_2' provide the freewheeling paths for the load current when S_1 and S_2 are off. An input capacitor filter may be provided to attenuate the high switching frequency currents drawn from the supply and also to improve the input power factor. Figure 18.10 shows the typical output voltage and load current waveform for a single-phase PWM ac chopper. It can be shown that the control characteristics of an ac chopper depend on the *modulation index, M* which theoretically varies from 0 to 1.

Three-phase PWM choppers consist of three single-phase choppers either connected in delta or four-wire star.

18.3 Three-phase AC–AC Voltage Controllers

18.3.1 Phase-controlled Three-phase AC Voltage Controllers

Various Configurations: Several possible circuit configurations for three-phase phase-controlled ac regulators with star or delta connected loads are shown in Fig. 18.11a–h.

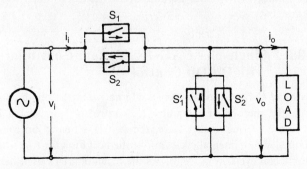

FIGURE 18.9 Single-phase PWM ac chopper circuit.

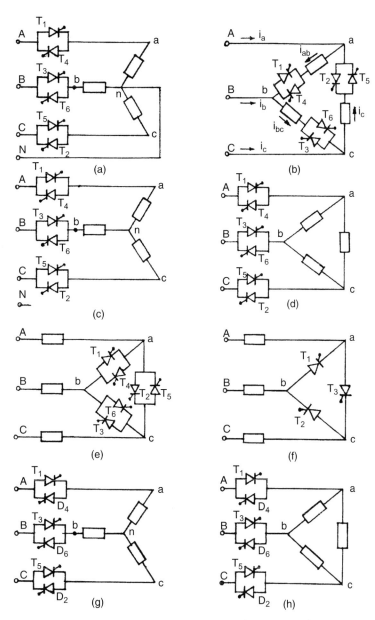

FIGURE 18.11 Three-phase ac voltage controller circuit configurations.

The configurations in (a) and (b) can be realized by three single-phase ac regulators operating independently of each other and they are easy to analyze. In (a), the SCRs are to be rated to carry line currents and withstand phase voltages whereas in (b) they should be capable to carry phase currents and withstand the line voltage. In (b), the line currents are free from triplen harmonics while these are present in the closed delta. The power factor in (b) is slightly higher. The firing angle control range for both these circuits is 0–180° for R-load.

The circuits in (c) and (d) are three-phase three-wire circuits and are complicated to analyze. In both these circuits, at least two SCRs, one in each phase, must be gated simultaneously to get the controller started by establishing a current path between the supply lines. This necessitates two firing pulses spaced at 60° apart per cycle for firing each SCR. The operation modes are defined by the number of SCRs conducting in these modes. The firing control range is 0–150°. The triplen harmonics are absent in both these configurations.

Another configuration is shown in (e) when the controllers are connected in delta and the load is connected between the supply and the converter. Here, current can flow between two lines even if one SCR is conducting so each SCR requires one firing pulse per cycle. The voltage and current ratings of SCRs are nearly the same as that of the circuit (b). It is also possible to reduce the number of devices to three SCRs in delta as shown in (f), connecting one source terminal directly to one load circuit terminal. Each SCR is provided with gate pulses in each cycle spaced at 120° apart. In both (e) and (f), each end of each phase must be accessible. The number of devices in (f) is less, but their current ratings must be higher.

As in the case of single-phase phase-controlled voltage regulator, the total regulator cost can be reduced by replacing six SCRs by three SCRs and three diodes, resulting in three-phase *half-wave controlled* unidirectional ac regulators as shown in (g) and (h) for star and delta connected loads. The main drawback of these circuits is the large harmonic content in the output voltage – particularly, the second harmonic because of the asymmetry. However, the dc components are absent in the line. The maximum firing angle in the half-wave controlled regulator is 210°.

18.3.2 Fully Controlled Three-phase Three-wire AC Voltage Controller

Star-connected Load with Isolated Neutral: The analysis of operation of the full-wave controller with isolated neutral as shown in Fig. 18.11c is, as mentioned, quite complicated in comparison to that of a single-phase controller, particularly for an RL or motor load. As a simple example, the operation of this controller is considered here with a simple star-connected R-load. The six SCRs are turned on in the sequence 1-2-3-4-5-6 at 60° intervals and the gate signals are sustained throughout the possible conduction angle.

The output phase voltage waveforms for $\alpha = 30°$, $75°$, and $120°$ for a balanced three-phase R-load are shown in Fig. 18.12. At any interval, either three SCRs or two SCRs, or no SCRs may be on and the instantaneous output voltages to the load are either a line-to-neutral voltage (three SCRs on), or one-half of the line-to-line voltage (two SCRs on), or zero (no SCR on).

Depending on the firing angle α, there may be *three* operating modes:

Mode I (also known as Mode 2/3): $0 \leq \alpha \leq 60°$; There are periods when *three* SCRs are conducting, one in each phase for either direction and periods when just *two* SCRs conduct.

For example, with $\alpha = 30°$ in Fig. 18.12a, assume that at $\omega t = 0$, SCRs T_5 and T_6 are conducting, and the current through the R-load in a-phase is zero making $v_{an} = 0$. At $\omega t = 30°$, T_1 receives a gate pulse and starts conducting; T_5 and T_6 remain on and $v_{an} = v_{AN}$. The current in T_5 reaches zero at 60°, turning T_5 off. With T_1 and T_6 staying on, $v_{an} = \frac{1}{2}v_{AB}$. At 90°, T_2 is turned on, the three SCRs T_1, T_2, and T_6 are then conducting and $v_{an} = v_{AN}$. At 120°, T_6 turns off, leaving T_1 and T_2 on, so $v_{an} = \frac{1}{2}v_{AC}$. Thus with the progress of firing in sequence till $\alpha = 60°$, the number of SCRs conducting at a particular instant alternates between two and three.

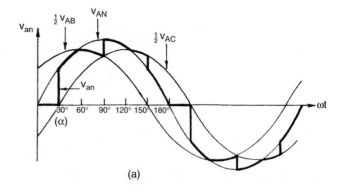

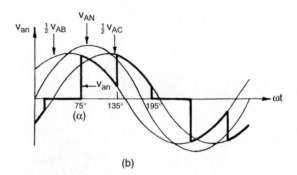

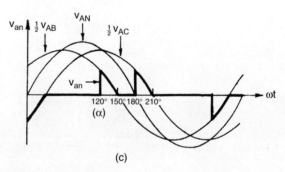

FIGURE 18.12 Output voltage waveforms for a three-phase ac voltage controller with star-connected R-load: (a) v_{an} for $\alpha = 30°$; (b) v_{an} for $\alpha = 75°$; and (c) $v_{an} = 120°$.

Mode II (also known as Mode 2/2): $60° \leq \alpha \leq 90°$; *Two* SCRs, one in each phase always conduct.

For $\alpha = 75°$ as shown in Fig. 18.12b, just prior to $\alpha = 75°$, SCRs T_5 and T_6 were conducting and $v_{an} = 0$. At 75°, T_1 is turned on, T_6 continues to conduct while T_5 turns off as v_{CN} is negative. $v_{an} = \frac{1}{2}v_{AB}$. When T_2 is turned on at 135°, T_6 is turned off and $v_{an} = \frac{1}{2}v_{AC}$. The next SCR to turn on is T_3 which turns off T_1 and $v_{an} = 0$. One SCR is always turned off when another is turned on in this range of α and the output voltage is either one-half line-to-line voltage or zero.

Mode III (also known as Mode 0/2): $90° \leq \alpha \leq 150°$; When *none* or *two* SCRs conduct.

For $\alpha = 120°$, Fig. 18.12c, earlier no SCRs were on and $v_{an} = 0$. At $\alpha = 120°$, SCR T_1 is given a gate signal while T_6 has a gate signal already applied. Since v_{AB} is positive,

18 AC–AC Converters

T_1 and T_6 are forward biased and they begin to conduct and $v_{an} = \frac{1}{2}v_{AB}$. Both T_1 and T_6 turn off, when v_{AB} becomes negative. When a gate signal is given to T_2, it turns on and T_1 turns on again.

For $\alpha > 150°$, there is no period when two SCRs are conducting and the output voltage is zero at $\alpha = 150°$. Thus, the range of the firing angle control is $0 \leq \alpha \leq 150°$.

For *star-connected R-load*, assuming the instantaneous phase voltages as

$$v_{AN} = \sqrt{2}V_s \sin \omega t$$

$$v_{BN} = \sqrt{2}V_s \sin(\omega t - 120°) \quad (18.17)$$

$$v_{CN} = \sqrt{2}V_s \sin(\omega t - 240°)$$

the expressions for the rms output phase voltage V_o can be derived for the three modes as:

$$0 \leq \alpha \leq 60° \quad V_o = V_s\left[1 - \frac{3\alpha}{2\pi} + \frac{3}{4\pi}\sin 2\alpha\right]^{\frac{1}{2}} \quad (18.18)$$

$$60° \leq \alpha \leq 90° \quad V_o = V_s\left[\frac{1}{2} + \frac{3}{4\pi}\sin 2\alpha + \sin(2\alpha + 60°)\right]^{\frac{1}{2}} \quad (18.19)$$

$$90° \leq \alpha \leq 150° \quad V_o = V_s\left[\frac{5}{4} - \frac{3\alpha}{2\pi} + \frac{3}{4\pi}\sin(2\alpha + 60°)\right]^{\frac{1}{2}} \quad (18.20)$$

For *star-connected pure L-load*, the effective control starts at $\alpha > 90°$ and the expressions for two ranges of α are:

$$90° \leq \alpha \leq 120° \quad V_o = V_s\left[\frac{5}{2} - \frac{3\alpha}{\pi} + \frac{3}{2\pi}\sin 2\alpha\right]^{\frac{1}{2}} \quad (18.21)$$

$$120° \leq \alpha \leq 150° \quad V_o = V_s\left[\frac{5}{2} - \frac{3\alpha}{\pi} + \frac{3}{2\pi}\sin(2\alpha + 60°)\right]^{\frac{1}{2}} \quad (18.22)$$

The control characteristics for these two limiting cases ($\phi = 0$ for R-load and $\phi = 90°$ for L-load) are shown in Fig. 18.13. Here also, like the single-phase case, the dead zone may be avoided by controlling the voltage with respect to the control angle or hold-off angle (γ) from the zero crossing of current in place of the firing angle α.

RL Load: The analysis of the three-phase voltage controller with star-connected RL load with isolated neutral is quite complicated since the SCRs do not cease to conduct at voltage zero, and the extinction angle β is to be known by solving the transcendental equation for the case. The Mode II operation, in this case, disappears [1] and the operation shift from

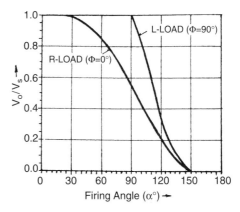

FIGURE 18.13 Envelope of control characteristics for a three-phase full-wave ac voltage controller.

Mode I to Mode III depends on the so-called critical angle α_{crit} [2, 3] which can be evaluated from a numerical solution of the relevant transcendental equations. Computer simulation either by PSPICE program [4, 5] or a switching-variable approach coupled with an iterative procedure [6] is a practical means of obtaining the output voltage waveform in this case. Figure 18.14 shows typical simulation results using the later approach [6] for a three-phase voltage controller fed RL load for $\alpha = 60°, 90°$, and $105°$ which agree with the corresponding practical oscillograms given in [7].

Delta-connected R-load: The configuration is shown in Fig. 18.11b. The voltage across an R-load is the corresponding line-to-line voltage when one SCR in that phase is on. Figure 18.15 shows the line and phase currents for $\alpha = 130°$ and $90°$ with an R-load. The firing angle α is measured from the zero crossing of the line-to-line voltage and the SCRs are turned on in the sequence as they are numbered. As in the single-phase case, the range of firing angle is $0 \leq \alpha \leq 180°$. The line currents can be obtained from the phase currents as

$$i_a = i_{ab} - i_{ca}$$
$$i_b = i_{bc} - i_{ab} \quad (18.23)$$
$$i_c = i_{ca} - i_{bc}$$

The line currents depend on the firing angle and may be discontinuous as shown. Due to the delta connection, the triplen harmonic currents flow around the closed delta and do not appear in the line. The rms value of the line current varies between the range

$$\sqrt{2}I_\Delta \leq I_{L,rms} \leq \sqrt{3}I_{\Delta,rms} \quad (18.24)$$

as the conduction angle varies from very small (large α) to $180°$ ($\alpha = 0$).

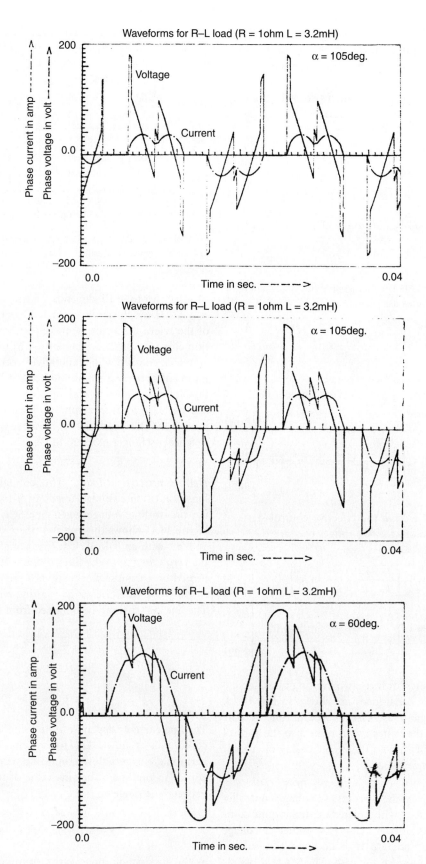

FIGURE 18.14 Typical simulation results for three-phase ac voltage controller-fed RL load ($R = 1$ ohm, $L = 3.2$ mH) for $\alpha = 60°$, $90°$, and $105°$.

18 AC–AC Converters

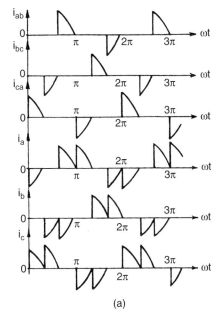

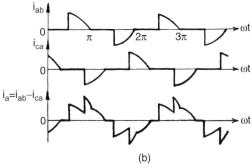

FIGURE 18.15 Waveforms of a three-phase ac voltage controller with a delta connected R-load: (a) $\alpha = 120°$ and (b) $\alpha = 90°$.

18.4 Cycloconverters

In contrast to the ac voltage controllers operating at constant frequency, discussed so far, a cycloconverter operates as a direct ac–ac frequency changer with inherent voltage control feature. The basic principle of this converter to construct an alternating voltage wave of lower frequency from successive segment of voltage waves of higher frequency ac supply by a switching arrangement was conceived and patented in 1920s. Grid-controlled mercury-arc rectifiers were used in these converters installed in Germany in 1930s to obtain $16\frac{2}{3}$ Hz single-phase supply for ac series traction motors from a three-phase 50 Hz system, while at the same time a cycloconverter using 18 thyratrons supplying a 400 hp synchronous motor was in operation for some years as a power station auxiliary drive in USA. However, the practical and commercial utilization of these schemes waited till the SCRs became available in 1960s. With the development of large power SCRs and microprocessor-based control, the cycloconverter today is a matured practical converter for application in large power, low speed variable-voltage variable-frequency (VVVF) ac drives in cement and steel rolling mills as well as in variable-speed constant-frequency (VSCF) systems in air-crafts and naval ships.

A cycloconverter is a naturally commuted converter with inherent capability of bi-directional power flow and there is no real limitation on its size unlike an SCR inverter with commutation elements. Here, the switching losses are considerably low, the regenerative operation at full power over complete speed range is inherent and it delivers a nearly sinusoidal waveform resulting in minimum torque pulsation and harmonic heating effects. It is capable of operating even with blowing out of individual SCR fuse (unlike inverter) and the requirements regarding turn-off time, current rise time, and dv/dt sensitivity of SCRs are low. The main limitations of a naturally commutated cycloconverter are (i) limited frequency range for sub-harmonic free and efficient operation, and (ii) poor input displacement/power factor, particularly at low-output voltages.

18.4.1 Single-phase to Single-phase Cycloconverter

Though rarely used, the operation of a single-phase to single-phase cycloconverter is useful to demonstrate the basic principle involved. Figure 18.16a shows the power circuit of a single-phase bridge type of cycloconverter which is the same arrangement as that of a single-phase dual converter. The firing angles of the individual two-pulse two-quadrant bridge converters are continuously modulated here, so that each ideally produces the same fundamental ac voltage at its output terminals as marked in the simplified equivalent circuit in Fig. 18.16b. Because of the unidirectional current carrying property of the individual converters, it is inherent that the positive half cycle of the current is carried by the P-converter and the negative half cycle of the current by the N-converter regardless of the phase of the current with respect to the voltage. This means that for a reactive load, each converter operates in both rectifying and inverting region during the period of the associated half cycle of the low-frequency output current.

Operation with R-load: Figure 18.17 shows the input and output voltage waveforms with a pure R-load for a 50–$16\frac{2}{3}$ Hz cycloconverter. The P- and N- converters operate for alternate $T_o/2$ periods. The output frequency ($1/T_o$) can be varied by varying T_o and the voltage magnitude by varying the firing angle α of the SCRs. As shown in the figure, three cycles of the ac input wave are combined to produce one cycle of the output frequency to reduce the supply frequency to one-third across the load.

If α_P is the firing angle of the P-converter, the firing angle of the N-converter α_N is $\pi - \alpha_P$ and the average voltage of the

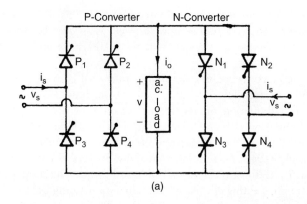

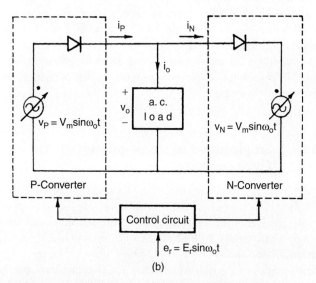

FIGURE 18.16 (a) Power circuit for a single-phase bridge cycloconverter and (b) simplified equivalent circuit of a cycloconverter.

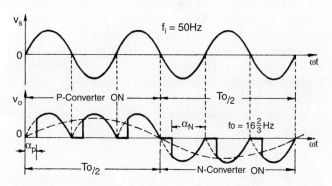

FIGURE 18.17 Input and output waveforms of a 50–$16\frac{2}{3}$ Hz cycloconverter with RL load.

P-converter is equal and opposite to that of the N-converter. The inspection of Fig. 18.17 shows that the waveform with α remaining fixed in each half cycle generates a square wave having a large low-order harmonic content. A near approximation to sine wave can be synthesized by a phase modulation of the

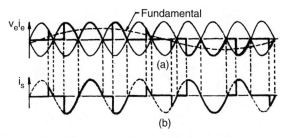

FIGURE 18.18 Waveforms of a single-phase/single-phase cycloconverter (50–10 Hz) with RL load: (a) load voltage and load current and (b) input supply current.

firing angles as shown in Fig. 18.18 for a 50–10 Hz cycloconverter. The harmonics in the load voltage waveform are less compared to earlier waveform. The supply current, however, contains a sub-harmonic at the output frequency for this case as shown.

Operation with RL Load: The cycloconverter is capable of supplying loads of any power factor. Figure 18.19 shows the idealized output voltage and current waveforms for a lagging power factor load where both the converters are operating as rectifier and inverter at the intervals marked. The load current lags the output voltage and the load current direction determines which converter is conducting. Each converter continues to conduct after its output voltage changes polarity and during this period, the converter acts as an inverter and the power is returned to the ac source. Inverter operation continues till the other converter starts to conduct. By controlling the frequency of oscillation and the *depth of modulation* of the firing angles of the converters (as shown later), it is possible to control the frequency and the amplitude of the output voltage.

The load current with RL load may be continuous or discontinuous depending on the load phase angle, ϕ. At light load inductance or for $\phi \leq \alpha \leq \pi$, there may be discontinuous load current with short zero-voltage periods. The current wave may contain even harmonics as well as sub-harmonic components. Further, as in the case of dual converter, though the mean output voltage of the two converters are equal and opposite, the

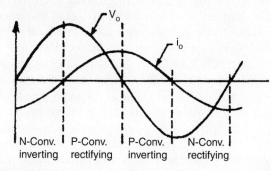

FIGURE 18.19 Idealized load voltage and current waveform for a cycloconverter with RL load.

instantaneous values may be unequal and a circulating current can flow within the converters. This circulating current can be limited by having a center-tapped reactor connected between the converters or can be completely eliminated by logical control similar to the dual converter case when the gate pulses to the converter remaining idle are suppressed, when the other converter is active. In practice, a zero-current interval of short duration is needed, in addition, between the operation of the P- and N- converters to ensure that the supply lines of the two converters are not short-circuited. With circulating current-free operation, the control scheme becomes complicated if the load current is discontinuous.

In the case of the circulating current scheme, the converters are kept in virtually continuous conduction over the whole range and the control circuit is simple. To obtain reasonably good sinusoidal voltage waveform using the line-commutated two quadrant converters and eliminate the possibility of the short circuit of the supply voltages, the output frequency of the cycloconverter is limited to a much lower value of the supply frequency. The output voltage waveform and the output frequency range can be improved further by using converters of higher pulse numbers.

18.4.2 Three-phase Cycloconverters

18.4.2.1 Three-phase Three-pulse Cycloconverter

Figure 18.20a shows the schematic diagram of a three-phase half-wave (three-pulse) cycloconverter feeding a single-phase load and Fig. 18.20b, the configuration of a three-phase half-wave (three-pulse) cycloconverter feeding a three-phase load. The basic process of a three-phase cycloconversion is illustrated in Fig. 18.20c at 15 Hz, 0.6 power factor lagging load from a 50 Hz supply. As the firing angle α is cycled from zero at "a" to 180° at "j", half a cycle of output frequency is produced (the gating circuit is to be suitably designed to introduce this oscillation of the firing angle). For this load, it can be seen that although the mean output voltage reverses at X, the mean output current (assumed sinusoidal) remains positive until Y. During XY, the SCRs A, B, and C in P-converter

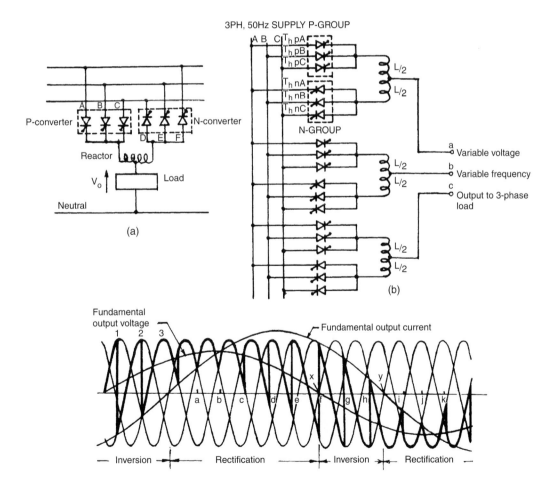

FIGURE 18.20 (a) Three-phase half-wave (three-pulse) cycloconverter supplying a single-phase load; (b) three-pulse cycloconverter supplying a three-phase load; and (c) output voltage waveform for one phase of a three-pulse cycloconverter operating at 15 Hz from a 50 Hz supply and 0.6 power factor lagging load.

are "inverting." A similar period exists at the end of the negative half cycle of the output voltage when D, E, and F SCRs in N-converter are "inverting." Thus the operation of the converter follows in the order of "rectification' and "inversion" in a cyclic manner, the relative durations being dependent on load power factor. The output frequency is that of the firing angle oscillation about a quiscent point of 90° (condition when the mean output voltage, given by $V_o = V_{do} \cos \alpha$, is zero). For obtaining the positive half cycle of the voltage, firing angle α is varied from 90° to 0° and then to 90° and for the negative half cycle, from 90° to 180° and back to 90°. Variation of α within the limits of 180° automatically provides for "natural" line commutation of the SCRs. It is shown that a complete cycle of low-frequency output voltage is fabricated from the segments of the three-phase input voltage by using the phase-controlled converters. The P-converter or N-converter SCRs receive firing pulses which are timed such that each converter delivers the same mean output voltage. This is achieved, as in the case of single-phase cycloconverter or the dual converter by maintaining the firing angle constraints of the two groups as $\alpha_P = (180° - \alpha_N)$. However, the instantaneous voltages of two converters are not identical and large circulating current may result unless limited by intergroup reactor as shown (*circulating-current cycloconverter*) or completely suppressed by removing the gate pulses from the non-conducting converter by an inter-group blanking logic (*circulating-current-free cycloconverter*).

Circulating-current Mode Operation: Figure 18.21 shows typical waveforms of a three-pulse cycloconverter operating with circulating current. Each converter conducts continuously with rectifying and inverting modes as shown, and the load is supplied with an average voltage of two converters reducing some of the ripple in the process, the inter-group reactor behaving as a potential divider. The reactor limits the circulating current, the value of its inductance to the flow of load current being one-fourth of its value to the flow of circulating current as the inductance is proportional to the square of the number of turns. The fundamental wave produced by both the converters are the same. The reactor voltage is the instantaneous difference between the converter voltages and the time integral of this voltage divided by the inductance (assuming negligible circuit resistance) is the circulating current. For a three-pulse cycloconverter, it can be observed that this current reaches its peak when $\alpha_P = 60°$ and $\alpha_N = 120°$.

Output voltage equation: A simple expression for the fundamental rms output voltage of the cycloconverter and the required variation of the firing angle α can be derived with the assumptions that (i) the firing angle α in successive half cycles is varied slowly resulting in a low-frequency output

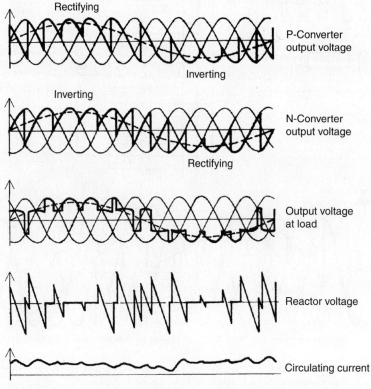

FIGURE 18.21 Waveforms of a three-pulse cycloconverter with circulating current.

(ii) the source impedance and the commutation overlap are neglected (iii) the SCRs are ideal switches and (iv) the current is continuous and ripple-free. The average dc output voltage of a p-pulse dual converter with fixed α is

$$V_{do} = V_{domax} \cos\alpha, \text{ where } V_{domax} = \sqrt{2}V_{ph}\frac{p}{\pi}\sin\frac{\pi}{p} \quad (18.25)$$

For the p-pulse dual converter operating as a cycloconverter, the average phase voltage output at any point of the low frequency should vary according to the equation

$$V_{o,av} = V_{o1,\max}\sin\omega_o t \quad (18.26)$$

where $V_{o1,\max}$ is the desired maximum value of the fundamental output of the cycloconverter.

Comparing Eq. (18.25) with Eq. (18.26), the required variation of α to obtain a sinusoidal output is given by

$$\alpha = \cos^{-1}[(V_{o1,\max}/V_{domax})\sin\omega_o t] = \cos^{-1}[r\sin\omega_o t] \quad (18.27)$$

where r is the ratio $(V_{o1,\max}/V_{domax})$, the *voltage magnitude control ratio*.

Equation (18.27) shows α as a non-linear function with r (≤ 1) as shown in Fig. 18.22.

However, the firing angle α_P of the P-converter cannot be reduced to 0° as this corresponds to $\alpha_N = 180°$ for the N-converter which, in practice, cannot be achieved because of allowance for commutation overlap and finite turn-off time of the SCRs. Thus the firing angle α_P can be reduced to a certain finite value $\alpha_{\min}$ and the maximum output voltage is reduced by a factor $\cos\alpha_{\min}$.

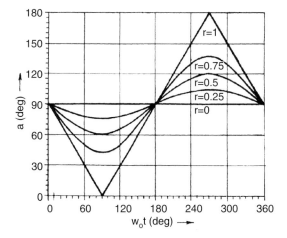

FIGURE 18.22 Variations of the firing angle (α) with r in a cycloconverter.

The fundamental rms voltage per phase of either converter is

$$V_{or} = V_{oN} = V_{oP} = rV_{ph}\frac{p}{\pi}\sin\frac{\pi}{p} \quad (18.28)$$

Though the rms value of the low-frequency output voltage of the P-converter and that of the N-converter are equal, the actual waveforms differ and the output voltage at the midpoint of the circulating current limiting reactor (Fig. 18.21) which is the same as the load voltage, is obtained as the mean of the instantaneous output voltages of the two converters.

Circulating Current-free Mode Operation: Figure 18.23 shows the typical waveforms for a three-pulse cycloconverter operating in this mode with RL load assuming *continuous current* operation. Depending on the load current direction, only one converter operates at a time and the load voltage is the same as the output voltage of the conducting converter. As explained earlier in the case of single-phase cycloconverter, there is a possibility of short-circuit of the supply voltages at the cross-over points of the converter unless taken care of in the control circuit. The waveforms drawn also neglect the effect of overlap due to the ac supply inductance. A reduction in the output voltage is possible by retarding the firing angle gradually at the points a, b, c, d, e in Fig. 18.23. (This can be easily implemented by reducing the magnitude of the reference voltage in the control circuit.) The circulating current is completely suppressed by blocking all the SCRs in the converter which is not delivering the load current. A current sensor is incorporated in each output phase of the cycloconverter which detects the direction of the output current and feeds an appropriate signal to the control circuit to inhibit or blank the gating pulses to the non-conducting converter in the same way as in the case of a dual converter for dc drives. The circulating current-free operation improves the efficiency and the displacement factor of the cycloconverter and also increases the maximum usable output frequency. The load voltage transfers smoothly from one converter to the other.

18.4.2.2 Three-phase Six-pulse and Twelve-pulse Cycloconverter

A six-pulse cycloconverter circuit configuration is shown in Fig. 18.24. Typical load voltage waveforms for 6-pulse (with 36 SCRs) and 12-pulse (with 72 SCRs) cycloconverters are shown in Fig. 18.25, the 12-pulse converter being obtained by connecting two 6-pulse configurations in series and appropriate transformer connections for the required phase-shift. It may be seen that the higher pulse numbers will generate waveforms closer to the desired sinusoidal form and thus permit higher frequency output. The phase loads may be isolated from each other as shown or interconnected with suitable secondary winding connections.

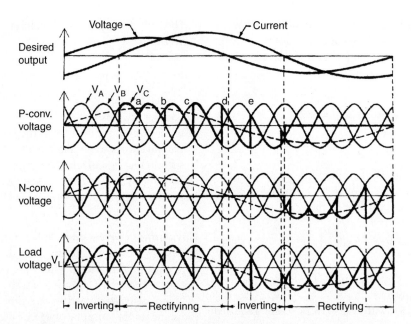

FIGURE 18.23 Waveforms for a three-pulse circulating current-free cycloconverter with RL load.

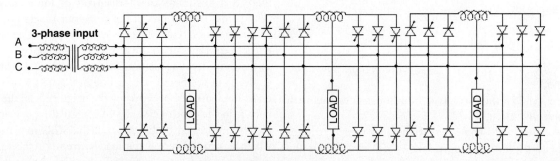

FIGURE 18.24 Three-phase six-pulse cycloconverter with isolated loads.

18.4.3 Cycloconverter Control Scheme

Various possible control schemes, analog as well as digital, for deriving trigger signals for controlling the basic cycloconverter have been developed over the years.

Out of the several possible signal combinations, it has been shown [8] that a sinusoidal reference signal ($e_r = E_r \sin \omega_o t$) at desired output frequency f_o and a cosine modulating signal ($e_m = E_m \cos \omega_i t$) at input frequency f_i is the best combination possible for comparison to derive the trigger signals for the SCRs (Fig. 18.26 [9]) which produces the output waveform with the lowest total harmonic distortion. The modulating voltages can be obtained as the phase-shifted voltages (B-phase for A-phase SCRs, C-phase voltage for B-phase SCRs, and so on) as explained in Fig. 18.27, where at the intersection point "a",

$$E_m \sin(\omega_i t - 120°) = -E_r \sin(\omega_o t - \phi)$$

or $\cos(\omega_i t - 30°) = (E_r/E_m) \sin(\omega_o t - \phi)$

From Fig. 18.27, the firing delay for A-phase SCR, $\alpha = (\omega_i t - 30°)$

So, $\cos \alpha = (E_r/E_m) \sin(\omega_o t - \phi)$

The cycloconverter output voltage for continuous current operation,

$$V_o = V_{do} \cos \alpha = V_{do}(E_r/E_m) \sin(\omega_o t - \phi) \quad (18.29)$$

in which the equation shows that the amplitude, frequency, and phase of the output voltage can be controlled by controlling corresponding parameters of the reference voltage, thus making the transfer characteristic of the cycloconverter linear. The derivation of the two complimentary voltage waveforms for the P-group or N-group converter "banks" in this way is illustrated in Fig. 18.28. The final cycloconverter output waveshape is composed of alternate half cycle segments of the complementary P-converter and the N-converter

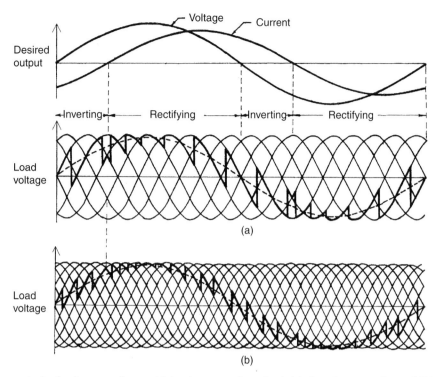

FIGURE 18.25 Cycloconverter load voltage waveforms with lagging power factor load: (a) six-pulse connection and (b) twelve-pulse connection.

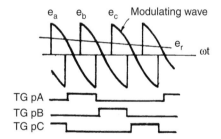

FIGURE 18.26 Deriving firing signals for one converter group of a three-pulse cycloconverter.

output voltage waveforms which coincide with the positive and negative current half cycles, respectively.

Control Circuit Block Diagram: Figure 18.29 [10] shows a simplified block diagram of the control circuit for a circulating current-free cycloconverter implemented with ICs in the early seventies in the Power Electronics Laboratory at IIT Kharagpur in India. The same circuit is also applicable to a circulating current cycloconverter with the omission of the *Converter Group Selection and Blanking* circuit.

The *Synchronizing circuit* produces the modulating voltages ($e_a = -Kv_b$, $e_b = -Kv_c$, $e_c = -Kv_a$), synchronized with the mains through step-down transformers and proper filter circuits.

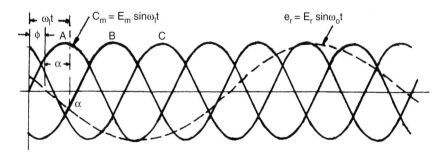

FIGURE 18.27 Derivation of the cosine modulating voltages.

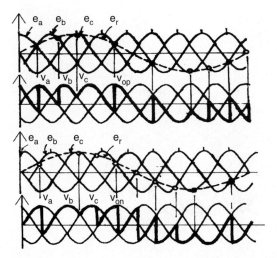

FIGURE 18.28 Derivation of P-converter and N-converter output voltages.

The *Reference Source* produces variable voltage variable frequency reference signal (e_{ra}, e_{rb}, e_{rc}) (three-phase for a three-phase cycloconverter) for comparison with the modulation voltages. Various ways, analog or digital, have been developed to implement this reference source as in the case of the PWM inverter. In one of the early analog schemes Fig. 18.30 [10], for a three-pulse cycloconverter, a variable-frequency UJT relaxation oscillator of frequency $6f_d$ triggers a ring counter to produce a three-phase square-wave output of frequency f_d which is used to modulate a single-phase fixed frequency (f_c) variable amplitude sinusoidal voltage in a three-phase full-wave transistor chopper. The three-phase output contains ($f_c - f_d$), ($f_c + f_d$), ($3f_d + f_c$), etc. frequency components from where the "wanted" frequency component ($f_c - f_d$) is filtered out for each phase using a low-pass filter. For example, with $f_c = 500$ Hz and frequency of the relaxation oscillator varying between 2820 and 3180 Hz, a three phase 0–30 Hz reference output can be obtained with the facility for phase sequence reversal.

The *Logic and Trigger Circuit* for each phase involves comparators for comparison of the reference and modulating voltages, and inverters acting as buffer stages. The outputs of the comparators are used to clock the flip-flops or latches

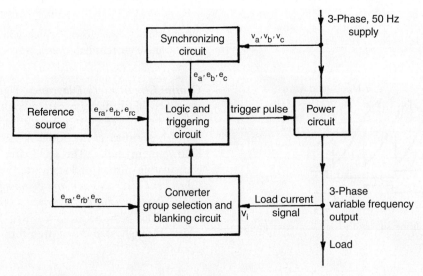

FIGURE 18.29 Block diagram for a circulating current-free cycloconverter control circuit.

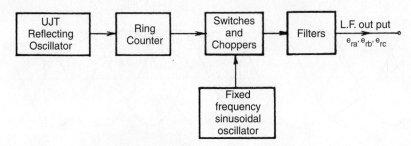

FIGURE 18.30 Block diagram of a variable-voltage variable-frequency three-phase reference source.

whose outputs in turn feed the SCR gates through AND gates and pulse amplifying and isolation circuits, the second input to the AND gates being from the *Converter Group Selection and Blanking Circuit*.

In the *Converter Group Selection and Blanking Circuit*, the zero crossing of the current at the end of each half cycle is detected and is used to regulate the control signals either to P-group or N-group converters depending on whether the current goes to zero from negative to positive or positive to negative, respectively. However, in practice, the current being discontinuous passes through multiple zero crossings while changing direction which may lead to undesirable switching of the converters. So, in addition to the current signal, the reference voltage signal is also used for the group selection, and a *threshold* band is introduced in the current signal detection to avoid inadvertent switching of the converters. Further, a delay circuit provides a blanking period of appropriate duration between the converter group switching to avoid line-to-line short circuits [10]. In some schemes, the delays are not introduced when a small circulating current is allowed during cross-over instants limited by reactors of limited size and this scheme operates in the so called "dual mode" – circulating current as well as circulating current-free mode for minor and major portions of the output cycle respectively. A different approach to the converter group selection, based on the closed-loop control of the output voltage where a bias voltage is introduced between the voltage transfer characteristics of the converters to reduce circulating current is discussed in [8].

Improved Control Schemes: With the development of microprocessors and PC-based systems, digital software control has taken over many tasks in modern cycloconverters, particularly in replacing the low-level reference waveform generation and analog signal comparison units. The reference waveforms can be easily generated in the computer, stored in the EPROMs and accessed under the control of a stored program and microprocessor clock oscillator. The analog signal voltages can be converted to digital signals by using analog-to-digital converters (ADCs). The waveform comparison can then be made with the comparison features of the microprocessor system. The addition of time delays and inter-group blanking can also be achieved with digital techniques and computer software. A modification of the cosine firing control, using communication principles like *regular sampling* in preference to the *natural sampling* (discussed so far) of the reference waveform, yielding a stepped sine wave before comparison with the cosine wave [11] has been shown to reduce the presence of *sub-harmonics* (discussed later) in the circulating current-cycloconverter, and facilitate microprocessor-based implementation, as in the case of PWM inverter.

For a six pulse non-circulating current cycloconverter-fed synchronous motor drive with a vector control scheme and a flux observer, a PC-based hybrid control scheme (a combination of analog and digital control) has been reported in [12]. Here the functions such as comparison, group selection, blanking between the groups and triggering signal generation, filtering and phase conversion are left to the analog controller and digital controller that takes care of more serious tasks like voltage decoupling for current regulation, flux estimation using observer, speed, flux and field current regulators using PI-controllers, position and speed calculation leading to an improvement of sampling time and design accuracy.

18.4.4 Cycloconverter Harmonics and Input Current Waveform

The exact waveshape of the output voltage of the cycloconverter depends on (i) the pulse number of the converter; (ii) the ratio of the output to input frequency (f_o/f_i); (iii) the relative level of the output voltage; (iv) load displacement angle; (v) circulating current or circulating current-free operation; and (vi) the method of control of the firing instants. The harmonic spectrum of a cycloconverter output voltage is different and more complex than that of a phase-controlled converter. It has been revealed [8] that because of the continuous "to-and-fro" phase modulation of the converter firing angles, the harmonic distortion components (known as *necessary distortion terms*) have frequencies which are sums and differences between multiples of output and input supply frequencies.

Circulating Current-free Operation: A derived general expression for the output voltage of a cycloconverter with circulating current-free operation [8] shows the following spectrum of harmonic frequencies for the 3-pulse, 6-pulse, and 12-pulse cycloconverters employing cosine modulation technique:

3-pulse: $f_{oH} = |3(2k-1)f_i \pm 2nf_o|$ and

$$|6kf_i \pm (2n+1)f_o|$$

6-pulse: $f_{oH} = |6kf_i \pm (2n+1)f_o|$

12-pulse: $f_{oH} = |6kf_i \pm (2n+1)f_o|$ \hfill (18.30)

where k is any integer from 1 to infinity and n is any integer from 0 to infinity.

It may be observed that for certain ratios of f_o/f_i, the order of harmonics may be less or equal to the desired output frequency. All such harmonics are known as *sub-harmonics*, since they are not higher multiples of the input frequency. These sub-harmonics may have considerable amplitudes (e.g. with a 50 Hz input frequency and 35 Hz output frequency, a sub-harmonic of frequency $3 \times 50 - 4 \times 35 = 10$ Hz is produced whose magnitude is 12.5% of the 35 Hz component [11]) and are difficult to filter and so objectionable. Their spectrum increase with the increase of the ratio f_o/f_i and so limits its value at which a tolerable waveform can be generated.

Circulating-current Operation: For circulating-current operation with continuous current, the harmonic spectrum in the output voltage is the same as that of the circulating current-free operation except that each harmonic family now terminates at a definite term, rather than having infinite number of components. They are

3-pulse: $f_{oH} = |3(2k-1)f_i \pm 2nf_o|$, $n \leq 3(2k-1)+1$

and $|6kf_i \pm (2n+1)f_o|$, $(2n+1) \leq (6k+1)$

6-pulse: $f_{oH} = |6kf_i \pm (2n+1)f_o|$, $(2n+1) \leq (6k+1)$

12-pulse: $f_{oH} = |6kf_i \pm (2n+1)f_o|$, $(2n+1) \leq (12k+1)$
(18.31)

The amplitude of each harmonic component is a function of the output voltage ratio for the circulating-current cycloconverter and the output voltage ratio as well as the load displacement angle for the circulating current-free mode.

From the point of view of maximum useful attainable output-to-input frequency ratio (f_i/f_o) with the minimum amplitude of objectionable harmonic components, a guideline is available in [8] for it as 0.33, 0.5, and 0.75 for 3-, 6-, and 12-pulse cycloconverter, respectively. However, with the modification of the cosine wave modulation timings like *regular sampling* [11] in the case of circulating-current cycloconverters only and using a *sub-harmonic detection and feedback control concept* [13, 14] for both circulating- and circulating-current-free cases, the sub-harmonics can be suppressed and useful frequency range for the naturally commutated cycloconverters can be increased.

Other Harmonic Distortion Terms: Besides the harmonics as mentioned, other harmonic distortion terms consisting of frequencies of integral multiples of desired output frequency appear, if the transfer characteristic between the output and reference voltages is not linear. These are called *unnecessary distortion terms* which are absent when the output frequencies are much less than the input frequency. Further, some *practical distortion terms* may appear due to some practical non-linearities and imperfections in the control circuits of the cycloconverter, particularly at relatively lower levels of output voltage.

Input Current Waveform: Although the load current, particularly for higher pulse cycloconverters can be assumed to be sinusoidal, the input current is more complex being made of pulses. Assuming the cycloconverter to be an ideal switching circuit without losses, it can be shown from the instantaneous power balance equation that in cycloconverter supplying a single-phase load, the input current has harmonic components of frequencies ($f_i \pm 2f_o$), called *characteristic harmonic frequencies* which are independent of pulse number and they result in an oscillatory power transmittal to the ac supply system. In the case of cycloconverter feeding a balanced three-phase load, the net instantaneous power is the sum of the three oscillating instantaneous powers when the resultant power is constant and the net harmonic component is much reduced compared to that of a single-phase load case. In general, the total rms value of the input current waveform consists of three components: in-phase, quadrature, and the harmonic. The in-phase component depends on the active power output while the quadrature component depends on the net average of the oscillatory firing angle and is always lagging.

18.4.5 Cycloconverter Input Displacement/ Power Factor

The input supply performance of a cycloconverter such as displacement factor or fundamental power factor, input power factor, and the input current distortion factor are defined similar to those of the phase-controlled converter. The harmonic factor for the case of a cycloconverter is relatively complex as the harmonic frequencies are not simple multiples of the input frequency but are sums and differences between multiples of output and input frequencies.

Irrespective of the nature of the load, leading, lagging, or unity power factor, the cycloconverter requires reactive power decided by the average firing angle. At low output voltage, the average phase displacement between the input current and the voltage is large and the cycloconverter has a low input displacement and power factor. Besides load displacement factor and output voltage ratio, another component of the reactive current arises due to the modulation of the firing angle in the fabrication process of the output voltage [8]. In a phase-controlled converter supplying dc load, the maximum displacement factor is unity for maximum dc output voltage. However, in the case of the cycloconverter, the maximum input displacement factor is 0.843 with unity power factor load [8, 15]. The displacement factor decreases with reduction in the output voltage ratio. The distortion factor of the input current is given by (I_1/I) which is always less than 1 and the resultant power factor (= distortion factor × displacement factor) is thus much lower (around 0.76 maximum) than the displacement factor and this is a serious disadvantage of the naturally commutated cycloconverter (NCC).

18.4.6 Effect of Source Impedance

The source inductance introduces commutation overlap and affects the external characteristics of a cycloconverter similar to the case of a phase-controlled converter with the dc output. It introduces delay in transfer of current from one SCR to another, results in a voltage loss at the output and a modified harmonic distortion. At the input, the source impedance causes "rounding off" of the steep edges of the input current waveforms resulting in reduction in the amplitudes of higher order harmonic terms as well as a decrease in the input displacement factor.

18.4.7 Simulation Analysis of Cycloconverter Performance

The non-linearity and discrete time nature of practical cycloconverter systems, particularly for discontinuous current conditions make an exact analysis quite complex and a valuable design and analytical tool is a digital computer simulation of the system. Two general methods of computer simulation of the cycloconverter waveforms for RL and induction motor loads with circulating current and circulating current-free operation have been suggested in [16] where one of the methods which is very fast and convenient is the *cross-over points method* that gives the cross-over points (intersections of the modulating and reference waveforms) and the conducting phase numbers for both P- and N-converters from which the output waveforms for a particular load can be digitally computed at any interval of time for a practical cycloconverter.

18.4.8 Power Quality Issues

Degradation of power quality (PQ) in a cycloconverter-fed system due to sub-harmonics/interharmonics in the input and the output has been a subject of recent studies [14, 17]. In [17], the study includes the impact of cycloconverter control strategies on the total harmonic distortion (THD), distribution transformers, and communication lines while in [14], the PQ indices are suitably defined and the effect on THD, input/output displacement factor and input/output power factor for a cycloconverter-fed synchronous motor drive are studied together with a development of a simple feedback method of reduction of subharmonics/low frequency interharmonics for improvement of the power quality. The implementation of this scheme, detailed in [14], requires a simple modification of the control circuit of the cycloconverter in contrast to the expensive power level active filters otherwise required for suppression of such harmonics [18].

18.4.9 Forced Commutated Cycloconverter

The naturally commutated cycloconverter (NCC) with SCRs as devices, so far discussed, is sometimes referred to as, a *restricted frequency changer* as in view of the allowance on the output voltage quality ratings, the maximum output voltage frequency is restricted ($f_o \ll f_i$), as mentioned earlier. With devices replaced by fully controlled switches like forced commutated SCRs, power transistors, IGBTs, GTOs, etc., a forced commutated cycloconverter can be built where the desired output frequency is given by $f_o = |f_s - f_i|$, where f_s = switching frequency which may be larger or smaller than the f_i. In the case when $f_o \geq f_i$, the converter is called *Unrestricted Frequency Changer (UFC)* and when $f_o \leq f_i$, it is called a *Slow Switching Frequency Changer (SSFC)*. The early FCC structures have been comprehensively treated in [15]. It has been shown that in contrast with the NCC, when the input displacement factor (IDF) is always lagging, in UFC it is leading when the load displacement factor is lagging and vice versa, and in SSFC, it is identical to that of the load. Further, with proper control in an FCC, the input displacement factor can be made unity displacement factor frequency changer (UDFFC) with concurrent composite voltage waveform or controllable displacement factor frequency changer (CDFFC), where P-converter and N-converter voltage segments can be shifted relative to the output current wave for control of IDF continuously from lagging via unity to leading.

In addition to allowing bilateral power flow, UFCs offer an unlimited output frequency range, offer good input voltage utilization, do not generate input current and output voltage sub-harmonics and require only nine bi-directional switches (Fig. 18.31) for a three-phase to three-phase conversion. The main disadvantage of the structures treated in [15] is that they generate large unwanted low-order input current and output voltage harmonics which are difficult to filter out, particularly for low output voltage conditions. This problem has largely been solved with an introduction of an imaginative PWM voltage control scheme in [19], which is the basis of the newly designated converter called the *Matrix Converter (also known as PWM Cycloconverter)* which operates as a *Generalized Solid-State Transformer* with significant improvement in voltage and input current waveforms resulting in sine-wave input and sine-wave output as discussed in the next section.

18.5 Matrix Converter

The matrix converter (MC) is a development of the FCC based on bi-directional fully controlled switches, incorporating PWM voltage control, as mentioned earlier. With the initial progress made by Venturini [19–21], it has received considerable attention in recent years as it provides a good alternative to the double-sided PWM voltage source rectifier–inverters having the advantages of being a single stage converter with only nine switches for three-phase to three-phase conversion and inherent bi-directional power flow, sinusoidal input/output waveforms with moderate switching frequency, possibility of a compact design due to the absence of dc link reactive components, and controllable input power factor independent of the output load current. The main disadvantages of the matrix converters developed so far are the inherent

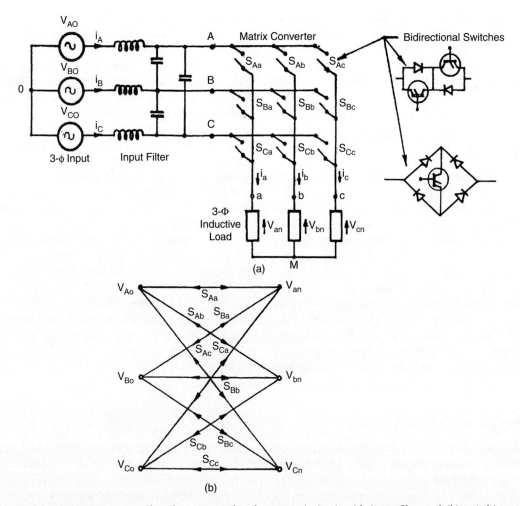

FIGURE 18.31 (a) 3φ-3φ Matrix converter (forced commutated cycloconverter) circuit with input filter and (b) switching matrix symbol for converter in (a).

restriction of the *voltage transfer ratio* (0.866), a more complex control, commutation and protection strategy, and above all the non-availability of a fully controlled bi-directional high frequency switch integrated in a silicon chip (triac, though bilateral, cannot be fully controlled).

The power circuit diagram of the most practical three-phase to three-phase (3φ–3φ) matrix converter is shown in Fig. 18.31a which uses nine bi-directional switches so arranged that any of the three input phases can be connected to any output phase as shown in the switching matrix symbol in Fig. 18.31b. Thus, the voltage at any input terminal may be made to appear at any output terminal or terminals while the current in any phase of the load may be drawn from any phase or phases of the input supply. For the switches, the inverse-parallel combination of reverse-blocking self-controlled devices like power MOSFETs or IGBTs or transistor embedded diode bridge as shown have been used so far. New perspective configuration of the bi-directional switch is to use two RB-IGBTs with reverse blocking capability in anti-parallel, eliminating the diodes reducing the conducting losses in the converter significantly. The circuit is called a *matrix converter* as it provides exactly one switch for each of the possible connections between the input and the output. The switches should be controlled in such a way that, at any time, one and only one of the three switches connected to an output phase must be closed to prevent "short circuiting" of the supply lines or interrupting the load current flow in an inductive load. With these constraints, it can be visualized that out of the possible 512 ($= 2^9$) states of the converter, only 27 switch combinations are allowed as given in Table 18.1 which includes the resulting output line voltages and input phase currents. These combinations are divided into three groups. Group-I consists of six combinations when each output phase is connected to a different input phase. In Group-II, there are three subgroups each having six combinations with two output phases short-circuited (connected to the same input phase). Group-III includes three combinations with all output phases short-circuited.

With a given set of input three-phase voltages, any desired set of three-phase output voltages can be synthesized by

TABLE 18.1 Three-phase/three-phase matrix converter switching combinations

Group	a	b	c	v_{ab}	v_{bc}	v_{ca}	i_A	i_B	i_C	S_{Aa}	S_{Ab}	S_{Ac}	S_{Ba}	S_{Bb}	S_{Bc}	S_{Ca}	S_{Cb}	S_{Cc}
	A	B	C	v_{AB}	v_{BC}	v_{CA}	i_a	i_b	i_c	1	0	0	0	1	0	0	0	1
	A	C	B	$-v_{CA}$	$-v_{BC}$	$-v_{AB}$	i_a	i_c	i_b	1	0	0	0	0	1	0	1	0
	B	A	C	$-v_{AB}$	$-v_{CA}$	$-v_{BC}$	i_b	i_a	i_c	0	1	0	1	0	0	0	0	1
I	B	C	A	v_{BC}	v_{CA}	v_{AB}	i_c	i_a	i_b	0	1	0	0	0	1	0	1	0
	C	A	B	v_{CA}	v_{AB}	v_{BC}	i_b	i_c	i_a	0	0	1	1	0	0	0	1	0
	C	B	A	$-v_{BC}$	$-v_{AB}$	$-v_{CA}$	i_c	i_b	i_a	0	0	1	0	1	0	1	0	0
	A	C	C	$-v_{CA}$	0	v_{CA}	i_a	0	$-i_a$	1	0	0	0	0	1	0	0	1
	B	C	C	v_{BC}	0	$-v_{BC}$	0	i_a	$-i_a$	0	1	0	0	0	1	0	0	1
	B	A	A	$-v_{AB}$	0	$-v_{AB}$	$-i_a$	i_a	0	0	1	0	1	0	0	1	0	0
II-A	C	A	A	v_{CA}	0	$-v_{CA}$	$-i_a$	0	i_a	0	0	1	1	0	0	1	0	0
	C	B	B	$-v_{BC}$	0	v_{BC}	0	$-i_a$	i_a	0	0	1	0	1	0	0	1	0
	A	B	B	v_{AB}	0	$-v_{AB}$	i_a	$-i_a$	0	1	0	0	0	1	0	0	1	0
	C	A	C	$-v_{CA}$	$-v_{CA}$	0	i_b	0	$-i_b$	0	0	1	1	0	0	0	0	1
	C	B	C	$-v_{BC}$	v_{BC}	0	0	i_b	$-i_b$	0	0	1	0	1	0	0	0	1
	A	B	A	v_{AB}	$-v_{AB}$	0	$-i_b$	i_b	0	1	0	0	0	1	0	1	0	0
II-B	A	C	A	$-v_{CA}$	v_{CA}	0	$-i_b$	0	i_b	1	0	0	0	0	1	1	0	0
	B	C	B	v_{BC}	$-v_{BC}$	0	0	$-i_b$	i_b	0	1	0	0	0	1	0	1	0
	B	A	B	$-v_{AB}$	v_{AB}	0	i_b	$-i_b$	0	0	1	0	1	0	0	0	1	0
	C	C	A	0	v_{CA}	$-v_{CA}$	i_c	0	$-i_c$	0	0	1	0	0	1	1	0	0
	C	C	B	0	$-v_{BC}$	v_{BC}	0	i_c	$-i_c$	0	0	1	0	0	1	0	1	0
	A	A	B	0	v_{AB}	$-v_{AB}$	$-i_c$	i_c	0	1	0	0	1	0	0	0	1	0
II-C	A	A	C	0	$-v_{CA}$	v_{CA}	$-i_c$	0	i_c	1	0	0	1	0	0	0	0	1
	B	B	C	0	v_{BC}	$-v_{BC}$	0	$-i_c$	i_c	0	1	0	0	1	0	0	0	1
	B	B	A	0	$-v_{AB}$	v_{AB}	i_c	$-i_c$	0	0	1	0	0	1	0	1	0	0
	A	A	A	0	0	0	0	0	0	1	0	0	1	0	0	1	0	0
III	B	B	B	0	0	0	0	0	0	0	1	0	0	1	0	0	1	0
	C	C	C	0	0	0	0	0	0	0	0	1	0	0	1	0	0	1

adopting a suitable switching strategy. However, it has been shown [21, 22] that regardless of the switching strategy, there are physical limits on the achievable output voltage with these converters as the maximum peak-to-peak output voltage cannot be greater than the minimum voltage difference between two phases of the input. To have complete control of the synthesized output voltage, the envelope of the three-phase reference or target voltages must be fully contained within the continuous envelope of the three-phase input voltages. Initial strategy with the output frequency voltages as references reported the limit as 0.5 of the input as shown in Fig. 18.32a. This can be increased to 0.866 by adding a third harmonic voltage of input frequency $(V_i/4) \cdot \cos 3\omega_i t$ to all target output voltages and subtracting from them a third harmonic voltage of output frequency $(V_o/6) \cdot \cos 3\omega_o t$ as shown in Fig. 18.32b [21, 22]. However, this process involves considerable amount of additional computations in synthesizing the output voltages. The other alternative is to use the space vector modulation (SVM) strategy as used in PWM inverters without adding third harmonic components but it also yields the maximum voltage transfer ratio as 0.866.

An ac input LC filter is used to eliminate the switching ripples generated in the converter and the load is assumed to be sufficiently inductive to maintain continuity of the output currents.

18.5.1 Operation and Control of the Matrix Converter

The converter in Fig. 18.31 connects any input phase (A, B, and C) to any output phase (a, b, and c) at any instant. When connected, the voltages v_{an}, v_{bn}, v_{cn} at the output terminals are related to the input voltages V_{Ao}, V_{Bo}, V_{Co} as

$$\begin{bmatrix} v_{an} \\ v_{bn} \\ v_{cn} \end{bmatrix} = \begin{bmatrix} S_{Aa} & S_{Ba} & S_{Ca} \\ S_{Ab} & S_{Bb} & S_{Cb} \\ S_{Ac} & S_{Bc} & S_{Cc} \end{bmatrix} \begin{bmatrix} v_{Ao} \\ v_{Bo} \\ v_{Co} \end{bmatrix} \quad (18.32)$$

where S_{Aa} through S_{Cc} are the switching variables of the corresponding switches shown in Fig. 18.31. For a balanced linear star-connected load at the output terminals, the input phase currents are related to the output phase currents by

$$\begin{bmatrix} i_A \\ i_B \\ i_C \end{bmatrix} = \begin{bmatrix} S_{Aa} & S_{Ab} & S_{Ac} \\ S_{Ba} & S_{Bb} & S_{Bc} \\ S_{Ca} & S_{Cb} & S_{Cc} \end{bmatrix} \begin{bmatrix} i_a \\ i_b \\ i_c \end{bmatrix} \quad (18.33)$$

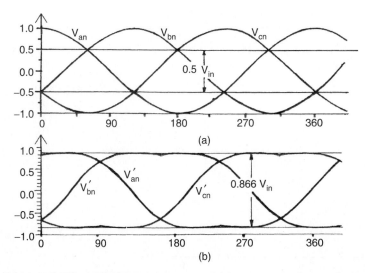

FIGURE 18.32 Output voltage limits for three-phase ac-ac matrix converter: (a) basic converter input voltages and (b) maximum attainable with inclusion of third harmonic voltages of input and output frequency to the target voltages.

Note that the matrix of the switching variables in Eq. (18.33) is a transpose of the respective matrix in Eq. (18.32). The matrix converter should be controlled using a specific and appropriately timed sequence of the values of the switching variables, which will result in balanced output voltages having the desired frequency and amplitude, while the input currents are balanced and in phase (for unity IDF) or at an arbitrary angle (for controllable IDF) with respect to the input voltages. As the matrix converter, in theory, can operate at any frequency, at the output or input, including zero, it can be employed as a three-phase ac–dc converter, dc/3-phase ac converter, or even a buck/boost dc chopper and thus as a *universal power converter*.

The control methods adopted so far for the matrix converter are quite complex and are subjects of continuing research [21–38]. Out of several methods proposed for independent control of the output voltages and input currents, two methods are of wide use and will be briefly reviewed here: (i) the *Venturini* method based on a mathematical approach of transfer function analysis and (ii) the *Space Vector Modulation* (SVM) approach (as has been standardized now in the case of PWM control of the dc link inverter).

Venturini Method: Given a set of three-phase input voltages with constant amplitude V_i and frequency $f_i = \omega_i/2\pi$, this method calculates a switching function involving the duty cycles of each of the nine bi-directional switches and generate the three-phase output voltages by sequential piecewise sampling of the input waveforms. These output voltages follow a predetermined set of reference or target voltage waveforms and with a three-phase load connected, a set of input currents I_i and angular frequency ω_i should be in phase for unity IDF or at a specific angle for controlled IDF.

A transfer function approach is employed in [29] to achieve the above mentioned features by relating the input and output voltages and the output and input currents as

$$\begin{bmatrix} V_{o1}(t) \\ V_{o2}(t) \\ V_{o3}(t) \end{bmatrix} = \begin{bmatrix} m_{11}(t) & m_{12}(t) & m_{13}(t) \\ m_{21}(t) & m_{22}(t) & m_{23}(t) \\ m_{31}(t) & m_{32}(t) & m_{33}(t) \end{bmatrix} \begin{bmatrix} V_{i1}(t) \\ V_{i2}(t) \\ V_{i3}(t) \end{bmatrix} \quad (18.34)$$

$$\begin{bmatrix} I_{i1}(t) \\ I_{i2}(t) \\ I_{i3}(t) \end{bmatrix} = \begin{bmatrix} m_{11}(t) & m_{21}(t) & m_{31}(t) \\ m_{12}(t) & m_{22}(t) & m_{32}(t) \\ m_{13}(t) & m_{23}(t) & m_{33}(t) \end{bmatrix} \begin{bmatrix} I_{o1}(t) \\ I_{o2}(t) \\ I_{o3}(t) \end{bmatrix} \quad (18.35)$$

where the elements of the modulation matrix, $m_{ij}(t)$ ($i, j = 1, 2, 3$) represent the duty cycles of a switch connecting output phase i to input phase j within a sample switching interval. The elements of $m_{ij}(t)$ are limited by the constraints

$$0 \leq m_{ij}(t) \leq 1 \quad \text{and} \quad \sum_{j=1}^{3} m_{ij}(t) = 1 \quad (i = 1, 2, 3)$$

The set of three-phase target or reference voltages to achieve the maximum voltage transfer ratio for unity IDF is

$$\begin{bmatrix} V_{o1}(t) \\ V_{o2}(t) \\ V_{o3}(t) \end{bmatrix} = V_{om} \begin{bmatrix} \cos \omega_o t \\ \cos(\omega_o t - 120°) \\ \cos(\omega_o t - 240°) \end{bmatrix} + \frac{v_{im}}{4} \begin{bmatrix} \cos(3\omega_i t) \\ \cos(3\omega_i t) \\ \cos(3\omega_i t) \end{bmatrix}$$

$$- \frac{V_{om}}{6} \begin{bmatrix} \cos(3\omega_o t) \\ \cos(3\omega_o t) \\ \cos(3\omega_o t) \end{bmatrix} \quad (18.36)$$

where V_{om} and V_{im} are the magnitudes of output and input fundamental voltages of angular frequencies ω_o and ω_i, respectively. With $V_{om} \leq 0.866\, V_{im}$, a general formula for the duty cycles $m_{ij}(t)$ is derived in [29]. For unity IDF condition, a simplified formula is

$$m_{ij} = \frac{1}{3}\left\{1 + 2q\cos(\omega_1 t - 2(j-1)60°)\right.$$

$$\left[\cos(\omega_o t - 2(i-1)60°) + \frac{1}{2\sqrt{3}}\cos(3\omega_i t) - \frac{1}{6}\cos(3\omega_o t)\right]$$

$$-\frac{2q}{3\sqrt{3}}\left[\cos(4\omega_i t - 2(j-1)60°)\right.$$

$$\left.\left. -\cos(2\omega_i t - 2(1-j)60°)\right]\right\} \qquad (18.37)$$

where $i, j = 1, 2, 3$ and $q = V_{om}/V_{im}$.

The method developed as above is based on a *Direct Transfer Function (DTF)* approach using a single modulation matrix for the matrix converter, employing the switching combinations of all the three groups in Table 18.1. Another approach called *Indirect Transfer Function (ITF)* approach [23, 24] considers the matrix converter as a combination of PWM voltage source rectifier–PWM voltage source inverter (VSR–VSI) and employs the already well established VSR and VSI PWM techniques for MC control utilizing the switching combinations of Group-II and Group-III only of Table 18.1. The drawback of this approach is that the IDF is limited to unity and the method also generates higher and fractional harmonic components in the input and the output waveforms.

SVM Method: The space vector modulation is a well documented inverter PWM control technique which yields high voltage gain and less harmonic distortion compared to the other modulation techniques. Here, the three-phase input currents and output voltages are represented as space vectors and SVM is simultaneously applied to the output voltage and input current space vectors, while the matrix converter is modeled as a rectifying and inverting stage by the *indirect modulation method* (Fig. 18.33). Applications of SVM algorithm to control of matrix converters have appeared extensively in the literature [27–37] and shown to have inherent capability to achieve full control of the instantaneous output voltage vector and the instantaneous current displacement angle even under supply voltage disturbances. The algorithm is based on the concept that the MC output line voltages for each

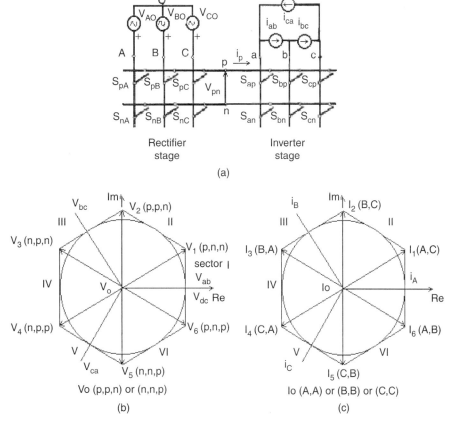

FIGURE 18.33 Indirect modulation model of a matrix converter: (a) VSR–VSI conversion; (b) output voltage switching vector hexagon; and (c) input current switching vector hexagon.

switching combination can be represented as a voltage space vector defined by

$$v_o = \frac{2}{3}\left[v_{ab} + v_{bc}\exp(j120°) + v_{ca}\exp(-j120°)\right] \quad (18.38)$$

Out of the three groups in Table 18.1, only the switching combinations of Group-II and Group-III are employed for the SVM method. Group-II consists of switching state voltage vectors having constant angular positions and are called *active* or *stationary* vectors. Each sub-group of Group-II determines the position of the resulting output voltage space vector and the six state space voltage vectors form a six-sextant hexagon used to synthesize the desired output voltage vector. Group-III comprises the *zero* vectors positioned at the center of the output voltage hexagon and these are suitably combined with the active vectors for the output voltage synthesis.

The modulation method involves selection of the vectors and their on-time computation. At each sampling period T_s, the algorithm selects four active vectors related to any possible combinations of output voltage and input current sectors in addition to the zero vector to construct a desired reference voltage. The amplitude and the phase angle of the reference voltage vector are calculated and the desired phase angle of the input current vector are determined in advance. For computation of the on-time periods of the chosen vectors, these are combined into two sets leading to two new vectors adjacent to the reference voltage vector in the sextant and having the same direction as the reference voltage vector. Applying the standard SVM theory, the general formulae derived for the vector on-times which satisfy, at the same time, the reference output voltage and input current displacement angle in [29] are

$$\begin{aligned}
t_1 &= \frac{2qT_s}{\sqrt{3}\cos\phi_i}\sin(60° - \theta_o)\cdot\sin(60° - \theta_i) \\
t_2 &= \frac{2qT_s}{\sqrt{3}\cos\phi_i}\sin(60° - \theta_o)\cdot\sin\theta_i \\
t_3 &= \frac{2qT_s}{\sqrt{3}\cos\phi_i}\sin\theta_o\cdot\sin(60° - \theta_i) \\
t_4 &= \frac{2qT_s}{\sqrt{3}\cos\phi_i}\sin\theta_o\cdot\sin\theta_i
\end{aligned} \quad (18.39)$$

where q = voltage transfer ratio, ϕ_i is the input displacement angle chosen to achieve the desired input power factor (with $\phi_i = 0$, a maximum value of $q = 0.866$ is obtained), θ_o and θ_i are the phase displacement angles of the output voltage and input current vectors, respectively, whose values are limited within the 0–60° range. The on-time of the zero vector is

$$t_o = T_s - \sum_{i=1}^{4} t_i \quad (18.40)$$

The integral value of the reference vector is calculated over one sample time interval as the sum of the products of the two adjacent vectors and their on-time ratios and the process is repeated at every sample instant.

Control Implementation and Comparison of the Two Methods: Both the methods need a digital signal processor (DSP) based system for their implementation. In one scheme [29] for the Venturini method, the programmable timers, as available, are used to time out the PWM gating signals. The processor calculates the six switch duty cycles in each sampling interval, converts them to integer counts and stores them in the memory for the next sampling period. In the SVM method, an erasable programmable read only memory (EPROM) is used to store the selected sets of active and zero vectors and the DSP calculates the on-times of the vectors. Then, with an identical procedure as in the other method, the timers are loaded with the vector on-times to generate PWM waveforms through suitable output ports. The total computation time of the DSP for the SVM method has been found to be much less that of the Venturini method. Comparison of the two schemes shows that while in the SVM method the switching losses are lower, the Venturini method shows better performance in terms of input current and output voltage harmonics.

A *direct control* method as used in conjunction with the voltage source converters has been developed recently and implemented with a 10 kVA matrix converter [38].

18.5.2 Commutation and Protection Issues in a Matrix Converter

As the matrix converter has no dc link energy storage, any disturbance in the input supply voltage will affect the output voltage immediately and a proper protection mechanism has to be incorporated, particularly against over-voltage from the supply and over-current in the load side. As mentioned, two types of bi-directional switch configurations have hitherto been used, one, the transistor (now IGBT) embedded in a diode bridge and the other, the two IGBTs in anti-parallel with reverse voltage blocking diodes. (shown in Fig. 18.31) In the latter configuration, each diode and IGBT combination operates in two quadrants only which eliminates the circulating currents otherwise built up in the diode bridge configuration that can be limited by only bulky commutation inductors in the lines.

Commutation: The MC does not contain freewheeling diodes which usually achieve safe commutation in the case of other converters. To maintain the continuity of the output current as each switch turns off, the next switch in sequence must be immediately turned on. In practice, with bi-directional switches, a momentary short circuit may develop between the input phases when the switches cross-over and one solution is

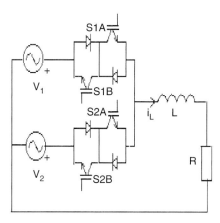

FIGURE 18.34 Safe commutation scheme.

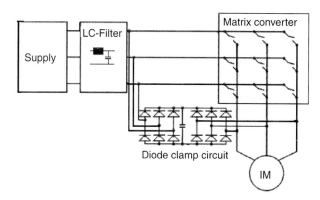

FIGURE 18.35 Diode clamp for matrix converter.

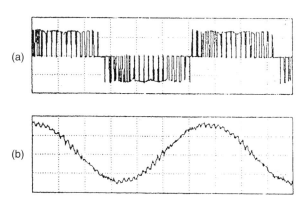

FIGURE 18.36 Experimental waveforms for a matrix converter at 30 Hz frequency from a 50 Hz input: (a) output line voltage and (b) output line current.

to use a *semi-soft current commutation* using a multi-stepped switching procedure to ensure safe commutation [39–41]. This method requires independent control of each two quadrant switches, sensing the direction of the load current and introducing a delay during the change of switching states. The switching rule for proper commutation from S1 to S2 of the arrangement shown in Fig. 18.34 for $i_L > 0$ with the two quadrant switches for *four-stepped commutation* [39, 41] is

(a) turn off S1B, (b) 2 turn on S2A, (c) turn off S1A, and (d) turn on S2B.

Analogously, for $i_L < 0$, the switching rule will be (a) turn off S1A, (b) turn on S2B, (c) turn off S1B, and (d) turn on S2A.

Typically, these commutation strategies are now implemented using programmable logic devices such as field programmable gate array (FPGA), programmable logic device (PLD) [41, 42], etc.

A robust voltage commutation scheme without sacrificing the line side current waveform quality and with minimal information requirement has been reported recently [43].

Protection strategies: A clamp capacitor (typically 2 μF for a 3 kW permanent magnet (PM) motor) connected through two three-phase full-bridge diode rectifiers involving additional 12 diodes as shown in Fig. 18.35 (a new configuration has been reported in [44] where the number of additional diodes are reduced to six using the anti-parallel switch diodes at the input and output lines of the MC) serves as a voltage clamp for possible voltage spikes under normal and fault conditions. A new passive protection strategy involving suppressor diodes and varistors for excellent over-voltage protection is discussed recently in [45] which allows the removal of the large and expensive diode clamp. A snubberless solution for over-voltage protection is presented in [46].

Input filter: A three-phase single stage LC filter at the input consisting of three capacitors in star and three inductors in the line is used to adequately attenuate the higher order harmonics and render sinusoidal input current. Typical values of L and C based on a 415 V converter with a maximum line current of 6.5 A and a switching frequency of 20 kHz are 3 mH and 1.5 μF only [47]. The filter may cause minor phase-shift in the input displacement angle which needs correction.

Figure 18.36 shows typical experimental waveforms of output line voltage and line current of an MC. The output line current is mostly sinusoidal except a small ripple, when the switching frequency is around 1 kHz only.

18.6 High Frequency Linked Single-phase to Three-phase Matrix Converters

Several kinds of single-phase to three-phase high frequency (typically, 20 kHz) cyclocoverters (actually, matrix converters with MCT switches) using "soft switching" have been reported recently for ac motor drive applications [48–50]. Figure 18.37 shows a typical configuration of such a converter where an H-bridge inverter produces a high frequency single-phase voltage which is fed to the cycloconverter through a high frequency transformer. The high frequency ac may be either sinusoidal generated by a resonant link inverter or quasi-square wave

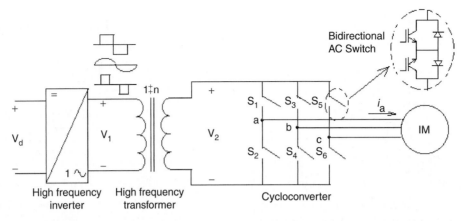

FIGURE 18.37 High frequency linked single-phase to three-phase matrix converter.

as shown. These systems have been developed at laboratories but not yet available commercially.

18.6.1 High Frequency Integral-pulse Cycloconverter [48]

The input to these converters may be sinusoidal or quasi-square wave and it is possible to use integral half-cycle pulse width modulation (IPM) principle to synthesize the output voltage waves. The advantage is that the devices can be switched at zero voltage reducing the switching loss. This converter can only work at output frequency which are multiples of the input frequency.

18.6.2 High Frequency Phase-controlled Cycloconverter [49]

Here, phase control principle as explained earlier is used to synthesize the output voltage. A sawtooth carrier wave is compared with the sine modulating wave to generate the firing instants of the switches. The phase control provides switching at zero current reducing switching loss but this scheme has more complex control circuit compared to the previous one. However, it can work at any frequency.

18.7 Applications of AC–AC Converters

18.7.1 Applications of AC Voltage Controllers

AC voltage controllers are used either for control of the rms value of voltage or current in lighting control, domestic and industrial heating, speed control of fan, pump or hoist drives, soft starting of induction motors, etc. or as *static ac switches (on/off control)* in transformer tap changing, temperature control, speed stabilization of high inertia induction motor drives like centrifuge, capacitor switching in static reactive power compensation, etc.

In *fan or pump drives* with induction motors, the torque varies as the square of the speed. So the speed control is required in a narrow range and an ac voltage controller is suitable for an induction motor with a full load slip of 0.1–0.2 in such applications. For these drives, braking or reverse operations are not needed but for the crane hoist drive, both motoring and braking are needed and a four quadrant ac voltage controller can be obtained by a modification of the ac voltage controller circuit as shown in Fig. 18.38. SCR pairs A, B, and C provide operation in quadrants I and IV and A' B, and C' in quadrants II and III. While changing from one set of SCR pairs to another, care should be taken to ensure that the incoming pair is activated only after the outgoing pair is fully turned off.

AC voltage controllers are being increasingly used for *soft-starting* of induction motors, as they have a number of advantages over the conventional starters such as smooth acceleration and deceleration, ease in implementation

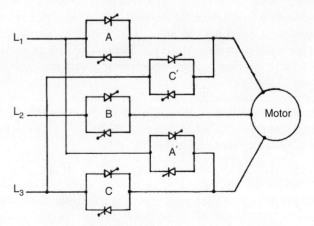

FIGURE 18.38 Four quadrant ac voltage controller.

of current control, simple protection against single-phasing or unbalanced operation, reduced maintenance and losses, absence of current inrush, etc. Even for the fixed-speed industrial applications, the voltage controllers can be used to provide a reduced stator voltage to an induction motor to improve its efficiency at light load and result in energy saving. Operation at an optimum voltage reduces the motor flux which, in turn, reduces the core-loss and the magnetizing component of the stator copper loss. Considerable savings in energy can be obtained in applications where a motor operates at no load for a significant time such as in drills, machine tools, woodworking machines, reciprocating air-compressors, etc. A popular approach to find an optimum operating voltage is to maximize the motor power factor by maintaining a minimum phase-shift between the voltage and current after sensing them.

The ac switch with on/off control used in driving a high inertia centrifuge involves switching on of the motor when the speed of the centrifuge drops below the minimum allowable level and switching off when the speed reaches the maximum allowable level – thus maintaining a constant average speed. An identical scheme of control is used with an ac switch for temperature control of an electric heater or air conditioner.

Integral cycle control is well suited to heating control while it may cause flicker in normal incandescent lighting control and speed fluctuations in motor control. However, with this control, less voltage distortion is produced in the ac supply system and less radio-frequency interference is propagated when compared with the phase-controlled system.

18.7.2 Applications of Cycloconverters

Cycloconverters as frequency changers essentially find well established applications in (i) high power low speed reversible ac motor drives with constant frequency input and (ii) constant frequency power supplies with a variable frequency input as in VSCF (variable-speed constant frequency) system, while they find potential applications in (iii) controllable VAR (volt-amperes reactive) generators for power factor correction, and (iv) ac system inter-ties linking two independent power systems as demonstrated in [15].

Variable Speed AC Drives: In this category, the applications of cycloconverter-controlled induction motor and synchronous motor drives have been adequately reviewed in [9]. Cycloconverter-fed synchronous motors are well suited for low speed drives with high torque at standstill and high capacity gear-less cement mills (tube or ball-mill above 5 MW) have been the first applications of these drives.

Since 1960s as developed by Siemens and Brown Boveri. one of the early installations employs a motor rating of 6.4 MW having a rotor diameter of 5 m in diameter and 18.5 m long while the stator construction is similar to that of a hydroelectric generator with 44 poles requiring 5.5 Hz for maximum speed of 15 rpm. The motor is flanged with the mill cylinder without additional bearings or "wrapped" directly around it, known as *ring motor* [51]. With the evolution of *field orientation* or *vector control*, cycloconverter-fed synchronous motors have replaced/are replacing the dc drives in the reversing rolling mills (2/4 MW) with extreme high dynamic requirements for torque and speed control [52–54], in mine winders and haulage [54, 55] of similar high ratings and in icebreakers and ships equipped with diesel generators with power ratings of about 20 MW per unit [56]. In these applications, the cycloconverter-fed synchronous motor is in *self-controlled* mode and known as *ac commutator-less motor* when the cycloconverter firing signals are derived from a rotor shaft position sensor, so that the frequency is slaved to the rotor speed and not vice versa with the results that the hunting and stability problems are eliminated and the torque is not limited to pull-out value. Further, with field control, the motor can be operated at leading power factor when the cycloconverter can operate with *load commutation* from the motor side at high speed in addition to the line commutation from the supply at low speed, thus providing speed control over a wide range. A cycloconverter-fed ac commutator-less motor is reported in [57] where the cycloconverter is operated both in sinusoidal and trapezoidal mode – the latter is attractive for better system power factor and higher voltage output at the cost of increased harmonic content [1]. A stator-flux oriented vector control scheme for a six-pulse circulating current-free cycloconverter-fed synchronous motor with a flux observer suitable for a rolling mill drive (300–0–300 rpm) is reported in [12]. A 12-pulse, 9.64 MVA, 120/33.1 Hz cycloconverter-linear synchronous motor combination for Maglev Vehicle ML-500, a high speed train (517 km/hr) is in the process of development in Japan since early 1980s [58]. Several recent projects involving very high (larger than 15 MW) power semi-autogenous (SAG) grinding mills with cycloconverter-fed synchronous motor drives in mining applications in Brazil are reported in [59].

Regarding cycloconverter-fed induction motors, early applications were for control of multiple run-out table motors of a hot strip mill, high performance servo drives, and controlled slip-frequency drive for diesel electric locomotives. Slip-power controlled drives in the form of static Scherbius drives with very high ratings using cycloconverter in the rotor of a doubly fed slip-ring induction motor have been in operation for high capacity pumps, compressors [60], and even in a proton–synchroton accelerator drive in CERN, Geneva [61]. Though synchronous motors have been preferred for very high capacity low speed drives because of their high rating, ability to control power factor, and precisely set speed independent of supply voltage and load variations, induction motors because of their absence of excitation control loop, simple structure, easy maintainability and quick response, have been installed for cycloconverter-fed drives in Japan. For example, a seemless tube piercing mill [62] where a squirrel cage 6-pole 3-MW,

188/300 rpm, 9.6/15.38 Hz, 2700 V motor is controlled by a cycloconverter bank of capacity 3750 KVA and output voltage 3190 volts.

Constant Frequency Power Supplies: Some applications such as aircraft and naval ships need a well regulated constant frequency power output from a variable frequency ac power source. For example, in aircraft power conversion, the alternator connected to the engine operating at a variable speed of 10,000–20,000 rpm provides a variable frequency output power over 1200–2400 Hz range which can be converted to an accurately regulated fixed frequency output power at 400 Hz through a cycloconverter with a suitable filter placed within a closed loop. The output voltages of the cycloconverter are proportional to the fixed frequency (400 Hz) sine wave reference voltage generator in the loop.

Both synchronous and induction motors can be used for VSCF generation. The static Scherbius system can be modified (known as Kramer drive) by feeding slip-power through a cycloconverter to a shaft mounted synchronous machine with a separate exciter for VSCF generation. A new application in very high power ratings of constant frequency variable speed motor generators with cycloconverter is in pumped storage schemes using reversible pump turbines for adaptation of the generated power to varying loads or keeping the ac system frequency constant. In 1993, a 400 MW variable speed pumped storage system was commissioned by Hitachi at Okhawachi hydropower plant [63] in Japan where the field windings of a 20-pole generator/motor are excited with three-phase low frequency ac current via slip-rings by a 72 MVA, 3-phase 12-pulse line-commuted cycloconverter. The armature terminals rated at 18 kV are connected to a 500 kV utility grid through a step-up transformer. The output frequency of the circulating current-free cycloconverter is controlled within ±5 Hz and the line frequency is 60 Hz. The variable speed system has a synchronous speed of 360 rpm with a speed range −330 to 390 rpm. The operational system efficiency in the pump mode is improved by 3% when compared to the earlier constant speed system.

Static VAR Generation: Cycloconverters with a high frequency (HF) base, either a HF generator or an oscillating LC tank, can be used for reactive power generation and control, replacing synchronous condensers or switched capacitors as demonstrated in [15]. If the cycloconverter is controlled to generate output voltage waves whose wanted components are in phase with the corresponding system voltages, reactive power can be supplied in either direction to the ac system by amplitude control of the cycloconverter output voltages. The cycloconverter will draw leading current from (that is, it will supply lagging current to) the ac system when its output voltage is greater than that of the system voltage and vice versa.

Inter-ties between AC Power Systems: The naturally commutated cycloconverter (NCC) was originally developed for this application to link a three-phase 50 Hz ac system with a single-phase $16\frac{2}{3}$ Hz railway supply system in Germany in the 1930s. Applications involve slip-power controlled motors with sub- and super-synchronous speeds. The stator of the motor is connected to 50 Hz supply which is connected to the rotor as well through a cycloconverter and the motor drives a single-phase synchronous generator feeding to $16\frac{2}{3}$ Hz system [64, 65]. A static asynchronous inter-tie between two different systems of different frequency can be obtained by using two NCCs in tandem, each with its input terminals connected to a common HF base. As long as the base frequency is appropriately higher than that of the either system, two system frequencies can be either same or different. The power factor at either side can be maintained at any desired level [15].

18.7.3 Applications of Matrix Converters

The practical applications of the matrix converters, as of now, are very limited. The main reasons are (i) non-availability of the bilateral fully controlled monolithic switches capable of high frequency operation; (ii) complex control law implementation; (iii) an intrinsic limitation of the output/input voltage ratio; and (iv) commutation and protection of the switches. To date, the switches are assembled from existing discrete devices resulting in increased cost and complexity and only experimental circuits of capacity upto 150 kVA [66] have been built. However, with the advances in device technology, it is hoped that the problems will be solved eventually and the MC will not only replace the NCCs in all the applications mentioned under Section 18.7.2, but will also takeover from the PWM rectifier–inverters as well.

Recently, reverse blocking IGBTs have opened up the possibility of bi-directional switch (BDS) construction for a practical matrix converter with just two back-to-back devices [67, 68]. A compact full matrix converter power module EconoMac [69] in a single package using 18 IGBT devices (35 amp, 1200 Volt) and diodes in the common collector configuration (Fig. 18.39) is available with Eupec/Siemens, Germany. This packaging minimizes the stray inductance in the current commutation paths. Fuji and Powerex have developed engineering samples for matrix converter output leg in a module using RB-IGBTs. A new power electronic building block (PEBB) configuration for a low cost MC has been proposed in [70]. Several new concepts in protection, commutation, switch design, and modulation strategy have been presented in [71].

Several new methods like overmodulation [72], adaptive rate regulation [73], two-side modulation control [74] have been introduced with experimental results to achieve *higher voltage transfer ratio*.

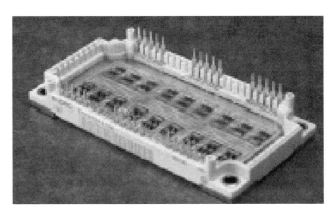

FIGURE 18.39 EconoMac matrix converter module by Eupec.

Design and loss comparison of matrix converters and dc-link voltage source converters (VSC) have been discussed in [75, 76]. It is shown that though MC requires 50% more semiconductors and gate drives excluding clamp circuit, the active silicon area and the number of gate unit power supplies are comparable to that of VSC of the same power rating. The losses of both converter systems are roughly the same at the typical range of 40–70% of rated torque and speed and a switching frequency of 10 kHz. The MC realizes a distinct better efficiency (92.5–96%) at 100% torque compared to VSC for the same IGBT modules. Also the maximum switching frequency of the MC (30 kHz at 250% rated torque) is also substantially higher. The low losses allow a reduction of current ratings (by 33%) of IGBT modules in MC.

Various potential applications of matrix converters have been proposed and experiments conducted in the field of VSCF systems such as wind-turbine and micro-turbine [77], switched mode power supplies [78], doubly fed induction motor [79, 80], and marine propulsion [81]. Few modern solutions for industrial matrix converter applications including a new integrated matrix converter motor (MCM) have been reported [82, 83]. The range of published practical implementations varies from a 2 kW matrix converter using silicon carbide devices and switching at 150 kHz for aerospace applications built at ETZ in Zurikh, Switzerland to a 150 kVA converter using 600 A IGBTs built at US army research labs in collaboration with the University of Nottingham, UK [42, 66]. A matrix converter using MCTs with enhanced commutation times is described in [84].

The complex control schemes of matrix converter demands higher test requirements and one of the modern means of testing the controllers before final integration on actual apparatus is to make hardware-in-the-loop (HIL) real time simulation tests as developed recently [85].

Matrix converter-fed adjustable speed drives (ASDs) have the advantages of inherent four-quadrant operation, absence of bulky dc-link electrolytic capacitors, clean input power characteristics with high input power factor and increased power density with the possibilities of operating at higher temperatures. However, due to the absence of the dc link, these are more susceptible to input power disturbances and a *ride-through module* is needed to be added for these drives under short-term power interruption. Such a module as developed with minimal addition of hardware and software into a matrix converter (230 V, 3 kVA) has been reported in [86].

References

1. C. Rombaut, G. Seguier, and R. Bausiere, *Power Electronic Converters—AC/AC Conversion*, McGraw-Hill, New York, 1987.
2. W. Shepherd, *Thyristor Control of AC Circuits*, Granada Publishing, Great Britain, 1975.
3. A. M. Trzynadlowski, *Introduction to Modern Power Electronics*, John Wiley, New York, 1998.
4. M. Rashid, *Power Electronics: Circuits, Devices and Applications*, Prentice Hall, N.J., 1993.
5. D. W. Hart, *Introduction to Power Electronics*, Prentice Hall, N.J. 1997.
6. A. K. Chattopadhyay, S. P. Das, and S. Karchowdhury, "Computer simulation of three-phase thyristor regulator-fed static and induction motor loads—A generalised approach", *Electric Machines and Power Systems*, vol.21, 1993, pp. 329–342.
7. S. B. Dewan and A. Straughen, *Power Semiconductor Circuits*, John Wiley, New York, 1975.
8. B. R. Pelly, *Thyristor Phase-Controlled Converters and Cycloconverters*, John Wiley, New York, 1971.
9. A. K. Chattopadhyay, "Cycloconverters and cycloconverter-fed drives—A Review", J. *Indian Inst. Sci.*, September/October 1997, vol.77, pp. 397–419.
10. A. K. Chattopadhyay and T. J. Rao, "Simplified control electronics for a practical cycloconverter", *Int. J. Electronics*, vol.43, 1977, pp. 137–150.
11. B. M. Bird and J. S. Ford, "Improvements in phase-controlled circulating current cycloconverters using communication principles", *Proc. IEE*, vol.121, no.10, 1974, pp. 1146–1149.
12. S. P. Das and A. K. Chattopadhyay, "Observer based stator flux oriented vector control of cycloconverter-fed synchronous motor drive", *IEEE Trans on Ind. Appl.*, vol.IA-33, no.4, July/August 1997, pp. 943–955.
13. P. Syam, P. K. Nandi, and A. K. Chattopadhyay, "An improvement feedback technique to suppress sub-harmonics in a naturally commutated cycloconverter", *IEEE Trans. Ind. Electron.*, vol.45, no.6, December 1998, pp. 950–952.
14. P. Syam, P. K. Nandi, and A. K. Chattopadhyay, "Improvement in power quality and a simple method of subharmonic suppression for a cycloconverter-fed synchronous motor drive", *IEE Proc. Elec. Power Appl.*, vol.149, no.4, July 2002.
15. L. Gyugi and B. R. Pelly, *Static Power Frequency Changers: Theory, Design and Applications*, Wiley-Interscience, New York, 1976.
16. A. K. Chattopadhyay and T. J. Rao, "Methods of digital computation of phase controlled cycloconverter performance", in *Conf. Rec. IEEE-IAS Ann. Meet.*, 1978, pp. 432–439.
17. Y. Lin, G. T. Heydt, and R. F. Chu, "The power quality impact of cycloconverter control strategies", *IEEE Trans. on Power Delivery*, vol.2, no.2, April 2005, pp. 1711–1718.

18. M. Loskarn, K. D. Tost, C. Unger, and R. Witzmann, "Mitigation of Interharmonics due to large cycloconverter-fed mill drives", *Proc. ICHQP'98*, Athens, Greece, October 1998, pp. 122–126.
19. M. Venturini, "A new sine-wave in sine-wave out converter technique eliminates reactor elements", *Proc. POWERCON'80*, vol.E3-1, 1980, pp. E3-1–E3-15.
20. A. Alesina and M. Venturini, "The generalised transformer: A new bidirectional waveform frequency converter with continuously adjustable input power factor", in *Proc. IEEE – PESC'80*, 1980 pp. 242–252.
21. ——, "Analysis and design of optimum amplitude nine-switch direct ac-ac converters", *IEEE Trans. Power Electron.*, vol.4, no.1, January 1989, pp. 101–112.
22. D. G. Holmes and T. A. Lipo, "Implementation of a controlled rectifier using ac-ac matrix converter theory", *IEEE Trans.on Power Electronics*, vol.7, no.1. January 1992, pp. 240–250.
23. P. D. Ziogas, S. I. Khan, and M. Rashid, "Some improved forced commutated cycloconverter structures", *IEEE Trans. Ind. Appl.* vol.Ia-21, July/August 1985, pp. 1242–1253.
24. ——, "Analysis and design of forced commutated cycloconverter structures and improved transfer characteristics", *IEEE Trans. Ind. Electron.* vol.IE-33, no.3, August 1986, pp. 271–280.
25. C. L. Neft and C. D. Schauder, "Theory and design of a 30 HP matrix converter", in *Conf. Rec IEEE-IAS Ann. Meet*, 1988, pp. 248–253.
26. A. Ishiguru, T. Furuhashi, and S. Okuma, "A novel control method of forced commutated cycloconverters using instantaneous values of input line voltages", *IEEE Trans. on Ind. Electron*, vol.38, no.3, June 1991, pp. 166–172.
27. L. Huber, D. Borojevic, and N. Burany, "Analysis, design and implementation of the space-vector modulator for forced commutated cycloconverters", *IEE Proc. B*, vol.139, no.2, March 1992, pp. 103–113.
28. L. Huber and D. Borojevic, "Space vector modulated three-phase to three-phase matrix converter with input power factor correction", *IEEE Trans. on Ind. Appl.*, vol.31, November/December 1995, pp. 1234–1246.
29. L. Zhang, C. Watthanasarn, and W. Shepherd, "Analysis and comparison of control strategies for ac-ac matrix converters", *IEE Proc. Power Elec. Appl.*, vol.145, no.4, July 1998, pp. 284–294.
30. D. Casadei, G. Serra, and A. Tani, "Reduction of the input current harmonic content in matrix converters under input/output unbalance", *IEEE Trans on Ind. Appl.*, vol.45, no.3, June 1998, pp. 401–411.
31. L. Zhang and C. Watthanasarn, "An efficient space vector modulation algorithm for ac-ac matrix converters: Analysis and implementation", *IEE Conf Proc. PEVD'96*, pp.108–111.
32. P. W. Wheeler, J. Rodriguez, J. C. Clare, L. Emperngen, and A. Wenstein, "Matrix converters: A technology review", *IEEE Trans. on Industrial Electronics*, vol.49, no.2, April 2002, pp. 276–288.
33. D. Casadei, G. Serra, A. Tani, and L. Zarri, "Matrix converter modulation strategies: A new general approach based on space vector representation of the switch state", *IEEE Trans on Ind. Electron.*, vol.49, no.2, April 2002.
34. H. J. Cha and P. N. Enjeti, "An approach to reduce common mode voltage in matrix converter", *IEEE Trans. on Ind. Appl.*, vol.39, July/August 2003, pp. 1151–1159.
35. M. Jussila, M. Salo, and H. Tuusa, "Realization of a three-phase indirect matrix converter with an indirect vector modulation method", *IEEE –PESC' 2003*, vol.2, pp. 689–694.
36. K. Sun, L. Huang, K. Matsuse, and T. Ishida, "Combined control of matrix converter fed induction motor drive system", *Proc. IEEE-IAS Ann. Meet*, 2003, vol.3, pp. 1723–1729.
37. R. Havrila, B. Dobrucky, and P. Balazovic, "Space vector modulated three-phase to three-phase matrix converter with unity power factor", *EPE-PEMC2000 Conf. Rec.*, Kosice, vol.2, pp. 103–108.
38. P. Mutschler and M. Marcks, "A direct control method for matrix converters", *IEEE Trans. on Ind. Electron.*, vol.49, no.2, April 2002, pp. 362–369.
39. N. Burany, "Safe control of four-quadrant switches," in *Conf. Rec IEEE-IAS Ann Meet*, 1989, pp. 1190–1194.
40. B. H. Kwon, B. D. Min, and J. H. Kim, "Novel commutatation techniques of ac-ac converters", *IEE Proc. Elec. Power Appl.*, vol.145, no.4, July 1998, pp. 295–300.
41. Y. Sun, F. Xu, and K. Sun, "Design of matrix converter with bidirectional switches", *Proc. IEEE-POWERCON'2002*, October 2002, vol.2, pp. 1034–1038.
42. J. Clare and P. Wheeler, "New technology: Matrix converters", *Industrial Electronics Society Newsletter*, vol.52, no.1, March 2005, pp. 10–12.
43. L. Wei, T. A. Lipo, and H. Chen, "Robust voltage commutation of conventional matrix converter", *Proc. IEEE-PESC'03*, vol.2, pp. 717–722.
44. P. Nielson, F. Blabjerg, and J. K. Pederson, "New protection issues of a matrix converter: Design considerations for adjustable-speed drives", *IEEE Trans. on Ind. Appl.*, vol.35, no.5, September/October 1999, pp. 1150–1161.
45. J. Mahlein, M. Bruckmann, and M. Braun, "Passive protection strategy for a drive system with a matrix converter and an induction machine", *IEEE Trans. on Ind. Electronics*, vol.49, no.2, April 2002, pp. 297–303.
46. O. Simon and M. Bruckmann, "Control and protection strategies for matrix converters", *Proc. PCIM Conf. Rec*, Nuremberg, Germany, 2000.
47. P. Wheeler and D. Grant, "Opimised input filter design and low-loss switching techniques for a practical matrix converter", *IEE Proc. Elec. Power Appl.*, vol.144, no.1, January 1997, pp. 53–59.
48. L. Hui, B. Ozpineci, and B. K. Bose, "A soft-switched high frequency non-resonant link integral pulse modulated dc-ac converter for ac motor drive", *Proc. IEEE-IECON*, Aachen, Germany, 1998, vol.2, pp. 726–732.
49. B. Ozpineei and B. K. Bose, "Soft-switched performance based high frequency non-resonant link phase-controlled converter for ac motor drive", *Proc. IEEE-IECON*, Aachen, Germany, 1998, vol.2, pp. 733–739.
50. B. K. Bose, "Modern power electronics and ac drives", *Pearson Education*, New Delhi, 2002.
51. E. Blauenstein, "The first gearless drive for a tube mill", *Brown Boveri Review*, vol.57, 1970, pp. 96–105.
52. W. Timpe, "Cycloconverter for rolling mill drives", *IEEE Trans. on Ind. Appl.*, vol.IA-18, no.4, 1982, pp. 401–404.
53. R. P. Pallman, "First use of cycloconverter-fed ac motors in an aluminium hot strip mill", *Siemens Engineering & Automation*, vol.XIV, no.2, 1992, pp. 26–29.
54. E. A. Lewis, "Cycloconverter drive systems", *IEE Conf. Proc. PEVD'96*, 1996, pp. 382–389.
55. K. Madisetti and M. A. Ramlu, "Trends in electronic control of mine hoists", *IEEE Trans.on Ind. Appl.*, vol.IA-22, no.6, 1986, pp. 1105–1112.

56. A. W. Hill, R. A. V. Turton, R. L. Dunzan, and C. L. Schwaln, "A vector controlled cycloconverter drive for an ice-breaker", *IEEE Trans. on Ind. Appl.*, vol.IA-23, November/December 1987, pp. 1036–1041.
57. S. P. Das and A. K. Chattopadhyay, "Comparison of simulation and test results for an ac commutatorless motor drive", *IEE Conf. Proc. PEVD'96*, 1996, pp. 294–299.
58. T. Saijo, S. Koike, and S. Takaduma. "Characteristics of linear synchronous motor drive cycloconverter for Maglev Vehicle ML-500 at Miyazaki test track", *IEEE Trans. on Ind. Appl.*, vol.IA-17, 1981, pp. 533–543.
59. J. R. Rodriguez, J. Pontt, P. Newman, R. Musalem, H. Miranda, L. Moran, and G. Azamora, "Technical evaluation and practical experience of high power grinding mill drives in mine applications", *IEEE Trans.on Ind. Appl.*, vol.41, no.3, May/June 2005, pp. 866–873.
60. H. W. Weiss, "Adjustable speed ac drive systems for pump and compressor applications", *IEEE Trans. on Ind. Appl.*, vol.IA-10, no.1, 1975, pp. 162–167.
61. R. Dirr, I. Neuffer, W. Schluter, and H. Waldmann, "New electronic control equipment for doubly-fed induction motors of high rating", *Siemens Review*, vol.3, 1972, pp. 121–126.
62. K. Sugi, Y. Naito, P. Kurosawa, Y. Kano, S. Katayama, and T. Yashida, "A microcomputer based high capacity cycloconverter drive for main rolling mill", *Proc. IPEC*, Tokyo, 1983, pp. 744–755.
63. S. Mori, E. Kita, H. Kojima, T. Sanematru, A. Shiboya, and A. Bando, "Commissioning of 400 MW adjustable speed pumped storage system at Okhawachi hydropower plant", in *Proc.1995 CIGRE Symp.*, 1995, no.520-04.
64. H. Stemmler, "High power industrial drives", *Proc. IEEE*, vol.82, no.8, August 1994, pp. 1266–1286
65. ———, "Active and reactive load control for converters interconnecting 50 and 16 2/3 Hz, using a static frequency changer cascade", *Brown Boveri Review*, vol.65, no.9, 1978, pp. 614–618.
66. T. F. Podlesak, D. C. Katsis, P. W. Wheeler, J. C. Clare, L. Empringham, and M. Bland, "A 150 -kVA vector controlled matrix converter induction motor drive", *IEEE Trans. on Ind. Appl.*, vol.41, no.3, May/June 2005, pp. 841–847.
67. C. Klumpner and F. Blaabjerg, "Using reverse blocking IGBTs in power converters for adjustable speed drives", *IEEE-IAS Conf. Rec 2003*, pp. 1516–1523.
68. J. Itoh, I. Sato, A. Odaka, H. Ohguchi, H. Kodatchi, and N. Eguchi, "A novel approach to practical matrix converter motor drive system with reverse blocking IGBT", *IEEE-PESC 2004*, Aachen, Germany, June 2004, vol.3, pp. 2380–2385.
69. M. Munzer, "Economac-The first all in one IGBT module for matrix converters", in *Proc. Drives and Control Conf.*, Sec.3, London, 2001, CD-ROM.
70. C. Klumpner, P. Nielson, I. Boldea, and F. Blaabjerg, "New solutions for a low cost power electronic building block for matrix converters", *IEEE Trans. on Ind. Electronics*, vol.49, no.2, April 2002, pp. 336–334.
71. J. Mahlein, J. Weigold, and O. Simon, "New concepts for matrix converter design", *Proc. IEEE-Int. Conf. on Ind.Electron., Control and Instrumentation*, Denver, 2001, vol.2, pp. 1044–1048.
72. J. Mahlein, O. Simon, and M. Braun, "A matrix converter with space vector control enabling overmodulation", *Conf. Proc. EPE'99*, Lussanne, September 1999, pp. 1–11.
73. L. Wang, K. Sun, and L. Huang, "A novel method to enhance the voltage transfer ratio of matrix converter", *Proc. 30th Ann. Meeting of IES*, November 2–6, 2004, Busan, Korea, pp. 723–727.
74. J. Chang, T. Sun, and A. Wang, "Highly compact ac-ac converter achieving a high voltage transfer ratio", *IEEE Trans. on Ind. Electronics*, vol.49, no.2, April 2002, pp. 345–352.
75. S. Bernett, S. Ponnaluri, and R. Teichmann, "Design and loss comparison of matrix converters and voltage-source converters for modern ac drives", *IEEE Trans. on Ind. Electronics*, vol.49, no.2, April 2002, pp. 304–314.
76. T. Matsuo, S. Bernet, and T. A. Lipo, "Application of matrix converter to field oriented induction motor drives", *WISPERC Report*, University of Wisconsin, January, 1996.
77. H. Cha and P. N. Enjeti, "A three-phase ac/ac high frequency link matrix converter for VSCF applications", *IEEE-PESC 2003*, vol.4, pp. 1971–1976.
78. S. Ratanapanaachote, H. J. Cha, and P. N. Enjeti, "A digitally controlled switch mode power supply", *IEEE-PESC 2004*, Aachen, Germany, pp. 2237–2243.
79. E. Chekhet, V. Sobolev, and I. Shapoval, "The steady state analysis of a doubly–fed induction motor (DIFM) with matrix converter", *Proc. EPE-PEMC 2000*, Kosice, vol.5, pp. 6–11.
80. A. K. Dalal, P. Syam, and A. K. Chattopadhyay, "Use of matrix converter as slip-power regulator in a doubly-fed induction motor drive for improvement of power quality", *Proc. IEEE Power India Conf. 2006*, 10–12 April, 2006, page(s) 6 pp, www.ieeexplore.ieee.org
81. K. M. Ciaramella and R. WG. Bucknall, "Potential use of matrix converters in marine propulsion applications", *IMAREST* 2003, Edinburgh, UK.
82. C. Klumpner, P. Nielson, I. Boldea, and F. Blaabjerg, "A new matrix converter motor (MCM) for industry applications, *IEEE Trans. on Ind. Electron.*, vol.49, no.2, April 2002, pp. 325–335.
83. O. Simon, J. Mahlein, M. N. Muenzer, and M. Bruckmann, "Modern solutions for industrial matrix-converter applications, *IEEE Trans. on Ind. Electron.*, vol.49, no.2, April 2002, pp. 401–406.
84. P. Wheeler, J. C. Chan, L. Empringham, M. Bland, and K. C. Kerris, "A vector controlled MCT matrix converter induction motor with minimized commutation times and enhanced waveform quality", *IEEE-Ind. Appl. Magazine*, vol.10, no.1, January/February 2004, pp. 59–65.
85. C. Dufour, L. Wei, and T. A. Lipo, "Real -time simulation of matrix converter drives", *WEMPEC Research Report*, University of Wisconsin, 2005-14.
86. H. J. Cha and P. N. Enjeti, "Matrix converter-fed ASDs", *IEEE Ind. Appl. Magazine*, vol.10, no.4, July/August 2004, pp. 33–39.

19
Power Factor Correction Circuits

Issa Batarseh, Ph.D. and Huai Wei, Ph.D.
School of Electrical Engineering and Computer Science, University of Central Florida, 4000 Central Florida Blvd., Orlando, Florida, USA

19.1 Introduction 517
19.2 Definition of PF and THD... 518
19.3 Power Factor Correction 520
   19.3.1 Energy Balance in PFC Circuits • 19.3.2 Passive Power Factor Corrector • 19.3.3 Basic Circuit Topologies of Active Power Factor Correctors • 19.3.4 System Configurations of PFC Power Supply
19.4 CCM Shaping Technique 525
   19.4.1 Current Mode Control • 19.4.2 Voltage Mode Control
19.5 DCM Input Technique 530
   19.5.1 Power Factor Correction Capabilities of the Basic Converter Topologies in DCM • 19.5.2 AC–DC Power Supply with DCM Input Technique • 19.5.3 Other PFC Techniques
19.6 Summary 538
   Further Reading 539

19.1 Introduction

Today, our society has become very aware of the necessity of the natural environment protection of our living plant in the face of a programmed utilization of natural resource. Like the earth, the utility power supply that we are now using was clean when it was invented in nineteenth century. Over hundred years, electrical power system has benefited people in every aspect. Meanwhile, due to the intensive use of this utility, the power supply condition becomes "polluted." However, public concern about the "dirty" environment in the power system has not been drawn until the mid 1980s [1–6].

Since ac electrical energy is the most convenient form of energy to be generated, transmitted, and distributed, ac power systems had been swiftly introduced into industries and residences since the turn of the twentieth century. With the proliferation of utilization of electrical energy, more and more heavy loads have been connected into the power system. During 1960s, large electricity consumers such as electrochemical and electrometallurgical industries applied capacitors as VAR compensator to their systems to minimize the demanded charges from utility companies and to stabilize the supply voltages. As these capacitors present low impedance in the system, harmonic currents are drawn from the line. Owing to the non-zero system impedance, line voltage distortion will be incurred and propagated. The contaminative harmonics can decline power quality and affect the system performance in several ways:

(a) The line *rms* current harmonics do not deliver any real power in Watts to the load, resulting in inefficient use of equipment capacity (i.e. low power factor).
(b) Harmonics will increase conductor loss and iron loss in transformers, decreasing transmission efficiency and causing thermal problems.
(c) The odd harmonics are extremely harmful to a three-phase system, causing overload of the unprotected neutral conductor.
(d) Oscillation in power system should be absolutely prevented in order to avoid endangering the stability of system operation.
(e) High peak harmonic currents may cause automatic relay protection devices to mistrigger.
(f) Harmonics could cause other problems such as electromagnetic interference to interrupt communication, degrading reliability of electrical equipment, increasing product defective ratio, insulation failure, audible noise, etc.

Perhaps the greatest impact of harmonic pollution appeared in early 1970s when static VAR compensators (SVCs) were extensively used for electric arc furnaces, metal rolling mills, and other high power appliances. The harmonic currents produced by partial conduction of SVC are odd order, which are especially harmful to three-phase power system. Harmonics can affect operations of other devices that are connected to the same system and, in some situation, the operations of themselves that generate the harmonics.

The ever deteriorated supply environment did not become a major concern until the early 1980s when the first technical standard IEEE519-1981 with respect to harmonic control at point of common coupling (PCC) was issued [7]. The significance of issuing this standard was not only that it provided the technical reference for design engineers and manufactures, but also that it opened the door of research area of harmonic reduction and power factor correction (PFC). Stimulated by the harmonic control regulation, researchers and industry users started to develop low-cost devices and power electronic systems to **reduce** harmonics since it is neither economical nor necessary to eliminate the harmonics.

Research on harmonic reduction and PFC has become intensified in the early nineties. With the rapid development in power semiconductor devices, power electronic systems have matured and expanded to new and wide application range from residential, commercial, aerospace to military and others. Power electronic interfaces, such as switch mode power supplies (SMPS), are now clearly superior over the traditional linear power supplies, resulting in more and more interfaces switched into power systems. While the SMPSs are highly efficient, but because of their non-linear behavior, they draw distorted current from the line, resulting in high total harmonic distortion (THD) and low power factor (PF). To achieve a smaller output voltage ripple, practical SMPSs use a large electrolytic capacitor in the output side of the single-phase rectifier. Since the rectifier diodes conduct only when the line voltage is higher than the capacitor voltage, the power supply draws high *rms* pulsating line current. As a result, high THD and poor PF (usually less than 0.67) are present in such power systems [6–10]. Even though each device, individually, does not present much serious problem with the harmonic current, utility power supply condition could be deteriorated by the massive use of such systems. In recent years, declining power quality has become an important issue and continues to be recognized by government regulatory agencies. With the introduction of compulsory and more stringent technical standard such as IEC1000-3-2, more and more researchers from both industries and universities are focusing in the area of harmonic reduction and PFC, resulting in numerous circuit topologies and control strategies. Generally, the solution for harmonic reduction and PFC are classified into passive approach and active approach. The passive approach offers the advantages of high reliability, high power handling capability, and easy to design and maintain. However, the operation of passive compensation system is strongly dependent on the power system and does not achieve high PF. While the passive approach can be still the best choice in many high power applications, the active approach dominates the low to medium power applications due to their extraordinary performance (PF and efficiency approach to 100%), regulation capabilities, and high density. With the power handling capability of power semiconductor devices being extended to megawatts, the active power electronic systems tend to replace most of the passive power processing devices [2–4].

Today's harmonic reduction and PFC techniques to improve distortion are still under development. Power supply industries are undergoing the change of adopting more and more PFC techniques in all off-line power supplies. This chapter presents an overview of various active harmonic reduction and PFC techniques in the open literature. The primary objective of writing this chapter is to give a brief introduction of these techniques and provide references for future researchers in this area. The discussion here includes definition of THD and PF, commonly used control strategies, and various types of converter topologies. Finally, the possible future research trends are stressed in the Summary Section.

19.2 Definition of PF and THD

Power factor is a very important parameter in power electronics because it gives a measure of how effective the real power utilization in the system is. It also represents a measure of distortion of the line voltage and the line current and phase shift between them. Referring to Fig. 19.1a, we define the input power factor (PF) at terminal a-a′ as the ratio of the average power and the apparent power measured at terminals a-a′ as described in Eq. (19.1) [7, 9, 10]

$$\text{Power Factor (PF)} = \frac{\text{Real Power (Average)}}{\text{Apparent Power}} \quad (19.1)$$

where, the apparent power is defined as the product of *rms* values of $v_s(t)$ and $i_s(t)$.

In a linear system, because load draws purely sinusoidal current and voltage, the PF is only determined by the phase difference between $v_s(t)$ and $i_s(t)$. Equation (19.1) becomes

$$PF = \frac{I_{s,rms} V_{s,rms} \cos\theta}{I_{s,rms} V_{s,rms}} = \cos\theta \quad (19.2)$$

where, $I_{s,rms}$ and $V_{s,rms}$ are *rms* values of line current and line voltage, respectively, and θ is the phase shift between line current and line voltage. Hence, in linear power systems, the PF is simply equal to the cosine of the phase angle between the current and voltage. However, in power electronic system, due to the non-linear behavior of active switching power devices, the phase-angle representation alone is not valid. Figure 19.1b shows that the non-linear load draws typical distorted line current from the line. Calculating PF for distorted waveforms is more complex when compared with the sinusoidal case. If both line voltage and line current are distorted, then

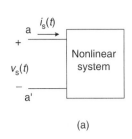
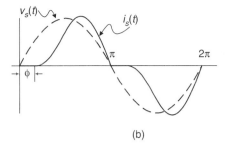

FIGURE 19.1 Non-linear load draws distorted line current.

Eqs. (19.3) and (19.4) give the Fourier expansion representations for the line current and line voltage, respectively

$$i_s(t) = I_{DC} + \sum_{n=1}^{\infty} I_{sn}\sin(n\omega t + \theta_{in})$$

$$= I_{DC} + I_{s1}\sin(\omega t + \theta_{i1}) + \sum_{n=2}^{\infty} I_{sn}\sin(n\omega t + \theta_{in}) \quad (19.3)$$

$$v_s(t) = V_{DC} + \sum_{n=1}^{\infty} V_{sn}\sin(n\omega t + \theta_{vn})$$

$$= V_{DC} + V_{s1}\sin(\omega t + \theta_{v1}) + \sum_{n=2}^{\infty} V_{sn}\sin(n\omega t + \theta_{vn}) \quad (19.4)$$

Applying the definition of PF given in Eq. (19.1) to the distorted current and voltage waveforms of Eqs. (19.3) and (19.4), PF may be expressed as

$$PF = \frac{\sum_{n=1}^{\infty} I_{sn,rms} V_{sn,rms} \cos\theta_n}{I_{s,rms} V_{s,rms}} = \frac{\sum_{n=1}^{\infty} I_{sn,rms} V_{sn,rms} \cos\theta_n}{\sqrt{\sum_{n=1}^{\infty} I_{sn,rms}^2 \sum_{n=1}^{\infty} V_{sn,rms}^2}} \quad (19.5)$$

where, $V_{sn,rms}$ and $I_{sn,rms}$ are the rms values of the nth harmonic voltage and current, respectively, and θ_n is the phase shift between the nth harmonic voltage and current.

Since most of power electronic systems draw their input voltage from a stable line voltage source $v_s(t)$, the above expression can be significantly simplified by assuming the line voltage is pure sinusoidal and distortion is only limited to $i_s(t)$, i.e.

$$v_s(t) = V_s \sin\omega t \quad (19.6)$$

$$i_s(t) = \text{distorted (non-sinusoidal)} \quad (19.7)$$

Then it can be shown that the PF can be expressed as

$$PF = \frac{I_{s1,rms}}{I_{s,rms}} \cos\theta_1 = k_{dist} \cdot k_{disp} \quad (19.8)$$

where,

θ_1: the phase angle between the voltage $v_s(t)$ and the fundamental component of $i_s(t)$;

$I_{s1,rms}$: rms value of the fundamental component in line current;

$I_{s,rms}$: total rms value of line current;

$k_{dist} = I_{s1,rms}/I_{s,rms}$: distortion factor;

$k_{disp} = \cos\theta_1$: displacement factor.

Another important parameter that measure the percentage of distortion is known as the current total harmonic distortion (THD_i) which is defined as follows

$$THD_i = \sqrt{\frac{\sum_{n=2}^{\infty} I_{sn,rms}^2}{I_{s1,rms}^2}} = \sqrt{\frac{1}{k_{dist}^2} - 1} \quad (19.9)$$

Conventionally SMPSs use capacitive rectifiers in front of the ac line which resulting in the capacitor voltage v_c and high rms pulsating line current $i_l(t)$ as shown in Fig. 19.2, when $v_l(t)$ is the line voltage. As a result, THD_i is as high as 70% and poor PF is usually less than 0.67.

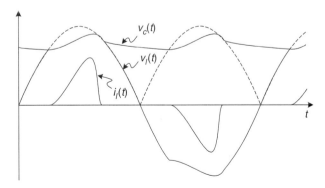

FIGURE 19.2 Typical waveforms in a poor PF system.

As we can see from Eqs. (19.8) and (19.9), PF and THD are related to distortion and displacement factors. Therefore, improvement in PF, i.e. power factor correction (PFC), also implies harmonic reduction.

19.3 Power Factor Correction

19.3.1 Energy Balance in PFC Circuits

Figure 19.3 shows a diagram of an ac–dc PFC unit. Let $v_l(t)$ and $i_l(t)$ be the line voltage and line current, respectively. For an ideal PFC unit (PF = 1), we assume

$$v_l(t) = V_{lm} \sin \omega_l t \quad (19.10\text{a})$$

$$i_l(t) = I_{lm} \sin \omega_l t \quad (19.10\text{b})$$

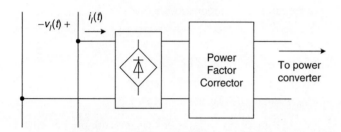

FIGURE 19.3 Block diagram of ac–dc PFC unit.

where, V_{lm} and I_{lm} are amplitudes of line voltage and line current, respectively, and ω_l is the angular line frequency. The instantaneous input power is given by

$$p_{in}(t) = V_{lm} I_{lm} \sin^2 \omega_l t = P_{in}(1 - \cos 2\omega_l t) \quad (19.11)$$

where, $P_{in} = 1/2 V_{lm} I_{lm}$ is the average input power.

As we can see from Eq. (19.11), the instantaneous input power contains not only the real power (average power) P_{in} component but also an alternative component with frequency $2\omega_l$ (i.e. 100 or 120 Hz), shown in Fig. 19.4. Therefore, the operation principle of a PFC circuit is to process the input power in a certain way that it stores the excessive input energy (area I in Fig. 19.4) when $p_{in}(t)$ is larger than $P_{in}(=P_o)$, and releases the stored energy when $p_{in}(t)$ is less than $P_{in}(=P_o)$ to compensate for area II.

The instantaneous excessive input energy, $w(t)$, is given by

$$w_{ex}(t) = \frac{P_o}{2\omega_l}(1 - \sin 2\omega_l t) \quad (19.12)$$

At $t = 3T_l/8$, the excessive input energy reaches the peak value

$$w_{ex,max} = \frac{P_o}{\omega_l} \quad (19.13)$$

The excessive input energy has to be stored in the dynamic components (inductor and capacitor) in the PFC circuit.

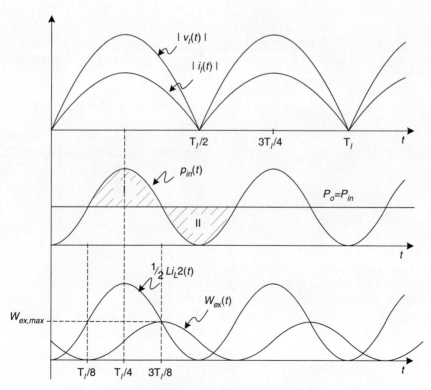

FIGURE 19.4 Energy balance in PF corrector.

19 Power Factor Correction Circuits

In most of the PFC circuits, an input inductor is used to carry the line current. For unity PF, the inductor current (or averaged inductor current in switch mode PFC circuit) must be a pure sinusoidal and in phase with the line voltage. The energy stored in the inductor ($1/2 L i_L^2(t)$) cannot completely match the change of the excessive energy as shown in Fig. 19.4. Therefore, to maintain the output power constant, another energy storage component (usually the output capacitor) is needed.

19.3.2 Passive Power Factor Corrector

Because of their high reliability and high power handling capability, passive power factor correctors are normally used in high power line applications. Series tuned LC harmonic filter is commonly used for heavy plant loads such as arc furnaces, metal rolling mills, electrical locomotives, etc. Figure 19.5 shows a connection diagram of harmonic filter together with line frequency switched reactor static VAR compensator. By tuning the filter branches to odd harmonic frequencies, the filter shunts the harmonic currents. Since each branch presents capacitive at line frequency, the filter also provides capacitive VAR for the system. The thyristor-controlled reactor keeps an optimized VAR compensation for the system so that higher PF can be maintained.

The design of the tuned filter PF corrector is particularly difficult because of the uncertainty of the system impedance and harmonic sources. Besides, this method involves too many expensive components and takes huge space.

For the applications where power level is less than 10 kW, the tuned filter PF corrector may not be a better choice. The most common off-line passive PF corrector is the inductive-input filter, shown in Fig. 19.6. Depending on the filter inductance, this circuit can give a maximum of 90% PF. For operation in continuous conduction mode (CCM), the PF is defined as [11]

$$PF = \frac{0.9}{\sqrt{1 + (0.075/K_1)^2}} \quad (19.14)$$

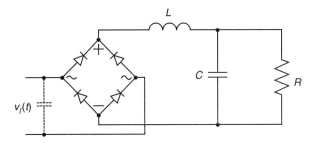

FIGURE 19.6 Inductive-input PF corrector.

where

$$K_1 = \frac{\omega_l L}{\pi R} \quad (19.15)$$

The PF corrector is simply a low pass inductive filter as shown in Fig. 19.7, whose transfer function and input impedance are given by

$$H(s) = \frac{1}{s^2 LC + sL/R + 1} \quad (19.16)$$

$$Z_{in}(s) = R \frac{s^2 LC + sL/R + 1}{sRC + 1} \quad (19.17)$$

The above equations show that the unavoidable phase displacement is incurred in the inductive-filter corrector. Because the filter frequency of operation is low (line frequency), large value inductor and capacitor have to be used. As a result, the following disadvantages are presented in most passive PF correctors:

(a) Only less than 0.9 PF can be achieved;
(b) THD is high;
(c) They are heavy and bulky;
(d) The output is unregulated;
(e) The dynamic response is poor;
(f) They are sensitive to circuit parameters;
(g) Optimization of the design is difficult.

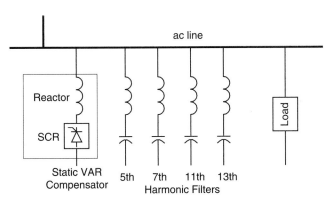

FIGURE 19.5 Series tuned LC harmonic filter PF corrector.

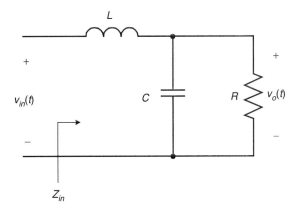

FIGURE 19.7 Low pass inductive filter.

19.3.3 Basic Circuit Topologies of Active Power Factor Correctors

In recent years, using the switched-mode topologies, many circuits and control methods are developed to comply with certain standard (such as IEEE Std 519 and IEC1000-3-2). To achieve this, high-frequency switching techniques have been used to shape the input current waveform successfully. Basically, the active PF correctors employ the six basic converter topologies or their variation versions to accomplish PFC.

A. The Buck Corrector

Figure 19.8a shows the buck PF corrector. By using PWM switch modeling technique [12], the circuit topology can be modeled by the equivalent circuit shown in Fig. 19.8b. It should be pointed out that the circuit model is a large signal model, therefore analysis of PF performance based on this model is valid. It can be shown that the transfer function and input impedance are given by

$$H(s) = \frac{d}{s^2 LC + sL/R + 1} \quad (19.18)$$

$$Z_{in}(s) = \frac{R}{d^2} \frac{s^2 LC + sL/R + 1}{sRC + 1} \quad (19.19)$$

where d is the duty ratio of the switching signal.

Notice that Eqs. (19.18) and (19.19) are different from Eqs. (19.16) and (19.17), in that they have introduced the control variable d. By properly controlling the switching duty ratio to modulate the input impedance and the transfer function, a pure resistive input impedance and constant output voltage can be approached. Thereby, unity PF and output regulation can both be achieved. These control techniques will be discussed in the next section.

Comparing with the other type of high frequency PFC circuits, the buck corrector offers inrush-current limiting, overload or short-circuit protection, and over-voltage protection for the converter due to the existence of the power switch in front of the line. Another advantage is that the output voltage is lower than the peak of the line voltage, which is usually the case normally desired. The drawbacks of using buck corrector may be summarized as follows:

(a) When the output voltage is higher than the line voltage, the converter draws no current from the line, resulting in significant line current distortion near the zero-across of the line voltage;
(b) The input current is discontinuous, leading to high differential mode EMI;
(c) The current stress on the power switch is high;
(d) The power switch needs a floating drive.

B. The Boost Corrector

The boost corrector and its equivalent PWM switch modeling circuit are shown in Figs. 19.9a and b. Its transfer function and

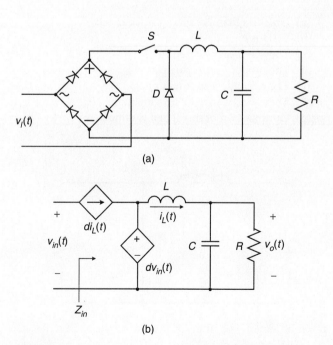

FIGURE 19.8 (a) Buck corrector and (b) PWM switch model for buck corrector.

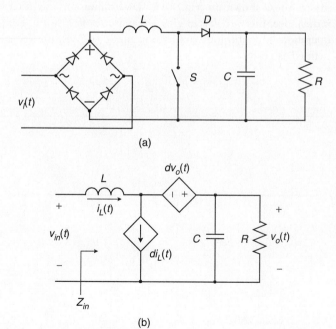

FIGURE 19.9 (a) Boost corrector and (b) PWM switch model of boost corrector.

input impedance are given by

$$H(s) = \frac{1/d'}{s^2(L/d'^2)C + s(L/d'^2)/R + 1} \quad (19.20)$$

$$Z_{in}(s) = d'^2 R \frac{s^2(L/d'^2)C + s(L/d'^2)/R + 1}{sRC + 1} \quad (19.21)$$

where $d' = 1 - d$.

Unlike in the buck case, it is interesting to note that in the boost case, the equivalent inductance is controlled by the switching duty ratio. Consequently, both the magnitude and the phase of the impedance, and both the dc gain and the pools of the transfer function are modulated by the duty ratio, which implies a tight control of the input current and the output voltage. Other advantages of boost corrector include less EMI and lower switch current and grounded drive. The shortcomings with the boost corrector are summarized as:

(a) The output voltage must be higher than the peak of line voltage;
(b) Inrush-current limiting, overload, and over-voltage protections are not available.

C. The Buck–Boost Corrector

The buck–boost corrector and its equivalent circuit are shown in Figs. 19.10a and b. The expressions for transfer function

and input impedance are

$$H(s) = \frac{d/d'}{s^2(L/d'^2)C + s(L/d'^2)/R + 1} \quad (19.22)$$

$$Z_{in}(s) = \left(\frac{d'}{d}\right)^2 R \frac{s^2(L/d'^2)C + s(L/d'^2)/R + 1}{sRC + 1} \quad (19.23)$$

The buck–boost corrector combines some advantages of the buck corrector and the boost corrector. Like a buck corrector, it can provide circuit protections and step-down output voltage, and like a boost corrector its input current waveform and output voltage can be tightly controlled. However, the buck–boost corrector has the following disadvantages:

(a) The input current is discontinued by the power switch, resulting in high differential mode EMI;
(b) The current stress on the power switch is high;
(c) The power switch needs a floating drive;
(d) The polarity of output voltage is reversed.

D. The Cuk, Sepic, and Zeta Correctors

Unlike the previous converters, the Cuk, Sepic, and Zeta converters are fourth-order switching-mode circuits. Their circuit topologies for PFC are shown in Figs. 19.11a, b, and c, respectively. Because there are four energy storage components available to handle the energy balancing involved in PFC, second harmonic output voltage ripples of these correctors are smaller when compared with the second-order buck, boost, and buck–boost topologies. These PF correctors are also able to provide overload protection. However, the increased count of components and current stress are undesired.

19.3.4 System Configurations of PFC Power Supply

The most common configurations of ac–dc power supply with PFC are two-stage scheme and one-stage (or single-stage) scheme. In two-stage scheme as shown in Fig. 19.12a, a non-isolated PFC ac–dc converter is connected to the line to create an intermediate dc bus. This dc bus voltage is usually full of second harmonic ripple. Therefore, followed by the ac–dc converter, a dc–dc converter is cascaded to provide electrical isolation and tight voltage regulation. The advantage of two-stage structure PFC circuits is that the two power stages can be controlled separately, and thus it makes it possible to have both converters optimized. The drawbacks of this scheme are lower efficiency due to twice processing of the input power, complex control circuits, higher cost, and low reliability. Although the two-stage scheme approach is commonly adopted in industry, it received limited attention by the common research, since the input stage and output stage can be studied independently. One-stage scheme combines the PFC circuit and power conversion circuit in one stage as shown in Fig. 19.12b. Due to its

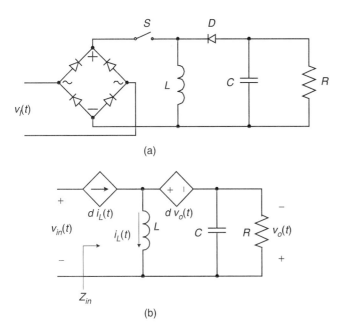

FIGURE 19.10 (a) Buck–boost corrector and (b) PWM switch model of buck–boost corrector.

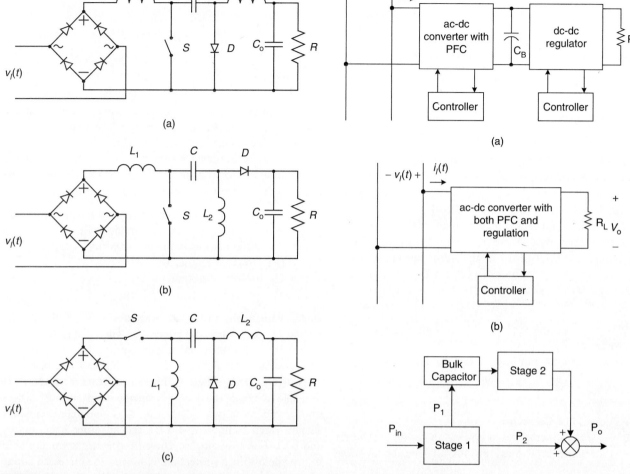

FIGURE 19.11 Fourth-order corrector: (a) Cuk corrector; (b) Sepic corrector; and (c) Zeta corrector.

FIGURE 19.12 System configurations of PFC power supply: (a) two-stage scheme; (b) one-stage scheme; and (c) parallel scheme.

simplified structure, this scheme is potentially more efficient and is very attractive in low to medium power level applications, particularly in those cost-sensitive applications. The one-stage scheme, therefore becomes the main stream of contemporary research due to the ever-increasing demands for inexpensive power supply in residential and office appliance.

For many single-stage PFC converters, one of the most important issues is the slow dynamic response under line and load changes. To remove the low frequency ripple caused by the line (120 Hz) from the output and keep a nearly constant operation duty ratio, a large volume output capacitor is normally used. Consequently, a low frequency pole (typically less than 20 Hz) must be introduced into the feedback loop. This results in very slow dynamic response of the system [13, 14].

To avoid twice power process in two-stage scheme, two converters can be connected in parallel to form so-called parallel PFC scheme [15]. In parallel PFC circuit, power from the ac main to the load flows through two parallel paths, shown in Fig. 19.12c. The main path is a rectifier, in which power is not processed twice for PFC, whereas the other path processes the input power twice for PFC purpose. It is shown that to achieve both unity PF and tight output voltage regulation, only the difference between the input and output power within a half cycle (about 32% of the average input power) needs to be processed twice [15]. Therefore, high efficiency can be obtained by this method.

The continuous research in improving system PF has resulted in countless circuit topologies and control strategies. Classified by their principles to realize PFC, they can be mainly categorized into discontinuous conduction mode (DCM) input technique and continuous conduction mode (CCM) shaping technique. The recent research interest in DCM input technique is focused on developing PFC circuit topologies with a single power switch, result in single-stage single-switch converter (so-called S^4-converter). The CCM shaping technique emphasizes on the control strategy to

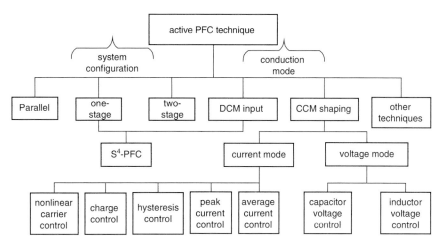

FIGURE 19.13 Overview of PFC techniques.

achieve unity input PF. The hot topics in this line of research are concentrated on degrading complexity of the control circuit and enhancing dynamic response of the system, resulting in some new control methods. Figure 19.13 shows an overview of these techniques based on conduction mode and system configuration types.

19.4 CCM Shaping Technique

Like other power electronic apparatus, the core of a PFC unit is its converter, which can operate either in DCM or in CCM. As shall be discussed in the next section, the benefit from DCM technique is that low-cost power supply can be achieved because of its simplified control circuit. However, the peak input current of a DCM converter is at least twice as high as its corresponding average input current, which causes higher current stresses on switches than that in a CCM converter, resulting in intolerable conduction and switching losses as well as transformer copper losses in high power applications. In practice, DCM technique is only suitable for low to medium level power application, whereas, CCM is used in high power cases. However, a converter operating in CCM does not have PFC ability inherently, i.e. unless a certain control strategy is applied, the input current will not follow the waveform of line voltage. This is why most of the research activities in improving PF under CCM condition have been focused on developing new current shaping control strategies. Depending on the system variable being controlled (either current or voltage), PFC control techniques may be classified as current control and voltage control. Current control is the most common control strategy since the primary objective of PFC is to force the input current to trace the shape of line voltage.

To achieve both PFC and output voltage regulation by using a converter operating in CCM, multiloop controls are generally used. Figure 19.14 shows the block diagram of ac–dc PFC

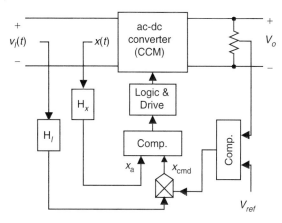

FIGURE 19.14 Block diagram of PFC converter with CCM shaping technique.

converter with CCM shaping technique, where, H_l is a line voltage compensator, H_x is a controlled variable compensator, and $x(t)$ is the control variable that can be either current or voltage.

Normally, in order to obtain a sinusoidal line current and a constant dc output voltage, line voltage $v_l(t)$, output voltage V_o, and a controlled variable $x(t)$ need to be sensed. Depending on whether the controlled variable $x(t)$ is a current (usually the line current or the switch current) or a voltage (related to the line current), the control technique is called "current mode control" or "voltage mode control," respectively. In Fig. 19.14, two control loops have been applied: the feedforward loop and the feedback loops. The feedforward loop is also called "inner loop" which keeps the line current to follow the line voltage in shape and phase, while the feedback loop (also called "outer loop") keeps the output voltage to be tightly controlled. These two loops share the same control command generated by the product of output voltage error signal and the line voltage (or rectified line voltage) signal.

19.4.1 Current Mode Control

Over many years, different current mode control techniques were developed. In this section, we will review several known methods.

A. Average Current Control

In average current control strategy, the average line current of the converter is controlled. It is more desired than the other control strategies because the line current in a SMPS can be approximated by the average current (per switching cycle) through an input EMI filter. The average current control is widely used in industries since it offers improved noise immunity, lower input ripple, and stable operation [13, 16–19].

Figure 19.15 shows a boost PFC circuit using average current control strategy. In the feedforward loop, a low value resistor R_s is used to sense the line current. Through the op-amp network formed by R_i, R_{imo}, R_f, C_p, C_z, and A_2, average line current is detected and compared with the command current signal, i_{cmd}, which is generated by the product of line voltage signal and the output voltage error signal.

There is a common issue in CCM shaping technique, i.e. when the line voltage increases, the line voltage sensor provides an increased sinusoidal reference for the feedforward loop. Since the response of feedback loop is much slow than the feedforward loop, both the line voltage and the line current increase, i.e. the line current is heading to wrong changing direction (with the line voltage increasing, the line current should decrease). This results in excessive input power, causing overshoot in the output voltage. The square block, x^2, in the line voltage-sensing loop shown in Fig. 19.15 provides a typical solution for this problem. It squares the output of the low-pass filter (LPF), which is in proportion to the amplitude of the line voltage, and provides the divider $(A*B)/C$ with a squared line voltage signal for its denominator. As a result, the amplitude of the sinusoidal reference i_{cmd} is negatively proportional to the line voltage, i.e. when the line voltage changes, the control circuit leads the line current to change in the opposite direction, which is the desired situation. The detailed analysis and design issues can be found in [16–18].

As it can be seen, the average current control is a very complicated control strategy. It requires sensing the inductor current, the input voltage, and the output voltage. An amplifier for calculating the average current and a multiplier are needed. However, because of today's advances made in IC technology, these circuits can be integrated in a single chip.

B. Variable Frequency Peak Current Control

Although the average current control is a more desired strategy, the peak current control has been widely accepted because it improves the converter efficiency and has a simpler control circuit [14, 20–24]. In variable frequency peak control strategy, shown in Fig. 19.16, the output error signal $k(t)$ is fed

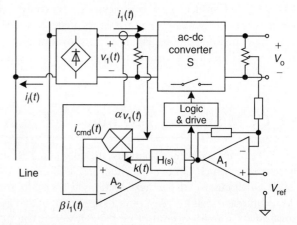

FIGURE 19.16 Block diagram for variable frequency peak current control.

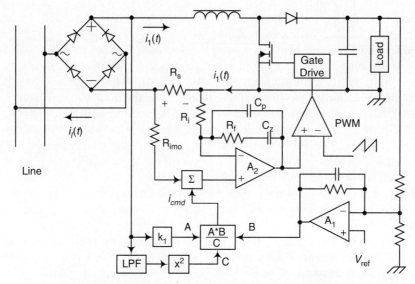

FIGURE 19.15 Boost corrector using average current control.

back through its outer loop. This signal is multiplied by the line voltage signal $\alpha v_1(t)$ to form a line current command signal $i_{cmd}(t)$ ($i_{cmd}(t) = \alpha k(t) \cdot v_1(t)$). The command signal $i_{cmd}(t)$ is the desired line current shape since it follows the shape of the line voltage. The actual line current is sensed by a transducer, resulting in signal $\beta i_1(t)$ that must be reshaped to follow $i_{cmd}(t)$ by feeding it back through the inner loop. After comparing the line current signal $\beta i_1(t)$ with the command signal $i_{cmd}(t)$, the following control strategies can be realized, depending on its logic circuit:

Constant on-time control: Its input current waveform is given in Fig. 19.17a. Letting the fixed on-time to be T_s, the control rules are:

- At $t = t_k$ when $\beta i_1(t_k) = i_{cmd}(t_k)$, S is turned on;
- At $t = t_k + T_{on}$, S is turned off.

Constant off-time control: The input current waveform is shown in Fig. 19.17b. Assuming the off-time is T_{off}, the control rules are:

- At $t = t_k$ when $\beta i_1(t_k) = i_{cmd}(t_k)$, S is turned off;
- At $t = t_k + T_{off}$, S is turned on.

C. Constant Frequency Peak Current Control

Generally speaking, to make it easier to design the EMI filter and to reduce harmonics, constant switching frequency ac–dc

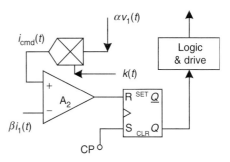

FIGURE 19.18 Logic circuit for constant frequency peak current control.

PFC converter is preferred. Based on the block diagram shown in Fig. 19.18, with T_s is the switching period, the following control rules can be considered to realize a constant frequency peak current control (shown in Fig. 19.19b):

- At $t = nT_s$, S is turned on;
- At $t = t_n$ when $\beta i_1(t_n) = i_{cmd}(t_n)$, S is turned off.

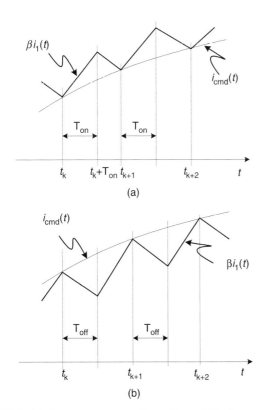

FIGURE 19.17 Input current waveforms for variable frequency peak current control: (a) constant on-time control and (b) constant off-time control.

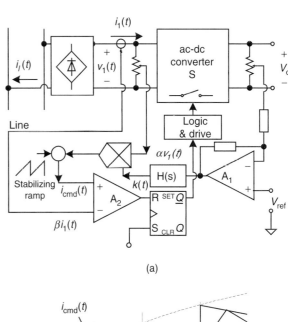

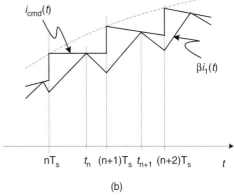

FIGURE 19.19 (a) Constant frequency peak current control with stabilizing ramp compensation and (b) line current waveform for constant frequency peak current control.

The logic circuit for the above control rules can be realized by using an R–S flip-flop with a constant frequency setting clock pulse (CP), as shown in Fig. 19.18. Unfortunately, this logic circuit will result in instability when the duty ratio exceeds 50%. This problem can be solved by subtracting a stabilizing ramp signal from the original command signal. Figure 19.19a shows a complete block diagram for typical constant frequency peak current control strategy. The line current waveform is shown in Fig. 19.19b.

It should be noticed that in both variable frequency and constant frequency peak current control strategies, either the input current or the switch current could be controlled. Thus it makes possible to apply these control methods to buck type converters. There are several advantages of using peak current control:

- The peak current can be sensed by current transformer, resulting in reduced transducer loss;
- The current-error compensator for average control method has been eliminated;
- Low gain in the feedforward loop enhances the system stability;
- The instantaneous pulse-by-pulse current limit leads to increased reliability and response speed.

However the three signals, line voltage, peak current, and output voltage signals, are still necessary to be sensed and multiplier is still needed in each of the peak current control method. Comparing with the average current control method, the input current ripple of these peak current control methods may be high when the line voltage is near the peak value. As a result, considerable line current distortion exists under high line voltage and light low operation conditions.

D. Hysteresis Control

Unlike the constant on-time and the constant off-time control, in which only one current command is used to limit either the minimum input current or the maximum input current, the hysteresis control has two current commands, $i_{hcmd}(t)$ and $i_{lcmd}(t)$ ($i_{lcmd}(t) = \delta i_{hcmd}(t)$), to limit both the minimum and the maximum of input current [25–28]. To achieve smaller ripple in the input current, we desire a narrow hyster-band. However, the narrower the hyster-band, the higher the switching frequency. Therefore, the hyster-band should be optimized based on circuit components such as switching devices and magnetic components. Moreover, the switching frequency varies with the change of line voltage, resulting in difficulty in the design of the EMI filter. The circuit diagram and input current waveform are given in Figs. 19.20a and b, respectively. When $\beta i_1(t) \geq i_{hcmd}(t)$, a negative pulse is generated by comparator A_1 to reset the R–S flip-flop. When $\beta i_1(t) \leq i_{lcmd}(t)$, a negative pulse is

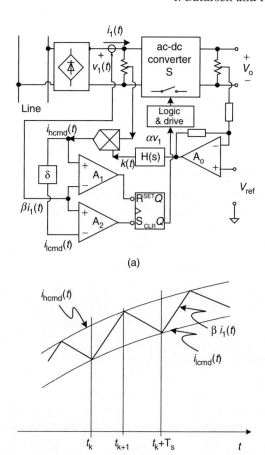

FIGURE 19.20 Hysteresis control: (a) block diagram for hysteresis control and (b) line current waveform of hysteresis control.

generated by comparator A_2 to set the R–S flip-flop. The control rules are:

- At $t = t_k$ when $\beta i_1(t_k) = i_{lcmd}(t)$, S is turned on;
- At $t = t_{k+1} \beta i_1(t_{k+1}) = i_{hcmd}(t)$, S is turned off;
- When $\beta i_1(t) = i_{hcmd}(t) = i_{lcmd}(t)$, S stays off or on.

Like the above mentioned peak current control methods, the hysteresis control method has simpler implementation, enhanced system stability, and increased reliability and response speed. In addition, it has better control accuracy than that the peak current control methods have. However, this improvement is achieved on the penalty of wide range of variation in the switching frequency. It is also possible to improve the hysteresis control in a constant frequency operation [29, 30], but usually this will increase the complexity of the control circuit.

E. Charge Control

In order to make the average control method to be applicable for buck-derived topologies where the switch current instead

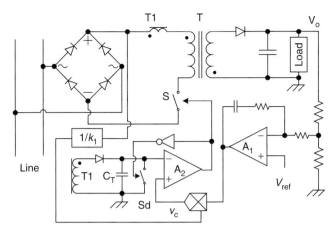

FIGURE 19.21 Flyback PFC converter using charge control.

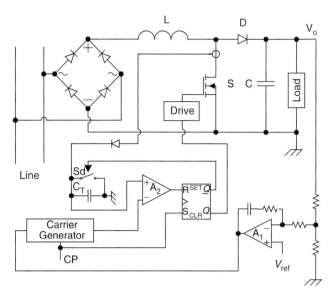

FIGURE 19.22 Boost PFC converter using NLC control.

of the inductor current needs to be controlled, an alternative method to realize average current control, namely, charge control was proposed in [31–33]. Since the total charge of the switch current per switching cycle is proportional to the average value of the switch current, the average current can be detected by a capacitor-switch network. Figure 19.21 shows a block diagram for charge control. The switch current is sensed by current transformer T1 and charges the capacitor C_T to form average line current signal. As the switch current increases, the charge on capacitor C_T also increases. When the voltage reaches the control command v_c, the power switch turns off. At the same time, the switch S_d turns on to reset the capacitor. The next switching cycle begins with the power switch turning on and the switch S_d turning off by a clock pulse.

The advantages of charge control are:

- Ability to control average switch current;
- Better switching noise immunity than peak current control;
- Good dynamic performance;
- Elimination of turn off failure in some converters (e.g. multiresonant converter) when the switch current reaches its maximum value.

The disadvantages are:

- Synthesis of the reference v_c still requires sensing both input and output voltage and use of a multiplier;
- Subharmonic oscillation may exist.

F. Non-linear-carrier (NLC) Control

To further simplify the control circuitry, non-linear-carrier control methods were introduced [34, 35]. In CCM operation, since the input voltage is related to the output voltage through the conversion ratio, the input voltage information can be recovered by the sensed output voltage signal. Thus the sensing of input voltage can be avoided, and therefore, the multiplier is not needed, resulting in significant simplification in the control circuitry. However, complicated NLC waveform generator and its designs are involved. Figure 19.22 shows the block diagram of the NLC charge control first introduced in [34].

19.4.2 Voltage Mode Control

Generally, current mode control is preferred in current source driven converters, as the boost converter. To develop controllers for voltage source driven converter, like the buck converter and to improve dynamic response, voltage mode control strategy was proposed [36, 37]. Figure 19.23 shows the

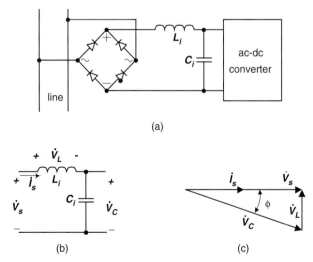

FIGURE 19.23 Input circuit and phasor diagram for voltage control: (a) input circuit of voltage control ac–dc converter; (b) simplified input circuit; and (c) phasor diagram.

input circuit of an ac–dc converter and its phasor diagram representation, where ϕ is the phase shift between the line current and the capacitor voltage. An LC network could be added to the input either before a switch mode rectifier (SMR) or after a passive rectifier to perform such kind of control. In boost type converter, the inductor L_i is the input inductor. It can be seen from the phasor diagram that to keep the line current in phase with the line voltage, we can either control the capacitor voltage or the inductor voltage. If the capacitor voltage is chosen as controlled variable, the control strategy is known as delta modulation control.

A. Capacitor Voltage Control

Figure 19.24 gives a SMR with PFC using capacitor voltage control [36]. The capacitor voltage $v_{c1}(t)$ is forced to track a sinusoidal command $v_{c1}^*(t)$ signal to indirectly adjust the line current in phase with the line voltage. The command signal is the product of the line voltage signal with a phase shift of ϕ and the feedback error signal. The phase shift ϕ is a function of the magnitudes of line voltage and line current, therefore the realization of a delta control is not really simple. In addition, since ϕ is usually very small, a small change in capacitor voltage will cause a large change in the inductor voltage, and hence in the line current. Thus it make the circuit very sensitive to parameter variations and perturbations.

B. Inductor Voltage Control

To overcome the above shortcomings, inductor voltage control strategies was reported in [37]. Figure 19.25 shows an SMR with PFC using inductor voltage control. As the phase difference between the line voltage and the inductor voltage is fixed at 90° ideally, the control circuit is simpler in implementation than that of capacitor voltage control. As the inductor voltage is sensitive to the phase shift ϕ, but not sensitive to the change in magnitude of reference, the inductor voltage control method is more effective in keeping the line current in phase

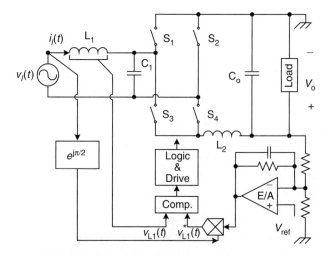

FIGURE 19.25 SMR using inductor voltage control.

with the line voltage. However, in the implementations of both the two kinds of voltage control methods hysteresis technique is normally used. Therefore, unlike the previous current mode control, variable frequency problem is encountered in these control methods.

Generally speaking, by using CCM shaping technique, the input current can trace the wave shape of the line voltage well. Hence the PF can be improved efficiently. However, this technique involves in the designing of complicated control circuits. Multiloop control strategy is needed to perform input current shaping and output regulation. In most CCM shaping techniques, current sensor, and multiplier are required, which results in higher cost in practical applications. In some cases, variable frequency control is inevitable, resulting in additional difficulties in its closedloop design. Table 19.1 gives a comparison among these control methods.

19.5 DCM Input Technique

To get rid of the complicated control circuit invoked by CCM shaping technique and reduce the cost of the electronic interface, DCM input technique can be adopted in low power to medium power level application.

In DCM, the inductor current of the core converter is no longer a valid state variable since its state in a given switching cycle is independent of the value in the previous switching cycle [38]. The peak of the inductor current is sampling the line voltage automatically, resulting in sinusoidal-like average input current (line current). This is why DCM input circuit is also called "voltage follower" or "automatic controller." The benefit of using DCM input circuit for PFC is that no feedforward control loop is required. This is also the main advantage over a CCM PFC circuit, in which multiloop control strategy is essential. However, the input inductor operating in DCM

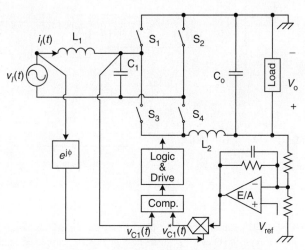

FIGURE 19.24 SMR using capacitor voltage control.

19 Power Factor Correction Circuits

TABLE 19.1 Comparison of CCM shaping techniques

	Average current	VF peak-current	CF peak-current	Hysteresis	Charge	Non-linear carrier	Capacitor voltage	Inductor voltage
Input ripple	Low	High	High	Low	Low	Low	Low	Low
Switching frequency	Constant	Variable	Constant	Variable	Constant	Constant	Variable	Variable
Dynamic response	Slow	Slow	Slow	Fast	Fast	Fast	Fast	Fast
Control signal sensed for inner loop	Input current & input voltage	Input (or switch) current & input voltage	Input (or switch) current & input voltage	Input current & input voltage	Input (or switch) current & input voltage	Input (or switch) current	Input voltage & capacitor voltage	Input voltage & inductor voltage
Inner loop E/A	Yes	No	No	Yes	No	No	No	No
Multiplier	Yes	Yes	Yes	Yes	Yes	No	Yes	Yes

cannot hold the excessive input energy because it must release all its stored energy before the end of each switching cycle. As a result, a bulky capacitor is used to balance the instantaneous power between the input and output. In addition, in DCM, the input current is normally a train of triangle pulses with nearly constant duty ratio. In this case, an input filter is necessary for smoothing the pulsating input current.

19.5.1 Power Factor Correction Capabilities of the Basic Converter Topologies in DCM

The DCM input circuit can be one of the basic dc–dc converter topologies. However, when they are applied to the rectified line voltage, they may draw different shapes of average line current. In order to examine the PFC capabilities of the basic converters, we first investigate their input characteristics. Because the input currents of these converters are discrete when they are operating in DCM, only averaged input currents are considered. Since switching frequency is much higher than the line frequency, let's assume the line voltage is constant in a switching cycle. In steady state operation, the output voltage is nearly constant and the variation in duty ratio is slight. Therefore, constant duty ratio is considered in deriving the input characteristics.

A. Buck Converter

The basic buck converter topology and its input current waveform when operating in DCM are shown in Figs. 19.26a and b, respectively. It can be shown that the average input current in one switching cycle is given by

$$i_{1,avg}(t) = \frac{1}{T_s}\left[\frac{1}{2} \cdot DT_s \cdot \frac{v_1(t) - V_o}{L}DT_s\right] \quad (19.24)$$

$$= \frac{D^2 T_s}{2L}v_1(t) - \frac{D^2 T_s}{2L}V_o$$

Figure 19.26c shows that the input voltage–input current I–V characteristic consists of two straight lines in quadrants I and III. It should be noted that these straight lines do not go through the origin. When the rectified line voltage $v_1(t)$ is less than the output voltage V_o, negative input current would occur. This is not allowed because the bridge rectifier will block the negative current. As a result, the input current is zero near the zero crossing of the line voltage, as shown in Fig. 19.26c. Actually, the input current is distorted simply because the buck converter can work only under the condition when the input voltage is larger than the output voltage. Therefore, the basic buck converter is not a good candidate for DCM input PFC.

B. Boost Converter

The basic boost converter and its input current waveform are shown in Figs. 19.27a and b, respectively. The input I–V characteristic can be found as follows

$$i_{1,avg}(t) = \frac{1}{T_s}\left[\frac{1}{2} \cdot (D + D_1)\, T_s \frac{v_1(t)}{L}DT_s\right] \quad (19.25)$$

$$= \frac{D^2 T_s}{2L} \frac{v_1(t)V_o}{V_o - v_1(t)}$$

where, $D_1 T_s$ is the time during which the inductor current decreases from its peak to zero.

By plotting Eq. (19.25), we obtain the input I–V characteristic curve as given in Fig. 19.27c. As we can see that as long as the output voltage is larger than the peak value of the line voltage in certain range, the relationship between $v_1(t)$ and $i_{1,avg}(t)$ is nearly linear. When the boost converter is connected to the line, it will draw almost sinusoidal average input current from the line, shown as in Fig. 19.27c. As one might notice from Eq. (19.25) that the main reason to cause the non-linearity is the existence of D_1. Ideally, if $D_1 = 0$, the input I–V characteristic will be a linear one.

Because of the above reasons, boost converter is comparably superior to most of the other converters when applied to do PFC. However, it should be noted that boost converter can operate properly only when the output voltage is higher

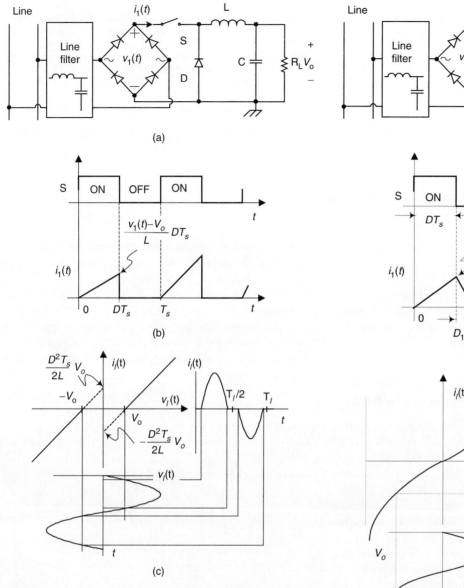

FIGURE 19.26 Input I–V characteristic of basic buck converter operating in DCM: (a) buck converter; (b) input current; and (c) input I–V characteristic.

than its input voltage. When low voltage output is needed, a stepdown dc–dc converter must be cascaded.

C. Buck–Boost Converter

Figure 19.28a shows a basic buck–boost converter. The averaged input current of this converter can be found according to its input current waveform, shown in Fig. 19.28b.

$$i_{1,avg}(t) = \frac{D^2 T_s}{2L} v_1(t) \qquad (19.26)$$

Equation (19.26) gives a perfect linear relationship between $i_{1,avg}(t)$ and $v_1(t)$, which proves that a buck–boost has an

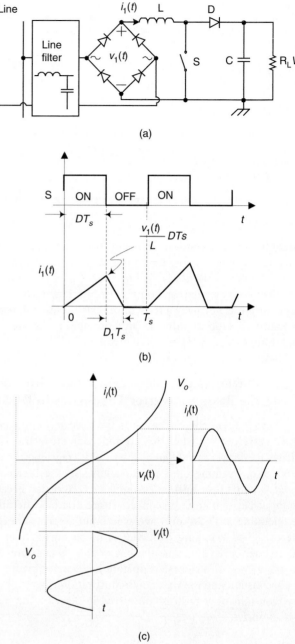

FIGURE 19.27 Input I–V characteristic of basic boost converter operating in DCM: (a) boost converter; (b) input current; and (c) input I–V characteristic.

excellent automatic PFC property. This is because the input current of buck–boost converter does not related to the discharging period D_1. Its input I–V characteristics and input voltage and current waveforms are shown in Fig. 19.28c. Furthermore, because the output voltage of buck–boost converter can be either larger or smaller than the input voltage, it demonstrates strong availability for DCM input technique to achieve PFC. So, theoretically buck–boost converter is a

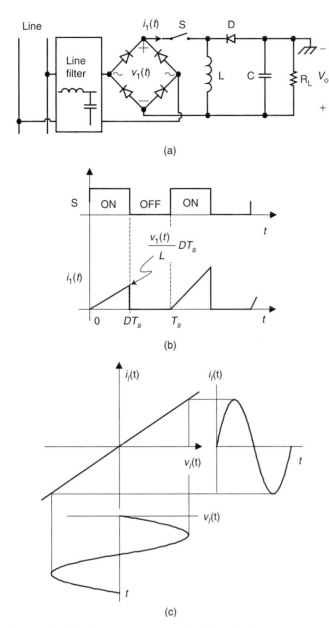

FIGURE 19.28 Input I–V characteristic of basic buck–boost converter operating in DCM: (a) buck–boost converter; (b) input current; and (c) input I–V characteristic.

perfect candidate. Unfortunately, this topology has two limitations: (1) the polarity of its output voltage is reversed, i.e. the input voltage and the output voltage don't have a common ground; and (2) it needs floating drive for the power switch. The first limitation circumscribes this circuit into a very narrow scope of applications. As a result, it is not widely used.

D. Flyback Converter

Flyback converter is an isolated converter whose topology and input current waveform are shown in Figs. 19.29a and b,

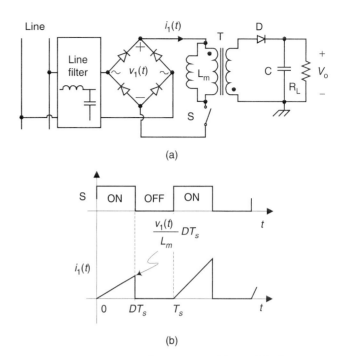

FIGURE 19.29 Input I–V characteristic of basic flyback converter operating in DCM: (a) flyback converter and (b) input current.

respectively. The input voltage–input current relationship is similar to that of buck–boost converter

$$i_{1,avg}(t) = \frac{D^2 T_s}{2L_m} v_1(t) \quad (19.27)$$

where, L_m is the magnetizing inductance of the output transformer.

Therefore, it has the same input I–V characteristic, and hence the same input voltage and input current waveforms as those the buck–boost converter has, shown in Fig. 19.29c.

Comparing with buck–boost converter, flyback converter has all the advantages of the buck–boost converter. What's more, input–output isolation can be provided by flyback converter. These advantages make flyback converter well suitable for PFC with DCM input technique. Comparing with boost converter, the flyback converter has better PFC and the output voltage can be either higher or lower than the input voltage. However, due to the use of power transformer, the flyback converter has high di/dt noise, lower efficiency, and lower density (larger size and heavier weight).

E. Forward Converter

The circuit shown in Fig. 19.30 is a forward converter. In order to avoid transformer saturation, it is well-known that forward converter needs the 3rd winding to demagnetize (reset) the transformer. When a forward converter is connected to the rectified line voltage, the demagnetizing current through the 3rd winding is blocked by the rectifier diodes.

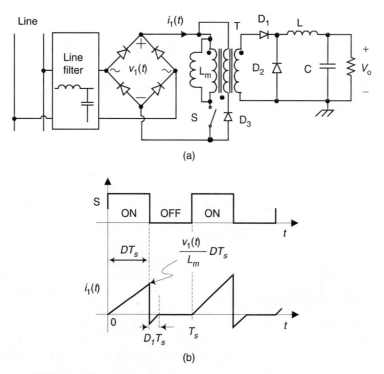

FIGURE 19.30 (a) Forward converter and (b) input current waveform.

Therefore, forward converter is not available for PFC purpose unless a certain circuit modification is applied.

F. Cuk Converter and Sepic Converter

It can be shown that Cuk converter and Sepic converter given in Figs. 19.31a and b, respectively, have the same input I–V characteristic. Each of these converter topologies has two inductors, with one located at its input and the other at its output. Let's consider the case when the input inductor operates in DCM while the output inductor operates in CCM. In this case, the capacitor C_1 can be designed with large value to balance the instantaneous input/output power, resulting in high PF in the input and low second harmonic ripple in the output voltage. To investigate the input characteristic of these converters, let's take the Cuk converter as an example. One should note that the results from the Cuk converter are also suitable for Sepic converter.

For the Cuk converter shown in Fig. 19.31a, the waveforms for input inductor current (the same as the input current), output inductor current, and the voltage across the output inductor are depicted in Fig. 19.31c. Assume that the capacitor C_1 is large enough to be considered as a voltage source V_c, in steady state, employing volt-second equilibrium principle on L_2, we obtain

$$V_C = \frac{1}{D} V_o \quad (19.28)$$

The input inductor current reset time ratio D_1 is given by

$$D_1 = \frac{D^2 v_1(t)}{V_o - D v_1(t)} \quad (19.29)$$

Therefore the averaged input current can be found as

$$i_{1,avg}(t) = \frac{1}{T_s} \left[\frac{1}{2} \cdot (D + D_1) T_s \frac{v_1(t)}{L} D T_s \right] \quad (19.30)$$

$$= \frac{D^2 T_s}{2L} \frac{v_1(t) V_o}{V_o - D v_1(t)}$$

It can be seen that Eq. (19.30) is very similar to Eq. (19.25) except that the denominator in the former equation is $(V_o - D v_1(t))$ instead of $(V_o - v_1(t))$. This will lead to some improvement in that I–V characteristic in Cuk converter. Referring to the I–V characteristic shown in Fig. 19.27c, Cuk converter has a curve more close to a straight line. Such improvement, however, is achieved at the expense of using more circuit components. It can be proved that the same results can be obtained by the Sepic converter.

G. Zeta Converter

Figure 19.32a gives a Zeta converter connected to the line. In DCM operation, the key waveforms are illustrated in Fig. 19.32b, where we presume the capacitor being equivalent to a voltage source V_c. As we can see that the converter input current waveform is exactly the same as that drawn by a

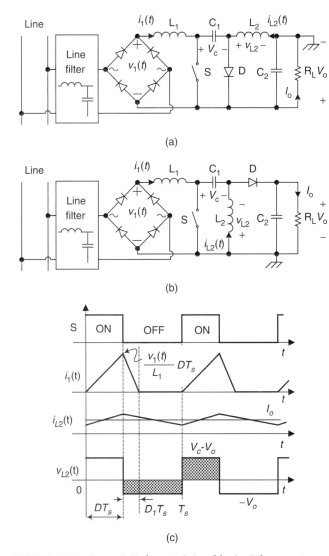

FIGURE 19.31 Input I–V characteristic of basic Cuk converter and Sepic converter operating in DCM: (a) Cuk converter; (b) Sepic converter; and (c) typical waveforms of Cuk converter with input inductor operating in DCM.

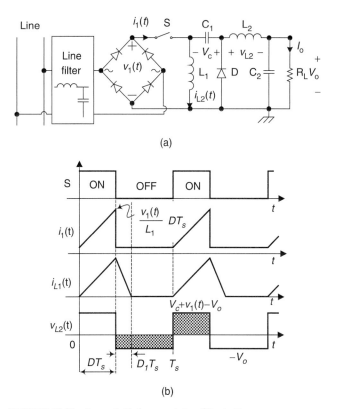

FIGURE 19.32 Input I–V characteristic of basic Zeta converter operating in DCM: (a) Zeta converter and (b) typical waveforms of Zeta converter with input inductor operating in DCM.

buck–boost converter. Thus, the average input current for the Zeta converter is identical to that for the buck–boost converter, which is given by Eq. (19.26). As a result, the Zeta converter has as good automatic PFC capability as the buck–boost converter. The improvement achieved here is the non-inverted output voltage. However, like the buck converter, floating drive is required for the power switch.

Based on the above discussion, we may conclude that all the eight basic converters except forward converter have good inherent PFC capability and are available for DCM PFC usage. Among them, boost converter and flyback converter are especially suitable for single-stage PFC scheme because they have minimum component count and grounded switch drive, and their power switches are easy to be shared with the output dc–dc converter. Hence, these two converters are most preferable by the designers for PFC purpose. The other converters could also be used to perform certain function such as circuit protection and small output voltage ripple. The characteristics of the eight basic converter topologies are summarized in Table 19.2.

19.5.2 AC–DC Power Supply with DCM Input Technique

In two-stage PFC power supply, the DCM converter is connected in front of the ac line to achieve high input PF and provide a roughly regulated dc bus voltage, as shown in Fig. 19.33. This stage is also known as "pre-regulator." The duty ratio of the pre-regulator should be maintained relatively stable so that high PF is ensured. To stabilize the dc bus voltage, a bank capacitor is used at the output of the pre-regulator. The second stage, followed by the pre-regulator, is a dc–dc converter, called post-regulator, with its output voltage being tightly controlled. This stage can operate either in DCM or in CCM. However, CCM is normally preferred to reduce the output voltage ripple.

DCM input technique has been widely used in one-stage PFC circuit configurations. Using a basic converter

TABLE 19.2 Comparison of basic converter topologies operating for DCM input technique

	Buck	Boost	Buck–boost	Flyback	Forward*	Cuk and sepic	Zeta
Line current waveform	(waveform)	(waveform)	(waveform)	(waveform)	–	(waveform)	(waveform)
Switch drive	Floating	Grounded	Floating	Grounded	Grounded	Grounded	Floating
Peak input current	High	Lower	High	High	–	Lower	High
Inrush and overload protection	Yes	No	Yes	Yes	–	Yes	Yes
Output voltage	$V_o < V_{l,m}$	$V_o > V_{l,m}$	Inverted	$V_o < V_{l,m}$ or $V_o > V_{l,m}$	–	$V_o < V_{l,m}$ or $V_o > V_{l,m}$; Inverted for Cuk	$V_o < V_{l,m}$ or $V_o > V_{l,m}$

*The standard forward converter is not recommended as a PF corrector since the rectifier at the input will block the demagnetizing current through the tertiary winding.

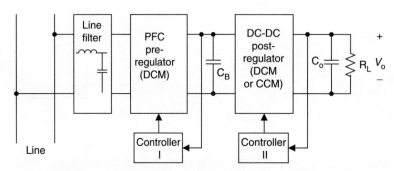

FIGURE 19.33 DCM input pre-regulator in two-stage ac–dc power supply.

(usually boost or flyback converter) operating in DCM, combining it with another isolation converter can form a one-stage PFC circuit. A storage capacitor is generally required to hold the dc bus voltage in these combinations. Unlike the two-stage PFC circuit, in which the bus voltage is controlled, the single-stage PFC converter has only one feedback loop from the output. The input circuit and the output circuit must share the same control signal. In [39–41] a number of combinations have been studied. Figures 19.34 and 19.35 show a few examples of successful combinations. Since the input circuit and the output circuit are in a single stage, it is possible for them to share the same power switch. Thus it results in single-stage single-switch PFC (S^4-PFC) circuit, as shown in Fig. 19.35 [39, 42, 43].

Due to the simplicity and low cost, DCM boost converter is most commonly used for unity PF operation. The main drawback of using boost converter is that it shows considerable distortion of the average line current owing to the slow discharging of the inductor after the switch is turned off.

The output dc–dc converter can operate either in DCM or in CCM if small output ripple is desired. If the output circuit operates in CCM, there exists a power unbalance in S^4-PFC converter when the load changes. Because the duty ratio is only sensitive to the output voltage in CCM operation, when the output power (output current) decreases, the duty ratio

will keep unchanged. As both the input and the output circuit share the power switch, the input circuit will draw an unchanged power from the ac source. As a result, the input power is higher than the output power. The difference between the input power and the output power has to be stored in the storage capacitor, and hence increase in the dc bus voltage occurs. With the dc bus voltage's rising, the duty ratio decreases. This process will be finished until a new power balance is built. As we can see, the new power balance is achieved at the penalty of increased voltage stress, resulting in high conduction losses in circuit components. Particularly, the high bus voltage causes difficulties in developing S^4-converter for universal input (input line voltage *rms* value from ac 90 to 260 V) application.

Recent research on solving this problem can be found in [44–52]. The circuit in [44] uses two bulk capacitors that share the dc bus voltage change, shown in Fig. 19.36a. As a result, lower voltage is present at each of the capacitor. Reference [45] proposed a modified boost–forward PFC converter, in which a negative current feedback is introduced to the input circuit by the coupled windings of forward transformer, shown in Fig. 19.36b. In [46] a series resonant circuit called charge pump circuit is introduced into S^4-PFC circuit, shown in Fig. 19.36c. As the load decrease, the charge pump circuit can suppress the dc bus voltage automatically.

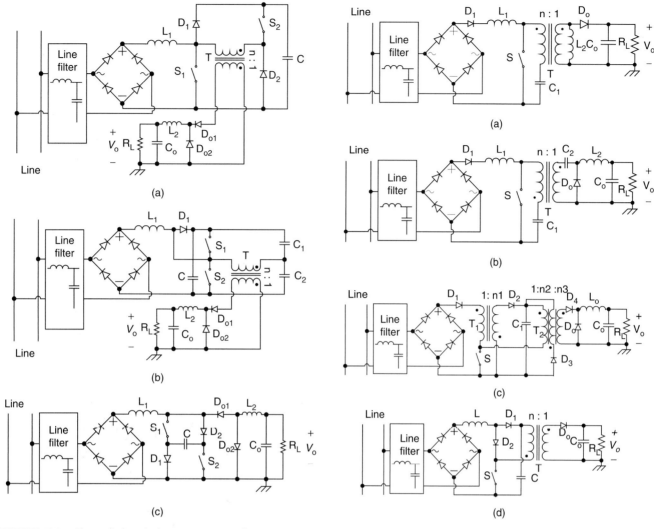

FIGURE 19.34 Two-switch single-stage power factor corrector: (a) boost–forward converter; (b) boost–half bridge converter; and (c) Sheppard–Taylor converter.

FIGURE 19.35 Single-stage single-switch PFC circuit: (a) boost–flyback combination circuit (BIFRED); (b) boost–buck combination circuit (BIBRED); (c) flyback–forward combination; and (d) boost–flyback combination.

19.5.3 Other PFC Techniques

Extensive research in PFC continues to yield countless new techniques [15, 53–63]. The research topics are mainly focused on improvements of the PFC circuit performs such as fast performs, high efficiency, low cost, small input current distortion, and output ripple. The classification of PFC techniques presented here can only cover those methods that are frequently documented in the open literature. There are still many PFC methods which do not fall into the specified categories. The following are some examples:

- **Second-harmonic-injected method** [56]: In DCM input technique, even the converter operates at constant duty ratio, current distortion still exists. The basic idea of second-harmonic-injected method is compensating the duty ratio by injecting a certain amount of second harmonic into the duty ratio to modify the input I–V characteristic of the input converter. However, the output voltage may be affected by the modified duty ratio.
- **Interleaved method** [57]: An interleaved PFC circuit composed of several input converters in parallel. The peak input current of these converters follow the line voltage and are interleaved. A sinusoidal total line current is obtained by superimposing all the input current of the converters. The advantage of this method is that the converter input current can be easily smoothed by input EMI filter.
- **Waveform synthesis method** [58]: This method combines passive and active PFC techniques. Since the rectifier in the passive inductive-input PF corrector has a limited

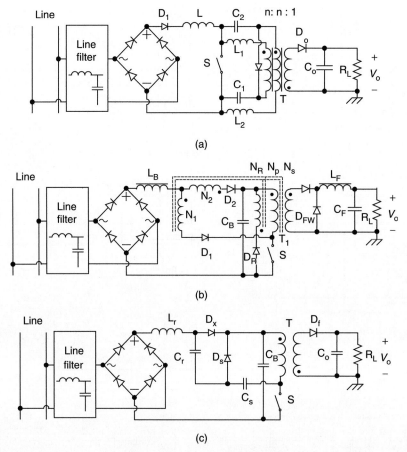

FIGURE 19.36 Improved S^4-PFC converter: (a) boost–forward PFC circuit using two bulk capacitors; (b) boost–forward PFC circuit with reduced bus voltage; and (c) boost–flyback PFC circuit with charge pump circuit.

conduction angle, the input current is a single pulse around the peak of the line voltage, whereas the boost converter draws a non-zero current around the zero-cross of the line voltage. By controlling the operation mode of the active switch (enable and disable the boost converter at certain line voltage), the waveforms of active and passive PFC circuits are tailored to extend the conduction angle of the rectifier. The resulting current waveform has a PF greater than 0.9 and a THD lower than 20%.

19.6 Summary

To reduce losses, and decrease weight and size associated with converting ac power to dc power in linear power supply, switch mode power supplies (SMPSs) were introduced. The high non-linearity of this kind of power electronic systems handicaps itself by providing the utility power system with low power factor (PF) and high total harmonic distortion (THD). These unwanted harmonics are commonly corrected by incorporating power factor correction (PFC) technique into the SMPS. This chapter gives a technical review of current research in high frequency PFC, including the definition of PF and THD, configuration of PFC circuit, DCM input technique, and CCM shaping technique. The common issue of these techniques is to properly process the power flow so that the constant power dissipation at the output is reflected into ac power dissipation with two times the line frequency. Technically, PFC techniques encounter the following tradeoffs:

(a) *Simplicity and accuracy*: Single-stage PFC circuit has simple topology and simple control circuit, but has less control accuracy while two-stage PFC circuit has the contrary performance;

(b) *Control simplicity and power handling capability*: DCM input technique requires no input current control, but has less power handling capability while CCM has multiloop control and has more power handling capability;

(c) *Switching frequency and conversion efficiency*: To reduce weight and size of the PFC converter, higher switching frequency is desired. However, the associated switching losses result in decrease in conversion efficiency;

(d) **Frequency response and bandwidth**: To have good dynamic response, wider bandwidth is desired, however to achieve high PF bulk storage capacitor and output capacitor has to be used.

In the past decades, research in PFC techniques has led to the development of more efficient circuits and control strategies in order to optimize the design without compromising the above tradeoffs. Moreover, since the growth in power electronics strongly relies on the development of semiconductor devices, the recent advent of higher rating power devices, it is believed that the switching mode PF correctors will completely replace the existing passive reactive compensators in power system. In the distributed power system (DPS) where small size and high efficiency are of extreme importance, a new soft-switching technique has been used in designing PFC circuits. With the ever increasing market demanding for ultra-fast computer, the need for low output voltage (typically less than 1 V!) with high output currents and high efficiency converters has never been greater. Research efforts in developing high frequency high efficiency PFC circuits will continue to grow.

Acknowledgment

I would like to thank my doctoral students *Guangyong Zhu, Shiguo Luo, and Wenkai Wu* for their valuable contribution to the area of power factor correction.

Further Reading

1. C. K. Duffey and R. P. Stratford, "Update of Harmonic Standard IEEE-519: IEEE Recommended Practices and Requirements for Harmonic Control in Electric Power Systems," IEEE Trans. on Industry Applications, vol. 25, no. 6, Nov. 1989, pp. 1025–1034.
2. B. K. Bose, "Power Electronics – A Technology Review," Proceedings of the IEEE, Aug. 1992, pp. 1303–1334.
3. H. Akagi, "Trends in Active Power Line Conditioners," IEEE Trans. on Power Electronics, vol. 9, no. 3, May 1994, pp. 263–268.
4. W. McMurray, "Power Electronics in The 1990's," Proceedings of IEEE-IECON'90, pp. 839–843.
5. A. McEachern, W. M. Grady, W. A. Moncrief, G. T. Heydt, and M. McGranaghan, "Revenue and Harmonics: An Evaluation of Some Proposed Rate Structures," IEEE Trans. on Power Delivery, vol. 10, no. 1, Jan. 1995, pp. 474–480.
6. R. Redl, P. Tenti, and J. D. Van WYK, "Power Electronics' Polluting Effects," IEEE Spectrum, May 1997, pp. 32–39.
7. J. S. Lai, D. Hurst, and T. Key, "Switch-Mode Power Supply Power Factor Improvement Via Harmonic Elimination Methods," Conference Record of IEEE-APEC'91, pp. 415–422.
8. IEEE Inc., "IEEE Guide for Harmonic Control and Reactive Compensation of Static Power Converters (IEEE Std. 519-1981)," ANSI/IEEE Inc., 1981.
9. IEEE Inc., "IEEE Recommended Practices and Requirements for Harmonic Control in Electrical Power systems (IEEE Std. 519-1992)," ANSI/IEEE Inc., 1993.
10. I. Batarseh, "Power Electronic Circuits," John Wiley & Sons Inc., (in press).
11. R. E. Tarter, "Solid-State Power Conversion Handbook," John Wiley & Sons Inc., 1993.
12. V. Vorperian, "Simplified Analysis of PWM Converters Using the Model of the PWM Switch: Parts I and II," IEEE Trans. on Aerospace and Electronic Systems, vol. 26, no. 3, 1990, pp. 490–505.
13. M. O. Eissa, S. B. Leeb, G. C. Verghese, and A. M. Stankovic, "A Fast Analog Controller for a Unity-Power-Factor AC/DC Converter," Conference Record of APEC'94, pp. 551–555.
14. R. Liu, I. Batarseh, and C. Q. Lee, "Resonant Power Factor Correction Circuits with Resonant Capacitor-Voltage and Inductor-Current-Programmed Controls," Conference Record of IEEE-APEC'93, pp. 675–680.
15. Y. Jiang and F. C. Lee, "Single-Stage Single-Phase Parallel Power Factor Correction Scheme," Conference Record IEEE-PESC'94, pp. 1145–1151.
16. L. Dixon, "Average Current Mode Control of Switching Power Supplies," Product & Applications Handbook, Unitrode Integrated Circuits Corporation, U140, 1993–94, pp. 9-457–9-470.
17. L. Dixon, "High Power Factor Switching Preregulator Design Optimization," Unitrode Power Supply Design Seminar, Sem-1000, 1994, pp. I3-1–I-12.
18. J. P. Noon and D. Dalal, "Practical Design Issues for PFC Circuits," Conference Record of APEC'97, pp. 51–58.
19. B. Mammano, "Average Current-Mode Control Provides Enhanced Performance for a Broad Range of Power Topologies," PCIM'92-Power Conversion, Sep. 1992, pp. 205–213.
20. J. P. Gegner and C. Q. Lee, "Linear Peak Current Mode Control: A Simple Active Power Factor Correction Control Technique for Continuous Conduction Mode," Conference Record of IEEE-PESC'96, 1996, pp. 196–202.
21. R. Redl and B. P. Erisman, "Reducing Distortion in Peak-Current-Controlled Boost Power-Factor Corrector," Conference Record of IEEE-APEC'94, pp. 576–583.
22. C. A. Caneson and I. Barbi, "Analysis and Design of Constant-Frequency Peak-Current-Controlled High-Power-Factor Boost Rectifier with Slope Compensation," Conference Record of IEEE-APEC'96, pp. 807–813.
23. A. R. Prasad, P. D. Ziogas, and S. Manias, "A New Active Power Factor Correction Method for Single-Phase Buck-Boost AC-DC Converter," Conference Record of IEEE-APEC'92, pp. 814–820.
24. A. V. Costa, C. H. G. Treviso, and L. C. Freitas, "A New ZCS-ZVS-PWM Boost Converter with Unit Power Factor Operation," Conference Record of IEEE-APEC'94, pp. 404–410.
25. J. Mahdavi, M. Tabandeh, and A. K. Shahriari, "Comparison of Conducted RFI Emission from Different Unity Power Factor AC/DC Converters," Conference Record of IEEE-PESC'96, pp. 1979–1985.
26. J. C. Salmon, "Techniques for Minimizing the Input Current Distortion of Current-Controlled Single-Phase Boost Rectifiers," IEEE Trans. on Power Electronics, vol. 8, no. 4, Oct. 1993, pp. 509–520.

27. R. Srinivasan and R. Oruganti, "A Unity Power Factor Converter Using Half-Bridge Boost Topology," IEEE Trans. on Power Electronics, vol. 13, no. 3, May 1998, pp. 487–500.
28. I. Barbi and S. A. Oliveira da Silva, "Sinusoidal Line Current Rectification at Unity Power Factor with Boost Quasi-Resonant Converters," Conference Record of APEC'90, pp. 553–562.
29. M. Kazerani, P. D. Ziogas, and G. Joos, "A Novel Active Current Waveshaping Technique for Solid-State Input Power Factor Conditioners," IEEE Tans. on IE, vol. 38, no. 1, Feb. 1991, pp. 72–78.
30. H. Y. Wu, C. Wang, K. W. Yao and J. F. Zhang, "High Power Factor Single-Phase AC/DC Converter with DC Biased Hysteresis Control Technique," Conference Record of APEC'97, pp. 88–93.
31. W. Tang, F. C. Lee, R. Ridley, and I. Cohen, "Charge Control: Modeling, Analysis and Design," Conference Record of IEEE-PESC'92, pp. 503–511.
32. W. Tang, C. S. Leu, and F. C. Lee, "Charge Control for Zero-Voltage-Switching Multiresonant Converter," IEEE Trans. on Power Electronics, vol. 11, no. 2, Mar. 1996, pp. 270–274.
33. R. Watson, G. C. Hua, and F. C. Lee, "Characterization of an Active Clamp Flyback Topology for Power Factor Correction Applications," Conference Record of APEC'94, pp. 412–418.
34. D. Maksimovic, Y. Jang, and R. Erickson, "Nonlinear-Carrier Control for High Power Factor Boost Rectifiers," Conference Record of APEC'95, pp. 635–641.
35. R. Zane and D. Maksimovic, "Nonlinear-Carrier Control for High-Power-Factor Rectifiers Based on Flyback, Cuk or Sepic Converters," Conference Record of APEC'96, pp. 814–820.
36. H. Y. Wu, X. M. Yuan, J. F. Zhang and W. X. Lin, "Single-Phase Unity Power Factor Current-Source Rectification with Buck-Type Input," Conference Record of IEEE-PESC'96, pp. 1149–1154.
37. R. Oruganti and M. Palaniapan, "Inductor Voltage Controlled Variable Power Factor Buck-Type Ac-Dc Converter," Conference Record of IEEE-PESC'96, pp. 230–237.
38. S. M. Cuk, "Modeling, and Design of Switching Converters," Ph.D. dissertation, California Institute of Technology, 1977.
39. R. Redl, L. Balogh, and N. O. Sokal, "A New Family of Single-Stage Isolated Power-Factor Correctors with Fast Regulation of the Output Voltage," Conference Record of IEEE-PESC'94, pp. 1137–1144.
40. T. F. Wu, T. H. Yu, and Y. C. Liu, "Principle of Synthesizing Single-Stage Converters for Off-Line Applications," Conference Record of IEEE-APEC'98, pp. 427–433.
41. M. Berg and J. A. Ferreira, "A Family of Low EMI, Unity Power Factor Correctors," Conference Record of IEEE-PECS'96, pp. 1120–1127.
42. M. Madigan, R. Erickson, and E. Ismail, "Integrated High-Quality Rectifier-Regulators," Conference Record of IEEE-PESC'92, pp. 1043–1051.
43. P. Kornetzky, H. Wei, and I. Batarseh, "A Novel One-Stage Power Factor Correction Converter," Conference Record IEEE-APEC'97, pp. 251–258.
44. P. Kornetzky, H. Wei, G. Zhu, and I. Bartarseh, "A Single-Switch Ac/Dc Converter with Power Factor Correction," Conference Record IEEE-PESC'97, pp. 527–535.
45. L. Huber and M. M. Jovanovic, "Single-Stage, Single-Switch, Isolated Power Supply Technique with Input-Current Shaping and Fast Output-Voltage Regulation for Universal Input-Voltage-Range Applications," Conference Record IEEE-APEC'97, pp. 272–280.
46. J. Qian, Q. Zhao, and F. C. Lee, "Single-Stage Single-Switch Power Factor Correction (S4-PFC) AC/DC Converters with DC Bus Voltage Feedback for Universal Line Applications," Conference Record of IEEE-Apec'98, pp. 223–229.
47. J. Qian and F. C. Lee, "A High Efficient Single Stage Single Switch High Power Factor AC/DC Converter with Universal Input," Conference Record IEEE-APEC'97, pp. 281–287.
48. R. Redl, "Reducing Distortion in Boost Rectifiers with Automatic Control," Conference Record of IEEE-APEC'97, pp. 74–80.
49. Y. S. Lee and K. W. Siu, "Single-Switch Fast-Response Switching Regulators with Unity Power Factor," Conference Record of APEC'96, pp. 791–796.
50. M. M. Jovanovic, D. M. C. Tsang, and F. C. Lee, "Reduction of Voltage Stress in Integrated High-Quality Rectifier-Regulators by Variable-Frequency Control," Conference Record of APEC'94, pp. 569–575.
51. M. M. Jovanovic, D. Tsang, and F. C. Lee, "Reduction of Voltage Stress in Integrated High-Quality Rectifier-Regulators by Variable-Frequency Control," Conference Record of APEC'94, pp. 569–575.
52. J. Wang, W. G. Dunford, and K. Mauch, "A Fixed Frequency, Fixed Duty Cycle Boost Converter with Ripple Free Input Inductor Current for Unity Power Factor Operation," Conference Record of IEEE-PESC'96, pp. 1177–1183.
53. J. Sebastian, P. Villegas, F. Nuno, and M. M. Hernando, "Very Efficient Two-Input DC-to-DC Switching Post-Regulators," Conference Record IEEE-PESC'96, pp. 874–880.
54. J. Sebastian, P. Villegas, M. M. Hernando, and S. Ollero, "Improving Dynamic Response of Power Factor Correctors by Using Series-Switching Post-Regulator," Conference Record IEEE-APEC'98, pp. 441–446.
55. J. Hwang and A. Chee, "Improving Efficiency of a Pre-/Post-Switching Regulator (PFC/PWM) at Light Loads Using Green-Mode Function," Conference Record IEEE-APEC'98, pp. 669–675.
56. D. Weng and S. Yuvarajan, "Constant-Switching Frequency AC-DC Second-Harmonic-Injected PWM," Conference Record IEEE-APEC'95, pp. 642–646.
57. A. Miwa, D. M. Otten, and M. F. Schlecht, "High Efficiency Power Factor Correction Using Interleaving Techniques," Conference Record IEEE-APEC'92, pp. 557–568.
58. M. S. Elmore, W. A. Peterson, and S. D. Sherwood, "A Power Factor Enhancement Circuit," Conference Record IEEE-APEC'91, pp. 407–414.
59. J. Sebastian, P. Villegas, M. M. Hernando, and S. Ollero, "High Quality Flyback Power Factor Correction Based on a Two-Input Buck Post-Regulator," Conference Record IEEE-APEC'97, pp. 288–294.
60. J. Rajagopalan, J. G. Cho, B. H. Cho, and F. C. Lee, "High Performance Control of Single-Phase Power Factor Correction Circuits Using a Discrete Time Domain Control Method," Conference Record IEEE-APEC'95, pp. 647–653.
61. L. J. Borle and C. V. Nayar, "Ramptime Current Control," Conference Record IEEE-APEC'96, pp. 828–834.
62. S. Z. Dai, N. L. Lujara, and B. T. Ooi, "A Unity Power Factor Current-Regulated SPWM Rectifier with a Notch Feedback for Stabilization and Active Filtering," IEEE Trans. on Power Electronics, vol. 7, no. 2, Apr. 1992, pp. 356–363.
63. T. Ohnuki, O. Miyashita, T. Haneyoshi, and E. Ohtsuji, "High Power Factor PWM Rectifiers with an Analog Pulsewidth Prediction

Controller," IEEE Trans. on Power Electronics, vol. 11, no. 3, May 1996, pp. 460–471.

64. D. Zaninelli and A. Domijan, Jr., "IEC and IEEE Standards on Harmonics and Comparisons," Proceedings of NSF Conf. on Unbundled Power Quality Services in Power Industry, Nov. 1996, pp. 46–52.

65. R. Redl, "Power Factor Correction in Single-Phase Switching-Mode Power Supplies – an Overview," International Journal of Electronics, vol. 77, no. 5, 1994, pp. 555–582.

66. A. Prasada, P. D. Ziogas, and S. Manias, "A Novel Passive Waveshaping Method for Single-Phase Diode Rectifiers," Conference Record of IEEE-PESC'89, pp. 99–105.

67. R. Redl and L. Balogh, "RMS, DC, Peak, and Harmonic Curent in High-frequency Power-Factor Correction with Capacitive Energy Storage," Conference Record of IEEE-APEC'92, pp. 533–540.

68. E. P. Nowicki, "A Comparison of Single-Phase Transformer-Less PWM Frequency Changers with Unity Power Factor," Conference record of IEEE-APEC'94, pp. 879–885.

69. M. M. Swamy and K. S. Bhat, "A Comparison of Parallel Resonant Converters Operating in Lagging Power Factor Mode," IEEE Trans. on Power Electronics, vol. 9, no. 2, Mar. 1994, pp. 181–195.

70. Z. Lai, K. M. Smedley, and Y. Ma, "Time Quantity One-Cycle Control for Power Factor Correctors," Conference Record IEEE-APEC'96, pp. 821–827.

71. J. Hong, D. Maksimovic, R. W. Erikson, and I. Khan, "Half-Cycle Control of the Parallel Resonant Converter Operated as a High Power Factor Rectifier," IEEE Trans. on Power Electronics, vol. 10, no. 1, Jan. 1995, pp. 1–8.

72. Z. Lai and K. M. Smedley, "A Family of Power-Factor-Correction Controllers," Conference Record of APEC'97, pp. 66–73.

73. J. Sun, W. C. Wu, and R. M. Bass, "Large-Signal Characterization of Single-Phase PFC Circuits with Different Types of Current Control," Conference Record of IEEE-Apec'98, pp. 655–661.

74. J. A. Pomilio and G. Spiazzi, "High-Precision Current Source Using Low-Loss, Single-Switch, Three-Phase AC/DC Converter," IEEE Trans. on Power Electronics, vol. 11, no. 4, July 1996, pp. 561–566.

75. M. S. Dawande and G. K. Dubey, "Programmable Input Power Factor Correction Method Rectifiers," IEEE Trans. on Power Electronics, vol. 11, no. 4, July 1996, pp. 585–591.

76. O. Garcia, C. P. Alou, R. Prieto, and J. Uceda, "A High Efficiency Low Output Voltage (3.3V) Single Stage AC/DC Power Factor Correction Converter," Conference Record of IEEE-Apec'98, pp. 201–207.

77. K. Tse and M. H. L. Chow, "Single Stage High Power Factor Converter Using the Sheppard-Taylor Topology," Conference Record of PESC'96, pp. 1191–1197.

78. M. Daniele, P. Jain, and G. Joos, "A Single Stage Single Switch Power Factor Correction AC/DC Converter," Conference Record of PECS'96, pp. 216–222.

79. A. Canesin and I. Barbi, "A Unity Power Factor Multiple Isolated Outputs Switching Mode Power Supply Using a Single Switch," Conference Record of APEC'91, pp. 430–436.

80. R. Erikson, M. Madigan, and S. Singer, "Design of a Simple High-Power-Factor Rectifier Based on the Flyback Converter," Conference Record of APEC'90, pp. 792–801.

81. A. Peres, D. C. Martins, and I. Barbi, "Zeta Converter Applied in Power Factor Correction," Conference Record of PESC'94, pp. 1152–1157.

82. J. Lazar and S. Cuk, "Feedback Loop Analysis for AC/DC Rectifiers Operating in Discontinuous Conduction Mode," Conference Record of APEC'96, pp. 797–806.

83. J. Sebastián, P. Villegas, M. M. Herando, J. Diáz and A. Fantán, "Input Current shaper Based on the Series Connection of a Voltage Source and a Loss-Free Resistor," Conference Record of IEEE-Apec'98, pp. 461–467.

84. A. Huliehel, F. C. Lee, and B. H. Cho, "Small-Signal Modeling of the Single-Phase Boost High Power Factor Converter with Constant Frequency Control," IEEE PESC'92, pp. 475–482.

85. G. Zhu, H. Wei, P. Kornetzky, and I. Batarseh, "Small-Signal Modeling of a Single-Switch AC/DC Power Factor Correction Circuit," IEEE PESC'98.

86. R. D. Middlebrook and S. Cuk, "A General Unified Approach to Modeling Switching Converter Stages," IEEE PESC'76, pp. 18–31.

87. S. Tsai, "Small-Signal and Transient Analysis of a Zero-Voltage-Switched, Phase-Controlled PWM Converter Using Averaging Switch Model," IEEE-IAS'91, pp. 1010–1016.

88. E. V. Dijk, H. J. N. Spruijt, D. M. O'Sullivan, and J. B. Klaassens, "PWM-Switch Modeling of DC-DC Converter," IEEE Trans. on Power Electronics, vol. 10, no. 6, Nov. 1995, pp. 659–665.

20
Gate Drive Circuitry for Power Converters

Irshad Khan,
MTech Eng, MSc
University of Cape Town, Department of Electrical Engineering, Cape Town, South Africa

20.1 Introduction to Gate Drive Circuitry 543
20.2 Semiconductor Drive Requirements 544
    20.2.1 Current-driven Devices • 20.2.2 Voltage-controlled Devices
20.3 Gate Drivers for Power Converters... 545
    20.3.1 Floating Supply • 20.3.2 Level Shifting
20.4 Gate Driver Circuit Implementation.. 549
    20.4.1 Isolated Gate Drivers • 20.4.2 Electronic Gate Drivers
20.5 Current Technologies 553
    20.5.1 Transformer Coupled Isolated Drivers • 20.5.2 Non-isolated Electronic Level Shifted Drivers • 20.5.3 High-speed Gate Drivers • 20.5.4 Resonant Gate Drivers
20.6 Current and Future Trends 558
20.7 Summary 558
    References 558

20.1 Introduction to Gate Drive Circuitry

Global trends towards energy efficiency over the last three decades has facilitated the need for technological advancements in the design and control of power electronic converters for energy processing. These power electronic converters find widespread applications in industry such as:

Consumer electronics

- Battery chargers for cellular telephones and cameras
- Computer power supplies

Automobile industries

- Electronic ignitions and lighting

Commercial sectors

- Variable speed motor drives for conveyor belt systems
- Induction heating installations for metals processing
- Uninterruptible power supplies (UPS)
- Industrial welding

Domestic electronics

- Fluorescent, compact fluorescent, and incandescent lighting, washing machines, cooking appliances, and dish washers

Utility applications

- DC transmission for electrification purposes

All of the above applications utilize similar power converter topologies. This allows for controlled and efficient power conversion from one form of energy to another, utilizing semiconductor technologies such as BJT transistors, SCRs (thyristors), IGBTs, and power MOSFETs.

It is true that "Power is nothing without control" (Pirelli Tyres). The gate drive circuitry of a power converter forms an important interface between the high power electronics and the intelligent control processing stages.

It is therefore of utmost importance that this interface between the control and power electronics is well designed, since it can have a substantial impact on the performance and reliability of a power electronic system. (See Fig. 20.1.)

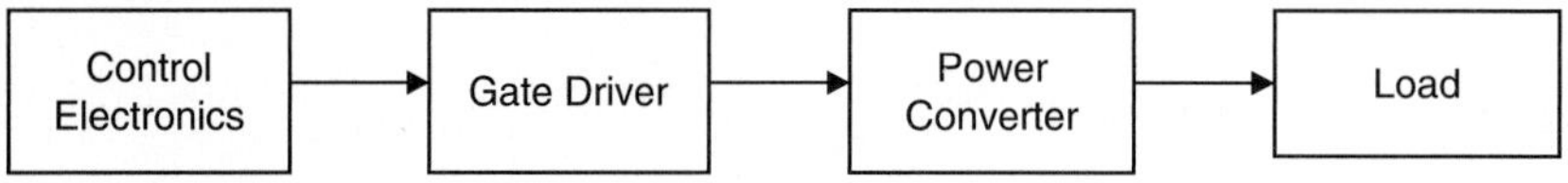

```
Control          Gate Driver        Power          Load
Electronics                         Converter
```

FIGURE 20.1 Generalized layout of a power electronic system showing the situation of the gate driver circuits.

20.2 Semiconductor Drive Requirements

Power semiconductor devices have three operating states commonly known as the cut-off mode, the active mode, and the saturation mode. In power electronic converters which utilize switch-mode operation, the aim is to operate these semiconductors in either the cut-off region or the saturation region, whilst making the transition through the active or linear region as short as possible in order to facilitate maximum power conversion efficiency. In order to achieve these fast transition times, a suitable gate driver circuit is required. This gate driver has to be able to supply the necessary charge to the power semiconductor device gate junction in order to achieve turn-on and turn-off.

Power semiconductors can be classified into two categories with respect to drive requirements, namely: current-driven devices and voltage-driven devices.

20.2.1 Current-driven Devices

A current-driven device is a device that requires a constant current drive for a period of time in order to initiate and/or remain in conduction. Two popular types of current controlled devices are the bipolar power transistor and the thyristor (SCRs). The SCR is mainly used in AC–DC converters such as controlled rectifiers where the input AC voltage helps to commutate (turn-off) the devices by polarity reversal. In DC–AC inverters, the gate turn-off thyristors (GTOs) are generally used due to their ability to be both turned on and off by a gate control signal. This eliminates the need for forced commutation circuitry, needed to turn-off the thyristor in these applications due to the absence of a reversing polarity AC input voltage. Thyristors and GTOs still dominate the high voltage and high current applications, like DC transmission, requiring converters up to the megawatt range. The evolution of the IGBT power semiconductor device has resulted in thyristors and GTOs being used less in conventional power converters these days. IGBTs are readily available with on-state currents of several hundred amperes and blocking voltages in excess of 1.7 kV. The IBGTs offer much faster switching speeds than thyristors and have much lower gate drive power requirements than SCRs or power bipolar junction transistors (BJT). The IGBTs have already replaced power BJTs in most applications due to their superior performance. The next generation of silicon carbide technology could realize IGBT devices with blocking voltages in excess of 4 kV and may one day even replace SCR and GTO devices completely.

20.2.2 Voltage-controlled Devices

These devices are semiconductors which require a constant voltage drive on the gate control terminal in order to remain in conduction. The input drive requirements of these devices are substantially lower than their current-driven counterparts and are the preferred choice in modern power electronics. Two such devices are the power MOSFET and the IGBT which are forced commutated switching devices being fully controlled at the gate terminal under normal operating conditions. These devices do not latch into conduction, and therefore do not require special commutation circuits. The gate input junction of a MOSFET and IGBT is purely capacitive, so, no gate drive current is needed in the steady state, unlike transistors. A minimum gate drive voltage however must be maintained (above the gate threshold voltage) at the device gate in order for it to remain in conduction. A high current low impedance drive circuit is needed to inject, or remove current, from the gate at high slew rates in order to switch the device rapidly. The gate drain capacitance, although small, can also require significant charge as at high drain voltage slew rates (the Miller effect) [1].

The process of supplying the necessary power for the efficient driving of voltage-controlled devices is an ongoing area of research. This chapter will however only focus on the gate driver circuitry required to drive power MOSFETs and IGBTs in the bridge circuit configuration. The basic symbolic representation of the power MOSFET and IGBT is shown below. Figure 20.2 represents the static model of these devices. Unlike the power MOSFET, the IGBT can be manufactured without the integral body diode. Often power MOSFETs are represented as shown below, omitting the integral body diode which is always present in the MOSFET device. Care should be taken when reading circuits, not to forget this extra unseen component.

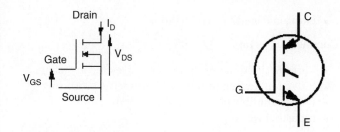

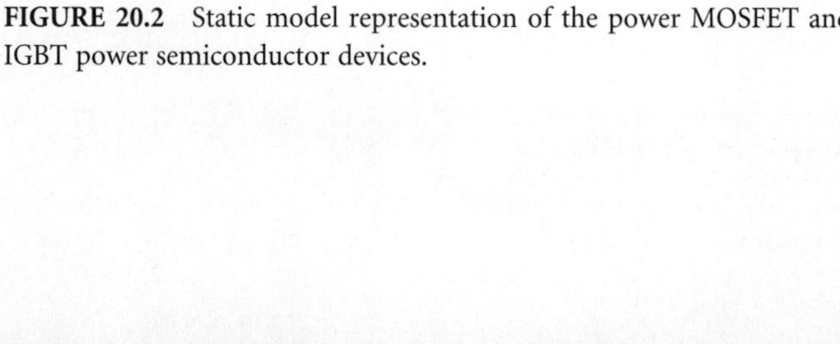

FIGURE 20.2 Static model representation of the power MOSFET and IGBT power semiconductor devices.

20 Gate Drive Circuitry for Power Converters

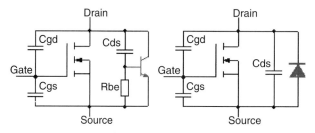

FIGURE 20.3 A dynamic model of the power MOSFET shows the presence of the parasitic BJT element found in the MOSFET structure [2, 1].

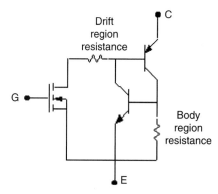

FIGURE 20.4 Dynamic model of an IGBT showing the location of the parasitic thyristor responsible for uncontrolled device latch up [1].

The dynamic model for a power MOSFET is shown in Fig. 20.3 [2, 1]. It can be noted that the integral body diode situated in the power MOSFET structure is actually a parasitic bipolar transistor. This parasitic element is the result of a fabrication process used when manufacturing power MOSFETs. The base of the BJT is usually floating, and a pull down resistor Rbe is included in the chip structure in order to keep the BJT base shorted to the MOSFET source terminal at all times. This is very necessary, since a floating BJT base will result in uncontrolled turn-on of the BJT device. By shorting the BJT base to the MOSFET's source terminal, a diode structure is now created, yielding the dynamic model with the integral body diode formed by the BJT base collector junction as shown in Fig. 20.3 [2, 1]. The turn-on of the BJT element during transients substantially degrades the performance of the power MOSFET. When the MOSFET is exposed to a fast changing transient voltage (*dv/dt*) across its drain to source terminals, enough current flows in Rbe, thereby biasing the parasitic bipolar into conduction. This transient can occur during the turn-on or turn-off of the device. To avoid transient susceptibility, the device *dv/dt* rating must be adhered to at all times. The input capacitance of the power MOSFET comprises the gate to source capacitance (Cgs) and the gate to drain capacitance (Cgd) also known as the Miller capacitance. The input gate structure for the IGBT is essentially the same as for the power MOSFET. A gate driver must be capable of supplying sufficient charge to both Cgs and Cgd when switching power MOSFETs and IGBTs in power converters.

A dynamic model of the IGBT is shown in Fig. 20.4. The IGBTs are essentially bipolar devices like BJTs with a MOSFET structured high impedance input. A parasitic transistor is also found in the IGBT chip structure as part of the manufacturing process. Combined with existing bipolar power transistor, it forms a parasitic thyristor, which once latched into conduction device, turn-off is not possible via the gate terminal. To avoid this uncontrolled behavior, the maximum allowable peak current and *dv/dt* ratings of the device must be adhered to at all times [1].

Both MOSFETs and IGBTs require sufficient charge deposited into their gate junctions, whilst maintaining a minimum gate threshold voltage in order to remain in conduction. When designing a gate driver, it is always important to understand both the static and dynamic behavior of the semiconductor device used as it aids the effectiveness of the design for a given gate driver system.

20.3 Gate Drivers for Power Converters

Power electronic converters are found in various configurations. Each has a complex switching function with the goal of achieving efficient power conversion. These converters make use of various floating potentials in order to achieve an AC or quasi-DC output characteristics. Gate drivers function as current buffers and signal converters. They convey both the switching state information and gate drive power required during the power semiconductor switching process.

20.3.1 Floating Supply

Power converters employing bridge configurations function by virtue of a high-side switch as shown in Fig. 20.5 below.

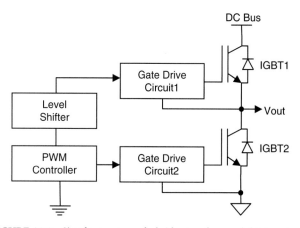

FIGURE 20.5 Simple structure of a bridge topology and driving circuit illustrating the concept of a high-side switch.

In order to drive a device like an IGBT or power MOSFET into conduction, the gate terminal must be made positive with respect to its source or emitter. A common misconception made by many newcomers to the field of power electronics is that, because the emitter or source terminal is usually at some ground potential, it must be made positive with respect to ground. It can be noted that the emitter terminal of IGBT1, in the circuit above, can be floating anywhere from ground up to the DC bus potential, depending on the operating states of IGBT1 and IGBT2, and is therefore not referenced to the system ground potential. A supply is therefore needed in order to provide power to any circuitry associated with this floating midpoint potential. This type of supply is commonly referred to as a floating supply. Three of the most common methods of generating a floating supply are described next.

- **Isolated Supply**
  The simplest way of generating a floating supply is to use a transformer isolated supply. Compared to other methods, this type of supply is able to supply a continuous, large amount of current. A mains frequency transformer supply is cheap, but is usually very bulky. By employing a high frequency isolated DC–DC converter, fed from an existing DC supply, an isolated floating supply can be generated employing a much smaller isolation transformer.
- **Charge-pump Supply**
  The charge-pump technique superimposes the voltage of one supply onto another. It is normally used for generating a boost voltage on top of the main high-voltage supply. The charge-pump supply is not suitable for generating a boost voltage for powering floating high-side circuitry. A benefit of employing the charge-pump technique is that a continuous supply to the circuit is maintained. Due to complexity and cost, this circuit is not commonly used for power converters [3, 4].
- **Bootstrap Supply**
  A very common technique employed to generate a floating supply, the bootstrap supply is a simple circuit using only one diode and a supply storage capacitor. This technique is commonly used for low cost solutions in converters up to several kilowatts. Typical applications include electronic ballasts and variable speed motor drives.

A simplified schematic of a bootstrap circuit is shown in Fig. 20.6. When the low-side switch M2 is on, the bootstrap diode, D, conducts and charges the storage capacitor. If we assume the saturation voltage drop of the low-side power device M2 and the forward volt drop of diode D to be negligible, the capacitor will charge to approximately the low voltage supply potential. When the high-side switch M1 is on low side switch M2 is off, D is reverse-biased and the high-side circuitry is powered from C. In this condition, the voltage on C droops as it discharges when supplying the high-side circuitry.

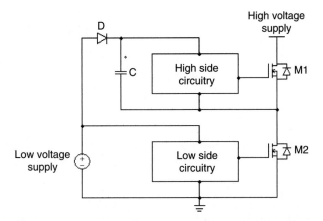

FIGURE 20.6 Bootstrap technique employed for creating a floating supply.

The amount of voltage droop is a function of the current drawn, the size of the capacitor, and the operating frequency of the converter [3]. Detailed design and application literature on bootstrap supplies can be found in [4].

A summary and comparison of the various methods of generating a floating supply is given in Table 20.1 [5].

20.3.2 Level Shifting

The second requirement for driving a high-side switch, is that the control signal fed from the PWM electronics needs to be conveyed to the floating driver circuitry. A level-shifting circuit is required in order to achieve this.

The GD1 and GD2 are the respective high- and low-side gate drive circuits. In order to signal the high-side gate driver circuitry to commence turn-on of the high-side IGBT switch, the control signal V1 which is referenced to the control circuit ground potential needs to be referenced to the floating potential V_{out} at the IGBT1 emitter. This implies that the control signal V1 will be level shifted to V_{g1} as shown in Fig. 20.7. A level shifter can be thought of as an isolating black-box that transfers a signal across a potential barrier [6]. The maximum level attained by V_{g1} will be $V_{\text{out}} + V1$. In power converters, operating off mains voltage, these levels can be in excess of 500 V. Level shifting is achieved in one of the following ways:

- **Transformer Level Shifting**
  Transformers are an obvious choice for providing a level shifted signal. They have excellent noise immunity when compared to opto-couplers. Level shifting transformers also have the benefit of providing both the control and gate drive power signal to the power semiconductor switch (Fig. 20.8), thereby eliminating. This eliminates the need for a floating power supply. When operating at high switching frequencies, above several hundred kilohertz, careful transformer design has to be applied in order to avoid the adverse effects of transformer

TABLE 20.1 Comparison of the various techniques of creating a gate driver floating supply [5]

Potential isolation	Transformers			None
System	50 Hz power supply	Switch-mode power supply		Bootstrap supply
Supplied by	Auxiliary voltage or mains voltage	Auxiliary voltage	DC bus voltage	Operating voltage on bottom side
AC-frequency	Low	Very high	Medium	Medium
Smoothing requirements	High	Very low	Low	Low
For power modules	1200 V	>1700 V	1700 V	1200 V
Output voltage	Positive and negative	Positive and negative	Positive and negative	Only positive
Duty cycle restriction	No	No	No	Yes
Coupling capacitance	High	Low	Medium	Low
HF interference emission	None	High	Low	None
Cost	Low	Low	High	Very low

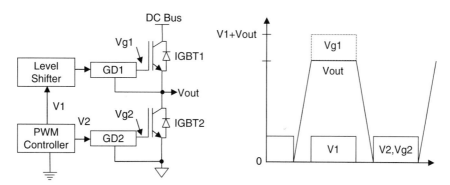

FIGURE 20.7 The concept of level shifting and the placement of the switching control signal on top of the inverter output voltage.

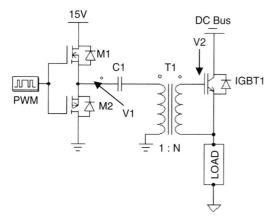

FIGURE 20.8 Combined transformer level shifter and gate driver. Capacitor C1 removes any DC offset in V1 thereby maintaining a zero average volt–time product across the transformer windings.

leakage inductance. Another limitation in the transformer gate driver, is when large currents are required at high speed for gate driving. The influence of the transformer parasitic components becomes significant. In the effort to deliver high peak currents at fast rise and fall times, the transformer turns are usually minimized, which leads to other transformer design limitations.

To avoid degradation of the gate drive waveform in the case of high speed and high current delivery, it is sometimes better to convey only the low power control signal via a transformer and also to employ a dedicated low-impedance output MOS-gate driver with a floating power supply [3].

A drawback of the gate drive transformer is that it cannot convey DC information (since the average volt–time product across any winding must be zero) and therefore has a limitation to its operation as shown in Fig. 20.9. When a high duty cycle is commanded, the gate drive capacitor will remove the DC offset in the signal, and could result in operation below the device threshold voltage $V_{ge}(th)$ as shown in Fig. 20.9c [3]. This condition will result in the device not turning on, or even operating in the linear region. This results in excessive semiconductor power dissipation.

- **Optical Level Shifting – Opto-couplers**
  Optical isolation is another technique used for achieving level shifting. The trade-off however, is that a separate floating supply is now required on the receiver end of the gate driver interface. Opto-couplers offer a cost effective, and easy solution, but are susceptible to noise and fast voltage transitions. This is common in gate drive circuits. This requirement places more demand on the power

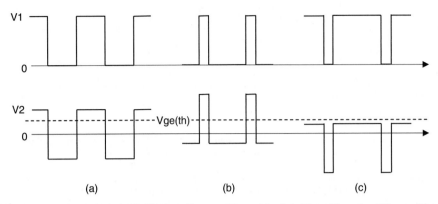

FIGURE 20.9 Operating waveforms for the pulse transformer circuit in Fig. 20.8 under different duty cycles.

supply filtering and PCB layout around the opto-coupler in order to achieve reliable operation. Other benefits of opto-coupler technology is the small footprint (6-8 pin dual in line or surface mount package), output enable pin, and compatibility with any logic level input since its input is current-driven (an input diode). Commercially available opto-couplers cover a wide range of operating speeds of up to 15 MBd, with rise and fall times of 10–20 ns and noise immunity levels (dv/dt rating) of 10–15 V/ns. Typical opto-coupler isolation voltages are in the region of 5 kV which is sufficient, since most semiconductors have breakdown voltages lower than this.

- **Optical Level Shifting – Fiber Optic Link**
Electrical conductors are susceptible to electromagnetic fields, and hence will radiate and pick up electromagnetic noise. Fiber optic links neither emit, nor receive electromagnetic noise, and pass through noisy environments unaffected. In addition, when dealing with very rapidly changing currents, ground noise can become a problem. A fiber optic link practically eliminates any ground loop or common-mode noise problems. A basic fiber optic communication system is given in Fig. 20.10.

The LED is driven by the PWM drive signal. The light emitted from the LED is sent down a length of fiber optic cable to the receiver. The receiver consists of a PIN photodiode and a trans-impedance amplifier. The output voltage of the amplifier is level detected by a comparator, and converted into a logic signal.

The galvanic isolation and dv/dt rating of this communication link can be increased to almost any desired value by simply lengthening the fiber optic cable. For a modest length of 10 cm of fiber optic cable, the isolation voltage is approximately 100 kV. The dv/dt rating is difficult to calculate, but as the coupling capacitance is practically zero, even the highest realizable dv/dt rating will have little effect. The bandwidth available in fiber optics systems makes operation at over 1 GHz possible [7].

- **Electronic Level Shifting**
Electronic level shifters work on the principle of having a current source referenced to a fixed potential. This reference potential is usually that of the low-side switches source, hence they are also called common source level shifters. The drive signal causes a certain current to be drawn through a current source regardless of the voltage

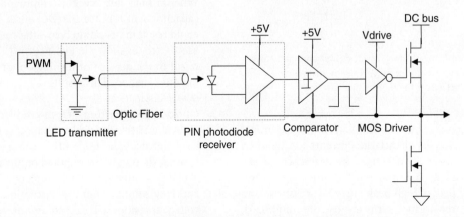

FIGURE 20.10 Fiber optic level shifting circuit used for high speed, extremely high noise immunity, and very high isolation voltage capability.

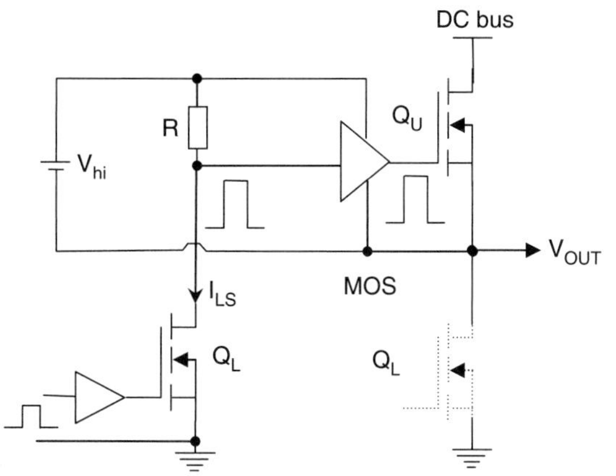

FIGURE 20.11 Electronic level shifting employed to convey the switching signal to the high-side gate drive circuitry. This technique does not offer any isolation but has widespread applications as an HV interface IC for power electronics applications.

across it. The current is sensed by the receiving side and turned into on/off information. Electronic level shifters do not provide isolation, however being an all electronic solution, allows it to be integrated onto a single chip. A simple common source non-inverting level shifter is shown in Fig. 20.11. The output capacitance and the power dissipation by the current source, limit the frequency and voltage it can feasibly work at. A better solution, that decreases the amount of power dissipated, is to use a pulsed current source, which is then latched on the receiving side. Topologies such as the single-ended and dual current pulsed latch are discussed in detail in reference [3].

A summary and comparison of the various methods of level shifting is given in Table 20.2 [5].

20.4 Gate Driver Circuit Implementation

20.4.1 Isolated Gate Drivers

Gate driver circuits incorporating electrical isolation, provide the benefit of good noise interference immunity between power and control circuits. This is as a result of the separated ground return paths. Several types of isolated driver circuits exist namely:

- **Isolated Power Supply with Opto-coupled Control Signal Inputs**
  A standard technique employed widely employed, is to generate floating supplies through the use of mains frequency transformer isolation (Fig. 20.12). Although not a very space-efficient solution (due to the mains frequency transformer size) compared to other methods, this technique is still well suited to almost any power electronic converter application. The power supply circuits can comprise standard low-voltage bridge rectifier modules. A regulated supply is easily realized using three-terminal voltage regulators, which can easily deliver continuous power to gate drive circuits in excess of 10 W. With each winding and its associated power supply, referenced to the IGBT or power MOSFET source, this system is a reliable and medium cost solution for almost any power converter up to switching frequencies well below 1 MHz. For higher operating frequencies, the inter-winding capacitance of the mains frequency transformer, leads to noise feed-through coupling. This causes spurious effects, like unscheduled turn-on of the power devices due to noise present in the driver circuit ground return path. The solution to this problem would then be to introduce a high frequency DC–DC converter, which incorporates a small HF isolation transformer with substantially reduced inter-winding capacitance.

  Level shifting of the switching control signal is achieved by means of optical isolation, (U3 and U4) with the input diodes (primary side of opto-coupler) referenced to the logic ground of the signal processing circuitry. The low impedance gate driver output is achieved by employing a high speed, high current buffer integrated circuit or a discrete bipolar or MOS complementary totem pole stage. Power to the opto-coupler and buffer U5 and U6 is derived from the respective floating power supply. The circuit above does not have any operating duty cycle limitations due to the floating power supply.

  The passive network comprising D1, R5, and R6 control the IGBT switching speed, and impacts on the

TABLE 20.2 Comparison of the various techniques of level shifting for gate drivers

Potential isolation	Transformers	Optical		None
System	Pulse transformer	Opto-coupler	Fiber optic link	Electronic level shifting
For power modules up to	>1700 V	1700 V	>1700 V	1200 V
Signal transmission	Bi-directional	Uni-directional	Uni-/bi-directional	Uni-directional
Duty cycle restriction	Yes	No	No	No
Coupling capacitance	5–20 pF	1–5 pF	<1 pF	>20 pF
dv/dt immunity	High	Low	High	Low
Costs	Medium	Low	High	Low

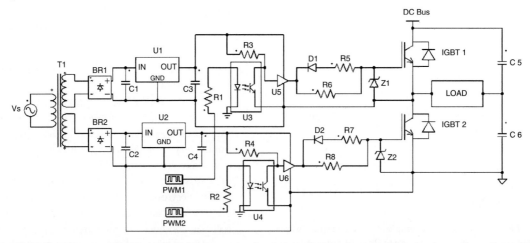

FIGURE 20.12 Mains frequency transformer employed to generate a floating supply. Simple and reliable, but larger than other solutions due to its mains frequency transformer. Works well for a half-bridge but would require more isolated power supplies if a full-bridge topology was employed. A 3-phase inverter would require either six separate floating supplies or three separate and one shared supply for the low-side devices.

performance and efficiency of the power converter. The R6 controls the turn-on switching speed of IGBT1. This controls the device switching loss as well as the turn-off dv/dt characteristics of the lower devices (IGBT2) free-wheeling diode for inductive loads. Diode D1 disconnects R5 from the circuit during the IGBT turn-on. The turn-off speed of IGBT1 is controlled by R5, provided that R5 is much smaller than R6. This is a desirable feature in voltage-fed inverters, since it ensures a minimum dead-time between device transitions as shown in Fig. 20.13.

The $U5_{out}$ and $U6_{out}$ in Fig. 20.13 represent the driver output signal at exactly 50% duty cycle. The passive gate network on each IGBT, alters the drive signals due to the RC time constant formed between the gate drive resistors and the IGBT gate capacitance. This is shown as Vg_{IGBT} which is measured directly on the IGBT gate terminal. This slewing action on the IGBT gate results in the IGBT having a delayed turn-on. Turn-on occurs when the IGBT gate voltage reaches its threshold level (Vg_{th}) and collector current starts to flow. The result is a dead-time created between switching transitions. This is required in any bridge circuit to avoid shoot through or cross-conduction of the upper and lower switches. This is depicted by the collector current trace (I_C) for a purely resistive load in Fig. 20.13.

- **Transformer Coupled Driver Supplying both Power and Control Signals**

  A transformer coupled gate driver is shown in Fig. 20.14. This system provides both the floating supply, as well as the level shifting of the switching signal. The push–pull driver on the primary side of the transformer T1, is used to supply a bi-directional current in the primary, without the use of a split power supply. The transformer T1 is rated for the PWM operating frequency, and can be realized by employing either a ferrite or iron powder core. The system operation is limited to a maximum PWM duty cycle of 50%. One of the benefits of this driver circuit, is its ability to generate a negative gate bias voltage when the device is off, due to transformer action. This feature is favorable since it reduces the IGBT dv/dt susceptibility by holding the gate terminal at a negative potential during turn-off transients, thereby avoiding uncontrolled turn-on or latch up in the IGBT. Back-to-back zener diodes placed across the gate-source terminal clamp the device gate voltage, thereby avoiding over voltages generated by the uncoupled transformer (T1) leakage inductance. The parallel resistor R5, acts as a gate pull down resistor holding the device in the off state during the initial power up of the gate driver circuit. The resistor – diode network in the gate drive circuit, (consisting of D1, R1, and R2) serves the same

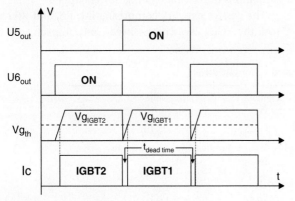

FIGURE 20.13 Switching waveforms for the circuit in Fig. 20.12.

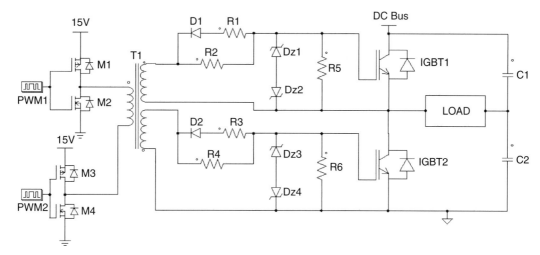

FIGURE 20.14 Transformer coupled gate driver used to supply both the control signal and the gate drive power to the device.

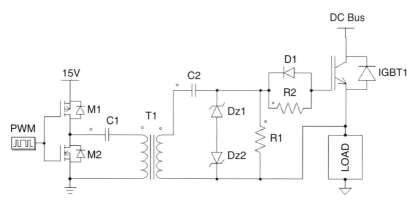

FIGURE 20.15 Transformer coupled gate driver with a large duty cycle operating range.

purpose as described in Fig. 20.12. The second function of this passive resistor network, is to dampen ringing effects. This is caused by the interaction between the IGBT or power MOSFET gate capacitance, and the gate drive transformer leakage inductance.

- **Transformer Coupled Gate Driver with Large Duty Cycle Capability**
  Transformers offer excellent noise immunity and provide simple and cost-effective gate drive solutions, whilst maintaining electrical isolation between the control and gate drive electronics. A drawback however is the limitation the transformer places on the maximum operating duty cycle. Figure 20.15 offers a simple but effective solution to the conventional limitations by the introduction of a DC restorer circuit which is formed by C2, Dz1, and Dz2. This system allows for removal of any DC information via C1, and restores the input waveform applied with the addition of a negative voltage bias needed for the IGBT gate drive. A small ferrite transformer core can be used for a MOSFET gate driver operational to several hundred kilohertz. This circuit can be redesigned for bridge topologies, but is also well suited for high voltage DC–DC converters requiring a high-side switch. The effective duty cycle range of this driver is from 5 to 95%. Operating waveforms are shown in Fig. 20.16.

  It should be noted that the gate drive voltage is clamped at fixed levels regardless of the duty cycle used, unlike the case in Fig. 20.9. This technique also supplies both the level shifted signal as well as the gate drive power, eliminating the need for an additional floating supply. The transformer turns ratio (T1) can also be adjusted to allow the circuit in Fig. 20.15 to operate from a 5 V supply, whilst generating an output voltage swing from +15 to −5 V at the IGBT gate.

- **Transformer Coupled Signal Modulated Gate Driver**
  The circuit in Fig. 20.17 employs a high-frequency carrier signal that is modulated by a lower frequency control signal (PWM). This is used to generate on/off switching instants of power IGBT1. By employing a high-frequency carrier, the transformer size is reduced, and

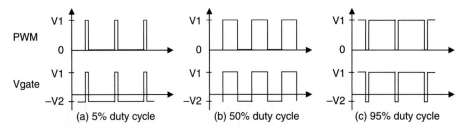

FIGURE 20.16 Operating waveforms of the transformer coupled gate driver in Fig. 20.15.

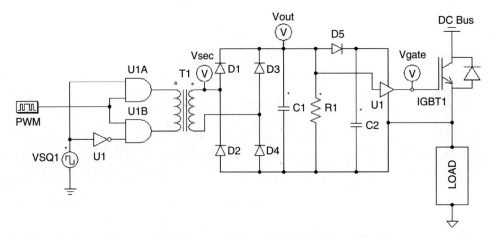

FIGURE 20.17 Signal modulated carrier used for level shifting and generation of a floating supply.

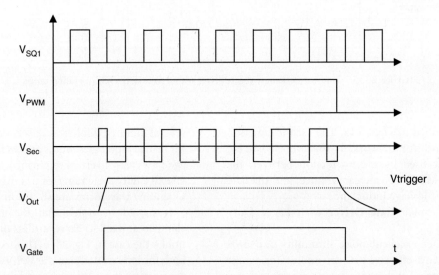

FIGURE 20.18 Signal modulated carrier used for level shifting and generation of a floating supply.

by modulating the time in which the carrier operates, it controls the energy delivered to the gate of the IGBT. A carrier frequency for VSQ1 should be chosen to be much higher than the frequency of the PWM control signal. When the PWM control signal is enabled, the carrier signal is transformed to the secondary of transformer T1, which is rectified and filtered to produce a DC signal V_{out}.

When the PWM control signal goes from the on- to the off-state, the charge stored in the filter capacitor C1 discharges by a time constant determined by R1. This is sometimes problematic when fast switching times are required, especially in inverter bridge configurations. A solution to this problem is to employ an active driver (U1) on the secondary side of the transformer. This will detect the carrier, and switch the IGBT

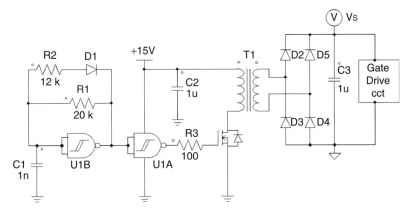

FIGURE 20.19 Low cost high frequency switch-mode floating supply.

gate accordingly. Operating waveforms for this circuit is shown in Fig. 20.18 [1].

- **High Frequency Floating Power Supply**
  Often gate driver systems require extra electronics which has to be referenced to the floating switch being driven. Extra functional electronics often leads to higher power consumption, resulting in the need for a small low cost floating power supply as shown Fig. 20.19. The input section consisting of U1B, forms an oscillator used to drive a MOSFET at high frequency. This MOSFET drives a high-frequency transformer, which forms the isolation medium between the common auxiliary power supply and the floating secondary circuit. Transformer T1 produces a secondary output voltage which is rectified to form a floating DC supply for the associated gate drive circuitry. These floating power supplies are also available as monolithic DC–DC converter ICs with isolated outputs.

20.4.2 Electronic Gate Drivers

These drivers utilize electronics in order to store energy in capacitors. This is used to produce floating referenced potentials. All circuits are referenced to a common ground potential. This technique provides a very cost-effective gate driver solution over the isolated versions, and are becoming increasingly popular in industries.

- **Gate Driver with Bootstrapped Floating Supply**
  The addition of a bootstrapped capacitor voltage allows for the generation of a floating supply in Fig. 20.20. Charging of C1 is achieved in the direction indicated during the period when MOSFET M2 is turned on. The turn-off of M2, results in the entrapment of charge in C1 which now acts as the power supply to OPTO1, U1, and the MOSFET M1's gate. As current is drawn from C1, the capacitor voltage tends to drop and needs to be replenished cyclically. The selection of the bootstrap capacitor value is critical for reliable operation of this circuit under extended duty cycle conditions. Diode D1 sees the full DC bus voltage when M2 is off, and therefore has to have a sufficiently high breakdown voltage. Applications with fast switching speed, require that D1 be a fast recovery diode in order to withstand the high dv/dt present across M2. Level shifting of the switching control signal is achieved by means of the opto-couplers in the circuit, providing the necessary control circuit isolation required.

- **Gate Driver with Floating Supply Derived from DC Bus**
  When the generation of a floating supply is required without the need for isolation, the gate drive power can be derived from the DC bus voltage. The circuit is charged during the period when IGBT1 is off, with the gate drive energy stored in C2. The circuit connected to the enable pin of the opto-coupler, forms an under-voltage lockout which inhibits the IGBT drive signal at low DC supply voltages, typically during the initial start-up of the inverter circuit. The control signals are level shifted through an opto-coupler which drives a totem pole complimentary buffer stage as shown in Fig. 20.21 [8].

20.5 Current Technologies

The evolution of power semiconductor technology has led to the development of smart driver IC modules. These modules offer an essentially single chip/package solution to the gate driver designer, with the inclusion of minimal external components. These smart driver ICs utilize sophisticated electronics, yielding an intelligent gate driver in a small module. Some of these functions include the electronics for generating the floating supply from the auxiliary power supplied to the IC, level shifting with or without isolation. Advanced on-board integrated functions include under-voltage lockout, thermal trip, overload and de-saturation detection, soft start, and shutdown. These features enhance the effective utilization of the power converter and also increase its reliability.

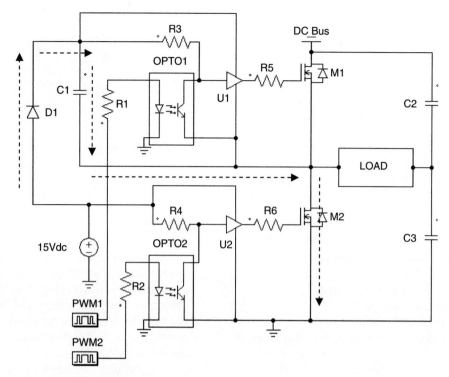

FIGURE 20.20 Optically isolated level shifter with an un-isolated bootstrapped floating supply derived from the low-side power supply.

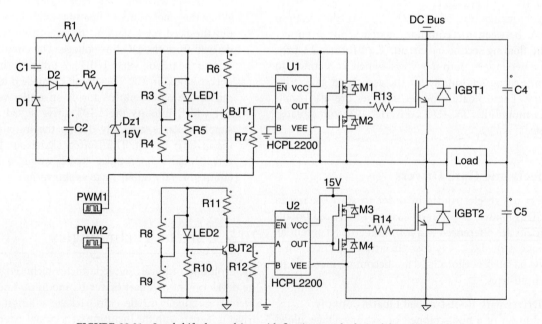

FIGURE 20.21 Level shifted gate driver with floating supply derived from the DC bus.

20.5.1 Transformer Coupled Isolated Drivers

Semikron Corporation has developed a gate driver module that employs integrated floating supplies (Fig. 20.22) [9]. The level shifting is accomplished through transformer coupling, which is used as a bi-directional communications link for the transmission of the gate drive and fault signals.

Each module is configured for a half-bridge inverter configuration and one additional driver module can be added for each inverter leg needed. The modules come in a variety of output current capabilities, depending on the IGBT drive requirements, with usable operating frequencies ranging from the low kHz range up to 30 kHz.

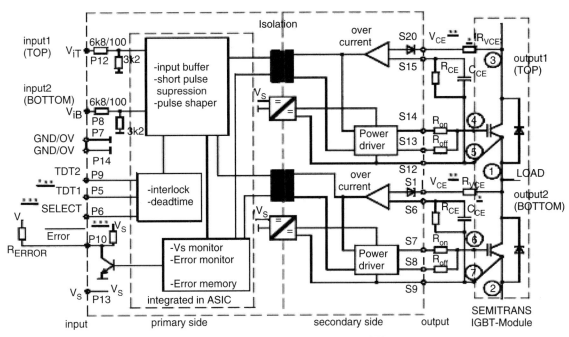

* When SKH122B is driving 1700V IGBTs, a 1kW/0,4W RVCE-resistor must be connected in series to the VCE-input.
** The VCE terminal is to be connected to the IGBT collector C. If the VCE-monitoring is not used, connect S1 to S9 or S20 to S12 respectively.
*** Terminals P5 and P6 are not existing for SKH122A/21A; internal pull-up resistor exists in SKH122A/21A only.
①-⑦ Connections to SEMITRANS GB-models

FIGURE 20.22 SKHI22 integrated half-bridge IGBT module capable of operation at 1200 V. A product of the Semikron Corporation.

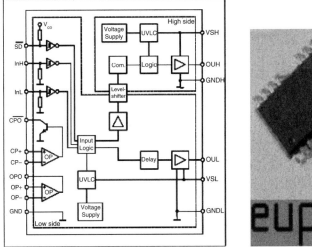

FIGURE 20.23 EUPEC EICE-Driver incorporating monolithic coreless transformer technology allowing for a substantial reduction in the device size.

Another manufacturer of isolated gate driver modules is Eupec semiconductor [10]. These modules employ a monolithically integrated coreless planar transformer design in a gate driver IC. This allows it to be substantially smaller and cheaper [6]. These modules are operational to 60 kHz with output current capabilities in excess of 2 A peak. The EUPEC EICE-Driver (2ED020I12-F) is shown in Fig. 20.23.

20.5.2 Non-isolated Electronic Level Shifted Drivers

The International Rectifier Corporation has developed a family of low cost gate driver ICs for the power electronics industry [11]. International Rectifiers, gate driver ICs (Fig. 20.24) utilize electronic level shifting for conveying the switching signal to

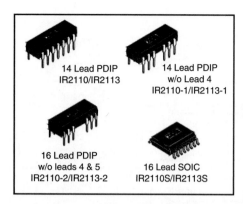 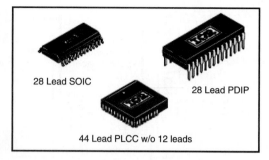

FIGURE 20.24 International Rectifiers IR2113 (Left) and the IR2233 (Right) high voltage gate driver ICs.

the high-side switch and a bootstrap technique for the generation of a floating supply. These ICs are relatively inexpensive and are being used mainly in the low cost consumer electronics industry. Application include Electronic Ballasts and low to medium power inverter applications such as variable speed drives and DC to single-phase AC backup inverter systems. Available in a single-channel, half-bridge, or three-phase inverter driver IC, these devices require a minimum amount of external components. Intelligent features like under voltage lockout, shut down, and current sensing inputs are included in the driver IC.

Functional diagrams of the IR2113 half-bridge driver and the IR2233 (Three-phase full-bridge) driver are shown in Fig. 20.25. The driver section of the IR2233 comprises a similar structure as the IR2113's bootstrap and level shifting circuitry. Even though the IR range of driver ICs do not provide isolation between the control and driver electronics, this device is a low cost and very compact solution for low to medium power applications where electrical noise and interference can be managed by PCB circuit layout techniques. The IR2113 half-bridge driver is capable of sourcing and sinking 2 A peak current to a capacitive load without external buffer circuitry. This feature coupled with low propagation delay times makes this device capable of operating at frequencies in excess of 100 kHz.

The IR2113 and IR2233 are capable of driving floating loads (high-side switches) at 600 and 1200 V DC bus voltages, respectively, making them suitable for most direct off-line single-phase and three-phase inverter applications.

Other driver ICs available include POWEREX's M57959L and INTERSIL's HIP2500 bridge driver.

20.5.3 High-speed Gate Drivers

In the quest for high conversion efficiency and high power densities, the trend in power electronics is the move towards high converter switching frequencies at increasingly higher power levels. The main reason for this is to reduce the size of the energy transfer and storage components such as transformers, capacitors, and inductors. This results in a substantial reduction in component size and to a certain extent also cost. Achieving the objectives of high speed and high power switching requires fast switching power semiconductors. These devices are readily available, but as the device die-size increases with voltage and current requirements, so does its gate drive power requirements. This is due to an increase in device input capacitance as the die-size is increased. Low impedance gate driver electronics capable of delivering gate peak drive currents to large die-size IGBTs and power MOSFETs in excess of 8 A are often required. In order to meet these requirements, driver ICs must be able to deliver these large pulse currents efficiently at high switching frequencies. This is mainly due to a limitation in the maximum allowable device power dissipation. Device package power dissipation is typically limited to around 1.5 W for a DIP-14 IC without heat sinking. Many manufacturers have high speed driver solutions which realize drivers with low output impedances such as the TC4427 and the TC4422. These devices are available as single or dual channel drivers capable of delivering up to 9 A peak to a 1–10 nF capacitive load with a typical rise time of 30–50 ns. A table of available high-speed drivers from TelCom Semiconductor Inc. is shown in Table 20.3 [12].

20.5.4 Resonant Gate Drivers

High speed gate driving requires the driver and power semiconductor device impedance to be as low as possible. It is also adversely affected by parasitic inductance and capacitance caused by incorrect layout of the gate driver circuit. Sometimes these unavoidable parasitic elements tend to increase the gate drive loop impedance, (between the driver and the power semiconductor device) resulting in a limitation of the peak current drawn by the device gate. This limits the maximum achievable switching speed of the device. It also results in uncontrolled voltage excursions on the device gate, making high-speed driver circuit design quite challenging.

Resonant gate drivers utilize these parasitic elements in the gate drive loop by virtue of a controlled series resonant

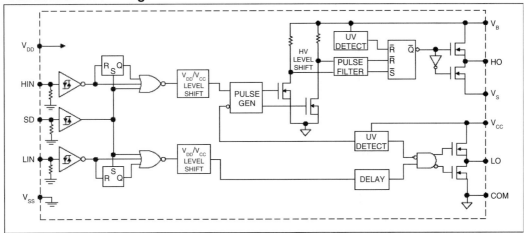

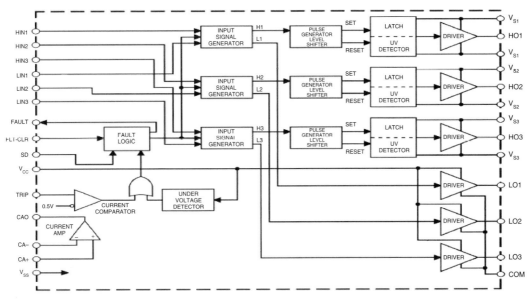

FIGURE 20.25 Functional diagrams of the IR2113 half-bridge driver and the IR2233 three-phase full-bridge driver.

mode of operation. The resonant circuit comprises the driver resistance (which is small), the device input capacitance, and its internal gate electrode resistance. By the addition of a defined amount of extra inductance to the gate drive loop, the voltage swing and switching speed of the gate driver can be controlled. A typical resonant gate driver for large die-power semiconductors has been developed by Turboswitchers Inc. A resonant gate driver called the TD-000, employs a patented low loss capacitance driver circuit topology, that reduces the driver power losses to less than half that of conventional high-speed drivers, at all switching frequencies. Operational from a 5 V supply, the driver is capable of high-switching speeds and generates a gate drive voltage swing of up to +24 to −19 V, depending on the inductance values used. Performance tests show that the TD-000 driver is capable of switching a 500 V/32 A power MOSFET supplied from a 400 V DC bus, in less than 3 ns. This switching speed allows for power electronic converters operating well into the MHz range. Detailed device application information can be found in [13].

TABLE 20.3 High current high-speed driver ICs available from TelCom semiconductor

Device no.	Drive current (Peak)	Output number and type		Rated load (pF)	Rise time @ rated load (ns)	Fall time @ rated load (ns)	Rising edge propogation delay (ns)	Falling edge propogation delay (ns)	Latch-up proof	Input protected to 5V below Gnd rail
		Inverting	Non-inverting							
TC1426	1.2 A	Dual	–	1000	30	20	55	80	Yes	No
TC1427	1.2 A	–	Dual	1000	30	20	55	80	Yes	No
TC1428	1.2 A	Single	Single	1000	30	20	55	80	Yes	No
TC4426	1.5 A	Dual	–	1000	25	25	33	38	Yes	Yes
TC4427	1.5 A	–	Dual	1000	25	25	33	38	Yes	Yes
TC4428	1.5 A	Single	Single	1000	25	25	33	38	Yes	Yes
TC4423	3.0 A	Dual	–	2200	25	25	33	38	Yes	Yes
TC4424	3.0 A	–	Dual	2200	25	25	33	38	Yes	Yes
TC4425	3.0 A	Single	Single	2200	25	25	33	38	Yes	Yes
TC4420	6.0 A	Single	Non-invert	4700	40	35	50	55	Yes	Yes
TC4429	6.0 A	Single	Inverting	4700	40	35	50	55	Yes	Yes
TC4421	9.0 A	Single	–	10,000	50	48	30	33	Yes	Yes
TC4422	9.0 A		Single	10,000	50	48	30	33	Yes	Yes
TC4469	1.2 A	– Quad –		1000	30	30	35	35	Yes	Yes
TC4468	1.2 A	– Quad AND –		1000	30	30	35	35	Yes	Yes
TC4467	1.2 A	– Quad NAND –		1000	30	30	35	35	Yes	Yes

20.6 Current and Future Trends

Integrated power electronic solutions have become the current trend for applications with the development of Smart Power modules. These modules contain the entire inverter semiconductor stack as well as fully integrated gate drive circuitry. This technology eliminates the challenges of inverter and gate drive design and allows for fast turnaround times in new product development through rapid prototyping. These devices are available for low to medium power applications and have an operating frequency limit of about 20 kHz.

Power semiconductor device manufacturers are constantly developing better die-structures for their power MOSFETs and IGBTs. This is to reduce device input capacitance, resulting in lower gate drive power requirements at high switching speeds.

20.7 Summary

The aim of this chapter was to expose the reader to the basic concepts, circuits, and technologies for gate drives in power converters. The main focus has been on voltage-controlled devices like power MOSFETs and IGBTs. Once the reader gains a basic understanding of the concepts and available solutions, detailed design information on gate drivers can be found on the device manufacturer's website. This is usually found under technical application notes or technical white papers. Other useful websites for power electronic design forums and application-specific information can be found in [2, 14]. The circuits presented can be adapted for the driving of other devices such as SCRs and power BJTs.

References

1. N. Mohan, T. Undeland, and W. Robbins, "Power Electronics: Converters, Applications and Design", Wiley, Brisbane, 1989.
2. www.powerdesigners.com
3. D. R. H. Carter, "Aspects of High Frequency Half-Bridge Circuits", PhD Thesis, Cambridge University, September 1996.
4. S. Clement and A. Dubhashi, "HV Floating MOS-Gate Driver IC", Integrated circuit designers manual.
5. Application Note, "Hints and Applications" Design manual, Chapter 3, Semikron Corporation.
6. M. Munzer, W. Ademmer, B. Strazalkowski, and K. T. Kaschani, "Coreless Transformer a New Technology for Half Bridge Driver IC's", application note, 2005, www.eupec.com.
7. I. de Vries, "High Power and High Frequency Class-DE Inverters", PhD Thesis, Department of Electrical Engineering, University of Cape Town, August 1999.
8. Application Note AN-937, "Gate Drive Characteristics and Requirements for HEXFET Power MOSFETs", www.irf.com.
9. Data sheet, SKHI22, www.semikron.com.
10. www.eupec.com
11. www.irf.com
12. Application Note 30, "Matching MOSFET drivers to MOSFETs", TelCom Semiconductor Inc.
13. I. de Vries, "Using Turbodriver-000", application note, February 2002, www.turboswitchers.com.
14. www.smpstech.com

21
Power Electronics in Capacitor Charging Applications

William C. Dillard, Ph.D.
*Archangel Systems, Incorporated
1635 Pumphrey Avenue Auburn,
Alabama, USA*

21.1 Introduction 559
21.2 High Voltage DC Power Supply with Charging Resistor 560
21.3 Resonance Charging 561
21.4 Switching Converters 562
     References 564

21.1 Introduction

Conventional dc power supplies operate at a given dc output voltage into a constant or near constant load. Pulse loads such as lasers, flashlamps, railguns, and radar, however, require short but intense bursts of energy. Typically, this energy is stored in a capacitor and then released into the load. The rate at which the capacitor is charged and discharged is called the repetition rate, T, and may vary from 0.01 Hz for large capacitor banks to a few kHz for certain lasers. Recharging the capacitor voltage to a specified voltage is tasked to a capacitor charging power supply (CCPS). The role of power electronics devices, topologies, and charging strategies for capacitor charging applications is presented in this chapter.

Figure 21.1 shows the voltage across the energy storage capacitor connected to the output of a CCPS. As seen in this figure, the CCPS has three modes of operation. The first mode is the charging mode in which the capacitor is charged from an initial voltage of zero to a specified final voltage. The duration of the charging mode is determined by the capacitance of the energy storage capacitor and the rate at which the CCPS delivers energy. The next mode of operation is the refresh mode, which can be considered a "standby mode" where the stored energy is simply maintained. When the output voltage drops below a predetermined value, the CCPS should turn on and deliver the energy necessary to compensate for capacitor leakage. Since refreshing is lost energy, the duration of the refresh mode should be as brief as possible. Issues that lead to non-zero refresh times include safety margins for worst-case charging and discharging mode times and SOA requirements of switching devices. The final mode of operation is the discharge mode where the load is actively discharging the capacitor. The CCPS does not supply any energy to the load in this mode. The amount of time the CCPS remains in this mode is determined by how quickly the load can discharge the capacitor.

The instantaneous output power for a CCPS varies over a wide range in comparison to a conventional dc power supply which supplies a near constant power to its load. This is shown in Fig. 21.2; the output power for the pulsed power load is drawn as linear for illustration purposes only. The charging mode is characterized by high peak power. At the beginning of this mode, the output power is zero (i.e. there is no voltage present but current is flowing). Thus, the load capacitor is equivalent to a short circuit. Additionally, at the end of the charging mode, the output power is again zero (i.e. there is an output voltage present but no current is flowing). Now the load capacitor is equivalent to an open circuit. The refresh mode is typically a low power mode, because the current required to compensate for capacitor leakage is small. The CCPS does not supply any power during the discharge mode when the energy storage capacitor is being discharged by the pulsed load.

The average output power for a CCPS depends on the discharge mode energy and the repetition rate of the load. It is maximum when the energy storage capacitor is discharged at the end of the charging mode (large voltage and current), which corresponds to operation without a refresh mode. Since the CCPS power is far from constant, the rating of a CCPS is often given in kJ/s instead of kW. The kJ/s rating can be

Copyright © 2007, 2001, Elsevier Inc.
All rights reserved.

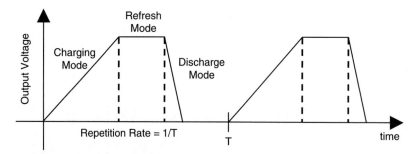

FIGURE 21.1 Three modes of operation of a capacitor charging power supply.

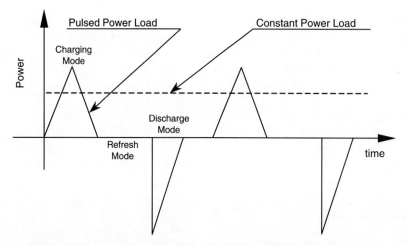

FIGURE 21.2 Power requirements for pulsed power and constant power loads.

written as,

$$\text{kJ/s} = \frac{W_{LOAD}}{T}$$

where W_{LOAD} is the energy delivered to the load per charging cycle and T is the repetition rate. In the optimum case with no refresh and instantaneous discharge, the kJ/s rating is limited to how fast a particular capacitor can be charged by its specified voltage.

21.2 High Voltage DC Power Supply with Charging Resistor

In this technique, the energy storage capacitor is charged by a high voltage dc power supply through a charging resistor as illustrated in Fig. 21.3. The charging mode ends when the capacitor voltage equals the output voltage of the power supply. The capacitor is continually refreshed by the power supply. During the discharge mode, the charging resistor isolates the

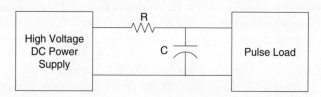

FIGURE 21.3 High voltage dc power supply and charging resistor.

power supply from the pulse load. The advantages of this technique are its simplicity, reliability, and low cost.

The major disadvantage of this technique is its poor efficiency. In the charging mode, the energy dissipated in the charging resistor is equal to the energy stored in the capacitor in the ideal case; therefore the maximum efficiency is 50%. As a result, this technique is utilized only in applications where the charge rate is low, i.e. 200 J/s. Another disadvantage of this technique is related to the charging time, which is determined by the RC time constant. Some laser applications require that the output voltage be within 0.1% of a target voltage. For this technique, more than five time constants are required for the capacitor voltage to meet this voltage specification.

21.3 Resonance Charging

The basic resonance charging technique is illustrated in Fig. 21.4. An ac input voltage is stepped up with a transformer, rectified, and filtered with capacitor C_2 to produce a high dc voltage V_0. In this circuit, C_2 is much greater than C_1. Thyristor T_1 is gated and current flows through the inductor and diode D_1 is transferring energy from C_2 to C_1. The voltage $v(t)$ and current $i(t)$ are described by the following equations assuming that $C_2 \gg C_1$. The charge time, t_c, for this circuit can be calculated by finding the time at which the current, described by Eq. (21.2), reaches zero and is given below. The voltage $v(t)$ has a value of $2V_0$ at the end of the charging mode.

$$v(t) = V_0(1 - \cos \omega_0 t) \qquad (21.1)$$

$$i(t) = V_0 \sqrt{\frac{C_1}{L}} \sin \omega_0 t \qquad (21.2)$$

$$\omega_0 = \frac{1}{\sqrt{LC_1}} \qquad (21.3)$$

$$t_c = \pi\sqrt{LC_1} = \frac{\pi}{\omega_0} \qquad (21.4)$$

Even though this technique is simple and efficient, it is not without its limitations. A high voltage capacitor with a large capacitance value is needed for C_2, which increases the cost. A single thyristor is shown in Fig. 21.4. Multiple thyristors connected in series or a thyratron may be required depending on the voltage level. The repetition rate of the pulse load should be such that C_1 is fully charged and $i(t)$ has reached zero before the load discharges to prevent latch up of T_1. It is not possible for this circuit to operate in the refresh mode because of the switch characteristics. Therefore, $v(t)$ will drift due to capacitor leakage. The charge time is a function of the circuit parameters and will drift as they change with temperature or due to aging.

Since all of the energy stored in C_1 is transmitted from C_2 in a single pulse, it can be difficult to achieve precise voltage regulation with the resonance charging technique. However, regulation can be improved with the addition of a de-queing circuit as shown in Fig. 21.5. The voltage $v(t)$ is monitored with a sensing network. Just before $v(t)$ reaches the desired level, thyristor T_2 is fired which terminates the charging mode. The remaining energy stored in the inductor is dissipated in R. The addition of the de-queing circuit reduces the circuit efficiency and increases the circuit complexity and cost, but still does not enable a refresh mode to compensate for capacitor leakage.

Boost charging, a variation on the resonance charging technique, is shown in Fig. 21.6 [1]. An extra switch is added to the circuit of Fig. 21.4 allowing energy to be stored in both C_2 and L. This can be modeled as an increase in the voltage gain of the CCPS. With switches S_1 and S_2 closed, the current $i(t)$ is given by

$$i(t) = \frac{V_0}{L} t \qquad (21.5)$$

At some time $t = t_{on}$, S_2 is opened, and the current is now described by

$$i(t) = I_0 \cos \omega_0 t + V_0 \sqrt{\frac{C_1}{L}} \sin \omega_0 t \qquad (21.6)$$

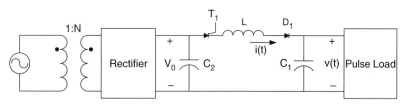

FIGURE 21.4 Resonance charging.

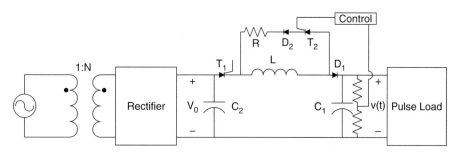

FIGURE 21.5 Resonance charging with de-queing.

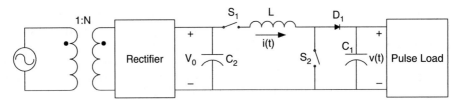

FIGURE 21.6 Boost charging.

where

$$I_0 = \frac{V_0}{L} t_{on} \quad (21.7)$$

is the inductor current initial value at t_{on}. The voltage $v(t)$ is then

$$v(t) = I_0 \sqrt{\frac{L}{C_1}} \sin \omega_0 t + V_0(1 - \cos \omega_0 t) \quad (21.8)$$

The time required for the current to reach zero and for the voltage $v(t)$ to reach its peak value can be calculated from

$$\tan(\omega_0 t_c) = -\frac{I_0 \sqrt{\frac{L}{C_1}}}{V_0} = -\omega_0 t_{on} \quad (21.9)$$

This is also the charge time, t_c, or the length of the charging mode. Note from equation (21.9) that the charge time depends on t_{on}, which is the on-time of switch S_2. In addition, the peak capacitor voltage is a function of t_{on}. The peak capacitor voltage is limited to $2V_0$ without S_2; voltage gains as high as 20 are possible with the addition of S_2 [1].

The switching elements in Figs. 21.4 and 21.5 are realized with thyristors. Simple switches are shown for the boost-charging technique in Fig. 21.6. Switch S_1 could be implemented with a thyristor. The boost capability provided by switch S_2 is best realized with a gate-controlled semiconductor device such as a GTO or an IGBT.

21.4 Switching Converters

The poor efficiency when charging a capacitor through a resistor from a high-voltage power supply limits its application to low charging rates. In the resonance-charging concepts, the energy is transferred to the load capacitor in a single pulse, and it is not possible to compensate for capacitor leakage. Energy storage capacitors may be charged utilizing the same power electronic technology that has been applied in switching converters for constant power loads. Instead of charging the energy storage capacitor with a single pulse, switching converters can charge the capacitor with a series of pulses or pulse train. The peak current is reduced when charging with a series of pulses, thus improving the efficiency of the charging process. In addition, soft-switching techniques may be employed in the switching converter to increase the efficiency. The regulation of the output voltage is also improved with the pulse train, because the energy is passed to the energy storage capacitor as small packets. Common control techniques such as pulse-width modulation can be used to control the size of the energy packet. This capability to control the size of the energy packet permits the CCPS to operate in the refresh mode and compensate for capacitor leakage. As a result, the CCPS may operate over a broad range of load repetition rates and still maintain tight output voltage regulation during refresh. During the refresh mode, energy lost due to capacitor leakage may be replaced in a burst fashion [2] or in a continuous fashion similar to trickle charging a battery [3].

In the switching converter, semiconductor switches may be operated on the low side of the transformer permitting the use of MOSFETs or IGBTs in the CCPS. Since the CCPS begins the charging mode with a short circuit across its output, the switching converter must be capable of operating under this severe load condition. This may require the implementation of a current limiting scheme in the converter control circuit.

One switching converter, topology candidate is the series resonant converter in Fig. 21.7. Note that the MOSFETs and resonant components L_r and C_r are connected on the low voltage side of the transformer. Only the rectifier diodes and energy storage capacitor must have high voltage ratings. When the output rectifier is conducting, the energy storage capacitor C_1 is connected in series with the resonant capacitor C_r. For a transformer turns ratio of 1:N, reflecting C_1 through the transformer yields a capacitance of $N^2 C_1$. Since N is typically large, this reflected capacitance is much larger than C_r, thus the resonant frequency, which is defined in Eq. (21.10), is not affected by C_1. For high voltage, high-frequency operation, the leakage inductance of the transformer may be utilized as L_r. Thus, the resonant frequency can be expressed as,

$$\omega_r = \frac{1}{\sqrt{L_r((NC_rC_1)/(C_r + NC_1))}} \quad (21.10)$$

One characteristic of this converter, which makes it attractive for capacitor charging, is the ability to operate under the short circuit conditions present at the beginning of the charging mode. The voltage across C_1 is zero at the beginning of

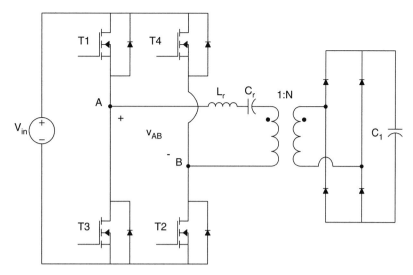

FIGURE 21.7 Series resonant converter.

this mode. The current in the switches is limited by the input voltage and impedance Z_0 as is defined in Eq. (21.11).

$$Z_0 = \sqrt{\frac{L_r}{C_r}} \qquad (21.11)$$

Another method for current limiting is to vary the ratio of f_s, the switching frequency of the MOSFETs, and the resonant frequency, f_r, which is $\omega_r/2\pi$. This effectively controls the flow of energy from the source to C_1. The ratio f_s/f_r may be set to a low value at the beginning of the charging mode and increased toward one as the voltage across C_1 builds up. This limits the current when the voltage across C_1 is low and allows increased energy transfer as the voltage approaches the target voltage. The disadvantage of this approach is that variable frequency operation complicates device and component selection and degrades EMI/EMC performance of the CCPS.

The flyback converter, shown in Fig. 21.8, also may be utilized for capacitor charging applications [4, 5]. When the MOSFET is turned on, current builds up in the primary winding storing energy in the magnetic field. When it reaches a specified level, the MOSFET is turned off and energy is transferred from the magnetic field to C_1. This energy transfer is terminated when the MOSFET is turned on again.

In cases where precise output regulation is not required and the packet energy is low, the diode in Fig. 21.8 can be replaced with a Zener diode with a Zener voltage of

$$V_Z = nV_{in} + V_{O,nom}$$

where $V_{O,nom}$ is the nominal value of V_O [6]. Once $V_{O,nom}$ is reached, the next energy packet will force the diode into a brief period of breakdown with the excess energy partially

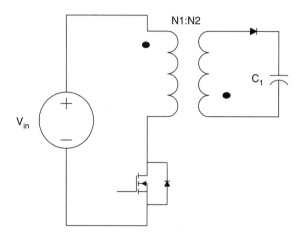

FIGURE 21.8 Flyback converter.

recycled to the input circuit. If the packet energy is small enough, the breakdown is not destructive and V_O is limited to

$$V_O = V_{O,nom} + \Delta V_{O,\Omega}$$

where $\Delta V_{O,\Omega}$ is the small excess voltage caused by the last energy packet.

Sokal and Redl [7] have investigated different control schemes for charging capacitors using the flyback converter. Their recommendation is to charge C_1 with current pulses which are nearly flat-topped. This strategy results in higher average current for a given peak current. The capacitor is charged faster because the charge delivered to it during a pulse is directly proportional to the average current. This desired pulse shape is achieved by turning on the MOSFET to terminate the transfer of energy to C_1 soon after the MOSFET is turned off, which increases the switching frequency. When the

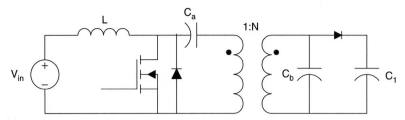

FIGURE 21.9 Ward converter.

primary current rises to a preset minimum level, the MOSFET is again turned off. This switching strategy is essentially hysteretic current mode control, where the switch current is limited between two preset bounds.

Another converter for capacitor charging applications is the Ward converter [8–10] shown in Fig. 21.9. When the MOSFET is on, energy is stored in the inductor and capacitor C_a transfers energy to the energy storage capacitor C_1 and capacitor C_b. The energy stored in the inductor is transferred to C_a when the MOSFET is off. The leakage inductance of the transformer and C_a resonate producing a sinusoidal current which flows in the primary winding of the transformer and the MOSFET. When the primary current reaches zero and starts negative, the diode turns on which allows the MOSFET to be turned off efficiently at zero current.

In some converter operating conditions, the voltage across C_a is very small because most of the energy has been transferred from C_a to C_1. The energy stored in C_a may be too small to ensure zero-current turn off of the MOSFET. In this case, the energy stored in C_b helps to ensure that the amplitude of the current is large enough for zero-current turn off of this device.

References

1. P. K. Bhadani, "Capacitor-Charging Power Supply for Laser Pulsers Using a Boost Circuit," Review of Scientific Instruments, Vol. 60, No. 4, April 1989, pp. 605–607.
2. A. C. Lippincott and R. M. Nelms, "A Capacitor-Charging Power Supply Using a Series-Resonant Topology, Constant On-Time/Variable Frequency Control, and Zero Current Switching," IEEE Transactions on Industrial Electronics, Vol. 38, No. 6, December 1991, pp. 438–447.
3. B. E. Strickland, M. Garbi, F. Cathell, S. Eckhouse, and M. Nelms, "2 kJ/sec 25-kV High-Frequency Capacitor-Charging Power Supply Using MOSFET Switches," Proceedings of the 1990 Nineteenth Power Modulator Symposium, June 1990.
4. R. L. Newsom, W. C. Dillard, and R. M. Nelms, "Digital Power-Factor Correction for a Capacitor-Charging Power Supply," IEEE Transaction on Industrial Electronics, Vol. 49, No. 5, October 2002, pp. 1146–1153.
5. F. P. Dawson and S. B. Dewan, "A Subresonant Flyback Converter for Capacitor Charging," Proceedings of the Second Annual IEEE Applied Power Electronics Conference, March 1987, pp. 274–283.
6. W. C. Dillard and R. M Nelms, "A Digitally-Controlled, Low-Cost Driver for Piezoceramic Flight Control Surfaces in Small Unmanned Aircraft and Munitions," Seventeenth Annual Applied Power Electronics Conference and Exposition, 2002, APEC '02, pp. 1154–1160.
7. N. O. Sokal and R. Redl, "Control Algorithms and Circuit Designs for Optimally Flyback-Charging an Energy-Storage Capacitor (e.g. for flash lamp or defibrillator)," IEEE Transactions on Power Electronics, Vol. 12, No. 5, September 1997, pp. 885–894.
8. M. A. V. Ward, "DC to DC Converter Current Pump," U.S. Patent Number 4,868,730, September 1989.
9. G. C. Chryssis, "High-Frequency Switching Power Supplies: Theory and Design," McGraw-Hill Publishing, New York, 1989.
10. R. M. Nelms and J. E. Schatz, "A Capacitor Charging Power Supply Utilizing a Ward Converter," IEEE Transactions on Industrial Electronics, Vol. 39, No. 5, October 1992, pp. 421–428.

22
Electronic Ballasts

J. Marcos Alonso, Ph.D.
University of Oviedo, DIEECS - Tecnologia Electronica, Campus de Viesques s/n, Edificio de Electronica, 33204 Gijon, Asturias, Spain

22.1 Introduction 565
    22.1.1 Basic Notions • 22.1.2 Discharge Lamps • 22.1.3 Electromagnetic Ballasts
22.2 High Frequency Supply of Discharge Lamps 571
    22.2.1 General Block Diagram of Electronic Ballasts • 22.2.2 Classification of Electronic Ballast Topologies
22.3 Discharge Lamp Modeling.. 575
22.4 Resonant Inverters for Electronic Ballasts 578
    22.4.1 Current-fed Resonant Inverters • 22.4.2 Voltage-fed Resonant Inverters • 22.4.3 Design Issues
22.5 High Power Factor Electronic Ballasts... 586
    22.5.1 Harmonic Limiting Standards • 22.5.2 Passive Solutions • 22.5.3 Active Solutions
22.6 Applications.. 589
    22.6.1 Portable Lighting • 22.6.2 Emergency Lighting • 22.6.3 Automotive Lighting • 22.6.4 Home and Industrial Lighting • 22.6.5 Microprocessor-based Lighting
    Further Reading 590

22.1 Introduction

Electronic ballasts, also called solid-state ballasts, are those power electronic converters used to supply discharge lamps. The modern age of electronic ballasts began with the introduction of power bipolar transistors with low storage time, allowing to supply fluorescent lamps at frequencies of several kilohertz and increasing lamp luminous efficacy by operating at these high frequencies. Later, electronic ballasts became very popular with the development of low cost power MOSFETs, whose unique features make them very attractive to implement solid-state ballasts. The main benefits of electronic ballasts are the increase in the lamp and ballast overall efficiency, increase in lamp life, reduction of ballast size and weight, and improvement in lighting quality. This chapter attempts to give a general overview about the more important topics related to this type of power converters.

22.1.1 Basic Notions

Discharge lamps generate electromagnetic radiation by means of an electric current passing through a gas or metal vapor. This radiation is discrete, as opposed to the continuous radiation emitted by an incandescent filament. Figure 22.1 shows the electromagnetic spectrum of an electric discharge, which consists of a number of separate spectral lines.

As can be seen in Fig. 22.1, only the electromagnetic radiation emitted within the visible region (380–780 nm) of the radiant energy spectrum is useful to provide lighting. The total power in watts emitted by an electric discharge can be obtained by integrating the spectral energy distribution. However, this is not a suitable parameter to measure the amount of light emitted by a discharge lamp.

The human eye presents different responses to the different types of electromagnetic waves within the visible range. As illustrated in Fig. 22.2, there exist two response curves. First, the photopic curve, also called $V(\lambda)$, is the characteristic used to represent the human eye behavior under normal illuminating level conditions or daylight vision. Second, the scotopic curve $V'(\lambda)$ is the response of the human eye for situations with low illuminating levels, also known as nocturnal vision. The reason for this different behavior is physiologic. The human eye consists of two types of photoreceptors: rods and cones. Rods are much more sensitive at low lighting levels than cones, but they are not sensitive to the different light colors. On the other hand, cones are responsible for normal color

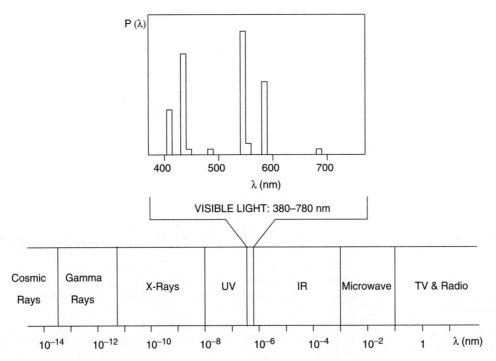

FIGURE 22.1 Spectral energy distribution of an arc discharge and radiant energy spectrum.

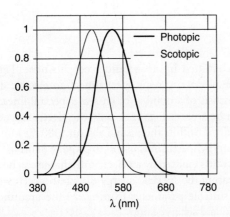

FIGURE 22.2 Spectral luminous efficiency functions for standard photopic and scotopic observers.

obtained by integrating the radiant power as follows:

$$\Phi = K_m \int_{380nm}^{780nm} P(\lambda) V(\lambda) d\lambda \qquad (22.1)$$

where K_m represents the maximal spectral luminous efficacy, which is equal to 683 lumens/watt (at $\lambda = 555$ nm) for photopic vision and 1700 lumens/watt (at $\lambda = 507$ nm) for scotopic vision. The standard photopic and scotopic functions were defined by the International Commission on Illumination (CIE) in 1924 and 1951, respectively.

The measurement of the lamp total luminous flux results is very useful to know whether or not the lamp is working properly. At the laboratory, the measurement of the lamp total luminous flux is performed by means of an integrating sphere and using the substitution method. The integrating sphere, also known as Ulbricht photometer, is internally coated with a perfectly diffusing material. Thus, the sphere performs the integral in Eq. (22.1) and the illuminance on the internal surface is proportional to the total luminous flux. A photometer with a $V(\lambda)$ filter is placed so that the internal illuminance can be measured, and a baffle is placed to avoid direct illumination of the photometer probe by the lamp. The measurement is made in two steps, one with the lamp under test in place and the other with a standard lamp of known total luminous flux. From the two measurements, the total luminous flux of the lamp under test is deduced by linear relationship. Figure 22.3 illustrates an integrating sphere photometer.

vision at higher lighting levels. Normally, only the photopic function is considered in lighting design and used to calibrate photometers.

Since, the human eye responds in different ways to the different wavelengths or colors, the output power of a lamp, measured in watts, is no longer applicable to represent the amount of light generated. Thus, a new unit is used to incorporate the human eye response, which is called *lumen*. The total light output of a lamp is then measured in lumens and it is known as lamp luminous flux. The lamp luminous flux is

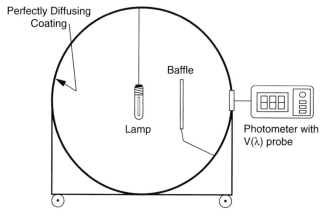

FIGURE 22.3 Integrating sphere used for measurement of the lamp total luminous flux.

An important parameter related to the supply of discharge lamps is the *luminous efficacy*. Luminous efficacy is defined as the rate of the lamp total luminous flux to the total electric power consumed by the lamp, usually expressed in lumens per watt. The luminous efficacy of a discharge lamp can be increased by proper designing of the electronic ballast, which finally results in energy saving.

22.1.2 Discharge Lamps

Basically, discharge lamps consists of a discharge tube inside which the electric energy is transformed into electromagnetic radiation. The discharge tube is made of a transparent or translucent material with two sealed-in electrodes placed at both ends, as shown in Fig. 22.4. The discharge tube is filled with an inert gas and a metal vapor. The electrodes generate free electrons, which are accelerated by the electrical field existing in the discharge. These accelerated electrons collide with the gas atoms, having both elastic and inelastic collisions depending on the electron kinetic energy. The basic processes inside the discharge tube are illustrated in Fig. 22.4. They are as follows:

1. Heat generation. When the kinetic energy of the electron is low, an elastic collision takes place and only a small part of the electron energy is transferred to the gas atom. The result of this type of collisions is an increase in the gas temperature. In this case, the electrical energy is consumed to produce heat dissipation. However, this is also an important process because the discharge has to set in its optimum operating temperature.

2. Gas atom excitation. Some electrons can have a high kinetic energy so that the energy transferred in the collision is used to send an electron of the gas atom to a higher orbit. This situation is unstable and the electron tends to recover its original level, then emitting the absorbed energy in the form of electromagnetic radiation. This radiation is used to directly generate visible light or, in other case, ultraviolet radiation is first generated and then transformed into visible radiation by means of a phosphor coating existing in the inside wall of the discharge tube.

3. Gas atom ionization. In some cases, electrons have gained such high kinetic energy that during the collision with the gas atom, an electron belonging to the gas atom is freed, resulting in a positive charged ion and a free electron. This freed electron can play the same roles as those generated by the electrodes. This process is specially important during both discharge ignition and normal running, because ionized atoms and electrons are necessary to maintain the electrical current through the lamp.

The number of free electrons in the discharge can increase rapidly due to continuing ionization, producing an unlimited current and finally a short-circuit. This is illustrated in Fig. 22.5, which shows how the voltage–current characteristics of a gas discharge exhibits a negative differential resistance. Therefore, in order to limit the discharge current, the use of an auxiliary supply circuit is mandatory. This circuit is called *ballast*.

Focusing on discharge lamps, the complete stabilization process consists of two main phases:

1. Breakdown phase. Most of the gases are very good insulators and an electric discharge is only possible if a sufficient

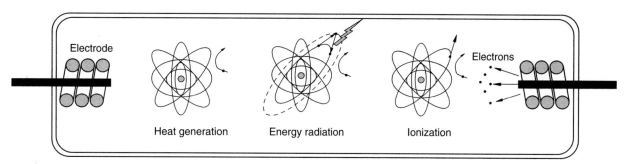

FIGURE 22.4 Basic processes inside the discharge tube.

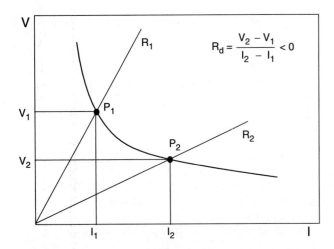

FIGURE 22.5 Voltage–current characteristics of a gas discharge.

concentration of charged particles is present. Normally, a high voltage is used to provide electricity carriers and to initiate the discharge. The minimum voltage applied to initiate the discharge is called *starting voltage*. The starting voltage mainly depends on the type of gas, gas pressure, and distance between electrodes. Figure 22.6a represents the starting voltage as a function of the product gas pressure by electrode distance, for different gases. These functions are known as Paschen curves.

Usually, auxiliary inert gases are used to decrease the starting voltage. Particularly, there exist some special inert gas mixtures, presenting a very low starting voltage, which are called *Penning mixtures*. These penning mixtures are very often used as initial starting gas. Figure 22.6a shows a typical penning mixture constituted by neon with 0.1% of argon.

2. Warm-up phase. Once the lamp is ignited, the collisions between free electrons and atoms generate heat and discharge temperature increases until reaching the normal operation conditions. During this phase, the heat is used to evaporate the metal atoms existing in the discharge tube and the emitted electromagnetic radiation assumes the character of a metal vapor discharge instead of an inert gas discharge. From the electrical point of view, the discharge warm-up phase shows initially low discharge voltage and high discharge current. As long as more and more metal atoms are evaporated, the discharge voltage increases and the discharge current decreases. Finally, an equilibrium is reached at steady-state operation with the normal values of voltage and current. The time constant of the warm-up phase strongly depends on the lamp type. It varies from seconds for fluorescent lamps to minutes for high-intensity discharge lamps.

Figure 22.6b illustrates some discharge waveforms during the warm-up phase.

The basic elements used in lamps to generate radiation in the visible part of the spectrum are sodium and mercury. The former generates radiation directly within the visible part of the spectrum; the latter generates radiation mainly in the ultraviolet region, but this radiation can be easily transformed in visible radiation by means of a phosphor coating on the internal wall of the discharge tube. Besides the element used, a very important parameter related to the efficiency and richness of the emitted radiation is the discharge pressure. For the sodium and mercury elements, there exist two pressure values around

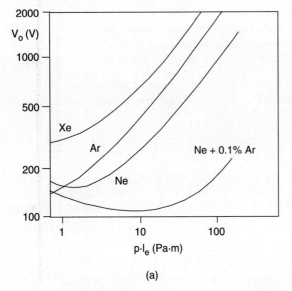

(a)

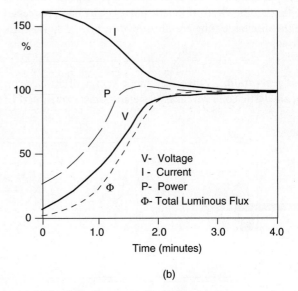

(b)

FIGURE 22.6 (a) Paschen curves for different inert gases and (b) stabilization curve of a discharge lamp.

which the luminous efficacy of the discharge is higher. The first is obtained at quite low-pressures, about 1 Pa, and the second at higher pressures, around 10^5 Pa (1 at). This is the reason why there exist two main types of discharge lamps:

1. *Low-pressure discharge lamps.* This type of lamps operates with pressures around 1 Pa and they feature low current density inside the discharge and low power per unit of discharge length. Therefore, these lamps present normally a quite large discharge volume with low power rating. Most representative examples are low-pressure mercury lamps, also known as fluorescent lamps and low-pressure sodium lamps.

2. *High-pressure discharge lamps.* The operating pressure in this type of lamps is around 10^5 Pa and higher, in order to achieve a considerable increase in the luminous efficacy of the discharge. These lamps present a high-current density in the discharge and a high power per discharge length ratio, thus showing much smaller discharge tubes. Examples of these lamps are high-pressure sodium lamps, high-pressure mercury lamps, and metal halide lamps.

Finally, to characterize the light produced by a discharge lamp, it is necessary to know two important concepts: the correlated color temperature (CCT) and the color rendering index (CRI).

The CCT is defined as the temperature of the blackbody radiator whose perceived color most closely resembles that of the discharge lamp. The color of an incandescent body changes as its temperature rises from deep red to orange, yellow, and finally white. Thus, a cool white fluorescent lamp has a CCT around 3500 K and it is perceived as a white source of light, whereas a high pressure-sodium lamp presents a CCT of about 2000 K and it appears as yellow.

The CRI of a light source is the effect that the source has on the color appearance of the objects under it, when compared to their appearance under a reference source of equal CCT. The measurement gives a value lower than 100, and the higher the CRI, the better the color rendition. For example, daylight and incandescent lamps has a CRI equal to 100.

To conclude this introduction on the discharge lamps, some comments regarding the most important types of discharge lamps will be given. Table 22.1 provides some additional data on the different discharge lamps for comparison.

1. *Fluorescent lamps.* These lamps belong to the category of low-pressure mercury vapor discharge lamps. The discharge generates two main lines at 185 and 253.7 nm and other weak lines in the visible range of the spectrum. A fluorescent powder on the inside wall of the discharge tube converts the ultraviolet radiation into visible radiation, resulting in a broadband spectral distribution and good color rendition. In these lamps, the optimum mercury vapor pressure (which gives the maximum luminous efficacy) is 0.8 Pa. For the tube diameters normally used, this pressure is reached at a wall temperature of about 40°C, not much higher than typical ambient temperature. The heat generated inside the discharge is sufficient to attain the required operating temperature without using an outer bulb. However, this structure causes a great variation of the lamp lumen output with the temperature, which is one important drawback of the fluorescent lamps. One solution to this problem is the addition of amalgams to stabilize the light output. This is specially used in compact fluorescent lamps.

2. *Low-pressure sodium lamps.* These lamps are the most efficient source of light. The reason is the almost monochromatic radiation that they generate, with two main lines at 589 and 589.6 nm, very close to the maximum human eye sensitivity. Therefore, color rendition of these lamps is very poor, however contrasts are seen more clearly under this light. This is the reason for using these lamps in situations where the recognition of objects and contours is essential for safety, such as motorway bridges, tunnels, intersections, and so on. The optimum pressure for the low-pressure sodium discharge is about 0.4 Pa, attained in normal discharge tubes at a temperature of 260°C. An outer bulb is normally used to reach and maintain this temperature.

3. *High-pressure mercury vapor lamps.* The increase in the pressure of the mercury vapor produces a radiation richer in spectral lines, some of them are in the visible part of the spectrum (405, 436, 546, and 577/579 nm). This leads

TABLE 22.1 Comparison of different discharge lamps

Lamp	Wattage (W)	Luminous efficacy (lm·W^{-1})	Life (h)	CCT (K)	CRI
Fluorescent	4–100	62	20,000	4200	62
Compact fluorescent	7–30	60–80	10,000	2700–5000	82
Low-pressure sodium	50–150	110–180	15,000	1800	<0
Mercury vapor	50–1000	40–70	24,000	4000–6000	15–50
Metal-halide	40–15,000	80–125	10,000	4000	65
HPS	35–1000	65–140	24,000	2000	22
HPS(Amalgam)	35–1000	45–85	10,000	2200	65

to an increase in the luminous efficacy, reaching values of 40–60 lm·W^{-1} at pressures 10^5–10^7 Pa (1–100 at). These lamps operate with unsaturated mercury vapor, which means that all the mercury in the discharge tube has evaporated and the number of mercury atoms per unit volume remains constant. Thus, the operation of this type of lamps is more independent of the temperature than most other discharge lamps. One drawback of these lamps is the lack of spectral lines in the long wavelengths (reds) of the spectrum, thus showing low CRI. An increase of the color rendition can be obtained by adding metal-halide compounds into the discharge tube, in order to generate radiation all over the visible spectrum. These lamps are known as *metal-halide lamps*.

4. High-pressure sodium lamps. This is a very popular source of light due to its high luminous efficacy and long life. The increase in the sodium vapor pressure produces a very widening spectrum, with good color rendition compared to the low-pressure sodium lamps. This also leads to a lower luminous efficacy but it is still higher than the other high intensity discharge lamps. Some of these lamps also incorporate mercury in the form of sodium amalgam to increase the field strength of the discharge, thus decreasing the discharge current. A lower lamp current and a higher lamp voltage allows to reduce the size and cost of the ballast. However, the addition of sodium amalgam strongly reduces the life of the lamp.

22.1.3 Electromagnetic Ballasts

Electromagnetic ballasts are commonly used to stabilize the lamp at the required operating point by limiting the discharge current. The operating point of the lamp is given by the intersection of both lamp and ballast characteristics, as shown in Fig. 22.7. The ballast line is the characteristic which shows the variation of the lamp power vs lamp voltage for a constant line voltage, and can be measured during the warm-up phase of the lamp. The lamp line is the characteristic which gives the variation of the lamp power as a function of the lamp voltage for different line voltages, and can be measured by varying the line voltage. Some lamps, like high-pressure sodium, exhibit a great variation of lamp voltage with changes in the lamp wattage. Because of this behavior, trapezoids have been established which define maximum and minimum permissible lamp wattage vs lamp voltage for purposes of ballast design, as shown in Fig. 22.7.

Figure 22.8 shows basic electromagnetic ballast used to supply low- and high-pressure lamps at line frequencies (50–60 Hz). Figure 22.8a illustrates the typical circuit used to supply fluorescent lamps with preheating electrodes, which basically uses a series inductor to limit the current through the discharge. Initially the glow switch is closed and the short-circuit current flows through the circuit, heating the electrodes.

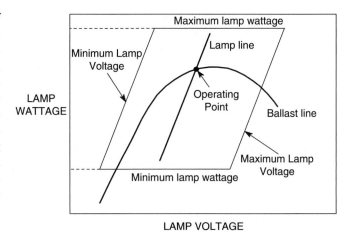

FIGURE 22.7 Lamp and ballast characteristics.

A fraction of a second later, the glow switch is opened and the energy stored in the ballast inductor causes a voltage spike between the lamp electrodes (about 800 V), which finally produces the discharge breakdown. Once ignited, the lamp voltage is lower than the line voltage and the glow switch remains open during the normal lamp operation. Typical glow switches are based on two bi-metal strips inside a small tube filled with an inert gas. An external capacitor of about 10 nF is used to enhance the glow-switch operation and also to reduce radio interference during lamp starting. Finally, in this type of inductive ballasts a capacitor placed across the line input is mandatory to achieve a reasonable value of the input power factor.

Starting voltages of high-pressure discharge lamps are normally higher than low-pressure discharge lamps and can range from 2500 V for a lamp at room temperature to 30–40 kV to re-ignite a hot lamp. Thus, the simple ignition system based on the glow switch is no longer applicable for these lamps. Figures 22.8b and c show two typical arrangements to supply high-intensity discharge lamps. A series inductor is also used to limit the lamp current at steady-state operation but autotransformers are used to attain higher voltage spikes for lamp ignition. For higher line voltages and short distances between starter and lamp, the inductor ballast can be used as ignition transformer as shown in Fig. 22.8b. In other cases, a separate igniting transformer is needed to provide higher voltage spikes and to avoid the effect of parasitic capacitance of connection cables (Fig. 22.8c).

The inductive ballast provides low lamp-power regulation against line voltage variation and therefore it is only recommended in those installations with low voltage fluctuations. When a good lamp power regulation is necessary the circuit shown in Fig. 22.8d is normally used. This circuit is commonly named as constant wattage autotransformer (CWA) and incorporates a capacitor in series with the lamp to limit the discharge current. Compared with the normal

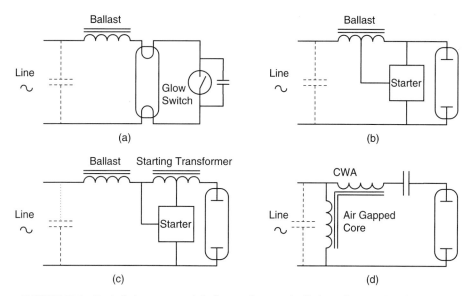

FIGURE 22.8 Typical electromagnetic ballast used to supply discharge lamps at low frequency.

inductive ballast, the CWA also exhibits higher input power factor, lower line extinguishing voltage and lower line starting currents.

The main advantage of electromagnetic ballasts is their simplicity, which in turn provides low cost and high reliability. However, since they operate at line frequencies, typically 50–60 Hz, they also feature high size and weight. Other important drawbacks of electromagnetic ballasts are as follows:

- Low efficiency, especially for those ballasts featuring good lamp power regulation against line voltage variation.
- Low reliability for ignition and re-ignition. If the voltage spike is not well located within the line period the ignition of the lamp can fail.
- Difficult to control the lamp luminous flux (dimming).
- Lamp operating point changes due to lamp aging process, thus reducing lamp life.
- Low input power factor and high harmonic distortion. Large capacitors are needed across the line input to increase power factor.
- Over-current risk due to ballast saturation caused by rectifying effect of some discharge lamps, specially at the end of their life.
- *Flickering* and *stroboscopic effect* due to low frequency supply. The energy radiated by the lamp is a function of the instantaneous input power. Therefore, when supplied from an AC line an instantaneous variation of the light output occurs, which is called *flicker*. For a line frequency of 60 Hz the resulting light frequency is 120 Hz. This variation is too fast for the human eye, but when rapidly moving objects are viewed under these lamps, the objects seem to move slowly or even halted. This is called *stroboscopic effect,* and can be very dangerous in industrial environments. A flicker index is defined with values from 0 to 1.0 [1]. The higher the flicker index, higher is the possibilities of noticeable stroboscopic effect.
- Unsuitable for DC applications (emergency lighting, automobile lighting, etc.).

22.2 High Frequency Supply of Discharge Lamps

22.2.1 General Block Diagram of Electronic Ballasts

Figure 22.9 shows the general block diagram of a typical electronic ballast. The main stages are the following:

- *EMI filter*. This filter is mandatory for commercial electronic ballast. Usually it consists of two coupled inductors and a capacitor. The input filter is used to attenuate the electro magnetic interference (EMI) generated by the high frequency stages of the ballasts. It also protects the ballast against possible line transients.
- *AC–DC converter*. This stage is used to generate a DC voltage level from the AC line. Normally a full-bridge diode rectifier followed by a filter capacitor is used. However, this simple rectifier provides low input power factor and poor voltage regulation. In order to obtain a higher power factor and a regulated bus voltage, active converters can be used as expounded later in this chapter.
- *DC–AC inverter and high frequency ballast*. These stages are used to supply the lamp at high frequency. The inverter generates a high frequency waveform and the

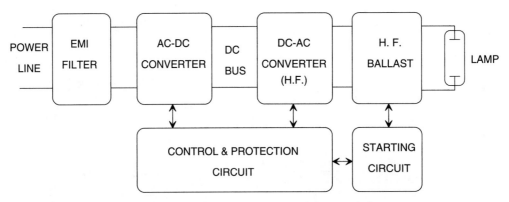

FIGURE 22.9 Block diagram of a typical electronic ballast.

ballast is used to limit the current through the discharge. Both inductors or capacitors can be used to perform this function, with the advantage of low size and weight, because they operate at high frequencies.
- *Starting circuit*. In most electronic ballast, specially those for low-pressure discharge lamps, the high frequency ballast is used to both ignite the lamp and limit the lamp current at steady state. Therefore, no extra starting circuit is necessary. However, when supplying high pressure discharge lamps, the starting voltages are quite higher and separate ignition circuits are needed, specially if hot re-ignition is pursued.
- *Control and protection circuit*. This stage includes the main oscillator, error amplifiers to regulate lamp current or power, output over-voltage protection, timers to control the ignition times, over-current protection, lamp failure protection, etc. It can vary from very simple circuits, those used in self-oscillating ballasts, to very complicated ones, which sometimes include a microprocessor-based control circuit.

There are several important topics when designing electronic ballasts:

- *Operating frequency*. The operating frequency should be high in order to take benefit of the lower size and weight of the reactive elements used to stabilize the discharge. Usually, the operating frequency should be higher than 20 kHz to avoid audible frequencies, which can produce annoying noises. On the other hand, a higher frequency produces higher switching losses and a practical limit of the switching frequency is about 100 kHz when using MOSFET switches. It is also important to avoid frequencies in the range 30–40 kHz, because these frequencies are normally used in IR remote controls and could generate some kind of interference.
- *Lamp current waveform*. In order to attain the maximum lamp life, it is important to drive the lamp with symmetrical alternating currents, thus making use of both the lamp electrodes alternately. Also, an important parameter is the *lamp current crest factor* (CF), which is the ratio of the peak value to the rms value of the lamp current. In case of electronic ballasts, the peak value of the low frequency modulated envelope to the rms value should be used. The higher the CF, the lower the lamp life. The ideal situation is to supply the lamp with a pure sinusoidal waveform. Usually, a CF lower than 1.7 is recommend to avoid early aging of the lamp.
- *Lamp starting procedure*. This is a very important issue when developing commercially available electronic ballasts. The reason is that the life of the lamp greatly depends on how well the lamp starting is performed, specially for hot-cathode fluorescent lamps. During the starting process, electrodes must be warmed up to the emission temperature, about 800°C, and no high voltage should be applied until their temperature is sufficiently high, thus avoiding sputtering damage. Once the electrodes reach the emission temperature, the starting voltage can be applied to ignite the lamp. For lamps with cold cathodes, the starting voltage must be applied rapidly to prevent harmful glow discharge and cathode sputtering. In any case, starting voltage must be limited to the minimum value to ignite the lamp, since higher voltages could provide undesirable starting conditions which would reduce the life of the lamp.
- *Dimming*. This is an important feature which allows the ballast to control lamp power and thereby lighting output. Usually the switching frequency is used in solid-state ballasts as a control parameter to provide dimming capability. Variations in frequency affect the high frequency ballast impedance and allow to change the discharge current. For example, if an inductor is used as high frequency ballast, a frequency increase yields an increase in the ballast impedance, thus decreasing lamp current. Dimming should be carried out smoothly, avoiding abrupt changes in lamp power when passing from one level to another. In an eventual power cutoff, the lamp should be re-started

at maximum lighting level and then slowly reduced to the required output level.

- *Acoustic resonance.* The HID lamps exhibit an unstable operation when they are supplied at high frequency. At a certain operating frequencies the arc fluctuates and becomes unstable, what can be observed as a high flicker due to important changes in the lamp power and thus in the lighting output. This can be explained by the dependence of the damping of acoustic waves on the plasma composition and pressure. More information about this topic can be found in [2]. The avoidance of acoustic resonance is mandatory to implement commercial electronic ballasts. This can be performed by selecting the operating frequency in a range free of acoustic resonances, typically below 1 kHz and over 100 kHz. Other methods are frequency modulation, square wave operation and sine-wave superposed with the third harmonic frequency [3, 4].

22.2.2 Classification of Electronic Ballast Topologies

Typical topologies used to supply discharge lamps at high frequency can be classified into two main groups: non-resonant ballasts and resonant ballasts.

22.2.2.1 Non-resonant Ballasts

These topologies are usually obtained by removing the output diode of DC-to-DC converters, in order to supply alternating current to the lamp. Current mode control is normally employed to limit the discharge lamp current. The lamp is supplied with a square current waveform, which can exhibit a DC level in some cases. A small capacitor is used to initially ignite the lamp, but its effect at steady-state operation can be neglected.

Examples of non-resonant electronic ballasts are shown in Fig. 22.10. Figure 22.10a, illustrates a boost-based and a flyback-based ballasts, respectively. Other topologies, which can supply symmetric alternating current through the lamp, are shown in Fig. 22.10c (symmetric boost) and 22.10d (push–pull).

These topologies present several drawbacks such as high-voltage spikes across the switch, which necessitates the use of high voltage transistors, and high switching losses due to hard switching, which gives low efficiency specially for high powers. Besides, since the ideal situation is the lamp being supplied with a sine wave, these circuits produce an early aging of the lamp. To conclude, typical applications of these topologies are portable and emergency equipment, where lamp power is low and the number of ignitions during its life is not very high. Some applications of these circuits can be found in [5–8].

22.2.2.2 Resonant Ballasts

These ballasts use a resonant tank circuit to supply the lamp. The resonant tank filters the high order harmonics, thus obtaining a sine current waveform through the lamp. Resonant ballasts can be classified into two categories:

A. Current-fed resonant ballasts

These ballasts are supplied with a DC current source, usually obtained by means of a choke inductor in series with the input DC voltage source. The DC current is transformed into an alternating square current waveform by switching power transistors. Typical topologies of this type of ballasts are shown in Fig. 22.11.

The topology shown in Fig. 22.11a corresponds to a class E inverter. Inductor Le is used to obtain a DC input current with low current ripple. This current supplies the resonant tank through the power switch formed by Q1–D1. The resonant tank used in this topology can vary from one ballast to another; the circuit shown in Fig. 22.11a is one which is normally used. The main advantage of this topology is that zero voltage switching (ZVS) can be attained in the power switch, thus reducing the switching losses and making possible the operation at very high frequencies, which can reach several megahertz. This allows to drastically reduce the size and weight of the ballast. However, the adjustment of the circuit parameters to obtain the optimum operation results is quite difficult, specially for mass production. Another important drawback is the high voltage stress across the switch, which can reach

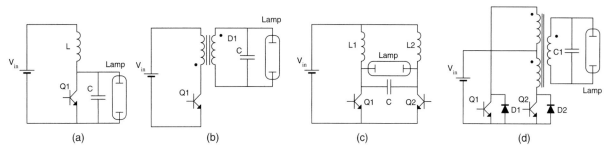

FIGURE 22.10 Non-resonant electronic ballasts.

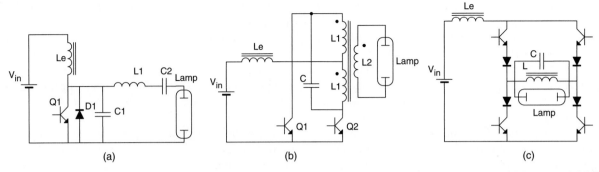

FIGURE 22.11 Two typical current-fed resonant inverters: (a) class E inverter; (b) current-fed push–pull inverter; and (c) current-fed full-bridge resonant inverter.

values of three times the DC input voltage. For these reasons, the main applications of this circuit are battery supply ballasts with low input voltage and low lamp power, as those used in emergency lighting and portable equipment. Typical power range of this ballast varies from 5 to 30 W. Applications of this circuit can be found in [9, 10].

Another typical topology in this group is the current-fed push-pull inverter shown in Fig. 22.11b. In this circuit, a DC input current is obtained by means of choke inductor Le. Transistors are operated with a 50% duty cycle, thus providing a current square wave, which supplies the current-fed parallel resonant circuit formed by the mutual inductance of the transformer and capacitor C. This circuit has the advantage of being relatively easy to implement in a self-oscillating configuration, avoiding the use of extra control circuits and thereby reducing the cost. Also, ZVS can be obtained in the power switches. However, the switches also present a high voltage stress, about three times the DC input voltage, which makes this topology unsuitable for power line applications. This circuit is also normally used in battery-operated applications in a self-oscillating arrangement. The typical power range is 4–100W. Applications based on this circuit can be found in [11, 12].

Finally, Fig. 22.11c shows a current fed full-bridge resonant inverter, which can be used for higher power rating. Also, this circuit allows to control the output power at constant frequency by switching the devices of the same leg simultaneously, generating a quasi-square current wave through the resonant tank [13].

B. Voltage-fed resonant ballasts

At present, electronic ballast manufactures mostly use voltage-fed resonant ballasts, specially for applications supplied from the AC mains. The circuit is fed from a DC voltage source, normally obtained by line voltage rectifying. A square wave voltage waveform is then obtained by switching the transistors with a 50% duty cycle, and used to feed a series resonant circuit. This resonant tank filters the high/order harmonics and supplies the lamp with a sine current waveform. One advantage of the voltage-fed series resonant circuit is that the starting voltage can be easily obtained without using extra ignition capacitors by operating close to resonant tank frequency. Figure 22.12 shows electrical diagrams of typical voltage-fed resonant ballasts.

The voltage-fed version of the push–pull inverter is illustrated in Fig. 22.12a. This inverter includes a transformer, which can be used to step up or down the input voltage in order to obtain an adequate rms value of the output square wave voltage. This provides higher design flexibility but also increases the cost. One disadvantage is that the voltage across transistors is twice the input voltage, what can be quite high for line applications. Therefore, this inverter is normally used for low voltage applications. Another important drawback of this voltage-fed inverter is that any asymmetry in the two primary windings (different number of turns) or in the switching times of power transistors would provide an undesirable DC level in the transformer magnetic flux, which in turn could saturate the core or decrease the efficiency due to the circulation of DC currents.

Figure 22.12b and 22.12c illustrates two possible arrangements for the voltage-fed half-bridge resonant inverter. The former is normally referred as asymmetric half-bridge, and uses one of the resonant tank capacitors ($C1$ in the figure) to block the DC voltage level of the square wave generated by the bridge. This means that capacitor $C1$ will exhibit a DC level equal to half the DC input voltage superimposed to its normal alternating voltage. A transformer can also be used in this inverter to step up or down the input voltage to the required level for each application. In this case, the use of the series capacitor $C1$ prevents from any DC current circulating through the primary winding, thus avoiding transformer saturation. This topology is widely used by ballast manufacturers to supply fluorescent lamps, especially in the self-oscillating version which allows to drastically reduce the cost. When supplying hot cathode fluorescent lamps, the parallel capacitor $C2$ is normally placed across two electrodes, as shown in Fig. 22.12b, in order to provide a preheating current for the electrodes and achieve soft ignition. Figure 22.12c shows other version of the half-bridge topology, using two bulk capacitors

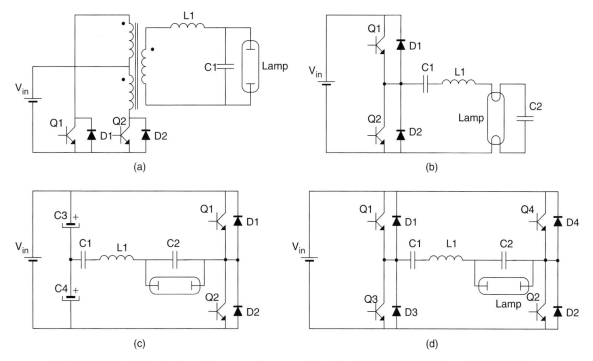

FIGURE 22.12 Typical voltage-fed resonant inverters: (a) push–pull; (b)–(c) half-bridge; and (d) full-bridge.

to provide a floating voltage level equal to half the input voltage. In this case, capacitor $C1$ is no longer used to block a DC voltage, thus showing lower voltage stress.

Finally, for the high power range (>200 W) the full-bridge topology shown in Fig. 22.12d is normally used. The transistors of each half-bridge are operated with a 50% duty cycle and their switching signals are phase-shifted by 180°. Thus, when switches Q1 and Q2 are activated, direct voltage V_{in} is applied to the resonant tank, and when switches Q3 and Q4 are activated, the reverse voltage $-V_{in}$ is obtained across the resonant circuit. One of the advantages of this circuit is that the switching signals of the two branches can be phase-shifted by angles between 0 and 180°, thus controlling the rms voltage applied to the resonant tank ranging from 0 to V_{in}. This provides an additional parameter to control the output power at constant frequency, the results are useful to implement dimming ballast.

22.3 Discharge Lamp Modeling

The low frequency of the mains is not an adequate power source for supplying discharge lamps. At these low frequencies, electrons and ionized atoms have enough time to recombine at each current reversal. For this reason, the discharge must be re-ignited twice within each line period. Figure 22.13a illustrates the current and voltage waveforms and the I–V characteristics of a 150 W HPS lamp operated with an inductive ballast at 50 Hz. As can be seen, the re-ignition voltage spike is nearly 50% higher than the normal discharge voltage, which is constant during the rest of the half-cycle.

When lamps are operated at higher frequencies (above 5 kHz), electrons and ions do not have enough time to recombine. Therefore, charge carrier density is sufficiently high at each current reversal and no extra power is needed to re-ignite the lamp. The result is an increase in the luminous flux compared to that at low frequencies, which is especially high for fluorescent lamps (10–15%).

Figure 22.13b shows the lamp waveforms and I–V characteristics for the same 150 W HPS when supplied at 50 kHz. It is shown how the re-ignition voltage spikes disappear and the lamp behavior is nearly resistive.

Figure 22.14 illustrates how the voltage waveforms change in a fluorescent lamp when increasing the supply frequency. As can be seen, at a frequency of 1 kHz, the voltage is already nearly sinusoidal and the lamp exhibits a resistive behavior.

Therefore, a resistor can be used to model the lamp at high frequencies for ballast design purposes. However, most lamp manufactures provide only lamp data for operating at low frequencies, where the lamp behaves as a square wave voltage source. Table 22.2 shows the low frequency electric data of different discharge lamps provided by the manufacturer and the measured values at high frequency for the same lamps. As can be seen, a power factor close to unity is obtained at high frequency.

The equivalent lamp resistance at high frequencies can be easily estimated from the low frequency data. Lamp power at

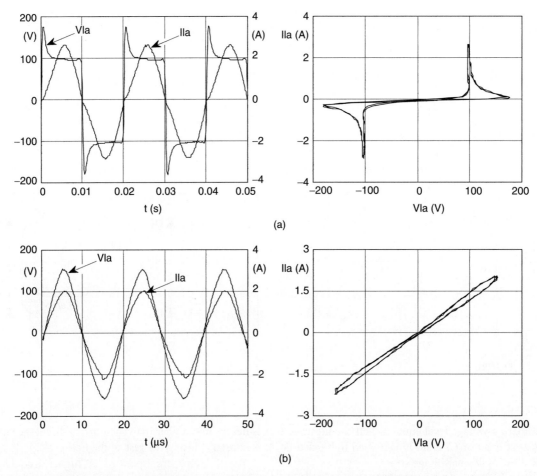

FIGURE 22.13 150 W HPS lamp waveforms and I–V characteristics at: (a) 50 Hz and (b) 50 kHz.

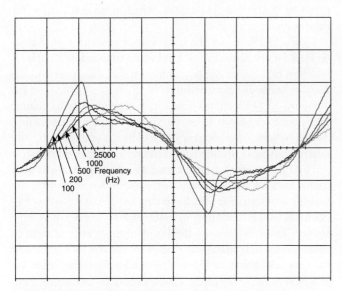

FIGURE 22.14 Voltage waveforms for a 36 W linear fluorescent lamp supplied through a resistive ballast at nominal power and different operating frequencies. Vertical scale: 100 V/DIV.

any operating frequency can be expressed as follows:

$$P_{LA} = V_{LA} I_{LA} FP_{LA} \qquad (22.2)$$

where V_{LA} and I_{LA} are the rms values of lamp voltage and current, and FP_{LA} is the lamp power factor.

At line frequencies, the lamp power factor is low (typically 0.8), due to the high distortion in the lamp voltage waveform. However, at high frequencies the lamp power factor reaches nearly 1.0. Then, lamp voltage and current at high frequency ($V_{LA,hf}$, $I_{LA,hf}$) can be estimated from the following equation:

$$I_{LA,hf} V_{LA,hf} = P_{LA} \qquad (22.3)$$

where P_{LA} is the nominal lamp power provided by the manufacturer.

As can be seen in Table 22.2, fluorescent lamps tend to maintain nearly the same rms current at low and high frequency, whereas high pressure discharge lamps tend to maintain nearly the same rms voltage. Based on these assumptions, the equivalent lamp resistance at high frequency estimated from the low frequency values are shown in Table 22.3.

22 Electronic Ballasts

TABLE 22.2 Electric data of different discharge lamps

Lamp*	Manufacturer @ 50 Hz				Measured @ H.F.			
	V (Vrms)	I (Arms)	P (W)	PF	V (Vrms)	I (Arms)	P (W)	PF
Fluorescent (TLD-36 W)	103	0.44	36	0.79	83.2	0.46	36	0.94
Compact fluorescent (PLC-26 W)	105	0.31	26	0.80	82	0.32	26	0.99
Low-pressure sodium (SOX-55 W)	109	0.59	55	0.86	75	0.76	56	0.98
Mercury vapor (HPLN-125 W)	125	1.15	125	0.87	132	0.92	120	0.99
Metal-halide (MHN-TD-150 W)	90	1.80	150	0.93	92	1.63	146	0.97
High-pressure sodium (SON-T-150 W)	100	1.80	150	0.83	105	1.42	148	0.99

*Lamps aged for 100 h.

TABLE 22.3 Estimated electric data of discharge lamps at high frequency

Lamp	$V_{LA,hf}$	$I_{LA,hf}$	$R_{LA,hf}$
Fluorescent lamps	$P_{LA}/I_{LA,lf}$	$I_{LA,lf}$	$P_{LA}/I_{LA,lf}^2$
High-pressure lamps	$V_{LA,lf}$	$P_{LA}/V_{LA,lf}$	$V_{LA,lf}^2/P_{LA}$

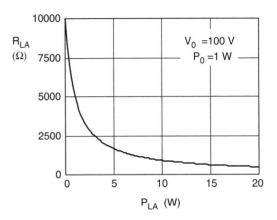

FIGURE 22.15 Lamp resistance vs lamp power characteristic.

Low-pressure sodium lamps neither maintain voltage nor current, constant at high frequency, and also they exhibit an equivalent resistance quite dependent on the frequency. Therefore their equivalent resistance can only be obtained by laboratory testing.

Note that the values given by Table 22.3 are only an approximation to the real value, which should be obtained by measurement at the laboratory. It can be used as the first starting point in the design of the electronic ballast, but final adjustments should be made at the laboratory.

Another important issue is that the lamp equivalent resistance is strongly dependent on power delivered to the lamp, which is specially important for designing electronic ballasts with dimming feature. The characteristic lamp resistance vs lamp power is different for each discharge lamp type and must be obtained by laboratory testing. One of the best possibilities to fit the lamp resistance vs power characteristic is the hyperbolic approximation. For example, Mader and Horn propose in [14] the following simple approximation:

$$R_{LA}(P_{LA}) = \frac{V_0^2}{P_{LA} + P_0} \quad (22.4)$$

where R_{LA} is the equivalent lamp resistance, P_{LA} is the average lamp power, and V_0 and P_0 are two parameters which depend on each lamp. This characteristic has been plotted in Fig. 22.15 for a particular lamp with $V_0 = 100$ V and $P_0 = 1$ W.

This model can be implemented very easily in circuit simulation programs, such as SPICE-based programs. Figure 22.16 shows the electric circuit and the description used to model the lamp behavior in a SPICE-based simulation program. The voltage-controlled voltage source EL is used to model the resistive behavior of the lamp. The voltage source VS is used to measure the lamp current so that the instantaneous and average lamp current can be calculated; for this reason its voltage value is equal to zero. GP is a voltage-controlled current source used to calculate the instantaneous lamp power, which is then filtered by RP and CP in order to obtain the averaged lamp power. Finally, the hyperbolic relationship between the lamp resistance and power is implemented by means of the voltage-controlled voltage source EK. The time constant $\tau = \text{RP} \cdot \text{CP}$ is related to the ionization constant of the discharge.

Figure 22.17 illustrates some simulation results at low frequency when the lamp is supplied from a sinusoidal voltage source and stabilized with an inductive ballast.

The Mader–Horn model can also be used at high frequencies obtaining a resistive behavior for the lamp. The equivalent lamp resistance at high frequency will also exhibit a hyperbolic variation with the averaged lamp power and with a time constant given by τ. This model is then useful to simulate electronic ballast with dimming feature.

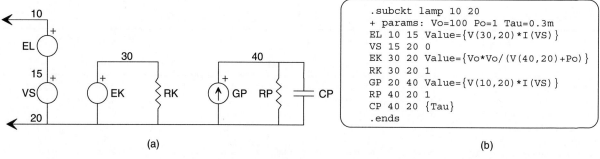

FIGURE 22.16 (a) Mader–Horn linear model for discharge lamps and (b) SPICE description of the model.

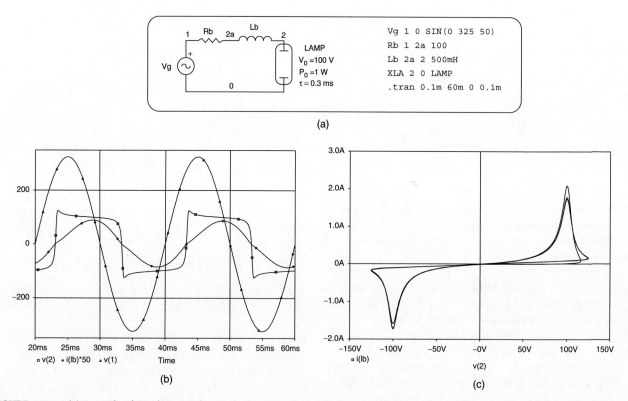

FIGURE 22.17 (a) Example of simulation with an inductive ballast at low frequency; (b) operating waveforms; and (c) lamp I–V characteristics.

Discharge lamp modeling has become an important subject, since its results are very useful to optimize the electronic ballast performance. Some improvements on the Mader–Horn model and other interesting models can be found in the literature [14–16].

22.4 Resonant Inverters for Electronic Ballasts

Most modern domestic and industrial electronic ballasts use resonant inverters to supply discharge lamps. They can be implemented in two basic ways: current-fed resonant inverters and voltage-fed resonant inverters.

22.4.1 Current-fed Resonant Inverters

One of the most popular topologies belonging to this category is the current-fed push–pull resonant inverter, previously shown in Fig. 22.11. For this reason, this inverter will be studied here to illustrate the operation of the current-fed resonant ballasts.

The current-fed push–pull inverter uses an input choke to obtain a DC input current with low current ripple. This current is alternatively conducted by the switches so that a parallel

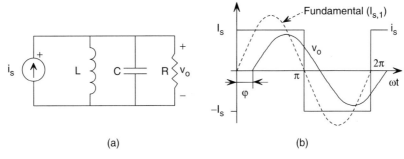

FIGURE 22.18 (a) Equivalent circuit of a current-fed parallel resonant inverter and (b) operating waveforms.

resonant tank can be supplied with a current square wave. Figure 22.18 shows the equivalent circuit and the operating waveforms of a current-fed parallel resonant inverter.

The input current can be expressed as a Fourier series in the following way:

$$i_s(t) = \sum_{n=1,3,5...} I_{S,n} \sin n\omega t = \sum_{n=1,3,5...} \frac{4I_S}{n\pi} \sin n\omega t \quad (22.5)$$

$I_{S,n}$ being the peak value of each current harmonic and I_S the DC input current of the inverter. The output voltage for each current harmonic is obtained by multiplying the input current $I_{S,n}$ by the equivalent parallel impedance $Z_{E,n}$, this is:

$$V_{0,n} = I_{S,n} Z_{E,n} = I_{S,n} \frac{1}{1/R + jn\omega C - j(1/n\omega L)} \quad (22.6)$$

Usually, normalized values are employed in order to provide more general results. Then, the output voltage can be expressed as follows:

$$V_{0,n} = I_{S,n} Z_B \frac{1}{1/Q + jn\Omega - j(1/n\Omega)} \quad (22.7)$$

where Z_B is the base impedance of the resonant tank, Q is the normalized load, Ω is the normalized frequency, and ω_0 is the natural frequency of the resonant circuit, given by:

$$Z_B = \sqrt{L/C} \quad Q = R/Z_B \quad \Omega = \omega/\omega_0 = \omega\sqrt{LC} \quad \omega_0 = 1/\sqrt{LC} \quad (22.8)$$

From Eq. (22.7) the peak output voltage $V_{0,n}$ and phase angle φ_n can be obtained for each harmonic:

$$V_{0,n} = I_{S,n} Z_B \frac{1}{\sqrt{1/Q^2 + (n\Omega - 1/n\Omega)^2}} \quad (22.9)$$

$$\varphi_n = -\tan^{-1} Q(n\Omega - 1/n\Omega) \quad (22.10)$$

The total harmonic distortion (THD) of the output voltage can be calculated as follows:

$$THD(\%) = \frac{\sqrt{\sum_{n=3,5,7...} V_{0,n}^2}}{V_{0,1}} \cdot 100 \quad (22.11)$$

Based on these equations, the analysis and design of the current-fed resonant inverter can be performed. Normally, the circuit operates close to the natural frequency ω_0 and the effect of the high frequency harmonics can be neglected. To probe this, Fig. 22.19a illustrates the THD of the output voltage as a function of the normalized load and frequency, obtained by plotting Eq. (22.11). As can be seen, for values of Q higher than 1 and for operation close to the natural frequency ($\Omega = 1$), the THD is low, which means that the output voltage is nearly a sinusoidal waveform. However, for low values of Q, the output voltage tends to be a square waveform and the THD tends to the value of about 48%, corresponding to the THD of a square waveform. Figure 22.19b illustrates the normalized output voltage for the fundamental component.

As stated previously, when used as lamp ballast, the current-fed parallel resonant inverter operates at the natural frequency of the resonant tank to both ignite the lamp and limit the current at normal running. Neglecting the effect of high-order harmonics, the rms output voltage is given by the fundamental component, and can be obtained using $\Omega = 1$ in Eq. (22.9) as follows:

$$V_{0(rms)} \approx V_{0,1(rms)} = I_{S,1(rms)} Z_B Q = \frac{4I_S R}{\pi\sqrt{2}} \quad (22.12)$$

In a current-fed resonant inverter, the DC input current I_S is supplied from a DC voltage source V_{in} with a series choke, as stated previously. Then, the DC input current can be obtained, assuming 100% efficiency, by equaling input and output power as follows:

$$P_{in} = V_{in} I_S = \frac{V_{0(rms)}^2}{R} \quad (22.13)$$

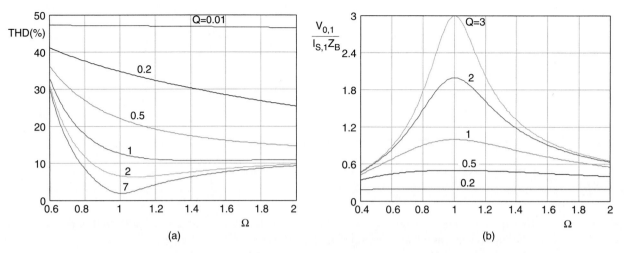

FIGURE 22.19 Characteristics of the current-fed parallel resonant inverter: (a) THD and (b) fundamental output voltage.

and then:

$$I_S = \frac{V_{0(rms)}^2}{V_{in}R} \qquad (22.14)$$

using Eq. (22.14) in Eq. (22.12) and solving for the output voltage:

$$V_{0(rms)} = \frac{\pi\sqrt{2}}{4}V_{in} = 1.1 V_{in} \qquad (22.15)$$

As can be seen, when operating at the natural frequency, the rms output voltage is independent of the resonant tank load. The peak output voltage is equal to $\pi V_{in}/2$. This value is directly related to the peak voltage stress in the switches. For a full-bridge topology, as shown in Fig. 22.11c, this value is equal to the switch voltage stress. However, for the current-fed push–pull inverter, the voltage stress is twice this value, this is πV_{in}, due to the presence of the transformer. This gives a very high voltage stress for the switches in this topology, this is the reason why the current-fed push–pull is mainly used to implement low input voltage ballasts.

On the other hand, lamp starting voltage can vary from 5 to 10 times the lamp voltage in normal discharge mode. This makes difficult the use of the current-fed parallel resonant inverter at constant frequency to both ignite the lamp and supply it at steady state, since the output voltage is independent of the resonant tank load.

One solution to this problem is to ignite the lamp at the resonant tank natural frequency and then change the frequency to decrease the output voltage and output current to the normal running values of the lamp. This solution makes necessary the use of extra circuitry to control the frequency, normally in closed loop to avoid lamp instabilities, which increases the ballast cost.

Another solution, very often used in low cost ballasts, is to design the parallel resonant tank to ignite the lamp, and limit the lamp current in discharge mode by using an additional reactive element in series with the lamp. Normally a capacitor is used to limit the lamp current in order to minimize the cost of the ballast. This solution is used in combination with the self-oscillating technique, which assures the operation at a constant frequency equal to the natural frequency of the resonant tank. Figure 22.20a illustrates this circuit. Normally the effect of the series capacitor is neglected and the resonant tank is assumed to behave as a sinusoidal voltage source during both ignition and normal operation, as shown in Fig. 22.20b. The high lamp starting voltage is obtained by means of a step-up transformer, this is why typically a push–pull topology is used. If V_{in} is the DC input voltage and V_{ig} is the lamp ignition voltage, then the necessary transformer turn ratio is given by the following expression:

$$\frac{N_2}{N_1} = \frac{V_{ig}}{\pi V_{in}/2} \qquad (22.16)$$

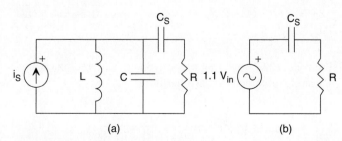

FIGURE 22.20 (a) Typical parallel resonant circuit used to supply discharge lamps and (b) equivalent circuit.

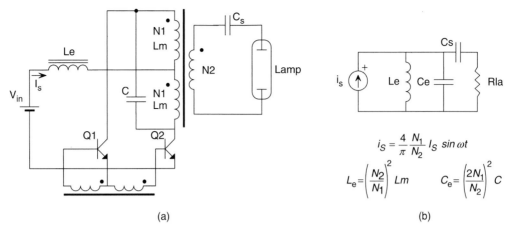

FIGURE 22.21 (a) Self-oscillating current-fed push–pull electronic ballast and (b) equivalent circuit.

and the rms lamp current in discharge mode can be approximated as follows:

$$I_{LA} = \frac{1.1(N_2/N_1)V_{in}}{\sqrt{R^2 + (1/2\pi f C_S)^2}} \quad (22.17)$$

where R is the equivalent resistance of the lamp. From Eq. (22.17) the necessary value of the series capacitor C_S, used to limit the lamp current to the nominal value I_{LA}, can be easily obtained.

Finally, Figs. 22.21a,b illustrate a typical ballast based on a current-fed resonant inverter and its equivalent circuit, respectively.

22.4.2 Voltage-fed Resonant Inverters

Some voltage-fed resonant inverters used to supply discharge lamps were previously shown in Fig. 22.12. Basically, they use two or more switches to generate a square voltage waveform. The different topologies are mainly given by the type of resonant tank used to filter this voltage waveform. There are three typical resonant tanks, the equivalent circuits are shown in Fig. 22.22. These circuits are the series-loaded resonant tank (Fig. 22.22a), the parallel-loaded resonant tank (Fig. 22.22b),
and the series–parallel-loaded resonant tank (Fig. 22.22c). The typical operating waveforms are shown in Fig. 22.22d.

Similar to the current-fed resonant inverter, the input voltage can be expressed as a Fourier series in the following way:

$$v_s(t) = \sum_{n=1,3,5...} V_{S,n} \sin n\omega t = \sum_{n=1,3,5...} \frac{4V_S}{n\pi} \sin n\omega t \quad (22.18)$$

$V_{S,n}$ being the peak value of each voltage harmonic and V_S the DC input voltage of the inverter. The same methodology as that used to analyze the current-fed resonant inverter will be used here to study the behavior of the three basic voltage-fed resonant inverters.

22.4.2.1 Series-loaded Resonant Circuit

The output voltage corresponding to the n-order harmonic is easily obtained as follows:

$$V_{0,n} = V_{S,n} \frac{1}{\sqrt{1 + \frac{1}{Q_S^2}\left(n\Omega - \frac{1}{n\Omega}\right)^2}} \quad (22.19)$$

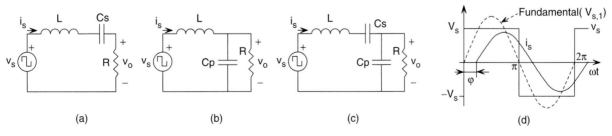

FIGURE 22.22 (a–c) Equivalent circuits of voltage-fed resonant inverters and (d) typical operating waveforms.

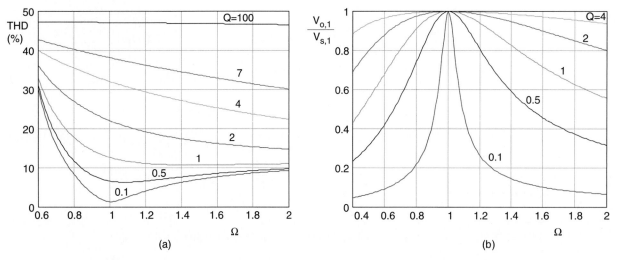

FIGURE 22.23 Characteristics of series-loaded resonant inverter: (a) THD and (b) fundamental output voltage.

where Q_S and Ω are the normalized load and switching frequency given by the following expressions:

$$Q_S = R/Z_B = R(L/C_S)^{-1/2} \quad \Omega = \omega/\omega_0 = \omega\sqrt{LC_S} \quad (22.20)$$

Figure 22.23a shows the THD of the series-loaded circuit and Fig. 22.23b shows the fundamental output voltage, which is normally considered for design purposes.

In this circuit the input and output current are equal, and can be calculated by dividing the output voltage by the load impedance. This current is also circulating through the inverter switches and therefore represents an important parameter for the design. Another important parameter is the phase angle of the input current, which defines the type of commutations in the inverter switches. For the fundamental harmonic, the phase angle of the current circulating through the resonant tank can be calculated as follows:

$$\varphi = -\tan^{-1}\frac{\Omega - 1/\Omega}{Q_S} \quad (22.21)$$

At the natural frequency (ω_0), the input voltage and current will be in phase, which means that there is no reactive energy handled by the circuit and all the input energy is transferred to the load at steady-state operation. For frequencies higher than ω_0, the current is lagged and some reactive energy will be handled. In this case the inverter switches will present zero voltage switching (ZVS) [17]. For frequencies lower than ω_0, the current is in advance and also some reactive energy will be handled. In this case the inverter switches will present ZVS.

As can be seen in Fig. 22.23a, the THD is lower for the lower values of the normalized load and for frequencies close to the natural frequency of the resonant tank. For the higher values of Q_S the THD tends to the value corresponding to a square wave. The output voltage is always lower than the input voltage, and for frequencies around the natural frequency, the circuit behaves as a voltage source, especially for the high values of Q_S. This means that the lamp cannot be ignited and supplied in discharge mode, maintaining a constant switching frequency. This behavior is similar to that encountered for the current-fed resonant inverter. The use of step-up transformers is mandatory to achieve lamp ignition, especially for the low input voltages. In order to maintain the operating frequency constant, a series element will be necessary to limit the lamp current at normal discharge operation. A capacitor can also be used as expounded in the previous section. As summary, this circuit is mainly used in high input voltage, low current applications, and it is not very frequently used to implement electronic ballasts.

22.4.2.2 Parallel-loaded Resonant Circuit

In this circuit, the output voltage corresponding to the n-order harmonic is given by the following expression:

$$V_{0,n} = V_{S,n}\frac{1}{\sqrt{\frac{n^2\Omega^2}{Q_P^2} + (n^2\Omega^2 - 1)^2}} \quad (22.22)$$

where

$$Q_P = R/Z_B = R(L/C_P)^{-1/2} \quad \Omega = \omega/\omega_0 = \omega\sqrt{LC_P} \quad (22.23)$$

The THD and fundamental output voltage are shown in Fig. 22.24. This circuit has useful characteristics very much useful to implement electronic ballasts than the series-loaded resonant circuit. First, the THD of the output voltage around

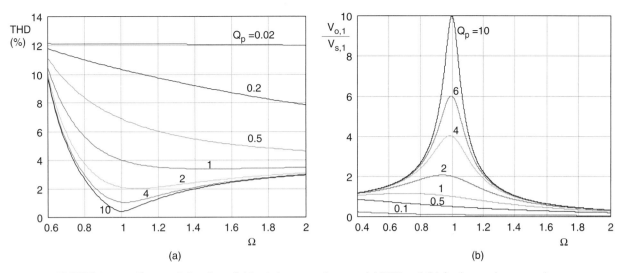

FIGURE 22.24 Characteristics of parallel-loaded resonant inverter: (a) THD and (b) fundamental output voltage.

the natural frequency is in general quite lower than that of the series circuit. For the lower values of Q_P, the THD tends to a value of 12%, which corresponds to the THD of a triangular wave. As a result, the lamp voltage and current waveforms will be very similar to a sine wave, which is an adequate waveform to supply the lamp. Second, the frequency response of the output voltage (Fig. 22.24b) makes it possible to both ignite the lamp and limit the lamp current at steady state, maintaining a constant operating frequency. During ignition, the lamp behaves as a very high impedance, thus giving a high value of Q_P. Under these conditions, the parallel circuit can generate a very high output voltage, provided that the operating frequency is close to the natural frequency. Once the lamp is ignited, the normalized load Q_P decreases and the circuit can limit the lamp current without changing the operating frequency. In fact, the parallel circuit operating close to the natural frequency behaves as a current source for the load, as will be shown later. This behavior results are adequate to supply discharge lamps because it assures a good discharge stability, avoiding the lamp being easily extinguished by transitory power fluctuations.

The maximum value of the voltage gain shown in Fig. 22.24 can be calculated to be equal to $Q_P/\sqrt{1 - 1/4Q_P^2}$, and it appears at a frequency $\Omega_m = \sqrt{1 - 1/2Q_P^2}$. This means that a maximum is only present if Q_P is greater than $1/\sqrt{2} \approx 0.71$. For the higher values of Q_P, the maximum gain can be approximated to Q_P and the frequency of maximum gain can be approximated to the natural frequency ω_0.

The input current of the parallel resonant circuit is another important parameter to calculate the current handled by the inverter switches. Since the operating frequency is normally around resonance, only the fundamental component is considered. The value of this fundamental current and its phase angle can be obtained as follows:

$$I_{S,1} = \frac{V_{S,1}}{Z_B} \sqrt{\frac{1 + Q_P^2 \Omega^2}{\Omega^2 + Q_P^2(\Omega^2 - 1)^2}} \qquad (22.24)$$

$$\varphi = \tan^{-1} \frac{-1}{\Omega Q_P} - \tan^{-1} Q_P \left(\Omega - \frac{1}{\Omega}\right) \qquad (22.25)$$

The condition for the input voltage being in phase with the input current can be obtained by equaling Eq. (22.25) to zero. This gives a value of the normalized frequency: $\Omega_{\varphi=0} = \sqrt{1 - 1/Q_P^2}$. For a frequency greater than that value, the input current will lag the input voltage and the inverter switches will present zero voltage switching. For frequencies lower than that value, the current will be in advance and zero current switching is obtained. The output voltage gain at that frequency is equal to Q_P.

Finally, it is very interesting to study the behavior of this circuit for frequencies close to the natural frequency ω_0 ($\Omega = 1$), since this is the normal region selected to operate for ballast applications. At this frequency, the output voltage gain is equal to Q_P and then the output current will be $V_{S,1}/Z_B$. This means that when operated at the natural frequency the parallel circuit behaves as a current source, whose value only depends on the input voltage. At the natural frequency, the input current is equal to $V_{S,1}\sqrt{1 + Q_P^2}/Z_B$ and the phase angle is equal to $\tan^{-1}(-1/Q_P)$. The behavior of the circuit is always inductive, with ZVS, and the phase angle decreases when increasing Q_P, which means that less reactive energy is handled by the circuit.

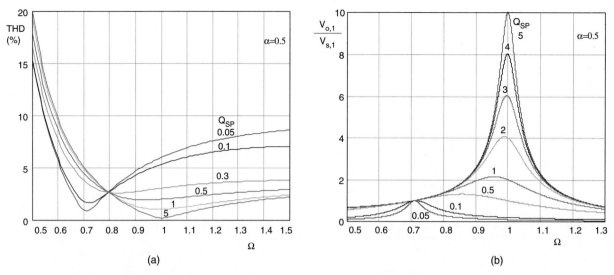

FIGURE 22.25 Characteristics of series–parallel-loaded resonant inverter: (a) THD and (b) fundamental output voltage.

22.4.2.3 Series–Parallel-loaded Resonant Circuit

This circuit is also very widely used to implement electronic ballasts. The fundamental output voltage is given by the following expression:

$$V_{0,1} = V_{S,1} \frac{1}{\sqrt{\frac{1}{Q_{SP}^2}\left(\Omega - \frac{1-\alpha}{\Omega}\right)^2 + \frac{1}{\alpha^2}\left(\Omega^2 - 1\right)^2}} \quad (22.26)$$

where

$$Q_{SP} = R/Z_B = R(L/C_E)^{-1/2} \quad \Omega = \omega/\omega_o = \omega\sqrt{LC_E}$$

$$\alpha = C_E/C_P = 1 - C_E/C_S \quad (22.27)$$

and $C_E = C_S C_P/(C_S + C_P)$ is the series equivalent of the two capacitors present in this resonant circuit.

The fundamental input current and its phase angle are the following:

$$I_{S,1} = \frac{V_{S,1}}{Z_B}\left[\frac{1+(\Omega^2 Q_{SP}^2/\alpha^2)}{\left(\Omega - (1-\alpha/\Omega)\right)^2 + (Q_{SP}^2/\alpha^2)\left(\Omega^2-1\right)^2}\right]^{1/2} \quad (22.28)$$

$$\varphi = \begin{cases} \tan^{-1}\frac{-\alpha}{Q_{SP}\Omega} - \tan^{-1}\left(\frac{Q_{SP}}{\alpha}\frac{\Omega^2-1}{\Omega-((1-\alpha)/\Omega)}\right) + 180° \\ \quad \text{if} \quad \Omega < \Omega_C = \sqrt{1-\alpha} \\ \tan^{-1}\frac{-\alpha}{Q_{SP}\Omega} - \tan^{-1}\left(\frac{Q_{SP}}{\alpha}\frac{\Omega^2-1}{\Omega-((1-\alpha)/\Omega)}\right) \\ \quad \text{if} \quad \Omega \geq \Omega_C = \sqrt{1-\alpha} \end{cases} \quad (22.29)$$

Figure 22.25 shows the characteristics corresponding to the THD of the output voltage and the fundamental output voltage for $\alpha = 0.5$. As can be seen, the THD is also very low around the natural frequency, especially for the higher values of the normalized load Q_{SP}. Regarding the output voltage, this circuit behaves as a parallel circuit around the natural frequency ω_0, with a maximum gain voltage of about Q_{SP}/α. Around the natural frequency of the series circuit given by L and $C_S \omega_{0S} = \omega_0 \sqrt{1-\alpha}$, the circuit behaves as a series-loaded circuit with a maximum voltage gain equal to unity.

The series–parallel circuit can also be used to both ignite and supply the lamp at constant frequency, since it also behaves as a current source at the natural frequency. Besides, this circuit allows to limit the ignition voltage by means of the factor α, thus avoiding sputtering damage of lamp electrodes. Also, the series capacitor can be used to block any DC component of the inverter square output voltage, as that existing in the asymmetric half-bridge. As summary, the series–parallel circuit combines the best features of the series-loaded and the parallel-loaded and this is why it is used widely to implement electronic ballasts.

When operated at frequency ω_0, the output voltage gain is equal to Q_{SP}/α, and then the circuit behaves as a current source equal to $V_{S,1}/\alpha Z_B$, which is independent of the load. As stated previously, this results are adequate to supply discharge lamps. The input current phase angle at this frequency is equal to $\tan^{-1}(-\alpha/Q_{SP})$, and the input current always lags the input voltage, thus showing zero voltage switching.

Finally, the condition to have input current in phase with the input voltage can be obtained by equaling Eq. (22.27) to zero. This gives the following value:

$$Q_{SP,\varphi=0} = \frac{\alpha}{\Omega}\sqrt{\frac{\Omega^2 - (1-\alpha)}{1-\Omega^2}} \quad (22.30)$$

Equation (22.30) defines the borderline between the ZVS and the zero current switching modes. The output voltage gain in this borderline can be obtained by using Eq. (22.30) in Eq. (22.26) and it is equal to $\alpha/\sqrt{\alpha(1-\Omega^2)}$.

22.4.3 Design Issues

The design methodology of a resonant inverter for discharge lamps supply can be very different depending on the lamp type and characteristics, inverter topology, design goals, etc. Some guidelines, especially focused for supplying fluorescent lamps with voltage-fed resonant inverters are presented in this section to illustrate the basic design methodology.

A typical starting process for a hot cathode fluorescent lamp is shown in Fig. 22.26. Initially lamp electrodes are heated up to the emission temperature. During this phase, the inverter should assure a voltage applied to the lamp not high enough to produce sputtering damage in lamp electrodes, thus avoiding a premature aging of the lamp. Once electrodes reach the emission temperature, the lamp can be ignited by applying the necessary starting voltage. Following this procedure, a soft starting is achieved and a long lamp life is assured.

The best method to perform this soft starting is to control the inverter switching frequency, so that the lamp voltage and current are always under control. During the heating process, the operating frequency is adjusted to a value above the natural frequency of the resonant tank. In this way, the heating current can be adjusted to the necessary value maintaining a lamp voltage quite lower than the starting voltage. Since normally MOSFET or bipolar transistors are used, the operation over the natural frequency is preferable, because it provides ZVS and the slow parasitic diodes existing in these transistors can be used. After a short period of time, the switching frequency is reduced until the starting voltage is obtained, then the lamp is ignited. Normally, the final operating point at steady-state operation is adjusted to be at a switching frequency equal to the natural frequency, so that a very stable operation for the lamp is assured.

Figure 22.27a shows a fluorescent lamp supplied using a series–parallel resonant tank. The input data for the design are normally the fundamental input voltage V_S, the switching frequency in normal discharge operation (running) f_S, the lamp voltage and current at high frequency: V_{LA}, I_{LA}, the electrode heating current I_H, the maximum lamp voltage during heating process: V_H, and the lamp starting voltage: V_{IG}.

The equivalent circuit during the heating and ignition phase is shown in Fig. 22.27b. The current circulating in this circuit is the electrode heating current, which can be calculated by using $Q_{SP} = \infty$ in Eq. (22.24):

$$I_H = \frac{V_S/Z_B}{\Omega - 1/\Omega} \qquad (22.31)$$

where the electrode resistance has been neglected for simplicity. For a given heating current I_H, the necessary switching frequency is obtained from Eq. (22.31) as follows:

$$\Omega_H = \frac{V_S}{2Z_B I_H} + \sqrt{1 + \left(\frac{V_S}{2Z_B I_H}\right)^2} \qquad (22.32)$$

Regarding the heating voltage, it can be calculated from Eq. (22.26) using $Q_{SP} = \infty$, and the following value is obtained:

$$V_H = \frac{\alpha V_S}{\Omega_H^2 - 1} \qquad (22.33)$$

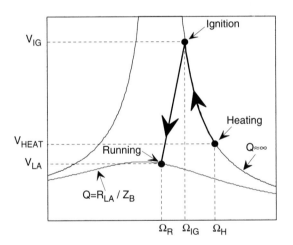

FIGURE 22.26 Typical starting process of discharge lamps.

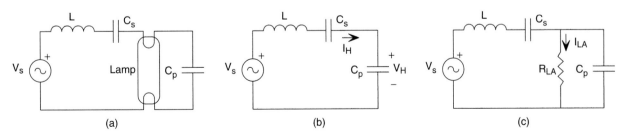

FIGURE 22.27 (a) Fluorescent lamp supplied with a series–parallel circuit; (b) equivalent circuit during heating and ignition; and (c) equivalent circuit during normal discharge mode.

The frequency at which the starting voltage is achieved can also be obtained using $Q_{SP}=\infty$ in Eq. (22.26), giving the following value:

$$\Omega_{IG}=\sqrt{1+\alpha\frac{V_S}{V_{IG}}} \qquad (22.34)$$

Once the lamp is ignited, the new operation circuit is shown in Fig. 22.27c. Normally the switching frequency is selected to be very close to the natural frequency, and the circuit characteristics can be obtained by using $\Omega=1$. Thus, as stated previously, the circuit behaves as a current source of the following value:

$$I_{LA}=\frac{V_S}{\alpha Z_B} \qquad (22.35)$$

From this equation, the impedance Z_B needed to provide a lamp current equal to I_{LA} is easily obtained:

$$Z_B=\frac{V_S}{\alpha I_{LA}} \qquad (22.36)$$

Using Eqs. (22.31)–(22.36), the design procedure can be performed as follows:

Step 1 Steady-state operation. Choose a value for the factor α; normally a value of 0.8–0.9 will be adequate for most applications. From Eq. (22.36) calculate the value of Z_B for the resonant tank. Since the natural frequency is equal to the switching frequency, the reactive elements of the resonant tank can be calculated as follows:

$$L=\frac{Z_B}{2\pi f_S} \quad C_P=\frac{1}{2\pi f_S \alpha Z_B} \quad C_S=\frac{1}{2\pi f_S(1-\alpha)Z_B} \qquad (22.37)$$

Step 2 Heating phase. From Eq. (22.32) calculate the switching frequency for a given heating current I_H. Then, calculate the value of the heating voltage V_H from Eq. (22.33).

Step 3 Check if the heating voltage is lower than the maximum value allowed to avoid electrode sputtering. If the voltage is too high, choose a higher value of α and repeat steps 1 and 2. Also, the maximum heating frequency can be limited to avoid excessive frequency variation. The lower the α, the lower is the frequency variation from heating to ignition, since the output voltage characteristics are narrower for the lower values of α.

The described procedure to achieve lamp soft ignition requires the use of a voltage-controlled oscillator to control the switching frequency. This can increase the cost of the ballast. A similar soft ignition can be achieved using the resonant circuit shown in Fig. 22.28. This circuit is very often used in self-oscillating ballasts, where the switches are driven from the resonant current using a current transformer [18].

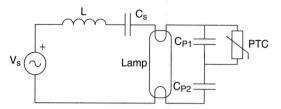

FIGURE 22.28 Resonant circuit using a PTC resistor.

In the circuit shown in Fig. 22.28, the PTC is initially cold and capacitor C_{P1} is practically short-circuited by the PTC. The resonant tank under these conditions is formed by $L-C_S-C_{P2}$, which can be designed to heat the lamp electrodes with a heating voltage low enough to avoid lamp cold ignition. Since the PTC is also heated by the circulating current, after a certain period of time it reaches the threshold temperature and trips. At this moment, the new resonant tank is formed by L, C_S, and the series equivalent of C_{P1} and C_{P2}. This circuit can be designed to generate the necessary ignition voltage and to supply the lamp in normal discharge mode. Once the lamp is ignited, the PTC maintains its high impedance since the dominant parallel capacitor in this phase is C_{P1} (usually $C_{P1} \ll C_{P2}$).

22.5 High Power Factor Electronic Ballasts

As commented in a previous section, when electronic ballasts are supplied from the AC line, an AC–DC stage is necessary to provide the DC input voltage of the resonant inverter (see Fig. 22.9). Since the introduction of harmonic regulations, as IEC 1000-3-2, the use of a full-bridge diode rectifier followed by a filter capacitor is no longer applicable for this stage due to the high harmonic content of the input current. Therefore, the use of an AC–DC stage showing a high input power factor (PF) and a low total harmonic distortion (THD) of the input current is mandatory. The inclusion of this stage can significantly increase the cost of the complete ballast, and therefore the search for low cost high power factor electronic ballasts is presently an important field of research.

Figure 22.29a shows a first possibility to increase the input power factor of the ballast by removing the filter capacitor across the diode rectifier. However, since there are no low frequency storage elements inside the resonant inverter, the output power instantaneously follows the input power, thus producing an annoying light flicker. Besides, the resulting lamp current crest factor is very high, which considerably decreases lamp life.

In order to avoid flicker and increase lamp current crest factor, continuous power must be delivered to the lamp. This can only be accomplished by using a PFC stage with a low frequency storage element. This solution is shown in Fig. 22.29b,

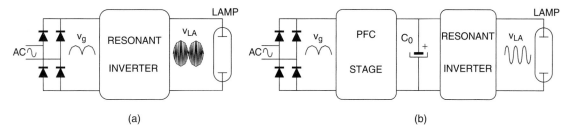

FIGURE 22.29 High power factor ballasts.

where capacitor C_0 is used as energy storage element. The main drawback of this solution is that the input power is handled by the two stages, which reduces the efficiency of the complete electronic ballast.

22.5.1 Harmonic Limiting Standards

The standards IEC 1000-3-2 and EN 61000-3-2 [19] are the most popular regulations regarding the harmonic pollution produced by electronic equipment connected to the mains. These standards are a new version of the previous IEC 555-2 regulation and they are applicable to the equipment with less than 16 A per phase and supplied from low voltage lines of 220/380 V, 230/400 V, and 240/415 V at 50 or 60 Hz. Limits for equipment supplied from voltages lower than 220 V have not yet been established.

This regulation divides the electrical equipment in several classes from class A to class D. Class C is especially for lighting equipment, including dimming devices; the harmonic limits for this class are shown in Table 22.4. As shown in Table 22.4, this regulation establishes a maximum amplitude for each input harmonic as a percentage of the fundamental harmonic component. The harmonic content established in Table 22.4 is quite restrictive, which means that the input current wave must be quite similar to a pure sine wave. For example, for a typical input power factor equal to 0.9, the THD calculated from Table 22.1 is only 32%.

TABLE 22.4 IEC 1000-3-2. Harmonic limits for class c equipment

Harmonic order n	Maximum permissible harmonic current expressed as a percentage of the input current at the fundamental frequency (%)
2	2
3	$30 \cdot \lambda^*$
5	10
7	7
9	5
$11 \leq n \leq 39$ (odd harmonics only)	3

*λ is the circuit power factor.

22.5.2 Passive Solutions

A first possibility to increase the ballast power factor and to decrease the harmonic content of the input current is the use of passive solutions. Figure 22.30 shows two typical passive solutions, which can be used to improve the input power factor of electronic ballast.

Figure 22.30a shows the most common passive solution based on a filter inductor L. Using a large inductance L a square input current can be obtained, with an input power factor of 0.9 and a THD of about 48%. A square input waveform does not satisfy the IEC 1000-3-2 requirements and then it is not a suitable solution. The addition of capacitor C across

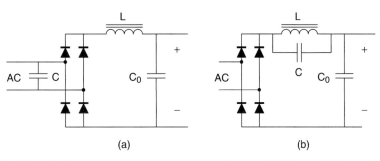

FIGURE 22.30 Passive circuits to improve input power factor (a) LC filter and (b) tuned LC filter.

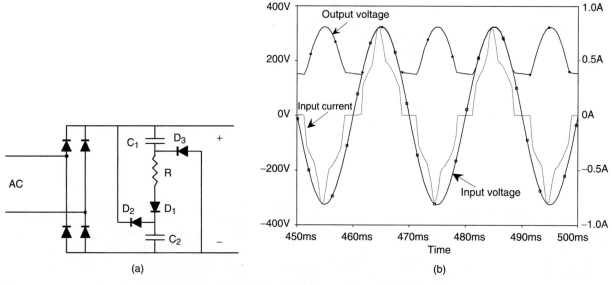

FIGURE 22.31 Valley-fill circuit: (a) electric diagram and (b) waveforms.

the AC terminals can increase the power factor up to 0.95, but still the standard requirements are difficult to fulfill.

A simple variation of this circuit is shown in Fig. 22.30b, where a parallel circuit tuned to the third harmonic of the line frequency is used to improve the shape of the line current. The input power factor obtained with this circuit can be close to unity.

A third possible solution, known as valley-fill circuit, is shown in Fig. 22.31a. The typical filter capacitor following the diode rectifier is split into two different capacitors which are alternately charged using three extra diodes. The addition of a small series resistor improves the power factor by about 2 points, maintaining a low cost for the circuit. An inductor in place of the resistor can also be used to improve the power factor but with a higher cost penalty. Figure 22.31b shows the output voltage and input current of the valley-fill circuit. The main disadvantage of this circuit is the high ripple of the output voltage, which produces lamp power and luminous flux fluctuation and high lamp current crest factor.

Passive solutions are reliable, rugged, and cheap. However, the size and weight of these solutions are high and their design to fulfill the IEC 1000-3-2 requirements is usually difficult. Therefore, they are normally applied in the lower power range.

22.5.3 Active Solutions

Active circuits are most popular solutions to implement high power factor electronic ballasts. They use controlled switches to correct the input power factor and in some cases to include galvanic isolation via high-frequency transformers. Active circuits normally used in electronic ballasts operate at a switching frequency well above the line frequency and over the audible range.

Some typical active circuits used in electronic ballasts are shown in Fig. 22.32. Buck–boost and flyback converters shown in Fig. 22.32a and 22.32b, respectively, can be operated in discontinuous conduction mode (DCM) with constant frequency and constant duty cycle in order to obtain an input power factor close to unity [20].

Fig. 22.32c shows the boost converter, which is one of the most popular active circuit used to correct the input power factor of electronic ballasts [21–23]. If the boost converter

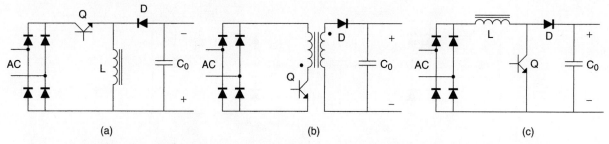

FIGURE 22.32 Power factor correction circuits for ballasts: (a) buck–boost; (b) flyback; and (c) boost.

is operated in DCM, an input power factor close to unity is obtained, provided that the output voltage is about twice the peak input voltage [21]. The main disadvantage of DCM operation, when compared to the CCM mode, is the high distortion of the input current (due to the discontinuous high frequency current) and the higher current and voltage stresses in the switches. Therefore, the DCM operation is only used for the lower power range.

For the medium power range, the operation of the boost converter at the DCM–CCM borderline is preferred. In this solution, the on-time of the controlled switch is maintained constant within the whole line period and the switching frequency is adjusted to allow the input current to reach zero at the end of the switching period. The typical control circuit used and the input current waveform are shown in Figs. 22.33a and b, respectively. The inductor current is sensed using a resistor in series with the switch and the peak inductor current is programmed to follow a sine wave using a multiplier. A comparator is employed to detect the zero-crossing of the inductor current in order to activate the switch. Most IC manufacturers provide a commercial version of this circuit to be used for electronic ballast applications.

The boost circuit operating with borderline control provides a continuous input current, which is easier to filter. Besides, it presents low switch turn-on losses and low recovery losses in the output diode. The main disadvantages are the variable switching frequency and the high output voltage, which must be higher than the peak line voltage.

For the higher power range, the boost converter can be operated in continuous conduction mode (CCM) to correct the input power factor. The input current in this scheme is continuous with very low distortion and easy to filter. The current stress in the switch is also lower, which means that more power can be handled maintaining a good efficiency. The normal efficiency obtained with a boost circuit operating either in the DCM–CCM borderline or in CCM can be as high as 95%.

22.6 Applications

Electronic ballasts are widely used in lighting applications as portable lighting, emergency lighting, automotive applications, home lighting, industrial lighting, and so on. They provide low volume and size, making it possible to reduce the luminaire size as well, which is a very interesting new trend in lighting design.

22.6.1 Portable Lighting

In this application, a battery is used as power source and then a low input voltage is available to supply the lamp. Examples are handlanterns and backlightings for laptop computers. Typical input voltages in these applications range from 1.5 to 48 V. Therefore, a step-up converter is necessary to supply the discharge lamp, and then electronic ballasts are the only suitable solution. Since the converter is supplied from a battery, the efficiency of the ballast should be as high as possible in order to optimize the use of the battery energy, thus increasing the operation time of the portable lighting. Typical topologies used are the class E inverter and the push–pull resonant inverter, obtaining efficiencies up to 95%.

22.6.2 Emergency Lighting

Emergency lightings are used to provide a minimum lighting level in case of a main supply cut-off. Batteries are employed to store energy from the mains and to supply the lamp in case of a main supply failure. A typical block diagram is shown in Fig. 22.34. An AC–DC converter is used as a

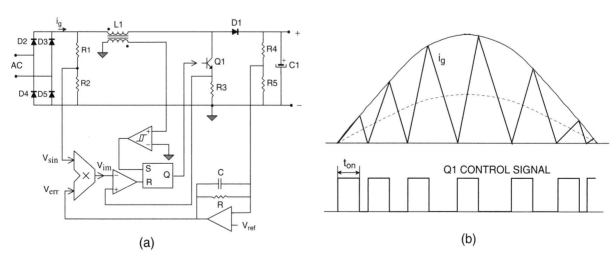

FIGURE 22.33 (a) Boost power factor corrector with borderline control and (b) input current waveform.

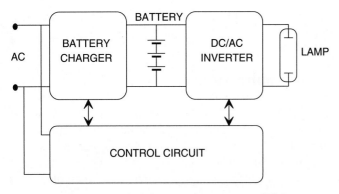

FIGURE 22.34 Block diagram of an emergency lighting.

battery charger to store energy during normal line operation. A control circuit continuously measures the line voltage and activates the inverter in case of a main supply failure. Normally a minimum operating time of one hour is required for the system in emergency state, thus the use of high efficient electronic ballasts is mandatory to reduce the battery size and cost. Typical topologies used include class E inverters, push–pull resonant inverters, and half-bridge resonant inverters. Fluorescent lamps are mainly used in emergency ballasts, but high intensity discharge lamps, such as metal-halide lamps or high-pressure sodium lamps, are also used in some special applications.

22.6.3 Automotive Lighting

Electronic ballasts are used in automotive applications such as automobiles, buses, trains, and aircrafts. Normally a low voltage DC bus is available to supply the lamps and then these applications are similar to portable and emergency lightings previously commented. In modern aircrafts, a 120/208 V, 400 Hz, 3-phase electrical system is also available and can be employed for lighting. Fluorescent lamps are typically used for automotive indoor lighting, whereas high intensity discharge lamps are preferred in the exterior lighting; for example, in automobile headlights.

22.6.4 Home and Industrial Lighting

Electronic ballasts, especially for fluorescent lamps, are also very often used in home and industrial applications. The higher efficiency of fluorescent lamps supplied at high frequency, provides an interesting energy saving when compared to incandescent lamps. A typical application is the use of compact fluorescent lamps with the electronic ballast inside the lamp base, which can directly substitute an incandescent lamp reducing the energy consumption four or five times. A self-oscillating half-bridge inverter is typically used in these energy saving lamps, since it allows to reduce the size and cost. The power of these lamps is normally below 25 W.

Other applications for higher power include more developed ballasts based on a power factor correction stage followed by a resonant inverter. Hot cathode fluorescent lamps are mostly used in these electronic ballasts. Also, with the development of modern HID lamps such as metal-halide lamps and very high-pressure sodium lamps (both showing very good color rendition), the use of HID lamps is being more and more frequent in home, commercial, and industrial lighting.

22.6.5 Microprocessor-based Lighting

The use of microprocessors in combination with electronic ballasts is very interesting from the point of view of energy saving [24–26]. The inclusion of microprocessor circuits allows to incorporate control strategies for dimming, such as scheduling, task tuning, daylighting, etc. [27]. Using these strategies, the achieved energy saving can be as high as 35–40%. Another advantage of using microprocessors is the possibility of detecting lamp failure or bad operation, thus increasing the reliability and decreasing the maintenance cost of the installation. Most advanced electronic ballasts can include a communication stage to send and receive information regarding the state of the lighting to or from a central control unit. In some cases, communications can be performed via power line, thus reducing the installation costs. Figure 22.35 shows the block diagram of a microprocessor-based electronic ballast.

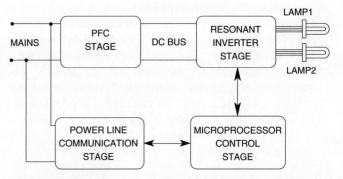

FIGURE 22.35 Block diagram of a microprocessor-based lighting.

Further Reading

1. Illuminating Engineering Society of North America: *IES Lighting Handbook,* 1984 Reference Volume, IESNA, New York, 1984.
2. De Groot, J. and Van Vliet, J.: *The High-Pressure Sodium Lamp.* Philips Technical Library, Macmillan Education, 1986.
3. Koshimura, Y., Aoike, N., and Nomura, O.: "Stable high frequency operation of high intensity discharge lamps and their ballast design," Proc. CIE 20th Session'83, 1983, pp. E36/1–E36/2.
4. Laskai, L., Enjeti, P. N., and Pitel, I. J.: "White-noise modulation of high-frequency high-intensity discharge lamp ballasts," IEEE Transactions on Industry Applications, Vol. 34, No. 3, May/June 1998, pp. 597–605.

5. Simoes, A. S., Silva, M. M., and Anunciada, A. V.: "A boost-type converter for DC-supply of fluorescent lamps," IEEE Transactions on Industrial Electronics, Vol. 41, No. 2, April 1994, pp. 251–255.
6. Vasiljevic, D. M.: "The design of a battery-operated fluorescent lamp," IEEE Transactions on Industrial Electronics, Vol. 36, No. 4, November 1989, pp. 499–503.
7. Hernando, M., Blanco, C., Alonso, J. M., and Rico, M.: "Fluorescent lamps supplied with dc current and controlled in current mode," Proc. of European Power Electronics Conference (EPE), Firenze, 1991, pp. 1/499–1/503.
8. Corominas, E. L., Alonso, J. M., Calleja, A. J., Ribas, J., and Rico, M.: "Analysis of tapped-inductor inverters as low-power fluorescent lamp ballast supplied from a very low input voltage," PESC Conf. Record, 1999, pp. 1103–1108.
9. Duarte, J. L., Wijntjens, J., and Rozenboom, J.: "Getting more from fluorescent lamps through resonant converters," Proc. of IECON, San Diego, 1992, pp. 560–563.
10. Ponce, M., Arau, J., Alonso, J. M., and Rico-Secades, M.: "Electronic ballast based on class E amplifier with a capacitive inverter and dimming for photovoltaic applications," Proc. of APEC, 1998, pp. 1156–1162.
11. Gulko, M. and Ben-Yaakov, S.: "Current-sourcing push-pull parallel-resonance inverter (CS-PPRI): theory and application as a discharge lamp driver," IEEE Transactions on Industrial Electronics, Vol. 41, No. 3, June 1994, pp. 285–291.
12. Lin, M.-S., Ho, W.-J., Shih, F.-Y., Chen, D.-Y., and Wu, Y.-P.: "A cold-cathode fluorescent lamp driver circuit with synchronous primary-side dimming control," IEEE Transactions on Industrial Electronics, Vol. 45, No. 2, April 1998, pp. 249–255.
13. Rashid, M. H.: *Power Electronics. Circuits, Devices, and Applications.* Prentice Hall. Second Edition, New Jersey, 1993.
14. Mader, U. and Horn, P.: "A dynamic model for the electrical characteristics of fluorescent lamps," IEEE Industry Applications Society Meeting, Conf. Records, 1992, pp. 1928–1934.
15. Sun, M. and Hesterman, B. L.: "Pspice high-frequency dynamic fluorescent lamp model," IEEE Trans. on Power Electronics, Vol. 13, No. 2, March 1998, pp. 261–272.
16. Tseng, K. J., Wang, Y., and Vilathgamuwa: "An experimentally verified hybrid Cassie-Mayr electric arc model for power electronics simulations," IEEE Trans. on Power Electronics, Vol. 12, No. 3, May 1997, pp. 429–436.
17. Alonso, J. M., Blanco, C., López, E., Calleja, A. J., and Rico, M.: "Analysis, design and optimization of the LCC resonant inverter as a high-intensity discharge lamp ballast," IEEE Trans. on Power Electronics, Vol. 13, No. 3, May 1998, pp. 573–585.
18. "Electronics ballasts for fluorescent lamps using BUL770/791 transistors." Application Report, Texas Instruments, 1992.
19. IEC 1000-3-2 (1995-03) standards on electromagnetic compatibility (EMC), Part 3, Section 2: Limits for harmonic current emissions. International Electrotechnical Commission, Geneva, Switzerland, April 1995.
20. Alonso, J. M., Calleja, A. J., López, E., Ribas, J., and Rico, M.: "A novel single-stage constant-wattage high-power-factor electronic ballast," IEEE Trans. on Industrial Electronics, Vol. 46, No 6, December 1999, pp. 1148–1158.
21. Liu, K.-H. and Lin, Y.-L.: "Current waveform distortion in power factor correction circuits employing discontinuous-mode boost converters," Power Electronics Specialists Conference proceedings, 1989, pp. 825–829.
22. Lai, J. S. and Chen, D.: "Design consideration for power factor correction boost converter operating at the boundary of continuous conduction mode and discontinuous conduction mode," Applied Power Electronics Conference proceedings, 1993, pp. 267–273.
23. Blanco, C., Alonso, J. M., López, E., Calleja, A. J., and Rico, M.: "A single-stage fluorescent lamp ballast with high power factor", IEEE APEC'96 proceedings, pp. 616–621.
24. Alling, W. R.: "The integration of microcomputers and controllable output ballasts - A new dimension in lighting control," IEEE Trans. on Industry Applications, September/October 1984, pp. 1198–1205.
25. Alonso, J. M., Díaz, J., Blanco, C., and Rico, M.: "A microcontroller-based emergency ballast for fluorescent lamps," IEEE Transactions on Industrial Electronics, Vol. 44, No. 2, April 1997, pp. 207–216.
26. Alonso, J. M., Ribas, J., Calleja, A. J., López, E., and Rico, M.: "An intelligent neuron-chip-based fluorescent lamp ballast for indoor applications", Conf. Proc. of the IEEE Industry Applications Society Annual Meeting (IAS'97), New Orleans, Louisiana, October 5–9, 1997, pp. 2388–2394.
27. Illuminating Engineering Society of North America: *IES Lighting Handbook,* 1984 Application Volume, IESNA, New York, 1987.
28. Meyer, Chr. and Nienhuis, H.: *Discharge Lamps,* Philips Technical Library, 1988.
29. Wyszecki, G. and Stiles, W. S.: *Color Science: Concepts and Methods, Quantitative Data and Formulae.* John Wiley & Sons, Second Edition, 1982.
30. Borton, J. A. and Daley, K. A.: "A comparison of light sources for the petrochemical industry," IEEE Industry Applications Magazine, July/August 1997, pp. 54–62.
31. Alling, W. R.: "Important design parameters for solid-state ballasts," IEEE Transactions on Industry Applications, Vol. 25, No. 2, March/April 1989, pp. 203–207.

23
Power Supplies

Y. M. Lai, Ph.D.
Department of Electronic and Information Engineering, The Hong Kong Polytechnic University, Hong Kong

23.1 Introduction 593
23.2 Linear Series Voltage Regulator 595
    23.2.1 Regulating Control • 23.2.2 Current Limiting and Overload Protection
23.3 Linear Shunt Voltage Regulator 598
23.4 Integrated Circuit Voltage Regulators 600
    23.4.1 Fixed Positive and Negative Linear Voltage Regulators • 23.4.2 Adjustable Positive and Negative Linear Voltage Regulators • 23.4.3 Applications of Linear IC Voltage Regulators
23.5 Switching Regulators 602
    23.5.1 Single-ended Isolated Flyback Regulators • 23.5.2 Single-ended Isolated Forward Regulators • 23.5.3 Half-bridge Regulators • 23.5.4 Full-bridge Regulators • 23.5.5 Control Circuits and Pulse-width Modulation
    Further Reading 618

23.1 Introduction

Power supplies are used in most electrical equipment. Their applications cut across a wide spectrum of product types, ranging from consumer appliances to industrial utilities, from milliwatts to megawatts, from hand-held tools to satellite communications.

By definition, a power supply is a device which converts the output from an ac power line to a steady dc output or multiple outputs. The ac voltage is first rectified to provide a pulsating dc, and then filtered to produce a smooth voltage. Finally, the voltage is regulated to produce a constant output level despite variations in the ac line voltage or circuit loading. Figure 23.1 illustrates the process of rectification, filtering, and regulation in a dc power supply. The transformer, rectifier, and filtering circuits are discussed in other chapters. In this chapter, we will concentrate on the operation and characteristics of the regulator stage of a dc power supply.

In general, the regulator stage of a dc power supply consists of a feedback circuit, a stable reference voltage, and a control circuit to drive a pass element (a solid-state device such as transistor, MOSFET, etc.). The regulation is done by sensing variations appearing at the output of the dc power supply. A control signal is produced to drive the pass element to cancel any variation. As a result, the output of the dc power supply is maintained essentially constant. In a transistor regulator, the pass element is a transistor, which can be operated in its active region or as a switch, to regulate the output voltage. When the transistor operates at any point in its active region, the regulator is referred to as a *linear voltage regulator*. When the transistor operates only at cutoff and at saturation, the circuit is referred to as a *switching regulator*.

Linear voltage regulators can be further classified as either series or shunt types. In a series regulator, the pass transistor is connected in series with the load as shown in Fig. 23.2. Regulation is achieved by sensing a portion of the output voltage through the voltage divider network R_1 and R_2, and comparing this voltage with the reference voltage V_{REF} to produce a resulting error signal that is used to control the conduction of the pass transistor. This way, the voltage drop across the pass transistor is varied and the output voltage delivered to the load circuit is essentially maintained constant.

In the shunt regulator shown in Fig. 23.3, the pass transistor is connected in parallel with the load, and a voltage-dropping

Copyright © 2007, 2001, Elsevier Inc.
All rights reserved.

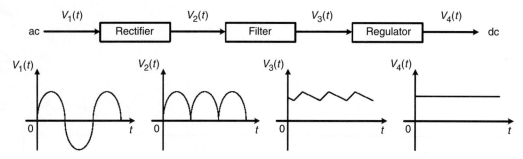

FIGURE 23.1 Block diagram of a dc power supply.

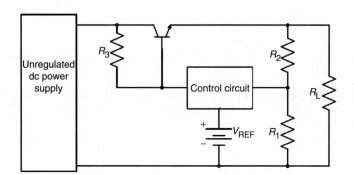

FIGURE 23.2 A linear series voltage regulator.

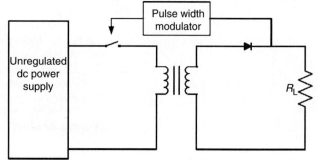

FIGURE 23.4 A simplified form of a switching regulator.

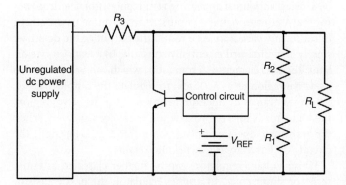

FIGURE 23.3 A linear shunt voltage regulator.

resistor R_3 is connected in series with the load. Regulation is achieved by controlling the current conduction of the pass transistor such that the current through R_3 remains essentially constant. This way, the current through the pass transistor is varied and the voltage across the load remains constant.

As opposed to linear voltage regulators, switching regulators employ solid-state devices, which operate as switches: either completely *on* or completely *off*, to perform power conversion. Because the switching devices are not required to operate in their active regions, switching regulators enjoy a much lower power loss than those of linear voltage regulators. Figure 23.4 shows a switching regulator in a simplified form. The high-frequency switch converts the unregulated dc voltage from one level to another dc level at an adjustable duty cycle. The output of the dc supply is regulated by means of a feedback control that employs a pulse-width-modulator (PWM) controller, where the control voltage is used to adjust the duty cycle of the switch.

Both linear and switching regulators are capable of performing the same function of converting an unregulated input into a regulated output. However, these two types of regulators have significant differences in properties and performances. In designing power supplies, the choice of using certain type of regulator in a particular design is significantly based on the cost and performance of the regulator itself. In order to use the more appropriate regulator type in the design, it is necessary to understand the requirements of the application and select the type of regulator that best satisfies those requirements. Advantages and disadvantages of linear regulators, as compared to switching regulators, are given below:

1. Linear regulators exhibit efficiency of 20–60%, whereas switching regulators have a much higher efficiency, typically 70–95%.
2. Linear regulators can only be used as a step-down regulator, whereas switching regulators can be used in both step-up and step-down operations.

3. Linear regulators require a mains-frequency transformer for off-the-line operation. Therefore, they are heavy and bulky. On the other hand, switching regulators use high-frequency transformers and can therefore be small in size.
4. Linear regulators generate little or no electrical noise at their outputs, whereas switching regulators may produce considerable noise if they are not properly designed.
5. Linear regulators are more suitable for applications of less than 20 W, whereas switching regulators are more suitable for large power applications.

In this chapter, we will examine the circuit operation, characteristics, and applications of linear and switching regulators. In Section 23.2, we will look at the basic circuits and properties of linear series voltage regulators. Some current-limiting techniques will be explained. In Section 23.3, linear shunt regulators will be covered. The important characteristics and uses of linear Integrated Circuit (IC) regulators will be discussed in Section 23.4. Finally, the operation and characteristics of switching regulators will be discussed in Section 23.5. Important design guidelines for switching regulators will also be given in this section.

23.2 Linear Series Voltage Regulator

A zener diode regulator can maintain a fairly constant voltage across a load resistor. It can be used to improve the voltage regulation and reduce the ripple in a power supply. However, the regulation is poor and the efficiency is low because of the non-zero resistance in the zener diode. To improve the regulation and efficiency of the regulator, we have to limit the zener current to a smaller value. This can be accomplished by using an amplifier in series with the load as shown in Fig. 23.5. The effect of this amplifier is to limit the variation of the current I_D through the zener diode D_z. This circuit is known as a linear series voltage regulator because the transistor is in series with the load.

Because of the current-amplifying property of the transistor, I_D is reduced by a factor of $(\beta + 1)$, where β is the dc current gain of the transistor. Hence there is a small voltage drop across the diode resistance and the zener diode approximates an ideal voltage source. The output voltage V_o of the regulator is

$$V_o = V_z + V_{BE} \qquad (23.1)$$

where V_z is the zener voltage and V_{BE} is the base-to-emitter voltage of the transistor. The change in output voltage is

$$\Delta V_o = \Delta V_z + \Delta V_{BE}$$
$$= \Delta I_D r_d + \Delta I_L r_e \qquad (23.2)$$

where r_d is the dynamic resistance of the zener diode and r_e is the output resistance of the transistor. Assume that V_i and V_z are constant. With $\Delta I_D \approx \Delta I_L/(\beta + 1)$, the change in output voltage is then

$$\Delta V_o \approx \Delta I_L r_e \qquad (23.3)$$

If V_i is not constant, then the current I will change with the input voltage. In calculating the change in output voltage, this current change must be absorbed by the zener diode.

In designing linear series voltage regulators, it is imperative that the series transistor must work within the rated Safe Operation Area (SOA) and be protected from excess heat dissipation because of current overload. The emitter-to-collector voltage V_{CE} of Q_1 is given by

$$V_{CE} = V_i - V_o \qquad (23.4)$$

Thus, with specified output voltage, the maximum allowable V_{CE} for a given Q_1 is determined by the maximum input

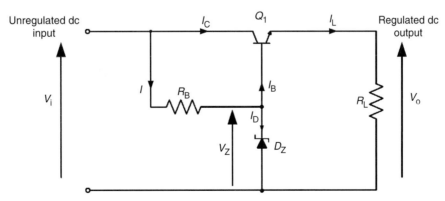

FIGURE 23.5 Basic circuit of a linear series regulator.

voltage to the regulator. The power dissipated by Q_1 can be approximated by

$$P_{Q1} \approx (V_i - V_o)I_L \qquad (23.5)$$

Consequently, the maximum allowable power dissipated in Q_1 is determined by the combination of the input voltage V_i and the load current I_L of the regulator. For a low output voltage and a high loading current regulator, the power dissipated in the series transistor is about 50% of the power delivered to the output.

In many high-current high-voltage regulator circuits, it is necessary to use a Darlington-connected transistor pair so that the voltage, current, and power ratings of the series element are not exceeded. The method is shown in Fig. 23.6. An additional desirable feature of this circuit is that the reference diode dissipation can be reduced greatly. The maximum base current I_{B1} is then $I_L/(\beta_1 + 1)(\beta_2 + 1)$, where β_1 and β_2 are the dc current gain of Q_1 and Q_2 respectively. This current is usually of the order of less than 1 mA. Consequently, a low-power reference diode can be used.

23.2.1 Regulating Control

The series regulators shown in Figs. 23.5 and 23.6 do not have feedback loop. Although they provide satisfactory performance for many applications, their output resistances and ripples cannot be reduced. Figure 23.7 shows an improved form of the series regulator, in which negative feedback is employed to improve the performance. In this circuit, transistors Q_3 and Q_4 form a single-ended differential amplifier, and the gain of this amplifier is established by R_6. Here D_z is a stable zener diode reference, biased by R_4. For higher accuracy, D_z can be replaced by an IC reference such as REF series from Burr-Brown. Resistors R_1 and R_2 form a voltage divider for output voltage sensing. Finally, transistors Q_1 and Q_2 form a Darlington pair output stage. The operation of the regulator can be explained as follows. When Q_1 and Q_2 are on, the output voltage increases, and hence the voltage V_A at the base of Q_3 also increases. During this time, Q_3 is off and Q_4 is on. When V_A reaches a level that is equal to the reference voltage V_{REF} at the base of Q_4, the base-emitter junction of Q_3 becomes forward-biased. Some Q_1 base current will divert into the collector of Q_3. If the output voltage V_o starts to rise above V_{REF}, Q_3 conduction increases to further decrease the conduction of Q_1 and Q_2, which will in turn maintain output voltage regulation. Figure 23.8 shows another improved series regulator that uses an operational amplifier to control the conduction of the pass transistor.

One of the problems in the design of linear series voltage regulators is the high-power dissipation in the pass transistors. If an excess load current is drawn, the pass transistor can be quickly damaged or destroyed. In fact, under short-circuit conditions, the voltage across Q_2 in Fig. 23.7, will be the input voltage V_i, and the current through Q_2 will be greater than the rated full-load output current. This current will cause Q_1 to exceed its rated SOA unless the current is reduced. In the next section, some current-limiting techniques will be presented to overcome this problem.

23.2.2 Current Limiting and Overload Protection

In some series voltage regulators, overloading causes permanent damage to the pass transistors. The pass transistors must be kept from excessive power dissipation under current overloads or short circuit conditions. A current-limiting mechanism must be used to keep the current through the transistors at a safe value as determined by the power rating of the transistors. The mechanism must be able to respond quickly to protect the transistor and yet permit the regulator to

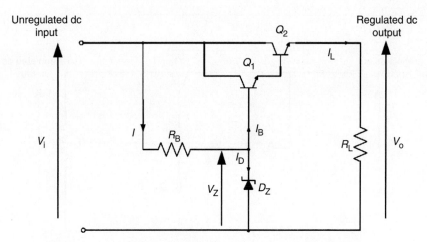

FIGURE 23.6 A linear series regulator with Darlington-connected amplifier.

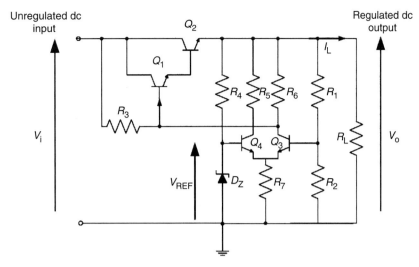

FIGURE 23.7 An improved form of discrete component series regulator.

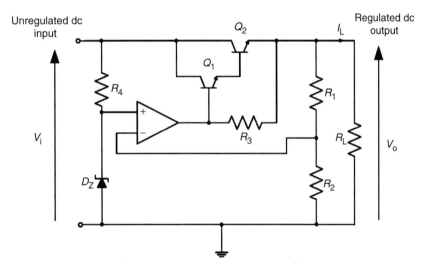

FIGURE 23.8 An improved form of op-amp series regulator.

return to normal operation as soon as the overload condition is removed. One of the current-limiting techniques to prevent current overload, called the constant current-limiting method, is shown in Fig. 23.9(a). Current limiting is achieved by the combined action of the components shown inside the dashed line. The voltage developed across the current-limit resistor R_3 and the base-to-emitter voltage of current-limit transistor Q_3 is proportional to the circuit output current I_L. During current overload, I_L reaches a predetermined maximum value that is set by the value of R_3 to cause Q_3 to conduct. As Q_3 starts to conduct, Q_3 shunts a portion of the Q_1 base current. This action, in turn, decreases and limits I_L to a maximum value $I_{L(max)}$. Since the base-to-emitter voltage V_{BE} of Q_3 cannot exceed above 0.7 V, the voltage across R_3 is held at this value and $I_{L(max)}$ is limited to

$$I_{L(max)} = \frac{0.7\,\text{V}}{R_3} \qquad (23.6)$$

Consequently, the value of the short-circuit current is selected by adjusting the value of R_3. The voltage–current characteristic of this circuit is shown in Fig. 23.9(b).

In many high-current regulators, foldback current limiting is always used to protect against excessive current. This technique is similar to the constant current-limiting method, except that as the output voltage is reduced as a result of load impedance moving toward zero, the load current is also reduced. Therefore, a series voltage regulator that includes

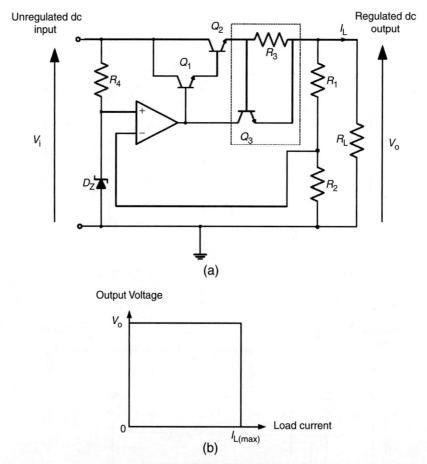

FIGURE 23.9 Series regulator with constant current limiting: (a) circuit and (b) voltage–current characteristic.

a foldback current-limiting circuit has the voltage–current characteristic shown in Fig. 23.10. The basic idea of foldback current limiting, with reference to Fig. 23.11, can be explained as follows. The foldback current-limiting circuit (in dashed outline) is similar to the constant current-limiting circuit, with the exception of resistors R_5 and R_6. At low output current, the current-limit transistor Q_3 is cutoff.

A voltage proportional to the output current I_L is developed across the current-limit resistor R_3. This voltage is applied to the base of Q_3 through the divider network R_5 and R_6. At the point of transition into current-limit, any further increase in I_L will increase the voltage across R_3 and hence across R_5, and Q_3 will progressively be turned on. As Q_1 conducts, it shunts a portion of the Q_1 base current. This action, in turn, causes the output voltage to fall. As the output voltage falls, the voltage across R_6 decreases and the current in R_6 also decreases, and more current is shunted into the base of Q_3. Hence, the current required in R_3 to maintain the conduction state of Q_3 is also decreased. Consequently, as the load resistance is reduced, the output voltage and current fall, and the current-limit point decreases toward a minimum when the output voltage is short-circuited. In summary, any regulator using foldback current limiting can have peak load current up to $I_{L(max)}$. But when the output becomes shorted, the current drops to a lower value to prevent overheating of the series transistors.

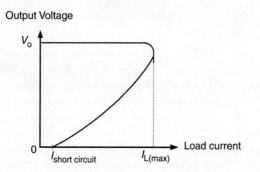

FIGURE 23.10 Voltage–current characteristic of foldback current-limit.

23.3 Linear Shunt Voltage Regulator

The second type of linear voltage regulator is the shunt regulator. In the basic circuit shown in Fig. 23.12, the pass transistor

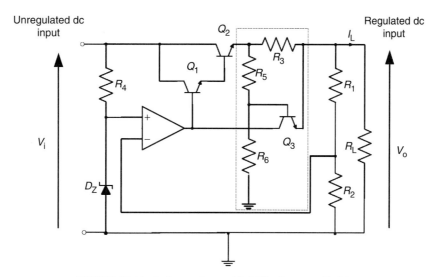

FIGURE 23.11 Series regulator with foldback current limiting.

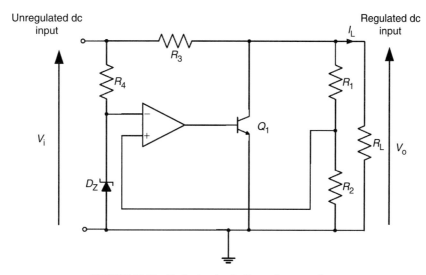

FIGURE 23.12 Basic circuit of a linear shunt regulator.

Q_1 is connected in parallel with the load. A voltage-dropping resistor R_3 is in series with this parallel network. The operation of the circuit is similar to that of the series regulator, except that regulation is achieved by controlling the current through Q_1. The operation of the circuit can be explained as follows. When the output voltage tries to increase because of a change in load resistance, the voltage at the non-inverting terminal of the operational amplifier also increases. This voltage is compared with a reference voltage and the resulting difference voltage causes Q_1 conduction to increase. With constant V_i and V_o, I_L will decrease and V_o will remain constant. The opposite action occurs when V_o tries to decrease. The voltage appearing at the base of Q_1 causes its conduction to decrease. This action offsets the attempted decrease in V_o and maintains it at an almost constant level.

Analytically, the current flowing in R_3 is

$$I_{R3} = I_{Q1} + I_L \qquad (23.7)$$

and

$$I_{R3} = \frac{V_i - V_o}{R_3} \qquad (23.8)$$

With I_L and V_o constant, a change in V_i will cause a change in I_{Q1}.

$$\Delta I_{Q1} = \frac{\Delta V_i}{R_3} \quad (23.9)$$

With V_i and V_o constant,

$$\Delta I_{Q1} = -\Delta I_L \quad (23.10)$$

Equation (23.10) shows that if I_{Q1} increases, I_L decreases, and vice versa. Although shunt regulators are not as efficient as series regulators for most applications, they have the advantage of greater simplicity. This topology offers inherently short-circuit protection. If the output is shorted, the load current is limited by the series resistor R_3 and is given by

$$I_{L(\max)} = \frac{V_i}{R_3} \quad (23.11)$$

The power dissipated by Q_1 can be approximated by

$$P_{Q1} \approx V_o I_C$$
$$= V_o(I_{R3} - I_L) \quad (23.12)$$

For a low value of I_L, the power dissipated in Q_1 is large and the efficiency of the regulator may drop to 10% under this condition.

To improve the power handling of the shunt transistor, one or more transistors connected in the common-emitter configuration in parallel with the load can be employed, as shown in Fig. 23.13.

23.4 Integrated Circuit Voltage Regulators

The linear series and shunt voltage regulators presented in the previous sections have been developed by various solid-state device manufacturers and are available in integrated circuit (IC) form. Like discrete voltage regulators, linear IC voltage regulators maintain an output voltage at a constant value despite variations in load and input voltage.

In general, linear IC voltage regulators are three-terminal devices that provide regulation of a fixed positive voltage, a fixed negative voltage, or an adjustable set voltage. The basic connection of a three-terminal IC voltage regulator to a load is shown in Fig. 23.14. The IC regulator has an unregulated input voltage V_i applied to the input terminals, a regulated voltage V_o at the output, and a ground connected to the third terminal. Depending on the selected IC regulator, the circuit can be operated with load currents ranging from milliamperes to tens of amperes and output power from milliwatts to tens

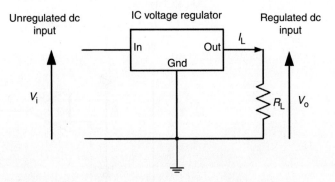

FIGURE 23.14 Basic connection of a three-terminal IC voltage regulator.

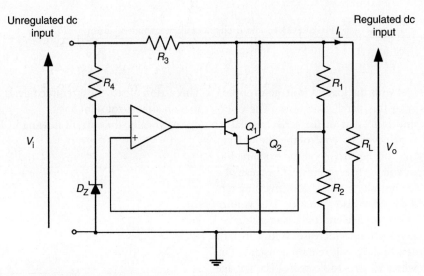

FIGURE 23.13 Linear shunt regulator with two transistors as shunt element.

23.4.1 Fixed Positive and Negative Linear Voltage Regulators

The 78XX series of regulators provide fixed regulated voltages from 5 to 24 V. The last two digits of the IC part number denote the output voltage of the device. For example, a 7824 IC regulator produces a +24 V regulated voltage at the output. The standard configuration of a 78XX fixed positive voltage regulator is shown in Fig. 23.15. The input capacitor C_1 acts as a line filter to prevent unwanted variations in the input line, and the output capacitor C_2 is used to filter the high-frequency noise that may appear at the output. In order to ensure proper operation, the input voltage of the regulator must be at least 2 V above the output voltage. Table 23.1 shows the minimum and maximum input voltages of the 78XX series fixed positive voltage regulator.

The 79XX series voltage regulator is identical to the 78XX series except that it provides negative regulated voltages instead of positive ones. Figure 23.16 shows the standard configuration of a 79XX series voltage regulator. A list of 79XX series regulators is provided in Table 23.2. The regulation of the circuit can be maintained as long as the output voltage is at least 2–3 V greater than the input voltage.

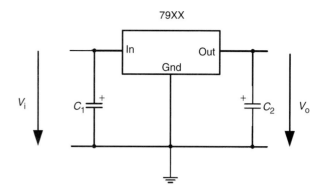

FIGURE 23.16 The 79XX series fixed negative voltage regulator.

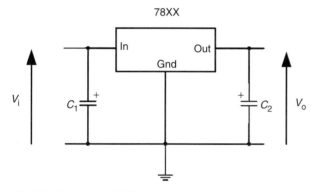

FIGURE 23.15 The 78XX series fixed positive voltage regulator.

TABLE 23.1 Minimum and maximum input voltages for 78XX series regulators

Type number	Output voltage V_o(V)	Minimum V_i(V)	Maximum V_i(V)
7805	+5	7	20
7806	+6	8	21
7808	+8	10.5	25
7809	+9	11.5	25
7812	+12	14.5	27
7815	+15	17.5	30
7818	+18	21	33
7824	+24	27	38

TABLE 23.2 Minimum and maximum input voltages for 79XX series regulators

Type number	Output voltage V_o(V)	Minimum V_i(V)	Maximum V_i(V)
7905	−5	−7	−20
7906	−6	−8	−21
7908	−8	−10.5	−25
7909	−9	−11.5	−25
7912	−12	−14.5	−27
7915	−15	−17.5	−30
7918	−18	−21	−33
7924	−24	−27	−38

23.4.2 Adjustable Positive and Negative Linear Voltage Regulators

The IC voltage regulators are also available in circuit configurations that allow the user to set the output voltage to a desired regulated value. The LM317 adjustable positive voltage regulator, for example, is capable of supplying an output current of more than 1.5 A over an output voltage range of 1.2–37 V. Figure 23.17 shows how the output voltage of an LM317 can be adjusted by using two external resistors R_1 and R_2. The capacitors C_1 and C_2 have the same function as those in the fixed linear voltage regulator.

As indicated in Fig. 23.17, the LM317 has a constant 1.25 V reference voltage, V_{REF}, across the output and the adjustment terminals. This constant reference voltage produces a constant current through R_1 regardless of the value of R_2. The output voltage V_o is given by

$$V_o = V_{REF}\left(1 + \frac{R_2}{R_1}\right) + I_{adj}R_2 \qquad (23.13)$$

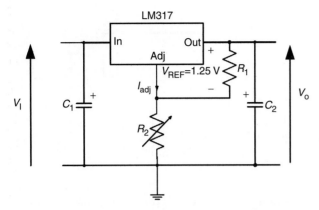

FIGURE 23.17 The LM317 adjustable positive voltage regulator.

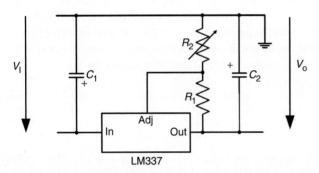

FIGURE 23.18 The LM337 adjustable negative voltage regulator.

where I_{adj} is a constant current into the adjustment terminal and has a value of approximately 50 μA for the LM317. As can be seen from Eq. (23.13), with fixed R_1, V_o can be adjusted by varying R_2.

The LM337 adjustable voltage regulator is similar to the LM317 except that it provides negative regulated voltages instead of positive ones. Figure 23.18 shows the standard configuration of a LM337 voltage regulator. The output voltage can be adjusted from −1.2 to −37 V, depending on the external resistors R_1 and R_2.

23.4.3 Applications of Linear IC Voltage Regulators

Most IC regulators are limited to an output current of 2.5 A. If the output current of an IC regulator exceeds its maximum allowable limit, its internal pass transistor will dissipate an amount of energy more than it can tolerate. As a result, the regulator will be shutdown.

For applications that require more than the maximum allowable current limit of a regulator, an external pass transistor can be used to increase the output current. Figure 23.19 illustrates such a configuration. This circuit has the capability of producing higher current to the load, but still preserving the thermal shutdown and short-circuit protection of the IC regulator.

A constant current-limiting scheme, as discussed in Section 23.2.2, is implemented by using the transistor Q_2 and the resistor R_2 to protect the external pass transistor Q_1 from excessive current under current-overload or short-circuit conditions. The value of the external current-sensing resistor R_1 determines the value of current at which Q_1 begins to conduct. As long as the current is less than the value set by R_1, the transistor Q_1 is off, and the regulator operates normally as shown in Fig. 23.15. But when the load current I_L starts to increase, the voltage across R_1 also increases. This turns on the external transistor Q_1 and conducts the excess current. The value of R_1 is determined by

$$R_1 = \frac{0.7 \text{ V}}{I_{max}} \qquad (23.14)$$

where I_{max} is the maximum current that the voltage regulator can handle internally.

23.5 Switching Regulators

The linear series and shunt regulators have control transistors that are operating in their linear active regions. Regulation is achieved by varying the conduction of the transistors to

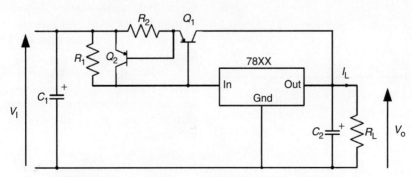

FIGURE 23.19 A 78XX series regulator with an external pass transistor.

maintain the output voltage at a desirable level. The switching regulator is different in that the control transistor operates as a switch, either in cutoff or saturation region. Regulation is achieved by adjusting the on-time of the control transistor. In this mode of operation, the control transistor does not dissipate as much power as that in the linear types. Therefore, switching voltage regulators have a much higher efficiency and can provide greater load currents at low voltage than linear regulators.

Unlike their linear counterparts, switching regulators can be implemented by many different topologies such as forward and flyback. In order to select an appropriate topology for an application, it is necessary to understand the merits and drawbacks of each topology and the requirements of the application. Basically, most topologies can work for various applications. Therefore, we have to determine from the factors such as cost, performances, and application that make one topology more desirable than the others. However, no matter which topology we decide to use, the basic building blocks of an off-the-line switching power supply are the same, as depicted in Fig. 23.1.

In this section, some popular switching regulator topologies, namely flyback, forward, half-bridge, and full-bridge topologies, are presented. Their basic operation is described, and the critical waveforms are shown and explained. The merits, drawbacks, and application areas of each topology are discussed. Finally, the control circuitry and PWM of the regulators will also be discussed.

23.5.1 Single-ended Isolated Flyback Regulators

An isolated flyback regulator consists of four main circuit elements: a power switch, a rectifier diode, a transformer, and a filter capacitor. The power switch, which can be either a power transistor or a MOSFET, is used to control the flow of energy in the circuit. A transformer is placed between the input source and the power switch to provide DC isolation between the input and the output circuits. In addition to being an energy storage element, the transformer also performs a stepping up or down function for the regulator. The rectifier diode and filter capacitor form an energy transfer mechanism to supply energy to maintain the output voltage of the supply. Note that there are two distinct operating modes for flyback regulators: *continuous* and *discontinuous*. However, both modes have an identical circuit. It is only the transformer magnetizing current that determines the operating mode of the regulator. Figure 23.20(a) shows a simplified isolated flyback regulator. The associated steady-state waveforms, resulting from a discontinuous-mode operation, is shown in Fig. 23.20(b). As shown in the figure, the voltage regulation of the regulator is achieved by a control circuit, which controls the conduction period or duty cycle of the switch, to keep the output voltage at a constant level. For clarity, the schematics and operation of the control circuit will be discussed in a separate section.

23.5.1.1 Discontinuous-mode Flyback Regulators

Under steady-state conditions, the operation of the regulator can be explained as follows. When the power switch Q_1 is on, the primary current I_p starts to build up and stores energy in the primary winding. Because of the opposite-polarity arrangement between the input and output windings of the transformer, the rectifier diode D_R is reverse-biased. In this period of time, there is no energy transferred from the input to the load R_L. The output voltage is supported by the load current I_L, which is supplied from the output filter capacitor C_F. When Q_1 is turned off, the polarity of the windings reverses as a result of the fact that I_p cannot change instantaneously. This causes D_R to turn on. Now D_R is conducting, charging the output capacitor C_F and delivering current to R_L. This charging action ends at the point where all the magnetic energy stored in the secondary winding during the first half-cycle is emptied. At this point, D_R will cease to conduct and R_L absorbs energy just from C_F until Q_1 is switched on again.

During the Q_1 on-time, the voltage across the primary winding of the transformer is V_i. The current in the primary winding I_p increases linearly and is given by

$$I_p = \frac{V_i t_{on}}{L_p} \qquad (23.15)$$

where L_p is the primary magnetizing inductance. At the end of the on-time, the primary current reaches a value equal to $I_{p(pk)}$ and is given by

$$I_{p(pk)} = \frac{V_i DT}{L_p} \qquad (23.16)$$

where D is the duty cycle and T is the switching period. Now when Q_1 turns off, the magnetizing current in the transformer forces the reversal of polarities on the windings. At the instant of turn off, the amplitude of the secondary current $I_{s(pk)}$ is

$$I_{s(pk)} = \left(\frac{N_p}{N_s}\right) I_{p(pk)} \qquad (23.17)$$

This current decreases linearly at the rate of

$$\frac{dI_s}{dt} = \frac{V_o}{L_s} \qquad (23.18)$$

where L_s is the secondary magnetizing inductance.

In the discontinuous-mode operation, $I_{s(pk)}$ will decrease linearly to zero before the start of the next cycle. Since the energy transfer from the source to the output takes place only in the first half cycle, the power drawn from V_i is then

$$P_{in} = \frac{L_p I_p^2}{2T} \qquad (23.19)$$

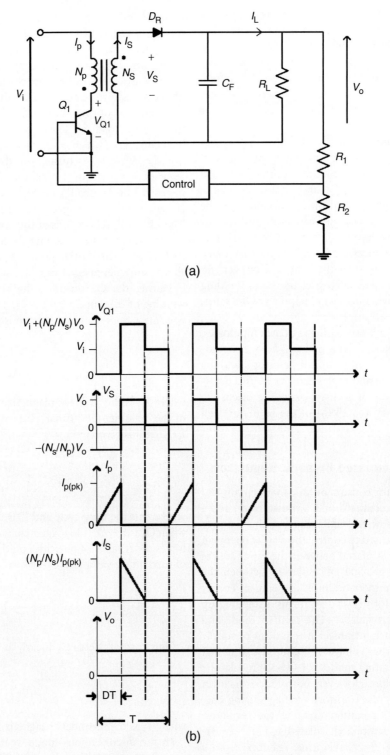

FIGURE 23.20 A simplified isolated flyback regulator: (a) circuit and (b) the associated waveforms.

Substituting Eq. (23.15) into Eq. (23.19), we have

$$P_{in} = \frac{(V_i t_{on})^2}{2TL_p} \quad (23.20)$$

The output power P_o may be written as

$$P_o = \eta P_{in}$$
$$= \frac{\eta(V_i t_{on})^2}{2TL_p} = \frac{V_o^2}{R_L} \quad (23.21)$$

where η is the efficiency of the regulator. Then, from Eq. (23.21), the output voltage V_o is related to the input voltage V_i by

$$V_o = V_i D \sqrt{\frac{\eta R_L T}{2L_p}} \quad (23.22)$$

Since the collector voltage V_{Q1} of Q_1 is maximum when V_i is maximum, the maximum collector voltage $V_{Q1(max)}$, as shown in the Fig. 23.20(b), is given by

$$V_{Q1(max)} = V_{i(max)} + \left(\frac{N_p}{N_s}\right) V_o \quad (23.23)$$

The primary peak current $I_{p(pk)}$ can be found in terms of P_o by combining Eq. (23.16) and Eq. (23.21) and then eliminating L_p as

$$I_{p(pk)} = \frac{2V_o^2}{\eta V_i D R_L}$$
$$= \frac{2P_o}{\eta V_i D} \quad (23.24)$$

The maximum collector current $I_{C(max)}$ of the power switch Q_1 at turn on is

$$I_{C(max)} = I_{p(pk)}$$
$$= \frac{2P_o}{\eta V_i D} \quad (23.25)$$

As can be seen from Eq. (23.21), V_o will be maintained constant by keeping the product $V_i t_{on}$ constant. Since maximum on-time $t_{on(max)}$ occurs at minimum supply voltage $V_{i(min)}$, the maximum allowable duty cycle for the discontinuous-mode can be found from Eq. (23.22) as

$$D_{max} = \frac{V_o}{V_{i(min)}} \sqrt{\frac{2L_p}{\eta R_L T}} \quad (23.26)$$

and V_o at D_{max} is then

$$V_o = V_{i(min)} D_{max} \sqrt{\frac{\eta R_L T}{2L_p}} \quad (23.27)$$

23.5.1.2 Continuous-mode Flyback Regulators

In the continuous-mode operation, the power switch is turned on before all the magnetic energy stored in the secondary winding empties itself. The primary and secondary current waveforms have a characteristic appearance as shown in Fig. 23.21. This mode produces a higher power capability without increasing the peak currents. During the on-time, the primary current I_p rises linearly from its initial value $I_p(0)$ and is given by

$$I_p = I_p(0) + \frac{V_i t_{on}}{L_p} \quad (23.28)$$

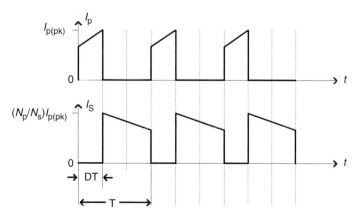

FIGURE 23.21 The primary and secondary winding currents of a flyback regulator operated in the continuous-mode.

At the end of the on-time, the primary current reaches a value equal to $I_{p(pk)}$ and is given by

$$I_{p(pk)} = I_p(0) + \frac{V_i DT}{L_p} \quad (23.29)$$

In general, $I_p(0) \gg V_i DT/L_p$; thus, Eq. (23.29) can be written as

$$I_{p(pk)} \approx I_p(0) \quad (23.30)$$

The secondary current I_s at the instant of turn off is given by

$$\begin{aligned} I_{s(pk)} &= \left(\frac{N_p}{N_s}\right) I_{p(pk)} \\ &= \left(\frac{N_p}{N_s}\right)\left(I_p(0) + \frac{V_i DT}{L_p}\right) \\ &\approx \left(\frac{N_p}{N_s}\right) I_p(0) \end{aligned} \quad (23.31)$$

This current decreases linearly at the rate of

$$\frac{dI_s}{dt} = \frac{V_o}{L_s} \quad (23.32)$$

The output power P_o is equal to V_o times the time-average of the secondary current pulses and is given by

$$\begin{aligned} P_o &= V_o I_s \frac{T - t_{on}}{T} \\ &\approx V_o I_{s(pk)} \frac{T - t_{on}}{T} \end{aligned} \quad (23.33)$$

or

$$I_{s(pk)} = \frac{P_o}{V_o(1 - t_{on}/T)} \quad (23.34)$$

For an efficiency of η, the input power P_{in} is

$$\begin{aligned} P_{in} &= \frac{P_o}{\eta} \\ &= V_i I_p \frac{t_{on}}{T} \\ &\approx V_i I_{p(pk)} \frac{t_{on}}{T} \end{aligned} \quad (23.35)$$

or

$$I_{p(pk)} = \frac{P_o}{\eta V_i (t_{on}/T)} \quad (23.36)$$

Combining Eqs. (23.31), (23.34), and (23.36) and solving for V_o, we have

$$V_o = \left(\frac{N_s}{N_p}\right) \frac{\eta V_i D}{1 - D} \quad (23.37)$$

The output voltage at D_{max} is then

$$V_o = \left(\frac{N_s}{N_p}\right) \frac{\eta V_{i(min)} D_{max}}{1 - D_{max}} \quad (23.38)$$

The maximum collector current $I_{C(max)}$ for the continuous-mode is then given by

$$\begin{aligned} I_{C(max)} &= I_{p(pk)} \\ &= \frac{P_o}{\eta V_i D_{max}} \end{aligned} \quad (23.39)$$

The maximum collector voltage of Q_1 is the same as that in the discontinuous-mode and is given by

$$V_{Q1(max)} = V_{i(max)} + \left(\frac{N_p}{N_s}\right) V_o \quad (23.40)$$

The maximum allowable duty cycle for the continuous-mode can be found from Eqs. (23.38) and is given by

$$D_{max} = \frac{1}{1 + \left(\frac{N_s}{N_p}\right) \frac{\eta V_{i(min)}}{V_o}} \quad (23.41)$$

At the transition from the discontinuous-mode to continuous-mode (or from continuous-mode to discontinuous-mode), the relationships in Eqs. (23.27) and (23.38) must hold. Thus, equating these two equations, we have

$$V_{i(min)} D_{max} \sqrt{\frac{\eta R_L T}{2 L_p}} = \frac{N_s}{N_p} \frac{D_{max}}{1 - D_{max}} \eta V_{i(min)} \quad (23.42)$$

Solving this equation for L_p to give the critical inductance $L_{p(limit)}$, at which the transition occurs, we have

$$\begin{aligned} L_{p(limit)} &= \frac{1}{2\eta} T R_L \left[(1 - D_{max})\frac{N_p}{N_s}\right]^2 \\ &= \frac{1}{2\eta} T \frac{V_o^2}{P_o} \left[(1 - D_{max})\frac{N_p}{N_s}\right]^2 \end{aligned} \quad (23.43)$$

Replacing V_o with Eq. (23.38) and solving for $L_{p(limit)}$, we have

$$L_{p(limit)} = \frac{1}{2} \eta T \frac{D_{max}^2 V_{i(min)}^2}{P_o} \quad (23.44)$$

Then, for a given D_{max}, input, and output quantities, the inductance value $L_{p(limit)}$ in Eq. (23.44) determines the mode of operation for the regulator. If $L_p < L_{p(\text{limit})}$, then the circuit is operated in the discontinuous-mode. Otherwise, if $L_P > L_{p(\text{limit})}$, the circuit is operated in the continuous-mode.

In designing flyback regulators, regardless of their operating modes, the power switch must be able to handle the peak collector voltage at turn off and the peak collector currents at turn on as shown in Eqs. (23.23), (23.25), (23.39), and (23.40). The flyback transformer, because of its unidirectional use of the B–H curve, has to be designed so that it will not be driven into saturation. To avoid saturation, the transformer needs a relatively large core with an air gap in it.

Although the continuous and discontinuous modes have an identical circuit, their operating properties differ significantly. As opposed to the discontinuous-mode, the continuous-mode can provide higher power capability without increasing the peak current $I_{p(pk)}$. It means that, for the same output power, the peak currents in the discontinuous-mode are much higher than those operated in the continuous-mode. As a result, a higher current rating and, therefore, a more expensive power transistor is needed. Moreover, the higher secondary peak currents in the discontinuous-mode will have a larger transient spike at the instant of turn off. However, despite all these problems, the discontinuous-mode is still much more widely used than its continuous-mode counterpart. There are two main reasons. First, the inherently smaller magnetizing inductance gives the discontinuous-mode a quicker response and a lower transient output voltage spike to sudden changes in load current or input voltage. Second, the continuous-mode has a right-half-plane zero in its transfer function, which makes the feedback control circuit more difficult to design.

The flyback configuration is mostly used in applications with output power below 100 W. It is widely used for high output voltages at relatively low power. The essential attractions of this configuration are its simplicity and low cost.

Since no output filter inductor is required for the secondary, there is a significant saving in cost and space, especially for multiple output power supplies. Since there is no output filter inductor, the flyback exhibits high ripple currents in the transformer and at the output. Thus, for higher power applications, the flyback becomes an unsuitable choice. In practice, a small LC filter is added after the filter capacitor C_F in order to suppress high-frequency switching noise spikes.

As mentioned previously, the collector voltage of the power transistor must be able to sustain a voltage as defined in Eq. (23.23). In cases where the voltage is too high for the transistor to handle, the double-ended flyback regulator shown in Fig. 23.22 may be used. The regulator uses two transistors that are switched on or off simultaneously. The diodes D_{R1} and D_{R2} are used to restrict the maximum collector voltage to V_i. Therefore, the transistors with a lower voltage rating can be used in the circuit.

23.5.2 Single-ended Isolated Forward Regulators

Although the general appearance of an isolated forward regulator resembles that of its flyback counterpart, their operations are different. The key difference is that the dot on the secondary winding of the transformer is so arranged that the output diode is forward-biased when the voltage across the primary is positive, that is, when the transistor is *on*. Energy is thus not stored in the primary inductance as it was for the flyback. The transformer acts strictly as a transformer. An inductive energy storage element is required at the output for proper and efficient energy transfer.

Unlike the flyback, the forward regulator is very suitable for working in the continuous-mode. In the discontinuous-mode, the forward regulator is more difficult to control because of a

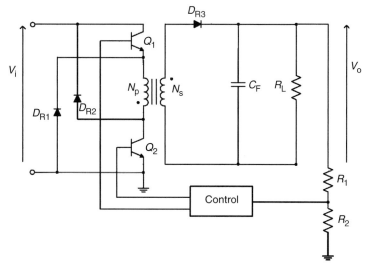

FIGURE 23.22 A double-ended flyback regulator.

double-pole at the output filter. Thus, it is not as much used as the continuous-mode. In view of it, only the continuous-mode will be discussed here.

Figure 23.23 shows a simplified isolated forward regulator and the associated steady-state waveforms for the continuous-mode operation. Again for clarity, the details of the control circuit are omitted from the figure. Under steady-state condition, the operation of the regulator can be explained as follows. When the power switch Q_1 turns on, the primary current I_p starts to build up and stores energy in the primary winding. Because of the same polarity arrangement of the primary and secondary windings, this energy is forward-transferred to the secondary and onto the $L_1 C_F$ filter and the load R_L through the rectifier diode D_{R2}, which is forward-biased. When Q_1 turns off, the polarity of the transformer winding voltage reverses. This causes D_{R2}, to turn off and D_{R1}, and D_{R3}, to turn on. Now D_{R3}, is conducting and delivering energy to R_L through the inductor L_1. During this period, the diode D_{R1}, and the tertiary winding provide a path for the magnetizing current returning to the input.

When the transistor Q_1 is turned on, the voltage across the primary winding is V_i. The secondary winding current is reflected into the primary, and the reaction current I_p, as shown in Fig. 23.24, is given by

$$I_p = \frac{N_s}{N_p} I_s \quad (23.45)$$

The magnetizing current in the primary has a magnitude of I_{mag} and is given by

$$I_{mag} = \frac{V_i t_{on}}{L_p} \quad (23.46)$$

The total primary current I'_p is then

$$I'_p = I_p + I_{mag}$$
$$= \frac{N_s}{N_p} I_s + \frac{V_i t_{on}}{L_p} \quad (23.47)$$

The voltage developed across the secondary winding V_s is

$$V_s = \frac{N_s}{N_p} V_i \quad (23.48)$$

Neglecting diode voltage drops and losses, the voltage across the output inductor is $V_s - V_o$. The current in L_1 increases linearly at the rate of

$$I_{L1} = \frac{(V_s - V_o) t_{on}}{L_1} \quad (23.49)$$

At the end of the on-time, the total primary current reaches a peak value equal to $I'_{p(pk)}$ and is given by

$$I'_{p(pk)} = I'_p(0) + \frac{V_i DT}{L_p} \quad (23.50)$$

The output inductor current I_{L1} is

$$I_{L1(pk)} = I_{L1}(0) + \frac{(V_s - V_o) DT}{L_1} \quad (23.51)$$

At the instant of turn on, the amplitude of the secondary current has a value of $I_{s(pk)}$ and is given by

$$I_{s(pk)} = \left(\frac{N_p}{N_s}\right) I'_{p(pk)}$$
$$= \left(\frac{N_p}{N_s}\right) \left[I'_p(0) + \frac{V_i DT}{L_p}\right] \quad (23.52)$$

During the off-time, the current I_{L1} in the output inductor is equal to the current I_{DR3} in the rectifier diode D_{R3} and both decrease linearly at the rate of

$$\frac{dI_{L1}}{dt} = \frac{dI_{DR3}}{dt}$$
$$= \frac{V_o}{L_1} \quad (23.53)$$

The output voltage V_o can be found from the time integral of the secondary winding voltage over a time equal to DT of the switch Q_1. Thus, we have

$$V_o = \frac{1}{T} \int_0^{DT} \frac{N_s}{N_p} V_i dt$$
$$= \frac{N_s}{N_p} V_i D \quad (23.54)$$

The maximum collector current $I_{C(max)}$ at turn on is equal to $I'_{p(pk)}$ and is given by

$$I_{C(max)} = I'_{p(pk)}$$
$$= \left(\frac{N_s}{N_p}\right) \left[I'_p(0) + \frac{V_i DT}{L_p}\right] \quad (23.55)$$

The maximum collector voltage $V_{Q1(max)}$ at turn-off is equal to the maximum input voltage $V_{i(max)}$ plus the maximum voltage $V_{r(max)}$ across the tertiary winding and is given by

$$V_{Q1(max)} = V_{i(max)} + V_{r(max)}$$
$$= V_{i(max)} \left(1 + \frac{N_p}{N_r}\right) \quad (23.56)$$

23 Power Supplies

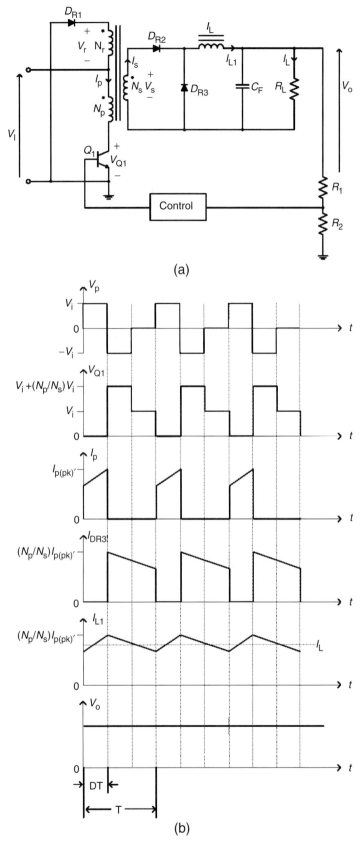

FIGURE 23.23 A simplified isolated forward regulator: (a) circuit and (b) the associated waveforms.

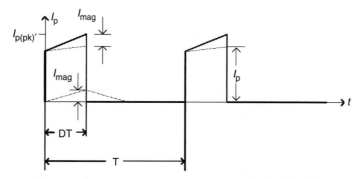

FIGURE 23.24 The current components in the primary winding.

The maximum duty cycle for the forward regulator operated in the continuous-mode can be determined by equating the time integral of the input voltage when Q_1 is on and the clamping voltage V_r when Q_1 is off.

$$\int_0^{DT} V_i dt = \int_{DT}^T V_r dt \quad (23.57)$$

which leads to

$$V_i DT = V_r (1-D)T \quad (23.58)$$

Grouping the D terms in Eq. (23.58) and replacing V_r/V_i with N_r/N_p, we have

$$D_{max} = \frac{1}{1 + N_r/N_p} \quad (23.59)$$

Thus, the maximum duty cycle depends on the turn ratio between the demagnetizing winding and the primary one.

In designing forward regulators, the duty cycle must be kept below the maximum duty cycle D_{max} to avoid saturating the transformer. It should also be noted that the transformer magnetizing current must be reset to zero at the end of each cycle. Failure to do so will drive the transformer into saturation, which can cause damage to the transistor. There are many ways of implementing the resetting function. In the circuit shown in Fig. 23.23(a), a tertiary winding is added to the transformer so that the magnetizing current will return to the input source V_i when the transistor turns off. The primary current always starts at the same value under the steady state condition.

Unlike flyback regulators, forward regulators require a minimum load at the output. Otherwise, excess output voltage will be produced. One commonly used method to avoid this situation is to attach a small load resistance at the output terminals. Of course, with such an arrangement, a certain amount of power will be lost in the resistor.

Because forward regulators do not store energy in their transformers, for the same output power level the transformer can be made smaller than for the flyback type. The output current is reasonably constant owing to the action of the output inductor and the flywheel diode; as a result, the output filter capacitor can be made smaller and its ripple current rating can be much lower than that required for the flybacks.

The forward regulator is widely used with output power below 200 W, though it can be easily constructed with a much higher output power. The limitation comes from the capability of the power transistor to handle the voltage and current stresses if the output power were to increase. In this case, a configuration with more than one transistor can be used to share the burden. Figure 23.25 shows a double-ended forward regulator. Like the double-ended flyback counterpart, the circuit uses two transistors which are switched on and off simultaneously. The diodes are used to restrict the maximum collector voltage to V_i. Therefore, the transistors with low voltage rating can be used in the circuit.

23.5.3 Half-bridge Regulators

The half-bridge regulator is another form of an isolated forward regulator. When the voltage on the power transistor in the single-ended forward regulator becomes too high, the half-bridge regulator is used to reduce the stress on the transistor. In a half-bridge regulator, the voltage stress imposed on the power transistors is subject to only the input voltage and is only half of that in a forward regulator. Thus, the output power of a half-bridge is double to that of a forward regulator for the same semiconductor devices and magnetic core.

Figure 23.26 shows the basic configuration of a half-bridge regulator and the associated steady state waveforms. As seen in Fig. 23.26(a), the half-bridge regulator can be viewed as two back-to-back forward regulators, fed by the same input voltage, each delivering power to the load at each alternate half cycle. The capacitors C_1 and C_2 are placed between the input and ground terminals. As such, the voltage across the primary winding is always half the input voltage. The power switches Q_1 and Q_2 are switched on and off alternatively to produce a square-wave ac at the input of the transformer.

23 Power Supplies

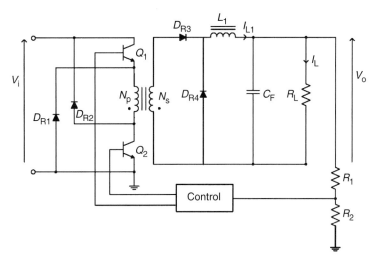

FIGURE 23.25 Double-ended forward regulator.

This square-wave is either stepped down or up by the isolation transformer and then rectified by the diodes D_{R1} and D_{R2}. Subsequently, the rectified voltage is filtered to produce the output voltage V_o.

Under steady state condition, the operation of the regulator can be explained as follows. When Q_1 is on and Q_2 off, D_{R1} conducts and D_{R2} is reverse-biased. The primary voltage V_p is $V_i/2$. The primary current I_p starts to build up and stores energy in the primary winding. This energy is forward-transferred to the secondary and onto the L_1C filter and the load R_L, through the rectifier diode D_{R1}. During the time interval Δ, when both Q_1 and Q_2 are off, D_{R1} and D_{R2} are forced to conduct to carry the magnetizing current that resulted in the interval during which Q_1 is turned on. The inductor current I_{L1} in this interval is equal to the sum of the currents in D_{R1} and D_{R2}. This interval terminates at half of the switching period T, when Q_2 is turned on. When Q_2 is on and Q_1 off, D_{R1} is reverse-biased and D_{R2} conducts. The primary voltage V_p is now $-V_i/2$. The circuit operates in a likewise manner as during the first half cycle.

With Q_1 on, the voltage across the secondary winding is

$$V_{s1} = \frac{N_{s1}}{N_p}\left(\frac{V_i}{2}\right) \quad (23.60)$$

Neglecting diode voltage drops and losses, the voltage across the output inductor is then given by

$$V_{L1} = \frac{N_{s1}}{N_p}\left(\frac{V_i}{2}\right) - V_o \quad (23.61)$$

In this interval, the inductor current I_{L1} increases linearly at the rate of

$$\frac{dI_{L1}}{dt} = \frac{V_{L1}}{L_1}$$
$$= \frac{1}{L_1}\left[\frac{N_{s1}}{N_p}\left(\frac{V_i}{2}\right) - V_o\right] \quad (23.62)$$

At the end of Q_1 on-time, I_{L1} reaches a value which is given by

$$I_{L1(pk)} = I_{L1}(0) + \frac{1}{L_1}\left[\frac{N_{s1}}{N_p}\left(\frac{V_i}{2}\right) - V_o\right]DT \quad (23.63)$$

During the interval Δ, I_{L1} is equal to the sum of the rectifier diode currents. Assuming the two secondary windings are identical, I_{L1} is given by

$$I_{L1} = 2I_{DR1} = 2I_{DR2} \quad (23.64)$$

This current decreases linearly at the rate of

$$\frac{dI_{L1}}{dt} = \frac{V_o}{L_1} \quad (23.65)$$

The next half cycle repeats with Q_2 on and for the interval Δ. The output voltage can be found from the time integral of the inductor voltage over a time equal to T. Thus, we have

$$V_o = 2 \times \frac{1}{T}\left[\int_0^{DT}\left(\frac{N_{s1}}{N_p}\left(\frac{V_i}{2}\right) - V_o\right)dt + \int_{T/2}^{T/2+DT} -V_o\, dt\right] \quad (23.66)$$

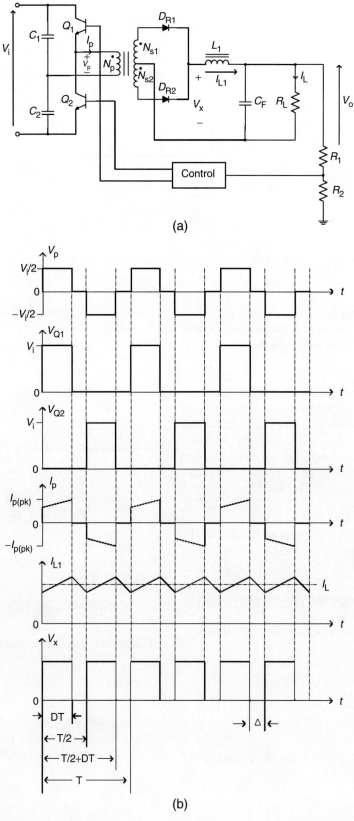

FIGURE 23.26 A simplified half-bridge regulator: (a) circuit and (b) the associated waveforms.

Note that the multiplier of 2 appears in Eq. (23.66) because of the two alternate half cycles. Solving Eq. (23.66) for V_o, we have

$$V_o = \frac{N_{s1}}{N_p} V_i D \qquad (23.67)$$

The output power P_o is given by

$$\begin{aligned} P_o &= V_o I_L \\ &= \eta P_{in} \\ &= \eta \frac{V_i I_{p(avg)} D}{2} \end{aligned} \qquad (23.68)$$

or

$$I_{p(avg)} = \frac{2 P_o}{\eta V_i D} \qquad (23.69)$$

where $I_{p(avg)}$ has the value of the primary current at the center of the rising or falling ramp. Assuming the reaction current I'_p reflected from the secondary is much greater than the magnetizing current, then the maximum collector currents for Q_1 and Q_2 are given by

$$\begin{aligned} I_{C(max)} &= I_{p(avg)} \\ &= \frac{2 P_o}{\eta V_i D_{max}} \end{aligned} \qquad (23.70)$$

As mentioned, the maximum collector voltages for Q_1 and Q_2 at turn off are given by

$$V_{C(max)} = V_{i(max)} \qquad (23.71)$$

In designing half-bridge regulators, the maximum duty cycle can never be greater than 50%. When both the transistors are switched on simultaneously, the input voltage is short-circuited to ground. The series capacitors C_1 and C_2 provide a dc bias to balance the volt–second integrals of the two switching intervals. Hence, any mismatch in devices would not easily saturate the core. However, if such a situation arises, a small coupling capacitor can be inserted in series with the primary winding. A dc bias voltage proportional to the volt–second imbalance is developed across the coupling capacitor. This balances the volt–second integrals of the two switching intervals.

One problem in using half-bridge regulators is related to the design of the drivers for the power switches. Specifically, the emitter of Q_1 is not at ground level, but is at a high ac level. The driver must therefore be referenced to this ac level. Typically, transformer-coupled drivers are used to drive both switches, thus solving the grounding problem and allowing the controller to be isolated from the drivers.

The half-bridge regulator is widely used for medium-power applications. Because of its core-balancing feature, the half-bridge becomes the predominant choice for output power ranging from 200 to 400 W. Since the half-bridge is more complex, for application below 200 W, the flyback or forward regulator is considered to be a better choice and more cost-effective. Above 400 W, the primary and switch currents of the half-bridge become very high. Thus, it becomes unsuitable.

23.5.4 Full-bridge Regulators

The full-bridge regulator is yet another form of isolated forward regulator. Its performance is improved over that of the half-bridge regulator because of the reduced peak collector current. The two series capacitors that appeared in half-bridge circuits are now replaced by another pair of transistors of the same type. In each switching interval, two of the switches are turned on and off simultaneously such that the full input voltage appears at the primary winding. The primary and the switch currents are only half that of the half-bridge for the same power level. Thus, the maximum output power of this topology is twice that of the half-bridge.

Figure 23.27 shows the basic configuration of a full-bridge regulator and the associated steady-state waveforms. Four power switches are required in the circuit. The power switches Q_1 and Q_4 turn on and off simultaneously in one of the half cycles. Q_2 and Q_3 also turn on and off simultaneously in the other half cycle, but with opposite phase as Q_1 and Q_4. This produces a square-wave ac with a value of $\pm V_i$ at the primary winding of the transformer. Like the half-bridge, this voltage is stepped down, rectified, and then filtered to produce a dc output voltage. The capacitor C_1 is used to balance the volt-second integrals of the two switching intervals and prevent the transformer from being driven into saturation.

Under steady-state conditions, the operation of the full-bridge is similar to that of the half-bridge. When Q_1 and Q_4 turn on, the voltage across the secondary winding is

$$V_{s1} = \frac{N_{s1}}{N_p} V_i \qquad (23.72)$$

Neglecting diode voltage drops and losses, the voltage across the output inductor is then given by

$$V_{L1} = \frac{N_{s1}}{N_p} V_i - V_o \qquad (23.73)$$

In this interval, the inductor current I_{L1} increases linearly at the rate of

$$\begin{aligned} \frac{dI_{L1}}{dt} &= \frac{V_{L1}}{L_1} \\ &= \frac{1}{L_1}\left[\frac{N_{s1}}{N_p} V_i - V_o\right] \end{aligned} \qquad (23.74)$$

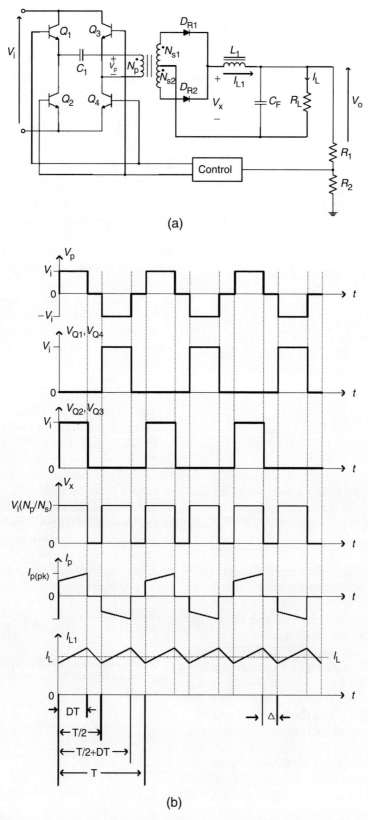

FIGURE 23.27 A simplified full-bridge regulator: (a) circuit and (b) the associated waveforms.

At the end of Q_1 and Q_4 on-time, I_{L1} reaches a value which is given by

$$I_{L1(pk)} = I_{L1}(0) + \frac{1}{L_1}\left[\frac{N_{s1}}{N_p}V_i - V_o\right]DT \qquad (23.75)$$

During the interval Δ, I_{L1} decreases linearly at the rate of

$$\frac{dI_{L1}}{dt} = \frac{V_o}{L_1} \qquad (23.76)$$

The next half cycle repeats with Q_2 and Q_3 on and the circuit operates in a similar manner as in the first half cycle.

Again, as in the half-bridge, the output voltage can be found from the time integral of the inductor voltage over a time equal to T. Thus, we have

$$V_o = 2 \times \frac{1}{T}\left[\int_0^{DT}\left(\frac{N_{s1}}{N_p}V_i - V_o\right)dt + \int_{T/2}^{T/2+DT} -V_o dt\right] \qquad (23.77)$$

Solving Eq. (23.77) for V_o, we have

$$V_o = \frac{N_{s1}}{N_p}2V_i D \qquad (23.78)$$

The output power P_o is given by

$$P_o = \eta P_{in} \qquad (23.79)$$
$$= \eta V_i I_{p(avg)} D$$

or

$$I_{p(avg)} = \frac{P_o}{\eta V_i D} \qquad (23.80)$$

where $I_{p(avg)}$ has the same definition as in the half-bridge case.

Comparing Eq. (23.80) with Eq. (23.69), we see that the output power of a full-bridge is twice that of a half-bridge with same input voltage and current. The maximum collector currents for Q_1, Q_2, Q_3, and Q_4 are given by

$$I_{C(max)} = I_{p(avg)}$$
$$= \frac{P_o}{\eta V_i D_{max}} \qquad (23.81)$$

Comparing Eq. (23.81) with Eq. (23.70), for the same output power, the maximum collector current is only half that of the half-bridge.

As mentioned, the maximum collector voltage for Q_1 and Q_2 at turn off is given by

$$V_{C(max)} = V_{i(max)} \qquad (23.82)$$

The design of the full-bridge is similar to that of the half-bridge. The only difference is the use of four power switches instead of two in the full-bridge. Therefore, additional drivers are required by adding two more secondary windings in the pulse transformer of the driving circuit.

For high power applications ranging from several hundred to thousand kilowatts, the full-bridge regulator is an inevitable choice. It has the most efficient use of magnetic core and semiconductor switches. The full-bridge is complex and therefore expensive to build, and is only justified for high-power applications, typically over 500 W.

23.5.5 Control Circuits and Pulse-width Modulation

In previous subsections, we presented several popular voltage regulators that may be used in a switching mode power supply. This section discusses the control circuits that regulate the output voltage of a switching regulator by constantly adjusting the conduction period t_{on} or duty cycle d of the power switch. Such adjustment is called *pulse-width modulation* (PWM).

The duty cycle is defined as the fraction of the period during which the switch is on, i.e.

$$d = \frac{t_{on}}{T}$$
$$= \frac{t_{on}}{t_{on} + t_{off}} \qquad (23.83)$$

where T is the switching period, i.e. $t_{off} = T - t_{on}$. By adjusting either t_{on} or t_{off}, or both, d can be modulated. Thus, PWM controlled regulators can operate at variable frequency as well as fixed frequency.

Among all types of PWM controllers, the fixed frequency controller is by far the most popular choice. There are two main reasons for their popularity. First, low-cost fixed-frequency PWM IC controllers have been developed by various solid-state device manufacturers, and most of these IC controllers have all the features that are needed to build a PWM switching power supply using a minimum number of components. Second, because of their fixed-frequency nature, fixed-frequency controllers do not have the problem of unpredictable noise spectrum associated with variable frequency controllers. This makes EMI control much easier.

There are two types of fixed-frequency PWM controllers, namely, *the voltage-mode controller* and the *current-mode controller*. In its simplified form, a voltage-mode controller consists of four main functional components: an adjustable clock for setting the switching frequency, an output voltage error amplifier for detecting deviation of the output from the nominal value, a ramp generator for providing a sawtooth signal that is synchronized to the clock, and a comparator that compares the output error signal with the sawtooth signal. The output of the comparator is the signal that drives the controlled switch. Figure 23.28 shows a simplified PWM voltage-mode controlled forward regulator operating at fixed frequency and its associated driving signal waveform. As shown, the duration of the on-time t_{on} is determined by the time between the reset of the ramp generator and the intersection of the error voltage with the positive-going ramp signal.

The error voltage v_e is given by

$$v_e = \left(1 + \frac{Z_2}{Z_1}\right) V_{REF} - \frac{Z_2}{Z_1} v_2 \qquad (23.84)$$

From Eq. (23.84), the small-signal term can be separated from the dc operating point by

$$\Delta v_e = -\frac{Z_2}{Z_1} \Delta v_2 \qquad (23.85)$$

The dc operating point is given by

$$V_e = \left(1 + \frac{Z_2}{Z_1}\right) V_{REF} - \frac{Z_2}{Z_1} V_2 \qquad (23.86)$$

Inspecting the waveform of the sawtooth and the error voltage shows that the duty cycle is related to the error voltage by

$$d = \frac{v_e}{V_p} \qquad (23.87)$$

where V_p is the peak voltage of the sawtooth.

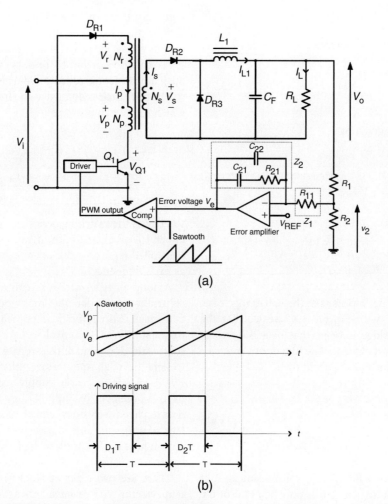

FIGURE 23.28 A simplified voltage-mode controlled forward regulator: (a) circuit and (b) the associated driving signal waveform.

Hence, the small-signal duty cycle is related to the small-signal error voltage by

$$\Delta d = \frac{\Delta v_e}{V_p} \tag{23.88}$$

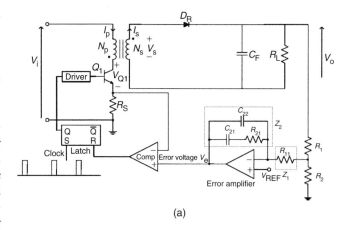

The operation of the fixed-frequency voltage-mode controller can be explained as follows. When the output is lower than the nominal dc value, a high error voltage is produced. This means that Δv_e is positive. Hence, Δd is positive. The duty cycle is increased to cause a subsequent increase in output voltage. The feedback dynamics (stability and transient response) is determined by the operational amplifier circuit that consists of Z_1 and Z_2. Some of the popular voltage-mode control ICs are SG1524/25/26/27, TL494/5, and MC34060/63.

The current-mode control makes use of the current information in a regulator to achieve output voltage regulation. In its simplest form, current-mode control consists of an inner loop that samples the inductance current value and turns the switches off as soon as the current reaches a certain value set by the outer voltage loop. In this way, the current-mode control achieves faster response than the voltage mode. There are two types of fixed-frequency PWM current-mode control, namely, the *peak current-mode control* and the *average current-mode control*.

In the peak current-mode control, no sawtooth generator is needed. In fact, the inductance current waveform is itself a sawtooth. The voltage analog of the current may be provided by a small resistance, or by a current transformer. Also, in practice, the switch current is used since only the positive-going portion of the inductance current waveform is required. Figure 23.29 shows a peak current-mode controlled flyback regulator.

In Fig. 23.29, the regulator operates at fixed frequency. Turn on is synchronized with the clock pulse, and turn off is determined by the instant at which the input current equals the error voltage V_e.

Because of its inherent peak current-limiting capability, the peak current-mode control can enhance reliability of power switches. The dynamic performance is improved because of the use of the additional current information. One main disadvantage of the peak current-mode control is that it is extremely susceptible to noise, since the current ramp is usually small compared to the reference signal. A second disadvantage is its inherent instability property at duty cycle exceeding 50%, which results in sub-harmonic oscillation. Typically, a compensating ramp is required at the comparator input to eliminate this instability. The third disadvantage is that it has a non-ideal loop response because of the use of the peak, instead of the average current sensing.

Figure 23.30 shows an average current-mode controlled flyback regulator. In the circuit, a PWM modulator (instead

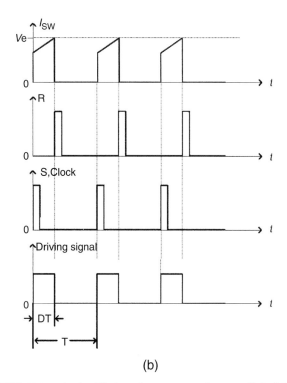

FIGURE 23.29 A simplified peak current-mode controlled flyback regulator: (a) circuit and (b) the associated waveforms.

of a clocked SR latch in the peak current-mode control) is employed to compare the current error to an externally generated sawtooth signal to formulate the desired control signal. The main advantages of this method over the peak current-control are that it has excellent noise immunity property; it is stable at duty cycle exceeding 50%; and it provides good tracking of average current. However, since there are three compensation networks (Z_1, Z_2, and Z_3), the analysis

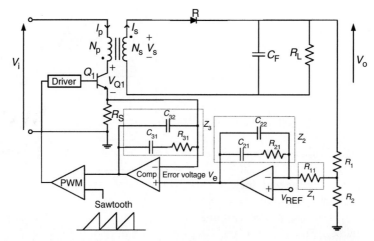

FIGURE 23.30 A simplified average current-mode controlled flyback regulator: (a) circuit and (b) the associated waveforms.

and optimal design of these networks are non-trivial. This is a major obstacle for adopting the average current mode control.

It should be noted that current-mode control is particularly effective for the flyback and boost-type regulators that have an inherent right-half-plane zero. Current-mode control effectively reduces the system to first-order by forcing the inductor current to be related to the output voltage, thus achieving faster response. In the case of the buck-type regulator, current-mode control presents no significant advantage because the current information can be derived from the output voltage, and hence faster response can still be achieved with a proper feedback network. Some of the popular current-mode control ICs are UC3840/2, UC3825, MC34129, and MC34065.

Further Reading

1. A.I. Pressman, *Switching Power Supply Design*, 2nd ed., New York: McGraw-Hill, 1999.
2. M. Brown, *Practical Switching Power Supply Design*, 2nd ed., New York: McGraw-Hill, 1999.
3. P.T. Krein, *Elements of Power Electronics*, 1st ed., Cambridge: Oxford University Press, 1998.
4. J.G. Kassakian, M.R. Schlecht, and G.C. Verghese, *Principles of Power Electronics*, 1st ed., MA Reading: Addison-Wesley, 1991.
5. G. Chryssis, *High-Frequency Switching Power Supplies*, 1st ed., New York: McGraw-Hill, 1984.
6. M.H. Rashid, *Power Electronics – Circuits, Devices, and Applications*, 2nd ed., New Jersey: Prentice-Hall, 1993.
7. T.L. Floyd and D. Buchla, *Fundamentals of Analog Circuits*, 1st ed., New Jersey: Prentice-Hall, 1999.

24

Uninterruptible Power Supplies

Adel Nasiri, Ph.D.
Power Electronics and Motor Drives Laboratory, University of Wisconsin-Milwaukee, 3200 North Cramer Street, Milwaukee Wisconsin, USA

24.1 Introduction 619
24.2 Classifications 619
    24.2.1 Standby UPS • 24.2.2 On-line UPS System • 24.2.3 Line-interactive UPS
    • 24.2.4 Universal UPS • 24.2.5 Rotary UPS • 24.2.6 Hybrid Static/Rotary UPS
    • 24.2.7 Comparison of UPS Configurations
24.3 Performance Evaluation 626
24.4 Applications 627
24.5 Control Techniques 628
24.6 Energy Storage Devices 630
    24.6.1 Battery • 24.6.2 Flywheel • 24.6.3 Fuel Cell
    Further Reading 632

24.1 Introduction

Power distortions such as power interruptions, voltage sags and swells, voltage spikes, and voltage harmonics can cause severe impacts on sensitive loads in the electrical systems. Uninterruptible power supply (UPS) systems are used to provide uninterrupted, reliable, and high quality power for these sensitive loads. Applications of UPS systems include medical facilities, life supporting systems, data storage and computer systems, emergency equipment, telecommunications, industrial processing, and on-line management systems [1–3]. The UPS systems are especially required in places where power outages and fluctuations occur frequently. A UPS provides a backup power circuitry to supply vital systems when a power outage occurs. In situations where short time power fluctuations or disturbed voltage occur, a UPS provides constant power to keep the important loads running. During extended power failures, a UPS provides backup power to keep the systems running long enough so that they can be gracefully powered down.

Most of the UPS systems also suppress line transients and harmonic disturbances. Generally, an ideal UPS should be able to simultaneously deliver uninterrupted power and provide the necessary power conditioning for the particular power application. Therefore, an ideal UPS should have the following features: regulated sinusoidal output voltage with low total harmonic distortion (THD) independent from the changes in the input voltage or in the load, on-line operation that means zero transition time from normal to back-up mode and vice versa, low THD sinusoidal input current and unity power factor, high reliability, high efficiency, low EMI and acoustic noise, electric isolation, low maintenance, low cost, weight, and size. Obviously, there is not a single configuration that can provide all of these features. Different configurations of UPS systems emphasize on some of the features mentioned above. Classifications of UPS systems are described in Section 24.2.

24.2 Classifications

24.2.1 Standby UPS

This configuration of UPS system is also known as "off-line UPS" or "line-preferred UPS" [4, 5]. Figure 24.1 shows the configuration of a typical standby UPS system. It consists of an AC/DC converter, a battery bank, a DC/AC inverter, and a static switch. A passive low pass filter may also be used at the output of the UPS or inverter to remove the switching frequency from the output voltage. The static switch is on during the normal mode of operation. Therefore, load is supplied from the AC line directly without any power conditioning. At the same time, the AC/DC rectifier charges the battery set. This converter is rated at a much lower power rating than the power demand of the load. When a power outage occurs or the primary power is out of a given preset tolerance, the static switch is opened and the DC/AC inverter provides power to

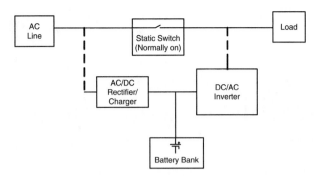

FIGURE 24.1 Configuration of a typical standby UPS system.

isolation of the load from the AC line, no output voltage regulation, long switching time, poor performance with non-linear loads, and no line conditioning are the main disadvantages of this configuration.

Different configurations of AC/DC rectifiers such as linear or switching may be used in this system. To reduce the cost, a simple diode-bridge rectifier with a capacitor at the front end is used. A full-bridge or half-bridge full controlled converter is also used to charge the battery bank. Two typical topologies for a single-phase UPS system are shown in Fig. 24.2. The full controlled topologies can provide power factor correction (PFC) to meet the corresponding standards. To optimize the charging process, the charging cycle is divided into "constant current" and "constant voltage" modes. In the constant current mode, the converter injects a constant current into the battery till the battery is charged up to about 95% of its capacity. After this mode, the constant voltage mode starts that applies a constant voltage on the battery. In this mode, the input current of the battery declines exponentially until it is fully charged.

the load from the battery set for the duration of the preset backup time or till the AC line is back again. This inverter is rated at 100% of the load power demand. It is connected in parallel to the load and stays standby during the normal mode of operation. The transition time from the AC line to DC/AC inverter is usually about one quarter of the line cycle, which is enough for most of the applications such as personal computers. The main advantages of this topology are simple design, low cost, and small size. On the other hand, lack of real

The purpose of the DC/AC inverter is to provide high quality AC power to the load when the static switch is opened. A full- or half-bridge topology is used for this inverter. Figure 24.3 shows two simple single-phase topologies for the DC/AC inverter.

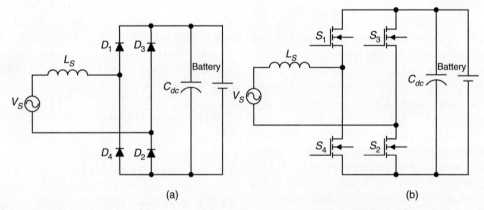

FIGURE 24.2 Two simple topologies of AC/DC rectifier: (a) full-bridge diode rectifier and (b) full-bridge full controlled topology.

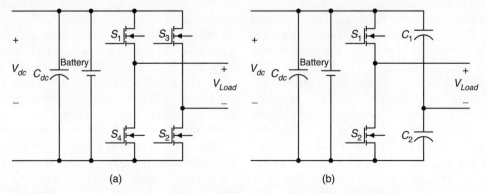

FIGURE 24.3 Two simple single-phase topologies for the DC/AC inverter: (a) full-bridge and (b) half-bridge.

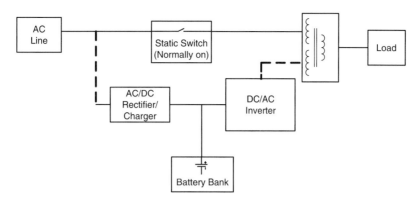

FIGURE 24.4 Typical configuration of ferroresonant standby UPS system.

In some topologies of standby UPS systems, an isolating transformer is used at the output stage of the UPS. This topology is called ferroresonant standby UPS system. The transformer also acts as a low pass filter that cancels out switching frequency from the output voltage of the DC/AC inverter. On the other hand, the transformer stores electromagnetic energy in the core and acts as a buffer when a power outage occurs. For a short time, the transformer provides power to the load and protects sensitive equipment from being affected during the transfer time from the input AC to the UPS. Figure 24.4 shows the configuration of a ferroresonant standby UPS system. Since the transformer is bulky and expensive, this configuration is more appropriate for high power applications.

24.2.2 On-line UPS System

Similar to standby UPS systems, on-line UPS systems also consist of a rectifier/charger, a battery set, an inverter, and a static switch (bypass). Other names for this configuration are "true UPS," "inverter preferred UPS," and "double-conversion UPS" [6, 7]. Figure 24.5 shows the block diagram of a typical on-line UPS. The rectifier/charger continuously supplies power to the DC bus. The power rating of this converter must be designed appropriately to supply power to the load and charge the battery bank at the same time. The batteries are rated in order to supply full power to the load during the backup time. The duration of this time varies in different applications. The inverter is rated at 100% of the load power since it must supply the load during the normal mode of operation as well as during the backup time. It is connected in series with the load; hence, there is no transfer time associated with the transition from normal mode to stored energy mode. This is the main advantage of on-line UPS systems. The static switch provides redundancy of the power source in the case of UPS malfunction or overloading. The AC line and load voltages must be in phase in order to use the static switch. This can be achieved easily by a phase-locked loop control. During the normal mode of operation, the power to the load is continuously supplied via the rectifier/charger and inverter. In fact, a double conversion from AC to DC and then from DC to AC takes place. This configuration of the UPS allows good power conditioning. The AC/DC converter charges the battery set and also supplies power to the load via the inverter. Therefore, it has the highest power rating in this topology, thereby

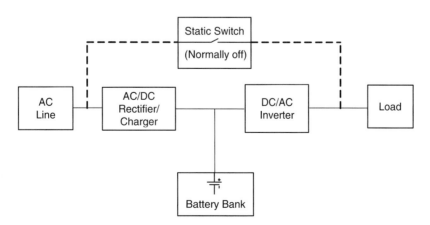

FIGURE 24.5 Block diagram of an on-line UPS system.

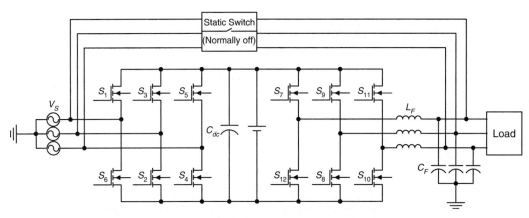

FIGURE 24.6 Configuration of a three-phase on-line UPS system.

increasing the cost. When the AC input voltage is outside the preset tolerance, the inverter and battery maintain continuity of power to the load. The duration of this mode is the duration of preset UPS backup time or till the AC line returns within the preset tolerance.

The main advantages of on-line UPS are very wide tolerance to the input voltage variation and very precise regulations of output voltage. In addition, there is no transfer time during the transition from normal to stored energy modes. It is also possible to regulate or change the output frequency [8]. The main disadvantages of this topology are low power factor, high THD at the input, and low efficiency. The input current is distorted by the rectifier unless an extra PFC circuit is added; but, this adds to the cost of the UPS system [9].

As mentioned for the standby UPS system, different topologies are employed for the AC/DC rectifier and DC/AC inverter. Unlike standby UPS system, in this system, these converters provide power to the load continuously. Therefore, more care should be given to the quality of the input current and output voltage as well as the efficiency of the system. Figure 24.6 shows the configuration of a three-phase on-line UPS system. The proper switching method such as PWM is employed for the AC/DC rectifier to minimize the input current harmonics and provide regulated DC bus voltage. A low pass filter at the output of the system removes the switching frequency from the output voltage.

power factor of the load or regulate the output voltage for the load. When the AC line is within the preset tolerance, it feeds the load directly. The AC/DC converter is connected in parallel with the load and charges the battery. This converter may also be used to improve the power factor of the system and compensate the load current harmonics. [10, 11]. Typical configuration of a line-interactive UPS is shown in Fig. 24.7.

When a power outage occurs or input voltage falls outside the preset tolerance, the system goes to bypass mode. In this mode, the bi-directional converter operates as a DC/AC inverter and supplies power to the load from the battery set. The static switch disconnects the AC line in order to prevent back feed from the inverter. The main advantages of the line-interactive UPS systems are simple design and, as a result, high reliability and lower cost compared to the on-line UPS systems. They also have good harmonic suppression for the input current. Since this is a single stage conversion topology, the efficiency is higher than on-line UPS system. The main disadvantage is the lack of effective isolation of the load from the AC line. Employing a transformer in the output can eliminate this; but, it will add to the cost, size, and weight of the UPS system. Furthermore, the output voltage conditioning is not good because the inverter is not connected in series with the load. In addition, since the AC line supplies the load directly during the normal mode of operation, there is no possibility for regulation of the output frequency.

24.2.3 Line-interactive UPS

Line-interactive UPS systems consist of a static switch, a series inductor, a bi-directional converter, and a battery bank. An optional passive filter can be added at the output of the bi-directional converter or at the input side of the load. A line-interactive UPS can operate either as an on-line UPS or as an off-line UPS. For an off-line line-interactive UPS, the series inductor is not required. However, most of the line-interactive UPS systems operate on-line in order to either improve the

24.2.4 Universal UPS

This type of UPS is also called "series-parallel" or "delta conversion." Its topology is derived from unified power quality conditioner (UPQC) topology and combines the advantages of both on-line and line-interactive UPS systems [12, 13]. It can achieve unity power factor, precise regulation of the output voltage, and high efficiency simultaneously. Its configuration is shown in Fig. 24.8. It consists of two bi-directional converters connected to a common battery set, static switch, and

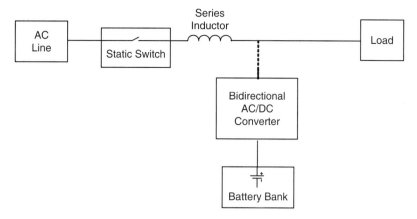

FIGURE 24.7 A typical configuration of a line-interactive UPS system.

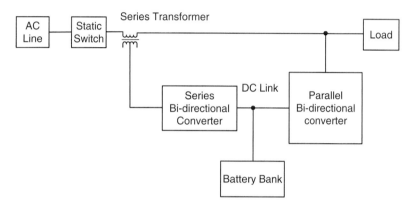

FIGURE 24.8 Block diagram of a universal UPS system.

a series transformer. The series bi-directional converter is rated at about 20% of the output power of the UPS system and it is connected via a transformer in series with the AC line. The second bi-directional converter is the usual inverter for a line-interactive UPS connected in parallel to the load and rated at 100% of the output power.

When the input voltage is in the acceptable range, the system is in the bypass mode. In this mode, parallel converter deals with current-based distortions. It mitigates load current harmonics and improves input power factor. At the same time, it charges the battery pack. Series converter deals with voltage-based distortions. It cancels input voltage harmonics and compensates voltage sags and swells. Most of the power is supplied directly from the AC line to the load. Only a small percentage of the input power is absorbed by parallel converter. This power is used to compensate the differences between input and reference voltages and to charge the battery pack. On the other hand, when the input voltage shuts down, the static switch separates the source and the load and the system goes to backup mode. In this situation, the parallel inverter acts as a DC/AC inverter and supplies power to the load. Since a large portion of the power flows without any conversion from the AC line to the load, the efficiency is higher than that of an on-line UPS system. Having eliminated the main drawback of double-conversion UPS systems, the universal UPS topology appears to be a strong competitor of on-line UPS systems in many applications. Figure 24.9 shows the topology of a three-phase universal UPS system.

24.2.5 Rotary UPS

Rotary UPS systems use the stored kinetic energy in the electrical machines to provide power to the load when a power outage occurs. There are different configurations for rotary UPS systems. The simplest topology consists of an AC motor and an AC generator, which are mechanically coupled. A flywheel is also used on the shaft of the machines to store more kinetic energy in the system. In normal operation, the input AC line provides power to the AC motor and this AC motor drives the AC generator. The configuration of this system is shown in Fig. 24.10a. In backup mode, the kinetic energy stored in the motor, flywheel, and generator is converted to electric power and supplies the load. This simple topology is designed to provide short time backup power to the

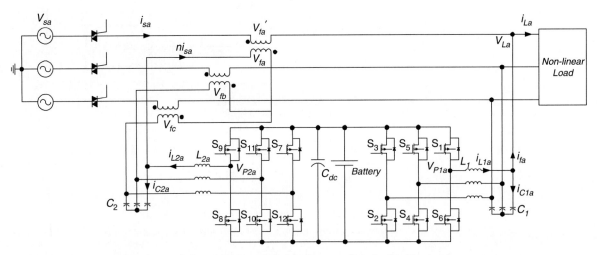

FIGURE 24.9 A universal UPS topology based on two three-leg bi-directional converters.

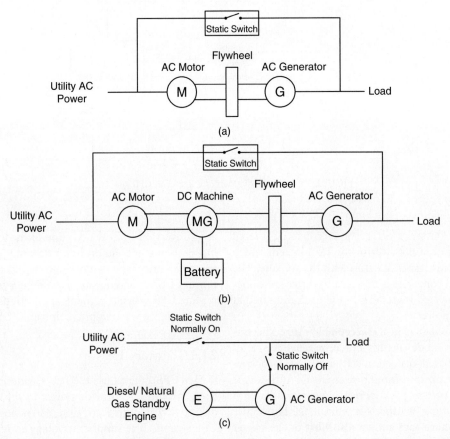

FIGURE 24.10 Different configuration of rotary UPS systems: (a) motor–generator set; (b) rotary UPS with battery backup; and (c) rotary UPS with standby diesel/natural gas engine.

load (typically less than 2 s) in case of power interruption. In another configuration of rotary UPS system which is shown in Fig. 24.10b, an AC motor, a DC machine, an AC generator, and a battery bank are used. During the normal mode of operation, the AC line supplies the AC motor, which drives the DC machine. The DC machine drives the AC generator, which supplies the load. During the backup mode of operation, the battery bank supplies the DC machine, which, in turn, drives the AC generator and the AC generator supplies the load. This system can provide long time backup power to the load depending on the capacity of the battery set. These two rotary UPS systems are much more reliable than the static UPS systems and provide complete electrical isolation between the load and input AC line. Yet, they require more maintenance and have much bigger size and weight. Therefore, they are usually used for high power applications [14, 15]. The configuration of a standby rotary UPS system is shown in Fig. 24.10c. This system does not provide electrical isolation between the load and input AC. There is also a transition delay for switching from main AC to backup AC generator. However, it can provide power to the load as long as needed.

24.2.6 Hybrid Static/Rotary UPS

Hybrid static/rotary UPS systems combine the main features of both static and rotary UPS systems. They have low output impedance, high reliability, excellent frequency stability, and low maintenance requirements [6]. Typical configurations of hybrid static/rotary UPS are depicted in Fig. 24.11. They are usually used in high power applications. In the system shown in Fig. 24.11a, during normal operation, the input AC power feeds the AC motor. The power is provided to the load from the AC generator, which is driven by the AC motor.

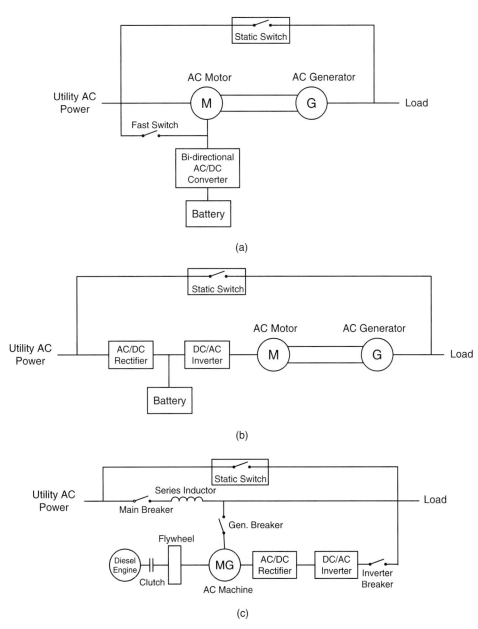

FIGURE 24.11 Three configurations of hybrid rotary-static UPS systems: (a) motor–generator set with battery backup; (b) motor–generator set with power conditioning at input side; and (c) battery-less hybrid UPS system.

In case of low input power quality or power interruption, the bidirectional AC/DC converter acts as an inverter and feeds the AC motor from battery pack. Configuration of a hybrid UPS system with power conditioning at the input is shown in Fig. 24.11b. Figure 24.11c shows the configuration of a more complicated hybrid UPS system. This system has three operation modes. In normal operation, the load is directly supplied by the main AC input and the AC motor is rotated at no-load. In the case of short power interruption, main breaker and generator breaker are opened and the inverter breaker is closed. The DC/AC inverter provides power to the load from the kinetic energy stored in the AC machine. If power is not restored in the short-term, the diesel engine is turned on, which provides power to the load through the AC generator. In this mode, the main breaker and inverter breaker remain open. One of the advantages of this topology is operation without a battery set to minimize cost, space, and required maintenance. The second advantage is avoiding double power conversion in long-term power interruption.

24.2.7 Comparison of UPS Configurations

Table 24.1 below provides the comparison between characteristics of different types of UPS systems.

24.3 Performance Evaluation

There are four criteria for evaluating the performance of a UPS system: quality of output voltage, input PFC and current harmonic cancellation, transition time, and efficiency. The quality of output voltage is the most important factor. The output voltage of a UPS system should be sinusoidal with low THD in different loading conditions even with non-linear loads. The control system should have small transient responses to provide appropriate line conditioning in different loading profiles. Typically, rotary UPS systems, which employ an AC generator at the load side, have better output voltage quality than static UPS systems. In these systems, there is no converter switching frequency present at the output voltage. Among the static UPS systems, on-line UPS configuration provides better output voltage quality. In this system, output voltage is provided by a DC/AC inverter regardless of input voltage quality. Usually, a pulse width modulation (PWM) method is used to regulate the output voltage. This kind of UPS should be designed to have minimum switching frequency at the output and provide pure sinusoidal voltage at different loading conditions. Followed by on-line UPS system are universal and line-interactive configurations. In universal topology, during normal mode of operation, the series converter provides voltage conditioning and regulates output voltage. In the backup mode, the parallel converter provides the load with sinusoidal voltage. In the line-interactive topology, during normal operation mode, input voltage directly supplies the load and no voltage conditioning is provided. In the backup mode, the DC/AC inverter provides the load with sinusoidal voltage.

The second criterion is transition time from normal mode of operation to stored energy mode. On-line rotary and static UPS systems have superior performance in this regard. The output voltage is always provided by the output generator or output DC/AC inverter and there is no transition time between operation modes of the systems. However, some of the rotary and hybrid configurations shown in Figs. 24.10 and 24.11 can only provide power to the load for a limited time. This time is determined by the amount of kinetic energy stored in the mechanical system. The transfer time in universal and line-interactive topologies depends on the time necessary for converting the power flow from the battery bank through the inverter to the load. Improved performance is achieved by choosing the DC bus capacitor voltage at the battery side to be slightly higher than the floating voltage of the batteries. Therefore, when the AC line fails, it is not necessary to sense the failure because the DC bus voltage will immediately fall under the floating voltage of the batteries and the power flow will naturally turn to the load. For off-line UPS systems, the transfer time is the longest. It depends upon the speed of sensing the failure of the AC line and starting the inverter.

The next important factor is the input power factor and the ability of the system to provide conditioning for load power. Universal UPS system has better performance followed

TABLE 24.1 Performance comparison of different configurations of UPS systems

Parameter	On-line	Line interactive	Off-line	Universal	Rotary	Hybrid
Surge protection	Excellent	Good	Good	Good	Excellent	Excellent
Transition time	Excellent	Good	Poor	Good	Excellent	Excellent
Line conditioning	Poor	Good	Poor	Excellent	Good	Good
Backup duration	Depends on battery	Depends on battery	Depends on battery	Depends on battery	Typically 0.1–0.5 s	Depends on battery
Efficiency	Low around 80%	High up to 95%	High	High up to 95%	High typically above 85%	High typically around 95%
Input/Output isolation	Poor	Poor	Poor	Poor	Perfect	Perfect
Cost	High	Medium-high	Low	High	Very high	Very high

by line-interactive and on-line UPS in this regard. During normal mode of operation, the parallel converter acts as an active filter and compensates reactive current and current harmonics generated by the load. In the line-interactive system, the bi-directional AC/DC converter performs this task. In an on-line UPS system, an additional system must be added to improve PFC and mitigate current harmonics.

The last criterion for performance evaluation is efficiency. To emphasize this factor, it should be noted that losses in UPS systems represent about 5–12% of all the energy consumed in data centers. Efficiency in rotary and hybrid configurations depends on the topology of the system but typically for low power application due to mechanical loss in the motor and generator, the efficiency is not very high. Among the static UPS systems, on-line UPS system has the poorest efficiency due to double conversion. Line-interactive and universal topologies provide higher efficiencies since most of the power directly flows from the input AC to the load during normal operation.

24.4 Applications

The UPS systems have wide applications in a variety of industries. Their common applications range from small power rating for personal computer systems to medium power rating for medical facilities, life supporting systems, data storage, and emergency equipment and high power rating for telecommunications, industrial processing, and on-line management systems. Different considerations should be taken into account for these applications. For emergency systems and lighting, the UPS should support the system for at least 90 minutes. Except for emergency systems, the UPS is designed to provide backup power to sensitive loads for 15–20 minutes. After this time, if the power is not restored, the system will be gracefully shut down. If a longer backup period is considered, a larger battery with higher cost and space is required. For process equipment and high power applications, some UPS systems are designed to provide enough time for the secondary power sources such as diesel generators to start up.

For industrial applications, it should be noted that UPS systems add to the complexity of the electrical system. They also add installation and ongoing maintenance costs. They may also add non-linearity to the system, decrease the efficiency, and deteriorate the input PFC mechanism. The power rating of the UPS should be appropriately selected considering the existing load and future extensions. For many applications, input voltage surges and spikes cause more damage than power outages. For these systems, another device instead of UPS can be utilized. Load characteristics should also be considered in UPS selection. For motor loads, the inrush current, which is sometimes 2.5 times of the rated current, should be considered. A good UPS for the motor loads is the one with higher transient overloads. For non-linear loads such as switching power supplies, the input current is not sinusoidal. Therefore, the instantaneous current is higher than the RMS current. This high instantaneous current should be considered in UPS selection.

For a power distribution network, two different approaches are taken to support sensitive loads. In a distributed approach, which is more suitable for highly proliferated loads such as medical equipment, data processing, and telecommunications many separate UPS units operate in parallel to supply critical loads. UPS units are placed flexibly in the system to form a critical load network. A typical on-line distributed UPS system is shown in Fig. 24.12. High flexibility and redundancy are the main advantages of distributed systems. Individual load increase can be supported by adding more UPS systems. Consideration for future extension can also be delayed until the loads are added. On the other side, this method has some disadvantages. The load sharing between different UPS units is a difficult task. Complicated digital control methods and communication between units are required to perform optimal load sharing. The second disadvantage is that the monitoring of the whole system is difficult and requires specially trained staff.

The other method to support distributed loads is to use a large UPS unit to supply all the critical loads in a centralized approach. This approach is more desirable for industrial and

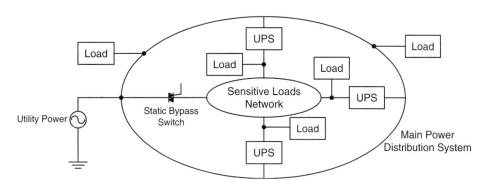

FIGURE 24.12 Typical configuration of a distributed UPS network.

utility applications. The advantage of this method is easier maintenance and troubleshooting. The disadvantages on the other side are lack of redundancy and high installation cost. In addition, consideration for system expansion should be taken into account when the original UPS unit is selected.

24.5 Control Techniques

The main task of the control system in a UPS unit is to minimize the output voltage total harmonic distortion in different loading profiles. In addition, it should provide the proper mechanism to recharge the battery set and maintain high input power factor and low total input current harmonic distortion. Other factors considered for a good control technique are nearly zero steady-state inverter output voltage error, good voltage regulation, robustness, fast transient response, and protection of the inverter against overload under linear/non-linear loads.

The most common switching technique is Sinusoidal PWM. This method can be utilized for both single-phase and three-phase systems. The advantage of this method is low output voltage harmonic and robustness. This strategy uses a single feedback loop to provide well-regulated output voltage with low THD. The feedback control can be continuous or discontinuous. Analog techniques are used in continuous approach. The sinusoidal PWM (SPWM) can be of natural sampling type, average type, or instantaneous type [17, 18].

In natural sampling type, the peak value of the output voltage is detected and compared with a reference voltage in order to obtain the error, which is used to control the reference to the modulator. The average approach is basically the same; but, the sensed voltage is converted to an average value and after that, is compared with a reference signal. These approaches control only the amplitude of the output voltage and are good only at high frequencies. In an instantaneous voltage feedback SPWM control, the output voltage is continuously compared with the reference signal improving the dynamic performance of the UPS inverter.

A typical block diagram of a three-phase DC/AC inverter for UPS systems and SPWM switching control technique is shown in Fig. 24.13. The disadvantage of this method is lack of flexibility for non-linear loads. Other programmed PWM techniques such as selective harmonic elimination, minimum THD, minimum loss, minimum current ripple, and reduced acoustic noise may be used for the inverter.

Better performance even with non-linear and step-changing loads can be achieved by multiple control loop strategies [19]. As shown in Fig. 24.14, there are two control loops: an outer and an inner. The outer control loop uses the output voltage as a feedback signal, which is compared with a reference signal. The error is compensated by a PI-integrator to achieve stable output voltage under steady-state operation. This error is also used as a reference signal for the inner current regulator loop, which uses the inductor or the capacitor output filter current as the feedback signal. The minor current loop ensures

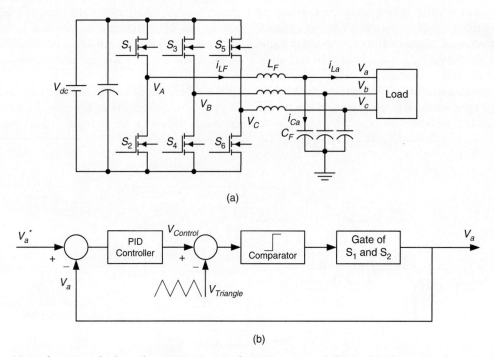

FIGURE 24.13 (a) Configuration of a three-phase DC/AC inverter for UPS systems and (b) simple voltage controller using PWM technique.

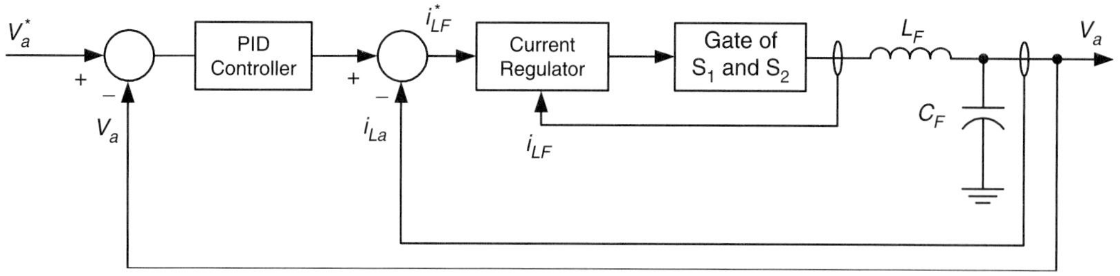

24 Uninterruptible Power Supplies

FIGURE 24.14 Typical current and voltage control loops for UPS inverter.

fast dynamic responses enabling good performance with non-linear or step-changing loads. The basic current regulators employed as minor current loop are: hysteresis regulators, sinusoidal PWM regulator, and predictive regulators. In a typical hysteresis regulator, the reference signal is compared with the feedback signal. The sign and predetermined amplitude of the error determine the output of the modulator. The duration between two successive levels is determined by the slope of the reference signal. The output voltage tracks the reference signal within the upper and lower boundary levels. This hysteresis control has fast transient response; but, the switching frequency varies widely [20].

In SPWM control technique, the output voltage feedback is compared with a sine reference signal and the error voltage is compensated by a PI-regulator to produce the current reference. The current through the inductor or the capacitor is sensed and compared with the reference signal. After being compensated by a PI-regulator, the error signal is compared with a triangular waveform to generate SPWM signal for switching control. The SPWM current control has a constant switching frequency and also provides fast dynamic responses. In predictive current control method, the switching instants are determined by suitable error boundaries. When the current vector touches the boundary line, the next switching state vector is determined by prediction and optimization in order to minimize the error. Predictive current control requires a good knowledge of the load parameters. All these current regulators are typically used as an inner loop to regulate the current in the filter inductor. The current reference for the current regulator is obtained by summing together the error in an outer voltage loop with the actual load current to yield the rated output voltage.

With the increase of speed and reliability of digital processors and a decrease in their cost, digital processors have been facing an enormous growth of popularity in control applications in the past few years. Many digital and discrete control techniques such as dead-beat control [21], dissipativity-based control [22], sliding-mode control [23], space vector-based control [24], and multiple-feedback loop [25] have been developed using digital signal processors (DSP).

In this section, fundamental analysis of a dead-beat control method is explained for the three-phase UPS configuration shown in Fig. 24.13a. The state space equations of one phase of this system in the continuous time-domain are as follows.

$$C_F \frac{dV_a}{dt} = i_{LF} - i_{La} \quad (24.1)$$

$$L_F \frac{di_{LF}}{dt} = V_A - V_a \quad (24.2)$$

Considering V_a and i_{LF} as state variables, the state space equation of the system is as follows:

$$\begin{bmatrix} V_a \\ i_{LF} \end{bmatrix}^\bullet = \begin{bmatrix} 0 & 1/C_F \\ -1/L_F & 0 \end{bmatrix} \begin{bmatrix} V_a \\ i_{LF} \end{bmatrix} + \begin{bmatrix} 0 \\ 1/L_F \end{bmatrix} V_A + \begin{bmatrix} -1/C_F \\ 0 \end{bmatrix} i_{La} \quad (24.3)$$

These continuous time-domain state space equations are converted to the discontinuous time domain with a sampling period of T_s [26].

$$\begin{bmatrix} V_a(k+1) \\ i_{LF}(k+1) \end{bmatrix} = \begin{bmatrix} \cos\omega_0 T_s & \dfrac{\sin\omega_0 T_s}{\omega_0 C_F} \\ -\dfrac{\sin\omega_0 T_s}{\omega_0 L_F} & \cos\omega_0 T_s \end{bmatrix} \begin{bmatrix} V_a(k) \\ i_{LF}(k) \end{bmatrix}$$

$$+ \begin{bmatrix} 1 - \cos\omega_0 T_s \\ \dfrac{1}{\omega_0^2 L_F} \sin\omega_0 T_s \end{bmatrix} V_A(k)$$

$$+ \begin{bmatrix} -\dfrac{1}{\omega_0 C_F} \sin\omega_0 T_s \\ 1 - \cos\omega_0 T_s \end{bmatrix} i_{La}(k) \quad (24.4)$$

Where ω_0 is the angular resonance frequency of L_F and C_F. The sampling frequency of the system is always considered much higher than the resonance frequency of L_F and C_F. With this assumption, Eq. (24.4) is simplified to Eq. (24.5). This conversion is valid for almost $f_s \geq 20 f_0$.

$$\begin{bmatrix} V_a(k+1) \\ i_{LF}(k+1) \end{bmatrix} = \begin{bmatrix} 1 & T_s/C_F \\ -T_s/L_F & 1 \end{bmatrix} \begin{bmatrix} V_a(k) \\ i_{LF}(k) \end{bmatrix}$$

$$+ \begin{bmatrix} 0 \\ T_s/L_F \end{bmatrix} V_A(k) + \begin{bmatrix} -T_s/C_F \\ 0 \end{bmatrix} i_{La}(k) \quad (24.5)$$

The current equation according to Eq. (24.5) is given by:

$$i_{LF}(k+1) = i_{LF}(k) + \frac{T_s}{L_F}[V_A(k) - V_a(k)] \quad (24.6)$$

Alternatively, this equation can be achieved by converting Eq. (24.2) from a differential equation to a difference equation. The same suggestion of $f_s \geq 20f_0$ has to be made for this conversion as well. If V_a and i_{LF}^* are considered constant over the next switching period, the output voltage of the inverter, which corrects the error of i_{LF} after two sampling periods, is described by:

$$V_A(k+1) = V_a(k+1) + \frac{L_F}{T_s}[i_{LF}^*(k+1) - i_{LF}(k+1)] \quad (24.7)$$

A linear estimation of $V_a(k+1)$ can be achieved from previous values:

$$V_a(k+1) = V_a(k) + [V_a(k) - V_a(k-1)] = 2V_a(k) - V_a(k-1) \quad (24.8)$$

By substituting Eqs. (24.8) and (24.10) in Eq. (24.9) and updating reference current for i_{LF} in every two sampling periods, the dead-beat digital control for series converter is described by:

$$V_A(k+1) = \frac{L_F}{T_s}[i_{LF}^*(k) - i_{LF}(k)] - V_A(k) + 3V_a(k) - V_a(k-1) \quad (24.9)$$

Equation (24.9) ensures that the current error between i_{LF} and i_{LF}^* at time $k+2$ goes to zero with a delay of two sampling periods. Avoiding interaction between voltage and current control loops, load voltage, V_a, is sampled at half of the current sampling frequency. The voltage equation according to Eq. (24.5) is as follows.

$$V_a(k+1) = V_a(k) + \frac{T_s}{C_F} i_{CF}(k) \quad (24.10)$$

$$V_a(k+2) = V_a(k+1) + \frac{T_s}{C_F} i_{CF}(k+1)$$

$$= V_a(k) + \frac{T_s}{C_F} i_{CF}(k) + \frac{T_s}{C_F} i_{CF}(k+1) \quad (24.11)$$

As current control is suggested to be dead-beat with a delay of two sampling periods, capacitor current at time k and $(k+1)$ are given by:

$$i_{CF}(k) = i_{CF}^*(k-2), i_{CF}(k+1) = i_{CF}^*(k-1) \quad (24.12)$$

Substituting Eq. (24.12) in Eq. (24.11) and updating the reference current at each of the two sampling periods, $V_a(k+2)$ is given by:

$$V_a(k+2) = V_a(k) + \frac{2T_s}{C_F} i_{CF}^*(k-2) \quad (24.13)$$

The current of i_{CF}^* at time k which corrects the voltage error of V_a at time $k+4$ is as follows.

$$i_{CF}^*(k) = \frac{C_F}{2T_s}[V_a^*(k) - V_a(k)] - i_{CF}^*(k-2) \quad (24.14)$$

A block diagram of the implementation of voltage and current control of the inverter is shown in Fig. 24.15. Block diagram of the current and voltage controller for the inverter is also shown in Fig. 24.16. Voltage regulator is a pure dead-beat controller with a delay of two sampling periods including the consumed time for calculation. G_1 is the time delay needed for calculations and analog to digital conversions. G_2 is the time delay caused by the PWM inverter and G_3 is the transfer function of the low pass filter. Current regulator is also considered as a pure delay. The output voltage of the inverter follows its reference with four sampling periods of delay. In practice, the dynamics of the current regulator is not a pure delay and shows some deviation from the dead-beat controller.

24.6 Energy Storage Devices

In this section, three dominant energy storage devices for the existing and future UPS systems are described. These energy storage devices are battery, flywheel, and fuel cell.

24.6.1 Battery

Battery is the energy storage component of current static UPS systems. It determines the capacity and run-time of the UPS.

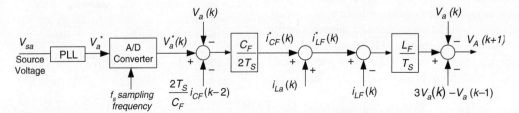

FIGURE 24.15 Implementation of the current and voltage control for the inverter shown in Figure 24.13a.

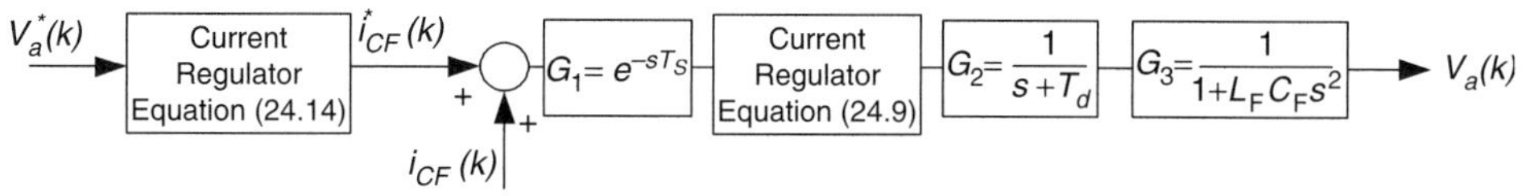

FIGURE 24.16 Block diagram of the current and voltage controller for the inverter shown in Figure 24.13a.

For small units, it is the size of battery that determines the size of the UPS. Different types of batteries are used in UPS systems but the most commonly used types are lead-acid, nickel–cadmium, and lithium ion. The lead acid batteries used in this application are the same as the ones used in the cars. However, there is one small difference. Car batteries generate electricity by the reaction of sulfuric acid on lead plates that are drowned under the liquid. These types of battery cells are not suitable for UPS applications because there is a chance of acid spillage from them. In addition, during the charging process, they release hydrogen that is explosive and dangerous in a closed environment. Lead acid batteries used in UPS systems are a special kind called sealed or valve-regulated. The nickel–cadmium batteries are another popular type of batteries used in UPS systems. They usually provide higher energy and power density compared to lead-acid batteries. The nominal voltage of nickel–cadmium cells is 1.2 V, which is smaller than 1.5 V of lead-acid batteries. However, the cell voltage variation throughout different charge levels is less than lead-acid batteries. These batteries also have less series resistance and can provide higher surge currents. Lithium-ion batteries have much higher energy density. This kind of battery can be molded into different shapes. They have a nominal voltage of 4.2 V. The main disadvantage of lithium-ion battery is that they lose their capacity from the time of manufacturing regardless of their charge level and conditions of use. Table 24.2 shows a comparison between different kinds of batteries for UPS application.

The traditional method of charging batteries is to apply constant current and constant voltage in two consecutive periods. Constant current is applied at the beginning of a typical full-charge cycle, when the battery voltage is low. When the battery voltage rises to a specified limit, the charger switches to constant voltage and continues in that mode until the charging current declines to nearly zero. At that time, the battery is fully charged. During the constant-voltage phase, the current drops exponentially due to the sum of battery resistance and any resistance in series with the battery (much like charging a capacitor through a resistor). Because current drops exponentially, a complete, full charge takes a long time.

24.6.2 Flywheel

Flywheel is simply a mechanical mass that is placed on the shaft of a motor–generator set and stores mechanical energy in the form of kinetic energy. When the electrical power is required, this kinetic energy is converted to electricity by the generator coupled with the flywheel. Flywheels are the oldest type of energy storage devices. The advantages of flywheel energy storage systems are high efficiency, high energy and power density, and long life. On the other hand, flywheels are more expensive and require more space than batteries and fuel cells. There are also some safety concerns about flywheels rotating at high speeds.

24.6.3 Fuel Cell

Due to high efficiency and low emissions, fuel cell systems have been gaining popularity in recent years. A fuel cell uses hydrogen as fuel and produces electricity, heat, and water from the reaction between hydrogen and oxygen. Each cell consists of an electrolyte and two electrodes as anode and cathode. Figure 24.17 shows the configuration of a typical fuel cell system. There are different kinds of fuel cell system depending on the types of electrolyte and hydrogen sources. Some fuel cell systems have an on-board fuel reformer and generate hydrogen from natural gas, methanol, and other hydrocarbons. Recent technology development in this field has made fuel cells a more reliable and cost-effective alternative for batteries. Fuel cells currently have a variety of applications in automotive, electric utility, and portable power industries. Table 24.3 provides a comparison between the most popular types of fuel cells.

TABLE 24.2 A comparison between different types of batteries for UPS systems

Battery type	Energy density (WH/kg)	Power density (W/kg)	Commercial availability
Lead-acid	35	300	Very mature and readily available
Nickel–cadmium	40	200	Mature and available
Lithium-ion	120	180	Available
Nickel hydride	70	200	Available
Zinc-air	350	60–225	Research stage
Aluminum-air	400	10	Research stage
Sodium chloride	110	150	Available
Sodium sulfur	170	260	Available
Zinc bromine	70	100	Available

TABLE 24.3 A comparison between different types of fuel cell system

Fuel cell type	Applications	Advantages	Disadvantages
Proton exchange membrane (PEM)	• Electric utility • Portable power • Automotive	• Solid electrolyte reduces corrosion and management problems • Low temperature • Quick start-up	• Expensive catalysts • High sensitivity to fuel impurities
Alkaline (AFC)	• Military • Space	• High performance	• Expensive removal of CO_2 from fuel and air streams required
Phosphoric acid (PAFC)	• Electric utility • Automotive	• Up to 85% efficiency in cogeneration of electricity and heat • Can use impure H_2 as fuel	• Expensive catalysts • Low power • Large size/weight
Molten carbonate (MCFC)	• Electric utility	• High efficiency • Fuel flexibility • Can use a variety of catalysts	• High temperature enhances corrosion and breakdown of cell components
Solid oxide fuel cell (SOFC)	• Electric utility	• High efficiency • Fuel flexibility • Can use a variety of catalysts • Solid electrolyte reduces corrosion and management problems • Low temperature • Quick start-up	• High temperature enhances the breakdown of cell components
Direct alcohol fuel cell (DAFC)	• Automotive • Portable power	• Compactness • High energy density	• Lower efficiency • Alcohol passing between electrodes without reacting

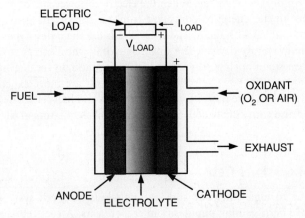

FIGURE 24.17 Configuration of a typical fuel cell system.

Further Reading

1. S. Karve, "Three of a kind," *IEE Review*, vol. 46, no. 2, pp. 27–31, March 2000.
2. R. H. Carle, "UPS applications: mill perspective," *IEEE Industry Application Magazine*, pp. 12–17, 1995.
3. R. Krishnan and S. Srinivasan, "Topologies for uninterruptible power supplies," in *Proc. IEEE International Symposium on Industrial Electronics*, Hungary, pp. 122–127, June 1993.
4. F. Kamran and T. G. Habetler, "A novel on-line UPS with universal filtering capabilities," *IEEE Transactions on Power Electronics*, vol. 13, no. 2, pp. 366–371, 1998.
5. J. H. Choi, J. M. Kwon, J. H. Jung, and B. H. Kwon, "High-performance online UPS using three-leg-type converter," *IEEE Transactions on Industrial Electronics*, vol. 52, no. 3, pp. 889–897, 2005.
6. H. Pinheiro, P. K. Jain, and G. Joos, "A comparison of UPS for powering hybrid fiber/coaxial networks," *IEEE Transactions on Power Electronics*, vol. 17, no. 3, pp. 389–397, 2002.
7. I. Youichi, I. Satoru, T. Isao, and H. Hitoshi, "New power conversion technique to obtain high performance and high efficiency for single-phase UPS," in *Proc. 36th IEEE Industry Applications Conference*, vol. 4, pp. 2383–2388, 2001.
8. H. Gueldner, H. Wolf, and N. Blacha, "Single phase UPS inverter with variable output voltage and digital state feedback control," in *Proc. IEEE International Symposium on Industrial Electronics*, vol. 2, pp. 1089–1094, 2001.
9. J. Lee, Y. Chang, and F. Liu, "A new UPS topology employing a PFC boost rectifier cascaded high-frequency tri-port converter," *IEEE Transactions on Industrial Electronics*, vol. 46, no. 4, pp. 803–813, 1999.
10. F. Kamran and T. G. Habetler, "A novel on-line UPS with universal filtering capabilities," *IEEE Transactions on Power Electronics*, vol. 13, no. 3, pp. 410–418, 1998.
11. B. Kwon, J. Choi, and T. Kim, "Improved single-phase line-interactive UPS," *IEEE Transactions on Industrial Electronics*, vol. 48, no. 4, pp. 804–811, 2001.

12. A. Nasiri, S. Bekiarov, and A. Emadi, "Reduced parts three-phase series-parallel UPS system with active filter capabilities," in *Proc. IEEE 38th Industry Applications Conference*, vol. 2, pp. 963–969, 2003.
13. S. da Silva, P. F. Donoso-Garcia, P. C. Cortizo, and P. F. Seixas, "A three-phase line-interactive UPS system implementation with series-parallel active power-line conditioning capabilities," *IEEE Transactions on Industry Applications*, vol. 38, no. 6, pp. 1581–1590, 2002.
14. A. Kuskoand and S. Fairfax, "Survey of rotary uninterruptible power supplies," in *Proc. 18th International Telecommunications Energy Conference*, pp. 416–419, 1996.
15. A. Windhorn, "A hybrid static/rotary UPS system," *IEEE Transactions on Industry Applications*, vol. 28, no. 3, pp. 541–545, 1992.
16. W. W. Hung and G. W. A. McDowell, "Hybrid UPS for standby power systems," *Power Engineering Journal*, vol. 4, no. 6, pp. 281–291, November 1990.
17. S. R. Bowes, "Advanced regular-sampled PWM control techniques for drives and static power converters," *IEEE Transactions on Industrial Electronics*, vol. 42, no. 4, pp. 367–373, 1995.
18. C. Rech, H. A. Grundling, and J. R. Pinheiro, "Comparison of discrete control techniques for UPS applications," in *Proc. IEEE Industry Applications Conference*, pp. 2531–2537, 2000.
19. J. Chen and C. Chu, "Combination voltage-controlled and current-controlled PWM inverters for UPS parallel operation," *IEEE Transactions on Power Electronics*, vol. 10, no. 5, pp. 547–558, 1995.
20. P. Mattavelli and W. Stefanutti, "Fully digital hysteresis modulation with switching time prediction," in *Proc. 19th Applied Power Electronics Conference and Exposition*, pp. 493–499, 2004.
21. P. Mattavelli, "An improved deadbeat control for UPS using disturbance observers," *IEEE Transactions on Industrial Electronics*, vol. 52, no. 1, pp. 206–212, 2005.
22. G. E. Valderrama, A. M. Stankovic, and P. Mattavelli, "Dissipativity-based adaptive and robust control of UPS in unbalanced operation," *IEEE Transactions on Power Electronics*, vol. 18, no. 4, pp. 1056–1062, 2003.
23. T. Tai and J. Chen, "UPS inverter design using discrete-time sliding-mode control scheme," *IEEE Transactions on Industrial Electronics*, vol. 49, no.1, pp. 67–75, 2002.
24. U. Burup, P. N. Enjeti, and F. Blaabjerg, "A new space-vector-based control method for UPS systems powering nonlinear and unbalanced loads," *IEEE Transactions on Industry Applications*, vol. 37, no. 6, pp. 1864–1870, 2001.
25. N. M. Abdel-Rahim and J. E. Quaicoe, "Analysis and design of a multiple feedback loop control strategy for single-phase voltage-source UPS inverters," *IEEE Transactions on Power Electronics*, vol. 11, no. 4, pp. 532–541, 1996.
26. A. Nasiri and A. Emadi, "Digital control of a three-phase series-parallel uninterruptible power supply/active filter system," in Proc. IEEE 35th *Annual Power Electronics Specialists Conference*, pp. 4115–4120, 2004.

25
Automotive Applications of Power Electronics

David J. Perreault
Massachusetts Institute of Technology, Laboratory for Electromagnetic and Electronic Systems, 77 Massachusetts Avenue, 10-039, Cambridge Massachusetts, USA

Khurram Afridi
Techlogix, 800 West Cummings Park, 1925, Woburn, Massachusetts, USA

Iftikhar A. Khan
Delphi Automotive Systems, 2705 South Goyer Road, MS D35 Kokomo, Indiana, USA

25.1 Introduction 635
25.2 The Present Automotive Electrical Power System 636
25.3 System Environment 636
    25.3.1 Static Voltage Ranges • 25.3.2 Transients and Electromagnetic Immunity • 25.3.3 Electromagnetic Interference • 25.3.4 Environmental Considerations
25.4 Functions Enabled by Power Electronics 641
    25.4.1 High Intensity Discharge Lamps • 25.4.2 Pulse-width Modulated Incandescent Lighting • 25.4.3 Piezoelectric Ultrasonic Actuators • 25.4.4 Electromechanical Engine Valves • 25.4.5 Electric Air Conditioner • 25.4.6 Electric and Electrohydraulic Power Steering Systems • 25.4.7 Motor Speed Control
25.5 Multiplexed Load Control 644
25.6 Electromechanical Power Conversion 646
    25.6.1 The Lundell Alternator • 25.6.2 Advanced Lundell Alternator Design Techniques • 25.6.3 Alternative Machines and Power Electronics
25.7 Dual/High Voltage Automotive Electrical Systems 652
    25.7.1 Trends Driving System Evolution • 25.7.2 Voltage Specifications • 25.7.3 Dual-voltage Architectures
25.8 Electric and Hybrid Electric Vehicles 655
25.9 Summary 657
    References 657

25.1 Introduction

The modern automobile has an extensive electrical system consisting of a large number of electrical, electromechanical, and electronic loads that are central to vehicle operation, passenger safety, and comfort. Power electronics is playing an increasingly important role in automotive electrical systems – conditioning the power generated by the alternator, processing it appropriately for the vehicle electrical loads, and controlling the operation of these loads. Furthermore, power electronics is an enabling technology for a wide range of future loads with new features and functions. Such loads include electromagnetic engine valves, active suspension, controlled lighting, and electric propulsion.

This chapter discusses the application and design of power electronics in automobiles. Section 25.2 provides an overview of the architecture of the present automotive electrical power system. The next section, Section 25.3, describes the environmental factors, such as voltage ranges, EMI/EMC requirements, and temperature, which strongly influence the design of automotive power electronics. Section 25.4 discusses a number of electrical functions that are enabled by power electronics, while Section 25.5 addresses load control via multiplexed remote switching architectures that can be implemented with power electronic switching. Section 25.6 considers the application of power electronics in automotive electromechanical energy conversion, including power generation. Section 25.7 describes the potential evolution of automotive electrical systems towards high- and dual-voltage systems, and provides an overview of the likely requirements of power electronics in such systems. Finally, the application of power electronics in electric and hybrid electric vehicles is addressed in Section 25.8.

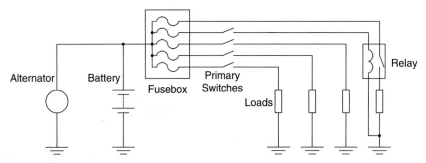

FIGURE 25.1 The 12-V point-to-point automotive electrical power system.

25.2 The Present Automotive Electrical Power System

Present-day automobiles can have over 200 individual electrical loads, with average power requirements in excess of 800 W. These include such functions as the headlamps, tail lamps, cabin lamps, starter, fuel pump, wiper, blower fan, fuel injector, transmission shift solenoids, horn, cigar lighter, seat heaters, engine control unit, cruise control, radio, and spark ignition. To power these loads, present day internal combustion engine (ICE) automobiles use an electrical power system similar to the one shown in Fig. 25.1. Power is generated by an engine-driven three-phase wound-field synchronous machine – a Lundell (claw-pole) alternator [1, 2]. The ac voltage of this machine is rectified and the dc output regulated to about 14 V by an electronic regulator that controls the field current of the machine. The alternator provides power to the loads and charges a 12 V lead-acid battery. The battery provides the high power needed by such loads as the starter, and supplies power when the engine is not running or when the demand for electrical power exceeds the output power of the alternator. The battery also acts as a large capacitor and smoothes out the system voltage.

Power is distributed to the loads via fuses and point-to-point wiring. The fuses, located in one or more fuseboxes, protect the wires against overheating and fire in the case of a short. Most of the loads are controlled directly by manually actuated mechanical switches. These primary switches are located in areas in easy reach of either the driver or the passengers, such as the dashboard, door panels, and the ceiling. Some of the heavy loads, such as the starter, are switched indirectly via electromechanical relays.

25.3 System Environment

The challenging electrical and environmental conditions found in the modern automobile have a strong impact on the design of automotive power electronic equipment. Important factors affecting the design of electronics for this application include static and transient voltage ranges, electromagnetic interference and compatibility requirements (EMI/EMC), mechanical vibration and shock, and temperature and other environmental conditions. This section briefly describes some of the factors that most strongly affect the design of power electronics for automotive applications. For more detailed guidelines on the design of electronics for automotive applications, the reader is referred to [1, 3–16] and the documents cited therein, from which much of the information presented here is drawn.

25.3.1 Static Voltage Ranges

In most present-day automobiles, a Lundell-type alternator provides dc electrical power with a lead-acid battery for energy storage and buffering. The nominal battery voltage is 12.6 V, which the alternator regulates to 14.2 V when the engine is on in order to maintain a high state of charge on the battery. In practice, the regulation voltage is adjusted for temperature to match the battery characteristics. For example, in [1], a 25°C regulation voltage of 14.5 V is specified with a −10 mV/°C adjustment. Under normal operating conditions, the bus voltage will be maintained in the range of 11–16 V [3]. Safety-critical equipment is typically expected to be operable even under battery discharge down to 9 V, and equipment operating during starting may see a bus voltage as low as 4.5–6 V under certain conditions.

In addition to the normal operating voltage range, a wider range of conditions is sometimes considered in the design of automotive electronics [3]. One possible condition is reverse-polarity battery installation, resulting in a bus voltage of approximately −12 V. Another static overvoltage condition can occur during jump starting from a 24-V system such as on a tow truck. Other static overvoltage conditions can occur due to failure of the alternator voltage regulator. This can result in a bus voltage as high as 18 V, followed by battery electrolyte boil-off and a subsequent unregulated bus voltage as high as 130 V. Typically, it is not practical to design the electronics for operation under such an extreme fault condition, but it should be noted that such conditions can occur. Table 25.1 summarizes the range of static voltages that can be expected in the automotive electrical system.

TABLE 25.1 Static voltage range for the automotive electrical system [3]

Static voltage condition	Voltage
Nominal voltage with engine on	14.2 V
Nominal voltage with engine off	12.6 V
Maximum normal operating voltage	16 V
Minimum normal operating voltage	9 V
Minimum voltage during starting	4.5 V
Jump start voltage	24 V
Reverse battery voltage	−12 V
Maximum voltage with alternator regulator failure followed by battery failure	130 V

25.3.2 Transients and Electromagnetic Immunity

Power electronic circuits designed for automotive applications must exhibit electromagnetic compatibility, i.e. the conducted and radiated emissions generated by the circuit must not interfere with other equipment on board the vehicle, and the circuit must exhibit immunity to radiated and conducted disturbances. The Society of Automotive Engineers (SAE) has laid out standards and recommended practices for the electromagnetic compatibility of automotive electronics in a set of technical reports [4]. These reports are listed in Table 25.2. Here we will focus on two of the basic requirements of automotive power electronics: immunity to power lead transients and limitation of conducted emissions.

A major consideration in the design of an automotive power electronic system is its immunity to the transients that can appear on its power leads. A number of transient sources exist in the vehicle [5] and procedures for validating immunity to these transients have been established in documents such as SAE J1113/11 [4, 6] and DIN 40389 [1]. Table 25.3 illustrates the transient test pulses specified in SAE J1113/11. Each test pulse corresponds to a different type of transient. The vehicle manufacturer determines which test pulses apply to a specific device.

Transients occur when inductive loads such as solenoids, motors, and clutches are turned on and off. The transients can be especially severe when the bus is disconnected from the battery, as is the case for the accessory loads when the ignition is switched off. Test pulse 1 in Table 25.3 simulates the transient generated when an inductive load is disconnected from the battery and the device under test remains in parallel with it. When the inductive load is a dc motor, it may briefly act as a generator after disconnection. This transient is simulated by test pulse 2b. Test pulse 2a models the transient when current in an inductive element in series with the device under test is interrupted. Test pulses 3a and 3b model switching spikes that appear on the bus during normal operation. Test pulse 4 models the voltage transient that occurs on starting.

Perhaps the best-known electrical disturbance is the so-called load dump transient that occurs when the alternator load current drops sharply and the battery is unable to properly buffer the change. This can occur when the battery becomes disconnected while drawing a large amount of current. To understand why a major transient can occur under this situation, consider that the Lundell-alternator has a very large leakage reactance. The high commutating reactance interacting with the diode rectifier results in a high degree of load regulation, necessitating the use of a large back emf to source rated current at high speed [7]. Back voltages as high as 120 V may be needed to generate rated current into a 14 V output at top speed. Analytical modeling of such systems is addressed in [8]. Two effects occur when the load on the alternator suddenly steps down. First, as the machine current drops, the energy in

TABLE 25.2 SAE J1113 electromagnetic compatibility technical reports

SAE specification	Type	Description
SAE J1113/1	Standard	Electromagnetic compatability measurement procedures and limits, 60 Hz–18 GHz
SAE J1113/2	Standard	Conducted immunity, 30 Hz–250 kHz
SAE J1113/3	Standard	Conducted immunity, direct injection of RF power, 250 kHz–500 MHz
SAE J1113/4	Standard	Conducted immunity, bulk current injection method
SAE J1113/11	Standard	Conducted immunity to power lead transients
SAE J1113/12	Recommended practice	Electrical interference by conduction and coupling – coupling clamp
SAE J1113/13	Recommended practice	Immunity to electrostatic discharge
SAE J1113/21	Information report	Electrical disturbances by narrowband radiated electromagnetic energy – component test methods
SAE J1113/22	Standard	Immunity to radiated magnetic fields from power lines
SAE J1113/23	Recommended practice	Immunity to radiated electromagnetic fields, 10 kHz–200 MHz, strip line method
SAE J1113/24		Immunity to radiated electromagnetic fields, 10 kHz–200 MHz, TEM cell method
SAE J1113/25	Standard	Immunity to radiated electromagnetic fields, 10 kHz–500 MHz, tri-plate line method
SAE J1113/26	Recommended practice	Immunity to ac power line electric fields
SAE J1113/27	Recommended practice	Immunity to radiated electromagnetic fields, reverberation method
SAE J1113/41	Standard	Radiated and conducted emissions, 150 kHz–1000 MHz
SAE J1113/42	Standard	Conducted transient emissions

TABLE 25.3 Transient pulse waveforms specified in SAE J1113/11

Pulse	Shape	Maximum excursion	Source impedance	Duration and repetition rate
1		−100 V	10 Ω	$T_{pulse} = 2$ ms 0.5 s $< T_{rep} < 5$ s
2a		100 V	10 Ω	$T_{pulse} = 50$ µs 0.5 s $< T_{rep} < 5$ s
2b		10 V	0.5–3 Ω	$T_{pulse} \geq 200$ ms
3a		−150 V	50 Ω	$T_{pulse} = 100$ ns $T_{rep} = 100$ µs
3b		100 V	50 Ω	$T_{pulse} = 100$ ns $T_{rep} = 100$ µs
4		−7 V	0.01 Ω	$T_{pulse} \leq 20$ s
5		84 A	0.6 Ω	$\tau = 115$ ms $T_{pulse} \sim 4\tau$

the alternator leakage reactances is immediately delivered to the alternator output, causing a voltage spike. The peak voltage reached depends on the electrical system impedance, and may be limited by suppression devices. Second, once the alternator current is reduced, the voltage drops across the leakage (commutating) reactances are reduced, and a much larger fraction of the machine back-emf is impressed across the dc output. The proper output voltage is only re-established as the voltage regulator reduces the field current appropriately. With conventional regulator circuits, this takes place on the time scale of the field winding time constant (typically 100 ms), and results in a major transient event. In systems without centralized protection, a load dump can generate a transient with a peak voltage in excess of 100 V lasting hundreds of milliseconds. Test pulse 5 in Table 25.3 (expressed as a current waveform in parallel with an output resistance) is designed to simulate such a load-dump transient; other load-dump tests are even more severe [1, 3].

25.3.3 Electromagnetic Interference

Strict limits also exist for the amount of electromagnetic interference (EMI) that an automotive electronic component can generate. Limits for both conducted and radiated emissions are specified in SAE standards J1113/41 and J1113/42 [4, 9, 10]. Here we will consider the conducted EMI specifications for power leads, since they directly impact the design of EMI filters for automotive power electronics. Meeting the conducted specifications is a major step towards achieving overall compliance.

The conducted EMI specifications in SAE J1113/41 limit the ripple that an electronic circuit can inject onto the voltage bus over the frequency range from 150 KHz to 108 MHz. The amount of ripple injected by a circuit usually depends on the bus impedance. To eliminate any variability due to this, EMI compliance testing is done using a line impedance stabilization network (LISN) between the bus and the device under test, as illustrated in Fig. 25.2. The LISN is also sometimes referred to as an artificial mains network (AN). Essentially, the LISN ensures that the equipment under test receives the proper dc voltage and current levels and also sees a controlled impedance for the ripple frequencies of interest. Figure 25.3 shows the magnitude of the LISN output impedance for a low-impedance input source; the effective impedance is 50 Ω over most of the frequency range of interest. The 50-Ω termination impedance of the LISN is typically provided by the measurement equipment. The EMI specifications are stated in terms of the allowable voltage ripple (in dB μV) appearing across the 50-Ω LISN resistance as a function of frequency.

There are a wide range of other technical considerations for EMI testing, including the arrangement of the equipment over a ground plane and the types and settings of the measuring devices. One characteristic to consider is that the EMI measurements are done across frequency with a spectrum analyzer having a prespecified receiver bandwidth (RBW). For frequencies between 150 kHz and 30 MHz, the receiver bandwidth is 9 kHz, resulting in spectral components within 9 kHz of one another being lumped together for purposes of the test. A full test procedure is defined in the SAE specifications, beginning with narrowband measurements and moving to wideband measurements if necessary. Figure 25.4 illustrates the narrowband conducted EMI limits for power leads in SAE J1113/41. It is interesting to note that for the commonly used Class 5 limits, the allowable ripple current into the LISN at 150 kHz is less than 100 μA!

As seen in the previous section, the transient disturbances generated by electrical and electronic equipment are an important consideration in automotive applications. Because power electronic circuits typically contain switches and magnetic elements, they are potential sources for such transients, especially when powered from the switched ignition line. SAE J1113/42 specifies methods for testing and evaluating the transients generated by automotive electrical components, and proposes transient waveform limits for different severity levels. The equipment under test is set up in a configuration similar to that in Fig. 25.2, but with a switching device on one side or the other of the LISN, depending on the application. The equipment under test is then evaluated for transient behavior at turn on, turn off, and across its operating range. The voltage transients at the input of the equipment are measured and

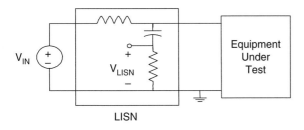

FIGURE 25.2 Conducted EMI test set up with LISN. $L_{LISN} = 5\,\mu H$, $C_{LISN} = 0.1\,\mu F$, and $R_{LISN} = 50\,\Omega$.

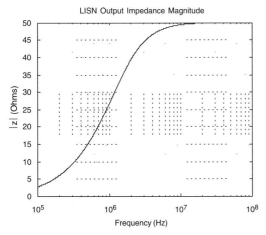

FIGURE 25.3 The LISN output impedance magnitude for a low impedance input source.

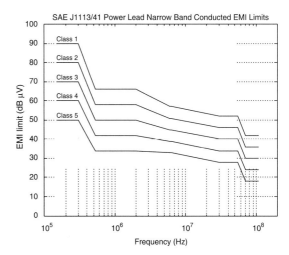

FIGURE 25.4 SAE J1113/41 narrowband conducted EMI limits for power leads. The specification covers the frequency range from 150 kHz to 108 MHz.

evaluated with respect to magnitude, duration, and rise and fall times. Specific limits for such transients are specified by the vehicle manufacturer, but SAE J1113/42 proposes a representative set of limits for four different transient severity levels.

Due to the tight conducted emissions limits, input EMI filter design is an important consideration in automotive power electronics. Single or multistage low-pass filters are typically used to attenuate converter ripple to acceptable levels [11–13]. When designing such filters, the parasitic behavior of the filter components, such as capacitor equivalent series resistance and inductance, and suitable filter damping are important considerations [14]. One must also ensure that the filter design yields acceptable transients at switch on and off, and does not result in undesired dynamic interactions with the power circuit [13]. Attention to appropriate filter design, coupled with proper circuit layout, grounding, and shielding goes a long way towards meeting electromagnetic interference specifications [14].

25.3.4 Environmental Considerations

The automobile is a very challenging environment for electronics. Environmental factors influencing the design of automotive electronics include temperature, humidity, mechanical shock, vibration, immersion, salt spray, and exposure to sand, gravel, oil, and other chemicals. In 1978, the SAE developed a recommended practice for electronic equipment design to address these environmental considerations [3, 4]. This document, SAE J1211, provides quantitative information about the automotive environment to aid the designer in developing environmental design goals for electronic equipment. Here, we briefly summarize a few of the most important factors affecting the design of power electronics for automotive applications. For more detailed guidelines, the reader is referred to [3] and the documents cited therein.

Perhaps the most challenging environmental characteristic is the extreme range of temperatures that can occur in the automobile. Table 25.4 summarizes some of the temperature extremes listed in SAE J1211 for different locations in the automobile. Ambient temperatures as low as −40°C may be found during operation, and storage temperatures as low as −50°C may be found for components shipped in unheated aircraft. Maximum ambient temperatures vary widely depending on vehicle location, even for small differences in position. Because ambient temperature has a strong impact on the design of a power electronic system it is important to work closely with the vehicle manufacturer to establish temperature specifications for a particular application. For equipment that is air-cooled, one must also consider that the equipment may be operated at altitudes up to 12,000 feet above sea level. This results in low ambient pressure (down to 9 psia), which can reduce the heat transfer efficiency [3]. For equipment utilizing the radiator-cooling loop, maximum coolant temperatures in the range of 105–120°C at a pressure of 1.4 bar are possible [15].

TABLE 25.4 Automotive temperature extremes by location [3]

Vehicle location	Min temp. (°C)	Max temp. (°C)
Exterior	−40	85
Chassis		
Isolated	−40	85
Near heat source	−40	121
Drive train high temperature location	−40	177
Interior		
Floor	−40	85
Rear deck	−40	104
Instrument panel	−40	85
Instrument panel top	−40	177
Trunk	−40	85
Under hood		
Near radiator support structure	−40	100
Intake manifold	−40	121
Near alternator	−40	131
Exhaust manifold	−40	649
Dash panel (normal)	−40	121
Dash panel (extreme)	−40	141

In addition to the temperature extremes in the automobile, thermal cycling and shock are also important considerations due to their effect on component reliability. Thermal cycling refers to the cumulative effects of many transitions between temperature extremes, while thermal shock refers to rapid transitions between temperature extremes, as may happen when a component operating at high temperature is suddenly cooled by water splash. The damaging effects of thermal cycling and shock include failures caused by thermal expansion mismatches between materials. Test methods have been developed which are designed to expose such weaknesses [3, 16]. The thermal environment in the automobile, including the temperature extremes, cycling, and shock, are challenging issues that must be addressed in the design of automotive power electronics.

A number of other important environmental factors exist in the automobile. Humidity levels as high as 98% at 38°C can exist in some areas of the automobile, and frost can occur in situations where the temperature drops rapidly. Salt atmosphere, spray, water splash, and immersion are also important factors for exterior, chassis, and underhood components. Failure mechanisms resulting from these factors include corrosion and circuit bridging. Dust, sand, and gravel bombardment can also be significant effects depending on equipment location. Mechanical vibration and shock are also important considerations in the design of automotive power electronic equipment. Details about the effects of these environmental factors, sample recorded data, and recommended test procedures can be found in [3].

25.4 Functions Enabled by Power Electronics

Over the past 20 years, power electronics has played a major role in the introduction of new functions such as the antilock breaking system (ABS), traction control, and active suspension, as well as the electrification of existing functions such as the engine-cooling fan, in the automobile. This trend is expected to continue, as a large number of new features being considered for introduction into automobiles require power electronics. This section discusses some of the new functions that have been enabled by power electronics, and some existing ones that benefit from it.

25.4.1 High Intensity Discharge Lamps

High intensity discharge (HID) lamps have started to appear in automobiles as low-beam headlights and fog lights. The HID lamps offer higher luminous efficacy, higher reliability, longer life, and greater styling flexibility than the traditional halogen lamps [17, 18]. The luminous efficacy of an HID lamp is over three times that of a halogen lamp and its life is about 2000 hours, compared to 300–700 hours for a halogen lamp. Therefore, HID lamps provide substantially higher road illumination while consuming the same amount of electrical power and, in most cases, should last the life of the automobile. The HID lamps also produce a whiter light than halogen lamps since their color spectrum is closer to that of the sun.

High intensity discharge lamps do not have a filament. Instead, light is generated by discharging an arc through a pressurized mixture of mercury, xenon, and vaporized metal halides – mercury produces most of the light, the metal halides determine the color spectrum, and xenon helps reduce the start-up time of the lamp [17, 19]. Unlike halogen lamps that can be powered directly from the 12-V electrical system, HID lamps require power electronic ballasts for their operation. Initially, a high voltage pulse of 10–30 kV is needed to ignite the arc between the electrodes and a voltage of about 85 V is needed to sustain the arc [4.3]. Figure 25.5 shows a simplified power electronic circuit that can be used to start and drive an HID lamp. A step-up dc–dc converter is used to boost the voltage from 12 V to the voltage needed for the steady-state operation of the HID lamp. Any dc-dc converter that can step up the voltage, such as the boost or flyback converter, can be used for this application. An H-bridge is then used to create the ac voltage that drives the lamp in steady state. The circuit to initiate the arc can be as simple as a circuit that provides an inductive voltage kick, as shown in Fig. 25.5.

25.4.2 Pulse-width Modulated Incandescent Lighting

Future automobiles may utilize a 42 V electrical system in place of today's 14 V electrical system (see Section 25.7). Because HID lamps are driven through a power electronic ballast, HID lighting systems operable from a 42 V bus can be easily developed. However, the high cost of HID lighting – as much as an order of magnitude more expensive than incandescent lighting – largely limits its usefulness to headlight applications. Incandescent lamps compatible with 42 V systems can also be implemented. However, because a much longer, thinner filament must be employed at the higher voltage, lamp lifetime suffers greatly. An alternative to this approach is to use pulse-width modulation to operate 12 V incandescent lamps from a 42 V bus [20].

In a pulse-width modulated (PWM) lighting system, a semiconductor switch is modulated to apply a periodic pulsed voltage to the lamp filament. Because of its resistive nature, the power delivered to the filament depends on the rms of the applied voltage waveform. The thermal mass of the system filters the power pulsations so that the filament temperature and light production are similar to that generated by a dc voltage with the same rms value. The PWM frequency is selected low enough to avoid lamp mechanical resonances and the need for EMI filtering, while being high enough to limit visible flicker; PWM frequencies in the range of 90–250 Hz are typical [20].

Ideally, a 11.1% duty ratio is needed to generate 14 V rms across a lamp from a 42 V nominal voltage source. In practice, deviations from this duty ratio are needed to adjust for input voltage variations and device drops. In some proposed systems, multiple lamps are operated within a single lighting module with phase staggered (interleaved) PWM waveforms to reduce the input rms current of the module.

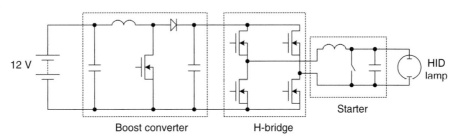

FIGURE 25.5 Simplified power electronic circuit for an HID lamp ballast.

Another issue with PWM lighting relates to startup. Even with operation from a 12 V dc source, incandescent lamps have an inrush current that is 6–8 times higher than the steady-state value, because of how filament resistance changes with temperature; this inrush impacts lamp durability. The additional increase in peak inrush current due to operating from a 42 V source can be sufficient to cause destruction of the filament, even when using conventional PWM soft-start techniques (a ramping up of duty ratio). Means for limiting the peak inrush current – such as operating the controlling MOSFET in current limiting mode during startup – are needed to make practical use of PWM lighting control.

While PWM incandescent lighting technology is still in the early stages of development, it offers a number of promising advantages in future 42 V vehicles. These include low-cost adaptation of incandescent lighting to high-voltage systems, control of lighting intensity independent of bus voltage, the ability to implement multiple intensities, flashing, dimming, etc. through PWM control, and the potential improvement of lamp durability through more precise inrush and operating control [20].

25.4.3 Piezoelectric Ultrasonic Actuators

Piezoelectric ultrasonic motors are being considered as actuators for window lifts, seat positioning, and head restraints in automobiles [21, 22]. These motors work on the principle of converting piezoelectrically induced ultrasonic vibrations in an elastic body into unidirectional motion of a moving part. Unidirectional motion is achieved by allowing the vibrating body to make contact with the moving part only during a half-cycle of its oscillation, and power is transferred from the vibrating body to the moving part through frictional contact. Ultrasonic motors have a number of attractive features, including high-torque density, large holding torque even without input power, low speed without gears, quiet operation, no magnetic fields, and high dynamics [21, 23]. These characteristics make ultrasonic motors an attractive alternative to electromagnetic motors for low-power high-torque applications.

Various types of ultrasonic motors have been developed. However, because of its compact design, the traveling wave type is the most popular ultrasonic motor [24]. Figure 25.6a shows the basic structure of such a motor. It consists of a metal stator and rotor, which are pushed against each other by a spring. The rotor is coated with a special lining material to increase friction and reduce wear at the contacting surfaces. A layer of piezoelectric material, such as lead zirconate titanate (PZT), is bonded to the underside of the stator. Silver electrodes are printed on both sides of the piezoceramic ring. The top electrode is segmented and the piezoceramic is polarized as shown in Fig. 25.6b. The number of segments is twice the order of the excited vibration mode.

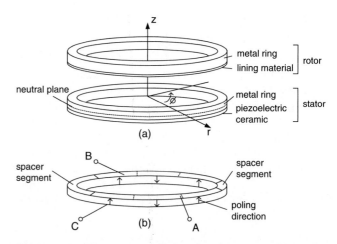

FIGURE 25.6 (a) Basic structure of a traveling wave piezoelectric ultrasonic motor and (b) structure of the piezoceramic ring and electrode for a four-wavelength motor. Arrows indicate direction of polarization. Dashed lines indicate segments etched in the electrode for poling but electrically connected during motor operation.

When a positive voltage is applied between terminals A and C, the downwards poled segment elongates and the upwards poled segments contract. This causes the stator to undulate, waving down at the elongated section and up at the contracted one. When the polarity of the voltage is inverted, the undulations are also inverted. Hence, when an ac voltage is applied a flexural standing wave is created in the stator. To get a large wave amplitude, the stator is driven at the resonance frequency of the flexural mode. An ac voltage between terminals B and C similarly produces another standing wave. However, because of the spacer segments in the piezoceramic ring, the second standing wave is 90° spatially out of phase from the first one. If the two standing waves are excited by ac voltages that are out of phase in time by 90°, a traveling wave is generated. As the traveling wave passes through a point along the neutral plane, that points simply exhibits axial (z-axis) motion. However, off-neutral plane points also have an azimuthal (ϕ-axis) component of motion. This azimuthal motion of the surface points propels the rotor. Ultrasonic motors require a power electronic drive. A power electronic circuit suitable for driving an ultrasonic motor is shown in Fig. 25.7. The two H-bridges are controlled to generate waveforms that are 90° out of phase with each other.

25.4.4 Electromechanical Engine Valves

Electromagnetic actuators are finding increasing application in automotive systems. These actuators are more desirable than the other types of actuators, such as the hydraulic and pneumatic actuators, because they can be more easily controlled by a microprocessor to provide more precise control. An application of electromagnetic actuators that is of particular interest is the replacement of the camshaft and tappet valve assembly

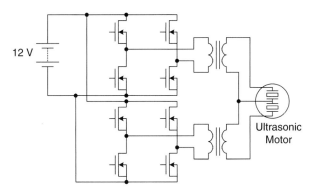

FIGURE 25.7 Drive circuit for an ultrasonic motor.

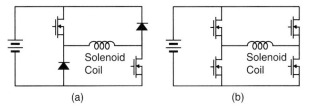

FIGURE 25.8 Power electronic circuits for driving solenoids.

by electromechanically driven engine valves [25]. The opening and closing of the intake and exhaust valves can be controlled to achieve optimum engine performance and improved fuel economy over a wide range of conditions determined by variables such as the speed, load, altitude, and temperature. The present cam system provides a valve profile that can give optimum engine performance and improved fuel economy only under certain conditions.

Two power electronic circuits suitable for driving the solenoids for valve actuation are shown in Fig. 25.8. The circuit of Fig. 25.8a is suitable for solenoids that require unidirectional currents through their coils, while the circuit of Fig. 25.8b is suitable for solenoids that require bidirectional currents through their coils.

25.4.5 Electric Air Conditioner

It is desirable to replace some of the engine-driven functions of a vehicle with electrically driven counterparts. The benefits of driving these functions electrically include the elimination of belts and pulleys, improved design and control due to independence from engine speed, and resulting increased efficiency and improved fuel economy. Furthermore, there is the opportunity for operation of the function in the engine-off condition.

The air conditioner is an example of an engine-driven function that could benefit from electrification. The engine drives the compressor of the air conditioner. Consequently, the speed of the compressor varies over a wide range and the compressor has to be over-sized to provide the desired performance at engine idle. Also, since the compressor speed is dependent on the engine speed, excessive cooling occurs at highway speeds requiring the cool air to be blended with the hot air to keep the temperature at the desired level. Furthermore, shaft seals and rubber hoses can lead to the loss of refrigerant (CFC) and pose an environmental challenge.

In an electric air conditioner, an electric motor is used to drive the compressor [26]. The motor is usually a three-phase brushless dc motor driven by a three-phase MOSFET bridge. The speed of the compressor in an electric air conditioner is independent of the engine speed. As a result, the compressor does not have to be over-sized and excessive cooling does not occur. Also, shaft seals and hoses can be replaced with a hermetically sealed system. Another benefit of an electric air conditioner is the flexibility in its location, since it does not have to be driven by the engine.

25.4.6 Electric and Electrohydraulic Power Steering Systems

The hydraulic power steering system of a vehicle is another example of an engine-driven accessory. This system can be replaced with an electric power steering (EPS) system in which a brushless dc motor is used to provide the steering power assist [27]. The electric power steering system is more efficient than the hydraulic power steering system because, unlike the engine-driven hydraulic steering pump, which is driven by the engine all the time, the motor operates only on demand. Another system that can replace the hydraulic power steering system is the electrohydraulic power steering (EHPS) system. In this case, a brushless dc motor and inverter can be employed to drive the hydraulic steering pump. The ability of the EPHS system to drive the pump only on demand leads to energy savings of as much as 80% as compared with the conventional hydraulic system. Challenges in implementing EPS and EPHS systems include meeting the required levels of cost and reliability for this critical vehicle subsystem.

25.4.7 Motor Speed Control

Some of the motors used in a vehicle require variable speed control. Consider, as an example, the blower motor used to provide air flow to the passenger compartment. This motor is typically a permanent magnet dc motor with a squirrel-cage fan. The speed of the motor is usually controlled by varying the resistance connected in series with the motor winding. This method of speed control leads to a significant power loss. A low-loss method of speed control employs semiconductor devices as shown in Fig. 25.9. In this case, the speed of the motor is controlled via PWM – that is, by switching the MOSFET on and off with different duty-ratios for different speed settings. An input filter is needed to reduce the EMI generated by the switching of the MOSFET. This method of speed control is equivalent to supplying power to the motor

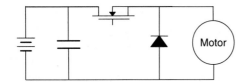

FIGURE 25.9 Low-loss circuit to control the speed of a motor.

through a variable-output dc-to-dc converter. The converter is located close to the motor and no filter is required between the converter output and motor winding.

Another low-loss method that can be used to control the speed of a motor employs a three-phase brushless dc motor. The speed in this case is controlled by controlling the MOSFETs in the dc-to-three-phase-ac converter that drives the motor.

25.5 Multiplexed Load Control

Another emerging application of power electronics in automobiles is in the area of load control. In the conventional point-to-point wiring architecture, most of the loads are controlled directly by the primary mechanical switches, as shown in Fig. 25.1. In a point-to-point wiring architecture, each load has a dedicated wire connecting it to the fuse box via the primary switch. Consequently, fairly heavy wires have to be routed all over the vehicle, as illustrated in Fig. 25.10a. The situation is made worse when multiple switches control the same load, as is the case with power windows and power door locks. The complete harness of a 1994 C-class Mercedes-Benz that uses point-to-point wiring has about 1000 wires, with a total length of 2 km, over 300 connectors and weighs 36 kg. The process of assembling the wiring harness is difficult and time consuming, leading to high labor costs. Retrofitting, fault tracing, and repairing are time consuming and expensive. The bulky harness also places constraints on the vehicle body design, and the large number of connectors compromise system reliability.

An alternative wiring technique is to control the loads remotely and multiplex the control signals over a communication bus, as shown in Fig. 25.10b and c. A control message is sent on the communication bus to switch a particular load on or off. This allows more flexibility in the layout of the power cables and could allow the pre-assembly of the harness to be more automated. Furthermore, with communication between the remote switches, it is practical to have a power management system than can turn off non-essential loads when there is a power shortage. One possibility is to group the remote switches into strategically located distribution boxes, as shown in Fig. 25.10b. A power and a communication bus connect the distribution boxes. Another possibility is to integrate the remote switches with the load, i.e. point-of-load switching,

as shown in Fig. 25.10c. In Fig. 25.10b the transceivers are also built into the distribution boxes, while in Fig. 25.10c each load and primary switch has an integrated transceiver. The point-of-load switching topology is attractive because of its simplicity, but raises cost and fusing challenges.

Multiplexed remote switching architectures have been under consideration since at least the early 1970s, when Ziomek investigated their application to various electrical subsystems [28]. The initial interest was dampened by cost and reliability concerns and the non-availability of appropriate remote switches. However, advances in semiconductor technology and rapid growth in the automotive electrical system revived interest in multiplexed architectures. The SAE Multiplexing Standards Committee has partitioned automotive communications into three classes: Class A for low data-rate (1–10 kbit/s) communication for the control of body functions, such as headlamps, windshield wipers, and power windows, Class B for medium data-rate (10–100 kbit/s) parametric data exchange, and Class C for high data-rate (1 Mbit/s) real-time communication between safety critical functions, such as between ABS sensors and brake actuators [29]. Although load control is categorized as Class A, lack of any widely accepted Class A communication protocols has lead to the application of Class B and Class C communication IC's to load control. Class B has received the most attention due to the California Air Resources Board mandated requirement for on-board diagnostics (OBD II) and a large number of competing protocols, including the French vehicle-area network (VAN), the ISO 9141 and the SAE J1850, have been developed [30]. Of these, the SAE J1850 is the most popular in the US. Another popular protocol is the controller area network (CAN) developed by Bosch [31]. Although designed for Class C with bit rates up to 1 Mbit/s, it is being applied for Class A and Class B applications due to the availability of inexpensive CAN ICs from a large number of semiconductor manufacturers.

Remote switching systems require remote power switches. An ideal remote switch must have a low on-state voltage, be easy to drive from a micro-controller, and incorporate current sensing. A low on-state voltage helps minimize the heatsinking requirements, while current sensing is needed for the circuit protection function to be incorporated into the switch. To withstand the harsh automotive environment the switch must also be rugged. Furthermore, if PWM control is required for the load, the switch must have short turn-on and turn-off times and a high cycle-life. The traditional means of remotely switching loads in an automobile is via electromechanical relays. Although relays offer the lowest voltage drop per unit cost, they require large drive current, are relatively large, are difficult to integrate with logic, and are not suitable for PWM applications [32–34]. Therefore, their use will be limited to very high current, non-PWM applications. The power levels of the individual loads in the automobile are too low for IGBTs and MCTs to be competitive. Bipolar transistors

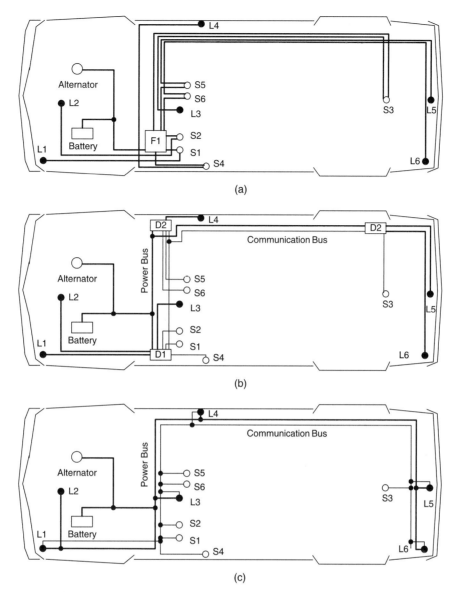

FIGURE 25.10 Alternative control strategies illustrated for a simple automotive electrical system with six loads (L1-6) and six primary switches (S1-6): (a) conventional direct switching architecture with a single fusebox (F1); (b) multiplexed remote switching architecture, with remote switches and transceivers in three distribution boxes (D1-3); and (c) multiplexed point-of-load switching with electronics integrated into the loads and the primary switches.

are also not very attractive because they are harder to drive than a MOS-gated device. Because of its fast switching speed, low voltage drop, relative immunity to thermal runaway, low drive requirements, and ease of integration with logic, the power MOSFET is the most attractive candidate for remote switching. Smart-power MOSFET devices with integrated logic interface and circuit protection have recently become available. Use of these devices for power electronic control of individual loads has become economically competitive in some subsystems, and may be expected to become more so with the advent of higher voltage electrical systems.

The benefits of remote switching electrical distribution systems have been demonstrated by Furuichi *et al.* [35]. The multiplexed architecture they implemented had 10 remote units (two power units with fuses, power drivers and signal inputs, five load control units with power drivers and signal inputs but no fuses, and three signal input units with only signal inputs). To increase system reliability, each power unit was connected to the battery via independently fused power cables. Although wiring cost decreased, the authors report an increase in overall system cost due to the additional cost of the remote units. Intel's CAN ICs with data rates of 20 kbit/s were used

TABLE 25.5 Comparison of a multiplexed and the conventional system, as reported by Furuichi *et al.* for a compact vehicle [35]. In the multiplexed system, the function of nine electronic control units (ECUs) was integrated into the remote units

	Point-to-point	Multiplexed	Change (%)
Harness weight (kg)	14.0	9.8	−30
ECU weight (kg)	1.2	0.0	N/A
Remote unit weight (kg)	0.0	3.5	N/A
Total weight (kg)	15.2	13.3	−12.5
Number of wires	743	580	−21.9
Number of terminals	1195	915	−23.4
Number of splices	295	246	−16.6
Length of wire (m)	809	619	−23.5

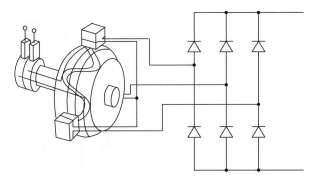

FIGURE 25.11 Structure and circuitry of the conventional Lundell alternator.

for the transmission and reception of control signals over an unshielded twisted-pair ring bus. Intelligent power MOSFETs were used as the remote switches and fusing was done with mini-fuses. The results of their work are shown in Table 25.5. Although weight of the wiring harness was reduced by 30%, the total system weight decreased by only 12.5% due to the added weight of the remote units.

25.6 Electromechanical Power Conversion

Power is generated in the automobile by an electrical machine driven by the engine. In the early days of the automobile, the electrical load was small and a dc generator was used for this purpose. As the electrical loads grew, the dc generator could not meet the growing demand of electrical power and was displaced by a three-phase alternator and diode rectifier. Continuously increasing power and performance requirements are driving further evolution in automotive power generation and control, and are motivating the introduction of power electronics and improved electrical machines in automobiles. In addition to high-power alternators, future applications of electromechanical power conversion may include integrated starter/alternators and propulsion systems. This section describes some of the machine and power electronic technologies that are useful for meeting the increasing challenges in the automobile.

25.6.1 The Lundell Alternator

The Lundell, or claw-pole, alternator is a three-phase wound-field synchronous machine that is almost universally used for power generation in present-day vehicles [1]. As illustrated in Fig. 25.11, the rotor is made of a pair of stamped pole pieces ("claw poles") fixed around a cylindrical field coil. The field winding is driven from the stator via a pair of slip rings and brushes, and causes the two pole pieces to become opposing magnetic poles. A full-bridge diode rectifier is traditionally used at the machine output, and a fan mounted on the rotor is typically used to cool the whole assembly.

The dc output voltage of the alternator system is regulated by controlling the field current. A switching field regulator applies a pulse width modulated voltage across the field. The steady-state field current is determined by the field-winding resistance and the average voltage applied by the regulator. Changes in the field current occur with an L/R field-winding time constant in the order of 100 ms or more. This long field-winding time constant and a large stator leakage reactance are characteristic of this type of alternator and tend to dominate its performance.

The alternator is driven by means of a belt, and is designed to operate over a wide speed ratio of about 10:1, though much of its operating lifetime is spent within a narrower 3:1 or 4:1 range. The gearing ratio provided by the belt is a design variable for the alternator; an alternator mechanical speed range from 1800 to 18,000 rpm for a 12-pole machine is typical.

A simple electrical model for the Lundell alternator is shown in Fig. 25.12. The armature of the alternator is modeled as a Y-connected set of leakage inductances L_s and back voltages v_{sa}, v_{sb}, and v_{sc}. The fundamental electrical frequency ω of

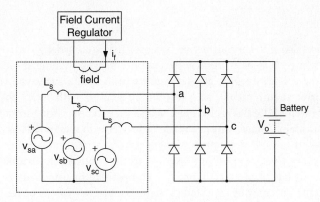

FIGURE 25.12 A simple Lundell alternator model.

the back-emfs is one-half of the product of the number of machine poles p and the mechanical speed ω_m. The magnitude of the back-emfs is proportional to the electrical frequency and the field current. For the sinusoidal case, the line-to-neutral voltage back-emf magnitude can be calculated as:

$$V_s = k\omega i_f \qquad (25.1)$$

where k is the machine constant and i_f is the field current. The diode bridge feeds a constant voltage V_o representing the battery and other loads. This simple model captures many of the important characteristics of the Lundell alternator, while remaining analytically tractable. Other effects, such as stator resistance, mutual coupling, magnetic saturation, and waveform harmonic content, can be incorporated into this model at the expense of simplicity. The constant-voltage battery load on the alternator makes the analysis of this system different from the classic case of a diode rectifier with a current-source load. Nevertheless, with reasonable approximations, the behavior of this system can be described analytically [8]. Using the results presented in [8], alternator output power vs operating point can be calculated as:

$$P_o = \frac{3V_o\sqrt{V_s^2 - (4V_o^2/\pi^2)}}{\pi\omega L_s} \qquad (25.2)$$

where V_o is the output voltage, V_s is the back-emf magnitude, ω is the electrical frequency, and L_s is the armature leakage inductance. Extensions of Eq. (25.2) that also include the effect of the stator resistance are given in [8].

As can be inferred from Eq. (25.2), alternator output power varies with speed, and is maximized when the back-emf magnitude of the machine is substantially larger than the output voltage. In a typical Lundell alternator, back voltages in excess of 80 V may be necessary to source-rated current into a 14 V output at high speed. Furthermore, as can be seen from Eq. (25.2), the armature leakage reactance limits the output power capability of the alternator. These characteristics are a result of the fact that significant voltage drops occur across the leakage reactances when current is drawn from the machine. These drops increase with speed and current, and cause the alternator to exhibit significant drop in output voltage with increasing current. Thus, an appropriate dc-side model for a Lundell alternator is a large open-circuit voltage (related to the back-emf magnitude) in series with a large current- and speed-dependent output impedance. This characteristic, coupled with the long field time constant, is the source of the undesirable load-dump transient characteristic of the Lundell alternator. In this transient, the large open-circuit voltage is transiently impressed across the alternator output when the load is suddenly reduced.

The efficiency of the conventional Lundell alternator is relatively poor. Typical efficiency values are in the order of 40–60%, depending on the operating point [1, 36, 37]. At low and medium speeds, losses tend to be dominated by stator copper losses. Iron losses become dominant only at very high speeds [1].

25.6.2 Advanced Lundell Alternator Design Techniques

The conventional diode-rectified Lundell alternator, though inefficient, has so far met vehicle electrical power requirements in a cost-effective manner. However, continuing increase in electrical power demand and growing interest in improved fuel economy is pushing the limits of conventional Lundell alternator technology. This section describes some established and emerging technologies that can be used to improve the performance of the Lundell alternator.

25.6.2.1 Third-harmonic Booster Diodes

One widely used approach for improving the high-speed output power capability of Lundell alternators is the introduction of third-harmonic booster diodes [1]. In this technique, the neutral point of the Y-connected stator winding is coupled to the output via a fourth diode leg, as illustrated in Fig. 25.13. While the fundamental components of the line-to-neutral back voltages are displaced by 120° in phase, any third-harmonic components will be exactly in phase. As a result, third-harmonic energy can be drawn from the alternator and transferred to the output by inducing and rectifying common-mode third-harmonic currents through the three windings. The booster diodes provide a means for achieving this. At high speed, the combination of the third-harmonic voltages at the main rectifier bridge (at nodes a, b, and c in Fig. 25.13), combined with the third-harmonic of the back voltages are large enough to forward bias the booster diodes and deliver third-harmonic energy to the output. In systems with significant (e.g. 10%) third-harmonic voltage content, up to 10% additional output power can be delivered at high speed. Additional power is not achieved at low speed, or in cases where the third-harmonic of the back voltage is small.

25.6.2.2 Lundell Alternator with Permanent Magnets

The structure of the rotor of the claw-pole alternator is such that the leakage flux is high. This reduces the output current capability of the alternator. The leakage flux can be reduced by placing permanent magnets on the pole faces or in the spaces between the adjacent poles of the rotor. This modification allows the alternator to deliver more output current. Placing the magnets in the spaces between adjacent poles is a better approach because it is simpler to implement and leads to a higher output current at engine idle [38].

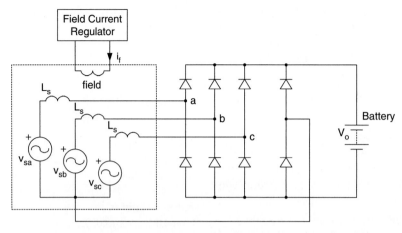

FIGURE 25.13 Lundell alternator with booster diodes.

25.6.2.3 Twin-rotor Lundell Alternator

The maximum power capability of the Lundell alternator is limited in part by the limit on its length-to-diameter ratio imposed by mechanical stresses on the stamped pole pieces. This prevents the Lundell alternator from being arbitrarily scaled up in size. The power capability of conventional designs is probably limited to 3 kW, which is likely to be unacceptable in the foreseeable future [39]. One way to retain the cost-effectiveness of the claw-pole alternator while achieving higher output power is to place two claw-pole rotors back-to-back on a common shaft inside a common stator [40]. This effectively increases the length of the claw-pole alternator without changing its diameter. This design allows higher power alternators to be built while retaining most of the cost benefits of the claw pole design.

25.6.2.4 Power Electronic Control

Another approach for improving the output power and efficiency of the Lundell alternator is through the use of more sophisticated power electronics. Power electronics technology offers tremendous value in this application. For example, replacing the conventional diode rectifier with a switched-mode rectifier provides an additional degree of design and control freedom, and allows substantially higher levels of power and efficiency to be attained from a given machine. One such design is shown in Fig. 25.14. It employs a simple switched-mode rectifier along with a special load-matching control technique to achieve dramatic improvement in alternator output power, efficiency, and transient performance [37]. The switched-mode rectifier provides improved control without the cost and complexity of a full active converter bridge. By controlling the duty ratio of the switched-mode rectifier based on available signals such as alternator speed, the alternator output power characteristic Eq. (25.2) can be altered and improved, particularly for speeds above idle [37]. Improvements in average power capability of a factor of two and average efficiency improvements on the order of 20% are possible with this technology. Furthermore, the

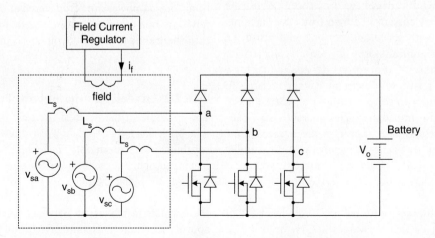

FIGURE 25.14 Lundell alternator with a switched-mode rectifier.

switched-mode rectifier can be employed to achieve greatly improved load-dump transient control.

25.6.3 Alternative Machines and Power Electronics

The demand for increased alternator power levels, efficiency, and performance also motivates the consideration of alternative electrical machines, power electronics, and design approaches. While no alternative machine has yet displaced the Lundell alternator in production vehicles, primarily due to cost considerations, some potential candidates are reviewed in this section. These include machines that are mounted directly on the engine rather than driven from a belt. These direct-driven machines become important as power levels rise. This section also addresses the more general case of the combined starter/alternators. While the use of a single machine to do both starting and generation functions is clearly possible, a separate (transient-rated) dc machine is presently used for starting. This is because the large mismatch in starting and generating requirements has made the combined starter/alternator approach unattractive. However, as alternator power ratings increase, the mismatch is reduced, and a single starter/alternator system becomes more practical. A combined system has the potential to eliminate the need for a separate flywheel, starter, solenoid switch, and pinion engaging drive. It also has the potential to allow regenerative braking and "light hybrid" operation, and to provide idle-stop capability (i.e. the ability to turn off the engine when the vehicle is stopped and seamlessly restart when the vehicle needs to move) for reduced fuel consumption. A move to this more sophisticated approach relies upon advanced electrical machines and power electronics.

25.6.3.1 Synchronous Machine with a Cylindrical Wound Rotor

The claw-pole rotor can be replaced with a cylindrical rotor to achieve better coupling between the stator and rotor. The cylindrical rotor is made from steel laminations and the field winding is placed in the rotor slots. The cylindrical rotor is similar to the armature of a dc machine except that the connection of the field winding to the external circuit is made through slip rings instead of a commutator. The cylindrical rotor structure leads to quiet operation and increased output power and efficiency. Unlike the claw-pole alternator, the length of the machine can be increased to get higher output power at a higher efficiency. The efficiency is higher since the effect of the end windings on the machine performance is less in a machine with a long length. It is also possible to build the machine with a salient-pole rotor instead of a cylindrical rotor. However, a machine with a salient-pole rotor is likely to produce more noise than a machine with a cylindrical rotor.

A machine with a cylindrical wound rotor has similar power electronics and control options as a claw-pole machine. If generation-only operation is required, a diode bridge and field current control is sufficient to regulate the output voltage. Better performance can be achieved by using a switched-mode rectifier in conjunction with field control [37]. If motoring operation is desired (e.g. for starting), or even better performance is desired, a full-bridge (active-switch) converter can be used, as shown in Fig. 25.15. Since this is a synchronous machine, some form of rotor position sensing or estimation is typically necessary. The full-bridge converter allows maximum performance and flexibility but carries a significant cost penalty.

25.6.3.2 Induction Machine

The stator of a three-phase induction machine is similar to that of a three-phase synchronous machine. The rotor is either a squirrel-cage or wound rotor. The machine with the squirrel-cage rotor is simpler in construction and more robust than the machine with a wound rotor in which the three-phase rotor winding is brought outside the rotor through slip rings. The rotor is cylindrical and is constructed from steel laminations. It is also possible to use a solid rotor instead of a laminated rotor. However, a solid rotor leads to higher losses as compared with a laminated rotor. The losses in a solid rotor can be

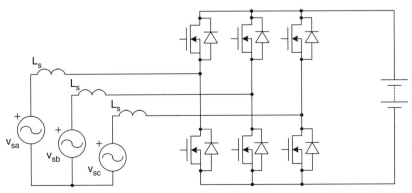

FIGURE 25.15 Model of an alternator with full-bridge converter.

reduced by cutting slots in the rotor surface, filling the stator slot openings with magnetic wedges to reduce the field ripple, and placing a copper cage on the rotor.

An induction machine requires a source that can provide the leading reactive power to magnetize the airgap. This means that a three-phase induction generator cannot supply power to a load through a three-phase diode bridge. Capacitor supply of the reactive energy is impractical because of the wide operating speed range. In the most general case (in which both motoring and generating operation can be achieved) a three-phase active bridge can be used. If only generating operation is desired, the power to the load can be supplied through a three-phase diode bridge and the reactive power can be obtained from a small three-phase active bridge provided for this purpose. This design requires a large number of devices and complex control.

25.6.3.3 Reluctance Machines

The switched reluctance machine is a doubly salient machine. Both the stator and rotor of the machine are made from steel laminations to reduce the iron losses. Only the stator carries windings; the rotor is constructed of steel laminations with a salient shape. The structure of a three-phase switched reluctance machine with six stator poles and four rotor poles is shown in Fig. 25.16a. A winding placed on diametrically opposite stator poles forms a phase winding. When a phase of the machine is excited, a pair of rotor poles tends to align with the excited stator poles to provide a path of minimum reluctance. If the rotor is moving towards alignment with the excited pair of stator poles, then the machine develops a positive torque and acts as a motor. If the rotor is moving away from the excited pair of stator poles, then the machine develops a negative torque and acts as a generator. The advantages of the switched reluctance machine include simple construction, fault-tolerant power electronic circuit, high reliability, unidirectional phase currents, and low cost. The drawbacks of the machine include high levels of torque ripple, vibration and acoustic noise, and a relatively high power electronics cost.

The synchronous reluctance machine is a singly salient machine. The stator of the machine is similar to that of a synchronous or induction machine. The rotor has a segmented structure with each segment consisting of a stack of axially laminated steel sheets sandwiched with a non-magnetic material. The structure of a four-pole synchronous reluctance machine is shown in Fig. 25.16b. A synchronous reluctance machine has less torque ripple, lower losses, and higher power density than a comparable switched reluctance machine. Inclusion of permanent magnets in the rotor structure allows both reluctance and magnet torque to be achieved. Such interior permanent magnet (IPM) machines can achieve very high performance and power density. When permanent magnets are included, however, careful attention must be paid to the effects of shutdown of the power electronics as an uncontrolled back-emf component will exist in this case [41].

The switched reluctance machine, like the induction machine, requires an external source to magnetize the airgap. Several circuits are available to excite the switched reluctance machine. A circuit that is suitable for the automotive application of this machine is shown in Fig. 25.17. A phase leg is needed for each stator phase of the machine. In this case, the switched reluctance machine obtains its excitation from the same bus that it generates into. Unlike the synchronous and induction machines in which the number of wires needed to connect the machines to the power converters is usually equal to the number of phases, the number of wires needed to connect the switched reluctance machine to a converter is equal to twice the number of phases. This is of no particular concern in a switched reluctance machine in which the power converter is integrated with the machine in the same housing. The synchronous reluctance machine also requires an external source to magnetize the airgap. The machine usually employs an active bridge similar to the one used with an induction machine for the desired power conversion. The machine can also employ the converters used with the switched reluctance machine. In this case, the currents through the stator windings are unidirectional. The relative complexity of the power electronics is a disadvantage of these machine types in the case where only generator operation is necessary.

25.6.3.4 Permanent Magnet and Hybrid Synchronous Machines

The permanent magnet synchronous machine designed with high-energy rare-earth magnets operates with high efficiency, high power density, low rotor inertia, and low acoustic noise. The excitation from the permanent magnets is fixed and, therefore, the regulation of the output voltage of the machine is not as straightforward as in a synchronous machine with a wound rotor. For generator operation, machines of this type can use switched-mode rectifiers to regulate the output voltage [42, 43]. The boost rectifier of Fig. 25.14 is one possible implementation of this approach. Alternatively, a diode rectifier followed by a dc/dc converter can be used to regulate the generator system output [44]. Another method proposed

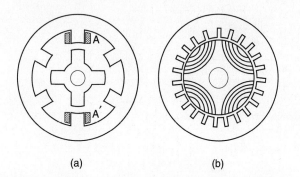

FIGURE 25.16 Structures of: (a) switched reluctance and (b) synchronous reluctance machines. AA′ represents phase A winding.

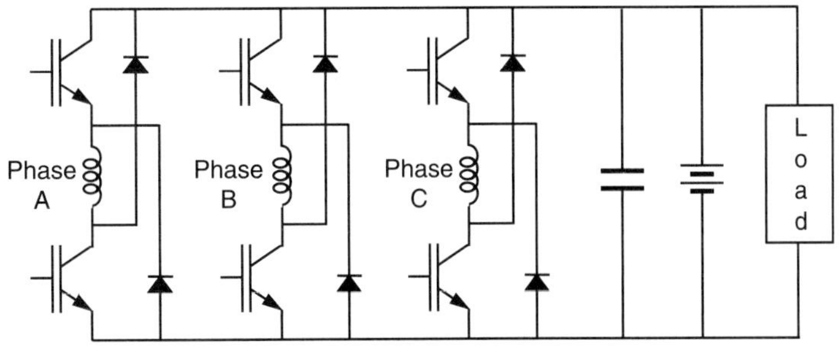

FIGURE 25.17 Circuit for a switched reluctance machine.

for this type of system involves the use of tapped windings and two three-phase SCR bridges [45]. The taps on the phase windings are connected to one bridge, while full phase windings are connected to the other bridge. The bridge connected to the full phase windings is used to supply power to the dc bus at low engine speeds, while the converter connected to the taps is used at high speed. The use of a tapped winding and dual bridges helps the system cope with the wide speed range of the alternator and limit the losses associated with the pulsating output currents. In the case when both motoring and generating modes are desired, a full-bridge converter can be used. Again, as this is a synchronous machine, some form of position sensing or estimation is necessary. Also, in all of these systems the effects of failure of the power electronics must be carefully considered as there is no possibility of regulating the back voltages by field control.

Attempts to develop a simpler voltage regulation scheme for permanent magnet synchronous machines have led to a permanent magnet/wound-rotor hybrid synchronous machine in which the rotor consists of two parts: a part with permanent magnets and a part with a field winding [46]. The two parts are placed next to each other on a common shaft. The rotor with the field winding can employ claw-pole, salient-pole, or cylindrical structure. The field current generates a flux that is used to either aid or oppose the permanent magnet flux and regulate the output voltage of the machine. One possible failure mode of this approach that can lead to catastrophic failure is if the field winding breaks while the machine is operating at high speed. In this case, the generated output voltage will become large and uncontrolled. Some means of mechanically disconnecting the alternator at the input or electrically disconnecting it at the output may be necessary to limit the impact of this failure mode.

25.6.3.5 Axial-airgap Machines

The principle of operation of an axial-airgap, or axial-flux, machine is the same as that of a radial-airgap machine. An axial-airgap machine is characterized by a short axial length and large diameter. The structure of an axial-airgap permanent magnet machine with surface magnets is shown in

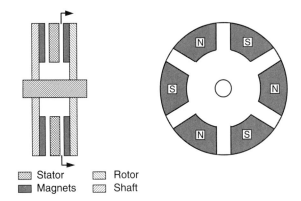

FIGURE 25.18 Structure of an axial-airgap permanent magnet machine.

Fig. 25.18 [47]. The stator of the machine can be slotless or slotted. Two different magnetic circuit configurations are possible. In the NN configuration, the magnetic polarities in one pole pitch on both sides of stator are the same so that there are two main fluxes with symmetrical distribution through the stator. In this case, the conductors can be wound into two back-to-back stator slots to make one coil. The machine has a large stator yoke dimension because the flux passes through the yoke, but less copper loss because of short end windings. In the NS configuration, the magnetic polarities in one pole pitch on the opposite sides of stator are the opposite of each other so that there is only one main axial flux through the stator. In this case, the stator yoke dimension is small, but the end windings are long because the direction of current in the back-to-back stator slots is the same. The iron losses are small due to small yoke dimension and the copper losses are high because of long end windings. Heat removal is more challenging due to small stator dimensions. The structure shown in Fig. 25.18 is that of an axial-airgap permanent magnet machine with surface magnets. In an axial-airgap machine with interior permanent magnets, the magnets are embedded in the steel of the rotor.

The axial airgap versions of other types of machines, such as the induction and switched reluctance machines, are also possible. The structure of an axial-airgap induction machine is similar to that of an axial-airgap permanent magnet machine except that windings are used instead of permanent magnets.

25.7 Dual/High Voltage Automotive Electrical Systems

The electrical system of a 1920s internal combustion engine (ICE) automobile had only a few loads: a starter, an ignition device, a horn, and some lamps [48]. The mean power consumption of these loads was less than 100 W. An engine-driven dc generator charged a 6 V lead-acid battery that provided electrical power. The power was distributed via point-to-point wiring, with most loads controlled directly by manually operated primary switches located within the reach of the driver. Only the starter was switched indirectly by an electromechanical relay. After the Second World War, the automotive electrical system started to grow rapidly in complexity and power consumption as additional features, including radios, multispeed windshield wipers, and power windows, were added. The introduction of higher compression engines stretched the 6-V system beyond its technological limits. The 8.5 to 1 compression ratio engines required 100–200% greater ignition voltages than the 6.4 to 1 engines. As a result, the primary side current of the ignition coil was doubled or tripled and the life of the distributor contacts was reduced to an unacceptable level. To overcome this problem, the battery voltage was increased to 12 V in the mid-1950s [49, 50].

Over the past four decades, the electrical power requirements of automobiles have increased even more rapidly. From a mere 400 W in 1955, the power rating of a luxury vehicle's generator has increased to over 1800 W [51, 52]. However, the electrical system of a modern automobile is architecturally identical to the 12-V point-to-point system of the 1950s. The only changes that have taken place have been at the component level, such as the replacement of the dc generator by a three-phase alternator-rectifier, the replacement of wound-field dc motors by permanent magnet ones, and an increased use of relays. The rapid growth in the electrical system is expected to continue due to environmental regulations, consumer demand for increased functionality, safety, security and comfort, and replacement of some mechanical actuators by electrical counterparts. The average electrical power requirement of a modern luxury vehicle is about 800 W. With the addition of such loads as electric power steering, engine-cooling fan, water pump, and electromechanical engine valves, the average power requirement could increase to 2.5 kW by 2005 [53]. The traditional solution of increasing the size of the alternator and the battery is not practical due to space limitations and fuel efficiency requirements. Furthermore, the peak power requirements of some of the anticipated loads – heated windshield (2.5 kW), heated catalyst (3 kW), electromechanical engine valves (2.4 kW at 3000 rpm), and active suspension (12 kW) – cannot be met economically using the present architecture. These factors have motivated the development of new dual/high voltage electrical architectures that incorporate a higher-voltage bus in addition to the standard 14 V bus [39, 54–56]. A dual/high voltage approach allows an efficient supply of power to many loads which benefit from operating at a higher voltage, while retaining the 14 V bus for loads (such as lamps and electronics) which do not benefit from a higher voltage. High-voltage architectures that do not retain the 14 V bus are also possible, but will require a substantial investment in the design and production of new high-voltage components. This section describes some of the characteristics and preliminary specifications of the new dual/high voltage electrical system architectures. It also discusses some of the widely considered implementation approaches.

25.7.1 Trends Driving System Evolution

The conventional 12-V automotive electrical power system has many defects, including a widely varying steady-state system voltage and large transients, which force the electrical functions to be over-designed. However, these limitations alone have not been a strong enough driver for automotive companies to seriously evaluate advanced alternatives. Now a number of new factors are changing this situation. The most important of these are future load requirements that cannot be met by the present 12-V architecture.

25.7.1.1 Future Load Requirements

Table 25.6 gives a list of electrical loads expected to be introduced into automobiles in the next ten years [53]. Some of these loads (electrohydraulic power steering, electric engine fan, electric water pump, and electromechanical valves) will replace existing mechanically or hydraulically driven loads. The remaining are new loads introduced to either meet government mandates or satisfy customer needs.

TABLE 25.6 Electrical loads expected to be introduced into automobiles in the next decade [53]

Load	Peak power (W)	Average power (W)
Exhaust air pump	300	10
Electrohydraulic power steering	1000	150
Electric engine fan	800	150
Heated catalytic converter	3000	90
Electric water pump	300	150
Heated windshield	2500	120
Electromechanical engine valves (6 cylinders at 6000 rpm)	2400	800
Active suspension	12,000	360
Total		1830

The average electrical power requirement of a present-day automobile is in the range of 500–900 W depending on whether it is an entry-level or a luxury vehicle. When the loads of Table 25.6 are introduced, the average electrical power requirement will increase by 1.8 kW. Furthermore, if the air-conditioning (A/C) pump were to ever become electrically driven, the peak and average power demands would increase by an additional 3.5 kW and 1.5 kW, respectively. Distributing such high power at a relatively low voltage will result in unacceptably bulky wiring harnesses and large distribution losses. Since the alternator has to generate both the power consumed by the loads and the power dissipated in the distribution network, its output rating (and hence size and power consumption) will be greater than in an architecture with lower distribution losses. With the large premium attached to the size of the alternator (due to space constraints in the engine compartment), an architectural change in the distribution and generation systems is essential before many of the future loads can be introduced.

There is also an increasing disparity in the voltage requirements of future electrical loads. High pulse-power loads, such as the heated windshield and electrically heated catalytic converter, become feasible only at voltages greater than the current 14 V [57]. On the other hand, incandescent lamps and electronic control units (ECUs) will continue to require low voltages. For example, present day ECUs have linear regulators which convert the 14 V distribution voltage to the 5 V needed by the integrated circuits. The efficiency of these regulators is equal to the ratio of output to input voltage, i.e. 35%. Furthermore, the next generation of higher speed lower power consumption integrated circuits operate at 3.3 V, making the regulators more inefficient. This inefficiency also means that larger heat sinks are required to remove the heat from the ECUs.

25.7.1.2 Higher Fuel Efficiency

A secondary motivating factor for the introduction of a higher system voltage is the challenge of achieving higher fuel economy. The average fuel economy of present-day automobiles in the United States is in the vicinity of 30 miles per gallon (mpg). There is little market incentive for automobile manufacturers to increase the fuel economy of vehicles for the US market where the price of fuel is relatively low. The price of gasoline in the US ($1.70 per gallon) is less than the price of bottled water ($4.00 per gallon when bought by the quart). Although market forces have not been a driver for the development of fuel-efficient vehicles, a number of new incentives have emerged over the past few years. One of these is the fine imposed on the automakers by the US government if the average fuel economy of their fleet falls below the mandated standard. The mandated standard for cars has increased from 24 mpg in 1982 to its 1997 level of 27.5 mpg, and will continue to increase. In Europe, the German Automotive Industry Association (VDA) plans to increase the average fleet fuel efficiency to 39.9 mpg by 2005 – compared to 31.4 mpg in 1990 [58].

Another driver behind the development of fuel-efficient vehicles is the partnership for a new generation of vehicles (PNGV). This ten-year research program, launched in September 1993, is a collaboration between the US Federal Government and the big three US automakers (General Motors, Ford, and DaimlerChrysler) that aims to strengthen national competitiveness in the automotive industry and reduce dependence on foreign oil. The PNGV has set a goal to develop an 80 mpg midsize vehicle by 2004 [59]. The German Automotive Industry Association is pursuing similar targets. The VDA has undertaken a pledge to introduce a 3 L/100 km (78 mpg) vehicle by the year 2000. This is complemented by the introduction of highly fuel-efficient vehicles (in excess of 50 mpg) in both the Japanese and American markets.

With the present alternator, 800 W of electrical power consumes 1.33 L of gasoline for every 100 km driven when the vehicle has an average speed of 33.7 km/h. This represents a 45% increase in fuel consumption for a 3 L/100 km vehicle. Hence, if future high fuel economy vehicles are going to have comfort, convenience, and safety features comparable to present-day vehicles, the efficiency of the electrical generation and distribution system will have to be substantially improved. Furthermore, as discussed in Section 25.8, one widely considered means of achieving high fuel economy is the use of a hybrid vehicle architecture. In practice, this approach necessitates the introduction of a higher voltage in the vehicle.

25.7.2 Voltage Specifications

A major issue when implementing a high or dual voltage system is the nominal voltage of the high-voltage bus, and the operating limits of both buses. While there are many possibilities, there is a growing consensus in the automotive industry for a nominal voltage of 42 V for the high-voltage bus (corresponding to a 36 V lead-acid storage battery) [39, 56, 60]. This voltage is gaining acceptance because it is as high as possible while remaining within acceptable safety limits for open wiring systems (once headroom is added for transients) and it provides substantial benefits in the power semiconductors and wiring harness [61]. Furthermore, this voltage is sufficient to implement starter/alternator systems and "light" hybrid vehicle designs [62, 63]. While no vehicles equipped at 42 V are in production at present, availability of 42 V components is rapidly increasing and 42 V equipped vehicles may be expected early in this decade.

The permissible static and transient voltage ranges in an electrical system are important design considerations for power electronic equipment. At present, no universally accepted specification exists for high or dual voltage automotive electrical systems. However, the preliminary specifications proposed by the European automotive working group, *Forum Bordnetz*, are under wide consideration by the

TABLE 25.7 Voltage limits for 14 and 42 V buses proposed in [56]

Voltage	Description	Value
$V_{42,OV-dyn}$	Maximum dynamic overvoltage on 42 V bus during fault conditions	55 V
$V_{42,OV-stat}$	Maximum static overvoltage on 42 V bus	52 V
$V_{42,E-max}$	Maximum operating voltage of 42 V bus while engine is running	43 V
$V_{42,E-nom}$	Nominal operating voltage of 42 V bus while engine is running	41.4 V
$V_{42,E-min}$	Minimum operating voltage of 42 V bus while engine is running	33 V
$V_{42,OP-min}$	Minimum operating, voltage on the 42 V bus. Also, lower limit operating voltage for all non-critical loads (i.e. loads not required for starting and safety)	33 V
$V_{42,FS}$	Failsafe minimum voltage: lower limit on operating voltage for all loads critical to starting and safety on the 42 V bus	25 V
$V_{14,OV-dyn}$	Maximum dynamic overvoltage on 14 V bus during fault conditions	20 V
$V_{14,OV-stat}$	Maximum static overvoltage on 14 V bus	16 V
$V_{14,E-max}$	Maximum operating voltage of 14 V bus while engine is running	14.3 V
$V_{14,E-nom}$	Nominal operating voltage of 14 V bus while engine is running	13.8 V
$V_{14,E-min}$	Minimum operating voltage of 14 V bus while engine is running	12 V
$V_{14,OP-min}$	Minimum operating voltage of 14 V bus. Also lower limit operating voltage for all non-critical loads	11 V
$V_{14,FS}$	Failsafe minimum voltage: lower limit on operating voltage for all critical loads on the 14 V bus	9 V

automotive industry [56]. These specifications, summarized in Table 25.7, impose tight static and transient limits on both the 42 and 14 V buses. The upper voltage limit on the 14 V bus is far lower than in the conventional 12-V system. The allowed upper limit on the 42 V bus is also proportionally tight. These strict limits facilitate the use of power semiconductor devices such as power MOSFETs and lower the cost of the protection circuitry needed in individual functions. However, they also require much more sophisticated means for limiting transients (such as load dump) than is found in conventional systems, which imposes a significant cost. Appropriate voltage range specifications for dual/high voltage electrical systems are thus a subject of ongoing investigation by vehicle manufacturers, and will likely continue to evolve for some time.

25.7.3 Dual-voltage Architectures

Conventional automotive electrical systems have a single alternator and battery. Dual-voltage electrical systems have two voltage buses and typically two batteries. Single-battery configurations are possible, but tend to be less cost effective [61]. A variety of different methods for generating and supplying energy to the two buses are under investigation in the automotive community. Many of these have power electronic circuits at their core. This section describes three dual-voltage electrical system architectures that have received broad attention. In all three cases the loads are assumed to be partitioned between the two buses with the starter and many of the other high-power loads on the 42 V bus and most of the lamps and electronics on the 14 V bus.

The dc/dc converter-based implementation of Fig. 25.19 is perhaps the most widely considered dual-voltage architecture.

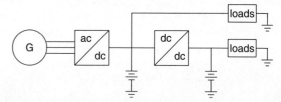

FIGURE 25.19 Dual-voltage architecture based on a dc/dc converter.

In this implementation, an alternator and associated battery provide energy to one bus (typically the 42 V bus), while the other bus is supplied via a dc/dc converter. If a battery is used at the dc/dc converter output, the converter needs to be rated for slightly above average power. Otherwise, the converter needs to be rated a factor of two to three higher to meet peak power requirements [61]. The architecture of Fig. 25.19 has a number of advantages. The dc/dc converter provides high-bandwidth control of energy flow between the two buses, thus enabling better transient control on the 14 V bus than is available in present-day systems or in most other dual-voltage architectures. Furthermore, in systems with batteries on both buses, the dc/dc converter can be used to implement an energy management system so that generated energy is always put to best use. If the converter is bidirectional it can even be used to recharge the high-voltage (starter) battery from the low-voltage battery, thus providing a *self jump start* capability. The major challenge presented by this architecture is the implementation of dc/dc converters having the proper functionality within the tight cost constraints dictated by the automotive industry. Some aspects of design and optimization of converters for this application are addressed in [64].

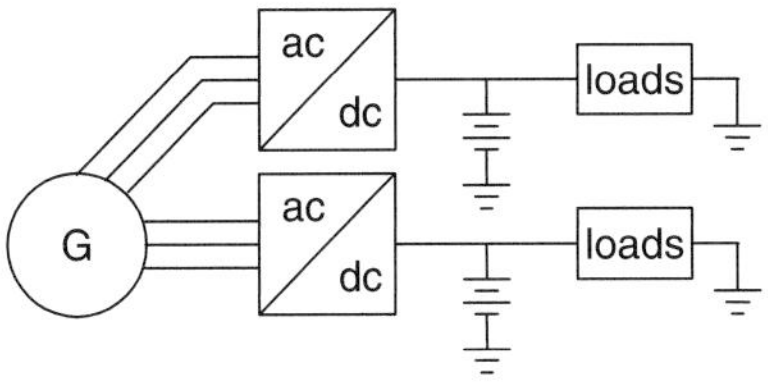

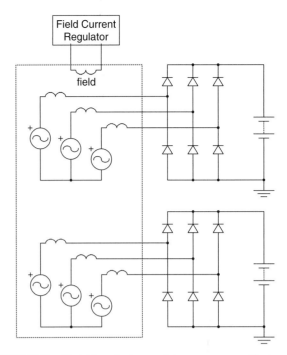

FIGURE 25.20 Dual-voltage architecture based on a dual-wound alternator.

The dual-stator alternator architecture of Fig. 25.20 is also often considered for dual-voltage automotive electrical systems [65, 66]. In this case, an alternator with two armature windings is used along with two rectifiers to provide energy to the buses and their respective batteries. Control of the bus voltages is achieved via a combination of controlled rectification and field control. Typically, field control is used to regulate one output, while the other output is regulated using a controlled rectifier. Figure 25.21 shows one possible implementation of this architecture. It should be noted that to achieve sufficient output power and power steering from the dual-wound alternator, the winding ratio between the two outputs must be carefully selected. For 42/14 V systems, a winding ratio of 2.5:1 is typical [66]. Advantages of this electrical architecture include low cost. However, it does not provide the bidirectional energy control that is possible in the dc/dc converter architecture. Furthermore, there are substantial issues of cross-regulation and transient control with this architecture that remain to be fully explored.

In a third architecture, a single-output alternator with a dual-output rectifier is employed. This approach is shown schematically in Fig. 25.22. As with the dual-stator alternator configuration, this architecture has the potential for low cost. One widely considered implementation of the dual-rectified alternator is shown in Fig. 25.23 [65, 67–69]. Despite its simplicity, this implementation approach provides less functionality than the dc/dc converter-based architecture, generates substantial low-frequency ripple which must be filtered, and has serious output power and control limitations [66]. An alternative implementation, proposed in [37] and shown in Fig. 25.24, seems to overcome these limitations, and may potentially provide the same capabilities as the dc/dc converter-based architecture at lower cost. Clearly, this architecture has promise for dual-voltage electrical systems, but remains to be fully explored.

25.8 Electric and Hybrid Electric Vehicles

Battery-powered electric vehicles were first introduced over one hundred years ago, and continue to incite great public interest because they do not generate tailpipe emissions.

FIGURE 25.21 Model for a dual-wound alternator system. The two output voltages are regulated through field control and phase control. For a 42/14 V system, a winding ratio between the two stator windings of 2.5:1 is typical.

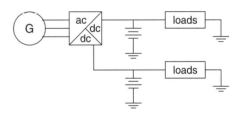

FIGURE 25.22 Dual-voltage architecture based on a dual-rectified alternator.

Nevertheless, the low energy storage density and the high cost of suitable batteries makes pure electric vehicles non-competitive with internal combustion engine vehicles in most applications. An alternative approach that is generating widespread attention is the hybrid electric vehicle (HEV). An HEV combines electrical propulsion with another energy source, such as an internal combustion engine, allowing the traditional range and performance limitations of pure electric vehicles to be overcome [70]. Alternative energy sources, such as fuel cells, are also possible in place of an internal combustion engine.

Hybrid electric vehicles can be classified as having either a parallel or series driveline configuration [71]. In a series HEV all of the propulsion force is produced from electricity; the engine is only used to drive a generator to produce electricity. In a parallel hybrid, propulsive force can come from either the

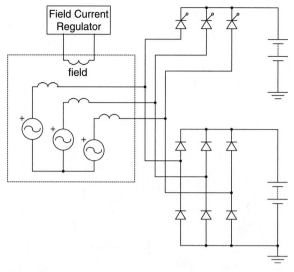

FIGURE 25.23 A dual-rectified alternator with a phase-controlled rectifier.

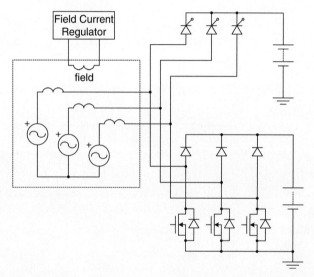

FIGURE 25.24 A dual-rectified alternator with a switched-mode rectifier.

engine or the electrical drive. In both cases, batteries or other electrical storage devices are used to buffer the instantaneous difference between the power needed for propulsion and that generated by the engine. The selection of a series or parallel driveline depends heavily on the performance requirements and mission of the vehicle.

In a series HEV, all power delivered to the wheels of the vehicle must be delivered through the electrical driveline. The electrical driveline components, including the batteries, power electronics, and machine(s), must all be rated for the *peak* traction power requirements, making these components relatively large and expensive if performance (e.g. acceleration) comparable to a conventional vehicle is to be achieved. To achieve the required power levels, the electrical driveline must operate at hundreds of volts, necessitating the electrical subsystem to be sealed from access by the user. The engine, on the other hand, need only be rated to deliver the *average* power required by the vehicle, which is much lower. In a system that does not require utility recharge of the batteries (i.e. can drive indefinitely on fuel alone), the engine size is set by the power requirements of the vehicle at maximum cruising speed. If utility recharge of the batteries and a battery-limited driving range is acceptable, engine power requirements can be reduced even further. Because the engine does not provide tractive power, it can be designed to run at a single optimized condition, thus maximizing engine efficiency and minimizing emissions. Furthermore, the need for a transmission is eliminated and there is a great deal of flexibility in the engine placement.

In a parallel HEV, traction power is split between the engine and the electrical driveline. One possible approach is to utilize a single machine mounted on the engine crankshaft to provide starting capability along with electrical traction power and regeneration [72–75]. This approach can be replaced or complemented with other approaches, such as use of a power-splitting device such as a planetary gear set [70, 76], or using different propulsion and generation techniques on different sets of wheels [71, 77, 78]. In all parallel hybrid approaches, some form of transmission is needed to limit the required speed range of the engine. A wide range of divisions between engine size and electrical system size is possible in the parallel hybrid case, depending on structure. Depending on this split, the necessary electrical driveline system voltage may be as low as 42 V (which is safe for an open wiring system) or as high as 300 V. Also because the electrical subsystem, the internal combustion engine subsystem, or both may provide tractive power under different conditions, there exists a wide range of possible operating approaches for a parallel hybrid system. Consequently, the control strategy for a parallel hybrid tends to be substantially more complex than for a series hybrid.

One parallel hybrid approach that is receiving a lot of attention for near-term vehicles is a "light" or "mild" hybrid. In this case, a somewhat conventional vehicle driveline is complemented with a relatively small starter/alternator machine mounted on the crankshaft [62, 63, 72–75, 79]. The electrical drive power is typically below 10 kW average and 20 kW peak. The starter/alternator can be used to provide rapid, clean restart of the vehicle so that the engine can be turned off at idling conditions and seamlessly restarted. This so-called "stop and go" operation of the engine is valuable for fuel economy and emissions. The starter/alternator can also be used to implement regenerative braking, to provide engine torque smoothing (replacing the flywheel and allowing different engine configurations to be used) and to provide boost power for short-term acceleration. At the low-power end, such systems can be integrated directly into the open wiring configuration of a 42 V electrical system, simplifying the vehicle electrical architecture. System-level control remains a major challenge in realizing the full benefits of such systems. Starter/alternator-based hybrids are expected to be a significant near-term application of power electronics and machines in automobiles.

25.9 Summary

Power electronics is playing an increasingly important role in automobiles. It is being used to enhance the output power capability and efficiency of the electrical power generation components. Power electronics is also an enabling technology for a wide range of new and improved functions that enhance vehicle performance, safety, and functionality.

The design of automotive power electronic systems is strongly influenced by the challenging electrical and environmental conditions found in automobiles. Important factors include the static and transient voltage ranges, electromagnetic interference and compatibility requirements, and temperature and other environmental conditions. Some of the most important design considerations for automotive power electronics were addressed in Section 25.3.

Section 25.4 described some of the vehicle functions that benefit from, or are enabled by, power electronics. These functions range from lighting to actuation and steering. Power electronic switches also play a central role in multiplexed electrical distribution systems. This role of power electronics was addressed in Section 25.5.

The rapid increase in electrical power demand in automobiles is motivating the introduction of new technologies for electrical power generation and control. Lundell alternators are presently used for power generation in automobiles, but are rapidly reaching their power limits. Section 25.6 reviewed the operating characteristics of the Lundell alternator. It also described several techniques for extending the power capabilities of this machine. To meet the growing demand for electrical power, alternative machine, and power electronic configurations may be necessary in the future. A number of candidate machine and power circuit configurations were reviewed in Section 25.6. Such configurations can also be applied towards the design of integrated starter/alternators and hybrid propulsion systems, as was discussed in Section 25.8.

The increasing electrical and electronic content of automobiles is beginning to stretch the capabilities of the conventional 12-V electrical system. Furthermore, there is a desire on the part of vehicle manufacturers to introduce new high-power loads, such as electromechanical engine valves, active suspension, and integrated starter/alternator. These are not likely to be practical within the present 12-V framework. These challenges are forcing the automotive industry to seriously consider high and dual voltage electrical systems. The ongoing developments in this area were reviewed in Section 25.7.

The increasing electrical content of vehicles both underscores the need for power electronics and reflects the benefits of their introduction. It is safe to say that power electronics will continue to play an important role in the evolution of automobiles far into the future.

References

1. *Automotive Electric/Electronic Systems*, 2nd Edition, Robert Bosch GmbH, Stuttgart, Germany, 1995.
2. T. Denton, *Automotive Electrical and Electronic Systems*, Arnold, London, UK, 1995.
3. SAE Electronic Systems Committee, "Recommended Environmental Practices for Electronic Equipment Design," *SAE Recommended Practice SAE J1211*, November, 1978.
4. *SAE Handbook*, Society of Automotive Engineers, Warrendale, PA, USA, 1999.
5. J. Alkalay, et al., "Survey of Conducted Transients in the Electrical System of a Passenger Automobile," *IEEE Symposium on Electromagnetic Compatability*, pp. 271–278, May, 1989.
6. SAE EMI Standards Committee, "Immunity to Conducted Transients on Power Leads," *SAE Standard SAE J1113/11*, June, 1995.
7. J.G Kassakian, M.F. Schlecht, and G.C. Verghese, *Principles of Power Electronics*, Addison-Wesley, New York, 1991.
8. V. Caliskan, D.J. Perreault, T.M. Jahns, and J.G. Kassakian, "Analysis of Three-Phase Rectifiers with Constant-Voltage Loads," *IEEE Power Electronics Specialists Conference*, pp. 715–720, Charleston, SC, USA, June, 1999.
9. SAE EMI Standards Committee, "Limits and Methods of Measurement of Radio Disturbance Characteristics of Components and Modules for the Protection of Receivers Used On Board Vehicles," *SAE Standard SAE J1113/41*, July, 1995.
10. SAE EMI Test Methods and Standards Committee, "Electromagnetic Compatability – Component Test Procedure – Part 42–Conducted Transient Emissions," *SAE Standard SAE J1113/42*, July, 1994.
11. T.K. Phelps and W.S. Tate, "Optimizing Passive Input Filter Design," *Proceedings of Powercon 6*, pp. G1-1–G1-10, May, 1979.
12. M.J. Nave, *Power Line Filter Design for Switched-Mode Power Supplies*, Van Nostrand Reinhold, New York, 1991.
13. R.D. Middlebrook, "Input Filter Considerations in Design and Application of Switching Regulators," *IEEE Industry Applications Society Annual Meeting*, 1976.
14. H.W. Ott, *Noise Reduction Techniques in Electronic Systems*, John Wiley, New York, 1988.
15. *Automotive Handbook*, 3rd Edition, Robert Bosch GmbH, Stuttgart, Germany, 1993.
16. R.K. Jurgen, Editor, *Automotive Electronics Handbook*, 2nd Edition, McGraw Hill, New York, 1999.
17. F.S. Schwartz, W. Hendrischk, and J. Jiao, "Intelligent Automotive Lighting," *Proceedings of International Congress on Transportation Electronics*, pp. 299–303, Dearborn, MI, USA, October, 1994.
18. T. Yamamoto and T. Futami, "Development of 2-Lamp type HID Headlighting System," SAE paper 900562, *Proceedings of the SAE International Congress and Exposition*, pp. 1–6, Detroit, MI, USA, February, 1990.
19. B. Woerner and R. Neumann, "Motor Vehicle Lighting Systems with High Intensity Discharge Lamps," SAE paper 900569, *Proceedings of the SAE International Congress and Exposition*, pp. 87–94, Detroit, MI, USA, February, 1990.

20. L. So, P. Miller, and M. O'Hara, "42V-PWM – Lighting the Way in the New Millenium," SAE paper 2000-01-3053, *Future Transportation Technology Conference*, Costa Mesa, CA, USA, 2000.
21. H.P. Schoner, "Piezo-Electric Motors and their Applications," *Ferroelectrics*, vol. 133, pp. 27–34, 1992.
22. J. Wallaschek, "Piezoelectric Ultrasonic Motors," *Journal of Intelligent Material Systems and Structures*, vol. 6, pp. 71–83, January, 1995.
23. T. Sashida and T. Kenjo, *An Introduction to Ultrasonic Motors*, Oxford University Press, New York, 1993.
24. T. Sashida, Japanese Patent No. 58-148682, February, 1982.
25. M.A. Theobald, B. Lequesne, and R. Henry, "Control of Engine Load via Electromagnetic Valve Actuators," *International Congress and Exposition*, SAE Paper 920447, Detroit, MI, USA, February, 1994.
26. J.L. Oldenkamp and D.M. Erdman, "Automotive Electrically Driven Air Conditioner System," *Proceedings of the IEEE Workshop on Automotive Power Electronics*, pp. 71–72, 1989.
27. R.J. Valentine, "Electric Steering Power Electronics," *IEEE Workshop on Power Electronics in Transportation*, pp. 105–110, Dearborn MI, USA, October, 1996.
28. J.F. Ziomek, "One-Wire Automotive Electrical Systems," SAE Paper 730132, *Proceedings of the SAE International Automotive Engineering Congress*, Detroit, MI, USA, January, 1973.
29. M. Leonard, "Multiplexed Buses Unravel Auto Wiring," *Electronic Design*, pp. 83–90, August 8, 1991.
30. *Class B Data Communication Network Interface*, SAE J1850, Rev. Aug 91, Society of Automotive Engineers, Warrendale, PA, USA, 1992.
31. CAN Specification, Robert Bosch GmbH, Stuttgart, Germany, 1991.
32. R. Frank, "Replacing Relays with Semiconductor Devices in Automotive Applications," SAE Paper 880177, pp. 47–53, 1988.
33. A. Marshall and K.G. Buss, "Automotive Semiconductor Switch Technologies," *Proceedings of IEEE Workshop on Electronic Applications in Transportation*, pp. 68–72, Dearborn, MI, USA, October, 1990.
34. F. Krieg, "Automotive Relays," *Automotive Engineering*, pp. 21–23, March, 1993.
35. K. Furuichi, K. Ishida, K. Enomoto, and K. Akashi, "An Implementation of Class A Multiplex Application," SAE Paper 920230, *SAE International Congress and Exposition, Multiplex Technology Applications to Vehicle Wire Harnesses*, pp. 125–133, Detroit, MI, USA, February, 1992.
36. S. Kuppers and G. Henneberger, "Numerical Procedures for the Calculation and Design of Automotive Alternators," *IEEE Transactions on Magnetics*, vol. 33, no. 2, pp. 2022–2025, March, 1997.
37. D. Perreault and V. Caliskan, "A New Design for Automotive Alternators," SAE Paper 2000-01-C084, *IEEE-SAE International Conference on Transportation Electronics (Convergence)*, Dearborn, MI, USA, October, 2000.
38. G. Henneberger and S. Kuppers, "Improvement of the Output Performance of Claw-Pole Alternators by Additional Permanent Magnets," *Proceedings of the International Conference on Electrical Machines*, vol. 2, pp. 472–476, 1994.
39. J. M. Miller, D. Goel, D. Kaminski, H.-P. Shoener, and T.M. Jahns, "Making the Case for a Next Generation Automotive Electrical System," *International Congress on Transportation Electronics (Convergence)*, pp. 41–51, Dearborn, MI, USA, October, 1998.
40. T. A. Radomski, "Alternating Current Generator," U.S. Patent No. 4,882,515, November 21, 1989.
41. T. M. Jahns and V. Caliskan, "Uncontrolled Generator Operation of Interior PM Synchronous Machines Following High-Speed Inverter Shutdown," *IEEE Transactions on Industry Applications*, vol. 35, no. 6, pp. 1347–1357, November/December, 1999.
42. G. Venkataramanan, B. Milovska, V. Gerez, and H. Nehrir, "Variable Speed Operation of Permanent Magnet Alternator Wind Turbines using a Single Switch Power Converter," *Journal of Solar Energy Engineering – Transactions of the ASME*, vol. 118, no. 4, pp. 235–238, November, 1996.
43. W.T. Balogh, "Boost converter regulated alternator," U.S. Patent No. 5,793,625, August 11, 1998.
44. H.-J. Gutt and J. Muller, "New Aspects for Developing and Optimizing Modern Motorcar Generators," *IEEE Industry Applications Society Annual Meeting*, Denver, CO, USA, October, 1994.
45. M. Naidu, N. Boules and R. Henry, "A High-Efficiency, High Power Generation System For Automobiles," *Proceedings of the IEEE Industry Applications Society*, pp. 709–716, 1995.
46. C.D. Syverson and W.P. Curtiss, "Hybrid Alternator with Voltage Regulator," U.S. Patent No. 5,502,368, March 26, 1996.
47. Z. Zhang, F. Profumo, and A. Tenconi, "Wheels Axial Flux Machines for Electric Vehicle Applications," *Proceedings of the International Conference on Electrical Machines*, vol. 2, pp. 7–12, 1994.
48. L. Givens, "A Technical History of the Automobile," *Automotive Engineering*, pp. 61–67, June, 1990.
49. S.M. Terry, "12 Volts Presents Its Case," *SAE Journal*, pp. 29–30, February, 1954.
50. H.L. Hartzell, "It's Still 12 Volts!," *SAE Journal*, pp. 63, June, 1954.
51. J.G.W. West, "Powering Up – a Higher System Voltage for Cars," *IEE Review*, pp. 29–32, January, 1989.
52. J.M. Miller, "Multiple Voltage Electrical Power Distribution System for Automotive Applications," *Intersociety Energy Conversion Engineering Conference*, Washington, DC, USA, August, 1996.
53. J.G. Kassakian, H.C. Wolf, J.M. Miller, and C.J. Hurton, "Automotive Electrical Systems circa 2005," *IEEE Spectrum*, pp. 22–27, August, 1996.
54. J.V. Hellmann and R.J. Sandel, "Dual/High Voltage Electrical Systems," *Future Transportation Technology Conference and Exhibition*, SAE 911652, August, 1991.
55. SAE J2232, "Vehicle System Voltage – Initial Recommendations," SAE Information Report J2232, June, 1992.
56. "Draft Specification of a Dual Voltage Vehicle Electrical Power System 42V/14V," *Forum Bordnetz*, Working Document, Germany, March 4, 1997.
57. *Heated Windshield*, PPG Industries Inc., Pittsburg, PA, USA.
58. "Automotive Industry Association Expects Production of Fuel-Efficient Cars to Begin by 2000," *The Week in Germany*, pp. 4, German Information Center, New York, NY, USA, September 15, 1995.
59. *Review of the Research Program of the Partnership for a New Generation of Vehicles*, Second Report, National Academy Press, Washington, DC, USA, 1996.
60. "MIT Prof Drives the Shift to 42-V Cars," *Electronic Engineering Times*, Issue 1053, pp. 115–116, March 22, 1999.
61. K.K. Afridi, *A Methodology for the Design and Evaluation of Advanced Automotive Electrical Power Systems*, PhD Thesis, Department of

Electrical Engineering and Computer Science, Massachusetts Institute of Technology, Cambridge, MA, USA, February, 1999.
62. J.M. Miller, K. Hampton, and R. Eriksson, "Identification of the Optimum Vehicle Class for the Application of 42V Integrated Starter Generator," SAE Paper 2000-01-C073, *IEEE-SAE International Conference on Transportation Electronics (Convergence)*, Dearborn, MI, October, 2000.
63. E.C. Lovelace, T.M. Jahns, and J.H. Lang, "Impact of Saturation and Inverter Cost on Interior PM Synchronous Machine Drive Optimization," *IEEE Transactions IAS*, vol. 36, no. 3, May/June, 2000.
64. T.C. Neugebauer and D.J. Perreault, "Computer-Aided Optimization of dc/dc Converters for Automotive Applications," *IEEE Power Electronics Specialists Conference*, pp. 689–695, Galway, Ireland, June, 2000.
65. J.C. Byrum, *Comparative Evaluation of Dual-Voltage Automotive Alternators*, S.M. Thesis, Department of Electrical Engineering and Computer Science, Massachusetts Institute of Technology, Cambridge, MA, USA, September, 2000.
66. V. Caliskan, *A Dual/High-Voltage Automotive Electrical Power System with Superior Transient Performance*, Ph.D. Thesis, Department of Electrical Engineering and Computer Science, Massachusetts Institute of Technology, Cambridge, MA, USA, September, 2000.
67. C.R. Smith, "Review of Heavy Duty Dual Voltage Systems," *International Off-Highway & Powerplant Congress and Exposition*, SAE Paper 911857, September, 1991.
68. J. Becker, M. Pourkermani, and E Saraie, "Dual-Voltage Alternators," *International Truck and Bus Meeting and Exposition*, SAE Paper 922488, Toledo, Ohio, USA, November, 1992.
69. J. O'Dwyer, C. Patterson, and T. Reibe, "Dual Voltage Alternator," *IEE Colloquium on Machines for Automotibe Applications*, pp. 4/1–4/5, London, UK, November, 1996.
70. D. Hermance and S. Sasaki, "Hybrid Electric Vehicles take to the Streets," *IEEE Spectrum*, pp. 48–52, November, 1998.
71. A.F. Burke, "Hybrid/Electric Vehicle Design Options and Evaluations," *Electric and Hybrid Vehicle Technology (SP-915) International Congress and Exposition*, SAE Paper 920447, Detroit, MI, USA, February, 1992.
72. K. Nakano and S. Ochiai, "Development of the Motor Assist System for the Hybrid Automobile – the Insight," *IEEE-SAE International Conference on Transportation Electronics (Convergence)*, SAE Paper 2000-01-C079, Dearborn, MI, USA, October, 2000.
73. R.L. Davis, T.L. Kizer, "Energy Management in DaimlerChrysler's PNGV Concept Vehicle," *IEEE-SAE International Conference on Transportation Electronics (Convergence)*, SAE Paper 2000-01-C063, Dearborn, MI, USA, October, 2000.
74. A. Gale and D. Brigham, "Starter/Alternator Design for Optimized Hybrid Fuel Economy," *IEEE-SAE International Conference on Transportation Electronics (Convergence)*, SAE Paper 2000-01-C061, Dearborn, MI, USA, October, 2000.
75. J.M. Miller, A.R. Gale, and V.A. Sankaran, "Electric Drive Subsystem for a Low-Storage Requirement Hybrid Electric Vehicle," *IEEE Transaction on Vehicular Technology*, vol. 48, no. 6, November, 1999.
76. S. Abe, "Development of the Hybrid Vehicle and its Future Expectation," *IEEE-SAE International Conference on Transportation Electronics (Convergence)*, SAE Paper 2000-01-C042, Dearborn, MI, USA, October, 2000.
77. W.L. Shepard, G.M. Claypole, M.G. Kosowski, and R.E. York, "Architecture for Robust Efficiency: GM's 'Precept' PNGV Vehicle," *Future Car Congress*, SAE Paper 2000-01-1582, Arlington, VA, USA, April, 2000.
78. M.G. Kosowski and P.H. Desai, "A Parallel Hybrid Traction System for GM's Precept PNGV Vehicle," *Future Car Congress*, SAE Paper 2000-01-1534, Arlington, VA, USA, April, 2000.
79. J.M. Miller, A.R. Gale, P.J. McCleer, F. Lonardi, and J.H. Lang, "Starter Alternator for Hybrid Electric Vehicle: Comparison of Induction and Variable Reluctance Machines and Drives," *IEEE IAS Annual Meeting*, St. Louis, MO, USA, October, 1998.

26
Solar Power Conversion

Lana Chaar, Ph.D.
*Electrical Engineering Department,
American University in Dubai,
P. O. Box 28282, Dubai,
UAE*

26.1 Introduction 661
26.2 How does a Solar Cell Work? 662
26.3 Solar Energy Conversion 663
26.4 Maximum Power Tracker 664
    26.4.1 Switch-mode Converter • 26.4.2 Controller • 26.4.3 MPPT Controller Algorithm
26.5 Photovoltaic Systems' Components 667
    26.5.1 Grid-connected Photovoltaic System • 26.5.2 Stand-alone Photovoltaic Systems
26.6 Factors Affecting Output 670
    26.6.1 Temperature • 26.6.2 Dirt and Dust • 26.6.3 DC–AC Conversion
26.7 System Design 671
    26.7.1 Criteria for a Quality PV System • 26.7.2 Design Procedures • 26.7.3 Power Conditioning System • 26.7.4 Battery Sizing
26.8 Summary 671
    References 671

26.1 Introduction

For many years, fossil fuel has been the primary source of energy. However, there is a limited supply of these fuels on Earth and they are being used much more rapidly than they are being created. Eventually, they will run out. In addition, because of safety concerns and waste disposal problems, renewable energy is definitely the solution since such technology is "clean" or "green" because they produce few if any, pollutants. The world trend nowadays is to find a non-depletable and clean source of energy. The most effective and harmless energy source is probably solar energy, which for many applications is so technically straightforward to use. With the exception of nuclear and geothermal, all forms of energy used on Earth originate from the sun's energy.

Solar energy is considered one of the most promising energy sources due to its infinite power. Thus modern solar technologies have been penetrating the market at faster rates. The solar technology that has the greatest impact on our lives is photovoltaic. Not in terms of the amount of electricity it produces, but because of the fact that photovoltaic cells – work silently, not polluting – can generate electricity wherever sun shines, even in places where no other form of electricity can be obtained [1]. The word Photovoltaic is a combination of the Greek word for light and the name of the physicist Alessandro Volta [2]. It identifies the direct conversion of sunlight into electricity by means of solar cells.

In this chapter, we will discuss how solar cells produce electricity and what are the components required for such system.

26.2 How does a Solar Cell Work?

Solar cells are composed of various semiconductor materials which become electrically conductive when supplied by heat or light. The majority of solar cells produced are composed of Silicon (Si) which exist in sufficient quantities and do not add any burden on the environment. However, over 95% of these cells have efficiency about 17% [3]. However solar cells, having power conversion efficiencies as high as 31% have been developed over the last decade in laboratory environment [4].

Doping technique is used to obtain a surplus of positive charge carriers (p-type) or a surplus of negative carriers (n-type). When two layers of different doping are in contact, then a p-n junction is formed on the boundary.

An internal electric field is built up which then causes the separation of charge carriers released by light as shown in Fig. 26.1. We all know that light is composed of small packets called photons. When these photons bombard our cell, many electrons are freed within the electric field proximity, which then pull the electrons from the p-side to n-side. Through metal contacts, an electric charge can be taped. If the outer circuit is closed, then direct current flows as illustrated in Fig. 26.2.

A solar cell is approximately 10×10 cm, protected by transparent antireflection film. Each of these cells produces around 0.5 V (for Silicon). The voltage across a solar cell is primarily dependent on the design and materials of the cell, whilst the electrical current depends primarily on the incident solar irradiance and the cell area.

The position of the sun in the sky varies dramatically with hour and season. The sun elevation angle $\theta_{sun}^{elevation}$ is expressed in degrees above the horizon. Sun azimuth $\theta_{sun}^{azimuth}$ angle is expressed in degrees from true north. Sun zenith angle θ_{sun}^{zenith} equals ninety degrees less than the sun elevation, or

$$\theta_{sun}^{zenith} = 90° - \theta_{sun}^{elevation} \qquad (26.1)$$

Azimuth, zenith, and elevation angle are illustrated in Fig. 26.3

The output from a typical solar cell which is exposed to the sun therefore increases from zero at sunrise to a maximum at midday and then falls again to zero at dusk. The I–V characteristics of a solar panel are nonlinear and follow the general shape and equation shown in Fig. 26.4.

$$I = I_{sc} - A\left(e^{BV} - 1\right) \qquad (26.2)$$

where A, B are constants

I_{sc} is the short circuit current, thus the greatest current value generated by the cell under short circuit conditions ($V = 0$).

In order to obtain the appropriate voltages and outputs for different applications, single solar cells are interconnected in series (for larger voltage) and in parallel (for larger current) to

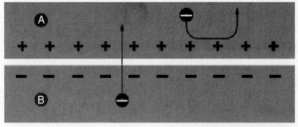

FIGURE 26.1 Effect of the electric field in a PV cell [1].

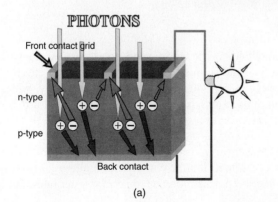

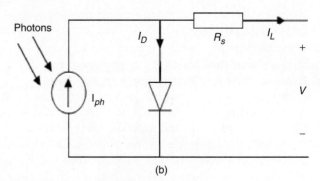

FIGURE 26.2 (a) Operation of a PV cell [2] and (b) solar cell equivalent circuit.

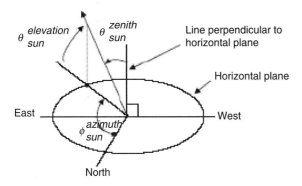

FIGURE 26.3 Azimuth, zenith, and elevation angles of a vector pointed toward the sun [5].

26.3 Solar Energy Conversion

Photovoltaic technology is used to produce electricity in areas where power lines do not reach. In developing countries, it is improving living conditions in rural areas especially in healthcare, education, and agriculture. In the industrialized countries, they have been used extensively and integrated with the utility grid.

Photovoltaic arrays are usually mounted in a fixed position and tilted toward the south to optimize the noontime and daily energy production. The orientation of fixed panels should be carefully chosen to capture the maximum energy for the season, or a year. Photovoltaic arrays have an optimum operating point called the maximum power point (MPP) as shown in Fig. 26.6 [8].

The curve in Fig. 26.6 shows that power increases as the voltage is increased, reaching a peak value before decreasing as the resistance increases to the point where current drops-off. According to the maximum power transfer theory, this point is where the load is matched to the solar panel's resistance at

form the photovoltaic module. Then several of these modules are connected to each other to form the photovoltaic array as shown in Fig. 26.5. This array is then fitted with an aluminum or stainless steel frame and covered with transparent glass on the front side.

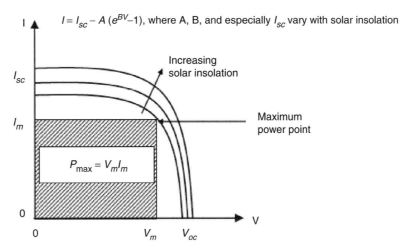

FIGURE 26.4 I–V characteristics of a solar cell [6].

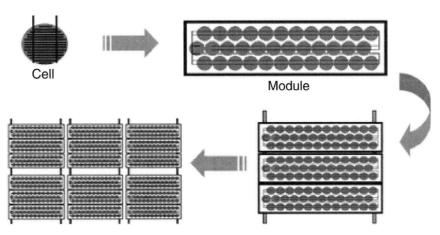

FIGURE 26.5 Photovoltaic cells, modules, panels, and array [7].

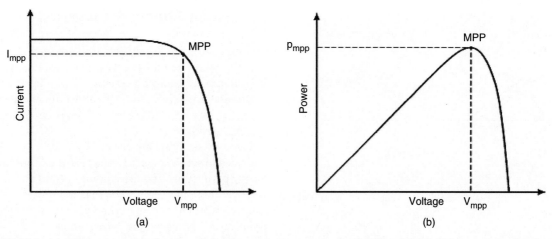

FIGURE 26.6 (a) *I–V* characteristics of a solar cell showing maximum power point (MPP) and (b) *P–V* characteristics showing MPP.

a certain level of temperature and insolation. The *I–V* curve changes as the temperature and insolation levels change as shown in Fig. 26.7, thus the MPP will vary accordingly [9].

It has shown that the open circuit voltage increases logarithmically while the short circuit increases linearly as the insolation level increases [10]. Moreover, increasing the cell's temperature decreases the open circuit voltage and increases slightly the short circuit current. This then makes the cell less efficient.

Since solar power is relatively expensive, it is important to operate panels at their maximum power conditions. Thus, PV systems will operate more efficiently with systems that can adjust automatically their loads to match the PV resistance. In addition, panels would change orientation to track the sun.

We need then to control either the operating voltage or the current to get maximum power from the PV panel at the prevailing temperature and insolation conditions using maximum power point tracker (MPPT) which should meet the following conditions [11]:

- Operate the PV system as close as possible to the MPP irrespective of the atmospheric changes.
- Have low cost and high conversion efficiency.
- Provide an output interface compatible with the battery-charging requirement.

26.4 Maximum Power Tracker

The MPPT maximizes the energy that can be transferred from the array to an electrical system. Its main function is to adjust the panel output voltage to a value at which the panel supplies the maximum energy to the load. Most current designs consist of three basic components: a switch-mode dc–dc converter, a control, and tracking section.

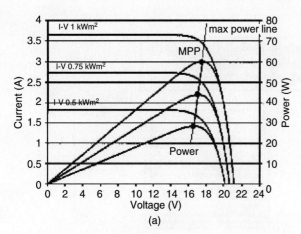
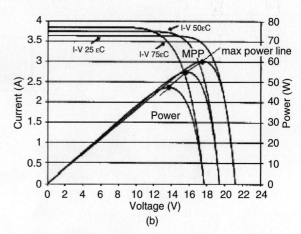

FIGURE 26.7 (a) *PV* panel insolation characteristics and (b) *PV* panel temperature characteristics [9].

26.4.1 Switch-mode Converter

The switch-mode converter is the core of the entire supply. It allows energy at one potential to be drawn, stored as magnetic energy in an inductor, and then released at a different potential. By setting up the switch-mode section in various different topologies, either high-to-low (Buck converter) or low-to-high (Boost converter), voltage converters can then be built. The main goal is to provide a fixed input voltage and/or current, such that the array is held at the maximum power point, while allowing the output to match the battery.

26.4.2 Controller

The controller should keep testing if the PV system is operating at the PV maximum power point. It should force the system to track this MPP. Continuous measuring of the voltage and current from the PV array, and then performing either voltage or power feedback control [12] is the method used.

26.4.2.1 Voltage Feedback Control

The control variable here is the PV array terminal voltage. The controller forces the PV array to operate at its MPP by changing the array terminal voltage. It neglects, however, the variation in the temperature and insolation level [12, 13].

26.4.2.2 Power Feedback Control

The control variable here is the power delivered to the load. To achieve maximum power the quantity dp/dv is forced to zero. This control scheme is not affected by the characteristics of the PV array, yet it maximizes power to the load and not power from the PV array [12, 13].

Fast shadows cause trackers to lose the MPP momentarily, and the time lost in seeking it again, because the point has moved away quickly and then moved back to the original position, equating to the energy lost while the array is off power point. On the other hand, if lighting conditions do change, the tracker needs to respond within a short amount of time to the change to avoid energy loss. Thus the controller should be capable of adjusting and keeping the PV at its MPPT.

Several algorithms were proposed to accomplish MPPT controller. Published MPPT methods include: (1) Perturb and Observe (PAO) [14], (2) Incremental Conductance Technique (ICT) [14], and (3) Constant Reference Voltage/Current [11, 14].

26.4.3 MPPT Controller Algorithm

26.4.3.1 Perturb and Observe (PAO)

The Perturb and Observe method has a simple feedback structure and few measured parameters. It operates by periodically perturbing (i.e. incrementing or decrementing) the duty cycle

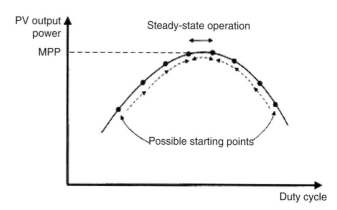

FIGURE 26.8 PAO technique [11].

controlling the array current as shown in Fig. 26.8 and comparing the PV output power with that of the previous perturbation cycle. If the perturbation leads to an increase (or decrease) in array power, the subsequent perturbation is made in the same (or opposite) direction. In this manner, the peak power tracker continuously seeks the peak power condition.

The PAO technique is easy to implement and costs the least among the other available techniques. It is considered to be a very efficient scheme in terms of power being extracted from the PV array [15]. However the PAO technique will be confused in catching the MPP under rapid varying solar radiation [14]. If the Insolation level increases ($I_2 > I_1$) then the controller will assume that the incremental step should keep moving in the same direction toward point ② when the new MPP is really in the other direction at point ③ as shown in Fig. 26.9 [14]. So for the PAO algorithm, the power has increased because the new MPP is toward the right whereas it has already been passed to point ③. In the following perturbation the PAO algorithm will increment the array operating voltage further right, point ②. In this way the PAO algorithm will continue to deviate from the actual MPP, with a corresponding power loss, until the solar radiation change slows or settles down [14].

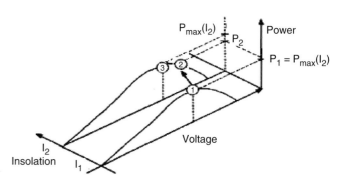

FIGURE 26.9 Deviation of the PAO technique from the MPP [14].

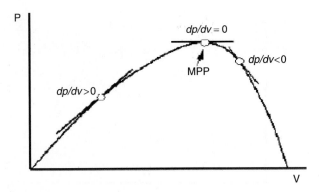

FIGURE 26.10 The slope "conductance" of the P-V curve [16].

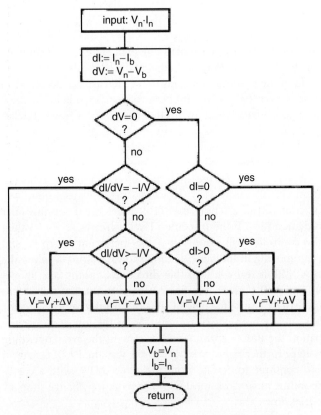

FIGURE 26.11 Flow chart of the ICT algorithm [14].

26.4.3.2 Incremental Conductance Technique (ICT)

The basic idea is that the derivative of the power with respect to the voltage (dp/dv) vanishes at the MPP since it is the maximum point on the curve as shown in Fig. 26.10.

The ICT algorithm checks for MPP by comparing dI/dV against $-I/V$ till it reaches the voltage operating point at which the incremental conductance is equal to the source conductance [14, 17]. The flow chart for the ICT algorithm is described in Fig. 26.11.

The algorithm starts by obtaining the present values of I and V, then using the corresponding values stored at the end of the preceding cycle, I_b and V_b, the incremental changes are approximated as: $dI = I - I_b$, and $dV = V - V_b$ and according to the result of this check, the control reference signal V_{ref} will be adjusted in order to move the array voltage toward the MPP voltage. At the MPP, $dI/dV = -I/V$, no control action is needed, therefore the adjustment stage will be bypassed and the algorithm will update the stored parameters at the end of the cycle as usual. Another check is included in the algorithm to detect whether a control action is required when the array was operating at the previous cycle MPP ($dV = 0$); in this case the change in weather condition will be detected using ($dI \neq 0$) [14].

This technique offers good performance under varying atmospheric conditions contrary to the PAO technique. However it requires complete mathematical model for the topology used and its complex circuitry adds to the cost of the MPPT controller [16].

26.4.3.3 Constant Reference Voltage

One very common MPPT technique is to compare the PV array voltage (or current) with a constant reference voltage (or current), which corresponds to the PV voltage (or current) at the maximum power point, under specific atmospheric conditions as shown in Fig. 26.12. The resulting difference signal (error signal) is used to drive a power conditioner, which interfaces the PV array to the load. Although the implementation of this method is simple, the method itself is not very accurate, since it does not take into account the effects of temperature and irradiation variations [11].

26.4.3.4 Other Techniques

Other techniques exist such as current-based maximum power point tracker "CMPPT" and voltage-based maximum power

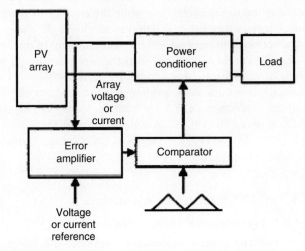

FIGURE 26.12 MPPT control system with constant voltage reference [11].

point tracker "**VMPPT**" [16]. Employed numerical methods show a linear dependence between the "cell currents corresponding to maximum power" and the "cell-short circuit currents". The current I_{MPP} operating at the MPP is calculated using the following equation:

$$I_{MPP} = M_C \, I_{SC} \qquad (26.3)$$

where M_C is called the "current factor". This factor M_C differs from one panel to another and is affected by the panel surface conditions, especially if partial shading covers the panel [18].

Similarly the MPP operating voltage is calculated directly from V_{OC}:

$$V_{MPP} = M_V \, V_{OC} \qquad (26.4)$$

where M_V is the "voltage factor".

The open circuit voltage V_{OC} is sampled by analog sampler, and then V_{MPP} is calculated by Eq. (26.4). This operating V_{MPP} voltage is the reference voltage for the voltage control loop as shown in Fig. 26.13. This method always, "results in a considerable power error because the output voltage of the PV module only follows the unchanged reference voltage during one sampling period" [19].

Others argue that these two techniques are considered to be "fast, practical, and powerful methods for MPP estimation of PV generators under all insolation and temperature conditions" [21].

26.4.3.5 Comparative Study

A comprehensive experimental comparison between different MPPT algorithms was prepared at South Dakota State University [22]. After presenting the advantages and disadvantages of each algorithm, an experiment for the same PV array setup was run. Results showed that the ICT method has the highest efficiency (98%) in terms of power extracted from the PV array, next is the PAO technique efficiency (96.5%), and finally the Constant Voltage method efficiency (88%).

The ICT method offers good performance under rapidly changing weather conditions and seems to provide the highest tracking efficiency, however four sensors are required to perform the measurements for computations and decision making [14]. If the system requires more conversion time in tracking the MPP, a large amount of power loss will occur [12]. On the contrary, if the sampling and execution speed of the perturbation and observation method is increased, then the system loss will be reduced. This technique requires only two sensors. This results in the reduction of hardware requirement and cost.

26.5 Photovoltaic Systems' Components

Once the PV array is controlled to perform efficiently, a number of other components are required to control, convert, distribute, and store the energy produced by the array. Such components may vary depending on the functional and operational requirements of the system. They may require battery banks and controller, dc–ac inverters, in addition to other components such as overcurrent, surge protection and other processing equipment. Figure. 26.14 shows a basic diagram of a photovoltaic system and the relationship with each component.

Photovoltaic systems are classified into two major classes: utility grid-connected photovoltaic systems and stand-alone photovoltaic systems.

26.5.1 Grid-connected Photovoltaic System

Grid-connected photovoltaic systems are designed to operate in parallel with the electric utility grid as shown. There are two general types of electrical designs for PV power systems: systems that interact with the utility power grid as shown in Fig. 26.15a and have no battery backup capability, and systems that interact and include battery backup as well, as shown in Fig. 26.15b. The latter type of system incorporates energy storage in the form of a battery to keep "critical load" circuits operating during utility outage. When an outage occurs, the unit disconnects from the utility and powers specific circuits of the load. If the outage occurs in daylight, the PV array is able to assist the load in supplying the loads.

The major component in both systems is the dc–ac inverter or also called the power conditioning system (PCS). Figure 26.16 shows the block diagram of such connection.

The inverter, used to convert photovoltaic dc energy to ac energy, is the key to the successful operation of the system, but it is also the most complex hardware. The most important

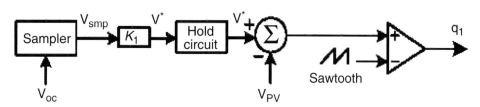

FIGURE 26.13 The conventional MPPT controller using open circuit voltage V_{oc} [20].

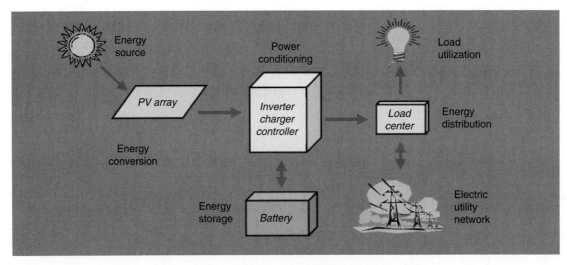

FIGURE 26.14 Major photovoltaic system components [6].

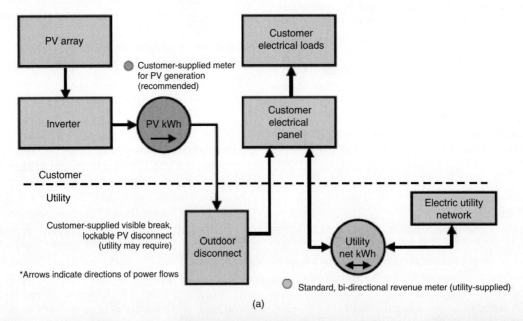

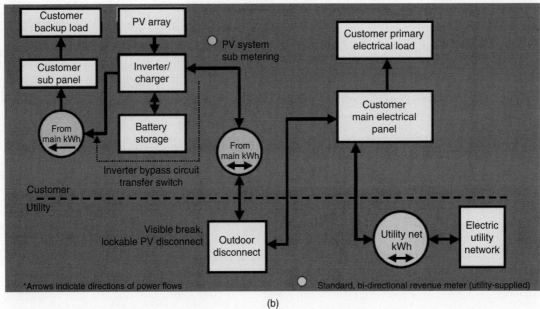

FIGURE 26.15 Grid-connected PV system: (a) without battery back-up [23] and (b) with battery storage [23].

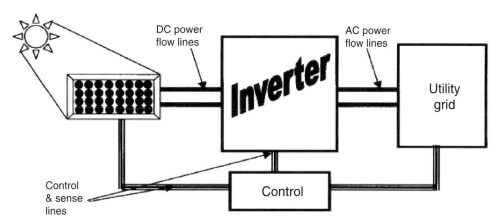

FIGURE 26.16 Diagram of grid-connected photovoltaic system [24].

inverter characteristics requirement are: operate over a wide range of voltages and currents, regulate output voltage and frequency, in addition to providing ac power with good power quality. Several interconnection circuits have been described in [25, 26].

For the last twenty years, researchers have been working on developing different inverter topologies that satisfy the above listed requirement. The evolution of solid state devices such as MOSFETs, IGBTs, microprocessors, PWM integrated circuits have allowed improvements on the inverter. However, more work is required to ensure quality control, reliability and lower cost since these are the key for a sustainable photovoltaic market.

A typical utility-interactive PV system is shown in Fig. 26.17 with metering to provide indication of system performance.

26.5.2 Stand-alone Photovoltaic Systems

Stand-alone photovoltaic systems are usually a utility power substitute. They generally include solar charging modules, storage batteries, and controls/regulator as shown in Fig. 26.18. Ground or roof mounted systems will require a mounting structure, and if 120/240 volt ac power is desired, a dc to ac inverter will also be required. They are especially used in remote places that are not connected to the electrical main utility grid. In many stand-alone PV systems, batteries are used for energy storage. A charge controller is then used to control the whole system and prevent the battery from overcharging and overdischarging. Photovoltaic modules charge the battery during the day and supplies power to the load as needed.

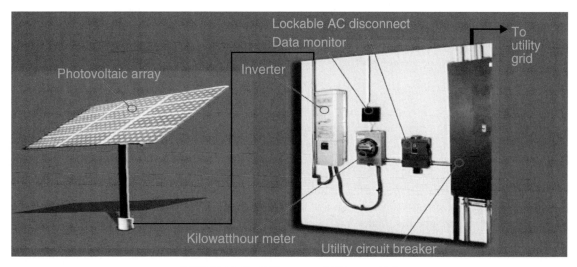

FIGURE 26.17 Grid-connected PV system [23].

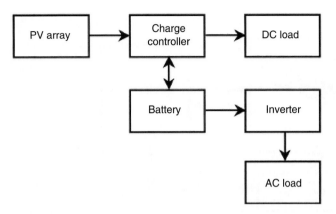

FIGURE 26.18 Diagram of stand-alone PV system with battery storage power dc and ac loads [6].

26.5.2.1 Batteries

Batteries are often used in PV systems for storing energy produced by the PV array during day time and supplying it to electrical loads as needed (during night time or cloudy weather). Moreover, batteries are also needed in the tracker systems to keep the operation at MPP in order to provide electrical loads with stable voltages. Nearly, most of the batteries used in PV systems are deep cycle lead-acid [27]. These batteries have thicker lead plates that make them tolerate deep discharges. The thicker the lead plates, the longer the life span. The heavier the battery for a given group size, the thicker the plates and the better the battery will tolerate deep discharges [28].

All deep cycle batteries are rated in ampere-hour where Ampere-hour (AH) capacity is a quantity of the amount of usable energy it can store at nominal voltage. For example an ampere-hour is one ampere for one hour or 10 A for one-tenth of an hour and so forth [29]. A good charge rate is approximately 10% of the total capacity of the battery per hour (i.e. 200 ampere-hour battery charged at 20A). This will reduce electrolyte loss and damage to the plates [28]. A PV system may have to be sized to store a sufficient amount of power in the batteries to meet power demand during several days of cloudy weather. This is known as "days of autonomy".

26.5.2.2 Charge Controller

The charge controller regulates the flow of electricity from the PV modules to the battery and the load. The controller keeps the battery fully charged without overcharging it. When the load is drawing power, the controller allows charge to flow from the modules into the battery, the load, or both. When the controller senses that the battery is fully charged, it stops the flow of charge from the modules. Many controllers will also sense when loads have taken too much electricity from batteries and will stop the flow until sufficient charge is restored to the batteries. This last feature can greatly extend the battery's lifetime. However, controllers in stand-alone photovoltaic system are more complex devices that depend on battery state-of-charge, which in turn depends on many factors and is difficult to measure. The controller must be sized to handle the maximum current produced.

Several characteristics should be considered before selecting a controller:

- Adjustable Setpoints.
    - High voltage disconnect.
    - Low voltage disconnect.
- Temperature compensation.
- Low voltage warning.
- Reverse current protection.

The controller should ensure that no current flows from the battery to the array at night.

26.6 Factors Affecting Output

PV systems produce power in proportion to the intensity of sunlight striking the solar array surface. Thus there are some factors that affect the overall output of the PV system.

26.6.1 Temperature

Output power of a PV system reduces as the module temperature increases. For Crystalline modules, a typical temperature reduction factor recommended by CEC is 89% in the middle of spring or a fall day, under full light conditions.

26.6.2 Dirt and Dust

Dirt and dust can accumulate on the solar module surface, blocking some of the sunlight and reducing output. A typical annual dust reduction factor to use is 93%. J. P. Thornton showed that sand and dust can cause erosion of the PV surface which affects the system's running performance by decreasing the output power to more than 10% [30].

26.6.3 DC–AC Conversion

Since the power from the PV array is converted back to ac as shown previously, some power is being lost in the conversion process, in addition to losses in the wiring. Common inverters used have peak efficiencies of about 88–90%.

Thus a 100 Watts module under well controlled conditions is actually a 95 Watts module under normal condition.

This power is then reduced due to the factors listed to:

Effect of Temperature: $95 \times 0.89 = 85$ Watts

Effect of Dirt and Dust: $85 \times 0.93 = 79$ Watts

Effect of Conversion: $79 \times 0.90 = 71$ Watts

26.7 System Design

The goal for a solar electric, or photovoltaic system is to provide high-quality, reliable renewable electrical power.

26.7.1 Criteria for a Quality PV System

- Be properly sized and oriented to provide electrical power and energy.
- Good control circuit to reduce electrical losses, overcurrent protection, switches and inverters.
- Good charge controller and battery management system, should the system contain batteries.

26.7.2 Design Procedures

The first task in designing a PV system is to estimate the system load. This is achieved by defining the power demand of all loads, the number of hours of use per day, and operating voltage [31].

From the load ampere-hours, and the given operating voltage for each load, the power demand is then calculated. For a stand-alone system, the system voltage is the potential required by the largest load. When ac loads dominate, the dc system voltage should be chosen to be compatible with the inverter input.

26.7.3 Power Conditioning System

The choice of the PCS has a great impact on the performance and economics of the system. The choice of PCS depends on the type of waveform produced which in turn depends on the method used for conversion as well as the filtering techniques of unwanted frequencies. Several factors must be considered when selecting or designing the inverter [32]:

- The power conversion efficiency.
- Rated power.
- Duty rating; the amount of time the inverter can supply maximum load.
- Input voltage.
- Voltage regulation.
- Voltage protection.
- Frequency requirement.
- Power factor.

26.7.4 Battery Sizing

The amount of battery storage needed depends on the load energy demand and on weather patterns at the site. There is always a trade-off between keeping cost low and meeting energy demand.

26.8 Summary

This chapter has discussed the conversion of solar energy into electricity using photovoltaic system. There are two types of PV systems: the grid-connected and the stand-alone. All major components for such systems have been discussed. Maximum power point tracking is the most important factor in PV systems to provide the maximum power. For this reason, several tracking systems have been described and compared. Factors affecting the output of such systems have been defined and steps for a good and reliable design have been considered.

References

1. www.worldenergy.org/
2. Photovoltaics: Solar Electricity and Solar Cells in Theory and Practice www.solarserver.de/wissen/photovoltaic-e.html
3. Moller H., Semiconductors for Solar Cells. London: Artech House, Inc., 1993.
4. Berkeley Lab; http://www.lbl.gov/msd/pis/walukiewicz/02\02_08_full_solar_spectrum.html
5. The Solar Sprint PV Panel http://chuck-wright.com/SolarSprintPV/SolarSprintPV.html
6. EE362L, Power Electronics, Solar Power, I-V Characteristics, Version October 14, 2005. http://www.ece.utexas.edu/~grady/EE362L_Solar.pdf
7. Photovoltaic Fundamentals http://www.fsec.ucf.edu/PVT/pvbasics/index.htm
8. Serhan M. A., "Maximum Power Point Tracking system: An Adaptive Algorithm for Solar Panels", Thesis, American University of Beirut, January 2005.
9. Johan H., Enslin R., Wolf M. S., Snyman. D. B., and Swiegers W., "Integrated Photovoltaic Maximum Power Point Tracking Converter", IEEE Transactions on Industrial Electronics, Vol. 44, No. 6, December 1997.
10. Hansen A., Sorensen P., Hansen L., and Bindner H., "Models for a Stand Alone PV System", Technical Report, Riso National Laboratory, Roskilde, Norway, December 2000. http://www.solenergi.dk/rapporter/sec-r-12.pdf
11. Kroutoulis E., Kalaitzakis K., and Voulgaris N.C., "Development of a Microcontroller-Based, Photovoltaic Maximum Power Tracking Control System", IEEE Transactions on Power Electronics, Vol. 16, No. 1, January 2001.
12. Hua C. and Shen C., "Comparative study of Peak Power tracking Techniques for Solar Storage System", IEEE Applied Power Electronics Conference and Exposition, Vol. 2, February 1998, pp. 679–685.

13. Hua C. and Lin J., "DSP-Based Controller in Battery Storage of Photovoltaic System", IEEE IECON 22nd International Conference on Industrial Electronics, Control and Instrumentation, Vol. 3, 1996, pp. 1705–1710.
14. Hussein K. H., Mutta I., Hoshino T., and Osakada M., "Maximum photovoltaic power tracking: An algorithm for rapidly changing atmospheric conditions", IEE Proceedings, Generation, Transmission, and Distribution, Vol. 142, No. 1, January 1995.
15. Yu G., Jung Y., Choy I., Song J., and Kim G., "A novel two Mode MPPT Control Algorithm Based on Comparative Study of Existing Algorithms", IEEE Photovoltaics Specialist Conference, May 2002, pp. 1531–1534.
16. Shengyi Liu, "Maximum Power Point Tracker Model", Control model, University of South Carolina, May 2000, Available from: http://vtb.engr.sc.edu/modellibrary_old
17. Tse K.K., Henry S. H., Chung, S. Y. R. Hui, and Ho M. T., "Novel Maximum Power Point Tracking Technique for PV Panels", IEEE Power Electronics Specialists Conference, Vol. 4, June 2001, pp. 1970–1975.
18. Noguchi T., Togashi S., and Nakamoto R., "Short-Current Pulse-Based Maximum Power Point Tracking Method for Multiple Photovoltaic-and-Converter Module System", IRRR Transactions on Industrial Electronics, Vol. 49, No. 1, February 2002.
19. Lee D., Noh H., Hyun D., and Choy I., " An Improved MPPT Converter Using Current Compensation Method for Small Scaled PV-Applications", Applied Power Electronics Conference and Exposition (APEC'03), Vol. 1, February 2003, pp. 540–545.
20. HowStuffworks http://science.howstuffworks.com/solar-cell5.htm
21. Masoum M., Dehbonei H., and Fuschs E., "Theoretical and Experimental Analyses of Photovoltaic Systems with Voltage-and Current-Based Maximum Power Point Tracking", IEEE Transactions Conversion, Vol. 17, No. 4, December 2002.
22. Hohm D. and Ropp M., "Comparative Study of Maximum Power Point Tracking Algorithms Using an Experimental, Programmable, Maximum Point Test Bed", IEEE Photovoltaic Specialists Conference, September 2000, pp. 1699–1702.
23. DER Road Show, "Solar Photovoltaic Systems", Overview and Standards for permitting, Installations, Code Compliance, and Inspections, http://www.eere.energy.gov/de/pdfs/road_shows/arlington_pv.pdf
24. Bower W., "Inverters–Critical Photovoltaic Balance-of-system Components: Status, Issues, and New-Millennium Opportunities", Progress in Photovoltaics: Research and Applications, Vol. 8, 2000, pp. 113–126.
25. Mohan N., Undeland T., and Robbins W., Power Electronics: Converter, Applications and Design, John Wiley and Sons, 3rd edition.
26. Rashid M. H., Power Electronics: Circuits, Devices, and Applications, Prentice Hall.
27. Enslin J. and Snyman D., "Combined Low Cost, High Efficient Inverter, Peak Power Tracker and Regulator for PV Applications", IEEE Transactions on Power Electronics, Vol. 6, No. 1, January 1991.
28. Linden D., Handbook of Batteries, New York: McGraw Hill, 1995.
29. Jian W., Jianzheng L., and Zhengming Z., " Optimal Control of Solar Energy Combined with MPPT and Battery Charging", Proceedings of IEEE International Conference on Electrical Machines and Systems, Vol. 1, November 2003, pp. 285–288.
30. Thornton J. P., "The Effect of Sand-Storm on Photovoltaic Array and Components", Solar energy Conference, 1992.
31. Sandia National Laboratories: Stand-Alone Photovoltaic Systems" A Handbook of Recommended Design Practices, National Technical Information Service, Springfield, VA, 1988.
32. Wolete J. N., "An interactive Menu-Driven Design Tool for Stand-Alone Photovoltaic Systems", Thesis, Virginia Polytechnic Institute. http://scholar.lib.vt.edu/theses/public/etd-11698-16389/materials

27
Power Electronics for Renewable Energy Sources

C. V. Nayar, S. M. Islam, H. Dehbonei, and K. Tan
Department of Electrical and Computer Engineering, Curtin University of Technology, GPO Box U1987, Perth, Western Australia 6845, Australia

H. Sharma
Research Institute for Sustainable Energy, Murdoch University, Perth, Western Australia, Australia

27.1 Introduction 673
27.2 Power Electronics for Photovoltaic Power Systems 674
   27.2.1 Basics of Photovoltaics • 27.2.2 Types of PV Power Systems • 27.2.3 Stand-alone PV Systems • 27.2.4 Hybrid Energy Systems • 27.2.5 Grid-connected PV Systems
27.3 Power Electronics for Wind Power Systems 700
   27.3.1 Basics of Wind Power • 27.3.2 Types of Wind Power Systems • 27.3.3 Stand-alone Wind Power Systems • 27.3.4 Wind–diesel Hybrid Systems • 27.3.5 Grid-connected Wind Energy Systems • 27.3.6 Control of Wind Turbines
References 714

27.1 Introduction

The Kyoto agreement on global reduction of greenhouse gas emissions has prompted renewed interest in renewable energy systems worldwide. Many renewable energy technologies today are well developed, reliable, and cost competitive with the conventional fuel generators. The cost of renewable energy technologies is on a falling trend and is expected to fall further as demand and production increases. There are many renewable energy sources (RES) such as biomass, solar, wind, mini hydro and tidal power. However, solar and wind energy systems make use of advanced power electronics technologies and, therefore the focus in this chapter will be on solar photovoltaic and wind power.

One of the advantages offered by (RES) is their potential to provide sustainable electricity in areas not served by the conventional power grid. The growing market for renewable energy technologies has resulted in a rapid growth in the need of power electronics. Most of the renewable energy technologies produce DC power and hence power electronics and control equipment are required to convert the DC into AC power.

Inverters are used to convert DC to AC. There are two types of inverters: (a) stand-alone or (b) grid-connected. Both types have several similarities but are different in terms of control functions. A stand-alone inverter is used in off-grid applications with battery storage. With back-up diesel generators (such as photovoltaic (PV)/diesel/hybrid power systems), the inverters may have additional control functions such as operating in parallel with diesel generators and bi-directional operation (battery charging and inverting). Grid interactive inverters must follow the voltage and frequency characteristics of the utility generated power presented on the distribution line. For both types of inverters, the conversion efficiency is a very important consideration. Details of stand-alone and grid-connected inverters for PV and wind applications are discussed in this chapter.

Section 27.2 covers stand-alone PV system applications such as battery charging and water pumping for remote areas. This section also discusses power electronic converters suitable for PV-diesel hybrid systems and grid-connected PV for rooftop and large-scale applications. Of all the renewable energy options, the wind turbine technology is maturing very fast. A marked rise in installed wind power capacity has been noticed worldwide in the last decade. Per unit generation cost of wind power is now quite comparable with the conventional generation. Wind turbine generators are used in stand-alone battery charging applications, in combination with fossil fuel generators as part of hybrid systems and as grid-connected systems. As a result of advancements in blade design, generators, power electronics, and control systems, it has been possible to increase dramatically the availability of large-scale wind power.

Copyright © 2007, 2001, Elsevier Inc.
All rights reserved.

Many wind generators now incorporate speed control mechanisms like blade pitch control or use converters/inverters to regulate power output from variable speed wind turbines. In Section 27.3, electrical and power conditioning aspects of wind energy conversion systems were included.

27.2 Power Electronics for Photovoltaic Power Systems

27.2.1 Basics of Photovoltaics

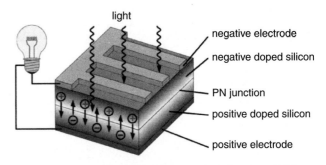

FIGURE 27.1 Principle of the operation of a solar cell [2].

The density of power radiated from the sun (referred as "solar energy constant") at the outer atmosphere is $1.373\,kW/m^2$. Part of this energy is absorbed and scattered by the earth's atmosphere. The final incident sunlight on earth's surface has a peak density of $1\,kW/m^2$ at noon in the tropics. The technology of photovoltaics (PV) is essentially concerned with the conversion of this energy into usable electrical form. Basic element of a PV system is the solar cell. Solar cells can convert the energy of sunlight directly into electricity. Consumer appliances used to provide services such as lighting, water pumping, refrigeration, telecommunication, television, etc. can be run from PV electricity. Solar cells rely on a quantum-mechanical process known as the "photovoltaic effect" to produce electricity. A typical solar cell consists of a p–n junction formed in a semiconductor material similar to a diode. Figure 27.1 shows a schematic diagram of the cross section through a crystalline solar cell [1]. It consists of a 0.2–0.3 mm thick monocrystalline or polycrystalline silicon wafer having two layers with different electrical properties formed by "doping" it with other impurities (e.g. boron and phosphorous). An electric field is established at the junction between the negatively doped (using phosphorous atoms) and the positively doped (using boron atoms) silicon layers. If light is incident on the solar cell, the energy from the light (photons) creates free charge carriers, which are separated by the electrical field. An electrical voltage is generated at the external contacts, so that current can flow when a load is connected. The photocurrent (I_{ph}), which is internally generated in the solar cell, is proportional to the radiation intensity.

A simplified equivalent circuit of a solar cell consists of a current source in parallel with a diode as shown in Fig. 27.2a. A variable resistor is connected to the solar cell generator as a load. When the terminals are short-circuited, the output voltage and also the voltage across the diode is zero. The entire photocurrent (I_{ph}) generated by the solar radiation then flows to the output. The solar cell current has its maximum (I_{sc}). If the load resistance is increased, which results in an increasing voltage across the p–n junction of the diode, a portion of the current flows through the diode and the output current decreases by the same amount. When the load resistor is open-circuited, the output current is zero and the entire photocurrent flows through the diode. The relationship between current and voltage may be determined from the diode characteristic equation

$$I = I_{ph} - I_0(e^{qV/kT} - 1) = I_{ph} - I_d \qquad (27.1)$$

where q is the electron charge, k is the Boltzmann constant, I_{ph} is photocurrent, I_0 is the reverse saturation current, I_d is diode current, and T is the solar cell operating temperature (°K). The current vs voltage (I–V) of a solar cell is thus equivalent to an "inverted" diode characteristic curve shown in Fig. 27.2b.

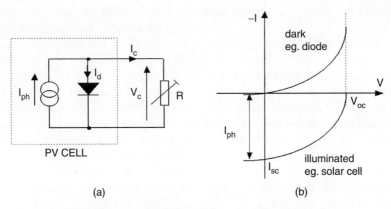

FIGURE 27.2 Simplified equivalent circuit for a solar cell.

A number of semiconductor materials are suitable for the manufacturing of solar cells. The most common types using silicon semiconductor material (Si) are:

- Monocrystalline Si cells.
- Polycrystalline Si cells.
- Amorphous Si cells.

A solar cell can be operated at any point along its characteristic current–voltage curve, as shown in Fig. 27.3. Two important points on this curve are the open-circuit voltage (V_{oc}) and short-circuit current (I_{sc}). The open-circuit voltage is the maximum voltage at zero current, while short-circuit current is the maximum current at zero voltage. For a silicon solar cell under standard test conditions, V_{oc} is typically 0.6–0.7 V, and I_{sc} is typically 20–40 mA for every square centimeter of the cell area. To a good approximation, I_{sc} is proportional to the illumination level, whereas V_{oc} is proportional to the logarithm of the illumination level.

A plot of power (P) against voltage (V) for this device (Fig. 27.3) shows that there is a unique point on the I–V curve at which the solar cell will generate maximum power. This is known as the maximum power point (V_{mp}, I_{mp}). To maximize the power output, steps are usually taken during fabrication, the three basic cell parameters: open-circuit voltage, short-circuit current, and fill factor (FF) – a term describing how "square" the I–V curve is, given by

$$\text{Fill Factor} = (V_{mp} \times I_{mp})/(V_{oc} \times I_{sc}) \qquad (27.2)$$

For a silicon solar cell, FF is typically 0.6–0.8. Because silicon solar cells typically produce only about 0.5 V, a number of cells are connected in series in a PV module. A panel is a collection of modules physically and electrically grouped together on a support structure. An array is a collection of panels (see Fig. 27.4).

The effect of temperature on the performance of silicon solar module is illustrated in Fig. 27.5. Note that I_{sc} slightly increases linearly with temperature, but, V_{oc} and the maximum power, P_m decrease with temperature [1].

Figure 27.6 shows the variation of PV current and voltages at different insolation levels. From Figs. 27.5 and 27.6, it can be seen that the I–V characteristics of solar cells at a given insolation and temperature consist of a constant voltage segment

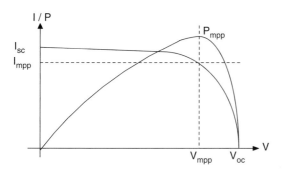

FIGURE 27.3 Current vs voltage (I–V) and current power (P–V) characteristics for a solar cell.

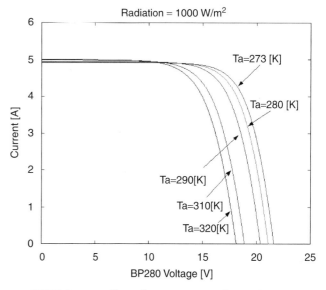

FIGURE 27.5 Effects of temperature on silicon solar cells.

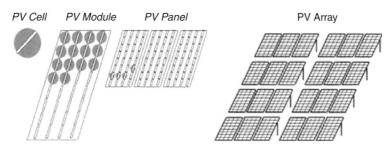

FIGURE 27.4 PV generator terms.

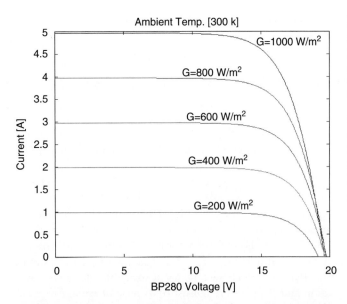

FIGURE 27.6 Typical current/voltage (*I–V*) characteristic curves for different insolation.

and a constant current segment [3]. The current is limited, as the cell is short-circuited. The maximum power condition occurs at the knee of the characteristic curve where the two segments meet.

27.2.2 Types of PV Power Systems

Photovoltaic power systems can be classified as:

- Stand-alone PV systems.
- Hybrid PV systems.
- Grid-connected PV systems.

Stand-alone PV systems, shown in Fig. 27.7, are used in remote areas with no access to a utility grid. Conventional power systems used in remote areas often based on manually controlled diesel generators operating continuously or for a few hours. Extended operation of diesel generators at low load levels significantly increases maintenance costs and reduces their useful life. Renewable energy sources such as PV can be added to remote area power systems using diesel and other fossil fuel powered generators to provide 24-hour power economically and efficiently. Such systems are called "hybrid energy systems." Figure 27.8 shows a schematic of a PV-diesel hybrid system. In grid-connected PV systems shown in Fig. 27.9, PV panels are connected to a grid through inverters without battery storage. These systems can be classified as small systems like the residential rooftop systems or large grid-connected systems. The grid-interactive inverters must be synchronized with the grid in terms of voltage and frequency.

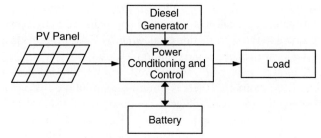

FIGURE 27.8 PV-diesel hybrid system.

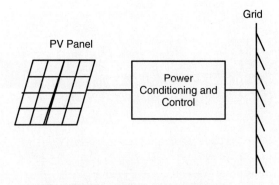

FIGURE 27.9 Grid-connected PV system.

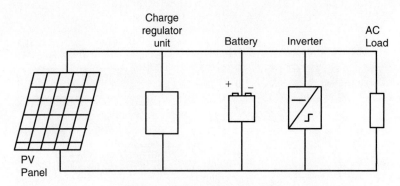

FIGURE 27.7 Stand-alone PV system.

27.2.3 Stand-alone PV Systems

The two main stand-alone PV applications are:

- Battery charging.
- Solar water pumping.

27.2.3.1 Battery Charging

27.2.3.1.1 Batteries for PV Systems Stand-alone PV energy system requires storage to meet the energy demand during periods of low solar irradiation and nighttime. Several types of batteries are available such as the lead acid, nickel–cadmium, lithium, zinc bromide, zinc chloride, sodium sulfur, nickel–hydrogen, redox, and vanadium batteries. The provision of cost-effective electrical energy storage remains one of the major challenges for the development of improved PV power systems. Typically, lead-acid batteries are used to guarantee several hours to a few days of energy storage. Their reasonable cost and general availability has resulted in the widespread application of lead-acid batteries for remote area power supplies despite their limited lifetime compared to other system components. Lead-acid batteries can be deep or shallow cycling gelled batteries, batteries with captive or liquid electrolyte, sealed and non-sealed batteries etc. [4]. Sealed batteries are valve regulated to permit evolution of excess hydrogen gas (although catalytic converters are used to convert as much evolved hydrogen and oxygen back to water as possible). Sealed batteries need less maintenance. The following factors are considered in the selection of batteries for PV applications [1]:

- Deep discharge (70–80% depth of discharge).
- Low charging/discharging current.
- Long duration charge (slow) and discharge (long duty cycle).
- Irregular and varying charge/discharge.
- Low self discharge.
- Long life time.
- Less maintenance requirement.
- High energy storage efficiency.
- Low cost.

Battery manufacturers specify the nominal number of complete charge and discharge cycles as a function of the depth-of-discharge (DOD), as shown in Fig. 27.10. While this information can be used reliably to predict the lifetime of lead-acid batteries in conventional applications, such as uninterruptable power supplies or electric vehicles, it usually results in an overestimation of the useful life of the battery bank in renewable energy systems.

Two of the main factors that have been identified as limiting criteria for the cycle life of batteries in PV power systems are incomplete charging and prolonged operation at a low state-of-charge (SOC). The objective of improved battery control strategies is to extend the lifetime of lead-acid batteries

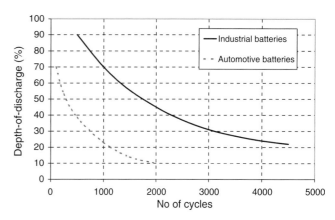

FIGURE 27.10 Nominal number of battery cycles vs DOD.

to achieve a typical number of cycles shown in Fig. 27.10. If this is achieved, an optimum solution for the required storage capacity and the maximum DOD of the battery can be found by referring to manufacturer's information. Increasing the capacity will reduce the typical DOD and therefore prolong the battery lifetime. Conversely, it may be more economic to replace a smaller battery bank more frequently.

27.2.3.1.2 PV Charge Controllers Blocking diodes in series with PV modules are used to prevent the batteries from being discharged through the PV cells at night when there is no sun available to generate energy. These blocking diodes also protect the battery from short circuits. In a solar power system consisting of more than one string connected in parallel, if a short circuit occurs in one of the strings, the blocking diode prevents the other PV strings to discharge through the short-circuited string.

The battery storage in a PV system should be properly controlled to avoid catastrophic operating conditions like overcharging or frequent deep discharging. Storage batteries account for most PV system failures and contribute significantly to both the initial and the eventual replacement costs. Charge controllers regulate the charge transfer and prevent the battery from being excessively charged and discharged. Three types of charge controllers are commonly used:

- Series charge regulators.
- Shunt charge regulators.
- DC–DC converters.

A. A Series Charge Regulators

The basic circuit for the series regulators is given in Fig. 27.11. In the series charge controller, the switch S_1 disconnects the PV generator when a predefined battery voltage is achieved. When the voltage reduces below the discharge limit, the load is disconnected from the battery to avoid deep discharge beyond the limit. The main problem associated with this type of controller is the losses associated with the switches. This extra

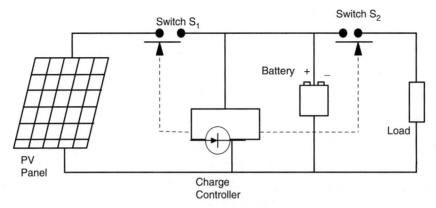

FIGURE 27.11 Series charge regulator.

power loss has to come from the PV power and this can be quite significant. Bipolar transistors, metal oxide semi conductor field effect transistors (MOSFETs), or relays are used as the switches.

B. Shunt Charge Regulators

In this type, as illustrated in Fig. 27.12, when the battery is fully charged the PV generator is short-circuited using an electronic switch (S_1). Unlike series controllers, this method works more efficiently even when the battery is completely discharged as the short-circuit switch need not be activated until the battery is fully discharged [1].

The blocking diode prevents short-circuiting of the battery. Shunt-charge regulators are used for the small PV applications (less than 20 A).

Deep discharge protection is used to protect the battery against the deep discharge. When the battery voltage reaches below the minimum set point for deep discharge limit, switch S_2 disconnects the load. Simple series and shunt regulators allow only relatively coarse adjustment of the current flow and seldom meet the exact requirements of PV systems.

C. DC–DC Converter Type Charge Regulators

Switch mode DC-to-DC converters are used to match the output of a PV generator to a variable load. There are various types of DC–DC converters such as:

- Buck (step-down) converter.
- Boost (step-up) converter.
- Buck–boost (step-down/up) converter.

Figures 27.13–27.15 show simplified diagrams of these three basic types converters. The basic concepts are an electronic switch, an inductor to store energy, and a "flywheel" diode, which carries the current during that part of switching cycle

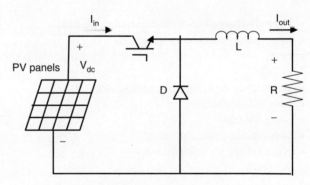

FIGURE 27.13 Buck converter.

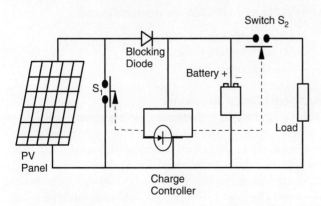

FIGURE 27.12 Shunt charge regulator.

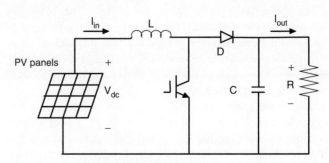

FIGURE 27.14 Boost converter.

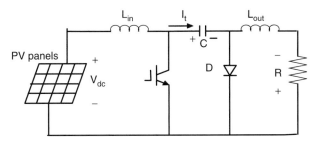

FIGURE 27.15 Boost–buck converter.

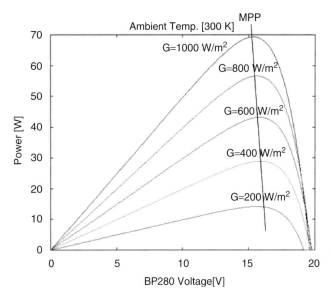

FIGURE 27.16 Typical power/voltage characteristics for increased insolation.

when the switch is off. The DC–DC converters allow the charge current to be reduced continuously in such a way that the resulting battery voltage is maintained at a specified value.

27.2.3.1.3 Maximum Power Point Tracking (MPPT)

A controller that tracks the maximum power point locus of the PV array is known as the MPPT. In Fig. 27.16, the PV power output is plotted against the voltage for insolation levels from 200 to 1000 W/m² [5]. The points of maximum array power form a curve termed as the maximum power locus. Due to high cost of solar cells, it is necessary to operate the PV array at its maximum power point (MPP). For overall optimal operation of the system, the load line must match the PV array's MPP locus.

Referring to Fig. 27.17, the load characteristics can be either curve OA or curve OB depending upon the nature of the load and it's current and voltage requirements. If load OA is considered and the load is directly coupled to the solar array, the array will operate at point A1, delivering only power P1. The maximum array power available at the given insolation is P2. In order to use PV array power P2, a power conditioner coupled between array and the load is needed.

There are generally two ways of operating PV modules at maximum power point. These ways take advantage of analog and/or digital hardware control to track the MPP of PV arrays.

27.2.3.1.4 Analog Control

There are many analog control mechanisms proposed in different articles. For instance, fractional short-circuit current (I_{SC}) [6–9], fractional open-circuit voltage (V_{OP}) [6, 7, 10–13], and ripple correlation control (RCC) [14–17].

Fractional open-circuit voltage (V_{OP}) is one of the simple analogue control method. It is based on the assumption that the maximum power point voltage, V_{MPP}, is a linear function

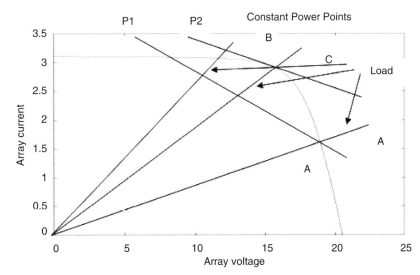

FIGURE 27.17 PV array and load characteristics.

of the open-circuit voltage, V_{OC}. For example $V_{MPP} = kV_{OC}$ where $k \approx 0.76$. This assumption is reasonably accurate even for large variations in the cell short-circuit current and temperature. This type of MPPT is probably the most common type. A variation to this method involves periodically open-circuiting the cell string and measuring the open-circuit voltage. The appropriate value of V_{MPP} can then be obtained with a simple voltage divider.

27.2.3.1.5 Digital Control
There are many digital control mechanisms that were proposed in different articles. For instance, perturbation and observation (P&O) or hill climbing [18–23], fuzzy logic [24–28], neural network [18, 29–31], and incremental conductance (IncCond) [32–35].

The P&O or hill climbing control involves around varying the input voltage around the optimum value by giving it a small increment or decrement alternately. The effect on the output power is then assessed and a further small correction is made to the input voltage. Therefore, this type of control is called a hill climbing control. The power output of the PV array is sampled at an every definite sampling period and compared with the previous value. In the event, when power is increased then the solar array voltage is stepped in the same direction as the previous sample time, but if the power is reduced then the array voltage is stepped in the opposite way and try to operate the PV array at its optimum/maximum power point.

To operate the PV array at the MPP, perturb and adjust method can be used at regular intervals. Current drawn is sampled every few seconds and the resulting power output of the solar cells is monitored at regular intervals. When an increased current results in a higher power, it is further increased until power output starts to reduce. But if the increased PV current results in lesser amount of power than in the previous sample, then the current is reduced until the MPP is reached.

27.2.3.2 Inverters for Stand-alone PV Systems
Inverters convert power from DC to AC while rectifiers convert it from AC to DC. Many inverters are bi-directional, i.e. they are able to operate in both inverting and rectifying modes. In many stand-alone PV installations, alternating current is needed to operate 230 V (or 110 V), 50 Hz (or 60 Hz) appliances. Generally stand-alone inverters operate at 12, 24, 48, 96, 120, or 240 V DC depending upon the power level. Ideally, an inverter for a stand-alone PV system should have the following features:

- Sinusoidal output voltage.
- Voltage and frequency within the allowable limits.
- Cable to handle large variation in input voltage.
- Output voltage regulation.
- High efficiency at light loads.
- Less harmonic generation by the inverter to avoid damage to electronic appliances like television, additional losses, and heating of appliances.
- Photovoltaic inverters must be able to withstand overloading for short term to take care of higher starting currents from pumps, refrigerators, etc.
- Adequate protection arrangement for over/under-voltage and frequency, short circuit etc.
- Surge capacity.
- Low idling and no load losses.
- Low battery voltage disconnect.
- Low audio and radio frequency (RF) noise.

Several different semiconductor devices such as metal oxide semiconductor field effect transistor (MOSFETs) and insulated gate bipolar transistors (IGBTs) are used in the power stage of inverters. Typically MOSFETs are used in units up to 5 kVA and 96 V DC. They have the advantage of low switching losses at higher frequencies. Because the on-state voltage drop is 2 V DC, IGBTs are generally used only above 96 V DC systems.

Voltage source inverters are usually used in stand-alone applications. They can be single phase or three phase. There are three switching techniques commonly used: square wave, quasi-square wave, and pulse width modulation. Square-wave or modified square-wave inverters can supply power tools, resistive heaters, or incandescent lights, which do not require a high quality sine wave for reliable and efficient operation. However, many household appliances require low distortion sinusoidal waveforms. The use of true sine-wave inverters is recommended for remote area power systems. Pulse width modulated (PWM) switching is generally used for obtaining sinusoidal output from the inverters.

A general layout of a single-phase system, both half bridge and full bridge, is shown in Fig. 27.18. In Fig. 27.18a, single-phase half bridge is with two switches, S_1 and S_2, the capacitors C_1 and C_2 are connected in series across the DC source. The junction between the capacitors is at the mid-potential. Voltage across each capacitor is $V_{dc}/2$. Switches S_1 and S_2 can be switched on/off periodically to produce AC voltage. Filter (L_f and C_f) is used to reduce high-switch frequency components and to produce sinusoidal output from the inverter. The output of inverter is connected to load through a transformer. Figure 27.18b shows the similar arrangement for full-bridge configuration with four switches. For the same input source voltage, the full-bridge output is twice and the switches carry less current for the same load power.

The power circuit of a three phase four-wire inverter is shown in Fig. 27.19. The output of the inverter is connected to load via three-phase transformer (delta/Y). The star point of the transformer secondary gives the neutral connection. Three phase or single phase can be connected to this system. Alternatively, a center tap DC source can be used to supply the converter and the mid-point can be used as the neutral.

27 Power Electronics for Renewable Energy Sources

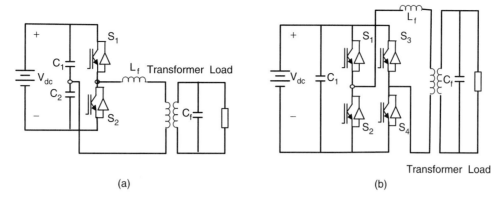

FIGURE 27.18 Single-phase inverter: (a) half bridge and (b) full bridge.

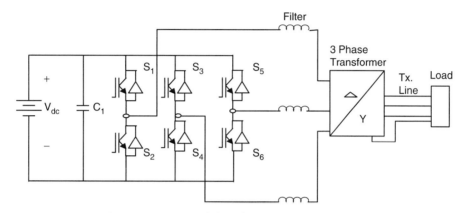

FIGURE 27.19 A stand-alone three-phase four wire inverter.

Figure 27.20 shows the inverter efficiency for a typical inverter used in remote area power systems. It is important to consider that the system load is typically well below the nominal inverter capacity P_{nom}, which results in low conversion efficiencies at loads below 10% of the rated inverter output power. Optimum overall system operation is achieved if the total energy dissipated in the inverter is minimized. The high conversion efficiency at low power levels of recently developed inverters for grid-connected PV systems shows that there is a significant potential for further improvements in efficiency.

Bi-directional inverters convert DC power to AC power (inverter) or AC power to DC power (rectifier) and are becoming very popular in remote area power systems [4, 5]. The principle of a stand-alone single-phase bi-directional inverter used in a PV/battery/diesel hybrid system can be explained by referring Fig. 27.21. A charge controller is used to interface the PV array and the battery. The inverter has a full-bridge configuration realized using four power electronic switches (MOSFET or IGBTs) S_1–S_4. In this scheme, the diagonally opposite switches (S_1, S_4) and (S_2, S_3) are switched using a sinusoidally PWM gate pulses. The inverter produces sinusoidal output voltage. The inductors X_1, X_2, and the AC output capacitor C_2 filter out the high-switch frequency components

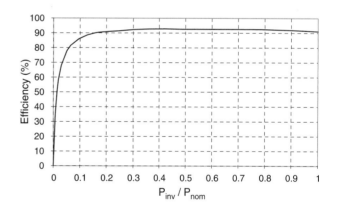

FIGURE 27.20 Typical inverter efficiency curve.

from the output waveform. Most inverter topologies use a low frequency (50 or 60 Hz) transformer to step up the inverter output voltage. In this scheme, the diesel generator and the converter are connected in parallel to supply the load. The voltage sources, diesel and inverter, are separated by the link inductor X_m. The bi-directional power flow between inverter and the diesel generator can be established.

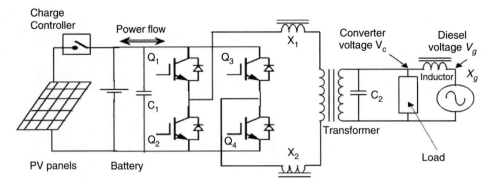

FIGURE 27.21 Bi-directional inverter system.

The power flow through the link inductor, X_m, is

$$S_m = V_m I_m^* \tag{27.3}$$

$$P_m = (V_m V_c \sin \delta)/X_m \tag{27.4}$$

$$Q_m = (V_m/X_m)(V_m - V_c \cos \delta) \tag{27.5}$$

$$\delta = \sin^{-1}[(X_m P_m)/(V_m V_c)] \tag{27.6}$$

where δ is the phase angle between the two voltages. From Eq. (27.4), it can be seen that the power supplied by the inverter from the batteries (inverter mode) or supplied to the batteries (charging mode) can be controlled by controlling the phase angle δ. The PWM pulses separately control the amplitude of the converter voltage, V_c, while the phase angle with respect to the diesel voltage is varied for power flow.

27.2.3.3 Solar Water Pumping

In many remote and rural areas, hand pumps or diesel driven pumps are used for water supply. Diesel pumps consume fossil fuel, affects environment, needs more maintenance, and are less reliable. Photovoltaic powered water pumps have received considerable attention recently due to major developments in the field of solar cell materials and power electronic systems technology.

27.2.3.3.1 Types of Pumps Two types of pumps are commonly used for the water pumping applications: positive and centrifugal displacement. Both centrifugal and positive displacement pumps can be further classified into those with motors that are (a) surface mounted and those which are (b) submerged into the water ("submersible").

Displacement pumps have water output directly proportional to the speed of the pump, but, almost independent of head. These pumps are used for solar water pumping from deep wells or bores. They may be piston type pumps, or use diaphragm driven by a cam, rotary screw type, or use progressive cavity system. The pumping rate of these pumps is directly related to the speed and hence constant torque is desired.

Centrifugal pumps are used for low-head applications especially if they are directly interfaced with the solar panels. Centrifugal pumps are designed for fixed-head applications and the pressure difference generated increases in relation to the speed of pump. These pumps are rotating impeller type, which throws the water radially against a casing, so shaped that the momentum of the water is converted into useful pressure for lifting [4]. The centrifugal pumps have relatively high efficiency but it reduces at lower speeds, which can be a problem for the solar water pumping system at the time of low light levels. The single-stage centrifugal pump has just one impeller whereas most borehole pumps are multistage types where the outlet from one impeller goes into the center of another and each one keeps increasing the pressure difference.

From Fig. 27.22, it is quite obvious that the load line is located relatively faraway from P_{max} line. It has been reported that the daily utilization efficiency for a DC motor drive is 87% for a centrifugal pump compared to 57% for a constant torque characteristics load. Hence, centrifugal pumps are more compatible with PV arrays. The system operating point

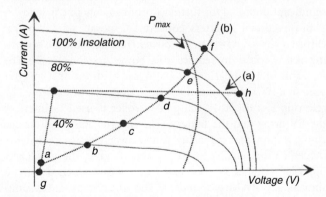

FIGURE 27.22 I–V characteristics of a PV array and two mechanical loads: (a) constant torque and (b) centrifugal pump.

is determined by the intersection of the *I–V* characteristics of the PV array and the motor as shown in Fig. 27.22. The torque-speed slope is normally large due to the armature resistance being small. At the instant of starting, the speed and the back emf are zero. Hence the motor starting current is approximately the short-circuit current of the PV array. By matching the load to the PV source through MPPT, the starting torque increases.

The matching of a DC motor depends upon the type of load being used. For instance, a centrifugal pump is characterized by having the load torque proportional to the square of speed. The operating characteristics of the system (i.e. PV source, permanent magnet (PM) DC motor and load) are at the intersection of the motor and load characteristics as shown in Fig. 27.23 (i.e. points a, b, c, d, e, and f for centrifugal pump). From Fig. 27.23, the system utilizing the centrifugal pump as its load tends to start at low solar irradiation (point a) level. However, for the systems with an almost constant torque characteristics in Fig. 27.22, the start is at almost 50% of one sun (full insolation) which results in short period of operation.

27.2.3.3.2 Types of Motors There are various types of motors available for the PV water pumping applications:

- DC motors.
- AC motors.

DC motors are preferred where direct coupling to PV panels is desired whereas AC motors are coupled to the solar panels through inverters. AC motors in general are cheaper than the DC motors and are more reliable but the DC motors are more

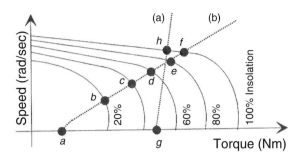

FIGURE 27.23 Speed torque characteristics of a DC motor and two mechanical loads: (a) helical rotor and (b) centrifugal pump.

efficient. The DC motors used for solar pumping applications are:

- Permanent magnet DC motors with brushes.
- Permanent DC magnet motors without brushes.

In DC motors with the brushes, the brushes are used to deliver power to the commutator and need frequent replacement due to wear and tear. These motors are not suitable for submersible applications unless long transmission shafts are used. Brush-less DC permanent magnet motors have been developed for submersible applications.

The AC motors are of the induction motor type, which is cheaper than DC motors and available, worldwide. However, they need inverters to change DC input from PV to AC power. A comparison of the different types of motors used for PV water pumping is given in Table 27.1.

TABLE 27.1 Comparison of the different types of motor used for PV water pumping

Types of motor	Advantages	Disadvantages	Main features
Brushed DC	Simple and efficient for PV applications. No complex control circuits is required as the motor starts without high current surge. These motors will run slowly but do not overheat with reduced voltage.	Brushes need to be replaced periodically (typical replacement interval is 2000–4000 hr or 2 years).	Requires MPPT for optimum performance. Available only in small motor sizes. Increasing current (by paralleling PV modules) increases the torque. Increasing voltage (by series PV modules) increases the speed.
Brush-less DC	Efficient. Less maintenance is required.	Electronic computation adds to extra cost, complexity, and increased risk of failure/malfunction. In most cases, oil cooled, can't be submerged as deep as water cooled AC units.	Growing trend among PV pump manufacturers to use brush-less DC motors, primarily for centrifugal type submersible pumps.
AC induction motors	No brushes to replace. Can use existing AC motor/pump technology which is cheaper and easily available worldwide. These motors can handle larger pumping requirements.	Needs an inverter to convert DC output from PV to AC adding additional cost and complexity. Less efficient than DC motor-pump units. Prone to overheating if current is not adequate to start the motor or if the voltage is too low.	Available for single or three supply. Inverters are designed to regulate frequency to maximize power to the motor in response to changing insolation levels.

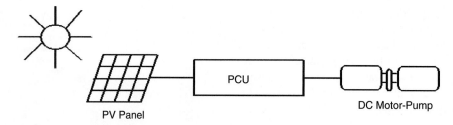

FIGURE 27.24 Block diagram for DC motor driven pumping scheme.

27.2.3.3.3 Power Conditioning Units for PV Water Pumping

Most PV pump manufacturers include power conditioning units (PCU) which are used for operating the PV panels close to their MPP over a range of load conditions and varying insolation levels and also for power conversion. DC or AC motor-pump units can be used for PV water pumping. In its simplest form, a solar water pumping system comprises of PV array, PCU, and DC water-pump unit as shown in Fig. 27.24.

In case of lower light levels, high currents can be generated through power conditioning to help in starting the motor-pump units especially for reciprocating positive displacement type pumps with constant torque characteristics, requiring constant current throughout the operating region. In positive displacement type pumps, the torque generated by the pumps depends on the pumping head, friction, and pipe diameter etc. and needs certain level of current to produce the necessary torque. Some systems use electronic controllers to assist starting and operation of the motor under low solar radiation. This is particularly important when using positive displacement pumps. The solar panels generate DC voltage and current. The solar water pumping systems usually has DC or AC pumps. For DC pumps, the PV output can be directly connected to the pump through MPPT or a DC–DC converter can also be used for interfacing for controlled DC output from PV panels. To feed the AC motors, a suitable interfacing is required for the power conditioning. These PV inverters for the stand-alone applications are very expensive. The aim of power conditioning equipment is to supply the controlled voltage/current output from the converters/inverters to the motor-pump unit.

These power-conditioning units are also used for operating the PV panels close to their maximum efficiency for fluctuating solar conditions. The speed of the pump is governed by the available driving voltage. Current lower than the acceptable limit will stop the pumping. When the light level increases, the operating point will shift from the MPP leading to the reduction of efficiency. For centrifugal pumps, there is an increase in current at increased speed and the matching of I–V characteristics is closer for wide range of light intensity levels. For centrifugal pumps, the torque is proportional to the square of speed and the torque produced by the motors is proportional to the current. Due to decrease in PV current output, the torque from the motor and consequently the speed of the pump is reduced resulting in decrease in back emf and the required voltage of the motor. Maximum power point tracker can be used for controlling the voltage/current outputs from the PV inverters to operate the PV close to maximum operating point for the smooth operation of motor-pump units. The DC–DC converter can be used for keeping the PV panels output voltage constant and help in operating the solar arrays close to MPP. In the beginning, high starting current is required to produce high starting torque. The PV panels cannot supply this high starting current without adequate power conditioning equipment like DC–DC converter or by using a starting capacitor. The DC–DC converter can generate the high starting currents by regulating the excess PV array voltage. DC–DC converter can be boost or buck converter.

Brush-less DC motor (BDCM) and helical rotor pumps can also be used for PV water pumping [36]. Brush-less DC motors are a self synchronous type of motor characteristics by trapezoidal waveforms for back emf and air flux density. They can operate off a low voltage DC supply which is switched through an inverter to create a rotating stator field. The current generation of BDCMs use rare earth magnets on the rotor to give high air gap flux densities and are well suited to solar application. The block diagram of such an arrangement is shown in Fig. 27.25 which consists of PV panels, DC–DC converter, MPPT, and BDCM.

The PV inverters are used to convert the DC output of the solar arrays to the AC quantity so as to run the AC motors driven pumps. These PV inverters can be variable frequency type, which can be controlled to operate the motors over wide range of loads. The PV inverters may involve impedance matching to match the electrical characteristics of the load and array. The motor-pump unit and PV panels operate at their maximum efficiencies. Maximum power point tracker is also used in the power conditioning. To keep the voltage stable for the inverters, the DC–DC converter can be used. The inverter/converter has a capability of injecting high-switch frequency components, which can lead to the overheating and the losses. So care shall be taken for this. The PV arrays are usually connected in series, parallel, or a combination of series parallel, configurations. The function of power electronic interface, as mentioned before, is to convert the DC power from the array to the required voltage and frequency to drive the AC motors.

27 Power Electronics for Renewable Energy Sources

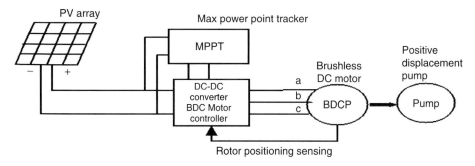

FIGURE 27.25 Block diagram for BDCM for PV application.

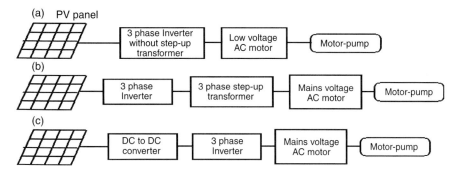

FIGURE 27.26 Block diagrams for various AC motor driven pumping schemes.

The motor-pump system load should be such that the array operates close to it's MPP at all solar insolation levels. There are mainly three types solar powered water pumping systems as shown in Fig. 27.26.

The first system shown in Fig. 27.26a is an imported commercially available unit, which uses a specially wound low voltage induction motor driven submersible pump. Such a low voltage motor permits the PV array voltage to be converted to AC without using a step-up transformer. The second system, shown in Fig. 27.26b makes use of a conventional "off-the-shelf" 415 V, 50 Hz, induction motor [6]. This scheme needs a step-up transformer to raise inverter output voltage to high voltage. Third scheme as shown in Fig. 27.26c comprises of a DC–DC converter, an inverter that switches at high frequency, and a mains voltage motor driven pump. To get the optimum discharge (Q), at a given insolation level, the efficiency of the DC–DC converter and the inverter should be high. So the purpose should be to optimize the output from PV array, motor, and the pump. The principle used here is to vary the duty cycle of a DC–DC converter so that the output voltage is maximum. The DC–DC converter is used to boost the solar array voltage to eliminate the need for a step-up transformer and operate the array at the MPP. The three-phase inverter used in the interface is designed to operate in a variable frequency mode over the range of 20–50 Hz, which is the practical limit for most 50 Hz induction motor applications. Block diagram for frequency control is given in Fig. 27.27.

This inverter would be suitable for driving permanent magnet motors by incorporating additional circuitry for position sensing of the motor's shaft. Also the inverter could be modified, if required, to produce higher output frequencies for high-speed permanent magnet motors. The inverter has a three-phase full-bridge configuration implemented by MOSFET power transistors.

27.2.4 Hybrid Energy Systems

The combination of RES, such as PV arrays or wind turbines, with engine-driven generators and battery storage, is widely recognized as a viable alternative to conventional remote area power supplies (RAPS). These systems are generally classified as hybrid energy systems (HES). They are used increasingly for electrification in remote areas where the cost of grid extension is prohibitive and the price for fuel increases drastically with the remoteness of the location. For many applications, the combination of renewable and conventional energy sources compares favorably with fossil fuel-based RAPS systems, both in regard to their cost and technical performance. Because these systems employ two or more different sources of energy, they enjoy a very high degree of reliability as compared to single-source systems such as a stand-alone diesel generator or a stand-alone PV or wind system. Applications of hybrid energy systems range from small power supplies for remote

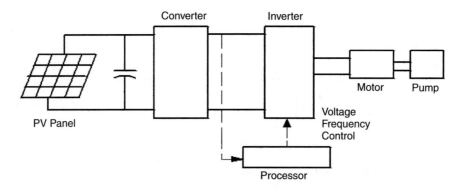

FIGURE 27.27 Block diagram for voltage/frequency control.

households, providing electricity for lighting and other essential electrical appliances, to village electrification for remote communities has been reported [37].

Hybrid energy systems generate AC electricity by combining RES such as PV array with an inverter, which can operate alternately or in parallel with a conventional engine-driven generator. They can be classified according to their configuration as [38]:

- Series hybrid energy systems.
- Switched hybrid energy systems.
- Parallel hybrid energy systems.

The parallel hybrid systems can be further divided to DC or AC coupling. An overview of the three most common system topologies is presented by Bower [39]. In the following comparison of typical PV-diesel system configurations are described.

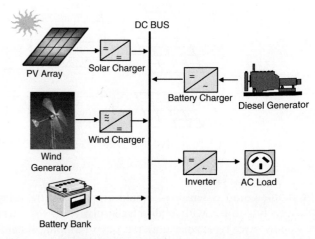

FIGURE 27.28 Series hybrid energy system.

27.2.4.1 Series Configuration

In the conventional series hybrid systems shown in Fig. 27.28, all power generators feed DC power into a battery. Each component has therefore to be equipped with an individual charge controller and in the case of a diesel generator with a rectifier.

To ensure reliable operation of series hybrid energy systems both the diesel generator and the inverter have to be sized to meet peak loads. This results in a typical system operation where a large fraction of the generated energy is passed through the battery bank, therefore resulting in increased cycling of the battery bank and reduced system efficiency. AC power delivered to the load is converted from DC to regulated AC by an inverter or a motor generator unit. The power generated by the diesel generator is first rectified and subsequently converted back to AC before being supplied to the load, which incurs significant conversion losses.

The actual load demand determines the amount of electrical power delivered by the PV array, wind generator, the battery bank, or the diesel generator. The solar and wind charger prevents overcharging of the battery bank from the PV generator when the PV power exceeds the load demand and the batteries are fully charged. It may include MPPT to improve the utilization of the available PV energy, although the energy gain is marginal for a well-sized system. The system can be operated in manual or automatic mode, with the addition of appropriate battery voltage sensing and start/stop control of the engine-driven generator.

Advantages:

- The engine-driven generator can be sized to be optimally loaded while supplying the load and charging the battery bank, until a battery SOC of 70–80% is reached.
- No switching of AC power between the different energy sources is required, which simplifies the electrical output interface.
- The power supplied to the load is not interrupted when the diesel generator is started.
- The inverter can generate a sine-wave, modified square-wave, or square-wave depending on the application.

Disadvantages:

- The inverter cannot operate in parallel with the engine-driven generator, therefore the inverter must be sized to supply the peak load of the system.
- The battery bank is cycled frequently, which shortens its lifetime.
- The cycling profile requires a large battery bank to limit the depth-of-discharge (DOD).
- The overall system efficiency is low, since the diesel cannot supply power directly to the load.
- Inverter failure results in complete loss of power to the load, unless the load can be supplied directly from the diesel generator for emergency purposes.

27.2.4.2 Switched Configuration

Despite its operational limitations, the switched configuration remains one of the most common installations in some developing countries. It allows operation with either the engine-driven generator or the inverter as the AC source, yet no parallel operation of the main generation sources is possible. The diesel generator and the RES can charge the battery bank. The main advantage compared with the series system is that the load can be supplied directly by the engine-driven generator, which results in a higher overall conversion efficiency. Typically, the diesel generator power will exceed the load demand, with excess energy being used to recharge the battery bank. During periods of low electricity demand the diesel generator is switched off and the load is supplied from the PV array together with stored energy. Switched hybrid energy systems can be operated in manual mode, although the increased complexity of the system makes it highly desirable to include an automatic controller, which can be implemented with the addition of appropriate battery voltage sensing and start/stop control of the engine-driven generator (Fig. 27.29).

Advantages:

- The inverter can generate a sine-wave, modified square-wave, or square-wave, depending on the particular application.
- The diesel generator can supply the load directly, therefore improving the system efficiency and reducing the fuel consumption.

Disadvantages:

- Power to the load is interrupted momentarily when the AC power sources are transferred.
- The engine-driven alternator and inverter are typically designed to supply the peak load, which reduces their efficiency at part load operation.

27.2.4.3 Parallel Configuration

The parallel hybrid system can be further classified as DC and AC couplings as shown in Fig. 27.30. In both schemes, a bi-directional inverter is used to link between the battery and an AC source (typically the output of a diesel generator). The bi-directional inverter can charge the battery bank (rectifier operation) when excess energy is available from the diesel generator or by the renewable sources, as well as act as a DC–AC converter (inverter operation). The bi-directional inverter may also provide "peak shaving" as part of a control strategy when the diesel engine is overloaded. In Fig. 27.30a, the renewable energy sources (RES) such as photovoltaic and wind are coupled on the DC side. DC integration of RES results in "custom" system solutions for individual supply cases requiring high costs for engineering, hardware, repair, and maintenance. Furthermore, power system expandability for covering needs of growing energy and power demand is also difficult. A better approach would be to integrate the RES on the AC side rather than on the DC side as shown in Fig. 27.30b.

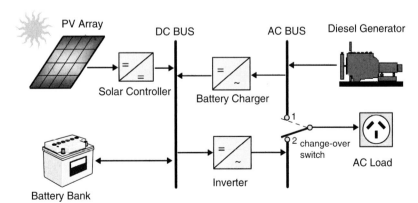

FIGURE 27.29 Switched PV-diesel hybrid energy system.

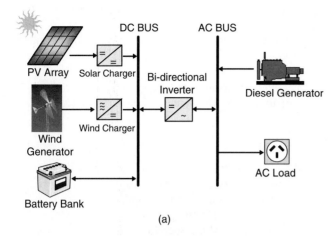

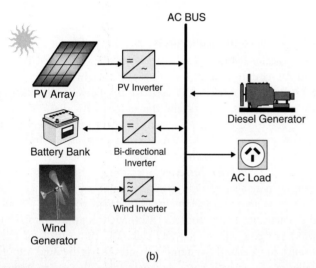

FIGURE 27.30 Parallel PV diesel hybrid energy system: (a) DC decoupling and (b) AC coupling.

Parallel hybrid energy systems are characterized by two significant improvements over the series and switched system configuration.

The inverter plus the diesel generator capacity rather than their individual component ratings limit the maximum load that can be supplied. Typically, this will lead to a doubling of the system capacity. The capability to synchronize the inverter with the diesel generator allows greater flexibility to optimize the operation of the system. Future systems should be sized with a reduced peak capacity of the diesel generator, which results in a higher fraction of directly used energy and hence higher system efficiencies.

By using the same power electronic devices for both inverter and rectifier operation, the number of system components is minimized. Additionally, wiring and system installation costs are reduced through the integration of all power-conditioning devices in one central power unit. This highly integrated system concept has advantages over a more modular approach to system design, but it may prevent convenient system upgrades when the load demand increases.

The parallel configuration offers a number of potential advantages over other system configurations. These objectives can only be met if the interactive operation of the individual components is controlled by an "intelligent" hybrid energy management system. Although today's generation of parallel systems include system controllers of varying complexity and sophistication, they do not optimize the performance of the complete system. Typically, both the diesel generator and the inverter are sized to supply anticipated peak loads. As a result most parallel hybrid energy systems do not utilize their capability of parallel, synchronized operation of multiple power sources.

Advantages:

- The system load can be met in an optimal way.
- Diesel generator efficiency can be maximized.
- Diesel generator maintenance can be minimized.
- A reduction in the rated capacities of the diesel generator, battery bank, inverter, and renewable resources is feasible, while also meeting the peak loads.

Disadvantages:

- Automatic control is essential for the reliable operation of the system.
- The inverter has to be a true sine-wave inverter with the ability to synchronize with a secondary AC source.
- System operation is less transparent to the untrained user of the system.

27.2.4.4 Control of Hybrid Energy Systems

The design process of hybrid energy systems requires the selection of the most suitable combination of energy sources, power-conditioning devices, and energy storage system together with the implementation of an efficient energy dispatch strategy. System simulation software is an essential tool to analyze and compare possible system combinations. The objective of the control strategy is to achieve optimal operational performance at the system level. Inefficient operation of the diesel generator and "dumping" of excess energy is common for many RAPS, operating in the field. Component maintenance and replacement contributes significantly to the lifecycle cost of systems. These aspects of system operation are clearly related to the selected control strategy and have to be considered in the system design phase.

Advanced system control strategies seek to reduce the number of cycles and the DOD for the battery bank, run the diesel generator in its most efficient operating range, maximize the utilization of the renewable resource, and ensure high reliability of the system. Due to the varying nature of the load demand, the fluctuating power supplied by the photovoltaic generator, and the resulting variation of battery SOC, the hybrid energy system controller has to respond

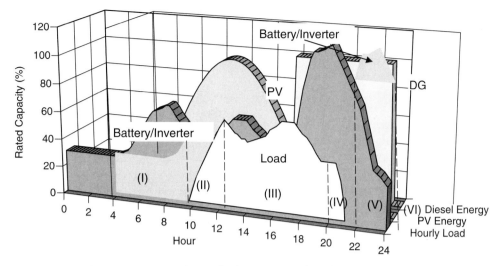

FIGURE 27.31 Operating modes for a PV single-diesel hybrid energy system.

to continuously changing operating conditions. Figure 27.31 shows different operating modes for a PV single-diesel system using a typical diesel dispatch strategy.

Mode (I): The base load, which is typically experienced at nighttime and during the early morning hours, is supplied by energy stored in the batteries. Photovoltaic power is not available and the diesel generator is not started.

Mode (II): PV power is supplemented by stored energy to meet the medium load demand.

Mode (III): Excess energy is available from the PV generator, which is stored in the battery. The medium load demand is supplied from the PV generator.

Mode (IV): The diesel generator is started and operated at its nominal power to meet the high evening load. Excess energy available from the diesel generator is used to recharge the batteries.

Mode (V): The diesel generator power is insufficient to meet the peak load demand. Additional power is supplied from the batteries by synchronizing the inverter AC output voltage with the alternator waveform.

Mode (VI): The diesel generator power exceeds the load demand, but it is kept operational until the batteries are recharged to a high SOC level.

In principle, most efficient operation is achieved if the generated power is supplied directly to the load from all energy sources, which also reduces cycling of the battery bank. However, since diesel generator operation at light loads is inherently inefficient, it is common practice to operate the engine-driven generator at its nominal power rating and to recharge the batteries from the excess energy. The selection of the most efficient control strategy depends on fuel, maintenance and component replacement cost, the system configuration, environmental conditions, as well as constraints imposed on the operation of the hybrid energy system.

27.2.5 Grid-connected PV Systems

The utility interactive inverters not only conditions the power output of the PV arrays but ensures that the PV system output is fully synchronized with the utility power. These systems can be battery less or with battery backup. Systems with battery storage (or flywheel) provide additional power supply reliability. The grid connection of PV systems is gathering momentum because of various rebate and incentive schemes. This system allows the consumer to feed its own load utilizing the available solar energy and the surplus energy can be injected into the grid under the energy by back scheme to reduce the payback period. Grid-connected PV systems can become a part of the utility system. The contribution of solar power depends upon the size of system and the load curve of the house. When the PV system is integrated with the utility grid, a two-way power flow is established. The utility grid will absorb excess PV power and will feed the house during nighttime and at instants while the PV power is inadequate. The utility companies are encouraging this scheme in many parts of the world. The grid-connected system can be classified as:

- Rooftop application of grid-connected PV system.
- Utility scale large system.

For small household PV applications, a roof mounted PV array can be the best option. Solar cells provide an environmentally clean way of producing electricity, and rooftops have always been the ideal place to put them. With a PV array on the rooftop, the solar generated power can supply residential load. The rooftop PV systems can help in reducing the peak summer load to the benefit of utility companies by feeding the household lighting, cooling, and other domestic loads. The battery storage can further improve the reliability of the system at the time of low insolation level, nighttime, or cloudy days. But the battery storage has some inherent problems like maintenance and higher cost.

For roof-integrated applications, the solar arrays can be either mounted on the roof or directly integrated into the roof. If the roof integration does not allow for an air channel behind the PV modules for ventilation purpose, then it can increase the cell temperature during the operation consequently leading to some energy losses. The disadvantage with the rooftop application is that the PV array orientation is dictated by the roof. In case, when the roof orientation differs from the optimal orientation required for the cells, then efficiency of the entire system would be suboptimal.

Utility interest in PV has centered on the large grid-connected PV systems. In Germany, USA, Spain, and in several other parts of the world, some large PV scale plants have been installed. The utilities are more inclined with large scale, centralized power supply. The PV systems can be centralized or distributed systems.

Grid-connected PV systems must observe the islanding situation, when the utility supply fails. In case of islanding, the PV generators should be disconnected from mains. PV generators can continue to meet only the local load, if the PV output matches the load. If the grid is re-connected during islanding, transient overcurrents can flow through the PV system inverters and the protective equipments like circuit breakers may be damaged. The islanding control can be achieved through inverters or via the distribution network. Inverter controls can be designed on the basis of detection of grid voltage, measurement of impedance, frequency variation, or increase in harmonics. Protection shall be designed for the islanding, short circuits, over/under-voltages/currents, grounding, and lightening, etc.

The importance of the power generated by the PV system depends upon the time of the day specially when the utility is experiencing the peak load. The PV plants are well suited to summer peaking but it depends upon the climatic condition of the site. PV systems being investigated for use as peaking stations would be competitive for load management. The PV users can defer their load by adopting load management to get the maximum benefit out of the grid-connected PV plants and feeding more power into the grid at the time of peak load.

The assigned capacity credit is based on the statistical probability with which the grid can meet peak demand [4]. The capacity factor during the peaks is very similar to that of conventional plants and similar capacity credit can be given for the PV generation except at the times when the PV plants are generating very less power unless adequate storage is provided. With the installation of PV plants, the need of extra transmission lines, transformers can be delayed or avoided. The distributed PV plants can also contribute in providing reactive power support to the grid and reduce burden on VAR compensators.

27.2.5.1 Inverters for Grid-connected Applications

Power conditioner is the key link between the PV array and mains in the grid-connected PV system. It acts as an interface that converts DC current produced by the solar cells into utility grade AC current. The PV system behavior relies heavily on the power-conditioning unit. The inverters shall produce good quality sine-wave output. The inverter must follow the frequency and voltage of the grid and the inverter has to extract maximum power from the solar cells with the help of MPPT and the inverter input stage varies the input voltage until the MPP on the I–V curve is found. The inverter shall monitor all the phases of the grid. The inverter output shall be controlled in terms of voltage and frequency variation. A typical grid-connected inverter may use a PWM scheme and operates in the range of 2–20 kHz.

27.2.5.2 Inverter Classifications

The inverters used for the grid interfacing are broadly classified as:

- Voltage source inverters (VSI).
- Current source inverters (CSI).

Whereas the inverters based on the control schemes can be classified as:

- Current controlled (CC).
- Voltage controlled (VC).

The source is not necessarily characterized by the energy source for the system. It is a characteristic of the topology of the inverter. It is possible to change from one source type to another source type by the addition of passive components. In the voltage source inverter (VSI), the DC side is made to appear to the inverter as a voltage source. The VSIs have a capacitor in parallel across the input whereas the CSIs have an inductor is series with the DC input. In the CSI, the DC source appears as a current source to the inverter. Solar arrays are fairly good approximation to a current source. Most PV inverters are voltage source even though the PV is a current source. Current source inverters are generally used for large motor drives though there have been some PV inverters built using a current source topology. The VSI is more popular with the PWM VSI dominating the sine-wave inverter topologies.

Figure 27.32a shows a single-phase full-bridge bi-directional VSI with (a) voltage control and phase-shift (δ) control – voltage-controlled voltage source inverter (VCVSI). The active power transfer from the PV panels is accomplished by controlling the phase angle δ between the converter voltage and the grid voltage. The converter voltage follows the grid voltage. Figure 27.32b shows the same VSI operated as a current controlled (CCVSI). The objective of this scheme is to control active and reactive components of the current fed into the grid using PWM techniques.

27 Power Electronics for Renewable Energy Sources

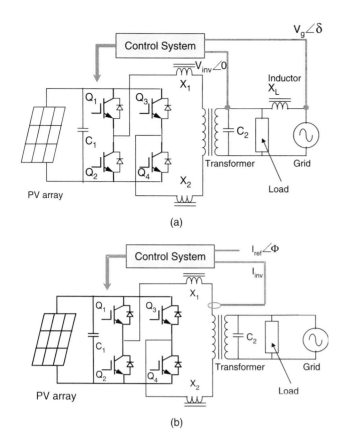

FIGURE 27.32 Voltage source inverter: (a) voltage control and (b) current control.

27.2.5.3 Inverter Types

Different types are being in use for the grid-connected PV applications such as:

- Line-commutated inverter.
- Self-commutated inverter.
- Inverter with high-frequency transformer.

27.2.5.3.1 Line-commutated Inverter
The line-commutated inverters are generally used for the electric motor applications. The power stage is equipped with thyristors. The maximum power tracking control is required in the control algorithm for solar application. The basic diagram for a single-phase line-commutated inverter is shown in the Fig. 27.33 [3].

The driver circuit has to be changed to shift the firing angle from the rectifier operation ($0 < \phi < 90$) to inverter operation ($90 < \phi < 180$). Six-pulse or 12-pulse inverter are used for the grid interfacing but 12-pulse inverters produce less harmonics. The thysistor type inverters require a low impedance grid interface connection for commutation purpose. If the maximum power available from the grid connection is less than twice the rated PV inverter power, then the line-commutated inverter should not be used [3]. The line-commutated inverters are cheaper but inhibits poor power quality. The harmonics injected into the grid can be large unless taken care of by employing adequate filters. These line-commutated inverters also have poor power factor, poor power quality, and need additional control to improve the power factor. Transformer can be used to provide the electrical isolation. To suppress the harmonics generated by these inverters, tuned filters are employed and reactive power compensation is required to improve the lagging power factor.

27.2.5.3.2 Self-commutated Inverter
A switch mode inverter using pulse width modulated (PWM) switching control, can be used for the grid connection of PV systems. The basic block diagram for this type of inverter is shown in the Fig. 27.34. The inverter bridges may consist of bipolar transistors, MOSFET transistors, IGBT's, or gate turn-off thyristor's (GTO's), depending upon the type of application. GTO's are used for the higher power applications, whereas IGBT's can be switched at higher frequencies i.e. 16 kHz, and are generally used for many grid-connected PV applications. Most of the present day inverters are self-commutated sine-wave inverters.

Based on the switching control, the voltage source inverters can be further classified based on the switching control as:

- PWM (pulse width modulated) inverters.
- Square-wave inverters.
- Single-phase inverters with voltage cancellations.
- Programmed harmonic elimination switching.
- Current controlled modulation.

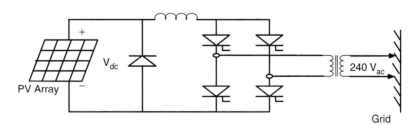

FIGURE 27.33 Line-commuted single-phase inverter.

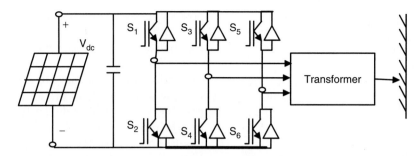

FIGURE 27.34 Self-commutated inverter with PWM switching.

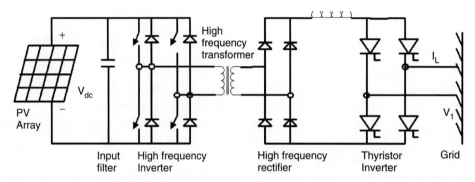

FIGURE 27.35 PV inverter with high frequency transformer.

27.2.5.3.3 Inverter with High-frequency Transformer The 50 Hz transformer for a standard PV inverter with PWM switching scheme can be very heavy and costly. While using frequencies more than 20 kHz, a ferrite core transformer can be a better option [3]. A circuit diagram of a grid-connected PV system using high frequency transformer is shown in the Fig. 27.35.

The capacitor on the input side of high frequency inverter acts as the filter. The high frequency inverter with PWM is used to produce a high frequency AC across the primary winding of the high frequency transformer. The secondary voltage of this transformer is rectified using high frequency rectifier. The DC voltage is interfaced with a thyristor inverter through low-pass inductor filter and hence connected to the grid. The line current is required to be sinusoidal and in phase with the line voltage. To achieve this, the line voltage (V_1) is measured to establish the reference waveform for the line current I_L^*. This reference current I_L^* multiplied by the transformer ratio gives the reference current at the output of high frequency inverter. The inverter output can be controlled using current control technique [40]. These inverters can be with low frequency transformer isolation or high frequency transformer isolation. The low frequency (50/60 Hz) transformer of a standard inverter with PWM is a very heavy and bulky component. For residential grid interactive rooftop inverters below 3 kW rating, high frequency transformer isolation is often preferred.

27.2.5.3.4 Other PV Inverter Topologies In this section, some of the inverter topologies discussed in various research papers have been discussed.

A. Multilevel Converters
Multilevel converters can be used with large PV systems where multiple PV panels can be configured to create voltage steps. These multilevel voltage-source converters can synthesize the AC output terminal voltage from different level of DC voltages and can produce staircase waveforms. This scheme involves less complexity, and needs less filtering. One of the schemes (half-bridge diode-clamped three level inverter [41]) is given in Fig. 27.36. There is no transformer in this topology. Multilevel converters can be beneficial for large systems in terms of cost and efficiency. Problems associated with shading and malfunction of PV units need to be addressed.

B. Non-insulated Voltage Source
In this scheme [42], string of low voltage PV panels or one high-voltage unit can be coupled with the grid through DC to DC converter and voltage-source inverter. This topology is shown in Fig. 27.37. PWM-switching scheme can be used to generate AC output. Filter has been used to reject the switching components.

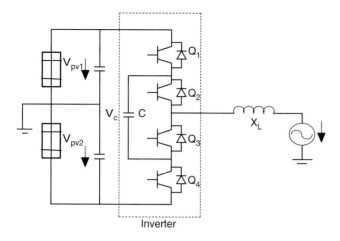

FIGURE 27.36 Half-bridge diode-clamped three-level inverter.

C. Non-insulated Current Source

This type of configuration is shown in Fig. 27.38. Non-insulated current-source inverters [42] can be used to interface the PV panels with the grid. This topology involves low cost which can provide better efficiency. Appropriate controller can be used to reduce current harmonics.

D. Buck Converter with Half-bridge Transformer Link

PV panels are connected to grid via buck converter and half bridge as shown in Fig. 27.39. In this, high-frequency PWM switching has been used at the low-voltage PV side to generate an attenuated rectified 100 Hz sine-wave current waveform [43]. Half-wave bridge is utilized to convert this output to 50 Hz signal suitable for grid interconnection. To step up the voltage, transformer has also been connected before the grid connection point.

E. Flyback Converter

This converter topology steps up the PV voltage to DC bus voltage. Pulse width modulation operated converter has been used for grid connection of PV system (Fig. 27.40). This scheme is less complex and has less number of switches. Flyback converters can be beneficial for remote areas due to less complex power conditioning components.

F. Interface Using Paralleled PV Panels

Low voltage AC bus scheme [44] can be comparatively efficient and cheaper option. One of the schemes is shown in Fig. 27.41. A number of smaller PV units can be paralleled together and then connected to combine single low-frequency transformer. In this scheme, the PV panels are connected in parallel rather than series to avoid problems associated with shading or malfunction of one of the panels in series connection.

27.2.5.4 Power Control through PV Inverters

The system shown in Fig. 27.42 shows control of power flow on to the grid [45]. This control can be an analog or a microprocessor system. This control system generates the waveforms and regulates the waveform amplitude and phase to control the power flow between the inverter and the grid. The grid-interfaced PV inverters, voltage-controlled VSI (VCVSI), or current-controlled VSI (CCVSI) have the potential of bi-directional power flow. They cannot only feed the local load but also can export the excess active and reactive power to the utility grid. An appropriate controller is required in order to avoid any error in power export due to errors in synchronization, which can overload the inverter.

There are advantages and limitations associated with each control mechanism. For instance, VCVSIs provide voltage support to the load (here the VSI operates as a voltage source), while CCVSIs provide current support (here the VSI operates as a current source). The CCVSI is faster in response compared to the VCVSI, as its power flow is controlled by the switching instant, whereas in the VCVSI the power flow is controlled by adjusting the voltage across the decoupling inductor. Active and reactive power are controlled independently in the CCVSI, but are coupled in the VCVSI. Generally, the advantages of one type of VSI are considered as a limitation of the other type [46].

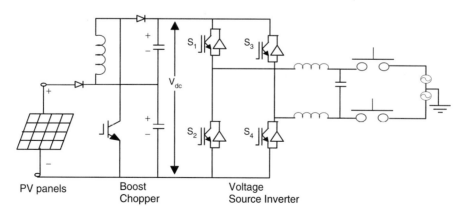

FIGURE 27.37 Non-insulated voltage source.

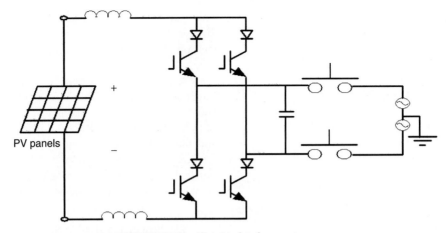

FIGURE 27.38 Non-insulated current source.

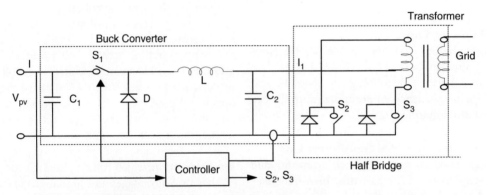

FIGURE 27.39 Buck converter with half-bridge transformer link.

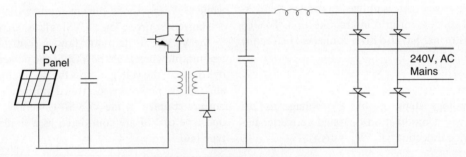

FIGURE 27.40 Flyback converter.

Figure 27.43 shows the simplified/equivalent schematic diagram of a VCVSI. For the following analysis it is assumed that the output low-pass filters (L_f and C_f) of VSIs will filter out high-order harmonics generated by PWMs. The decoupling inductor (X_m) is an essential part of any VCVSI as it makes the power flow control possible. In a VCVSI, the power flow of the distributed generation system (DGS) is controlled by adjusting the amplitude and phase (power angle (δ)) of the inverter output voltage with respect to the grid voltage. Hence, it is important to consider the proper sizing of the decoupling inductor and the maximum power angle to provide the required power flow when designing VCVSIs. The phasor diagram of a simple grid-inverter interface with a first-order filter are shown in Fig. 27.44.

Referring to Fig. 27.43, the fundamental grid current (I_g) can be expressed by Eq. (27.7):

$$I_g = \frac{V_g < 0 - V_c < \delta}{jX_m} = -\frac{V_c \sin\delta}{X_m} - j\frac{V_g - V_c \cos\delta}{X_m} \tag{27.7}$$

where V_g and V_c are respectively the grid and the VCVSI's fundamental voltages, and X_m is the decoupling inductor impedance. Using per unit values ($S_{base} = V_{base}^2/Z_{base}$,

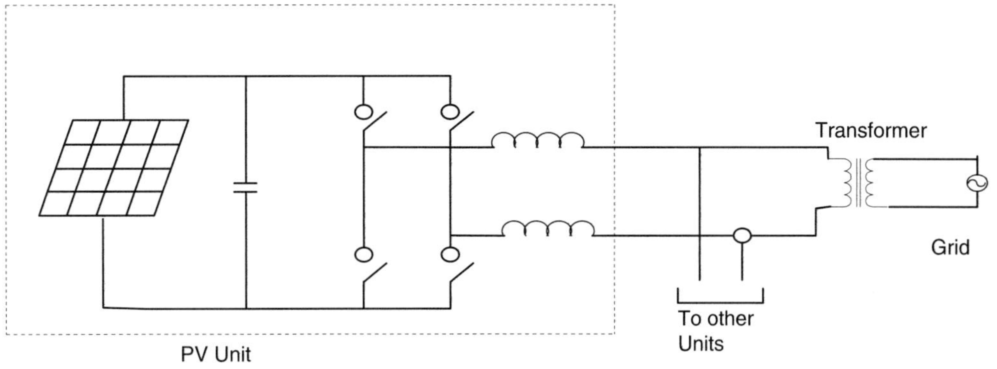

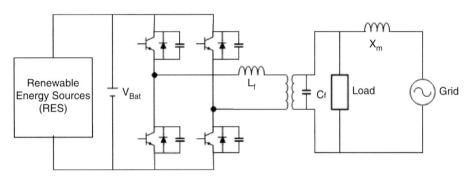

FIGURE 27.41 Converter using parallel PV units.

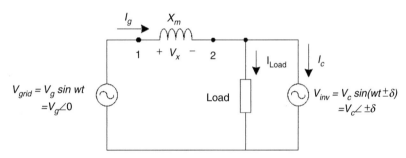

FIGURE 27.42 Schematic diagram of a parallel processing DGS.

FIGURE 27.43 The equivalent circuit diagram of a VCVSI.

$V_{base} = V_c$, and $Z_{base} = X_m$) where V_{base}, Z_{base}, and S_{base} are the base voltage, impedance and complex power values respectively. The grid apparent power can be expressed as Eq. (27.8).

$$S_{gpu} = -V_{gpu}\sin\delta + j(V_{gpu}^2 - V_{gpu}\cos\delta) \qquad (27.8)$$

Using per unit values, the complex power of the VCVSI and decoupling inductor are

$$S_{cpu} = -V_{gpu}\sin\delta + j(V_{gpu}\cos\delta - 1) \qquad (27.9)$$

$$S_{xpu} = j(V_{gpu}^2 - 2V_{gpu}\cos\delta + 1) \qquad (27.10)$$

where S_{gpu}, S_{cpu}, and S_{xpu} are per unit values of the grid, VCVSI, and decoupling inductor apparent power respectively, and V_{gpu} is the per unit value of the grid voltage.

Figure 27.45 shows the equivalent schematic diagram of a CCVSI. As a CCVSI controls the current flow using the VSI switching instants, it can be modeled as a current source and there is no need for a decoupling inductor (Fig. 27.45). As the current generated from the CCVSI can be controlled independently from the AC voltage, the active and reactive power controls are decoupled. Hence, unity power factor operation is possible for the whole range of the load. This is one of the main advantages of CCVSIs.

As the CCVSI is connected in parallel to the DGS, it follows the grid voltage. Figure 27.46 shows the phasor diagram of a

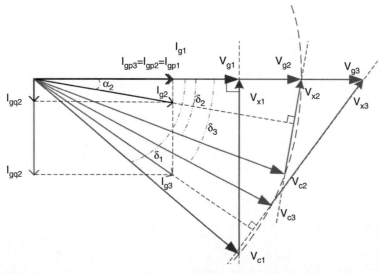

FIGURE 27.44 Phasor diagram of a VCVSI with resistive load and assuming the grid is responsible for supplying the active power [46].

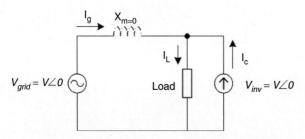

FIGURE 27.45 The equivalent circuit of a CCVSI.

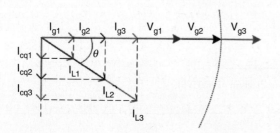

FIGURE 27.46 Phasor diagram of a CCVSI with inductive load and assuming grid is responsible for supplying the active power [46].

CCVSI based DGS in the presence of an inductive load (considering the same assumption as VCVSI section). Figure 27.49 shows that when the grid voltage increases, the load's active power consumption, which supplied by the grid increases and the CCVSI compensates the increase in the load reactive power demand. In this case, the CCVSI maintains grid supply at unity power factor, keeping the current phase delay with respect to the grid voltage at a fixed value (θ). Therefore, the CCVSI cannot maintain the load voltage in the presence of a DGS without utilizing extra hardware and control mechanisms. This limitation on load voltage stabilization is one of the main drawbacks of CCVSI based DGS.

Assuming the load active current demand is supplied by the grid (reactive power support function), the required grid current can be rewritten as follows

$$I_g^* = \text{Re}\,[I_L] = \text{Re}\left[\frac{S_L}{V_g}\right] \quad (27.11)$$

where, S_L is the demanded load apparent power. For grid power conditioning, it is preferred that the load operate at unity power factor. Therefore, the CCVSI must provide the remainder of the required current Eq. (27.12)

$$I_c = I_L - I_g^* \quad (27.12)$$

For demand side management (DSM), it is desirable to supply the active power by the RES, where excess energy from the RES is injected into the DGS. The remaining load reactive power will be supplied by the CCVSI. Hence Eq. (27.12) can be rewritten as Eq. (27.13).

$$I_g^* = \text{Re}\,[I_L] - \text{Re}\,[I_c] = \text{Re}\left[\frac{S_L - P_{RES}}{V_g}\right] \quad (27.13)$$

When using a voltage controller for grid-connected PV inverter, it has been observed that a slight error in the phase of synchronizing waveform can grossly overload the inverter whereas a current controller is much less susceptible to voltage phase shifts [45]. Due to this reason, the current controllers are better suited for the control of power export from the PV inverters to the utility grid since they are less sensitive to errors in synchronizing sinusoidal voltage waveforms.

A prototype current-controlled type power conditioning system has been developed by the first author and tested on a weak rural feeder line at Kalbarri in Western Australia [47]. The choice may be between additional conventional generating capacity at a centralized location or adding smaller distributed generating capacities using RES like PV. The latter option can have a number of advantages like:

- The additional capacity is added wherever it is required without adding additional power distribution infrastructure. This is a critical consideration where the power lines and transformers are already at or close to their maximum ratings.
- The power conditioning system can be designed to provide much more than just a source of real power, for minimal extra cost. A converter providing real power needs only a slight increase in ratings to handle significant amounts of reactive or even harmonic power. The same converter that converts DC PV power to AC power can simultaneously provide the reactive power support to the week utility grid.

The block diagram of the power conditioning system used in the Kalbarri project has been shown in the Fig. 27.47. This CCVSI operates with a relatively narrow switching frequency band near 10 kHz. The control diagram indicates the basic operation of the power conditioning system. The two outer control loops operate to independently control the real and reactive power flow from the PV inverter. The real power is controlled by an outer MPPT algorithm with an inner DC link voltage control loop providing the real current magnitude request I_p^* and hence the real power export through PV converter is controlled through the DC link voltage regulation. The DC link voltage is maintained at a reference value by a PI control loop, which gives the real current reference magnitude as it's output. At regular intervals, the DC link voltage is scanned over the entire voltage range to check that the algorithm is operating on the absolute MPP and is not stuck around a local MPP. During the night, the converter can still be used to regulate reactive power of the grid-connected system although it cannot provide active power. During this time, the PI controller maintain a minimum DC link voltage to allow the power conditioning system to continue to operate, providing the necessary reactive power.

The AC line voltage regulation is provided by a separate reactive power control, which provides the reactive current magnitude reference I_Q^*. The control system has a simple transfer function, which varies the reactive power command in response to the AC voltage fluctuations. Common to the outer real and reactive power control loops is an inner higher bandwidth zero average current error (ZACE) current control loop. I_p^* is in phase with the line voltages, and I_Q^* is at 90° to the line voltage. These are added together to give one (per phase) sinusoidal converter current reference waveform (I_{ac}^*). The CCVSI control consists of analog and digital circuitry which acts as a ZACE transconductance amplifier in converting I_{ac}^* into AC power currents [48].

27.2.5.5 System Configurations

The utility compatible inverters are used for power conditioning and synchronization of PV output with the utility power. In general, four types of battery-less grid-connected PV system configurations have been identified:

- Central plant inverter.
- Multiple string DC/DC converter with single output inverter.
- Multiple string inverter.
- Module integrated inverter.

27.2.5.5.1 Central Plant Inverter In the central plant inverter, usually a large inverter is used to convert DC power output of PV arrays to AC power. In this system, the PV modules are serially stringed to form a panel (or string) and several such panels are connected in parallel to a single DC bus. The block diagram of such a scheme is shown in Fig. 27.48.

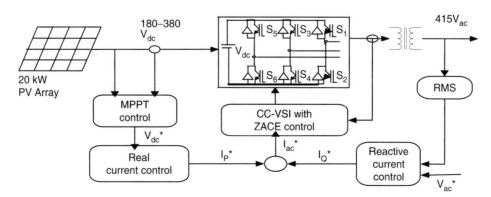

FIGURE 27.47 Block diagram of Kalbarri power conditioning system.

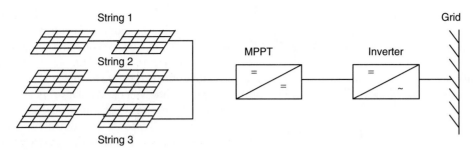

FIGURE 27.48 Central plant inverter.

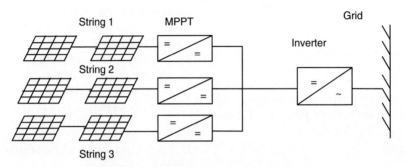

FIGURE 27.49 Multiple string DC/DC converter.

27.2.5.5.2 Multiple String DC/DC Converter In multiple string DC/DC converter, as shown in Fig. 27.49, each string will have a boost DC/DC converter with transformer isolation. There will be a common DC link, which feeds a transformer-less inverter.

27.2.5.5.3 Multiple String Inverters Figure 27.50 shows the block diagram of multiple string inverter system. In this scheme, several modules are connected in series on the DC side to form a string. The output from each string is converted to AC through a smaller individual inverter. Many such inverters are connected in parallel on the AC side. This arrangement is not badly affected by the shading of the panels. It is also not seriously affected by inverter failure.

27.2.5.5.4 Module Integrated Inverter In the module integrated inverter system (Fig. 27.51), each module (typically 50–300 W) will have a small inverter. No cabling is required. It is expected that high volume of small inverters will bring down the cost.

27.2.5.6 Grid-compatible Inverters Characteristics

The characteristics of the grid-compatible inverters are:

- Response time.
- Power factor.
- Frequency control.
- Harmonic output.
- Synchronization.
- Fault current contribution.

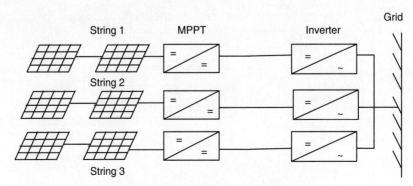

FIGURE 27.50 Multiple string inverter.

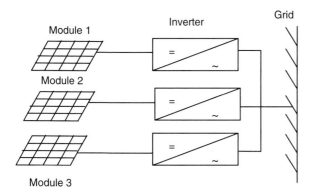

FIGURE 27.51 Module integrated inverter.

- DC current injection.
- Protection.

The response time of the inverters shall be extremely fast and governed by the bandwidth of the control system. Absence of rotating mass and use of semiconductor switches allow inverters to respond in millisecond time frame. The power factor of the inverters is traditionally poor due to displacement power factor and the harmonics. But with the latest development in the inverter technology, it is possible to maintain the power factor close to unity. The converters/inverters have the capability of creating large voltage fluctuation by drawing reactive power from the utility rather than supplying [49]. With proper control, inverters can provide voltage support by importing/exporting reactive power to push/pull towards a desired set point. This function would be of more use to the utilities as it can assist in the regulation of the grid system at the domestic consumer level.

Frequency of the inverter output waveshape is locked to the grid. Frequency bias is where the inverter frequency is deliberately made to run at 53 Hz. When the grid is present, this will be pulled down to the nominal 50 Hz. If the grid fails, it will drift upwards towards 53 Hz and trip on over frequency. This can help in preventing islanding.

Harmonics output from the inverters have been very poor traditionally. Old thyristor-based inverters are operated with slow switching speeds and could not be pulse width modulated. This resulted in inverters known as six-pulse or twelve-pulse inverters. The harmonics so produced from the inverters can be injected into the grid, resulting in losses, heating of appliances, tripping of protection equipments, and poor power quality. The number of pulses being the number of steps in a sine-wave cycle. With the present advent in the power electronics technology, the inverter controls can be made very good. Pulse width modulated inverters produce high quality sine waves. The harmonic levels are very low, and can be lower than the common domestic appliances. If the harmonics are present in the grid voltage waveform, harmonic currents can be induced in the inverter. These harmonic currents, particularly those generated by a voltage-controlled inverter, will in fact help in supporting the grid. These are good harmonic currents. This is the reason that the harmonic current output of inverters must be measured onto a clean grid source so that the only harmonics being produced by the inverters are measured.

Synchronization of inverter with the grid is performed automatically and typically uses zero crossing detection on the voltage waveform. An inverter has no rotating mass and hence has no inertia. Synchronization does not involve the acceleration of a rotating machine. Consequently the reference waveforms in the inverter can be jumped to any point required within a sampling period. If phase-locked loops are used, it could take up a few seconds. Phase-locked loops are used to increase the immunity to noise. This allows the synchronization to be based on several cycles of zero crossing information. The response time for this type of locking will be slower.

Photovoltaic panels produce a current that is proportional to the amount of light falling on them. The panels are normally rated to produce $1000\,W/m^2$ at $25°\,C$. Under these conditions, the short-circuit current possible from these panels is typically only 20% higher than the nominal current whereas it is extremely variable for wind. If the solar radiation is low then the maximum current possible under short-circuit is going to be less than the nominal full load current. Consequently PV systems cannot provide the short-circuit capacity to the grid. If a battery is present, the fault current contribution is limited by the inverter. With the battery storage, it is possible for the battery to provide the energy. However, inverters are typically limited between 100 and 200% of nominal rating under current limit conditions. The inverter needs to protect itself against the short circuits because the power electronic components will typically be destroyed before a protection device like circuit breaker trips.

In case of inverter malfunction, inverters have the capability to inject the DC components into the grid. Most utilities have guidelines for this purpose. A transformer shall be installed at the point of connection on the AC side to prevent DC from being entering into the utility network. The transformer can be omitted when a DC detection device is installed at the point of connection on the AC side in the inverter. The DC injection is essentially caused by the reference or power electronics device producing a positive half cycle that is different from the negative half cycle resulting in the DC component in the output. If the DC component can be measured, it can then be added into the feedback path to eliminate the DC quantity.

27.2.5.6.1 Protection Requirements A minimum requirement to facilitate the prevention of islanding is that the inverter energy system protection operates and isolates the inverter energy system from the grid if:

- Over voltage.
- Under voltage.

- Over frequency.
- Under frequency exists.

These limits may be either factory set or site programmable. The protection voltage operating points may be set in a narrower band if required, e.g. 220–260 V. In addition to the passive protection detailed above, and to prevent the situation where islanding may occur because multiple inverters provide a frequency reference for one another, inverters must have an accepted active method of islanding prevention following grid failure, e.g. frequency drift, impedance measurement, etc. Inverter controls for islanding can be designed on the basis of detection of grid voltage, measurement of impedance, frequency variation, or increase in harmonics. This function must operate to force the inverter output outside the protection tolerances specified previously, thereby resulting in isolation of the inverter energy system from the grid. The maximum combined operation time of both passive and active protections should be 2 s after grid failure under all local load conditions. If frequency shift is used, it is recommended that the direction of shift be down. The inverter energy system must remain disconnected from the grid until the reconnection conditions are met. Some inverters produce high voltage spikes, especially at light load, which can be dangerous for the electronic equipment. IEEE P929 gives some idea about the permitted voltage limits.

If the inverter energy system does not have the above frequency features, the inverter must incorporate an alternate anti-islanding protection feature that is acceptable to the relevant electricity distributor. If the protection function above is to be incorporated in the inverter it must be type tested for compliance with these requirements and accepted by the relevant electricity distributor. Otherwise other forms of external protection relaying are required which have been type tested for compliance with these requirements and approved by the relevant electricity distributor. The inverter shall have adequate protection against short circuit, other faults, and overheating of inverter components.

27.3 Power Electronics for Wind Power Systems

In rural USA, the first wind mill was commissioned in 1890 to generate electricity. Today, large wind generators are competing with utilities in supplying clean power economically. The average wind turbine size has been 300–600 kW until recently. The new wind generators of 1–3 MW have been developed and are being installed worldwide, and prototype of even higher capacity is under development. Improved wind turbine designs and plant utilization have resulted in significant reduction in wind energy generation cost from 35 cents per kWh in 1980 to less than 5 cents per kWh in 1999, in locations where wind regime is favorable. At this generation cost, wind energy has become one of the least cost power sources. Main factors that have contributed to the wind power technology development are:

- High strength fiber composites for manufacturing large low-cost blades.
- Variable speed operation of wind generators to capture maximum energy.
- Advancement in power electronics and associated cost.
- Improved plant operation and efficiency.
- Economy of scale due to availability of large wind generation plants.
- Accumulated field experience improving the capacity factor.
- Computer prototyping by accurate system modeling and simulation.

The Table 27.2 is for wind sites with average annual wind speed of 7 m/s at 30 m hub height. Since 1980s, wind technology capital costs have reduced by 80% worldwide. Operation and maintenance costs have declined by 80% and the availability factor of grid-connected wind plants has increased to 95%. At present, the capital cost of wind generator plants has dropped to about $600 per kW and the electricity generation cost has reduced to 6 cents per kWh. It is expected to reduce the generation cost below 4 cents per kWh. Keeping this in view, the wind generation is going to be highly competitive with the conventional power plants. In Europe, USA, and Asia the wind power generation is increasing rapidly and this trend is going to continue due to economic viability of wind power generation.

TABLE 27.2 Wind power technology developments

	1980	1999	Future
Cost per kWh	$0.35–0.40	$0.05–0.07	<$0.04
Capital cost per kW	$2000–3000	$500–700	<$400
Operating life	5–7 Years	20 Years	30 Years
Capacity factor (average)	15%	25–30%	>30%
Availability	50–65%	95%	>95%
Wind turbine unit size range	50–150 kW	300–1000 kW	500–2000 kW

The technical advancement in power electronics is playing an important part in the development of wind power technology. The contribution of power electronics in control of fixed speed/variable speed wind turbines and interfacing to the grid is of extreme importance. Because of the fluctuating nature of wind speed, the power quality and reliability of the wind based power system needs to be evaluated in detail. Appropriate control schemes require power conditioning.

27.3.1 Basics of Wind Power

The ability of a wind turbine to extract power from wind is a function of three main factors:

- Wind power availability.
- Power curve of the machine.
- Ability of the machine to respond to wind perturbations.

The mechanical power produced by a wind turbine is given by

$$P_m = 0.5\rho C_p A U^3 \text{ W} \quad (27.14)$$

The power from the wind is a cubic function of wind speed. The curve for power coefficient C_p and λ is required to infer the value of C_p for λ based on wind speed at that time.

Where tip speed ratio, $\lambda = \frac{r\omega_A}{U}$, ρ = Air density, Kg m^{-3}, C_p = power coefficient, A = wind turbine rotor swept area, m^2, U = wind speed in m/s.

The case of a variable speed wind turbine with a pitch mechanism that alters the effective rotor dynamic efficiency, can be easily considered if an appropriate expression for C_p as a function of the pitch angle is applied. The power curve of a typical wind turbine is given in Fig. 27.52 as a function of wind speed.

The C_p–λ curve for 150 kW windmaster machine is given in Fig. 27.53, which has been inferred from the power curve of the machine. The ratio of shaft power to the available power in the wind is the efficiency of conversion, known as the power coefficient C_p

$$C_p = \frac{P_m}{(1/2\rho A U^3)} \quad (27.15)$$

The power coefficient is a function of turbine blade tip speed to wind speed ratio (β). A tip speed ratio of 1 means the blade tips are moving at the same speed as the wind and when β is 2 the tips are moving at twice the speed of the wind and so on [51]. Solidity (σ) is defined as the ratio of the sum of the width of all the blades to the circumference of the rotor. Hence,

$$\sigma = Nd/(2\pi R) \quad (27.16)$$

where N = number of blades and d = width of the blades.

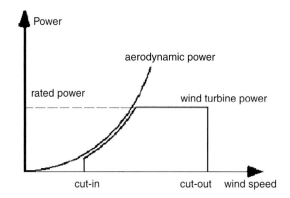

FIGURE 27.52 Power curve of wind turbine as a function of wind speed [50].

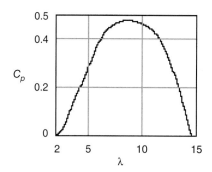

FIGURE 27.53 Cp–λ curve of wind machine [50].

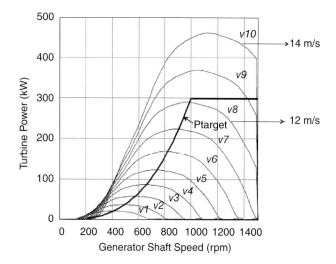

FIGURE 27.54 Turbine power vs shaft speed curves.

The power from a wind turbine doubles as the area swept by the blades doubles. But doubling of the wind speed increases the power output eight times. Figure 27.54 gives a family of power curves for a wind turbine. If the loading of the turbine is controlled such that the operating point is along the maximum power locus at different wind speeds, then the wind energy system will be more efficient.

27.3.1.1 Types of Wind Turbines

There are two types of wind turbines available Fig. 27.55:

- Horizontal axis wind turbines (HAWTs).
- Vertical axis wind turbines (VAWTs).

Vertical axis wind turbines (VAWTs) have an axis of rotation that is vertical, and so, unlike the horizontal wind turbines, they can capture winds from any direction without the need to reposition the rotor when the wind direction changes (without a special yaw mechanism). Vertical axis wind turbines were also used in some applications as they have the advantage that they do not depend on the direction of the wind. It is possible to extract power relatively easier. But there

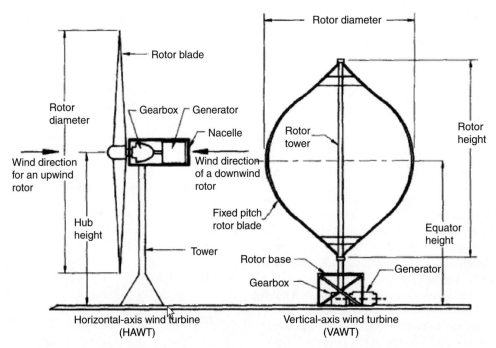

FIGURE 27.55 Typical diagram of HAWTs and VAWTs.

are some disadvantages such as no self starting system, smaller power coefficient than obtained in the horizontal axis wind turbines, strong discontinuation of rotations due to periodic changes in the lift force, and the regulation of power is not yet satisfactory.

The horizontal axis wind turbines are generally used. Horizontal axis wind turbines are, by far, the most common design. There are a large number of designs commercially available ranging from 50 W to 4.5 MW. The number of blades ranges from one to many in the familiar agriculture windmill. The best compromise for electricity generation, where high rotational speed allows use of a smaller and cheaper electric generator, is two or three blades. The mechanical and aerodynamic balance is better for three bladed rotor. In small wind turbines, three blades are common. Multiblade wind turbines are used for water pumping on farms.

Based on the pitch control mechanisms, the wind turbines can also be classified as:

- Fixed pitch wind turbines.
- Variable pitch wind turbines.

Different manufacturers offer fixed pitch and variable pitch blades. Variable pitch is desirable on large machines because the aerodynamic loads on the blades can be reduced and when used in fixed speed operation they can extract more energy. But necessary mechanisms require maintenance and for small machines, installed in remote areas, fixed pitch seems more desirable and economical. In some machines, power output regulation involves yawing blades so that they no longer point into the wind. One such system designed in Western Australia has a tail that progressively tilts the blades in a vertical plane so that they present a small surface to the wind at high speeds.

The active power of a wind turbine can be regulated by either designing the blades to go into an aerodynamic stall beyond the designated wind speed, or by feathering the blades out of the wind, which results in reducing excess power using a mechanical and electrical mechanism. Recently, an active stall has been used to improve the stability of wind farms. This stall mechanism can prevent power deviation from gusty winds to pass through the drive train [52].

Horizontal axis wind turbines can be further classified into fixed speed (FS) or variable speed (VS). The FS wind turbine generator (FSWT) is designed to operate at maximum efficiency while operating at a rated wind speed. In this case, the optimum tip-speed ratio is obtained for the rotor airfoil at a rated wind speed. For a VS wind turbine generator (VSWT), it is possible to obtain optimum wind speed at different wind speeds. Hence this enables the VS wind turbine to increase its energy capture. The general advantages of a VSWT are summarized as follows:

- VSWTs are more efficient than the FSWTs.
- At low wind speeds the wind turbines can still capture the maximum available power at the rotor, hence increasing the possibility of providing the rated power for wide speed range.

27.3.1.2 Types of Wind Generators

Schemes based on permanent magnet synchronous generators (PMSG) and induction generators are receiving close attention in wind power applications because of their qualities such as ruggedness, low cost, manufacturing simplicity, and low maintenance requirements. Despite many positive features over the conventional synchronous generators, the PMSG was not being used widely [23]. However, with the recent advent in power electronics, it is now possible to control the variable voltage, variable frequency output of PMSG. The permanent magnet machine is generally favored for developing new designs, because of higher efficiency and the possibility of a rather smaller diameter. These PMSG machines are now being used with variable-speed wind machines.

In large power system networks, synchronous generators are generally used with fixed-speed wind turbines. The synchronous generators can supply the active and reactive power both, and their reactive power flow can be controlled. The synchronous generators can operate at any power factor. For the induction generator, driven by a wind turbine, it is a well-known fact that it can deliver only active power, while consuming reactive power

Synchronous generators with high power rating are significantly more expensive than induction generators of similar size. Moreover, direct connected synchronous generators have the limitation of rotational speed being fixed by the grid frequency. Hence, fluctuation in the rotor speed due to wind gusts lead to higher torque in high power output fluctuations and the derived train. Therefore in grid-connected application, synchronous generators are interfaced via power converters to the grid. This also allows the synchronous generators to operate wind turbines in VS, which makes gear-less operation of the VSWT possible.

The squirrel-cage induction generators are widely used with the fixed-speed wind turbines. In some applications, wound rotor induction generators have also been used with adequate control scheme for regulating speed by external rotor resistance. This allows the shape of the torque-slip curve to be controlled to improve the dynamics of the drive train. In case of PMSG, the converter/inverter can be used to control the variable voltage, variable frequency signal of the wind generator at varying wind speed. The converter converts this varying signal to the DC signal and the output of converter is converted to AC signal of desired amplitude and frequency.

The induction generators are not locked to the frequency of the network. The cyclic torque fluctuations at the wind turbine can be absorbed by very small change in the slip speed. In case of the capacitor excited induction generators, they obtain the magnetizing current from capacitors connected across its output terminals [51, 53, 54].

To take advantage of VSWTs, it is necessary to decouple the rotor speed and the grid frequency. There are different approaches to operate the VSWT within a certain operational range (cut-in and cut-out wind speed). One of the approaches is dynamic slip control, where the slip is allowed to vary upto 10% [55]. In these cases, doubly-fed induction generators (DFIG) are used (Fig. 27.56). One limitation is that DFIG require reactive power to operate. As it is

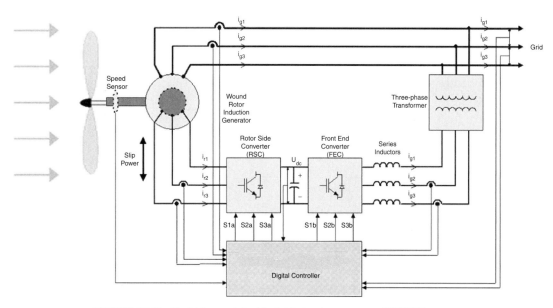

FIGURE 27.56 Variable speed doubly-fed induction generator (VSDFIG) system.

not desired that the grid supply this reactive power, these generators are usually equipped with capacitors. A gear box forms an essential component of the wind turbine generator (WTG) using induction generators. This results in the following limitations:

- Frequent maintenance.
- Additional cost.
- Additional losses.

With the emergence of large wind power generation, increased attention is being directed towards wound rotor induction generators (WRIG) controlled from the rotor side for variable speed constant frequency (VSCF) applications. A wound rotor induction generator has a rotor containing a 3-phase winding. These windings are made accessible to the outside via slip rings. The main advantages of a wound rotor induction generator for VSCF applications are:

- Easier generator torque control using rotor current control.
- Smaller generator capacity as the generated power can be accessed from the stator as well as from the rotor. Usually the rotor power is proportional to the slip speed (shaft speed–synchronous speed). Consequently smaller rotor power converters are required. The frequency converter in the rotor (inverter) directly controls the current in the rotor winding, which enables the control of the whole generator output. The power electronic converters generally used are rated at 20–30% of the nominal generator power.
- Fewer harmonics exist because control is in the rotor while the stator is directly connected to the grid.

If the rotor is short-circuited (making it the equivalent of a cage rotor induction machine), the speed is primarily determined by the supply frequency and the nominal slip is within 5%. The mechanical power input ($P_{TURBINE}$) is converted into stator electrical power output (P_{STATOR}) and is fed to the AC supply. The rotor power loss, being proportional to the slip speed, is commonly referred to as the slip power (P_{ROTOR}).

The possibility of accessing the rotor in a doubly-fed induction generator makes a number of configurations possible. These include slip power recovery using a cycloconverter, which converts the ac voltage of one frequency to another without an intermediate DC link [56–58], or back-to-back inverter configurations [59, 60].

Using voltage-source inverters (VSIs) in the rotor circuit, the rotor currents can be controlled at the desired phase, frequency, and magnitude. This enables reversible flow of active power in the rotor and the system can operate in sub-synchronous and super-synchronous speeds, both in motoring and generating modes. The DC link capacitor acts as a source of reactive power and it is possible to supply the magnetizing current, partially or fully, from the rotor side. Therefore, the stator side power factor can also be controlled. Using vector control techniques, the active and reactive powers can be controlled independently and hence fast dynamic performance can also be achieved.

The converter used at the grid interface is termed as the line-side converter or the front end converter (FEC). Unlike the rotor side converter, this operates at the grid frequency. Flow of active and reactive powers is controlled by adjusting the phase and amplitude of the inverter terminal voltage with respect to the grid voltage. Active power can flow either to the grid or to the rotor circuit depending on the mode of operation. By controlling the flow of active power, the DC bus voltage is regulated within a small band. Control of reactive power enables unity power factor operation at the grid interface. In fact, the FEC can be operated at a leading power factor, if it is so desired. It should be noted that, since the slip range is limited, the DC bus voltage is less in this case when compared to the stator side control. A transformer is therefore necessary to match the voltage levels between the grid and the DC side of the FEC. With a PWM converter in the rotor circuit, the rotor currents can be controlled at the desired phase, frequency, and magnitude. This enables reversible flow of active power in the rotor and the system can operate in sub-synchronous and super-synchronous speeds, both in motoring and generating modes (Fig. 27.57).

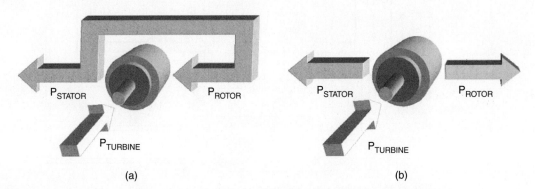

FIGURE 27.57 Doubly-fed induction generator power flow in generating mode: (a) sub-synchronous and (b) super-synchronous.

27.3.2 Types of Wind Power Systems

Wind power systems can be classified as:

- Stand-alone.
- Hybrid.
- Grid-connected.

27.3.3 Stand-alone Wind Power Systems

Stand-alone wind power systems are being used for the following purposes in remote area power systems:

- Battery charging.
- Household power supply.

27.3.3.1 Battery Charging with Stand-alone Wind Energy System

The basic elements of a stand-alone wind energy conversion system are:

- Wind generator.
- Tower.
- Charge control system.
- Battery storage.
- Distribution network.

In remote area power supply, an inverter and a diesel generator are more reliable and sophisticated systems. Most small isolated wind energy systems use batteries as a storage device to level out the mismatch between the availability of the wind and the load requirement. Batteries are a major cost component in an isolated power system.

27.3.3.2 Wind Turbine Charge Controller

The basic block diagram of a stand-alone wind generator and battery charging system is shown in Fig. 27.58.

The function of charge controller is to feed the power from the wind generator to the battery bank in a controlled manner. In the commonly used permanent magnet generators, this is usually done by using the controlled rectifiers [61]. The controller should be designed to limit the maximum current into the battery, reduce charging current for high battery SOC, and maintain a trickle charge during full SOC periods.

27.3.4 Wind–diesel Hybrid Systems

The details of hybrid systems are already covered in Section 27.2.4. Diesel systems without batteries in remote area are characterized by poor efficiency, high maintenance, and fuel costs. The diesel generators must be operated above a certain minimum load level to reduce cylinder wear and tear due to incomplete combustion. It is a common practice to install dump loads to dissipate extra energy. More efficient systems can be devised by combining the diesel generator with a battery inverter subsystem and incorporating RES, such as wind/solar where appropriate. An integrated hybrid energy system incorporating a diesel generator, wind generator, battery or flywheel storage, and inverter will be cost effective at many sites with an average daily energy demand exceeding 25 kWh [62]. These hybrid energy systems can serve as a mini grid as a part of distributed generation rather than extending the grid to the remote rural areas. The heart of the hybrid system is a high quality sine-wave inverter, which can also be operated in reverse as battery charger. The system can cope with loads ranging from zero (inverter only operation) to approximately three times greater capacity (inverter and diesel operating in parallel).

Decentralized form of generation can be beneficial in remote area power supply. Due to high cost of PV systems, problems associated with storing electricity over longer periods (like maintenance difficulties and costs), wind turbines can be a viable alternative in hybrid systems. Systems with battery storage although provide better reliability. Wind power penetration can be high enough to make a significant impact on the operation of diesel generators.

High wind penetration also poses significant technical problems for the system designer in terms of control and transient stability [30]. In earlier stages, wind diesel systems were installed without assessing the system behavior due to lack of design tools/software. With the continual research in this area, there are now software available to assist in this process. Wind diesel technology has now matured due to research and development in this area. Now there is a need to utilize this

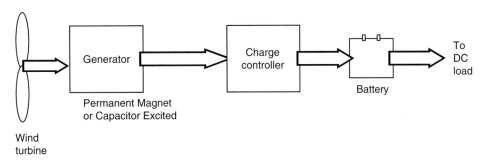

FIGURE 27.58 Block diagram for a stand-alone wind generator and battery charging system.

knowledge into cost effective and reliable hybrid systems [63]. In Western Australia, dynamic modeling of wind diesel hybrid system has been developed in Curtin/MUERI, supported by the Australian Cooperative Research Centre for Renewable Energy (ACRE) program 5.21.

27.3.5 Grid-connected Wind Energy Systems

Small scale wind turbines, connected to the grid (weak or strong grid), have been discussed here. Wind diesel systems have been getting attention in many remote parts of the world lately. Remote area power supplies are characterized by low inertia, low damping, and poor reactive power support. Such weak power systems are more susceptible to sudden change in network operating conditions [64]. In this weak grid situation, the significant power fluctuations in the grid would lead to reduced quality of supply to users. This may manifest itself as voltage and frequency variations or spikes in the power supply. These weak grid systems need appropriate storage and control systems to smooth out these fluctuations without sacrificing the peak power tracking capability. These systems can have two storage elements. The first is the inertia of the rotating mechanical parts, which includes the blades, gearbox, and the rotor of the generators. Instead of wind speed fluctuation causing large and immediate change in the electrical output of the generator as in a fixed speed machine, the fluctuation will cause a change in shaft speed and not create a significant change in generator output. The second energy storage element is the small battery storage between the DC–DC converter and the inverter. The energy in a gust could be stored temporarily in the battery bank and released during a lull in the wind speed, thus reducing the size of fluctuations.

In larger scale wind turbines, the addition of inverter control further reduces fluctuation and increases the total output power. Thus the total output of the wind energy system can be stabilized or smoothed to track the average wind speed and can omit certain gusts. The system controller should track the peak power to maximize the output of the wind energy system. It should monitor the stator output and adjust the inverter to smooth the total output. The amount of smoothing would depend on SOC of the battery. The nominal total output would be adjusted to keep the battery bank SOC at a reasonable level. In this way, the total wind energy system will track the long-term variations in the wind speed without having fluctuations caused by the wind. The storage capacity of the battery bank need only be several minutes to smooth out the gusts in the wind, which can be easily handled by the weak grid. In the cases, where the weak grid is powered by diesel generators, the conventional wind turbine can cause the diesel engines to operate at low capacity. In case of strong wind application, the fluctuations in the output of the wind energy generator system can be readily absorbed by the grid. The main aim here is to extract the maximum energy from the wind. The basic block layout of such a system [65] is shown in the Fig. 27.59.

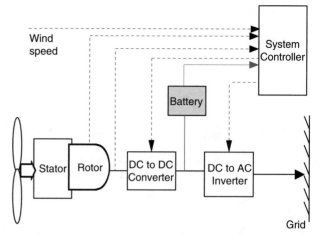

FIGURE 27.59 System block diagram of grid-connected wind energy system.

The function of the DC–DC converter will be to adjust the torque on the machine and hence ensure by measurement of wind speed and shaft speed that the turbine blades are operating so as to extract optimum power. The purpose of the inverter is to feed the energy gathered by the rotor and DC–DC converter, in the process of peak power tracking, to the grid system. The interaction between the two sections would be tightly controlled so as to minimize or eliminate the need for a battery bank. The control must be fast enough so that the inverter output power set point matches the output of the DC–DC converter. For a wound rotor induction machine operating over a two to one speed range, the maximum power extracted from the rotor is equal to the power rating of the stator. Thus the rating of the generator from a traditional point of view is only half that of the wind turbine [65]. Since half the power comes from the stator and half from the rotor, the power electronics of the DC–DC converter and inverter need to handle only half the total wind turbine output and no battery would be required.

Power electronic technology also plays an important role in both system configurations and in control of offshore wind farms [66]. Wind farms connect in various configurations and control methods using different generator types and compensation arrangements. For instance, wind farms can be connected to the AC local network with centralized compensation or with a HVDC transmission system, and DC local network. Decentralized control with a DC transmission system has also been used [67].

27.3.5.1 Soft Starters for Induction Generators

When an induction generator is connected to a load, a large inrush current flows. This is something similar to the direct online starting problem of induction machines. It has been observed that the initial time constants of the induction

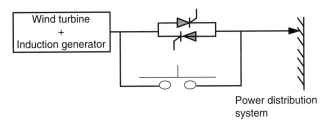

FIGURE 27.60 Soft starting for wind turbine coupled with induction generator.

machines are higher when it tries to stabilize initially at the normal operating conditions. There is a need to use some type of soft starting equipment to start the large induction generators. A simple scheme to achieve this is shown in the Fig. 27.60.

Two thyristors are connected in each phase, back-to-back. Initially, when the induction generator is connected, the thyristors are used to control the voltage applied to the stator and to limit the large inrush current. As soon as the generator is fully connected, the bypass switch is used to bypass the soft starter unit.

27.3.6 Control of Wind Turbines

Theory indicates that operation of a wind turbine at fixed tip speed ratio (C_{pmax}) ensures enhanced energy capture [50]. The wind energy systems must be designed so that above the rated wind speed, the control system limit the turbine output. In normal operation, medium to large-scale wind turbines are connected to a large grid. Various wind turbine control policies have been studied around the world. Grid-connected wind turbines generators can be classified as:

- Fixed speed wind turbines.
- Variable speed wind turbines.

27.3.6.1 Fixed Speed Wind Turbines

In case of a fixed speed wind turbine, synchronous or squirrel-cage induction generators are employed and is characterized by the stiff power train dynamics. The rotational speed of the wind turbine generator in this case is fixed by the grid frequency. The generator is locked to the grid, thereby permitting only small deviations of the rotor shaft speed from the nominal value. The speed is very responsive to wind speed fluctuations. The normal method to smooth the surges caused by the wind is to change the turbine aerodynamic characteristics, either passively by stall regulation or actively by blade pitch regulation. The wind turbines often subjected to very low (below cut in speed) or high wind speed (above rated value). Sometimes they generate below rated power. No pitch regulation is applied when the wind turbine is operating below rated speed, but pitch control is required when the machine is operating above rated wind speed to minimize the stress. Figure 27.61 shows the effect of blade pitch angle on the torque speed curve at a given wind speed.

Blade pitch control is a very effective way of controlling wind turbine speed at high wind speeds, hence limiting the power and torque output of the wind machine. An alternative but cruder control technique is based on airfoil stall [50]. A synchronous link maintaining fixed turbine speed in combination with an appropriate airfoil can be designed so that, at higher than rated wind speeds the torque reduces due to

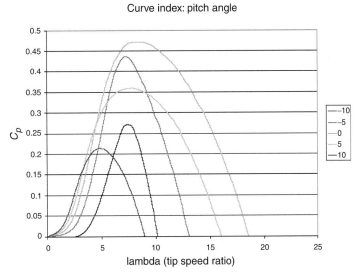

FIGURE 27.61 C_p/λ curves for different pitch settings.

airfoil stall. This method does not require external intervention or complicated hardware, but it captures less energy and has greater blade fatigue.

The aims of variable pitch control of medium- and large-scale wind turbines were to help in start-up and shutdown operation, to protect against overspeed and to limit the load on the wind turbine [68]. The turbine is normally operated between a lower and an upper limit of wind speed (typically 4.5–26 m/s). When the wind speed is too low or too high, the wind turbine is stopped to reduce wear and damage. The wind turbine must be capable of being started and run up to speed in a safe and controlled manner. The aerodynamic characteristics of some turbines are such that they are not self starting. The required starting torque may be provided by motoring or changing the pitch angle of the blade. In case of grid-connected wind turbine system, the rotational speed of the generator is locked to the frequency of the grid. When the generator is directly run by the rotor, the grid acts like an infinite load. When the grid fails, the load rapidly decreases to zero resulting in the turbine rotor to accelerate quickly. Overspeed protection must be provided by rapid braking of the turbine. A simple mechanism of one of blade pitch control techniques is shown in Fig. 27.62.

In this system, the permanent magnet synchronous generator (PMSG) has been used without any gearbox. Direct connection of generator to the wind turbine requires the generator to have a large number of poles. Both induction generators and wound filed synchronous generators of high pole number require a large diameter for efficient operation. Permanent magnet synchronous generators allow a small pole pitch to be used [69]. The power output, P_{mech}, of any turbine depends mainly upon the wind speed, which dictates the rotational speed of the wind turbine rotor. Depending upon the wind speed and rotational speed of turbine, tip speed ratio λ is determined. Based on computed λ, the power coefficient C_p is inferred. In the control strategy above, the torque output, T_{actual}, of the generator is monitored for a given wind speed and compared with the desired torque, T_{actual}, depending upon the load requirement. The generator output torque is passed through the measurement filter. The pitch controller then infers the modified pitch angle based on the torque error. This modified pitch angle demand and computed λ decides the new C_p resulting in the modified wind generator power and torque output. The controller will keep adjusting the blade pitch angle till the desired power and torque output are achieved.

Some of the wind turbine generator includes the gearbox for interfacing the turbine rotor and the generator. The general drive train model [68] for such a system is shown in Fig. 27.63. This system also contains the blade pitch angle control provision.

The drive train converts the input aerodynamic torque on the rotor into the torque on the low-speed shaft. This torque on the low-speed shaft is converted to high-speed shaft torque using the gearbox and fluid coupling. The speed of the wind turbine here is low and the gear box is required to increase the speed so as to drive the generator at rated rpm e.g. 1500 rpm. The fluid coupling works as a velocity-in-torque-out device and transfer the torque [68]. The actuator regulates the tip angle based on the control system applied. The control system here is based on a pitch regulation scheme where the blade pitch angle is adjusted to obtain the desired output power.

27.3.6.2 Variable Speed Wind Turbines

The variable speed constant frequency turbine drive trains are not directly coupled to the grid. The power-conditioning device is used to interface the wind generator to the grid.

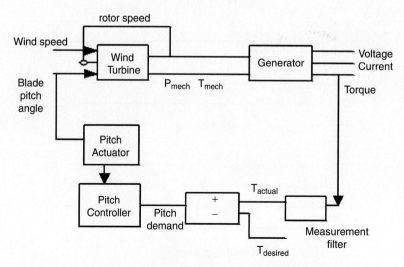

FIGURE 27.62 Pitch control block diagram of a PMSG.

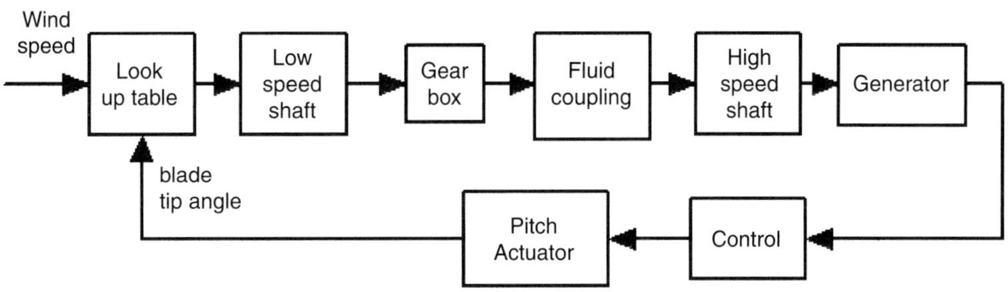

FIGURE 27.63 Block diagram of drive train model.

The output of the wind generator can be variable voltage and variable frequency, which is not suitable for grid integration and appropriate interfacing is required. The wind turbine rotor in this case is permitted to rotate at any wind speed by power generating unit.

A number of schemes have been proposed in the past which allow wind turbines to operate with variable rotor speed while feeding the power to a constant frequency grid. Some of the benefits that have been claimed for variable speed constant frequency wind turbine configuration is as follow [65]:

- The variable speed operation results in increased energy capture by maintaining the blade tip speed to wind speed ratio near the optimum value.
- By allowing the wind turbine generator to run at variable speed, the torque can be fixed, but the shaft power allowed to increase. This means that the rated power of the machine can be increased with no structural changes.
- A variable speed turbine is capable of absorbing energy in wind gusts as it speeds up and gives back this energy to the system as it slows down. This reduces turbulence induced stresses and allows capture of a large percentage of the turbulent energy in the wind.
- More efficient operation can be achieved by avoiding aerodynamic stall over most of operating range.
- Better grid quality due to support of grid voltage.

Progress in the power electronics conversion system has given a major boost to implementing the concept of variable speed operation. The research studies have shown that the most significant potential advancement for wind turbine technology was in the area of power electronic controlled variable speed operation. There is much research underway in the United States and Europe on developing variable speed wind turbine as cost effective as possible. In United States, the NASA MOD-0 and MOD-5B were operated as variable speed wind turbines [65]. Companies in United States and Enercon (Germany) made machines incorporate a variable speed feature. Enercon variable speed wind machine is already in operation in Denham, Western Australia.

The ability to operate at varying rotor speed, effectively adds compliance to the power train dynamics of the wind turbine. Although many approaches have been suggested for variable speed wind turbines, they can be grouped into two main classes: (a) discretely variable speed and (b) continuously variable speed [65, 70].

27.3.6.3 Discretely Variable Speed Systems

The discretely variable speed category includes electrical system where multiple generators are used, either with different number of poles or connected to the wind rotor via different ratio gearing. It also includes those generators, which can use different number of poles in the stator or can approximate the effect by appropriate switching. Some of the generators in this category are those with consequent poles, dual winding, or pole amplitude modulation. A brief summary of some of these concepts is presented below.

27.3.6.3.1 Pole Changing Type Induction Generators These generators provide two speeds, a factor of two apart, such as four pole/eight pole (1500/750 rpm at a supply frequency of 50 Hz or 1800/900 rpm at 60 Hz). They do this by using one-half the poles at the higher speed. These machines are commercially available and cost about 50% more than the corresponding single speed machines. Their main disadvantage, in comparison with other discretely variable machines is that the two to one speed range is wider than the optimum range for a wind turbine [71].

27.3.6.3.2 Dual Stator Winding Two Speed Induction Generators These machines have two separate stator windings, only one of which is active at a time. As such, a variety of speed ranges can be obtained depending on the number of poles in each winding. As in the consequent pole machines only two speeds may be obtained. These machines are significantly heavier than single speed machines and their efficiency is less, since one winding is always unused which leads to increased losses. These machines are commercially available. Their cost is approximately twice that of single speed machines [71].

27.3.6.3.3 Multiple Generators
This configuration is based on the use of a multiple generator design. In one case, there may simply be two separate generators (as used on some European wind turbines). Another possibility is to have two generators on the same shaft, only one of which is electrically connected at a time. The gearing is arranged such that the generators reach synchronous speed at different turbine rotor speeds.

27.3.6.3.4 Two Speed Pole Amplitude Modulated Induction Generator (PAM)
This configuration consists of an induction machine with a single stator, which may have two different operating speeds. It differs from conventional generators only in the winding design. Speed is controlled by switching the connections of the six stator leads. The winding is built in two sections which will be in parallel for one speed and in series for the other. The result is the superposition of one alternating frequency on another. This causes the field to have an effectively different number of poles in the two cases, resulting in two different operating speeds. The efficiency of the PAM is comparable to that of a single speed machine. The cost is approximately twice that of conventional induction generators.

The use of a discretely variable speed generator will result in some of the benefits of continuously variable speed operation, but not all of them. The main effect will be in increased energy productivity, because the wind turbine will be able to operate close to its optimum tip speed ratio over a great range of wind speeds than will a constant speed machine. On the other hand, it will perform as single speed machine with respect to rapid changes in wind speed (turbulence). Thus it could not be expected to extract the fluctuating energy as effective from the wind as would be continuously variable speed machine. More importantly, it could not use the inertia of the rotor to absorb torque spikes. Thus, this approach would not result in improved fatigue life of the machine and it could not be an integral part of an optimized design such as one using yaw/speed control or pitch/speed control.

27.3.6.4 Continuously Variable Speed Systems
The second main class of systems for variable speed operation are those that allow the speed to be varied continuously. For the continuously variable speed wind turbine, there may be more than one control, depending upon the desired control action [72–76]:

- Mechanical control.
- Combination of electrical/mechanical control.
- Electrical control.
- Electrical/power electronics control.

The mechanical methods include hydraulic and variable ratio transmissions. An example of an electrical/mechanical system is one in which the stator of the generator is allowed to rotate. All the electrical category includes high-slip induction generators and the tandem generator. The power electronic category contains a number of possible options. One option is to use a synchronous generator or a wound rotor induction generator, although a conventional induction generator may also be used. The power electronics is used to condition some or all the power to form a appropriate to the grid. The power electronics may also be used to rectify some or all the power from the generator, to control the rotational speed of the generator, or to supply reactive power. These systems are discussed below.

27.3.6.4.1 Mechanical Systems

A. Variable Speed Hydraulic Transmission
One method of generating electrical power at a fixed frequency, while allowing the rotor to turn at variable speed, is the use of a variable speed hydraulic transmission. In this configuration, a hydraulic system is used in the transfer of the power from the top of the tower to ground level (assuming a horizontal axis wind turbine). A fixed displacement hydraulic pump is connected directly to the turbine (or possibly gearbox) shaft. The hydraulic fluid is fed to and from the nacelle via a rotary fluid coupling. At the base of the tower is a variable displacement hydraulic motor, which is governed to run at constant speed and drive a standard generator.

One advantage of this concept is that the electrical equipment can be placed at ground level making the rest of the machine simpler. For smaller machines, it may be possible to dispense with a gearbox altogether. On the other hand, there are a number of problems using hydraulic transmissions in wind turbines. For one thing, pumps and motors of the size needed in wind turbines of greater than about 200 kW are not readily available. Multiples of smaller units are possible but this would complicate the design. The life expectancy of many of the parts, especially seals, may well be less than five years. Leakage of hydraulic fluid can be a significant problem, necessitating frequent maintenance. Losses in the hydraulics could also make the overall system less efficient than conventional electric generation. Experience over the last many years has not shown great success with the wind machines using hydraulic transmission.

B. Variable Ratio Transmission
A variable ratio transmission (VRT) is one in which the gear ratio may be varied continuously within a given range. One type of VRT suggested for wind turbines is using belts and pulleys, such as are used in some industrial drives [65, 77]. These have the advantage of being able to drive a conventional fixed speed generator, while being driven by a variable speed turbine rotor. On the other hand, they do not appear to be commercially available in larger sizes and those, which do exist, have relatively high losses.

27.3.6.4.2 Electrical/Mechanical Variable Speed Systems – Rotating Stator Induction Generator

This system uses a conventional squirrel-induction generator whose shaft is driven by a wind turbine through a gearbox [50, 77]. However, the stator is mounted to a support, which allows bi-directional rotation. This support is in turn driven by a DC machine. The armature of the DC machine is fed from a bi-directional inverter, which is connected to the fixed frequency AC grid. If the stator support allowed to turn in the same direction as the wind turbine, the turbine will turn faster. Some of the power from the wind turbine will be absorbed by the induction generator stator and fed to the grid through the inverter. Conversely, the wind turbine will turn more slowly when the stator support is driven in the opposite direction. The amount of current (and thus the torque) delivered to or from the DC machine is determined by a closed loop control circuit whose feedback signal is driven by a tachometer mounted on the shaft of the DC machine.

One of the problems with this system is that the stator slip rings and brushes must be sized to take the full power of the generator. They would be subjected to wear and would require maintenance. The DC machine also adds to cost, complexity, and maintenance.

27.3.6.4.3 Electrical Variable Speed Systems

A. High Slip Induction Generator

This is the simplest variable speed system, which is accomplished by having a relatively large amount of resistance in the rotor of an induction generator. However, the losses increase with increased rotor resistance. Westwind Turbines in Australia investigated such a scheme on a 30 kW machine in 1989.

B. Tandem Induction Generator

A tandem induction generator consists of an induction machine fitted with two magnetically independent stators, one fixed in position and the other able to be rotated, and a single squirrel-cage rotor whose bars extend to the length of both stators [65, 77]. Torque control is achieved by physical adjustment of the angular displacement between the two stators, which causes a phase shift between the induced rotor voltages.

27.3.6.4.4 Electrical/Power Electronics

The general configuration is shown in the Fig. 27.64. It consists of the following components:

- Wind generator.
- Rectifier.
- Inverter.

The generator may be DC, synchronous (wound rotor or permanent magnet type), squirrel-cage wound rotor, or brushless doubly-fed induction generator. The rectifier is used to convert the variable voltage variable frequency input to a DC voltage. This DC voltage is converted into AC of constant voltage and frequency of desired amplitude. The inverter will also be used to control the active/reactive power flow from the inverter. In case of DC generator, the converter may not be required or when a cycloconverter is used to convert the AC directly from one frequency to another.

27.3.6.5 Types of Generator Options for Variable Speed Wind Turbines Using Power Electronics

Power electronics may be applied to four types of generators to facilitate variable speed operation:

- Synchronous generators.
- Permanent magnet synchronous generators.
- Squirrel-cage induction generators.
- Wound rotor induction generators.

27.3.6.5.1 Synchronous Generator In this configuration, the synchronous generator is allowed to run at variable speed, producing power of variable voltage and frequency. Control may be facilitated by adjusting an externally supplied field current. The most common type of power conversion uses a bridge rectifier (controlled/uncontrolled), a DC link, and inverter as

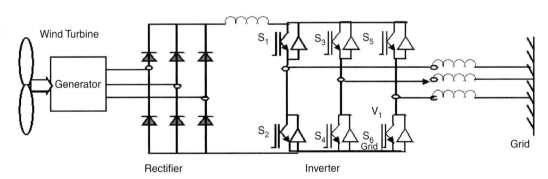

FIGURE 27.64 Grid-connected wind energy system through AC/DC/AC converter.

shown in Fig. 27.64. The disadvantage of this configuration include the relatively high cost and maintenance requirements of synchronous generators and the need for the power conversion system to take the full power generated (as opposed to the wound rotor system).

27.3.6.5.2 Permanent Magnet Synchronous Generators
The permanent magnet synchronous generator (PMSG) has several significant advantageous properties. The construction is simple and does not required external magnetization, which is important especially in stand-alone wind power applications and also in remote areas where the grid cannot easily supply the reactive power required to magnetize the induction generator. Similar to the previous externally supplied field current synchronous generator, the most common type of power conversion uses a bridge rectifier (controlled/uncontrolled), a DC link, and inverter as shown in Fig. 27.65 [78–80].

Figure 27.66 shows a wind energy system where a PMSG is connected to a three-phase rectifier followed by a boost converter. In this case, the boost converter controls the electromagnet torque and the supply side converter regulates the DC link voltage as well as controlling the input power factor. One drawback of this configuration is the use of diode rectifier that increases the current amplitude and distortion of the PMSG. As a result, this configuration have been considered for small size wind energy conversion systems (smaller than 50 kW).

The advantage of the system in Fig. 27.65 with regardant to the system showed in Fig. 27.66 is, it allows the generator to operate near its optimal working point in order to minimize the losses in the generator and power electronic circuit. However, the performance is dependent on the good knowledge of the generator parameter that varies with temperature and frequency. The main drawbacks, in the use of PMSG, are the cost of permanent magnet that increase the price of machine, demagnetization of the permanent magnet material, and it is not possible to control the power factor of the machine

To extract maximum power at unity power factor from a PMSG and feed this power (also at unity power factor) to the grid, the use of back-to-back connected PWM voltage source converters are proposed [81]. Moreover, to reduce the overall cost, reduced switch PWM voltage source converters (four switch) instead of conventional (six switch) converters for variable speed drive systems can be used. It is shown that by using both rectifier and inverter current control or flux based control, it is possible to obtain unity power factor operation both at the WTG and the grid. Other mechanisms can also be included to maximize power extraction from the VSWT (i.e. MPPT techniques) or sensor-less approaches to further reduce cost and increase reliability and performance of the systems.

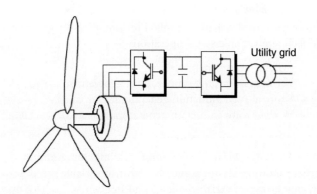

FIGURE 27.65 Grid-connected PMSG wind energy system through DC/AC converter.

27.3.6.5.3 Squirrel-cage Induction Generator
Possible architecture for systems using conventional induction generators which have a solid squirrel-cage rotor have many similarities to those with synchronous generators. The main difference is that the induction generator is not inherently self-exciting and it needs a source of reactive power. This could be done by a generator side self-commutated converter operating in the rectifier mode. A significant advantage of this configuration is the low cost and low maintenance requirements of induction generators. Another advantage of using the self-commutated double converter is that it can be on the ground, completely separate from the wind machine. If there is a problem in the converter, it could be switched out of the circuit for repair and the wind machine could continue to run at constant speed. The main disadvantage with this configuration is that, as with the synchronous generator, the power conversion system would have to take the full power generated and could be relatively costly compared to some other configurations. There would also be additional complexities associated with the supply of reactive power to the generator.

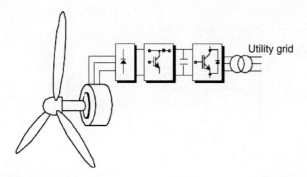

FIGURE 27.66 Grid-connected PMSG wind energy system through DC/AC converter with a boost chopper.

27.3.6.5.4 Wound Rotor Induction Generator
A wound rotor induction rotor has three-phase winding on the rotor,

accessible to the outside via slip rings. The possibility of accessing the rotor can have the following configurations:

- Slip power recovery.
- Use of cycloconverter.
- Rotor resistance chopper control.

A. Slip Power Recovery (Static Kramer System)

The slip power recovery configuration behaves similarly to a conventional induction generator with very large slip, but in addition energy is recovered from the rotor. The rotor power is first carried out through slip rings, then rectified and passed through a DC link to a line-commutated inverter and into the grid. The rest of the power comes directly from the stator as it normally does. A disadvantage with this system is that it can only allow super-synchronous variable speed operation. Its possible use in the wind power was reported by Smith and Nigim [82].

In this scheme shown in Fig. 27.67, the stator is directly connected to the grid. Power converter has been connected to the rotor of wound rotor induction generator to obtain the optimum power from variable speed wind turbine. The main advantage of this scheme is that the power-conditioning unit has to handle only a fraction of the total power so as to obtain full control of the generator. This is very important when the wind turbine sizes are increasing for the grid-connected applications for higher penetration of wind energy and the smaller size of converter can be used in this scheme.

B. Cycloconverter (Static Scherbius System)

A cycloconverter is a converter, which converts AC voltage of one frequency to another frequency without an intermediate DC link. When a cycloconverter is connected to the rotor circuit, sub- and super-synchronous operation variable speed operation is possible. In super-synchronous operation, this configuration is similar to the slip power recovery. In addition, energy may be fed into the rotor, thus allowing the machine to generate at sub-synchronous speeds. For that reason, the generator is said to be doubly fed [83]. This system has a limited ability to control reactive power at the terminals of the generator, although as a whole it is a net consumer of reactive power. On the other hand, if coupled with capacitor excitation, this capability could be useful from the utility point of view. Because of its ability to rapidly adjust phase angle and magnitude of the terminal voltage, the generator can be resynchronized after a major electrical disturbance without going through a complete stop/start sequence. With some wind turbines, this could be a useful feature.

C. Rotor Resistance Chopper Control

A fairly simple scheme of extracting rotor power as in the form of heat has been proposed in [44].

27.3.6.6 Isolated Grid Supply System with Multiple Wind Turbines

The isolated grid supply system with a wind park is shown in Fig. 27.68. Two or more wind turbines can be connected to this system. A diesel generator can be connected in parallel. The converters, connected with wind generators will work in parallel and the supervisory control block will control the output of these wind generators in conjunction with the diesel generator. This type of decentralized generation can be a better option where high penetration of wind generation is sought. The individual converter will control the voltage and frequency of the system. The supervisory control system will play an important part in co-ordination between multiple power generation systems in a remote area power supply having weak grid.

27.3.6.7 Power Electronics Technology Development

To meet the needs of future power generation systems, power electronics technology will need to evolve on all levels, from devices to systems. The development needs are as follows:

- There is a need for modular power converters with plug-and-play controls. This is particularly important for high power utility systems, such as wind power. The power

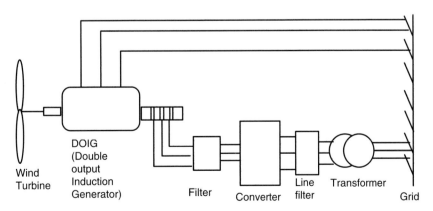

FIGURE 27.67 Schematic diagram of doubly-fed induction generator.

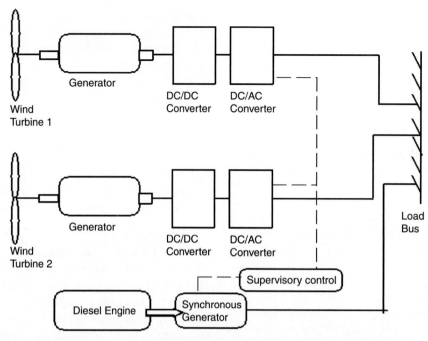

FIGURE 27.68 Schematic diagram of isolated grid system having a wind park.

electronics equipment used today is based on industrial motor drives technology. Having dedicated, high power density, modular systems will provide flexibility and efficiency in dealing with different energy sources and large variation of generation systems architectures.

- There is a need for new packaging and cooling technologies, as well as integration with PV and fuel cell will have to be addressed. The thermal issues in integrated systems are complex, and new technologies such as direct fluid cooling or microchannel cooling may find application in future systems. There is large potential for advancement in this area.
- There is a need for new switching devices with higher temperature capability, higher switching speed, and higher current density/voltage capability. The growth in alternative energy markets will provide a stronger pull for further development of these technologies.

References

1. G. Hille, W. Roth, and H. Schmidt, "Photovoltaic system," in *Fraunhofer Institute for Solar Energy Systems*. Freiburg, Germany, 1995.
2. D. P. Hodel, "Photovoltaics-electricity from sunlight," U.S. Department of Energy Report, DOE/NBMCE.
3. O. H. Wilk, "Utility connected photovoltaic systems," presented at International Energy Agency (IEA): Solar Heating and Cooling Program, Montreux, 1992.
4. R. S. Wenham, M. A. Green, and M. E. Watt, "Applied photovoltaic," in *Centre for Photovoltaic Device and Systems*, University of New Southwales, Sydney, 1998.
5. W. B. Lawrance and H. Dehbonei, "A versatile PV array simulation tools," presented at ISES 2001 Solar World Congress, Adelaide, South Australia, 2001.
6. M. A. S. Masoum, H. Dehbonei, and E. F. Fuchs, "Theoretical and experimental analyses of photovoltaic systems with voltage and current-based maximum power-point tracking," IEEE Trans. on Energy Conversion, vol. 17, pp. 514–522, Dec 2002.
7. B. Bekker and H. J. Beukes, "Finding an optimal PV panel maximum power point tracking method," presented at 7th AFRICON Conf., Africa, 2004.
8. T. Noguchi, S. Togashi, and R. Nakamoto, "Short-current pulse based adaptive maximum-power-point tracking for photovoltaic power generation system," presented at Proc. 2000 IEEE International Symp. on Ind. Electronics, 2000.
9. N. Mutoh, T. Matuo, K. Okada, and M. Sakai, "Prediction-data-based maximum-power-point-tracking method for photovoltaic power generation systems," presented at 33rd Annual IEEE Power Electronics Specialists Conf., 2002.
10. G. W. Hart, H. M. Branz, and C. H. Cox, "Experimental tests of open-loop maximum-power-point tracking techniques," Solar Cells, vol. 13, pp. 185–195, 1984.
11. D. J. Patterson, "Electrical system design for a solar powered vehicle," presented at 21st Annual IEEE Power Electron. Specialists Conf., 1990.
12. H. J. Noh, D. Y. Lee, and D. S. Hyun, "An improved MPPT converter with current compensation method for small scaled PV-applications," presented at 28th Annual Conf. of the Ind. Electronics Society, 2002.
13. K. Kobayashi, H. Matsuo, and Y. Sekine, "A novel optimum operating point tracker of the solar cell power supply system," presented at 35th Annual IEEE Power Electronics Specialists Conf., 2004.
14. P. Midya, P. T. Krein, R. J. Turnhull, R. Reppa, and J. Kimball, "Dynamic maximum power point tracker for pholovoltaic

applications," presented at 27th Annual IEEE Power Electron. Specialists Conf., 1996.
15. V. Arcidiacono, S. Corsi, and L. Larnhri, "Maximum power point tracker for photovoltaic power plants," presented at Proc. IEEE Photovotaic Specialists Conf., 1982.
16. Y. H. Lim and D. C. Hamill, "Synthesis, simulation and experimental verification of a maximum power point tracker from nonlinear dynamics," presented at 32nd Annual IEEE Power Electronics Specialists Conf., 2001.
17. Y. H. Lim and D. C. Hamill, "Simple maximum power point tracker for photovoltaic arrays," Electron. Lett, vol. 36, pp. 997–999, May 2000.
18. L. Zhang, Y. Bai, and A. Al-Amoudi, "GA-RBF neural network based maximum power point tracking for grid-connected photovoltaic systems," presented at International Conf. on Power Electronics Machines and Drives, 2002.
19. C. C. Hua and J. R. Lin, "Fully digital control of distributed photovoltaic power systems," presented at IEEE International Symp. on Ind. Electronics, 2001.
20. M. L. Chiang, C. C. Hua, and J. R. Lin, "Direct power control for distributed PV power system," presented at Proc. Power Conversion Conf., 2002.
21. S. Jain and V. Agarwal, "A new algorithm for rapid tracking of approximate maximum power point in photovoltaic systems," IEEE Power Electronics Lett., vol. 2, pp. 16–19, Mar 2004.
22. N. Femia, G. Petrone, G. Spagnuolo, and M. Vitelli, "Optimization of perturb and observe maximum power point tracking method," IEEE Trans. Power Electronics, vol. 20, pp. 963–973, July 2005.
23. N. S. D'Souza, L. A. C. Lopes, and X. Liu, "An intelligent maximum power point tracker using peak current control," 36th Annual IEEE Power Electronics Specialists Conf., pp. 172–177, 2005.
24. M. G. Simoes, N. N. Franceschetti, and M. Friedhofer, "A fuzzy logic based photovoltaic peak power tracking control," Proc. 1998 IEEE International Symp. on Ind. Electron, pp. 300–305, 1998.
25. A. M. A. Mahmoud, H. M. Mashaly, S. A. Kandil, H. E. Khashab, and M. N. F. Nashed, "Fuzzy logic implementation for photovoltaic maximum power tracking," presented at Proc. 9th IEEE International Workshop on Robot and Human Interactive Commum., 2000.
26. N. Patcharaprakiti and S. Premrudeepreechacharn, "Maximum power point tracking using adaptive fuzzy logic control for grid-connected photovoltaic system," presented at IEEE Power Eng. Society Winter Meeting, 2002.
27. B. M. Wilamowski and X. Li, "Fuzzy system based maximum power point tracking for PV system," presented at 28th Annual Conf. of the IEEE Industrial Electronics Society, 2002.
28. M. Veerachary, T. Senjyu, and K. Uezato, "Neural-network-based maximum-power-point tracking of coupled-inductor interleaved-boost-converter-supplied PV system using fuzzy controller," IEEE Trans. Industry Electronics, vol. 50, pp. 749–758, Aug 2003.
29. K. Ro and S. Rahman, "Two-loop controller for maximizing performance of a grid-connected photovoltaic-fuel cell hybrid power plant," IEEE Trans. on Energy Conversion, vol. 13, pp. 276–281, Sept 1998.
30. A. Hussein, K. Hirasawa, J. Hu, and J. Murata, "The dynamic performance of photovoltaic supplied dc motor fed from DC–DC converter and controlled by neural networks," presented at Proc. 2002 International Joint Conf. on Neural Networks, 2002.
31. X. Sun, W. Wu, X. Li, and Q. Zhao, "A research on photovoltaic energy controlling system with maximum power point tracking," presented at Proc. Power Conversion Conf., 2002.
32. O. Wasynczuk, "Dynamic behavior of a class of photovoltaic power systems," IEEE Trans. Power App. Syst., vol. 102, pp. 3031–3037, Sept 1983.
33. Y. C. Kuo, T. J. Lian, and J. F. Chen, "Novel maximum-power-point-tracking controller for photovoltaic energy conversion system," IEEE Trans. Ind. Electron., vol. 48, pp. 594–601, June 2001.
34. G. J. Yu, Y. S. Jung, J. Y. Choi, I. Choy, J. H. Song, and G. S. Kim, "A novel two-mode MPPT control algorithm based on comparative study of existing algorithms," presented at Conf. Record of the Twenty-Ninth IEEE Photovoltaic Specialists Conf., 2002.
35. K. Kohayashi, I. Takano, and Y. Sawada, "A study on a two stage maximum power point tracking control of a photovoltaic system under partially shaded insolation conditions," presented at IEEE Power Enç. Society General Meeting, 2003.
36. D. Langridge, W. Lawrance, and B. Wichert, "Development of a photo-voltaic pumping system using a brushless D.C. motor and helical rotor pump," Solar Energy, vol. 56, pp. 151–160, 1996.
37. H. Dehbonei and C. V. Nayar, "A new modular hybrid power system," presented at IEEE International Symposium on Industrial Electronics, Rio de Janeiro, Brazil, 2003.
38. C. V. Nayar, S. J. Phillips, W. L. James, T. L. Pryor, and D. Remmer, "Novel wind/diesel/battery hybrid system," Solar Energy, vol. 51, pp. 65–78, 1993.
39. W. Bower, "Merging photovoltaic hardware development with hybrid applications in the U.S.A," presented at Proceedings Solar '93 ANZSES, Fremantle, Western Australia, 1993.
40. N. Mohan, M. Undeland, and W. P. Robbins, *Power Electronics*: John Wiley and Sons, Inc., New York, 1995.
41. M. Calais, V. G. Agelidis, and M. Meinhardt, "Multilevel converters for single phase grid-connected photovoltaic systems an overview," Solar Energy, vol. 66, pp. 525–535, 1999.
42. K. Hirachi, K. Matsumoto, M. Yamamoto, and M. Nakaoka, "Improved control implementation of single phase current fed PWM inverter for photovoltaic power generation," presented at Seventh International Conference on Power Electronics and Variable Speed Drives (PEVD'98), 1998.
43. U. Boegli and R. Ulmi, "Realisation of a new inverter circuit for direct photovoltaic energy feedback into the public grid," IEEE Trans. on Industry Application, vol. 22, Mar/Apr 1986.
44. B. Lindgren, "Topology for decentralised solar energy inverters with a low voltage A-bus," presented at EPE99 -European Power Electronics Conf., 1999.
45. K. Masoud and G. Ledwich, "Aspects of grid interfacing: current and voltage controllers," presented at Proceedings of AUPEC 99, 1999.
46. H. K. Sung, S. R. Lee, H. Dehbonei, and C. V. Nayar, "A comparative study of the voltage controlled and current controlled voltage source inverter for the distributed generation system," presented at Australian Universities Power Engineering Conference(AUPEC), Hobart, Australia, 2005.
47. L. J. Borle, M. S. Dymond, and C. V. Nayar, "Development and testing of a 20 kW grid interactive photovoltaic power conditioning system in Western Australia," IEEE Trans. on Industry Applications, vol. 33, pp. 1–7, 1999.

48. L. J. Borle and C. V. Nayar, "Zero average current error controlled power flow for ac-dc power converters," IEEE Trans. on Power Electronics, vol. 10, pp. 725–732, 1995.
49. H. Sharma, "Grid integration of photovoltaics," Australia: The University of Newcastle, 1998.
50. L. L. Freris, *Wind Energy Conversion Systems*: Prentice Hall, New York, 1990.
51. C. V. Nayar, J. Perahia, and F. Thomas., "Small scale wind powered electrical generators," The Minerals and Energy Research Institute of Western Australia, 1992.
52. R. D. Richardson and G. M. McNerney, "Wind energy systems," Proceedings of the IEEE, vol. 81, pp. 378–389, 1993.
53. J. Arillaga and N. Watson, "Static power conversion from self excited induction generators," Proc of Institution of Electrical Engineers, vol. 125, no. 8, pp. 743–746.
54. C. V. Nayar, J. Perahia, F. Thomas, S. J. Phillips, T. L. Pryor, and W. L. James, "Investigation of capacitor excited induction generators and permanent magnet alternators for small scale wind power generation," Renewable Energy, vol. 125 1991.
55. T. Ackermann and L. Sörder, "An overview of wind energy status-2002," Renewable & Sustainable Energy Reviews, pp. 67–128, June 2002.
56. S. Peresada, A. Tilli, and A. Tonielli, "Robust active-reactive power control of a doubly-fed induction generator," presented at IECON '98, 1998.
57. W. E. Long and N. L. Schmitz, "Cycloconverter control of the doubly fed induction motor," IEEE Trans. Ind. and Gen. Appl., vol. 7, pp. 162–167, 1971.
58. A. Chattopadhyay, "An adjustable-speed induction motor drive with a thyristor-commutator in the rotor," IEEE Trans. Ind. Appl., vol. 14, pp. 116–122, 1978.
59. P. Pena, J. C. Clare, and G. M. Asher, "Doubly fed induction generator using back-to-back PWM converters and its application to variable speed wind-energy generation," IEE Proceedings Electric Power Applications, vol. 143, 1996.
60. H. Azaza, "On the dynamic and steady state performances of a vector controlled DFM drive," presented at IEEE International Conference on Systems, Man and Cybernetics, 2002.
61. Bergey Windpower User Manual "10 kW Battery charging wind energy generating system," Bergey Windpower Co., Oklahoma, USA, 1984.
62. J. H. R. Enslin and F. W. Leuschner., "Integrated hybrid energy systems for isolated and semi-isolated users," presented at Proc. of Renewable Energy Potential in Southern African Conference, UCT, South Africa, 1986.
63. D. G. Infield, "Wind diesel systems technology and modelling-a review," International Journal of Renewable Energy Engineering, vol. 1, no. 1, pp. 17–27, 1999.
64. H. Sharma, S. M. Islam, C. V. Nayar, and T. Pryor, "Dynamic response of a remote area power system to fluctuating wind speed," presented at Proceedings of IEEE Power Engineering Society (PES 2000) Winter Meeting, 2000.
65. W. L. James, C. V. Nayar, F. Thomas, and M. Dymond, "Variable speed asynchronous wind powered generator with dynamic power conditioning," Murdoch University Energy Research Institute (MUERI), Australia, 1993.
66. F. Blaabjerg, Z. Chen, and S. B. Kjaer, "Power electronics as efficient interface in dispersed power generation systems," IEEE Trans. on Power Electronics, vol. 19, pp. 1184–1194, 2004.
67. F. Blaahjerg, Z. Chen, and P. H. Madsen, "Wind power technology status, development and trends," presented at Proc. Workshop on Wind Power and Impacts on Power Systems, Oslo, Norway, 2002.
68. J. Wilkie, W. E. Leithead, and C. Anderson, "Modelling of wind turbines by simple models," Wind Engineering, vol. 14, pp. 247–273, 1990.
69. A. L. G. Westlake, J. R. Bumby, and E. Spooner, "Damping the power angle oscillations of a permanent magnet synchronous generator with particular reference to wind turbine applications," IEE Proc. Electrical Power Applications, vol. 143, pp. 269–280, 1996.
70. J. F. Manwell, J. G. McGowan, and B. H. Bailey, "Electrical/mechanical options for variable speed turbines," Solar Energy, vol. 46, pp. 41–51, 1991.
71. T. S. Andersen and H. S. Kirschbaum, "Multi-speed electrical generator applications in wind turbines," presented at AIAA/SERI Wind Energy Conference Proc., Boulder, 1980.
72. E. Muljadi and C. P. Butterfield, "Pitch-controlled variable-speed wind turbine generation," IEEE Transactions on Industry Applications, vol. 37, no. 1, pp. 240–246, Jan/Feb 2001.
73. G. Riahy and P. Freere, "Dynamic controller to operate a wind turbine in stall region," presented at Proceeding of Solar'97-Australia and New Zealand Solar Energy Society, 1997.
74. K. Tan and S. Islam, "Optimum control strategies in energy conversion of PMSG wind turbine system without mechanical sensors," IEEE Transactions on Energy Conversion, vol. 19, no. 2, pp. 392–399, June 2004.
75. K. Tan and S. Islam, "Mechanical sensorless robust control of permanent magnet synchronous generator for maximum power operation," presented at Australia University Power Engineering Conference, Australia, 2001.
76. Q. Wang and L. Chang, "An independent maximum power extraction strategy for wind energy conversion system," presented at IEEE Canadian Conference on Electrical and Computer Engineering, Canada, Alberta, 1999.
77. J. Perahia and C. V. Nayar, "Power controller for a wind-turbine driven tandem induction generator," Electric Machines and Power Systems, vol. 19, pp. 599–624, 1991.
78. K. Tan, S. Islam, and H. Tumbelaka, "Line commutated inverter in maximum wind energy conversion," International Journal of Renewable Energy Engineering, vol. 4, no. 3, pp. 506–511, Dec 2002.
79. E. Muljadi, S. Drouilhet, R. Holz, and V. Gevorgian, "Analysis of permanent magnet generator for wind power battery charging," presented at Thirty-First IAS Annual Meeting, IAS '96, San Diego, CA, USA, 1996.
80. B. S. Borowy and Z. M. Salameh, "Dynamic response of a stand-alone wind energy conversion system with battery energy storage to a wind gust," IEEE Transactions on Energy Conversion, vol. 12, no. 1, pp. 73–78, Mar 1997.
81. A. B. Raju, "Application of power electronic interfaces for grid-connected variable speed wind energy conversion systems," in *Department of Electrical Engineering*. Bombay: Indian Institute of Technology, 2005.
82. G. A. Smith and K. A. Nigim., "Wind energy recovery by static Scherbius induction generator," IEE Proceedings, Part 'C', Generation, Transmission and Distribution, vol. 128, pp. 317–324, 1981.
83. T. S. Anderson and P. S. Hughes, "Investigation of doubly fed induction machine in variable speed applications," Westinghouse Electric Corporation, 1983.

28

Fuel-cell Power Electronics for Distributed Generation

S. K. Mazumder, Ph.D.
Department of Electrical and Computer Engineering, Director Laboratory for Energy and Switching-Electronics Systems (LESES), University of Illinois, Chicago, Illinois, USA

28.1 Distributed Generation 717
28.2 Fuel-cell Based Energy Systems for DG 719
28.3 Power-electronic Topologies for Residential Stationary Fuel-cell Energy Systems 720
28.4 Towards Power-electronic Topologies for High Power Fuel-cell Based DG 731
References 736

28.1 Distributed Generation

On-site power generation (often called as distributed generation (DG) [1, 2]) using alternative/renewable-energy technologies, as illustrated in Fig. 28.1, can minimize environmental pollution and reduce our dependence on fossil fuels. Distributed energy resources (DER) are parallel and stand-alone electric generation units located within the electric distribution system near the end user. The distributed energy resources, if properly integrated can be beneficial to electricity consumers and energy utilities, providing energy independence and increased energy security. Each home and commercial unit with DER equipments can be a micro-power station, generating much of the electricity it needs on-site and sell the excess power to the national grid.

Table 28.1 provides information regarding commercially available equipment for DER [1, 2] and for those technologies still undergoing development. Some of these technologies are listed in both categories because they are commercially available and undergoing further research and development as well. There are several customer DG applications including (i) allowing customers to continuously generate their own electricity, with or without grid backup; (ii) permitting customers to generate power while serving their thermal and/or cooling loads; (iii) generating a portion of electricity on-site to reduce the amount of electricity purchased during peak-price periods; (iv) licensing customers to sell excess generation back onto the grid when their own demand is low, especially during peak-pricing periods; (v) using standby or emergency power to backup grid-based power; (vi) improving customer power quality and reliability; (vii) serving niche applications, such as "green" power or remote power; and (viii) meeting continuous power, premium power, or cogeneration needs of the residential market.

Customer Generation

Continuous customer generation applications produce power on a nearly continuous basis, running at least 6,000 hours per year. When evaluating the usage of DG technologies in this capacity, customers consider competing grid price, as well as the installed cost of the unit and fuel costs. Maintenance costs, power quality, and reliability of grid power are other critical components. In non-attainment areas, emissions can provide a strong barrier to these applications.

Cogeneration

Cogeneration, also known as combined heat and power (CHP), utilizes otherwise wasted exhaust heat as useful thermal output, typically steam. The steam may be used either for space heating or space cooling. Again, CHP applications are driven by grid price and installed cost, but emissions can provide a strong barrier to implementation, especially in non-attainment areas. As with continuous power applications, these

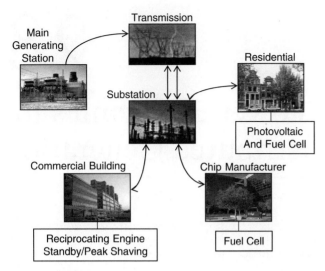

FIGURE 28.1 Illustration of a DG-based power generation.

TABLE 28.1 Attributes of commercially available and emerging DER technologies

DER Technologies	Commercially Available	Emerging Technology	Size	Efficiency (%)
Microturbines	X	X	10–300 kW	20–30
Reciprocating engines	X	–	50 kW–6 MW	33–37
Stirling engines	–	X	10–200 kW	<40
Solid-oxide fuel cells (SOFC)	X	X	5 kW–3 MW	45–65
Proton-exchange-membrane fuel cells	X	–	1 kW–3 MW	34–36
Photovoltaic systems	X	–	1 kW–1 MW	<22
Wind systems	X	X		<30
Hybrid systems	–	X	<1 kW–1 MW	40–50
Combined heat and power (CHP)	X	X	5 kW–25 MW	50–80

units will run on a nearly continuous basis, typically at least 6,000 hours per year.

Peak Shaving

Driven primarily by high utility demand charges, peak-shaving applications (sometimes called peak clipping) are also affected by installed cost, perceived unit reliability, and fuel prices. These units operate much less frequently than do continuous power or CHP applications, often running as few as several hundred hours annually.

Selling Power to the Grid under Net Metering

Depending on what state they live in, some customers may be eligible for net metering. If eligible, net metering effectively allows a customer to sell its excess generation back to the grid at the same retail price as the customer buys power from the grid during other periods. With this financial incentive, the market for small commercial and residential DG installations, especially those based on the renewable energy technologies, should increase.

Standby/Emergency Generation

Applications providing standby or emergency power are typically driven by the reliability (perceived or real) of the grid and the cost to the customer of outage. Although the DG unit may only operate a few hours a year, it is used to power critical devices whose failure would result in property damage and/or threatened health and safety. Some customers like hospitals and airports may be required by code to install and maintain these units. For others, the high cost of outage drives the application. The choice of DG technologies for these emergency and standby applications is determined by installed cost, time required to start (i.e. black-start response), fuel access/storage, and size/weight of the unit.

Premium Power

On-site generation can improve both power quality and power reliability, especially when backed up with grid-based power. This application requires a DG technology that can operate continuously. In an era of both increasing power outages and rising demand for premium power, many businesses may install DG units to protect against the risk and cost of power outages. These customers include banks, semiconductor manufacturers, grocery stores, hospitals, and many other industrial and commercial market sites.

Green Power

Many of the renewable technologies and fuel cells have very low emissions. With the addition of emission-reducing technologies, microturbines and miniturbines also emit low levels of regulated gases. Customers, who are environmentally inclined, may purchase these DG applications for this reason, even if they have to pay a slight premium for green power compared with grid-based power purchases.

Remote Power

Residences and small commercial establishments (such as ranches, dairy farms, and flower growers) that are located well away from the transmission and distribution (T&D) system may prefer to generate their power on-site. By doing so, they eliminate both the cost of connecting to the grid and any problems associated with their position at the end of a long T&D line. The elimination of these problems, which include power outages and lower quality power, can produce the compelling economics necessary to the further use of DG technologies in a remote power capacity.

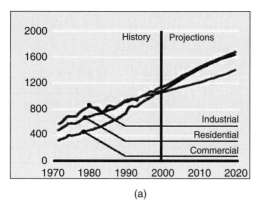
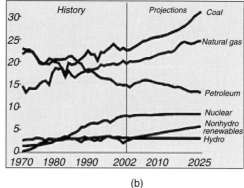

FIGURE 28.2 (a) Current and projected electricity sales in terawatt-hours (TWh) by sector and (b) energy production (quadrillion BTU) by fuel in USA [3].

28.2 Fuel-cell Based Energy Systems for DG

Hydrogen-based fuel-cell[1] energy is one of the two frontrunner alternative-energy solutions to address and alleviate the imminent and critical problems of rapid depletion of existing fossil-fuels (as illustrated in Fig. 28.2) and environmental pollution due to high emission. The framework for integrating these "zero-emission" alternative-energy sources to the existing energy infrastructure has been provided by the concept of DG, which provide an additional advantage: reduced reliance on existing and new centralized power generation, thereby saving significant capital cost. However, to achieve the projected worldwide target of $50 billion by 2015, the fuel-cell energy systems (as illustrated in Fig. 28.5) have to address cost, durability and reliability, and energy efficiency.

Currently, the applications of fuel cells (especially SOFC) for stationary applications are primarily for (i) lower power residential with a power range of <5 kW and (ii) for very high power reactive-power and harmonic compensations and high temperature coal-power applications. In Sections 28.3 and 28.4, we outline some established and possible power-electronic topologies for low- and high-power stationary applications, respectively.

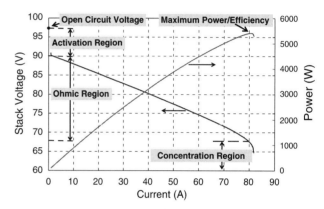

FIGURE 28.3 Illustration of a SOFC operation in activation, ohmic, and concentration/mass-transport regions. The simulated current–voltage characteristics and the corresponding power output for a 100-cell planar SOFC stack are shown.

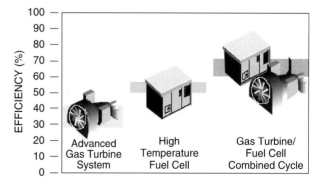

FIGURE 28.4 An illustration showing the vision of combined SOFC and gas-turbine operation to improve energy efficiency.

---

[1]Table 28.1 shows that SOFC (electrical characteristics illustrated in Fig. 28.3) is the most energy efficient among the listed DG equipment. The efficiencies of the SOFC systems can be further enhanced by utilizing the quality of waste heat derived from the fuel-cell reactions for CHP and combined cycle applications (as illustrated in Fig. 28.4).

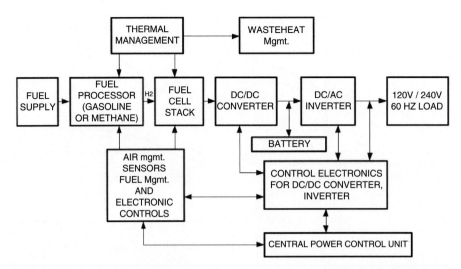

FIGURE 28.5 A typical fuel-cell energy system architecture comprising: (a) the fuel processor/balance-of-plant (BOP); (b) the fuel-cell stack (see Fig. 28.3 for a typical characteristic); and (c) the power-electronic subsystem (PES). We note that single-stage PES design is also feasible.

28.3 Power-electronic Topologies for Residential Stationary Fuel-cell Energy Systems

The choice of power-electronic topologies can be broadly categorized as push–pull and full-bridge based topologies. In [4–7], low-cost push–pull topology for power levels >1 kW is proposed. Push–pull based topology, owing to its low part count, is a good candidate for a low-cost fuel-cell converter and can achieve an efficiency of 90% for low and medium power. However, at higher power it can suffer from transformer flux imbalance and core-saturation problems. For current-fed topologies, the main limitation is encountered at greater than 3 kW, where, leakage inductance of the transformer poses a problem, unless soft switching is used. Further, the limited available range of switch duty cycle also makes it difficult to track variations in the input voltage. In [8–14], full-bridge based fuel-cell inverter topologies are discussed. Because of the symmetrical transformer flux and equal electrical stress distribution, several variations of full-bridge inverter topologies have been found to be useful from the cost and efficiency point of view, especially when implemented for power levels greater than 3 kW. A 1.8 kW prototype of a boost converter followed by a two-stage dc/ac converter is discussed in [9]. The two-stage dc/ac converter comprises a full-bridge high-frequency inverter and a cycloconverter, both operating at the same switching frequency, is based on a novel multi-carrier pulse-width modulation (PWM). A similar two-stage dc/ac converter is discussed in [13], which incorporates a zero-current switched cycloconverter; while the full-bridge high-frequency inverter is switched with a fixed 50% duty cycle. An improved version of the prototype in [13], comprising a soft-switched sine-wave pulse-width modulation (SPWM) full-bridge HF inverter followed by a cycloconverter (operating at line frequency) is proposed in [14]. Among all the low-cost fuel-cell inverters [4–14], which aim to achieve an efficiency of greater than 90%, a maximum full-load efficiency of 87 and 89% is demonstrated in [4] and [14], respectively. Also, the power-electronic topologies in [4–14] are not designed to improve efficiency of the stack.

In this section, we describe three low-power stationary fuel-cell power-conversion schemes being pursued at the University of Illinois, Chicago with focus low cost, energy efficiency, and reliability. The first PES comprises a zero-ripple boost converter followed by a two-stage dc/ac converter comprising a soft-switched phase-shifted SPWM multilevel high-frequency inverter and a line-frequency-switched cycloconverter. The second PES comprises a front-end phase-shifted high-frequency full-bridge inverter followed by a step-up high-frequency transformer and a dual forced cycloconverter for universal power conditioning at high efficiency. The final PES comprises a single-stage isolated differential Ĉuk inverter [15] and has very low switch count and reduced cost.

A. Ripple-mitigating Power Conditioner [16–18]

The power conditioner described in this section comprises a fuel-cell powered dc/dc zero-ripple boost converter (ZRBC), which generates a high voltage dc at its output, followed by a soft-switched, transformer isolated dc/ac inverter, which generates a 110 V ac. The high-frequency (HF) inverter switches are arranged in a multilevel fashion and are modulated by a fully rectified sine wave to create a HF, three-level ac voltage as shown in Fig. 28.6. Multilevel arrangement of the switches is particularly useful when the intermediate

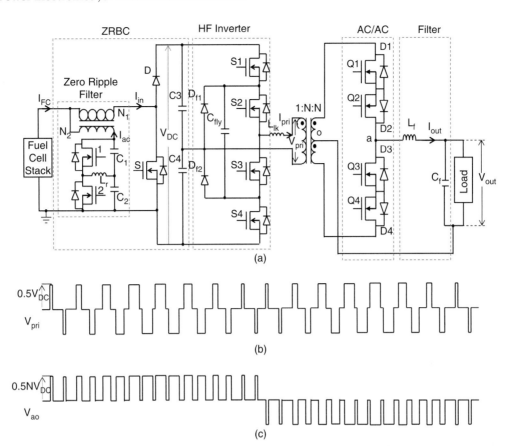

FIGURE 28.6 Schematic of the ripple-mitigating fuel cell power conditioner [16–18].

dc voltage >500 V. The HF inverter is followed by the ac/ac cycloconverter which converts the three-level ac to a voltage that carries the line-frequency sinusoidal information.

Zero Ripple Boost Converter (ZRBC)

The ZRBC is a standard non-isolated boost converter, with the conventional inductor replaced by a hybrid zero-ripple filter (ZRF). The ZRF (shown in Fig. 28.6) is viewed as a combination of a coupled inductor (shown in Fig. 28.6) and a half-bridge active power filter (APF) (shown in Fig. 28.6). Such a hybrid structure serves the dual purpose of reducing the high- and the low-frequency current ripples. The coupled inductor minimizes the high-frequency ripple from the fuel-cell current ($I_{FC} + i_2 = i_1$) and the APF minimizes the low-frequency ripple from the fuel-cell current ($I_{FC} + i_{ac} = i_{in}$). I_{FC} is the dc current supplied by the fuel cell, i_2 is the high-frequency ac current supplied by the series combination of identical capacitors C_1 and C_2 (in Fig. 28.6), and i_{ac} is the low-frequency ac current supplied by the APF storage reactor L_r. For effective reduction of the high-frequency current from the fuel-cell output, the value of the capacitors C_1 and C_2 should be as large as possible. However, the series combination should be small enough to provide a high impedance path to the low-frequency current i_{ac}. Therefore, for a chosen value of capacitor, the values of the following expression hold true [18]:

$$C_1 = C_2 = 2C \qquad (28.1a)$$

$$f_{HF} = \frac{1}{\sqrt{L_2 C}} \qquad (28.1b)$$

$$f_{LF} = \frac{1}{\sqrt{4 L_r C}} \qquad (28.1c)$$

where

f_{HF} is the switching frequency of the converter,
f_{LF} is the lowest frequency component in i_{ac}.

Assuming the switching frequency is approximately twenty times the lowest frequency component, the value of ZRF

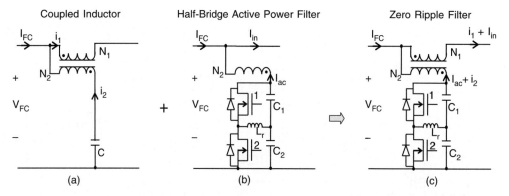

FIGURE 28.7 Schematic diagrams for: (a) coupled-inductor structure for reducing the HF current ripple; (b) half-bridge active filter, which compensates for the low frequency harmonic current ripple demand by the inverter; and (c) the proposed hybrid zero ripple filter structure.

passive components L_2 and L_r can be determined as follows [18]:

$$f_{HF} \geq 20 f_{LF} \quad (28.1\text{d})$$

$$\frac{1}{\sqrt{L_2 C}} \geq \frac{10}{\sqrt{L_r C}} \quad (28.1\text{e})$$

$$L_r \geq 100 L_2 \quad (28.1\text{f})$$

Therefore, the value of L_2 should be small in order to limit the value of L_r and also to minimize the phase shift in the injected low frequency current i_{ac}.

In the following sub-sections, the high- and low-frequency ac-current reduction mechanisms (Fig. 28.7) and the conditions to achieve it are discussed. In addition to this, the effect of coupled inductor parameters on the bandwidth of the open loop system will be discussed. For the purpose of analysis, the value of the capacitors C_1 and C_2 are assumed to be large. Hence, the dynamics of the APF is assumed to have minimal effect on the coupled inductor analysis.

High-frequency Current-ripple Reduction
In this section, the inductance offered by the coupled inductor and the ripple reduction achievable is discussed. For that purpose we need to derive an expression for the effective inductance of the coupled inductor. Because, the value of the capacitors C_1 and C_2 are large and that of L_{22} is small, the dynamics of the APF is assumed to have minimal effect on the coupled-inductor analysis. In the pi-model for the zero ripple coupled inductor and the excitation voltage and the current for the primary and the secondary windings are shown in the Fig. 28.8. The currents i_{1HF} and i_{2HF} are, respectively, the primary and the secondary AC currents shown in Fig. 28.8.

$$v_{FC} = (L_1 + L_M)\frac{di_{1HF}}{dt} + nL_M \frac{di_{2HF}}{dt} \quad (28.2\text{a})$$

$$v_C = (L_1 + L_M)\frac{di_{1HF}}{dt} + (L_2 + nL_M)\frac{di_{2HF}}{dt} \quad (28.2\text{b})$$

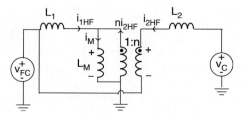

FIGURE 28.8 AC model for the coupled inductor shown in Fig. 28.7a.

$$n = \frac{N_2}{N_1} \cong \sqrt{\frac{L_{22}}{L_{11}}} \quad (28.2\text{c})$$

where L_{11} is the self inductance of the primary winding with N_1 turns. Solving Eqs. (28.2a and b), the expressions for di_{1HF}/dt and di_{2HF}/dt are obtained using

$$\frac{di_{1HF}}{dt} = \frac{(L_2 + nL_M)v_{FC} - nL_M v_C}{(L_1 + L_M)L_2}$$

$$= \frac{v_{FC}}{L_1 + L_M} + \frac{nL_M(v_{FC} - v_C)}{(L_1 + L_M)L_2} \quad (28.3\text{a})$$

$$\frac{di_{2HF}}{dt} = \frac{v_{FC} - v_C}{-L_2} \quad (28.3\text{b})$$

By substituting Eq. (28.3a) in (28.3b), we obtain the following expression:

$$\frac{di_{1HF}}{dt} = \frac{(L_2 + nL_M)v_{FC} - nL_M v_C}{(L_1 + L_M)L_2}$$

$$= \frac{v_{FC}}{L_1 + L_M} + \frac{nL_M}{(L_1 + L_M)}\frac{di_{2HF}}{dt} \quad (28.3\text{c})$$

To reduce the AC component of the fuel-cell-stack current to zero, the following condition should hold true:

$$\frac{di_{1HF}}{dt} = \frac{di_{2HF}}{dt} \quad (28.4)$$

Therefore, using the above condition and Eq. (28.3c), we obtain

$$\frac{di_{1HF}}{dt} = \frac{v_{FC}}{L_{11}\left[1 + \frac{(1+n)}{n}k\right]} = \frac{V_{FC}}{L_{eff}} \quad (28.5)$$

The denominator of Eq. (28.5) is the effective inductance L_{eff} offered by the coupled-inductor structure of the hybrid filter. The effective inductance depends on the turns ratio n, the coupling coefficient k, and the self inductance L_{11} of the primary winding. For very small values of turns ratio ($n \ll 1$), significantly large values of effective inductances can be obtained. Figure 28.9 shows the effective inductance curves and the corresponding reduction in the ripple. Figure 28.9a shows the dependence of normalized L_{eff} on n as a function of k. For the values of effective inductance shown in Fig. 28.9a, the corresponding values of achievable ripple current in both the coupled-inductor windings are shown in Fig. 28.9b. Using Fig. 28.9b, a designer can decide on a value of high-frequency current ripple and using the corresponding values of n and k, the normalized effective inductance can be chosen from Fig. 28.9a. While deciding the value of high-frequency ripple, one should choose a small value for n (<0.25), to ensure that L_{22} is small enough to prevent significant variations in the voltage across capacitors C_1 and C_2. Also, the effective inductance should be chosen to meet the bandwidth requirements of the ZRBC. Increase in the effective input inductance has a two-pronged effect on the open loop frequency response of the ZRBC. First, the bandwidth is reduced and second, the RHP zero is drawn closer to the imaginary axis, resulting in a reduction in the available phase margin and thereby the ZRBC stability.

Active Power Filter

The input current of the inverter comprises as dc component and a 120 Hz ac component and is expressed as

$$I_{dc} + I_{ac} = \frac{V_{out}I_{out}}{V_{dc}}\cos(\theta) - \frac{V_{out}I_{out}}{V_{dc}}\cos(2\omega t - \theta) \quad (28.6)$$

where

V_{out} inverter output voltage,

I_{out} are inverter output current,

V_{dc} is the average value of voltage across the series capacitors C_1 and C_2,

θ is the load power factor angle.

Here, we derive the condition for low frequency current ripple elimination from the PCS input current. For the APF shown in Fig. 28.6, the voltage across the storage reactor L_r of the APF is expressed as

$$V_{ab} = V_a - \frac{V_{dc}}{2} = V_{dc}\left(S_a - \frac{1}{2}\right) \quad (28.7)$$

where S_a is the modulating signal. The reactor current i_r is

$$i_r = \frac{V_{dc}(S_a - 1/2)}{j\omega L_r} \quad (28.8)$$

where

$$S_a = 0.5 + \sum_{n=1}^{\infty} B_n \sin n(\omega t + \phi)$$

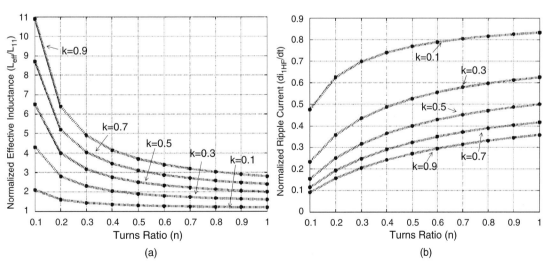

FIGURE 28.9 Normalized: (a) effective inductance and (b) ripple current of the coupled inductor.

$$i_r = \frac{V_{dc}}{\omega L_r} B_n \sin\left(\omega t + \phi - \frac{\pi}{2}\right) \quad \text{(considering only the fundamental component)}$$

The current injected by the APF is

$$i_{ac} = \left(S_a - \frac{1}{2}\right) i_r \quad (28.9a)$$

$$i_{ac} = \frac{V_{dc}}{\omega L_r} B_n^2 \sin(\omega t + \phi) \sin\left(\omega t + \phi - \frac{\pi}{2}\right) \quad (28.9b)$$

$$i_{ac} = \frac{V_{dc}}{2\omega L_r} B_n^2 \left[\cos\left(\frac{\pi}{2}\right) - \cos\left(2\omega t + 2\phi - \frac{\pi}{2}\right)\right] \quad (28.9c)$$

In order to reduce the 2nd harmonic in the input current to zero, $i_{ac} = I_{ac}$

$$\frac{V_{dc}}{2\omega L_r} B_n^2 \left[\cos\left(2\omega t + 2\phi - \frac{\pi}{2}\right)\right] = \frac{V_o I_o}{V_{dc}} \cos(2\omega t - \theta) \quad (28.10a)$$

which yields

$$B_n = \frac{\sqrt{2\omega L_r V_o I_o}}{V_{dc}} \quad (28.10b)$$

$$\phi = \frac{\pi}{4} - \frac{\theta}{2} \quad (28.10c)$$

DC/AC Converter

A two-stage dc/ac converter (shown in Fig. 28.10) comprises a soft-switched phase-shifted synchronized pulse-width modulation (SPWM) multilevel HF inverter and a line-frequency switched ac/ac cycloconverter. The multilevel arrangement of the HF inverter switches reduces the voltage stress and the cost of the high-frequency semiconductor switches. The ac/ac stage comprises a single-phase cycloconverter and an output LC filter. The cycloconverter has two bidirectional switch pairs Q_1 and Q_2 and Q_3 and Q_4 for a single-phase output. In order to achieve a 60-Hz sine-wave ac at the output, the sine wave modulation can be performed either on the HF inverter or on the cycloconverter. Therefore, two different modulation strategies are possible for the dc/ac inverter. Both schemes result in the soft switching of the HF inverter while the cycloconverter is hard switched.

In the first modulation scheme the cycloconverter switches follow SPWM, while the HF inverter switches are switched at fixed 50% duty pulse. The HF inverter switches in this scheme undergo zero-voltage turn-on. In the second modulation scheme, the switches of the multilevel HF inverter follow SPWM and the cycloconverter switches are switched based on the power-flow information. Unlike the first modulation scheme, which modulates the cycloconverter switches at high frequency, in the second modulation scheme, cycloconverter operates at line frequency. The switches are commutated at high frequency only when the polarities of output current and voltage are different. Usually this duration is very small and therefore, the switching loss of the ac/ac cycloconverter is considerably reduced compared to the conventional control method. Therefore, the heat-sinking requirement of the cycloconverter switches is significantly reduced. The HF inverter switches in this scheme undergo zero-current turn-off. Control signals for the second modulation scheme are shown in Fig. 28.10.

B. Universal Power Conditioner [19][2]

This approach achieves a direct power conversion and does not require any intermediate energy storage components. As shown in Fig. 28.11, the final approach has a HF inverter followed by a HF transformer and a forced cycloconverter. Switches (Q1–Q4) on the primary of the HF transformer are sine-wave modulated to create a HF three-level ac voltage. The three-level ac at the output of the HF transformer is converted to 60/50 Hz line-frequency ac by the cycloconverter and the LC filter. For an input of 30 V, the transformation ratio of the HF transformer is calculated to be $N = 13$. Fabrication of a 1:13 transformer is relatively difficult. Furthermore, high turns ratio yields enhanced secondary leakage inductance and secondary winding resistance, which result in measurable loss of duty cycle and secondary copper losses, respectively. Higher leakage also leads to the higher voltage spike, which added to the high nominal voltage of the secondary that necessitate the use of high-voltage power devices. Such devices have higher on-resistance and slower switching speeds.

Therefore, a combination of two transformers and two cycloconverters on the secondary side of the HF transformer is identified to be an optimum solution. For an input voltage in the range of 30–42 V, we use $N = 6.5$, while for an input voltage above 42 V, we use $N = 4.3$. To change the transformation ratio of the HF transformer, we use a single-pole-double-throw (SPDT) relay, as shown in Figs. 28.11a, b. Such an arrangement not only improves the efficiency of the transformer but also significantly improves the utilization of the cycloconverter switches for operation at 120/240 V ac and 60/50 Hz. For 120 V ac output, the two cycloconverter filter capacitors are paralleled (as shown in Fig. 28.11a) while for 240 V ac output, the voltage of the filter capacitor are connected in opposition (as shown in Fig. 28.11b).

Modes of Operation

In this section, we discuss the modes of operation of the inverter in Fig. 28.11 for 120 V ac output and for an input voltage in the range of 42–60 V (i.e. $N = 4.3$). The modes of operation below 42 V (i.e. $N = 6.5$) remains the same.

---

[2]University of Illinois, Chicago secured the first position among U.S.A and 3rd position among worldwide university competition for this topology, which was developed as a part of 2005 IEEE Future Energy Challenge Competition.

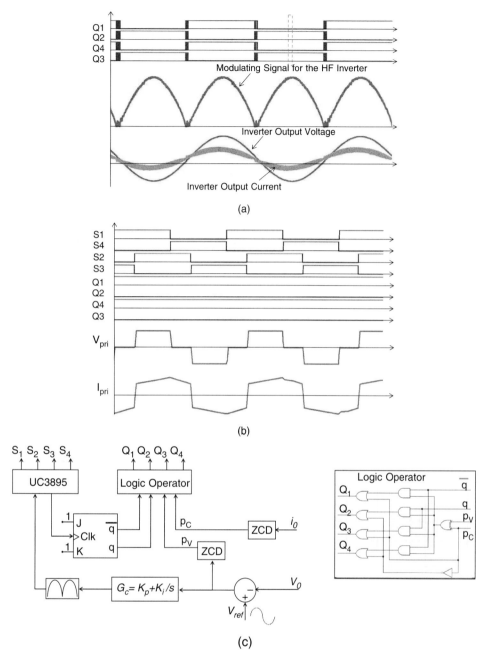

FIGURE 28.10 (a) Schematic waveforms for the inverter operation; (b) schematic waveforms for the HF inverter; and (c) overall control scheme for the HF inverter.

Figure 28.10 shows the waveforms of the five operating modes of the phase-shifted HF inverter and a positive primary and a positive filter-inductor current. Modes 2 and 4 show the zero-voltage switching (ZVS) turn-on mechanism for switches Q_3 and Q_4, respectively. Unlike conventional control scheme for cycloconverter [13], which modulates the switches at high frequency, the proposed cycloconverter operates at line frequency. The switches are commutated at high frequency only when the polarities of the output current and voltage are different [13]. For unity power-factor operation, this duration is negligibly small and therefore, the switching loss of the ac/ac cycloconverter is considerably reduced compared to the conventional control method [14, 16].

Five modes of the PES operation are discussed for positive primary current. A set of 5 modes exists for a negative primary current as well. A similar set of five modes of operation for the 240 V ac exists [19] for input voltage above 42 V (N = 4.5).

Again, the mode of operation for input voltage below 42 V ($N = 6.5$) remains the same.

Mode 1 (Fig. 28.12a): During this mode, switches Q_1 and Q_2 of the HF inverter are on and the transformer primary current I_{p1} and I_{p2} is positive. The load current splits equally between the two cycloconverter modules. For the top cycloconverter module, the load current $I_{out}/2$ is positive and flows through the switches pair S_1 and S'_1, the output filter L_{f1} and C_{f1}, switches S_2 and S'_2, and the transformer secondary. Similarly, for the bottom cycloconverter module, the load current

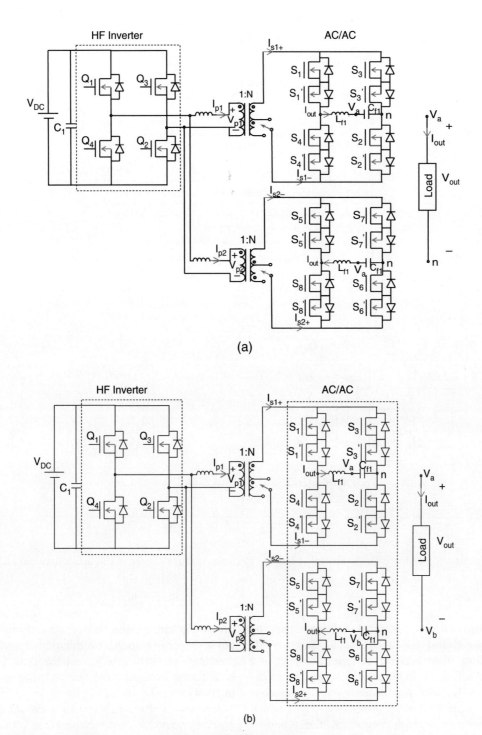

FIGURE 28.11 (a) and (b) Schematics for converter operation, respectively, at 120 V ac and 60 Hz and 240 V ac and 50 Hz; (c) and (d) control schemes of the power electronic system (PES) in stand-alone and grid-connected modes [19].

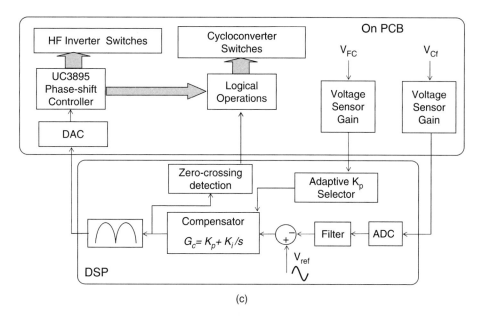

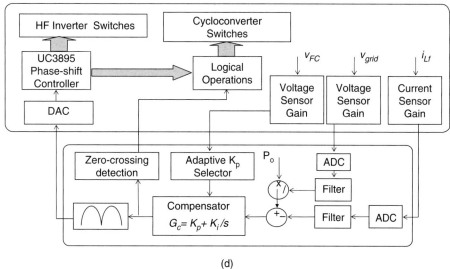

FIGURE 28.11 *continued*

$0.5 \times I_{out}$ is positive and flows through the switches pair S_5 and S'_5, the output filter L_{f2} and C_{f2}, switches S_6 and S'_6, and the transformer secondary.

Mode 2 (Fig. 28.12b): At the beginning of this interval the gate voltage of the switch Q_1 undergoes a high-to-low transition. As a result, the output capacitance of Q_1 begins to accumulate charge and at the same time the output capacitance of switch Q_4 begins to discharge. Once the voltage across Q_4 goes to zero, it is can be turned on under ZVS. The transformer primary currents I_{p1} and I_{p2} and the load current I_{out} continue to flow in the same direction. This mode ends, when the switch Q_1 is completely turned off and its output capacitance is charged to V_{DC}.

Mode 3 (Fig. 28.12c): This mode initiates when Q_1 turns off. The transformer primary currents I_{p1} and I_{p2} are still positive, and freewheels through Q_4 as shown in Fig. 28.12c. Also the load current continues to flow in the same direction as in Mode 2. Mode 3 ends at the commencement of turn off Q_2.

Mode 4 (Fig. 28.12d): At the beginning of this interval, the gate voltage of Q_2 undergoes a high-to-low transition. As a result of this, the output capacitance of Q_2 begins to accumulate charge and at the same time, the output capacitance of switch Q_3 begins to discharge as shown in the Fig. 28.12d. The charging current of Q_2 and the discharging current of Q_3 together add up to the primary current I_{p1} and I_{p2}.

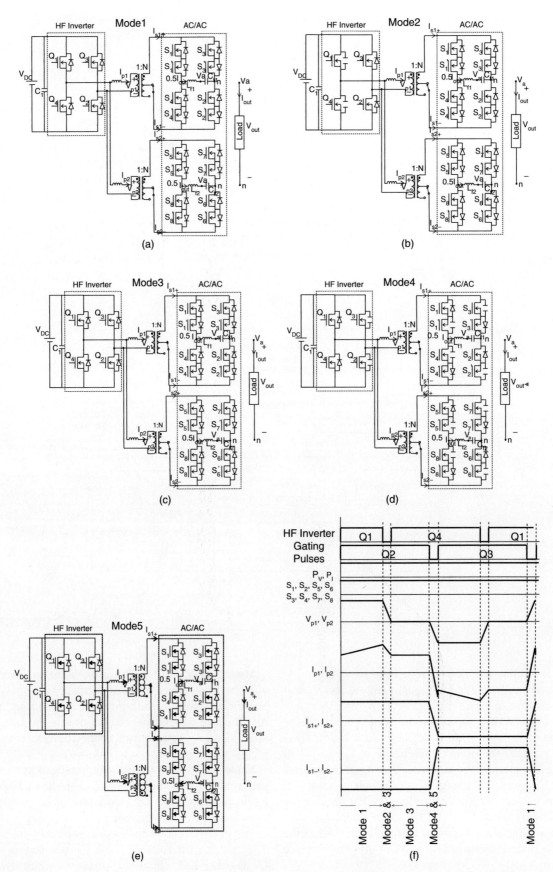

FIGURE 28.12 Modes of operation for 120 V ac for input voltage in the range of 42–60 V ($N = 4.3$): (a–e) topologies corresponding to the five operating modes of the overall dc/ac converter for positive primary current and for power flow from the input to the load and (f) schematic waveforms show the operating modes of the HF inverter when primary currents are positive. *The modes of operation below 42 V (i.e. $N = 6.5$) remains the same.*

The transformer current makes a transition from positive to negative. Once the voltage across Q_3 goes to zero, it is turned on under ZVS. The load current flows in the same direction as in Mode 3, but makes a rapid transition from the bidirectional switches S_1 and S'_1 and S_2 and S'_2 to S_3 and S'_3 and S_4 and S'_4 and during this process $I_{out}/2$ splits between the two legs of the cycloconverter modules as shown in Fig. 28.12d. Mode 4 ends when the switch Q_2 is completely turned off and its output capacitance is charged to V_{DC}. At this point, it is necessary to note that since S_1 and S_2 are off simultaneously, each of them support a voltage of V_{DC}.

Mode 5 (Fig. 28.12e): This mode starts when Q_2 is completely turned off. The primary current I_{p1} and I_{p2} is negative, while the load current is positive as shown in Fig. 28.12e.

C. Low-cost Fuel-cell Power Conditioner

Low cost power electronics system (PES) that converts a fuel cell stack's low voltage (typically 30–60 V) output into a commercial ac output is critical for the success of fuel-cell energy system, especially for the low power applications ($\leq$3 kW). According to [20], the cost of stationary residential PES should be less than \$40/kW for high volume of production. Many types of configurations have been tried to achieve fuel-cell power conversion at lower cost without compromising efficiency [5, 7, 11–13, 19, 21–23]. Based on where the transformer is inserted for isolating the fuel cell from the load, they can be broadly categorized into two types of topologies.

The first type, usually called "dc/dc type converter," uses a dc/dc at front end to boost the low input voltage. The transformer is inserted into the dc/dc stage which usually is a push–pull or full-bridge converter. A diode rectifier is required at the secondary to obtain a higher dc voltage. Following the dc/dc stage, a non-isolated dc/ac stage is used to get low frequency ac. Since this type of topology has three power conversion stages (dc-high frequency ac–dc-low frequency ac), it may not be the best choice from efficiency and cost point of view because each stage has to have a very high efficiency to make the overall efficiency high, and usually more than six active components (switches and diodes) are required.

The second type of topology, referred as "cycloconverter type converter," reduce the system complexity by removing the dc/dc stage. The galvanic isolation is achieved by embedding the transformer into a dc/ac converter which usually composes of a three-level or full-bridge inverter followed by a cycloconverter. Although it has lesser power conversion stage, the number of active components may not necessary be less than that of the first type mainly due to the bidirectional switches, which are required for the cycloconverter.

A single-stage isolated dc/ac inverter, which was originally proposed in [15] as "push–pull amplifier" achieves direct dc/ac conversion by connecting load differentially across two bidirectional dc/dc Cuk converters and modulating them sinusoidally with 180° phase difference (Fig. 28.13). Since only four main switches are used, it would potentially reduce system complexity, costs, improve reliability, and increase efficiency. Furthermore, the common source connection between two devices both at primary and secondary sides (Q_a and Q_c and Q_b and Q_d in Fig. 28.13) makes the gate drive circuit relatively simple. In addition, the possibility of coupling inductors or integrated magnetics [24] will further reduce the overall volume and weight thereby achieving lesser material and space usage. Therefore, it would be a better alternative for fuel-cell application and will eventually lead to a very low cost, high density power conversion system.

Another advantage of this inverter is the reduction of turns ratio of the step-up transformer, which is usually required to achieve rated ac from low fuel-cell stacks dc voltage. The inherent voltage boosting capability of the dc/ac Cuk inverter can reduce the transformer turns ratio requirement by at least half. Low transformer turns ratio yields less leakage inductance

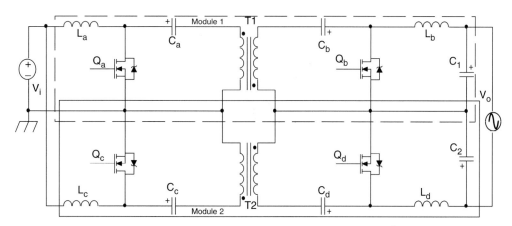

FIGURE 28.13 Schematic of the single-stage dc/ac differential isolated Ĉuk inverter.

and secondary winding resistance, which reduces the loss of duty cycle and secondary copper losses, respectively.

Although the non-isolated dc/ac inverter has already been proposed [15], analysis of the isolated version has not appeared in any literature, nor have the issues involved in the design towards higher power rating (>1 kW) for fuel-cell application. Reference [25] presents the analysis and design of a 1 kW fuel cell PES using the isolated differential Cuk inverter.

The output of the proposed dc/ac inverter is the difference between two "sine-wave modulated PWM controlled" isolated Cuk inverter (module 1 and module 2), with their primary sides connected in parallel. The two diagonal switches of two modules are triggered by a same signal ($Q_a = Q_d$, $Q_b = Q_c$), while the two switches in each module have complementary gate signals ($Q_a = /Q_b$, $Q_c = /Q_d$). As we know, the output voltage of an isolated Cuk inverter can be expressed as

$$V_o = V_i \cdot \frac{D}{N(1-D)} \quad (28.11)$$

where D is the duty ratio, N is the transformer turns ratio, V_i is the input voltage. Since duty ratios for module 1 and module 2 are complementary, the output difference between two modules is:

$$V_o = Vc1 - Vc2 = V_i \cdot \left[\frac{D}{N \cdot (1-D)} - \frac{1-D}{N \cdot D} \right] \quad (28.12)$$

The curves corresponding to the terms in Eq. (28.12) with respect to the duty ratio D (assuming $N = 1$) are plotted in Fig. 28.14. The figure shows that although the gain-duty ratio curves of modules 1 and 2 are not linear, their difference is almost linear. Therefore, if a sine-wave modulated duty ratio D is used as a control signal for inverter in Fig. 28.13, its output voltage will be a sine wave with small distortion [15].

In order to understand how the current flows and energy transfers during the switching and to help select the device rating, four different modes of the inverter are analyzed and shown in Fig. 28.15. Figures 28.15a, b show the direction of the current when the load current flows from top to the bottom.

Mode 1: Figure 28.15a shows the current flow for the case when switch Q_a, Q_d are ON and Q_b, Q_c are OFF. During this time, the current flowing through the input inductor L_a increases and the inductor stores energy. At the same time, the capacitor C_a discharges through Q_a, and thus, there is transfer of energy from primary side to the secondary side through the transformer T_1. The capacitor C_b is discharged to the circuit formed by L_b, C_1, and the load R. Meanwhile, the inductor L_d stores energy and its current increases. The capacitor C_d discharges through Q_d. The power flows in opposite direction in the module 2 from secondary to the primary side. The capacitor C_c is also discharged to provide the power.

Mode 2: When Q_a, Q_d are turned off, and Q_b, Q_c are ON (Fig. 28.15b) C_a, C_d and C_b, C_c are charged using the energy, which was stored in the inductors L_a and L_d while Q_a, Q_d were on. During this time, L_b and L_c will release their energy.

Figures 28.15c, d show the current direction when the load current flows in the opposite direction. The description for these two modes is omitted due to the similarity with Figs. 28.15a, b.

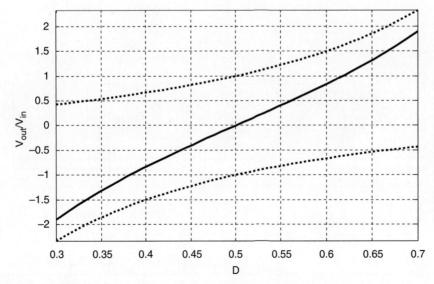

FIGURE 28.14 Voltage gain vs D for module 1 (top), module 2 (bottom), and their difference (middle).

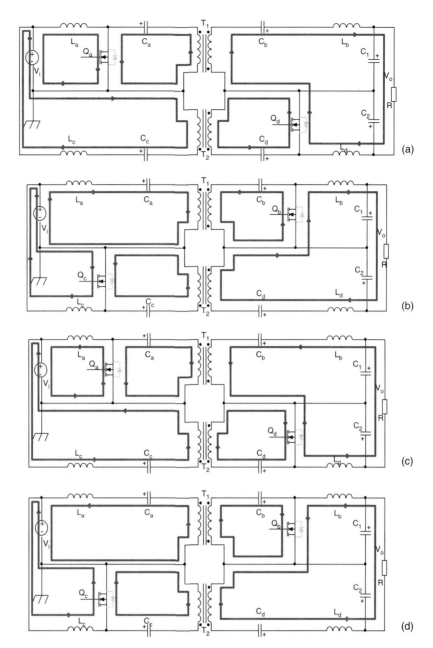

FIGURE 28.15 (a) and (b) Direction of the current flow in two modes of operation for positive load current: (a) Mode 1, when Q_a, Q_d are ON, Q_b, Q_c are OFF and (b) Mode 2 when Q_a, Q_d are OFF Q_b, Q_c are ON. Figures 28.15c, d are the corresponding current flow when load current flows in the opposite direction.

28.4 Towards Power-electronic Topologies for High Power Fuel-cell Based DG

In addition to residential use, fuel cells are being increasingly considered for high-power (MW-power) applications. In that context, currently, Department of Energy's (DOEs) focus is on reactive-power and harmonic compensations and high-temperature coal-power applications. Due to the practical manufacture limitation and reliability concern, the maximal output voltage and output power of one fuel-cell stack cannot be too high. To provide the desired amount of voltage and/or power, multiple fuel-cell stacks are required to form a fuel-cell module. These modules along with their power conditioning system can be connected in several different configurations, which will be presented in this section.

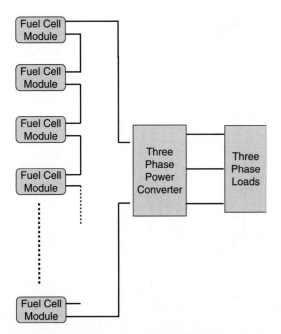

FIGURE 28.16 A general diagram of multiple fuel-cell modules connected in series to generate dc bus followed by a three-phase inverter. Copyright © IEEE, 2003, Ozpineci etc. [26].

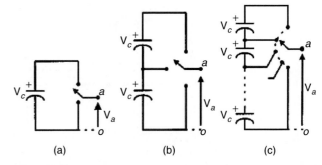

FIGURE 28.17 One phase multilevel diagram: (a) two levels; (b) three levels; and (c) n-levels. Copyright © IEEE, 2002, Rodriguez etc. [27].

The advantages and disadvantages of each approach are also summarized.

One of the most common configurations of fuel-cell modules and PCS for high power application is to connect multiple fuel-cell modules directly in series and apply the obtained dc bus to a three-phase dc/ac inverter, as shown in Fig. 28.16. For each module, fuel-cell stacks can also be connected in parallel to provide the required power, if necessary.

For the inverter, recently, multilevel power conversion has received increasing attention for medium-voltage and high power applications. Among the multilevel topologies, diode-clamped inverter, capacitor-clamped (flying capacitor) inverter, and cascaded multi-cell inverter with separate dc sources are the three basic architectures. Other emerging novel topologies include asymmetric hybrid cells and soft-switched multilevel inverters [27].

Multilevel inverters include an array of power semiconductor devices and voltage sources. They generate output voltage with stepped waveforms. Figure 28.17 shows a schematic diagram of one phase leg multilevel converter with different number of levels. Note the actual power devices are replaced by ideal switches with several positions. Figure 28.17a is a two-level inverter since the output voltage V_a has only two possible values, while Fig. 28.17b is a three-level inverter since its output can have three different values. If m is the number of possible output voltage levels, it is called m-level inverter (Fig. 28.17c). By increasing the number of levels, the output voltage waveforms will have more steps and thus have a reduced harmonic distortion. However, a high number of levels will increase the complexity and introduce voltage imbalance problems.

Figure 28.18 show the diagrams of one phase leg fuel-cell power conditioner for high power applications using diode-clamped multilevel inverter. Its operational principle was explained in [27]. The number of switches equals $2 \times (m-1)$. The key components that make the circuit different are clamp diodes, which clamp the switch voltage to the corresponding level of the dc bus voltage. Assuming all the clamp diodes are the same, the number of diodes required for each leg will be $(m-1) \times (m-2)$. This represents a quadratic increase in m (number of levels). Another problem caused by diode clamping is if the inverter operates under high frequency, the diode reverse recovery will become a major design challenges.

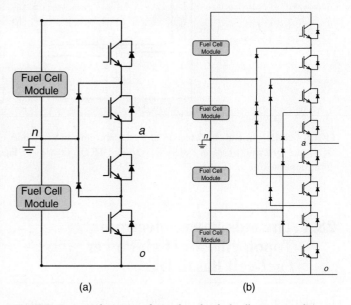

FIGURE 28.18 Schematics of one phase leg fuel-cell power conditioner for high power applications using diode-clamped multilevel inverter: (a) three-level and (b) five-level inverter. Copyright © IEEE 2002, Rodriguez etc. [27] and 2003, Ozpineci etc. [26].

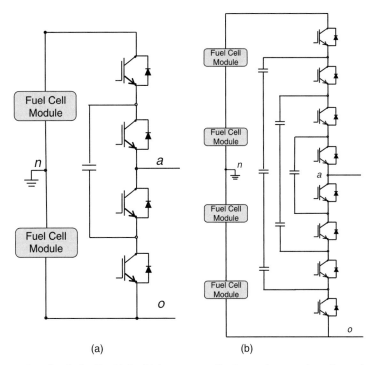

FIGURE 28.19 Schematics of one phase leg fuel-cell PCS for high power applications using capacitor-clamped multilevel inverter: (a) three-level and (b) five-level inverter. Copyright © IEEE, 2002, Rodriguez etc. [27].

Figure 28.19 show diagrams of one phase leg multilevel capacitor-clamped inverter. The operational principle is similar to diode-clamped type multilevel except the clamping of device voltages is achieved by the flying capacitors [27]. Similar to diode-clamping, capacitor clamping requires a large number of bulk capacitors to clamp the voltage. For a m-level inverter, a total of $(m-1) \times (m-2)/2$ clamping capacitors are needed per phase leg.

Figure 28.20 shows a 24 MVA NPC inverter using integrated gate commutated thyristor (IGCT) series connection for medium-voltage applications [28]. Topology used here is a diode clamped three-level inverter with series connection of semiconductors. The overall input dc bus voltage V_{dc} is 14.4 kV. To sustain this voltage, each switch is composed of three 4.5 kV IGCTs and each diode is made up of three 4.5 kV diodes. The RC snubbers are also used for the purpose of voltage balancing.

As compared to the other configurations, the series connection of fuel-cell modules followed by an inverter has less complexity and less number of semiconductor devices.

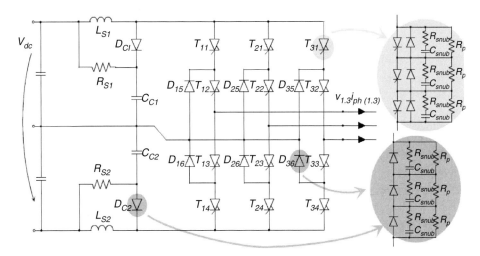

FIGURE 28.20 A 24 MVA inverter using IGCT series connection for medium-voltage applications. Copyright © IEEE, 2001 Nagel etc. [28].

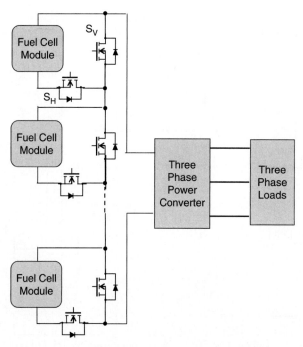

FIGURE 28.21 Level reduction configuration of multiple fuel-cell modules to generate dc bus followed by a three-phase inverter [29, 30].

But the system has potential reliability problem since failure of one fuel cell can potentially make the whole system inoperable. So far, the power conditioning systems aforementioned have to sustain the variation of fuel-cell stack voltage caused by the change of loads. Since power semiconductor devices have to be selected for the highest input voltage, this degrades the device utilization and increases the cost by using more expensive higher voltage devices.

The problem can be alleviated by cascading the fuel-cell modules and dc/dc converters [29, 30]. As shown in Fig. 28.21, each fuel-cell module has an associated vertical (S_V) and a horizontal (S_H) switches. They are complementarily controlled. When S_H is on and S_V is off, the fuel-cell module supplies power. If S_V is on and S_H is off, the fuel-cell module is inhibited. With this configuration, the need for degrading power semiconductors in fuel-cell systems is reduced. By inhibiting some of the fuel cells and using the inhibited fuel cells in other applications, like charging batteries, the system efficiency and the fuel-cell utilization increases. It also has the advantages of increasing reliability. This means that if a fuel cell fails, the system will continue to operate. A three-phase inverter is still required for the voltage inversion. This inverter will require a variable modulation index to compensate for the varying dc bus voltage but this variation will not be as extreme as the one in Fig. 28.16.

A different converter configuration is introduced in [31], which is based on the series connection of single-phase inverters with separate dc sources (Fig. 28.22). Figure 28.22a shows a phase leg of a seven-level inverter with three cells in each phase. The output is formed by the combination of the ac voltages generated by each single-phase full-bridge inverter cell, which outputs three voltage levels: $+V_{dc}$, $-V_{dc}$, and 0 (Fig. 28.22b). Cascade multilevel converter can have several types of variation.

Figure 28.23 shows another way to achieve a voltage source multilevel inverter by voltage addition with transformers, which has been realized for high power applications [32, 33]. The step voltage waveforms are obtained by adding the each output of the inverter at the transformer secondary side. In this configuration, only one fuel-cell module is required. The isolation and voltage boost can be achieved by transformer. However, the transformer in this configuration requires special design and may saturate if the primary-side current has a dc component.

Figure 28.24a shows a mixed-level hybrid multilevel cell, obtained by replacing the full bridge of the cascade multilevel cell in Fig. 28.24 by diode-clamped or capacitor-clamped multilevel inverter. The number of the fuel cells is reduced by half. The dc voltage of fuel-cell modules in this configuration can be different which produces asymmetric hybrid multilevel cells. It is also possible to modulate different cells of cascade multilevel converter with different frequencies and implement it using different types of power devices, as shown in Fig. 28.24b.

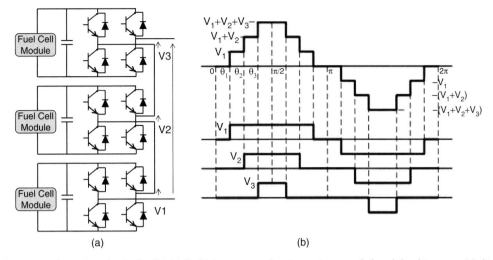

FIGURE 28.22 Schematics of one phase leg fuel-cell PCS for high power applications using cascaded multilevel inverter: (a) diagram of seven-level cascaded multilevel inverter and (b) its corresponding waveforms. Copyright © IEEE, 2003, Ozpineci etc. [26].

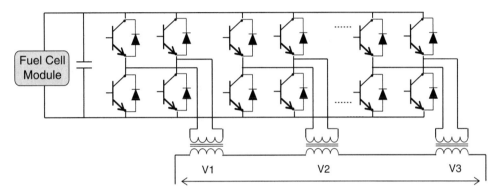

FIGURE 28.23 Diagram of one phase leg of a voltage addition with transformer [32].

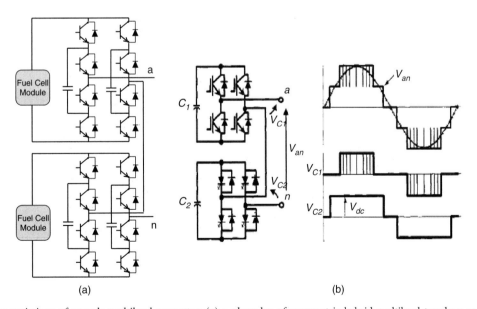

FIGURE 28.24 Two variations of cascade multilevel converter: (a) a phase leg of asymmetric hybrid multilevel topology and (b) a phase leg of asymmetric cascaded multilevel topology. Copyright © IEEE, 2002, Rodriguez etc. [27].

Acknowledgments

The work described in this paper is supported in parts by the California Energy Commission (CEC) under Award No. 53422A/03-02 and by the Department of Energy (DOE) under Award No. DE-FC2602NT41574. However, any opinions, findings, conclusions, or recommendations expressed herein are those of the authors and do not necessarily reflect the views of the CEC or DOE.

References

1. http://www.pbworld.com/library/technical_papers/pdf/61_DieselEngine.pdf
2. http://www.eere.energy.gov/de/pdfs/ams_program_plan.pdf
3. http://www.eia.doe.gov/
4. G.K. Andersen, C. Klumpner, S.B. Kjaer, and F. Blaabjerg, "A new green power inverter for fuel cells", *IEEE Power Electronics Specialists Conference*, 2002, pp. 727–733.
5. R. Gopinath, S.S. Kim, and P. Enjeti, "Development of a low cost fuel cell inverter with DSP control", *IEEE Power Electronics Specialists Conference*, 2002, pp. 1256–1262.
6. G.W. Pradeep, H. Mohammad, and P. Famouri, "High efficiency low cost inverter system for fuel cell application", *Fuel Cell Seminar*, 2003. (Also available at http://www.energychallenge.org/FuelCellSeminar.pdf.)
7. J. Mazumdar, I. Batarseh, and N. Kutkut, "High frequency low cost dc-ac inverter design with fuel cell source for home applications", *IEEE Industry Applications Conference*, 2002, pp. 789–794.
8. A.M. Tuckey and J.N. Krase, "A low-cost inverter for domestic fuel cell applications", *IEEE Power Electronics Specialists Conference*, 2002, pp. 339–346.
9. P.T. Krein and R. Balog, "Low cost inverter suitable for medium-power fuel cell sources", *IEEE Power Electronics Specialists Conference*, 2002, pp. 321–326.
10. H. Ertl, J.W. Kolar, and F.C. Zach, "A novel multicell dc-ac converter for applications in renewable energy systems", *IEEE Transactions on Industrial Electronics*, 2002, pp. 1048–1057.
11. J. Wang and F.Z. Peng, "A new low cost inverter system for 5 kW fuel cell", *Fuel Cell Seminar*, 2003. (Also available at http://www.energychallenge.org/FuelCellSeminar.pdf.)
12. T.P. Bohn and R.D. Lorenz, "A low-cost inverter for domestic fuel cell applications", *Fuel Cell Seminar*, 2003. (Also available at http://www.energychallenge.org/FuelCellSeminar.pdf.)
13. T. Kawabata, H. Komji, and K. Sashida, "High frequency link dc/ac converter with PWM cycloconverter", *IEEE Industrial Application Society Conference*, 1990, pp. 1119–1124.
14. S. Deng and H. Mao, "A new control scheme for high-frequency link inverter design", *IEEE Applied Power Electronics Conference and Exposition*, 2003, pp. 512–517.
15. S. Ćuk and R.W. Erickson, "A conceptually new high-frequency switched-mode amplifier technique eliminates current ripple", *Proceedings of Fifth National Solid-State Power Conversion Conference*, May 1978, pp. G3.1–G3.22.
16. S.K. Mazumder, R.K. Burra, and K. Acharya, *A novel efficient and reliable dc-ac converter for fuel cell power conditioning*, USPTO Patent Application# 20050141248, 2005.
17. S.K. Mazumder and R.K. Burra, "Fuel cell power conditioner for stationary power system: Towards optimal design from reliability, efficiency, and cost standpoint", Keynote Lecture, *ASME Proceedings on Third International Conference on Fuel Cell Science, Engineering and Technology*, 2005, FUELCELL2005-74178.
18. R.K. Burra, *A high performance power conditioner for solid-oxide fuel cell based stationary power generation*, Doctoral Dissertation, University of Illinois at Chicago, Chicago, Illinois, 2005.
19. S.K. Mazumder et al., *Topic B: Design and development of a 1 kW fuel cell grid connected inverter*, Final Report, IEEE Future Energy Challenge Competition, 2005.
20. http://www.eren.doe.gov/distributedpower
21. G.W. Pradeep, H. Mohammad, and P. Famouri, et al., "High efficiency low cost inverter system for fuel cell application", *Fuel Cell Seminar*, 2003.
22. P.T. Krein and R. Balog, "Low cost inverter suitable for medium-power fuel cell sources", *IEEE Power Electronics Specialists Conference*, 2002, pp. 321–326.
23. A.M. Tuckey and J.N. Krase, "A low-cost inverter for domestic fuel cell applications", *IEEE Power Electronics Specialists Conference*, 2002, pp. 339–346.
24. R.D. Middlebrook and S. Ćuk, *Advances in switched-mode power conversion*, vols. I, II, and III, TESLACO, 1983.
25. S.K. Mazumder et al., "A low-cost single-stage isolated differential Cuk inverter for fuel cell application", submitted for review, *IEEE Power Electronics Specialists Conference*, 2005.
26. B. Ozpineci, Z. Du, L.M. Tolbert, D.J. Adams, and D. Collins, "Integrating multiple solid oxide fuel cell modules", *IEEE Industrial Electronics Conference*, 2003, pp. 1568–1573.
27. J. Rodriguez, J-S. Lai, and F.Z. Peng, "Multilevel inverters: A survey of topologies, controls, and applications", *IEEE Transactions on Industrial Electronics*, vol. 49, no. 4, 2002, pp. 724–738.
28. A. Nagel, S. Bernet, P.K. Steimer, and O. Apeldoorn, "A 24 MVA inverter using IGCT series connection for medium voltage applications", *IEEE Thirty-Sixth IAS Annual Meeting Conference Record*, 2001.
29. B. Ozpineci, L.M. Tolbert, and D. Zhong, "Optimum fuel cell utilization with multilevel inverters", *IEEE Power Electronics Specialists Conference*, 2004, pp. 4798–4802.
30. B. Ozpineci, L.M. Tolbert, G-J. Su, and Z. Du, "Optimum fuel cell utilization with multilevel dc-dc converters", *IEEE Applied Power Electronics Conference*, 2004, pp. 1572–1576.
31. P.W. Hammond, "A new approach to enhance power quality for medium voltage ac drives", *IEEE Transactions on Industry Applications*, vol. 33, no. 1, 1997, pp. 202–208.
32. S. Mariethoz and A. Rufer, "Design and control of asymmetrical multi-level inverters", *IEEE Industrial Electronics Conference*, 2002 pp. 840–845.
33. N.P. Schibli, T. Nguyen, and A.C. Rufer, "A three-phase multilevel converter for high-power induction motors", *IEEE Transactions on Power Electronics*, vol. 13, no. 5, 1998, pp. 978–986.

29
Wind Turbine Applications

Juan M. Carrasco,
Eduardo Galván, and
Ramón Portillo
Department of Electronic Engineering, Engineering School, Seville University, Spain

29.1 Wind Energy Conversion Systems 737
    29.1.1 Horizontal-axis Wind Turbine • 29.1.2 Simplified Model of a Wind Turbine • 29.1.3 Control of Wind Turbines
29.2 Power Electronic Converters for Variable Speed Wind Turbines 743
    29.2.1 Introduction • 29.2.2 Full Power Conditioner System for Variable Speed Turbines • 29.2.3 Rotor Connected Power Conditioner for Variable Speed Wind Turbines • 29.2.4 Grid Connection Standards for Wind Farms
29.3 Multilevel Converter for Very High Power Wind Turbines 757
    29.3.1 Multilevel Topologies • 29.3.2 Diode Clamp Converter (DCC) • 29.3.3 Full Converter for Wind Turbine Based on Multilevel Topology • 29.3.4 Modeling • 29.3.5 Control • 29.3.6 Application Example
29.4 Electrical System of a Wind Farm 761
    29.4.1 Electrical Schematic of a Wind Farm • 29.4.2 Protection System • 29.4.3 Electrical System Safety: Hazards and Safeguards
29.5 Future Trends 762
    29.5.1 Semiconductors • 29.5.2 Power Converters • 29.5.3 Control Algorithms • 29.5.4 Offshore and Onshore Wind Turbines
    Nomenclature 765
    References 766

29.1 Wind Energy Conversion Systems

Wind energy has matured to a level of development where it is ready to become a generally accepted utility generation technology. Wind turbine technology has undergone a dramatic transformation during the last 15 years, developing from a fringe science in the 1970s to the wind turbine of the 2000s using the latest in power electronics, aerodynamics, and mechanical drive train designs [1, 2].

Most countries have plans for increasing their share of energy produced by wind power. The increased share of wind power in the electric power system makes it necessary to have grid-friendly interfaces between the wind turbines and the grid in order to maintain power quality.

In addition, power electronics is undergoing a fast evolution, mainly due to two factors. The first one is the development of fast semiconductor switches, which are capable of switching quickly and handling high powers. The second factor is the control area, where the introduction of the computer as a real-time controller has made it possible to adapt advanced and complex control algorithms. These factors together make it possible to have cost-effective and grid-friendly converters connected to the grid [3, 4].

29.1.1 Horizontal-axis Wind Turbine

A horizontal-axis wind turbine is the most extensively used method for wind energy extraction. The power rating varies from a few watts to megawatts on large grid-connected wind turbines.

In relation to the position of the rotor regarding the tower, the rotors are classified as leeward (rotor downstream the tower) or windward (rotor upstream the tower), this last configuration being the most widely used.

These turbines consist of a rotor, a gearbox, and a generator. The group is completed with a nacelle that includes the mechanisms, as well as a tower holding the whole system and hydraulic subsystems, electronic control devices, and electric infrastructure as it is shown in Fig. 29.1 [1]. A photograph of a real horizontal-axis wind turbine is shown in Fig. 29.2. We will briefly explain the above-mentioned devices.

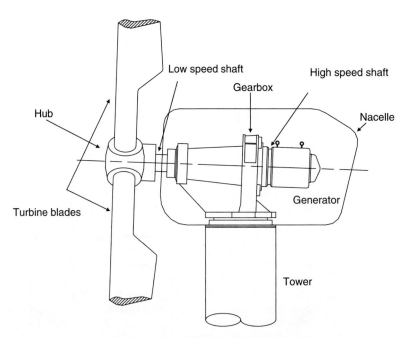

FIGURE 29.1 View of horizontal-axis wind turbine.

FIGURE 29.2 Wind turbine in Monteahumada (Spain). Made S.A. AE-41PV.

29.1.1.1 The Rotor

The rotor is the part of the wind turbine that transforms the energy from the wind into mechanical energy [1]. The area swept by the rotor is the area that captures the energy from the wind. The parameter measuring the influence of the size of the capturing area is the ratio area/rated power. Thus, for the same installed power, more energy will be delivered if this ratio is greater, and so, more equivalent hours (kWh/kW). Values for this ratio close to 2.2 m^2/kW are found today in locations with high average wind speed (>7 m/s), but there is a trend to elevate this ratio above 2.5 m^2/kW for certain locations of medium and low potential. In this case, the technical limits are the high tangential speed at the tip of the blade, that force to lower the speed of the rotors, hence the variable speed and the technology used are most important. Making a bigger rotor for a certain wind turbine involves the possibility of using it for a lower wind speed location by compensating wind loss with a bigger capturing area. The rotor consists of a shaft, blades, and a hub, which holds the fastening system of the blades to the shaft. The rotor and the gearbox form the so-called drive train.

A basic classification of the rotors is between constant pitch and variable-pitch machines, according to whether the type of tie of the blade to the hub is constant or whether it allows rotation to the rotor axis.

The pitch control of a wind turbine makes it possible to regulate energy extraction at high speed wind condition. On the other hand, the use of variable speed makes the systems more expensive to build and maintain.

The use of variable-speed generators (other than 50 Hz of the grid), allows the reduction of sudden load surges.

This condition differentiates between constant speed and variable speed generators. The hub includes the blade pitch controller in case of variable pitch, and the hydraulic brake system in case of constant pitch. The axis to which the hub is tied to the so-called low speed shaft is usually hollow which allows for the hydraulic conduction for regulation of the power by varying the blade pitch or by acting on the aerodynamic brakes in case of constant pitch.

29.1.1.2 The Gearbox

The function of the gearbox, shown in Fig. 29.1, is to adapt a low rotation speed of the rotor axis to a higher one in the electric generator [1, 2]. The gearbox may have parallel or planetary axis. It consists of a system of gears that connect the low speed shaft to the high speed shaft connected to the electric generator by a coupler. In some cases, using multi-pole, the gearbox is not necessary.

29.1.1.3 The Generator

The main objective of the generator is to transform the mechanical energy captured by the rotor of the wind turbine into electrical energy that will be injected into the utility grid.

Asynchronous generators are commonly used in wind turbine applications with fixed speed or variable speed control strategies. Also, in large power wind turbine applications synchronous machines are used. In the asynchronous generator, the electric energy is produced in the stator when the rotating speed of the rotor is higher than the speed of the rotary field of the stator. The asynchronous generator needs to take energy from the grid to create the rotary field of the stator. Because of this, the power factor is decreased and so a capacitor bank is needed. The synchronous generator with an excitation system includes electromagnets in the rotor that generate the rotating field. The rotor electromagnets are fed back with a DC current by rectifying part of the electricity generated. Another kind of generator recently used is the permanent magnet [5]. This type of machine does not need an excitation system, and it is used mainly for low power wind turbine applications. The advantages of using an asynchronous generator are low cost, robustness, simplicity, and easier coupling to the grid, yet its main disadvantage is the necessity of a power factor compensator and a lower efficiency.

29.1.1.3.1 Induction Machine The induction generator, as can be deducted from its torque/speed characteristic, has a nearly constant speed in a wide working torque range, as they are positive (working as a motor) or negative (working as a generator). This characteristic curve is very useful for machines with constant speed, as the machine is auto-regulated to keep the synchronous frequency. But the situation is very different when we proceed to change the speed of the generator. It is then necessary to use power converters in order to adapt the generator frequency to the frequency of the grid [3, 6, 7]. The general principles applicable to change the speed of an induction generator can be deduced from the following equation:

$$N_r = N_1 \cdot (1 - s) = \frac{60 \cdot f_1}{p} \cdot (1 - s) \quad (29.1)$$

where N_r is the generator speed (rpm), N_1 is the generator synchronous speed, s is the induction generator slip, p is the pole pair number, and f_1 is the excitation stator frequency (Hz).

From Eq. (29.1), it is immediately inferred that the speed can be controlled in either ways; one way is changing the synchronous speed and the other is changing the slip. The speed is deduced from the number of pole pairs p and the supplying frequency into the machine f_1. The slip can be easily changed when modifying the torque/slip characteristic curve. This modification can be achieved as follows: first, by changing the input voltage of the generator; second, by changing the resistance of the rotor circuit; and third, by injecting a voltage into the rotor so that it has the same frequency as the electromotive force induced in it and an arbitrary magnitude and phase. The techniques used to vary the supplying frequency permit a wide range of variation of the speed, from 0 to 100% or even greater than the synchronous speed. Another variable-speed technique is achieved by changing the number of poles which permit a regulation of the speed in discrete steps. If we proceed to vary the slip, then the range of variation of the speed is within a narrow margin of regulation.

Among all these techniques, only the variation of the voltage can be actually implemented using a squirrel cage machine with a short-circuited rotor. The rest are implemented by means of a wound-rotor machine.

The stator voltage can be varied by means of a power converter [4, 8, 9]. This converter should be connected in series to the generator and to the grid. Since it is only necessary to vary the voltage of the generator and not its frequency, an AC–AC converter can be used. Furthermore, the power converter bears all the power of the generator so it deals with all the disadvantages of the other wide-range of control techniques.

For the speed to be varied by changing the slip, it is necessary to work with wound rotor induction machines.

29.1.1.3.2 Synchronous Machine with Excitation System As it is well-known, the general principle to change the speed of a synchronous machine is summarized in the following equation [3, 6, 7]:

$$N_r = N_1 = \frac{60 \cdot f_1}{p} \quad (29.2)$$

The only way to control the speed is by changing the number of pole pairs or by supplying frequency into the machine, f_1. Therefore, wide range or discrete steps are permitted.

The synchronous machine will always be controlled in a wide range of the rotor speed, ω_r. In this kind of system, the excitation current permits an easier torque and power control.

29.1.1.3.3 Permanent Magnet Synchronous Machine As with the synchronous generator with excitation system, the permanent magnet synchronous machine can be controlled in a wide range of rotor speeds ω_r. In this case, a magnetic field control has to be made from the power converter. The advantage of this machine is better performance and less complexity [3, 6, 7, 10].

29.1.1.4 Power Electronic Conditioner

The power electronic conditioner is a converter that is mainly used in variable speed applications. This converter is connected between the generator machine and the utility grid by an isolating transformer and permits different frequency and voltage levels in its input and output. The power converter is connected to the stator voltage or to the rotor of a wound rotor machine. This system includes large power switches that can be GTOs, Thyristors, IGCTs, or IGBTs arranged in different topologies.

29.1.2 Simplified Model of a Wind Turbine

The mechanical power P_m in the low speed shaft can be expressed as a function of the available power in the wind P_v by the Eq. (29.3):

$$P_m = C_p(\lambda, \beta) \cdot P_w \tag{29.3}$$

where $C_p(\lambda, \beta)$ is the power coefficient, which is a function of the blade angle β and the dimensionless variable $\lambda = \omega_L R / v_w$ (where ω_L is the angular speed on the low speed shaft, R is

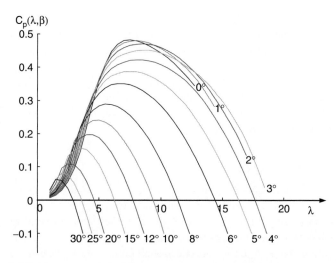

FIGURE 29.3 Analytical approximation of $C_p(\lambda, \beta)$ characteristic (blade pitch angle β as parameter).

the turbine radius, and v_w the wind speed). In Fig. 29.3 an analytical approximation of the power coefficient $C_p(\lambda, \beta)$ is shown.

In Fig. 29.4 the power characteristic of a wind turbine P_m is shown.

The power of the wind can be expressed by the following equations [1, 2]:

$$P_w = \frac{1}{2} \rho \pi R^2 v_w^3 \tag{29.4}$$

where ρ is the air density.

Substitution of Eq. (29.4) in Eq. (29.3) and including λ in such expression, the following can be obtained:

$$Q_L = \frac{C_p}{2\lambda^3} \rho \pi R^5 \omega_L^2 \tag{29.5}$$

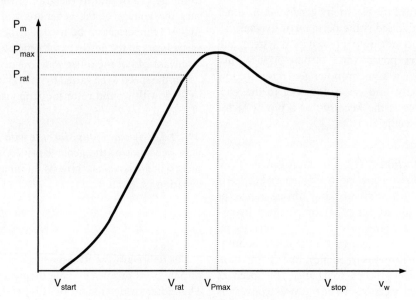

FIGURE 29.4 Power characteristic of the wind turbine.

29 Wind Turbine Applications

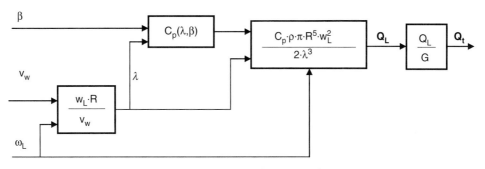

FIGURE 29.5 Torque calculation block diagram.

where Q_L is the torque in the low speed shaft that the wind turbine draws from the wind.

This Eq. (29.5) is represented in Fig. 29.5.

Neglecting mechanical losses, the total torque on the high speed shaft, Q_t is equal to the torque in the low speed shaft, Q_L, divided by the gearbox ratio, G.

$$Q_t = \frac{Q_L}{G} \qquad (29.6)$$

Equation (29.7) shows the differential equation for the dynamics of the rotational speed that depends on the difference of load and generator torque.

$$Q_t - Q_e = J\frac{d\omega_r}{dt} \qquad (29.7)$$

In Eq. (29.7) J is the total inertia of the system referred to the high speed shaft.

Figure 29.6 shows the block diagram of the simplified mechanical model of a wind turbine. Also, it has been represented by the electrical power P_e, obtained by multiplying the electrical torque Q_e by the rotor speed ω_r and the electrical performance η.

29.1.3 Control of Wind Turbines

Many horizontal axes, grid-connected, and medium- to large-scale wind turbines are regulated by pitch control, and most of the wind turbines built so far have practically constant speed, since they use an AC generator, directly connected to the distribution grid, which determines its speed of rotation.

In the last few years, variable speed control has been added to pitch-angle control design in order to improve the performance of the system [11]. Variable speed operation of a wind turbine has important advantages vs the constant speed ones. The main advantages of variable speed wind tunnel are the reduction of electric power fluctuations by changes in kinetic energy of the rotor, the potential reduction of stress loads on the blades and the mechanical transmissions, and the possibility to tune the turbine to local conditions by adjusting the control parameters. On the other hand, variable speed control is normally used with fixed pitch angle and very few applications using both controls have been reported [12, 13].

In short, four different wind turbine types are provided depending on the controller [14]:

- No control. The generator is directly connected to a constant frequency grid, and the aerodynamics of the blade is used to regulate power in high winds.
- Fixed speed pitch regulated. In this case, the generator is also directly connected to a constant frequency grid, and pitch control is used to regulate power in high winds.
- Variable speed stall regulated. A frequency converter decouples the generator from the grid, allowing the rotor

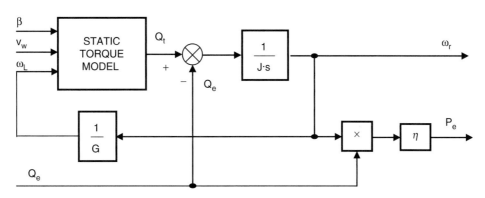

FIGURE 29.6 Mechanical model of a wind turbine.

speed to be varied by controlling the generator reaction torque. In high winds, this speed control capability is used to slow the rotor down until aerodynamic stall limits the power to a desired level [15].
- Variable speed pitch regulated. A frequency converter decouples the generator from the grid, allowing the rotor speed to be varied by controlling the generator reaction torque. In high winds, the torque is held at a rated level, and pitch control is used to regulate the rotor speed, and hence, also the power [13].

A power converter will be mainly used in variable speed applications. In fixed speed control, a power converter could be used for a better system performance, for example, smooth transition during turn on, harmonics, and flicker reduction, etc. Next, the operation of the most general controller, namely, the variable speed pitch regulator controller is explained. Another controller can be obtained from this control scheme, but will not be presented here.

29.1.3.1 Variable Speed Variable Pitch Wind Turbine

Objectives of variable speed control systems are summarized by the following general goals [12, 16, 17]:

- to regulate and smooth the power generated
- to maximize the energy captured
- to alleviate the transient loads throughout the wind turbine
- to achieve unity power factor on the line side with no low frequency harmonics current injection
- to reduce the machine rotor flux at light load reducing core losses

Objectives for the pitch-angle control are similar to the variable speed. If pitch-angle control is used together with variable speed, better performance in the system is obtained. For instance, to permit starting, blade pitch angle differs from the operation pitch angle, allowing an easier starting and optimum running. Moreover, the power and speed can be limited through rotor pitch regulation.

The control diagram is shown schematically in Fig. 29.7. The generator torque Q_e and the pitch angle β control the wind turbine. The control system acquires the actual generated electric power P_e and the generator speed, ω_r, and calculates the reference generator torque Q_e^{ref} and the reference pitch angle β^{ref}, using two control loops [14].

In low winds it is possible to maximize the energy captured by following a constant tip speed ratio λ load line which corresponds to operating at the maximum power coefficient. This load line is a quadratic curve in the torque-speed plane as it is shown in Fig. 29.8. During that time, the pitch angle is adjusted to a constant value, the maximum power pitch angle.

If there is a minimum allowed operating speed, then it is not possible to follow this curve in very low winds, and the turbine is then operated at a constant speed N_{min} shown in Fig. 29.8. On the other hand, in high wind speed, it is necessary to limit the torque Q_{rate} or power P_{rate} of the generator to a constant value.

The control parameters are: the minimum speed ω_r^{min}, the maximum speed in constant tip speed ratio mode ω_r^{max}, the nominal steady-state operating speed ω_r^{rat}, and the parameter K_λ which defines the constant tip speed ratio line $Q_e = K_\lambda \omega_r^2$. K_λ is given by (29.8):

$$K_\lambda = \frac{\pi \rho R^5 C_p(\lambda, \beta)}{2\lambda^3 G^3} \quad (29.8)$$

When the generator torque demand is set to $K_\lambda \omega_r^2$ where ω_r is the measured generator speed, this ensures that in the steady state the turbine will maintain the tip speed ratio λ_{opt} and the corresponding maximum power coefficient $C_p(\lambda, \beta)$.

Figure 29.9 shows the simplified control loops used to generate pitch and torque demands. When operating below rated power, the torque controller is active, and the pitch demand

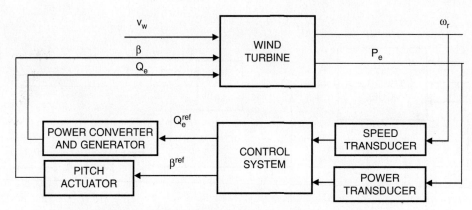

FIGURE 29.7 Block control diagram of the variable speed pitch regulated wind turbine.

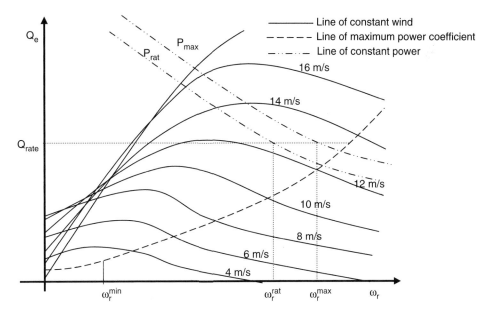

FIGURE 29.8 Variable speed pitch regulated operating curve.

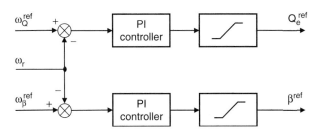

FIGURE 29.9 Pitch regulated variable speed control loops.

loop is active when operating above rated power. Below rated, the speed set-point, ω_Q^{ref}, is the optimum speed given by the optimal tip speed ratio curve and pitch angle is held at zero. Above rated, the reference generator torque is hold rated constant value Q_{rate} and the pitch angle controller is achieving the reference nominal speed ω_β^{ref}. During this control interval, the captured power is constant because the reference torque is maintained at a rated torque of the machine and the rotor speed is controlled to maintain a rated value.

29.2 Power Electronic Converters for Variable Speed Wind Turbines

29.2.1 Introduction

Power electronic converters can operate the stator of synchronous or asynchronous machines. In other applications, the power converter can be connected to the rotor of a wound rotor induction machine. In the first case, the converter handles the overall power of the machine and it operates in a wide speed range. In the wound rotor machine case, the converter handles a fraction of the rated power but it does not allow a very low speed to obtain higher energy from the wind. So, the advantage is that the power converter is smaller and cheaper than the stator converter.

29.2.2 Full Power Conditioner System for Variable Speed Turbines

In this section, the different topologies of power electronic converters that are currently used for wide range speed control of generators will be presented. A variable speed wind turbine control method within a wide range has the following advantage and disadvantage compared to those for narrow-range speed control. The advantage of a variable speed wind turbine control method in a wind range is that it allows for very low speed to obtain higher energy from the wind. On the other hand, the disadvantage is that the power converter must be rated to the 100% of the nominal generation power.

The power conditioners, next to be considered in this section, may be used for synchronous as well as asynchronous generators. For both cases the control block to be employed is defined. The main objective of power converters to be used for wind energy applications is to handle the energy captured from the wind and the injection of this energy into the grid. The characteristics of the generator to be connected to the grid and where to inject the electric energy are decisive when designing the power converter. To attain this design, it is necessary to consider the type of semiconductor to be used, components and subsystems.

By using cycloconverters (AC/AC) or frequency converters based on double frequency conversion, normally AC/DC–DC/AC, and connected by a DC link, a rapid control of the active and reactive power can be accomplished along with a low incidence in the distribution electric grid. The commutation frequency of the power semiconductors is also an important factor for the control of the wind turbine because it allows not only to maximize the energy captured from the wind but also to improve the quality of the energy injected into the electrical grid. Because of this, the semiconductors required are those that have a high power limit and allow a high commutation frequency. The insulated gate bipolar transistor (IGBTs) are commonly used because of their high breakdown voltage and because they can bear commutation frequencies within the range of 3–25 kHz, depending on the power handled by the device. Other semiconductors such as gate turn-off thyristor (GTOs) are used for high power applications allowing lower commutation frequencies, and thus, worsening not only the control of the generator but also the quality of the energy injected into the electric grid. [4, 8, 9, 10, 18].

The different topologies used for a wide-range rotor speed control are described next. Advantages and disadvantages for using these topologies, as power electronics is concerned, are:

Advantages:

- Wide-range speed control
- Simple generator-side converter and control
- Generated power and voltage increased with speed
- VAR-reactive power control possible

Disadvantages:

- One or two full-power converter in series
- Line-side inductance of 10–15% of the generated power
- Power loss up to 2–3% of the generated power
- Large DC link capacitors

29.2.2.1 Double Three Phase Voltage Source Converter Connected by a DC-link

Figure 29.10 shows the scheme of a power condition for a wind turbine. The three phase inverter on the left side of the power converter works as a driver controlling the torque generator by using a vectorial control strategy. The three phase inverter on the right side of the figure permits the injection of the energy extracted from the wind into the grid, allowing a control of the active and reactive power injected into the grid. It also keeps the total harmonic distortion coefficient as low as possible improving the quality of the energy injected into the public grid. The objective of the DC-link is to act as an energy storage, so that the captured energy from the wind is stored as a charge in the capacitors and is instantaneously injected into the grid. The control signal is set to maintain a constant reference to the voltage of the capacitors battery V_{dc}. The control strategy for the connection to the grid will be described in Section 29.2.3.

The power converter shown in Fig. 29.10 can be used for a variable speed control in generators of wind turbines, either for synchronous or asynchronous generators.

29.2.2.1.1 Asynchronous Generator
Next to be considered is the case of an asynchronous generator connected to a wind turbine. The control of a variable speed generator requires a torque control, so that for low speed winds the control is required with optimal tip speed ratio, λ_{opt}, to allow maximum captured wind energy from low speed winds. The generator speed is adjusted to the optimal tip speed ratio λ_{opt} by setting a reference speed. For high speed winds the pitch or stall regulation of the blade limits the maximum power generated by the wind turbine. For low winds it is necessary to develop a control strategy, mentioned in Section 29.2.

The adopted control strategy is an algorithm for indirect vector control of an induction machine [3, 6, 7], which is described next and shown in Fig. 29.11.

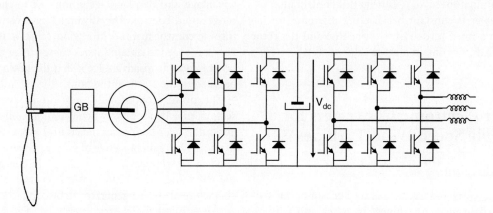

FIGURE 29.10 Double three phase voltage source inverter connected by a DC link used in wind turbine applications.

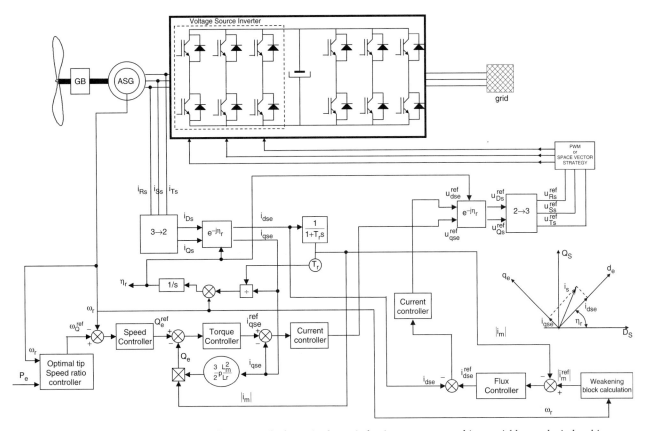

FIGURE 29.11 Schematic of the rotor flux-oriented of a squirrel cage induction generator used in a variable speed wind turbine.

A reference speed, ω_Q^{ref}, has been obtained from the control strategy used in order to achieve optimal speed ratio working conditions of the wind turbine to capture the maximum energy from the wind. In Fig. 29.11, the calculation block to obtain ω_Q^{ref} is shown, that is fed by the actual rotor speed, ω_r, and the electrical power generated, P_e, by the asynchronous generator. Using this ω_Q^{ref} and the actual rotor speed, ω_r, which is measured by the machine, the reference for the electric torque, Q_e^{ref}, is obtained from the speed regulator which is necessary to set the reference torque in the machine shaft in order to achieve the control objectives.

In Fig. 29.11 the induction generator is driven by a voltage-source pulse width moducation (PWM) inverter, which is connected by a second voltage-source PWM inverter to the public grid through a DC link battery capacitors. The output voltage of the inverter is controlled by a PWM technique in order to follow the voltage references, u_{Rs}^{ref}, u_{Ss}^{ref}, u_{Ts}^{ref}, provided by the control algorithm in each phase. There are many types of modulation techniques which are not discussed in detail here.

A flux model has been used to obtain the angular speed of the rotor flux and the modulus of magnetizing current, $|i_m|$, that has also been used to calculate the electromagnetic torque, Q_e, as shown in Fig. 29.11, using the Eq. (29.9).

$$Q_e = \frac{3}{2} \cdot p \cdot \frac{L_m^2}{L_r} \cdot |i_m| \cdot i_{qse} \qquad (29.9)$$

The speed controller provides the reference of the torque, Q_e^{ref}, and the torque controller gives the reference value of the quadrature-axis stator current in the rotor flux-oriented reference frame i_{qse}^{ref}.

In Fig. 29.11 the field weakening block, used to obtain the reference value of the modulus of the rotor magnetizing current space phasor $\left|\vec{i}_m^{ref}\right|$, is rotor speed-dependent. This reference signal is then compared with the actual value of the rotor magnetizing current, $|i_m|$, and the error generated is used as input to the flux controller. The output of this controller is the direct-axis stator current reference expressed in the rotor-flux-oriented reference frame i_{dse}^{ref}.

The difference in values of the direct and quadrature-axis stator current references (i_{dse}^{ref}, i_{qse}^{ref}) and their actual values (i_{dse}, i_{qse}) are given as inputs to the respective PI current controllers. The outputs of these PI controllers are values of the direct and quadrature-axis stator voltage reference expressed in the rotor-flux-oriented reference frame (u_{dse}^{ref}, u_{qse}^{ref}). After this they are transformed into the steady reference frame (u_{Ds}^{ref}, u_{Qs}^{ref}), using the $e^{-j\eta r}$ transformation. This is followed by the $2 \to 3$ block and finally, the reference values of the three phase stator voltage (u_{Rs}^{ref}, u_{Ss}^{ref}, u_{Ts}^{ref}) are obtained. These signals are used to control the pulse-width modulator, which transforms these reference signals into appropriate on–off switching signals to command the inverter phase.

29.2.2.1.2 Synchronous Generator

When the generator, used to transform the mechanical energy into electrical energy in the wind turbine, is a salient-pole synchronous machine with an electrically excited rotor, the control criteria are the same as the one applied in the induction generator case, so as to minimize the angular speed error in order to obtain an optimum tip speed ratio performance of the wind turbine. In this section, a drive control based on magnetizing field-oriented control is described and applied to a wind turbine with a salient-pole synchronous machine. This control method can be applied to the synchronous generator using a voltage-source inverter or a cycloconverter as it is shown in Figs. 29.10 and 29.13 respectively, using the same block control diagram.

In both cases, a controllable three-phase rectifier supplies the excitation winding on the rotor of the synchronous machine. As it is well-known, the cycloconverter is a frequency converter which converts power directly from a fixed frequency to a lower frequency. Each of the motor phases is supplied through a three-phase transformer, an antiparallel thyristor bridge, and the field winding is supplied by another three-phase transformer and a three-phase-rectifier using the bridge connection.

In the control block described in Fig. 29.12, the rotor speed and monitored current have been used in order to control the relationship between the magnetizing flux and currents of the machines by modifying the voltages of the power converter and the excitation current in the field winding.

The reference rotor speed, ω_Q^{ref}, and the measured rotor speed, ω_r, are compared and the error is introduced into the speed controller. The output voltage is proportional to the electromagnetic torque PI, of the synchronous machine and the reference torque, Q_e^{ref}, is obtained. Dividing the reference torque by the modulus of the magnetizing flux-linkage space

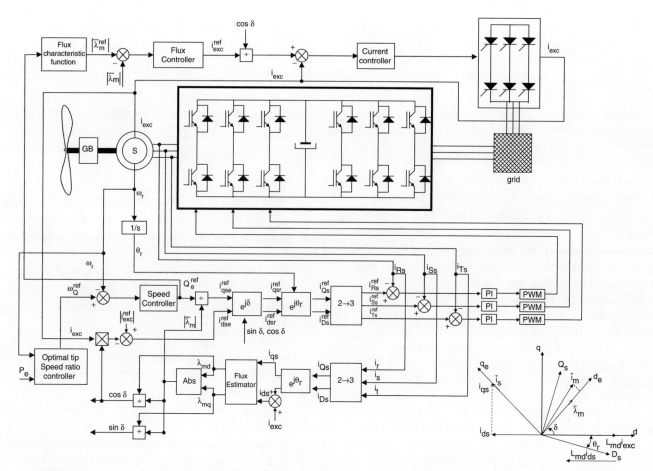

FIGURE 29.12 Block diagram of a field-oriented control of a salient-pole synchronous machine used in wind turbine applications.

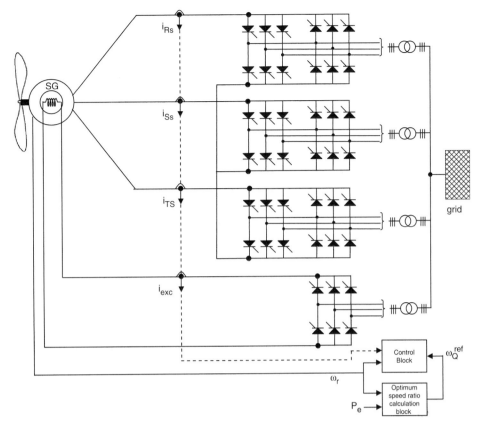

FIGURE 29.13 Schematic of the cycloconverter synchronous generator used in variable speed wind turbines.

phasor $|\lambda_m|$, the reference value of the torque-producing stator current, $|\lambda_m|$, component is obtained.

Using a characteristic function of the magnetizing flux reference and the actual rotor speed, ω_r, the magnetizing flux reference is obtained, $\left|\lambda_m^{ref}\right|$. Below base speed, this function yields a constant value of the magnetizing flux reference, λ_m^{ref}; above base speed, this flux is reduced. The magnetizing flux controller, $|\bar{\lambda}_m|$, is introduced and compared with the estimated magnetizing flux of the synchronous machine, $\left|\lambda_m^{ref}\right|$, obtaining the error, which is fed into the flux controller as shown in Fig. 29.12. The flux controller maintains the magnetizing linkage flux to a pre-set value independent of the load. As an output of this controller the reference excitation current, i_{exc}^{ref} is obtained.

The value of the reference magnetizing stator current, i_{exc}^{ref}, is obtained using a steady state in which there is no reactive current drawn from the stator [3]. In this case the power factor is unity and the stator current value is the optimal. The zero reactive power condition can be fulfilled by controlling the magnetizing current stator component that is shown in Fig. 29.12.

The stator current components (i_{dsc}^{ref}, i_{qse}^{ref}) are first transformed into the stator current components established in the rotor reference frame (i_{dsr}^{ref}, i_{qsr}^{ref}). After these components are transformed into the stationary-axes current components by a similar transformation, but taking into account that the phase displacement between the stator direct axis of the rotor is θ_r. The obtained two-axis stator current references (i_{Qs}^{ref}, i_{Ds}^{ref}), are transformed into the three phase stator current references (i_{Rs}^{ref}, i_{Ss}^{ref}, i_{Ts}^{ref}) by the application of the three phase to two phase transformation. The reference stator currents are compared with their respective measured currents, and their errors are fed into the respective stator current controllers.

29.2.2.2 Step-up Converter and Full Power Converter

An alternative for the power conditioning system of a wind turbine is to choose a synchronous generator and a three phase diode rectifier, as shown in Fig. 29.14. Such a choice is based on low cost compared with an induction generator connected to a voltage source inverter used as a rectifier. When the speed of the synchronous generator alters, the voltage value on the DC-side of the diode rectifier will change. A step-up chopper is used to adapt the rectifier voltage to the DC-link voltage of the inverter, see Fig. 29.14. When the inverter system is analyzed, the generator/rectifier system can be modeled as an ideal current source. The step-up chopper used as a rectifier utilizes

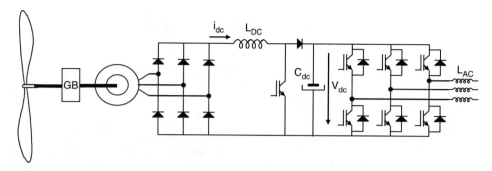

FIGURE 29.14 Step-up converter in the rectifier circuit and full power inverter topology used in wind turbine applications.

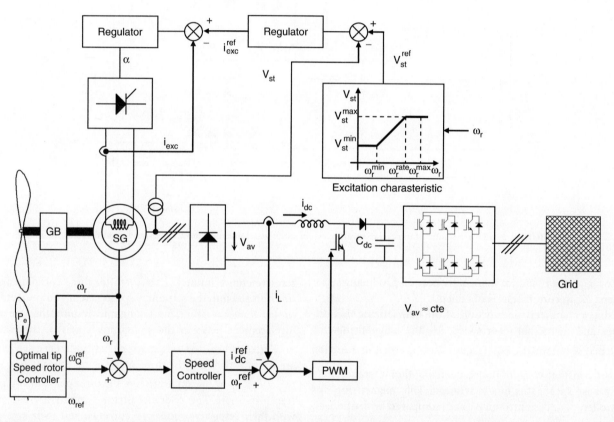

FIGURE 29.15 Control block diagram of a step-up converter in the rectifier circuit and full power inverter used in wind turbine applications.

a high switching frequency so the bandwidth of these components is much higher than the bandwidth of the generator. The generator/rectifier current is denoted as i_{dc} and is independent of the value of the DC-link voltage. The inductor of the step-up converter is denoted as LDC. The capacitor of the DC-link has the value C_{dc}. The inductors on the AC-side of the inverter have the inductance L_{AC}. The three-phase voltage system of the grid has phase voltages e_R, e_S, and e_T and the phase currents are denoted as i_{Rg}, i_{Sg}, and i_{Tg}. The DC-link voltage value is denoted as V_{dc}.

The control system, shown in Fig. 29.15, is based on the measurement of the rotor speed ω_r of the synchronous generator by means of a speed transducer. This value is compared with the reference rotor speed obtained by the control algorithm of the variable speed wind turbine used in the application in order to achieve optimal speed ratio, and therefore, to capture the maximum energy from the wind.

The objective of the synchronous machine excitation system is to keep the stator voltage V_{st} following the excitation characteristic of the generator, shown in Fig. 29.15. This excitation characteristic is linear in the range of the minimum of the rotor speed ω_r^{min} and the rated value of the rotor speed ω_Q^{rate}. Outside this range, the stator voltage is saturated to V_{st}^{min} or V_{st}^{max}. For rotor speeds ω_r in the range between ω_r^{min} and ω_r^{rate}, the control is carried out using an inductor current i_{dc} proportional to the shaft torque of the generator. Above the rated rotor speed, this current is proportional to the power because the stator voltage V_{st} is constant.

29.2.2.3 Grid Connection Conditioning System

Injection into the public grid is accomplished by means of PWM voltage source interver. This requires the control of the current of each phase of the inverter, as shown in Fig. 29.16. There are several methods of generating the reference current to be injected into any of the phases of the inverter. A very useful method to calculate this current was proposed by Professor Akagi [19], that is applied to the active power filters referred to as "Instantaneous Reactive Power Theory." The control block shown in Fig. 29.16 is based on the comparison of the actual capacitor array voltage V_{dc} with a reference voltage V_{dc}^{ref}. Subtracting the actual capacitor array voltage V_{dc}^{ref} from the reference voltage V_{dc}^{ref}, the error signal voltage is obtained. This error signal is fed into a compensator, usually a PI compensator, that transforms the output to an active instant power signal that, after being injected into the electric grid, is responsible for the error of the regulator of the voltage in the capacitor array to be zero.

Transformation of the phase voltage e_{rg}, e_{sg}, and e_{tg} into the α–β orthogonal coordinates is given by the following expression:

$$\begin{bmatrix} e_\alpha \\ e_\beta \end{bmatrix} = \sqrt{\frac{2}{3}} \cdot \begin{bmatrix} 1 & -1/2 & -1/2 \\ 0 & \sqrt{3}/2 & -\sqrt{3}/2 \end{bmatrix} \cdot \begin{bmatrix} e_{rg} \\ e_{sg} \\ e_{tg} \end{bmatrix} \quad (29.10)$$

Using the inverse transformation equations from the control algorithm [19] and the reference of real power of the capacitor array, it is possible to derive the phase sinusoidal reference current to be injected by the converter into the grid.

$$\begin{bmatrix} i_{rg}^{ref} \\ i_{sg}^{ref} \\ i_{tg}^{ref} \end{bmatrix} = \sqrt{\frac{2}{3}} \cdot \begin{bmatrix} 1 & 0 \\ -1/2 & \sqrt{3}/2 \\ -1/2 & -\sqrt{3}/2 \end{bmatrix} \cdot \begin{bmatrix} e_\alpha & e_\beta \\ -e_\beta & e_\alpha \end{bmatrix}^{-1} \cdot \begin{bmatrix} p^{ref} \\ q^{ref} \end{bmatrix} \quad (29.11)$$

As can be seen in the above equation, the reference of the reactive power appears which is normally set at zero so that the current is being injected into the public grid with unity power factor.

There is another type of wind energy extraction system where the objective is to compensate the reactive power

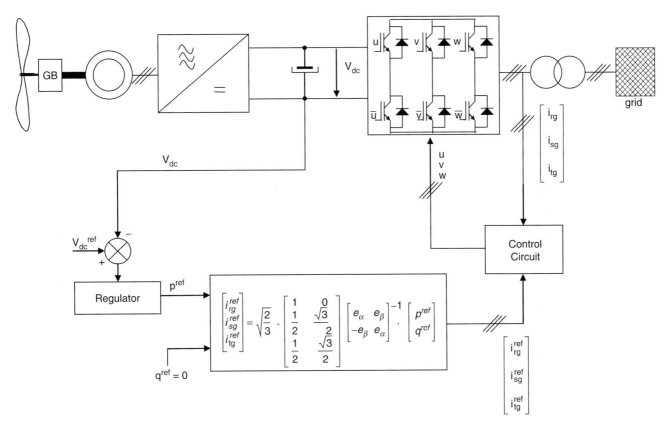

FIGURE 29.16 Control block diagram of the grid connection conditioning system.

generated by non-linear loads of the grid. In this case, the reactive reference power is set to the corresponding value, for either reactive or capacitive power factor "leading or lagging."

29.2.3 Rotor Connected Power Conditioner for Variable Speed Wind Turbines

As it was introduced before, variable speed can also be obtained by slip change of induction generator using a wound rotor machine.

Since it is necessary to have an electric connection to the rotor winding, rotation is achieved by using a slip ring. The power delivered by the rotor through the slip rings is equal to the product of the slip by the electrical power that flows into the stator P_S, Eq. (29.12). This is the so-called "slip power" P_{slip}.

$$P_{slip} = s \cdot P_S \qquad (29.12)$$

The slip power can be handled as follows:

- It can be dissipated in a resistor (Fig. 29.17).
- Using a single doubly fed scheme, the slip power is returned to the electrical grid or to the machine stator (Fig. 29.18).
- Using a cascaded scheme. This is accomplished by connecting a second machine. Part of the power is transferred as mechanical power through the shaft, and the other part is transferred to the grid by a power converter (Fig. 29.19).
- A single frame cascaded or brushless doubly fed induction machine [20] can be used in the same way as before, but using only one machine instead of two (Fig. 29.20).

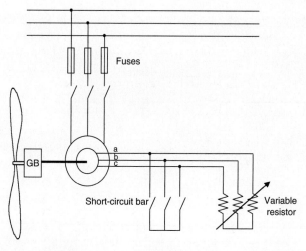

FIGURE 29.17 Slip power dissipation in a resistor.

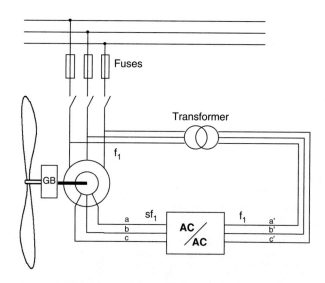

FIGURE 29.18 Single doubly fed induction machine.

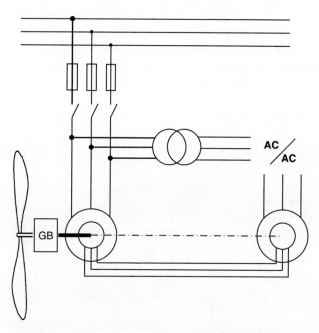

FIGURE 29.19 Wound rotor cascaded induction machines.

29.2.3.1 Slip Power Dissipation

Figure 29.17 shows a system in which the power delivered by the rotor is dissipated in a resistor.

The variable resistor can be substituted by the power converter in Fig. 29.21. The power converter controls the power delivered to the resistor using an uncontrolled rectifier and a parallel DC–DC chopper. This design has the disadvantage of the current having a higher harmonic content. This is caused by the rectangular rotor current waveform in the case of a three phase uncontrolled rectifier. This disadvantage can be avoided using a six IGBT's controlled rectifier, however this topology increases the cost significantly.

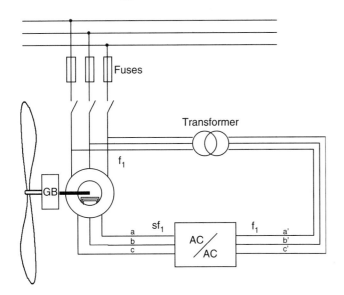

FIGURE 29.20 Brushless doubly fed induction machine.

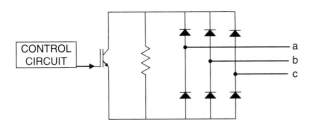

FIGURE 29.21 Variable resistor using a power converter.

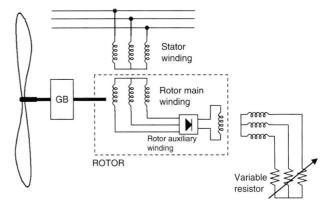

FIGURE 29.22 Slip power dissipation in an external resistor using brushless machine.

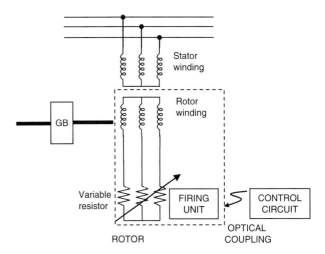

FIGURE 29.23 Slip power dissipation in an internal resistor using brushless machine.

The variation of the rotor resistance is not a recommended technique due to the high copper losses in the regulation resistance and so, the generator system efficiency is lower. It only can be efficient within a very narrow range of the rotor speed. Another disadvantage is that this technique is applicable only to wound-rotor machines and so, slip rings and brushes are needed. In order to solve this problem some brushless schemes are proposed. A solution is to use a rotor auxiliary winding which couples the power to an external variable resistor. The scheme can be observed in Fig. 29.22.

Another solution is to dissipate the energy within a resistor placed in the rotor as it is shown in Fig. 29.23. This method is currently used in generators for wind conversion systems, but as the efficiency of the system decreases with increasing the slip, the speed control is limited to a narrow margin. This scheme includes the power converter and the resistors in the rotor. Trigger signals to the power switches are accomplished by optical coupling.

29.2.3.2 Single Doubly Fed Induction Machine

In Fig. 29.18 the connection scheme shows that slip power is injected into the public grid by a power converter and a transformer. The power converter changes the frequency and controls the slip power. In some cases a transformer is used due to public grid voltages which can be higher than rotor voltages.

Disregarding losses, the simplified scheme of Fig. 29.24 shows real power flux in all different connection points of the diagram. In this figure, the electrical power in the stator machine P_S, the mechanical power $(1-s) \cdot P_S$, and the slip power and power converter $s \cdot P_S$ are represented.

In generation mode, the power is positive when the arrow direction shown in Fig. 29.24 is considered. The power handled on the power converter depends on the sign of the machine slip. When this slip is positive, i.e. subsynchronous mode of operation, the slip power goes through the converter from the grid to the rotor of the machine. On the other hand, when the slip is negative, i.e. hypersynchronous mode of operation, the slip power comes out of the rotor to the power converter. Since the slip power is the real power through the converter, this power is determined directly by the maximum slip or by the speed range of the machine. For instance, if the speed range

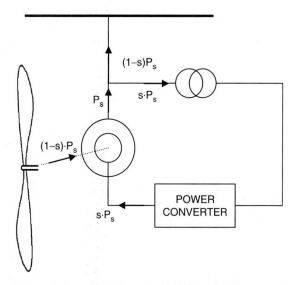

FIGURE 29.24 Simplified scheme of a single doubly fed induction machine.

used is 20% of the synchronous speed, the power rating of the converter is 20% of the main power.

29.2.3.3 Power Converter in Wound-rotor Machines

The power converter used in wound-rotor machines can be a force-commutated converter connected to a line-commutated converter by an inductor as shown in Fig. 29.25. In this case, the power can only flow from the rotor to the grid, and the induction generator works above synchronous speed. Main disadvantages of this scheme are a low power factor and a high content of low frequency harmonics whose frequency depends on the speed.

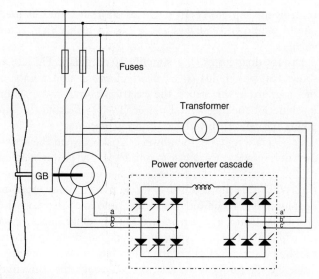

FIGURE 29.25 Single doubly fed induction machine with a force-commutated converter connected to a line-commutated converter.

Since the feeding frequency of the rotor is much lower than the grid's, a cycloconverter, as shown in Fig. 29.26, can be used. In this case the controllability of the system is greatly improved. The cycloconverter is an AC–AC converter based on the use of two three-phase thyristor bridges connected in parallel, one for each phase. This scheme allows working with speed above and below the synchronous speed.

The two schemes mentioned before (Figs. 29.25 and Fig. 29.26) are based on line switched converters. A disadvantage of this scheme is that voltages in the rotor decrease when the machine is working in frequencies close to the synchronous frequency. This fact makes the line-commutated converter not to commute satisfactorily, and we need to use a forced-commutated converter. Also, when a forced-switched converter is used, quality of the voltage and current injected into the public grid is improved. The forced-switched power converter scheme is shown in Fig. 29.27. The converter includes two three-phase AC–DC converters linked by a DC capacitor battery. This scheme allows, on one hand, a vector control of the active and reactive power of the machine, and on the other hand, a decrease by a high percentage of the harmonic content injected into the grid by the power converter.

29.2.3.4 Control of Wound-rotor Machines

The vector control of the rotor flux can be accomplished very easily in a wound-rotor machine. Power converters that can be used for vector control applications are either a controlled rectifier in series with an inverter or a cycloconverter.

In this system, the slip power can flow in both directions, from the rotor to the grid (subsynchronous) or from the grid to the rotor (hyper-synchronous). In both working modes, the machine must be working as a generator. When the speed is hyper-synchronous, the converter connected to the rotor will work as a rectifier and the converter connected to the grid as an inverter, as was deduced before from Fig. 29.24.

The wound-rotor induction machine can be modeled as follows (see nomenclature page):

$$u_S = R_S \cdot i_S + \frac{d\lambda_S}{dt} \qquad (29.13)$$

$$u'_r = n^2 \cdot R_r \cdot i'_r + \frac{d\lambda'_r}{dt} - j \cdot \omega_r \cdot \lambda'_r \qquad (29.14)$$

$$\lambda_S = L_S \cdot i_S + n \cdot L_m \cdot i'_r \qquad (29.15)$$

$$\lambda'_r = n^2 \cdot L_S \cdot i'_r + n \cdot L_m \cdot i_S \qquad (29.16)$$

$$Q_e = \frac{3}{2} \cdot \frac{p}{2} \cdot I_m \{\lambda_r \cdot i^*_r\} \qquad (29.17)$$

$$P_S = \frac{3}{2} \cdot R_e \{u_S \cdot i^*_s\} \qquad (29.18)$$

29 Wind Turbine Applications

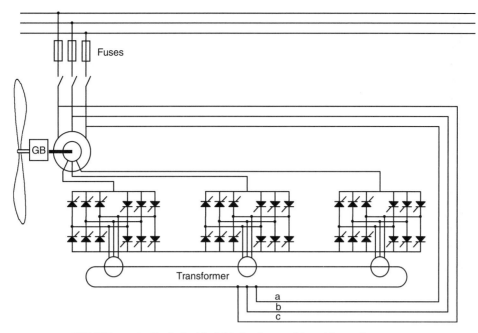

FIGURE 29.26 Single doubly fed induction machine with a cycloconverter.

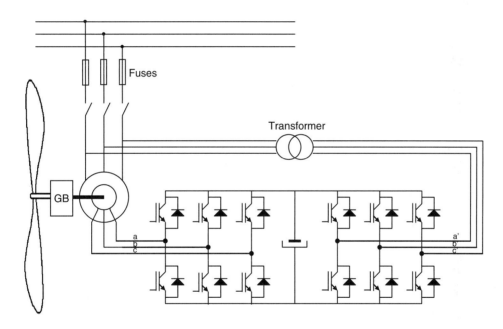

FIGURE 29.27 Single doubly fed induction machine with two fully controlled AC–DC power converters.

$$Q_S = \frac{3}{2} \cdot I_m \left\{ u_S \cdot i_S^* \right\} \qquad (29.19)$$

In the induction machine model, rotor magnitudes, the rotor voltage u_r, the rotor flux λ_r, and the rotor current i_r are referred to the stator, and so, the rotor voltage u'_r, the rotor flux λ'_r, and the rotor current i'_r are defined as:

$$u'_r = n \cdot u_r \cdot e^{j \cdot \theta_r} \qquad (29.20)$$

$$\lambda'_r = n \cdot \lambda_r \cdot e^{j \cdot \theta_r} \qquad (29.21)$$

$$i'_r = \frac{i_r \cdot e^{j \cdot \theta_r}}{n} \qquad (29.22)$$

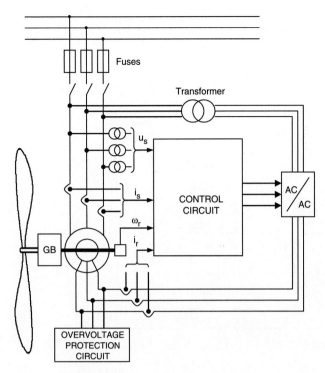

FIGURE 29.28 Wind turbine control block.

The stator magnetizing current i_m is defined as:

$$i_m = \frac{\lambda_S}{n \cdot L_m} = \frac{L_S}{n \cdot L_m} \cdot i_S + i'_r \qquad (29.23)$$

Figure 29.28 shows the control block of a wound-rotor induction machine. The wound-rotor induction machine is controlled by the rotor using a power converter that controls the rotor current i'_r by changing the rotor voltage u'_r. The control of the stator current via the rotor current makes sense only if the converter power is kept lower than the rated power of the machine. The AC stator voltage generates a rotating magnetic field with angular frequency ω_e. Relative to the rotor, this magnetic field rotates only with the angular slip frequency. The frequency of voltages induced in the rotor is low, so voltages of the power converter are low too. Active and reactive power of the induction generator, or a certain percentage of them, can be controlled by the rotor current.

The machine model can be referred to the reference axes that move with the magnetizing current. This system of coordinates rotates with an angle θ_e relative to the stator. In these axes $i_{qm} = 0$ as shown in Fig. 29.29.

Equations of the rotor and stator voltage become:

$$u_S = R_S \cdot i_S + \frac{d\lambda_S}{dt} + j \cdot \omega_e \cdot \lambda_S \qquad (29.24)$$

$$u'_r = n^2 \cdot R_r \cdot i'_r + \frac{d\lambda'_r}{dt} + j \cdot (\omega_e - \omega_r) \cdot \lambda'_r \qquad (29.25)$$

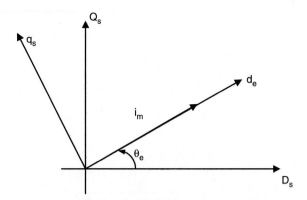

FIGURE 29.29 Stator and rotor reference frames.

$$\omega_e = \frac{d\theta_e}{dt} \qquad (29.26)$$

Supposing steady-state conditions and disregarding the resistors in the stator and the rotor, because the voltage drop is very low in comparison to the stator voltage, the stator and rotor voltages can be determined as:

$$u_S = j\omega_e \lambda_S \qquad (29.27)$$

$$u'_r = j\omega_{s1} \lambda'_r \qquad (29.28)$$

The flux is determined from the stator voltage u_S and the angular frequency ω_e of the AC system. Since both are constant, the flux linkage and magnetizing current are constant too.

$$i_m = \frac{\lambda_S}{nL_m} = \frac{u_{ds} + ju_{qs}}{j\omega_e nL_m} \qquad (29.29)$$

As i_m is constant, the stator current can be controlled at any time, by means of controlling the rotor current that can be deduced from Eq. (29.23). From Eq. (29.29) we can also deduce that the direct component of the stator voltage u_{ds} is zero due to the quadrature component of the stator magnetizing current i_{qm} is zero and so, the active and reactive power can be obtained by the following equations:

$$P_S = \frac{3}{2} u_{ds} i_{dsm} \qquad (29.30)$$

$$Q_S = \frac{3}{2} u_{qs} i_{qsm} \qquad (29.31)$$

Figure 29.30 shows a block diagram of a vector control of active and reactive power for a wound-rotor machine. Vector control of cascaded doubly fed machine is presented in [21].

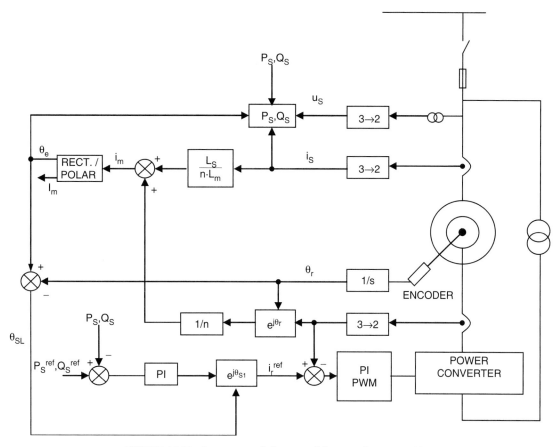

FIGURE 29.30 Power control diagram of the wound-rotor machine.

29.2.4 Grid Connection Standards for Wind Farms

29.2.4.1 Voltage Dip Ride-through Capability of Wind Turbines

As wind capacity increases, network operators have to ensure that consumer power quality is not compromised. To enable large-scale application of wind energy without compromising power system stability, the turbines should stay connected and contribute to the grid in case of a disturbance such as a voltage dip. Wind farms should generate similar to conventional power plants supplying active and reactive power for frequency and voltage recovery, immediately after the fault has been produced.

Thus, several utilities have introduced special grid connection codes for wind farm developers, covering reactive power control, frequency response, and fault ride-through, especially in places where wind turbines provide for a significant part of the total power. Examples are Spain [22], Denmark [23], and part of Northern Germany [24].

The correct interpretation of these codes is crucial for wind farm developers, manufacturers, and network operators. They define the operational boundary of a wind turbine connected to the network in terms of frequency range, voltage tolerance, power factor, and fault ride-through. Among all these requirements, fault ride-through is regarded as the main challenge to the wind turbine manufacturers. Though the definition of fault ride-through varies, the E.ON (German Transmission and Distribution Utility) regulation is likely to set the standard [24]. This stipulates that a wind turbine should remain stable and connected during the fault, while voltage at the point of connection drops to 15% of nominal (i.e. a drop of 85%) for a period of 150 ms: see Fig. 29.31.

Only when the grid voltage drops below the curve, the turbine is allowed to disconnect from the grid. When the voltage is in the shaded area, the turbine should also supply reactive power to the grid in order to support grid voltage restoration.

A major drawback of variable-speed wind turbines, especially for turbines with doubly fed induction generators (DFIGs), is their operation during grid faults [25, 26]. Faults in the power system, even far away from the location of the turbine, can cause a voltage dip at the connection point of the wind turbine. The dip in the grid voltage will result in an increase of the current in the stator windings of the DFIG. Because of the magnetic coupling between the stator

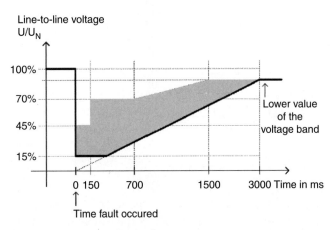

FIGURE 29.31 E. On Netz requirements for wind farm behavior during faults.

and rotor, this current will also flow in the rotor circuit and the power electronic converter. This can lead to the permanent damage of the converter. It is possible to try to limit the current by current-control on the rotor side of the converter; however, this will lead to high voltages at the converter terminals, which might also lead to the destruction of the converter. A possible solution that is sometimes used is to short-circuit the rotor windings of the generator with the so-called crowbars.

The key of the protection technique is to limit the high currents and to provide a bypass for it in the rotor circuit via a set of resistors that are connected to the rotor windings (Fig. 29.32). This should be done without disconnecting the converter from the rotor or from the grid. Thyristors can be used to connect the resistors to the rotor circuit. Because the generator and converter stay connected, the synchronism of operation remains established during and after the fault.

The impedance of the bypass resistors is of importance but not critical. They should be sufficiently low to avoid excess voltage on the converter terminals. On the other hand, they should be high enough to limit the current. A range of values can be found that satisfies both conditions. When the fault in the grid is cleared, the wind turbine is still connected to the grid. The resistors can be disconnected by inhibiting the gating signals and the generator resumes normal operation.

29.2.4.2 Power Quality Requirements for Grid-connected Wind Turbines

The grid interaction and grid impact of wind turbines has been focused in the past few years. The reason behind this interest is that wind turbines are among utilities considered to be potential sources of bad power quality. Measurements show that the power quality impact of wind turbines has been improved in recent years. Especially variable-speed wind turbines have some advantages concerning flicker. But a new problem is faced with variable-speed wind turbines. Modern forced-commutated inverters used in variable-speed wind turbines produce not only harmonics but also inter-harmonics.

The IEC initiated the standardization on power quality for wind turbines in 1995 as a part of the wind turbine standardization in TC88. In 1998, the IEC issued a draft IEC-61400-21 standard for "Power Quality Requirements for Grid Connected Wind Turbines" [27]. The methodology of that IEC standard consists on three analyses. The first one is the flicker analysis. IEC-61400-21 specifies a method that uses current and voltage time series measured at the wind turbine terminals to simulate the voltage fluctuations on a fictitious grid with no source of voltage fluctuations other than the wind turbine switching operation. The second one is switching operations.

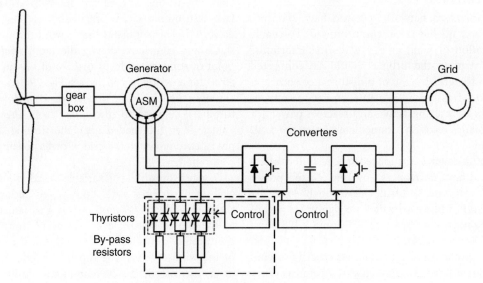

FIGURE 29.32 DFIG bypass resistors in the rotor circuit.

Voltage and current transients are measured during the switching operations of the wind turbine (start-up at cut wind speed and start-up at rated wind speed). The last one is the harmonic analysis which is carried out by the fast fourier transform (FFT) algorithm. Rectangular windows of eight cycles of fundamental frequency width, with no gap and no overlapping between successive windows are applied. Furthermore, the current total harmonic distortion (THD) is calculated up to 50th harmonic order [28, 29].

Recently, high frequency harmonics and inter-harmonics are treated in the IEC 61000-4-7 and IEC 61000-3-6 [30, 31]. The methods for summing harmonics and inter-harmonics in the IECE61000-3-6 are applicable to wind turbines. In order to obtain a correct magnitude of the frequency components, the use of a well-defined window width, according to the IEC 61000-4-7, Amendment 1 is of a great importance, as it has been reported in ref. [32]

Wind turbines not only produce harmonics, they also produce inter-harmonics, i.e. harmonics which are not a multiple of 50 Hz. Since the switching frequency of the inverter is not constant but varies, the harmonics will also vary. Consequently, since the switching frequency is arbitrary, the harmonics are also arbitrary. Sometimes they are a multiple of 50 Hz and sometimes they are not. Figure 29.33 shows the total harmonics spectrum from a variable-speed wind turbine. As can be seen in the figure, at lower frequencies there are only pure harmonics but at higher frequencies there are a whole range of harmonics and inter-harmonics. This whole range of harmonics and inter-harmonics represents variations in the switching frequency of that wind turbine.

29.3 Multilevel Converter for Very High Power Wind Turbines

29.3.1 Multilevel Topologies

In 1980s, power electronics concerns were focused on the increase of the power converters by increasing the voltage or current to fulfill the requirements of the emerging applications. There were technological drawbacks, that endure nowadays, which make impossible to increase the voltage or current in the individual power devices, so researchers were developing new topologies based on series and parallel association of individual power devices in order to manage higher levels of current and voltage, respectively. Due to the higher number of individual power devices on such topologies it is possible to obtain more than the classical two levels of voltage at the output of the converter hence the multilevel denomination for this converter.

29.3.2 Diode Clamp Converter (DCC)

In 1981, Nabae *et al.* presented a new neutral-point-clamped PWM inverter (NPC-PWM) [33]. This converter was based on a modification of the two-level converter topology. In the two-level case, each power switch must support at the most a voltage equal to DC-link total voltage so the switches should be dimensioned to support such voltage.

The proposed modification adds two new switches and two clamp diodes in each phase. In this converter each transistor support at the most a half of the total DC-link voltage; hence,

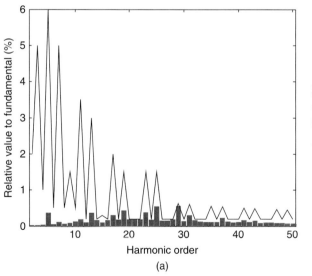
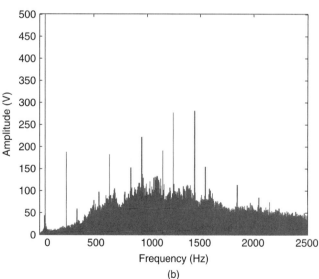

FIGURE 29.33 Typical results of a variable-speed wind turbine with a synchronous generator and full converter. (a) Harmonic content and the comparison with the maximum level of IEC 1000-3-6 standard and (b) harmonic and inter-harmonic content in voltage.

if the used power devices have the same characteristics of those used in the two-level case, the DC-link can be doubled and hence, the power which the converter can manage. Figure 29.34 shows one phase of a three-level DCC with the capacitors voltage divider and the additional switches and diodes.

The analysis of the DCC converter states shows that there are three different switching configurations. These possible configurations are shown in Table 29.1.

When transistors S_3 and S_4 are switched on, the phase is connected to the lowest voltage in the DC-link. In the same manner, when the transistors S_1 and S_2 are switched on, the phase is connected to the highest voltage in the DC-link, and when the transistors S_2 and S_3 are switched on, the phase is connected to the mid DC-link voltage through one of the transistors and clamping diodes.

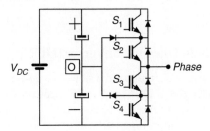

FIGURE 29.34 Three-level DCC.

TABLE 29.1 Switching configurations for the three-level DCC

State	S_1	S_2	S_3	S_4	Phase-O voltage
0	Off	Off	On	On	$-V_{DC}/2$
1	Off	On	On	Off	0
2	On	On	Off	Off	$V_{DC}/2$

29.3.3 Full Converter for Wind Turbine Based on Multilevel Topology

In order to decrease the cost per megawatt and to increase the efficiency of the wind energy conversion, the nominal power of wind turbines has been continuously growing in the last years. The limitations of the two-level converters power ratings versus three-level ones and the capacity of this to reduce the harmonic distortion and electromagnetic interferences (EMI) make the multilevel converters suitable for modern high power wind turbine applications.

Figure 29.35 shows the diagram of high power wind turbine directly connected to the utility grid, with a full converter based on two coupled three-level DCC. The converter connected to the generator acts like an AC–DC converter and its main function is to extract the energy from the generator and to deliver it to the DC-link. The converter connected to the grid acts like a DC–AC converter and its main function is to collect the energy at the DC-link and to deliver it to the utility grid.

29.3.4 Modeling

The use of multilevel converters is limited by the following drawbacks: typically very complex, control and voltage imbalance problems at the DC-link capacitors. An analytical model of the whole system is necessary to study this dynamic and to develop control algorithms that meet with the design specifications.

In [34], a general modeling strategy is proposed to obtain the equations that describe the dynamics of the currents and the capacitors voltages as functions of the control signals that represent the voltage in each phase. Based on the nomenclature that can be seen in Fig. 29.36, this modeling strategy yields in the next mathematical model for the currents dynamics Eqs. (29.32), (29.33):

$$\begin{bmatrix} v_{sr1} \\ v_{sr2} \\ v_{sr3} \end{bmatrix} = L_r \begin{bmatrix} di_{r1}/dt \\ di_{r2}/dt \\ di_{r3}/dt \end{bmatrix} + \frac{1}{3} \begin{bmatrix} 2 & -1 & -1 \\ -1 & 2 & -1 \\ -1 & -1 & 2 \end{bmatrix} \begin{bmatrix} v_{r1} \\ v_{r2} \\ v_{r3} \end{bmatrix} \quad (29.32)$$

$$\begin{bmatrix} v_{si1} \\ v_{si2} \\ v_{si3} \end{bmatrix} = -L_i \begin{bmatrix} di_{i1}/dt \\ di_{i2}/dt \\ di_{i3}/dt \end{bmatrix} + \frac{1}{3} \begin{bmatrix} 2 & -1 & -1 \\ -1 & 2 & -1 \\ -1 & -1 & 2 \end{bmatrix} \begin{bmatrix} v_{i1} \\ v_{i2} \\ v_{i3} \end{bmatrix} \quad (29.33)$$

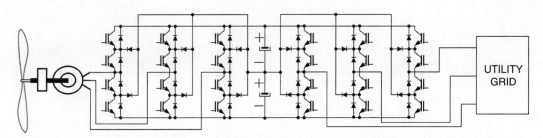

FIGURE 29.35 Diagram of a high power wind turbine with a full converter directly connected to the utility grid.

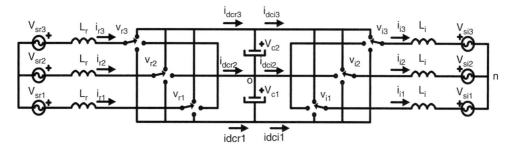

FIGURE 29.36 Nomenclature criterion for the modeling of the full DCC converter.

And the next ones for capacitors voltages dynamics Eq. (29.34):

$$2C\frac{dx_1}{dt} = (\delta_{r1}i_{r1} + \delta_{r2}i_{r2} + \delta_{r3}i_{r3}) - (\delta_{i1}i_{i1} + \delta_{i2}i_{i2} + \delta_{i3}i_{i3})$$

$$x_1 = \frac{v_{c1} + v_{c2}}{2}$$

$$2C\frac{dx_2}{dt} = (\delta_{r1}^2 i_{r1} + \delta_{r2}^2 i_{r2} + \delta_{r3}^2 i_{r3}) - (\delta_{i1}^2 i_{i1} + \delta_{i2}^2 i_{i2} + \delta_{i3}^2 i_{i3})$$

$$x_2 = \frac{v_{c2} - v_{c1}}{2}$$

(29.34)

where, x_1 and x_2 are chosen as variables to facilitate the controller design and represent the dynamics of the sum and the difference of the capacitors voltages, respectively.

As indicated in [34], it is useful to represent the system in $\alpha\beta\gamma$-coordinates because after the transformation appears the γ control signal as a third freedom degree of the control, moreover this transformation shows the direct relation between this control signal and the capacitors voltage balance. To change to the $\alpha\beta\gamma$-coordinates, an invariant power transformation has been used. The voltages and currents, which are vectors originally in abc-coordinates, are transformed into $\alpha\beta\gamma$-coordinates according to the following matrix transformation shown in Eq. (29.35):

$$T = \sqrt{\frac{2}{3}}\begin{bmatrix} 1 & -1/2 & -1/2 \\ 0 & \sqrt{3}/2 & -\sqrt{3}/2 \\ 1/\sqrt{2} & 1/\sqrt{2} & 1/\sqrt{2} \end{bmatrix}$$

(29.35)

The transformed equations are:

$$L_r \begin{bmatrix} \frac{di_{r\alpha}}{dt} \\ \frac{di_{r\beta}}{dt} \end{bmatrix} = \begin{bmatrix} v_{sr\alpha} \\ v_{sr\beta} \end{bmatrix} - x_1 \begin{bmatrix} \delta_{r\alpha} \\ \delta_{r\beta} \end{bmatrix}$$

$$- x_2 \begin{bmatrix} \frac{\delta_{r\alpha}^2 - \delta_{r\beta}^2}{\sqrt{6}} + \frac{2\delta_{r\alpha}\delta_{r\gamma}}{\sqrt{3}} \\ -\sqrt{\frac{2}{3}}\delta_{r\alpha}\delta_{r\beta} + \frac{2}{\sqrt{3}}\delta_{r\beta}\delta_{r\gamma} \end{bmatrix}$$

(29.36)

$$L_i \begin{bmatrix} \frac{di_{i\alpha}}{dt} \\ \frac{di_{i\beta}}{dt} \end{bmatrix} = -\begin{bmatrix} v_{si\alpha} \\ v_{si\beta} \end{bmatrix} + x_1 \begin{bmatrix} \delta_{i\alpha} \\ \delta_{i\beta} \end{bmatrix}$$

$$+ x_2 \begin{bmatrix} \frac{\delta_{i\alpha}^2 - \delta_{i\beta}^2}{\sqrt{6}} + \frac{2\delta_{i\alpha}\delta_{i\gamma}}{\sqrt{3}} \\ -\sqrt{\frac{2}{3}}\delta_{i\alpha}\delta_{i\beta} + \frac{2}{\sqrt{3}}\delta_{i\beta}\delta_{i\gamma} \end{bmatrix}$$

(29.37)

$$2C\frac{dx_1}{dt} = \begin{bmatrix} \delta_{r\alpha} & \delta_{r\beta} \end{bmatrix}\begin{bmatrix} i_{r\alpha} \\ i_{r\beta} \end{bmatrix} - \begin{bmatrix} \delta_{i\alpha} & \delta_{i\beta} \end{bmatrix}\begin{bmatrix} i_{i\alpha} \\ i_{i\beta} \end{bmatrix}$$

$$2C\frac{dx_2}{dt} = \frac{2}{\sqrt{3}}\begin{bmatrix} \delta_{r\alpha} & \delta_{r\beta} \end{bmatrix}\begin{bmatrix} i_{r\alpha} \\ i_{r\beta} \end{bmatrix}\delta_{r\gamma}$$

$$+ \begin{bmatrix} \frac{\delta_{r\alpha}^2 - \delta_{r\beta}^2}{\sqrt{6}}, -\sqrt{\frac{2}{3}}\delta_{r\alpha}\delta_{r\beta} \end{bmatrix}\begin{bmatrix} i_{r\alpha} \\ i_{r\beta} \end{bmatrix} \ldots$$

$$- \frac{2}{\sqrt{3}}\begin{bmatrix} \delta_{i\alpha} & \delta_{i\beta} \end{bmatrix}\begin{bmatrix} i_{i\alpha} \\ i_{i\beta} \end{bmatrix}\delta_{i\gamma}$$

$$- \begin{bmatrix} \frac{\delta_{i\alpha}^2 - \delta_{i\beta}^2}{\sqrt{6}}, -\sqrt{\frac{2}{3}}\delta_{i\alpha}\delta_{i\beta} \end{bmatrix}\begin{bmatrix} i_{i\alpha} \\ i_{i\beta} \end{bmatrix}$$

(29.38)

In these final equations, it is important to point out the relation between the γ control signal and the input and output power of the DCC full converter.

29.3.5 Control

As it can be observed in Eqs. (29.36) and (29.37) the rectifier and inverter currents $i_{\alpha\beta}^r$, $i_{\alpha\beta}^i$ can be controlled separately due to the decoupling of these equations. Also it can be seen in Eq. (29.38) that the control objective on x_1 can be achieved using the normalized voltage references $\delta_{\alpha\beta}^r$ or $\delta_{\alpha\beta}^i$, and x_2 can be controlled using δ_γ^r or δ_γ^i. The implemented control consists basically of independently controlling the inverter and the rectifier. The inverter controls the voltage balance in the DC-link, whereas the rectifier controls the active and reactive power extracted from the generator.

29.3.5.1 Rectifier Control

Figure 29.37 shows the control scheme proposed for the rectifier. The objective of this controller is to make the currents of the generator such that the active and reactive power achieve the reference ones.

It is necessary to notice that the rectifier γ component of the normalized voltage is imposed to be equal to zero $\delta_\gamma^r = 0$ for not affecting on voltage balance, because this balancing is implemented on the inverter control.

29.3.5.2 Inverter Control

Inverter control is divided in two parts. The first part controls the sum of the capacitor voltages x_1, while the second part makes the difference between the capacitor voltages x_2 as small as possible.

29.3.5.3 Sum of the Capacitor Voltages Control

The controller scheme, which can be seen in Fig. 29.38, has been described before in [35], and it is appropriated for this application due to the similarities found in the equations.

The main objective of the controller is to achieve a desired value of the total DC-link voltage. Additionally, the controller can take a reactive power reference to control the power factor of the energy delivered to the utility grid.

29.3.5.4 Difference of the Capacitor Voltages Control

Avoiding the quadratic terms in δ from the equation of the difference of the capacitor voltages in Eq. (29.38), expression (29.39) is obtained:

$$\frac{dx_2}{dt} = K \cdot P_{ref}^i \cdot \delta_\gamma^i \quad (29.39)$$

where K is a constant. With this equation, the following control scheme (Fig. 29.39) is proposed:

The objective of the controller is to add a voltage reference in γ direction that depends on the sign of the power and the imbalance of the capacitors voltages.

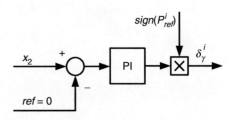

FIGURE 29.39 Proposed capacitors voltages balancing control.

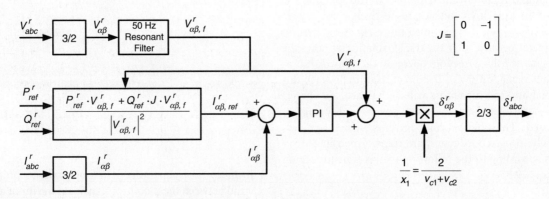

FIGURE 29.37 Control diagram of the rectifier.

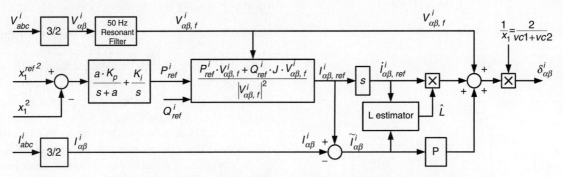

FIGURE 29.38 Inverter control scheme for the sum of the capacitors voltages.

29.3.5.5 Modulation

Finally, the normalized voltage references $\delta^r_{\alpha\beta\gamma}$ and $\delta^i_{\alpha\beta\gamma}$, obtained from the whole controllers, are translated to *abc*-coordinates and the 3D-space vector modulation algorithm [36] is used to generate the duty cycles and the switching times of power semiconductors.

29.3.6 Application Example

As it was explained before, the standards on energy quality related to renewable energy are focusing to request the plants to contribute to the general stability of the electrical system. To show that the exposed modeling strategy and control scheme can be used to meet the design specification, the electrical system of a wind turbine has been modeled. It consists of an asynchronous induction motor connected to the utility grid through a full DCC converter. The parameters of the example are: nominal power: 3 MW, switching frequency: 2.5 kHz, DC-link nominal voltage: 5 kV, and utility grid line voltage: 2.6 kV. The experiment consists of studying the behavior of the system when there is a voltage dip in the utility grid due to a short-circuit. Figure 29.40 shows the envelope of the voltage dip that has been used to carry out the results.

Figure 29.41 shows the results obtained under the voltage dip condition. Good behavior of the currents on both the sides of the full DCC converter, DC-link voltage, and energy extracted from the generator illustrates the suitability of the control scheme and the model to study the system.

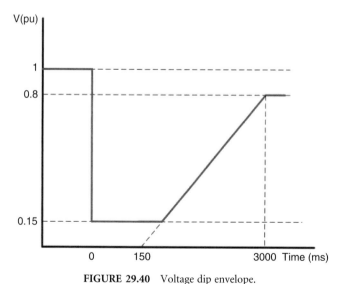

FIGURE 29.40 Voltage dip envelope.

29.4 Electrical System of a Wind Farm

29.4.1 Electrical Schematic of a Wind Farm

A wind farm is integrated by wind turbines and the substation that connect the farm to the utility grid to evacuate the electrical energy. The wind farm is arranged by string of wind turbines. Figure 29.42 shows a string compounded by several aerogenerators. These wind turbines are connected by manual switch breakers which isolate a wind turbine or it isolates the whole string. In variable-speed applications, an AC/AC power converter is used. This power converter is connected by a manual switch to the machine. The power converter includes a remote controlled switch breaker which isolates from the power transformer. The switch breaker is used for automatic reconnection after a fault. Figure 29.42 shows the transformer connection.

A schematic diagram of a typical substation is shown in Fig. 29.43. A large transformer, depicted in the figure, or several transformer connected in parallel, changes from the medium voltage to a higher voltage level. A typical voltage levels in Europe could be 20 kV/320 kV. The substation also incorporates bus bar, protections systems, measurement instrumentation, and auxiliary services circuit. Bus bar voltage measurement is made by voltage transformer. Each branch current, including several wind turbines, is measured by current transformer.

Some farms with lower rated power or connected to an isolated grid, e.g. wind-diesel systems, do not use this large transformer. The schematic of an isolated wind-diesel installation is represented in Fig. 29.44. Every power generator and load are connected to a medium voltage bus bar, in the typical range of 10–20 kV. The transformers are protected by circuit breakers that connect the lines directly to ground when open. A measurement system is used for power consumption and electrical quality control. Also auxiliaries' power supply feeds the substation equipment.

29.4.2 Protection System

Protection of wind power systems requires an understanding of system faults and their detection, as well as their safe disconnection. The protection system of a wind farm is mainly included in the substation. Circuit breaker and switchgear [37] are extensively used for overcurrent protection. New type of relay has been designed for the protection of wind farms that incorporate fixed-speed induction generators as described in [38]. A protection relay can be installed in the medium-voltage collecting line at the common point connection to the utility grid. This relay provides short-circuit protection for the collecting line and the medium-voltage (MV) and low-voltage (LV) circuits. Consequently, the relay allows wind farms to be constructed and adequately protected without the need to include fuses on the MV side of each generator–transformer.

The variable speed generator also includes digital relay protection and can be programmed for complex coordination and selectivity. This modern protection system can be used for voltage gap or sag function protection. Moreover, they can implement modern stabilization programs [39].

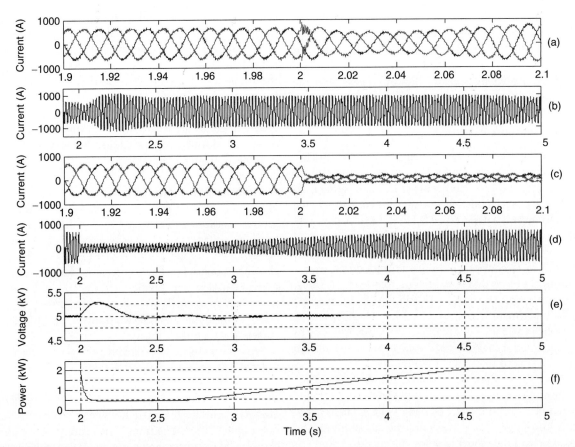

FIGURE 29.41 Response of the system to a voltage dip: (a) inverter side currents (detail); (b) inverter one phase current; (c) rectifier side currents (detail); (d) rectifier one phase current; (e) DC-link voltage; and (f) active power extracted from the generator.

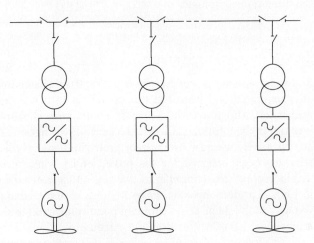

FIGURE 29.42 Typical branch composed by several wind turbines.

29.4.3 Electrical System Safety: Hazards and Safeguards

It is important to understand the hazards of electricity at the power system supply level. The safety of wind farm includes a good knowledge of electrical blast, electrocution, short circuits, overloads, ground faults, fires, lifting and pinching injuries.

It is also recommended to review the principles, governmental regulations, work practices, and specialized equipment relating to electrical safety. Installers and maintenance personnel have to know the different types of "Personal Protective Equipment" through demonstrations of locking and tagging devices, protective clothing, and specialized equipment. The isolation and "Lockout Practices Procedures" for the lockout and isolation of electrical equipment can also be implemented into the existing site regulations and policies. A common practise is to use isolation transformer and grounding circuit breaker as they are being operated.

29.5 Future Trends

Future trends relating to power electronics used in wind turbine applications can be summarized in the following points:

29.5.1 Semiconductors

Improvements in the performance of power electronics variable frequency drives for wind turbine applications have been directly related to the availability of power semiconductor

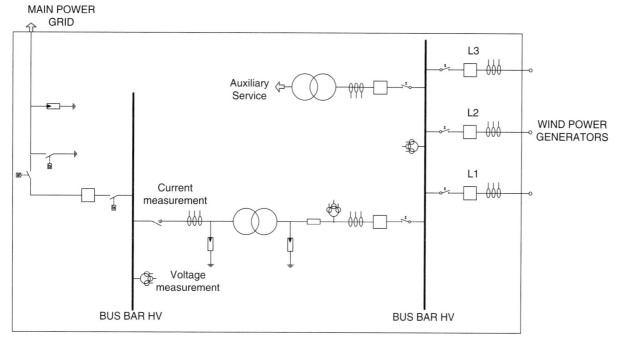

FIGURE 29.43 Schematic diagram of a typical wind farm substation.

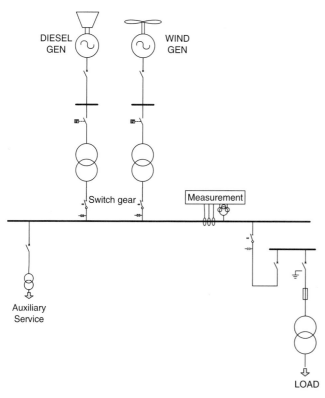

FIGURE 29.44 Schematic diagram of a wind–diesel generation system.

devices with better electrical characteristics and lower prices because the device performance determines the size, weight, and cost of the entire power electronics used as interfaces in wind turbines [8, 18, 9].

The thyristor is the component that started power electronics. It is an old device with decreasing use in medium power applications, which was replaced by turn-off components like insulated gate bipolar transistor (IGBTs). The IGBT, which can be considered as an MOS bipolar Darlington, is now the main component for power electronics, and also for wind turbine applications. They are now mature technology turn-on components adapted to very high power (6 kV–1.2 kA), and they are in competition with gate turn-off thyristor (GTO) for high power applications [40].

Recently, the integrated gated control thyristor (IGCT) has been developed, consisting of the mechanical integration of a GTO plus a delicate hard drive circuit that transforms the GTO into a modern high performance component with a large safe operation area (SOA), lower switching losses, and a short storage time [41–43].

The comparison between IGCT and IGBT for frequency converters, used especially in wind turbines is explained below:

- IGBTs have higher switching frequency than IGCTs so they introduced less distortion in the grid. Accordingly if we use two three-phase systems in parallel, it is possible to double the resulting switching frequency without increasing the power loss, hence it is possible to have a total harmonic distortion (THD) of less than 2% without special harmonic filters.

- IGCTs are made like disk devices. They have to be cooled with a cooling plate by electrical contact on the high voltage side. This is a problem because high electromagnetic emission will occur. Another point of view is the number of allowed load cycles. Heating and cooling the device will always bring mechanical stress to the silicon chip and can be destroyed. This is a serious problem, especially in wind turbine applications. On the other hand, IGBTs are built like module devices. The silicon is isolated to the cooling plate and can be connected to ground for low electromagnetic emission even with higher switching frequency. The base plate of this module is made of a special material which has exactly the same thermal behavior as silicon, so nearly no thermal stress occurs. This increases the lifetime of the device by 10-fold approximately.
- The main advantage of IGCTs versus IGBTs is that they have a lower on state voltage drop, which is about 3.0 V for a 4500 V device. In this case, the power dissipation due to a voltage drop for a 1500 kW converter will be 2400 W per phase. On the other hand, in the case of IGBT, the voltage drop is higher than IGCTs. For a 1700 V device having a drop of 5 V, and in this case the power dissipation due to the voltage drop for a 1500 kW condition will be 5 kW per phase.

In conclusion, with the present semiconductor technology, IGBTs present better characteristics, for frequency converters in general and especially for wind turbine applications.

29.5.2 Power Converters

The technology of power electronic interfaces for variable-speed wind turbines is focused on the following points:

- Development of high efficiency/high quality voltage source AC/DC/AC converter for a main connection of variable wind turbines, operating with either a permanent magnet, a synchronous or an asynchronous generator.
- Operation at a power factor around one with higher-harmonic voltage distortion less than international standards.
- The power quality of the electrical output of the wind farms may be improved by the use of advanced static var compensators STATCOM or active power filters using power semiconductors like IGBTs, IGCTs, or GTOs. These kind of power conditioning systems are a new emerging family of FACTS (flexible AC transmission system) converters, which allow improved utilization of the power network. These systems will allow wind farms to reduce voltage drops and electrical losses in the network without the possibility of transient over voltage at islanding due to self-excitation of wind generators. Moreover, power conditioning systems equipment with different control algorithm can be used to control the network voltage, which will fluctuate in response to the wind farm output if the distribution network is weak [44, 45].
- For large power wind turbine applications where it is necessary to increase the voltage level of the semiconductor of the power electronic interface, multilevel power converter technology is emerging as a new breed of power converter options for high-power applications. The general structure of the multilevel converter is to synthesize a sinusoidal voltage from several levels of voltages, typically obtained from capacitor voltage sources. Additionally, these converters have better performance and controllability because they use more than two voltage levels [46–48].

29.5.3 Control Algorithms

A variable pitch and speed wind turbine is a very complex non-lineal system. The control problem is more difficult to solve because some performance objectives, such as maximum power captured, minimum mechanical stress, constant speed, and power constant counteract each other. To solve this problem, a Fuzzy Logic control has recently been proposed. The Fuzzy Logic controller implements a rule-based structure [49] that can be easily adapted in order to optimize performance control objectives and has been widely used in introduction motor control applications [50–53]. In [54], a Fuzzy Logic controller is used to optimize the power captured using maximum power tracking algorithms. Other original structures have been proposed in [55]. The structure implements different control strategies depending on the rotor speed and generates a current torque control action. Presently, more complex control structures are being researched.

29.5.4 Offshore and Onshore Wind Turbines

One of the main trends in wind turbine technology is offshore installations. There are great wind resources at sea for installing wind turbines in many areas where the depth of the sea is relatively shallow. There are several demonstration plants that have had extremely positive results, so interest has increased in installing offshore wind farms, because of the development of large commercial power MW wind turbines. Offshore wind turbines may have slightly more favorable energy balance than onshore turbines, depending on local wind conditions. In places where onshore wind turbines are typically placed on flat terrain, offshore wind turbines will generally yield some 50% more energy than a turbine placed on a nearby onshore site. The reason is the low roughness of the sea surface. On the other hand, the construction and installation of a foundation requires 50% more energy than onshore turbines. It should be remembered, however, that offshore wind turbines have a longer life expectancy, around 25 to 30 years, than onshore turbines. The reason is that the low turbulence at sea gives lower fatigue loads on the wind turbine.

From a power electronics point of view, offshore wind turbines are interesting because, under certain circumstances, they become desirable to transmit the generated power to the load center over DC transmission lines (HVDC). This alternative becomes economically attractive versus AC transmission when a large amount of power is to be transmitted over a long distance from a remote wind farm to the load center [8]. Moreover, the transient stability and the dynamic damping of the electrical system oscillations can be improved by HVDC transmissions.

Nomenclature

Symbol	Description
C_{dc}	DC link capacitor.
$C_p(\lambda, \beta)$	Power coefficient at tip speed ratio λ and pitch β.
e_{rg}, e_{sg}, e_{tg}	Instantaneous values of grid voltages.
e_α, e_β	Grid voltages expressed in an orthogonal reference frame.
f_1	Excitation frequency (the same as the grid frequency) in hertz.
f_2	Frequency of the voltage supplied to the rotor of machine 2 in hertz.
G	Gearbox ratio.
i_{dc}	DC inductor current of the step-up converter.
i_{dc}^{ref}	Desired DC inductor current of the step-up converter.
i_{dse}, i_{qse}	Instantaneous values of the direct and quadrature-axis stator current components, respectively, and expressed in rotor-flux-oriented reference frame.
$i_{dse}^{ref}, i_{qse}^{ref}$	Desired instantaneous values of the direct and quadrature-axis stator current components, respectively, and expressed in rotor-flux-oriented reference frame.
i_{dsr}, i_{qsr}	Stator current components established in the rotor reference frame.
$i_{dsr}^{ref}, i_{qsr}^{ref}$	Desired stator current components established in the rotor reference frame.
i_{ds}, i_{qs}	Instantaneous values of the direct and quadrature-axis stator current components, respectively, and expressed in the rotor reference phase.
i_{Ds}, i_{Qs}	Instantaneous values of the direct and quadrature-axis stator current components, respectively, and expressed in the stator reference frame.
$i_{Ds}^{ref}, i_{Qs}^{ref}$	Desired direct and quadrature-axis stator current components expressed in the stator reference frame.
i_{exc}	Synchronous generator excitation current.
i_{exc}^{ref}	Desired synchronous generator excitation current.
$i_{R_g}^{ref}, i_{S_g}^{ref}, i_{T_g}^{ref}$	Desired stator current.
$i_{R_r}, i_{S_r}, i_{T_r}$	Rotor current.
$i_{R_s}, i_{S_s}, i_{T_s}$	Stator current.
$i_{R_s}^{ref}, i_{S_s}^{ref}, i_{T_s}^{ref}$	Desired stator current.
i_{rg}, i_{sg}, i_{tg}	Public grid phase currents.
i_m^{ref}	Desired magnetizing current.
i_m	Magnetizing current.
i_{dsm}, i_{qsm}	Instantaneous values of the direct and quadrature-axis magnetizing stator current components, respectively, and expressed in the magnetizing current-oriented reference frame.
J	Total inertia of the system referred to the high speed shaft.
k_{ws}, k_{wr}	Winding factors of rotor and stator.
L_m	Coupled inductance.
L_{AC}	Inductance of inductors at the AC side of the inverter.
L_S, L_r	Stator and rotor windings inductances.
L_{DC}	Inductance of the step-up chopper.
n	Turns ratio of the machine. $n = k_{ws} \cdot n_s / k_{wr} \cdot n_r$
n_s, n_r	Number of turns of each rotor and stator phase.
N_1	Angular speed of the magnetic field (synchronous speed) expressed in rpm.
N_r	Angular speed of the generator rotor expressed in rpm.
p	Number of pole pairs.
p_1, p_2	Number of pole pairs of machine number 1 and 2.
P_1	Electrical power in the stator of principal machine number 1.
P_2	Electrical power in the stator of auxiliary machine number 2.
P_e	Electrical generated power.
P_m	Mechanical power in the low speed shaft.
P_{max}	Generated maximum power.
P_{rate}	Generator rate power.
P_S, Q_S	Active and reactive power through the stator.
P_S^{ref}, Q_S^{ref}	Desired active and reactive power through the stator.
P_{slip}	Slip power.
P_w	Available wind power.
p^{ref}, q^{ref}	Desired real power and desired reactive power on the grid side.
Q_e	Electromagnetic torque of the machine.
Q_e^{ref}	Desired electromagnetic torque of the machine.
Q_L	Torque in the low speed shaft.
Q_{rate}	Generator rate torque.
Q_t	Torque in the high speed shaft.
R	Rotor radius.
R_S, R_r	Stator and rotor winding resistors.
s	Slip.
$s_1 = \omega_1 - \omega_r / \omega_1$	Slips of machine principal number 1.
$s_2 = \omega_2 - \omega_r / \omega_2$	Slips of machine principal number 2.
T_s, T_r	Stator and rotor time constants, respectively.
$u_{dse}^{ref}, u_{qse}^{ref}$	Desired direct and quadrature-axis stator voltage expressed in the rotor-flux-oriented reference frame.
$u_{Ds}^{ref}, u_{Qs}^{ref}$	Desired direct and quadrature-axis stator voltage expressed in the stator reference frame.

$u_{R_r}, u_{S_r}, u_{T_r}$	Rotor voltage.
$u_{R_s}, u_{S_s}, u_{T_s}$	Stator voltage.
$u_{R_s}^{ref}, u_{S_s}^{ref}, u_{T_s}^{ref}$	Desired stator voltage.
u_r, i_r, λ_r	Rotor voltage, current, and flux, respectively, referred to a reference frame that rotates with the rotor.
u'_r, i'_r, λ'_r	Rotor voltage, current, and flux, respectively, referred to a reference frame fixed with the stator.
u_s, i_s, λ_s	Stator voltage, current, and flux, respectively, referred to a reference frame fixed with the stator.
u_{ds}, u_{qs}	Direct and quadrature-axis stator voltage expressed in the magnetizing current reference frame.
v_{rat}	Rated wind speed.
$v_{p_{max}}$	Maximum power wind speed.
v_{start}	Start wind speed.
v_{stop}	Stop wind speed.
v_w	Wind speed.
V_{dc}	DC-Link capacitor voltage.
V_{dc}^{ref}	Desired DC-Link capacitor voltage.
V_{st}^{max}	Maximum RMS stator voltage of the synchronous generator.
V_{st}^{min}	Maximum RMS stator voltage of the synchronous generator.
β	Pitch angle.
β^{ref}	Desired pitch angle.
δ	Load angle.
$\lambda = \omega \cdot R/V_v$	Tip speed ratio.
λ_{opt}	Optimal tip speed ratio.
λ_m	Magnetizing flux linkage vector.
$\lambda_{md}, \lambda_{mq}$	Instantaneous values of the direct and quadrature axis magnetizing flux linkage components expressed in the rotor reference frame.
$|\lambda_m|$	Estimated modulus of magnetizing flux linkage vector.
$\left|\lambda_m^{ref}\right|$	Desired modulus of magnetizing flux linkage vector.
η	Electrical performance.
η_r	Phase angle of the rotor flux linkage space phasor with respect to the direct axis of the stator reference frame.
θ_e	Magnetizing current angle.
θ_{sl}	Angle corresponding to the angular slip frequency.
θ_r	Rotor angle.
ρ	Air density.
$\omega_1 = 2\pi f_1/p_1$	Angular speed of the rotating magnetic flux produced in the stator of machine 1 relative to the stator.
$\omega_2 = 2\pi f_2/p_2$	Angular speed of the rotating magnetic flux produced in the rotor of machine 2 relative to the rotor.
ω_L	Low-speed shaft angular speed.
ω_r	Angular speed of the generator rotor.
ω_r^{ref}	Reference rotor speed.
ω_r^{max}	Maximum angular speed of the synchronous generator.
ω_r^{min}	Minimum angular speed of the synchronous generator.
ω_Q^{ref}	Desired angular speed for the torque controller.
ω_β^{ref}	Desired angular speed for the pitch controller.
ω_r^{rate}	Rated value of the rotor speed.
ω_e	Electrical angular speed of the magnetizing current reference frame.

References

1. S. Heier, "Grid Integration of Wind Energy Conversion Systems". Chichester, Sussex (UK): John Wiley & Sons, 1998.
2. G.L. Johnson, "Wind Energy Systems". Englewood Cliffs, NJ (US): Prentice-Hall, INC., 1985.
3. P. Vas, "Vector Control of AC Machines", NY (US): Oxford Clarendon Press, 1990.
4. V. Subrahmanyam, "Electric Drives. Concepts and Applications". NY (US): MacGraw-Hill, 1996.
5. M. Alatalo, M. Sc, and T. Svensson, "Variable Speed Direct-Driven PM-Generator With a PWM Controlled Current Source Inverter". *European Community Wind Energy Conference*. March 1993. Lùbeck-Travemùnde, Germany.
6. P. Vas, "Sensorless Vector and Direct Torque Control". NY (US): Oxford University Press, 1998.
7. Chee-Mun Ong. "Dynamic Simulation of Electric Machinery Using Matlab/Simulink". Prentice Hall PTR, 1998.
8. N. Mohan, T.M. Undeland, and W.P. Robbins, "Power Electronics, Converters, Applications, and Design". Second edition, John Wiley & Sons, INC., 1995.
9. R.E. Tarter. "Solid-State Power Conversion Handbook". John Wiley & Sons, Inc., 1993.
10. B.K. Bose, "Power Electronics and Variable Frequency Drives. Technology and Applications". IEEE Press, 1997.
11. S.A. Papathanassiou and M.P. Papadopoulos, "A Comparison of Variable Speed Wind Turbine Configurations". *Wind Energy Conference*. March 1999, France.
12. D.S. Zinger and E. Muljadi. "Annualized Wind Energy Improvement Using Variable Speeds". *IEEE Transactions on Industry Applications*, Vol. 33, No. 6, Nov–Dec 1997.
13. K. Pierce, "Control Method for Improved Energy Capture below Rated Power". *Third edition ASME/JSME Joint Fluids Engineering Conference*. July, 1999. San Francisco, California.
14. *Theory Manual*. E.A Bossanyi, "Bladed for Windows". Garrad Hassan and Partners Limited. September 1997.
15. E. Muljadi, K. Pierce, and P. Migliore, "Control Strategy for Variable-Speed, Stall-Regulated Wind Turbines". *American Controls Conference*. June 1998. Philadelphia, NREL.
16. A.D. Simmons, L.L. Freris, and J.A.M. Bleijs, "Comparison of Energy Capture and Structural Implementaions of Various Policies of Controlling Wind Turbines". *Wind Energy: Technology and Implementation*. 1991. (Amsterdam EWEC'91).
17. W.E. Leithead, S. de la Salle, and D. Reardon, "Wind Turbine Control Objectives and Design". *European Community Wind Energy Conference*. September 1990. Madrid, Spain.

18. M.H. Rashid, *Power Electronics. Circuits, Devices, and Applications*. Second edition. Englewood Cliffs, NJ(US): Prentice Hall, 1993.
19. H. Akagi, A. Nabae, and S. Atoh, "Control Strategy of Active Power Filters Using Multiple Voltage-Source PWM Converters". *IEEE Trans. Ind. Applications*, Vol. IA-22, No. 3, pp. 460–465, May–June 1986.
20. S. Bhowmik, R. Spée, and J.H.R. Enslin, "Performance Optimization for Doubly Fed Wind Power Generation Systems". *IEEE Transactions on Industry Applications*, Vol. 35, No. 4, July–August 1999.
21. B. Hopfensperger, D.J. Atkinson, and R.A. Lakin, "Application of Vector Control to the Cascaded Induction Machine for Wind Power Generation Schemes". *7th IEE European on Power Electronics, EPE'97*. September 1997. Trondheim (Norway).
22. P.O. 12.3 "Propuesta sobre requisitos de respuesta frente a huecos de tensión de las instalaciones eólica". Red Eléctrica de España, S.A. October 2005.
23. C. Rasmussen, P. Jorgensen, and J. Havsager, "Integration of Wind Power in the Grid in Eastern Denmark". In *Proc 4th International Workshop on Large-scale Integration of Wind Power and Transmission Network for Offshore Wind Farm*, 20–21 October 2003, Billund, Denmark.
24. E.ON Netz Grid Code, Bayreuth; E.ON Netz GmbH. Germany, 1 August 2003.
25. M. Johan and de Hann Sjoerd W.H. "Ridethrough of Wind Turbines with Doubly-fed Induction Generator During a Voltage Dip". *IEEE Transaction on Energy Conversion*, Vol. 20, No. 2. June 2005.
26. X. Bing, F. Brendan, and F. Damian, "Study of Fault Ride Through for DFIG Wind Turbines". *IEEE Transaction on Energy Conversion*, Vol. 20, No. 2. June 2005.
27. International Electrotechnical Commission. Draft IEC 61400-21: Power Quality Requirements for Grid Connected Wind Turbines. Committee Draft (CD). December 1998.
28. D. Foussekis, F. Kokkalidis, S. Tentzevakis, and D. Agoris, "Power Quality Measurement on Different Type of Wind Turbines Operating in the Same Wind Farm". EWEC 2003.
29. S. Poul, G. Gert, S. Fritz, R. Niel, D. Willie, K. Maria, M. Evangelis, and L. Ake, "Standards for Measurements and Testing of Wind Turbine Power Quality". EWEC 1999.
30. International Electrotechnical Commission, IEC Standard, Amendment 1 to Publication 61000-4-7, Electromagnetic Compatibility, General Guide on Harmonics and Inter-harmonics Measurements and Instrumentation, 1997.
31. International Electrotechnical Commission, IEC Standard, Publication 61000-3-6, Electromagnetic Compatibility, Assessment of Emission Limits for Distorting Loads in MV and HV Power Systems, 1996.
32. L. Ake, S. Poul, and S. Fritz, *Grid Impact of Variable Speed Wind Turbines*. EWEC' 1999.
33. A. Nabae, H. Akagi, and I. Takahashi, "A New Neutral-Point-Clamped PWM Inverter". *IEEE Transactions on Industry Applications*, Vol. IA-17, No. 5, pp. 518–523, September/October 1981.
34. G. Escobar, J. Leyva, J.M. Carrasco, E. Galvan, R. Portillo, M.M. Prats, and L.G. Franquelo, "Modeling of a Three Level Converter Used in a Synchronous Rectifier Application", in *Proc. Power Electronics Specialists Conference, PESC'04*, Aachen, Germany, Vol. 6, pp. 4306–4311, 2004.
35. G. Escobar, J. Leyva-Ramos, J. M. Carrasco, E. Galvan, R. Portillo, M.M. Prats, and L.G. Franquelo, "Control of a Three Level Converter Used as a Synchronous Rectifier". *2004 IEEE 35th Annual Power Electronics Specialists Conference, PESC'04*, Vol. 5, pp. 3458–3464, June 20–25, 2004.
36. M.M. Prats, L.G. Franquelo, R. Portillo, J.I. León, E. Galván, and J.M. Carrasco, "A Three Dimensional Space Vector Modulation Generalized Algorithm for Multilevel Converters". *IEEE Power Electronics Letters*, Vol. 1, pp. 110–114, 2003.
37. W.D. Goodwin, "High-voltage Auxiliary Switchgear for Power Stations". *Power Engineering Journal* [see also Power Engineer], Vol. 3, No. 3, pp. 145–154, May 1989.
38. S.J. Haslam, P.A. Crossley, and N. Jenkins, "Design and Evaluation of a Wind Farm Protection Relay". Generation, Transmission and Distribution, *IEE Proceedings*, Vol. 146, No. 1, pp. 37–44, January 1999, Digital Object Identifier 10.1049/ip-gtd:19990045.
39. M.P. Palsson, T. Toftevaag, K. Uhlen, and J.O.G. Tande, "Large-scale Wind Power Integration and Voltage Stability Limits in Regional Networks". Power Engineering Society Summer Meeting, *IEEE Proceedings*, Vol. 2, pp. 762–769, 2002. Digital Object Identifier 10.1109/PESS.2002.1043417.
40. J.M. Peter, "Main Future Trends for Power Semiconductors from the State of the Art to Future Trends". *Power Conversion Intelligent Motion (PCIM'99)*. Nürnberg, June 1999.
41. H. Grüning, B. Ødegård, J. Rees, A. Weber, E. Carroll, and S. Eicher, "High Power Hard-Driven GTO Module for 4.5kV/3kA Snubberless Operations". *PCI Europe Proceedings*, Nürnberg, 1996.
42. A. Jaecklin, "Integration of Power Components – State of the Art and Trends". *European Power Electronic Conference EPE'97*. Trondheim, 1997.
43. H.R. Zeller, "High Power Components from the State of the Art to Future Trends". *PCIM Europe Proceedings*, Nürnberg, 1998.
44. J.M. Carrasco, M. Perales, B. Ruiz, E. Galván, L.G. Franquelo, S. Gutiérrez, and E. Gonzalez "DSP Control of an Active Power Line Conditioning System". *7th IEE European Conference on Power Electronics, EPE'97*. Trondheim (Norway). September 1997.
45. J. Balcells, M. Lamich, and D. González, "Parallel Active Filter Based on a Three Level Inverter". *European Power Electronic Conference EPE'99*. Laussane, 1999.
46. R. Teodorescu, F. Blaabjerg, J.K. Pedersen, E. Cengelci, S.U. Sulistijo, B.O. Woo, and P. Enjeti, "Multilevel Converters – A Survey", *European Power Electronic Conference EPE'99*. Lausanne, September 1999.
47. K. Oguchi, T. Karaki, and N. Hoshi, "Space Vector of Output Voltages of Reactor-Coupled Three Phase Multilevel Voltage-Source Inverters". *European Power Electronic Conference EPE´99*. Lausanne, September 1999.
48. J.-S. Lai and F.Z. Peng, "Multilevel Converters – A New Breed of Power Converters". *IEEE Transactions on Industry Application*, Vol. 32, No. 3. May–June 1996.
49. M. Sugeno, "An Introductory Survey of Fuzzy Control". *Information Sciences*, 36, pp. 59–83, 1985.
50. E. Galván, A. Torralba, F. Barrero, M.A. Aguirre, and L.G. Franquelo, "Fuzzy-Logic Based Control of an Induction Motor". *Industrial Fuzzy Control and Applications*. Tarrasa (Spain), March 1993.
51. E. Galván, A. Torralba, F. Barrero, M.A. Aguirre, and L.G. Franquelo, "A Robust Speed Control of AC Motor Drives based on Fuzzy

Reasoning". *Industry Application Society (IAS)*. Toronto (Canada), October 1993.

52. F. Barrero, E. Galván, A. Torralba, and L.G. Franquelo, "Fuzzy Selftuning System for Induction Motor Controllers". *IEE European Conference on Power Electronics, EPE'95*. Seville (Spain), September 1995.

53. F. Barrero, A. Torralba, E. Galván and L.G. Franquelo, "A Switching Fuzzy Controller for Induction Motors with Self-tunning Capability". *Int. Con. on Industrial Electronics, IECON'95*. Orlando, Florida (USA), November 1995.

54. M. Godoy Simoes, B.K. Bose, and R.J. Spiegel "Fuzzy Logic Based Intelligent Control of a Variable Speed Cage Machine Wind Generation System". *IEEE Transaction on Power Electronics*, Vol. 12, No. 1, pp 87–95, January 1997.

55. M. Perales, J. Pérez, F. Barrero, J.L. Mora, E. Galván, J.M Carrasco, L.G. Franquelo, D. de la Cruz, L. Fernández, and A. Zazo, "A New Fuzzy Based Approach for a Variable Speed, Variable Pitch Wind Turbine". *8th International Fuzzy Systems Association World Congress. IFSA'99*.

30
HVDC Transmission

Vijay K. Sood, Ph.D.
Hydro-Quebec (IREQ), 1800 Lionel Boulet, Varennes, Quebec, Canada

30.1 Introduction 769
    30.1.1 Comparison of AC–DC Transmission • 30.1.2 Evaluation of Reliability and Availability Costs • 30.1.3 Applications of DC Transmission • 30.1.4 Types of HVDC Systems

30.2 Main Components of HVDC Converter Station 775
    30.2.1 Converter Unit • 30.2.2 Converter Transformer • 30.2.3 Filters • 30.2.4 Reactive Power Source • 30.2.5 DC Smoothing Reactor • 30.2.6 DC Switchgear • 30.2.7 DC Cables

30.3 Analysis of Converter Bridge 778

30.4 Controls and Protection 778
    30.4.1 Basics of Control for a Two-terminal DC Link • 30.4.2 Control Implementation • 30.4.3 Control Loops • 30.4.4 Hierarchy of DC Controls • 30.4.5 Monitoring of Signals • 30.4.6 Protection against Overcurrents • 30.4.7 Protection against Overvoltages

30.5 MTDC Operation 786
    30.5.1 Series Tap • 30.5.2 Parallel Tap • 30.5.3 Control of MTDC System

30.6 Application 787
    30.6.1 HVDC Interconnection at Gurun (Malaysia)

30.7 Modern Trends 789
    30.7.1 Converter Station Design of the 2000s

30.8 HVDC System Simulation Techniques 792
    30.8.1 DC Simulators and TNAs [9] • 30.8.2 Digital Computer Analysis

30.9 Concluding Remarks 794

    References 795

30.1 Introduction

High voltage direct current (HVDC) transmission [1–3] is a major user of power electronics technology. The HVDC technology first made its mark in the early undersea cable interconnections of Gotland (1954) and Sardinia (1967), and then in long distance transmission with the Pacific Intertie (1970) and Nelson River (1973) schemes using mercury arc valves. A significant milestone development occurred in 1972 with the first back-to-back (BB) asynchronous interconnection at Eel River between Quebec and New Brunswick; this installation also marked the introduction of thyristor valves to the technology and replaced the earlier mercury arc valves.

Until 2005, a total transmission capacity of 70,000 MW HVDC is installed in some 95 projects all over the world. To understand the rapid growth of dc transmission (Table 30.1) [4] in the past 50 years, it is first necessary to compare it to conventional ac transmission.

30.1.1 Comparison of AC–DC Transmission

Making a planning selection between either ac or dc transmission is based on an evaluation of transmission costs, technical considerations, and the reliability/availability offered by the two power transmission alternatives.

30.1.1.1 Evaluation of Transmission Costs

The cost of a transmission line comprises of the capital investment required for the actual infrastructure (i.e. right-of-way (ROW), towers, conductors, insulators, and terminal equipment) and costs incurred for operational requirements (i.e. losses). Assuming similar insulation requirements for peak voltage levels for both ac and dc lines, a dc line can carry as much power, with two conductors (having positive/negative polarities with respect to ground), as an ac line with three conductors of the same size. Therefore, for a given power level, a dc line requires a smaller ROW, simpler and cheaper towers

TABLE 30.1 Listing of HVDC installations

HVDC link	Supplier	Year	Power (MW)	DC voltage (kV)	Length (km)	Location
Gotland I#	A	1954	20	±100	96	Sweden
English channel	A	1961	160	±100	64	England–France
Volgograd–Donbass*	Unknown Russian manufacturer	1965	720	±400	470	Russia
Inter-island	A	1965	600	±250	609	New Zealand
Konti-Skan I	A	1965	250	250	180	Denmark–Sweden
Sakuma	A	1965	300	2 × 125	B-B***	Japan
Sardinia	I	1967	200	200	413	Italy
Vancouver I	A	1968	312	260	69	Canada
Pacific intertie	JV	1970	1440	±400	1362	U.S.A.
Pacific intertie	JV	1982	1600	±400	1362	U.S.A
Nelson River I**	I	1972	1620	±450	892	Canada
Kingsnorth	I	1975	640	±260	82	England
Gotland	A	1970	30	±150	96	Sweden
Eel River	C	1972	320	2 × 80	B-B	Canada
Skagerrak I	A	1976	250	250	240	Norway–Denmark
Skagerrak II	A	1977	500	±250		Norway–Denmark
Skagerrak III	A	1993	440	350	240	Norway–Denmark
Vancouver II	C	1977	370	−280	77	Canada
Shin-Shinano	D	1977	300	2 × 125	B-B	Japan
Shin-Shinano	D	1992	600	3 × 125	B-B	Japan
Square Butte	C	1977	500	±250	749	U.S.A.
David A. Hamil	C	1977	100	50	B-B	U.S.A.
Cahora Bassa	J	1978	1920	±533	1360	Mozambique-S. Africa
Nelson River II	J	1978	900	±250	930	Canada
Nelson River II	J	1985	1800	±500	930	Canada
CU Project	A	1979	1000	±400	710	U.S.A.
Hokkaido–Honshu	E	1979	150	125	168	Japan
		1980	300	250		
		1993	600	±250		
Acaray	G	1981	50	25.6	B-B	Paraguay
Vyborg	F	1981	355	1 × 170(±85)	B-B	Russia (tie w/Finland)
Vyborg	F	1982	710	2 × 170		
Vyborg	F		1065	3 × 170		
Duernrohr	J	1983	550	145	B-B	Austria
Gotland II	A	1983	130	150	100	Sweden
Gotland III	A	1987	260	±150	103	Sweden
Eddy County	C	1983	200	82	B-B	U.S.A.
Chateauguay	J	1984	1000	2 × 140	B-B	Canada
Oklaunion	C	1984	200	82	B-B	U.S.A.
Itaipu	A	1984	1575	±300	785	Brazil
Itaipu	A	1985	2383		785	Brazil
Itaipu	A	1986	3150	±600	785	Brazil
Inga-Shaba	A	1982	560	±500	1700	DR Congo
Pacific Intertie upgrade	A	1984	2000	±500	1362	U.S.A.
Blackwater	B	1985	200	57	B-B	U.S.A.
Highgate	A	1985	200	±56	B-B	U.S.A.
Madawaska	C	1985	350	140	B-B	Canada
Miles City	C	1985	200	±82	B-B	U.S.A.
Broken Hill	A	1986	40	2 × 17(±8.33)	B-B	Australia
Intermountain power project	A	1986	1920	±500	784	U.S.A.
Cross-channel:						
(Les Mandarins)	H	1986	1000	±270	72	France–England
(Sellindge)	I	1986	2000	2 × ±270		
Des Cantons-Comerford	C	1986	690	±450	172	Canada-U.S.A.
Sacoi##	H	1986	200	200	415	Corsica Island, Italy
Sacoi###	H	1992	300			

TABLE 30.1—Contd

HVDC link	Supplier	Year	Power (MW)	DC Voltage (kV)	Length (km)	Location
Itaipu II	A	1987	3150	±600	805	Brazil
Sidney (Virginia Smith)	G	1988	200	55.5	B-B	U.S.A.
Gezhouba–Shanghai	B+G	1989	600	500	1000	China
		1990	1200	±500		
Konti-Skan II	A	1988	300	285	150	Sweden–Denmark
Vindhyachal	A	1989	500	2 × 69.7	B-B	India
Pacific Intertie Exp.	B	1989	1100	±500	1362	U.S.A.
McNeill	I	1989	150	42	B-B	Canada
Fenno-Skan	A	1989	500	400	200	Finland–Sweden
Sileru–Barsoor	K	1989	100	+100	196	India
			200	+200		
			400	±200		
Rihand-Delhi	A	1991		1500	± 500	India
Quebec-New England	A	1990	2000****	±450	1500	Canada-U.S.A.
Nicolet Tap	A	1992	1800			Canada
DC Hybrid Link	AB	1992	992	+270/−350	617	New Zealand
Etzenricht	G	1993	600	160	B-B	Germany (tie w/Czech)
Vienna-South east	G	1993	600	145	B-B	Austria (tie w/Hungary)
Haenam–Cheju	I	1993	300	±180	100	South Korea
Baltic Cable Project	AB	1994	600	450		Sweden-Germany
Welch–Monticello	AB	1995	600	450	B-B	U.S.A.
Kontek Interconnection	G	1995	600	400	170	Denmark–Germany
Scotland–N. Ireland	G	1996	250	250		United Kingdom
Chandrapur–Ramagundum	I	1996	1000	2 × 500	B-B	India
Chandrapur–Padghe	AB	1997	1500	±500	900	India
Greece–Italy	AB	1997	500	400	200, sea	Italy
Gazuwaka–Jeypore	AB	1997	500		B-B	India
Leyte-Luzun	AB	1997	1600	400	440	Philippines
Cahora Bassa	G	1998	1920	±533	1456	Mozambique–S.Africa
TSQ-Beijao	G	2000	1800	±500	903	China
Thailand–Malaysia	G	2001	300	±300	B-B	Thailand–Malaysia
Moyle	G	2001	2 × 250	250	64	Ireland–Scotland
East-South Intercon.	G	2003	2000	±500	1450	India
Rapid City DC tie	AB	2003	2 × 100	±13	B-B	S.Dakota, U.S.A.
Three Gorges-Changzhou	AB	2003	3000	±500	890	China
Three Gorges-Quangdong	AB	2004	3000	±500	940	China
Guizhou-Guangdong	G	2004	3000	±500	940	China
Celilo Conv. station	AB	2004	2000	±500	upgrade	U.S.A.
Nelson River Bipole II	G	2004	1000	450	Pole 1	Canada
Basslink	G	2005	500	400	350	Australia–Tasmania
Lamar	G	2005	210	63.6	B-B	Colorado, U.S.A.
Vizag II	AB	2005	500	±88	B-B	India
Estlink	AB	2006	350	±150	HVDC Light	Estonia – Finland
Three Gorges-Shanghai	AB	2007	3000	±500	1059	China
NorNed	AB	2007	700	±450	560	Norway – Netherlands
Valhall offshore	AB	2009	950		HVDC Light	Norway

A – ASEA; B – Brown Boveri; C – General Electric; D – Toshiba; E – Hitachi; F – Russian; G – Siemens; H – CGEE Alsthom; I – GEC (Formerly English Electric); J – HVDC Working Group. (AEG, BBC, Gmens); K – (Independent); AB – ABB (ASEA Brown Boveri); JV – Joint Venture (GE and ASEA); *two valve groups replaced with thyristors in 1977; **two valve groups in Pole 1 replaced with thyristors by GEC in 1991; ***Back-to-back HVDC System; ****Multiterminal system. Largest terminal is rated 2250 MW; #Retired from service; ##50 MW thyristor tap; ###Uprated w/thyristor valves.

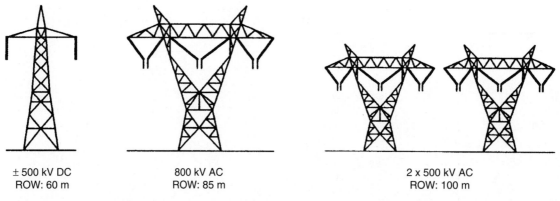

FIGURE 30.1 Comparison of ROW for ac and dc transmission systems.

and reduced conductor and insulator costs. As an example, Fig. 30.1 shows the comparative case of ac and dc systems carrying 2000 MW.

With the dc option, since there are only two conductors (with the same current capacity of three ac conductors), the power transmission losses are also reduced to about two-thirds of the comparable ac system. The absence of skin effect with dc is also beneficial in reducing power losses marginally, and the dielectric losses, in case of power cables is also very much less for dc transmission.

Corona effects tend to be less significant on dc than for ac conductors. The other factors that influence line costs are the costs of compensation and terminal equipment. DC lines do not require reactive power compensation but the terminal equipment costs are increased due to the presence of converters and filters.

Figure 30.2 shows the variation of infrastructure costs with distance for ac and dc transmission. AC tends to be more economical than dc for distances less than the "breakeven distance" but is more expensive for longer distances. This is due to a combination of the terminal equipment costs and line costs for the two types of transmission. The breakeven distances can vary from about 500 to 800 km in overhead lines depending on the per unit line costs. With a cable system, this breakeven distance approaches 50 km.

30.1.1.2 Evaluation of Technical Considerations

Due to its fast controllability, a dc transmission system has full control over transmitted power, an ability to enhance transient and dynamic stability in associated ac networks and can limit fault currents in the dc lines. Furthermore, dc transmission overcomes some of the following problems associated with ac transmission:

Stability limits

The power transfer in an ac line is dependent on the angular difference between the voltage phasors at the two line ends. For a given power transfer level, this angle increases with distance. The maximum power transfer is limited by the considerations of steady state and transient stability. The power carrying capability of an ac line is inversely proportional to transmission distance whereas the power carrying ability of dc lines is unaffected by the distance of transmission.

Voltage control

Voltage control in ac lines is complicated by the line charging requirements and voltage drops. The voltage profile in an ac line is relatively flat only for a fixed level of power transfer, corresponding to its surge impedance loading (SIL). The voltage profile varies with the line loading. For constant voltage at the line ends, the midpoint voltage is reduced for line loadings higher than SIL and increased for loadings lesser than SIL.

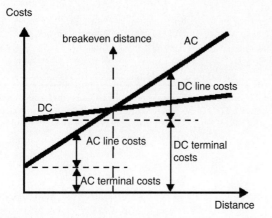

FIGURE 30.2 Comparison of ac and dc transmission system costs.

The maintenance of constant voltage at the two ends requires reactive power control as the line loading is increased. The reactive power requirements increase with line length.

Although dc converter stations require reactive power related to the power transmitted, the dc line itself does not require any reactive power.

The steady-state charging currents in ac cables pose serious problems and make the breakeven distance for cable transmission around 50 km.

Line compensation

Line compensation is necessary for long distance ac transmission to overcome the problems of line charging and stability limitations. The increase in power transfer and voltage control is possible through the use of shunt inductors, series capacitors, static var compensators (SVCs), and lately, the new generation static compensators (STATCOMs).

In the case of dc lines, such compensation is not needed.

Problems of ac interconnection

The interconnection of two power systems through ac ties requires the automatic generation controllers of both systems to be coordinated using tie line power and frequency signals. Even with coordinated control of interconnected systems, the operation of ac ties can be problematic due to (a) the presence of large power oscillations which can lead to frequent tripping, (b) increase in fault level, and (c) transmission of disturbances from one system to the other.

The fast controllability of power flow in dc lines eliminates all of the above problems. Furthermore, the asynchronous interconnection of two power systems can only be achieved with the use of dc links.

Ground impedance

In ac transmission, the existence of ground (zero sequence) current cannot be permitted in steady state due to the high magnitude of ground impedance that will not only affect efficient power transfer, but also result in telephonic interference.

The ground impedance is negligible for dc currents and a dc link can operate using one conductor with ground return (monopolar operation). The ground return is objectionable only when buried metallic structures (such as pipelines) are present and are subject to corrosion with dc current flow. It is to be noted that even while operating in the monopolar mode, the ac network, feeding the dc converter station operates with balanced voltages and currents. Hence, single-pole operation of dc transmission systems is possible for extended periods, while in ac transmission single-phase operation (or any) unbalanced operation is not feasible for more than a second.

Problems of dc transmission

The application of dc transmission is limited by factors such as:

1. High cost of conversion equipment.
2. Inability to use transformers to alter voltage levels.
3. Generation of harmonics.
4. Requirement of reactive power.
5. Complexity of controls.

Over the years, there have been significant advances in dc technology, which have tried to overcome the disadvantages listed above, except for item 2. These are:

1. Increase in the ratings of a thyristor cell that makes up a valve.
2. Modular construction of thyristor valves.
3. Twelve-pulse operation of converters.
4. Use of force-commutation.
5. Application of digital electronics and fiber optics in the control of converters.

Some of the above advances have resulted in improving the reliability and reduction of conversion costs in dc systems.

30.1.2 Evaluation of Reliability and Availability Costs

Statistics on the reliability of HVDC links are maintained by CIGRE and IEEE Working Groups. The reliability of dc links has been very good and is comparable with ac systems. The availability of dc links is quoted in the upper 90%.

30.1.3 Applications of DC Transmission

Due to their costs and special nature, most applications of dc transmission generally fall into one of the following four categories:

Underground or underwater cables

In the case of long-cable connections over the breakeven distance of about 40–50 km, the dc cable transmission system has a marked advantage over the ac cable connections. Examples of this type of applications were the Gotland (1954) and Sardinia (1967) schemes.

The recent development of voltage source converters (VSCs) and the use of rugged polymer dc cables, with the so-called "HVDC Light" option, are being increasingly considered. An example of this type of application is the 180 MW Directlink connection (2000) in Australia.

Long distance bulk power transmission

Bulk power transmission over long distances is an application ideally suited for dc transmission and is more economical than ac transmission whenever the breakeven distance is exceeded.

Examples of this type of application abound from the earlier Pacific Intertie to the recent links in China and India.

The breakeven distance is being effectively decreased with the reduced costs of new compact converter stations possible due to the recent advances in power electronics (discussed in a later section).

Asynchronous interconnection of ac systems

In terms of an asynchronous interconnection between two ac systems, the dc option reigns supreme. There are many instances of BB connections where two ac networks have been tied together for the overall advantage to both ac systems. With recent advances in control techniques, these interconnections are being increasingly made at weak ac systems. The growth of BB interconnections is best illustrated with the example of North America where the four main independent power systems are interconnected with 12 BB links.

In the future, it is anticipated that these BB connection will also be made with VSCs offering the possibility of full four-quadrant operation and the total control of active/reactive power coupled with the minimal generation of harmonics.

Stabilization of power flows in integrated power system

In large interconnected systems, power flow in ac ties (particularly under disturbance conditions) can be uncontrolled and lead to overloading and stability problems, thus endangering system security. Strategically placed dc lines can overcome this problem due to the fast controllability of dc power and provide much needed damping and timely overload capability. The planning of dc transmission in such applications requires detailed study to evaluate the benefits. Examples are the IPP link in the USA and the Chandrapur–Padghe link in India.

Presently the number of dc lines in a power grid is very small compared to the number of ac lines. This indicates that dc transmission is justified only for specific applications. Although advances in technology and introduction of multiterminal dc (MTDC) systems are expected to increase the scope of application of dc transmission, it is unlikely that the ac grid will be replaced by a dc power grid in the future. There are two major reasons for this. First, the control and protection of MTDC systems is complex and the inability of voltage transformation in dc networks imposes economic penalties. Second, the advances in power electronics technology have resulted in the improvement of the performance of ac transmissions using FACTS devices, for instance through introduction of static var systems, static phase shifters, etc.

30.1.4 Types of HVDC Systems

Three types of dc links are considered.

30.1.4.1 Monopolar Link

A monopolar link (Fig. 30.3a) has one conductor and uses either ground- and/or sea-return. A metallic return can also be used where concerns for harmonic interference and/or corrosion exist. In applications with dc cables (i.e. HVDC light), a cable return is used. Since the corona effects in a dc line are substantially less with negative polarity of the conductor as compared to the positive polarity, a monopolar link is normally operated with negative polarity.

30.1.4.2 Bipolar Link

A bipolar link (Fig. 30.3b) has two conductors, one with positive and the other with negative polarity. Each terminal has two sets of converters of equal rating, in series on the dc side. The junction between the two sets of converters is grounded at one or both ends by the use of a short electrode line. Since both poles operate with equal currents under normal operation, there is zero ground current flowing under these conditions. Monopolar operation can also be used in the early stages of the development of a bipolar link. Alternatively, under faulty converter conditions, one dc line may be temporarily used as a metallic return with the use of suitable switching.

30.1.4.3 Homopolar Link

In this type of link (Fig. 30.3c), two conductors having the same polarity (usually negative) can be operated with ground or metallic return.

Due to the undesirability of operating a dc link with ground return, bipolar links are mostly used. A homopolar link has the advantage of reduced insulation costs, but the disadvantages of earth return outweigh the advantages.

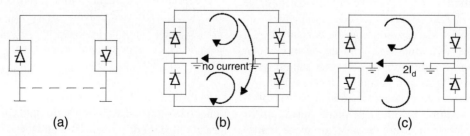

FIGURE 30.3 Types of HVDC links: (a) monopolar link; (b) bipolar link; and (c) homopolar link.

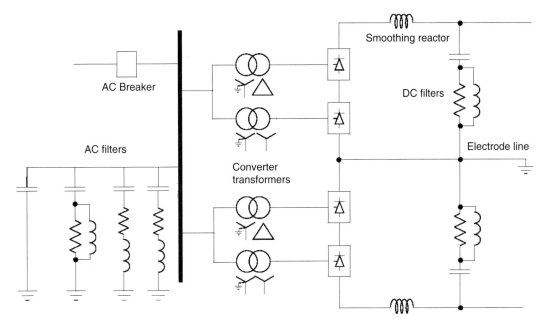

FIGURE 30.4 Typical HVDC converter station equipment.

30.2 Main Components of HVDC Converter Station

The major components of a HVDC transmission system are the converter stations at the ends of the transmission system. In a typical two-terminal transmission system, both a rectifier and an inverter are required. The role of the two stations can be reversed, as controls are usually available for both functions at the terminals. The major components of a typical 12-pulse bipolar HVDC converter station (Fig. 30.4) are discussed below.

30.2.1 Converter Unit

This usually consists of two three-phase converter bridges connected in series to form a 12-pulse converter unit. The design of valves is based on a modular concept where each module contains a limited number of series-connected thyristor levels. The valves can be packaged either as a single-valve, double-valve, or quadruple-valve arrangement. The converter is fed by two converter transformers, connected in star/star and star/delta arrangement, to form a 12-pulse pair. The valves may be cooled by air, oil, water, or freon. However, the cooling using deionized water is more modern and considered efficient and reliable. The ratings of a valve group are limited more by the permissible short-circuit currents than by the steady-state load requirements.

Valve firing signals are generated in the converter control at ground potential and are transmitted to each thyristor in the valve through a fiber-optic light guide system. The light signal received at the thyristor level is converted to an electrical signal using gate drive amplifiers with pulse transformers. Recent trends in the industry indicate that direct optical firing of the valves with light triggered thyristors (LTTs) is also feasible.

The valves are protected using snubber circuits, protective firing, and gapless surge arresters.

30.2.1.1 Thyristor Valves

Many individual thyristors are connected in series to build up an HVDC valve. To distribute the off-state valve voltage uniformly across each thyristor level and protect the valve from di/dt and dv/dt stresses, special snubber circuits are used across each thyristor level (Fig. 30.5).

The snubber circuit is composed of the following components:

- A saturating reactor is used to protect the valve from di/dt stresses during turn-on. The saturating reactor offers a high inductance at low current and a low inductance at high current.
- A dc grading resistor, RG distributes the direct voltage across the different thyristor levels. It is also used as a voltage divider to measure the thyristor level voltage.
- The RC snubber circuits are used to damp out voltage oscillations from power frequency to a few kilohertz.
- A capacitive grading circuit, CFG is used to protect the thyristor level from voltage oscillations at a much higher frequency.

A thyristor is triggered ON by a firing pulse sent via a fiber-optic cable from the valve base electronics (VBE) unit at earth potential. The fiber-optic signal is amplified by a gate

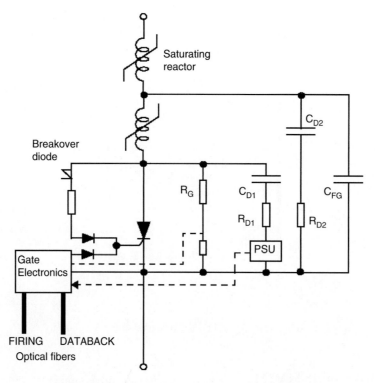

FIGURE 30.5 Electrical circuit of the thyristor level [2].

electronic unit (GEU), which receives its power from the RC snubber circuit during the valve's OFF period. The GEU can also affect the protective firing of the thyristor independent of the central control unit. This is achieved by a breakover diode (BOD) via a current limiting resistor that triggers the thyristor when the forward voltage threatens to exceed the rated voltage for the thyristor. This may arise in a case when some thyristors may block forward voltage while others may not.

It is normal to include some extra redundant thyristor levels to allow the valve to remain in service after the failure of some thyristors. A metal oxide surge arrestor is also used across each valve for overvoltage protection.

The thyristors produce considerable heat loss, typically 24–40 W/cm^2 (or over 1 MW for a typical quadruple valve), and so an efficient cooling system is essential.

30.2.2 Converter Transformer

The converter transformer (Fig. 30.6) can have different configurations: (a) three phase, two-winding, (b) single-phase, three-winding, and (c) single-phase, two-winding. The valve-side windings are connected in star and delta with neutral point ungrounded. On the ac side, the transformers are connected in parallel with the neutral grounded. The leakage impedance of the transformer (typical value between 15 and 18%) is chosen to limit the short-circuit currents through any valve.

FIGURE 30.6 Spare converter transformer in the switchyard of an HVDC station.

The converter transformers are designed to withstand dc voltage stresses and increased eddy current losses due to harmonic currents. One problem that can arise is due to the dc magnetization of the core due to unsymmetrical firing of valves.

30.2.3 Filters

Due to the generation of characteristic and non-characteristic harmonics by the converter, it is necessary to provide suitable filters on the ac–dc sides of the converter to improve the power quality and meet telephonic and other requirements. Generally, three types of filters are used for this purpose.

30.2.3.1 AC Filters

AC Filters (Fig. 30.7) are passive circuits used to provide low impedance, shunt paths for ac harmonic currents. Both tuned and damped filter arrangements are used. In a typical 12-pulse station, filters at 11th, 13th harmonics are required as tuned filters. Damped filters (normally tuned to the 23rd harmonic) are required for the higher harmonics. In recent years, C-type filters have also been used since they provide more economic designs. Double- or even triple-tuned filters exist to reduce the cost of the filter (see Fig. 30.23).

The availability of cost-effective active ac filters will change the scenario in the future.

30.2.3.2 DC Filters

These are similar to ac filters and are used for the filtering of dc harmonics. Usually a damped filter at the 24th harmonic is utilized. Modern practice is to use active dc filters (see also the application example system presented later). Active dc filters are increasingly being used for efficiency and space saving purposes.

30.2.3.3 High Frequency (RF/PLC) Filters

These are connected between the converter transformer and the station ac bus to suppress any high-frequency currents. Sometimes such filters are provided on the high-voltage dc bus connected between the dc filter and dc line and also on the neutral side.

30.2.4 Reactive Power Source

Converter stations consume reactive power that is dependent on the active power loading (typically about 50–60% of the active power). The ac filters provide a part of this reactive power requirement. In addition, shunt (switched) capacitors and static var systems are also used.

30.2.5 DC Smoothing Reactor

A sufficiently large series reactor is used on the dc side of the converter to smooth the dc current and for the converter protection from line surges. The reactor (Fig. 30.8) is usually designed as a linear reactor and may be connected on the line side, neutral side, or at an intermediate location. Typical values of the smoothing reactor are in the range 240–600 mH for long distance transmission and about 24 mH for a BB connection.

30.2.6 DC Switchgear

This is usually a modified ac equipment and used to interrupt only small dc currents (i.e. employed as disconnecting switches). DC breakers or metallic return transfer breakers (MRTB) are used, if required, for the interruption of rated load currents.

In addition to the equipment described above, ac switchgear and associated equipment for protection and measurement are also part of the converter station.

FIGURE 30.7 Installation of an ac filter in the switchyard.

FIGURE 30.8 Installation of an air-cooled smoothing reactor.

30.2.7 DC Cables

Contrary to the use of ac cables for transmission, dc cables do not have a requirement for continuous charging current. Hence the length limit of about 50 km does not apply. Moreover, dc voltage gives less aging and hence a longer lifetime for the cable. The new design of HVDC light cables from ABB are based on extruded polymeric insulating material instead of classic paper-oil insulation that has a tendency to leak. Due to their rugged mechanical design, flexibility, and low weight, polymer cables can be installed in underground cheaply with a plowing technique, or in submarine applications, it can be laid in very deep waters and on rough sea-bottoms. Since dc cables are operated in bipolar mode, one cable with positive polarity and one cable with negative polarity, very limited magnetic fields result from the transmission. HVDC light cables have successfully achieved operation at a stress of 20 kV/mm.

30.3 Analysis of Converter Bridge

To consider the theoretical analysis of a conventional 6-pulse bridge (Fig. 30.9), the following assumptions are made:

- DC current I_d is considered constant.
- Valves are ideal switches.
- AC system is strong (infinite).

Due to the leakage impedance of the converter transformer, commutation from one valve to the next is not instantaneous. An overlap period is necessary and, depending on the leakage, either two, three, or four valves may conduct at any time. In the general case, with a typical value of converter transformer leakage impedance of about 13–18%, either two or three valves conduct at any one time.

The analysis of the bridge gives the following dc output voltages:

For a rectifier:

$$V_{dr} = V_{dor} \cdot \cos\alpha - R_{cr} \cdot I_d \quad (30.1)$$

where

$$V_{dor} = \frac{3}{\pi} \cdot \sqrt{2} \cdot V_{LL}$$

$$R_{cr} = \frac{3}{\pi} \cdot \omega L_{cr}$$

where

$$\omega = 2 \cdot \pi \cdot f$$

where "f" is the power frequency.

For an inverter:

There are two options possible depending on choice of the delay angle or extinction angle as the control variable

$$-V_{di} = V_{doi} \cdot \cos\beta - R_{ci} \cdot I_d \quad (30.2)$$

$$-V_{di} = V_{doi} \cdot \cos\gamma - R_{ci} \cdot I_d \quad (30.3)$$

where

V_{dr} and V_{di} – dc voltage at the rectifier and inverter respectively.

V_{dor} and V_{doi} – open circuit dc voltage at the rectifier and inverter respectively.

R_{cr} and R_{ci} – equivalent commutation resistance at the rectifier and inverter respectively.

L_{cr} and L_{ci} – leakage inductance of converter transformer at rectifier and inverter respectively.

I_d – dc current.

α – delay angle.

β – advance angle at the inverter, $(\beta = \pi - \alpha)$.

γ – extinction angle at the inverter, $\gamma = \pi - \alpha - \mu$.

30.4 Controls and Protection

In a typical two-terminal dc link connecting two ac systems (Fig. 30.10), the primary functions of the dc controls are to:

- Control the power flow between the terminals.
- Protect the equipment against the current/voltage stresses caused by faults.

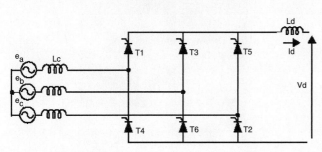

FIGURE 30.9 Six-pulse bridge circuit.

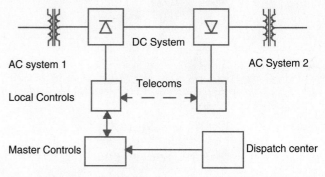

FIGURE 30.10 Typical HVDC system linking two ac systems.

- Stabilize the attached ac systems against any operational mode of the dc link.

The dc terminals each have their own local controllers. A centralized dispatch center will communicate a power order to one of the terminals that will act as a master controller and has the responsibility to coordinate the control functions of the dc link. Besides the primary functions, it is desirable that the dc controls have the following features:

- Limit the maximum dc current: Due to a limited thermal inertia of the thyristor valves to sustain overcurrents, the maximum dc current is usually limited to less than 1.2 pu for a limited period of time.
- Maintain a maximum dc voltage for transmission: This reduces the transmission losses, and permits optimization of the valve rating and insulation.
- Minimize reactive power consumption: This implies that the converters must operate at a low firing angle. A typical converter will consume reactive power between 50 and 60% of its MW rating. This amount of reactive power supply can cost about 15% of the station cost, and comprise about 10% of the power loss.

The desired features of the dc controls are indicated below:

1. Limit maximum dc current: Since the thermal inertia of the converter valves is quite low, it is desirable to limit the dc current to prevent failure in the valves.
2. Maintain maximum dc voltage for transmission purposes to minimize losses in the dc line and converter valves.
3. Keep the ac reactive power demand low at either converter terminal: This implies that the operating angles at the converters must be kept low. Additional benefits of doing this are to reduce the snubber losses in the valves and reduce the generation of harmonics.
4. Prevent commutation failures at the inverter station and hence improve the stability of power transmission.
5. Other features, i.e. the control of frequency in an isolated ac system or to enhance power system stability.

In addition to the above desired features, the dc controls will have to cope with the steady-state and dynamic requirements of the dc link, as shown in Table 30.2.

TABLE 30.2 Requirements of the dc link

Steady-state requirements	Dynamic requirements
Limit the generation of non-characteristic harmonics	Step changes in dc current or power flow
Maintain the accuracy of the controlled variable, i.e. dc current and/or constant extinction angle	Start-up and fault induced transients
Cope with the normal variations in the ac system impedances due to topology changes	Reversal of power flow
	Variation in frequency of attached ac system

30.4.1 Basics of Control for a Two-terminal DC Link

From converter theory, the relationship between the dc voltage V_d and dc current I_d is given by Eqs. (30.1)–(30.3). These three characteristics represent straight lines on the V_d–I_d plane. Notice that Eq. (30.2), i.e. beta characteristic, has a positive slope while the Eq. (30.3), i.e. gamma characteristic, has a negative slope. The choice of the control strategy for a typical two-terminal dc link is made according to the conditions in the Table 30.3.

Condition 1 implies the use of the rectifier in constant current control mode and condition 3 implies the use of the inverter in constant extinction angle (CEA) control mode. Other control modes may be used to enhance the power transmission during contingency conditions depending upon applications. This control strategy is illustrated in Fig. 30.11.

The rectifier characteristic is composed of two control modes: alpha-min (line AB) and constant current (line BC). The alpha-min mode of control at the rectifier is imposed by the natural characteristics of the rectifier ac system, and the ability of the valves to operate when alpha is equal to zero, i.e. in the limit the rectifier acts a diode rectifier. However, since a minimum positive voltage is desired before firing of the valves to ensure conduction, an alpha-min limit of about 2–5° is typically imposed.

TABLE 30.3 Choice of control strategy for two-terminal dc link

Condition no.	Desirable features	Reason	Control implementation
1	Limit the maximum dc current, I_d	For the protection of valves	Use constant current control at the rectifier
2	Employ the maximum dc voltage, V_d	For reducing power transmission losses	Use constant voltage control at the inverter
3	Reduce the incidence of commutation failures	For stability purposes	Use minimum extinction angle control at inverter
4	Reduce reactive power consumption at the converters	For voltage regulation and economic reasons	Use minimum firing angles

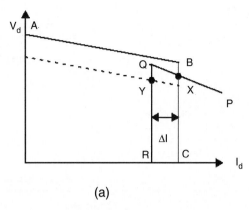
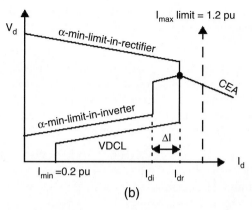

FIGURE 30.11 Static V_d–I_d characteristic for a two-terminal link: (a) unmodified and (b) modified.

The inverter characteristic is composed of two modes: gamma-min (line PQ) and constant current (line QR). The crossover point X of the two characteristics defines the operating point for the dc link. In addition, a constant current characteristic is also used at the inverter. However, the current demanded by the inverter I_{di} is usually less than the current demanded by the rectifier I_{dr} by the current margin ΔI which is typically about 0.1 pu; its magnitude is selected to be large enough so that the rectifier and inverter constant current modes do not interact due to any current harmonics which may be superimposed on the dc current. This control strategy is termed as the current margin method.

The advantage of this control strategy becomes evident if there is a voltage decrease at the rectifier ac bus. The operating point then moves to point Y. In this way, the current transmitted will be reduced to 0.9 pu of its previous value and voltage control will shift to the rectifier. However, the power transmission will be largely maintained near to 90% of its original value.

The control strategy usually employs the following other modifications to improve the behavior during system disturbances:

At the rectifier:
1. **Voltage dependent current limit, VDCL**
    This modification is made to limit the dc current as a function of either the dc voltage or, in some cases, the ac voltage. This modification assists the dc link to recover from faults. Variants of this type of VDCL do exist. In one variant, the modification is a simple fixed value instead of a sloped line.
2. **I_d-min limit**
    This limitation (typically 0.2–0.3 pu) is to ensure a minimum dc current to avoid the possibility of dc current extinction caused by the valve current dropping below the hold-on current of the thyristors; an eventuality that could arise transiently due to harmonics superimposed on the low value of the dc current. The resultant current chopping would cause high overvoltages to appear on the valves. The magnitude of I_d-min is affected by the size of the smoothing reactor employed.

At the inverter:
1. **Alpha-min limit at inverter**
    The inverter is usually not permitted to operate inadvertently in the rectifier region, i.e. a power reversal occurring due to an inadvertent current margin sign change. To ensure this, an alpha-min-limit in inverter mode of about 100–110° is imposed.
2. **Current error region**
    When the inverter operates into a weak ac system, the slope of the CEA control mode characteristic is quite steep and may cause multiple crossover points with the rectifier characteristic. To avoid this possibility, the inverter CEA characteristic is usually modified into either a constant beta characteristic or constant voltage characteristic within the current error region.

30.4.2 Control Implementation

30.4.2.1 Historical Background

The equidistant pulse firing control systems used in modern HVDC control systems were developed in the mid-1960s [5, 6]; although improvements have occurred in their implementation since then, such as the use of microprocessor based equipment, their fundamental philosophy has not changed much. The control techniques described in [5, 6] are of the pulse frequency control (PFC) type as opposed to the now-out-of-favor pulse phase control (PPC) type. All these controls use an independent voltage controlled oscillator (VCO) to decouple the direct coupling between the firing pulses and the commutation voltage, V_{com}. This decoupling was necessary to eliminate the possibility of harmonic instability detected in the converter operation when the ac system capacity became nearer to the power transmission capacity of the HVDC link, i.e. with the use of weak ac systems. Another advantage of the equidistant firing pulse control was the elimination of non-characteristic harmonics during steady-state operation. This was a prevalent feature during the use of the earlier individual phase control (IPC) system where the firing pulses were directly coupled to the commutation voltage, V_{com}.

30.4.2.2 Firing Angle Control

To control the firing angle of a converter, it is necessary to synchronize the firing pulses emanating from the ring counter to the ac commutation voltage that has a frequency of 60 Hz in steady state. However, it was noted quite early on (early 1960s), that the commutation voltage (system) frequency is not a constant, neither in frequency nor in amplitude, during a perturbed state. However, it is the frequency that is of primary concern for the synchronization of firing pulses. For strong ac systems, the frequency is relatively constant and distortion free to be acceptable for most converter type applications. But, as converter connections to weak ac systems became required more often than not, it was necessary to devise a scheme for synchronization purposes which would be decoupled from the commutation voltage frequency for durations when there were perturbations occurring on the ac system. The most obvious method is to utilize an independent oscillator at 60 Hz that can be synchronously locked to the ac commutation voltage frequency. This oscillator would then provide the (phase) reference for the generation of firing pulses to the ring counter during the perturbation periods, and would use the steady-state periods for locking in step with the system frequency. The advantage of this independent oscillator would be to provide an ideal (immunized and clean) sinusoid for synchronizing and timing purposes. There are two possibilities for this independent oscillator:

- Fixed frequency operation.
- Variable frequency operation.

Use of a fixed frequency oscillator (although feasible, and called the PPC oscillator) is not recommended, since it is known that the system frequency does drift, between 55 and 65 Hz, due to the rotating machines used to generate electricity. Therefore, it is preferable to use a variable frequency oscillator (called the PFC oscillator) with a locking range of between 50 and 70 Hz and the center frequency of 60 Hz. This oscillator would then need to track the variations in the system frequency and a control loop of some sort would be used for this tracking feature; this control loop would have its own gain and time parameters for steady-state accuracy and dynamic performance requirements.

The control loop for frequency tracking purposes would also need to consider the mode of operation for the dc link. The method widely adopted for dc link operation is the so-called current margin method.

30.4.3 Control Loops

Control loops are required to track the following variables:

- Ordered current I_{or} at the rectifier and the inverter.
- Ordered extinction angle (γ_o) at the inverter.

30.4.3.1 Current Control Loops

In conventional HVDC systems, a proportional integral (PI) regulator is used (Fig. 30.12) for the rectifier current controller. The rectifier plant system is inherently non-linear and has a relationship given in Eq. (30.1). For constant I_d and for small changes in α, we have

$$\frac{\Delta V_d}{\Delta \alpha} = -V_{dor} \cdot \sin \alpha \tag{30.4}$$

It is obvious from Eq. (30.4) that the maximum gain ($\Delta V_d/\Delta \alpha$) occurs when $\alpha = 90°$. Thus the control loop must be stabilized for this operation point, resulting in slower dynamic properties at normal operation within the range 12–18°. Attempts have been made to linearize this gain and have met with some limited success. However, in practical terms, it is not always possible to have the dc link operating with the rectifier at 90° due to harmonic generation and other protection elements coming into operation also. Therefore, optimizing the gains of the PI regulator can be quite arduous and take a long time. For this reason, the controllers are

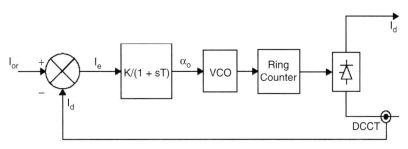

FIGURE 30.12 Control loop for the rectifier.

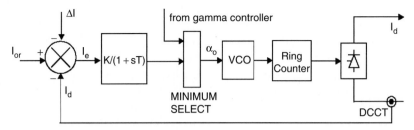

FIGURE 30.13 Current controller at the inverter.

often pre-tested in a physical simulator environment to obtain approximate settings. Final (often very limited) adjustments are then made on site.

Other problems with the use of a PI regulator are listed below:

- It is mostly used with fixed gains, although some possibility for gain scheduling exists.
- It is difficult to select optimal gains, and even then they are optimal over a limited range only.
- Since the plant system is varying continually, the PI controller is not optimal.

A similar current control loop is used at the inverter (Fig. 30.13). Since the inverter also has a gamma controller, the selection between these two controllers is made via a MINIMUM SELECT block. Moreover, in order to bias the inverter current controller off, a current margin signal ΔI is subtracted from the current reference I_{or} received from the rectifier via a communication link.

Telecommunication requirements

As was discussed above, the rectifier and inverter current orders must be coordinated to maintain a current margin of about 10% between the two terminals at all times, otherwise there is a risk of loss of margin and the dc voltage could run down. Although, it is possible to use slow voice communication between the two terminals, and maintain this margin, the advantage of fast control action possible with converters may be lost for protection purposes. For maintaining the margin during dynamic conditions, it is prudent to raise the current order at the rectifier first followed by the inverter; in terms of reducing the current order, it is necessary to reduce at the inverter first and then at the rectifier.

30.4.3.2 Gamma Control Loop

At the inverter end, there are two known methods for the gamma control loop. The two variants are different only due to the method of determining the extinction angle:

- Predictive method for the indication of extinction angle (gamma).
- Direct method for actual measurement of extinction angle (gamma).

In either case, a delay of one cycle occurs from the indication of actual gamma and the reaction of the controller to this measurement. Since the avoidance of a commutation failure often takes precedence at the inverter, it is normal to use the minimum value of gamma measured for the 6- or 12-inverter valves for the converter(s).

1. Predictive method of measuring gamma

The predictive measurement tries to maintain the commutation voltage–time area after commutation larger than a specified minimum value. Since the gamma prediction is only approximate, the method is corrected by a slow feedback loop that calculates the error between the predicted value and actual value of gamma (one cycle later) and feeds it back.

The predictor calculates continuously, by a triangular approximation, the total available voltage–time area that remains after commutation is finished. Since an estimate of the overlap angle m is necessary, it is derived from a well-known fact in converter theory that the overlap commutation voltage–time area is directly proportional to direct current and the leakage impedance (assumed constant and known) value of the converter transformer.

The prediction process is inherently of an individual phase firing character. If no further measures were taken, each valve would fire on the minimum margin condition. To counteract this undesired property, a special firing symmetrizer is used; when one valve has fired on the minimum margin angle, the following two or five valves fire equidistantly. (The choice of either two or five symmetrized valves is mainly a stability question.)

2. Direct method of measuring gamma

In this method, the gamma measurement is derived from a measurement of the actual valve voltage. Waveforms of the gamma measuring circuit are shown in the Fig. 30.14. An internal timing waveform, consisting of a ramp function of fixed slope, is generated after being initiated from the instant of zero anode current. This value corresponds to a direct voltage proportional to the last value of gamma. From the gamma values of all (either 6 or 12) valves, the smallest value is selected to

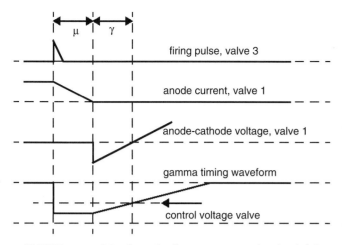

FIGURE 30.14 Waveforms for the gamma measuring circuit [5].

produce the indication of measured gamma for use with the feedback regulator.

This value is compared to a gamma-ref value and a PI regulator defines the dynamic properties for the controller (Fig. 30.15). Inherently, this method has an individual phase control characteristic. One version of this type of control implementation overcomes this problem by using a symmetrizer for generating equidistant firing pulses. The 12-pulse circuit generates 12 gamma measurements; the minimum gamma value is selected and then used to derive the control voltage for the firing pulse generator with symmetrical pulses.

30.4.4 Hierarchy of DC Controls

Since HVDC controls are hierarchical in nature, they can be subdivided into blocks that follow the major modules of a converter station (Fig. 30.16). The main control blocks are:

1. The bipole (master) controller is usually located at one end of the dc link, and receives its power order from a centralized system dispatch center. The bipole controller derives a current order for the pole controller using a local measurement of either ac or dc voltage. Other inputs, i.e. frequency measurement, may also be used by the bipole controller for damping or modulation purposes. A communication to the remote terminal of the dc link is also necessary to coordinate the current references to the link.

2. The pole controller then derives an alpha order for the next level. This alpha order is sent to both the positive and negative poles of the bipole.

3. The valve group controllers generate the firing pulses for the converter valves. Controls also receive measurements of dc current, dc voltage, and ac current into

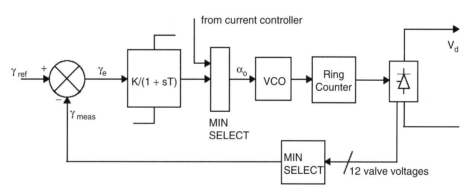

FIGURE 30.15 Gamma feedback controller.

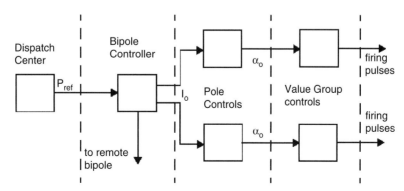

FIGURE 30.16 Hierarchy of controllers.

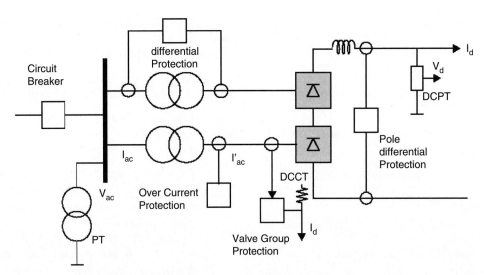

FIGURE 30.17 Monitoring points for the protection circuits.

the converter transformer. These measurements assist in the rapid alteration of firing angle for protection of the valves during perturbations. A slow loop for control of tap changer position as a function of alpha is also available at this level.

30.4.5 Monitoring of Signals

As described earlier, monitoring of the following signals is necessary for the controls (Fig. 30.17) to perform their functions and assist in the protection of the converter equipment:

V_d, I_d – dc voltage and dc current respectively.
I_{ac}, I'_{ac} – ac current on line side and converter side of the converter transformer respectively.
V_{ac} – ac voltage at the ac filter bus.

30.4.6 Protection against Overcurrents

Faults and disturbances can be caused by malfunctioning equipment or insulation failures due to lightening and pollution. First, these faults need to be detected with the help of monitored signals. Second, the equipment must be protected by control or switching actions. Since dc controls can react within one cycle, control action is used to protect equipment against overcurrent and overvoltage stresses, and minimize loss of transmission. In a converter station, the valves are the most critical (and most expensive) equipment that needs to be protected rapidly due to their limited thermal inertia.

The basic types of faults that the converter station can experience are:

Current extinction (CE)

Current extinction can occur if the valve current drops below the holding current of the thyristor. This can happen at low-current operation accompanied by a transient leading to current extinction. Due to the phenomena of current chopping of an inductive current, severe overvoltages may result. The size of the smoothing reactor and the rectifier minimum current setting I_{min} helps to minimize the occurrence of CE.

Commutation failure (CF) or misfire

In line-commutated converters, the successful commutation of a valve requires that the extinction angle γ-nominal be maintained more than the minimum value of the extinction angle γ-min. Note that γ-nominal $= 180 - \alpha - \mu$. The overlap angle, μ is a function of the commutation voltage and the dc current. Hence, a decrease in commutation voltage or an increase in dc current can cause an increase in μ, resulting in a decrease in γ-nominal. If γ-nominal $<$ γ-min, a CF may result. In this case, the outgoing valve will continue to conduct current and when the incoming valve is fired in sequence, a short circuit of the bridge will occur.

A missing firing pulse can also lead to a misfire (at a rectifier) or a CF (at an inverter). The effects of a single misfire are similar to those of a single CF. Usually a single CF is self-clearing, and no special control actions are necessary. However, a multiple CF can lead to the injection of ac voltages into the dc system. Control action may be necessary in this case.

The detection of a CF is based on the differential comparison of dc current and the ac currents on the valve side of the converter transformer. During a CF, the two valves in an arm of the bridge are conducting. Therefore, the ac current goes to zero while the dc current continues to flow.

The protection features employed to counteract the impact of a CF are indicated in Table 30.4.

Short circuits – internal or dc line

An internal bridge fault is rare as the valve hall is completely enclosed and is air-conditioned. However, a bushing can fail,

30 HVDC Transmission

TABLE 30.4 Protection against overcurrents

Fault type	Occurrence	Fault current level	Protection method
Internal faults	Infrequent	10 pu	Valve is rated to withstand this surge
DC line faults	Frequent	2–3 pu	– Forced retard of firing angle
			– Dynamic VDCL deployment
			– Trip ac breaker CB after third attempt
Commutation failures (single or multiple)	Very frequent	1.5–2.5 pu	Single CF:
			– Self-clearing
			Multiple CF:
			–Beta angle advanced in stages
			– Static VDCL deployment

or valve-cooling water may leak resulting in a short circuit. The ac breaker may have to be tripped to protect against bridge faults.

The protection features employed to counteract the impact of short circuits are indicated in Table 30.4.

The fast-acting HVDC controls (which operate within one cycle) are used to regulate the dc current for protection of the valves against ac and dc faults.

The basic protection (Fig. 30.18) is provided by the VGP differential protection that compares the rectified current on the valve side of the converter transformer with the dc current measured on the line side of the smoothing reactor. This method is applied because of its selectivity possible due to high impedances in the smoothing reactor and converter transformer.

The over current protection (OCP) is used as a back-up protection in case of malfunction in the VGP. The level of overcurrent setting is set higher than the differential protection.

The pole differential protection (PDP) is used to detect ground faults, including faults in the neutral bus.

30.4.7 Protection against Overvoltages

The typical arrangement of metal oxide surge arrestors for protecting equipment in a converter pole is shown in the

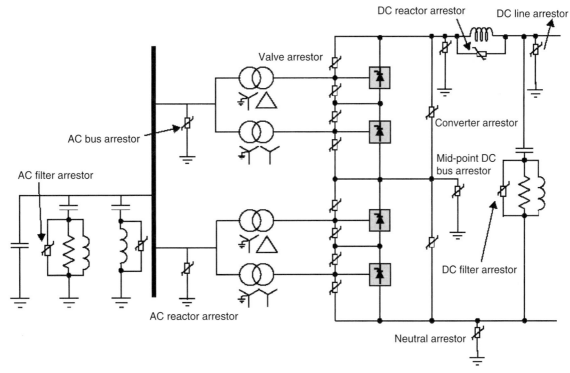

FIGURE 30.18 Typical arrangement of surge arrestors for a converter pole.

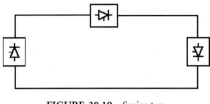

FIGURE 30.19 Series tap.

Fig. 30.18. In general, overvoltages entering from the ac bus are limited by the ac bus arresters; similarly, overvoltages entering the converter from the dc line are limited by the dc arrestor. The ac and dc filters have their respective arrestors also. Critical components such as the valves have their own arrestors placed close to these components. The protective firing of a valve is used as a back-up protection for overvoltages in the forward direction. Owing to their varied duty, these arrestors are rated accordingly for the location used. For instance, the converter arrestor for the upper bridge is subjected to higher energy dissipation than for the lower bridge.

Since the evaluation of insulation coordination is quite complex, detailed studies are often required with dc simulators to design an appropriate insulation coordination strategy.

30.5 MTDC Operation

Most HVDC transmission systems are two-terminal systems. A multiterminal dc system (MTDC) has more than two terminals and there are two existing installations of this type. There are two possible ways of tapping power from an HVDC link, i.e. with series or parallel taps.

30.5.1 Series Tap

A monopolar version of a three-terminal series dc link is shown in Fig. 30.19. The system is grounded at only one suitable location. In a series dc system, the dc current is set by one terminal and is common to all terminals; the other terminals are operated at a constant delay angle for a rectifier and CEA for inverter operation with the help of transformer tap changers.

Power reversal at a station is done by reversing the dc voltage with the aid of angle control.

No practical installation of this type exists in the world at present. From an evaluation of ratings and costs for series taps, it is not practical for the series tap to exceed 20% of the rating for a major terminal in the MTDC system.

30.5.2 Parallel Tap

A monopolar version of a three-terminal parallel dc link is shown in Fig. 30.20. In a parallel MTDC system, the system voltage is common to all terminals. There are two variants possible for a parallel MTDC system: radial or mesh. In a radial system, disconnection of one segment of the system will interrupt power from one or more terminals. In a mesh system, the removal of one segment will not interrupt power flow providing the remaining links, the capability of carrying the required power.

Power reversal in a parallel system will require mechanical switching of the links as the dc voltage cannot be reversed.

From an evaluation of ratings and costs for parallel taps, it is not practical for the parallel tap to be less than 20% of the rating for a major terminal in the MTDC system.

There are two installations of parallel taps existing in the world. The first is the Sardinia–Corsica–Italy link where a 50 MW parallel tap at Corsica is used. Since the principal terminals are rated at 200 MW, a commutation failure at Corsica can result in very high overcurrents (typically 7 pu); for this reason, large smoothing reactors (2.5 H) are used in this link.

The second installation is the Quebec–New Hampshire 2000 MW link where a parallel tap is used at Nicolet. Since the rating of Nicolet is at 1800 MW, the size of the smoothing reactors was kept to a modest size.

30.5.3 Control of MTDC System

Although several control methods exist for controlling MTDC systems, the most widely utilized method is the so-called

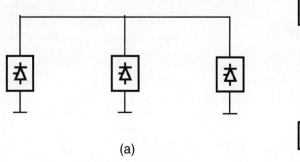
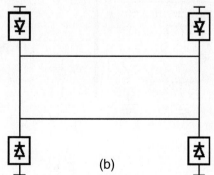

FIGURE 30.20 Parallel tap: (a) radial type and (b) mesh type.

current margin method, which is an extension of the control method, used for two-terminal dc systems.

In this method, the voltage setting terminal (VST) operates at the angle limit (minimum alpha or minimum gamma) while the remaining terminals are controlling their respective currents. The control law that is used sets the current reference at the voltage setting terminal according to

$$\sum I_{jref} = \Delta I \quad (30.5)$$

where ΔI is known as the current margin.

The terminal with the lowest voltage ceiling acts as the VST.

An example of the control strategy is shown in Fig. 30.21 for a three-terminal dc system with one rectifier, REC and two inverters, INV 1 at 40% and INV2 at 60% rating. The REC 1 and INV 2 terminals are maintained in current control, while INV 1 is the VST operating in CEA mode.

Due to the requirement to maintain current margin for the MTDC system at all times, a centralized current controller, known as the current reference balancer (CRB) (Fig. 30.22), is required. With this technique, reliable two-way telecommunication links are required for current reference coordination purposes. The current orders, $I_{ref1} - I_{ref3}$, at the terminals must satisfy the control law according to Eq. (30.5). The weighting factors (K_1, K_2, and K_3) and limits are selected as a function of the relative ratings of the terminals.

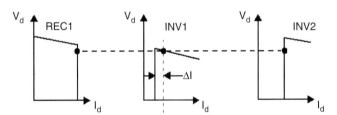

FIGURE 30.21 Current margin method of control for MTDC system.

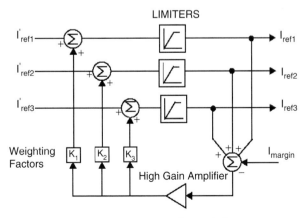

FIGURE 30.22 Current reference balancer.

30.6 Application

30.6.1 HVDC Interconnection at Gurun (Malaysia)

The 240/600 MW HVDC interconnection between Malaysia and Thailand is a major first step in implementing electric power network interconnections in the ASEAN region (Fig. 30.23). It is jointly undertaken by two utilities, Tenaga Nasional Berhad (TNB) of Malaysia and Electricity Generating Authority of Thailand (EGAT). This will be the first HVDC project for both these utilities. The HVDC interconnection consists of a 110 km HVDC line (85 km owned by TNB and 25 km owned by EGAT) with the dc converter stations at Gurun in the Malaysian side and Khlong Ngae in the Thailand side. The scheme entered into commercial operation in 2000. The interconnection provides a range of diverse benefits such as:

- Spinning reserve sharing.
- Economic power exchange – commercial transactions.
- Emergency assistance to either ac system.
- Damping of ac system oscillations.
- Reactive power support (voltage control).
- Deferment of generation plant up.

30.6.1.1 Power Transmission Capacity

The converter station is presently constructed for monopolar operation for power transfer of 240 MW in both directions with provisions for future extensions to a bipole configuration giving a total power transfer capability of 600 MW. The HVDC line is constructed with two pole conductors to cater for the second 240 MW pole. Full-length neutral conductors are used instead of ground electrodes because of high land costs and inherently high values of soil resistivity.

The monopole is rated for a continuous power of 240 MW (240 kV, 1000 A) at the dc terminal of the rectifier station. In addition, there is a 10 minutes overload capability of up to 450 MW, which may be utilized once per day when all redundant cooling equipment is in service (Tables 30.5 and 30.6).

The HVDC interconnection scheme is capable of continuous operation at a reduced dc voltage of 210 kV (70%) over the whole load range up to the rated dc current of 1000 A with all redundant cooling equipment in service.

30.6.1.2 Performance Requirements

A high degree of energy availability was a major design objective. Guaranteed targets for both stations are:

Energy availability	> 99.5%.
Forced energy unavailability	< 0.43%.
Forced outage rate	< 5.4 outages per year.

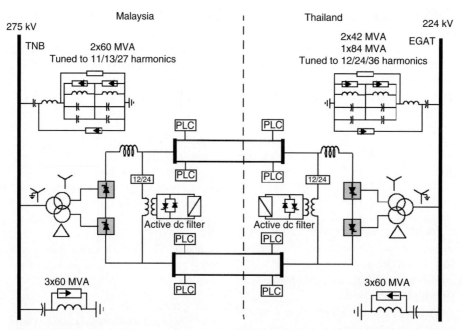

FIGURE 30.23 One-line diagram of the HVDC interconnection between Malaysia and Thailand.

30.6.1.3 Technical Information

TABLE 30.5 Main data 1

Monopolar operation		Rectifier (Malaysia)	Inverter (Thailand)
DC power, P_{dn}	Rated	240 MW (Stage 1)	240 MW
	Maximum overload (10 min).	450 MW	
	Minimum (10% of rated)	24 MW	
DC current, I_{dn}		1000 A	1000 A
DC voltage, U_{dn}		240 kV	293.5 kV
Firing angle, alpha		15.0°	–
Extinction angle, gamma		–	19.6°
Transformer secondary voltage, U_v		122.2 kV	122.2 kV
Converter reactive power, Q_{dc}		133.3 MVAr	150.9 MVAr
Reduced dc voltage operation	210 kV (70% dc voltage)		
AC system voltage		275 kV, 50 Hz	224 kV, 50 Hz
DC transmission line distance	110 kms		

TABLE 30.6 Main data 2

Main equipment	GURUN converter station
Smoothing reactor	100 mH
DC filter	Active type dc filters including passive type (12th/24th harmonics)
AC filter	2 × 60 MVAr (11/13/27)
Reactive power compensation	3 × 60 MVAr C-shunt
Power thyristor:	Type T1501 N75T-S34
Diameter	100 mm
Blocking voltage	7.5 kV repetitive, 8.0 kV non-repetitive
Maximum dc current	1550 A. 120° electrical. $T_c = 60°C$
Maximum effective current	3240 A
Converter transformer	Single-phase, three winding
	Rated power = 275 kV/116 MVA
Transmission line:	
Pole conductor	Cardinal ACSR 546 mm^2, twin conductors per pole
Neutral conductor	Hen ACSR 298 mm^2

30.6.1.4 Major Technical Features

The station incorporates the latest state-of-the-art technology in power electronics and control equipment. The features include:

- A fully decentralized control and protection system.
- Active dc filter technology.
- Hybrid dc current shunt measuring devices.
- A triple-tuned ac filter.

30.7 Modern Trends

30.7.1 Converter Station Design of the 2000s

The 1500 MW, ±500 kV Rihand–Delhi HVDC transmission system (commissioned in 1991) serves as an example of a typical design of the last decade. Its major design aspects were:

- Stations employ water-cooled valves of indoor design with the valves arranged as three quadruple valves suspended from the ceiling of the valve hall.
- Converter transformers are one-phase, three-winding units situated close to the valve hall with their bushings protruding into the valve hall.
- AC filtering is done with conventional, passive, double-tuned, and high-pass units employing internally-fused capacitors and air-cored reactors; these filters are mechanically switched for reactive power control.
- DC filtering is done with conventional, passive units employing a split smoothing reactor consisting of both an oil-filled reactor and an air-cored reactor.
- DC current transformer (DCCT) measurement is based on a zero-flux principle.
- Control and protection hardware is located in a control room in the service building in between the two valve halls. The controls still employed some analog parts, particularly for the protection circuits. The controls were duplicated with an automatic switchover to hot standby for reliability reasons.

This design was done in the 1980s to meet the requirements of the day for increased reliability and performance requirements, i.e. availability, reduced losses, higher overload capability, etc. However, these requirements led to increased costs for the system.

The next generation equipment is now being spearheaded by a desire to reduce costs and make HVDC as competitive as ac transmission. This is being facilitated by the major developments of the past decade that have taken place in power electronics. Therefore, the following will influence the next generation HVDC equipment.

30.7.1.1 Thyristor Development

The HVDC thyristors are now available with a diameter of 150 mm at a rating of 8–9 kV and power-handling capacity of 1500 kW, which will lead to a dramatic decrease in the number of series connected components for a valve with consequential cost reductions and improvement in reliability.

The development of the LTT (Figs. 30.24 and 30.25) is likely to eliminate the electronic unit (with their high number of electronic components) for generating the firing pulses for the thyristor (Fig. 30.26). The additional functional requirements of monitoring and protection of the device are being incorporated also.

Both of the above developments present the possibility of achieving compact valves that can be packaged for outdoor construction thereby reducing the overall cost and reliability of the station.

30.7.1.2 Higher DC Transmission Voltages

For long distance transmission, there is a tendency to use a high voltage to minimize the losses. The typical transmission voltage has been ±500 kV, although the Itaipu project in Brazil uses 600 kV. The transmission voltage has to be balanced against the cost of insulation. The industry is considering raising the voltage to ±800 kV.

FIGURE 30.24 Silicon wafer and construction of the LTT. The light guides appear in the bottom right-hand corner.

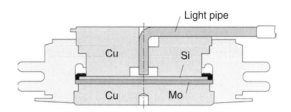

FIGURE 30.25 Cross section of the LTT with the light pipe entry.

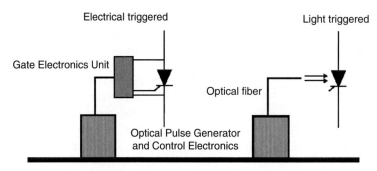

FIGURE 30.26 Conventional firing vs light triggered firing.

30.7.1.3 Controls

It is now possible to operate an HVDC scheme into an extremely weak ac system, even with short-circuit capacity down to unity.

Using programmable digital signal processors (DSPs) has resulted in a compact and modular design that is low cost; furthermore, the number of control cubicles has decreased by a factor of almost 10 times in the past decade. Now all control functions are implemented on digital platforms. The controls are fully integrated having monitoring, control and protection features. The design incorporates self-diagnostic and supervisory characteristics. The controls are optically coupled to the control room for reliable operation. Redundancy and duplication of controls is resulting in very high reliability and availability of the equipment.

30.7.1.4 Outdoor Valves

The introduction of air-insulated, outdoor valves will reduce the cost requirements for a valve hall. The use of a modular, compact design that can be preassembled in the manufacturing plant will save installation time and provide for a flexible station layout. This design has been feasible because of the development of a composite insulator for dc applications that is used as a communications channel for fiber-optics, cooling water, and ventilation air between the valve-unit and ground.

30.7.1.5 Active DC Filters

This development, made possible due to advances in power electronics and microprocessors, has resulted in a more efficient dc filter operating over a wide spectrum of frequencies and provides a compact design (Fig. 30.27).

30.7.1.6 AC Filters

Conventional ac filters used passive components for tuning out certain harmonic frequencies. Due to the variations in frequency and capacitance, physical aging and thermal characteristics of these tuned filters, the quality factors for the filters were typically about 100–125, which meant that the filters could not be too sharply tuned for efficient harmonic filtering. The advent of electronically tunable inductors based on the transductor principal, means that the Q-factors could now approach the natural one of the inductor. This will lead to much more enhanced filtering capacities (Fig. 30.28).

30.7.1.7 AC–DC Current Measurements

The new optical current transducers utilize a precision shunt at high potential. A small optical fiber link between ground and high potential is utilized resulting in a lower probability of a flashover. The optical power link transfers power to high potential for use in the electronics equipment. The optical data link transfers data to ground potential. This transducer results

FIGURE 30.27 Compact outdoor container with an active dc filter.

FIGURE 30.28 Installation of the triple-tuned filter at the Gurun station in Malaysia.

30 HVDC Transmission

FIGURE 30.29 The optical current transducer.

in high reliability, compact design, and efficient measurement (Fig. 30.29).

30.7.1.8 New Topologies

Two new topological changes to the converter design will impact greatly on future designs.

1. **Series compensated commutation:** There are two variants to this topology: the capacitor-commutated converter (CCC) (Fig. 30.30a) and the controlled series capacitor converter (CSCC) (Fig. 30.30b). Essentially, the behavior of the two variants is very similar. The insertion of a capacitor in series with the converter transformer leakage reactance causes a major reduction in the commutation impedance of the converter resulting in a reduction in the reactive power requirement of the converter. An increase in the size of the series capacitor can even result in the converter operation at a leading power factor if so desired. However, the negative impact of lower commutation impedance results in additional stresses on the valves and transformers and additional cost implications [7].

   The first CCC back-back converter station (rated at 2×550 MW), has been put into service at Garabi, on the Brazilian side of the Uruguay river. The system interconnects the electrical systems of Argentina and Brazil.

2. **Voltage source converters (VSC):** The use of self-commutation with the new generation switching devices (i.e. gate turn off thyristors (GTOs) and insulated gate bipolar transistors (IGBTs)) has resulted in the topology of a VSC (Fig. 30.31) as opposed

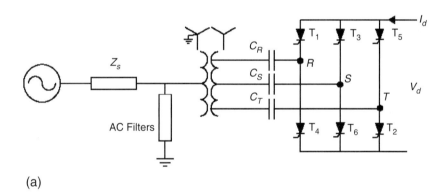

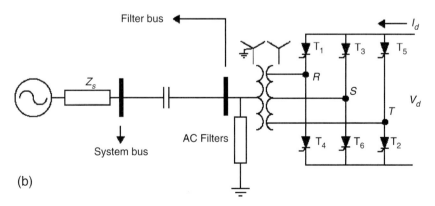

FIGURE 30.30 (a) Capacitor-commutated converter circuit and (b) controlled series capacitor converter circuit.

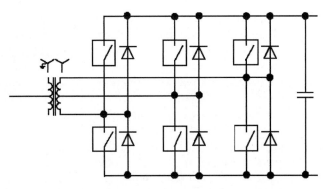

FIGURE 30.31 Voltage source converter topology.

to the conventional converter using line-commutated thyristors and current source converter (CSC) topology. The VSC, being self-commutated, can control active/reactive power and, with PWM techniques, control harmonic generation as well. The application of such circuits is presently limited by the switching losses and ratings of available switching devices. The ongoing advances in power electronic devices are expected to have a major impact on the future application of this type of converter in HVDC transmission. New application areas, particularly in distribution systems, are being actively investigated with this topology. Table 30.7 provides a partial list of the HVDC links in operation using this technique.

One major difficulty for the use of the VSC is the threat posed to the valves from a short circuit on the dc line. Unlike the CSC where the valves are inherently protected against short-circuit currents by the presence of the smoothing reactor, the VSC is relatively unprotected. For this reason, the VSC applications are almost always used with dc cables where the risk of a dc line short circuit is greatly reduced.

30.7.1.9 Compact Station Layout

The advances discussed above have resulted in marked improvement of the footprint requirement of the compact station [8] of the year 2000 which has about 24% space requirement of the comparable HVDC station designed in the past decade (Fig. 30.32).

30.8 HVDC System Simulation Techniques

Modern HVDC systems incorporate complex control and protection features. The testing and optimization of these features require powerful tools that are capable of modeling all facets of the system and have the flexibility to do the evaluation in a rapid, effective, and cost efficient manner.

30.8.1 DC Simulators and TNAs [9]

For decades, this has been achieved with the aid of physical power system simulators or transient network analyzers (TNAs) which incorporate scaled physical models of all power system elements (three-phase ac network lines/cables, sources as e.m.f. behind reactances, model circuit breakers for precisely timed ac system disturbances, transformers (system and convertor transformers with capacity to model saturation characteristics), filter capacitors, reactors, resistors, arrestors, and machines). Until the 1970s, these were built with analog components. However, with the developments in microprocessors, it is now feasible to utilize totally digital simulators operating in real time for even the most complex HVDC system studies. Most simulators operating scale is in the range 20–100 V dc, 0.2–1 A ac and at power frequency of 50 or 60 Hz. The stray capacitances and inductances are, however, not normally represented since the simulator is primarily used to assess control system behavior and temporary overvoltages of frequencies below 1000 Hz. Due to the developments of flexible ac transmission systems (FACTS) application, most modern simulators now include similarly scaled models of HVDC converters, static compensators, and other thyristor-controlled equipments. The controls of these equipments are usually capable of realistic performance during transients such as the ac faults and commutation failure. The limited

TABLE 30.7 Applications of HVDC light technology

No.	Project	Rating		Distance (km)	Application	Commissioned
		MVA	kV			
1	Hellsjon	3	±10	10	AC–DC conversion	Mar. 1997
2	Gotland	50	±80	70	Feed from wind power generation	June 1999
3	Tjaereborg	7	±10	4	Feed from wind power generation	Aug. 1999
4	Directlink	180	±140	65	Asynchronous interconnection	Dec. 1999
5	Murraylink	220	±140	180	Asynchronous interconnection	2002
6	Shoreham	330	±140	40	Cross sound cable link	2002
7	Troll A	2 × 42	±60	70	Gas production offshore platform	2005

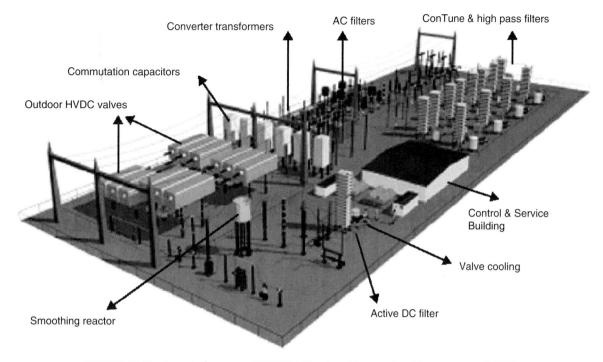

FIGURE 30.32 Layout of a compact HVDC station (graphic reproduced here courtesy of ABB).

availability of adequate models of some of the system elements restricts the scope of the studies which can be completed entirely by means of the simulator. Due to the scaling problems, the losses in the simulator may be disproportionately high and need to be partly compensated by electronic circuits (negative resistances) to simulate appropriate damping of overvoltages and other phenomena.

30.8.2 Digital Computer Analysis

The main type of program employed for studies is an electromagnetic transients program (EMTP) that solves sets of differential equations by step-by-step integration methods. The digital program must allow for the modeling of both the linear and non-linear components (single- and three-phase) and of the topological changes caused; for example, by valve firing or by circuit-breaker operation. Detailed modeling of the converter control system is necessary depending on the type of study.

The EMTP has become an industrial standard analysis tool for power systems and is widely used. The program has had a checkered history and numerous variants have appeared. Initially, the development of the program was supported by the Bonneville Power Administration (BPA). Some of the drawbacks in the capabilities of EMTP became more pronounced as the modeling of FACTS with power electronic switches and VSCs became more desirable.

Some of the drawbacks of the original EMTP version were:

1. The use of a fixed timestep that did not take into account the relatively long periods of inaction during non-switching events. This results in unnecessarily long simulation times and huge amounts of data to be manipulated. This is particularly problematic for simulating power electronic converters.
2. The use of a fixed timestep results in the modeled switches chopping inductive current that causes numerical oscillations. The use of artificial RC "snubbers" helped to alleviate some of these problems. The choice of the snubber capacitor was a function of the magnitude of the current to be chopped and the timestep size.
3. The use of the trapezoidal integration method results in numerical oscillations when the network admittance matrix to be inverted becomes singular. This is the direct result of modeling switches as truly either ON or OFF without representation of their intermediate non-linear characteristics.
4. The requirement of a one-timestep delay between the main program and the transient analysis of control systems (TACS) subroutine for controls simulation.
5. The use of new VSCs with multiple switching per cycle made the problem of switching "jitter" much more evident.
6. The lack of user-friendly input and output processors.

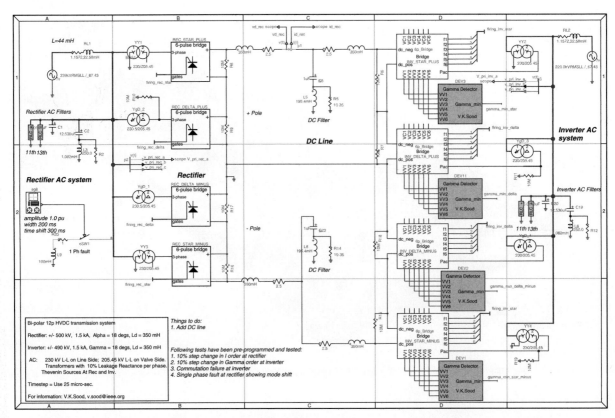

FIGURE 30.33 A sample of the graphical input file of EMTP RV.

In recent years, considerable effort has been made by the EMTP Development Coordination Group (DCG) to restructure the program. This has resulted in the latest version called the EMTP RV (restructured version) (see www.emtp.com). The entire code of the program has been re-written and graphical input and output processors have been added. A sample of the graphical input file is shown in Fig. 30.33.

The main advantage of digital studies is the possibility of correct representation of the damping present in the system. This feature permits more accurate evaluation of the nature and rate of decay of transient voltages following their peak levels in the initial few cycles, and also a more realistic assessment of the peak current and total energy absorption of the surge arresters. The digital program also allows modeling of stray inductances and capacitances and can be used to cover a wider frequency range of transients than the dc simulator.

The main disadvantage of the digital studies is the lack of adequate representation of commutation failure phenomena with the use of power electronic converters. However, with the increasing capacity of computers, this is likely to be overcome in the future.

The models used in the simulators and digital programs depend on the assumptions made and on the proper understanding of the component and system characteristics; therefore, they require care in their usage to avoid unrealistic results in inexperienced hands.

30.9 Concluding Remarks

The HVDC technology is now mature, reliable, and accepted all over the world. From its modest beginning in the 1950s, the technology has advanced considerably and maintained its leading edge image. The encroaching technology of flexible ac transmission systems (FACTS) has learned and gained from the technological enhancements made initially by HVDC systems. The FACTS technology may challenge some of the traditional roles for HVDC applications since the deregulation of the electrical utility business will open up the market for increased interconnection of networks [7]. HVDC transmission has unique characteristics, which will provide it with new opportunities. Although the traditional applications of HVDC transmission will be maintained for bulk power transmission in places like China, India, South America, and Africa, the increasing desire for the exploitation of renewable resources will provide both a challenge and an opportunity for innovative solutions in the following applications:

- Connection of small dispersed generators to the grid.
- Alternatives to local generation.
- Feeding to urban city centers.

Acknowledgments

The author pays tribute to the many pioneers whose vision of HVDC transmission has led to the rapid evolution of the power industry. It is not possible here to name all of them individually.

A number of photographs of equipment have been included in this chapter, and I thank the suppliers (Mr. P. Lips of Siemens and Mr. R. L. Vaughan from ABB) for their assistance.

I also thank my wife Vinay for her considerable assistance in the preparation of this manuscript.

References

1. E.W. Kimbark, Direct Current Transmission – Volume I, Wiley Interscience, USA, 1971, ISBN 0-471-47580-7.
2. J. Arrillaga, High Voltage Direct Current Transmission, 2nd Edition, The Institution of Electrical Engineers, UK, 1998, ISBN 0-85296-941-4.
3. K.R. Padiyar, HVDC Power Transmission Systems - Technology and System Interactions, John Wiley & Sons, India, 1990, ISBN 0-470-21706-5.
4. D. Melvold, HVDC Projects Listing, Prepared by IEEE DC and Flexible AC Transmission Subcommittee.
5. J. Ainsworth, "The Phase-Locked Oscillator – A New Control System for Controlled Static Converters," IEEE Transactions on Power Apparatus and Systems, Vol. PAS-87, No. 3, March 1968, pp. 859–865.
6. A. Ekstrom and G. Liss, "A Refined HVDC Control System, " IEEE Trans. on Power Apparatus and Systems, Vol. PAS-89, No. 5/6, May/June 1970, pp. 723–732.
7. V. K. Sood, HVDC and FACTS Controllers – Applications of Static Converters in Power Systems, Kluwer Academic Publishers, Canada, April 2004, ISBN 1-4020-7890-0.
8. L. Carlsson, G. Asplund, H. Bjorklund, and H. Stomberg, "Recent and Future Trends in HVDC Converter Station Design," IEE 2nd International Conference on Advances in Power System Control, Operation and Management, Hong Kong, December 1993, pp. 221–226.
9. C. Gagnon, V.K. Sood, J. Belanger, A. Vallee, M. Toupin, and M. Tetreault, "Hydro-Québec Power System Simulator", IEEE Canadian Review, No. 19, Spring-Summer 1994, pp. 6–9.

31
Flexible AC Transmission Systems

E. H. Watanabe
Electrical Engineering Department, COPPE/Federal University of Rio de Janeiro, Brazil, South America

M. Aredes
Electrical Engineering Department, Polytechnic School and COPPE/ Federal University of Rio de Janeiro, Brazil, South America

P. G. Barbosa
Electrical Engineering Department, Federal University of Juiz de Fora, Brazil, South America

G. Santos Jr.
Electrical Engineering Department, COPPE/Federal University of Rio de Janeiro, Brazil, South America

F. K. de Araújo Lima
Electrical Engineering Department, COPPE/Federal University of Rio de Janeiro, Brazil, South America

R. F. da Silva Dias
Electrical Engineering Department, COPPE/Federal University of Rio de Janeiro, Brazil, South America

31.1 Introduction 797
31.2 Ideal Shunt Compensator 798
31.3 Ideal Series Compensator 799
31.4 Synthesis of FACTS Devices 802
    31.4.1 Thyristor-based FACTS Devices
    31.4.2 FACTS Devices Based on Self-commutated Switches
    References 820

This chapter presents the basic operation principles of FACTS devices. Starting with a brief introduction on the concept and its origin, the text then focuses on the ideal behavior of each basic shunt and series FACTS device. Guidelines on the synthesis of the first generation of these devices, based on thyristors, are presented, followed by the newer generations of FACTS devices based on self-commutated semiconductor switches.

31.1 Introduction

In 1988, Hingorani [1] published a paper entitled "Power Electronics in Electric Utilities: Role of Power Electronics in Future Power Systems," which proposed the extensive use of power electronics for the control of AC systems [2]. The basic idea was to obtain AC systems with a high level of control flexibility, just as in high voltage direct current (HVDC) systems [3], based on the use of the thyristor, as well as on self-commutated (controllable turn-on and turn-off) semiconductor devices like gate turn-off thyristors (GTOs), insulated gate bipolar transistors (IGBTs), and integrated gate controlled thyristors (IGCTs) [4, 5], which were not developed at that time yet.

The switching characteristics of thyristors – controlled turn-on and natural turn-off – are appropriate for using in line-commutated converters, like in conventional HVDC transmission systems with a current source in the DC side. In this latter application, the technology for series connection of thyristors is very important due to the high-voltage characteristics of the transmission voltage. This is a well-known technology. Maximum breakdown voltage and current conduction capabilities are around 8 kV and 4 kA, respectively. These are some features that make thyristors important for very high-power applications, although they also present some serious drawbacks: the lack of controlled turn-off capability and low switching speeds.

Self-commutated switches are adequate for use in converters where turn-off capability is necessary. The device with highest ratings in this group was, for a long time, the GTO, with maximum switching capability of 6 kV and 6 kA. At present, there are IGBTs with ratings in the range of 6.5 kV and 3 kA and IGCTs with switching capability of about 6 kV and 4 kA. Other semiconductors switches – like the injection enhanced gate transistor (IEGT), faster than IGBTs and with high ratings – can be also found in the market. The GTOs and IGCTs are devices that need turn-on current rate of change (di/dt) limitation, normally achieved with a small inductor. Normally, GTOs also need a snubber circuit for voltage rate of change (dv/dt) limitation.

The GTOs, IGCTs, and IGBTs are the most used options for self-commutated high-power converters. Because the switching time of these devices is in the microsecond range (or below), their series connection is more complicated than in the case of thyristors. However, there are examples of series connections of various GTOs or IGCTs and, in the case of IGBTs, the number of series connected devices can go as high as 32 [6].

Because of the commutation nature of the thyristors, the converters used in HVDC systems are of the current source type [7]. On the other hand, the force commutated converters using the self-commutated devices are basically of the voltage source type. More details about current source and voltage source converters can be found in many power electronics text books, e.g. [3, 7].

31.2 Ideal Shunt Compensator

A simple and lossless AC system is composed of two ideal generators and a short transmission line, as shown in Fig. 31.1, is considered as basis to the discussion of the operating principles of a shunt compensator [8]. The transmission line is modeled by an inductive reactance X_L. In the circuit, a continuously controlled voltage source is connected in the middle of the transmission line. It is assumed that the voltage phasors V_S and V_R have the same magnitude and are phase-shifted by δ. The subscript "S" stands for "Source" and "R" stands for "Receptor." Figure 31.2 shows the phasor diagram of the system in Fig. 31.1, for the case in which the compensation voltage phasor V_M has also the same magnitude as V_S and V_R and its phase is exactly $(-\delta/2)$ with respect to V_S and $(+\delta/2)$ with respect to V_R.

In this situation, the current I_{SM} flows from the source and the current I_{MR} flows into the receptor. The phasor I_M is the resulting current flowing through the ideal shunt compensator, figure shows that this current I_M, in this case, is orthogonal to the voltage V_M, which means that the ideal shunt compensator voltage source does not have to generate or absorb active power and have only reactive power in its terminals.

From Fig. 31.2 and knowing that no active power flows to or from the ideal shunt compensator, it is possible to calculate the power transferred from V_S to V_R which is given by:

$$P_S = \frac{2V^2}{X_L} \sin(\delta/2) \qquad (31.1)$$

where, P_S is the active power flowing from the source, V is the magnitude of the voltages V_S and V_R.

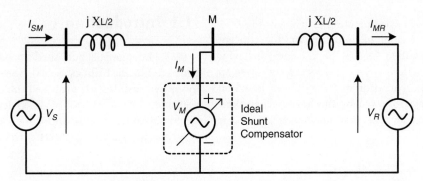

FIGURE 31.1 Ideal shunt compensator connected in the middle of a transmission line.

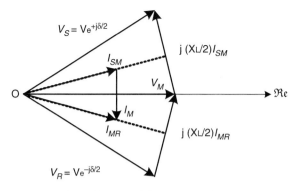

FIGURE 31.2 Phasor diagram of the system with shunt reactive power compensation.

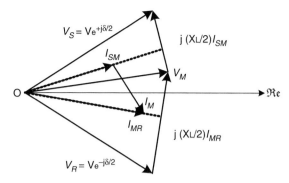

FIGURE 31.3 Phasor diagram of the system with shunt reactive and active power compensation.

If the ideal shunt compensator were not present, the transferred power would be given by:

$$P_S = \frac{V^2}{X_L} \sin \delta \qquad (31.2)$$

Since $2\sin(\delta/2)$ is always greater than $\sin \delta$ for δ in the range of $[0, 2\pi]$, the ideal shunt compensator does improve the power transfer capability of the transmission line. This voltage source is in fact operating as an ideal reactive power shunt compensator.

If the phase angle between V_M and V_S is different from $\delta/2$ (as shown in Fig. 31.3), the power flowing through V_M has both active and reactive components.

With the characteristics of the ideal shunt compensator presented above it is possible to synthesize power electronics-based devices to operate as active or reactive power compensators. This is discussed in the following sections. It will be seen that the requirements of the device synthesis with actual semiconductor switches for the situations of reactive or active power compensation are different, due to the need of energy storage element or energy source if active power is to be drained/generated by the shunt compensator.

31.3 Ideal Series Compensator

Similar to the previous section, the ideal series compensator is modeled by a voltage source for which the phasor is V_C, connected in the middle of a lossless transmission line as shown in Fig. 31.4.

The current flowing through the transmission line is given by:

$$I = (V_{SR} - V_C)/jX_L \qquad (31.3)$$

where $V_{SR} = V_S - V_R$.

If the ideal series compensator voltage is generated in such a way that its phasor V_C is in quadrature with line current I, this series compensator does not supply neither absorb active power. As previously discussed, power at the series source is only reactive and the voltage source may, in this particular case,

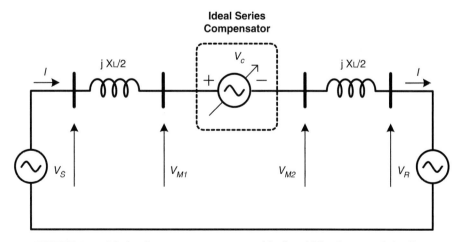

FIGURE 31.4 Ideal series compensator connected in the middle of a transmission line.

be replaced by capacitive or inductive equivalent impedance. The equivalent impedance would then given by

$$X_{eq} = X_L(1+s) \quad (31.4)$$

where,

$$s = \frac{X_{Comp}}{X_L}; \quad (0 \leq |s| \leq 1) \quad (31.5)$$

is the compensation factor and X_{Comp} is the series equivalent compensation reactance, negative if capacitive and positive if inductive. In this case the compensation voltage is given by

$$V_C = I_L X_{eq} \quad (31.6)$$

and the transmitted power is equal to

$$P_s = \frac{V^2}{X_L(1-s)} \sin\delta \quad (31.7)$$

Equation (31.6) shows that the transmitted power can be considerably increased by series compensation, choosing a proper compensation factor s. The reactive power at the series source is given by

$$Q_{CS} = \frac{2V^2}{X_L} \frac{s}{(1-s)^2}(1-\cos\delta) \quad (31.8)$$

The left-hand side of Fig. 31.5a shows the phasor diagram of the system in Fig. 31.4 without the ideal series compensator. The voltage phasor V_L on the line reactance X_L and the compensator voltage phasor V_C are shown for a given compensation level, assuming that this voltage V_C corresponds to a capacitive compensation. In this case, the line current phasor leads voltage phasor V_C by 90° and the total voltage drop in the line V_Z ($=V_S - V_R - V_C$) is larger than the original voltage drop V_L. The current flowing in the line is larger after compensation than before. This situation shows the case where the series compensator is used to increase power flow.

The left-hand side of Fig. 31.5b shows the same non-compensated situation as in the previous case. In the middle is shown the case of inductive compensation. In this case, the compensation voltage V_C is in phase with the line drop voltage V_L, producing an equivalent total voltage drop V_Z smaller than in the original case. As a result, the current phasor I flowing in the line is smaller than before compensation. This kind of compensation may be interesting in the case that the power flowing through the line has to be decreased. In either capacitive or inductive compensation modes, no active power is absorbed or generated by the ideal series compensator.

Figure 31.6 shows an AC system with an ideal generic series compensation voltage source V_C for the general case where it may not be in quadrature with the line current. In this case, the compensator is able to fully control the phase difference between the two systems, thus controlling also the active and reactive power exchanged between them. However, in this case, the compensation source V_C may have to absorb or generate active power (P_C), as well as to control the reactive power (Q_C).

Figure 31.7 shows the phasor diagram for the case of this ideal generic series compensator. This figure shows also a dashed-line circle with the locus of all the possible position

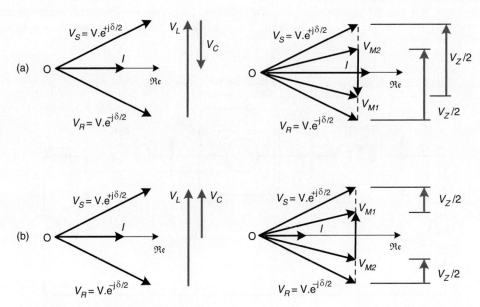

FIGURE 31.5 Phasor diagram of the series reactive compensator: (a) capacitive and (b) reactive mode.

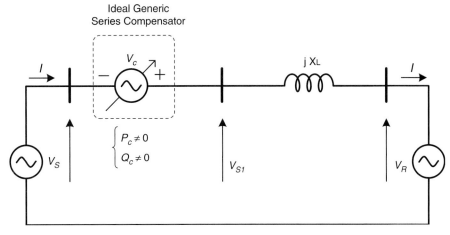

FIGURE 31.6 Ideal generic series compensator.

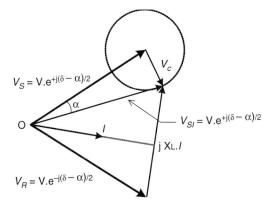

FIGURE 31.7 Phasor diagram of the AC system compensated with an ideal generic series compensator.

that the compensation voltage V_C could take, assuming that the magnitude shown for this voltage is its maximum. Naturally, if the sum of the compensation voltage and the source voltage V_S is on the circle, the magnitude of V_{S1} may be smaller or larger than the magnitude of V_S.

If the compensation voltage V_C added in series with V_S produces a voltage V_{S1} that has the same magnitude as V_S but is phase-shifted by an angle α, the power flowing through the transmission line in Fig. 31.6 is given by:

$$P_S = \frac{V^2}{X_L} \sin(\delta - \alpha) \qquad (31.9)$$

Equation (31.8) shows that transmitted power increases as the phase difference $(\delta - \alpha)$ reaches $90°$. However, its maximum value is the same as in the case of no compensation. The difference is that with this compensator the angle between the two voltage sources at the terminals of the line can be controlled.

In Fig. 31.7, voltage V_C may have any phase angle with respect to line current. Therefore, it may have to supply or absorb active power, as well as control reactive power. As in the case of the shunt device, this feature must also be taken into account in the synthesis of the actual devices. As a first approximation, when the goal is to control active power flow through the transmission line, compensator location seems to be just a question of convenience.

Figure 31.8 summarizes the active power transfer characteristics in a transmission line as a function of the phase difference δ between its sending and receiving ends, as shown in Figs. 31.1 and 31.2, for the cases of the line without compensation, line with series or shunt compensation, as well as line with a phase-shift compensation. These characteristics are drawn on the assumption that the source voltages V_S and V_R (see Fig. 31.2) have the same magnitude, which is a conventional situation. A 50% series compensation ($s = 0.5$ as defined in Eq. (31.4) presents a significant increase in the line power transfer capability.

In general, series compensation is the best choice for increasing power transfer capability. The phase-shifter compensator is important to connect two systems with excessive or uncontrollable phase difference. It does not increase power transfer capability significantly; however, it may allow the adjustment of large or highly variable phase differences. The shunt compensator does not increase power transfer capability in a significant way in its normal operating region, where the angle δ is naturally below $90°$ and in general around $30°$. The great importance of the shunt compensator is the increase in the stability margin, as explained in Fig. 31.9.

Figure 31.9 shows the power transfer P_δ characteristics of a transmission line, which is first assumed to be transmitting power P_{S0} at phase angle δ_0. If a problem happens in the line (a fault, for example) the turbine that drives the generator cannot change its input mechanical power immediately even if there is no power transmission for a short time. This situation

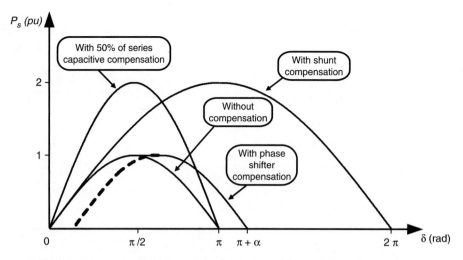

FIGURE 31.8 Power transfer characteristics for the case of shunt, series, and no compensation.

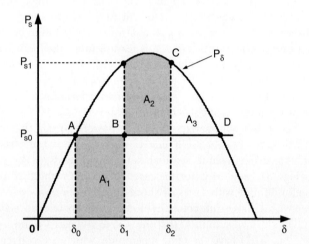

FIGURE 31.9 Stability margin characteristics – stable situation.

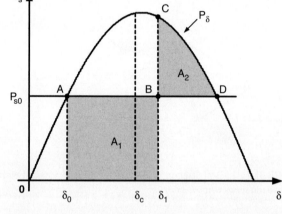

FIGURE 31.10 Stability margin characteristics – unstable situation.

accelerates the generator, increasing its frequency and leading to an increase of the phase angle δ to δ_1. If the line restarts operation at the instant corresponding to this phase angle δ_1, the transmitted power will be P_1, which is larger than P_0 and decelerates the turbine/generator. The area A_1 corresponds to the energy that accelerated the turbine. As the frequency gets higher than the rated frequency at (P_{S1}, δ_1), the phase angle will increase up to δ_2, where the area A_2 is equal to the area A_1. If the area given by the A_2 plus A_3 is larger than A_1, the system is said to be dynamically stable. On the contrary, if it is not possible to have an area A_2 equal to A_1, the system is said to be unstable. An unstable situation is shown in Fig. 31.10, where the system is the same as in Fig. 31.9 but with a longer interval with no power transmission. In this case, the turbine/generator accelerates more than in the case in Fig. 31.9 and the phase angle δ goes over its critical value δ_c reaching δ_1. Therefore, the area below the P_δ curve to decelerate the system is not enough leading to an unstable system because A_2 is smaller than A_1.

Looking at Fig. 31.8 it is possible to see that, depending on the operating point, all three compensation methods increase the stability margin as the area under the curve of transmitted power P_δ versus phase angle δ is increased. However, the ideal shunt compensator is the one that most increases this area, this is the reason why it is said to be the best option to increase the stability margin.

31.4 Synthesis of FACTS Devices

It has been stated that the synthesis of the compensators presented in Sections 31.2 and 31.3 may be achieved with thyristors or self-commutated switches like GTO, IGBT, or IGCT. Each type of switches leads to devices with different operating principles and synthesis concepts, and that is a reason why they should be discussed separately. Terms and definitions for most of the FACTS devices are given in [9].

Thyristor-based FACTS devices use line or natural commutation together with large energy storage elements (capacitors or reactors). On the other hand, devices based on self-commutating switches like GTOs, IGCTs, or IGBTs uses gate-controlled commutation. In general, it is said that the first generation of FACTS devices is based on conventional line commutated thyristors and the subsequent generations are based on gate-controlled devices. The most important FACTS devices based on thyristors and self-commutating devices are presented in the next sections.

31.4.1 Thyristor-based FACTS Devices

31.4.1.1 Thyristor-controlled Reactor

The most used thyristor-based FACTS device is the thyristor-controlled reactor (TCR) shown in Fig. 31.11a. This is a shunt compensator which produces an equivalent continuous variable inductive reactance by using phase-angle control. Figure 31.11b shows the voltage and current waveforms of the TCR. The current is controlled by the firing-angle α – its fundamental component can be larger or smaller depending on the angle α which may vary from 90 to 180°, measured from the zero-crossing of the voltage. At $\alpha = 90°$, the reactor is fully inserted in the circuit and for $\alpha = 180°$, the reactor is completely out of the circuit. Figure 31.12 shows the equivalent admittance of the TCR as function of the firing-angle α. Naturally, this admittance is always inductive.

31.4.1.2 Thyristor-switched Capacitor

Figure 31.13 shows the thyristor-switched capacitor (TSC). In this device the word "controlled" used in the case of the reactor is substituted by "switched," because the thyristor is turned-on only when zero-voltage switching (ZVS) condition is achieved. This means that the voltage across the thyristor terminals has to be zero at the turn-on instant. In practical cases, it may be slightly positive, since thyristors need positive anode–cathode voltages to be triggered (large anode–cathode voltage during turn-on, however, may produce a large current spike that may damage the thyristors). Therefore, due to this switching characteristic, the thyristors can only connect the capacitor to the grid or disconnect it. Consequently, only step-like control is possible and, therefore, a continuous control is not possible. The capacitor connection to or disconnection from the grid is normally done at very low frequencies and the harmonics, when they appear, are not a serious concern.

31.4.1.3 Static Var Compensator

The use of the TCR shown in Fig. 31.11 or the TSC shown in Fig. 31.13 allows only continuous inductive compensation or capacitive discontinuous compensation. However, in most applications, it is desirable to have continuous capacitive or inductive compensation. The static var compensator (SVC) is generally designed to operate in both inductive and capacitive continuous compensation [10, 11]. The TCR serves as the controller basis for the conventional SVC used for reactive power compensation, for either voltage regulation or power factor correction.

Figure 31.14 shows a single-line diagram of a SVC, where the TCRs are Δ connected and the capacitors are Y connected.

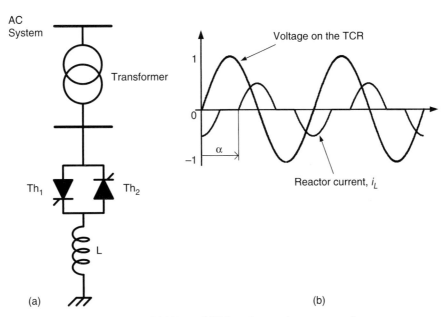

FIGURE 31.11 (a) TCR and (b) its voltage and current waveforms.

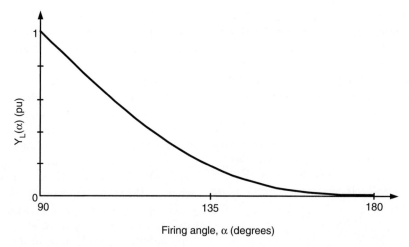

FIGURE 31.12 Equivalent admittance of a TCR as function of the firing-angle α.

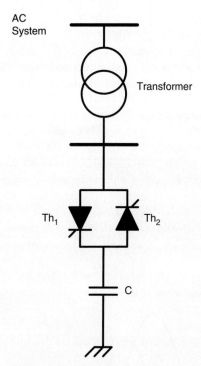

FIGURE 31.13 Thyristor-switched capacitor.

The circuit does not show the filters that are normally needed due to the switching-generated harmonics. In some cases, the fixed capacitor can be replaced by a TSC to get more flexibility in terms of control range.

The capacitor of the SVC is calculated in such a way as to generate the maximum capacitive reactive power that it has to control. This condition is achieved when the thyristors are turned-off ($\alpha = 180°$). On the other hand, the TCR inductor maximum reactive power has to be greater than the reactive power of the capacitor bank. In this way, the SVC is able to control the reactive power from capacitive to inductive.

The maximum inductive reactive power is given for the case when the thyristors are turned-on at minimum firing-angle ($\alpha = 90°$). Thus, the SVC can control reactive power from maximum capacitive for $\alpha = 180°$ to maximum inductive for $\alpha = 90°$. In this sense, the SVC represents an adjustable fundamental frequency susceptance to the AC network, controlled by the firing-angle of the TCR thyristors ($90° < \alpha < 180°$).

The SVC is well-known and many examples of successful applications can be found around the world.

Due to the once-per-cycle thyristor firing with phase-angle control, current with low-order harmonic components appears and Y–Δ transformers and passive filters may be needed to eliminate them. Three sets of TCRs connected in the Δ side of Y–Δ transformers form a conventional 6-pulse TCR. To minimize harmonic generation, it is common to have two sets of transformers connected in Y–Δ and Δ–Δ with the TCR connected in the Δ side forming a 12-pulse TCR.

31.4.1.4 Thyristor-switched Series Capacitor

Figure 31.15 shows the thyristor-switched series capacitor (TSSC). In this device, the thyristors should be kept untriggered so as to connect the capacitors in series with the transmission line. If the thyristors are turned-on, the capacitor is bypassed. Thyristors turn-on must be at a zero-voltage condition (ZVS), as it occurs in the case of the TSC, to avoid current spikes in the switches. An example of an application based on this concept is presented in [12].

This compensation system has the advantage of being very simple. However, it does not allow continuous control. If the connection/disconnection of the capacitors is to be made at sporadic switching, no harmonic problem occurs. Depending on the frequency the thyristors are switched, harmonic or sub-harmonics may appear. In this arrangement, it is interesting to choose the value of the capacitors in such a way that many different combinations can be achieved. For example, if the

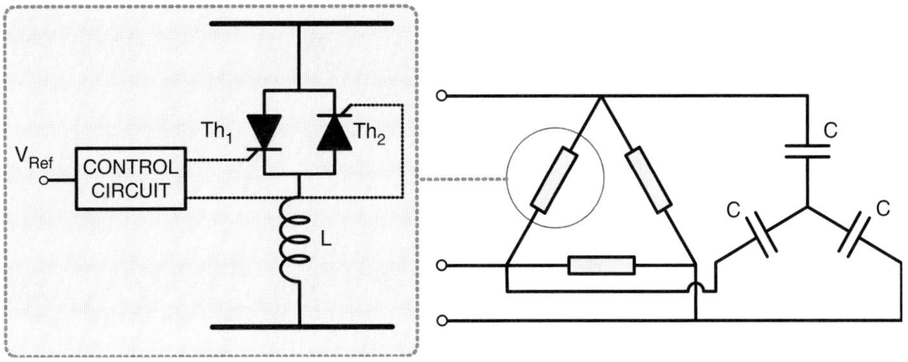

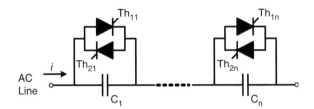

FIGURE 31.14 6-Pulse SVC.

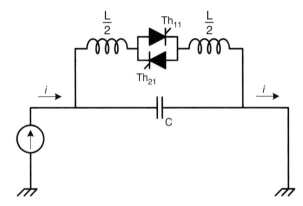

FIGURE 31.15 Thyristor-switched series capacitors.

FIGURE 31.16 Thyristor-controlled series capacitors.

total number of capacitors is three, they could have values proportional to 1, 2, and 3. Therefore, by combining these values it is possible to obtain equivalent capacitor proportional to 1, 2, 3, 4, 5, and 6.

31.4.1.5 Thyristor-controlled Series Capacitor (TCSC)

Figure 31.16 shows the thyristor-controlled series capacitor (TCSC). In this figure, the transmission line and the voltage sources in its ends are represented by a current source because this is the actual behavior of most of the transmission system. This compensator is also based on the TCR which was first developed for shunt connection. When the TCR is used connected in series with the line, it has to be always connected in parallel with a capacitor because it is not possible to control the current if the equivalent of the transmission line and the sources is a current source. This circuit is similar to the conventional SVC, with the difference that the TCSC is connected in series with the line. In this compensator, the equivalent value of the series connected reactor can be continuously controlled by adjusting the firing-angle of the thyristors. As a consequence, this device presents a continuously controllable series capacitor. Various practical system based on this concept is under operation around the world [12–14]. This device has been used for power flow control and power oscillation damping.

Figure 31.17 presents the current and voltage waveforms in the TCSC, showing that although there is a large amount of harmonics in the capacitor and reactor currents, capacitor voltage is somehow almost sinusoidal. In actual applications, these harmonics are not a serious concern and they are filtered by the capacitor itself and the transmission line impedance.

Figure 31.18 shows the equivalent impedance of the TCSC as a function of the firing-angle α. This figure shows that this device has both capacitive and inductive characteristic regions. It also has a resonance for α around 145°, in this example. In normal operation, the TCSC is controlled in the capacitive compensation region where its impedance varies from its minimum value Z_{min} for firing-angle $\alpha = 180°$ and its maximum safe value Z_{max} for α around 150°. Operation with α close to the resonance region is not safe. This device can operate also in the inductive region, but in this case, normally it is used only with $\alpha = 90°$ to decrease power transfer capability of the transmission line.

31.4.1.6 Thyristor-controlled Phase Angle Regulator

The thyristor-controlled phase angle regulator (TCPAR), shown in Fig. 31.19 as an example, may improve considerably the controllability of a utility power transmission system. In this figure, only control of phase "a" is detailed. The series

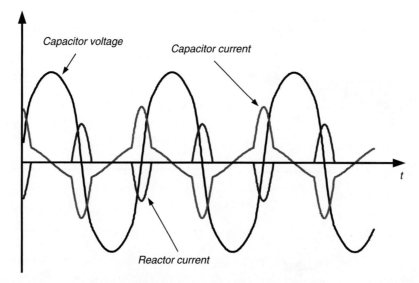

FIGURE 31.17 TCSC voltage and current waveforms.

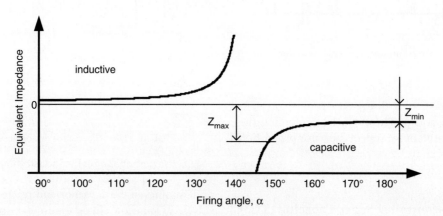

FIGURE 31.18 TCSC equivalent reactance.

voltage generated in phase "a" comes from three secondary windings of a transformer whose primary side is connected between phases "b" and "c." Each of the three secondary windings can be connected in series with the line through the thyristors switching. The thyristors are connected in antiparallel, forming bidirectional naturally commutated switches. By turning on a set of thyristors, a voltage whose magnitude can be controlled by phase control is connected in series with the transmission line. The number of secondary windings is chosen as to decrease harmonic content of the series compensation voltage.

The TCPAR in Fig. 31.19 has some peculiarities that should be pointed out. One of them is that active power can only flow from the shunt to the series windings – therefore reverse power flow is not possible. The compensation voltage phasor, as shown in Fig. 31.20, has a limited range of variation: in the case of phase "a," its locus is along a line orthogonal to V_a, because the injected voltage is in phase with the voltage $(V_b - V_c)$. As a consequence, it is not possible for the TCPAR to generate a compensating voltage phasor whose locus is a circle, as shown in Fig. 31.7, for the case of an ideal generic series compensator.

Other configuration of phase-shifters can be found in the literature, e.g. [15, 16].

31.4.2 FACTS Devices Based on Self-commutated Switches

There are various different types of FACTS devices based on self-commutated switches and the names used here are in accordance with the names published in [9]. Some of them are newer devices and are not in that reference. In this case, the name used is as it appears in the literature. In [2] it is possible to find most of the details of FACTS devices.

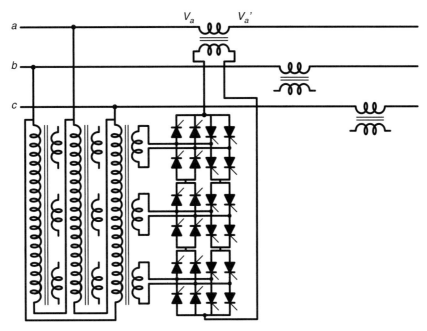

FIGURE 31.19 Thyristor-controlled phase angle regulator (TCPAR).

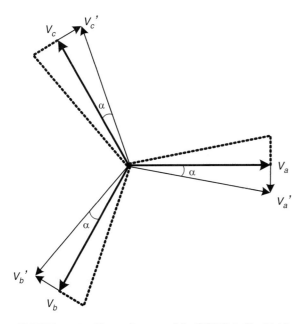

FIGURE 31.20 Phasor diagram of the TCPAR in Fig. 31.19.

31.4.2.1 The Static Synchronous Compensator

The development of high-power self-commutated devices like GTOs, IGBTs, and IGCTs has led to the development of high-power voltage source inverters (VSI), as the 6-pulse 2-level VSI shown in Fig. 31.21a [4] or the 3-level neutral point clamped VSI shown in Fig. 31.21b [17]. In the figure, the switches are GTOs (they could be IGBTs, IGCTs, or other self-commutating switches) with an anti-parallel-connected diode, which operates with a unidirectional voltage blocking capability and a bidirectional current flow. In contrast, the current source inverters (CSI) used in HVDC transmission systems use switches (thyristors) operating with unidirectional current flow and bidirectional voltage blocking capabilities.

In a conventional VSI, for industrial applications, a voltage source is connected at the inverter DC side. However, in the case that only reactive power has to be controlled, the DC voltage source may be replaced by a small capacitor. If active power has to be absorbed or generated by the compensator, an energy storage system has to be connected at the DC side of the VSI.

In practical applications, small reactors (L) are necessary to connect the VSI to the AC network. This is necessary to avoid currents peaks during switching transients. In most cases, these small reactors are just the leakage inductance of the coupling transformers.

The first high rating STATCOM is under operation since 1991 [18] in Japan and uses three single-phase VSI to form one 3-phase, 6-pulses, 10 MVA converter. To guarantee low losses, the switching frequency is equal to the line frequency and a total of eight sets of 3-phase converters are used to form a 48-pulse converter. All the converters have a common DC capacitor in their DC side. In the AC side, the converters are connected in series through a zigzag transformer to eliminate low frequency voltage harmonics. The device, whose compensation capability is 80 MVA, was developed to increase the transient stability margin of a transmission line and has allowed a 20% increase of the transmitted power above the previous stability limit. Since it was developed for improving

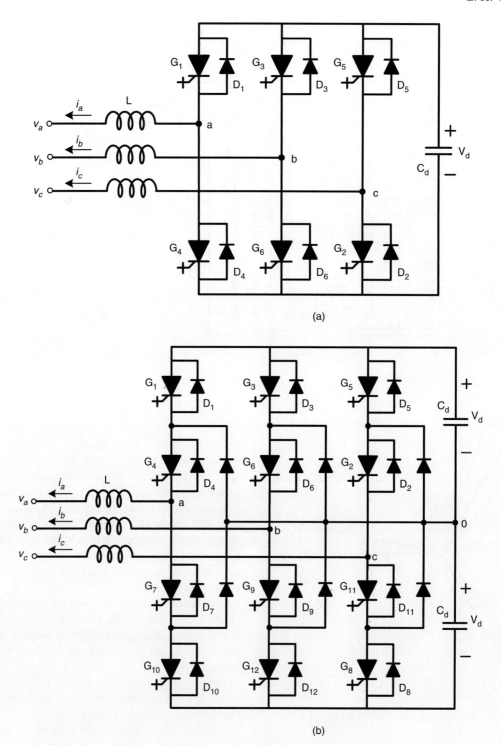

FIGURE 31.21 (a) Basic 6-pulse VSI 2-level var compensator and (b) basic 3-level var compensator.

transient stability margin, it normally operates in standby mode, without reactive compensation and, consequently, low losses. During transient situation, this STATCOM operates for a short time until the system is stable.

The development of a ±100 Mvar STATCOM, in USA, was reported in [19]. It is based on eight sets of 3-phase bridge converters, similar to that shown in Fig. 31.21b and was developed for reactive power control, so it can operate continuously with acceptable losses. The switching frequency is equal to line frequency and the number of pulses is 48 and, therefore, the output voltage waveform is almost sinusoidal and harmonic filters are not used in both cases referred in [18] and [19].

31.4.2.1.1 Basic Switching Control Techniques
In FACTS applications the power ratings of the converters are in the range of some MW to hundreds of MW and the switching frequency is lower, if compared to the switching frequency used in industrial application converters, to avoid excessive switching losses. However, there are various switching control types, being the most known so far:

– Multi-pulse converters switched at line frequency, as in [18] and [19];
– Pulse width modulation (PWM) with harmonic elimination technique [20];
– Sinusoidal PWM [6];
– Cascade converters [21].

31.4.2.1.2 Multi-pulse Converters Switched at Line Frequency
The multi-pulse converter was the first choice for STATCOM application as it presents low losses and low harmonic content [18, 19]. Figure 31.22 shows a 24-pulse converter based on 3-phase, 2-level 6-pulse converters. In this case, the zigzag transformers are connected in such a way as to produce phase differences of 15, 30, 45, and 60°. With this arrangement, the resulting output voltage and its harmonic spectrum are as shown in Fig. 31.23. The first two harmonics components are the 23rd and 25th order harmonics. Figure 31.24 shows the voltage waveform for a 48-pulse converter and its respective harmonic spectrum. In this case, the first two harmonic components are the 47th and 49th order harmonics. The total harmonic distortion (THD) for the 24- and 48-pulse converters are 7 and 3.3%, respectively. These converters can also be built by using 3-level converters. However, one drawback of the multi-pulse converter is the complexity of the transformers, which have to operate with high harmonic content in their voltage as well as various different turns ratio.

31.4.2.1.3 PWM (Pulse Width Modulation) with Harmonic Elimination Technique
One way to avoid the complexity of the multi-pulse converters is to use PWM with harmonic elimination technique [20]. With this approach, it is possible to use relatively low switching frequency and, consequently, have low switching losses. The PWM modulation is obtained by offline calculation of the switches "on" and "off" instants in such a way as to eliminate the low frequency harmonics. Figure 31.25a shows an example of voltage waveform with "on" and "off" instants calculated in such a way as to eliminate the 5th, 7th, 11th, 13th order harmonics. This voltage corresponds to the voltage between one phase of the converter and the negative terminal of the DC side. Figure 31.25b shows the control angle as a function of the modulation index m_a. Figure 31.26 shows the harmonic spectrum for the phase-to-phase voltage waveform corresponding to that shown in Fig. 31.25a. Here it is considered that the RMS value of the fundamental component of the voltage in Fig. 31.25 is equal to unity. In Fig. 31.26 the magnitude of the fundamental component is equal to $\sqrt{3}$. The higher order harmonics in the voltage waveform can be eliminated by a relatively small passive filter, so the voltage and current at the converter terminals are practically harmonic-free and, therefore, the transformer to connect a PWM-controlled STATCOM to the grid may be a conventional transformer designed for sinusoidal operation.

31.4.2.1.4 Sinusoidal PWM
The sinusoidal PWM control technique is possibly the simplest to implement and can be synthesized by comparing a sinusoidal reference voltage with a triangular carrier [4]. This switching control method needs a relatively high switching frequency which is in the range of 1–2 kHz and consequently produces higher switching losses if compared with multi-pulse STATCOM. The harmonic content at low frequencies is negligible; however, there is a relatively high harmonic content at the switching frequency which is eliminated by a passive filter.

31.4.2.1.5 Cascade Converter
The basic cascade converter [21] topology is shown in Fig. 31.27. Only two single-phase full bridge converters are shown, the first and the nth. However, in actual application, several of them are connected in series and switched at line frequency. The resulting voltage waveform can be similar to the multi-pulse converter waveform with the advantage that there is no need of transformers to sum up the converters output voltage. Due to the line frequency switching, the switching losses are very low. The resulting voltage waveform can be almost sinusoidal depending on the number of series converters and the transformer used to connect them to the grid can be a conventional sinusoidal waveform transformer, if necessary. One drawback of this converter topology is that it is not possible to have a back-to-back connection. The need to have one DC capacitor for each single-phase converter has two consequences: the number of capacitors is equal to the number of single-phase converters; and the capacitance of each capacitor has to be much higher, if compared with three-phase converters. This is because the instantaneous power in the single-phase converter has a large oscillating component at double of the line frequency and it would force a large voltage ripple in the capacitor if they were small.

31.4.2.1.6 STATCOM Control Techniques
The control of reactive power in the STATCOM is done by controlling its terminal voltage. Figure 31.28 shows a simplified circuit where the AC grid is represented by a voltage source V_S behind an impedance X_L and the STATCOM is represented by its fundamental voltage V_I. Figure 31.29a shows the case when the AC grid phasor voltage V_S is in phase with the STATCOM voltage V_I and both have the same magnitude. In this case, the line current I_L is zero. Figure 31.28b shows the case when V_S is a little larger than V_I. In this case, the line current I_L,

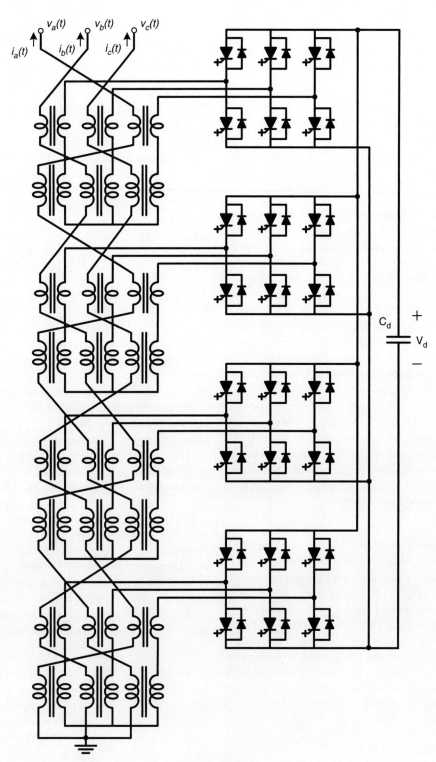

FIGURE 31.22 6-Pulse 2-level VSI-based 24-pulse var compensator.

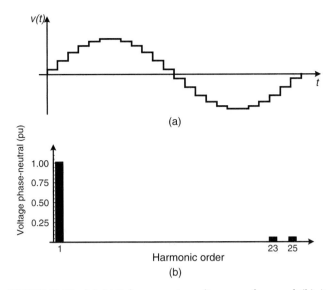

FIGURE 31.23 (a) 24-Pulse converter voltage waveform and (b) its harmonic spectrum.

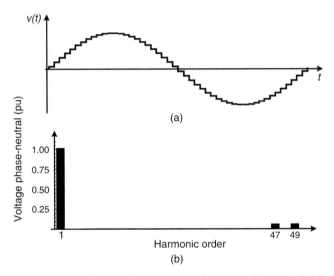

FIGURE 31.24 (a) 48-Pulse converter voltage waveform and (b) its harmonic spectrum.

which lags the voltage V_L by 90°, is also lagging the AC grid phasor voltage V_S, and therefore the STATCOM is producing an inductive reactive power. On the other hand, Fig. 31.29(c) shows the case when V_S is a little smaller than V_I. As a result, the line current I_L, which lags the voltage V_L by 90°, leads the voltage V_S, and therefore the STATCOM is producing a capacitive reactive power. In summary, the STATCOM reactive power can be controlled if the magnitude of its V_I voltage is controlled, assuming that it is in phase with V_S.

- if V_S is equal to V_I there is no reactive power and no active power in the STATCOM;
- if V_S is larger than V_I the STATCOM reactive power is inductive;
- if V_S is smaller than V_I the STATCOM reactive power is capacitive.

The reactive power control in a STATCOM is, therefore, a problem of how to control the magnitude of its voltage V_I. There are two basic principles: in the case of multi-pulse converters, the output voltage magnitude can only be controlled by controlling the DC side voltage that is the DC capacitor voltage; in the case of PWM control, the DC capacitor voltage can be kept constant and the voltage can be controlled by the PWM controller itself.

Figure 31.30a shows the phasor diagram for the case when a phase difference δ between V_S and V_I is positive. The resulting line current is in such a way that produces an active power flowing into the converter, charging the DC capacitor. Figure 31.30b shows the phasor diagram for a negative phase angle δ. In this case the DC capacitor is discharged. Therefore, by controlling the phase angle δ it is possible to control the DC capacitor voltage.

In general, STATCOM based on multi-pulse converter without PWM has to control its voltage by charging or discharging the DC capacitor and this voltage has to be variable. On the other hand, STATCOM based on a PWM-controlled converter has to control DC side capacitor voltage only to keep it constant. In both cases, the principle shown in Fig. 31.30 is valid.

The STATCOM control technique presented above illustrates the basic scalar control concepts. However, this compensator can be also controlled by a vector technique [22]. In this case, the three-phase voltages are transformed to a synchronous reference frame where they can be controlled in such a way as to regulate the quadrature component of the current, which controls reactive power. The direct component of the current are used to control the DC capacitor voltage as it represents the active power.

Another way to control the STATCOM is by using the instantaneous power theory [23, 24]. This theory was first proposed for controlling active power filters and is used in the design of the compensators operating with unbalanced systems. If a high frequency PWM converter is used, this theory allows the design of active filters to compensate for harmonic components or fundamental reactive component.

31.4.2.1.7 STATCOM DC Side Capacitor Theoretically the DC side capacitor of a STATCOM based on three-phase converters operating in a balanced system and controlling only reactive power could have a capacitance equal to zero Farad. However, in actual STATCOMs, a finite capacitor has to be used with the objective of keeping constant or controlled DC voltage as it tends to vary due to the converter switching. One parameter commonly used in synchronous machine is

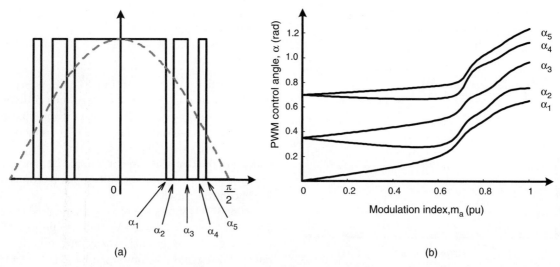

FIGURE 31.25 (a) Example of one-phase voltage with the harmonic elimination technique and (b) the control angle as function of the modulation index.

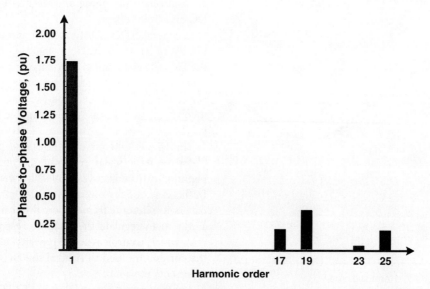

FIGURE 31.26 Line voltage harmonic spectrum for the voltage waveform in Fig. 31.25a.

the inertia constant H defined by

$$H = \frac{J\omega^2/2}{S} \quad (31.10)$$

where J and ω are the rotor moment of inertia and angular speed and S is the machine apparent power. A similar parameter for the STATCOM, H_{ST}, can be defined by

$$H_{ST} = \frac{CV_{DC}^2/2}{S} \quad (31.11)$$

where C and V_{DC} are the DC capacitance and its voltage and S is the STATCOM apparent power.

In both cases, the constants H and H_{ST} are values in time units corresponding to the relation of the amount of energy stored in the rotor inertia or in the capacitor and the machine or STATCOM apparent power. In the case of synchronous machines, the value of H is in the range of few seconds (generally, in the range of 1–3 s) and in the case of the STATCOM, H_{ST} is in the range of milliseconds or below (0.5–5 ms) if only reactive power is to be controlled. These numbers show that the STATCOM based on three-phase converters and designed for reactive power control only (which is the general case), has almost no stored energy in its DC capacitor. On the other hand, STATCOM based on single-phase converters without a common DC link may have larger capacitors, as in the case of

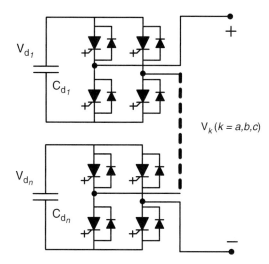

FIGURE 31.27 Cascade converter basic topology.

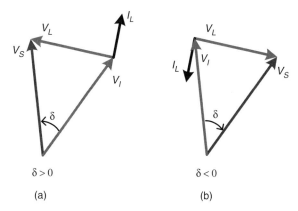

FIGURE 31.30 Active power control in a STATCOM.

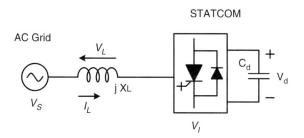

FIGURE 31.28 Simplified circuit for the AC grid and the STATCOM.

cascade converters due to the power oscillations at double of the line frequency.

There are STATCOMs (in some cases with different names) that are designed for operation with unbalanced loads. In this case, the DC capacitor has to be also much larger than in the case of balanced systems to avoid large voltage ripple on the DC voltage due to power oscillations at twice the line frequency, which appears naturally in unbalanced systems [25, 26] or unbalanced voltages [27] or system with flicker problem [28, 29]. In this case, the STATCOM compensates reactive power as well as the instantaneous oscillating active power due to the negative sequence currents components. In fact, this is an extension of the shunt active power filter application where the goal is the current harmonic compensation, which includes negative sequence currents even at the fundamental frequency. If sub-harmonics are present, the device is able to filter them out as well.

31.4.2.1.8 STATCOM with Energy Storage In general, the STATCOM is designed for reactive power compensation and it does not need large energy storage elements. However, there are some applications where it may be interesting to have some energy stored in the DC side, for example to compensate for active power for a short time. In these applications, the DC side capacitor has to be substituted by a voltage source energy storage device like a battery [30] or

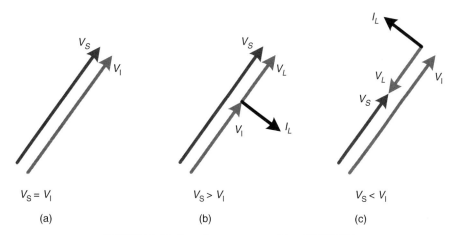

FIGURE 31.29 Reactive power control in a STATCOM.

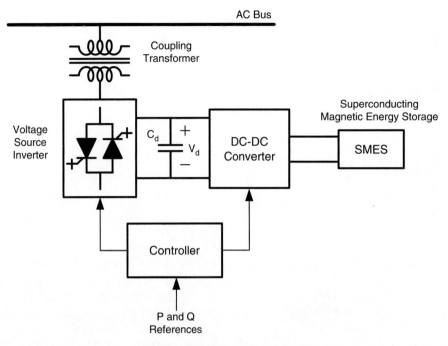

FIGURE 31.31 STATCOM with SMES.

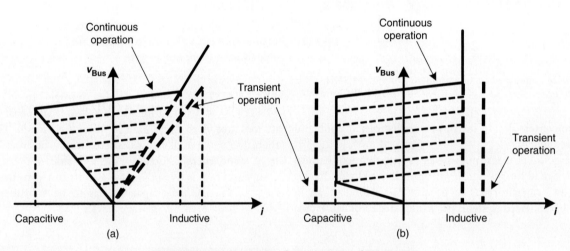

FIGURE 31.32 Comparison between SVC and STATCOM.

a double layer capacitor (super capacitor). Another possibility is to store energy in superconducting magnetic energy storage (SMES) systems [31, 32]. A natural solution for the use of the superconducting reactor would be the connection to the AC grid through a current source inverter (CSI) instead of the voltage source inverter. However, this has not been the case found in the literature. Figure 31.31 shows a block diagram of a typical STATCOM/SMES system, where the SMES is connected to the DC side of the STATCOM through a DC–DC converter which converts the direct current in the superconductor magnet to DC voltage in the STATCOM DC side and vice versa. This STATCOM is able to control reactive power continuously, as well as active power for a short time, depending on the amount of energy stored in the superconducting magnet.

31.4.2.1.9 Comparison between SVC and STATCOM Figure 31.32a shows the steady-state volt–ampere characteristics for the SVC shown in Fig. 31.14, while Fig. 31.32b shows the same characteristics for a STATCOM. For operation at rated voltage, both devices can present similar characteristics in terms of control range. However, SVC current compensation capability for lower voltages becomes smaller, while in the STATCOM it does not change significantly for voltages lower than rated (but approximately above 0.2 pu). This is explained by the fact that the SVC is based on impedance

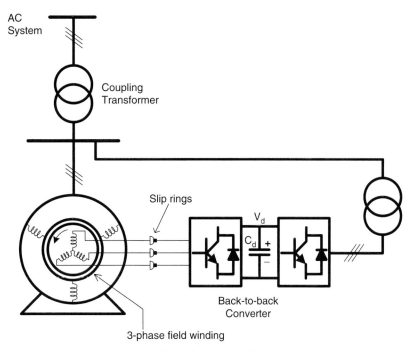

FIGURE 31.33 Adjustable speed synchronous condenser.

control, while the STATCOM is based on voltage source control. Therefore, while in the SVC the current decreases with a corresponding voltage decrease, in the STATCOM the current capability of the converters depends only on the switching device used, so the maximum current can be kept unchanged even for a low voltage condition. This is an important characteristic, especially in applications where the voltage may drop (as in most cases) where the STATCOM presents a better performance.

31.4.2.2 Adjustable Speed Synchronous Condenser

The adjustable speed synchronous condenser is not exactly a FACTS device, as it contains an electrical machine. However, it may be an interesting shunt device to compensate reactive power continuously and relatively large amounts of active power for a short time. The basic topology of this device is shown in Fig. 31.33. It is based on a double-fed induction machine with a conventional 3-phase winding in the stator and a 3-phase winding in the rotor. The latter is supplied by a 3-phase converter connected back-to-back to a second converter, which is connected to the grid.

This configuration allows the generation of a rotating magnetic flux in the rotor, which depends on the rotor converter frequency. When the machine is rotating at synchronous speed, the rotor converter operates at zero frequency, and the magnetic flux in the rotor is stationary, with respect to the rotor itself. In this case, the compensator operates as a conventional synchronous condenser.

However, when the rotor speed is lower or higher than the synchronous speed (normally during transients), the rotor converter generates a field current with the necessary frequency to keep the stator and rotor fluxes synchronized – if the synchronous frequency is 60 Hz and the rotor is running at 58 Hz, the rotor converter has to supply voltage or current at 2 Hz, so as to synchronize the fluxes. Naturally, it would be more interesting to use field-oriented control [33] instead of scalar control to get a better performance.

This hybrid compensator may supply energy to the AC system, if rotor speed is decreased. This machine is designed to have relatively large rotor inertia so as to present a large inertia constant (which may be in the range of more than 10 s). It is also called adjustable speed rotary condenser [34]. Operation at speeds higher than the synchronous speed is also possible, if it is necessary to absorb energy from the grid.

One of the advantages of this device is that a compensator with power in the range of 400 MVA may be synthesized with power electronics converter rated at a small fraction of this power and with a large capability to supply both active (for a few seconds) and reactive power (continuously) [34].

31.4.2.3 Static Synchronous Series Compensator

In contrast with the STATCOM, which is a shunt FACTS device, it is possible to build a converter-based compensator for series compensation. Figure 31.34 shows the basic diagram of a static synchronous series compensator (SSSC) based on voltage source inverter (VSI) with a capacitor in its DC side

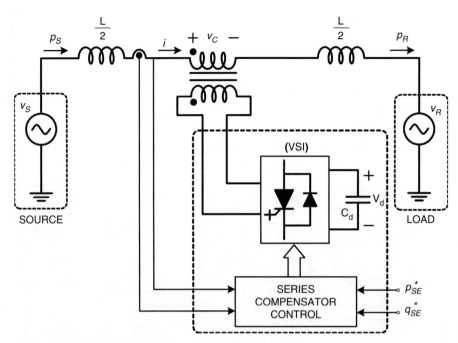

FIGURE 31.34 Static synchronous series compensator (SSSC).

and connected in series with the transmission line through a transformer [35]. The inputs to the SSSC controller shown are the line current and voltage, as well as the active and reactive power references p_{SE}^* and q_{SE}^*, respectively. In general, only reactive power is compensated and, in this case, the active power reference p_{SE}^* is zero and q_{SE}^* is chosen as to control power flow. Naturally, in the case of power flow control, it is necessary to have another control loop for this purpose and this is not shown in the figure.

One should note that if current is flowing in the transmission line, the SSSC controls reactive power by generating voltage v_C in quadrature with the line current. The device then behaves as capacitive or inductive equivalent impedance increasing or decreasing the power flow, respectively. The compensation characteristic is as shown in Fig. 31.8, for the case of series compensation where the transmitted power is always positive for $0 < \delta < 180°$. That is, with reactive power control it is only possible to transmit in one direction. However, if instead of controlling q_{SE}^* voltage v_C is controlled, it is possible to have power flow reversion. Figure 31.35 shows the power flow characteristics of a transmission line with an SSSC using constant voltage control. The voltage is in quadrature with the current and its magnitude is kept constant. The figure shows that it is possible to have power flow reversion for small values of δ with a constant compensation voltage v_C.

It should be noted that the discussion presented with respect to the converters for the case of the STATCOM is also valid for the case of the SSSC. The SSSC can be used for power flow control and for power oscillation damping as well.

31.4.2.4 Gate-controlled Series Capacitor

Figure 31.16 showed the TCSC, which is basically a TCR in parallel with a capacitor and both connected in series with a transmission line. The combination is effective in continuously controlling the equivalent capacitive reactance presented to the system, mainly for power flow control and oscillation damping purposes. It was also pointed out that the device has the disadvantage of a resonance area due to the association capacitor/TCR (see Fig. 31.18).

In [36], the continuously regulated series capacitor using GTO thyristors to directly control capacitor voltage is presented. Figure 31.36 shows the GTO thyristor-controlled series capacitor (GCSC) [37], hereafter renamed as the gate-controlled series capacitor, since it may also be built using other self-commutated switches such as IGBTs or IGCTs.

The GCSC circuit consists of a capacitor and a pair of self-commutated switches in anti-parallel. As the switches operate under AC voltage, they must be able to block both direct and reverse voltage, as well as allow current control in both directions.

Figure 31.37 shows the voltage and current waveforms for the GCSC, where the current in the transmission line is assumed to be sinusoidal. If the switches are kept turned-on, the capacitor is bypassed and does not present any compensation effect. If they are kept off, the capacitor is fully inserted in the line. On the other hand, if the switches are conducting and are turned-off at a given blocking angle γ counted from the zero-crossings of the line current, the capacitor voltage v_C appears as a result of the integration of the line current passing

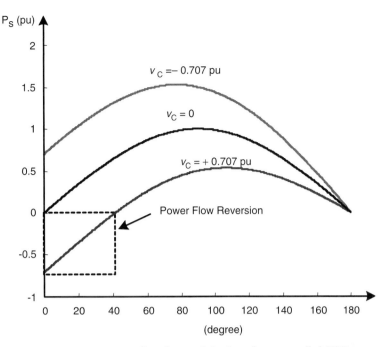

FIGURE 31.35 Power flow characteristics for voltage controlled SSSC.

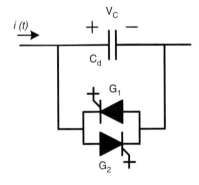

FIGURE 31.36 Gate-controlled series capacitor.

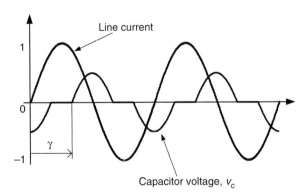

FIGURE 31.37 GCSC voltage and current waveforms.

through it. The next time the capacitor voltage crosses zero, the switches are turned-on again, to be turned-off at the next turn-off angle γ. With this switching control sequence, it must be clear that the switches always switch at zero voltage. This is an interesting feature for the series connection of the switches under high voltage operation [38].

The GCSC has some advantages when compared to the TCSC – the blocking angle can be continuously varied, which in turn varies the fundamental component of the voltage v_C. Also, it can be smaller than the TCSC [39]. Moreover, the dynamic response of the GCSC is generally better than that of the TCSC [40].

The fundamental impedance of the GCSC as a function of the blocking angle γ is shown in Fig. 31.38. A blocking angle of 90° means the capacitor is fully inserted in the circuit, while a value of 180° corresponds to a situation where the capacitor is bypassed and no compensation occurs.

31.4.2.5 Unified Power Flow Controller

The unified power flow controller (UPFC) is a more complete transmission line compensator [41], shown as a simplified block diagram in Fig. 31.39. This device can be understood as a STATCOM and an SSSC with a common DC link. The energy storing capacity of the DC capacitor is generally small, and so the shunt converter has to draw (or generate) active power from the grid in the exactly same amount as the active power being generated (or drawn) by the series converter. If this is not followed, the DC link voltage may increase or decrease

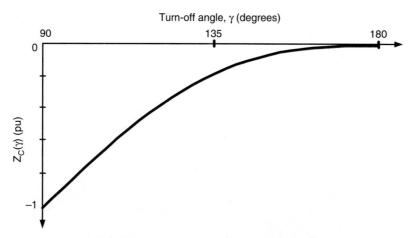

FIGURE 31.38 GCSC equivalent fundamental impedance.

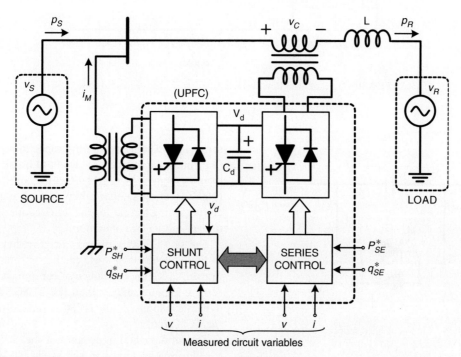

FIGURE 31.39 UPFC block diagram.

with respect to the rated voltage, depending on the net power being absorbed or delivered by both converters. On the other hand, the reactive power in the shunt or series converter can be controlled independently, giving a great flexibility to the power flow control.

The phasor diagram in Fig. 31.40 shows that the UPFC can be controlled in such a way as to produce any voltage phasor in series with the transmission line that fits inside the dashed line circle on top of the phase voltage phasors. The maximum radius of the circle is limited by the voltage limitation of the series converter. The fact that the locus of v_C is a circle is one of the greatest advantage of the UPFC when compared with the thyristor-based phase shifter. If the UPFC injects or absorbs reactive power in parallel with the system, the magnitude of voltage V_S will be increased or decreased, respectively. This extra-characteristic increases the locus of the series voltage v_C. Of course this effect is only possible if an inductive impedance is present in series with the voltage source V_S, which is normally the case.

The shunt compensator of the UPFC is normally used with two objectives. The first is to control the reactive power in the point of connection and therefore control the voltage at this point. The second is to control active power in such a way as to control the DC link voltage. The control technique to be used can be similar to the one used in STATCOM. The series compensator can be controlled in the same way as the SSSC,

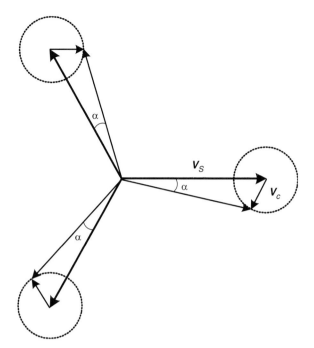

FIGURE 31.40 Phasor diagram for a system with a UPFC.

with the difference that in this case it may control active and reactive power. Naturally, the active power control in the series compensator would change the capacitor voltage that should be controlled by the shunt compensator.

31.4.2.6 Interline Power Flow Controller

The interline power flow controller (IPFC) [42] is a UPFC-derived device with the objective of controlling power flow between lines instead of one line as in the UPFC. Figure 31.41 shows a basic block diagram of an IPFC with two converters to control the power flow in lines 1 and 2. The minimum number of converters connected in back-to-back is two, but there may be more. Each converter should be connected in series with a different transmission line and should control power flow in this line with the following conditions:

- the reactive power control can be totally independent in each converter; and
- the active power flowing into or out of each converter has to be coordinated in such a way that the DC link voltage is kept controlled.

The DC link voltage control can be achieved in a similar way as in the case of the UPFC. In this case, one of the series converters can control the compensation voltage freely and it may produce active power flowing into or from the DC link, which would charge or discharge the DC link capacitor. Therefore, the other converter has to be controlled to regulate this DC voltage. If there are n converters with n greater than two, $(n-1)$ converters can absorb or generate active power while one converter has to control the DC link voltage. Anyway, all n converters can control reactive power freely. For instance, this concept allows the control of the power flow in n lines and it is possible to transfer active power from one line to another being an interesting device to balance power flow in n parallel transmission lines.

31.4.2.7 Convertible Static Converter

Following the IPFC, a more generic concept is the convertible static converter (CSC) [43, 44], which is based on the connection of a voltage source converter in various different topologies. Considering one simple case of two transmission lines and two converters with apparent power each equal to S, it is possible to have the following topologies:

- Two converters connected in shunt operating as a STATCOM rated at 2S apparent power;
- Two converters connected in series with one transmission line forming an SSSC with 2S apparent power;
- One converter connected in shunt and the other in series forming a UPFC;
- One converter connected in series with one line and the other connected in series with the other line forming an IPFC.

Other topologies are possible depending on how the converters are connected in the system. The basic control of each converter depends on how it is being used if, as a STATCOM, SSSC, UPFC, or IPFC.

31.4.2.8 Voltage Source Inverter-based HVDC Transmission

The VSI can be used for DC transmission with a circuit configuration exactly dual to the conventional HVDC transmission system, which is based on the thyristor-controlled CSI. The duality can be explained by the fact that the thyristor-controlled HVDC system controls the DC link current, while the system based on VSI controls the DC link voltage. The basic circuit configuration for a two converter VSI-HVDC system is shown in Fig. 31.42. This concept can be used for the connection of asynchronous systems, systems with different frequencies or located in places where cable transmission is more applicable than conventional transmission lines (as in congested urban areas or underwater transmission). The number of VSIs can be two for point-to-point transmission or more for multipoint transmission.

One great advantage of the VSI-HVDC system is that it allows independent active and reactive power control in each terminal. While the CSI-HVDC transmission can be synthesized for higher power as compared to the VSI-HVDC system, it can only control active power and may have large reactive power consumption, which means that reactive power compensation equipment is normally necessary.

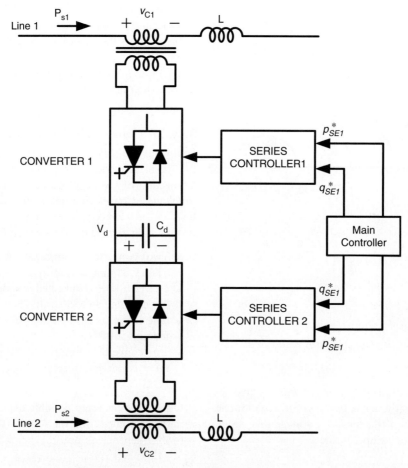

FIGURE 31.41 Block diagram of IPFC with two converters.

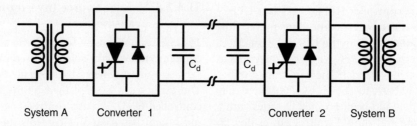

FIGURE 31.42 Voltage source inverter HVDC system.

References

1. N. G. Hingorani, "Power Electronics in Electric Utilities: Role of Power Electronics in Future Power Systems," Proceedings of the IEEE, Vol. 76, No. 4, April, 1988, pp. 481–482.
2. N. G. Hingorani and L. Gyugyi, "Understanding FACTS – Concepts and Technology of Flexible AC Transmission Systems," IEEE Press, New York, 1999.
3. E. W. Kimbark, "Direct Current Transmission," Wiley-Interscience, New York, 1971.
4. N. Mohan, T. M. Undeland and W. P. Robins, "Power Electronics – Converters, Applications and Design," John Wiley & Sons, Inc. New York, 2003.
5. M. H. Rashid, "Power Electronics, Circuits, Devices, and Applications," 3rd Edition, Prentice Hall, 2003.
6. E. M. John, A. Oskoui and A. Petersson, "Using a STATCOM to retire urban generation," IEEE PES - Power Systems Conference and Exposition, Vol. 2, 2004, pp. 693–698.
7. B. K. Bose, "Power Electronics and AC Drives," Prentice-Hall, 1986.
8. L. Gyugyi, "Solid-State Synchronous Voltage Sources for Dynamic Compensation and Real Time Control AC Transmission Lines," Emerging Practices in Technology, IEEE-Transmission Lines, IEEE Standards Press, Piscataway, USA, 1993.
9. A. A. Edris, R. Adapa, M. H. Baker, L. Bhomann, K. Clark, K. Habashi, L. Gyugyi, J. Lemay, A. S. Mehraban, A. K. Myers, J. Reeve, F. Sener, D. R. Togerson and R. R. Wood, "Proposed Terms and

Definitios for Flexibe AC Transmission Systems (FACTS)," IEEE Transactions on Power Delivery, Vol. 12, No. 4, October, 2002, pp. 1848–1853.
10. C. Wei-Nan and C.-J. Wu, "Developing Static Reactive Power Compensators in a Power System Simulator for Power Education," IEEE Transactions on Power Systems, Vol. 10, No. 4, November, 1995, pp. 1734–1741.
11. T. Hasegawa, Y. Aoshima and T. Sato, "Development of 60 MVA SVC (Static Var Compensator) Using Large Capacity 8 kV and 3.5 kA Thyristors," IEEE Power Converter Conference – PCC-Nagaoka, 1997, pp. 725–730.
12. A. J. F. Deri, B. J. Ware, R. A. Byron, A. S. Mehraban, M. Chamia, P. Halvarsson and L. Ängquist, "Improving Transmission System Performance Using Controlled Series Capacitors," Cigré Joint Session 14/37/38–07, Paris, France, 1992.
13. N. Christl, R. Hedin, K. Sadek, P. Lützelberger, P. E. Krause, S. M. McKenna, A. H. Montoya and D. Ttorgerson, "Advanced Series Compensation (ASC) with Thyristor Controlled Impedance," Cigré Joint Session 14/37/38–05, Paris, France, 1992.
14. C. Gama, "Brazilian North–South Interconnection Control Application and Operating Experience with a TCSC," IEEE – PES Summer Meeting, Vol. 2, 1999, pp. 1103–1108.
15. K. Xing and G. L. Kusic, "Damping Sub-synchronous Resonance by Phase Shifters," IEEE Transactions on Energy Conversion, Vol. 4, No. 3, September, 1989, pp. 344–350.
16. S. Nyati, M. Eitzmann, J. Kappenman, D. VanHouse, N. Mohan and A. Edris, "Design Issues for a Single Core Transformer Thyristor Controlled Phase-Angel Regulator," IEEE Transactions on Power Delivery, Vol. 10, No. 4, October, 1995, pp. 2013–2019.
17. D. G. Holmes and T. A. Lipo, "Pulse Width Modulation for Power Converters: Principles and Practice", Wiley-IEEE Press, New Jersey, 2003.
18. S. Mori, K. Matsuno, M. Takeda and M. Seto, "Development of a Large Static Var Generator Using Self-commutated Inverters for Improving Power System Stability," IEEE Transactions on Power Delivery, Vol. 8, No. 1, February, 1993, pp. 371–377.
19. C. Schauder, M. Gernhardt, E. Stacey, T. Lemak, L. Gyugyi, T. W. Cease and A. Edris, "Development of a ±100 Mvar Static Condenser for Voltage Control of Transmission System," IEEE Transactions on Power Delivery, Vol. 10, No. 3, July, 1995, pp. 1486–1496.
20. G. Reed, J. Paserba, T. Croasdaile, M. Takeda, Y. Hamasaki, T. Aritsuka, N. Morishima, S. Jochi et al., "The VELCO STATCOM Based Transmission System Project," IEEE Power Engineering Society Winter Meeting, 2001, Vol. 3, 2001, pp. 1109–1114.
21. C. Qian and M. L. Crow, "A Cascade Converter-Based STATCOM with Energy Storage," IEEE PES Winter Meeting, 2002, pp. 544–549.
22. C. Schauder and H. Mehta, "Vector Analysis and Control of Advanced Static VAr Compensators," IEE Proceedings on Generation, Transmission and Distribution, Vol. 140, No. 4, July, 1993, pp. 299–306.
23. H. Akagi, Y. Kanazawa and A. Nabae, "Instantaneous Reactive Power Compensators Comprising Switching Devices without Energy Storage Components," IEEE Transactions on Industrial Application, Vol. IA-20, No. 3, 1984.
24. E. H. Watanabe, R. M. Stephan and M. Aredes, "New Concepts of Instantaneous Active and Reactive Power for Three Phase System and Generic Loads," IEEE Transactions on Power Delivery, Vol. 8, No. 2, April, 1993.
25. F. Ichikawa et al., "Operating Experience of a 50 MVA Self-commutated SVC at the Shin–Shimano Substation," Proceedings of the International Power Electronics Conference – IPEC-Yokohama'95, Yokohama, Japan, April, 1995, pp. 597–602.
26. M. Takeda et al., "Development of an SVG Series for Voltage Control over Three-Phase Unbalance Caused by Railway Load," Proceedings of the International Power Electronics Conference – IPEC-Yokohama'95, Yokohama, Japan, April, 1995, pp. 603–608.
27. C. Hochgraf and R. H. Lasseter, "STATCOM Controls for Operation with Unbalanced Voltages," IEEE Transactions on Power Delivery, Vol. 13, No. 2, April, 1998, pp. 538–544.
28. G. F. Reed, J. E. Greaf, T. Matsumoto, Y. Yonehata, M. Takeda, T. Aritsuka, Y. Hamasaki, F. Ojima, A. P. Sidell, R. E. Chervus and C. K. Nebecker, "Application of a 5 MVA, 4.16 kV D-STATCOM System for Voltage Flicker Compensation at Seattle Iron and Metals," IEEE Power Engineering Society Summer Meeting 2000, Vol. 3, 2000, pp. 1605–1611.
29. C. Schauder, "STATCOM for Compensation of Large Electric Arc Furnace Installations," IEEE Power Engineering Society Summer Meeting 1999, Vol. 2, 1999, pp. 1109–1112.
30. C. Shen, Z. Yang, M. L. Crow and S. Atcitty, "Control of STATCOM with Energy Storage Device," IEEE Power Engineering Society Winter Meeting 2000, Vol. 4, 2000, pp. 2722–2728.
31. A. B. Arsoy, Y. Liu, P. F. Ribeiro and F. Wang, "STATCOM-SMES," Industry Applications Magazine, IEEE Vol. 9, No. 2, March–April, 2003, pp. 21–28.
32. M. G. Molina, P. E. Mercado and E. H. Watanabe, "Dynamic Performance of a Static Synchronous Compensator with Superconducting Magnetic Energy Storage," IEEE Power Electronics Specialist Conference – PESC'2005, June, 2005, pp. 224–230.
33. W. Leonhard, "Control of Electrical Drives," 3rd Edition, Springer, 2001.
34. H. Akagi, "The State-of-the-Art of Power Electronics in Japan," IEEE Transactions on Power Electronics, Vol. 13, No. 2, March, 1998, pp. 345–356.
35. L. Gyugyi, C. D. Schauder and K. S. Kalyan, "Static Synchronous Series Compensator: A Solid-State Approach to the Series Compensation of Transmission Lines," IEEE Transactions on Power Delivery, Vol. 12, No. 1, January, 1997, pp. 406–417.
36. G. G. Karady, T. H. Ortmeyer, B. R. Pilvelait and D. Maratukulam, "Continuous Regulated Series Capacitor," IEEE Transactions on Power Delivery, Vol. 8, No. 3, July, 1993, pp. 1348–1354.
37. L. F. W. de Souza, E. H. Watanabe and M. Aredes, "GTO Controlled Series Capacitors: Multi-module and Multi-pulse Arrangements," IEEE Transactions on Power Delivery, Vol. 15, No. 2, April, 2000, pp. 725–731.
38. E. H. Watanabe, M. Aredes, L. F. W. de Souza and M. D. Bellar, "Series Connection of Power Switches for Very High-Power Applications and Zero Voltage Switching," IEEE Transactions on Power Electronics, Vol. 15, No. 1, January, 2000, pp. 44–50.
39. L. F. W. de Souza, E. H. Watanabe, J. E. R. Alves and L. A. S. Pilotto, "Thyristor and Gate Controlled Series Capacitors: Comparison of Components Rating," IEEE Power Engineering Society General Meeting, Toronto, Canada, July, 2003, pp. 2542–2547.
40. S. Banerjee, J. K. Chatterjee and S. C. Triphathy, "Application of Magnetic Energy Storage Unit as Continuous VAR Controller," IEEE Transactions on Energy Conversion, Vol. 5, No. 1, March, 1990.

41. L. Gyugyi, "Unified Power-Flow Control Concept for Flexible AC Transmission Systems," IEE-Proceedings-C, Vol. 139, No. 4, July, 1992, pp. 323–331.
42. L. Gyugyi, C. D. Schauder and K. S. Kalyan, "The Interline Power Flow Controller Concept: A New Approach to Power Flow Management in Transmission Systems," IEEE Transactions on Power Delivery, Vol. 14, No. 3, July, 1999, pp. 1115–1123.
43. A. A. Edris, S. Zelinger, L. Gyugyi and L. J. Kovalsky, "Squeezing More Power from the Grid," IEEE Power Engineering Review, Vol. 22, No. 6, June, 2002, pp. 4–6.
44. E. Uzunovic, B. Fardanesh, L. Hopkins, B. Shperling, S. Zelingher and A. Schuff, "NYPA Convertible Static Compensator (CSC) Application Phase I: STATCOM," IEEE - PES Transmission and Distribution Conference and Exposition. Vol. 2, 2001, pp. 1139–1143.

32
Drives Types and Specifications

Yahya Shakweh, Ph.D.
Technical Director, FKI Industrial Drives & Controls, England, UK

32.1 An Overview 823
  32.1.1 Introduction • 32.1.2 Historical Review • 32.1.3 Advantages of VSD
  • 32.1.4 Disadvantages of VSD
32.2 Drives Requirements & Specifications 827
  32.2.1 General Market Requirements • 32.2.2 Drive Specifications
32.3 Drive Classifications and Characteristics 829
  32.3.1 Classification by Applications • 32.3.2 Classification by Type of Power Device
  • 32.3.3 Classification by the Type of Converter
32.4 Load Profiles and Characteristics 835
  32.4.1 Load Profile Types • 32.4.2 Motor Drive Duty
32.5 Variable Speed Drive Topologies 837
  32.5.1 DC Motor Drives • 32.5.2 Induction Motor Drive • 32.5.3 Synchronous Motor Drives
  • 32.5.4 Special Motors
32.6 PWM-VSI DRIVE 843
  32.6.1 Drive Comparison • 32.6.2 Medium Voltage PWM-VSI • 32.6.3 Control Strategies
  • 32.6.4 Communication in VSDs • 32.6.5 PWM Techniques • 32.6.6 Impact of PWM
  Waveform • 32.6.7 Techniques Used to Reduce the Effect of PMP Voltage Waveform
  • 32.6.8 Supply Front-end for PWM-VSI Drives
32.7 Applications 851
  32.7.1 VSD Applications • 32.7.2 Applications by Industry
  • 32.7.3 Examples of Modern VSD Systems
32.8 Summary 854
  Further Reading 855

32.1 An Overview

32.1.1 Introduction

In every industry there are industrial processes of some form, which require adjustments either for normal operation or optimum performance. Such adjustments are usually accomplished with a variable speed drive (VSD) system. They are an important part of automation. They help to optimize the process, reduce investment costs, energy consumption, and energy cost.

There are three basic types of VSD systems: electrical drives, hydraulic drives and finally mechanical drives. This chapter focuses mainly on electrical drives.

A typical electric VSD system consists of three basic components. The electric motor, the power converter, and the control system, as illustrated in Fig. 32.1. The electric motor is connected directly or indirectly (through gears) to the load. The power converter controls the power flow from an AC supply (often via a supply transformer), to the motor by appropriate control of power semiconductor switches (part of the power converter).

With recent advances of power semiconductor and converter topologies, electric variable speed drives are witnessing a revolution in applications including computer peripheral drives, machine tools and robotic drives, test benches, fan pumps and compressors, paper mill drives, automation, traction and ship propulsion, cement mill and rolling mill drives.

For a proper control system, the VSD system variables, both mechanical and electrical, are required for control and protection. Signals are usually derived from sensors, whose outputs are very much dependent on the control strategy employed and the functionality required.

This chapter introduces electric variable speed drives, and briefly describes their benefits. It examines their classifications

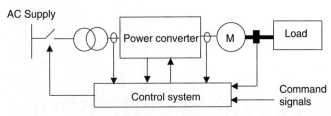

FIGURE 32.1 VSD schematic diagram.

from different perspectives. Their specification requirement to meet applications of different industries is briefly outlined. Various VSD topologies have been carefully examined and compared with each other. A selection of modern VSD applications are examined and briefly commented upon.

32.1.2 Historical Review

In order to appreciate electric VSDs, significant dates in the evolution of electric drives are summarized in Table 32.1 [1].

TABLE 32.1 Historical review

Year	Key advancement
1886	The birth of the electric variable speed drive system represented by Ward Leonard system
1889	The invention of squirrel cage induction motor
1890	The slip ring induction motor drive – speed control via rotor resistance control
1904	Kramer drives – introduce a DC link between the slip rings and the AC supply.
1911	Variable speed system based on induction motor with a commutator on the rotor
1923	Ignitron made controlled rectification possible
1928	The invention of thyratron and grid controlled mercury arc rectifiers
1930	DC to AC power inversion
1931	AC to AC power conversion by cyclo-converters
1950	Silicon based power switches
1960	Thyristors (SCRs) became available and variable speed drives began
1961	Back-to-back reversing DC drive introduced
1960s	Power semiconductor voltage and current ratings grew and performance characteristics improved.
1970	The concept of packaging industrial drives was introduced
1972	First integrated motors with DC converter
1973	Isolated thyristors packages
1970s	The principle of vector control (field-oriented control) evolved.
1983	Plastic molding made their first significant impact on VSDs
1985	Direct torque control as a concept
1990	Integrated power modules
1992	A new packaging trend emerged
1996	Universal drives (a general purpose open loop vector drive, a closed loop flux vector drive and a servo drive)
1998	Complete AC/AC integral converter up to 15 kW
1998	Medium voltage pulse width modulated voltage source inverter drives – became a commercial product

The increased popularity of electric VSD systems, witnessed in recent years, may be explained by the many advantages a VSD can offer. Such advantages include operation at speeds significantly different from the synchronous speed, energy saving, reduced mechanical shock, improved process performance, improved efficiency, reduced mechanical wear, increased plant life, reduced total ownership costs, reduced system fault levels, reduced AC disturbances in certain applications. Furthermore modern electric drives are equipped with many features, including serial communication, remote control, diagnostics, trip history, etc. In the low voltage low power arena, packaged electric drives are becoming a commodity product. The disadvantages of such a system are also recognized. They include the need for extra space to accommodate the equipment, cooling, capital cost, noise, and power system harmonic effects. The following is a brief review of some of the benefits and drawbacks of VSDs.

32.1.3 Advantages of VSD

The author suggests that VSDs benefit most industrial processes with some form of drive. The challenge has often been how to quantify these benefits. The energy saving potential of VSD can be easily quantified, particularly for fan and pump drive applications.

32.1.3.1 Energy Saving

Electric VSD provides savings in two ways: (a) directly by consuming less energy and (b) indirectly by improving the product quality. The latter is often more difficult to quantify.

Direct energy saving is possible only with centrifugal loads such as centrifugal pumps and fans. Such loads are often run at fixed speeds. Traditionally, an automatic valve, or some other mechanical means is used to vary fluid flow rates in pumps. However, if a VSD is used, then the motor speeds can be controlled electronically to obtain a desired flow rate and can result in significant energy savings.

On the basis of the laws of affinity for centrifugal loads,

- The volume of flow is directly proportional to speed
- Pressure is proportional to the square of the speed
- Input power is proportional to the cube of the speed

The affinity law states that the power consumption is proportional to the cube of the motor speed. This implies that if the speed is halved, then the power consumption is reduced to one-eighth. So, energy savings occur as the requirement for volume decreases. If, for example, a cooling system calls for operation at 50% airflow volume, it requires only 12.5% of the power needed to run the system at 100% volume. Because power requirements decrease faster than the reduction in volume, there is a potential for significant energy reduction at lower volume.

Generally, centrifugal pumps and fans are sized to handle peak volume requirements that typically occur for short periods. As a result, centrifugal pumps and fans mostly operate at reduced volumes.

Opening or closing of a damper allows the airflow of fans to be controlled. Restricting the airflow causes the motor to work hard even with a low throughput.

With a variable speed drive, the speed of the fan can be reduced, thus giving the opportunity to reduce energy consumption. Adjusting the speed of the motor regulates the airflow. The control can be achieved by monitoring humidity, temperature, flow, etc. The lower the required throughput, the greater the energy saved.

It has been estimated that the payback period of a 50 kW fan or pump VSD equipment, operating 2000 hours/year is 1.9 years for operation at 75% speed, and 1.23 years for 50% speed. It has been assumed that the cost of the VSD is £5.5k and the cost of power is £0.05/kW.

32.1.3.2 Improved Process Control

Using VSDs to improve process control results in more efficient operating systems. The throughput rates of most industrial processes are functions of many variables. For example, throughput in continuous metal annealing depends on, amongst other factors, the material characteristics, the cross-sectional area of the material being processed and the temperature of one or more heat zones. If constant speed motors are used to run conveyors on the line, it must either run without material during the time required to change temperature in a heat zone or produce scrap during this period. Both choices waste energy or material.

With VSDs, however, the time needed to change speed is significantly less than the time it takes to change heat-zone temperature. By adjusting the material flow continuously to match the heat zone conditions, a production line can operate continuously. The results are less energy use and less scrap metal.

32.1.3.3 Reduced Mechanical Stress (Soft Starts)

Starting a motor on line-power increases stress on the mechanical system e.g. belts and chains. Direct on-line start-up of an induction motor is always associated with high inrush current with poor power factor.

VSD can improve the operating conditions for a system by giving a smooth, controlled start and by saving some energy during starting and running. Smoother start-up operation will prolong life and reduce maintenance, but it is difficult to do more than make an estimate of the cost-advantages of these. The benefits of soft start, inherent in VSD, is that it eliminates the uncontrolled inrush of current that occurs when stationary motor is connected to full line voltage, and also the inevitable suddenly applied high start-up torque. Benefits are that the power wasted by current inrush is eliminated and that the life of the motor and the driven machine are prolonged by the gentle, progressive application of torque.

32.1.3.4 Improved Electrical System Power Factors

When a diode supply bridge is used for rectification, electric variable speed drives operate at near unity power factor over the whole speed range (the supply delivers mostly real power). When a fully controlled thyristor supply bridge is used (as in DC, Cyclo and current source drives) the power factor starts at around 0.9 at full speed, and proportionately worsens as speed declines due to front-end thyristors (typically 0.45 at 50% speed and 0.2 at 25% speed).

Modern pulse width modulated (PWM) drives convert the three phases AC line voltage to a fixed-level DC voltage. They do this regardless of inverter output speed and power. The PWM inverters, therefore, provide a constant power factor regardless of the power factor of the load machine and the controller installation configuration, for example, by adding a reactor or output filter between the VSD and the motor.

32.1.4 Disadvantages of VSD

The cost of VSD is generally space, cooling, and capital cost. Some of the drawbacks are:

- Acoustic noise
- Motor derating
- Supply harmonics

The PWM voltage source inverter (VSI) drives, equipped with fast switching devices, add other possible problems such as (a) premature motor insulation failures, (b) bearing/earth current, and (c) electromagnetic compatibility (EMC).

32.1.4.1 Acoustic Noise

In some installations, placing a VSD on a motor increases the motor's acoustic noise level. The noise occurs when the drive's non-sinusoidal (current and voltage) waveforms produce vibration in the motor's laminations. The non-sinusoidal current and voltage waveforms produced by the VSD are the result of the transistor switching frequency and modulation in the DC-to-AC inverter. The switching frequency, fixed or variable determines the audible motor noise. In general, the higher the carrier frequency, the closer the output waveform is to a pure sine wave. One method of reducing audible motor noise is full-spectrum switching (random switching frequency). The VSD manufacturers accomplish full-spectrum switching by an algorithm within the VSD controller. The motor performance is optimized by evaluating motor characteristics, including motor current, voltage, and the desired output frequency. The resulting frequency band, though audible to humans, produces a family of tones across a wide frequency band. So, the perceived motor noise is considerably less than it would be with a single switching frequency.

Motor noise may not present a problem. Relevant factors include motor locations and the amount of noise produced by other equipment. Traditionally motor noise level is reduced by adding a LC filter between the VSD and the motor, i.e. reducing the high frequency component of the motor voltage waveform. Modern PWM inverter drives run at very high switching frequency and with random switching frequency thus reducing the noise level too. Various methods have been proposed to reduce the magnetically generated noise, which is radiated from inverter-fed induction motors.

32.1.4.2 Motor Heating

Most motor manufacturers design their products according to NEMA standards to operate on utility supplied power. Designers base their motors' heating characteristics and cooling methods on power supplied at fixed voltage and frequency.

For many drive applications, particularly those requiring relatively low power, inverters with a high switching speed can produce variable voltage and variable frequency with little significant harmonic content. With these, either standard or high efficiency induction motors can be used with little or no motor derating. However, the inverters used in larger drives have limits on switching rate that cause their output voltages to contain substantial harmonics of orders 5, 7, 11, 13, and so on. These, in turn, cause harmonic currents and additional heating (copper & iron losses) in the stator and rotor windings. These harmonic currents are limited mainly by the leakage inductance. For simple six-step inverters, the additional power losses, particularly those in the rotor, may require derating of the motor by 10–15%.

Existing constant speed drives often have an oversized induction motor. These can usually be converted to variable speed operation using the original induction motor. Most of the subsequent operation will be at lower load and lower loss than that for which the motor was designed.

Modern PWM VSI drives produce a voltage wave with negligible lower-order harmonics. The wave consists of pulses formed by switching at relatively high frequency between the positive and negative sides of the DC link voltage supply. With larger motors that operate from AC supplies up to 6600 V, the rapid rate of change of the voltage applied to the winding may cause deterioration and failure in the insulation on the entry turns of standard motors.

On self-ventilated (fan-cooled) motors, reducing the motor shaft speed decreases the available cooling airflow. Operating a motor at full torque and reduced speed results in inadequate airflow. This consequently results in increased motor insulation temperature. This potentially can be damaging and can reduce the life of the motor's insulation or cause the motor to fail. One potential solution is to add a constant speed, separately driven cooling fan to the motor. This approach ensures adequate stator cooling over the whole speed range. However the rotor will run hotter than designed as internal airflow remains a function of speed. As there are no windings in the rotor, insulation failure is not an issue but bearings may run hotter and require more frequent lubrication.

Fan-cooled motors with centrifugal loads present less of a problem. Pumps and fans, for example, do not require full torque at reduced speeds. So, in these cases, there is less thermal stress on motors at reduced speeds. Centrifugal load does not cause the motor to exceed thermal limits defined by the insulation system.

32.1.4.3 Supply Harmonics

Current and voltage harmonics in the AC supply are created by VSD (as a non-linear load) connected on the power distribution system. Such harmonics pollute the electric plant, which could cause problems if harmonic level increases beyond a certain level. The effect of harmonics can be overheating of transformers, cables, motors, generators, and capacitors connected to the same power supply with the devices generating the harmonics.

The IEEE 519 recommends practices and requirements for harmonic control in electrical power systems. The philosophy of such regulations is to limit the harmonics injection from customers so that they will not cause unacceptable voltage distortion levels for normal system characteristics and to limit the overall total harmonic distortion of the system voltage supplied by the utility.

In order to reduce supply harmonics that are generated by VSDs, equipped with a 6-pulse diode bridge rectifier, VSD equipment manufacturers adopt various techniques. Table 32.2 summarizes the most common methods and their advantages and disadvantages [2].

Reference [2] quantifies the cost of these options as a percentage of the cost of a basic system with 6-pulse Diode Bridge. For low power VSDs, the cost of a drive with a line reactor is estimated to be 120% of that without. A VSD with a 12-pulse diode bridge with a polygon transformer is 200% while for a double wound transformer is 210%. The most expensive solution is that with active front-end, estimated at 250%.

For 6-pulse converter $n6p \pm 1$ (5, 7, 11, 13, 17, 19, etc) order harmonics are generated. To minimize the effects on the supply network, recommendations are laid down by IEEE 519 as to the acceptable harmonic limits. For higher drive powers, therefore either harmonic filtering or use of a higher converter pulse number is necessary. It is generally true that the use of a higher pulse number is the cheaper alternative. Reference [2] also quantifies the harmonic levels generated by each of the above method, refer to Table 32.3 for a direct comparison.

TABLE 32.2 Techniques used to reduce supply harmonics

Topology	Advantage	Disadvantage
6-pulse bridge with a choke	• Least expensive – Low cost • Known technology • Simple to apply	• Bulky • Too large a value can reduce available torque • Only applies to the drive • Least effective method of filtering
12-pulse bridge	• Eliminates the 5, 7, 17, 19 harmonics • Known technology • Simple to apply	• Bulky and expensive • Only applies to the drive • A lot of 12-pulse drives on one site will shift the problem to the 11th and 13th harmonics
6-pulse, fully controlled active front-end	• Comprehensive filtering for the drive • Cancels all low order harmonics	• Very expensive • Not widely available • New technology
Harmonic filters	• Filters the installation • Reduces the harmonics at the point of common coupling • Least expensive filter to install	• Needs a site survey • Only sized to the existing load
Active filter	• Intelligent filter • Extremely efficient • Can be used globally or locally • More than one device can be installed on the same supply	• Very expensive

TABLE 32.3 Supply harmonics for different supply bridge configurations

Harmonic order number	5th	7th	11th	13th	17th	19th
6-pulse	54%	36%	10%	6.7%	7%	5%
6-pulse with inductor	30%	12%	9%	6%	4%	4%
12-pulse with polygon transformer	11%	6%	6%	5%	2%	1%
12-pulse with double wound transformer	4%	3%	8%	5%	1%	1%
24-pulse 250% cost	4%	3%	1%	1%	1%	1%
Active front-end	3%	3%	3%	0%	2%	2%

32.2 Drives Requirements & Specifications

32.2.1 General Market Requirements

Some of the most common requirements of VSDs are: high reliability, low initial and running costs, high efficiency across speed range, compactness, satisfactory steady-state and dynamic performance, compliance with applicable national and international standards (e.g. EMC, shock, and vibration), durability, high availability, ease of maintenance, and repairs.

The order and priority of such requirements may vary from one application to another and from one industry to another. For example, for low performance drives such as fans and pumps, the initial cost and efficiencies are paramount, as the main reason for employing variable speed drives are energy saving. However, in other industries such as Marine, the compactness of the equipment (high volumetric power densities) is priority requirement due to shortage in space. In such environments direct raw water-cooling is the preferred choice as water is plentiful, and forced water-cooling results in a more compact drive solution.

In critical VSD applications, such as Military Marine Propulsion, reliability, availability and physical size are very critical requirements. Cost is relatively less critical. However, achieving these requirements adds to the cost of the basic drive unit. Series and parallel redundancy of components enable the VSD equipment to continue operation even with failed components. These are usually repaired during regular maintenance. In other critical applications (such as hot mill strips or sub-sea drives) the cost of drive failures could be many times more expensive than the drive itself. For example accessing a drive down on the seabed, many kilometers below the sea-water level could be very difficult.

This section identifies the VSD requirements in various drive applications in different industries.

32.2.1.1 The Mining Industry

The majority of early generation large mine-winders are DC Drives. Modern plants and retrofits generally employ cyclo-converters with AC motors. However, small mine-winders (below 1 MW) tend to remain DC.

The main requirements are:

- High reliability & availability
- Fully regenerative
- Small number requiring single quadrant operation
- High range of speeds
- High starting torque required
- High torque required continuously during slow speed running
- Low torque ripple required
- Low supply harmonics
- Low audible noise emissions
- Flameproof packaging

32.2.1.2 The Marine Industry

The requirements of this industry are:

- Initial purchase price
- Reliability
- Ease of maintenance, i.e. minimum component count, simple design
- Size and weight of equipment
- Transformerless, water-cooled VSD equipment is always preferred

Other desirable features include:

- A requirement for the integration of Power Management functions
- High volumetric power density (the smallest possible)
- Remote diagnostics, to allow faultfinding by experts onshore in critical situations

Drive powers are commonly in the range of 0.75 to 5.8 MW for thrusters, and 6 to 24 MW for propulsion. The evolution in the commercial market is towards powers from 1 to 10 MW for propulsion. Higher powers are required for naval applications. The package drive efficiency must be equal to, or better than 96%. Noise and harmonics problems are to be considered when using PWM inverters. The supply side harmonics produced must be capable of being filtered. Above 1 MW, power converters are usually equipped with a 12-pulse supply bridge, given today's technology.

Two-quadrant operation is required in general, hence, a diode supply bridge is adequate. Occasional requirement for crash stops force use of dynamic brake chopper. DC Bus – can be advantageous for supply to wharf loading equipment, but the drive power ranges are such that commercially available products already adequately serve this application.

The use of standard AC machines is desirable; however, if motors matched to the inverter prove to be cheaper their use could be preferred. Low-noise emission (acoustic and electromagnetic) is very important. There is no requirement for high torque at low speed. Programming and expanded input and output capabilities are required to avoid the need for additional Programmable Logic Control (PLC).

32.2.1.3 The Process Industries

The main requirements of this market are:

- Initial purchase price (long-term cost of ownership does not generally influence purchasing decision)
- Efficiency in continuous processes
- Reliability
- Ease of maintenance
- Bypass facility

The industry preference is for air-cooled drives. It is perceived that air-cooled drives are less costly than their water-cooled equivalents. Customers often have the belief that water and electricity does not mix well, and are wary of problems with leaks. The exception is the offshore industry where equipment size is paramount, and therefore, water-cooling is standard. In general there is no perceived requirement for space-saving in majority of process plants. The desirable features often requested by customers are ease of maintenance and good diagnostic facilities.

The market requirement is for cost-effective, stand-alone drives at various power level from a fraction of a kW up to 30 MW. The use of standard AC machines is desirable. However, if non-standard, but simpler & cheaper machines can be offered an advantage could be gained.

- Two-quadrant operation for fans, pumps, and compressors
- Four-quadrant operation for some Test Benches
- Control must allow additional functions such as temperature protection, motor bearing temperature, flow and pressure control etc.
- There is no requirement, in general, for field weakening
- The harmonics produced by the drive, imposed on the power system should not require a harmonic filter. Harmonics must be minimized

In the Low Voltage (LV) arena, the PWM VSI is dominating the market. In the Medium Voltage (MV) arena, there are a number of viable drive solutions – Load Commutated Inverters (LCI's) and cyclo-converters. However, there is a developing market for MV PWM VSI drives.

32.2.1.4 The Metal Industries

The requirements of this industry are:

- Reliability – high availability
- Efficiency of the equipment – long-term costs of ownership
- Low maintenance costs – (this has been a key factor in the move from DC to AC)

- Power supply system distortion – more onerous regulations from the supply authorities
- Initial purchase cost – very competitive market, and large drive costs have a big impact on total project costs
- Confidence in the supplier and their solution

The following is a list of desirable features:

- Programmable system drives with powerful programming tools
- Preference for air-cooled stacks, but water-cooled is acceptable if a water-to-air heat exchanger is used
- Powerful maintenance and diagnostic tools
- Low EMC noise signature
- Ability to interface to existing automation system via network, Fieldbus or serial link
- Physical size of equipment is often not an important consideration
- Fire protection systems integral to drive equipment

The main market concerns are: (a) EMC regulations, (b) effects on motor insulation of higher voltage levels, and (c) cooling with "Dirty" Mill water is not acceptable. The maintenance of deionized water circuits is a big issue.

32.2.2 Drive Specifications

Failure to properly specify an electric VSD can result in a conflict between the equipment's supplier and the end user. Often the cost can be delayed project completion and/or loss of revenue.

In order to avoid such a problem, requirement specifications should reflect the operating and environmental conditions (Table 32.4). The equipment supplier and the customer need to work as partners and cooperate from the beginning of the project until successful commissioning and hand over. It is advisable that the end user procures the complete drive system, including system engineering, commissioning, engineering support, from one competent supplier.

It is one of the first priorities to identify applicable national and international standards on issues related to EMC, harmonics, safety, noise, smoke emissions during faults, dust, and vibration. Over specifying the requirements could often result in a more expensive solution than necessary. Under specifying the requirements result in poor performance and disappointment.

As far as the end user is concerned, they need to specify the drive interfaces – the AC input voltage, shaft mechanical power, and shaft speed. The torque and current are calculated from these. Frequency and power factor depends on the choice of motor.

For a high-power drive, it is always recommended to carry out a "harmonic survey." Such a survey will reveal the existing level of harmonics, and quantify the impact of the new drive on the harmonic levels.

TABLE 32.4 Typical example of VSD specifications

Variable	Specification
Application	Dynamometer application for a test bench
Motor type	Induction motor
Duty cycle	Continuous at full rating. 150% overload for 1 minute every 60 minutes
Power rating	100 kW
Supply voltage	690 V±5%
Supply frequency	50±0.05 Hz
Speed range	1000:1
Accuracy	0.1%
Min/Max. speed	0/1500 rpm
Torque dynamic response	<10 ms from 100% positive torque to 100% negative torque
Power factor	>96% lagging at all speeds
Efficiency	>98% at full load
Performance	Fully regenerative
	Full torque at zero speed
Ambient temperature	0–40°C
Supply harmonics	G5/3, IEEE519
Life expectancy	>5 years
MTBF	>50,000 hours
MTTR	<2 hours
IP rating	IP45
IEEE 519	IEEE recommended practices and requirements for harmonic control in electrical power systems
IEC 60146	Semiconductor converters. Specifications of basic requirements
IEC 61800	Adjustable speed electrical power drives systems

32.3 Drive Classifications and Characteristics

Table 32.5 illustrates the most commonly used classifications of electric VSDs. In this section, particular emphases will be given to classification by applications and by converter types.

Other classifications, not listed in Table 32.5, include:

- Working voltage: Low-voltage <690 V or Medium Voltage (MV) 2.4–11 kV
- Current type: Unipolar or bipolar drive
- Mechanical coupling: Direct (via a gearbox) or indirect mechanical coupling
- Packaging: Integral motors as opposed to separate motor inverter
- Movement: Rotary movement, vertical, or linear
- Drive configuration: Stand-alone, system, DC link bus
- Speed: High speed and low speed
- Regeneration mode: Regenerative or non-regenerative
- Cooling method: Direct and indirect air, direct water (raw water and deionized water)

Section 32.2 deals with the subject of drives requirement and specification from applications point of view, while

TABLE 32.5 Classifications of electric VSD

By application	By devices	By converter	By motors	By industry	By rating
• Appliances	• Thyristor	• AC/DC (chopper)	• DC	• Power generation	• Fraction kW power < 1 kW
• Low performance (2Q)	• Transistor	• AC/AC direct (cyclo- and matrix-converter)	• Induction motor (squirrel cage and wound rotor)	• Metal	• Low power (1 < P < 5 kW)
• High performance (4Q)	• Gate Turn-off Thyristor (GTO)	• AC/AC via a DC link Voltage source	• Synchronous motor	• Petrochemical	• Medium Power <500 kW
	• Integrated Gate Commutated Thyristor (IGCT)				
• Servo	• Insulated Gate Bipolar Transistor (IGBT)	• AC/AC via a DC link current source	• *Special motors:* SRM, BDCM, Stepper, Actuators, Linear motor	• Process industry	• High power 1-50 MW
	• MOSFET			• Mining	
				• Marine	

Section 32.5 deals with drive topologies from the point of view of motor classifications.

32.3.1 Classification by Applications

Under this classification there are four main groups:

- Appliances (white goods)
- General purpose drives
- System drives
- Servo drives

Table 32.6 describes the main features of these groups and lists typical applications.

32.3.2 Classification by Type of Power Device

The Silicon Controlled Rectifier (SCR), also known as the Thyristor, is the oldest controllable solid-state power device and still the most widely used power device for MV – AC voltages between 2.4 kV and 11 kV – high power drive applications. Such devices are available at high voltages and currents, but the maximum switching frequency is limited and requires a complex commutation circuit for VSI drive. The SCRs are therefore most popular in applications where natural commutation is possible (e.g. cyclo-converters and LCI current source converters).

The Gate Turn-Off Thyristor (GTO) has made PWM VSI drives viable in LV drive applications. The traction industry was one of the first to benefit from such a device on a large scale. Complex gate drive and limited switching performance, combined with the need for a snubber circuit, limited this device to high performance applications where the SCR-based drives could not give the required performance.

The main power devices available in the market can be divided into two groups as shown in Table 32.7.

Bipolar/MOSFET type transistors have witnessed significant popularity in the late eighties, however, they have been replaced by the IGBT which combines the characteristics of both devices – the current handling capability of the bipolar transistor and the ease of drive of the MOSFET.

Traction inverters are designed for DC link voltages between 650 V DC and 3 kV DC with ratings up to 3 MW. The first generation of widely used traction inverter equipment was GTO-based while the latest generation is almost exclusively IGBT-based. Conversion to IGBT has enabled a 30% to 50% reduction in cost, weight, and volume of the equipment.

Early attempts to use GTOs in MV applications failed because of their high cost, snubber requirements, and associated snubber energy loss, which is proportional to the square of the supply voltage. Energy recovery circuitry enables recovery of most of the snubber energy but adds to the cost and complexity of the converter. With high voltage IGBT and IGCT, MV PWM VSI have become commercially available with supply voltage up to 6.6 kV, and power rating in excess of 19 MW.

32.3.3 Classification by the Type of Converter

The power converter is capable of changing both its output voltage magnitude and frequency. However, in many applications these two functions are combined into a single converter by the use of the appropriate switching function; e.g. PWM. By appropriate control of the stator frequency of AC machines, the speed of rotation of the magnetic field in the machine's air gap and thus output speed of the mechanical drive shaft can be adjusted. As the magnetic flux density in the machine must be kept constant under normal operation, the ratio of motor voltage over stator frequency must be kept constant.

The input power of the majority of VSD systems is obtained from sources with constant frequency (e.g. AC supply grid or

32 Drives Types and Specifications

TABLE 32.6 Classification of electric VSD by application

Type of drive	Appliances	General purpose	System	Servo
Performance	Low	Low	High	Very high
Power rating	Very low	Whole range	Whole range	Low
Motor	Universal and induction motor. Recently: PM & SRM are being used	DC motor, induction motor, and synchronous motor	DC motors, induction motors, and synchronous motors	DC motors, brushless DC motors, induction motor, stepper motors, and actuators
Converter	Simple, low cost	AC and DC drives with open loop controller	PWM drives with DC bus, cyclo converter, good quality control with closed loop control, and needs encoder or an observer	DC drive, AC drive, and special motor drives. Tendency towards brushless DC motors
Typical industry	Home	Process	Metal	Automation
Feature	Mass production, low cost, price sensitive, and very low power	Non-regenerative, cost sensitive, low or no overload, low start-up, low performance, and stand-alone	Accuracy with encoders $\lll 0.1\%$ in steady state and dynamic, good precision and linearity of I/O and control, flexible with operations capability, and set up and configuration Communication and feedback	Closed loop, PM motor, >1000 Hz torque response, precise and rapid response, and frequent full speed reversal High precision and linearity of I/Os
Applications	Home appliances e.g. washing machines, dishwasher, temple-dryers, freezers	Fans, pumps, and compressors, Mixers, and simple elevator	Test benches, winders, sectional process line, elevator, cranes, and hoists	Positioning, pick and place, robotics, coordinate control, and machine tools

TABLE 32.7 Power devices used in the VSD converters

Group 1: THYRISTORS	Group 2: TRANSISTORS
This group covers devices having a four-layer, three-junction monolithic structure. They are characterized by low conduction losses and high surge and current carrying capabilities. They operate as an on/off switch. The most popular types of devices listed under this group. • Silicon Controlled Rectifier (SCR) • Gate Turn Off Thyristor (GTO) • MOSFET Controlled Thyristor (MCT) • Field Controlled Thyristor (FCT) • Emitter Switched Thyristor (EST) • MOS Turn Off Thyristor (MTO) • Integrated Gate Commutated Thyristor (IGCT).	Switches listed under this group are basically three-layer two-junction structure devices, which operate in switching and linear modes. They are best recognized for ruggedness of their turn-off capabilities. • Bipolar Junction Transistor (BJT) • Darlington Transistor • MOSFET • Injection Enhanced Gate Transistor (IEGT) • Carrier Stored Trench-Gate Bipolar Transistor (CSTBT) • Insulated Gate Bipolar Transistor (IGBT)

AC generator). In order to achieve a variable frequency output energy an AC/AC converter is needed. Some converters achieve direct power conversion from AC/AC without an intermediate step (e.g. cyclo-converters and matrix-converters). Other converters require DC link (as current source or voltage source).

In all AC variable speed drives the direction of shaft rotation is reversed by simply changing the phase rotation of the inverter through the sequence of driving the switches.

32.3.3.1 DC Static Converter

This drive employs the simplest static converter. It is easily configured to be a regenerative drive with a wide speed range. Table 32.8 summarizes its key features.

High torque is available throughout the speed range with excellent dynamic performance. Unfortunately, the motor requires regular maintenance and the top speed often is a limiting factor. Commutator voltage is limited to around 1000 V and this limits the maximum power available. The continuous stall-torque rating is very limited due to the motor's commutator.

32.3.3.2 Direct AC/AC Converters

Cyclo-Converter

A typical cyclo-converter comprises the equivalent of 3 anti-parallel 6-pulse bridges (for regenerative converter) whose output may be operated in all four quadrants with natural commutation. The main features of cyclo-converters are listed

TABLE 32.8 Converter topologies

Converter	Schematic	Features
(a) Controlled Rectifier		• DC motor • Fully controlled SCR converter • Controlled DC voltage • Simple converter topology • Power factor is a function of speed
(b) Cyclo		• Induction motor & synchronous motor • Direct AC/AC power conversion • 3 × 6-pulse SCR-based fully controlled converters – APT for fully regen • Natural commutation • Low supply harmonics, 18-pulse • Power factor is a function of speed
(c) Matrix		• Squirrel cage induction motor • Synchronous motor • Direct AC/AC power conversion • Forced commutated, reverse conducting switches • Four-quadrant operation inherent PWM in/PWM out • Controlled Power factor
(d) LCI		• Synchronous motor • Simple converter arrangement • Power factor is a function of speed • Load commutated SCRs • Synchronous motor requires excitation • Suffer from torque pulsation at low speeds
(e) FCI		• Squirrel cage induction motor • Similar to LCI • Requires output capacitors for commutation • Requires a diverter commutation circuit for commutation at low speeds • Torque pulsation and resonance
(f) VSI		• Synchronous and squirrel cage induction motors • 6-pulse diode front-end • Good Power factor across speed range • DC link voltage source • PWM output voltage
(g) Kramer		• Wound rotor induction motor with slip rings • Small energy recovery converter • Any type converter may be used between slip ring & AC input

in Table 32.8. This type of drive is best suited for high performance high power >2 MW drives where the maximum motor frequency is less than 33% of the mains frequency.

Matrix-converter

The force-commutated cyclo-converter (better known as a matrix-converter) represents possibly the most advanced state of the art at present, enabling a good input and output current waveform, as well as eliminating the DC link components with very little limitation in input to output frequency ratio. This type of converter is still at its early stages of development. The main advantage of this drive is the ability to convert AC fixed frequency supply input to AC output without DC bus. It is ideal for integrated motor drives with relatively low power ratings. Major drawbacks include: (a) the increased level of silicon employed (bi-directional switches), (b) its output voltage is always less than its input voltage and (c) complexity of commutation and protection.

Matrix-converters provide direct AC/AC power conversion without an intermediate DC link and the associated reactive components. They have substantial benefits for integrated drives as outlined below:

- Reduced volume due to the absence of DC link components
- Ability to operate at the higher thermal limit imposed by the power devices
- Reduced harmonic input current compared to diode bridge
- Ability to regenerate into the supply without dumping heat in dynamic braking resistors
- Matrix converters have not been commercially exploited because of voltage ratio limitation, device count, and difficulties with current commutation control and circuit protection

32.3.3.3 Current Source Inverter (CSI)

The output of this inverter is rectangular blocks of current from the motor bridge supplied from a supply converter whose output is kept at constant current by a DC link reactor and current servo. This type of inverter is typically based on fast thyristors.

Load Commutated Inverter (LCI)

Natural commutations of thyristors is usually achieved with Synchronous Machines at speeds >10%. Natural commutation is induced as a result of the presence of the motors Electromotive Force (EMF), this is called Load Commutation hence the drives other name of LCI. At low speeds the motor voltage is too low to give motor bridge commutations. This is achieved by using the supply converter. Induction motor LCI drives can be supplied by adding a large capacitor on the motor terminals.

The LCI drive covers a wider speed range (up to 10,000 rpm) with power rating up to 100 MW. It gives full load torque throughout the speed range with moderate dynamic performance. Its simple converter design combined with a maintenance free motor design (both induction and synchronous) has increased the popularity of these drives. It is still a popular solution for high power drives (e.g. conveyors, pumps, fans, compressors, and marine propulsion).

The LCI drive has limited performance at low speeds. It also suffers from torque pulsation at 6 and 12 times motor's frequency and beat frequencies. Critical speeds can excite mechanical resonance. Its AC power factor varies with speed. Torque pulsations can be reduced in 12-pulse systems if required.

Forced Commutated Inverter (FCI)

Externally commutated current source converters with an induction motor is also a viable solution. To compensate for the inductive component in the motor current a bank of capacitors is usually used at the motor terminals. The capacitor current is proportional to the motor voltage and frequency. Load commutation at high speed where the compensation current is high enough. Forced commutation at lower speed where the capacitive current is too low for compensation. Forced commutation is achieved using various techniques. The one shown above is based on DC link diverter which consists of a GTO, loading equipment in parallel with the diverting/compensating capacitor. Modern drives employ forced commutated devices, such as reverse blocking GTOs and IGCTs.

32.3.3.4 Slip Power Recovery (Kramer)

In this type of converter, which is described in Table 32.8, the rotor current of a slip-ring wound-rotor induction motor is rectified and the power then reconverted to AC at fixed frequency and fed back into the supply network. For traditional designs the low frequency slip ring currents are rectified with a diode bridge and the DC power is then inverted into AC power at mains frequency.

The traditional designs had poor AC mains dip immunity, high torque pulsation and high levels of low frequency AC supply harmonics. The latest generation of this type of drive is called the Rotor Drive and uses PWM-VSI inverters for the rotor and AC supply bridges.

This keeps sine wave currents in the AC rotor circuits and the drive has many advantages over traditional circuits including:

- No torque pulsation
- Low AC harmonics
- Very high immunity to AC supply dips
- Very cost-effective if a limited speed range is required, but still requires a separate starter

- Inherent ability to run at rated speed without electronic circuits
- Converter cost reduced by 2:1 if uses the ± speed ability to give a speed range

32.3.3.5 PWM-VSI Converter

The availability of power electronic switches with turn-off capability; e.g. FETs, BJTs, IGBTs, and GTOs have currently favored drives with voltage-fed PWM converters on induction.

The PWM VSI drives offer the highest possible performance of all variable speed drives; refer to Table 32.9. Recent improvements in switching technology and the use of microcontrollers have greatly advanced this type of drive. The inverters are now able to operate with an infinite speed range. The supply power factor is always near unity. Additional hardware is easily added if there is a requirement to regenerate power back into the mains supply. Motor ripple current is related to the switching frequency and in large drives the motor may be derated by less than 3%.

TABLE 32.9 Drives features

Type	DC DRIVE			AC DRIVE		
	DC	Cyclo	CSI (FCI)	CSI (LCI)	Kramer	PWM-VSI
Motor type	• DC motor	• Induction and synchronous motors	• Induction motor	• Synchronous motor	• Slip-ring wound rotor induction motor	• Induction or synchronous
Power	• Up to 10 MW	• 2 to 30 MW	• 1 to 10 MW	• 1 to 100 MW	• 0.5 to 50 MW	• 0.5 to 2 MW
• Speed range	• 1000:1	• 1000:1	• 10:1	• 10:1	• 0.8:1.2	• 1000:1
• Accuracy	• 0.01%	• ±0.01%	• ±1%	• ±0.01%	• 0.1%	• 0.01%
• Maximum speed	• Limited by motor capability	• 1000 rpm	• 6000 rpm	• 10,000 rpm	• <1200 rpm	• 10,000 rpm
Performance	• High torque over speed range • High dynamic performance	• High torque over speed range • High dynamic performance	• Poor dynamic response • Low starting torque	• High torque over speed range • Reasonable dynamic performance	• High torque over speed range • High dynamic performance	• High torque over speed range • High dynamic performance
Advantages	• Simple regenerative	• High stall torque (induction) • Inherently regenerative • Robust motors • Low maintenance motor • High over-load capacity	• Standard robust maintenance-free motor • Minimal derating	• Simple • Inherently regenerative • Maintenance-free motor	• Regenerative (new) • Robust • Slip ring wound rotor • High over-load capacity	• Good Power factor • Tolerant to supply dips • Standard robust maintenance-free motor • Minimal derating
Disadvantages	• Stall torque rating • Motor maintenance • Custom motor design	• Motor custom design • Low AC supply Power factor	• Complex • Poor dynamic performance • Torque pulsation & resonance	• Motor custom design • Torque pulsation	• Complex • Motor custom design	• Complex • Expensive • Regeneration at extra cost
Applications	• Mill drives (ball and sag) • Marine propulsion • Mine winders • Process lines • Conveyors	• Mill drives (ball and sag) • Marine propulsion • Mine winders • Conveyors	• Pumps, fans, and compressors • Soft-starter	• Pumps, fans, and compressors • Soft-starter • Marine propulsion • Conveyors • Mill drives	• Pumps, fans, and compressors • Power generation • Mills (ball and sag)	• Process lines • Paper machines • Traction

32.3.3.6 Comparison

Table 32.9 summarizes the main features of all types of converter drives discussed above and assesses their merits and drawbacks. It also illustrates typical applications.

32.4 Load Profiles and Characteristics

The way the drive performs is very much dependent on the load characteristics. Here are four load characteristics described.

32.4.1 Load Profile Types

In the literature, four different load profiles have been described, e.g. Reference [3] (Table 32.10). These are:

1. Torque proportional to the square of the shaft speed (Variable torque)
2. Torque linearly proportional to speed (Linear torque)
3. Torque independent of speed (Constant torque)
4. Torque inversely proportional to speed (Inverse torque)

32.4.2 Motor Drive Duty

32.4.2.1 Duty Cycle

The size of the driven motors is generally chosen for continuous operation at rated output, yet a considerable proportion of motor drives are used for duties other than continuous. As the output attainable under such deviating conditions may differ from the continuous rating, fairly accurate specification of the duty is an important prerequisite for proper planning. There is hardly a limit to the number of possible duty types.

In high performance applications, such as traction and robotics, the load and speed demands vary with time. During acceleration of traction equipment, a higher start-up torque (typically twice the nominal torque) is required; this is usually followed by cruising and deceleration intervals. As the torque varies with time so does the motor current (and motor flux linkage level). The electric, magnetic, and thermal loading of the motor and the electric and thermal loading of the power electronics converter are definite constraints in a drive specification.

Table 32.11 categorizes operating duties into eight major types, Reference [4].

32.4.2.2 Mean Output

Variation of the required motor output during the periods of loaded operation is among the most frequent deviations from the duty types defined in Table 32.11. In such cases the load (defined as current or torque) is represented by the mean load. This represents the root mean square(RMS) value, calculated from the load versus time characteristics. The maximum torque must not exceed 80% of the breakdown torque of an induction motor.

TABLE 32.10 Load characteristics

Type I	Type II	Type III	Type IV
• $T = f(n^2)$ • $P = f(n^3)$ • Low start-up torque • Best suited for energy saving • Torque-speed curve is required when specifying a drive	• $T = f(n)$ • $P = f(n^2)$ Information about process is needed (e.g. density, consistency, viscosity, temperature)	• T = Constant • $P = f(n)$ At start-up the torque may be higher than nominal. Examples static friction with conveyor belts. Vertical and horizontal forces need to be taken into consideration for inclined conveyors.	• $T = f(1/n)$ • P = Constant Mostly dominated by DC drives, but modern PWM VSI is taking over. Certain loads such as winding and reeling machinery require closed loop controls
• Axial and centrifugal pumps • Axial and centrifugal ventilators • Screw and centrifugal compressors • Centrifugal mixers • Agitators	• Mixers • Stirrers	• Extrusions, draw-benches • Paper and printing continuous machines • Volumetric gear pumps / pistons pumps etc. • Piston compressors • Conveyor machines • Lift-machine	• Lift-machines • Reciprocating rolling mill • Winding machines • Lathes • Winders • Reelers • Wire drawers • Web-feed printing machines

TABLE 32.11 Definition of load cyclic duties – VDE0530, in accordance with IEC 34-1

Duty type	Representation	Description
S1: Continuous running duty		Operation at constant load of sufficient duration for the thermal equilibrium to be reached. Specify by indicating "S1" and required output.
S2: Short-time duty		Operation at constant load during a given time, less than required to reach thermal equilibrium, followed by a rest and de-energized period of sufficient duration to re-establish machine temperatures within 2°C of the coolant.
S3: Intermittent periodic duty with a high start-up torque		A sequence of identical duty cycles, each including a period of operation at constant load and a rest and de-energized period. In this duty type the cycle is such that the starting current does not significantly affect the temperature rise.
S4: Intermittent periodic with a high start-up torque		A sequence of identical duty cycles, each cycle including a significant period of starting, a period of operation at constant load and a rest and de-energized period.
S5: Intermittent periodic duty with high start-up torque and electric braking		A sequence of identical cycles, each cycle consisting of a period of starting, a period of operation at constant load, a period of rapid electric braking and a rest and de-energized period.
S6: Continuous operation periodic duty		A sequence of identical duty cycles, each cycle consisting of a period of operation at constant load and a period of operation at no-load. There is no rest and de-energized period.
S7: Continuous operation periodic duty with high start-up torque and electric braking		A sequence of identical duty cycles, each cycle consisting of a period of starting, a period of operation at constant load and a period of electric braking. There is no rest and de-energized period.
S8: Continuous operation periodic duty with related load/speed changes		A sequence of identical duty cycles, each cycle consisting of a period of operation at constant load corresponding to a predetermined speed of rotation, followed by one or more periods of operation at other constant loads corresponding to different speeds of rotation. There is no rest and de-energized period.

If the ratio of the peak torque to the minimum power requirements is greater than 2:1, the error associated with using the root mean square (RMS) output becomes excessive and the mean current has to be used instead. No such mean value approximation is possible with duty type S2, which therefore necessitates special enquiry.

Careful assessment of duty types S2 to S8 reveal that there exist two distinct groups:

1. Duties S2, S3, and S6 permit up rating of motors relative to the output permissible in continuous running duty (S1).

2. Duties S4, S5, S7, and S8 requiring derating relative to the output permissible in continuous running duty (S1).

32.4.2.3 Thermal Cycling

The drive duty cycle also affects the reliability and the life expectancy of power devices. Repetitive load cyclic duty results in additional thermal stresses on power devices. Frequent acceleration and deceleration of drives results in repetitive junction temperature rise and falls at the cyclic duty. The life expectancy of devices is often determined by the maximum allowed number of cycles for a given power device junction temperature rise.

Although this is true for all types of power devices, it is more critical for IGBTs where wire bonds and solder layers are used.

In modern IGBT-based converter design, the maximum junction temperature rise of the IGBTs is limited to a level, which ensures a conservative number of thermal cycles over the lifetime of the drive. Typical junction temperature rise is 30°C for a repetitive cyclic duty (e.g. steel mill) and 40°C for non-repetitive cyclic duty (e.g. fan pumps).

32.4.2.4 Multi-quadrant Operation

Fully regenerative electric VSDs offer a rapid regenerative dynamic braking in both forward and reverse directions. Operation in motoring implies that torque and speed are in the same direction (QI, III). In regenerative braking the torque is opposite to the speed direction (QII, IV) and the electric power flow in the motor is reversed. (See Fig. 32.2.)

Positive power flow of electric energy means that electric power is drawn from the power supply via the power electronics converter by the motor while negative power flow refers to electric power delivered by the motor in the generator mode to the power electronics converter. This could be regenerated back to the supply or dissipated, as a heat in the dynamic brake dissipative mechanism.

For regenerative drive, the power electronics converter has to be designed to be able to handle bidirectional power flow. In low and medium power converter (say <500 kW) with slow dynamic braking demands, the generated power during the braking period is interchanged with the strong filter capacitor of power electronics converter, or DC (dynamic) braking is used.

32.4.2.5 Dynamic Braking Energy

There exist two types of energy stored in VSD, which need to be dealt with during dynamic braking:

- *Inertia or kinetic energy loads:* Typically moving (rotating or linear) machines. These would decelerate naturally to rest. Braking can speed up the process cycle for the sake of productivity.

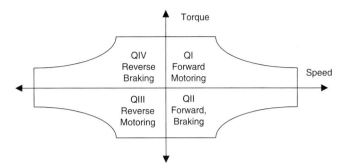

FIGURE 32.2 Operating regions of electric VSD.

- *Mass or potential energy loads:* Typically hoists or lifts – which would run on or even accelerate. Braking must apply full power to maintain constant speed while the load is lowered.

The drive losses, mechanical resistance, and the transmission efficiency, work in favor of deceleration, reducing the braking power demand. The energy regenerated by potential energy loads depend on maximum power and both the overrun time and the decelerating time.

The braking time and the duty cycle time are decided by the requirements of the process system, but note particularly the effect of varying the braking duty cycle time and the deceleration time.

For DC injection braking the kinetic energy of the motor - load system is converted to heat in the motor rotor. For fast and frequent generator braking the power electronics converter has to handle the generated power either by a controlled dynamic brake chopper (with braking resistor) or through bidirectional power flow. The power losses in the converter can assist in dynamic braking.

For a fast speed response, modern variable speed drives may develop a maximum transient torque up to base speed and maximum transient power up to maximum speed, provided that both the motor and the power electronics converter can handle these powers. For a 200 kW dynamometer drive application, a rapid change of torque from full positive torque to full negative torque is required in less than 10 ms.

32.5 Variable Speed Drive Topologies

In this section drive topologies are classified according to the motor they employ. Various publications dealt with this subject e.g. References [3, 5]. The most common motors are illustrated in Fig. 32.3.

32.5.1 DC Motor Drives

Until recently, the DC motor drive was the most commonly used type of electric variable speed drive, with only very few

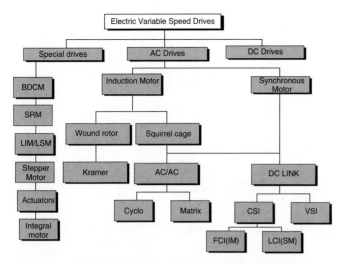

FIGURE 32.3 Classification of electric VSD.

exceptions is the least expensive. The mechanical commutator is an electromechanical DC to AC bidirectional power flow power converter, as the currents in the rotor armature coils are AC while the brush-current is DC. The DC drive is well-known, well-proven and widely applied; yet its popularity is in relative decline due to the emergence of the more robust, lower cost squirrel cage induction motor drive.

Unfortunately the mechanical commutator though not bad in terms of losses and power density has serious commutation current and speed limits and thus limits the power per unit to 1–2 MW at 1000 rpm and may not be accepted at all in chemically aggressive or explosion-prone environments. The application of the DC drive has been restricted to hazardous areas due to the very limited availability of flameproof DC machines. Commutator and brush maintenance is difficult in such environments. Furthermore, continuous sparking at the brushes is virtually inevitable at full load output.

Due to the inherent ease of speed control of the separately excited DC machine, DC drives found popularity in early electric drive applications, by varying the applied armature voltage. This variable armature voltage is simply generated by phase-controlled rectification and this technique has now almost entirely replaced the Ward-Leonard systems previously used.

The AC/DC converter offers a variable DC voltage, which is capable of four-quadrant operation (positive and negative DC voltage and DC current output). Permanent Magnet-excited brushed motors have been used in numerous applications for sometime, particularly in non-regenerative drive applications.

Motor output torque is approximately proportional to armature current and motor speed is approximately proportional to converter output voltage. Speed control by sensing armature voltage is therefore feasible giving an accuracy of around 5%.

Provided the motor excitation is kept constant, the DC drive power factor is proportional to motor speed. Since most pumps, compressors, and fans demand a torque proportional to the square of speed, constant excitation systems are used and so the above relationship applies.

A typical power factor, at maximum rated speed for a DC drive is 0.85. This relationship applies to many other types of electric drives.

If a slow dynamic response is satisfactory, regeneration to the mains supply is achieved either by reversing the motor field or armature connection. Alternatively regeneration with faster response is achieved by connecting another Thyristor Bridge in anti-parallel with the main bridge. In this case fast response is possible with changeover time of <15 ms between full torque motoring to full torque regenerating. The 6-pulse drive configuration is acceptable for powers up to 1 MW. This limitation arises not from any semiconductor device limitation, but is due to AC line current harmonics the converter generates.

A force-commutated or "chopper" converter for DC motors uses the principle of variable mark-space control using a Thyristor or transistor solid-state switch. With a diode front-end converter a fixed, smoother DC supply is derived from the mains by uncontrolled rectification and rapidly applied, removed, and reapplied to the machine for adjustable intervals, thus applying a variable mean DC voltage to the DC motor, refer to Fig. 32.4b.

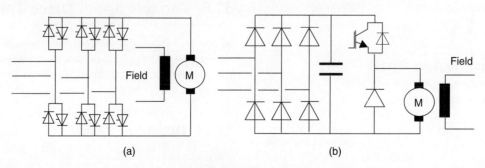

FIGURE 32.4 DC drive: (a) with fully controlled anti-parallel supply bridge and (b) diode rectifier with DC chopper.

This type of DC drive has the advantage of high (near unity) power factor at all motor speeds and much reduced harmonic spectrum.

32.5.2 Induction Motor Drive

32.5.2.1 Squirrel Cage Induction Motor

Squirrel cage induction motors are simpler in structure than DC motors and are most commonly used in the VSD industry. They are robust and reliable. They require little maintenance and are available at very competitive prices. They can be designed with totally enclosed motors to operate in dirty and explosive environments. Their initial cost is substantially less than that of commutator motors and their efficiency is comparable. All these features make them attractive for use in industrial drives.

The three-stator windings develop a rotating magnetic flux rotating at synchronous speed. This speed depends on the motor pole number and supply frequency: The rotating flux intersects the rotor windings and induces an EMF in the rotor winding, which in turn results in circulating current. The rotor currents produce a second magnetic flux, which interacts with the stator flux to produce torque to accelerate the machine. As the rotor accelerates, the induced rotor voltage falls in magnitude and frequency until an equilibrium speed is reached. At this point the induced rotor current is sufficient to produce the torque demanded by the load. The rotor speed is slightly lower than the synchronous speed by the slip frequency, typically 3%.

In order to ensure constant excitation of the machine, and to maximize torque production up to the base speed, the ratio of stator voltage to frequency needs to be kept approximately constant.

Induction motor drive has three distinct operating regions:

(a) *Constant Torque:* The inverter voltage is controlled up to a maximum value limited by the supply voltage. As the motor speed and the voltage are increased in proportion, constant V/F, the rated flux linkage is maintained up to the base speed. Values of torque up to the maximum value can be produced at speeds up to about this base value. The maximum available torque is proportional to the square of the flux linkage. Typically, the induction motor is designed to provide a continuous torque rating of about 40–50% of its maximum torque.

(b) *Constant Power:* For higher speed, the frequency of the inverter can be increased, but the supply voltage has to be kept constant at the maximum value available in the supply. This causes the stator flux linkage to decrease in inverse proportion to the frequency. Constant power can be achieved up to the speed at which the peak torque available from the motor is just sufficient to reach the constant power curve. A constant power speed range of 2–2.5 can usually be achieved. Within this range, the motor frequency is increased until, at maximum speed.

(c) *Machine Limit (Pullout Torque):* Once the machine limit has been reached the torque falls off in proportion to the square of motor frequency. Operation at the higher end of this speed range may not be feasible as the motor power factor worsens. This in turn results in a higher stator current than the rated value. The motor heating may be excessive unless the duty factor is low.

Induction motors are used in applications requiring fast and precise control of torque, speed, and shaft position.

The control method widely used in this type of application is known as Vector control, a transient response at least equivalent to that of a commutator motor can be achieved.

The voltage, current, and flux linkage variables in this circuit are space vectors from which the instantaneous values of the phase quantities can be obtained by projecting the space vector on three radial axes displaced 120° from each other. The real and imaginary components of the space vectors are separated, resulting in separate direct and quadrature axis equivalent circuits but with equal parameters in the two axes.

Changes in the rotor flux linkage can be made to occur only relatively slowly because of the large value of the magnetizing inductance of the induction motor. Vector control is based on keeping the magnitude of the instantaneous magnetizing current space vector constant so that the rotor flux linkage remains constant. The motor is supplied from an inverter that provides an instantaneously controlled set of phase currents that combine to form the space vector, which is controlled to have constant magnitude to maintain constant rotor flux linkage. The second component is a space vector, which is in space quadrature with the instantaneous magnetizing current space vector. This component is instantaneously controlled to be proportional to the demand torque.

To the extent that the inverter can supply instantaneous stator currents meeting these two requirements, the motor is capable of responding without time delay to a demand for torque. This feature, combined with the relatively low inertia of the induction motor rotor, makes this drive attractive for high-performance control systems.

Vector control requires a means of measuring or estimating the instantaneous magnitude and angle of the space vector of the rotor flux linkage. Direct measurement is generally not feasible. Rapid advances are being made in devising control configurations that use measured electrical terminal values for estimation.

32.5.2.2 Slip-ring (Wound-rotor) Induction Motor Drive

Wound rotor induction motors with three rotor slip rings have been used in adjustable speed drives for many years.

In an induction motor, torque is equal to the power crossing the air gap divided by the synchronous mechanical speed. In early slip-ring induction motor drives, power was transferred through the motor to be dissipated in external resistances, connected to the slip ring terminals of the rotor. This resulted in an inefficient drive over most of the speed ranges. More modern slip ring drives use an inverter to recover the power from the rotor circuit, feeding it back to the supply system.

The speed of slip-ring induction motor can be controlled by:

- Stator frequency control as with a cage rotor machine
- Rotor frequency control
- Rotor resistance control
- Slip energy recovery (Kramer system). For capital cost reasons the last two are commonly used

Addition of rotor resistance especially for starting large induction motors is well-known. The basic effect produced by adding rotor resistance is to alter the speed at which maximum motor torque is developed. Unfortunately, power dissipation as heat in the rotor resistance bank takes place, earlier means to overcome this shortcoming were to convert the rotor power to DC, and feed a DC motor on the same shaft. The rotor slip energy, when running at reduced speed, is therefore reconverted to mechanical power. This is the "Kramer" system. The disadvantages of this approach were the extra maintenance and capital costs.

The static Kramer system overcomes these shortcomings by replacing the DC machine with a line commutated inverter which returns the slip-energy directly to the AC line, either directly (on lower power systems) or via a transformer. A key advantage of the Kramer drive system is that the slip energy recovery equipment (DC machines or static inverter) needs only be rated for a fraction of the maximum motor rating. This is true when a small speed range is required and provided that a separate means is provided of starting the motor. This is because the motor rotor current is proportional to torque and the rotor voltage inversely proportional to speed.

Naturally if the slip energy recovery network can be rated to withstand full rotor voltage (developed at standstill) a controlled speed range of zero to maximum could be achieved. However this is generally only feasible on smaller motors (below 2000 kW) where the rotor voltage is sufficiently low for an economic inverter package. Secondly if a full speed range is needed the slip energy recovery network has to be rated at full motor power, so static Kramer drives become uneconomic for wide speed ranges. The overall system power factor would be very low for a wide speed range system.

For the above reasons Kramer drives are very suitable for high power drives (>200 kW) where a small speed range is required. Pump and fan drives present therefore good economic applications. Kramer drives have also been used for low speed range endurance dynos using the recovery system to control torque of induction generator. As with all line-commutated converters and inverters, current harmonics are produced and these can be reduced to acceptable values. However, as the slip-energy recovery network is only power-rated in direct proportion to the speed reduction required (assuming constant load torque), the magnitudes of the harmonic currents generated are proportionally less than with drives where the solid-state converters have to handle the whole drive power. Harmonics of the rotor rectifiers are transmitted through the rotor and appear as non-integer harmonics in the main supply.

The main disadvantages of the slip-ring induction motor drive are: (a) the increased cost of the motor in comparison with a squirrel cage, (b) the need for slip ring maintenance, (c) difficulty in operating in hazardous environments, (d) the need for switchable start-up resistors, and (e) the poor power factor compared with other types of drive.

32.5.3 Synchronous Motor Drives

To understand the way the synchronous machine operates, let us assume that the induction motor were to rotate at the synchronous speed by an external means. Under this condition the frequency and magnitude of the rotor currents would become zero. If an external DC power supply were connected to the rotor winding, then the rotor would become polarized in a similar way to a permanent magnet. The rotor would pull into step with the air-gap-rotating magnetic field, generated by the stator but lagging it by a small constant angle, referred to as the load angle. The load angle is proportional to the torque applied to the shaft, and the rotor keeps rotating at synchronous speed, provided that the DC supply is maintained to the rotor field winding. The magnetic flux produced by the rotor winding intersects the stator windings and generates a back EMF, which makes the synchronous motor significantly different from the induction motor.

As with the induction motor drive, the requirement is to keep the ratio V/F constant (i.e. varies both the stator frequency and applied voltages in proportion to the desired motor speed).

The supply bridge converter is phase controlled generating an adjustable DC current in the DC link choke. To generate maximum torque from the synchronous motor this current is switched into the motor stator windings at the correct phase position with respect to rotor angular position as detected by position sensor by the Inverter Bridge. When running above about 10% speed, the back EMF generated by the synchronous motor is sufficient to commutate the current into the next arm of Inverter Bridge. So, as this type of inverter is machine (motor) commutated, the inverter configuration is merely that of a conventional DC drive. The complexity, expense, and limited power capability of the force-commutated circuitry is therefore avoided.

The motor back EMF is insufficient for Thyristor commutation at low speeds. The technique here, therefore, is to rapidly phase back the supply converter bridge to reduce the DC link current to zero and after a short delay (to ensure that all thyristors in machine bridge are turned off) reapply DC current when the correct Thyristor trigger pattern has been reestablished. As the motor speed and thus back EMF, increase to a value sufficient for machine commutation, changeover to continuous DC link current operation is effected.

During the starting mode the correct Inverter Bridge firing instant is determined by rotor position sensor, which is mounted on the motor shaft whose angular position is detected by opto or magnetic probes. When in the machine-commutated mode, sensing of stator voltage is used. To develop maximum torque in the low speed or pulsed mode, angular rotor position sensing is necessary. However if less than full load torque availability at low speed can be tolerated, the inverter system can be set to produce a low fixed frequency in the pulsed mode. This frequency is then increased, as motor rotation is detected (either in steps or on a pre-set ramp rate) until sufficient back EMF is generated to facilitate changeover to the voltage-sensing mode.

As previously stated the key advantage of this type of drive is that all Thyristor devices are line or machine commutated. Expensive and complex forced commutation circuitry is avoided and fast turn-off thyristors are unnecessary. Inverter systems of this type can therefore be built at very high powers, up to 100 MW. Also, as a result of avoiding force commutation, converter efficiency is high.

The thyristors in the machine Inverter Bridge must be triggered at such an angle to give sufficient time for commutation from one device to the next. This results in the synchronous motor operating at a high leading power factor of around 0.85. However, as far as the mains supply is concerned, the total drive has the characteristics of a DC drive where power factor is proportional to speed.

Another important characteristic of this type of drive is that it is inherently reversible and regenerative. For regenerative operation the Inverter Bridge is triggered in the fully advanced position so, in effect, it becomes a plain diode bridge. A DC output voltage, approximately proportional to motor speed, is therefore generated at the DC side of the supply Converter Bridge. This converter bridge is now triggered in the regenerative mode thus returning power to the supply system. Reversing operation is achieved by altering the sequence in which the thyristors in Inverter Bridge are triggered.

This type of drive is widely applied over a wide power range as it embodies an efficient brushless motor and relatively simple and efficient converter. At lower powers, say below 30 kW, permanent magnet synchronous motors is more common.

Unlike the induction motor, the synchronous type requires two types of converter. The first for main power conversion while the second is low power for field excitation. The field converter feeds the rotor exciter winding through slip rings and brushes or alternatively a brushless exciter can be used. A coordinated control of the two converters provide for active power and reactive power control and for efficient wide speed range control in high power applications.

For high power applications, synchronous motors are preferred because of the ability to control reactive power flow through appropriate control of excitation. Synchronous motors tend to have wider speed range and higher efficiency. However, synchronous motors are generally more expensive than induction motors.

With modern high power PWM-VSI drives, synchronous motor can be driven for same inverter with vector control methods.

32.5.4 Special Motors

Motors under this category employ power electronics converters for normal operation. Generally, this type of motor has a large number of phases in order to limit torque pulsation and self-start from any rotor initial position. This is a new breed of motors, which can be fed through a unipolar or bipolar current. Also they have singly salient or doubly salient magnetic structures with or without permanent magnets on the rotor.

32.5.4.1 Brushless DC Motor (BDCM) Drive

This type of machine has a similar construction to a standard synchronous machine, but the rotor magnetic field is produced by permanent magnet material. A position sensor is used to ensure synchronism between the rotor position and the stator magneto motive force (MMF) via drive signals to the inverter. The use of new magnet materials characterized by high coercive force levels has reduced magnet sizes, and largely overcome the demagnetization problem. The absence of the field copper losses improves the machine efficiency.

As the permanent magnet is the source for excitation, the BDCM can be viewed as a constant flux motor. A limited amount of flux weakening can be achieved by increasing the load angle of the stator current. Achieving a useful constant power range is not usually practical with this type of motor. A large demagnetizing component of stator current would be required to produce a significant reduction in magnet flux, and this would increase the stator loss substantially.

The required base torque determines the motor size and the losses are essentially independent of the number of stator turns. At speeds up to the base speed of the constant power range, the efficiency of the motor is essentially the same as for one designed for rated voltage at base speed. For operation above base speed, the stator current from the inverter is reduced in inverse proportion to the speed. This mode of operation in the high-speed range reduces the dominant stator winding losses relative to a machine in which the flux is reduced and the current kept constant. The losses in the inverter are, however, increased due to its higher current rating. For an electric road vehicle that must carry its energy store, the net energy saving may be sufficiently valuable to overcome the additional cost of the larger inverter. A further advantage of this approach is that, if the DC supply to the inverter is lost, the open circuit voltage applied to the inverter switches will be within their normal ratings.

The BDCM has higher volumetric power density compared to other types of motors (induction or synchronous). They are particularly suited for the high values of acceleration required in drives (e.g. machine tools). They are often operated with high acceleration for a short time followed by a longer period of low torque. At such low values of load factor, the cooling capability is frequently not a limitation. The major interest is in obtaining the maximum acceleration from the motor. The short-term stator current of a BDCM is limited to the value required for magnet protection. These values of acceleration are significantly higher than that can be achieved with either induction or DC motors of similar maximum torque rating.

32.5.4.2 Switched Reluctance Motor (SRM) Drive

This motor can be regarded as a special case of a salient synchronous machine in which the field MMF is zero and the torque is produced by reluctance or saliency action only. The rotor has no winding. The SRM drive needs an inverter whose frequency is locked to the shaft speed, but since the torque is linearly proportional to the square of the stator current, the use of unidirectional current involves little sacrifice in performance.

Generally, the use of position sensors in the SRM and BDCM is something of a disadvantage in both cases. The SRM does not require permanent magnets, which can cost and may involve demagnetization risks and limit top speeds due to centrifugal forces. The SRM hence has a simpler construction and is more robust. However the need to magnetize the motor from the AC side adds to inverter costs and may increase peak current levels significantly, hence raising stator copper losses.

Switched reluctance synchronous motors have a cylindrical stator with three AC windings and a solid rotor (without any winding) with a moderate orthogonal axis magnetic saliency up to 4 (6) to 1. High magnetic saliency is obtained with multiple flux barriers. The conventional SRMs are to some extent (up to 100 kW) used in low dynamics variable speed drives with open loop speed control, as the speed does not decrease with load. Consequently the control is simpler than with induction motors.

The main drawback of the conventional SRD is the low motor power factor and the relatively poor torque density, which leads to a higher kVA rating of the power converter (approximately 20%). The main advantage of the synchronous reluctance motor over the induction motor of similar rating is the higher efficiency. Compared to the squirrel cage induction motor, the rotor loss is small or negligible in synchronous reluctance machines. If the saliency ratio is sufficient to produce a power factor equal to that of the induction motor, the stator winding loss will be the same. Also, the stator iron losses will be similar for the two motors.

The reluctance motor is capable of operation in the constant power mode of operation. As for all AC drives, when the supply voltage limit is reached above the base speed, the flux linkage is reduced in inverse proportion to the shaft speed, and the torque is inversely proportional to speed squared.

32.5.4.3 Linear Motors

There are applications in which linear motion, as opposed to rotational, is required. A linear machine has the same operating principles as those applied to all other rotating machines. The PWM-VSI converters and motor control principles discussed in this chapter are also applicable to this type of motor.

There are two types of linear motors:

- LIM – Linear Induction Motor
- LSM – Linear Synchronous Motor with permanent magnetic excitation

The LSM type has the following advantages over the LIM:

- Better power factor
- More responsive control
- Higher efficiency

The disadvantages of LSM are:

- Very accurate position feedback is required
- The use of PM – expensive and heavy

Transport, material handling, and extrusion processes are a few examples in which linear motors have successfully been employed.

32.5.4.4 Stepper Motors

Stepper motors are either built in a similar manner to BDCM, with permanent magnets embedded in or bonded to the rotor or a rotor with no magnets. The latter type is made of a ferrite magnetic material and its circumference is cut to form a number of slots, forming teeth lengthwise to the rotor axis.

Torque production can be based on (a) magnetic reluctance (as in SRM), (b) magnetic attraction (as in BDCM), or (c) both magnetic reluctance and attraction.

Stepper drives do not offer dynamic speed control, and the main action is to accelerate at full torque to full speed, maintain the speed and decelerate at full torque. In comparison to the reluctance type stepper motor, the permanent magnet type offers greater torque for a given speed, particularly at start and low speeds.

Most drives incorporate controllers with connections for a communications link for supervisory control by PLC, hard-wiring connectors for analog/digital inputs and outputs, and some are equipped with software for communications with

32 Drives Types and Specifications

TABLE 32.12 Control features for servo and stepper motor drives, Reference [6]

Control features	Servo drive	Stepper drive
Acceleration/deceleration time	Adjustable	Accelerate at maximum torque, time is dependent on maximum torque and inertia
Maximum speed	Part of the motor specification	Part of the motor specification
Speed control	Permit a range of speed settings	Not necessarily available
Torque control	Many offer speed & torque control	Always operate at maximum torque
Auto tuning	A feature of some servo drive	Not applicable
Reversing	Commonly available by digital control signal	Commonly available by digital control signal
Zero speed clamp	Applies full torque to hold the position constant	Applies full torque to hold the position constant
Dynamic braking	Controlled deceleration, may require dissipative brake resistor	Usually standard
Regenerative braking	Dedicated circuit for controlled braking	Not applicable
Travel limits	Definition of travel limits in the forward and reverse directions	Standard
Jog or inch	Digital command to "jog" one step (with defined distance)	Optional feature
Closed loop configuration	Most drives accept external signals for closed loop control	Most drives accept external signals for closed loop control
Programming functions	Many drives incorporate programming functions as in PLCs, reset all functions to default states, return to a home position, enable or disable repetition or a pre-set sequence, select a particular set of control inputs, increased or decreased speed, change the torque boost, etc.	

a computer or handheld keypad. Table 32.12 lists typical options.

Unlike above motor drives, the stepper motor can achieve precise position control without the need for any external feedback.

32.5.4.5 Actuators

Actuators are widely used in industry, primarily for positioning tasks. Their designs are based on all sorts of force producing principles. Reference [7] describes several types of direct drive electric actuators, including (a) the DC actuator (Moving coil type), (b) induction actuators, (c) synchronous actuators (moving magnet DC type), (d) reluctance actuators, and (e) inductor actuators (polarized reluctance type).

Electric actuators are used increasingly in control systems and automated electromechanical equipment. Typical specification factors include: range of motion, type of motion (linear or rotary, stepwise or continuous), resolution needs, speed of response, environmental conditions, supply conditions, allowable electromagnetic noise emission level, need for integrated position and velocity sensors, maintenance needs, eligibility, cost, peak, and continuos torque, etc.

The main demands of industry for high performance systems are:

(a) A convenient supply and low power consumption
(b) Reliability and robustness
(c) Low initial cost and maintenance
(d) Fast response
(e) Linear "torque-excitation" characteristics

32.5.4.6 Integrated Motors

The Integral Motor consists of a standard AC motor with an integrated frequency inverter and EMC filter. It is robust and specified for reliable operation, and often designed to handle rough working conditions, including ambient temperature −25–40°C and dusty, corrosive as well as humid environments (Enclosure IP55). This type of drive uses a standard induction motor with the AC/AC converter integrated in the motor frame often as a separate converter box mounted directly above the motor frame in place of the terminal box. The power popularity of this type of drive is limited to 0.5–7.5 kW.

This type of motor offers the following advantages:

- Save space by eliminating the need for a separate controller
- Reduce installed costs because cabling between motor and converter is eliminated
- Eliminate motor problems caused by high voltage transient due to output cable capacitance
- Minimize EMC due to high dV/dt

The integrated drive includes most features including start, stop, forward, reverse, speed and torque controls, controlled acceleration and deceleration, etc.

32.6 PWM-VSI DRIVE

In recent years, the popularity of PWM VSI has increased beyond recognition. Its dynamic performance and controllability is better than the DC drive. Its power range has extended

TABLE 32.13 Drive comparison

Control features	Cyclo	LCI	PWM-VSI	Matrix
Speed	Limited	Wide	Wide	Wide
Dynamic response	Excellent	Good	Excellent	Excellent
Torque pulsation	Low	High	Very low	Very good
Power factor at low speed	Poor	Poor	Very good	Very good
Stability	Good	Moderate	Very good	Very good
Motor	Custom	Custom	Standard	Standard*
Regeneration	Inherent	Inherent	Needs extra hardware	Inherent
Volumetric power density	Moderate	Good	Very good	Excellent

*The AC output voltage of the matrix-converter is always less than the input voltage – derating is expected or a larger frame size is required.

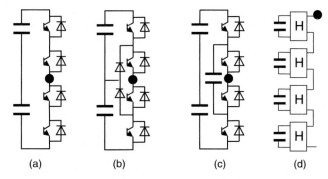

FIGURE 32.5 MV stack topologies.

to areas dominated for years by traditional solutions such as the cyclo-converter and LCI drives.

32.6.1 Drive Comparison

Table 32.13 shows a direct comparison between the cyclo-converter, LCI, and PWM-VSI drives. The DC drive and the Slip Power Recovery converter type are not listed because AC drives have already replaced DC drives in most applications due to low maintenance and better reliability of AC motors. Slip recovery is only suitable for applications with a limited speed range and requires a slip ring wound rotor.

In comparison with the cyclo-converter and LCI current source converter drives, the PWM-VSI drive offers the following advantages:

- Excellent dynamic response
- Smooth torque/speed control over full speed range (0–200 Hz)
- High volumetric power density
- Ride through of dips in supply voltage
- Use of standard motors (squirrel cage induction motor or synchronous motor)
- Improved AC supply power factor over full speed range
- Reduced cabling and transformer size and cost in comparison with cyclo-converters
- No significant torque pulsation
- Lower noise level
- Low maintenance

32.6.2 Medium Voltage PWM-VSI

The maximum power rating of LV VSD is limited by practical current ratings of power components such as motor, cable and transformer (typically 1500 A), giving a limit of about 2 MVA at 600 V. At this rating, motor manufacturers always prefer a MV machine design – significant saving and improved thermal performance of power components can be achieved by operating at medium voltages instead of low voltages. Many variable-speed drive applications will benefit from the availability of economic MV alternatives.

When adequately rated high blocking voltage devices are available, a simple 2-level inverter or alternatively 3-level Neutral Point Clamped (NPC) has always been the choice to meet required output voltages. These topologies offer a simple and cost-effective solution.

Series connection of power devices is the traditional solution for high power high voltage Thyristor-based drives. This approach is perceived to be complex with fast switching IGBTs because of simultaneous switching and correct static and dynamic voltage sharing of series devices.

The "Multi-level" inverter drive is seen to offer a better solution for high power, high voltage inverter drive. The output waveform is high quality, even at very high modulation frequencies, which inherently results in lower harmonic content in the output voltage waveform (less losses, less torque pulsation, and lower insulation voltage stresses).

Reference [8] and Fig. 32.5 categorizes MV converter topologies as follows:

(a) Series Connected 2-Level (SC2L)
(b) 3 Level Neutral Point Clamped (3LNPC)
(c) Multi-level: Diode Clamped Multi-Level (DCML), Capacitor Clamped Multi-Level (CCML)
(d) Isolated Series H-Bridge (ISHB)

32.6.3 Control Strategies

Several control techniques can be found in the VSD industry, refer to Fig. 32.6. These are:

- Open loop inverter with fixed V/Hz control
- Open loop inverter with flux vector control
- Closed loop inverter with flux vector control (induction motor)

Table 32.15 summarizes the main features, advantages, and disadvantages of each technique.

32 Drives Types and Specifications

TABLE 32.14 Comparison between different MV converter stack topologies

Topologies	Advantages	Disadvantages
2-level with series devices (SC2L)	• Simple & proven technology • Same converter design over supply voltage range • Standard fully developed PWM control • Provision for series redundancy of power switches per inverter phase arm (n+1)	• Static and dynamic voltage sharing of series devices • High dV/dt due to synchronous commutation of series devices • High switching frequency harmonic content in inverter output voltage
3-level NPC (3LNPC)	• Well-proven • Reduced harmonic content • Better utilization of switches • Reduced dV/dt (half the SC2L equivalent)	• Series redundancy is difficult to achieve • More complex PWM control is needed, than 2 level • Requires extra clamping diodes • Requires split DC link • Requires mid-point voltage balance control • Even number of power devices per arm is always needed • Switches requires snubbers
Diode clamped multi-level (DCML)	• Reduced harmonic contents • Reduced dV/dt	• Series redundancy is very difficult to achieve • Very complex PWM control is needed • Requires many steering diodes • Requires split DC link • Requires voltage balance control of split DC link capacitors • Uneven current stresses on power devices • Requires snubbers
Capacitor clamped multi-level (CCML)	• This configuration has all the advantages of a multi-level converter plus Simpler arrangement, modular building block • Less components • Snubberless operation is possible • Easier capacitor voltage balance than 3LNPC	• Possible parasitic resonance between decoupling capacitors • Complex to provide series redundancy • More complex PWM control strategy than for 2-level • Voltage redistribution of capacitors during supply voltage surges • Too many capacitors (bulky stack design & poor capacitor utilization at high ratings) • Complex converter arrangement (for low stray inductance) • Inverter rating is limited by the load current flowing through the capacitors
Series connected isolated h-bridges (ISHB)	• Modular design of the converter power modules • The basic building block is based on a DC supply bridge, decoupling capacitor and a H-bridge arrangement • In the AC supply the combined diode bridge rectifiers act like a multi-pulse bridge (18p for 4-level and 24p for 5-level), reducing harmonic injection into the AC supply. • Its output has very low harmonic contents in spite of the low switching frequency	• Employs a special (bulky and expensive) transformer • Complex to achieve series redundancy • Different supply transformer designs are required for applications operating at different AC line voltages • Power pulsation for poor power factor loading • Poor utilization of capacitors • Not suitable for common DC bus applications • Dynamic Braking difficult

32.6.4 Communication in VSDs

The use of a high speed advanced digital communication (Fieldbus) to build industrial automation system for real-time control or simply for data logging has become well-established in modern industries. Digital communication resulted in replacing wiring looms with a digital serial network, this resulted in a lower cost installation and a more reliable solution. Over the last few years many industrial Fieldbuses emerged and endusers, system integrators and original equipment manufacturers (OEMs) chose the optimum system for their applications.

A Fieldbus is a digital communication system that allows a control system to exchange data with remote sensors, actuators and drives, using a single communication link. The major benefits seen are (a) reduced installation and cabling cost, and better overall immunity of the system. Both factors result in more reliable operation and reduced maintenance costs.

There exist two main types of network:

(a) Centralized network – requires a network master controller, typically a PLC. The master device is entirely responsible for controlling communications over the network, while the slave devices tend to be "dumb" devices with no local intelligence.

(b) Decentralized network – which require some local intelligence at each node, but no overall master device. This is ideal for real time application

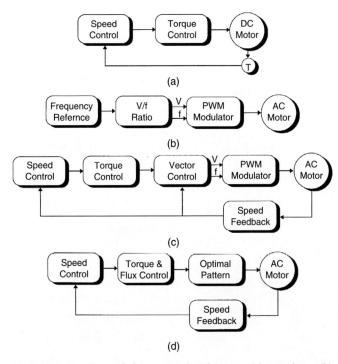

FIGURE 32.6 Electrical drive control techniques. (a) DC drive; (b) frequency control (PWM scalar control); (c) flux vector control (field-oriented control); and (d) direct torque control.

environment, as all nodes are effectively running in parallel.

Most modern VSDs are equipped with hardware and software, which enable local and remote communication with plant automated system via a Fieldbus system. The most popular Fieldbuses are Profibus, Interbus, Ctnet, Sercos, Worldfib, and Devicenet.

32.6.5 PWM Techniques

Different PWM techniques have been employed in PWM-VSD converters. Figure 32.7 identifies the most commonly used techniques.

32.6.6 Impact of PWM Waveform

32.6.6.1 PWM Voltage Waveform

Fast switching of IGBTs (typically $<1\,\mu s$) results in high dV/dt, typically 3–5 kV/μs, and possible voltage overshoot at turn off which can last for a few microseconds. The fast rate of rise/fall of voltage combined with high peak voltage at the turn off, result in a premature failure of motors as well as EMC. References [9–11] deal with the effect of PWM waveforms of VSD.

The following is a brief summary of the effect of the unfiltered waveforms.

TABLE 32.15 Comparison between various control methods used in VSD

Drive type	DC drive	AC drive		
Control method	• Field-oriented	• Frequency control	• Flux vector control	• Direct torque control
Features	• Field orientation via mechanical commutator • Controlling variables are armature current and field current • Torque control is direct • Typical response 10–20 ms	• Voltage and frequency control • Simulation of variable speed drive using modulator • Flux provided with a constant V/F ratio • Open loop drive • Load dictate torque level • Typical torque dynamic response 100 ms	• Field-oriented control – similar to DC drive • Motor electrical characteristics are modelled (observer) • Closed loop drive • Torque controlled indirectly • Typical torque dynamic response 10–20 ms	• Use advance control theory • Controlled variables are magnetizing flux and motor control • Typical torque dynamic response is <5 ms
Advantages	• Accurate and fast torque control • High dynamic speed response • Simple to control	• Low cost • No feedback devices are required • Simple	• Good torque response • Accurate speed control • Full torque at zero speed • Performance approaching DC drive	• Simple • No feedback requirements • No need for an observe
Disadvantages	• Reduced motor reliability • Regular maintenance • Motor costly to purchase • Needs encoder for feedback	• Field orientation not used • Motor status ignored • Torque is not controllable • Delaying modulator used	• Feedback is needed • Costly • Modulator is needed	

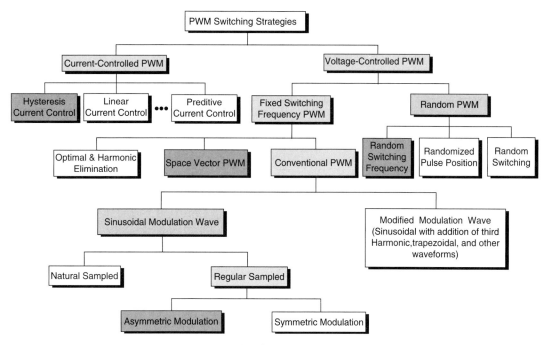

FIGURE 32.7 Classifications of PWM techniques.

32.6.6.2 Effect on Motor

- Premature insulation failure due to partial discharge as a result of peak voltage, high dV/dt and high frequency.
- Motor shaft voltage, which forces current into the shaft bearing, leading to early bearing failure.
- Motor stray capacitance (between windings and earthed frame) leads to earth-current flow caused by high dV/dt.
- High dV/dt creates nonuniform distribution of voltage across the winding, with high voltage-drop across the first few turns and consequential failures.
- In a large motor, voltage differential on the frame is likely to develop in spite of protective earthing of the motor. More than one earthing point is needed.

32.6.6.3 Effect on Cables

- Voltage doubling effect at the rising/falling edges of voltage waveform due to wave propagation in long cables [9].
- Earth-current flows in cable stray capacitance due to dV/dt.
- Restriction on cable type used and earthing methods employed.
- Cable type (armored, screen, multi-core).
- Likelihood of cross talk with other surrounding cables running in parallel.
- For PWM drives, the cost of cabling is likely to be significant due to special requirements of cables, and termination methods employed.

32.6.6.4 Effect on EMC/Insulation/Earthing

- Inductive and capacitive couplings between live components and earth result in common-mode and differential-mode noise. This could lead to malfunctioning of nearby sensitive equipment.
- The voltage to earth applied on drive components pulsates at the switching frequency, adding voltage stresses (worst at low speeds, low modulation index). This poses additional insulation requirements on main power components (motor, cable, output filter, and transformer).
- A 4th wire may be required between the motor frame and the converter virtual earth so that a low impedance path is provided for the motor earth current.
- Strict rules must be observed when cabling and earthing.

32.6.6.5 Motor Insulation

High peak voltages can be experienced at the motor terminals especially when long cable is employed (10–100m depending on the size of motor). This is usually caused by voltage doubling phenomena of a transmission line with unequal line and load impedance. Motor line voltage can reach twice the DC link voltage with long cables.

Fast voltage rise times of 5000 V/μs can be measured at the motor terminals. Under this condition the motor insulation becomes stressed and can lead to a premature breakdown of a standard motor insulation. When motor fails due to insulation stress caused by high peak voltage and fast voltage rise times, failure occurs in the first turn as phase-to-phase short or phase to stator short. The highest voltage is normally seen by the first turn of the winding.

Standard motor capabilities, established by the National Electric Manufacturers Association (NEMA) and expressed in the MG-I standard (part 30), indicate that standard NEMA type B motors can withstand 1000 V peak at a minimum rise time of 2 µs (500 V/µs). Reference [10] describes the effect of PWM inverter waveform on motor insulation in more detail.

Partial Discharge

The phenomena which starts deterioration of the motor insulation is called Partial Discharge (PD). When electric stresses in insulation voids exceed the breakdown voltage of the air, a partial discharge occurs. Successive PDs destroy the insulation slowly.

Voltage Strength between Phase to Phase and Phase to Frame

Both NEMA and IEC are proposing: (a) maximum 1000 V at rise time less than 2 µs and (b) a maximum rate of rise of 500 V/µs. It is believed that low voltage standard motors can withstand a lot larger voltage stresses than specified by NEMA and IEC, possibly up to 1300 V, almost regardless of the rise time.

Voltage Strength between Turn to Turn

In low voltage AC motors, the conductor insulation is designed for 245 V RMS (350 V peak). The insulation strength is however higher depending on the impregnation method.

32.6.6.6 Bearing Current

Bearing current and shaft voltages under 50/60 Hz sine-wave operation has been recognized since 1924. The bearing impedance characteristics largely determine the resulting bearing current that will flow for a given shaft voltage [11].

The rotating machines have three basic sources of shaft voltage. These are:

- Electromagnetic induction from the stator winding to the rotor shaft (due to small asymmetries of the magnetic field in the air gap that is inherent in a practical machine design. The design limit is <1 V RMS.
- Electrostatic coupled from internal sources: such a voltage in motors where rotor charge accumulation may occur (belt-driven coupling, ionized air passing over rotor fan blades).
- Electrostatic coupled from external sources such as PWM inverter. The presence of high dV/dt across the stator neutral to frame ground causes a portion of the voltage to ground due to capacitor divider action. The presence of PWM related voltage components is undesirable and lead to a premature bearing failure.

- The fundamental cause of the shaft voltage is magnetic asymmetry between the stator and the rotor or possibly a phase shift of the motor voltage waveform. System ground may also contribute to this condition through unbalance system voltage.
- NEMA-500 recommends the consideration of insulated bearing for motor frame of certain sizes.

32.6.6.7 EMC

The main sources of electromagnetic emission of PWM-VSI drives are described in [12] as follows:

AC/DC Converter: Supply harmonics caused by supply bridge rectifier (100 Hz–2.5 kHz): As already explained the input bridge circuit with a SCR or diode bridge is a source of supply harmonics in the input current.

DC/AC Inverter: Harmonics caused by the switching of the inverter bridge (3 kHz–20 MHz): the inverter bridge uses fast switching devices to create PWM voltage output. The inverter is a source of a wide band of frequencies, typically extending from the basic switching frequency (usually several kilohertz) to the radio high-frequency bands at 20 MHz. The radio frequency current spreads out into both the supply and motor connections. An EMC filter is often used to limit spread of high frequency harmonics into the supply.

Control Electronics: The control circuit employs a microprocessor with clock frequency of several megahertz, typically 20 MHz. The clock wave produces frequencies, which are multiple of 20 MHz up to 300 MHz.

32.6.7 Techniques Used to Reduce the Effect of PMP Voltage Waveform

32.6.7.1 Output Line Reactor

A reactor increases the rise time but the benefit of its connection may be negated as follows:

- Beneficial connection if cable length is short enough for reflections to be superimposed within rise time, i.e. if rise time is increased beyond critical value of cable length.
- Harmful connection if cable length is too long, the reactor may have negligible effect on peak voltage (theoretically its presence is insignificant in this case) or ringing period but it will increase the duration of each overshoot, thus increasing the probability of partial discharge.

Adding a series line reactor between the motor and inverter is not as simple as illustrated above because the reactor adds or adjusts other resonant modes where the reactor rings with lumped capacitance's. These resonant modes are pure transmission line modes and can double voltage. Some line inductance helps short circuit protection. If earth current is limited by other means, then the coupled reactors may be helpful.

TABLE 32.16 An overview of techniques used as a counter measure to EMI

Effect	Frequency range (f)	Counter measure	
		At source	At load
Mains	≤100 Hz	• Avoid circulating currents	• Balanced signal circuits • Avoid earth loops in signal paths • Screening (electric field only)
Mains harmonics	100 < f ≤ 2.5 kHz	• Line and/or DC link reactor on rectifiers. • Higher pulse number rectifier (e.g. 12, 18, or 24) • Low impedance supply • Harmonic filters	• Balanced signal circuits • Avoid earth loops in signal paths • Filtering
Intermediate	2.5 < f ≤ 150 kHz	• Filters	• Filtering • Screening • Balanced signal circuits
Low-frequency	150 kHz < f ≤ 30 MHz	• Filters – one per apparatus • Cable screening	• Filtering • Screening
High frequency	30 MHz < f ≤ 1 GHz	• Screening • Internal filtering	• Screening

32.6.7.2 Sine-wave Filter

This mechanism filters the PWM carrier frequency; thus the converter output voltages are sinusoidal. This type of filter is best suited for low performance drives and/or retrofit applications (old or standard motors). Reference [13] and Table 32.17 illustrates the filtering options for high power VSDs.

Employing a filter at the inverter output has some practical consequences:

- Cost and weight of filter
- Filter power losses, voltage drop
- A small derating of power switches due to circulating current between filter L, C, and DC link capacitor

TABLE 32.17 Filtering options for PWM-VSI drives, Reference [13]

Option	No filter	dV/dt filter	Sine-wave filter
Motor dV/dt	High	Acceptable	Low
Motor insulation	Must be increased	Normal	Normal
EMC ground noise	Very high	Low	Very low
PWM carrier at motor	100%	100%	Very low
Motor audible noise	Higher	Decreased a little	Minimum
Motor derating	Approx. 13%	Approx. 3%	0%
Torque response	Fast	Fast	Suits most applications
Motor cost	Typically +10% cost	Normally no extra cost	No extra cost
Conclusions	Impractical	Suitable only for high dynamic torque response	Best choice for most drives

- Reduced torque response due to time delay in the filter, sine-wave type
- Potential oscillations which have to be electronically dampened
- Potential induction motor self excitation

32.6.7.3 PWM (dV/dt) Filter

This reduces the dV/dt seen by the motor to a level, which does not compromise the motor or EMC. It is ideal for high performance drives with custom-built motors.

32.6.7.4 RC Filter at Motor Terminals

A simple RC network is used at the motor terminal; the capacitor would represent a short circuit for the high frequency components (sharp dV/dt). Wave reflection will not happen if the resistor value is similar to the cable characteristic impedance. Resistor losses are generally small, as current flow will only occur at the rising and falling edges of the PWM waveform.

32.6.7.5 Common Mode Reactor

The presence of capacitive current due to the high dV/dt can be improved by employing a common mode reactor. It is well-established that such a choke is not effective to reduce the RMS and mean values of the leakage current, but only effective to reduce the peak value. The presence of such a choke in the circuit, increases the inductance and resistance of the zero sequence impedance.

TABLE 32.18 Types of supply front-end bridges of PWM-VSI drives

Type	Power device	Motor speed reversal	Regenerative capability	Regenerative with AC supply loss	Comment
I	Diode	Yes	No	No	• Good power factor across speed range • Needs pre-charge circuit • Lack of protection
II	Diode	Yes	Dissipative	Yes	• Ditto
III	SCR	Yes	No	No	• Power factor is function of speed • Fully controlled DC link • Phase back when (a) supply voltage rises, (b) fault on DC bus side • Needs gate drivers for SCRs
IV	SCR	Yes	Dissipative	Yes	• Ditto
V	SCR	Yes	Regenerative into supply	No	• Ditto
VI	Forced commutated devices (e.g. IGBT/IGCT)	Yes	Regenerative into supply	No	• Can operate with controlled power factor (unity, lagging, leading) • High frequency harmonics • DC link voltage higher than the crest of the supply voltage • Fully controlled DC link, even during a supply dip • Output voltage equals to input voltage • Requires a pre-charge circuit

32.6.8 Supply Front-end for PWM-VSI Drives

There are many types of PWM voltage source drive depending on the supply front-end type and regenerative technique employed (Table 32.18)

  (i) PWM-VSI with a diode supply front-end
 (ii) As above, but with a dynamic brake chopper
(iii) Fully controlled thyristor front-end
(iv) As above but with a dynamic brake chopper
 (v) Fully controlled anti-parallel thyristor supply bridge
(vi) PWM supply front-end

The use of a higher pulse number than 6-pulse would necessitate the use of a supply transformer. This is always considered to be an unnecessary "evil" because of additional cost, losses, and the need for extra space to accommodate this component. For MV applications, this is considered to be a necessity for isolation and protection.

32.6.8.1 Regenerative Braking

Several techniques are usually used for regenerative braking.

A simple diode front-end supply bridge will operate in two quadrants (positive and negative speeds). There is no regenerative power capability as any regeneration of power would result in an increase in the DC link voltage, and the drive will trip on over-voltage.

If a small amount of regeneration is required, during stopping, or speed reversal, then a dynamic brake chopper may be used. This is a simple chopper with a dynamic brake resistor. The size of the resistor is very much dependent on the regenerative brake energy, its magnitude, and repetition rate.

Full power regeneration is possible by employing a fully controlled anti-parallel thyristor front-end. This is similar to that used on DC drives or cyclo-converters.

A more modern approach is to use pulse converter front-end (fully controlled bridge). This is a four-quadrant converter with the ability to control the power factor and the DC link. Such an option necessitates the use of a pre-charge circuit for the DC link, and smoothing inductance on the AC side.

For fully regenerative drives, the supply needs to be receptive.

By using a PWM rectifier as a primary converter in this composite structure both the problems of regeneration and line current distortion are successfully solved – with the penalty of having a much more complicated converter structure and control system.

With the modern PWM-VSI VSD controller, the supply bridge can be fully controlled. Such an option offers the following benefits:

- Fully regenerative drive
- Unity power factor all time
- Sine wave input voltage and current
- Can operate with controlled power factor (e.g. leading power factor)
- Can operate as an active filter while supplying power to the load. Possible elimination of low order supply harmonics (5th & 7th)
- Output voltage equals the input voltage

TABLE 32.19 Application analysis of VSDs

Industry	Current drive topology	Preferences	Applications
Power generation	Direct On Line (DOL) Soft start CSI	• 6.6–11 kV	Boiler feed pump, start-up converter, coal mills
Petrochemical	LCI DOL CSI	• Air-cooled, stand-alone. • Induction motors up to 10 MW • Synchronous above 10 MW	Petrochemical and derivatives, gas liquefaction, pipelines and storage, oil on/off shore and pipelines
Mining	Cyclo-converters	• Low maintenance • Reliability • Low power supply distortion	Mine winders, conveyor belts, coal mills, ventilation fans, underground machinery
Stand-alone and process industries		• Low cost • Efficiency • Ease of repair and maintenance	Water and sewage pumps, wind mills, material handling (extruders), test benches, paper and plastic machines
Metals	Mill drives – cyclo-converters	• Air-cooled • High dynamic performance • Low maintenance motor	Hot mills, medium section mills, finishing section mills, cold mills
Marine	LCI cyclo-converters	• Small size • Low maintenance • Water-cooled	Warships, drilling vessels (mono-hulls), chemical tankers shuttle tankers, cruise liners, icebreakers, semi-submersibles, fishing vessels, cable layers, floating exploration rigs, ferries, research vessels, container vessels

32.7 Applications

32.7.1 VSD Applications

Table 32.19 summarizes main industries and applications.

Present solution of drives, and electric drive application examples from various industries have been described in this section.

32.7.2 Applications by Industry

32.7.2.1 Deep Mining

Reference [14] lists various high-power MV VSI inverter drive applications for the mining industry. In deep mine conveyor belt applications a PWM-VSI drive offers significant advantages over other conventional alternatives. The following benefits have been identified for deep mine conveyor belts applications:

- Improved drive starting and stopping
- Improved reliability
- Matching belt speed to production
- Easier belt inspection
- Reduced belt wear, increased belt life
- Lower specification belt material may be used
- Low speed running to reduce coal removal by windage
- Manpower saving – less coal spillage
- Unity power factor with low harmonic content
- Reduced AC supplies disturbances

For hoist applications, the PWM-VSI drives can also be used to replace DC and cyclo-converter drives for Mine Hoist applications. The benefits are:

- Improved drive control, with 100% continuous stall torque available with induction motors
- Reduced AC supplies disturbances
- Very unlikely to need reactive MVAR correction even at high ratings
- Improved immunity to AC supplies dips

The use of electrically coupled Mine Hoist systems have many advantages especially for deep mines and set to become an essential feature of many new mine shaft systems. The circuit is shown in Fig. 32.8.

The power flows naturally from the motor 1 to motor 2 such that at the point of balance the AC supply current is virtually zero and at near unity power factor.

This technology is the natural successor to the DC electrically coupled winders and totally solves the poor AC power factor that would result if twin drive cyclo motors were used.

32.7.2.2 Industrial Processes

In this industry, there are a number of viable drive solutions available for the major market power ranges, from LCIs

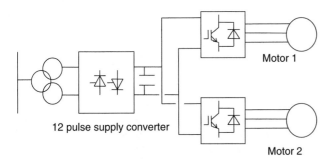

FIGURE 32.8 PWM-VSI electrically coupled Mine Hoist drive.

to FCI. However, there is a developing market for MV variable speed drives. The PWM-VSI using new high power IGBTs or IGCTs appear to be the best solution for the future. Benefits include better power factor, no limit on frequency, and higher voltages.

PWM-VSI converter cost is likely to be higher than equivalent other well-established technologies (e.g. LCI). Hence, the flexibility in choice of motors, and improved control must be exploited. The advantage of offering a MV solution may prove significant. Possible means of reducing motor costs are:

1. Higher frequencies are achievable, allowing the use of high-speed motors and gearboxes
2. Higher pole number machines can be used, giving a cost saving
3. Better power factor over the speed range giving power supply saving
4. Induction motors with rotors adapted for use with VSDs can be used with resultant cost savings over standard DOL, fixed speed motors
5. Higher voltages, smaller conductors
6. However, at low powers the relative cost of the machines is less significant versus the cost of the converters. Hence, the viability of this technology in this market requires close examination

The cyclo-converter drive with a synchronous motor is used when four-quadrant operation is required. Particularly for high power rating with high torque at low speed and at standstill but with a rather low maximum speed are drive requirements. Gear-less cement mill drives were the first applications of cyclo-converters. The mill tube is driven from low-speed wraparound motor with a high number of poles.

32.7.2.3 Metal Industry

The majority of installed hot mill drives are cyclo-converters. A few LCI drives have been used, but applications are limited for such technology on Mill Main Drives. Most early generation plants are equipped with DC drives. The trend is to replace DC with AC.

Direct current drive applications were universally used in the first generation of Rolling Mills. The market for New Mills requiring this technology is declining as the Steel Industry moves to AC as a preference. On early generation Mills where motors are retained, DC drives are likely to be required. Customers in their enquiries, some requesting AC alternatives, are still requesting DC drive solutions.

DC drives are probably still the most economic for the power range 750–1500 kW. The number of manufacturers however, producing DC motors, is declining, particularly in the case of large DC motor manufacture. The lower price of DC solutions is offset by the advantage of use of AC motors in AC solutions, making AC the more popular choice.

Current source inverter LCI can be applied to Roughing stands of Rod & Bar Mills. Technical limitations include the risk of torque pulsation and a minimum drive output frequency of 8 Hz.

Cyclo-converter is the solution most often used. However, it is relatively expensive compared to alternative technology. Major cost penalties arise from supply transformers, cabling and bridge configuration. In some cases, active power supply compensation equipment may be required, taking the costs even higher. Cyclo-converter solutions will still be cost-effective for medium to high power, low speed, low frequency (say below 21 Hz maximum operating frequency) applications. This would include hot reversing mills, with direct drive, for the primary rolling processes; albeit a declining application area, and possibly for direct drive, low speed, high torque roller table applications. Technical limitations include limited output frequency (typically 29 Hz for 12-pulse, at 60 Hz supplies), which can necessitate the use of 2-pole motors to reach application speeds.

The high power PWM-VSI using new power devices (IGBT/IGCT) appears to be the best solution for the future. Benefits include better power factor, no limit on frequency, and higher voltages. Potentially either the 2-level or the multi-level solution will meet the market requirements.

In some applications, like coilers/uncoilers, the system is composed of several drives, which have different power cycles, when some drives are furnishing power, other are braking. A common DC bus system will allow that the energy fed from drives operating in the regenerative braking mode will be utilized by other drives connected to the same DC bus, but operating in the motoring mode. The supply bridge, i.e. rectifier, feeding the DC bus system, will only be rated for the total system power.

The benefits of the DC bus systems include:

- Good operating power factor
- Low harmonics (lowest when using 12-pulse, or 18-pulse front ends)
- Possibility of energy transfer on the common DC link solutions (reducing front-end converter and transformer sizes with attendant energy saving, possibility of using kinetic energy to allow controlled stopping)

32.7.2.4 Marine and Offshore

Drive powers are commonly in the range of 0.75–5.8 W for thrusters, and 6–24 MW for propulsion. The evolution in the commercial market is towards powers from 1 to 10 MW for propulsion. Higher powers are required for naval applications with package drive efficiency better than 96%.

The PWM inverters at these powers would allow the use of induction machines, rather than the more expensive synchronous alternatives required for LCI drive. This could give savings in the price of the motor.

Current source inverter drive LCI is used for all applications except for icebreakers where cyclo-converter drives are used. The PWM (voltage source) inverter using new force commutated driven appears to be the best solution for the future. Benefits include better power factor, no limit on frequency, and higher voltages. Many icebreakers and some other ships are equipped with diesel generator fed cyclo-converter synchronous motors with power ratings up to about 20 MW per unit.

32.7.3 Examples of Modern VSD Systems

32.7.3.1 Integrated Power System for All Electric Ship

This is a full-scale main propulsion drive for the US Navy [15]. It consists of a main propulsion 19 MW induction motor drive system. The power converter consists of three 6-pulse rectifier stages, three 6 kV DC links and 15 IGBT-based H-bridges feeding a 15-phase induction motor (Fig. 32.9).

This drive demonstrates the potential of modern power electronics over more traditional solutions such as cyclo-converter and LCI. The volumetric power density of the new converter is reported to be 905 kW/m^3, compared to 455 kW/m^3 for cyclo and 313 kW/m^3 for LCI.

32.7.3.2 Sub-sea Separation and Injection System

This is a full-scale pilot plant developed to increase recovery and improve the economics of offshore oil and gas fields. The system comprises several VSD units, typically 500 kW oil pump. 1 MW multi-phase booster and 1–2.5 MW water injection drive unit and such a system is called "SUBSIS" [16]. The main task for such a system is to separate the bulk water from the well stream and treat it either for discharge into the sea or re-injection into the reservoir.

This system employs sub-sea based rotating machinery for pumping, boosting, and compression. The sub-sea Electrical Power Distribution System (SEPDIS) is an innovative and cost-effective sub-sea processing (Fig. 32.10). The pump motors are mounted in a pressurized vessel and positioned on the seabed. Reference [16] identifies the benefits of sub-sea drives as follows:

- 3–6% increase in oil and gas recovery
- improved pipeline transportation conditions by removing water from the well stream
- reduced environmental impact due to lower energy consumption and reduction in chemicals used to inhibit corrosion
- reduced size and cost of new platforms
- cost-effective development of marginal fuels through reuse of existing infrastructure

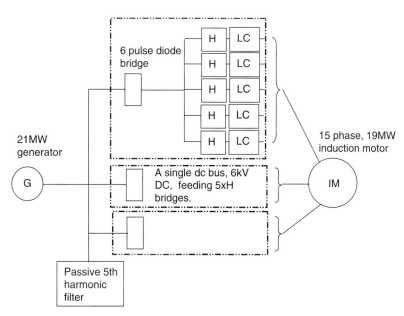

FIGURE 32.9 Schematic diagram of the IPS drive system [15].

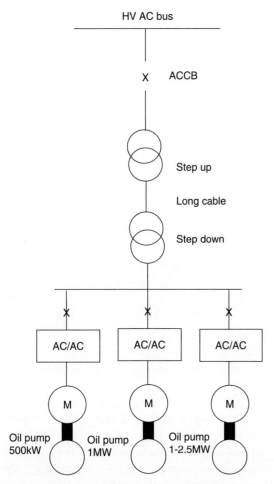

FIGURE 32.10 Schematic diagram of SEPDIS.

32.7.3.3 Shaft-generator for Marine Application

During cruising at sea, up to 3.5 MW of electric power is extracted from the ship's main diesel engine/propeller shaft (90,000 hp per ship) via a salient pole shaft generator, which is fitted to the main propeller shaft. The converter output voltage (set at 60 Hz) is stepped up to 6.6 kV [17].

The converter is based on 24-pulse converter (LCI) technology, which is traditionally used as main ship propulsion. The shaft generator output frequency varied between 14 and 25.7 Hz, 6-phase generator and the output stage is configured as 24-pulse, via a step-up transformer. At the output a passive LC filter is employed, with a synchronous condenser, started by a pony motor.

This type of application is likely to significantly benefit from the higher volumetric power density of the PWM VSI with fully controlled front-end. Such a system will eliminate the need for a passive filter at the output stage, and use a standard 3-phase generator.

The same technology is also applicable to high-speed generators and windmill energy. The ability of the active front-end to sustain fixed DC link voltage over a relatively wide shaft speed range, results in a very good control of the output voltage, irrespective of the shaft speed.

In wind power plants the optimal efficiency of the wind turbine depends on the speed when the wind conditions change. It is, therefore, advantageous to vary the speed of the generator and link it via a frequency converter to the AC system.

For high-speed generators, driven by diesel engine or gas turbine, the fully controlled converters enable direct power conversion from AC high frequency (hundreds of Hz) to fixed power frequency (50/60) Hz fixed output voltage. The magnitude of the output voltage is kept constant irrespective of speed variation of the generator.

32.7.3.4 Linear Motor Drive for Roller Coaster

This drive involves a fully regenerative PWM VSI. The supply front-end is made of anti-parallel thyristors front-end while the machine-bridge is based on PWM IGBT VSI. The Escape' has been developed for Six Flags California at Magic Mountain. The inverter output frequency is 0–230 Hz, and 525 V AC RMS. The power rating is 1.8 MW. The duty cycle is 1.8 MW for 7 seconds, followed by 16 seconds at zero power, and 1.3 MW for 5 seconds, and a stop period of 32 seconds. This ride involves acceleration at 4.5 g, speed and free-fall (6.5 seconds of weightlessness, during which a height of 415 ft is achieved [18]).

The same concept employed in this application could be used for aeroplane launcher on aircraft carriers, instead of the conventional catapult.

32.8 Summary

The benefits of VSD are there to be quantified, and energy saving has been the prime reason for employing a VSD in stand-alone drive applications. Other benefits such as improved process control or increase life expectancy are often difficult to quantify in real terms.

There is a large selection of VSD systems to meet a wide range of applications. In the low and medium power, the induction motor and PWM-VSI are supreme. At higher power ratings, MV PWM-VSIs are gaining popularity, but LCI and cyclo-converter drives would remain key technologies with very high power applications.

Modern drives are becoming more available at competitive prices with good reliability record. However, there are concerns with regard to the impact of fast switching on the motor and the environment.

To ensure successful implementation of a VSD system, both the supplier and end-users need to work in partnership. Ideally, one competent supplier should supply the full drive package, with some after sale service support. Understanding the nature of the load plays an important role in specifying

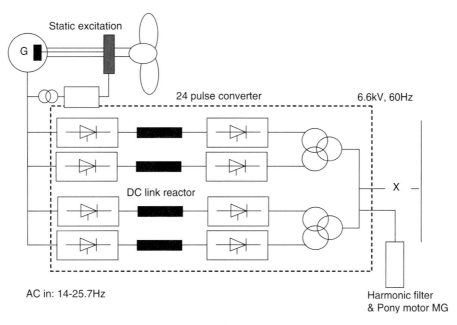

FIGURE 32.11 Schematic diagram of shaft generator.

the power rating of VSD correctly to meet performance requirements and required life expectancy.

New areas of VSD applications are emerging as the power electronics advances and become more reliable at affordable prices. Packaged drives up to several hundred kWs are becoming a commodity product, and end-users do not need to involve a third party during specification, installation, and commissioning. Integrated motors are likely to increase their popularity, possibly with new types of power converters, e.g. matrix.

Further Reading

1. Dury, W., "Electrical Variable Speed Drives Mature Consumable or Radical Infant," Power Engineering Journal, April 1999, Vol.13, No.2, IEE, London.
2. Guide to Harmonics with AC drives. ABB Technical Guide No.6.
3. Bose, B.K., "Power Electronics and Variable Frequency Drives: Technology and Applications," Piscataway, NJ: IEEE Press 1997.
4. Siemens, Complete Guide Series to Variable Speed Drives, "Knowledge of Motor Duty – Key to Proper Planning of Drives," 3.6.
5. Hobbs, P.J., "Electrical Variable Speed Drives Saves Energy," Conference on Development in Variable Speed Drives for Fluid Machinery, ImechE 1981, C101/81.
6. Richmond, A.W., Apprentice Engineers' Handbook No. 2, "Servos and Steppers," Drives and Controls Publication, Kamtech Publishing Ltd, Croydon, Surrey.
7. Shakweh, Y., "Aspects of Limited Motion Actuators and Sub-kW Unipolar Drives," PhD thesis, August 1989, London University.
8. Shakweh, Y. and Lewis, E., "Assessment of MV converter stack topologies," Power Electronics Specialist Conference 99.
9. Jouanne, et al., "Application Issues for PWM Adjustable Speed AC Motor Drives," IEEE Industry Application Magazine, September/October 1998, pp. 10–18.
10. Manz, L., "Motor Insulation System Quality for IGBT Drives," IEEE Industry Application Magazine, January/February 1997, pp. 51–55.
11. Chen, S. and Lipo, T.A., "Circulating type Motor Bearing Current in Inverter Drives," IEEE Industry Application Magazine, January/February 1998, pp. 32–38.
12. Hargis, C., "Electro-Magnetic Compatibility – a Basic Guide for Power Engineers," Control Techniques Technical Publications, Powys, UK.
13. Shakweh, Y. and Aufleger, P., "Multi-Megawatts, Medium Voltage, PWM Voltage Source Sine-Wave Converter For Industrial Drive Applications," Power Electronics & Variable Speed Drives Conference (PEVD'98), UK, London, 21–23 September 98.
14. Shakweh, Y., Lewis, E.A., and Gent, A., "High-power drives for mining applications," Minmech 98, South Africa, September 1998.
15. Crane, A. and McCoy, T.J., "*EMC design for a 19MW PWM motor drive.*" 1999 IEEE Industry Applications Society Annual Meeting'99, Vol.3, pp. 1590–1995.
16. Stromquist, R. and Gustafson, S., "SUBSIS – World's First Separation and Injection System," ABB Review, 6/1998.
17. Clegg, B., et al., "The Application of Drives and Generator Technology to a Modern Container Ship," IEE, PEVD98, London, UK.
18. Elliott, N.J., "Novel Application of a Linear Synchronous Motor Drive," IEE Colloquium on "Update on New Power Electronics Techniques," IEE, London, May 23 1997.
19. Shakweh, Y., "Power Devices for MV PWM VSI Converters", Power Engineering Journal, IEE, U.K., December 1999.
20. Richmond, A.W., 'A Practical Engineer's handbook, Industrial Electric Drives', Drives & Controls Publications, Kamtech Publishing Ltd, Croydon, Surrey, UK.
21. Direct Torque Control, ABB Technical Guide No.1.

33
Motor Drives

M. F. Rahman
School of Electrical Engineering and Telecommunications, The University of New South Wales, Sydney, New South Wales 2052, Australia

D. Patterson
Northern Territory Centre for Energy Research, Faculty of Technology, Northern Territory University, Darwin, Northern Territory 0909, Australia

A. Cheok
Department of Electrical and Computer Engineering, National University of Singapore, 10 Kent Ridge Crescent, Singapore

R. Betz
Department of Electrical and Computer Engineering, University of Newcastle, Callaghan, New South Wales, Australia

33.1 Introduction 858
33.2 DC Motor Drives 859
  33.2.1 Introduction • 33.2.2 DC Motor Representation and Characteristics • 33.2.3 Converters for DC Drives • 33.2.4 Drive System Integration • 33.2.5 Converter-DC Drive System Considerations
33.3 Induction Motor Drives 865
  33.3.1 Introduction • 33.3.2 Steady-state Representation • 33.3.3 Characteristics and Methods of Control • 33.3.4 Vector Controls
33.4 Synchronous Motor Drives 877
  33.4.1 Introduction • 33.4.2 Steady-state Equivalent-circuit Representation of the Motor • 33.4.3 Performance with Voltage-source Drive • 33.4.4 Characteristics under Current-source Inverter (CSI) Drive • 33.4.5 Brushless DC Operation of the CSI-driven Motor • 33.4.6 Operating Modes • 33.4.7 Vector Controls
33.5 Permanent-magnet AC Synchronous Motor Drives 885
  33.5.1 Introduction • 33.5.2 The Surface-magnet Synchronous Motor • 33.5.3 The PM Sinewave Motor
33.6 Permanent-magnet Brushless DC Motor Drives 891
  33.6.1 Machine Background • 33.6.2 Electronic Commutation • 33.6.3 Current/Torque Control • 33.6.4 The Signal Processing for Producing Switch Drive Signals from Hall Effect Sensors and Current Sensors • 33.6.5 Summary
33.7 Servo Drives 902
  33.7.1 Introduction • 33.7.2 Servo Drive Performance Criteria • 33.7.3 Servo Motors, Shaft Sensors, and Coupling • 33.7.4 The Inner Current/Torque Loop • 33.7.5 Sensors for Servo Drives • 33.7.6 Servo Control-loop Design Issues
33.8 Stepper Motor Drives 907
  33.8.1 Introduction • 33.8.2 Motor Types and Characteristics • 33.8.3 Mechanism of Torque Production • 33.8.4 Single- and Multi step Responses • 33.8.5 Drive Circuits • 33.8.6 Micro Stepping • 33.8.7 Open-loop Acceleration–Deceleration Profiles
33.9 Switched-reluctance Motor Drives 915
  33.9.1 Introduction • 33.9.2 Advantages and Disadvantages of Switched-reluctance Motors • 33.9.3 Switched-reluctance Motor Variable-speed Drive Applications • 33.9.4 SR Motor and Drive Design Options • 33.9.5 Operating Theory of the Switched-reluctance Motor: Linear Model • 33.9.6 Operating Theory of the SR Motor (II): Magnetic Saturation and Nonlinear Model • 33.9.7 Control Parameters of the SR Motor • 33.9.8 Position Sensing
33.10 Synchronous Reluctance Motor Drives 926
  33.10.1 Introduction • 33.10.2 Basic Principles • 33.10.3 Machine Structure • 33.10.4 Basic Mathematical Modeling • 33.10.5 Control Strategies and Important Parameters • 33.10.6 Practical Considerations • 33.10.7 A Syncrel Drive System • 33.10.8 Conclusion
References 932
Further Reading 932

33.1 Introduction

The widespread proliferation of power electronics and ancillary control circuits into motor control systems in the past two or three decades have led to a situation where motor drives, which process about two-thirds of the world's electrical power into mechanical power, are on the threshold of processing all of this power via power electronics. The days of driving motors directly from the fixed ac or dc mains via mechanical adjustments are almost over.

The marriage of power electronics with motors has meant that processes can now be driven much more efficiently with a much greater degree of flexibility than previously possible. Of course, certain processes are more favorable to certain types of motors, because of the more favorable match between their characteristics. Historically, this situation was brought about by the demands of the industry. Increasingly, however, power electronic devices and control hardware are becoming able to easily tailor the rigid characteristics of the motor (when driven from a fixed dc or ac supply source) to the requirements of the load. Development of novel forms of machines and control techniques therefore has not abated, as recent trends would indicate.

It should be expected that just as power electronics equipment has tremendous variety, depending on the power level of the application, motors also come in many different types, depending on the requirements of application and power level. Often the choice of a motor and its power electronic drive circuit for application are forced by these realities, and the application engineer therefore needs to have a good understanding of the application, the available motor types, and the suitable power electronic converter and its control techniques. Table 33.1 gives a rough guide of combinations of suitable motors and power electronic converters for a few typical applications.

For many years, the brushed dc motor has been the natural choice for applications requiring high dynamic performance. Drives of up to several hundred kilowatts have used this type of motor. In contrast, the induction motor was considered for low-performance, adjustable-speed applications at low and medium power levels. At very high power levels, the slip-ring induction motor or the synchronous motor drive were the natural choices. These boundaries are increasingly becoming blurred, especially at the lower power levels.

Another factor for motor drives was the consideration for servo performance. The ever-increasing demand for greater productivity or throughput and higher quality of most of the industrial products that we use in our everyday lives means that all aspects of dynamic response and accuracy of motor drives have to be increased. Issues of energy efficiency and harmonic proliferation into the supply grid are also increasingly affecting the choices for motor-drive circuitry.

A typical motor-drive system is expected to have some of the system blocks indicated in Fig. 33.1. The load may be a conveyor system, a traction system, the rolls of a mill drive, the cutting tool of a numerically controlled machine tool, the compressor of an air conditioner, a ship propulsion system, a control valve for a boiler, a robotic arm, and so on. The power electronic converter block may use diodes, metal-oxide semiconductor field effect transistors (MOSFETS), gate turn-off thyristors (GTOs), insulated gate bipolar transistors (IGBTs), or thyristors. The controllers may consist of several control loops, for regulating voltage, current, torque, flux, speed, position, tension, or other desirable conditions of the load. Each of these may have their limiting features purposely placed in order to protect the motor, the converter, or the load. The input commands and the limiting values to these controllers would normally come from the supervisory control systems that produce the required references for a drive. This supervisory control system is normally more concerned

TABLE 33.1 Typical motor, converter, and application guides

Motor	Type of converter	Type of control	Applications
Brushed dc motor	Thyristor ac–dc converter	Phase control, with inner current loop	Process rolling mills, winders, locomotives, large cranes, extruders, and elevators
	GTO/IGBT/MOSFET chopper	Pulse-width modulator(PWM) control with inner current loop	Drives for transportation, machine tools, and office equipment
Induction motor (cage)	Back-back thyristor	Phase control	Pumps, and compressors
	IGBT/GTO inverter/ cycloconverter	PWM V-f control	General-purpose industrial drive such as for cranes, pumps, fans, elevators, material transport and handling, extruders, and subway trains
	IGBT/GTO	Vector control	High-performance ac drives in transportation, motion control, and automation
Induction motor (slip-ring)	Thyristor ac–dc converter	Phase control with dc-link current loop	Large pumps, fans, and cement kilns
Synchronous motor (excited)	Thyristor ac–dc converter	DC-link current loop	Large pumps, fans, blowers, compressors, and rolling mills
Synchronous motor (PM)	IGBT/MOSFET inverter	PWM current control	High-performance ac servo drives for office equipment, machine tools, and motion control

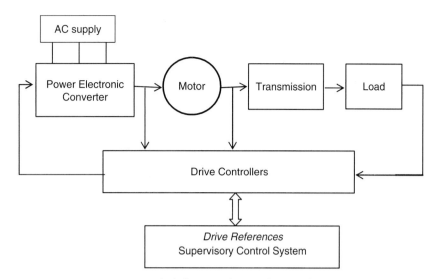

FIGURE 33.1 Block diagram of a typical motor-drive system.

with the overall operation of the process rather than the drive.

Consequently, a vast array of choices and technologies exist for a motor-drive application. Against this background, this chapter gives a brief description of the dominant forms of motor drives in current usage. The interested reader is expected to consult the further reading material listed at the end of this chapter for more detailed coverage.

33.2 DC Motor Drives

33.2.1 Introduction

Direct-current motors are extensively used in variable-speed drives and position-control systems where good dynamic response and steady-state performance are required. Examples are in robotic drives, printers, machine tools, process rolling mills, paper and textile industries, and many others. Control of a dc motor, especially of the separately excited type, is very straightforward, mainly because of the incorporation of the commutator within the motor. The commutator-brush allows the motor-developed torque to be proportional to the armature current if the field current is held constant. Classical control theories are then easily applied to the design of the torque and other motion loops of a drive system.

The mechanical commutator limits the maximum applicable voltage to about 1500 V and the maximum power capacity to a few hundred kilowatts. Series or parallel combinations of more than one motor are used, when dc motors are applied in applications that handle larger loads. The maximum armature current and its rate of change are also limited by the commutator.

33.2.2 DC Motor Representation and Characteristics

The dc motor has two separate sources of fluxes that interact to develop torque. These are the field and the armature circuits. Because of the commutator action, the developed torque is given by

$$T = K i_f i_a \tag{33.1}$$

where i_f and i_a are the field and the armature currents, respectively, and K is a constant relating motor dimensions and parameters of the magnetic circuits.

The dynamic and the steady-state responses of the motor and load are given by

Dynamic   *Steady-state*

$$v_a = R_a i_a + L_a \frac{di_a}{dt} + e \qquad V_a = R_a I_a + E \tag{33.2}$$

$$v_f = R_f i_f + L_f \frac{di_f}{dt} \qquad V_f = R_f I_f \tag{33.3}$$

$$e = K i_f \omega \qquad E = K I_f \omega \tag{33.4}$$

$$T = J \frac{d\omega}{dt} + D\omega + T_L \qquad T = J\omega + T_L \tag{33.5}$$

where J, D, and T_L are the moment of inertia, damping factor, and load torque, respectively, referred to the motor, and the subscripts a and f refer to the armature and field circuits, respectively. R, L, I, and E refer to resistance, inductance, current, and back-emf of the motor in associated circuits referred by the subscripts.

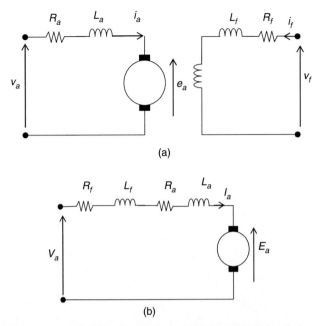

FIGURE 33.2 Representation of the separate and series excited motor circuits: (a) separately excited dc motor circuit and (b) series excited dc motor circuit.

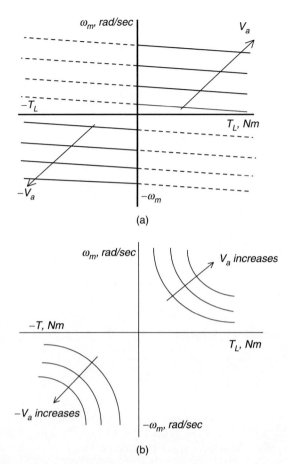

FIGURE 33.3 Torque–speed characteristics of the (a) separately and (b) series excited motors.

Small servo-type dc motors normally have (PM) permanent-magnet excitation for the field, whereas larger size motors tend to have separate field-supply, V_f, for excitation. The separately excited dc motors represented in Fig. 33.2a, have fixed field excitation, and these motors are very easy to control via the armature current that is supplied from a power electronic converter. Thyristor ac–dc converters with phase angle control are popular for the larger size motors, whereas the duty-cycle controlled pulse-width modulated switching dc–dc converters are popular for servo motor drives. The series-excited dc motor has its field circuit in series with the armature circuit as shown in Fig. 33.2b. Such a connection gives high torque at low speed and low torque at high speed, a pseudo constant-power-like characteristic that may match traction-type loads well.

Torque–speed characteristics of the separately and series excited dc motors are indicated in Figs. 33.3a and b, respectively. The speed of the separately excited dc motor drops with load, the net drop being about 5–10% of the base speed at full load. The voltage drop across the armature resistance and the armature reaction are responsible for this. Operation of the motor above the base speed at which the armature voltage reaches for the rated field excitation is by means of field weakening, whereby the field current is reduced in order to increase speed beyond the base speed. The armature voltage is now maintained at the rated value actively, by overriding the field control if required. Note that the range of field control is limited because of the magnetic nonlinearity of the field circuit and the problem of good commutation at weak field.

Usually, the top speed is limited to about three times the base speed. Note also that field weakening results in reduced torque production per ampere of armature current. Depending on the type of load, the armature current, and speed change as dictated by Eqs. (33.1)–(33.5).

For the separately excited dc motor, assuming that the field excitation is held constant, the transfer characteristic between the shaft speed and the applied voltage to the armature can be expressed as indicated in the block diagram of Fig. 33.4. If we ignore the load torque T_L, the transfer characteristic is

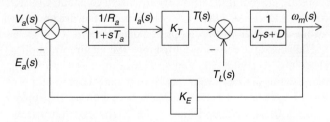

FIGURE 33.4 Transfer characteristic block diagram of a separately excited motor.

given by

$$\frac{\omega_m}{V_a} = \frac{K_T}{(sL_a + R_a)(Js + D) + K_E K_T} \quad (33.6)$$

the characteristic roots of which are given by

$$\left(s + \frac{1}{T_a}\right)\left(s + \frac{D}{J}\right) + \frac{1}{T_a T_m} = 0 \quad (33.7)$$

If we compare this with a standard second-order system, the undamped natural frequency and damping factor are given by

$$\omega_n = \sqrt{\frac{1}{T_a}\left(\frac{1}{T_m} + \frac{D}{J}\right)} \quad (33.8)$$

and

$$\sigma = \xi \omega_n = \frac{1}{2}\left[\frac{1}{T_a} + \frac{D}{J}\right] \quad (33.9)$$

where, T_m = mechanical time constant = $\frac{R_a J}{K_E K_T}$ in which $K_E = K i_f = K_T$ in SI units, and T_a = electrical time constant = $\frac{L_a}{R_a}$.

The speed response of the motor around an operating speed to the application of load torque on the shaft is given by

$$\frac{\Delta \omega_m}{\Delta T_L} = \frac{1 + sT_a}{(1 + sT_a)(Js + D) + \frac{K_E K_T}{R_a}} \quad (33.10)$$

33.2.3 Converters for DC Drives

Depending on the application requirements, the power converter for a dc motor may be chosen from a number of topologies. For example, a half-controlled thyristor converter or a single-quadrant pulse-width modulated (PWM) switching converter may be adequate for a drive that does not require controlled deceleration with regenerative braking. On the other hand, a full four-quadrant thyristor or transistor converter for the armature circuit and a two-quadrant converter for the field circuit may be required for a high-performance drive with a wide speed range.

The frequency at which the power converter is switched, e.g. 100 Hz for a single-phase thyristor bridge converter supplied from a 50 Hz ac source (or 300 Hz for a three-phase thyristor bridge converter), 20 kHz for a PWM MOSFET H-bridge converter, and so on, has a profound effect on the dynamics achievable with a motor drive. Low-power switching devices tend to have faster switching capability than high-power devices. This is convenient for low-power motors since these are normally required to be operated with high dynamic response and accuracy.

33.2.3.1 Thyristor Converter Drive

Consider the dc drive of Fig. 33.5 for which the armature supply voltage v_a to the motor is given by

$$v_a = \frac{2V_{max}}{\pi} \cos \alpha \quad (33.11)$$

where V_{max} is the peak value of the line-line ac supply voltage to the converter and α is the firing angle. The dc output voltage v_a is controllable via the firing angle α, which in turn is controlled by the control voltage e_c as the input to the firing control circuit (FCC). The FCC is synchronized with the mains ac supply and drives individual thyristors in the ac–dc converter according to the desired firing angle. Depending on the load and the speed of operation, the conduction of the current may become discontinuous as indicated in Fig. 33.6a. When this happens, the converter output voltage does not change with the control voltage as proportionately as with the continuous conduction. The motor speed now drops much more with the load as indicated by Fig. 33.6b. The consequent loss of gain of the converter may have to be avoided or compensated if good control over speed is desired.

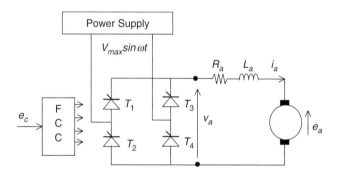

FIGURE 33.5 A two-quadrant single-phase thyristor bridge converter drive.

The output voltage of the simple, two-pulse ac–dc converter of Fig. 33.5 is rich in ripples of frequency nf, where n is an even integer starting with two and f is the frequency of the ac supply. Such low-frequency ripple may derate the motor considerably. Converters with higher pulse number, such as the 6- or 12-pulse converters deliver much smoother output voltage and may be desirable for more demanding applications.

A high-performance dc drive for a rolling mill drive may consist of such converter circuits connected for bidirectional operation of the drive, as indicated in Fig. 33.7. The interfacing of the FCC to other motion-control loops, such as speed and position controllers, for the desired motion is also indicated.

Two fully controlled bridge ac–dc converter circuits are used back-to-back from the same ac supply. One is for forward and the other is for reverse driving of the motor. Since each is a two-quadrant converter, either may be used for regenerative braking of the motor. For this mode of operation, the braking converter, which operates in the inversion mode, sinks the

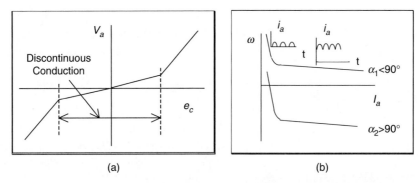

FIGURE 33.6 Converter output voltage and motor torque–speed characteristic with discontinuous conduction.

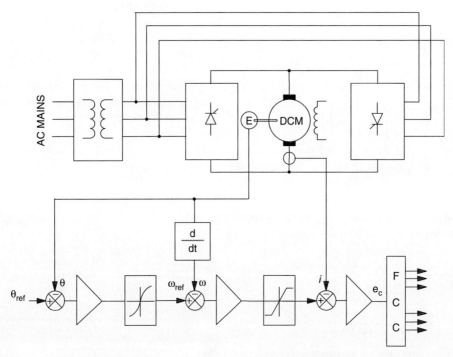

FIGURE 33.7 Bidirectional speed and position control system with a back-to-back (dual) thyristor converter.

motor current aided by the back-emf of the motor. The energy of the overhauling motor now returns to the ac source.

It may be noted that the braking converter may be used to maintain the braking current at the maximum allowable level right down to zero speed. A complete acceleration–deceleration cycle of such a drive is indicated in Fig. 33.8. During braking, the firing angle is maintained at an appropriate value at all times so that the controlled and predictable deceleration takes place at all times.

The innermost control loop indicated in Fig. 33.7 is for torque, which translates to an armature current loop for a dc drive. Speed and position control loops are usually designed as hierarchical control loops. Operation of each loop is sufficiently decoupled from each other so that each stage can be designed in isolation and operated with its special limiting features.

33.2.3.2 PWM Switching Converter Drive

Pulse-width modulated switching converters have traditionally been referred to as choppers in many traction, and forklift-type drives. These are essentially PWM dc–dc converters operating from rectified dc or battery mains. These converters can also operate in one, two, or four quadrants, offering a few choices to meet application requirements. Servo drive systems normally use the full four-quadrant converter of Fig. 33.9 which allows the bidirectional drive and regenerative braking capabilities.

For forward driving, the transistors T_1, T_4, and diode D_2 are used as a buck converter which supplies a variable voltage, v_a, to the armature given by

$$v_a = \delta V_{DC} \tag{33.12}$$

where V_{DC} is the dc supply voltage to the converter, and δ is the duty cycle of the transistor T_1.

33 Motor Drives

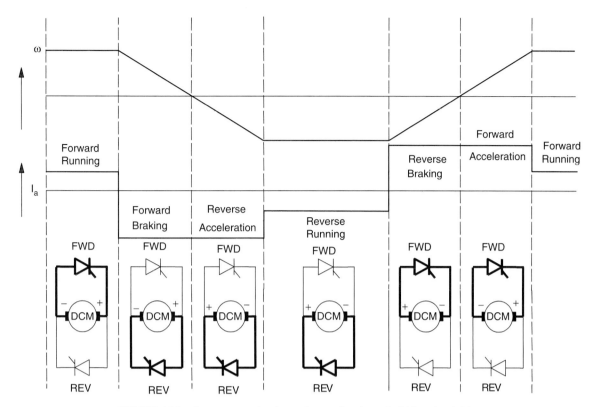

FIGURE 33.8 Converter conduction and operating duty of a bidirectional drive.

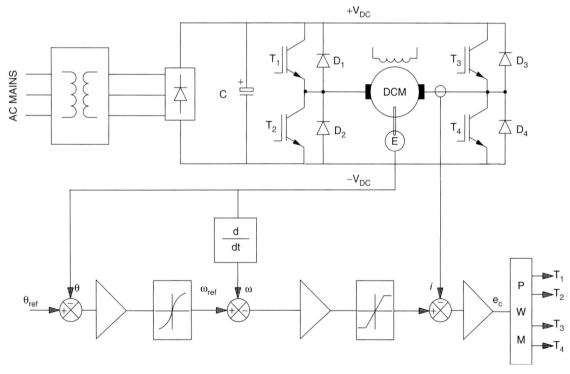

FIGURE 33.9 Bidirectional speed and position control system with a PWM transistor bridge drive.

The duty cycle δ is defined as the duration of the ON-time of the modulating (switching transistor) as a fraction of the switching period. The switching frequency is normally dictated by the application and the type of switching devices selected for the application.

During regenerative braking in the forward direction, transistors T_2 and diode D_4 are used as a boost converter that regulates the braking current through the motor by automatically adjusting the duty cycle of T_2. The energy of the overhauling motor now returns to the dc supply through the diode D_1, aided by the motor back-emf and the dc supply. Again, note that the braking converter, comprising T_2 and D_1, may be used to maintain the regenerative braking current at the maximum allowable level right down to zero speed.

Figure 33.10 shows a typical acceleration–deceleration cycle of such a drive under the action of the control loops indicated in Fig. 33.9.

Four-quadrant PWM converter drives such as that in Fig. 33.10 are widely used for motion-control equipment in the automation industry. Because of the significant development of power switching devices, switching frequencies of 10–20 kHz are easily attainable. At such frequencies, virtually no derating of the motor is necessary.

In order to satisfy the requirements of a drive application, simpler versions of the drive circuits indicated in Figs. 33.7 and 33.9 may be used. For instance, for a unidirectional drive, half of the converter circuits indicated earlier may be used. Further simplification of the drive circuit is possible, if regenerative braking is not required.

33.2.4 Drive System Integration

A complete drive system has a torque controller (armature current controller for a dc drive) as its inner most loop, followed by a speed controller as indicated in Fig. 33.11.

The inner current loop is often regulated with a proportional plus integral (PI) type controller of high gain. The rest of the inner loop consists of the converter, the motor armature, which is essentially an R–L circuit with the armature back-emf as disturbance, and the current sensor. The current sensor is typically an isolated circuit, such as a Hall sensor or a direct current transformer (DCT). A well-designed torque (armature current) loop behaves essentially as a first-order lag system. Together with the mechanical inertia load, this loop can be indicated as the middle Bode plot of Fig. 33.12, in which $1/T_i$ represents the current-controlled system bandwidth. (Note that the damping factor D and the load torque T_L indicated in Fig. 33.11 have been neglected in this description for simplicity.) The current loop is normally designed by analyzing the block diagram of Fig. 33.11, the converter and the PI controller for the current loop using Bode analysis or other control-system design tools.

The next step is usually the design of the speed controller. The 0-db intercept of $1/Js(1 + T_i s)$ is normally too low. Again, if a PI controller is selected for the speed loop, its Bode plot

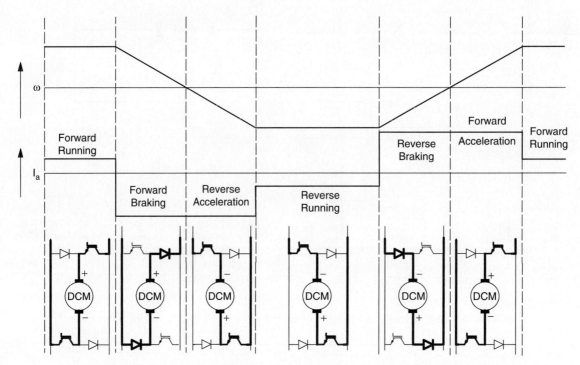

FIGURE 33.10 Converter conduction and operating duty with a PWM bridge transistor drive.

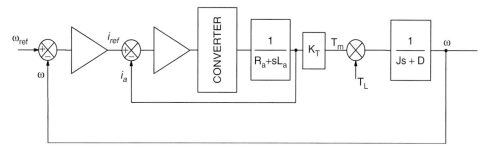

FIGURE 33.11 Structure of a closed-loop speed-control system with a dc drive.

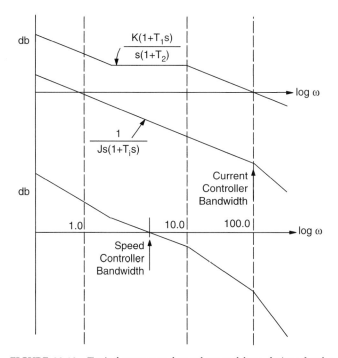

FIGURE 33.12 Typical current and speed control-loop designs for the system of Fig. 33.11.

is superimposed on the current-controlled system as indicated in Fig. 33.12 to obtain the desired speed-control bandwidth.

33.2.5 Converter-DC Drive System Considerations

Several operational factors need to be considered in applying a dc drive. Some of the important ones are as follows:

(1) The armature current may be rich in harmonics. This is particularly true for thyristor converters. The feedback of these current ripples into the FCC may cause overloading of individual switches and tripping. Adequate filtering is necessary to avoid such problems.

(2) Since the converter is switched at regular intervals while the current controller operates continuously, current overshoot may occur because of the delay in the FCC. The current controller gains must be limited to limit this overshoot.

(3) The switching frequency of the converter should be selected according to the desired motor-current ripple, supply input current harmonics, and the dynamic performance of the drive.

(4) Ripple in the speed-sensor output limits the performance of the speed controller. Analog tachogenerator output is particularly noisy and defines the upper limit of the speed-control bandwidth. Digital speed sensors, such as encoders and resolvers, alleviate this limit significantly.

33.3 Induction Motor Drives

33.3.1 Introduction

The ac induction motor is by far the most widely used motor in the industry. Traditionally, it has been used in constant and variable-speed drive applications that do not cater for fast dynamic processes. Because of the recent development of several new control technologies, such as vector and direct torque controls, this situation is changing rapidly. The underlying reason for this is the fact that the cage induction motor is much cheaper and more rugged than its competitor, the dc motor, in such applications. This section starts with the induction motor drives that are based on the steady-state equivalent circuit of the motor, followed by vector-controlled drives that are based on its dynamic model.

33.3.2 Steady-state Representation

The traditional methods of variable-speed drives are based on the equivalent circuit representation of the motor shown in Fig. 33.13.

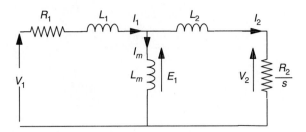

FIGURE 33.13 Steady-state equivalent circuit of an induction motor.

From this representation, the following power relationships in terms of motor parameters and the rotor slip can be found.

$$\text{Power in the rotor circuit, } P_2 = 3I_2^2 \frac{R_2}{s}$$

$$= \frac{3sR_2E_1^2}{R_2^2 + (s\omega_1 L_2)^2} \quad (33.13)$$

$$\text{Output power, } P_o = P_2 - 3I_2^2 R_2 = (1-s)P_2$$

$$= \omega_0 T = \frac{(1-s)\omega_1}{P} T \quad (33.14)$$

where

$$\text{slip, } s = \frac{\omega_1 - \omega_r}{\omega_1} = \frac{\omega_1 - p\omega_o}{\omega_1} \quad (33.15)$$

$P =$ number of pole pairs

$\omega_o = \dfrac{2\pi N}{60}$ rad/s; N is the rotor speed in rev/min,

$\omega_r =$ rotor speed in electrical rad/s,

and

$\omega_1 = 2\pi f_1$ rad/s (electrical), f_1 being the supply frequency.

$$\text{The developed torque, } T = \frac{P_2}{\omega_1/P} \text{ Nm} \quad (33.16)$$

The slip frequency, sf_1, is the frequency of the rotor current and the airgap voltage E_1 is given by

$$E_1 = \omega_1 L_m I_m = \omega_1 \lambda_m \quad (33.17)$$

where λ_m is the stator flux linkage due to the airgap flux. If the stator impedance is negligible compared to E_1, which is true when f_1 is near the rated frequency f_o,

$$V_1 \approx E_1 = 2\pi f_1 \lambda_m \quad (33.18)$$

and

$$T = \frac{3P}{\omega_1} \frac{sR_2 V_1^2}{R_2^2 + (s\omega_1 L_2)^2} = \frac{3P}{\omega_1} \frac{sR_2 (2\pi f_1)^2 \lambda_m^2}{R_2^2 + (s\omega_1 L_2)^2} \quad (33.19)$$

33.3.3 Characteristics and Methods of Control

The above analysis suggests several speed-control methods. The following are the widely used methods:

(1) Stator voltage control,
(2) Slip power control,
(3) Variable-voltage, variable-frequency (V–f) control,
(4) Variable-current, variable-frequency (I–f) control.

These methods are sometimes called *scalar controls* to distinguish them from *vector controls*, which are described in Section 33.3.4. The torque–speed characteristics of the motor differ significantly under different types of control, as will be evident in the following sections.

33.3.3.1 Stator Voltage Control

In this method of control, back-to-back thyristors are used to supply the motor with variable ac voltage, as indicated in the converter circuit diagram of Fig. 33.14a.

The analysis of Section 33.3.1 implies that the developed torque varies inversely as the square of the input root mean square (RMS) voltage to the motor, as indicated in Fig. 33.14b. This makes such a drive suitable for fan- and impeller-type loads for which the torque demand rises faster with speed. For other type of loads, the suitable speed range is very limited. Motors with high rotor resistance may offer an extended speed range. It should be noted that this type of drive with back-to-back thyristors with the firing-angle control suffers from poor power and harmonic distortion factors when operated at low speed.

If unbalanced operation is acceptable, the thyristors in one or two supply lines to the motor may be bypassed. This offers the possibility of dynamic braking or plugging, desirable in some applications.

33.3.3.2 Slip Power Control

Variable-speed, three-phase, wound-rotor (or slip-ring) induction motor drives with slip power control may take several forms. In a passive scheme, the rotor power is rectified and dissipated in a liquid resistor or in a multi-tapped resistor that may be adjustable and forced cooled. In a more popular scheme, which is widely used in medium- to large-capacity pumping installations, the rectified rotor power is returned to the ac mains by a thyristor converter operating in a naturally commutated inversion mode. This static Scherbius scheme is indicated in Fig. 33.15. In this scheme, the rotor terminals are connected to a three-phase diode bridge which rectifies the rotor voltage. This rotor output is then inverted into mains frequency ac by a fully controlled thyristor converter operating from the same mains as the motor stator.

The converter in the rotor circuit handles only the rotor slip power, so that the cost of the power converter circuit can be much less than that of an equivalent inverter drive, albeit at

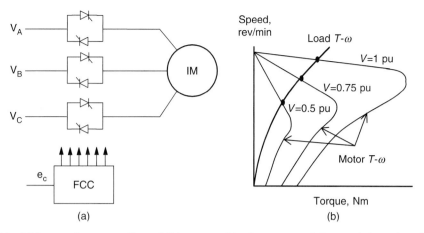

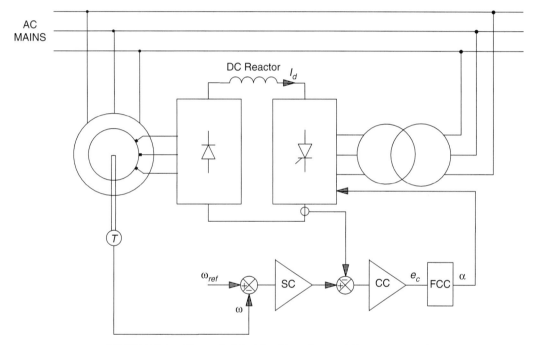

FIGURE 33.14 (a) Stator voltage controller and (b) motor and load torque–speed characteristics under voltage control.

FIGURE 33.15 The static Scherbius drive scheme of slip power control.

the cost of the more expensive motor. The dc link current, smoothed by a reactor, may be regulated by controlling the firing angle of the converter in order to maintain the developed torque at the level required by the load. The current controller (CC) and speed controller (SC) are also indicated in Fig. 33.15. The current controller output determines the converter firing angle α from the firing control circuit (FCC).

From the equivalent circuit of Fig. 33.13 and ignoring the stator impedance, the RMS voltage per phase in the rotor circuit is given by

$$V_R = \frac{V_s}{n}\frac{\omega_r}{\omega_s} = \frac{V_s}{n}\frac{s\omega_s}{\omega_s} = \frac{V_s s}{n} \quad (33.20)$$

where ω_s and ω_r are the angular frequencies of the voltages in the stator and rotor circuits, respectively, and n is the ratio of the equivalent stator to rotor turns. The dc-link voltage at the rectifier terminals of the rotor, v_d, is given by

$$v_d = \frac{3\sqrt{6}V_R}{\pi}$$

Assuming that the transformer interposed between the inverter output is and the ac supply has the same turns ratio n as the effective stator-to-rotor turns of the motor,

$$v_d = -\frac{3\sqrt{6}}{\pi}\frac{V_s}{n}\cos\alpha \quad (33.21)$$

The negative sign arises because the thyristor converter develops negative dc voltage in the inverter mode of operation. The dc-link inductor is mainly to ensure continuous current through the converter so that the expression (33.21) holds for all conditions of operation. Combining the preceding three equations gives

$$s\omega_s = -\omega_s \cos\alpha \text{ so that, } s = -n\cos\alpha$$

and the rotor speed

$$\omega_o = \frac{1}{P}(1-s)\omega_s = \frac{1}{P}\omega_s(1+n\cos\alpha) \text{ rad/s} \quad (33.22)$$

Thus, the motor speed can be controlled by adjusting the firing angle α. By varying α between 180° and 90°, the speed of the motor can be varied from zero to full speed, respectively. For a motor with low rotor resistance and with the assumptions taken earlier, it can be shown that the developed torque of the motor is given by

$$T = 3P\frac{V_s}{\omega_s}i_d \approx 3P\lambda_m i_d \text{ Nm} \quad (33.23)$$

where i_d is the dc-link current. Thus, the inner torque control loop of a variable-speed drive using the Scherbius scheme normally employs a dc-link current loop as the innermost torque loop. Figure 33.16 shows the transient responses of the dc-link, rotor and stator currents of such a drive when the motor is accelerated between two speeds.

The drive is normally started with a short-time-rated liquid resistor, and the thyristor speed controller is started when the drive reaches a certain speed.

By replacing the diode rectifier of Fig. 33.16 with another thyristor bridge, power can be made to flow to and from the rotor circuit, allowing the motor to operate at a rate higher than synchronous speed. For very large drives, a cycloconverter may also be used in the rotor circuit with direct conversion of frequency between the ac supply and the rotor and driving the motor above and below synchronous speed.

33.3.3.3 Variable-voltage, Variable-frequency (V–f) Control

33.3.3.3.1 SPWM Inverter Drive

When an induction motor is driven from an ideal ac voltage source, its normal operating speed is less than 5% below the synchronous speed, which is determined by the ac source frequency and the number of motor poles. With a sinusoidally modulated (SPWM) inverter, indicated in Fig. 33.17, the supply frequency to the motor can be easily adjusted for variable speed. Equation (33.18) implies that, if rated airgap flux is to be maintained at its rated value at all speeds, the supply voltage V_1 to the motor should be varied in proportion to the frequency f_1. The block diagram of Fig. 33.18a shows how the frequency f_1 and the output voltage V_1 of the SPWM inverter are proportionately adjusted with the speed reference. The speed reference signal is normally passed through a filter that only allows a gradual change in the frequency f_1. This type of control is widely referred to as the V–f inverter drive. Control of the stator input voltage V_1 as a function of the frequency f_1 is readily arranged within the inverter by modulating the switches T_1–T_6. At low speed, however, where the input voltage V_1 is low, most of the input voltage may drop across the stator impedance, leading to a reduction in airgap flux and loss of torque.

Compensation for the stator resistance drop, as indicated in Fig. 33.18b, is often employed. However, if the motor becomes lightly loaded at low speed, the airgap flux may exceed the rated value, causing the motor to overheat.

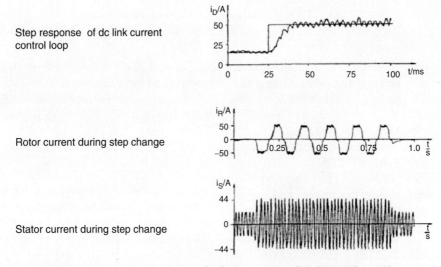

FIGURE 33.16 Transient responses of a slip power controlled drive under acceleration.

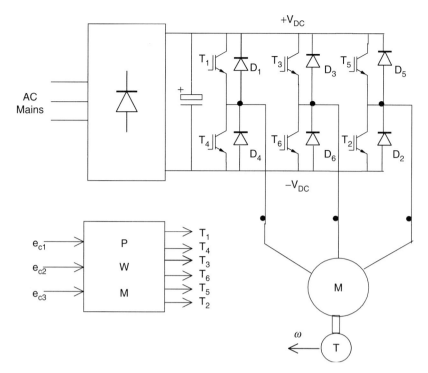

FIGURE 33.17 V–f drive with SPWM inverter.

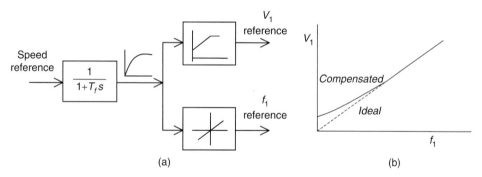

FIGURE 33.18 (a) Input reference filter and voltage and frequency reference generation for the V–f inverter drive and (b) voltage compensation at low speed.

From the equivalent circuit of Fig. 33.13 and neglecting the rotor leakage inductance, the developed torque T and the rotor current I_2 are given by

$$I_2 = \frac{E_1 s\omega_1}{R_2 \omega_1} = \frac{\lambda_m}{R_2} s\omega_1 \qquad (33.24)$$

and

$$T = 3P\frac{R_2}{\omega_r}I_2^2 \qquad (33.25)$$

where $s\omega_1$ is the slip frequency, which is also the frequency of the voltages and currents in the rotor. Equation (33.24) implies that by limiting the slip s, the rotor current can be limited, which in turn limits the developed torque Eq. (33.25). Consequently, a slip-limited drive is also a torque-limited drive. Note that this is true only in steady state. A speed-control system with such a slip limiter is shown in Fig. 33.19. In this scheme, the motor speed is sensed and added to a limited speed error (or limited slip speed) to obtain the frequency (or speed reference for the V–f drive).

Many applications of the V–f controller, however are open-loop schemes, in which any demanded variation in V_1 is passed through a ramp limiter (or filter) so that sudden changes in the slip speed ω_r are precluded, thereby allowing the motor to follow the change in the supply frequency without exceeding the rotor current and torque limits.

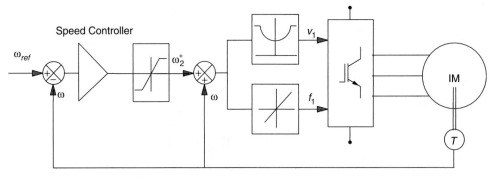

FIGURE 33.19 Closed-loop speed controller with an inner slip loop.

From the foregoing analyses, it is obvious that the V–f inverter drive essentially operates in all four quadrants, with rotor speed dropping slightly with the load, and developing full torque at the same slip speed at all speeds. This assumes that the stator input voltage is properly compensated, so that the motor is operated with constant (or rated) airgap flux at all speed. The motor can be operated above the base speed by keeping the input voltage V_1 constant, while increasing the stator frequency above base frequency in order to run the motor at speeds higher than the base speed. The airgap flux and hence the maximum developed torque now fall with speed, leading to constant power type characteristic. Figure 33.20 depicts the T–ω characteristics of such a voltage- and frequency-controlled drive for various operating frequencies. In this figure, the T–ω characteristic for base speed has been drawn in full, indicating the maximum developed torque T_{max} and the rated torque. Below base speed, the V_1–f_1 ratio is maintained to keep the airgap flux constant. Above base speed, V_1 is kept constant, while f_1 increases with speed, thus weakening the airgap flux. Forward driving in quadrant 1 takes place with the inverter output voltage sequence of a–b–c, whereas reverse driving in quadrant 3 takes place with the sequence a–c–b. Regenerative braking while forward driving takes place by adjusting the input frequency f_1 in such a way that the motor operates in quadrant 2 (quadrant 4 for reverse braking) with the desired braking characteristic.

Note that the characteristics in Fig. 33.20 are based on the steady-state equivalent circuit model of the motor. Such a drive suffers from the poor torque response during transient operation because of time-dependent interactions between the stator and rotor fluxes. Figure 33.21 indicates the machine airgap flux during acceleration with V-f control obtained from a dynamic model. Clearly, the airgap flux does not remain constant during the dynamic operation.

33.3.3.3.2 Cycloconverter Drive
For large-capacity induction motor drives, the variable-frequency supply at variable voltage is effectively obtained from a cycloconverter in which a back-to-back thyristor converter pairs are used, one for each phase of the motor, as indicated in Fig. 33.22. Each thyristor block in this figure represents a fully controlled thyristor ac–dc converter. The maximum output frequency of such a converter can be as high as about 40% of the supply frequency. In view of the large number of thyristor switches required, cycloconverter drives are suitable for large capacity but low-speed applications.

33.3.3.4 Variable Current–Variable Frequency (I-f) Control

In this scheme, medium- to large-capacity induction motors are driven from a variable but stiff current supply that may be obtained from a thyristor converter and a dc-link inductor as indicated in Fig. 33.23. The frequency of the current supply to the motor is adjusted by a thyristor converter with

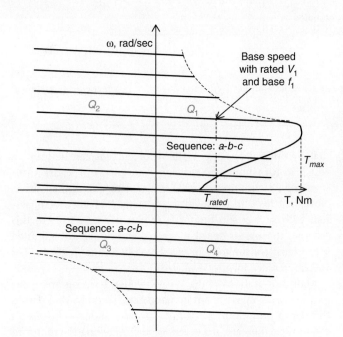

FIGURE 33.20 Typical T–ω characteristics of V–f drive with the input frequency f_1 and voltage V_1 below and above base speed.

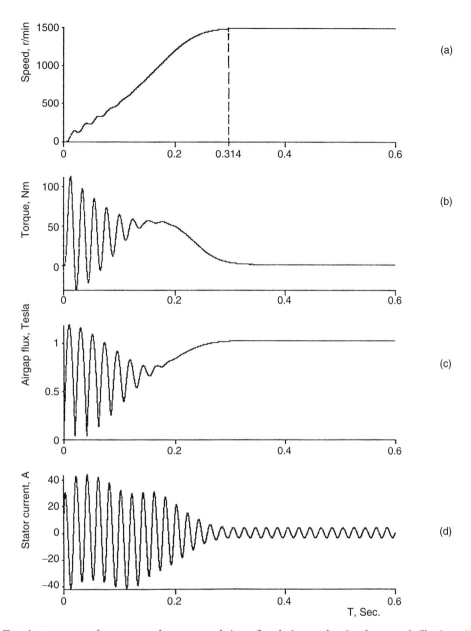

FIGURE 33.21 Transient response of torque, speed, current, and airgap flux during acceleration from standstill using a V–f inverter drive.

auxiliary diodes and capacitors. The diodes in each inverter leg and the capacitors across them are needed for turning off the thyristors when current is to be commutated from one to the next in sequence. The motor current waveforms are normally six-step, or quasi-square, as indicated in Fig. 33.24. The switching states of the inverter thyristors are also indicated in this figure. The motor voltage waveforms are determined by the load. These waveforms are more nearly sinusoidal than the current waveforms.

The thyristor converter supplying the quasi-square current waveforms to the motor has firing-angle control, in order to regulate the dc-link current to the inverter. The dynamics of the dc-link current control is such that this current may be considered to be constant during the time, the inverter switches commutate the dc-link current from one switch to the next. Such a current-source drive offers four-quadrant operation, with independent control of the dc-link current and output frequency. One drawback is that the motor voltage waveforms have voltage spikes due to commutation.

From the analysis of Section 33.3.2, if the higher order harmonics of the current waveforms in Fig. 33.24 are neglected, and it is assumed that the motor voltage and current waveforms are taken to be sinusoidal, the magnetizing current I_m in Fig. 33.13 can be kept constant (for constant-airgap flux

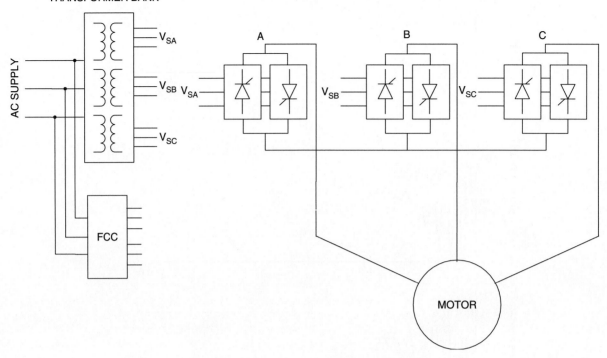

FIGURE 33.22 V–f drive with cycloconverter drive.

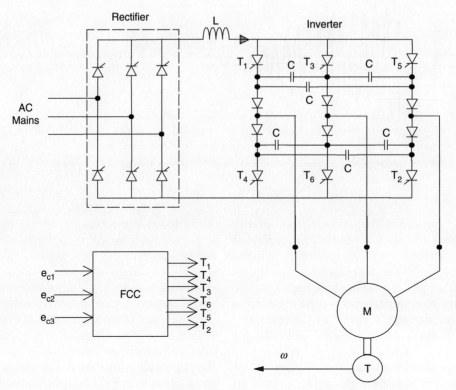

FIGURE 33.23 DC-link current-source thyristor inverter drive.

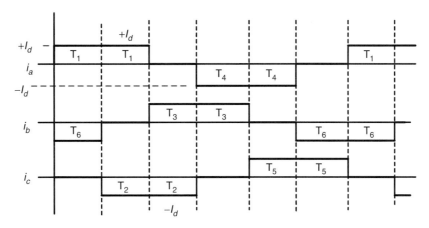

FIGURE 33.24 Motor-current waveforms and the thyristor switching states for a current-source drive.

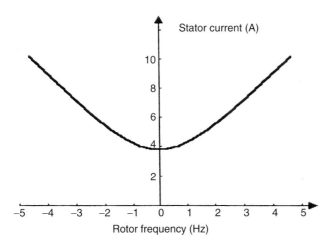

FIGURE 33.25 Stator current vs rotor (slip) frequency for constant-airgap flux operation.

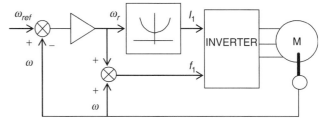

FIGURE 33.26 Variable-current–variable-frequency inverter drive scheme.

operation) if the RMS value of the stator supply current I_1 is defined according to Eq. (33.26). This relationship is also shown in graphical form in Fig. 33.25.

$$I_m = I_1 \left[\frac{R_2^2 + (2\pi f_1 s L_2)^2}{R_2^2 + 2\pi f_1 (L_2 + L_m)^2} \right]^{1/2} = \text{constant} \quad (33.26)$$

The control scheme for variable-speed operation with a current-source drive is indicated in the block diagram of Fig. 33.26. The speed reference defines the stator current reference according to Eq. (33.26) and the frequency reference is obtained by adding the rotor frequency to the actual speed of the motor. The inverter drive may consist of the thyristor current source and the inverter of Fig. 33.23 or a diode rectifier supplied SPWM transistor inverter of Fig. 33.17 with independent current regulators, one for each phase.

The dynamic performance of such current-controlled induction motor drives is not very satisfactory, just as for the voltage-source inverters. Furthermore, the current-source inverter drive cannot normally be operated open-loop, like the V–f inverter drive. For high dynamic performance, vector-controlled drives are becoming popular.

33.3.4 Vector Controls

The foregoing scalar control methods are only suitable for adjustable speed applications in which the load speed or position is not controlled like in a servo system. Instantaneous frequency control with a view to control motor speed or position cannot be defined, and therefore, instantaneous torque control cannot be addressed by scalar control methods. Vector control technique allows a squirrel-cage induction motor to be driven with high dynamic performance, comparable to that of a dc motor. In vector control, machine dynamic model rather than steady-state model (on which scalar controllers are based on), is used to design the controller. For this, the controller needs to know the rotor speed (in the indirect method) or the airgap flux vector (in the direct method) accurately, using sensors. The latter method is not practical because of the requirement of attaching airgap flux sensors. The indirect method, which is being widely accepted in recent years, requires the controller to be matched with the motor being driven. This is because the controller also needs to know some rotor parameter(s), which may vary according to the conditions of operation, continuously.

33.3.4.1 Basic Principles

The methods of vector control are based on the dynamic equivalent circuit of the induction motor. There are at least three fluxes (rotor, airgap, and stator) and three currents or mmfs (stator, rotor, and magnetizing) in an induction motor. For high dynamic response, interactions among current, fluxes, and speed, must be taken into account in determining appropriate control strategies. These interactions are understood only via the dynamic model of the motor.

All fluxes rotate at synchronous speed. The three-phase currents create mmfs (stator and rotor) that also rotate at synchronous speed. Vector control aligns axes of an mmf and a flux orthogonally at all times. It is easier to align the stator current mmf orthogonally to the rotor flux.

Any three-phase sinusoidal set of quantities in the stator can be transformed to an orthogonal reference frame by

$$\begin{bmatrix} f_{\alpha s} \\ f_{\beta s} \\ f_o \end{bmatrix} = \frac{2}{3} \begin{bmatrix} \cos\theta & \cos(\theta - \frac{2\pi}{3}) & \cos(\theta - \frac{4\pi}{3}) \\ \sin\theta & \sin(\theta - \frac{2\pi}{3}) & \sin(\theta - \frac{4\pi}{3}) \\ \frac{1}{2} & \frac{1}{2} & \frac{1}{2} \end{bmatrix} \begin{bmatrix} f_{as} \\ f_{bs} \\ f_{cs} \end{bmatrix} \quad (33.27)$$

where θ is the angle of the orthogonal set α–β–0 with respect to any arbitrary reference. If the α–β–0 axes are stationary and the α-axis is aligned with the stator a-axis, then $\theta = 0$ at all times, Thus

$$\begin{bmatrix} f_{\alpha s} \\ f_{\beta s} \\ f_{os} \end{bmatrix} = \frac{2}{3} \begin{bmatrix} 1 & -\frac{1}{2} & -\frac{1}{2} \\ 0 & \frac{\sqrt{3}}{2} & \frac{\sqrt{3}}{2} \\ \frac{1}{2} & \frac{1}{2} & \frac{1}{2} \end{bmatrix} \begin{bmatrix} f_{as} \\ f_{bs} \\ f_{cs} \end{bmatrix} \quad (33.28)$$

If the orthogonal set of reference rotates at the synchronous speed ω_1, its angular position at any instant is given by

$$\theta = \int_0^t \omega_1 t + \theta_o \quad (33.29)$$

The orthogonal set is then referred to as d–q–0 axes. The three-phase rotor variables, transformed to the synchronously rotating frame, are

$$\begin{bmatrix} f_{dr} \\ f_{qr} \\ f_o \end{bmatrix} = \frac{2}{3} \begin{bmatrix} \cos(\omega_e - \omega_r)t & \cos\left((\omega_e - \omega_r)t - \frac{2\pi}{3}\right) & \cos\left((\omega_e - \omega_r)t - \frac{4\pi}{3}\right) \\ \sin(\omega_e - \omega_r)t & \sin\left((\omega_e - \omega_r)t - \frac{2\pi}{3}\right) & \sin\left((\omega_e - \omega_r)t - \frac{4\pi}{3}\right) \\ \frac{1}{2} & \frac{1}{2} & \frac{1}{2} \end{bmatrix}$$

$$\times \begin{bmatrix} f_{ar} \\ f_{br} \\ f_{cr} \end{bmatrix} \quad (33.30)$$

It should be noted that the difference $\omega_e - \omega_r$ is the relative speed between the synchronously rotating reference frame and the frame attached to the rotor. This difference is also the slip frequency, ω_{sl}, which is the frequency of the rotor variables. By applying these transformations, voltage equations of the motor in the synchronously rotating frame reduce to

$$\begin{bmatrix} v_{qs} \\ v_{ds} \\ v_{qr} \\ v_{dr} \end{bmatrix} = \begin{bmatrix} R_s + pL_s & \omega_e L_s & pL_m & \omega_e L_m \\ -\omega_e L_s & R_s + pL_s & -\omega_e L_m & pL_m \\ pL_m & (\omega_e - \omega_r)L_m & R_r + pL_r & (\omega_e - \omega_r)L_r \\ -(\omega_e - \omega_r) & pL_m & -(\omega_e - \omega_r)L_r & R_r + pL_r \end{bmatrix}$$

$$\times \begin{bmatrix} i_{qs} \\ i_{ds} \\ i_{qr} \\ i_{dr} \end{bmatrix} \quad (33.31)$$

where the speed of the reference frame, ω_e, is equal to ω_1 and $L_s = L_{ls} + L_m$, $L_r = L_{lr} + L_m$.

Subscripts l and m stand for leakage and magnetizing, respectively, and p represents the differential operator d/dt. The equivalent circuits of the motor in this reference frame are indicated in Figs. 33.27a and b.

The stator flux linkage equations are:

$$\lambda_{qs} = L_{ls} i_{qs} + L_m(i_{qs} + i_{qr}) = L_s i_{qs} + L_m i_{qr} \quad (33.32)$$

$$\lambda_{ds} = L_{ls} i_{ds} + L_m(i_{ds} + i_{dr}) = L_s i_{ds} + L_m i_{dr} \quad (33.33)$$

$$\hat{\lambda}_s = \sqrt{\left(\lambda_{qs}^2 + \lambda_{ds}^2\right)} \quad (33.34)$$

The rotor flux linkages are given by

$$\lambda_{qr} = L_{lr} i_{qr} + L_m(i_{qs} + i_{qr}) = L_r i_{qr} + L_m i_{qs} \quad (33.35)$$

$$\lambda_{dr} = L_{lr} i_{dr} + L_m(i_{ds} + i_{dr}) = L_r i_{dr} + L_m i_{ds} \quad (33.36)$$

$$\hat{\lambda}_r = \sqrt{\left(\lambda_{qr}^2 + \lambda_{dr}^2\right)} \quad (33.37)$$

The airgap flux linkages are given by

$$\lambda_{mq} = L_m \left(i_{qs} + i_{qr}\right) \quad (33.38)$$

$$\lambda_{md} = L_m (i_{ds} + i_{dr}) \quad (33.39)$$

$$\hat{\lambda}_m = \sqrt{\left(\lambda_{mqs}^2 + \lambda_{mds}^2\right)} \quad (33.40)$$

The torque developed by the motor is given by

$$T = \frac{3P}{2} \left[\lambda_{ds} i_{qs} - \lambda_{qs} i_{ds}\right] \quad (33.41)$$

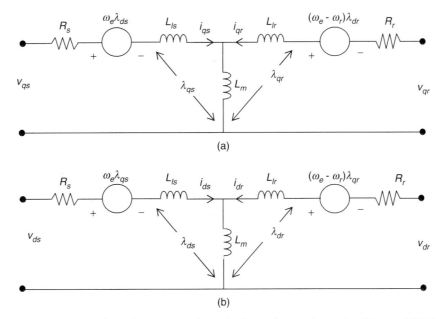

FIGURE 33.27 Motor dynamic equivalent circuits in the synchronously rotating: (a) q- and (b) d-axes.

From Eq. (33.31), the rotor voltage equations are

$$v_{qr} = 0 = L_m \frac{di_{qs}}{dt} + (\omega_e - \omega_r) L_m i_{ds}$$
$$+ \left(R_r i_r + L_r \frac{di_{qr}}{dt}\right) + (\omega_e - \omega_r) L_r i_{dr} \quad (33.42)$$

$$v_{dr} = 0 = L_m \frac{di_{ds}}{dt} + (\omega_e - \omega_r) L_m i_{qs}$$
$$+ \left(R_r i_r + L_r \frac{di_{dr}}{dt}\right) + (\omega_e - \omega_r) L_r i_{qr} \quad (33.43)$$

Using Eqs. (33.35) and (33.36),

$$\frac{d\lambda_{qr}}{dt} + R_r i_{qr} + (\omega_e - \omega_r)\lambda_{dr} = 0 \quad (33.44)$$

and

$$\frac{d\lambda_{dr}}{dt} + R_r i_{dr} + (\omega_e - \omega_r)\lambda_{qr} = 0 \quad (33.45)$$

Also from Eqs. (33.35) and (33.36),

$$i_{qr} = \frac{1}{L_r}\lambda_{qr} - \frac{L_m}{L_r} i_{qs} \quad (33.46)$$

$$i_{dr} = \frac{1}{L_r}\lambda_{dr} - \frac{L_m}{L_r} i_{ds} \quad (33.47)$$

The rotor currents i_{qr} and i_{dr} can be eliminated from Eqs. (33.44) and (33.45) by using Eqs. (33.46) and (33.47). Thus

$$\frac{d\lambda_{qr}}{dt} + \frac{L_r}{R_r}\lambda_{qr} - \frac{L_m}{L_r} R_r i_{qs} + (\omega_e - \omega_r)\lambda_{dr} = 0 \quad (33.48)$$

$$\frac{d\lambda_{dr}}{dt} + \frac{R_r}{L_r}\lambda_{dr} - \frac{L_m}{L_r} R_r i_{ds} + (\omega_e - \omega_r)\lambda_{qr} = 0 \quad (33.49)$$

The elimination of transients in rotor flux and the coupling between the two axes occurs when

$$\lambda_{qr} = 0 \quad \text{and} \quad \hat{\lambda}_r = \lambda_{dr} \quad (33.50)$$

The rotor flux should also remain constant so that

$$\frac{d\lambda_{dr}}{dt} = 0 = \frac{d\lambda_{qr}}{dt} \quad (33.51)$$

From Eq. (33.48) to Eq. (33.51),

$$\omega_e - \omega_r = \omega_{sl} = \frac{L_m}{\hat{\lambda}_r} \frac{R_r}{L_r} i_{qs} \quad (33.52)$$

and

$$\frac{L_r}{R_r} \frac{d\hat{\lambda}_r}{dt} + \hat{\lambda}_r = L_m i_{ds} \quad (33.53)$$

Substituting the expressions for i_{qr} and i_{dr} into Eqs. (33.35) and (33.36),

$$\lambda_{qs} = \left(L_s - \frac{L_m^2}{L_r}\right) i_{qs} + \frac{L_m}{L_r} \lambda_{qr} \quad (33.54)$$

$$\lambda_{ds} = \left(L_s - \frac{L_m^2}{L_r}\right) i_{ds} + \frac{L_m}{L_r} \lambda_{dr} \quad (33.55)$$

Substituting λ_{qs} and λ_{ds} from Eqs. (33.54) and (33.55) into the torque equation of Eq. (33.41),

$$T = \frac{3P}{2} \frac{L_m}{L_r} (\lambda_{dr} i_{qs} - \lambda_{qr} i_{ds}) = \frac{3P}{2} \frac{L_m}{L_r} \hat{\lambda}_r i_{qs} \quad (33.56)$$

It is clear from Eq. (33.53) that the rotor flux $\hat{\lambda}_{dr}$ is determined by i_{ds}, subject to a time delay T_r which is the rotor time constant (L_r/R_r). The Current i_{qs}, according to Eq. (33.56), controls the developed torque T without delay. Currents i_{ds} and i_{qs} are orthogonal to each other and are called the flux- and torque-producing currents, respectively. This correspondence between the flux- and torque-producing currents is subject to maintaining the conditions in Eqs. (33.50) and (33.51). Normally, i_{ds} would remain fixed for operation up to the base speed. Thereafter, it is reduced in order to weaken the rotor flux so that the motor may be driven with a constant-power-like characteristic.

Based on how the rotor flux is detected and regulated, two methods of control are available. One is the more popular indirect rotor flux oriented control (IFOC) method and the other is the direct vector control method; both are described hereafter.

33.3.4.2 Indirect Rotor Flux Oriented (IFOC) Vector Control

In the indirect scheme, the relationship between the slip frequency and current i_{qs} given by Eq. (33.52) is used to relate the compensated speed error $\omega_1 - \omega_r$ to i_{qs}. The i_{qs}, in turn, is used to develop to the demanded torque T^* according to the Eq. (33.56). The rotor flux is maintained at the base value for operation below speed, and it may be reduced to a lower value for field weakening above base speed. The orthogonal relationship between the torque-producing stator current i_{qs} and the flux-producing stator current i_{ds} is maintained at all times by generating the stator current references in the synchronously rotating dq reference frame, using sine and cosine functions of angle θ_1. This angle is obtained as indicated in Fig. 33.28.

The compensated speed error produces the current reference i_{qs}^* according to Eq. (33.56). The current reference i_{qs}^* also gives the slip speed ω_{sl}, according to Eq. (33.53). The slip speed ω_{sl} is added to the rotor speed ω_r to obtain the stator frequency ω_1. This frequency is integrated with respect to time to produce the required angle θ_1 of the stator mmf relative to the rotor flux vector. This angle is used to transform the stator currents to the dq reference frame. Two independent current controllers are used to regulate the i_q and i_d currents to their reference values. The compensated i_q and i_d errors are then inverse transformed into the stator a–b–c reference frame for obtaining switching signals for the inverter via PWM or hysteresis comparators.

It is clear that this scheme uses a feedforward scheme, or a machine model, in which the current reference for i_{qs} is also determined by the rotor time constant T_r. This is also indicated in Fig. 33.28. The rotor time constant T_r cannot be expected to remain constant for all conditions of operation.

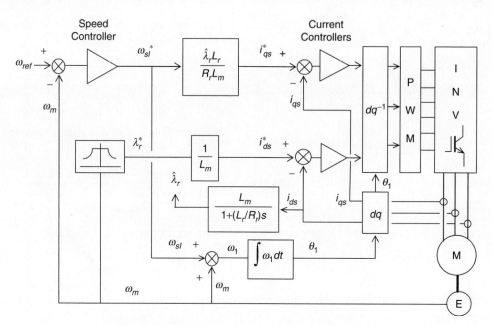

FIGURE 33.28 Indirect rotor flux oriented vector control scheme.

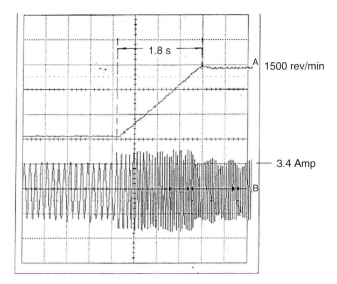

FIGURE 33.29 Speed and current responses of an induction motor drive under the IFOC scheme of Fig. 33.28.

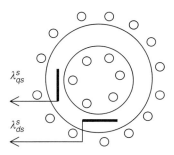

FIGURE 33.30 Quadrature sensors for airgap flux for direct vector control.

Its considerable variation with operating conditions means that the slip speed ω_{sl}, which directly affects the developed torque and the rotor flux vector position, may vary widely. Many rotor time-constant identification schemes have been developed in recent years to overcome the problem.

The mandatory requirement for a rotor-speed sensor is also a significant drawback, because its presence reduces the reliability of the IFOC drive. Consequently, sensorless schemes of identifying the rotor flux position have also drawn considerable interest in recent years.

Figure 33.29 shows the transient response of an induction motor under the IFOC drive scheme of Fig. 33.28. In this case, the drive accelerates a large inertia load from standstill to the base speed of 1500 rev/min. It is clear that the overcurrent transients of Fig. 33.21 are eliminated, while the motor accelerates under a constant torque (implied by the constant rate at which the speed increases) and settles at the final speed with little over- or under-shoot in speed. Clearly, rotor and airgap fluxes remain constant at all times.

33.3.4.3 Direct Vector Control with Airgap Flux Sensing

In the direct scheme, use is made of the airgap flux linkages in the stator d- and q-axes, which are then compensated for the respective leakage fluxes in order to determine the rotor flux linkages in the stator reference frame. The airgap flux linkages are measured by installing quadrature flux sensors in the airgap, as indicated in Fig. 33.30.

By returning to Eqs. (33.32)–(33.40) in the stator reference frame and using some simplifications, it can be shown that

$$\lambda_{qr}^s = \frac{L_r}{L_m}\lambda_{qm}^s - L_{lr} i_{qs}^s \qquad (33.57)$$

$$\lambda_{dr}^s = \frac{L_r}{L_m}\lambda_{dm}^s - L_{lr} i_{ds}^s \qquad (33.58)$$

where the superscript s stands for the stator reference frame. Since the rotor flux rotates at a synchronous speed with respect to the stator reference frame, the angle θ_1 used for the co-ordinate transformations in Fig. 33.27 can be obtained from

$$\cos\theta_1 = \frac{\lambda_{dr}^s}{\hat{\lambda}_r} \qquad (33.59)$$

$$\text{and} \quad \sin\theta_1 = \frac{\lambda_{qr}^s}{\hat{\lambda}_r} \qquad (33.60)$$

where $\left|\hat{\lambda}_r\right| = \sqrt{\left(\lambda_{dr}^s\right)^2 + \left(\lambda_{qr}^s\right)^2}$

The control of torque via i_{qs} and the rotor flux via i_{ds}, subject to satisfying conditions (33.50) and (33.51), remain as indicated in Fig. 33.28, according to the basic principle of vector control.

The requirement of airgap flux sensors is rather restrictive. Such fittings also reduce reliability. Even though this method of control offers better low-speed performance than the IFOC, this restriction has practically precluded the adoption of this scheme. In an alternative method, the d- and q-axes stator flux linkages of the motor may be computed from integrating the stator input voltages.

33.4 Synchronous Motor Drives

33.4.1 Introduction

Variable-speed synchronous motors have been widely used in very large-capacity (>MW) pumping and centrifuge-type applications using naturally commutated current-source thyristor converters. At the low-power end, the current-source SPWM inverter-driven synchronous motors have become very popular in recent years in the form of permanent-magnet brushless dc and ac synchronous motor drives in servo-type applications. There are certain features of three-phase synchronous motors that have allowed them, especially the lower capacity motors, to be controlled with high dynamic performance using cheaper control hardware than is required for the induction motor of similar capacity. Since the average speed of the synchronous motor is precisely related to the supply frequency, which can be precisely controlled, multi-motor drives with a fixed speed ratio among them are also good candidates for synchronous motor drives. This section begins with the performance of the variable-speed nonsalient-pole and salient-pole synchronous motor drive using the steady-state equivalent circuit followed by the dynamics of the vector-controlled synchronous motor drive.

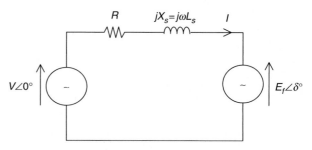

FIGURE 33.31 Equivalent circuit of a nonsalient pole motor.

33.4.2 Steady-state Equivalent-circuit Representation of the Motor

Some of the operating characteristics of variable-frequency voltage- and current-source driven synchronous motors can be readily obtained from their steady-state equivalent representation, as was the case with the scalar controls of induction motor drives. Assuming balanced, sinusoidal distribution of stator and rotor mmfs and an uniform airgap (nonsalient-pole motor), the per phase equivalent circuit of Fig. 33.31 represents the motor at a constant speed.

The representation in Fig. 33.31 is in terms of the RMS voltage V applied to the motor phase winding which consists of the phase resistance R, synchronous reactance X_s (in Ω/phase), and the per phase induced voltage E_f. The back-emf E_f develops in the stator phase winding as a result of rotor excitation supplied from an external dc source via slip rings or by permanent magnets in the rotor. The phasor E_f has an arbitrary phase angle δ with respect to the input voltage V. This is the load angle of the motor.

Unlike the induction motor, a synchronous motor may derive part or all of its excitation from the rotor via rotor excitation. For small synchronous motors that are used in the brushless dc and ac servo applications, this excitation is derived from permanent magnets in the rotor. This is readily and economically obtained with the modern permanent magnets, thus dispensing with the slip-ring-brush assembly. The i^2R losses in the rotor windings with the external excitation are also eliminated. These magnets also allow considerable reduction in space requirement for the rotor excitation. For large synchronous motors, this excitation is supplied more economically from an external dc source via slip rings or via an exciter.

The phasor diagrams of Fig. 33.32 may be used to analyze characteristics of the nonsalient-pole synchronous motor drive. Since the rotor magnetic field may be such that the motor may develop a back-emf that is smaller or larger than the ac supply voltage to the stator windings, the motor may accordingly be under- or overexcited, respectively. The overexcited motor will normally operate at a leading power factor, as is the case in Fig. 33.32b. This is desirable in high-power applications.

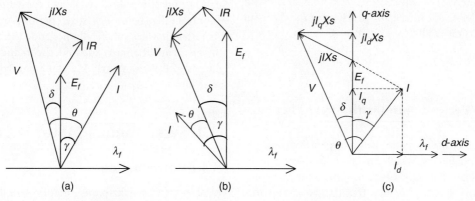

FIGURE 33.32 Phasor diagrams of synchronous motors: (a) under-excited nonsalientpole motor; (b) overexcited nonsalient-pole motor; and (c) under-excited salient-pole motor.

In the phasor diagrams of Fig. 33.32c, the stator current I has been resolved into two components I_d and I_q, which are current phasors responsible for developing mmfs in the rotor d- and q-axes. These representations are in the stator reference frame, and hence are sinusoidal quantities at the frequency of the stator supply.

If the voltage drop across the stator resistance is neglected, which may be acceptable when the stator frequency is near the base frequency or higher, the developed torque T of the motor can be found from the phasor relationships of Fig. 33.32. Thus

$$T = \frac{3}{\omega_r}\frac{EV}{X_s}\sin\delta = \frac{3P}{\omega}\frac{EV}{\omega L_s}\sin\delta = \frac{3PK_\phi}{L_s}\frac{V}{\omega}\sin\delta \text{ Nm} \quad (33.61)$$

for the nonsalient-pole motor (also for the sine wave PM ac motor with rotor magnets at the surface of the rotor) and

$$T = \frac{3P}{\omega_r}\left[\frac{EV}{\omega L_d}\sin\delta + \frac{V^2}{2}\left(\frac{L_d - L_q}{\omega L_d L_q}\right)\sin 2\delta\right]$$
$$= 3P\left[\frac{K_\phi}{L_d}\frac{V}{\omega}\sin\delta + \frac{V^2}{2\omega^2}\left(\frac{L_d - L_q}{L_d L_q}\right)\sin 2\delta\right] \text{ Nm} \quad (33.62)$$

for the salient-pole motor (also for the interior-magnet motor in which the rotor magnets are buried inside the rotor). Here the flux constant of the motor, K_ϕ, is the ratio of the RMS value of the phase voltage E_f induced in the stator only due to the rotor excitation and speed. Note that for a given rotor excitation, the ratio $K_\phi = E_f/\omega$ remains constant at all speed.

33.4.3 Performance with Voltage-source Drive

Equations (33.61) and (33.62) imply that if the motor is driven from a voltage-source supply and if the input voltage to frequency ratio, V–f, is kept constant, the motor will develop the same maximum torque at all speeds. For the nonsalient-pole motor, this maximum torque will occur for a load angle $\delta = 90°$. For the salient-pole motor, this will occur for a load angle which is also influenced by the relative values of L_d and L_q.

At low speed, where the supply voltage V is small, the voltage drop in the stator resistance may become significant compared to E_f. This may lead to a significant drop in the maximum available torque, as given by the torque Eqs. (33.63) and (33.64), which are derived from the phasor diagrams of Fig. 33.32. The stator per phase resistance R is now included in the analysis. The developed torque is then given by

$$T = \frac{3PE_{fo}V_1}{\omega_o}\frac{\lambda X_{so}\sin\delta + R\left(\cos\delta - \frac{\lambda E_{fo}}{V_1}\right)}{R^2 + (\lambda X_{so})^2} \text{ Nm} \quad (33.63)$$

for the nonsalient-pole motor and

$$T = \frac{3P}{\omega_o}\left[\frac{\frac{V_1^2}{\lambda}R + VE_{fo}\lambda X_{qo}\sin\delta + \frac{V^2}{2}(X_{do} - X_{qo})\sin 2\delta - VRE_{fo}\cos\delta}{R^2 + \lambda^2 X_d X_q}\right] \text{ Nm} \quad (33.64)$$

for the salient-pole motor where the subscript o refers to the base quantities and λ refers to the per unit input frequency f_1/f_0.

Equations (33.63) and (33.64) indicate that the maximum torque that the motor can develop diminishes at low speed because of the voltage drop across the stator resistance R. This drop in maximum available torque at low speed can be avoided by boosting the input voltage at low speed, as indicated in Fig. 33.33a. This voltage boosting is similar to the IR compensation applied to a variable-frequency induction motor drive.

Figure 33.33 indicates an open-loop V–f inverter drive scheme, similar to the scheme for an induction motor drive in which the RMS stator input voltage is made proportional to frequency. The speed reference is passed through a first-order filter, as shown in Fig. 33.33a, so that a large and abrupt change in the input frequency command to the inverter is avoided. The filtered speed reference is translated into a proportional frequency reference f_1. The voltage reference V_1 is also proportional to frequency reference, but with a zero frequency bias. The variable RMS input voltage V_1 to the motor may be obtained from an inverter by SPWM methods or from a cycloconverter with phase angle control.

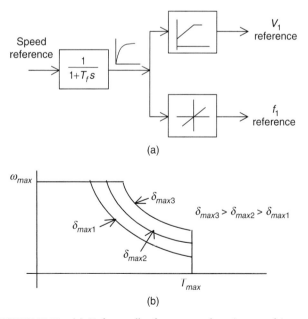

FIGURE 33.33 (a) V–f controller for an open-loop inverter drive and (b) T–ω characteristics with a limited maximum δ for constant-power operation.

The available input voltage V_1 is normally limited by the available dc-link voltage to the inverter or the ac supply voltage to a cycloconverter. This limit is normally arranged to occur at base speed. Above this speed, the stator flux drops leading to field weakening and constant-power-like operation, as indicated in Fig. 33.33b. In some control scheme, the maximum load angle δ is not allowed to exceed a certain limiting value δ_{max}. By selecting δ_{max}, constant-power operation at various power levels is possible.

33.4.4 Characteristics under Current-source Inverter (CSI) Drive

A CSI-driven synchronous motor drive generally gives higher dynamic response. It also gives better reliability because of the automatic current-limiting feature. In a variable-speed application, the synchronous motor is normally driven from a stiff current source. A rotor position sensor is used to place the phase current phasor I of each phase at a suitable angle with respect to the back-emf phasor (E_f) of the same phase. The rotor position sensor is thus mandatory.

Two converter schemes have generally been used. In one scheme, as indicated in Fig. 33.34, a large dc-link reactor (inductor) makes the current source to the inverter stiff. The scheme is suitable for large synchronous motors for which thyristor switches are used in the inverter. A current loop may also be established by sensing the dc-link current and by using a closed-loop current controller that continuously regulates the firing angle of the controllable rectifier in order to supply the inverter with the desired dc-link current. It can be shown that the motor-developed torque is proportional to the level of the dc-link current.

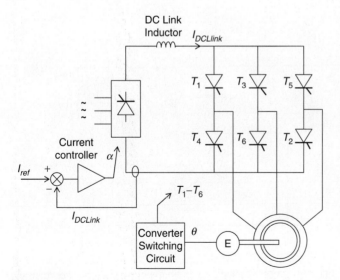

FIGURE 33.34 Schematic of a current-source inverter (CSI) driven synchronous motor.

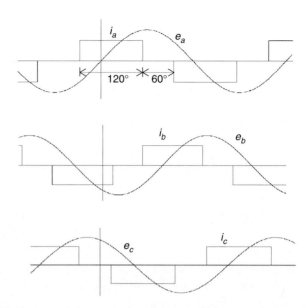

FIGURE 33.35 Current waveforms in the dc-link current-source-driven motor.

The inverter drives the motor with quasi-squarewave current waveforms as indicated in Fig. 33.35. The current waveforms are switched according to the measured rotor position information, such that the current waveform in each phase has a fixed angular displacement, γ, with respect to the induced emf of the corresponding phase. Because of this, the drive is sometimes referred to as self-controlled. The angular displacement of these current waveforms (or their fundamental components) with the respective back-emf waveforms is indicated in Fig. 33.35. Because of the large dc-link inductor, the phase currents may be considered to remain essentially constant between the switching intervals. The quasi-square current waveforms contain many harmonics, and are responsible for large torque pulsations that may become troublesome at low speed.

In the forgoing scheme, the motor can be reversed easily by reversing the sequence of switching of the inverter. It can also be braked regeneratively by increasing the firing angle of the input rectifier beyond 90° while maintaining the dc-link current at the desired braking level until braking is no longer required. The rectifier now returns the energy of the overhauling load to the ac mains regeneratively.

In another scheme, which is preferred for lower capacity drives for which higher dynamic response is frequently sought, phase currents are regulated within the inverter. The inverter typically employs gate turn-off switches, such as the IGBT, and pulse-width modulation techniques, as indicated in Fig. 33.36. Motor currents are sensed and used to close independent current controllers for each phase. Normally, two current controllers suffice for a balanced star-connected motor. Three-phase sinusoidal ac currents are supplied to the motor, the

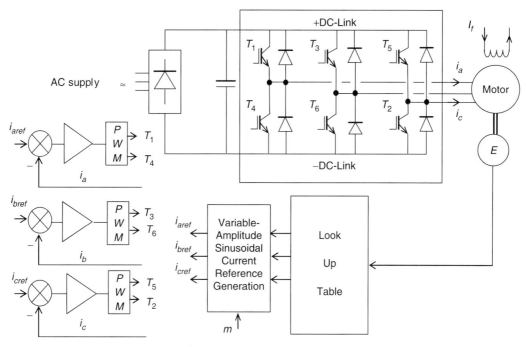

FIGURE 33.36 Control scheme of the SPWM current-source drive.

amplitude and phase angle of which can be independently controlled as required. The references for the current controllers are obtained from a three-phase current reference generator that is addressed by the feedback of the rotor position. The rotor position is continuously measured by a high-resolution encoder. In this way, the current references, and hence the actual stator currents are synchronized to the rotor.

33.4.5 Brushless DC Operation of the CSI-driven Motor

The torque characteristic of the CSI drive scheme of the foregoing section can be easily analyzed using the phasor diagrams shown earlier, if the harmonics in the motor current waveform are neglected or if the motor current waveforms are indeed sinusoidal as in the second scheme described earlier. In the following analysis, it is assumed that the supply current waveforms are sinusoidal. It is also assumed that the phase angle of these current sources with respect to the induced voltage in each phase can be arbitrarily chosen.

The phase back-emf and the current waveforms and the phasor diagram of the nonsalient-pole motor are shown in Fig. 33.37. The phase angle γ between the E_f and I phasors and the RMS value (or amplitude) of I are determined according to the desired torque and power factor considerations. The developed torque is found from the phasor diagram to be

$$T = \frac{3 E_f I \cos \gamma}{\omega_r} = K \phi I \cos \gamma \text{ Nm} \quad (33.65)$$

If the angle $\gamma = 0°$ is chosen, the familiar dc-motor-like torque characteristic is obtained. It should be noted from the forgoing that the developed torque at any speed is independent of R since a high-gain (stiff) current-source drive is used. Note also that the ratio E_f/ω at any operating speed is proportional to the amplitude of the stator flux linkage, λ_f, due to rotor excitation. For fixed rotor excitation this ratio is a constant.

Equation (33.65) indicates that the developed torque of a nonsalient-pole synchronous motor can be controlled by controlling the amplitude of the rotor field (field control), or more conveniently, by controlling the amplitude of the stator phase current. The highest torque per ampere characteristic is achieved when $\gamma = 0°$. Note that the operation with a fixed γ angle is key to this dc-motor-like torque characteristic.

33.4.5.1 Operation with Field Weakening

If the stator impedance drop is neglected, the maximum E_f is largely determined by the dc-link voltage and $E_f = K \lambda_f \omega$ implies that speed ω can be increased by decreasing λ_f. Consequently, the operation above base speed is normally achieved with field weakening. In this speed range, because of the limited dc-link voltage, the rotor field must be weakened; otherwise, the amplitude of the phase-induced emf will exceed the dc-link voltage and current control will not be effective. Field weakening is a means of keeping this voltage at the rated level for speeds higher than the base speed.

The flux linkage due to the rotor excitation, λ_f, can be adjusted when a variable rotor supply is available. This may also be achieved by demagnetizing the rotor mmf by using

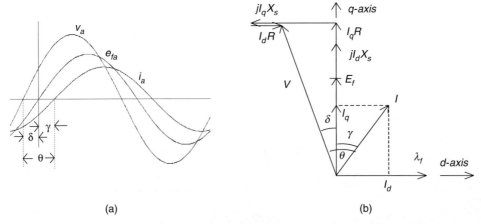

FIGURE 33.37 Phasor relationships of CSI-driven motor: (a) back-emf and current waveforms and (b) the phasor diagram.

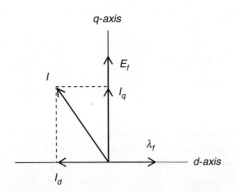

FIGURE 33.38 Field weakening using stator mmf.

the mmf produced by the stator currents. For motors with permanent-magnet excitation, the latter is the only means of weakening the λ_f. Referring to the phasor diagram of Fig. 33.38, if I is made to lead E_f, the d-axis component of I, i.e. I_d, will lead E_f by 90°. The mmf due to I_d then opposes the rotor d-axis mmf. The net rotor flux linkage along the q-axis is then given by

$$\lambda_f' = \lambda_f + L_d I_d \quad (33.66)$$

where I_d is negative when I leads E_f. If the airgap is small, the d-axis component of the armature current may reduce the rotor flux to the required extent.

33.4.6 Operating Modes

A synchronous motor may be driven with a view to achieving various operating characteristics, such as power factor compensation, maximum torque per ampere characteristic, and field weakening. The power factor at which a synchronous motor operates is an important issue, especially for a large drive. A large angle θ between the input voltage and current phasors of the motor results in a poor overall power factor. Operation of the synchronous motor with a CSI which delivers the stator current waveforms with phase angle with respect to the respective phase back-emf waveforms that allows interesting power factor compensation possibilities. Consider the following three cases.

33.4.6.1 Case 1: Operation with I Lagging E_f

In this case, the motor is under-excited and I lags E_f, by an angle γ, as indicated in Fig. 33.39. The overall power factor in this case is lagging, since I lags V by an angle θ. The power-factor angle θ is larger than γ. Note that I_d now magnetizes the rotor field.

33.4.6.2 Case 2: I is in phase with E_f; Maximum Torque per Ampere Operation ($\gamma = 0°$)

If $\gamma = 0°$ is used, the motor input current i_a is in phase with the back-emf e_a, as indicated in the waveforms of Fig. 33.40a. From Eq. (33.65), the developed torque is given by

$$T = K\phi I \quad (33.67)$$

Thus, for a fixed rotor excitation, the developed torque is the highest that can be achieved per ampere of stator current I. In other words, if $\gamma = 0°$ is chosen, the drive operates with its maximum torque per ampere characteristic. From the phasor diagram of Fig. 33.40b, it is clear that the input current I phasor now lags the voltage V phasor at the motor terminals. (see the phasor diagram in the figure). Note that the level of E_f, which is determined by the level of excitation, also determines the angle θ to some extent. Clearly, when maximum torque per ampere characteristic is required, a power factor less than unity has to be accepted.

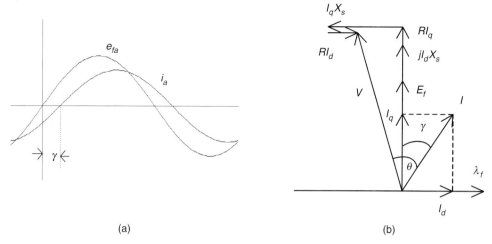

FIGURE 33.39 (a) Phase back-emf and current waveforms and (b) the phasor diagram with I lagging E_f.

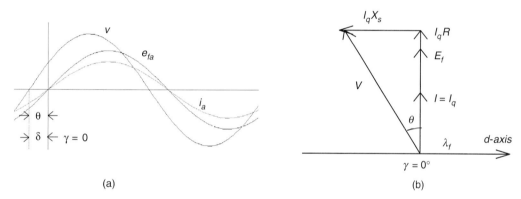

FIGURE 33.40 (a) Phase back-emf and current waveforms and (b) phasor diagram for I in phase with E_f.

33.4.6.3 Case 3: Operation with I Leading E_f

If I is chosen to lead E_f, the overall power factor can be higher, including unity, as is indicated in Fig. 33.41. Note that the motor now operates with less than maximum torque per ampere characteristic. Note also that the d-axis component of I now tends to demagnetize the rotor and that the operation with field weakening is implied.

With a CSI-driven motor, the amplitude and the angle of the phase current relative to the back-emf can be selected according to one of the desirable operating characteristics mentioned above. Additionally, other operational limits such as the inverter/motor current limit, the maximum stator voltage limit, and the maximum power limit can also be addressed. The amplitude of the stator current I clearly determines the developed torque of the motor. Consequently, the error of the speed controller is used to determine the amplitude of I. The overall control system with an inner torque loop can be described by Fig. 33.42.

33.4.7 Vector Controls

The foregoing controls were based on the steady-state equivalent circuit of the motor. Even though the torque Eq. (33.65) for a current-source drive evokes vector-control-like relationships, they do not address the dynamics of the current controls as is possible in an orthogonal reference frame. Using an orthogonal set of reference attached to the rotor, a simple set of decoupled, dc-motor-like torque control relationships is readily obtained. Following the transformation technique used in Section 33.3.4, the stator voltage equations of a synchronous motor with fixed rotor excitation in the rotor reference frame are

$$\begin{bmatrix} v_q \\ v_d \end{bmatrix} = \begin{bmatrix} R + pL_q & \omega L_d \\ -\omega L_q & R + pL_d \end{bmatrix} \begin{bmatrix} i_q \\ i_d \end{bmatrix} + \begin{bmatrix} \omega \lambda_f \\ 0 \end{bmatrix} \quad (33.68)$$

$$T = \frac{3}{2} P \left(\lambda_f i_q + (L_d - L_q) i_d i_q \right) \quad (33.69)$$

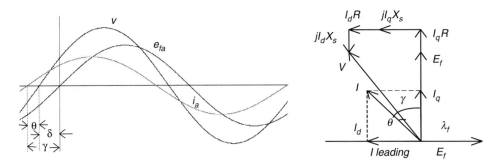

FIGURE 33.41 Back-emf and current waveforms and phasor diagram for I leading E_f.

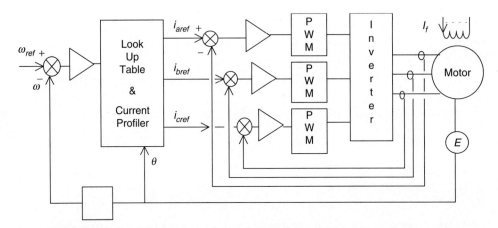

FIGURE 33.42 Structure of a speed-control system with a CSI-driven synchronous motor.

where all the quantities in lower case represent instantaneous quantities in the rotor dq-frame. The λ_f is the flux linkage per phase due to the rotor excitation, ω is the electrical angular velocity in rad/sec, and P is the number of pole pairs. Here, p is the time derivative operator d/dt. Assuming the magnetic linearity, the stator flux linkages are

$$\lambda_d = L_d i_d + \lambda_f$$
$$\lambda_q = L_q i_q \tag{33.70}$$

Note that the Eq. (33.68) can be written down directly from Eq. (33.31), taking into account the fixed rotor excitation so that the third and fourth rows and columns of Eq. (33.31) may be dropped. Since the reference frame now rotates at the speed of the rotor, $\omega_e = \omega$. The induced back-emf due to the fixed rotor excitation occurs in the rotor q-axis and is included in Eq. (33.68), as a separate term. Similarly, the torque expression of Eq. (33.69) may also be written down from Eq. (33.41), using the flux linkages of Eq. (33.70).

Equations (33.67)–(33.69) are for a salient-pole motor for which $L_d \neq L_q$. For a nonsalient-pole motor, L_d is equal to L_q and the developed torque is proportional to i_q only. In either case, the inner torque loop consist of two separate current loops; one for i_d and the other for i_q, as indicated in Fig. 33.43. The i_q current loop generally derives its reference signal from the output of the speed controller and constitutes the inner torque loop. The reference for the i_d current loop is normally specified by the extent of field weakening for which a negative i_d reference is used. Otherwise, the d–axis current is maintained at zero. Note that for large synchronous motors with variable external excitation, field weakening is normally applied through adjustment of the rotor excitation, using a spillover signal from the output of the speed controller.

From Eq. (33.68), it is clear that the couplings of q- and d-axes voltages exist through the d- and q-axes currents, respectively, and the back-emf. During dynamic operation, such coupling effects become undesirable. The coupling effects of d- and q-axes currents and the back-emf into q- and d-axes voltages, respectively, can be removed by the feedforward terms shown in the shaded part of the block the diagram of Fig. 33.44 which also shows the two current-control loops. The two outputs v_d^* and v_q^* from the decoupled current controllers are transformed to the stator reference frame before being subjected to pulse-width modulators or hysteresis comparators. Note that the current references i_d^* and i_q^* are obtained with

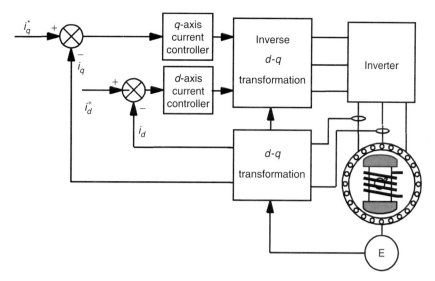

FIGURE 33.43 Inner torque loop of a vector-controlled synchronous motor drive.

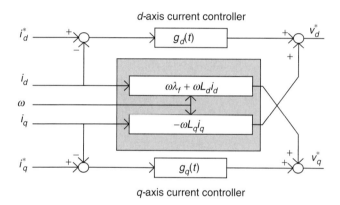

FIGURE 33.44 The d- and q-axis decoupling compensation.

due regard for the desired operating and limiting conditions as described in Section 33.4.5.

Under this type of control, which is exercised in the rotor reference frame, the d- and q-axes currents are separately regulated. The purpose of the q-axis current controller is primarily to control the developed torque, especially for the nonsalient-pole motor. For this motor, the d-axis current is normally maintained at zero, when the motor is operated below the base speed. The d-axis current may be used to weaken the airgap flux so that the motor operates with a constant power-like characteristic. It should be noted that the field weakening may also be carried out more directly by adjusting the rotor excitation current by using a spillover signal from the speed controller, when the base speed is exceeded.

The dynamic response of the drive under the vector control scheme indicated in Fig. 33.44 is the highest possible with a CSI drive. It should also be noted that the dq currents in the rotor reference frame vary only the mechanical dynamics of the rotor. In fact, they are dc quantities when the motor runs at a constant speed. Consequently, the following error (or lag) associated with tracking a sinusoidally time varying current reference, which is the case when current control is exercised in the stator a–b–c reference frame, can be reduced easily by using an integral-type current controllers.

33.5 Permanent-magnet AC Synchronous Motor Drives

33.5.1 Introduction

Since the introduction of samarium cobalt and neodymium–iron–boron magnetic materials in the 1970s and 1980s, synchronous motors with permanent-magnet excitation in the rotor have been displacing the dc motor in many high-performance applications. The trend is more noticeable in applications requiring high-performance motors of up to a few kilowatts. In low- to medium-power applications, the superior dynamics, smaller size, and higher efficiency of motors with PM excitation in the rotor, compared to all other motors, are well known. Prior to this development, the ferrite and alnico magnets were routinely used in small servomotors, with the magnetic excitation in the outer stator. Such motors need a brush-commutator assembly to supply power to the armature, that is a problem. Nevertheless, interesting low-inertia, low-armature inductance designs are possible that are desirable for servo applications. One such design is the pancake ironless armature with the commutator-brush assembly directly located on the printed armature. These shortcomings of the commutator-brush are now avoided by locating the magnets in the rotor and by having three-phase windings in the stator that are supplied from an inverter.

The arrangement just mentioned is very similar to a conventional synchronous motor. This is particularly so when the stator windings and the rotor mmf are sinusoidally distributed.

Several rotor configurations of the PM synchronous motor have been developed, of which the important ones are indicated in Figs. 33.45–33.47.

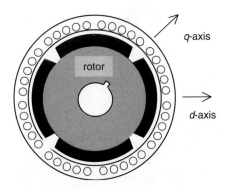

FIGURE 33.45 Schematic of the cross section of a surface-magnet synchronous motor.

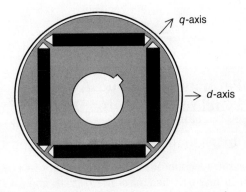

FIGURE 33.46 Schematic of the rotor cross section of an interior-magnet motor.

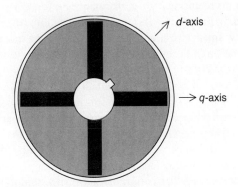

FIGURE 33.47 Schematic of the rotor cross section of a circumferential-magnet motor.

33.5.2 The Surface-magnet Synchronous Motor

The surface-magnet motor comes with the rotor magnets glued onto the surface of the rotor. An additional stainless steel (i.e. nonmagnetic) cylindrical shell may be used to cover the rotor in order to keep the magnets in place against centrifugal force in high-speed applications. Since the relative permeability of the magnetic material is very close to unity, the effective airgap is uniform and large. The airgap is normally about 8 mm. Consequently, the synchronous inductances along the rotor d- and q-axes, as indicated in Fig. 33.45, are equal and small (i.e. $L_d = L_q = L_s$). The armature reaction in this type of motor is small.

The three-phase winding in the stator has sinusoidal distributed windings. In another form, the motor may have trapezoidal distributed winding, and the rotor mmf is also uniformly distributed. Such a motor, often called a brushless dc (BLDC) motor because of its similarity to the inside-out conventional brushed PM DC motor, develops a trapezoidal back-emf waveform as indicated in Fig. 33.48, when it is driven at a constant speed. The back-emf waveforms have flat tops for nearly 120° in each half-cycle followed by 60° of transition from positive to negative polarity of voltage and vice versa. These motors are very suitable for variable-speed applications such as spindle drives in machine-tool and disk drives.

33.5.2.1 Control of the Trapezoidal-wave Motor

Neglecting higher-order harmonic terms, the back-emf in the motor phases may be as indicated in Fig. 33.49. Each back-emf has a constant amplitude (or flat top) for 120° (electrical) followed by 60° of transition in each half-cycle. The developed torque at any instant is given by

$$T = \frac{e_{an}i_a + e_{bn}i_b + e_{cn}i_c}{\omega} \text{ Nm} \quad (33.71)$$

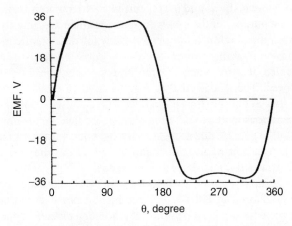

FIGURE 33.48 Back-emf of a trapezoidal waveform motor.

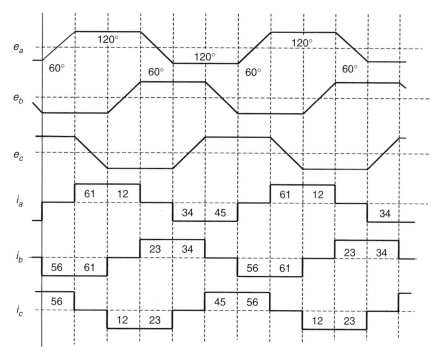

FIGURE 33.49 Back-emf and current waveforms and on-states of transistor switches in the trapezoidal-waveform motor.

It is readily seen that the ideal current waveform in each phase needs to be a quasi-square waveform of 120° of conduction angle in each half-cycle. The conduction of current in each phase winding coincides with the flat part of the back-emf waveforms, which guarantees that the developed torque, i.e. $\left(\sum_{x=a}^{c} \frac{e_{xn} \cdot i_{xn}}{\omega_r}\right)$, is constant or ripple-free at all times. With such quasi-square current waveforms, a simple set of six opto-couplers or Hall-effect sensors would be required to drive the six inverter switches indicated in Fig. 33.36. The output current waveforms for the three-phase inverter and the switching devices that conduct during the six switching intervals per cycle are also indicated in Fig. 33.49. Since only six discrete outputs per electrical cycle are required from the rotor position sensors, the requirement of a high-resolution position sensor is dispensed with. Continuous current control for each phase of the motor, by hysteresis or PWM control, to regulate the amplitude of the motor current in each phase is normally employed. The operation of the BLDC motor drive is described in detail in Section 33.6.

Even though careful electromagnetic design is employed in order to have perfect trapezoidal back-emf waveforms as indicated in Fig. 33.49, the back-emf waveforms in a practical brushless dc motor exhibit some harmonics, as indicated in Fig. 33.48. If ripples in the back-emf waveforms are significant, then torque ripples will also exist when the motor is driven with quasi-square phase current waveforms. These ripples may become troublesome when the motor is operated at low speed, when the motor load inertia may not filter out the torque ripples adequately.

A second source of torque ripple in the permanent-magnet brushless dc (PM BLDC) motor is from the commutation of current in the inverter. Since the actual phase currents cannot have the abrupt rise and fall times as indicated in Fig. 33.49, torque spikes, one for each switching, may exist.

Even though the PM BLDC motor does not have sinusoidal back-emf and inductance variations with rotor angle, the analysis of Section 33.4 in terms of the fundamental quantities will often suffice. The switching of the inverter using the six rotor-position sensors guarantees that the current waveform in each phase always remains in synchronism with the back-emf of the respective phase. Since the quasi-square phase current waveform in each phase coincides with the flat part of the back-emf waveform of the same phase, the angle γ is clearly zero for such operation. Thus, considering fundamental quantities, the motor back-emf and the torque characteristics can be expressed by

$$E = K_E \omega \text{ V} \qquad (33.72)$$

$$T = K_T I \cos \gamma = K_T I \text{ Nm} \qquad (33.73)$$

where K_E and K_T are equal in consistent units. Note that E and I are now RMS values of the fundamental components of these quantities.

33.5.2.2 Sensorless Operation of the PM BLDC Motor

In spindle and other variable-speed drive applications, where the lowest speed of operation is not less than a few hundred revs/min, it may be possible to obtain the switching signals for the inverter from the motor back-emf, thus dispensing with rotor-position sensors. The method consists of integrating the back-emf waveforms, which are the same as the applied phase voltage if the other voltage drops are neglected, and comparing the integrated outputs with a fixed reference. These comparator outputs determine the switching signals for inverter. It may be noted that the amplitude of the back-emf waveforms is proportional to the operating speed, so that the frequency of the comparator outputs increases automatically with speed. In other words, the angle γ and the current waveform relative to the back-emf waveform in each phase remains the same regardless of the operating speed. Integrated circuits are available from several suppliers, that perform this task of sensorless BLDC operation satisfactorily, covering a reasonable speed range.

33.5.3 The PM Sinewave Motor

The PM sinewave motor, which may also have magnets on the rotor surface or buried inside the rotor (as in the interior-magnet motor of Fig. 33.46), has sinusoidally distributed windings. The airgap flux distribution produced by the rotor magnets is also sinusoidal, arranged through magnet shaping. Consequently, the back-emf waveform of each phase is also a sinusoidal waveform, as indicated in Fig. 33.50, when the motor is driven at a constant speed.

It should be noted, however, that with magnets mounted on the rotor surface, the effective airgap is large and uniform so that $L_d = L_q = L_s$. Because of the large equivalent airgap, these inductances are also small, and consequently, the armature reaction is small. As a result, this motor essentially operates with fixed excitation, and there is hardly any scope for altering the operating power factor or the rotor mmf, once the motor and its drive voltage and current ratings have been selected.

The PM sinewave interior-magnet motor, which comes with magnets buried inside the rotor as indicated in Figs. 33.46 and 33.47, has an easier magnetization path along the rotor q-axis, so that $L_q > L_d$. The small airgap implies that the inductances L_d and L_q may not be small, and hence may allow considerable scope for field weakening.

If the sinewave motor is supplied from an SPWM inverter, the analyses and vector diagrams of this motor are not different from those in Section 33.4 for the nonsalient-pole and salient-pole synchronous motor drives. The surface-magnet motor is more akin to the nonsalient-pole motor, since $L_d = L_q = L_s$, whereas the interior-magnet motor is more akin to the salient-pole motor because of the d- and q-axes inductances being unequal. Thus, the equations in Section 33.4 will apply equally well for this motor, both in the steady state and dynamically and for both voltage- and current-source supply.

33.5.3.1 Control of the PM Sinewave Motor

The sinusoidal back-emf waveforms imply that the three-phase motor must be supplied with a sinusoidal three-phase currents if a dc-motor-like (or vector-control-like) torque characteristic is desired, as was found in Section 33.4. For such operation, the phase currents must also be synchronized with the respective phase back-emf waveforms, in other words, with the rotor dq-reference frame. This implies the operation with a specified γ angle. Two implementations are possible.

In one scheme, the stator currents are regulated in the stator reference frame and the stator current references are produced with reference to the rotor position, as indicated in Fig. 33.51. The rotor position θ is continuously sensed and used to produce three sinusoidal current references of unit amplitude. The phase angle γ of the references relative to the respective back-emf waveforms is usually derived from the speed controller, which defines the field-weakening regime of the drive. The operating power factor of the drive may also be addressed using the γ angle control, as explained in Section 33.4.5. The amplitudes of these references are then multiplied by the error of the speed controller in order to produce the desired torque reference. In this way, both the RMS value I of the phase current and its angle γ with respect to E_f can be adjusted independently (which is equivalent to independent dq-axes current control). Three PWM current controllers are indicated, but two suffice for a balanced motor. Other types of current controllers, such as hysteresis controllers, may also be used.

In the second scheme, the motor currents are first transformed into the rotor dq-reference frame using continuous rotor position feedback and the measured rotor d- and q-axes currents are then regulated in the rotor reference frame, as indicated in Fig. 33.52. The main advantage of regulating

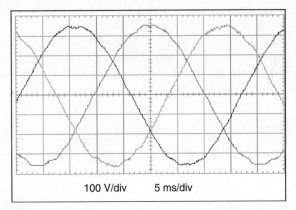

FIGURE 33.50 Back-emf waveforms of a sinewave PM motor. Speed = 1815 rev/min.

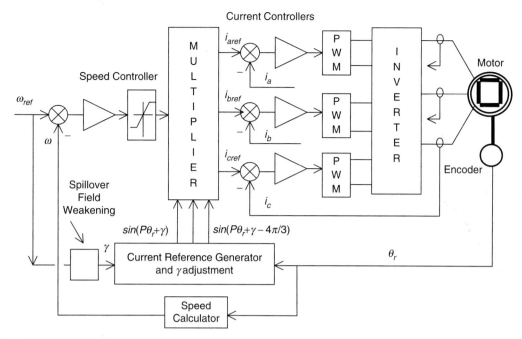

FIGURE 33.51 Speed-controlled drive with the inner torque loop via current control in the stator reference frame.

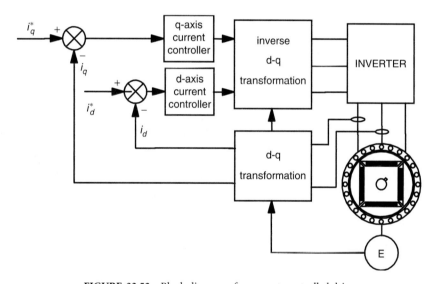

FIGURE 33.52 Block diagram of a current-controlled drive.

the stator currents in the rotor reference frame, compared to the stator reference frame as in the first scheme, is that current references now vary more slowly and hence suffer from lower tracking error. At constant speed, the stator current references are in fact dc quantities as opposed to ac quantities, and hence the inevitable tracking errors of the former scheme are easily removed by an integral-type controller. The other advantage is that the current references may now include feedforward decoupling so as to remove the cross-coupling of the d- and q-axes variables as shown in the motor representation of Fig. 33.52. The latter scheme is currently preferred, since it allows more direct control of the currents in the rotor reference frame.

It should be noted that continuous transformations to and from rotor dq-axes are required; hence the rotor-position sensor is a mandatory requirement. Rotor-position sensors of 10–12 bit accuracy are normally required. This is generally viewed as a weakness for this type of drive.

The general torque relationship of a sinewave motor supplied from a current-source SPWM inverter is given by

Eq. (33.69) which is rewritten here.

$$T = \frac{3}{2} P \left(\lambda_f i_q + (L_d - L_q) i_d i_q\right) \quad (33.74)$$

Precise control of the developed-torque control requires that both i_d and i_q be actively controlled in order to regulate it to the level determined by the error of the speed controller. For the surface-magnet synchronous motor, the large effective airgap means that $L_d \approx L_q$, so that i_d should be maintained at zero level. This implies that the i_d current reference in Fig. 33.52 should be kept zero. For the interior-magnet motor, $L_q > L_d$, so that i_d and i_q have to be controlled simultaneously to develop the desired torque. A few modes of operation are possible.

33.5.3.2 Operating Modes

33.5.3.2.1 Operation with Maximum Torque per Ampere (MTPA) Characteristic
In this mode of control, the developed torque T per ampere of stator current is the highest for a given rotor excitation. The combination of i_d and i_q that develops the maximum torque per ampere of stator current is indicated in Fig. 33.53. The reference signal for the inner torque loop normally determines the i_q reference; while the i_d reference is given by

$$i_d = \frac{\lambda_f}{2(L_q - L_d)} - \sqrt{\frac{\lambda_f^2}{4(L_q - L_d)^2} + i_q^2} \quad (33.75)$$

It should be noted that with $i_d = 0$, this mode of operation is achieved for the surface-magnet motor.

33.5.3.2.2 Operation with Voltage and Current Limits
The maximum current limit of the motor/inverter, I_{smax}, is normally imposed by setting appropriate limits on i_d and i_q such that

$$i_d^2 + i_q^2 \leq I_{s\,max}^2 \quad (33.76)$$

Equation (33.75) defines a circle around the origin of the i_d–i_q plane.

The available dc-link voltage to the inverter places an upper limit of the motor phase voltage, V_{smax}, given by

$$V^2 = V_d^2 + V_q^2 \leq V_{max}^2 \quad (33.77)$$

$$(L_q i_q)^2 + (L_d i_d + \lambda_f)^2 \leq \left(\frac{V_{am}}{\omega}\right)^2 \quad (33.78)$$

where the stator voltages v_d and v_q are expressed in the rotor reference frame, as in Eq. (33.68). Equation (33.78) is obtained from the voltage equation of Eq. (33.68) when the phase resistance R is neglected. Equation (33.78) defines elliptical trajectories that contract as speed increases. All the three modes of operation are included in Fig. 33.53.

33.5.3.2.3 Operation with Field Weakening
This mode of operation is normally required for operating the motor above the base speed. A constant-power characteristic is normally desired over the full field-weakening range. For a given rotor excitation of λ_f and developed power of P_0, the developed torque T and the net rotor flux linkage required for constant-power operation can be determined. From this, the limiting values for i_d and i_q, which further constrain the allowable i_d–i_q trajectory is also indicated in Fig. 33.53, can be determined.

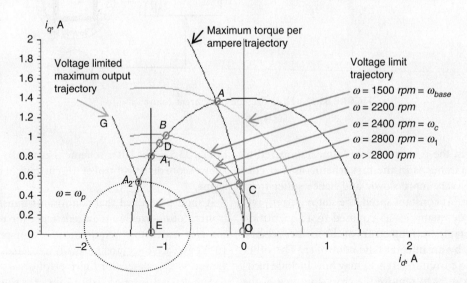

FIGURE 33.53 i_d–i_q Trajectories for the maximum torque per ampere characteristic and for maximum voltage and current limits of an interior PM motor.

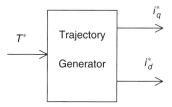

FIGURE 33.54 i_d and i_q trajectory generation satisfying the desired operating modes.

The operating modes described in the preceding section are normally included in a trajectory controller that generates the references for i_d and i_q continuously, as indicated in Fig. 33.54.

33.6 Permanent-magnet Brushless DC Motor Drives

33.6.1 Machine Background

Anyone studying electric machines would be aware that the classical synchronous machine offers the possibility of the very highest efficiencies. These machines have the power-carrying conductors in the stationary part, therefore there is no need to transfer high power through a brush system. The field is provided on the rotating part either by an electromagnet, which does require some very low-power brushes or at best, by a permanent magnet with no power requirements at all.

This is clearly better than the traditional dc machine with high-power brushes and rotating conductors, with those conductors having a very poor path for heat to get away under overload conditions. You might ask, "why did we use so many of them?" It was of course because the dc machine offered the ability to control the speed fully and easily, which the synchronous machine, when running from the mains at a fixed frequency, was definitely not capable of. Thus, for example, all electric trains and trams built up to the 1980s, many of them still in service, use brushed dc motors.

The induction machine, the work horse of industry is quite efficient, but because of the need to carry magnetizing current in the stator as well as load current, and the existence of slip, it cannot be as good as the synchronous machine. It also innately has a very limited speed range.

All this changed with the advent of inexpensive, reliable power electronics. Quite suddenly power electronics engineers had the ability to vary both the frequency and the amplitude of an ac supply and thus to provide the holy grail of variable-speed ac motors. This was first applied to induction machines, with the simplest form of open-loop control. The complexity of control for synchronous machines added substantially to the cost and was not so appealing. Today the thought of putting a digital signal processor (DSP) or a microcontroller in a small motor controller is no longer as daunting as it was ten years ago, so the distinction between the complexity of a controller for induction machines and synchronous machines has almost disappeared.

Whilst power electronics radically changed our attitudes to electric machine design and operation, there was a second very important development which arrived on the scene at almost the same time; the appearance of very powerful rare earth permanent-magnet materials.

The first rare earth magnet material was Samarium Cobalt, used in very special applications such as space and defence, but it was then, and still is, very expensive. Neodymium–Iron–Boron (Nd–Fe–B) came next with even higher energy product. Energy product is a measure used to quantify the effectiveness of magnets. Nd–Fe–B began expensively, but has continued to drop in price at a spectacular rate and recently became cheaper than ferrites in term of dollars per unit of energy product.

Thus all the machines discussed here are permanent–magnet (PM) machines, in general using Nd–Fe–B magnets.

33.6.1.1 Clarifying Torque Versus Speed Control

As is well understood, given a fixed magnetic flux level, the magnitude of the current in a motor determines the magnitude of the torque out of the machine. In a general application if a torque is set, the speed is determined by the load. This is what happens when driving an automobile. The throttle or accelerator varies the torque. The speed can be very slow or very high for the same foot position, depending on the road conditions and vehicle motion history, even if the vehicle is in the same gear. This principle of torque control rather than speed control is fundamental to the operation of electric machines. When speed control is required in PMBDCMs, it is usual to implement torque control, with an extra outside control loop, like a cruise control in an automobile, to use the torque control to deliver speed control. So from here on the discussion will describe torque control for the variable-speed motor.

33.6.1.2 Permanent-magnet AC Machines

There are two quite different types of permanent-magnet ac machines, actually looking very similar in their physical realization but dramatically different in their electrical characteristics and in the way in which they are controlled.

33.6.1.2.1 Permanent-magnet Synchronous Machines

These are simple extensions of the classical synchronous machine, where all of the voltages and currents are designed to be sinusoidal functions of time, as they are in the synchronous machine when used as a supply generator. These machines are known as permanent-magnet synchronous machines (PMSMs) or brushless ac machines and are discussed elsewhere in this book. Whilst the application of sinusoidal waveforms everywhere is comforting to many and does result in less acoustic noise, less "stray losses" etc. there is in general a requirement to know the exact rotor angular position at every instant of time, with an accuracy in the order of one degree. This knowledge is then used to shape the current waveforms to be sure that they are in phase with the back-emf of the windings. Whilst much research is going on to run these machines without position sensors, the reality in the workplace today is that you need to include a relatively expensive shaft position encoder with your PMSM, if you wish to control it electronically.

33.6.1.2.2 Permanent-magnet Brushless DC Machines A very much simpler possibility has emerged, which also gives the benefits of smooth torque and rapid controllability. This results in smaller minimum machine size, yielding maximum machine power density.

For this variant, the current waveforms and the back-emf waveforms are trapezoidal rather than sinusoidal. In this case, the machine is known as a permanent-magnet brushless dc machine (PMBDCM), with waveforms as shown in Fig. 33.55.

33.6.1.3 Brief Tutorial on Electric Machine Operation

The operation of electric machines can be explained in a variety of ways. Two possible and different methods are by:

(1) Using an understanding of the interaction of magnetic fields and the tendency of magnetic fields to align. This tendency to align is what provides the force that makes a compass needle swing around, until it aligns with the earth's magnetic field.

(2) Using the physical principle that "a current carrying conductor in a magnetic field has a force exerted on it." This force is commonly known as the Lorentz force.

The difference between the two explanations is that the first method gives better "physical pictures." However when one gets a little more serious, wanting to put in numbers, the force equations, whilst not greatly more difficult, are a little more challenging as they involve vector cross products.

With the second method, once the principle is accepted, there is a particularly simple scalar version of the force equation that very rapidly and simply gives numeric answers. The directions of the force, current, and motion are all orthogonal, i.e. at right angles and a formal exposition would again use a vector cross product. However, provided the rules learned in high school science are applied, the scalar version, $F = BLI$ can be used, where F is the force in newtons, B is the flux density in Tesla, L is the length of the conductor in meters, and I is the current in amperes.

Let us try both the methods, on the simple machine of Fig. 33.56.

The rotor is a magnet with a north and a south pole, and there is a coil of wire in the slots in the stator. Current is shown going into the conductor called **a** at the top of the stator and coming out of the conductor called **a'** at the bottom. To simulate a real machine more closely, one should imagine putting a one turn "coil" of wire into the slot, with a connection made at the back of the machine to connect the top wire to the bottom, so that by applying a positive voltage to the top wire and a negative voltage to the end of the coil (the bottom wire), current will flow in at the top and out at the bottom as discussed previously.

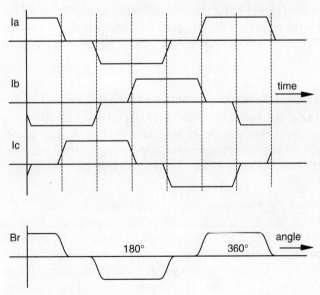

FIGURE 33.55 Phase currents as functions of time and the flux density on the surface of the rotor as a function of angular position for a typical PMBDCM.

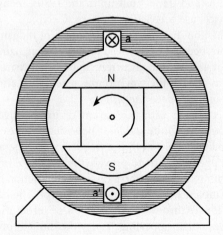

FIGURE 33.56 A simple motor with a single coil in the stator and a permanent-magnet rotor.

Method 1 If the current is put into the top and comes out of the bottom as marked, then according to the right-hand rule (if you forget, then look at the Institution of Electrical and Electronic Engineers (IEEE) logo), magnetic flux will be forced to go through the center of the coil from right to the left, creating a north pole on the left. The flux will then double back through the iron frame of the motor to arrive back at the South pole on the right.

Figure 33.57 shows a computer generated magnetic field plot. This was produced using finite element analysis and modeling the stator as electrical steel, the volume in the center as air and injecting current into the coil.

If the permanent-magnet rotor is then put inside the stator of Fig. 33.57, the rotor will align by rotating 90° counterclockwise.

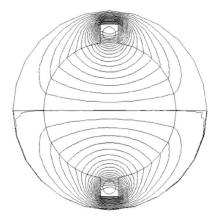

FIGURE 33.57 A computer generated magnetic field for current in the windings and a stator made of electrical steel.

Method 2 In this case, looking at Fig. 33.58, the conductor **a** will be immersed in a quite dense field produced by the magnets, with the flux lines going up the page in the rotor, across the airgap into the steel of the stator, heading down the stator to jump back across the lower airgap to return to the south pole. If a current is directed into the upper conductor and out of the bottom as before, it will produce force. Applying the direction rules associated with the Lorentz force, the force on the conductor will be to the right at the top and to the left at the bottom. Invoking Newton's third law, that action and reaction are equal and opposite, it is clear that if the wire is fixed, as it is, then there will be a force on the rotor to the left at the top and to the right at the bottom, thus the rotor will move counterclockwise as before.

33.6.2 Electronic Commutation

If the rotor of Fig. 33.58 moves, then to keep it rotating in the same direction, sooner or later the current in the conductors must be reversed. The time at which that needs to be done is when the rotor has moved to an almost aligned position, so it is not really a matter of time, but a matter of rotor position. Since the rotor is a permanent magnet, it is a very simple matter to determine at least where the physical pole edges are, using simple, reliable and inexpensive Hall effect (HE) sensors. These are small semiconductor devices which respond to magnetic flux density. (There are many ways to sense the position of the rotor; however the remainder of this exposition will stay with the HE sensor for simplicity.)

A very simple motor, relying on the inertia of the rotor for continuous rotation, could have a control circuit which sensed when the rotor magnet was horizontal and then reversed the current in **aa′** of Fig. 33.58.

The traditional "H" bridge switching circuit shown in Fig. 33.59 does that very effectively. Switch drive logic must ensure that either S1 and S4 are on, or S2 and S3 are on and never S1 and S2 simultaneously, or S3 and S4 simultaneously. This simultaneous operation would provide a short circuit. When this happens due to errors in the switch drive signal it is called "shoot through." There are generally important parts

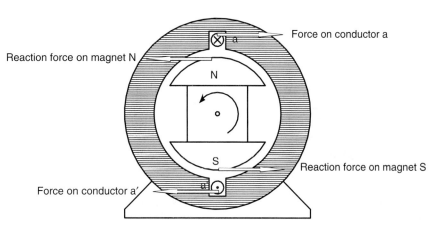

FIGURE 33.58 Lorentz forces on conductor and rotor of a simple machine.

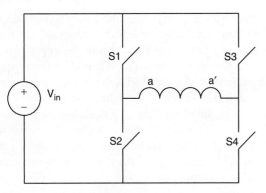

FIGURE 33.59 An "H" bridge circuit that can reverse the current through aa', as drawn with mechanical switches.

of the switch drive signal logic, discussed later, which try to make this impossible in normal operation.

In real motors, the permanent-magnet rotor field does not change instantaneously from north to south as the rotor rotates, so there would be reasonably large angles over which the torque could not be effectively produced. Also more coils are put in so that the space of the machine frame is used more efficiently.

The use of three "phases," as in Fig. 33.60, is very common. There are many things which balance naturally in a three-phase system and it is the simplest system with which it is possible to develop constant and unidirectional torque at all times and positions.

A very common method has developed, known as "six step switching." In this system, three phases are used. They are connected in star and there is a space between each of the magnet poles on the rotor and the two phases are activated at any one time, the third "resting" as the space between the magnet poles passes over it.

While the motor in Fig. 33.60 looks just like a three-phase synchronous machine, its operation is rather different.

As stated above, the windings are invariably star connected, as shown in Fig. 33.61.

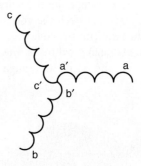

FIGURE 33.61 The schematic connection of the windings shown in Fig. 33.60.

Six Step Switching Explained

Consider the coil **aa'**, as shown in Fig. 33.60 and also diagrammatically in Fig. 33.61. Current driven into the **a** terminal, coming out of the **a'** terminal, will produce, as discussed before, flux from left to right, as shown here.

Obviously if the current direction in **aa'** is reversed, the flux will be

Similarly if current is driven into **b** and out of **b'** flux will result

And if it is reversed,

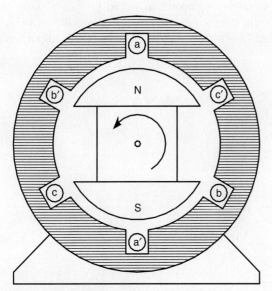

FIGURE 33.60 The physical layout of a three-phase PMBDCM.

Finally, if current is driven into **c** and out of **c′**, flux will result

And if it is reversed,

The permanent-magnet rotor would tend to align with the flux arrow shown in each of the states above, as discussed under machine operation, Method 1.

Now if the motor is wired in star as shown in Fig. 33.61, and a steady current is driven into **a** and out of **b**, the first and the fourth cases above occur at the same time. They add to produce a resultant,

and the rotor will tend to align with this resultant.

The Hall effect sensors are positioned so that just as the rotor gets to within 60° of being aligned, another arrangement of the windings is energized, and the field jumps forward another 60°, so the rotor is 120° away from being aligned. The rotor is always an average of 90° away from its desired position!

If a controller has input power from positive and negative of a dc supply and outputs to the terminals **a, b**, and **c** only, then the sequence of connections shown in Table 33.2 will produce the resultant flux directions as shown, providing continuous rotation.

The switching of Table 33.2 is simply achieved by a variant of the "H" bridge, as shown in Fig. 33.62. This process of switching current into different windings is called commutation and is the equivalent of the sliding brush contacts in a traditional brushed dc machine.

33.6.3 Current/Torque Control

You may have noted that the discussion has been about current in the windings rather than the voltage across them. In very simple small brushless DC machines, like the one in the muffin fan in your computer power supply, voltages are connected directly to the windings. For these small motors, the resistance of the windings is relatively high and this helps limit the actual current that flows and swamps inductive effects.

TABLE 33.2 The sequence of connections which results in counterclockwise continuous rotation of the rotor of the PMBDCM of Fig. 33.60

When rotor gets to this position, connections are made as at right	Terminal **a**	Terminal **b**	Terminal **c**	Flux directions
↑	Current in	Current out	Zero	↗
↖	Zero	Current out	Current in	↑
↗	Current out	Zero	Current in	↖
↓	Current out	Current in	Zero	↙
↘	Zero	Current in	Current out	↓
↗	Current in	Zero	Current out	↘

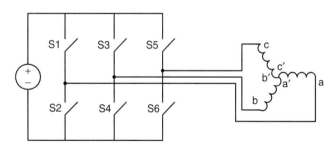

FIGURE 33.62 The switching circuit used for commutation of a PMBDCM.

There is an extra difficulty that must be addressed with high-performance, high-efficiency, well-made machines, and it adds another layer to the control of the motor. Such machines can easily be designed with very low resistance windings. It is not uncommon to have windings for a 200 V machine at the 20 kw level with the winding resistances less than 0.1 Ω.

When starting, or at a low speed, the current in a winding is limited only by the very low resistance and, for the machine above by Ohm's law, $I = E/R$ would result in more than 2000 amperes.

The most common requirement for a steady current in the windings is to provide a steady torque. There is always a back-emf generated in the windings whenever the motor is rotating, which is proportional to speed and subtracts from the applied voltage. Thus currents cannot be determined just by terminal voltage. The winding does however have inductance. Whenever the copper conductors are put in coils in an iron structure, particularly if there are low reluctance magnetic paths with only small airgaps, the creation of quite large inductances cannot be avoided. These are used to very good effect.

The nature of inductance is that when a voltage is applied to an inductor, instantaneous current does not result, rather the current begins to increase and ramps up in a quite controlled fashion. If the voltage across the inductor is reversed the current does not immediately reverse, rather it ramps down, will go through zero and reverses if the reversed voltage is left there long enough. However, if the voltage is alternated by switching rapidly, as can be done with power electronics, the current can be controlled to ramp-up and ramp-down either side of a desired current, staying within any determined tolerance of that desired current.

Figure 33.63 looks very much like the simple "H" bridge commutation circuit, but is performing a very different function. It is controlling the current amplitude to stay within a desired band. If S1 and S4 are turned on then the current will begin to increase from left to right in the winding. The current sensor in the circuit detects when the current reaches a value of half the hysteresis band of the comparator above the desired current level and initiates turn-off of S1 and S4 and turn-on of S2 and S3. (If they were all turned off at once, the inductive nature of the circuit would produce very high voltages which would cause arcing in mechanical switches, or breakdown and failure of semiconductor switches, see later.) The current then begins to reduce. It reduces a small amount, down to half the hysteresis band of the comparator below the desired current level and then the switches reverse again. Thus a desired current level is achieved, with an arbitrarily small triangular ripple superimposed, as shown in Fig. 33.64.

The general process of controlling by switching a voltage fully on or fully off at high speed is called pulse width modulation (PWM) and the specific method of current control achieved above with PWM, is called hysteresis band current control (HBCC).

Of course to keep this current ripple small, the switching may need to be very fast, but with the modern semiconductor switches there is no great problem up to 100 kHz for small machines and typically above 15 kHz for acoustic noise reasons, for machines rated up to several hundred kilowatts.

A perceived "drawback" of HBCC is that the switching frequency is determined by the circuit inductance, the width of hysteresis band, the back-emf, and the applied voltage, ranging very widely in normal operation. It is not difficult, but it is a little more complicated, to use a fixed frequency, and a linear analog of the current error to modify the pulse width of the PWM signal.

33.6.3.1 Switching Losses

There is a practical limit to how often semiconductor switches can be operated. At every change of state, if the switch is carrying current as it is opened, then as the voltage rises across the

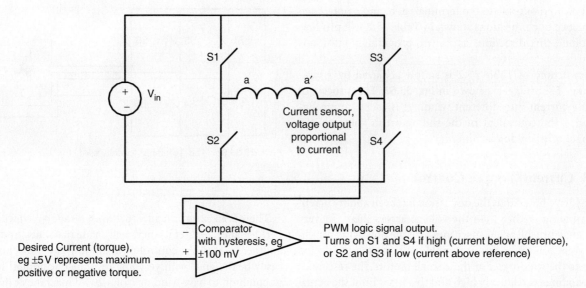

FIGURE 33.63 Hysteresis band current control using pulse-width modulation (PWM).

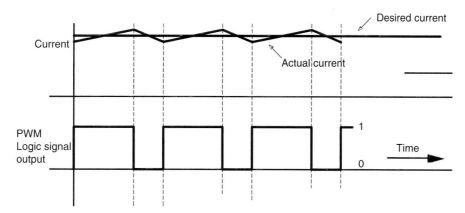

FIGURE 33.64 Hysteresis band current control and PWM waveforms.

switch and the current through it falls, there is a short pulse of power dissipated in the switch. Similarly, as the switch is closed, the voltage will take some time to fall and the current will take some time to rise, again producing a pulse of power dissipation. This loss is called switching loss. Fairly obviously it will represent a power loss proportional to the switching frequency and so the switching frequency is generally set as low as it can be without impinging on the effective operation of the circuit. "Effective operation" might well include criteria for acoustic noise and levels of vibration.

33.6.3.2 High Efficiency Method of Managing the Switching in the H Bridge

A very common way to control the current with the smallest number of switching transitions is to combine HBCC with, for example, alternating only S1 and S2 in Fig. 33.65 leaving S4 on all the time, on the understanding that there will be a back-emf in the winding and the current can still be increased or decreased as desired.

Thus when the motor is rotating and the back-emf is somewhere between zero and the rail voltage, alternating two switches rather than four will still allow current control in the coil, using for example HBCC, exactly as before.

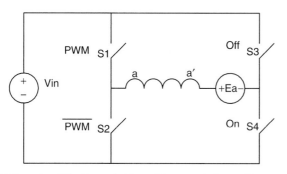

FIGURE 33.65 H-bridge switching with one switch steadily on and a back-emf.

This is a very common control scheme and will need some extra logic to reverse the direction of rotation of the motor, by either turning S4 off and S3 on continuously, or by swapping the control signals to the left and right "legs." For full-servo operation normal H-bridge switching can be used and the logic is slightly different, but not significantly more complicated. However, following the discussion above, the switching losses will be higher.

33.6.3.3 Combining Commutation and PWM Current Control

The real break-through is that one set of six switches can be used for both PWM and commutation. That is the clever part and also the confusing part when one first tries to understand what is going on.

Thus, in a controller there are two control loops. The first is an inner current loop switching at, for example, 15 kHz to control carefully and exactly the current in two of the coils. Then at a much lower rate, for example at 50 times per second at 3000 rpm, the two coils doing the work are changed according to Table 33.2, controlled by an outer commutation loop, using information from the Hall effect shaft position sensors.

A complete controller is shown in the block diagram form in Fig. 33.66. Various aspects of this block diagram will now be examined and explained in detail.

33.6.3.3.1 Hardware Details – Semiconductor Switches The three most likely semiconductor switches for a six step controller are the bipolar junction transistor (BJT), the metal-oxide silicon field effect transistor (MOSFET) and the insulated gate bipolar transistor (IGBT). Older controllers used BJTs, however contemporary controllers tend to use MOSFETs for lower voltages and powers and IGBTs for higher voltages and powers. Both of these devices are controlled by a gate signal and will turn-on when the voltage of the gate above

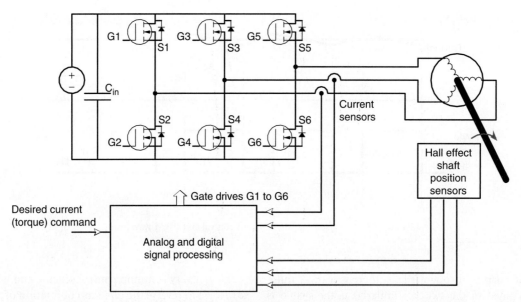

FIGURE 33.66 A complete controller showing the two feedback paths, one for the position sensors and one for the current sensors.

the source or emitter is greater than a threshold, which is typically about 5 V. Use of about 10 V is common. The devices are thus off when the gate voltage below the threshold. Systems typically use zero volts for the off state. The controller of Fig. 33.66 shows MOSFETs used for the six switches.

The trick is that the voltage at terminal **a**, also S1's source, is either ground or the positive potential of the battery, depending on which switches are on. Driving S2, S4, and S6 is easy since the MOSFET sources are all at the potential of the negative rail and the lower gate drive signals are referred to this rail.

There is a range of dedicated integrated circuits which can drive the switches S1, S3, and S5, and which use a "charge pump" principle to generate the drive signal and the drive power internally, all related to the MOSFET source potential. Various approaches to this technical challenge of providing a floating gate drive are commonly discussed under the generic heading of "high side drives."

For the most sophisticated drives, transformer coupling is used to provide a tiny power supply especially for the isolated gate drive and send the control signals either through an optocoupler or a separate transformer coupling. The high-side drive problems here are exactly the same as those encountered in the traditional buck converter, or in drives for induction motors and PMSMs.

33.6.3.3.2 Dead Time and Flyback Diodes
Two issues have been mentioned above that must be addressed when using high-speed electronic switches in inductive circuits.

The first, in Section 33.6.2 "*Electronic Commutation*," was that care should be taken to ensure that the upper and lower switches in the same "leg," (e.g. S1 and S2) are never turned on at the same time. If the controller attempts to turn one off and the other on at the same instant and switch turn-off is slower than turn-on (as it is with BJTs and IGBTs), then a short circuit will result for a brief time. The bus capacitor is usually very large to provide ripple current (see later) and usually of very high quality being fabricated especially for power-electronic applications, and can easily provide thousands of amperes for a few microseconds, which is enough to destroy the semiconductor switches.

The second issue, discussed in Section 33.6.3 "*Current/ Torque control*," is that one cannot turn-off both switches in a leg at the same time, even for a few nanoseconds, since the voltages resulting from attempting to interrupt current in an inductor will cause avalanche breakdown and failure of the semiconductors. This sounds like quite a dilemma.

There is actually a very simple and effective solution. At any transition, the control circuitry ensures that the active switches are all turned off before any switch is turned on, usually for a few microseconds. This is known as "dead time" and its provision is an essential part of most of the dedicated integrated circuits in use. Then a "flyback"/"freewheel" diode is put in anti-parallel with each semiconductor switch and this provides the current path during dead time. These diodes are shown in Fig. 33.66.

The diode has a little more loss than the switch, since the diode forward drop is more than the switch drop. This is significant in a low voltage controller. However, as stated above, for a low voltage controller, MOSFETs are the device of choice. They have a lesser known property, that when gated on, they can carry current in both directions. Intriguingly the "on" state resistance is lower in reverse than in the forward direction!

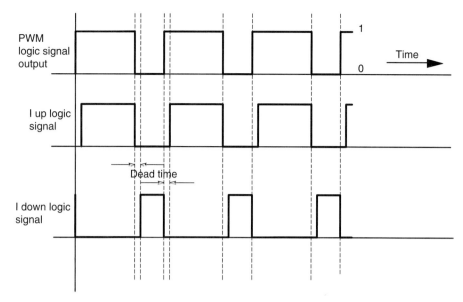

FIGURE 33.67 Dead time introduced into the PWM logic signal, for switch drive.

Thus for low voltage controllers, when the switch forward drop represents a significant contribution to losses, the MOSFET is turned on after dead time for both the current directions. Thus the higher loss in the diode is only for a few microseconds.

It is not difficult to produce from the PWM signal an "I up" logic signal which is used to cause the current to increase and an "I down" logic signal used to decrease it, with the timing as shown in Fig. 33.67. The function can be executed in sequential logic, or with the simple analog timing circuits.

A. Semiconductor Detail

In MOSFETs and most IGBTs, there is a diode already within the device; it is unavoidable and results from the fabrication processes. In modern power semiconductors, this intrinsic diode is optimized to be a good switching diode. The serious designer, however, will check the specifications for reverse recovery of this diode, since in highly optimized controller designs, reverse recovery losses in the intrinsic diode can be significant and are very difficult to control. In low voltage controllers, you can put a Schottky diode in parallel with the intrinsic body diode. The Schottky diode, with its lower forward voltage drop will tend to take the current and has no reverse recovery problems, but the current must commutate to it from the semiconductor die and the inductance of the connections is critical.

B. The Smoothing Capacitor on the Input to the Controller

This is a substantial capacitor, often very expensive, (it is shown in Fig. 33.66 as C in) and its design is quite challenging. The issue of smoothing is quite serious. If there are high-frequency or sudden current changes in the leads from the dc supply to the controller, they will radiate electromagnetic energy. Good design will limit the length of conductors in which the current is changing rapidly. Thus, a very large capacitor is placed physically as close to the positive bus of the switches as possible, aiming to have a steady current in the longer conductors from the dc supply, up to the capacitor. When the motor is running at, say, half speed and providing large torques, a very high level of ripple current is carried by this capacitor. Kirchhoff's current law (KCL) must be applied at node A, as shown in Fig. 33.68. Good capacitors have a ripple current maximum buried away in their specification sheet. It turns out that in general, the size of the capacitor in a given design has very little to do with how much voltage ripple you can tolerate at the bus, but rather is determined by the ability to carry the ripple current without the capacitor heating up and failing. It has been known that the small electrolytics in prototype controllers mysteriously explode. On searching, it is found that they are in parallel with the main capacitor and quite close to it, so they carry a lot of ripple current, then heat up and explode!

33.6.4 The Signal Processing for Producing Switch Drive Signals from Hall Effect Sensors and Current Sensors

33.6.4.1 Operation of the Hall Sensors

The flux density directly under a magnet pole can be anywhere from 500 to 800 millitesla with Nd–Fe–B magnets. Hall effect (HE) sensors with a digital output, called Hall effect switches, change state at very close to zero flux density. Thus, they will change state when the north and south pole are equidistant

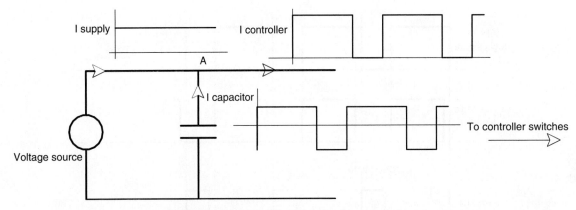

FIGURE 33.68 Kirchhoff's current law at node A.

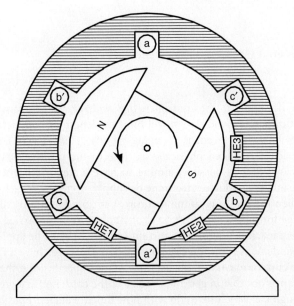

FIGURE 33.69 Possible Hall effect switch positions in a three-phase machine.

TABLE 33.3 Hall effect switch outputs for rotor positions as shown, HE switches placed as in Fig. 33.69

When the center of the rotor north pole is in this sector	HE1 outputs	HE2 outputs	HE3 outputs
	1	0	0
	1	1	0
	1	1	1
	0	1	1
	0	0	1
	0	0	0

from them, so that for example, in Fig. 33.69, the switch HE1 is just changing state with the rotor as shown.

In practice, a motor designer needs to consider what magnetomotive force comes from the current in the windings which might result in flux which would trip the HE sensor at a slightly different time. If you follow all the above logic about six step switching, you will see that you only need the magnet poles to have a span of a bit more than 120°. Using 180° magnet poles can add considerably to the cost, as well as having an impact on such things as cogging torque. Actual designs often add extra "sense" magnets to cover 180° just at the circumferential strip where the HE switches are located, adding minimally to the magnet mass and ensuring good and accurate triggering.

However, if the switches operate as above, then Table 33.3 will result for the HE switches located as shown in Fig. 33.69.

33.6.4.2 Sensing the Current in the Motor Windings

Figure 33.66 shows two current sensors. In the most sophisticated systems, there are two current sensors, one in each of the two motor phases. The current sensing is done at the winding and isolated with either an HE sensor in a soft ferromagnetic magnetic core surrounding the conductor (commercial items are available), or by using a resistive shunt sensor and some accurate analog signal isolation/coupling through transformers or opto-couplers. The isolation is necessary since the potential at points **a**, **b**, and **c** is either the dc bus voltage or zero, depending on which switches are on, so that any current measure such as the small voltage across a shunt is superimposed on these very large voltage changes. This is a very similar problem to that for the high side gate drives discussed earlier. The current in the third winding is determined by the algebraic application of KCL, given the other two readings.

Simple controllers sometimes avoid the complexities of isolated current measurement and instead measure the current in the return negative supply, for example from the bottom of the three lower switches to the bottom of the supply smoothing capacitor. This arrangement senses current when an upper and a lower switch is on, but not when the current is being carried in flyback diodes or by two lower switches. While it is inexpensive, it does not provide fully accurate control. The system works because the current should be decreasing when a measure is not available, heading towards zero, so switch or system failure due to over-current should not occur.

33.6.4.3 Management of Current Sensing

The controller must select the right current to increase or decrease, dependent on rotor position. The following convention is adopted. Positive current provides torque in the counterclockwise direction and therefore goes into winding **a**, **b**, or **c**.

All systems are capable of regeneration, which implies that negative torque can be commanded (without reversing the direction of rotation) to make the machine operate as a generator, developing retarding torque.

Thus for the above sequence of sector determinations, referring back to Table 33.2, the output of current sensors should be directed to the current controller as shown in Table 33.4.

The addition and negation required can be carried out with the standard operational-amplifier circuitry. The three required analog measures are then fed to a three to one analog multiplexer, gated from the HE switch signals suitably processed in combinational logic. The resulting single analog output is fed to the current comparator.

33.6.4.4 Distribution of Control Signals to the Switches

Given that the dead time is introduced elsewhere in sequential logic, or with timing circuits, it is a simple matter to develop

TABLE 33.4 Current sensors to use as input to the current controller, for each of the six rotor position sectors

Hall effect switch outputs (HE1, HE2, HE3)	Monitor current as read by
100	Negative of (sensor a+ sensor b)
110	Negative of (sensor a+ sensor b)
111	Sensor b
011	Sensor b
001	Sensor a
000	Sensor a

TABLE 33.5 Distribution of control signals to the switches using "High Efficiency Method of Managing the Switching in the H Bridge"

HE states(1,2,3)	S1	S2	S3	S4	S5	S6
100	0	0	0	I up	I up	I down
011	0	I up	0	0	I up	I down
001	0	I up	I up	I down	0	0
000	0	0	I up	I down	0	1
100	I up	I down	0	0	0	1
110	I up	I down	0	1	0	0

the combinational logic for directing, or steering, the switching signals to the right switches. A typical scheme for a specific controller is shown in Table 33.5.

It is usual also to include some shutdown logic from dedicated protection circuits, for example sensing over-current, over-bus voltage, under voltage for gate drive and over-temperature both in the motor and in the controller power stage. For simplicity, this is not shown in the table.

33.6.5 Summary

What is Discussed in the above

The physical principles of the operation of a PMBDCM have been discussed which lead to the development of the necessary parts of a power electronic controller. One specific type of current control, hysteresis band current control was explained in detail, and one specific type of switch logic pattern was developed. The exposition has included many of the issues that can cause difficulties for controller designers if they are not careful.

What is Not Discussed in the above

Many PMBDCMs have more than one pair of poles. The arguments above can all be extended to higher pole count machines, by taking any mention of degrees to be electrical degrees rather than mechanical degrees. The controller discussed in detail only manages one direction of rotation. It is an excellent exercise, and straightforward, but not trivial, to repeat the above steps, preparing the tables for clockwise rotation of the simple machine discussed above. Then, following the discussion in the first part of Section 33.6.3, *Current/Torque Control* about H bridge switching, prepare the logic tables again for full bidirectional control, using the I up and I down logic signals exactly as above, but applying them to both "legs" determined by the rotor position. Only one form of current sensing was discussed in detail. There are many simpler schemes in use which do not have quite the flexibility and accuracy of the above, but which can suit certain applications. Similarly, there are other forms of current control such as the constant-frequency linear method briefly discussed. Shaft position sensors take many forms. Adherence to the HE sensor was for simplicity, and to reinforce the magnetic field aspects of the machine operation.

33.7 Servo Drives

33.7.1 Introduction

Servo drives are motor drives that operate with high dynamic response. Historically, servo drives have implied motion-control systems in which sophisticated motor design, drive, and control techniques have been employed to obtain very much shorter positioning times than is possible with conventional drive systems. Examples are in machine tool drives, robotic actuators, computer disk drives, and so on. The power range for these drives has typically been in the range of a few kilowatts or less. This range has steadily increased in recent years as a result of advances in magnetic materials, machine design, power and signal electronic devices, and sensors.

Apart from the fast positioning times, "high dynamic response" also means that the drive operates with the following:

1. Very smooth torque up to a very low speed,
2. Very high reliability and little maintenance,
3. Immunity from load disturbances.

The last of the foregoing items is brought about by robust and intelligent control algorithms; the first two items are brought about by innovative and often costly motor and controller designs. As a result of these, the cost of a servo motor drive is usually much higher than the equivalent power rated industrial drives.

The distinctions just mentioned may be easily recognized by noting, for example, that the drives that bring material to a mill may not require high performance, but the drives that take part in shaping, milling, or reducing the material should have high dynamic response in order to increase throughput and meet the accuracy requirements of the final product.

33.7.2 Servo Drive Performance Criteria

The performance of a servo drive can be expressed in terms of a number or factors such as servo bandwidth, accuracy, percentage regulation, and stiffness. While servo bandwidth indicates the ability of the drive to track a moving or cyclic reference, the percentage regulation and stiffness stipulates the drive's static holding performance for speed or position, in the face of disturbances from the load and in the supply conditions. The servo bandwidth, specified as a frequency in Hertz or rad/sec, is often found from the system frequency response plot, such as the Bode diagram.

The percentage regulation of a speed-controlled system often refers to the percentage change in speed from no load to full load. In a type-zero system, this figure will have a finite value. Many systems are type zero, albeit with a high gain so that the regulation is acceptably low. For such systems, the regulation is often necessary for operational reasons. In some applications, zero percentage error is required, which calls for type 1 or integral type control system.

The servo stiffness is similar to the percentage error mentioned earlier, but it applies mainly for the position servo. It specifies the deflection of the load from its reference position, when full load torque is applied. It is usually the slope of the deflection versus the applied load torque in rad/Nm around the reference position.

33.7.3 Servo Motors, Shaft Sensors, and Coupling

Servo drives use motors which allow the desired goals of high dynamic response to be achieved. The important parameters/attributes of a servo motor are:

1. High torque-to-inertia ratio,
2. High torque-to-volume ratio,
3. Low inductance of the motor windings,
4. Low cogging torque at low speed,
5. Efficient heat dissipation,
6. Low coefficient of shaft compliance,
7. Direct-coupled, high-resolution, shaft-mounted sensors for position and speed.

High torque-to-inertia ratio allows fast acceleration or deceleration of the drive when motion references are changed. This is often achieved through innovative low-inertia rotor design and low inductance in the stator winding. One example of a dc servo motor is the pancake printed armature dc motor with no iron in the rotor, as indicated in Fig. 33.70. The rotor is sandwiched between axially mounted stator poles. The commutator is also on the printed armature. Another example is the disk rotor stepping motor, also without iron in the rotor, as indicated in Fig. 33.71.

The PM ac synchronous motors with modern high-energy-density magnets in the rotor, as described in Section 33.6, are also examples where the motor designer strives to minimize the rotor inertia. Modern permanent magnets allow the required airgap flux to be developed with a much reduced volume of the magnets, consequently reducing the diameter of the rotor.

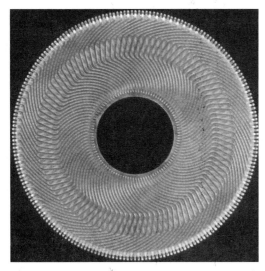

FIGURE 33.70 Pancake armature of a dc servo motor. Courtesy: Printed Motors Ltd., UK.

FIGURE 33.71 A disk rotor stepping motor with ironless rotor for low inertia and inductance. Courtesy: Escap Motors.

It is well-known that the moment of inertia of a motor increases as the fourth power of its outer radius!

Another benefit of the modern permanent-magnet material is that the motor volume is also reduced. Servo motors often have to be located in a very confined space, and this reduction in volume is an important attribute.

The ironless designs mentioned earlier bring other benefits in the form of reduced inductance and cogging torque. Brushed pancake ironless motors are available with armature inductance as low as $100\,\mu H$.

From Section 33.2.2, the mechanical and electrical time constants of a brushed dc motor are given by

$$\tau_m = \text{mechanical time constant} = \frac{R_a J}{K_E K_T}\,s$$

and $\quad \tau_a = \text{electrical time constant} = \frac{L_a}{R_a}\,s.$

It is well known that for the highest load acceleration, the load inertia referred to the motor should be equal to the rotor inertia. Thus, in a matched system, the total inertia the motor accelerates is twice its own inertia. In other words, the motor inertia should also be minimized.

For a good servo motor, the ratio between the mechanical and the electrical time constants is often of the order of five or more. This allows the speed and the current-control loops to be decoupled and noninteracting. The electrical time constant of a motor determines how quickly the motor current may be changed and hence how quickly the torque can be changed. As also mentioned in Section 33.2, drives with a reasonable dynamic performance should have an inner torque loop. This torque loop is built around current loops, for the armature for the brushed dc motor, or for the d- and q-axes currents for the induction and synchronous motor drives. Having a low inductance in the winding allows these currents to be followed dynamically changing current or the torque references with higher accuracy and bandwidth.

The cogging torque, if appreciable, causes the rotor to have preferential positions. As a result, the position accuracy of the motor may suffer. Another problem is the ripple in speed as the motor is operated at low speed. At high speed, these ripples due to cogging torque may be filtered out by the motor inertia; however, the extra loss due to cogging remains. The ironless or toothless rotor obviously produce very small cogging torque because of the absence of preferential paths for the airgap flux to establish through the rotor iron of the brushed dc motor. The surface-magnet synchronous motor also has this feature. The interior-magnet motor normally has skewed stator slots to avoid the production of cogging torque.

Servo motors often operate with frequent start-and-stop duty, with the fastest allowable acceleration and deceleration during which the motor current is allowed to reach about 2–3 times the continuously rated current. The increased $I^2 R$

loss in such duty must be dissipated. This calls for adequate cooling measures to be incorporated in the motor housing. With such operation, it is sometimes possible to excite the mechanical resonance due to shaft compliance. This is avoided through proper arrangement of the shaft position/speed sensor and the coupling between the motor and the sensor. A belt-driven speed sensor may be acceptable for an industrial drive; however, for servo applications, a rigid, direct-coupled sensor mounted as close as possible to the motor armature is preferable. Additionally, the speed sensor is also required to have negligible noise. Speed signal from analog tachogenerators, which were used for speed sensing until recently, invariably needed to be filtered to remove the cyclic ripple/noise that existed. Such filtering often limits the maximum speed-control bandwidth of a drive.

33.7.4 The Inner Current/Torque Loop

The inner current loop(s) in a servo motor drive play a more important role than just limiting the current in case of overload. These loops operate continuously to regulate the motor-developed torque so as to meet the load demand, and for meeting the speed trajectory specified by the motion controller. Motor drives of high dynamic response currently employ PWM current sources. These sources use MOSFET or IGBT switching devices that allow the modulator to be operated with a switching frequency between 10 and 25 kHz. At these frequencies, the inherent switching delay, which is equivalent to half of the PWM switching period, is made rather small for the bandwidth of the torque control loop. The bandwidth of the current control loops closely represents the bandwidth of the torque control. This is because the motor-developed torque generally is proportional to these currents. Servo drives up to a few kilowatts presently have torque/current control-loop bandwidths in excess of 1 kHz.

For higher power, fast-response drives, such as those used in the metal-processing industries, thyristor converters have been used for many years. The switching frequencies of these converters are rather low, being some multiple of the mains frequency, according to the converter chosen. Fortunately, the larger mechanical time constant of the larger power motor and the nature of the applications have allowed the 300 Hz (360 Hz in the United States) switch frequency of the three-phase thyristor bridge converter to be used satisfactorily in many applications requiring high dynamic response. The growing availability of faster and higher power IGBT devices is continually enhancing the dynamic performance of larger drives.

Fast-response, inner torque control loops have in recent years been extended to ac induction and synchronous motors. These motors were hitherto considered only for industrial drives. The vector methods described in Sections 33.3 and 33.4, which employ inner quadrature axis current controllers in the synchronous (for the induction motor) or the rotor (for the synchronous motor) reference frame, have transformed the prospects of ac motor drives in servo applications.

Because of the fast dynamic response requirement of servo drives, the servo motor is nearly always driven with the maximum torque per ampere (MTPA) characteristic. Field weakening is normally not used. In other words, field control either directly for a brushed dc motor or a synchronous motor or indirectly through armature reaction (i.e. through i_d current control) for induction or PM ac synchronous motors is not used for field weakening. It is nevertheless used for regulating the field at the desired level. Field weakening are mainly used for drives where the operation at higher than base speed with constant-power characteristic is desirable.

33.7.5 Sensors for Servo Drives

Servo drives require high-bandwidth current sensors for the inner torque loop and high-accuracy, noise-free speed and position sensors for the outer loops. The current sensor is often a Hall device with an amplifier, which can have bandwidths as high as 100 kHz. The inner current loop both limits and continuously regulates the motor current in all operating modes of the drive, including acceleration and deceleration. About 2–3 times, the continuous rated current of the motor is tolerated during acceleration and deceleration. This entails limiting the speed controller output to the level corresponding to the current sensor output for the limiting values of motor currents.

The current-sensor output has to be filtered to adequately remove the switching frequency noise. Otherwise, certain switching devices in converter may be overloaded. This task is more important for the thyristor converters for dc drives for which the switching frequency is rather low. This filtering of the current-sensor output limits the bandwidth of the current-control system, i.e. the inner torque-control loop.

Performance of servo-motor drives depends critically on the noise and accuracy of the speed and position sensors. Synchro-resolvers with 12 bit or higher digital accuracy were used in many servo-drive systems until recently. The advent of cheaper incremental and absolute optical encoders has altered this situation completely. These digital sensors are actually position sensors. The speed information is derived from positions measured by discrete differentiation. Such differentiation is not feasible with analog position sensors, because of the noise.

Analog tachogenerators are also avoided for speed servo systems. This is because of the tachogenerator ripples inherent in the sensor.

Modern discrete position sensors provide for virtually noise-free speed and position sensing. This allows very fast dynamic response to be achieved if the switching frequency of the converter allows it.

33.7.6 Servo Control-loop Design Issues

33.7.6.1 Typical Controllers

33.7.6.1.1 Proportional Controller A proportional controller provides for a straight gain to amplify the error signal. It has no discriminatory properties. With the input and feed-resistance values indicated in Fig. 33.72, the total gain of the controller is $(K + 1)$.

For this controller, $\quad v_c = (K + 1)(\theta_i - \theta) \quad$ (33.79)

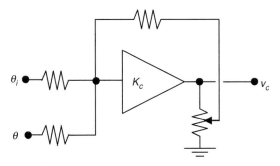

FIGURE 33.72 Proportional controller.

33.7.6.1.2 Transient Velocity Feedback Controller It is well-known that a following error will exist in the preceding system when a moving or ramp reference is tracked. If a rate feedback, such as the speed feedback in a position-control system, is used to damp the system, this error is further increased. To overcome this following error due to velocity feedback, transient velocity feedback can be used as indicated in Fig. 33.73.

The speed (velocity) signal is passed through an RC circuit at the input of the amplifier circuit. An input current occurs only when the speed signal changes. In the steady state, the capacitor is fully charged, so that no following error in the steady state due to the velocity feedback can exist.

In the steady state when the velocities are equal, the output may lag or lead depending on the relative values of R_1 and R_2. It can be shown that in the absence of frictional load torque,

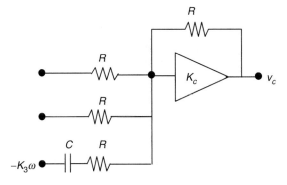

FIGURE 33.73 Transient velocity feedback controller.

as is often the case in servo applications, no following error is introduced if $R_1 = R_2$.

33.7.6.1.3 Integral Controller In the transient-velocity and error-rate feedback schemes, the following error will exist if viscous friction and load torque are present. If such loads are present, the system gain has to be infinity to have a zero error. Very large gains will make any physical systems unstable, unless bandwidth limitations exist. One way to employ infinite gain in the steady state is to use an integrator. This amplifies the steady-state error until it is eliminated.

Normally, a proportional plus integral (PI) action is used. A derivative term is normally not used in the control system of a drive system, since the drive feedback signals are very noisy. Instead, derivative signals are obtained through sensors such as tachogenerators. The structure of a PI controller is indicated in Fig. 33.74.

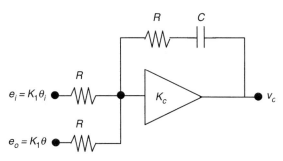

FIGURE 33.74 Integral controller.

It can be shown that there will be no steady-state error even in the presence of frictional or other load torque.

Many types of more complex controllers are available, such as the variable structure controller. Drives with fuzzy controllers have also been in the marketplace for some years.

The controller circuits just described are usually implemented in analog circuits using operational amplifiers. Digital implementations are also being gradually introduced using the embedded microcontrollers and digital signal processors.

33.7.6.2 Simplified Drive Representations and Control

Consider the block diagram of Fig. 33.75 in which the individual elements (blocks) are represented in terms of their transfer functions in terms of the Laplace operators.

Here, $G_A(s)$, $G_C(s)$, $G_L(s)$, $H_T(s)$, and $H_F(s)$ represent the transfer functions of the power converter plus the motor, the controller, the load, the sensor (of speed in this example), and the filter following the sensor respectively. The reference input for speed and the feedback signal are connected to a summing junction of an operational amplifier through resistors R_i and R_f, respectively.

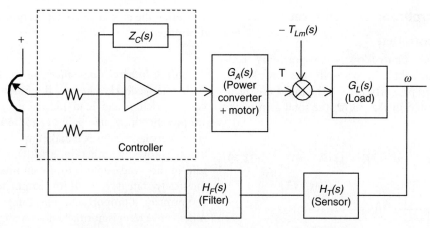

FIGURE 33.75 Block diagram of a speed-control system.

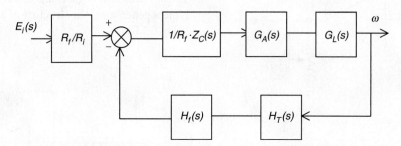

FIGURE 33.76 Simplified representation of Fig. 33.75.

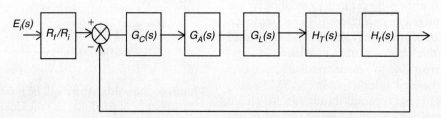

FIGURE 33.77 Further simplified representation of Fig. 33.75.

The preceding system can be simplified to that shown in Fig. 33.76, and further to that in Fig. 33.77.

In general, if the individual control blocks are approximated as first-order systems and are mutually decoupled, meaning that each block operates in a frequency band that is far outside the frequency bands of all other blocks, then the foregoing systems can be represented by a transfer function of the form

$$G_1(s) = \frac{K}{(1+sT_1)(1+sT_2)(1+sT_3)(1+sT_4)\cdots} \tag{33.80}$$

When T_3 and T_4 are much smaller time constants than T_1 and T_2, the preceding may be approximated by

$$G_1(s) \approx \frac{K}{(1+sT_1)(1+sT_2)(1+sT_s)} \tag{33.81}$$

where $T_s = T_3 + T_4 + \cdots$ etc. A dc-motor speed-control system with current and speed sensors falls in this category. For such a system there exist two dominant time constants (poles).

For such a system, a proportional plus integral controller is of the form

$$G_c(s) = \frac{(1+s\tau_1)(1+s\tau_2)}{s\tau_0(1+s\tau_{F1})(1+s\tau_{F2})} \tag{33.82}$$

One optimization criterion (Kessler's) stipulates that $\tau_1 \approx T_1$, $\tau_2 \approx T_2$, and $\tau_o \approx 2KT_s$. With this stipulation, the transfer function of the complete system is given by

$$G(s) = G_1(s)G_c(s) = \frac{1}{2sT_s(1+T_s)} \tag{33.83}$$

and

$$\frac{V(s)}{V_i(s)} = \frac{G}{1+G} = \frac{1}{1+2sT_s+2s^2T_s} \tag{33.84}$$

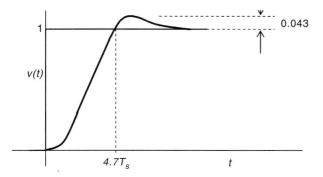

FIGURE 33.78 Response of the optimized system of Fig. 33.75.

Note that the two filter time constants τ_{F1} and τ_{F2} are included in $G_c(s)$ for the sake of its realizability. These can be relegated to frequencies far higher than the range of interest and can be ignored for further analysis of the system. For an unit step input of V_i, the output V is given by

$$v(t) = 1 - \sqrt{2}\, e^{-1/2T_s} \sin\left(\frac{t}{2T_s} + \frac{\pi}{4}\right) \text{ for } t \geq 0. \quad (33.85)$$

A typical output is sketched Fig. 33.78.

If the transfer function $G_1(s)$ has one dominant time constant $T_1(s)$, as for the field current control of a dc motor, a suitable controller is the form

$$G_c(s) = \frac{(1 + sT_1)}{s\tau_0 (1 + sT_F)} \quad (33.86)$$

In some cases, the transfer function $G_1(s)$ is of the form

$$G_1(s) = \frac{K}{sT_1(1 + sT_2)(1 + sT_s)} \quad (33.87)$$

where T_s is the sum of a number of short time constants, associated with sensors, switching frequency, and so on. The current controller of the dc motor with back-emf has such a characteristic. A suitable PI controller for this system is

$$G_c(s) = \frac{(1 + s\tau_1)(1 + s\tau_2)}{s\tau_0(1 + s\tau_{F1})(1 + s\tau_{F2})} \quad (33.88)$$

For this system, Kessler's optimization criterion stipulates that

$$\tau_1 = 4T_s, \tau_2 = T_2, \text{ and } \tau_0 = \frac{8KT_s}{T_1}$$

The transfer function of the complete system is then

$$\frac{V(s)}{V_i(s)} = \frac{1 + 4sT_s}{1 + 4sT_s + 8s^2T_s^2 + 8s^3T_s^3} \quad (33.89)$$

The peak overshoot of this system to a unit step unit is usually unacceptable, as indicated by the response of Fig. 33.79. This overshoot is usually reduced by inserting a first-order filter in the reference circuit. The filter network and the responses are given in Fig. 33.80.

33.8 Stepper Motor Drives

33.8.1 Introduction

A stepper motor is a positioning device that increments its shaft position in direct proportion to the number of current pulses supplied to its windings. A digital positioning system without any position or speed feedback is thus easily implemented at a much lower cost than with the other types of motors, simply by delivering a counted number of switching signals to the motor. Typically, a 200 steps-per-revolution stepper motor with 5% stepping accuracy will be equivalent to a dc motor with a 12-bit (or 4000 counts/rev) encoder plus the closed-loop speed and position controllers for obtaining similar positioning resolution. This advantage, however, is obtained at a cost of increased complexity of the drive circuits. A disadvantage of the motor is perhaps its inability to reach an absolute position, since the final position reached is only relative to its arbitrary initial position. Nevertheless, the true digital nature of this motor makes it a very suitable candidate for digital positioning systems in many manufacturing, automation, and indexing systems.

The working principle of stepper motors is based on the tendency of the rotor to align with the position where the stator flux becomes maximum (i.e. seeking of the minimum-reluctance position, also called the detent position). The rotor

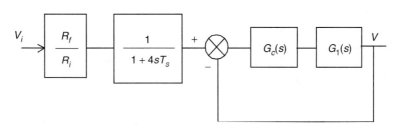

FIGURE 33.79 Block diagram representation of a typical current controller.

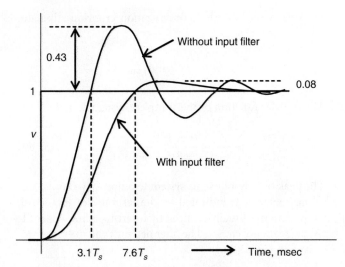

FIGURE 33.80 Optimized response of the system of Fig. 33.80.

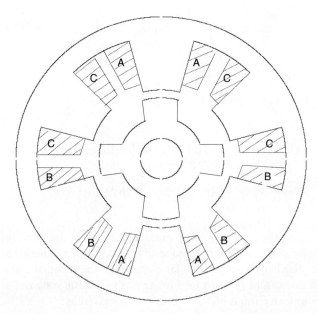

FIGURE 33.81 Cross section of a single-stack variable-reluctance stepper motor.

and the stator are both toothed structures, and the stator normally has more than two windings to step the rotor in the desired direction when they are energized in certain combinations with current. Some motors additionally have permanent magnets embedded in the rotor that accentuate an already existing, zero-excitation detent torque. These motors hold their positions even when the stator excitations are removed completely, a feature desirable for some applications.

In addition to the point-to-point stepping action, these motors can also be operated at high slewing speed, simply by increasing the pulsing rate of phase currents. Since the motor is inherently a synchronous actuator, the pulsing rate has to be increased and decreased properly, so that the rotor may follow it. At the end of a complete run, the motor always stops at the desired incremental position or angle without any accumulated error. The only error that may be encountered is mainly due to the machining accuracy of the teeth in the stator and rotor. This error is of the order of about 5% of one step position/angle and it is nonaccumulative.

33.8.2 Motor Types and Characteristics

33.8.2.1 Single-stack Variable-reluctance Stepper Motor

Single-stack motors are normally of the variable-reluctance type with no excitation in the rotor. The cross section of a three-phase motor with two stator poles/phase and four rotor poles are indicated in Fig. 33.81. The motor can be stepped clock or anticlockwise by energizing the phase winding in the ABCA or ACBA sequence, respectively. The step angle, i.e. the angle moved by the rotor for each change in excitation sequence, of the motor is given by

$$\theta_s = \frac{360}{NP} \text{ degrees} \tag{33.90}$$

where N is the number of phases in the stator and P is the number of poles in the stator. Single stack motors typically has larger step angles than other types because of limitations of space for the windings. The step angle of these motor tend to be larger than the multistack and hybrid stepper motors.

For each excited winding, the motor develops a torque angle (T–θ) characteristic as indicated in Fig. 33.82. Note that there are two equilibrium positions of the rotor, namely, X and Y, where the motor develops zero torque.

The position X is referred to as the stable detent position, around which the rotor develops a restoring torque when displaced. The restoring torque increases as the rotor is moved from its detent position, becoming a maximum T_{max} on either side of this position. The slope of the T–θ characteristic around this detent position and the maximum torque, both of which depend on the level of excitation, indicate how far the rotor will be displaced under load torque. This means that the level of excitation also affects the position holding accuracy of the motor.

The motor may also be excited in the sequence: AB–BC–CA or AB–CA–BC for forward and reverse stepping, respectively. The two phases-on scheme develops more torque around the detent positions at the expense of twice the resistive losses.

Yet another excitation scheme is AB–B–BC–C–CA–A–AB for forward stepping and AB–A–AC–C–BC–B–AB for reverse stepping. In this scheme, the step size is halved as opposed to the full-step size of the previous sequences. Two different levels of torque is produced for alternate detent positions. However, the reduced step size and the more damped nature of each step may outweigh this disadvantage.

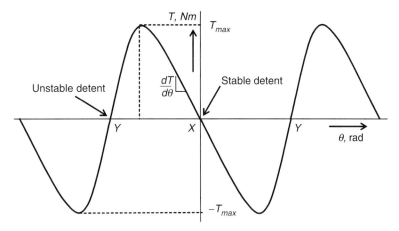

FIGURE 33.82 Static torque characteristic of a stepper motor.

33.8.2.2 Multi Stack Variable-Reluctance Stepping Motor

In a multi stack variable-reluctance motor, the stator windings are stacked along the shaft. Each stack section now has the same number of poles in the stator and the rotor. Normally each stator stack is staggered with respect to its neighbor by *one/Nth* of a pole pitch, where N is the number of stator/rotor phases or sections. The cut out view of Fig. 33.83 shows some internal details of a six-phase multi stack motor, in which each stack has a phase winding between two rings, each with 32 stator and rotor poles. The step size of this motor is

$$\theta_s = \frac{360°}{Np} = \frac{360°}{6 \times 32} = 1.875° \quad (33.91)$$

The excitation sequence of this motor is similar to the ones mentioned in Section 33.8.2.1, except that more excitation sequences are available. When a stator winding is energized, the rotor poles of that section tend to align with those defined by the stator excitation. The stator and rotor teeth in the other sections are not aligned. By changing the combination of excited phases to the next in sequence, the rotor is made to move by one step angle.

33.8.2.3 Hybrid Stepping Motor

A hybrid stepper motor has an axially oriented permanent magnet sandwiched between two sections of the stator and rotor, as indicated in Fig. 33.84. The magnetic flux distributes radially through the two stator and rotor sections, both of which are toothed, and axially through the back iron of the stator and the shaft. The stator has two phase windings, each of which creates alternate polarities of magnetic poles in both sections of the stator. Stator windings are excited with bipolar

FIGURE 33.83 Cut out view of a six-phase, multi stack, variable-reluctance stepper motor. Courtesy: Pratt Hydraulics, UK.

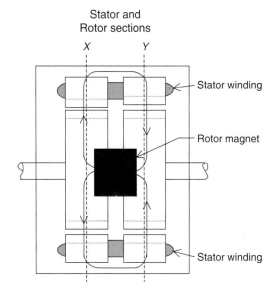

FIGURE 33.84 Axial section of the hybrid motor.

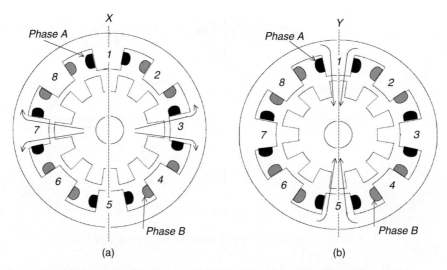

FIGURE 33.85 Cross section of the hybrid motor: (a) section X and (b) section Y.

currents, as opposed to the unipolar currents in the variable-reluctance motors of the two preceding sections. The magnetic flux produced by the stator windings is circumferential in each stator and rotor section, but also crosses the airgap radially. It does not, however, pass through the rotor magnet. The two rotor sections are offset by half its tooth pitch.

The rotor magnet causes to the stator and rotor teeth to settle at the minimum reluctance position with a modest amount of detent torque to keep the rotor in position, when the stator windings are not energized. The rotor magnetic flux distributes outward through stator poles 3 and 7 in section X and inward through poles 1 and 5 in section Y, as shown in Figs. 33.85a and b. When the stator windings A and B (indicated as dark and faint shaded, respectively) are energized with positive and negative currents, respectively, the resulting stator flux also distributes through these same poles, so that the rotor then develops a much higher detent torque (T–θ) characteristic. The motor can be stepped forward or backward by energizing windings in sequence $A\bar{B} - AB - \bar{A}B - \bar{A}\bar{B}$ or $A\bar{B} - \bar{A}\bar{B} - \bar{A}B - AB$ respectively, where the over bar indicates the polarity of currents in phases A and B.

The stepping angle of a hybrid stepper motor is given by

$$\theta_s = \frac{90°}{P} \qquad (33.92)$$

where P is the number of rotor poles.

33.8.2.4 Permanent-magnet Stepping Motor

Permanent-magnet stepper motors have alternate polarities of permanent magnets on the rotor surface while the rotor iron, if it is used, has no teeth. In one type of construction, the rotor has no iron, and the stator consists of two windings that

FIGURE 33.86 Rotor of a PM stepper motor. Courtesy: Escap Motors.

setup alternate poles when energized, just as in the case of the hybrid motor. The rotor consists of permanent magnets, alternately polarized, attached to the surface of a nonmagnetic disk, as shown in Fig. 33.86. The stator and rotor fluxes cross the airgap, one on either side of the disk, axially.

33.8.3 Mechanism of Torque Production

33.8.3.1 Variable-reluctance Motor

If it is assumed that the current in the excited winding remains constant, the production of static torque of a variable-reluctance motor around a detent position is given by

$$T = \frac{dW_m}{d\theta} \qquad (33.93)$$

This torque expression may also be expressed as in Eq. 33.94, when it is further assumed that the inductance of the excited

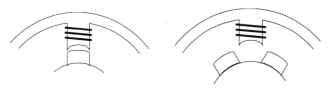

FIGURE 33.87 Stator and rotor teeth alignment: (a) aligned position, $\theta = X$ and (b) unaligned position, $\theta = Y$.

winding at any given position remains constant for all currents.

$$T = \frac{1}{2} i^2 \frac{dL}{d\theta} \qquad (33.94)$$

The developed torque is due to the variation of inductance (or reluctance) with position. Note that the direction of current has no bearing on the developed torque. When the stator and rotor poles are perfectly aligned, as indicated in Fig. 33.87a, the inductance L changes little with a small change in θ. The developed torque is thus very small around this position, corresponding to the position X in Fig. 33.82. When the stator and rotor teeth are unaligned, as in Fig. 33.87b, L changes more significantly with θ, and the restoring torque becomes much larger. As θ increases, $dL/d\theta$ goes through a maximum, producing T_{max}. It should be noted that around a stable detent, L reduces as θ increases, so that the slope of the $T-\theta$ characteristic is negative at the origin. Beyond the position where T_{max} is developed, L increases as a result of the next set of rotor teeth coming under the stator teeth. This explains the drop in T_{max} and the positive slope of the $T-\theta$ characteristic in the region between where T_{max} is developed and Y in Fig. 33.82.

If stepper motors are operated in magnetically linear region where L remains constant with the current for a given angular position, the developed torque per unit volume is small. Because of this, steppers motors are normally driven far into saturation. Equation (33.94) then does not represent the torque characteristic adequately.

For a saturated stepper motor, the calculation of the $T-\theta$ characteristic for any given current involves complex computation of stored energy, or coenergy, for each position of the rotor. This requires the magnetization characteristics of the motor for different levels of stator currents and rotor positions to be known. Reference [33] may be consulted for further reading on this.

33.8.3.2 Hybrid and PM Motors

In hybrid stepper motors, most of the developed torque is contributed by the variable-reluctance principle explained earlier. The rest is developed by the rotor magnet in striving to find the minimum-reluctance position. It should be noted that the alternate polarities of the magnetic poles created by each winding may be reversed by the direction of its current. Consequently, the polarity of the winding currents also determines the direction in which the developed torque increases positively around a detent position.

33.8.4 Single- and Multi-step Responses

When the rotor is at a detent position and phase currents are changed to a new value, the detent position is moved and the rotor proceeds towards it and settles down at the new detent position. The movement of the rotor is influenced by the shape of the $T-\theta$ characteristic and the load friction. The rotor stepping is normally quite under-damped. The final positioning error is also determined largely by the load torque. For instance, if the $T-\theta$ characteristic is assumed to be a sinusoidal function of θ, the error in stepping is given by Eq. (33.95), where T_{max} is the peak of the $T-\theta$ characteristic and T_L is the load friction torque.

$$\theta_e = \left|\sin^{-1}(T_L/T_{\max})\right| \qquad (33.95)$$

However, this error does not accumulate as further stepping is performed. If the phase currents are switched in succession, the rotor makes multiple steps. Typical single and multi step responses are as indicated in Fig. 33.88.

The maximum rate at which the rotor can be moved depends on several factors. The rise and fall times of the winding currents, which are largely determined by the electrical parameters of the windings and the type of drive circuits used, and the combined inertia and friction parameters of the motor and load are important factors.

The discrete signals to step the motor in the forward or reverse direction are translated into current-switching signals for the drive circuits. This translator is a simple logical operation that is embedded in most of the integrated circuits available for driving stepper motors.

In many applications, the stepper motor is operated at far higher speeds than which it can start/stop from. The performance of a stepper motor at high speed is normally given in terms of its pull-out torque-speed $(T-\omega)$ characteristic.

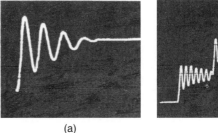
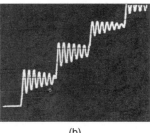

(a)   (b)

FIGURE 33.88 Typical step responses of a stepper motor: (a) single step response and (b) multi step response.

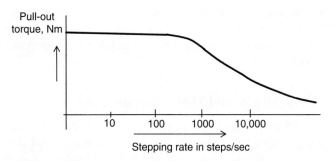

FIGURE 33.89 Typical pull-out torque characteristic of a stepper motor.

This characteristic indicates the maximum average torque, the motor may develop while stepping continuously at a given rate. This torque is also largely determined by the parameters of the motor and its drive circuits. Figure 33.89 indicates the typical shape of the pull-out T–ω characteristic of a stepper motor drive.

At low speed, the pull-out torque is roughly equal to the average value of the positive half-cycle of the T–θ waveforms of Fig. 33.82. At high speed, the finite but fixed rise and fall times of the currents and the back-emf of the winding reduces the extent to which the windings are energized during each switching period. Consequently, the pull-out torque of the motors falls as the stepping rate (speed) increases.

For operation at high speed, the stepping rate is gradually increased and decreased from one speed to another. Without careful acceleration and deceleration to and from a high speed, the motor will not be able to follow the stepping commands and will lose its synchronism with the stepping pulses or winding excitations. The acceleration and deceleration rates of a stepper motor are also determined largely by the pull-out torque characteristic.

Stepper motors are known to suffer from mechanically induced resonance and consequent mis-stepping when its switching rate falls within certain bands, which are largely determined by the way the developed torque varies with time, as the motor steps. Careful selection of stepping rate is normally employed to overcome the problem. Some shaft-mounted external damping measures may also be used when the stepping rate needs to be continuously varied, such as in the case of machine-tool profile following.

33.8.5 Drive Circuits

Two types of drive circuits are in general use for stepper motors. The unipolar drive is suitable for variable-reluctance stepper motors, for which the developed torque is determined by the level of current, not its polarity. For hybrid and permanent-magnet motors, the direction of current is also important, so that the bipolar drive circuits are more suitable.

33.8.5.1 Unipolar Drive Circuits

In its simplest form, the unipolar drive circuits, one for each winding, are as indicated in Fig. 33.90. The transistor (MOSFET) is turned on to energize the winding, with a current that is limited either by the winding resistance or by hysteresis or PWM current controllers. The freewheeling diode allows the winding current a circulating path when the transistor is turned off.

The drive circuit of Fig. 33.90a is a basic one. A better drive circuit is shown in Fig. 33.90b, which includes a zener diode in the freewheeling path. A pulse-width modulator is also included in the gate driving circuit. The pulse-width modulator allows a higher dc supply voltage (typically 5–10 times the voltage for the resistance-limited drive) to be used, thereby reducing the rise time of current at switch-on by 5–10 times. The zener diode allows a fast fall time for the current when the transistor is turned off by dissipating the trapped energy of the winding at switch-off faster. Yet another scheme is shown in Fig. 33.90c which allows the trapped energy of the winding at switch-off to be returned to the dc source when the transistor is turned off, rather than being dissipated in the winding or the freewheeling circuits. This circuit is by far the most efficient, and at the same time gives the fastest possible rise and fall times for the winding currents.

33.8.5.2 Bipolar Drive Circuits

The bipolar drive allows the motor windings to be driven with bidirectional currents. The four-transistor bridge drive circuit of Fig. 33.91, one for each winding, is the most popular. The circuit can cater to the required rise and fall times of the winding by properly selecting the dc supply voltage V_{dc}, the pulse-width modulator, and the current controller gains.

Some hybrid and PM motors come with four windings, two for each phase. These may be connected in series or parallel, depending on the torque characteristics desired. In any case, only two drive circuits of the type indicated in Fig. 33.91 are required.

33.8.5.3 Drive Circuits for Bifilar Wound Motors

Hybrid stepping motors may also come with bifilar windings, which allow the simpler unipolar drive circuits to be used. These motors have two tightly coupled windings for each phase. Figure 33.92 illustrates two bifilar windings on stator pole and their unipolar drives. The two windings on each pole have opposite sense, so that the magnetic polarity is reversed by simply switching the other winding. Since only unidirectional current is involved, the unipolar drive circuits of Fig. 33.90a or b may be used at a considerable savings in terms of the drive circuits. This benefit is, however, derived at the cost of extra winding space, and hence larger volume, for the same torque.

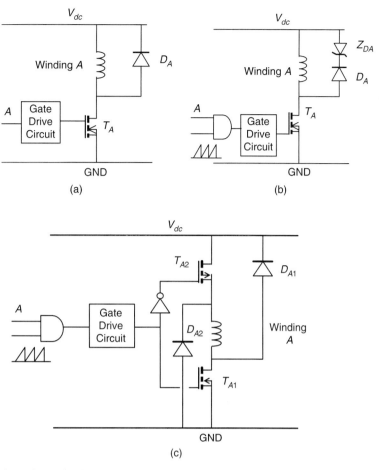

FIGURE 33.90 Three commonly used unipolar drive circuits: (a) the basic unipolar drive; (b) unipolar drive with PWM current limiting and zener diode turn-off; and (c) unipolar drive with regenerative turn-off.

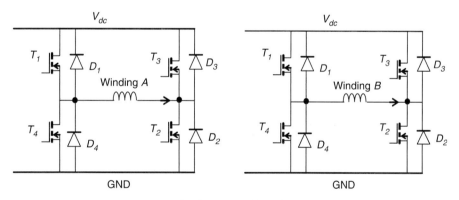

FIGURE 33.91 Bipolar drive circuit (gate-drive circuits omitted).

33.8.6 Micro Stepping

The drive sequences mentioned in Section 33.8.2 normally switch rated current through the motor windings. These produce regular step angles. The half-stepping operation also uses rated motor currents. Halving of the step angle is arranged mainly through the selection of the windings switched.

In micro stepping, the regular step angle of the motor is subdivided further by a factor, typically from 10 to 100, by energizing the windings partially, with combinations of currents ranging from zero to full rated value in more than one windings simultaneously. This does not lead to any sacrifice of the developed torque, since the phase currents are so selected that the peak of total torque contributed by two partially energized windings is not lower than the peak detent torque T_{max} obtained in regular stepping.

The idea behind micro stepping is readily understood when it is considered that by increasing the current in phase A of a two-phase hybrid in 10 equal steps to full value and decreasing the current in phase B in a similar manner, the motor step size may be divided by a factor of 10. If the closed-loop current controllers are added to the two drive circuits of Fig. 33.91 and distinct current references are obtained from a reference generator, a complete micro stepping drive is realized.

In micro stepping, the two current references must have values such that the motor does the following:

1. Develops the same T_{max} for every combination of winding currents.
2. Develops the same torque slope, i.e. $dt/d\theta$ at every micro stepping detent position.
3. Dissipates no more than the rated power loss ($I^2 R$) for every combination of winding currents.

The preceding conditions are necessary if the motor is to retain its static accuracy, maximum torque, and power dissipation characteristics.

The static torque characteristics (Fig. 33.92) of stepper motors are close to, but not exactly, sinusoidal functions of angle θ. The required current references for all windings of a stepper motor, including the variable-reluctance motor of three or more phases, can easily be calculated from the data of the T–θ characteristics of the motor for each phase for various currents and rotor positions. A typical set of T–θ data for a three-phase variable-reluctance motor is shown in Fig. 33.93. The application of the three conditions mentioned earlier leads to an unique set of current references for each phase of the motor for each micro step. Figure 33.94 shows the current references for this motor for micro stepping.

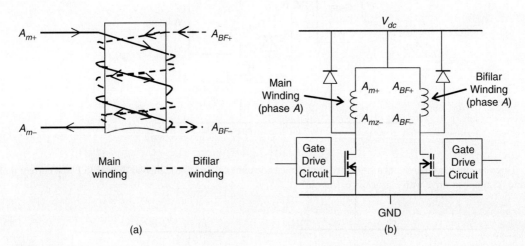

FIGURE 33.92 Drive circuits for one phase of a bifilar-wound motor: (a) bifilar pole windings and (b) drive circuits.

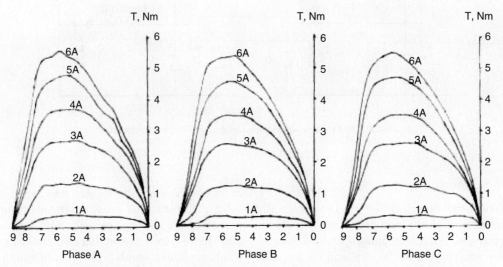

FIGURE 33.93 T–θ characteristics of a three-phase variable-reluctance motor.

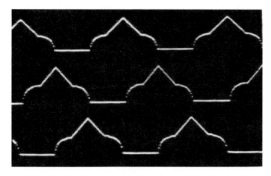

FIGURE 33.94 Micro stepping current references for the VR motor of Fig. 33.93. Stepping rate: 28,800 steps/s., $I = 6$ A (maximum).

In multi stepping operation, these micro stepping current references have to be issued to the current controllers for each phase, at a rate determined by the commanded stepping rate.

Care has to be taken in designing the phase-current controllers so that the actual winding currents match the current references in both single and multi stepping operation up to the maximum stepping rate desired. Since the current references are time varying, high-bandwidth current controllers are normally required to cover the desired speed range.

33.8.7 Open-loop Acceleration–Deceleration Profiles

As mentioned in Section 33.8.4, many applications require the stepper motors to be driven far above the stepping rates to and from which the motor can start and stop abruptly without losing or gaining any step. This calls for carefully designed acceleration–deceleration profiles that the stepping pulse rate must not exceed.

The number of steps the motor is to be stepped and its direction are normally under the control of the motion controller. Once this reference is known, a digital timer/counter circuit can be used in the controller to progressively adjust the time between the stepping pulses such that a prescribed acceleration–deceleration profile, as indicated in Fig. 33.95, is followed. The timer/counter and the pulsing sequence controller (the translator) need to be managed in realtime to execute the motion-control task at hand.

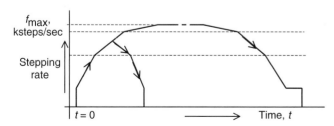

FIGURE 33.95 Typical acceleration–deceleration profiles.

The fastest acceleration–deceleration profile, a stepper motor is capable of is largely determined by its pull-out (T–ω) characteristic, which in turn is determined by the motor winding parameters and the drive circuit. An optimized stepping profile to and from the top speed may have a number of segments as indicated in Fig. 33.95. These profiles are easily computed from the pull-out (T–ω) characteristic by integrating the dynamic torque balance equation of the drive. For a large positioning angle, the entire profile, including some constant-speed running at the top speed, may be used. For short positioning angles, only part of the profile may be traversed. In general, a single segment acceleration–deceleration profile is used in commercial stepper motor controllers, so as to avoid a great deal of realtime number crunching by the profile controller.

The overall stepper motor controller thus consists of the blocks depicted in Fig. 33.96.

33.9 Switched-reluctance Motor Drives

33.9.1 Introduction

The switched-reluctance (SR) motor is a doubly salient electric machine with salient-poles on both the stator and rotor. The machine is operated by switching current pulses to each stator winding on and off in a continuous switching sequence. The rotor poles have no excitation. Figure 33.97 shows the physical topology of a typical SR motor. The diagram illustrates a motor with eight salient stator poles (numbered A1 to D2) and six salient rotor poles (numbered 1 to 6). Although many combinations of the number of stator and rotor poles are possible, this particular type has found widespread use.

The phase windings on the stator of the SR motor consist of concentrated windings wrapped around the stator poles. In the conventional arrangement, each stator pole winding is connected with that of the diametrically opposite pole to form a stator phase. In Fig. 33.97, the connected stator pole pairs are indicated by the same prefix letter.

The general principle of operation of the SR motor is the same as all types of reluctance machines, i.e. the stator and the rotor poles seek the minimum-reluctance position, so that the stator excited flux becomes maximum. Hence, when current flows in an SR motor stator phase and produces a magnetic field, the nearest rotor pole will tend to position itself with the direction of the developed magnetic field. This position, which is termed the *aligned position*, is reached when the rotor pole center axis is aligned with the stator pole center axis (assuming symmetrical poles). The aligned position also corresponds to the position of minimum reluctance, and hence the position of maximum inductance.

It should be noted that the *unaligned* position is defined as the position when the *inter-pole axis*, or the axis of the center of the inter-polar space in the rotor, is aligned with a

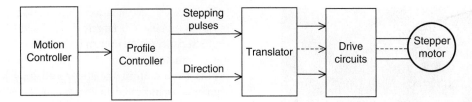

FIGURE 33.96 Structure of an open-loop motion controller for a stepper motor.

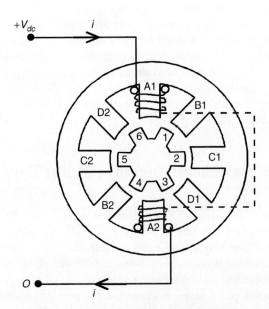

FIGURE 33.97 Four-phase SR motor topology.

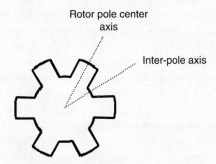

FIGURE 33.98 Rotor pole axis positions.

stator pole axis. This position corresponds to the position of minimum inductance. These rotor axis positions are illustrated in Fig. 33.98.

To achieve continuous rotation, the stator phase currents are switched on and off in each phase in a sequence according to the position of the rotor. Consider the motor schematic illustrated in Fig. 33.97. If coils A1 and A2 of phase A are excited and produce a magnetic field in a vertical direction, then poles 1 and 4 on the rotor will align themselves with the stator poles of phase A. If the coils of phase A now have their current switched off, and coils B1 and B2 of phase B are now excited, then in a similar fashion the rotor will move so that the poles 2 and 5 are aligned with stator poles B1 and B2. Exciting phases A, B, C, and D in sequence will produce rotor rotation in the counterclockwise direction.

From the preceding discussion, one may see that the switching on and off of excitation current to the motor phases is related to the rotor pole positions. This means that some form of position sensor is essential for the effective operation of the SR motor.

33.9.2 Advantages and Disadvantages of Switched-reluctance Motors

The SR motor has a number of inherent advantages that makes it suitable for use in certain variable-speed drive applications. Nevertheless, the motor also has some inherent disadvantages that must be considered before choosing the motor for a particular application. In Table 33.6, the main advantages and disadvantages of the SR motor drive are summarized.

33.9.3 Switched-reluctance Motor Variable-speed Drive Applications

The main application for SR motors is in variable-speed drive systems. One application area has been general-purpose industrial drives where speed, acceleration, and torque control are desired. SR-motor-based industrial drives provide the advantages of a very wide range of operating speeds as well as

TABLE 33.6 Advantages and disadvantages of SR drives

Advantages	Disadvantages
Low cost motor.	Need for position measurement.
Robust motor construction.	Higher torque ripple than other machine types.
Absence of brushes.	Higher noise than other machine types.
No motor short-circuit fault.	Nonlinear and complex characteristics.
No shoot-through faults.	
Ability to operate with faulted phase.	
High torque to inertia ratio.	
Unidirectional currents.	
High efficiency.	

high efficiency and robustness. Other applications of the SR drive include automotive applications, where the SR motor has advantages of robustness and fault tolerance. The SR motor in this application can also be easily controlled for acceleration, steady speed, and regenerative braking.

The SR motor is also well suited to aerospace applications where the ability to operate under faulted conditions and its suitability for operation under harsh environments are critical. Additionally, the very high-speed capability and high-power density also make these motors well suited in the aerospace field. There are also many domestic appliances where cost is of primary concern. In these products, the SR motor can provide a low-cost solution for a brushless fully controllable motor drive. In addition, the motor can be used in battery-powered applications, where the motor-high efficiency and ability to use a dc supply are important.

33.9.4 SR Motor and Drive Design Options

The main components of the drive system are shown in Fig. 33.99. It is important to design the motor and drive together in an integrated manner. The main criteria that need to be considered in designing the components of the SR drive system will be discussed later. It will be seen that certain design choices, which may be advantageous for one component of the drive system, may bring about disadvantages in another component. This highlights the need for a careful, integrated system approach to be taken when designing the drive system.

33.9.4.1 Number of Motor Phases

There are many possibilities in choosing the number of stator phases and rotor poles in SR motors. The simplest SR motor may consist of only one phase; however, to operate the motor in four quadrants (motoring or generating in both forward or reverse directions), at least three phases are required. The most common configuration to date has been the four-phase SR motor, which has eight rotor poles and six stator poles, as was shown in Fig. 33.97.

33.9.4.2 Maximum Speed

The SR motor is capable of operating at very high speeds because of its robust rotor construction, and in most applications the maximum speed is limited by the inverter switching speed and not limited by the motor itself. The maximum speed of the SR motor is itself normally greater than 15,000 rpm for a standard SR motor.

However, to determine the maximum drive speed, the controller and motor must be considered together. This is because the power-electronic device switching speed is directly proportional to the commutation frequency, which is in turn proportional to the motor speed. The maximum switching frequency of the power devices must therefore be taken into account in the SR drive design.

33.9.4.3 Number of Power Devices

In general, the number of switches per phase in SR motor drives will vary according to the inverter topology. A wide range of different SR drive circuits are available for SR drives, and these are detailed below. Circuits with only one switch per phase are possible; however, these have various disadvantages such as control restrictions, a need for extra windings, or higher switch voltages. However, with two switches per phase, the motor is fully controllable in four quadrants and has completely independent motor phase control. Therefore, the maximum number of power switches required for the motor operation is normally $2q$, where q is the number of phases.

33.9.4.4 Inverter Topology Types for SR Motors

As was mentioned, the torque produced in the SR motor is independent of the direction of current flow in each motor phase. This means the inverter is only required to supply unidirectional currents into the stator windings. The three major circuit topology types that have been used for each winding of an SR motor drives are shown in Fig. 33.100. As indicated in this figure, these are commonly termed the *bifilar*, *split dc supply*, and *two-switch* type inverter circuits.

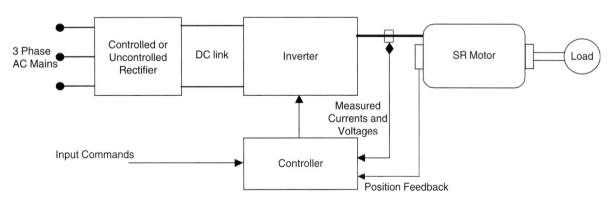

FIGURE 33.99 Main components of an SR drive.

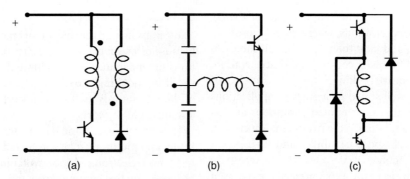

FIGURE 33.100 Major SR inverter topology types: (a) bifilar type; (b) split dc supply type; and (c) two switch type.

In the circuits shown in Fig. 33.100, only one or two switching components per phase are required. Other circuit topology types that use shared components between the motor phases have limitations in control flexibility.

33.9.4.4.1 Bifilar Type Inverter Circuit

In Fig. 33.100a, a drive circuit for a bifilar-wound SR motor is shown. The bifilar windings are closely coupled, with one winding being connected to a switching device while the other is connected to a freewheeling diode. Current is increased in the winding when the switching device closes. At turn-off, the current transfers to the secondary winding through transformer action, and the inductive energy flows back into the supply via the freewheeling diode. If perfect coupling is assumed, then the voltage across the switching device will rise to twice the dc supply voltage during turn-off. However, in practice this would be higher. This is because there will be some uncoupled inductance in the primary that will cause high induced voltages when the current in the winding collapses to zero. Thus, snubbing circuits would almost certainly be required to protect the switching components from over-voltage.

The advantage of the bifilar circuit is that it requires only one switching device per phase. However, with the advent of modern power electronic devices, which have both low cost and low losses, this advantage quickly disappears.

33.9.4.4.2 Split DC Supply Inverter Circuit

The split dc supply type inverter circuit is in Fig. 33.100b. As in the bifilar circuit, this configuration also uses only one switching device and one diode per phase. However, a center-tapped dc source is required. When the switching device is turned on, current increases in the phase winding because of the positive capacitor voltage being applied. At turn-off, the current is forced to flow through the diode and thus decays to zero more quickly because of the connection to the negative voltage. It is usual for the dc center tap to be implemented using a split capacitor in the dc-link. The voltages across each capacitor must remain balanced, which means that there can be no significant power-flow difference between the two capacitors.

Upon examination of the circuit, it can be seen that because of the split capacitor bank, only half the available dc voltage can be switched across the phase winding. Thus, for the same voltage across the motor phases that is supplied by the bifilar circuit described earlier, the dc supply voltage must be doubled with respect to the bifilar circuit supply. This means that the voltage rating of the devices would effectively be the same as in the bifilar circuit.

This is inherently inefficient. The configuration also has the need for balanced split capacitive components. In addition, it will be seen that the *soft-chopping* form of control described in Section 33.9.7 is not available in this drive.

33.9.4.4.3 Two-switch Inverter Circuit

The two-switch inverter type circuit, which is shown in Fig. 33.100c, uses two switching devices and two diodes per phase. Unlike the previous two circuits, three modes of operation are possible:

Mode 1: Positive phase voltage
A positive phase voltage can be applied by turning both switching devices on. This will cause the current to increase in the phase winding.

Mode 2: Zero phase voltage
A zero-voltage loop can be imposed on the motor phases when one of the two switches is turned off while current is flowing through the phase winding. This results in current flow through a freewheeling loop consisting of one switching device and one diode, with no energy being supplied by or returned to the dc supply. The current will decay slowly because of the small resistance of the semiconductors and connections, which leads to small conduction losses. This mode of operation is used in soft-chopping control, as described in Section 33.9.7.

Mode 3: Negative phase voltage
When both switches in a motor phase leg are turned off, the third mode of operation occurs. In this mode, the motor phase current will transfer to both of the freewheeling diodes and

return energy to the supply. When both of the diodes in the phase circuit are conducting, a negative voltage with amplitude equal to the dc supply voltage level is imposed on the phase windings.

In this circuit, the switching devices and diodes must be able to block the dc supply voltage amplitude when they are turned off, in addition to any switching transient voltages. However, because the circuit contains two devices in series, the blocking voltage is essentially half the value seen in the previous two circuit types for the same applied motor phase voltage amplitude. Another advantage of the two-switch inverter circuit is that it offers greater control flexibility with its three modes of voltage control.

A disadvantage of this inverter type, as compared to the bifilar and split dc supply types, is that it contains twice as many switching components per phase. However, with the current wide availability and economy of power semiconductors, in most applications, the advantages of the two-switch circuit outweigh the cost of an extra switching device per phase.

33.9.5 Operating Theory of the Switched-reluctance Motor: Linear Model

If a linear magnetic circuit is assumed, the flux linkage is proportional to phase current for any rotor position θ. This is demonstrated in Fig. 33.101, where the magnetization curves for the linear SR motor for various rotor positions and currents are shown. In this linear case, the inductance L at any position θ, which is the slope of these curves, is *constant* and independent of current.

As the motor rotates, each stator phase undergoes a cyclic variation of inductance. As can be seen in Fig. 33.101, in the fully aligned position (when a rotor pole axis is directly aligned with the stator pole axis) the reluctance of the magnetic circuit through the stator and rotor poles will be at a minimum, and thus the inductance of the stator winding will be at a maximum. The opposite will occur in the fully unaligned position (when the rotor inter-pole axis is aligned with the stator pole). Thus, the inductance becomes a function of position only and is not related to the current level. If it is also assumed that mutual inductance between the phases is zero, then a typical inductance variation $L(\theta)$ with respect to the rotor position similar to that shown in Fig. 33.102 arises. Although this is an idealized inductance variation, it is helpful in the understanding of key operating principles of the machine. One should note that in the idealized inductance variation there are sharp corners, which can only arise if flux fringing is completely ignored.

Four distinct regions can be identified in the plot of the linear inductance variation shown in Fig. 33.102. These distinct regions correspond to a ranges of rotor pole positions relative to the stator pole positions as described below:

Region A

This region begins at rotor angle θ_1, where the first edge of the rotor, with respect to the direction of rotation, just meets the first edge of the stator pole. The inductance will then rise in a linear fashion until the poles of the stator and rotor are completely overlapped at angle θ_2. At this point, the magnetic reluctance is at a minimum and the phase inductance is at a maximum. These rotor positions are illustrated in Figs. 33.103a and b, for example, four phase motor with rotor pole 1 approaching the stator pole of phase A.

Region B

This region spans from rotor positions θ_2 to θ_3. In this region, the inductance remains constant because the rotor pole is completely overlapped by the stator pole (i.e. the overlap area of the poles remains constant). At rotor angle θ_3 the edge of the rotor pole leaves the stator pole overlap region, and thus the area of overlap will again begin to decrease. The position at which this occurs is illustrated in Fig. 33.103c.

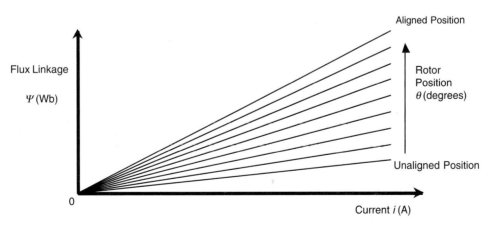

FIGURE 33.101 Magnetization characteristics of linear SR motor.

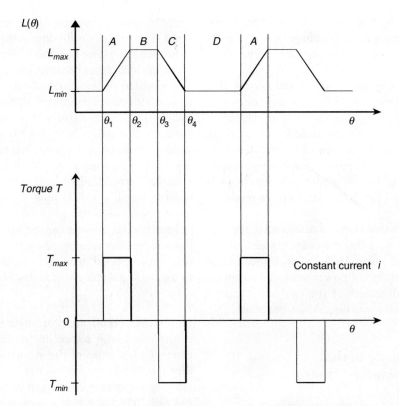

FIGURE 33.102 Typical linear inductance variations and corresponding torque variations for constant phase current.

Region C
When the rotor moves past θ_3, the rotor pole leading edge begins to leave the pole overlap region, and region C begins. At this point, the inductance begins to linearly decrease, until at θ_4, the rotor pole has completely left the stator pole face overlap region. At this point, the inductance is at its minimum once more. The rotor position at which the rotor pole has completely left the overlap is indicated in Fig. 33.103d.

Region D
In this region, the rotor and stator have no overlap, and thus the inductance remains constant at the minimum level, until region A is reached once again.

It was mentioned earlier that when a stator phase is excited, the rotor poles will tend to move toward the maximum-inductance region. Thus, a motoring torque is produced when a stator phase is provided with a current pulse during the angles when the inductance is rising (assuming motoring rotation is in the direction of increasing θ in Fig. 33.102). This means that if positive torque is desired, excitation should be arranged such that the current flows between the appropriate rotor angles when the inductance is rising.

Conversely, if current flows during the decreasing inductance region, a negative torque would result. This is because the rotor will be attracted to the stator pole in such a way that it rotates in the opposite direction to the motoring rotation, or in other words, the rotor experiences a torque opposite to the direction of rotation.

It should be noted that this reluctance-machine torque always acts to decrease the reluctance. The direction of current flowing into the stator winding is irrelevant. This signifies that unidirectional current excitation is possible in the SR motor drive.

The variation of torque with rotor angle for a constant phase winding current is as shown in Fig. 33.102. It can be seen that the torque is constant in the increasing and decreasing inductance regions, and is zero when the inductance remains constant.

The preceding physical explanation of the developed torque is also given by the familiar torque Eq. (33.96) for a variable-reluctance machine.

$$T = \frac{1}{2} i^2 \frac{dL(\theta)}{d\theta} \quad (33.96)$$

From Eq. (33.96), it is evident that the magnitude of the instantaneous torque developed in the SR motor is proportional to both i^2 and $dL/d\theta$. If the inductance is increasing with respect to the angle, and current flows in the phase winding, then the torque will be positive and the machine will operate in motoring mode. Hence, from Eq. (33.96), it can be seen that when the motor phase is excited during a rising inductance region, part of the energy from the supply is

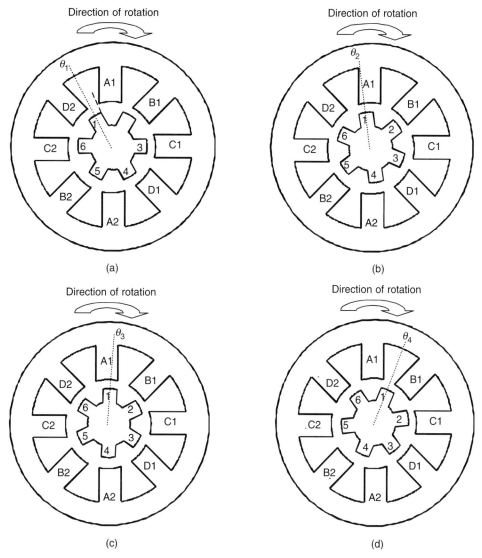

FIGURE 33.103 Rotor pole 1 positions: (a) meeting edge of stator pole A; (b) overlapped by stator pole A; (c) edge of rotor pole leaving overlap region; and (d) rotor pole completely leaving overlap region. (Note: Airgap space is exaggerated for clarity.)

converted to mechanical energy to produce the torque, and another part is stored in the magnetic field. If the supply is turned off during this region, then any stored magnetic energy is partly converted to mechanical energy and partly returned to the supply.

However, a negative, or braking torque will be developed by the motor if the inductance is decreasing with respect to the rotor angle and current flows in the phase winding. In this case energy flows back to the supply from both the stored magnetic energy and the mechanical load, which acts as a generator.

It can also be seen from Eq. (33.96) that the sign (or direction) of the torque is independent of the direction of the current and is only dependent on the sign of $dL/d\theta$. This explains the torque waveforms that were seen in Fig. 33.102, where for constant current (and constant $dL/d\theta$ magnitude), the magnitude of the torque was constant in the rising or decreasing inductance regions. However, it was seen that the torque changes from positive to negative according to the sign of $dL/d\theta$.

Hence, the ideal waveform for the production of motoring torque would be a square wave pulse of current (with magnitude equal to the maximum possible supply current) flowing only during the increasing inductance period (Fig. 33.104). This current waveform is illustrated in Fig. 33.104b. However, in practice this type of current waveform is difficult to produce in a motor phase. This is because the motor phase current is supplied from a finite dc voltage source, and thus inductance of the stator phase winding would delay the rise and fall of current at the pulse edges. Instead, a more practical current waveform is normally used as is illustrated in Fig. 33.104c.

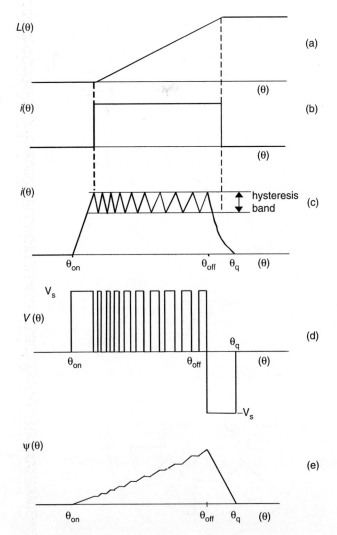

FIGURE 33.104 (a) Linear phase inductance variation; (b) ideal square wave phase current; (c) chopping-mode phase current; (d) chopping-mode phase voltage; and (e) flux linkage waveform corresponding to chopping-mode current.

It can be seen that in this waveform, the ideal square waveform is closely approximated by the use of hysteresis current control. At higher speeds, hysteresis current control can no longer be used and a current waveform similar to that shown in Fig. 33.104b is seen in the phase winding. These two types of practical current waveforms, which approximate the ideal square pulse waveform *(a) at low to medium speeds* and *(b) at high speeds*, will be discussed next.

33.9.5.1 Low to Medium-speed Approximation to Square-pulse Current Waveform

At low to medium motor speeds, the ideal square-pulse current waveform is approximated in the practical motor drive using hysteresis current control, as is shown in Fig. 33.104c. The hysteresis method of controlling the current is termed the *chopping-mode control* method in SR motor drives. During the time of conduction (between the turn-on and turn-off angles), the current is maintained within the hysteresis band by the switching off and on of the phase voltage by the inverter when the phase current reaches the maximum and minimum hysteresis band. An example of the voltage waveform used for the hysteresis current control is shown Fig. 33.104d, where a constant inverter dc supply voltage of magnitude V_s is used. It can be seen that the switching frequency of the voltage waveform decreases as the angle increases. This is due to the fact that the phase inductance is linearly increasing with angle, which has the effect of increasing the current rise and fall time within the hysteresis band.

In the chopping-mode control method, the *turn-on region* is defined as the angle between the *turn-on angle* θ_{on} and the *turn-off angle* θ_{off}, and is chosen to occur during the rising inductance region for motoring torque. In the practical chopping current waveform, the current turn-on angle θ_{on} is placed somewhat before the rising-inductance region. This is to ensure that the current can quickly rise to the maximum level in the minimum-inductance region before the rising- inductance, or torque-producing, region. Similarly the turn-off angle θ_{off} is placed a little before the maximum-inductance region so that the current has time to decay before the negative-torque, or decreasing-inductance, region. The angle at which the current decays to zero after turn-off is labeled as θ_q in Fig. 33.104c.

33.9.5.2 High-speed Approximation to Square-pulse Current Waveform

The chopping-mode of operation cannot be used at higher speeds, as at these speeds the hysteresis band current level will not be reached. This is because at high speeds, the back-emf of the motor becomes equal to or larger than the voltage supply in the rising-inductance region (Fig. 33.105), which limits the increase of the motor phase current. In addition, the rise time of the current will correspond to an ever-increasing angle as the speed is increased. Eventually, at high speeds, the rise-time angle will be so large that the turn-off angle θ_{off} will be reached before the hysteresis current level has been exceeded. Thus, at high speeds, the current is switched on and off only once per cycle. In SR motor drive control, this is called the *single-pulse mode* of operation. An example of the single-pulse mode current is illustrated in Fig. 33.105b.

In the single-pulse mode of operation, the inverter power switches turn-on at rotor angle θ_{on}, which places the dc voltage supply V_s across the phase winding, as is shown for the example single-pulse voltage waveform in Fig. 33.105c.

As for the chopping-mode case, in order to maximize torque, θ_{on} must usually be located prior to the rising-inductance region. This is so that, while the inductance is low, the current has a chance to rise rapidly to a substantial value before the torque-producing region begins and the motor

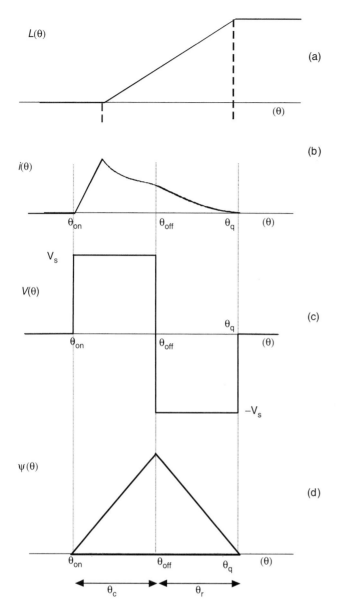

FIGURE 33.105 (a) Linear phase inductance variation; (b) single-pulse mode phase current; (c) single-pulse mode phase voltage, and (d) flux linkage waveform corresponding to single-pulse mode current.

back-emf increases. At rotor angle θ_{off}, the power switches are turned off, and the phase will have a negative voltage (typically $-V_s$) thrown across it. The current will then decay until it becomes zero at rotor angle θ_q.

33.9.6 Operating Theory of the SR Motor (II): Magnetic Saturation and Nonlinear Model

In the linear model described earlier, it was assumed that the inductance of a phase winding is independent of current. However, in a real SR motor, significant saturation of the magnetic circuit normally occurs as the phase current increases, and thus the phase inductance is related to both the phase current level and position. Because of the magnetic saturation effect, the actual phase inductances at a given rotor position can be reduced significantly compared to the inductance given by linear magnetization characteristics. In addition, the effect of magnetic saturation becomes larger as the motor current level increases.

The effects of saturation in an SR motor can be observed in a plot of its magnetization curves. This shows the relationship of flux linkage vs current, at rotor positions varying between the fully aligned and unaligned angles. A typical set of SR motor magnetization curves is shown in Fig. 33.106, where it can be seen that there is a nonlinear relationship between the flux linkage and current for each curve.

Due to the magnetic saturation effect discussed earlier, the instantaneous torque Eq. (33.96) which was derived assuming linear conditions, will not be generally valid for calculating the torque in SR motors. Therefore, for accurate calculations, the torque must take into account the dependence of phase inductance with current and position.

If one considers the phase-inductance saturation, the expression for instantaneous torque production of an SR motor phase can be written as

$$T = \left[\frac{\partial W'}{\partial \theta}\right]_{i=constant} \quad (33.97)$$

where the coenergy W' is defined as

$$W' = \int_0^i \Psi \, di \quad (33.98)$$

33.9.7 Control Parameters of the SR Motor

A variety of performance characteristics can be obtained in the SR motor by controlling various parameters. These parameters include the chopping-mode control hysteresis level at low to medium- speeds, and the turn-on and turn-off angles θ_{on} and θ_{off} at all motor speeds. By controlling these parameters, it is possible to produce any desired characteristic such as constant torque, constant power, or some other particular characteristic in between.

As discussed in Section 33.9.5, two distinct modes of operation apply in the SR motor depending on the nature of the current waveform. These modes are the chopping-mode control, which can be used at low to medium motor speeds, and single-pulse mode of control, which is used at high speeds. Both of these modes of operation will be further detailed hereafter, with an explanation of the corresponding inverter switching operation.

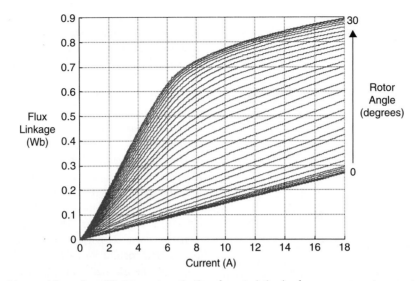

FIGURE 33.106 Measured four-phase SR motor magnetization characteristics (each curve represents a constant rotor position).

33.9.7.1 Chopping-mode Control

In the chopping-mode control region, the turn-on and turn-off angles are controlled together with the current level. As described in Section 33.9.5, the turn-on angle and turn-off angle are controlled, so that the current flows during the rising-inductance, or positive torque-producing, region. This normally means that the turn-on angle is placed shortly before the place where the rising-inductance angle begins, and the turn-off angle is placed shortly before this region ends.

In the chopping mode, the current level is controlled to remain below the maximum allowable level. This involves switching the voltage across the phase on and off in such a manner that the current is maintained between some chosen upper and lower hysteresis current levels. An example of this form of current chopping control was shown in Fig. 33.104c.

The actual torque production of the motor in the chopping mode is set by the control turn-on and turn-off angles and the current hysteresis level. Within the chopping-mode of operation, two current hysteresis control schemes can be used. These are termed *soft* and *hard* chopping. Soft chopping can only be used in some circuit configurations, such as that shown in Fig. 33.107. For soft-chopping control, one switching device remains on during the entire conduction period, while the other is switched on and off to maintain the desired current level. This can be seen in Figs. 33.107a and b, where the two conduction modes during chopping are shown. When both switches are on, the phase winding receives the full positive supply, whereas when only one switching device is on, the phase experiences a zero-voltage freewheeling loop that will decrease the current.

In the hard-chopping scheme, both devices are switched simultaneously and have the same switching state at all times. If both switching devices are turned on, the phase winding

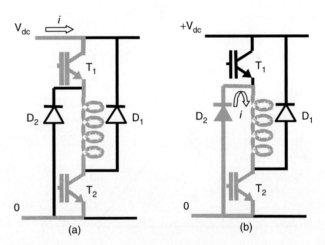

FIGURE 33.107 Soft-chopping mode conduction paths: (a) both devices on: positive voltage applied to motor phase and (b) T1 turned off: zero voltage freewheeling loop applied to motor phase.

sees the full positive supply. To decrease the current, the full negative supply is applied by turning both devices off as shown in Fig. 33.108. In circuit configurations with fewer than two switches per phase, only hard chopping can be used.

Soft chopping is more advantageous than hard chopping. This is because of a smaller dc ripple current in the supply, which can substantially minimize the ripple-current rating of the dc-link capacitor, as well as lower the hysteresis loss in the motor. It has also been found that the soft chopping lowers acoustic noise and electromagnetic radiation.

33.9.7.2 Single-pulse Mode Control

At higher speeds, the back-emf of the SR motor eventually becomes greater than or equal to the supply voltage during

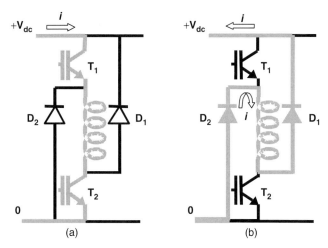

FIGURE 33.108 Hard-chopping mode conduction paths: (a) both devices on: positive voltage applied to motor phase and (b) T1 and T2 turned off: negative voltage applied to motor phase.

the rising-inductance region. This means that even if a phase is excited, the current in the motor phase will not increase in the rising-inductance region. Therefore, at higher speeds, the turn-on angle must be placed before the beginning of the increasing inductance region, so that the phase current will have an adequate time to increase before the back-emf becomes high.

In addition, the time available for the current to rise after turn-on becomes less and less as the speed of the motor increases. This is due to the fact that the available conduction time is lower for constant switching angles as the speed of rotation increases. This can be seen by considering that speed is the time rate of change of angle. Thus, as the speed increases, there will be a point when the current level never rises to the chopping level. At this point, the single-pulse mode of operation will come into effect and the current will decrease or remain constant throughout the increasing-inductance zone. An example of a single-pulse mode current waveform was seen in Fig. 33.105b.

As the current is not commutated in the single-pulse mode, the control in this mode consists only of controlling the on and off angles. The turn-on angle θ_{on} can be placed at some point in advance of the rising-inductance region where the phase inductance is low, so that the current can increase at a faster rate before the increasing-inductance region. The angle can be advanced up until maximum allowable current occurs at the peak of the waveform (this may even mean switching on in the previous decreasing-inductance zone). The actual control turn-on and turn-off angles for the single-pulse mode, for a given load torque and speed, can be determined by simulating the motor equations.

The speed at which a changeover between single-pulse and chopping-mode occurs is called the *base speed*. Base speed is defined as the highest speed at which the chopping mode can be maintained at the rated voltage and with fixed on and off angles. Below the base speed, the current increases during the rising-inductance region, unless it is maintained at the maximum or a lower level by chopping.

Therefore, it can be seen that at lower speeds that are below the base speed, the motor is controlled using chopping-mode control, whereas at speeds above the base speed, the single pulse mode of control is used. In both the control modes, the control turn-on and turn-off angles are chosen so that the motor provides the required load torque.

33.9.8 Position Sensing

It can be seen from the preceding discussion that to control the SR motor satisfactorily, the motor phases are excited at the rotor angles determined by the control method. It is therefore essential to have knowledge of the rotor position. Furthermore, the rotor-angle information must be accurate and have high resolution to allow implementation of the more sophisticated nonlinear control schemes that can minimize torque ripple and optimize the motor performance.

This means that the performance of an SR drive depends on the accurate position sensing. The efficiency of the drive and its torque output can be greatly decreased by the inaccurate position sensing, and the corresponding inaccurate excitation angles. It has been demonstrated that at high motor speeds, an error of only 1° may decrease the torque production by almost 8% of the maximum torque output.

Traditionally, the rotor-position information has been measured using some form of mechanical angle transducer or encoder. The position-sensing requirements are in fact similar to those for brushless PM motors. However, although position sensing is required for the motor operation, the position-measurement sensors are often undesirable. The disadvantages of the electromechanical sensors include the following:

(a) The position sensors have a tendency to be unreliable because of environmental factors such as dust, high temperature, humidity, and vibration.
(b) The cost of the sensors rises with the position resolution. Hence, if high-performance control is required, an expensive high-resolution encoder needs to be employed.
(c) There is an additional manufacturing expense and inconvenience due to the sensor installation on the motor shaft. In addition, consideration must be given to maintenance of the motor because of the mechanical mounting of the sensors, which also adds to the design time and cost.
(d) Mechanical position sensors entail extra electrical connections to the motor. This increases the quantity of electrical wiring between the motor and the motor drive. This wire normally needs to be shielded

from electromagnetic noise and thus further adds to the expense of the drive system.

(e) The allocation of space for the mounting of the position sensor may be a problem for small applications (such as for motors used in consumer products).

Hence, to overcome the problems induced by rotor-position transducers, researchers have developed a number of methods to eliminate the electromechanical sensor for deriving position information. This is achieved by indirectly determining the rotor position. Such methods are commonly termed *sensorless* rotor-position estimation methods. The term sensorless seems to imply that there are no sensors at all. However, there must be some form of sensor used to measure the rotor position. In fact, the term sensorless position estimation in reality implies that there are no additional sensors required to determine position apart from those that measure the motor electrical parameters to control the motor. These are normally current- or voltage-measuring circuits.

Hence, all sensorless position estimation methods for the SR motor use some form of processing on electrical waveforms of the motor windings. In essence, the major difference between sensorless position detection, and the electromechanical sensors mentioned above, is that there is no mechanical connection of the sensor to the motor shaft. Therefore, sensorless position detection involves electrical measurements only.

33.10 Synchronous Reluctance Motor Drives

33.10.1 Introduction

In recent years, there has been a revival of interest in reluctance machines. Two main machines have been the focus of this interest: the switched-reluctance machine (SRM) and the synchronous reluctance machine (Syncrel). The SRM is a machine that does not have sine-wave spatial distributed windings, but instead has concentrated coils and a doubly salient rotor and stator structure. The operation of this machine is highly nonlinear in character, and normal ac machine modeling techniques cannot be applied in a straightforward manner to describe its operation. The SRM drive has been considered in detail in an earlier section.

The Syncrel, on the other hand, has conventional three-phase sinusoidally distributed windings on the stator. The word "synchronous" in the machine's name emphasizes the fact that the stator windings generate a spatial sinusoidally distributed magnetomotive force (mmf) in the airgap between the stator and the rotor, and under steady-state conditions the rotor rotates in synchronism with this field. Therefore, the stator winding configuration of this machine is virtually exactly the same as that of the induction machine or the conventional synchronous machine. The major difference between the Syncrel and conventional synchronous and induction machines is in the rotor structure. In both the induction machine and the synchronous machine, there is a source of flux in the rotor itself. In the case of the induction machine, this flux is produced by currents resulting from an induction mechanism, and for the synchronous machine, there is a field winding wound on the rotor that is fed with the dc current to produce flux. The permanent-magnet synchronous machine replaces the wound field on the rotor with a permanent magnet. The Syncrel, on the other hand, does not have any source of flux on the rotor, but instead the rotor is designed to distort the flux density distribution produced by the sinusoidally distributed mmf.

Sinusoidally wound reluctance machines were traditionally used in the fiber-spinning industry because of their synchronous nature. This made it simple to keep a large number of machines running at the same speed using just the frequency of the supply to the machines. These machines were direct-on-line-start machines. This was facilitated by the presence of an induction machine starting cage on the rotor. This cage was also essential to damp out oscillations in the rotor speed when running at synchronous speed. It should be pointed out that these machines are not considered to be Syncrels – *a Syncrel does not have an induction machine cage on the rotor*. A Syncrel is absolutely dependent on an intelligent inverter drive in order to start the machine and to stabilize it when running. The lack of a requirement for a starting cage means that the rotor design can be optimized for best torque and power performance.

The revival of interest in the Syncrel in the early 1980s was motivated by the development of low-cost microprocessors and reliable power electronics, coupled with the perception that the Syncrel may be more efficient and simpler to control in variable-speed applications compared to the induction machine. The control simplicity is achieved in practice, mainly because one does not have to locate the flux vector in order to implement vector control. The potential for improved efficiency and torque density compared to the induction machine is very much dependent on the rotor design. The Syncrel has the advantage over the switched-reluctance machine; in that it produces relatively smooth torque naturally, and it uses a conventional three-phase inverter. Therefore, inverter technology developed for the induction machine can be applied directly.

33.10.2 Basic Principles

Reluctance machines are one of the oldest electric machine structures, since they are based on the basic physical fact that a magnet attracts a piece of iron. In fact, Syncrel structures were being published in the early 1920s [1]. The essential idea behind the operation of all reluctance machines is that the windings of the machine produce magnetic poles that are used to attract the reluctance rotor. If the magnetic poles are moved around the periphery of the machine at the rate at which the

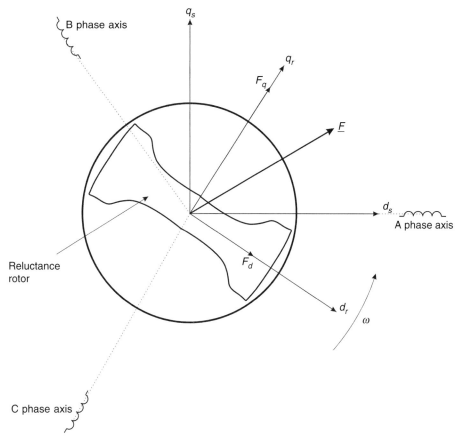

FIGURE 33.109 Conceptual diagram of a synchronous reluctance machine.

rotor is moving, then sustained torque and rotation can be achieved.

A conceptual diagram of a Syncrel is shown in Fig. 33.109. In this figure, the rotor is represented as a simple "dumbbell"-type rotor. The axes of the three-phase sinusoidally distributed windings are indicated by the dashed lines. If these windings are being fed with currents, then a spatial sinusoidally distributed, mmf results. Because this mmf is sinusoidally distributed, it can be represented by a "space vector" (similar to sinusoidal time-varying quantities being represented by a time phasor). In Fig. 33.109, this resultant space vector is indicated by F, F_d, and F_q are the components of this vector that lie along the high-permeance and low-permeance axes of the machine, respectively.

If one considers the situation shown in Fig. 33.109, then the rotor would tend to rotate in the direction indicated. This rotation would continue until the high permeance d_r axis (i.e. the least-reluctance axis) of the machine aligns with the mmf vector. When this alignment occurs, the flux produced by the stator mmf vector would be maximized. If the vector F also rotates as the rotor rotates, then as mentioned previously, the angle between F and d_r will remain constant and the rotor will continue to chase the F vector, continuous rotation being the result.

One can ask even more fundamental questions such as "Why does the rotor rotate to the position that maximizes the flux density?" This is essentially asking, why does a magnet attract a piece of iron. To completely answer this question one has to delve into the field of quantum physics, which is beyond the scope of this presentation. A less complicated explanation is based on the fact that the stator flux density tends to align the domains in the ferromagnetic rotor material, which produces an effect similar to having a current-carrying winding wrapped around the rotor. This effective current then interacts with the stator flux density to produce a force that has a component that is oriented radially around the periphery of the machine. It is this component that produces the torque on the rotor that causes the alignment with the stator mmf vector.

It was mentioned previously that in order to have a continuous motion, the mmf vector F must rotate at the same angular velocity as the rotor, so that the angle between the mmf vector and the d_r axis of the rotor is kept at a constant value. The rotation of the mmf vector is achieved by feeding the three-phase windings of the machine with time-varying currents.

It can be shown that if these form a balanced 120° temporally phase-shifted set of sinusoidal currents, then the resultant mmf vector will rotate at a constant velocity related to the frequency of the input current waveforms and the number of pole pairs in the machine.

In order to get a more precise figure for the torque produced by a Syncrel, one has to develop techniques of modeling the machine. Because the Syncrel is a reluctance machine, the coenergy technique for developing the torque expressions can be used [2, 3]. The coenergy technique is a very accurate way of determining the torque as it explicitly takes into account, the saturation nonlinearities in the iron of the machine. However, the technique does not lend itself to mathematical analysis and is not a good way of understanding the basic dynamic properties of the machine. The coenergy approach will not be pursued any further in this presentation.

33.10.3 Machine Structure

The essential difference between the Syncrel and, say, the induction machine is the design of the rotor. Most experimental Syncrel systems that have been built use the stator of an induction machine, including the same windings. The rotor designs can take on a number of different forms, from the very simple and basic dumbbell-shaped rotor (such as that sketched in Fig. 33.109) to more complex designs. Unfortunately, the designs that are simple to manufacture (such as the dumbbell design) do not give good performance; therefore, one is forced into more complex designs. The design of the rotor in a Syncrel is the key to whether it is economic to manufacture and has competitive performance with similar machines.

The design of Syncrel rotors fall into four main categories of increasing manufacturing complexity and performance: dumbbell or higher pole-number equivalent designs, flux-barrier designs, radially laminated flux-barrier designs, and axially laminated designs. The first two of these design methodologies are old and lead to designs with poor to modest performance. Therefore, they will not be considered any further. The latter two, however, lead to machine designs with performance comparable to that of the induction machine.

Figures 33.110 and 33.111 show the cross section of a four-pole machine with a radial lamination flux barrier designed rotor and an axially laminated rotor. The radial lamination design allows the rotor to be built using similar techniques to standard radial laminations for other machines. The flux barriers can be punched for mass production, or wire-eroded for low production numbers. These laminations are simply stacked onto the shaft to form the rotor. The punched areas can be filled with plastic or epoxy materials for extra strength, if required. The iron bridges at the outside of the rotor are designed to saturate under normal flux levels and therefore do not adversely affect the performance of the machine. They are there to provide mechanical strength.

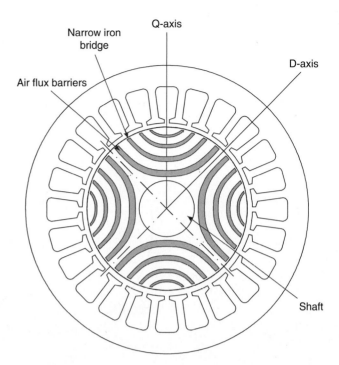

FIGURE 33.110 Cross section of a radially laminated Syncrel.

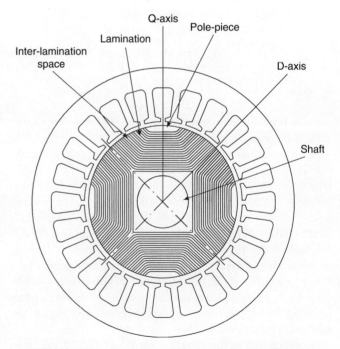

FIGURE 33.111 Cross section of an axially laminated Syncrel.

The axially laminated rotor is constructed with laminations running the length of the rotor (i.e. into the page on Fig. 33.111). In between the laminations, a nonmagnetic packing material is used. This can be aluminum or bronze, for example, but a nonconductive material such as slot insulation is better since eddy currents can be induced in conductive materials. The ratio of the steel laminations to nonmagnetic material is usually about 1:1. The axial laminations are all stacked on top of each other, and a nonmagnetic pole piece is bolted on top of the stack to hold the laminations to the shaft. The strength of these bolts is usually the main limitation on the mechanical strength and hence the speed of rotation of this rotor. If more or thicker bolts are used to increase strength, the magnetic properties of the rotor are compromised because of the amount of lamination that has to be cut out to make room for them.

Radial and axial laminated rotors are usually limited to four-pole or higher machines because of the difficulty of accommodating the shaft in two-pole designs. An axially laminated two-pole rotor has been built with the shafts effectively bonded onto the end of the rotor. Another design was constructed of a block of alternating steel and bronze laminations, the whole structure being brazed together and the resultant stack then being machined into a round rotor and shafts (this rotor was used for high-speed generator applications).

Of the two rotor designs, the radially laminated one has the best potential for economic production. The axially laminated rotor in general gives the best performance, but the mass production difficulties with folding and assembling the laminations make its adoption by industry unlikely. On the other hand, improved designs for radially laminated rotors mean that they can now produce performance very close to that of the axial-laminated designs, and the ease of manufacture would indicate that these rotors are the future of Syncrel rotors.

33.10.4 Basic Mathematical Modeling

In order to give a more quantitative understanding of the machine, a basic mathematical dynamic model of the machine will be introduced.[1] This model will assume that the iron material in the machine does not saturate. This means that the flux density and flux linkage of the windings in the machine are linear functions of the currents in the machine.

To derive the electrical dynamic model of a machine, one usually uses Faraday's flux linkage expressions. In the case of the Syncrel, the self and mutual flux linkage between the phases is obviously a function of the angular position of the rotor; therefore, one needs to have expressions for these inductances in terms of rotor position. The fundamental assumption used to make this mathematically tractable is that the inductances vary as a sinusoidal function of the rotor position.[2] The other major part of the modeling process is the conversion from a three-phase model to a two-phase model. This is a process that is carried out for most sinusoidally wound machines, since it allows a variety of machines to be represented by very similar models.

A further complication in this process is that the two-phase model is derived in a "rotating reference frame," as opposed to a stationary reference frame. Developing the equations in a rotating reference frame has the advantage that the normal sinusoidal currents feeding the machine are transformed into dc currents in steady state, and the angular dependence of the machine's inductances disappear.

One way of heuristically understanding the effect of the rotating frame transformation is to imagine that we are observing the machine's behavior from the vantage point of the rotor. Because the sinusoidal flux density waveform is rotating around the machine in synchronism with the rotor, it appears from the rotor that the flux density is not changing with time – i.e. it is a flux density created by dc currents flowing in a single sinusoidally distributed winding. This single sinusoidal winding is effectively rotating with the rotor. It should be noted that the transformation process of the fluxes, currents, voltages, and machine parameters to the two-phase rotating frame is an invertible process; therefore one can apply the inverse transformation to ascertain what is happening in the original three-phase machine.

The models derived using the three-phase to two-phase transformations are known as dq models, the d and q referring to the two axes of the machine in the two-phase model (both stationary and rotating frame dq axes are shown in Fig. 33.109). The linear[3] dq equations for the Syncrel can be derived as [4]

$$v_d = Ri_d + L_d \frac{di_d}{dt} - \omega L_q i_q \quad (33.99)$$

$$v_q = Ri_q + L_q \frac{di_q}{dt} - \omega L_d i_d \quad (33.100)$$

where

$v_d i_d$ = the d-axis voltage and current,
$v_q i_q$ = the q-axis voltage and current,
L_d, L_q = the d-and q-axes inductances, respectively,
ω = the electrical angular velocity of the rotor.

Thus far we have concentrated on the electrical dynamics of the machine. The other very important aspect is the torque produced by the machine. It is possible to derive the torque of the Syncrel using the principle of virtual work based on coenergy as:

$$T_e = \frac{3}{2} p_p (L_d - L_q) i_d i_q \quad (33.101)$$

---

[1] Note that the model is not derived but instead just stated. The structure of the model will be heuristically explained.

[2] The sinusoidal variation of inductance with rotor position turns out to be very accurate because of the fact that the stator windings are sinusoidally wound. This forces the flux linkage to behave in a sinusoidal fashion.

[3] Linear refers to the fact that the equations are derived assuming that the iron circuit behaves linearly in relation to applied mmf and the flux produced.

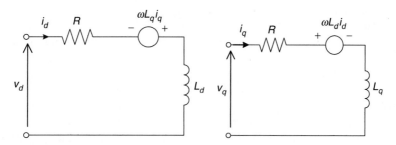

FIGURE 33.112 Two-phase equivalent circuit of the Syncrel in a rotating frame.

The 3/2 factor is to account for the fact that the two-phase machine produces two-thirds the torque of the three-phase machine.[4]

The only other remaining equation is the mechanical equation for the system:

$$J\dot{\omega}_r + D\omega_r + T_F = T_e \qquad (33.102)$$

where

$J \equiv$ the rotational inertia of the rotor/load,
$D \equiv$ the friction coefficient for the load,
$T_F \equiv$ the fixed load torque of the load,
$\omega_r \equiv$ the rotor mechanical angular velocity ($= \omega/p_p$).

Remark

Equation (33.101) shows that the machine must be designed so that $L_d - L_q$ is as large as possible. This will maximize the torque that is produced by the machine for given d- and q-axis currents. To lower L_q, one must design the q-axis so that it has as much air obstructing the flow of flux as possible, and the d-axis must be designed so that it has as much iron as possible. In practice these quantities cannot be varied independently.

Figure 33.112 shows the equivalent circuit for the Syncrel corresponding to Eqs. (33.99) and (33.100). One can see that the dynamic equations for the Syncrel are intrinsically simple. In contrast, the induction machine electrical equations consist of a set of four complex coupled differential equations.

33.10.5 Control Strategies and Important Parameters

The dq model captured in Eqs. (33.99)–(33.101) can be used to explain a number of control strategies for the Syncrel. It is beyond the scope of this section to present the derivation of

---

[4]The 3/2 conversion factor is required if the transformations are power-variant transformations, as opposed to the power-invariant transformations. The power-variant transformations are the most common ones used because the single-phase machine parameters can be used directly in the resultant models, and the two-phase voltages and currents are identical in magnitude to their three-phase counterparts.

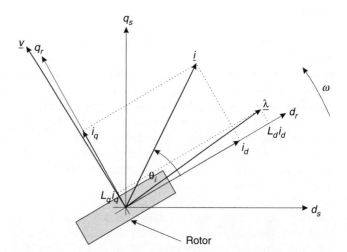

FIGURE 33.113 Space phasor diagram of a Syncrel.

these. The interested reader should consult references [4–6] cited at the end of this chapter.

One of the most common control strategies for any electrical machine is to maximize the torque per ampere of input current. The following discussion should be considered in conjunction with Fig. 33.113, which shows the relationship of the various vectors in the machine to the $d_r q_r$- and $d_s q_s$-axes.

It turns out that one of the critical parameters for the control of the Syncrel is the angle of the resultant current vector in the machine in relation to the d-axis of the machine. It is possible to write the torque expression for the Syncrel in terms of this as

$$T_e = \frac{3}{2} p \left(L_d - L_q \right) |i|^2 \sin 2\delta \qquad (33.103)$$

It is obvious that this expression is maximized for a given value of i if $\delta = \frac{\pi}{4}$. Therefore, one should control the currents so that δ stays at this angle, if maximum torque per ampere is desired.

Another control objective for the Syncrel is to maximize the power factor for the machine. This is important to minimize the kVA for the inverter. It can be shown that the current angle to maximize the power factor is [4]

$$\delta = \tan^{-1} \sqrt{\xi} \qquad (33.104)$$

where

$\xi = L_d/L_q$ (the inductance ratio).

Remark

Equation (33.104) indicates that ξ is the important parameter in relation to power factor. In order to obtain a power factor of 0.8, one requires an inductance ratio of approximately 10.

Finally, we shall consider another control objective – maximize the rate of change of torque with a fixed-current-angle control strategy. In effect, this means that one is maximizing the rate of change of the currents in the machine for a given voltage applied to it. The analysis of this requirement results in [4]

$$\delta = \tan^{-1} \xi \qquad (33.105)$$

Remark

As with the maximum-power-factor case, ξ is the most important parameter in relation to the rate of change of torque. Because this control effectively optimizes the current into the machine for a given voltage and angular velocity, this angle also corresponds to that required to maximize the field-weakening range of the machine.

Other control strategies for the machine can be devised, as well as the current angles required to obtain the maximum power from the machine during field-weakening operation.

Remark

If one carries out a thorough analysis of all the control properties of the Syncrel, then it emerges that all performance measures for the machine are enhanced by a large value of the ξ ratio.

33.10.6 Practical Considerations

The control strategies discussed in the previous section were all derived assuming that the machine does not exhibit saturation and there are no iron losses. The q-axis of the machine does not have any saturation, as the flux path on this axis is dominated by air. However, the d-axis of the machine does exhibit substantial saturation under operational flux levels, and this effect must be accounted for to optimize the drive performance.

The effect that saturation has on the ideal current angles is to increase them. This increase is most pronounced for the maximum torque per ampere control strategy, since this strategy results in a larger component of current in the d-axis, and consequently more saturation. Maximum power factor and maximum rate of change of torque are not affected as much. In order to get the correct current angle for maximum torque per ampere, a lookup table of the saturation characteristic of the machine must be stored in the controller, which is consulted in order to calculate the desired current angle [7].

It has been found that iron losses in the stator and the rotor also affect the optimal current angles. However, usually saturation effects dominate, and the effects of iron losses can be ignored.

33.10.7 A Syncrel Drive System

The basic structure of a variable-speed drive system based on using the Syncrel is shown in Fig. 33.114. Many components of this drive are very similar to those found in an induction machine drive system. One notable exception is the L_d lookup table block and the current reference generator. The L_d lookup stores the current vs d-axis inductance table for the machine, thereby allowing the inductance to be determined for various current levels. This table is also used to generate the incremental d-axis inductance. The inductance values generated from this table are used in the state feedback block and the torque estimator.

The state feedback block effectively generates an offset voltage to the PWM generator so that the voltage it produces is at least enough to counter the back-emf. This technique effectively eliminates the back-emf disturbance from the current-control loops.

The current reference generator takes the desired torque as an input and generates the required d- and q-axis currents at the output. This block uses a lookup-table technique together with an inverse of the torque equation to generate these currents and takes into account the saturation characteristics of the machine.

The three-to-two-phase block converts the currents from a three-phase stationary frame to a two-phase rotating frame. This is a standard block in induction machine drives, and as with induction machine drives, this means that the Syncrel control algorithm is implemented in a rotating reference frame. The conversion from this frame back to the stationary frame occurs implicitly in the space vector PWM generator.

The Syncrel control algorithm is essentially a simplified vector controller, and consequently the computational requirements are not high. This means that a Syncrel controller can be implemented on a modest microprocessor. As far as input and output hardware is concerned, the requirements are basically the same as those for an induction machine system – i.e. sample two of the phase currents, the link voltage, and the rotor position.

33.10.8 Conclusion

The Syncrel-based drive system offers simplicity in control, excellent performance for variable speed and position-control

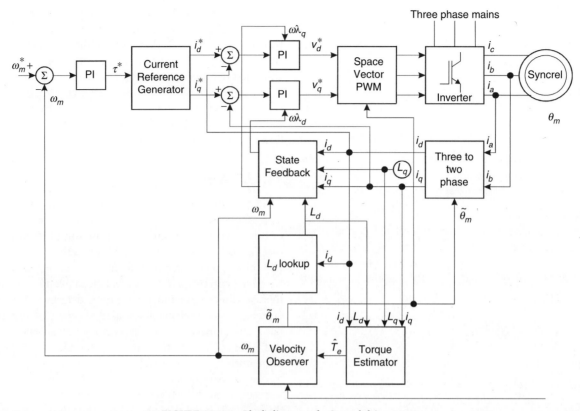

FIGURE 33.114 Block diagram of a Syncrel drive system.

applications, good torque and power density, and efficiency that is more than competitive with that of the induction machine drive systems. To date very few commercial drive systems are available using Syncrels, this being mainly due to the slow emergence of easy-to-manufacture rotors that give good performance, and the conservatism of the motor-drive industry. One commercial application that has emerged is ac-servo applications, where the Syncrel offers low torque ripple (with appropriate rotor design) together with a small moment of inertia. Other applications under consideration are in the area of drives for electric vehicles and generators for flywheel energy storage systems. It remains to be seen whether the Syncrel can ever challenge the supremacy of the vector-controlled induction machine in mainstream industrial applications.

The interested reader who wishes to pursue Syncrel drives in more detail can find a good coverage of the control and motor design issues in [8].

References

1. J. Kostko, "Polyphase reaction synchronous motors," *J. Amer. Inst. Elec. Eng.*, 42, 1162–1168, 1923.
2. D. O'Kelly and S. Simmons, *Introduction to Generalised Electrical Machine Theory*, McGraw Hill, U.K., 1968.
3. D. Staton, W. Soong, and T.J. Miller, "Unified theory of torque production in switched reluctance motors," *IEEE Trans. on Industry Applications*, IA-31, 329–337, 1995.
4. R.E. Betz, "Theoretical aspects of control of synchronous reluctance machines," *IEE Proc. B*, 139, 355–364, 1992.
5. A. Chiba and T. Fukao, "A closed-loop operation of super high speed reluctance motor for quick torque response," *IEEE Trans. on Industry Applications*, IA-28, 600–606, 1992.
6. M.G. Jovanovic and R.E. Betz, "Theoretical aspects of the control of synchronous reluctance machine including saturation ad iron losses," *Proceedings of IEEE-IAS Annual Meeting*, Houston, October, 1992.
7. M.G. Jovanovic, "*Sensorless Control of Synchronous Reluctance Machines*," Ph.D. thesis, University of Newcastle, Australia, 1997.
8. I. Boldea, *Reluctance Synchronous Machines and Drives*. Oxford University Press, 1996.

Further Reading

DC Motor Drives

9. G. K. Dubey, *Power Semiconductor Controlled Drives*. Prentice Hall, 1989.
10. W. Leonard, *Control of Electric Drives*. Springer-Verlag, 1985.
11. V. Subrahmanyam, *Electric Drives; Concepts and Applications*. McGraw Hill, 1994.

12. M. A. El-Sharkawi, *Fundamentals of Electric Drives*. Thompson Learning, 2000.

Induction Motor Drives

13. B. K. Bose, *Power Electronics and AC Drives*. Prentice Hall, 1987.
14. J. M. D. Murphy and G. G. Turnbull, *Power Electronic Control of AC Motors*. Pergamon Press, 1988.
15. D. W. Novotny and T. A. Lipo, *Vector Control and Dynamics of Drives*. Oxford Science Publications, 1996.
16. P. Vas, *Electrical Machines and Drives*. a Space Vector Theory Approach, Clarendon Press, 1992.
17. G. K. Dubey, *Power Semiconductor Controlled Drives*. Prentice Hall, 1989.

Synchronous Motor Drives

18. W. Leonard, *Control of Electrical Drives*. Springer-Verlag, 1985.
19. J. M. D. Murphy and G. G. Turnbull, *Power Electronic Control of AC Drives*. Pergamon Press, 1988.
20. G. R. Slemon, *Electric Machines and Drives*. Addison-Wesley, 1992.
21. W. W. Novotny and T. A. Lipo, *Vector Control and Dynamics of AC Drives*. Oxford Science Publications, 1996.

Permanent-magnet AC Synchronous Motor Drives

22. T. J. E. Miller, *Brushless Permanent Magnet and Reluctance Motor Drives*. Oxford Science Publications, 1989.
23. "Performance and Design of Permanent Magnet AC Motor Drives", Tutorial course, *IEEE Industry Application Society*, 1989.
24. R. M. Crowder, *Electric Drives and Their Controls*. Oxford Science Publications, 1995.

Permanent-magnet Brushless DC (BLDC) Motor Drives

25. T. J. E. Miller, *Brushless Permanent-magnet and Reluctance Motor Drives*. Oxford University press, 1989. For "square-wave brushless dc motors."
26. T. Kenjo and S. Nagamori, *Permanent-magnet and Brushless DC Motors*. Oxford University Press, 1985. For "small machines used particularly in consumer electronics."
27. D. C. Hanselman, *Brushless Permanent-magnet Motor Design*. Mc Graw Hill, 1994. For "the possible range of controllers in the last chapter."
28. Application notes from the various integrated circuit companies, particularly those that specialize in motor control.

Servo Drives

29. G. W. Younkin, *Industrial Servo Control Systems: Fundamentals and Applications*. Marcel Dekker, 1996.
30. B. C. Kuo and J. Tal, *Incremental Motion Control 1: DC motors and systems*. SRL Publishing, 1978.
31. D. Shetty and R. Kolk, *Mechatronics Systems Design*. PWS Publishing, 1997.
32. A. Fransua and R. Magureanu, *Electric machines and Drive Systems*. Techical Press, Oxford, 1984.

Stepper Motor Drives

33. H. B. Ertan, A. Hughes, and P. J. Lawrenson, "Efficient numerical method for predicting the torque-displacement curve of saturated VR stepping motors," *Proc. IEE*, 133, Part B, 1980.
34. T. Kenjo and A. Sugawara, *Stepping Motors and Their Microprocessor Controls*, Oxford University Press, 1994.
35. P. P. Acarnley, *Stepping Motors: A Guide to Modern Theory and Practice. IEE Control Engineering Series*, 19, Peter Peregrinus, Ltd., 1982.
36. Application Notes AN235, "Stepper Motor Driving," Discrete Semiconductor Handbook. SGS Thompson Microelectronics, 1995.
37. M. F. Rahman, C. S. Chang, and A. N. Poo, "Approaches to ministepping step motor controllers and their accuracy considerations", *IEEE Transactions on Industrial Electronics*, IE32, No. 3, 1985.
38. *Proceedings of Symposium on Incremental Motion Controls Systems and Devices*. 1972–1992. University of Illinois, Urbana-Champaign, IL.

Switched-reluctance Motor Drives

39. J. R Hendershot, "Application of SR drives, "*IEEE Industry Application Society Tutorial on Switched Reluctance Drives*, 60–90, 1990.
40. P. J. Lawrenson, "Switched reluctance motor drives," *Electronics and Power*, 144–147, February, 1983.
41. E. Richter, "Switched reluctance machines for high performance operations in a harsh environment – a review paper," *Int. Conf. Electrical machines (ICEM '90)*, Part 1, 18–24, August, 1990.
42. P. J. Lawrenson, "Switched Reluctance Drives – A Fast Growing Technology," *Electric Drives and Controls*, April/May, 1985.
43. T. J. E. Miller, *Switched Reluctance Motors and Their Control*. Clarendon, Oxford, 1993.
44. S. R. MacMinn, "Control of the Switched Reluctance Machine," Switched reluctance drives tutorial, *IEEE Industry Applications Society Conference, 25th Industry Application Society Annual Meeting*, Seattle, October 12, 36–59, 1990.
45. A. D. Cheok and N. Ertugrul, "High robustness of an sr motor angle estimation algorithm using fuzzy predictive filters and heuristic knowledge based rules," *IEEE Trans. Ind. Electr.*, V. 46, October, 1999.

34
Control Methods for Switching Power Converters

J. Fernando Silva, Ph.D. and Sónia Ferreira Pinto, Ph.D.
Instituto Superior Técnico, DEEC, A.C. Energia, Laboratório de Máquinas Eléctricas e Electrónica de Potência, Centro de Automática da Universidade Técnica de Lisboa, AV. Rorisco Pais 1, 1049-001 Lisboa, Portugal

34.1 Introduction 935
34.2 Switching Power Converter Control Using State-space Averaged Models 936
  34.2.1 Introduction • 34.2.2 State-space Modeling • 34.2.3 Converter Transfer Functions • 34.2.4 Pulse Width Modulator Transfer Functions • 34.2.5 Linear Feedback Design Ensuring Stability • 34.2.6 Examples: Buck–Boost DC/DC Converter, Forward DC/DC Converter, 12 Pulse Rectifiers, Buck–Boost DC/DC Converter in the Discontinuous Mode (Voltage and Current Mode), Three-phase PWM Inverters
34.3 Sliding-mode Control of Switching Converters 955
  34.3.1 Introduction • 34.3.2 Principles of Sliding-mode Control • 34.3.3 Constant-frequency Operation • 34.3.4 Steady-state Error Elimination in Converters with Continuous Control Inputs • 34.3.5 Examples: Buck–Boost DC/DC Converter, Half-bridge Inverter, 12-pulse Parallel Rectifiers, Audio Power Amplifiers, Near Unity Power Factor Rectifiers, Multilevel Inverters, Matrix Converters
34.4 Fuzzy Logic Control of Switching Converters 993
  34.4.1 Introduction • 34.4.2 Fuzzy Logic Controller Synthesis • 34.4.3 Example: Near Unity Power Factor Buck–Boost Rectifier
34.5 Conclusions 996
  References 997

34.1 Introduction

Switching power converters must be suitably designed and controlled in order to supply the voltages, currents, or frequency ranges needed for the load and to guarantee the requested dynamics [1–4]. Furthermore, they can be designed to serve as "clean" interfaces between most loads and the electrical utility system. Thereafter, the set switching converter plus load behaves as an almost pure electrical utility resistive load.

This chapter provides basic and some advanced skills to control electronic power converters, taking into account that the control of switching power converters is a vast and interdisciplinary subject. Control designers for switching converters should know the static and dynamic behavior of the electronic power converter and how to design its elements for the intended operating modes. Designers must be experts on control techniques, especially the nonlinear ones, since switching converters are nonlinear, time-variant, discrete systems, and designers must be capable of analog or digital implementation of the derived modulators, regulators, or compensators. Powerful modeling methodologies and sophisticated control processes must be used to obtain stable-controlled switching converters, not only with satisfactory static and dynamic performance, but also with low sensitivity against load or line disturbances or, preferably, robustness.

In Section 34.2, the techniques to obtain suitable nonlinear and linear state-space models, for most switching converters, are presented and illustrated through examples. The derived linear models are used to create equivalent circuits, and to design linear feedback controllers for converters operating in the continuous or discontinuous mode. The classical linear time-invariant systems control theory, based on Laplace transform, transfer function concepts, Bode plots or root locus, is best used with state-space averaged models, or derived circuits, and well-known triangular wave modulators for generating the switching variables or the trigger signals for the power semiconductors.

Nonlinear state-space models and sliding-mode controllers, presented in Section 34.3, provide a more consistent way of handling the control problem of switching converters, since sliding mode is aimed at variable structure systems, as are switching power converters. Chattering, a characteristic of sliding mode, is inherent to switching power converters, even if they are controlled with linear methods. Chattering is very hard to remove and is acceptable in certain converter variables. The described sliding-mode methodology defines exactly the variables that need to be measured, while providing the necessary equations (control law and switching law) whose implementation gives the robust modulator and compensator low-level hardware (or software). Therefore, the sliding-mode control integrates the design of the switching converter modulator and controller electronics, reducing the needed designer expertise. This approach requires measurement of the state variables, but eliminates conventional modulators and linear feedback compensators, enabling better performance and robustness. It also reduces the converter cost, control complexity, volume, and weight (increasing power density). The so-called main drawback of sliding mode, variable switching frequency, is also addressed, providing fixed-frequency auxiliary functions and suitable augmented control laws to null steady-state errors due to the use of constant switching frequency.

Fuzzy control of switching converters (Section 34.4) is a control technique needing no converter models, parameters, or operating conditions, but only an expert knowledge of the converter dynamics. Fuzzy controllers can be used in a diverse array of switching converters with only small adaptations, since the controllers, based on fuzzy sets, are obtained simply from the knowledge of the system dynamics, using a model reference adaptive control philosophy. Obtained fuzzy control rules can be built into a decision-lookup table, in which the control processor simply picks up the control input corresponding to the sampled measurements. Fuzzy controllers are almost immune to system parameter fluctuations, since they do not take into account their values. The steps to obtain a fuzzy controller are described, and the example provided compares the fuzzy controller performance to the current-mode control.

34.2 Switching Power Converter Control Using State-space Averaged Models

34.2.1 Introduction

State-space models provide a general and strong basis for dynamic modeling of various systems including switching converters. State-space models are useful to design the needed linear control loops, and can also be used to computer simulate the steady state, as well as the dynamic behavior, of the switching converter, fitted with the designed feedback control loops and subjected to external perturbations. Furthermore, state-space models are the basis for applying powerful nonlinear control methods such as sliding mode. State-space averaging and linearization provides an elegant solution for the application of widely known linear control techniques to most switching converters.

34.2.2 State-space Modeling

Consider a switching converter with sets of power semiconductor structures, each one with two different circuit configurations, according to the state of the respective semiconductors, and operating in the continuous mode of conduction. Supposing the power semiconductors as controlled ideal switches (zero on-state voltage drops, zero off-state currents, and instantaneous commutation between the on- and off-states), the time (t) behavior of the circuit, over period T, can be represented by the general form of the state-space model (34.1):

$$\dot{\mathbf{x}} = \mathbf{A}\mathbf{x} + \mathbf{B}\mathbf{u}$$
$$\mathbf{y} = \mathbf{C}\mathbf{x} + \mathbf{D}\mathbf{u} \qquad (34.1)$$

where $\mathbf{x}$ is the state vector, $\dot{\mathbf{x}} = d\mathbf{x}/dt$, $\mathbf{u}$ is the input or control vector, and $\mathbf{A}$, $\mathbf{B}$, $\mathbf{C}$, $\mathbf{D}$ are respectively the dynamics (or state), the input, the output, and the direct transmission (or feedforward) matrices.

Since the power semiconductors will either be conducting or blocking, a time-dependent switching variable $\delta(t)$ can be used to describe the allowed switch states of each structure (i.e. $\delta(t) = 1$ for the on-state circuit and $\delta(t) = 0$ for the off-state circuit). Then, two subintervals must be considered: subinterval 1 for $0 \leq t \leq \delta_1 T$, where $\delta(t) = 1$ and subinterval 2 for $\delta_1 T \leq t \leq T$ where $\delta(t) = 0$. The state equations of the circuit, in each of the circuit configurations, can be written as:

$$\dot{\mathbf{x}} = \mathbf{A}_1\mathbf{x} + \mathbf{B}_1\mathbf{u}$$
$$\mathbf{y} = \mathbf{C}_1\mathbf{x} + \mathbf{D}_1\mathbf{u} \qquad \text{for} \quad 0 \leq t \leq \delta_1 T \quad \text{where} \quad \delta(t) = 1$$
$$(34.2)$$

$$\dot{\mathbf{x}} = \mathbf{A}_2\mathbf{x} + \mathbf{B}_2\mathbf{u}$$
$$\mathbf{y} = \mathbf{C}_2\mathbf{x} + \mathbf{D}_2\mathbf{u} \qquad \text{for} \quad \delta_1 T \leq t \leq T \quad \text{where} \quad \delta(t) = 0$$
$$(34.3)$$

34.2.2.1 Switched State-space Model

Given the two binary values of the switching variable $\delta(t)$, Eqs. (34.2) and (34.3) can be combined to obtain the nonlinear and time-variant switched state-space model of the switching converter circuit, Eq. (34.4) or (34.5):

$$\dot{x} = [A_1\delta(t) + A_2(1-\delta(t))]x + [B_1\delta(t) + B_2(1-\delta(t))]u$$

$$y = [C_1\delta(t) + C_2(1-\delta(t))]x + [D_1\delta(t) + D_2(1-\delta(t))]u \quad (34.4)$$

$$\dot{x} = A_S x + B_S u$$
$$y = C_S x + D_S u \quad (34.5)$$

where $A_S = [A_1\delta(t) + A_2(1-\delta(t))]$, $B_S = [B_1\delta(t) + B_2(1-\delta(t))]$, $C_S = [C_1\delta(t) + C_2(1-\delta(t))]$, and $D_S = [D_1\delta(t) + D_2(1-\delta(t))]$.

34.2.2.2 State-space Averaged Model

Since the state variables of the **x** vector are continuous, using Eq. (34.4), with the initial conditions $x_1(0) = x_2(T)$, $x_2(\delta_1 T) = x_1(\delta_1 T)$, and considering the duty cycle δ_1 as the average value of $\delta(t)$, the time evolution of the converter state variables can be obtained, integrating Eq. (34.4) over the intervals $0 \leq t \leq \delta_1 T$ and $\delta_1 T \leq t \leq T$, although it often requires excessive calculation effort. However, a convenient approximation can be devised, considering λ_{max}, the maximum of the absolute values of all eigenvalues of **A** (usually λ_{max} is related to the cutoff frequency f_c of an equivalent low-pass filter with $f_c \ll 1/T$). For $\lambda_{max}T \ll 1$, the exponential matrix (or state transition matrix) $e^{At} = I + At + A^2t^2/2 + \cdots + A^n t^n/n!$, where **I** is the identity or unity matrix, can be approximated by $e^{At} \approx I + At$. Therefore, $e^{A_1\delta_1 t} \cdot e^{A_2(1-\delta_1)t} \approx I + [A_1\delta_1 + A_2(1-\delta_1)]t$. Hence, the solution over the period T, for the system represented by Eq. (34.4), is found to be:

$$x(T) \cong e^{[A_1\delta_1 + A_2(1-\delta_1)]T} x_1(0)$$
$$+ \int_0^T e^{[A_1\delta_1 + A_2(1-\delta_1)](T-\tau)} [B_1\delta_1 + B_2(1-\delta_1)] u d\tau \quad (34.6)$$

This approximate response of Eq. (34.4) is identical to the exact response obtained from the nonlinear continuous time-invariant state-space model (34.7), supposing that the average values of **x**, denoted $\bar{x}$, are the new state variables, and considering $\delta_2 = 1 - \delta_1$. Moreover, if $A_1 A_2 = A_2 A_1$, the approximation is exact.

$$\dot{\bar{x}} = [A_1\delta_1 + A_2\delta_2]\bar{x} + [B_1\delta_1 + B_2\delta_2]\bar{u}$$
$$\bar{y} = [C_1\delta_1 + C_2\delta_2]\bar{x} + [D_1\delta_1 + D_2\delta_2]\bar{u} \quad (34.7)$$

For $\lambda_{max}T \ll 1$, the model (34.7), often referred to as the state-space averaged model, is also said to be obtained by "averaging" Eq. (34.4) over one period, under small ripple and slow variations, as the average of products is approximated by products of the averages. Comparing Eq. (34.7) to Eq. (34.1), the relations (34.8), defining the state-space averaged model, are obtained.

$$A = [A_1\delta_1 + A_2\delta_2]; \quad B = [B_1\delta_1 + B_2\delta_2];$$
$$C = [C_1\delta_1 + C_2\delta_2]; \quad D = [D_1\delta_1 + D_2\delta_2] \quad (34.8)$$

EXAMPLE 34.1 State-space models for the buck–boost dc/dc converter

Consider the simplified circuitry of the buck–boost converter of Fig. 34.1 switching at $f_s = 20\,\text{kHz}$ ($T = 50\,\mu s$) with $V_{DCmax} = 28\,\text{V}$, $V_{DCmin} = 22\,\text{V}$, $V_o = 24\,\text{V}$, $L_i = 400\,\mu\text{H}$, $C_o = 2700\,\mu\text{F}$, $R_o = 2\,\Omega$.

The differential equations governing the dynamics of the state vector $x = [i_L, v_o]^T$ (^{T} denotes the transpose of vectors or matrices) are:

$$L_i \frac{di_L}{dt} = V_{DC}$$
$$C_o \frac{dv_o}{dt} = -\frac{v_o}{R_o}$$
for $0 \leq t \leq \delta_1 T$ ($\delta(t) = 1$, Q_1 is on and D_1 is off) \quad (34.9)

$$L_i \frac{di_L}{dt} = -v_o$$
$$C_o \frac{dv_o}{dt} = i_L - \frac{v_o}{R_o}$$
for $\delta_1 T \leq t \leq T$ ($\delta(t) = 1$, Q_1 is off and D_1 is on) \quad (34.10)

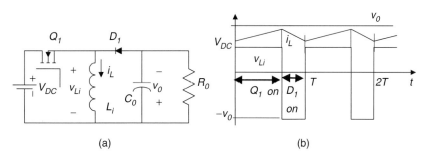

FIGURE 34.1 (a) Basic circuit of the buck–boost dc/dc converter and (b) ideal waveforms.

Comparing Eqs. (34.9) and (34.10) to Eqs. (34.2) and (34.3) and considering $\mathbf{y} = [v_o, i_L]^T$, the following matrices can be identified:

$$\mathbf{A}_1 = \begin{bmatrix} 0 & 0 \\ 0 & -1/(R_o C_o) \end{bmatrix}; \quad \mathbf{A}_2 = \begin{bmatrix} 0 & -1/L_i \\ 1/C_o & -1/(R_o C_o) \end{bmatrix};$$

$$\mathbf{B}_1 = [1/L_i, 0]^T; \quad \mathbf{B}_2 = [0, 0]^T; \quad \mathbf{u} = [V_{DC}];$$

$$\mathbf{C}_1 = \begin{bmatrix} 0 & 1 \\ 1 & 0 \end{bmatrix}; \quad \mathbf{C}_2 = \begin{bmatrix} 0 & 1 \\ 1 & 0 \end{bmatrix};$$

$$\mathbf{D}_1 = [0, 0]^T; \quad \mathbf{D}_2 = [0, 0]^T$$

From Eqs. (34.4) and (34.5), the switched state-space model of this switching converter is

$$\begin{bmatrix} \dot{i}_L \\ \dot{v}_o \end{bmatrix} = \begin{bmatrix} 0 & -(1-\delta(t))/L_i \\ (1-\delta(t))/C_o & -1/(R_o C_o) \end{bmatrix} \begin{bmatrix} i_L \\ v_o \end{bmatrix}$$

$$+ \begin{bmatrix} \delta(t)/L_i \\ 0 \end{bmatrix} V_{DC}$$

$$\begin{bmatrix} v_o \\ i_L \end{bmatrix} = \begin{bmatrix} 0 & 1 \\ 1 & 0 \end{bmatrix} \begin{bmatrix} i_L \\ v_o \end{bmatrix} + \begin{bmatrix} 0 \\ 0 \end{bmatrix} [V_{DC}] \quad (34.11)$$

Now, applying Eq. (34.7), Eqs. (34.12) and (34.13) can be obtained:

$$\begin{bmatrix} \dot{\bar{i}}_L \\ \dot{\bar{v}}_o \end{bmatrix} = \left[\begin{bmatrix} 0 & 0 \\ 0 & -1/R_o C_o \end{bmatrix} \delta_1 + \begin{bmatrix} 0 & -1/L_i \\ 1/C_o & -1/R_o C_o \end{bmatrix} \delta_2 \right]$$

$$\times \begin{bmatrix} \bar{i}_L \\ \bar{v}_o \end{bmatrix} + \left[\begin{bmatrix} 1/L_i \\ 0 \end{bmatrix} \delta_1 + \begin{bmatrix} 0 \\ 0 \end{bmatrix} \delta_2 \right] [\bar{V}_{DC}]$$

$$(34.12)$$

$$\begin{bmatrix} \bar{v}_o \\ \bar{i}_L \end{bmatrix} = \left[\begin{bmatrix} 0 & 1 \\ 1 & 0 \end{bmatrix} \delta_1 + \begin{bmatrix} 0 & 1 \\ 1 & 0 \end{bmatrix} \delta_2 \right] \begin{bmatrix} \bar{i}_L \\ \bar{v}_o \end{bmatrix}$$

$$+ \left[\begin{bmatrix} 0 \\ 0 \end{bmatrix} \delta_1 + \begin{bmatrix} 0 \\ 0 \end{bmatrix} \delta_2 \right] [\bar{V}_{DC}] \quad (34.13)$$

From Eqs. (34.12) and (34.13), the state-space averaged model, written as a function of δ_1, is

$$\begin{bmatrix} \dot{\bar{i}}_L \\ \dot{\bar{v}}_o \end{bmatrix} = \begin{bmatrix} 0 & -1-\delta_1/L_i \\ 1-\delta_1/C_o & -1/R_o C_o \end{bmatrix} \begin{bmatrix} \bar{i}_L \\ \bar{v}_o \end{bmatrix} + \begin{bmatrix} \delta_1/L_i \\ 0 \end{bmatrix} [\bar{V}_{DC}]$$

$$(34.14)$$

$$\begin{bmatrix} \bar{v}_o \\ \bar{i}_L \end{bmatrix} = \begin{bmatrix} 0 & 1 \\ 1 & 0 \end{bmatrix} \begin{bmatrix} \bar{i}_L \\ \bar{v}_o \end{bmatrix} + \begin{bmatrix} 0 \\ 0 \end{bmatrix} [\bar{V}_{DC}] \quad (34.15)$$

The eigenvalues $s_{bb_{1,2}}$, or characteristic roots of $\mathbf{A}$, are the roots of $|s\mathbf{I} - \mathbf{A}|$. Therefore:

$$s_{bb_{1,2}} = \frac{-1}{2R_o C_o} \pm \sqrt{\frac{1}{4(R_o C_o)^2} - \frac{(1-\delta_1)^2}{L_i C_o}} \quad (34.16)$$

Since λ_{max} is the maximum of the absolute values of all the eigenvalues of $\mathbf{A}$, the model (34.14, 34.15) is valid for switching frequencies f_s ($f_s = 1/T$) that verify $\lambda_{max} T \ll 1$. Therefore, as $T \ll 1/\lambda_{max}$, the values of T that approximately verify this restriction are $T \ll 1/max(|s_{bb_{1,2}}|)$. Given this buck–boost converter data, $T \ll 2$ ms is obtained. Therefore, the converter switching frequency must obey $f_s \gg max(|s_{f1,2}|)$, implying switching frequencies above, say, 5 kHz. Consequently, the buck–boost switching frequency, the inductor value, and the capacitor value were chosen accordingly.

This restriction can be further used to discuss the maximum frequency ω_{max} for which the state-space averaged model is still valid, given a certain switching frequency. As λ_{max} can be regarded as a frequency, the preceding constraint brings $\omega_{max} \ll 2\pi f_s$, say $\omega_{max} < 2\pi f_s/10$, which means that the state-space averaged model is a good approximation at frequencies under one-tenth of the power converter switching frequency.

The state-space averaged model (34.14, 34.15) is also the state-space model of the circuit represented in Fig. 34.2. Hence, this circuit is often named "the averaged equivalent circuit" of the buck–boost converter and allows the determination, under small ripple and slow variations, of the average equivalent circuit of the converter switching cell (power transistor plus diode).

The average equivalent circuit of the switching cell (Fig. 34.3a) is represented in Fig. 34.3b and emerges directly from the state-space averaged model (34.14, 34.15). This equivalent circuit can be viewed as the model of an "ideal transformer" (Fig. 34.3c), whose primary to secondary ratio (v_1/v_2) can be calculated applying Kirchhoff's voltage law to obtain $-v_1 + v_s - v_2 = 0$. As $v_2 = \delta_1 v_s$, it follows that $v_1 = v_s(1 - \delta_1)$, giving $(v_1/v_2) = (1 - \delta_1)/\delta_1$. The same ratio could be obtained beginning with $i_L = i_1 + i_2$, and $i_1 = \delta_1 i_L$ (Fig. 34.3b) which gives $i_2 = i_L(1 - \delta_1)$ and $(i_2/i_1) = \delta_2/\delta_1$.

The average equivalent circuit concept, obtained from Eq. (34.7) or Eqs. (34.14) and (34.15), can be applied to other switching converters, with or without a similar switching cell, to obtain transfer functions or to computer simulate

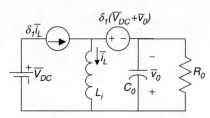

FIGURE 34.2 Equivalent circuit of the averaged state-space model of the buck–boost converter.

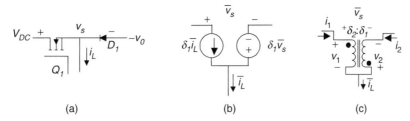

FIGURE 34.3 Average equivalent circuit of the switching cell: (a) switching cell; (b) average equivalent circuit; and (c) average equivalent circuit using an ideal transformer.

the converter average behavior. The average equivalent circuit of the switching cell can be applied to converters with the same switching cell operating in the continuous conduction mode. However, note that the state variables of Eq. (34.7) or Eqs. (34.14) and (34.15) are the mean values of the converter instantaneous variables and, therefore, do not represent their ripple components. The inputs of the state-space averaged model are the mean values of the converter inputs over one switching period.

34.2.2.3 Linearized State-space Averaged Model

Since the converter outputs $\bar{y}$ must be regulated actuating on the duty cycle $\delta(t)$, and the converter inputs $\bar{u}$ usually present perturbations due to the load and power supply variations. State variables are decomposed in small ac perturbations (denoted by "~") and dc steady-state quantities (represented by uppercase letters). Therefore:

$$\bar{x} = X + \tilde{x}$$
$$\bar{y} = Y + \tilde{y}$$
$$\bar{u} = U + \tilde{u} \quad (34.17)$$
$$\delta_1 = \Delta_1 + \tilde{\delta}$$
$$\delta_2 = \Delta_2 - \tilde{\delta}$$

Using Eq. (34.17) in Eq. (34.7) and rearranging terms, we obtain:

$$\dot{\tilde{x}} = [\mathbf{A}_1 \Delta_1 + \mathbf{A}_2 \Delta_2] \mathbf{X} + [\mathbf{B}_1 \Delta_1 + \mathbf{B}_2 \Delta_2] \mathbf{U}$$
$$+ [\mathbf{A}_1 \Delta_1 + \mathbf{A}_2 \Delta_2] \tilde{x} + [(\mathbf{A}_1 - \mathbf{A}_2) \mathbf{X} + (\mathbf{B}_1 - \mathbf{B}_2) \mathbf{U}] \tilde{\delta}$$
$$+ [\mathbf{B}_1 \Delta_1 + \mathbf{B}_2 \Delta_2] \tilde{u} + [(\mathbf{A}_1 - \mathbf{A}_2) \tilde{x} + (\mathbf{B}_1 - \mathbf{B}_2) \tilde{u}] \tilde{\delta}$$
$$(34.18)$$

$$\mathbf{Y} + \tilde{y} = [\mathbf{C}_1 \Delta_1 + \mathbf{C}_2 \Delta_2] \mathbf{X} + [\mathbf{D}_1 \Delta_1 + \mathbf{D}_2 \Delta_2] \mathbf{U}$$
$$+ [\mathbf{C}_1 \Delta_1 + \mathbf{C}_2 \Delta_2] \tilde{x} + [(\mathbf{C}_1 - \mathbf{C}_2) \mathbf{X} + (\mathbf{D}_1 - \mathbf{D}_2) \mathbf{U}] \tilde{\delta}$$
$$+ [\mathbf{D}_1 \Delta_1 + \mathbf{D}_2 \Delta_2] \tilde{u} + [(\mathbf{C}_1 - \mathbf{C}_2) \tilde{x} + (\mathbf{D}_1 - \mathbf{D}_2) \tilde{u}] \tilde{\delta}$$
$$(34.19)$$

The terms $[\mathbf{A}_1 \Delta_1 + \mathbf{A}_2 \Delta_2] \mathbf{X} + [\mathbf{B}_1 \Delta_1 + \mathbf{B}_2 \Delta_2] \mathbf{U}$ and $[\mathbf{C}_1 \Delta_1 + \mathbf{C}_2 \Delta_2] \mathbf{X} + [\mathbf{D}_1 \Delta_1 + \mathbf{D}_2 \Delta_2] \mathbf{U}$, respectively from Eqs. (34.18) and (34.19), represent the steady-state behavior of the system. As in steady state $\dot{\mathbf{X}} = \mathbf{0}$, the following relationships hold:

$$\mathbf{0} = [\mathbf{A}_1 \Delta_1 + \mathbf{A}_2 \Delta_2] \mathbf{X} + [\mathbf{B}_1 \Delta_1 + \mathbf{B}_2 \Delta_2] \mathbf{U} \quad (34.20)$$
$$\mathbf{Y} = [\mathbf{C}_1 \Delta_1 + \mathbf{C}_2 \Delta_2] \mathbf{X} + [\mathbf{D}_1 \Delta_1 + \mathbf{D}_2 \Delta_2] \mathbf{U} \quad (34.21)$$

Neglecting higher order terms ($[(\mathbf{A}_1 - \mathbf{A}_2) \tilde{x} + (\mathbf{B}_1 - \mathbf{B}_2) \tilde{u}] \tilde{\delta} \approx 0$) of Eqs. (34.18) and (34.19), the linearized small-signal state-space averaged model is

$$\dot{\tilde{x}} = [\mathbf{A}_1 \Delta_1 + \mathbf{A}_2 \Delta_2] \tilde{x} + [(\mathbf{A}_1 - \mathbf{A}_2) \mathbf{X} + (\mathbf{B}_1 - \mathbf{B}_2) \mathbf{U}] \tilde{\delta}$$
$$+ [\mathbf{B}_1 \Delta_1 + \mathbf{B}_2 \Delta_2] \tilde{u}$$
$$\tilde{y} = [\mathbf{C}_1 \Delta_1 + \mathbf{C}_2 \Delta_2] \tilde{x} + [(\mathbf{C}_1 - \mathbf{C}_2) \mathbf{X} + (\mathbf{D}_1 - \mathbf{D}_2) \mathbf{U}] \tilde{\delta}$$
$$+ [\mathbf{D}_1 \Delta_1 + \mathbf{D}_2 \Delta_2] \tilde{u}$$
$$(34.22)$$

or

$$\dot{\tilde{x}} = \mathbf{A}_{av} \tilde{x} + \mathbf{B}_{av} \tilde{u} + [(\mathbf{A}_1 - \mathbf{A}_2) \mathbf{X} + (\mathbf{B}_1 - \mathbf{B}_2) \mathbf{U}] \tilde{\delta}$$
$$\tilde{y} = \mathbf{C}_{av} \tilde{x} + \mathbf{D}_{av} \tilde{u} + [(\mathbf{C}_1 - \mathbf{C}_2) \mathbf{X} + (\mathbf{D}_1 - \mathbf{D}_2) \mathbf{U}] \tilde{\delta}$$
$$(34.23)$$

with

$$\mathbf{A}_{av} = [\mathbf{A}_1 \Delta_1 + \mathbf{A}_2 \Delta_2]$$
$$\mathbf{B}_{av} = [\mathbf{B}_1 \Delta_1 + \mathbf{B}_2 \Delta_2]$$
$$\mathbf{C}_{av} = [\mathbf{C}_1 \Delta_1 + \mathbf{C}_2 \Delta_2]$$
$$\mathbf{D}_{av} = [\mathbf{D}_1 \Delta_1 + \mathbf{D}_2 \Delta_2]$$
$$(34.24)$$

34.2.3 Converter Transfer Functions

Using Eq. (34.20) in Eq. (34.21), the input $\mathbf{U}$ to output $\mathbf{Y}$ steady-state relations (34.25), needed for open-loop and feedforward control, can be obtained.

$$\frac{\mathbf{Y}}{\mathbf{U}} = -\mathbf{C}_{av}\mathbf{A}_{av}^{-1}\mathbf{B}_{av} + \mathbf{D}_{av} \quad (34.25)$$

Applying Laplace transforms to Eq. (34.23) with zero initial conditions, and using the superposition theorem, the small-signal duty-cycle $\tilde{\delta}$ to output $\tilde{\mathbf{y}}$ transfer functions (34.26) can be obtained considering zero line perturbations ($\tilde{u} = 0$).

$$\frac{\tilde{\mathbf{y}}(s)}{\tilde{\delta}(s)} = \mathbf{C}_{av}\left[s\mathbf{I} - \mathbf{A}_{av}\right]^{-1}\left[(\mathbf{A}_1 - \mathbf{A}_2)\mathbf{X} + (\mathbf{B}_1 - \mathbf{B}_2)\mathbf{U}\right]$$
$$+ \left[(\mathbf{C}_1 - \mathbf{C}_2)\mathbf{X} + (\mathbf{D}_1 - \mathbf{D}_2)\mathbf{U}\right] \quad (34.26)$$

The line to output transfer function (or audio susceptibility transfer function) (34.27) is derived using the same method, considering now zero small-signal duty-cycle perturbations ($\tilde{\delta} = 0$).

$$\frac{\tilde{\mathbf{y}}(s)}{\tilde{\mathbf{u}}(s)} = \mathbf{C}_{av}\left[s\mathbf{I} - \mathbf{A}_{av}\right]^{-1}\mathbf{B}_{av} + \mathbf{D}_{av} \quad (34.27)$$

EXAMPLE 34.2 Buck–Boost dc/dc converter transfer functions

From Eqs. (34.14) and (34.15) of Example 34.1 and Eq. (34.23), making $\mathbf{X} = [I_L, V_o]^T$, $\mathbf{Y} = [V_o, I_L]^T$, and $\mathbf{U} = [V_{DC}]$, the linearized state-space model of the buck–boost converter is

$$\begin{bmatrix} \dot{\tilde{i}}_L \\ \dot{\tilde{v}}_o \end{bmatrix} = \begin{bmatrix} 0 & -1-\Delta_1/L_i \\ 1-\Delta_1/C_o & -1/R_oC_o \end{bmatrix} \begin{bmatrix} \tilde{i}_L \\ \tilde{v}_o \end{bmatrix} + \begin{bmatrix} \Delta_1/L_i \\ 0 \end{bmatrix} [\tilde{v}_{DC}]$$
$$+ \begin{bmatrix} 0 & \tilde{\delta}/L_i \\ -\tilde{\delta}/C_o & 0 \end{bmatrix} \begin{bmatrix} I_L \\ V_o \end{bmatrix} + \begin{bmatrix} V_{DC}/L_i \\ 0 \end{bmatrix} [\tilde{\delta}]$$

$$\begin{bmatrix} \tilde{v}_o \\ \tilde{i}_L \end{bmatrix} = \begin{bmatrix} 0 & 1 \\ 1 & 0 \end{bmatrix} \begin{bmatrix} \tilde{i}_L \\ \tilde{v}_o \end{bmatrix} + \begin{bmatrix} 0 \\ 0 \end{bmatrix} [\tilde{v}_{DC}]$$

(34.28)

From Eqs. (34.24) and (34.28), the following matrices are identified:

$$\mathbf{A}_{av} = \begin{bmatrix} 0 & -(1-\Delta_1)/L_i \\ 1-\Delta_1/C_o & -1/R_oC_o \end{bmatrix}; \quad \mathbf{B}_{av} = \begin{bmatrix} \Delta_1/L_i \\ 0 \end{bmatrix};$$

$$\mathbf{C}_{av} = \begin{bmatrix} 0 & 1 \\ 1 & 0 \end{bmatrix}; \quad \mathbf{D}_{av} = \begin{bmatrix} 0 \\ 0 \end{bmatrix}$$

(34.29)

The averaged linear equivalent circuit, resulting from Eq. (34.28) or from the linearization of the averaged equivalent circuit (Fig. 34.2) derived from Eqs. (34.14) and (34.15), now includes the small-signal current source $\tilde{\delta}I_L$ in parallel with the current source $\Delta_1\tilde{i}_L$, and the small-signal voltage source $\tilde{\delta}(V_{DC} + V_o)$ in series with the voltage source $\Delta_1(\tilde{v}_{dc} + \tilde{v}_o)$. The supply voltage source $\bar{V}_{DC}$ is replaced by the voltage source $\tilde{v}_{DC}$.

Using Eq. (34.29) in Eq. (34.25), the *input U to output Y steady-state relations* are:

$$\frac{I_L}{V_{DC}} = \frac{\Delta_1}{R_o(\Delta_1-1)^2} \quad (34.30)$$

$$\frac{V_o}{V_{DC}} = \frac{\Delta_1}{1-\Delta_1} \quad (34.31)$$

These relations are the well-known steady-state transfer relationships of the buck–boost converter [2, 5, 6]. For open-loop control of the V_o output, knowing the nominal value of the power supply V_{DC} and the required V_o, the value of Δ_1 can be off-line calculated from Eq. (34.31) ($\Delta_1 = V_o/(V_o + V_{DC})$). A modulator such as that described in Section 34.2.4, with the modulation signal proportional to Δ_1, would generate the signal $\delta(t)$. The open-loop control for fixed output voltages is possible, if the power supply V_{DC} is almost constant and the converter load does not change significantly. If the V_{DC} value presents disturbances, then the feedforward control can be used, calculating Δ_1 on-line, so that its value will always be in accordance with Eq. (34.31). The correct V_o value will be attained at steady state, despite input-voltage variations. However, because of converter parasitic reactances, not modeled here (see Example 34.3), in practice a steady-state error would appear. Moreover, the transient dynamics imposed by the converter would present overshoots, being often not suited for demanding applications.

From Eq. (34.27), the *line to output transfer functions* are:

$$\frac{\tilde{i}_L(s)}{\tilde{v}_{DC}(s)} = \frac{\Delta_1(1+sC_oR_o)}{s^2L_iC_oR_o + sL_i + R_o(1-\Delta_1)^2} \quad (34.32)$$

$$\frac{\tilde{v}_o(s)}{\tilde{v}_{DC}(s)} = \frac{R_o\Delta_1(1-\Delta_1)}{s^2L_iC_oR_o + sL_i + R_o(1-\Delta_1)^2} \quad (34.33)$$

From Eq. (34.26), *the small-signal duty-cycle $\tilde{\delta}$ to output $\tilde{y}$ transfer functions* are:

$$\frac{\tilde{i}_L(s)}{\tilde{\delta}(s)} = \frac{V_{DC}(1+\Delta_1+sC_oR_o)/(1-\Delta_1)}{s^2L_iC_oR_o + sL_i + R_o(1-\Delta_1)^2} \quad (34.34)$$

$$\frac{\tilde{v}_o(s)}{\tilde{\delta}(s)} = \frac{V_{DC}\left(R_o - sL_i\Delta_1/(1-\Delta_1)^2\right)}{s^2L_iC_oR_o + sL_i + R_o(1-\Delta_1)^2} \quad (34.35)$$

These transfer functions enable the choice and feedback-loop design of the compensation network. Note the positive zero in $\tilde{v}_o(s)/\tilde{\delta}(s)$, pointing out a

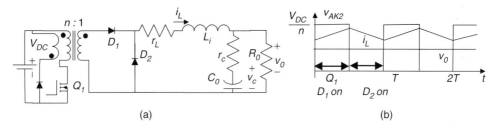

FIGURE 34.4 (a) Basic circuit of the forward dc/dc converter and (b) circuit main waveforms.

nonminimum-phase system. These equations could also be obtained using the small-signal equivalent circuit derived from Eq. (34.28), or from the linearized model of the switching cell Fig. 34.3b, substituting the current source $\delta_1 \tilde{i}_L$ by the current sources $\Delta_1 \tilde{i}_L$ and $\tilde{\delta} I_L$ in parallel, and the voltage source $\delta_1 \tilde{v}_s$ by the voltage sources $\Delta_1 (\tilde{v}_{DC} + \tilde{v}_o)$ and $\tilde{\delta}(V_{DC} + V_o)$ in series.

EXAMPLE 34.3 Transfer functions of the forward dc/dc converter

Consider the forward (buck derived) converter of Fig. 34.4 switching at $f_s = 100$ kHz ($T = 10\,\mu$s) with $V_{DC} = 300$ V, $n = 30$, $V_o = 5$ V, $L_i = 20\,\mu$H, $r_L = 0.01\,\Omega$, $C_o = 2200\,\mu$F, $r_C = 0.005\,\Omega$, $R_o = 0.1\,\Omega$.

Assuming $\mathbf{x} = [i_L, v_C]^T$, $\delta(t) = 1$ when both Q_1, D_1 are on and D_2 is off ($0 \le t \le \delta_1 T$), $\delta(t) = 0$ when both Q_1, D_1 are off and D_2 is on ($\delta_1 T \le t \le T$), the switched state-space model of the forward converter, considering as output vector $\mathbf{y} = [i_L, v_o]^T$, is

$$\frac{di_L}{dt} = -\frac{(R_o r_C + R_o r_L + r_L r_C)}{L_i (R_o + r_C)} i_L$$
$$\quad - \frac{R_o}{L_i (R_o + r_C)} v_C + \frac{\delta(t)}{n} V_{DC}$$
$$\frac{dv_C}{dt} = \frac{R_o}{(R_o + r_C) C_o} i_L - \frac{1}{(R_o + r_C) C_o} v_C \qquad (34.36)$$
$$v_o = \frac{r_C}{1 + r_C/R_o} i_L + \frac{1}{1 + r_C/R_o} v_C$$

Making $r_{cm} = r_C/(1 + r_C/R_o)$, $R_{oc} = R_o + r_C$, $k_{oc} = R_o/R_{oc}$, $r_P = r_L + r_{cm}$ and comparing Eq. (34.36) to Eqs. (34.2) and (34.3), the following matrices can be identified:

$$\mathbf{A}_1 = \mathbf{A}_2 = \begin{bmatrix} -r_P/L_i & -k_{oc}/L_i \\ k_{oc}/C_o & -1/(R_{oc} C_o) \end{bmatrix};$$

$$\mathbf{B}_1 = [1/(nL_i), 0]^T; \quad \mathbf{B}_2 = [0, 0]^T; \quad \mathbf{u} = [V_{DC}]$$

$$\mathbf{C}_1 = \mathbf{C}_2 = \begin{bmatrix} 1 & 0 \\ r_{cm} & k_{oc} \end{bmatrix}; \quad \mathbf{D}_1 = \mathbf{D}_2 = [0, 0]^T$$

Now, applying Eq. (34.7), the exact (since $\mathbf{A}_1 = \mathbf{A}_2$) state-space averaged model (34.37, 34.38) is obtained:

$$\begin{bmatrix} \dot{\bar{i}}_L \\ \dot{\bar{v}}_C \end{bmatrix} = \begin{bmatrix} -r_P/L_i & -k_{oc}/L_i \\ k_{oc}/C_o & -1/(R_{oc} C_o) \end{bmatrix} \begin{bmatrix} \bar{i}_L \\ \bar{v}_C \end{bmatrix} + \begin{bmatrix} \frac{\delta_1}{nL_i} \\ 0 \end{bmatrix} [\bar{V}_{DC}]$$
(34.37)

$$\begin{bmatrix} \bar{i}_L \\ \bar{v}_o \end{bmatrix} = \begin{bmatrix} 1 & 0 \\ r_{cm} & k_{oc} \end{bmatrix} \begin{bmatrix} \bar{i}_L \\ \bar{v}_C \end{bmatrix} + \begin{bmatrix} 0 \\ 0 \end{bmatrix} [\bar{V}_{DC}] \qquad (34.38)$$

Since $\mathbf{A}_1 = \mathbf{A}_2$, this model is valid for $\omega_{max} < 2\pi f_s$. The converter eigenvalues $s_{f1,2}$ are:

$$s_{f1,2} = -\frac{L_i + C_o R_{oc} r_P \pm \sqrt{-4 R_{oc} L_i C_o (R_{oc} k_{oc}^2 + r_P) + (L_i + C_o R_{oc} r_P)^2}}{2 R_{oc} L_i C_o}$$
(34.39)

The equivalent circuit arising from Eqs. (34.37) and (34.38) is represented in Fig. 34.5. It could also be obtained with the concept of the switching cell equivalent circuit Fig. 34.3 of Example (34.1).

Making $\mathbf{X} = [I_L, V_C]^T$, $\mathbf{Y} = [I_L, V_o]^T$ and $\mathbf{U} = [V_{DC}]$, from Eq. (34.23) the small-signal state-space averaged model is

$$\begin{bmatrix} \dot{\tilde{i}}_L \\ \dot{\tilde{v}}_C \end{bmatrix} = \begin{bmatrix} -r_P/L_i & -k_{oc}/L_i \\ k_{oc}/C_o & -1/(R_{oc} C_o) \end{bmatrix} \begin{bmatrix} \tilde{i}_L \\ \tilde{v}_C \end{bmatrix}$$
$$\quad + \begin{bmatrix} \Delta_1/nL_i \\ 0 \end{bmatrix} [\tilde{v}_{DC}] + \begin{bmatrix} V_{DC}/nL_i \\ 0 \end{bmatrix} [\tilde{\delta}] \quad (34.40)$$

$$\begin{bmatrix} \tilde{i}_L \\ \tilde{v}_o \end{bmatrix} = \begin{bmatrix} 1 & 0 \\ r_{cm} & k_{oc} \end{bmatrix} \begin{bmatrix} \tilde{i}_L \\ \tilde{v}_C \end{bmatrix} + \begin{bmatrix} 0 \\ 0 \end{bmatrix} [\tilde{v}_{DC}] \quad (34.41)$$

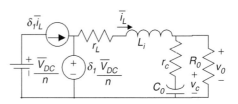

FIGURE 34.5 Equivalent circuit of the averaged state-space model of the forward converter.

From Eq. (34.25), the input **U** to output **Y** steady-state relations are:

$$\frac{I_L}{V_{DC}} = \frac{\Delta_1}{n\left(k_{oc}^2 R_{oc} + r_P\right)} \quad (34.42)$$

$$\frac{V_o}{V_{DC}} = \frac{\Delta_1 \left(k_{oc}^2 R_{oc} + r_{cm}\right)}{n\left(k_{oc}^2 R_{oc} + r_P\right)} \quad (34.43)$$

Making $r_C = 0$, $r_L = 0$ and $n = 1$, the former relations give the well-known dc transfer relationships of the buck dc/dc converter. Relations (34.42, 34.43) allow the open-loop and feedforward control of the converter, as discussed in Example 34.2, provided that all the modeled parameters are time-invariant and accurate enough.

From Eq. (34.27), the line to output transfer functions are derived:

$$\frac{\tilde{i}_L(s)}{\tilde{v}_{DC}(s)} = \frac{(\Delta_1/n)(1 + sC_o R_{oc})}{s^2 L_i C_o R_{oc} + s(L_i + C_o R_{oc} r_P) + k_{oc}^2 R_{oc} + r_P} \quad (34.44)$$

$$\frac{\tilde{v}_o(s)}{\tilde{v}_{DC}(s)} = \frac{(\Delta_1/n)\left(k_{oc}^2 R_{oc} + r_{cm} + sC_o R_{oc} r_{cm}\right)}{s^2 L_i C_o R_{oc} + s(L_i + C_o R_{oc} r_P) + k_{oc}^2 R_{oc} + r_P} \quad (34.45)$$

Using Eq. (34.26), the small-signal duty-cycle $\tilde{\delta}$ to output $\tilde{y}$ transfer functions are:

$$\frac{\tilde{i}_L(s)}{\delta(s)} = \frac{(V_{DC}/n)(1 + sC_o R_{oc})}{s^2 L_i C_o R_{oc} + s(L_i + C_o R_{oc} r_P) + k_{oc}^2 R_{oc} + r_P} \quad (34.46)$$

$$\frac{\tilde{v}_o(s)}{\delta(s)} = \frac{(V_{DC}/n)\left(k_{oc}^2 R_{oc} + r_{cm} + sC_o R_{oc} r_{cm}\right)}{s^2 L_i C_o R_{oc} + s(L_i + C_o R_{oc} r_P) + k_{oc}^2 R_{oc} + r_P} \quad (34.47)$$

The real zero of Eq. (34.47) is due to r_C, the equivalent series resistance (ESR) of the output capacitor. A similar zero would occur in the buck–boost converter (Example 34.2), if the ESR of the output capacitor had been included in the modeling.

34.2.4 Pulse Width Modulator Transfer Functions

In what is often referred to as the pulse width modulation (PWM) voltage mode control, the output voltage $u_c(t)$ of the error (between desired and actual output) amplifier plus regulator, processed if needed, is compared to a repetitive or carrier waveform $r(t)$, to obtain the switching variable $\delta(t)$ (Fig. 34.6a). This function controls the power switch, turning it on at the beginning of the period and turning it off when the ramp exceeds the $u_c(t)$ voltage. In Fig. 34.6b the opposite occurs (turn-off at the end of the period, turn-on when the $u_c(t)$ voltage exceeds the ramp).

Considering $r(t)$ as represented in Fig. 34.6a ($r(t) = u_{cmax} t/T$), δ_k is obtained equating $r(t) = u_c$ giving $\delta_k = u_c(t)/u_{cmax}$ or $\delta_k/u_c(t) = G_M$ ($G_M = 1/u_{cmax}$). In Fig. 34.6b, the switching-on angle α_k is obtained from $r(t) = u_{cmax} - 2u_{cmax}\omega t/\pi$, $u_c(t) = u_{cmax} - 2u_{cmax}\alpha_k/\pi$, giving $\alpha_k = (\pi/2) \times (1 - u_c/u_{cmax})$ and $G_M = \partial \alpha_k/\partial u_c = -\pi/(2u_{cmax})$.

Since, after turn-off or turn-on, any control action variation of $u_c(t)$ will only affect the converter duty cycle in the next period, a time delay is introduced in the control loop. For simplicity, with small-signal perturbations around the operating point, this delay is assumed almost constant and equal to its mean value ($T/2$). Then, the transfer function of the PWM

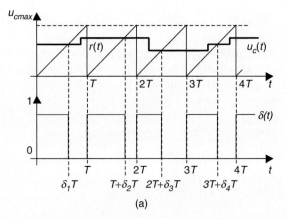

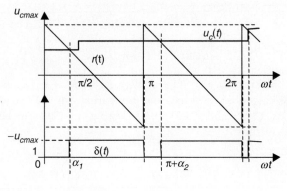

FIGURE 34.6 Waveforms of pulse width modulators showing the variable time delays of the modulator response: (a) $r(t) = u_{cmax} t/T$ and (b) $r(t) = u_{cmax} - 2u_{cmax}\omega t/\pi$.

modulator is

$$\frac{\tilde{\delta}(s)}{\tilde{u}_c(s)} = G_M e^{-sT/2} = \frac{G_M}{e^{s(T/2)}}$$

$$= \frac{G_M}{1 + s\frac{T}{2} + \frac{s^2}{2!}\left(\frac{T}{2}\right)^2 + \cdots + \frac{s^j}{j!}\left(\frac{T}{2}\right)^2 + \cdots} \approx \frac{G_M}{1 + s\frac{T}{2}} \quad (34.48)$$

The final approximation of Eq. (34.48), valid for $\omega T/2 < \sqrt{2}/2$, [7] suggests that the PWM modulator can be considered as an amplifier with gain G_M and a dominant pole. Notice that this pole occurs at a frequency doubling the switching frequency, and most state-space averaged models are valid only for frequencies below one-tenth of the switching frequency. Therefore, in most situations this modulator pole can be neglected, being simply $\delta(s) = G_M u_c(s)$, as the dominant pole of Eq. (34.48) stays at least one decade to the left of the dominant poles of the converter.

34.2.5 Linear Feedback Design Ensuring Stability

In the application of classical linear feedback control to switching converters, Bode plots and root locus are, usually, suitable methods to assess system performance and stability. General rules for the design of the compensated open-loop transfer function are as follows:

(i) The low-frequency gain should be high enough to minimize output steady-state errors;
(ii) The frequency of 0 dB gain (unity gain), ω_{0dB}, should be placed close to the maximum allowed by the modeling approximations ($\lambda_{max}T \ll 1$), to allow fast response to transients. In practice, this frequency should be almost an order of magnitude lower than the switching frequency;
(iii) To ensure stability, the phase margin, defined as the additional phase shift needed to render the system unstable without gain changes (or the difference between the open-loop system phase at ω_{0dB} and $-180°$), must be positive and in general greater than $30°$ ($45°-70°$ is desirable). In the root locus, no poles should enter the right-half of the complex plane;
(iv) To increase stability, the gain should be less than -30 dB at the frequency where the phase reaches $-180°$ (gain margin greater than 30 dB).

Transient behavior and stability margins are related: the obtained damping factor is generally 0.01 times the phase margin (in degrees), and overshoot (in percent) is given approximately by $75°$ minus the phase margin. The product of the rise time (in seconds) and the closed-loop bandwidth (in rad/s) is close to 2.8.

To guarantee gain and phase margins, the following series compensation transfer functions (usually implemented with operational amplifiers) are often used [8]:

34.2.5.1 Types of Compensation

Lag or lead compensation

Lag compensation should be used in converters with good stability margin but poor steady-state accuracy. If the frequencies $1/T_p$ and $1/T_z$ of Eq. (34.49) with $1/T_p < 1/T_z$ are chosen well below the unity gain frequency, lag–lead compensation lowers the loop gain at high frequency but maintains the phase unchanged for $\omega \gg 1/T_z$. Then, the dc gain can be increased to reduce the steady-state error without significantly decreasing the phase margin.

$$C_{LL}(s) = k_{LL}\frac{1 + sT_z}{1 + sT_p} = k_{LL}\frac{T_z}{T_p}\frac{s + 1/T_z}{s + 1/T_p} \quad (34.49)$$

Lead compensation can be used in converters with good steady-state accuracy but poor stability margin. If the frequencies $1/T_p$ and $1/T_z$ of Eq. (34.49) with $1/T_p > 1/T_z$ are chosen below the unity gain frequency, lead–lag compensation increases the phase margin without significantly affecting the steady-state error. The T_p and T_z values are chosen to increase the phase margin, fastening the transient response and increasing the bandwidth.

Proportional–Integral compensation

Proportional–integral (PI) compensators (34.50) are used to guarantee null steady-state error with acceptable rise times. The PI compensators are a particular case of lag–lead compensators, therefore suitable for converters with good stability margin but poor steady-state accuracy.

$$C_{PI}(s) = \frac{1 + sT_z}{sT_p} = \frac{T_z}{T_p} + \frac{1}{sT_p} = K_p + \frac{K_i}{s} = K_p\left(1 + \frac{K_i}{K_p s}\right)$$

$$= K_p\left(1 + \frac{1}{sT_z}\right) = \frac{1 + sT_z}{sT_z/K_p} \quad (34.50)$$

Proportional–Integral plus high-frequency pole compensation

This integral plus zero-pole compensation (34.51) combines the advantages of a PI with lead or lag compensation. It can be used in converters with good stability margin but poor steady-state accuracy. If the frequencies $1/T_M$ and $1/T_z$ ($1/T_z < 1/T_M$) are carefully chosen, compensation lowers the loop gain at high frequency, while only slightly lowering the phase to achieve the desired phase margin.

$$C_{ILD}(s) = \frac{1 + sT_z}{sT_p(1 + sT_M)} = \frac{T_z}{T_p T_M}\frac{s + 1/T_z}{s(s + 1/T_M)}$$

$$= W_{cp}\frac{s + \omega_z}{s(s + \omega_M)} \quad (34.51)$$

Proportional–Integral derivative (PID), plus high-frequency poles

The PID notch filter type (34.52) scheme is used in converters with two lightly damped complex poles, to increase the response speed, while ensuring zero steady-state error. In most switching converters, the two complex zeros are selected to have a damping factor greater than the converter complex poles and slightly smaller oscillating frequency. The high-frequency pole is placed to achieve the needed phase margin [9]. The design is correct if the complex pole loci, heading to the complex zeros in the system root locus, never enter the right half-plane.

$$C_{PIDnf}(s) = T_{cp} \frac{s^2 + 2\xi_{cp}\omega_{0cp}s + \omega_{0cp}^2}{s(1+s/\omega_{p1})}$$

$$= \frac{T_{cp}s}{1+s/\omega_{p1}} + \frac{2T_{cp}\xi_{cp}\omega_{0cp}}{1+s/\omega_{p1}} + \frac{T_{cp}\omega_{0cp}^2}{s(1+s/\omega_{p1})}$$

$$= \frac{T_{cp}s}{1+s/\omega_{p1}} + \frac{T_{cp}\omega_{0pc}^2(1+2s\xi_{cp}/\omega_{0cp})}{s(1+s/\omega_{p1})} \quad (34.52)$$

For systems with a high-frequency zero placed at least one decade above the two lightly damped complex poles, the compensator (34.53), with $\omega_{z1} \approx \omega_{z2} < \omega_p$, can be used. Usually, the two real zeros present frequencies slightly lower than the frequency of the converter complex poles. The two high-frequency poles are placed to obtain the desired phase margin [9]. The obtained overall performance will often be inferior to that of the PID type notch filter.

$$C_{PID}(s) = W_{cp} \frac{(1+s/\omega_{z1})(1+s/\omega_{z2})}{s(1+s/\omega_p)^2} \quad (34.53)$$

34.2.5.2 Compensator Selection and Design

The procedure to select the compensator and to design its parameters can be outlined as follows:

1. Compensator selection: In general, since V_{DC} perturbations exist, null steady-state error guarantee is needed. High-frequency poles are usually necessary, if the transfer function shows a −6 dB/octave roll-off due to high frequency left plane zeros. Therefore, in general, two types of compensation schemes with integral action (34.51 or 34.50), and (34.52 or 34.53) can be tried. Compensator (34.52) is usually convenient for systems with lightly damped complex poles;
2. Unity gain frequency ω_{0dB} choice:
   - If the selected compensator has no complex zeros, it is better to be conservative, choosing ω_{0dB} well below the frequency of the lightly damped poles of the converter (or the frequency of the right half plane zeros if lower). However, because of the resonant peak of most converter transfer functions, the phase margin can be obtained at a frequency near the resonance. If the phase margin is not enough, the compensator gain must be lowered;
   - If the selected compensator has complex zeros, ω_{0dB} can be chosen slightly above the frequency of the lightly damped poles;
3. Desired phase margin (ϕ_M) specification $\phi_M \geq 30°$ (preferably between 45° and 70°);
4. Compensator zero-pole placement to achieve the desired phase margin:
   - With the integral plus zero-pole compensation type (34.51), the compensator phase ϕ_{cp}, at the maximum frequency of unity gain (often ω_{0dB}), equals the phase margin (ϕ_M) minus 180° and minus the converter phase ϕ_{cv}, ($\phi_{cp} = \phi_M - 180° - \phi_{cv}$). The zero-pole position can be obtained calculating the factor $f_{ct} = tg(\pi/2 + \phi_{cp}/2)$ being $\omega_z = \omega_{0dB}/f_{ct}$ and $\omega_M = \omega_{0dB}f_{ct}$.
   - With the PID notch filter type (34.52) controller, the two complex zeros are placed to have a damping factor equal to two times the damping of the converter complex poles, and oscillating frequency ω_{0cp} 30% smaller. The high-frequency pole ω_{p1} is placed to achieve the needed phase margin ($\omega_{p1} \approx (\omega_{0cp} \cdot \omega_{0dB})^{1/2} f_{ct}^2$ with $f_{ct} = tg(\pi/2 + \phi_{cp}/2)$ and $\phi_{cp} = \phi_M - 180° - \phi_{cv}$ [5]).
5. Compensator gain calculation (the product of the converter and compensator gains at the ω_{0dB} frequency must be one).
6. Stability margins verification using Bode plots and root locus.
7. Results evaluation. Restarting the compensator selection and design, if the attained results are still not good enough.

34.2.6 Examples: Buck–Boost DC/DC Converter, Forward DC/DC Converter, 12 Pulse Rectifiers, Buck–Boost DC/DC Converter in the Discontinuous Mode (Voltage and Current Mode), Three-phase PWM Inverters

EXAMPLE 34.4 Feedback design for the buck–boost dc/dc converter

Consider the converter output voltage v_o (Fig. 34.1) to be the controlled output. From Example 34.2 and Eqs. (34.33) and (34.35), the block diagram of Fig. 34.7 is obtained. The modulator transfer function is considered a pure gain ($G_M = 0.1$). The magnitude and phase of the

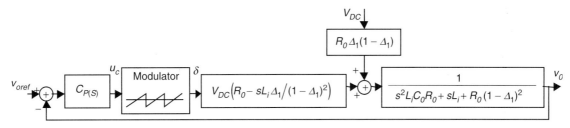

FIGURE 34.7 Block diagram of the linearized model of the closed loop buck–boost converter.

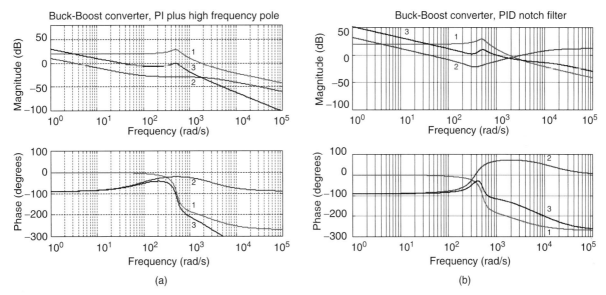

FIGURE 34.8 Bode plots for the buck–boost converter. Trace 1 – switching converter magnitude and phase; trace 2 – compensator magnitude and phase; trace 3 – resulting magnitude and phase of the compensated converter: (a) PI plus high-frequency pole compensation with 60° phase margin, $\omega_{0dB} = 500$ rad/s and (b) PID notch filter compensation with 65° phase margin, $\omega_{0dB} = 1000$ rad/s.

open-loop transfer function v_o/u_c (Fig. 34.8a trace 1), shows a resonant peak due to the two lightly damped complex poles and the associated -12 dB/octave roll-off. The right half-plane zero changes the roll-off to -6 dB/octave and adds $-90°$ to the converter phase (nonminimum-phase converter).

Compensator selection. As V_{DC} perturbations exist null steady-state error guarantee is needed. High-frequency poles are needed given the -6 dB/octave final slope of the transfer function. Therefore, two compensation schemes (34.51 and 34.52) with integral action are tried here. The buck–boost converter controlled with integral plus zero-pole compensation presents, in closed-loop, two complex poles closer to the imaginary axes than in open-loop. These poles should not dominate the converter dynamics. Instead, the real pole resulting from the open-loop pole placed at the origin should be almost the dominant one, thus slightly lowering the calculated compensator gain. If the ω_{0dB} frequency is chosen too low, the integral plus zero-pole compensation turns into a pure integral compensator ($\omega_z = \omega_M = \omega_{0dB}$).

However, the obtained gains are too low, leading to very slow transient responses.

Results showing the transient responses to v_{oref} and V_{DC} step changes, using the selected compensators and converter Bode plots (Fig. 34.8), are shown (Fig. 34.9). The compensated real converter transient behavior occurs in the buck and in the boost regions. Notice the nonminimum-phase behavior of the converter (mainly in Fig. 34.9b), the superior performance of the PID notch filter compensator and the unacceptable behavior of the PI with high-frequency pole. Care should be taken with load changes, when using this compensator, since instability can easily occur.

The compensator critical values, obtained with the root-locus studies, are $W_{cpcrit} = 700$ s^{-1} for the integral plus zero-pole compensator, $T_{cpcrit} = 0.0012$ s for the PID notch filter, and $W_{Icpcrit} = 18$ s^{-1} for the integral compensation derived from the integral plus zero-pole compensator ($\omega_z = \omega_M$). This confirms the Bode-plot design and allows stability estimation with changing loads and power supply.

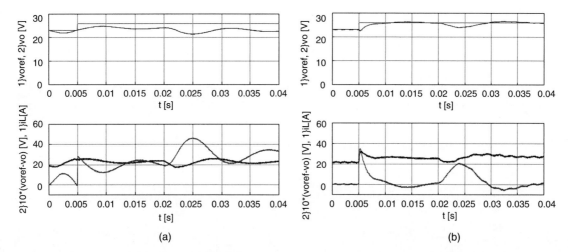

FIGURE 34.9 Transient responses of the compensated buck–boost converter. At $t = 0.005$ s, v_{oref} step from 23 to 26 V. At $t = 0.02$ s, V_{DC} step from 26 to 23 V. Top graphs: step reference v_{oref} and output voltage v_o. Bottom graphs: trace starting at 20 is i_L current; trace starting at zero is $10 \times (v_{oref} - v_o)$: (a) PI plus high-frequency pole compensation with 60° phase margin and $\omega_{0dB} = 500$ rad/s and (b) PID notch filter compensation with 64° phase margin and $\omega_{0dB} = 1000$ rad/s.

EXAMPLE 34.5 Feedback design for the forward dc/dc converter

Consider the output voltage v_o of the forward converter (Fig. 34.4a) to be the controlled output. From Example 34.3 and Eqs. (34.45) and (34.47), the block diagram of Fig. 34.10 is obtained. As in Example 34.4, the modulator transfer function is considered as a pure gain ($G_M = 0.1$). The magnitude and phase of the open-loop transfer function v_o/u_c (Fig. 34.11a, trace 1), shows an open-loop stable system. Since integral action is needed to have some disturbance rejection of the voltage source V_{DC}, the compensation schemes used in Example 34.4, obtained using the same procedure (Fig. 34.11), were also tested.

Results, showing the transient responses to v_{oref} and V_{DC} step changes, are shown (Fig. 34.12). Both compensators (34.51) and (34.52) are easier to design than the ones for the buck–boost converter, and both have acceptable performances. Moreover, the PID notch filter presents a much faster response.

Alternatively, a PID feedback controller such as Eq. (34.53) can be easily hand-adjusted, starting with the proportional, integral, and derivative gains all set to zero. In the first step, the proportional gain is increased until the output presents an oscillatory response with nearly 50% overshoot. Next, the derivative gain is slowly increased until the overshoot is eliminated. Finally, the integral gain is increased to eliminate the steady-state error as quickly as possible.

EXAMPLE 34.6 Feedback design for phase controlled rectifiers in the continuous mode

Phase controlled, p pulse ($p > 1$), thyristor rectifiers (Fig. 34.13a), operating in the continuous mode, present an output voltage with p identical segments within the mains period T. Given this cyclic waveform, the **A**, **B**, **C**, and **D** matrices for all these p intervals can be written with the same form, inspite of the topological variation. Hence, the state-space averaged model is obtained simply by averaging all the variables within the period T/p. Assuming small variations, the mean value of the rectifier output voltage U_{DC} can be written [10]:

$$U_{DC} = U_p \frac{p}{\pi} \sin\left(\frac{\pi}{p}\right) \cos\alpha \qquad (34.54)$$

where α is the triggering angle of the thyristors, and U_p the maximum peak value of the rectifier output voltage, determined by the rectifier topology and the ac supply voltage. The α value can be obtained ($\alpha = (\pi/2) \times (1 - u_c/u_{cmax})$) using the modulator of Fig. 34.6b, where $\omega = 2\pi/T$ is the mains frequency. From Eq. (34.54), the incremental gain K_R of the modulator plus rectifier yields:

$$K_R = \frac{\partial U_{DC}}{\partial u_c}$$

$$= U_p \frac{p}{2u_{cmax}} \sin\left(\frac{\pi}{p}\right) \cos\left(\frac{\pi u_c}{2u_{cmax}}\right) \qquad (34.55)$$

For a given rectifier, this gain depends on u_c, and should be calculated for a certain quiescent point. However, for feedback design purposes, keeping in mind that the rectifier could be required to be stable in all operating

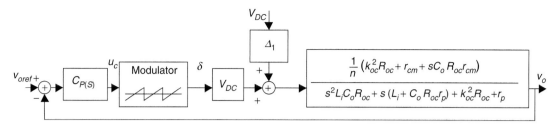

FIGURE 34.10 Block diagram of the linearized model of the closed-loop controlled forward converter.

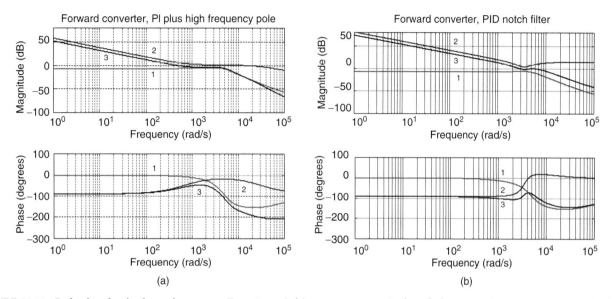

FIGURE 34.11 Bode plots for the forward converter. Trace 1 – switching converter magnitude and phase; trace 2 – compensator magnitude and phase; trace 3 – resulting magnitude and phase of the compensated converter: (a) PI plus high-frequency pole compensation with 115° phase margin, $\omega_{0dB} = 500$ rad/s and (b) PID notch filter compensation with 85° phase margin, $\omega_{0dB} = 6000$ rad/s.

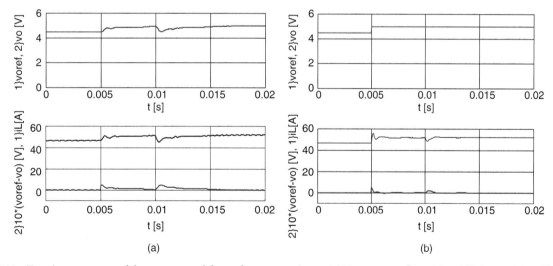

FIGURE 34.12 Transient responses of the compensated forward converter. At $t = 0.005$ s, v_{oref} step from 4.5 to 5 V. At $t = 0.01$ s, V_{DC} step from 300 to 260 V. Top graphs: step reference v_{oref} and output voltage v_o. Bottom graphs: top traces i_L current; bottom traces $10\times (v_{oref} - v_o)$; (a) PI plus high-frequency pole compensation with 115° phase margin and $\omega_{0dB} = 500$ rad/s and (b) PID notch filter compensation with 85° phase margin and $\omega_{0dB} = 6000$ rad/s.

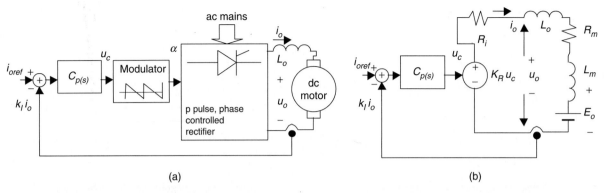

FIGURE 34.13 (a) Block diagram of a p pulse phase controlled rectifier feeding a separately excited dc motor and (b) equivalent averaged circuit.

points, the maximum value of K_R, denoted K_{RM}, can be used:

$$K_{RM} = U_p \frac{p}{2u_{cmax}} \sin\left(\frac{\pi}{p}\right) \quad (34.56)$$

The operation of the modulator, coupled to the rectifier thyristors, introduces a non-neglectable time delay, with mean value $T/2p$. Therefore, from Eq. (34.48) the modulator-rectifier transfer function $G_R(s)$ is

$$G_R(s) = \frac{U_{DC}(s)}{u_c(s)}$$

$$= K_{RM}\, e^{-s(T/2p)} \approx \frac{K_{RM}}{1 + s\,(T/2p)} \quad (34.57)$$

Considering zero U_p perturbations, the rectifier equivalent averaged circuit (Fig. 34.13b) includes the loss-free rectifier output resistance R_i, due to the overlap in the commutation phenomenon caused by the mains inductance. Usually, $R_i \approx p\omega l/\pi$ where l is the equivalent inductance of the lines paralleled during the overlap, half of the line inductance for most rectifiers, except for single-phase bridge rectifiers where l is the line inductance. Here, L_o is the smoothing reactor and R_m, L_m, and E_o are respectively the armature internal resistance, inductance, and back electromotive force of a separately excited dc motor (typical load). Assuming the mean value of the output current as the controlled output, making $L_t = L_o + L_m$, $R_t = R_i + R_m$, $T_t = L_t/R_t$ and applying Laplace transforms to the differential equation obtained from the circuit of Fig. 34.13b, the output current transfer function is

$$\frac{i_o(s)}{U_{DC}(s) - E_o(s)} = \frac{1}{R_t(1 + sT_t)} \quad (34.58)$$

The rectifier and load are now represented by a perturbed (E_o) second-order system (Fig. 34.14). To achieve zero steady-state error, which ensures steady-state insensitivity to the perturbations, and to obtain closed-loop second-order dynamics, a PI controller (34.50) was selected for $C_p(s)$ (Fig. 34.14). Canceling the load pole ($-1/T_t$) with the PI zero ($-1/T_z$) yields:

$$T_z = L_t/R_t \quad (34.59)$$

The rectifier closed-loop transfer function $i_o(s)/i_{oref}(s)$, with zero E_o perturbations, is

$$\frac{i_o(s)}{i_{oref}(s)} = \frac{2pK_{RM}k_I/(R_t T_p T)}{s^2 + (2p/T)s + 2pK_{RM}k_I/(R_t T_p T)} \quad (34.60)$$

The final value theorem enables the verification of the zero steady-state error. Comparing the denominator of

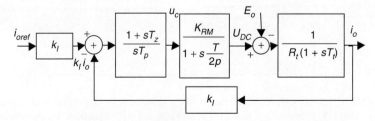

FIGURE 34.14 Block diagram of a PI controlled p pulse rectifier.

Eq. (34.60) to the second-order polynomial $s^2 + 2\zeta\omega_n s + \omega_n^2$ yields:

$$\omega_n^2 = 2pK_{RM}k_I/(R_t T_p T)$$
$$4\zeta^2\omega_n^2 = (2p/T)^2 \quad (34.61)$$

Since only one degree of freedom is available (T_p), the damping factor ζ is imposed. Usually $\zeta = \sqrt{2}/2$ is selected, since it often gives the best compromise between response speed and overshoot. Therefore, from Eq. (34.61), Eq. (34.62) arises:

$$T_p = 4\zeta^2 K_{RM} k_I T/(2pR_t) = K_{RM}k_I T/(pR_t) \quad (34.62)$$

Note that both T_z (34.59) and T_p (34.62) are dependent upon circuit parameters. They will have the correct values only for dc motors with parameters closed to the nominal load value. Using Eq. (34.62) in Eq. (34.60) yields Eq. (34.63), the second-order closed-loop transfer function of the rectifier, showing that, with loads close to the nominal value, the rectifier dynamics depend only on the mean delay time $T/2p$.

$$\frac{i_o(s)}{i_{oref}(s)} = \frac{1}{2(T/2p)^2 s^2 + sT/p + 1} \quad (34.63)$$

From Eq. (34.63) $\omega_n = \sqrt{2}p/T$ results, which is the maximum frequency allowed by $\omega T/2p < \sqrt{2}/2$, the validity limit of Eq. (34.48). This implies that $\zeta \geq \sqrt{2}/2$, which confirms the preceding choice. For $U_p = 300$ V, $p = 6$, $T = 20$ ms, $l = 0.8$ mH, $R_m = 0.5$ Ω, $L_t = 50$ mH, $E_o = -150$ V, $u_{cmax} = 10$ V, $k_I = 0.1$, Fig. 34.15a shows the rectifier output voltage u_{oN} ($u_{oN} = u_o/U_p$) and the step response of the output current i_{oN} ($i_{oN} = i_o/40$) in accordance with Eq. (34.63). Notice that the rectifier is operating in the inverter mode. Fig. 34.15b shows the effect, in the i_o current, of a 50% reduction in the E_o value. The output current is initially disturbed but the error vanishes rapidly with time.

This modeling and compensator design are valid for small perturbations. For large perturbations either the rectifier will saturate or the firing angles will originate large current overshoots. For large signals, antiwindup schemes (Fig. 34.16a) or error ramp limiters (or soft starters) and limiters of the PI integral component (Fig. 34.16b) must be used. These solutions will also work with other switching converters.

To use this rectifier current controller as the inner control loop of a cascaded controller for the dc motor speed regulation, a useful first-order approximation of Eq. (34.63) is $i_o(s)/i_{oref}(s) \approx 1/(sT/p + 1)$.

Although allowing a straightforward compensator selection and precise calculation of its parameters, the rectifier modeling presented here is not suited for stability studies. The rectifier root locus will contain two complex conjugate poles in branches parallel to the imaginary axis. To study the current controller stability, at least the second-order term of Eq. (34.48) in Eq. (34.57) is needed. Alternative ways include the first-order Padé approximation of $e^{-sT/2p}$, $e^{-sT/2p} \approx (1 - sT/4p)/(1 + sT/4p)$, or the second-order approximation, $e^{-sT/2p} \approx (1 - sT/4p + (sT/2p)^2/12)/(1 + sT/4p + (sT/2p)^2/12)$. These approaches introduce zeros in the right half-plane (nonminimum-phase systems), and/or extra poles, giving more realistic results. Taking a first-order approximation and root-locus techniques, it is found that the rectifier is stable for $T_p > K_{RM}k_I T/(4pR_t)$ ($\zeta > 0.25$). Another approach uses the conditions of magnitude and angle of the delay function $e^{-sT/2p}$ to obtain the

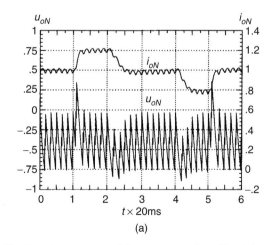
(a)

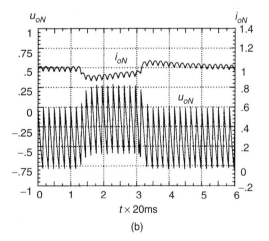
(b)

FIGURE 34.15 Transient response of the compensated rectifier: (a) step response of the controlled current i_o and (b) the current i_o response to a step chance to 50% of the E_o nominal value during 1.5 T.

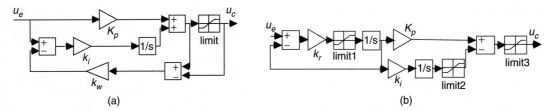

FIGURE 34.16 (a) PI implementation with antiwindup (usually $1/K_p \leq k_w \leq K_i/K_p$) to deal with rectifier saturation and (b) PI with ramp limiter/soft starter ($k_r \gg K_p$) and integral component limiter to deal with large perturbations.

system root locus. Also, the switching converter can be considered as a sampled data system, at frequency p/T, and Z transform can be used to determine the critical gain and first frequency of instability $p/(2T)$, usually half the switching frequency of the rectifier.

EXAMPLE 34.7 Buck–Boost dc/dc converter feedback design in the discontinuous mode

The methodologies just described do not apply to switching converters operating in the discontinuous mode. However, the derived equivalent averaged circuit approach can be used, calculating the mean value of the discontinuous current supplied to the load, to obtain the equivalent circuit. Consider the buck–boost converter of Example 34.1 (Fig. 34.1) with the new values $L_i = 40\,\mu\text{H}$, $C_o = 1000\,\mu\text{F}$, $R_o = 15\,\Omega$. The mean value of the current i_{Lo}, supplied to the output capacitor and resistor of the circuit operating in the discontinuous mode, can be calculated noting that, if the input V_{DC} and output v_o voltages are essentially constant (low ripple), the inductor current rises linearly from zero, peaking at $I_P = (V_{DC}/L_i)\delta_1 T$ (Fig. 34.17a). As the mean value of i_{Lo}, supposed linear, is $I_{Lo} = (I_P\delta_2 T)/(2T)$, using the steady-state input–output relation $V_{DC}\delta_1 = V_o\delta_2$ and the above I_P value, I_{Lo} can be written:

$$I_{Lo} = \frac{\delta_1^2 V_{DC}^2 T}{2 L_i V_o} \tag{34.64}$$

This is a nonlinear relation that could be linearized around an operating point. However, switching converters in the discontinuous mode seldom operate just around an operating point. Therefore, using a quadratic modulator (Fig. 34.18), obtained integrating the ramp $r(t)$ (Fig. 34.6a) and comparing the quadratic curve to the term $u_{cPI}v_o/V_{DC}^2$ (which is easily implemented using the Unitrode UC3854 integrated circuit), the duty cycle δ_1 is $\delta_1 = \sqrt{u_{cPI}V_o/(u_{cmax}V_{DC}^2)}$, and a constant incremental factor K_{CV} can be obtained:

$$K_{CV} = \frac{\partial I_{Lo}}{\partial u_{cPI}} = \frac{T}{2 u_{cmax} L_i} \tag{34.65}$$

Considering zero-voltage perturbations and neglecting the modulator delay, the equivalent averaged circuit (Fig. 34.17b) can be used to derive the output voltage to input current transfer function $v_o(s)/i_{Lo}(s) = R_o/(sC_o R_o + 1)$. Using a PI controller (34.50), the closed-loop transfer function is

$$\frac{v_o(s)}{v_{oref}(s)} = \frac{K_{CV}(1+sT_z)/C_o T_p}{s^2 + s(T_p + T_z K_{CV} k_v R_o)/C_o R_o T_p + K_{CV} k_v/C_o T_p} \tag{34.66}$$

Since two degrees of freedom exist, the PI constants are derived imposing ζ and ω_n for the second-order denominator of Eq. (34.66), usually $\zeta \geq \sqrt{2}/2$ and $\omega_n \leq 2\pi f_s/10$. Therefore:

$$\begin{aligned} T_p &= K_{CV} k_v/(\omega_n^2 C_o) \\ T_z &= T_p (2\zeta\omega_n C_o R_o - 1)/(K_{CV} k_v R_o) \end{aligned} \tag{34.67}$$

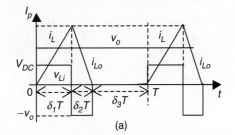 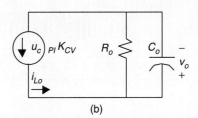

FIGURE 34.17 (a) Waveforms of the buck–boost converter in the discontinuous mode and (b) equivalent averaged circuit.

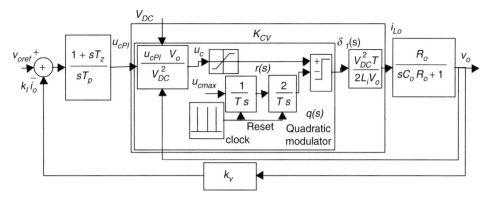

FIGURE 34.18 Block diagram of a PI controlled (feedforward linearized) buck–boost converter operating in the discontinuous mode.

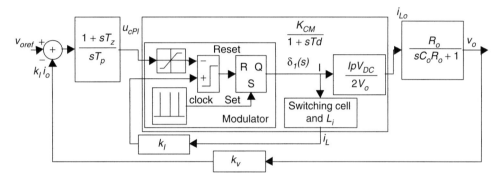

FIGURE 34.19 Block diagram of a current-mode controlled buck–boost converter operating in the discontinuous mode.

The transient behavior of this converter, with $\zeta = 1$ and $\omega_n \approx \pi f_s/10$, is shown in Fig. 34.20a. Compared to Example 34.2, the operation in the discontinuous conduction mode reduces, by 1, the order of the state-space averaged model and eliminates the zero in the right-half of the complex plane. The inductor current does not behave as a true state variable, since during the interval $\delta_3 T$ this current is zero, and this value is always the i_{Lo} current initial condition. Given the differences between these two examples, care should be taken to avoid the operation in the continuous mode of converters designed and compensated for the discontinuous mode. This can happen during turn-on or step load changes and, if not prevented, the feedback design should guarantee stability in both modes (Example 34.8, Fig. 34.19a).

EXAMPLE 34.8 Feedback design for the buck–boost dc/dc converter operating in the discontinuous mode and using current-mode control

The performances of the buck–boost converter operating in the discontinuous mode can be greatly enhanced if a *current-mode control* scheme is used, instead of the voltage mode controller designed in Example 34.7. Current-mode control in switching converters is the simplest form of state feedback. Current mode needs the measurement of the current i_L (Fig. 34.1) but greatly simplifies the modulator design (compare Fig. 34.18 to Fig. 34.19), since no modulator linearization is used. The measured value, proportional to the current i_L, is compared to the value u_{cPI} given by the output voltage controller (Fig. 34.19). The modulator switches off the power semiconductor when $k_I I_P = u_{cPI}$.

Expressed as a function of the peak i_L current I_P, I_{Lo} becomes (Example 34.7) $I_{Lo} = I_P \delta_1 V_{DC}/(2V_o)$, or considering the modulator task $I_{Lo} = u_{cPI} \delta_1 V_{DC}/(2k_I V_o)$. For small perturbations, the incremental gain is $K_{CM} = \partial I_{Lo}/\partial u_{cPI} = \delta_1 V_{DC}/(2k_I V_o)$. An I_{Lo} current delay $T_d = 1/(2f_s)$, related to the switching frequency f_s can be assumed. The current mode control transfer function $G_{CM}(s)$ is

$$G_{CM}(s) = \frac{I_{Lo}(s)}{u_{cPI}(s)} \approx \frac{K_{CM}}{1+sT_d} \approx \frac{\delta_1 V_{DC}}{2k_I V_o(1+sT_d)} \quad (34.68)$$

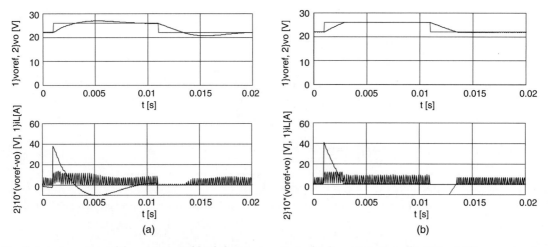

FIGURE 34.20 Transient response of the compensated buck–boost converter in the discontinuous mode. At $t = 0.001$ s v_{oref} step from 23 to 26 V. At $t = 0.011$ s, v_{oref} step from 26 to 23 V. Top graphs: step reference v_{oref} and output voltage v_o. Bottom graphs: pulses, i_L current; trace peaking at 40, $10\times (v_{oref} - v_o)$: (a) PI controlled and feedforward linearized buck–boost converter with $\zeta = 1$ and $\omega_n \approx \pi f_s/10$ and (b) Current-mode controlled buck–boost with $\zeta = 1$ and maximum value $I_{pmax} = 15$ A.

Using the approach of Example 34.6, the values for T_z and T_p are given by Eq. (34.69).

$$T_z = R_o C_o$$
$$T_p = 4\zeta^2 K_{CM} k_v R_o T_d \quad (34.69)$$

The transient behavior of this converter, with $\zeta = 1$ and maximum value for I_p, $I_{pmax} = 15$ A, is shown in Fig. 34.19b. The output voltage step response presents no overshoot, no steady-state error, and better dynamics, compared to the response (Fig. 34.19a) obtained using the quadratic modulator (Fig. 34.18). Notice that, with current mode control, the converter behaves like a reduced order system and the right half-plane zero is not present.

The current-mode control scheme can be advantageously applied to converters operating in the continuous mode, guarantying short-circuit protection, system order reduction, and better performances. However, for converters operating in the step-up (boost) regime, a stabilizing ramp with negative slope is required, to ensure stability, the stabilizing ramp will transform the signal u_{cPI} in a new signal $u_{cPI} - \text{rem}(k_{sr} t/T)$ where k_{sr} is the needed amplitude for the compensation ramp and the function rem is the remainder of the division of $k_{sr} t$ by T. In the next section, current control of switching converters will be detailed.

Closed-loop control of resonant converters can be achieved using the outlined approaches, if the resonant phases of operation last for small intervals compared to the fundamental period. Otherwise, the equivalent averaged circuit concept can often be used and linearized, now considering the resonant converter input–output relations, normally functions of the driving frequency and input or output voltages, to replace the δ_1 variable.

EXAMPLE 34.9 Output voltage control in three-phase voltage-source inverters using sinusoidal wave PWM (SWPWM) and space vector modulation (SVM)

Sinusoidal wave PWM
Voltage-source three-phase inverters (Fig. 34.21) are often used to drive squirrel cage induction motors (IM) in variable speed applications.

Considering almost ideal power semiconductors, the output voltage $u_{bk}(k \in \{1, 2, 3\})$ dynamics of the inverter is negligible as the output voltage can hardly be considered a state variable in the time scale describing the motor behavior. Therefore, the best known method to create sinusoidal output voltages uses an open-loop modulator with low-frequency sinusoidal waveforms $\sin(\omega t)$, with the amplitude defined by the modulation index m_i ($m_i \in [0, 1]$), modulating high-frequency triangular waveforms $r(t)$ (carriers), Fig. 34.22, a process similar to the one described in Section 34.2.4.

This sinusoidal wave PWM (SWPWM) modulator generates the variable γ_k, represented in Fig. 34.22 by the rectangular waveform, which describes the inverter k leg state:

$$\gamma_k = \begin{cases} 1 \rightarrow \text{when } m_i \sin(\omega t) > r(t) \\ 0 \rightarrow \text{when } m_i \sin(\omega t) < r(t) \end{cases} \quad (34.70)$$

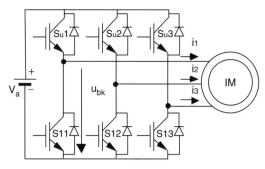

FIGURE 34.21 IGBT-based voltage-sourced three-phase inverter with induction motor.

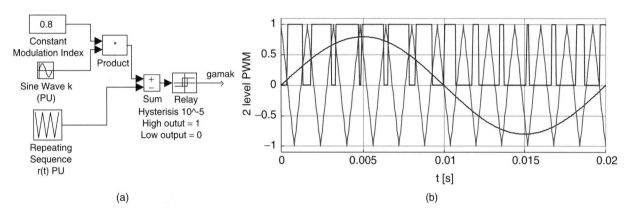

FIGURE 34.22 (a) SWPWM modulator schematic and (b) main SWPWM signals.

The turn-on and turn-off signals for the k leg inverter switches are related with the variable γ_k as follows:

$$\gamma_k = \begin{cases} 1 \rightarrow \text{then } Su_k \text{ is on and } sl_k \text{ is off} \\ 0 \rightarrow \text{then } Su_k \text{ is off and } sl_k \text{ is on} \end{cases} \quad (34.71)$$

This applies constant-frequency sinusoidally weighted PWM signals to the gates of each insulated gate bipolar transistor (IGBT). The PWM signals for all the upper IGBTs (Su_k, $k \in \{1, 2, 3\}$) must be 120° out of phase and the PWM signal for the lower IGBT Sl_k must be the complement of the Su_k signal. Since transistor turn-on times are usually shorter than turn-off times, some dead time must be included between the Su_k and Sl_k pulses to prevent internal short-circuits.

Sinusoidal PWM can be easily implemented using a microprocessor or two digital counters/timers generating the addresses for two lookup tables (one for the triangular function, another for supplying the per unit basis of the sine, whose frequency can vary). Tables can be stored in read only memories, ROM, or erasable programmable ROM, EPROM. One multiplier for the modulation index (perhaps into the digital-to-analog (D/A) converter for the sine ROM output) and one hysteresis comparator must also be included.

With SWPWM, the first harmonic maximum amplitude of the obtained line-to-line voltage is only about 86% of the inverter dc supply voltage V_a. Since it is expectable that this amplitude should be closer to V_a, different modulating voltages (for example, adding a third-order harmonic with one-fourth of the fundamental sine amplitude) can be used as long as the fundamental harmonic of the line-to-line voltage is kept sinusoidal. Another way is to leave SWPWM and consider the eight possible inverter output voltages trying to directly use then. This will lead to space vector modulation.

Space vector modulation

Space vector modulation (SVM) is based on the polar representation (Fig. 34.23) of the eight possible base output voltages of the three-phase inverter (Table 34.1, where v_α, v_β are the vector components of vector $\vec{V}_g$, $g \in \{0, 1, 2, 3, 4, 5, 6, 7\}$, obtained with Eq. (34.72). Therefore, as all the available voltages can be used, SVM does not present the voltage limitation of SWPWM.

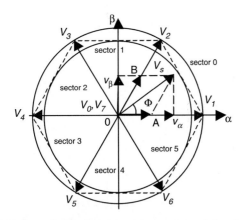

FIGURE 34.23 α, β space vector representation of the three-phase bridge inverter leg base vectors.

TABLE 34.1 The three-phase inverter with eight possible γ_k combinations, vector numbers, and respective α, β components

γ_1	γ_2	γ_3	u_{bk}	$u_{bk} - u_{bk+1}$	v_α	v_β	Vector
0	0	0	0	0	0	0	$\vec{V}_0$
1	0	0	$\gamma_k V_a$	$(\gamma_k - \gamma_{k+1}) V_a$	$\sqrt{2/3} V_a$	0	$\vec{V}_1$
1	1	0	$\gamma_k V_a$	$(\gamma_k - \gamma_{k+1}) V_a$	$V_a/\sqrt{6}$	$V_a/\sqrt{2}$	$\vec{V}_2$
0	1	0	$\gamma_k V_a$	$(\gamma_k - \gamma_{k+1}) V_a$	$-V_a/\sqrt{6}$	$V_a/\sqrt{2}$	$\vec{V}_3$
0	1	1	$\gamma_k V_a$	$(\gamma_k - \gamma_{k+1}) V_a$	$-\sqrt{2/3} V_a$	0	$\vec{V}_4$
1	1	1	V_a	0	0	0	$\vec{V}_7$
1	0	1	$\gamma_k V_a$	$(\gamma_k - \gamma_{k+1}) V_a$	$V_a/\sqrt{6}$	$-V_a/\sqrt{2}$	$\vec{V}_6$
0	0	1	$\gamma_k V_a$	$(\gamma_k - \gamma_{k+1}) V_a$	$-V_a/\sqrt{6}$	$-V_a/\sqrt{2}$	$\vec{V}_5$

Furthermore, being a vector technique, SVM fits nicely with the vector control methods often used in IM drives.

$$\begin{bmatrix} v_\alpha \\ v_\beta \end{bmatrix} = \sqrt{\frac{2}{3}} \begin{bmatrix} 1 & -1/2 & -1/2 \\ 0 & \sqrt{3}/2 & -\sqrt{3}/2 \end{bmatrix} \begin{bmatrix} \gamma_1 \\ \gamma_2 \\ \gamma_3 \end{bmatrix} V_a \quad (34.72)$$

Consider that the vector $\vec{V}_s$ (magnitude V_s, angle Φ) must be applied to the IM. Since there is no such vector available directly, SVM uses an averaging technique to apply the two vectors, $\vec{V}_1$ and $\vec{V}_2$, closest to $\vec{V}_s$. The vector $\vec{V}_1$ will be applied during $\delta_A T_s$ while vector $\vec{V}_2$ will last $\delta_B T_s$ (where $1/T_s$ is the inverter switching frequency, δ_A and δ_B are duty cycles, $\delta_A, \delta_B \in [0, 1]$). If there is any leftover time in the PWM period T_s, then the zero vector is applied during time $\delta_0 T_s = T_s - \delta_A T_s - \delta_B T_s$. Since there are two zero vectors ($\vec{V}_0$ and $\vec{V}_7$) a symmetric PWM can be devised, which uses both $\vec{V}_0$ and $\vec{V}_7$, as shown in Fig. 34.24. Such a PWM arrangement minimizes the power semiconductor switching frequency and IM torque ripples.

The input to the SVM algorithm is the space vector $\vec{V}_s$, into the sector s_n, with magnitude V_s and angle Φ_s. This vector can be rotated to fit into sector 0 (Fig. 34.23) reducing Φ_s to the first sector, $\Phi = \Phi_s - s_n \pi/3$. For any $\vec{V}_s$ that is not exactly along one of the six nonnull inverter base vectors (Fig. 34.23), SVM must generate an approximation by applying the two adjacent vectors during an appropriate amount of time. The algorithm can be devised considering that the projections of $\vec{V}_s$, onto the two closest base vectors, are values proportional to δ_A and δ_B duty cycles. Using simple trigonometric relations in sector 0 ($0 < \Phi < \pi/3$) Fig. 34.23, and considering K_T the proportional ratio, δ_A and δ_B are, respectively, $\delta_A = K_T \overline{OA}$ and $\delta_B = K_T \overline{OA}$, yielding:

$$\delta_A = K_T \frac{2 V_s}{\sqrt{3}} \sin\left(\frac{\pi}{3} - \Phi\right)$$
$$\delta_B = K_T \frac{2 V_s}{\sqrt{3}} \sin \Phi \quad (34.73)$$

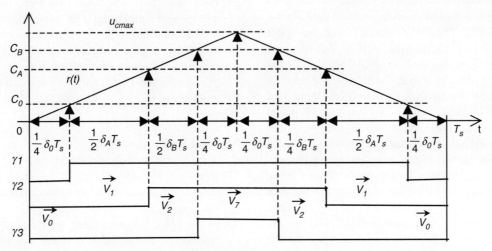

FIGURE 34.24 Symmetrical SVM.

The K_T value can be found if we notice that when $\vec{V}_s = \vec{V}_1$, $\delta_A = 1$, and $\delta_B = 0$ (or when $\vec{V}_s = \vec{V}_2$, $\delta_A = 0$, and $\delta_B = 1$). Therefore, since when $\vec{V}_s = \vec{V}_1$, $V_s = \sqrt{v_\alpha^2 + v_\beta^2} = \sqrt{2/3}V_a$, $\Phi = 0$, or when $\vec{V}_s = \vec{V}_2$, $V_s = \sqrt{2/3}V_a$, $\Phi = \pi/3$, the K_T constant is $K_T = \sqrt{3}/(\sqrt{2}V_a)$. Hence:

$$\delta_A = \frac{\sqrt{2}\,V_s}{V_a}\sin\left(\frac{\pi}{3} - \Phi\right)$$

$$\delta_B = \frac{\sqrt{2}\,V_s}{V_a}\sin\Phi \qquad (34.74)$$

$$\delta_0 = 1 - \delta_A - \delta_B$$

The obtained resulting vector $\vec{V}_s$ cannot extend beyond the hexagon of Fig. 34.23. This can be understood if the maximum magnitude V_{sm} of a vector with $\Phi = \pi/6$ is calculated. Since, for $\Phi = \pi/6$, $\delta_A = 1/2$, and $\delta_B = 1/2$ are the maximum duty cycles, from Eq. (34.74) $V_{sm} = V_a/\sqrt{2}$ is obtained. This magnitude is lower than that of the vector $\vec{V}_1$ since the ratio between these magnitudes is $\sqrt{3}/2$. To generate sinusoidal voltages, the vector $\vec{V}_s$ must be inside the inner circle of Fig. 34.23, so that it can be rotated without crossing the hexagon boundary. Vectors with tips between this circle and the hexagon are reachable, but produce nonsinusoidal line-to-line voltages.

For sector 0, (Fig. 34.23) SVM symmetric PWM switching variables (γ_1, γ_2, γ_3) and intervals (Fig. 34.24) can be obtained by comparing a triangular wave with amplitude u_{cmax}, (Fig. 34.24, where $r(t) = 2u_{cmax}t/T_s$, $t \in [0, T_s/2]$) with the following values:

$$C_0 = \frac{u_{c\max}}{2}\delta_0 = \frac{u_{c\max}}{2}(1 - \delta_A - \delta_B)$$

$$C_A = \frac{u_{c\max}}{2}\left(\frac{\delta_0}{2} + \delta_A\right) = \frac{u_{c\max}}{2}(1 + \delta_A - \delta_B)$$

$$C_B = \frac{u_{c\max}}{2}\left(\frac{\delta_0}{2} + \delta_A + \delta_B\right) = \frac{u_{c\max}}{2}(1 + \delta_A + \delta_B)$$
$$(34.75)$$

Extension of Eq. (34.75) to all six sectors can be done if the sector number s_n is considered, together with the auxiliary matrix Ξ:

$$\Xi^T = \begin{bmatrix} -1 & -1 & 1 & 1 & 1 & -1 \\ -1 & 1 & 1 & 1 & -1 & -1 \end{bmatrix} \qquad (34.76)$$

Generalization of the values C_0, C_A, and C_B, denoted C_{0sn}, C_{Asn}, and C_{Bsn} are written in Eq. (34.77), knowing that, for example, $\Xi_{((sn+4)\bmod 6+1)}$ is the Ξ matrix row with number $(s_n + 4)_{\bmod 6} + 1$.

$$C_{0sn} = \frac{u_{c\max}}{2}\left(1 + \Xi_{((Sn)\bmod 6+1)}\begin{bmatrix}\delta_A\\\delta_B\end{bmatrix}\right)$$

$$C_{Asn} = \frac{u_{c\max}}{2}\left(1 + \Xi_{((Sn+4)\bmod 6+1)}\begin{bmatrix}\delta_A\\\delta_B\end{bmatrix}\right) \qquad (34.77)$$

$$C_{Bsn} = \frac{u_{c\max}}{2}\left(1 + \Xi_{((Sn+2)\bmod 6+1)}\begin{bmatrix}\delta_A\\\delta_B\end{bmatrix}\right)$$

Therefore, γ_1, γ_2, γ_3 are:

$$\gamma_1 = \begin{cases} 0 \to \text{when } r(t) < C_{0sn} \\ 1 \to \text{when } r(t) > C_{0sn} \end{cases}$$

$$\gamma_2 = \begin{cases} 0 \to \text{when } r(t) < C_{Asn} \\ 1 \to \text{when } r(t) > C_{Asn} \end{cases} \qquad (34.78)$$

$$\gamma_3 = \begin{cases} 0 \to \text{when } r(t) < C_{Bsn} \\ 1 \to \text{when } r(t) > C_{Bsn} \end{cases}$$

Supposing that the space vector $\vec{V}_s$ is now specified in the orthogonal coordinates $\alpha, \beta (\vec{V}_\alpha, \vec{V}_\beta)$, instead of magnitude V_s and angle Φ_s, the duty cycles δ_A, δ_B can be easily calculated knowing that $v_\alpha = V_s\cos\Phi$, $v_\beta = V_s\sin\Phi$ and using Eq. (34.74):

$$\delta_A = \frac{\sqrt{2}}{2V_a}\left(\sqrt{3}v_\alpha - v_\beta\right)$$
$$(34.79)$$
$$\delta_B = \frac{\sqrt{2}}{V_a}v_\beta$$

This equation enables the use of Eqs. (34.77) and (34.78) to obtain SVM in orthogonal coordinates.

Using SVM or SWPWM, the closed-loop control of the inverter output currents (induction motor stator currents) can be performed using an approach similar to that outlined in Example 34.6 and decoupling the currents expressed in a d, q rotating frame.

34.3 Sliding-mode Control of Switching Converters

34.3.1 Introduction

All the designed controllers for switching power converters are in fact variable structure controllers, in the sense that the control action changes rapidly from one to another of, usually, two possible $\delta(t)$ values, cyclically changing the converter topology. This is accomplished by the modulator (Fig. 34.6), which creates the switching variable $\delta(t)$ imposing $\delta(t) = 1$ or $\delta(t) = 0$, to turn on or off the power semiconductors. As a consequence of this discontinuous control action, indispensable for efficiency reasons, state trajectories move back and forth around a certain average surface in the state-space, and variables present some ripple. To avoid the effects of this ripple in the modeling and to apply linear control methodologies to time-variant systems, average values of state variables and state-space averaged models or circuits were presented (Section 34.2). However, a nonlinear approach to the modeling and control problem, taking advantage of the inherent ripple and variable structure behavior of switching converters, instead of just trying to live with them, would be desirable, especially if enhanced performances could be attained.

In this approach switching converters topologies, as discrete nonlinear time-variant systems, are controlled to switch from one dynamics to another when just needed. If this switching occurs at a very high frequency (theoretically infinite), the state dynamics, described as in Eq. (34.4), can be enforced to slide along a certain prescribed state-space trajectory. The converter is said to be in sliding mode, the allowed deviations from the trajectory (the ripple) imposing the practical switching frequency.

Sliding mode control of variable structure systems, such as switching converters, is particularly interesting because of the inherent robustness [11, 12], capability of system order reduction, and appropriateness to the on/off switching of power semiconductors. The control action, being the control equivalent of the management paradigm "Just in Time" (JIT), provides timely and precise control actions, determined by the control law and the allowed ripple. Therefore, the switching frequency is not constant over all operating regions of the converter.

This section treats the derivation of the control (sliding surface) and switching laws, robustness, stability, constant-frequency operation, and steady-state error elimination necessary for sliding-mode control of switching converters, also giving some examples.

34.3.2 Principles of Sliding-mode Control

Consider the state-space switched model Eq. (34.4) of a switching converter subsystem, and input–output linearization or another technique, to obtain, from state-space equations, one Eq. (34.80), for each controllable subsystem output $y = x$. In the controllability canonical form [13] (also known as input–output decoupled or companion form), Eq. (34.80) is:

$$\frac{d}{dt}[x_h, \ldots, x_{j-1}, x_j]^T = [x_{h+1}, \ldots, x_j, -f_h(\mathbf{x}) - p_h(t) + b_h(\mathbf{x})u_h(t)]^T \quad (34.80)$$

where $\mathbf{x} = [x_h, \ldots, x_{j-1}, x_j]^T$ is the subsystem state vector, $f_h(\mathbf{x})$ and $b_h(\mathbf{x})$ are functions of $\mathbf{x}$, $p_h(t)$ represents the external disturbances, and $u_h(t)$ is the control input. In this special form of state-space modeling, the state variables are chosen so that the x_{i+1} variable ($i \in \{h, \ldots, j-1\}$) is the time derivative of x_i, that is $\mathbf{x} = \left[x_h, \dot{x}_h, \ddot{x}_h, \ldots, \overset{m}{x}_h\right]^T$, where $m = j - h$ [14].

34.3.2.1 Control Law (Sliding Surface)

The required closed-loop dynamics for the subsystem output vector $\mathbf{y} = \mathbf{x}$ can be chosen to verify Eq. (34.81) with selected k_i values. This is a model reference adaptive control approach to impose a state trajectory that advantageously reduces the system order $(j - h + 1)$.

$$\frac{dx_j}{dt} = -\sum_{i=h}^{j-1} \frac{k_i}{k_j} x_{i+1} \quad (34.81)$$

Effectively, in a single-input single-output (SISO) subsystem the order is reduced by unity, applying the restriction Eq. (34.81). In a multiple-input multiple-output (MIMO) system, in which ν independent restrictions could be imposed (usually with ν degrees of freedom), the order could often be reduced in ν units. Indeed, from Eq. (34.81), the dynamics of the jth term of $\mathbf{x}$ is linearly dependent from the $j - h$ first terms:

$$\frac{dx_j}{dt} = -\sum_{i=h}^{j-1} \frac{k_i}{k_j} x_{i+1} = -\sum_{i=h}^{j-1} \frac{k_i}{k_j} \frac{dx_i}{dt} \quad (34.82)$$

The controllability canonical model allows the direct calculation of the needed control input to achieve the desired dynamics Eq. (34.81). In fact, as the control action should enforce the state vector $\mathbf{x}$, to follow the reference vector $\mathbf{x}_r = \left[x_{h_r}, \dot{x}_{h_r}, \ddot{x}_{h_r}, \ldots, \overset{m}{x}_{h_r}\right]^T$, the tracking error vector will be $\mathbf{e} = [x_{h_r} - x_h, \ldots, x_{j-1r} - x_{j-1}, x_{jr} - x_j]^T$ or $\mathbf{e} = [e_{x_h}, \ldots, e_{x_{j-1}}, e_{x_j}]^T$. Thus, equating the sub-expressions for dx_j/dt of Eqs. (34.80) and (34.81), the necessary control input $u_h(t)$ is

$$u_h(t) = \frac{p_h(t) + f_h(\mathbf{x}) + \frac{dx_j}{dt}}{b_h(\mathbf{x})}$$

$$= \frac{p_h(t) + f_h(\mathbf{x}) - \sum_{i=h}^{j-1} \frac{k_i}{k_j} x_{i+1_r} + \sum_{i=h}^{j-1} \frac{k_i}{k_j} e_{x_{i+1}}}{b_h(\mathbf{x})} \quad (34.83)$$

This expression is the required closed-loop control law, but unfortunately it depends on the system parameters, on external perturbations and is difficult to compute. Moreover, for some output requirements, Eq. (34.83) would give extremely high values for the control input $u_h(t)$, which would be impractical or almost impossible.

In most switching converters $u_h(t)$ is discontinuous. Yet, if we assume one or more discontinuity borders dividing the state-space into subspaces, the existence and uniqueness of the solution is guaranteed out of the discontinuity borders, since in each subspace the input is continuous. The discontinuity borders are subspace switching hypersurfaces, whose order is the space order minus one, along which the subsystem state slides, since its intersections with the auxiliary equations defining the discontinuity surfaces can give the needed control input.

Within the sliding-mode control (SMC) theory, assuming a certain dynamic error tending to zero, one auxiliary equation (sliding surface) and the equivalent control input $u_h(t)$ can be obtained, integrating both sides of Eq. (34.82) with null initial conditions:

$$k_j x_j \sum_{i=h}^{j-1} k_i x_i = \sum_{i=h}^{j} k_i x_i = 0 \qquad (34.84)$$

This equation represents the discontinuity surface (hyperplane) and just defines the necessary sliding surface $S(x_i, t)$ to obtain the prescribed dynamics of Eq. (34.81):

$$S(x_i, t) = \sum_{i=h}^{j} k_i x_i = 0 \qquad (34.85)$$

In fact, by taking the first time derivative of $S(x_i, t)$, $\dot{S}(x_i, t) = 0$, solving it for dx_j/dt, and substituting the result in Eq. (34.83), the dynamics specified by Eq. (34.81) is obtained. This means that the control problem is reduced to a first-order problem, since it is only necessary to calculate the time derivative of Eq. (34.85) to obtain the dynamics (34.81) and the needed control input $u_h(t)$.

The sliding surface Eq. (34.85), as the dynamics of the converter subsystem, must be a Routh–Hurwitz polynomial and verify the sliding manifold invariance conditions, $S(x_i, t) = 0$ and $\dot{S}(x_i, t) = 0$. Consequently, the closed-loop controlled system behaves as a stable system of order $j - h$, whose dynamics is imposed by the coefficients k_i, which can be chosen by pole placement of the poles of the order $m = j - h$ polynomial. Alternatively, certain kinds of polynomials can be advantageously used [15]: Butterworth, Bessel, Chebyshev, elliptic (or Cauer), binomial, and minimum integral of time absolute error product (ITAE). Most useful are Bessel polynomials $B_E(s)$ Eq. (34.88), which minimize the system response time t_r, providing no overshoot, the polynomials $I_{TAE}(s)$ Eq. (34.87), that minimize the ITAE criterion for a system with desired natural oscillating frequency ω_o, and binomial polynomials $B_I(s)$ Eq. (34.86). For $m > 1$, ITAE polynomials give faster responses than binomial polynomials.

$$B_I(s)_m = (s+\omega_o)^m$$

$$= \begin{cases} m=0 \Rightarrow B_I(s)=1 \\ m=1 \Rightarrow B_I(s)=s+\omega_o \\ m=2 \Rightarrow B_I(s)=s^2+2\omega_o s+\omega_o^2 \\ m=3 \Rightarrow B_I(s)=s^3+3\omega_o s^2+3\omega_o^2 s+\omega_o^3 \\ m=4 \Rightarrow B_I(s)=s^4+4\omega_o s^3+6\omega_o^2 s^2+4\omega_o^3 s+\omega_o^4 \\ \ldots \end{cases}$$

$$(34.86)$$

$$I_{TAE}(s)_m = \begin{cases} m=0 \Rightarrow I_{TAE}(s)=1 \\ m=1 \Rightarrow I_{TAE}(s)=s+\omega_o \\ m=2 \Rightarrow I_{TAE}(s)=s^2+1.4\omega_o s+\omega_o^2 \\ m=3 \Rightarrow I_{TAE}(s)=s^3+1.75\omega_o s^2+2.15\omega_o^2 s+\omega_o^3 \\ m=4 \Rightarrow I_{TAE}(s)=s^4+2.1\omega_o s^3+3.4\omega_o^2 s^2 \\ \qquad\qquad +2.7\omega_o^3 s+\omega_o^4 \\ \ldots \end{cases}$$

$$(34.87)$$

$$B_E(s)_m = \begin{cases} m=0 \Rightarrow B_E(s)=1 \\ m=1 \Rightarrow B_E(s)=st_r+1 \\ m=2 \Rightarrow B_E(s)=\frac{(st_r)^2+3st_r+3}{3} \\ m=3 \Rightarrow B_E(s)=\frac{((st_r)^2+3.678st_r+6.459)(st_r+2.322)}{15} \\ \qquad\qquad = \frac{(st_r)^3+6(st_r)^2+15st_r+15}{15} \\ m=4 \Rightarrow B_E(s)=\frac{(st_r)^4+10(st_r)^3+45(st_r)^2+105(st_r)+105}{105} \\ \ldots \end{cases}$$

$$(34.88)$$

These polynomials can be the reference model for this model reference adaptive control method.

34.3.2.2 Closed-loop Control Input–Output Decoupled Form

For closed-loop control applications, instead of the state variables x_i, it is worthy to consider, as new state variables, the errors e_{x_i}, components of the error vector $\mathbf{e} = \left[e_{x_h}, \dot{e}_{x_h}, \ddot{e}_{x_h}, \ldots, \overset{m}{e}_{x_h}\right]^T$ of the state-space variables x_i, relative to a given reference x_{i_r} Eq. (34.90). The new controllability canonical model of the system is

$$\frac{d}{dt}[e_{x_h}, \ldots, e_{x_{j-1}}, e_{x_j}]^T = [e_{x_{h+1}}, \ldots, e_{x_j}, -f_e(\mathbf{e}) + p_e(t) \\ - b_e(\mathbf{e})u_h(t)]^T \qquad (34.89)$$

where $f_e(\mathbf{e})$, $p_e(t)$, and $b_e(\mathbf{e})$ are functions of the error vector $\mathbf{e}$.

As the transformation of variables

$$e_{x_i} = x_{i_r} - x_i \quad \text{with} \quad i = h, \ldots, j \quad (34.90)$$

is linear, the Routh–Hurwitz polynomial for the new sliding surface $S(e_{x_i}, t)$ is

$$S(e_{x_i}, t) = \sum_{i=h}^{j} k_i e_{x_i} = 0 \quad (34.91)$$

Since $e_{x_{i+1}}(s) = s e_{x_i}(s)$, this *control law*, from Eqs. (34.86–34.88) can be written as $S(\mathbf{e}, s) = e_{x_i}(s + \omega_0)^m$, does not depend on circuit parameters, disturbances, or operating conditions, but only on the imposed k_i parameters and on the state variable errors e_{x_i}, which can usually be measured or estimated. The control law Eq. (34.91) enables the desired dynamics of the output variable(s), if the semiconductor switching strategy is designed to guarantee the system stability. In practice, the finite switching frequency of the semiconductors will impose a certain dynamic error ε tending to zero. The control law Eq. (34.91) is the required controller for the closed-loop SISO subsystem with output $\mathbf{y}$.

34.3.2.3 Stability

Existence condition. The existence of the operation in sliding mode implies $S(e_{x_i}, t) = 0$. Also, to stay in this regime, the control system should guarantee $\dot{S}(e_{x_i}, t) = 0$. Therefore, the semiconductor switching law must ensure the stability condition for the system in sliding mode, written as

$$S(e_{x_i}, t)\dot{S}(e_{x_i}, t) < 0 \quad (34.92)$$

The fulfillment of this inequality ensures the convergence of the system state trajectories to the sliding surface $S(e_{x_i}, t) = 0$, since

– if $S(e_{x_i}, t) > 0$ and $\dot{S}(e_{x_i}, t) < 0$, then $S(e_{x_i}, t)$ will decrease to zero,
– if $S(e_{x_i}, t) < 0$ and $\dot{S}(e_{x_i}, t) > 0$, then $S(e_{x_i}, t)$ will increase toward zero.

Hence, if Eq. (34.92) is verified, then $S(e_{x_i}, t)$ will converge to zero. The condition (34.92) is the manifold $S(e_{x_i}, t)$ invariance condition, or the sliding-mode existence condition.

Given the statespace model Eq. (34.89) as a function of the error vector $\mathbf{e}$ and, from $\dot{S}(e_{x_i}, t) = 0$, the equivalent average control input $U_{eq}(t)$ that must be applied to the system in order that the system state slides along the surface Eq. (34.91), is given by

$$U_{eq}(t) = \frac{k_h \frac{de_{x_h}}{dt} + k_{h+1} \frac{de_{x_{h+1}}}{dt} + \cdots + k_{j-1} \frac{de_{x_{j-1}}}{dt} + k_j(-f_e(\mathbf{e}) + p_e(t))}{k_j b_e(\mathbf{e})} \quad (34.93)$$

This control input $U_{eq}(t)$ ensures the converter subsystem operation in the sliding mode.

Reaching condition. The fulfillment of $S(e_{x_i}, t)\dot{S}(e_{x_i}, t) < 0$, as $S(e_{x_i}, t)\dot{S}(e_{x_i}, t) = (1/2)\dot{S}^2(e_{x_i}, t)$, implies that the distance between the system state and the sliding surface will tend to zero, since $S^2(e_{x_i}, t)$ can be considered as a measure for this distance. This means that the system will reach sliding mode. Additionally, from Eq. (34.89) it can be written:

$$\frac{de_{x_j}}{dt} = -f_e(\mathbf{e}) + p_e(t) - b_e(\mathbf{e})u_h(t) \quad (34.94)$$

From Eq. (34.91), Eq. (34.95) is obtained.

$$S(e_{x_i}, t) = \sum_{i=h}^{j} k_i e_{x_i} = k_h e_{x_h} + k_{h+1} \frac{de_{x_h}}{dt} + k_{h+2} \frac{d^2 e_{x_h}}{dt^2}$$

$$+ \cdots + k_j \frac{d^m e_{x_h}}{dt^m} \quad (34.95)$$

If $S(e_{x_i}, t) > 0$, from the Routh–Hurwitz property of Eq. (34.91), then $e_{x_j} > 0$. In this case, to reach $S(e_{x_i}, t) = 0$ it is necessary to impose $-b_e(\mathbf{e})u_h(t) = -U$ in Eq. (34.94), with U chosen to guarantee $de_{x_j}/dt < 0$. After a certain time, e_{x_j} will be $e_{x_j} = d^m e_{x_h}/dt^m < 0$, implying along with Eq. (34.95) that $\dot{S}(e_{x_i}, t) < 0$, thus verifying Eq. (34.92). Therefore, every term of $S(e_{x_i}, t)$ will be negative, which implies, after a certain time, an error $e_{x_h} < 0$ and $S(e_{x_i}, t) < 0$. Hence, the system will reach sliding mode, staying there if $U = U_{eq}(t)$. This same reasoning can be made for $S(e_{x_i}, t) < 0$, it is now being necessary to impose $-b_e(\mathbf{e})u_h(t) = +U$, with U high enough to guarantee $de_{x_j}/dt > 0$.

To ensure that the system always reaches sliding-mode operation, it is necessary to calculate the maximum value of $U_{eq}(t)$, U_{eqmax}, and also impose the reaching condition:

$$U > U_{eqmax} \quad (34.96)$$

This means that the power supply voltage values U should be chosen high enough to additionally account for the maximum effects of the perturbations. With step inputs, even with $U > U_{eqmax}$, the converter usually loses sliding mode, but it will reach it again, even if the U_{eqmax} is calculated considering only the maximum steady-state values for the perturbations.

34.3.2.4 Switching Law

From the foregoing considerations, supposing a system with two possible structures, the semiconductor switching strategy must ensure $S(e_{x_i}, t)\dot{S}(e_{x_i}, t) < 0$. Therefore, if $S(e_{x_i}, t) > 0$, then $\dot{S}(e_{x_i}, t) < 0$, which implies, as seen, $-b_e(\mathbf{e})u_h(t) = -U$ (the sign of $b_e(\mathbf{e})$ must be known). Also, if $S(e_{x_i}, t) < 0$, then $\dot{S}(e_{x_i}, t) > 0$, which implies $-b_e(\mathbf{e})u_h(t) = +U$. This imposes the switching between two structures at infinite frequency. Since power semiconductors can switch only at finite frequency, in practice, a small enough error for $S(e_{x_i}, t)$ must be allowed $(-\varepsilon < S(e_{x_i}, t) < +\varepsilon)$. Hence, the switching law between the two possible system structures might be

$$u_h(t) = \begin{cases} U/be(\mathbf{e}) & \text{for } S(e_{x_i}, t) > +e \\ -U/be(\mathbf{e}) & \text{for } S(e_{x_i}, t) < -e \end{cases} \quad (34.97)$$

The condition Eq. (34.97) determines the control input to be applied and therefore represents the semiconductor switching strategy or switching function. This law determines a two-level pulse width modulator with JIT switching (variable frequency).

34.3.2.5 Robustness

The dynamics of a system, with closed-loop control using the control law Eq. (34.91) and the switching law Eq. (34.97), does not depend on the system operating point, load, circuit parameters, power supply, or bounded disturbances, as long as the control input $u_h(t)$ is large enough to maintain the converter subsystem in sliding mode. Therefore, it is said that the switching converter dynamics, operating in sliding mode, is robust against changing operating conditions, variations of circuit parameters, and external disturbances. The desired dynamics for the output variable(s) is determined only by the k_i coefficients of the control law Eq. (34.91), as long as the switching law (34.97) maintains the converter in sliding mode.

34.3.3 Constant-frequency Operation

Prefixed switching frequency can be achieved, even with the sliding-mode controllers, at the cost of losing the JIT action. As the sliding-mode controller changes the control input when needed, and not at a certain prefixed rhythm, applications needing constant switching frequency (such as thyristor rectifiers or resonant converters), must compare $S(e_{x_i}, t)$ (hysteresis width 2ι much narrower than 2ε) with auxiliary triangular waveforms (Fig. 34.25a), auxiliary sawtooth functions (Fig. 34.25b), three-level clocks (Fig. 34.25c), or phase locked loop control of the comparator hysteresis variable width 2ε[16]. However, as illustrated in Fig. 34.25d, steady-state errors do appear. Often, they should be eliminated as described in Section 34.3.4.

34.3.4 Steady-state Error Elimination in Converters with Continuous Control Inputs

In the ideal sliding mode, state trajectories are directed toward the sliding surface (34.91) and move exactly along the discontinuity surface, switching between the possible system structures, at infinite frequency. Practical sliding modes cannot switch at infinite frequency, and therefore exhibit phase plane trajectory oscillations inside a hysteresis band of width 2ε, centered in the discontinuity surface.

The switching law Eq. (34.91) permits no steady-state errors as long as $S(e_{x_i}, t)$ tends to zero, which implies no restrictions on the commutation frequency. Control circuits operating at constant frequency, or needed continuous inputs, or particular limitations of the power semiconductors, such as minimum on or off times, can originate $S(e_{x_i}, t) = \varepsilon_1 \neq 0$. The steady-state error (e_{x_h}) of the x_h variable, $x_{h_r} - x_h = \varepsilon_1/k_h$, can be eliminated, increasing the system order by 1. The new state-space controllability canonical form, considering the error e_{x_i}, between the variables and their references, as the state vector, is

$$\frac{d}{dt}\left[\int e_{x_h} dt, e_{x_h}, \ldots, e_{x_{j-1}}, e_{x_j}\right]^T$$
$$= [e_{x_h}, e_{x_{h+1}}, \ldots, e_{x_j}, -f_e(\mathbf{e}) - p_e(t) - b_e(\mathbf{e})u_h(t)]^T \quad (34.98)$$

The new sliding surface $S(e_{x_i}, t)$, written from Eq. (34.91) considering the new system Eq. (34.98), is

$$S(e_{x_i}, t) = k_0 \int e_{x_h} dt + \sum_{i=h}^{j} k_i e_{x_i} = 0 \quad (34.99)$$

This sliding surface offers zero-state error, even if $S(e_{x_i}, t) = \varepsilon_1$ due to the hardware errors or fixed (or limited)

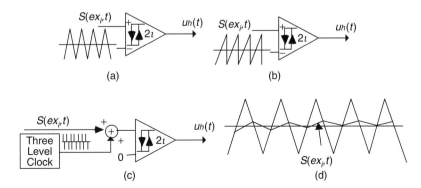

FIGURE 34.25 Auxiliary functions and methods to obtain constant switching frequency with sliding-mode controllers.

frequency switching. Indeed, at the steady state, the only nonnull term is $k_0 \int e_{x_h} dt = \varepsilon_1$. Also, like Eq. (34.91), this closed-loop control law does not depend on system parameters or perturbations to ensure a prescribed closed-loop dynamics similar to Eq. (34.81) with an error approaching zero.

The approach outlined herein precisely defines the control law (sliding surface (34.91) or (34.99)) needed to obtain the selected dynamics, and the switching law Eq. (34.97). As the control law allows the implementation of the system controller, and the switching law gives the PWM modulator, there is no need to design linear or nonlinear controllers, based on linear converter models, or devise offline PWM modulators. Therefore, sliding-mode control theory, applied to switching converters, provides a systematic method to generate both the controller(s) (usually nonlinear) and the modulator(s) that will ensure a model reference robust dynamics, solving the control problem of switching converters.

In the next examples, it is shown that the sliding-mode controllers use (nonlinear) state feedback, therefore, needing to measure the state variables and often other variables, since they use more system information. This is a disadvantage since more sensors are needed. However, the straightforward control design and obtained performances are much better than those obtained with the averaged models, the use of more sensors being really valued. Alternatively to the extra sensors, state observers can be used [13, 14].

34.3.5 Examples: Buck–Boost DC/DC Converter, Half-bridge Inverter, 12-pulse Parallel Rectifiers, Audio Power Amplifiers, Near Unity Power Factor Rectifiers, Multilevel Inverters, Matrix Converters

EXAMPLE 34.10 Sliding-mode control of the buck–boost dc/dc converter

Consider again the buck–boost converter of Fig. 34.1 and assume the converter output voltage v_o to be the controlled output. From Section 34.2, using the switched state-space model of Eq. (34.11), making $dv_o/dt = \theta$, and calculating the first time derivative of θ, the controllability canonical model (34.100), where $i_o = v_o/R_o$, is obtained:

$$\frac{dv_o}{dt} = \theta = \frac{1-\delta(t)}{C_o} i_L - \frac{i_o}{C_o}$$

$$\frac{d\theta}{dt} = -\frac{(1-\delta(t))^2}{L_i C_o} v_o - \frac{C_o \theta + i_o}{C_o(1-\delta(t))} \frac{d\delta(t)}{dt}$$

$$- \frac{1}{C_o} \frac{di_o}{dt} + \frac{\delta(t)(1-\delta(t))}{C_o L_i} V_{DC} \quad (34.100)$$

This model, written in the form of Eq. (34.80), contains two state variables, v_o and θ. Therefore, from Eq. (34.91) and considering $e_{v_o} = v_{o_r} - v_o$, $e_\theta = \theta_r - \theta$, the control law (sliding surface) is

$$S(e_{x_i}, t) = \sum_{i=h}^{2} k_i e_{x_i} = k_1(v_{o_r} - v_o) + k_2 \frac{dv_{o_r}}{dt} - k_2 \frac{dv_o}{dt}$$

$$= k_1(v_{o_r} - v_o) + k_2 \frac{dv_{o_r}}{dt} - \frac{k_2}{C_o}(1-\delta(t))i_L$$

$$+ \frac{k_2}{C_o} i_o = 0 \quad (34.101)$$

This sliding surface depends on the variable $\delta(t)$, which should be precisely the result of the application, in Eq. (34.101), of a switching law similar to Eq. (34.97). Assuming an ideal up–down converter and slow variations, from Eq. (34.31) the variable $\delta(t)$ can be averaged to $\delta_1 = v_o/(v_o + V_{DC})$. Substituting this relation in Eq. (34.101), and rearranging, Eq. (34.102) is derived:

$$S(e_{x_i}, t) = \frac{C_o k_1}{k_2} \left(\frac{v_o + V_{DC}}{v_o} \right)$$

$$\times \left((v_{o_r} - v_o) + \frac{k_2}{k_1} \frac{dv_{o_r}}{dt} + \frac{k_2}{k_1} \frac{1}{C_o} i_o \right) - i_L = 0 \quad (34.102)$$

This control law shows that the power supply voltage V_{DC} must be measured, as well as the output voltage v_o and the currents i_o and i_L.

To obtain the switching law from stability considerations (34.92), the time derivative of $S(e_{x_i}, t)$, supposing $(v_o + V_{DC})/v_o$ almost constant, is

$$\dot{S}(e_{x_i}, t) = \frac{C_o k_1}{k_2} \left(\frac{v_o + V_{DC}}{v_o} \right)$$

$$\times \left(\frac{de_{v_o}}{dt} + \frac{k_2}{k_1} \frac{d^2 v_{o_r}}{dt^2} + \frac{k_2}{k_1 C_o} \frac{di_o}{dt} \right) - \frac{di_L}{dt} \quad (34.103)$$

If $S(e_{x_i}, t) > 0$ then, from Eq. (34.92), $\dot{S}(e_{x_i}, t) < 0$ must hold. Analyzing Eq. (34.103), we can conclude that, if $S(e_{x_i}, t) > 0$, $\dot{S}(e_{x_i}, t)$ is negative if, and only if, $di_L/dt > 0$. Therefore, for positive errors $e_{v_o} > 0$ the current i_L must be increased, which implies $\delta(t) = 1$. Similarly, for $S(e_{x_i}, t) < 0$, $di_L/dt < 0$ and $\delta(t) = 0$. Thus, a switching law similar to Eq. (34.97) is obtained:

$$\delta(t) = \begin{cases} 1 & \text{for } S(e_{x_i}, t) > +e \\ 0 & \text{for } S(e_{x_i}, t) < -e \end{cases} \quad (34.104)$$

The same switching law could be obtained from knowing the dynamic behavior of this nonminimum-phase up-down converter: to increase (decrease) the output voltage, a previous increase (decrease) of the i_L current is mandatory.

Equation (34.101) shows that, if the buck–boost converter is into the sliding mode ($S(e_{x_i}, t) = 0$), the dynamics of the output voltage error tends exponentially to zero with time constant k_2/k_1. Since during step transients, the converter is in the reaching mode, the time constant k_2/k_1 cannot be designed to originate error variations larger than the one allowed by the self-dynamics of the converter excited by a certain maximum permissible i_L current. Given the polynomials (34.86–34.88) with $m = 1$, $k_1/k_2 = \omega_o$ should be much lower than the finite switching frequency ($1/T$) of the converter. Therefore,

the time constant must obey $k_2/k_1 \gg T$. Then, knowing that k_2 and k_1 are both imposed, the control designer can tailor the time constant as needed, provided that the above restrictions are observed.

Short-circuit-proof operation for the sliding-mode controlled buck–boost converter can be derived from Eq. (34.102), noting that all the terms to the left of i_L represent the set point for this current. Therefore, limiting these terms (Fig. 34.26, saturation block, with $i_{Lmax} = 40$ A), the switching law (34.104) ensures that the output current will not rise above the maximum imposed limit. Given the converter nonminimum-phase behavior, this i_L current limit is fundamental

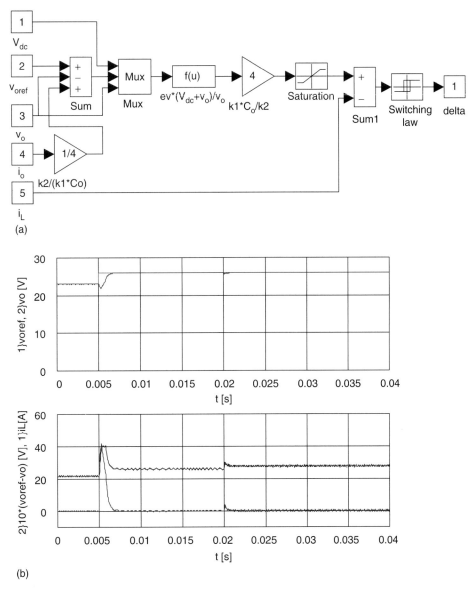

FIGURE 34.26 (a) Block diagram of the sliding-mode nonlinear controller for the buck–boost converter and (b) transient responses of the sliding-mode controlled buck–boost converter. At $t = 0.005$ s, v_{oref} step from 23 to 26 V. At $t = 0.02$ s, V_{DC} step from 26 to 23 V. Top graph: step reference v_{oref} and output voltage v_o. Bottom graph: trace starting at 20 is i_L current; trace starting at zero is $10\times(v_{oref} - v_o)$.

to reach the sliding mode of operation with step disturbances.

The block diagram (Fig. 34.26a) of the implemented control law Eq. (34.102) (with $C_o k_1/k_2 = 4$) and switching law (34.103) (with $\varepsilon = 0.3$) does not included the time derivative of the reference (dv_{o_r}/dt) since, in a dc/dc converter its value is considered zero. The controller hardware (or software), derived using just the sliding-mode approach, operates only in a closed-loop.

The resulting performance (Fig. 34.26b) is much better than that obtained with the PID notch filter (compare to Example 34.4, Fig. 34.9b), with a higher response speed and robustness against power-supply variations.

EXAMPLE 34.11 Sliding mode control of the single-phase half-bridge converter
Consider the half-bridge four quadrant converter of Fig. 34.27 with the output filter and the inductive load ($V_{DCmax} = 300$ V; $V_{DCmin} = 230$ V; $R_i = 0.1\,\Omega$; $L_o = 4$ mH; $C_o = 470\,\mu$F; inductive load with nominal values $R_o = 7\,\Omega$, $L_o = 1$ mH).

Assuming that power switches, output filter capacitor, and power supply are all ideal, and a generic load with allowed slow variations, the switched state-space model of the converter, with state variables v_o and i_L, is

$$\frac{d}{dt}\begin{bmatrix} v_o \\ i_L \end{bmatrix} = \begin{bmatrix} 0 & 1/C_o \\ -1/L_o & -R_i/L_o \end{bmatrix}\begin{bmatrix} v_o \\ i_L \end{bmatrix} + \begin{bmatrix} -1/C_o & 0 \\ 0 & 1/L_o \end{bmatrix}\begin{bmatrix} i_o \\ \delta(t)V_{DC} \end{bmatrix} \quad (34.105)$$

where i_o is the generic load current and $v_{PWM} = \delta(t)V_{DC}$ is the extended PWM output voltage ($\delta(t) = +1$ when one of the upper main semiconductors of Fig. 34.27 is conducting and $\delta(t) = -1$ when one of the lower semiconductors is on).

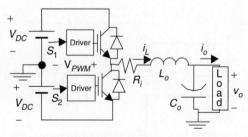

FIGURE 34.27 Half-bridge power inverter with insulated gate bipolar transistors, output filter, and load.

34.3.5.1 Output Current Control (Current-mode Control)

To perform as a v_{i_L} voltage controlled i_L current source (or sink) with transconductance g_m ($g_m = i_L/v_{i_L}$), this converter must supply a current i_L to the output inductor, obeying $i_L = g_m v_{i_L}$. Using a bounded v_{i_L} voltage to provide output short-circuit protection, the reference current for a sliding-mode controller must be $i_{L_r} = g_m v_{i_L}$. Therefore, the controlled output is the i_L current and the controllability canonical model (34.106) is obtained from the second equation of (34.105), since the dynamics of this subsystem, being governed by $\delta(t)V_{DC}$, is already in the controllability canonical form for this chosen output.

$$\frac{di_L}{dt} = -\frac{R_i}{L_o}i_L - \frac{1}{L_o}v_o + \frac{\delta(t)V_{DC}}{L_o} \quad (34.106)$$

A suitable sliding surface (34.107) is obtained from Eq. (34.91), making $e_{i_L} = i_{L_r} - i_L$.

$$S(e_{i_L}, t) = k_p e_{i_L} = k_p (i_{L_r} - i_L) = k_p (g_m v_{i_L} - i_L) = 0 \quad (34.107)$$

The switching law Eq. (34.108) can be devised calculating the time derivative of Eq. (34.107) $\dot{S}(e_{i_L}, t)$, and applying Eq. (34.92). If $S(e_{i_L}, t) > 0$, then $di_L/dt > 0$ must hold to obtain $\dot{S}(e_{i_L}, t) < 0$, implying $\delta(t) = 1$.

$$\delta(t) = \begin{cases} 1 & \text{for } S(e_{i_L}, t) > +\varepsilon \\ -1 & \text{for } S(e_{i_L}, t) < -\varepsilon \end{cases} \quad (34.108)$$

The k_p value and the allowed the ripple ε define the instantaneous value of the variable switching frequency. The sliding-mode controller is represented in Fig. 34.28a. Step response (Fig. 34.29a) shows the variable-frequency operation, a very short rise time (limited only by the available power supply) and confirms the expected robustness against supply variations.

For systems where fixed-frequency operation is needed, a triangular wave, with frequency (10 kHz) slightly greater than the maximum variable frequency, can be added (Fig. 34.28b) to the sliding-mode controller, as explained in Section 34.3.3. Performances (Fig. 34.29b) are comparable to those of the variable-frequency sliding-mode controller (Fig. 34.29a). Fig. 34.29b shows the constant switching frequency, but also a steady-state error dependent on the operating point.

To eliminate this error, a new sliding surface Eq. (34.109), based on Eq. (34.99), should be used. The constants k_p and k_0 can be calculated, as discussed in Example 34.10.

$$S(e_{i_L}, t) = k_0 \int e_{i_L} dt + k_p e_{i_L} = 0 \quad (34.109)$$

The new constant-frequency sliding-mode current controller (Fig. 34.30a), with added antiwindup techniques (Example 34.6), since a saturation (errMax) is needed to keep the frequency constant, now presents no steady-state error (Fig. 34.30b). Performances are comparable to those of the

34 Control Methods for Switching Power Converters

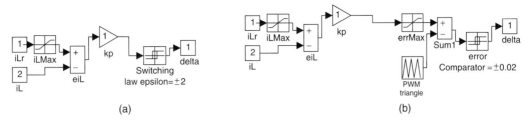

FIGURE 34.28 (a) Implementation of short-circuit-proof sliding-mode current controller (variable frequency) and (b) implementation of fixed frequency, short-circuit-proof sliding-mode current controller using a triangular waveform.

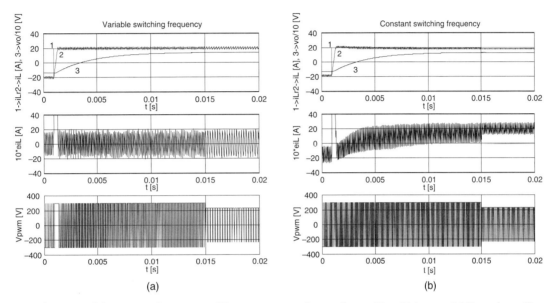

FIGURE 34.29 Performance of the transconductance amplifier; response to a i_{Lr} step from -20 to 20 A at $t = 0.001$ s and to a V_{DC} step from 300 to 230 V at $t = 0.015$ s: (a) variable-frequency sliding-mode controller and (b) fixed-frequency sliding-mide controller.

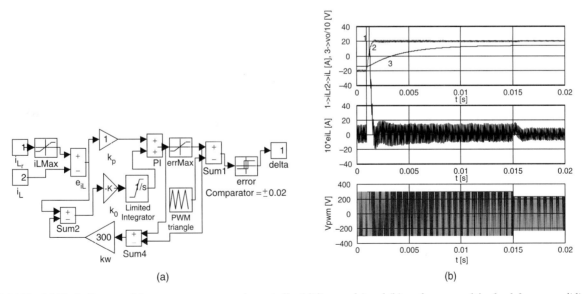

FIGURE 34.30 (a) Block diagram of the average current-mode controller (sliding mode) and (b) performance of the fixed-frequency sliding-mode controller with removed steady-state error: response to a i_{L_r} step from -20 to 20 A at $t = 0.001$ s and to a V_{DC} step from 300 to 230 V at $t = 0.015$ s.

variable-frequency controller, and no robustness loss is visible. The applied sliding-mode approach led to the derivation of the known average current-mode controller.

34.3.5.2 Output Voltage Control

To obtain a power operational amplifier suitable for building uninterruptible power supplies, power filters, power gyrators, inductance simulators, or power factor active compensators, v_o must be the controlled converter output. Therefore, using the input–output linearization technique, it is seen that the first time derivative of the output $(dv_o/dt) = (i_L - i_o)/C_o = \theta$, does not explicitly contain the control input $\delta(t)V_{DC}$. Then, the second derivative must be calculated. Taking into account Eq. (34.105), as $\theta = (i_L - i_o)/C_o$, Eq. (34.110) is derived.

$$\frac{d^2 v_o}{dt^2} = \frac{d}{dt}\theta = \frac{d}{dt}\left(\frac{i_L - i_o}{C_o}\right)$$

$$= -\frac{R_i}{L_o}\theta - \frac{1}{L_o C_o}v_o - \frac{R_i}{L_o C_o}i_o - \frac{1}{C_o}\frac{di_o}{dt} + \frac{1}{L_o C_o}\delta(t)V_{DC} \quad (34.110)$$

This expression shows that the second derivative of the output depends on the control input $\delta(t)V_{DC}$. No further time derivative is needed, and the state-space equations of the equivalent circuit, written in the phase canonical form, are

$$\frac{d}{dt}\begin{bmatrix} v_o \\ \theta \end{bmatrix} = \begin{bmatrix} \theta \\ -\frac{R_i}{L_o}\theta - \frac{1}{L_o C_o}v_o - \frac{R_i}{L_o C_o}i_o - \frac{1}{C_o}\frac{di_o}{dt} + \frac{1}{L_o C_o}\delta(t)V_{DC} \end{bmatrix} \quad (34.111)$$

According to Eqs. (34.91), (34.111), and (34.105), considering that e_{v_o} is the feedback error $e_{v_o} = v_{o_r} - v_o$, a sliding surface $S(e_{v_o}, t)$, can be chosen:

$$S(e_{v_o}, t) = k_1 e_{v_o} + k_2 \frac{de_{v_o}}{dt} = e_{v_o} + \frac{k_2}{k_1}\frac{de_{v_o}}{dt}$$

$$= e_{v_o} + \beta \frac{de_{v_o}}{dt} = \frac{C_o}{\beta}(v_{o_r} - v_o) + C_o \frac{dv_{o_r}}{dt} + i_o - i_L = 0 \quad (34.112)$$

where β is the time constant of the desired first-order response of output voltage ($\beta \gg T > 0$), as the strong relative degree [14] of this system is 2, and the sliding-mode operation reduces by one, the order of this system (the strong relative degree represents the number of times the output variable must be time differentiated until a control input explicitly appears).

Calculating $\dot{S}(e_{v_o}, t)$, the control strategy (switching law) Eq. (34.113) can be devised since, if $S(e_{v_o}, t) > 0$, then di_L/dt must be positive to obtain $\dot{S}(e_{i_L}, t) < 0$, implying $\delta(t) = 1$. Otherwise, $\delta(t) = -1$.

$$\delta(t) = \begin{cases} 1 & \text{for } S(e_{v_o}, t) > 0 (v_{PWM} = +V_{DC}) \\ -1 & \text{for } S(e_{v_o}, t) < 0 (v_{PWM} = -V_{DC}) \end{cases} \quad (34.113)$$

In the ideal sliding-mode dynamics, the filter input voltage v_{PWM} switches between V_{DC} and $-V_{DC}$ with the infinite frequency. This switching generates the equivalent control voltage V_{eq} that must satisfy the sliding manifold invariance conditions, $S(e_{v_o}, t) = 0$ and $\dot{S}(e_{v_o}, t) = 0$. Therefore, from $\dot{S}(e_{v_o}, t) = 0$, using Eqs. (34.112) and (34.105), (or from Eq. (34.110)), V_{eq} is

$$V_{eq} = L_o C_o \left[\frac{d^2 v_{o_r}}{dt^2} + \frac{1}{\beta}\frac{dv_{o_r}}{dt} + \frac{v_o}{L_o C_o} \right. $$
$$\left. + \frac{(\beta R_i - L_o)i_L}{\beta L_o C_o} + \frac{i_o}{\beta C_o} + \frac{1}{C_o}\frac{di_o}{dt} \right] \quad (34.114)$$

This equation shows that only smooth input v_{o_r} signals ("smooth" functions) can be accurately reproduced at the inverter output, as it contains derivatives of the v_{o_r} signal. This fact is a consequence of the stored electromagnetic energy. The existence of the sliding-mode operation implies the following necessary and sufficient condition:

$$-V_{DC} < V_{eq} < V_{DC} \quad (34.115)$$

Equation (34.115) enables the determination of the minimum input voltage V_{DC} needed to enforce the sliding-mode operation. Nevertheless, even in the case of $|V_{eq}| > |V_{DC}|$, the system experiences only a saturation transient and eventually reaches the region of sliding-mode operation, except if, in the steady state, operating point and disturbances enforce $|V_{eq}| > |V_{DC}|$.

In the ideal sliding mode, at infinite switching frequency, state trajectories are directed toward the sliding surface and move exactly along the discontinuity surface. Practical switching converters cannot switch at infinite frequency, so a typical implementation of Eq. (34.112) (Fig. 34.31a) with neglected $\dot{v}_{o_r}$ features a comparator with hysteresis 2ε, switching occurring at $|S(e_{v_o}, t)| > \varepsilon$ with frequency depending on the slopes of i_L. This hysteresis causes phase-plane trajectory oscillations of width 2ε around the discontinuity surface $S(e_{v_o}, t) = 0$, but the V_{eq} voltage is still correctly generated, since the resulting duty cycle is a continuous variable (except for error limitations in the hardware or software, which can be corrected using the approach pointed out by Eq. (34.98)).

The design of the compensator and the modulator is integrated with the same theoretical approach, since the signal $S(e_{v_o}, t)$ applied to a comparator generates the pulses for the power semiconductors drives. If the short-circuit-proof operation is built into the power semiconductor drives, there is the possibility to measure only the capacitor current $(i_L - i_o)$.

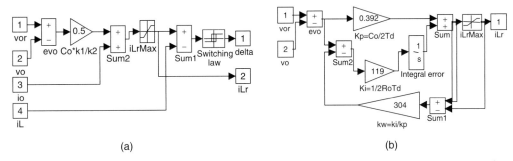

FIGURE 34.31 (a) Implementation of short-circuit-proof, sliding-mode output voltage controller (variable frequency) and (b) implementation of antiwindup PI current-mode (fixed frequency) controller.

34.3.5.3 Short-circuit Protection and Fixed-frequency Operation of the Power Operational Amplifier

If we note that all the terms to the left of i_L in Eq. (34.112) represent the value of i_{L_r}, a simple way to provide short-circuit protection is to bound the sum of all these terms (Fig. 34.31a with $i_{L_{rmax}} = 100$ A). Alternatively, the output current controllers of Fig. 34.28 can be used, comparing Eq. (34.107) to Eq. (34.112), to obtain $i_{L_r} = S(e_{v_o}, t)/k_p + i_L$. Therefore, the block diagram of Fig. 34.31a provides the i_{L_r} output (for $k_p = 1$) to be the input of the current controllers (Fig. 34.28 and Fig. 34.31a). As seen, the controllers of Fig. 34.28b and Fig. 34.30a also ensure fixed-frequency operation.

For comparison purposes a proportional–integral (PI) controller, with antiwindup (Fig. 34.31b) for output voltage control, was designed, supposing the current-mode control of the half bridge ($i_{L_r} = g_m v_{i_L}/(1 + sT_d)$ considering a small delay T_d), a pure resistive load R_o, and using the approach outlined in Examples 34.6 and 34.8 ($k_v = 1$, $g_m = 1$, $\zeta^2 = 0.5$, $T_d = 600\,\mu$s). The obtained PI (34.50) parameters are

$$T_z = R_o C_o$$
$$T_p = 4\zeta^2 g_m k_v R_o T_d \tag{34.116}$$

Both variable frequency (Fig. 34.32) and constant frequency (Fig. 34.33) sliding-mode output voltage controllers present excellent performance and robustness with nominal loads. With loads much higher than the nominal value (Fig. 34.32b and Fig. 34.33b), the performance and robustness are also excellent. The sliding-mode constant-frequency PWM controller presents the additional advantage of injecting lower ripple in the load.

As expected, the PI regulator presents lower performance (Fig. 34.34). The response speed is lower and the insensitivity to power supply and load variations (Fig. 34.34b) is not as high as with the sliding mode. Nevertheless, the PI performances are acceptable, since its design was carried considering a slow

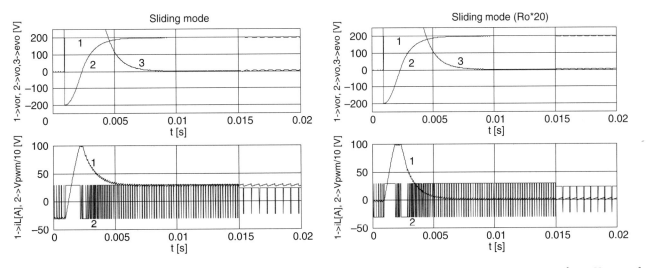

FIGURE 34.32 Performance of the power operational amplifier; response to a v_{o_r} step from -200 to 200 V at $t = 0.001$ s and to a V_{DC} step from 300 to 230 V at $t = 0.015$ s: (a) variable-frequency sliding mode (nominal load) and (b) variable-frequency sliding mode ($R_o \times 20$).

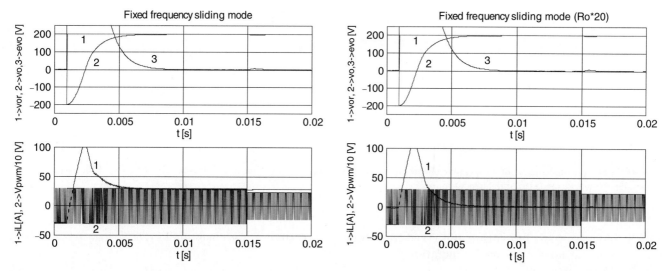

FIGURE 34.33 Performance of the power operational amplifier; response to a v_{o_r} step from −200 to 200 V at $t = 0.001$ s and to a V_{DC} step from 300 to 230 V at $t = 0.015$ s: (a) fixed-frequency sliding mode (nominal load) and (b) fixed-frequency sliding mode ($R_o \times 20$).

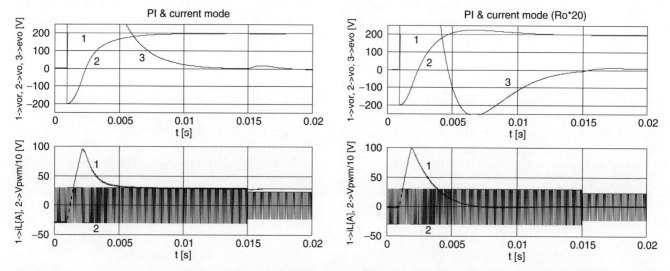

FIGURE 34.34 Performance of the PI controlled power operational amplifier; response to a v_{o_r} step from −200 to 200 V at $t = 0.001$ s and to a V_{DC} step from 300 to 230 V at $t = 0.015$ s: (a) PI current-mode controller (nominal load) and (b) PI current-mode controller ($R_o \times 20$).

and fast manifold sliding-mode approach: the fixed-frequency sliding-mode current controller (34.109) for the fast manifold (the i_L current dynamics) and the antiwindup PI for the slow manifold (the v_o voltage dynamics, usually much slower than the current dynamics).

EXAMPLE 34.12 Constant-frequency sliding-mode control of p pulse parallel rectifiers

This example presents a new paradigm to the control of thyristor rectifiers. Since p pulse rectifiers are variable-structure systems, sliding-mode control is applied here to 12-pulse rectifiers, still useful for very high-power applications [3]. The design determines the variables to be measured and the controlled rectifier presents robustness, and much shorter response times, even with the parameter uncertainty, perturbations, noise, and non-modeled dynamics. These performances are not feasible using linear controllers, obtained here for comparison purposes.

34.3.5.4 Modeling the 12-pulse Parallel Rectifier

The 12-pulse rectifier (Fig. 34.35a) is built with four three-phase half-wave rectifiers, connected in parallel with current-sharing inductances l and l' merged with capacitors C', C_2, to

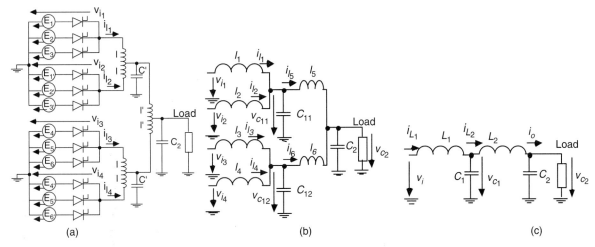

FIGURE 34.35 (a) 12-pulse rectifier with interphase reactors and intermediate capacitors; (b) rectifier model neglecting the half-wave rectifier dynamics; and (c) low-order averaged equivalent circuit for the 12-pulse rectifier with the resulting output double LC filter.

obtain a second-order LC filter. This allows low-ripple output voltage and continuous mode of operation (laboratory model with $l = 44$ mH; $l' = 13$ mH; $C' = C_2 = 10$ mF; star-delta connected ac sources with $E_{RMS} \approx 65$ V and power rating 2.2 kW, load approximately resistive $R_o \approx 3\text{--}5\,\Omega$).

To control the output voltage v_{C_2}, given the complexity of the whole system, the best approach is to derive a low-order model. By averaging the four half-wave rectifiers, neglecting the rectifier dynamics and mutual couplings, the equivalent circuit of Fig. 34.35b is obtained ($l_1 = l_2 = l_3 = l_4 = l$; $l_5 = l_6 = l'$; $C_{11} = C_{12} = C'$). Since the rectifiers are identical, the equivalent 12-pulse rectifier model of Fig. 34.35c is derived, simplifying the resulting parallel associations ($L_1 = l/4$; $L_2 = l'/2$; $C_1 = 2C'$).

Considering the load current i_o as an external perturbation and v_i the control input, the state-space model of the equivalent circuit of Fig. 34.35c is

$$\frac{d}{dt}\begin{bmatrix} i_{L_1} \\ i_{L_2} \\ v_{C_1} \\ v_{C_2} \end{bmatrix} = \begin{bmatrix} 0 & 0 & -1/L_1 & 0 \\ 0 & 0 & 1/L_2 & -1/L_2 \\ 1/C_1 & -1/C_1 & 0 & 0 \\ 0 & 1/C_2 & 0 & 0 \end{bmatrix} \begin{bmatrix} i_{L_1} \\ i_{L_2} \\ v_{C_1} \\ v_{C_2} \end{bmatrix}$$

$$+ \begin{bmatrix} 1/L_1 & 0 \\ 0 & 0 \\ 0 & 0 \\ 0 & -1/C_2 \end{bmatrix} \begin{bmatrix} v_i \\ i_o \end{bmatrix} \quad (34.117)$$

34.3.5.5 Sliding-mode Control of the 12-pulse Parallel Rectifier

Since the output voltage v_{C_2} of the system must follow the reference $v_{C_2 r}$, the system equations in the phase canonical (or controllability) form must be written, using the error $e_{v_{C_2}} = v_{C_2 r} - v_{C_2}$ and its time derivatives as new state error variables, as done in Example 34.11.

$$\frac{d}{dt}\begin{bmatrix} e_{v_{C_2}} \\ e_\theta \\ e_\gamma \\ e_\beta \end{bmatrix} = \begin{bmatrix} e_\theta \\ e_\gamma \\ e_\beta \\ -\left(\frac{1}{C_1 L_1} + \frac{1}{C_1 L_2} + \frac{1}{C_2 L_2}\right)e_\gamma - \frac{e_{v_{C_2}}}{C_1 L_1 C_2 L_2} - \left(\frac{1}{C_1 L_1 C_2}\right) \\ + \frac{1}{C_1 C_2 L_2}\right)\frac{di_o}{dt} - \frac{1}{C_2}\frac{d^3 i_o}{dt^3} - \frac{v_i}{C_1 L_1 C_2 L_2} \end{bmatrix}$$

$$(34.118)$$

The sliding surface $S(e_{x_i}, t)$, designed to reduce the system order, is a linear combination of all the phase canonical state variables. Considering Eqs. (34.118) and (34.117), and the errors $e_{v_{C_2}}$, e_θ, e_γ, and e_β, the sliding surface can be expressed as a combination of the rectifier currents, voltages, and their time derivatives:

$$S(e_{x_i}, t) = e_{v_{C_2}} + k_\theta e_\theta + k_\gamma e_\gamma + k_\beta e_\beta$$

$$= v_{C_2 r} + k_\theta \theta_r + k_\gamma \gamma_r + k_\beta \beta_r - \left(1 - \frac{k_\gamma}{C_2 L_2}\right) v_{C_2}$$

$$- \frac{k_\gamma}{C_2 L_2} v_{C_1} + \left(\frac{k_\theta}{C_2} - \frac{k_\beta}{C_2^2 L_2}\right) i_o + \frac{k_\gamma}{C_2}\frac{di_o}{dt} + \frac{k_\beta}{C_2}\frac{d^2 i_o}{dt^2}$$

$$- \frac{k_\beta}{C_1 C_2 L_2} i_{L_1} - \left(\frac{k_\theta}{C_2} - \frac{k_\beta}{C_1 C_2 L_2} - \frac{k_\beta}{C_2^2 L_2}\right) i_{L_2} = 0$$

$$(34.119)$$

Equation (34.119) shows the variables to be measured (v_{C_2}, v_{C_1}, i_o, i_{L_1}, and i_{L_2}). Therefore, it can be concluded that the output current of each three-phase half-wave rectifier must be measured.

The existence of the sliding mode implies $S(e_{x_i}, t) = 0$ and $\dot{S}(e_{x_i}, t) = 0$. Given the state models (34.117, 34.118), and from $\dot{S}(e_{x_i}, t) = 0$, the available voltage of the power supply v_i must exceed the equivalent average dc input voltage V_{eq} (34.120), which should be applied at the filter input, in order that the system state slides along the sliding surface (34.119).

$$V_{eq} = \frac{C_1 L_1 C_2 L_2}{k_\beta}(\theta_r + k_\theta \gamma_r + k_\gamma \beta_r + k_\beta \dot{\beta}_r) + v_{c_2} - \frac{C_1 L_1 C_2 L_2}{k_\beta}$$
$$\times (\theta + k_\gamma \beta) + \left(C_2 L_2 + C_2 L_1 + C_1 L_1 - C_1 L_1 C_2 L_2 \frac{k_\theta}{k_\beta}\right)\gamma$$
$$+ (L_1 + L_2)\frac{di_o}{dt} + C_1 L_1 L_2 \frac{d^3 i_o}{dt^3} \quad (34.120)$$

This means that the power supply root mean square (RMS) voltage values should be chosen high enough to account for the maximum effects of the perturbations. This is almost the same criterion adopted when calculating the RMS voltage values needed with linear controllers. However, as the V_{eq} voltage contains the derivatives of the reference voltage, the system will not be able to stay in sliding mode with a step as the reference.

The switching law would be derived, considering that, from Eq. (34.118) $b_e(\mathbf{e}) > 0$. Therefore, from Eq. (34.97), if $S(e_{x_i}, t) > +\varepsilon$, then $v_i(t) = V_{eqmax}$, else if $S(e_{x_i}, t) < -\varepsilon$, then $v_i(t) = -V_{eqmax}$. However, because of the lack of gate turn-off capability of the rectifier thyristors, power rectifiers cannot generate the high-frequency switching voltage $v_i(t)$, since the statistical mean delay time is $T/2p(T = 20\,\text{ms})$ and reaches $T/2$ when switching from $+V_{eqmax}$ to $-V_{eqmax}$. To control mains switched rectifiers, the described constant-frequency sliding-mode operation method is used, in which the sliding surface $S(e_{x_i}, t)$ instead of being compared to zero, is compared to an auxiliary constant-frequency function $r(t)$ (Fig. 34.6b) synchronized with the mains frequency. The new switching law is

$$\left.\begin{array}{l}\text{If } k_p S(e_{x_i}, t) > r(t) + \iota \Rightarrow \text{Trigger the next thyristor}\\ \text{If } k_p S(e_{x_i}, t) < r(t) - \iota \Rightarrow \text{Do not trigger any}\\ \qquad\qquad\qquad\qquad\qquad\qquad\text{thyristor}\end{array}\right\} \Rightarrow v_i(t)$$
$$(34.121)$$

Since now $S(e_{x_i}, t)$ is not near zero, but around some value of $r(t)$, a steady-state error $e_{v_{c_{2av}}}$ appears ($\min[r(t)]/k_p < e_{v_{c_{2av}}} < \max[r(t)]/k_p$), as seen in Example 34.11. Increasing the value of k_p (toward the ideal saturation control) does not overcome this drawback, since oscillations would appear even for moderate k_p gains, because of the rectifier dynamics. Instead, the sliding surface (34.122), based on Eq. (34.99), should be used. It contains an integral term, which, given the canonical controllability form and the Routh–Hurwitz property, is the only nonzero term at steady state, enabling the complete elimination of the steady-state error.

$$S_i(e_{x_i}, t) = \int e_{v_{c_2}} dt + k_{1v} e_{v_{c_2}} + k_{1\theta} e_\theta + k_{1\gamma} e_\gamma + k_{1\beta} e_\beta$$
$$(34.122)$$

To determine the k constants of Eq. (34.122) a pole-placement technique is selected, according to a fourth-order Bessel polynomial $B_E(s)_m$, $m = 4$, from Eq. (34.88), in order to obtain the smallest possible response time with almost no overshoot. For a delay characteristic as flat as possible, the delay t_r is taken inversely proportional to a frequency f_{ci} just below the lowest cutoff frequency ($f_{ci} < 8.44\,\text{Hz}$) of the double LC filter. For this fourth-order filter, the delay is $t_r = 2.8/(2\pi f_{ci})$. By choosing $f_{ci} = 7\,\text{Hz}$ ($t_r \approx 64\,\text{ms}$), and dividing all the Bessel polynomial terms by st_r, the characteristic polynomial (34.123) is obtained:

$$S_i(e_{x_i}, s) = \frac{1}{st_r} + 1 + \frac{45}{105} st_r + \frac{10}{105} s^2 t_r^2 + \frac{1}{105} s^3 t_r^3$$
$$(34.123)$$

This polynomial must be applied to Eq. (34.122) to obtain the four sliding functions needed to derive the thyristor trigger pulses of the four three-phase half-wave rectifiers. These sliding functions will enable the control of the output current (i_{l_1}, i_{l_2}, i_{l_3}, and i_{l_4}) of each half-wave rectifier, improving the current sharing among them (Fig. 34.35b). Supposing equal current share, the relation between the i_{L_1} current and the output currents of each threephase rectifier is $i_{L_1} = 4i_{l_1} = 4i_{l_2} = 4i_{l_3} = 4i_{l_4}$. Therefore, for the nth half-wave three-phase rectifier, since for $n = 1$ and $n = 2$, $v_{c_1} = v_{c_{11}}$ and $i_{L_2} = 2i_{l_5}$ and for $n = 3$ and $n = 4$, $v_{c_1} = v_{c_{12}}$ and $i_{L_2} = 2i_{l_6}$, the four sliding surfaces are ($k_{1v} = 1$):

$$S_i(e_{x_i}, t)_n = \left[k_{1v} v_{c_{2r}} + \frac{45 t_r}{105}\theta_r + \frac{10 t_r^2}{105}\gamma_r + \frac{t_r^3}{105}\beta_r\right.$$
$$+ \frac{1}{t_r}\int v_{c_{2r}} - v_{c_2} dt - \left(\frac{k_{1v}}{C_2 L_2} - \frac{10 t_r^2}{105 C_2 L_2}\right)v_{c_2}$$
$$- \frac{10 t_r^2}{105 C_2 L_2} v_{c_{11_2}} + \left(\frac{45 t_r}{105 C_2} - \frac{t_r^3}{105 C_2^2 L_2}\right)i_o$$
$$\left.+ \left(\frac{10 t_r^2}{105 C_2}\right)\frac{di_o}{dt} + \left(\frac{t_r^3}{105 C_2}\right)\frac{d^2 i_o}{dt^2}\right]/4$$
$$- \left[\left(\frac{45 t_r}{105 C_2} - \frac{t_r^3}{105 C_2^2 L_2} - \frac{t_r^3}{105 C_1 L_2 C_2}\right)i_{l_{56}}\right]/2$$
$$- \left(\frac{t_r^3}{105 C_1 L_2 C_2}\right)i_{l_n} \quad (34.124)$$

If an inexpensive analog controller is desired, the successive time derivatives of the reference voltage and the output current of Eq. (34.124) can be neglected (furthermore, their calculation is noise prone). Nonzero errors on the first, second, and third-order derivatives of the controlled variable will appear, worsening the response speed. However, the steady-state error is not affected.

To implement the four equations (34.124), the variables v_{c_2}, $v_{c_{11}}$, $v_{c_{12}}$, i_o, i_{l_5}, i_{l_6}, i_{l_1}, i_{l_2}, i_{l_3}, and i_{l_4} must be measured. Although this could be done easily, it is very convenient to further simplify the practical controller, keeping its complexity and cost at the level of linear controllers, while maintaining the advantages of sliding mode. Therefore, the voltages $v_{c_{11}}$ and $v_{c_{12}}$ are assumed almost constant over one period of the filter input current, and $v_{c_{11}} = v_{c_{12}} = v_{c_2}$, meaning that $i_{l_5} = i_{l_6} = i_o/2$. With these assumptions, valid as the values of C' and C_2 are designed to provide an output voltage with very low ripple, the new sliding-mode functions are

$$S_i(e_{x_i},t)_n \approx \frac{\frac{1}{t_r}\int vc_{2r} - vc_2\, dt + k_{1v}(v_{c_{2r}} - v_{c_2}) + \frac{t_r^3}{105}\frac{1}{C_1 C_2 L_2} i_o}{4}$$

$$- \left(\frac{t_r^3}{105 C_1 L_2 C_2}\right) i_{ln} \quad (34.125)$$

These approximations disregard only the high-frequency content of $v_{c_{11}}$, $v_{c_{12}}$, i_{l_5}, and i_{l_6}, and do not affect the rectifier steady-state response, but the step response will be a little slower, although still much faster (150 ms, Fig. 34.39) than that obtained with linear controllers (280 ms, Fig. 34.38). Regardless of all the approximations, the low switching frequency of the rectifier would not allow the elimination of the dynamic errors. As a benefit of these approximations, the sliding-mode controller (Fig. 34.36a) will need only an extra current sensor (or a current observer) and an extra operational amplifier in comparison with linear controllers derived hereafter (which need four current sensors and six operational amplifiers). Compared to the total cost of the 12-pulse rectifier plus output filter, the control hardware cost is negligible in both the cases, even for medium-power applications.

34.3.5.6 Average Current-mode Control of the 12-pulse Rectifier

For comparison purposes a PI-based controller structure is designed (Fig. 34.36b), taking into account, that small mismatches of the line voltages or of the trigger angles can completely destroy the current share of the four paralleled rectifiers, inspite of the current equalizing inductances (l and l'). Output voltage control sensing only the output voltage is, therefore, not feasible. Instead, the slow and fast manifold approach is selected. For the fast manifold, four internal current control loops guarantee the same dc current level in each three-phase rectifier and limit the short-circuit currents. For the output slow dynamics, an external cascaded output voltage control loop (Fig. 34.36b), measuring the voltage applied to the load, is the minimum.

For a straightforward design, given the much slower dynamics of the capacitor voltages compared to the input current, the PI current controllers are calculated as shown in Example 34.6, considering the capacitor voltage constant during a switching period, and $r_t \approx 1\,\Omega$ the intrinsic resistance of the transformer windings, thyristor overlap, and inductor l. From Eq. (34.59), $T_z = l/r_t \approx 0.044$ s. From, Eq. (34.62), with the common assumptions, $T_p \approx 0.16 k_I$ s ($p = 3$). These values guarantee a small overshoot ($\approx 5\%$) and a current rise time of approximately $T/3$.

To design the external output voltage control loop, each current-controlled rectifier can be considered a voltage-controlled current source $i_{L_1}(s)/4$, since each half-wave rectifier current response will be much faster than the filter output voltage response. Therefore, in the equivalent circuit

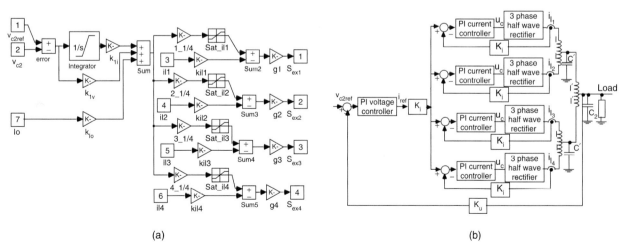

FIGURE 34.36 (a) Sliding-mode controller block diagram and (b) linear control hierarchy for the 12-pulse rectifier.

of Fig. 34.35b, the current source $i_{L_1}(s)$ substitutes the input inductor, yielding the transfer function $v_{c_2}(s)/i_{L_1}(s)$:

$$\frac{V_{C_2}(s)}{i_{L_1}(s)} = \frac{R_o}{C_2 C_1 L_2 R_o s^3 + C_1 L_2 s^2 + (C_2 R_o + C_1 R_o)s + 1} \quad (34.126)$$

Given the real pole ($p_1 = -6.7$) and two complex poles ($p_{2,3} = -6.65 \pm j140.9$) of Eq. (34.126), the PI voltage controller zero ($1/T_{zv} = p_1$) can be chosen with a value equal to the transfer function real pole. The integral gain T_{pv} can be determined using a root-locus analysis to determine the maximum gain, that still guarantees the stability of the closed-loop controlled system. The critical gain for the PI was found to be $T_{zv}/T_{pv} \approx 0.4$, then $T_{pv} > 0.37$. The value $T_{pv} \approx 2$ was selected to obtain weak oscillations, together with almost no overshoot.

The dynamic and steady-state responses of the output currents of the four rectifiers ($i_{l_1}, i_{l_2}, i_{l_3}, i_{l_4}$) and the output voltage v_{c_2} were analyzed using a step input from 2 to 2.5 A applied at $t = 1.1$ s, for the currents, and from 40 to 50 V for the v_{c_2} voltage. The PI current controllers (Fig. 34.37) show good sharing of the total current, a slight overshoot ($\zeta = 0.7$) and response time 6.6 ms (T/p).

The open-loop voltage v_{c_2} presents a rise time of 0.38 s. The PI voltage controller (Fig. 34.38) shows a response time of 0.4 s, no overshoot. The four three-phase half-wave rectifier output currents ($i_{l_1}, i_{l_2}, i_{l_3}$, and i_{l_4}) present nearly the same transient and steady-state values, with no very high current peaks. These results validate the assumptions made in the PI design.

The closed-loop performance of the fixed-frequency sliding-mode controller (Fig. 34.39) shows that all the $i_{l_1}, i_{l_2}, i_{l_3}$, and i_{l_4} currents are almost equal and have peak values only slightly higher than those obtained with the PI linear controllers. The output voltage presents a much faster response time (150 ms) than the PI linear controllers, negligible or no steady-state error, and no overshoot. From these waveforms

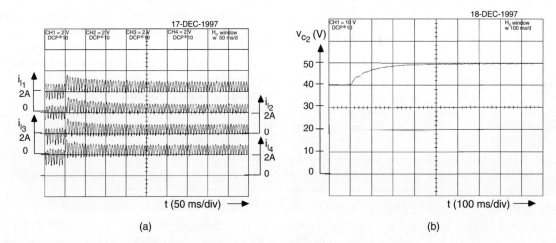

FIGURE 34.37 PI current controller performance: (a) $i_{l_1}, i_{l_2}, i_{l_3}, i_{l_4}$ closed-loop currents and (b) open-loop output voltage V_{C_2}.

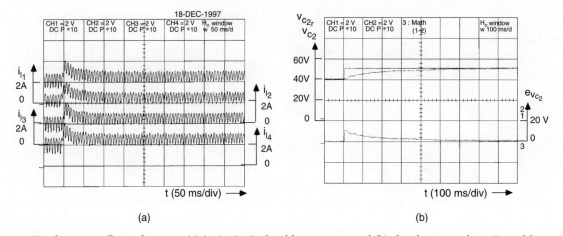

FIGURE 34.38 PI voltage controller performance: (a) $i_{l_1}, i_{l_2}, i_{l_3}, i_{l_4}$ closed-loop currents and (b) closed output voltage V_{C_2} and $I_{v_{C_2}}$ output voltage error.

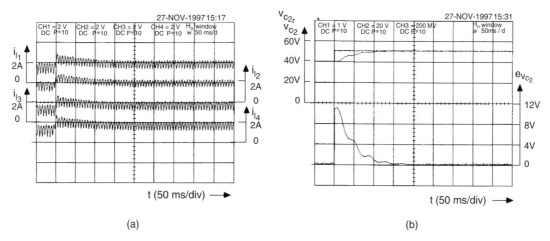

FIGURE 34.39 Closed-loop constant-frequency sliding-mode controller performance: (a) i_{l_1}, i_{l_2}, i_{l_3}, i_{l_4} closed-loop currents and (b) closed output voltage V_{C_2} and $e_{v_{C_2}}$ output voltage error.

it can be concluded that the sliding-mode controller provides a much more effective control of the rectifier, as the output voltage response time is much lower than the obtained with PI linear controllers, without significantly increasing the thyristor currents, overshoots, or costs. Furthermore, sliding mode is an elegant way to know the variables to be measured, and to design all the controller and the modulator electronics.

EXAMPLE 34.13 Sliding-mode control of pulse width modulation audio power amplifiers

Linear audio power amplifiers can be astonishing, but have efficiencies as low as 15–20% with speech or music signals. To improve the efficiency of audio systems while preserving the quality, PWM switching power amplifiers, enabling the reduction of the power-supply cost, volume, and weight and compensating the efficiency loss of modern loudspeakers, are needed. Moreover, PWM amplifiers can provide a complete digital solution for audio power processing.

For high-fidelity systems, PWM audio amplifiers must present flat passbands of at least 16–20 kHz (± 0.5 dB), distortions less than 0.1% at the rated output power, fast dynamic response, and signal-to-noise ratios above 90 dB. This requires fast power semiconductors (usually metal-oxide semiconductor field effect transistor (MOSFET) transistors), capable of switching at frequencies near 500 kHz, and fast nonlinear controllers to provide the precise and timely control actions needed to accomplish the mentioned requirements and to eliminate the phase delays in the LC output filter and loudspeakers.

A low-cost PWM audio power amplifier, able to provide over 80 W to 8 Ω loads ($V_{dd} = 50$ V), can be obtained using a half-bridge power inverter (switching at $f_{PWM} \approx 450$ kHz), coupled to an output filter for high-frequency attenuation (Fig. 34.40). A low-sensitivity,

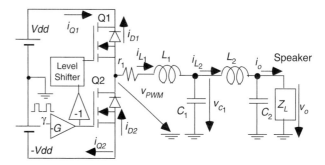

FIGURE 34.40 PWM audio amplifier with fourth-order Chebyshev low-pass output filter and loudspeaker load.

doubly terminated passive ladder (double LC), low-pass filter using fourth-order Chebyshev approximation polynomials is selected, given its ability to meet, while minimizing the number of inductors, the following requirements: passband edge frequency 21 kHz, passband ripple 0.5 dB, stopband edge frequency 300 kHz and 90 dB minimum attenuation in the stopband ($L_1 = 80\,\mu$H; $L_2 = 85\,\mu$H; $C_1 = 1.7\,\mu$F; $C_2 = 820$ nF; $R_2 = 8\,\Omega$; $r_1 = 0.47\,\Omega$).

34.3.5.7 Modeling the PWM Audio Amplifier

The two half-bridge switches must always be in complementary states, to avoid power supply internal short-circuits. Their state can be represented by the time-dependent variable γ, which is $\gamma = 1$ when Q1 is on and Q2 is off, and is $\gamma = -1$ when Q1 is off and Q2 is on.

Neglecting switch delays, on state semiconductor voltage drops, auxiliary networks, and supposing small dead times, the half-bridge output voltage (v_{PWM}) is $v_{PWM} = \gamma V_{dd}$. Considering the state variables and circuit components of

Fig. 34.40, and modeling the loudspeaker load as a disturbance represented by the current i_o (ensuring robustness to the frequency dependent impedance of the speaker), the switched state-space model of the PWM audio amplifier is

$$\frac{d}{dt}\begin{bmatrix} i_{L_1} \\ v_{C_1} \\ i_{L_2} \\ v_o \end{bmatrix} = \begin{bmatrix} -r_1/L_1 & -1/L_1 & 0 & 0 \\ 1/C_1 & 0 & -1/C_1 & 0 \\ 0 & 1/L_2 & 0 & -1/L_2 \\ 0 & 0 & 1/C_2 & 0 \end{bmatrix} \begin{bmatrix} i_{L_1} \\ v_{C_1} \\ i_{L_2} \\ v_o \end{bmatrix}$$
$$+ \begin{bmatrix} 1/L_1 & 0 \\ 0 & 0 \\ 0 & 0 \\ 0 & -1/C_2 \end{bmatrix} \begin{bmatrix} \gamma V_{dd} \\ i_o \end{bmatrix} \quad (34.127)$$

This model will be used to define the output voltage v_o controller.

34.3.5.8 Sliding-mode Control of the PWM Audio Amplifier

The filter output voltage v_o, divided by the amplifier gain ($1/k_v$), must follow a reference v_{o_r}. Defining the output error as $e_{v_o} = v_{o_r} - k_v v_o$, and also using its time derivatives ($e_\theta, e_\gamma, e_\beta$) as a new state vector $\mathbf{e} = [e_{v_o}, e_\theta, e_\gamma, e_\beta]^T$, the system equations, in the phase canonical (or controllability) form, can be written in the form

$$\frac{d}{dt}[e_{v_o}, e_\theta, e_\gamma, e_\beta]^T = [e_\theta, e_\gamma, e_\beta, -f(e_{v_o}, e_\theta, e_\gamma, e_\beta) + p_e(t)$$
$$- \gamma V_{dd}/C_1 L_1 C_2 L_2]^T \quad (34.128)$$

Sliding-mode control of the output voltage will enable a robust and reduced-order dynamics, independent of semiconductors, power supply, filter, and load parameters. According to Eqs. (34.91) and (34.128), the sliding surface is

$$S(e_{v_o}, e_\theta, e_\gamma, e_\beta, t) = e_{v_o} + k_\theta e_\theta + k_\gamma e_\gamma + k_\beta e_\beta$$
$$= v_{o_r} - k_v v_o + k_\theta \frac{d(v_{o_r} - k_v v_o)}{dt}$$
$$+ k_\gamma \frac{d}{dt}\left(\frac{d(v_{o_r} - k_v v_o)}{dt}\right)$$
$$+ k_\beta \frac{d}{dt}\left[\frac{d}{dt}\left(\frac{d(v_{o_r} - k_v v_o)}{dt}\right)\right] = 0 \quad (34.129)$$

In sliding mode, Eq. (34.129) confirms the amplifier gain ($v_o/v_{o_r} = 1/k_v$). To obtain a stable system and the smallest possible response time t_r, a pole placement according to a third-order Bessel polynomial is used. Taking t_r inversely proportional to a frequency just below the lowest cutoff frequency (ω_1) of the double LC filter ($t_r \approx 2.8/\omega_1 \approx 2.8/(2\pi \times 21\,\text{kHz}) \approx 20\,\mu\text{s}$) and using Eq. (34.88) with $m = 3$, the characteristic polynomial Eq. (34.130), verifying the Routh–Hurwitz criterion is obtained.

$$S(\mathbf{e}, s) = 1 + st_r + \frac{6}{15}(st_r)^2 + \frac{1}{15}(st_r)^3 \quad (34.130)$$

From Eq. (34.97) the switching law for the control input at time t_k, $\gamma(t_k)$, must be

$$\gamma(t_k) = \text{sgn}\{S(\mathbf{e}, t_k) + \varepsilon \, \text{sgn}[S(\mathbf{e}, t_{k-1})]\} \quad (34.131)$$

To ensure reaching and existence conditions, the power supply voltage V_{dd} must be greater than the maximum required mean value of the output voltage in a switching period $V_{dd} > (\overline{v_{PWMmax}})$. The sliding-mode controller (Fig. 34.41) is obtained from Eqs. (34.129–34.131) with $k_\theta = t_r$, $k_\gamma = 6t_r^2/15$, $k_\beta = t_r^3/15$. The derivatives can be approximated by the block diagram of Fig. 34.41b, were h is the oversampling period.

Fig. 34.42a shows the v_{PWM}, v_{o_r}, $v_o/10$, and the error $10 \times (v_{o_r} - v_o/10)$ waveforms for a 20 kHz sine input. The overall behavior is much better than the obtained with the sigma-delta controllers (Figs. 34.43 and 34.44) explained below for comparison purposes. There is no 0.5 dB loss or phase delay over the entire audio band; the Chebyshev filter behaves as a maximally flat filter, with higher stopband attenuation. Fig. 34.42b shows v_{PWM}, v_{o_r}, and $10 \times (v_{o_r} - v_o/10)$ with a 1 kHz square input. There is almost no steady-state error and almost no overshoot on the speaker voltage v_o, attesting to the speed of response ($t \approx 20\,\mu\text{s}$ as designed, since, in contrast to Example 34.12, no derivatives were neglected). The stability, the system order reduction, and the sliding-mode controller usefulness for the PWM audio amplifier are also shown.

34.3.5.9 Sigma Delta Controlled PWM Audio Amplifier

Assume now the fourth-order Chebyshev low-pass filter, as an ideal filter removing the high-frequency content of the v_{PWM} voltage. Then, the v_{PWM} voltage can be considered as the amplifier output. However, the discontinuous voltage $v_{PWM} = \gamma V_{dd}$ is not a state variable and cannot follow the almost continuous reference v_{PWM_r}. The new error variable $e_{vPWM} = v_{PWM_r} - k_v \gamma V_{dd}$ is always far from the zero value. Given this nonzero error, the approach outlined in Section 34.3.4 can be used. The switching law remains Eq. (34.131), but the new control law Eq. (34.132) is

$$S(e_{vPWM}, t) = \kappa \int (v_{PWM_r} - k_v \gamma V_{dd}) dt = 0 \quad (34.132)$$

34 Control Methods for Switching Power Converters

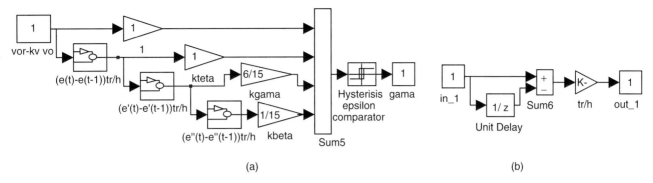

FIGURE 34.41 (a) Sliding-mode controller for the PWM audio amplifier and (b) implementation of the derivative blocks.

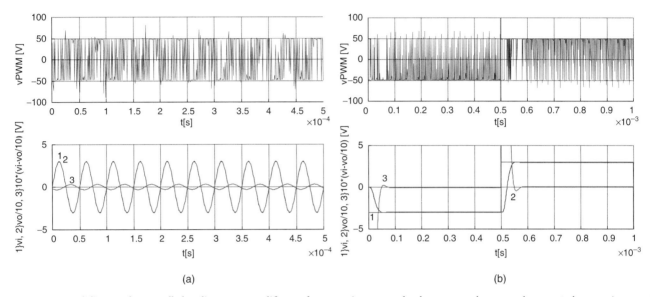

FIGURE 34.42 Sliding-mode controlled audio power amplifier performance (upper graphs show v_{PWM}, lower graph traces 1 show v_{o_r} ($v_{o_r} \equiv v_i$), lower graph traces 2 show $v_o/10$, and lower graph traces 3 show $10 \times (v_{o_r} - v_o/10)$): (a) response to a 20 kHz sine input, at 55 W output power and (b) response to 1 kHz square wave input, at 100 W output power.

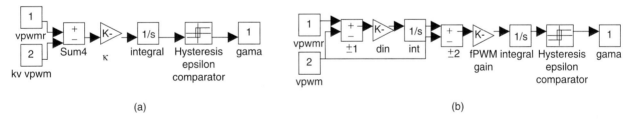

FIGURE 34.43 (a) First-order sigma delta modulator and (b) second-order sigma delta modulator.

The κ parameter is calculated to impose the maximum switching frequency f_{PWM}. Since $\kappa \int_0^{1/2f_{PWM}} (v_{PWM_{rmax}} + k_v V_{dd}) dt = 2\varepsilon$, we obtain

$$f_{PWM} = \kappa(v_{PWM_{rmax}} + k_v V_{dd})/(4\varepsilon) \quad (34.133)$$

Assuming that v_{PWM_r} is nearly constant over the switching period $1/f_{PWM}$, Eq. (34.132) confirms the amplifier gain, since $\overline{v_{PWM}} = v_{PWM_r}/k_v$.

Practical implementation of this control strategy can be done using an integrator with gain κ ($\kappa \approx 1800$), and a comparator with hysteresis ε ($\varepsilon \approx 6$ mV), Fig. 34.43a. Such an arrangement is called a first-order sigma-delta ($\Sigma\Delta$) modulator.

Fig. 34.44a shows the v_{PWM}, v_{o_r}, and $v_o/10$ waveforms for a 20 kHz sine input. The overall behavior is as expected, because the practical filter and loudspeaker are not ideal, but notice the

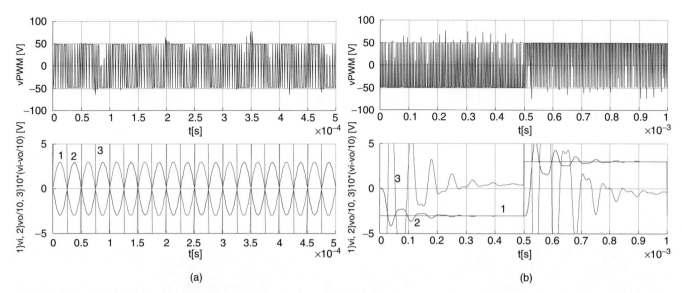

FIGURE 34.44 First-order sigma-delta audio amplifier performance (upper graphs show v_{PWM}, lower graphs trace 1 show $v_{o_r} \equiv v_i$, lower graphs trace 2 shows $v_o/10$, and lower graphs trace 3 show $10 \times (v_{o_r} - v_o/10)$): (a) response to a 20 kHz sine input, at 55 W output power and (b) response to 1 kHz square wave input, at 100 W output power.

0.5 dB loss and phase delay of the speaker voltage v_o, mainly due to the output filter and speaker inductance. In Fig. 34.44b, the v_{PWM}, v_{o_r}, $v_o/10$, and error $10 \times (v_{o_r} - v_o/10)$ for a 1 kHz square input are shown. Note the oscillations and steady-state error of the speaker voltage v_o, due to the filter dynamics and double termination.

A second-order sigma-delta modulator is a better compromise between circuit complexity and signal-to-quantization noise ratio. As the switching frequency of the two power MOSFET (Fig. 34.40) cannot be further increased, the second-order structure named "cascaded integrators with feedback" (Fig. 34.43b) was selected, and designed to eliminate the step response overshoot found in Fig. 34.44b.

Fig. 34.45a, for 1 kHz square input, shows much less overshoot and oscillations than Fig. 34.44b. However, the v_{PWM}, v_{o_r}, and $v_o/10$ waveforms, for a 20 kHz sine input presented

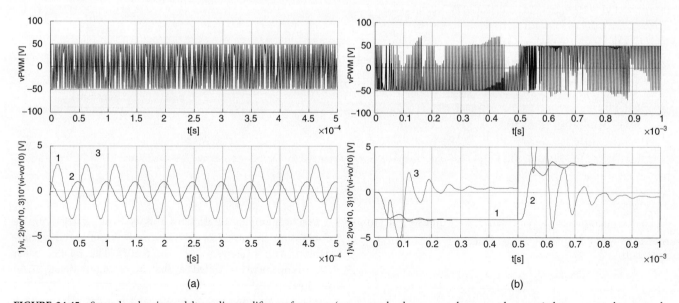

FIGURE 34.45 Second-order sigma-delta audio amplifier performance (upper graphs show v_{PWM}, lower graphs trace 1 show $v_{o_r} \equiv v_i$, lower graphs trace 2 show $v_o/10$, and lower graphs trace 3 show $10 \times (v_{o_r} - v_o/10)$): (a) response to 1 kHz square wave input, at 100 W output power and (b) response to a 20 kHz sine input, at 55 W output power.

in Fig. 34.45b, show increased output voltage loss, compared to the first-order sigma-delta modulator, since the second-order modulator was designed to eliminate the v_o output voltage ringing (therefore reducing the amplifier bandwidth). The obtained performances with these and other sigma-delta structures are inferior to the sliding-mode performances (Fig. 34.42). Sliding mode brings definite advantages as the system order is reduced, flatter passbands are obtained, power supply rejection ratio is increased, and the nonlinear effects, together with the frequency-dependent phase delays, are cancelled out.

EXAMPLE 34.14 Sliding-mode control of near unity power factor PWM rectifiers

Boost-type voltage-sourced three-phase rectifiers (Fig. 34.46) are multiple-input multiple-output (MIMO) systems capable of bidirectional power flow, near unity power factor operation, and almost sinusoidal input currents, and can behave as ac/dc power supplies or power factor compensators.

The fast power semiconductors used (usually MOSFETs or IGBTs) can switch at frequencies much higher than the mains frequency, enabling the voltage controller to provide an output voltage with fast dynamic response.

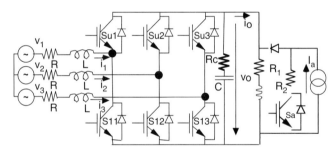

FIGURE 34.46 Voltage-sourced PWM rectifier with IGBTs and test load.

34.3.5.10 Modeling the PWM Boost Rectifier

Neglecting switch delays and dead times, the states of the switches of the kth inverter leg (Fig. 34.46) can be represented by the time-dependent nonlinear variables γ_k, defined as

$$\gamma_k = \begin{cases} 1 > \text{if } Su_k \text{ is on and } Sl_k \text{ is off} \\ 0 > \text{if } Su_k \text{ is off and } Sl_k \text{ is on} \end{cases} \quad (34.134)$$

Consider the displayed variables of the circuit (Fig. 34.46), where L is the value of the boost inductors, R their resistance, C the value of the output capacitor, and R_c its equivalent series resistance (ESR). Neglecting semiconductor voltage drops, leakage currents, and auxiliary networks, the application of Kirchhoff laws (taking the load current i_o as a time-dependent perturbation) yields the following switched state-space model of the boost rectifier:

$$\frac{d}{dt}\begin{bmatrix} i_1 \\ i_2 \\ i_3 \\ v_o \end{bmatrix} = \begin{bmatrix} -R/L & 0 & 0 & -2\gamma_1+\gamma_2+\gamma_3/3L \\ 0 & -R/L & 0 & -2\gamma_2+\gamma_3+\gamma_1/3L \\ 0 & 0 & -R/L & -2\gamma_3+\gamma_1+\gamma_2/3L \\ A_{41} & A_{42} & A_{43} & A_{44} \end{bmatrix}\begin{bmatrix} i_1 \\ i_2 \\ i_3 \\ v_o \end{bmatrix}$$

$$+ \begin{bmatrix} 1/L & 0 & 0 & 0 & 0 \\ 0 & 1/L & 0 & 0 & 0 \\ 0 & 0 & 1/L & 0 & 0 \\ \gamma_1 R_c/L & \gamma_2 R_c/L & \gamma_3 R_c/L & -1/C & -R_c \end{bmatrix}\begin{bmatrix} v_1 \\ v_2 \\ v_3 \\ i_o \\ di_o/dt \end{bmatrix}$$

(34.135)

where $A_{41} = \gamma_1\left(\frac{1}{C} - \frac{RR_c}{L}\right); A_{42} = \gamma_2\left(\frac{1}{C} - \frac{RR_c}{L}\right); A_{43} = \gamma_3\left(\frac{1}{C} - \frac{RR_c}{L}\right); A_{44} = \frac{-2R_c(\gamma_1(\gamma_1-\gamma_2)+\gamma_2(\gamma_2-\gamma_3)+\gamma_3(\gamma_3-\gamma_1))}{3L}$.

Since the input voltage sources have no neutral connection, the preceding model can be simplified, eliminating one equation. Using the relationship (34.136) between the fixed frames $x_{1,2,3}$ and $x_{\alpha,\beta}$, in Eq. (34.135), the state-space model (34.137), in the α, β frame, is obtained.

$$\begin{bmatrix} x_1 \\ x_2 \end{bmatrix} = \begin{bmatrix} \sqrt{2/3} & 0 \\ -\sqrt{1/6} & \sqrt{1/2} \end{bmatrix}\begin{bmatrix} x_\alpha \\ x_\beta \end{bmatrix} \quad (34.136)$$

$$\frac{d}{dt}\begin{bmatrix} i_\alpha \\ i_\beta \\ v_o \end{bmatrix} = \begin{bmatrix} -R/L & \omega & -\gamma_\alpha/L \\ 0 & -R/L & -\gamma_\beta/L \\ A_{31}^\alpha & A_{32}^\alpha & A_{33}^\alpha \end{bmatrix}\begin{bmatrix} i_\alpha \\ i_\beta \\ v_o \end{bmatrix}$$

$$+ \begin{bmatrix} 1/L & 0 & 0 & 0 \\ 0 & 1/L & 0 & 0 \\ \gamma_\alpha R_c/L & \gamma_\beta R_c/L & -1/C & -R_c \end{bmatrix}\begin{bmatrix} v_\alpha \\ v_\beta \\ i_o \\ di_o/dt \end{bmatrix}$$

(34.137)

where $A_{31}^\alpha = \gamma_\alpha\left(\frac{1}{C} - \frac{RR_c}{L}\right); A_{32}^\alpha = \gamma_\beta\left(\frac{1}{C} - \frac{RR_c}{L}\right); A_{33}^\alpha = \frac{-R_c(\gamma_\alpha^2+\gamma_\beta^2)}{L}$.

34.3.5.11 Sliding-mode Control of the PWM Rectifier

The model (34.137) is nonlinear and time-variant. Applying the Park transformation (34.138), using a frequency ω rotating reference frame synchronized with the mains (with the q component of the supply voltages equal to zero), the nonlinear, time-invariant model (34.139) is written:

$$\begin{bmatrix} i_a \\ i_b \end{bmatrix} = \begin{bmatrix} \cos(\omega t) & -\sin(\omega t) \\ \sin(\omega t) & \cos(\omega t) \end{bmatrix} \begin{bmatrix} i_d \\ i_q \end{bmatrix} \quad (34.138)$$

$$\frac{d}{dt}\begin{bmatrix} i_d \\ i_q \\ v_o \end{bmatrix} = \begin{bmatrix} -R/L & \omega & -\gamma_d/L \\ -\omega & -R/L & -\gamma_q/L \\ A_{31}^d & A_{32}^d & A_{33}^d \end{bmatrix} \begin{bmatrix} i_d \\ i_q \\ v_o \end{bmatrix}$$

$$+ \begin{bmatrix} 1/L & 0 & 0 & 0 \\ 0 & 1/L & 0 & 0 \\ \gamma_d R_c/L & \gamma_q R_c/L & -1/C & -R_c \end{bmatrix} \begin{bmatrix} v_d \\ v_q \\ i_o \\ di_o dt \end{bmatrix}$$

$$(34.139)$$

where $A_{31}^d = \gamma_d \left(\frac{1}{C} - \frac{RR_c}{L} \right)$; $A_{32}^d = \gamma_q \left(\frac{1}{C} - \frac{RR_c}{L} \right)$; $A_{33}^d = \frac{-R_c \left(\gamma_d^2 + \gamma_q^2 \right)}{L}$.

This state-space model can be used to obtain the feedback controllers for the PWM boost rectifier. Considering the output voltage v_o and the i_q current as the controlled outputs and γ_d, γ_q the control inputs (MIMO system), the input–output linearization of Eq. (34.72) gives the state-space equations in the controllability canonical form (34.140):

$$\frac{di_q}{dt} = -\omega i_d - \frac{R}{L} i_q - \frac{\gamma_q}{L} v_o + \frac{1}{L} v_q$$

$$\frac{dv_o}{dt} = \theta$$

$$\frac{d\theta}{dt} = \frac{R + R_c \left(\gamma_d^2 + \gamma_q^2 \right)}{L} \theta - \frac{\gamma_d^2 + \gamma_q^2}{LC} v_o$$

$$+ \frac{\gamma_d v_d + \gamma_q v_q}{LC} - \frac{R i_o}{LC} - \left(\frac{1}{C} + \frac{RR_c}{L} \right) \frac{di_o}{dt} \quad (34.140)$$

$$+ \omega \left(\frac{1}{C} - \frac{RR_c}{L} \right) (\gamma_d i_q - \gamma_q i_d) - R_c \frac{d^2 i_o}{dt^2}$$

where

$$\theta = \left(\frac{1}{C} - \frac{RR_c}{L} \right) (\gamma_d i_d + \gamma_q i_q) - \frac{R_c \left(\gamma_d^2 + \gamma_q^2 \right)}{L} v_o$$

$$+ \frac{R_c}{L} (\gamma_d v_d + \gamma_q v_q) - \frac{i_o}{C} - R_c \frac{di_o}{dt}.$$

Using the rectifier overall power balance (from Tellegen's theorem, the converter is conservative, i.e. the power delivered to the load or dissipated in the converter intrinsic devices equals the input power), and neglecting the switching and output capacitor losses, $v_d i_d + v_q i_q = v_o i_o + R i_d^2$. Supposing unity power factor ($i_{qr} \approx 0$), and the output v_o at steady state, $\gamma_d i_d + \gamma_q i_q \approx i_o$, $v_d = \sqrt{3} V_{RMS}$, $v_q = 0$, $\gamma_q \approx v_q/v_o$, $\gamma_d \approx (v_d - R i_d)/v_o$. Then, from Eqs. (34.140) and (34.91), the following two sliding surfaces can be derived:

$$S_q(e_{i_q}, t) = k_{e_{i_q}}(i_{qr} - i_q) = 0 \quad (34.141)$$

$$S_d(e_{v_o}, e_\theta, t) \approx \left[\beta^{-1}(v_{o_r} - v_o) + \frac{dv_{o_r}}{dt} + \frac{1}{C} i_o + R_c \frac{di_o}{dt} \right]$$

$$\times \frac{LC}{L - CRR_c} \frac{v_o}{\sqrt{3} V_{RMS} - R i_d} - i_d = i_{d_r} - i_d = 0 \quad (34.142)$$

where β^{-1} is the time constant of the desired first-order response of output voltage v_o ($\beta \gg T > 0$). For the synthesis of the closed-loop control system, notice that the terms of Eq. (34.142) inside the square brackets can be assumed as the i_d reference current i_{d_r}. Furthermore, from Eqs. (34.141) and (34.142) it is seen that the current control loops for i_d and i_q are needed. Considering Eqs. (34.138) and (34.136), the two sliding surfaces can be written

$$S_\alpha(e_{i_\alpha}, t) = i_{\alpha_r} - i_\alpha = 0 \quad (34.143)$$

$$S_\beta(e_{i_\beta}, t) = i_{\beta_r} - i_\beta = 0 \quad (34.144)$$

The switching laws relating the sliding surfaces (34.143, 34.144) with the switching variables γ_k are

$$\begin{cases} \text{If } S_{\alpha\beta}(e_{i_{\alpha\beta}}, t) > \varepsilon \text{ then } i_{\alpha\beta_r} > i_{\alpha\beta} \text{ hence choose } \gamma_k \text{ to} \\ \qquad \text{increase the } i_{\alpha\beta} \text{ current} \\ \text{If } S_{\alpha\beta}(e_{i_{\alpha\beta}}, t) < -\varepsilon \text{ then } i_{\alpha\beta_r} < i_{\alpha\beta} \text{ hence choose } \gamma_k \text{ to} \\ \qquad \text{decrease the } i_{\alpha\beta} \text{ current} \end{cases} \quad (34.145)$$

The practical implementation of this switching strategy could be accomplished using three independent two-level hysteresis comparators. However, this might introduce limit cycles as only two line currents are independent. Therefore, the control laws (34.143, 34.144) can be implemented using the block diagram of Fig. 34.47a, with $d, q/\alpha, \beta$ (from Eq. (34.138)) and $1,2,3/\alpha,\beta$ (from Eq. (34.136)) transformations and two three-level hysteretic comparators with equivalent hysteresis ε and ρ to limit the maximum switching frequency. A limiter is included to bound the i_d reference current to i_{dmax}, keeping the input line currents within a safe value. This helps to eliminate the nonminimum-phase behavior (outside sliding mode) when large transients are present, while providing short-circuit proof operation.

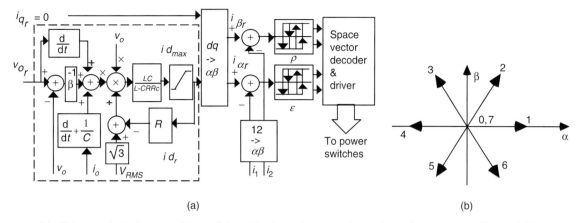

FIGURE 34.47 (a) Sliding-mode PWM controller modulator for the unity power factor three-phase PWM rectifier and (b) α, β space vector representation of the PWM bridge rectifier leg voltages.

34.3.5.12 α, β Space Vector Current Modulator

Depending on the values of γ_k, the bridge rectifier leg output voltages can assume only eight possible distinct states represented as voltage vectors in the α, β reference frame (Fig. 34.47b), for sources with isolated neutral.

With only two independent currents, two three-level hysteresis comparators, for the current errors, must be used in order to accurately select all eight available voltage vectors. Each three-level comparator can be obtained by summing the outputs of two comparators with two levels each. One of these two comparators ($\delta_{L\alpha}$, $\delta_{L\beta}$) has a wide hysteresis width and the other ($\delta_{N\alpha}$, $\delta_{N\beta}$) has a narrower hysteresis width. The hysteresis bands are represented by ε and ρ. Table 34.1 represents all possible output combinations of the resulting four two-level comparators, their sums giving the two three-level comparators (δ_α, δ_β), plus the voltage vector needed to accomplish the current tracking strategy $(i_{\alpha,\beta_r} - i_{\alpha,\beta}) = 0$ (ensuring $(i_{\alpha,\beta_r} - i_{\alpha,\beta}) \times d(i_{\alpha,\beta_r} - i_{\alpha,\beta})/dt < 0$), plus the γ_k variables and the α, β voltage components.

From the analysis of the PWM boost rectifier it is concluded that, if, for example, the voltage vector 2 is applied ($\gamma_1 = 1$, $\gamma_2 = 1$, $\gamma_3 = 0$), in boost operation, the currents i_α and i_β will both decrease. Oppositely, if the voltage vector 5 ($\gamma_1 = 0$, $\gamma_2 = 0$, $\gamma_3 = 1$) is applied, the currents i_α and i_β will both increase. Therefore, vector 2 should be selected when both i_α and i_β currents are above their respective references, that is for $\delta_\alpha = -1$, $\delta_\beta = -1$, whereas vector 5 must be chosen when both i_α and i_β currents are under their respective references, or for $\delta_\alpha = 1$, $\delta_\beta = 1$. Nearly all the outputs of Table 34.2 can be filled using this kind of reasoning.

TABLE 34.2 Two-level and three-level comparator results, showing corresponding vector choice, corresponding γ_k and vector α, β component voltages; vectors are mapped in Fig. 34.47b

$\delta_{L\alpha}$	$\delta_{N\alpha}$	$\delta_{L\beta}$	$\delta_{N\beta}$	δ_α	δ_β	**Vector**	γ_1	γ_2	γ_3	v_α	v_β
−0.5	−0.5	−0.5	−0.5	−1	−1	2	1	1	0	$v_o/\sqrt{6}$	$v_o/\sqrt{2}$
0.5	−0.5	−0.5	−0.5	0	−1	2	1	1	0	$v_o/\sqrt{6}$	$v_o/\sqrt{2}$
0.5	0.5	−0.5	−0.5	1	−1	3	0	1	0	$-v_o/\sqrt{6}$	$v_o/\sqrt{2}$
−0.5	0.5	−0.5	−0.5	0	−1	3	0	1	0	$-v_o/\sqrt{6}$	$v_o/\sqrt{2}$
−0.5	0.5	0.5	−0.5	0	0	0 or 7	0 or 1	0 or 1	0 or 1	0	0
0.5	0.5	0.5	−0.5	1	0	4	0	1	1	$-\sqrt{2/3}v_o$	0
0.5	−0.5	0.5	−0.5	0	0	0 or 7	0 or 1	0 or 1	0 or 1	0	0
−0.5	−0.5	0.5	−0.5	−1	0	1	1	0	0	$\sqrt{2/3}v_o$	0
−0.5	−0.5	0.5	0.5	−1	1	6	1	0	1	$v_o/\sqrt{6}$	$-v_o/\sqrt{2}$
0.5	−0.5	0.5	0.5	0	1	6	1	0	1	$v_o/\sqrt{6}$	$-v_o/\sqrt{2}$
0.5	0.5	0.5	0.5	1	1	5	0	0	1	$-v_o/\sqrt{6}$	$-v_o/\sqrt{2}$
−0.5	0.5	0.5	0.5	0	1	5	0	0	1	$-v_o/\sqrt{6}$	$-v_o/\sqrt{2}$
−0.5	0.5	−0.5	0.5	0	0	0 or 7	0 or 1	0 or 1	0 or 1	0	0
0.5	0.5	−0.5	0.5	1	0	4	0	1	1	$-\sqrt{2/3}v_o$	0
0.5	−0.5	−0.5	0.5	0	0	0 or 7	0 or 1	0 or 1	0 or 1	0	0
−0.5	−0.5	−0.5	0.5	−1	0	1	1	0	0	$\sqrt{2/3}v_o$	0

The cases where $\delta_\alpha = 0$, $\delta_\beta = -1$, the vector is selected upon the value of the i_α current error (if $\delta_{L\alpha} > 0$ and $\delta_{N\alpha} < 0$ then vector 2, if $\delta_{L\alpha} < 0$ and $\delta_{N\alpha} > 0$ then vector 3). When $\delta_\alpha = 0$, $\delta_\beta = 1$, if $\delta_{L\alpha} > 0$ and $\delta_{N\alpha} < 0$ then vector 6, else if $\delta_{L\alpha} < 0$ and $\delta_{N\alpha} > 0$ then vector 5. The vectors 0 and 7 are selected in order to minimize the switching frequency (if two of the three upper switches are on, then vector 7, otherwise vector 0). The space-vector decoder can be stored in a lookup table (or in an EPROM) whose inputs are the four two-level comparator outputs and the logic result of the operations needed to select between vectors 0 and 7.

34.3.5.13 PI Output Voltage Control of the Current-mode PWM Rectifier

Using the α, β current-mode hysteresis modulators to enforce the i_d and i_q currents to follow their reference values, i_{d_r}, i_{qr} (the values of L and C are such that the i_d and i_q currents usually exhibit a very fast dynamics compared to the slow dynamics of v_o), a first-order model (34.146) of the rectifier output voltage can be obtained from Eq. (34.73).

$$\frac{dv_o}{dt} = \left(\frac{1}{C} - \frac{RR_c}{L}\right)(\gamma_d i_{d_r} + \gamma_q i_{qr}) - \frac{R_c\left(\gamma_d^2 + \gamma_q^2\right)}{L}v_o$$
$$+ \frac{R_c}{L}(\gamma_d v_d + \gamma_q v_q) - \frac{i_o}{C} - R_c\frac{di_o}{dt} \quad (34.146)$$

Assuming now a pure resistor load $R_1 = v_o/i_o$, and a mean delay T_d between the i_d current and the reference i_{d_r}, continuous transfer functions result for the i_d current ($i_d = i_{dr}(1 + sT_d)^{-1}$) and for the v_o voltage ($v_o = k_A i_d/(1 + sk_B)$ with k_A and k_B obtained from Eq. (34.146)). Therefore, using the same approach as Examples 34.6, 34.8, and 34.11, a linear PI regulator, with gains K_p and K_i (34.147), sampling the error between the output voltage reference v_{o_r} and the output v_o, can be designed to provide a voltage proportional (k_I) to the reference current i_{d_r} ($i_{d_r} = (K_p + K_i/s)k_I(v_{o_r} - v_o)$).

$$K_p = \frac{R_1 + R_c}{4\zeta^2 T_d R_1 K_1 \gamma_d(1/C - RR_c/L)}$$
$$K_i = \frac{\left(R_c(\gamma_d^2 + \gamma_q^2)/L\right) + (1/R_1 C)}{4\zeta^2 T_d K_1 \gamma_d(1/C - RR_c/L)} \quad (34.147)$$

These PI regulator parameters depend on the load resistance R_1, on the rectifier parameters (C, R_c, L, R), on the rectifier operating point γ_d, on the mean delay time T_d, and on the required damping factor ζ. Therefore, the expected response can only be obtained with the nominal load and input voltages, the line current dynamics depending on the K_p and K_i gains.

Results (Fig. 34.48) obtained with the values $V_{RMS} \approx 70$ V, $L \approx 1.1$ mH, $R \approx 0.1\,\Omega$, $C \approx 2000\,\mu$F with equivalent series resistance ESR$\approx 0.1\,\Omega$ ($R_c \approx 0.1\,\Omega$), $R_1 \approx 25\,\Omega$, $R_2 \approx 12\,\Omega$, $\beta = 0.0012$, $K_p = 1.2$, $K_i = 100$, $k_I = 1$, show that the α, β space vector current modulator ensures the current tracking needed (Fig. 34.48) [17]. The v_o step response reveals a faster sliding-mode controller and the correct design of the current mode/PI controller parameters. The robustness property of the sliding-mode controlled output v_o, compared to the current mode/PI, is shown in Fig. 34.49.

EXAMPLE 34.15 Sliding-mode controllers for multilevel inverters

Diode clamped multilevel inverters (Fig. 34.50) are the converters of choice for high-voltage high-power dc/ac or ac/ac (with dc link) applications, as the active semiconductors (usually gate turn-off thyristors (GTO) or IGBT transistors) of n-level power conversion systems, must withstand only a fraction (normally $U_{cc}/(n-1)$) of the total supply voltage U_{cc}. Moreover, the output voltage of multilevel converters, being staircase-like

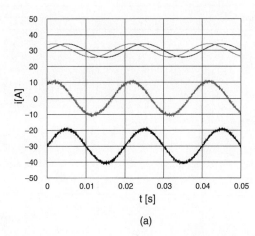
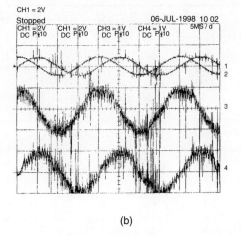

(a)   (b)

FIGURE 34.48 α, β space vector current modulator operation at near unity power factor: (a) simulation result ($i_{1r} + 30$; $i_{2r} + 30$; $2 \times i_1$; $2 \times i_2 - 30$) and (b) experimental result (1→i_{1r}, 2→i_{2r} (10 A/div); 3→i_1, 4→i_2 (5 A/div)).

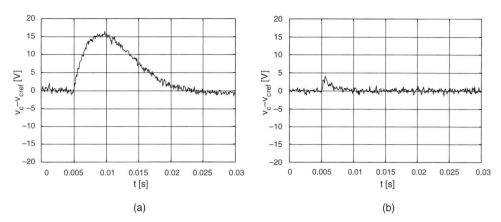

FIGURE 34.49 Transition from rectifier to inverter operation (i_o from 8 to -8 A) obtained by switching off IGBT S_a (Fig. 34.46) and using $I_a = 16$ A: (a) $v_o - v_{or}$ [v] with sliding-mode control and (b) $v_o - v_{or}$ [v] with current-mode /PI control.

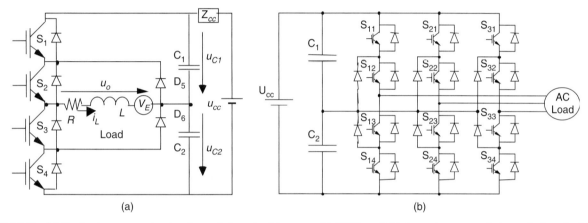

FIGURE 34.50 (a) Single-phase, neutral point clamped, three-level inverter with IGBTs and (b) three-phase, neutral-clamped, three-level inverter.

waveforms with n steps, features lower harmonic distortion compared to the two-level waveforms with the same switching frequency.

The advantages of multilevel converters are paid into the price of the capacitor supply voltage dividers (Fig. 34.51) and voltage equalization circuits, into the cost of extra power supply arrangements (Fig. 34.51c), and into increased control complexity. This example shows how to extend the two-level switching law (34.97) to n-level converters, and how to equalize the voltage of the capacitive dividers.

Considering single-phase three-level inverters (Fig. 34.50a), the open-loop control of the output voltage can be made using three-level SWPWM. The two-level modulator, seen in Example 34.9, can be easily extended (Fig. 34.52a) to generate the γ_{III} command (Fig. 34.52b) to three-level inverter legs, from the two-level γ_{II} signal, using the following relation:

$$\gamma_{III} = \gamma_{II}(m_i \sin(\omega t) - \text{sgn}(m_i \sin(\omega t))/2 - r(t)/2)$$
$$- 1/2 + \text{sgn}(m_i \sin(\omega t))/2 \qquad (34.148)$$

The required three-level SWPWM modulators for the output voltage synthesis seldom take into account the semiconductors and the capacitor voltage divider non-ideal characteristics. Consequently, the capacitor voltage divider tends to drift, one capacitor being overcharged, the other discharged, and an asymmetry appears in the currents of the power supply. A steady-state error in the output voltage can also be present. Sliding-mode control can provide the optimum switching timing between all the converter levels, together with robustness to supply voltage disturbances, semiconductor non-idealities, and load parameters.

A. Sliding-mode switching law For a variable-structure system where the control input $u_i(t)$ can present n levels, consider the n values of the integer variable γ, being $-(n-1)/2 \leq \gamma \leq (n-1)/2$ and $u_i(t) = \gamma U_{cc}/(n-1)$, dependent on the topology and on the conducting semiconductors. To ensure the sliding-mode manifold invariance condition (34.92) and the reaching mode behavior, the switching strategy $\gamma(t_{k+1})$

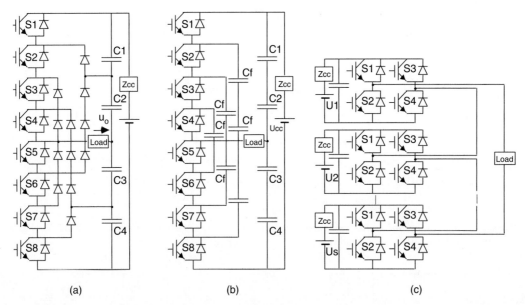

FIGURE 34.51 (a) Five-level ($n = 5$) diode clamped inverter with IGBTs; (b) five-level ($n = 5$) flying capacitor converter; and (c) multilevel converter based on cascaded full-bridge inverters.

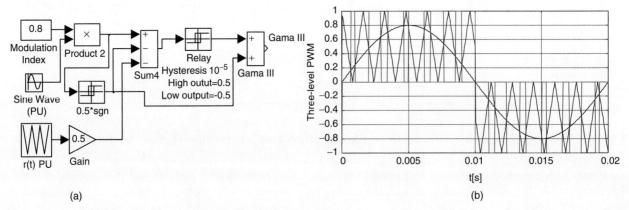

FIGURE 34.52 (a) Three-level SWPWM modulator schematic and (b) main three-level SWPWM signals.

for the time instant t_{k+1}, considering the value of $\gamma(t_k)$ must be

$$\gamma(t_{k+1}) = \begin{cases} \gamma(t_k) + 1 & \text{if } S(e_{x_i}, t) > \varepsilon \wedge \dot{S}(e_{x_i}, t) \\ & > \varepsilon \wedge \gamma(t_k) < (n-1)/2 \\ \gamma(t_k) - 1 & \text{if } S(e_{x_i}, t) < -\varepsilon \wedge \dot{S}(e_{x_i}, t) \\ & < -\varepsilon \wedge \gamma(t_k) > -(n-1)/2 \end{cases} \quad (34.149)$$

This switching law can be implemented as depicted in Fig. 34.53.

34.3.5.14 Control of the Output Voltage in Single-phase Multilevel Converters

To control the inverter output voltage, in closed-loop, in diode-clamped multilevel inverters with n levels and supply voltage U_{cc}, a control law similar to Eq. (34.132), $S(e_{uo}, t) = \kappa \int \left(u_{o_r} - k_v \gamma(t_k) U_{cc}/(n-1) \right) dt = 0$, is suitable.

Figure 34.54a shows the waveforms of a five-level sliding-mode controlled inverter, namely the input sinus voltage, the generated output staircase wave, and the sliding-surface instantaneous error. This error is always within a band centered around the zero value and presents zero mean value, which is not the case of sigma-delta modulators followed by n-level quantizers, where the error presents an offset mean value in each half period.

Experimental multilevel converters always show capacitor voltage unbalances (Fig. 34.54b) due to small differences between semiconductor voltage drops and circuitry offsets. To obtain capacitor voltage equalization, the voltage error $(v_{C_2} - U_{cc}/2)$ is fed back to the controller (Fig. 34.55a) to counteract the circuitry offsets. Experimental results (Fig. 34.54c)

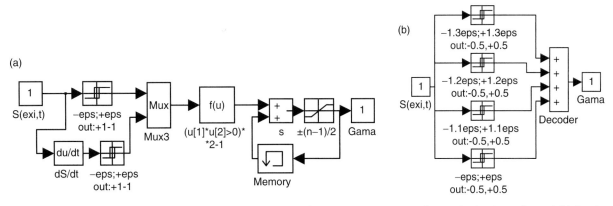

FIGURE 34.53 (a) Multilevel sliding-mode PWM modulator with n-level hysteresis comparator with quantization interval ε and (b) four hysteresis comparator implementation of a five-level switching law.

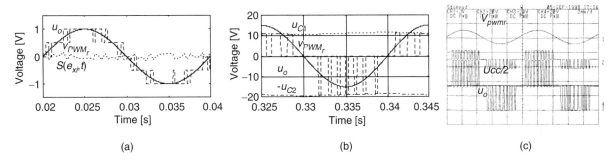

FIGURE 34.54 (a) Scaled waveforms of a five-level sliding-mode controlled single-phase converter, showing the input sinus voltage v_{PWM_r}, the generated output staircase wave u_o and the value of the sliding surface $S(e_{x_i}, t)$; (b) scaled waveforms of a three-level neutral point clamped inverter showing the capacitor voltage unbalance (shown as two near flat lines touching the tips of the PWM pulses); and (c) experimental results from a laboratory prototype of a three-level single-phase power inverter with the capacitor voltage equalization described.

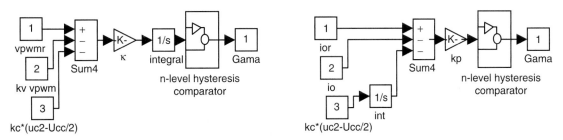

FIGURE 34.55 (a) Multilevel sliding-mode output voltage controller and PWM modulator with capacitor voltage equalization and (b) sliding-mode output current controller with capacitor voltage equalization.

clearly show the effectiveness of the correction made. The small steady-state error, between the voltages of the two capacitors, still present, could be eliminated using an integral regulator (Fig. 34.55b).

Figure 34.56 confirms the robustness of the sliding-mode controller to power supply disturbances.

34.3.5.15 Output Current Control in Single-phase Multilevel Converters

Considering an inductive load with current i_L, the control law (34.107) and the switching law of (34.159), should be used for single-phase multilevel inverters. Results obtained using the capacitor voltage equalization principle just described are shown in Fig. 34.57.

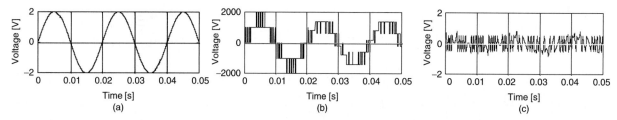

FIGURE 34.56 Simulated performance of a five-level power inverter, with a U_{cc} voltage dip (from 2 to 1.5 kV). Response to a sinusoidal wave of frequency 50 Hz: (a) v_{PWM_r} input; (b) PWM output voltage u_o; and (c) the integral of the error voltage, which is maintained close to zero.

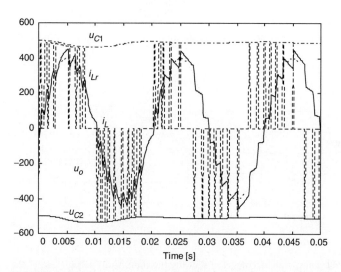

FIGURE 34.57 Operation of a three-level neutral point clamped inverter as a sinusoidal current source: Scaled waveforms of the output current sine wave reference i_{Lr}, the output current i_L, showing ripple, together with the PWM-generated voltage u_o, with nearly equal pulse heights, corresponding to the equalized dc capacitor voltages u_{C_1} and u_{C_2}.

EXAMPLE 34.16 Sliding-mode controllers for three-phase multilevel inverters

Three-phase n-level inverters (Fig. 34.58) are suitable for high-voltage, high-power dc/ac applications, such as modern high-speed railway traction drives, as the controlled turn-off semiconductors must block only a fraction (normally $U_{dc}/(n-1)$) of the total supply voltage U_{dc}.

This example presents a real-time modulator for the control of the three output voltages and capacitor voltage equalization, based on the use of sliding mode and space vectors represented in the α, β frame. Capacitor voltage equalization is done with the proper selection of redundant space vectors.

34.3.5.16 Output Voltage Control in Multilevel Converters

To guarantee the topological constraints of this converter and the correct sharing of the U_{dc} voltage by the semiconductors, the switching strategy for the k leg ($k \in \{1, 2, 3\}$) must ensure complementary states to switches S_{k1} and S_{k3}. The same restriction applies for S_{k2}, S_{k4}. Neglecting switching delays, dead times, on-state semiconductor voltage drops, snubber networks, and power supply variations, supposing small dead times and equal capacitor voltages $U_{C1} = U_{C2} = U_{dc}/2$, and using the time-dependent switching variable $\gamma_k(t)$, the leg output voltage U_k (Fig. 34.58) will be $U_k = \gamma_k(t)U_{dc}/2$, with

$$\gamma_k(t) = \begin{cases} 1 & \text{if } S_{k1} \wedge S_{k2} \text{ are ON } \wedge S_{k3} \wedge S_{k4} \text{ are OFF} \\ 0 & \text{if } S_{k2} \wedge S_{k3} \text{ are ON } \wedge S_{k1} \wedge S_{k4} \text{ are OFF} \\ -1 & \text{if } S_{k3} \wedge S_{k4} \text{ are ON } \wedge S_{k1} \wedge S_{k2} \text{ are OFF} \end{cases} \quad (34.150)$$

The converter output voltages U_{Sk} of vector $\mathbf{U}_S$ can be expressed

$$\mathbf{U}_S = \begin{bmatrix} 2/3 & -1/3 & -1/3 \\ -1/3 & 2/3 & -1/3 \\ -1/3 & -1/3 & 2/3 \end{bmatrix} \begin{bmatrix} \gamma_1 \\ \gamma_2 \\ \gamma_3 \end{bmatrix} \frac{U_{dc}}{2} \quad (34.151)$$

The application of the Concordia transformation $U_{S1,2,3} = [\mathbf{C}]\, U_{S\alpha,\beta,o}$ (Eq. (34.152)) to Eq. (34.151))

$$\begin{bmatrix} U_{S1} \\ U_{S2} \\ U_{S3} \end{bmatrix} = \sqrt{\frac{2}{3}} \cdot \begin{bmatrix} 1 & 0 & 1/\sqrt{2} \\ -1/2 & \sqrt{3}/2 & 1/\sqrt{2} \\ -1/2 & -\sqrt{3}/2 & 1/\sqrt{2} \end{bmatrix} \begin{bmatrix} U_{S\alpha} \\ U_{S\beta} \\ U_{So} \end{bmatrix} \quad (34.152)$$

gives the output voltage vector in the α, β coordinates $\mathbf{U}_{S\alpha,\beta}$:

$$\mathbf{U}_{S\alpha,\beta} = \begin{bmatrix} U_{S\alpha} \\ U_{S\beta} \end{bmatrix} = \sqrt{2/3} \begin{bmatrix} 1 & -1/2 & -1/2 \\ 0 & \sqrt{3}/2 & -\sqrt{3}/2 \end{bmatrix} \begin{bmatrix} \Gamma_1 \\ \Gamma_2 \\ \Gamma_3 \end{bmatrix} \frac{U_{dc}}{2} = \begin{bmatrix} \Gamma_\alpha \\ \Gamma_\beta \end{bmatrix} \frac{U_{dc}}{2} \quad (34.153)$$

where

$$\Gamma_1 = \frac{2}{3}\gamma_1 - \frac{1}{3}\gamma_2 - \frac{1}{3}\gamma_3; \quad \Gamma_2 = \frac{2}{3}\gamma_2 - \frac{1}{3}\gamma_3 - \frac{1}{3}\gamma_1;$$

$$\Gamma_3 = \frac{2}{3}\gamma_3 - \frac{1}{3}\gamma_1 - \frac{1}{3}\gamma_2$$

$$(34.154)$$

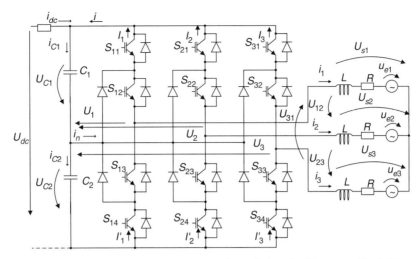

FIGURE 34.58 Three-phase, neutral point clamped, three-level inverter with IGBTs.

The output voltage vector in the α, β coordinates $\mathbf{U}_{S_{\alpha,\beta}}$ is discontinuous. A suitable state variable for this output can be its average value $\bar{\mathbf{U}}_{S_{\alpha,\beta}}$ during one switching period:

$$\bar{\mathbf{U}}_{S_{\alpha,\beta}} = \frac{1}{T}\int_0^T \mathbf{U}_{S_{\alpha,\beta}} dt = \frac{1}{T}\int_0^T \Gamma_{\alpha,\beta}\frac{U_{dc}}{2} dt \quad (34.155)$$

The controllable canonical form is

$$\frac{d}{dt}\bar{\mathbf{U}}_{S_{\alpha,\beta}} = \frac{\mathbf{U}_{S_{\alpha,\beta}}}{T} = \frac{\Gamma_{\alpha,\beta}}{T}\frac{U_{dc}}{2} \quad (34.156)$$

Considering the control goal $\bar{\mathbf{U}}_{S_{\alpha,\beta}} = \bar{\mathbf{U}}_{S_{\alpha,\beta_{ref}}}$ and Eq. (34.91), the sliding surface is

$$\mathbf{S}(\mathbf{e}_{\alpha,\beta},t) = \sum_{o=1}^{\varphi} k_{\alpha\beta_o}\mathbf{e}_{\alpha,\beta_o} = k_{\alpha,\beta_1}\mathbf{e}_{\alpha,\beta_1} = k_{\alpha,\beta_1}\left(\bar{\mathbf{U}}_{S_{\alpha,\beta_{ref}}} - \bar{\mathbf{U}}_{S_{\alpha,\beta}}\right)$$

$$= \frac{k_{\alpha,\beta}}{T}\int_0^T \left(\mathbf{U}_{S_{\alpha,\beta_{ref}}} - \mathbf{U}_{S_{\alpha,\beta}}\right) dt = 0 \quad (34.157)$$

To ensure reaching mode behavior, and sliding-mode stability (34.92), as the first derivative of Eq. (34.157), $\dot{\mathbf{S}}(\mathbf{e}_{\alpha,\beta},t)$, is

$$\dot{\mathbf{S}}(\mathbf{e}_{\alpha,\beta},t) = \frac{k_{\alpha,\beta}}{T}\left(\mathbf{U}_{S_{\alpha,\beta_{ref}}} - \mathbf{U}_{S_{\alpha,\beta}}\right) \quad (34.158)$$

The switching law is

$$\mathbf{S}(\mathbf{e}_{\alpha,\beta},t) > 0 \Rightarrow \dot{\mathbf{S}}(\mathbf{e}_{\alpha,\beta},t) < 0 \Rightarrow \mathbf{U}_{S_{\alpha,\beta}} > \mathbf{U}_{S_{\alpha,\beta_{ref}}}$$
$$\mathbf{S}(\mathbf{e}_{\alpha,\beta},t) < 0 \Rightarrow \dot{\mathbf{S}}(\mathbf{e}_{\alpha,\beta},t) > 0 \Rightarrow \mathbf{U}_{S_{\alpha,\beta}} < \mathbf{U}_{S_{\alpha,\beta_{ref}}} \quad (34.159)$$

This switching strategy must select the proper values of $\mathbf{U}_{S_{\alpha,\beta}}$ from the available outputs. As each inverter leg (Fig. 34.58) can deliver one of the three possible output voltages ($U_{dc}/2$; 0; $-U_{dc}/2$), all the 27 possible output voltage vectors listed in Table 34.3 can be represented in the α, β frame of Fig. 34.59 (in per units, 1 p.u. = U_{dc}). There are nine different levels for the α space vector component and only five for the β component. However, considering any particular value of α (or β) component, there are at most five levels available in the remaining orthogonal component. From the load viewpoint, the 27 space vectors of Table 34.3 define only 19 distinct space positions (Fig. 34.59).

To select one of these 19 positions from the control law (34.157) and the switching law of Eq. (34.159), two five-level hysteretic comparators (Fig. 34.53b) must be used ($5^2 = 25$). Their outputs are the integer variables λ_α and λ_β, denoted $\lambda_{\alpha,\beta}$ ($\lambda_\alpha, \lambda_\beta \in \{-2; -1; 0; 1; 2\}$) corresponding to the five selectable levels of Γ_α and Γ_β. Considering the sliding-mode stability, $\lambda_{\alpha,\beta}$, at time step $j+1$, is given by Eq. (34.160), knowing their previous values at step j. This means that the output level is increased (decreased) if the error and its derivative are both positive (negative), provided the maximum (minimum) output level is not exceeded.

$$\begin{cases} (\lambda_{\alpha,\beta})_{j+1} = (\lambda_{\alpha,\beta})_j + 1 & \text{if } \mathbf{S}(\mathbf{e}_{\alpha,\beta},t) > \varepsilon \wedge \dot{\mathbf{S}}(\mathbf{e}_{\alpha,\beta},t) \\ & > \varepsilon \wedge (\lambda_{\alpha,\beta})_j < 2 \\ (\lambda_{\alpha,\beta})_{j+1} = (\lambda_{\alpha,\beta})_j - 1 & \text{if } \mathbf{S}(\mathbf{e}_{\alpha,\beta},t) < -\varepsilon \wedge \dot{\mathbf{S}}(\mathbf{e}_{\alpha,\beta},t) \\ & < -\varepsilon \wedge (\lambda_{\alpha,\beta})_j > -2 \end{cases}$$
(34.160)

The available space vectors must be chosen not only to reduce the mean output voltage errors, but also to guarantee transitions only between the adjacent levels, to minimize the capacitor voltage unbalance, to minimize the switching frequency, to observe minimum on or off times if applicable, and to equally stress all the semiconductors.

TABLE 34.3 Vectors of the three-phase three-level converter, switching variables γ_k, switch states s_{kj}, and the corresponding output voltages, line to neutral point, line-to-line, and α, β components in per units

Vector	γ_1	γ_2	γ_3	S_{11}	S_{12}	S_{13}	S_{14}	S_{21}	S_{22}	S_{23}	S_{24}	S_{31}	S_{32}	S_{33}	S_{34}	U_1	U_2	U_3	U_{12}	U_{23}	U_{31}	$U_{s\alpha}/U_{dc}$	$U_{s\beta}/U_{dc}$
1	1	1	1	1	1	0	0	1	1	0	0	1	1	0	0	$U_{dc}/2$	$U_{dc}/2$	$U_{dc}/2$	0	0	0	0.00	0.00
2	1	1	0	1	1	0	0	1	1	0	0	0	1	1	0	$U_{dc}/2$	$U_{dc}/2$	0	0	$U_{dc}/2$	$-U_{dc}/2$	0.20	0.35
3	1	1	−1	1	1	0	0	1	1	0	0	0	0	1	1	$U_{dc}/2$	$U_{dc}/2$	$-U_{dc}/2$	0	U_{dc}	$-U_{dc}$	0.41	0.71
4	1	0	−1	1	1	0	0	0	1	1	0	0	0	1	1	$U_{dc}/2$	0	$-U_{dc}/2$	$U_{dc}/2$	$U_{dc}/2$	$-U_{dc}$	0.61	0.35
5	1	0	0	1	1	0	0	0	1	1	0	1	1	0	0	$U_{dc}/2$	0	0	$U_{dc}/2$	0	$-U_{dc}/2$	0.41	0.00
6	1	0	1	1	1	0	0	0	1	1	0	1	1	0	0	$U_{dc}/2$	0	$U_{dc}/2$	$U_{dc}/2$	$-U_{dc}/2$	0	0.20	−0.35
7	1	−1	1	1	1	0	0	0	0	1	1	1	1	0	0	$U_{dc}/2$	$-U_{dc}/2$	$U_{dc}/2$	U_{dc}	$-U_{dc}$	0	0.41	−0.71
8	1	−1	0	1	1	0	0	0	0	1	1	0	1	1	0	$U_{dc}/2$	$-U_{dc}/2$	0	U_{dc}	$-U_{dc}/2$	$-U_{dc}/2$	0.61	−0.35
9	1	−1	−1	1	1	0	0	0	0	1	1	0	0	1	1	$U_{dc}/2$	$-U_{dc}/2$	$-U_{dc}/2$	U_{dc}	0	$-U_{dc}$	0.82	0.00
10	0	−1	−1	0	1	1	0	0	0	1	1	0	0	1	1	0	$-U_{dc}/2$	$-U_{dc}/2$	$U_{dc}/2$	0	$-U_{dc}/2$	0.41	0.00
11	0	−1	0	0	1	1	0	0	0	1	1	0	1	1	0	0	$-U_{dc}/2$	0	$U_{dc}/2$	$-U_{dc}/2$	0	0.20	−0.35
12	0	−1	1	0	1	1	0	0	0	1	1	1	1	0	0	0	$-U_{dc}/2$	$U_{dc}/2$	$U_{dc}/2$	$-U_{dc}$	$U_{dc}/2$	0.00	−0.71
13	0	0	1	0	1	1	0	0	1	1	0	1	1	0	0	0	0	$U_{dc}/2$	0	$-U_{dc}/2$	$U_{dc}/2$	−0.20	−0.35
14	0	0	0	0	1	1	0	0	1	1	0	0	1	1	0	0	0	0	0	0	0	0.00	0.00
15	0	0	−1	0	1	1	0	0	1	1	0	0	0	1	1	0	0	$-U_{dc}/2$	0	$U_{dc}/2$	$-U_{dc}/2$	0.20	0.35
16	0	1	−1	0	1	1	0	1	1	0	0	0	0	1	1	0	$U_{dc}/2$	$-U_{dc}/2$	$-U_{dc}/2$	U_{dc}	$-U_{dc}/2$	0.00	0.71
17	0	1	0	0	1	1	0	1	1	0	0	0	1	1	0	0	$U_{dc}/2$	0	$-U_{dc}/2$	$U_{dc}/2$	0	−0.20	0.35
18	0	1	1	0	1	1	0	1	1	0	0	1	1	0	0	0	$U_{dc}/2$	$U_{dc}/2$	$-U_{dc}/2$	0	$U_{dc}/2$	−0.41	0.00
19	−1	1	1	0	0	1	1	1	1	0	0	1	1	0	0	$-U_{dc}/2$	$U_{dc}/2$	$U_{dc}/2$	$-U_{dc}$	0	U_{dc}	−0.82	0.00
20	−1	1	0	0	0	1	1	1	1	0	0	0	1	1	0	$-U_{dc}/2$	$U_{dc}/2$	0	$-U_{dc}$	$U_{dc}/2$	$U_{dc}/2$	−0.61	0.35
21	−1	1	−1	0	0	1	1	1	1	0	0	0	0	1	1	$-U_{dc}/2$	$U_{dc}/2$	$-U_{dc}/2$	$-U_{dc}$	U_{dc}	0	−0.41	0.71
22	−1	0	−1	0	0	1	1	0	1	1	0	0	0	1	1	$-U_{dc}/2$	0	$-U_{dc}/2$	$-U_{dc}/2$	$U_{dc}/2$	0	−0.20	0.35
23	−1	0	0	0	0	1	1	0	1	1	0	0	1	1	0	$-U_{dc}/2$	0	0	$-U_{dc}/2$	0	$U_{dc}/2$	−0.41	0.00
24	−1	0	1	0	0	1	1	0	1	1	0	1	1	0	0	$-U_{dc}/2$	0	$U_{dc}/2$	$-U_{dc}/2$	$-U_{dc}/2$	U_{dc}	−0.61	−0.35
25	−1	−1	1	0	0	1	1	0	0	1	1	1	1	0	0	$-U_{dc}/2$	$-U_{dc}/2$	$U_{dc}/2$	0	$-U_{dc}/2$	U_{dc}	−0.41	−0.71
26	−1	−1	0	0	0	1	1	0	0	1	1	0	1	1	0	$-U_{dc}/2$	$-U_{dc}/2$	0	0	$-U_{dc}/2$	$U_{dc}/2$	−0.20	−0.35
27	−1	−1	−1	0	0	1	1	0	0	1	1	0	0	1	1	$-U_{dc}/2$	$-U_{dc}/2$	$-U_{dc}/2$	0	0	0	0.00	0.00

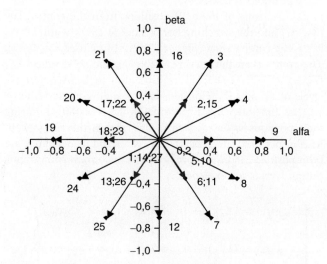

FIGURE 34.59 Output voltage vectors (1 to 27) of three-phase, neutral-clamped three-level inverters, in the α, β frame.

TABLE 34.4 Switching table to be used if $(U_{C1} - U_{C2}) > \varepsilon_{eU}$ in the inverter mode, or $(U_{C1} - U_{C2}) < -\varepsilon_{eU}$ in the regenerative mode, showing vector selection upon the variables $\lambda_\alpha, \lambda_\beta$

$\lambda_\beta \backslash \lambda_\alpha$	−2	−1	0	1	2
−2	25	25	12	7	7
−1	24	13	13;6	6	8
0	19	18	1;14;27	5	9
1	20	17	17;2	2	4
2	21	21	16	3	3

TABLE 34.5 Switching table to be used if $(U_{C1} - U_{C2}) > \varepsilon_{eU}$ in the regenerative mode, or $(U_{C1} - U_{C2}) < -\varepsilon_{eU}$ in the inverter mode, showing vector selection upon the variables $\lambda_\alpha, \lambda_\beta$

$\lambda_\beta / \lambda_\alpha$	−2	−1	0	1	2
−2	25	25	12	7	7
−1	24	26	26;11	11	8
0	19	23	1;14;27	10	9
1	20	22	22;15	15	4
2	21	21	16	3	3

Using Eq. (34.160) and the control laws $S(\mathbf{e}_{\alpha,\beta,t})$ Eq. (34.157), Tables 34.4 and 34.5 can be used to choose the correct voltage vector in order to ensure stability, output voltage tracking, and DC capacitor voltage equalization. The vector with α, β components corresponding to the levels of the pair $\lambda_\beta, \lambda_\alpha$ is selected, provided that the adjacent transitions on inverter legs are obtained. If there is no directly corresponding vector, then the nearest vector guaranteeing adjacent transitions is selected. If a zero vector must be applied, then, one of the three zero vectors (1, 14, 27) is selected, to minimize the switching frequency. If more than one vector is the nearest, then, one of them is selected to equalize the capacitor voltages, as shown next.

34.3.5.17 DC Capacitor Voltage Equalization

The discrete values of $\lambda_{\alpha,\beta}$ allow 25 different combinations. As only 19 are distinct from the load viewpoint, the extra ones can be used to select vectors that are able to equalize the capacitor voltages ($U_{C1} = U_{C2} = U_{dc}/2$).

Considering the control goal $U_{C1} = U_{C2}$, since the first derivatives of U_{C1} and U_{C2} Eq. (34.161) directly depend on the $\gamma_k(t)$ control inputs, from Eq. (34.91) the sliding surface is given by Eq. (34.162), where k_U is a positive gain.

$$\frac{d}{dt}\begin{bmatrix}U_{C1}\\U_{C2}\end{bmatrix}=\begin{bmatrix}-\frac{\gamma_1(1+\gamma_1)}{2C_1} & -\frac{\gamma_2(1+\gamma_2)}{2C_1} & -\frac{\gamma_3(1+\gamma_3)}{2C_1} & \frac{1}{C_1}\\-\frac{\gamma_1(1-\gamma_1)}{2C_2} & -\frac{\gamma_2(1-\gamma_2)}{2C_2} & -\frac{\gamma_3(1-\gamma_3)}{2C_2} & \frac{1}{C_2}\end{bmatrix}\begin{bmatrix}i_1\\i_2\\i_3\\i_{dc}\end{bmatrix}$$
(34.161)

$$S(e_{Uc},t) = k_U e_{Uc}(t) = k_U(U_{C1} - U_{C2}) = 0 \quad (34.162)$$

The first derivative of $U_{C1} - U_{C2}$ (the sliding surface) is (Fig. 34.58 with $C_1 = C_2 = C$):

$$\frac{d}{dt}e_{Uc} = \frac{i_{C1}}{C_1} - \frac{i_{C2}}{C_2} = \frac{i_n}{C} = \frac{(\gamma_3^2 - \gamma_1^2)i_1 + (\gamma_3^2 - \gamma_2^2)i_2}{C}$$
(34.163)

To ensure reaching mode behavior and sliding-mode stability, from Eq. (34.92), considering a small enough $e_{Uc}(t)$ error, ε_{eU}, the switching law is

$$\begin{aligned}S(e_{Uc},t) > \varepsilon_{eU} &\Rightarrow \dot{S}(e_{Uc},t) < 0 \Rightarrow i_n < 0\\ S(e_{Uc},t) < -\varepsilon_{eU} &\Rightarrow \dot{S}(e_{Uc},t) > 0 \Rightarrow i_n > 0\end{aligned}$$
(34.164)

From circuit analysis, it can be seen that vectors {2, 5, 6, 13, 17, 18} result in the discharge of capacitor C_1, if the converter operates in inverter mode, or in the charge of C_1, if the converter operates in boost-rectifier (regenerative) mode. Similar reasoning can be applied for vectors {10, 11, 15, 22, 23, 26} and capacitor C_2, since this vector set give i_n currents with opposite sign relatively to the set {2, 5, 6, 13, 17, 18}. Therefore, considering the vector $[\Upsilon_1, \Upsilon_2] = [(\gamma_1^2 - \gamma_3^2), (\gamma_2^2 - \gamma_3^2)]$

the switching law is:

IF $(U_{C1} - U_{C2}) > \varepsilon_{eU}$

THEN $\begin{cases}\text{IF the candidate vector from } \{2, 5, 6, 13, 17, 18\}\\ \text{gives } (\Upsilon_1 i_1 + \Upsilon_2 i_2) > 0, \text{THEN choose the vector}\\ \text{according to } \lambda_{\alpha,\beta} \text{ on Table 34.4;}\\ \text{ELSE, the candidate vector of } \{10, 11, 15, 22, 23,\\ 26\} \text{ gives } (\Upsilon_1 i_1 + \Upsilon_2 i_2) > 0, \text{the vector being}\\ \text{chosen according to } \lambda_{\alpha,\beta} \text{ from (table 34.5)}\end{cases}$

IF $(U_{C1} - U_{C2}) < -\varepsilon_{eU}$

THEN $\begin{cases}\text{IF the candidate vector from } \{2, 5, 6, 13, 17, 18\}\\ \text{gives } (\Upsilon_1 i_1 + \Upsilon_2 i_2) < 0, \text{THEN choose the vector}\\ \text{according to } \lambda_{\alpha,\beta} \text{ on Table 34.4;}\\ \text{ELSE, the candidate vector of } \{10, 11, 15, 22, 23,\\ 26\} \text{ gives } (\Upsilon_1 i_1 + \Upsilon_2 i_2) < 0, \text{the vector being}\\ \text{chosen according to } \lambda_{\alpha,\beta} \text{ from (table 34.5)}\end{cases}$

For example, consider the case where $U_{C1} > U_{C2} + \varepsilon_{eU}$. Then, the capacitor C_2 must be charged and Table 34.4 must be used if the multilevel inverter is operating in the inverter mode or Table 34.5 for the regenerative mode. Additionally, when using Table 34.4, if $\lambda_\alpha = -1$ and $\lambda_\beta = -1$, then vector 13 should be used.

Experimental results shown in Fig. 34.61 were obtained with a low-power, scaled down laboratory prototype (150 V, 3 kW) of a three-level inverter (Fig. 34.60), controlled by two four-level comparators, plus described capacitor voltage equalizing procedures and EPROM-based lookup Tables 34.3–34.5. Transistors IGBT (MG25Q2YS40) were switched at frequencies near 4 kHz, with neutral clamp diodes 40HFL, $C_1 \approx C_2 \approx 20$ mF. The load was mainly inductive (3×10 mH, $2\,\Omega$).

The inverter number of levels (three for the phase voltage and five for the line voltage), together with the adjacent transitions of inverter legs between levels, are shown in Fig. 34.61a and, in detail, in Fig. 34.62a.

The performance of the capacitor voltage equalizing strategy is shown in Fig. 34.62b, where the reference current of phase 1 and the output current of phase 3, together with the power supply voltage ($U_{dc} \approx 100$ V) and the voltage of capacitor $C_2 (U_{C2})$, can be seen. It can be noted that the U_{C2} voltage is nearly half of the supply voltage. Therefore, the capacitor voltages are nearly equal. Furthermore, it can be stated that without this voltage equalization procedure, the three-level inverter operates only during a brief transient, during which one of the capacitor voltages vanishes to nearly zero volt and the other is overcharged to the supply voltage. Figure 34.61b shows the harmonic spectrum of the output voltages, where the harmonics due to the switching frequency (≈ 4.5 kHz) and the fundamental harmonic can be seen.

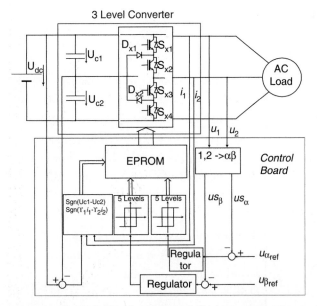

FIGURE 34.60 Block diagram of the multilevel converter and control board.

34.3.5.18 On-line Output Current Control in Multilevel Inverters

Considering a standard inductive balanced load (R, L) with electromotive force (u) and isolated neutral, the converter output currents i_k can be expressed

$$U_{Sk} = Ri_k + L\frac{di_k}{dt} + u_{ek} \qquad (34.165)$$

Now analyzing the circuit of Fig. 34.58, the multilevel converter switched state-space model can be obtained:

$$\frac{d}{dt}\begin{bmatrix} i_1 \\ i_2 \\ i_3 \end{bmatrix} = -\begin{bmatrix} R/L & 0 & 0 \\ 0 & R/L & 0 \\ 0 & 0 & R/L \end{bmatrix}\begin{bmatrix} i_1 \\ i_2 \\ i_3 \end{bmatrix} - \begin{bmatrix} 1/L & 0 & 0 \\ 0 & 1/L & 0 \\ 0 & 0 & 1/L \end{bmatrix}\begin{bmatrix} u_{e1} \\ u_{e2} \\ u_{e3} \end{bmatrix}$$

$$+ \begin{bmatrix} \Gamma_1/L \\ \Gamma_2/L \\ \Gamma_3/L \end{bmatrix}\frac{U_{dc}}{2} \qquad (34.166)$$

The application of the Concordia matrix Eq. (34.152) to Eq. (34.166), reduces the number of the new model equations (Eq. (34.167)) to two, since an isolated neutral is assumed.

$$\frac{d}{dt}\begin{bmatrix} i_\alpha \\ i_\beta \end{bmatrix} = -\begin{bmatrix} R/L & 0 \\ 0 & R/L \end{bmatrix}\begin{bmatrix} i_\alpha \\ i_\beta \end{bmatrix} - \begin{bmatrix} 1/L & 0 \\ 0 & 1/L \end{bmatrix}\begin{bmatrix} u_{e\alpha} \\ u_{e\beta} \end{bmatrix}$$

$$+ \begin{bmatrix} 1/L & 0 \\ 0 & 1/L \end{bmatrix}\begin{bmatrix} U_{S\alpha} \\ U_{S\beta} \end{bmatrix} \qquad (34.167)$$

The model Eq. (34.167) of this multiple-input multiple-output system (MIMO) with outputs i_α, i_β reveals the control inputs $U_{S\alpha}$, $U_{S\beta}$, dependent on the control variables $\gamma_k(t)$.

From Eqs. (34.167) and (34.91), the two sliding surfaces $S(e_{\alpha,\beta}, t)$ are

$$S(e_{\alpha,\beta}, t) = k_{\alpha,\beta}(i_{\alpha,\beta ref} - i_{\alpha,\beta}) = k_{\alpha,\beta}e_{\alpha,\beta} = 0 \qquad (34.168)$$

The first derivatives of Eq. (34.167), denoted $\dot{S}(e_{\alpha,\beta,t})$, are

$$\dot{S}(e_{\alpha,\beta}, t) = k_{\alpha,\beta}(\dot{i}_{\alpha,\beta ref} - \dot{i}_{\alpha,\beta})$$
$$= k_{\alpha,\beta}\left[\dot{i}_{\alpha,\beta ref} + RL^{-1}i_{\alpha,\beta} + u_{e\alpha,\beta}L^{-1} - U_{S\alpha,\beta}L^{-1}\right] \qquad (34.169)$$

Therefore, the switching law is

$$S(e_{\alpha,\beta}, t) > 0 \Rightarrow \dot{S}(e_{\alpha,\beta}, t) < 0 \Rightarrow U_{S\alpha,\beta} > L\dot{i}_{\alpha,\beta ref} + Ri_{\alpha,\beta} + u_{e\alpha,\beta}$$

$$S(e_{\alpha,\beta}, t) < 0 \Rightarrow \dot{S}(e_{\alpha,\beta}, t) > 0 \Rightarrow U_{S\alpha,\beta} < L\dot{i}_{\alpha,\beta ref} + Ri_{\alpha,\beta} + u_{e\alpha,\beta} \qquad (34.170)$$

These switching laws are implemented using the same α, β vector modulator described above in this example.

Figure 34.63a shows the experimental results. The multilevel converter and proposed control behavior are obtained for step inputs (4 to 2A) in the amplitude of the sinus references with frequency near 52 Hz ($U_{dc} \approx 150$ V). Observe the tracking ability, the fast transient response, and the balanced

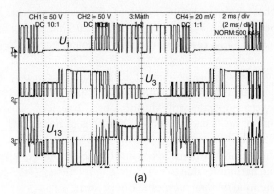
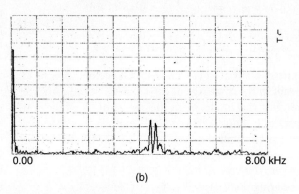

FIGURE 34.61 (a) Experimental results showing phase and line voltages and (b) harmonic spectrum of output voltages.

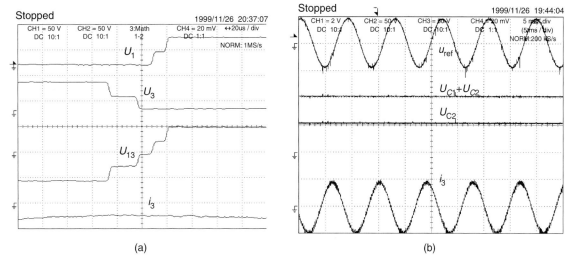

FIGURE 34.62 Experimental results showing (a) the transitions between adjacent voltage levels (50 V/div; time 20 μs/div) and (b) performance of the capacitor voltage equalizing strategy; from top trace to bottom: 1 is the voltage reference input; 2 is the power supply voltage; 3 is the mid-point capacitor voltage, which is maintained close to $U_{dc}/2$; 4 is the output current of phase 3 (2 A/div; 50 V/div; 5 ms/div).

three-phase currents. Figure 34.63b shows almost the same test (step response from 2 to 4 A at the same frequency), but now the power supply is set at 50 V and the inductive load was unbalanced (±30% on resistor value). The response remains virtually the same, with tracking ability, almost no current distortions due to dead times or semiconductor voltage drops. These results confirm experimentally that the designed controllers are robust concerning these nonidealities.

EXAMPLE 34.17 Sliding-mode vector controllers for matrix converters

Matrix converters are all silicon ac/ac switching converters, able to provide variable amplitude almost sinusoidal output voltages, almost sinusoidal input currents, and controllable input power factor [18]. They seem to be very attractive to use in ac drives speed control as well as in applications related to power-quality enhancement. The lack of an intermediate energy storage link, their

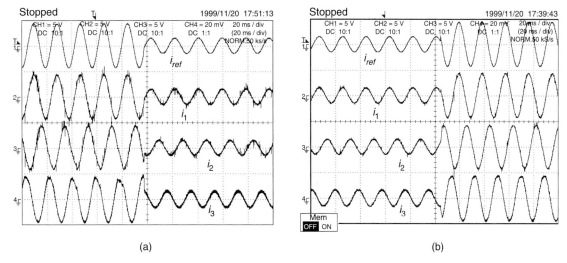

FIGURE 34.63 Step response of the current control method: (a) step from 4 to 2 A. Traces show the reference current for phase 1 and the three output currents with 150 V power supply (5 A/div; time scale 20 ms/div) and (b) step from 2 to 4 A in the reference amplitude at 52 Hz. Traces show the reference current for phase 1 and the three output currents with 50 V power supply.

main advantage, implies an input/output coupling which increases the control complexity.

This example presents the design of sliding-mode controllers considering the switched state-space model of the matrix converter (nine bidirectional power switches), including the three-phase input filter and the output load (Fig. 34.64).

34.3.5.19 Output Voltage Control

Ideal three-phase matrix converters are obtained by assembling nine bidirectional switches, with the turn-off capability, to allow the connection of each one of the input phases to any one of the output phases (Fig. 34.64). The states of these switches are usually represented as a nine-element matrix $\mathbf{S}$ (Eq. (34.171)), in which each matrix element, $S_{kj}\ k,j \in \{1,2,3\}$, has two possible states: $S_{kj} = 1$ if the switch is closed (ON) and $S_{kj} = 0$ if it is open (OFF). Only 27 switching combinations are possible (Table 34.6), as a result of the topological constraints (the input phases should never be short-circuited and the output inductive currents should never be interrupted), which implies that the sum of all the S_{kj} of each one of the matrix, k rows must always equal 1 (Eq. (34.171)).

$$\mathbf{S} = \begin{bmatrix} S_{11} & S_{12} & S_{13} \\ S_{21} & S_{22} & S_{23} \\ S_{31} & S_{32} & S_{33} \end{bmatrix} \quad \sum_{j=1}^{3} s_{kj} = 1 \quad k,j \in \{1,2,3\} \quad (34.171)$$

Based on the matrix $\mathbf{S}$, the output phase v_A, v_B, v_C and line voltages v_{AB}, v_{BC}, v_{CA}, can be expressed in terms of the input phase voltages v_a, v_b, v_c. The input currents i_a, i_b, i_c can be expressed as a function of the output currents i_A, i_B, i_C:

$$\begin{bmatrix} v_A \\ v_B \\ v_C \end{bmatrix} = \mathbf{S} \begin{bmatrix} v_a \\ v_b \\ v_c \end{bmatrix};$$

$$\begin{bmatrix} v_{AB} \\ v_{BC} \\ v_{CA} \end{bmatrix} = \begin{bmatrix} S_{11} - S_{21} & S_{12} - S_{22} & S_{13} - S_{23} \\ S_{21} - S_{31} & S_{22} - S_{32} & S_{23} - S_{33} \\ S_{31} - S_{11} & S_{32} - S_{12} & S_{33} - S_{13} \end{bmatrix} \begin{bmatrix} v_a \\ v_b \\ v_c \end{bmatrix};$$

$$\begin{bmatrix} i_a \\ i_b \\ i_c \end{bmatrix} = \mathbf{S}^T \begin{bmatrix} i_A \\ i_B \\ i_C \end{bmatrix}$$

(34.172)

The application of the Concordia transformation $[X_{\alpha,\beta,0}]^T = \mathbf{C}^T [X_{a,b,c}]^T$ to Eq. (34.172) results in the output voltage vector:

$$v_{o\alpha\beta} = \begin{bmatrix} v_{o\alpha} \\ v_{o\beta} \end{bmatrix} = \sqrt{\frac{2}{3}} \begin{bmatrix} 1 & -1/2 & -1/2 \\ 0 & \sqrt{3}/2 & -\sqrt{3}/2 \end{bmatrix}$$

$$\times \begin{bmatrix} S_{11} - S_{21} & S_{12} - S_{22} & S_{13} - S_{23} \\ S_{21} - S_{31} & S_{22} - S_{32} & S_{23} - S_{33} \\ S_{31} - S_{11} & S_{32} - S_{12} & S_{33} - S_{13} \end{bmatrix} \begin{bmatrix} v_a \\ v_b \\ v_c \end{bmatrix}$$

$$= \begin{bmatrix} \rho_{v\alpha\alpha} & \rho_{v\alpha\beta} \\ \rho_{v\beta\alpha} & \rho_{v\beta\beta} \end{bmatrix} \begin{bmatrix} v_{c\alpha} \\ v_{c\beta} \end{bmatrix} \quad (34.173)$$

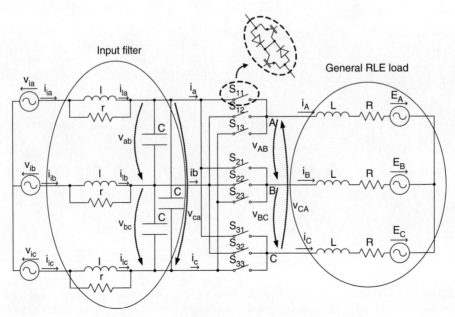

FIGURE 34.64 AC/AC matrix converter with input lCr filter.

TABLE 34.6 Switching combinations and output voltage/input current state-space vectors

Group	Name	A	B	C	v_{AB}	v_{BC}	v_{CA}	i_a	i_b	i_c	V_o	δ_o	I_i	μ_i
I	1g	a	b	c	v_{ab}	v_{bc}	v_{ca}	i_A	i_B	i_C	v_i	δ_i	i_o	μ_o
	2g	a	c	b	$-v_{ca}$	$-v_{bc}$	$-v_{ab}$	i_A	i_C	i_B	$-v_i$	$-\delta_i + 4\pi/3$	i_o	$-\mu_o$
	3g	b	a	c	$-v_{ab}$	$-v_{ca}$	$-v_{bc}$	i_B	i_A	i_C	$-v_i$	$-\delta_i$	i_o	$-\mu_o + 2\pi/3$
	4g	b	c	a	v_{bc}	v_{ca}	v_{ab}	i_C	i_A	i_B	v_i	$\delta_i + 4\pi/3$	i_o	$\mu_o + 2\pi/3$
	5g	c	a	b	v_{ca}	v_{ab}	v_{bc}	i_B	i_C	i_A	v_i	$\delta_i + 2\pi/3$	i_o	$\mu_o + 4\pi/3$
	6g	c	b	a	$-v_{bc}$	$-v_{ab}$	$-v_{ca}$	i_C	i_B	i_A	$-v_i$	$-\delta_i + 2\pi/3$	i_o	$-\mu_o + 4\pi/3$
II	+1	a	b	b	v_{ab}	0	$-v_{ab}$	i_A	$-i_A$	0	$2/\sqrt{3}v_{ab}$	$\pi/6$	$2/\sqrt{3}i_A$	$-\pi/6$
	−1	b	a	a	$-v_{ab}$	0	v_{ab}	$-i_A$	i_A	0	$-2/\sqrt{3}v_{ab}$	$\pi/6$	$-2/\sqrt{3}i_A$	$-\pi/6$
	+2	b	c	c	v_{bc}	0	$-v_{bc}$	0	i_A	$-i_A$	$2/\sqrt{3}v_{bc}$	$\pi/6$	$2/\sqrt{3}i_A$	$\pi/2$
	−2	c	b	b	$-v_{bc}$	0	v_{bc}	0	$-i_A$	i_A	$-2/\sqrt{3}v_{bc}$	$\pi/6$	$-2/\sqrt{3}i_A$	$\pi/2$
	+3	c	a	a	v_{ca}	0	$-v_{ca}$	$-i_A$	0	i_A	$2/\sqrt{3}v_{ca}$	$\pi/6$	$2/\sqrt{3}i_A$	$7\pi/6$
	−3	a	c	c	$-v_{ca}$	0	v_{ca}	i_A	0	$-i_A$	$-2/\sqrt{3}v_{ca}$	$\pi/6$	$-2/\sqrt{3}i_A$	$7\pi/6$
	+4	b	a	b	$-v_{ab}$	v_{ab}	0	i_B	$-i_B$	0	$2/\sqrt{3}v_{ab}$	$5\pi/6$	$2/\sqrt{3}i_B$	$-\pi/6$
	−4	a	b	a	v_{ab}	$-v_{ab}$	0	$-i_B$	i_B	0	$-2/\sqrt{3}v_{ab}$	$5\pi/6$	$-2/\sqrt{3}i_B$	$-\pi/6$
	+5	c	b	c	$-v_{bc}$	v_{bc}	0	0	i_B	$-i_B$	$2/\sqrt{3}v_{bc}$	$5\pi/6$	$2/\sqrt{3}i_B$	$\pi/2$
	−5	b	c	b	v_{bc}	$-v_{bc}$	0	0	$-i_B$	i_B	$-2/\sqrt{3}v_{bc}$	$5\pi/6$	$-2/\sqrt{3}i_B$	$\pi/2$
	+6	a	c	a	$-v_{ca}$	v_{ca}	0	$-i_B$	0	i_B	$2/\sqrt{3}v_{ca}$	$5\pi/6$	$2/\sqrt{3}i_B$	$7\pi/6$
	−6	c	a	c	v_{ca}	$-v_{ca}$	0	i_B	0	$-i_B$	$-2/\sqrt{3}v_{ca}$	$5\pi/6$	$-2/\sqrt{3}i_B$	$7\pi/6$
	+7	b	b	a	0	v_{ab}	v_{ab}	i_C	$-i_C$	0	$2/\sqrt{3}v_{ab}$	$3\pi/2$	$2/\sqrt{3}i_C$	$-\pi/6$
	−7	a	a	b	0	v_{ab}	v_{ab}	$-i_C$	i_C	0	$-2/\sqrt{3}v_{ab}$	$3\pi/2$	$-2/\sqrt{3}i_C$	$-\pi/6$
	+8	c	c	b	0	$-v_{bc}$	v_{bc}	0	i_C	$-i_C$	$2/\sqrt{3}v_{bc}$	$3\pi/2$	$2/\sqrt{3}i_C$	$\pi/2$
	−8	b	b	c	0	v_{bc}	$-v_{bc}$	0	$-i_C$	i_C	$-2/\sqrt{3}v_{bc}$	$3\pi/2$	$-2/\sqrt{3}i_C$	$\pi/2$
	+9	a	a	c	0	$-v_{ca}$	v_{ca}	$-i_C$	0	i_C	$2/\sqrt{3}v_{ca}$	$3\pi/2$	$2/\sqrt{3}i_C$	$7\pi/6$
	−9	c	c	a	0	v_{ca}	$-v_{ca}$	i_C	0	$-i_C$	$-2/\sqrt{3}v_{ca}$	$3\pi/2$	$-2/\sqrt{3}i_C$	$7\pi/6$
III	z_a	a	a	a	0	0	0	0	0	0	0	-	0	-
	z_b	b	b	b	0	0	0	0	0	0	0	-	0	-
	z_c	c	c	c	0	0	0	0	0	0	0	-	0	-

where $v_{c\alpha\beta}$ is the input filter capacitor voltage and $\rho_{v\alpha\alpha}, \rho_{v\alpha\beta}, \rho_{v\beta\alpha}, \rho_{v\beta\beta}$ are functions of the ON/OFF state of the nine S_{kj} switches:

$$\begin{bmatrix} \rho_{v\alpha\alpha} & \rho_{v\alpha\beta} \\ \rho_{v\beta\alpha} & \rho_{v\beta\beta} \end{bmatrix} = \begin{bmatrix} 1/2(S_{11} - S_{21} - S_{12} + S_{22}) & \sqrt{3}/2(S_{11} - S_{21} + S_{12} - S_{22}) \\ 1/2\sqrt{3}(S_{11} + S_{21} - 2S_{31} - S_{12} - S_{22} + 2S_{32}) & 1/2(S_{11} + S_{21} - 2S_{31} + S_{12} + S_{22} - 2S_{32}) \end{bmatrix} \quad (34.174)$$

The average value $\overline{v_{o\alpha,\beta}}$ of the output voltage vector, in $\alpha\beta$ coordinates, during one switching period is the output variable to be controlled (since $v_{o\alpha,\beta}$ is discontinuous).

$$\overline{v_{o\alpha\beta}} = \frac{1}{T_s} \int_{nT_s}^{(n+1)T_s} v_{o\alpha\beta} \, dt \quad (34.175)$$

Considering the control goal $\overline{v_{o\alpha\beta}} = \overline{v_{o\alpha\beta_{ref}}}$, the sliding surface $S(e_{\alpha\beta}, t)$ ($k_{\alpha\beta} > 0$) is:

$$S(e_{\alpha\beta}, t) = \frac{k_{\alpha\beta}}{T} \int_0^T (v_{o\alpha\beta_{ref}} - v_{o\alpha\beta}) dt = 0 \quad (34.176)$$

The first derivative of Eq. (34.176) is:

$$\dot{S}(e_{\alpha\beta}, t) = k_\alpha (v_{o\alpha\beta_{ref}} - v_{o\alpha\beta}) \quad (34.177)$$

As the sliding-mode stability is guaranteed if $S_{\alpha\beta}(e_{\alpha\beta}, t) \dot{S}_{\alpha\beta}(e_{\alpha\beta}, t) < 0$, the criterion to choose the state-space vectors is:

$$\begin{aligned} S_{\alpha\beta}(e_{\alpha\beta}, t) < 0 &\Rightarrow \dot{S}_{\alpha\beta}(e_{\alpha\beta}, t) > 0 \Rightarrow v_{o\alpha\beta} < v_{o\alpha\beta_{ref}} \\ S_{\alpha\beta}(e_{\alpha\beta}, t) > 0 &\Rightarrow \dot{S}_{\alpha\beta}(e_{\alpha\beta}, t) < 0 \Rightarrow v_{o\alpha\beta} > v_{o\alpha\beta_{ref}} \end{aligned} \quad (34.178)$$

This implies that the sliding mode is reached only when the vector applied to the converter has the desired amplitude and angle.

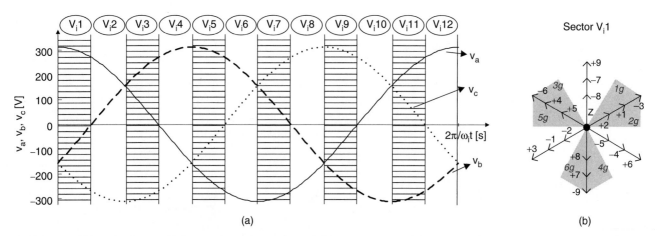

FIGURE 34.65 (a) Input voltages and their corresponding sector and (b) representation of the output voltage state-space vectors when the input voltages are located at sector V_i1.

According to Table 34.6, the 6 vectors of group I have fixed amplitude but time varying phase, the 18 vectors of group II have variable amplitude and vectors of group III are null. Therefore, from the load viewpoint, the 18 highest amplitude vectors (6 vectors from group I and 12 vectors from group II) and one null vector are suitable to guarantee the sliding-mode stability.

Therefore, if two three-level comparators ($C_{\alpha\beta} \in \{-1, 0, 1\}$) are used to quantize the deviations of Eq. (34.178) from zero, the nine output voltage error combinations (3^3) are not enough to guarantee the choice of all the 19 available vectors. The extra vectors may be used to control the input power factor. As an example, if the output voltage error is quantized as $C_\alpha = 1$, $C_\beta = 1$, at sector V_i1 (Fig. 34.65), the vectors -3, $+1$, or $1g$ might be used to control the output voltage. The final choice would depend on the input current error.

34.3.5.20 Input Power Factor Control

Assuming that the source is a balanced sinusoidal three-phase voltage supply with frequency ω_i, the switched state-space model equations of the converter input filter is obtained in abc coordinates.

$$\begin{cases} \frac{di_{l_a}}{dt} = \frac{1}{3l} v_{bc} + \frac{2}{3l} v_{ca} + \frac{1}{l} v_{i_a} \\ \frac{di_{l_b}}{dt} = -\frac{2}{3l} v_{bc} - \frac{1}{3l} v_{ca} + \frac{1}{l} v_{i_b} \\ \frac{dv_{bc}}{dt} = \frac{1}{3C} i_{l_a} + \frac{2}{3C} i_{l_b} - \frac{1}{3Cr} v_{bc} + \frac{1}{3Cr} v_{i_a} + \frac{2}{3Cr} v_{i_b} \\ \qquad - \frac{1}{3C}(S_{11} - S_{31} + 2S_{12} - S_{32}) i_A \\ \qquad - \frac{1}{3C}(S_{21} - S_{31} + 2S_{22} - S_{32}) i_B \\ \frac{dv_{ca}}{dt} = -\frac{2}{3C} i_{l_a} - \frac{1}{3C} i_{l_b} - \frac{1}{3Cr} v_{ca} - \frac{2}{3Cr} v_{i_a} - \frac{1}{3Cr} v_{i_b} \\ \qquad + \frac{1}{3C}(2S_{11} - 2S_{31} + S_{12} - S_{32}) i_A \\ \qquad + \frac{1}{3C}(2S_{21} - 2S_{31} + S_{22} - S_{32}) i_B \end{cases}$$

(34.179)

To control the input power factor, a reference frame synchronous with one of the input voltages v_{i_a}, may be used applying the Blondel–Park transformation to the matrix converter switched state-space model (Eq. (34.179)), where ($\rho_{i_{dd}}, \rho_{i_{dq}}, \rho_{i_{qd}}, \rho_{i_{qq}}$) are functions of the ON/OFF states of the nine S_{kj} switches):

$$\begin{cases} \frac{di_{l_d}}{dt} = \omega_i i_{l_q} - \frac{1}{2l} v_{c_d} - \frac{1}{2\sqrt{3}l} v_{c_q} + \frac{1}{l} v_{i_q} \\ \frac{di_{l_q}}{dt} = -\omega_i i_{l_d} + \frac{1}{2\sqrt{3}l} v_{c_d} - \frac{1}{2l} v_{c_q} + \frac{1}{l} v_{i_q} \\ \frac{dv_{c_d}}{dt} = \frac{1}{2C} i_{l_d} - \frac{1}{2\sqrt{3}C} i_{l_q} - \frac{1}{3Cr} v_{c_d} + \omega_i v_{c_q} + \frac{-\rho_{i_{dd}}+\left(\rho_{i_{qd}}/\sqrt{3}\right)}{2C} i_{o_d} \\ \qquad + \frac{-\rho_{i_{dq}}+\left(\rho_{i_{qq}}/\sqrt{3}\right)}{2C} i_{o_q} + \frac{1}{2Cr} v_{i_d} - \frac{1}{2\sqrt{3}Cr} v_{i_q} \\ \frac{dv_{c_q}}{dt} = \frac{1}{2\sqrt{3}C} i_{l_d} + \frac{1}{2C} i_{l_q} - \omega_i v_{c_d} - \frac{1}{3Cr} v_{c_q} + \frac{-\left(\rho_{i_{dd}}/\sqrt{3}\right)-\rho_{i_{qd}}}{2C} i_{o_d} \\ \qquad + \frac{-\left(\rho_{i_{dq}}/\sqrt{3}\right)-\rho_{i_{qq}}}{2C} i_{o_q} + \frac{1}{2\sqrt{3}Cr} v_{i_d} + \frac{1}{2Cr} v_{i_q} \end{cases}$$

(34.180)

As a consequence, neglecting ripples, all the input variables become time-invariant, allowing a better understanding of the sliding-mode controller design, as well as the choice of the most adequate state-space vector. Using this state-space model, the input i_{i_d} and i_{i_q} currents are:

$$\begin{cases} i_{i_d} = i_{l_d} + \frac{1}{r}\left(\frac{di_{l_d}}{dt} - \omega i_{l_q}\right) \\ i_{i_q} = i_{l_q} + \frac{1}{r}\left(\frac{di_{l_q}}{dt} + \omega i_{l_d}\right) \end{cases} \Leftrightarrow \begin{cases} i_{i_d} = i_{l_d} - \frac{1}{2r} v_{c_d} - \frac{1}{2\sqrt{3}r} v_{c_q} + \frac{1}{r} v_{i_d} \\ i_{i_q} = i_{l_q} + \frac{1}{2\sqrt{3}r} v_{c_d} - \frac{1}{2r} v_{c_q} + \frac{1}{r} v_{i_q} \end{cases}$$

(34.181)

The input power factor controller should consider the input–output power constraint (Eq. (34.182)) (the converter losses and ripples are neglected), obtained as a function of the input and output voltages and currents (the input voltage v_{i_q} is equal to zero in the chosen dq rotating frame). The choice of one output voltage vector automatically defines the instantaneous value of the input $i_{i_d}(t)$ current.

$$v_{i_d} i_{i_d} \approx \frac{1}{3}\left(\frac{\sqrt{3}}{2}v_{o_d} + \frac{1}{2}v_{o_q}\right) i_{o_d} + \frac{1}{3}\left(-\frac{1}{2}v_{o_d} + \frac{\sqrt{3}}{2}v_{o_q}\right) i_{o_q} \tag{34.182}$$

Therefore, only the sliding surface associated to the $i_{i_q}(t)$ current is needed, expressed as a function of the system state variables and based on the state-space model determined in Eq. (34.180):

$$\begin{cases} \dfrac{di_{i_q}}{dt} = -\omega i_{l_d} + \dfrac{1}{3Cr} i_{l_q} + \left(-\dfrac{1}{6\sqrt{3}Cr^2} + \dfrac{\omega}{2r}\right) v_{c_d} \\ \quad + \left(\dfrac{1}{6Cr^2} + \dfrac{\omega}{2\sqrt{3}r}\right) v_{c_q} + \dfrac{1}{2\sqrt{3}l} v_{c_d} - \dfrac{1}{2l} v_{c_q} \\ \quad + \dfrac{1}{3Cr}\left(\rho_{i_{qd}} i_{o_d} + \rho_{i_{qq}} i_{o_q}\right) \dfrac{1}{r} \dfrac{dv_{i_q}}{dt} - \dfrac{1}{3Cr^2} v_{i_q} + \dfrac{1}{l} v_{i_q} \end{cases} \tag{34.183}$$

As the derivative of the input i_{i_q} current depends directly on the control variables $\rho_{i_{qd}}$, $\rho_{i_{qq}}$, the sliding function $S_{i_q}(e_{i_q}, t)$ will depend only on the input current error $e_{i_q} = i_{q\text{ref}} - i_{i_q}$.

$$S_{i_q}(e_{i_q}, t) = k_{i_q}\left(i_{q\text{ref}} - i_{i_q}\right) \tag{34.184}$$

As the sliding-mode stability is guaranteed if $S_{\alpha\beta}(e_{\alpha\beta}, t)$ $\dot{S}_{\alpha\beta}(e_{\alpha\beta}, t) < 0$, the criterion to choose the state-space vectors is:

$$S_{i_q}(e_{i_q}, t) > 0 \Rightarrow \dot{S}_{i_q}(e_{i_q}, t) < 0 \Rightarrow \frac{di_q}{dt} > \frac{di_{q\text{ref}}}{dt} \Rightarrow i_{i_q} \uparrow$$

$$S_{i_q}(e_{i_q}, t) < 0 \Rightarrow \dot{S}_{i_q}(e_{i_q}, t) > 0 \Rightarrow \frac{di_q}{dt} < \frac{di_{q\text{ref}}}{dt} \Rightarrow i_{i_q} \downarrow \tag{34.185}$$

Also, to choose the adequate input current vector it is necessary: (a) to know the location of the output currents, as the input currents depend on the output currents location (Table 34.6); (b) to know the dq frame location. As in the chosen frame (synchronous with the v_{i_a} input voltage), the dq-axis location depends on the v_{i_a} input voltage location, the sign of the input current vector i_{i_q} component can be determined knowing the location of the input voltages and the location of the output currents (Fig. 34.66).

Considering the previous example, at sector $V_i 1$ (Fig. 34.65), for an error of $C_\alpha = 1$ and $C_\beta = 1$, vectors -3, $+1$ or $1g$ might be used to control the output voltage. When compared, at sector $I_o 1$ (Fig. 34.66b), these three vectors have positive i_d components and, as a result, will have a similar effect on the input i_d current. However, they have a different effect on the i_q current: vector -3 has a positive i_q component, vector $+1$ has a negative i_q component and vector $1g$ has a nearly zero i_q component. As a result, if the output voltage errors are $C_\alpha = 1$ and $C_\beta = 1$, at sectors $V_i 1$ and $I_o 1$, vector -3 should be chosen if the input current error is quantized as $C_{iq} = 1$ (Fig. 34.66b), vector $+1$ should be chosen if the input current error is quantized as $C_{iq} = -1$ and if the input current error is $C_{iq} = 0$, vector $1g$ or -3 might be used.

When the output voltage errors are quantized as zero $C_{\alpha\beta} = 0$, the null vectors of group III should be used only if the input current error is $C_{iq} = 0$. Otherwise (being $C_{iq} \neq 0$), the lowest amplitude voltage vectors ({$+2, -8, +5, -2, +8, -5$} at sector $V_i 1$ at Fig. 34.65b), that were not used to control the output voltages, might be chosen to control the input i_q current as these vectors may have a strong influence on the input i_q current component (Fig. 34.66b).

To choose one of these six vectors, only the vectors located as near as possible to the output voltages sector (Fig. 34.67) are chosen (to minimize the output voltage ripple), and a five level comparator is enough. As a result, there will be $9 \times 5 = 45$ error combinations to select 27 space vectors. Therefore, the same vector may have to be used for more than one error combination.

With this reasoning, it is possible to obtain Table 34.7 for sector $V_i 1$, $I_o 1$, and $V_o 1$ and generalize it for all the other sectors.

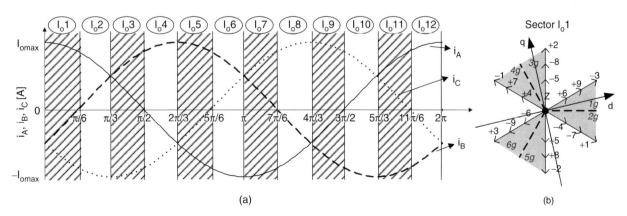

FIGURE 34.66 (a) Output currents and their corresponding sector and (b) representation of input current state-space vectors, when the output currents are located at sector $I_o 1$. The dq-axis is represented considering that the input voltages are located in zone $V_i 1$.

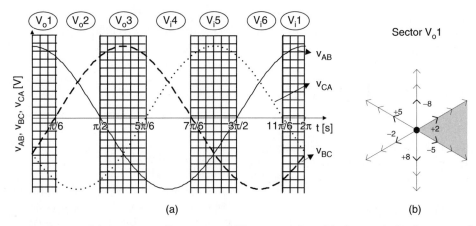

FIGURE 34.67 (a) Output voltages and their corresponding sector and (b) representation of the lowest amplitude output voltage vectors, when the input voltages are located at sector $V_o 1$.

TABLE 34.7 State-space vectors choice at sector $V_i 1$, $I_o 1$, and $V_o 1$

		C_{iq}				
C_α	C_β	−2	−1	0	1	2
−1	−1	+3	+3	+3	−1	−1
−1	0	5g	+3	−6	−1	−1
−1	1	−6	−6	−6	+4	3g
0	−1	6g	−9	−9	+7	4g
0	0	+8	+8	0	−5	2
0	1	−7	−7	+9	+9	+9
1	−1	−4	−4	+6	+6	+6
1	0	+1	+1	+6	−3	−3
1	1	+1	+1	1g	−3	−3

The experimental results shown in (Fig. 34.68) were obtained with a low-power prototype (1 kW), with two three-level comparators and one five-level comparator, associated to an EPROM lookup table. The transistors IGBT were switched at frequencies near 10 kHz.

The results show the response to a step on the output voltage reference (Fig. 34.68a) and on the input reference current (Fig. 34.68b), for a three-phase output load ($R = 7\,\Omega$, $L = 15\,\text{mH}$), with $k_{\alpha\beta} = 100$ and $k_{iq} = 2$. These results show that the matrix converter may operate with a near unity input power factor (Fig. 34.68a – $f_o = 20\,\text{Hz}$), or with lead/lag power factor (Fig. 34.68b), guaranteeing very low ripple on the output currents, a good tracking capability and fast transient response times.

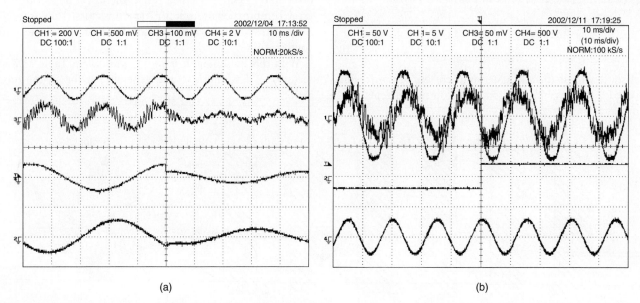

FIGURE 34.68 Dynamic responses obtained with a three-phase load: (a) output reference voltage step ($R = 7\,\Omega$, $L = 15\,\text{mH}$, $f_o = 20\,\text{Hz}$): input voltage $v_{i_a}(t)$ (CH1), input current $i_{i_a}(t)$ (CH3), output reference voltage $v_{BC_{ref}}(t)$ (CH4), and output current $i_A(t)$ (CH2) and (b) input reference current $i_{i_{qref}}(t)$ step: input voltage $v_{i_a}(t)$ (CH1), input current $i_{i_a}(t)$ (CH3), input reference current $i_{i_{qref}}(t)$ (CH2), and output current $i_A(t)$ (CH4).

34.4 Fuzzy Logic Control of Switching Converters

34.4.1 Introduction

Fuzzy logic control is a heuristic approach that easily embeds the knowledge and key elements of human thinking in the design of nonlinear controllers [19–21]. Qualitative and heuristic considerations, which cannot be handled by conventional control theory, can be used for control purposes in a systematic form, and applying fuzzy control concepts [22]. Fuzzy logic control does not need an accurate mathematical model, can work with imprecise inputs, can handle nonlinearity, and can present disturbance insensitivity greater than the most nonlinear controllers. Fuzzy logic controllers usually outperform other controllers in complex, nonlinear, or undefined systems for which a good practical knowledge exists.

Fuzzy logic controllers are based on fuzzy sets, i.e. classes of objects in which the transition from membership to nonmembership is smooth rather than abrupt. Therefore, boundaries of fuzzy sets can be vague and ambiguous, making them useful for approximation models.

The first step in the fuzzy controller synthesis procedure is to define the input and output variables of the fuzzy controller. This is done accordingly with the expected function of the controller. There are no general rules to select those variables, although typically the variables chosen are the states of the controlled system, their errors, error variation and/or error accumulation. In switching power converters, the fuzzy controller input variables are commonly the output voltage or current error, and/or the variation or accumulation of this error. The output variables $u(k)$ of the fuzzy controller can define the converter duty cycle (Fig. 34.60), or a reference current to be applied in an inner current-mode PI or a sliding-mode controller.

The fuzzy controller rules are usually formulated in linguistic terms. Thus, the use of linguistic variables and fuzzy sets implies the fuzzification procedure, i.e. the mapping of the input variables into suitable linguistics values.

Rule evaluation or decision-making infers, using an inference engine, the fuzzy control action from the knowledge of the fuzzy rules and the linguistic variable definition.

The output of a fuzzy controller is a fuzzy set, and thus it is necessary to perform a defuzzification procedure, i.e. the conversion of the inferred fuzzy result to a nonfuzzy (crisp) control action, that better represents the fuzzy one. This last step obtains the crisp value for the controller output $u(k)$ (Fig. 34.69).

These steps can be implemented on-line or off-line. On-line implementation, useful if an adaptive controller is intended, performs real-time inference to obtain the controller output and needs a fast enough processor. Off-line implementation employs a lookup table built according to the set of all possible combinations of input variables. To obtain this lookup table, the input values in a quantified range are converted (fuzzification) into fuzzy variables (linguistic). The fuzzy set output, obtained by the inference or decision-making engine according to linguistic control rules (designed by the knowledge expert), is then, converted into numeric controller output values (defuzzification). The table contains the output for all the combinations of quantified input entries. Off-line process can actually reduce the controller actuation time since the only effort is limited to consulting the table at each iteration.

This section presents the main steps for the implementation of a fuzzy controller suitable for switching converter control. A meaningful example is provided.

34.4.2 Fuzzy Logic Controller Synthesis

Fuzzy logic controllers consider neither the parameters of the switching converter or their fluctuations, nor the operating conditions, but only the experimental knowledge of the switching converter dynamics. In this way, such a controller can be used with a wide diversity of switching converters implying only small modifications. The necessary fuzzy rules are simply obtained considering roughly the knowledge of the switching converter dynamic behavior.

34.4.2.1 Fuzzification

Assume, as fuzzy controller input variables, an output voltage (or current) error, and the variation of this error. For the output, assume a signal $u(k)$, the control input of the converter.

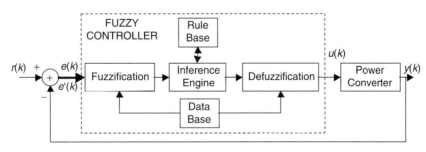

FIGURE 34.69 Structure of a fuzzy logic controller.

A. Quantization Levels Consider the reference $r(k)$ of the converter output kth sample, $y(k)$. The tracking error $e(k)$ is $e(k) = r(k) - y(k)$ and the output error change $\Delta_e(k)$, between the samples k and $k-1$, is determined by $\Delta_e(k) = e(k) - e(k-1)$.

These variables and the fuzzy controller output $u(k)$, usually ranging from -10 to 10 V, can be quantified in m levels $\{-(m-1)/2, +(m-1)/2\}$. For off-line implementation, m sets a compromise between the finite length of a lookup table and the required precision.

B. Linguistic Variables and Fuzzy Sets The fuzzy sets for x_e, the linguistic variable corresponding to the error $e(k)$, for $x_{\Delta e}$, the linguistic variable corresponding to the error variation $\Delta_e(k)$, and for x_u the linguistic variable of the fuzzy controller output $u(k)$, are usually defined as positive big (*PB*), positive medium (*PM*), positive small (*PS*), zero (*ZE*), negative small (*NS*), negative medium (*NM*), and negative big (*NB*), instead of having numerical values.

In most cases, the use of these seven fuzzy sets is the best compromise between accuracy and computational task.

C. Membership Functions A fuzzy subset, for example $\mathbf{S}_i$ ($\mathbf{S}_i$ = (NB, NM, NS, ZE, PS, PM, or PB)) of a universe E, collection of $e(k)$ values denoted generically by $\{e\}$, is characterized by a membership function μ_{Si}: $E \rightarrow [0,1]$, associating with each element e of universe E, a number $\mu_{Si}(e)$ in the interval $[0,1]$, which represents the grade of membership of e to E. Therefore, each variable is assigned a membership grade to each fuzzy set, based on a corresponding membership function (Fig. 34.70). Considering the m quantization levels, the membership function $\mu_{Si}(e)$ of the element e in the universe of discourse E, may take one of the discrete values included in $\mu_{Si}(e) \in \{0; 0.2; 0.4; 0.6; 0.8; 1; 0.8; 0.6; 0.4; 0.2; 0\}$. Membership functions are stored in the database (Fig. 34.69).

Considering $e(k) = 2$ and $\Delta_e(k) = -3$, taking into account the staircase-like membership functions shown in Fig. 34.70, it can be said that x_e is *PS* and also *ZE*, being equally *PS* and *ZE*. Also, $x_{\Delta e}$ is *NS* and *ZE*, being less *ZE* than *NS*.

D. Linguistic Control Rules The generic linguistic control rule has the following form: "IF $x_e(k)$ is membership of the set $\mathbf{S}_i$ = (NB, NM, NS, ZE, PS, PM, or PB) AND $x_{\Delta e}(k)$ is

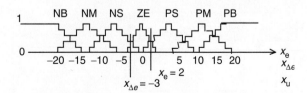

FIGURE 34.70 Membership functions in the universe of discourse.

membership of the set $\mathbf{S}_j$ = (NB, NM, NS, ZE, PS, PM, or PB), THEN the output control variable is membership of the set $\mathbf{S}_u$ = (NB, NM, NS, ZE, PS, PM, or PB)."

Usually, the rules are obtained considering the most common dynamic behavior of switching converters, the second-order system with damped oscillating response (Fig. 34.71). Analyzing the error and its variation, together with the rough linguistic knowledge of the needed control input, an expert can obtain linguistic control rules such as the ones displayed in Table 34.8. For example, at point 6 of Fig. 34.71 the rule is "if $x_e(k)$ is NM AND $x_{\Delta e}(k)$ is ZE, THEN $x_u(k+1)$ should be NM."

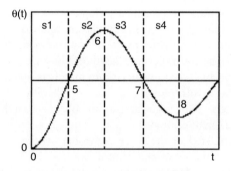

FIGURE 34.71 Reference dynamic model of switching converters: second-order damped oscillating error response.

TABLE 34.8 Linguistic control rules

$x_e(k)x_{\Delta e}(k)$	NB	NM	NS	ZE	PS	PM	PB
NB	NB	NB	NB	NM	NM	PS	PM
NM	NB	NB	NM	NS	NM	PM	PB
NS	NB	NB	NM	NS	NS	PM	PB
ZE	NB	NM	NS	ZE	PS	PM	PB
PS	NB	NM	PS	PS	PM	PB	PB
PM	NB	NM	PM	PS	PM	PB	PB
PB	NM	NS	PM	PM	PB	PB	PB

Table 34.8, for example, states that:

IF $x_e(k)$ is NB AND $x_{\Delta e}(k)$ is NB, THEN $x_u(k+1)$ must be NB, or

IF $x_e(k)$ is PS AND $x_{\Delta e}(k)$ is NS, THEN $x_u(k+1)$ must be NS, or

IF $x_e(k)$ is PS AND $x_{\Delta e}(k)$ is ZE, THEN $x_u(k+1)$ must be PS, or

IF $x_e(k)$ is ZE AND $x_{\Delta e}(k)$ is NS, THEN $x_u(k+1)$ must be NS, or

IF $x_e(k)$ is ZE AND $x_{\Delta e}(k)$ is ZE, THEN $x_u(k+1)$ must be ZE, or

IF...

These rules (rule base) alone do not allow the definition of the control output, as several of them may apply at the same time.

34.4.2.2 Inference Engine

The result of a fuzzy control algorithm can be obtained using the control rules of Table 34.8, the membership functions, and an inference engine. In fact, any quantified value for $e(k)$ and $\Delta_e(k)$ is often included into two linguistic variables. With the membership functions used, and knowing that the controller considers $e(k)$ and $\Delta_e(k)$, the control decision generically must be taken according to four linguistic control rules.

To obtain the corresponding fuzzy set, the min–max inference method can be used. The minimum operator describes the "AND" present in each of the four rules, that is, it calculates the minimum between the discrete value of the membership function $\mu_{Si}(x_e(k))$ and the discrete value of the membership function $\mu_{Sj}(x_{\Delta e}(k))$. The "THEN" statement links this minimum to the membership function of the output variable. The membership function of the output variable will therefore include trapezoids limited by the segment $\min(\mu_{Si}(x_e(k)), \mu_{Sj}(x_{\Delta e}(k)))$.

The OR operator linking the different rules is implemented by calculating the maximum of all the (usually four) rules. This mechanism to obtain the resulting membership function of the output variable is represented in Fig. 34.72.

34.4.2.3 Defuzzification

As shown, the inference method provides a resulting membership function $\mu_{Sr}(x_u(k))$, for the output fuzzy variable x_u (Fig. 34.72). Using a defuzzification process, this final membership function, obtained by combining all the membership functions, as a consequence of each rule, is then converted into a numerical value, called $u(k)$. The defuzzification strategy can be the center of area (COA) method. This method generates one output value $u(k)$, which is the abscissa of the gravity center of the resulting membership function area, given by the following relation:

$$u(k) = \left(\sum_{i=1}^{m} \mu_{Sr}(x_u(k))x_u(k)\right) \bigg/ \sum_{i=1}^{m} \mu_{Sr}(x_u(k)) \quad (34.186)$$

This method provides good results for output control. Indeed, for a weak variation of $e(k)$ and $\Delta_e(k)$, the center of the area will move just a little, and so does the controller output value. By comparison, the alternative defuzzification method, mean of maximum strategy (MOM) is advantageous for fast response, but it causes a greater steady-state error and overshoot (considering no perturbations).

34.4.2.4 Lookup Table Construction

Using the rules given in Table 34.8, the min–max inference procedure and COA defuzzification, all the controller output values for all quantified $e(k)$ and $\Delta_e(k)$, can be stored in an array to serve as the decision-lookup table. This lookup table usually has a three-dimensional representation similar to Fig. 34.73. A microprocessor-based control algorithm just picks up output values from the lookup table.

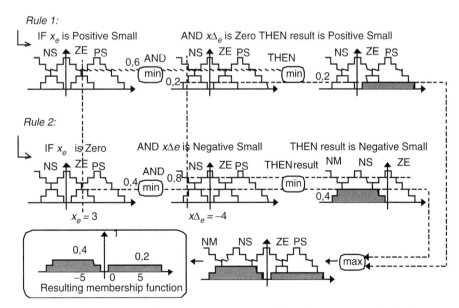

FIGURE 34.72 Application of the min–max operator to obtain the output membership function.

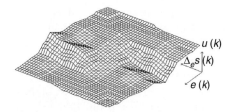

FIGURE 34.73 Three-dimensional view of the lookup table.

34.4.3 Example: Near Unity Power Factor Buck–Boost Rectifier

EXAMPLE 34.18 Fuzzy logic control of unity power factor buck–boost rectifiers

Consider the near unity power factor buck–boost rectifier of Fig. 34.74.

The switched state-space model of this converter can be written:

$$\begin{cases} \dfrac{di_s}{dt} = -\dfrac{R_f}{L_f}i_s - \dfrac{1}{L_f}v_{C_f} + \dfrac{1}{L_f}v_s \\ \dfrac{dv_{C_f}}{dt} = \dfrac{1}{C_f}i_s - \dfrac{\gamma_p}{C_f}i_{L_o} \\ \dfrac{di_{L_o}}{dt} = \dfrac{\gamma_p}{L_o}v_{C_f} - \dfrac{\gamma(1-|\gamma_p|)}{L_o}V_{C_o} \\ \dfrac{dV_{C_o}}{dt} = \dfrac{1-|\gamma_p|}{C_o}i_{L_o} - \dfrac{1}{R_o C_o}V_o \end{cases} \quad (34.187)$$

where $\gamma_p = \begin{cases} 1, & (\text{switch 1 and 4 are ON}) \text{ and} \\ & (\text{switch 2 and 3 are OFF}) \\ 0, & \text{all switches are OFF} \\ -1, & (\text{switch 2 and 3 are ON}) \text{ and} \\ & (\text{switch 1 and 4 are OFF}) \end{cases}$

and $\gamma = \begin{cases} 1, & i_{L_o} > 0 \\ 0, & i_{L_o} \leq 0 \end{cases}$

For comparison purposes, a PI output voltage controller is designed considering that a current-mode PWM modulator enforces the reference value for the i_s current (which usually exhibits a fast dynamics compared with the dynamics of V_{C_o}). A first-order model, similar to Eq. (34.146) is obtained. The PI gains are similar to Eq. (34.116) and load-dependent ($K_p = C_o/(2T_d)$, $K_i = 1/(2T_d R_o)$).

A fuzzy controller is obtained considering the approach outlined, with seven membership functions for the output voltage error, five for its change, and three membership functions for the output. The linguistic control rules are obtained as the ones depicted in Table 34.8 and the lookup table gives a mapping similar to Fig. 34.73. Performances obtained for the step response show a fuzzy controlled rectifier behavior close to the PI behavior. The advantages of the fuzzy controller emerge for perturbed loads or power supplies, where the low sensitivity of the fuzzy controller to system parameters is clearly seen (Fig. 34.75). Therefore, the fuzzy controllers can be advantageous for switching converters with changing loads, power supply voltages, and other external disturbances.

34.5 Conclusions

Control techniques for switching converters were reviewed. Linear controllers based on state-space averaged models or circuits are well established and suitable for the application of linear systems control theory. Obtained linear controllers are useful, if the converter operating point is almost constant and the disturbances are not relevant. For changing operating points and strong disturbances, linear controllers can be enhanced with nonlinear, antiwindup, soft-start, or saturation techniques. Current-mode control will also help to overcome the main drawbacks of linear controllers.

Sliding mode is a nonlinear approach well adapted for the variable structure of the switching converters. The critical problem of obtaining the correct sliding surface was highlighted, and examples were given. The sliding-mode control law allows the implementation of the switching converter controller, and the switching law gives the PWM modulator. The system variables to be measured and fed back are identified. The obtained reduced-order dynamics is not dependent on system parameters or power supply (as long as it is high enough), presents no steady-state errors, and has a faster response speed (compared with linear controllers), as the system order is reduced and non-idealities are eliminated. Should the measure of the state variables be difficult, state observers may be used, with steady-state errors easily corrected. Sliding-mode controllers provide robustness against bounded disturbances and an elegant way to obtain the controller and modulator, using just the same theoretical approach. Fixed-frequency operation was addressed and solved, together with the short-circuit-proof operation.

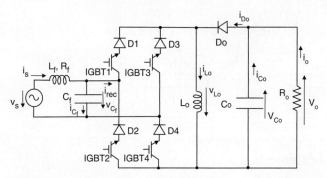

FIGURE 34.74 Unity power factor buck–boost rectifier with four IGBTs.

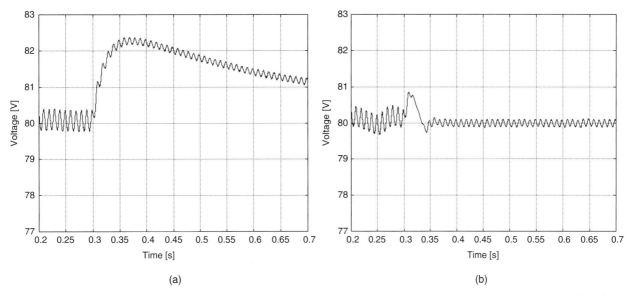

FIGURE 34.75 Simulated result of the output voltage response to load disturbances ($R_o = 50$–$150\,\Omega$ at time 0.3 s): (a) PI control and (b) fuzzy logic control.

Presently, fixed-frequency techniques were applied to converters that can only operate with fixed frequency. Sliding-mode techniques were successfully applied to MIMO switching power converters and to multilevel converters, solving the capacitor voltage divider equalization. Sliding-mode control needs more information from the controlled system than do the linear controllers, but is probably the most adequate tool to solve the control problem of switching power converters.

Fuzzy logic controller synthesis was briefly presented. Fuzzy logic controllers are based on human experience and intuition and do not depend on system parameters or operating points. Fuzzy logic controllers can be easily applied to various types of power converters having the same qualitative dynamics. Fuzzy logic controllers, like sliding-mode controllers, show robustness to load and power supply perturbations, semiconductor non-idealities (such as switch delays or uneven conduction voltage drops), and dead times. The controller implementation is simple, if based on the off-line concept. On-line implementation requires a fast microprocessor but can include adaptive techniques to optimize the rule base and/or the database.

Acknowledgments

J. Fernando Silva thanks all the researchers whose works contributed to this chapter, namely Professors S. Pinto, V. Pires, J. Quadrado, T. Amaral, M. Crisóstomo, Engineers J. Costa, N. Rodrigues, and the suggestions of Professor M. P. Kazmierkowski. The authors also thank FCT, POSI, POCTI, FEDER for funding the projects enabling the presented results.

References

1. Bose, B. K. *Power Electronics and AC Drives*, Prentice-Hall, New Jersey, 1986.
2. Kassakian, J. Schlecht, M. and Verghese, G. *Principles of Power Electronics*, Addison Wesley, 1992.
3. Rashid, M. *Power Electronics: Circuits, Devices and Applications*, 2nd ed., Prentice-Hall International, 1993.
4. Thorborg, K. *Power Electronics*, Prentice-Hall, 1988.
5. Mohan, N. Undeland, T. and Robins, W. *Power Electronics: Converters, Applications and Design*, 2nd ed., John Wiley & Sons, 1995.
6. Sum, K. K. *Switched Mode Power Conversion*, Marcel Dekker Inc., 1984.
7. Bühler, H. *Electronique de Réglage et de Commande*, Traité D'électricité, vol. XV, éditions Georgi, 1979.
8. Irwin, J. D. ed. *The Industrial Electronics Handbook*, CRC/IEEE Press, 1996.
9. Chryssis, G. *High Frequency Switching Power Supplies*, McGraw-Hill, 1984.
10. Labrique, F. and Santana, J. *Electrónica de Potência*, Fundação Calouste Gulbenkian, Lisboa, 1991.
11. Utkin, V. I. *Sliding Modes and Their Application on Variable Structure Systems*, MIR Publishers Moscow, 1978.
12. Utkin, V. I. *Sliding Modes in Control Optimization*, Springer-Verlag, 1981.
13. Ogata, K. *Modern Control Engineering*, 3rd ed., Prentice-Hall International, 1997.
14. Levine, W. S. ed. *The Control Handbook*, CRC/IEEE Press, 1996.
15. Fernando Silva, J. *Electrónica Industrial*, Fundação Calouste Gulbenkian, Lisboa, 1998.
16. Fernando Silva, J. *Sliding Mode Control Design of Control and Modulator Electronics for Power Converters*, Special Issue on Power Electronics of Journal on Circuits, Systems and Computers, vol. 5, no. 3, pp. 355–371, 1995.
17. Fernando Silva, J. *Sliding Mode Control of Boost Type Unity Power Factor PWM Rectifiers*, IEEE Trans. on Industrial Electronics,

Special Section on High-Power-Factor Rectifiers I, vol. 46, no. 3, pp. 594–603, June 1999. ISSN 0278-0046.
18. José Rodriguez, Special Section on Matrix Converters, *IEEE Transactions on Industrial Electronics*, vol. 49, no. 2, April 2002.
19. Zadeh, L. A. "Fuzzy Sets", *Information and Control*, vol. 8, pp. 338–353, 1965.
20. Zadeh, L. A. "Outline of a New Approach to the Analysis of Complex Systems and Decision Process," *IEEE Trans. Syst. Man Cybern.*, vol. SMC-3, pp. 28–44, 1973.
21. Zimmermann, H. J. *Fuzzy Sets: Theory and its Applications*, Kluwer-Nijhoff, 1995.
22. Candel, A. and Langholz, G. *Fuzzy Control Systems*, CRC Press, 1994.

35
Fuzzy Logic in Electric Drives

Ahmed Rubaai, Ph.D.
Department of Electrical Engineering, Howard University, Washington, D.C., USA

35.1 Introduction 999
35.2 The Fuzzy Logic Concept 999
    35.2.1 The Fuzzy Inference System (FIS) • 35.2.2 Fuzzification • 35.2.3 The Fuzzy Inference Engine • 35.2.4 Defuzzification
35.3 Applications of Fuzzy Logic to Electric Drives 1005
    35.3.1 Fuzzy Logic-based Microprocessor Controller • 35.3.2 Fuzzy Logic-based Speed Controller • 35.3.3 Fuzzy Logic-based Position Controller
35.4 Hardware System Description 1008
    35.4.1 Experimental Results
35.5 Conclusion 1011
    Further Reading 1013

35.1 Introduction

Over the years, we have increasingly been on the search to understand the human ability to reason and make decisions, often in the face of only partial knowledge. The ability to generalize from limited experience into areas as yet encountered is one of the fascinating abilities of the human mind. Traditionally, our attempt to understand the world and its functions has been limited to finding mathematical models or equations for the systems under study. This approach has proven extremely useful, particularly in an age when very fast computers are available to most of us with only a minimum amount of capital outlay. And even when these computers are not fast enough, many researchers can gain access to super computers capable of giving numerical solutions to multiorder differential equations that are capable of describing most of the industrial processes.

This analytical enlightenment, however, has come at the cost of realizing just how complex the world is. At this point, we have come to realize that no matter how simple the system is, we can never hope to model it completely. So instead, we select suitable approximations that give us answers that we think, are sufficiently precise. Because our models are incomplete, we are faced with one of the following choices:

1. Use the approximate model and introduce probabilistic representations to allow for the possible errors.

2. Seek to develop an increasingly complex model in the hope that we can find one, that completely describes the systems while being solvable in real time.

This dilemma has led a few, most notably Zadeh [1], to return the decision-making process employed by our brilliant minds when confronted with incomplete information. The approach taken in those cases makes allowances for the imprecision caused by incomplete knowledge and actually embracing the imprecision in forming an analytical framework. This approach involved artificial intelligence using approximate reasoning or fuzzy logic as it now commonly known. As a result, artificial intelligence using fuzzy logic has proven extremely useful in ascribing a logic mechanism to a wide range of topics from economic modeling and prediction to biology analysis to control engineering. In this chapter, an examination of the principles involved in artificial intelligence using fuzzy logic and its application to electric drives is discussed.

35.2 The Fuzzy Logic Concept

Fuzzy logic arose from a desire to incorporate logical reasoning and the intuitive decision making of an expert operator into an automated system [1]. The aim is to make decisions based on a number of learned or predefined rules, rather than numerical calculations. Fuzzy logic incorporates rule-base structure in attempting to make decisions [1–5]. However, before the rule-base can be used, the input data should be represented in such a way as to retain meaning, while, still allowing for manipulation. Fuzzy logic is an aggregation of rules, based on the input state variables condition with a corresponding desired output. A mechanism must exist to decide on which output, or combination of the different outputs, will be used since each rule could conceivably result in a different output action.

Fuzzy logic can be viewed as an alternative form of input/output mapping. Consider the input premise, x, and a particular qualification of the input x represented by, A_i. Additionally, the corresponding output, y, can be qualified by expression C_i. Thus, a fuzzy logic representation of the relationship between the input x and the output y could be described with the following:

$$R_1: \quad \text{IF} \quad x \text{ is } A_1 \quad \text{THEN } y \text{ is } C_1$$
$$\ldots \quad \ldots \quad \ldots \quad \ldots$$
$$R_2: \quad \text{IF} \quad x \text{ is } A_2 \quad \text{THEN } y \text{ is } C_2$$
$$\ldots \quad \ldots \quad \ldots \quad \ldots$$
$$R_n: \quad \text{IF} \quad x \text{ is } A_n \quad \text{THEN } y \text{ is } C_n \quad (35.1)$$

where

x is the input (state variable).
y is the output of the system.
A_i are the different fuzzy variables used to classify the input x.
C_i are the different fuzzy variables used to classify the output y.

The fuzzy rule representation is based on linguistic [1, 3]. Thus, the input x is a linguistic variable that corresponds to the state variable under consideration. Furthermore, the elements A_i are fuzzy variables that describe the input x. Correspondingly, the elements C_i are the fuzzy variables used to describe the output y. In fuzzy logic control, the term "linguistic variable" refers to whatever state variables the system designer is interested in [1]. Linguistic variables that are often used in control applications include speed, speed error, position, and derivative of position error. The fuzzy variable is perhaps better described as a fuzzy linguistic qualifier. Thus the fuzzy qualifier performs classification (qualification) of the linguistic variables. The fuzzy variables frequently employed include negative large, positive small, and zero. Several papers in the literature use the term "fuzzy set" instead of "fuzzy variable," however, the concept remains the same. Table 35.1 illustrates the difference between fuzzy variables and linguistic variables.

TABLE 35.1 Fuzzy and linguistic variables

Linguistic variables	Fuzzy variables (linguistic qualifiers)
Speed error (SE)	Negative large (NL)
Position error (PE)	Zero (ZE)
Acceleration (AC)	Positive medium (PM)
Derivative of position error (DPE)	Positive very small (PVS)
Speed (SP)	Negative medium small (NMS)

Once the linguistic and fuzzy variables have been specified, the complete inference system can be defined. The fuzzy linguistic universe, U, is defined as the collection of all the fuzzy variables used to describe the linguistic variables [6–8], i.e. the set U for a particular system could be comprised of NS, ZE, and PS. Thus, in this case the set U is equal to the set of [NS, ZE, PS]. For the system described by Eq. (35.1), the linguistic universe for the input x would be the set $U_x = [A_1 \ A_2 \ldots A_n]$. Similarly, the linguistic universe for the output y would be the set $U_y = [C_1 \ C_2 \ldots C_n]$.

35.2.1 The Fuzzy Inference System (FIS)

The basic fuzzy inference system (FIS) can be classified as:

Type 1   fuzzy input fuzzy output (FIFO)
Type 2   fuzzy input crisp output (FICO)

Type 2 differs from the first in that the crisp output values are predefined and, thus, built into the inference engine of the FIS. On the contrary, Type 1 produces linguistic outputs. Type 1 is more general than Type 2 as it allows redefinition of the response without having to redesign the entire inference engine. One draw back is the additional step required converting the fuzzy output of the FIS to a crisp output.

Developing a FIS and applying it to a control problem involves several steps:

1. Fuzzification.
2. Fuzzy rule evaluation (fuzzy inference engine).
3. Defuzzification.

The total FIS is a mechanism that relates the inputs to a specific output or set of outputs. First, the inputs are categorized linguistically (fuzziffication), then the linguistic inputs are related to outputs (fuzzy inference), and finally, all the different outputs are combined to produce a single output (defuzziffication). Figure 35.1 shows a block diagram of the fuzzy inference system.

35.2.2 Fuzzification

Fuzzification is the conversion of crisp numerical values into fuzzy linguistic quantifiers' [7, 8]. Fuzzification is performed using membership functions. Each membership function evaluates how well the linguistic variable may be described by

35 Fuzzy Logic in Electric Drives

FIGURE 35.1 Fuzzy inference system.

a particular fuzzy qualifier. In other words, the membership function derives a number that is representative of the suitability of the linguistic variable to be classified by the fuzzy variable (set). This suitability is often described as the degree of membership. In order to maintain a relationship to traditional binary logic, the membership values must range from 0 to 1 inclusive. Figure 35.2 shows the mechanism involved in the fuzzification of crisp inputs when multiple inputs are involved. Since each input has a number of membership functions (one for each fuzzy variable), the outputs of all the membership functions for a particular crisp numerical input are combined to form a fuzzy vector.

Any number of normalizing expressions can perform fuzzification. Two of the more common functions are the linear and Gaussian [4]. In both cases there is one parameter, μ, that indicates the midpoint of the region and another, σ, that defines the width of the membership functions. For the linear function, the width is specified by, σ_L, and the midpoint by, μ_L. Similarly for the gaussian function, the width is specified by, σ_G, and the midpoint by, μ_G. Equations (35.2a) and (35.2b) define the linear and gaussian membership functions, respectively.

Linear function:

$$\begin{cases} 1 - \left|\frac{x-\mu_L}{\sigma_L}\right| & \text{if } x \in [(\mu_L - \sigma_L), (\mu_L + \sigma_L)] \\ 0 & \text{Otherwise} \end{cases} \quad (35.2a)$$

Gaussian function:

$$\exp\left(-\frac{(x-\mu_G)^2}{2(\sigma_G)^2}\right) \quad (35.2b)$$

where

$$\mu_L = \mu_G \quad (35.3a)$$

$$\sigma_L = 3\sigma_G \quad (35.3b)$$

The relations expressed by equations (35.3a) and (35.3b) are made because of the characteristics of the gaussian function. Because a gaussian membership function may never have a membership value of zero, some appropriate value close to zero must be chosen as the cut-off point. At a distance of $3\sigma_G$ from the mean, the gaussian membership function results in a membership value of 0.05. Thus the width of the gaussian function is chosen as $3\sigma_G$.

As previously mentioned, fuzzification of the input has resulted in a fuzzy vector where each component of this vector represents the degree of membership of the linguistic variables into a specific fuzzy variable's category. The number of components of the fuzzy vector is equal to the number of fuzzy variables used to categorize specific linguistic variable. For illustrative purposes, we consider an example with a linguistic variable x and three fuzzy variables PV, ZE, and NV. If we describe the membership function (fuzzifier) as X, then we will have three membership functions: X_{PV}, X_{ZE}, and X_{NV}. The fuzzy linguistic universe for the input x can be described by the set U_x that is defined in Eq. (35.4).

$$U_x = [X_{NV} \; X_{ZE} \; X_{PV}] \quad (35.4)$$

where

X_{PV} is the membership function for the positive fuzzy variables.

X_{ZE} is the membership function for the zero fuzzy variables.

X_{NV} is the membership function for the negative fuzzy variables.

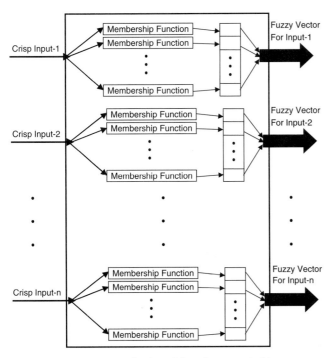

FIGURE 35.2 Fuzzification of the crisp numerical inputs.

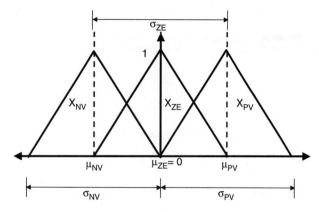

FIGURE 35.3 Linear membership functions.

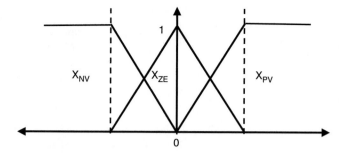

FIGURE 35.4 Linear-trapezoidal membership functions.

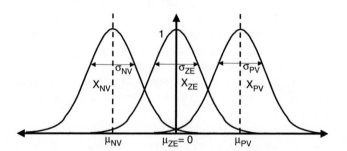

FIGURE 35.5 Gaussian membership functions.

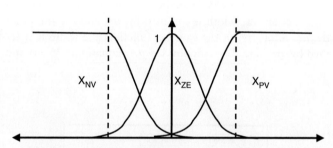

FIGURE 35.6 Gaussian-trapezoidal membership functions.

Thus, the fuzzy vector, which is the output of the fuzzification step of the inference system, can be denoted by $\underline{x}$.

$$\underline{x} = [x_{NV}\ x_{ZE}\ x_{PV}] = [X_{NV}(x)\ X_{ZE}(x)\ X_{PV}(x)] \quad (35.5)$$

where

x_{NV} = the membership value of x into the fuzzy region denoted by negative.

x_{ZE} = the membership value of x into the fuzzy region denoted by zero.

x_{PV} = the membership value of x into the fuzzy region denoted by positive.

Equation (35.2a) represents the linear membership functions, which are illustrated in Fig. 35.3. The linear function can be modified to form the linear-trapezoidal function. Under this modification, if the input x falls between zero and the mean, μ_L, of the respective region, then Eq. (35.2a) is used, otherwise, the membership value is equal to one. Thus, we arrive at the membership functions shown in Fig. 35.4. Each region of the linear and trapezoidal-linear membership functions is distinguished from another by the different values of σ_L and μ_L. One important criterion that should be taken into consideration is that the union of the domain of all membership functions for a given input must cover the entire range of the input [9]. Thus the trapezoidal modification is often employed to ensure coverage of the entire input space.

The gaussian membership function is characterized by Eq. (35.2b). The gaussian function can also be modified to form the trapezoidal-gaussian function. In this case, if the input falls between the mean, μ_G and zero, Eq. (35.2b) is used to find the membership value. Otherwise the membership value becomes one. The gaussian function is shown in Fig. 35.5 and the modified version is shown in Fig. 35.6.

A third type of membership function known as the fuzzy singleton is also considered. The fuzzy singleton is a special function in which the membership value is one for only one particular value of the linguistic input variable, and zero otherwise [4]. Thus, the fuzzy singleton is a special case of the membership function with a width, σ, of zero. Therefore, the only parameter that needs to be defined is the mean, μ_s, of the singleton. Thus, if the input is equal to μ_s, then the membership value is one. Otherwise it is zero. We can denote the singleton membership function as $S(\mu_s)$.

The fuzzy singleton function is quite useful in defining some special membership functions. If we would like to dispense with the need for a continuous degree of membership and prefer a binary valued function, the fuzzy singleton is an ideal candidate. We can form the membership function representing the fuzzy variable as a collection of fuzzy singletons ranging within the regions denoted by $[\mu+\sigma/2,\ \mu-\sigma/2]$. A graphical representation using three fuzzy variables (membership functions) is shown in Fig. 35.7 (in Fig. 35.7 the corners are only slanted so that the regions are easier to distinguish from each other). Thus, the membership functions shown in Fig. 35.7

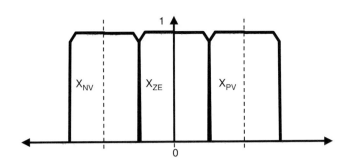

FIGURE 35.7 Membership functions comprised of fuzzy singletons.

would be best defined as an integral of the fuzzy singleton with respect to the mean over the width of the function. Consequently, the membership function $X(\sigma,\mu)$ would be defined as follows:

$$X(\sigma,\mu) = \int_{\mu-\sigma/2}^{\mu+\sigma/2} S(\mu_s)d\mu_s \qquad (35.6)$$

where

$S(\mu_s)$ is the singleton function.

35.2.3 The Fuzzy Inference Engine

The fuzzy inference engine uses the fuzzy vectors to evaluate the fuzzy rules and produce an output for each rule. Figure 35.8 shows a block diagram of the fuzzy inference engine. Note that the rule-based system takes the form found in Eq. (35.1). This form could be applied to traditional logic as well as fuzzy logic albeit with some modification. A typical rule R would be:

$$R_i: \quad \text{IF} \quad x_i \quad \text{THEN} \quad y = C_i \qquad (35.7)$$

where

x_i is the result of some logic expression.

The logical expression used in the case of fuzzy inference Eq. (35.7) is of the form

$$x \in X_i \qquad (35.8)$$

where

x is the input.

X_i is the linguistic variable.

In binary logic, the expression in Eq. (35.8) results in either true or false. However, in fuzzy logic we often require a continuum of truth-values. Figures 35.9 and 35.10 illustrate the difference between binary logic and fuzzy logic. In traditional logic, there is a single point representing the boundary between true and false. While in fuzzy logic, there is an entire region over which there is a continuous variation between truth and falsehood. The second part of Eq. (35.7), $y = C_i$, is the action prescribed by the particular rule. This portion indicates what value will be assigned to the output. This value could be either a fuzzy linguistic description or a crisp numerical value.

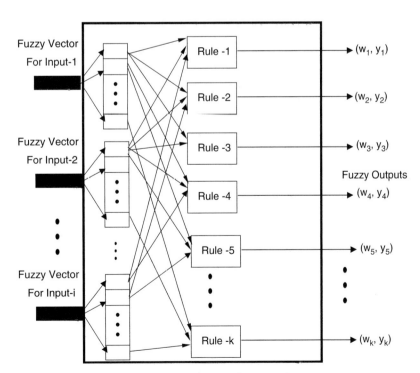

FIGURE 35.8 Fuzzy inference engine.

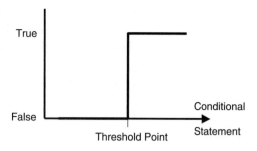

FIGURE 35.9 Binary logic statement evaluation.

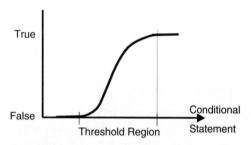

FIGURE 35.10 Fuzzy logic statement evaluation.

The logical expression that dictates whether the result of a particular rule is carried out could involve multiple criteria. Multiple conditions imply multiple input, as is most often the case in many applications of fuzzy logic to dynamic systems. Let us describe a system with two inputs x^1 and x^2. For simplicity of explanation and without loss of generality, we will use three fuzzy variables, namely, PV, ZE, and NV. Although each linguistic variable x^1 and x^2 uses the same fuzzy qualifiers, each input must have its own membership functions since they belong to different spaces. Thus, we will have two fuzzy vectors $\underline{x}^1$ and $\underline{x}^2$, for the first and second input, respectively.

$$\underline{X}^1 = \lfloor X_{NV}^1(x^1) \quad X_{ZE}^1(x^1) \quad X_{PV}^1(x^1) \rfloor = \lfloor x_1^1 \quad x_2^1 \quad x_3^1 \rfloor \tag{35.9a}$$

$$\underline{X}^2 = \lfloor X_{NV}^2(x^2) \quad X_{ZE}^2(x^2) \quad X_{PV}^2(x^2) \rfloor = \lfloor x_1^2 \quad x_2^2 \quad x_3^2 \rfloor \tag{35.9b}$$

where

X_{NV}^n = Membership function for the negative fuzzy variable for input n.

X_{ZE}^n = Membership function for the zero fuzzy variable for input n.

X_{PV}^n = Membership function for the positive fuzzy variable for input n.

x^1 = First linguistic variable (input-1).

x^2 = Second linguistic variable (input-2).

x_i^1 = The degree of membership of input-1 into the ith fuzzy variable's category.

x_j^2 = The degree of membership of input-2 into the jth fuzzy variable's category.

The inference mechanism in this case would be specified by the rule R_{ij}:

$$R_{ij}: \quad \text{IF} \quad (x_i^1 \text{ AND } x_j^2) \text{ THEN } y = C_{ij} \tag{35.10}$$

The specification R_{ij} is made so as to emphasize that all combinations of the components of the fuzzy vectors should be used in separate rules. A specific example of one of the rules can be described as follows:

$$R_{13} \text{ IF } (x_1^1 \text{ AND } x_3^2 \text{ THEN } y = C_{13}$$

This rule can be written linguistically as:

$$R_{13} \text{ IF}((x^1 \text{ is Negative}) \text{ AND } (x^2 \text{ is Positive})) \text{THEN } y = C_{13}$$

where

x^1 is the first linguistic variable.

x^2 is the second linguistic variable.

C_{13} is the output action to be defined by the system designer.

The AND in Eq. (35.10) can be interpreted and evaluated in two different ways. First, the AND could be evaluated as the product of x_i^1 and x_j^2 [7]. Thus,

$$x_i^1 \text{ AND } x_j^2 = x_i^1 x_j^2$$

The second method is by taking the minimum of the term's [7]. In this case the result is the minimum value of the membership values. Therefore,

$$x_i^1 \text{ AND } x_j^2 = \min(x_i^1, x_j^2)$$

The most commonly used method in the literature is the product method. Therefore, the product method is used in this chapter to evaluate the AND function. Equation (35.10) can be expanded to multiple input with multiple fuzzy variables. In the most general case of n inputs and k linguistic qualifiers, we would have the rule R_i:

$$R_i: \quad \text{IF } [x_i^1 \text{ AND } x_i^2 \text{ AND } \ldots \text{ AND } x_i^n] \text{ THEN } y = C_i \tag{35.11}$$

Recalling that all combinations of input vector components must be taken between fuzzy vector components the system designer could have up to $n*k$ rules. Where n is the number of fuzzy variables used to describe the inputs and k is the number of inputs. The number of rules used by the fuzzy inference engine could be reduced if the designer could eliminate some combinations of input conditions.

35.2.4 Defuzzification

The fuzzy inference engine as described previously often has multiple rules, each with possibly a different output. Defuzzification refers to the method employed to combine these many outputs into a single output. Using Eq. (35.11) where multiple inputs ($x^1\ x^2\ \ldots\ x^n$) should be evaluated, the product due to the evaluation of the premise conditions (defined by the components of the fuzzy vectors) determines the strength of the overall rule evaluation, w_i.

$$w_i = x_i^1 \text{ AND } x_i^2 \text{ AND } \ldots \text{ AND } x_i^n$$

$$w_i = (x_i^1)(x_i^2)\ldots(x_i^n) \quad (35.12)$$

where, x_i^k is the membership value of the kth input into the ith fuzzy variable's category.

This value, w_i, becomes extremely important in defuzzification. Ultimately, defuzzification involves both the set of outputs C_i and the corresponding rule strength w_i.

There are a number of methods used for defuzzification, including the center of gravity (COG) and mean of maxima (MOM) [10]. The COG method otherwise known as the fuzzy centroid is denoted by y^{COG}.

$$y^{\text{COG}} = \frac{\sum_i w_i C_i}{\sum_i w_i} \quad (35.13)$$

where

$$w_i = \prod_k x_i^k$$

C_i: is the corresponding output.

The mean of maxima method selects the outputs C_i that have the corresponding highest values of w_i.

$$y^{\text{MOM}} = \sum_{C_i \in G} C_i \Big/ \text{Card}(G) \quad (35.14)$$

where, G denotes a subset of C_i consisting of these values that have the maximum value of w_i. Out of the two methods of defuzzification, the most common method is the fuzzy centroid and is the one employed in this chapter.

35.3 Applications of Fuzzy Logic to Electric Drives

High performance drives requires that the shaft speed and the rotor position follow preselected tracks (trajectories) at all times [11, 12]. To accomplish this, two fuzzy control systems were designed and implemented. The goal of fuzzy control system is to replace an experienced human operator with a fuzzy rule-based system. The fuzzy logic controller provides an algorithm that converts the linguistic control maneuvering, based on expert knowledge, into an automatic control approach. In this section, a fuzzy logic controller (FLC) is proposed and applied to high performance speed and position tracking of a brushless DC (BLDC) motors. The proposed controller provides the high degree of accuracy required by high performance drives without the need for detailed mathematical models. A laboratory implementation of the fuzzy logic-tracking controller using the Motorola MC68HC11E9 microprocessor is described in this chapter. Additionally, in this experiment a bang–bang controller is compared to the fuzzy controller.

35.3.1 Fuzzy Logic-based Microprocessor Controller

The first step in designing a fuzzy controller is to decide which state variables representative of system dynamic performance can be taken as the input signals to the controller. Further, choosing the proper fuzzy variables, formulating the fuzzy control rules are also significant factors in the performance of the fuzzy control system. Empirical knowledge and engineering intuition play an important role in choosing fuzzy variables and their corresponding membership functions. The motor drive's state variables and their corresponding errors are usually used as the fuzzy controller's inputs including, rotor speed, rotor position, and rotor acceleration. After choosing proper linguistic variables as input and output of the fuzzy controller, it is required to decide on the fuzzy variables to be used. These variables transform the numerical values of the input of the fuzzy controller, to fuzzy quantities. The number of these fuzzy variables specifies the quality of the control, which can be achieved using the fuzzy controller. As the number of the fuzzy variable increases, the management of the rules is more involved and the tuning of the fuzzy controller is less straightforward. Accordingly, a compromise between the quality of control and computational time is required to choose the number of fuzzy variables. For the BLDC motor drive under study, two inputs are usually required. After specifying the fuzzy sets, it is required to determine the membership functions for these sets. Finally, the fuzzy logic control (FLC) is implemented by using a set of fuzzy decision rules. After the rules are evaluated, a fuzzy centroid is used to determine the fuzzy control output. Details of the design of the proposed controllers are given in the following sections.

35.3.2 Fuzzy Logic-based Speed Controller

In the case of shaft speed control to achieve optimal tracking performance, the motor speed error (ω_e) and the motor acceleration error (α_e) are used as inputs to the proposed controller. The controller output is the change in the motor voltage. For the fuzzy logic-based speed controller, the two inputs required are defined as

$$\omega_e = \omega_{\text{ref}} - \omega_{\text{act}} \quad (35.15)$$

$$\alpha_e = \alpha_{\text{ref}} - \alpha_{\text{act}} \quad (35.16)$$

where

ω_{ref} = the desired speed (rad/s).

ω_{act} = the measured speed (rad/s).

$\alpha_{\text{ref}} = 0$, because we want to minimize the acceleration to zero.

α_{act} = the calculated acceleration in (rad/s²).

For both the speed and acceleration errors, three regions of operation are established according to the fuzzy variables. These regions are positive error, zero, and negative error. The proposed controller uses these regions to determine the required motor voltage, which enables the motor speed to follow a desired reference trajectory. Examples of the broad fuzzy decisions are:

IF speed error is positive, THEN decrease the output.
IF speed error is zero, THEN maintain the output.
IF speed error is negative, THEN increase the output.
IF acceleration error is positive, THEN decrease the output.
IF acceleration error is zero, THEN maintain the output.
IF acceleration error is negative, THEN increase the output.

To achieve a sufficiently good quality of control, the three basic variables must be further refined. Thus the linguistic variable "acceleration error" has seven fuzzy variables: negative large (NL), negative medium (NM), negative small (NS), zero (Z), positive small (PS), positive medium (PM), and positive large (PL). The values associated with the fuzzy variables for the acceleration error are shown in Table 35.2.

The linguistic variable "speed error" has nine fuzzy variables: negative large (NL), negative medium (NM), negative medium small (NMS), negative small (NS), zero (Z), positive small (PS), positive medium (PM), positive medium small (PMS), and positive large (PL). Two fuzzy sets, namely, negative medium small (NMS) and positive medium small (PMS) are added to enhance the tracking performance. The

TABLE 35.2 Fuzzy variables for the acceleration error (α_e) for speed control

Fuzzy variable	Acceleration error (rad/s²)
(NL)	$\alpha_e \leq -738$
(NM)	$-738 < \alpha_e \leq -369$
(NS)	$-369 < \alpha_e < 0$
(ZE)	$\alpha_e = 0$
(PS)	$0 < \alpha_e < 369$
(PM)	$369 \leq \alpha_e < 738$
(PL)	$738 \leq \alpha_e$

TABLE 35.3 Fuzzy variables for the speed error (ω_e) for speed control

Fuzzy variable	Speed error (rad/s)
(NL)	$\omega_e \leq -209$
(NM)	$-209 < \omega_e \leq -104$
(NMS)	$-104 < \omega_e \leq -49$
(NS)	$-49 < \omega_e < 0$
(Z)	$\omega_e = 0$
(PS)	$0 < \omega_e < 49$
(PMS)	$49 \leq \omega_e < 104$
(PM)	$104 \leq \omega_e < 209$
(PL)	$209 \leq \omega_e$

regions defined for each fuzzy variable for the speed error is summarized in Table 35.3.

After specifying the fuzzy sets, it is required to determine the membership functions for these sets. The membership function for the fuzzy variable-representing ZERO is a fuzzy singleton. Additionally, the other membership functions are of the type described in Eq. (35.6) and are composed of fuzzy singletons within the region defined for each particular fuzzy variable. Figures 35.11 and 35.12 show the resulting membership function for the acceleration and speed errors, respectively.

The two fuzzy sets illustrated in Figs. 35.11 and 35.12 result in 63 linguistic rules for the BLDC drive system under study. The conditional rules listed in Table 35.4 are clearly implied, and the physical meanings of some rules are briefly explained as follows:

Rule 1: IF speed error is PL AND acceleration is PS, THEN change in control voltage (output of fuzzy controller) is PL. This rule implies a general condition when the measured speed is far from the desired reference speed. Accordingly, it requires a large increase in the control voltage to force the shaft speed to the desired reference speed quickly.

Rule 2: IF speed error is PS AND acceleration is ZERO, THEN change in control voltage is positive very very small PVVS. This rule implements the conditions when the error starts to decrease and the measured speed is approaching the desired reference speed. Consequently, a very small increase in the control voltage is applied.

Rule 3: IF speed error is ZERO AND acceleration is NS, THEN change in control voltage is PVVS. This rule deals with the circumstances when overshoot does occur. A very small decrease in the control voltage is required, which brings the motor speed to the desired reference speed.

These rules comprise the decision mechanism for the fuzzy speed controller. The decision table, Table 35.4, consists of values showing the different situations experienced by the drive system and the corresponding control input functions. It is clear that each entry in Table 35.4 represents a particular rule.

Now it is necessary to find the fuzzy output (change in control voltage). In this experiment the fuzzy centroid is used.

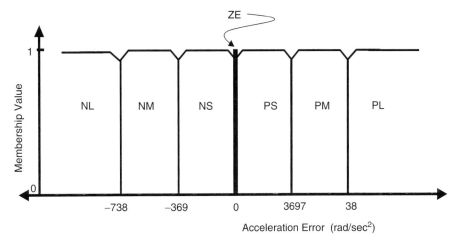

FIGURE 35.11 Membership functions for the acceleration error.

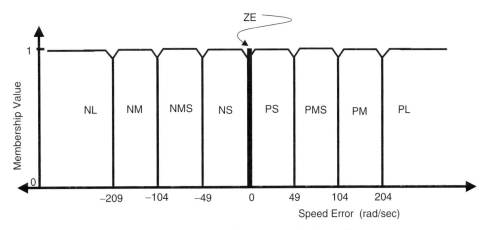

FIGURE 35.12 Membership functions for the speed error.

TABLE 35.4 Decision table for speed control

		\multicolumn{9}{c}{Velocity error}
		NL	NM	NMS	NS	ZERO	PS	PMS	PM	PL
Acceleration error	PL	NS	NVVS	PVVS	PVS	PS	PMS	PM	PL	PVVL
	PM	NMS	NVS	ZERO	PVVS	PVS	PS	PMS	PML	PVL
	PS	NMS	NS	NVVS	ZERO	PVVS	PVS	PS	PM	PL
	ZERO	NML	NMS	NVS	NVVS	ZERO	PVVS	PVS	PMS	PML
	NS	NL	NM	NS	NVS	NVVS	ZERO	PVVS	PS	PML
	NM	NVL	NML	NMS	NS	NVS	NVVS	ZERO	PVS	PMS
	NL	NVVL	NL	NM	NMS	NS	NVS	NVVS	PVVS	PS

Equation (35.17) shows the fuzzy centroid used to compute the final output of the controller.

$$\text{Output} = \left(\frac{w_1 o_1 + w_2 o_2 + w_3 o_3 + \cdots + w_{63} o_{63}}{w_1 + w_2 + w_3 + \cdots + W_{63}} \right) \quad (35.17)$$

The weights w_i will be the strength of each particular rule's evaluation. The rule strength is evaluated as the product of the membership values associate with the speed and acceleration errors for the particular fuzzy variables involved in that rule. Since the membership functions were comprised solely of fuzzy singletons and the AND operator is evaluated as a product, the strength of each rules evaluation (w_i) will be either 0 or 1. Additionally, since the membership functions do not overlap, only one rule will be evaluated as true at each sample time. Thus all the weights w_i will be zero except one. So the actual implementation can be simplified.

35.3.3 Fuzzy Logic-based Position Controller

The task of the position control algorithm is to force the rotor position of the motor to follow a desired reference track without overshoot. The same method applied to the fuzzy speed controller is applied to the position controller. The inputs to the fuzzy position controller are, angular position error (θ_e) and motor speed error (ω_e). The output from the controller is

TABLE 35.5 Fuzzy variables of the speed error for position control

Fuzzy variable	Speed error (rad/s)
(NL)	$x^2 \leq -209$
(NM)	$-209 < x^2 \leq -104$
(NS)	$-104 < x^2 < 0$
(ZE)	$x^2 = 0$
(PS)	$0 < x^2 < 104$
(PM)	$104 \leq x^2 < 209$
(PL)	$209 \leq x^2$

TABLE 35.6 Fuzzy variables of position error (θ_e) for position control

Fuzzy variable	Position error (rad)
(NL)	$\theta_e \leq -2.21$
(NM)	$-2.21 < \theta_e \leq -1.05$
(NMS)	$-1.05 < \theta_e \leq -0.53$
(NS)	$-0.53 < \theta_e < 0$
(Z)	$\theta_e = 0$
(PS)	$0 < \theta_e < 0.53$
(PMS)	$0.53 \leq \theta_e < 1.05$
(PM)	$1.05 \leq \theta_e < 2.21$
(PL)	$2.21 \leq \theta_e$

a change in the motor voltage. The two inputs are defined as follows

$$\theta_e = \theta_{\text{ref}} - \theta_{\text{act}} \qquad (35.18)$$

$$\omega_e = \omega_{\text{ref}} - \omega_{\text{act}} \qquad (35.19)$$

where

θ_{ref} = the desired position in radians.

θ_{act} = the measured position in radians.

$\omega_{\text{ref}} = 0$, because we want to minimize the speed to zero.

ω_{act} = the measured speed in radians/second.

Nine fuzzy variables were defined for the position error and seven for the speed error. The fuzzy variables for the position error and the speed error are defined in Tables 35.5 and 35.6, respectively. After specifying the fuzzy sets for the position controller, the membership functions can be fully defined, it is required to determine the membership functions for these sets. The membership function for the fuzzy variable ZERO is a fuzzy singleton. Additionally, the other membership functions are of the type described in Eq. (35.6) and are composed of fuzzy singletons within the region defined for each particular fuzzy variable. Figures 35.13 and 35.14 show the resulting membership function for the speed and position errors respectively.

A corresponding output to the motor based on the speed and the rotor position error detected in each sampling interval must be assigned and thereby specifying the fuzzy inference engine. This is done by the fuzzy rules. For example, IF the position error is PL AND the speed is PS, THEN a large positive change in driving effort is used. The crisp (defuzzified) output signal of the fuzzy controller is obtained by calculating the centriod of all fuzzy output variables. Fuzzy rule-base, on which the required change in the motor voltage is generated, is illustrated in Table 35.7.

35.4 Hardware System Description

The BLDC drive system under control consists of the components illustrated in Fig. 35.15. The drive system is made up of several distinct subsystems: the motor, a personal computer (PC), the driving circuit, and the microprocessor evaluation board. The motor is 1/4 HP, 3000-rev/min BLDC motor, and was manufactured by Pittman Company. The BLDC motor is equipped with a hall-effect sensor and an incremental optical encoder. The hall-effect sensors detect the rotor position to indicate which of the three phases of the motor is to be excited as the motor spins. The optical position encoder with

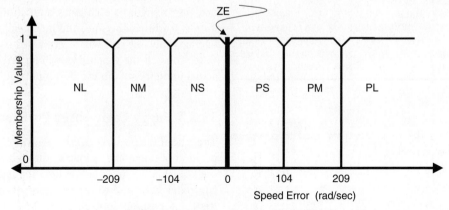

FIGURE 35.13 Membership functions for the speed error.

35 Fuzzy Logic in Electric Drives

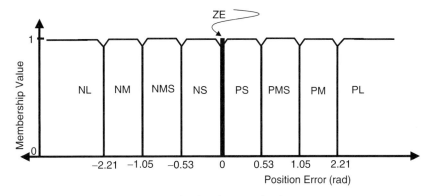

FIGURE 35.14 Membership functions for the position error.

TABLE 35.7 Decision table for position control

		Position error							
		NL	NM	NMS	NS	ZERO	PS	PMS	PM	PL
Speed error	PL	NS	NVVS	PVVS	PVS	PS	PMS	PM	PL	PVVL
	PM	NMS	NVS	ZERO	PVVS	PVS	PS	PMS	PML	PVL
	PS	NMS	NS	NVVS	ZERO	PVVS	PVS	PS	PM	PL
	ZERO	NML	NMS	NVS	NVVS	ZERO	PVVS	PVS	PMS	PML
	NS	NL	NM	NS	NVS	NVVS	ZERO	PVVS	PS	PML
	NM	NVL	NML	NMS	NS	NVS	NVVS	ZERO	PVS	PMS
	NL	NVVL	NL	NM	NMS	NS	NVS	NVVS	PVVS	PS

resolution of 500 pulses/revolution is used to give speed and position feedback.

The controller is implemented by software and executed using a microprocessor. The control algorithm is written and loaded into the microprocessor using the PC. The computer used is an IBM 486 PC. The driving circuit is constructed using two major components: (1) the integrated circuit (UDN2936W-120) designed for BLDC Drives [13] and (2) the digital to analog converter (DAC) and power amplifier. The inputs to the integrated circuit are the rotor position (which is obtained from hall-effect sensors), the direction of desired

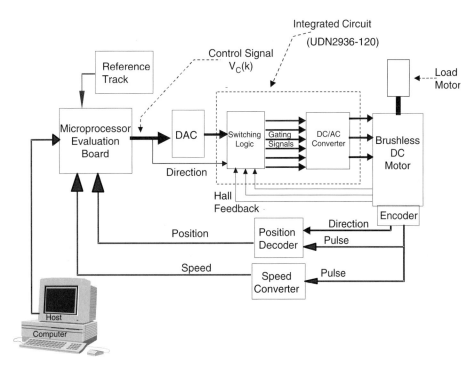

FIGURE 35.15 Block diagram of the laboratory setup.

rotation, and the magnitude of the control voltage. The output of the switching logic section is a sequence of gating signals that are pulse width modulated (PWM). These signals are used to drive the power converter portion of the chip. The power converter is a DC/AC inverter utilizing six MOSFETs. The output of the power converter is a chopped three-phase AC waveform.

The microprocessor used was the Motorola MC68HC11E9 microprocessor [14]. The on board memory system includes 8k bytes of read only memory (ROM), 512 bytes of electrically erasable programmable (ROM) (EEPROM), and 256 bytes of random access memory (RAM). The processor also has four eight bit parallel input/output ports, namely, A, B, C, and E. The program is completely contained in the microprocessor and the computer is only required to load new programs into the processor. Additionally, it has an internal analog-to-digital converter, and can accept up to four analog inputs. Figure 35.15 includes the integrated circuit (UDN2936W-120) designed for operating the BLDC motor. The range of the input voltage (control voltage) is between zero and 25 V. The motor speed for any control voltage less than 7 V is zero. This indicates that the motor requires a minimum of 7 V to start. This limitation is actually a protective feature of the circuit to prevent malfunctions [13]. Thus, the actual output of the microprocessor (the control signal $V_C(k)$) is added to a seven volt DC offset, using a summing amplifier. The current output of the summing amplifier must be first increased before it can be used to drive the motor. The current amplification is accomplished by using a 40 V power transistor (TIP31A). Additionally, when a sudden change in speed or direction occurs, the back EMF produced by the motor can, sometimes, cause large currents. Since the output stage of the integrated circuit (UDN2936W-120) has transistors with a current rating of ± 3 A, some limiting resistors were needed to prevent damage. Figure 35.16 shows a snapshot of the laboratory experiment.

Speed measurements were taken using a frequency to voltage converter LM2907. The LM2907 produces a voltage proportional to the frequency of the pulses it receives at its input. In this experiment, the pulses were sourced from the encoder signal-A. The circuit was constructed such that the maximum speed of the motor was corresponded to 5 V. This value was the input to the microprocessor, via the on chip A/D converter, as an 8-bit word. The direction of rotation was observed using a D-flip-flop. This direction signal was used to indicate the sign of the speed. The two signals, speed and direction, were combined within the software environment to produce an integer. Thus internally the speed could range from +256 to −256. These limits represent +3000 and −3000 R.P.M. respectively. Additionally, the digital output of the microprocessor

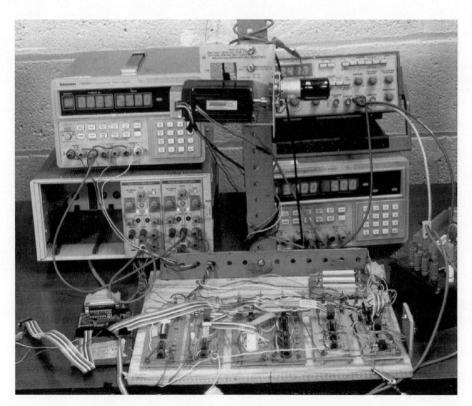

FIGURE 35.16 Photograph of the hardware implementation.

was limited to the resolution provided by seven bits. Therefore, if the desired output voltage was 22 V, then the maximum byte output was 127. Thus each increment in control signal, numerically within the microprocessor, would produce a change of 0.17 V.

35.4.1 Experimental Results

A square-wave followed by sinusoidal reference track was considered. In this experiment, there is a weight attached to the motor via a cable and pulley assembly. Figure 35.17 shows the speed tracking performance of the fuzzy logic controller. The motor is under constant load. The actual motor speed is superimposed on the desired reference speed in order to compare tracking accuracy. High tracking accuracy is observed at all speeds. The corresponding position tracking performance is displayed in Fig. 35.18. Reasonable position tracking accuracy is displayed. One can see from these figures that the results were very successful. For comparison purposes, Figs. 35.19 and 35.20 exhibit cases in which the fuzzy logic controller was replaced by a bang–bang controller. However, when bang–bang controller was applied, the controller could not maintain tracking accuracy. That is, the response using the bang–bang controller was unsatisfactory.

35.5 Conclusion

This chapter describes a fuzzy logic-based microprocessor controller, which incorporates attractive features such as

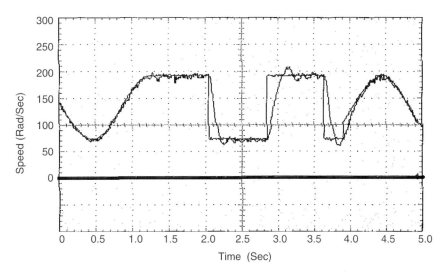

FIGURE 35.17 Fuzzy speed tracking under loading condition.

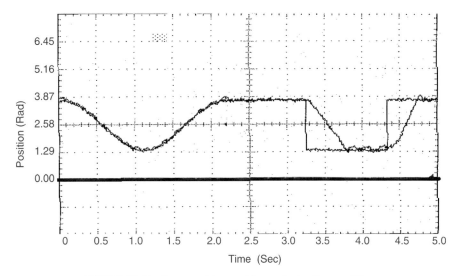

FIGURE 35.18 Fuzzy position tracking under loading condition.

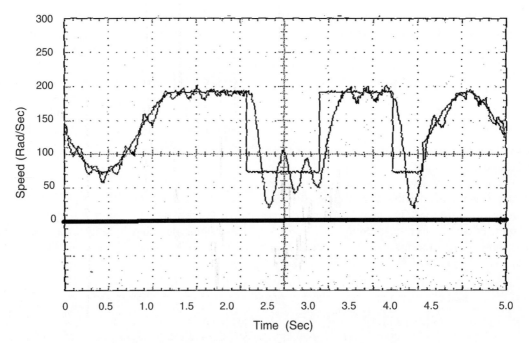

FIGURE 35.19 Bang–bang speed tracking under loading condition.

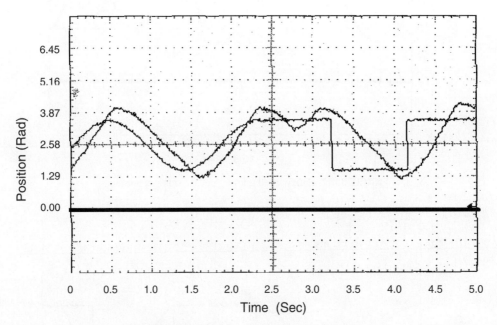

FIGURE 35.20 Bang–bang position tracking with load.

simplicity, good performance, and automation while utilizing a low cost hardware and software implementation. The test results indicate that the fuzzy logic-tracking controllers follow the trajectories successfully. Additionally, experimental results show that the effective smooth speed/position control and tracking of brushless DC (BLDC) motors can be achieved by the fuzzy logic controler (FLC), thus making it suitable for high performance motor drive applications. The controller design does not require explicit knowledge of the motor/load dynamics. This is a useful feature when dealing with parameter and load uncertainties. The advantage of using a fuzzy logic-tracking controller in this application is that it is well suited to the control of unknown or ill-defined non-linear dynamics.

Acknowledgments

The author would like to acknowledge the assistance of Mr. Daniel O. Ricketts, a graduate student at Howard University in clarifying these concepts and obtaining the experimental results. The author would also like to acknowledge the financial support by NASA Glenn Research Center, in the form of grant no. NAG3-2287 technically monitored by Mr. Donald F. Noga.

Further Reading

1. Ross, T. J., Fuzzy Logic with Engineering Applications, McGraw-Hill, 1995.
2. Mitra, S. and Pal. K. S., "Fuzzy Self-Organization, Inferencing, and Rule Generation," IEEE Trans. Syst., Man, Cybern.– Part A: Syst. Humans, Vol. 26, No. 5, pp. 608–619, September 1996.
3. Jamshidi, M., Vadiee, N. and Ross, T. Fuzzy Logic and Control, Prentice Hall, 1993.
4. Lee, C. C., "Fuzzy Logic in Control Systems: Fuzzy Logic Controller. Part I," IEEE Trans. Syst., Man, Cybern, Vol. 20, No. 2, pp. 404–418, March/April 1990.
5. Lee, J., "On Methods for Improving Performance of PI-Type Fuzzy Logic Controllers," IEEE Trans. Fuzzy Systems, Vol. 1, No. 4, November 1993.
6.
7. Lofti, A. and Tsoi, A. C. "Learning Fuzzy Inference Systems Using and Adaptive Membership Function Scheme," IEEE Trans. Syst., Man, Cybern., Vol. 26, No. 2, pp. 326–331, April 1996.
8. Lofti, A., Andersen, H. C., and Tsoi, A. C. "Matrix Formulation of Fuzzy Rule-Based Systems," IEEE Trans. Syst., Man, Cybern., Vol. 26 No. 2, pp. 332–339, April 1996.
9. Motorola, MC68HC11E9 Technical Data Manual, 1990.
10. Rubaai, A., Kotaru, R. and David Kankam, M. "A Continually Online-Trained Neural Network Controller for Brushless DC Motor Drives," IEEE Trans. Ind. Appl., Vol. 36, No. 2, pp. 475–483, March/April 2000.
11. Ross, T. J., Fuzzy Logic with Engineering Applications, McGraw-Hill, 1995.
12. Rubaai, A. and Kotaru, R. "Neural Net-Based Robust Controller Design for Brushless DC Motor Drives," IEEE Trans. Syst., Man, Cybern., Part C: Applications and Review, Vol. 29, No. 3, pp. 460–474, August 1999.
13. Allegro, UDN2936W-120 Technical Data Manual, 1993.
14. Marks II, R. J. (Editor), Fuzzy Logic Technology and Applications, IEEE Press, 1994.
15. Zadeh, L. "Outline of a New Approach to the Analysis of Complex Systems and Decision Processes," IEEE Trans. Syst., Man, Cybern., Vol. 3, No. 1, pp. 28–44, January 1973.

36
Artificial Neural Network Applications in Power Electronics and Electrical Drives

B. Karanayil, Ph.D. and
M.F. Rahman, Ph.D.
*School of Electrical Engineering and Telecommunications,
The University of New South Wales, Sydney, New South Wales 2052, Australia*

36.1 Introduction 1015
36.2 Conventional and Neural Function Approximators 1016
  36.2.1 Conventional Approximator • 36.2.2 Neural Function Approximator
36.3 ANN-based Estimation in Induction Motor Drives 1017
  36.3.1 Speed Estimation • 36.3.2 Flux and Torque Estimation • 36.3.3 Rotor Resistance Identification Using ANN • 36.3.4 Stator Resistance Estimation Using ANN
36.4 ANN-based Controls in Motor Drives 1025
  36.4.1 Induction Motor Current Control • 36.4.2 Induction Motor Control • 36.4.3 Efficiency Optimization in Electric Drives
36.5 ANN-based Controls in Power Converters 1029
  Further Reading 1030

36.1 Introduction

In classical control systems, knowledge of the controlled system (plant) is required in the form of a set of algebraic and differential equations, which analytically relate inputs and outputs. However, these models can become complex, rely on many assumptions, may contain parameters which are difficult to measure or may change significantly during operation as in the case of the rotor flux oriented control (RFOC) induction motor drive. Classical control theory suffers from some limitations due to the assumptions made for the control system such as linearity, time-invariance, etc. These problems can be overcome by using artificial intelligence-based control techniques, and these techniques can be used, even when the analytical models are not known. Such control systems can also be less sensitive to parameter variation than classical control systems.

The main advantages of using artificial intelligence-based controllers and estimators are:

- Their design does not require a mathematical model of the plant.
- They can lead to improved performance, when properly tuned.
- They can be designed exclusively on the basis of linguistic information available from experts or by using clustering or other techniques.
- They may require less tuning effort than conventional controllers.
- They may be designed on the basis of data from a real system or a plant in the absence of necessary expert knowledge.
- They can be designed using a combination of linguistic and response-based information.

Generally, the following two types of intelligence-based systems are used for estimation and control of drives, namely:

(a) *Artificial Neural Networks (ANNs)*
(b) *Fuzzy Logic Systems (FLSs)*

In different applications in power electronics and electrical drives, there are occasions where an output y has to be estimated for an input x. This is generally accomplished with the help of mathematical equations of the system under consideration. Sometimes it may not be possible to have an accurate mathematical model, or there is no conventional model at all which can be used. In these circumstances, model-free estimators required which can be either using ANNs or fuzzy logic systems. A mathematical model-based system can operate smoothly when there is no noise in the inputs but they might fail when the input has noise or if the inputs themselves are uncertain. Also, depending on the complexity of the mathematical model, the computation times can be excessive leading to difficulties for practical implementation.

A conventional function approximator, a neural function estimator, and how these estimators are arrived at are discussed briefly in Section 36.2. It will be shown how neural estimators are used in the estimation of speed, flux, torque, etc. for an induction motor. Some of these estimators are briefly reviewed in Section 36.3. Also, the ANNs have been used for identification and control of stator current in induction motor drives and these are briefly described in Section 36.4. In addition to their applications to motor drives, they are also used in control of power converters. Some of these applications are discussed in Section 36.5.

36.2 Conventional and Neural Function Approximators

36.2.1 Conventional Approximator

A conventional function approximator should give a good prediction of output data, when a system is presented with a new input data. A conventional function approximator uses a mathematical model of the system as shown in Fig. 36.1a. Sometimes it is not possible to have an accurate mathematical model of the system, in such a case; mathematical model free approximators are required, as shown in Fig. 36.1b. A conventional function approximator can easily be replaced by a neural network-based function approximator.

36.2.2 Neural Function Approximator

The function $y = f(x_1, x_2, \ldots, x_n)$ is the function to be approximated and $x_1, x_2, x_3, \ldots x_n$ are the n variables (n inputs) and the approximator uses the sum of non-linear functions $g_1, g_2, g_3, g_4, \ldots g_i$, where each of the g are non-linear functions of a single variable. Thus

$$y = f(x_1, x_2, \ldots, x_n) = g_1 + g_2 + g_3 + \cdots + g_{2n+1}$$
$$= \sum_{i=1}^{2n+1} g_i \quad (36.1)$$

where g_i is a real and continuous non-linear function which depends only on a single variable z_i. The output y can also be represented as;

$$y = \sum_{j=1}^{n} w_{Mj} x_j = w_{M1} x_1 + w_{M2} x_2 + \cdots w_{Mn} x_n \quad (36.2)$$

These equations can be represented by a network shown in Fig. 36.2. There are n input nodes in the input layer, M hidden nodes in the hidden layer.

Figure 36.3 shows the schematic diagram of a neural approximation and Fig. 36.4 shows the technique of its training. The error (e) between the desired non-linear function (y) and the non-linear function ($\hat{y}$) obtained by the neural estimator

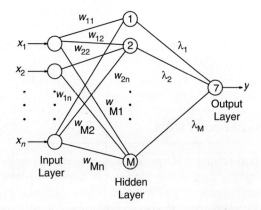

FIGURE 36.2 Artificial neural network to generate the required function.

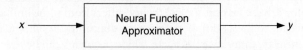

FIGURE 36.3 Schematic of a neural function approximator.

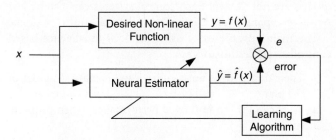

FIGURE 36.4 Training of a neural function approximator.

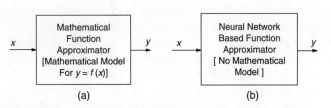

FIGURE 36.1 Conventional and neural network-based function approximators.

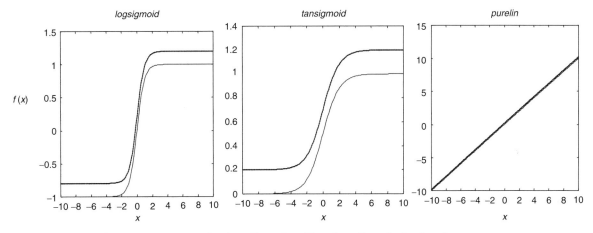

FIGURE 36.5 Logsigmoid, tansigmoid, and purelin activation functions.

is the input to the learning algorithm for the artificial neural estimator. This type of learning is known as back propagation.

The output of a single neuron can be represented as

$$a_i = f_i \left\{ \sum_{j=1}^{n} w_{ij} x_j(t) + b_i \right\} \quad (36.3)$$

where f_i is the activation function and b_i is the bias. Figure 36.5 shows a number of possible activation functions in a neuron. The simplest of all is the linear activation function, where the output varies linearly with the input but saturates at ± 1 as shown with a large magnitude of the input. The most commonly used activation functions are non-linear, continuously varying types between two asymptotic values 0 and 1 or -1 and $+1$. These are respectively, the sigmoidal function also called logsigmoid and the hyperbolic tan function also called tansigmoid.

The learning process of an ANN is based on the training process. One of the most widely used training techniques is the error back-propagation technique; a scheme which is illustrated in Fig. 36.4. When this technique is employed, the ANN is provided with input and output training data and the ANN configures its weights. The training process is then followed by supplying with the real input data and the ANN then produces the required output data.

The total network error (sum of squared errors) can be expressed as

$$E = \frac{1}{2} \sum_{k=1}^{P} \sum_{j=1}^{K} \left(d_{kj} - o_{kj} \right)^2 \quad (36.4)$$

where E is the total error, P is the number of patterns in the training data, k is the number of outputs in the network, d_{kj} is the target (desired) output for the pattern K and $o_{kj} (= y_{kj})$ is the jth output of the kth pattern. The minimization of the error can be arranged with different algorithms such as the gradient descent with momentum, Levenberg–Marquardt, reduced memory Levenberg–Marquardt, Bayesian regularization, etc.

36.3 ANN-based Estimation in Induction Motor Drives

Artificial neural networks have found widespread use in function approximation. It has been shown that, theoretically, a three layer ANN can approximate arbitrarily closely, any non-linear function, provided it is non-singular. Some of the ANN-based estimators reported in the literature for rotor flux, torque and rotor speed of induction motor drive are discussed in the following section.

36.3.1 Speed Estimation

In general, steady-state and transient analysis of induction motors is done using space vector theory, with the mathematical model having the parameters of the motor. To estimate the various machine quantities such as stator and rotor flux linkages, rotor speed, electromagnetic torque, etc., the above mathematical model is normally used. However, these machine quantities could be estimated without the mathematical model by using an ANN. Here no assumptions have to be made about any type of non-linearity.

As an example, the rotor speed of an induction motor can be estimated from the direct and quadrature axis stator voltages and currents in the stationary reference frame, as shown in Fig. 36.6. A three layer feedforward neural network structure with $8 \times 7 \times 1$ (8 input, 7 hidden, and 1 output) is used in this case. The input nodes were selected as equal to the number of input signals and the output nodes as equal to the number of output signals. The number of hidden layer neurons is generally taken as the mean of the input and output

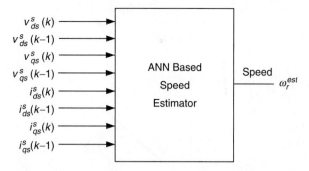

FIGURE 36.6 ANN-based speed estimator for induction motor.

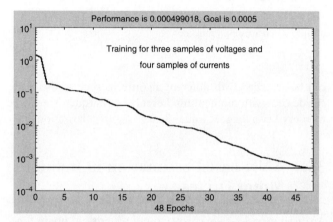

FIGURE 36.7 The error plot of $8 \times 7 \times 1$ network training for speed estimation.

nodes. In this speed estimator ANN, 7 hidden neurons were selected.

The results of an experiment conducted on a 1.1 kW, 415 V, 3 phase, 4 pole, 50 Hz induction motor is explained in this section. In order to investigate the case of speed estimation using ANN, a rotor flux oriented induction motor drive was set up in the laboratory, where the speed reference was changed in steps of 100 rpm and reversed every time the speed reached 1000 rpm. The load torque on the motor was kept constant at its full load rating. The stator voltages, stator currents, and the rotor speed were measured for 5 s and a data file was generated. The neural network was then trained using the *trainlm* algorithm with this data file. The training of the neural network converged after 48 epochs and the error plot for this network training is shown in Fig. 36.7. The estimated speed was predicted with the trained neural network, and the result is shown in Fig. 36.8.

The noisy data in the plot is the estimated speed and the continuous line is the speed measured with the encoder. The speed estimation was found to fail for speeds less than 100 rpm. If the trained neural network has to predict the speed under the complete range of operation of the drive, the data for training the neural network also has to be taken for the whole range. From this example investigated, it was found that the off-line training of the neural network could not produce satisfactory results, and it can be concluded that these off-line methods are not most suitable for these applications.

Artificial neural networks was also used for the estimation of the rotor speed of an induction motor together with the help of induction motor dynamic model. Though the technique gives a fairly good estimate of the speed, this technique lies more in the adaptive control area than in neural networks. The speed is not obtained at the output of a neural network; instead, the magnitude of one of the weights corresponds to the speed. The four quadrant operation of the drive was not possible for speeds less than 500 rpm. The motor was not able to follow the speed reference during the reversal for speeds less than 500 rpm. The drive worked satisfactorily for speeds above 500 rpm. Even though this method does not fall into a true neural network estimator, the results achieved with this type of implementation were very good except for lower speeds.

Alternately, the estimated speed can be made available at the output of a neural network as shown in Fig. 36.9. This speed estimator used a three layer neural network with five input nodes, one hidden layer, and one output layer to give the estimated speed $\hat{\omega}_r(k)$ as shown in Fig. 36.9. The three inputs to the ANN are a reference model flux λ_r^*, an adjustable model flux $\hat{\lambda}_r$, and $\hat{\omega}_r(k-1)$, the time delayed estimated speed. The multilayer and recurrent structure of the network makes it robust to parameter variations and system noise. The main advantage of their ANN structure lies in the fact that they have used a recurrent structure which is robust to parameter variations and system noise. These authors were able to achieve a speed control error of 0.6% for a reference speed of 10 rpm. The speed control error dropped to 0.584% for a reference speed of 1000 rpm.

36.3.2 Flux and Torque Estimation

The same principle as described in Section 36.3.1 can also be extended for simultaneous estimation of more quantities such as torque and stator flux. When more quantities or variables have to be estimated, the complex ANN has to implement a complex non-linear mapping.

The four feedback signals required for a direct field oriented induction motor drive can be estimated using ANNs. A $4 \times 20 \times 4$ multilayer network has been used for the estimation of the rotor flux magnitude, the electromagnetic torque, and the sine/cosine of the rotor flux angle. It has been demonstrated both by modeling and experimental results that the above estimated quantities were almost equal to the same quantities computed by a DSP-based estimator. Both the estimated torque and rotor flux signals using neural network was found to have higher ripple content compared to the DSP-based estimated quantities. It could be concluded that a properly trained ANN could totally eliminate the machine model equations as is evident from the results reported.

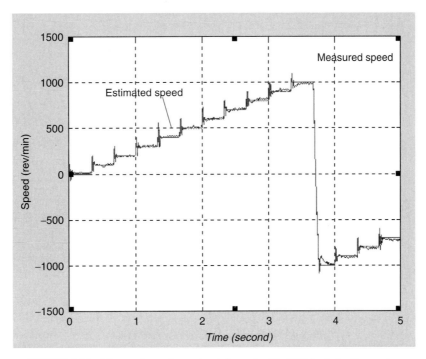

FIGURE 36.8 Measured speed vs estimated speed with ANN for induction motor.

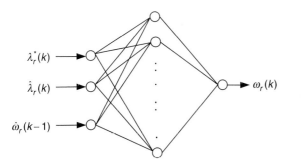

FIGURE 36.9 Structure of the neural network for the induction motor speed estimation.

In another application, an ANN with $5 \times 8 \times 8 \times 2$ structure has been used to estimate the stator flux using the measured induction motor stator variables quantities. After successful training, the ANN was used in a direct field oriented controlled drive. The rotor flux is computed from the stator flux estimate provided by the ANN and the stator current. This particular implementation also included an ANN-based decoupler which was used for the indirect field oriented (IFO) drive. An ANN with a structure of $2 \times 8 \times 8 \times 1$ was used for implementing the mapping between the flux and torque references and the stator current references. The estimated rotor flux using an ANN and a conventional FOC controller was shown to be equal. These authors have used experimental data for the ANN training and thus the effect of motor parameters was reduced. The authors were unable to use these estimated fluxes for controlling the induction motor in the experiment. The structure of the ANN used for this estimation is only that of a static ANN. It is preferable to have a dynamic neural network for this purpose.

36.3.3 Rotor Resistance Identification Using ANN

36.3.3.1 Off-line Trained ANN

The artificial neural network can also be used for the rotor time constant adaptation in indirect field oriented controlled drives. One of the implementation reported in the literature is shown in Fig. 36.10. There are five inputs to the T_r estimator, namely v_{ds}^s, v_{qs}^s, i_{ds}^s, i_{qs}^s, ω_r. The training signals are generated with step variations in rotor resistance for different torque reference T_e^* and flux command λ_r^* and the final network is connected in the IFO controller as shown in Fig. 36.10. The rotor time constant was tracked by a proportional integral (PI) regulator that corrects any errors in the slip calculator. The output of this regulator is summed with that of the slip calculator and the result constitutes the new slip command that is required to compensate for the rotor time constant variation. The major drawback of this scheme is that the final neural network is only an off-line trained neural network with a limited data file obtained from the modeling.

36.3.3.2 On-line Trained ANN

The limitations of off-line trained ANN are overcome by using an on-line trained ANN configuration. The method discussed

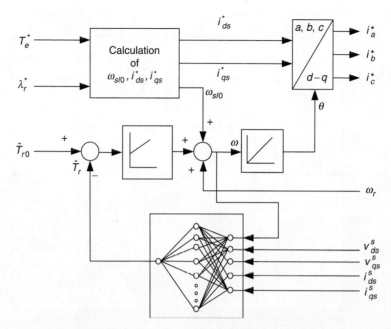

FIGURE 36.10 Principle of rotor time constant adaptation.

in this section has used an on-line trained ANN for adaptation of R_r of an induction motor in the RFOC induction motor drive. The error between the desired state variable of an induction motor and the actual state variable of a neural model is back propagated to adjust the weights of the neural model, so that the actual state variable tracks the desired value.

The principle of on-line estimation of rotor resistance (R_r) with multilayer feedforward artificial neural networks using on-line training has been described in Section 36.3.3.2.1. This technique was then investigated with the help of modeling studies with a 1.1 kW squirrel-cage induction motor (SCIM), described in Section 36.3.3.2.2. The modeling results are presented in Section 36.3.3.2.3. In order to validate the modeling studies, modeling results were compared with those from an experimental set-up with a SCIM under RFOC. These experimental results are presented in Section 36.3.3.2.4.

36.3.3.2.1 Multilayer Feedforward ANN Multilayer feedforward neural networks are regarded as universal approximations and have the capability to acquire non-linear input–output relationships of a system by learning via the back-propagation algorithm. It should be possible that a simple two-layer feedforward neural network trained by the back-propagation technique can be employed in the rotor resistance identification. The modified technique using ANN proposed in this section can be implemented in real time so that the resistance updates are available instantaneously and there is no convergence issues related to the learning algorithm. The two-layered neural network based on a back-propagation technique is used to estimate the rotor resistance.

Two models of the state variable estimation are used, one provides the actual induction motor output and the other one gives the neural model output. The total error between the desired and actual state variables is then back propagated as shown in Fig. 36.11, to adjust the weights of the neural model, so that the output of this model coincides with the actual output. When the training is completed, the weights of the neural network should correspond to the parameters in the actual motor.

36.3.3.2.2 Rotor Resistance Estimation for RFOC Using ANN
The basic structure of an adaptive scheme described by Fig. 36.11 is extended for rotor resistance estimation of an induction motor as illustrated in Fig. 36.12. Two independent observers are used to estimate the rotor flux vectors of the induction motor. If the rotor flux linkages are estimated using the stator voltages and stator currents, they are referred to as *voltage model* and if the rotor flux linkages are estimated using the stator currents and rotor speed they are referred to as *current model*.

The stator flux linkages based on the neural network model in Fig. 36.12 can be represented using Eq. (36.5), which is derived from the sample data models of the combined voltage and current model equations of the induction motor.

$$\overrightarrow{\lambda_r^{s\,nm}}(k) = W_1 X_1 + W_2 X_2 + W_3 X_3 \qquad (36.5)$$

The neural network model represented by Eq. (36.5) is shown in Fig. 36.13, where W_1, W_2, and W_3 represent the weights of the networks and X_1, X_2, X_3 are the three inputs to

36 ANN Applications in Power Electronics and Electrical Drives

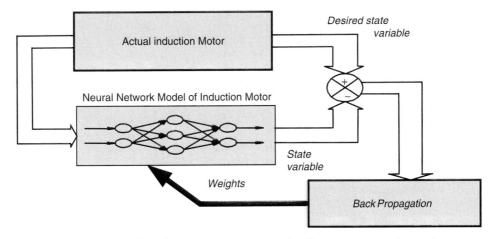

FIGURE 36.11 Block diagram of rotor resistance identification using neural networks.

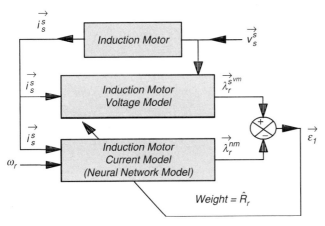

FIGURE 36.12 Structure of the neural network system for R_r estimation.

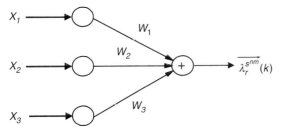

FIGURE 36.13 Two-layered neural network model for rotor flux linkage estimation.

the network. If the network shown in Fig. 36.13 has to be used to estimate R_r, where W_2 is already known, then W_1 and W_3 need to be updated.

The weights of the network, W_1 and W_3 are found from training, so as to minimize the cumulative error function E_1,

$$E_1 = \frac{1}{2}\vec{\varepsilon}_1^{\,2}(k) = \frac{1}{2}\left\{\overrightarrow{\lambda_r^{svm}}(k) - \overrightarrow{\lambda_r^{sim}}(k)\right\}^2 \quad (36.6)$$

The weight of the neural network W_1 has to be adjusted using generalized delta rule.

To accelerate the convergence of the error back-propagation learning algorithm, the current weight adjustment has to be supplemented with a fraction of the most recent weight adjustment, as indicated in Eq. (36.7).

$$W_1(k) = W_1(k-1) - \eta_1 \vec{\delta} X_2 + \alpha_1 \Delta W_1(k-1) \quad (36.7)$$

where α_1 is a user-selected positive momentum constant and η_1 is the training coefficient.

The rotor resistance R_r can now be calculated from W_3 from Eq. (36.8) as follows:

$$\hat{R}_r = \frac{L_r W_3}{L_m T_s} \quad (36.8)$$

36.3.3.2.3 Modeling Results A schematic diagram showing the implementation of a rotor resistance estimator in RFOC controller for induction motor is shown in Fig. 36.14. The stator voltages and currents are measured to estimate the rotor flux linkages using the voltage model as shown in this figure. The inputs to the rotor resistance estimator (RRE) are the stator currents, rotor flux linkages λ_{dr}^{svm}, λ_{qr}^{svm}, and the rotor speed ω_r. The estimated rotor resistance $\hat{R}_r$ will then be used in the RFOC controllers for the flux model. The response of the drive together with the rotor resistance estimator OFF is shown in Fig. 36.15 for an abrupt change in R_r of motor, from 6.03 to 8.5 Ω at 0.8 s. The possible changes in the estimated motor torque T_e, the rotor flux linkage λ_{rd} with this estimator were noted.

Subsequently the results of the drive were looked at with RRE ON, so that the rotor resistance in the controller R_r' was updated with the estimated rotor resistance $\hat{R}_r$ as shown in Fig. 36.16. The estimated rotor resistance $\hat{R}_r$ has converged to the rotor resistance of the motor R_r within 50 ms.

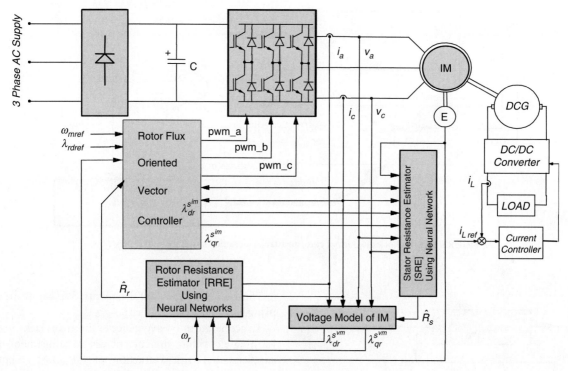

FIGURE 36.14 Block diagram of the RFOC induction motor drive with on-line rotor and stator resistance tracking using ANN.

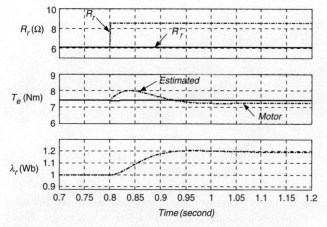

FIGURE 36.15 Effect of rotor resistance variation without rotor resistance estimator for 40% step change in R_r – modeling results.

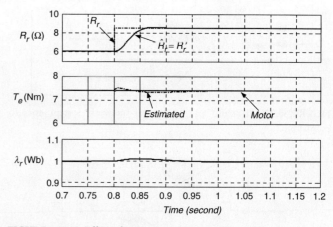

FIGURE 36.16 Effect of rotor resistance variation with rotor resistance estimator using ANN for 40% step change in R_r – modeling results.

The response of the drive system with the RRE using ANN are also shown for a practical profile in load torque, speed, and change in R_r. The R_r was increased from 6.03 to 8.5 Ω over a period of 8 s and the torque and rotor flux linkage are shown in Fig. 36.17 when the RRE is OFF. For rapid reversals of the drive and variations in load torque, there are significant errors in rotor flux linkages and thus errors in estimated torques. Subsequently, the RRE block was switched ON and it can be observed that the rotor resistance estimator was tracking very well even during the worst dynamics in the motor speed and load torques as shown in Fig. 36.18. The estimated rotor resistance $\hat{R}_r$ tracked the actual rotor resistance of the motor R_r throughout except there was some small error during the reversal of the motor. The rotor flux linkage λ_{rd} was found to remain constant at 1.0 Wb during this transient condition. The current i_{qs} also remained constant as the torque is now perfectly decoupled.

36.3.3.2.4 Experimental Results

The practical implementation of the above rotor resistance estimation was found

36 ANN Applications in Power Electronics and Electrical Drives

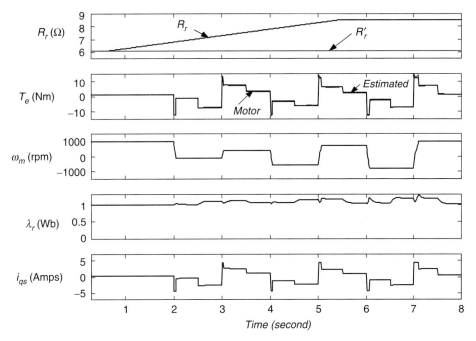

FIGURE 36.17 Effect of rotor resistance variation without rotor resistance estimator for 40% ramp change in R_r – modeling results.

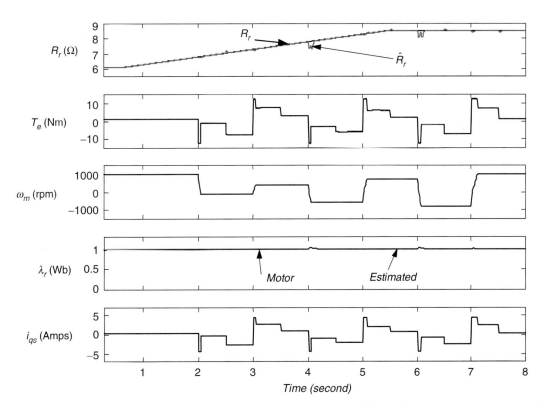

FIGURE 36.18 Effect of rotor resistance variation with rotor resistance estimator using ANN for 40% ramp change in R_r – modeling results.

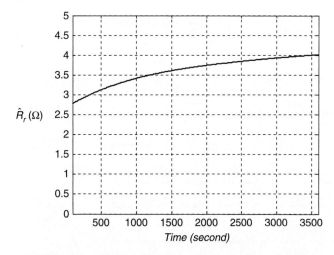

FIGURE 36.19 Estimated R_r using ANN – experimental results.

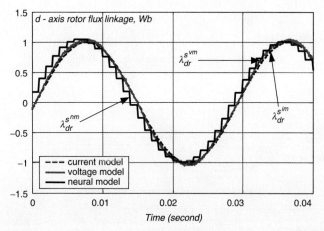

FIGURE 36.20 Rotor fluxes estimated during R_r estimation using ANN – experimental results.

functioning very well in the experimental drive set-up. The results of the R_r estimation obtained from the experiment is shown in Fig. 36.19 taken from a heat-run test conducted on an induction motor. As the RRE calculated, the rotor resistance using variables in stationary reference frame, the d-axis rotor flux linkages of the current model $\left(\lambda_{dr}^{im}\right)$, the voltage model $\left(\lambda_{dr}^{vm}\right)$, and the neural model $\left(\lambda_{dr}^{nm}\right)$, taken at the end of heat run are also recorded as shown in Fig. 36.20.

36.3.4 Stator Resistance Estimation Using ANN

In this section, the capability of a neural network has been deployed to have on-line estimator for stator resistance in an RFOC induction motor drive. The stator resistance observer was realized with a recurrent neural network with feedback loops trained using the standard back-propagation learning algorithm. Such architecture with recurrent neural network is known to be a more desirable approach and the implementation reported in this section confirms this.

The d-axis stator current in the stationary reference frame in the discrete form can be represented using the voltage and current model equations of the induction motor,

$$i_{ds}^{s*}(k) = W_4 i_{ds}^{s*}(k-1) + W_5 \lambda_{dr}^{s\,im}(k-1) + W_6 \omega_r \lambda_{qr}^{s\,im}(k-1) + W_7 v_{ds}^{s}(k-1) \quad (36.9)$$

where, $W_4 = 1 - \dfrac{T_s}{\sigma L_s} \dfrac{L_m^2}{L_r T_r} - \dfrac{T_s}{\sigma L_s} R_s;\quad W_5 = \dfrac{T_s}{\sigma L_s} \dfrac{L_m}{L_r T_r}$

$W_6 = \dfrac{T_s}{\sigma L_s} \dfrac{L_m}{L_r};\quad W_7 = \dfrac{T_s}{\sigma L_s}$

The weights W_5, W_6, and W_7, are calculated using the induction motor parameters, rotor speed ω_r, and the sampling interval used in the estimator T_s.

The relationship between stator current and stator resistance is non-linear which could be easily mapped using a neural network.

When this Eq. (36.9) is represented graphically, it resembles a recurrent neural network as shown in Fig. 36.21. The standard back-propagation learning rule can then be employed for training this neural network. The weight W_4 is the result of training so as to minimize the cumulative error function E_2,

$$E_2 = \frac{1}{2}\vec{\varepsilon}_2^{\,2}(k) = \frac{1}{2}\left\{i_{ds}^{s}(k) - i_{ds}^{s*}(k)\right\}^2 \quad (36.10)$$

To accelerate the convergence of the error back-propagation learning algorithm, the current weight adjustment is supplemented with a fraction of the most recent weight adjustment, as indicated in Eq. (36.11).

$$W_4(k) = W_4(k-1) + \eta_2 \Delta W_4(k) + \alpha_2 \Delta W_4(k-1) \quad (36.11)$$

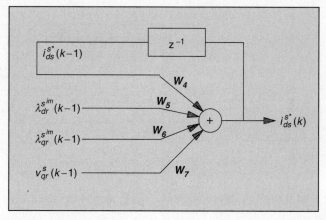

FIGURE 36.21 d-Axis stator current estimation using recurrent neural network based on Eq. (36.9).

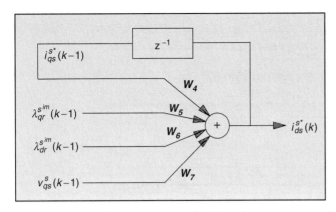

FIGURE 36.22 q-Axis stator current estimation using recurrent neural network based on Eq. (36.12).

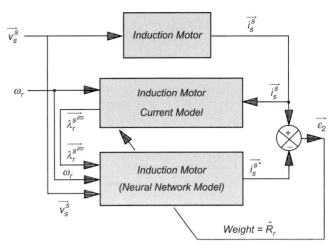

FIGURE 36.23 Block diagram of R_s estimation using artificial neural network.

where η_2 is the training coefficient, α_2 is a user-selected positive momentum constant.

Following a similar procedure, the q-axis stator current of the induction motor can be estimated using the discrete form as shown in Eq. (36.12),

$$i_{qs}^{s*}(k) = W_4 i_{qs}^{s*}(k-1) + W_5 \lambda_{qr}^{s^{im}}(k-1) - \omega_r W_6 \lambda_{dr}^{s^{im}}(k-1) + W_7 v_{qs}^{s}(k-1) \quad (36.12)$$

Equation (36.12) can be represented by a neural network as shown in Fig. 36.22. The weight W_4 is updated with the training based on Eq. (36.12).

The stator resistance $\hat{R}_s$ of the induction motor can now be calculated using Eq. (36.13) as follows:

$$\hat{R}_s = \left\{ 1 - W_4 - \frac{T_s}{\sigma L_s} \frac{L_m^2 \hat{R}_r}{L_r^2} \right\} \frac{\sigma L_s}{T_s} \quad (36.13)$$

The stator resistance of an induction motor can be thus estimated from the stator current using the neural network system as indicated in Fig. 36.23.

36.3.4.1 Modeling Results of Stator Resistance Estimation Using ANN

The block diagram of a rotor flux oriented induction motor drive together with stator resistance identification is already shown in Fig. 36.14, where the stator resistance estimation is implemented by the stator resistance estimator (SRE) block.

The stator resistance estimation results are as shown in Fig. 36.24. It has three results (1) without both rotor and stator resistance estimators, (2) with only rotor resistance estimation, and (3) with both rotor and stator resistance estimations. As shown in the figure, both R_r and R_s were increased abruptly by 40% at 1.5 s. It can be seen that the estimated stator resistance $\hat{R}_s$ converges to $\hat{R}_s$ within 200 ms. It can be noted from this figure that the convergence of the stator resistance estimation is not affected by the convergence of the rotor resistance estimator.

36.3.4.2 Experimental Results of Stator Resistance Estimation Using ANN

The stator resistance estimation algorithm was tested together with the rotor flux oriented induction motor drive of Fig. 36.14 implemented in the laboratory. To test the stator resistance estimation, an additional 3.4 Ω per phase was added in series with the induction motor stator, with the motor running at 1000 rev/min and with a load torque of 7.4 Nm.

The estimated stator resistance together with the actual stator resistance is shown in Fig. 36.25. The estimated stator resistance converges to 9.4 Ω within less than 200 ms.

Figure 36.26 shows both the measured d-axis stator current and the one estimated by the neural network model. The neural network model output $i_{ds}^{s*}(k)$ follows the measured values $i_{ds}^{s}(k)$, due to the on-line training of the neural network.

36.4 ANN-based Controls in Motor Drives

36.4.1 Induction Motor Current Control

Another application of ANN is to identify and control the stator current of an induction motor. In one study, Burton *et al.* have used a current control strategy outlined by Wishart and Harley to train an ANN to control the induction motor stator currents. They have used a training algorithm named random weight change (RWC) which is reported to be slightly faster than back-propagation. In RWC algorithm, the weights are perturbed by a fixed step-size and a random sign.

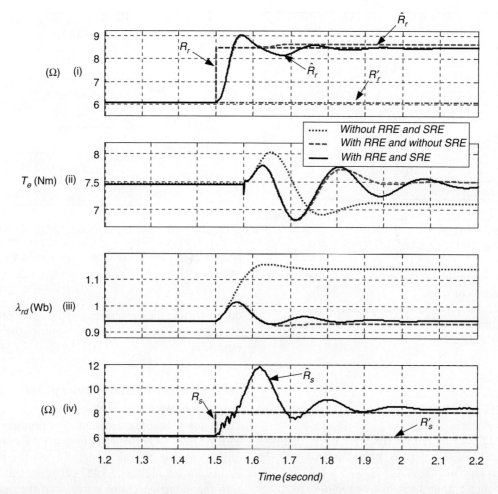

FIGURE 36.24 Performance of the drive with and without RRE and SRE using ANN for 40% step change in R_r and R_s, R'_r and R'_s compensated – modeling results.

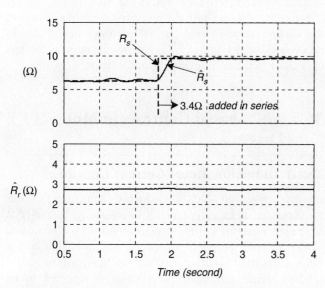

FIGURE 36.25 Estimated stator resistance R_s using ANN – experimental results.

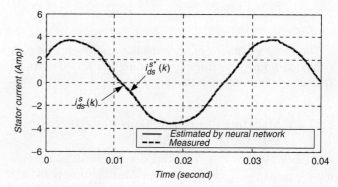

FIGURE 36.26 Stator currents in R_s estimation – experimental results.

This is done for fixed number of trials and after each trial, the error with the desired output is computed. Finally, the set of weight changes which result in the least error are chosen and the whole process is repeated till convergence is reached. The modeling results reported in this scheme was excellent. Later, Burton *et al.* presented their practical implementation of this proposed stator current controller with a transputer controller card.

36.4.2 Induction Motor Control

Narendra and Parthasarathy proposed methods for identification and control of dynamical systems using ANNs. Wishart and Harley used the above basic principles to identify and control induction machines. A block diagram of the control scheme is shown in Fig. 36.27. For the induction motor, the Non-linear AutoRegressive Moving Average with eXogenous inputs (NARMAX) model for the stationary frame stator current is derived and used for the identification of electromagnetic model. In its general form, the NARMAX model represents a system in terms of its delayed inputs and outputs. Random steps in the stator voltage are given for the purpose of identification. The neural network used is of the multilayer back-propagation type, and a quantity based on the rotor time constant is also computed as an extra weight. As opposed to the regular ANN architecture, this ANN has non-linearity in the output layer and the weighted sum of the inputs is used as the output. This gives an estimate of the rotor time constant and makes the system robust against variation of the parameters. Once, the identification is over, the ANN is used for current control. The stator currents predicted by the ANN are used to compute the input voltage for the induction motor, and the ANN output is made to track the reference currents by back-propagating the error.

The rotor speed is also controlled in this system by identifying a NARMAX model for the speed increment rather than the absolute value of speed. To simplify the NARMAX model, the load torque is assumed to be a function of the motor speed, as is the case in a fan or pump type of load. For the current control case, the relationship between the control variable (voltage) and the controlled quantity (current) was linear. In the speed control case, this relationship is non-linear, thus necessitating two ANNs, one for identification of speed and the other for control.

The identification ANN (N_i) predicts the value for the speed increment, which is compared with the actual speed increment and the error (ε_i) is back-propagated through the ANN. A PI controller is used for basic speed control, and the control ANN (N_c) produces the slip frequency, and the difference between the desired speed increment and actual speed increment (ε_c) is back-propagated through the ANN. The induction motor drive therefore employs three ANNs as shown in Fig. 36.28.

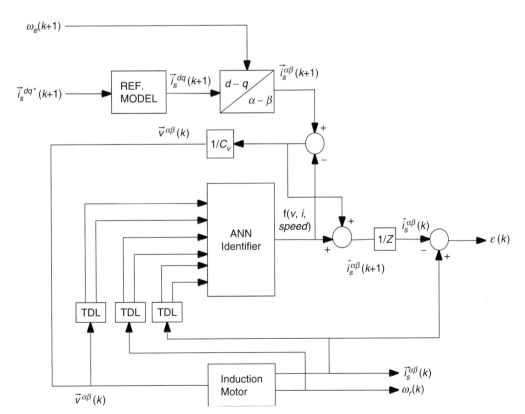

FIGURE 36.27 Adaptive current control using ANN proposed by Harley.

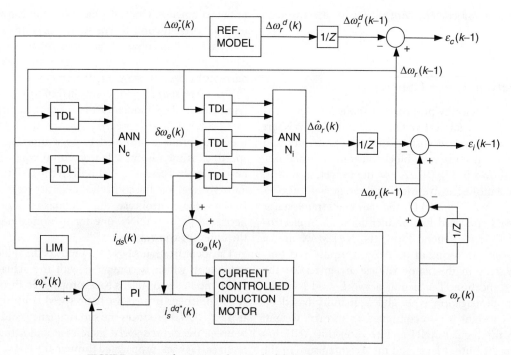

FIGURE 36.28 Adaptive speed control of the induction motor using ANN.

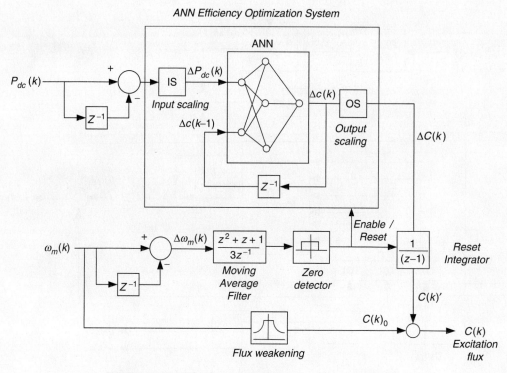

FIGURE 36.29 Induction motor efficiency optimizer using ANN.

36.4.3 Efficiency Optimization in Electric Drives

The efficiency improvement of induction motor drives via flux control can be classified into three groups: pre-computed flux programs, real-time computation of losses, and on-line input–output efficiency optimization control. All of these methods target choosing a value of the motor excitation that optimizes the motor-converter losses. The main requirement of an input–output optimization is to achieve the flux optimization in a minimum number of search steps. A constant search step may take too long if the step is too small, or it may bypass the minimum power input if the step is too large. For each mechanical operating point of the motor, it is possible to find a combination of rotor flux linkage and torque producing current i_q at which the dc link power P_{dc} to the drive system is minimum.

One of the possible ANN efficiency optimizer reported is shown in Fig. 36.29. An ANN-based search algorithm is employed to operate as an efficiency optimizer. The inputs to the system are the dc power fed into the drive system at instant k, and the change in the control variable at the previous instant $\Delta c(k-1)$. The only output is the actual change in control variable $\Delta c(k)$. Appropriate scaling from engineering units to the normalized interval $[-1, 1]$ are implemented by the input and output interfaces, input scaling (IS) and output scaling (OS). The speed signal is used to generate the reference flux $C(k)_0$ corresponding to each speed. The mechanical steady-state is detected by applying a moving average filtering to the speed variation $\Delta \omega_m$. The ANN efficiency optimizer is enabled only during the mechanical steady-state.

36.5 ANN-based Controls in Power Converters

A feedforward ANN can implement a non-linear input–output mapping. A feedforward carrier-based pulse-width modulation (PWM) technique, such as space vector modulator (SVM), can be looked at as a non-linear mapping where the command phase voltages are sampled at the input and the corresponding pulse-width patterns are established at the output. Figure 36.30 shows the block diagram of an open-loop V/f-controlled induction motor drive incorporating the proposed

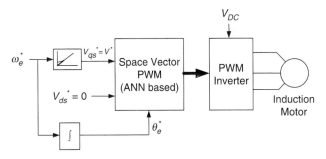

FIGURE 36.30 V/f control of induction motor using ANN-based SVM.

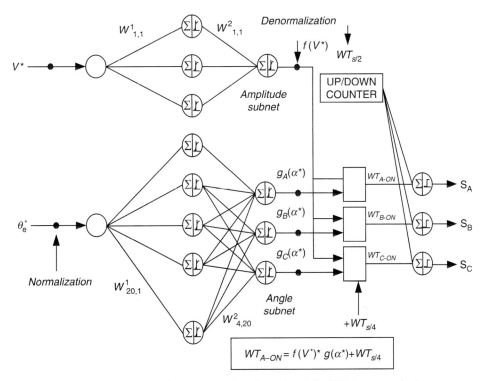

FIGURE 36.31 Neural network topology (2 × 20 × 4) for PWM wave synthesis.

ANN-based SVM controller. The command voltage $V_{qs}^*(=V^*)$ is generated from the frequency or speed command, and the angle command θ_e^* is obtained by integrating the frequency, as shown. The output of the modulator generates the PWM patterns for the inverter switches.

The ANN can be conveniently trained off-line with the data generated by calculation of the SVM algorithm. The ANN has inherent learning capability that can give improved precision by interpolation unlike the standard lookup table method. Figure 36.31 shows such an SVM that can operate during both undermodulation and overmodulation regions linearly extending smoothly up to a square wave. The SVM is implemented using two subnets: angle subnet and amplitude subnet. The subnets use a multiplayer perceptron-type network with sigmoidal-type transfer function. The bias is not shown in the figure. The composite network uses two neurons at the input, 20 neurons in the hidden layer, and four output neurons. The input signal to the angle subnet is θ_e angle which is normalized and then pulse-width functions at unit amplitude are solved (or mapped) at the output for three phases, as indicated. The amplitude subnet implements the $f(V^*)$ function. The digital words corresponding to the turn-on time are generated by multiplying the angle subnet output with that of the amplitude subnet and then adding the $T_s/4$ bias signal, as shown. The PWM signals are then generated using a single timer. The angle subnet is trained with an angle interval of 2.16° in the range of 0–360°. Due to learning or interpolation capability, both the subnets will operate higher signal resolution. A sampling interval T_s of 50 μs corresponds to a switching frequency of 20 kHz and a 100 μs that corresponds to 10 kHz.

Further Reading

1. P. Vas, *Artificial Intelligence-Based Electrical Machines and Drives: application of fuzzy, neural, fuzzy-neural and genetic-algorithm – based techniques*, Oxford University Press, New York, 1999.
2. B.K. Bose, *Modern Power Electronics and AC Drives*, Prentice Hall, New Jersey, 2002.
3. L. Ben-Brahim, S. Tadakuma, and A. Akdag, "Speed control of induction motor without rotational transducers," *IEEE Transactions on Industry Applications*, vol.35, no.4, pp. 844–850, July/August 1999.
4. M.G. Simoes and B.K. Bose, "Neural network based estimator of feedback signals for a vector controlled induction motor drive," *IEEE Transactions on Industry Applications*, vol.31, pp. 620–629, May/June 1995.
5. K. Funahashi, "On the approximate realization of continuous mappings by neural networks," *Neural Networks*, vol.2, pp. 183–192, 1989.
6. D.T. Pham and X. Liu, *Neural Networks for Identification, Prediction and Control*, Springer–Verlag, New York, 1995.
7. M. Wishart and R.G. Harley, "Identification and control of induction machines using artificial neural networks," *IEEE Transaction on Industry Applications*, vol.31, no.3, pp. 612–619, May/June 1995.
8. K.S. Narendra and K. Parthasarathy, "Identification and control of dynamical systems using neural networks," *IEEE Transactions on Neural Networks*, vol.1, no.1, pp. 4–27, March 1990.
9. J.O. Pinto, B.K. Bose, L.E.B. De Silva, and M.P. Kazmierkowski, "A Neural-Network based Space-Vector PWM Controller for Voltage-fed Inverter induction motor drive," *IEEE Transactions on Industry Applications*, vol.36, no.6, pp. 1628–1636, November/December 2000.

37
DSP-based Control of Variable Speed Drives

Hamid A. Toliyat, Ph.D.
Electrical and Computer Engineering Department, Texas A&M University, 3128 Tamus, 216g Zachry Engineering Center, College Station, Texas, USA

Mehdi Abolhassani, Ph.D.
Black & Decker (US) Inc., 701 E Joppa Rd., TW100, Towson, Maryland, USA

Peyman Niazi, Ph.D.
Maxtor Co., 333 South St., Shrewsbury, Massachusetts, USA

Lei Hao, Ph.D.
Wavecrest Laboratories, 1613 Star Batt Drive, Rochester Hills, Michigan, USA

37.1 Introduction 1031
37.2 Variable Speed Control of AC Machines 1032
37.3 General Structure of a Three-phase AC Motor Controller 1032
    37.3.1 Pulse Width Modulation Generation • 37.3.2 Analog-to-Digital Conversion Requirements • 37.3.3 Position Sensing and Encoder Interface Units • 37.3.4 The PI regulator
37.4 DSP-based Control of Permanent Magnet Brushless DC Machines 1037
    37.4.1 Mathematical Model of the BLDC Motor • 37.4.2 Torque Generation • 37.4.3 BLDC Motor Control Topology • 37.4.4 DSP Controller Requirements • 37.4.5 Implementation of the BLDC Motor Control Algorithm Using LF2407
37.5 DSP-based Control of Permanent Magnet Synchronous Machines 1041
    37.5.1 Mathematical Model of PMSM • 37.5.2 Mathematical Model of PMSM in Rotor Reference Frame • 37.5.3 PMSM Control Topology • 37.5.4 DSP Controller Requirements • 37.5.5 Implementation of the PMSM Algorithm Using the LF2407
37.6 DSP-based Vector Control of Induction Motors 1046
    37.6.1 Induction Motor Field-oriented Control • 37.6.2 DSP Controller Requirements • 37.6.3 Implementation of Field-oriented Speed Control of Induction Motor

37.1 Introduction

High-performance motor drives are characterized by the need for smooth rotation down to stall, full control of torque at stall, and fast accelerations and decelerations. In the past, variable speed drives employed predominantly dc motors because of their excellent controllability. However, modern high-performance motor drive systems are usually based on three-phase ac motors, such as the ac induction motor (ACIM) or the permanent magnet synchronous motor (PMSM). These machines have supplanted the dc motor as the machine of choice for variety of applications because of their simple robust construction, low inertia, high power density, high torque density, and good performance at high speeds of rotation.

The vector-control techniques established for controlling these ac motors; and most modern high-performance drives now implement digital closed-loop current control. In such systems, the achievable closed-loop bandwidths are directly related to the rate at which the computationally intensive vector-control algorithms and associated vector rotations can be implemented in real time. Because of this computational burden, many high-performance drives now use digital signal processors (DSPs) to implement the embedded motor- and vector-control schemes. The DSPs are special microprocessors used where real-time manipulation of large amounts of digital data is required in order to implement complicated control algorithms. The inherent computational power of the DSP permits very fast cycle times and closed-loop current control bandwidths to be achieved.

The complete current control scheme for these machines also requires a high-precision pulse-width modulation (PWM) voltage-generation scheme and high-resolution analog-to-digital (A/D) conversion (ADC) for measurement of the motor currents. In order to maintain a smooth control of torque to zero speed, rotor position feedback is essential for modern vector controllers. Therefore, many systems include rotor-position transducers, such as resolvers and incremental encoders.

The Texas Instruments TMS320LF2407 DSP Controller (referred to as the LF2407 in this chapter) is a programmable digital controller with a C2xx DSP central processing unit (CPU) as the core processor. The LF2407 contains the DSP core processor and useful peripherals integrated onto a single piece of silicon. The LF2407 combines the powerful CPU with on-chip memory and peripherals. With the DSP core and control-oriented peripherals integrated into a single chip, users can design very compact and cost-effective digital control systems.

The LF2407 DSP controller offers 40 million instructions per second (MIPS) performance. This high processing speed of the C2xx CPU allows users to compute parameters in real time rather than look up approximations from tables stored in memory. This fast performance is well suited for processing control parameters in applications such as notch filters or sensorless motor control algorithms where a large amount of calculations must be computed quickly.

While the "brain" of the LF2407 DSP is the C2xx core, the LF2407 contains several control-orientated peripherals onboard (see Fig. 37.1). The peripherals on the LF2407 make virtually any digital control requirement possible. Their applications range from analog to digital conversion to pulse width modulation (PWM) generation. Communication peripherals make possible the communication with external peripherals, personal computers, or other DSP processors. Below is a graphical listing of the different peripherals onboard the LF2407 depicted in Fig. 37.1.

We describe here the fundamental principles behind the implementation of high-performance controllers for three-phase ac motors – combining an integrated DSP controller, LF2407, flexible PWM generation, high-resolution A/D conversion, and an embedded encoder interface.

37.2 Variable Speed Control of AC Machines

Efficient variable speed control of three-phase ac machines requires the generation of a balanced three-phase set of variable voltages with variable frequency. The variable-frequency supply is typically produced by conversion from dc using power-semiconductor devices (typically MOSFETs or IGBTs) as solid-state switches. A commonly used converter configuration is shown in Fig. 37.2a. It is a two-stage circuit, in which the fixed-frequency 50 or 60 Hz ac supply is first rectified to provide the dc link voltage, V_d, stored in the dc link capacitor. This voltage is then supplied to an inverter circuit that generates the variable-frequency ac power for the motor. The power switches in the inverter circuit permit the motor terminals to be connected to either V_d or ground.

This mode of operation gives high efficiency because, ideally, the switch has zero loss in both the open and closed positions. By rapid sequential opening and closing of the six switches (Fig. 37.2a), a three-phase ac voltage with an average sinusoidal waveform can be synthesized at the output terminals. The actual output voltage waveform is a pulse-width modulated (PWM) high-frequency waveform, as shown in Fig. 37.2b. In practical inverter circuits using solid-state switches, high-speed switching of about 20 kHz is possible. Therefore sophisticated PWM waveforms with fundamental frequencies, nominally in the range of 0–250 Hz can be generated. The inductive reactance of the motor increases with frequency. Thus, higher-order harmonic currents are very small and near-sinusoidal currents flow in the stator windings. The fundamental voltage and output frequency of the inverter, as indicated in Fig. 37.2b, are adjusted by changing the PWM waveform using an appropriate controller. When controlling the fundamental output voltage, the PWM process inevitably modifies the harmonic content of the output voltage waveform. A proper choice of modulation strategy can minimize these harmonic voltages and in result, harmonic losses in the motor.

37.3 General Structure of a Three-phase AC Motor Controller

Accurate control of any motor-drive process may ultimately be reduced to the problem of accurate control of both the torque and speed of the motor. In general, motor speed is controlled directly by measuring the motor's speed or position using appropriate transducers, and torque is controlled indirectly by suitable control of the motor-phase currents. Figure 37.3 shows a block diagram of a typical synchronous frame current controller for a three-phase motor. The figure also shows the proportioning of tasks between software code modules and the dedicated motor-control peripherals of a motor controller such as the LF2407. The controller consists of two proportional-plus-integral-plus-differential (PID) current regulators that are used to control the motor current vector in a reference frame that rotates synchronously with the measured rotor position.

Sometimes it may be desirable to implement a decoupling between voltage and speed that removes the speed dependencies and associated axes cross coupling from the control loop. The reference voltage components are then synthesized on the inverter using a suitable PWM strategy, such as space vector modulation (SVM). It is also possible to incorporate some compensation schemes to overcome the distorting effects of the inverter switching dead time, finite inverter device on-state voltages, and dc-link voltage ripple. The two components of the stator current vector are known as the direct-axis and quadrature-axis components. The direct-axis current controls the motor flux and is usually controlled to be zero with permanent magnet machines. The motor torque may then be controlled directly by regulation of the quadrature-axis component. Fast, accurate torque control is essential for high-performance drives in order to ensure rapid acceleration and deceleration – and smooth rotation down to zero speed under all load conditions.

37 DSP-based Control of Variable Speed Drives

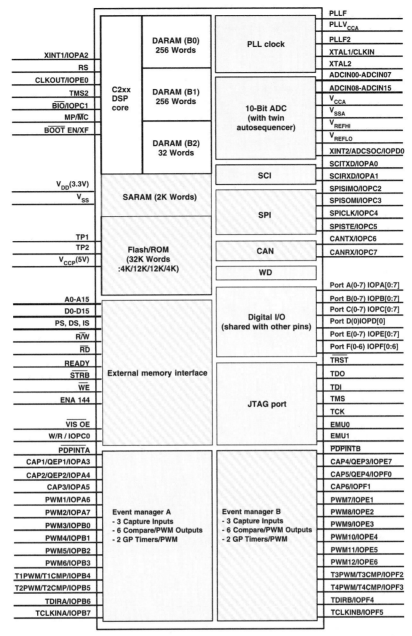

FIGURE 37.1 Graphical overview of DSP core and peripherals on the LF2407. (Courtesy of Texas Instruments)

The actual direct and quadrature current components are obtained by first measuring the motor phase currents with suitable current-sensing transducers and converting them to digital, using an on-chip ADC system. It is usually sufficient to simultaneously sample just two of the motor line currents: since the sum of the three currents is zero, the third current can, when necessary, be deduced from simultaneous measurements of the other two currents The controller software makes use of mathematical vector transformations, known as Park Transformations that ensure that the three-phase set of currents applied to the motor is synchronized to the actual rotation of the motor shaft, under all operating conditions. This synchronism ensures that the motor always produces the optimal torque per ampere – i.e. operates at optimal efficiency. The vector rotations require real-time calculation of the sine and cosine of the measured rotor

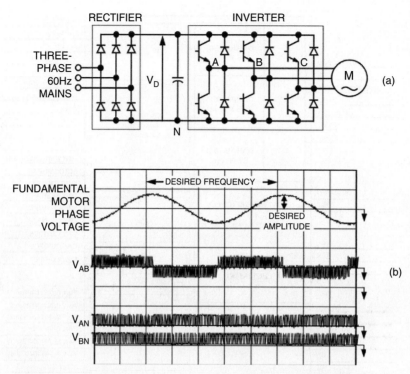

FIGURE 37.2 (a) Typical configuration of power converter used to drive three-phase ac motors and (b) typical PWM waveforms in the generation of a variable-voltage, variable-frequency supply for the motor.

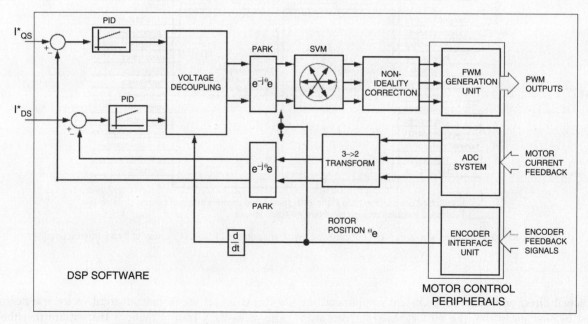

FIGURE 37.3 Configuration of typical control system for three-phase ac motor.

angle, plus a number of multiply-and-accumulate operations. The overall control-loop bandwidth depends on the speed of implementation of the closed-loop control calculations – and the resulting computation of new duty-cycle values. The inherent fast computational capability of the 40-MIPS, 16-bit fixed-point DSP core makes it the ideal computational engine for these embedded motor-control applications.

37.3.1 Pulse Width Modulation Generation

In typical ac motor-controller design, both hardware and software considerations are involved in the process of generating the PWM signals that are ultimately used to turn on or off the power devices in the three-phase inverter. In typical digital control environments, the controller generates a regularly timed interrupt at the PWM switching frequency (nominally 10–20 kHz). In the interrupt service routine, the controller software computes new duty-cycle values for the PWM signals used to drive each of the three legs of the inverter. The computed duty cycles depend on both the measured state of the motor (torque and speed) and the desired operating state. The duty cycles are adjusted on a cycle-by-cycle basis in order to make the actual operating state of the motor follow the desired trajectory.

Once the desired duty cycle values have been computed by the processor, a dedicated hardware PWM generator is needed to ensure that the PWM signals are produced over the next PWM and controller cycle. The PWM generation unit typically consists of an appropriate number of timers and comparators that are capable of producing very accurately timed signals. Typically, 10-to-12 bit performance in the generation of the PWM timing waveforms is desirable.

Figure 37.4 shows a typical PWM waveform for a single leg inverter. In general, there is a small delay required between turning off one power device like A-phase lower device and turning on the complementary power device A-phase upper device. This dead-time is required to ensure the device being turned off has sufficient time to regain its blocking capability before the other device is turned on. Otherwise a short circuit of the dc voltage could result.

In LF2407, the compare units have been used to generate the PWM signals. The PWM output signal is high when the output of current PI regulation matches the value of T1CNT and is set to low when the timer underflow occurs. The switch-states are controlled by the ACTR register. In order to minimize the switching losses, the lower switches are always kept on and the upper switches are chopped on/off to regulate the phase current.

37.3.2 Analog-to-Digital Conversion Requirements

For control of high-performance motor drives, fast, high-accuracy, simultaneous-sampling A/D conversion of the measured current values is required. The drives have a rated operation range – a certain power level that they can sustain continuously, with an acceptable temperature rise in the motor and power converter. They also have a peak rating – the ability to handle a current far in excess of the rated current for short periods of time. This allows a large torque to be applied transiently, to accelerate or decelerate the drive very quickly, and then to revert to the continuous range for normal operation. This also means that in the normal operating mode of the drive, only a small percentage of the total input range is being used.

At the other end of the scale, in order to achieve the smooth and accurate rotations desired in these machines, it is wise to compensate for small offsets and non-linearities such as core saturation and parameter detuning. In any current-sensor electronics, the analog signal processing is often subject to gain and offset errors. Gain mismatches, for example, can exist between the current-measuring systems for different windings. These effects combine to produce undesirable oscillations in the torque. To meet both of these conflicting resolution requirements, modern motor-drives use 10-bit A/D converters, depending on the cost/performance trade-off required by the application.

The bandwidth of the system is essentially limited by the amount of time it takes to input information and then perform the calculations. The A/D converters that take many

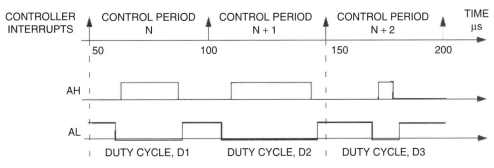

FIGURE 37.4 Typical PWM waveforms for a single inverter leg.

microseconds to convert can produce intolerable delays in the system. A delay in a closed-loop system will degrade the achievable bandwidth of the system, and bandwidth is one of the most important figures of merit in these high-performance drives. Therefore, fast analog-to-digital conversion is a necessity for these applications.

A third important characteristic of the A/D converter used in these applications is timing. In addition to high resolution and fast conversion, simultaneous sampling is needed. In any three-phase motor, it is necessary to measure the currents in the three windings of the motor at exactly the same time in order to get an instantaneous "snapshot" of the torque in the machine. Any time skew (time delay between the measurements of the different currents) is an error factor that is artificially inserted by the means of measurement. Such a nonideality translates directly into a ripple of the torque – a very undesirable characteristic.

The analog-to-digital converter (ADC) on the LF2407 allows the DSP to sample analog or "real-world" voltage signals. The output of the ADC is an integer number which represents the voltage level sampled. The integer number may be used for calculations in an algorithm. The resolution of the ADC is 10 bits, meaning that the ADC will generate a 10-bit number for every conversion it performs. However, the ADC stores the conversion results in registers that are 16-bit wide. The 10 most significant bits are the ADC result, while the least significant bits (LSBs) are filled with "0"s. There are a total of 16 input channels to the single input ADC. The control logic of the ADC consists of auto-sequencers, which control the sampling of the 16 input channels to the ADC. The auto-sequencers not only control which channels (input channels) will be sampled by the ADC, but also the order of the channels that the ADC performs conversions on. The two 8-conversion auto-sequencers can operate independently or cascade together as a "virtual" 16-conversion ADC.

37.3.3 Position Sensing and Encoder Interface Units

Usually the motor position is measured through the use of an encoder mounted on the rotor shaft. The incremental encoder produces a pair of quadrature outputs (A and B), each with a large number of pulses per revolution of the motor shaft. For a typical encoder with 1024 lines, both signals produce 1024 pulses per revolution. Using a dedicated quadrature counter, it is possible to count both the rising and falling edges of both the A and B signals so that one revolution of the rotor shaft may be divided into 4096 different values. In other words, a 1024 line encoder allows the measurement of rotor position to 12-bit resolution. The direction of rotation may also be inferred from the relative phasing of quadrature signals A and B.

Figure 37.5 shows the structure of an optical encoder. It consists of a light source, a radially slotted disk, and photoelectric

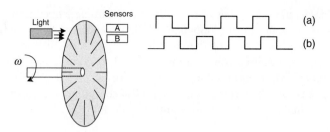

FIGURE 37.5 The structure of an encoder.

sensors. The disk rotates with the rotor. The two photosensors detect the light passing through the slots in the disk. When the light is hidden, a logic "0" is generated by the sensors. When the light passes through the slots of the disk, a logic "1" is produced. These logic signals are shown in Fig. 37.5. By counting the number of pulses, the motor speed can be calculated. The direction of rotation can be determined by detecting the leading signal between signals A and B.

This is all very well, but there is an increasing class of cost-sensitive motor drive applications with lower performance demands that can afford neither the cost nor the space requirements of the rotor position transducer. In these cases, the same motor-control algorithms can be implemented with estimated rather than measured rotor position.

The DSP core is quite capable of computing rotor position using sophisticated rotor-position estimation algorithms, such as extended Kalman estimators that extract estimates of the rotor position from measurements of the motor voltages and currents. These estimators rely on the real-time computation of a sufficiently accurate model of the motor in the DSP. In general, these sensorless algorithms can be made to work as well as the sensored algorithms at medium to high-speeds of rotation. But as the speed of the motor decreases, the extraction of reliable speed-dependent information from voltage and current measurements becomes more difficult. In general, sensorless motor control is applicable principally to applications such as compressors, fans and pumps, where continuous operation at zero or low speeds is not required.

37.3.4 The PI regulator

An electrical drive based on the field-orientated control (FOC) needs two constants as control parameters: the torque component reference i_{qs}^{e*} and the flux component reference i_{ds}^{e*}. The classical PI regulator is well suited to regulate the torque and flux feedback to the desired values. This is because it is able to reach constant references by correctly setting both the proportional term (K_p) and the integral term (K_i), which are, respectively, responsible for the error sensibility and for the steady-state error. The numerical expression of the PI regulator

37 DSP-based Control of Variable Speed Drives

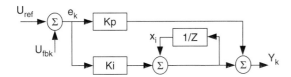

FIGURE 37.6 Classical PI regulator structure in discrete domain.

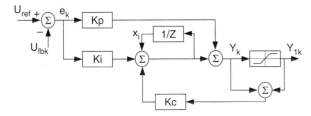

FIGURE 37.7 Numerical PI regulator with correction.

is as follows:

$$Y_{(k)} = K_p e_{(k)} + K_i e_{(k)} + \sum_{n=0}^{k-1} e_{(n)} \qquad (37.1)$$

which is represented in Fig. 37.6.

During normal operation, large reference value variations or disturbances may occur, that result in the saturation and overflow of the regulator variables and output. To solve this problem, one solution is to add a correction of the integral component as depicted in Fig. 37.7.

The constants K_p, K_i, K_c, proportional, integral, and integral correction components, are selected based on the sampling period and on the motor parameters. After defining the DSP-controlled motor drives requirements, in the following, we describe the digital control algorithms for permanent magnet motors and induction motors.

37.4 DSP-based Control of Permanent Magnet Brushless DC Machines

Permanent magnet alternating current (PMAC) motors are synchronous motors that have permanent magnets mounted on the rotor and poly-phase, usually three-phase, armature windings located on the stator. Since the field is provided by the permanent magnets, the PMAC motor has higher efficiency than induction or switched reluctance motors. The advantages of PMAC motors, combined with a rapidly decreasing cost of permanent magnets, have led to their widespread use in many variable speed drives such as robotic actuators, computer disk drives, appliances, automotive applications, and air conditioning (HVAC) equipment.

In general, PMAC motors are categorized into two types. The first type of motor is referred to as PM synchronous motor (PMSM). These motors produce a sinusoidal back-EMF, shown in Fig. 37.8a, and should be supplied with sinusoidal current/voltage. The PMSM's electronic control and drive system uses continuous rotor position feedback and PWM to supply the motor with the sinusoidal voltage or current. With this, constant torque is produced with very little ripple.

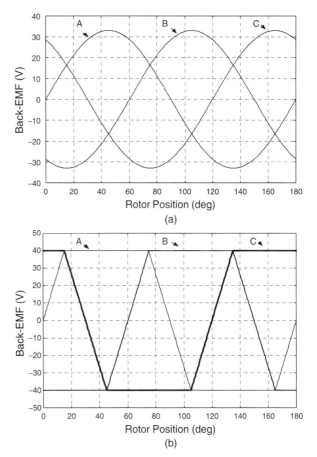

FIGURE 37.8 The back-EMF of PMAC motors: (a) three-phase back-EMF of PMSM and (b) three-phase back-EMF of BLDC motors.

The second type of PMAC motor has a trapezoidal back-EMF and is referred to as the brushless DC (BLDC) motor. The back-EMF of the BLDC motor is shown in Fig. 37.8b. The BLDC motor requires that quasi-rectangular-shaped currents are fed into the machine. Alternatively, the voltage may be applied to the motor every 120°, with a current limit to hold the currents within the motor's capabilities.

37.4.1 Mathematical Model of the BLDC Motor

The phase variables are used to model the BLDC motor due to its non-sinusoidal back-EMF and phase current. The terminal voltage equation of the BLDC motor can be written as

$$\begin{bmatrix} v_a \\ v_b \\ v_c \end{bmatrix} = \begin{bmatrix} R + pL_s & 0 & 0 \\ 0 & R + pL_s & 0 \\ 0 & 0 & R + pL_s \end{bmatrix} \cdot \begin{bmatrix} i_a \\ i_b \\ i_c \end{bmatrix} + \begin{bmatrix} e_a \\ e_b \\ e_c \end{bmatrix} \quad (37.2)$$

where v_a, v_b, v_c are the phase voltages, i_a, i_b, i_c are the phase currents, e_a, e_b, e_c are the phase back-EMF voltages, R is the phase resistance, L_s is the synchronous inductance per phase and includes both leakage and armature reaction inductances, and p represents d/dt. The electromagnetic torque is given by

$$T_e = \frac{(e_a i_a + e_b i_b + e_c i_c)}{\omega_m} \quad (37.3)$$

where ω_m is the mechanical speed of the rotor. The equation of motion is

$$\frac{d}{dt}\omega_m = \frac{(T_e - T_L - B\omega_m)}{J} \quad (37.4)$$

where T_L is the load torque, B is the damping constant, and J is the moment of inertia of the rotor shaft and the load.

37.4.2 Torque Generation

From Eq. (37.3), the electromagnetic torque of the BLDC motor is related to the product of the phase back-EMF and current. The back-EMFs in each phase are trapezoidal in shape and are displaced by 120 electrical degrees with respect to each other in a three-phase machine. A rectangular current pulse is injected into each phase so that current coincides with the crest of the back-EMF waveform; hence the motor develops an almost constant torque. This strategy, commonly called six-step current control is shown Fig. 37.9. The amplitude of each phase's back-EMF is proportional to the rotor speed, and is given by

$$E = k\phi\omega_m \quad (37.5)$$

where k is a constant and depends on the number of turns in each phase, ϕ is the permanent magnet flux, and ω_m is the mechanical speed. In Fig. 37.9, during any 120° interval, the instantaneous power converted from electrical to mechanical, P_o, is the sum of the contributions from two phases in series, and is given by

$$P_o = \omega_m T_e = 2EI \quad (37.6)$$

where T_e is the output torque and I is the amplitude of the phase current. From Eqs. (37.4) and (37.6), the expression for

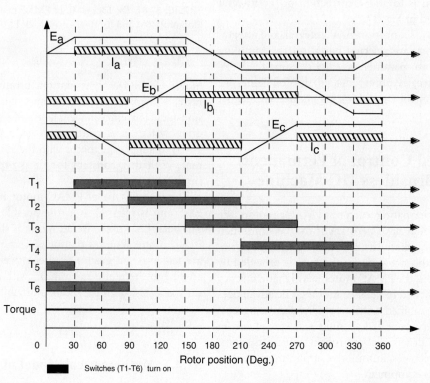

FIGURE 37.9 The principle of the six-step current control algorithm. T_1–T_6 are the gate signals, E_a, E_b, and E_c are the motor phase back-EMF, I_a, I_b, and I_c are the motor phase currents.

output torque can be written as

$$T_e = 2k\phi I = k_t I \quad (37.7)$$

where K_t is the torque constant. Since the electromagnetic torque is only proportional to the amplitude of the phase current in Eq. (37.7), torque control of the BLDC motor is essentially accomplished by phase current control.

37.4.3 BLDC Motor Control Topology

Based on the previously discussed concept, a BLDC motor-drive system is shown in Fig. 37.10. It can be seen that the total drive system consists of the BLDC motor, power electronics converter, sensor, and controller.

The BLDC motors are predominantly surface-magnet machines with wide magnet pole arcs. The stator windings are usually concentrated windings, which produce a square waveform distribution of flux density around the air-gap. The design of the BLDC motor is based on the crest of each half-cycle of the back-EMF waveform. In order to obtain a smooth output torque, the back-EMF waveform should be wider than 120° electrical degrees. A typical BLDC motor with 12 stator slots and 4 poles on the rotor is shown in Fig. 37.11. The inverter is usually responsible for the electronic commutation and current regulation. For the six-step current control, if the motor windings are Y-connected without the neutral connection, only two of the three phase currents flow through the inverter in series. This results in the amplitude of the DC link current always being equal to that of the phase currents. The PWM current controllers are typically used to regulate the actual machine currents in order to match the rectangular current reference waveforms shown in Fig. 37.9. For example, during one 60° interval, when switches T_1 and T_6 are active, phases A and B conduct. The lower switch T_6 is always turned on and the upper switch T_1 is chopped on/off using either a hysteresis current controller with variable switch frequency or a PI controller with fixed switch frequency.

When T_1 and T_6 are conducting, current builds up in the path as shown in Fig. 37.12a with dashed line. When switch T_1 is turned off, the current decays through diode D_4 and switch T_6 as depicted in Fig. 37.12b. In the next interval, switch T_2 is

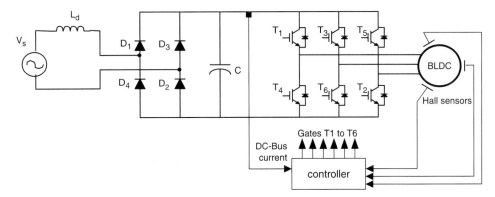

FIGURE 37.10 BLDC motor control system.

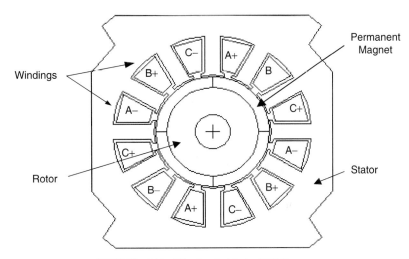

FIGURE 37.11 The 4-pole 12-slot BLDC motor.

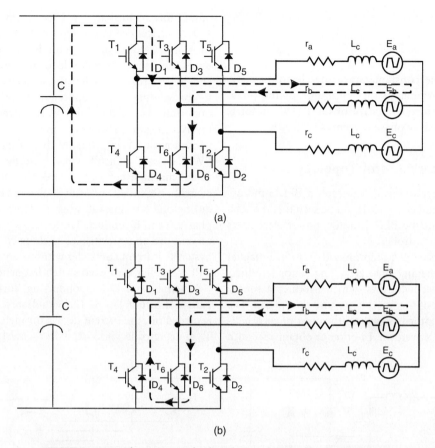

FIGURE 37.12 The current path when the switch T_1 turns on and turns off.

on, and T_1 is chopped so that phase A and phase C conduct. During the commutation interval, the phase B current rapidly decreases through the freewheeling diode D_3 until it becomes zero and the phase C current builds up.

From the above analysis, each of the upper switches is always chopped for one 120° interval and the corresponding lower switch is always turned on per interval. The freewheeling diodes provide the necessary paths for the currents to circulate when the switches are turned off and during the commutation intervals. There are two types of sensors for the BLDC drive system: a current sensor and a position sensor. Since the amplitude of the dc link current is always equal to the motor phase current in six-step current control, the dc link current is measured instead of the phase current. Thus, a shunt resistor, which is in series with the inverter, is usually used as the current sensor. Hall-effect position sensors typically provide the position information needed to synchronize the stator excitation with rotor position in order to produce constant torque. Hall-effect sensors detect the change in magnetic field. The rotor magnets are used as triggers for the Hall sensors. A signal conditioning circuit is needed for noise cancellation in Hall-effect sensors circuits. In six-step current control algorithm, rotor position needs to be detected at only six discrete points in each electrical cycle. The controller tracks these six points so that the proper switches are turned on or off for the correct intervals. Three Hall-effect sensors, spaced 120 electrical degrees apart, are mounted on the stator frame. The digital signals from the Hall sensors are then used to determine the rotor position and switch gating signals for the inverter switches.

37.4.4 DSP Controller Requirements

The controller of BLDC drive systems reads the current and position feedback, implements the speed or torque control algorithm, and finally generates the gate signals. The connectivity of the LF2407 in this application is illustrated in Fig. 37.13. Three capture units in the LF2407 are used to detect both the rising and falling edges of Hall-effect signals. Hence, every 60 electrical degrees of motor rotation, one capture unit interrupt is generated which ultimately causes a change in the gating signals and the motor to move to the next position. One input channel of the 10-bit A/D converter reads the dc link current. The output pins PWM1–PWM6 are used to supply the gating signals to the inverter.

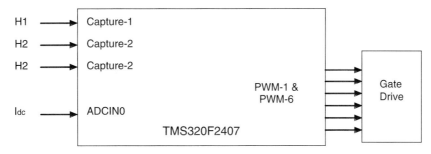

FIGURE 37.13 The interface of LF2407.

37.4.5 Implementation of the BLDC Motor Control Algorithm Using LF2407

A block diagram of the BLDC motor control system is shown in Fig. 37.14. The dashed line separates the software from the hardware components introduced in the previous section. It is necessary to choose hardware components carefully in order to ensure high processing speed and precision in the overall control system. The overall control algorithm of the BLDC motor consists of nine modules:

- Initialization procedure
- Detection of Hall-effect signals
- Speed control subroutine
- Measurement of current
- Speed profiling
- Calculation of actual speed
- PID regulation
- PWM generation
- DAC output

The flowchart of the overall control algorithm is illustrated in Fig. 37.15.

37.5 DSP-based Control of Permanent Magnet Synchronous Machines

As previously described, the permanent magnet synchronous motor (PMSM) is a PM motor with a sinusoidal back-EMF. Compared to the BLDC motor, it has less torque ripple because the torque pulsations associated with current commutation do not exist. A carefully designed machine in combination with a good control technique can yield a very low level of torque ripple (<2% rated), which is attractive for high-performance motor control applications such as machine tool and servo applications.

In this section, following the same procedures used in the previous section, the principles of the PMSM drive system will

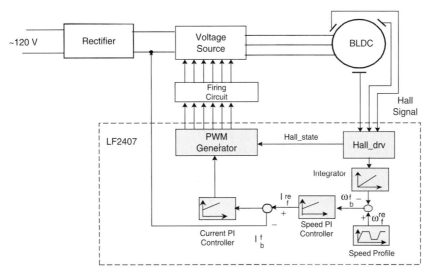

FIGURE 37.14 The block diagram of BLDC motor control algorithm.

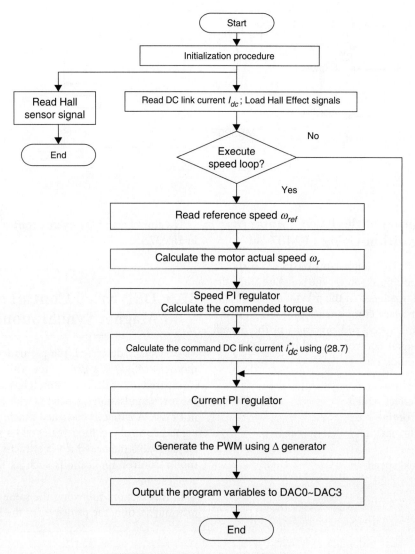

FIGURE 37.15 BLDC algorithm flowchart.

be introduced. Later, the control implementation using the LF2407 DSP will be described in detail.

37.5.1 Mathematical Model of PMSM

Figure 37.16 depicts the simplified three-phase surface mounted PMSM motor for our discussion. The stator windings, as–as', bs–bs', and cs–cs', are shown as lumped windings for simplicity, but are actually distributed around the stator. The rotor has 2 poles. Mechanical rotor speed and position are denoted as ω_{rm} and θ_{rm}, respectively. Electrical rotor speed and position, ω_r and θ_r, are defined as $P/2$ times the corresponding mechanical quantities, where P is the number of poles.

Based on the above motor definition, the voltage equation in the abc stationary reference frame is given by

$$V_{abcs} = R_s i_{abcs} + \frac{d}{dt}\lambda_{abcs} \qquad (37.8)$$

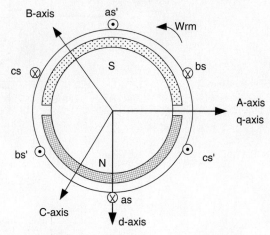

FIGURE 37.16 The cross section of PMSM.

where

$$f_{abcs} = [f_{as}\ f_{bs}\ f_{cs}]^T \quad (37.9)$$

and the stator resistance matrix is given by

$$R_s = \text{diag}[r_s\ r_s\ r_s] \quad (37.10)$$

The flux linkages equation can be expressed by

$$\lambda_{abcs} = L_s i_{abcs} + \lambda'_m \begin{bmatrix} \sin\vartheta_r \\ \sin(\vartheta_r - \frac{2\pi}{3}) \\ \sin(\vartheta_r - \frac{4\pi}{3}) \end{bmatrix} \quad (37.11)$$

where λ'_m denotes the amplitude of the flux linkages established by the permanent magnet as viewed from the stator phase windings. Note that in Eq. (37.11) the back-EMFs are sinusoidal waveforms that are 120° apart from each other.

The stator self-inductance matrix, L_s, is given as

$$L_s = \begin{bmatrix} L_{ls} + L_A - L_B \cos 2\theta_r & -\frac{1}{2}L_A - L_B \cos 2(\theta_r - \pi/3) & -\frac{1}{2}L_A - L_B \cos 2(\theta_r + \pi/3) \\ -\frac{1}{2}L_A - L_B \cos 2(\theta_r - \pi/3) & L_{ls} + L_A - L_B \cos 2(\theta_r - 2\pi/3) & -\frac{1}{2}L_A - L_B \cos 2(\theta_r + \pi) \\ -\frac{1}{2}L_A - L_B \cos 2(\theta_r + \pi/3) & -\frac{1}{2}L_A - L_B \cos 2(\theta_r + \pi) & L_{ls} + L_A - L_B \cos 2(\theta_r + 2\pi/3) \end{bmatrix} \quad (37.12)$$

The torque and speed are related by the electromechanical motion equation

$$J\frac{d}{dt}\omega_{rm} = \frac{P}{2}(T_e - T_L) - B_m\omega_{rm} \quad (37.15)$$

where J is the rotational inertia, B_m is the approximated mechanical damping due to friction, and T_L is the load torque.

37.5.2 Mathematical Model of PMSM in Rotor Reference Frame

The voltage and torque equations can be expressed in the rotor reference frame in order to transform the time-varying variables into steady-state constants. The transformation of the three-phase variables in the stationary reference frame to the rotor reference frame is defined as

$$f_{qd0r} = K_r f_{abcs} \quad (37.16)$$

where

$$k_r = \frac{2}{3}\begin{bmatrix} \cos\theta_r & \cos(\theta_r - \frac{2\pi}{3}) & \cos(\theta_r + \frac{2\pi}{3}) \\ \sin\theta_r & \sin(\theta_r - \frac{2\pi}{3}) & \sin(\theta_r + \frac{2\pi}{3}) \\ \frac{1}{2} & \frac{1}{2} & \frac{1}{2} \end{bmatrix}$$

The electromagnetic torque may be written as

$$T_e = \frac{P}{2}\left\{\lambda'_m\left[(i_{as} - \frac{1}{2}i_{bs} - \frac{1}{2}i_{cs})\cos\theta_r - \frac{\sqrt{3}}{2}(i_{bs} - i_{cs})\sin\vartheta_r\right]\right.$$

$$+ \frac{L_{md} - L_{mq}}{3}\left[(i_{as}^2 - \frac{1}{2}i_{bs}^2 - \frac{1}{2}i_{cs}^2 - i_{as}i_{bs} - i_{as}i_{cs} + 2i_{bs}i_{cs})\right.$$

$$\left.\left.\sin 2\theta_r + \frac{\sqrt{3}}{2}(i_{bs}^2 i_{cs}^2 - 2i_{as}i_{bs} + 2i_{as}i_{cs})\cos 2\theta_r\right]\right\} + T_{cog}(\theta_r) \quad (37.13)$$

In Eq. (37.13), $T_{cog}(\theta_r)$ represents the cogging torque and the d- and q-axes magnetizing inductances are defined by

$$L_{md} = \frac{3}{2}(L_A - L_B)$$

and

$$L_{md} = \frac{3}{2}(L_A + L_B) \quad (37.14)$$

If the applied stator voltages are given by

$$\begin{cases} V_{as} = \sqrt{2}V_s \cos\theta_{ev} \\ V_{bs} = \sqrt{2}V_s \cos(\theta_{ev} - \frac{2\pi}{3}) \\ V_{cs} = \sqrt{2}V_s \cos(\theta_{ev} + \frac{2\pi}{3}) \end{cases} \quad (37.17)$$

Then applying (37.16) to (37.8), (37.11), and (37.17) yields

$$v_{qs}^r = r_s i_{qs}^r + \omega_r \lambda_{ds}^r + \frac{d}{dt}\lambda_{qs}^r \quad (37.18)$$

$$v_{ds}^r = r_s i_{ds}^r - \omega_r \lambda_{qs}^r + \frac{d}{dt}\lambda_{ds}^r \quad (37.19)$$

$$\lambda_{qs}^r = L_{qs} i_{qs}^r \quad (37.20)$$

$$\lambda_{ds}^r = L_{ds} i_{ds}^r + \lambda'^r_m \quad (37.21)$$

where the q- and d-axes self-inductances are given by $L_{qs} = L_{ls} + L_{mq}$ and $L_{ds} = L_{ls} + L_{md}$, respectively. The electromagnetic torque can be written as

$$T_e = \frac{3}{2}\frac{P}{2}[\lambda'^r_m i_{qs}^r + (L_{ds} - L_{qs})i_{qs}i_{ds}] \quad (37.22)$$

From Eq. (37.22), it can be seen that torque is related only to the d- and q-axes currents. Since $L_q \geq L_d$ (for surface mount PMSM both inductances are equal), the second item contributes a negative torque if the flux weakening control has been used. In order to achieve the maximum torque/current ratio, the d-axis current is set to zero during the constant torque control so that the torque is proportional only to the q-axis current. Hence, this results in the control of the q-axis current for regulating the torque in rotor reference frame.

37.5.3 PMSM Control Topology

Based on the above analysis, a PMSM drive system is developed as shown in Fig. 37.17. The total drive system looks similar to that of the BLDC motor and consists of a PMSM, power electronics converter, sensors, and controller. These components are discussed in detail in the following sections.

The design consideration of the PMSM is to first generate the sinusoidal back-EMF. Unlike the BLDC, which needs concentrated windings to produce the trapezoidal back-EMF, the stator windings of PMSM are distributed in as many slots per pole as deemed practical to approximate a sinusoidal distribution. To reduce the torque ripple, standard techniques such as skewing and chorded windings are applied to the PMSM. With the sinusoidally excited stator, the rotor design of the PMSM becomes more flexible than the BLDC motor where the surface-mount permanent magnet is a favorite choice. Besides the common surface-mount non-salient pole PM rotor, the salient pole rotor, like inset and buried magnet rotors, are often used because they offer appealing performance characteristics during the flux weakening region. A typical PMSM with 36 stator slots in stator and 4 poles on the rotor is shown in Fig. 37.18.

Due to the sinusoidal nature of the PMSM, control algorithms such as V/f and vector control, developed for other AC motors, can be directly applied to the PMSM control system. If the motor windings are Y-connected without a neutral connection, three-phase currents can flow through the inverter at any

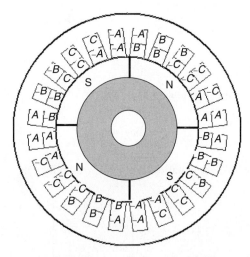

FIGURE 37.18 A 4-pole 24-slot PMSM.

moment. With respect to the inverter switches, three switches, one upper and two lower in three different legs conduct at any moment as shown in Fig. 37.19. The PWM current control is still used to regulate the actual machine current. Either a hysteresis current controller, a PI controller with sine-triangle, or an SVPWM strategy is employed for this purpose. Unlike the BLDC motor, the three switches are switched at any time.

37.5.4 DSP Controller Requirements

The LF2407 is used as the controller to implement speed control of the PMSM system. The interface of the LF2407 is illustrated in Fig. 37.20. Similar to the BLDC motor control system, three input channels are selected to read the two-phase currents and resolver signal. Because a resolver is used in one case, the quadrature-encoder-pulse (QEP) inputs are not used. The QEP inputs work only with a QEP signal that a rotary encoder supplies. The DSP output pins PWM1–PWM6 used to supply the gating signals to the switches and form the output of the control part of the system.

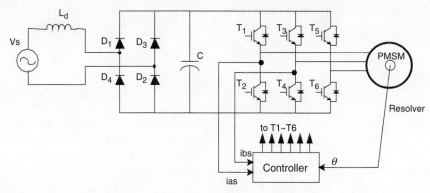

FIGURE 37.17 The PMSM speed control system.

37 DSP-based Control of Variable Speed Drives

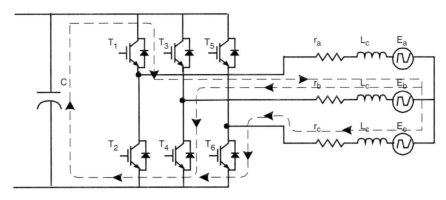

FIGURE 37.19 The current path when the three phases are chopped.

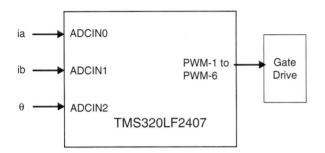

FIGURE 37.20 The interface of LF2407.

37.5.5 Implementation of the PMSM Algorithm Using the LF2407

The block diagram of the PMSM drive system is displayed in Fig. 37.21.

The flowchart of the developed software is shown in Fig. 37.22. The control program of the PMSM has one main routine and includes four modules:

- Initialization procedure
- DAC module
- ADC module
- Speed control module

In the following section speed control module is disscussed.

37.5.5.1 The Speed Control Algorithm

The requirement for speed control algorithm can be itemized as:

- Reading the current and position signal, then generating the commanded speed profile
- Calculating the actual motor speed, transferring the variables in the abc model to the d–q model and reverse
- Regulating the motor speed and currents using the vector-control strategy

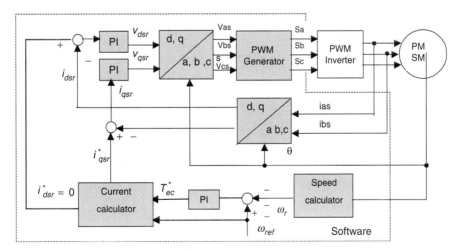

FIGURE 37.21 Block diagram of PMSM speed control system.

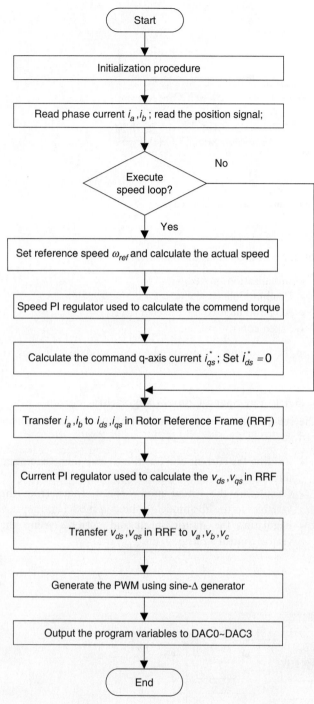

FIGURE 37.22 The flowchart of PMSM control system.

- Generating the PWM signal based on the calculated motor phase voltages

The PWM frequency is determined by the time interval of the interrupt, with the controlled phase voltages being recalculated every interrupt.

37.6 DSP-based Vector Control of Induction Motors

For many years, induction motors have been preferred for a variety of industrial applications because of their robust and rugged construction. Compared to dc motors, induction motors are not as easy to control. They typically draw large starting currents, about six to eight times their full load values, and operate with lagging power factor when loaded. However, with the advent of the vector-control concept for motor control, it is possible to decouple the torque and the flux, thus making the control of the induction motor very similar to that of the dc motor.

The most popular type of induction motor used is the squirrel cage induction motor. The rotor consists of a laminated core with parallel slots for carrying the rotor conductors, which are usually heavy bars of copper, aluminum, or alloys. One bar is placed in each slot; or rather, the bars are inserted from the end when the semi-closed slots are used. The rotor bars are brazed, electrically welded, or bolted to two heavy and stout short-circuiting end-rings, thus completing the squirrel cage construction. The rotor bars are permanently short-circuited on themselves. The rotor slots are usually not parallel to the shaft, but are given a slight angle, called a skew, which increases the rotor resistance due to increased length of rotor bars and an increase in the slip for a given torque. The skew is also advantageous because it reduces the magnetic hum while the motor is operating and reduces the locking tendency, or cogging, of the rotor teeth.

When the three-phase stator windings are fed by a three-phase supply, a magnetic flux of a constant magnitude rotating at synchronous speed is created in the air-gap. Due to the relative speed between the rotating flux and the stationary conductors, an electromagnetic force (EMF) is induced in the rotor in accordance with Faraday's laws of electromagnetic induction. The frequency of the induced EMF is the same as the supply frequency, and the magnitude is proportional to the relative velocity between the flux and the conductors. Therefore, the rotor current develops in the same direction as the flux and tries to catch up with the rotating flux.

37.6.1 Induction Motor Field-oriented Control

The term "vector" control refers to the control technique that controls both the amplitude and the phase of ac excitation voltage. Vector control controls the spatial orientation of the EMF in the machine. This has led to the coining of the term FOC, which is used for controllers that maintain a 90° spatial orientation between the critical field components.

The required 90° of spatial orientation between key field components can be compared to the dc motor, where the armature winding magnetic field and the filed winding magnetic filed are always in quadrature. The objective is to force the control of the induction machine to be similar to the control of a dc motor, i.e., torque control. In dc machines, the field and the armature winding axes are orthogonal to one another, making the magnetomotive forces (MMFs) established orthogonal. If the iron saturation is ignored, then the orthogonal fields can be considered to be completely decoupled.

It is important to maintain a constant field flux for proper torque control. It is also important to maintain an independently controlled armature current in order to overcome the effects of the detuning of resistance of the armature winding, and leakage inductance. A spatial angle of 90° between the flux and MMF axes has to be maintained in order to limit interaction between the MMF and the flux. If these conditions are met at every instant of time, the torque will always follow the current.

With vector control, the mechanically robust induction motors can be used in high-performance applications where dc motors were previously used. The key feature of the control scheme is the orientation of the synchronously rotating q–d–0 frame to the rotor flux vector. The d-axis component is aligned with the rotor flux vector and regarded as the flux-producing current component. On the other hand, the q-axis current, which is perpendicular to the d-axis, is solely responsible for torque production.

In order to apply a rotor flux field-orientation condition, the rotor flux linkage is aligned with the d-axis, so the q-axis rotor flux in excitation reference frame λ_{qr}^e will be zero and the d-axis rotor flux in the excitation reference frame will be the rotor flux; $\lambda_{dr}^e = \hat{\lambda}_r$. Therefore we have:

$$i_{ds}^e = \frac{\lambda_{dr}^e}{L_m} \quad (37.23)$$

$$\omega_{slip} = \frac{r_r}{\hat{\lambda}_r}\left(\frac{L_m}{L_r}\right)i_{qs}^e = \frac{L_m i_{qs}^e}{\tau_r \lambda_r} \quad (37.24)$$

$$T_e = \frac{3}{2}\frac{P}{2}\frac{L_m}{L_r}\hat{\lambda}_d i_{qs}^e \quad (37.25)$$

where τ_r is rotor time constant, L_m is magnetizing inductance, L_r rotor leakage inductance, r_r rotor resistance, i_{qs}^e q-axis stator current in excitation frame, i_{ds}^e d-axis stator current in excitation frame, and ω_{slip} angular frequency of slip. We can find out that in this case i_{ds}^e controls the rotor flux linkage and i_{qs}^e controls the electromagnetic torque. The reference currents of the q–d–0 axis (i_{qs}^{e*}, i_{ds}^{e*}) are converted to the reference phase voltages (v_{ds}^{e*}, v_{qs}^{e*}) as the commanded voltages for the control loop. Given the position of the rotor flux and two-phase currents, this generic algorithm implements the instantaneous direct torque and flux control by means of coordinate transformations and PI regulators, thereby achieving accurate and efficient motor control.

It is clear that for implementing vector control we have to determine the rotor flux position. This usually is performed by measuring the rotor position and utilizing the slip relation to compute the angle of the rotor flux relative to the rotor axis.

Equations (37.23) and (37.24) show that we can control torque and field by i_{ds} and i_{qs} in the excitation frame. However, in the implementation of FOC, we need to know i_{ds} and i_{qs} in the stationary reference frame. So, we have to know the angular position of the rotor flux to transform i_{ds} and i_{qs} from the excitation frame to the stationary frame. By using ω_{slip}, which is shown in Eq. (37.24) and using actual rotor speed, the rotor flux position is obtained.

$$\int_0^t \omega_{slip} dt + \theta_{re}(t) = \theta_r(t) \quad (37.26)$$

Where $\theta_{re}(t)$ is electrical angular rotor position, and $\theta_r(t)$ angular rotor flux position.

The Current Model takes i_{ds} and i_{qs} as inputs as well as the rotor mechanical speed and gives the rotor flux position as an output. Figure 37.23 shows the block diagram of the vector-control strategy in which speed regulation is possible using a control loop.

As shown in Fig. 37.23, two-phase currents are measured and fed to the Clarke transformation block. These projection outputs are indicated as i_{ds}^s and i_{qs}^s. These two components of the current provide the inputs to Park's transformation, which gives the currents in the qds^e excitation reference frame. The i_{ds}^e and i_{qs}^e components, which are outputs of the Park transformation block, are compared to their reference values i_{ds}^{e*}, the flux reference, and i_{qs}^{e*}, the torque reference. The torque command, i_{qs}^{e*}, comes from the output of the speed controller. The flux command, i_{ds}^{e*}, is the output of the flux controller which indicates the right rotor flux command for every speed reference. For i_{ds}^{e*}, we can use the fact that the magnetizing current is usually between 40 and 60% of the nominal current. For operating in speeds above the nominal speed, a field weakening section should be used in the flux controller section. The current regulator outputs, v_{ds}^{e*} and v_{qs}^{e*}, are applied to the inverse Park transformation. The outputs of this projection are v_{ds}^s and v_{qs}^s, which are the components of the stator voltage vector in the dqs^s orthogonal reference frame. They form the inputs of the space-vector PWM block. The outputs of this block are the signals that drive the inverter.

Note that both the Park and the inverse Park transformations require the exact rotor flux position, which is given by the Current Model block. This block needs the rotor resistance or rotor time constant as a parameter. Accurate knowledge of the rotor resistance is essential to achieve the highest possible efficiency from the control structure. Lack of this knowledge results in the detuning of the FOC. In Fig. 37.23, a space-vector PWM has been used to emulate v_{ds}^s and v_{qs}^s in order to implement current regulation.

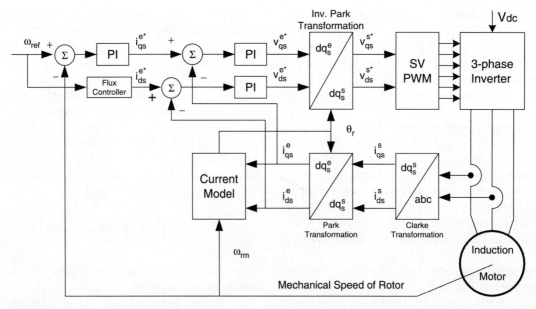

FIGURE 37.23 Vector-control algorithm for induction motor.

37.6.2 DSP Controller Requirements

The controller of the induction motor control system is used to read the feedback current and position signals, to implement the speed or torque control algorithm, and to generate the gate signals based on the control signal. Analog controllers or digital signal processors, such as LF2407 can perform these tasks.

The interface of the LF2407 is illustrated in Fig. 37.24. Two quadrature counters detect the rising and falling edges of the encoder signals. Two input channels related to the 10-bit ADC are selected to read the two-phase currents. The pins PWM1 to PWM6 output the gating signals to the gate drive circuitry.

37.6.3 Implementation of Field-oriented Speed Control of Induction Motor

Some practical aspects of implementing the block diagram of Fig. 37.24 are discussed in this section and subsections. The software organization, the utilization of different variables, and the handling of the DSP controller resources are described. In addition, the control structure for the per-unit model is presented. Next, some numerical considerations have been made in order to address the problems inherent within the fixed-point calculation. As described, current model is one of the most important blocks in the block diagram depicted in Fig. 37.23. The inputs of this block are the currents and mechanical speed of rotor. In the next sections, technical

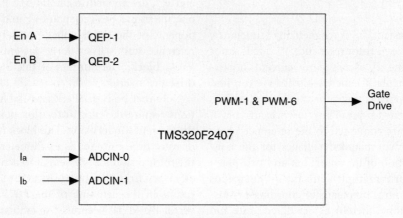

FIGURE 37.24 The interface of LF2407.

37.6.3.1 Software Organization

The body of the software consists of two main modules: the initialization module and the PWM interrupt service routine (ISR) module. The initialization model is executed only once at start-up. The PWM ISR module interrupts the waiting infinite loop when the timer underflows. When the underflow interrupt flag is set, the corresponding ISR is served. Figure 37.25 shows the general structure of the software. The complete FOC algorithm is executed within the PWM ISR so that it runs at the same frequency as the switching frequency or at a fraction of it. The wait loop could be easily replaced with a user interface.

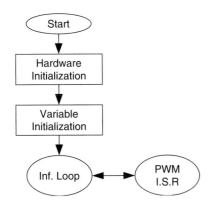

FIGURE 37.25 General structure of software.

37.6.3.2 Base Values and Per-unit Model

It is often convenient to express machine parameters and variables of per-unit quantities. Moreover, the LF2407 is a fixed point DSP, so using a normalized per-unit model of the induction motor is easier than using real parameters. In this model, all quantities refer to the base values. Base power and base voltage are selected, and all parameters and variables are normalized using these base quantities. Although one might violate this convention from time to time when dealing with instantaneous quantities, the rms values of the rated phase voltage and current are generally selected as the base voltage for the a–b–c variables while the peak value is generally selected as the base voltage for d–q variables. The base values are determined from the nominal values by using Eq. (37.27), where I_n, V_n, f_n are the nominal phase current, the nominal phase to neutral voltage, and the nominal frequency in a star-connected induction motor, respectively. The base value definitions are as follows:

$$I_b = \sqrt{2}I_n$$
$$V_b = \sqrt{2}V_n$$
$$\omega_b = 2\pi f_n \quad (37.27)$$
$$\psi_b = \frac{V_b}{\omega_b}$$

I_b and V_b are the maximum values of the nominal phase current and voltage, ω_b is the electrical nominal rotor flux speed, and ψ_b is the base flux.

37.6.3.3 Speed Estimation during High-speed Region

As previously mentioned, this method is based on counting the number of encoder pulses in a specified time interval. The QEP assigned timer counts the number of pulses and records it in the timer counter register (TxCNT). As the mechanical time constant is much slower than the electrical one, the speed regulation loop frequency might be lower than the current loop frequency. The speed regulation loop frequency is obtained in this algorithm by means of a software counter. This counter accepts the PWM interrupt as input clock and its period is the software variable called SPEEDSTEP. The counter variable is named speedstep. When speedstep is equal to SPEEDSTEP, the number of pulses counted is stored in another variable called n_p and thus the speed can be calculated. The scheme depicted in Fig. 37.26 shows the structure of the speed feedback generator.

Assuming that n_p is the number of encoder pulses in one SPEEDSTEP period when the rotor turns at the nominal speed, a software constant K_{speed} should be chosen as follows:

$$01000h = K_{speed} \cdot n_p$$

The speed feedback can then be transformed into a Q4.12 format, which can be used in the control software. In the proposed control system, the nominal speed is 1800 rpm and SPEEDSTEP is set to 125. The n_p can be calculated as follows:

$$n_p = \frac{1800 \times 64 \times 4}{60} \times \text{SPEEDSTEP} \times T_p = 288 \quad (37.28)$$

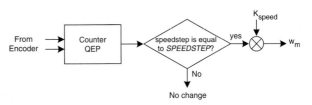

FIGURE 37.26 Block diagram of speed feedback calculator.

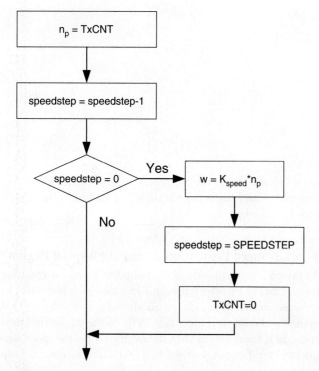

FIGURE 37.27 Complete flowchart of speed measurement block during high-speed region.

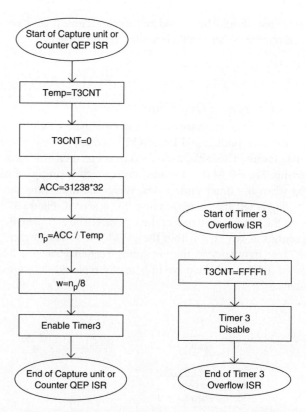

FIGURE 37.28 Flowchart of speed measurement at low speed.

where $T_p = \frac{3}{f_{pwm}} = 3 \times 10^{-4}$ (PWM frequency is 10 kHz but the program is running at 3333 Hz) and hence K_{speed} is given by:

$$K_{speed} = \frac{4096}{288} = 14.22 \Leftrightarrow 0E38h \quad Q8.8$$

Note that K_{speed} is out of the Q4.12 format range. The most appropriate format to handle this constant is the Q8.8 format. The speed feedback in Q4.12 format is then obtained from the encoder by multiplying n_p by K_{speed}. The flowchart of speed measurement is presented in Fig. 37.27.

37.6.3.4 Speed Measurement during Low-speed Region

To detect the edges of two successive encoder pulses, the developed program can use either the QEP counter or the capture unit input pins. The program has to measure the time between two successive pulses, therefore it must utilize another GP timer. In this program, Timer 3 has been dedicated to the time measurement. During the interrupt service routine of the capture unit or counter QEP, speed can be calculated. To obtain the actual speed of the motor, the appropriate number is divided by the value in the count register of Timer 3.

As it can be inferred, at very low speeds an overflow may occur in Timer 3. The counter would then reset itself to zero and start counting up again. This event results in a large error in speed measurement. To avoid this event, Timer 3 will be disabled in the overflow interrupt service routine. However, this timer is enabled in the capture unit (counter QEP) interrupt.

The flowchart of this implementation is presented in Fig. 37.28

37.6.3.5 The Current Model

The Current Model is used to find the rotor flux position. This module takes i_{ds} and i_{qs} as inputs plus the rotor electrical speed and then calculates the rotor flux position. The current model is based on Eqs. (37.23) and (37.24). Equation (37.23) in transient form can be written as:

$$\frac{L_r}{r_r L_m} \frac{d\lambda_{dr}}{dt} + \frac{\lambda_{dr}}{L_m} = i_{ds} \quad (37.29)$$

Assume $\lambda_{dr}/L_m = i_m$ where i_m is the magnetizing current, therefore Eq. (37.29) can be written as follows:

$$T_r \frac{d}{dt} i_m + i_m = i_{ds} \quad (37.30)$$

Rotor flux speed in a per-unit system can be shown by:

$$f_s = \frac{1}{\omega_b} \frac{d\theta}{dt} = \omega_{re} + \frac{i_{qs}}{T_r i_m \omega_b} \quad (37.31)$$

where θ is the rotor flux position and $T_r = L_r/r_r$ and ω_{re} are the rotor time constant and rotor electrical speed, respectively. The rotor time constant is critical to the correct functionality of the Current Model. This system outputs the rotor flux speed, which in turn will be integrated to get the rotor flux position. Assuming that $i_{qs(k+1)} \approx i_{qs(k)}$, Eqs (37.30) and (37.31) can be discretized as follows:

$$i_{mr(k+1)} = i_{mr(k)} + \frac{T_p}{T_r}\left(i_{ds(k)} - i_{mr(k)}\right)$$

$$f_{s(k+1)} = n_{(K+1)} + \frac{1}{T_r\omega_b}\frac{i_{qs(k)}}{i_{mr(k+1)}}$$

(37.32)

For example, let the constants T_p/T_r and $1/T_r\omega_b$ be renamed to K_t and K_r, respectively. Here $L_r = 73.8$ mH, $r_r = 0.73\,\Omega$, and $f_n = 60$ Hz. So for K_t and K_r we have:

$$K_r = \frac{T_p}{T_r} = \frac{(10000/3)^{-1}}{101.09 \times 10^{-3}}$$

$$= 2.967 \times 10^{-3} \Leftrightarrow 000Ch \quad Q4.12$$

$$K_t = \frac{1}{T_r\omega_b} = \frac{1}{30.232 \times 10^{-3} \times 377}$$

$$= 26.237 \times 10^{-3} \Leftrightarrow 006Bh \quad Q4.12$$

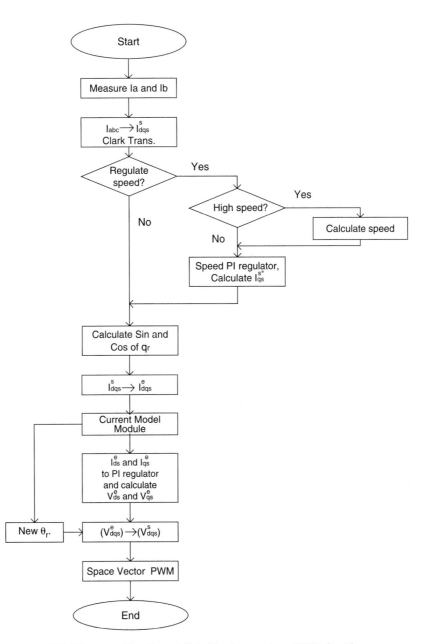

FIGURE 37.29 Flowchart of digital implementation of FOC algorithm.

By knowing the rotor flux speed (f_s), the rotor flux position (θ_{cm}) is computed by the integration formula in the per-unit system.

$$\theta_{cm(k+1)} = \theta_{cm(k)} + \omega_b \cdot f_{s(k)} \cdot T \quad (37.33)$$

In Eq. (37.33), let $\omega_b f_s T$ be called θ_{inc}. This variable is the rotor angle variation within one sampling period. Thus, the Current Model Module has three input variables i_{ds}, i_{ds}, and ω_{re} and one output, which is the rotor flux position θ_{cm} represented as a 16-bit integer value. The flowchart of the field-oriented speed control of induction motor is presented in Fig. 37.29. This routine is placed inside the PWM interrupt service routine.

38
Power Quality

S. Mark Halpin and Angela Card
Department of Electrical and Computer Engineering, Auburn University, Alabama, USA

38.1 Introduction 1053
38.2 Power Quality 1054
    38.2.1 Steady-state Voltage Frequency and Magnitude • 38.2.2 Voltage Sags
    • 38.2.3 Grounding • 38.2.4 Harmonics • 38.2.5 Voltage Fluctuations and Flicker
    • 38.2.6 Transients • 38.2.7 Monitoring and Measurement
38.3 Reactive Power and Harmonic Compensation... 1061
    38.3.1 Typical Harmonics Produced by Equipment • 38.3.2 Resonance • 38.3.3 Harmonic Filters
38.4 IEEE Standards... 1063
38.5 Conclusions 1065
    Further Reading 1066

38.1 Introduction

Power electronics and power quality are irrevocably linked together as we strive to advance both broad areas. With the dramatic increases over the last 20 years in energy conversion systems utilizing power electronic devices, we have seen the emergence of "power quality" as a major field of power engineering. The power electronic technology has played a major role in creating "power quality," and simple control algorithm modifications *to this same technology* can often play an equally dominant role in enhancing overall quality of electrical energy available to end-users.

Power electronics has given us, as a industrial society, a plethora of new ways to manufacture products, provide services, and utilize energy. From a power quality impact viewpoint, applications such as

1. Switched-mode power supplies,
2. DC arc furnaces,
3. Electronic fluorescent lamp ballasts,
4. Adjustable speed drives, and
5. Flexible ac transmission components.

are often cause for concern. From the viewpoint of a utility supply system, these converter-based systems can lead to operational and life expectancy problems for other equipment, possibly not owned or operated by the same party. It was from this initial perspective that the field of power quality emerged.

In most cases, the same devices and systems that create power quality problems can also be used to solve power quality problems. "Problem solving" applications such as

1. Active harmonic filters,
2. Static and adaptive var compensators, and
3. Uninterruptable power supplies.

all utilize the same switching device technology as the "problem causing" applications.

As the number of potentially problematic power electronic-based loads has increased over time, so attention has given to enhanced converter control to maximize power quality. Perfect examples of these improvements include

1. Unity power factor converters,
2. Dip-proof inverters, and
3. Limited-distortion electronic lamp ballasts.

While these direct product enhancements are not mandatory in North America, today's global economy necessitates consideration of power quality standards and limits in order to conduct business in the European Union.

While many studies suggest increases in power electronic-based energy utilization as high as 70–80% (of all energy consumed), it is equally clear that we are beginning to realize the total benefit of such end-use technologies. Power quality problems associated with grounding, sags, harmonics, and transients will continue to increase because of the sheer number of sensitive electronic loads expected to be placed in service. At the same time, we are only now beginning to realize the total benefits that such loads can offer.

38.2 Power Quality

The term "power quality" means different things to different people. To utility suppliers, power quality initially referred to the quality of the service delivered as "measured" by the consumer's ability to use the energy delivered in the desired manner. This conceptual definition included such conventional utility planning topics as voltage and frequency regulation and reliability. The end-user's definition of power quality also centers around their ability to use the delivered energy in the desired manner, but the topics considered can be much more specific and include magnitude and duration of different events as well as waveshape concerns. Fortunately, a good working definition of power quality has not been a point of contention, and most parties involved consider "power quality" to be that, which allows the user to meet their end-use goals. The working definition is not complicated by particular issues; engineers are well aware that topics from many aspects of power engineering may be important.

Power quality can be roughly broken into categories as follows:

1. Steady-state voltage magnitude and frequency,
2. Voltage sags,
3. Grounding,
4. Harmonics,
5. Voltage fluctuations and flicker,
6. Transients, and
7. Monitoring and measurement.

The remainder of this section discusses each of the major categories in turn.

38.2.1 Steady-state Voltage Frequency and Magnitude

In most areas of North America, steady-state frequency regulation is not a significant issue due to the sufficient levels of generating capacity and the strong interconnections among generating companies and control areas. In other parts of the world, and North America under extreme conditions, frequency can deviate from 1/4 to 1/2 Hz during periods of insufficient generating capacity. Under transient conditions, frequency can deviate up to 1–2 Hz.

Frequency deviations can affect power electronic equipment that use controlled switching devices unless the control signals are derived from a signal that is phase-locked with the applied voltage. In most cases, phase locks are used, or the converters consist of uncontrolled rectifiers. In either case, frequency deviations are not a major cause of problems. In most cases, frequency deviations have more impacts on conventional equipment that does not use electronics or in very inexpensive electronic devices. Clocks can run fast (or slow), motor speeds can drop (or rise) by a few revolutions per minute, etc. In most cases, these effects have minimal economic impact and are not considered a real power quality problem.

TABLE 38.1 ANSI C84.1 Voltage ranges

	Service voltage (%)	Utilization voltage (%)
Range A	114–125	108–125
Range B	110–127	104–127

Range A is for normal conditions and Range B is for emergency or short-time conditions.

Steady-state voltage regulation is a much more pronounced issue that can impact a wide range of end-use equipment. In most cases, utility supply companies do a very effective job of providing carefully regulated voltage within permissible ranges. In North America, ANSI Standard C84.1 suggests steady-state voltage ranges both at the utility service entrance and at the point of connection of end-use equipment. Furthermore, equipment manufacturers typically offer equipment that is tolerant of steady-state voltage deviations in the range of ±10%. Table 38.1 shows the voltage ranges suggested by ANSI C84.1, with specific mention of normal (Range A) and contingency (Range B) allowable voltages, expressed in percent.

Virtually all equipment, especially sensitive electronic equipment, can be effected by deviating voltage outside the ±10% range. In most cases, overvoltages above +10% lead to loss of life, usually over time; excessive overvoltages can immediately fail equipment. Undervoltages below −10% usually lead to excessive current demands, especially for equipment that has a controlled output like an adjustable speed drive controlling a motor to a constant speed/torque point. The impacts of these prolonged excessive currents can be greater voltage drop, temperature rise in conductors, etc. In the extreme, undervoltages of greater than 15–20% can cause equipment to immediately trip. In most cases, such extreme undervoltages are associated with system faults and the associated protection system. These extreme undervoltages are so important that they are classified in a power quality category of their own called voltage sags.

38.2.2 Voltage Sags

Other than improper grounding, voltage sags are probably the most problematic of all power quality problems. At this time, a number of standards-making bodies, including IEEE, ANSI, and IEC, are working on standards related to sags. In most cases, sags are generally agreed to be more severe and outside of the scope of ANSI C84.1 and they are temporary in nature due to the operation of system protection elements. Because the electrical system is a continuous electrical circuit, faults in any location will have some impact on voltages throughout the network. Of course, areas closer to the faulted area will see a greater voltage sag due to the fault than other, more (electrically) remote areas. Sags can originate anywhere in a system, but are more pronounced in utility distribution systems because of the greater exposure of low-voltage systems to the causes of short circuits.

Most utility companies implement distribution system protection in what is known as a "fuse saving" methodology. Figure 38.1 shows a typical overhead distribution system with two feeders being supplied from the same substation transformer. Each primary circuit has its own automatic circuit recloser (ACR) and shows one fused tap.

With the protection system set up based on fuse-saving methodology, any fault downstream of a fault will be cleared first by the substation recloser followed by a reclosing operation (re-energization of the circuit) 1/2–2 later. If the fault is still present, the closest fuse should blow to permanently isolate the fault. (Note that in some cases, multiple reclosing attempts are made prior to the clearing of the fuse.)

For a fault on the load side of fused tap #2 in Fig. 38.1, customers on feeder #1 will see a voltage sag determined by the system and transformer impedance at the substation. Because this impedance is typically on the same order (or larger) as the feeder circuit impedance, a sag in substation bus voltage of 50% is common. This sag will persist until feeder #2 is cleared by the recloser opening. When the recloser re-energizes the circuit, a permanent fault will still be present and the substation bus will again experience a voltage sag. Of course, any sag in substation bus voltage will be delivered directly to all customers on feeder #1, even though there is no electrical problem on that feeder. Figure 38.2 shows a possible rms voltage profile that might be supplied to the customers on feeder #1 for

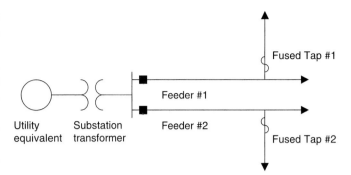

FIGURE 38.1 Overhead distribution system.

a permanent fault on the load side of fused tap #2. Only one recloser operation is shown prior to fuse clearing.

Just based on the voltage information shown in Fig. 38.2, it is impossible to tell if the end-use loads on feeder #1 will experience a problem. Equipment tolerance curves are required to assess the vulnerability of equipment to voltage deviations, including sags, and all equipment is different. Figure 38.3 shows the lower portions of two equipment tolerance curves, the (older) CBEMA and the (newer) ITIC curves for computer equipment. Most, but not all, power electronic-based equipment has a similar shape. Voltage sags with a duration that correspond to a point that is "below and to the right" of the tolerance curve will result in loss of equipment function, while sags of duration that plot "above and to the left" of the tolerance curve will not effect equipment performance. Note that only the lower portion of the curve has been shown; an upper tolerance curve also exists that is often used in transient (overvoltage) studies.

Voltage sags are probably the most common power quality problem that is "given" to the end-user by the supplying utility. However, improper equipment grounding is responsible for

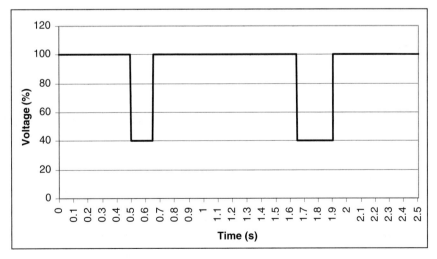

FIGURE 38.2 The rms voltage supplied to feeder #1 customers.

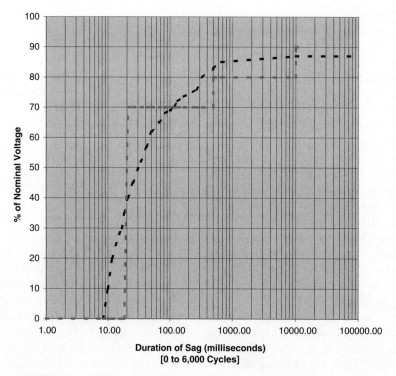

FIGURE 38.3 CBEMA (curved line) and ITIC (square shape) tolerance curves.

the vast majority of power quality problems on the customer's side of the meter.

38.2.3 Grounding

Grounding of equipment was originally conceived as a personnel safety issue. However, the presence of an electrical conductor that is at zero potential has been widely used in many power electronic and microprocessor-controlled loads. In the United States, electrical systems in residential, commercial, and industrial facilities fall under the purview of the National Electric Code (NEC) which establishes specific criteria for grounding of equipment. While it was once thought that proper grounding according to the NEC was detrimental to power quality concerns, these opinions have gradually faded over time.

From a power quality perspective, improper grounding can be considered in three broad categories

1. Ground loops,
2. Improper neutral-to-ground connections, and
3. Excessive neutral-to-ground voltage.

The ground loop problem is a significant issue when power, communications, and control signals all originate in different locations, but come together at a common electrical point. Transients induced in one location can travel through the created ground loop, damaging equipment along the way.

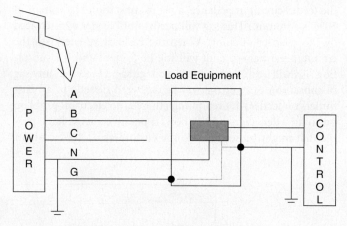

FIGURE 38.4 Powering and control ground loop.

Improper neutral-to-ground connections will create a "noisy" ground reference that may interfere with low-voltage communications and control devices. Excessive neutral-to-ground voltage may damage equipment that is not properly insulated or that has an inexpensive power supply.

Figure 38.4 shows a common wye-connected service (assumed at the terminals of a transformer) that supplies power to equipment that is also remotely monitored and controlled from another location with a separate ground reference.

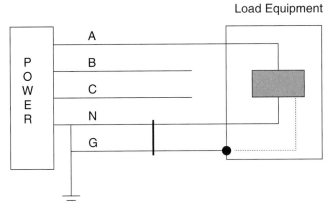

FIGURE 38.5 Improper neutral-to-ground connections.

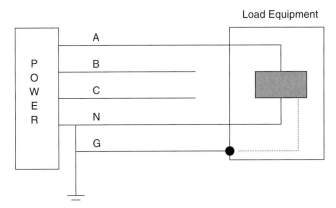

FIGURE 38.6 Excessive neutral-to-ground voltage.

For any shift in ground potential for the power circuit, often caused by lightning as shown in Fig. 38.4, potentially large currents can flow through the grounding circuits and through the sensitive electronic equipment. Such currents can easily lead to equipment damage. Situations like these are common in

1. Residential areas, if power and CATV or telephone grounds are not the same and
2. Commercial and industrial complexes consisting of multiple buildings with linking communications, computer, or control circuits, when each building has its own power service (and therefore ground).

Figure 38.5 shows an example of an improper neutral-to-ground connection, and how this connection can create power quality problems.

Load current returning in the neutral conductor will, at the point of improper connection to ground, divide between neutral and ground. This current flow in the ground conductor will produce a voltage at the load equipment, which can easily disrupt equipment operation.

Figure 38.6 shows an example of the possibility for excessive neutral-to-ground voltage and how this can lead to power quality problems.

For load equipment that produces significant voltage drop in the neutral, such as laser printers and copying machines when the thermal heating elements are on, the voltage from the neutral-to-ground reference inside the equipment can exceed several volts. In many cases, this voltage is sufficient to damage printed circuit boards, disrupt control logic, and fail components.

38.2.4 Harmonics

In most cases, power electronic equipment is considered to be the "cause" of harmonics. While switching converters of all types produce harmonics because of the non-linear relationship between the voltage and current across the switching device, harmonics are also produced by a large variety of "conventional" equipment including

1. Power generation equipment (slot harmonics),
2. Induction motors (saturated magnetics),
3. Transformers (overexcitation leading to saturation),
4. Magnetic-ballast fluorescent lamps (arcing), and
5. AC electric arc furnaces (arcing).

All these devices will cause harmonic currents to flow and some devices, actually, directly produce voltage harmonics.

Any ac current flow through any circuit at any frequency will produce a voltage drop at that same frequency. Harmonic currents, which are produced by power electronic loads, will produce voltage drops in the power supply impedance at those same harmonic frequencies. Because of this inter-relationship between current flow and voltage drop, harmonic currents created at any location will distort the voltage in the entire supply circuit.

In most cases, equipment is not overly sensitive to the direct impacts of harmonic current flow. Note, however, that equipment heating is a function of the rms value of the current, which can significantly exceed the fundamental frequency value when large harmonic components are present. It is because harmonic currents produce harmonic voltages that there is a real power quality concern.

Most equipment can operate satisfactorily as long as the voltage distortion at the equipment terminals does not exceed around 5%. Exceptions to this general rule include ripple-control systems for converters (which are impacted by small even-order harmonics) and small harmonics at sufficiently high frequency to produce multiple zero crossing in a waveform. (Note that voltage notching due to simultaneous commutation of switching devices can also create multiple zero crossings.) Such a multiple crossing scenario is shown in Fig. 38.7 and represents a 60 Hz waveform plus a 1% voltage harmonic at 3000 Hz.

Converters that have a time-limited firing signal can directly suffer from excessive voltage distortion. For a six-pulse

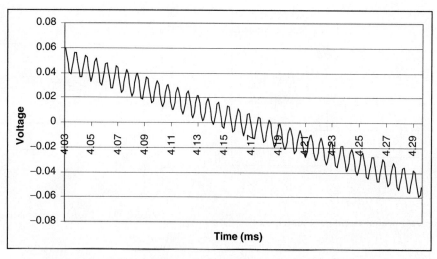

FIGURE 38.7 Multiple zero crossings.

converter, a maximum time of $1/(6\times60)$ seconds is available to turn on a switching device. Similarly, for a 12-pulse converter, a maximum of $1/(12\times60)$ is available to turn on a switching device. Considering that all switching devices have a short (but non-zero) turn-on time, manufacturers tend to design drive circuits that bring up the firing pulse for a limited amount of time. If, for example, a firing pulse is maintained for $100\,\mu s$, the device must begin conduction in that time. In situations where voltage distortion is excessive, the device to be switched could be reverse biased during the first several milliseconds of the time available for device firing during which time conduction cannot begin. If the firing signal is removed before the certain classes of switching devices are correctly biased, conduction will not begin at all. This situation, commonly called a "misfire," can lead to equipment mis-operation and failure.

Because some switching devices can conduct in both directions when the firing signal is applied (but only one direction is intended to carry appreciable current), applying the firing pulse at a time when the voltage is of the wrong polarity can destroy the device. Excessive voltage distortion can certainly lead to such a situation, and manufacturers typically design products to function only under limited-distortion conditions.

Because of the numerous potential problems with harmonic currents, standards exist for their control. The IEC goes as far as to limit the harmonic currents produced by certain individual pieces of equipment, while the IEEE takes more "system-level" point of view and prescribes limits for harmonic currents for a facility as a whole, including one of more harmonic producing loads. Harmonic standards will be further discussed in Section 38.4.

38.2.5 Voltage Fluctuations and Flicker

Voltage flicker is not directly caused by electronic loads except in the largest of applications. Voltage fluctuations, and the corresponding light flicker due to them, are usually created by large power fluctuations at frequencies less than about 30 Hz. In most applications, only

1. Large dc arc furnaces and welders,
2. Reactive power compensators, and
3. Cycloconverters

are potentially problematic. Each of these types of end-use devices can create large, low-frequency (about 30 Hz or less) variations in the system voltage, and can therefore lead to voltage flicker complaints. At this time, the IEEE prescribes a "flicker curve" based originally on research conduction by General Electric. The IEC, however, has adopted a different methodology that can consider voltage fluctuations and flicker that are more complex than those considered by the IEEE flicker curve.

Most equipment is not sensitive to the voltage fluctuations that cause flicker complaints. The change in output of incandescent lamps as viewed by human observers becomes objectionable at levels of change around 0.3%, but electronic equipment will not be affected at all. Because most utility supply companies limit voltage fluctuations, regardless of the frequency of repetition, to less than a few percent, equipment malfunction or damage due to flicker is very rare. Figures 38.8 and 38.9 show plots of single-cycle rms voltage fluctuations due to large dc welders and arc furnaces, respectively; it is clear that the magnitude of the fluctuations are well above the level that could impact equipment. The waveform in Fig. 38.8 probably would generate numerous light-flicker complaints, whereas the waveform in Fig. 38.9 probably would not. Neither would disrupt equipment.

Due to the advances in power electronics that have offered devices with higher power ratings, reactive compensation systems have been developed to compensate for voltage fluctuations by adding or removing reactive power from

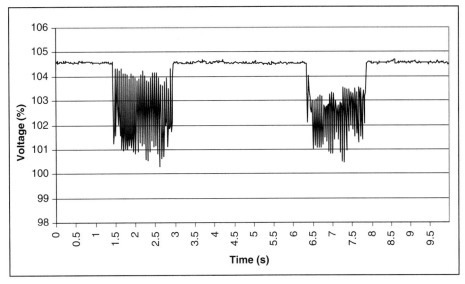

FIGURE 38.8 Single-cycle rms voltage fluctuations due to a large dc welder.

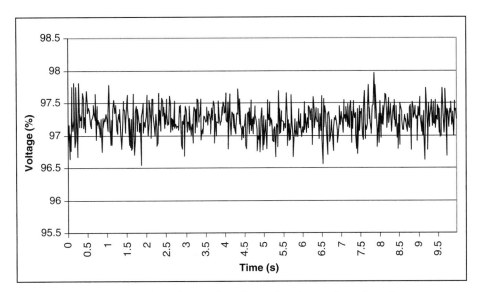

FIGURE 38.9 Single-cycle rms voltage fluctuations due to a large dc arc furnace.

the supply circuit. These devices have allowed large flicker-producing loads like arc furnaces to be served from utility circuits that, without the compensator, could not serve the load. However, because the compensators can so directly impact system voltage, they can create flicker problems if they are not properly applied and controlled.

38.2.6 Transients

Transients, especially in the voltage supply, can create numerous power quality problems. The major sources of transients are

1. Lightning,
2. Utility circuit switching and fault clearing,
3. Capacitor switching, and
4. Load switching.

Lightning events can create the most severe overvoltages, but these transients decay rapidly. A typical lightning transient has decayed to zero in a few hundred microseconds, but it can reach a peak magnitude of several hundred percent if not controlled with surge suppression devices. Other categories of transients associated with power system switching are much smaller in magnitude (typically less than 200%), but last in the order of several hundred milliseconds. Considering the energy available in a transient, therefore, there is a considerable overlap in the range of severity of lightning and switching transients. It is the available energy that typically determines

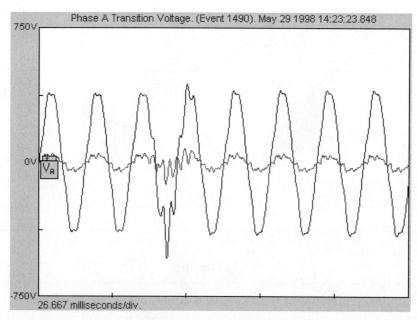

FIGURE 38.10 Capacitor switching transient.

whether or not an equipment will be affected or damaged. Figure 38.10 shows a capacitor switching transient on a low-voltage (480 V) circuit. The magnitude and duration of the event are quite clear.

Transients such as those shown in Fig. 38.10 are generally sufficient to cause nuisance trips of electronic loads like adjustable speed drives. For these types of loads, the protection system settings are usually very tight due to the use of sensitive switching devices. Overcurrent and overvoltage settings of 120% are not uncommon. For a transient similar to that shown in Fig. 38.10, there is sufficient overvoltage for very large currents to flow through any conducting switching device to the drive's dc bus. The device's protection system see these overcurrents as a fault, and trip the drive. Similarly, the overvoltage at the terminals can be passed through to the dc bus and accumulate, where the drive may trip due to overvoltge on the dc bus.

38.2.7 Monitoring and Measurement

To consider or be able to diagnose power quality related problems, it is imperative to be able to measure various power quality parameters. Several different categories of monitoring and measurement equipment exist for these purposes, with costs ranging from a few hundred dollars to $10,000–20,000 for fully equipped disturbance analyzer.

The most basic category of power quality measurement tool is the handheld voltmeter. It is important that the voltmeter be a true-rms meter, or erroneous readings will be obtained that incorrectly suggest low or high voltage when harmonics are present in the signal. It is especially important to have true rms capability when measuring currents; voltage distortion is not typically severe enough to create large errors in the readings of non-true rms meters. Virtually all major measurement equipment vendors offer true rms meters, with the costs starting around $100.

The next step up from the basic voltmeter is a class of instruments that have come to be called "power quality analyzers." These instruments are handheld and battery powered. These instruments can measure and display various power quality indices, especially those that relate to harmonics like THD, etc. and can also display the input waveform. Newer models feature 20 MHz (and higher) bandwidth oscilloscopes, inrush measurements, time trending, and other useful features. Manufacturers such as Fluke, Dranetz, BMI, and Tektronix offer these types of instruments for around $2000.

In most power quality investigations, it is not possible to use handheld equipment to collect sufficient data to solve the problem. Most power quality problems are intermittent in nature, so some type of long-term monitoring is usually required. Various recorders are available that can measure and record voltage, current, and power over user-defined time period. Such recorders typically cost in the order of $3000–10,000. More advanced long-term monitors can record numerous power quality events and indices, including transients, harmonics, sags, flicker, etc. These devices, often called "line disturbance analyzers", typically cost between $10,000 and $20,000.

It is important to use the right instrument to measure the phenomenon that is suspected of causing the problem. Some meters record specific parameters, while others are more flexible. With this flexibility comes an increased learning curve for the user, so it is important to spend time on them before going out to monitor, to make sure all aspects and features of the equipment are understood.

It is equally important to measure in the correct location. The best place to measure power quality events is at the equipment terminals that is experiencing problems. With experience, an engineer can evaluate the waveforms recorded at the equipment terminals and correlate them to events and causes elsewhere in the power system. In general, the farther away from the equipment location the monitoring takes place, the more difficult is to diagnose a problem.

38.3 Reactive Power and Harmonic Compensation

The previous section specifically identified harmonics as a potential power quality problem. In that discussion, it was pointed out that non-linear loads such as adjustable speed drives create harmonic currents, and when these currents flow through the impedances of the power supply system, harmonic voltages are produced. While harmonic currents have secondary (in most cases) negative impacts, it is these harmonic voltages that can be supplied to other load equipment (and disrupt operation) that are of primary concern. Having parallel or series resonant conditions present in the electrical supply system can quickly exacerbate the problem.

TABLE 38.2 Typical harmonic spectra of load equipment

Harmonic no.	Switched-mode power supply	Fluorescent lamp	Six-pulse dc drive	Six-pulse ac drive
1	100.0	100.0	100.0	100.0
2	0.7	1.0	4.8	1.1
3	91.9	12.6	1.2	3.9
4	1.0	0.3	1.5	0.5
5	80.2	1.8	33.6	82.8
6	1.3	0.1	0.0	1.7
7	64.8	0.7	1.6	77.5
8	1.4	0.1	1.7	1.2
9	47.7	0.5	0.4	7.6
10	1.0	0.1	0.3	0.7
11	30.8	0.2	8.7	46.3
12	0.8	0.1	0.0	1.0
13	16.0	0.2	1.2	41.2
14	0.4	0.0	1.3	0.2
15	5.0	0.1	0.3	5.7
16	0.1	0.1	0.2	0.3
17	4.0	0.2	4.5	14.2
18	0.3	0.1	0.0	0.4
19	7.2	0.1	1.3	9.7
20	0.4	0.2	1.1	0.4
21	7.7	0.2	0.3	2.3
22	0.4	0.1	0.3	0.5
23	6.2	0.1	2.8	1.5
24	0.2	0.0	0.0	0.5
25	4.0	0.1	1.2	2.5

38.3.1 Typical Harmonics Produced by Equipment

In theory, most harmonic currents follow the "$1/n$" rule where n is the harmonic order ($180\,\text{Hz} = 3 \times 60$; $n = 3$). Also in theory, most harmonic currents in three-phase systems are not integer multiples of three. Finally, in theory, harmonic currents are not usually even-order integer multiples of the fundamental. In practice, none of these statements are completely true and using any of them "exactly" could lead to either over- or under-conservatism depending on many factors. Consider the following examples:

1. Switched-mode power supplies, such as found in televisions, personal computers, etc. often produce a third-harmonic current that is nearly as large (80–90%) as the fundamental frequency component.
2. Unbalance in voltages supplied to a three-phase converter load will lead to the production of even-order harmonics and, in some extreme cases, establish a positive feedback situation leading to stability problems.
3. Arcing loads, particularly in the steel industry, generate significant harmonics of all orders, including harmonics that are not integer multiples of the power frequency.
4. Cycloconverters produce dominant harmonics that are integer multiples of the power frequency, but they also produce sideband components at frequencies that are not integer multiples of the power frequency. In some control schemes, the amplitudes of the sideband components can reach damaging levels.

Table 38.2 gives the magnitudes, in percent of fundamental, of the first 25 (integer) harmonics for a single-phase switched-mode power supply, a single-phase fluorescent lamp, a three-phase (six-pulse) dc drive, and a three-phase (six-pulse, no input choke) ac drive. Together, these load types represent the range of harmonic sources in power systems. Note that seemingly minor changes in parameter values and control methods can have significant impacts on harmonic current generation; the values given here are on the conservative side of "typical."

38.3.2 Resonance

Considering only the harmonic current spectra given in Table 38.2, it would appear that a large number of harmonic-related power quality problems are on the verge of appearing. In reality, most current drawn by many residential, commercial, and industrial customers is of the fundamental frequency;

the amplitudes of the individual harmonic currents, in percent of the *total* fundamental current, are often much less than that shown in Table 38.2. For this reason, end-use locations employing non-linear loads often do not lead directly to significant voltage distortion problems. A parallel resonance in the power supply system, however, changes the picture entirely.

Series and parallel resonance exist in any ac power supply network that contains inductance(s) and capacitance(s). The following simple (but usable) definitions apply:

1. Series resonance occurs when the impedance to current flow at a certain frequency (or frequencies) is low.
2. Parallel resonance occurs when the impedance to current flow at a certain frequency (or frequencies) is high.

For a given harmonic-producing load that generates harmonics at frequencies that correspond to parallel resonance in the supply system, even small currents at the resonant frequencies can produce excessive voltages at these same frequencies. The principle of series resonance, however, is usually exploited to reduce harmonics in power systems by providing intentionally low impedance paths to ground.

In many cases, end-users will install power factor correction capacitors in order to minimize reactive power charges by the supply utility. In most cases, these capacitors are located on the customer's low-voltage supply buses and are therefore in parallel with the service transformer. In most cases where the power factor correction capacitors are sized to provide a net power factor (at the service entrance) to 0.85(lag)–1.0, the parallel resonance occurs somewhere between the fifth and nineth harmonic. Considering Table 38.2, it is apparent that a large number of loads produce harmonics at these frequencies, and the amplitudes can be significant. Even a small increase in impedance at these frequencies due to resonance can lead to unacceptable voltages being produced at these same frequencies. Figure 38.11 shows and example plot of impedance looking into a utility supply system when a typically-sized capacitor bank has been installed to improve overall plant power factor.

From Fig. 38.11, it is clear that 1.0 A at the 9th harmonic will produce at least 150 times more voltage drop than would be produced by the same 1.0 A if it were at the fundamental frequency. A look back at the Table 38.2 shows the clear potential for problems. Fortunately, the series resonance principle can be used to provide a low-impedance path to ground for the harmonic currents and thus reduce the potential for problematic voltage distortion.

Since capacitors are required to produce parallel resonance, it is often a "cheap fix" to slightly modify the capacitor to include a properly sized series reactor and create filter. This filter approach, designed based on the series resonance concept, is usually the most cost-effective means to control harmonic voltage distortion.

38.3.3 Harmonic Filters

Harmonic filters come in many "shapes and sizes." In general, harmonic filters are "shunt" filters because they are connected in parallel with the power system and provide low impedance paths to ground for currents at one or more harmonic frequencies. For power applications, shunt filters are almost always more economical than series filters (like those found in many communications applications) for the following reasons:

1. Series components must be rated for the full current, including the power frequency component. Such a requirement leads to larger component sizes and therefore costs.

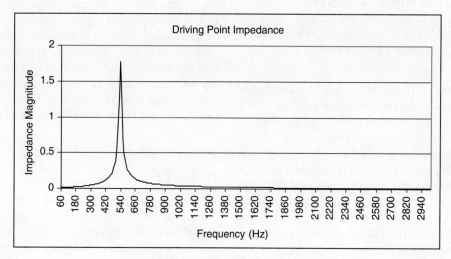

FIGURE 38.11 Driving point impedance.

2. Shunt filter components generally must be rated for only part of the system voltage (usually with respect to ground). Such requirements lead to smaller component sizes and therefore costs.

Shunt filters are designed (or can be purchased) in three basic categories as follows:

1. Single-tuned filters,
2. Multiple- (usually limited to double) tuned filters, and
3. Damped filters (of first-, second-, or third- order, or newer "c-type").

The single- and double-tuned filters are usually used to filter specific frequencies, while the damped filters are used to filter a wide range of frequencies. In applications involving small harmonic producing loads, it is often possible to use one single-tuned filter (usually tuned near the fifth harmonic) to eliminate problematic harmonic currents. In large applications, like those associated with arc furnaces, multiple tuned filters and a damped filter are often used. Equivalent circuits for single- and double-tuned filters are shown in Fig. 38.12. Equivalent circuits for first, second, third, and "c-type" damped filters are shown in Fig. 38.13.

A plot of the impedance as a function of frequency for a single-tuned filter is shown in Fig. 38.14. The filter is based on a 480 V, 300 kvar (three-phase) capacitor bank and is tuned to the 4.7th harmonic with a quality factor, Q, of 150. Note that the quality factor is a measure of the "sharpness" of the tuning and is defined as X/R where X is the inductive reactance for the filter inductor at the (undamped) resonant frequency; typically $50 < Q < 150$ for tuned filters.

A plot of impedance as a function of frequency for a second-order damped filter is shown in Fig. 38.15. This filter is based on a 480 V, 300 kvar capacitor bank and is tuned to the 12th harmonic. The quality factor is chosen to be 1.5. Note that the quality factor for damped filters is the inverse of the definition

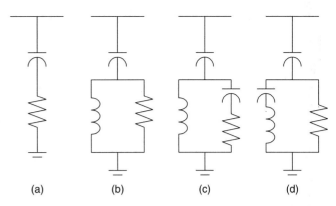

FIGURE 38.13 Damped filters: (a) first-order; (b) second-order; (c) third-order; and (d) C-type.

for tuned filters; $Q = R/X$ where X is the inductive reactance at the (undamped) resonant frequency. Typically, $0.5 < Q < 1.5$ for damped filters.

In most cases, it is common to tune single-tuned filter banks to slightly below (typically around 5%) the frequency of the harmonic to be removed. The reasons for this practice are as follows:

1. For a low-resistance series resonance filter that is exactly tuned to a harmonic frequency, the filter bank will act as a sink to all harmonics (at the tuned frequency) in the power system, regardless of their source(s). This action can quickly overload the filter.
2. All electrical components have some non-zero temperature coefficient, and capacitors are the most temperature sensitive component in a tuned filter. Because most capacitors have a negative temperature coefficient (capacitance decreases and therefore tuned frequency increases with temperature), tuning slightly lower than the desired frequency is desirable.

Damped filters are typically used to control higher-order harmonics as a group. In general, damped filters are tuned in between the corresponding pairs of harmonics (11th and 13th, 17th and 19th, etc.) to provide the maximum harmonic reduction at those frequencies while continuing to serve as a (not quite as effective) filter bank for frequencies higher than the tuned frequency. Because damped filters have significantly higher resistance than single- or double-tuned filters, they are usually not used to filter harmonics near the power frequency so that filter losses can be maintained at low values.

38.4 IEEE Standards

The IEEE has produced numerous standards relating to the various power quality phenomena discussed in Section 38.2. Of these many standards, the one most appropriate to power

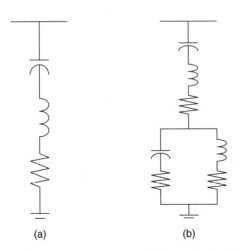

FIGURE 38.12 Harmonic filters: (a) single-tuned and (b) double-tuned.

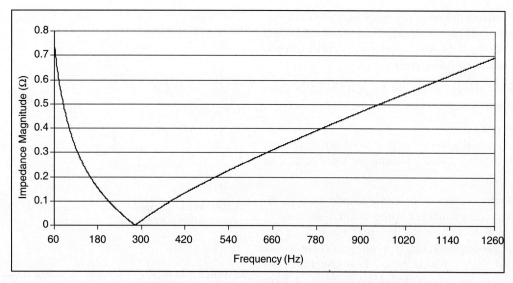

FIGURE 38.14 Single-tuned filter frequency response.

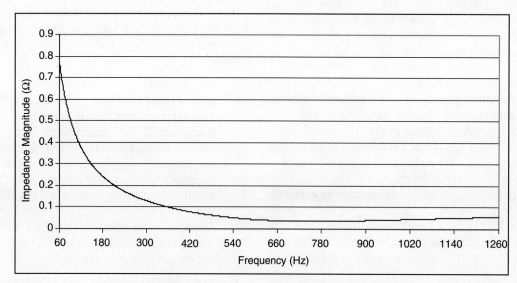

FIGURE 38.15 Second-order damped filter frequency response.

electronic equipment is IEEE Standard 519-1992. This standard is actually a "recommended practice," which means that the information contained within represents a set of "recommendations," rather than a set of "requirements." In practice, this seemingly small difference in wording means that the harmonic limits prescribed are merely suggested values; they are not (nor were they ever intended) to be absolute limits that could not be exceeded.

Harmonic control via IEEE 519-1992 is based on the concept that all parties use and pay for the public power supply network. Due to the nature of utility company rate structures, end-users that have a higher demand pay more of the total infrastructure cost through higher demand charges. In this light, IEEE 519-1992 allows these larger end-users to produce a greater percentage of the maximum level of harmonics that can be absorbed by the supply utility before voltage distortion problems are encountered. Because the ability for a harmonic source to produce voltage distortion is directly dependent on the supply system impedance upon the point where distortion is to be evaluated, it is necessary to consider both

1. The size of the end-user and
2. The strength (impedance) of the system

at the same time in order to establish meaningful limits for harmonic emissions. Furthermore, it is necessary to establish tighter limits in higher voltage supply systems than lower voltage because the potential for more widespread problems associated with high-voltage portions of the supply system.

Unlike limits set forth in various IEC Standards, IEEE 519-1992 established the "point of common coupling," or PCC as the point at which harmonic limits shall be evaluated. In most cases (recall that IEEE 519-1992 is a "recommended practice"), this point will be:

1. In the supply system owned by the utility company,
2. The closest electrical point to the end-user's premises, and
3. As in (2), but further restricted to points where other customers are (or could be in the future) provided with electric service.

In this context, IEEE 519-1992 harmonic limits are designed for an entire facility and should not be applied to individual pieces of equipment without great care.

Because the PCC is used to evaluate harmonic limit compliance, the system strength (impedance) is measured at this point and is described in terms of available (three-phase) short-circuit current. Also, the end-user's maximum average demand current is evaluated at this point. Maximum demand is evaluated based on one of the following:

1. The maximum value of the 15 or 30 minute average demand, usually considering the previous 12 month's billing history or
2. The connected kVA or horsepower, perhaps multiplied by a diversity factor.

The ratio of I_{SC} to I_L, where I_{SC} is the available fault current and I_L is the maximum demand current, implements the founding concept of IEEE 519-1992: larger end-users can create more harmonic currents, but the specific level of current that any end-user may produce is dependent on the strength of the system at the PCC. Tables 38.3–38.5 show the harmonic current limits in IEEE 519-1992 for various voltage levels.

In general, it is the responsibility of the end-user to insure that their net harmonic currents at the PCC do not exceed the values given in the appropriate table. In some cases, usually associated with parallel resonance involving a utility-owned capacitor bank, it is possible that all customers will be within the prescribed limits, but voltage distortion problems exist. In these cases, it is generally the responsibility of the supply utility to insure that excessive voltage distortion levels are not present. The harmonic voltage limits that are recommended for utility companies are given in Table 38.6.

TABLE 38.3 Current distortion limits for general distribution systems, 120 V–69 kV

	Maximum harmonic current distortion in percent of I_L					
I_{SC}/I_L	<11	$11 \leq h < 17$	$17 \leq h < 23$	$23 \leq h < 35$	$h \geq 35$	TDD
<20[1]	4.0	2.0	1.5	0.6	0.3	5.0
20<50	7.0	3.5	2.5	1.0	0.5	8.0
50<100	10.0	4.5	4.0	1.5	0.7	12.0
100<1000	12.0	5.5	5.0	2.0	1.0	15.0
≥1000	15.0	7.0	6.0	2.5	1.4	20.0

Individual harmonic order h (odd harmonics).
Even harmonics are limited to 25% of the odd harmonic limits above.
Current distortions that result in a dc offset are not allowed.
[1] All power generation equipment is limited to these values of current distortion regardless of the value of I_{SC}/I_L.

TABLE 38.4 Current distortion limits for general subtransmission systems, 69.001–161 kV

	Maximum harmonic current distortion in percent of I_L					
I_{SC}/I_L	<11	$11 \leq h < 17$	$17 \leq h < 23$	$23 \leq h < 35$	$h \geq 35$	TDD
<20[1]	2.0	1.0	0.75	0.3	0.15	2.5
20<50	3.5	1.75	1.25	0.5	0.25	4.0
50<100	5.0	2.25	2.0	0.75	0.35	6.0
100<1000	6.0	2.75	2.5	1.0	0.5	7.5
≥1000	7.5	3.5	3.0	1.25	0.7	10.0

Individual harmonic order h (odd harmonics).
Even harmonics are limited to 25% of the odd harmonic limits above.
Current distortions that result in a dc offset are not allowed.
[1] All power generation equipment is limited to these values of current distortion regardless of the value of I_{SC}/I_L.

TABLE 38.5 Current distortion limits for general transmission systems, >161 kV

	Maximum harmonic current distortion in percent of I_L					
I_{SC}/I_L	<11	$11 \leq h < 17$	$17 \leq h < 23$	$23 \leq h < 35$	$h \geq 35$	TDD
<50[1]	2.0	1.0	0.75	0.3	0.15	2.5
≥50	3.0	1.5	1.15	0.45	0.22	3.75

Individual harmonic order h (odd harmonics).
Even harmonics are limited to 25% of the odd harmonic limits above.
Current distortions that result in a dc offset are not allowed.
[1] All power generation equipment is limited to these values of current distortion regardless of the value of I_{SC}/I_L.

TABLE 38.6 Voltage distortion limits

Bus voltage at PCC	Individual harmonic magnitude (%)	Total voltage distortion (THD in %)
≤69 kV	3.0	5.0
69.001–161 kV	1.5	2.5
>161 kV	1.0	1.5

38.5 Conclusions

In this chapter, various power quality phenomena have been described, with particular focus on the implications on power electronic converters and equipment. While one popular opinion "blames" power electronic equipment for "causing" most power quality problems, it is quite clear that power electronic converter systems can play an equally-important role in reducing the impact of power quality problems. While it is true that power electronic converters and systems are the major cause of harmonic-related problems, the application (in general terms) of IEEE 519-1992 limits for current and voltage harmonics has led to the reduction, elimination, and prevention of most harmonics problems. Other power quality phenomena, like grounding, sags, and voltage flicker, are most often completely unrelated to power electronic systems. In reality, advances in power electronic circuits and control algorithms are making it more possible to control these events and minimize the financial impacts of the majority of power quality problems.

Further Reading

1. ANSI Std C84.1-1995, *Electric Power Systems and Equipment – Voltage Ratings (60 Hz)*.
2. IEEE Std 493-1997, *IEEE Recommended Practice for the Design of Reliable Industrial and Commercial Power Systems* (IEEE Gold Book), © IEEE 1998.
3. IEEE Std 142-1991, *IEEE Recommended Practice for Grounding of Industrial and Commercial Power Systems* (IEEE Green Book), © IEEE 1992.
4. National Fire Protection Association 70-1999, *National Electrical Code*, 1999.
5. IEEE Std 519-1992, *IEEE Recommended Practices and Requirements for Harmonic Control in Electrical Power Systems*, © IEEE 1993.
6. IEEE Std 141-1993, *IEEE Recommended Practice for Electric Power Distribution for Industrial Plants* (IEEE Red Book), © IEEE 1994.
7. J.L. Gutierrez Iglesias, Chairman UIE Working Group WG2, "Part 5: Flicker and Voltage Fluctuations," 1999.
8. International Electrotechnical Commission, IEC Technical Report 61000-4-15, *Flickermeter – Functional and Design Specifications*, 1997.
9. IEEE Std 1100-1999, *IEEE Recommended Practice for Powering and Grounding Electronic Equipment* (IEEE Emerald Book), © IEEE 1999.
10. IEEE Std 1159-1995, *IEEE Recommended Practice for Monitoring Electric Power Quality*, © IEEE 1995.
11. E.W. Kimbark, *Direct Current Transmission, Volume I*. © John Wiley & Sons, 1948.

39
Active Filters

Luis Morán
*Electrical Engineering Dept.,
Universidad de Concepción
Concepción, Chile*

Juan Dixon
*Electrical Engineering Dept.,
Universidad Católica de Chile
Santiago, Chile*

39.1 Introduction 1067
39.2 Types of Active Power Filters 1068
39.3 Shunt Active Power Filters 1069
    39.3.1 Power Circuit Topologies • 39.3.2 Control Scheme • 39.3.3 Power Circuit Design
    • 39.3.4 Technical Specifications
39.4 Series Active Power Filters 1085
    39.4.1 Power Circuit Structure • 39.4.2 Principles of Operation • 39.4.3 Power Circuit Design
    • 39.4.4 Control Issues • 39.4.5 Control Circuit Implementation • 39.4.6 Experimental
    Results
39.5 Hybrid Active Power Filters 1094
    39.5.1 Principles of Operation • 39.5.2 The Hybrid Filter Compensation Performance
    • 39.5.3 Experimental Results
    Further Reading 1101

39.1 Introduction

The growing number of power electronics-based equipment has produced an important impact on the quality of electric power supply. Both high power industrial loads and domestic loads cause harmonics in the network voltages. At the same time, much of the equipment causing the disturbances is quite sensitive to deviations from the ideal sinusoidal line voltage. Therefore, power quality problems may originate in the system or may be caused by the consumer itself. Moreover, in the last years the growing concern related to power quality comes from:

- Consumers that are becoming increasingly aware of the power quality issues and being more informed about the consequences of harmonics, interruptions, sags, switching transients, etc. Motivated by deregulation, they are challenging the energy suppliers to improve the quality of the power delivered.
- The proliferation of load equipment with microprocessor-based controllers and power electronic devices which are sensitive to many types of power quality disturbances.
- Emphasis on increasing overall process productivity, which has led to the installation of high-efficiency equipment, such as adjustable speed drives and power factor correction equipment. This in turn has resulted in an increase in harmonics injected into the power system, causing concern about their impact on the system behavior.

For an increasing number of applications, conventional equipment is proving insufficient for mitigation of power quality problems. Harmonic distortion has traditionally been dealt with the use of passive LC filters. However, the application of passive filters for harmonic reduction may result in parallel resonances with the network impedance, over compensation of reactive power at fundamental frequency, and poor flexibility for dynamic compensation of different frequency harmonic components.

The increased severity of power quality in power networks has attracted the attention of power engineers to develop dynamic and adjustable solutions to the power quality problems. Such equipment, generally known as active filters, are also called active power line conditioners, and are able to compensate current and voltage harmonics, reactive power, regulate terminal voltage, suppress flicker, and to improve voltage balance in three-phase systems. The advantage of active filtering is that it automatically adapts to changes in the network and load fluctuations. They can compensate for several harmonic orders, and are not affected by major changes in network characteristics, eliminating the risk of resonance between the filter and network impedance. Another plus is that they take up very little space compared with traditional passive compensators.

39.2 Types of Active Power Filters

The technology of active power filter has been developed during the past two decades reaching maturity for harmonics compensation, reactive power, and voltage balance in ac power networks. All active power filters are developed with pulse width modulated (PWM) converters (current-source or voltage-source inverters). The current-fed PWM inverter bridge structure behaves as a non-sinusoidal current source to meet the harmonic current requirement of the non-linear load. It has a self-supported dc reactor that ensures the continuous circulation of the dc current. They present good reliability, but have important losses and require higher values of parallel capacitor filters at the ac terminals to remove unwanted current harmonics. Moreover, they cannot be used in multilevel or multistep modes configurations to allow compensation in higher power ratings.

The other converter used in active power filter topologies is the PWM voltage-source inverter (PWM-VSI). This converter is more convenient for active power filtering applications since it is lighter, cheaper, and expandable to multilevel and multistep versions, to improve its performance for high power rating compensation with lower switching frequencies. The PWM-VSI has to be connected to the ac mains through coupling reactors. An electrolytic capacitor keeps a dc voltage constant and ripple free.

Active power filters can be classified based on the type of converter, topology, control scheme, and compensation characteristics. The most popular classification is based on the topology such as shunt, series, or hybrid. The hybrid configuration is a combination of passive and active compensation. The different active power filter topologies are shown in Fig. 39.1.

Shunt active power filters (Fig. 39.1a) are widely used to compensate current harmonics, reactive power, and load current unbalanced. It can also be used as a static var generator in power system networks for stabilizing and improving voltage profile. Series active power filters (Fig. 39.1b) is connected

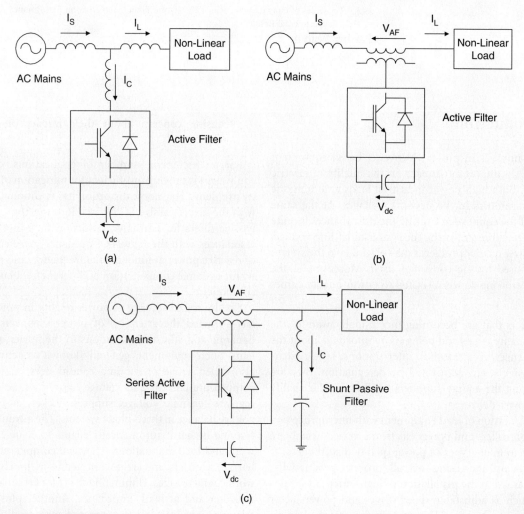

FIGURE 39.1 Active power filter topologies implemented with PWM-VSI: (a) shunt active power filter; (b) series active power filter; and (c) hybrid active power filter.

before the load in series with the ac mains, through a coupling transformer to eliminate voltage harmonics and to balance and regulate the terminal voltage of the load or line. The hybrid configuration is a combination of series active filter and passive shunt filter (Fig. 39.1c). This topology is very convenient for the compensation of high power systems, because the rated power of the active filter is significantly reduced (about 10% of the load size), since the major part of the hybrid filter consists of the passive shunt LC filter used to compensate lower-order current harmonics and reactive power at fundamental frequency.

Due to the operation constraint, shunt or series active power filters can compensate only specific power quality problems. Therefore, the selection of the type of active power filter to improve power quality depends on the source of the problem as can be seen in Table 39.1.

The principles of operation of shunt, series, and hybrid active power filters are described in the following sections.

TABLE 39.1 Active filter solutions to power quality problems

Active filter connection	Load on ac supply	AC supply on load
Shunt	Current harmonic filtering	
	Reactive current compensation	
	Current unbalance	
	Voltage flicker	
Series	Current harmonic filtering	Voltage sag/swell
	Reactive current compensation	Voltage unbalance
	Current unbalance	Voltage distortion
	Voltage flicker	Voltage interruption
	Voltage unbalance	Voltage flicker
		Voltage notching

39.3 Shunt Active Power Filters

Shunt active power filters compensate current harmonics by injecting equal but opposite harmonic compensating current. In this case, the shunt active power filter operates as a current source injecting the harmonic components generated by the load but phase shifted by 180°. As a result, components of harmonic currents contained in the load current are cancelled by the effect of the active filter, and the source current remains sinusoidal and in phase with the respective phase-to-neutral voltage. This principle is applicable to any type of load considered as an harmonic source. Moreover, with an appropriate control scheme, the active power filter can also compensate the load power factor. In this way, the power distribution system sees the non-linear load and the active power filter as an ideal resistor. The compensation characteristics of the shunt active power filter is shown in Fig. 39.2.

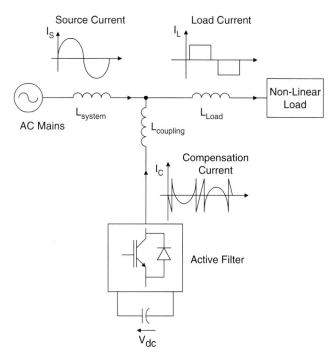

FIGURE 39.2 Compensation characteristics of a shunt active power filter.

39.3.1 Power Circuit Topologies

Shunt active power filters are normally implemented with PWM-VSIs. In this type of application, the PWM-VSI operates as a current-controlled voltage source. Traditionally, two level PWM-VSI have been used to implement such system connected to the ac bus through a transformer. This type of configuration is aimed to compensate non-linear load rated in the medium power range (hundreds of kVA) due to semiconductors rated values limitations. However, in the last years multilevel PWM-VSIs have been proposed to develop active power filters for medium voltage and higher rated power applications. Also, active power filters implemented with multiple VSIs connected in parallel to a dc bus but in series through a transformer or in cascade has been proposed in the technical literature. The different power circuit topologies are shown in Fig. 39.3.

The use of VSI connected in cascade is an interesting alternative to compensate high power non-linear loads. The use of two PWM-VSI with different rated power allows the use of different switching frequencies, reducing switching stresses, and commutation losses in the overall compensation system. The power circuit configuration of such a system is shown in Fig. 39.4.

The VSI connected closer to the load compensates for the displacement power factor and lower frequency current harmonic components (Fig. 39.5b), while the second compensates only high-frequency current harmonic components. The first converter requires higher rated power than the second and

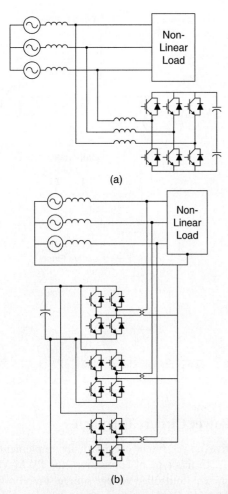

FIGURE 39.3 Shunt active power filter topologies implemented with PWM-VSIs: (a) a three-phase PWM unit and (b) three single-phase units in parallel to a common dc bus.

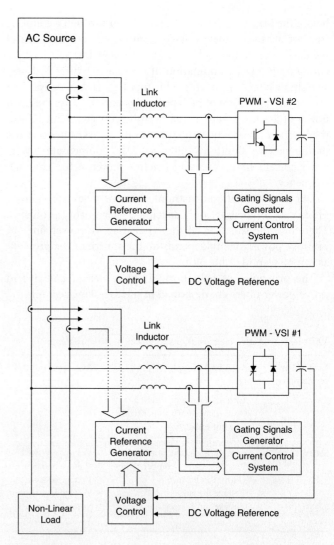

FIGURE 39.4 A shunt active power filter implemented with two PWM-VSI connected in cascade.

can operate at lower switching frequency. The compensation characteristics of the cascade shunt active power filter is shown in Fig. 39.5.

In recent years, there has been an increasing interest in using multilevel inverters for high power energy conversion, especially for drives and reactive power compensation. The use of neutral-point-clamped (NPC) inverters (Fig. 39.6) allows equal voltage shearing of the series-connected semiconductors in each phase. Basically, multilevel inverters have been developed for applications in medium voltage ac motor drives and static var compensation. For these types of applications, the output voltage of the multilevel inverter must be able to generate an almost sinusoidal output current. In order to generate a near sinusoidal output current, the output voltage should not contain low-frequency harmonic components.

However, for active power filter applications, the three-level NPC inverter output voltage must be able to generate an output current that follows the respective reference current containing the harmonic and reactive component required by the load. Current and voltage waveforms obtained for a shunt active power filter implemented with a three-level NPC-VSI are shown in Fig. 39.7.

39.3.2 Control Scheme

The control scheme of a shunt active power filter must calculate the current reference waveform for each phase of the inverter, maintain the dc voltage constant, and generate the inverter gating signals. The block diagram of the control scheme of a shunt active power filter is shown in Fig. 39.8.

The current reference circuit generates the reference currents required to compensate the load current harmonics and reactive power, and also try to maintain constant the dc voltage across the electrolytic capacitors. There are many possibilities to implement this type of control, and the most popular of them will be explained in this chapter. Also, the compensation

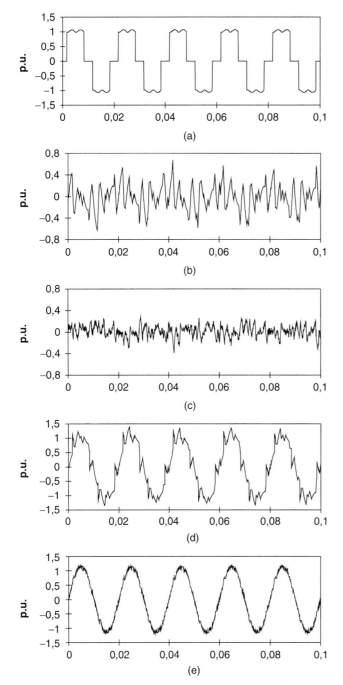

FIGURE 39.5 Current waveforms of active power filter implemented with two PWM-VSI in cascade: (a) load current waveform; (b) current waveform generated by PWM-VSI no. 1; (c) current waveform generated by PWM-VSI no. 2; (d) power system current waveform between the two inverters ($THD_i = 13.7\%$); and (e) power system current waveform ($THD_i = 4.5\%$).

effectiveness of an active power filter depends on its ability to follow with a minimum error and time delay, the reference signal calculated to compensate the distorted load current. Finally, the dc voltage control unit must keep the total dc bus voltage constant and equal to a given reference value. The dc voltage control is achieved by adjusting the small amount of real power absorbed by the inverter. This small amount of real power is adjusted by changing the amplitude of the fundamental component of the reference current.

39.3.2.1 Current Reference Generation

There are many possibilities to determine the reference current required to compensate the non-linear load. Normally, shunt active power filters are used to compensate the displacement power factor and low-frequency current harmonics generated by non-linear loads. One alternative to determine the current reference required by the VSI is the use of the instantaneous reactive power theory, proposed by Akagi [1], the other one is to obtain current components in *d–q* or synchronous reference frame [2], and the third one to force the system line current to follow a perfectly sinusoidal template in phase with the respective phase-to-neutral voltage.

39.3.2.1.1 Instantaneous Reactive Power Theory This concept is very popular and useful for this type of application, and basically consists of a variable transformation from the *a, b, c,* reference frame of the instantaneous power, voltage, and current signals to the α, β reference frame. The transformation equations from the *a, b, c,* reference frame to the α, β coordinates can be derived from the phasor diagram shown in Fig. 39.9.

The instantaneous values of voltages and currents in the α, β coordinates can be obtained from the following equations:

$$\begin{bmatrix} v_\alpha \\ v_\beta \end{bmatrix} = [A] \cdot \begin{bmatrix} v_a \\ v_b \\ v_c \end{bmatrix} \quad \begin{bmatrix} i_\alpha \\ i_\beta \end{bmatrix} = [A] \cdot \begin{bmatrix} i_a \\ i_b \\ i_c \end{bmatrix} \quad (39.1)$$

where *A* is the transformation matrix, derived from Fig. 39.9 and is equal to

$$[A] = \sqrt{\frac{2}{3}} \begin{bmatrix} 1 & -1/2 & -1/2 \\ 0 & \sqrt{3}/2 & -\sqrt{3}/2 \end{bmatrix} \quad (39.2)$$

This transformation is valid if and only if $v_a(t) + v_b(t) + v_c(t)$ is equal to zero, and also if the voltages are balanced and sinusoidal. The instantaneous active and reactive power in the α, β coordinates are calculated with the following expressions:

$$p(t) = v_\alpha(t) \cdot i_\alpha(t) + v_\beta(t) \cdot i_\beta(t) \quad (39.3)$$

$$q(t) = -v_\alpha(t) \cdot i_\beta(t) + v_\beta(t) \cdot i_\alpha(t) \quad (39.4)$$

It is evident that *p(t)* becomes equal to the conventional instantaneous real power defined in the *a, b, c* reference frame.

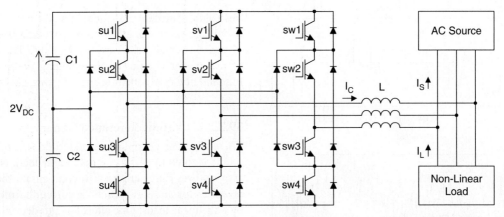

FIGURE 39.6 A shunt active power filter implemented with a three-level neutral point-clamped VSI.

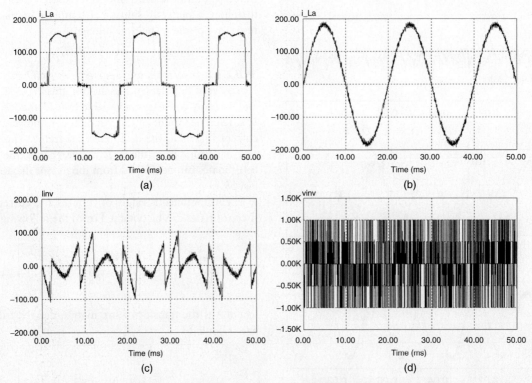

FIGURE 39.7 Current and voltage waveforms for a shunt active power filter implemented with a three-level NPC-VSI: (a) load current; (b) compensated system current ($THD = 3.5\%$); (c) current generated by the shunt active power filter; and (d) inverter output voltage.

However, in order to define the instantaneous reactive power, Akagi introduces a new instantaneous space vector defined by expression (39.4) or by the vector equation:

$$q = v_\alpha x i_\beta + v_\beta x i_\alpha \qquad (39.5)$$

The vector q is perpendicular to the plane of α, β coordinates, to be faced in compliance with a right-hand rule, v_α is perpendicular to i_β, and v_β is perpendicular to i_α. The physical meaning of the vector q is not "instantaneous power" because of the product of the voltage in one phase and the current in the other phase. On the contrary, $v_\alpha i_\alpha$ and $v_\beta i_\beta$ in Eq. (39.3) obviously mean "instantaneous power" because of the product of the voltage in one phase and the current in the same phase. Akagi named the new electrical quantity defined in Eq. (39.5) "instantaneous imaginary power," which is represented by the product of the instantaneous voltage and current in different axes, but cannot be treated as a conventional quantity.

39 Active Filters

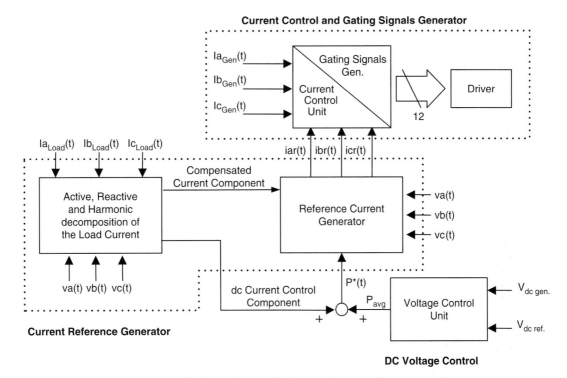

FIGURE 39.8 The block diagram of a shunt active power filter control scheme.

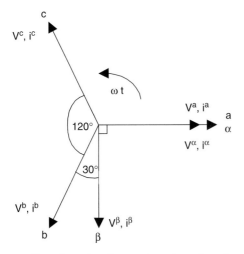

FIGURE 39.9 Transformation diagram from the *a, b, c* reference frame to the α, β coordinates.

The expression of the currents in the α–β plane, as a function of the instantaneous power is given by the following equation:

$$\begin{bmatrix} i_\alpha \\ i_\beta \end{bmatrix} = \frac{1}{v_\alpha^2 + v_\beta^2} \cdot \left\{ \begin{bmatrix} v_\alpha & v_\beta \\ v_\beta & -v_\alpha \end{bmatrix} \cdot \begin{bmatrix} p \\ 0 \end{bmatrix} + \begin{bmatrix} v_\alpha & v_\beta \\ v_\beta & -v_\alpha \end{bmatrix} \cdot \begin{bmatrix} 0 \\ q \end{bmatrix} \right\}$$

$$\equiv \begin{bmatrix} i_{\alpha p} \\ i_{\beta p} \end{bmatrix} + \begin{bmatrix} i_{\alpha q} \\ i_{\beta q} \end{bmatrix} \qquad (39.6)$$

and the different components of the currents in the α–β plane are shown in the following expressions:

$$i_{\alpha p} = \frac{v_\alpha p}{v_\alpha^2 + v_\beta^2} \qquad (39.7)$$

$$i_{\alpha q} = \frac{v_\beta q}{v_\alpha^2 + v_\beta^2} \qquad (39.8)$$

$$i_{\beta p} = \frac{v_\beta p}{v_\alpha^2 + v_\beta^2} \qquad (39.9)$$

$$i_{\beta q} = \frac{-v_\alpha q}{v_\alpha^2 + v_\beta^2} \qquad (39.10)$$

From Eqs. (39.3) and (39.4), the values of *p* and *q* can be expressed in terms of the dc components plus the ac components, that is:

$$p = \bar{p} + \tilde{p} \qquad (39.11)$$

$$q = \bar{q} + \tilde{q} \qquad (39.12)$$

where,

$\bar{p}$ dc component of the instantaneous power *p*, and is related to the conventional fundamental active current.

$\tilde{p}$ is the ac component of the instantaneous power *p*, it does not have average value, and is related to the harmonic currents caused by the ac component of the instantaneous real power.

- $\bar{q}$ is the dc component of the imaginary instantaneous power q, and is related to the reactive power generated by the fundamental components of voltages and currents.
- $\tilde{q}$ is the ac component of the instantaneous imaginary power q, and it is related to the harmonic currents caused by the ac component of instantaneous reactive power.

In order to compensate reactive power (displacement power factor) and current harmonics generated by non-linear loads, the reference signal of the shunt active power filter must include the values of $\tilde{p}$, $\bar{q}$, and $\tilde{q}$. In this case the reference currents required by the shunt active power filters are calculated with the following expression:

$$\begin{bmatrix} i^*_{c,\alpha} \\ i^*_{c,\beta} \end{bmatrix} = \frac{1}{v_\alpha^2 + v_\beta^2} \cdot \begin{bmatrix} v_\alpha & v_\beta \\ v_\beta & -v_\alpha \end{bmatrix} \cdot \begin{bmatrix} \tilde{p}_L \\ \bar{q}_L + \tilde{q}_L \end{bmatrix} \quad (39.13)$$

The final compensating currents including the zero sequence components in a, b, c reference frame are the following:

$$\begin{bmatrix} i^*_{c,a} \\ i^*_{c,b} \\ i^*_{c,c} \end{bmatrix} = \sqrt{\frac{2}{3}} \cdot \begin{bmatrix} \frac{1}{\sqrt{2}} & 1 & 0 \\ \frac{1}{\sqrt{2}} & \frac{-1}{2} & \frac{\sqrt{3}}{2} \\ \frac{1}{\sqrt{2}} & \frac{-1}{2} & \frac{-\sqrt{3}}{2} \end{bmatrix} \cdot \begin{bmatrix} -i_0 \\ i^*_{c,\alpha} \\ i^*_{c,\beta} \end{bmatrix} \quad (39.14)$$

where the zero sequence current component i_0 is equal to $1/\sqrt{3}$ $(i_a + i_b + i_c)$. The block diagram of the circuit required to generate the reference currents defined in Eq. (39.14) is shown in Fig. 39.10.

The advantage of instantaneous reactive power theory is that real and reactive power associated with fundamental components are dc quantities. These quantities can be extracted with a low-pass filter. Since the signal to be extracted is dc, filtering of the signal in the α–β reference frame is insensitive to any phase shift errors introduced by the low-pass filter, improving compensation characteristics of the active power filter. The same advantage can be obtained by using the *synchronous reference frame method*, proposed in [2]. In this case, transformation from a, b, c axes to d–q synchronous reference frame is done.

Effects of the Low-pass Filter Design Characteristics in Compensation Performance

A Butterworth filter is normally used due to the adequate frequency response. A second-order filter offers an appropriate relation between the transient response and the required attenuation characteristic. Higher-order filters achieve better filtering characteristic, but the settling time is increased. Since the low-pass filter cannot eliminate completely the low-frequency harmonic contained in the p and q signals, the shunt active power filter cannot compensate the entire low-frequency harmonic contained in the load current. Normally the cut-off frequency is equal to 127 Hz with an attenuation factor of 15 dB for the first ac component to be eliminated, which means an 82.2% attenuation of the fifth and seventh harmonic components (Fig. 39.11). The expression that relates the system line current harmonic distortion with the LPF cut-off frequency and load power factor is shown in Eq. (39.15).

$$THD_{isys} = \frac{\sqrt{\sum_{h=6k}^{\infty} \frac{1}{(f_h/f_c)^4 + 1} \left[((h^2+1)/(h^2-1)) - (1/(h^2-1))\cos(2\varphi) \right]}}{\cos(\varphi)} \quad (39.15)$$

with $k = 1, 2, 3, \ldots$ and f_h is the frequency of the harmonic component of order h, and f_c is the LPF cut-off frequency.

Figure 39.12 shows the total harmonic distortion (THD) in the line currents introduced by the second-order Butterworth filtering characteristics, as a function of the cut-off frequency, and considering a 50 Hz ac mains frequency. The harmonic distortion of the compensated line current depends on the

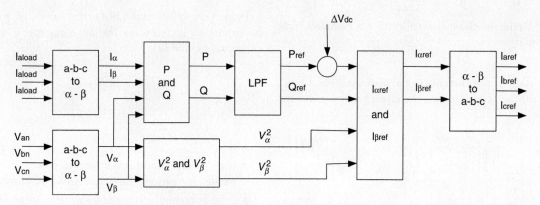

FIGURE 39.10 The block diagram of the current reference generator using p–q theory.

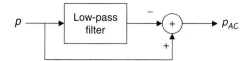

FIGURE 39.11 The low-pass filter block diagram used to extract the ac component of p.

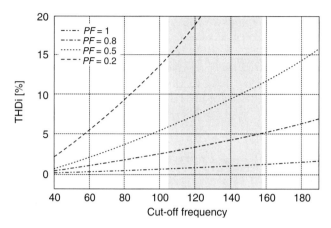

FIGURE 39.12 Harmonic distortion of the compensated line current as a function of the low-pass filter cut-off frequency. $PF = \cos(\varphi)$.

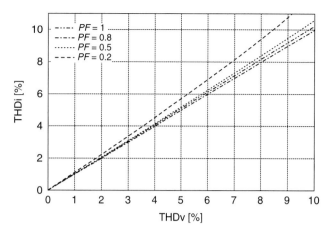

FIGURE 39.13 Harmonic distortion in the compensated line current as a function of the harmonic distortion in the system voltages. $PF = \cos(\varphi)$.

load displacement power factor, as shown in Eq. (39.15). If the cut-off frequency of the low-pass filter is changed, the active power filter compensation performance is affected as well as the transient response of the control scheme.

Effects of the Supply Voltage Distortion in Compensation Performance

One of the most important characteristics of the instantaneous imaginary power concept is that in order to obtain the current reference signal required to compensate reactive and harmonic current components, the system phase-to-neutral voltages are used. In general, purely sinusoidal voltages are considered in previously reported analysis. In case voltage is purely sinusoidal, the dc component of p and q in the α–β plane are related with the fundamental components in the real a, b, c reference system. This is not the case if the system voltages are distorted or unbalanced, as it is demonstrated below.

It is assumed that the supply voltages have harmonic distortion, and these are represented by:

$$v_a(t) = V_1\cos(\omega t) + V_h\cos[h(\omega t - \delta_h)]$$

$$v_b(t) = V_1\cos\left(\omega t - \frac{2\pi}{3}\right) + V_h\cos\left[h\left(\omega t - \delta_h - \frac{2\pi}{3}\right)\right]$$

$$v_c(t) = V_1\cos\left(\omega t + \frac{2\pi}{3}\right) + V_h\cos\left[h\left(\omega t - \delta_h + \frac{2\pi}{3}\right)\right] \quad (39.16)$$

Since the harmonic voltage component introduces a dc component in p and q, compensation performance of the shunt active power filter is reduced, as shown in Fig. 39.13. The larger the harmonic distortion in the system voltage is, the active power filter performance is more affected.

Effects of the Supply Voltage Unbalance in Compensation Performance

Voltage unbalance also affects the active power filter compensation performance. In this analysis, the phase-to-neutral voltages of the ac supply are equal to:

$$v_a(t) = V_1\cos(\omega t)$$
$$v_b(t) = V_1(1+m)\cos\left(\omega t - \frac{2\pi}{3}\right) \quad (39.17)$$
$$v_c(t) = V_1(1-m)\cos\left(\omega t + \frac{2\pi}{3}\right)$$

with $0 < m < 1$.

If the low-pass filter is considered as ideal, and the active filter follows exactly the current references, the compensated supply current (phase a) is:

$$i_{Sa} = I_1\cos(\varphi)\cos(\omega t) + \frac{\sqrt{3}}{3}m\,[I_1\sin(\varphi)\cos(3\omega t)$$
$$+ I_1\cos(3\omega t - \varphi)] \quad (39.18)$$

In this case, the active power filter is not able to fully compensate the system line current, since compensation performance depends on the voltage unbalance magnitude.

If the unbalance is defined as a function of the positive and negative sequence component as:

$$V_{a1} = \frac{1}{3}[V_a + a^2 V_b + a V_c]$$

$$V_{a2} = \frac{1}{3}[V_a + a V_b + a^2 V_c], \quad a = 1\angle 120° \quad (39.19)$$

$$\text{Unbalance} = \frac{|V_{a2}|}{|V_{a1}|} = \frac{\sqrt{3}}{3} m$$

The harmonic distortion is equal to:

$$THD_i = \frac{\sqrt{3}}{3} m \quad (39.20)$$

The relation between line current harmonic distortion and voltage unbalance is shown in Fig. 39.14.

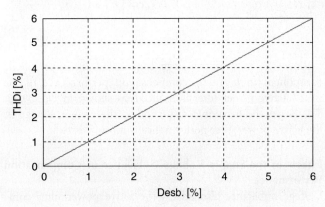

FIGURE 39.14 Harmonic distortion in the compensated line current as a function of the system voltages unbalance.

Effects of the Time Delay Introduced by the DSP in Compensation Performance

If a DSP is used to derive the reference generation signals, a time delay, T, associated with the processing time is introduced in the calculation of the reference signals. Considering a simultaneous sampling, the ac mains compensated line current with an ideal low-pass filter in the control scheme (phase a) is:

$$i_{Sa} = I_1 \cos(\omega t)[\cos(\varphi) + \sin(\omega T)] + \Delta V_{DC} \cos(\omega t)$$

$$+ \sum_{n=2k-1}^{\infty} I_n [1 - \cos(n\omega T)] \cos[n(\omega t - \varphi)]$$

$$+ \sum_{n=2k-1}^{\infty} I_n \sin(n\omega T) \sin[n(\omega t - \varphi)] \quad (39.21)$$

The time delay introduced by the DSP affects the compensation performance, especially when fast changes are present in the load current. Nevertheless, when T is small ($T > 50\,\mu s$), the effect in compensation performance is negligible.

In four-wire systems, unbalances in the load current also affect the generation of the adequate current reference signal that will assure high performance active power filter compensation. This assumption can be proved by considering the following load currents:

$$i_{La} = I_1 \cos(\omega t - \varphi) + \sum_{n=2k-1}^{\infty} I_n \cos[n(\omega t - \varphi)]$$

$$i_{Lb} = I_1 (1+l) \cos\left(\omega t - \varphi - \frac{2\pi}{3}\right)$$

$$+ \sum_{n=2k-1}^{\infty} I_n (1+l) \cos\left[n\left(\omega t - \varphi - \frac{2\pi}{3}\right)\right] \quad (39.22)$$

$$i_{Lc} = I_1 (1-l) \cos\left(\omega t - \varphi + \frac{2\pi}{3}\right)$$

$$+ \sum_{n=2k-1}^{\infty} I_n (1-l) \cos\left[n\left(\omega t - \varphi + \frac{2\pi}{3}\right)\right]$$

$$i_{LN} = i_{La} + i_{Lb} + i_{Lc}$$

with $0 < l < 1$, $k = 1, 2, 3, \ldots$

Equation (39.23) shows the presence of low-frequency even harmonics in the active power expression introduced by the load current unbalanced. These low frequency current components forces to reduce the cut-off frequency of the low-pass filter, close to 100 Hz, which affects the transient response of the active power filter.

$$p = \frac{3}{2} V_1 \left[I_1 \cos(\varphi) + \sum_{n=2k-1}^{\infty} I_n \cos[(n+1)\omega t - n\varphi] \right.$$

$$\left. + \sum_{m=2k-1}^{\infty} I_m \cos[(m-1)\omega t - m\varphi] \right] \text{ with } k = 1, 2, 3, \ldots$$

$$(39.23)$$

39.3.2.1.2 Synchronous Reference Frame Algorithm The block diagram of a current reference generator that uses the synchronous reference frame algorithm is shown in Fig. 39.15.

In this case, the real currents are transformed into a synchronous reference frame [2]. The reference frame is synchronized with the ac mains voltage, and is rotating at the

39 Active Filters

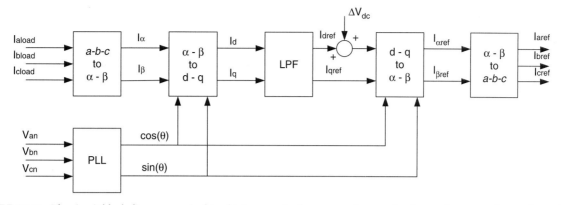

FIGURE 39.15 The circuit block diagram required to obtain current reference waveforms using the synchronous reference frame theory.

same frequency. The transformation is defined by:

$$\begin{bmatrix} i_d \\ i_q \end{bmatrix} = \begin{bmatrix} \cos(\omega t) & \sin(\omega t) \\ \sin(\omega t) & \cos(\omega t) \end{bmatrix} \begin{bmatrix} i_\alpha \\ i_\beta \end{bmatrix} \quad (39.24)$$

As for the instantaneous reactive power theory, d and q terms are composed by a dc and multiple ac components, such as $i_d = i_{ddc} + i_{dac}$ and $i_q = i_{qdc} + i_{qac}$. The compensation reference signals are obtained from the following expressions: $i_{dref} = -i_{dac}$ and $i_{qref} = -i_{qdc} - i_{qac}$. The compensated currents generated by the shunt active power filter are obtained from Eq. (39.25).

$$\begin{bmatrix} i_{aref} \\ i_{bref} \\ i_{cref} \end{bmatrix} = \sqrt{\frac{2}{3}} \begin{bmatrix} \frac{1}{\sqrt{2}} & 1 & 0 \\ \frac{1}{\sqrt{2}} & -\frac{1}{2} & \frac{\sqrt{3}}{2} \\ \frac{1}{\sqrt{2}} & -\frac{1}{2} & -\frac{\sqrt{3}}{2} \end{bmatrix} \begin{bmatrix} -1 & 0 & 0 \\ 0 & \cos(\omega t) & -\sin(\omega t) \\ 0 & \sin(\omega t) & \cos(\omega t) \end{bmatrix} \begin{bmatrix} i_0 \\ i_{dref} \\ i_{qref} \end{bmatrix}$$

(39.25)

One of the most important characteristics of this algorithm is that the reference currents are obtained directly from the loads currents without considering the source voltages. This is an important advantage since the generation of the reference signals is not affected by voltage unbalance or voltage distortion, therefore increasing the compensation robustness and performance. However, in order to transform from the α–β plane to the d–q synchronous reference frame, sine and cosine signals synchronized with the respective phase-to-neutral voltages are required. A phase-locked loop (PLL) per each phase, as the one shown in Fig. 39.16 must be used.

Since the algorithm used to obtain the reference current presents the same mathematical procedure and operations that the ones required in the instantaneous reactive power concept, the effects introduced by the filter and DSP time delay are similar. Unbalanced load currents generate a different harmonic spectrum in the synchronous reference frame, and low-order harmonic components appear in the reference signal. In order

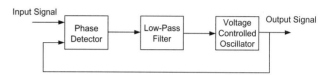

FIGURE 39.16 The PLL circuit block diagram.

to separate these uncharacteristic low-frequency current components, the cut-off frequency of the low-pass filter must be reduced.

39.3.2.1.3 Peak Detection Method There are other possibilities to generate the current reference signal required to compensate reactive power and current harmonics. Basically, all the different schemes try to obtain the current reference signals that include the reactive components required to compensate the displacement power factor and the current harmonics generated by the non-linear load. Figure 39.17 shows another scheme used to generate the current reference signals required by a shunt active power filter. In this case, the ac current generated by the inverter is forced to follow the reference signal obtained from the current reference generator. In this circuit, the distorted load current is filtered, extracting the fundamental component, i_{f1}. The band-pass filter is tuned at the fundamental frequency (50 or 60 Hz), so that the gain attenuation introduced in the filter output signal is zero and the phase-shift angle is 180°. Thus, the filter output current is exactly equal to the fundamental component of the load current but phase shifted by 180°. If the load current is added to the fundamental current component obtained from the second-order band-pass filter, the reference current waveform required to compensate only harmonic distortion is obtained. In order to provide the reactive power required by the load, the current signal obtained from the second-order band-pass filter I_{f1} is synchronized with the respective phase-to-neutral source voltage (see Fig. 39.17) so that the inverter ac output current is forced to lead the respective inverter output

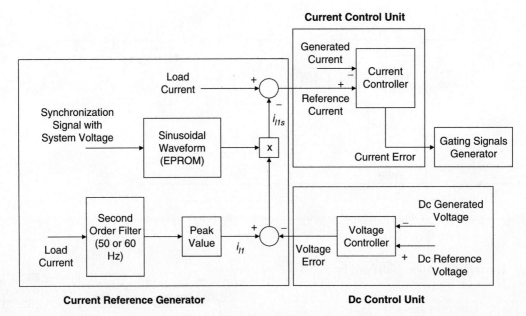

FIGURE 39.17 The block diagram of an active power filter control scheme that does not use the instantaneous reactive power concept.

voltage, thereby generating the required reactive power and absorbing the real power necessary to supply the switching losses and also to maintain the dc voltage constant.

The real power absorbed by the inverter is controlled by adjusting the amplitude of the fundamental current reference waveform, I_{l1}, obtained from the reference current generator. The amplitude of this sinusoidal waveform is equal to the amplitude of the fundamental component of the load current plus or minus the error signal obtained from the dc voltage control unit. In this way, the current signal allows the inverter to supply the current harmonic components, the reactive power required by the load, and to absorb the small amount of active power necessary to cover the switching losses and to keep the dc voltage constant (Fig. 39.18).

The main characteristic of this method is the direct derivation of the compensating component from the load current, without the use of any reference frame transformation [1, 2]. Nevertheless, this technique presents a low frequency oscillation problem in the active power filter dc bus voltage. To improve this technique, a modification of the previous scheme (Fig. 39.17) is shown in Fig. 39.19. The scheme is necessary for each phase. The expression for i_{Ma} is:

$$i_{Ma} = \frac{I_1 \cos(\varphi)}{2} + \frac{I_1 \cos(2\omega t - \varphi)}{2}$$

$$+ \sum_{n=2k-1}^{\infty} \frac{I_n}{2} \{\cos[(n-1)\omega t - n\varphi]$$

$$+ \cos[(n+1)\omega t - n\varphi]\} \quad (39.26)$$

with $k = 1, 2, 3, \ldots$

Figure 39.20 shows the current harmonic distortion introduced by the low-pass filter used in the control scheme and the associated cut-off frequency. The current distortion of the compensated current depends on the phase angle of the fundamental load current component.

The supply voltage has no effect on the reference current generation. Synchronization with the ac mains voltage is the important issue in this scheme as well as in the synchronous reference frame theory. Unbalanced loads do not affect the reference generation. Nevertheless, the method cannot achieve active power balance in four-wire systems. The control circuit implementation of the peak detection method is simple and does not require complex calculation, so the processing time on a DSP is lower than the required in the two previous implementations, ($T < 10\,\mu s$). The use of this method minimizes the distortion introduced on current harmonics.

Technical Comparison of the Three Different Techniques

In order to validate the effectiveness of the proposed analysis, a common industrial system is considered. The results are obtained using the previous equations and Matlab simulations results. The DSP delay introduced is using the processing time according the DSP ADSP2187. The parameters considered are:

$$THD_{iL} = 29\%$$

$$THD_V = 5\%$$

$$\cos(\varphi) = 0.87$$

$$\text{Unbalance} = 3\%$$

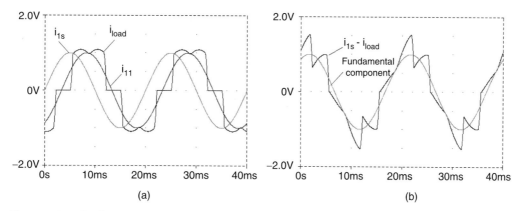

FIGURE 39.18 The procedure for the generation of the current reference waveform: (a) the load current i_{load}, its fundamental component, i_{l1}, and the fundamental current component synchronized with the respective phase-to-neutral source voltage, i_{ls} and (b) the synchronized fundamental current signal minus the load current, $i_{ls} - i_{load}$, and its fundamental component.

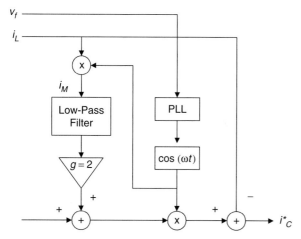

FIGURE 39.19 Modified version of the original peak detection method.

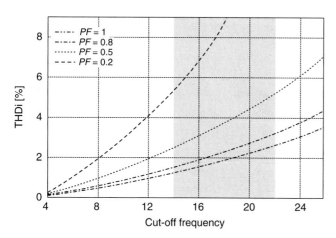

FIGURE 39.20 Harmonic distortion of the system line current as a function of the cut-off frequency of the low-pass filter and the load displacement power factor.

Table 39.2 shows the numeric results:

TABLE 39.2 Results of the considered case

Technique	THD_{iS} [%]	FP	Transient delay t_d [ms]
PQ	8.2	0.99	8.1
DQ	3.1	0.99	8.1
DPVM	2.3	0.99	56.8

Table 39.3 shows a comparison of the three different techniques analyzed. The effects of the non-ideal conditions are described for each technique.

In conclusion, it can be mentioned that the compensation performance of the different techniques is similar under

TABLE 39.3 Comparison of the techniques

Parameter	PQ	SRF	PDM
Load PF required to achieve full compensation	PF > 0.3	PF > 0.3	PF > 0.2
Harmonic distortion voltage effect on the compensated current.	$THD_i \approx THD_v$	0	0
Unbalanced voltage effect	$THD_i \approx$ Unbalance	0	0
Dynamic response under load changes (in function of f_c, $t_d = 5\tau$)	Fast (load balanced)	Fast (load balanced)	Slow
Capability of load balance	Yes (It is necessary to reduce cut-off frequency f_c)	Yes (It is necessary to reduce cut-off frequency f_c)	No
DSP delay time introduced	Minimum	Minimum	≈ 0

ideal conditions, but under the presence of unbalanced and voltage distortion, the synchronous reference frame algorithm presents the best performance, since it is insensitive to voltage perturbations. It is fundamental to consider adequately the cut-off frequency in the filter used to extract the ac component in the different techniques. If the frequency is changed, the compensation performance is affected as well as the transient response of the control scheme. In four-wire systems, unbalance in the load current also affects the generations of the adequate current reference. Unbalanced load currents generate a different harmonic spectrum in the dc reference frame, and low-order harmonic components appear in the reference signal; then the cut-off frequency of the filter must be reduced.

39.3.2.2 Current Modulator

The effectiveness of an active power filter depends basically on the design characteristics of the current controller, the method implemented to generate the reference template and the modulation technique used. Most of the modulation techniques used in active power filters are based on PWM strategies. In this chapter, four of these methods, whose characteristics are their simplicity and effectiveness, are analyzed: periodical sampling control, hysteresis band control, triangular carrier control, and vector control. The first three methods have been tested with different waveform templates sinusoidal, quasisquare, and rectifier compensation current and were compared in terms of the harmonic content and distortion at the same switching frequency [3]. The analysis shows that for sinusoidal current generation the best method is triangular carrier, followed by hysteresis band and periodical sampling. For other types of references, however, one strategy may be better than the others. Also it was shown that each control method is affected in a different way by the switching time delays present in the driving circuitry and in the power semiconductors.

39.3.2.2.1 Periodical Sampling The periodical sampling method switches the power transistors of the converter during the transitions of a square wave clock of fixed frequency (the sampling frequency). As shown in Fig. 39.21, this type of control is very simple to implement since it requires a comparator and a D-type flip-flop per phase. The main advantage of this method is that the minimum time between switching transitions is limited to the period of the sampling clock. However, the actual switching frequency is not clearly defined.

39.3.2.2.2 Hysteresis band The hysteresis band method switches the transistors when the current error exceeds a fixed magnitude: the hysteresis band. As can be seen in Fig. 39.22, this type of control needs a single comparator with hysteresis per phase. In this case the switching frequency is not determined, but it can be estimated.

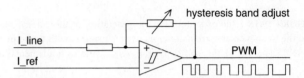

FIGURE 39.22 Control modulator block for hysteresis band.

39.3.2.2.3 Triangular Carrier The triangular carrier method, shown in Fig. 39.23, compares the current error with fixed amplitude and fixed frequency triangular wave (the triangular carrier). The error is processed through a proportional-integral (PI) gain stage before the comparison with the triangular carrier takes place. As can be seen, this control scheme is more complex than the periodical sampling and hysteresis band. The values for the PI control gain k_p and k_i determine the transient response and steady-state error of the triangular carrier method. It was found empirically that the values for k_p and k_i shown in Eqs. (39.27) and (39.28) give a good dynamic performance under transient and steady-state operating conditions.

$$k_p^* = \frac{(L + L_o) \cdot \omega_c}{2 V_{dc}} \quad (39.27)$$

$$k_i^* = \omega_c k_p^* \quad (39.28)$$

where $L+L_o$ is the total series inductance seen by the converter, ω_c is the triangular carrier frequency, whose amplitude is one volt peak-peak, and V_{dc} is the dc supply voltage of the inverter.

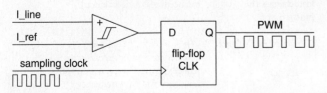

FIGURE 39.21 Control modulator block for periodical sampling method.

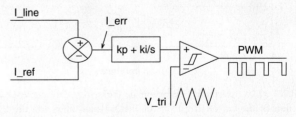

FIGURE 39.23 Control modulator block for triangular carrier method.

39.3.2.2.4 Vector Control Technique

This current control technique proposed in [4] divides the α–β reference frame of currents and voltages in six regions, phase shifted by 30° (Fig. 39.24), identifies the region where the current vector error Δi is located, and selects the inverter output voltage vector V_{inv} that will force Δi to change in the opposite direction, keeping the inverter output current close to the reference signal.

Figure 39.25 shows the single-phase equivalent circuit of the shunt active power filter connected to a non-linear load and to the power supply.

The equation that relates the active power filter currents and voltages is obtained by applying Kirchhoff law to the equivalent circuit shown in Fig. 39.25:

$$V_{inv} = L\frac{di_{gen}}{dt} + E_0 \quad (39.29)$$

The current error vector Δi is defined by the following expression:

$$\Delta i = i_{ref} - i_{gen} \quad (39.30)$$

where i_{ref} represents the inverter reference current vector defined by the instantaneous reactive power concept. By replacing Eq. (39.30) in Eq. (39.29):

$$V_{inv} = L\frac{d}{dt}(i_{ref} - \Delta i) + E_0 \Rightarrow L\frac{d\Delta i}{dt} = L\frac{di_{ref}}{dt} + E_0 - V_{inv} \quad (39.31)$$

If $E = L(di_{ref}/dt) + E_0$ then Eq. (39.31) becomes

$$L\frac{d\Delta i}{dt} = E - V_{inv} \quad (39.32)$$

Equation (39.32) represents the active power filter state equation and shows that the current error vector variation $d\Delta i/dt$ is defined by the difference between the fictitious voltage vector E and the inverter output voltage vector V_{inv}. In order to keep $d\Delta i/dt$ close to zero, V_{inv} must be selected near E_0.

The selection of the inverters' gating signals is defined by the region in which Δi is located and by its amplitude. In order to improve the current control accuracy and associate time response, depending on the amplitude of Δi the following actions are defined:

- if $\Delta i \leq \delta$ the gating signals of the inverter are not changed,
- if $h \leq \Delta i \leq \delta$, the inverter gating signals are defined following Mode a,
- if $\Delta i > h$, the inverter gating signals are defined following Mode b;

where δ and h are reference values that define the accuracy and the hysteresis window of the current control scheme.

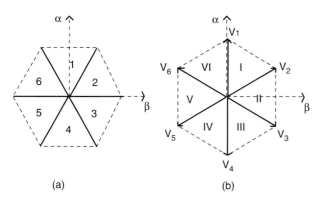

FIGURE 39.24 The hexagons defined in the α–β reference frame by the current control scheme: (a) the hexagon defined by the inverter output current vector and (b) the hexagon defined by the inverter output voltage vector.

39.3.2.2.5 Mode a: Small Changes in Δi ($h \leq \Delta i \leq \delta$)

The selection of the inverter switching Mode a can be explained with the following example. Assuming that the voltage vector E is located in region I (Fig. 39.26a) and the current error vector Δi is in region 6 (Fig. 39.26b), the inverter voltage vectors, V_{inv}, located closest to E are V_1 and V_2. The vectors $E-V_2$ and $E-V_1$ define two vectors $Ld\Delta i/dt$, located in region III and V respectively, as shown in Fig. 39.26a, so in order to reduce the current vector error Δi, $Ld\Delta i/dt$ must be located in region III. Thus the inverter output voltage has to be equal to V_1. In this way Δi will be forced to change in the opposite direction reducing its amplitude faster. By doing the same analysis for all the possible combinations, the inverter switching modes for each location of Δi and E can be defined (Table 39.4).

V_k represents the inverter switching functions defined in Table 39.5.

39.3.2.2.6 Mode b: Large Changes in Δi ($\Delta i > h$)

If Δi becomes larger than h in a transient state, it is necessary to choose the switching mode in which the $d\Delta i/dt$ has the largest

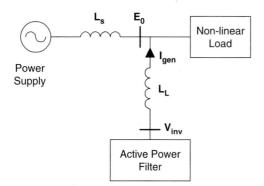

FIGURE 39.25 The single-phase equivalent circuit of a shunt active power filter connected to the power system.

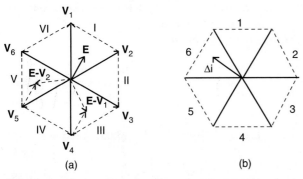

FIGURE 39.26 Selection of the inverter switching pattern according to the region where Δi and E are located: (a) the regions where $Ld\Delta i/dt$ are located (b) the region where Δi is located.

TABLE 39.4 Inverter switching modes

E region	Δi region					
	1	2	3	4	5	6
I	V_1	V_2	V_2	$V_0 - V_7$	$V_0 - V_7$	V_1
II	V_2	V_2	V_3	V_3	$V_0 - V_7$	$V_0 - V_7$
III	$V_0 - V_7$	V_3	V_3	V_4	V_4	$V_0 - V_7$
IV	$V_0 - V_7$	$V_0 - V_7$	V_4	V_4	V_5	V_5
V	V_6	$V_0 - V_7$	$V_0 - V_7$	V_5	V_5	V_6
VI	V_1	V_1	$V_0 - V_7$	$V_0 - V_7$	V_6	V_6

TABLE 39.5 Relationship between switching function and inverter output voltage

k	Switch on phase a	Switch on phase b	Switch on phase c	Inverter output voltage V_k
0	4	6	2	0
1	1	6	2	$\frac{2}{3}V_{dc}$
2	1	3	2	$\frac{2}{3}V_{dc}\,e^{j\pi/3}$
3	4	3	2	$\frac{2}{3}V_{dc}\,e^{j2\pi/3}$
4	4	3	5	$\frac{2}{3}V_{dc}\,e^{j\pi}$
5	4	6	5	$\frac{2}{3}V_{dc}\,e^{j4\pi/3}$
6	1	6	5	$\frac{2}{3}V_{dc}\,e^{j5\pi/3}$
7	1	3	5	0

opposite direction to Δi. In this case the best inverter output voltage V_{inv} corresponds to the value located in the same region of Δi.

The switching frequency may be fixed by controlling the time between commutations and not applying a new switching pattern if the time between two successive commutations is lower than a selected value ($t = 1/2f_c$).

Figure 39.27 shows the block diagram of the inverter vector current control scheme implemented in a microcontroller. In Fig. 39.27 $^*E$ represents the region where the vector E is located, $^*\Delta i$, the region of Δi, k_1 keeps the same value of k (no commutation in the inverter), k_2 selects the new inverter output voltage from Table 39.4, and k_3 selects V_{inv} in the same region of Δi.

39.3.2.3 Control Loop Design

Active power filters based on self-controlled dc bus voltage requires two control loops, one to control the inverter output current and the other to regulate the inverter dc voltage. Different design criteria have been presented in the technical literature; however, a classic design procedure using a PI controller will be presented in this chapter. In general, the design procedure for the current and voltage loops is based on the respective time response requirements. Since the transient response of the active power is determined by the current control loop, its time response has to be fast enough to follow the current reference waveform closely. On the other hand, the time response of the dc voltage does not need to be fast and is selected to be at least 10 times slower than the current control loop time response. Thus, these two control systems can be decoupled and designed as two independent systems.

A PI controller is normally used for the current and the voltage control loops since it contributes to zero steady-state error in tracking the reference current and voltage signals, respectively.

39.3.2.3.1 Design of the Current Control Loop The design of the current control loop gains depends on the selected current modulator. In the case of selecting the triangular carrier technique, to generate the gating signals, the error between the generated current and the reference current is processed through a PI controller, then the output current error is compared with a fixed amplitude and fixed frequency triangular wave (Fig. 39.23) The advantage of this current modulator technique is that the output current of the converter has well-defined spectral line frequencies for the switching frequency components.

Since the active power filter is implemented with a voltage-source inverter, the ac output current is defined by the inverter ac output voltage. The block diagram of the current control loop for each phase is shown in Fig. 39.28 where

E phase-to-neutral source voltage,
$Z(s)$ impedance of the link reactor,
K_s gain of the converter, and
$Gc(s)$ gain of the controller.

The values of K_s and $Gc(s)$ are given in Eqs. (39.33) and (39.34).

$$K_s = \frac{V_{dc}}{2\xi} \qquad (39.33)$$

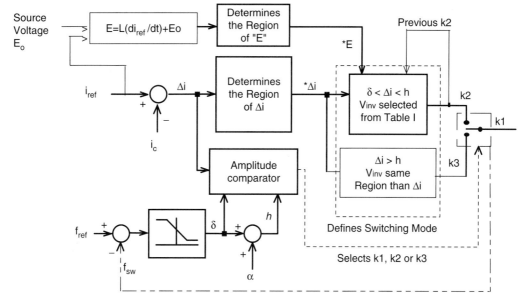

FIGURE 39.27 The current control block diagram.

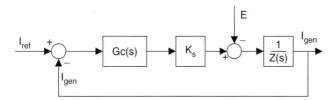

FIGURE 39.28 The block diagram of the current control loop.

$$G_c(s) = K_p + \frac{K_i}{s} \qquad (39.34)$$

From Fig. 39.27 and Eq. (39.34), the following expression is obtained:

$$I_{gen} = \frac{[K_s(K_p + (K_i/s))]/(R_r + sL_r)}{1 + [K_s(K_p + (K_i/s))/(R_r + sL_r)]} I_{ref}$$
$$- \frac{1/(R_r + sL_r)}{1 + (K_s((K_p + K_i/s))/(R_r + sL_r))} E \qquad (39.35)$$

The characteristic equation of the current loop is given by

$$1 + \frac{(K_p s + K_i/s)}{s(R_r + sL_r)} \qquad (39.36)$$

The analysis of the characteristic equation proves that the current control loop is stable for all values of K_p and K_i. Also, this analysis shows that K_p determines the speed response and K_i defines the damping factor of the control loop. If K_p is too big, the error signal can exceed the amplitude of the triangular waveform, affecting the inverter switching frequency, and if K_i is too small, the gain of the PI controller decreases, which means that the generated current will not be able to follow the reference current closely. The active filter transient response can be improved by adjusting the gain of the proportional part (K_p) to equal one and the gain of the integrator (K_i) to equal the frequency of the triangular waveform.

39.3.2.3.2 DC Voltage Control Loop Voltage control of the dc bus is performed by adjusting the small amount of real power flowing into the dc capacitor, thus compensating for the conduction and switching losses. The voltage loop is designed to be at least 10 times slower than the current loop, hence the two loops can be considered decoupled. The dc voltage control loop need not to be fast, since it only responds for steady-state operating conditions. Transient changes in the dc voltage are not permitted and are taken into consideration with the selection criteria of the appropriate electrolytic capacitor value.

39.3.3 Power Circuit Design

The selection of the ac link reactor and the dc capacitor values affects directly the performance of the active power filter. Static var compensators implemented with voltage-source inverters present the same power circuit topology, but for this type of application, the criteria used to select the values of L_r and C are different. For reactive power compensation, the design of the synchronous link inductor, L_r, and the dc capacitor, C, is performed based on harmonic distortion constrain. That is, L_r must reduce the amplitude of the current harmonics generated by the inverter, while C must keep the dc voltage ripple factor below a given value. This design criteria cannot be applied in the active power filter since it must be able to generate distorted current waveforms. However, L_r must be specified so that it keeps the high-frequency switching ripple of the inverter ac output current smaller than a defined value.

39.3.3.1 Design of the Synchronous Link Reactor

The design of the synchronous link reactor depends on the current modulator used. The design criteria presented in this section is based considering that the triangular carrier modulator is used. The design of the synchronous reactor is performed with the constraint that for a given switching frequency the minimum slope of the inductor current is smaller than the slope of the triangular waveform that defines the switching frequency. In this way, the intersection between the current error signal and the triangular waveform will always exist. In the case of using another current modulator, the design criteria must allow an adequate value of L_r in order to ensure that the di/dt generated by the active power filter will be able to follow the inverter current reference closely. In the case of the triangular carrier technique, the slope of the triangular waveform, λ, is defined by

$$\lambda = 4\xi f_t \quad (39.37)$$

where ξ is the amplitude of the triangular waveform, which has to be equal to the maximum permitted amount of ripple current, and f_t is the frequency of the triangular waveform (i.e. the inverter switching frequency). The maximum slope of the inductor current is equal to

$$\frac{di_L}{dt} = \frac{V_{an} + 0.5 V_{dc}}{L_r} \quad (39.38)$$

Since the slope of the inductor current (di_L/dt) has to be smaller than the slope of the triangular waveform (λ), and the ripple current is defined, from Eqs. (39.37) and (39.38), as

$$L_r = \frac{V_{an} + 0.5 V_{dc}}{4\xi f_t} \quad (39.39)$$

39.3.3.2 Design of the DC Capacitor

Transient changes in the instantaneous power absorbed by the load, generate voltage fluctuations across the dc capacitor. The amplitude of these voltage fluctuations can be controlled effectively with an appropriate dc capacitor value. It must be noticed that the dc voltage control loop stabilizes the capacitor voltage after a few cycles, but is not fast enough to limit the first voltage variations. The capacitor value obtained with this criteria is bigger than the value obtained based on the maximum dc voltage ripple constraint. For this reason, the voltage across the dc capacitor presents a smaller harmonic distortion factor.

The maximum overvoltage generated across the dc capacitor is given by

$$V_{C\max} = \frac{1}{C}\int_{\theta_1/\omega}^{\theta_2/\omega} i_C(t)dt + V_{dc} \quad (39.40)$$

where V_{Cmax} is the maximum voltage across the dc capacitor, V_{dc} is the steady-state dc voltage, and $i_C(t)$ is the instantaneous dc bus current. From Eq. (39.40)

$$C = \frac{1}{\Delta V}\int_{\theta_1/\omega}^{\theta_2/\omega} i_C(t)dt \quad (39.41)$$

Eq. (39.41) defines the value of the dc capacitor, C, that will maintain the dc voltage fluctuation below ΔV p.u. The instantaneous value of the dc current is defined by the product of the inverter line currents with the respective switching functions. The mean value of the dc current that generates the maximum overvoltage can be estimated by

$$\int_{\theta_1/\omega}^{\theta_2/\omega} i_C(t)dt = I_{inv}\int_{\theta_1/\omega}^{\theta_2/\omega} \left[\sin(\omega t) + \sin(\omega t + 120°)\right] dt \quad (39.42)$$

In this expression the inverter ac current is assumed to be sinusoidal. These operating conditions represent the worst case.

39.3.4 Technical Specifications

The standard specifications of shunt active power filters are the following:

- Number of phases: three-phase and three wires or three-phases and four wires (in case neutral currents need to be compensated).
- Input voltage: 200, 210, 220 ±10%, 400, 420, 440 ±10%, 6600 ±10%.
- Frequency: 50/60 Hz ±5%.
- Number of restraint harmonic orders: 2–25th.
- Harmonic restraint factor: 85% or more at the rated output.
- Type of rating: continuous.
- Response: 1 ms or less.

For shunt active power filter the harmonic restraint factor is defined as $\left[1 - (I_{H_2}/I_{H_1})\right] \times 100\%$, where I_{H_1} are the harmonic currents flowing on the source side when no measure are taken for harmonic suppression, and I_{H_2} are the harmonic currents flowing on the source side when harmonics are suppressed using an active filter.

39.4 Series Active Power Filters

Series active power filters were introduced by the end of the 1980s [5], and operate mainly as a voltage regulator and harmonic isolator between the non-linear load and the utility system. The series-connected active power filter is more preferable to protect the consumer from an inadequate supply voltage quality. This type of approach is specially recommended for compensation of voltage unbalances, voltage distortion, and voltage sags from the ac supply, and for low power applications represents an economically attractive alternative to UPS, since no energy storage (battery) is necessary and the overall rating of the components is smaller. The series active power filter injects a voltage component in series with the supply voltage and therefore can be regarded as a controlled voltage source, compensating voltage sags and swells on the load side (Fig. 39.29).

If passive LC filters are connected in parallel to the load, the series active power filter operates as an harmonic isolator forcing the load current harmonics to circulate mainly through the passive filter rather than the power distribution system (hybrid topology) (Fig. 39.30). The main advantage of this scheme is that the rated power of the series active power filter is a small fraction of the load kVA rating, typically 5%. However, the rated apparent power of the series active power filter may increase, in case voltage compensation is required.

39.4.1 Power Circuit Structure

The topology of the series active power filter is shown in Fig. 39.31. In most cases, the power circuit configuration is based on a three-phase PWM voltage-source inverter connected in series with the power lines through three single-phase coupling transformers. For certain type of applications, the three-phase PWM voltage-source converter can be replaced by three single-phase PWM inverters. However, this type of approach requires more power components, which increases the cost.

In order to operate as an harmonic isolator, a parallel LC filter must be connected between the non-linear loads and the coupling transformers (Fig. 39.30). Current harmonic and voltage compensation are achieved by generating the appropriate voltage waveforms with the three-phase PWM voltage-source inverter, which are reflected in the power system through three coupling transformers. With an adequate control scheme, series active power filters can compensate for current harmonics generated by non-linear loads, voltage unbalances, voltage distortion, and voltage sags or swells at the load terminals. However, it is very difficult to compensate the load power factor with this type of topology. In four-wire power distribution systems, series active power filters with the power topology can also compensate the current harmonic components that circulate through the neutral conductor.

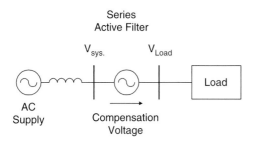

FIGURE 39.29 The series active power filter operating as a voltage compensator.

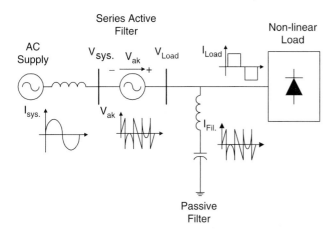

FIGURE 39.30 Combination of series active power filter and passive filter for current harmonic compensation.

39.4.2 Principles of Operation

Series active power filters compensate current system distortion caused by non-linear loads by imposing a high impedance path to the current harmonics, which forces the high-frequency currents to flow through the LC passive filter connected in parallel to the load (Fig. 39.30). The high impedance imposed by the series active power filter is created by generating a voltage of the same frequency that the current harmonic component needs to be eliminated. Voltage regulation or voltage unbalance can be corrected by compensating the fundamental frequency positive, negative, and zero sequence voltage components of the power distribution system (Fig. 39.29) In this case, the series active power filter injects a voltage component in series with the supply voltage and therefore can be regarded as a controlled voltage source, compensating voltage regulation on the load side (sags or swells), and voltage unbalance. Voltage injection of arbitrary phase with respect to the load current implies active power transfer capabilities which increases the rating of the series active power filter, and in most cases requires an energy storage element connected in the dc bus. Voltage and current waveforms

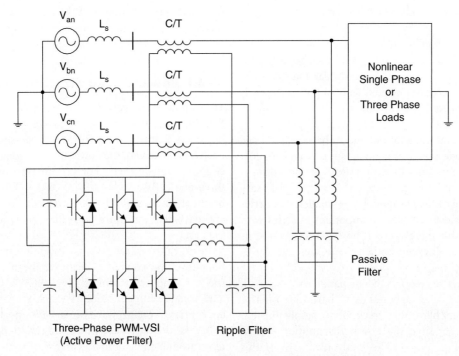

FIGURE 39.31 The series active power filter topology.

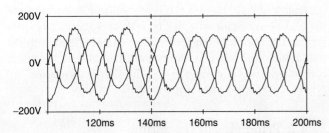

FIGURE 39.32 Load voltage waveforms for voltage unbalance compensation. Phase-to-neutral voltages at the load terminals before and after series compensation. (Compensation starts at 140 ms, current harmonic compensator not operating.)

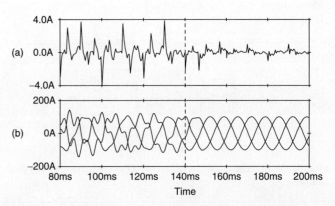

FIGURE 39.33 System current waveforms for current harmonic compensation: (a) neutral current flowing to the ac mains before and after compensation and (b) line currents flowing to the ac mains before and after compensation. (Voltage unbalance compensator not operating.)

shown in Figs. 39.32, 39.33, and 39.34 illustrate the compensation characteristics of a series active power filter operating with a shunt passive filter.

39.4.3 Power Circuit Design

The power circuit topology of the series active power filter is composed by the three-phase PWM voltage-source inverter, the second-order resonant LC filters, the coupling transformers, and the secondary ripple frequency filter (Fig. 39.30). The design characteristics for each of the power components are described below.

39.4.3.1 PWM Voltage-source Inverter

Since series active power filter can compensate voltage unbalance and current harmonics simultaneously, the rated power of the PWM voltage-source inverter increases compared with other approaches that compensate only current harmonics, since voltage injection of arbitrary phase with respect to the load current implies active power transfer from the inverter to the system. Also, the transformer leakage inductance entails fundamental voltage drop and apparent power, which has to be supported by the inverter, reducing the series active filter

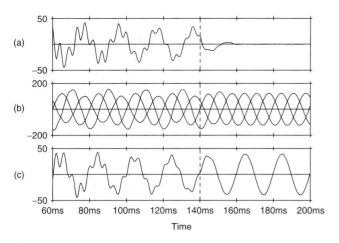

FIGURE 39.34 Load voltages and system currents for voltage unbalance and current harmonic compensation, before and after compensation: (a) power system neutral current; (b) phase-to-neutral load voltages; and (c) power system line current.

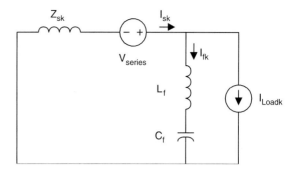

FIGURE 39.35 The equivalent circuit of the series active power filter for harmonic components.

inverter rating available for harmonic and voltage compensation. The rated apparent power required by the inverter can be obtained by calculating the apparent power generated in the primary of the coupling transformers. The voltage reflected across the primary winding of each coupling transformer is defined in Eq. (39.43).

$$V_{series} = \left[K_1^2 \left\{ \sum_{k \neq 1} I_{sk}^2 \right\}^{1/2} + K_2^2 \{V_2 + V_0\}^2 \right]^{1/2} \quad (39.43)$$

where V_{series} is the rms voltage across the primary winding of the coupling transformer. Equation (39.43) shows that the voltage across the primary winding of the transformer is defined by two terms. The first one is inversely proportional to the quality factor of the passive LC filter, while the second one depends on the voltage unbalance that needs to be compensated. K_1 depends on the LC filter values while K_2 is equal to one. The current flowing through the primary winding of the coupling transformer, due to the harmonic currents (Eq. (39.44)), can be obtained from the equivalent circuit shown in Fig. 39.35.

$$I_{sk} = \frac{Z_{fk} I_{lk}}{Z_{fk} + Z_{sk} + K_1} \quad (39.44)$$

where $V_{series} = -K_1 I_{sk}$. The fundamental component of the primary current depends on the amplitude of the negative and zero sequence component of the source voltage due to the system unbalance.

39.4.3.2 Coupling Transformer

The purpose of the three coupling transformers is not only to isolate the PWM inverters from the source but also to match the voltage and current ratings of the PWM inverters with those of the power distribution system. The total apparent power required by each coupling transformer is one-third the total apparent power of the inverter. The turn ratio of the current transformer is specified according to the inverter dc bus voltage, K_1 and V_{ref}. The correct value of the turn ratio "a" must be specified according to the overall series active power filter performance. The turn ratio of the coupling transformer must be optimized through the simulation of the overall active power filter, since it depends on the values of different related parameters. In general, the transformer turn ratio must be high in order to reduce the amplitude of the inverter output current and to reduce the voltage induced across the primary winding. Also, the selection of the transformer turn ratio influences the performance of the ripple filter connected at the output of the PWM inverter. Taking into consideration all these factors, in general, the transformer turn ratio is selected equal to 1:20.

39.4.3.3 Secondary Ripple Filter

The design of the ripple filter connected in parallel to the secondary winding of the coupling transformer is performed following the method presented by Akagi in [6]. However, it is important to notice that the design of the secondary ripple filter depends mainly on the coupling transformer turn ratio and the current modulator used to generate the inverter gating signals. If the triangular carrier is used, the frequency of the triangular waveform has to be considered in the design of the ripple filter. The ripple filter connected at the output of the inverter avoid the induction of the high-frequency ripple voltage generated by the PWM inverter switching pattern at the terminals of the primary winding of the coupling transformer. In this way, the voltage applied in series to the power system corresponds to the components required to compensate voltage unbalanced and current harmonics. The single-phase equivalent circuit is shown in Fig. 39.36.

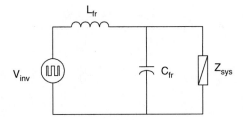

FIGURE 39.36 The single-phase equivalent circuit of the inverter output ripple filter.

The voltage reflected in the primary winding of the coupling transformer has the same waveform as that of the voltage across the filter capacitor. For low-frequency components, the inverter output voltage must be almost equal to the voltage across C_{fr}. However, for high-frequency components, most of the inverter output voltage must drop across L_{fr}, in which case the voltage at the capacitor terminals is almost zero. Moreover, C_{fr} and L_{fr} must be selected in order not to exceed the burden of the coupling transformer. The ripple filter must be designed for the carrier frequency of the PWM voltage-source inverter. To calculate C_{fr} and L_{fr} the system equivalent impedance at the carrier frequency, Z_{sys}, reflected in the secondary must be known. This impedance is equal to

$$Z_{sys(secondary)} = a^2 \cdot Z_{sys(primary)} \qquad (39.45)$$

For the carrier frequency, the following design criteria must be satisfied:

(i) $X_{Cfr} \ll X_{Lfr}$ – to ensure that at the carrier frequency most of the inverter output voltage will drop across L_{fr}.
(ii) X_{Cfr} and $X_{Lfr} \ll Z_{sys}$ – to ensure that the voltage divider is between L_{fr} and C_{fr}.

The small-rated LC passive filter exhibits a high quality factor circuit because of the high impedance on the output side. Oscillation between the small-rated inductor and capacitor may occur, causing undesirable high-frequency voltage across the ripple filter capacitor, which is reflected in the primary winding of the coupling transformer generating high-frequency current to flow through the power distribution system. It is important to note that this oscillation is very difficult to eliminate through the design and selection of the L_{fr} and C_{fr} values. However, it can be eliminated with the addition of a new control loop. The cause of the output voltage oscillation is explained with the help of Fig. 39.37. The transfer function $G_p(s)$ between the input voltage $V_i(s)$ and the output voltage $V_C(s)$ is given by the equation:

$$G_p(s) = \frac{\omega_n^2}{s^2 + 2\xi\omega_n s + \omega_n^2} \qquad (39.46)$$

where $\omega_n = \sqrt{1/L_{fr}C_{fr}}$ and $\xi = r_L/2\sqrt{C_{fr}/L_{fr}}$.

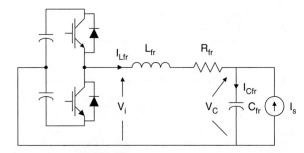

FIGURE 39.37 Single-phase equivalent circuit of series active power filter connected to the ripple filter.

Normally, the damping factor ξ is smaller than one, causing the voltage oscillation across the capacitor ripple filter, C_{fr}. Generally, relatively low impedance loads are connected to the output terminals of voltage-source PWM inverters. In these cases, the quality factor of the LC filter can be low, and the oscillation between the inductor L_{fr} and the capacitor C_{fr} is avoided [7].

39.4.3.4 Passive Filters

Passive filters connected between the non-linear load and the series active power filter play an important role in the compensation of the load current harmonics. With the connection of the passive filters the series active power filter operates as an harmonic isolator. The harmonic isolation feature reduces the need for precise tuning of the passive filters and allows their design to be insensitive to the system impedance and eliminates the possibility of filter overloading due to supply voltage harmonics. The passive filter can be tuned to the dominant load current harmonic and can be designed to correct the load displacement power factor.

However, for industrial loads connected to stiff supply, it is difficult to design passive filters that can absorb a significant part of the load harmonic current and therefore its effectiveness deteriorates. Specially, for compensation of diode rectifier type of loads, where a small kVA passive filter is required, it is difficult to achieve the required tuning to absorb significant percentage of the load harmonic currents. For this type of application, the passive filter cannot be tuned exactly to the harmonic frequencies because they can be overloaded due to the system voltage distortion and/or system current harmonics.

The single-phase equivalent circuit of a passive LC filter connected in parallel to a non-linear current source and to the power distribution system is shown in Fig. 39.38. From this figure, the design procedure of this filter can be derived. The harmonic current component flowing through the passive filter i_{fh} and the current component flowing through the source

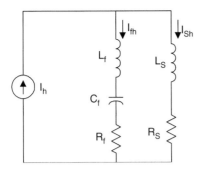

FIGURE 39.38 The single-phase equivalent circuit of the passive filter connected to a non-linear load.

i_{sh} are given by the following expressions:

$$i_{fh} = \frac{Z_s}{Z_s + Z_f} i_h; \quad i_{sh} = \frac{Z_f}{Z_s + Z_f} i_h \quad (39.47)$$

At the resonant frequency the inductive reactance of the passive filter is equal to the capacitive reactance of the filter, that is:

$$2\pi fr L = \frac{1}{2\pi fr C} \quad (39.48)$$

Therefore, the resonant frequency of the passive filter is equal to:

$$fr = \frac{1}{2\pi \sqrt{LC}} \quad (39.49)$$

The passive filter band width is defined by the upper and lower cut-off frequency, at which values the filter current gain is -3 dB, as shown in Fig. 39.39.

The magnitude of the passive filter impedance as a function of the frequency is shown in Fig. 39.40.

At the resonant frequency the passive filter magnitude is equal to the resistance. If the resistance is zero, the filter is in

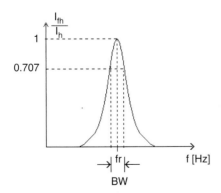

FIGURE 39.39 The passive filter bandwidth.

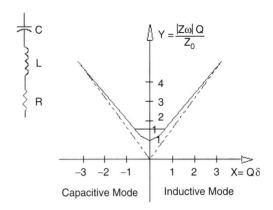

FIGURE 39.40 The frequency response of the passive LC filter.

short circuit. The quality factor of the passive filter is defined by the following expression:

$$Q = \frac{\omega_n L}{R} \quad (39.50)$$

It is important to note that the operation of the series active power filter with off-tuned passive filter has an adverse impact on the inverter power rating compared to the normal case. The more off-tuned the passive filter, the more rated apparent power is required by the series active power filter.

39.4.3.5 Protection Requirements

Short circuits in the power distribution system generate large currents that flow through the power lines until the circuit breaker operates clearing the fault. The total clearing time of a short circuit depends on the time delay imposed by the protection system. The clearing time cannot be instantaneous due to the operating time imposed by the overcurrent relay and by the total interruption time of the power circuit breaker. Although power system equipment, such us power transformers, cables, etc. are designed to withstand short-circuit currents during at least 30 cycles, the active power filter may suffer severe damage during this short time. The withstand capability of the series active power filter depends mainly on the inverter power semiconductor characteristics.

Since the most important feature of series active power filters is the small rated power required to compensate the power system, typically 10–15% of the load rated apparent power, the inverter semiconductors are rated for low values of blocking voltages and continuous currents. This makes series active power filters more vulnerable to power system faults.

The block diagram of the protection scheme described in this section is shown in Fig. 39.41. It consists of a varistor connected in parallel to the secondary winding of each coupling transformer, and a couple of antiparallel thyristors [8]. A special circuit detects the current flowing through the varistors and generates the gating signals of the antiparallel thyristors.

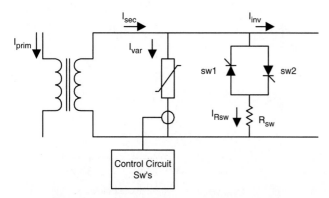

FIGURE 39.41 The series active power filter protection scheme.

The protection circuit of the series active power filter must protect only the PWM voltage-source inverter connected to the secondary of the coupling transformers and must not interfere with the protection scheme of the power distribution system. Since the primary of the active power filter coupling transformers are connected in series to the power distribution system, they operate as current transformers, so that their secondary windings cannot operate in open circuit. For this reason, if a short circuit is detected in the power distribution system, the PWM voltage-source inverter cannot be disconnected from the secondary of the current transformer. Therefore, the protection scheme must be able to limit the amplitude of the currents and voltages generated in the secondary circuits. This task is performed by the varistors and by the magnetic saturation characteristic of the transformers.

The main advantages of the series active power filter protection scheme described in this section are the following:

(i) it is easy to implement and has a reduced cost,
(ii) it offers full protection against power distribution short-circuit currents, and
(iii) it does not interfere with the power distribution system.

When short-circuit currents circulate through the power distribution system, the low saturation characteristic of the transformers increases the current ratio error and reduces the amplitude of the secondary currents. The larger secondary voltages induced by the primary short-circuit currents are clamped by the varistors, reducing the amplitude of the PWM voltage-source inverter currents. After a few cycles of duration of the short circuit, the PWM voltage-source inverter is bypassed through a couple of antiparallel thyristors, and at the same time the gating signals applied to the PWM voltage-source inverter are removed. In this way, the PWM voltage-source inverter can be turned off. The principles of operation and the effectiveness of the protection scheme are shown in Fig. 39.42.

The secondary short-circuit currents will circulate through the antiparallel thyristors and the varistors until the fault is cleared by the protection equipment of the power distribution system.

By using the protection scheme described in this subsection, the voltage and currents reflected in the secondary of the coupling transformers are significantly reduced. When short-circuit currents circulate through the power distribution system, the low saturation characteristic of the coupling transformers increases the current ratio error and reduces the amplitude of the secondary voltages and currents. Moreover, the saturated high secondary voltages induced by the primary short-circuit currents are clamped by the varistors, reducing the amplitude of the PWM voltage-source inverter ac currents. Once the secondary current exceeds a predefined reference value, the PWM voltage-source inverter is bypassed through a couple of antiparallel thyristors, and then the gating signals applied to the PWM voltage-source inverter are removed. The effectiveness of the protection scheme is shown in Fig. 39.43.

By increasing the current ratio error due to the magnetic saturation, the energy dissipated in the secondary of the coupling transformer is significantly reduced. The total energy dissipated in the varistor for the different protection conditions is shown in Fig. 39.44.

39.4.4 Control Issues

The block diagram of a series active power filter control scheme that compensates current harmonics and voltage unbalance simultaneously is shown in Fig. 39.45.

Current and voltage reference waveforms are obtained by using the Park transformation (instantaneous reactive power theory) voltage unbalance is compensated by calculating the negative and zero sequence fundamental components of the system voltages. These voltage components are added to the source voltages through the coupling transformers compensating the voltage unbalance at the load terminals. In order to reduce the amplitude of the current flowing through the neutral conductor, the zero sequence components of the line currents are calculated. In this way, it is not necessary to sense the current flowing through the neutral conductor.

39.4.4.1 Reference Signals Generator

The compensation characteristics of series active power filters are defined mainly by the algorithm used to generate the reference signals required by the control system. These reference signals must allow current and voltage compensation with minimum time delay. Also it is important that the accuracy of the information contained in the reference signals allows the elimination of the current harmonics and voltage unbalance present in the power system. Since the voltage and

39 Active Filters

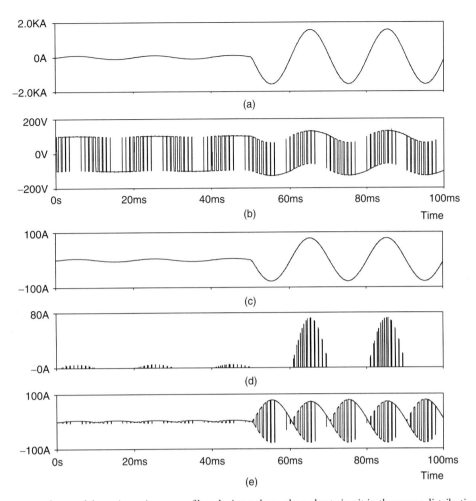

FIGURE 39.42 Current waveforms of the series active power filter during a three-phase short circuit in the power distribution system set at 50 ms: (a) power distribution system line current; (b) current transformer secondary voltage; (c) inverter ac current; (d) current through an inverter leg; and (e) inverter dc bus current.

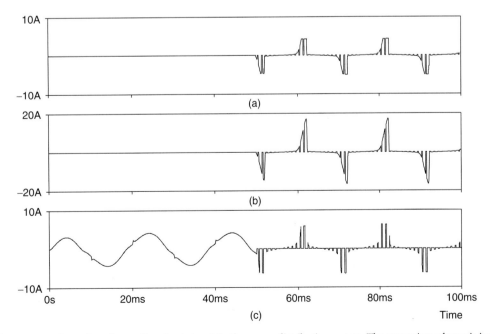

FIGURE 39.43 Current waveforms for a line-to-line short circuit in the power distribution system. The protection scheme is implemented with the varistor, a couple of antiparallel thyristors, and a coupling transformer with low saturation characteristic: (a) current through the varistor; (b) current through the thyristors; and (c) inverter ac current.

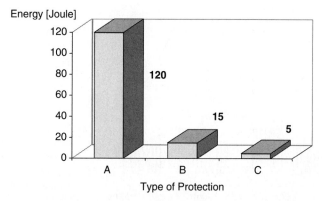

FIGURE 39.44 The total energy dissipated in the varistor during the power system short circuit for different protection schemes implementations: (a) only with the varistor used; (b) the varistor is connected in parallel to a bidirectional switch; and (c) with the complete protection scheme operating (including the coupling transformer with low saturation characteristics).

current control scheme are independent, the equations used to calculate the voltage reference signals are the following:

$$\begin{bmatrix} v_{a0} \\ v_{a1} \\ v_{a2} \end{bmatrix} = \frac{1}{\sqrt{3}} \begin{bmatrix} 1 & 1 & 1 \\ 1 & a & a^2 \\ 1 & a^2 & a \end{bmatrix} \cdot \begin{bmatrix} v_a \\ v_b \\ v_c \end{bmatrix} \quad (39.51)$$

The voltages v_a, v_b, and v_c correspond to the power system phase-to-neutral voltages before the current transformer. The reference voltage signals are obtained by making the positive sequence component, v_{a1} zero and then applying the inverse of the Fortescue transformation. In this way the series active power filter compensates only voltage unbalance and not voltage regulation. The reference signals for the voltage unbalance control scheme are obtained by applying the following equations:

$$\begin{bmatrix} v_{refa} \\ v_{refb} \\ v_{refc} \end{bmatrix} = \frac{1}{\sqrt{3}} \begin{bmatrix} 1 & 1 & 1 \\ 1 & a^2 & a \\ 1 & a & a^2 \end{bmatrix} \cdot \begin{bmatrix} -v_{a0} \\ 0 \\ -v_{a2} \end{bmatrix} \quad (39.52)$$

In order to compensate current harmonics generated by the non-linear loads, the following equations are used:

$$\begin{bmatrix} i_{aref} \\ i_{bref} \\ i_{cref} \end{bmatrix} = \sqrt{\frac{2}{3}} \cdot \begin{bmatrix} 1 & 0 \\ -\frac{1}{2} & \frac{\sqrt{3}}{2} \\ -\frac{1}{2} & -\frac{\sqrt{3}}{2} \end{bmatrix} \cdot \begin{bmatrix} v_\alpha & v_\beta \\ -v_\beta & v_\alpha \end{bmatrix}^{-1} \begin{bmatrix} p_{ref} \\ q_{ref} \end{bmatrix} + \frac{1}{\sqrt{3}} \begin{bmatrix} i_0 \\ i_0 \\ i_0 \end{bmatrix} \quad (39.53)$$

where i_0 is the fundamental zero sequence component of the line current and is calculated using the Fortescue transformation Eq. (39.50).

$$i_0 = \frac{1}{\sqrt{3}} (i_a + i_b + i_c) \quad (39.54)$$

In Eq. (39.40) p_{ref}, q_{ref}, v_α, and v_β are defined according to the instantaneous reactive power theory. In order to avoid the compensation of the zero sequence fundamental frequency current the reference signal i_0 must be forced to circulate through a high-pass filter generating a current i_0' which

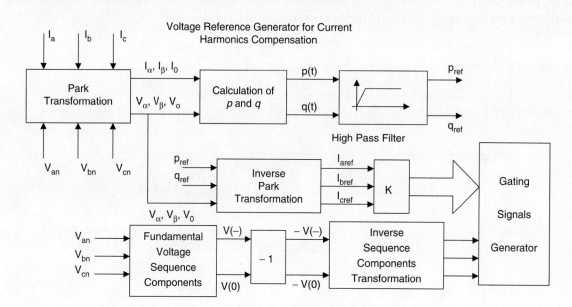

FIGURE 39.45 Compensation scheme for voltage unbalance correction.

is used to create the reference signal required to compensate current harmonics. Finally, the general equation that defines the references of the PWM voltage-source inverter required to compensate voltage unbalance and current harmonics is the following:

$$\begin{bmatrix} V_{refa} \\ V_{refb} \\ V_{refc} \end{bmatrix} = K_1 \left\{ \sqrt{\frac{2}{3}} \begin{bmatrix} 1 & 0 \\ \frac{-1}{2} & \frac{\sqrt{3}}{2} \\ \frac{-1}{2} & \frac{-\sqrt{3}}{2} \end{bmatrix} \cdot \begin{bmatrix} v_\alpha & v_\beta \\ -v_\beta & v_\alpha \end{bmatrix}^{-1} \begin{bmatrix} p_{ref} \\ q_{ref} \end{bmatrix} \right.$$

$$\left. + \frac{1}{\sqrt{3}} \begin{bmatrix} i'_0 \\ i'_0 \\ i'_0 \end{bmatrix} \right\} + K_2 \left\{ \frac{1}{\sqrt{3}} \begin{bmatrix} 1 & 1 & 1 \\ 1 & a^2 & a \\ 1 & a & a^2 \end{bmatrix} \cdot \begin{bmatrix} -v_{a0} \\ 0 \\ -v_{a2} \end{bmatrix} \right\}$$

(39.55)

where K_1 is the gain of the coupling transformer which defines the magnitude of the impedance for high-frequency current components, and K_2 defines the degree of compensation for voltage unbalance, ideally K_2 equals 1.

39.4.4.2 Gating Signals Generator

This circuit provides the gating signals of the three-phase PWM voltage-source inverter required to compensate voltage unbalance and current harmonic components. The current and voltage reference signals are added and then the amplitude of the resultant reference waveform is adjusted in order to increase the voltage utilization factor of the PWM inverter for steady-state operating conditions. The gating signals of the inverter are generated by comparing the resultant reference signal with a fixed frequency triangular waveform. The triangular waveform helps to keep the inverter switching frequency constant (Fig. 39.46).

A higher voltage utilization of the inverter is obtained if the amplitude of the resultant reference signal is adjusted for the steady-state operating condition of the series active power filter. In this case, the reference current and reference voltage waveforms are smaller. If the amplitude is adjusted for transient operating conditions, the required reference signals will have a larger value, creating a higher dc voltage in the inverter and defining a lower voltage utilization factor for steady-state operating conditions.

The inverter switching frequency must be higher in order not to interfere with the current harmonics that need to be compensated.

39.4.5 Control Circuit Implementation

Since the control scheme of the series active power filter must translate the current harmonics components that need to be compensated in voltage signals, a proportional controller is used. The use of a PI controller is not recommended since it would modify the reference waveform and generate new current harmonic components. The gain for proportional controller depends on the load characteristics and its value fluctuates between one and two.

Another important element used in the control scheme is the filter that allows to generate p_{ref} and q_{ref} (Fig. 39.35). The frequency response of this filter is very important and must not introduce any phase-shift or attenuation to the low-frequency harmonic components that must be compensated. A high-pass first-order filter tuned at 15 Hz is adequated. This corner frequency is required when single-phase non-linear loads are compensated. In this case, the dominant current harmonic is the third. An example of the filter that can be used is shown in Fig. 39.47.

The operator "a" required to calculate the sequence components of the system voltages (Fortescue transformation) can be obtained with the all-pass filter shown in Fig. 39.48.

The phase-shift between V_o and V_i is given by the following expression:

$$\theta = 2\arctan(2\pi f R_2 C) \tag{39.56}$$

Since the phase-shift is negative, the operator "a" is obtained by using an inverter (180°) and then by tuning $\theta = -60°$.

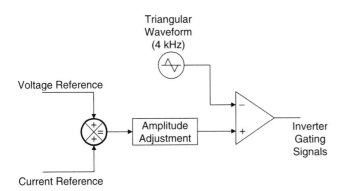

FIGURE 39.46 The block diagram of the gating signals generator.

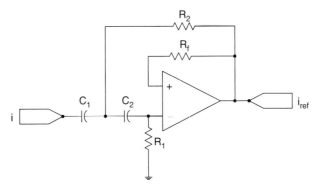

FIGURE 39.47 The first-order high-pass filter implemented for the calculation of p_{ref} and q_{ref} ($C_1 = C_2 = 0.1\,\mu F$, $R_1 = R_f = 150\,k\Omega$, $R_2 = 50\,k\Omega$).

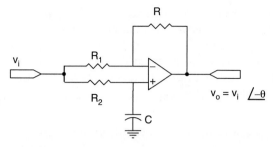

FIGURE 39.48 The all-pass filter used as a phase shifter.

The operator "a^2" is obtained by phase-shifting V_i by $-120°$. V_i is synchronized with the system phase-to-neutral voltage V_{an}.

39.4.6 Experimental Results

The viability and effectiveness of series active power filters used to compensate current harmonics was proved in an experimental setup of 5 kVA. Current waveforms generated by a non-linear load and using only passive filters and the combination of passive and the series active power filter are shown in Figs. 39.49 and 39.50.

These figures show the effectiveness of the series active power filter, which reduces the overall *THD* of the current that flows through the power system from 28.9% (with the operation of only the passive filters) to 9% with the operation of the passive filters with the active power filter connected in series. The compensation of the current that flows through the neutral conductor is shown in Fig. 39.51.

39.5 Hybrid Active Power Filters

Hybrid topologies composed of passive LC filters connected in series to an active power filter have already been proposed by the end of the eighties [2, 5, 6, 9, 10, 11]. Hybrid topology improves significantly the compensation characteristics of simple passive filters, making the use of active power filter available for high power applications, at a relatively lower cost. Moreover, compensation characteristics of already installed passive

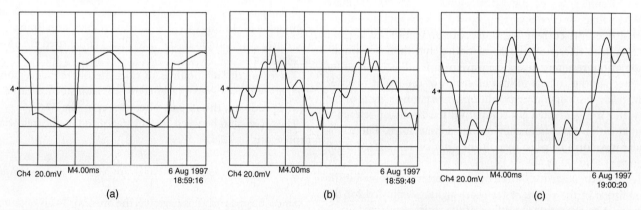

FIGURE 39.49 Experimental results without the operation of the series active power filter: (a) load current; (b) current flowing through the passive filter; and (c) current through the power supply.

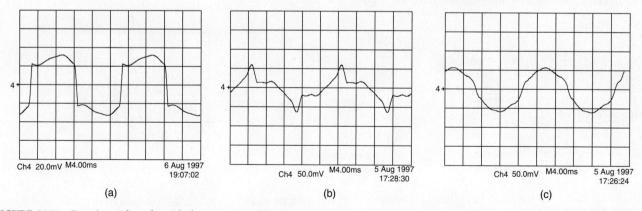

FIGURE 39.50 Experimental results with the operation of the series active power filter: (a) load current; (b) current flowing through the passive filter; and (c) current through the power supply.

39 Active Filters

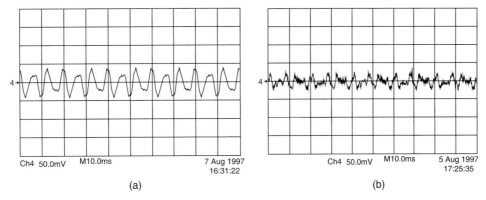

FIGURE 39.51 Experimental results of neutral current compensation: (a) current flowing through the neutral without the operation of the series active power filter and (b) current flowing through the neutral with the operation of the active power filter.

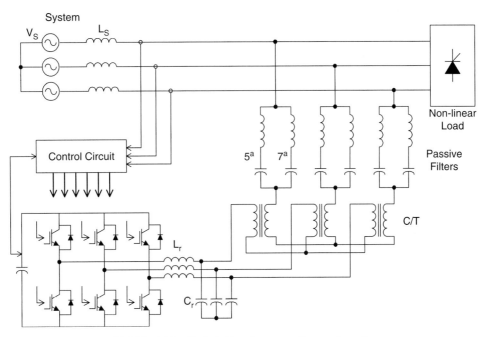

FIGURE 39.52 The hybrid active power filter configuration.

filters can be significantly improved by connecting a series active power filter at its terminals, giving more flexibility to the compensation scheme. Most of the technical disadvantages of passive filters described before can be effectively attenuate if an active power filter is connected in series to the passive approach as shown in Fig. 39.52.

The hybrid active power filter topology is shown in Fig. 39.52. The active power filter is implemented with a three-phase PWM voltage-source inverter operating at fixed switching frequency, and connected in series to the passive filter through a coupling transformer. Basically, the active power filter forces the utility line currents to become sinusoidal and in phase with the respective phase-to-neutral voltage, improving the compensation characteristics of the passive filter.

39.5.1 Principles of Operation

Since the active power filter is connected in series to the passive filter through a coupling transformer, it imposes a voltage signal at its primary terminals that forces the circulation of current harmonics through the passive filter, improving its compensation characteristics, independently of the variations in the selected resonant frequency of the passive filter. The block diagram of the proposed control scheme shown in Fig. 39.53 consists of three modules: the dc voltage control, the voltage reference generator, and the inverter gating signal generator.

The voltage reference waveform required by the inverter control scheme is obtained by adjusting the amplitude of a

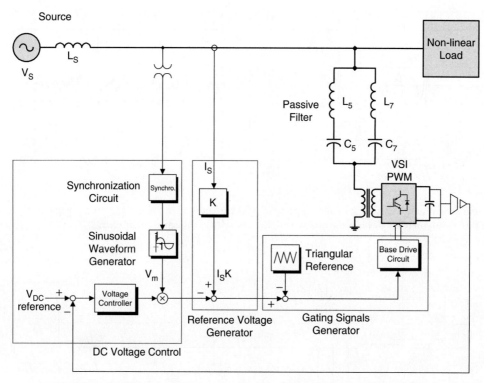

FIGURE 39.53 The hybrid active power filter topology and associated control scheme.

sinusoidal reference waveform in phase with the respective phase-to-neutral voltage and then subtracting the respective ac line current (Fig. 39.53). The sinusoidal reference signal can be obtained from the voltage system (in case of low voltage distortion) or it can be generated from an EPROM synchronized with the respective phase-to-neutral voltage. The amplitude of this reference waveform controls the inverter dc voltage and the ac mains displacement power factor. The inverter dc voltage varies according with the amount of real power absorbed by the inverter, while the ac mains power factor depends on the amount of reactive power generated by the hybrid filter, which can be controlled by changing the amplitude of the fundamental component of the inverter output voltage.

The principles of operation for current harmonic and power factor compensation are explained with the help of the single-phase equivalent circuit shown in Fig. 39.54. In the current harmonic compensation mode, the active filter improves the filtering characteristic of the passive filter by imposing a voltage harmonic waveform at its terminals with an amplitude value equals to:

$$V_{Ch} = K \cdot I_{Sh} \quad (39.57)$$

where I_{Sh} is the harmonic content of the line current to be compensated, and K is the active power filter gain. If the ac mains voltage is purely sinusoidal, the ratio between

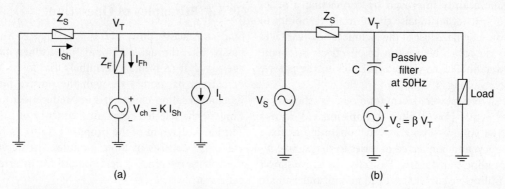

FIGURE 39.54 Single-phase equivalent circuits of the hybrid active power filter scheme: (a) for current harmonic compensation and (b) for displacement power factor compensation.

the harmonic component of the non-linear load current and the harmonic component of the ac line current (attenuation factor) is obtained from Fig. 39.54a and is equal to:

$$\frac{I_{Sh}}{I_{Lh}} = \frac{Z_F}{K + Z_F + Z_S} \quad (39.58)$$

Equation (39.58) shows that the filtering characteristics of the hybrid topology (I_{Sh}/I_{Lh}) depends on the value of the passive filter equivalent impedance, Z_F. Moreover, since the tuned factor, δ, and the quality factor, Q, can modify the filter band width and the passive filter harmonic equivalent impedance (Fig. 39.55), their values must be carefully selected in order to maintain the compensation effectiveness of the hybrid topology. In particular, a high value of the quality factor, Q, defines a large band width of the passive filter, improving the compensation characteristics of the hybrid topology. On the other hand, a low value in the quality factor and/or a large value in the tuned factor increases the required voltage generated by the active power filter necessary to keep the same compensation effectiveness, which increases the active power filter rated power.

Figure. 39.55 shows how the active power filter gain, K, in Eq. (39.57) affects the harmonic attenuation factor of the line currents. The attenuation factor of the line current harmonics expressed in percentage is obtained from Eq. (2), and is shown in Fig. 39.55, for a power distribution system with two passive filters tuned at fifth and seventh harmonics.

Also, the K factor affects the THD of the line current, as it is shown in Eq. (39.59).

$$THD\,i = \frac{\sqrt{\sum_{h=2} [I_{Lh} \cdot (Z_F/(Z_S + Z_F + K))]^2}}{I_{S1}} \quad (39.59)$$

Equation (39.59) indicates that the total harmonic distortion of the line current decreases if K increases. In other words, a better hybrid filter compensation is obtained for

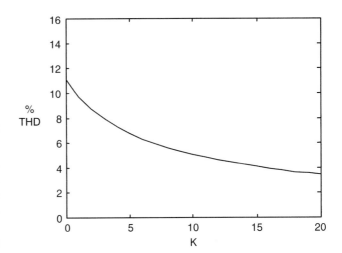

FIGURE 39.56 System line current THD vs K factor.

larger values of voltage harmonic components generated by the active power filter. Also, it is shown that the compensation capability of the hybrid filter depends on the compensation characteristic of the passive filter, that is the filter impedance value and tuned frequency will affect the active filter rated power required to satisfy the system line current compensation requirements. Figure 39.56 shows how the line current THD is affected for different values of K, for a power distribution system connected to a high power six-pulses rectifier and passive filters tuned at the fifth and seventh harmonics.

Displacement power factor correction can be achieved by controlling the voltage drop across the passive filter capacitor. In order to do that a voltage at fundamental frequency is generated at the inverter ac terminals, with an amplitude equals to:

$$V_C = \beta \cdot V_T \quad (39.60)$$

Displacement power factor control can be achieve since at fundamental frequency the passive filter equivalent impedance is capacitive. The reactive power generated by the passive filter is obtained by changing the voltage imposed by the active power filter across the passive filter capacitor terminals. The passive filter fundamental current component is defined by the following expression:

$$i_F = C\frac{d}{dt}(v_T - \beta v_T) = (1 - \beta) C\frac{dv_T}{dt} = C_\gamma \beta \frac{dv_T}{dt} \quad (39.61)$$

Equation (39.61) proves that the equivalent capacitance at fundamental frequency C_γ, can be modified by changing β. The reactive power generated by the active filter is β times the reactive power generated by the passive filter and can be defined by:

$$Q_\gamma = V_C \cdot I_F = \beta V_T I_F \quad (39.62)$$

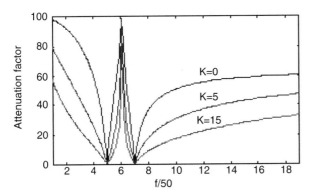

FIGURE 39.55 The attenuation factor of the line current harmonics for different frequency values.

Equation (39.62) shows that if $\beta > 0$ the active power filter generates a voltage at fundamental frequency in phase with V_T, reducing the reactive power that flows to the load. If $\beta < 0$ the active power filter generates a voltage at fundamental frequency phase shifted by 180° with respect to V_T, increasing the reactive power that flows to the load. In other words, by selecting a β positive or negative the hybrid topology can generate or absorb reactive power at fundamental frequency, compensating for leading or lagging displacement power factor of the non-linear load. This is achieved by changing the voltage across the passive filter capacitor C_F.

39.5.2 The Hybrid Filter Compensation Performance

The design characteristics of the passive filters have an important influence on the compensation performance of the hybrid topology. The tuned frequency and quality factor of the passive filter directly influence the compensation characteristics of the hybrid topology. If these two factors are not properly selected, the rated power of the active filter must be increased in order to get the required compensation performance. The tuned factor, δ, in per unit with respect to the resonant frequency is defined by Eq. (39.63).

$$\delta = \frac{\omega - \omega_n}{\omega_n} \quad (39.63)$$

Here, δ, defines the magnitude in which the passive filter resonant frequency changes due to the variations in the power system frequency and modifications in the values of the passive filter parameters L and C. The values of L and C can change due to aging conditions, temperature, or design tolerances. The tuning factor, δ, can also be defined as:

$$\delta = \frac{\Delta f}{f_n} + \frac{1}{2}\left(\frac{\Delta L}{L_n} + \frac{\Delta C}{C_n}\right) \quad (39.64)$$

The quality factor, Q, of the passive filter is defined by Eq. (39.65).

$$Q = \frac{X_0}{R} \quad (39.65)$$

where X_0 is equal to:

$$X_0 = \omega_n L = \frac{1}{\omega_n C} = \sqrt{\frac{L}{C}} \quad (39.66)$$

The passive filter quality factor, Q, defines the ratio between the passive filter reactance with respect to its resistance. In order to have a passive filter low impedance value in a limited frequency band width, defined by the maximum expected changes of the rated power system frequency, it is necessary to reduce X_0 or increase the passive filter resistance R,

reducing the value of Q. However, if R increases, the equivalent impedance of the passive filter at the resonant frequency will increase, with the associated power losses, increasing the overall passive filter operational cost. On the other hand, the larger the value of Q, the lower is the passive filter equivalent impedance at the resonant frequency, increasing the currents harmonic components across the filter, at the resonant operating point.

39.5.2.1 Effects of the Power System Equivalent Impedance

The influence of the power system equivalent impedance on the hybrid filter compensation performance is related with its effects on the passive filter, since if the system equivalent impedance is lower compared to the passive filter equivalent impedance at the resonant frequency, most of the load current harmonics will flow mainly to the power distribution system. In order to compensate this negative effect on the hybrid filter compensation performance, K must be increased, as shown in Eq. (39.57), increasing the active power filter rated power.

Figure 39.57 shows how the system equivalent impedance affects the relation between the system current THD with the active filter gain, K, in a power distribution system with passive filters tuned at the fifth and seventh harmonics. If Z_s decreases, the current system THD increases, so in order to keep the same compensation performance of the hybrid scheme, the active power filter gain, K, must be increased. On the other hand, if Z_s is high, it is not necessary to increase K in order to ensure a low THD value in the system current.

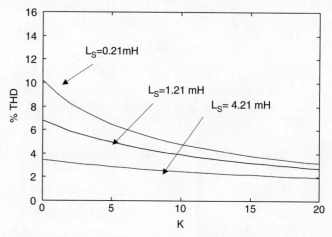

FIGURE 39.57 Relation between the THD of the line current vs the active power filter gain, K, for different values of the system equivalent impedance.

39.5.2.2 Effects of the Passive Filter Quality Factor

The quality factor, Q, defines the passive filter bandwidth. A passive filter with a high quality factor, Q, presents a larger bandwidth and better compensation characteristics. Equation (39.67) defines the quality factor value in terms of the passive filter parameters, L, R, and C. This equation shows that if R or C decreases, Q increases, and the filter bandwith becomes larger improving the hybrid scheme compensation characteristics.

$$Q = \frac{1}{R}\sqrt{\frac{L}{C}} \qquad (39.67)$$

Although by decreasing R or C the passive filter quality factor is improved, each element produces a different effect in the hybrid filtering behavior. For example, by increasing R, the filter equivalent impedance at the resonant frequency becomes bigger, affecting the current harmonic compensation characteristics at this specific frequency.

Figure 39.58 illustrates how the system current THD changes with different values of C, while keeping the filter tuned factor, δ, constant. It is important to note that the larger the value of C, the better is the compensation characteristic of the hybrid scheme. An appropriate value of C must be selected, since C cannot be increased too much. If C is too big, the fundamental component of the filter current will increase, overloading the hybrid filter and generating a large amount of reactive power to the power distribution system.

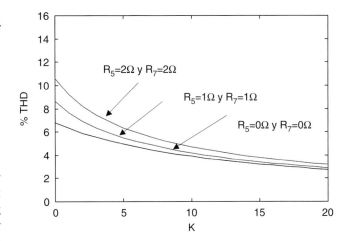

FIGURE 39.59 THD of the system line current vs the K factor for different values of filter resistor.

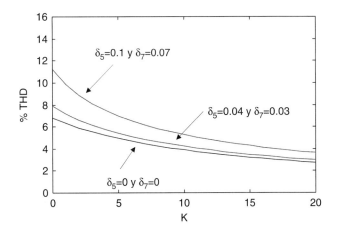

FIGURE 39.60 THD of the system line current vs the K factor for different values of tuned factor.

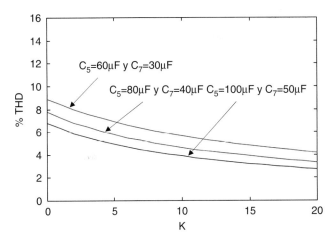

FIGURE 39.58 THD of the system line current vs the K factor for different values of filter capacitor.

of δ, the filtering performance of the hybrid filter becomes worst.

39.5.3 Experimental Results

A laboratory prototype using IGBT switches was implemented and tested in the compensation of a six-pulses controlled rectifier. The inverter was operated at 4 kHz switching frequency. Steady-state experimental results are illustrated in Figs. 39.61 and 39.62.

In case, resonance is generated between the passive filter and system equivalent reactance, the system current THD increases to 60%. By connecting the active power filter, the line current THD is reduced to 4.9%, as shown in Figs. 39.63 and 39.64.

39.5.2.3 Effects of the Passive Filter Tuned Factor

The tuned factor, δ, is defined in Eq. (39.63). This factor affects the hybrid scheme performance especially at the passive filter resonant frequency, since δ defines the changes in the system frequency, affecting the value of the passive filter resonant frequency. Figure 39.60 shows how the system current THD changes with respect to the active power filter gain, K, for different values of passive filter tuned factors. The larger the value

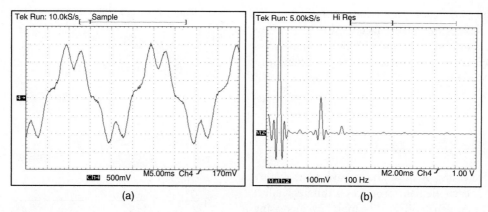

FIGURE 39.61 Experimental ac line current waveform with passive filtering compensation: (a) line current waveform ($THD = 24\%$) and (b) line current frequency spectrum.

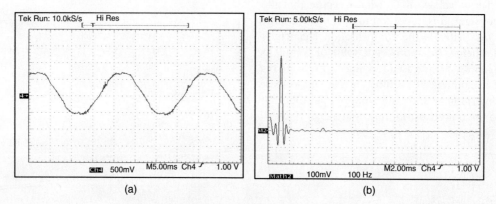

FIGURE 39.62 Experimental ac line current waveform with hybrid filter compensation: (a) line current waveform ($THD = 6.3\%$) and (b) associated frequency spectrum.

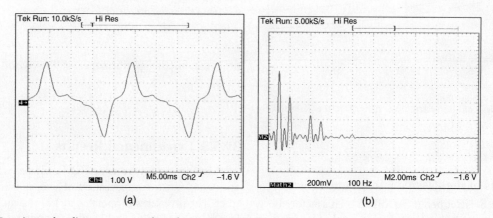

FIGURE 39.63 Experimental ac line current waveform for resonant compensation: (a) line current waveform ($THD = 60\%$) and (b) associated frequency spectrum.

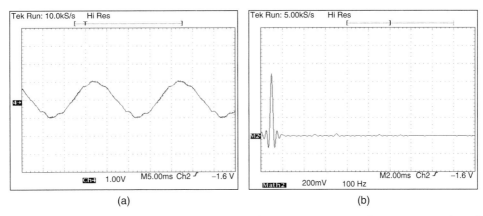

FIGURE 39.64 Experimental ac line current waveform with hybrid topology compensation: (a) line current waveform ($THD = 4.9\%$) and (b) associated frequency spectrum.

Acknowledgment

The authors would like to acknowledge the financial support from "FONDECYT" through the 1050067 project. The collaboration of Dr. Víctor Manuel Cárdenas from the University of San Luis Potosí and Pedro Ruminot from Universidad de Concepción are also recognized.

Further Reading

1. H. Akagi, Y. Kanzawa, and A. Nabae (1984) Instantaneous reactive power compensators comprising switching devices without energy components. *IEEE Trans. Ind. Appl.*, **20** (3), pp. 625–630.
2. S. Bhattacharaya and D. Divan (1995) Design and implementation of a hybrid series active filter system. *IEEE Conference Record PESC 1995*, pp. 189–195.
3. J. Dixon, S. Tepper, and L. Morán (1996) Practical evaluation of different modulation techniques for current controlled voltage-source inverters, *IEE Proc. on Electric Power Applications*, **143** (4), pp. 301–306.
4. A. Nabae, S. Ogasawara, and H. Akagi (1986) A novel control scheme for current controlled PWM inverters, *IEEE Trans. on Ind. Appl.*, **22** (4), pp. 312–323.
5. F.Z. Peng, H. Akagi, and A. Nabae (1990) A new approach to harmonic compensation in power system – A combined system of shunt passive and series active filter, *IEEE Trans. on Ind. Appl.*, **26** (6), pp. 983–989.
6. F.Z. Peng, M. Kohata, and H. Akagi (1993) Compensation characteristics of shunt passive and series active filters, *IEEE Trans. on Ind. Appl.*, **29** (1), pp. 144–151.
7. T. Tanaka, K. Wada, and H. Akagi (1995) A new control scheme of series active filters, *Conference Record IPEC-95*, pp. 376–381.
8. L. Morán, I. Pastorini, J. Dixon, and R. Wallace (1999) A fault protection scheme for series active power filters, *IEEE Trans. on Power Electronics*, **14** (5), pp. 928–938.
9. S. Fukuda and T. Endoh (1995) Control method for a combined active filter system employing a current-source converter and high pass filter, *IEEE Trans. Ind. Appl.*, **31** (3), pp. 590–597.
10. Weimin Wu, Liqing Tong, MingYue Li, Z.M. Qian, ZhengYu Lu, and F.Z. Peng (2004) A novel series hybrid active power filter, *IEEE 35th Annual Power Electronics Specialists Conference PESC 04*, **4**, pp. 3045–3049.
11. H. Fujita, T. Yamasaki, and H. Akagi (2000) A hybrid active filter for damping of harmonic resonance in industrial power systems, *IEEE Trans. Power Electronics*, **15**, pp. 215–222.
12. A. Campos, G. Joos, P.D. Ziogas, and J. Lindsay (1994) Analysis and design of a series voltage unbalance compensator based on a three-phase VSI operating with unbalanced switching functions, *IEEE Trans. on Power Electronics*, **9** (3), pp. 269–274.
13. G. Joos and L. Morán (1998) Principles of active power filters, *IEEE – IAS 98 Tutorial Course Notes*, October.
14. H. Akagi (1997) Control strategy and site selection of a shunt active filter for damping harmonics propagation in power distribution system, *IEEE Trans. Power Delivery*, **12** (1), pp. 17–28.
15. H. Akagi (1994) Trends in active power line conditioners, *IEEE Trans. Power Electronics*, **9** (3), pp. 263–268.
16. L. Morán, P. Ziogas, and G. Goos (1993) A solid-state high performance reactive-power compensator, *IEEE Trans. Ind. Appl.*, **29** (5), pp. 969–978.
17. L. Morán, J. Dixon, and R. Wallace (1995) A three-phase active power filter operating with fixed switching frequency for reactive power and current harmonic compensation, *IEEE Trans. Industrial Electronics*, **42** (4), pp. 402–408.
18. H. Fujita, S. Tominaga, and H. Akagi (1996) Analysis and design of a dc voltage-controlled static var compensator using quad-series voltage-source inverters, *IEEE Trans. Ind. Appl.*, **32** (4), pp. 970–978.
19. V. Bhavaraju and P. Enjeti (1996) An active line conditioner to balance voltages in a three-phase system, *IEEE Trans. Ind. Appl.*, **32** (2), pp. 287–292.
20. T. Tanaka and H. Akagi (1995) A new method of harmonic power detection based on the instantaneous active power in three-phase circuit, *IEEE Trans. Power Delivery*, **10** (4), pp. 1737–1742.

21. E. Watanabe, R. Stephan, and M. Aredes (1993) New concepts of instantaneous active and reactive power in electrical system with generic loads, *IEEE Trans. Power Delivery*, **8** (2), pp. 697–703.
22. M. Aredes and E. Watanabe (1995) New control algorithms for series and shunt three-phase four-wire active power filter, *IEEE Trans. Power Delivery*, **10** (3), pp. 1649–1656.
23. T. Tanaka, K. Wada, and H. Akagi (1995) A new control scheme of series active filters, *Conference Record IPEC-95*, pp. 376–381.
24. T. Thomas, K. Haddad, G. Joos, and A. Jaafari (1998) Design and performance of active power filters, *IEEE Industry Applications Magazine*, **4** (5), pp. 38–46.
25. S. Bhattacharya, T. M. Frank, D. Divan, and B. Banerjee (1998) Active filter implementation, *IEEE Industry Applications Magazine*, **4** (5), pp. 47–63.
26. Sangsun Kim and P.N. Enjeti (2002) A new hybrid active power filter (APF) topology, *IEEE Trans. Power Electronics*, **17**, pp. 48–54.

40
EMI Effects of Power Converters

Andrzej M. Trzynadlowski, Ph.D.
*University of Nevada,
Electrical Engineering Dept.,
260 Reno, Nevada, USA*

40.1 Introduction 1103
40.2 Power Converters as Sources of EMI 1103
40.3 Measurements of Conducted EMI... 1106
40.4 EMI Filters 1108
40.5 Random Pulse Width Modulation 1113
40.6 Other Means of Noise Suppression 1115
40.7 EMC Standards 1117
      References 1118

40.1 Introduction

The term "electromagnetic interference" (EMI) refers to the disturbance of operation of an engineering system due to an electromagnetic impact of the same system or other systems. Conducted electric noise and radiated electromagnetic noise are typical culprits. Watching TV in industrial centers of Western Europe in late 1960s was virtually impossible during work hours, due to operation of large dc drives in the factories and inadequate efforts to fight the EMI threat. Today, proliferation of power electronic converters and communication devices makes EMI mitigation as important as ever, because of the high, and growing, number of both the potential perpetrators and victims of EMI. To ensure electromagnetic compatibility (EMC), various EMI norms have been established. The issue of EMI is closely linked to that of susceptibility of an electronic circuit to external disturbances. Considering a multicircuit system, if each constituent circuit does not emit above-limit EMI and is immune to below-limit EMI, then each circuit is electromagnetically compatible in the system.

The institution historically entrusted with development of international EMC standards is Comité International Special des Perturbations Radioelectriques (CISPR), founded in 1933, and now forming part of International Electrotechnical Commission (IEC). Drawing on CISPR expertise, numerous other international and national organizations issue norms pertaining to various aspects of EMC. Specific EMI limits are set for internal use by individual countries and organizations, such as armed forces or large corporations. Manufacturers of electric apparatus and systems must comply with the EMC standards to have an access to the market.

In this chapter, EMI effects of power electronic converters are illustrated, and basic terms and units are introduced. EMI measurement techniques and preventive measures are described, and EMC standardization efforts are examined.

40.2 Power Converters as Sources of EMI

The switch-mode operation of power electronic converters results in current ripple mixed with the fundamental currents. Also, rapid changes of the switched voltages induce currents in parasitic (stray) capacitances that couple the converter with other closely spaced circuits. Distorted currents drawn from the power grid produce corresponding voltage drops, generating pervasive voltage noise. In general, the voltage disturbances can be divided into three classes: (1) the noise, (2) short-duration impulses superimposed on the mains voltage, and (3) transients.

Electromagnetic disturbances travel by conduction on wiring (conducted EMI), radiation in space (radiated EMI), and inter-circuit capacitive or inductive coupling. Below some 10 MHz, the EMI spreads mainly by conduction, while at higher frequencies mostly by radiation. The frequency range of 0.15–30 MHz occupies an important place in EMC norms, because it covers the radio and TV broadcast. In that range, according to those norms, it is the conducted EMI that must be measured and suppressed, while measurements of the radiated EMI are required only under special circumstances. Power electronic equipment typically produces EMI that is broadband and contained within a frequency range from the operating frequency to several megahertz. The very

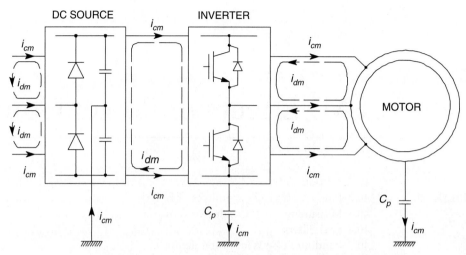

FIGURE 40.1 AC drive system with a PWM voltage-source inverter.

low frequency range of 10–150 kHz has been recently receiving an increased attention too. Consequently, with respect to most power electronic converters, it is the conducted EMI that is considered the major nuisance.

Conducted EMI appears in the form of common-mode (CM) and differential-mode (DM) voltages and currents. These two modes are illustrated in Fig. 40.1, which depicts an adjustable-speed ac drive with a PWM voltage-source inverter. Common-mode currents, i_{cm}, close to ground through parasitic capacitances, C_p, while DM currents, i_{dm}, flow only in the wires of the system. Considering the two-wire dc line supplying the inverter, the CM voltage, $V_{cm(2)}$, is defined as

$$V_{cm(2)} = \frac{V_1 + V_2}{2} \quad (40.1)$$

and the DM voltage, $V_{dm(2)}$, as

$$V_{dm(2)} = V_1 - V_2 \quad (40.2)$$

where V_1 and V_2 are line-to-ground voltages of the two wires. The CM and DM currents, $I_{cm(2)}$ and $I_{dm(2)}$, are defined as

$$I_{cm(2)} = I_1 + I_2 \quad (40.3)$$

$$I_{dm(2)} = \frac{I_1 - I_2}{2} \quad (40.4)$$

For the three-wire ac lines, the CM voltage, $V_{cm(3)}$, and current, $I_{cm(3)}$, are defined as

$$V_{cm(3)} = \frac{V_1 + V_2 + V_3}{3} \quad (40.5)$$

$$I_{cm(3)} = I_1 + I_2 + I_3 \quad (40.6)$$

where subscript "3" applies to the third wire. The DM voltage and current, if measured with respect to wires 1 and 2, are defined by Eqs. (40.2) and (40.4) as for the two-wire system.

To illustrate EMI generation by PWM power converters, a typical waveform of output current of the inverter in Fig. 40.1 is shown in Fig. 40.2. The fundamental sine wave dominates, but the high-frequency current ripple is easy to observe too. As seen in Fig. 40.3, numerous clusters of higher harmonics appear in the frequency spectrum of the current. Their locations coincide with multiples of the switching frequency (here, 2 kHz), which, as in all PWM converters is much higher than the fundamental frequency (here, 9 Hz) of the ac output (or input) voltage.

The range of frequency in Fig. 40.3 is narrow (24 kHz). A wide-range frequency spectrum of another inverter current is shown in Fig. 40.4. It can be seen that the harmonics become progressively less prominent when the frequency increases. This observation justifies the already mentioned focus on the below 30 MHz frequency range.

As already mentioned, the fast ON–OFF and OFF–ON state transition of switches, whose switching times are typically in the order of a fraction of microsecond, causes EMI too. It is due to the high dv/dt ratios, which generate transient charge currents in inter-wire capacitances of cables, intra-winding capacitances of transformers, and p–n junctions of power diodes, as well as transient circuit-to-ground currents through parasitic capacitances. If not filtered out, the transient currents appear in current waveforms as spikes on top of the fundamental sinusoid. The circuit-to-ground current spikes constitute CM emissions, while the DM emissions comprise of the ripple current and periodic current spikes coincident with the switching pattern of the converter.

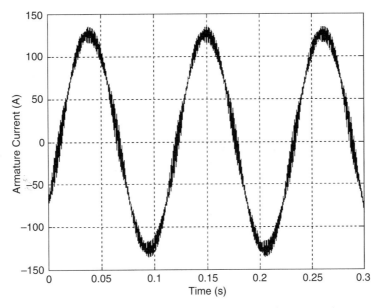

FIGURE 40.2 Waveform of output current of a PWM voltage-source inverter.

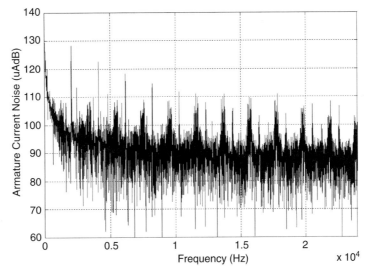

FIGURE 40.3 Frequency spectrum of the current in Fig. 40.2.

Power electronic converters, such as rectifiers and inverters, tend to generate current-related interference at their input, injecting noise to the power grid, and voltage-related interference at their outputs, which may disturb operation of communication and control systems in proximity to the converter. Considering the popular low-power dc-to-dc converters, the flyback and buck converters produce particularly noisy input currents, because the semiconductor switch is directly in series with the input power line. The power line conducted EMI from dc-to-dc converters, in conjunction with the finite source impedance, can cause disturbance of operation of system operating from a common power bus. Disturbances can also be caused by emissions radiated from converters with very high switching frequencies. The radiated EMI is most often a magnetic field, due to the typically low voltages and high currents involved, but electric fields can sometimes be a nuisance too, especially with respect to the own control circuitry of the converter. Power lines can also radiate EMI, whose source is the current noise conducted from converters.

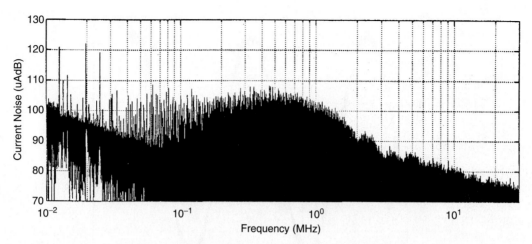

FIGURE 40.4 Wide-range frequency spectrum of an inverter current.

40.3 Measurements of Conducted EMI

It has been already mentioned that in power electronic systems it is the conducted EMI, whose mitigation is of major practical importance. Therefore, the topic of EMI measurements in this chapter is limited to that type of EMI, within the 10 kHz–30 MHz frequency range. The related terms, units, instruments, and techniques are subsequently described. Concerning the radiated interference, the reader is referred to one of many publications covering the radio-frequency (RF) aspects of EMC, e.g. [1–5].

The basic quantity of signal is power. Because of the potentially wide magnitude ranges involved, the decibel (dB) system is often used, in which 1 W serves as the reference. In the dB system, signal power, $P_{(dBW)}$, expressed in dBW, is calculated from the power in watts, $P_{(W)}$, as

$$P_{(dBW)} = 10 \log P_{(W)} \text{ dBW} \quad (40.7)$$

If milliwatts, mW, often abbreviated as "m," are considered, then

$$P_{(dBm)} = P_{(dBW)} + 30 \text{ dBm} \quad (40.8)$$

However, in EMI measurements, it is voltage that is commonly used, and voltage probes are employed. Power is proportional to the square of voltage. Thus, taking 1 V as reference, voltage in the logarithmic scale, $V_{(dbV)}$, is defined as

$$V_{(dBV)} = 20 \log V_{(V)} \text{ dBV} \quad (40.9)$$

In practice, the reference voltage is usually 1 μV, and

$$V_{(dB\mu V)} = 20 \log V_{(\mu V)} \text{ dB}\mu V = V_{(dBV)} + 120 \text{ dB}\mu V \quad (40.10)$$

If the commonly used resistance of 50 Ω is employed in the measurements, conversions of $P_{(dBm)}$ into $V_{(dB\mu V)}$ and vice versa can be made from the equation

$$V_{(dB\mu V)} + P_{(dBm)} = 107 \quad (40.11)$$

Current probes for EMI current measurements are used when limits on conducted EMI are expressed in units of current. Denoting the terminal-pair transfer impedance of a current probe by Z_T, current expressed in dBμA is given by

$$I_{(dB\mu A)} = V_{(dB\mu V)} - Z_{T(dB\Omega)} \quad (40.12)$$

where

$$Z_{(dB\Omega)} = 20 \log Z_{(\Omega)} \text{ dB}\Omega \quad (40.13)$$

To characterize broadband coherent (continuous spectrum) signals, an amplitude density function, A, is used in place of a discrete amplitude measure. The so-called broadband unit is 1 V/Hz. In EMC techniques, the 1 μV/MHz unit is more convenient. The conversion equations are

$$A_{(dB\mu V/MHz)} - A_{(dBV/Hz)} = 240 \quad (40.14)$$

and

$$V_{(dB\mu V)} - A_{(dB\mu V/MHz)} = 20 \log B_{(MHz)} \quad (40.15)$$

where $B_{(MHz)}$ is the measurement bandwidth, usually 9 kHz, that is, 0.009 MHz.

Conducted emissions are measured using EMI receivers, which usually measure voltage from an appropriate probe. Basically, the EMI receiver is a tunable voltmeter, whose block diagram is shown in Fig. 40.5. Modern EMI receivers cover extremely wide frequency ranges, e.g. from 9 kH to 18 GHz.

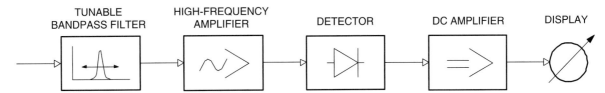

FIGURE 40.5 Block diagram of the EMI receiver.

To allow comparison of EMI measurement results, precise specifications for the detector have been established. The most common types are peak, quasi-peak, and average detectors. They are based on RC circuits having different charging and discharging time constants. It must be mentioned that the displays are usually calibrated to the rms value of an equivalent sinusoidal signal, while the actual detector often measures a different characteristic of the culprit noise.

Measurement of the peak EMI is important because of the ubiquity of impulsive sources and sensitivity of many types of electronic circuits to impulses. The peak detectors have a short charging time constant, in the order of a fraction of microsecond, and a very long discharging time constant, in the order of tens of seconds. The quasi-peak detectors, strongly recommended by CISPR, have been primarily developed to emulate the human ear acting as a broadcast disturbance sensor. The charging and discharging time constants are 1 and 160 ms, respectively. The average detector is characterized by a long, in the order of a second, integration time constant. For the 0.15–30 MHz frequency range, CISPR requires that an EMI measuring instrument have a 9 kHz bandwidth of the filter at 6 dB (from −5 to +4 kHz from the central frequency value) and that the measurement accuracy for a sinusoidal signal be <2 dB.

To increase the accuracy within wide ranges of the frequency and magnitude of signals, modern EMI receivers have several measuring channels. Each channel contains an independent quasi-peak detector, and the saturation and threshold levels of individual channels differ from each other. Output signals of the detectors are added in a circuit operating the display.

For comparability of EMI measurements, an interface circuit is inserted between the EMI source and the mains. This circuit is called a line impedance stabilization network (LISN) or artificial mains. The LISN produces a standard load impedance for the EMI source and filters out high-frequency disturbances on the mains, which could distort measurement results. An example of single-phase LISN, recommended by CISPR 16, a publication devoted to proper EMI testing, is shown in Fig. 40.6. In the equipment under test (EUT) block, the positive and negative line terminals are denoted by P and N, respectively, while G is the grounded terminal. The load resistance, R_L, is specified by CISPR 16 to be 50 Ω. The switch, S, allows transition from the CM to DM measurements. Three-phase LISNS are also available. Refer to the manual of a given LISN for operating instructions.

For quick qualitative EMI measurements, e.g. while developing an EMI filter, spectrum analyzers can be employed. A direct view of the interference levels over a wide frequency

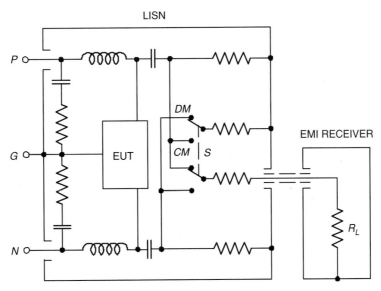

FIGURE 40.6 Simplified circuit diagram of a LISN.

range is very convenient. The frequency axis can be switched to a linear or logarithmic scale, while the magnitude is usually displayed in dBm. True peak and average detectors are available. Spectrum analyzers are rms calibrated, meaning that the reading is correct for sinusoidal waveforms only, such as discrete harmonics of the measured noise. For impulse disturbances, correct peak values are obtained. For broadband measurements, corrections must be made to convert the true peak to quasi-peak, and to account for the difference in measurement bandwidths. For help, refer to the manual of a given spectrum analyzer.

Technical details of EMI measurements for EMC compliance are too vast a topic to be fully covered here. The reader is referred to such authoritative sources as the already mentioned CISPR Publication 16 and books [1, 6].

40.4 EMI Filters

EMI generated in power electronic systems usually exceeds the allowable levels, and it must be reduced. The most common means of EMI mitigation are low-pass filters, often called radio-frequency (RF) filters. They are installed on the input and output sides of power converters. The filters are simple capacitive (C) or inductive–capacitive (LC) circuits. Resistors are sometimes added to dampen possible resonances caused by converters with high switching frequencies.

Choke coils are employed as the inductive components of the filters. They are usually single-layer solenoid structures with relatively low coefficients of inductance. In the case of CM EMI filters, two or three windings are placed on the same closed iron core. The windings are so arranged that the coil's impedance for the CM noise is high, but that for the load current and DM noise is low. A single-phase CM choke coil is shown in Fig. 40.7. Letters P, N, and G again denote the positive, negative, and ground terminals of the mains. It can be seen that magnetomotive forces $\mathscr{F}_L$ and $\mathscr{F}_{dm}$, produced by the load current i_L and DM noise current i_{dm} in the upper and lower parts of the coil, cancel each other. As a result, the magnetic flux Φ_{cm12} in the core is generated only by magnetomotive forces $\mathscr{F}_{cm1}$ and $\mathscr{F}_{cm2}$ produced by the CM noise currents i_{cm1} and i_{cm2}.

Capacitors employed in EMI filters must have low parasitic series inductance, low dielectric and ohmic losses, and stable capacitance vs frequency characteristics. Paper, metalized paper, polystyrene, and ceramic capacitors are widely used. The so-called multifunction ceramic (MFC) capacitors can suppress the high-frequency noise and absorb impulse-type transients. According to their electrical connection mode, capacitors in power line filters can be classified as X- and Y-capacitors, the former posing no threat of electrical shock to personnel, even in the case of a breakdown. Y-capacitors are typically used for attenuation of the CM noise in power line filters, but require stricter safety standards and higher reliability than X-capacitors.

To reduce the parasitic inductance, contributed mostly by the leads, the so-called feedthrough capacitors have been developed for use in EMI filters. The wire conducting the load current passes through the capacitor structure, instead of the capacitor being connected between it and the ground. This is illustrated in Figs. 40.8 and 40.9. Figure 40.8a shows the structure of the feedthrough capacitor and its electric symbol is shown in Fig. 40.8b. Figure 40.9a shows that in a conventional capacitor the parasitic inductances, L_p, of the leads are in series with the capacitance, C. The total inductance is thus $2L_p$. In contrast, as seen in Fig. 40.9b, in the feedthrough capacitor, lead inductances are connected in parallel to each other, so that the total inductance is $L_p/2$.

From the very beginning of modern power electronics, capacitors have been employed in the so-called snubbers. In the simplest embodiment, a snubber circuit consists of a capacitor with a resistor connected in series to dampen

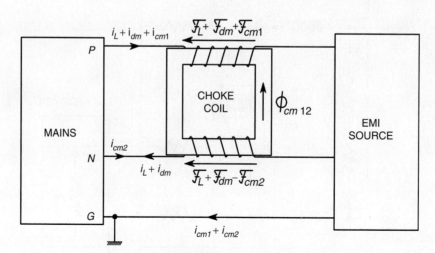

FIGURE 40.7 Single-phase CM choke coil.

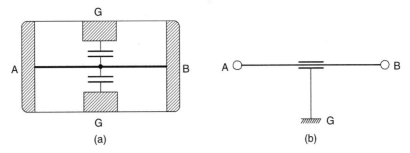

FIGURE 40.8 Feedthrough capacitor: (a) structure and (b) electric symbol.

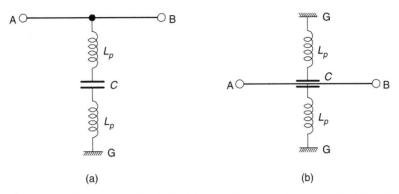

FIGURE 40.9 Parasitic inductances of the leads: (a) conventional capacitor and (b) feedthrough capacitor.

possible resonance with parasitic inductances. The snubber is placed across a semiconductor power switch to improve the switching conditions and keep the switch within its safe operating area. Snubbers reduce the EMI, which is particularly strong when the switch turns off, interrupting the load current.

Noise suppression capacitors, connected to various other points in a power electronic system, are also widely used, although they do not mitigate the EMI as effectively as LC filters with choke coils. Capacitive filters most often appear at the input and output of power electronic converters. They can be connected between lines or line-to-ground.

Basic topologies of low-pass LC filters are illustrated in Fig. 40.10. It must be stressed that EMI filters must be placed in all wires of a power electronic system. Most common configurations are shown in Fig. 40.11 for a two-wire system. The filters attenuate both the CM and DM conducted EMI. An extension of the T-type filter on a three-wire system is shown in Fig. 40.12. Similar extension can be made with respect to the other topologies of filters.

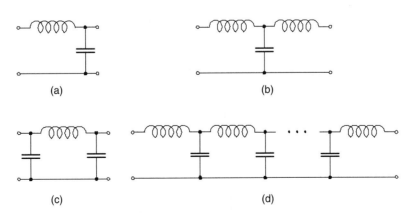

FIGURE 40.10 Basic topologies of low-pass LC filters: (a) inverted-gamma (Γ'); (b) pi (Π); (c) tee (T); and (d) multistage.

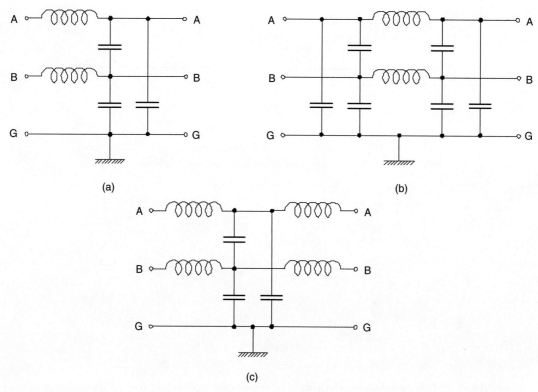

FIGURE 40.11 EMI filters in a two-wire power electronic system: (a) inverted-gamma (Γ'); (b) pi (Π); and (c) tee (T).

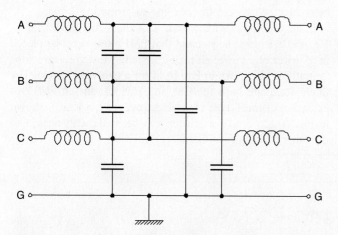

FIGURE 40.12 T-type EMI filter in a three-wire power electronic system.

An example arrangement of the CM and DM EMI filters in the ac drive system of Fig. 40.1 is shown in Figs. 40.13 and 40.14, respectively (both types of filters are installed in the actual system, and the two separate circuit diagrams have only been used for an instructional purpose).

The insertion loss, IL, is the main parameter of EMI filters. For a given frequency, f, it can be determined by measuring the voltage, V_G, of a sinusoidal signal generator, and then connecting the filter to the generator and measuring the voltage, V_F, at the filter's output. Then,

$$\text{IL} = 20 \log\left(\frac{V_G}{V_F}\right) \text{ dB} \quad (40.16)$$

The insertion loss is usually given as an IL(f) graph with logarithmic coordinates. For filter design, the two-port network theory is employed, using the transmission, A, parameters expressing the dynamic relations between the input and output variables of a two-port network. Specifically,

$$\begin{bmatrix} V_1(s) \\ I_1(s) \end{bmatrix} = \begin{bmatrix} A_{11}(s) & A_{12}(s) \\ A_{21}(s) & A_{22}(s) \end{bmatrix} \begin{bmatrix} V_2(s) \\ I_2(s) \end{bmatrix} \quad (40.17)$$

where V_1 denotes the input (EMI-source side) voltage of the network, I_1 is the input current, V_2 is the output voltage, and I_2 is the output current, while s denotes the Laplace variable (complex frequency). The transmission parameters are defined as

$$A_{11}(s) = \frac{V_1(s)}{V_2(s)} \Big|_{I_2 = 0} \quad (40.18)$$

$$A_{12}(s) = \frac{V_1(s)}{I_2(s)} \Big|_{V_2 = 0} \quad (40.19)$$

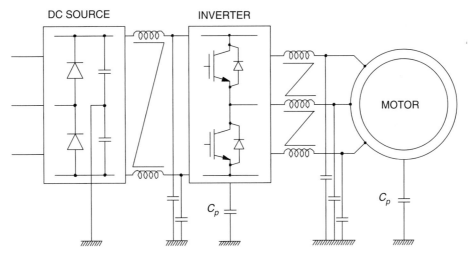

FIGURE 40.13 Placement of CM EMI filters in the drive system of Fig. 40.1.

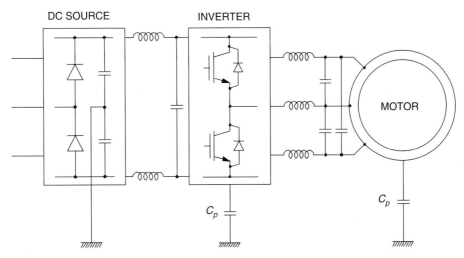

FIGURE 40.14 Placement of DM EMI filters in the drive system of Fig. 40.1.

$$A_{21}(s) = \frac{I_1(s)}{V_2(s)} \bigg|_{I_2 = 0} \quad (40.20)$$

$$A_{22}(s) = \frac{I_1(s)}{I_2(s)} \bigg|_{V_2 = 0} \quad (40.21)$$

Once the transmission parameters for a given filter are determined using the transient analysis of its circuit, the insertion loss can be found from the equation

$$IL(f)$$
$$= 20\log \left| \frac{[A_{21}(j\omega)Z_L(j\omega) + A_{22}(j\omega)]Z_S(j\omega) + A_{11}(j\omega)Z_L(j\omega) + A_{11}(j\omega)}{Z_S(j\omega) + Z_L(j\omega)} \right|$$
$$(40.22)$$

where $\omega = 2\pi f$. Then, the IL(f) graph can be plotted, as illustrated in Fig. 40.15 for an actual commercial EMI filter.

The insertion loss is roughly proportional to the total inductance and capacitance (total LC), and square of the frequency. Because of the cost and size associated with the total LC value of the filter, an optimal filter is the one with minimum total LC value, which still mitigates the EMI sufficiently to satisfy the pertaining norm. A difficulty in optimizing the EMI filter design consists in the mismatched impedance conditions under which the filters operate. From the point of view of a manufacturer of the filters, the source and load impedances are simply unknown. However, from the user's viewpoint, estimating the source and load impedances allows for better selection of the filter configuration. Specifically, for low source and load impedance, the "T" filter is recommended. The "Π" configuration is best for high source and load impedance, while for low source and high load impedance, the "Γ" (LC) filter is preferred. Finally, for high source and low load impedance, the "Γ" (CL) topology is

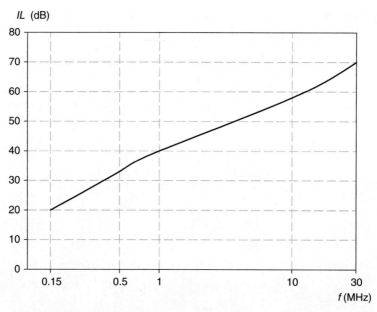

FIGURE 40.15 An example insertion loss vs frequency graph for an EMI filter.

favored. If possible, multiple-stage filter topologies should be employed.

Practical design of EMI filters is complicated by the fact that there are no ideal inductors and capacitors. Considering the equivalent electric circuit of an inductor in Fig. 40.16a and that of a capacitor in Fig. 40.16b, it can be seen that at very high frequencies, the impedance of the parasitic components, marked by the subscript "p," may dominate that of the main component. Thus, the reactance of the parasitic capacitance, C_p, of the inductor can make the overall impedance to decrease with the frequency, and the reactance of the parasitic inductance, L_p, of the capacitor can make the overall impedance of the capacitor to increase with the frequency. That is why manufacturers of EMI filters and their components go to such lengths to minimize the parasitics.

In the common case of power electronic converters fed from the grid, the so-called power line filter is installed between the grid and converter. It protects the converter and its load from external disturbances and, vice versa, screens out the grid from EMI generated in the converter. Voltage impulses originated within the grid are particularly hazardous for the converter, and varistors are used in the filter to attenuate the impulses. An example of commercial power line filter is shown in Fig. 40.17. In the drive system in Fig. 40.1, the filter would be connected to the three-phase input terminals of the dc source (rectifier).

Numerous companies offer a variety of power line and EMI filters with various voltage and current ratings. Thus, designers of power converters and converter-fed systems have a choice of quality products, without the need for designing their own filters. Useful details of the process of EMI filters design can be found in [6–8].

Active EMI filters, although very effective, are less common than the passive filters described because of their increased complexity and cost. They are based on the feedback principle, generating currents that neutralize the noise currents. In dc-to-dc converters, they are also used to limit inrush currents, for example during the so-called hot swapping when a faulty electronic board is replaced without interrupting operation of the system. In certain three-phase systems with power converters such as ac drives, the CM voltage noise, apart from producing external interference, causes internal problems,

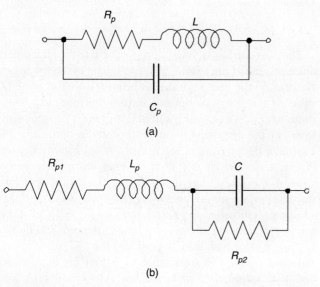

FIGURE 40.16 High-frequency equivalent circuits of: (a) capacitor and (b) inductor.

40 EMI Effects of Power Converters

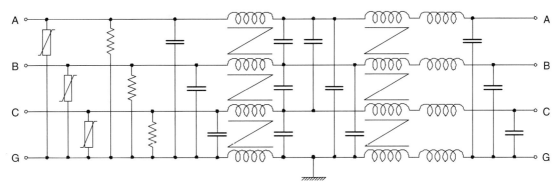

FIGURE 40.17 Power line filter.

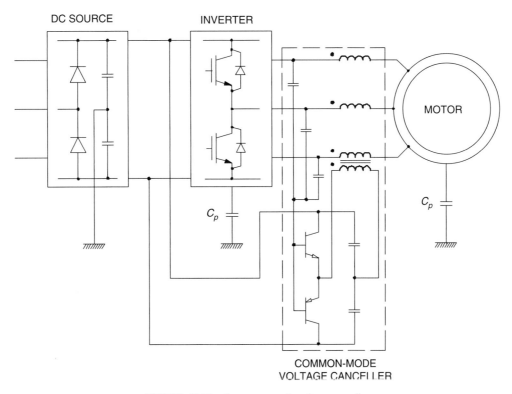

FIGURE 40.18 Common-mode voltage canceller.

e.g. accelerated wear of bearings. The so-called CM voltage cancellers are one of the means to mitigate those problems. Such a voltage canceller in the drive system of Fig. 40.1 is shown in Fig. 40.18.

40.5 Random Pulse Width Modulation

Suppression of the low-frequency noise, originating from the PWM operating mode of most power converters, is particularly difficult. As already pointed out in Section 40.1, the noise, mostly harmonics of the voltages and currents, appears in the frequency spectra as harmonic clusters around multiples of the switching frequency, f_{sw}. As seen in Fig. 40.1, the highest magnitudes of the harmonics appear at the lower end of the frequency range, while, as shown in Fig. 40.15, the insertion loss of EMI filters progressively increases with the frequency. Thus, at low frequencies, effectiveness of the filters decreases. Increasing the total LC value helps, but it results in corresponding increase of the cost and bulk of the filters. Therefore, mitigation of the harmonics by means of special PWM strategies is the best solution for filter minimization.

Most of the commercial PWM power converters are characterized by a fixed switching frequency, f_{sw}, defined here as a reciprocal of the switching period, T_{sw}. The switching period constitutes the length of a switching interval, which houses pulses of switching variables of the converter, one pulse per variable. Duty ratio of a given pulse varies from 0 to 1, that is, the width of the pulse is in the range of $0-T_{sw}$. It has been demonstrated in several publications that if individual switching periods are randomly varied, then the discrete harmonic power (watts) of spectra of the voltages and currents of the converter is transferred to continuous power spectral density (watts/hertz) [9]. This strategy of random pulse width modulation (RPWM) results in significant mitigation of both the acoustic and electromagnetic noise associated with the current harmonics [10, 11]. Accumulated experience and theoretical considerations show that varying T_{sw} from 50 to 150% of the average switching period, $T_{sw,ave}$, is sufficient. Thus, the nth switching period is determined as

$$T_{sw,n} = (r_n + 0.5)T_{sw,ave} \quad (40.23)$$

where r_n is a uniform-probability random number in the 0–1 range.

For convenience, in practical digital modulators for PWM converters, the switching cycles coincide with the sampling cycles of the modulator, that is, $f_{sw} = f_{smp}$, where f_{smp} denotes the sampling frequency (not to be confused with the much higher clock frequency). Consequently, the varying switching rate is associated with identically varying sampling rate. When a single digital system performs more tasks than just PWM, the random sampling rate is a distinct disadvantage. For instance, in a control system, the sampling rate defines the control bandwidth and it is selected at a specific trade-off level. Therefore, an RPWM technique with a fixed sampling rate but variable switching frequency is more practical. Figure 40.19 illustrates: (a) the most common, non-random PWM technique with fixed-period and coinciding sampling and switching cycles, (b) RPWM with randomly varied and coinciding sampling and switching cycles, and (c) RPWM with fixed sampling periods and randomly varied switching periods, subsequently referred to as variable-delay RPWM (VD RPWM).

As seen in Fig. 40.19c, the switching cycles in the VD RPWM method are delayed with respect to the corresponding sampling cycles by a randomly varied time delay, d. The value of d for the nth switching cycle is calculated as

$$d_n = r_n T_{smp} \quad (40.24)$$

where $T_{smp} = 1/f_{smp}$ denotes the sampling period. When in two consecutive switching cycles, the kth and $k+1$th, r_k is close to 1 and r_{k+1} is close to 0, the second, $k+1$th, switching cycle may be too short, that is, its length, $T_{sw,k+1}$, may be lower than the minimum allowable length, $T_{sw,min}$. Therefore, in case of such occurrence, $T_{sw,k+1}$ is set to $T_{sw,min}$, or another value of r_k is selected. As a result, the switching periods vary from T_{min} to $2T_{smp}$. The average switching period, $T_{sw,ave}$, equals the sampling period, T_{smp}.

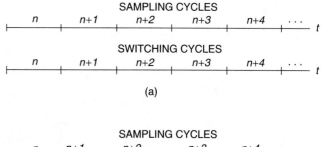
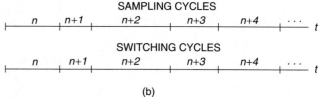
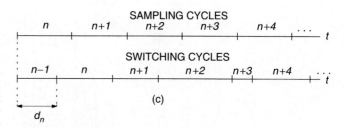

FIGURE 40.19 Illustration of PWM techniques: (a) fixed-period and coinciding sampling and switching cycles; (b) randomly varied and coinciding sampling and switching cycles; and (c) fixed-period sampling cycles and variable-period switching cycles (VD RPWM).

Any PWM strategy based on the concept of switching cycles can be employed within the RPWM method. For instance, the popular space-vector modulation can be used in three-phase voltage-source inverters. As long as the average switching period is sufficiently short, the quality of output current is similar to that in a converter with a fixed switching frequency having the same period. An example spectrum of the output current of a voltage-source inverter controlled using the VD RPWM is shown in Fig. 40.20. It is the same converter whose current spectrum, with fixed-period PWM, was shown in Fig. 40.3. The noise suppression by some 10 dBµA is easily observable. A spectrum corresponding to that in Fig. 40.4 is shown in Fig. 40.21. Here, the EMI mitigation, by about 20 dBµA, is even stronger.

Although RPWM is not a perfect tool for EMI mitigation, it effectively eliminates high harmonics of the input and output currents and output voltages in PWM power converters. The EMI filters are usually still needed for the suppression of transients, impulses, and high-frequency noise. However, their total LC value can be greatly reduced. The RPWM technique with random switching and sampling periods is a little more effective than the VD RPWM because of the somewhat bigger freedom of randomization, and it is

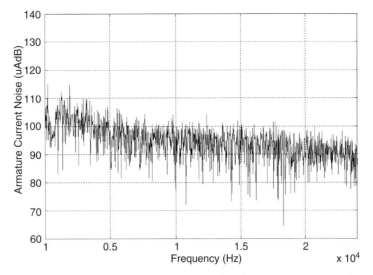

FIGURE 40.20 Frequency spectrum of output current of a voltage-source inverter with VD RPWM.

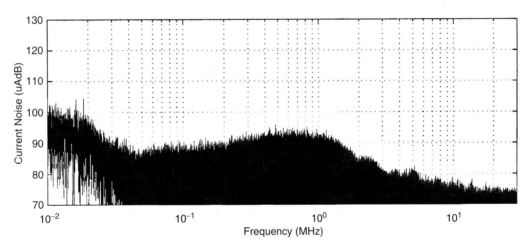

FIGURE 40.21 Wide-range frequency spectrum of inverter current with VD RPWM.

recommended for converters whose pulse width modulator operates independently of the rest of the system.

It is worth mentioning that PWM techniques have been developed that reduce the CM noise in three-phase inverters by elimination (or significant reduction of use) of the so-called zero states, in which all the three-phase terminals are clamped by the inverter switches to one of the dc buses. Other solutions involve addition of a fourth leg to the inverter, or use of multilevel inverters.

40.6 Other Means of Noise Suppression

In addition to filters and PWM methods, means of mitigation of EMI effects in systems with power converters include grounding, shielding, and reduction of electromagnetic coupling. When skillfully used, they can significantly reduce the EMI and circuit susceptibility at a low extra cost.

Effects of grounding on the EMI must be considered in the context of the whole system employing power electronic converters. Grounding of electric equipment has historically been based on the requirements of safe and reliable power distribution, maximum protection against overvoltages (including lightning strikes and electrostatic buildup), and safe-touch conditions for personnel during ground faults. The so-called safety ground (green wire in the USA) protects personnel and equipment by conducting fault current, operating a circuit breaker or fuse, and limiting the voltage to ground during faults. This type of grounding can be termed low-frequency grounding. The earth ground is made using buried rods, plates, or grids. Note that the ground wiring for lightning protection must be isolated from other grounding conductors, and the

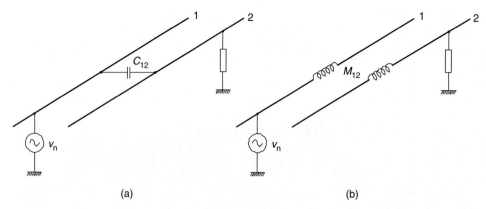

FIGURE 40.22 Electromagnetic coupling: (a) capacitive and (b) inductive.

lightning current may not pass through any circuit breaker or fuse. Considering the drive system in Fig. 40.1 and assuming the "wye" connection of the secondary windings of the supplying transformer, the neutral of the transformer can be ungrounded, directly grounded, or connected to the ground through a resistor. Each of these arrangements has advantages and disadvantages, but the resistance-grounded system has gained the greatest popularity, because the line-to-ground faults do not cause immediate shutdown of the system. Proper selection of the grounding resistance allows sufficient limitation of the ground fault currents and the voltage-to-ground to avoid equipment failures.

The high-frequency grounding, which should be considered separately from the low-frequency scheme, has a strong impact on the CM EMI. The general rule for proper high-frequency grounding design is to understand the CM noise path and to reroute the noise away from the sensitive electronics. A single-point grounding scheme is theoretically the best, although the existence of many stray system-to-ground capacitances makes it practically a multipoint grounding above certain frequencies. Physical capacitors are often used to augment the stray capacitances, for example, between the chassis and the grounding wire, to reduce the CM noise voltage. High-power parts of a system should be connected the closest to the single-point ground, with the most sensitive circuits the furthest from it.

Most practical systems with power converters include cables, e.g. those connecting the inverter to the motor in the drive in Fig. 40.1. The cables act as antennas for the radiated EMI and are the receptors in the electromagnetic coupling. Shielding the cables significantly reduces those unwanted phenomena. Such types of cables as the aluminum-armor or braided-shield ones offer very good protection from EMI. In particular, they prevent interference with sensitive equipment by providing an isolated and predictable metallic path for the CM noise. The shield is usually bonded with the ground wire at both ends of the cable. For industrial installations, such as drive systems in a factory, use of grounded steel conduits to carry cables is a convenient and effective solution. Shielding can also be applied to other than cables parts of the electronic circuitry, such as circuit enclosures.

As illustrated in Fig. 40.22, the electromagnetic noise coupling can be capacitive (electric field) or inductive (magnetic field). The noise source-voltage is denoted by v_n, and the stray capacitance, C_{12}, and mutual inductance, M_{12} link wires 1 and 2 transmitting noise from the source. The coupling capacitances and inductances should therefore be minimized. The simplest, but not always practical, approach is to increase the distances between noise sources and receivers. Shielding is a good protection from electromagnetic noise coupling, and the unshielded leads extending beyond the shield should be possibly short.

Electromagnetic noise coupling can also be minimized by proper circuit geometry. The potential noise source should not be placed in parallel to the potential noise receiver. Perpendicular placement is best. Twisting two interconnecting wires helps as well, because currents in individual wires flow in opposite directions, producing magnetic fields that cancel each other.

EMI reduction techniques should also be considered at the printed circuit board (PCB) level. Proper grounding, power distribution, and interconnection techniques, depending on whether the PCB contains analog or digital circuitry, improve immunity to both the internal and external noise sources. This highly specialized topic extends the tutorial scope of this publication.

A thorough compendium of various means of EMI mitigation in inverter-fed ac drive systems can be found in [12]. It should be stressed that EMI suppression techniques should be considered in early stages of the design. Otherwise, the designed system is very likely to have serious and expensive to solve noise problems.

40.7 EMC Standards

The number of institutions involved in regulations and recommendations that concern EMI is quite large, and to an average person the issue of EMC standards can be somewhat confusing. EMC is defined as the ability of equipment to function satisfactorily in its electromagnetic environment without introducing intolerable disturbances to anything in that environment. EMC requirements entail two major items: emissions and susceptibility, or its opposition, the immunity. Electromagnetic disturbance is any phenomenon that may degrade the performance of a device or system, or adversely affect the living and inert matter. The term "EMI" pertains to that performance degradation. It is worth mentioning that in the colloquial engineering language, EMI is often meant as emissions and EMC as immunity, which is inexact. In the important U.S. military standard MIL-STD 461, the emissions and immunity requirements are referred to as conducted/radiated emissions (CE/RE) and conducted/radiated susceptibility (CS/RS).

In the USA, it is the Federal Communication Commission (FCC) that sets the general EMC requirements (medical products are regulated by the Food and Drug Administration). The FCC Rules and Regulations, Title 47, Part 15, Subpart B concerns "*any unintentional radiator (device or system) that generates and uses timing pulses at a rate in excess of 9000 pulses (cycles) per second and uses digital techniques.*" Clearly, that mandatory requirement applies to almost every product that employs a microprocessor. It is illegal to sell or advertise for sale any product regulated under Part 15, Sub-part B until its emissions have been measured and found to be in compliance. Products regulated by Part 15, Subpart B are divided in two classes. Class A devices are those marketed for use in commercial applications, while the domestic applications belong in Class B. As illustrated in Fig. 40.23, Class B limits are more stringent than those for Class A products, and the Class B administrative *certification* process is more rigorous than the Class A *verification* process. The American National Standards Institute (ANSI) standard C63.4 defines the required emission test procedures. However, there are no FCC regulations pertaining to product immunity to electromagnetic fields.

EMC requirements for products used by the U.S. military are listed in the already mentioned MIL-STD-461. This standard is applicable to a wide range of systems, from power tools to computer workstations. Unlike the FCC Regulations, MIL-STD-461 specifies limits for both the emissions and immunity. The ANSI and Institute of Electrical and Electronics Engineers (IEEE) also publish EMC standards, as do, for internal use, such private organizations as Society of Automotive Engineers (SAE) and automobile manufacturers.

Under the General Trade Agreement on Tariffs and Trade (GATT) and its successor, World Trade Organization (WTO) agreements, member countries are obliged to adopt international standards for national use. With respect to EMC, international standards are primarily developed by International Electrotechnical Commission (IEC) and its International

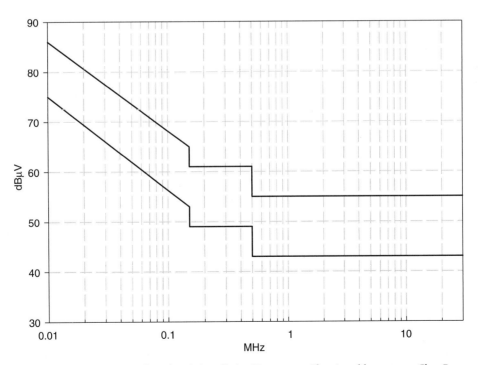

FIGURE 40.23 FCC conducted emissions limits. Upper trace: Class A and lower trace: Class B.

Special Committee on Radio Interference (CISPR), both already mentioned in the introduction to this chapter. The series of IEC standards, IEC 61000-1 through IEC 61000-6, covers all aspects of EMC. The FCC standards are harmonized to IEC, as are the European Norms (EN) used in the European Union (EU). Some specific EMC standards have also been published by International Organization for Standardization (ISO).

In 1992, the EU eliminated internal borders, necessitating a common system for establishing EMC standards and accreditation, testing, and certification procedures. The "New Approach" and "Global Approach" have been initiated, whose goal is to include EU directives into national laws of EU states, harmonize national standards with the European standards, and ensure validity of test reports and conformity certificates between all Member States.

According to the "New Approach," technical contents have been removed from the European Directives and entrusted to the European Standardization Bodies, which are Comitè Europèen de Normalization (CEN), Comitè Europèen de Normalization Electro-Technique (CENELEC), and European Telecommunication Standards Institute (ETSI). The key European Directive 89/336/EEC "Electromagnetic Compatibility" gives only a general definition of the *essential protection requirements* for all electric and electronic equipment and systems, while referring to CENELEC and ETSI standards for technical details. The "Global Approach" requires every product on the European market to have the permanent "CE" marking, which indicates that the affixer declares, and takes full responsibility of, the conformity to all applicable European directives.

The EMC Directive 89/336/EEC sets up emission and immunity requirements. It defines (a) components, (b) systems, and (c) installations. The Directive applies only to components performing direct function and to the systems. Standards referred to in the Directive are divided into basic standards, generic standards, and product standards. The basic standards define general EMC requirements and testing procedures, without specifying any limit values or assessment criteria. Generic standards specify the requirements for products in specific electromagnetic environments. For instance, EN 50081-1 applies to emissions in residential, commercial, and light industrial equipment, the latter including power supplies for industrial equipment, and EN 50081-2 to emissions in industrial environments. Respective immunity norms are EN 50082-1 and EN 50082-2. Product standards address EMC requirements for certain products and product families, such as household appliances, information technology equipment, or generic light industrial equipment, Various European Norms are employed as the product standards.

The IEC 555-2 (EN 60555-2), IEC 1000-3-2 (EN 61000-3-2), and IEC 1000-3-4 are emission standards for low-frequency harmonics, closely associated with operation of power electronic converters. Voltage fluctuations and flicker (*impression of unsteadiness of visual sensation induced by a light stimulus whose luminance or spectral distribution fluctuates in time*) emission limits are defined in IEC 555-3 (EN 555-3), IEC 1000-3-3 (EN 61000-3-3), and IEC 1000-3-5. IEC 61000-4-7 and IEC 61000-4-15 define the required instrumentation for EMI measurements. Radio-frequency conducted and radiated emissions are dealt with in EN 55011, EN 55014, and EN 55022, while CISPR 16 is the basic standard for radio-interference measurements. Finally, IEC 1000-4-1 through IEC 1000-4-12 are immunity standards.

EMC standards are continuously being developed and revised. Therefore, it is important to keep track of standards' publication dates (DoP) and dates of withdrawal (DoW) of conflicting earlier standards. Temporary EN standards are called ENV. Numerous Internet resources are available, in particular:

- American National Standards Institute (ANSI).
- Canadian Standards Association.
- Electronic Industries Association (EIA).
- European Telecommunications Standards Institute (ETSI).
- Federal Communications Commission (FCC).
- IEEE Standards Association.
- International Electrotechnical Commission (IEC).
- International Organization for Standardization (ISO).
- National Institute of Standards and Technology (NIST).
- NSSN, A National Resource for Global Standards.
- Society of Automotive Engineers (SAE).
- Standards Australia.
- VCCI (Japanese EMC Regulation and Certification).
- Verband Deutscher Elektrotechniker e. V. (VDE - German standards).

For more EMC information links go to http://www.dbtechnology.co.uk/links.htm. A comprehensive treatment of contemporary EMC issues can be found in [13].

References

1. D. A. Weston, "Electromagnetic Compatibility: Principles and Applications," 2nd Ed., *Marcel Dekker*, New York, 2001.
2. J. J. Carr, "The Technician's EMI Handbook," *Nevnes Press*, Woburn, 2000.
3. M. Mardiguian, "EMI Troubleshooting Techniques," *McGraw Hill*, New York, 1999.
4. C. R. Paul, "Introduction to Electromagnetic Compatibility," *John Wiley*, New York, 1992.
5. H. W. Ott, "Noise Reduction Techniques in Electronic Systems," 2nd Ed., *John Wiley*, New York, 1988.
6. L. Tihanyi, "Electromagnetic Compatibility in Power Electronics," *IEEE Press*, New York, 1995.
7. R. L. Ozenbaugh, "EMI Filter Design," 2nd Ed., *Marcel Dekker*, New York, 2001.

8. M. J. Nave, "Power Line Filter Design for Switched-Mode Power Supplies," *Van Nostrand Reinhold*, New York, 1991.
9. R. L. Kirlin, M. M. Bech, and A. M. Trzynadlowski, "Analysis of Power and Power Spectral Density in PWM Inverters with Randomized Switching Frequency," *IEEE Transactions on Industrial Electronics*, vol. 49, no. 2, pp. 486–499, Apr. 2002.
10. M. Vilathgamuwa, J. Deng, and K. J. Tseng, "EMI Suppression with Switching Frequency Modulated DC-DC Converters," *IEEE Industry Applications Magazine*, vol. 5, no. 6, pp. 27–33, 1999.
11. A. M. Trzynadlowski, M. Zigliotto, and M. M. Bech, "Random Pulse Width Modulation Quiets Motors, Reduces EMI," *PCIM Power Electronics Systems Magazine*, pp. 55–58, Feb. 1999.
12. G. L. Skibinski, R. J. Kerkman, and D. Schlegel, "EMI Emissions of Modern PWM AC Drives," *IEEE Industry Applications Magazine*, vol. 5, no. 6, pp. 47–80, 1999.
13. L. Rosetto, P. Tenti, and A. Zuccato, "Electromagnetic Compatibility Issues," *IEEE Industry Applications Magazine*, vol. 5, no. 6, pp. 34–46, 1999.

41
Computer Simulation of Power Electronics and Motor Drives

Michael Giesselmann, P. E.
Center for Pulsed Power and Power Electronics, Department of Electrical and Computer Engineering, Texas Tech University, Lubbock, Texas, USA

41.1 Introduction 1121
41.2 Use of Simulation Tools for Design and Analysis 1121
41.3 Simulation of Power Electronics Circuits with PSpice® 1122
41.4 Simulations of Power Electronic Circuits and Electric Machines 1128
41.5 Simulations of AC Induction Machines Using Field Oriented (Vector) Control 1131
41.6 Simulation of Sensorless Vector Control Using PSpice® 1135
41.7 Simulations Using Simplorer® 1141
41.8 Conclusions 1144
    References 1145

41.1 Introduction

This chapter shows how power electronics circuits, electric motors, and drives, can be simulated with modern simulation programs. The main focus will be on PSpice®, which is one of the most widely used general-purpose simulation programs and Simplorer®, which is more specialized towards the power electronics and motor drives application area. Ali Ricardo Buendia, who obtained his M.S.E.E. degree from Texas Tech University, has created the examples for Simplorer®. The PSpice® examples have been developed for the free student version of OrCAD Capture 9.1 from Cadence. The author found the use of examples that can be run on the student version very beneficial in an educational environment, since such examples can be shared with students to enhance their understanding of the lecture material. This shall by no means lead to the conclusion, that the programs and simulations presented here cannot be used for serious professional work. In fact, the author has used these tools with great success in many research and consulting projects. In addition to the programs mentioned above, MathCAD® has been used to derive and present the underlying equations. The advantage of using MathCAD® for this purpose is that in MathCAD® it is possible to check equations by actually executing them.

The examples have been developed to illustrate advanced techniques for simulation of systems from the power electronics and drives area but not to teach the basic features of the individual programs. It is assumed, that the reader will familiarize themself with the basics on how to run the programs using the accompanying documentation. In addition, it is assumed, that the reader is familiar with the basics of power electronics and electric machines, specifically AC induction machines. For a review the reader shall be referred to [1] for power electronics and [2, 3] for induction machines.

41.2 Use of Simulation Tools for Design and Analysis

It is appropriate to reflect upon the value of simulations and its place in the design and analysis process before any in-depth discussion of specific simulation examples. Computer simulations enable engineers to study the behavior of complex and powerful systems without actually building or operating them. Simulations therefore have a place in the analysis of existing equipment as well as the design of new systems. In addition, computer simulations enable engineers to safely study abnormal operating or fault conditions without actually creating such conditions in the real environment.

However, the reader should be reminded that even the most modern simulation programs cannot perfectly represent all parameters and aspects of real equipment. The accuracy of the

Copyright © 2007, 2001, Elsevier Inc.
All rights reserved.

simulation results depends on the accuracy of the component models and the proper identification and inclusion of parasitic circuit elements such as parasitic inductance, capacitance, and mutual coupling. Accuracy of component models in this context shall not mean that the model is actually faulty but rather that the limitations of the model are exceeded. For example, if the transformer inrush phenomenon were to be studied using a linear model for a transformer, the simulation would not yield useful results.

In particular, the precise prediction of voltage and current traces during fast switching transitions in power electronics circuits has been proven to be difficult. To obtain useful results, extensive experimental validation, advanced device models (and the values for their parameters!), and detailed knowledge of parasitic elements, including the ones of the packaging of the circuit elements, are necessary. In addition, numerical convergence is often a problem, if gate-drive signals, with rise and fall times as steep as in real circuits, are applied. Therefore, the exact prediction of waveforms during switching transitions shall be excluded from the discussions in this chapter. Consequently, the author prefers to measure parameters such as voltage rise and fall times, over and undershoot etc. on actual circuitry in the laboratory.

Sometimes, users of PSpice® claim that the convergence problems are so severe, that it's use for simulations of power electronics circuits is just not possible or worth the effort. However, this is absolutely not true and with the proper techniques of gate signal generation, we can simulate just about any given circuit with little or no convergence problems. In addition, if convergence problems are avoided, simulations run much faster and larger numbers of individual transitions can be studied. This is achieved by generating gate signals that are slightly less steep than in real circuits using analog behavioral elements. This gives a lot of insight into the cycle-by-cycle as well as the system level behavior of a power electronics circuit. In this fashion, the function of an existing, as well as the expected performance of a new proposed circuit can be studied. An excellent application for these cycle-by-cycle simulations is the development and verification of control strategies for the power semiconductors.

Analog behavioral modeling (ABM) techniques included in PSpice® can be used to study large and complex systems like the control of induction machines using field oriented (also called vector) control techniques. Examples are given that replace the power electronics inverter with an ABM source that produces voltages, which represent the short-term average (filtering away the voltage components of the switching frequency and above) of the output of a three-phase inverter. These examples represent pure system level simulations, which could have also been done using programs like MatLab/Simulink®. However, circuit simulation programs provide the option of studying actual circuit level details in complex systems. To demonstrate this capability, the start-up of an induction motor, fed by a three-phase metal oxide semiconductor field effect transistor (MOSFET) inverter, is presented.

In all modeling cases, the user needs to define the goal of the simulation effort. In other words, the user must answer the question "What information shall be obtained through the simulation of the circuit or system?" The user must then select the appropriate simulation software and the appropriate models. This process requires a detailed understanding of the properties and limitations of the device models and the sensitivity of the results to the model limitations. In order to obtain such an understanding, it is often recommended and necessary to perform numerous simulation test runs, carefully scrutinize the results and compare them with measured data, results from other simulation packages or otherwise known facts.

41.3 Simulation of Power Electronics Circuits with PSpice®

The first example of a power electronics circuit is a step-down (also called buck) converter with synchronous rectification. For the purpose of synchronous rectification, the diode, which connects the inductor to ground in the regular circuit, is replaced with a power MOSFET transistor. The benefit of this circuit is that the power MOSFET represents a purely resistive channel in the on state. This channel does not have a residual, current independent, voltage drop like the p–n junction of a diode. Therefore, the voltage drop across the MOSFET can be made lower than what can be achieved with diodes. The results are reduced losses and increased efficiency. To achieve this, the lower MOSFET must be turned on whenever the upper MOSFET is turned off and the current in the inductor is positive. If the current in the inductor is continuous, the drive signal for the lower MOSFET is simply the inverted drive signal for the upper MOSFET. However, if the current in the inductor is discontinuous, the drive signal for the lower MOSFET must be cut off as soon as the current in the inductor goes to zero.

Figure 41.1 shows a simulation setup for a synchronous buck converter that can operate correctly for continuous as well as discontinuous inductor current. As mentioned before, the key element of this example is the circuit for the generation of the gate-drive signals for the MOSFETs. For clarity, this circuit has been realized using only standard elements from the libraries of the evaluation version. The basic principle of the operation of the gate-drive circuit is the well-known carrier based scheme, where a control voltage is compared with a triangular carrier with fixed amplitude. An "ABM" block, shown in the lower left part of Fig. 41.1, generates the triangular carrier. Equation (41.1) shown below gives the closed form equation for the triangular carrier wave. The output

41 Computer Simulation of Power Electronics and Motor Drives

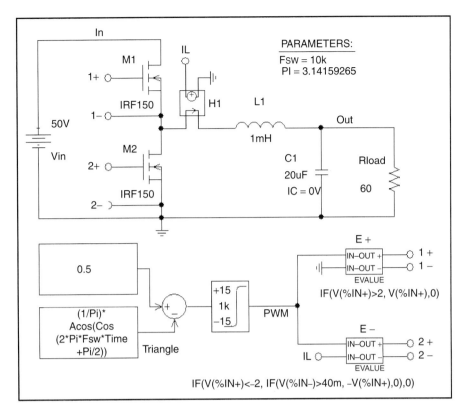

FIGURE 41.1 Simulation setup for a synchronous buck converter.

range of the function shown in Eq. (41.1) is between 0.0 and 1.0.

$$E_{Tri}(t) := \frac{1}{\pi} a\cos\left(\cos\left(2\pi Fsw\, t + \frac{\pi}{2}\right)\right) \quad (41.1)$$

For the generation of the gate-drive signals, the carrier wave is compared with a control signal that can have values between 0.0 and 1.0, corresponding to a duty cycle input between 0 and 100%. Feeding the difference of the carrier wave and the control signal into a soft comparator generates the primary PWM signal. Careful inspection of the implementation of the soft-limiter element provided in the evaluation version of PSpice® shows that it uses a scaled hyperbolic tangent function. Figure 41.2 shows a plot of a hyperbolic tangent function. It can easily be seen, that the result of the soft-limiter is an output signal with smooth transitions, which is crucial to avoid convergence problems in PSpice®. The soft-limiter used here has an upper and lower limit of ±15 V and a gain (steepness control for the tanh function) of 50.

Figure 41.3 shows the output of the simulation run for the synchronous buck converter. The top-level graph shows the generated triangle carrier. It has a frequency (Fsw, see parameter statement in Fig. 41.1) of 10 kHz. This frequency has been chosen rather low to improve the readability of Fig. 41.3. In the graph of the triangle voltage in Fig. 41.3, the gate-drive signals are shown for both MOSFET transistors. Please note, that the gate-drive signal for the lower MOSFET is vertically shifted by 30 V in order to separate the traces for readability. The graph below the gate-drive signals shows the inductor current. It can be seen that the current is discontinuous after the initial inrush peak. The inrush peak is caused by the fact that the capacitor is initially discharged (IC = 0 V). It is evident, that the gate-drive signal of the lower (synchronous rectification) MOSFET is appropriate for the inductor current. The bottom trace shows the capacitor voltage, which has a steady-state value of slightly more than 0.5 × 40 V (0.5 = 50% being the duty cycle and 40 V being the input voltage) due to the fact that the inductor current is slightly discontinuous even at steady-state conditions. To test the gate-drive circuit for the lower MOSFET, the load has been chosen such that the steady-state current would be discontinuous. Following the soft-limiter are two voltage-controlled voltage sources that generate isolated gate-source voltages of 15 V for the on condition and 0 V for the off condition of the MOSFETs. To enable operation with discontinuous inductor current, the source "E−" in Fig. 41.1 also monitors the polarity of the inductor current through a current-controlled voltage source "H1" with unity gain.

In addition to the cycle-by-cycle simulation of a DC–DC converter, it is also possible to use a time-averaged replacement for the MOSFET transistors used in the circuit in Fig. 41.1. In fact, a common time-averaged model can be used for the buck, the boost, the buck–boost, and the Cuk converter as

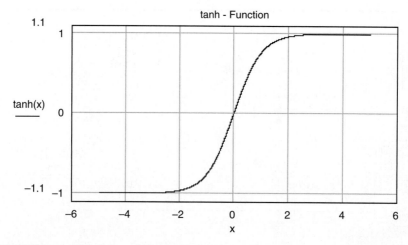

FIGURE 41.2 Plot of a hyperbolic tangent function used to generate smooth PWM signals.

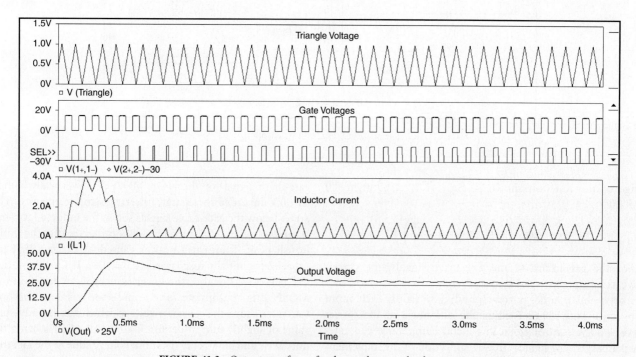

FIGURE 41.3 Output waveforms for the synchronous buck converter.

long as they operate with continuous inductor current. The time-averaged model has the advantage that it can run much faster since it does not have to follow each switching transition. It is also possible to perform DC and AC sweep analyses. A DC sweep would sweep the duty cycle over a wide range and show the output voltage as a function of the duty cycle. An AC sweep analysis would sweep the frequency of an AC signal, which is superimposed on top of the duty cycle bias signal. The AC sweep allows the study of the behavior of the converter, including a feedback control system, in the frequency domain for traditional stability analysis and system tuning. A detailed description of this time-averaged modeling technique, including detailed examples is given in [4].

To illustrate the capabilities of the PSpice® simulation program, the next example shows a complete three-phase inverter bridge using six power MOSFETs. This circuit is shown in Fig. 41.4. Note that free-wheeling diodes are an integral part of every power MOSFET and are not shown separately. The inverter drives a three-phase load, which could represent an induction motor for a singular operating point. The load is connected to the inverter output terminals with so-called connection bubbles. Due to the number of elements involved, the

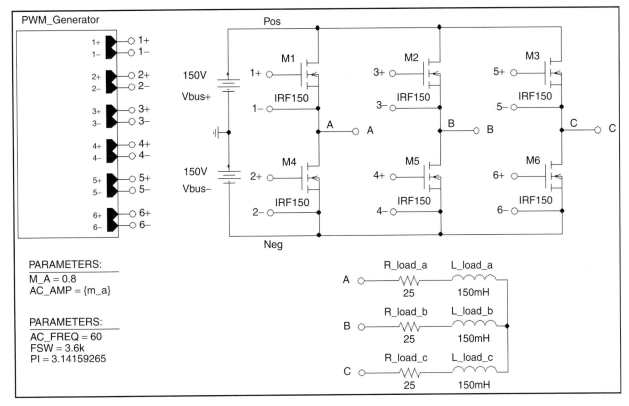

FIGURE 41.4 Circuit for a three-phase inverter with MOSFETs.

circuit for the gate drive-signal generation is contained in a hierarchical block. Blocks like this are available from the main toolbar in the schematic editor. Selecting "Descend Hierarchy" for the block called "PWM_Generator" reveals the subcircuit which is shown in Fig. 41.5.

The hexagonal shaped symbols named "1+," "1−," "2+," etc. are called interface ports. These interface ports provide the connection between the subcircuit and the ports of the hierarchical block above. Here the connection is to the ports (dots) on the "PWM_Generator" block. The interface ports are created by simply drawing a wire up to the boundary of the block. The name of the port is initially generic, "Px," where x is a running number, but can be easily edited by double-clicking on the generic name. After drawing a block and creating all the ports, right-clicking the "Descend Hierarchy" will open up a schematic page for the subcircuit which has all the appropriately named interface ports already in it. Additional details on hierarchical techniques can be found in [5].

The circuit shown in Fig. 41.5 is similar to the gate-drive generation circuit discussed before. Circuits like the circuit shown in Fig. 41.1, compare a triangular carrier with one or more reference signals. In this case, three reference signals, one for each phase, are used. The triangular carrier signal is symmetrical with respect to the time axis. The values cover the range from −1.0 to 1.0. The equation for the triangular carrier for PWM modulation for AC reference signals is given by the Eq. (41.2).

$$E_{Tri}(t) := \frac{2}{\pi} a \sin\left(\sin\left(2\pi f_s t + \frac{\pi}{2}\right)\right) \quad (41.2)$$

The three reference signals are sinusoidal signals with equal amplitude and a relative phase shift of 120°. For linear modulation, the amplitude range of the reference signals is limited to the amplitude of the triangular carrier, e.g. 1 V. The ratio of the reference wave amplitude and the (fixed) carrier amplitude is called amplitude modulation ratio "m_a." In the circuit shown in Fig. 41.4 "m_a" has a value of 0.8. This value is defined by a parameter symbol and represents a global parameter, which is visible throughout all levels of the hierarchy. The phase to neutral voltage amplitude of each inverter leg is equal to "Vbus+" (shown in Fig. 41.4) multiplied with the amplitude modulation ratio. The frequency and waveshape of the phase to neutral voltage of each phase leg is equal to the reference waveform, if the high-frequency components resulting from the carrier wave are filtered away. This way, each inverter leg can be viewed as a linear power amplifier for its reference voltage. In fact, in drive applications, inverters are often called "servo-amplifiers." The load typically reacts

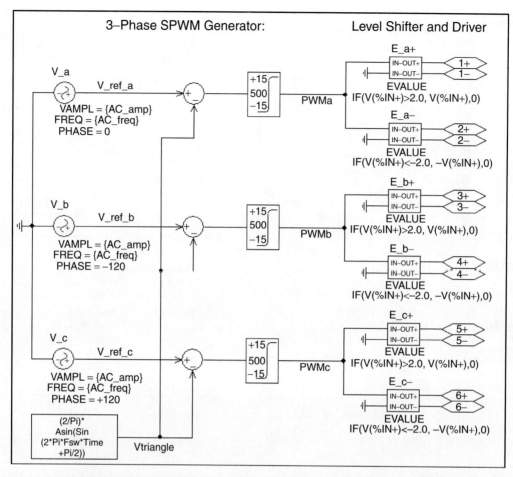

FIGURE 41.5 PWM generation subcircuit for a three-phase MOSFET inverter.

only to the low-frequency components of the inverter output voltage. The high-frequency components, which include the triangular carrier frequency (also called switching frequency) and its harmonics, are typically "just a blur" for the load. This is especially true in recent times, where switching frequencies of 20 kHz and above are possible. As an added benefit, audible noise is avoided at these frequency levels.

The circuit involving the soft-limiter and level-shifter/high-side driver in Fig. 41.5 is very similar to the circuit for the synchronous buck converter, except for the fact that the load current is not monitored. The control functions for the "E_x+, E_x-" sources, where x denotes the phase, are chosen such that the activation voltage levels are ± 2 V. If the output voltage of the soft-limiter is between -2 and $+2$ V, no MOSFET is activated, and shoot-through, meaning a short circuit between the positive and negative bus, is avoided.

Figure 41.6 shows the simulation results for the three-phase inverter. The time scale is slightly stretched, to better show the details of the PWM signals. The line-to-line voltage V_{AB} and the load current in all three phases are shown. Due to the inductors contained in the load, the current cannot instantaneously change and follow the PWM signal. Therefore the load currents are almost pure sinusoids with very little ripple. This is representative of the real line currents in induction motors.

Figure 41.7 shows an example where the MOSFETs in Fig. 41.4 have been replaced with insulated gate bipolar transistor (IGBT). The particular IGBT shown here is included in the library of the evaluation version. Note that free-wheeling diodes are needed, if IGBTs are used. The free-wheeling diodes carry the load current when the IGBTs are turned off to provide a continuous path for the current. This is very important, since the load can have a substantial inductive component. Whenever the diodes are conducting, energy flows momentarily back to the source. In the case of power MOSFETs, the diodes (often called body diodes) are an integral part of the device. In the symbol graphic of PSpice® these body diodes are not shown for MOSFETs. For this circuit, the gate-drive circuit and the results are the same as for the three-phase bridge with power MOSFETs.

41 Computer Simulation of Power Electronics and Motor Drives

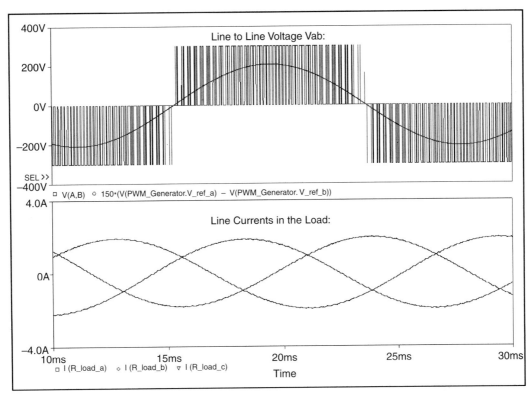

FIGURE 41.6 Output waveforms of the three-phase inverter with MOSFETs.

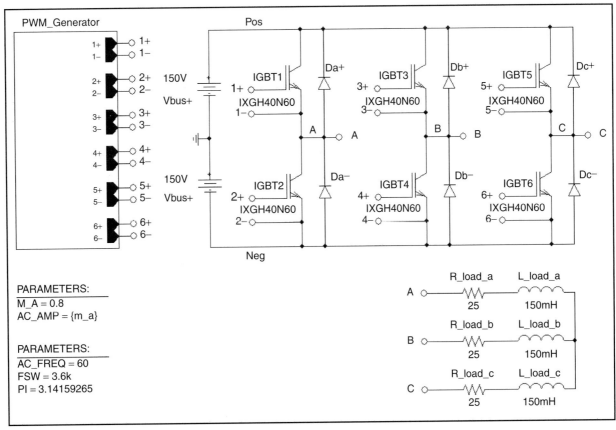

FIGURE 41.7 Three-phase inverter circuit with IGBTs.

41.4 Simulations of Power Electronic Circuits and Electric Machines

In the following, the start-up of an induction motor, fed by the three-phase inverter shown in Fig. 41.4, is presented. For this purpose, the simple passive load in Fig. 41.4 is replaced by an induction motor. For this and further discussions, it is assumed that the reader is familiar with the theory of induction machines. A number of excellent references are given at the end of this chapter [2, 3, 6, 7]. The induction motor symbol represents the electro-mechanical model of an induction motor. The model is suitable for studies of electrical and mechanical transients as well as steady-state conditions. The output pin on the motor shaft represents the mechanical output. The voltage on this pin represents the mechanical angular velocity using the relation $1\,V = 1\,\text{rad/s}$. In addition, any current drawn from or fed into this terminal represents applied motor or generator torque according to the relation $1\,A = 1\,\text{Nm}$. Due to these definitions, the electrical power associated with the voltage of the motor shaft (with respect to ground) is identical to the mechanical power. Following the well-known theory, the induction motor model has been derived for a two-phase (direct and quadrature, D, Q) equivalent motor. Attached to the motor is a bidirectional two-phase to three-phase converter module. This module is voltage and current invariant. This means that the voltage and current levels in the two-phase and the three-phase machine are equal. Consequently, the power in the two-phase machine is only two third of the power in the three-phase circuit. This is accounted for in the calculation of the electromagnetic torque (factor 3/2; see Eq. (41.4)). The internally generated torque can be monitored on the output labeled "Torque" on top of the sleeve around the motor shaft. The linear load in Fig. 41.8 is a symbol that represents an appropriately sized resistor to ground.

In Fig. 41.8 the motor is represented by a custom symbol called "Motor1." A simple hierarchical block could have been used for the motor, but a custom symbol has been created to achieve a more realistic and pleasing graphical representation. The symbol can be easily created with the symbol editor, which is built into the regular schematics (capture) editor. The editor provides standard graphical elements (lines, rectangles, circles etc.) so that professional looking symbols can easily be created. More details on this are shown in Figs. 41.19 and 41.20 and the accompanying discussion. Parameters are passed onto the subcircuit by using the name of the parameter preceded by a '@' symbol. The advantage of passing parameters to subcircuits

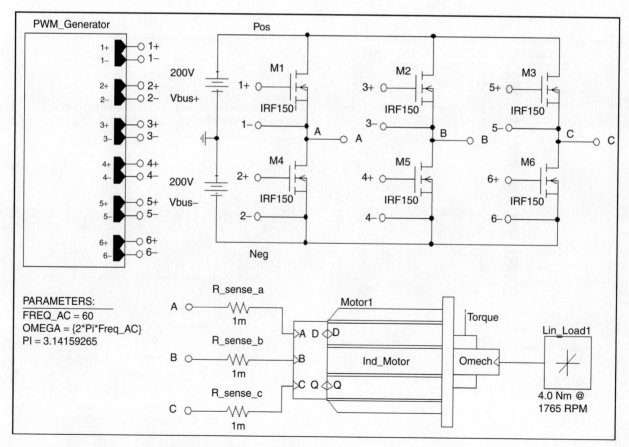

FIGURE 41.8 Induction motor start-up with three-phase inverter circuit.

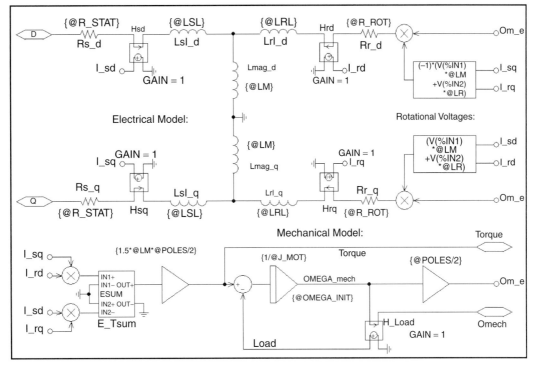

FIGURE 41.9 Subcircuit for induction motor model.

in this way is that several symbols can call one set of subcircuits with different parameters.

Right-clicking on the motor symbol and selecting "Descend Hierarchy" reveals the associated subcircuit that implements its function. This subcircuit is shown in Fig. 41.9. The upper portion of this subcircuit represents the electrical model. The task of the electrical model is to calculate the stator and rotor currents, with the stator voltages and the mechanical speed of the machine being input parameters. However, it is also possible without any changes, to feed stator currents (with controlled current sources) into the D and Q inputs and have the model to calculate the appropriate stator voltages. This option is useful for vector control applications, which are discussed later.

The equation system for the electrical model of a two-phase induction machine is given by Eq. (41.3). The theory for this equation system is derived in [2, 3, 6, 7]. The equation system and the model are formulated for the stationary reference frame. This reference frame assumes that the frame of the machine is stationary (which is hopefully the case in a real machine!) and the voltages and currents of the rotor are equivalent AC values with stator frequency. From machine theory we know that the actual rotor currents have slip frequency. Another reference frame is the synchronous (also called excitation) reference frame. In this reference frame, the stator of a **fictitious** machine is assumed to rotate with synchronous speed. The advantage of this reference frame is that the input frequency is zero (DC), which makes it easy to explain the principle of vector control by extending the theory of DC machines to AC machines.

$$\begin{bmatrix} V_d \\ V_q \\ 0 \\ 0 \end{bmatrix} = \begin{bmatrix} R_{stat}+pL_s & 0 & pL_m & 0 \\ 0 & R_{stat}+pL_s & 0 & pL_m \\ pL_m & \omega_e L_m & R_{rot}+pL_r & \omega_e L_r \\ -\omega_e L_m & pL_m & -\omega_e L_r & R_{rot}+pL_r \end{bmatrix} \begin{bmatrix} I_{sd} \\ I_{sq} \\ I_{rd} \\ I_{rq} \end{bmatrix}$$

$$L_s = L_m + L_{s1} \quad L_r = L_m + L_{r1} \quad p = \tfrac{d}{dt}$$

(41.3)

In typical implementations of vector control using digital signal processors (DSPs), the synchronous reference frame is used internally to calculate the reference values for the currents in the D and Q axis. These values are then transformed to the stationary reference frame in an additional step. Sometimes still other reference frames are used and it is possible to generate a universal electrical model with a reference frame speed input. This model could then be used for any reference frame.

The electrical model in Fig. 41.9 implements the equation system shown in Eq. (41.3). The circuit closely resembles the well-known T-equivalent circuit for the steady-state analysis of induction machines. Two instances of the T-equivalent circuit are necessary to implement the two-phase (D, Q) model. The two instances of the circuit are almost mirror images of each other (and actually drawn that way) except for some differences in the circuit elements that calculate the voltages, which are generated due to the rotation of the rotor. The bottom of

Fig. 41.9 represents the mechanical model. This circuitry calculates the internally generated electro-magnetic torque, using the rotor and stator currents as input values. The equation for the internal electromagnetic torque of the induction machine is given by Eq. (41.4) [3, 6, 7]. The factor 3/2 accounts for the fact that the real motor is a three-phase machine. Using the generated torque, the load torque, and the moment of inertia, the angular acceleration can be calculated. Integration of the angular acceleration yields the rotor speed, which is used in the electrical model. To avoid clutter and to improve readability, connection bubbles are used to connect the various parts of the model together.

$$T = \frac{3}{2}\frac{P}{2}L_m \left(I_{sq}I_{rd} - I_{sd}I_{rq}\right) \qquad (41.4)$$

Since typical induction machines are three-phase machines, it is often desirable to have a machine model with a three-phase input. Therefore, a bidirectional two-phase to three-phase converter module, which can be attached to the motor, has been developed. A subcircuit for this module is shown in Fig. 41.10. This circuit is truly bidirectional, meaning that the circuit can be fed with voltage or current sources from either side. The equation system for this voltage and current invariant transformation is given by Eq. (41.5). This transformation is sometimes called "Clark" or "ABC–DQ" transformation. Note that V_0 denotes a zero-sequence voltage, which is assumed to be zero. This voltage would only have non-zero values for unbalanced conditions. An interesting detail of the subcircuit is the three-phase switch on the input. This switch is necessary to ensure a stable initialization of the simulator in case the machine is fed with a controlled current source. The switch provides an initial shunt resistor from the three-phase input to ground. Soon after the simulation has started, the switch opens and leaves only a negligible shunt conductance to ground.

$$\begin{bmatrix} V_d \\ V_q \\ V_o \end{bmatrix} = \frac{1}{3}\begin{bmatrix} 2 & -1 & -1 \\ 0 & \sqrt{3} & -\sqrt{3} \\ 1 & 1 & 1 \end{bmatrix} \begin{bmatrix} V_a \\ V_b \\ V_c \end{bmatrix}$$
$$\begin{bmatrix} V_a \\ V_b \\ V_c \end{bmatrix} = \frac{1}{2}\begin{bmatrix} 2 & 0 & 2 \\ -1 & \sqrt{3} & 2 \\ -1 & -\sqrt{3} & 2 \end{bmatrix} \begin{bmatrix} V_d \\ V_q \\ V_o \end{bmatrix} \qquad (41.5)$$

Figure 41.11 shows the result for the start-up of the induction motor for the circuit of Fig. 41.8. The motor is a half (Attention LE(½)) hp, 208 V, 4-pole machine. The detailed parameters are shown in Table 41.1. The PWM generation was identical to the example shown in Fig. 41.4, except that the switching frequency was 4 kHz and 21.1% of the third harmonic has been added to each of the reference sinusoids in order to increase the linear modulation range. The resulting reference waveform was then multiplied with 1.14, which represented the maximum voltage for linear modulation. The top trace in Fig. 41.11 shows the developed electromagnetic torque, the level for the rated steady-state torque (4 Nm as commanded by the load in Fig. 41.8) and the zero level. This graph shows the typical oscillatory torque production of the induction machine for an uncontrolled line start. The scale for this graph is 1 V = 1 Nm. The graph below shows the mechanical angular velocity with a scale of 1 V = 1 rad/s. Below the graph for the rotor speed, all three input currents are shown. It is evident, that the current traces are almost perfect sinusoids, despite the fact that the input voltage is the

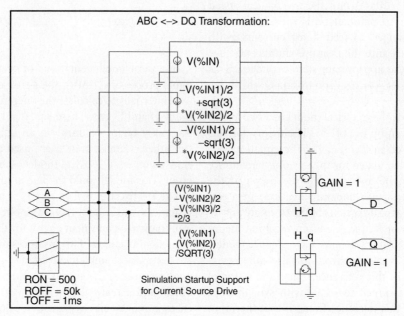

FIGURE 41.10 Subcircuit for the ABC–DQ transformation module.

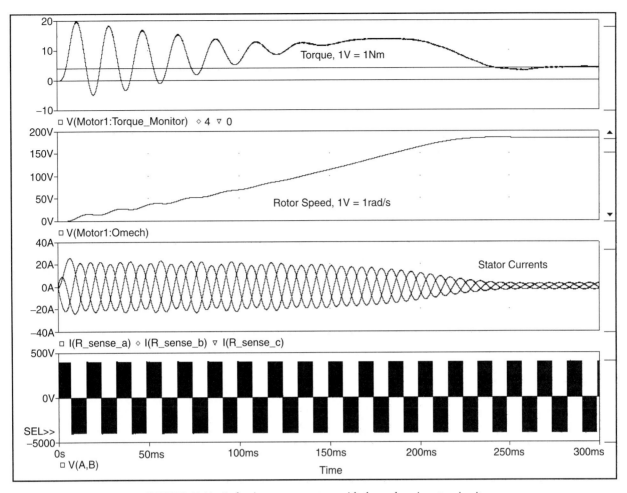

FIGURE 41.11 Induction motor start-up with three-phase inverter circuit.

TABLE 41.1 List of all attributes used for the half (**Attention LE:(1/2)**)hp, 208 V, four-pole induction motor in Fig. 41.8

Attributes:

PART = Induction_Servo
MODEL = Ind_Motor
TEMPLATE =
J_mot = 0.01
Omega_init = 0.0
Ls = (@Lm + @Lsl)
Lr = (@Lm + @Lrl)
Poles = 4
Tau_r = (@Lr/@R_Rot)
REFDES = Motor?
Lsl = 14.96mH
Lrl = 8.79mH
R_Stat = 3.60
R_Rot = 1.90
Lm = 424.41mH

PWM waveform shown in the bottom graph. Also, the trace for the torque shows no discernable high frequency ripple. The reason is, of course, that the motor windings are inductive, and represent a low-pass filter for the applied voltages. Nevertheless, recent research suggests that filtering the output voltage of the inverter is advantageous anyway, because it significantly reduces the voltage stress on the windings and suppresses displacement currents through the bearings [8].

41.5 Simulations of AC Induction Machines Using Field Oriented (Vector) Control

The following examples will demonstrate the use of PSpice® for simulations of AC induction machines using field oriented control (FOC). Again, it is assumed that the reader is familiar with the basics of induction machine theory. Often times, FOC is also called vector control, and both expressions can be used interchangeably. The FOC was proposed in the 1960s by Hasse and Blaschke, working at the Technical University of Darmstadt [9]. The basic idea of FOC is to inject currents into the stator of an induction machine such that the magnetic flux level and the production of electromagnetic torque can be independently controlled and the dynamics of the machine resembles that of a separately excited DC machine (without armature reaction; no cross coupling). The previously discussed two-phase model for the induction machine is very helpful for studies of vector control and shall be used in all examples. If a two-phase induction motor model for the synchronous (or excitation) reference frame is used, the similarities between the control of a separately excited DC machine and vector control of an AC induction machine would be most evident. In this case, the D input would correspond to the field excitation input of the DC machine and both inputs would be fed with DC current. Assuming unsaturated machines, the current into the D input of the induction machine or the field current in the DC machine would control the flux level. The Q input of the induction machine would correspond to the armature winding input of the DC machine and again both inputs would receive DC current. These currents would directly control the production of electromagnetic torque with a linear relation (constant k_T) between the current level and the torque level. Furthermore, the Q component of the current would not change the flux level established by the D component (no cross coupling). To make such a simulation work, it would finally be necessary to calculate the slip value that corresponds to the commanded torque and supply this DC value to the D, Q (synchronous reference frame) machine model.

Of course this is very interesting from an academic standpoint and the author uses this example in a semester long lecture on FOC. However, it should again be noted, that a machine represented by a model with a synchronous reference frame would have a **stator**, which rotates with synchronous speed. Of course this is not realistic and therefore it is more interesting to generate a simulation example that uses the previously discussed motor with a model for the stationary reference frame. This motor must be supplied with AC voltages and currents with a frequency determined mostly by the rotor speed to a small extent by the commanded torque. We still supply DC values representing the commanded flux and torque but we transform these DC values to appropriate AC values. In the following example, we will assume that we can measure the actual rotor speed with a sensor. This can, in effect, be easily accomplished and many types of sensors are available on the market. If we add the slip speed, that we determine mathematically from the torque command, to the measured rotor speed, we obtain the synchronous speed for the given operating point. With this synchronous speed we can transform the DC flux and torque command values from the synchronous reference frame to the stationary reference frame. We accomplish this by using a rotational transformation according to the matrix equations in Eq. (41.6). This rotational transformation is also called "Park" transformation.

$$\begin{bmatrix} V_{Dout} \\ V_{Qout} \end{bmatrix} = \begin{bmatrix} \cos(\rho) & -\sin(\rho) \\ \sin(\rho) & \cos(\rho) \end{bmatrix} \begin{bmatrix} V_{Din} \\ V_{Qout} \end{bmatrix}$$

$$\begin{bmatrix} V_{Dout} \\ V_{Qout} \end{bmatrix} = \begin{bmatrix} \cos(\rho) & \sin(\rho) \\ -\sin(\rho) & \cos(\rho) \end{bmatrix} \begin{bmatrix} V_{Din} \\ V_{Qout} \end{bmatrix} \quad (41.6)$$

$$\begin{bmatrix} \cos(\rho) & -\sin(\rho) \\ \sin(\rho) & \cos(\rho) \end{bmatrix} \begin{bmatrix} \cos(\rho) & \sin(\rho) \\ -\sin(\rho) & \cos(\rho) \end{bmatrix} = \begin{bmatrix} 1 & 0 \\ 0 & 1 \end{bmatrix}$$

As shown in Eq. (41.6), the transformation is bidirectional and the product of the transformation matrices yields the unity matrix. For the following discussion we shall define the transformation, which produces AC values from DC inputs as a positive vector transformation and the reverse operation consequently a negative vector transformation. The matrix equation for the positive vector transformation is shown on the left side of Eq. (41.6). The negative transformation is very useful to extract DC values from AC voltages and currents for diagnostic and feedback control purposes. We will also make use of it for sensorless vector control, which is discussed below. Both rotational transformations use the angle, ρ, in the equations. This angle can be interpreted as the momentary rotational displacement angle between two Cartesian coordinate systems; one containing the input values and the other, the output values. This angle is obtained by integration of the angular velocity with which the coordinate systems are rotating (typically the synchronous speed).

In summary, we replaced a theoretical motor model using a synchronous reference frame by a reference frame transformation of the supply voltages and currents. In fact, modern DSPs like the TMS320C2000™ Digital Signal Controller series from Texas Instruments are capable to perform both the Clark and the Park transformation in both directions at very high speeds [10]. These DSPs are well supported with proven reference designs, including free software examples.

Figure 41.12 shows the top level of a simulation example that implements vector control for an induction machine with a stationary reference frame model. In fact, the motor model and the associated subcircuits are identical to the ones used for the circuit shown in Fig. 41.8. However, a more powerful motor is used here, specifically a 3 hp, 4-pole 208 V motor with circuit parameters shown in Table 41.2. As discussed above, the actual speed of the rotor is used as an input signal for the control unit. This scheme is known as indirect vector control and represents one of the most often used arrangements. The symbol for the controller has the same parameters as the motor. This is necessary to achieve correct field orientation. In real systems, the controller also must know or somehow determine the machine parameters. The machine parameters could have also been established globally using "PARAM" symbols, but if the parameters for the controller can be set separately as it is the case here, the influence of parameter mismatch on the

41 Computer Simulation of Power Electronics and Motor Drives

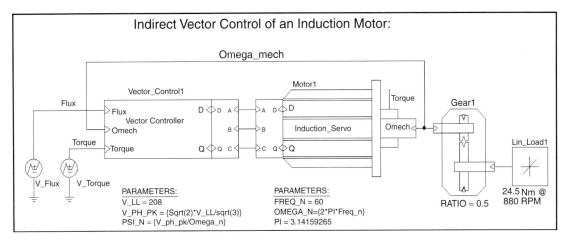

FIGURE 41.12 Top level circuit for indirect vector control of induction motors.

TABLE 41.2 List of all attributes for the induction motor symbol for vector control

Attributes:

PART = Induction_Servo
MODEL = Ind_Motor
TEMPLATE =
J_mot = 0.1
Omega_init = 0.0
Ls = (@Lm + @Lsl)
Lr = (@Lm + @Lrl)
Poles = 4
Tau_r = (@Lr/@R_Rot)
REFDES = Motor?
Lsl = 2.18mH
Lrl = 2.89mH
R_Stat = 0.48
R_Rot = 0.358
Lm = 51.25mH

performance of the control can be easily studied. An example of this is given in Fig. 41.12.

Figure 41.12 also shows a symbol for a mechanical gear, which is attached to the output of the induction motor. Let us recall that the voltage on the mechanical output terminal represents the angular velocity and the current represents the torque. We also know that the product of the angular velocity and the torque is the mechanical power. Therefore it is easily understood that the electrical representation of an ideal gear is an ideal transformer. The ideal transformer changes speed (voltage) and torque (current) just like an ideal gear. Also there are no power losses in an ideal transformer as well as in an ideal gear.

In this fashion, many more mechanical properties and devices can be modeled. For example, a mechanical flywheel is simply represented by a capacitor to ground. Due to the scaling factors for the angular velocity and the torque, a flywheel with $J = 1 \text{ kgm}^2$ would be a capacitor with $C = 1$ F. The compliance of a drive-shaft (elastic twisting by the applied torque) can be modeled by a series inductor. By including both capacitors and inductors, effects like mechanical resonance can be included in a model.

Figure 41.13 shows the subcircuit for the vector control unit. The central part is a vector rotator for positive direction. This element transforms the DC reference values for the flux (D-axis) and the torque (Q-axis) to the stationary reference frame. The input angle for the vector rotator is the integral of the synchronous angular velocity. The signal called "Omega_o" is the measured rotor speed.

This speed is multiplied with the number of pole-pairs (poles/2) to obtain the electrical angular velocity. Then the slip value (see Eq. (41.7) for slip frequency calculation for vector control) appropriate to the torque command is added and the resulting signal is routed through an integrator to generate the input angle for the vector rotator.

In the D-axis, a differentiator function "DDT()" is used in a compensation (see Eq. (41.8) for D-axis reference current for vector control) element which assures that the actual flux in the machine follows the commanded signal without delay. The input and output values of the vector rotator are voltage signals which correspond 1:1 to current signals. (In a real controller the currents are typically scaled values in the memory of a DSP.) In fact, the vector controller calculates the appropriate currents that need to be injected into the machine to perform as desired. Two voltage-controlled current sources with unity gain are connected to the output of the vector rotator to generate these currents. In a real system, the controlled current sources are realized by an inverter with current feedback. In the most realistic case, this would be a three-phase inverter and the ABC–DQ transformation would be performed before the current controlled inverter. In this example, the ABC–DQ transformation has been performed outside the controller and

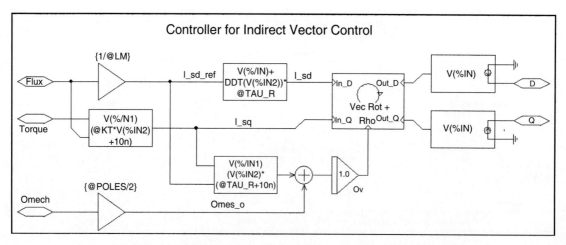

FIGURE 41.13 Subcircuit for indirect vector control of induction motors.

the motor. This way, it is possible to study vector control principles using a DQ controller and a DQ motor by eliminating the ABC–DQ transformation elements. An example is given in Fig. 41.14.

$$\omega_{slip} = L_m \frac{I_{sq}}{\tau_r \lambda_{rd}} \tag{41.7}$$

$$I_{sd} = \frac{(\lambda_{rd} + \tau_r \lambda_{rd} p)}{L_m} = \frac{1 + \tau_r p}{L_m} \lambda_{rd} \quad p = \frac{d}{dt} \tag{41.8}$$

Figure 41.15 shows the results obtained for the circuit shown in Figs. 41.12 and 41.14 with perfect tuning of the vector controller. Perfect tuning means that the controller precisely knows all motor parameters at all times (including resistance changes due to heating of the windings). The two traces in the diagram on top of Fig. 41.15 represent the traces for the D and Q input signals of the vector rotator. The graph below shows the reference value for the flux. It is evident, that the flux level is being changed at the same time when 10 Nm of torque is commanded (and produced). This is done to check if the torque and the flux can be independently controlled, which is true for correct FOC. Below the flux reference is the graph for the mechanical angular velocity. It can be seen, that the machine accelerates whenever torque is developed and slows down due to the load when the torque command is driven to zero. The graph on the bottom of Fig. 41.15 provides the easiest way to judge the quality of the correct field orientation. This graph shows the traces of the commanded and the actually produced torque and in this case they are perfectly on top of each other at all times.

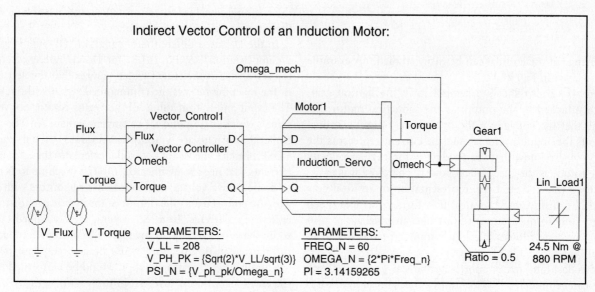

FIGURE 41.14 Circuit for indirect vector control without ABC–DQ transformations.

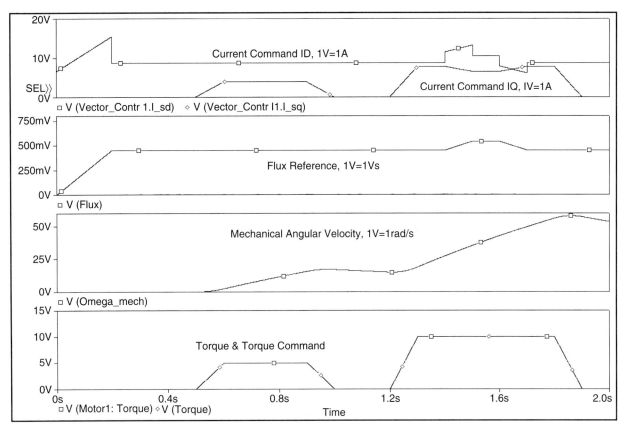

FIGURE 41.15 Results for indirect vector control with perfect tuning.

Figure 41.16 is an example of the results obtained from a de-tuned vector controller. The circuit is identical to the circuit in Fig. 41.12, except for the fact that the rotor resistance value in the controller was increased to 125%, which is thought to be attributed to heating of the rotor bars. It is obvious that the traces for the commanded and the actually produced torque are no longer identical. This is especially true, during times when the flux is changing. De-tuning is actually a real problem in industrial vector control applications. De-tuning is caused by the fact that the machine parameters are not precisely known to begin with and/or, are changing during the operation of the machine. The values of the winding resistance are most likely to change due to heating of the machine.

41.6 Simulation of Sensorless Vector Control Using PSpice®

In the previous example, the advantages of vector control have been shown. However, for the implementation of the control scheme a sensor for the mechanical speed was necessary. This could pose a problem for applications, where a vector control unit is to be retrofitted into existing equipment. The motor installation may not easily allow the installation of a mechanical speed sensor. Therefore engineers have thought to replace the mechanical speed sensor with a speed observer, which is a mathematical model that is evaluated by the control processor (typically a DSP), which is performing the standard vector control computations anyway. The algorithm for the observer would use the measured stator voltages and currents for the D- and Q-axes as input parameters. It would also rely on the knowledge of the machine model and on the correct machine parameters (rotor and stator resistance, mutual and leakage inductance, etc.). The following example shows such an arrangement. It could be derived from the previous example with the only difference being that the speed sensor signal is replaced by a speed observer. However, careful examination of the derivation [2] of the speed observer reveals, that it is easier to calculate the synchronous angular velocity, which is ultimately desired anyway, than the angular velocity of the rotor. Therefore, the observer was modified and the calculation of the rotor speed and the subsequent addition of the slip speed was foregone. The speed observer used here basically solves the D, Q equation system of the induction machine shown in Eq. (41.3), with the only difference that some of the dependent variables are now independent and vice versa. In addition to the synchronous angular velocity, the observer

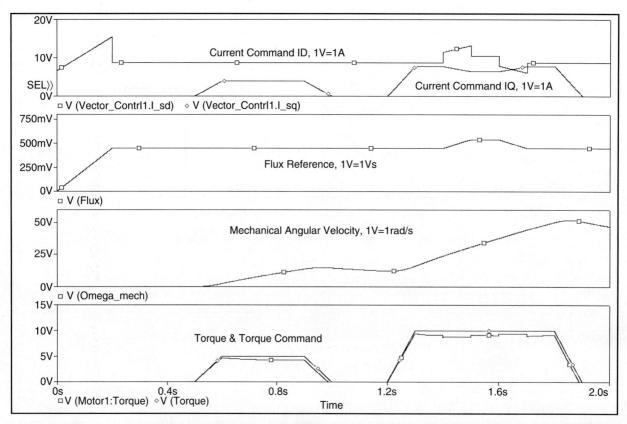

FIGURE 41.16 Results for indirect vector control with 125% rotor resistance.

provides the values of the rotor flux, which are used in the D-axis signal path. The advantage of this is that the compensator with the differentiation function, which is problematic from a numerical stability standpoint, can be eliminated.

Figure 41.17 shows the top level of a simulation project for sensorless vector control. The top view of this circuit is very similar to the circuit for the indirect vector control represented by Figs. 41.12 and 41.14 except for the missing motor-speed feedback. The model for the motor and the motor's parameters are precisely the same as in the example for the indirect vector control. Selecting the "Sensorless Vector Control" block and choosing "Descend Hierarchy" reveals the associated subcircuit, which is shown in Fig. 41.18. This subcircuit is similar to the subcircuit for the indirect vector control with two exceptions: first and foremost, the motor-speed feedback signal is replaced by the speed observer. In this case the speed observer directly provides the synchronous angular velocity. Therefore, it is not necessary to calculate the slip speed and add it to the rotor speed to obtain the synchronous speed. Second, the values of the rotor flux, which are available from the speed observer, are used to calculate the reference value for the torque producing current (Q) component. Therefore, the compensation term, which contains a differentiator in the D-axis of the controller shown in Fig. 41.13 can be eliminated. The purpose of the compensation term is to ensure that the flux is equal to the flux command at all times with no delay. If that is assured, the command signal can be chosen in place of the real flux to calculate how much torque producing current is necessary to obtain the desired torque.

In order to demonstrate how to obtain a more compact motor model, the extensive subcircuit of the induction motor model (see Fig. 41.9) has been replaced by a number of additional attributes which have been added to the motor symbol using the symbol editor.

However, if the actual flux is known (observed), this value can be used instead. Since the flux observer is fed by the D- and Q-axes voltages and currents for the stationary reference frame, the flux components need to be transformed back into the synchronous reference frame. This is done with the "Negative Vector Rotator" located above and to the left of the speed observer in Fig. 41.18.

The easiest way to start the symbol editor and to open the appropriate library is to select the symbol by clicking into it and then select the "Edit Symbol" function. The additional attributes of the modified motor symbol will enter the equivalent to the subcircuit represented by Fig. 41.9 into the netlist. The netlist is a compilation of the schematic pages into a textual description in ASCII format. From a usability standpoint, a "self-contained" symbol like this is a very elegant solution, since the file for the subcircuit is no longer required.

41 Computer Simulation of Power Electronics and Motor Drives

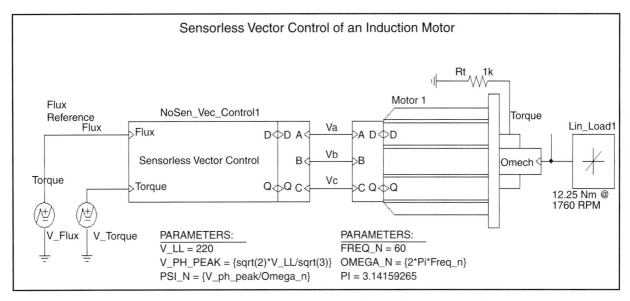

FIGURE 41.17 Top level of simulation circuit for sensorless vector control.

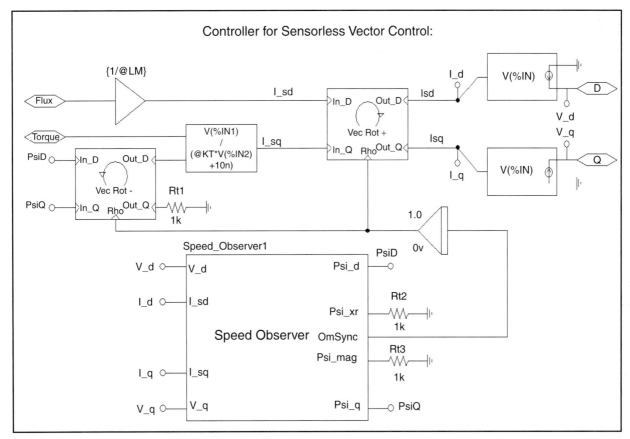

FIGURE 41.18 Subcircuit for sensorless vector controller.

Since the PSpice® simulation engine always uses the netlist as the input, the circuit works identically. For the PSpice® simulation it makes no difference, where the netlist or part of it comes from. This also means that even the most recent release of PSpice® can still simulate legacy files that have been created before the introduction of schematic editors. The introduction of netlist entries is done via the "TEMPLATE" attribute. This attribute is a system attribute and a part of every symbol. Therefore, the TEMPLATE attribute is of course present in the attribute list for the motor symbols in the previous examples. These attribute lists are shown in Tables 41.1 and 41.2. In these tables the TEMPLATE attribute has no value since the netlist entries are made by the symbols in the associated subcircuit. To give a reader a better understanding of the self-contained machine symbol, the format of some common netlist entries and the syntax of the value of the TEMPLATE attribute is discussed. It is also very helpful to examine the TEMPLATE attributes of existing symbols.

Figure 41.19 shows the screen view of the symbol for the induction motor in the symbol editor in PSpice®/Cadence Release 9.1, which was used for the development of this part. Figure 41.20 shows a similar view; however, here the window for entry and editing of attributes is also visible. This screen can be invoked using the "Options/Part Properties..." dialog.

The format of the netlist entries for some common elements is:
[ ] denotes space holder for name extension specific for a symbol to avoid duplicate names.

Resistors, (R devices):
Generic: Rname[]   +Node[]   −Node[]   Value   ;Optional Comment
Example: Rsd[]   Rsd+[]   Rsd−[]   1.0k   ;Resistor, fixed Value
Example: Rsd[]   Rsd+[]   Rsd−[]   {@Rs}   ;Resistor, Value Rs passed on
Example: Rsd[]   %D   0   1.0k   ;Resistor, 1k, Pin 'D' to Gnd

Inductors, (L devices):
Generic: Lname[]   +Node[]   −Node[]   Value   ;Optional Comment
Example: Lsd[]   Lsd+[]   Lsd−[]   1.0u   ;Inductor, 1.0 µH fixed
Example: Lsd[]   Lsd+[]   Lsd−[]   {@Lsl}   ;Inductor, Value Lsl passed on

Voltage-controlled voltage sources (E devices):
Generic: Ename[]   +Out[]   −Out[]   VALUE { Control Function }
Example: ETorque[]   %Torque   0   VALUE { 1.5*(Vt1[] − Vt2[]) }
;E source, output between pin "Torque" and Gnd, Control function as shown

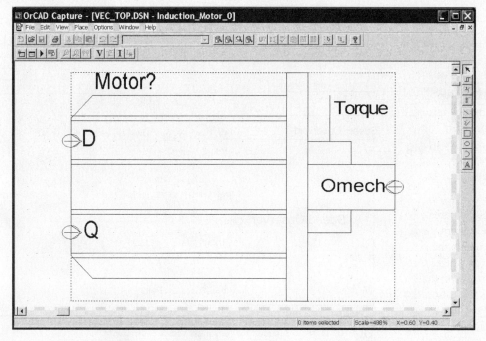

FIGURE 41.19 Screen view of motor symbol in the symbol editor.

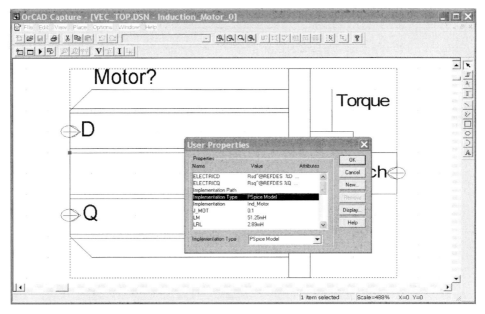

FIGURE 41.20 Screen view of symbol editor with user properties input window.

The attributes, that create the netlist entries, which insert the previously discussed model for the motor are entered here. A complete list of all attributes, extracted from a working example, is given in Table 41.3. Therefore, the reader should be able to enter the attributes exactly as printed and obtain a working model.

According to Table 41.3, the value for the TEMPLATE attribute is as follows:

TEMPLATE = @ElectricD\n\n@ElectricQ\n\n@Mechanical

This statement will insert the expression for "ElectricD," i.e. the T-equivalent circuit for the D-axis, two carriage returns "\n\n," the expression for "ElectricQ," i.e. the T-equivalent circuit for the Q-axis, two more carriage returns "\n\n," and finally the expression for "Mechanical," i.e. the mechanical model into the netlist. The expressions for "ElectricD," "ElectricQ," and "Mechanical" are defined in separate attributes. Careful examination of the values of the "ElectricD," "ElectricQ," and "Mechanical" attributes reveals a number of repetitive terms. The meaning of these terms are explained below:

@Name — Substitutes what is defined for "@Name" at the current place.

^@REFDES — Inserts the path "^" and the reference designator "@REFDES" to create a unique node or part name, that does not repeat. The path is the concatenation of the names of the symbols and subcircuits in the hierarchy above the part. The reference designator is the name of the part on a particular schematic page.

\n — new line (carriage return) is inserted into the netlist, however the value of the attribute does **not** have a carriage return in it. This means everything listed after "ElectricD" until "ElectricQ" goes on one line. If the symbol definition is printed however, it is shown as in Table 41.3.

\n+ — new line and continuation of expression.

\n+ + — new line, continuation of expression, numerical operator "+."

Figure 41.21 shows a subcircuit for the speed observer. This schematic shows the structure and all the details of the implementation in the form of a block diagram. The speed observer uses the stator voltages and currents for the D- and the Q-axis as input variables. After subtracting the voltage drop across the winding resistance and the leakage inductance of the stator from the stator input voltage and scaling the result by L_r/L_m, the observer calculates the D and Q components of the rotor flux by integration. Since the input values are in the stationary reference frame, so are the results. The observer also calculates the magnitude of the rotor flux (Eq. (41.9)) and then calculates the synchronous angular velocity (Eq. (41.10)) by evaluating the rate of change of the ratio of the D and Q components. The mathematical relationships are given by the Eqs. (41.9) and 41.10 [2].

$$p\begin{bmatrix}\lambda_{rd}\\ \lambda_{rq}\end{bmatrix} = \frac{L_r}{L_m}\left[\begin{bmatrix}V_{sd}\\ V_{sq}\end{bmatrix} - \begin{bmatrix}R_s + \sigma L_s p & 0\\ 0 & R_s + \sigma L_s p\end{bmatrix}\begin{bmatrix}I_{sd}\\ I_{sq}\end{bmatrix}\right]$$

$$\sigma = \left[1 - \frac{L_m^2}{(L_r L_s)}\right] \quad \sigma = \text{Leakage factor}$$

(41.9)

TABLE 41.3 List of all attributes used for the self-contained induction motor symbol

---

Attributes:

REFDES = Motor?
PART = Ind_Motor
MODEL = Ind_Motor
TEMPLATE = @ElectricD\n\n @ElectricQ\n\n@Mechanical
R_Stat = 0.48
R_Rot = 0.358
Lm = 51.25mH
Lsl = 2.18mH
Lrl = 2.89mH
J_mot = 0.1
Omega_init = 0.0
Poles = 4

ElectricD = Rsd^@REFDES %D 1^ @REFDES @R_Stat
\nLsld^@REFDES 1^@REFDES Vmd^@REFDES @Lsl
\nLmd^@REFDES Vmd^@REFDES 0 @Lm
\nLrld^@REFDES Vmd^@REFDES 2^@REFDES @Lrl
\nRrd^@REFDES 2^@REFDES ErotD^@REFDES @R_Rot
\nErotd^@REFDES ErotD^@REFDES 0 VALUE {-(Ome^@REFDES)*((V(%Q,3^@REFDES)/@R_Stat)*@Lm
\n+ +(V(ErotQ^@REFDES,4^@REFDES)/@R_Rot)*(@Lm+@Lrl)) }

ElectricQ=Rsq^@REFDES %Q 3^@REFDES @R_Stat
\nLslq^@REFDES 3^@REFDES Vmq^@REFDES @Lsl
\nLmq^@REFDES Vmq^@REFDES 0 @Lm
\nLrlq^@REFDES Vmq^@REFDES 4^@REFDES @Lrl
\nRrq^@REFDES 4^@REFDES ErotQ^@REFDES @R_Rot
\nErotq^@REFDES ErotQ^@REFDES 0 VALUE {V(Ome^@REFDES)*((V(%D,1^@REFDES)/@R_Stat)*@Lm
\n+ +(V(ErotD^@REFDES,2^@REFDES)/@R_Rot)*(@Lm+@Lrl)) }

Mechanical=ETorque^@REFDES %Torque 0
\n+ VALUE { (1.5*@Lm*@Poles/2) * (
\n+ ((V(%Q,3^@REFDES)/@R_Stat)*(V(ErotD^@REFDES,2^@REFDES)/@R_Rot)) -
\n+ ((V(%D,1^@REFDES)/@R_Stat)*(V(ErotQ^@REFDES,4^@REFDES)/@R_Rot))) }
\nEOme^@REFDES Ome^@REFDES 0 VALUE { V(Om^@REFDES)*@Poles/2 }
\nVLd^@REFDES Om^@REFDES %Omech 0V \nEom^@REFDES Om^@REFDES 0 VALUE {SDT((V(%Torque)-I(VLd^@REFDES))/@J_mot) + @Omega_init }

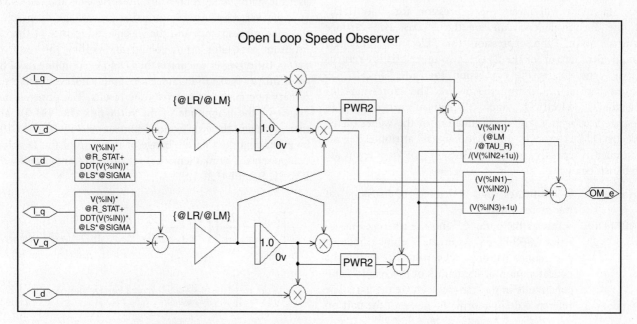

FIGURE 41.21 Subcircuit for speed observer for sensorless vector controller.

TABLE 41.4 List of all attributes for the speed observer symbol

---
Attributes:

REFDES = Speed_Observer?
PART = Speed_Obs
MODEL = Speed_Obs

TEMPLATE = Ed^@REFDES %Psi_d 0 VALUE { @EXP_d1
\n+ @EXP_d2 \n+ @EXP_d3 }
\nEq^@REFDES %Psi_q 0 VALUE { @EXP_q1
\n+ @EXP_q2 \n+ @EXP_q3 }
\nEmag^@REFDES %Psi_mag 0 VALUE { V(%Psi_d)*V(%Psi_d) + V(%Psi_q)*V(%Psi_q) }
\nExr^@REFDES %Psi_xr 0 VALUE { @EXP_x1
\n+ @EXP_x2 \n+ @EXP_x3 \n+ @EXP_x4 }
\nEOmSync^@REFDES %OmSync 0 VALUE { @EXP_om1 }

SIMULATIONONLY =
EXP_d1 = @Ini_d +
EXP_d2 = SDT((@Lr/@Lm)*(V(%V_d) - V(%I_sd)*@R_Stat
EXP_d3 = -DDT(V(%I_sd))*@Ls*@Sigma))
EXP_q1 = @Ini_q +
EXP_q2 = SDT((@Lr/@Lm)*(V(%V_q) - V(%I_sq)*@R_Stat
EXP_q3 = -DDT(V(%I_sq))*@Ls*@Sigma))
Ini_d = 0.0V
Ini_q = 0.0V
EXP_x1 = (V(%Psi_d)*(@Lr/@Lm)*
EXP_x2 = (V(%V_q) - V(%I_sq)*@R_Stat - DDT(V(%I_sq))*@Ls*@Sigma))
EXP_x3 = -(V(%Psi_q)*(@Lr/@Lm)*
EXP_x4 = (V(%V_d) - V(%I_sd)*@R_Stat - DDT(V(%I_sd))*@Ls*@Sigma))
EXP_om1 = (V(%Psi_xr) / (V(%Psi_mag)+1u))

$$\frac{d}{dt}atan\left(\frac{\lambda_{rq}(t)}{\lambda_{rd}(t)}\right) = \omega_s = \frac{\left(\frac{d}{dt}\lambda_{rq}(t)\lambda_{rd}(t) - \lambda_{rq}(t)\frac{d}{dt}\lambda_{rd}(t)\right)}{\left(\lambda_{rd}(t)^2 + \lambda_{rq}(t)^2\right)}$$
(41.10)

Since the observer shown in Fig. 41.21 has a large number of elements, this subcircuit has also been integrated into a custom part by creating an appropriate TEMPLATE and other supporting attributes. A complete listing, which was extracted from thoroughly tested part, is given in Table 41.4. Using this table, the reader should be able to create this very complex part with relative ease.

Figure 41.22 shows the output for the speed sensorless vector control project as it presents itself in the PSpice® 9.1/ORCAD evaluation version.

A comparison of the traces for the torque and the torque command (they are perfectly on top of each other) shows that the scheme works extremely well. It even works for a start from zero speed, which is typically not the case in real systems. The reason is, that the uncertainties of the winding resistance values (note that they change with temperature) create errors in observers like this one. The problem is worst at low speeds, because at low speeds and associated low stator frequencies, the uncertain winding resistance represents the largest part of the total machine impedance.

41.7 Simulations Using Simplorer®

In the following, some examples of simulations with the evaluation version of Simplorer® [11], Release 4.1 are shown. It should be acknowledged at this point that these examples have been created by Ali Ricardo Buendia, who obtained M.S.E.E degree from Texas Tech University. The advantages of Simplorer® are that it has a lot of models for power electronics devices and machines built-in, since it is specialized for this type of simulation. Also, no numerical convergence problems have been noticed thus far. Figure 41.23 shows the schematic for a three-phase diode rectifier. The input is a balanced three-phase source with a reactive source impedance. An exponential characteristic, closely resembling real diodes, has been chosen as a model for the diodes.

Figure 41.24 shows a circuit, which transforms the load currents from a three-phase to a two-phase system. Here the

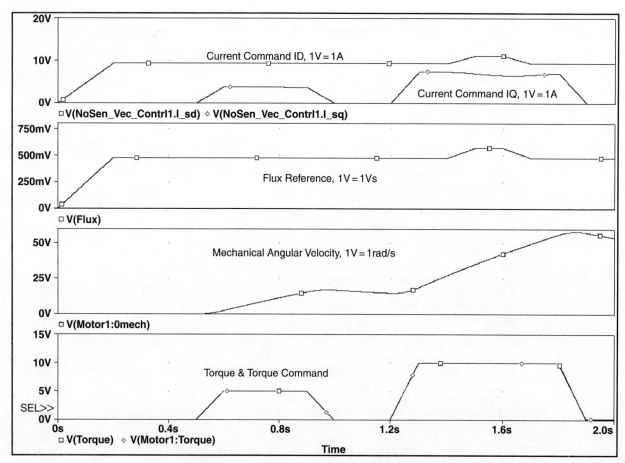

FIGURE 41.22 Simulation results for sensorless vector control.

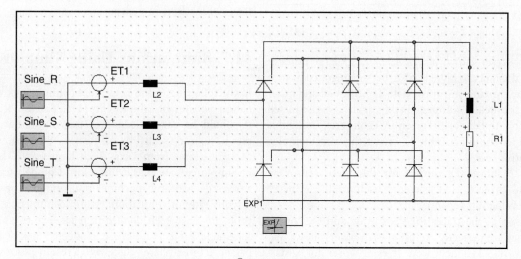

FIGURE 41.23 Simplorer® schematics for three-phase rectifier.

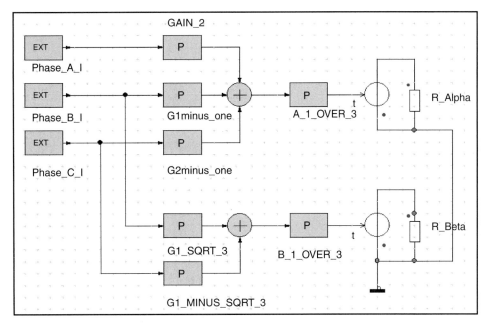

FIGURE 41.24 Three-phase to two-phase conversion for load currents.

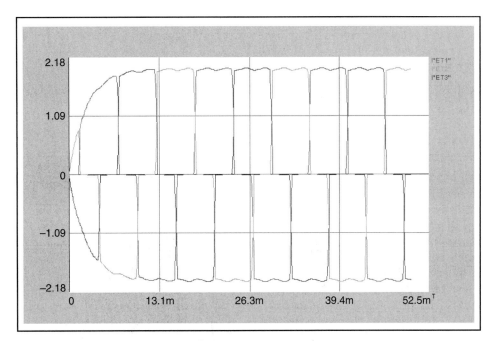

FIGURE 41.25 Simplorer® plot of load currents for the three-phase rectifier.

D and Q components are called alpha and beta components respectively. The results of the simulation are shown on the same page that is used for the schematic diagrams shown by the previous two figures. Figure 41.25 shows a plot of the line currents during the start-up of the rectifier, where the initial load current is zero. The next graph shows a plot of the line currents that have been transformed by the circuit shown in Fig. 41.26. The components of the line currents are the variables of the axes. The plot shows the typical hexagonal trace that can be expected for this type of line-commutated rectifier (six step operation). If the converted source voltages were plotted in this fashion, a perfect circle would be the result.

The last two graphs demonstrate a Simplorer® simulation of the start of an induction motor. As previously mentioned, the

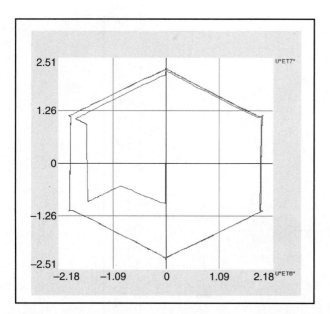

FIGURE 41.26 Plot of converted load currents for the three-phase rectifier.

the machine model. Like the plot of the current components for the rectifier, the flux components are the variables of the axes in the graphs of Fig. 41.28.

41.8 Conclusions

In this chapter, the capabilities of PSpice® [12] and Simplorer® [11] have been used to simulate a number of projects from the power electronics, machines, and drives area. The advantage of PSpice® is that it is based on the almost universal Spice simulation language, which can be seen as the worldwide de facto standard. On the other hand, Simplorer® has the advantage of built-in machine models. If both programs are used, comparisons and mutual validations of models can be performed. The reader should always validate any model before it is used for critical engineering decisions. It was pointed out in the introduction, that model validation often means verification that the limitations of the model are not exceeded.

In addition to the detailed examples, some general guidelines on the uses of simulation for analysis and design have been developed. All tested programs yielded excellent results. The simulation time for each project shown is typically less than a minute on a typical (2005) PC. The author hopes, that the reader has gotten an insight and appreciation of the power of these modern simulation codes and some useful ideas and inspirations for projects of his own.

symbol and the model for the induction motor is built into the code. Figure 41.27 shows the schematic with the source and the induction motor as well as a graph that shows the rotational speed as a function of time. The graph below shows plots of the stator and rotor flux components, which are available from

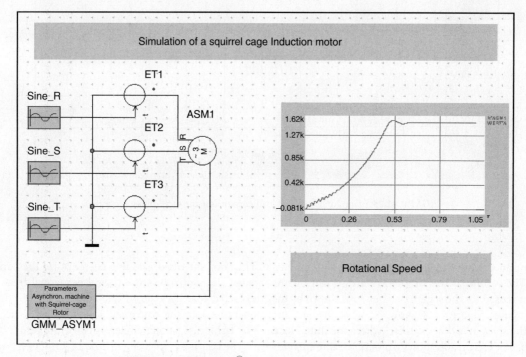

FIGURE 41.27 Simplorer® schematics for induction motor start.

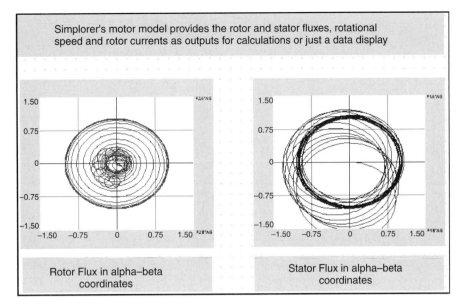

FIGURE 41.28 Simplorer® plots of machine fluxes for induction motor start.

References

1. N. Mohan, T. Undeland, and W. Robbins, "Power Electronics, Converters, Applications, and Design," 2nd edition, John Wiley & Sons, New York, 1995, ISBN: 0-471-58408-8.
2. A. M. Trzynadlowski, "The Field Orientation Principle in Control of Induction Motors," Kluwer Academic Publishers, Inc., Boston, MA 1994, ISBN: 0-7923-9420-8.
3. P. C. Krause, O. Wasynczuk, and S. D. Sudhoff, "Analysis of Electric Machinery," IEEE Press, Hoboken, NJ 1995, ISBN: 0-7803-1101-9.
4. M. G. Giesselmann, "Averaged and Cycle by Cycle Switching Models for Buck, Boost, Buck-Boost and Cuk Converters with common Average Switch Model," Proceedings of the 32nd Intersociety Energy Conversion Engineering Conference, IECEC-97, Honolulu, HI, July 27–Aug. 01, 1997.
5. M. Giesselmann, "A PSpice® Tutorial for Demonstrating Digital Logic," IEEE Transactions on Education, Nov. 1999, http://coeweb.engr.unr.edu/eee/59/CDROM/Begin.htm.
6. D. W. Novotny and T. A. Lipo, "Vector Control and Dynamics of AC Drives," Oxford Science Publications, New York, NY 1996, ISBN: 0-19-856439-2.
7. P. Vas, "Vector Control of AC Machines," Oxford Science Publications, 1990 ISBN: 0-19-859370-8.
8. A. Von Jouanne, D. Rendusara, P. Enjeti, and W. Gray, "Filtering Techniques to Minimize the Effect of Long Motor Leads on PWM Inverter Fed AC Motor Drive Systems," IEEE Transactions on Industry Applications, July/Aug. 1996, pp. 919–926.
9. K. Hasse, "Zur Dynamik Drehzahlgeregelter Antriebe mit Stromrichter gespeisten Asynchron-Kurzschlussläufermachinen," (On the Dynamics of Adjustable Speed Drives with converter-fed Squirrel Cage Induction Machines), Ph.D. Dissertation, Technical University of Darmstadt, 1969.
10. *ACI3_4: Sensor-less Direct Flux Vector Control of 3-phase ACI Motor*, Data Sheet, Texas Instruments Incorporated, 12500 TI Boulevard, Dallas, TX 75243-4136, 800-336-5236.
11. Simplorer®, Technical Documentation, Ansoft Corporate Headquarters, 225 West Station Square Drive, Suite 200, Pittsburgh, PA 15219, USA, (412) 261 3200, http://www.ansoft.com/products/em/simplorer/.
12. PSpice® Documentation, 2655 Seely Avenue, San Jose, California 95134, USA, (408) 943-1234, http://www.cadence.com/.

42
Packaging and Smart Power Systems

Douglas C. Hopkins, Ph.D.
Dir.—Electronic Power and Energy Research Laboratory, University at Buffalo, 332 Bonner Hall, Buffalo, New York, USA

42.1 Introduction 1147
42.2 Background 1148
42.3 Functional Integration 1148
    42.3.1 Steps to Partitioning
42.4 Assessing Partitioning Technologies 1149
    42.4.1 Levels of Packaging • 42.4.2 Technologies • 42.4.3 Semiconductor Power Integrated Circuits
42.5 Cost-driven Partitioning [5] 1153
42.6 Technology-driven Partitioning 1154
42.7 Example 2.2 kW Motor Drive Design 1155
    42.7.1 User Requirements (Constraints) • 42.7.2 Component Characterization Map • 42.7.3 Component Grouping • 42.7.4 Strategic Partitioning with Constraints • 42.7.5 Optimization within Partitions
42.8 High Temperature (HT) Packaging [6] 1157
    42.8.1 HT Materials Selection • 42.8.2 Module Construction
    Further Reading 1158

42.1 Introduction

A continual endeavor in power electronics is to increase power density. This is achieved by shrinking component size, moving components closer, and reducing component count. During the last two decades, circuit frequencies increased sharply to shrink component dimensions. Improved thermal management and physical packaging materials brought components closer, and finally, increased integration of functions at the semiconductor and package levels reduced component count. This has been marked in the microelectronics world by "system on chip" (SOC), "system in package" (SIP), and "system on package" (SOP) with subsystems including "stacked die" and "multichip modules" (MCMs), all addressing higher densities and all applicable to lower power, power electronic systems.

The approach of "functional integration" has been ongoing for decades. Until the 1980s, nearly all such integration was done at the packaging level melding control and power processing. The term "smart power" (within the context of power electronic conditioning) applied in the 1960s–1970s to the integration of computers and microprocessors into large rectifier and converter cabinets. With the advent of high-voltage-silicon integrated circuits, more functionality was brought directly to the power semiconductors, and in the 1980s–1990s the term applied mostly to smart power semiconductors. In the late 1990s, there was a move back to hybrid integration following the trend to SOP. During the 1980s–1990s "smart power" also became associated with digital control of higher power systems, such as motor drives and uninterruptible power supply (UPS) systems and became commercially associated with power management circuits. The first decade of the twenty-first century has ushered in "digital power" for direct control of high-frequency inner control loops in power supplies. Smart power is now more generically used since the cost of digital controllers, such as microcontrollers and programmable ICs (pics), are low cost and easily used throughout the power electronic systems.

From a designer's perspective, "functional integration" exists in a *packaging continuum* with "smart power" as a subset dependent on the definition in vogue. To take advantage of "functional integration" the designer, in reverse thinking, partitions or modularizes circuits, and functions to achieve the most cost-effective approach that meets a set of required performance specifications. This chapter provides background information, framework, and procedures to produce partitioning and functional integration.

42.2 Background

Circuits are typically designed based upon a pre-determined set of packaging technologies ranging from silicon integration of sub-circuit functions to multiple boards in a rack. Partitioning a circuit for packaging in one technology, such as all silicon, is straightforward. Partitioning for multiple technologies is much more difficult since higher performing technologies duplicate the aspects of lower technologies. The duplication geometrically increases parameter trade-offs and complicates design. A study on the status on power electronics packaging (STATPEP) [1] identifies metrics to evaluate the relative technical merits of the technologies.

To optimize the use of multiple technologies in functional integration, a structured method should be used. A full-cost model for various technologies is used as a basis to produce a comparative cost diagram. The diagram allows intermixing of high and low performance technologies based on surface density, which is interpreted as circuit area and, hence a partition. An example is given in Section 42.7 to demonstrate the method using a 2.2 kW motor-drive module product.

The method is also applied to product modularization, i.e. system partitioning where a specific function is used across several products. A module can represent functional integration within a packaging technology or use multiple packaging technologies to create integrated power modules (IPMs) or power electronic building blocks (PEBBs). The importance of modularization is to increase the product volume to lower cost. The cost model includes variations based on volume.

This partitioning approach matches user requirements to "Levels of Packaging" as defined in the "Framework for Power Electronics Packaging"[2] and provides optimum integration of packaging levels for a product. The framework also identifies critical technical issues that need to be considered in evaluating technical performance. This partitioning approach looks at electrical, magnetic, thermal, and mechanical issues (multiple energy forms).

42.3 Functional Integration

Figure 42.1 shows a 2.2 kW ac motor drive. Functional integration requires that the system should be partitioned both electrically and physically. The systems integrator is usually an electrical designer and the first partitioning is usually electrical. The *electrical* partitions and distributed power losses are also shown in the figure. The physical partitioning, or packaging, involves different components with different functions ranging from fine-line control to high-current, high-loss power processing. Several partitions can be pursued. The line-communications and motor-control blocks can use a signal-level packaging approach, such as all-silicon application-specific integrated circuit (ASIC), or discrete components on an epoxy-glass flame resistant 4 (FR-4) or insulated-metal substrates (IMS). If the power supply and control blocks are to be combined, a surface mount technology (SMT) approach cannot accommodate bulky storage components in the power supply. Hence, a through-hole approach is considered for part or both blocks. Regardless, such trade-offs can be nearly endless.

A structured method needs to be used to establish essential requirements and guide circuit and system partitioning. The method described here is based on characterizing and grouping the components, evaluating the cost and technical constraints, and then, matching packaging technologies to the groupings. All this is set against a set of comprehensive user requirements.

42.3.1 Steps to Partitioning

A first step to partitioning is creation of a comprehensive categorized list of electrical, mechanical, and thermal, technical user requirements. The second step is creation of a simple component characterization map that identifies dominant attributes of the components. The block diagram of a 2.2 kW motor drive is shown in Fig. 42.1 and a partial characterization map is given in Table 42.1. The map is divided into metrics by energy form to categorize and record extreme operating values for each component. Not all blocks need to be completed or components included, only those that most impact the technology selection. For example, any 5 V, <0.1 W resistor in the control circuit need not be listed since it is accommodated by nearly all technologies (e.g. as 0806, SMT, plated through holes (PTH), thick film, etc.). For each of the remaining components, all the mechanical package formats should be listed under the delivery form.

The third step is to strategically group components by delivery form taking into consideration limits on electrical and thermal operating points. This first-cut grouping brings a high level of packaging integration to the system and is a critical step. Similar components from all parts of the circuit become associated.

The fourth step uses the user requirements as constraints along with the engineering experience to re-associate components into different groupings. Not all components are easily regrouped. The unassociated components become dominant factors during technology selection. As an example, the high-voltage components of a bootstrap gate-drive supply can be associated with the gate-drive circuit board or the high-voltage power inverter components. Interestingly, most unassociated components reside at the interfaces between functional blocks (as shown in Fig. 42.1).

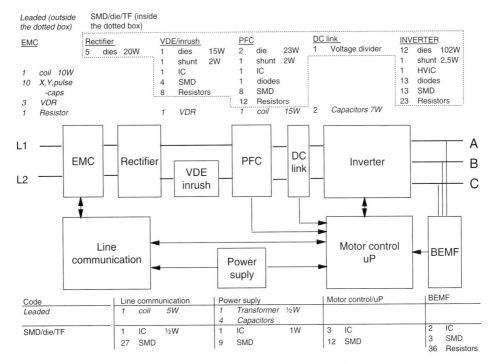

FIGURE 42.1 Block diagram of a 2.2 kW motor-drive module.

TABLE 42.1 Component characterization map (partial listing)

Functional block	Function	Component	Quantity	Mechanical		Electrical			Thermal	
				Delivery form	Size	Voltage V	Current A/comp	Constraint	Loss W/comp	Max temp °C
Rectifier	Bridge	Diode	4	die	3.5 × 2.5	600	11 rms		5	125
	Clamp	Diode	4	die	3.5 × 2.5	600			<1	125
Inrush/ VDE	Switch	IGBT	1	die	6 × 4.3	1,200	11 rms		15	125
	Current sense	Shunt	1	TF			11 rms		2	
	Controller	IC	1	die		<18			<1	125
	Support	C R	4 8	SMD TF	0603					
	Transient clamp	VDR	3	leaded	$\phi 21 \times 5$	300 ac		Low L		
PFC	Switch	MOS	1	die	7.5×7.5	500	26 peak		16	125
	Freewheel	Diode	1	die	3×4	500			7	125
	Current sense	Shunt	1	TF			11 rms		2	
	Controller	IC	1	die			10 m		<1	125
	Support	C R	8 12	SMD TF	0603	<70 V				
	Choke	L	1	leaded		500	11 rms		15	130
DC-link	DC-cap	E-lytics	2	leaded	$\phi 26 \times 50$	500	1.25 rms		3.5	75
	Voltage sense	R	2	TF						

The fifth step is to map the groupings of components to the packaging technologies. This was partially performed in the previous step as engineering judgment guided the regrouping. Refinement of the selection comes when the unassociated components are incorporated. Steps four and five become iterative to provide optimum partition(s).

42.4 Assessing Partitioning Technologies

To better understand the correlation between electrical and physical circuits (*partitioning and packaging*), consider the morphology of a generic circuit. A circuit has three partitions: components, topologies, and controls. Components are active or passive, such as ICs or heat sinks. Topologies are the positioning of the components to provide a function, such as a buck converter or the thermal structure of silicon soldered on copper clad to ceramic. Controls provide a preferred set of rules for operation of those components, such as voltage regulation or a thermostatically controlled fan. Hence, a "circuit" design involves electrical and physical design. The physical design is packaging and can be defined as

> "... the art (design) of arranging components to provide a function or characteristic"

Note that *packaging* is a design function, whereas manufacturing embodies the *processes* to fabricate the designed arrangement. Although packaging and manufacturing are strongly interrelated, they are not synonymous.

The term "physical" should be further defined. The term "Electrical" identifies the form of energy being processed. Hence, "physical" represents other forms, such as mechanical, thermal, chemical, photonic, etc. The power electronics designer is mostly interested in electric, magnetic, mechanical, and thermal. Discussion will be limited to these four energy forms.

An integrated design problem example relating the four energy forms of interest is as follows. A high-frequency magnetic core couples the radiated field into a copper conductor on a printed wiring board (PWB) and causes eddy current heating, increasing the skin-effect resistance. Higher resistance loss further increases conductor heating which increases the mechanical stresses between the conductor and PWB leading to early failure. Who would notice the problem first? The electrical designer through circuit loss measurements; the thermal designer through a thermograph of that specific spot, or the packaging engineer who first notices the conductors are lifting off the board and assumes the conductor adhesion is poor because of faulty chemistry?

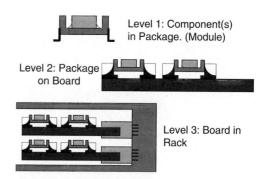

FIGURE 42.2 Levels of packaging.

42.4.1 Levels of Packaging

The *Levels of Packaging* divides a system, top–down, into lower and lower subassemblies with the boundary drawn between assembly and subassemblies as shown in Fig. 42.2.

Each level is defined and numbered, bottom–up, in a micro-to-macro manner. Three traditional levels in electronic packaging [3] are also applicable to power packaging. Note that levels are not easily defined. Some packages may be categorized in either of the two levels depending on the application.

Level-0: Component(s). This is the base level for a component that a designer can obtain and may be a passive, discrete semiconductor or integrated circuit, including a "smart power" circuit. The semiconductors and ICs are typically Silicon (Si), though there is significant development in SiC and GaAs for higher operating temperature and speed.

Level-1: Component(s) in Package. This is basic component packaging. Examples include mount-down and lead-attach of a component or semiconductor in a discrete package, or multiple components in a module. Traditional "chip and wire" hybrid circuits mounted in a housing (often, hermetic) are Level-1 packages. The package provides a "self contained environment" that allows the components to be tested, transported, and used at the next higher level of packaging while buffering electrical, mechanical, and chemical discontinuities from the next level. This package becomes a subassembly to the next higher level.

Level-2: Package on Board or Board-Level Assembly. These boards carry mixed-technology components (capacitors, resistors, inductors, and packaged discretes) that are usually coated and terminated with a connector. Examples are printed circuit board (PCB), IMS, and SMT boards. They differ from Level-1 in the lesser sophistication in fabrication. The board provides a functional partition and is a subassembly to the next packaging level.

Level-1.5 *(half-level)*: Chip on Board (COB). This mounts "chip and wire" semiconductors directly to a PCB or on IMS. A driver in packaging is to combine levels. An objective of the road map will be the development of a direction to combine levels.

Level-3: Board in Rack or Box-Level Assembly. At this level the rack or case is considered. Each board or module is a subassembly connected to a rack backplane, motherboard, or free-wired together. An example is the output modules in a multi-output power supply. The sub-module approach provides flexibility and fast assembly time.

Definition of levels can continue into Level-4: Rack in Cabinet and Level-5: Cabinet in Room or Multiple Cabinet Level. Note that all the technical issues of packaging, embodied in passing energy across an interface or along a pathway, are the same for all levels of packaging.

42.4.2 Technologies

The delivery form, i.e. mechanical support structure for integrating functions, divides the technologies. The simplest delivery form is a mono-material approach, such as a silicon integrated circuit. A thick-film hybrid uses ceramic-glass structures to create functions. A glass-epoxy board (FR-4) allows the attachment of discrete components to "integrate functions." Functions are not restricted to electrical, but may include magnetic, mechanical, and thermal.

The delivery form is important since the size of the mounted components greatly limits the choices in technologies. The greater the mass of the components, the more mechanically robust the technology needs to be. The technologies reviewed below belong to packaging Level-0 through Level-2 and, generally, sequentially range from fully imbedded components as in silicon ICs to modestly robust for surface mounted components, to very robust for clamped, screwed, and axially leaded. The transition from PTH to SMT occurs within FR-4 and partly explains the greater acceptance of this versatile technology.

Semiconductor Power Integrated Circuits – This is considerably different from the remaining packaging technologies in which it approaches a "mono-material system." Multiple functions can be produced in one material, usually silicon. This is expanded in the following section.

Thick Film on Ceramic (TFC) – Glass-based pastes or inks are loaded with electrically conductive materials, such as copper, gold, or silver, to form interconnects, loaded with resistive materials to form components, or used unloaded as dielectrics. The pastes are screen printed on ceramic and fired at ~900°C. Vias are formed as holes in dielectric layers and discrete components are surface mounted with solder or adhesives. Only two types of air-fired thick film are considered here: multilayer thick film (TF-multilayer) for control circuity and thick thick-film(TTF) where silver is printed to form up to 160 μm conductors for power.

Cu Plated on Ceramic (CuPC) – Patterns are imaged or transferred to the surface of ceramic. Copper is then plated to a thickness <125 μm (5 mils). Discrete components are attached or full thick-film processing is placed on the plated copper with screen-printed components imbedded or discrete components attached.

Glass-epoxy with Surface Mount Pads (FR-4, SMT) – A fiberglass mesh is impregnated with epoxy and metalized with copper. Interconnect patterns are etched into the foil. The patterned copper clad mesh can be laminated and vias formed by drilled and plated holes. Chip components are "surface mounted" with solder attachment or conductive polymer. Components can also be imbedded using loaded polymers similar to the TFC process, but with low temperature curing. (SMT is "surface mount technology.")

Insulated Metal Substrate – Polymer on Metal (IMS-PM) – A polymer is used to isolate and attach a conductive interconnect to a metal plate which provides mechanical support. Vias can be placed between the interconnect and plate, and a layer of polymer and interconnect can be attached to the interconnect layer.

Insulated Metal Substrate – Steel Corded (IMS-PS) – A high temperature (HT) glass (~900°C) coats a steel plate and a thick-film conductive cermet interconnect is applied upon the glass. The structure is similar to traditional thick film. Vias are processed as in multilayer thick film.

Direct Bonded Aluminum (DBA) – Aluminum foil is bonded to, or cast onto ceramic, patterned, and etched. Low-resolution patterns can be direct cast. Discrete components are surface mounted with solder or adhesive. There are no vias. This is an excellent approach for Al bearing substrate materials.

Direct Bonded Copper (DBC) – Copper foil is applied to ceramic, bonded at ~1063°C, and a pattern is etched. Discrete components are surface mounted with solder or adhesive. There are no vias.

Glass-epoxy with Plated through Holes (FR-4, PTH) – Same as above FR-4 except leaded components are solder attached with leads placed through holes. (PTH is "plated through holes.")

Molded Interconnect Device (MID) – A HT plastic or polymer structure hosting electrical interconnects is fabricated by 1-shot, 2-shot, or insert molding. The interconnections are formed by hot-stamping copper foil, imaging and metal

Comparison of technologies

	TFC/IMS-PS	CuPC	IMS-PM	DBC	DBA	FR-4-PTH	MID
Conductor material	Ag/Cu	Cu	Cu	Cu	Al	Cu	Al
Thickness (μm)	15	25–125	35–140	100–1000	100–1000	17.5–140	5–15
Line width (μm)	100–150	50–100	75–125	75–125	75–125	75–125	75–225
Line pitch (μm)	250–350	50–100	150–250	150–500	150–500	150–250	150–250
Bond pad pitch (μm)	250–350	200	200	200	200	200	200
Max # layers	5–10	2	2	2	2	36	2
Sheet resistance (mΩ/sq)	3–1.2	0.14–0.69	0.14–0.69	0.034–0.135	0.068–0.270	1.1–0.14	9–3
Dielectric material	Glass/ceramic	Ceramic	Polymer	(Coating)	(Coating)	Epoxy/glass	Polymer
Dielectric constant	6–9	9.5	6.4	n–a	n–a	4.8	
Thickness/layer(μm)	35–50	n–a	75–150	n–a	n–a	120	
Min. via dia. (μm)	200	50–150	300	n–a	n–a	300	
Substrate material	Al_2O_3AlN	Al_2O_3	Al, Cu	Al_2O_3,	Al_2O_3, AlN	Epoxy	Polymer
Thermal (W/m-C)	20–35	20–35		26	26, 150–270	0.17	
TCE (ppm/K)	7.1	7.1	23	7.3	7.3	13–18	

plating the polymer, or insert molding of structured metal. The MID lends itself to high volume, 3D, net shape packaging and is extensively overlooked in the power electronics area (excluding automotive). Components can be surface mounted or through-hole with moderate to course line resolution. Only the hot embossing is considered here.

Laminated Bus-bar – A polymer, such as epoxy, glues together thick conductor bars while providing electrical isolation. The bars can be free-floating laminated interconnects or, if sufficiently thick, be the metal carrier. Vias between layers are metal posts or fasteners placed through drilled or stamped holes. These are used in high-current systems and can accommodate very large components. These were not considered in this development.

42.4.3 Semiconductor Power Integrated Circuits

As noted in the Introduction, the term "smart power" has been used for several decades to describe the imbedding of control into power processing systems. One approach integrates control and power into a monolithic circuit, such as silicon, and takes on two forms. One is the integration of analog and digital circuitry with discrete power devices. The second applies to high-voltage ICs used for power monitoring and fault control. The term "smart power" has become synonymous with power integrated circuits (Power ICs) or application-specific power ICs (Power ASICs). Motorola trademarked the term "SMARTpower" circa 1980.

A designer typically is a user of power ICs and seldom influences the chip design. Systems partitioning, as described throughout this chapter, is not directly applicable. However, once the chip is available, the designer is armed with a more functionally integrated component. A background to power ICs is given below to aid the designer in better understanding the technology. An excellent reference noting the beginning of high-voltage ICs is an IEEE Press Book by B. J. Baliga [4].

Power ICs can be divided into four groups resulting from a matrix of low and high voltage, and low and high current capabilities as identified in Table 42.2. The low-voltage, low-current ICs are readily available for the control and monitoring of power processing functions. These smart chips control power supplies, battery chargers, motor drives, etc. and are often referred to as "power controllers." These chips are produced from standard IC processes and limited to the voltages of the process. The cost follows typical IC cost structures.

TABLE 42.2 Examples of power ICs (smart power)

	Low current	High current
Low voltage	Power control ICs PWM controllers	Bipolar drivers Automotive actuators
High voltage	Bridge gate drivers Gas-display drivers	(limited application)

Low-voltage, *low-current* ICs can be further subdivided into "dedicated" and "programmable" chips. In the late 1990s and early 2000s, the incorporation of imbedded control expanded the definition of "power controllers." Sophisticated control algorithms that were implemented in digital signal processors (DSPs) were incorporated into programmable power controllers. The role of the power electronics designer further changed to become adept at high-performance programming.

Low-voltage, *high-current* power ICs again use standard IC processes for fabrication. The higher current requirement is met by creating effectively large device areas that maintain current densities consistent with process characteristics. In the 1970s and 1980s, bipolar processing was dominant and large area devices were fabricated. Typically, processes were limited to 40 V and pushed to 60 V for actuator and transistor driver applications. As a side note, the most successful power metal-oxide semiconductor FET (MOSFET) driver in the 1980s used a commercially available digital "line driver" IC. Driver chips were later developed with FET processes that paralleled many low-power FET cells. Again, the required area was determined by the maximum current density of the allowed process.

Dedicated chips of the 1990s used power MOSFET technology to create driver and actuator chips. Applications of the low-voltage, high-current ICs fall mostly in the areas of power conditioning for photovoltaic systems, actuators for computer hard drives, actuators and motor drives for automotive and appliance applications, and driver applications in power semiconductors circuits.

Since mostly all IC technology was created for computer and telecommunications applications, creation of "higher voltage" ICs for power was slow to develop. Lack of market size in power did not support substantial technology development, but rather incremental product development. However, high-voltage ICs were developed early on for the gas-tube display market (circa. 1980s). Other significant developments slowly occurred mostly in drivers for power-bridge circuits as used in motor drives and "application specific ICs."

High-voltage ICs are processed with either dielectric isolation or junction isolation. In the 1980s, dielectric isolation was used extensively by Dionics Incorporated for display drivers which had ratings of several hundred volts. Dielectric isolation utilizes silicon-dioxide wells. Devices, such as bipolar transistors, are fabricated in the wells, which serve as functional islands. The devices are then interconnected at the surface.

Limitation of the dielectric isolation process is the higher cost. However, dielectric isolation does provide for more reliable isolation with greater circuit flexibility. Both power rating and current capacity are low relative to junction isolation because of the planar nature of the structure and interconnects.

Junction isolation became the preferred method starting in mid-1990s with the developments from General Electric and Harris companies followed by power-ASICs from power semiconductor manufacturers. The isolation method used multiple levels of p–n junctions to form wells. A cross section of several

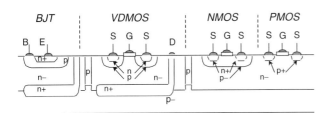

FIGURE 42.3 Junction isolation used to separate devices or circuits.

basic technologies is shown in Fig. 42.3. Note the p-type sinkers connecting to the p-material of the substrate to provide cell isolation. There is also a combination of structures used to produce a BiCMOS process. The complementary metal-oxide semiconductor (CMOS) provides the control circuitry while the bipolar structures provide high current-density transistors for power processing.

Junction isolation has several limitations, most significant of which is possible "layer inversion." Inversion occurs when the reference substrate, or portions thereof, becomes reversed biased. Relatively large currents can flow and biasing of four layer structures can cause latch-up. Manufacturers have paid significant attention to minimizing this problem. However, designers must always be cautious that a fault condition or capacitive current from a high-frequency transient does not induce an inversion.

42.5 Cost-driven Partitioning [5]

This is the one issue seldom discussed in open literature, yet is the greatest driver to the selection of circuit design approaches and determination of partitions. Unfortunately, a designer often limits cost estimation to only component cost, i.e. the bill of materials (BOM). The greatest cost is often not the component, but the handling, mounting, and testing of a component. An excellent example is the selection of output filter capacitors in dc–dc supplies. The use of a multitude of smaller ceramic chip capacitors, which can be automatically surface mounted, is often less expensive than larger electrolytic through-hole-mounted capacitors, and provide much greater reliability over time. (This applies to larger volume production.)

The use of cost-driven partitioning is also dependent on the company structure. A vertically structured company with captive manufacturing has the advantage of increasing volume by *modularizing* their circuits to be used across several product lines. The following procedure uses a *Full-cost Model* applicable to both captive and out-sourced manufacturing.

When discussing cost, it is necessary to define centers of cost for both the product and the business. The terms are defined as follows.

1. *Materials cost* represent direct costs of packaging, and include the minimum packaged component (e.g. silicon chip), component packaging materials (e.g. plastic housing on a TO-220), and packaging materials for manufacture (e.g. solder or adhesives for mount down). If the manufacturer can mount bare die, then these quantities are determined separately. If manufacturing is out-sourced, then the "pre-packaged" component cost accounts for the first two costs and the manufacturer (or assembler) determines the third cost. The variation in cost by volume must also be included. Volume dependency is greatest for custom products at low volume and lowest for standard high-volume products. A typical volume cost factor is 20% decrease in cost per 10-fold increase in volume.

2. *Production cost* includes factors for wages and product volume, but are independent of material costs, (which is not often assumed when assessing overhead). Production cost can be characterized as a function of technology and quantity. To reflect this into a design tool, it is necessary to describe production cost as a function of simple information, such as the number of surface mount device (SMD) and leaded components, and square inches of substrate board. Assessment is as follows for captive production:

    1. Determine the total wages, equipment and facility depreciation, and other facility overhead.
    2. Determine the number of production technologies in the facility, both in place and available with minimal extension.
    3. Determine technology costs by a ratio of the above two parameters.
    4. Add scaling factors for volume dependency.

In Fig. 42.4, the relative production costs for various technologies (circa. 1999) are shown for fixed volume. Note that the chip and wire is less expensive

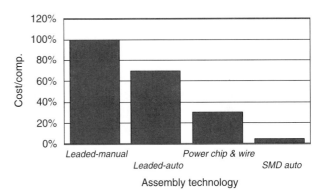

FIGURE 42.4 Relative production costs.

than handling a leaded component and is typical for captive facilities. Including the scaling factors in your calculations to give a volume dependency for a highly automated production technology. Depreciation is for production equipment and buildings, whereas other overhead covers the significant cost involved in purchasing, management, production technology, etc.

3. *Partitioning cost* is incurred for each technology used. From the previous technology descriptions, it appears straightforward to choose "this technology for these components and those technologies for those components" based on *technical* performance attributes. However, there is a drawback to this partitioning. Each partition adds one circuit to be handled through production with an additional interconnect and assembly process. This means the additional incremental costs. Assembling sub-circuits into a product is similar to assembling components on boards and is modeled as cost in wages modified by a different overhead factor. For chip and wire, costs for protecting (encapsulating) chips are included if necessary.

4. *Full cost* combines material costs and production costs as shown in Fig. 42.5. A minimum-packaged-component system is chosen to highlight the possibility of buying non-packaged components, but the model is valid for any level of packaging. If there is not a captive circuit fabricator, then the cost is obtained through competitive quotes or experience with the manufacturer. A mixture of in-house and out-sourced costs can be included in the model.

5. *Product business cost*, i.e. returns on investment for development of one product, is an investment in future payback. The total cash flow from development until the end of production determines the business costs for a product.

6. *Company business cost*, i.e. return on investment for cross products reflects the cost of sub-optimization within one single product. The value of reusing the same packaging technologies, designs (diagrams), and even physical circuits (building blocks) across different products should be measured at the company level. The value of building blocks becomes obvious through savings in repetitive development costs and maintenance of function. Development and maintenance costs are saved since the function is developed only once and unilaterally maintained across all products.

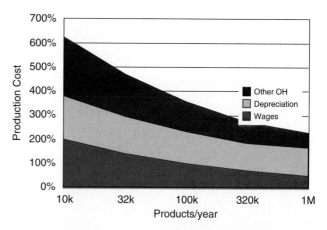

FIGURE 42.6 Cost variation due to volume.

The impact of volume on building-block cost applied to three motor drive products is shown in Fig. 42.6. At low volumes, the main savings are in development and maintenance costs, while at high volumes only savings in full cost matters. The overall conclusion is that if a partition is necessary to meet requirements, then, the partition must be guided by strategic choices in order to optimize cost on a company business level and the relative cost diagrams should be used only for optimizing within partitions.

42.6 Technology-driven Partitioning

There are several natural aids to partitioning. Ordering common packaging technologies by *technical performance*, orders most other attributes. As one moves down the list of technologies described in Section 42.4.2, one finds increasing current-carrying capacity, decreasing voltage isolation and operating frequency, lower thermal performance and density, less sophisticated processing, and lower cost for lower volumes (except for MID). These monotonic trends allow rich engineering judgment to effectively group components (step four in Section 42.3.1) for optimized partitioning.

Partitioning proceeds by following a sequence of first grouping and matching the most challenging components with higher performance technologies. The next challenging component grouping is matched with the next technology of lower performance and typically lower cost. Starting with the highest performance technology, often lower performing components

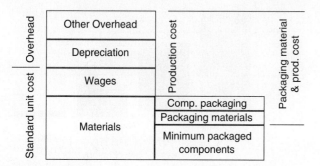

FIGURE 42.5 Full-cost model for circuit partitioning.

to be included at little increase in cost. For example, if thick film ceramic (TFC) is used for chip and wire power die and current-sense resistors, the inclusion of thick-film control circuits comes with little added real estate (cost).

Mixing of technologies and technology selection is misunderstood because a typical perspective is to look at the "substrate area cost," i.e. the famous "dollars per square inch" costing of technologies. This is as limiting as using only a BOM for cost-driven decisions. A better understanding is required and is aided by the graphical perspective in Fig. 42.7. View the curves right to left as density decreases. The falling curves represent relative *Full Cost* (Section 42.5) of each technology as substrate area changes. The starting and ending points are the practical limits in the use of the technology at certain densities.

As an example, assume a given circuit is designed with only one technology, such as thick film (TF), and as dense as possible. As the board area increases (becoming less dense), the components grow in size (0603–0805) with larger interconnect traces; the cost increases, following the curves up and to the left. A point is reached in an area when a less costly technology may be suitable, such as SMT FR-4. This other technology would decrease the cost for the same area. Hence, not only do the cost and density decrease (with lower technology), but the performance also decreases. Within a range near maximum density, the higher performance TF technology with an added area is still *less costly*. This is due to the *packaging* and *production costs*, and is often overlooked by designers who look at the "cost per square area of boards" without looking at the full-cost model.

A more generalized set of curves is shown in Fig. 42.8. This graph, in essence, is created for a specific production facility. The circuit designer, or design team, would follow the steps to partitioning, letting the costs of the technologies drive where partitions are best drawn. Remember that the overall circuit

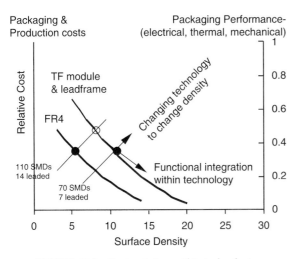

FIGURE 42.7 Cost variation within technologies.

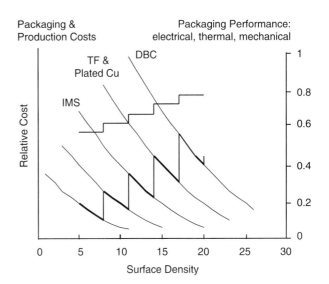

FIGURE 42.8 Generalized relationship of cost and technology.

is composed of electric, magnetic, mechanical, and thermal circuits.

42.7 Example 2.2 kW Motor Drive Design

A 2.2 kW motor drive, consisting of electronics, motor and pump encased in one housing, is used as an example product. The block diagram of the electronics is shown in Fig. 42.1. For a (planar) electrical-mesh circuit, the physical assembly pattern would closely follow the electrical schematic layout and one packaging technology, such as FR-4, could be used, though not efficiently. The design would, then follow a single line up and to the left in Fig. 42.7. Using mixed packaging technologies provide multiple assembly levels, and the assembly pattern more closely follows grouping of the physical Delivery Forms of the components. The steps outlined in Section 42.3.2 are followed to determine the proper partitioning of the system to meet the performance requirements and provide maximum business profit. The steps are summarized as:

1. User requirements,
2. Component characterization,
3. Component grouping,
4. Strategic partitioning with constraints,
5. Optimizing within partitions.

42.7.1 User Requirements (Constraints)

Many user requirements direct the system design as outlined in Section 42.3.1. However, several requirements place specific constraints on the packaging of the 2.2 kW drive as noted below.

Mechanical: Built in a 65 mm dia. stainless steel tube; short as possible.
Thermal: Cooling through tube with non-flow of water at 30°C.
Environment: Potting electronics is disallowed.
Regulatory: UL, CE.
Reliability: 1,000,000 quick start/stop,
30,000 maximum gradient start/stop,
40,000 h lifetime @ 10°C water.

42.7.2 Component Characterization Map

A component characterization map is performed on all the components to identify the technical and physical attributes that dominate it, and is illustrated for part of the circuit as in Table 42.1. In this component characterization map, components are listed for each electrical functional block.

42.7.3 Component Grouping

An overview of possible groupings into packaging partitions is obtained by attaching main components and key attributes to the functional block diagram of Fig. 42.1.

42.7.4 Strategic Partitioning with Constraints

A major constraint is the limited space available (65-mm diameter). This makes it obvious that some miniaturization is very valuable, but what should be miniaturized? Packaging cannot miniaturize leaded components. These components require either through-hole PCB (FR-4, for soldering) or some form of lead frame (MID for welding). Power die are top candidates for miniaturization because the die can be grouped into a power module that is much smaller than discrete power components. Also, high-power losses do not allow the same packaging technologies to be used as for leaded components.

The remaining non-power die and associated components are prime candidates for modularization. Highest value is reached if a building block can be reused across different products. Therefore, as much control circuitry, as possible, should be integrated without violating the possibility for reuse in other products. For this product, the line communications bus and motor control circuitry would be excluded, but the control for VDE/inrush and PFC would be integrated together with the driver and all-sense resistors. This integrates 82% of all power losses for easier cooling, integrates all power-component-dependent control circuitry, and enables product-independent maintenance and power die optimization.

At this stage, there are usually new requirements added for cross-product reuse. This application requires 125°C baseplate temperature.

42.7.5 Optimization within Partitions

Optimization requires choosing optimum technologies to meet cost and performance requirements. In Fig. 42.9, the relative cost of various substrates is shown together with the cost of suitable production technologies. Note that the substrate cost is for equal substrate area but different performance. For example, IMS requires more space for control circuitry than TF multilayer because IMS has only one conductor layer.

Figure 42.9 should be used together with Fig. 42.1, which shows that the module includes both power chip and wire (PC&W), and low-power control circuitry (SMT). The DBC, IMS, TTF, and CuPC can accommodate the PC&W and FR-4, IMS, TF multilayer, and CuPC can accommodate fine-line SMT. This should initially lead to the conclusion that DBC, IMS, or TTF should be used for power, excluding CuPC due to cost; and FR-4 for control, excluding the others due to cost.

Are all cost issues taken into account and all requirements met? Not necessarily. Packaging approaches influences component cost. Power sense resistors, which are typically in SMT form, can be integrated in TF multilayer at near-zero incremental cost. Also, less-expensive integrated circuits can be chosen when the packaging approach allows active trimming of associated components. Besides cost, technical issues limit packaging choices for certain circuit partitions. Reliability and temperature requirement (125°C) rule out FR-4.

There are fewer and fewer choices. If power die were available as known good die, then power and control could be combined on one substrate with IMS or CuPC. The IMS has drawbacks, such as lower power cycling capability due to a high thermal coefficient of expansion (CTE) and is only a one-layer technology, which means more area and less noise immunity. The CuPC has neither of these problems, but due to the lack of known good power die that was not chosen. Also, CuPC

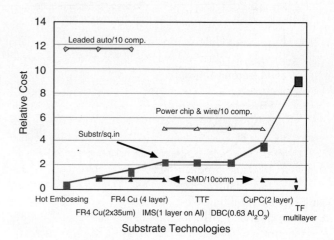

FIGURE 42.9 Substrate costs (1999).

FIGURE 42.10 Final module combining several packaging approaches.

does not allow component integration at the cost indicated in Fig. 42.9. A two-substrate solution was needed.

The power DBC was chosen as the obvious highest performing technology among the comparable low cost power substrates. The DBC is soldered onto a low cost copper base for thermal management and extends to form a mounting base for the control substrate.

Multilayer thick film was chosen for control circuitry despite the apparently high substrate cost. In the motor module, this substrate is the optimum cost choice because of the high component integration, such as the three buried power current-sense resistors and many printed resistors for accurate active trimming of functions associated with the integrated circuits. Partitioning cost is minimized by combining interconnections of substrates with interconnection to I/O terminals in one technology – heavy wire bonding. This has been possible by designing a MID interconnection component with terminals that are wire bondable on one end and solderable on the other. The resulting module is shown in Fig. 42.10.

Other components, both SMT and leaded, not best accommodated in the module. Therefore, a two layer FR-4 is chosen as the lowest cost technology suitable for both Delivery Forms and used for the module and components. Mechanical stability and cooling is achieved by using a patented structure of extruded aluminum profiles.

Using bare die, higher cost substrates and partitioning with different technologies allows the product to surpass cost targets. The partitioning in packaging Levels 1 and 2 address optimization of *product business cost* as defined in Section 42.5. Designing the module building block as a component for reuse across other products increases volume and reduces cost. More importantly, relative low-volume products can benefit from the building block by faster development cycles, lower development cost, lower Level-3 packaging cost, and lower maintenance cost. The building block value addresses optimization of *company business cost*.

42.8 High Temperature (HT) Packaging [6]

Enabling issues for many power electronics applications are size, weight, and efficiency. Silicon carbide (SiC) semiconductor technology has the potential to provide up to a 5-fold reduction in converter volume if the high-temperature, high-frequency power electronics can be implemented. Higher frequency operation reduces the size of the passive components and, thus, the system volume. Higher operating temperatures allow a larger temperature difference between the heat sink and cooling fluid, which increases radiator effectiveness and decreases size. Silicon devices are limited to 150°C junction temperature prior to de-rating; whereas SiC devices can operate in excess of 400°C. In addition, SiC devices offer the potential for incorporating power electronics at point-of-load (POL), e.g. at the motor or actuator housing, thus greatly reducing system cabling and volume, and provide increased flexibility in equipment arrangement. Inherent in all this is the requirement for very reliable HT electronics packaging.

42.8.1 HT Materials Selection

A paramount requirement of HT packaging is to minimize dissimilarity in material interfaces and can be achieved with a "nearly all" Al (aluminum) approach. This includes Al backed SiC JFETs, AlN substrate, Al_2O_3 (anodized) electrical insulators, Al interconnects (in place of copper), and AlSiC (aluminum silicon carbide) heat sinks [7]. Aluminum is adequately ductile to act as an excellent stress relief during temperature cycling. Also, Al provides a common metallurgical bonding medium.

42.8.2 Module Construction

A proposed module structure consists of Al conductors on AlN substrate on AlSiC. In a one- to two-step casting process, an Al conductor pattern is formed on AlN, which is captured into a netshape-cast AlSiC heat sink. A test sample was created first with multiple pads for JFETs and wirebonds, shown in Fig. 42.11 (Courtesy of PCC-AFT Inc.). Also, using AlSiC for

FIGURE 42.11 Al/AlN/AlSiC casting.

TABLE 42.3 Material properties

	CTE	Thermal conductance	Electrical resistance	Young's modulus	Flexural Strength
	ppm/°C	W/m°C	mW-cm	GPa	(MPa)$_{avg}$
Al/SiC	8	175		220	369
AlN	4.5	100–180		320	300
SiC	3.7	120–490			550
Al	23	240	4.3		
Copper	17	393	1.7		
Gold	14.2	297	2.2		
Silver	19.7	418	1.6		11

the housing and heat sink allows a ceramic substrate, connectors, and other hardware to be integrated into the mold, or directly cast into the structure. Connectors that require solder are nickel and gold plated. Other parts of the module follow below.

Aluminum Interconnect – Material properties are given in Table 42.3. Electrical resistance of Al is 2.5 × higher than Cu and 40% lower in thermal conductance. The performance reduction is offset with the thicker conductor greatly mitigating stresses between the components. Both Al and Cu approximately double the resistance every 100°C.

Die Attachment – A significant challenge is die attach. The SiC semiconductor must be bonded to the Al interconnect with a material electrically conductive, having a high physical resistance to temperature excursion, and imparting little stress on substrate or die during power and temperature cycling. One approach uses a silver–glass composite, which can be cured and operate at or above 350°C. One material (QMI3555R) has electrical conductivity of 15 $\mu\Omega$-cm and thermal conductivity 80 W/m-K, and is qualified replacement for Si/Au Eutectic.

Cover Coating and Sealing – If required, an inert HT cover coating, such as an alumina refractory cement (Cotronics 920), can be used across die and wire bonds. The material has the properties of alumina ceramic, devoid of outgassing, completely inert, and good to a service temperature of up to 1634°C. The coefficient of thermal expansion (CTE) is 4.5 ppm/°C, dielectric strength 270 V/mil, and volume resistivity 1011 W-cm.

Acknowledgment

The author wish to thank Mr. John B. Jacobsen, Technical Manager, Electronic Engineering and Packaging Department of Grundfos A/S, Denmark, for supplying most of the application data used in the module design.

About the Author

Dr. Hopkins received BS and MS degrees from the State University of New York at Buffalo (UB), and Ph.D. from Virginia Tech (VPI&SU) where his primary study was in megahertz-frequency power supplies using high-density packaging techniques. He is an Associate Research Professor, Director of the Electronic Power and Energy Research Laboratory, Assistant Director of the Electronic Packaging Laboratory at UB, and and IEEE and IMAPS senior member. He chairs the power electronics packaging technical committees for IEEE-CPMT, IEEE-PELS, and IMAPS. He has authored over 50 journal and conference publications. He also has over 15 years of industrial experience at GEs and Carrier Air Conditioning Companys' R&D centers, was visiting fellow at several national labs and is president of DCHopkins & Associates.

Further Reading

1. M. Meinhardt, P. Cheasty, S. Eckert, J. Flannery, S. C. ÓMathúna, and A. Alderman, "STATPEP-Current Status of Power Electronics Packaging for Power Supplies - Methodology," *Proc. of the 14th Annual Applied Power Electronics Conference and Exposition*, pp. 16–22, March 14–18, 1999.
2. D. C. Hopkins, S. C. ÓMathúna, A. Alderman, and J. Flannery, "A Framework for Developing Power Electronics Packaging," *Proc. of the 14th Annual Applied Power Electronics Conference and Exposition*, pp. 9–15, February 15–19, 1998.
3. R. R. Tummala and E. J. Rymaszewski, *Microelectronics Packaging Handbook*, Van Nostrand Reinhold, New York, 1989.
4. "High Voltage Integrated Circuits," ed. B. J. Baliga, IEEE Press, New York, IEEE order number: PC0232-9, ISBN 0-87942-242-4, 1988.
5. J. B. Jacobsen and D. C. Hopkins, "Optimally Selecting Packaging Technologies and Circuit Partitions based on Cost and Performance," *Applied Power Electronics Conference*, New Orleans, LA, February 6–10, 2000.
6. D. C. Hopkins, D. W. Kellerman, R. A. Wunderlich, C. Basaran, and J. Gomez, "High-Temperature, High-Density Packaging of a 60 kW Converter for >200°C Embedded Operation," *IEEE Applied Power Electronics Conference*, Dallas, TX, March 19–24, 2006.
7. D. C. Hopkins, J. M. Pitarressi, and J. A. Karker, "Systems Design Considerations for Using a Direct-Attached-Ceramic MMC Power Package," PCIM '96 EUROPE. *Proc. 32nd Int'l Power Conversion Conf. 1996*, pp. 683–9 Nurnberg, Germany.

Index

78XX series of voltage regulator, 601
79XX series voltage regulator, 601

ABM techniques, 1122
AC induction motor (ACIM), 683, 865, 1031
AC voltage controllers, 483, 510
AC/AC converters, 483
   applications, 510
Active power filter, 749, 1078, 1608
   application, 811
   concept from, 813
   control loop design, 1082
   control scheme, 1070
   efficiency, 1071, 1080
   impact on inverter power, 1089
   technical specifications, 1084
   types, 1068–1069
Actuators, 335, 642, 644, 652, 843, 845, 902, 1037, 1152
   electromagnetic actuators, 642
   piezoelectric ultrasonic, 642
Adjustable speed drives (ASDs), 353, 390, 480, 513–515, 839, 1053, 1061, 1067, 1145
   advantages, 513
Adjustable speed synchronous condenser, 815
AH capacity, 670
Analog tachogenerator, 865, 904
Anode shorts, 98, 101, 117
ANSI standard, 1054, 1117
Antilock breaking system, 641
Application specific integrated circuits (ASIC), 257
Applications of DC transmission, 773
Artificial intelligence-based controllers, 1015
Artificial neural networks (ANNs), 639, 1015, 1017, 1018
   Ann-based controls in power converters, 1029
   controls in motor drives, 1025–1029
   estimation in induction motor drives, 1017–1025
Asea Brown Boveri (ABB), 236
Automatic circuit recloser (ACR), 1055
Auxiliary inert gases, 568
Axial-airgap/axial-flux machine, 651, 652

Back-to-back diode-clamped converter, 458
Baker's clamp, 34, 126
Ballast, 197, 546, 556, 565, 567, 570, 586, 641, 1053, 1057
   advantages, 571
   applications, 589
   block diagram, 571
   classification, 573
   drawbacks, 571
   factors affecting design, 572
Base resistance-controlled thyristor (BRT), 127
Basic converter topologies, 522–525
   characteristics, 535, 536
   correction capabilities, 531–535
Batteries, 4, 451, 453, 589, 621, 630, 656, 670, 671, 677
   cadmium, 631
   car, 631
   charging, 631, 677, 734
   comparison among types, 631
   lithium-ion, 631
   nickel–cadmium, 631
   sealed, 677
   storage, 669
Bell Telephone Laboratories, 27
Bi-phase half-wave rectifier, 180, 181
Bipolar drive circuits, 912
Bipolar junction transistor (BJT), 10, 27, 41, 71, 138, 544, 897
   applications, 37
   base–emitter junction, 2–4, 6, 8, 9
   collector–base junction, 4
   Darlington connection, 30
   operation, 28
   spice simulation, 36
   structure, 28
Bipolar mode operation of SIT (BSIT), 136
Blade pitch control, 674, 707, 708, 739
Blanking time, 255
BLDC motor, 841, 933, 1005, 1012, 1037, 1041, 1044
   control topology, 1039
Bonneville power administration (BPA), 793

Boost corrector, 522
    advantages of, 523
Bridge rectifier, 22, 189, 262, 549, 711, 712, 826, 845, 848
Brushed DC motor, 858, 891, 903, 904, 1005
Buck corrector, 522, 523
    advantages of, 522
    drawbacks of, 522
Buck–boost converter, 251, 532, 533, 938, 940, 951, 960, 961
    basic converter, 251
Burst-firing, 487
Butterworth filter, 1074

Capacitive snubbers, 125
Capacitor–inductor stored energy ratio (CIR), 346, 347
Carrier frequency gating, 487
Carrier lifetime, 73, 76, 78, 81, 86, 91, 97, 99
Cascade converter, 457, 809, 813
Cascaded H-bridges, 452, 466
Cascaded inverters, 453
Cathode shorts, 97, 98, 115
CBEMA tolerance curves, 1055
CCM shaping technique, 189, 247, 525, 530, 536, 544, 545, 589
    modes of operation, 559
Characteristic harmonic frequencies, 502
Chattering, 936
Chopper devices, 37, 39, 111, 126, 246, 247, 258, 319, 341, 488, 500, 713, 838, 850, 862
Chopping-mode control method, 922
Clock pulse (CP), 528, 529, 617
Closed-loop operation of inverters, 379
    feedback techniques in current source inverters, 385
    feedback techniques in voltage source inverters, 382
    feedforward techniques in current source inverters, 382
    feedforward techniques in voltage source inverters, 379
Cogeneration/combined heat and power (CHP), 717–719
Cogging torque, 900, 903, 1043
Collector current, 30, 34, 73, 75–81, 136
Color rendering index (CRI), 569
Combining commutation and PWM current control, 897
Comitè Europèen de Normalization (CEN), 1118
Comitè Europèen de Normalization Electro-Technique (CENELEC), 1118
Comité International Special des Perturbations Radioelectriques (CISPR), 1103
Common control techniques, 562
Common mode reactor, 849
Common source level shifters, 548
Communication principles, 501
Commutation, 45, 210, 582, 773, 779, 895
    commutation failure (CF) or misfire, 784
    electronic commutation, 893
    load commutation, 833

Company business cost, 1154
Comparison of CCM shaping techniques, 531
Complementary metal-oxide semiconductor, 1153
Component characteristic map, 1149
Continuous conduction mode (CCM), 189, 247, 524–526, 844
Continuously variable speed systems, 710–711
Control schemes
    current-mode control, 256, 257, 618, 936, 951, 964, 965, 969, 996
    hysteretic (or bang-bang) control, 257, 564, 976, 983
    voltage-mode control, 246, 256, 615, 617
Controllable displacement factor frequency changer, 503
Controller, 665, 667, 670, 671, 677, 684, 686, 687, 705, 741, 760, 888, 901, 965, 968, 993, 994, 1005, 1134
Controller area network (CAN) IC, 644, 645
Conventional approximator, 1016
Converter arrangements, 4, 5
Converter group selection and blanking circuit, 501
Converters, resonant, soft-switching, 405–447
    classification, 407
    converter, active-clamp flyback, 421, 424
        ZVS-MRC, clamp mode, 421
    converters, power width modulated (PWM), 405
    DC–AC power inverters, soft-switching, 436–447
        control techniques, 628
        digital signal processor (DSP), 444, 445
        intensity pulse density modulation (IPDM), 437
        inverter, quasi-resonant soft-switched, 442–445
        inverter, resonant dc link, 438, 439
        inverter, resonant dc link, active clamped, 439
        inverter, resonant dc link, low voltage stress, 440–442
        resonant pole inverter (RPI), 445–447
        uninterrupted power supplies (UPS), 438
        ZACE concept, 444
    electromagnetic interference (EMI), 406
    EMI suppression, soft-switching, 434, 435
    extended-period quasi-resonant (EP-QR) converters, 427–434
        DC–DC bidirectional flyback converter, 434, 435
        DC–DC flyback converter, soft-switched, 434, 435
        EP-QR boost type, 427, 429, 431, 432
        PFC converter, hard-switched boost type, 427, 431
    frequency modulation (FM), 355, 406, 573
    high power devices, soft-switching, snubbers, 435, 436
        safe operating area (SOA), 435
        snubber circuits, Mcmurray, 435–437
        snubber, Undeland, 435, 437
    ICs, customized control, 406
    load resonant converters (LRCS), 421–425
        parallel resonant converters (PRCS), 425, 427
        series resonant converters (SRC), 421, 426
        series-parallel resonant converters (SPRC), 425, 429

multi-resonant converters (MRC), 415–421, 425
  constant-frequency multi-resonant (CF-MR), 420, 421
  zero-current multi-resonant (ZC-MR), 418
  zero-voltage multi-resonant (ZV-MR), 418
  ZVS-MRC, buck, 418–420
power switches, 405, 406
quasi-resonant converters (QRC), 408–415, 425
  ZCS-QRCs, 408–411
  ZVS-QRC, 409
resonant converters, control circuits, 425–427
  UCC3895, PWM controller, 427, 431
  ZVS-MR forward converter, 427, 430
resonant switch, 407, 408
  gate turn-off (GTO), 407
  insulated gate bipolar transistor (IGBT), 407
  zero current (ZC), 407
switching trajectory, 405, 406
zero-current switching (ZCS), 406, 407
zero-voltage switching (ZVS), 406, 407
ZVS, high frequency, applications, 412–415
  clamped voltage, 412
  full-bridge (FB), 412, 417
  half-bridge (HB), 412, 416
  zero voltage transition, 414, 418
ZVT converters, 421, 422
ZVT-PWM, 421, 422
Convertible static converter (CSC), 792, 805, 816, 817, 819
  topologies, 819
Correlated color temperature (CCT), 569, 789
Cost-driven partitioning, 1153
Coupling transformers, 807, 1085–1087, 1090
Cross-over points method, 497, 503
CSI-driven synchronous motor, 880, 883, 884
  brushless dc operation of, 881
Cu plated on ceramic (CuPC), 1151
Cuk converter, 246, 252, 523, 534
Current control loops, 630, 781, 884, 903, 904, 931, 969, 976
Current distortion limits – subtransmission systems, 1065
Current distortion limits – transmission systems, 1065
Current mode control techniques, 526–529
Current modulator, 977, 978, 1080–1082, 1084, 1087
Current rating of active devices, 3, 18, 27, 29, 48, 177, 443, 453, 473, 1112
Current source inverter (CSI), 353, 354, 371–373, 376, 390, 690, 819, 833, 857, 885
  carrier-based PWM techniques in CSIs, 355, 372, 459
  normalized sampling frequency, 369, 379
  selective harmonic elimination (SHE) in three-phase CSIs, 376, 452
  space-vector-based modulating techniques in CSIs, 376, 379, 382, 397, 403, 978, 1047, 1114
  square-wave operation of three-phase CSIs, 364, 376, 379
Current-based maximum power point tracker (CMPPT), 666
Customer generation, 717

Cycloconverter, 752, 831, 844, 870, 1058, 1061
  application of, 493, 510
  circulating current-free mode operation, 496, 497, 499, 501–503, 511, 512
  circulating-current mode operation, 496
  limitations of, 493
  operation with RL load, 494
  operation with R-load, 493
  principle, 493
  single-phase to single-phase, 493
  three-phase, 495
  three-phase six-pulse, 498
  three-phase twelve-pulse, 499
Cycloconverter drive, 870

Darlington pair output stage, 596
Days of autonomy, 670
DC choppers, 246, 247, 258
DC controls, 778, 779, 783, 784
DC link capacitor specifications, 226, 227, 354, 466, 477, 819, 924, 1032
DC motor drives, 262, 837
  applications of, 859
  converter-dc drive system considerations, 865
  converters for, 861
  representation and characteristics of, 859
DC ripple reinjection, 214
DC–DC converters, 1, 247, 248, 253, 254, 256, 257, 259, 418, 553, 678, 862
  applications of, 258, 343
  control principles, 255
  converter family tree, 264
  converter topologies, 247–253
  generations of, 262, 263
  hard-switching PWM converters, 245
  resonant and soft-switching converters, 245
  seven self-lift dc/dc converters, 269, 270
  synchronous and bidirectional converters, 254
DCM input technique, 247, 524, 530, 532, 535, 537, 538
Dead time, 219, 255, 427, 550, 898, 901, 953, 975, 982, 997, 1032, 1035
Defuzzification, 956, 993, 995, 997, 1000, 1005
Delta modulation control, 530
Device static ratings, 29
De-queing circuit, 561
Digital signal processors (DSP), 444, 508, 891, 1018, 1031, 1032, 1036, 1037, 1040, 1044, 1046, 1076, 1078, 1152
  definition, 1031
  DSP controller requirements, 1044, 1048
  DSP-based control of permanent magnet synchronous machines, 1041
  DSP-based vector control of induction motors, 1046
Diode clamp converter (DCC), 757, 758, 789

Diode-clamped multilevel inverter, 454, 980
   advantages and disadvantages of, 455
Diodes, 919, 1104, 1126, 1141
   ac diode parameters, 16
   applications, 22, 23
   characteristics, 15, 16, 24
   circuit symbol, 15, 16
      forward bias, 15
      reverse bias, 15, 16
   dc diode parameters, 15
   PSPICE simulation, 22
   ratings
      current ratings, 18
      voltage ratings, 17
   series and parallel connection, 19
   snubber circuits, 19
   standard datasheet for selection, 24
   types, 17
Direct bonded aluminum (DBA), 1151
Direct bonded copper (DBC), 1151
Direct transfer function (DTF) approach, 507
Direct vector control method, 876, 877
Direct vector control with airgap flux sensing, 877
Discharge lamps, 565, 567–570
Discretely variable speed systems, 709, 710
Distributed energy resource (DER), 717, 718
Distributed generation/power generation, 694, 705, 717
   applications, 717
DMOS, 72, 140, 141
Doping technique, 29, 73, 91, 662, 674
Double output Luo-converters, 263, 284, 301
   five circuits of, 284–288
Doubly-fed induction generators (DFIG), 703, 704, 711
Drive circuits for bifilar wound motors, 912, 914
Dual converters, 218
Dual-voltage electrical systems, 324, 327, 654
Duty ratio, 246, 536, 648, 1114

Ebers–Moll model, 36
Electric machine operation, 892, 893
Electric power conversion, 2
Electric power steering (EPS) system, 643
Electrohydraulic power steering (EHPS) system, 643
Electromagnetic actuators, 642
Electromagnetic ballasts, 570, 571
   classification of, 573–575
   factors influencing design, 572
   merits and demerits, 571
Electromagnetic compatibility (EMC), 194, 262, 324, 636, 637, 829, 848, 1103, 1106, 1117
Electromagnetic interference (EMI), 81, 86, 87, 95, 262, 307, 327, 336, 434, 445, 522, 1103, 1105, 1108
Electromagnetic transients program (EMTP), 793, 794

Electromechanical power conversion system, 646
Electronic converters, 543–545, 565, 673, 743, 1103
   applications of, 543
Electronic gate drivers, 553
Electronic power supplies
   high ends, 4
   low ends, 4
   switch-mode power supply, 4
EMC standards, 1103, 1117, 1118
Emission standards for low frequency harmonics, 1118
Energy factor, 263, 345–347
Energy storage devices, 630, 631
EnergyStar® program, 4
Equivalent series resistance (ESR), 162, 253, 942, 975
EUPEC EICE-driver, 555
Eupec semiconductor, 555
European Telecommunication Standards Institute (ETSI), 1118

FACTS devices, 802, 803, 806, 809, 815
Fault diagnosis system, 478, 480
Ferroresonant standby UPS system, 621
Fiber optic links, 548
Field-orientated control (FOC), 1036, 1047, 1049, 1052, 1131, 1132
   implementation of field-oriented speed control of induction motor, 1048
Filtering systems, 145, 151, 158
   capacitive-input dc filters, 160–162
   inductive-input dc filters, 158–159
Finite element models, 83–85
Fixed-frequency PWM controllers, 615, 617
Flexible ac transmission systems (FACTS), 764, 774, 792–794, 797, 802, 803, 806, 809, 815
Floating supply, 545, 550, 551, 553
   methods of generation, 546
Fluorescent lamps, 197, 543, 565, 568, 569, 572, 574, 576, 585, 590, 1053, 1057, 1061
Flyback converter, 37, 162, 172, 251, 252, 258, 421, 434, 533, 535, 536, 563, 641, 693
Flyback regulators, 603, 607, 610, 617
   continuous mode, 605
   discontinuous mode, 603
   operating properties, 607
Flyback transformer, 251, 252, 255, 258, 434, 607
Flying-capacitor multilevel inverter, 455, 456
   advantages and disadvantages, 456
Flywheel, 38, 162, 163, 167, 623, 630, 631, 649, 656, 678, 689, 1133
Force commutated cycloconverters (FCC), 483, 503, 861, 865, 867, 1117, 1118
Force-commutated pulse width modulated (PWM) rectifiers, 201

Index

Force-commutated rectifiers, 221, 223, 233, 237
   new technologies, 233
      active power filter, 233
      frequency link systems, 233
      topologies for high power applications, 233
   three-phased, 221
Forced commutated inverter (FCI), 833, 834, 852
Fortescue transformation, 1092, 1093
Forward two-quadrant (F 2Q) Luo-converter, 302
Four-quadrant dc/dc Luo-converter, 305, 322
   four modes of operation, 310–315
Four-quadrant ZVS quasi-resonant dc/dc Luo-converter, 327, 331
Freewheeling diode, 19, 76, 78, 125, 126, 160, 441, 446, 488, 508, 912, 918, 924, 1040
Front end converter (FEC), 389, 391, 393, 704, 852
Fuel cell, 2, 258, 259, 451, 453, 478, 480, 630–632, 655, 714, 718–721, 729, 730–734
Fuel-cell based energy systems for DG, 719, 731
Full cost, 1148, 1153–1155
Full-bridge based topology, 720
Full-bridge regulators, 613, 615
Full-bridge resonant inverter, 574
Functional integration, 1147, 1148
Fuzzification, 993, 1000–1003
Fuzzy and linguistic variables, 1000
Fuzzy inference engine, 1000, 1003–1005, 1008
Fuzzy inference system (FIS), 1000
   classification, 1000
   steps in development of, 1000
Fuzzy logic, 1011, 1012, 1015, 1016
Fuzzy logic control of switching converters, 764, 993, 996, 997, 1000, 1005, 1011, 1012
   fuzzy logic controller synthesis, 993
Fuzzy logic controller (FLC), 996, 997, 1000, 1005, 1011, 1012
Fuzzy logic based microprocessor controller, 1005
Fuzzy logic-based speed controller, 1005

Gamma control loop, 782
Gas atom excitation, 567
Gas atom ionization, 567
Gate control Luo-resonator, 342
Gate drive circuit, 80–82, 93, 101, 105, 196, 412, 452, 543, 546, 547, 549, 550, 553, 558, 729, 1048, 1122, 1123, 1126, 1148
   conventional model, 80
   new gate drive circuits, 81
   significance of, 79–81
Gate drive requirements, 71, 80, 90, 106
   gate circuits, 106
   snubber circuits, 106

Gate turn-off (GTO) thyristor, 10, 11, 13, 82, 85, 90, 92, 93, 96, 99, 101–103, 109, 118, 119, 126, 483, 544, 744, 816, 830, 858, 978
   applications of, 121
   basic structure and operation, 115, 116
   definition, 115
   GTO thyristor model, 116
   power devices used in the VSD converters, 831
   process of, 120
   PSPICE model, 109
   SPICE GTO model, 120
      SPICE sub-circuit model, 121
   static characteristics
      gate triggering characteristics, 118
      off-state characteristics, 118
      on-state characteristics, 117
      rate of rise of off-state voltage, 118
   structure, 116
   switching phases, 118–120
   thyristor models, 116, 117
   turn-off circuit arrangement, 120
   types, 115
Gate-controlled series capacitor, 816
Gating signals, 1010, 1040, 1044, 1081, 1093
   control scheme, 1070
   generator, 1093
Gearbox, 706, 708, 710, 711, 737–739, 829, 852
General structure of a three-phase ac motor controller, 1032
General Trade Agreement on Tariffs and Trade (GATT), 1117
Generators, 704, 705, 711, 739
   analog tachogenerators, 904, 905
   asynchronous generators, 739, 744–746
   constant frequency generators, 240
   conventional fuel generators, 673
   diesel generators, 511, 627, 673, 676, 705, 706
   high slip induction generator, 711
   induction generator, 240, 703, 706, 707, 712, 739
   permanent magnet synchronous generator (PMSG), 712
   PV generators, 690
   squirrel-cage induction generator, 712
   synchronous generator, 703, 708–712, 739, 746, 747
   tandem induction generator, 711
   wind generators, 674, 700, 703, 713
   wound rotor induction generator (WRIG), 704, 712
Glass-epoxy with plated through holes (FR-4, PTH), 1151
Glass-epoxy with surface mount pads (FR-4, SMT), 1151
Graetz bridge, 206, 208
Green power, 717, 718
Grid-compatible inverters, 698
Grid-connected PV systems, 676, 681, 689, 690, 692, 697
Grid-controlled mercury-arc rectifiers, 493
Grounding, 33, 613, 640, 690, 762, 1053, 1054, 1056, 1057, 1066, 1115, 1116
GTO thyristor-controlled series capacitor (GCSC), 816, 817

H-bridge, 8, 236, 241, 242, 452–454, 457, 509, 641, 642, 844, 853, 861, 893, 895–897, 901
   high efficiency method, 897
Half controlled bridge converter, 181, 186, 208, 209, 861
Half-bridge regulator, 610, 613
Hall effect (HE) sensors, 37, 887, 895, 897, 899, 901, 1008, 1040, 1041
   operation of, 899
Handheld voltmeter, 1060
Hardware-in-the-loop (HIL), 513
Harmonic distortion, 758, 866, 1067, 1074, 1077, 1079
   recommended practices, 1065
   special configurations, 213
   standards, 219
Harmonic filters, 233, 521, 790, 808, 826, 827, 1053, 1062
Harmonics, 1069–1071, 1074, 1078, 1083, 1085–1088, 1090, 1093, 1095, 1097, 1104, 1113, 1114
Harris PMCT, 127
Harris semiconductors (Intersil), 124, 127
High intensity discharge (HID) lamps, 568, 570, 590, 641
High power factor electronic ballasts, 586, 588
High temperature (HT) packaging, 1157
High voltage dc power supply, 560
   advantages of, 560
   disadvantages of, 560
High voltage direct current (HVDC) systems, 706, 765, 769, 770, 773–776, 778, 781, 785, 787, 789, 792, 794, 795, 807, 819
High-frequency current-ripple reduction, 722
High-frequency diode rectifier circuits,
   clamping, 162
   design consideration, 172
   flywheeling, 162
   guidelines, selection of diodes, 177
   rectifying, 162
High-pressure mercury vapor lamps, 569
High-pressure sodium lamps, 569, 570
Home and industrial lighting, 590
Hot spots, 31, 126
HVAC equipment, 1037
HVDC technology, 769, 794
   applications of, 787
   main components of HVDC converter station, 775–778
   modern trends of, 789–792
   types of HVDC systems, 774, 775
HVDC transmission, 769
Hybrid active power filters, 1069, 1094, 1095
Hybrid electric vehicle (HEV), 635, 655, 656
Hybrid energy systems (HES), 676, 685, 687, 688, 705
   classification, 686
Hybrid PV systems, 676
Hybrid static/rotary UPS, 625, 626
Hybrid stepper motor, 908–911

Hybrid topologies, 1085, 1094, 1097, 1098, 1101
   compensation performance, 1074–1076, 1079, 1080, 1098
   effects of the passive filter quality factor, 1098
   effects of the passive filter tuned factor, 1099
   influence of the power system equivalent impedance on, 1098
   principles of operation, 1095
Hydrogen-based fuel-cell energy, 719
Hysteresis band, 227, 896, 901, 922, 959, 977, 1080
Hysteresis band current control (HBCC), 896, 901

I down logic signal, 899, 902
I up logic signal, 899
Ideal series compensator, 799, 800
IEEE standards, 826, 829, 893, 1058, 1063, 1064, 1117, 1118, 1158
IGCT vs IGBT, 763
Incremental conductance technique (ICT), 665, 666
Indirect rotor flux oriented control (IFOC) method, 876, 877
Indirect transfer function (ITF) approach, 507
Induction motor (cage), 510–513, 858, 865
Induction motor (slip-ring), 511, 512, 833, 834, 839, 840, 858, 866, 878
Induction motor control, 1027, 1048
Induction motor drive, 86, 197, 510, 685, 838–840
   characteristics and methods of control, 866–873
   steady-state representation, 865
Input capacitance, 27, 53, 76, 79, 257, 545, 556, 558
   components, 53
Input impedance, 27, 123, 257, 522, 523
In-quadrature term, 523
Instantaneous reactive power theory, 749, 1071, 1074, 1077, 1078, 1081, 1090, 1092
Insulated gate bipolar transistor (IGBT), 10, 11, 13, 27, 28, 41, 71–73, 123, 543–545
   applications, 85
   basic structure and operation, 72
   characteristics, 28
   dynamic switching characteristics, 76
      turn-off characteristics, 76, 77
      turn-on characteristics, 76
   forward conduction characteristics, 73, 74
   NPT IGBT, 73
   operation of, 27
   performance parameters, 78–80
      clamped inductive load current, 78
      collector–emitter blocking voltage, 78
      collector–emitter leakage current, 79
      collector–emitter saturation voltage, 79
      continuous collector current, 78
      emitter–collector blocking voltage, 78
      fall time, 79, 127, 547, 548, 640, 887, 911, 912, 1122

Index

forward transconductance, 79
gate–emitter threshold voltage, 79
gate–emitter voltage, 27, 73, 76, 78, 79
input capacitance, 27, 51, 53, 76, 79, 257, 545, 556–558
junction temperature, 17, 24, 29, 50, 78, 92, 96, 100, 102–104, 117, 118, 1157
maximum power dissipation, 78, 100
output capacitance, 51, 53, 79, 84, 254, 412, 415, 418, 549, 727, 729
peak collector repetitive current, 78
reverse transfer capacitance, 51, 79
rise time (*tr*), 32, 76, 79, 104, 126, 493, 556, 847, 848, 912, 922, 943, 962, 969, 970
safe operating area (*SOA*), 25, 27, 31, 79, 125, 126, 129, 435, 1109
total gate charge, 79
turn-off delay time, 79
turn-on delay time, 76, 79, 125
saber IGBT model, 84
static characteristics, 15, 29, 74, 92, 117
transfer characteristics, 52, 56, 59, 73, 498, 501, 502, 801
Insulated metal substrate – polymer on metal (IMS-PM), 1151
Insulated metal substrate – steel corded (IMS-PS), 1151
Insulation test bench, 343
Integral cycle control, 484, 487, 488, 511
Integrated motors, 824, 843, 855
Intelligence-based systems
artificial neural networks (ANN), 1015, 1017–1020, 1025, 1030
fuzzy logic systems (FLS), 1015, 1016
Interior permanent magnet (IPM) machines, 650, 651
Interline power flow controller (IPFC), 819, 822
Internal combustion engine (ICE) automobiles, 636, 652
International Commission on Illumination, 566
International Electrotechnical Commission (IEC), 1103, 1118
International Special Committee on Radio Interference (CISPR), 1103, 1107, 1108, 1118
Interphase transformer, 205, 206, 208
INTERSIL's HIP2500 bridge driver, 556
Inverters, 578, 581, 585, 590, 667, 670, 673, 676, 680, 681, 683, 684, 691, 698–700, 734
battery-less grid-connected PV system configurations, 697
classification, 690
multilevel inverters, 980–982, 986, 1070, 1083, 1114
other PV inverter topologies, 692
types of, 673
Islanding control, 690
ITIC tolerance curves, 1055

Junction field effect transistor (JFET), 75, 133

Kirchhoff laws, 975
Kirchhoff's voltage law, 8
Kramer, 840

Laminated bus-bar, 1152
Lamp luminous flux, 566, 571
Latching, 10, 50, 76, 81, 89, 92, 93, 98, 107, 125, 126
Lateral punch-through transistor (LPTT), 133, 137
Lead acid batteries, 631, 677
Levels of packaging, 1148, 1150
Level-shifting circuit, 546
Levenberg–Marquardt algorithm, 1017
LF2407, 1031–1033, 1035, 1036, 1040–1042, 1044, 1045, 1048, 1049
applications of, 1032
Lighting control circuits, 114
Line commutated rectifiers, 179, 181, 188, 189, 212, 214, 215, 221, 233
Line-commutated controlled rectifiers (thyristor rectifiers), 201
double star, 152–154, 161, 204–206, 208
six-pulse or double star, 201, 204
three-phase half-wave, 151, 201, 202, 495, 967, 968, 970
Line compensation, 773
Line-interactive UPS, 622, 623
Line impedance stabilization network (LISN), 639, 1107
Line regulation, 44, 256, 257
Linear feedback control, 935, 943, 944
types of compensation, 943
compensator selection and design, 944
Linear IC voltage regulators, 600
applications, 602
Linear motors, 830, 842, 854
Linear vs switching regulators, 594, 595
Lithium-ion batteries, 631
LM317 regulator, 601, 602
Load average voltage, 179, 180, 201, 205, 206
Load commutated inverter (LCI), 828, 833
Logic and trigger circuit, 500
Lossless filters, 8, 12, 13
Low drop-out voltage (LDO) regulators, 245
Low-cost fuel-cell inverters, 720
Low-cost fuel-cell power conditioner, 729
Low-pass filter (LPF), 526, 1074, 1077
Low-pressure sodium lamps, 569, 570
Luminous efficacy, 565, 567, 569, 570, 641
Lundell alternator, 637, 646–649, 657
Luo-converter triplelift circuit, 343

Macro models, 83–85
Mader–Horn model, 577, 578
Main converter, 241

Materials cost, 1153
MathCAD®, 1121
Matrix converter motor (MCM), 513
   applications of, 512, 513
   demerits, 503
   merits, 503
Matrix converters, 8, 483, 484, 503–510, 512, 513, 831, 833, 844, 960, 987, 988, 990, 992
Maximum power point tracker (MPPT), 664–667, 679, 680, 683–686, 690, 697, 698, 712
Medium voltage PWM-VSI, 844
Metal-halide lamps, 570, 590
Metal oxide field effect transistor (MOSFET), 3, 4, 10, 12, 13, 27, 28, 41, 48–56, 58–69, 71–76, 78–80, 82–85, 90, 95, 104, 107, 108, 123–130, 232, 246, 255, 306, 307, 332, 343, 347, 412, 427, 472, 504, 543–546, 549, 551, 553, 556–558, 562–565, 572, 585, 593, 603, 642–646, 654, 669, 678, 680, 681, 685, 691, 830, 831, 858, 861, 897–899, 904, 912, 971, 974, 975, 1010, 1032, 1122–1127, 1152
   advantages of, 123
   current MOSFET performance, 66
      levels, 62
   MOSFET large-signal model, 65
   MOSFET PSPICE model, 62
   operation of, 27
   regions of operation, 48, 51
   SOA for, 61
   structure of, 48, 90, 104
   turn-off characteristic of, 55, 61
   types, 48, 49, 62
Method of energy balance, 6, 7
Micro stepping, 913–915
Miller capacitance, 79, 83, 545
Miller effect, 53, 81, 138, 544
MIMO, 956, 975, 976, 986, 997
Mine hoist systems, 851
Mixed-level hybrid multilevel converter, 457
Mixed mode circuit simulators, 72
Modulating techniques, 353–355, 357, 359, 360, 364, 368, 372, 376, 380, 382, 390, 391, 394, 396–398, 400, 401
Molded interconnect device (MID), 1151, 1152, 1154, 1156, 1157
MOS controlled thyristor (MCT)
   applications of, 128, 129
   C-MCT, 124
   comparison of MCT with other power devices, 125
   gate drive for, 126
   generation-1 MCTs, 125, 127, 129, 130
   generation-2 MCTs, 125, 127, 129, 130
   MCT chopper, 126
   NMCT, 126, 127, 130
   paralleling of, 126
   P-channel MCT, 124, 447
   simulation model of, 127
   snubbers of, 126
MOS turn-off (MTO) thyristor, 128, 831
Motor drive duty, 835
Motorola MC68HC11E9 microprocessor, 1005, 1010
MPPT controller algorithm, 665
Multi stack variable-reluctance motor, 909
Multicell topologies, 391
Multichip modules (MCMs), 1147
Multi-element resonant power converters, 262, 263, 335
   bipolar current voltage sources, 341, 342
   four energy-storage elements resonant power converters, 336
   three energy-storage elements resonant power converters, 336
   two energy-storage elements resonant power converters, 335
   two groups, 193, 496, 830
Multifunction ceramic (MFC) capacitors, 1108
Multilayer feedforward ANN, 1020
Multilevel cascaded H-bridge converters
   advantages and disadvantages of, 453, 454
Multilevel converter design example, 470
Multilevel converter PWM modulation strategies, 459
Multilevel converters, 451, 452, 456–459, 470, 478, 480, 692, 758, 978–982, 997
Multilevel inverter drive (MLID), 478
Multilevel topologies, 240, 391, 394, 398, 456, 732, 757
Multilevel universal power conditioner (MUPC), 470, 473, 474, 478
Multi-quadrant operation, 262, 307, 318, 324, 837
Multi-quadrant ZCS quasi-resonant Luo-converters, 263, 323, 324
   four-quadrant ZCS quasi-resonant Luo-converter, 327
   two-quadrant ZCS quasi-resonant Luo-converter in forward operation, 324
   two-quadrant ZCS quasi-resonant Luo-converter in Reverse Operation, 325
Multi-quadrant ZVS quasi-resonant Luo-converters, 263, 327
   two-quadrant ZVS quasi-resonant DC/DC Luo-converter in forward operation, 327, 328
Multistage inverters, 390
Multiterminal dc system (MTDC), 774, 786, 787
Multiple-lift push–pull switched-capacitor Luo-converters
   P/O multiple-lift (ML) push–pull (PP) switched capacitor (SC) DC/DC Luo-converter, 315–317
   N/O multiple-lift push–pull switched-capacitor DC/DC Luo-converter, 317, 318
Multiple-quadrant operating Luo-converters, 301

NARMAX model, 1027
National electric code (NEC), 1056
Naturally commutated cycloconverter (NCC), 502, 503, 512

Near unity power factor, 825, 839, 851, 960, 975, 978, 996
Necessary distortion terms, 501, 502
Neodymium–iron–boron magnetic materials, 885, 891
Net metering, 718
Neural function approximator, 1016
Neutral-point-clamped (NPC) inverters, 733, 844, 845, 1070, 1072
Neutral-point-clamped PWM inverter (NPC-PWM), 757
NMOSFET, 124, 125, 127
Non-isolated electronic level shifted drivers, 543, 555–556

Off-line trained ANN, 1019
On-line output current control in multilevel inverters, 986
On-line trained ANN, 1019, 1020
On-line UPS systems, 621–623
Operational amplifiers (OA), 36, 342, 905, 943, 969
Opto–couplers, 546–549, 553, 901
Overlap angle, 210, 211, 782, 784
Overlap time, 210
Overmodulation, 231, 355, 357, 360, 364, 375, 376, 396, 399, 462, 464, 512, 1030

P2 topology, 456
Pancake ironless armature, 885, 903
Parallel hybrid energy systems, 686, 688
Parallel inverter power rating, 474
PARAM symbols, 1132
Parasitic capacitance, 50, 51, 65, 137, 140, 162, 167, 418, 570, 1104, 1112
Parasitic thyristor, 77–79, 545
Paschen curves, 568
Passive power factor corrector, 521
Peak current-mode control, 617
Peak inverse voltage (PIV), 17, 146
Peak rating, 1035
Peak shaving, 718
Penning mixtures, 568
Performance evaluation, 626
Periodical sampling (PS), 227, 1080
Permanent dc magnet motors without brushes, 683
Permanent-magnet ac synchronous motor drives, 885, 993
Permanent magnet alternating current (PMAC) motors, 1037
Permanent-magnet brushless dc (PM BLDC) motor, 887, 891, 892, 993
   machine background, 891
   permanent-magnet ac machines, 891
   sensorless operation of, 888
Permanent magnet dc motors with brushes, 683
Permanent-magnet stepper motors, 910

Permanent magnet synchronous machine (PMSM), 650, 651, 703, 708, 711, 712, 740, 841, 891, 926, 1031, 1041
   implementation of the PMSM algorithm using the LF2407, 1045
   mathematical model, 1042, 1043
   PMSM control topology, 1044
Perturb and observe (PAO) method, 665
PES operation, 725
Phase-to-neutral voltages, 207, 223, 224, 1069, 1071, 1075, 1077, 1092, 1094–1096
Photometers, 566
Photopic curve, 565
Photovoltaic arrays, 258, 259, 663, 669, 672, 715
Photovoltaic system, 667, 669–671, 718, 1152
PI regulator, 629, 781–783, 965, 978, 1019, 1036, 1047
Piezoelectric ultrasonic motors, 642, 658
Pittman company, 1008
PM sinewave motors, 888–891
   surface-magnet motors, 886
PMOSFET, 124, 125
p–n junctions, 49, 89, 90, 94, 101, 137, 662, 674, 1104, 1122, 1152
Point of common coupling (PCC), 518, 1065
Poisson equation, 135
PolarPAK device, 68
Poly-phase diode rectifiers, 150, 155
   poly-phase design parameters, 157
   six-phase parallel bridge rectifier, 157, 158
   six-phase series bridge rectifier, 156, 157
   six-phase star rectifier, 155, 156
Potential saddle, 135
Power conditioning system (PCS), 667, 671, 697, 731, 734, 747, 764
Power conditioning units (PCU), 684
Power devices, 10, 11, 41, 48, 68, 69, 83, 87, 94, 105, 112, 123, 125, 405–407, 412, 435–440, 443, 446, 518, 539, 732, 837, 1035
   maximum switching frequency, 917
   power devices available, 830
   series connection, 844
   types, 734
Power electronic building block (PEBB), 86, 87, 512, 1148
Power electronic circuits, 8, 72, 82, 86, 257, 452, 637, 639, 641, 642, 650, 654, 712, 1066, 1128
   advancements, 13
   classification, 8
   key characteristics, 2
   method of energy balance, 6
   reliability objective, 3, 4
   resonant Switch, 407
   switching, 42, 44
Power electronic conditioner, 740
Power electronic converter block, 858

Power electronic converters, 16, 543–545, 557, 565, 673, 704, 743, 793, 794, 858, 1066, 1103–1105, 1109, 1112, 1115, 1118
Power electronic system, 2, 3, 8, 13, 36, 42, 72, 82, 85, 86, 518, 519, 538, 543, 544, 637, 640, 657, 682, 726, 1066, 1106, 1109, 1110, 1147
  challenges, 8, 9, 11–13
Power electronics, 699, 700, 703, 706, 709, 710, 711, 713, 729, 737, 744, 762, 763, 789, 790, 797, 799, 837, 841, 853, 926, 1108, 1122
  applications, 1, 2
  characteristics, 2
  definitions, 1
Power electronics for photovoltaic power systems
  basics of photovoltaics, 674
  types of PV power systems, 676
Power electronics laboratory, 499
Power factor (PF), 128, 149, 160, 161, 179, 212, 221, 222, 224, 227, 230, 259, 359, 370, 390, 398, 429, 431, 432, 488, 494, 495, 499, 502, 503, 508, 518, 570, 571
  correction stage, 590
  definition of, 518
  disadvantages of, 521
  grid connection conditioning system, 749, 755
  harmonic limiting standards, 587
  high power factor, 586
  improved electrical system power factors, 825, 829
  input power factor, 588, 589
  power factor and harmonics, 487
  unity power factor, 712, 725, 739, 742, 747
  universal UPS, 622
Power factor correction (PFC) technique, 179, 190, 196, 259, 427, 429, 431, 518, 520, 521, 523–525, 530, 531, 533, 535, 536, 538, 1156
  classification of, 525, 537
Power quality, 622, 626, 669, 691, 699, 700, 717, 718, 737, 755, 756, 764, 777, 987, 1053, 1054, 1056
  diagnosis, 1060
  power quality analyzers, 1060
  power quality problems, 1066, 1067
Power rectifiers using multilevel topologies, 240
Power semiconductors, 544, 653, 734, 959, 964, 971, 975, 1080, 1122, 1147, 1152
  categories, 544
  diepower semiconductors, 557
  features, 10, 13, 27,
  power losses in power semiconductors, 193
  sliding-mode control, 971
  switching law, 958
Power-electronic topologies, 720, 731
POWEREX's M57959L bridge driver, 512, 556
Power supply, 161, 189, 198, 201, 440, 448, 517, 523, 535, 547, 549, 559, 560, 562, 593, 595, 619, 690, 705, 826
  switch-mode, 4, 37, 51, 125

Premium power, 717, 718
Product business cost, 1154, 1157
Production cost, 1153
Programmable logic control (PLC), 777, 828, 842, 845
PSpice®, 1121, 1123, 1124, 1126, 1131, 1135, 1138, 1141, 1144
Public utility board (PUB), 345
Pulse density modulation (PDM), 130, 1079
  advantages of, 130
Pulse width modulation (PWM), 23, 353, 355, 357, 359, 360, 364, 365, 379, 395, 1035, 1085–1088, 1090, 1093, 1104, 1105
Pulse width modulator transfer functions, 942
Pumping energy (PE), 346, 347
Punch-through emitter (PTE), 133, 137
Push–pull based topology, 720
PV charge controllers, 677
  types of, 677–678
PWM AC chopper, 488
PWM flyback converter, 251
PWM generator, 931, 1035, 1125
PWM phase-to-phase voltages
PWM switching converter drive/choppers, 861, 862
PWM techniques, 369, 372, 396, 399, 439, 440, 459, 462, 464, 507, 628, 690, 792, 846, 1115
PWM-VSI converter, 834, 842, 852

Radio-frequency (RF) filters, 1, 511, 1106, 1108
Rbe, 543
RC snubbers, 733
Rectification, filtering, and regulation in a dc power supply, 593
Regeneration, 124, 128, 196, 240, 354, 386–390, 829
Regeneration in inverters, 386–390
Regenerative braking, 39, 112, 198, 215, 239, 307, 319, 324, 327, 453, 649, 837, 850, 861, 862
Regulator circuits, 596, 601, 638
  transistor, 44
  voltage divider, 42
  zener, 42
Remote area power supplies (RAP), 685, 688
Remote power, 644, 717, 718
Renewable energy, 717, 761
Renewable energy interface, 478
Renewable energy sources (RES), 259, 451, 453, 480, 673, 676, 687
Repetition rate, 559–562, 850
Resonance, 573, 583, 641, 642, 805, 816, 833, 904, 912, 944, 1061–1063, 1067, 1099, 1109
Resonance charging technique, 561
Resonant inverters, 335, 437, 581, 585, 590
  types, 578–585
Resonant power converter (RPC), 262, 263, 335, 336, 339

Restricted switches, 11
 FCRB, 11
 types of, 11
Reverse recovery process, 19
Reverse recovery time, 16, 17, 19, 24, 25, 145, 162, 167
Reverse two-quadrant operating (R 2Q) Luo-converter, 303, 304
Ridethrough module, 513
Ripple-mitigating power conditioner, 720
Root mean square (RMS), 5, 100, 148, 183, 201, 835, 836, 866, 968
Rotary UPS, 623–626,
Rotor connected power conditioner, 750–754
Rotor flux oriented control (RFOC) induction motor drive, 876, 1015
Rotor resistance estimator (RRE), 1021, 1022, 1025

SAE standards J1113/41 and J1113/42, 639
Safe operating area (SOA), 25, 31, 71, 79, 125, 126, 129, 435, 1109
 FBSOA, 31
 RBSOA, 31, 35
Samarium cobalt, 885, 891
Sawtooth carrier wave, 510
Scalar control methods, 873
Schmitt trigger comparator, 126
Schottky diode, 134, 137, 138, 162, 165, 167, 177, 254, 899
Schottky emitters, 137
Scotopic curve, 565
SCR (thyristor), 543, 830, 848
 applications, 110
 basic structure and operation, 90
 limitations of, 123
 major types of, 90
 static characteristics, 92–95
 structure, 142
 thyristor parameters, 99–101
 types of, 101–106
Sealed batteries, 677
Secondary ripple filter, 1087
Selective harmonic elimination (SHE), 354, 357, 360, 367, 376, 452, 469, 470, 628
Semi-autogenous (SAG) grinding mills, 511
Semiconductor power integrated circuits, 1151, 1152
Semiconductor-switching network, 42
 characteristics, 44, 45
 efficiency of, 42
Semiconductors, 10, 899, 918, 919, 936, 952, 956, 958, 959, 962, 964, 971, 975, 982, 1069, 1147, 1152
 characteristics of power devices, 10
 groups, 10

Semiconductors switches, 27, 798
 characteristics of, 27
 operating regions, 27, 28
Semiconverter, 209
Semikron Corporation, 554, 555
SEPDIS, 853, 854
Sepic converter, 534
Series active power filters, 1068, 1069, 1085, 1089, 1090
 advantages of, 1090
 characteristics, 1086
 compensation characteristics, 1090
 power circuit structure, 1085
 power circuit topology, 1086
 principles of operation, 1085
Series hybrid energy systems, 686
Series-parallel/delta conversion UPS, 622
Servo drives, 511, 830, 902, 904
Shoot through, 81, 550, 893, 916, 1126
Silicon carbide technology, 544
Silicon control rectifier (SCR), 142, 483–487, 489, 491, 493, 496, 497, 503, 510, 543, 544, 651, 830
Silicon Power Corporation (SPCO), 124, 128, 130
Silicon semiconductor material, 675
Simplorer®, 1141
Simulation of sensorless vector control using PSpice®, 1135
Simulation tools for design and analysis, 1121
Simulations of power electronic circuits, 1128–1131
Simulations using Simplorer®, 1141
Single-ended isolated flyback regulator, 603
Single-ended isolated forward regulators, 607
Single-phase ac–ac voltage controller, 484, 487
 phase-controlled, 485
 with on/off control, 487
Single-phase diode rectifiers, 145
 design parameters, 150
 full-wave rectifiers, 145
  bridge rectifiers, 146
 half-wave rectifiers, 146
 performance parameters
  current relationships, 148
  form factor, 148
  harmonics, 149
  rectification ratio, 148
  ripple factor, 149
  transformer utilization factor, 149
  voltage relationships, 147
Single-phase rectifiers, 179, 186, 188, 189, 198
 bridge rectifier, 189, 262, 331
 classification, 179
Single-phase voltage source inverters, 355
Single-pulse mode of operation, 922, 925
Sinusoidal pulse width modulation, 229
Sinusoidal wave PWM, 452, 952

Six step switching, 894, 895, 900
Sliding-mode control of switching converters, 955, 956
    principles of, 956
Slow switching frequency changer (SSFC), 503
Society of automotive engineers (SAE), 637, 657, 658, 1117, 1118
Soft starters for induction generators, 706
Soft-switched multilevel converter, 457
Soft-switching converters, 86, 245, 262, 323, 406
Solar cells, 478, 662, 674, 675, 679, 680, 690
Solar energy constant, 674
Solar energy conversion, 663
Solar water pumping, 667, 682, 684
    types of pumps, 682
Solidtron, 130
Space charge limiting load (SCLL), 133, 140
Space vector modulation (SVM), 369, 378, 452, 460, 505–507, 952, 953, 955, 1029, 1030, 1032
Special motors, 830, 841
Speed control algorithm, 1045
Spice, 1144
SPWM inverter drive, 868, 878
Square-wave operation, 364, 376
Squirrel cage induction motor, 824, 838, 839, 842, 844, 873, 952, 1020, 1046
Stacked die, 1147
Staircase waveform quality, 451
Stand-alone PV systems, 669, 673, 677, 680
    features of, 680
Standby UPS, 619–622
Standby/emergency generation, 718
Star-connected R-load, 490, 491
Starting voltage, 568, 570, 572, 574, 580, 585, 586
STATCOM control techniques, 809, 811
Static induction diode (SID), 137
Static induction MOS transistor (SIMOS), 133, 139
    advantages of, 140
        short switching time, 137
        thermal stability, 137
Static induction transistor (SIT), 133–135, 137
    bipolar mode operation of SI devices (BSIT), 136
    characteristics of, 134
        linear scales, 135
        logarithmic scales, 135
    junction field effect transistor (JFET), 133
    theory, 133, 134
Static induction transistor logic (SITL), 133, 137, 138
Static Kramer system, 713, 840
Static Scherbius system, 713
Static switching systems, 112
Static synchronous compensator, 807
Static synchronous series compensator, 815
Static var compensators (SVC), 517
Static var compensators STATCOM, 764

Steady-state frequency regulation, 1054
Steady-state voltage regulation, 1054
Step-down (buck) converter
    basic converter, 247
    scheme, 248
    transformer versions of
        forward converter, 248
        full-bridge converter, 249
        half-bridge converter, 249
        push–pull converter, 249
    waveforms, 248
Stepper motors, 842
Step-up (boost) converter, 250
Super-lift (SL) Luo-converters, 288
    in CCM, 349
    N/O cascade boost-converters (CBC), 297
    N/O super-lift Luo-converters, 292
    P/O cascade boost-converters (CBC), 294
    P/O super-lift Luo-converters, 288
Surface-magnet synchronous motor, 886
SVC vs STATCOM, 814
Switch, 2
    ideal switches, 9, 497, 732, 778, 936
    power handling rating, 3
    real switches, 9
Switch matrices, 8
Switch-mode converter, 665
Switch mode power supplies (SMPS), 518, 519, 526
Switched-capacitor converters, 318, 322
Switched-capacitor (SC) DC/DC converters, 262
Switched-capacitor multi-quadrant Luo-converters, 306, 307
    two modes of, 307–315
Switched hybrid energy systems, 687
Switched-inductor (SI) DC/DC converters, 262
Switched mode power supply (SMPS), 125, 1061
Switched reluctance motor (SRM) drive, 842, 857, 915, 916, 919, 1037
    advantages and disadvantages of, 916
    applications of, 916
    drive design options of, 917–919
    operating theory of, 919
    position sensing, 925
    principle of operation, 915
Switching functions, 11–13, 1081, 1084
    parameter of, 12
Switching power converters, 2, 13, 140, 935, 936, 993, 997
    linearized state-space averaged model, 939
    state-space averaged model, 935–939, 943, 946, 956, 996
    state-space modeling, 936, 956
    switched state-space model, 936
Switching regulators, 245, 594, 595, 602, 603
Synchronizing circuit, 499
Synchronous link reactor, 1084
Synchronous motor (excited), 858

Synchronous motor (PM), 858
Synchronous motor drives, 511, 840, 877, 878, 885, 888, 903
    CSI-driven, 880
    operating modes, 882
    steady-state equivalent-circuit representation, 878
Synchronous-rectifier (SR) DC/DC converters, 262, 331
Synchronous reluctance motor drives, 926
    basic mathematical modeling, 929
    basic principles, 926
    machine structure, 928–930
Syncrel, 926–932
Syncrel drive system, 931
System in package (SIP), 1147
System on chip (SOC), 1147
System on package (SOP), 1147

Tail-current amplitude, 129
Tapped inductor (Watkins–Johnson) converters, 271
Techniques of level shifting, 549
Technology-driven partitioning, 1154, 1155
TelCom Semiconductor Inc., 556
The Comité Européen de Normalisation Electrotechnique (CENELEC), 219
The Lundell/claw-pole alternator, 646
Thick film on ceramic (TFC), 1151
Three-phase ac–ac voltage controllers, 488
    fully controlled three-wire, 490
        modes, 490, 491
    phase-controlled, 488
Three-phase diode rectifiers, 150
    bridge rectifiers
        high power applications, 153
    design parameters, 150
    operations, 155
    star rectifiers
        basic circuit, 151, 152
        double-star rectifier, 153
        inter-star or zig-zag rectifier circuit, 152, 153
Three-phase voltage source inverters, 363, 1114
    DC Link current in three-phase VSI, 369
    load-phase voltages in three-phase VSI, 371
    selective harmonic elimination, 367
    sinusoidal PWM with zero sequence signal injection, 364, 365
    space-vector (SV)-based modulating techniques, 368
    square-wave operation of three-phase VSI, 364
Thyristor-controlled inductor (TCI), 113
Thyristor-controlled phase angle regulator (TCPAR), 805, 807
Thyristor-controlled reactor (TCR), 521, 803
Thyristor-controlled series capacitor (TCSC), 805, 816
Thyristor converter drive, 861

Thyristor-switched series capacitor (TSSC), 804, 805
Thyrode controller, 804
Topologies converter, 1150
    developed, 266, 267
    fundamental, 263–266
Total FIS, 1000
Total harmonic distortion (THD), 160, 184, 188, 190, 221, 262, 346, 453, 462, 469, 498, 503, 518, 619, 628, 744, 757, 763, 809, 1074
Transients, 815, 1053, 1054, 1056, 1059, 1060, 1067, 1108, 1114, 1128
Transistors, 3, 10, 13, 27, 28, 30, 32, 33, 36, 37, 71, 138, 140, 141, 227
    dynamic switching characteristics, 32, 76, 95
    power-npn and -pnp bipolar transistors, 27–29
    SPICE simulation of, 36, 37
Trapezoidal back-emf waveform, 886, 887
    control of, 886
Triangular carrier, 227, 809, 1080, 1082, 1084, 1087, 1122, 1125, 1126
True UPS, 621
Turboswitchers Inc., 557

Ultra-lift (UL) Luo-converters, 261, 263, 299
    continuous conduction mode, 299
    discontinuous conduction mode, 301
    negative output Luo-converters
        five circuits of, 279–284
    positive output Luo-converters, 271
        five circuits of, 271–276
    simplified positive output Luo-converters
        four circuits of, 279
Unified power quality conditioner (UPQC) topology, 622
Uninterruptible power supply (UPS), 619, 1085, 1147
    classification, 619–626
    performance comparison of, 626
Unipolar capacitors, 477
Unipolar drive circuits, 912
Unity displacement factor frequency changer (UDFFC), 503
Universal power conditioner, 458, 459, 470, 474, 724
Universal power converter, 506
Universal UPS, 622, 623, 626
Unnecessary distortion terms, 502
Unrestricted frequency changer (UFC), 503

VAR compensators, 27, 101, 105, 112, 353, 354, 517, 690, 764, 773, 1053
Variable ratio transmission (VRT), 710
Variable speed constant frequency (VSCF), 240, 493, 511, 704, 708, 709
Variable speed control of AC machines, 1032

Variable speed drive (VSD), 218, 455, 556, 712, 823–825, 827, 831, 842, 852, 859, 865, 868, 888, 916, 931, 1031, 1037
    advantages of, 824
    applications, 851–854
    classification by applications, 830
    classification by the type of converter, 830
    classification by type of power device, 830
    common requirements, 827
    communication in, 845
    control strategies, 844
    disadvantages of, 825
    drive classifications and characteristics, 829
    drive specifications, 827–829
    improved process control, 825
    modern VSD systems, 853
Vector control methods, 841, 954
    basic principles of, 874
Vector control technique, 704, 873, 1031, 1081, 1122
Venturini method, 503, 506, 508
Venturini vs SVM method, 508
VMOS, 140, 141
Voltage–current characteristic curves, 30
Voltage distortion limits, 1065
Voltage flicker, 1058, 1066, 1069
Voltage lift Luo-converters, 271
Voltage-lift (VL) technique, 262, 271, 288
Voltage mode control, 246, 256, 525, 529, 615–617, 942, 951
Voltage sags, 474, 1054, 1055, 1085
Voltage source converters (VSC), 513, 774, 791–793
Voltage source inverter (VSI), 353–357, 360, 364, 367, 369, 376, 379, 382, 386, 390, 690, 691
Voltage-source voltage-controlled PWM rectifier, 225, 228, 229
VSI-multilevel topologies, 394–398

West German Standards (VDE), 219
Wind turbine generators, 673
Wind turbine technology, 673, 709, 737, 764
    basics of wind power, 701
    classification, 702
    control of, 741
    control of wind turbines, 707
    electrical system of a wind farm, 761, 762
    fixed speed generator (FSWT), 702, 707, 708
    grid connection standards, 755–757
    horizontal-axis wind turbine, 737
    offshore and onshore wind turbines, 764
    types of wind generators, 703
    types of wind power systems, 705
    types of wind turbines, 701
    variable speed pitch wind turbine, 742
    variable speed wind turbine generator, 702, 707, 708

x-Element RPC, 335

Zener diode regulator, 17, 595
Zener diodes, 17, 34, 550
Zero-ripple boost converter (ZRBC), 720, 721, 723
Zeta converter, 269, 523, 534, 535